Plant Physiology and Development

SIXTH EDITION

Plant Physiology and Development

SIXTH EDITION

Lincoln Taiz
Professor Emeritus, University of California, Santa Cruz

Eduardo Zeiger
Professor Emeritus, University of California, Los Angeles

Ian Max Møller
Associate Professor, Aarhus University, Denmark

Angus Murphy
Professor, University of Maryland

Sinauer Associates, Inc • Publishers
Sunderland, Massachusetts U.S.A.

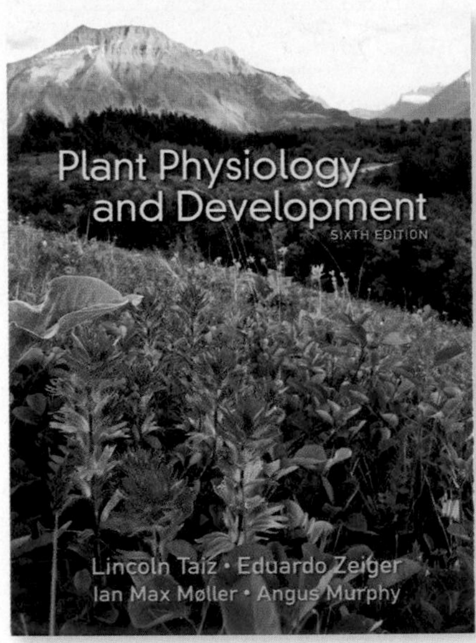

Front Cover

Red Indian Paintbrush (*Castilleja miniata*) grows on hillside in Waterton Lakes National Park, Alberta, Canada. © All Canada Photos/Corbis.

About the Book

Editor: Andrew D. Sinauer
Project Editors: Laura Green and Stephanie Bonner
Production Manager: Christopher Small
Book and Cover Design: Jefferson Johnson
Photo Researcher: David McIntyre
Copyeditor: Liz Pierson
Indexer: Grant Hackett
Illustrator: Elizabeth Morales
Book and Cover Manufacturer: Courier Corporations, Inc.

Plant Physiology and Development, Sixth Edition

For information, address
Sinauer Associates, Inc., P.O. Box 407, Sunderland, MA 01375 U.S.A.
FAX: 413-549-1118
E-mail: publish@sinauer.com
Internet: www.sinauer.com

Library of Congress Cataloging-in-Publication Data
Plant physiology
Plant physiology and development / editors, Lincoln Taiz, University of California, Santa Cruz, Eduardo Zeiger, University of California, Los Angeles. -- Sixth edition.
pages cm
Revised edition of: Plant physiology. Fifth edition. c2010.
ISBN 978-1-60535-255-8 (casebound)
1. Plant physiology. 2. Plants--Development. I. Taiz, Lincoln. II. Zeiger, Eduardo. III. Title.
QK711.2.T35 2014
571.2--dc23 2014030480

Printed in U.S.A.
10 9 8 7 6 5 4 3 2

Brief Contents

Editors

Lincoln Taiz is Professor Emeritus of Molecular, Cellular, and Developmental Biology at the University of California at Santa Cruz. He received his Ph.D. in Botany from the University of California at Berkeley. Dr. Taiz's main research focus has been on the structure, function, and evolution of vacuolar H^+-ATPases. He has also worked on gibberellins, cell wall mechanical properties, metal transport, auxin transport, and stomatal opening. (Chapters 15, 16, 18, 19, 20, 21, 22, and 23)

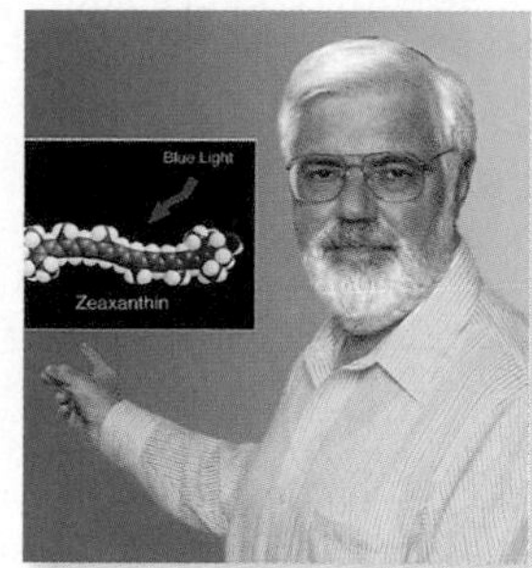

Eduardo Zeiger is Professor Emeritus of Biology at the University of California at Los Angeles. He received a Ph.D. in Plant Genetics at the University of California at Davis. His research interests include stomatal function, the sensory transduction of blue-light responses, and the study of stomatal acclimations associated with increases in crop yields. (Chapter 10)

Sub-Editors

Ian Max Møller is Associate Professor at Department of Molecular Biology and Genetics at Aarhus University, Denmark. He received his Ph.D. in Plant Biochemistry from Imperial College, London, UK. He has worked at Lund University, Sweden and, more recently, at Risø National Laboratory and the Royal Veterinary and Agricultural University in Copenhagen, Denmark. Professor Møller has investigated plant respiration throughout his career. His current interests include turnover of reactive oxygen species and the role of protein oxidation in plant cells. (Chapter 12)

Angus Murphy is Professor and Chair of the Department of Plant Science and Landscape Architecture at the University of Maryland. He earned his Ph.D. in Biology from the University of California, Santa Cruz in 1996. Dr. Murphy studies ATP-binding cassette transporters, auxin transport proteins, and the role of auxin transport in programmed and plastic growth. (Chapters 15, 16, 17, 18, and 19)

Principle Contributors

Sarah M. Assmann is a Professor in the Biology Department at the Pennsylvania State University. She received a Ph.D. in the Biological Sciences at Stanford University. Dr. Assmann studies how plants respond to environmental stresses, with a focus on abiotic stress regulation of RNA structure, heterotrimeric G-protein signaling, and guard cell systems biology. (Chapter 6)

Christine Beveridge is a Professor in the School of Biological Sciences at the University of Queensland. She received a Ph.D. in Plant Sciences at the University of Tasmania in 1994. Her research focuses on shoot architecture and hormonal control of development, especially strigolactones, and involves genetic approaches, molecular physiology and plant modelling. (Chapter 19)

Robert E. Blankenship is a Professor of Biology and Chemistry at Washingon University in St. Louis. He received his Ph.D. in Chemistry from the University of California at Berkeley in 1975. His professional interests include mechanisms of energy and electron transfer in photosynthetic organisms, and the origin and early evolution of photosynthesis. (Chapter 7)

Arnold J. Bloom is a Professor in the Department of Sciences at the University of California at Davis. He received a Ph.D. in Biological Sciences at Stanford University in 1979. His research focuses on plant-nitrogen relationships, especially the differences in plant responses to ammonium and nitrate as nitrogen sources. He is the co-author with Emanuel Epstein of the textbook, *Mineral Nutrition of Plants* and author of the textbook, *Global Climate Change: Convergence of Disciplines*. (Chapters 5 and 13)

Eduardo Blumwald is a Professor of Cell Biology and the Will W. Lester Endowed Chair at the Department of Plant Sciences, University of California at Davis. He received his Ph.D. in Bioenergetics from the Hebrew University of Jerusalem in 1984. His research focuses on the adaptation of plants to environmental stress and the cellular and molecular bases of fruit quality. (Chapter 24)

John Browse is a Professor in the Institute of Biological Chemistry at Washington State University. He received his Ph.D. from the University of Aukland, New Zealand, in 1977. Dr. Browse's research interests include the biochemistry of lipid metabolism and the responses of plants to low temperatures. (Chapter 12)

Bob B. Buchanan is a Professor of Plant and Microbial Biology at the University of California at Berkeley. He continues to work on thioredoxin-linked regulation in photosynthesis, seed germination, and related processes. His findings with cereals hold promise for societal application. (Chapter 8)

Victor Busov is a Professor at Michigan Technological University. His work is focused on understanding the molecular mechanisms that regulate growth and development of woody perennial species. He is interested in how these mechanisms are important for adaptation to environment, evolution of different life forms and applications to tree improvement and biotechnology. (Chapter 19)

John Christie holds an undergraduate degree in Biochemistry and Ph.D. from the University of Glasgow and is currently Professor of Photobiology at the University. During his postdoc with Winslow Briggs at Stanford, he contributed to uncovering the molecular identity of higher plant phototropins. He established his own research group at the University of Glasgow in 2002 and continues to investigate the molecular basis of plant UV/blue light receptor function and signaling. His research also extends to developing new technologies derived from photoreceptor characterization. (Chapter 16)

Daniel J. Cosgrove is a Professor of Biology at the Pennsylvania State University at University Park. His Ph.D. in Biological Sciences was earned at Stanford University. Dr. Cosgrove's research interest is focused on plant growth, specifically the biochemical and molecular mechanisms governing cell enlargement and cell wall expansion. His research team discovered the cell wall loosening proteins called expansins and is currently studying the structure, function, and evolution of this gene family. (Chapter 14)

Susan Dunford is an Associate Professor of Biological Sciences at the University of Cincinnati. She received her Ph.D. from the University of Dayton in 1973 with a specialization in plant and cell physiology. Dr. Dunford's research interests include long-distance transport systems in plants, especially translocation in the phloem, and plant water relations. (Chapter 11)

James Ehleringer is at the University of Utah where he is a Distinguished Professor of Biology and serves as Director of both the Global Change and Sustainability Center and of the Stable Isotope Ratio Facility for Environmental Research (SIRFER). His research focuses on understanding terrestrial ecosystem processes through stable isotope analyses, gas exchange and biosphere–atmosphere interactions, and water relations. (Chapter 9)

Jürgen Engelberth is an Associate Professor of Plant Biochemistry at the University of Texas at San Antonio. He received his Ph.D. in Plant Physiology at the Ruhr-University Bochum, Germany in 1995 and did postdoctoral work at the Max Planck Institute for Chemical Ecology, at USDA, ARS, CMAVE in Gainesville, and at Penn State University. His research focuses on signaling involved in plant–insect and plant–plant interaction. (Chapter 23)

Lawrence Griffing is an Associate Professor in the Biology Department at Texas A&M University. He received his Ph.D. in Biological Sciences at Stanford University. Dr. Griffing's research mainly focuses on plant cell biology, concentrating on the interaction between the endoplasmic reticulum and other membranes and the dynamics of endomembranes through their interactions with the cytoskeleton. His teaching focuses on incorporating authentic inquiry and scientific discovery into undergraduate courses. (Chapter 1)

N. Michele Holbrook is a Professor in the Department of Organismic and Evolutionary Biology at Harvard University. She received her Ph.D. from Stanford University in 1995. Dr. Holbrook's research group focuses on water relations and long-distance transport through xylem and phloem. (Chapters 3 and 4)

Andreas Madlung is a Professor in the Department of Biology at the University of Puget Sound. He received a Ph.D. in Molecular and Cellular Biology from Oregon State University in 2000. Research in his laboratory addresses fundamental questions concerning the influence of genome structure on plant physiology and evolution, especially with respect to polyploidy. (Chapter 2)

Ron Mittler is a Professor in the Department of Biological Sciences at the University of North Texas. He got his Ph.D. in biochemistry from Rutgers the State University of New Jersey. His current research is focused on plant responses to abiotic stress and reactive oxygen signaling and metabolism in plant and cancer cells. (Chapter 24)

Gabriele B. Monshausen is an Assistant Professor of Biology at the Pennsylvania State University. She received her PhD in plant biology at the University of Bonn, Germany. Dr. Monshausen's research focuses on mechanisms of cellular ion signaling in plant hormone responses and plant responses to mechanical forces. (Chapter 15)

Wendy Peer is an Assistant Professor in the Department of Environmental Science and Technology and an affiliate in the Department of Plant Science and Landscape Architecture at the University of Maryland, College Park. Wendy Peer's research focuses on seedling establishment and the integration of developmental and environmental signals that lead to successful seedling establishment. (Chapters 15, 18, and 19)

Allan G. Rasmusson is Professor in Plant Physiology at Lund University in Sweden. He received his Ph.D. in plant physiology at Lund University in 1994, and made a postdoc at IGF Berlin. Dr. Rasmusson's current research centers on redox control in respiratory metabolism and on peptide-membrane interactions. (Chapter 12)

Darren R. Sandquist is a Professor of Biological Science at California State University, Fullerton. He received his Ph.D. from the University of Utah. His research focuses on plant ecophysiological responses to disturbance, invasion, and climate change in arid and semi-arid ecosystems. (Chapter 9)

Graham B. Seymour is Professor of Plant Biotechnology and Head of the Plant and Crop Science Division at the University of Nottingham in the UK. His major research interests are the mechanistic basis of fruit quality traits and understanding the role of the epigenome in regulating the ripening process. (Chapter 21)

Sally Smith is an Emeritus and Adjunct Professor in the Soils Group, School of Agriculture, Food and Wine, the University of Adelaide, Australia. She is a fellow of the Australian Academy of Science and co-author of a major research text on mycorrhizas. Her research interests include interactions between arbuscular mycorrhizal fungi and plants, especially roles of the symbiosis in plant phosphate nutrition and growth. (Chapter 5)

Joe H. Sullivan is a Professor in the department of Plant Science and Landscape Architecture at the University of Maryland. He received his Ph.D. in Plant Physiology at Clemson University in 1985. His research interests include Plant Physiological Ecology in natural and urban ecosystems with particular interest in the response of plants to ultraviolet radiation and other parameters of global climate change. (Chapter 16)

Heven Sze is a Professor at the University of Maryland at College Park. She earned a Ph.D. in plant physiology at Purdue University, and was a postdoctoral fellow at Harvard Medical School. Her research has focused on the mechanism and regulation of ion transport and how ion and pH homeostasis are integrated with growth, development and reproduction. (Chapter 21)

Bruce Veit is a senior scientist at AgResearch in Palmerston North, New Zealand. He received his Ph.D. in Genetics from University of Washington, Seattle in 1986 before undertaking postdoctoral research at the Plant Gene Expression Center in Albany, California. Dr. Veit's current research interests focus on mechanisms that influence the determination of cell fate. (Chapter 17)

Philip A. Wigge is a Principal Investigator at the Sainsbury Laboratory, Cambridge University, UK. He received his Ph.D in Cell Biology from the University of Cambridge, UK, in 2001. Dr. Wigge has studied how florigen controls plant development at the Salk Institute, CA, in the laboratory of Detlef Weigel. His research group is fascinated by how plants are able to sense and respond to climate change. (Chapter 20)

Ricardo A. Wolosiuk is Professor at the University of Buenos Aires and senior scientist at Instituto Leloir (Buenos Aires). He received his Ph.D. in Chemistry from the University of Buenos Aires in 1974. His current research centers on the modulation of photosynthetic CO2 assimilation and the structure and function of plant proteins. (Chapter 8)

Reviewers

Javier Abadía
Aula Dei Experimental Station, Spanish Council for Scientific Research
Chapter 5

Elizabeth A. Ainsworth
USDA Agricultural Research Service
Chapter 24

Richard Amasino
University of Wisconsin
Chapter 20

Diane Bassham
Iowa State University
Chapter 22

Tom Beeckman
VIB/Ghent University
Chapter 17

J. Derek Bewley
Emeritus, University of Guelph
Chapters 18 and 21

Winslow Briggs
Carnegie Institution for Science, Stanford
Chapter 16

Alice Y. Cheung
University of Massachusetts, Amherst
Chapter 20

Karl-Josef Dietz
Bielefeld University
Chapter 24

Anna Dobritsa
The Ohio State University
Chapter 21

Xinnian Dong
Duke University
Chapter 23

Anna F. Edlund
Lafayette College
Chapter 20

Christian Fankhauser
University of Lausanne
Chapter 16

Ruth Finkelstein
University of California, Santa Barbara
Chapter 15

James J. Giovannoni
Cornell University
Chapter 20

Heiner E. Goldbach
University of Bonn
Chapter 5

Sigal Savaldi Goldstein
The Technion – Israel Institute of Technology
Chapter 15

Michael Gutensohn
Purdue University
Web Appendix 4

Philip J. Harris
University of Auckland, New Zealand
Chapter 14

George Haughn
University of British Columbia
Chapters 18 and 21

J. S. (Pat) Heslop-Harrison
University of Leicester
Chapter 2

Joseph Kieber
The University of North Carolina
Chapter 15

Kenneth L. Korth
University of Arkansas
Chapter 23

Clark Lagarias
University of California, Davis
Chapter 16

Jane Langdale
University of Oxford
Chapter 19

Andrew R. Leitch
Queen Mary University of London
Chapter 2

Gerhard Leubner-Metzger
Royal Holloway, University of London
Chapter 18

David Macherel
University of Angers
Chapter 12

Massimo Maffei
University of Turin
Chapter 23

Ján A. Miernykj
University of Missouri
Chapter 12

June B. Nasrallah
Cornell University
Chapter 20

Lars Østergaard
The John Innes Centre
Chapter 20

Jarmila Pittermann
University of California, Santa Cruz
Chapters 3 and 4

Jerry Roberts
University of Nottingham
Chapter 22

John Roden
Southern Oregon University
Chapter 9

Jocelyn K. C. Rose
Cornell University
Chapter 14

Rowan F. Sage
University of Toronto
Chapters 8 and 9

Pill-Soon Song
Jeju National University
Chapter 16

Valerie Sponsel
The University of Texas at San Antonio
Chapter 15

Venkatesan Sundaresan
University of California, Davis
Chapter 21

Dan Szymanski
Purdue University
Chapter 1

Lawrence D. Talbott
University of California, Los Angeles
Chapter 10

Paolo Trost
University of Bologna
Chapter 8

Miltos Tsiantis
University of Oxford
Chapter 19

Robert Turgeon
Cornell University
Chapter 11

David Twell
University of Leicester
Chapter 21

Michael Udvardi
The Samuel Roberts Noble Foundation
Chapter 13

Luis Vidali
Worcester Polytechnic Institute
Chapter 1

Rick Vierstra
University of Wisconsin, Madison
Chapter 16

John M. Ward
University of Minnesota
Chapter 6

John C. Watson
Indiana University–Purdue University Indianapolis
Chapter 18

Dolf Weijers
Wageningen University
Chapter 17

Ramin Yadegari
University of Arizona
Chapter 21

Preface

Readers of previous editions of this text will notice a significant new feature of the Sixth Edition from the cover alone: the title has been changed from *Plant Physiology* to *Plant Physiology and Development*. The new title reflects a major reorganization of Unit III (*Growth and Development*) along developmental lines. Instead of separate chapters on the structure and function of individual photoreceptors and hormones, the interactions of photoreceptors and hormones are described in the context of the plant life cycle, from seed to seed. This change in approach has been facilitated by the virtual explosion of information on the interactions of signaling pathways and gene networks during the past four years. Among the many new topics that are being covered for the first time in the Sixth Edition are seed dormancy, germination, seedling establishment, root and shoot architecture, gametophyte development, pollination, seed development, fruit development, biotic interactions, and plant senescence. The resulting up-to-date, comprehensive, and meticulously illustrated presentation of plant development will provide students with an unprecedented appreciation of the integration of light, hormones, and other signaling agents that regulate the various stages of the plant life cycle.

The chapters in Units I and II covering traditional plant physiological topics such as water relations, mineral nutrition, transport, photosynthesis, and respiration, have also been extensively updated for the Sixth Edition. These processes function more or less continuously throughout the life of the plant and, in our view, attempting to insert them arbitrarily into a particular stage of the life cycle is not only misleading, it disrupts the flow of the developmental narrative. Therefore, for pedagogical reasons, we have maintained the integrity of the physiological chapters at the front end of the book. After mastering the basic physiological processes discussed in Units I and II, students are fully prepared to focus their attention on the signaling pathways and gene networks that govern the temporal changes that occur during the plant life cycle, as described in Unit III.

Besides the title change, a second important novel feature of the Sixth Edition can be gleaned from the cover: the addition of two new editors, Ian Max Møller, Associate Professor at the Department of Molecular Biology and Genetics at Aarhus University, Denmark, and Angus Murphy, Professor and Chair, Department of Plant Science and Landscape Architecture at the University of Maryland in College Park. Max Møller served as a Developmental Editor for the text as a whole, assessing every chapter for level, consistency, and pedagogy. Angus Murphy spearheaded the reorganization of Unit III and was a contributing author on several of the chapters. Both new editors have been invaluable during the preparation of the Sixth Edition, and their presence ensures that continuity will be preserved for many more editions of the text. In addition, Wendy Peer, Assistant Professor in the Department of Environmental Science and Technology at the University of Maryland, also made important contributions to the redesign of Unit III as well as serving as a contributing author to several chapters.

Editors	**Sub-editors**
L. T.	I. M. M.
E. Z.	A. M.

Media and Supplements

to accompany *Plant Physiology and Development*, Sixth Edition

For the Student

Companion Website (www.plantphys.net)

Available free of charge, this website supplements the coverage provided in the textbook with additional and more advanced material on selected topics of interest and current research. In-text references to Web Topics and Essays are included throughout the textbook, and the end of each chapter includes a complete list of Topics and Essays for that chapter. The site includes the following:

- *Web Topics:* Additional coverage of selected topics
- *Web Essays:* Articles on cutting-edge research, written by the researchers themselves
- *Study Questions:* A set of short-answer questions for each chapter
- *References*: A set of chapter-specific references, categorized by section heading.
- *Appendices*: New for the Sixth Edition, four complete appendices are available online:
    - Appendix 1: Energy and Enzymes
    - Appendix 2: The Analysis of Plant Growth
    - Appendix 3: Hormone Biosynthetic Pathways
    - Appendix 4: Secondary Metabolites

For the Instructor

Instructor's Resource Library

(Available to qualified adopters)

The *Plant Physiology and Development,* Sixth Edition Instructor's Resource Library includes a collection of visual resources from the textbook for use in preparing lectures and other course materials. The textbook figures have all been sized and formatted for optimal legibility when projected. The IRL includes all textbook figures and tables in JPEG (both high- and low-resolution) and PowerPoint formats.

Value Options

eBook

Plant Physiology and Development is available as an eBook, in several different formats, including VitalSource CourseSmart, Yuzu, and BryteWave. The eBook can be purchased as either a 180-day rental or a permanent (non-expiring) subscription. All major mobile devices are supported. For details on the eBook platforms offered, please visit www.sinauer.com/ebooks.

Looseleaf Textbook (ISBN 978-1-60535-353-1)

Plant Physiology and Development is available in a three-hole punched, looseleaf format. Students can take just the sections they need to class and can easily integrate instructor material with the text.

Table of Contents

UNIT II Biochemistry and Metabolism 169

CHAPTER 7 Photosynthesis: The Light Reactions 171

CHAPTER 8

Photosynthesis: The Carbon Reactions 203

CHAPTER 9

Photosynthesis: Physiological and Ecological Considerations 245

UNIT III Growth and Development 377

CHAPTER 14 Cell Walls: Structure, Formation, and Expansion 379

CHAPTER 15 Signals and Signal Transduction 407

CHAPTER 16 Signals from Sunlight 447

CHAPTER 21 Gametophytes, Pollination, Seeds, and Fruits 625

1 Plant and Cell Architecture

Plant physiology is the study of plant *processes*—how plants grow, develop, and function as they interact with their physical (abiotic) and living (biotic) environments. Although this book will emphasize the physiological, biochemical, and molecular functions of plants, it is important to recognize that, whether we are talking about gas exchange in the leaf, water conduction in the xylem, photosynthesis in the chloroplast, ion transport across membranes, signal transduction pathways involving light and hormones, or gene expression during development, all of these functions depend entirely on structures.

Function derives from structures interacting at every level of scale. It occurs when tiny molecules recognize and bind each other to produce a complex with new functions. It occurs as a new leaf unfolds, as cells and tissues interact during the process of plant development. It occurs when huge organisms shade, nourish, or mate with each other. At every level, from molecules to organisms, structure and function represent different frames of reference of a biological unity.

The fundamental organizational unit of plants, and of all living organisms, is the cell. The term *cell* is derived from the Latin *cella*, meaning "storeroom" or "chamber." It was first used in biology in 1665 by the English scientist Robert Hooke to describe the individual units of the honeycomb-like structure he observed in cork under a compound microscope. The cork "cells" Hooke observed were actually the empty lumens of dead cells surrounded by cell walls, but the term is an apt one, because cells are the basic building blocks that define plant structure.

Moving outward from the cell, groups of specialized cells form specific tissues, and specific tissues arranged in particular patterns are the basis of three-dimensional organs. Just as plant anatomy, the study of the macroscopic arrangements of cells and tissues within organs, received its initial impetus from improvements to the light microscope in the seventeenth century, so plant cell biology, the study of the interior of cells, was stimulated by the first application of the electron microscope to biological material in the mid-twentieth century. Subsequent improvements in both light and electron microscopy have revealed astonishing variety and dynamics in the components that make up cells—the cellular organelles, whose combined activities are required for the wide range of cellular and physiological functions that characterize biological organisms.

This chapter provides an overview of the basic anatomy and cell biology of plants, from the macroscopic structure of organs and tissues to the microscopic ultrastructure of cellular organelles. Subsequent chapters will treat these structures in greater detail from the perspective of their physiological and developmental functions at different stages of the plant life cycle.

Plant Life Processes: Unifying Principles

The spectacular diversity of plant size and form is familiar to everyone. Plants range in height from less than 1 cm to more than 100 m. Plant morphology, or form, is also surprisingly diverse. At first glance, the tiny plant duckweed (*Lemna*) seems to have little in common with a giant saguaro cactus or a redwood tree. No single plant shows the entire spectrum of adaptations to the range of environments that plants occupy on Earth, so plant physiologists often study **model organisms**, plants with short generation times and small **genomes** (the sum of their genetic information) (see WEB TOPIC 1.1). These models are useful because all plants, regardless of their specific adaptations, carry out fundamentally similar processes and are based on the same architectural plan.

We can summarize the major unifying principles of plants as follows:

- As Earth's primary producers, plants and green algae are the ultimate solar collectors. They harvest the energy of sunlight by converting light energy to chemical energy, which they store in bonds formed when they synthesize carbohydrates from carbon dioxide and water.
- Other than certain reproductive cells, plants do not move from place to place; they are sessile. As a substitute for motility, they have evolved the ability to grow toward essential resources, such as light, water, and mineral nutrients, throughout their life span.
- Plants are structurally reinforced to support their mass as they grow toward sunlight against the pull of gravity.
- Plants have mechanisms for moving water and minerals from the soil to the sites of photosynthesis and growth, as well as mechanisms for moving the products of photosynthesis to nonphotosynthetic organs and tissues.
- Plants lose water continuously by evaporation and have evolved mechanisms for avoiding desiccation.
- Plants develop from embryos that derive nutrients from the mother plant, and these additional food stores facilitate the production of large self-supporting structures on land.

Plant Classification and Life Cycles

Based on the principles listed above, we can define plants generally as sessile, multicellular organisms derived from embryos, adapted to land, and able to convert carbon dioxide into complex organic compounds through the process of photosynthesis. This broad definition includes a wide spectrum of organisms, from the mosses to the flowering plants, as illustrated in the diagram, or cladogram, depicting evolutionary lineage as branches, or clades, on a tree (**Figure 1.1**). The relationships of current and past plant identification systems, classification systems (taxonomies), and evolutionary thought are discussed in WEB TOPIC 1.2. Plants share with (mostly aquatic) green algae the primitive trait that is so important for photosynthesis in both clades: their chloroplasts contain the pigments chlorophyll *a* and *b* and β-carotene. **Plants**, or **embryophytes**, share the evolutionarily derived traits for surviving on land that are absent in the algae. Plants include the **nonvascular plants**, or **bryophytes** (mosses, hornworts, and liverworts), and the **vascular plants**, or **tracheophytes**. The vascular plants, in turn, consist of the **non-seed plants** (ferns and their relatives) and the **seed plants** (gymnosperms and angiosperms). The characteristics of many of these plant clades are in the descriptions of their representative model species (see WEB TOPIC 1.1).

Because plants have many agricultural, industrial, timber, and medical uses, as well as an overwhelming dominance in terrestrial ecosystems, most research in plant biology has focused on the plants that have evolved in the last 300 million years, the seed plants (see Figure 1.1). The **gymnosperms** (from the Greek for "naked seed") include the conifers, cycads, ginkgo, and gnetophytes (which include *Ephedra*, a popular medicinal plant). About 800 species of gymnosperms are known. The largest group of gymnosperms is the **conifers** ("cone-bearers"), which include such commercially important forest trees as pine, fir, spruce, and redwood. The **angiosperms** (from the Greek for "vessel seed") evolved about 145 million years ago and include three major groups: the **monocots**, **eudicots**, and so-called basal angiosperms, which include the Magnolia family and its relatives. Except in the great coniferous forests of Canada, Alaska, and northern Eurasia, angiosperms dominate the landscape. About 120,000 species are known, with an additional 17,000 undescribed species predicted by taxonomists using computer models. Most of the predicted species are imperiled because they occur primarily in regions of rich biodiversity where habitat destruction is common. The major anatomical innovation of the angiosperms is the

Figure 1.1 Cladogram showing the evolutionary relationships among the various members of the plants and their close relatives, the algae. The sequence of evolutionary innovations given on the right side of the figure eventually gave rise to the angiosperms. Mya, million years ago.

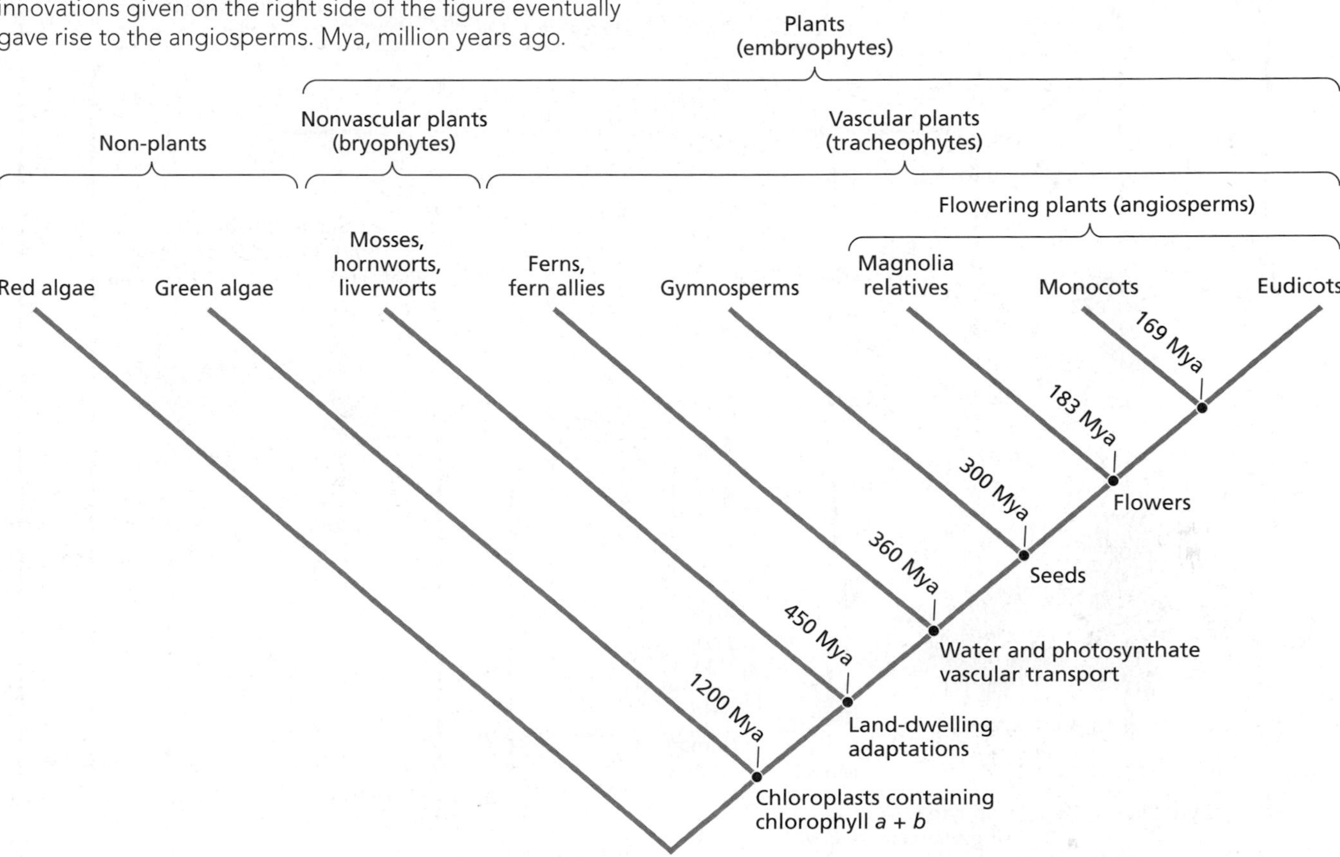

flower; hence they are referred to as **flowering plants**. WEB TOPIC 1.3 discusses the relationship between flower anatomy and the plant life cycle.

Plant life cycles alternate between diploid and haploid generations

Plants, unlike animals, alternate between two distinct multicellular generations to complete their life cycle. This is called **alternation of generations**. One generation has **diploid** cells, cells with two copies of each chromosome and abbreviated as having **2*N*** chromosomes, and the other generation has **haploid** cells, cells with only one copy of each chromosome, abbreviated as **1*N***. Each of these multicellular generations may be more or less physically dependent on the other, depending on their evolutionary grouping.

When diploid (2*N*) animals, as represented by humans on the inner cycle in **Figure 1.2**, produce haploid **gametes**, egg (1*N*) and sperm (1*N*), they do so directly by the process of **meiosis**, cell division resulting in a reduction of the number of chromosomes from 2*N* to 1*N*. In contrast, the products of meiosis in diploid plants are **spores**, and diploid plant forms are therefore called **sporophytes**. Each spore is capable of undergoing **mitosis**, cell division that doesn't change the number of chromosomes in the daughter cells, to form a new haploid multicellular individual, the **gametophyte**, as shown by the outer cycles in Figure 1.2. The haploid gametophytes produce gametes, egg and sperm, by simple mitosis, whereas haploid gametes in animals are produced by meiosis. This is a fundamental difference between plants and animals and gives the lie to some stories about "the birds and the bees"—bees don't carry around sperm to fertilize female flowers, they carry around the male gametophyte, the **pollen**, which is a multicellular structure that produces sperm cells. When placed on receptive sporophytic tissue, the pollen grain germinates to form a pollen tube that must grow through sporophytic tissue until it reaches the female gametophyte. The male gametophyte penetrates the female gametophyte and releases sperm to fertilize the egg. This hidden nature of sex in plants, where it occurs deep inside sporophytic tissue, made its discovery difficult, and when discovered, was so "shocking" that it was frequently denied.

Once the haploid gametes fuse and **fertilization** takes place to create the 2*N* zygote, the life cycles of animals and plants are similar (see Figure 1.2). The 2*N* zygote undergoes a series of mitotic divisions to produce the embryo, which eventually grows into the mature diploid adult.

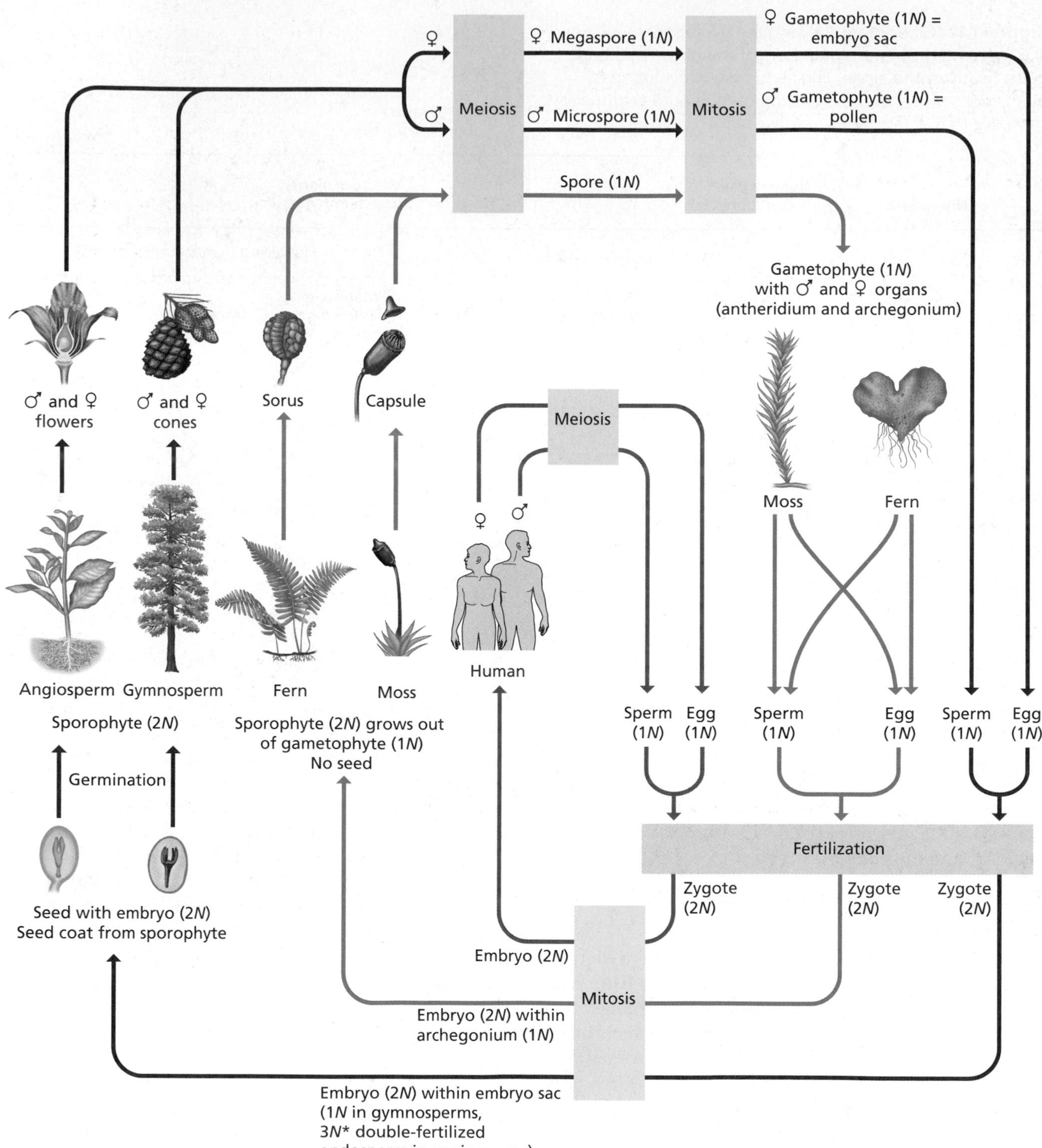

Figure 1.2 Diagram of the generalized life cycles of plants and animals. In contrast to animals, plants exhibit alternation of generations. Rather than producing gametes directly by meiosis as animals do, plants produce vegetative spores by meiosis. These 1*N* (haploid) spores divide to produce a second multicellular individual called the gametophyte. The gametophyte then produces gametes (sperm and egg) by mitosis. Following fertilization, the resulting 2*N* (diploid) zygote develops into the mature sporophyte generation, and the cycle begins again. In angiosperms, the process of double fertilization produces a 3*N* (triploid) or higher ploidy level (*; see Chapter 21) feeding tissue called the endosperm.

Thus, all plant life cycles encompass two separate generations: the diploid, spore-producing **sporophyte generation** and the haploid, gamete-producing **gametophyte generation**. A line drawn between fertilization and meiosis divides these two separate stages of the generalized plant life cycle (see Figure 1.2). Increasing the number of mitoses between fertilization and meiosis increases the size of the sporophyte generation and the number of spores that can be produced. Having more spores per fertilization event could compensate for low fertility when water becomes scarce on land. This could explain the marked tendency for the increase in size of the sporophyte generation, relative to the gametophyte generation, during the evolution of plants.

The sporophyte generation is dominant in the seed plants, the gymnosperms and angiosperms, and gives rise to different spores: the **megaspores**, which develop into the female gametophyte, and the **microspores**, which develop into the male gametophyte (see Figure 1.2). The way the resulting male and female gametophytes are separated is quite diverse. In angiosperms, a single individual in a **monoecious** (from the Greek for "one house") species has flowers that produce both male and female gametophytes; both can occur in the single "perfect" flower as in tulips, or they can occur in separate male (staminate) and female (pistillate) flowers as in maize (corn; *Zea mays*). If male and female flowers occur on separate individuals, as in willow or poplar trees, then the species is **dioecious** (from the Greek for "two houses"). In gymnosperms, ginkos and cycads are dioecious, while conifers are monoecious. Conifers produce female cones, **megastrobili** (from the Greek for "large cones"; singular *megastrobilus*), usually higher up on the plant than the male cones, **microstrobili** (from the Greek for "small cones"; singular *microstrobilus*). Both megaspores and microspores produce gametophytes with only a few cells, compared with the sporophyte.

Sperm and egg production, as well as the dynamics of fertilization, differs among gametophytes of the seed plants (see WEB TOPIC 1.3). In angiosperms there is the amazing process of **double fertilization**, whereby two sperm are produced, only one of which fertilizes the egg. The other sperm fuses with two nuclei in the female gametophyte to produce the 3*N* (three sets of chromosomes) endosperm, the storage tissue for the angiosperm seed. (Some angiosperms produce endosperm of higher ploidy levels; see Chapter 21.) The storage tissue for the seed in gymnosperms is 1*N* gametophytic tissue because there is no double fertilization (see Figure 1.2). So the seed of seed plants is not at all a spore (defined as a cell that produces the gametophyte generation), but it does contain gametophytic (1*N*) storage tissue in gymnosperms and gametophyte-derived 3*N* storage tissue in angiosperms.

In the lower plants, the ferns and mosses, the sporophyte generation gives rise to spores that grow into adult gametophytes that then have regions that differentiate into male and female structures, the male **antheridium** and the female **archegonium**. In ferns the gametophyte is a small monoecious **prothallus**, which has antheridia and archegonia that divide mitotically to produce motile sperm and egg cells, respectively. The dominant leafy gametophyte generation in mosses contains antheridia and archegonia on the same (monoecious) or different (dioecious) individuals. The motile sperm then enters the archegonium and fertilizes the egg, to form the 2*N* zygote, which develops into an embryo enclosed in the gametophytic tissue, but no seed is formed. The embryo directly develops into the adult 2*N* sporophyte.

Overview of Plant Structure

Despite their apparent diversity, all seed plants have the same basic body plan (**Figure 1.3**). The vegetative body is composed of three organs—the stem, the root, and the leaves—each with a different direction, or polarity, of growth. The **stem** grows upward and supports the aboveground part of the plant. The **root**, which anchors the plant and absorbs nutrients and water, grows down below the ground. The **leaves**, whose primary function is photosynthesis, grow out laterally from the stem at the **nodes**. Variations in leaf arrangement can give rise to many different forms of **shoots**, the term for the leaves and stem together. For example, leaf nodes can spiral around the stem, rotating by a fixed angle between each **internode** (the region between two nodes). Alternatively, leaves can arise oppositely or alternating on either side of the stem.

Organ shape is defined by directional patterns of growth. The polarity of growth of the **primary plant axis** (the main stem and taproot) is vertical, whereas the typical leaf grows laterally at the margins to produce the flattened **leaf blade**. The growth polarities of these organs are adapted to their functions: leaves function in light absorption, stems elongate to lift the leaves toward sunlight, and roots elongate in search of water and nutrients from the soil. The cellular component that directly determines growth polarity in plants is the cell wall.

Plant cells are surrounded by rigid cell walls

The outer fluid boundary of the living cytoplasm of plant cells is the **plasma membrane** (also called **plasmalemma**), similar to the situation in animals, fungi, and bacteria. The **cytoplasm** is defined as all of the organelles and cytoskeleton suspended within the **cytosol**, the water-soluble and colloidal phase, residing within the plasma membrane, but which excludes the nucleoplasm, the internal compartment of the membrane-bounded nucleus in eukaryotes. However, plant cells, unlike animal cells, are further enclosed by a rigid, cellulosic **cell wall** (**Figure 1.4**). Because of the absence of cell walls in animals, embryonic cells are able to migrate from one location to another; developing tissues

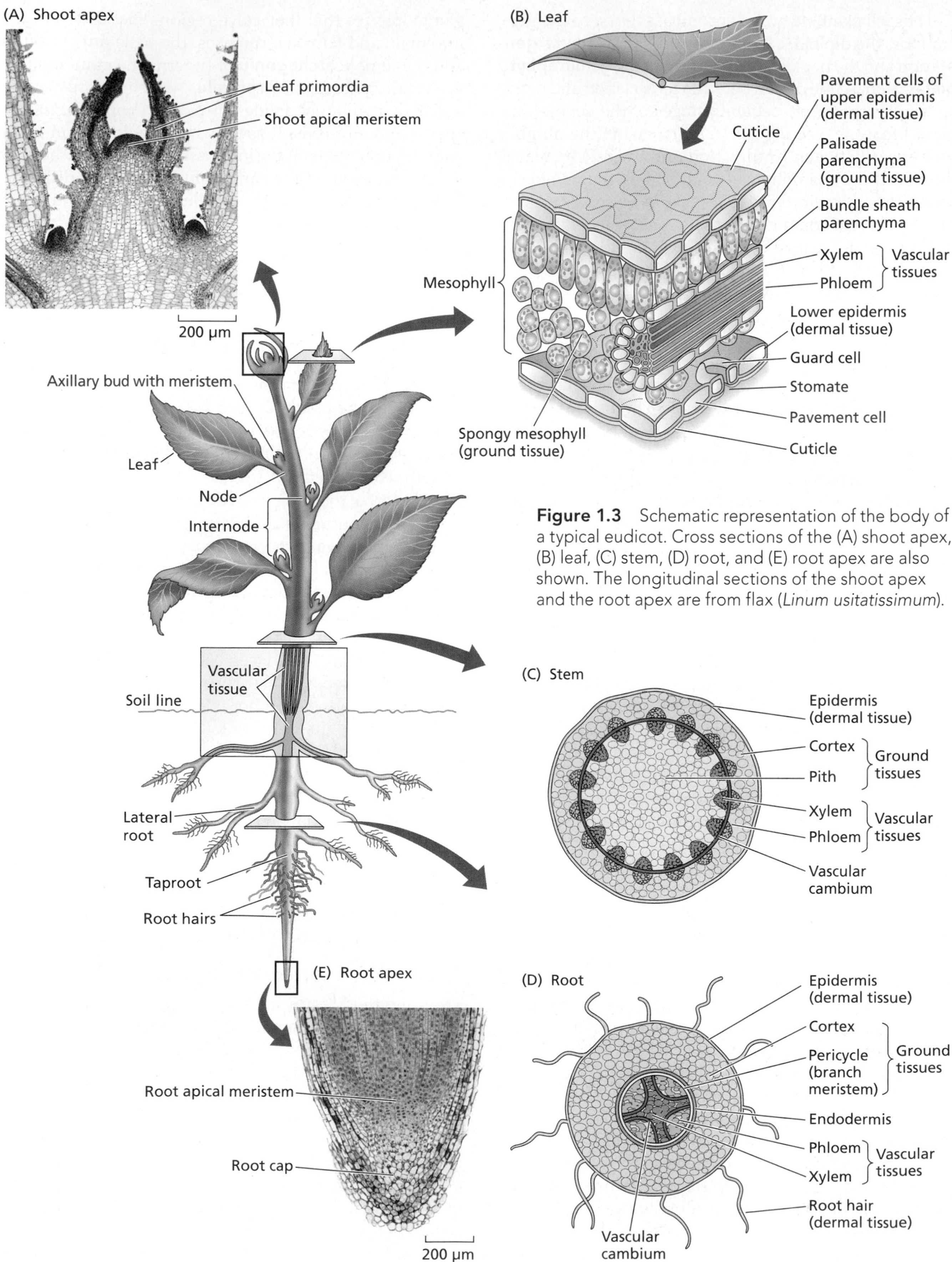

Figure 1.3 Schematic representation of the body of a typical eudicot. Cross sections of the (A) shoot apex, (B) leaf, (C) stem, (D) root, and (E) root apex are also shown. The longitudinal sections of the shoot apex and the root apex are from flax (*Linum usitatissimum*).

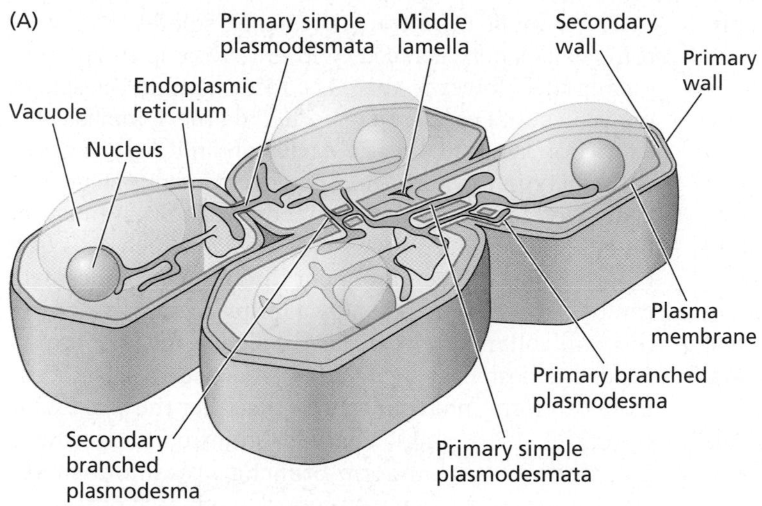

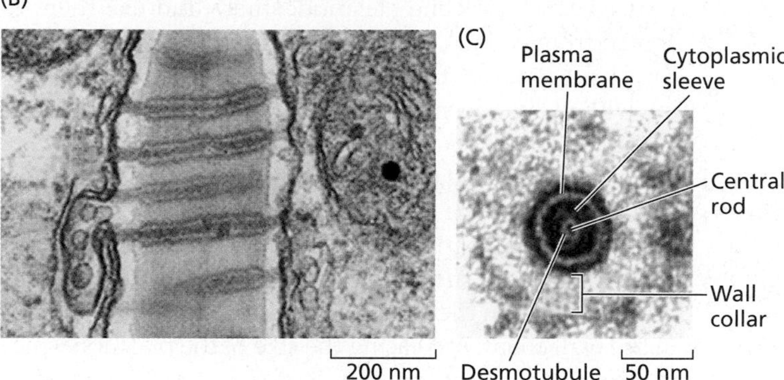

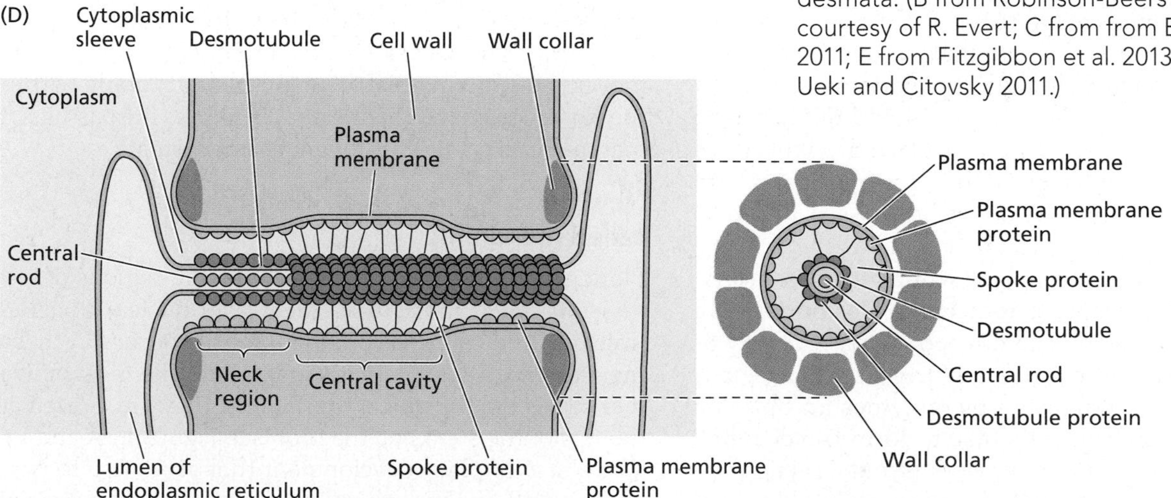

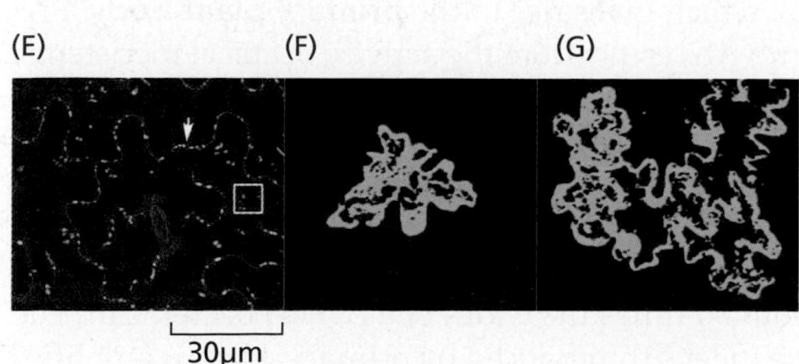

Figure 1.4 Plant cell walls and their associated plasmodesmata. (A) Diagrammatic representation of the cell walls surrounding four adjacent plant cells. Cells with only primary walls and with both primary and secondary walls are illustrated. The secondary walls form inside the primary walls. The cells are connected by both simple (unbranched) and branched plasmodesmata. Plasmodesmata formed during cell division are primary plasmodesmata. (B) Electron micrograph of a wall separating two adjacent cells, showing simple plasmodesmata in longitudinal view. (C) Tangential section through a cell wall showing a plasmodesma. (D) Schematic surface and cross-section views of a plasmodesma. The pore consists of a central cavity down which the desmotubule runs, connecting the endoplasmic reticulum of the adjoining cells. (E) Epidermal cells of an Arabidopsis leaf imaged with fluorescence microscopy showing the cell wall in red and complex plasmodesmata in green. The arrow points to the high number of plasmodesmata at cell three-way junctions, and the rectangle outlines plasmodesmata that connect the epidermal cells to cells beneath them, the mesophyll cells. (F) A single tobacco leaf epidermal cell expressing a green fluorescent viral movement protein imaged with fluorescence microscopy. (G) After a single tobacco leaf epidermal cell expresses the gene for a viral movement protein, several tobacco leaf epidermal cells express green fluorescent viral movement protein because it has moved to them through plasmodesmata. (B from Robinson-Beers and Evert 1991, courtesy of R. Evert; C from from Bell and Oparka 2011; E from Fitzgibbon et al. 2013; F and G from Ueki and Citovsky 2011.)

and organs may thus contain cells that originated in different parts of the organism. In plants such cell migrations are prevented, because each walled cell is cemented to its neighbors by a **middle lamella**. As a consequence, plant development, unlike animal development, depends solely on patterns of cell division and cell enlargement.

Plant cells have two types of walls: primary and secondary (see Figure 1.4A). **Primary cell walls** are typically thin (less than 1 μm) and are characteristic of young, growing cells. **Secondary cell walls** are thicker and stronger than primary walls and are deposited on the inner surface of the primary wall after most cell enlargement has ended. Secondary cell walls owe their strength and toughness to **lignin**, a brittle, gluelike material (see Chapter 14). The evolution of lignified secondary cell walls provided plants with the structural reinforcement necessary to grow vertically above the soil and to colonize the land. Bryophytes, which lack lignified cell walls, are unable to grow more than a few centimeters above the ground.

Plasmodesmata allow the free movement of molecules between cells

The cytoplasm of neighboring cells is usually connected by means of **plasmodesmata** (singular *plasmodesma*), tubular channels 40 to 50 nm in diameter and formed by the connected plasma membranes of adjacent cells (see Figure 1.4A–D). They facilitate intercellular communication during plant development, enabling cytoplasmic exchange of vital developmental signals in the form of proteins, nucleic acids, and other macromolecules (see Chapters 18–20). Plant cells interconnected in this way form a cytoplasmic continuum referred to as the **symplast**. Intercellular transport of small molecules through plasmodesmata is called **symplastic transport** (see Chapters 4 and 6). Transport through the wall spaces, which constitute the apoplast, is called **apoplastic transport**. Both forms of transport are important in the vascular system of plants (see Chapter 6).

Primary plasmodesmata are created as the primary cell wall assembles during and following cell division (discussed later in the chapter). **Secondary plasmodesmata** form after cell division is completed, across primary or secondary cell walls (see Figure 1.4A), when small regions of the cell walls are digested by enzymes and plasma membranes of adjacent cells fuse to form the channel. The endoplasmic reticulum network (see the section *The Endomembrane System*, below) of adjacent cells is also connected, forming the **desmotubule** (see Figure 1.4C and D) that runs through the center of the channel. Proteins line the outer surface of the desmotubule and the inner surface of the plasma membrane (see Figure 1.4D); the two surfaces are thought to be connected by filamentous proteins (**spokes**), which divide the **cytoplasmic sleeve** into microchannels. Valvelike **wall collars**, composed of the polysaccharide callose, surround the necks of the channel at either end and serve to restrict the size of the pore.

The symplast can transport water, solutes, and macromolecules between cells without crossing the plasma membrane. However, there is a restriction on the size of molecules that can be transported via the symplast; this restriction is called the **size exclusion limit**, which varies with cell type, environment, and developmental stage. The transport can be followed by studying the movement of fluorescently labeled proteins or dyes between cells (see Figure 1.4E–G). The movement through plasmodesmata can be regulated, or gated, by altering the dimensions of the wall collars, the cytoplasmic sleeve, and the lumen inside the desmotubule. In addition, adjacent plasmodesmata can form interconnections that alter the size exclusion limit. Thus, single channels, referred to as **simple plasmodesmata**, can form **branched plasmodesmata** (see Figure 1.4A) when they connect with each other.

In a situation that occurs all too frequently, plant viruses can hijack the plasmodesmata and use them to spread from cell to cell. **Movement proteins**, encoded by the virus genome, facilitate viral movement by interacting with plasmodesmata through one of two mechanisms. Movement proteins from some viruses coat the surface of the viral genome (typically RNA), forming ribonucleoprotein complexes. The 30-kDa movement protein of tobacco mosaic virus acts in this way. It can move between cells in leaves that are susceptible to the virus, where it recruits other proteins in the cell that reduce the amount of callose in the wall collar, increasing the size of the plasmodesmatal pore. As a result, even virus-sized particles can readily move through the plasmodesmata to a neighboring cell (see Figure 1.4F and G). Other viruses, such as cowpea mosaic virus and tomato spotted wilt virus, encode movement proteins that form a transport tubule within the plasmodesmatal channel that enhances the passage of mature virus particles through plasmodesmata.

New cells originate in dividing tissues called meristems

Plant growth is concentrated in localized regions of cell division called **meristems**. Nearly all nuclear division (mitosis) and cell division (cytokinesis) occurs in these meristematic regions. In a young plant, the most active meristems are the **apical meristems**; they are located at the tips of the stem and the root (see Figure 1.3A and E). The phase of plant development that gives rise to new organs and to the basic plant form is called **primary growth**, which gives rise to the **primary plant body**. Primary growth results from the activity of apical meristems. Cell division in the meristem produces cuboidal cells about 10 μm on each side. Division is followed by progressive cell enlargement, typically elongation, whereby cells become much longer than they are wide (30–100 μm long, 10–25 μm wide—about half the width of a baby's fine hair and about 50 times the width of a typical bacterium). The increase in length produced by primary growth amplifies the plant's axial (top-to-bottom) polarity, which is established in the embryo.

Cell differentiation into specialized tissues follows cell enlargement (**Figure 1.5**, see also Figure 1.3). There are three major tissue systems present in all plant organs: dermal tissue, ground tissue, and vascular tissue (see Figure 1.3B–D). **Dermal tissue** forms the outer protective layer of the plant and is called the **epidermis** in the primary plant body; **ground tissue** fills out the three-dimensional bulk of the plant and includes the **pith** and **cortex** of primary stems and roots, and the **mesophyll** in leaves. **Vascular tissue**, which moves, or **translocates**, water and solutes throughout the length of the plant, consists of two types of tissues: **xylem** and **phloem**, each of which consists of conducting cells, generalized parenchyma cells, and thick-walled fibers. Some of the different cell types that

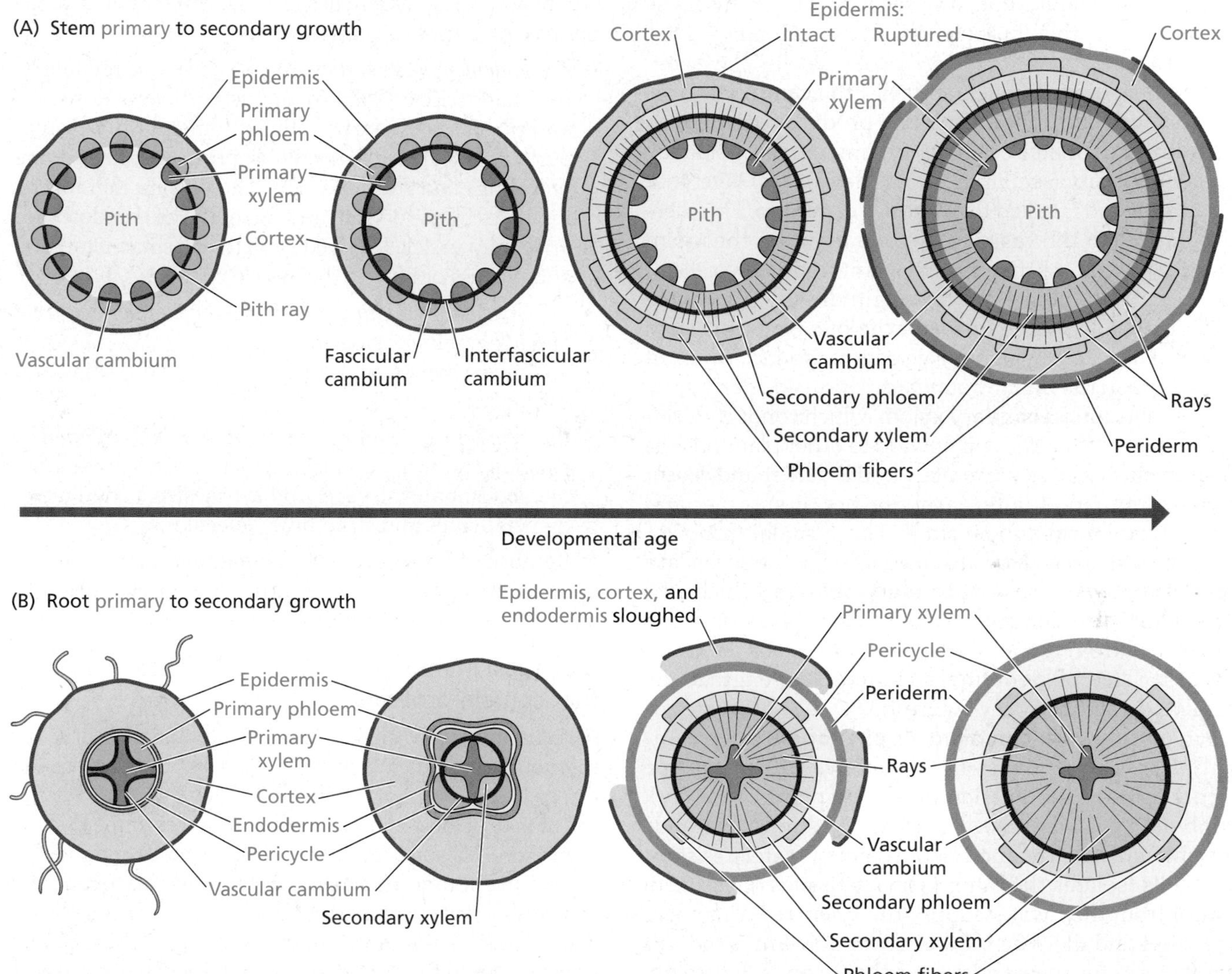

Figure 1.5 Secondary growth in stems and roots. (A) Stem primary to secondary growth. Primary growth is labeled in green, while secondary growth is labeled in red. The vascular cambium starts as separated growth regions in the vascular bundles, or fascia, of primary xylem and phloem. As the plant grows, the bundled, fascicular cambium becomes connected by interfascicular cambium between the bundles. Once the vascular cambium forms a continuous ring, it divides inward to generate secondary xylem and it divides outward to generate the secondary phloem. Regions in the cortex develop into phloem fibers and the periderm, which contains the phellogen, or cork cambium, and the outer phelloderm. With growth, the epidermis ruptures and rays connect the inner and outer vasculature. (B) Root primary to secondary growth. The central vascular cylinder contains the primary phloem and primary xylem. As in the stem, the vascular cambium becomes connected and grows outward, generating secondary phloem and rays. As roots increase in girth, the pericycle generates the root periderm, while the outer epidermis, cortex, and endodermis are sloughed off. The pericycle produces the phloem fibers and rays as well as lateral roots (not shown). The vascular cambium produces secondary phloem and rings of secondary xylem.

make up these tissues will be covered in detail at the end of the chapter, when we discuss the interplay of their different organelles that accompanies differentiation.

Meristematic tissue is also found along the length of the root and shoot. **Axillary buds** are meristems that develop in the node, or axil—the region between the leaf and the shoot. Axillary buds become the apical meristems of branches. The branches of roots, the **lateral roots**, arise from meristematic cells in the **pericycle**, or root branch meristem (see Figure 1.5B; also see Figure 1.3). This meristematic tissue then becomes the apical meristem of the lateral root.

Another set of meristematic cells, the **cambium**, gives rise to **secondary growth**, which produces an increase in width or diameter of plants, having radial (inside-to-outside) polarity (see Figure 1.5). The cambial layer that produces wood is called the **vascular cambium**. This meristem arises in the vascular system, between the xylem and the phloem of the primary plant body. The cells of the vascular cambium divide longitudinally to produce derivatives toward the inside or the outside of the stem or root. They also divide transversely to produce **rays** that transmit material radially outward. The inside derivatives differentiate into **secondary xylem**, which conducts water and nutrients from the soil upward to other plant organs. In temperate climates, summer wood is darker and denser than spring wood; alternating layers of summer and spring wood form annual rings. The vascular cambium derivatives displaced toward the outside of the secondary stem or root give rise to **secondary phloem**, which, like primary phloem, conducts the products of photosynthesis downward from the leaves to other organs of the plant. The associated **phloem fibers** add tensile strength to the stem, as do all fibers (see Figure 1.37).

Finally, the **cork cambium**, or **phellogen**, is the cambial layer that produces the protective **periderm** (see Figure 1.5) on the outside of woody plants. The cork cambium typically arises each year within the secondary phloem. The production of layers of water-resistant cork cells isolates the outer primary tissues of the stem or root from their water supply, the xylem, causing them to shrivel and die. The **bark** of a woody plant is the collective term for several tissues—the secondary phloem, secondary phloem fibers, cortex (in stems), pericycle (in roots) and periderm—that can be peeled off as a unit at the soft layer of vascular cambium.

Plant Cell Organelles

All plant cells have the same basic eukaryotic organization: They contain a nucleus, a cytoplasm, and subcellular organelles; and they are enclosed in a plasma membrane that defines their boundaries, as well as a cellulosic cell wall (**Figure 1.6**). Small changes in these components can produce large changes in the evolution and development of plants. In addition to the lignification of secondary cell walls (mentioned above) that allows plants to produce large, sturdy stems, xylem development is accompanied by the loss of the nucleus and many other organelles, enabling the cell to form a pipelike conduit for movement of water. The end of the chapter will include more examples of how organelles change during differentiation to produce the 40 or so different cell types in plants. But first, all plant cells *begin* with a similar complement of organelles. These organelles fall into three main categories based on how they arise:

- *The endomembrane system*: the endoplasmic reticulum, nuclear envelope, Golgi apparatus, vacuole, endosomes, and plasma membrane. The endomembrane system plays a central role in secretory processes, membrane recycling, and the cell cycle. The plasma membrane regulates transport into and out of the cell. Endosomes arise from vesicles derived from the plasma membrane and process or recycle the vesicle contents.
- *Independently dividing or fusing organelles derived from the endomembrane system*: the oil bodies, peroxisomes, and glyoxysomes, which function in lipid storage and carbon metabolism.
- *Independently dividing, semiautonomous organelles*: plastids and mitochondria, which function in energy metabolism and storage, and synthesize a wide range of metabolites and structural molecules.

Because all of these cellular organelles are membranous compartments, we'll begin with a description of membrane structure and function.

Biological membranes are phospholipid bilayers that contain proteins

All cells are formed enclosed in a membrane that serves as their outer boundary, separating the cytoplasm from the external environment. This plasma membrane allows the cell to take up and retain certain substances while excluding others. Various transport proteins embedded in the plasma membrane are responsible for this selective traffic of solutes—water-soluble ions and small, uncharged molecules—across the membrane. The accumulation of ions or molecules in the cytosol through the action of transport proteins consumes metabolic energy. In eukaryotic cells, membranes enshroud the genetic material, delimit the boundaries of other specialized internal organelles of the cell, and regulate the fluxes of ions and metabolites into and out of these compartments.

According to the **fluid-mosaic model**, all biological membranes have the same basic molecular organization. They consist of a double layer (*bilayer*) of lipid in which proteins are embedded (**Figure 1.7**). Each layer is called a *leaflet* of the bilayer. In most membranes, proteins make up about half of the membrane's mass. However, the composition of the lipid components and the properties of the

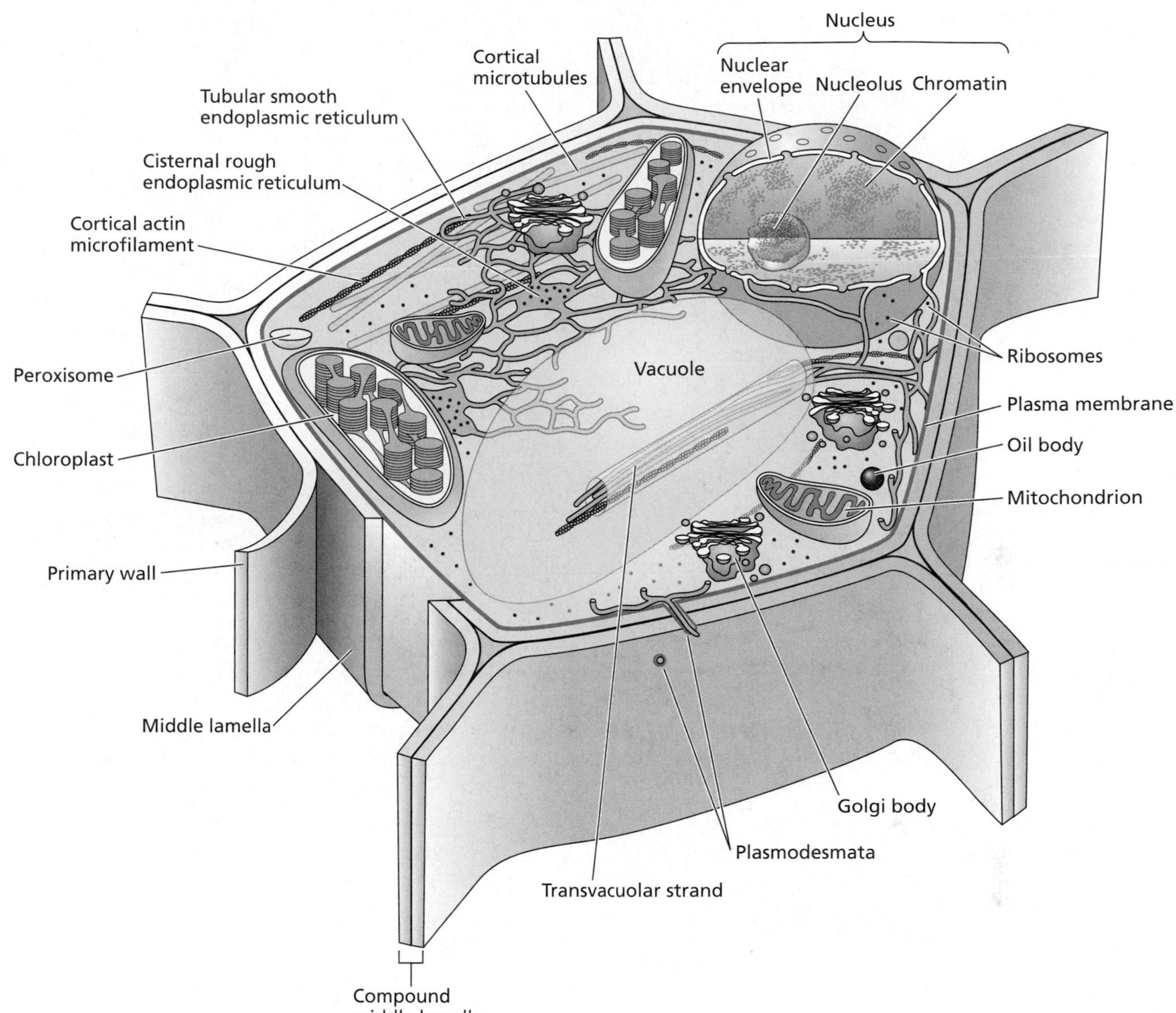

Figure 1.6 Diagrammatic representation of a plant cell. Various intracellular compartments are defined by their respective membranes, such as the tonoplast, the nuclear envelope, and the membranes of the other organelles. The two adjacent primary walls, along with the middle lamella, form a composite structure called the compound middle lamella.

proteins vary from membrane to membrane, conferring on each membrane its unique functional characteristics.

LIPIDS The most prominent membrane lipids found in plants are phospholipids, a class of lipids in which two fatty acids are covalently linked to glycerol, which is covalently linked to a phosphate group. Attached to the phosphate group in the phospholipid is a variable component, called the *head group*, such as serine, choline, glycerol, or inositol (see Figure 1.7C). The nonpolar hydrocarbon chains of the fatty acids form a region that is hydrophobic—that is, that excludes water. In contrast to the fatty acids, the head groups are highly polar; consequently, phospholipid molecules display both hydrophilic and hydrophobic properties (i.e., they are *amphipathic*). Various phospholipids are distributed asymmetrically across the plasma membrane, giving the membrane sidedness; in terms of phospholipid composition, the outside leaflet of the plasma membrane that faces the outside of the cell is different from the inside leaflet that faces the cytosol.

The membranes of specialized plant organelles called **plastids**, the group of membrane-bound organelles to which chloroplasts belong, are unique in that their lipid component consists almost entirely of **glycosylglycerides**, the glycosyl polar head groups of which are galactose derivatives. These **galactolipids** may contain galactose (see Figure 1.7C), digalactose, or sulfated galactose,

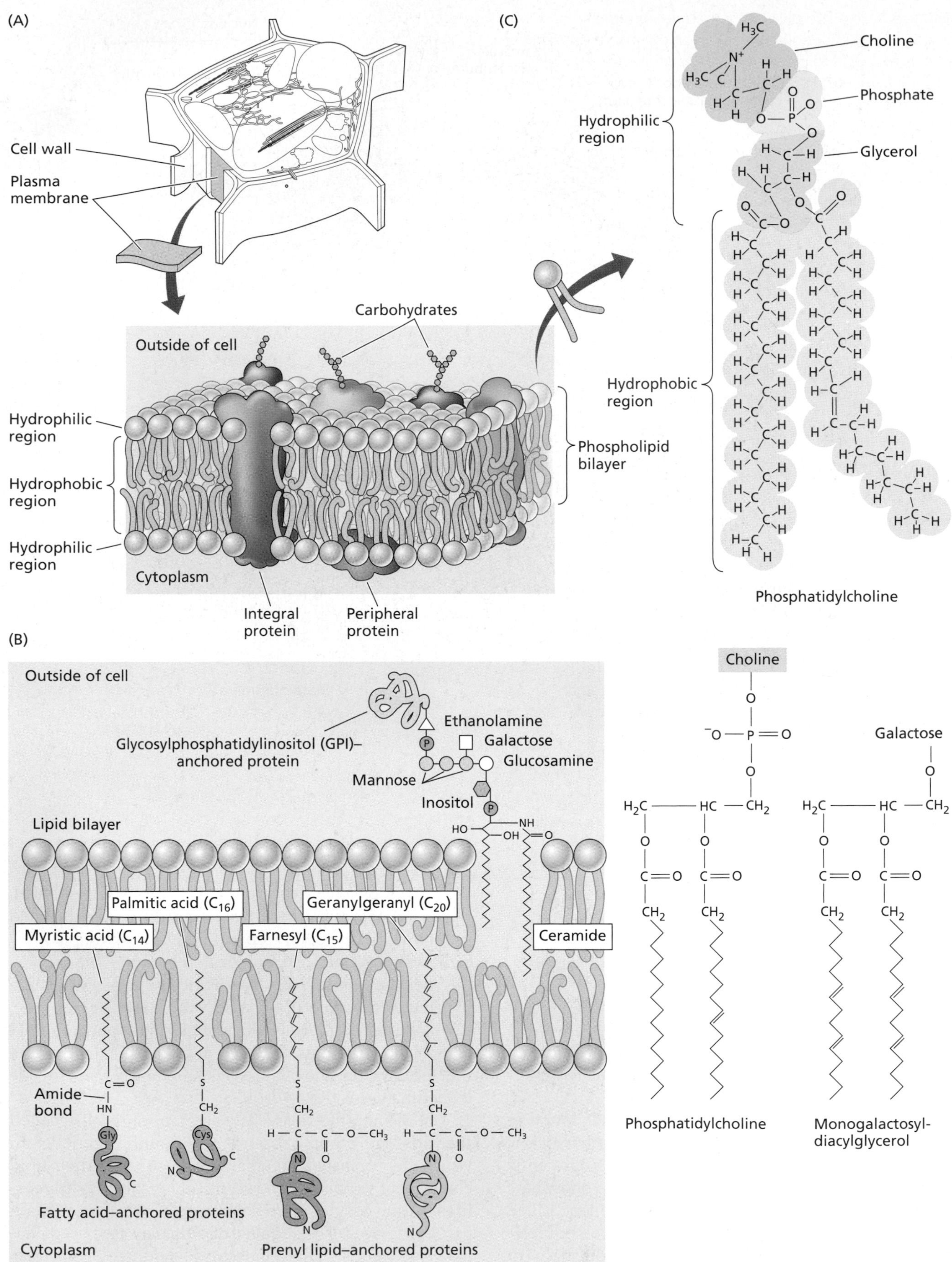

(A)
Cell wall
Plasma membrane
Carbohydrates
Outside of cell
Hydrophilic region
Hydrophobic region
Hydrophilic region
Cytoplasm
Phospholipid bilayer
Integral protein
Peripheral protein
(C)
Choline
Phosphate
Glycerol
Hydrophilic region
Hydrophobic region
Phosphatidylcholine
(B)
Outside of cell
Ethanolamine
Galactose
Glycosylphosphatidylinositol (GPI)–anchored protein
Glucosamine
Mannose
Inositol
Lipid bilayer
Palmitic acid (C_{16})
Geranylgeranyl (C_{20})
Myristic acid (C_{14})
Farnesyl (C_{15})
Ceramide
Amide bond
Fatty acid–anchored proteins
Cytoplasm
Prenyl lipid–anchored proteins
Choline
Galactose
Phosphatidylcholine
Monogalactosyl-diacylglycerol

◀ **Figure 1.7** (A) The plasma membrane, endoplasmic reticulum, and other endomembranes of plant cells consist of proteins embedded in a phospholipid bilayer, while plastid membranes have a galactolipid bilayer. (B) Various anchored membrane proteins, attached to the membrane via GPI, fatty acids, and prenyl groups, enhance the sidedness of membranes. (C) Chemical structures of typical phospholipids: phosphatidylcholine and monogalactosyldiacylglycerol. (B after Buchanan et al. 2000.)

in their head group, but have no phosphate. They are products of a prokaryotic pathway for lipid biosynthesis that both plastids and mitochondria inherited from their endosymbiotic ancestors. As described later in the chapter, there is some exchange of lipids between these organelles and the rest of the cell (also see Chapter 12).

The fatty acid chains of phospholipids and glycosylglycerides are variable in length but usually consist of 16 to 24 carbons. If the carbons are linked by single bonds, the fatty acid chain is *saturated* (with hydrogen atoms), but if the chain includes one or more double bonds, it is *unsaturated*.

Double bonds in a fatty acid chain create a kink in the chain that prevents tight packing of the phospholipids in the bilayer (i.e., the bonds adopt a kinked *cis* configuration, as opposed to an unkinked *trans* configuration). The kinks promote membrane fluidity, which is critical for many membrane functions. Membrane fluidity is also strongly influenced by temperature. Because plants generally cannot regulate their body temperature, they are often faced with the problem of maintaining membrane fluidity under conditions of low temperature, which tends to decrease membrane fluidity. To maintain membrane fluidity at cold temperatures, plants can produce a higher percentage of unsaturated fatty acids, such as *oleic acid* (one double bond), *linoleic acid* (two double bonds), and *linolenic acid* (three double bonds) (see also Chapter 12).

Another lipid component of plant cells is the family of **sterols**. Although foods from plants are often labeled as "cholesterol-free," plants do have cholesterol, but in quantities that are small enough to permit this sort of food labeling (animals have at least 5 g cholesterol/kg total lipid, whereas plants have about 50 mg cholesterol/kg total lipid). Plants have up to 250 other sterols and sterol derivatives (usually sterol esters), the most common of which are β-sitosterols. Sterols contribute to the formation and assembly of membranes and the waxy cuticles on the surfaces of plants. In fact, most of the cholesterol in plants is on the surfaces of various organs (e.g., seed pod, leaves). Some families of sterols, such as the brassinosteroids, also serve as hormones, as they do in animals (e.g., estrogens and testosterone).

PROTEINS The proteins associated with the lipid bilayer are of three main types: integral, peripheral, and anchored. Proteins and lipids can also combine in transient aggregates in the membrane called *lipid rafts*.

Integral proteins are embedded in the lipid bilayer (see Figure 1.7A). Most integral proteins span the entire width of the phospholipid bilayer, so one part of the protein interacts with the outside of the cell, another part interacts with the hydrophobic core of the membrane, and a third part interacts with the interior of the cell, the cytosol. Proteins that serve as ion channels (see Chapter 6) are always integral membrane proteins, as are certain receptors that participate in signal transduction pathways (see Chapter 15). Some receptor-like proteins on the outer surface of the plasma membrane recognize and bind tightly to cell wall constituents, effectively cross-linking the membrane to the cell wall.

Peripheral proteins are bound to the membrane surface (see Figure 1.7A) by noncovalent bonds, such as ionic bonds or hydrogen bonds, and can be dissociated from the membrane with high-salt solutions or chaotropic agents, which break ionic and hydrogen bonds, respectively. Peripheral proteins serve a variety of functions in the cell. For example, some are involved in interactions between the membranes and the major elements of the cytoskeleton, the microtubules and actin microfilaments (see Figure 1.6 and the section *The Plant Cytoskeleton*, below).

Anchored proteins are bound to the membrane surface via lipid molecules, to which they are covalently attached. These lipids include fatty acids (myristic acid and palmitic acid), prenyl groups derived from the isoprenoid pathway (farnesyl and geranylgeranyl groups), and glycosylphosphatidylinositol (GPI)-anchored proteins (see Figure 1.7B). These lipid anchors make the two sides of the plasma membrane even more distinct, with the fatty acid and prenyl anchors occurring on the cytoplasm-facing leaflet of the membrane and the GPI linkages occurring on the leaflet facing the outside of the cell.

The Endomembrane System

The endomembrane system of eukaryotic cells is the collection of related internal membranes that divides the cell into functional and structural compartments and that distributes membranes and proteins via vesicular traffic among cellular organelles. Our discussion of the endomembrane system will begin with the nucleus, where the genetic information for organelle biogenesis is mainly stored. This will be followed by a description of the independently dividing or fusing endomembrane organelles and the semiautonomous organelles.

The nucleus contains the majority of the genetic material

The **nucleus** (plural *nuclei*) is the organelle that contains the genetic information primarily responsible for regulating the metabolism, growth, and differentiation of the cell.

Collectively, these genes and their intervening sequences are referred to as the **nuclear genome**. The size of the nuclear genome in plants is highly variable, ranging from about 1.2×10^8 base pairs for the mustard relative *Arabidopsis thaliana* to 1×10^{11} base pairs for the lily *Fritillaria assyriaca*. The remainder of the genetic information of the cell is contained in the two semiautonomous organelles—the plastid and the mitochondrion—which we will discuss later in this chapter.

(A)

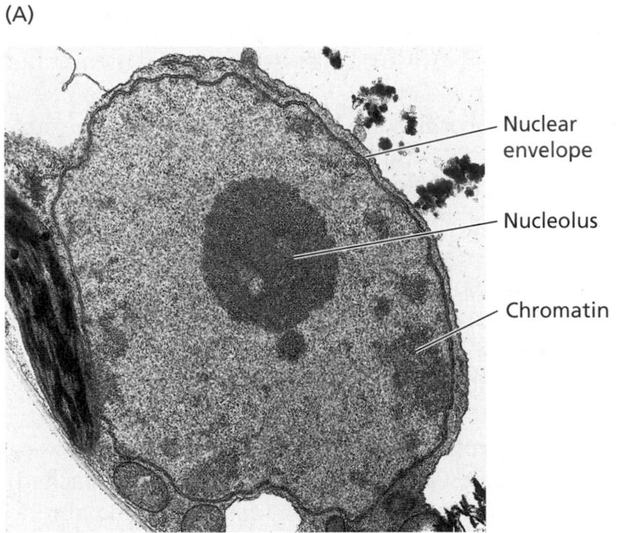

(B)

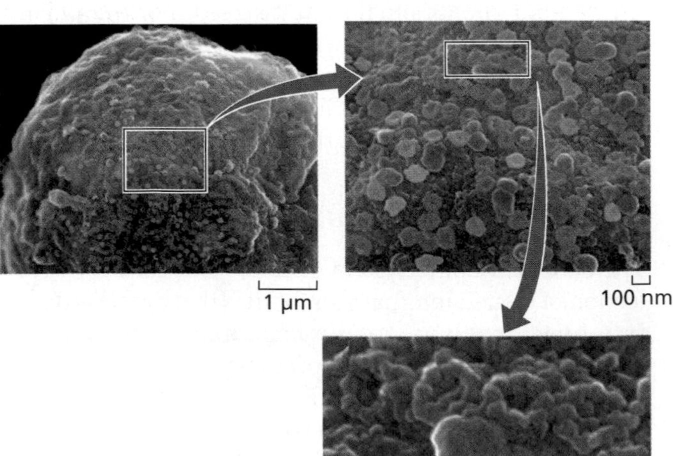

Figure 1.8 (A) Transmission electron micrograph of a plant cell, showing the nucleolus and the nuclear envelope. (B) Organization of the nuclear pore complexes (NPCs) on the nuclear surface in tobacco culture cells. The NPCs that touch each other are colored brown; the rest are colored blue. The first inset (top right) shows that most of the NPCs are closely associated, forming rows of 5–30 NPCs. The second inset (bottom right) shows the tight associations of the NPCs. (A courtesy of R. Evert; B from Fiserova et al. 2009.)

The nucleus is surrounded by a double membrane called the **nuclear envelope** (**Figure 1.8A**), which is a subdomain of the endoplasmic reticulum (ER; see below). **Nuclear pores** form selective channels across both membranes, connecting the nucleoplasm (the region inside the nucleus) with the cytoplasm (**Figure 1.8B**). There can be very few to many thousands of nuclear pores on an individual nuclear envelope, and they can be arranged into higher-order aggregates.

The nuclear "pore" is actually an elaborate structure composed of more than 100 different **nucleoporin** proteins arranged octagonally to form a 105-nm **nuclear pore complex** (**NPC**). The nucleoporins lining the 40-nm channel of the NPC form a meshwork that acts as a supramolecular sieve. Several proteins required for nuclear import and export have been identified (see WEB TOPIC 1.4). A specific amino acid sequence called the **nuclear localization signal** is required for a protein to gain entry into the nucleus (see WEB TOPIC 1.5).

The nucleus is the site of storage and replication of the **chromosomes**, composed of DNA and its associated proteins (**Figure 1.9**). Collectively, this DNA–protein complex is known as **chromatin**. The linear length of the entire DNA in any plant genome is usually millions of times greater than the diameter of the nucleus in which it is found. To solve the problem of packaging this chromosomal DNA within the nucleus, segments of the linear double helix of DNA are coiled twice around a solid cylinder of eight **histone** protein molecules, forming a **nucleosome**. Nucleosomes are arranged like beads on a string along the length of each chromosome. When the nucleus is not dividing, the chromosomes maintain their spatial independence. Although they move around inside the nucleus, they do not "tangle" and remain quite discrete (**Figure 1.10**).

During mitosis, the chromatin condenses, first by coiling tightly into a **30-nm chromatin fiber**, with six nucleosomes per turn, followed by further folding and packing processes that depend on interactions between proteins and nucleic acids (see Figure 1.9). At interphase, two types of chromatin are distinguishable, based on their degree of condensation: heterochromatin and euchromatin. **Heterochromatin** is a highly compact and transcriptionally inactive form of chromatin and accounts for about 10% of the DNA. Most of the heterochromatin is concentrated along the periphery of the nuclear membrane and is associated with regions of the chromosome containing few genes, such as telomeres and centromeres. The rest of the DNA consists of **euchromatin**, the dispersed, transcriptionally active form. Only about 10% of the euchromatin is transcriptionally active at any given time. The remainder exists in a state of condensation intermediate between that of the transcriptionally active euchromatin and that of heterochromatin. The

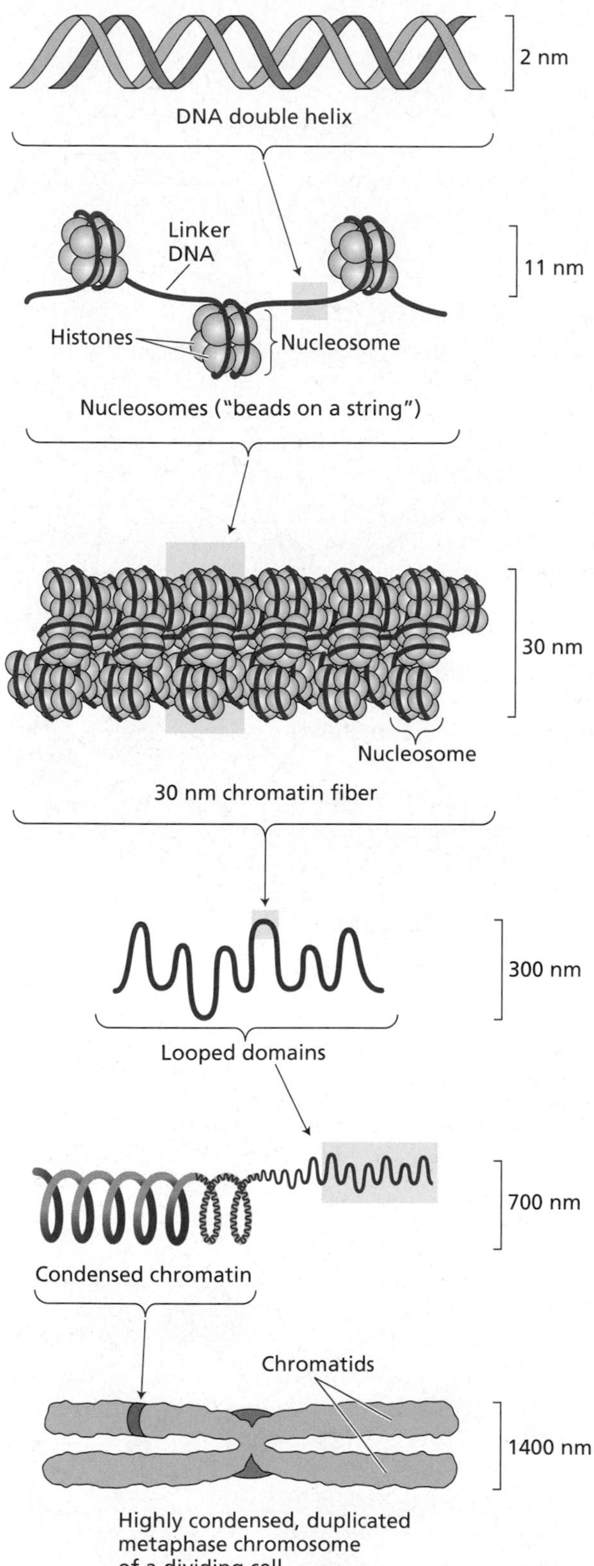

Figure 1.9 Packaging of DNA in a metaphase chromosome. The DNA is first aggregated into nucleosomes and then wound to form the 30-nm chromatin fibers. Further coiling leads to the condensed metaphase chromosome. (After Alberts et al. 2002.)

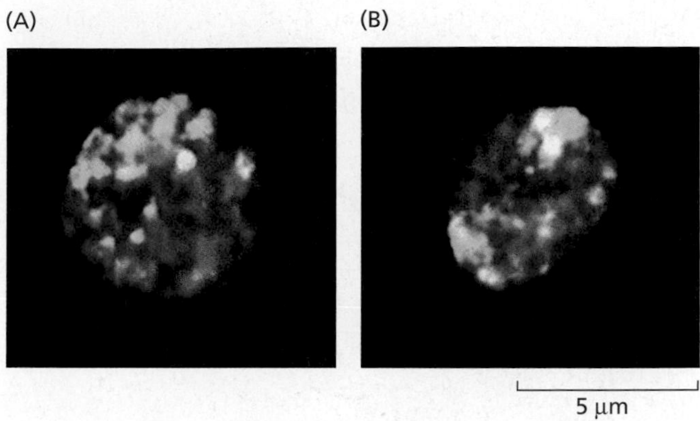

Figure 1.10 With the use of fluorescent probes for chromosome 1 and chromosome 2 of the model grass plant *Brachypodium distachyon*, these two chromosomes can be visualized in interphase (nondividing) cells with a technique called chromosome painting. (A) Both of the homologous copies of chromosome 1 (light blue) can be seen on one side of the nucleus, while both copies of chromosome 2 (magenta) are on the other side of the nucleus. (B) A nucleus at a different time in interphase. Although the homologous chromosomes are no longer together, they still occupy unique and separate regions of the nucleus. (From Idziak et al. 2011.)

chromosomes reside in specific regions of the nucleoplasm, each in its own separate space, giving rise to the possibility of separate regulation of each chromosome.

During the cell cycle, chromatin undergoes dynamic structural changes. In addition to transient local changes required for transcription, heterochromatic regions can be converted to euchromatic regions, and vice versa, by the addition or removal of functional groups on the histone proteins (see Chapter 2). Such global changes in the genome can give rise to stable changes in gene expression. In general, stable changes in gene expression that occur without changes in the DNA sequence are referred to as *epigenetic regulation*.

Nuclei contain a densely granular region called the **nucleolus** (plural *nucleoli*), which is the site of **ribosome** synthesis. Typical cells have one nucleolus per nucleus; some cells have more. The nucleolus includes portions of one or more chromosomes where ribosomal RNA (rRNA) genes are clustered to form a structure called the **nucleolar organizer region** (**NOR**). Even though chromosomes remain largely separate within the nucleus, parts of several may come together in their middle to help form the nucleolus. The nucleolus assembles the proteins and rRNA of the ribosome into a large and a small subunit, each exiting the nucleus separately through the nuclear pores. The two subunits unite in the cytoplasm to form a complete ribosome (**Figure 1.11A**). Assembled ribosomes are protein-synthesizing machines. Those produced by the nucleus for cytoplasmic, "eukaryotic" protein synthesis,

(A)

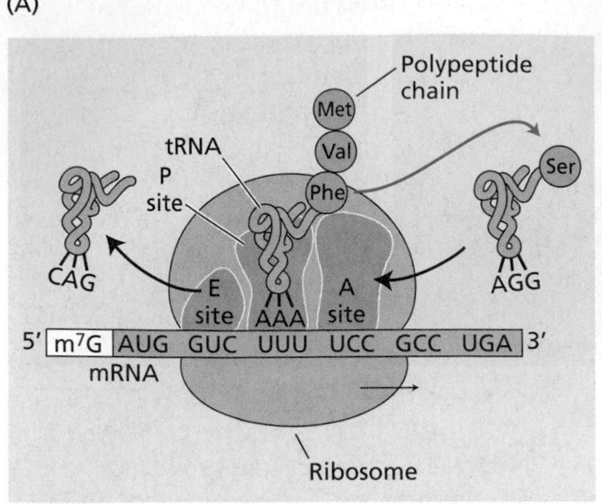

(B)

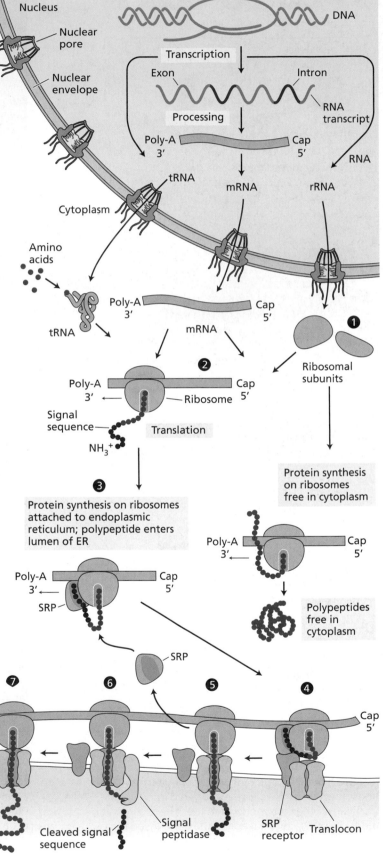

Figure 1.11 (A) Amino acids are polymerized on the ribosome, with the help of tRNA, to form the elongating polypeptide chain. (B) Basic steps in gene expression, including transcription, processing, export to the cytoplasm, and translation. (1–2) Proteins may be synthesized on free or bound ribosomes. (3) Proteins destined for secretion are synthesized on the rough endoplasmic reticulum and contain a hydrophobic signal sequence. A signal recognition particle (SRP) binds the signal peptide to the ribosome, interrupting translation. (4) SRP receptors associate with protein-transporting channels called translocons. The ribosome–SRP complex binds to the SRP receptor on the ER membrane, and the ribosome docks with the translocon. (5) The translocon pore opens, the SRP particle is released, and the elongating polypeptide enters the lumen of the ER. (6) Translation resumes. Upon entering the lumen of the ER, the signal sequence is cleaved off by a signal peptidase on the membrane. (7–8) After carbohydrate addition and chain folding, the newly synthesized polypeptide is shuttled to the Golgi apparatus via vesicles.

the 80S ribosomes, are larger than the ribosomes assembled in and remaining in mitochondria and plastids for their separate program of "prokaryotic" protein synthesis, the 70S ribosomes.

Gene expression involves both transcription and translation

The nucleus is the site of read-out, or **transcription**, of the cell's DNA. Some of the DNA is transcribed as messenger RNA (mRNA), which codes for proteins. Ribosomes read mRNA in one direction, from the 5′ to the 3′ end (see Figure 1.11A). Other regions of the DNA are transcribed into transfer RNA (tRNA) and rRNA to be used in **translation**. The RNA moves through nuclear pores to the cytoplasm (**Figure 1.11B**), where the polyribosomes (groups of ribosomes translating a single RNA strand) that are "free" in the cyctoplasm (not membrane-bound) translate the RNA for proteins destined for the cytosol and organelles that receive proteins independently of the endomembrane pathway. Endomembrane and secreted proteins are inserted during the process of translation, or co-translationally, on polyribosomes bound to the ER. The mechanism of **co-translational insertion** of proteins into the ER is complex, involving the ribosomes, the mRNA that codes for the secretory protein, and a special protein-translocating pore, the **translocon**, in the ER membrane, as described below. The proteins synthesized on cytosolic ribosomes that are targeted to membrane organelles after translation have **posttranslational** insertion. The process of translation on either cytosolic or membrane-bound polysomes produces the primary protein sequence of the protein, which includes not only the sequence involved in protein function, but also sequence information required to "target" the protein to different destinations within the cell (see **WEB TOPIC 1.5**).

The endoplasmic reticulum is a network of internal membranes

The ER is composed of an extensive network of tubules that is continuous with the nuclear envelope (**Figure 1.12**). The tubules join together to form a network of polygons and flattened saccules called **cisternae** (singular *cisterna*) (see Figure 1.12 and **Figure 1.13**). The tubules spread throughout the cell, forming very close associations with other organelles (**Figure 1.14**). The ER network may therefore be a communication network between organelles within a cell, while also serving as a synthesis and delivery system for protein or lipid. The ER that lies just under, and is probably attached to, the plasma membrane resides in the outer layer of cytoplasm called the *cell cortex* or **cortical ER**

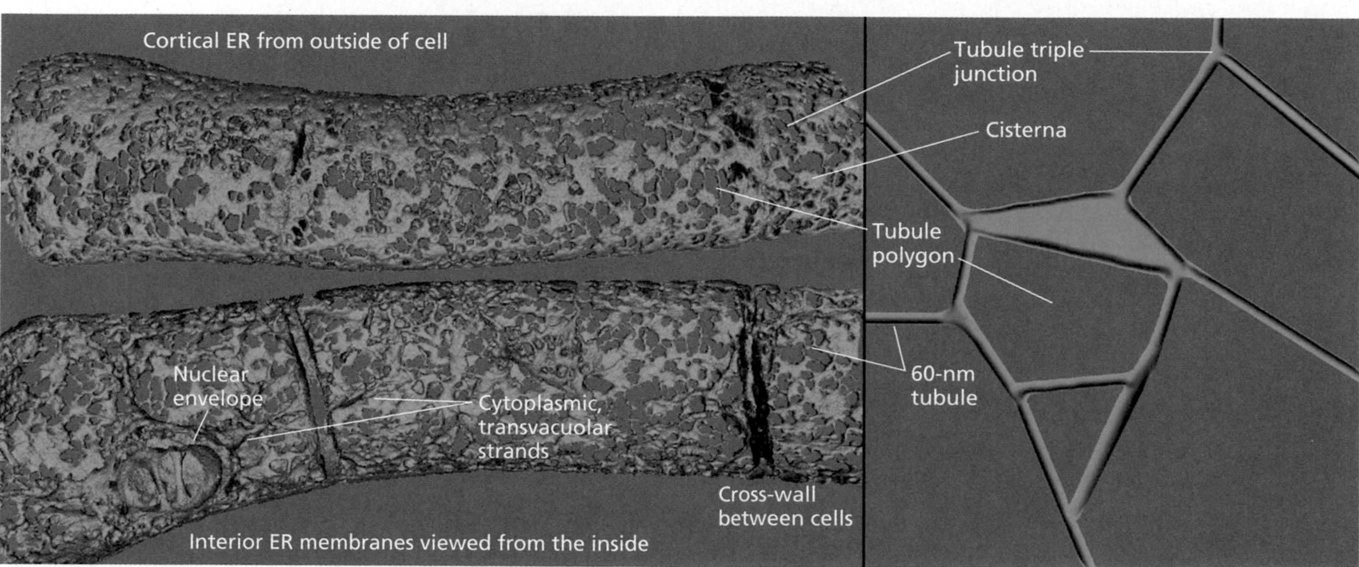

Figure 1.12 Three-dimensional reconstruction of ER in tobacco suspension culture cells. (A) When the cells are viewed from the outside looking inward (top), the cortical network of the ER is clearly made up of cisternal domains and polygonal tubule domains. When the cells are viewed from the inside looking outward (bottom), transvacuolar strands containing ER tubules can be seen, as well as the nuclear envelope, a subdomain of the ER. The nuclei have trans-nuclear channels and invaginations in the nuclear envelope. (B) Diagrammatic representation of tubules and cisternae arranged in the network of polygons typical of cortical ER. (Courtesy of L. R. Griffing.)

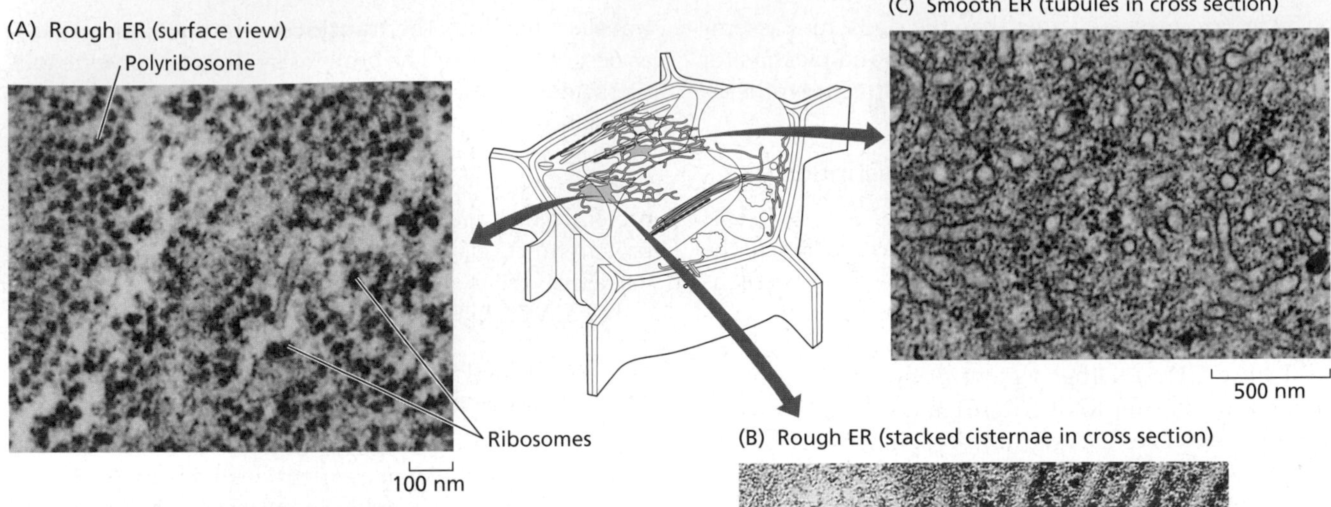

Figure 1.13 The endoplasmic reticulum. (A) Rough ER from the alga *Bulbochaete* can be seen in surface view in this micrograph. The polyribosomes (strings of ribosomes attached to mRNA) in the rough ER are clearly visible. (B) Cross-sections of stacks of regularly arranged rough cisternal ER (white arrow) in glandular trichomes of *Coleus blumei*. The plasma membrane is indicated by the black arrow, and the material outside the plasma membrane is the cell wall. (C) Smooth ER often forms a tubular network, as shown in this transmission electron micrograph from a young petal of *Primula kewensis*. (Micrographs from Gunning and Steer 1996.)

(see Figure 1.14). In expanded or elongated cells, cortical ER forms a *polygonal network* of tubules (see Figure 1.12) that is traversed by dynamic, flowing tubule bundles. Individual tubules and tubule bundles can also detach from the cortex, becoming **internal ER** in the inner layer of the cytoplasm, and can traverse the cell via *transvacuolar strands*—strands of cytoplasm that extend through the central vacuole (see Figures 1.6 and 1.14), wrapped in vacuole membrane. In unexpanded, meristematic cells, the ER is predominantly cisternal. As the cell develops and expands, tubular and cisternal forms of the ER rapidly transition between each other. The transition may be controlled by a class of proteins called **reticulons**, which form tubules from membrane sheets. The actomyosin cytoskeleton, which we'll discuss later in the chapter, is also involved in this transition, while participating in tubule rearrangement, cisterna formation, and the flow of proteins through the network in expanding, nondividing cells.

The region of the ER that has many membrane-bound ribosomes is called **rough ER** because the bound ribosomes give the ER a rough appearance in electron micrographs (see Figure 1.13A and B). ER without bound ribosomes is called **smooth ER** (see Figure 1.13C). Most ER has the capacity to bind ribosomes, since most ER contains translocons. The distinction between rough and smooth ER is sometimes correlated with changes in the form of the ER, rough ER being cisternal, and smooth ER being tubular. This classical distinction applies best to certain cell types, such as in floral glands producing nectar (see the section *Plant Cell Types* and Figure 1.33), which, if secreting lipid, contain much smooth tubular ER and, if secreting protein, contain rough cisternal ER.

The ER is the major source of membrane phospholipids and provides membrane proteins and protein cargo for the other compartments in the endomembrane pathway: the nuclear envelope, the Golgi apparatus, vacuoles, the plasma membrane, and the endosomal system. It even transports some proteins to the chloroplast. Most of this transport occurs via specialized vesicles moving between the endomembrane organelles. However, specialized regions of the ER can apparently exchange lipids and other molecules with "partnering" organelles, such as the plasma membrane, chloroplasts, and mitochondria, when in close association (see Figure 1.14) without the involvement of transport vesicles.

There is an intrinsic sidedness or asymmetry to membrane bilayers because the enzyme that initiates phospholipid synthesis on the ER adds new phospholipid precursors exclusively to the cytosolic leaflet of the bilayer (i.e., the side of the membrane facing the cytosol). The enzymes involved in synthesizing the phospholipid head groups (serine, choline, glycerol, or inositol) are also on the cytosolic leaflet. This causes intrinsic lipid asymmetry in the membranes of endomembranes, with the cytoplas-

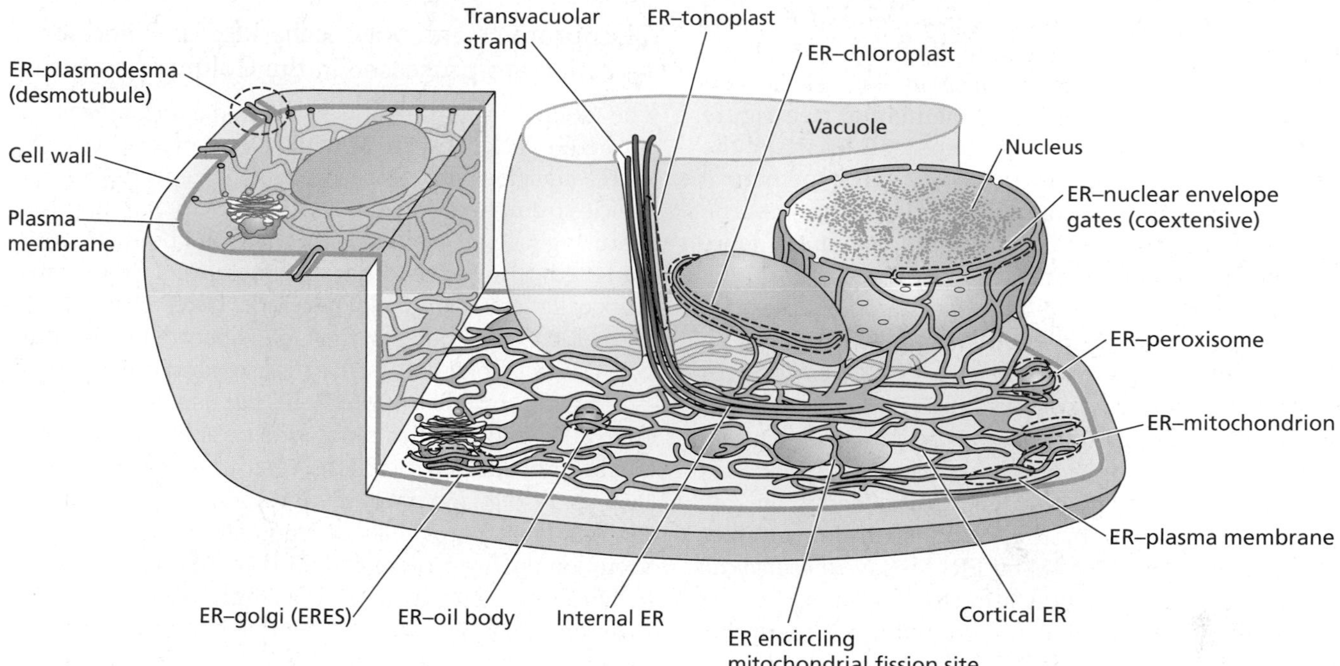

Figure 1.14 ER–organelle associations. Cortical ER is attached to the plasma membrane and traverses the plasmodesmata. More internal ER bundles move along the cortex and through transvacuolar channels. The ER is continuous with the nuclear envelope, but there are restrictions that allow only certain proteins to be shared. The ER associates with the mitochondria during mitochondrial division, using a select set of proteins that are part of the ER–mitochondrial encounter structure (ERMES). The mitochondrion is also tethered to the plasma membrane through a complex that involves the ER. ER tubules also surround the chloroplast. The peroxisome and oil bodies are closely associated with the ER during their formation (see movie, **WEB TOPIC 1.9**). The Golgi apparatus is associated with the ER at the ER exit sites (ERES) (see movie, **WEB TOPIC 1.7**).

mic leaflet of the organelles having a different composition from the lumenal (inside) leaflet of the organelles. The lumenal leaflet eventually becomes the leaflet of the membrane that faces the outside of the cell on the plasma membrane. Further asymmetrical modifications of lipid head groups and posttranslational modification of proteins by covalent addition of lipids and carbohydrates amplify the sidedness of membranes (see Figure 1.7). Membrane asymmetry can be counteracted by enzymes called **flippases**, which "flip" newly synthesized phospholipids across the bilayer to the inner leaflet.

The ER, plastids, and mitochondria are able to add new membrane directly through lipid and protein synthesis. However, for endomembrane organelles "downstream" from the ER, including the Golgi apparatus, vacuole, oil bodies, peroxisomes, and plasma membrane, the addition of new membrane occurs primarily through the process of **fusion** of transport tubules or vesicles with these membranes. Because membranes are fluid, new membrane constituents can be transferred to an existing membrane even if the new membrane subsequently separates from the existing membrane by **fission**. These cycles of membrane fusion and fission are the basis for the growth and division of all endomembrane organelles that are derived directly or indirectly from the ER. Selective fusion and fission of vesicles and tubules that serve as transporters between the compartments of the endomembrane system are achieved by means of a special class of targeting recognition proteins called **SNAREs** and **Rabs** (see **WEB TOPIC 1.6**).

Secretion of proteins from cells begins with the rough ER

Secretory proteins are inserted into the ER as they are being translated, a process called co-translational insertion. All secretory proteins and most integral membrane proteins have been shown to have a **signal peptide**, a hydrophobic leader sequence of 18 to 30 amino acid residues at the amino-terminal end of the chain (see Figure 1.11). Early in translation, a **signal recognition particle** (**SRP**), made up of protein and RNA, binds both to this hydrophobic leader and to the ribosome, interrupting translation. The rough ER membrane contains **SRP receptors** that can associate with the translocons, through which the newly synthesized protein is threaded. During co-translational insertion into the ER, the mRNA–ribosome–SRP complex in the cytosol binds to the SRP receptor on the ER membrane, and the ribosome docks with the translocon. Docking opens the translocon pore, the SRP particle is released, translation resumes, and the elongating polypeptide enters the lumen of the ER. For secretory proteins, the signal sequence is cleaved off by a signal peptidase on the ER membrane (see Figure 1.11). For integral membrane proteins, some parts of the polypeptide chain are translocated across the membrane while others are not. Completed integral membrane proteins are anchored to the membrane by one or more hydrophobic membrane-spanning domains.

Many of the proteins found in the lumen of the endomembrane system are **glycoproteins**—proteins with small sugar chains covalently attached—destined for secretion from the cell or delivery to the other endomembranes. In the vast majority of cases, a branched oligosaccharide chain made up of *N*-acetylglucosamine (GlcNAc), mannose (Man), and glucose (Glc) is attached to the free amino group of one or more specific asparagine residues of the secretory protein in the ER. This *N-linked glycan* ("N" is the one-letter abbreviation for asparagine) is first assembled on a lipid molecule, **dolichol diphosphate**, which is embedded in the ER membrane (see Chapter 12). The completed 14-sugar glycan is then transferred to the nascent polypeptide as it enters the lumen. As in animal cells, these **N-linked glycoproteins** are then transported to the Golgi apparatus (discussed next) via small vesicles or tubules. However, in the Golgi apparatus the glycans are further processed in a plant-specific way, causing potential problems for the production of plant-based injectable vaccines or antibodies for medical use. The plant-specific modifications make the proteins highly antigenic (recognized as foreign) in vertebrate immune systems.

Glycoproteins and polysaccharides destined for secretion are processed in the Golgi apparatus

The Golgi body (in plants, also called a *dictyosome*) is a polarized stack of cisternae, with fatter cisternae occurring on the ***cis*** side, or forming face, which accepts tubules and vesicles from the ER (**Figures 1.15 and 1.16**). The opposite, maturing face, or ***trans*** side, of the Golgi body has more flattened, thinner cisternae and includes a tubular network called the ***trans* Golgi network** (**TGN**). There may be up to 100 Golgi bodies in the Golgi apparatus (the entire collection of Golgi bodies) of a meristematic cell; other cell types differ in their Golgi content but usually have fewer than 100. Golgi bodies can divide by splitting, and they can be built up from the ER in a *cis*-to-*trans* fashion in cells lacking a Golgi apparatus. By controlling their number of Golgi bodies, plant cells can regulate their capacity for secretion during growth and differentiation.

Different cisternae within a single Golgi body have different enzymes, and different biochemical functions depending on the type of polymer being processed—whether polysaccharides for the cell wall or glycoproteins for the cell wall or the vacuole. For example, as N-linked glycoproteins pass from the *cis* to the *trans* Golgi cisternae, they are successively modified by the specific sets of enzymes localized in the different cisternae. Certain sugars, such as mannose, are removed from the oligosaccharide chains, and other sugars are added. In addition to these modifications, glycosylation of the –OH groups of hydroxyproline, serine, threonine, and tyrosine residues (**O-linked oligosaccharides**) also occurs in the Golgi. The enzymes involved in polysaccharide biosynthesis in the Golgi bodies are substantially different but occur side by side with glycoprotein modification enzymes. The different enzymes for polysaccharide biosynthesis are also

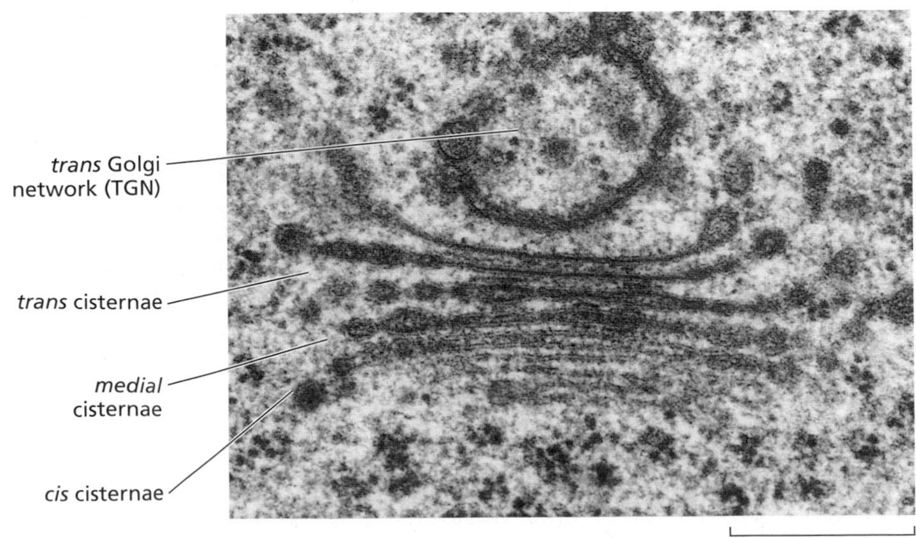

Figure 1.15 Electron micrograph of a Golgi apparatus in a tobacco (*Nicotiana tabacum*) root cap cell. The *cis*, *medial*, and *trans* cisternae are indicated. The *trans* Golgi network is associated with the *trans* cisternae. (From Gunning and Steer 1996.)

(A)

1. COPII-coated vesicles bud from the ER and are transported to the *cis* face of the Golgi apparatus.

2. Cisternae progress through the Golgi stack in the anterograde direction, carrying their cargo with them.

3. Retrograde movement of COPI-coated vesicles maintains the correct distribution of enzymes in the *cis*, *medial*, and *trans* cisternae of the stack.

4. Uncoated vesicles bud from the *trans* Golgi membrane and fuse with the plasma membrane.

5. Endocytotic clathrin-coated vesicles fuse with the prevacuolar compartment.

6. Uncoated vesicles bud off from the prevacuolar compartment and carry their cargo to a lytic vacuole.

7. Proteins destined for lytic vacuoles are secreted from the *trans* Golgi to the prevacuolar compartment via clathrin-coated vesicles, and then repackaged for delivery to the lytic vacuole.

8. Endocytic clathrin-coated vesicles can also uncoat and recycle via the early recycling endosome. Vesicles produced by the early recycling endosome either directly fuse with the plasma membrane or fuse with the *trans* Golgi.

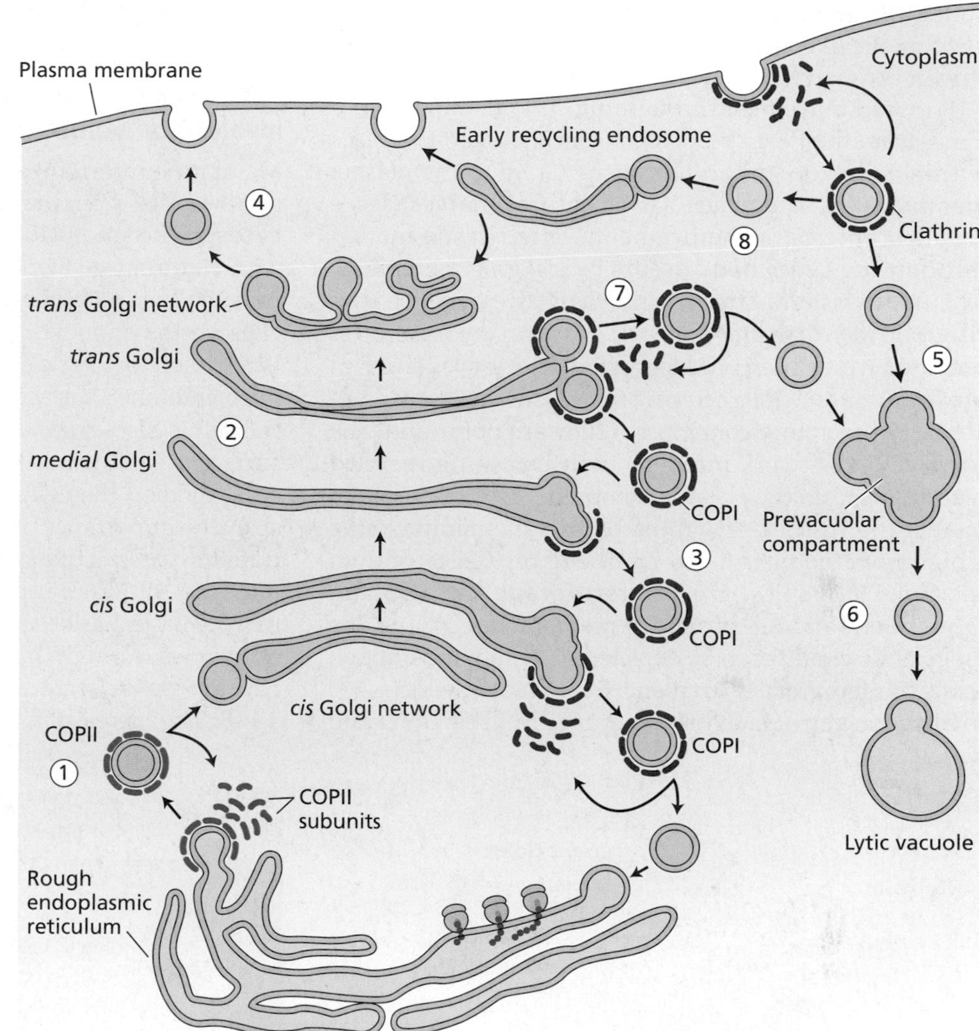

(B)

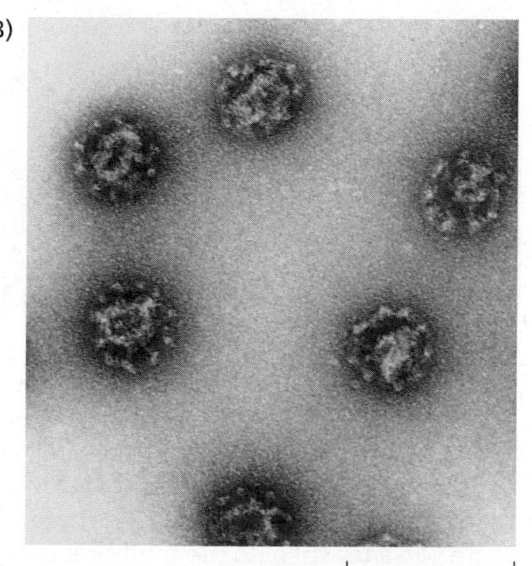

Figure 1.16 Vesicular traffic along the secretory and endocytotic pathways. (A) Diagram of vesicular traffic mediated by three types of coat proteins. COPII is indicated in green, COPI in blue, and clathrin in red. (B) Electron micrograph of clathrin-coated vesicles isolated from bean leaves. (B courtesy of D. G. Robinson.)

found in different parts of the Golgi body, with pectic polysaccharides, for example, being assembled in the late *cis*, *medial*, and *trans* Golgi and xyloglucan primarly in just the *trans* Golgi.

Delivery of membrane and its contents to the Golgi body from the ER occurs at specialized **ER exit sites** (**ERES**). The ER exit sites are determined by the presence of a **coat protein** called **COPII** (see Figure 1.16A). This surface protein associates with the transmembrane receptors that bind the specific cargo destined for the Golgi. These membrane regions then bud off to form coated vesicles or tubules that lose their COPII coats prior to fusion with the target *cis* Golgi membranes. By using fluorescent tags for both ERES and Golgi, it has been possible to show that the ERES move in concert with the Golgi bodies as the latter stream through the cell (see movie in **WEB TOPIC 1.7**).

Forward movement through the endomembrane system—from the ER to the Golgi, from the *cis* to *trans* face within a Golgi body, and from the Golgi to the plasma membrane or to prevacuolar structures via vesicles—is called anterograde movement. Anterograde motion through the Golgi body occurs by *cisternal maturation*, whereby *cis* cisternae mature into *trans* cisternae. The *trans* cisternae then "fall off" the stack and form the *trans* Golgi network, which then produces secretory vesicles (see Figure 1.16A). The "sloughed off" *trans* cisternae sometimes are full of secretory content, but they are not round. The *trans* Golgi network interacts with incoming recycled membrane from the plasma membrane. This **recycling** of membrane vesicles—from the plasma membrane to the Golgi, from the *trans* to the *cis* face of the Golgi, or from the Golgi to the ER—is called **retrograde**, or backward, movement. Without retrograde membrane recycling, the Golgi body would soon be depleted of membrane through loss by anterograde movement. **COPI**-coated vesicles are involved in retrograde movement within the Golgi and from the Golgi to the ER. The retrograde movement of membrane from the plasma membrane, discussed next, uses a different set of coat proteins.

The plasma membrane has specialized regions involved in membrane recycling

Membrane internalization by retrograde movement of small vesicles from the plasma membrane is called **endocytosis**. The small (100 nm) vesicles are initially coated with **clathrin** (see Figure 1.16), but they quickly lose that coat and fuse with other tubules and vesicles; the organelles of this endocytic pathway are called **endosomes**. When secretory vesicles fuse with the plasma membrane, the membrane's surface area necessarily increases. Unless the cell is also expanding to keep pace with the added surface area, it needs some method of membrane recycling to keep the cell's surface area in balance with its size. The importance of membrane recycling is best illustrated by cells active in secretion, such as root cap cells (**Figure 1.17**). Root cap cells secrete copious amounts of mucopolysaccharides (slime), which lubricate the root tip as it grows through the soil; this slime can be seen as the electron-dense secretory material in the Golgi in Figure 1.17B. The increase in membrane surface area caused by

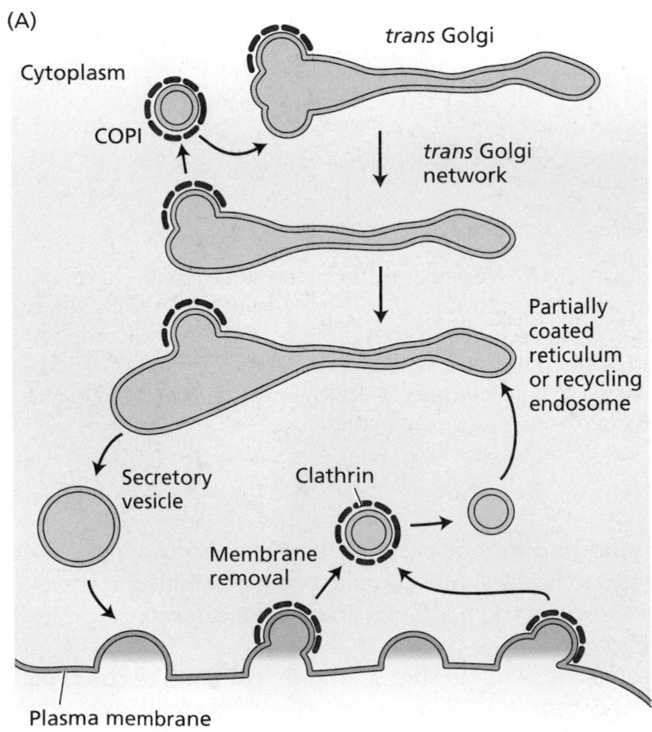

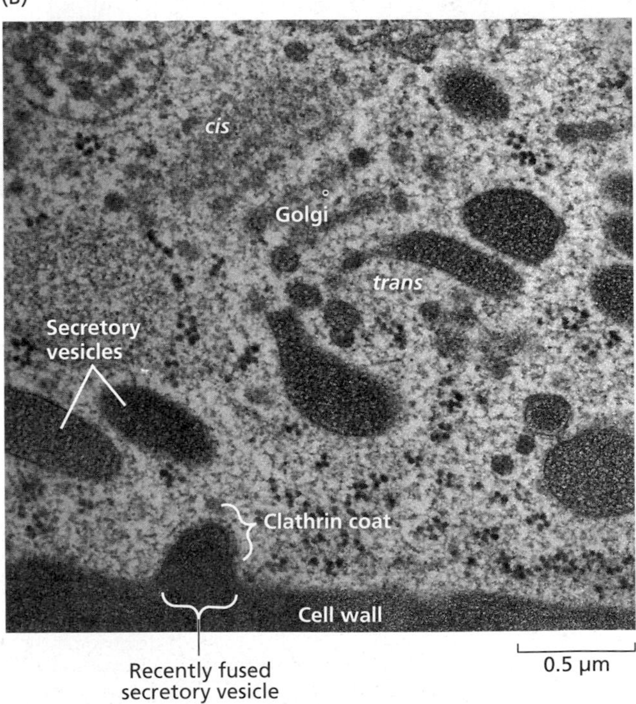

Figure 1.17 Clathrin-coated pits are associated with secretion of slime in maize root cap. (A) Diagram of the recycling of membrane using clathrin-coated vesicles from recent secretion sites on the plasma membrane. (B) Recent secretion site showing a secretory vesicle that has just deposited its contents into the cell wall and a clathrin-coated invagination, which recycles membrane from the secretion site. There are 20 times more coated pits on secretion sites than on the membrane in general. (B micrograph by H. H. Mollenhauer, courtesy of L. R. Griffing.)

the fusion of large slime-filled vesicles with the plasma membrane would become excessive if it were not for the process of endocytosis, which constantly recycles plasma membrane back to an organelle called the **early endosome**. The endosome can then be targeted either back to the *trans* Golgi network for secretion or to a structure called the **prevacuolar compartment** for hydrolytic degradation (see Figure 1.16A). Movement of these membranes is generally guided by the actomyosin cytoskeleton (described later in the chapter), but some endosomes also travel along the microtubule cytoskeleton.

Endocytosis and endocytotic recycling take place in a wide variety of plant cells. The control of endocytosis at the plasma membrane differentially regulates the abundance of ion channels (see Chapter 6), such as the potassium ion channel in stomatal guard cells and the borate transporter in roots. During gravitropism, the differential internalization of transporters for the growth hormone auxin causes a change in the concentration of hormone across the root, resulting in a bending of the root (see Chapter 15).

Vacuoles have diverse functions in plant cells

The plant vacuole was originally defined by its appearance in the microscope—a membrane-enclosed compartment without cytoplasm. Instead of cytoplasm, it contains **vacuolar sap** composed of water and solutes. The increase in volume of plant cells during growth takes place primarily through an increase in the volume of vacuolar sap. A large central vacuole occupies up to 95% of the total cell volume in many mature plant cells; sometimes there are two or more large central vacuoles, as in the case of certain flower petals with both pigmented and unpigmented vacuoles (see images and movies in WEB TOPIC 1.8). The fact that vacuoles can differ in size and appearance suggests how diverse in form and function the vacuolar compartment can be. Some of the variations are probably due to differences in degree of vacuole maturation. For example, meristematic cells have no large central vacuole, but rather many small vacuoles or a highly convoluted system of vacuolar membranes. Some of these probably fuse together and remodel themselves to form the large central vacuole as the cell matures (see Figure 1.35).

The vacuolar membrane, the **tonoplast**, contains proteins and lipids that are synthesized initially in the ER. In addition to its role in cell expansion, the vacuole can also serve as a storage compartment for secondary metabolites involved in plant defense against herbivores (see Chapter 23). Inorganic ions, sugars, organic acids, and pigments are just some of the solutes that can accumulate in vacuoles, thanks to the presence of a variety of specific membrane transporters (see Chapter 6). Protein-storing vacuoles, called **protein bodies**, are abundant in seeds.

Vacuoles also play a role in protein turnover, analogous to the animal lysosome, as in the case of the **lytic vacuoles** that accumulate in senescing leaves. During senescence-associated programmed cell death (see Chapter 22), cellular constituents are degraded by specialized lytic vacuoles called **autophagosomes**. Membrane sorting to plant vacuoles and animal lysosomes occurs by different mechanisms. Although in both cases the sorting to the prevacuolar compartment occurs in the Golgi, the recognition processes used in sorting receptors and lytic proteins to the vacuole versus the lysosome are different. In mammalian cells, many lysosomal proteins are recognized by an ER enzyme that then adds mannose 6-phosphate to them; this modification is subsequently recognized by a receptor in the Golgi that sorts the lysosomal proteins into vesicles destined for the lysosomes. This sorting pathway apparently is missing in plants. Instead, some plant lytic vacuoles are derived directly from the ER, bypassing the Golgi entirely via a pathway apparently missing in mammals.

The delivery of some Golgi-derived vesicles to the vacuole is indirect. As described above, there are multiple vacuolar compartments in the cell, not all of which serve as targets for Golgi vesicles. Those vacuoles that do receive Golgi-derived vesicles do so via an intermediate, prevacuolar compartment that also serves as a sorting organelle for membrane endocytosed from the plasma membrane (see Figure 1.16A). This prevacuolar sorting compartment includes the **multivesicular body**, which in some cases is also a *postvacuolar compartment* that functions in the degradation of vacuoles and their membranes. The multivesicular body is a specialized organelle that has a distinct structure of an outer limiting membrane 0.3–0.5 mm in diameter and contains 50-nm internal vesicles. The internal vesicles take up ubiquitinated proteins by the ESCRT (for *endosomal sorting complexes required for transport*) pathway. These proteins include endocytosed surface receptors that are subsequently selectively broken down in the vacuolar system.

Independently Dividing or Fusing Organelles Derived from the Endomembrane System

Several organelles are able to grow, proliferate, or fuse independently even though they are derived from the endomembrane system. These organelles include oil bodies, peroxisomes, and glyoxysomes.

Oil bodies are lipid-storing organelles

Many plants synthesize and store large quantities of oil during seed development. These oils accumulate in organelles called oil bodies (also known as oleosomes, lipid bodies, lipid droplets, or spherosomes) (**Figure 1.18**). Oil bodies are unique among the organelles in that they are surrounded by a "half–unit membrane"—that is, a phospholipid monolayer—derived from the ER. The phospholipids in the half–unit membrane are oriented with their polar head groups toward the aqueous phase of the cytosol

(A)

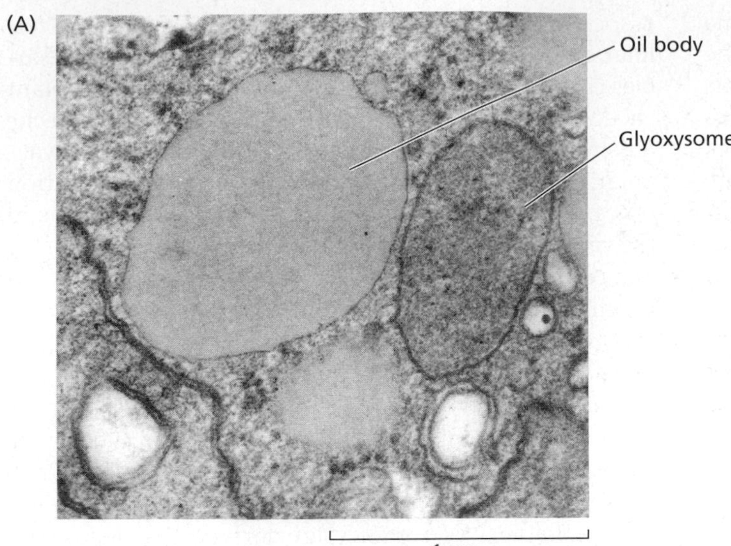

(B)

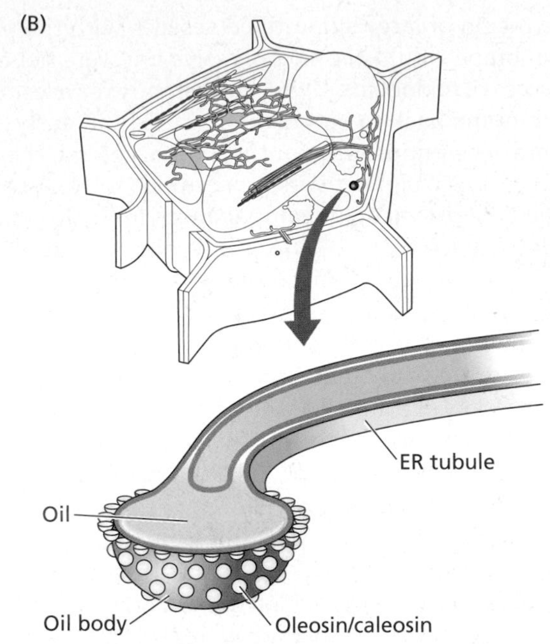

Figure 1.18 (A) Electron micrograph of an oil body beside a peroxisome. (B) Diagram showing the formation of oil bodies by the synthesis and deposition of oil within the phospholipid bilayer of the ER. After budding off from the ER, the oil body is surrounded by a phospholipid monolayer containing specific oil body proteins, such as oleosin. (A from Huang 1987; B after Buchanan et al. 2000.)

and their hydrophobic fatty acid tails facing the lumen, dissolved in the stored lipid.

Oil bodies initially form as regions of differentiation within the ER. The nature of the storage product, **triglycerides** (three fatty acids covalently linked to a glycerol backbone), dictates that this storage organelle will have a hydrophobic lumen. Consequently, as triglyceride is stored, it appears to be initially deposited in the hydrophobic region between the outer and inner leaflets of the ER membrane (see Figure 1.18B). Triglycerides do not have the polar head groups of membrane phospholipids, so are not exposed to the hydrophilic cytoplasm. Although the nature of the budding process that gives rise to the oil body is not yet fully understood, when the oil body separates from the ER it contains a single outer leaflet of phospholipid containing special proteins that coat oil bodies: **oleosin**, **caleosin**, and **steroleosin**. These proteins are synthesized on ER polysomes and are inserted into an oil body–forming region of the ER co-translationally. The proteins consist of a central, hairpin-like hydrophobic region, which inserts itself inside the oil-containing lumen, and two hydrophilic ends, which remain on the outside of the oil body. Oil body size is regulated by the abundance of these proteins. Once they have budded from the ER, oil bodies may increase in size by fusion of smaller oil bodies into larger ones. When oil bodies break down during seed germination, they associate with other organelles that contain the enzymes for lipid oxidation, the glyoxysomes.

Microbodies play specialized metabolic roles in leaves and seeds

Microbodies are a class of spherical organelles surrounded by a single membrane and specialized for one of several metabolic functions. **Peroxisomes** and **glyoxysomes** are microbodies specialized for the **β-oxidation** of fatty acids and the metabolism of **glyoxylate**, a two-carbon acid aldehyde (see Chapter 12). Microbodies lack DNA and are intimately associated with other organelles, with which they exchange intermediate metabolites. The glyoxysome is associated with mitochondria and oil bodies, while the peroxisome is associated with mitochondria and chloroplasts (**Figure 1.19**).

Initially, it was thought that peroxisomes and glyoxysomes were independent organelles, produced separately from the ER. However, experiments using antibodies specific for each type of organelle have supported a model in which peroxisomes develop directly from glyoxysomes, at least in greening cotyledons. For example, in cucumber seedlings, the nongreen cotyledon cells initially contain glyoxysomes, but after greening only peroxisomes are present. At intermediate stages of greening, the microbodies have both glyoxysomal and peroxisomal proteins, demonstrating that the glyoxysomes are converted to peroxisomes during the greening process.

In the peroxisome, glycolate, a two-carbon oxidation product of the photorespiratory cycle in an adjacent chloroplast, is oxidized to the acid aldehyde glyoxylate (see Chapter 8). During this conversion, hydrogen peroxide is generated, which can easily oxidize and destroy other compounds. However, the most abundant protein of the peroxisome is **catalase**, an enzyme that splits hydrogen

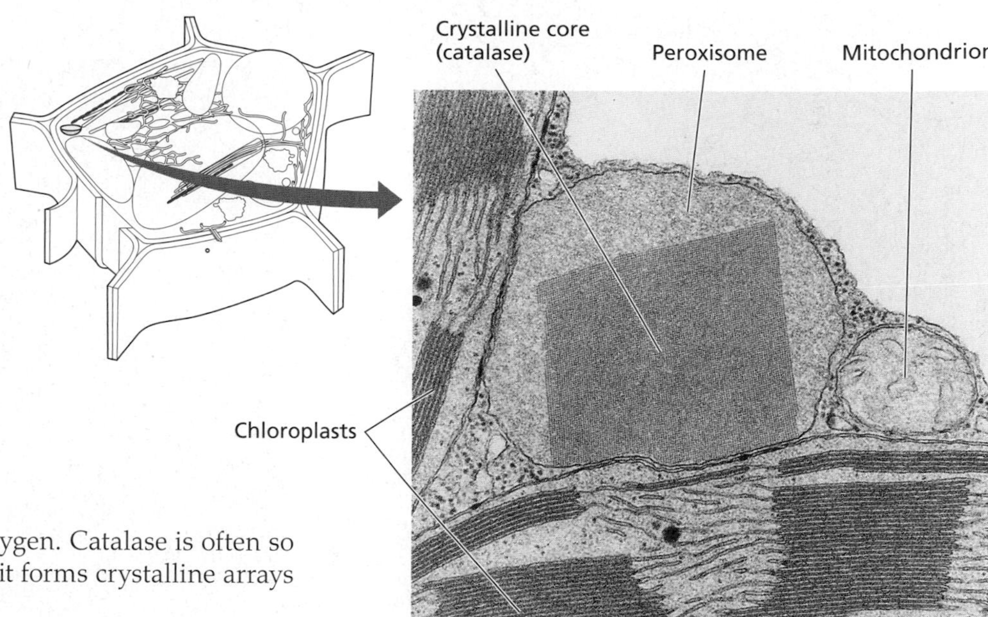

Figure 1.19 Catalase crystal in a peroxisome of a mature leaf of tobacco. Note the close association of the peroxisome with two chloroplasts and a mitochondrion, organelles that exchange metabolites with peroxisomes. (Micrograph by S. E. Frederick, courtesy of E. H. Newcomb.)

peroxide to water, releasing oxygen. Catalase is often so abundant in peroxisomes that it forms crystalline arrays of the protein (see Figure 1.19).

The observation that glyoxysomes can mature into peroxisomes explains the appearance of peroxisomes in developing cotyledons, but it does not explain how peroxisomes arise in other tissues. If they are inherited during cell division, peroxisomes can grow and divide separately from other organelles, using proteins similar to those involved in the division of mitochondria. In fact, about 20 proteins, including some involved in division, are dual-targeted to both the peroxisome and mitochondrion. Many proteins enter peroxisomes directly from the cytosol posttranslationally by means of a specific targeting signal, consisting of serine-lysine-leucine at the carboxyl termini of the proteins (see **WEB TOPIC 1.5**). Other peroxisomal proteins are made in the ER and traffic to the peroxisomes by a pathway that has yet to be fully elucidated. One possibility is that protein transfer could occur as peroxisomes and their tubular extensions, called **peroxules**, directly associate with the ER without fusing (see images and movies in **WEB TOPIC 1.9**). Although the peroxisome can divide independently, it is still dependent on the ER for some of its protein, and in this way the organelle could be considered semiautonomous. But unlike the mitochondria and chloroplasts that we'll describe next, peroxisomes have only a single outer membrane and do not contain their own DNA and ribosomes.

Independently Dividing, Semiautonomous Organelles

A typical plant cell has two types of energy-producing organelles: mitochondria and chloroplasts. Both types are separated from the cytosol by a double membrane (an outer and an inner membrane) and contain their own DNA and ribosomes.

Mitochondria (singular *mitochondrion*) are the cellular sites of respiration, a process in which the energy released from sugar metabolism is used for the synthesis of ATP (adenosine triphosphate) from ADP (adenosine diphosphate) and inorganic phosphate (P_i) (see Chapter 12).

Mitochondria are highly dynamic structures that can undergo both fission and fusion. Mitochondrial fusion can result in long, tubelike structures that may branch to form mitochondrial networks. Regardless of shape, all mitochondria have a smooth outer membrane and a highly convoluted inner membrane (**Figure 1.20**). The inner membrane contains an **ATP synthase** that uses a proton gradient to synthesize ATP for the cell. The **proton gradient** is generated through the cooperation of electron transporters called the **electron transport chain**, which is embedded in, and peripheral to, the inner membrane (see Chapter 12).

The infoldings of the inner membrane are called **cristae** (singular *crista*). The compartment enclosed by the inner membrane, the mitochondrial **matrix**, contains the enzymes of the pathway of intermediary metabolism called the citric acid cycle.The matrix also contains a special region, the nucleoid, that contains the mitochondrion's DNA.

Mitochondria change during seed germination and plant development. In the dry seed, they start as promitochondria that have no cristae. Within 6 hours of soaking the seed (imbibition; see Chapter 18), the genes for ATP synthase are activated and transcribed, and within 12 hours the mitochondria contain cristae.

Chloroplasts (**Figure 1.21A**) belong to another group of double-membrane–enclosed organelles called plastids, and are the sites of photosynthesis. Chloroplast membranes are rich in galactolipids (e.g., monogalactosyl glyc-

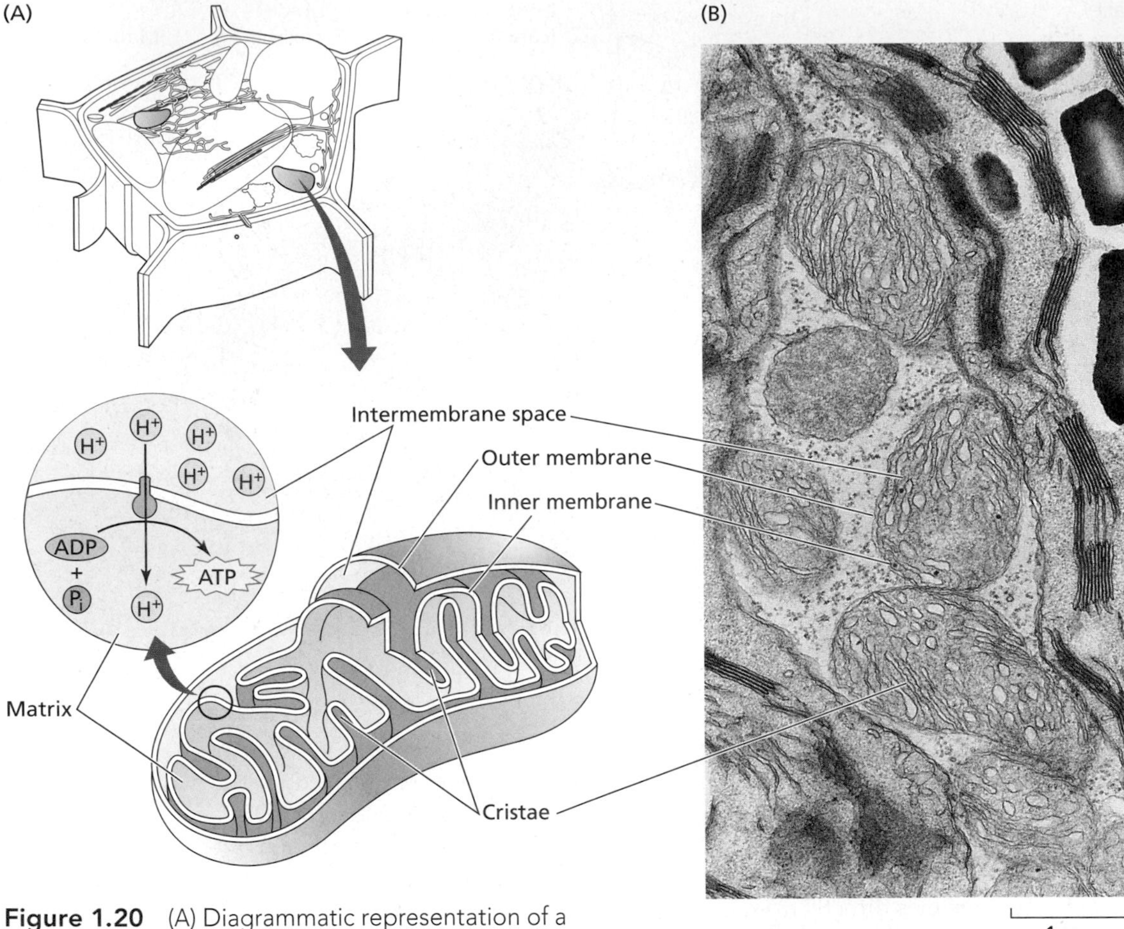

Figure 1.20 (A) Diagrammatic representation of a mitochondrion, including the location of the H^+-ATPases involved in ATP synthesis on the inner membrane. (B) Electron micrograph of mitochondria from a leaf cell of the Bermuda grass *Cynodon dactylon*. (Micrograph by S. E. Frederick, courtesy of E. H. Newcomb.)

erol; see Figure 1.7C). In addition to their inner and outer envelope membranes, chloroplasts possess a third system of membranes called **thylakoids**. A stack of thylakoids forms a **granum** (plural *grana*) (**Figure 1.21B**). Proteins and pigments (chlorophylls and carotenoids) that function in the photochemical events of photosynthesis are embedded in the thylakoid membrane. Adjacent grana are connected by unstacked membranes called **stroma lamellae** (singular *lamella*). The fluid compartment surrounding the thylakoids, called the **stroma**, is analogous to the matrix of the mitochondrion and contains what may be Earth's most abundant protein, **rubisco**, the protein involved in converting the carbon from carbon dioxide into organic acids during photosynthesis (see Chapter 8). The large subunit of rubisco is encoded by the chloroplast genome, while the small subunit is encoded by the nuclear genome. Concerted expression of each subunit (and other proteins) by each genome is necessary as chloroplasts grow and divide.

The various components of the photosynthetic apparatus are localized in different areas of the grana and the stroma lamellae. The ATP synthases of the chloroplast are located on the thylakoid membranes (**Figure 1.21C**). During photosynthesis, light-driven electron transfer reactions result in a proton gradient across the thylakoid membrane (**Figure 1.21D**) (see Chapter 7). As in the mitochondria, ATP is synthesized when the proton gradient is dissipated via ATP synthase. In the chloroplast, however, the ATP is not exported to the cytosol, but is used for many stromal reactions, including the fixation of carbon from carbon dioxide in the atmosphere, as described in Chapter 8.

Plastids that contain high concentrations of carotenoid pigments, rather than chlorophyll, are called **chromoplasts**. Chromoplasts are responsible for the yellow, orange, or red colors of many fruits and flowers, as well as of autumn leaves (**Figure 1.22**; also see Figure 21.35).

Nonpigmented plastids are called **leucoplasts**. Leucoplasts in specialized secretory tissues, such as the nectary, make monoterpenoids (see Figure 1.33), volatile molecules (in essential oils) that often have a strong smell. The most

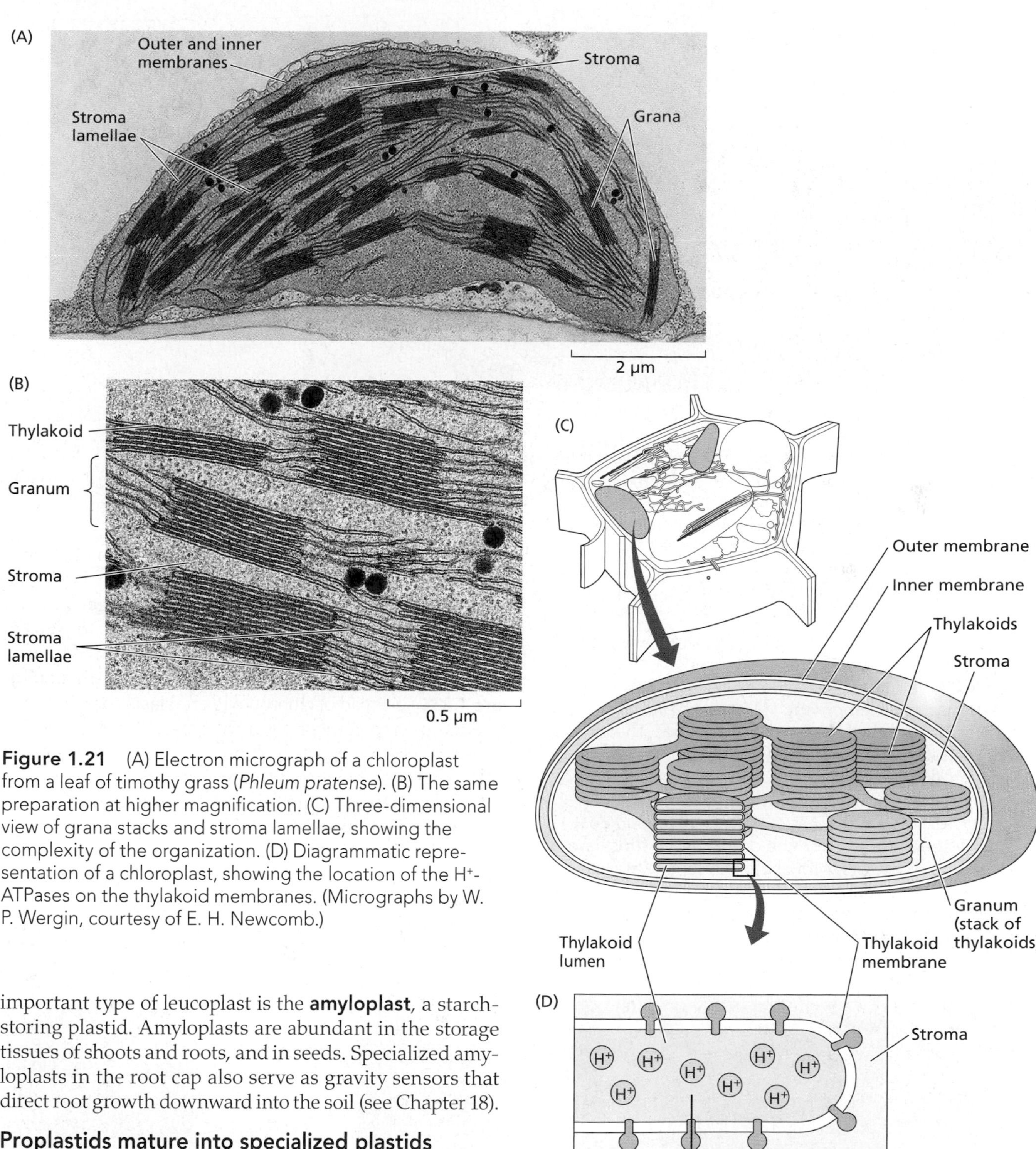

Figure 1.21 (A) Electron micrograph of a chloroplast from a leaf of timothy grass (*Phleum pratense*). (B) The same preparation at higher magnification. (C) Three-dimensional view of grana stacks and stroma lamellae, showing the complexity of the organization. (D) Diagrammatic representation of a chloroplast, showing the location of the H^+-ATPases on the thylakoid membranes. (Micrographs by W. P. Wergin, courtesy of E. H. Newcomb.)

important type of leucoplast is the **amyloplast**, a starch-storing plastid. Amyloplasts are abundant in the storage tissues of shoots and roots, and in seeds. Specialized amyloplasts in the root cap also serve as gravity sensors that direct root growth downward into the soil (see Chapter 18).

Proplastids mature into specialized plastids in different plant tissues

Meristem cells contain **proplastids**, which have few or no internal membranes, no chlorophyll, and an incomplete complement of the enzymes necessary to carry out photosynthesis (**Figure 1.23A**). In angiosperms and some gymnosperms, chloroplast development from proplastids is triggered by light. Upon illumination, enzymes are formed inside the proplastid or imported from the cytosol;

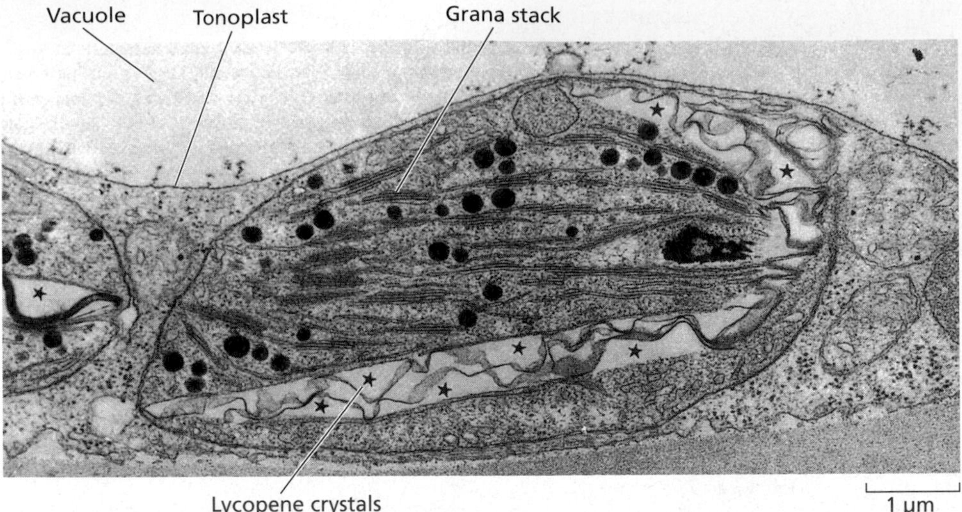

Figure 1.22 Electron micrograph of a chromoplast from tomato (*Solanum esculentum*) fruit at an early stage in the transition from chloroplast to chromoplast. Small grana stacks are still visible. Stars indicate crystals of the carotenoid lycopene. (From Gunning and Steer 1996.)

Figure 1.23 Electron micrographs illustrating several stages of plastid development. (A) Proplastid from the root apical meristem of the broad bean (*Vicia faba*). The internal membrane system is rudimentary, and grana are absent. (B) Mesophyll cell of a young oat (*Avena sativa*) leaf grown in the light, at an early stage of differentiation. The plastids are developing grana stacks. (C) Cell of a young oat leaf from a seedling grown in the dark. The plastids have developed as etioplasts, with elaborate semicrystalline lattices of membrane tubules called prolamellar bodies. When exposed to light, the etioplast can convert to a chloroplast by the disassembly of the prolamellar body and the formation of grana stacks. (From Gunning and Steer 1996.)

light-absorbing pigments are produced; and membranes proliferate rapidly, giving rise to stroma lamellae and grana stacks (**Figure 1.23B**).

Seeds usually germinate in the soil in the dark, and their proplastids mature into chloroplasts only when the young shoot is exposed to light. If, instead, germinated seedlings are kept in the dark, the proplastids differentiate into **etioplasts**, which contain semicrystalline tubular arrays of membrane known as **prolamellar bodies** (**Figure 1.23C**). Instead of chlorophyll, etioplasts contain a pale yellow-green precursor pigment, protochlorophyllide.

Within minutes after exposure to light, an etioplast differentiates, converting the prolamellar body into thylakoids and stroma lamellae, and the protochlorophyllide into chlorophyll (for a discussion of chlorophyll biosynthesis, see **WEB TOPIC 7.11**). The maintenance of chloroplast structure depends on the presence of light; mature chloroplasts can revert to etioplasts during extended periods of darkness. Likewise, under different environmental

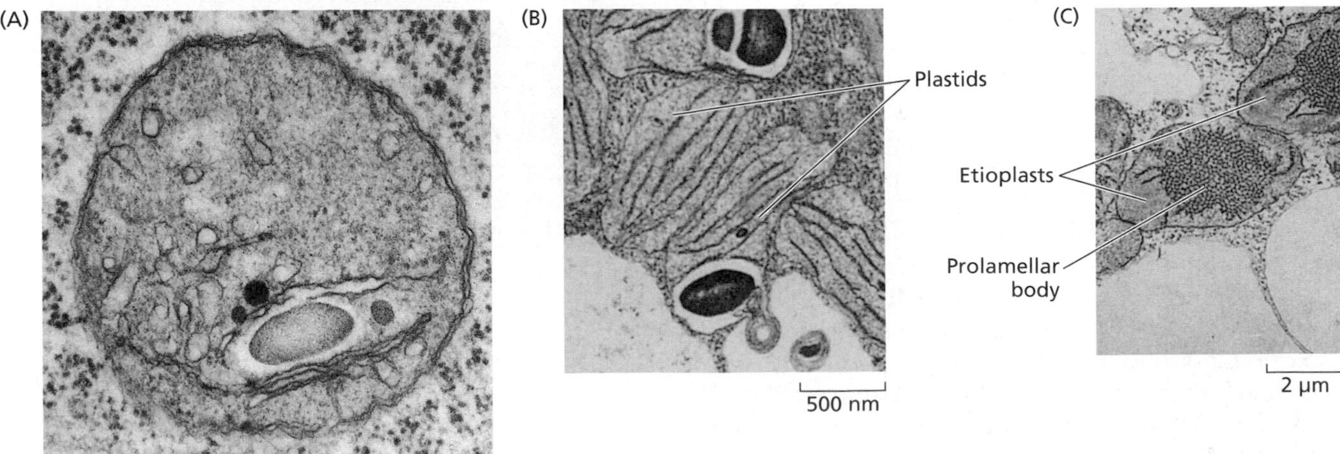

conditions, chloroplasts can be converted to chromoplasts (see Figure 1.22), as in autumn leaves and ripening fruit.

Chloroplast and mitochondrial division are independent of nuclear division

As noted earlier, plastids and mitochondria divide by the process of fission, consistent with their prokaryotic origins. Fission and organellar DNA replication are regulated independently of nuclear division. For example, the number of chloroplasts per cell volume depends on the cell's developmental history and its local environment. Accordingly, there are many more chloroplasts in the mesophyll cells in the interior of a leaf than in the epidermal cells forming the outer layer of the leaf.

Although the timing of the fission of chloroplasts and mitochondria is independent of the timing of cell division, these organelles require nuclear-encoded proteins to divide. In both bacteria and the semiautonomous organelles, fission is facilitated by proteins that form rings on the inner membrane at the site of the future division plane. In plant cells, the genes that encode these proteins are located in the nucleus. The proteins may be delivered to the site through associated ER, which forms a ring around the dividing organelle. Mitochondria and chloroplasts can also increase in size without dividing, in order to meet energy or photosynthetic demand. If, for example, the proteins involved in mitochondrial division are experimentally inactivated, the fewer mitochondria become larger, allowing the cell to meet its energy needs.

Protrusions of the outer and inner membrane occur in both mitochondria and chloroplasts. In chloroplasts these protrusions are called **stromules** because they contain stroma, but no thylakoids (see stromules in WEB TOPIC 1.9). In mitochondria they are called **matrixules**. Although there is little evidence at this time for the function of matrixules and stromules, they may function by exchanging materials with other organelles that they encounter.

Both plastids and mitochondria can move around plant cells. In some plant cells the chloroplasts are anchored in the outer, cortical cytoplasm of the cell, but in others they are mobile. The movement of chloroplasts in response to light is shown in WEB TOPIC 1.9. Like Golgi bodies and peroxisomes, mitochondria are motorized by plant myosins that move along actin microfilaments (see WEB TOPIC 1.9). Actin microfilament networks are among the main components of the plant cytoskeleton, which we will describe next.

The Plant Cytoskeleton

The cytoplasm is organized into a three-dimensional network with filamentous proteins called the **cytoskeleton**. This network provides the spatial organization for the organelles and serves as scaffolding for the movements of organelles and other cytoskeletal components. It also plays fundamental roles in mitosis, meiosis, cytokinesis, wall deposition, the maintenance of cell shape, and cell differentiation.

The plant cytoskeleton consists of microtubules and microfilaments

Two major types of cytoskeletal elements have been demonstrated in plant cells: microtubules and microfilaments. Each type is filamentous, having a fixed diameter and a variable length, up to many micrometers. Microtubules and microfilaments are macromolecular assemblies of globular proteins. Another class of cytoskeletal proteins found in animal cells, the *intermediate filaments*, is not found in the plant genome. This is not surprising, since intermediate filaments (e.g., keratin) are used to produce skin, hair, scales, feathers, and claws—none of which are found in plants. However, there are some places *inside* the animal cell where intermediate filaments have important functions, such as at the inner surface of the nuclear envelope. In these places plants have structural proteins with large interacting regions that coil around each other, the so-called *coiled-coil* domains, and these proteins may serve a function similar to that served by animal intermediate filaments.

Microtubules are hollow cylinders with an outer diameter of 25 nm; they are composed of polymers of the protein **tubulin**. The tubulin monomer of microtubules is a heterodimer composed of two similar polypeptide chains (α- and β-**tubulin**) (**Figure 1.24A**). A single microtubule consists of hundreds of thousands of tubulin monomers arranged in columns called **protofilaments**.

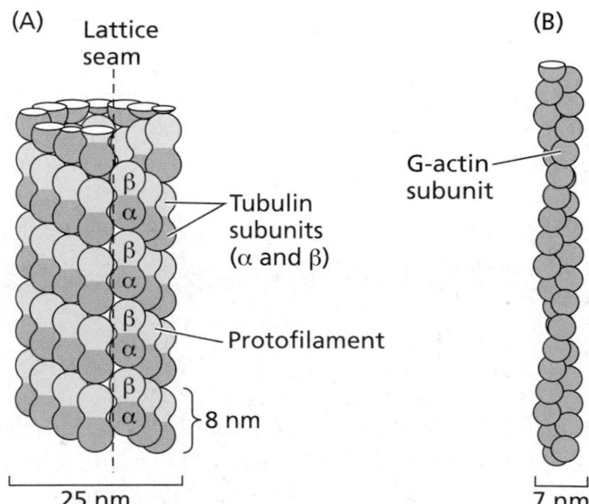

Figure 1.24 (A) Drawing of a microtubule in longitudinal view. Each microtubule is typically composed of 13 protofilaments (but varies with species and cell type). The organization of the α and β subunits is shown. (B) Diagrammatic representation of a microfilament, showing an F-actin strand (protofilament) with a helical pitch based on the asymmetry of the monomers, the G-actin subunits.

(A) Kinetics of actin polymerization without actin-binding proteins

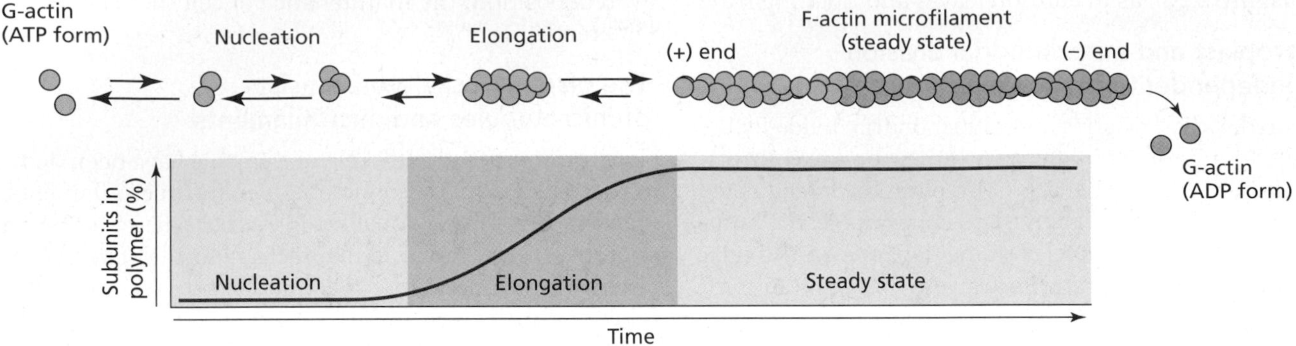

(B) Interaction with profilin and production of bundled actin cables by villin

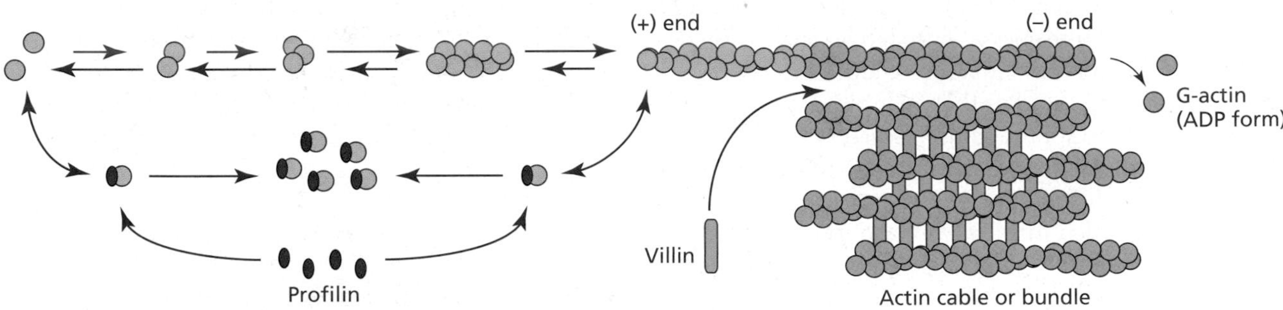

(C) Creation of single and bundled actin microfilaments by formin and fimbrin

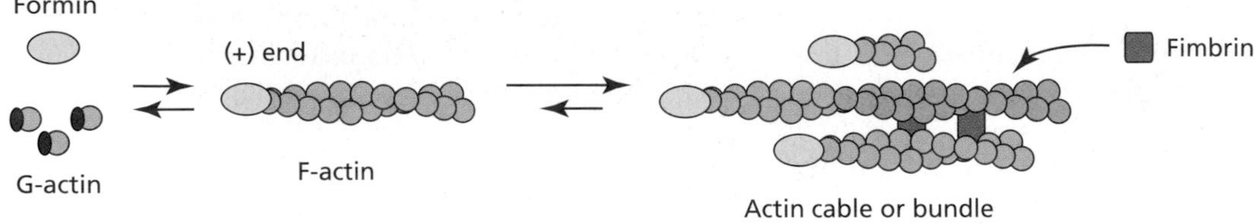

(D) Creation of branched actin filaments by the Arp 2/3 complex

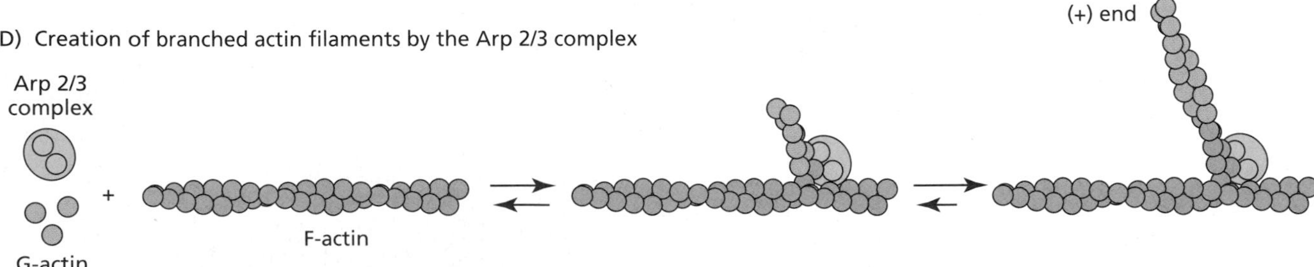

Figure 1.25 Models for the assembly of actin microfilaments. (A) Polymerization of actin filaments occurs at the plus end with actin that has ATP bound to it (green). There is a time lag for polymerization that is dependent on the nucleation of G-actin into small F-actin primers. Once a critical size of the primer has been achieved, the rate of polymerization increases until it reaches a steady state where the rate of addition to the plus end is balanced by the depolymerization at the minus end. ATP hydrolysis to ADP occurs after the ATP-charged units are polymerized and the G-actin that comes off the minus end has ADP (orange). (B) Profilin helps keep a pool of readily available ATP-charged actin in the cell which can then be added to the plus end. Growing tubules can be stabilized by cross-linking the actin with the bundling protein villin. (C) Formins can nucleate actin monomers, thereby overcoming the lag phase seen in the absence of accessory proteins (see A above). The newly formed F-actin associated with formin can be stabilized and cross-linked with the ABP fimbrin. (D) Besides formins, the other protein that can nucleate the formation of new actin filaments is the Arp 2/3 complex. This complex initiates actin branching at a certain angle from preexisting microfilaments.

Microfilaments are solid, with a diameter of 7 nm. They are composed of the monomeric form of the protein **actin**, called globular actin, or **G-actin**. Monomers of G-actin polymerize to form a single chain of actin subunits, also called a *protofilament*. The actin in the polymerized protofilament is referred to as filamentous actin, or **F-actin**. A microfilament is helical, a shape resulting from the polarity of association of the G-actin monomers (**Figure 1.24B**). As described below, this polarity dictates microfilament dynamics.

Actin, tubulin, and their polymers are in constant flux in the living cell

In the cell, actin and tubulin subunits exist as pools of free proteins that are in dynamic equilibrium with their polymerized forms. The cycle of polymerization–depolymerization is essential for cell life; drugs that halt this cycle will eventually kill the cell. Each of the monomers contains a bound nucleotide: ATP or ADP in the case of actin, GTP or GDP (guanosine tri- or diphosphate) in the case of tubulin. Both microtubules and microfilaments are polarized; that is, the two ends are different. The polarity is displayed by the different rates of growth of the two ends, with the more active end being the **plus end** and the less active end the **minus end**. In microfilaments, the polarity arises from the polarity of the actin monomer itself; the ATP/ADP binding cleft is exposed at the minus end, while the side opposite the cleft is exposed at the plus end. In microtubules, the polarity arises from the polarity of the α- and β-tubulin heterodimer. In microtubules, the α-tubulin monomer exists only in the GTP form and is exposed on the minus end, and the β-tubulin can bind either GTP or GDP and is exposed on the plus end. Microfilaments and microtubules have half-lives, usually counted in minutes, determined by accessory proteins: *actin-binding proteins* (*ABPs*) in microfilaments and *microtubule-associated proteins* (*MAPs*) in microtubules. ABPs and MAPs perform diverse functions which can regulate the dynamics of microfilaments and microtubules.

The polymerization of G-actin in the absence of ABPs in a test tube is not only concentration-dependent, whereby it must reach a critical concentration in order to polymerize, but also time-dependent, requiring time for nucleation of monomers into a size stable enough to allow further elongation (**Figure 1.25A**). During elongation and in the steady state, the plus end polymerizes rapidly, while the other end of the microfilament, the minus end, polymerizes slowly. F-actin slowly hydrolyzes ATP to ADP (green subunit to orange subunit transition in Figure 1.25). **Profilins** regulate the balance between G- and F-actin (**Figure 1.25B**). New F-actin is initiated in two ways: by filament growth activated by proteins called **formins** (**Figure 1.25C**), and by branching from filaments at junctions formed by the actin filament nucleator **Arp 2/3** (**Figure 1.25D**). There are also proteins involved in shearing the actin filament into pieces, such as **actin depolymerizing factor** (**ADF**). Turnover of actin in the living cell involves extensive shearing. But microfilaments can be stabilized into bundles through association with the proteins **villin** and **fimbrin** (see Figure 1.25B and C). These bundles form the center of transvacuolar strands and cytoplasmic thickenings in the cell cortex where fast lanes of cytoplasmic streaming occur (see Figure 1.6).

The assembly of microtubules from free tubulin in a test tube follows a similar time-dependent pattern as that of actin, involving nucleation, elongation, and steady state phases (see Figure 1.25A). Microtubule nucleation and initiation of growth occurs inside the cell at *microtubule-organizing centers* (*MTOCs*), also called *initiation complexes*, but the nature of the initiation complex is still something of a mystery. One type of initiation complex contains a much less abundant type of tubulin called *γ-tubulin*. γ-Tubulin, along with accessory proteins, can form a ring from which microtubules grow. These γ-tubulin ring complexes are present in the cortical cytoplasm, sometimes associated with microtubule branches (**Figure 1.26A–C**), similar to how Arp 2/3 is present at branches of microfilaments. γ-Tubulin ring complexes prime the polymerization of α- and β-tubulin heterodimers into short longitudinal protofilaments. Next, the protofilaments (the number varies with species) associate laterally to form a flat sheet (see Figure 1.26A). The sheet curls into a cylindrical microtubule as GTP is hydrolyzed (see Figure 1.26B). In most other organisms, γ-tubulin ring complexes are involved in the initiation of microtubule growth, but in plants some initiation complexes do not contain γ-tubulin ring complexes. The principal sites of initiation complexes include the cortical cytoplasm in cells at interphase, the periphery of the nuclear envelope, and the spindle poles in dividing cells.

Each tubulin heterodimer contains two GTP molecules, one on the α-tubulin monomer and the other on the β-tubulin monomer. GTP on the α-tubulin monomer is tightly bound and nonhydrolyzable, while the GTP on the β-tubulin site is hydrolyzed to GDP at some time after the heterodimer assembles onto the plus end of a microtubule. The hydrolysis of GTP to GDP on the β-tubulin subunit causes the dimer to bend slightly, and if the rate of GTP hydrolysis "catches up" with the rate of addition of new heterodimers, the GTP-charged cap of tubulin vanishes and the protofilaments come apart from each other, initiating a *catastrophic* depolymerization that is much more rapid than the rate of polymerization (see Figure 1.26C). These catastrophes can also occur when a microtubule runs into another at an angle greater than 40°. Such catastrophes can be rescued (depolymerization stopped and polymerization resumed) if the increase in the local free tubulin concentration (with GTP) caused by the catastrophe once again favors polymerization. This process is called **dynamic instability** (see Figure 1.26A–C). The minus end, or slowly growing

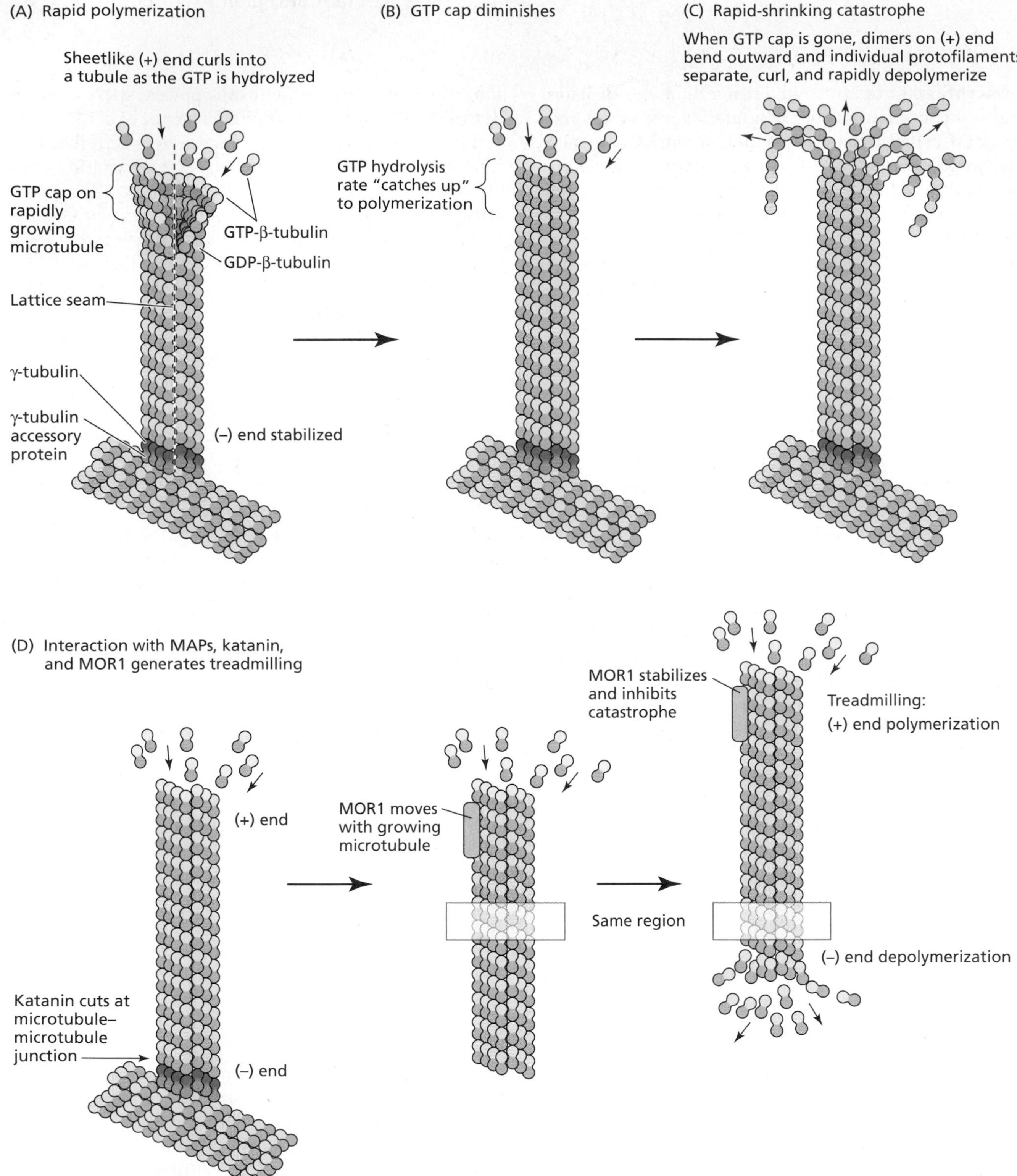

(A) Rapid polymerization
Sheetlike (+) end curls into a tubule as the GTP is hydrolyzed
GTP cap on rapidly growing microtubule
GTP-β-tubulin
GDP-β-tubulin
Lattice seam
γ-tubulin
γ-tubulin accessory protein
(–) end stabilized
(B) GTP cap diminishes
GTP hydrolysis rate "catches up" to polymerization
(C) Rapid-shrinking catastrophe
When GTP cap is gone, dimers on (+) end bend outward and individual protofilaments separate, curl, and rapidly depolymerize
(D) Interaction with MAPs, katanin, and MOR1 generates treadmilling
(+) end
Katanin cuts at microtubule–microtubule junction
(–) end
MOR1 moves with growing microtubule
Same region
MOR1 stabilizes and inhibits catastrophe
Treadmilling:
(+) end polymerization
(–) end depolymerization

Figure 1.26 Models for microtubule dynamic instability and treadmilling. (A) The minus ends of microtubules at new initiation sites can be stabilized by γ-tubulin ring complexes, some of which are found along the sides of previously existing microtubules. Microtubule plus ends grow rapidly, producing a "cap" of tubulin that has GTP bound to the β subunit. The newly added cap has a sheetlike structure that curls into a tubule as the GTP is hydrolyzed. (B) As the rate of growth slows or GTP hydrolysis increases, the GTP cap is diminished. (C) When the GTP cap is gone, the microtubule protofilaments peal apart because the heterodimer with GDP bound to the β subunit of tubulin is slightly curved. The protofilaments are unstable, and rapid, catastrophic depolymerization ensues. (D) If the microtubule is severed at the branch point by the ATPase katanin, the minus end becomes unstable and can depolymerize. If the stability of the microtubule is stabilized against dynamic instability at the plus end by MOR1 (microtubule organizing 1 protein), a MAP, then the rate of addition on the plus can match the depolymerization at the minus end and treadmilling ensues.

microtubule end, does not depolymerize if it is capped by γ-tubulin. However, plant microtubules can be released from γ-tubulin ring complexes by an ATPase, **katanin** (from the Japanese word *katana*, "samurai sword"), which severs the microtubule at the point where the growing microtubule branches off another (see Figure 1.26D). Once the microtubules are released by katanin, they travel, snakelike, through the cell cortex by a mechanism called treadmilling.

Cortical microtubules move around the cell by treadmilling

Microtubules in the cortical cytoplasm can migrate laterally around the cell periphery by a process called **treadmilling**. During treadmilling, tubulin heterodimers are added to the growing plus end at about the same rate that they are removed from the shrinking minus end (see Figure 1.26D). The actual tubulin subunits do not move relative to the cell once they are polymerized into the microtubule (see shaded region in Figure 1.26D), because the microtubule is usually bound to a membrane through a variety of MAPs. But the microtubule does move along, just under the plasma membrane, as it adds more subunits to the plus end and removes them from the minus end. Such movement would stop if the plus end were to shrink rapidly with a catastrophe. However, the microtubule is stabilized against catastrophes by MAPs, most notably MOR1 ("*m*icrotubule *or*ganization *1*"), which can move down a microtubule while it is treadmilling. As we will discuss in Chapter 19 and in the section *Cell Cycle Regulation* below (see Figures 1.37 and 1.39), the transverse orientation of the cortical microtubules determines the orientation of the newly synthesized cellulose microfibrils in the cell wall. The presence of transverse cellulose microfibrils in the cell wall reinforces the wall in the transverse direction, promoting growth along the longitudinal axis. In this way microtubules play an important role in the polarity of plant growth.

Cytoskeletal motor proteins mediate cytoplasmic streaming and directed organelle movement

As shown in the movies in WEB TOPIC 1.9, mitochondria, peroxisomes, and Golgi bodies are extremely dynamic in plant cells. These approximately 1 μm particles move at rates of about 1–10 μm s^{-1} in seed plants. This movement is quite fast; if you proportionally scale up ($\times 10^6$ in size), it is equivalent to a 1 m object moving at 10 m sec^{-1}, approximately the speed of the world's fastest human. It is much faster than animal cells crawling across a substrate (0.01–0.2 μm s^{-1}), and is about the same speed as pigment particles moving in the melanocytes of fish or octopi when they rapidly change color to hide or attack. But plant cells "move it" almost all the time! Actin and its motor protein, myosin, act together within the plant endoplasm to generate this movement and are therefore frequently referred to as the actomyosin cytoskeleton.

This movement of individual organelles can be part of **cytoplasmic streaming**, the coordinated bulk flow of the cytosol and cytoplasmic organelles inside the cell. But it is perhaps better called **directed organelle movement** because organelles can frequently move past each other in opposite directions (see WEB TOPIC 1.9). If directed organelle movement exerts sufficient viscous drag on the surrounding cytosol and the organelles in it, then it will drive cytoplasmic streaming. As cells enlarge, movement rates tend to increase. In the giant cells of the green algae *Chara* and *Nitella*, cytoplasmic streaming occurs in a helical path down one side of a cell and up the other side, occurring at speeds of up to 75 μm s^{-1}. Directed organelle movement works in concert with **organelle tethering**, the anchoring of organelles to each other, the cytoskeleton, or the plasma membrane, to organize the cytoplasm of plant cells. For example, the chloroplasts of *Chara* and *Nitella* are tethered so that they do not move, even though there is extremely active movement of other organelles in the inner cytoplasm. Likewise, the cortical ER is tethered to the plasma membrane, while internal ER (e.g., in transvacuolar strands; see Figure 1.14) is more dynamic.

Molecular motors participate in both directed organelle movements and tethering. Plants have two types of motors, the myosins and the kinesins. The myosins are ABPs, reversibly binding to actin microfilaments. There are two families of plant myosins, myosin VIII, which may serve primarily as tethers of organelles during plant development, and myosin XI, members of which are primarily responsible for most organelle movement in vegetative nondividing cells. The **kinesins** are MAPs and bind microtubules. When they move along the cytoskeleton, they do so in a particular direction along the polar cytoskeletal polymers. Myosins generally move toward the plus end of the actin filaments (only myosin VI in animals is known to move toward the minus end). Of the 61 members of the kinesin family, two-thirds move toward the plus end of microtubules and one-third move toward the minus end. Although members of the kinesin family interact with some organelle membranes, they tend to tether the organelles rather than mediate movement along microtubules. Kinesins can bind to chromatin or to other microtubules, and help organize the spindle apparatus during mitosis (see below and WEB TOPIC 1.10). *Dyneins*, the predominantly minus end–directed microtubule motors in animals and protists, are absent in plants but are present in flagellated green algae such as *Chlamydomonas*.

How is it that motors can both move and tether organelles? All of the above motors have separate head, neck, and tail domains, as does myosin XI (**Figure 1.27**). The globular head domain binds to the cytoskeleton reversibly, depending on the energy state of ATP in the ATPase active site. The neck domain changes angle upon hydrolysis of ATP, flexing the head relative to the tail. The tail domain usually contains coiled-coil regions for dimerization, and the end

(A) Unfolded sequence of domains in myosin XI

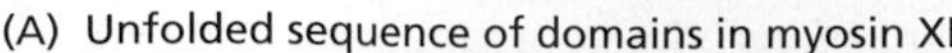

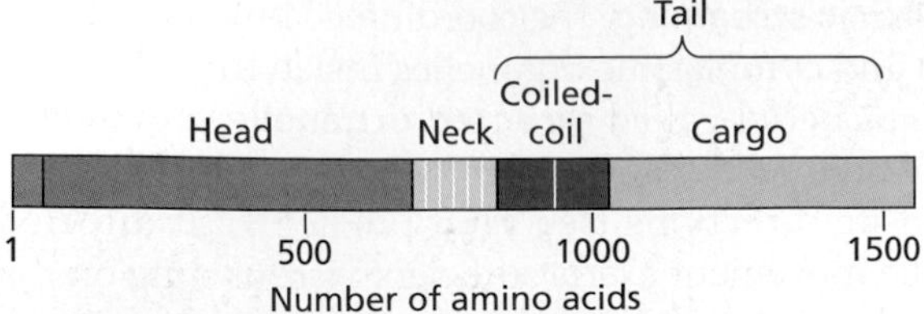

(B) Folded dimer configuration of myosin XI

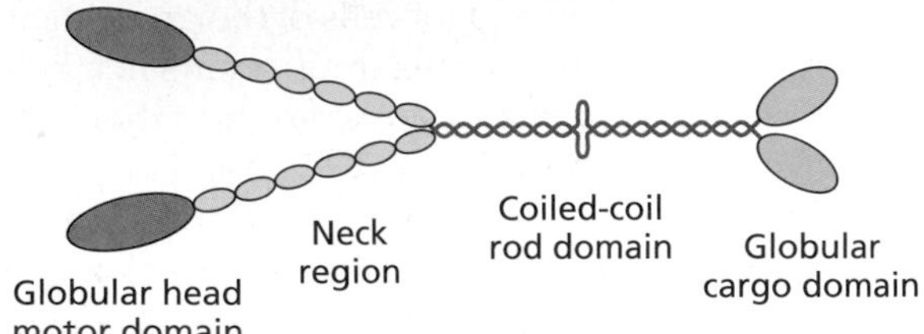

(C) Movement of cargo and powerstroke of myosin XI

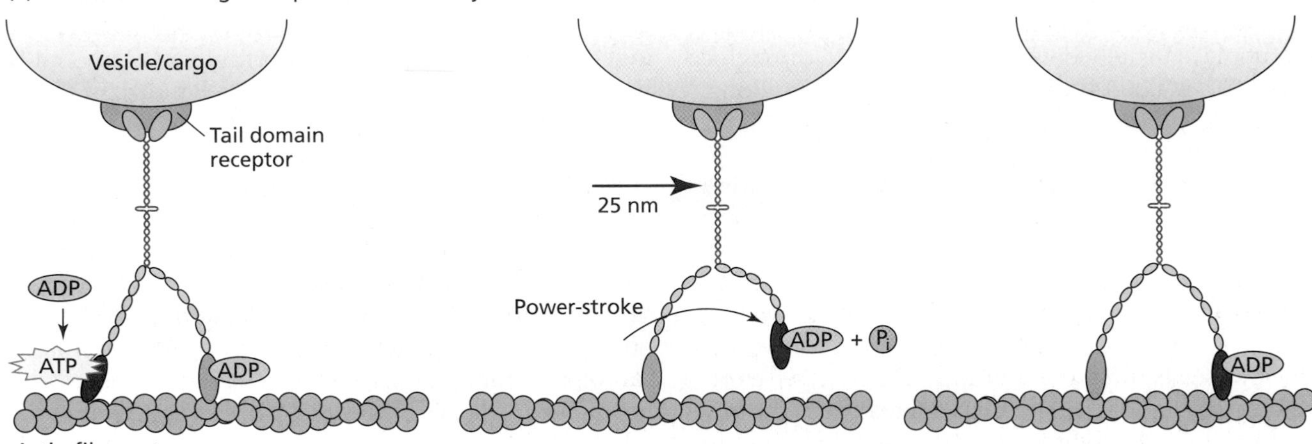

Figure 1.27 Myosin-driven movement of organelles. (A) Extended amino acid domains of the myosin motor. The tail domain includes a coiled-coil region for dimerization and a cargo domain for interacting with membranes. (B) The head domain folds to become globular. Near the extended neck domain, ATP/ADP binds to the head domain. The neck consists of regions with a particular amino acid composition (IQ motif) that can interact with modulating proteins. (C) Movement and power-stroke of myosin XI. The tail binds to the organelle via its cargo domains and a receptor complex on the membrane. The two heads, shown in red and pink, have ATPase and motor activity such that a change in the configuration of the neck region adjacent to the head produces a "walking," processive movement along the actin filament during the power-stroke of the motor, when ATP is hydrolyzed to ADP and inorganic phosphate (P_i). The cargo moves about 25 nm with each step. When the phosphate is released, the dimer is "reset" to the pre–power-stroke state.

of the globular tail domain binds to specific organelles or "cargo" and is called the *cargo domain* (see Figure 1.27B). In order for a motor to tether an organelle via the cytoskeleton to the plasma membrane, the motor head binds to the cytoskeleton, which is bound to the organelle, while the cargo domain binds to a protein in the plasma membrane. Often tethering motors are monomeric, so when the myosin-bound ATP hydrolyses, the head domain releases from the cytoskeleton and the organelle bound to the cytoskeleton via the motor is released. In order for a motor to move an organelle, the motor dimerizes; the two molecules interact at the coiled-coil tail domain and bind to the organelle at the cargo domain. The two heads of the dimer then alternately bind to the cytoskeleton and "walk" forward while the neck flexes as ATP is hydrolyzed (see Figure 1.27C). This way, the organelle is moved along the cytoskeleton.

Some movements are differentially regulated. For example, the chloroplasts of plants reorient themselves under certain lighting conditions (see **WEB TOPIC 1.9**; also see Chapter 9). These movements can be "superimposed" on other directed organelle movements because the mechanism of chloroplast movement is different from the movement of Golgi bodies, peroxisomes, and mitochondria. Chloroplast movement occurs on small actin bundles and uses some kinesin-like proteins to move along the actin (instead of microtubules!) and other kinesin-like proteins to tether it to the plasma membrane.

Organelle movement serves the dynamic changes that accompany growth and development. Differential tethering and movement can set up a polar distribution of organelles inside the cell, as in pollen tubes, where the tube cell grows only at the tip, not along the shank, by secreting membrane and wall materials only at the tip (see Chapter 21). In the moss *Physcomytrella patens,* which has a reduced number of myosin XI isoforms, myosin XI is involved not only in vesicle transport but also in polarized cell growth. Directed organelle movement and cytoplasmic streaming provide a rapid response system which plant cells need when responding to the external biotic and abiotic environment. For example, a local insult, such as a fungal infection or wounding, can lead to rapid reorganization of the cell, and a change in the direction of light or gravity can necessitate repositioning of organelles. Finally, organelle movement serves the choreography of cell division, which we will discuss next.

Cell Cycle Regulation

The cell division cycle, or cell cycle, is the process by which cells reproduce themselves and their genetic material, the nuclear DNA (**Figure 1.28**). The cell cycle consists of four phases: G_1, S, G_2, and M. **G_1** is the phase when a newly formed daughter cell has not yet replicated its DNA. DNA is replicated during the **S phase**. **G_2** is the phase when a cell with replicated DNA has not yet proceeded to mitosis. Collectively, the G_1, S, and G_2 phases are referred to as **interphase**. The M phase is mitosis. In vacuolated cells, the vacuole enlarges throughout interphase, and the plane of cell division bisects the vacuole during mitosis (see Figure 1.28).

Each phase of the cell cycle has a specific set of biochemical and cellular activities

Nuclear DNA is prepared for replication in G_1 by the assembly of a prereplication complex at the origins of replication along the chromatin. DNA is replicated during the S phase, and G_2 cells prepare for mitosis.

The whole architecture of the cell is altered as it enters mitosis. If the cell has a large central vacuole, this vacuole must first be divided in two by a coalescence of cytoplasmic transvacuolar strands that contain the nucleus; this becomes the region where nuclear division will occur. (Compare Figure 1.28, division in a vacuolated cell, with

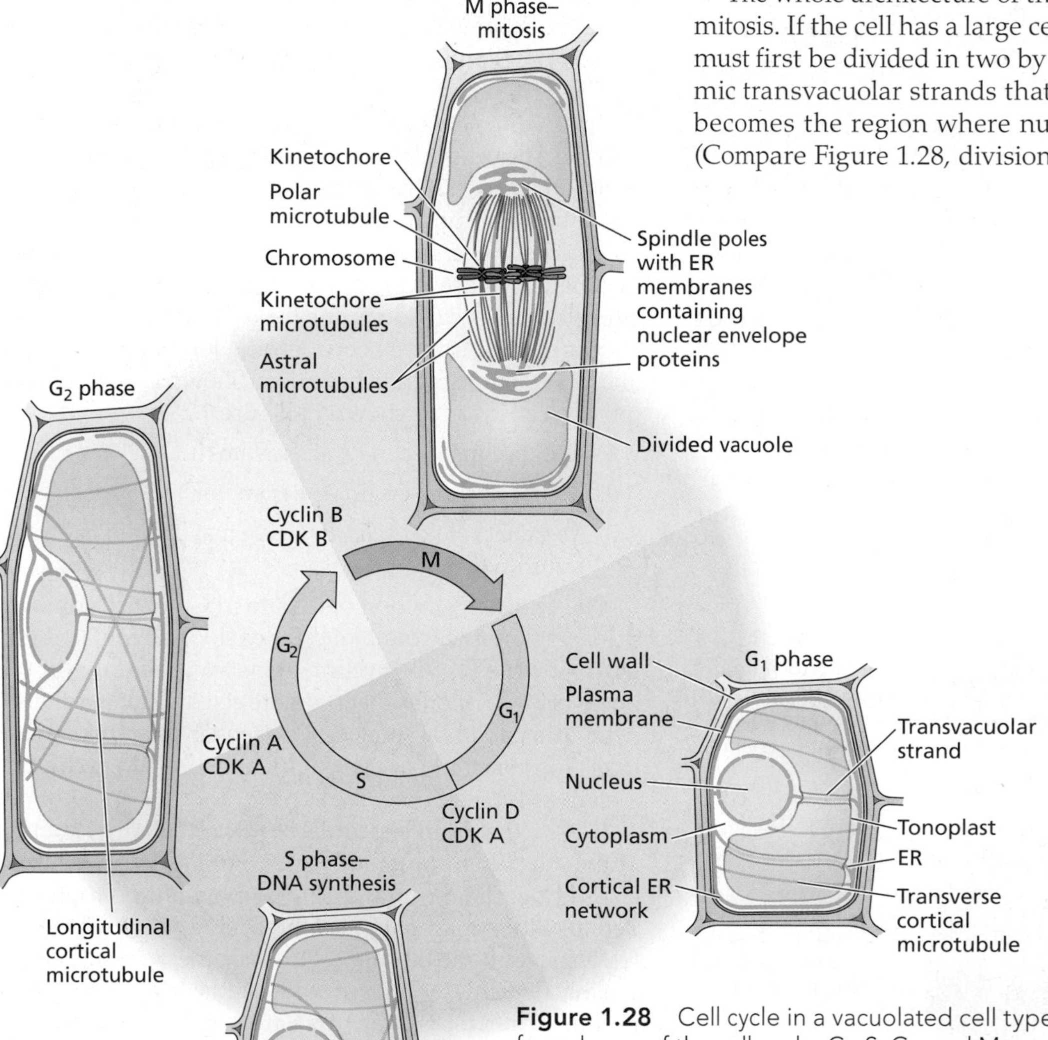

Figure 1.28 Cell cycle in a vacuolated cell type (a tobacco cell). The four phases of the cell cycle, G_1, S, G_2, and M, are shown in relation to the elongation and division of a vacuolated cell. Various cyclins and cyclin-dependent kinases (CDKs) regulate the transitions from one phase to the next. Cyclin D and cyclin-dependent kinase A (CDK A) are involved in the transition from G_1 to S. Cyclin A and CDK A are involved in the transition from S into G_2. Cyclin B and cyclin-dependent kinase B (CDK B) regulate the transition from G_2 into M. The kinases phosphorylate other proteins in the cell, causing major reorganization of the cytoskeleton and the membrane systems. The cyclin/CDK complexes have a finite lifetime, usually regulated by their own phosphorylation state; the decrease in their abundance toward the end of each phase allows progression to the next stage of the cell cycle.

Figure 1.30, division in a nonvacuolated cell.) Golgi bodies and other organelles divide and partition themselves equally between the two halves of the cell. As described below, the endomembrane system and cytoskeleton are extensively rearranged.

As a cell enters mitosis, the chromosomes change from their interphase state of organization within the nucleus and begin to condense to form the metaphase chromosomes (**Figure 1.29**; also see Figure 1.9). The metaphase chromosomes are held together by special proteins called *cohesins*, which reside in the centromeric region of each chromosome pair. In order for the chromosomes to separate, these proteins must be cleaved by the enzyme *separase*, which first has to be activated. This occurs when the kinetochore attaches to spindle microtubules (described in the next section).

At a key regulatory point, or **checkpoint**, early in G_1 of the cell cycle, the cell becomes committed to the initiation of DNA synthesis. In mammalian cells, DNA replication and mitosis are linked—the cell division cycle, once begun, is not interrupted until all phases of mitosis have been completed. In contrast, plant cells can leave the cell division cycle either before or after replicating their DNA (i.e., during G_1 or G_2). As a consequence, whereas most animal cells are diploid (having two sets of chromosomes), plant cells are frequently tetraploid (having four sets of chromosomes) or sometimes even polyploid (having many sets of chromosomes) after going through additional cycles of nuclear DNA replication without mitosis, a process called **endoreduplication**. The role of polyploidy in plant evolution will be discussed in Chapter 2. The cell cycle can regulate differentiation in some cell types, with "giant" cells forming as a result of cycles of endoreduplication, while "small" cells remain mitotically active.

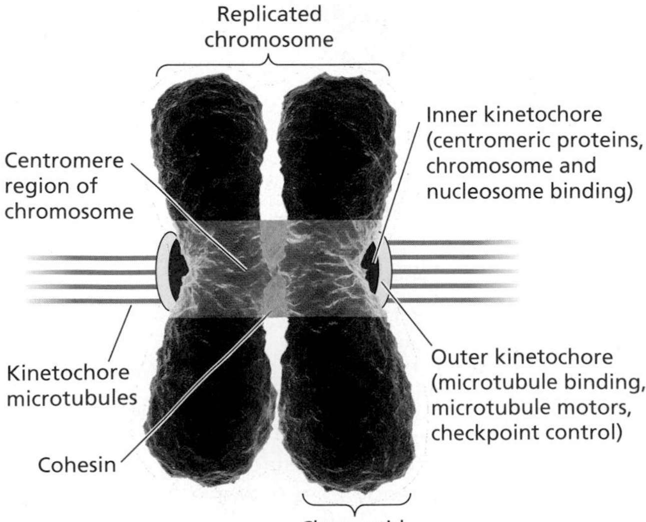

Figure 1.29 Structure of a metaphase chromosome. The centromeric DNA is highlighted, and the region where cohesion molecules bind the two chromosomes together is shown in orange. The kinetochore is a layered structure (inner layer purple, outer layer yellow) that contains microtubule-binding proteins, including kinesins that help depolymerize the microtubules during shortening of kinetochore microtubules in anaphase.

The cell cycle is regulated by cyclins and cyclin-dependent kinases

The biochemical reactions governing the cell cycle are evolutionarily highly conserved in eukaryotes, and plants have retained the basic components of this mechanism. Progression through the cycle is regulated mainly at three checkpoints: during the late G_1 phase (as mentioned above), late S phase, and at the G_2/M boundary.

The key enzymes that control the transitions between the different phases of the cell cycle, and the entry of nondividing cells into the cell cycle, are the **cyclin-dependent kinases**, or **CDKs**. Protein kinases are enzymes that phosphorylate proteins using ATP. Most multicellular eukaryotes have several protein kinases that are active in different phases of the cell cycle. All depend on regulatory subunits called **cyclins** for their activities. Several classes of cyclins have been identified in plants, animals, and yeast. Three cyclins have been shown to regulate the tobacco cell cycle, as shown in Figure 1.28:

1. G_1/S cyclins, cyclin D, active late in G_1.
2. S-type cyclins, cyclin A, active in late S phase.
3. M-type cyclins cyclin B, active just prior to the mitotic phase.

The critical restriction point late in G_1, which commits the cell to another round of cell division, is regulated primarily by the D-type cyclins. As we will see later in the book, plant hormones that promote cell division, including cytokinins and brassinosteroids (see Chapter 15), appear to do so at least in part through an increase in cyclin D3, a plant D-type cyclin.

CDK activity can be regulated in various ways, but two of the most important mechanisms are (1) cyclin synthesis and degradation, and (2) the phosphorylation and dephosphorylation of key amino acid residues within the CDK protein. In the first regulatory mechanism, CDKs are inactive unless they are associated with a cyclin. Most cyclins turn over rapidly; they are synthesized and then actively degraded (using ATP) at specific points in the cell cycle. Cyclins are degraded in the cytosol by a large proteolytic complex called the **26S proteasome** (see Chapter 2). Before being degraded by the proteasome, the cyclins are marked for destruction by the attachment of a small protein called *ubiquitin*, a process that requires ATP. Ubiquitination is a general mechanism for tagging cellular proteins destined for turnover (see Chapter 2).

The second mechanism regulating CDK activity is phosphorylation and dephosphorylation. CDKs possess two tyrosine phosphorylation sites: One causes activation of the enzyme, and the other causes inactivation. Specific kinases activate the CDKs, while other specific kinases inactivate them. Similarly, protein phosphatases can remove phosphates from CDKs, either stimulating or inhibiting their activity depending on the position of the phosphate. The addition or removal of phosphate groups from CDKs is highly regulated and an important mechanism for the control of cell cycle progression. Further control of the pathway is exerted by the presence of CDK inhibitors (ICKs) that can influence the G_1/S transition.

Mitosis and cytokinesis involve both microtubules and the endomembrane system

Mitosis is the process by which previously replicated chromosomes are aligned, separated, and distributed in an orderly fashion to daughter cells (**Figure 1.30**; also see Figure 1.28). Microtubules are an integral part of mitosis. The period immediately prior to prophase is called **preprophase**. During preprophase, the G_2 microtubules are completely reorganized into a **preprophase band** of microtubules around the nucleus at the site of the future cell plate—the precursor of the cross-wall (see Figure 1.30). The position of the preprophase band, the underlying *cortical division site*, and the partition of cytoplasm that divides the central vacuoles determine the plane of cell division in plants, and thus play a crucial role in development (see Chapters 17–19).

At the start of prophase, microtubules, polymerizing on the surface of the nuclear envelope, begin to gather at two foci on opposite sides of the nucleus, initiating **spindle formation** (see Figure 1.30). Although not associated with the centrosomes (which plants, unlike animal cells, lack), these foci serve the same function in organizing microtubules. During **prophase**, the nuclear envelope remains intact, but it breaks down as the cell enters **metaphase** in a process that involves reorganization and reassimilation of the nuclear envelope into the ER (see Figure 1.30). During division, ER tubules "switch tracks," jumping off of the actin cytoskeleton and jumping onto the microtubules of the mitotic spindle. Two polar regions of aggregated ER then reside on either side of the spindle and individual ER tubules pervade the spindle. Throughout the division cycle, cell division kinases interact with microtubules to help reorganize the spindle by phosphorylating MAPs and kinesins.

As the chromosomes condense, the nucleolar organizer regions (NORs) of the different chromosomes disassociate, causing the nucleolus to fragment. The nucleolus completely disappears during mitosis and gradually reassembles after mitosis, as the chromosomes decondense and reestablish their positions in the daughter nuclei.

In early metaphase, or **prometaphase**, the preprophase band is gone and new microtubules polymerize to complete the **mitotic spindle**. The mitotic spindles of plant cells, which lack centrosomes, are more boxlike in shape than those in animal cells. The spindle microtubules in a plant cell arise from a diffuse zone consisting of multiple foci at opposite ends of the cell and extend toward the middle in nearly parallel arrays.

Metaphase chromosomes are completely condensed through close packing of the histones into nucleosomes, which are further wound into condensed fibers (see Figures 1.9 and 1.29). The **centromere**, the region where the two chromatids are attached near the center of the chromosome, contains repetitive DNA, as does the **telomere**, which forms the chromosome ends that protect it from degradation. Some of the spindle microtubules bind to the chromosomes in the special region of the centromere called the **kinetochore**, and the condensed chromosomes align at the metaphase plate (see Figures 1.28 and 1.30). Some of the unattached microtubules overlap with microtubules from the opposite polar region in the spindle midzone.

Just as there are checkpoints that control the four phases of the cell cycle, there are also checkpoints that operate *within* mitosis. For example, a **spindle attachment checkpoint** stops cells from proceeding into anaphase if spindle microtubules have incorrectly interacted with the kinetochores. The cyclin B–CDK B complex plays a central role in regulating this process. If the spindle microtubules have correctly attached to their kinetochores, the **anaphase promoting complex** brings about the proteolytic degradation of an inhibitor of separase, thereby activating separase, which cleaves the cohesin that binds the two chromosomes together (see Figure 1.29); this allows the chromatids aligned at the metaphase plate to segregate to their respective poles. The anaphase promoting complex also promotes the ubiquitination and subsequent proteolytic degradation of cyclin B. Without cyclin B, the cyclin B–CDK B complex can no longer form and the spindle dissembles, the chromosomes decondense, and the nuclear envelope re-forms. (Note that each chromatid has undergone a round of DNA replication and contains the diploid [2*N*] amount of DNA. Thus, as soon as separation occurs, the chromatids become chromosomes.)

The mechanism of chromosome separation during **anaphase** has two components:

- **anaphase A** or early anaphase, during which the sister chromatids separate and begin to move toward their poles; and
- **anaphase B** or late anaphase, during which the polar microtubules slide relative to each other and elongate to push the spindle poles farther apart. At the same time, the sister chromosomes are pushed to their respective poles.

In plants, the spindle microtubules are apparently not anchored to the cell cortex at the poles, and so the chromosomes cannot be pulled apart. Instead they are probably pushed apart by kinesins in the overlapping polar microtubules of the spindle (see **WEB TOPIC 1.10**).

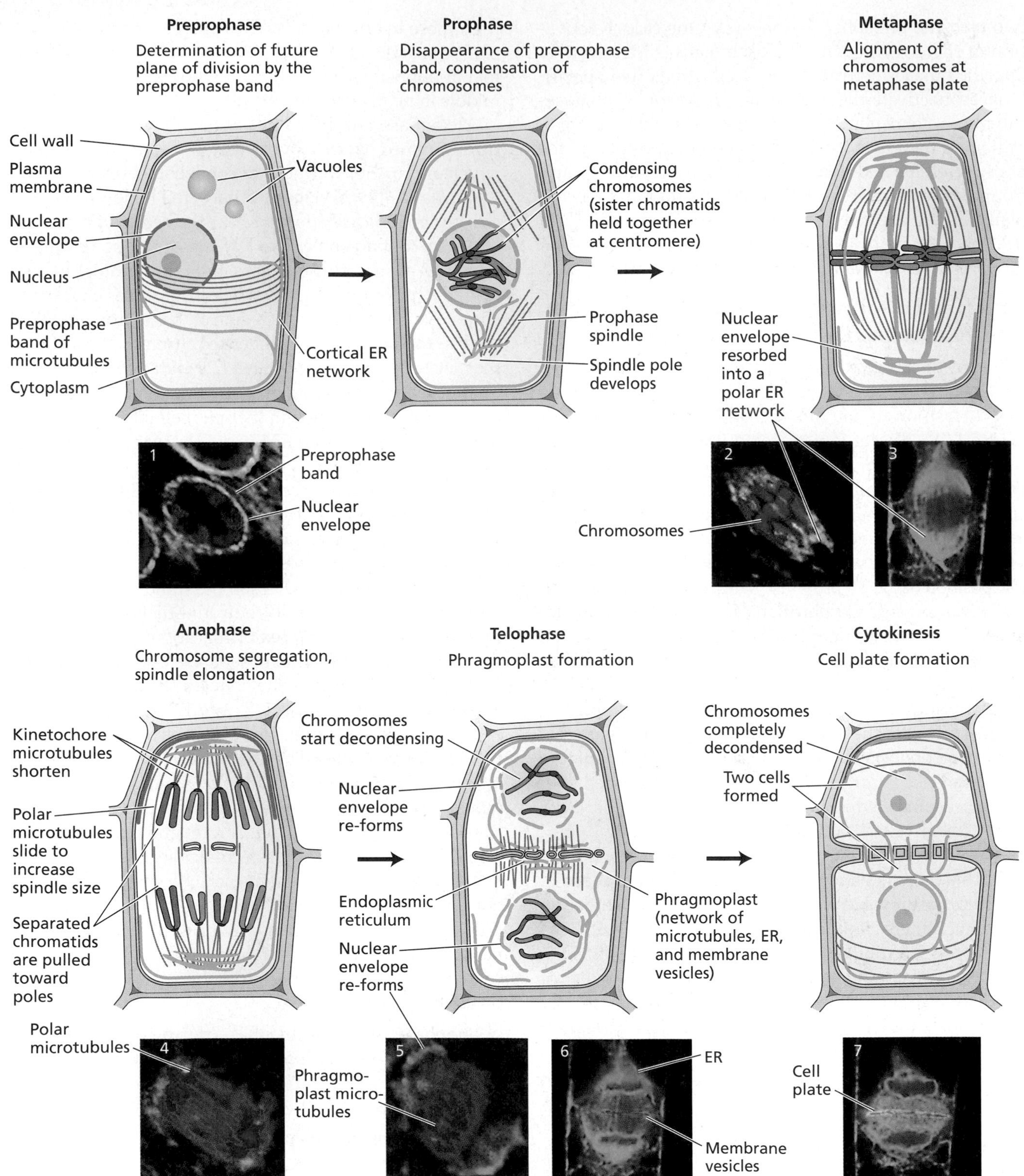

Figure 1.30 Changes in cellular organization that accompany mitosis in a meristematic (nonvacuolated) plant cell. (1, 2, 4, and 5) Red fluorescence is from anti-α-tubulin antibody (microtubules), green fluorescence is from WIP–GFP (green fluorescent protein fused to a nuclear envelope protein), and blue fluorescence is from DAPI (a DNA-binding dye). (3, 6, and 7) The ER is labeled with green-fluorescing HDEL–GFP, and the cell plate is labeled with red-fluorescing FM4-64. (1, 2, 4, and 5 from Xu et al. 2007; 3, 6, and 7 from Higaki et al. 2008.)

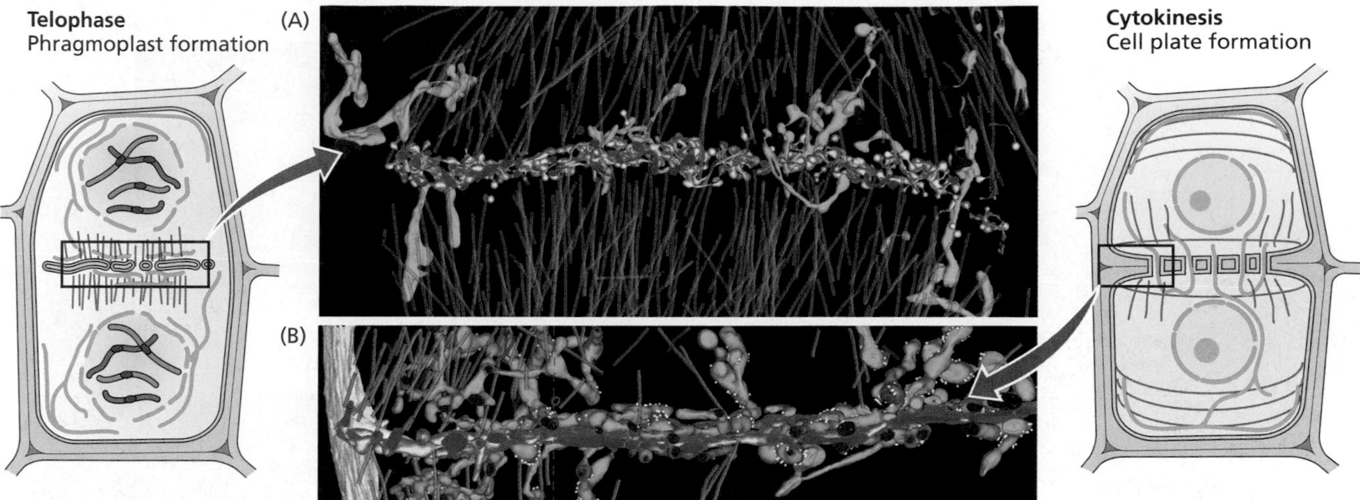

Figure 1.31 Changes in the organization of the phragmoplast and ER during cell plate formation. (A) The forming cell plate (yellow, seen from the side) in early telophase has only a few sites where it interacts with the tubulovesicular network of the ER (blue). The solid phragmoplast microtubules (purple) also have few ER cisternae among them. (B) Side view of the forming peripheral cell plate (yellow), showing that although many cytoplasmic ER tubules (blue) intermingle with microtubules (purple) in the peripheral growth zone, there is little direct contact between ER tubules and cell plate membranes. The small white dots are ER-bound ribosomes. (Three-dimensional tomographic reconstruction of electron microscopy of the phragmoplast from Seguí-Simarro et al. 2004.)

At **telophase**, a new network of microtubules and F-actins called the **phragmoplast** appears (**Figure 1.31**). The phragmoplast organizes the region of the cytoplasm where cytokinesis takes place. The microtubules have now lost their spindle shape but retain polarity, with their minus ends still pointed toward the separated, now decondensing, chromosomes, where the nuclear envelope is in the process of re-forming (see Figure 1.31, "Telophase"). The plus ends of the microtubules point toward the midzone of the phragmoplast, where small vesicles accumulate, partly derived from endocytotic vesicles from the parent cell plasma membrane. These vesicles have long tethers that may aid in the formation of the cell plate in the next cell cycle stage: cytokinesis.

Cytokinesis is the process that establishes the **cell plate**, a precursor of the cross-wall that will separate the two daughter cells (see Figure 1.30). This cell plate, with its enclosing plasma membrane, forms as an island in the center of the cell that grows outward toward the parent wall by vesicle fusion. The targeting recognition protein **KNOLLE**, which belongs to the SNARE family of proteins involved in vesicle fusion (see **WEB TOPIC 1.6**), is present at the forming cell plate, as is **dynamin**, a noose-shaped GTPase that is involved in vesicle formation. There are also several molecular motors and tethering factors involved in the assembly. The site at which the forming cell plate fuses with the parental plasma membrane is determined by the location of the preprophase band (which disappeared earlier) and specific microtubule-associated proteins (MAPs). As the cell plate assembles, it traps ER tubules in plasmalemma-lined membrane channels spanning the plate, thus connecting the two daughter cells (see Figure 1.31). The cell plate–spanning ER tubules establish the sites for the primary plasmodesmata (see Figure 1.4B–D). After cytokinesis, microtubules re-form in the cell cortex. The new cortical microtubules have a transverse orientation relative to the cell axis, and this orientation determines the polarity of future cell extension.

Plant Cell Types

As described at the beginning of the chapter, the basic plant body plan has three groups of plant tissues: dermal tissue, ground tissue, and vascular tissue. In this section we will discuss examples of the particular cell types found in of each of these tissues, including their subcellular organization and the differentiation of their organelles.

Dermal tissues cover the surfaces of plants

We show five examples of dermal tissue here: three in leaves (**Figure 1.32**) and two in floral nectaries (**Figure 1.33**). Leaves have an upper and lower epidermis with different cell types in each (see Figure 1.32A). The epidermis includes the pavement cells, which are puzzle-piece shaped (or shaped like certain bricks used in decorative paving) in many flowering plants. In Figure 1.32B, the

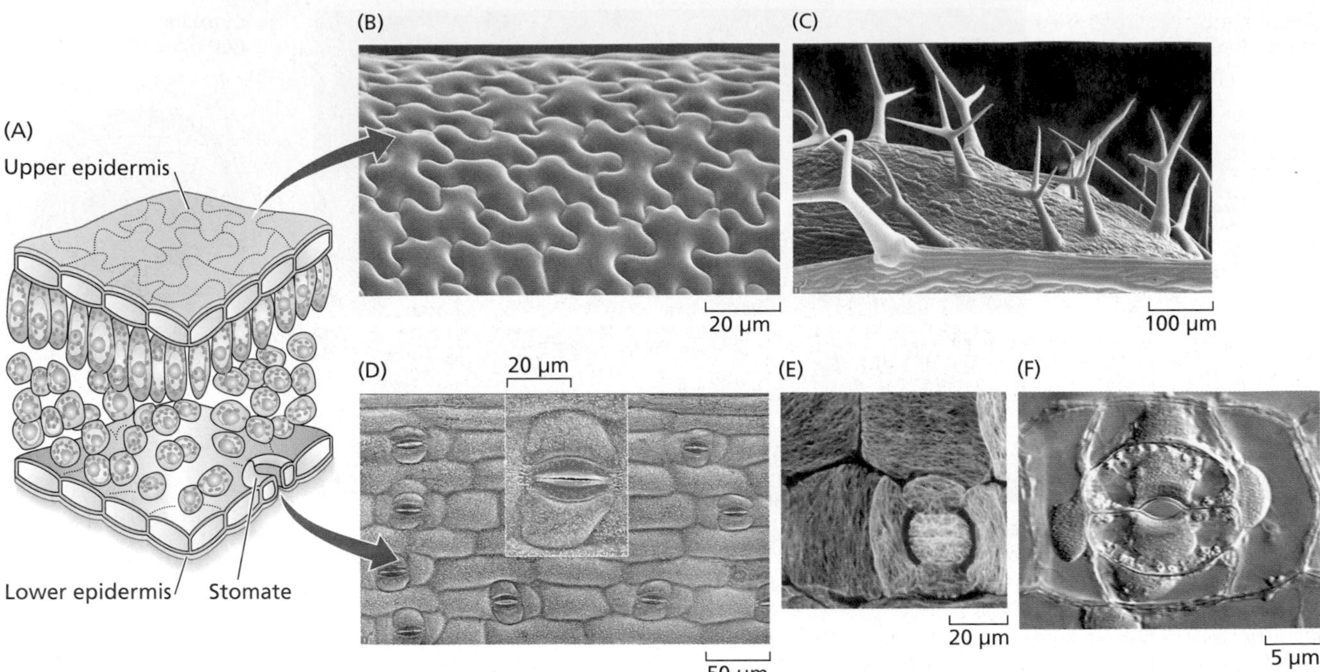

Figure 1.32 Dermal tissue of the leaf of a typical eudicot plant. (A) Generalized view of leaf structure. (B) Scanning electron micrograph of the epidermal cells of a *Galium aparine* leaf showing the puzzle-piece arrangement of pavement cells in the epidermis. (C) Scanning electron micrograph of an epidermis of a true leaf of Arabidopsis. The three-branched trichomes arise from a field of pavement cells and guard cell complexes. (D) Scanning electron micrograph with slightly higher magnification inset of the stomatal complexes in a *Tradescantia* sepal. (E) Fluorescence micrograph of microtubule distribution in the developing stomatal complex of the *Tradescantia* sepal. (F) Light micrograph of the stomatal complex of the *Tradescantia* sepal. There is a movie of cytoplasmic streaming in these cells in **WEB TOPIC 1.9**. (Micrographs from Gunning 2009; B and D courtesy of B. Gunning; C courtesy of R. Heady; E and F courtesy of A. Cleary.)

pavement cells of a seed leaf, or cotyledon, of Arabidopsis are the only cells in evidence, but when the first primary, true leaf appears, its epidermis differentiates into more cell types, becoming covered in three-pronged leaf hairs, or trichomes (see Figure 1.32C). In roots, root hairs differentiate from the epidermis. Many plants have relatively few chloroplasts in the epidermal tissue of true leaves, perhaps because chloroplast division is shut down. The exception to this is the guard cells—the amazing cells that form the "lips" of the mouths, or stomata, of the leaf—which contain many chloroplasts (see Figure 1.32D–F). Guard cells develop from the same precursor cells as the rest of the leaf epidermis, but the microtubule cytoskeleton, which helps determine the guard cell shape, is radially oriented, centered on the stomatal aperture, whereas the surrounding tissue has a longitudinal or lateral orientation (see Figure 1.32E). Once made, guard cells remain cytoplasmically isolated from the rest of the leaf because during the last division that forms them, no plasmodesmata are made in the forming cell plate (note the absence of the green spots that indicate plasmodesmata in the guard cells in Figure 1.4E). Hence, unlike the rest of the leaf, the guard cells are not part of the symplast. As we will describe in Chapters 6 and 10, the plasma membrane of the guard cell is quite dynamic, regulating the abundance of ion-regulating K^+ channels on the cell surface through endocytosis.

The floral nectary of the *Abutilon* flower (see Figure 1.33A) has epidermal pavement cells and multicellular secretory trichomes (see Figure 1.33B), which end in a final hemispherical cap cell. Ultrastructurally, this cap cell is quite interesting (see Figure 1.33C), being connected to the underlying cells by many plasmodesmata (see Figure 1.33D and E) and containing considerable endoplasmic reticulum, both smooth ER in tubular form and rough ER in cisternal form. The smooth ER is probably involved in the synthesis of lipid included in the cap cell secretions (see Figure 1.33B), while the rough ER supports the synthesis of special proteins for the secretions. Lipid and terpenoid synthesis also occurs in the leucoplasts found in the tip cell of the trichome. The walls surrounding the trichome are cut off from the rest of the wall spaces, or apoplast, of the tissue by a thick ring of cutin at the base of the trichome, thereby isolating the secretions from the apoplast of the rest of the tissue (see Figure 1.33F).

Ground tissues form the bodies of plants

The leaf has ground tissue of two different types: the upper elongated *palisade mesophyll* and the irregularly

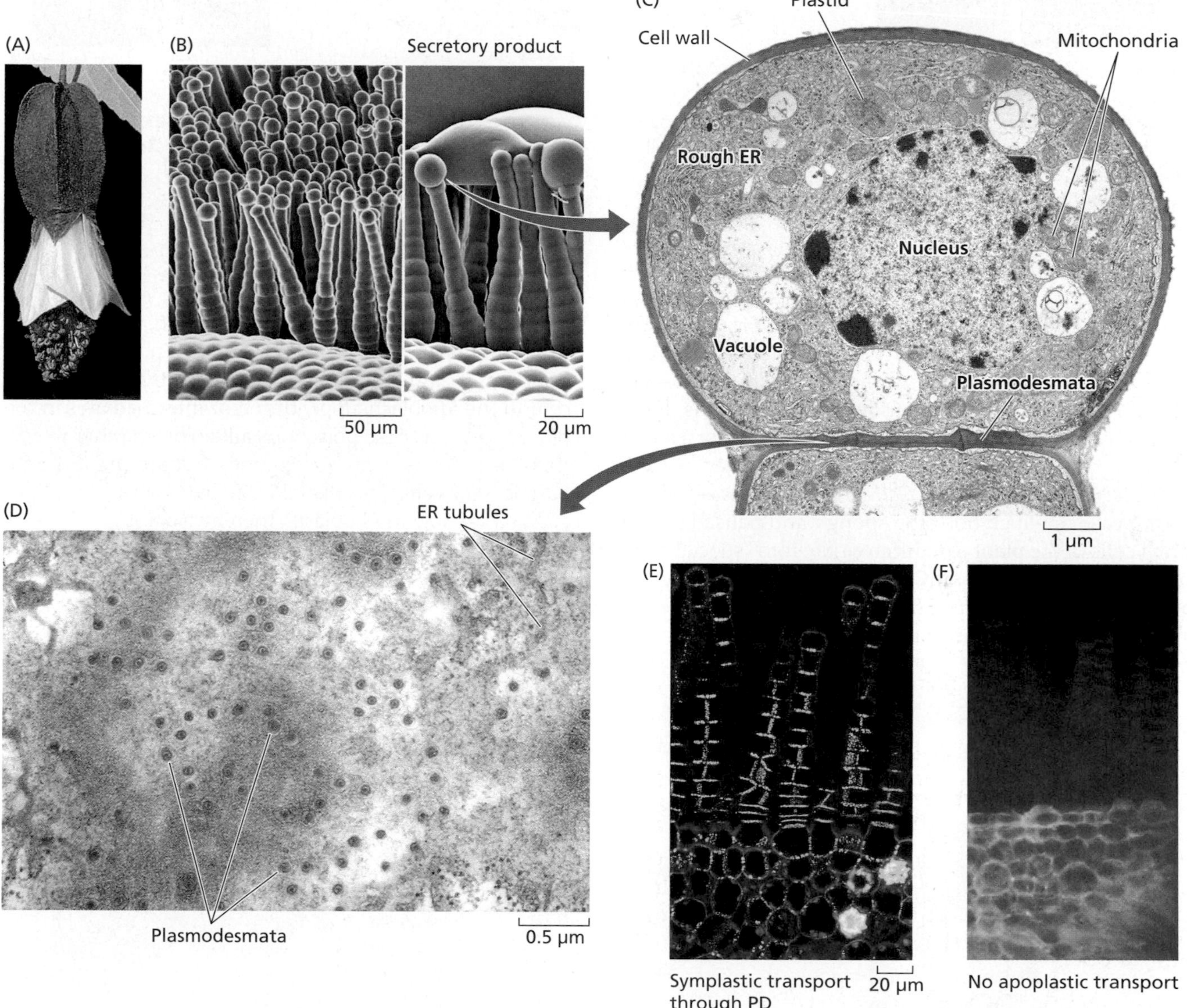

Figure 1.33 Structure of secretory trichomes in the epidermis of the floral nectary of *Abutilon*. (A) Flower of *Abutilon*. (B) Scanning electron micrographs of fields of the secretory trichomes in the nectary. The micrograph on the right shows secreted product accumulated on the surface of the trichomes. (C) Cap cell of secretory trichome. (D) Field of plasmodesmata in the cross-wall below the cap cell. (E) Fluorescence micrograph of secretory trichomes stained with aniline blue to label the callose associated with the plasmodesmata. (F) Absence in the secretory trichomes of a fluorescent marker labeling the apoplast of the rest of the leaf suggests that all transport through the trichomes occurs via the symplast. (From Gunning 2009; A, B, D, and F courtesy of B. Gunning; C courtesy of C. H Busby; E courtesy of J. E. Hughs.)

shaped *spongy mesophyll* (**Figure 1.34**). The spongy mesophyll has large air spaces between the cells—they are not cemented all around their periphery with middle lamellae. This allows exchange of gasses (carbon dioxide and oxygen) through the air spaces of the leaf during photosynthesis and respiration. Both cell types have many chloroplasts (see Figure 1.34B and C), oriented at the cell periphery but able to move (see **WEB TOPIC 1.9**) when the light changes. Both cell types also have plentiful spherical peroxisomes (see Figure 1.34B) and elongated, branched mitochondria that can be many micrometers long (see Figure 1.34D). Mitochondria and the network of ER in mesophyll cells (see Figure 1.34B) are tethered near the chloroplasts in a way that presumably allows efficient exchange of the intermediate metabolites of respiration and photosynthesis (see Figure 1.14). This tethering is the focus of an emerging area of active research.

Mesophyll cells of the leaf can differentiate into a variety of other cell types under the appropriate environment, and mesophyll is therefore considered a form of **parenchyma**,

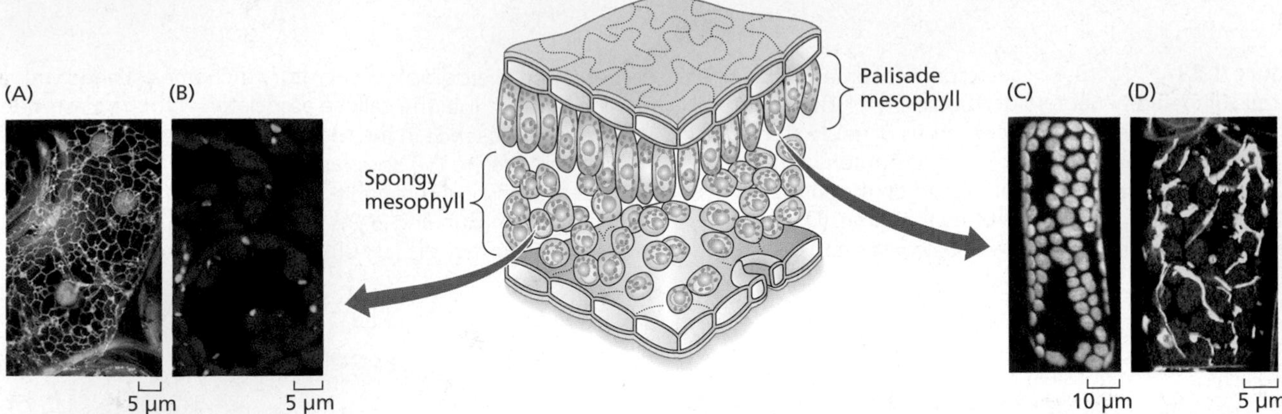

Figure 1.34 (A) Fluorescence micrograph of fluorescent endoplasmic reticulum network in the spongy mesophyll cells. (Oparka in Gunning 2009.) (B) Fluorescence micrograph of spongy mesophyll cells showing peroxisomes (green) and chloroplasts (red). (C) Three-dimensional stereo view of chloroplast distribution in a palisade cell of a leaf. (D) Three-dimensional stereo view of the distribution of the mitochondria in a palisade cell. (Micrographs from Gunning 2009; A courtesy of K. Oparka; B–D courtesy of B. Gunning.)

a ground tissue with thin primary walls (**Figure 1.35A**). Parenchyma have the capacity to continue division and can differentiate into a variety of other ground tissues and vascular tissues, after being produced by meristems. Young parenchyma cells have multiple small vacuoles, or a reticulated vacuole network (**Figure 1.35B**), which then develops into a large central vacuole. *Catharanthus roseus* is an interesting case in which both the spongy and palisade mesophyll cells of the plant can differentiate into "special cells," or **idioblasts** (**Figure 1.35C**), which contain the important anti-cancer drugs vincristine and vinblastine and are presumably used for defense against herbivory. Vinblastine interferes with microtubule polymerization, thereby inhibiting cell division, by causing tubulin to polymerize into nontubular aggregates. Different parenchyma cells adjacent to the idioblast supply different intermediates in the biosynthesis of these poisonous alkaloids; in other words, there is a biochemical division of labor among different parenchyma cells. The alkaloids are partitioned into a specialized vacuole in idioblasts, thereby isolating the rest of the cell from the toxic effects of the chemicals.

Parenchyma can differentiate into ground tissue that has thickened primary walls which can nevertheless continue to elongate (**Figure 1.36**). **Collenchyma** in the

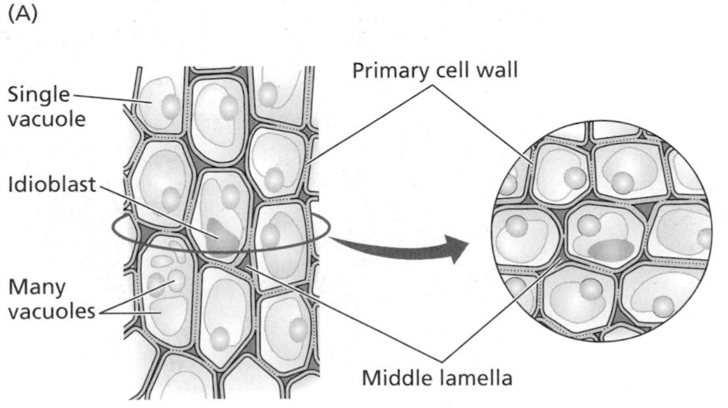

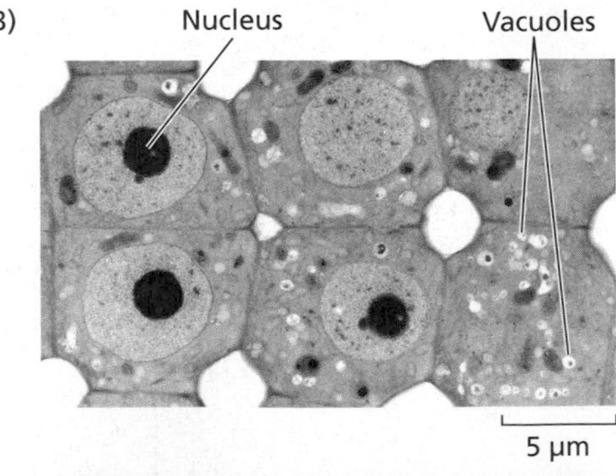

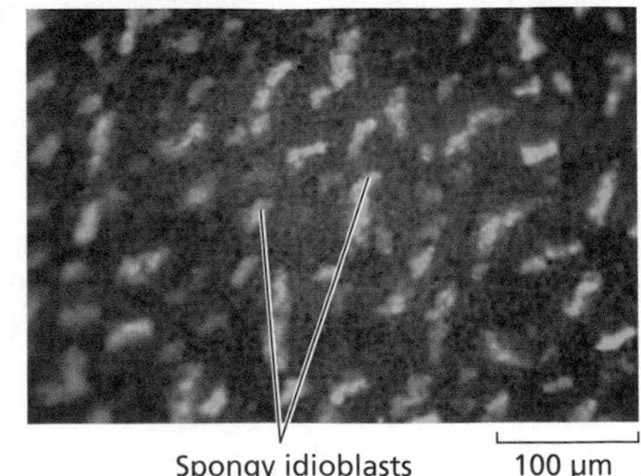

Figure 1.35 Ground tissues with thin primary walls. (A) Diagram of parenchyma cells showing change from multiple small vacuoles to a single large central vacuole and the differentiation of one of the parenchyma cells into an idioblast. (B) Electron micrograph of *Cyperus* root parenchyma. (C) Fluorescence micrograph of spongy mesophyll parenchyma of *Catharanthus roseus* with differentiated idioblasts, which contain green-yellow fluorescent alkaloids. Red fluorescence is from the chloroplasts in the other spongy parenchyma cells. (B from Gunning 2009, courtesy of B. Gunning; C from St. Pierre et al. 1999.)

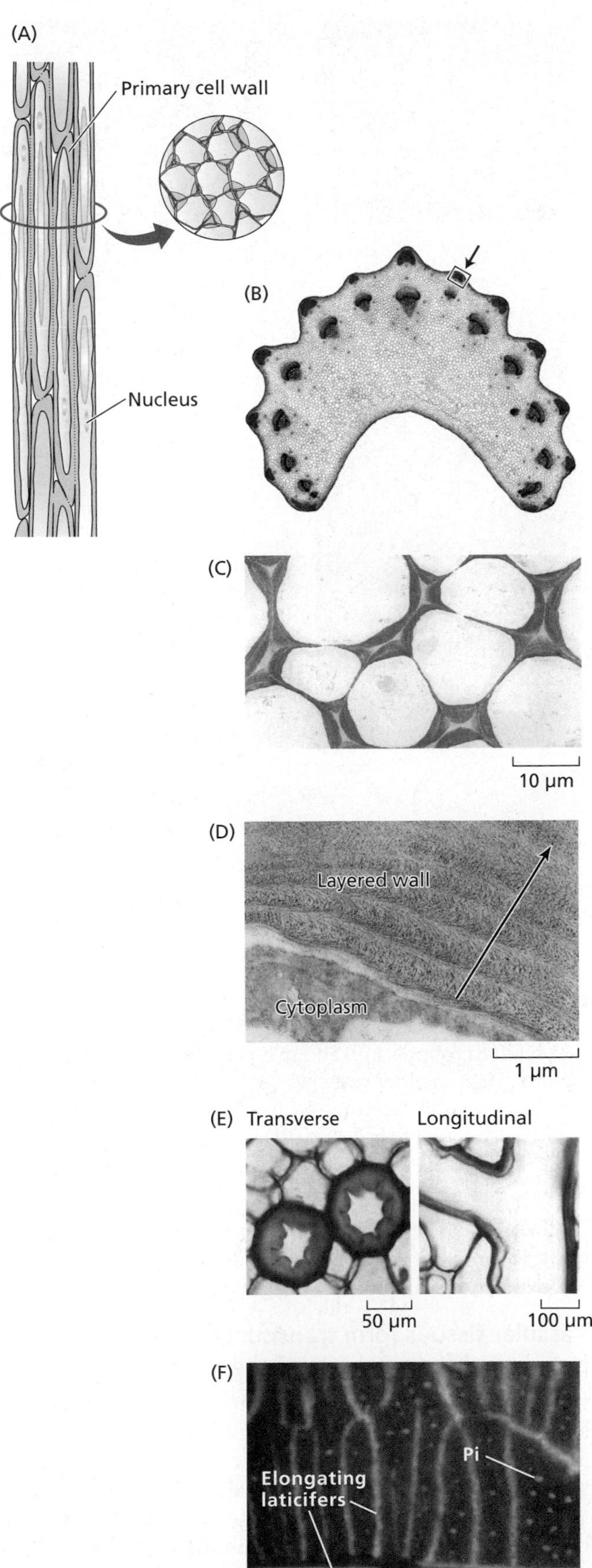

Figure 1.36 Ground tissue with thick primary walls. (A) Diagram of the collenchyma of celery in longitudinal view and cross section. (B) Light micrograph cross section of celery showing collenchyma clusters. Arrow shows region enriched in collenchyma. (C) Electron micrograph of cross section of celery collenchyma. (D) Electron micrograph showing layered primary wall of the collenchyma of celery. The arrow crosses five layers of primary wall celery. (E) Light micrographs of transverse and longitudinal sections of a *Euphorbia* laticifer showing thickened primary walls. (F) Elongated laticifers in the leaf of *Catharanthus roseus*, which contain green-yellow fluorescent alkaloids, as do idioblasts associated with the palisade parenchyma (denoted as Pi in the figure). (B–D from Leroux 2012; E from Rudall 1987; F from St. Pierre et al. 1999.)

ribs of celery stalks have very thick, layered walls (see Figure 1.36A–D) and a tasty crunch! The network of **laticifers** that carry the milky white latex in rubber trees, poppy plants, leaf-lettuce, and dandelions have irregularly thickened primary walls and continue to elongate (see Figure 1.36E and F). Laticifers come in two forms, articulated and non-articulated (see Figure 23.12). *Articulated* laticifers recruit adjacent parenchyma cells to form laticifers, and when the adjacent laticifer cells digest their connecting walls, latex moves through them. *Non-articulated* laticifers have relatively thinner walls but grow and branch and divide without forming cross-walls, thereby forming a multinucleate conduit for latex. The latex particles are contained in differentiated vacuoles; *Catharanthus roseus* laticifer vacuoles, like the idioblast vacuoles, contain toxic alkaloids (see Figure 1.36F).

Parenchyma can also differentiate into the **schlerenchyma**, which has thick secondary walls (**Figure 1.37**). *Sclereids* come from parenchyma in leaves, fruit (e.g., pear), and flowers (e.g., camellia; see Figure 1.37A). They often have an irregularly branched shape (see Figure 1.37B). Their development in some tissues is dependent on their exposure to environmental stress, such as wind and rain (see Figure 1.37C and D). **Fibers** develop from parenchyma and form elongated support structures with thickened secondary walls both in ground tissue (see Figure 1.37E) and in vascular tissues (see the phloem fibers in Figure 1.5). They can elongate into the longest cells of higher plants; for example, individual fiber cells of the *Ramie* plant can be 25 cm long! Because the walls are hardened with lignin after elongation, the cells have very high tensile strength, and it is no wonder that humans extensively use such fibers, called bast fibers. During elongation, the fibers are characterized by many Golgi (see Figure 1.37F), actively secreting polysaccharide, and large parallel arrays of microtubules involved in aligning the cellulose deposition at the plasma membrane (see Figure 1.37G). The plasma membrane has a peculiar structure in growing fiber cells, but the role this unique structure plays

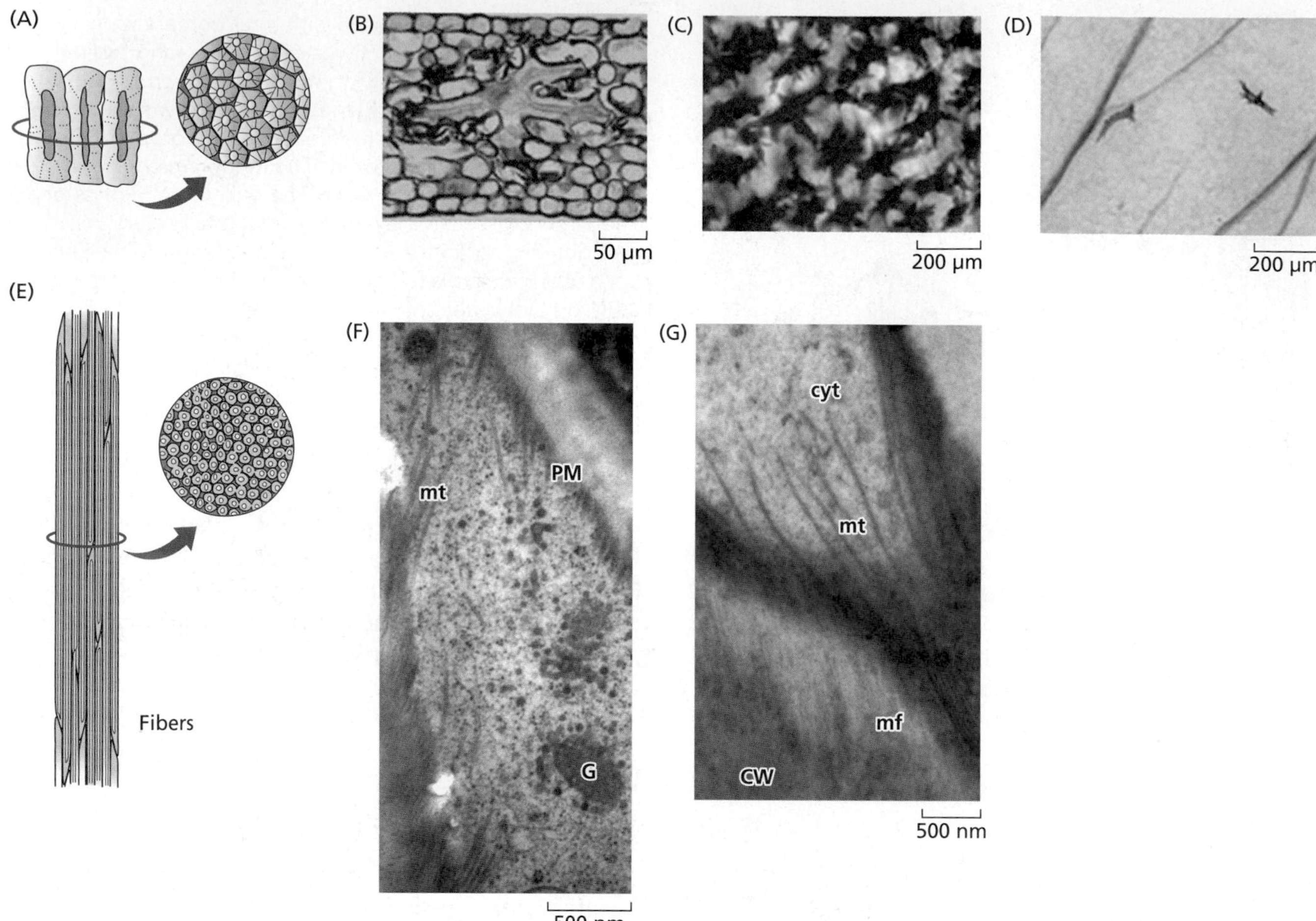

Figure 1.37 Ground tissue with thick primary and secondary walls. (A) Diagram of sclereid cluster in longitudinal and cross section views. (B) Sclereid that has developed from a parenchyma cell in the mesophyll of a *Camellia sinensis* petal. (C) Light micrograph of the pigmented sclereids in a *C. japonica* petal grown in the wild, where it is subject to wind and rain. (D) Light micrograph of much-reduced number of pigmented sclereids from a *C. japonica* petal grown in the greenhouse. (E) Diagram of fibers in longitudinal and cross section views. (F) Electron micrograph of Golgi (G; face view) and microtubule (mt) cluster in a plasma membrane (PM) glancing section of a fiber cell actively engaged in secondary wall deposition in poplar. (G) Electron micrograph showing alignment of cell wall (CW) cellulose microfibrils (mf) and microtubules (mt) in the cytoplasm (cyt) of another glancing section of a fiber engaged in secondary wall biosynthesis. (B–D from Zhang et al. 2011; F and G from Kaneda et al. 2010.)

in wall deposition or elongation is unknown. Specialized holes in the secondary wall, called **pits**, connect the living fiber cells with each other. They are often the site of fields of plasmodesmata.

The presence of fibers in both ground and vascular tissues brings up the topic of how the different tissues are separated. In the stem, the vascular cylinder can be filled with ground tissue, the pith, in addition to the vascular tissue (see Figure 1.5A). In the root, the ground tissue is between the dermal tissue and the vascular tissue and is called the root *cortex* (see Figure 1.5B). The boundary between the cortex cells and the vascular tissue is a specialized cell type called the **endodermis**, which has a suberin-impregnated **Casparian strip**. As we will describe in Chapter 6, the Casparian strip, like the cutin band on the trichomes of *Abutilon*, separates the apoplast of the cortex from the apoplast of the vascular tissue.

Vascular tissues form transport networks between different parts of the plant

Phloem cells, which conduct the products of photosynthesis from the leaves to the root (see Chapter 11), are living at maturity and have nonlignified cell walls. They include **sieve cells** in gymnosperms, and **sieve tube elements**—which stack to form **sieve tubes**—in angiosperms (**Figure 1.38**). Specialized proteins are made in the phloem cells, such as the P-protein (see Figure 1.38B). The P-protein network forms a reticulum centered on the cross-

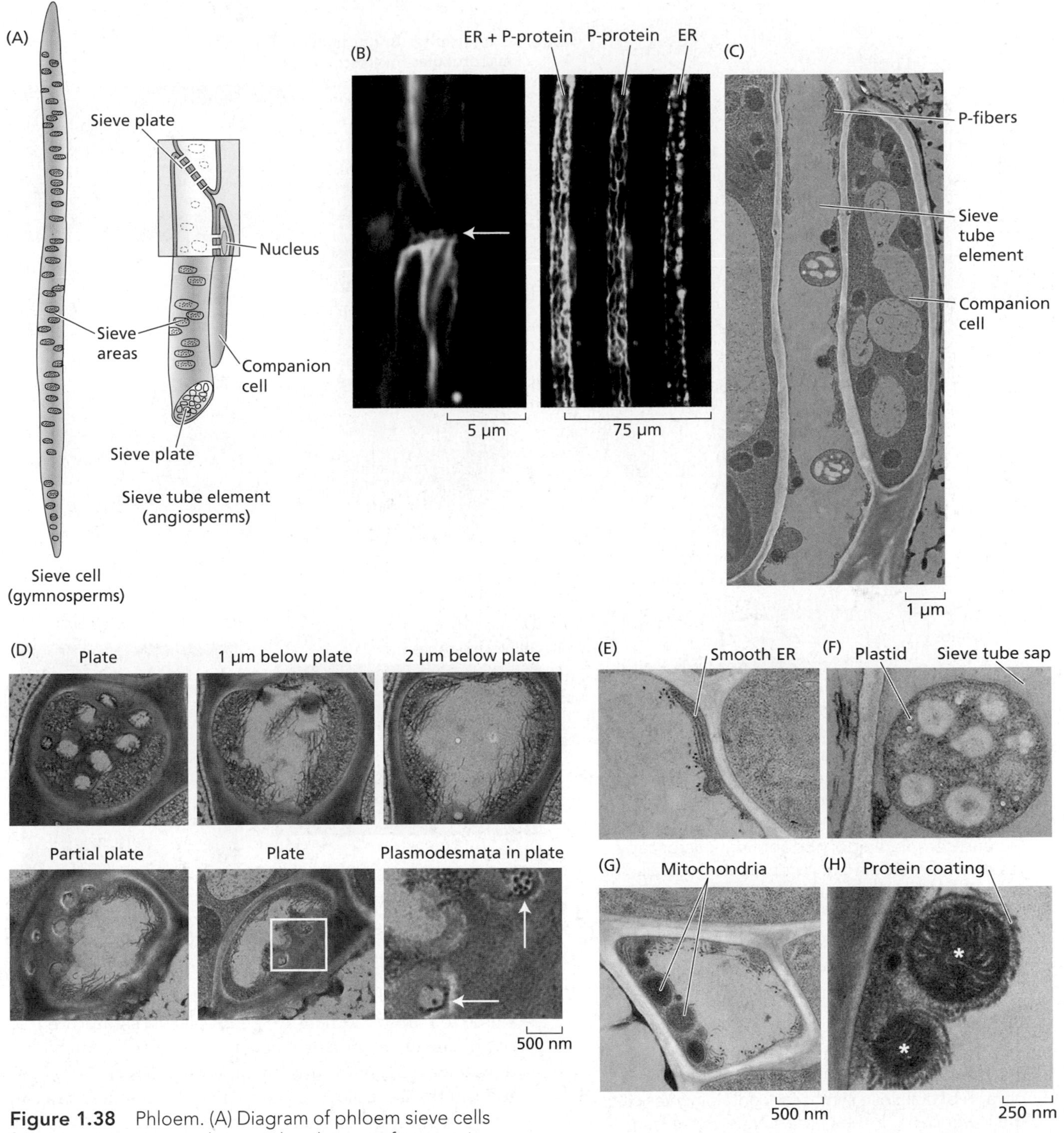

Figure 1.38 Phloem. (A) Diagram of phloem sieve cells from gymnosperms and sieve tube elements from angiosperms. (B) Fluorescence micrograph of P-protein (SERB2; blue) and ER (green) in a mature living sieve tube element. Note that the P-protein is focused on the sieve plate (arrow in left panel), but experiments reveal that it does not occlude the plate. (C) Electron micrograph of a sieve tube element and a companion cell. Note the fibrillar P-protein aggregates. (D) Cross section of the sieve plate of Arabidopsis. Top row: Serial sections through a sieve plate in 1-μm steps. There are several open pores in the center of the plate (left panel), then the lumen fills with P-protein (middle and right panels). Second row: When the sieve tube element is sectioned at an angle to the plate, multiple open pores are revealed (left and middle panels), some of which contain multiple plasmodesmata (white arrows in the right panel, which is a higher magnification of the boxed area in the middle panel). (E) Smooth ER in a sieve tube element. (F) Specially differentiated plastid in a mature sieve tube element in direct contact with sieve tube sap. (G) Disposition of mitochondria in relation to the rest of the cytoplasm in a cross section of mature sieve tube element. (H) Ultrastructure of mitochondria (asterisks), with filamentous protein ring around them. (B–H from Froelich et al. 2011.)

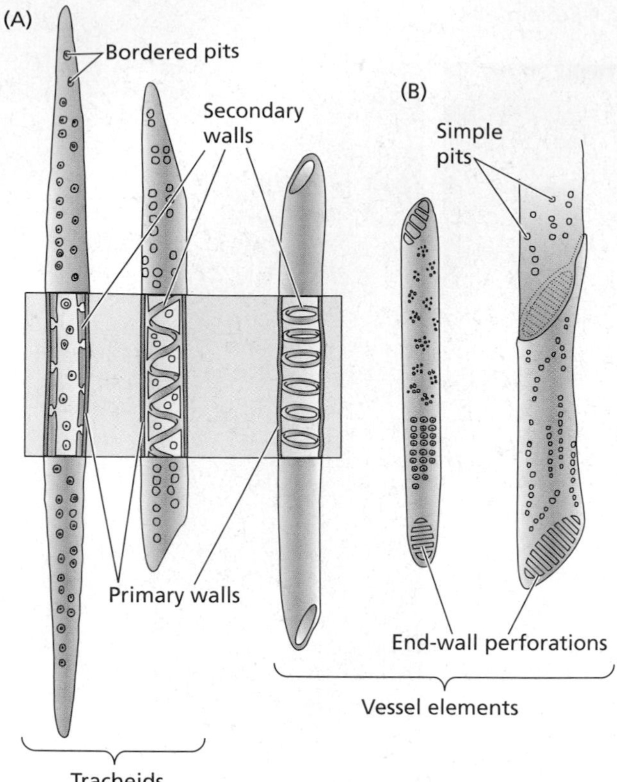

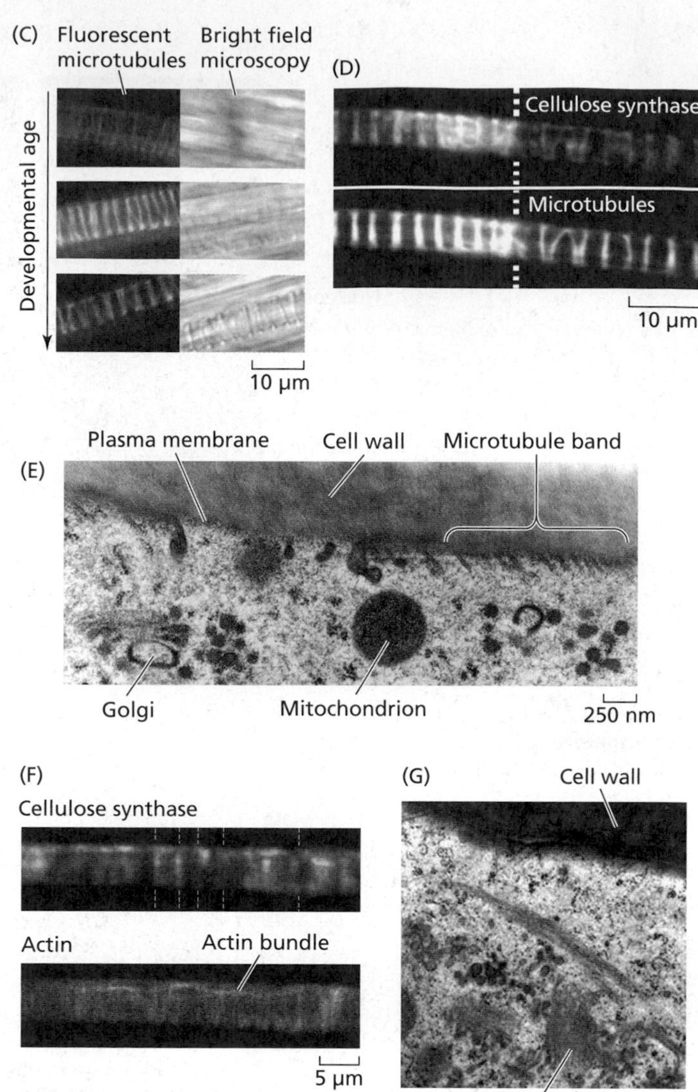

Figure 1.39 Xylem. (A) Diagram of two tracheids and a vessel element. The cross sections (in the blue window) reveal thickenings of secondary wall in spiral (helical) and annular (ring-shaped) arrangements. (B) Diagram of two vessel elements, showing pits and end-wall perforations. (C) Developing xylem in poplar as seen in fluorescence microscopy images of microtubules and complementary bright field microscopy images of the developing rings of secondary wall. The microtubules align with the developing rings. (D) Co-localization of microtubules and cellulose synthase complexes in a developing vessel element in poplar. The upper panel shows fluorescent cellulose synthase complexes; the lower panel shows fluorescent microtubules in the same cell. (E) Electron micrograph of a microtubule patch and Golgi in developing xylem in pine. (F) Trafficking of cellulose synthase in developing poplar xylem. The upper panel is a fluorescence micrograph of cellulose synthase–containing vesicles, and the lower panel is the fluorescence signal from actin bundles in the same cell. There is some co-localization along the longitudinal actin strands, indicating that cellulose synthase complexes might cycle through the cytoplasm in vesicles guided by actin. (G) Electron micrograph of actin microfilament bundles, vesicles, tubules (arrowhead), and Golgi in developing xylem in pine. (C, D, and F from Wightman and Turner 2008; E and G from Samuels et al. 2002.)

walls, or sieve plates (see Figure 1.38B and D), which is next to the reticulum of the ER. The ER plays a significant role in phloem development, whereby cisternal ER centered on a plasmodesma marks the site of future cross-wall changes to form the large sieve plate opening. Plasmodesmata can be seen in tangential sections of the forming sieve plate (see Figure 1.38E). As with other plasmodesmatal arrays, callose is desposited at the sieve plate. The living sieve tube element is connected by plasmodesmata fields, or sieve areas, to adjacent cells, **companion cells** (in angiosperms) and **albuminous cells** (in gymnosperms). As shown in Figure 1.38C, organelles in the mature sieve tube element and companion cell are quite different. In the sieve tube element, there is abundant smooth ER, condensed mitochondria with a specialized surrounding structure, and altered plastids (see Figure 1.38E–G). Also associated with phloem in some plants are fibers and storage parenchyma.

The xylem cells that conduct water and minerals from the root, the **tracheary elements**, are nonliving at maturity and consist of **tracheids** (in all vascular plants) and shorter **vessel elements** (mostly in angiosperms) (**Figure 1.39**). Vessel elements stack end-to-end to form wide (up to 0.7 mm) columns called **vessels**. Protoxylem cells with primary walls begin to differentiate into mature tracheary elements by laying down secondary cell walls with spiral bands of cellulose and strengthened with lignin. The cellulose is aligned in bands by bands of microtubules (see Figure 1.39C, D, and F) which connect with cellulose synthase complexes on the plasma membrane (see Figure 1.39D). Actin filaments also participate in the wall deposition, guiding actively secreting Golgi to sites of polysaccharide deposition (see Figure 1.39E–G). As with fiber cells, secondary wall deposition is characterized by abundant and active Golgi (see Figure 1.39F). Once elongation has stopped, the top and bottom end walls get large perforations in them. On the side walls, secondary walls continue to thicken, except in spots containing pits, which start as plasmodesmata fields and ultimately become channels in the walls shared between adjacent cells. The cells of the tracheids and vessels die, undergoing a process called programmed cell death (see Chapter 19), leaving behind the hardened bundle of pipes made by the secondary walls and connected side-to-side by pits. The pits are important because flow through these narrow pipes depends on having a continuous liquid stream (see Chapters 4 and 6). If an air bubble, or embolism, forms in a pipe, the stream can be diverted around the embolism through the pits into adjacent cells.

SUMMARY

Despite their great diversity in form and size, all plants carry out similar physiological processes. All plant tissues, organs, and whole organisms show a growth polarity, being derived from axial or radial polarity of cell division of meristems.

Plant Life Processes: Unifying Principles

- All plants convert solar energy to chemical energy. They use growth instead of motility to obtain resources, have vascular systems, have rigid structures, and have mechanisms to avoid desiccation on land. They develop from embryos sustained and protected by tissues from the mother plant.

Plant Classification and Life Cycles

- Plant classification is based on evolutionary relationships (**Figure 1.1**).
- Plant life cycles alternate between diploid and haploid generations (**Figure 1.2**).

Overview of Plant Structure

- All plants chave a common body plan (**Figure 1.3**).
- Dermal tissue, ground tissue, and vascular tissue are the three major tissue systems present in all plant organs (**Figure 1.3**).
- Because of plants' rigid cell walls, plant development depends solely on patterns of cell division and cell enlargement (**Figure 1.4**).
- The cytoplasm of clonally derived cells is connected by plasmodesmata, forming the symplast, which allows water and small molecules to move between cells without crossing the outer membrane (**Figure 1.4**).
- Nearly all mitosis and cytokinesis occurs in meristems.
- Secondary growth results in the increase in girth of roots and shoots through the action of the specialized meristems, the vascular and cork cambium. (**Figure 1.5**).

Plant Cell Organelles

- In addition to cell walls and plasma membrane, plant cells contain compartments derived from the endomembrane system (**Figure 1.6**).
- The endomembrane system plays a central role in secretory processes, membrane recycling, and the cell cycle.
- Plastids and mitochondria are semiautonomous, independently dividing organelles that are not derived from the endomembrane system.
- The composition and fluid-mosaic structure of all plasma membranes permits regulation of transport into and out of the cell and between subcellular compartments (**Figure 1.7**).

The Endomembrane System

- The endomembrane system conveys both membrane and cargo proteins to diverse organelles.
- The specialized membranes of the nuclear envelope are derived from the endoplasmic reticulum (ER), a component of the endomembrane system (**Figures 1.8, 1.12, 1.13**).
- The nucleus is the site of storage, replication, and transcription of the DNA in the chromatin, as well as the site for the synthesis of ribosomes (**Figures 1.9–1.11**).

- The ER is a system of membrane-bound tubules that form a complex and dynamic structure (**Figure 1.12**).
- The rough ER is involved in synthesis of proteins that enter the lumen of the ER; the smooth ER is the site of lipid biosynthesis (**Figure 1.13**).
- The ER forms close membrane associations with the other organelles in the cell and may help organize the cell by that means (**Figure 1.14**).
- The ER provides the membrane protein and protein cargo for the other compartments of the endomembrane system.
- Secretion of proteins from cells begins with the rough ER (**Figures 1.11, 1.16**).
- Glycoproteins and polysaccharides destined for secretion are processed in the Golgi apparatus (**Figures 1.15, 1.16**).
- During endocytosis, membrane is removed from the plasma membrane by formation of small clathrin-coated vesicles (**Figures 1.16, 1.17**).
- Endocytosis from the plasma membrane provides membrane recycling (**Figure 1.17**).
- Vacuoles serve multiple functions and can be generated from multiple endomembrane pathways.

Independently Dividing or Fusing Organelles Derived from the Endomembrane System

- Oil bodies, peroxisomes, and glyoxysomes can grow, proliferate, or fuse independently (**Figures 1.18, 1.19**).

Independently Dividing, Semiautonomous Organelles

- Mitochondria and chloroplasts each have an inner and an outer membrane (**Figures 1.20, 1.21**).
- Chloroplasts have an additional internal thylakoid membrane system which contains chlorophylls and carotenoids.
- Plastids may contain high concentrations of pigments or starch (**Figure 1.22**).
- Proplastids pass through distinct developmental stages to form specialized plastids (**Figure 1.23**).
- In plastids and mitochondria, fission and organellar DNA replication are regulated independently of nuclear division.

The Plant Cytoskeleton

- A three-dimensional network of polymerizing and depolymerizing tubulin to form microtubules and actin to form microfilaments organizes the cytosol and is required for life (**Figure 1.24**).
- Microfilament and microfilament bundle formation and breakdown are regulated by a variety of accessory proteins (**Figure 1.25**).
- Microtubules show dynamic instability but can stabilize and treadmill through the cell with accessory proteins (**Figure 1.26**).
- Molecular motors reversibly bind the cytoskeleton and can tether organelles or direct organelle movement (**Figure 1.27**).
- During cytoplasmic streaming, bulk flow of the cytosol is driven by the viscous drag in the wake of motor-driven organelles.

Cell Cycle Regulation

- The cell cycle, during which cells replicate their DNA and reproduce themselves, consists of four phases (**Figure 1.28**).
- Cyclins and cyclin-dependent kinases (CDKs) regulate the cell cycle, including the separation of paired metaphase chromosomes (**Figure 1.28**).
- Successful mitosis (**Figures 1.29, 1.30**) and cytokinesis (**Figure 1.31**) require the participation of microtubules and the endomembrane system.

Plant Cell Types

- Dermal tissue includes the epidermis, which has multiple cell types, including pavement cells, guard cells, and trichomes (**Figure 1.32**).
- Epidermal differentiation includes changes in ER and plastids, and plasmodesmata between adjacent cells (**Figure 1.33**).
- Ground tissue is made up of parenchyma cells that can differentiate into a variety of cell types that can be distinguished by the nature of the primary and secondary cell walls and vacuolar content (**Figures 1.34–1.37**).
- Vascular tissue has thickened secondary cell walls, perforated cell ends, and pit fields generated by the interplay of organelles involved in cell wall formation during development (**Figures 1.38, 1.39**).

WEB MATERIAL

- **WEB TOPIC 1.1 Model Organisms** Certain plant species are used extensively in the lab to study their physiology.
- **WEB TOPIC 1.2 Plant Identification, Classification, and Evolutionary Thought** Our organization of how plants are identified started from utilitarian motives but now relies on evolutionary relationships.
- **WEB TOPIC 1.3 Flower Anatomy and the Angiosperm Life Cycle** The cellular differentiation of the gametophytic and sporophytic generations and their contributions to flower structure are considered.
- **WEB TOPIC 1.4 The Nuclear Pore and Proteins Involved in Nuclear Import and Export** The nuclear pore is believed to be lined by a meshwork of unstructured nucleoporin proteins and GTP-charged proteins that mediate transfer into and out of the nucleoplasm.
- **WEB TOPIC 1.5 Protein Signals Used to Sort Proteins to Their Destinations** The primary sequence of a protein can include a ticket to its final destination.
- **WEB TOPIC 1.6 SNARES, Rabs, and Coat Proteins Mediate Vesicle Formation, Fission, and Fusion** Models are presented for the mechanisms of vesicle fission and fusion.
- **WEB TOPIC 1.7 ER Exit Sites (ERES) and Golgi Bodies Are Interconnected** The co-migration of ERES and Golgi bodies during cytoplasmic streaming is shown in movies.
- **WEB TOPIC 1.8 Specialized Vacuoles in Plant Cells** Plant cells contain diverse types of vacuoles, which are dynamic and motile, as shown by movies.
- **WEB TOPIC 1.9 Directed Organelle Movement and Cytoplasmic Streaming** The movement and interaction of plant cell organelles are shown with movies.
- **WEB TOPIC 1.10 Microtubule Movement and Microtubule-mediated Movement in Plants** Kinesin motors, MAPs, and how they regulate microtubule dynamics through the cell cycle are discussed.

available at plantphys.net

Suggested Reading

Albersheim, P., Darvill, A., Roberts, K., Sederoff, R., and Staehelin, A. (2011) *Plant Cell Walls: From Chemistry to Biology*. Garland Science, Taylor and Francis Group, New York.

Bell, K., and Oparka, K. (2011) Imaging plasmodesmata. *Protoplasma* 248: 9–25.

Burch-Smith, T. M., Stonebloom, S., Xu, M., and Zambryski, P. C. (2011) Plasmodesmata during development: Re-examination of the importance of primary, secondary, and branched plasmodesmata structure versus function. *Protoplasma* 248: 61–74.

Burgess, J. (1985) *An Introduction to Plant Cell Development*. Cambridge University Press, Cambridge.

Carrie, C., Murcha, M. W., Giraud, E., Ng, S., Zhang, M. F., Narsai, R., and Whelan, J. (2013) How do plants make mitochondria? *Planta* 237: 429–439.

Chapman, K. D., Dyer, J. M., and Mullen, R. T. (2012) Biogenesis and functions of lipid droplets in plants: Thematic Review Series: Lipid droplet synthesis and metabolism: from yeast to man. *J. Lipid Res.* 53: 215–226.

Griffing, L. R. (2010) Networking in the endoplasmic reticulum. *Biochem. Soc. Trans.* 38: 747–753.

Gunning, B. E. S. (2009) *Plant Cell Biology on DVD*. Springer, New York, Heidelberg.

Henty-Ridilla, J. L., Li, J., Blanchoin, L., and Staiger, C. J. (2013) Actin dynamics in the cortical array of plant cells. *Curr. Opinion Plant Biol.* 16: 678–687.

Hu, J., Baker, A., Bartel, B., Linka, N., Mullen, R. T., Reumann, S., and Zolman, B. K. (2012) Plant peroxisomes: biogenesis and function. *Plant Cell* 24: 2279–2303.

Jones, R., Ougham, H., Thomas, H., and Waaland, S. (2013) *The Molecular Life of Plants*. Wiley-Blackwell, Oxford.

Joppa, L. N., Roberts, D. L., and Pimm, S. L. (2011) How many species of flowering plants are there? *Proc. R. Soc. B* 278: 554–559.

Leroux, O. (2012) Collenchyma: a versatile mechanical tissue with dynamic cell walls. *Ann. Bot.* 110: 1083–1098.

McMichael, C. M., and Bednarek, S. Y. (2013) Cytoskeletal and membrane dynamics during higher plant cytokinesis. *New Phytol.*197: 1039–1057.

Müller, S., Wright, A. J., and Smith, L. G. (2009) Division plane control in plants: new players in the band. *Trends Cell Biol.* 19: 180–188.

Williams, M. E. (July 16, 2013). How to be a plant. Teaching tools in plant biology: Lecture notes. *Plant Cell* 25(7): DOI 10.1105/tpc.113.tt0713.

Wasteneys, G. O., and Ambrose, J. C. (2009) Spatial organization of plant cortical microtubules: Close encounters of the 2D kind. *Trends Cell Biol.* 19: 62–71.

2 Genome Structure and Gene Expression

A plant's phenotype is the result of three major factors: its genotype (all the genes, or alleles, that determine the plant's traits), the pattern of epigenetic chemical modifications of its DNA (chemical groups attached to some of the DNA's nitrogenous bases that can affect gene activity), and the environment in which it lives. In Chapter 1 we reviewed the fundamental structure and function of DNA, its packaging into chromosomes, and the two major phases of gene expression: transcription and translation. In this chapter we will discuss how the composition of the genome beyond its genes influences the physiology and evolution of the organism. First we will look at the structure and organization of the nuclear genome and the non-gene elements it contains. Then we will turn to the cytoplasmic genomes, which are contained inside the mitochondria and plastids. Next we will discuss the cellular machinery needed to transcribe and translate the genes into functional proteins, and we will see how gene expression is regulated both transcriptionally and posttranscriptionally. We will then introduce some of the tools used to study gene function, and we will finish up with a discussion of the use of genetic engineering in research and agriculture.

Nuclear Genome Organization

As discussed in Chapter 1, the nuclear genome contains most of the genes required for the plant's physiological functions. The first plant genome to be fully sequenced, in 2000, was that of a small dicotyledonous angiosperm called thale cress, or *Arabidopsis thaliana*. The genome of Arabidopsis is made up of only about 157 million base pairs (Mbp), which are distributed over five chromosomes. By contrast, the genome of the monocot Japanese canopy plant (*Paris japonica*), the plant species with the largest known genome, contains about 150,000 Mbp. Within its nuclear genome, Arabidopsis holds 27,416 protein-coding genes and another 4827 genes that are either **pseudogenes** (nonfunctional genes) or parts of transposons (mobile DNA elements). The Arabidopsis genome also contains 1359 genes that produce non-protein-coding RNAs (ncRNAs).

Some of these ncRNAs include ribosomal and transfer RNAs, others are probably involved in gene regulation. We will describe both transposons and ncRNAs in more detail later in this chapter.

The plant genome, however, consists of much more than genes. In this section we will examine the organization and chemical makeup of the genome, then see how certain regions of the genome correspond to specific functions.

The nuclear genome is packaged into chromatin

The nuclear genome consists of DNA molecules that are wrapped around histone proteins to form beadlike structures called **nucleosomes** (see Chapter 1). DNA and histones, together with other proteins that bind to the DNA, are referred to as **chromatin** (see Figure 1.10). Two types of chromatin can be distinguished: euchromatin and heterochromatin. Historically, these two types were differentiated based on their appearance in light microscopy when stained with specific dyes. **Heterochromatin** is usually more tightly packaged and thus appears darker than the less condensed **euchromatin**. Most genes that are actively transcribed in a plant are located within the euchromatic regions of a chromosome, while genes located in heterochromatic regions are either inactive or silent, at least in most tissues. Complete silencing of genes will eventually lead to the accumulation of mutations that incur no evolutionary cost (i.e., the mutation does not help or hurt the individual) and render the gene defunct. Such genes are examples of pseudogenes. Compared with euchromatin, heterochromatin is relatively gene poor. Heterochromatic regions include the centromeres, several so-called knobs, and the regions immediately adjacent to the telomeres, or chromosome ends, known as the **subtelomeric regions**.

Heterochromatic structures often consist of highly repetitive DNA sequences, or **tandem repeats**: blocks of nucleotide motifs of about 150 to 180 bp that are repeated over and over. A second class of repeats is the **dispersed repeats**. One type of dispersed repeat is known as **simple sequence repeats** (**SSRs**), or **microsatellites**. These repeats consist of sequence motifs that are usually between two and six nucleotides long and that are repeated hundreds or even thousands of times. Another dominant group of dispersed repeats found in heterochromatin is the transposons.

Centromeres, telomeres, and nucleolar organizer regions contain repetitive sequences

The most prominent structural landmarks on chromosomes are centromeres, telomeres, and nucleolar organizer regions. These regions contain repetitive DNA sequences that can be made visible by fluorescent in situ hybridization (FISH), a technique that uses fluorescently labeled molecular probes—usually fragments of DNA—that bind specifically to the sequence to be identified (**Figure 2.1**). **Centromeres** are constrictions of the chromosomes where spindle fibers attach during cell division. The attachment of fibers to the centromere is mediated by the kinetochore, a protein complex surrounding the centromere (see Chapter 1). Usually centromeres consist of highly repetitive DNA regions including tandem repeats and/or inactive transposons. Although these repetitive sequences are often only between 150 and 180 bp long, the total size of plant centromeres can range from hundreds of kilobase pairs to several megabase pairs in length. Because of the length and repetitiveness of centromeres, determining their exact sequence has been difficult for genome scientists, even in the era of routine whole-genome sequencing. **Telomeres** are sequences located at the ends of each chromosome. Telomeres act as caps on the chromosome ends, prevent loss of DNA during DNA replication, and inhibit end-to-end fusion of chromosomes mediated via double-strand break repair mechanisms.

The RNA molecules that make up ribosomes (rRNA) are transcribed from **nucleolar organizer regions** (**NORs**). Because ribosomes are composed mostly of rRNA and proteins, and because many ribosomes are needed for translation, it is not sur-

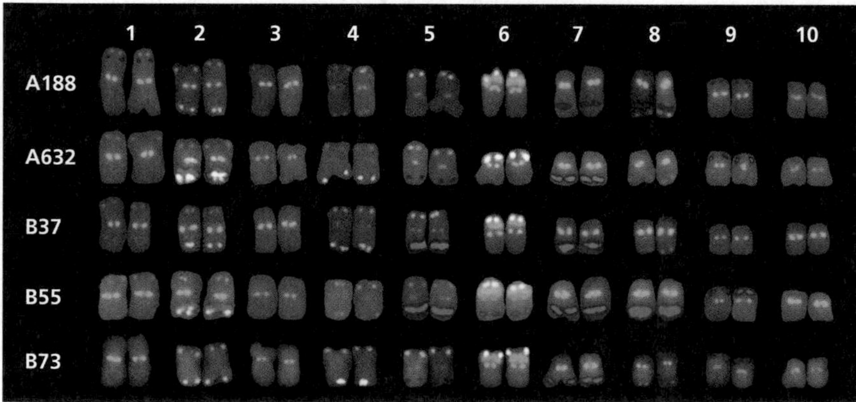

Figure 2.1 Chromosomal landmarks, including centromeres, telomeres, and nucleolar organizer regions (NORs), can be used to identify individual chromosomes. Each row shows the ten chromosome pairs of a different inbred line of maize (*Zea mays*; five common lines are shown here, from A188 to B73). DNA sequences (probes) complementary to certain chromosomal landmarks were labeled with fluorescent dyes and hybridized to chromosome preparations. The centromeres can be seen as green dots near the middle of the chromosomes, the NOR as a larger green area on chromosome 6, and the telomeres as faint red dots, most clearly visible at the top of chromosomes 2–4. The larger blue areas are specific heterochromatic regions. (From Kato et al. 2004.)

prising that NORs contain hundreds of repeated copies of each rRNA gene. Depending on the plant species, one or several NORs are present within the genome (maize has one, on chromosome 6; see Figure 2.1). Due to their repetitive nature and high GC content, NORs can be seen through a light microscope (after staining) and thus can serve as chromosome-specific markers. Visible chromosomal markers such as these were used by early geneticists to map phenotypic traits to specific chromosomal regions. Despite its repetitive nature, rDNA (DNA that encodes rRNA) is actively transcribed. The prominent nuclear structure called the nucleolus (see Figure 1.4) consists of the rDNA of the NOR, the proteins that transcribe the rDNA and process the rRNA primary transcripts for assembly into ribosomes, and the immature ribosomes just being assembled.

Transposons are mobile sequences within the genome

One dominant type of repetitive DNA within the heterochromatic regions of the genome is the transposon. **Transposons**, or **transposable elements**, are also known as *jumping genes* because some of them have the ability to insert a copy of themselves into a new location within the genome.

There are two large classes of transposons: the retroelements, or retrotransposons (Class 1), and the DNA transposons (Class 2). These two classes are distinguished by their mode of replication and insertion into a new location (**Figure 2.2**). **Retrotransposons** make an RNA copy of themselves, which is then reverse-transcribed into DNA before it is inserted elsewhere in the genome (see Figure 2.2A). Because they do not normally leave their original location, but rather generate additional copies of themselves, active retrotransposons tend to multiply within the genome. Retrotranspososon-derived genome content varies widely among species. In Norway spruce (*Picea abies*), retrotransposons make up an estimated 58% of the genome, while in the carnivorous bladderwort *Utricularia gibba*, retrotransposons occupy no more than about 2.5% of the genome. **DNA transposons**, by contrast, move from one position to another using a cut-and-paste mechanism catalyzed by an enzyme that is encoded within the transposon sequence. This enzyme, **transposase**, splices out the transposon and inserts it elsewhere in the genome, in most cases keeping the total transposon copy number the same (see Figure 2.2B).

Transposition into a gene can result in mutations. If a transposon lands within a coding region, the gene may be inactivated. Insertion of a transposon close to a gene can also alter that gene's expression pattern. For example, the transposon may disrupt the gene's normal regulatory elements, preventing transcription or, since transposons often carry promoters, increasing transcription of the gene. The mutagenic capability of transposons may play an important role in the evolution of the host genome. A low level of mutagenesis can lead to novel variation in an individual that can be passed on to the next generation. However, if the transposition rate is high, resulting in individuals with many mutations, at least some of those mutations are likely to be deleterious and to decrease the overall fitness of the individual.

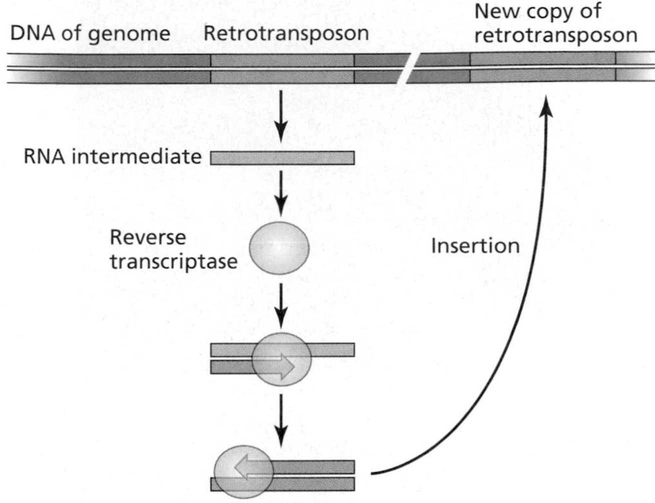

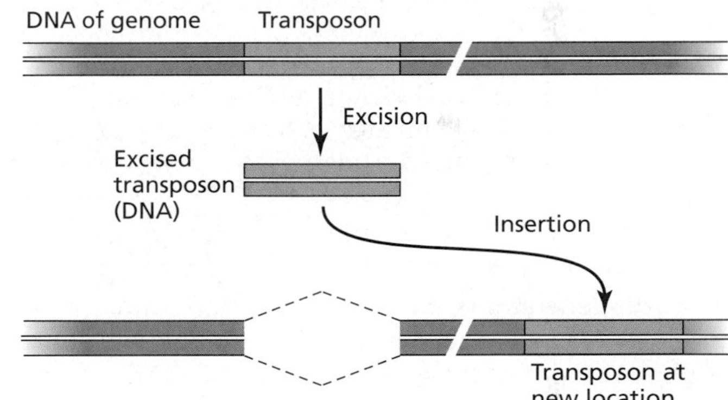

Figure 2.2 The two major classes of transposons differ in their mode of transposition. (A) Retrotransposons move via an RNA intermediate. (B) DNA transposons move using a cut-and-paste mechanism.

Plants and other organisms seem to be able to regulate the activity of transposons through the methylation of DNA and histones. As we will see later in this chapter, these same processes are used to repress transcription in heterochromatic regions of the genome. As more genomic DNA sequences have become available, scientists have noticed large numbers of highly methylated transposons in heterochromatic regions. Did methylation of transposons cause heterochromatin to form there? Or did the transposons become methylated because they happened to land in heterochromatic regions?

Studies of mutants that are unable to maintain genome methylation have shown that slow loss of methyla-

(A)

(B)

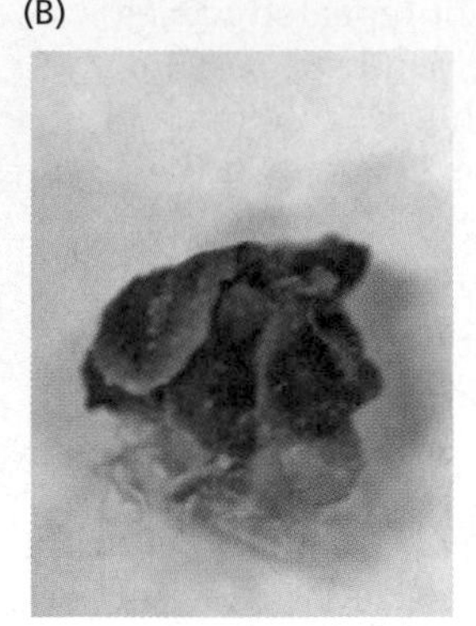

Figure 2.3 Loss of methylation can lead to mutations as unmethylated transposons become active. A mutation called *decrease in dna methylation* (*ddm1*) causes hypomethylation (decreased methylation) of endogenous transposons. The *clam* mutation, which arose in a *ddm1* mutant, is the result of insertion of a transposon into the *DWARF4* (*DWF4*) gene, which is required for the biosynthesis of the growth hormone brassinosteroid. (A) The transposon-induced *clam* mutant (left) next to wild-type Arabidopsis. (B) The *clam* mutant without (left) and with (right) a sector that has reverted to the wild-type phenotype after the transposon jumped back out of the *DWF4* gene. (From Miura et al. 2001.)

tion over generations can activate dormant transposons and increase the frequency of transpositional mutations (**Figure 2.3**). Such transposon activity may considerably decrease the fitness of offspring. Therefore, methylation and the formation of heterochromatin appear to play important roles in the stability of the genome.

Chromosome organization is not random in the interphase nucleus

During interphase of the cell cycle, the chromosomes decondense. However, interphase chromosomes are not randomly arranged or entangled with each other like a plate of spaghetti; rather, each chromosome occupies a discrete location within the nucleus called the **chromosome territory**. Chromosomes in species with larger genomes orient their chromosomes so that centromeres and telomeres of each chromosome are on opposite poles of the nucleus, a conformation known as the **Rabl configuration**, after the Austrian scientist Carl Rabl, who first proposed such arrangement as early as 1885 (**Figure 2.4A**). However, chromosomes in plants with smaller genomes, such as Arabidopsis, do not adopt a Rabl configuration, but instead seem to cluster their telomeres around the nucleolus in a rosette-like formation (**Figure 2.4B**). For a long time, scientists have tried to determine whether the arrangement of chromosomes during interphase affects gene expression, but to date, this question is unresolved.

Meiosis halves the number of chromosomes and allows for the recombination of alleles

In Chapter 1 we discussed the events during mitotic cell division. During the production of gametes, cells are divided in a similar fashion as during mitosis but with several important differences. During the first meiotic division, DNA is exchanged between homologous chromosomes before the chromosomes are separated into daughter cells, resulting in the recombination of genetic material (meiotic recombination) (**Figure 2.5**). The second meiotic division segregates the sister chromatids, resulting in four daughter cells per original cell. Because meiosis involves two cell divisions but only one round of DNA replication, the daughter cells each have half as much genetic

(A)

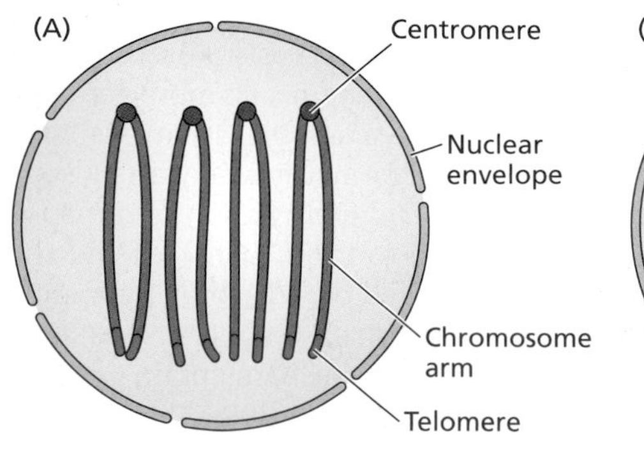

(B)

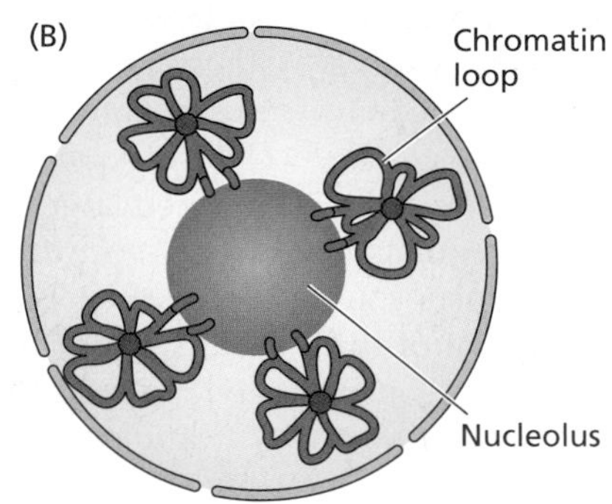

Figure 2.4 Chromosome arrangements in the interphase nucleus. (A) Rabl configuration of chromosomes, where centromeres and telomeres of all chromosomes point away from each other. (B) Rosette configuration of chromosomes, where the telomeres are oriented toward the nucleolus. (After Tiang et al. 2012.)

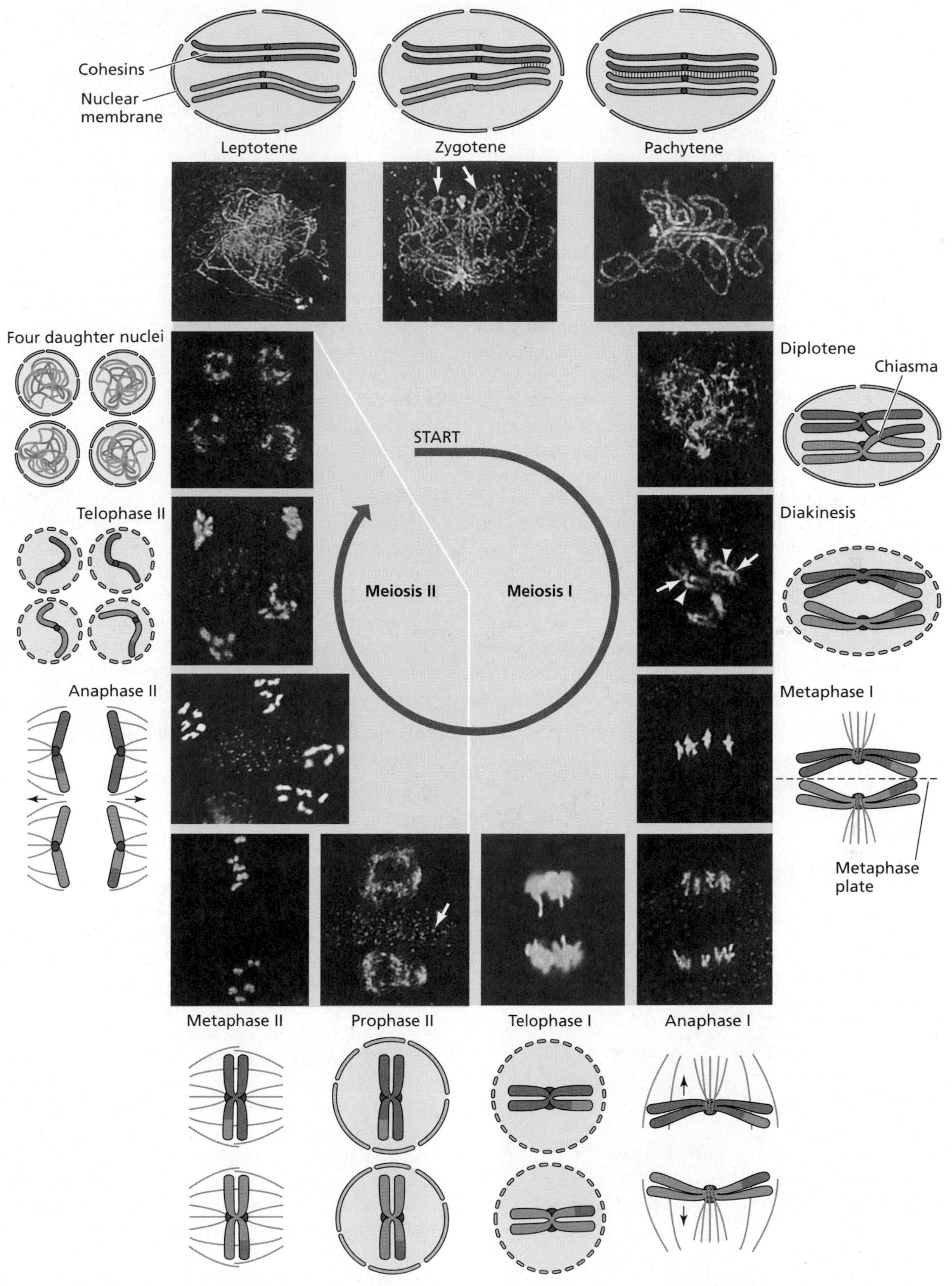

Figure 2.5 Male meiosis in Arabidopsis. The illustration shows the chromosomal state at each stage for one chromosome only. Prophase 1 starts at the Leptotene stage and goes through Diakinesis. See text for details. Arrows in zygotene of prophase I indicate visible areas of chromosome pairing; in diakinesis of prophase I, arrows indicate chiasmata and arrowheads indicate centromeres. Cohesins are proteins that hold sister chromatids together. Synaptonemal complexes (green lines) are protein complexes that form between homologs. (Micrographs by Wuxing Li, from Ma 2005; diagrams after Grandont et al. 2013.)

material as the original cell: a diploid (2*N*) plant produces haploid (1*N*) gametes. This is called meiotic reduction.

The first major phase in meiosis, prophase, is divided into five stages: leptotene, zygotene, pachytene, diplotene, and diakinesis. During leptotene, homologous regions within pairs of homologous chromosomes begin to associate with each other and meiotic recombination is initiated with the help of several specific proteins. Once homologous regions are identified, homologous chromosomes (homologs) begin to pair during zygotene and form **synaptonemal complexes** (**Figure 2.6**), which eventually run continuously along the length of each chromosome pair. Paired chromosomes are also referred to as bivalents. Toward the end of pachytene, chromosomes have condensed enough to be seen in the microscope as distinguishable threads (see Figure 2.5). Crossing over (the exchange of DNA between homologs) starts during pachytene, and in diplotene visible junctions, also called chiasmata, can be seen between homologous chromosomes. Chiasmata are resolved (i.e., DNA exchange is completed) toward the end of diplotene and into diakinesis. At this point, chromosomes further condense and the centromeres appear to move away from each other, while the chromosome ends still maintain contact between homologs. The nuclear membrane breaks down at the end of diakinesis. During metaphase I, the still paired homologs line up along the metaphase plate, where spindle fibers attach to each centromere via the proteinaceous kinetochore. In anaphase I, the homologous pairs are separated with the help of the spindle fibers pulling the homologs to opposite poles. Sister chromatids remain attached to each other during anaphase I. During telophase I, chromosomes may decondense somewhat, as is the case in Arabidopsis, or as in some plants, stay condensed and move quickly through prophase II into metaphase II. During these phases the chromosomes again align at the metaphase plate, and spindle fibers attach at the centromeres. In anaphase II, sister chromatids are separated and pulled toward the poles. Chromosomes begin to decondense during telophase II, and four haploid nuclei form. Cytokinesis then produces four separate cells. In the case of male meiosis in angiosperms, these four cells are the microspores, which stay together in a cluster called a **tetrad**. The tetrad later releases the four microspores, which undergo mitosis to produce mature pollen (the male gametophytes). In the case of female meiosis in angiosperms, only one of the four daughter cells survives, giving rise to the megaspore. The megaspore eventually undergoes mitosis and produces eight haploid nuclei, which form the female gametophyte.

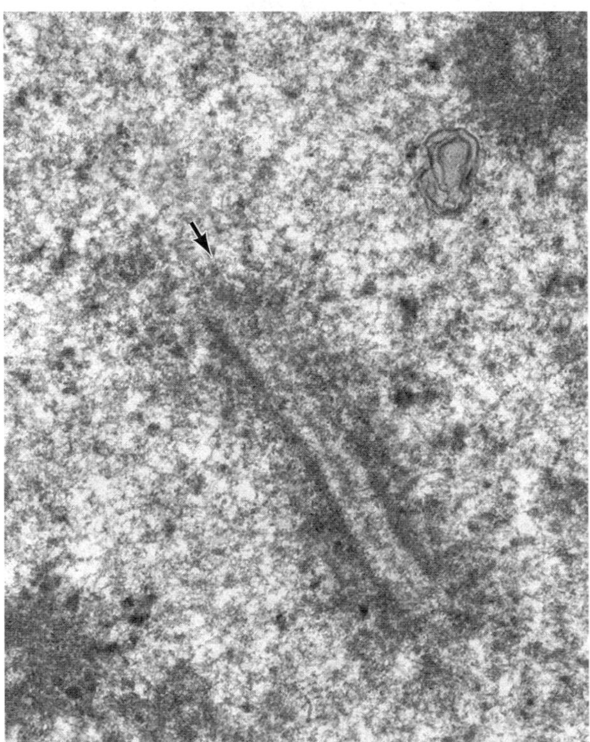

Figure 2.6 Synaptonemal complex in Arabidopsis. The arrowhead points toward a partial synaptonemal complex during pachytene. (Micrograph by L. Timofejeva, from Ma 2005.)

Polyploids contain multiple copies of the entire genome

Ploidy level—the number of copies of the entire genome in a cell—is another important aspect of genome structure that may have implications for both physiology and evolution. In many organisms, but especially in plants, the entire diploid (2*N*) genome can undergo one or more additional rounds of replication without undergoing cytokinesis (see Chapter 1) to become **polyploid**. If polyploidy is restricted to somatic tissues, the terminology used to describe this state is **endopolyploidy**. Examples of endopolyploidy are salivary glands in *Drosophila* and liver cells in humans. In plants, endopolyploidy occurs frequently in fully differentiated leaf cells.

If whole-genome duplication in a somatic cell is carried into the germ line (gametes), a uniformly polyploid next generation can result. Polyploidy is not a rare event, nor is it normally associated with mutation or disease. In fact, polyploidy is a common phenomenon that has occurred at least once in all angiosperm lineages. Evidence for multiple polyploidization events can be found in many plant genomes, but interestingly, whole-genome duplications appear to be less common in gymnosperms. Two forms of polyploidy are distinguished: autopolyploidy and allopolyploidy. **Autopolyploids** contain multiple complete genomes of a single species, while **allopolyploids** contain multiple complete genomes derived from two or more separate species.

Both types of polyploidy can result from incomplete meiosis during gametogenesis. During normal meiosis,

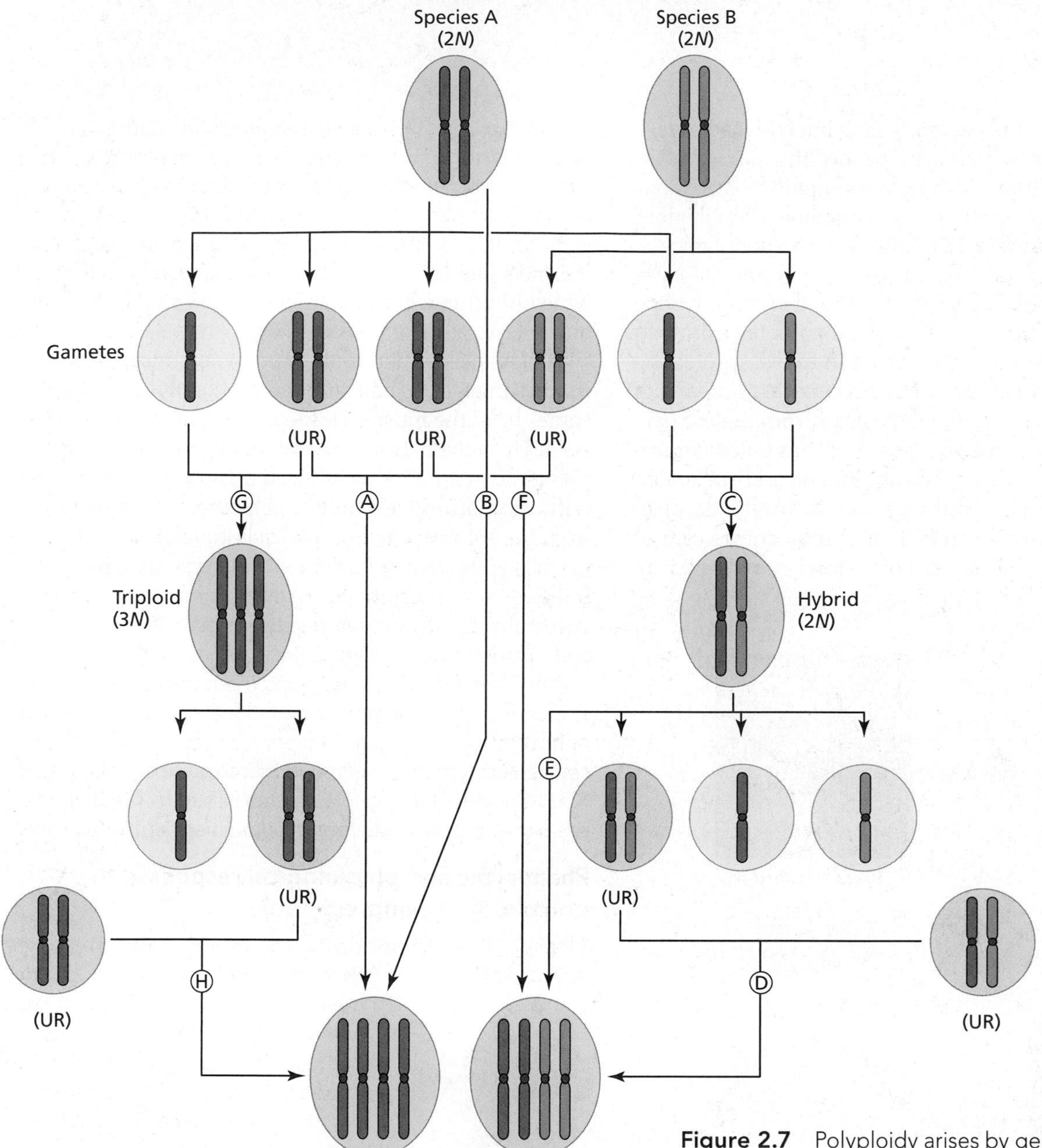

Figure 2.7 Polyploidy arises by genome duplication. (A) Fusion of unreduced, diploid gametes from the same diploid species results in an autotetraploid. (B) Spontaneous genome duplication in a diploid can also result in an autotetraploid. (C) Fusion of a haploid sperm from one species and a haploid egg from another species results in a diploid interspecies hybrid. (D) In rare cases, an interspecies hybrid may produce unreduced diploid hybrid gametes, which, if fused with another unreduced hybrid gamete, would produce an allopolyploid. (E) An interspecies hybrid also may in rare cases undergo spontaneous genome duplication in somatic cells, leading to an allopolyploid. (F) Fusion of diploid gametes from two different species results in an allopolyploid. (G) The fusion of a normal haploid gamete with an accidentally formed diploid gamete first leads to a triploid individual, which (if fertile at all) may produce both haploid and diploid (unreduced) gametes. Fusion of two diploid gametes then results in an autotetraploid, as shown in (H). This pathway to polyploidy is called the triploid bridge. (UR: unreduced gametes.) (After Bomblies and Madlung 2014)

the chromosomes of a diploid germ cell undergo DNA replication followed by two rounds of division (meiosis I and meiosis II), producing four haploid cells (see Figure 2.5). If chromosome duplication is not followed by the normal two rounds of cell division during meiosis, diploid **unreduced gametes** may result. Within a species, or in a self-fertilizing individual, if a diploid egg is fertilized by a diploid sperm, the resulting zygote contains four copies of each chromosome and is said to be *autotetraploid* (**Figure 2.7A**). Likewise, if cell division does not occur after chromosome duplication during mitosis, cells become autotetraploid (**Figure 2.7B**). Both types of errors during meiosis or mitosis occur spontaneously in plants at variable frequencies depending on the species.

Allopolyploids may also form in one of two ways: (1) A haploid sperm from one species and a haploid egg from another species may form a diploid interspecies hybrid (**Figure 2.7C**).

Meiosis in these plants generally fails but can lead to rare duplicated gametes, which can produce the allopolyploid (**Figure 2.7D**). Additionally, if hybrid somatic cells accidentally omit cell division, these cells become allopolyploid spontaneously (**Figure 2.7E**). This type of allopolyploidization is called somatic doubling and can happen as early as in a hybrid zygote, or later in vegetative or reproductive tissues of the hybrid plant. (2) Diploid gametes from two different species may join to form a tetraploid zygote. The diploid gametes can come either from tetraploid parents that have undergone normal meiosis or from diploid parents in which normal reduction meiosis has failed (**Figure 2.7F**). The latter event is known as gametic non-reduction.

Diploid interspecies hybrids occur naturally, but they are frequently sterile because their chromosomes cannot pair properly during prophase I of meiosis (see Figure 2.5). Another pathway to stable polyploidy is via the so-called **triploid bridge**, which requires a two-step process. First, a reduced gamete fuses with an unreduced gamete, resulting in a triploid individual (**Figure 2.7G**). This individual is more likely to produce unreduced gametes itself (some $2N$ and some $1N$), and thus is more likely to produce tetraploid offspring when its gametes fuse with another unreduced ($2N$) gamete (**Figure 2.7H**).

A classic example from nature where multiple species in the same genus have produced allopolyploid offspring comes from the mustard family, Brassicaceae (**Figure 2.8**). Aside from the natural occurrence of genome duplication, polyploidy can also be induced artificially by treatment with the natural cell toxin colchicine, which is derived from the autumn crocus (*Colchicum autumnale*). Colchicine inhibits spindle fiber formation and prevents cell division, but does not interfere with DNA replication. Treatment with colchicine therefore results in an undivided nucleus containing multiple copies of the genome.

The lack of fertility in diploid interspecies hybrids is in stark contrast to the phenomenon known as **hybrid vigor** or **heterosis**: the increased vigor often observed in the offspring of crosses between two inbred varieties of the same plant species. Heterosis can contribute to larger plants, greater biomass, and higher yields in agricultural crops.

Phenotypic and physiological responses to polyploidy are unpredictable

The widely held notion that autopolyploids are larger than their diploid progenitors does not always hold true. For example, when individual maize plants with the same genetic background, but differing in ploidy level, were compared, it was found that plant height increased from haploidy to diploidy, but *decreased* with further increases in ploidy level (**Figure 2.9A**). One hypothesis to explain the greater vigor of some autopolyploids compared with their diploid progenitors is that plant vigor increases with

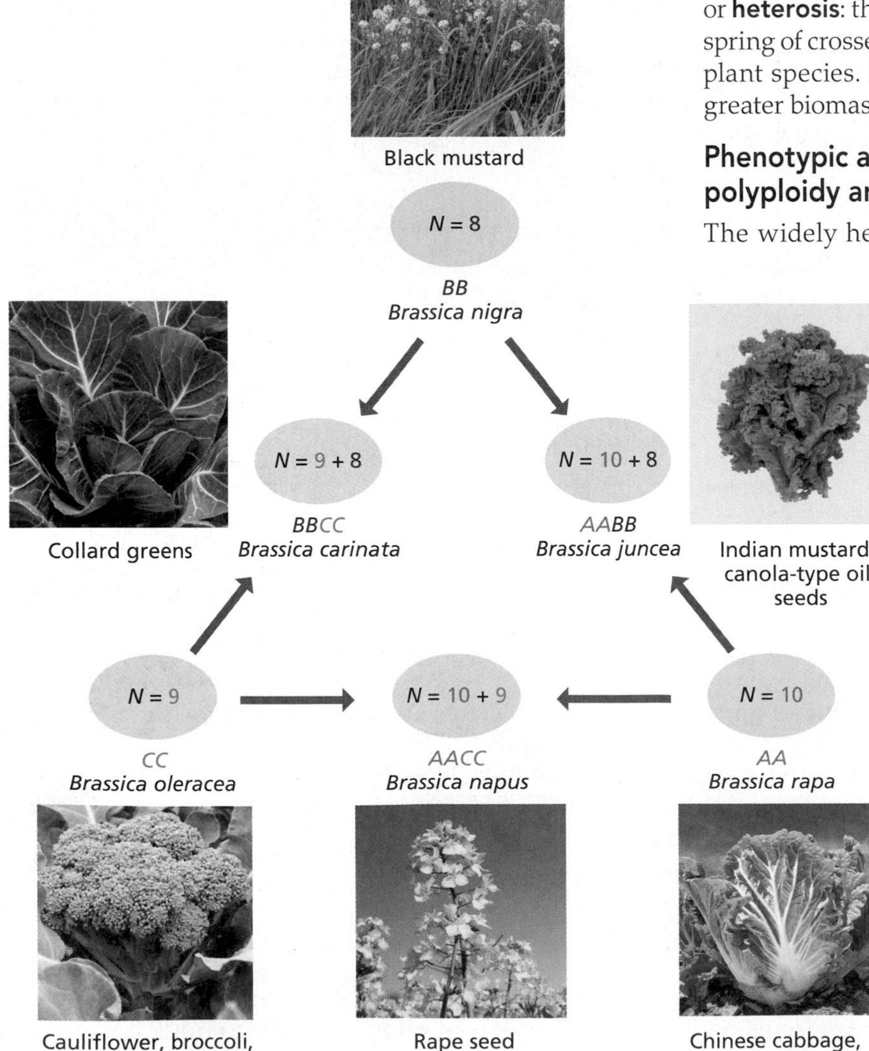

Figure 2.8 Three common species of plants in the mustard family (Brassicaceae) have interbred with one another in nature to form new allotetraploid species. Their relationship is depicted in the so-called triangle of U, named after the Korean scientist Nagaharu U. The three corners of the triangle show diploid species of *Brassica*. Each of the three species can interbreed with the two others to form new allopolyploids.

(A)

(B)

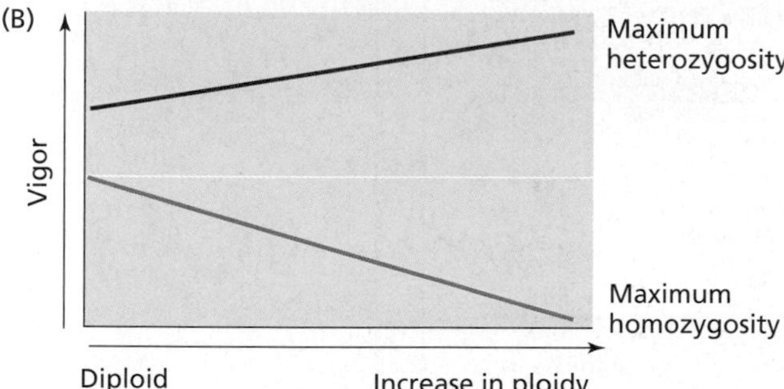

Figure 2.9 (A) Ploidy series in maize. Plants of the same age are shown from left to right: haploid, diploid, triploid, and tetraploid. In inbred maize, autopolyploidy correlates with reduced vigor compared with the diploid. Each black or white section on the scale bar measures 20 cm. (B) A generalized view of the relationship between plant vigor and ploidy level. As ploidy increases, plant vigor increases only in plants whose overall level of heterozygosity also increases due to a greater number of different alleles per genome (red line). By contrast, increasing ploidy in homozygous or inbred plants is correlated with decreasing overall plant vigor (blue line). (A courtesy of E. Himelblau; B adapted from a diagram courtesy of J. Birchler.)

increasing ploidy only if hybridity (heterozygosity) also increases. If instead, the level of *homozygosity* increases in plants with increasing ploidy level (through inbreeding), their vigor decreases (**Figure 2.9B**).

Allopolyploids differ from their parental diploid progenitor species in two major ways:

1. Their genomes, like those of autopolyploids, are duplicated.
2. They are hybrids between two different species.

When comparing allopolyploids with their progenitors, it is therefore difficult to determine whether the observed phenotypic differences are due to genome duplication or to hybridization. Current data suggest that hybridization makes a greater contribution than does genome duplication to the divergence of allopolyploid offspring from their parents. Allopolyploids are frequently more vigorous or higher yielding than their parent species and are very common among agriculturally important plants. Examples of such allopolyploids include canola and collard, coffee, cotton, wheat, rye, oat, and sugarcane.

Regardless of how allopolyploids arise, the fusion of two divergent genomes has many consequences, although it is not yet clear if there is a common set of responses across all species during or immediately after allopolyploidization. Some of the genetic changes that have been observed in newly formed allopolyploids compared with their parent species are the following:

- Reorganization of the genome, including loss or gain of DNA sequences
- Changes in epigenetic modifications
- Changes in gene transcriptional activity
- Variability in exon use (alternative splicing)
- Increase in meiotic recombination frequency
- Activation of previously dormant transposable elements through loss of gene silencing

Transcriptional activity changes between parent and allopolyploid species at the whole-genome level (as opposed to one gene at a time) have been studied using microarray analysis and high-throughput RNA sequencing, two techniques we will discuss later in this chapter. It is likely that epigenetic modifications, including DNA and histone methylation and histone acetylation, are responsible for many of these changes. Due to the unpredictable nature of genomic changes in polyploids, physiological responses to polyploidy can vary among individuals from the same cross. In contrast to the mostly unpredictable phenotypic changes due to allopolyploidy, some phenotypes that are generally associated with autopolyploidy are greater flower diameter, larger stomata size, and in Arabidopsis, greater resistance to salt stress.

Polyploidy leads to multiple, redundant copies of genes in the genome. When evolution acts on the duplicate genes, one copy may be either lost or changed in function, while the other retains its original function. Duplicate gene copies may also take on expression patterns that are tissue-specific. These types of processes are known as **subfunctionalization**. Genome analysis shows that even in many diploid species there is clear evidence of genome duplication in the species' evolutionary history. In such cases, a subsequent gradual loss of DNA has led back to a diploid-like state (**Figure 2.10**). Species that show signs of ancient genome duplications followed by DNA loss are known as **paleopolyploids** and include Arabidopsis, maize, and *Brassica* species.

Polyploidy is in striking contrast to a condition called **aneuploidy**. Aneuploids are organisms whose genomes contain more or fewer individual chromosomes (not entire chromosome sets) than normal. Such states are known as

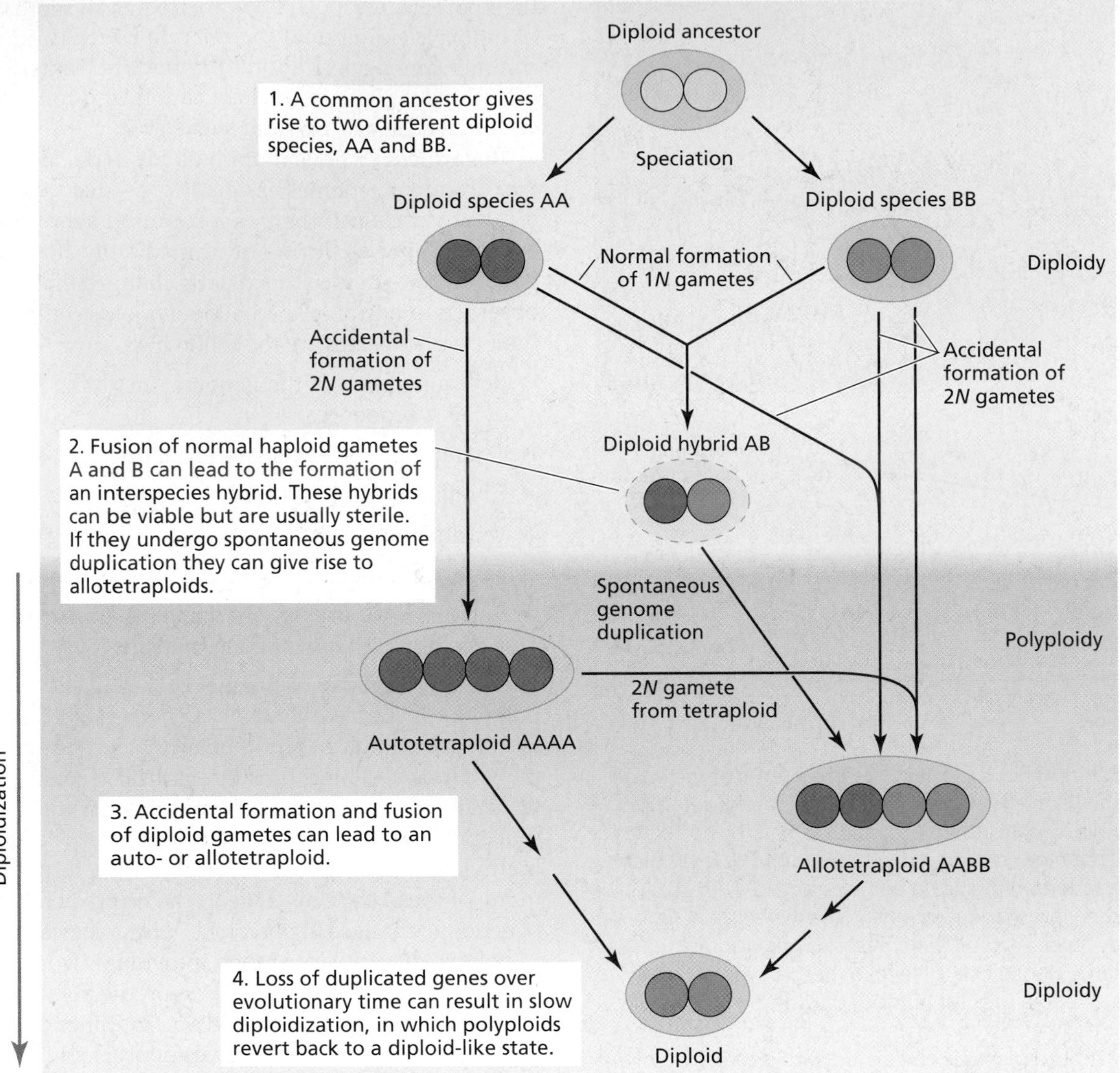

Figure 2.10 Continuum in the evolution of polyploid species. Diploids can give rise to autopolyploids or allopolyploids by the mechanisms outlined in Figure 2.5. Polyploids can revert to a diploid-like state via the gradual loss of DNA, including duplicated genes, over evolutionary time. Light purple underlying ovals represent the nuclei of a species; colored circles inside nuclei represent entire genomes. (After Comai 2005.)

trisomies if one type of chromosome is tripled or as **monosomies** if only one chromosome of a given type is present. In humans and many animals, aneuploidy usually leads to death or to severe physiological problems, such as Down syndrome (trisomy 21). Aneuploid plants, although often phenotypically distinct from normal (euploid) plants, are generally viable. In polyploids, the effects of aneuploidy can be masked by the additional chromosomes in the genome.

The role of polyploidy in evolution is still unclear

It is now known that all angiosperm lineages have undergone at least one genome duplication event in their evolutionary history, and new cases of polyploidization are frequent in nature. Given the breadth of genomic and epigenetic changes found in newly formed polyploids (discussed above), polyploidy might be expected to have enormous potential for shaping plant evolution. The genetic novelty that arises from polyploidy-induced changes provides new genetic material on which selection can act. Yet phylogenetic analysis has also shown that polyploidy is found more often in young species than in older species. This observation has led to the hypothesis that polyploidy might not be as advantageous as previously presumed, and that many new polyploid species are little more than an evolutionary dead end. Genomic changes and complex chromosome pairing interactions during meiosis might explain why, in some cases, polyploids have reduced fitness and go extinct. Nonetheless, the prominence of polyploidy as a process shaping plant evolution suggests that in the long term the advantages of polyploidy outweigh its dangers.

Plant Cytoplasmic Genomes: Mitochondria and Plastids

In addition to the nuclear genome, plant cells contain two additional genomes: the **mitochondrial genome**, which they share with animal cells, and the **plastid genome**. In this section we will see where these genomes come from and what roles they play. We will then look at their organization and discuss some important differences from the nuclear genome in the way their genetic information is transmitted.

The endosymbiotic theory describes the origin of cytoplasmic genomes

Cytoplasmic genomes are probably the evolutionary remnants of the genomes of bacterial cells that were engulfed by another cell. The **endosymbiotic theory**, championed by Lynn Margulis in the 1980s, postulates that the original mitochondrion was an oxygen-using (aerobic) bacterium that was absorbed by another prokaryotic organism. Over time, this original endosymbiont evolved into an organelle that was no longer able to live on its own. The host cell, along with its endosymbiont, gave rise to a lineage of cells that were able to use oxygen in aerobic metabolism; these cells, in turn, eventually gave rise to all animal cells. Plant cells, according to this theory, arose when a second endosymbiosis event took place. This time, a mitochondrion-containing cell engulfed a photosynthetic cyanobacterium, which over time evolved inside the cell into the plastid.

Two main lines of evidence are often cited in support of the endosymbiotic theory. First, both mitochondria and plastids are enclosed by an outer and an inner membrane. This observation is consistent with the idea that engulfment of the original aerobic or photosynthetic cell through invagination of the plasma membrane of the prokaryotic host cell left a double membrane around the new organelle. Second, both organellar genomes show sequence similarity to prokaryotic genomes. The organellar genomes, like those of prokaryotes, are not enclosed in a nuclear envelope and are called **nucleoids**.

Organellar genomes vary in size

Plastid genomes generally range in size from about 120 to 160 kilobase pairs (kbp) and encode genes that are needed for photosynthesis and the expression of plastid genes. The mitochondrial genome is much more variable in size than the plastid genome. Plant mitochondrial genomes range between approximately 180 kbp and nearly 11 Mbp—much larger than mitochondrial genomes of animals or fungi, many of which are only 15 to 50 kbp. Much of the difference in size between these genomes is made up by noncoding repeat DNA. Plant mitochondrial DNA contains genes that encode proteins needed in the electron transport chain, or that are involved in providing cofactors for electron transport. Additionally, plant mitochondrial DNA carries genes for proteins required for the organelle's own gene expression, such as ribosomal proteins, tRNAs, and rRNAs. In both organelles, most of the genes required for proper plastid or mitochondrial function are no longer encoded in the organellar genome itself but over evolutionary time have been transferred to the nucleus of modern plants. The proteins encoded by these genes are synthesized in the cytoplasm and then imported into the organelles.

For many years organellar chromosomes had been thought to contain a genome-sized DNA molecule in circular form, similar to the circular plasmids of bacteria. Recent data, however, show that most of the DNA in both plant mitochondria and plastids is found in linear molecules that may contain more than one copy of the genome. These copies are connected to one another in head-to-tail orientation, and the chromosomal DNA molecules can be highly branched, resembling a bush or tree, unlike the simpler structures of linear nuclear chromosomes. While nuclear chromosomes are of constant size generation after generation, the chromosome size in mitochondria and plastids can vary. Nonetheless, each organellar chromosome contains at least one complete genome. (For photographs of plant plastid genomes, see WEB TOPIC 2.1.)

Organellar genetics do not obey Mendelian principles

The genetics of organellar genes are governed by two principles that distinguish them from Mendelian genetics. First, both mitochondria and plastids generally show **uniparental inheritance**, which means that sexual offspring (via pollen and eggs) inherit organelles from only one parent. Among the gymnosperms, the conifers normally inherit their plastids from the paternal parent. Among angiosperms, the general rule is that the plastids come from the maternal parent. However, there are a few angiosperms in which plastids are inherited either biparentally or paternally. Mitochondrial inheritance is usually maternal in the majority of plants, but again a few exceptions can be found; for example, some types of conifers, such as the cypresses, show paternal inheritance of mitochondria. (For a discussion on how uniparental inheritance is achieved during development, see WEB TOPIC 2.1.)

The second major feature of organellar inheritance is the fact that both plastids and mitochondria can exhibit **vegetative segregation**. This means that a vegetative cell (as opposed to a gamete) can give rise to another vegetative cell via mitosis that is genetically different. For example, consider a vegetative cell that contains a mixture of two genetically distinct types of plastids. During mitosis, the

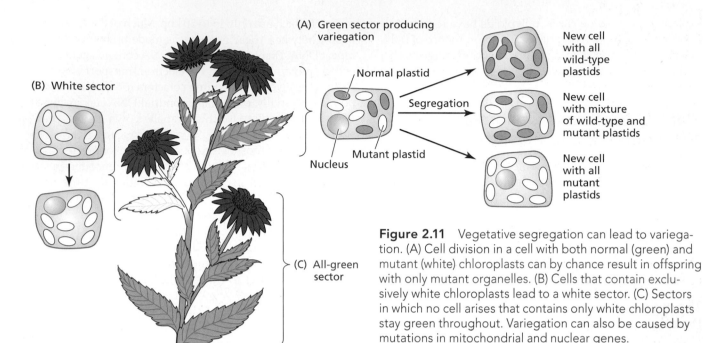

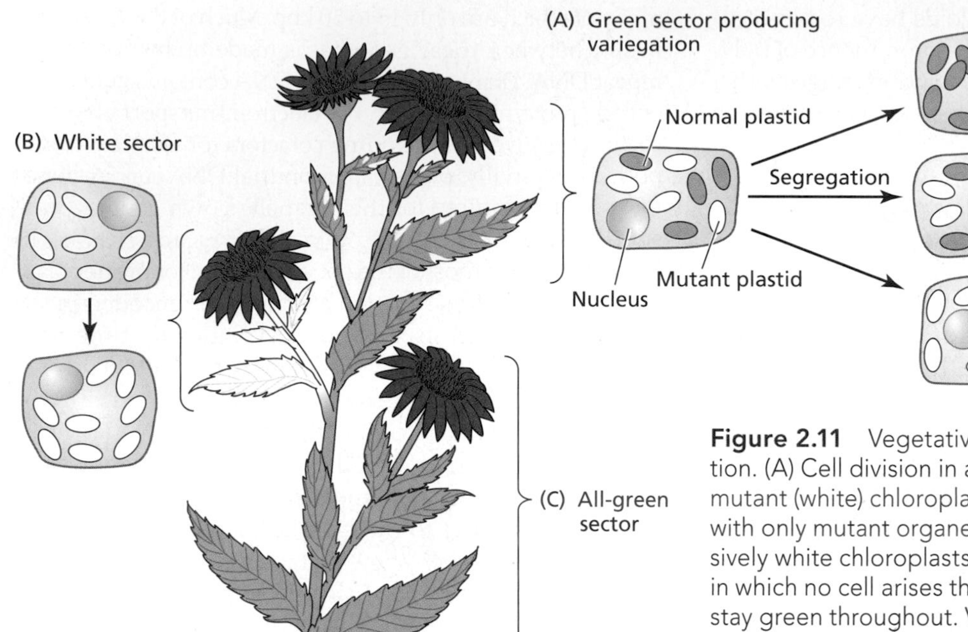

Figure 2.11 Vegetative segregation can lead to variegation. (A) Cell division in a cell with both normal (green) and mutant (white) chloroplasts can by chance result in offspring with only mutant organelles. (B) Cells that contain exclusively white chloroplasts lead to a white sector. (C) Sectors in which no cell arises that contains only white chloroplasts stay green throughout. Variegation can also be caused by mutations in mitochondrial and nuclear genes.

plastids are distributed randomly to the daughter cells. By chance, one daughter cell may receive plastids with one type of genome, while the other daughter cell may receive plastids with different genetic information, perhaps containing one or more mutations. Vegetative segregation, which is also referred to as **sorting-out**, can result in the formation of phenotypically different sectors within a tissue (**Figure 2.11**). The presence of such sectors in leaves may result in what horticulturists often refer to as **variegation**. Leaf variegation can also be caused by mutations in nuclear and mitochondrial genes.

Now that we have looked at the organization of nuclear and cytoplasmic genomes in plants, we will turn to the structure of the nuclear genome and how that structure influences the expression of the genes the genome contains. The basic mechanisms of gene transcription will be reviewed first, followed by a description of the transcriptional regulation of gene expression.

Transcriptional Regulation of Nuclear Gene Expression

As we introduced in Chapter 1, the path from gene to protein is a multistep process catalyzed by many enzymes (see Figure 1.11). Each step is subject to regulation by the plant to control the amount of protein that is produced by each gene. Regulation of the first step, transcription, determines when and whether an mRNA is made. This level of regulation, which is referred to as **transcriptional regulation**, includes the control of transcription initiation, maintenance of transcription, and termination. The next level in the regulation of gene expression, known as **posttranscriptional regulation**, occurs after transcription. This level, which we will cover later in this chapter, includes the control of mRNA stability, translation efficiency, and degradation. Finally, **protein stability** (posttranslational regulation) plays an important role in determining the overall activity of a gene or its product.

RNA polymerase II binds to the promoter region of most protein-coding genes

Gene transcription is facilitated by an enzyme called **RNA polymerase**, which binds to the DNA to be transcribed and makes an mRNA transcript complementary to the DNA sequence (**Figure 2.12**). There are several types of RNA polymerase. RNA polymerase II is the polymerase that transcribes most protein-coding genes.

The region of the gene that recruits the transcription machinery, including RNA polymerase, is called the **promoter**. The structure of the eukaryotic promoter can be divided into two parts: the **core promoter** or **minimum promoter**, consisting of the minimum sequence required for gene expression, and the **regulatory promoter sequences**, which control the activity of the core promoter. For protein-coding genes, the core promoter generally occupies the approximately 80 bp surrounding the transcription start site.

Before transcription of a gene can begin, several steps have to occur to allow the RNA polymerase to gain access to the gene's nucleotide sequence. Nuclear DNA is wrapped around histones, forming beadlike nucleosomes. As we will discuss in more detail later in this section, the

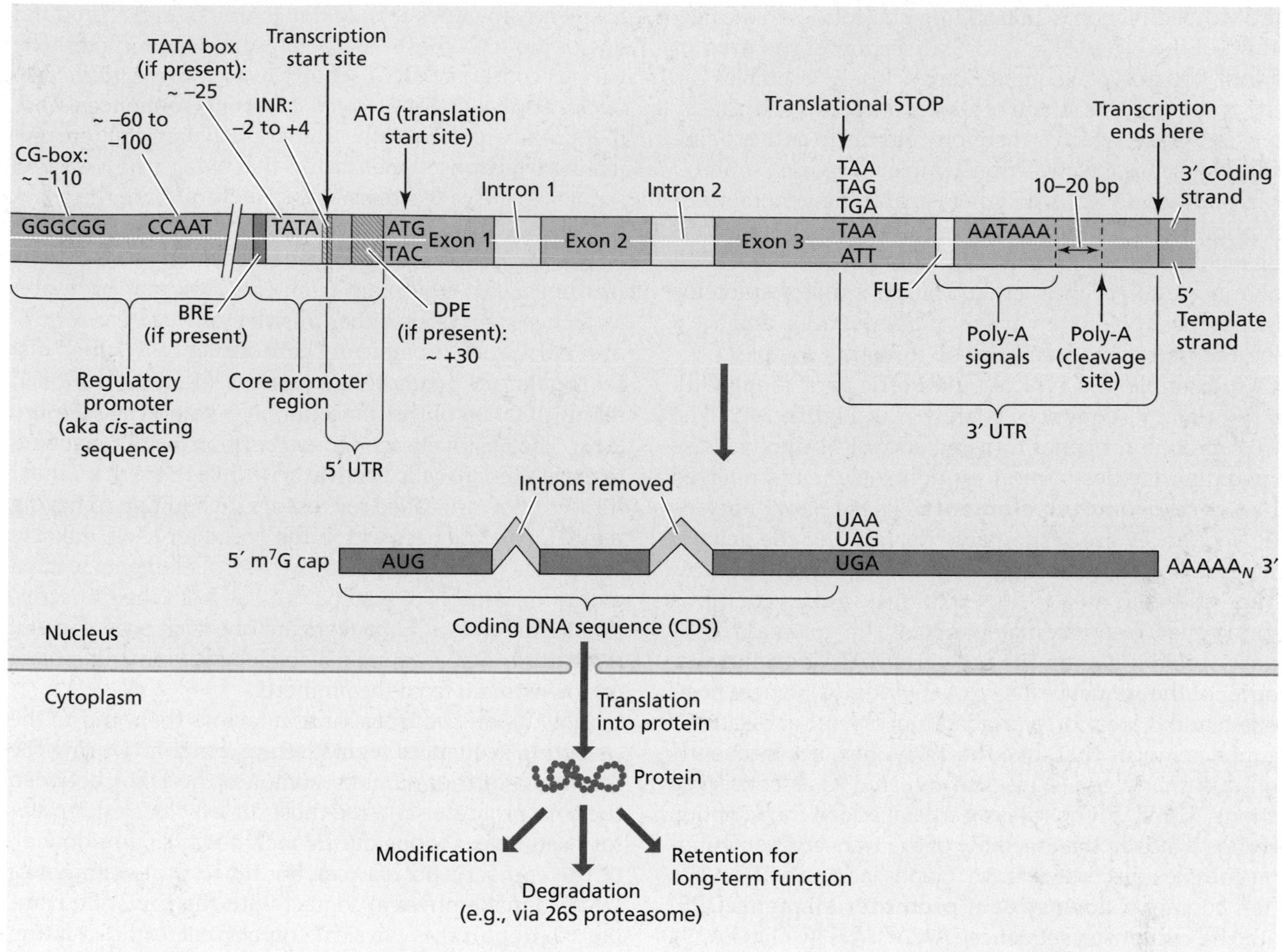

Figure 2.12 Gene expression in eukaryotes. RNA polymerase II binds to the promoters of genes that encode proteins. Unlike prokaryotic genes, eukaryotic genes are usually not clustered in operons, and each is divided into introns and exons. Transcription from the template strand proceeds in the 3′-to-5′ direction at the transcription start site, and the growing RNA chain extends one nucleotide at a time in the 5′-to-3′ direction. Translation begins with the first AUG encoding methionine, as in prokaryotes, and ends with a stop codon. The pre-mRNA transcript is first "capped" by the addition of 7-methylguanylate (m7G) to the 5′ end. The 3′ end is shortened slightly by cleavage at a specific site, and a poly-A tail is added. The capped and polyadenylated pre-mRNA is then spliced by a protein complex called the spliceosome, and the introns are removed. The mature mRNA exits the nucleus through the nuclear pores and initiates translation on ribosomes in the cytosol. As each ribosome progresses toward the 3′ end of the mRNA, new ribosomes attach at the 5′ end and begin translating, leading to the formation of polysomes. After translation, some proteins are modified by the addition of chemical groups to the protein chain. The released polypeptides have characteristic half-lives, which are regulated by the ubiquitin pathway and a large proteolytic complex called the 26S proteasome. Eukaryotic genes typically contain binding sites for the RNA polymerase, such as the TATA box within the core promoter region, as well as sites binding general and specific transcription factors in the proximal and distal regulatory promoter.

histones are subject to modifications, and only if those modifications are favorable to transcription will RNA polymerase be able to bind to the DNA. To be functional, the RNA polymerases of eukaryotes require additional proteins called **general transcription factors** to position the polymerase at the transcription start site. These general transcription factors, together with the RNA polymerase, make up a large, multi-subunit **transcription initiation complex**. Transcription is initiated when the final transcription factor to join the complex phosphorylates the RNA polymerase. The RNA polymerase then separates from the initiation complex and proceeds along the antisense strand (also sometimes referred to as the noncoding, template, negative, or Watson strand) of the DNA in the 3′-to-5′ direction, while adding nucleotides to the new strand of mRNA in the 5′-to-3′ direction of the nascent strand. The mRNA sequence most closely resembles the code of the opposite strand of DNA not used as template by the polymerase, which is therefore referred to as the coding strand (or sense, positive, non-template, or Crick strand).

In addition to RNA polymerase and the general transcription factors, many genes require specific transcription factors (also called gene regulatory proteins) for RNA polymerase to become active. These regulatory proteins bind to the DNA, often at specific sequences, and are a required part of the transcription initiation complex.

An example of a typical eukaryotic gene along with its regulatory sequences is shown in Figure 2.12. The core promoter for genes transcribed by RNA polymerase II usually includes several sequence elements referred to as **core promoter elements**. These short nucleotide sequences are responsible for binding the general transcription factors and the RNA polymerase. Many, although by no means all, eukaryotic genes contain a short sequence approximately 25 to 30 bp upstream of the transcription initiation site called the **TATA box** and consisting of the sequence TATA(A/T)AA(G/A), where positions 5 and 8 are more variable than the other positions. Sequence motifs that, like the TATA box, are frequently found in many genes are also referred to as conserved regions. The TATA box plays a crucial role in transcription because it aids in the assembly of the transcription initiation complex discussed above. Genes lacking a TATA box often contain a **downstream promoter element** (**DPE**) with the consensus sequence (A/G)G(A/T)(C/T)(G/A/C), which is located at nucleotides 28 to 32 downstream of the transcription initiation site. A third important part of the core promoter is the **initiator element** (**INR**). This nucleotide sequence also binds to general transcription factors and can be found both in TATA box–containing and TATA box–lacking genes around the transcription start site from nucleotide position –2 to +4. The fourth binding element in the core promoter is the so-called **TFIIB recognition element** (**BRE**). This sequence recognizes a different general transcription factor than the other elements. The BRE is located within the six nucleotides immediately adjacent to and upstream of the TATA box (see Figure 2.12).

Further upstream of the core promoter sequence, many eukaryotic genes also contain two additional conserved sequences: the **CCAAT box** and the **GC box** (see Figure 2.12). The region containing these sequences is called the **regulatory promoter** or the **proximal** (close) **promoter**. This part of the promoter does not bind the RNA polymerase and its associated general transcription factors required for all genes, but instead binds to gene-specific transcription factors. The CCAAT box, if present, is usually located 60 to 100 bp upstream of the transcription start site. It is important to note that not all genes contain all conserved elements. For example, GC boxes are found more frequently in genes that do not contain a TATA box, and one or more GC boxes can be present in a promoter. The various conserved DNA sequences we have described so far are also collectively termed ***cis*-acting sequences**, since they are adjacent (*cis*) to the transcription units they are regulating. The transcription factors that bind to the *cis*-acting sequences are called ***trans*-acting factors**, since the genes that encode them are located elsewhere in the genome.

Numerous *cis*-acting sequences located more distally (farther away) upstream from the proximal promoter sequences can exert either positive or negative control over eukaryotic promoters. These sequences, termed **distal regulatory promoter sequences**, are usually located within 1000 bp of the transcription initiation site (**Figure 2.13**). The positively acting transcription factors that bind to these sites are called **activators**, while those that inhibit transcription are called **repressors**. In addition to having regulatory sequences within the promoter itself, eukaryotic genes can be regulated by control elements located tens of thousands of base pairs away from the transcription initiation site. **Enhancers** are one such type of distal regulatory sequence and may be located either upstream or downstream from the promoter.

How do all the transcription factors that bind to the *cis*-acting sequences regulate transcription? During the formation of the initiation complex, the DNA between the core promoter and the most distally located regulatory sequences loops out in such a way as to allow all of the transcription factors bound to that segment of DNA to make physical contact with the initiation complex. Through this physical contact, each transcription factor exerts its control, either positive or negative, over transcription.

Conserved nucleotide sequences signal transcriptional termination and polyadenylation

As RNA polymerase II reaches the 3′ region of the gene, it first passes the DNA sequence that codes for the stop codon in the mRNA (see Figure 2.12). The stop codon is part of the mRNA and indicates to the ribosomes where the mRNA region that should be translated into protein ends. The 3′ untranslated region lies 3′ of the sequence for the stop codon. Signals for termination of transcription in plants, fungi, and animals have both similarities and differences. Before terminating transcription, plant RNA polymerase II encounters three conserved DNA sequences that aid in the termination of transcription and the addition of the poly-A tail, which helps stabilize the mRNA. The first of these somewhat conserved DNA sequences is the **far upstream element** (**FUE**), which is six nucleotides long and found somewhere between 30 and 170 bp before the poly-A addition site. Soon after the FUE many—but by no means all—plant genes contain a conserved AAUAAA sequence. This exact sequence seems to be strictly required for polyadenylation in ani-

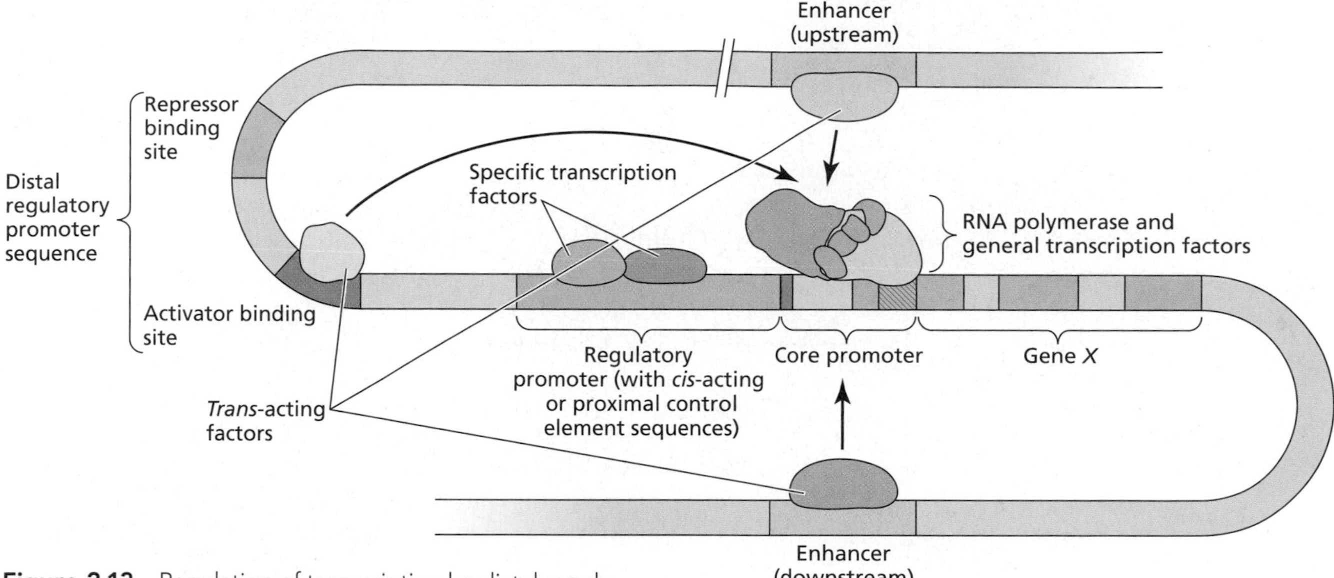

Figure 2.13 Regulation of transcription by distal regulatory promotor sequences, enhancers, and *trans*-acting factors. The *trans*-acting factors can act in concert with the distal regulatory promotor sequences to which they are bound, to activate transcription by making direct physical contact with the transcription initiation complex.

mals, but in plants variations of this sequence with similarity to the AAUAAA element are sufficient for proper function. Both the FUE and the AAUAAA site are also referred to as poly-A signals. The poly-A cleavage site is the DNA sequence that codes for the region on the mRNA where the nascent mRNA strand is cleaved and the poly-A tail added (see Figure 2.12). Together, these three conserved sequences on the DNA strand also promote termination of transcription by RNA polymerase II.

Epigenetic modifications help determine gene activity

As mentioned earlier, transcription can be initiated only if the DNA is accessible to the RNA polymerase and other required binding proteins. In order to make the DNA accessible, its packaging has to be "loosened," a process mediated by covalent modifications of both DNA and histones. Because these modifications can change a gene's behavior without changing the DNA sequence of the gene itself, they are referred to as **epigenetic modifications** (from the Greek word *epi*, meaning "over" or "on top of").

One common type of DNA modification is the **methylation** of cytosine residues (**Figure 2.14A**). DNA sequences that are frequently methylated in plants are CG, CHG, and CHH (where H can be any base except guanine). In contrast, cytosine methylation in mammals occurs mostly on CG. Cytosine methylation is catalyzed by one of several methyltransferases, whereas DNA demethylation is catalyzed by glycosylases that replace methylcytosine with unmethylated cytosine.

Epigenetic modifications can also occur on histones, which, together with the DNA wrapped around them, make up the nucleosomes. Each histone has a "tail," which is made up of the first part of the histone's amino acid chain and protrudes outward from the nucleosome. Histone modifications occur on these tails, usually within the outermost 40 or so amino acids. These modifications can influence the conformation of the nucleosomes and thereby gene activity in the associated DNA.

One of the histone modifications that influences gene activity is methylation, especially at specific lysine residues (single-letter amino acid abbreviation K) in the tail of histone type H3. These residues are K4, K9, K27, and K36, counting from the outermost amino acid inward to the center of the histone coil. One, two, or three methyl groups can be added to a single lysine (**Figure 2.14B**). Histones dimethylated at position H3K4 are generally associated with active genes, whereas dimethylation at position H3K9 is often associated with inactive genes and elements, such as silent transposons. Methyl groups can be removed by histone demethylases.

Another form of modification that occurs on the histone tail is **acetylation**, which is catalyzed by enzymes called histone acetyltransferases (HATs). In general, acetylated histones are associated with genes that are actively transcribed. Histone deacetylases (HDACs) can reverse this activation by removing acetyl groups.

Both methylation and acetylation change the architecture of the chromatin complex, which may result in condensation or relaxation of the chromatin. These changes occur when multiprotein chromatin remodeling complexes bind to the modified histones. Using the energy released by ATP hydrolysis to drive the reaction, these complexes open up the chromatin by displacing

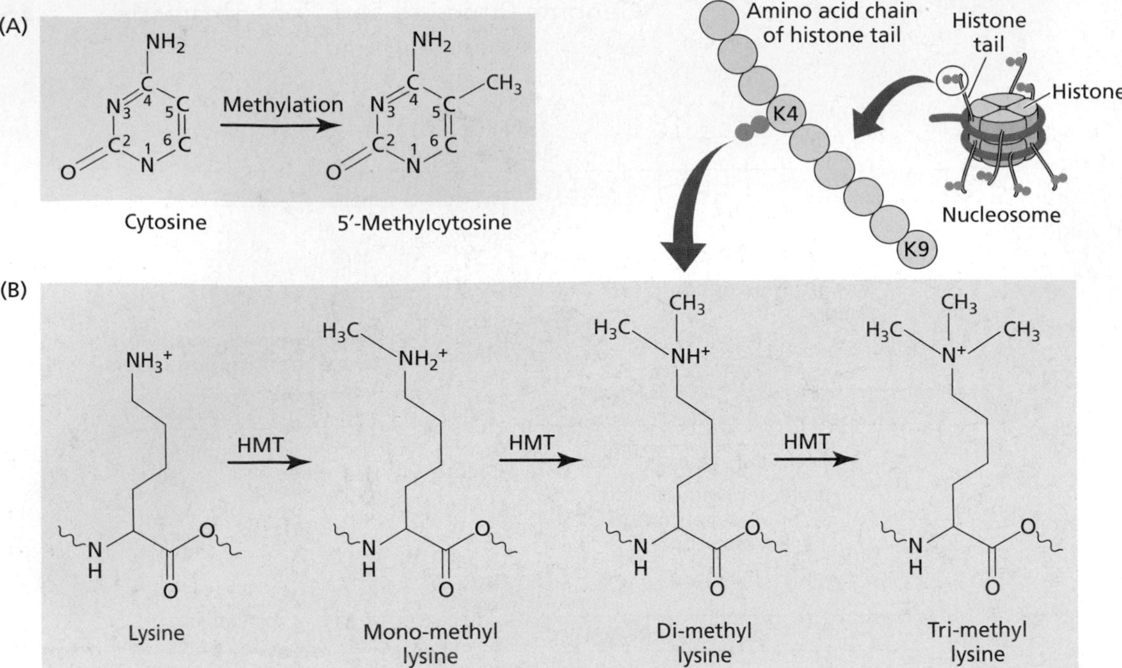

(C) ATP-dependent chromatin remodeling protein

RNA polymerase II and transcription factors

Transcription

GENE INDUCTION

Activating histone modifications (e.g., acetyl groups or dimethylated H3K4)

Histone deacetylase

Histone acetyltransferase, histone methylase

Histone

DNA

Nucleosome

GENE REPRESSION

H3K9 histone methyltransferase

Histone demethylase/ glycosylase

Deactivating histone modifications (e.g., dimethylated H3K9)

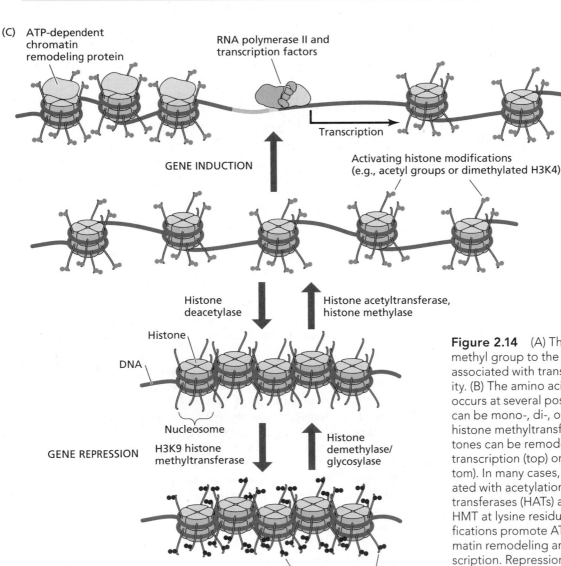

Figure 2.14 (A) The addition of a methyl group to the C5 in cytosine is associated with transcriptional inactivity. (B) The amino acid lysine (K), which occurs at several positions in histones, can be mono-, di-, or trimethylated by histone methyltransferase (HMT). (C) Histones can be remodeled to activate gene transcription (top) or to repress it (bottom). In many cases, activation is associated with acetylation by histone acetyltransferases (HATs) and methylation by HMT at lysine residue H3K4. These modifications promote ATP-dependent chromatin remodeling and stimulate transcription. Repression of transcription can be achieved by methylation of H3K9 and deacetylation by histone deacetylases.

the nucleosomes in the area toward the 5′ or 3′ direction of the remodeling complex. The resulting gap between nucleosomes is now wide enough for RNA polymerase to bind and initiate transcription (**Figure 2.14C**). Alternatively, histone modifications may present novel binding sites for regulatory proteins that affect gene activity. Scientists are only beginning to understand the effects of specific chemical modifications on each of the first 40 or so amino acids of the histone tails. In addition to methylation and acetylation, other types of histone modifications, including phosphorylation and ubiquitination, can influence the transcriptional activity of a given gene. The entirety of histone modifications on a specific nucleosome is sometimes called the *histone code* to underline the strong link between nucleosome constitution and gene activity.

Posttranscriptional Regulation of Nuclear Gene Expression

Immediately after transcription, the resulting mRNAs are processed: their introns are removed by splicing, and 5′ caps and 3′ poly-A tails are added. The transcripts are then exported to the cytoplasm for translation (see Figure 2.12).

An organism often produces mRNA in response to a specific situation. In order to remain useful as a specific response to a specific situation, individual mRNAs must have a finite lifetime. For example, in order to cope with a transient environmental stress, a plant may need to briefly produce specific enzymes. After the stress subsides, it would be wasteful, maybe even detrimental, to continue to produce those enzymes. Thus, mRNA production, activity, and stability are all regulated. Differential degradation of mRNA species can change the level of molecules available for translation and thus have an influence on overall gene activity. We discussed the regulation of transcription (mRNA production) in the previous section. Now we turn to mechanisms of posttranscriptional regulation (regulation of mRNA activity and stability).

All RNA molecules are subject to decay

Eukaryotic mRNA molecules can be degraded by exonucleases after removal of the poly-A tail (deadenylation) or removal of the 5′ cap (decapping). These processes are guided by environmental cues and other cellular pathways. One mechanism by which mRNA stability is regulated depends on the presence of certain sequences within the mRNA molecule itself, called ***cis*-elements**—an unfortunate choice of terms, since the same term is used for the DNA regions that influence transcriptional activity. These *cis*-elements can be bound by RNA-binding proteins, which may either stabilize the mRNA or promote its degradation by nucleases. Depending on the types of *cis*-elements present, the stability of a given mRNA molecule can vary widely.

Noncoding RNAs regulate mRNA activity via the RNA interference (RNAi) pathway

Another mechanism for regulating mRNA stability is the **RNA interference (RNAi) pathway**. This pathway involves several types of small RNA molecules that do not code for proteins and are thus called **noncoding RNAs (ncRNAs)**. The RNAi pathway has important roles in gene regulation and genomic defense.

The RNAi pathway is a set of cellular reactions to the presence of double-stranded RNA (dsRNA) molecules. Recall that mRNA is usually a single-stranded molecule (ssRNA). In plant cells, dsRNAs usually occur as a result of one of three types of events:

1. The presence of **microRNAs (miRNAs)**, which are involved in normal developmental processes (see Figure 2.15)
2. The production of **short interfering RNAs (siRNAs)**, which silence certain genes (see Figure 2.16)
3. The introduction of foreign RNAs, either by viral infection or via transformation by a foreign gene (see Figure 2.17).

Regardless of how the dsRNAs are produced, the cell mounts the RNAi response. The dsRNAs are chopped up, or "diced," into small 21- to 24-nucleotide RNAs, which bind to complementary single-stranded RNAs (e.g., mRNAs) from endogenous genes, viruses, or transgenes and promote their degradation or translational inhibition. In some cases, the RNAi pathway can also lead to gene silencing or **heterochromatization** of endogenous DNA or introduced foreign genes. To explore RNAi in more detail, we will first take a look at events leading to dsRNA accumulation in the cell. Then we will discuss the molecular components and downstream events of the RNAi process.

MicroRNAs REGULATE MANY DEVELOPMENTAL GENES POSTTRANSCRIPTIONALLY Plants contain hundreds of genes encoding miRNAs, which act by repressing the translation of mRNAs into protein or by targeting specific mRNAs for degradation. miRNAs are involved in many developmental processes, such as reproduction, cell division, embryogenesis, the formation of new organs (including leaves and flowers), and the transition from the vegetative to the floral phase. miRNAs arise from RNA polymerase II–mediated transcription of a specific locus that codes for the primary miRNA transcripts (pri-miRNAs), which can vary in length from hundreds to thousands of nucleotides (**Figure 2.15**). This primary transcript is capped at the 5′ end, polyadenylated at the 3′ end, and forms a double-stranded stem, whose base-paired arms border a single-stranded loop. Next, pri-miRNAs are processed into pre-miRNAs, which are usually 60 to 80 nucleotides long in animals but can be up to several hundred nucleotides long in plants. In plants, pri-miRNAs are converted into

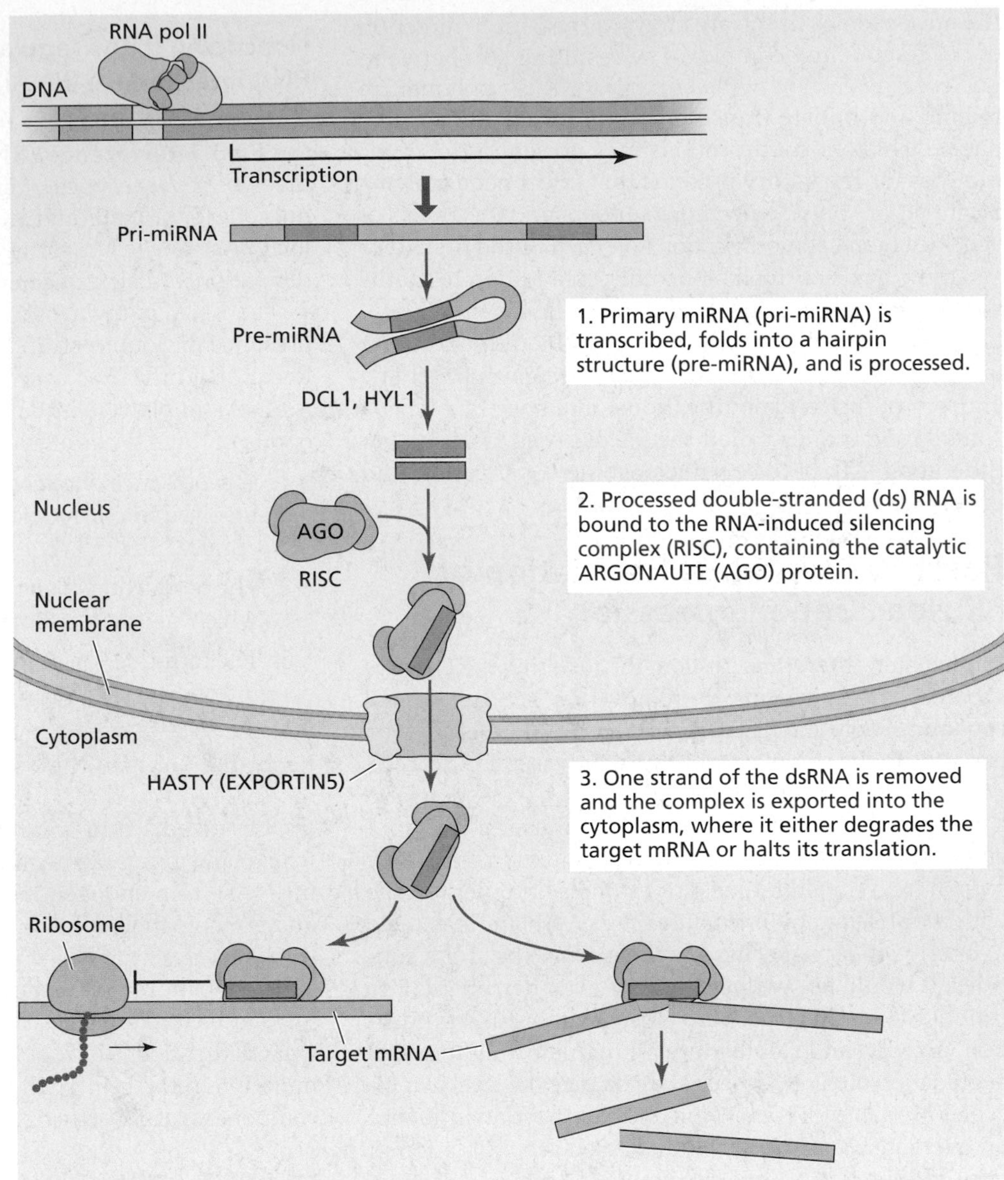

RNA pol II
DNA
Transcription
Pri-miRNA
Pre-miRNA
1. Primary miRNA (pri-miRNA) is transcribed, folds into a hairpin structure (pre-miRNA), and is processed.
DCL1, HYL1
Nucleus
AGO
RISC
2. Processed double-stranded (ds) RNA is bound to the RNA-induced silencing complex (RISC), containing the catalytic ARGONAUTE (AGO) protein.
Nuclear membrane
Cytoplasm
HASTY (EXPORTIN5)
3. One strand of the dsRNA is removed and the complex is exported into the cytoplasm, where it either degrades the target mRNA or halts its translation.
Ribosome
Target mRNA

Figure 2.15 The RNAi pathway in plants: microRNAs. MicroRNAs (miRNAs) are part of many genetic pathways that are active during plant development.

miRNAs inside the nucleus by the proteins **DICER-LIKE1 (DCL1)** and the double-stranded RNA-binding domain protein (dsRBP) HYPONASTIC LEAVES 1 (HYL1), both of which are involved in processing the primary transcripts into the mature miRNA duplexes. In animals, this process is carried out by the RNase III endonuclease Drosha aided by various dsRBPs. After processing, the miRNA is transported through the nuclear pores with the help of the nuclear export protein called EXPORTIN5 in animals, or its homolog in plants, called HASTY. Once in the cytoplasm, mature miRNAs are ready to be used in RNAi.

SHORT INTERFERING RNAS ORIGINATE FROM REPETITIVE DNA Mature siRNAs are structurally and functionally similar to miRNAs and also lead to the initiation of RNAi. However, siRNAs differ from miRNAs in the way they are generated. siRNAs can be produced in three ways. First, they can arise by transcription from opposing promoters that produce mRNA from opposite strands of a single segment of DNA (**Figure 2.16A**). Simultaneous transcription from such promoters generates two fully or partially complementary single-stranded RNA (ssRNA) molecules that can subsequently form one double-stranded molecule. The second way siRNAs can be formed is by transcription of a duplicated sequence in opposite directions (**Figure 2.16B**). This generates the sense strand of one copy and the antisense strand of the other copy. Short interfering RNA can also be produced from DNA sequences arrayed in such a way that their continuous transcription results in a message that contains at the end of its sequence a mirror image of the beginning of its sequence (a palindrome) and thus can fold back on itself to produce a dou-

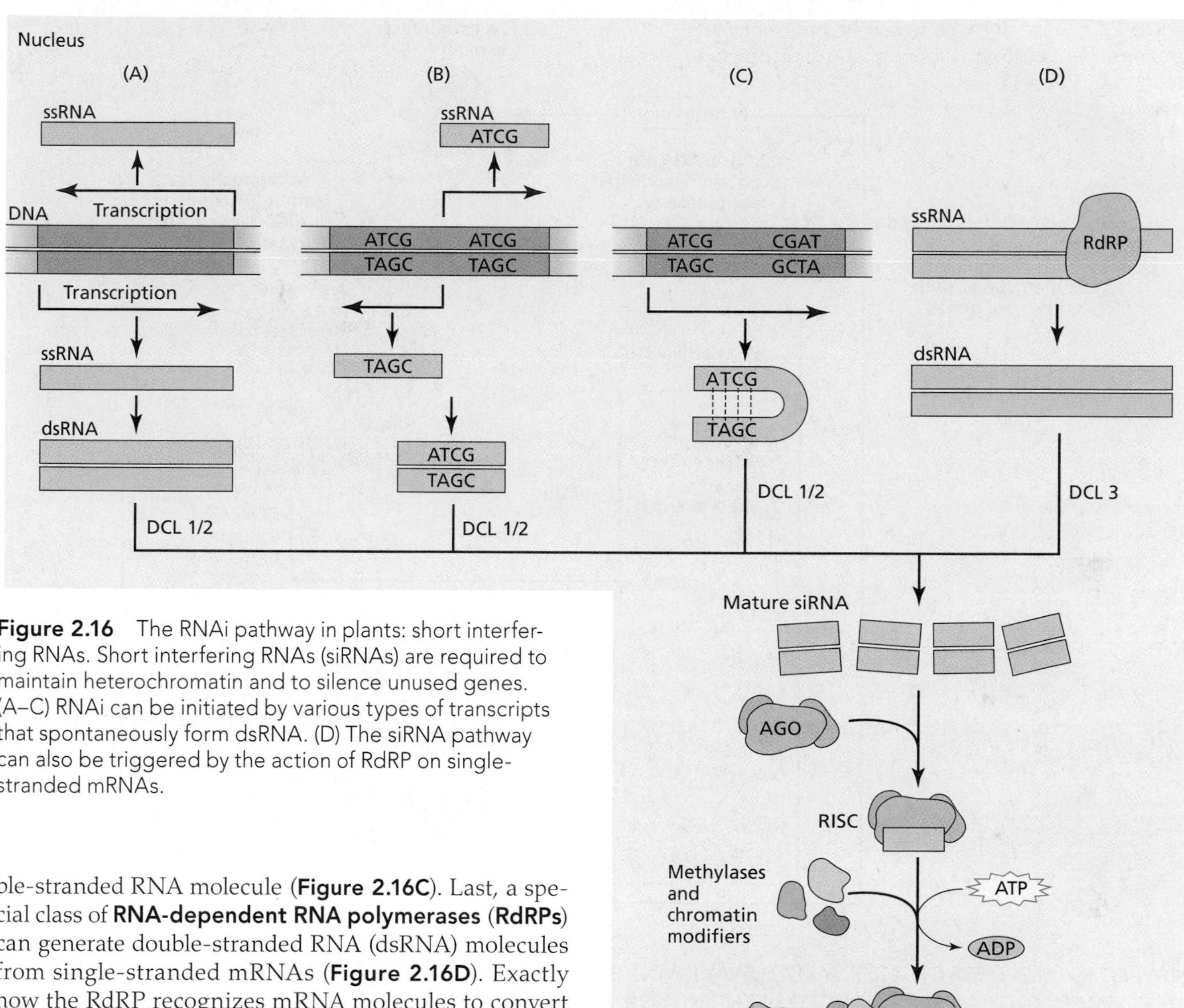

Figure 2.16 The RNAi pathway in plants: short interfering RNAs. Short interfering RNAs (siRNAs) are required to maintain heterochromatin and to silence unused genes. (A–C) RNAi can be initiated by various types of transcripts that spontaneously form dsRNA. (D) The siRNA pathway can also be triggered by the action of RdRP on single-stranded mRNAs.

ble-stranded RNA molecule (**Figure 2.16C**). Last, a special class of **RNA-dependent RNA polymerases (RdRPs)** can generate double-stranded RNA (dsRNA) molecules from single-stranded mRNAs (**Figure 2.16D**). Exactly how the RdRP recognizes mRNA molecules to convert to dsRNA is at this time still under investigation. While most protein-coding genes and miRNA-coding sequences are transcribed by RNA polymerase II, transcription of siRNAs is accomplished by RNA polymerases IV and V.

Not only the biogenesis of siRNAs described above differs from that of miRNAs. Unlike miRNAs, endogenous siRNAs are transcribed from chromosomal regions that have in the past often been regarded as transcriptionally inactive: repetitive DNA, transposons, and centromeric regions. Indeed, siRNAs originating from such repeat regions are sometimes referred to as **repeat-associated silencing RNAs (ra-siRNAs)**. As you will see below, this may not be coincidental: it appears that the formation of siRNAs and the induction of RNAi actually cause these regions to become heterochromatic and largely transcriptionally silent. Once a dsRNA is produced, either by direct transcription or by conversion of ssRNA into dsRNA via RdRPs, it is cut into 21- to 24-nucleotide RNA duplexes by members of the DICER-LIKE (DCL) protein family (see Figure 2.16). This process takes place in the nucleus in plants, but in some animals, such as *Caenorhabditis elegans*, it occurs in the cytoplasm.

In addition to these siRNAs of endogenous origin, exogenous RNAs can also trigger the formation of siRNAs. Sources for such exogenous RNAs include artificially introduced transgenes and viral RNA. In both cases, RdRPs and DCL proteins are involved in producing mature siRNAs (**Figure 2.17**).

In addition to miRNAs and siRNAs, a third class of small RNAs, called PIWI-associated RNAs (piRNAs), is found in animal germ cells. This class of small RNAs specifically targets transposon transcripts to keep these genetic elements inactive.

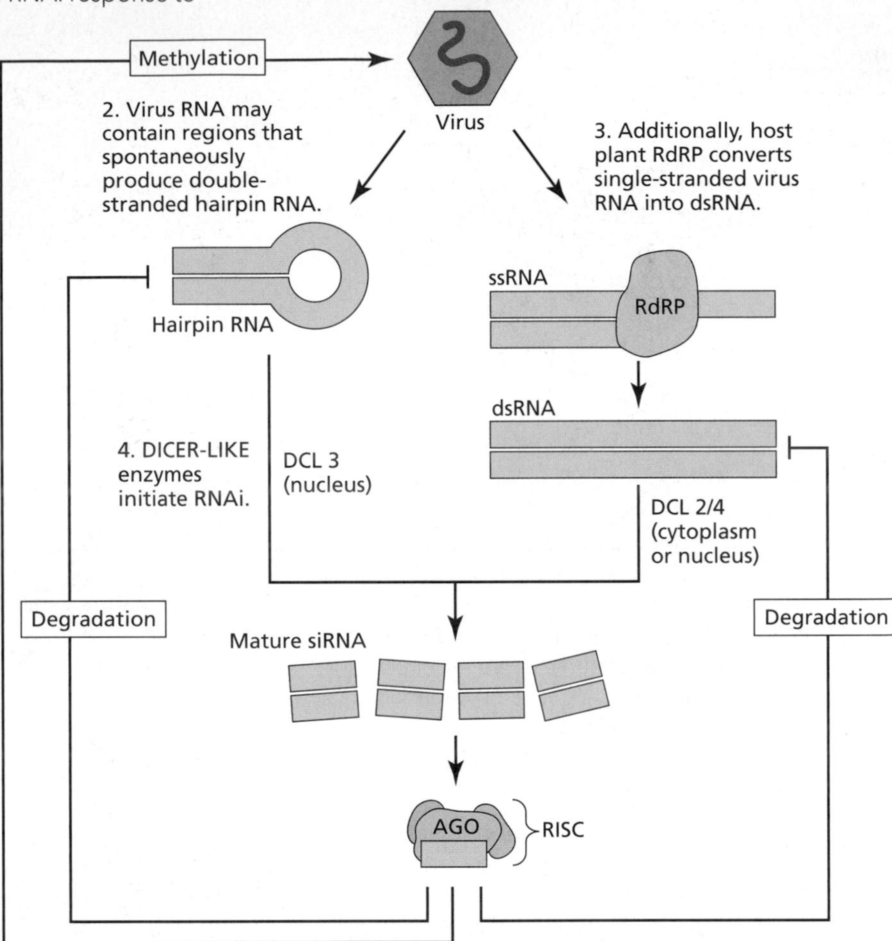
1. A virus infects the plant cell.
Methylation
Virus
2. Virus RNA may contain regions that spontaneously produce double-stranded hairpin RNA.
3. Additionally, host plant RdRP converts single-stranded virus RNA into dsRNA.
5. RISC binds DCL-produced siRNAs as guides to identify the origins of the RNAs, recruits methylases, and silences the viral genes.
Hairpin RNA
ssRNA
RdRP
dsRNA
4. DICER-LIKE enzymes initiate RNAi.
DCL 3 (nucleus)
DCL 2/4 (cytoplasm or nucleus)
Degradation
Degradation
Mature siRNA
AGO
RISC

Figure 2.17 The RNAi pathway in plants: antiviral defenses. Plant cells can mount an RNAi response to infection by viruses.

DOWNSTREAM EVENTS OF THE RNAI PATHWAY INVOLVE THE FORMATION OF AN RNA-INDUCED SILENCING COMPLEX For miRNAs, siRNAs, and RNAs of exogenous origin, the end result of the RNAi process is similar: the inactivation or silencing of their complementary mRNAs or DNA sequences. After 21- to 24-nucleotide miRNAs or siRNAs have been formed by the DICER-LIKE proteins, one strand of the short RNA duplex associates with a ribonuclease complex called the **RNA-induced silencing complex** (**RISC**) (see Figures 2.15 through 2.17). In both animals and plants, RISC contains at least one catalytic **ARGONAUTE** (**AGO**) protein. In some cases, RISC can recruit additional proteins to the complex. In Arabidopsis, ten different members of the AGO gene family are known. After the diced duplex miRNA or siRNA binds to AGO, one of the two RNA strands is removed. Upon this removal, RISC is active. In the case of miRNAs, the small ssRNA strand that binds to AGO now guides RISC to a complementary mRNA. Upon binding of RISC and target mRNA, the target mRNA is cleaved by the "slicer" activity of AGO. The resulting fragments are released into the cytoplasm, where they are further degraded. Instead of slicing the target, the association of RISC with an mRNA molecule may also simply inhibit translation of the mRNA into protein.

While RISC-bound miRNAs primarily target the expression of protein-coding genes, RISC-bound siRNAs also facilitate methylation of DNA and associated histones at sequences complementary to the siRNA. This allows the organism to silence certain genes permanently, and to form heterochromatin, predominantly in the centromeric and subtelomeric regions. Although the mechanism is unclear, RISC with its siRNA somehow guides DNA-modifying enzymes to the genomic sequence to be silenced. The chromatin structure is then "remodeled" in an ATP-requiring reaction and subsequently methylated, resulting in tighter condensation and heterochromatization of the DNA region involved (see Figure 2.14).

RNA-INTERFERENCE MAY HELP RESET EPIGENETIC MARKS IN THE GERM LINE By now you may be wondering why the transcriptional machinery would go through the cellular expense of transcribing transposons and other heterochromatic genes just to silence them again at the end of the siRNA pathway. Epigenetic marks such as DNA or histone methylation can change during the lifetime of the organism and are reset to a certain baseline of marks in the germ line. This ensures that epigenetic marks don't simply accumulate in successive generations but can have a regulatory function during the life of an individual. In plants, this is particularly important for some genes regulating the onset of flowering (described in more detail in Chapter 20).

The question of how selective epigenetic resetting in the germ line is accomplished at the molecular level is still unresolved. However, recent findings in Arabidopsis suggest a model in which genetic elements that are normally suppressed, such as transposons, are allowed to become active in non-gamete cells in both the male and female gametophyte. It is assumed that the transcription of these genetic elements leads to the formation of siRNAs complementary to the genes that should remain silent in the next generation. According to this model, the siRNAs generated in non-gamete cells (the central cell in the female and the vegetative nucleus in the male) would then travel within the gametophyte to the germ cells (egg and sperm), where the siRNAs could direct the methylation of those DNA sequences to which they are complementary. This process would essentially accept the detrimental consequences of increased transposon-induced mutagenesis in those cells of the gametophyte not being transmitted to the next generation, while at the same time maintaining proper methylation in the germ line.

SMALL RNAs AND RNAi COMBAT VIRAL INFECTION In addition to the processing of miRNAs and endogenous siRNAs, plants have also adopted the RNAi pathway as a type of molecular immune response against infection by viruses. (For other types of plant pathogen defenses aside from RNAi, see Chapter 23.) The genomic structures of plant viruses are quite diverse. Some plant viruses inject double-stranded DNA into plant cells, but the majority of plant-infecting viruses use single- or double-stranded RNA. Plants use the siRNA pathway to produce siRNA molecules against the viral genome. Scientists propose three possible ways to generate viral siRNA: (1) via the formation of double-stranded hairpin loops from viral ssRNA, (2) via the generation of complementary sense and antisense viral RNA molecules by host or virus RNA polymerase, and (3) via one of the plant's RdRPs. Regardless of their origin, once dsRNA is recognized by the plant's DCL proteins, siRNAs are produced, loaded onto AGO, and assembled into a RISC (see Figure 2.17). Virus-derived siRNAs can then both degrade viral RNA and methylate the virus genome inside the host cell.

In the process of cutting invasive RNA into 21- to 24-nucleotide siRNAs, the plant generates a pool of "memory" molecules that can travel via the plasmodesmata throughout the entire plant body, effectively immunizing the plant before the virus can spread. Not to be outdone by the plant's defenses, viruses have evolved a variety of molecular pathways to circumvent the plant's siRNA mechanism. Some of these countermechanisms include the inhibition of RISC formation, degradation of AGO, and indirect destabilization of the siRNA molecule itself.

CO-SUPPRESSION IS A GENE-SILENCING PHENOMENON MEDIATED BY RNA One of the first experiments leading to the discovery of RNAi involved an unexpected plant response to the introduction of transgenes. In the early 1990s, Richard Jorgensen and his colleagues were working with the petunia gene for chalcone synthase, a key enzyme in the pathway that produces purple pigment molecules in petunia flowers. When they inserted a highly active copy of the gene into petunia plants, they expected to see a deepening of the purple color in the flowers of the offspring. To their surprise, the flowers ranged in petal color from dark purple (as expected) to completely white (as if chalcone synthase levels had gone *down* instead of up). This phenomenon—decreased expression of a gene when extra copies are introduced—was termed **co-suppression**. With our present understanding of RNAi, we now know that in some cells, overexpression of chalcone synthase triggered an RNA-dependent RNA polymerase to produce dsRNA molecules, which initiated the RNAi response. This response eventually led to posttranscriptional silencing and to methylation of both the introduced and the endogenous copies of the chalcone synthase gene. Interestingly, posttranscriptional silencing did not occur in all cells. The cells in which gene silencing took place gave rise to white sectors, explaining why some of the transgenic petunia plants had variegated purple and white flowers.

In summary, RNAi is a process in which dsRNA elicits a posttranscriptional response that leads to the silencing of specific transcripts. miRNAs aid in the posttranscriptional regulation of genes in the cytoplasm, while siRNAs act in the nucleus to keep heterochromatin transcriptionally inactive or function as a molecular immune response against viruses.

Posttranslational regulation determines the life span of proteins

As we have seen, mRNA stability plays an important role in the ability of a gene to produce a functional protein. We turn next to the stability of proteins and the mechanisms that regulate a protein's longevity. A protein, once synthesized, has a finite life span in the cell, ranging from a few minutes to hours or days. Thus, steady-state levels of cellular enzymes reflect equilibrium between the synthesis and the degradation of those proteins, known as **turnover**. In both plant and animal cells, there are two distinct pathways of protein turnover, one in specialized lytic vacuoles (called lysosomes in animal cells) and the other in the cytoplasm (see also Chapter 1).

The cytoplasmic pathway of protein turnover involves the ATP-dependent formation of a covalent bond between

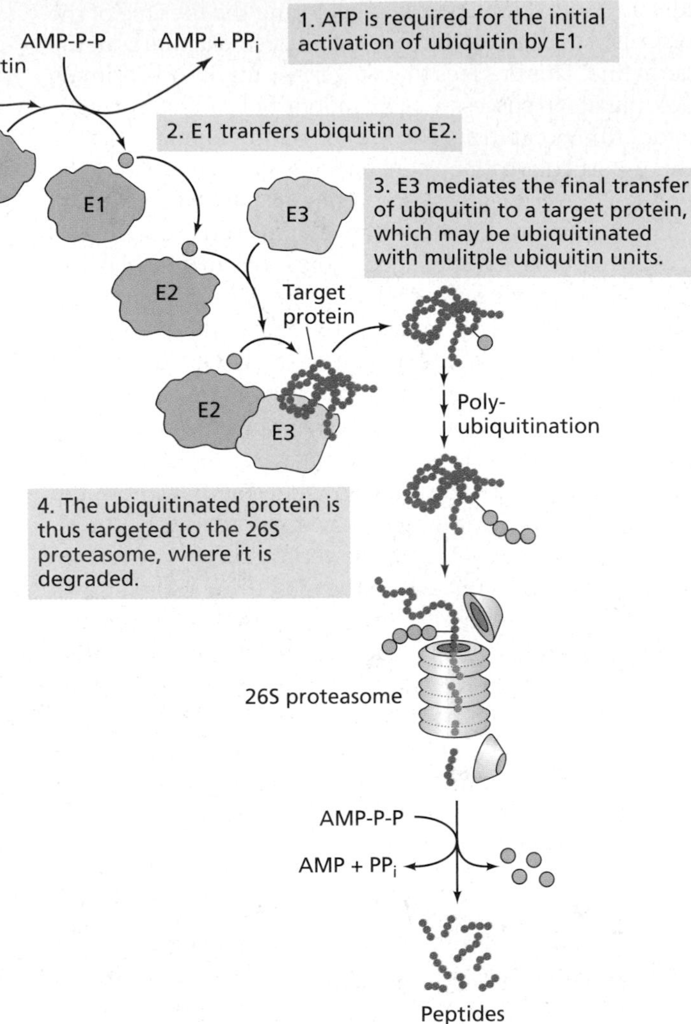
1. ATP is required for the initial activation of ubiquitin by E1.
AMP-P-P
AMP + PP_i
Ubiquitin
2. E1 tranfers ubiquitin to E2.
E1
E1
E3
3. E3 mediates the final transfer of ubiquitin to a target protein, which may be ubiquitinated with mulitple ubiquitin units.
E2
Target protein
E2
E3
Poly-ubiquitination
4. The ubiquitinated protein is thus targeted to the 26S proteasome, where it is degraded.
26S proteasome
AMP-P-P
AMP + PP_i
Peptides

Figure 2.18 Generalized diagram of the cytoplasmic pathway of protein degradation.

the protein that is to be degraded and a small, 76–amino acid polypeptide called **ubiquitin**. Addition of one or many molecules of ubiquitin to a protein is called *ubiquitination*. Ubiquitination marks a protein for destruction by a large, ATP-dependent proteolytic complex called the **26S proteasome**, which specifically recognizes such "tagged" molecules (**Figure 2.18**). More than 90% of the short-lived proteins in eukaryotic cells are degraded via the ubiquitin pathway.

Ubiquitination is initiated when the **ubiquitin-activating enzyme** (**E1**) catalyzes the ATP-dependent adenylation of the C terminus of ubiquitin. The adenylated ubiquitin is then transferred to a cysteine residue on a second enzyme, the **ubiquitin-conjugating enzyme** (**E2**). Proteins destined for degradation are bound by a third type of protein, a **ubiquitin ligase** (**E3**). The E2–ubiquitin complex then transfers its ubiquitin to a lysine residue of the protein bound to E3. This process can occur multiple times to form a polymer of ubiquitin. The ubiquitinated protein is then targeted to the proteasome for degradation.

As we now know, there are a multitude of protein-specific ubiquitin ligases that regulate the turnover of specific target proteins (see Chapter 14). We will discuss an example of this pathway in more detail when we cover developmental regulation by the plant hormone auxin in Chapter 19.

Tools for Studying Gene Function

Individuals that contain specific changes in their DNA sequence are called **mutants**. The analysis of mutants is an extremely powerful tool that can help scientists infer the function of a gene or map its location on the chromosomes. In this section we will discuss how mutants are generated and how they can be used in genetic analysis. We will also discuss some modern biotechnological tools that allow researchers to study or manipulate the expression of genes.

Mutant analysis can help elucidate gene function

Throughout this book we will discuss the genes and genetic pathways involved in physiological functions at length, often referring to certain types of mutants that allowed researchers to understand the genes and pathways under discussion. Why is a mutant gene a more powerful tool for elucidating gene function than the normal, wild-type gene on its own?

The use of mutants for gene identification relies on the ability to distinguish a mutant from a normal individual, so the change in the mutant's nucleotide sequence must result in an altered phenotype. If a mutant can be restored to the normal phenotype with a wild-type version of a candidate gene, the researcher knows that a mutation in that gene was responsible for conferring the originally observed mutant phenotype. This method is called **complementation**. For example, assume that a plant with a single-gene mutation shows a delay in producing flowers compared with the wild type. If we can determine the sequence and location of the gene responsible, we will presumably learn something about the mechanisms involved in floral development. Let's now assume that a researcher is able to find a gene in the mutant genome that differs from the wild-type gene in its DNA sequence. If the researcher is able to show that transferring the wild-type gene into the mutant restores the normal phenotype, we can be reasonably certain that the candidate gene plays a role in the initiation of flowering.

In the 1920s, H. J. Muller and L. J. Stadler independently experimented with the effects of X-rays on the stability of chromosomes in flies and in barley, respectively. Both researchers reported heritable changes in the treated organisms. In the following years, other techniques for inducing mutations were developed. These techniques include the use of ultraviolet or fast-neutron radiation and of mutagenic chemicals. For example, treatment with the chemical **ethylmethanesulfonate** (**EMS**) causes the addition of an ethyl group to a nucleotide base, usually guanine. Ethylated guanine pairs with thymine instead of cytosine. The cell's DNA repair machinery then replaces the ethylated guanine with adenine, causing a permanent mutation from G/C to A/T at that site. Mutagenesis with either radiation or chemicals induces nucleotide changes randomly throughout the genome.

There are several ways to map a mutation to its chromosome and ultimately clone the affected gene. WEB TOPIC 2.2 explains a method called **map-based cloning**, which uses crosses between a mutant and a wild-type plant and genetic analysis of the offspring to narrow down the location of the mutation to a short segment of the chromosome, which is then sequenced.

Another method of mutagenesis is the random insertion of transposons into genes. This technique involves crossing the plant of interest with a plant carrying a single active transposon and screening their offspring for mutant phenotypes caused by random insertion of the transposon into new locations. Because the sequence of the transposon is known, these mutations are "tagged"; thus, the DNA sequences adjacent to the transposon can be readily found and analyzed to identify the mutated gene. This technique is called **transposon tagging** and is explained in detail in WEB TOPIC 2.3.

Molecular techniques can measure the activity of genes

Once a gene of interest has been identified, scientists are usually interested in where and when the gene is expressed. For example, a gene may be expressed only in reproductive tissues, or only in vegetative ones. Likewise, a gene may encode general cell functions (so-called housekeeping functions) and be expressed continuously, or it may encode special functions and be expressed only in response to a certain stimulus, such as a hormone or an environmental cue. In the past, transcriptional analysis (the determination of the amount of mRNA produced from a gene at a given time) has been performed mainly on single genes. Tools developed for this type of analysis include Northern blotting, reverse transcription or quantitative polymerase chain reaction (RT-PCR or qPCR), and in situ hybridization. You can find applications of each of these techniques throughout this book. The increasing availability of whole-genome sequences has added two new methods to the arsenal of RNA techniques: **microarray analysis** and high-throughput **RNA sequencing** (**RNA-seq**). Both techniques have allowed the investigation of a sample's **transcriptome**, which is the sum of all transcribed genes at a given time. You can find more on the microarray technique in WEB TOPIC 2.4.

Transcriptome sequencing—or RNA-seq—is a technique that essentially relies on sequencing every mRNA molecule in a sample, counting up the number of mRNA molecules for each gene, and comparing these mRNA abundances with those obtained for a different sample (**Figure 2.19**). To do this, mRNAs are reverse transcribed into complementary DNA (cDNA) pools. Using one of several specific techniques available, each cDNA molecule is sequenced, producing a "read." Genes that are more highly expressed in one sample than in another will produce more mRNA molecules, and thus lead to more cDNA molecules and ultimately more sequence reads. To most efficiently compare mRNA abundance between two samples, the researcher must have the sequenced genome of the organism at hand or at least the DNA sequences corresponding to the transcribed regions of the genome (the transcriptome). The reads from the two mRNA samples to be com-

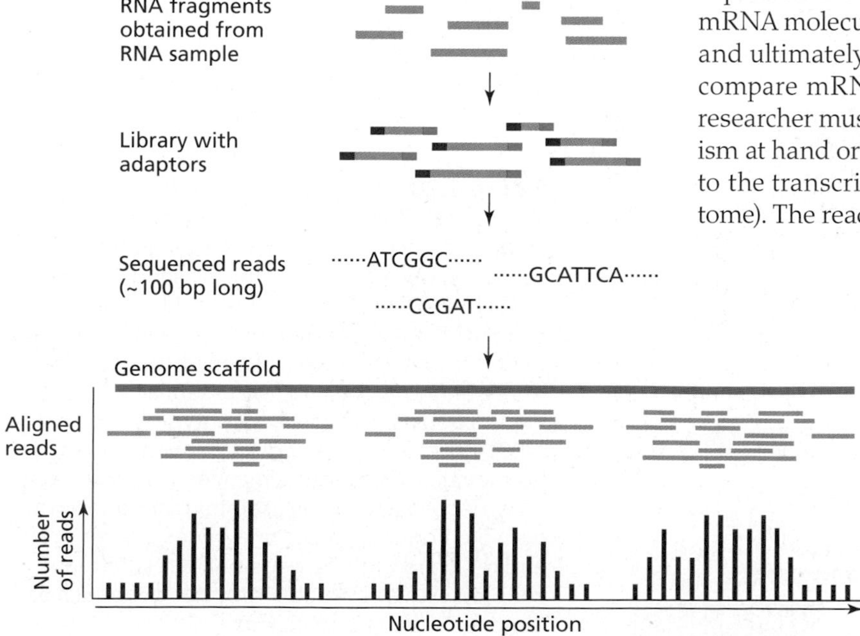

Figure 2.19 Workflow for RNA-seq analysis of gene expression. RNA fragments are reverse transcribed and the resulting cDNA fragments are ligated with adaptors and sequenced. Each sequence is computationally aligned with the known genome sequence of the organism. The greater the number of RNA fragments (reads) per coding region, the greater is the expression of that gene.

pared can then be aligned with this "reference genome." Using computational and statistical analysis, it can then be determined if the number of reads found in one sample is different from the number of reads in the other sample. With the cost for sequencing declining rapidly, RNA-seq is fast becoming the method of choice for genome-wide transcriptional analysis.

RNA abundance is currently much easier to measure than protein abundance for any given gene. Although the transcriptional activity of a gene is a relatively good indication of its protein levels, it is not a given that changes in RNA and protein levels are linearly related. Genome-wide protein analysis is still labor-intensive and costly, but techniques, including mass spectrometry, are being refined and employed to analyze samples for their sum of all expressed proteins, called the **proteome**. Just as the study of the transcriptome is called **transcriptomics**, the study of the proteome is called **proteomics**. Because of technological innovations, it is now possible to study many aspects of an organism, not just the transcriptome or the proteome, in their entirety, rather than one molecule at a time. Such comprehensive analyses are often referred to as *-omics* analyses. For example, the study of all metabolites in physiological pathways is called **metabolomics**, and the examination of all epigenetic modifications in the genome of a cell is referred to as **epigenomics**.

Gene fusions can introduce reporter genes

The identification of a gene containing a mutation provides information about that gene's location in the genome and about the effect of its altered function on the plant's phenotype. From the sequence of a gene alone, scientists can make inferences about the gene's cellular function by comparing the gene's structure with those of other known genes. For example, certain regions within the gene—so-called **domains**—might have similarity to domains found in certain families of genes, such as those encoding kinases, phosphatases, or membrane receptors. However, sequence information alone does not give direct evidence of the gene's cellular function, nor does it indicate where in the plant, or under what conditions, the gene is active.

One way to find out where and when a certain gene is expressed within a plant or cell is to measure the abundance of its mRNA by one of the methods described above. Another way is to make a gene fusion. A **gene fusion** is an artificial construct that combines part of the gene of interest—for example, the promoter—with another gene—referred to as a **reporter gene**—that produces a readily detectable protein. One such reporter gene is the green fluorescent protein (GFP) gene, which produces a fluorescent protein that can be observed in an intact plant or cell by fluorescence microscopy (for example, see Figure 19.18B). Recall that not all genes are transcribed in every cell of the plant at all times. A gene's expression is regulated by transcription factors that fine-tune its activity and allow it to be transcribed only where and when it is needed. If a plant carries a promoter–*GFP* gene fusion in all of its cells, *GFP* will be expressed only in those cells that would normally express the gene whose promoter was fused with *GFP*. In other words, green fluorescence will be visible wherever and whenever the gene under investigation is expressed. Another frequently used reporter gene is *β-glucuronidase*, usually called *GUS*. The *GUS* reporter system does not require fluorescent light to be visible, but the downside of this system is that in order to visualize β-glucuronidase the tissue first has to be fixed (killed) and then bathed in a substrate solution, which causes the formation of a blue color in those tissues that express *GUS* (for an example, see Figure 19.24).

To transform plants with gene fusions, such as reporter genes or those for complementation analysis, scientists

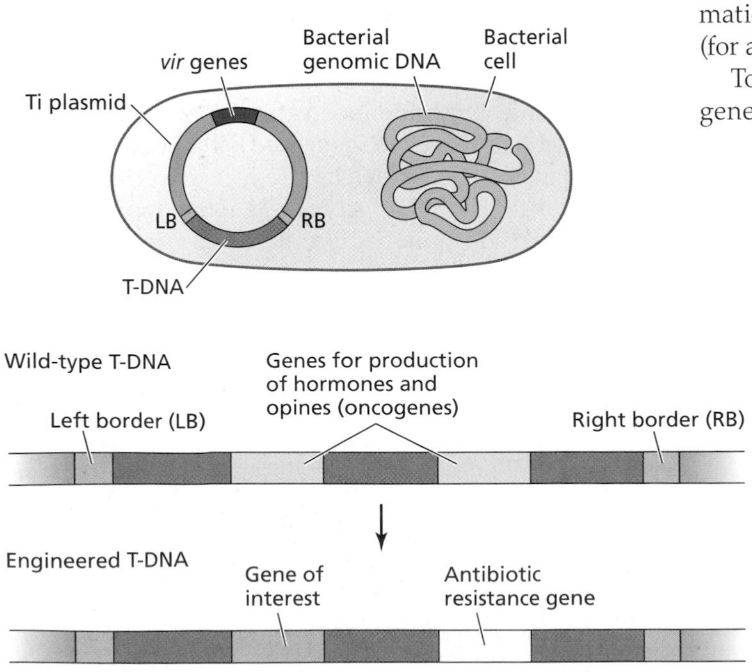

Figure 2.20 *Agrobacterium*'s tumor-inducing (Ti) plasmid. The Ti plasmid is a circular extrachromosomal piece of DNA contained inside the bacterial cell. A portion of this plasmid, the transfer DNA (T-DNA), is transferred to the infected plant, where it is inserted into the plant's nuclear genome. The virulence (*vir*) genes, located elsewhere on the Ti plasmid, are essential for initiating the transfer of T-DNA. Wild-type Ti plasmid T-DNA contains genes for production of plant hormones and non-protein amino acids (opines). When *Agrobacterium* is used for plant transformation, the hormone and opine genes are removed and replaced with the gene of interest, often coupled with a selectable reporter gene such as a gene for antibiotic resistance.

have harnessed the power of the microbial plant pathogen *Agrobacterium tumefaciens*. This bacterium causes infected plants to overproduce growth hormones, which induce the formation of a tumor called a **crown gall** (see Figure 15.11B). Crown gall disease is a serious problem for certain agricultural crops, such as fruit trees, as it can reduce crop yield and decrease overall plant health.

Agrobacterium is sometimes referred to as the natural genetic engineer because it has the ability to transform plant cells with a small subset of its own genes. The genes transferred to the plant genome are part of a circular extrachromosomal piece of DNA called the tumor-inducing (Ti) plasmid (**Figure 2.20**). The Ti plasmid contains a number of virulence (*vir*) genes as well as a region called the transfer DNA (T-DNA). The *vir* genes are required for initiating and conducting the transfer of the T-DNA into the plant cell. Once transferred, the T-DNA inserts itself randomly into the plant nuclear genome. It carries genes for two general functions: first, the induction of the crown gall, which will provide housing for the bacterium, and second, the production of non-protein amino acids called *opines*, which are used by the bacterium as a source of metabolic energy. An overview of the steps involved in the transformation of plant cells by *Agrobacterium* is shown in **Figure 2.21**. A more detailed description of the mechanism of transformation can be found in WEB TOPIC 2.5.

Given that *Agrobacterium* is usually a plant pathogen, how can it be a useful biotechnological tool? When *Agrobacterium* is used in the laboratory, scientists use a strain that has a modified Ti plasmid. The hormone and opine genes are removed from the T-DNA, and a gene of interest is inserted in their place (see Figure 2.20). Often a gene that confers resistance to an antibiotic is added as a selectable marker gene. The engineered Ti plasmid is then inserted into *Agrobacterium*. Any gene now contained within the T-DNA will be transferred into a plant cell infected with the engineered bacterium. The antibiotic resistance gene allows the researcher to easily screen for transformed cells since only transformed cells will survive when grown in the presence of the antibiotic.

Plants can be infected with the engineered bacterium in several ways. Small leaf segments can be cut from a plant and co-cultivated with a solution of the bacterium, before washing the plant cells and culturing them on tissue culture medium. The plant hormones auxin and cytokinin are then used to stimulate the generation of roots and shoots from the tissue, respectively. This technique eventually produces a transformed adult plant. Some plants, including Arabidopsis, are so easily transformed that just dipping the flowers in a suspension of the bacterium is sufficient to result in transformed embryos in the next generation.

Besides *Agrobacterium*-mediated transformation, several other techniques have been developed to incorporate foreign genes into plant genomes. One such technique is the fusion of two plant cells with different genomic information, called **protoplast fusion**. Another is **biolistics**, sometimes referred to as the **gene gun** technique, in which tiny particles of gold coated with the genetic construct of interest are shot into cells growing in culture

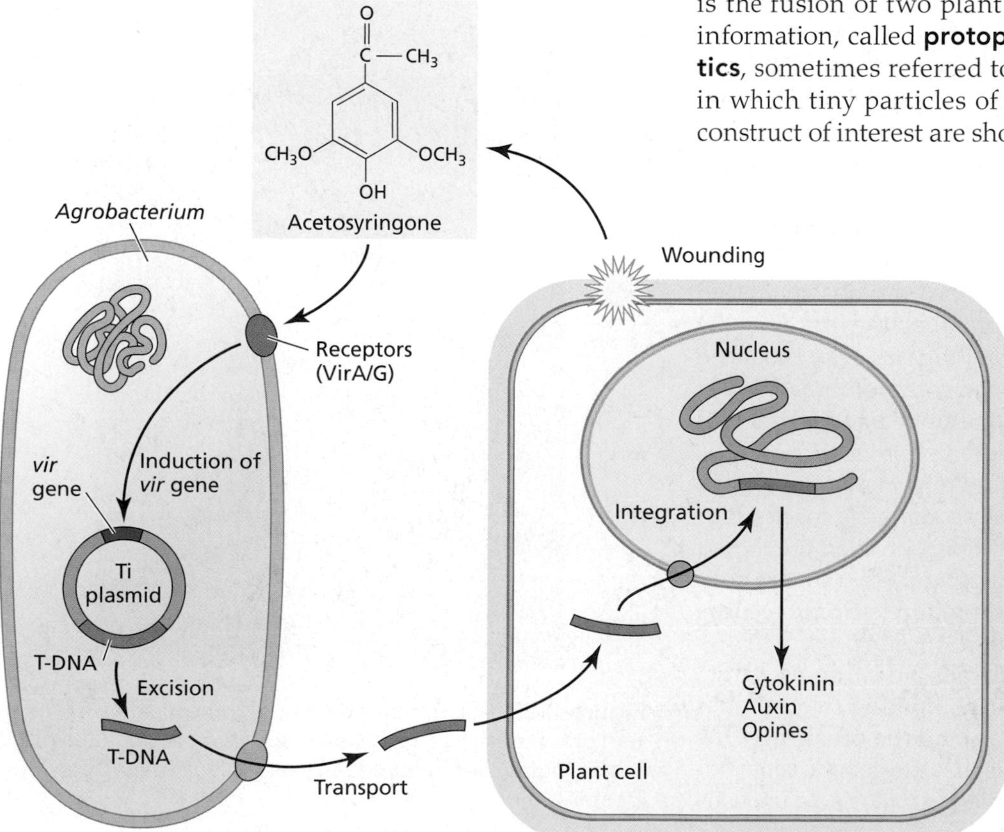

Figure 2.21 Infection of plant cells with *Agrobacterium*.

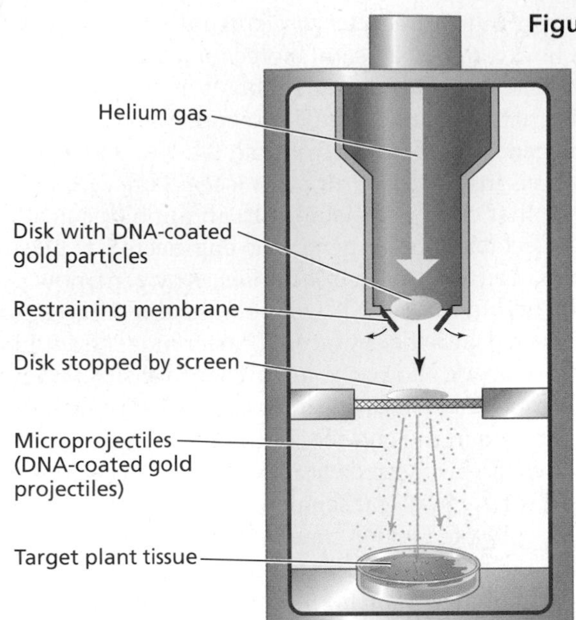

Figure 2.22 Plant cell transformation using the "gene gun."

dishes (**Figure 2.22**). The genetic material is randomly incorporated into the cells' genomes. The cells can then be transferred to solid culture medium and grown into mature transgenic individuals.

Genetic Modification of Crop Plants

Humans have modified crop plants through selective breeding for many centuries to produce varieties that have higher yields, are better adapted to specific climates, or are resistant to plant pathogens. For example, modern maize varieties are the domesticated descendants of wild relatives in the genus *Zea*, known as teosinte (**Figure 2.23**). As is apparent from the photo, breeding and domestication have modified this crop plant substantially from its original form. Likewise, selective breeding has produced crop tomatoes that are much larger than the fruit from the original progenitor species. Breeding has even produced entirely new species, such as the common bread wheat *Triticum aestivum*, which is allohexaploid and arose from the cross-pollination of three different progenitor species. While classic breeding techniques rely on random genetic recombination of traits in sexually compatible species, biotechnology allows the transfer of a controlled number of genes between species that cannot be crossed successfully. Let's discuss how classic breeding differs from breeding using biotechnological tools.

In classic breeding, desirable traits are introgressed into elite agricultural lines by cross-pollinating two varieties and selecting for those traits among the offspring. One disadvantage of this approach is that the genetic contributions of both parents are reshuffled in meiosis, so undesirable traits can be introduced into the recipient line along with the desirable ones. The undesirable traits must then be bred out again by repeated, often time-consuming backcrosses with the elite line while retaining the desirable traits. Biotechnological tools circumvent this problem by allowing insertion of only the desired genes into the recipient plant, most often either by *Agrobacterium*-mediated transformation or by biolistics. Plants produced in this way are commonly referred to as genetically modified organisms (GMOs).

There are three essential differences between GMOs and conventionally bred varieties of crops:

1. Gene transfer into GMOs occurs in the laboratory and does not require crossbreeding.
2. The donor genes of GMOs can be derived from any organism, not just those with which the recipient can be successfully crossed.
3. GMOs may carry gene constructs that are the product of splicing a variety of genetic components together to produce genes with new uses (for example, the promoter–*GFP* fusion genes we described earlier).

We will turn next to some examples of genes commonly used to modify crops.

Figure 2.23 Classic breeding and domestication of the wild grass teosinte (left) have led to the crop plant *Zea mays* (maize, right) over hundreds of years. (Courtesy of John Doebly.)

Transgenes can confer resistance to herbicides or plant pests

Any gene artificially transferred into an organism is usually referred to as a **transgene**. Most often transgenes are introduced from one species into another species. However, some researchers prefer to distinguish gene transfer between sexually compatible species that could also exchange genetic material via classic breeding (**cisgenics**) from gene transfer between species that cannot be naturally crossed, for which these researchers reserve the term **transgenics**. Two of the types of transgenes most commonly used in commercial crops today are genes that allow plants to resist herbicide applications and genes that allow plants to withstand attack by certain insects. Weed invasion and insect infestation are two of the main causes of crop yield reductions in agriculture.

Plants carrying a transgene for **glyphosate resistance** will survive a field application of glyphosate (the commercial herbicide Roundup), which kills weeds but does not harm resistant crop plants. Glyphosate inhibits the enzyme enolpyruvalshikimate-3-phosphate synthase (EPSPS), which catalyzes a key reaction in the shikimic acid pathway, a plant-specific metabolic pathway necessary for the production of many secondary compounds, including auxin and aromatic amino acids (see WEB APPENDIX 4). Glyphosate-resistant plants carry either a gene encoding a bacterial form of EPSPS that is insensitive to the herbicide, or transgene constructs that fuse high-activity promoters with the wild-type EPSPS gene to achieve herbicide resistance by overproduction of the enzyme.

Another commonly used transgene encodes an insecticidal toxin from the soil bacterium ***Bacillus thuringiensis*** (**Bt**). Bt toxin interferes with a receptor found only in the larval gut of certain insects, eventually killing them. Plants expressing Bt toxin are toxic to susceptible insects but harmless to most other organisms, including nontarget insect species.

Transgenic plants are also being developed that have enhanced nutritional value. Every year, according to the World Health Organization, dietary vitamin A deficiency causes as many as 500,000 children in developing nations to go blind. Many of these children live in Southeast Asia, where rice is the main part of the diet. Although rice plants synthesize abundant levels of β-carotene (provitamin A) in their leaves, rice endosperm, which makes up the bulk of the grain, does not normally express the genes required for three of the steps in the β-carotene biosynthesis pathway. To overcome this block, Ingo Potrykus, Peter Beyer, and their colleagues developed new strains of rice that carry genes from other species that can complete the β-carotene biosynthesis pathway (**Figure 2.24**). The most efficient strain uses two transgenes: a phytoene synthase gene from maize and a bacterial carotene desaturase gene. Together, these two genes allow the rice plant to accumulate large amounts of β-carotene. Facing many regulatory and intellectual property hurdles, this "golden rice" has been field tested but not released for public use at the time of this printing. This was not the first time that the β-carotene content of a crop plant was altered by agriculturalists. Carrots, for example, were either red or yellow before the seventeenth century, when a Dutch horticulturist selected the first orange-colored varieties.

Other researchers are developing transgenic plants that express vaccines in their edible fruit as an alternative, more convenient way to vaccinate people in parts of the world where medical facilities are insufficient for the administration of conventional vaccines.

Genetically modified organisms are controversial

The development of GMOs has not been greeted with universal enthusiasm and support. In spite of the enormous humanitarian potential of GMOs, many individuals, as well as the governments of some countries, look on them with suspicion and concern.

Opponents of the use of biotechnology in agriculture cite, for example, the possibility of inadvertently producing crops that express allergens introduced from another spe-

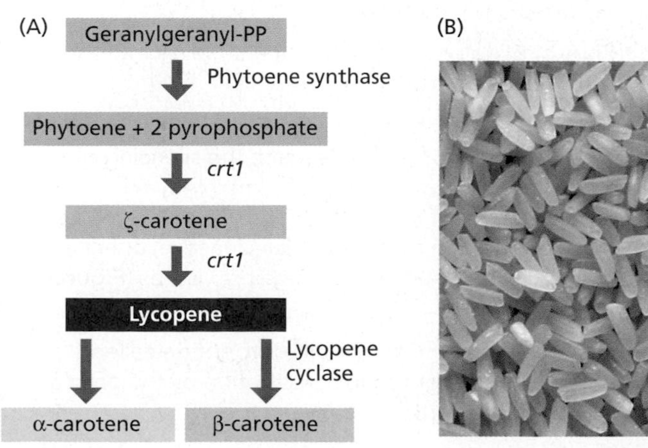

Figure 2.24 Golden rice was produced by inserting two foreign genes involved in β-carotene synthesis into rice. (A) The β-carotene biosynthesis pathway in golden rice. (B) Normal white rice (left) compared with golden rice (right). (Photo courtesy of the Golden Rice Humanitarian Board, www.goldenrice.org)

cies. They also worry that the overuse of genes encoding Bt toxin might select for insects that develop resistance to the toxin or that windblown pollen from transgenic herbicide-resistant crops could cross-pollinate nearby wild species, thereby producing weeds with herbicide resistance or contaminating organic crops with transgenes. The concern that the ingestion of currently grown and marketed GMO foods poses risks to human health has so far been unsubstantiated according to the World Health Organization.

While many concerns regarding GMOs have been addressed by proponents of plant biotechnology, research is ongoing to monitor the effects of the new technologies on human health and the environment (**WEB ESSAY 2.1**). In the end, the controversy may come down to this question: How much risk is acceptable in an attempt to satisfy the needs of an ever-increasing world population for food, clothing, and shelter?

SUMMARY

Genotype, epigenetic chemical modifications of its DNA, and the environment in which it lives determine the phenotype of a plant. To thoroughly understand a plant's physiology requires understanding how genotype (nuclear, mitochondrial, and plastid) is translated into phenotype.

Nuclear Genome Organization

- The most prominent structural landmarks on chromosomes are centromeres, telomeres, and nucleolar organizer regions (NORs) (**Figure 2.1**).
- Heterochromatin (often highly repetitive DNA sequences) is transcriptionally less active than euchromatin.
- Transposons are mobile DNA sequences within the nuclear genome. Some can insert themselves into new places along the chromosomes (**Figure 2.2**).
- Active transposons can significantly damage their host, but most mobile elements are inactivated by epigenetic modifications, such as methylation (**Figure 2.3**).
- Epigenetic modifications are controlled by methylation of DNA, and methylation and acetylation of histones.
- Meiosis allows for recombination of genes and the organized reduction of the genome by one half of its chromosomes (**Figure 2.5**).
- All angiosperm lineages have experienced genome duplication at least once in their evolutionary history. Many modern plant species are polyploid, either because of genome multiplication within a species (autopolyploidy) or because of genome duplication in association with hybridization of two or more species (allopolyploidy) (**Figure 2.7**).
- Genomic signatures of ancient polyploidy (paleopolyploidy) can be detected in many modern plant genomes.
- Phenotypic and physiological responses to polyploidy are variable and often unpredictable.
- Polyploids have multiple complete genomes; this altered genomic balance can phenotypically distinguish polyploids, especially allopolyploids, from their parents, and may lead to speciation (**Figure 2.10**).

Plant Cytoplasmic Genomes: Mitochondria and Plastids

- Organellar genomes are mostly found in multiple copies of the genome on the same DNA molecule.
- Organellar genetics do not obey Mendelian laws, but usually show uniparental inheritance and vegetative segregation (**Figure 2.11**).

Transcriptional Regulation of Nuclear Gene Expression

- Gene expression is regulated on several levels: transcriptional, posttranscriptional, and posttranslational.
- For protein-coding genes, RNA polymerase II binds to the promoter region and requires general transcription factors and other regulatory proteins to initiate gene transcription (**Figures 2.12, 2.13**).
- Epigenetic modifications, such as methylation of DNA and methylation and acetylation of histone proteins, help determine gene activity (**Figure 2.14**).

Posttranscriptional Regulation of Nuclear Gene Expression

- RNA-binding proteins may either stabilize mRNA or promote its degradation.
- The RNA interference (RNAi) pathway is a posttranscriptional response that leads to the silencing of specific transcripts. MicroRNAs (miRNAs) aid in gene regulation. Short interfering RNAs (siRNAs) help keep heterochromatin transcriptionally inactive, or act as a molecular immune system against viruses (**Figures 2.15–2.17**).
- Proteins tagged with a small polypeptide called ubiquitin are targeted for destruction by the proteasome (**Figure 2.18**).

Tools for Studying Gene Function

- Tools developed for transcriptional analysis of single genes include Northern blotting, reverse transcription or quantitative polymerase chain reaction (RT-PCR or qPCR), and in situ hybridization.
- Microarray and RNA-seq technologies use information gained from genome sequencing for high-throughput analysis of gene expression (**Figure 2.19**).
- Reporter gene fusions contain part of a gene of interest (e.g., the promoter) fused to a reporter gene that encodes a protein that can be readily detected when expressed. Such constructs can be used to monitor where and when a particular gene is active.
- *Agrobacterium* can transform plant cells when targeted genes are transferred from the bacterial plasmid called the tumor-inducing (Ti) plasmid (**Figures 2.20, 2.21**).

Genetic Modification of Crop Plants

- In contrast to classic selective breeding, bioengineering allows the transfer of specific genes between species that cannot be crossed successfully, or between crossable species as a means for more precise gene transfer than is possible by traditional breeding.
- Artificially transferred genes can confer resistance to herbicides or plant pests, or provide enhanced nutrition.

WEB MATERIAL

- **WEB TOPIC 2.1 Inheritance Patterns of Plastid Genomes** Plastid genomes are inherited in a non-Mendelian fashion.
- **WEB TOPIC 2.2 Recombination Mapping and Gene Cloning: Overview** Map-based cloning can be used to isolate the gene(s) involved in a phenotype of interest.
- **WEB TOPIC 2.3 Transposon Tagging** Mutagenesis using transposable elements is another approach to gene identification.
- **WEB TOPIC 2.4 Microarray Technology** Microarray technology enables genome-scale measure ments of gene expression and other genome characteristics.
- **WEB TOPIC 2.5 Transformation by *Agrobacterium*** *Agrobacterium*, a plant pathogen that naturally transforms its host plant, has become an important tool for biotechnology.

- **WEB ESSAY 2.1 Agriculture, Population Growth and the Challenge of Climate Change** The current scientific consensus is that improving crop yields via genetically-engineered crops, as well as enhancing soil organic carbon sequestration, are essential components of an overall solution to the problems of population growth and climate change.

available at plantphys.net

Suggested Reading

Allen, J. F. (2003) The function of genomes in bioenergetic organelles. *Phil. Trans. R. Soc. B.* 358: 19–37.

Bendich, A. (2013) DNA abandonment and the mechanisms of uniparental inheritance of mitochondria and chloroplasts. *Chromosome Res.* 21: 287–296.

Birchler, J. A., Gao, Z., Sharma, A., Presting, G. G., and Han, F. (2011) Epigenetic aspects of centromere function in plants. *Curr. Opin. Plant Biol.* 14: 217–222.

Chen, X. (2012) Small RNAs in development—insights from plants. *Curr. Opin. Genet. Develop.* 22: 361–367.

Chen, Z. J. (2007) Genetic and epigenetic mechanisms for gene expression and phenotypic variation in plant polyploids. *Annu. Rev. Plant Biol. Mol. Biol.* 58: 377–406.

Ghildiyal, M., and Zamore, P. D. (2009) Small silencing RNAs: an expanding universe. *Nat. Rev. Genet.* 10: 94–108.

Gill, N., Hans, C. S., and Jackson, S. (2008) An overview of plant chromosome structure. Cytogenet. *Genome Res.* 120: 194–201.

Grandont, L., Jenczewski, E., and Lloyd, A. (2013) Meiosis and its deviations in polyploid plants. *Cytogenet. Genome Res.* 140:171–84.

Jiao, Y., Wickett, N. J., Ayyampalayam, S., Chanderbali, A. S., Landherr, L., Ralph, P. E., Tomsho, L. P., Hu, Y., Liang, H., Soltis, P. S., Soltis, D. E., Clifton, S. W., Schlarbaum, S. E., Schuster, S. C., Ma, H., Leebens-Mack, J., and dePamphilis, C. W. (2011) Ancestral polyploidy in seed plants and angiosperms. *Nature* 473: 97–100.

Leitch, A. R., and Leitch, I. J. (2008) Genomic plasticity and the diversity of polyploid plants. *Science* 320: 481–483.

Liu, C., Lu, F., Cui, X., and Cao, X. (2010) Histone methylation in higher plants. *Annu. Rev. Plant Biol.* 61: 395–420.

Madlung, A., and Wendel, J. F. (2013) Genetic and epigenetic aspects of polyploid evolution in plants. *Cytogenet. Genome Res.* 140: 270–285.

Mogensen, L. (1996) The hows and whys of cytoplasmic inheritance in seed plants. *Am. J. Bot.* 83: 383–404.

Parisod, C., Alix, K., Just, J., Petit, M., Sarilar, V., Mhiri, C., Ainouche, M., Chalhoub, B., and Grandbastien, M.-A. (2010) Impact of transposable elements on the organization and function of allopolyploid genomes. *New Phytol.* 186: 37–45.

UNIT I

Transport and Translocation of Water and Solutes

UNIT I

Chapter 3
Water and Plant Cells

Chapter 4
Water Balance of Plants

Chapter 5
Mineral Nutrition

Chapter 6
Solute Transport

3 Water and Plant Cells

Water plays a crucial role in the life of the plant. Photosynthesis requires that plants draw carbon dioxide from the atmosphere, and at the same time exposes them to water loss and the threat of dehydration. To prevent leaf desiccation, water must be absorbed by the roots and transported through the plant body. Even slight imbalances between the uptake and transport of water and the loss of water to the atmosphere can cause water deficits and severe malfunctioning of many cellular processes. Thus, balancing the uptake, transport, and loss of water represents an important challenge for land plants.

A major difference between plant and animal cells, which has a large impact on their respective water relations, is that plant cells have cell walls and animal cells do not. Cell walls allow plant cells to build up large internal hydrostatic pressures, called **turgor pressure**. Turgor pressure is essential for many physiological processes, including cell enlargement, stomatal opening, transport in the phloem, and various transport processes across membranes. Turgor pressure also contributes to the rigidity and mechanical stability of nonlignified plant tissues. In this chapter we will consider how water moves into and out of plant cells, emphasizing the molecular properties of water and the physical forces that influence water movement at the cell level.

Water in Plant Life

Of all the resources that plants need to grow and function, water is the most abundant and yet often the most limiting. The practice of crop irrigation reflects the fact that water is a key resource limiting agricultural productivity (**Figure 3.1**). Water availability likewise limits the productivity of natural ecosystems (**Figure 3.2**), leading to marked differences in vegetation type along precipitation gradients.

The reason that water is frequently a limiting resource for plants, but much less so for animals, is that plants use water in huge amounts. Most (~97%) of the water absorbed by a plant's roots is carried through the plant and evaporates from leaf surfaces. Such

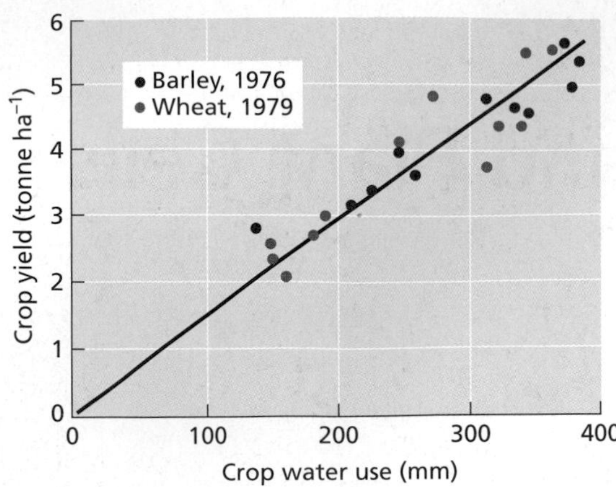

Figure 3.1 Grain yield as a function of water used under a range of irrigation treatments for barley in 1976 and wheat in 1979 in southeastern England. (After Jones 1992; data from Day et al. 1978 and Innes and Blackwell 1981.)

water loss is called **transpiration**. In contrast, only a small amount of the water absorbed by roots actually remains in the plant to supply growth (~2%) or to be consumed in the biochemical reactions of photosynthesis and other metabolic processes (~1%).

Water loss to the atmosphere appears to be an inevitable consequence of carrying out photosynthesis on land. The uptake of CO_2 is coupled to the loss of water through a common diffusional pathway: as CO_2 diffuses into leaves, water vapor diffuses out. Because the driving gradient for water loss from leaves is much larger than that for CO_2 uptake, as many as 400 water molecules are lost for every CO_2 molecule gained. This unfavorable exchange has had a major influence on the evolution of plant form and function and explains why water plays such a key role in the physiology of plants.

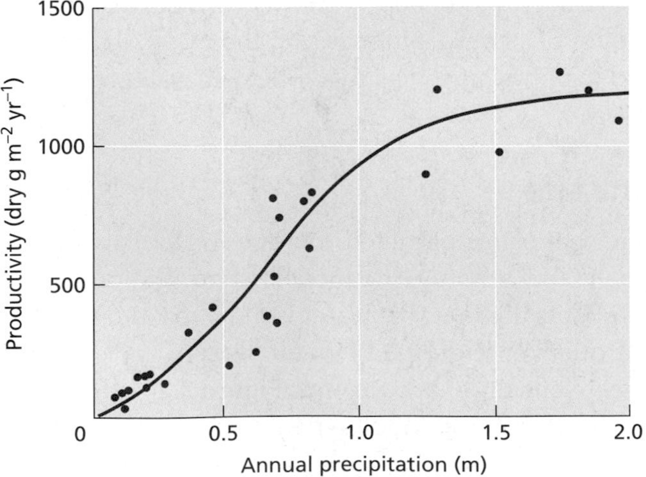

Figure 3.2 Productivity of various ecosystems as a function of annual precipitation. Productivity was estimated as net aboveground accumulation of organic matter through growth and reproduction. (After Whittaker 1970.)

We will begin our study of water by considering how its structure gives rise to some of its unique physical properties. We will then examine the physical basis for water movement, the concept of water potential, and the application of this concept to cell–water relations.

The Structure and Properties of Water

Water has special properties that enable it to act as a wide-ranging solvent and to be readily transported through the body of the plant. These properties derive primarily from the hydrogen bonding ability and polar structure of the water molecule. In this section we will examine how the formation of hydrogen bonds contributes to the high specific heat, surface tension, and tensile strength of water.

Water is a polar molecule that forms hydrogen bonds

The water molecule consists of an oxygen atom covalently bonded to two hydrogen atoms (**Figure 3.3A**). Because the oxygen atom is more **electronegative** than hydrogen, it tends to attract the electrons of the covalent bond. This attraction results in a partial negative charge at the oxygen end of the molecule and a partial positive charge at each hydrogen, making water a **polar** molecule. These partial charges are equal, so the water molecule carries no *net* charge.

Water molecules are tetrahedral in shape. At two points of the tetrahedron are the hydrogen atoms, each with a partial positive charge. The other two points of the tetrahedron contain lone pairs of electrons, each with a partial negative charge. Thus, each water molecule has two positive poles and two negative poles. These opposite partial charges create electrostatic attractions between water molecules, known as **hydrogen bonds** (**Figure 3.3B**).

Hydrogen bonds take their name from the fact that effective electrostatic bonds are formed only when highly electronegative atoms such as oxygen are covalently bonded to hydrogen. The reason for this is that the small size of the hydrogen atom allows the partial positive charges to be more concentrated, and thus more effective in bonding electrostatically.

Hydrogen bonds are responsible for many of the unusual physical properties of water. Water can form up to four hydrogen bonds with adjacent water molecules, resulting in very strong intermolecular interactions. Hydrogen bonds can also form between water and other molecules that contain electronegative atoms (O or N), especially when these are covalently bonded to H.

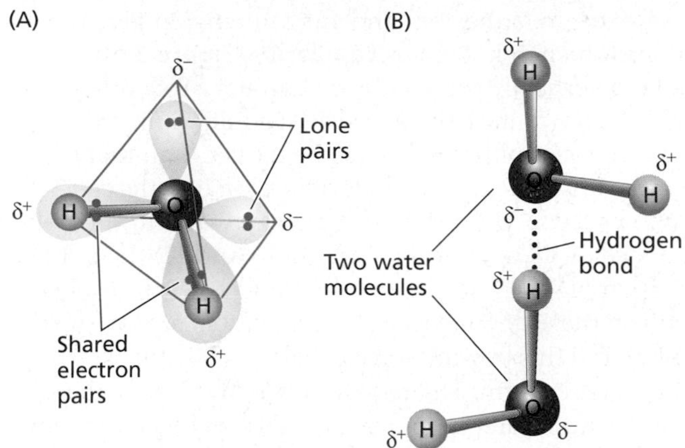

Figure 3.3 Structure of the water molecule. (A) The strong electronegativity of the oxygen atom means that the two electrons that form the covalent bond with hydrogen are unequally shared, such that each hydrogen atom has a partial positive charge. Each of the two lone pairs of electrons of the oxygen atom produces a partial negative charge. (B) The opposite partial charges (δ^- and δ^+) on the water molecule lead to the formation of intermolecular hydrogen bonds with other water molecules. Oxygen has six electrons in the outer orbitals; each hydrogen has one.

Water is an excellent solvent

Water dissolves greater amounts of a wider variety of substances than do other related solvents. Its versatility as a solvent is due in part to the small size of the water molecule. However, it is the hydrogen bonding ability of water and its polar structure that make it a particularly good solvent for ionic substances and for molecules such as sugars and proteins that contain polar —OH or —NH_2 groups.

Hydrogen bonding between water molecules and ions, and between water and polar solutes, effectively decreases the electrostatic interaction between the charged substances and thereby increases their solubility. Similarly, hydrogen bonding between water and macromolecules such as proteins and nucleic acids reduces interactions between macromolecules, thus helping draw them into solution.

Water has distinctive thermal properties relative to its size

The extensive hydrogen bonding between water molecules results in water having both a high specific heat capacity and a high latent heat of vaporization.

Specific heat capacity is the heat energy required to raise the temperature of a substance by a set amount. Temperature is a measure of molecular kinetic energy (energy of motion). When the temperature of water is raised, the molecules vibrate faster and with greater amplitude. Hydrogen bonds act like rubber bands that absorb some of the energy from applied heat, leaving less energy available to increase motion. Thus, compared with other liquids, water requires a relatively large heat input to raise its temperature. This is important for plants, because it helps buffer temperature fluctuations.

Latent heat of vaporization is the energy needed to separate molecules from the liquid phase and move them into the gas phase—a process that occurs during transpiration. The latent heat of vaporization decreases as temperature increases, reaching a minimum at the boiling point (100°C). For water at 25°C, the heat of vaporization is 44 kJ mol^{-1}—the highest value known for any liquid. Most of this energy is used to break hydrogen bonds between water molecules.

Latent heat does not change the temperature of water molecules that have evaporated, but it does cool the surface from which the water has evaporated. Thus, the high latent heat of vaporization of water serves to moderate the temperature of transpiring leaves, which would otherwise increase due to the input of radiant energy from the sun.

Water molecules are highly cohesive

Water molecules at an air–water interface are attracted to neighboring water molecules by hydrogen bonds, and this interaction is much stronger than any interaction with the adjacent gas phase. As a consequence, the lowest-energy (i.e., most stable) configuration is one that minimizes the surface area of the air–water interface. To increase the area of an air–water interface, hydrogen bonds must be broken, which requires an input of energy. The energy required to increase the surface area of a gas–liquid interface is known as **surface tension**.

Surface tension can be expressed in units of energy per area (J m^{-2}) but is generally expressed in the equivalent, but less intuitive, units of force per length (J m^{-2} = N m^{-1}). A joule (J) is the SI unit of energy, with units of force × distance (N m); a newton (N) is the SI unit of force, with units of mass × acceleration (kg m s^{-2}). If the air–water interface is curved, surface tension produces a net force perpendicular to the interface (**Figure 3.4**). As we will see later, surface tension and adhesion (defined below) at the evaporative surfaces in leaves generate the physical forces that pull water through the plant's vascular system.

The extensive hydrogen bonding in water also gives rise to the property known as **cohesion**, the mutual attraction between molecules. A related property, called **adhesion**, is the attraction of water to a solid phase such as a cell wall or glass surface, which again is due primarily to the formation of hydrogen bonds. The degree to which water is attracted to the solid phase versus to itself can be quantified by measuring the **contact angle** (**Figure 3.5A**). The contact angle describes the shape of the air–water interface and thus the effect that the surface tension has on the pressure in the liquid.

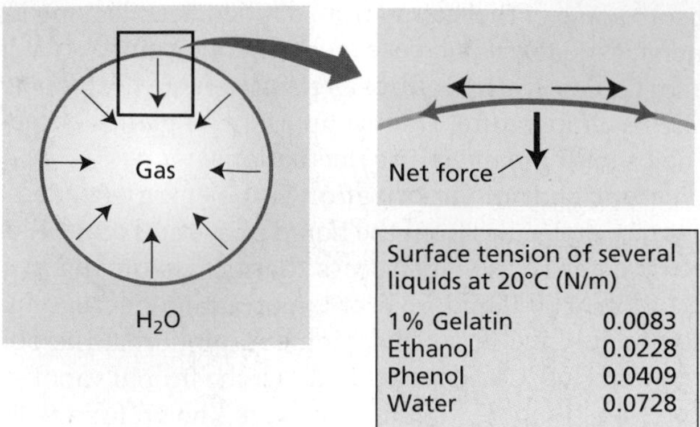

Surface tension of several liquids at 20°C (N/m)	
1% Gelatin	0.0083
Ethanol	0.0228
Phenol	0.0409
Water	0.0728

Figure 3.4 A gas bubble suspended within a liquid assumes a spherical shape such that its surface area is minimized. Because surface tension acts along the tangent to the gas–liquid interface, the resultant (net) force is inward, leading to compression of the bubble. The magnitude of the pressure (force/area) exerted by the interface is equal to $2T/r$, where T is the surface tension of the liquid (N/m) and r is the radius of the bubble (m). Water has an extremely high surface tension compared with other liquids at the same temperature.

Cohesion, adhesion, and surface tension give rise to a phenomenon known as **capillarity** (**Figure 3.5B**). Consider a vertically oriented glass capillary tube with wettable walls (contact angle < 90°). At equilibrium, the water level in the capillary tube will be higher than that of the water supply at its base. Water is drawn into the capillary tube because of (1) the attraction of water to the polar surface of the glass tube (adhesion) and (2) the surface tension of water. Together, adhesion and surface tension pull on the water molecules, causing them to move up the tube until this upward force is balanced by the weight of the water column. The narrower the tube, the higher the equilibrium water level. For calculations related to capillarity, see WEB TOPIC 3.1.

Water has a high tensile strength

Hydrogen bonding gives water a high **tensile strength**, defined as the maximum force per unit area that a continuous column of water can withstand before breaking. We do not usually think of liquids as having tensile strength; however, such a property is evident in the rise of a water column in a capillary tube.

We can demonstrate the tensile strength of water by placing it in a clean glass syringe (**Figure 3.6**). When we *push* on the plunger, the water is compressed, and a positive **hydrostatic pressure** builds up. Pressure is measured in units called *pascals* (Pa) or, more conveniently, *megapascals* (MPa). One MPa equals approximately 9.9 atmospheres. Pressure is equivalent to a force per unit area (1 Pa = 1 N m^{-2}) and to an energy per unit volume (1 Pa = 1 J m^{-3}). **Table 3.1** compares units of pressure.

If instead of pushing on the plunger we *pull* on it, a tension, or *negative hydrostatic pressure*, develops as the

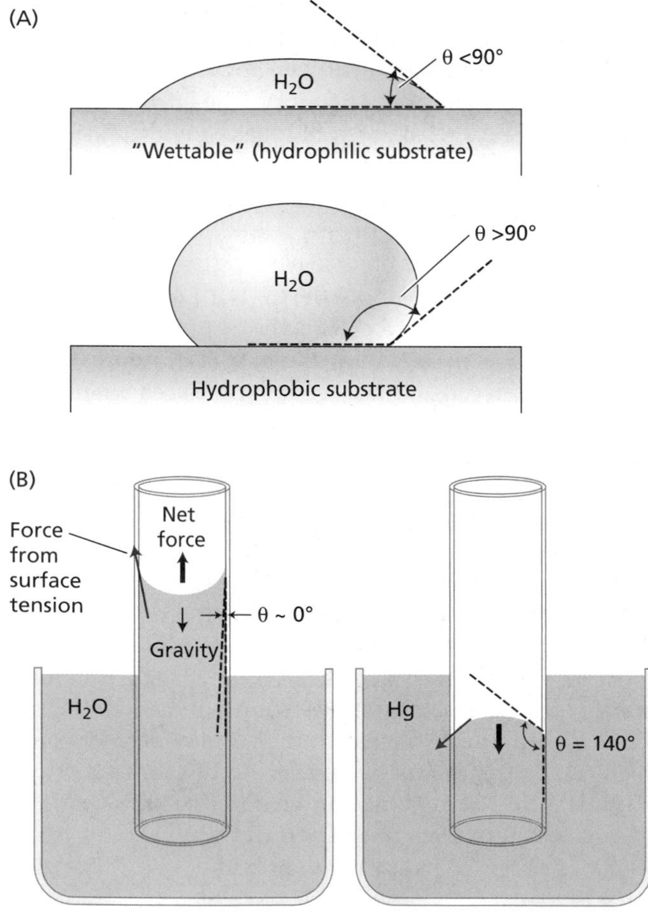

Figure 3.5 (A) The shape of a droplet placed on a solid surface reflects the relative attraction of the liquid to the solid versus to itself. The contact angle (θ), defined as the angle from the solid surface through the liquid to the gas–liquid interface, is used to describe this interaction. "Wettable" surfaces have contact angles of less than 90°; a highly wettable (i.e., hydrophilic) surface (such as water on clean glass or primary cell walls) has a contact angle close to 0°. Water spreads out to form a thin film on highly wettable surfaces. In contrast, nonwettable (i.e., hydrophobic) surfaces have contact angles greater than 90°. Water "beads" up on such surfaces. (B) Capillarity can be observed when a liquid is supplied to the bottom of a vertically oriented capillary tube. If the walls are highly wettable (e.g., water on clean glass), the net force is upward. The water column rises until this upward force is balanced by the weight of the water column. In contrast, if the liquid does not "wet" the walls (e.g., Hg on clean glass has a contact angle of approximately 140°), the meniscus curves downward, and the force resulting from surface tension lowers the level of the liquid in the tube.

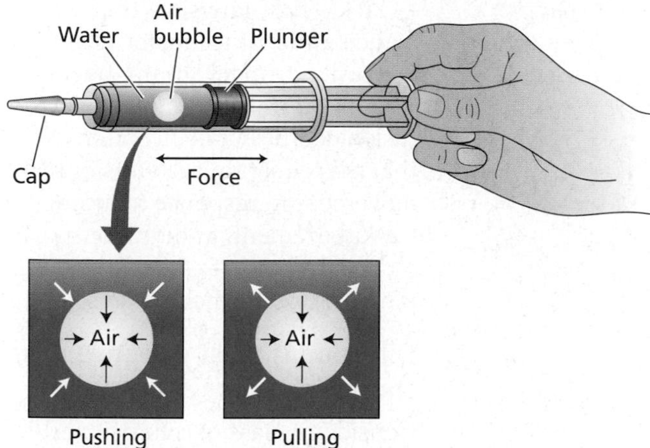

Figure 3.6 A sealed syringe can be used to create positive and negative pressures in fluids such as water. Pushing on the plunger causes the fluid to develop a positive, hydrostatic pressure (white arrows) that acts in the same direction as the inward force resulting from the surface tension of the gas–liquid interface (black arrows). Thus, a small air bubble trapped within the syringe will shrink as the pressure increases. Pulling on the plunger causes the fluid to develop a tension, or negative pressure. Air bubbles in the syringe will expand if the outward force on the bubble exerted by the fluid (white arrows) exceeds the inward force resulting from the surface tension of the gas–liquid interface (black arrows).

water molecules resist being pulled apart. Negative pressures develop only when molecules are able to pull on one another. Strong hydrogen bonds between water molecules allow tensions to be transmitted through water, even though it is a liquid. In contrast, gases cannot develop negative pressures because the interactions between gas molecules are limited to elastic collisions.

How hard must we pull on the plunger before the water molecules are torn away from each other and the water column breaks? Careful studies have demonstrated that water can resist tensions greater than 20 MPa. The water column in a syringe (see Figure 3.6), however, cannot sustain such large tensions because of the presence of microscopic gas bubbles. Because gas bubbles can expand, they interfere with the ability of the water in the syringe to resist the pull exerted by the plunger. The expansion of gas bubbles due to tension in the surrounding liquid is known as **cavitation**. As we will see in Chapter 4, cavitation can have a devastating effect on water transport through the xylem.

TABLE 3.1 Comparison of units of pressure

1 atmosphere	= 14.7 pounds per square inch
	= 760 mm Hg (at sea level, 45° latitude)
	= 1.013 bar
	= 0.1013 MPa
	= 1.013×10^5 Pa

A car tire is typically inflated to about 0.2 MPa.

The water pressure in home plumbing is typically 0.2–0.3 MPa.

The water pressure under 30 feet (10 m) of water is about 0.1 MPa.

Diffusion and Osmosis

Cellular processes depend on the transport of molecules both to the cell and away from it. **Diffusion** is the spontaneous movement of substances from regions of higher to lower concentration. At the scale of a cell, diffusion is the dominant mode of transport. The diffusion of water across a selectively permeable barrier is referred to as **osmosis**.

In this section we will examine how the processes of diffusion and osmosis lead to the net movement of both water and solutes.

Diffusion is the net movement of molecules by random thermal agitation

The molecules in a solution are not static; they are in continuous motion, colliding with one another and exchanging kinetic energy. The trajectory of any particular molecule after a collision is considered to be a random variable. Yet these random movements can result in the net movement of molecules.

Consider an imaginary plane dividing a solution into two equal volumes, A and B. As all the molecules undergo random motion, at each time step there is some probability that any particular solute molecule will cross our imaginary plane. The number we expect to cross from A to B in any particular time step will be proportional to the number at the beginning of the time step on side A, and the number crossing from B to A will be proportional to the number starting on side B.

If the initial concentration on side A is higher than that on side B, more solute molecules will be expected to cross from A to B than B to A, and we will observe a net movement of solutes from A to B. Thus, diffusion results in the net movement of molecules from regions of high concentration to regions of low concentration, even though each molecule is moving in a random direction. The independent motion of each molecule explains why the system will evolve toward an equal number of A and B molecules on each side (**Figure 3.7**).

This tendency for a system to evolve toward an even distribution of molecules can be understood as a consequence of the second law of thermodynamics, which tells us that spontaneous processes evolve in the direction of increasing entropy, or disorder. Increasing entropy is synonymous with a decrease in free energy. Thus, diffusion

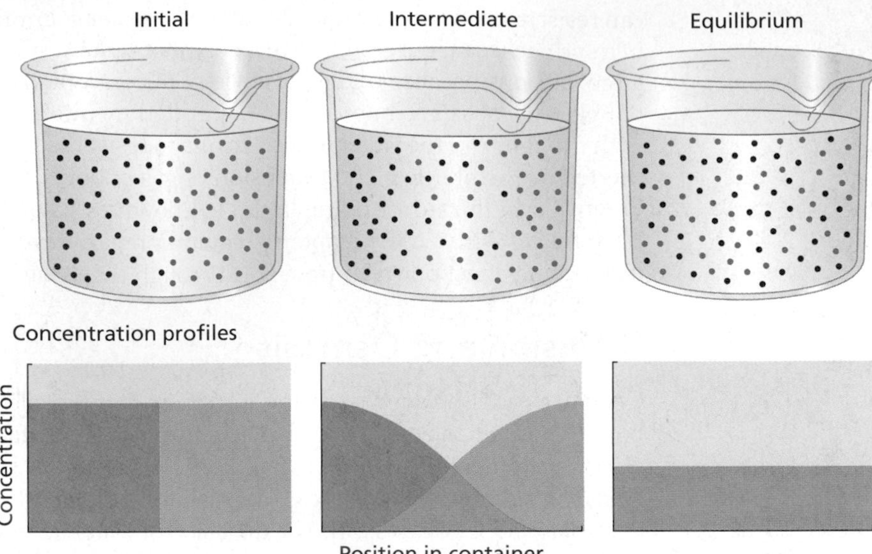

Figure 3.7 Thermal motion of molecules leads to diffusion—the gradual mixing of molecules and eventual dissipation of concentration differences. Initially, two materials containing different molecules are brought into contact. The materials may be gas, liquid, or solid. Diffusion is fastest in gases, slower in liquids, and slowest in solids. The initial separation of the molecules is depicted graphically in the upper panels, and the corresponding concentration profiles are shown in the lower panels as a function of position. The purple color indicates an overlap in the concentration profiles of red and blue solutes. With time, the mixing and randomization of the molecules diminish net movement. At equilibrium the two types of molecules are randomly (evenly) distributed. Note that at all points and times the *total* concentration of solutes (i.e., both red and blue solutes) remains constant.

represents the natural tendency of systems to move toward the lowest possible energy state.

It was Adolf Fick who first noticed in the 1850s that the rate of diffusion is directly proportional to the concentration gradient ($\Delta c_s/\Delta x$)—that is, to the difference in concentration of substance *s* (Δc_s) between two points separated by a very small distance Δx. In symbols, we write this relation as Fick's first law:

$$J_s = -D_s \frac{\Delta c_s}{\Delta x} \tag{3.1}$$

The rate of transport, expressed as the **flux density** (J_s), is the amount of substance *s* crossing a unit cross-sectional area per unit time (e.g., J_s may have units of moles per square meter per second [mol m^{-2} s^{-1}]). The **diffusion coefficient** (D_s) is a proportionality constant that measures how easily substance *s* moves through a particular medium. The diffusion coefficient is a characteristic of the substance (larger molecules have smaller diffusion coefficients) and depends on both the medium (diffusion in air is typically 10,000 times faster than diffusion in a liquid, for example) and the temperature (substances diffuse faster at higher temperatures). The negative sign in the equation indicates that the flux moves down a concentration gradient.

Fick's first law says that a substance will diffuse faster when the concentration gradient becomes steeper (Δc_s is large) or when the diffusion coefficient is increased. Note that this equation accounts only for movement in response to a concentration gradient, and not for movement in response to other forces (e.g., pressure, electric fields, and so on).

Diffusion is most effective over short distances

Consider a mass of solute molecules initially concentrated around a position $x = 0$ (**Figure 3.8A**). As the molecules undergo random motion, the concentration front moves away from the starting position, as shown for a later time point in **Figure 3.8B**.

Comparing the distribution of the solutes at the two times, we see that as the substance diffuses away from the starting point, the concentration gradient becomes less steep (Δc_s decreases); that is, the number of solute molecules that happen to step "backward" (i.e., toward $x = 0$) relative to those that step "forward" (away from $x = 0$) increases, and thus net movement becomes slower. Note that the average position of the solute molecules remains at $x = 0$ for all time, but that the distribution slowly flattens out.

As a direct consequence of the fact that each molecule is undergoing its own random walk, and thus is as likely to step toward some point of interest as away from it, the average time needed for a particle to diffuse a distance L grows as L^2/D_s. In other words, the average time required for a substance to diffuse a given distance increases as the *square* of that distance.

The diffusion coefficient for glucose in water is about 10^{-9} m^2 s^{-1}. Thus, the average time required for a glucose molecule to diffuse across a cell with a diameter of 50 μm is 2.5 s. However, the average time needed for the same glucose molecule to diffuse a distance of 1 m in water is approximately 32 years. These values show that diffusion in solutions can be effective within cellular dimensions but is far too slow to be effective over long distances. For additional calculations on diffusion times, see **WEB TOPIC 3.2**.

Osmosis describes the net movement of water across a selectively permeable barrier

Membranes of plant cells are **selectively permeable**; that is, they allow water and other small, uncharged substances to move across them more readily than larger sol-

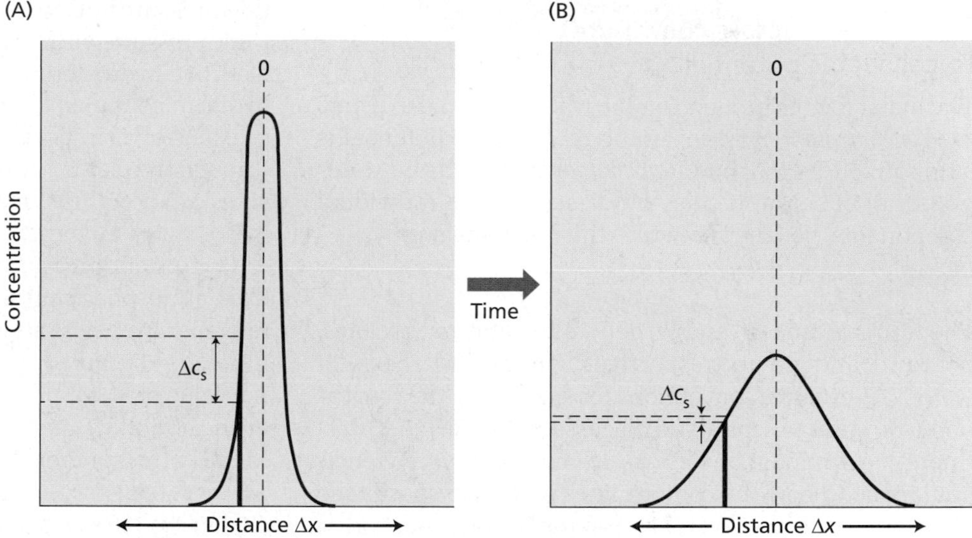

Figure 3.8 Graphical representation of the concentration gradient of a solute that is diffusing according to Fick's first law. The solute molecules were initially located in the plane indicated on the *x*-axis ("0"). (A) The distribution of solute molecules shortly after placement at the plane of origin. Note how sharply the concentration drops off as the distance, *x*, from the origin increases. (B) The solute distribution at a later time point. The average distance of the diffusing molecules from the origin has increased, and the slope of the gradient has flattened out. (After Nobel 1999.)

utes and charged substances. If the concentration of solutes is greater in the cell than in the solution surrounding it, water diffuses into the cell, but the solutes are unable to diffuse out of the cell. The net movement of water across a selectively permeable barrier is referred to as *osmosis*.

We saw earlier that the tendency of all systems to evolve toward increasing entropy results in solutes spreading out through the entire available volume. In osmosis, the volume available for solute movement is restricted by the membrane, and thus entropy maximization is realized by the volume of solvent diffusing through the membrane to dilute the solutes. Indeed, in the absence of any countervailing force, *all* the available water will flow to the solute side of the membrane.

Let's imagine what happens when we place a living cell in a beaker of pure water. The presence of a selectively permeable membrane means that the net movement of water will continue until one of two things happens: (1) the cell expands until the selectively permeable membrane ruptures, allowing solutes to diffuse freely, or (2) the expansion of the cell volume is mechanically constrained by the presence of a cell wall such that the driving force for water to enter the cell is balanced by a pressure exerted by the cell wall.

The first scenario describes what would happen to an animal cell, which lacks a cell wall. The second scenario is relevant to plant cells. The plant cell wall is very strong. The resistance of cells walls to deformation creates an inward force that raises the hydrostatic pressure within the cell. The word *osmosis* derives from the Greek word for "pushing"; it is an expression of the positive pressure generated when solutes are confined.

We will soon see how osmosis drives the movement of water into and out of plant cells. First, however, let's discuss the concept of a composite or total driving force, representing the free-energy gradient of water.

Water Potential

All living things, including plants, require a continuous input of free energy to maintain and repair their highly organized structures, as well as to grow and reproduce. Processes such as biochemical reactions, solute accumulation, and long-distance transport are all driven by an input of free energy into the plant. (For a detailed discussion of the thermodynamic concept of free energy, see WEB APPENDIX 1.) In this section we will examine how concentration, pressure, and gravity influence free energy.

The chemical potential of water represents the free-energy status of water

Chemical potential is a quantitative expression of the free energy associated with a substance. In thermodynamics, free energy represents the potential for performing work, force × distance. The unit of chemical potential is energy per mole of substance (J mol^{-1}). Note that chemical potential is a relative quantity: It represents the difference between the potential of a substance in a given state and the potential of the same substance in a standard state.

The chemical potential of water represents the free energy associated with water. Water flows spontaneously, that is, without an input of energy, from regions of higher chemical potential to ones of lower chemical potential.

Historically, plant physiologists have most often used a related parameter called **water potential**, defined as the chemical potential of water divided by the partial molal volume of water (the volume of 1 mol of water): 18 $\times 10^{-6}$ m^3 mol^{-1}. Water potential is thus a measure of the free energy of water per unit volume (J m^{-3}). These units are equivalent to pressure units such as the pascal, which is the common measurement unit for water potential. Let's look more closely at the important concept of water potential.

Three major factors contribute to cell water potential

The major factors influencing the water potential in plants are *concentration, pressure,* and *gravity*. Water potential is symbolized by Ψ (the Greek letter psi), and the water potential of solutions may be dissected into individual components, usually written as the following sum:

$$\Psi = \Psi_s + \Psi_p + \Psi_g \quad (3.2)$$

The terms Ψ_s and Ψ_p and Ψ_g denote the effects of solutes, pressure, and gravity, respectively, on the free energy of water. (Alternative conventions for expressing the components of water potential are discussed in **WEB TOPIC 3.3**.) Energy levels must be defined in relation to a reference, analogous to how the contour lines on a map specify the distance above sea level. The reference state most often used to define water potential is pure water at ambient temperature and standard atmospheric pressure. The reference height is generally set either at the base of the plant (for whole plant studies) or at the level of the tissue under examination (for studies of water movement at the cellular level). Let's consider each of the terms on the right-hand side of Equation 3.2.

SOLUTES The term Ψ_s, called the **solute potential** or the **osmotic potential**, represents the effect of dissolved solutes on water potential. Solutes reduce the free energy of water by diluting the water. This is primarily an entropy effect; that is, the mixing of solutes and water increases the disorder or entropy of the system and thereby lowers the free energy. This means that *the osmotic potential is independent of the specific nature of the solute*. For dilute solutions of nondissociating substances such as sucrose, the osmotic potential may be approximated by:

$$\Psi_s = -RTc_s \quad (3.3)$$

where R is the gas constant (8.32 J mol^{-1} K^{-1}), T is the absolute temperature (in degrees Kelvin, or K), and c_s is the solute concentration of the solution, expressed as **osmolarity** (moles of total dissolved solutes per volume of water [mol L^{-1}]). The minus sign indicates that dissolved solutes reduce the water potential of a solution relative to the reference state of pure water.

Equation 3.3 is valid for "ideal" solutions. Real solutions frequently deviate from the ideal, especially at high concentrations—for example, greater than 0.1 mol L^{-1}. Temperature also affects water potential (see **WEB TOPIC 3.4**). In our treatment of water potential, we will assume that we are dealing with ideal solutions.

PRESSURE The term Ψ_p, called the **pressure potential**, represents the effect of hydrostatic pressure on the free energy of water. Positive pressures raise the water potential; negative pressures reduce it. Both positive and negative pressures occur within plants. The positive hydrostatic pressure within cells is referred to as *turgor pressure*. Negative hydrostatic pressures, which frequently develop in xylem conduits, are referred to as **tension**. As we will see, tension is important in moving water long distances through the plant. The question of whether negative pressures can occur in living cells is considered in **WEB TOPIC 3.5**.

Hydrostatic pressure is often measured as the deviation from atmospheric pressure. Remember that water in the reference state is at atmospheric pressure, so by this definition $\Psi_p = 0$ MPa for water in the standard state. Thus, the value of Ψ_p for pure water in an open beaker is 0 MPa, even though its absolute pressure is approximately 0.1 MPa (1 atmosphere).

GRAVITY Gravity causes water to move downward unless the force of gravity is opposed by an equal and opposite force. The **gravitational potential** (Ψ_g) depends on the height (h) of the water above the reference-state water, the density of water (ρ_w), and the acceleration due to gravity (g). In symbols, we write the following:

$$\Psi_g = \rho_w gh \quad (3.4)$$

where $\rho_w g$ has a value of 0.01 MPa m^{-1}. Thus, raising water a distance of 10 m translates into a 0.1 MPa increase in water potential.

The gravitational component (Ψ_g) is generally omitted in considerations of water transport at the cell level, because differences in this component among neighboring cells are negligible compared with differences in the osmotic potential and the pressure potential. Thus, in these cases Equation 3.2 can be simplified as follows:

$$\Psi = \Psi_s + \Psi_p \quad (3.5)$$

Water potentials can be measured

Cell growth, photosynthesis, and crop productivity are all strongly influenced by water potential and its components. Plant scientists have thus expended considerable effort in devising accurate and reliable methods for evaluating the water status of plants.

The principal approaches for determining Ψ use psychrometers, of which there are two types, or the pressure chamber. Psychrometers take advantage of water's large latent heat of vaporization, which allows accurate measurements of (1) the vapor pressure of water in equilibrium with the sample or (2) the transfer of water vapor between the sample and a solution of known Ψ_s. The pressure chamber measures Ψ by applying external gas pressure to an excised leaf until water is forced out of the living cells.

In some cells, it is possible to measure Ψ_p directly by inserting a liquid-filled microcapillary that is connected to a pressure sensor into the cell. In other cases, Ψ_p is estimated as the difference between Ψ and Ψ_s. Solute concentrations (Ψ_s) can be determined using a variety of methods, including psychrometers and instruments that measure freezing point depression. A detailed explanation of the instruments that have been used to measure Ψ, Ψ_s, and Ψ_p can be found in WEB TOPIC 3.6.

In discussions of water in dry soils and plant tissues with very low water contents, such as seeds, one often finds reference to the **matric potential**, Ψ_m. Under these conditions, water exists as a very thin layer, perhaps one or two molecules deep, bound to solid surfaces by electrostatic interactions. These interactions are not easily separated into their effects on Ψ_s and Ψ_p, and are thus sometimes combined into a single term, Ψ_m. The matric potential is discussed further in WEB TOPIC 3.7.

Water Potential of Plant Cells

Plant cells typically have water potentials of 0 MPa or less. A negative value indicates that the free energy of water within the cell is less than that of pure water at ambient temperature, atmospheric pressure, and equal height. As the water potential of the solution surrounding the cell changes, water will enter or leave the cell via osmosis. In this section we will illustrate the osmotic behavior of water in plant cells with some numerical examples.

Water enters the cell along a water potential gradient

First imagine an open beaker full of pure water at 20°C (**Figure 3.9A**). Because the water is open to the atmosphere, the pressure potential of the water is the same as atmospheric pressure (Ψ_p = 0 MPa). There are no solutes in the water, so Ψ_s = 0 MPa. Finally, because we focus here on transport processes that take place within the beaker, we define the reference height as equal to the level of the beaker, and thus Ψ_g = 0 MPa. Therefore, the water potential is 0 MPa ($\Psi = \Psi_s + \Psi_p$).

Now imagine dissolving sucrose in the water to a concentration of 0.1 *M* (**Figure 3.9B**). This addition lowers the osmotic potential (Ψ_s) to –0.244 MPa and decreases the water potential (Ψ) to –0.244 MPa.

Next consider a flaccid plant cell (i.e., a cell with no turgor pressure) that has a total internal solute concentration of 0.3 *M* (**Figure 3.9C**). This solute concentration gives an osmotic potential (Ψ_s) of –0.732 MPa. Because the cell is flaccid, the internal pressure is the same as atmospheric pressure, so the pressure potential (Ψ_p) is 0 MPa and the water potential of the cell is –0.732 MPa.

What happens if this cell is placed in the beaker containing 0.1 *M* sucrose (see Figure 3.9C)? Because the water potential of the sucrose solution (Ψ = –0.244 MPa; see Figure 3.9B) is greater (less negative) than the water potential of the cell (Ψ = –0.732 MPa), water will move from the sucrose solution to the cell (from high to low water potential).

(A) Pure water

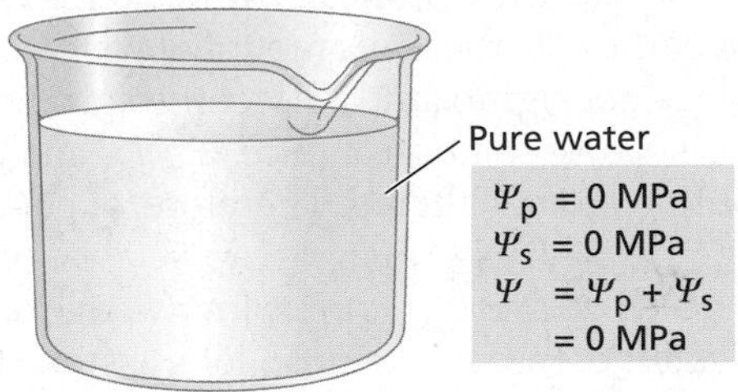

(B) Solution containing 0.1 *M* sucrose

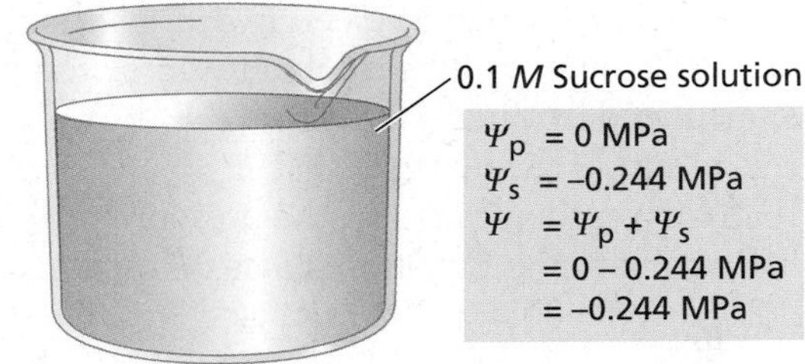

(C) Flaccid cell dropped into sucrose solution

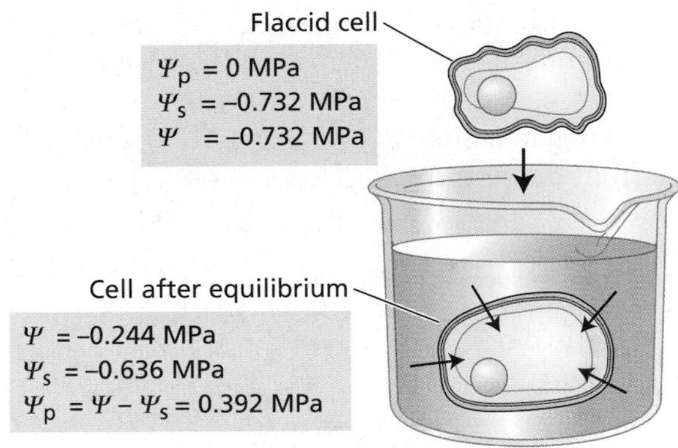

Figure 3.9 Water potential gradients can cause water to enter a cell. (A) Pure water. (B) A solution containing 0.1 *M* sucrose. (C) A flaccid cell (in air) is dropped in a 0.1 *M* sucrose solution. Because the starting water potential of the cell is less than the water potential of the solution, the cell takes up water. After equilibration, the water potential of the cell equals the water potential of the solution, and the result is a cell with a positive turgor pressure.

As water enters the cell, the plasma membrane begins to press against the cell wall. The wall stretches a little but also resists deformation by pushing back on the cell. This increases the pressure potential (Ψ_p) of the cell. Consequently, the cell water potential (Ψ) increases, and the difference between inside and outside water potentials ($\Delta\Psi$) is reduced.

Eventually, cell Ψ_p increases enough to raise the cell Ψ to the same value as the Ψ of the sucrose solution. At this point, equilibrium is reached ($\Delta\Psi$ = 0 MPa), and net water transport ceases.

At equilibrium, water potential is equal everywhere: $\Psi_{(cell)} = \Psi_{(solution)}$. Because the volume of the beaker is much larger than that of the cell, the tiny amount of water taken up by the cell does not significantly affect the solute concentration of the sucrose solution. Hence, Ψ_s, Ψ_p, and Ψ of the sucrose solution are not altered. Therefore, at equilibrium, $\Psi_{(cell)} = \Psi_{(solution)} = -0.244$ MPa.

Calculation of cell Ψ_p and Ψ_s requires knowledge of the change in cell volume. In this example, let's assume that we know that the cell volume increased by 15%, such that the volume of the turgid cell is 1.15 times that of the flaccid cell. If we assume that the number of solutes within the cell remains constant as the cell hydrates, the final concentration of solutes will be diluted by 15%. The new Ψ_s can be calculated by dividing the initial Ψ_s by the relative increase in size of the hydrated cell: $\Psi_s = -0.732/1.15 = -0.636$ MPa. We can then calculate the pressure potential of the cell by rearranging Equation 3.5 as follows: $\Psi_p = \Psi - \Psi_s = (-0.244) - (-0.636) = 0.392$ MPa (see Figure 3.9C).

Water can also leave the cell in response to a water potential gradient

Water can also leave the cell by osmosis. If we now remove our plant cell from the 0.1 *M* sucrose solution and place it in a 0.3 *M* sucrose solution (**Figure 3.10A**), $\Psi_{(solution)}$ (–0.732 MPa) is more negative than $\Psi_{(cell)}$ (–0.244 MPa), and water will move from the turgid cell to the solution.

As water leaves the cell, the cell volume decreases. As the cell volume decreases, cell Ψ_p and Ψ decrease until $\Psi_{(cell)} = \Psi_{(solution)} = -0.732$ MPa. As before, we assume that the number of solutes within the cell remains constant as water flows from the cell. If we know that the cell volume decreases by 15%, the concentration of solutes will increase by 15%. Thus, we can calculate the new Ψ_s by multiplying the initial Ψ_s by the relative amount the cell volume has decreased: $\Psi_s = -0.636 \times 1.15 = -0.732$ MPa. This allows us to calculate that $\Psi_p = 0$ MPa using Equation 3.5.

If, instead of placing the turgid cell in the 0.3 *M* sucrose solution, we leave it in the 0.1 *M* solution and slowly squeeze it by pressing the cell between two plates (**Figure 3.10B**), we effectively raise the cell Ψ_p, consequently raising the cell Ψ and creating a $\Delta\Psi$ such that water now flows *out* of the cell. This is analogous to the industrial process of reverse osmosis in which externally applied pressure is used to separate water from dissolved solutes by forcing it across a semipermeable barrier. If we continue squeezing until half the cell's water is removed and then hold the cell in this condition, the cell will reach a new equilibrium. As in the previous example, at equilibrium, $\Delta\Psi = 0$ MPa, and the amount of water added to the external solution is so small that it can be ignored. The cell will thus return to the Ψ value that it had before

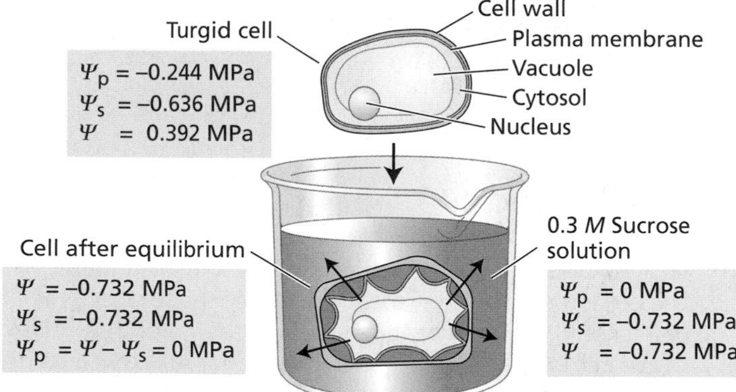

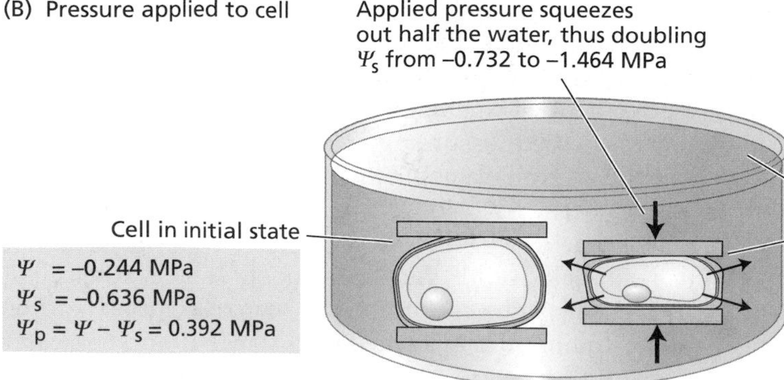

Figure 3.10 Water potential gradients can cause water to leave a cell. (A) Increasing the concentration of sucrose in the solution makes the cell lose water. The increased sucrose concentration lowers the solution water potential, draws water out of the cell, and thereby reduces the cell's turgor pressure. In this case, the protoplast pulls away from the cell wall (i.e., the cell plasmolyzes), because sucrose molecules are able to pass through the relatively large pores of the cell walls. When this occurs, the difference in water potential between the cytoplasm and the solution is entirely across the plasma membrane, and thus the protoplast shrinks independently of the cell wall. In contrast, when a cell desiccates in air (e.g., as in the flaccid cell in Figure 3.9C), plasmolysis does not occur. Instead, the cell (cytoplasm + wall) shrinks as a unit, resulting in the cell wall being mechanically deformed as the cell loses volume. (B) Another way to make the cell lose water is to squeeze it slowly between two plates. In this case, half of the cell water is removed, so cell osmotic potential increases by a factor of 2.

the squeezing procedure. However, the components of the cell Ψ will be quite different.

Because half of the water was squeezed out of the cell while the solutes remained inside the cell (the plasma membrane is selectively permeable), the cell solution is concentrated twofold, and thus Ψ_s is lower (–0.636 MPa × 2 = –1.272 MPa). Knowing the final values for Ψ and Ψ_s, we can calculate the pressure potential, using Equation 3.5, as $\Psi_p = \Psi - \Psi_s$ = (–0.244 MPa) – (–1.272 MPa) = 1.028 MPa.

In our example we used an external force to change cell volume without a change in water potential. In nature, it is typically the water potential of the cell's environment that changes, and the cell gains or loses water until its Ψ matches that of its surroundings.

One point common to all these examples deserves emphasis: *Water flow across membranes is a passive process. That is, water moves in response to physical forces, toward regions of low water potential or low free energy.* There are no known metabolic "pumps" (e.g., reactions driven by ATP hydrolysis) that can be used to drive water across a semipermeable membrane against its free-energy gradient.

The only situation in which water can be said to move across a semipermeable membrane against its water potential gradient is when it is coupled to the movement of solutes. The transport of sugars, amino acids, or other small molecules by various membrane proteins can "drag" up to 260 water molecules across the membrane per molecule of solute transported.

Such transport of water can occur even when the movement is against the usual water potential gradient (i.e., toward a higher water potential), because the loss of free energy by the solute more than compensates for the gain of free energy by the water. The net change in free energy remains negative. The amount of water transported in this way is generally quite small compared with the passive movement of water down its water potential gradient.

Water potential and its components vary with growth conditions and location within the plant

In leaves of well-watered plants, Ψ ranges from –0.2 to about –1.0 MPa in herbaceous plants and to –2.5 MPa in trees and shrubs. Leaves of plants in arid climates can have much lower Ψ, down to below –10 MPa under the most extreme conditions.

Just as Ψ values depend on the growing conditions and the type of plant, so too, the values of Ψ_s can vary considerably. Within cells of well-watered garden plants (examples include lettuce, cucumber seedlings, and bean leaves), Ψ_s may be as high as –0.5 MPa (low cell solute concentration), although values of –0.8 to –1.2 MPa are more typical. In woody plants, Ψ_s tends to be lower (higher cell solute concentration), allowing the more negative midday Ψ typical of these plants to occur without a loss in turgor pressure.

Although Ψ_s *within* cells may be quite negative, the apoplastic solution surrounding the cells—that is, in the cell walls and in the xylem—is generally quite dilute. The Ψ_s of the apoplast is typically –0.1 to 0 MPa, although in certain tissues (e.g., developing fruits) and habitats (e.g., high salinity environments) the concentration of solutes in the apoplast can be large.

Values for Ψ_p within cells of well-watered plants may range from 0.1 to as much as 3 MPa, depending on the value of Ψ_s inside the cell. A plant **wilts** when the turgor pressure inside the cells of such tissues falls toward zero. As more water is lost from the cell, the walls become mechanically deformed, and the cell may be damaged as a result. WEB TOPIC 3.8 contrasts the situation in which a cell is dehydrated osmotically due to the presence of apoplastic solutes that can diffuse freely through the cell wall to that in which water is withdrawn from the cell due to lower (more negative) water potentials in the apoplast.

Cell Wall and Membrane Properties

Structural elements make important contributions to the water relations of plant cells. Cell wall elasticity defines the relation between turgor pressure and cell volume, while the permeability of the plasma membrane and the tonoplast to water influences the rate at which cells exchange water with their surroundings. In this section we will examine how wall and membrane properties influence the water status of plant cells.

Small changes in plant cell volume cause large changes in turgor pressure

Cell walls provide plant cells with a substantial degree of volume homeostasis relative to the large changes in water potential that they experience every day as a consequence of the transpirational water losses associated with photosynthesis (see Chapter 4). Because plant cells have fairly rigid walls, a change in cell Ψ is generally accompanied by a large change in Ψ_p, with relatively little change in cell (protoplast) volume, as long as Ψ_p is greater than 0.

This phenomenon is illustrated by the *pressure–volume curve* shown in **Figure 3.11**. As Ψ decreases from 0 to –1.2 MPa, the relative or percent water content is reduced by only slightly more than 5%. Most of this decrease is due to a reduction in Ψ_p (by about 1.0 MPa); Ψ_s decreases by less than 0.2 MPa as a result of increased concentration of cell solutes.

Measurements of cell water potential and cell volume can be used to quantify how wall properties influence the water status of plant cells. Turgor pressure in most cells approaches zero as the relative cell volume decreases by 10 to 15%. However, for cells with very rigid cell walls, the volume change associated with turgor loss can be much smaller. In cells with extremely elastic walls, such as the water-storing cells in the stems of many cacti, this volume change may be substantially larger.

The volumetric elastic modulus, symbolized by ε (the Greek letter epsilon), can be determined by examining

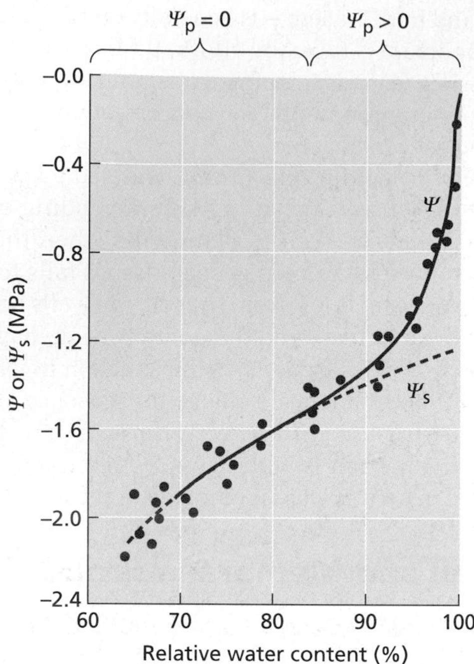

Figure 3.11 The relation between water potential (Ψ), solute potential (Ψ_s), and relative water content ($\Delta V/V$) in cotton (*Gossypium hirsutum*) leaves. Note that water potential (Ψ) decreases steeply with the initial decrease in relative water content. In comparison, osmotic potential (Ψ_s) changes little. As cell volume decreases below 90% in this example, the situation reverses: Most of the change in water potential is due to a drop in cell Ψ_s, accompanied by relatively little change in turgor pressure. (After Hsiao and Xu 2000.)

the relationship between Ψ_p and cell volume: ε is the change in Ψ_p for a given change in relative volume ($\varepsilon = \Delta\Psi_p/\Delta$[relative volume]). Cells with a large ε have stiff cell walls and thus experience larger changes in turgor pressure for the same change in cell volume than cells with a smaller ε and more elastic walls. The mechanical properties of cell walls vary among species and cell types, resulting in significant differences in the extent to which water deficits affect cell volume.

A comparison of the cell water relations within stems of cacti illustrates the important role of cell wall properties. Cacti are stem succulent plants, typically found in arid regions. Their stems consist of an outer, photosynthetic layer that surrounds nonphotosynthetic tissues that serve as a water storage reservoir (**Figure 3.12**). During drought, water is lost preferentially from these inner cells, despite the fact that the water potential of the two cell types remains in equilibrium (or very close to equilibrium). How does this happen?

Detailed studies of *Opuntia ficus-indica* demonstrate that the water storage cells are larger and have thinner walls than the photosynthetic cells, and are thus more flexible (have lower ε). For a given decrease in water potential, a water storage cell will lose a greater fraction of its water content than a photosynthetic cell. In addition, the solute concentration of the water storage cells decreases during drought, in part due to the polymerization of soluble sugars into insoluble starch granules. A more typical plant response to drought is to accumulate solutes, in part to prevent water loss from cells. However, in the case of cacti, the combination of more flexible cell walls and a decrease in solute concentration during drought allows water to be withdrawn preferentially from the water storage cells, thus helping maintain the hydration of the photosynthetic tissues.

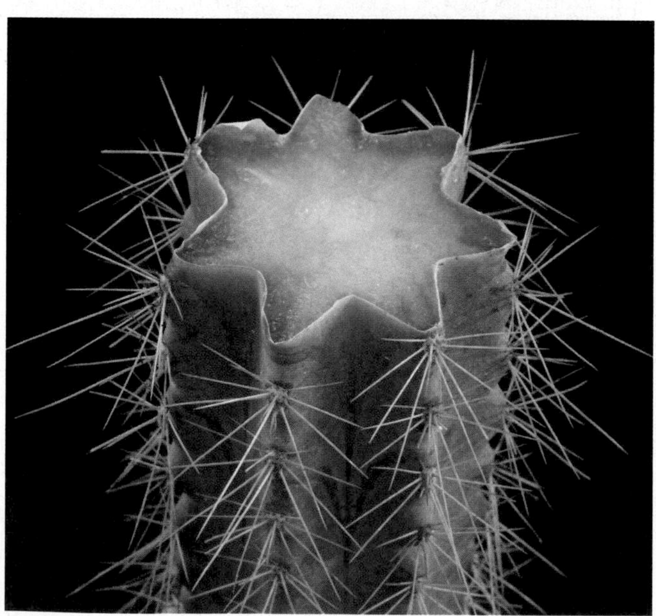

Figure 3.12 Cross section of a cactus stem, showing an outer, photosynthetic layer, and an inner, nonphotosynthetic tissue that functions in water storage. During drought, water is lost preferentially from nonphotosynthetic cells, so the water status of the photosynthetic tissue is maintained.

The rate at which cells gain or lose water is influenced by plasma membrane hydraulic conductivity

So far, we have seen that water moves into and out of cells in response to a water potential gradient. The direction of flow is determined by the direction of the Ψ gradient, and the rate of water movement is proportional to the magnitude of the driving gradient. However, for a cell that experiences a change in the water potential of its surroundings (e.g., see Figures 3.9 and 3.10), the movement of water across the plasma membrane will decrease with time as the internal and external water potentials converge

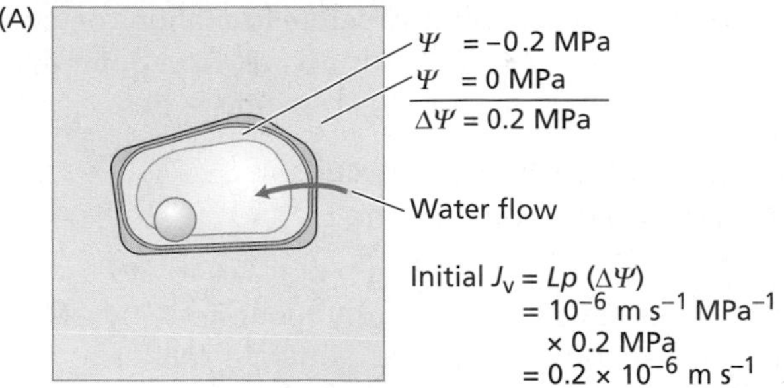

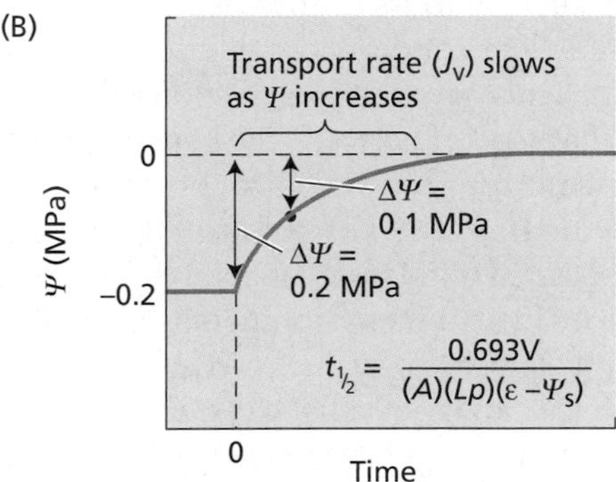

Figure 3.13 The rate of water transport into a cell depends on the magnitude of the water potential difference (ΔΨ) and the hydraulic conductivity of the plasma membranes (*Lp*). (A) In this example, the magnitude of the initial water potential difference is 0.2 MPa and *Lp* is 10^{-6} m s^{-1} MPa^{-1}. These values give an initial transport rate (J_v) of 0.2 × 10^{-6} m s^{-1}. (B) As water is taken up by the cell, the water potential difference decreases with time, leading to a slowing in the rate of water uptake. This effect follows an exponentially decaying time course with a half-time ($t_{1/2}$) that depends on the following cell parameters: volume (*V*), surface area (*A*), conductivity (*Lp*), volumetric elastic modulus (ε), and cell osmotic potential (Ψ_s).

(**Figure 3.13**). The rate approaches zero in an exponential manner. The time it takes for the rate to decline by half—its half-time, or $t_{1/2}$—is given by the following equation:

$$t_{1/2} = \left(\frac{0.693}{(A)(Lp)}\right)\left(\frac{V}{\varepsilon - \Psi_s}\right) \qquad (3.6)$$

where *V* and *A* are, respectively, the volume and surface of the cell, and *Lp* is the **hydraulic conductivity** of the plasma membrane. Hydraulic conductivity describes how readily water can move across a membrane; it is expressed in terms of volume of water per unit area of membrane per unit time per unit driving force (i.e., $m^3\ m^{-2}\ s^{-1}\ MPa^{-1}$). For additional discussion on hydraulic conductivity, see **WEB TOPIC 3.9**.

A short half-time means fast equilibration. Thus, cells with large surface-to-volume ratios, high membrane hydraulic conductivity, and stiff cell walls (large ε) will come rapidly into equilibrium with their surroundings. Cell half-times typically range from 1 to 10 s, although some are much shorter. Because of their short half-times, single cells come to water potential equilibrium with their surroundings in less than 1 min. For multicellular tissues, the half-times may be much longer.

Aquaporins facilitate the movement of water across plasma membranes

For many years, plant physiologists were uncertain about how water moves across plant membranes. Specifically, it was unclear whether water movement into plant cells was limited to the diffusion of water molecules across the plasma membrane's lipid bilayer or if it also involved diffusion through protein-lined pores (**Figure 3.14**). Some studies suggested that diffusion directly across the lipid bilayer was not sufficient to account for observed rates of water movement across membranes, but the evidence in support of microscopic pores was not compelling.

This uncertainty was put to rest in 1991 with the discovery of **aquaporins** (see Figure 3.14). Aquaporins are integral membrane proteins that form water-selective channels across the membrane. Because water diffuses much faster through such channels than through a lipid bilayer, aquaporins facilitate water movement into plant cells.

Although aquaporins may alter the *rate* of water movement across the membrane, they do not change the *direction of transport* or the *driving force* for water movement. However, aquaporins can be reversibly "gated" (i.e., transferred between an open and a closed state) in response to physiological parameters such as intercellular pH and Ca^{2+}

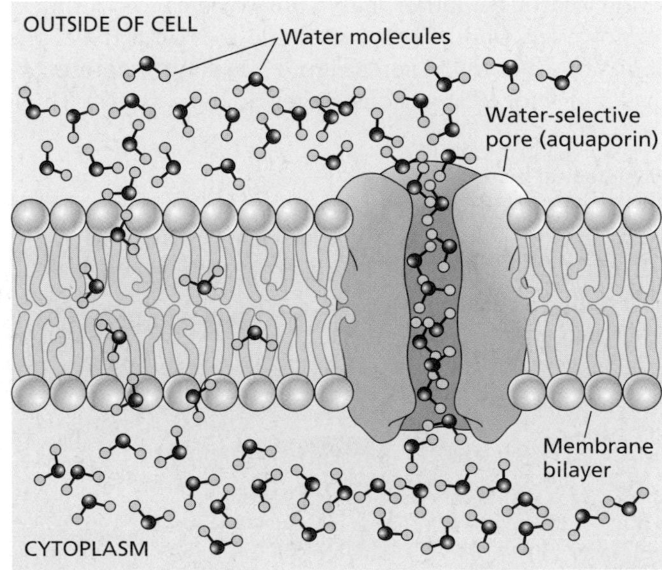

Figure 3.14 Water can cross plant membranes by diffusion of individual water molecules through the membrane bilayer, as shown on the left, and by the linear diffusion of water molecules through water-selective pores formed by integral membrane proteins such as aquaporins.

levels. As a result, plants have the ability to regulate the permeability of their plasma membranes to water.

Plant Water Status

The concept of water potential has two principal uses: First, water potential governs transport across plasma membranes, as we have described. Second, water potential is often used as a measure of the *water status* of a plant. In this section we will discuss how the concept of water potential helps us evaluate the water status of a plant.

Physiological processes are affected by plant water status

Because of transpirational water loss to the atmosphere, plants are seldom fully hydrated. During periods of drought, they suffer from water deficits that lead to inhibition of plant growth and photosynthesis. **Figure 3.15** lists some of the physiological changes that occur as plants experience increasingly drier conditions.

The sensitivity of any particular physiological process to water deficits is, to a large extent, a reflection of that plant's strategy for dealing with the range of water availability that it experiences in its environment. According to Figure 3.15, the process that is most affected by water deficit is cell expansion. In many plants, reductions in water supply inhibit shoot growth and leaf expansion but *stimulate* root elongation. A relative increase in roots relative to leaves is an appropriate response to reductions in water availability, and thus the sensitivity of shoot growth to decreases in water availability can be seen as an adaptation to drought rather than a physiological constraint.

However, what plants cannot do is to alter the availability of water in the soil. (Figure 3.15 shows representative values for Ψ at various stages of water stress.) Thus, drought does impose some absolute limitations on physiological processes, although the actual water potentials at which such limitations occur vary with species.

Solute accumulation helps cells maintain turgor and volume

The ability to maintain physiological activity as water becomes less available typically incurs some costs. The plant may spend energy to accumulate solutes to maintain turgor pressure, invest in the growth of nonphotosynthetic organs such as roots to increase water uptake capacity, or build xylem conduits capable of withstanding large tensions. Thus, physiological responses to water availability reflect a tradeoff between the benefits accrued by being able to carry out physiological processes (e.g., growth) over a wider range of environmental conditions and the costs associated with such capability.

Plants that grow in saline environments, called **halophytes**, typically have very low values of Ψ_s. A low Ψ_s lowers cell Ψ enough to allow root cells to extract water from saline water without allowing excessive levels of salts to enter at the same time. Plants may also exhibit quite negative Ψ_s under drought conditions. Water stress typically leads to an accumulation of solutes in the cytoplasm and vacuole of plant cells, thus allowing the cells to maintain turgor pressure despite low water potentials.

A positive turgor pressure ($\Psi_p > 0$) is important for several reasons. First, growth of plant cells requires turgor pressure to stretch the cell walls. The loss of turgor under water deficits can explain in part why cell growth is so sensitive to water stress, as well as why this sensitivity can be modified by varying the cell's osmotic potential (see Chapter 24). The second reason positive turgor is important is that turgor pressure increases the mechanical rigidity of cells and tissues.

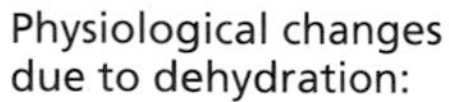

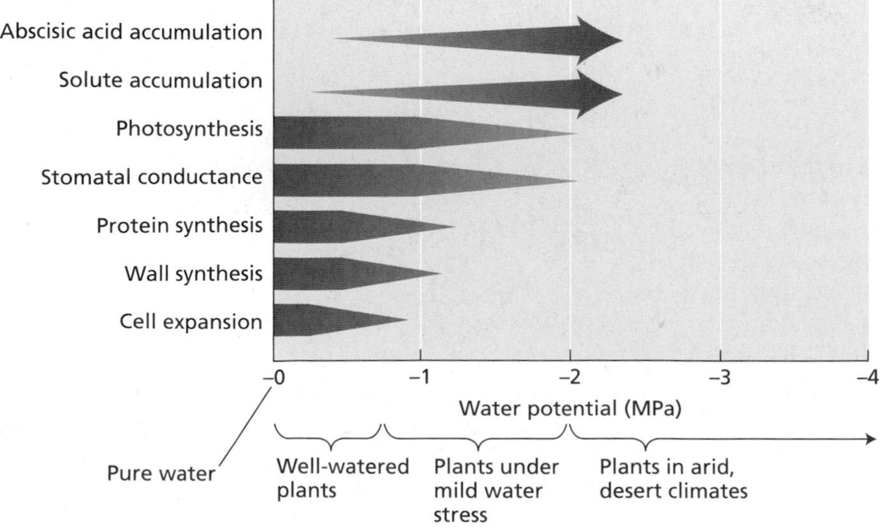

Figure 3.15 Sensitivity of various physiological processes to changes in water potential under various growing conditions. The thickness of the arrows corresponds to the magnitude of the process. For example, cell expansion decreases as water potential falls (becomes more negative). Abscisic acid is a hormone that induces stomatal closure during water stress (see Chapter 24). (After Hsiao and Acevedo 1974.)

Finally, although some physiological processes may be influenced directly by turgor pressure, it is likely that many more are affected by changes in cell volume. The existence of stretch-activated signaling molecules in the plasma membrane suggests that plant cells may sense changes in their water status via changes in volume, rather than by responding directly to turgor pressure.

SUMMARY

Photosynthesis exposes plants to water loss and the threat of dehydration. To prevent desiccation, water must be absorbed by the roots and transported through the plant body.

Water in Plant Life

- Cell walls allow plant cells to build up large internal hydrostatic pressures (turgor pressure). Turgor pressure is essential for many plant processes.
- Water limits both agricultural and natural ecosystem productivity (**Figures 3.1, 3.2**).
- About 97 percent of the water absorbed by roots is carried through the plant and is lost by transpiration from the leaf surfaces.
- The uptake of CO_2 is coupled to the loss of water through a common diffusional pathway.

The Structure and Properties of Water

- The polarity and tetrahedral shape of water molecules permit them to form hydrogen bonds that give water its unusual physical properties: it is an excellent solvent and has a high specific heat, an unusually high latent heat of vaporization, and a high tensile strength (**Figures 3.3, 3.6**).
- Cohesion, adhesion, and surface tension give rise to capillarity (**Figures 3.4, 3.5**).

Diffusion and Osmosis

- The random thermal motion of molecules results in diffusion (**Figures 3.7, 3.8**).
- Diffusion is important over short distances. The average time for a substance to diffuse a given distance increases as the square of that distance.
- Osmosis is the net movement of water across a selectively permeable barrier.

Water Potential

- Water's chemical potential measures the free energy of water in a given state.
- Concentration, pressure, and gravity contribute to water potential (Ψ) in plants.
- Ψ_s, the solute potential or osmotic potential, represents solutes' dilution of water and the reduction of the free energy of water.
- Ψ_p, the pressure potential represents the effect of hydrostatic pressure on the free energy of water. Positive pressure (turgor pressure) raises the water potential; negative pressure (tension) reduces it.
- The gravitational potential (Ψ_g) is generally omitted when calculating cell water potential. Thus, $\Psi = \Psi_s + \Psi_p$.

Water Potential of Plant Cells

- Plant cells typically have negative water potentials.
- Water enters or leaves a cell according to the water potential gradient.
- When a flaccid cell is placed in a solution that has a water potential greater (less negative) than the cell's water potential, water will move from the solution into the cell (from high to low water potential) (**Figure 3.9**).
- As water enters, the cell wall resists being stretched, increasing the turgor pressure (Ψ_p) of the cell.
- At equilibrium [$\Psi_{(cell)} = \Psi_{(solution)}$; $\Delta\Psi_w = 0$], the cell Ψ_p has increased sufficiently to raise the cell Ψ to the same value as the Ψ of the solution, and net water movement ceases.
- Water can also leave the cell by osmosis. When a turgid plant cell is placed in a sucrose solution that has a water potential more negative than the water potential of the cell, water will move from the turgid cell to the solution (**Figure 3.10**).
- If a cell is squeezed, its Ψ_p is raised, as is cell Ψ, resulting in a $\Delta\Psi$ such that water flows out of the cell (**Figure 3.10**).

Cell Wall and Membrane Properties

- Cell wall elasticity defines the relation between turgor pressure and cell volume, while the water permeability of the plasma membrane and tonoplast determines how fast cells exchange water with their surroundings.
- Because plant cells have fairly rigid walls, small changes in plant cell volume cause large changes in turgor pressure (**Figure 3.11**).

- For any non-zero initial $\Delta\Psi$, the net movement of water across the membrane will decrease with time as the internal and external water potentials converge (**Figure 3.13**).
- Aquaporins are water-selective membrane channels (**Figure 3.14**).

Plant Water Status

- During drought, photosynthesis and growth are inhibited, while concentrations of abscisic acid and solutes increase (**Figure 3.15**).
- During drought, plants must use energy to maintain turgor pressure by accumulating solutes, as well as to support root and vascular growth.
- Stretch-activated signaling molecules in the plasma membrane may permit plant cells to sense changes in their water status via changes in volume.

WEB MATERIAL

- **WEB TOPIC 3.1 Calculating Capillary Rise** Quantification of capillary rise allows us to assess its functional role in water movement of plants.
- **WEB TOPIC 3.2 Calculating Half-Times of Diffusion** The assessment of the time needed for a molecule such as glucose to diffuse across cells, tissues, and organs shows that diffusion has physiological significance only over short distances.
- **WEB TOPIC 3.3 Alternative Conventions for Components of Water Potential** Plant physiologists have developed several conventions to define water potential of plants. A comparison of key definitions in some of these convention systems provides us with a better understanding of the water relations literature.
- **WEB TOPIC 3.4 Temperature and Water Potential** Variation in temperature between 0°C and 30°C has a relatively minor effect on osmotic potential.
- **WEB TOPIC 3.5 Can Negative Turgor Pressures Exist in Living Cells?** It is assumed that Ψ_p is 0 or greater in living cells; is this true for living cells with lignified walls?
- **WEB TOPIC 3.6 Measuring Water Potential** Several methods are available to measure water potential in plant cells and tissues.
- **WEB TOPIC 3.7 The Matric Potential** Matric potential is used to quantify the chemical potential of water in soils, seeds, and cell walls.
- **WEB TOPIC 3.8 Wilting and Plasmolysis** Plasmolysis is a major structural change resulting from major water loss by osmosis.
- **WEB TOPIC 3.9 Understanding Hydraulic Conductivity** Hydraulic conductivity, a measurement of the membrane permeability to water, is one of the factors determining the velocity of water movements in plants.

available at plantphys.net

Suggested Reading

Bartlett, M. K., Scoffoni, C., and Sack, L. (2012) The determinants of leaf turgor loss point and prediction of drought tolerance of species and biomes: A global meta-analysis. *Ecol. Lett.* 15: 393–405.

Chaumont, F., and Tyerman, S. D. (2014) Aquaporins: Highly regulated channels controlling plant water relations. *Plant Physiol.* 164: 1600–1618.

Goldstein, G., Ortega, J. K. E., Nerd, A., and Nobel, P. S. (1991) Diel patterns of water potential components for the crassulacean acid metabolism plant *Opuntia ficus-indica* when well-watered or droughted. *Plant Physiol.* 95: 274–280.

Kramer, P. J., and Boyer, J. S. (1995) *Water Relations of Plants and Soils.* Academic Press, San Diego.

Maurel, C., Verdoucq, L., Luu, D.-T., and Santoni, V. (2008) Plant aquaporins: Membrane channels with multiple integrated functions. *Annu. Rev. Plant Biol.* 59: 595–624.

Munns, R. (2002) Comparative physiology of salt and water stress. *Plant Cell Environ.* 25: 239–250.

Nobel, P. S. (1999) *Physicochemical and Environmental Plant Physiology.* 2nd ed. Academic Press, San Diego.

Tardieu, F., Parent, B., Caldeira, C. F., and Welcker, C. (2014) Genetic and physiological controls of growth under water deficit. *Plant Physiol.* 164: 1628–1635.

Wheeler, T. D., and Stroock, A. D. (2008) The transpiration of water at negative pressures in a synthetic tree. *Nature* 455: 208–212.

4 Water Balance of Plants

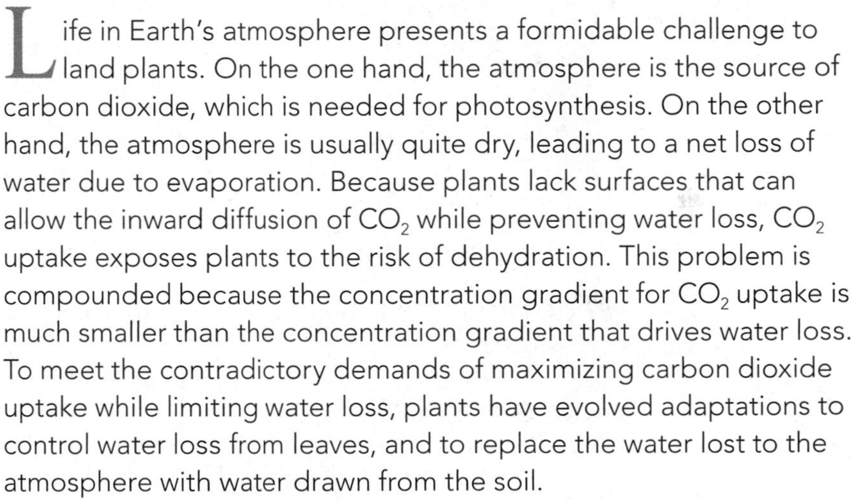

Life in Earth's atmosphere presents a formidable challenge to land plants. On the one hand, the atmosphere is the source of carbon dioxide, which is needed for photosynthesis. On the other hand, the atmosphere is usually quite dry, leading to a net loss of water due to evaporation. Because plants lack surfaces that can allow the inward diffusion of CO_2 while preventing water loss, CO_2 uptake exposes plants to the risk of dehydration. This problem is compounded because the concentration gradient for CO_2 uptake is much smaller than the concentration gradient that drives water loss. To meet the contradictory demands of maximizing carbon dioxide uptake while limiting water loss, plants have evolved adaptations to control water loss from leaves, and to replace the water lost to the atmosphere with water drawn from the soil.

In this chapter we will examine the mechanisms and driving forces operating on water transport within the plant and between the plant and its environment. We will begin our examination of water transport by focusing on water in the soil.

Water in the Soil

The water content and the rate of water movement in soils depend to a large extent on soil type and soil structure. At one extreme is sand, in which the soil particles may be 1 mm or more in diameter. Sandy soils have a relatively low surface area per gram of soil and have large spaces or channels between particles.

At the other extreme is clay, in which particles are smaller than 2 μm in diameter. Clay soils have much greater surface areas and smaller channels between particles. With the aid of organic substances such as humus (decomposing organic matter), clay particles may aggregate into "crumbs," allowing large channels to form that help improve soil aeration and infiltration of water.

When a soil is heavily watered by rain or by irrigation (see **WEB TOPIC 4.1**), the water percolates downward by gravity through the spaces between soil particles, partly displacing, and in some cases trapping, air in these channels. Because water is pulled into the spaces between soil particles by capillarity, the smaller channels

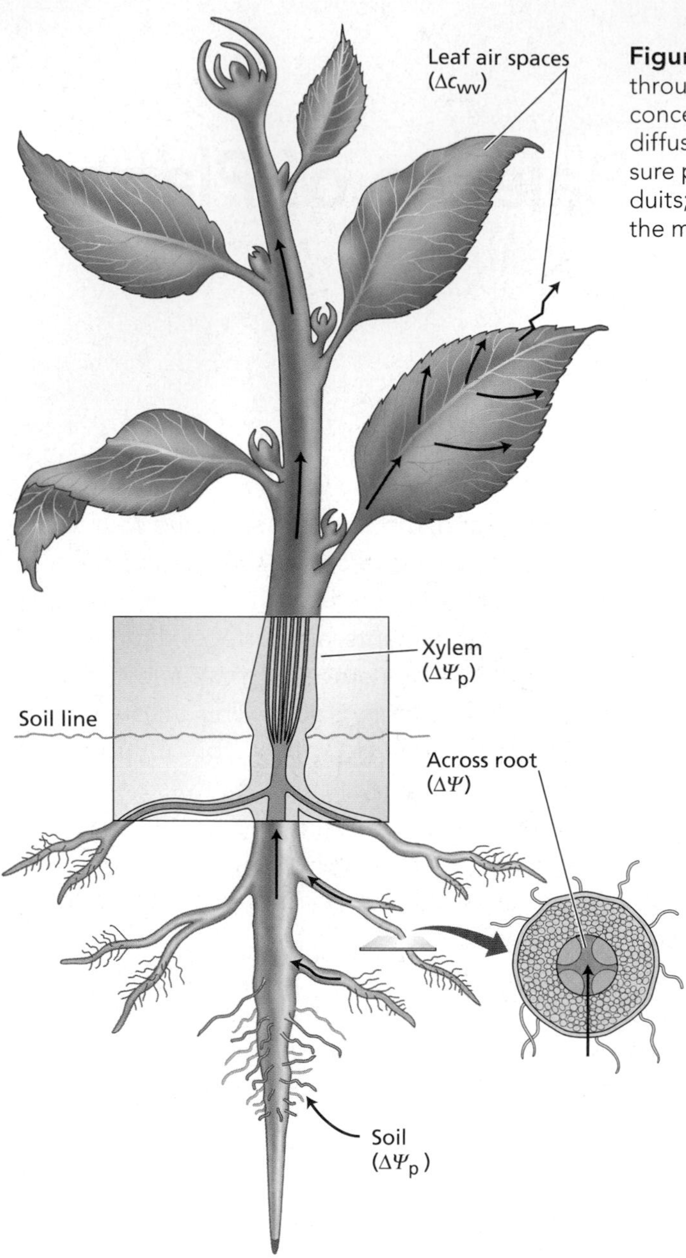

Figure 4.1 Main driving forces for water flow from the soil through the plant to the atmosphere: differences in water vapor concentration (Δc_{wv}) between leaf and air are responsible for the diffusion of water vapor from the leaf to the air; differences in pressure potential ($\Delta\Psi_p$) drive the bulk flow of water through xylem conduits; and differences in water potential ($\Delta\Psi$) are responsible for the movement of water across the living cells in the root.

become filled first. Depending on the amount of water available, water in the soil may exist as a film adhering to the surface of soil particles, it may fill the smaller but not the larger channels, or it may fill all of the spaces between particles.

In sandy soils, the spaces between particles are so large that water tends to drain from them and remain only on the particle surfaces and in the spaces where particles come into contact. In clay soils, the spaces between particles are so small that much water is retained against the force of gravity. A few days after a soaking rainfall, a clay soil might retain 40% water by volume. In contrast, sandy soils typically retain only about 15% water by volume after thorough wetting.

In the following sections we will examine how the physical structure of the soil influences soil water potential, how water moves in the soil, and how roots absorb the water needed by the plant.

A negative hydrostatic pressure in soil water lowers soil water potential

Like the water potential of plant cells, the water potential of soils may be dissected into three components: the osmotic potential, the pressure potential, and the gravitational potential. The **osmotic potential** (Ψ_s; see Chapter 3) of soil water is generally negligible, because except in saline soils, solute concentrations are low; a typical value might be –0.02 MPa. In soils that contain a substantial concentration of salts, however, Ψ_s can be significant, perhaps –0.2 MPa or lower.

The second component of soil water potential is the **pressure potential** (Ψ_p) (**Figure 4.1**). For wet soils, Ψ_p is very close to zero. As soil dries out, Ψ_p decreases and can become quite negative. Where does the negative pressure potential in soil water come from?

Recall from our discussion of capillarity in Chapter 3 that water has a high surface tension that tends to minimize air–water interfaces. However, because of adhesive forces, water also tends to cling to the surfaces of soil particles (**Figure 4.2**).

As the water content of the soil decreases, the water recedes into the channels between soil particles, forming air–water surfaces whose curvature represents the balance between the tendency to minimize the surface area of the air–water interface and the attraction of the water for the soil particles. Water under a curved surface develops a negative pressure that may be estimated by the following formula:

$$\Psi_p = \frac{-2T}{r} \tag{4.1}$$

where T is the surface tension of water (7.28×10^{-8} MPa m) and r is the radius of curvature of the air–water interface. Note that this is the same capillarity equation discussed in **WEB TOPIC 3.1** (see also Figure 3.5), where here the soil particles are assumed to be fully wettable (contact angle $\theta = 0$; $\cos\theta = 1$).

As soil dries out, water is first removed from the largest spaces between soil particles and subsequently from successively smaller spaces between and within soil particles. In this process, the value of Ψ_p in soil water can become quite negative due to the increasing curvature of air–water surfaces in pores of successively smaller diam-

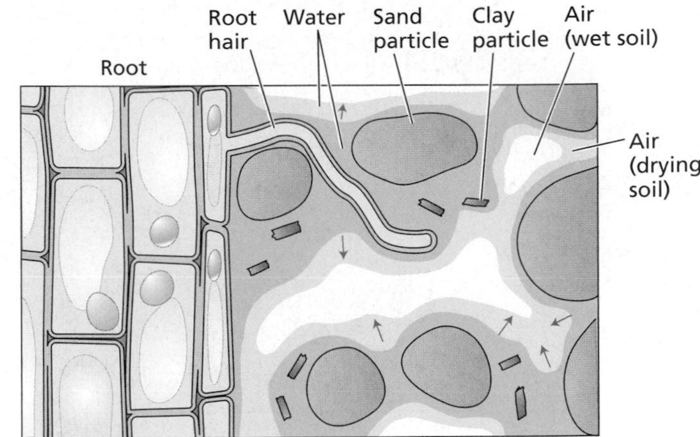

Figure 4.2 Root hairs make intimate contact with soil particles and greatly amplify the surface area used for water absorption by the plant. The soil is a mixture of particles (sand, clay, silt, and organic material), water, dissolved solutes, and air. Water is adsorbed to the surface of the soil particles. As water is absorbed by the plant, the soil solution recedes into smaller pockets, channels, and crevices between the soil particles. At the air–water interfaces, this recession causes the surface of the soil solution to develop concave menisci (curved interfaces between air and water, marked in the figure by arrows), and brings the solution into tension (negative pressure) by surface tension. As more water is removed from the soil, the curvature of the air–water menisci increases, resulting in greater tensions (more negative pressures).

eter. For instance, a curvature of $r = 1$ µm (about the size of the largest clay particles) corresponds to a Ψ_p value of -0.15 MPa. The value of Ψ_p may easily reach -1 to -2 MPa as the air–water interface recedes into the smaller spaces between clay particles.

The third component is **gravitational potential** (Ψ_g). Gravity plays an important role in drainage. The downward movement of water is due to the fact that Ψ_g is proportional to elevation: higher at higher elevations, and vice versa.

Water moves through the soil by bulk flow

Bulk or mass flow is the concerted movement of molecules en masse, most often in response to a pressure gradient. Common examples of bulk flow are water moving through a garden hose or down a river. The movement of water through soils is predominantly by bulk flow.

Because the pressure in soil water is due to the existence of curved air–water interfaces, water flows from regions of higher soil water content, where the water-filled spaces are larger and thus Ψ_p is less negative, to regions of lower soil water content, where the smaller size of the water-filled spaces is associated with more curved air–water interfaces and a more negative Ψ_p. Diffusion of water vapor also accounts for some water movement, which can be important in dry soils.

As plants absorb water from the soil, they deplete the soil of water near the surface of the roots. This depletion reduces Ψ_p near the root surface and establishes a pressure gradient with respect to neighboring regions of soil that have higher Ψ_p values. Because the water-filled pore spaces in the soil are interconnected, water moves down the pressure gradient to the root surface by bulk flow through these channels.

The rate of water flow in soils depends on two factors: the size of the pressure gradient through the soil, and the hydraulic conductivity of the soil. **Soil hydraulic conductivity** is a measure of the ease with which water moves through the soil, and it varies with the type of soil and its water content. Sandy soils, which have large spaces between particles, have a large hydraulic conductivity when saturated, whereas clay soils, with only minute spaces between their particles, have an appreciably smaller hydraulic conductivity.

As the water content (and hence the water potential) of a soil decreases, the hydraulic conductivity decreases dramatically. This decrease in soil hydraulic conductivity is due primarily to the replacement of water in the soil by air. When air moves into a soil channel previously filled with water, water movement through that channel is restricted to the periphery of the channel. As more of the soil spaces become filled with air, water flow is limited to fewer and narrower channels, and the hydraulic conductivity falls. (WEB TOPIC 4.2 shows how soil texture influences both the water-holding capacity of soils and their hydraulic conductivity.)

Water Absorption by Roots

Contact between the surface of the root and the soil is essential for effective water absorption by the root. This contact provides the surface area needed for water uptake and is maximized by the growth of the root and of root hairs into the soil. **Root hairs** are filamentous outgrowths of root epidermal cells that greatly increase the surface area of the root, thus providing greater capacity for absorption of ions and water from the soil. When 3-month-old wheat plants were examined, their root hairs were found to constitute more than 60% of the surface area of the roots (see Figure 5.7).

Water enters the root most readily near the root tip. Mature regions of the root are less permeable to water because they have developed a modified epidermal layer that contains hydrophobic materials in its walls. Although it might at first seem counterintuitive that any portion of the root system should be impermeable to water, the older regions of the root must be sealed off if there is to be water uptake (and thus bulk flow of nutrients) from the regions of the root system that are actively exploring new areas in the soil (**Figure 4.3**).

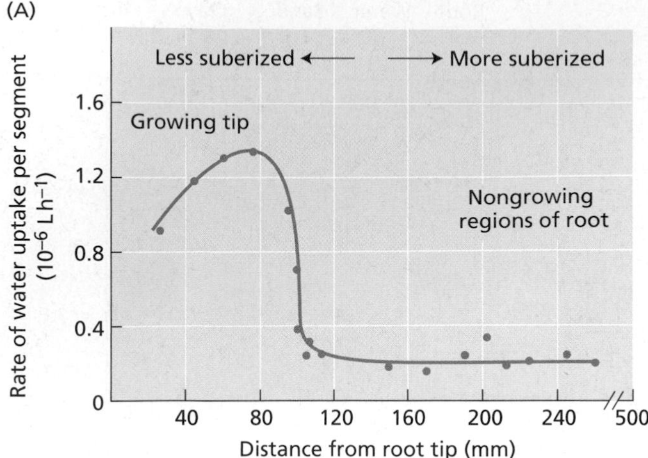

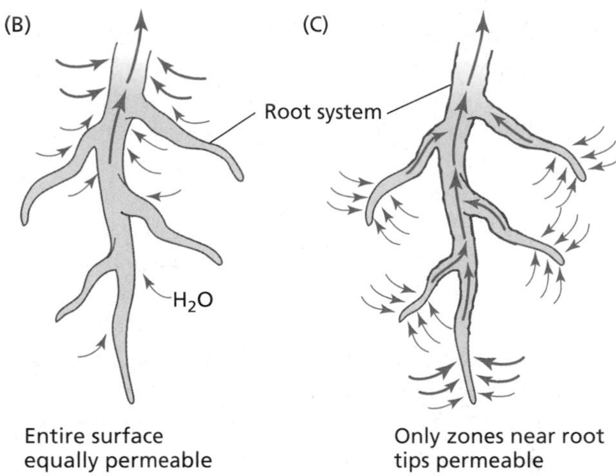

Figure 4.3 Rate of water uptake by short segments (3–5 mm) at various positions along an intact pumpkin (*Cucurbita pepo*) root (A). Diagram of water uptake in which the entire root surface is equally permeable (B) or is impermeable in older regions due to the deposition of suberin, a hydrophobic polymer (C). When root surfaces are equally permeable, most of the water enters near the top of the root system, with more distal regions being hydraulically isolated as the suction in the xylem is relieved due to the inflow of water. Decreasing the permeability of older regions of the root allows xylem tensions to extend further into the root system, allowing water uptake from distal regions of the root system. (A after Kramer and Boyer 1995.)

The contact between the soil and the root surface is easily ruptured when the soil is disturbed. It is for this reason that newly transplanted seedlings and plants need to be protected from water loss for the first few days after transplantation. Thereafter, new root growth into the soil reestablishes soil–root contact, and the plant can better withstand water stress.

Let's consider how water moves within the root, and the factors that determine the rate of water uptake into the root.

Water moves in the root via the apoplast, symplast, and transmembrane pathways

In the soil, water flows between soil particles. However, from the epidermis to the endodermis of the root, there are three pathways through which water can flow (**Figure 4.4**): the apoplast, the symplast, and the transmembrane pathway.

1. The apoplast is the continuous system of cell walls, intercellular air spaces, and the lumens of nonliving cells (e.g., xylem conduits and fibers). In this pathway, water moves through cell walls and extracellular spaces without crossing any membranes as it travels across the root cortex.
2. The symplast consists of the entire network of cell cytoplasm interconnected by plasmodesmata. In this pathway, water travels across the root cortex via the plasmodesmata (see Chapter 1).
3. The transmembrane pathway is the route by which water enters a cell on one side, exits the cell on the other side, enters the next in the series, and so on. In this pathway, water crosses the plasma membrane of each cell in its path twice (once on entering and once on exiting). Transport across the tonoplast may also be involved.

Although the relative importance of the apoplast, symplast, and transmembrane pathways has not yet been fully established, experiments with the pressure probe technique (see WEB TOPIC 3.6) indicate an important role for plasma membranes, and thus the transmembrane pathway, in the movement of water across the root cortex. And though we can define three pathways, it is important to remember that water moves not according to a single chosen path, but wherever the gradients and resistances direct it. A particular water molecule moving in the symplast may cross the membrane and move in the apoplast for a moment, and then move back into the symplast again.

At the endodermis, water movement through the apoplast pathway is obstructed by the Casparian strip (see Figure 4.4). The **Casparian strip** is a band within the radial cell walls of the endodermis that is impregnated with suberin and/or **lignin**, hydrophobic polymers. The Casparian strip forms in the nongrowing part of the root, several millimeters to several centimeters behind the root tip, at about the same time that the first xylem elements mature. The Casparian strip breaks the continuity of the apoplast pathway, forcing water and solutes to pass through the plasma membrane in order to cross the endodermis.

The requirement that water move symplastically across the endodermis helps explain why the permeability of roots to water depends strongly on the presence of aquaporins. Down-regulating the expression of aquaporin genes markedly reduces the hydraulic conductivity of

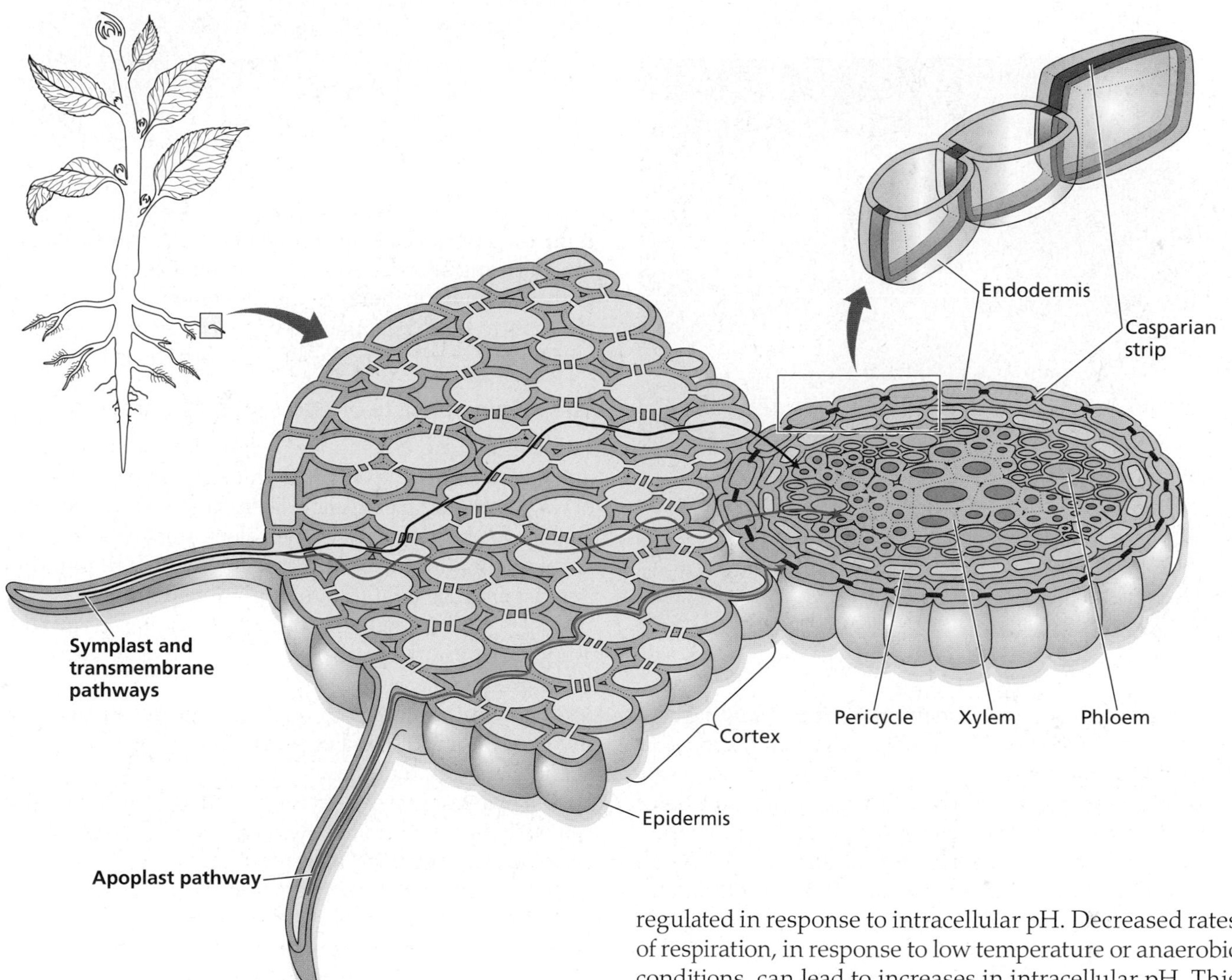

Figure 4.4 Pathways for water uptake by the root. Through the cortex, water may travel via the apoplast pathway, the transmembrane pathway, and the symplast pathway. In the symplast pathway, water flows between cells through the plasmodesmata without crossing the plasma membrane. In the transmembrane pathway, water moves across the plasma membranes, with a short visit to the cell wall space. At the endodermis, the apoplast pathway is blocked by the Casparian strip. Note that while these are drawn as three distinct pathways, in actuality water molecules will move between the symplast and apoplast as directed by gradients in water potential and hydraulic resistances.

roots and can result in plants that wilt easily or that compensate by producing larger root systems.

Water uptake decreases when roots are subjected to low temperature or anaerobic conditions, or treated with respiratory inhibitors. Until recently, there was no explanation for the connection between root respiration and water uptake, or for the enigmatic wilting of flooded plants. We now know that the permeability of aquaporins can be regulated in response to intracellular pH. Decreased rates of respiration, in response to low temperature or anaerobic conditions, can lead to increases in intracellular pH. This increase in cytosolic pH alters the conductance of aquaporins in root cells, resulting in roots that are markedly less permeable to water. Thus, maintaining membrane permeability to water requires energy expenditure by root cells that is supplied by respiration.

Solute accumulation in the xylem can generate "root pressure"

Plants sometimes exhibit a phenomenon referred to as **root pressure**. For example, if the stem of a young seedling is cut off just above the soil, the stump will often exude sap from the cut xylem for many hours. If a manometer is sealed over the stump, positive pressures as high as 0.2 MPa (and sometimes even higher) can be measured.

When transpiration is low or absent, positive hydrostatic pressure builds up in the xylem because roots continue to absorb ions from the soil and transport them into the xylem. The buildup of solutes in the xylem sap leads to a decrease in the xylem osmotic potential (Ψ_s) and thus a decrease in the xylem water potential (Ψ). This lowering of the xylem Ψ provides a driving force for water absorption, which in turn leads to a positive hydrostatic pressure in the xylem. In effect, the multicellular root tis-

Figure 4.5 Guttation in a leaf from lady's mantle (*Alchemilla vulgaris*). In the early morning, leaves secrete water droplets through the hydathodes, located at the margins of the leaves.

sue behaves as an osmotic membrane does, building up a positive hydrostatic pressure in the xylem in response to the accumulation of solutes.

Root pressure is most likely to occur when soil water potentials are high and transpiration rates are low. As transpiration rates increase, water is transported through the plant and lost to the atmosphere so rapidly that a positive pressure resulting from ion uptake never develops in the xylem.

Plants that develop root pressure frequently produce liquid droplets on the edges of their leaves, a phenomenon known as **guttation** (**Figure 4.5**). Positive xylem pressure causes exudation of xylem sap through specialized pores called *hydathodes* that are associated with vein endings at the leaf margin. The "dewdrops" that can be seen on the tips of grass leaves in the morning are actually guttation droplets exuded from hydathodes. Guttation is most noticeable when transpiration is suppressed and the relative humidity is high, such as at night. It is possible that root pressure reflects an unavoidable consequence of high rates of ion accumulation. However, the existence of positive pressures within the xylem at night can help dissolve gas bubbles, and thus play a role in reversing the deleterious effects of cavitation described in the next section.

Water Transport through the Xylem

In most plants, the xylem constitutes the longest part of the pathway of water transport. In a plant 1 m tall, more than 99.5% of the water transport pathway through the plant is within the xylem, and in tall trees the xylem represents an even higher percentage of the pathway. Compared with the movement of water through layers of living cells, the xylem is a simple pathway of low resistivity. In the following sections we will examine how the structure of the xylem contributes to the movement of water from the roots to the leaves, and how negative pressures generated by transpiration pull water through the xylem.

The xylem consists of two types of transport cells

The conducting cells in the xylem have a specialized anatomy that enables them to transport large quantities of water with great efficiency. There are two main types of water-transporting cells in the xylem: tracheids and vessel elements (**Figure 4.6**). Vessel elements are found in angiosperms, a small group of gymnosperms called the Gnetales, and some ferns. Tracheids are present in both angiosperms and gymnosperms, as well as in ferns and other groups of vascular plants.

The maturation of both tracheids and vessel elements involves the production of secondary cell walls and the subsequent death of the cell—the loss of the cytoplasm and all of its contents. What remain are the thick, lignified cell walls, which form hollow tubes through which water can flow with relatively little resistance.

Tracheids are elongated, spindle-shaped cells (see Figure 4.6A) that are arranged in overlapping vertical files (**Figure 4.7**). Water flows between tracheids by means of the numerous **pits** in their lateral walls (see Figure 4.6B). Pits are microscopic regions where the secondary wall is absent and only the primary wall is present (see Figure 4.6C). Pits of one tracheid are typically located opposite pits of an adjoining tracheid, forming **pit pairs**. Pit pairs constitute a low-resistance path for water movement between tracheids. The water-permeable layer between pit pairs, consisting of two primary walls and a middle lamella, is called the **pit membrane**.

Pit membranes in tracheids of conifers have a central thickening, called a **torus** (plural *tori*), surrounded by a porous and relatively flexible region known as the **margo** (see Figure 4.6C). The torus acts like a valve: When it is centered in the pit cavity, the pit remains open; when it is lodged in the circular or oval wall thickenings bordering the pit, the pit is closed. Such lodging of the torus effectively prevents gas bubbles from spreading into neighboring tracheids (we will discuss this formation of bubbles, a process called cavitation, shortly). With very few exceptions, the pit membranes in all other plants, whether in tracheids or vessel elements, lack tori. But because the water-filled pores in the pit membranes of nonconifers are very small, they also serve as an effective barrier against the movement of gas bubbles. Thus, pit membranes of

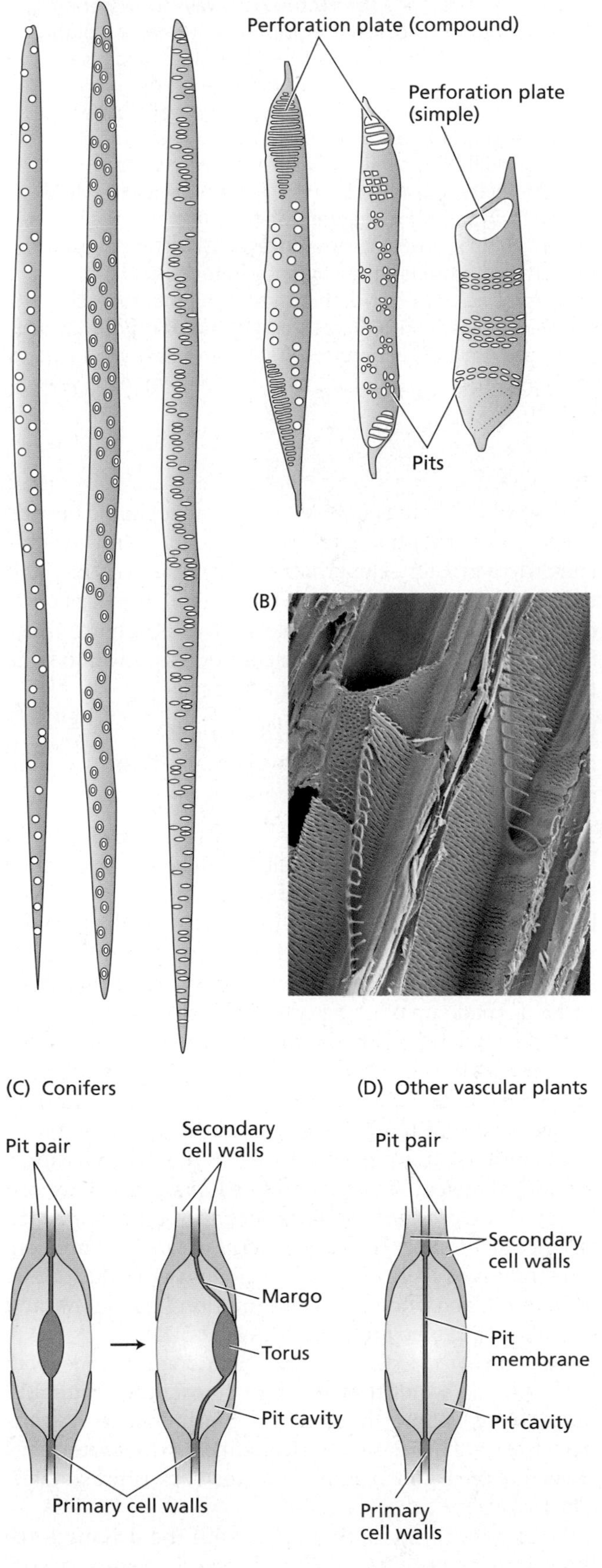

Figure 4.6 Xylem conduits and their interconnections. (A) Structural comparison of tracheids and vessel elements. Tracheids are elongated, hollow, dead cells with highly lignified walls. The walls contain numerous pits—regions where secondary wall is absent but primary wall remains. The shapes of pits and the patterns of wall pitting vary with species and organ type. Tracheids are present in all vascular plants. Vessels consist of a stack of two or more vessel elements. Like tracheids, vessel elements are dead cells and are connected to one another by perforation plates—regions of the wall where pores or holes have developed. Vessels are connected to other vessels and to tracheids through pits. Vessels are found in most angiosperms and are lacking in most gymnosperms. (B) Scanning electron micrograph showing two vessels (running diagonally from lower left to upper right). Pits are visible on the side walls, as are the scalariform end walls between vessel elements. (200×) (C) Diagram of a coniferous bordered pit with the torus centered in the pit cavity (left) or lodged to one side of the cavity (right). When the pressure difference between two tracheids is small, the pit membrane lies close to the center of the bordered pit, allowing water to flow through the porous margo region of the pit membrane; when the pressure difference between two tracheids is large, such as when one has cavitated and the other remains filled with water under tension, the pit membrane is displaced such that the torus becomes lodged against the overarching walls, thereby preventing the embolism from propagating between tracheids. (D) In contrast, the pit membranes of angiosperms and other nonconiferous vascular plants are relatively homogeneous in their structure. These pit membranes have very small pores compared with those of conifers, which prevents the spread of embolism but also imparts a significant hydraulic resistance. (C after Zimmermann 1983.)

both types play an important role in preventing the spread of gas bubbles, called *emboli,* within the xylem.

Vessel elements tend to be shorter and wider than tracheids and have perforated end walls that form a **perforation plate** at each end of the cell. Like tracheids, vessel elements have pits on their lateral walls (see Figure 4.6B). Unlike in tracheids, the perforated end walls allow vessel elements to be stacked end to end to form a much longer conduit called a **vessel** (see Figure 4.7). Vessels are multicellular conduits that vary in length both within and among species. Vessels range from a few centimeters in length to many meters. The vessel elements found at the extreme ends of a vessel lack perforations in their end walls and are connected to neighboring vessels via pits.

Water moves through the xylem by pressure-driven bulk flow

Pressure-driven bulk flow of water is responsible for long-distance transport of water in the xylem. It also accounts for much of the water flow through the soil and through the cell walls of plant tissues. In contrast to the diffusion of

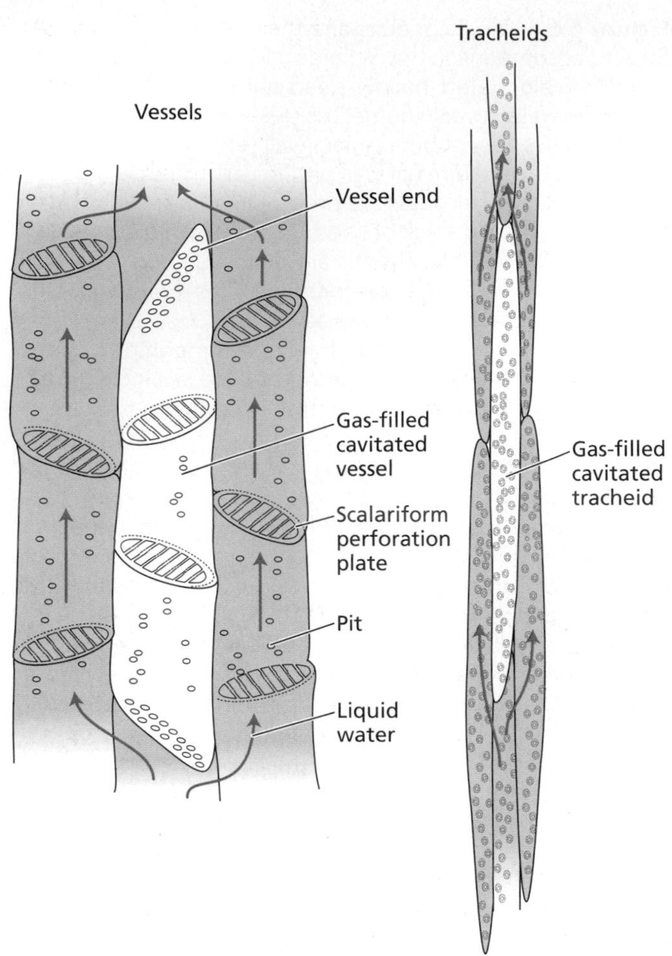

Figure 4.7 Vessels (left) and tracheids (right) form a series of parallel, interconnected pathways for water movement. Cavitation blocks water movement because of the formation of gas-filled (embolized) conduits. Because xylem conduits are interconnected through openings ("bordered pits") in their thick secondary walls, water can detour around the blocked vessel by moving through adjacent conduits. The very small pores in the pit membranes help prevent embolisms from spreading between xylem conduits. Thus, in the diagram on the right, the gas is contained within a single cavitated tracheid. In the diagram on the left, gas has filled the entire cavitated vessel, shown here as being made up of three vessel elements, each separated by scalariform (resembling the rungs of a ladder) perforation plates. In nature, vessels can be very long (up to several meters in length) and thus made up of many vessel elements.

water across semipermeable membranes, pressure-driven bulk flow is independent of solute concentration gradients, as long as viscosity changes are negligible.

If we consider bulk flow through a tube, the rate of flow depends on the radius (r) of the tube, the viscosity (η) of the liquid, and the pressure gradient ($\Delta\Psi_p/\Delta x$) that drives the flow. Jean Léonard Marie Poiseuille (1797–1869) was a French physician and physiologist, and the relation just described is given by one form of Poiseuille's equation:

$$\text{Volume flow rate} = \left(\frac{\pi r^4}{8\eta}\right)\left(\frac{\Delta\Psi_p}{\Delta x}\right) \tag{4.2}$$

expressed in cubic meters per second ($m^3\ s^{-1}$). This equation tells us that pressure-driven bulk flow is extremely sensitive to the radius of the tube. If the radius is doubled, the volume flow rate increases by a factor of 16 (2^4). Vessel elements up to 500 μm in diameter, nearly an order of magnitude greater than the largest tracheids, occur in the stems of climbing species. These large-diameter vessels permit vines to transport large amounts of water despite the slenderness of their stems.

Equation 4.2 describes water flow through a cylindrical tube and thus does not take into account the fact that xylem conduits are of finite length, such that water must cross many pit membranes as it flows from the soil to the leaves. All else being equal, pit membranes should impede water flow through single-celled (and thus shorter) tracheids to a greater extent than through multicellular (and thus longer) vessels. However, the pit membranes of conifers are much more permeable to water than are those found in other plants, allowing conifers to grow into large trees despite producing only tracheids.

Water movement through the xylem requires a smaller pressure gradient than movement through living cells

The xylem provides a pathway of low resistivity for water movement. Some numerical values will help us appreciate the extraordinary efficiency of the xylem. We will calculate the driving force required to move water through the xylem at a typical velocity and compare it with the driving force that would be needed to move water through a pathway made up of living cells at the same rate.

For the purposes of this comparison, we will use a value of 4 mm s^{-1} for the xylem transport velocity and 40 μm as the vessel radius. This is a high velocity for such a narrow vessel, so it will tend to exaggerate the pressure gradient required to support water flow in the xylem. Using a version of Poiseuille's equation (see Equation 4.2), we can calculate the pressure gradient needed to move water at a velocity of 4 mm s^{-1} through an *ideal* tube with a uniform inner radius of 40 μm. The calculation gives a value of 0.02 MPa m^{-1}. Elaboration of the assumptions, equations, and calculations can be found in **WEB TOPIC 4.3**.

Of course, *real* xylem conduits have irregular inner wall surfaces, and water flow through perforation plates and pits adds resistance to water transport. Such deviations from the ideal increase the frictional drag: Measurements show that the actual resistance is greater by approximately a factor of 2.

Let's now compare this value with the driving force that would be necessary to move water at the same velocity from cell to cell, crossing the plasma membrane each time. As calculated in **WEB TOPIC 4.3**, the driving force needed to move water through a layer of cells at 4 mm s^{-1} is 2×10^8 MPa m^{-1}. This is ten orders of magnitude greater than the driving force needed to move water through our 40-μm-radius xylem vessel. Our calculation clearly shows that water flow through the xylem is vastly more efficient than water flow across living cells. Nevertheless, the xylem can make a significant contribution to the total resistance to water flow through the plant.

What pressure difference is needed to lift water 100 meters to a treetop?

With the foregoing example in mind, let's see what pressure gradient is needed to move water up to the top of a very tall tree. The tallest trees in the world are the coast redwoods (*Sequoia sempervirens*) of North America and the mountain ash (*Eucalyptus regnans*) of Australia. Individuals of both species can exceed 100 m.

If we think of the stem of a tree as a long pipe, we can estimate the pressure difference that is needed to overcome the frictional drag of moving water from the soil to the top of the tree by multiplying the pressure gradient needed to move the water by the height of the tree. The pressure gradients needed to move water through the xylem of very tall trees are on the order of 0.01 MPa m^{-1}, smaller than in our previous example. If we multiply this pressure gradient by the height of the tree (0.01 MPa m^{-1} × 100 m), we find that the total pressure difference needed to overcome the frictional resistance to water movement through the stem is equal to 1 MPa.

In addition to frictional resistance, we must consider gravity. As described by Equation 3.4, for a height difference of 100 m, the difference in Ψ_g is approximately 1 MPa. That is, Ψ_g is 1 MPa higher at the top of the tree than at the ground level. So the other components of water potential must be 1 MPa more negative at the top of the tree to counter the effects of gravity.

To allow transpiration to occur, the pressure gradient due to gravity must be added to that required to cause water movement through the xylem. Thus, we calculate that a pressure difference of roughly 2 MPa, from the base to the top branches, is needed to carry water up the tallest trees.

The cohesion–tension theory explains water transport in the xylem

In theory, the pressure gradients needed to move water through the xylem could result from the generation of positive pressures at the base of the plant or negative pressures at the top of the plant. We mentioned previously that some roots can develop positive hydrostatic pressure in their xylem. However, root pressure is typically less than 0.1 MPa and disappears with transpiration or when soils are dry, so it is clearly inadequate to move water up a tall tree. Furthermore, because root pressure is generated by the accumulation of ions in the xylem, relying on this for transporting water would require a mechanism for dealing with these solutes once the water evaporates from the leaves.

Instead, the water at the top of a tree develops a large tension (a negative hydrostatic pressure), and this tension *pulls* water through the xylem. This mechanism, first proposed toward the end of the nineteenth century, is called the *cohesion–tension theory of sap ascent* because it requires the cohesive properties of water to sustain large tensions in the xylem water columns. One can readily demonstrate xylem tension by puncturing intact xylem through a drop of ink on the surface of a stem from a transpiring plant. When the tension in the xylem is relieved, the ink is drawn instantly into the xylem, resulting in visible streaks along the stem.

The xylem tensions needed to pull water from the soil develop in leaves as a consequence of transpiration. How does the loss of water vapor through open stomata result in the flow of water from the soil? When leaves open their stomata to obtain CO_2 for photosynthesis, water vapor diffuses out of the leaves. This causes water to evaporate from the surface of cell walls inside the leaves. In turn, the loss of water from the cell walls causes the water potential in the walls to decrease (**Figure 4.8**). This creates a gradient in water potential that causes water to flow toward the sites of evaporation.

One hypothesis for how a loss of water from cell walls results in a decrease in water potential is that as water evaporates, the surface of the remaining water is drawn into the interstices of the cell wall (see Figure 4.8) where it forms curved air–water interfaces. Because water adheres to the cellulose microfibrils and other hydrophilic components of the cell wall, the curvature of these interfaces induces a negative pressure in the water. As more water is removed from the wall, the curvature of these air–water interfaces increases and the pressure of the water becomes more negative (see Equation 4.1), a situation analogous to what occurs in the soil. An alternative hypothesis for how transpiration causes the water potential in cell walls to decrease focuses on the properties of the pectin component of the cell wall and is discussed in **WEB ESSAY 4.1**.

Some of the water that flows toward the sites of evaporation comes from the protoplasts of adjacent cells. However, because leaves are connected to the soil via a low-resistance pathway—the xylem—most of what replaces water lost from the leaves due to transpiration comes from the soil. Water will flow from the soil when the water potential of the leaves is low enough to overcome the Ψ_p of the soil, as well as the resistance associated with moving water through the plant. Note that for water to be pulled from the soil requires that there be a continuous liquid-filled pathway extending from the sites of evaporation, down through the plant, and out into the soil.

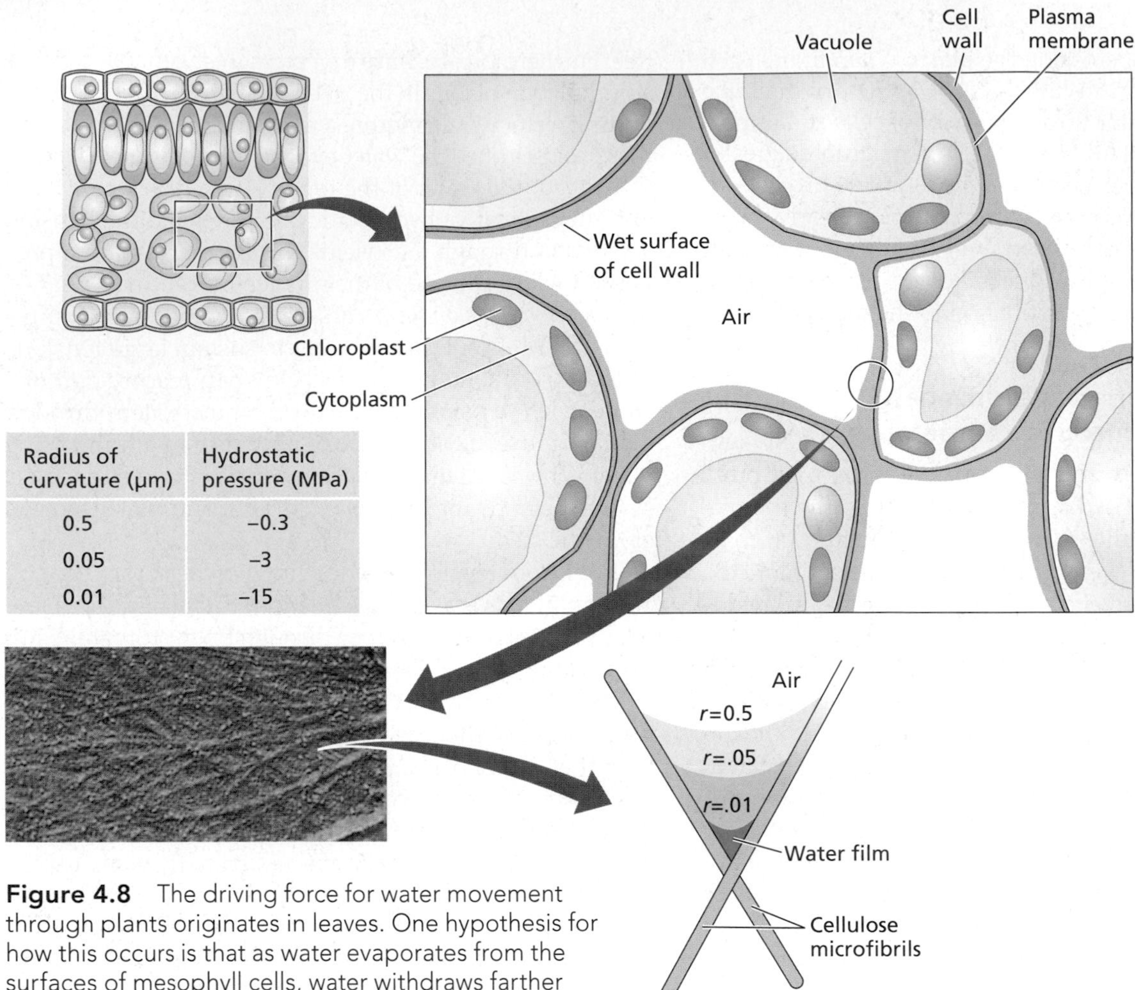

Radius of curvature (μm)	Hydrostatic pressure (MPa)
0.5	–0.3
0.05	–3
0.01	–15

Figure 4.8 The driving force for water movement through plants originates in leaves. One hypothesis for how this occurs is that as water evaporates from the surfaces of mesophyll cells, water withdraws farther into the interstices of the cell wall. Because cellulose is hydrophilic (contact angle = 0°), the force resulting from surface tension causes a negative pressure in the liquid phase. As the radius of curvature of the air–water interfaces decreases, the hydrostatic pressure becomes more negative, as calculated from Equation 4.1. (Micrograph from Gunning and Steer 1996.)

The cohesion–tension theory explains how the substantial movement of water through plants can occur without the direct expenditure of metabolic energy: The energy that powers the movement of water through plants comes from the sun, which, by increasing the temperature of both the leaf and the surrounding air, drives the evaporation of water. However, water transport through the xylem is not "free." The plant must build xylem conduits capable of withstanding the large tensions needed to pull water from the soil. In addition, plants must accumulate enough solutes in their living cells that they are able to remain turgid even as water potentials decrease due to transpiration.

The cohesion–tension theory has been a controversial subject for more than a century and continues to generate lively debate. The main controversy surrounds the question of whether water columns in the xylem can sustain the large tensions (negative pressures) necessary to pull water up tall trees. Recently, water transport through a microfluidic device designed to function as a synthetic "tree" demonstrated the stable flow of liquid water at pressures lower (more negative) than –7.0 MPa. For details on the history of research on water transport in the xylem, including the controversy surrounding the cohesion–tension theory, see **WEB ESSAYS 4.2 and 4.3**.

Xylem transport of water in trees faces physical challenges

The large tensions that develop in the xylem of trees (see **WEB ESSAY 4.4**) and other plants present significant physical challenges. First, the water under tension transmits an inward force to the walls of the xylem. If the cell walls were weak or pliant, they would collapse under this tension. The secondary wall thickenings and lignification of tracheids and vessels are adaptations that offset this tendency to collapse. Plants that experience large xylem tensions tend to have dense wood, reflecting the mechanical stresses imposed on the wood by water under tension.

A second challenge is that water under such tensions is in a *physically metastable state*. Water is stable as a liquid when its hydrostatic pressure exceeds its saturated vapor pressure. When the hydrostatic pressure in liquid water becomes equal to its saturated vapor pressure, the water will undergo a phase change. We are all familiar with the idea of vaporizing water by increasing its temperature (raising its saturated vapor pressure). Less familiar, but still easily observed, is the fact that water can be made to boil at room temperature by placing it in a vacuum chamber (lowering the hydrostatic pressure of the liquid phase by reducing the pressure of the atmosphere).

In our earlier example, we estimated that a pressure gradient of 2 MPa would be needed to supply water to leaves at the top of a 100-m-high tree. If we assume that the soil surrounding this tree is fully hydrated and lacks significant concentrations of solutes (i.e., $\Psi = 0$), the cohesion–tension theory predicts that the hydrostatic pressure of water in the xylem at the top of the tree will be –2 MPa. This value is substantially below the saturated vapor pressure (absolute pressure of ~0.002 MPa at 20°C), raising the question of what maintains the water column in its liquid state.

Water in the xylem is described as being in a metastable state because despite the existence of a thermodynamically lower energy state—the vapor phase—it remains a liquid. This situation occurs because (1) the cohesion and adhesion of water make the free-energy barrier for the liquid-to-vapor phase change very high, and (2) the structure of the xylem minimizes the presence of *nucleation sites*—sites that lower the energy barrier separating the liquid from the vapor phase.

The most important nucleation sites are gas bubbles. When a gas bubble grows to a sufficient size that the inward force resulting from surface tension is less than the outward force due to the negative pressure in the liquid phase, the bubble will expand. Furthermore, once a bubble starts to expand, the inward force due to surface tension decreases, because the air–water interface has less curvature. Thus, a bubble that exceeds the critical size for expansion will expand until it fills the entire conduit.

The absence of gas bubbles of sufficient size to destabilize the water column when under tension is, in part, due to the fact that in the roots, water must flow across the endodermis to enter the xylem. The endodermis serves as a filter, preventing gas bubbles from entering the xylem. Pit membranes also function as filters as water flows from one xylem conduit to another. However, when pit membranes are exposed to air on one side—due to injury, leaf abscission, or the existence of a neighboring gas-filled conduit—pit membranes can serve as sites of entry for air. Air enters when the pressure difference across the pit membrane is sufficient either to allow air to penetrate the cellulose microfibriller matrix of structurally homogeneous pit membranes (see Figure 4.6D), or to dislodge the torus of a coniferous pit membrane (see Figure 4.6C). This phenomenon is called *air seeding*.

A second mode by which bubbles can form in xylem conduits is freezing of the xylem tissues. Because water in the xylem contains dissolved gases and the solubility of gases in ice is very low, freezing of xylem conduits can lead to bubble formation.

The phenomenon of bubble expansion is known as *cavitation*, and the resulting gas-filled void is referred to as an *embolism*. Its effect is similar to that of a vapor lock in the fuel line of an automobile or an embolism in a blood vessel. Cavitation breaks the continuity of the water column and prevents the transport of water under tension.

Such breaks in the water columns in plants are not unusual. When plants are deprived of water, sound pulses or clicks can be detected. The formation and rapid expansion of air bubbles in the xylem, such that the pressure in the water is suddenly increased by perhaps 1 MPa or more, results in high-frequency acoustic shock waves through the rest of the plant. These breaks in xylem water continuity, if not repaired, would be disastrous to the plant. By blocking the main transport pathway of water, such embolisms would increase flow resistance and ultimately cause the dehydration and death of the leaves and other organs.

Vulnerability curves (**Figure 4.9**) provide a way of quantifying a species' susceptibility to cavitation and the impact of cavitation on flow through the xylem. A vulnerability curve plots the measured hydraulic conductivity (usually as a percent of maximum) of a branch, stem, or root segment versus the experimentally imposed level of xylem tension. Due to cavitation, xylem hydraulic conductivity decreases with increasing tensions until flow ceases entirely. However, the decrease in xylem hydraulic conductivity occurs at much lower tensions in species in

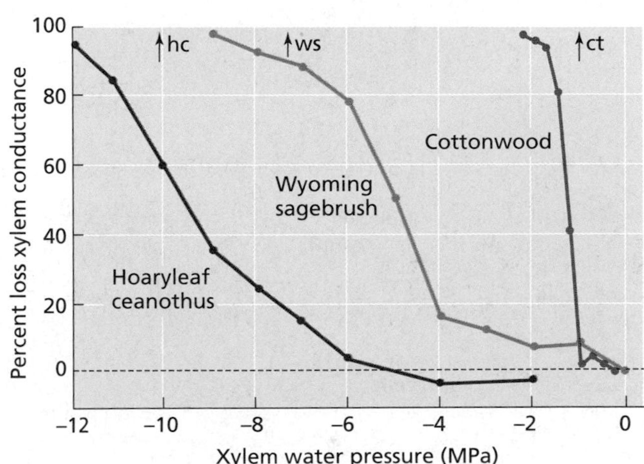

Figure 4.9 Xylem vulnerability curves represent the percentage loss of hydraulic conductance in stem xylem versus xylem water pressure, here in three species of contrasting drought tolerance. Data were obtained from excised branches subjected experimentally to increasing levels of xylem tension using a centrifugal force technique. Arrows on the upper axis indicate the minimum xylem pressure measured in the field for each species. (After Sperry 2000.)

moist habitats, such as birch, than in species from more arid regions, such as sagebrush.

Plants minimize the consequences of xylem cavitation

The impact of xylem cavitation on the plant can be minimized by several means. Because the water-transporting conduits in the xylem are interconnected, one gas bubble might, in principle, expand to fill the whole network. In practice, gas bubbles do not spread far, because an expanding gas bubble cannot easily pass through the small pores of the pit membranes. Because the capillaries in the xylem are interconnected, one gas bubble does not completely stop water flow. Instead, water can detour around the embolized conduit by traveling through neighboring, water-filled conduits (see Figure 4.7). Thus, the finite length of the tracheid and vessel conduits of the xylem, while resulting in an increased resistance to water flow, also provides a way to restrict the impact of cavitation.

Gas bubbles can also be eliminated from the xylem. As we have seen, some plants develop positive pressures (root pressures) in the xylem. Such pressures shrink bubbles and cause the gases to dissolve. Recent studies suggest that cavitation may be repaired even when the water in the xylem is under tension. A mechanism for such repair is not yet known and remains the subject of active research (see WEB ESSAY 4.5).

Finally, many plants have secondary growth in which new xylem forms each year. The production of new xylem conduits allows plants to replace losses in water-transport capacity due to cavitation.

Water Movement from the Leaf to the Atmosphere

On its way from the leaf to the atmosphere, water is pulled from the xylem into the cell walls of the mesophyll, where it evaporates into the air spaces of the leaf (**Figure 4.10**). The water vapor then exits the leaf through the stomatal pore. The movement of liquid water through the living tissues of the leaf is controlled by gradients in water potential. However, transport in the vapor phase is by diffusion, so the final part of the transpiration stream is controlled by the *concentration gradient of water vapor.*

The waxy cuticle that covers the leaf surface is an effective barrier to water movement. It has been estimated that only about 5% of the water lost from leaves escapes through the cuticle. Almost all of the water lost from leaves is lost by diffusion of water vapor through the tiny stomatal pores. In most herbaceous species, stomata are present on both the upper and lower surfaces of the leaf, usually

Figure 4.10 Water pathway through the leaf. Water is pulled from the xylem into the cell walls of the mesophyll, where it evaporates into the air spaces within the leaf. Water vapor then diffuses through the leaf air space, through the stomatal pore, and across the boundary layer of still air found next to the leaf surface. CO_2 diffuses in the opposite direction along its concentration gradient (low inside, higher outside).

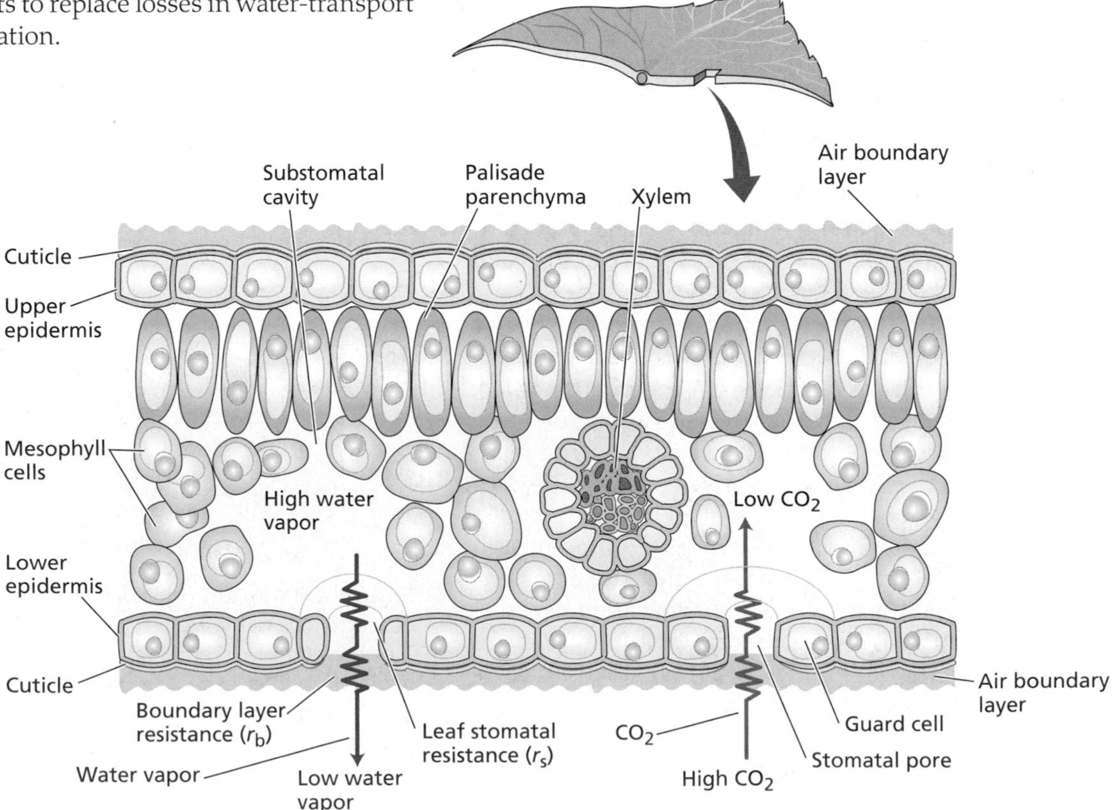

more abundant on the lower surface. In many tree species, stomata are located only on the lower surface of the leaf.

We will now examine the movement of liquid water through the leaf, the driving force for leaf transpiration, the main resistances in the diffusion pathway from the leaf to the atmosphere, and the anatomical features of the leaf that regulate transpiration.

Leaves have a large hydraulic resistance

Although the distances that water must traverse within leaves are small relative to the entire soil-to-atmosphere pathway, the contribution of the leaf to the total hydraulic resistance is large. On average, leaves constitute 30% of the total liquid-phase resistance, and in some plants their contribution is much larger. This combination of short path length and large hydraulic resistance also occurs in roots, reflecting the fact that in both organs, water transport takes place across highly resistive living tissues as well as through the xylem.

Water enters into leaves and is distributed across the leaf lamina in xylem conduits. Water must exit through the xylem walls and pass through multiple layers of living cells before it evaporates. Leaf hydraulic resistance thus reflects the number, distribution, and size of xylem conduits, as well as the hydraulic properties of leaf mesophyll cells. The hydraulic resistance of leaves of diverse vein architectures varies as much as 40-fold. A large part of this variation appears to be due to the density of veins within the leaf and their distance from the evaporative leaf surface. Leaves with closely spaced veins tend to have lower hydraulic resistance and higher rates of photosynthesis, suggesting that the proximity of leaf veins to sites of evaporation exerts a significant impact on the rates of leaf gas exchange.

The hydraulic resistance of leaves varies in response to growth conditions and exposure to low leaf water potentials. For example, leaves of plants growing in shaded conditions exhibit greater resistance to water flow than do leaves of plants grown in higher light. Leaf hydraulic resistance also typically increases with leaf age. Over shorter time scales, decreases in leaf water potential lead to marked increases in leaf hydraulic resistance. The increase in leaf hydraulic resistance may result from decreases in the membrane permeability of mesophyll cells, cavitation of xylem conduits in leaf veins, or in some cases, the physical collapse of xylem conduits under tension.

The driving force for transpiration is the difference in water vapor concentration

Transpiration from the leaf depends on two major factors: (1) the **difference in water vapor concentration** between the leaf air spaces and the external bulk air (Δc_{wv}) and (2) the **diffusional resistance** (r) of this pathway. The difference in water vapor concentration is expressed as $c_{wv(leaf)} - c_{wv(air)}$. The water vapor concentration of air ($c_{wv[air]}$) can be readily measured, but that of the leaf ($c_{wv[leaf]}$) is more difficult to assess.

Whereas the volume of air space inside the leaf is small, the wet surface from which water evaporates is large. Air space volume is about 5% of the total leaf volume in pine needles, 10% in maize (corn; *Zea mays*) leaves, 30% in barley, and 40% in tobacco leaves. In contrast to the volume of the air space, the internal surface area from which water evaporates may be from 7 to 30 times the external leaf area. This high surface-to-volume ratio makes for rapid vapor equilibration inside the leaf. Thus, we can assume that the air space in the leaf is close to water potential equilibrium with the cell wall surfaces from which liquid water is evaporating.

Within the range of water potentials experienced by transpiring leaves (generally greater than –2.0 MPa), the equilibrium water vapor concentration is within 2 percentage points of the saturation water vapor concentration. This allows one to estimate the water vapor concentration within a leaf from its temperature, which is easy to measure. Because the saturated water vapor content of air increases exponentially with temperature, leaf temperature has a marked impact on transpiration rates. (**WEB TOPIC 4.4** shows how we can calculate the water vapor concentration in the leaf air spaces and discusses other aspects of the water relations within a leaf.)

The concentration of water vapor, c_{wv}, changes at various points along the transpiration pathway. We see from **Table 4.1** that c_{wv} decreases at each step of the pathway from the cell wall surface to the bulk air outside the leaf. The important points to remember are that (1) the driving force for water loss from the leaf is

TABLE 4.1 Representative values for relative humidity, absolute water vapor concentration, and water potential for four points in the pathway of water loss from a leaf

		Water vapor	
Location	Relative humidity	Concentration (mol m^{-3})	Potential (MPa)[a]
Inner air spaces (25°C)	0.99	1.27	–1.38
Just inside stomatal pore (25°C)	0.97	1.21	–7.04
Just outside stomatal pore (25°C)	0.47	0.60	–103.7
Bulk air (20°C)	0.50	0.50	–93.6

Source: Adapted from Nobel 1999.

Note: See Figure 4.10.

[a]Calculated using Equation 4.5.2 in **WEB TOPIC 4.4**, with values for $RT/\overline{V}_w$ of 135 MPa at 20°C and 137.3 MPa at 25°C.

the *absolute* concentration difference (difference in c_{wv}, in mol m^{-3}), and (2) this difference is markedly influenced by leaf temperature.

Water loss is also regulated by the pathway resistances

The second important factor governing water loss from the leaf is the diffusional resistance of the transpiration pathway, which consists of two varying components (see Figure 4.10):

1. The resistance associated with diffusion through the stomatal pore, the **leaf stomatal resistance** (r_s).
2. The resistance due to the layer of unstirred air next to the leaf surface through which water vapor must diffuse to reach the turbulent air of the atmosphere. This second resistance, r_b, is called the leaf **boundary layer resistance**. We will discuss this type of resistance before considering stomatal resistance.

The thickness of the boundary layer is determined primarily by wind speed and leaf size. When the air surrounding the leaf is very still, the layer of unstirred air on the surface of the leaf may be so thick that it is the primary deterrent to water vapor loss from the leaf. Increases in stomatal apertures under such conditions have little effect on transpiration rate (**Figure 4.11**), although closing the stomata completely will still reduce transpiration.

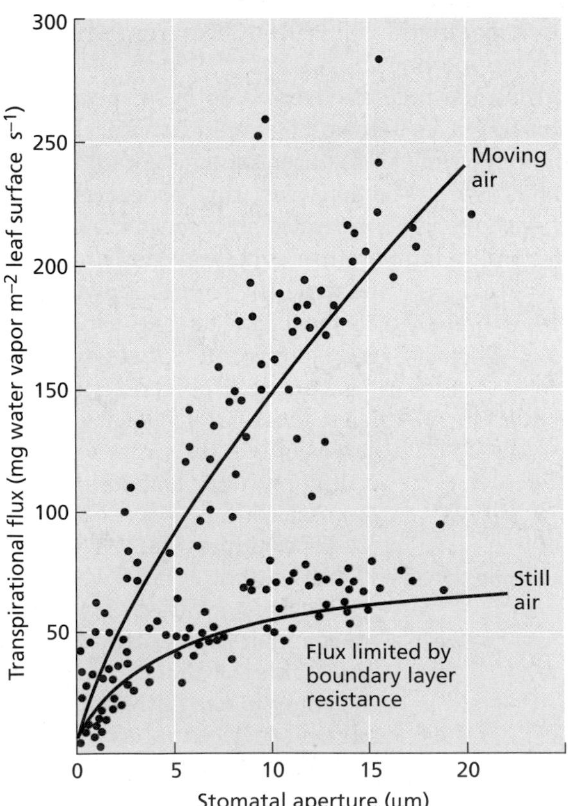

Figure 4.11 Dependence of transpiration flux on the stomatal aperture of zebra plant (*Zebrina pendula*) in still air and in moving air. The boundary layer is thicker and more rate-limiting in still air than in moving air. As a result, the stomatal aperture has less control over transpiration in still air. (After Bange 1953.)

When wind velocity is high, the moving air reduces the thickness of the boundary layer at the leaf surface, reducing the resistance of this layer. Under such conditions, stomatal resistance will largely control water loss from the leaf.

Various anatomical and morphological aspects of the leaf can influence the thickness of the boundary layer. Hairs on the surface of leaves can serve as microscopic windbreaks. Some plants have sunken stomata that provide a sheltered region outside the stomatal pore. The size and shape of leaves and their orientation relative to the wind direction also influence the way the wind sweeps across the leaf surface. Most of these factors, however, cannot be altered on an hour-to-hour or even a day-to-day basis. For short-term regulation of transpiration, control of stomatal apertures by the guard cells plays a crucial role in the regulation of leaf transpiration.

Some species are able to change the orientation of their leaves and thereby influence their transpiration rates. For example, when plants orient their leaves parallel to the sun's rays, leaf temperature is reduced and with it the driving force for transpiration, Δc_{wv}. Many grass leaves roll up as they experience water deficits, in this way increasing their boundary layer resistance. Even wilting can help ameliorate high transpiration rates by reducing the amount of radiation intercepted, resulting in lower leaf temperatures and a decrease in Δc_{wv}.

Stomatal control couples leaf transpiration to leaf photosynthesis

Because the cuticle covering the leaf is nearly impermeable to water, most leaf transpiration results from the diffusion of water vapor through the stomatal pore (see Figure 4.10). The microscopic stomatal pores provide a *low-resistance pathway* for diffusional movement of gases across the epidermis and cuticle. Changes in stomatal resistance are important for the regulation of water loss by the plant and for controlling the rate of carbon dioxide uptake necessary for sustained CO_2 fixation during photosynthesis.

When water is abundant, the functional solution to the leaf's need to limit water loss while taking in CO_2 is the *temporal* regulation of stomatal apertures—open during the day, closed at night. At night, when there is no photosynthesis and thus no demand for CO_2 inside the leaf, stomatal apertures are kept small or closed, preventing unnecessary loss of water. On a sunny morning when the supply of

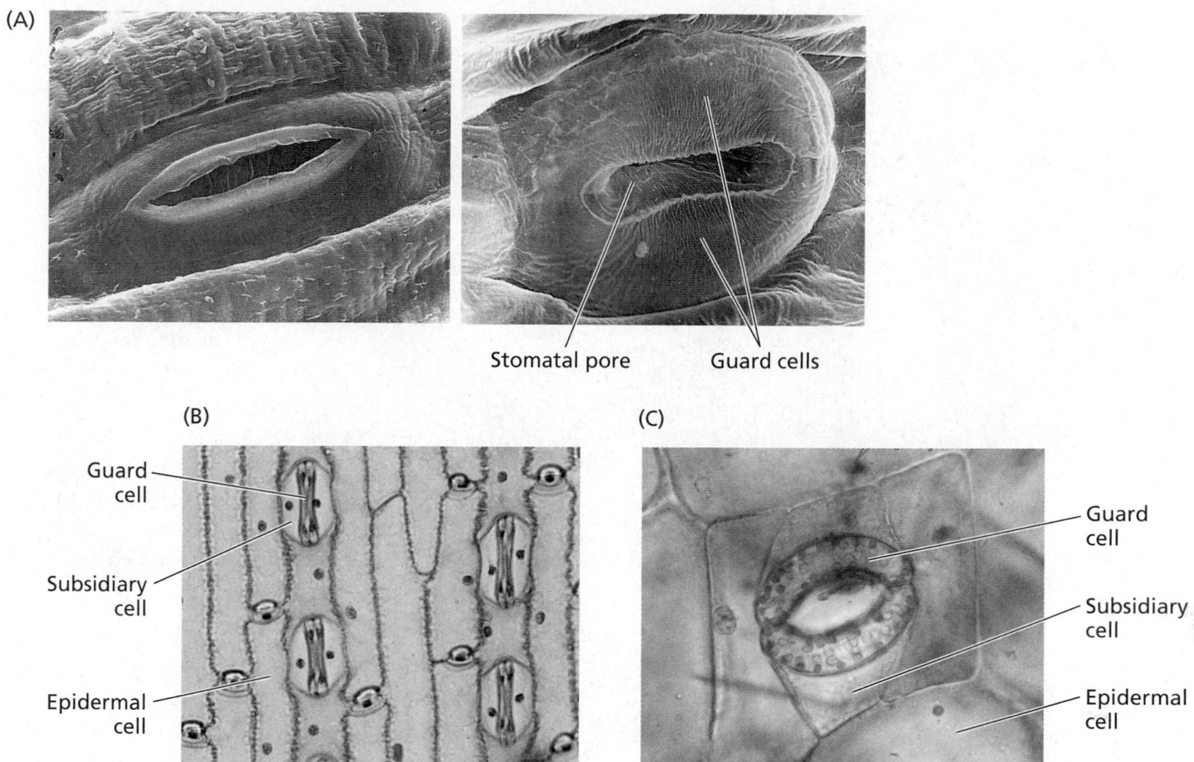

Figure 4.12 Stomata. (A) Scanning electron micrographs of onion epidermis. The left panel shows the outside surface of the leaf, with a stomatal pore inserted in the cuticle. The right panel shows a pair of guard cells facing the stomatal cavity, toward the inside of the leaf. (1640×) (B) Stomata of corn (*Zea mays*), showing the dumbbell-shaped guard cells typical of grasses. (C) Most other plants have kidney-shaped guard cells, as seen in this open stoma of *Tradescantia zebrina*. (A from Zeiger and Hepler 1976 [left] and E. Zeiger and N. Burnstein [right].)

water is abundant and the solar radiation incident on the leaf favors high photosynthetic activity, the demand for CO_2 inside the leaf is large, and the stomatal pores open wide, decreasing the stomatal resistance to CO_2 diffusion. Water loss by transpiration is substantial under these conditions, but since the water supply is plentiful, it is advantageous for the plant to trade water for the products of photosynthesis, which are essential for growth and reproduction.

On the other hand, when soil water is less abundant, the stomata will open less or even remain closed on a sunny morning. By keeping its stomata closed in dry conditions, the plant avoids dehydration. The leaf cannot control $c_{wv(air)}$ or r_b. However, it can regulate its stomatal resistance (r_s) by opening and closing of the stomatal pore. This biological control is exerted by a pair of specialized epidermal cells, the **guard cells**, which surround the stomatal pore (**Figure 4.12**).

The cell walls of guard cells have specialized features

Guard cells are found in leaves of all vascular plants, and they are also present in some nonvascular plants, such as hornworts and mosses. Guard cells show considerable morphological diversity, but we can distinguish two main types: One is typical of grasses, while the other is found in most other flowering plants, as well as in mosses, ferns, and gymnosperms.

In grasses (see Figure 4.12B), guard cells have a characteristic dumbbell shape, with bulbous ends. The pore proper is a long slit located between the two "handles" of the dumbbells. These guard cells are always flanked by a pair of differentiated epidermal cells called **subsidiary cells**, which help the guard cells control the stomatal pore. The guard cells, subsidiary cells, and pore are collectively called the **stomatal complex**.

In most other plants, guard cells have an elliptical contour (often called "kidney-shaped") with the pore at their center (see Figure 4.12C). Although subsidiary cells are common in species with kidney-shaped stomata, they can be absent, in which case the guard cells are surrounded by ordinary epidermal cells.

A distinctive feature of guard cells is the specialized structure of their walls. Portions of these walls are substantially thickened (**Figure 4.13**) and may be up to 5

(A)

Cytosol and vacuole

Pore

Heavily thickened guard cell wall

Subsidiary cells

Guard cells

Atmosphere

Interior of leaf

(B)

ATMOSPHERE

Pore

Vacuole

Plastid

Inner cell wall

SUBSTOMATAL CAVITY

2 μm

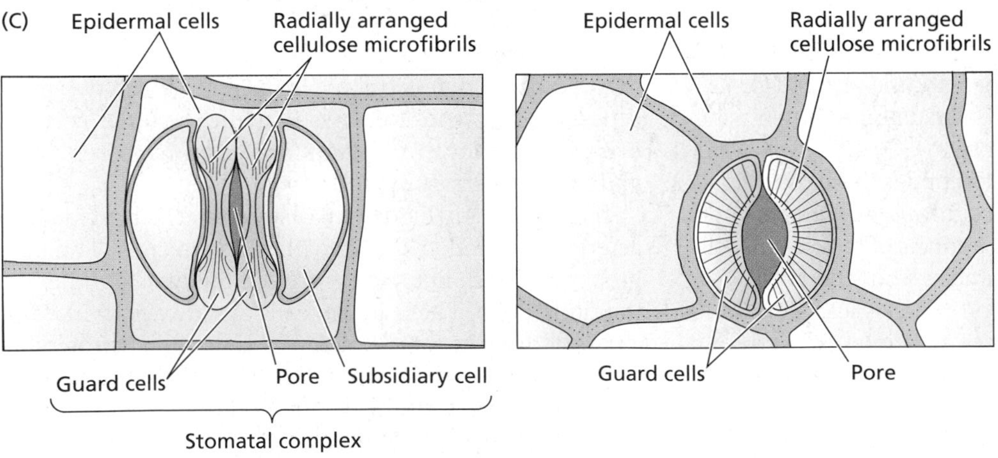

Figure 4.13 Guard cell wall structure. (A) Electron micrograph of a stoma from a grass (*Phleum pretense*). The bulbous ends of each guard cell show their cytosolic content and are joined by the heavily thickened walls. The stomatal pore separates the two midportions of the guard cells. (2560×) (B) Electron micrograph showing a pair of guard cells from tobacco (*Nicotiana tabacum*). The section was made perpendicular to the main surface of the leaf. The pore faces the atmosphere; the bottom faces the substomatal cavity inside the leaf. Note the uneven thickening pattern of the walls, which determines the asymmetric deformation of the guard cells when their volume increases during stomatal opening. (C) Radial alignment of the cellulose microfibrils in guard cells and epidermal cells of a grasslike stoma (left) and a kidney-shaped stoma (right). (A from Palevitz 1981, courtesy of B. Palevitz; B from Sack 1987, courtesy of F. Sack; C after Meidner and Mansfield 1968.)

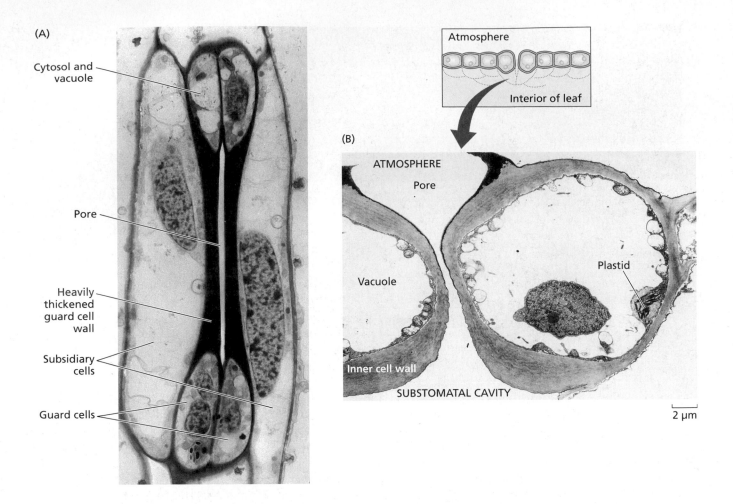

μm across, in contrast to the 1 to 2 μm typical of epidermal cells. In kidney-shaped guard cells, a differential thickening pattern results in very thick inner and outer (lateral) walls, a thin dorsal wall (the wall in contact with epidermal cells), and a somewhat thickened ventral (pore) wall. The portions of the wall that face the atmosphere often extend into well-developed ledges, which form the pore proper.

The alignment of **cellulose microfibrils**, which reinforce all plant cell walls and are an important determinant of cell shape (see Chapter 14), plays an essential role in the opening and closing of the stomatal pore. In ordinary, cylindrically shaped cells, cellulose microfibrils are oriented transversely to the long axis of the cell. As a result, the cell expands in the direction of its long axis, because the cellulose reinforcement offers the least resistance at right angles to its orientation.

In guard cells the microfibril organization is different. Kidney-shaped guard cells have cellulose microfibrils fanning out radially from the pore (see Figure 4.13C). As a result, the inner wall (facing the pore) is much stronger than the outer wall. Thus, as a guard cell increases in volume, the outer wall expands more than the inner wall. This causes the guard cells to bow apart and the pore to open. In grasses, the dumbbell-shaped guard cells function like beams with inflatable ends. The orientation of the cellulose microfibrils is such that as the bulbous ends of the cells increase in volume, the beams are separated from each other and the slit between them widens (see Figure 4.13C).

An increase in guard cell turgor pressure opens the stomata

Guard cells function as multisensory hydraulic valves. Environmental factors such as light intensity and quality, temperature, leaf water status, and intracellular CO_2 concentrations are sensed by guard cells, and these signals are integrated into well-defined stomatal responses. If leaves kept in the dark are illuminated, the light stimulus is perceived by the guard cells as an opening signal, triggering a series of responses that result in opening of the stomatal pore.

The early aspects of this process are ion uptake and other metabolic changes in the guard cells, which we will discuss in detail in Chapter 24. Here we will note the effect of decreases in osmotic potential (Ψ_s) resulting from ion uptake and from biosynthesis of organic molecules in the guard cells. Water relations in guard cells follow the same rules as in other cells. As Ψ_s decreases, the water potential decreases, and water consequently moves into the guard cells. As water enters the cell, turgor pressure increases and the stomata open (**Figure 4.14**).

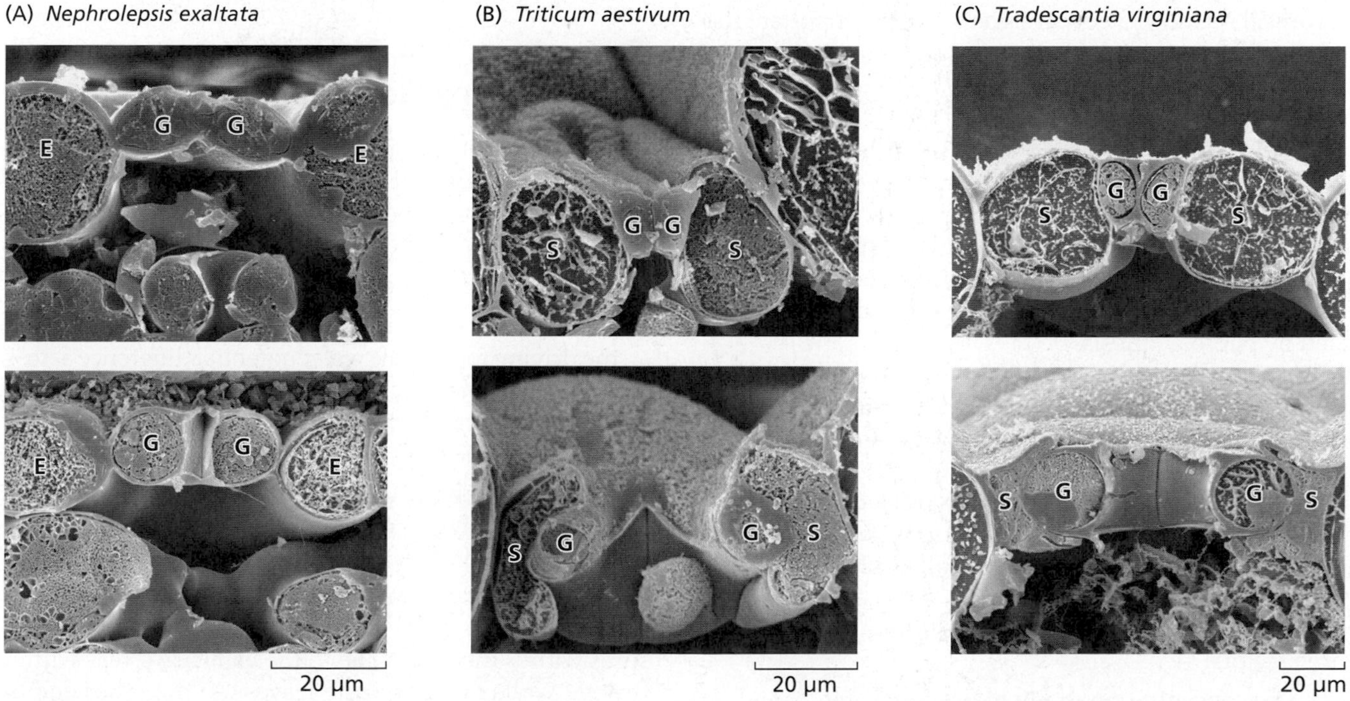

(A) *Nephrolepsis exaltata* (B) *Triticum aestivum* (C) *Tradescantia virginiana*

Figure 4.14 Cross section of stomata sampled by snap freezing of intact leaves of (A) *Nephrolepsis exaltata*, a fern, (B) *Triticum aestivum*, a grass, and (C) *Tradescantia virginiana*, a nongrass angiosperm. Closed stomata (upper figures) are from leaves sampled at night; open stomata (lower figures) are from leaves exposed to full sun with very high humidity for several hours. G, guard cell; S, subsidiary cell; E, epidermal cell. (From Franks and Farquhar 2007.)

In some plants, for example ferns, stomatal opening and closing involves changes in the volume and turgor pressure of only the guard cells (see Figure 4.14A). When leaves are well hydrated and thus leaf Ψ is high, guard cell turgor pressure is large and the stomata open. Conversely, when water availability decreases and leaf Ψ falls, guard cell turgor pressure also decreases and the stomata close, conserving water.

In angiosperms, stomatal opening and closing involves changes in the volume and turgor pressure of both the guard cells and the subsidiary (or adjacent epidermal) cells (see Figures 4.14B and C). At the same time that the uptake of solutes into guard cells causes them to increase in volume and turgor pressure, the subsidiary (or adjacent epidermal) cells release solutes into the apoplast. The transfer of solutes out of subsidiary cells and into the guard cells causes the former to decrease in both turgor pressure and size, facilitating the expansion of guard cells in the direction away from the stomatal pore. Conversely, the transfer of solutes from guard cells to the subsidiary cells increases the size and turgor pressure of the latter, thus pushing the guard cells together and causing the stomata to close.

Subsidiary cells appear to play an important role in allowing angiosperm stomata to open quickly and to achieve large apertures. One consequence of these interactions is that decreases in leaf water potential are not passively linked to stomatal closure. The subsidiary cells must increase in volume and turgor pressure for the stomata to close. In Chapter 24 we will see how chemical signals play an important role in controlling stomatal aperture during drought.

The transpiration ratio measures the relationship between water loss and carbon gain

The effectiveness of plants in moderating water loss while allowing sufficient CO_2 uptake for photosynthesis can be assessed by a parameter called the **transpiration ratio**. This value is defined as the amount of water transpired by the plant divided by the amount of carbon dioxide assimilated by photosynthesis.

For plants in which the first stable product of carbon fixation is a three-carbon compound (C_3 plants; see Chapter 8), as many as 400 molecules of water are lost for every molecule of CO_2 fixed by photosynthesis, giving a transpiration ratio of 400. (Sometimes the reciprocal of the transpiration ratio, called the *water use efficiency*, is cited. Plants with a transpiration ratio of 400 have a water use efficiency of 1/400, or 0.0025.)

The large ratio of H_2O efflux to CO_2 influx results from three factors:

1. The concentration gradient driving water loss is about 50 times larger than that driving the influx of CO_2. In large part, this difference is due to the low concentration of CO_2 in air (about 0.04%) and the relatively high concentration of water vapor within the leaf.
2. CO_2 diffuses about 1.6 times more slowly through air than water does (the CO_2 molecule is larger than H_2O and has a smaller diffusion coefficient).
3. CO_2 must cross the plasma membrane, the cytoplasm, and the chloroplast envelope before it is assimilated in the chloroplast. These membranes add to the resistance of the CO_2 diffusion pathway.

Some plants use variations in the usual photosynthetic pathway for fixation of carbon dioxide that substantially reduce their transpiration ratio. Plants in which a four-carbon compound is the first stable product of photosynthesis (C_4 plants; see Chapter 8) generally transpire less water per molecule of CO_2 fixed than C_3 plants do; a typical transpiration ratio for C_4 plants is about 150. This is largely because C_4 photosynthesis results in a lower CO_2 concentration in the intercellular air space (see Chapter 8), thus creating a larger driving force for the uptake of CO_2 and allowing these plants to operate with smaller stomatal apertures and thus lower transpiration rates.

Desert-adapted plants with crassulacean acid metabolism (CAM) photosynthesis, in which CO_2 is initially fixed into four-carbon organic acids at night, have even lower transpiration ratios; values of about 50 are not unusual. This is possible because their stomata have an inverted diurnal rhythm, opening at night and closing during the day. Transpiration is much lower at night, because the cool leaf temperature gives rise to only a very small Δc_{wv}.

Overview: The Soil–Plant–Atmosphere Continuum

We have seen that movement of water from the soil through the plant to the atmosphere involves different mechanisms of transport:

- In the soil and the xylem, liquid water moves by bulk flow in response to a pressure gradient ($\Delta\Psi_p$).
- When liquid water is transported across membranes, the driving force is the water potential difference across the membrane. Such osmotic flow occurs when cells absorb water and when roots transport water from the soil to the xylem.
- In the vapor phase, water moves primarily by diffusion, at least until it reaches the outside air, where convection (a form of bulk flow) becomes dominant.

However, the key element in the transport of water from the soil to the leaves is the generation of negative pressures within the xylem due to the capillary forces within the cell walls of transpiring leaves. At the other end of the plant, soil water is also held by capillary forces. This results in a "tug-of-war" on a rope of water by capillary forces at both ends. As a leaf loses water due to transpiration, water moves up the plant and out of the soil driven by physical forces, without the involvement of any metabolic pump. The energy for the movement of water is ultimately supplied by the sun.

This simple mechanism makes for tremendous efficiency energetically—which is critical when as many as 400 molecules of water are being transported for every CO_2 molecule being taken up in exchange. Crucial elements that allow this transport system to function are a low-resistivity xylem flow path that is protected from cavitation and a high-surface-area root system for extracting water from the soil.

SUMMARY

There is an inherent conflict between a plant's need for CO_2 uptake and its need to conserve water, resulting from water being lost through the same pores that let CO_2 in. To manage this conflict, plants have evolved adaptations to control water loss from leaves, and to replace the water that is lost.

Water in the Soil

- The water content and rate of movement in soils depend on soil type and structure, which influence the pressure gradient in the soil and its hydraulic conductivity.
- In soil, water may exist as a surface film on soil particles, or it may partially or completely fill the spaces between particles.
- Osmotic potential, pressure potential, and gravitational potential influence the movement of water from the soil through the plant to the atmosphere (**Figure 4.1**).
- The intimate contact between root hairs and soil particles greatly increases the surface area for water absorption (**Figure 4.2**).

Water Absorption by Roots

- Water uptake is mostly confined to regions near root tips (**Figure 4.3**).
- In the root, water may move via the apoplast, the symplast, or the transmembrane pathway (**Figure 4.4**).
- Water movement through the apoplast is obstructed by the Casparian strip in the endodermis, which forces water to move symplasmically before it enters the xylem (**Figure 4.4**).
- When transpiration is low or absent, the continued transport of solutes into the xylem fluid leads to a decrease in Ψ_s and a decrease in Ψ, providing the force for water absorption and a positive Ψ_p, which yields a positive hydrostatic pressure in the xylem (**Figure 4.5**).

Water Transport through the Xylem

- Xylem conduits, which can be either single-celled tracheids or multicellular vessels, provide a low-resistance pathway for the transport of water (**Figure 4.6**).
- Elongated, spindle-shaped tracheids and stacked vessel elements have pits in lateral walls (**Figure 4.7**).
- Pressure-driven bulk flow moves water long distances through the xylem.
- The ascent of water through plants results from the decrease in water potential at the sites of evaporation within leaves (**Figure 4.8**).
- Cavitation breaks the continuity of the water column and prevents the transport of water under tension (**Figure 4.9**).

Water Movement from the Leaf to the Atmosphere

- Water is pulled from the xylem into the cell walls of leaf mesophyll before evaporating into the leaf's air spaces (**Figure 4.10**).
- The hydraulic resistance of leaves is large and varies in response to growth conditions and exposure to low leaf water potentials.
- Transpiration depends on the difference in water vapor concentration between the leaf air spaces and the external air and on the diffusional resistance of this pathway, which consists of leaf stomatal resistance and boundary layer resistance (**Figure 4.11**).
- Opening and closing of the stomatal pore is accomplished and controlled by guard cells (**Figures 4.12–4.14**).
- The effectiveness of plants in limiting water loss while allowing CO_2 uptake is given by the transpiration ratio.

Overview: The Soil–Plant–Atmosphere Continuum

- Physical forces, without the involvement of any metabolic pump, drive the movement of water from soil to plant to atmosphere, with the sun being the ultimate source for the energy.

WEB MATERIAL

- **WEB TOPIC 4.1 Irrigation** Irrigation has a dramatic impact on crop yield and soil salinity.
- **WEB TOPIC 4.2 Physical Properties of Soils** The size distribution of soil particles influences the soil's ability to hold and conduct water.
- **WEB TOPIC 4.3 Calculating Velocities of Water Movement in the Xylem and in Living Cells** Water flows more easily through the xylem than across living cells.
- **WEB TOPIC 4.4 Leaf Transpiration and Water Vapor Gradients** Leaf transpiration and stomatal conductance affect leaf and air water vapor concentrations.

- **WEB ESSAY 4.1 Transpiration and Cell Walls** An alternative hypothesis for how partial dehydration results in a decrease in the water potential of cell walls.
- **WEB ESSAY 4.2 A Brief History of the Study of Water Movement in the Xylem** The history of our understanding of sap ascent in plants, especially in trees, is a beautiful example of how knowledge about plants is acquired.
- **WEB ESSAY 4.3 The Cohesion–Tension Theory at Work** The cohesion–tension theory has withstood a number of challenges.
- **WEB ESSAY 4.4 How Water Climbs to the Top of a 112-Meter-tall Tree** Measurements of photosynthesis and transpiration in 112-meter-tall trees show that some of the conditions experienced by the top foliage are comparable to those of extreme deserts.
- **WEB ESSAY 4.5 Cavitation and Refilling** A possible mechanism for cavitation repair is under active investigation.

available at **plantphys.net**

Suggested Reading

Bramley, H., Turner, N. C., Turner, D. W., and Tyerman, S. D. (2009) Roles of morphology, anatomy and aquaporins in determining contrasting hydraulic behavior of roots. *Plant Physiol.* 150: 348–364.

Brodribb, T. J., and McAdam, S. A. M. (2011) Passive origins of stomatal control in vascular plants. *Science* 331: 582–585.

Dainty, J. (1976) Water relations of plant cells. In *Transport in Plants*, Vol. 2, Part A: *Cells* (Encyclopedia of Plant Physiology, New Series, Vol. 2), U. Luttge and M. Pitman, eds., Springer, Berlin, pp. 12–35.

Franks, P. J., and Farquhar, G. D. (2007) The mechanical diversity of stomata and its significance in gas-exchange control. *Plant Physiol.* 143: 78–87.

Hacke, U. G., Sperry, J. S., Pockman, W. T., Davis, S. D., and McCulloh, K. (2001) Trends in wood density and structure are linked to prevention of xylem implosion by negative pressure. *Oecologia* 126: 457–461.

Milburn, J. A. (1979) *Water Flow in Plants*. Longman, London.

Nobel, P. S. (1991) *Physicochemical and Environmental Plant Physiology*. Academic Press, San Diego.

Pittermann, J., Sperry, J. S., Hacke, U. G., Wheeler, J. K., and Sikkema, E. H. (2005) Torus-margo pits help conifers compete with angiosperms. *Science* 310: 1924.

Smith, J. A. C., and Griffiths, H. (1993) *Water Deficits: Plant Responses from Cell to Community*. BIOS Scientific, Oxford.

Steudle, E., and Frensch, J. (1996) Water transport in plants: Role of the apoplast. *Plant Soil* 187: 67–79.

Tyree, M. T., and Jarvis, P. G. (1982) Water in tissues and cells. In *Physiological Plant Ecology II: Water Relations and Carbon Assimilation* (Encyclopedia of Plant Physiology, New Series, Vol. 12B), O. L. Lange, P. S. Nobel, C. B. Osmond, and H. Ziegler, eds., Springer, Berlin, pp. 35–77.

Weatherly, P. E. (1982) Water uptake and flow in roots. In *Physiological Plant Ecology II: Water Relations and Carbon Assimilation* (Encyclopedia of Plant Physiology, New Series, Vol. 12B), O. L. Lange, P. S. Nobel, C. B. Osmond, and H. Ziegler, eds., Springer, Berlin, pp. 79–109.

Zeiger, E. (1983) The biology of stomatal guard cells. *Annu. Rev. Plant Physiol.* 34: 441–475.

Zeiger, E., Farquhar, G., and Cowan, I., eds. (1987) *Stomatal Function*. Stanford University Press, Stanford, CA.

Ziegler, H. (1987) The evolution of stomata. In *Stomatal Function*, E. Zeiger, G. Farquhar, and I. Cowan, eds., Stanford University Press, Stanford, CA, pp. 29–57.

Zwieniecki, M. A., Thompson, M. V., and Holbrook, N. M. (2002) Understanding the hydraulics of porous pipes: Tradeoffs between water uptake and root length utilization. *J. Plant Growth Regul.* 21: 315–323.

5 Mineral Nutrition

Mineral nutrients are elements such as nitrogen, phosphorus, and potassium that plants acquire primarily in the form of inorganic ions from the soil. Although mineral nutrients continually cycle through all organisms, they enter the **biosphere** predominantly through the root systems of plants, so in a sense plants act as the "miners" of Earth's crust. The large surface area of roots and their ability to absorb inorganic ions at low concentrations from the soil solution increase the effectiveness of mineral acquisition by plants. After being absorbed by roots, the mineral elements are translocated to the different parts of the plant, where they serve in numerous biological functions. Other organisms, such as mycorrhizal fungi and nitrogen-fixing bacteria, often participate with roots in the acquisition of mineral nutrients.

The study of how plants obtain and use mineral nutrients is called **mineral nutrition**. This area of research is central for improving modern agricultural practices and environmental protection, as well as for understanding plant ecological interactions in natural ecosystems. High agricultural yields depend on fertilization with mineral nutrients. In fact, yields of most crop plants increase linearly with the amount of fertilizer they absorb. To meet increased demand for food, annual world consumption of the primary mineral elements used in fertilizers—nitrogen, phosphorus, and potassium—rose steadily from 30 million metric tons in 1960 to 143 million metric tons in 1990. For a decade after that, consumption remained relatively constant as fertilizers were used more judiciously in an attempt to balance rising costs. During the past few years, however, annual consumption has climbed to 180 million metric tons (**Figure 5.1**).

Over half of the energy used in agriculture is expended on the production, distribution, and application of nitrogen fertilizers. Moreover, production of phosphorus fertilizers depends on nonrenewable resources that are likely to reach peak production during this century. Crop plants, however, typically use less than half of the fertilizer applied to the soils around them. The remaining minerals may leach into surface waters or groundwater, become associated with soil particles, or contribute to air pollution or climate change.

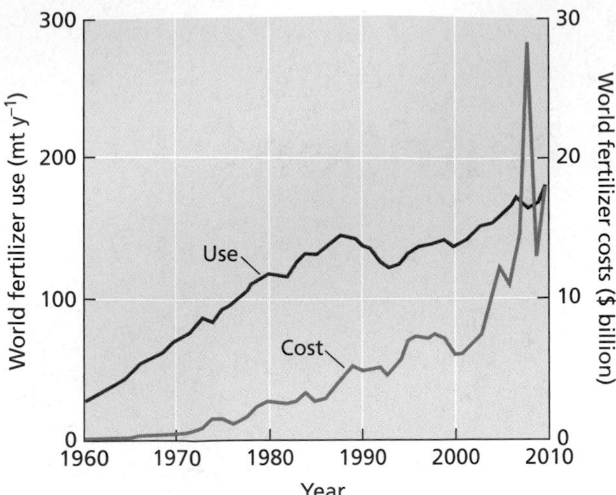

Figure 5.1 Worldwide fertilizer consumption and costs over the past five decades. (After http://faostat3.fao.org/faostat-gateway/go/to/download/R/*/E.)

As a consequence of fertilizer leaching, many water wells in the United States now exceed federal standards for nitrate (NO_3^-) concentrations in drinking water, and the same problem occurs in many agricultural areas in the rest of the world. Enhanced nitrogen availability through nitrate (NO_3^-) and ammonium (NH_4^+) released to the environment from human activities and deposited in the soil by rainwater, a process known as atmospheric nitrogen deposition, is altering ecosystems throughout the world.

On a brighter note, plants are the traditional means of recycling animal wastes and are proving useful for removing deleterious materials, including heavy metals, from toxic-waste dumps. Because of the complex nature of plant–soil–atmosphere relationships, studies of mineral nutrition involve atmospheric chemists, soil scientists, hydrologists, microbiologists, and ecologists as well as plant physiologists.

In this chapter we will discuss the nutritional needs of plants, the symptoms of specific nutritional deficiencies, and the use of fertilizers to ensure proper plant nutrition. Then we will examine how soil structure (the arrangement of solid, liquid, and gaseous components) and root morphology influence the transfer of inorganic nutrients from the environment into a plant. Finally, we will introduce the topic of symbiotic mycorrhizal associations, which play key roles in nutrient acquisition in the majority of plants. Chapters 6 and 13 will address additional aspects of solute transport and nutrient assimilation, respectively.

Essential Nutrients, Deficiencies, and Plant Disorders

Only certain elements have been determined to be essential for plants. An **essential element** is defined as one that is an intrinsic component in the structure or metabolism of a plant or whose absence causes severe abnormalities in plant growth, development, or reproduction and may prevent a plant from completing its life cycle. If plants are given these essential elements, as well as water and energy from sunlight, they can synthesize all the compounds they need for normal growth. **Table 5.1** lists the elements that

TABLE 5.1 Tissue levels of essential elements required by most plants

Element	Chemical symbol	Concentration in dry matter (% or ppm)[a]	Relative number of atoms with respect to molybdenum
Obtained from water or carbon dioxide			
Hydrogen	H	6	60,000,000
Carbon	C	45	40,000,000
Oxygen	O	45	30,000,000
Obtained from the soil			
Macronutrients			
Nitrogen	N	1.5	1,000,000
Potassium	K	1.0	250,000
Calcium	Ca	0.5	125,000
Magnesium	Mg	0.2	80,000
Phosphorus	P	0.2	60,000
Sulfur	S	0.1	30,000
Silicon	Si	0.1	30,000
Micronutrients			
Chlorine	Cl	100	3,000
Iron	Fe	100	2,000
Boron	B	20	2,000
Manganese	Mn	50	1,000
Sodium	Na	10	400
Zinc	Zn	20	300
Copper	Cu	6	100
Nickel	Ni	0.1	2
Molybdenum	Mo	0.1	1

Source: Epstein 1972, 1999.

[a]The values for the nonmineral elements (H, C, O) and the macronutrients are percentages. The values for micronutrients are expressed in parts per million.

are considered to be essential for most, if not all, higher plants. The first three elements—hydrogen, carbon, and oxygen—are not considered mineral nutrients because they are obtained primarily from water or carbon dioxide.

Essential mineral elements are usually classified as *macronutrients* or *micronutrients* according to their relative concentrations in plant tissue. In some cases the differences in tissue concentration between macronutrients and micronutrients are not as large as those indicated in Table 5.1. For example, some plant tissues, such as the leaf mesophyll, contain almost as much iron or manganese as they do sulfur or magnesium. Often elements are present in concentrations greater than the plant's minimum requirements.

Some researchers have argued that a classification into macronutrients and micronutrients is difficult to justify physiologically. Konrad Mengel and Ernest Kirkby have proposed that the essential elements be classified instead according to their biochemical role and physiological function. **Table 5.2** shows such a classification, in which plant nutrients have been divided into four basic groups:

1. Nitrogen and sulfur constitute the first group of essential elements. Plants assimilate these nutrients via biochemical reactions involving oxidation

TABLE 5.2 Classification of plant mineral nutrients according to biochemical function

Mineral nutrient	Functions
Group 1	**Nutrients that are part of carbon compounds**
N	Constituent of amino acids, amides, proteins, nucleic acids, nucleotides, coenzymes, hexosamines, etc.
S	Component of cysteine, cystine, methionine. Constituent of lipoic acid, coenzyme A, thiamine pyrophosphate, glutathione, biotin, 5′-adenylylsulfate, and 3′-phosphoadenosine.
Group 2	**Nutrients that are important in energy storage or structural integrity**
P	Component of sugar phosphates, nucleic acids, nucleotides, coenzymes, phospholipids, phytic acid, etc. Has a key role in reactions that involve ATP.
Si	Deposited as amorphous silica in cell walls. Contributes to cell wall mechanical properties, including rigidity and elasticity.
B	Complexes with mannitol, mannan, polymannuronic acid, and other constituents of cell walls. Involved in cell elongation and nucleic acid metabolism.
Group 3	**Nutrients that remain in ionic form**
K	Required as a cofactor for more than 40 enzymes. Principal cation in establishing cell turgor and maintaining cell electroneutrality.
Ca	Constituent of the middle lamella of cell walls. Required as a cofactor by some enzymes involved in the hydrolysis of ATP and phospholipids. Acts as a second messenger in metabolic regulation.
Mg	Required by many enzymes involved in phosphate transfer. Constituent of the chlorophyll molecule.
Cl	Required for the photosynthetic reactions involved in O_2 evolution.
Zn	Constituent of alcohol dehydrogenase, glutamic dehydrogenase, carbonic anhydrase, etc.
Na	Involved with the regeneration of phosphoenolpyruvate in C_4 and CAM plants. Substitutes for potassium in some functions.
Group 4	**Nutrients that are involved in redox reactions**
Fe	Constituent of cytochromes and nonheme iron proteins involved in photosynthesis, N_2 fixation, and respiration.
Mn	Required for activity of some dehydrogenases, decarboxylases, kinases, oxidases, and peroxidases. Involved with other cation-activated enzymes and photosynthetic O_2 evolution.
Cu	Component of ascorbic acid oxidase, tyrosinase, monoamine oxidase, uricase, cytochrome oxidase, phenolase, laccase, and plastocyanin.
Ni	Constituent of urease. In N_2-fixing bacteria, constituent of hydrogenases.
Mo	Constituent of nitrogenase, nitrate reductase, and xanthine dehydrogenase.

Source: After Evans and Sorger 1966 and Mengel and Kirkby 2001.

and reduction to form covalent bonds with carbon and create organic compounds (e.g., amino acids, nucleic acids, and proteins).

2. The second group is important in energy storage reactions or in maintaining structural integrity. Elements in this group are often present in plant tissues as phosphate, borate, and silicate esters in which the elemental group is covalently bound to an organic molecule (e.g., sugar phosphate).
3. The third group is present in plant tissue as either free ions dissolved in the plant water or ions electrostatically bound to substances such as the pectic acids present in the plant cell wall. Elements in this group have important roles as enzyme cofactors, in the regulation of osmotic potentials, and in controlling membrane permeability.
4. The fourth group, comprising metals such as iron, has important roles in reactions involving electron transfer.

Please keep in mind that this classification is somewhat arbitrary because many elements serve several functional roles. For example, manganese is listed in group 4 as a metal involved in several key electron transfer reactions, yet it is a mineral element that remains in ionic form, which would place it in group 3.

Some naturally occurring elements, such as aluminum, selenium, and cobalt, are not essential elements yet can also accumulate in plant tissues. Aluminum, for example, is not considered to be an essential element, although plants commonly contain from 0.1 to 500 μg aluminum per g dry matter, and the addition of low amounts of aluminum to a nutrient solution may stimulate plant growth. Many species in the genera *Astragalus, Xylorhiza,* and *Stanleya* accumulate selenium, although plants have not been shown to have a specific requirement for this element. Cobalt is part of cobalamin (vitamin B_{12} and its derivatives), a component of several enzymes in nitrogen-fixing microorganisms; thus, cobalt deficiency blocks the development and function of nitrogen-fixing nodules, but plants that are not fixing nitrogen do not require cobalt. Crop plants normally contain only relatively small amounts of such nonessential elements.

The following sections describe the methods used to examine the roles of nutrient elements in plants.

Special techniques are used in nutritional studies

To demonstrate that an element is essential requires that plants be grown under experimental conditions in which only the element under investigation is absent. Such conditions are extremely difficult to achieve with plants grown in a complex medium such as soil. In the nineteenth century, several researchers, including Nicolas-Théodore de Saussure, Julius von Sachs, Jean-Baptiste-Joseph-Dieudonné Boussingault, and Wilhelm Knop, approached this problem by growing plants with their roots immersed in a **nutrient solution** containing only inorganic salts. Their demonstration that plants could grow normally with no soil or organic matter proved unequivocally that plants can fulfill all their needs from only mineral nutrient elements, water, air (CO_2), and sunlight.

The technique of growing plants with their roots immersed in a nutrient solution without soil is called **solution culture** or **hydroponics**. Successful hydroponic culture (**Figure 5.2A**) requires a large volume of nutrient solution or frequent adjustment of the nutrient solution to prevent nutrient uptake by roots from producing large changes in the mineral nutrient concentrations and pH of the solution. A sufficient supply of oxygen to the root system is also critical and may be achieved by vigorous bubbling of air through the solution.

Hydroponics is used in the commercial production of many greenhouse and indoor crops, such as tomato (*Solanum lycopersicum*), cucumber (*Cucumis sativus*), and hemp (*Cannabis sativa*). In one form of commercial hydroponic culture, plants are grown in a supporting material such as sand, gravel, vermiculite, rockwool, polyurethane foams, or expanded clay (i.e., kitty litter). Nutrient solutions are then flushed through the supporting material, and old solutions are removed by leaching. In another form of hydroponic culture, plant roots lie on the surface of a trough, and nutrient solutions flow in a thin layer along the trough over the roots. This **nutrient film growth system** ensures that the roots receive an ample supply of oxygen (**Figure 5.2B**).

Another technique, which has sometimes been heralded as the medium of the future for scientific investigations, is to grow the plants in **aeroponics**. In this technique plants are grown with their roots suspended in air while being sprayed continuously with a nutrient solution (**Figure 5.2C**). This approach provides easy manipulation of the gaseous environment around the roots, but it requires higher concentrations of nutrients than hydroponic culture does to sustain rapid plant growth. For this reason and other technical difficulties, the use of aeroponics is not widespread.

An ebb-and-flow system (**Figure 5.2D**) is yet another approach to solution culture. In such systems, the nutrient solution periodically rises to immerse plant roots and then recedes, exposing the roots to a moist atmosphere. Like aeroponics, ebb-and-flow systems require higher concentrations of nutrients than do other hydroponic or nutrient film systems.

Nutrient solutions can sustain rapid plant growth

Over the years, many formulations have been used for nutrient solutions. Early formulations developed by Knop in Germany included only KNO_3, $Ca(NO_3)_2$, KH_2PO_4, $MgSO_4$, and an iron salt. At the time, this nutrient solution was believed to contain all the minerals required by

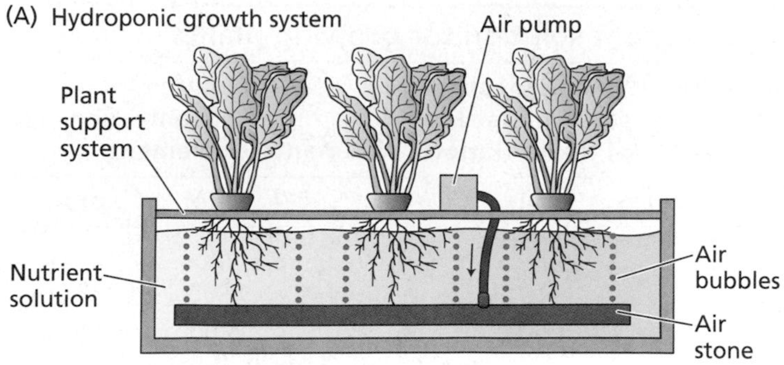

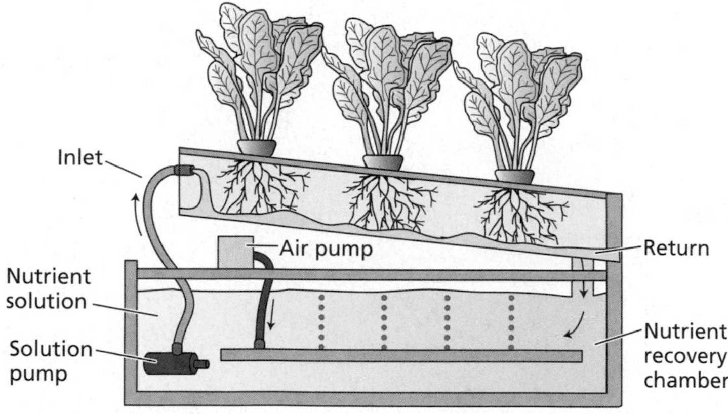

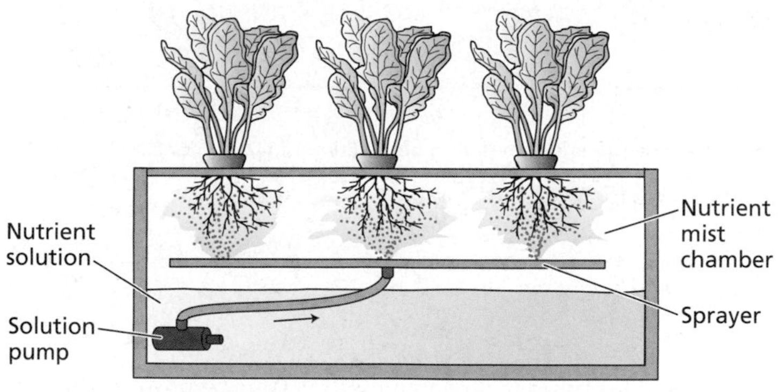

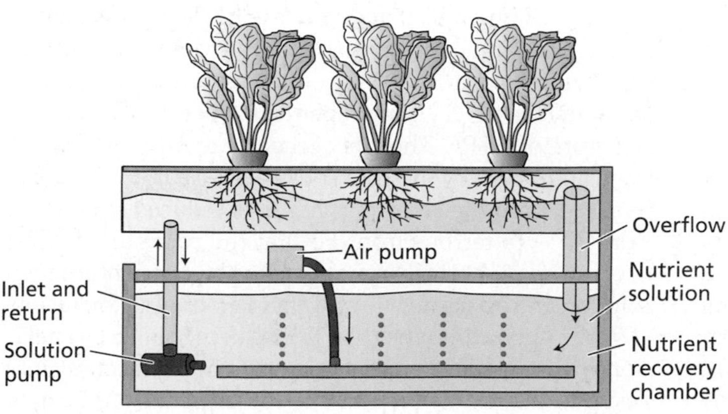

Figure 5.2 Various types of solution culture systems. (A) In a standard hydroponic culture, plants are suspended by the base of the stem over a tank containing a nutrient solution. The pumping of air through an air stone, a porous solid that generates a stream of small bubbles, keeps the solution fully saturated with oxygen. (B) In the nutrient film technique, a pump drives nutrient solution from a main reservoir along the bottom of a tilted tank, and down a return tube back to the reservoir. (C) In one type of aeroponics, a high-pressure pump sprays nutrient solution on roots enclosed in a tank. (D) In an ebb-and-flow system, a pump periodically fills an upper chamber containing the plant roots with nutrient solution. When the pump is turned off, the solution drains back through the pump into a main reservoir. (After Epstein and Bloom 2005.)

plants, but these experiments were carried out with chemicals that were contaminated with other elements that are now known to be essential (such as boron or molybdenum). **Table 5.3** shows a more modern formulation for a nutrient solution. This formulation is called a modified **Hoagland solution**, named after Dennis R. Hoagland, a researcher who was prominent in the development of modern mineral nutrition research in the United States.

A modified Hoagland solution contains all known mineral elements needed for rapid plant growth. The concentrations of these elements are set at the highest possible concentrations without producing toxicity symptoms or salinity stress, and thus may be several orders of magnitude higher than those found in the soil around plant roots. For example, whereas phosphorus is present in the soil solution at concentrations normally less than 0.06 $\mu g\ g^{-1}$ or 2 μM, here it is offered at 62 $\mu g\ g^{-1}$ or 2 mM. Such high initial concentrations permit plants to be grown in a medium for extended periods without replenishment of the nutrient, but they may injure young plants. Therefore, many researchers dilute their nutrient solutions severalfold and replenish them frequently to minimize fluctuations of nutrient concentration in the medium and in plant tissue.

Another important property of this modified Hoagland formulation is that nitrogen is supplied as both ammonium (NH_4^+) and nitrate (NO_3^-). Supplying nitrogen in a balanced mixture of cations (positively charged ions) and anions (negatively charged ions) tends to decrease the rapid rise in the pH of the medium that is commonly observed when the nitrogen is supplied solely as nitrate anion. Even when the pH of the medium is kept neutral, most plants grow better if they have

TABLE 5.3 Composition of a modified Hoagland nutrient solution for growing plants

Compound	Molecular weight	Concentration of stock solution	Concentration of stock solution	Volume of stock solution per liter of final solution	Element	Final concentration of element	
	g mol^{-1}	mM	g L^{-1}	mL		μM	ppm
Macronutrients							
KNO_3	101.10	1,000	101.10	6.0	N	16,000	224
$Ca(NO_3)_2 \cdot 4H_2O$	236.16	1,000	236.16	4.0	K	6,000	235
$NH_4H_2PO_4$	115.08	1,000	115.08	2.0	Ca	4,000	160
$MgSO_4 \cdot 7H_2O$	246.48	1,000	246.49	1.0	P	2,000	62
					S	1,000	32
					Mg	1,000	24
Micronutrients							
KCl	74.55	25	1.864		Cl	50	1.77
H_3BO_3	61.83	12.5	0.773		B	25	0.27
$MnSO_4 \cdot H_2O$	169.01	1.0	0.169	2.0 (KCl through H_2MoO_4 combined)	Mn	2.0	0.11
$ZnSO_4 \cdot 7H_2O$	287.54	1.0	0.288		Zn	2.0	0.13
$CuSO_4 \cdot 5H_2O$	249.68	0.25	0.062		Cu	0.5	0.03
H_2MoO_4 (85% MoO_3)	161.97	0.25	0.040		Mo	0.5	0.05
NaFeDTPA	468.20	64	30.0	0.3–1.0	Fe	16.1–53.7	1.00–3.00
Optional[a]							
$NiSO_4 \cdot 6H_2O$	262.86	0.25	0.066	2.0	Ni	0.5	0.03
$Na_2SiO_3 \cdot 9H_2O$	284.20	1,000	284.20	1.0	Si	1,000	28

Source: After Epstein and Bloom 2005.

Note: The macronutrients are added separately from stock solutions to prevent precipitation during preparation of the nutrient solution. A combined stock solution is made up containing all micronutrients except iron. Iron is added as sodium ferric diethylenetriaminepentaacetate (NaFeDTPA, trade name Ciba-Geigy Sequestrene 330 Fe; see Figure 5.3); some plants, such as maize, require the higher concentration of iron shown in the table.

[a]Nickel is usually present as a contaminant of the other chemicals, so it may not need to be added explicitly. Silicon, if included, should be added first and the pH adjusted with HCl to prevent precipitation of the other nutrients.

access to both NH_4^+ and NO_3^- because absorption and assimilation of the two nitrogen forms promotes inorganic cation–anion balance in the plant.

A significant problem with nutrient solutions is maintaining the availability of iron. When supplied as an inorganic salt such as $FeSO_4$ or $Fe(NO_3)_2$, iron can precipitate out of solution as iron hydroxide, particularly under alkaline conditions. If phosphate salts are present, insoluble iron phosphate will also form. Precipitation of the iron out of solution makes it physically unavailable to plants, unless iron salts are added at frequent intervals. Earlier researchers solved this problem by adding iron together with citric acid or tartaric acid. Compounds such as these are called **chelators** because they form soluble complexes with cations such as iron and calcium in which the cation is held by ionic forces rather than by covalent bonds. Chelated cations thus remain physically available to plants.

More modern nutrient solutions use the chemicals ethylenediaminetetraacetic acid (EDTA), diethylenetriaminepentaacetic acid (DTPA, or pentetic acid), or ethylenediamine-N,N′-bis(o-hydroxyphenylacetic) acid (*o,o*EDDHA) as chelating agents. **Figure 5.3** shows the structure of DTPA. The fate of the chelation complex during iron uptake by the root cells is not clear; iron may be released from the chelator when it is reduced from ferric iron (Fe^{3+}) to ferrous iron (Fe^{2+}) at the root surface. The chelator may then diffuse back into the nutrient (or soil) solution and associate with another Fe^{3+} or other metal ion.

After uptake into the root, iron is kept soluble by chelation with organic compounds present in plant cells. Citric acid may play a major role as such an organic iron chelator,

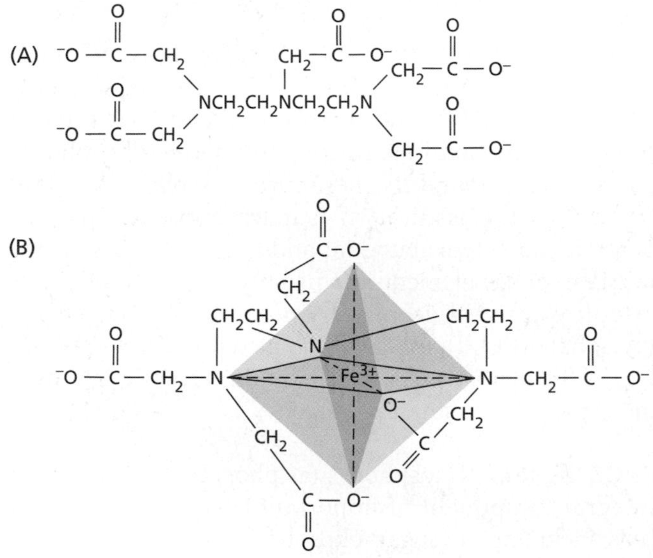

(A)

(B)

Figure 5.3 Chelator and chelated cation. Chemical structure of the chelator diethylenetriaminepentaacetic acid (DTPA) by itself (A) and chelated to Fe^{3+} (B). Iron binds to DTPA through interactions with three nitrogen atoms and the three ionized oxygen atoms of the carboxylate groups. The resulting ring structure clamps the metallic ion and effectively neutralizes its reactivity in solution. During the uptake of iron at the root surface, Fe^{3+} appears to be reduced to Fe^{2+}, which is released from the DTPA–iron complex. The chelator can then associate with other available Fe^{3+}. (After Sievers and Bailar 1962.)

and long-distance transport of iron in the xylem appears to involve an iron–citric acid complex.

Mineral defi[…]iencies disrupt plant metabolism and f[…]

[…] of an essential element results in […]anifested by characteristic defi- […] a[…]ponic culture, withholding […] readily correlated with a […]ple, a particular defi- […] of leaf discoloration. […] more complex for […]simultane- […]ant

• Some vi[…] toms similar t[…] Nutrient deficiency sy[…] sion of metabolic disorders […] supply of an essential element. Th[…] to the roles played by essential eleme[…] metabolism and function (listed in Table 5.2)

TABLE 5.4 Mineral elements classified on the basis of their mobility within a plant and their tendency to retranslocate during deficiencies

Mobile	Immobile
Nitrogen	Calcium
Potassium	Sulfur
Magnesium	Iron
Phosphorus	Boron
Chlorine	Copper
Sodium	
Zinc	
Molybdenum	

Note: Elements are listed in the order of their abundance in the plant.

Although each essential element participates in many different metabolic reactions, some general statements about the functions of essential elements in plant metabolism are possible. In general, the essential elements function in plant structure, metabolism, and cellular osmoregulation. More specific roles may be related to the ability of divalent cations such as Ca^{2+} or Mg^{2+} to modify the permeability of plant membranes. In addition, research continues to reveal specific roles for these elements in plant metabolism; for example, calcium ions act as a signal to regulate key enzymes in the cytosol. Thus, most essential elements have multiple roles in plant metabolism.

An important clue in relating an acute deficiency symptom to a particular essential element is the extent to which an element can be recycled from older to younger leaves. Some elements, such as nitrogen, phosphorus, and potassium, can readily move from leaf to leaf; others, such as boron, iron, and calcium, are relatively immobile in most plant species (**Table 5.4**). If an essential element is mobile, deficiency symptoms tend to appear first in older leaves. Conversely, deficiencies of immobile essential elements become evident first in younger leaves. Although the precise mechanisms of nutrient mobilization are not well understood, plant hormones such as cytokinins appear to be involved (see Chapter 15). In the discussion that follows, we will describe the particular deficiency symptoms and functional roles of the essential elements as they are grouped in Table 5.2. Please keep […]mind that many symptoms are highly dependent on […]species.

[…] DEFICIENCIES IN MINERAL NUTRI[…]T ARE PART OF CARBON COM[…]

[…] first group consists of nitrogen and sul- […]lability in soils limits plant productivity in most natural and agricultural ecosystems. By contrast, soils generally contain sulfur in excess. Despite this difference, nitrogen and sulfur are similar chemically in that their oxidation–reduction states range widely (see Chapter 13). Some of the most energy-intensive reactions in life convert highly oxidized inorganic forms, such as nitrate and sulfate, that roots absorb from the soil into highly reduced organic compounds, such as amino acids, within plants.

NITROGEN Nitrogen is the mineral element that plants require in the greatest amounts (see Table 5.1). It serves as a constituent of many plant cell components, including chlorophyll, amino acids, and nucleic acids. Therefore, nitrogen deficiency rapidly inhibits plant growth. If such a deficiency persists, most species show leaf **chlorosis** (yellowing of the leaves), especially in the older leaves near the base of the plant (for pictures of nitrogen deficiency and the other mineral deficiencies described in this chapter, see WEB TOPIC 5.1). Under severe nitrogen deficiency, these leaves become completely yellow (or tan) and fall off the plant. Younger leaves may not show these symptoms initially because nitrogen can be mobilized from older leaves. Thus, a nitrogen-deficient plant may have light green upper leaves and yellow or tan lower leaves.

When nitrogen deficiency develops slowly, plants may have markedly slender and often woody stems. This woodiness may be due to a buildup of excess carbohydrates that cannot be used in the synthesis of amino acids or other nitrogen-containing compounds. Carbohydrates not used in nitrogen metabolism may also be used in anthocyanin synthesis, leading to accumulation of that pigment. This condition is revealed as a purple coloration in leaves, petioles, and stems of nitrogen-deficient plants of some species, such as tomato and certain varieties of maize (corn; *Zea mays*).

SULFUR Sulfur is found in certain amino acids (i.e., cystine, cysteine, and methionine) and is a constituent of several coenzymes and vitamins, such as coenzyme A, *S*-adenosylmethionine, biotin, vitamin B_1, and pantothenic acid, which are essential for metabolism.

Many of the symptoms of sulfur deficiency are similar to those of nitrogen deficiency, including leaf chlorosis, stunting of growth, and anthocyanin accumulation. This similarity is not surprising, since sulfur and nitrogen are both constituents of proteins. The chlorosis caused by sulfur deficiency, however, generally arises initially in young and mature leaves, rather than in old leaves as in nitrogen deficiency, because sulfur, unlike nitrogen, is not easily remobilized to the younger leaves in most species. Nonetheless, in some plant species sulfur chlorosis may occur simultaneously in all leaves, or even initially in older leaves.

GROUP 2: DEFICIENCIES IN MINERAL NUTRIENTS THAT ARE IMPORTANT IN ENERGY STORAGE OR STRUCTURAL INTEGRITY This group consists of phosphorus, silicon, and boron. Phosphorus and silicon are found at concentrations in plant tissue that warrant their classification as macronutrients, whereas boron is much less abundant and is considered a micronutrient. These elements are usually present in plants as ester linkages between an inorganic acid group such a phosphate (PO_4^{3-}) and a carbon alcohol (i.e., X–O–C–R, where the element X is attached to a carbon-containing molecule C–R via an oxygen atom O).

PHOSPHORUS Phosphorus (as phosphate, PO_4^{3-}) is an integral component of important compounds in plant cells, including the sugar–phosphate intermediates of respiration and photosynthesis as well as the phospholipids that make up plant membranes. It is also a component of nucleotides used in plant energy metabolism (such as ATP) and in DNA and RNA. Characteristic symptoms of phosphorus deficiency include stunted growth of the entire plant and a dark green coloration of the leaves, which may be malformed and contain small areas of dead tissue called **necrotic spots** (for a picture, see WEB TOPIC 5.1).

As in nitrogen deficiency, some species may produce excess anthocyanins under phosphorus deficiency, giving the leaves a slight purple coloration. Unlike in nitrogen deficiency, the purple coloration of phosphorus deficiency is not associated with chlorosis. In fact, the leaves may be a dark greenish purple. Additional symptoms of phosphorus deficiency include the production of slender (but not woody) stems and the death of older leaves. Maturation of the plant may also be delayed.

SILICON Only members of the family Equisetaceae—called *scouring rushes* because at one time their ash, rich in gritty silica, was used to scour pots—require silicon to complete their life cycle. Nonetheless, many other specie[illegible] accumulate substantial amounts of silicon in their tis[illegible] and show enhanced growth, fertility, and stress r[illegible] when supplied with adequate amounts of sili[illegible]

Plants deficient in silicon are more sus[illegible] ing (falling over) and fungal infectio[illegible] ited primarily in the endoplasm[illegible] and intercellular spaces as h[illegible] ($SiO_2 \cdot nH_2O$). It also form[illegible] and thus serves as a co[illegible] ment of cell walls [illegible] ity of many m[illegible]

BORO[illegible] pl[illegible]

roles in cell elongation, nucleic acid synthesis, hormone responses, membrane function, and cell cycle regulation. Boron-deficient plants may exhibit a wide variety of symptoms, depending on the species and the age of the plant.

A characteristic symptom is black necrosis of young leaves and terminal buds. The necrosis of the young leaves occurs primarily at the base of the leaf blade. Stems may be unusually stiff and brittle. Apical dominance may also be lost, causing the plant to become highly branched; however, the terminal apices of the branches soon become necrotic because of inhibition of cell differentiation. Structures such as the fruits, fleshy roots, and tubers may exhibit necrosis or abnormalities related to the breakdown of internal tissues (see WEB ESSAY 5.1).

GROUP 3: DEFICIENCIES IN MINERAL NUTRIENTS THAT REMAIN IN IONIC FORM This group includes some of the most familiar mineral elements: the macronutrients potassium, calcium, and magnesium and the micronutrients chlorine, zinc, and sodium. These elements may be found as ions in solution in the cytosol or vacuoles, or they may be bound electrostatically or as ligands to larger, carbon-containing compounds.

POTASSIUM Potassium, present in plants as the cation K^+, plays an important role in regulation of the osmotic potential of plant cells (see Chapters 3 and 6). It also activates many enzymes involved in respiration and photosynthesis.

The first observable symptom of potassium deficiency is mottled or marginal chlorosis, which then develops into necrosis primarily at the leaf tips, at the margins, and between veins. In many monocotyledons, these necrotic lesions may initially form at the leaf tips and margins and then extend toward the leaf base. Because potassium can be mobilized to the younger leaves, these symptoms appear initially on the more mature leaves toward the base of the plant. The leaves may also curl and crinkle. The stems of potassium-deficient plants may be slender and weak, with abnormally short internodal regions. In potassium-deficient maize, the roots may have an increased susceptibility to root-rotting fungi present in the soil, and this susceptibility, together with effects on the stem, results in an increased tendency for the plant to be easily lodged (bent to the ground).

CALCIUM Calcium ions (Ca^{2+}) have two distinct roles in plants: (1) a structural/apoplastic role whereby Ca^{2+} binds to acidic groups of membrane lipids (phospho- and sulfolipids) and cross-links pectins, particularly in the middle lamellae that separate newly divided cells; and (2) a signaling role whereby Ca^{2+} acts as a second messenger that initiates plant responses to environmental stimuli. In its function as a second messenger, Ca^{2+} may bind to **calmodulin**, a protein found in the cytosol of plant cells. The calmodulin–Ca^{2+} complex then binds to several different types of proteins, including kinases, phosphatases, second messenger signaling proteins, and cytoskeletal proteins, and thereby regulates many cellular processes ranging from transcription control and cell survival to release of chemical signals (see Chapter 15).

Characteristic symptoms of calcium deficiency include necrosis of young meristematic regions such as the tips of roots or young leaves, where cell division and cell wall formation are most rapid. Necrosis in slowly growing plants may be preceded by a general chlorosis and downward hooking of young leaves. Young leaves may also appear deformed. The root system of a calcium-deficient plant may appear brownish, short, and highly branched. Severe stunting may result if the meristematic regions of the plant die prematurely.

MAGNESIUM In plant cells, magnesium ions (Mg^{2+}) have a specific role in the activation of enzymes involved in respiration, photosynthesis, and the synthesis of DNA and RNA. Mg^{2+} is also part of the ring structure of the chlorophyll molecule (see Figure 7.6A). A characteristic symptom of magnesium deficiency is chlorosis between the leaf veins, occurring first in older leaves because of the high mobility of this cation. This pattern of chlorosis occurs because the chlorophyll in the vascular bundles remains unaffected longer than that in the cells between the bundles. If the deficiency is extensive, the leaves may become yellow or white. An additional symptom of magnesium deficiency may be senescence and premature leaf abscission.

CHLORINE The element chlorine is found in plants as the chloride ion (Cl^-). It is required for the water-splitting reaction of photosynthesis through which oxygen is produced (see Chapter 7). In addition, chlorine may be required for cell division in leaves and roots. Plants deficient in chlorine develop wilting of the leaf tips followed by general leaf chlorosis and necrosis. The leaves may also exhibit reduced growth. Eventually, the leaves may take on a bronzelike color ("bronzing"). Roots of chlorine-deficient plants may appear stunted and thickened near the root tips.

Chloride ions are highly soluble and are generally available in soils because seawater is swept into the air by wind and delivered to soil when it rains. Therefore, chlorine deficiency is only rarely observed in plants grown in native or agricultural habitats. Most plants absorb chlorine at concentrations much higher than those required for normal functioning.

ZINC Many enzymes require zinc ions (Zn^{2+}) for their activity, and zinc may be required for chlorophyll biosynthesis in some plants. Zinc deficiency is characterized by a reduction in internodal growth, and as a result plants display a rosette habit of growth in which the leaves form a circular cluster radiating at or close to the ground. The leaves may also be small and distorted, with leaf margins having a puckered appearance. These symptoms may result from loss of the ability to produce sufficient amounts of the auxin indole-3-acetic acid (IAA). In some species (e.g., maize, sorghum, and beans), older leaves may show chlorosis between the leaf veins and then develop white necrotic spots. This chlorosis may be an expression of a zinc requirement for chlorophyll biosynthesis.

SODIUM Species using the C_4 and crassulacean acid metabolism (CAM) pathways of carbon fixation (see Chapter 8) may require sodium ions (Na^+). In these plants, Na^+ appears vital for regenerating phosphoenolpyruvate, the substrate for the first carboxylation in the C_4 and CAM pathways. Under sodium deficiency, these plants exhibit chlorosis and necrosis, or even fail to form flowers. Many C_3 species also benefit from exposure to low concentrations of Na^+. Sodium ions stimulate growth through enhanced cell expansion and can partly substitute for potassium ions as an osmotically active solute.

GROUP 4: DEFICIENCIES IN MINERAL NUTRIENTS THAT ARE INVOLVED IN REDOX REACTIONS This group of five micronutrients consists of the metals iron, manganese, copper, nickel, and molybdenum. All of these can undergo reversible oxidations and reductions (e.g., $Fe^{2+} \leftrightarrow Fe^{3+}$) and have important roles in electron transfer and energy transformation. They are usually found in association with larger molecules such as cytochromes, chlorophyll, and proteins (usually enzymes).

IRON Iron has an important role as a component of enzymes involved in the transfer of electrons (redox reactions), such as cytochromes. In this role, it is reversibly oxidized from Fe^{2+} to Fe^{3+} during electron transfer.

As in magnesium deficiency, a characteristic symptom of iron deficiency is intervenous chlorosis. This symptom, however, appears initially on younger leaves because iron, unlike magnesium, cannot be readily mobilized from older leaves. Under conditions of extreme or prolonged deficiency, the veins may also become chlorotic, causing the whole leaf to turn white. The leaves become chlorotic because iron is required for the synthesis of some of the chlorophyll–protein complexes in the chloroplast. The low mobility of iron is probably due to its precipitation in the older leaves as insoluble oxides or phosphates. The precipitation of iron diminishes subsequent mobilization of the metal into the phloem for long-distance translocation.

MANGANESE Manganese ions (Mn^{2+}) activate several enzymes in plant cells. In particular, decarboxylases and dehydrogenases involved in the citric acid (Krebs) cycle are specifically activated by manganese ions. The best-defined function of Mn^{2+} is in the photosynthetic reaction through which oxygen (O_2) is produced from water (see Chapter 7). The major symptom of manganese deficiency is intervenous chlorosis associated with the development of small necrotic spots. This chlorosis may occur on younger or older leaves, depending on plant species and growth rate.

COPPER Like iron, copper is associated with enzymes involved in redox reactions, through which it is reversibly oxidized from Cu^+ to Cu^{2+}. An example of such an enzyme is plastocyanin, which is involved in electron transfer during the light reactions of photosynthesis. The initial symptom of copper deficiency in many plant species is the production of dark green leaves, which may contain necrotic spots. The necrotic spots appear first at the tips of young leaves and then extend toward the leaf base along the margins. The leaves may also be twisted or malformed. Cereal plants exhibit a white leaf chlorosis and necrosis with rolled tips. Under extreme copper deficiency, leaves may drop prematurely and flowers may be sterile.

NICKEL Urease is the only known nickel-containing (Ni^{2+}) enzyme in higher plants, although nitrogen-fixing microorganisms require nickel (Ni^+ through Ni^{4+}) for the enzyme that reprocesses some of the hydrogen gas generated during fixation (hydrogen uptake hydrogenase) (see Chapter 13). Nickel-deficient plants accumulate urea in their leaves and consequently show leaf tip necrosis. Nickel deficiency in the field has been found in only one crop, pecan trees in the southeastern United States, because plants require only minuscule amounts of nickel (see Table 5.1).

MOLYBDENUM Molybdenum ions (Mo^{4+} through Mo^{6+}) are components of several enzymes, including nitrate reductase, nitrogenase, xanthine dehydrogenase, aldehyde oxidase, and sulphite oxidase. Nitrate reductase catalyzes the reduction of nitrate to nitrite during its assimilation by the plant cell; nitrogenase converts nitrogen gas to ammonia in nitrogen-fixing microorganisms (see Chapter 13). The first indication of a molybdenum deficiency is general chlorosis between veins and necrosis of older leaves. In some plants, such as cauliflower or broccoli, the leaves may not become necrotic but instead may appear twisted and subsequently die (whiptail disease). Flower formation may be prevented, or the flowers may abscise prematurely.

Because molybdenum is involved with both nitrate assimilation and nitrogen fixation, a molybdenum deficiency may bring about a nitrogen deficiency if the nitrogen source is primarily nitrate or if the plant depends on symbiotic nitrogen fixation. Although plants require only tiny amounts of molybdenum (see Table 5.1), some soils (for example, acidic soils in Australia) supply inadequate concentrations. Small additions of molybdenum to such soils can greatly enhance crop or forage growth at negligible cost.

Analysis of plant tissues reveals mineral deficiencies

Requirements for mineral elements change as a plant grows and develops. In crop plants, nutrient concentrations at certain stages of growth influence the yield of the economically important tissues (tuber, grain, and so on). To optimize yields, farmers use analyses of nutrient concentrations in soil and in plant tissue to determine fertilizer schedules.

Soil analysis is the chemical determination of the nutrient content in a soil sample from the root zone. As we will discuss later in the chapter, both the chemistry and the biology of soils are complex, and the results of soil analyses vary with sampling methods, storage conditions for the samples, and nutrient extraction techniques. Perhaps more important is that a particular soil analysis reflects the amount of nutrients *potentially* available to the plant roots from the soil, but soil analysis does not tell us how much of a particular mineral nutrient the plant actually needs or is able to absorb. This additional information is best determined by plant tissue analysis.

Proper use of **plant tissue analysis** requires an understanding of the relationship between plant growth (or yield) and the concentration of a nutrient in plant tissue samples. Bear in mind that the tissue concentration of a nutrient depends on the balance between nutrient absorption and dilution of the amount of nutrient through growth. **Figure 5.4** identifies three zones (deficiency, adequate, and toxic) in the response of growth to increasing tissue concentrations of a nutrient. When the nutrient concentration in a tissue sample is low, growth is reduced. In this **deficiency zone** of the curve, an increase in nutrient availability and absorption is directly related to an increase in growth or yield. As nutrient availability and absorption continue to increase, a point is reached at which further addition of the nutrient is no longer related to increases in growth or yield, but is reflected only in increased tissue concentrations. This region of the curve is called the **adequate zone**.

The point of transition between the deficiency and adequate zones of the curve reveals the **critical concentration** of the nutrient (see Figure 5.4), which may be defined as the minimum tissue concentration of the nutrient that is correlated with maximum growth or yield. As the nutrient concentration of the tissue increases beyond the adequate zone, growth or yield declines because of toxicity. This region of the curve is the **toxic zone**.

To evaluate the relationship between growth and tissue nutrient concentration, researchers grow plants in soil or a nutrient solution in which all the nutrients are present

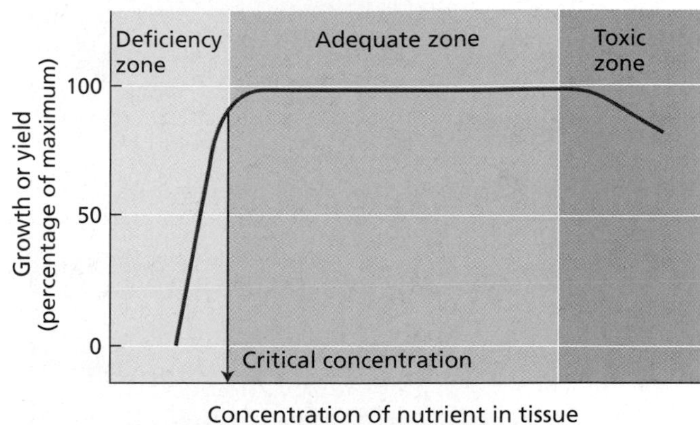

Figure 5.4 Relationship between yield (or growth) and the nutrient concentration of the plant tissue defines zones of deficiency, adequacy, and toxicity. Yield or growth may be expressed in terms of shoot dry weight or height. To obtain data of this type, plants are grown under conditions in which the concentration of one essential nutrient is varied while all others are in adequate supply. The effect of varying the concentration of this nutrient during plant growth is reflected in the growth or yield. The critical tissue concentration for that nutrient is the concentration below which yield or growth is reduced.

in adequate amounts except the nutrient under consideration. At the start of the experiment, the limiting nutrient is added in increasing concentrations to different sets of plants, and the concentrations of the nutrient in specific tissues are correlated with a particular measure of growth or yield. Several curves are established for each element, one for each tissue and tissue age.

Because agricultural soils are often limited in the elements nitrogen, phosphorus, and potassium (NPK), many farmers routinely take into account, at a minimum, growth or yield responses for these elements. If a nutrient deficiency is suspected, steps are taken to correct the deficiency before it reduces growth or yield. Plant analysis has proved useful in establishing fertilizer schedules that sustain yields and ensure the food quality of many crops.

Treating Nutritional Deficiencies

Many traditional and subsistence farming practices promote the recycling of mineral elements. Crop plants absorb nutrients from the soil, humans and animals consume locally grown crops, and crop residues and manure from humans and animals return the nutrients to the soil. The main losses of nutrients from such agricultural systems ensue from leaching that carries dissolved ions, especially nitrate, away with drainage water. In acidic soils, leaching of nutrients other than nitrate may be decreased by the addition of lime—a mix of CaO, $CaCO_3$, and $Ca(OH)_2$—to

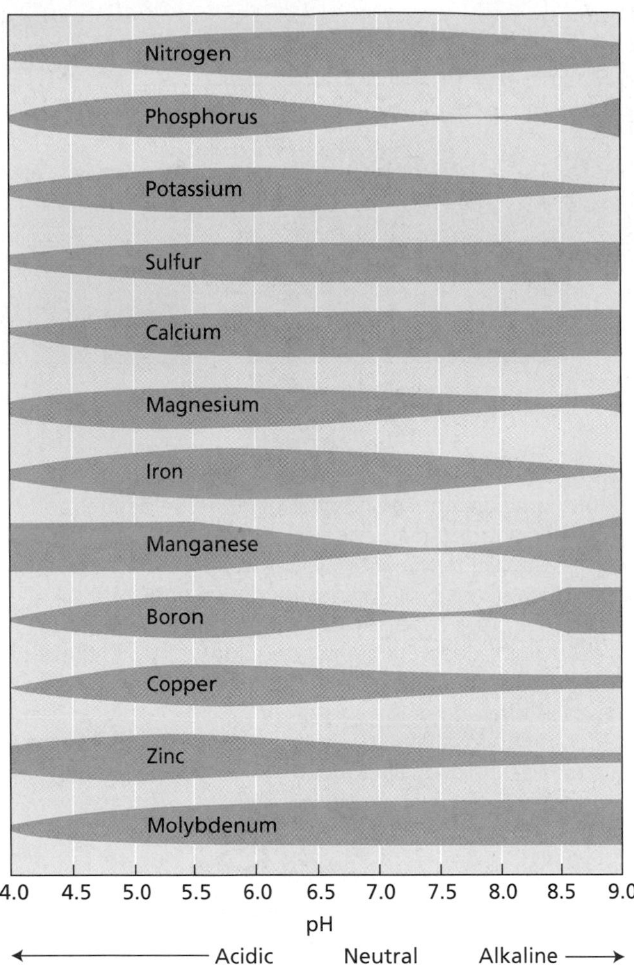

Figure 5.5 Influence of soil pH on the availability of nutrient elements in organic soils. The thickness of the horizontal bars indicates the degree of nutrient availability to plant roots. All of these nutrients are available in the pH range of 5.5 to 6.5. (After Lucas and Davis 1961.)

make the soil more alkaline because many mineral elements form less soluble compounds when the pH is higher than 6 (**Figure 5.5**). This decrease in leaching, however, may be gained at the expense of decreased availability of some nutrients, especially iron.

In the high-production agricultural systems of industrialized countries, a large proportion of crop biomass leaves the area of cultivation, and returning crop residues to the land where the crop was produced becomes difficult at best. This unidirectional removal of nutrients from agricultural soils makes it important to restore the lost nutrients to these soil through the addition of fertilizers.

Crop yields can be improved by the addition of fertilizers

Most **chemical fertilizers** contain inorganic salts of the macronutrients nitrogen, phosphorus, and potassium (see Table 5.1). Fertilizers that contain only one of these three nutrients are termed *straight fertilizers*. Some examples of straight fertilizers are superphosphate, ammonium nitrate, and muriate of potash (potassium chloride). Fertilizers that contain two or more mineral nutrients are called *compound fertilizers* or *mixed fertilizers*, and the numbers on the package label, such as "10-14-10," refer to the percentages of N, P, and K, respectively, in the fertilizer.

With long-term agricultural production, consumption of micronutrients by crops can reach a point at which they too must be added to the soil as fertilizers. Adding micronutrients to the soil may also be necessary to correct a preexisting deficiency. For example, many acidic, sandy soils in humid regions are deficient in boron, copper, zinc, manganese, molybdenum, or iron and can benefit from nutrient supplementation.

Chemicals may also be applied to the soil to modify soil pH. As Figure 5.5 shows, soil pH affects the availability of all mineral nutrients. Addition of lime, as mentioned previously, can raise the pH of acidic soils; addition of elemental sulfur can lower the pH of alkaline soils. In the latter case, microorganisms absorb the sulfur and subsequently release sulfate and hydrogen ions that acidify the soil.

Organic fertilizers are those approved for organic agricultural practices. In contrast to chemical fertilizers, they originate from natural rock deposits such as sodium nitrate and rock phosphate (phosphorite) or from the residues of plant or animal life. Natural rock deposits are chemically inorganic, but are acceptable for use in 'organic' agriculture. Plant and animal residues contain many nutrient elements in the form of organic compounds. Before crop plants can acquire the nutrient elements from these residues, the organic compounds must be broken down, usually by the action of soil microorganisms through a process called **mineralization**. Mineralization depends on many factors, including temperature, water and oxygen availability, pH, and the type and number of microorganisms present in the soil. As a consequence, rates of mineralization are highly variable, and nutrients from organic residues become available to plants over periods that range from days to months to years. This slow rate of mineralization hinders efficient organic fertilizer use, so farms that rely solely on organic fertilizers may require the addition of substantially more nitrogen or phosphorus and may suffer even higher nutrient losses than farms that use chemical fertilizers. Residues from organic fertilizers do improve the physical structure of most soils, enhancing water retention during drought and increasing drainage in wet weather. In some developing countries, organic fertilizers are all that is available or affordable.

Some mineral nutrients can be absorbed by leaves

In addition to absorbing nutrients added to the soil as fertilizers, most plants can absorb mineral nutrients applied to their leaves as a spray, a process known as **foliar application**. In some cases this method has agronomic advantages over the application of nutrients to the soil. Foliar application can reduce the lag time between application and uptake by the plant, which could be important during a phase of rapid growth. It can also circumvent the problem of restricted uptake of a nutrient from the soil. For example, foliar application of mineral nutrients such as iron, manganese, and copper may be more efficient than application through the soil, where these ions are adsorbed on soil particles and hence are less available to the root system.

Nutrient uptake by leaves is most effective when the nutrient solution is applied to the leaf as a thin film. Production of a thin film often requires that the nutrient solutions be supplemented with surfactant chemicals, such as the detergent Tween 80 or newly developed organosilicon surfactants, that reduce surface tension. Nutrient movement into the plant seems to involve diffusion through the cuticle and uptake by leaf cells, although uptake through the stomatal pores may also occur.

For foliar nutrient application to be successful, damage to the leaves must be minimized. If foliar sprays are applied on a hot day, when evaporation is high, salts may accumulate on the leaf surface and cause burning or scorching. Spraying on cool days or in the evening helps alleviate this problem. Addition of lime to the spray diminishes the solubility of many nutrients and limits toxicity. Foliar application has proved economically successful mainly with tree crops and vines such as grapes, but it is also used with cereals. Nutrients applied to the leaves can save an orchard or vineyard when soil-applied nutrients would be too slow to correct a deficiency. In wheat (*Triticum aestivum*), nitrogen applied to the leaves during the later stages of growth enhances the protein content of seeds.

Soil, Roots, and Microbes

Soil is complex physically, chemically, and biologically. It is a heterogeneous mixture of substances distributed in solid, liquid, and gaseous phases (see Chapter 4). All of these phases interact with mineral elements. The inorganic particles of the solid phase provide a reservoir of potassium, phosphorus, calcium, magnesium, and iron. Also associated with this solid phase are organic compounds containing nitrogen, phosphorus, and sulfur, among other elements. The liquid phase of soil constitutes the soil solution that is held in pores between the soil particles. It contains dissolved mineral ions and serves as the medium for ion movement to the root surface. Gases such as oxygen, carbon dioxide, and nitrogen are dissolved in the soil solution, but roots exchange gases with soils predominantly through the air-filled pores between soil particles.

From a biological perspective, soil constitutes a diverse ecosystem in which plant roots and microorganisms interact. Many microorganisms play key roles in releasing (mineralizing) nutrients from organic sources, some of which are then made directly available to plants. Under some soil conditions, free-living microbes compete with the plants for these mineral nutrients. In contrast, some specialized microorganisms, including mycorrhizal fungi and nitrogen-fixing bacteria, can form alliances with plants for their mutual benefit (**symbioses**, singular *symbiosis*). In this section we will discuss the importance of soil properties, root structure, and symbiotic mycorrhizal relationships to plant mineral nutrition. Chapter 13 will address the symbiotic relationships of plants with nitrogen-fixing bacteria.

Negatively charged soil particles affect the adsorption of mineral nutrients

Soil particles, both inorganic and organic, have predominantly negative charges on their surfaces. Many inorganic soil particles are crystal lattices that are tetrahedral arrangements of the cationic forms of aluminum (Al^{3+}) and silicon (Si^{4+}) bound to oxygen atoms, thus forming aluminates and silicates. When cations of lesser charge replace Al^{3+} and Si^{4+} in the crystal lattice, these inorganic soil particles become negatively charged.

Organic soil particles originate from dead plants, animals, and microorganisms that soil microorganisms have decomposed to various degrees. The negative surface charges of organic particles result from the dissociation of hydrogen ions from the carboxylic and phenolic acid groups present in this component of the soil. Most of the world's soils are composed of aggregates formed from organic and inorganic particles.

Soils are categorized by particle size:

- Gravel consists of particles larger than 2 mm.
- Coarse sand consists of particles between 0.2 and 2 mm.
- Fine sand consists of particles between 0.02 and 0.2 mm.
- Silt consists of particles between 0.002 and 0.02 mm.
- Clay consists of particles smaller than 0.002 mm (2 μm).

The silicate-containing clay materials are further divided into three major groups—kaolinite, illite, and montmorillonite—based on differences in their structure and physical properties (**Table 5.5**). The kaolinite group is generally found in well-weathered soils; the montmorillonite and illite groups are found in less-weathered soils.

TABLE 5.5 Comparison of properties of three major types of silicate clays found in the soil

Property	Type of clay		
	Montmorillonite	Illite	Kaolinite
Size (μm)	0.01–1.0	0.1–2.0	0.1–5.0
Shape	Irregular flakes	Irregular flakes	Hexagonal crystals
Cohesion	High	Medium	Low
Water-swelling capacity	High	Medium	Low
Cation exchange capacity (milliequivalents 100 g^{-1})	80–100	15–40	3–15

Source: After Brady 1974.

Mineral cations such as ammonium (NH_4^+) and potassium (K^+) are adsorbed to the negative surface charges of inorganic and organic soil particles or adsorbed within the lattices formed by soil particles. This cation adsorption is an important factor in soil fertility. Mineral cations adsorbed on the surface of soil particles are not readily leached when the soil is infiltrated by water and thus provide a nutrient reserve available to plant roots. Mineral nutrients adsorbed in this way can be replaced by other cations in a process known as **cation exchange** (**Figure 5.6**). The degree to which a soil can adsorb and exchange ions is termed its *cation exchange capacity* (*CEC*) and is highly dependent on the soil type. A soil with higher cation exchange capacity generally has a larger reserve of mineral nutrients.

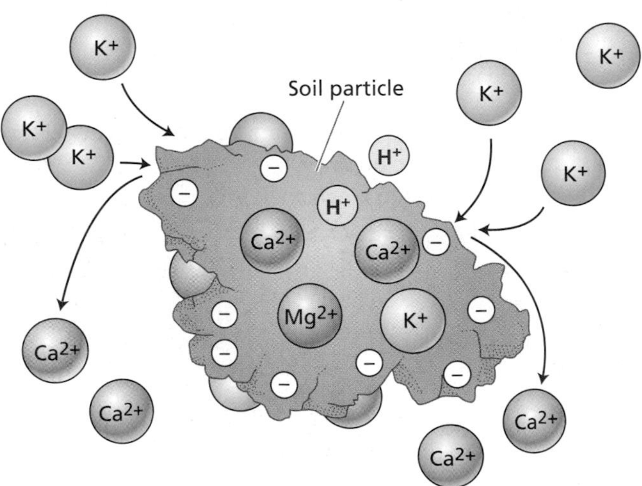

Figure 5.6 The principle of cation exchange on the surface of a soil particle. Cations are adsorbed on the surface of a soil particle because that surface is negatively charged. Addition of one cation, such as potassium (K^+), to the soil can displace other cations, such as calcium (Ca^{2+}), from the surface of the soil particle and make it available for uptake by roots.

Mineral anions such as nitrate (NO_3^-) and chloride (Cl^-) tend to be repelled by the negative charge on the surface of soil particles and remain dissolved in the soil solution. Thus, the anion exchange capacity of most agricultural soils is small compared with their cation exchange capacity. Nitrate in particular remains mobile in the soil solution, where it is susceptible to leaching by water moving through the soil.

Phosphate ions ($H_2PO_2^-$) may bind to soil particles containing aluminum or iron because the positively charged iron and aluminum ions (Fe^{2+}, Fe^{3+}, and Al^{3+}) are associated with hydroxyl ion (OH^-) groups that are exchanged for phosphate. Phosphate ions also react strongly with Ca^{2+}, Fe^{3+}, and Al^{3+} to form insoluble inorganic compounds. As a result, phosphate is frequently tightly bound at both low and high pH (see Figure 5.5), and its lack of mobility and availability in soil can limit plant growth. Formation of mycorrhizal symbioses (which we'll discuss later in this section) helps overcome this lack of mobility. Additionally, the roots of some plants, such as those of lupine (*Lupinus albus*) and members of the Proteaceae (e.g., *Macadamia, Banskia, Protea*), secrete large amounts of organic anions or protons into the soil that release phosphate from iron, aluminum, and calcium phosphates.

Sulfate (SO_4^{2-}) in the presence of Ca^{2+} forms gypsum ($CaSO_4$). Gypsum is only slightly soluble, but it releases sufficient sulfate to support plant growth. Most nonacidic soils contain substantial amounts of Ca^{2+}; consequently, sulfate mobility in these soils is low, and sulfate is not highly susceptible to leaching.

Soil pH affects nutrient availability, soil microbes, and root growth

Hydrogen ion concentration (pH) is an important property of soils because it affects the growth of plant roots and soil microorganisms. Root growth is generally favored in slightly acidic soils, at pH values between 5.5 and 6.5. Fungi generally predominate in acidic (pH below 7) soils; bacteria become more prevalent in alkaline (pH above 7) soils. Soil pH determines the availability of soil nutrients (see Figure 5.5). Acidity promotes the weathering of rocks that releases K^+, Mg^{2+}, Ca^{2+}, and Mn^{2+} and increases the solubility of carbonates, sulfates, and some phosphates. Increasing the solubility of nutrients enhances their availability to roots as concentrations increase in the soil solution.

Major factors that lower the soil pH are the decomposition of organic matter, ammonium assimilation by plants and microbes, and the amount of rainfall. Carbon dioxide is produced as a result of the decomposition of organic matter and equilibrates with soil water in the following reaction:

$$CO_2 + H_2O \leftrightarrow H^+ + HCO_3^-$$

This releases hydrogen ions (H^+), lowering the pH of the soil. Microbial decomposition of organic matter also produces ammonia/ammonium (NH_3/NH_4^+) and hydrogen sulfide (H_2S) that can be oxidized in the soil to form the strong acids nitric acid (HNO_3) and sulfuric acid (H_2SO_4), respectively. As plant roots absorb ammonium ions from the soil and assimilate them into amino acids, the roots generate hydrogen ions that they excrete into the surrounding soil (see Chapter 13). Hydrogen ions also displace K^+, Mg^{2+}, Ca^{2+}, and Mn^{2+} from the surfaces of soil particles. Leaching may then remove these ions from the upper soil layers, leaving a more acidic soil. By contrast, the weathering of rock in arid regions releases K^+, Mg^{2+}, Ca^{2+}, and Mn^{2+} into the soil, but because of the low rainfall, these ions do not leach from the upper soil layers, and the soil remains alkaline.

Excess mineral ions in the soil limit plant growth

When excess mineral ions are present in soil, the soil is said to be *saline*, and such soils may inhibit plant growth if the mineral ions reach concentrations that limit water availability or exceed the adequate levels for a particular nutrient (see Chapter 24). Sodium chloride and sodium sulfate are the most common salts in saline soils. Excess mineral ions in soils can be a major problem in arid and semiarid regions because rainfall is insufficient to leach them from the soil layers near the surface.

Irrigated agriculture fosters soil salinization if the amount of water applied is insufficient to leach the salt below the root zone. Irrigation water can contain 100 to 1000 g of mineral ions per cubic meter. An average crop requires about 10,000 m^3 of water per hectare. Consequently, 1000 to 10,000 kg of mineral ions per hectare may be added to the soil per crop, and over a number of growing seasons, high concentrations of mineral ions may accumulate in the soil.

In saline soils, plants encounter **salt stress**. Whereas many plants are affected adversely by the presence of relatively low concentrations of salt, other plants can survive (**salt-tolerant plants**) or even thrive (**halophytes**) at high salt concentrations. The mechanisms by which plants tolerate high salinity are complex (see Chapter 24), involving biochemical synthesis, enzyme induction, and membrane transport. In some plant species, excess mineral ions are not taken up, being excluded by the roots; in others, they are taken up but excreted from the plant by salt glands associated with the leaves. To prevent toxic buildup of mineral ions in the cytosol, many plants sequester them in the vacuole. Efforts are under way to bestow salt tolerance on salt-sensitive crop species using both classic plant breeding and biotechnology, as detailed in Chapter 24.

Another important problem with excess mineral ions is the accumulation of heavy metals in the soil, which can cause severe toxicity in plants as well as humans (see **WEB ESSAY 5.2**). These heavy metals include zinc, copper, cobalt, nickel, mercury, lead, cadmium, silver, and chromium.

Some plants develop extensive root systems

The ability of plants to obtain both water and mineral nutrients from the soil is related to their ability to develop an extensive root system and to various other traits such as ability to secrete organic anions or develop mycorrhizal symbioses. In the late 1930s, H. J. Dittmer examined the root system of a single winter rye plant after 16 weeks of growth. He estimated that the plant had 13 million primary and lateral root axes, extending more than 500 km in length and providing 200 m^2 of surface area. This plant also had more than 10^{10} root hairs, providing another 300 m^2 of surface area. The total surface area of roots from a single rye plant equaled that of a professional basketball court. Other plant species may not develop such extensive root systems, which may limit their nutrient absorption capacity and increase their reliance on mycorrhizal symbioses (discussed below).

In the desert, the roots of mesquite (genus *Prosopis*) may extend downward more than 50 m to reach groundwater. Annual crop plants have roots that usually grow between 0.1 and 2.0 m in depth and extend laterally to distances of 0.3 to 1.0 m. In orchards, the major root systems of trees planted 1 m apart reach a total length of 12 to 18 km per tree. The annual production of roots in natural ecosystems may easily surpass that of shoots, so in many respects the aboveground portions of a plant represent only "the tip of the iceberg." Nonetheless, making observations on root systems is difficult and usually requires special techniques (see **WEB TOPIC 5.2**).

Plant roots may grow continuously throughout the year if conditions are favorable. Their proliferation, however, depends on the availability of water and minerals in the immediate microenvironment surrounding the root, the so-called **rhizosphere**. If the rhizosphere is poor in nutrients or too dry, root growth is slow. As rhizosphere conditions improve, root growth increases. If fertilization

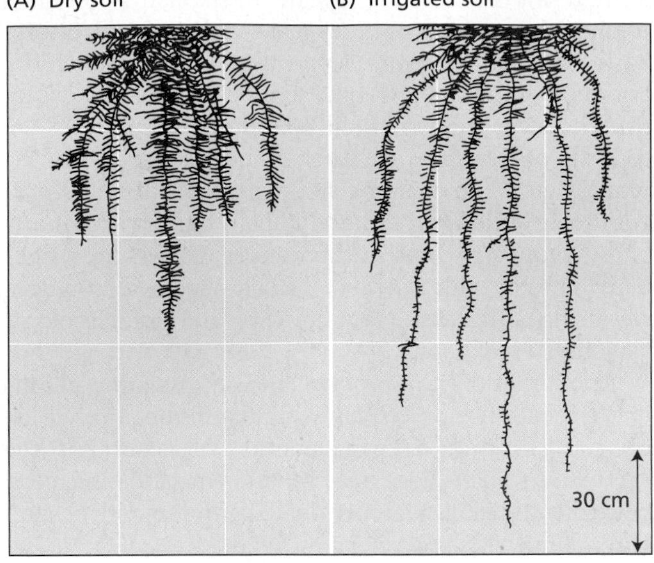

Figure 5.7 Fibrous root systems of wheat (a monocotyledon). (A) The root system of a mature (3-month-old) wheat plant growing in dry soil. (B) The root system of a mature wheat plant growing in irrigated soil. It is apparent that the morphology of the root system is affected by the amount of water present in the soil. In a mature fibrous root system, the primary root axes are indistinguishable. (After Weaver 1926.)

and irrigation provide abundant nutrients and water, root growth may not keep pace with shoot growth. Plant growth under such conditions becomes carbohydrate-limited, and a relatively small root system meets the nutrient needs of the whole plant. In crops for which we harvest aboveground parts, fertilization and irrigation cause greater allocation of resources to the shoot and reproductive structures than to roots, and this shift in allocation patterns often results in higher yields.

Root systems differ in form but are based on common structures

The *form* of the root system differs greatly among plant species. In monocotyledons, root development starts with the emergence of three to six **primary** (or *seminal*) **root** axes from the germinating seed. With further growth, the plant extends new adventitious roots, called **nodal roots** or *brace roots*. Over time, the primary and nodal root axes grow and branch extensively to form a complex *fibrous root system* (**Figure 5.7**). In fibrous root systems, all the roots generally have similar diameters (except where environmental conditions or pathogenic interactions modify the root structure), so it is impossible to distinguish a main root axis.

In contrast to monocotyledons, dicotyledons develop root systems with a main single root axis, called a **taproot**, which may thicken as a result of secondary cambial activity. From this main root axis, *lateral roots* develop to form an extensively branched root system (**Figure 5.8**).

The development of the root system in both monocotyledons and dicotyledons depends on the activity of the root apical meristem and the production of lateral root meristems. **Figure 5.9** is a generalized diagram of the apical region of a plant root and identifies three zones of activity: the meristematic, elongation, and maturation zones.

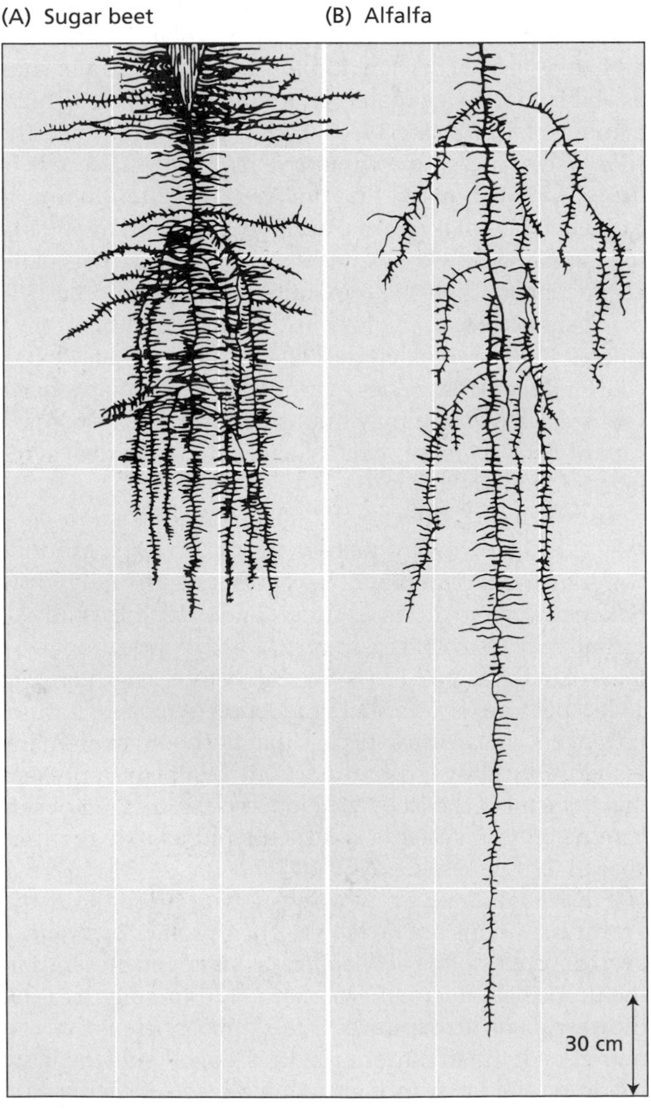

Figure 5.8 Taproot system of two adequately watered dicotyledons: sugar beet (A) and alfalfa (B). The sugar beet root system is typical of 5 months of growth; the alfalfa root system is typical of 2 years of growth. In both dicotyledons, the root system shows a major vertical root axis. In the case of sugar beet, the upper portion of the taproot system is thickened because of its function as storage tissue. (After Weaver 1926.)

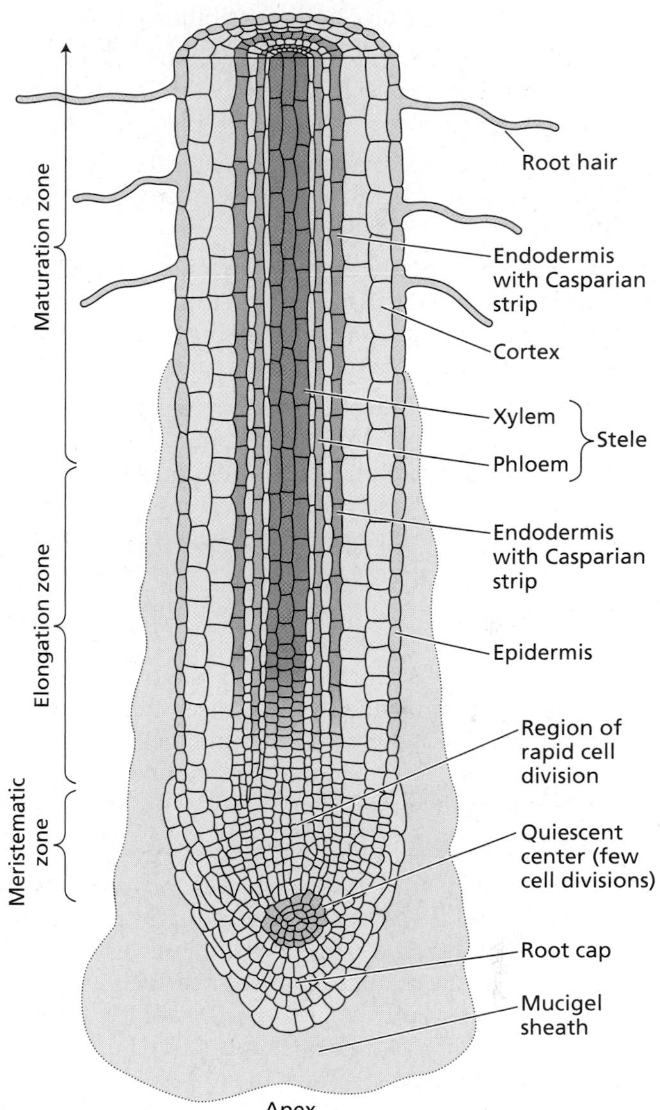

Figure 5.9 Diagrammatic longitudinal section of the apical region of the root. The meristematic cells are located near the tip of the root. These cells generate the root cap and the upper tissues of the root. In the elongation zone, cells differentiate to produce xylem, phloem, and cortex. Root hairs, formed in epidermal cells, first appear in the maturation zone.

In the **meristematic zone**, cells divide both in the direction of the root base to form cells that will differentiate into the tissues of the functional root and in the direction of the root apex to form the **root cap**. The root cap protects the delicate meristematic cells as the root expands into the soil. It commonly secretes a gelatinous material called *mucigel*, which surrounds the root tip. The precise function of mucigel is uncertain, but it may provide lubrication that eases the root's penetration of the soil, protect the root apex from desiccation, promote the transfer of nutrients to the root, and affect interactions between the root and soil microorganisms. The root cap is central to the perception of gravity, the signal that directs the growth of roots downward. This process is termed the **gravitropic response** (see Chapter 18).

Cell division in the root apex proper is relatively slow; thus this region is called the **quiescent center**. After a few generations of slow cell divisions, root cells displaced from the apex by about 0.1 mm begin to divide more rapidly. Cell division again tapers off at about 0.4 mm from the apex, and the cells expand equally in all directions.

The **elongation zone** begins approximately 0.7 to 1.5 mm from the apex (see Figure 5.9). In this zone cells elongate rapidly and undergo a final round of divisions to produce a central ring of cells called the **endodermis**. The walls of this endodermal cell layer become thickened, and suberin is deposited on the radial walls and forms the **Casparian strip**, a hydrophobic structure that prevents apoplastic movement of water or solutes across the root (see Figure 4.4).

The endodermis divides the root into two regions: the **cortex** toward the outside and the **stele** toward the inside. The stele contains the vascular elements of the root: the **phloem**, which transports metabolites from the shoot to the root and to fruits and seeds, and the **xylem**, which transports water and solutes to the shoot.

Phloem develops more rapidly than xylem, attesting to the fact that phloem function is critical near the root apex. Large quantities of carbohydrates must flow through the phloem to the growing apical zones in order to support cell division and elongation. Carbohydrates provide rapidly growing cells with an energy source and with the carbon skeletons required to synthesize organic compounds. Six-carbon sugars (hexoses) also function as osmotically active solutes in the root tissue. At the root apex, where the phloem is not yet developed, carbohydrate movement depends on symplastic transport and is relatively slow. The low rates of cell division in the quiescent center may result from the fact that insufficient carbohydrates reach this centrally located region or that this area is kept in an oxidized state.

Root hairs, with their large surface area for absorption of water and solutes and for anchoring the root to the soil, first appear in the **maturation zone** (see Figure 5.9), and here the xylem develops the capacity to translocate substantial quantities of water and solutes to the shoot.

Different areas of the root absorb different mineral ions

The precise point of entry of minerals into the root system has been a topic of considerable interest. Some researchers have claimed that nutrients are absorbed only at the apical regions of the root axes or branches; others claim that nutrients are absorbed over the entire root surface. Experimental evidence supports both possibilities, depending on the plant species and the nutrient being investigated:

- Root absorption of calcium ions in barley (*Hordeum vulgare*) appears to be restricted to the apical region.
- Iron may be taken up either at the apical region, as in barley and other species, or over the entire root surface, as in maize.
- Potassium ions, nitrate, ammonium, and phosphate can be absorbed freely at all locations of the root surface, but in maize the elongation zone has the maximum rates of potassium ion accumulation and nitrate absorption.
- In maize and rice and in wetland species, the root apex absorbs ammonium more rapidly than the elongation zone does. Ammonium and nitrate uptake by conifer roots varies significantly across different regions of the root and may be influenced by rates of root growth and maturation.
- In several species, the root apex and root hairs are the most active in phosphate absorption. For species with poorly developed root hairs, hyphae of arbuscular mycorrhizal fungi may play a significant role in uptake of phosphate and other nutrients, and the development of this symbiosis can change the regions of the root involved in uptake.

The high rates of nutrient absorption in the apical root zones result from the strong demand for nutrients in these tissues and the relatively high nutrient availability in the soil surrounding them. For example, cell elongation depends on the accumulation of solutes such as potassium, chloride, and nitrate ions to increase the osmotic pressure within the cell (see Chapter 14). Ammonium is the preferred nitrogen source to support cell division in the meristem because meristematic tissues are often carbohydrate-limited and because assimilation of ammonium into organic nitrogen compounds consumes less energy than assimilation of nitrate (see Chapter 13). The root apex and root hairs grow into fresh soil, where nutrients have not yet been depleted.

Within the soil, nutrients can move to the root surface both by bulk flow and by diffusion (see Chapter 3). In bulk flow, nutrients are carried by water moving through the soil toward the root. The amounts of nutrients provided to the root by bulk flow depend on the rate of water flow through the soil toward the plant, which itself depends on transpiration rates and on nutrient concentrations in the soil solution. When both the rate of water flow and the concentrations of nutrients in the soil solution are high, bulk flow can play an important role in nutrient supply. Hence, highly soluble nutrients such as nitrate are largely carried by bulk flow, but this process is less important for nutrients with low solubility, such as phosphate and zinc ions.

In diffusion, mineral nutrients move from a region of higher concentration to a region of lower concentration. Nutrient uptake by roots lowers the concentrations of nutrients at the root surface, generating concentration gradients in the soil solution surrounding the root. Diffusion of nutrients down their concentration gradients, along with bulk flow resulting from transpiration, can increase nutrient availability at the root surface.

When the rate of absorption of a nutrient by roots is high and the nutrient concentration in the soil solution is low, bulk flow can supply only a small fraction of the total nutrient requirement. Under these conditions, plant nutrient absorption becomes independent of plant transpiration rates, and diffusion rates limit the movement of the nutrient to the root surface. When diffusion is too slow to maintain high nutrient concentrations near the root, a **nutrient depletion zone** forms adjacent to the root surface (**Figure 5.10**). This zone may extend from about 0.2 to 2.0 mm from the root surface, depending on the mobility of the nutrient in the soil. The nutrient depletion zone is particularly important for phosphate.

The formation of a depletion zone tells us something important about mineral nutrition. Because roots deplete the mineral supply in the rhizosphere, their effectiveness in mining minerals from the soil is determined not only by the rate at which they can remove nutrients from the soil solution, but by their continuous growth into undepleted soil. Without continuous growth, roots would rapidly deplete the soil adjacent to their surfaces. Optimal nutrient acquisition therefore depends both on the root system's capacity for nutrient uptake and on its ability to grow into fresh soil. The ability of the plant to form a mycorrhizal symbiosis is also critical in overcoming the effects of

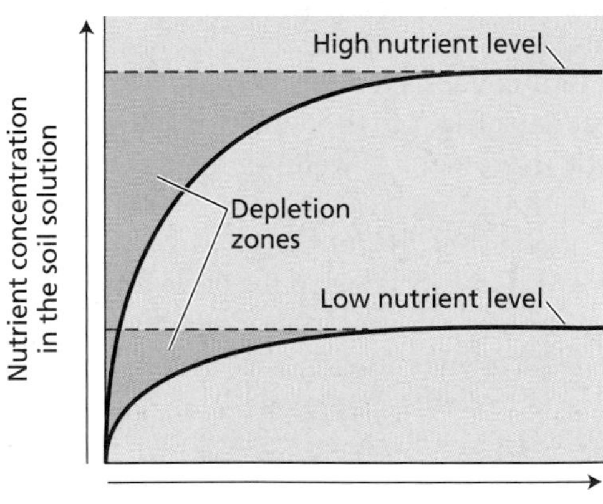

Figure 5.10 Formation of a nutrient depletion zone in the region of the soil adjacent to the plant root. A nutrient depletion zone forms when the rate of nutrient uptake by the cells of the root exceeds the rate of replacement of the nutrient by bulk flow and diffusion in the soil solution. This depletion causes a localized decrease in the nutrient concentration in the area adjacent to the root surface. (After Mengel and Kirkby 2001.)

depletion zones, because the hyphae of the fungal symbionts grow beyond the depletion zone. These fungal structures take up nutrients far from the root (up to 25 cm in the case of arbuscular mycorrhizas) and translocate them rapidly to the roots, overcoming the slow diffusion in soil.

Nutrient availability influences root growth

Plants, which have limited mobility for most of their lives, must deal with changes in their local environment because they cannot move away from unfavorable conditions. Above the ground, light intensity, temperature, and humidity may fluctuate substantially during the day and across the canopy, but CO_2 and O_2 concentrations remain relatively uniform. In contrast, soil buffers the roots from temperature extremes, but the belowground concentrations of CO_2 and O_2, water, and nutrients are extremely heterogeneous, both spatially and temporally. For example, inorganic nitrogen concentrations in soil may range a thousandfold over a distance of centimeters or the course of hours. Given such heterogeneity, plants seek the most favorable conditions within their reach.

Roots sense the belowground environment—through gravitropism, thigmotropism, chemotropism, and hydrotropism—to guide their growth toward soil resources. Some of these responses involve auxin (see Chapter 18). The extent to which roots proliferate within a soil patch varies with nutrient concentrations (**Figure 5.11**). Root growth is minimal in poor soils because the roots become nutrient-limited. As soil nutrient availability increases, roots proliferate.

Where soil nutrients exceed an optimal concentration root growth may become carbohydrate-limited and eventually cease. With high soil nutrient concentrations, a few roots—3.5% of the root system in spring wheat and 12% in lettuce—are sufficient to supply all the nutrients required, so the plant may diminish the allocation of its resources to roots while increasing its allocation to the shoot and reproductive structures. This resource shifting is one mechanism through which fertilization stimulates crop yields.

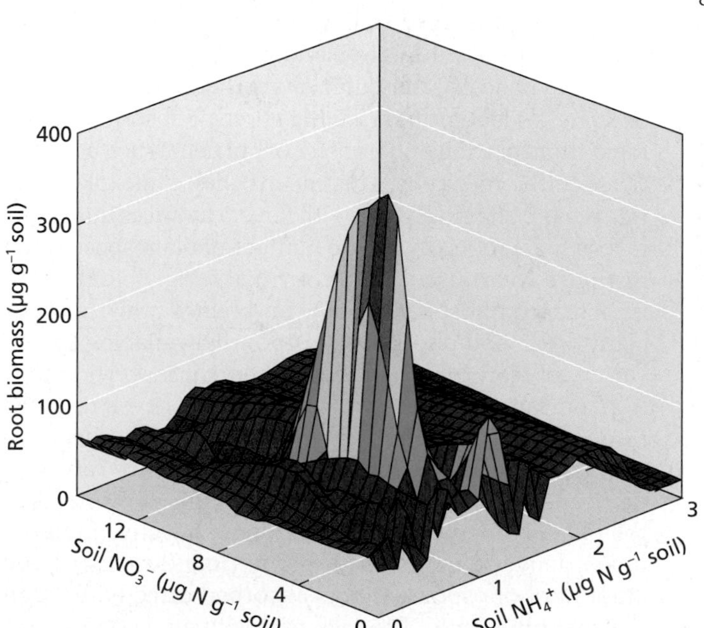

Figure 5.11 Root biomass as a function of extractable soil NH_4^+ and NO_3^-. The root biomass is shown (µg root dry weight g^{-1} soil) plotted against extractable soil NH_4^+ and NO_3^- (µg extractable N g^{-1} soil) for tomato (*Solanum lycopersicum* cv T-5) growing in an irrigated field that had been fallow the previous 2 years. The colors emphasize the differences among biomasses, ranging from low (purple) to high (red). (After Bloom et al. 1993.)

Mycorrhizal symbioses facilitate nutrient uptake by roots

Our discussion so far has centered on the direct acquisition of mineral elements by roots, but this process is usually modified by the association of **mycorrhizal fungi** with the root system to form a **mycorrhiza** (from the Greek words for "fungus" and "root"; plural *mycorrhizas* or *mycorrhizae*). The host plant supplies associated mycorrhizal fungi with carbohydrates and in return receives nutrients from the fungi. There are indications that drought and disease tolerance may also be improved in the host plant.

Mycorrhizal symbioses of two main types—**arbuscular mycorrhizas** and **ectomycorrhizas**—are widespread in nature, occurring in about 90% of terrestrial plant species, including most major crops. The majority, perhaps 80%, are arbuscular mycorrhizas, which are symbioses between a newly described phylum of fungi, the Glomeromycota, and a broad range of angiosperms, gymnosperms, ferns, and liverworts. Their importance in herbaceous species and in fruit trees of many types makes arbuscular mycorrhizas vital to agricultural production, particularly in nutrient-poor soils. This is the most ancient type of mycorrhiza, occurring in fossils of the earliest land plants. This symbiosis was probably important in facilitating plant establishment on land over 450 million years ago, because early land plants had poorly developed underground organs.

By contrast, ectomycorrhizal symbioses evolved more recently. They are formed by far fewer plant species, notably trees in the families Pinaceae (pines, larches, Douglas fir), Fagaceae (beech, oak, chestnut), Salicaceae (poplar, aspen), Betulaceae (birch), and Myrtaceae (*Eucalyptus*). The fungal partners belong to either the Basidomycota or, less frequently, the Ascomycota. These symbioses play major roles in the nutrition of trees and therefore in the productivity of vast areas of boreal forest.

Some plant species, particularly those in the families Salicaceae (*Salix* [willow] and *Populus* [poplar and aspen]) and Myrtaceae (*Eucalyptus*), can form both arbuscular and ectomycorrhizal symbioses. Other plants prove unable to form any kind of mycorrhiza. These include members of the families Brassicaceae, such as cabbage (*Brassica oleracea*) and the model plant *Arabidopsis thaliana*; Chenopodiaceae, such as spinach (*Spinacea oleracea*); and Proteaceae, such as macadamia nut (*Macadamia integrifolia*).

Certain agricultural practices may reduce or eliminate mycorrhiza formation in plants that normally form them. These practices include flooding (paddy rice does not form mycorrhizas, whereas upland rice does), extensive soil disturbance caused by plowing, application of high concentrations of fertilizer, and of course soil fumigation and application of some fungicides. Such practices may decrease yields in crops such as maize that are very dependent on mycorrhizas for nutrient uptake. Mycorrhizas also do not form in solution culture or in hydroponic cultivation. Nonetheless, for the majority of plants, mycorrhiza formation is the normal situation and the non-mycorrhizal state is essentially an artifact, brought about by particular agricultural practices.

Mycorrhizas modify the plant root system and influence plant mineral nutrient acquisition, but the way they do so varies between types. Arbuscular mycorrhizal fungi develop, outside the root of their host, a highly branched system (mycelium) of hyphae (fine filamentous structures 2 to 10 μm in diameter) that explores the soil (**Figure 5.12**). Different arbuscular mycorrhizal fungi vary considerably in their distance and intensity of soil exploration, but phosphate transfer from as far as 25 cm away from the root has been measured. The mycelium in soil also helps stabilize aggregates of soil particles, promoting good soil structure. The hyphae extend into soil well beyond the zone of depletion that develops around a root and thus can absorb an immobile nutrient such as phosphate from beyond the depletion zone. Hyphae also penetrate soil pores that are much narrower than those available to roots.

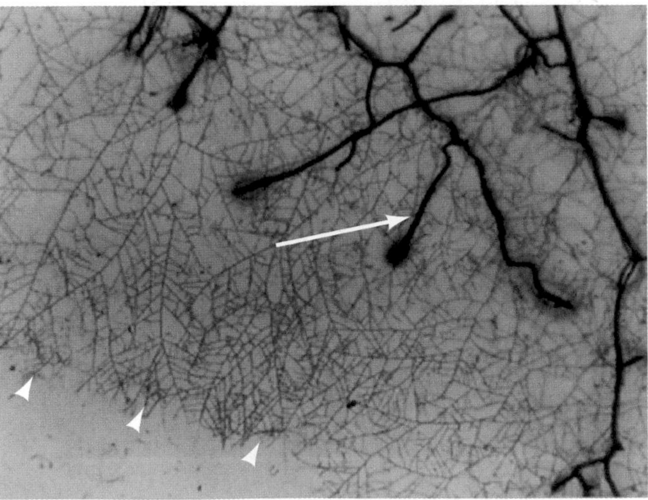

Figure 5.12 Visualization of the intact extraradical mycelium of *Glomus mosseae* spreading from colonized roots of cherry plum (*Prunus cerasifera*). The advancing front of the extraradical mycelium is indicated by arrowheads and the plant roots by an arrow. Note the differences in lengths and diameters of roots and hyphae. (From Smith and Read 2008.)

The root of the arbuscular mycorrhizal host plant looks almost the same as a non-mycorrhizal root, and the presence of the fungi can only be detected by staining and microscopy. Hyphae of arbuscular mycorrhizal fungi, growing from spores in the soil or roots of another plant, penetrate the root epidermis and colonize the root cortex, extending through intercellular spaces and invading the cortical cells to form either highly branched structures called **arbuscules** (Arum-type colonization; **Figure 5.13A**) or complex **hyphal coils** (Paris-type colonization; **Figure 5.13B**). The fungi are restricted to the cortex and never penetrate the endodermis or colonize the stele of the root. These structures increase the area of contact between the symbionts and remain surrounded by a plant membrane that is involved in transferring nutrients from fungal to plant cells. The penetration process is genetically controlled by a pathway that millions and millions of years later was partially co-opted for the colonization of legume roots by nitrogen-fixing bacteria (see Chapter 13).

Phosphate is delivered by the fungi directly to the root cortex. After export from the fungal arbuscules or coils, this phosphate is taken up by the plant cells. Some of the suite of plant **phosphate transporters** (see Chapter 6) are specifically or preferentially expressed only in the plant membrane surrounding the arbuscules and coils in the root cortex and are not expressed in non-mycorrhizal roots. The transporters play a key role in the transfer of phosphate from fungus to plant.

The hyphae of arbuscular mycorrhizal fungi have the capacity for fast growth, highly efficient absorption, and rapid translocation and transfer of nutrients such as phosphate to the root cells. This means they can explore soil much more effectively and with fewer resources than non-mycorrhizal roots. In a large number of plant species, the response to arbuscular mycorrhizal fungi colonization is increased phosphate uptake and hence growth, especially when soil phosphorus is poorly available. A wide variety of responses has been observed, however, ranging from large positive responses to zero or even negative responses. The conventional explanation for the negative responses is that the fungi consume excessive carbohydrate and fail to deliver adequate amounts of nutrients to the plant. Nonetheless, the fungi remain active in phosphate delivery, while at the same time decreasing the amount of phosphate that is absorbed directly through the root epidermis. The lack of positive responses may

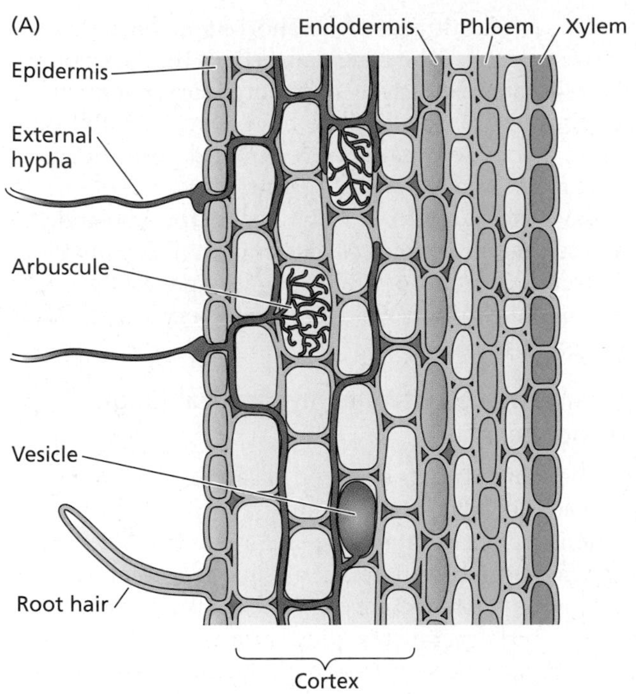

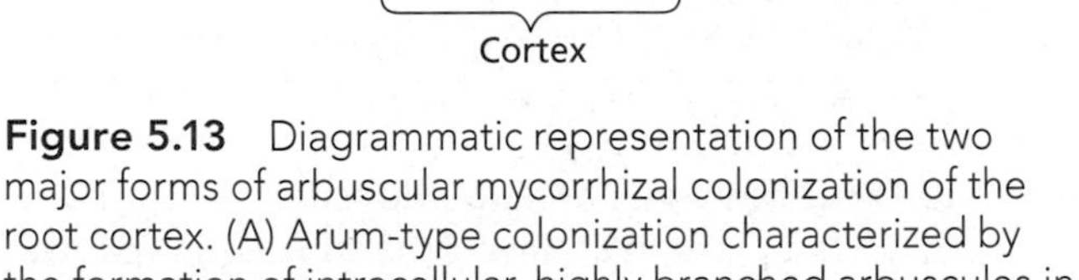

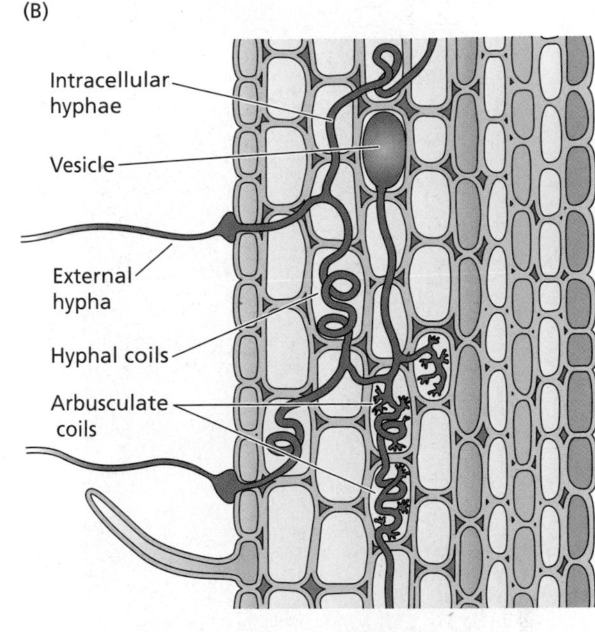

Figure 5.13 Diagrammatic representation of the two major forms of arbuscular mycorrhizal colonization of the root cortex. (A) Arum-type colonization characterized by the formation of intracellular, highly branched arbuscules in root cortical cells. (B) Paris-type colonization characterized by the formation of intracellular hyphal coils in root cortical cells, some of which (called arbusculate coils) bear small arbuscule-like branches.

therefore derive from "cross talk" between the plant and fungal symbionts that interferes with the way the roots absorb nutrients. High phosphate availability in soil tends to decrease the stimulatory effect that arbuscular mycorrhiza formation has on plant phosphorus uptake, growth, and yield, but substantive evidence for specific plant control of fungal colonization and activity by phosphate is still lacking.

Harnessing arbuscular mycorrhizal symbiosis to optimize crop nutrition as fertilizers become increasingly expensive will depend on understanding how the symbiotic partners interact to influence nutrient acquisition. At present, arbuscular mycorrhizal fungi are known to be important in uptake of immobile nutrients such as phosphate and zinc. Their role in increasing nitrogen uptake remains to be established.

Roots colonized by ectomycorrhizal fungi can be clearly distinguished from non-mycorrhizal roots; they grow more slowly and often appear thicker and highly branched. The fungi typically form a thick sheath, or *mantle*, of mycelium around roots, and some of the hyphae penetrate between the epidermal and sometimes (in the case of conifers) the cortical cells (**Figure 5.14**). The root cells themselves are not penetrated by the fungal hyphae, but instead are surrounded by a network of hyphae called the **Hartig net**, which provides a large area of contact between the symbionts that is involved in nutrient trans-

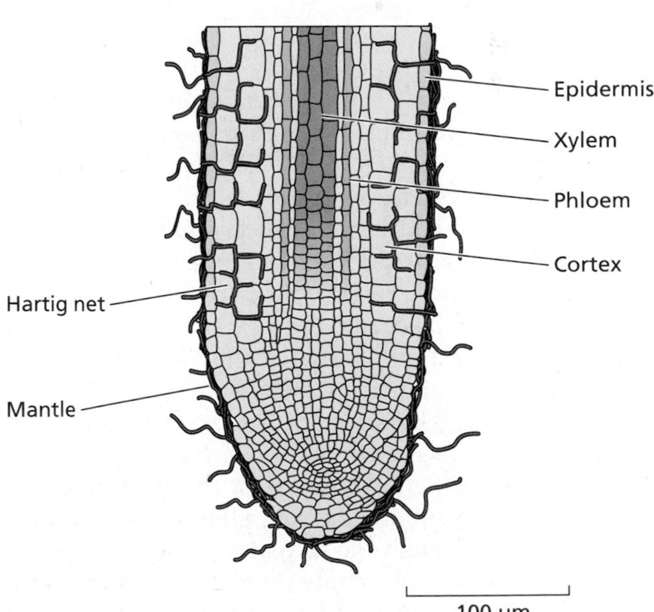

Figure 5.14 Diagrammatic representation of a longitudinal section of an ectomycorrhizal root. Fungal hyphae (shown in brown) form a dense fungal mantle over the surface of the root and penetrate between epidermal cells, or epidermal and cortical cells, to form the Hartig net. Hyphae also grow extensively in soil, forming dense mycelium and/or mycelial strands. (After Rovira et al. 1983.)

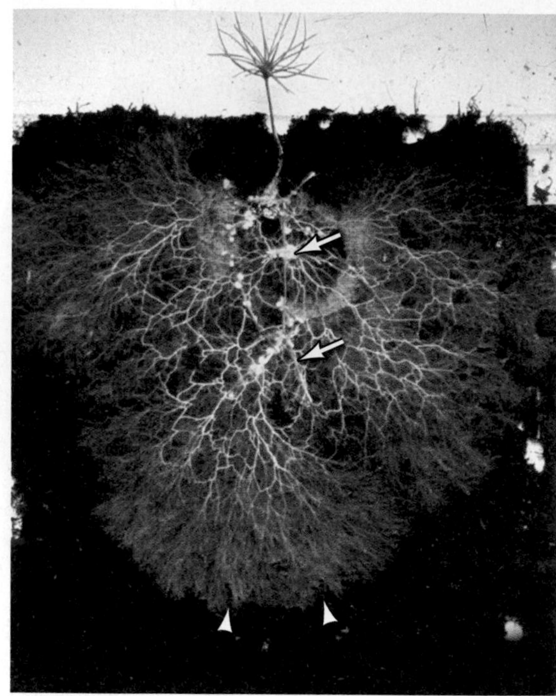

Figure 5.15 Seedling of pine (*Pinus*) showing mycorrhizal rootlets (upper arrow) colonized by an ectomycorrhizal fungus and grown in an observation chamber in forest soil. Note the differences between the mycelial front of dense hyphae advancing into the soil (arrowheads) and aggregated mycelial strands (lower arrow). (Courtesy of D. J. Read.)

fers. The fungal mycelium also extends into the soil, away from the compact sheath, where it is present as individual hyphae, mycelial fans (**Figure 5.15**), or mycelial strands. The fans in particular play important roles in obtaining nutrients from the soil, especially soil organic matter.

Ectomycorrhizal fungi produce many of the toadstools, puffballs, and truffles found in forests. Often the amount of fungal mycelium is so extensive that its total mass is much greater than that of the roots themselves. The arrangement and biochemical activities of the fungal structures in relation to the root tissues determine important aspects of nutrient acquisition by ectomycorrhizal roots and the form in which the nutrients pass from fungus to plant. In addition, all nutrients from the soil must pass through the fungal mantle covering the root epidermis before reaching the root cells themselves, giving the fungus a major role in uptake of all nutrients from the soil solution, including phosphate and inorganic forms of nitrogen (nitrate and ammonium). To what extent the fungi are actually involved in uptake of inorganic nitrogen and to what extent they may compete with the roots when nitrogen is in short supply are matters of active research. The fungal mycelium that develops in soil proliferates extensively in soil organic matter patches (see Figure 5.15). The hyphae have a marked ability to convert insoluble organic nitrogen and phosphorus to soluble forms and to pass these nutrients to the plants. In this way, ectomycorrhizal fungi enable their host plants to access organic sources of nutrients, avoid competition with free-living mineralizing organisms, and grow in highly organic forest soils that contain very low amounts of inorganic nutrients.

Nutrients move between mycorrhizal fungi and root cells

Movement of nutrients from soil via a mycorrhizal fungus to root cells involves complex integration of structure and function in both fungus and plant symbionts. The interfaces where fungus and plant are juxtaposed are critical zones for transport and are composed of the plasma membranes of both organisms, plus variable amounts of cell wall material. Therefore, nutrient movements from fungus to plant are potentially under the control of these two membrane types and subject to regulatory transport processes as described in Chapter 6. Movement of nutrients from soil to the plant via a mycorrhizal fungus requires (at least) uptake of a nutrient from soil by the fungus, long-distance translocation of the nutrient through fungal hyphae (and mycelial strands when present), release (or efflux) from the fungus to the apoplastic zone between the two membranes of the interface, and uptake by the plant plasma membrane. Important issues to be resolved include the form of nutrient that is transferred and the mechanism and amounts of transfers. Mechanisms promoting efflux from the fungus to the interfacial apoplastic zone are poorly understood, but uptake into the plant has received more attention. In the case of phosphate, the plant uptake step is an active process requiring energy and the presence of phosphate transporters in the plant membrane surrounding the intracellular fungal structures, which are specifically or preferentially expressed when the roots are mycorrhizal.

Transfer of nitrogen is more complex and more controversial. In ectomycorrhizas, for which a major role in plant nitrogen nutrition has long been accepted, organic nitrogen may move from fungus to plant, with the form (glutamine, glutamine and alanine, or glutamate) varying with the distribution of enzymes involved in inorganic nitrogen assimilation and the identity of the plant and fungal symbionts. Some transfer of nitrogen as ammonium or ammonia may also occur. As mentioned above, the involvement of arbuscular mycorrhizas in enhancing nitrogen uptake and transfer to host plants is not well established.

SUMMARY

Plants are autotrophic organisms capable of using the energy from sunlight to synthesize all their components from carbon dioxide, water, and mineral elements. Although mineral nutrients continually cycle through all organisms, they enter the biosphere predominantly through the root systems of plants. After being absorbed by the roots, the mineral elements are translocated to the various parts of the plant, where they serve in numerous biological functions.

Essential Nutrients, Deficiencies, and Plant Disorders

- Studies of plant nutrition have shown that specific mineral elements are essential for plant life (**Tables 5.1, 5.2**).
- These elements are classified as macronutrients or micronutrients, depending on the relative amounts found in plant tissue (**Table 5.1**).
- Certain visual symptoms are diagnostic for deficiencies in specific nutrients in higher plants. Nutritional disorders occur because nutrients have key roles within plants. They serve as components of organic compounds, in energy storage, in plant structures, as enzyme cofactors, and in electron transfer reactions.
- Mineral nutrition can be studied through the use of solution culture, which allows the characterization of specific nutrient requirements (**Figure 5.2; Table 5.3**).
- Soil and plant tissue analysis can provide information on the nutritional status of the plant–soil system and can suggest corrective actions to avoid deficiencies or toxicities (**Figure 5.4**).

Treating Nutritional Deficiencies

- When crop plants are grown under modern high-production conditions, substantial amounts of nutrients are removed from the soil.
- To prevent the development of deficiencies, nutrients can be added back to the soil in the form of fertilizers, particularly nitrogen, phosphorus, and potassium.
- Fertilizers that provide nutrients in inorganic forms are called chemical fertilizers; those that derive from plant or animal residues or from natural rock deposits are considered organic fertilizers. In both cases, plants absorb the nutrients primarily as inorganic ions. Most fertilizers are applied to the soil, but some are sprayed on leaves.

Soil, Roots, and Microbes

- Soil is a complex substrate—physically, chemically, and biologically. The size of soil particles and the cation exchange capacity of the soil determine the extent to which a soil provides a reservoir for water and nutrients (**Table 5.5; Figure 5.6**).
- Soil pH also has a large influence on the availability of mineral elements to plants (**Figure 5.5**).
- If mineral elements, especially sodium or heavy metals, are present in excess in the soil, plant growth may be adversely affected. Certain plants are able to tolerate excess mineral elements, and a few species—for example, halophytes in the case of sodium—may thrive under these extreme conditions.
- To obtain nutrients from the soil, plants develop extensive root systems (**Figures 5.7, 5.8**), form symbioses with mycorrhizal fungi, and produce and secrete protons or organic anions into the soil.
- Roots continually deplete the nutrients from the immediate soil around them (**Figure 5.10**)
- The majority of plants have the ability to form symbioses with mycorrhizal fungi.
- The fine hyphae of mycorrhizal fungi extend the reach of roots into the surrounding soil and facilitate the acquisition of nutrients (**Figures 5.12, 5.14, 5.15**). Arbuscular mycorrhizas increase uptake of mineral nutrients, particularly phosphorus, whereas ectomycorrhizas play a significant role in obtaining nitrogen from organic sources.
- In return, plants provide carbohydrates to the mycorrhizal fungi.

WEB MATERIAL

- **WEB TOPIC 5.1 Symptoms of Deficiency in Essential Minerals** Deficiency symptoms are characteristic of each essential element and can be diagnostic for the deficiency. The color photographs in this topic illustrate deficiency symptoms for each essential element in tomato.
- **WEB TOPIC 5.2 Observing Roots below the Ground** The study of roots growing under natural conditions requires a means to observe roots belowground. State-of-the-art techniques are described in this topic.

• **WEB ESSAY 5.1 Boron Functions in Plants: Looking beyond the Cell Wall** Presents one long list of "postulated roles of B essentiality" for microorganisms and for higher plant growth and development.

• **WEB ESSAY 5.2 From Meals to Metals and Back** Heavy metal accumulation is toxic to plants. Understanding the molecular process involved in toxicity is helping researchers develop better phytoremediation crops.

available at plantphys.net

Suggested Reading

Armstrong, F. A. (2008) Why did nature choose manganese to make oxygen? *Philos. Trans. R. Soc. Lond., B, Biol. Sci.* 363: 1263–1270.

Baker, A. J. M., and Brooks, R. R. (1989) Terrestrial higher plants which hyperaccumulate metallic elements—A review of their distribution, ecology and phytochemistry. *Biorecovery* 1: 81–126.

Berry, W. L., and Wallace, A. (1981) Toxicity: The concept and relationship to the dose response curve. *J. Plant Nutr.* 3: 13–19.

Bucher, M. (2007) Functional biology of plant phosphate uptake at root and mycorrhiza interfaces. *New Phytol.* 173: 11–26.

Burns, I. G. (1991) Short- and long-term effects of a change in the spatial distribution of nitrate in the root zone on N uptake, growth and root development of young lettuce plants. *Plant Cell Environ.* 14: 21–33.

Connor, D. J., Loomis, R. S., and Cassman, K. G. (2011) *Crop Ecology: Productivity and Management in Agricultural Systems,* 2nd ed. Cambridge University Press, Cambridge.

Cordell, D., Drangerta, J.-O., and White, S. (2009) The story of phosphorus: Global food security and food for thought. *Glob. Environ. Change* 19: 292–305.

Epstein, E., and Bloom, A. J. (2005) *Mineral Nutrition of Plants: Principles and Perspectives,* 2nd ed. Sinauer Associates, Sunderland, MA.

Fageria, N., Filho, M. B., Moreira, A., and Guimaraes, C. (2009) Foliar fertilization of crop plants. *J. Plant Nutr.* 32: 1044–1064.

Feldman, L. J. (1998) Not so quiet quiescent centers. *Trends Plant Sci.* 3: 80–81.

Fox, T. C., and Guerinot, M. L. (1998) Molecular biology of cation transport plants. *Annu. Rev. Plant Physiol. Plant Mol. Biol.* 49: 669–696.

Jackson, R. B., Canadell, J., Ehleringer, J. R., Mooney, H. A., Sala, O. A., and Schulze, E.-D. (1996) A global analysis of root distributions for terrestrial biomes. *Oecologia* 108: 389–411.

Jeong, J., and Guerinot, M. L. (2009) Homing in on iron homeostasis in plants. *Trends Plant Sci.* 14: 280–285.

Jones, M. D. and Smith, S. E. (2004) Exploring functional definitions of mycorrhizas: Are mycorrhizas always mutualisms? *Can. J. Bot.* 82: 1089–1109.

Kochian, L.V. (2000) Molecular physiology of mineral nutrient acquisition, transport and utilization. In *Biochemistry and Molecular Biology of Plants,* B. Buchanan, W. Gruissem and R. Jones, eds. American Society of Plant Physiologists, Rockville, Md.

Larsen, M. C., Hamilton, P. A., and Werkheiser, W. H. (2013) Water quality status and trends in the United States. In: *Monitoring Water Quality: Pollution Assessment, Analysis, and Remediation,* 1st ed. Ahuja, S., ed., Elsevier, Amsterdam. pp. 19–57.

Loomis, R. S., and Conner, D. J. (1992) *Crop Ecology: Productivity and Management in Agricultural Systems.* Cambridge University Press, Cambridge.

Mengel, K., and Kirkby, E. A. (2001) *Principles of Plant Nutrition,* 5th ed. Kluwer Academic Publishers, Dordrecht, Netherlands.

Sattelmacher, B. (2001) The apoplast and its significance for plant mineral. *New Phytol.* 149: 167–192.

Smith, S. E., and Read, D. J. (2008) *Mycorrhizal Symbiosis,* 3rd ed. Academic Press and Elsevier, Oxford.

Smith, F. A., Smith, S. E., and Timonen, S. (2003) Mycorrhizas. In: *Root Ecology,* H. de Kroon and E. J. W. Visser, eds. Springer, Berlin. pp. 257–295.

Zegada-Lizarazu, W., Matteucci, D. and Monti, A. (2010) Critical review on energy balance of agricultural systems. *Biofuels, Bioprod. Biorefin.* 4: 423–446.

6 Solute Transport

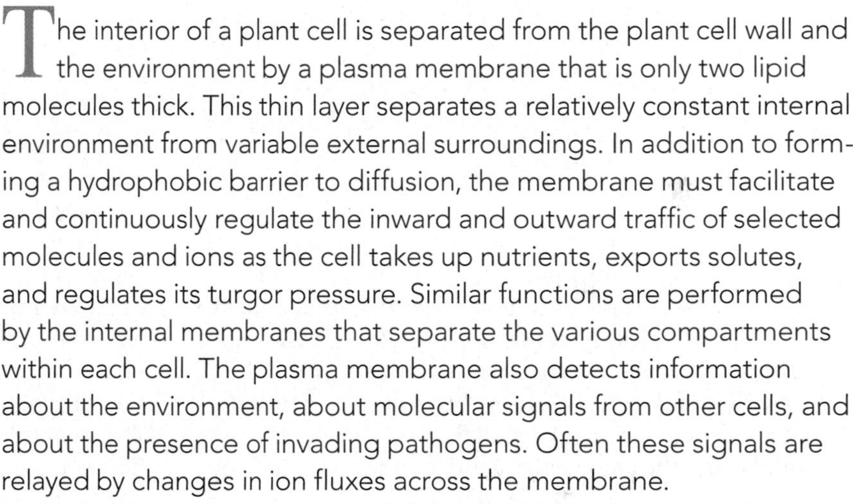

The interior of a plant cell is separated from the plant cell wall and the environment by a plasma membrane that is only two lipid molecules thick. This thin layer separates a relatively constant internal environment from variable external surroundings. In addition to forming a hydrophobic barrier to diffusion, the membrane must facilitate and continuously regulate the inward and outward traffic of selected molecules and ions as the cell takes up nutrients, exports solutes, and regulates its turgor pressure. Similar functions are performed by the internal membranes that separate the various compartments within each cell. The plasma membrane also detects information about the environment, about molecular signals from other cells, and about the presence of invading pathogens. Often these signals are relayed by changes in ion fluxes across the membrane.

Molecular and ionic movement from one location to another is known as **transport**. Local transport of solutes into or within cells is regulated mainly by membrane proteins. Larger-scale transport between plant organs, or between plant and environment, is also controlled by membrane transport at the cellular level. For example, the transport of sucrose from leaf to root through the phloem, referred to as **translocation**, is driven and regulated by membrane transport into the phloem cells of the leaf and from the phloem to the storage cells of the root (see Chapter 11).

In this chapter we will consider the physical and chemical principles that govern the movements of molecules in solution. Then we will show how these principles apply to membranes and biological systems. We will also discuss the molecular mechanisms of transport in living cells and the great variety of membrane transport proteins that are responsible for the particular transport properties of plant cells. Finally, we will examine the pathways that ions take when they enter the root as well as the mechanism of xylem loading, the process whereby ions are released into the tracheary elements of the stele. Because transported substances, including carbohydrates, amino acids, and metals such as iron and zinc, are vital for human nutrition, understanding and manipulating solute transport in plants can contribute solutions to sustainable food production.

Passive and Active Transport

According to Fick's first law (see Equation 3.1), the movement of molecules by diffusion always proceeds spontaneously, down a gradient of free energy or chemical potential, until equilibrium is reached. The spontaneous "downhill" movement of molecules is termed **passive transport**. At equilibrium, no further net movements of solutes can occur without the application of a driving force.

The movement of substances against a gradient of chemical potential, or "uphill," is termed **active transport**. It is not spontaneous, and it requires that work be done on the system by the application of cellular energy. One common way (but not the only way) of accomplishing this task is to couple transport to the hydrolysis of ATP.

Recall from Chapter 3 that we can calculate the driving force for diffusion or, conversely, the energy input necessary to move substances against a gradient by measuring the potential-energy gradient. For uncharged solutes this gradient is often a simple function of the difference in concentration. Biological transport can be driven by four major forces: concentration, hydrostatic pressure, gravity, and electric fields. (However, recall from Chapter 3 that in small-scale biological systems, gravity seldom contributes substantially to the force that drives transport.)

The **chemical potential** for any solute is defined as the sum of the concentration, electrical, and hydrostatic potentials (and the chemical potential under standard conditions). *The importance of the concept of chemical potential is that it sums all the forces that may act on a molecule to drive net transport.*

$$\underset{\substack{\text{Chemical}\\\text{potential}\\\text{for a given}\\\text{solute, } j}}{\tilde{\mu}_j} = \underset{\substack{\text{Chemical}\\\text{potential}\\\text{of } j \text{ under}\\\text{standard}\\\text{conditions}}}{\mu_j^*} + \underset{\substack{\text{Concentration}\\\text{(activity)}\\\text{component}}}{RT \ln C_j} + \underset{\substack{\text{Electric-}\\\text{potential}\\\text{component}}}{z_j FE} + \underset{\substack{\text{Hydrostatic-}\\\text{pressure}\\\text{component}}}{\bar{V}_j P} \tag{6.1}$$

Here $\tilde{\mu}_j$ is the chemical potential of the solute species j in joules per mole (J mol^{-1}), μ_j^* is its chemical potential under standard conditions (a correction factor that will cancel out in future equations and so can be ignored), R is the universal gas constant, T is the absolute temperature, and C_j is the concentration (more accurately the activity) of j.

The electrical term, z_jFE, applies only to ions; z is the electrostatic charge of the ion (+1 for monovalent cations, –1 for monovalent anions, +2 for divalent cations, and so on), F is Faraday's constant (96,500 Coulombs, equivalent to the electric charge on 1 mol of H^+), and E is the overall electrical potential of the solution (with respect to ground). The final term, $\bar{V}_jP$, expresses the contribution of the partial molal volume of j ($\bar{V}_j$) and pressure (P) to the chemical potential of j. (The partial molal volume of j is the change in volume per mole of substance j added to the system, for an infinitesimal addition.)

This final term, $\bar{V}_jP$, makes a much smaller contribution to $\tilde{\mu}_j$ than do the concentration and electrical terms, except in the very important case of osmotic water movements. As discussed in Chapter 3, when considering water movement at the cellular scale the chemical potential of water (i.e., the

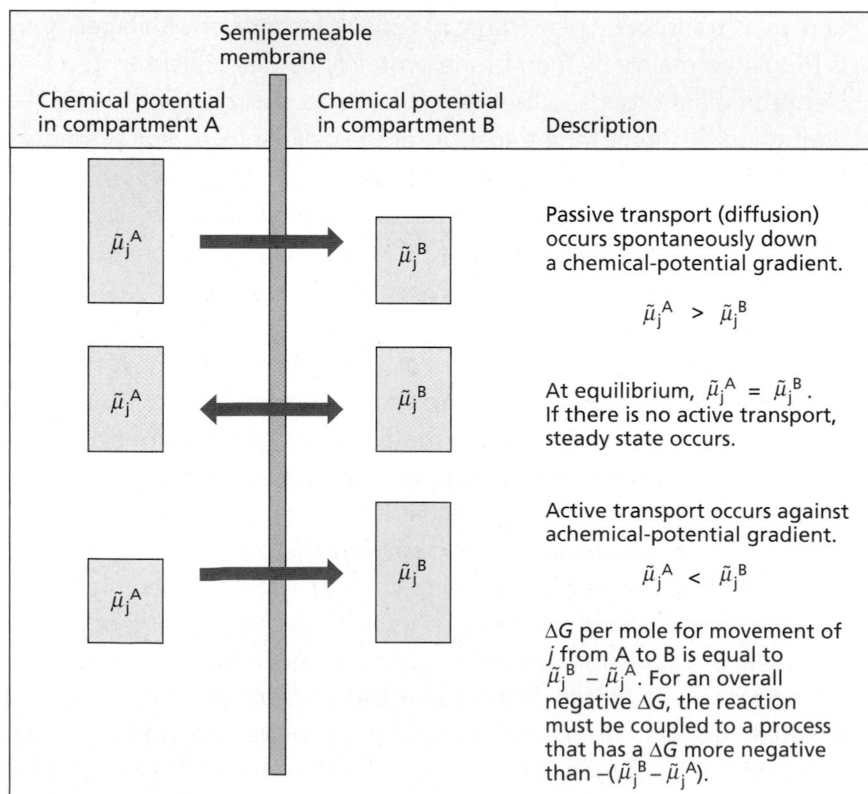

Figure 6.1 Relationship between chemical potential, $\tilde{\mu}$, and the transport of molecules across a permeability barrier. The net movement of molecular species j between compartments A and B depends on the relative magnitude of the chemical potential of j in each compartment, represented here by the size of the boxes. Movement down a chemical gradient occurs spontaneously and is called passive transport; movement against, or up, a gradient requires energy and is called active transport.

water potential) depends on the concentration of dissolved solutes and the hydrostatic pressure on the system.

In general, diffusion (passive transport) always moves molecules energetically downhill from areas of higher chemical potential to areas of lower chemical potential. Movement against a chemical-potential gradient is indicative of active transport (**Figure 6.1**).

If we take the diffusion of sucrose across a plasma membrane as an example, we can accurately approximate the chemical potential of sucrose in any compartment by the concentration term alone (unless a solution is concentrated, causing hydrostatic pressure to build up within the plant cell). From Equation 6.1, the chemical potential of sucrose inside a cell can be described as follows (in the next three equations, the subscript *s* stands for sucrose and the superscripts *i* and *o* stand for inside and outside, respectively):

$\tilde{\mu}_s^i$	=	μ_s^*	+	$RT \ln C_s^i$	
Chemical potential of sucrose solution inside the cell		Chemical potential of sucrose solution under standard conditions		Concentration component	(6.2)

The chemical potential of sucrose outside the cell is calculated as follows:

$$\tilde{\mu}_s^o = \mu_s^* + RT \ln C_s^o \quad (6.3)$$

We can calculate the difference in the chemical potential of sucrose between the solutions inside and outside the cell, $\Delta\tilde{\mu}_s$, regardless of the mechanism of transport. To get the signs right, remember that for inward transport, sucrose is being removed (–) from outside the cell and added (+) to the inside, so the change in free energy in joules per mole of sucrose transported will be as follows:

$$\Delta\tilde{\mu}_s = \tilde{\mu}_s^i - \tilde{\mu}_s^o \quad (6.4)$$

Substituting the terms from Equations 6.2 and 6.3 into Equation 6.4, we get the following:

$$\begin{aligned} \Delta\tilde{\mu}_s &= \left(\mu_s^* + RT \ln C_s^i\right) - \left(\mu_s^* + RT \ln C_s^o\right) \\ &= RT\left(\ln C_s^i - \ln C_s^o\right) \\ &= RT \ln \frac{C_s^i}{C_s^o} \end{aligned} \quad (6.5)$$

If this difference in chemical potential is negative, sucrose can diffuse inward spontaneously (provided the membrane has a permeability to sucrose; see the next section). In other words, the driving force ($\Delta\tilde{\mu}_s$) for solute diffusion is related to the magnitude of the concentration gradient (C_s^i/C_s^o).

If the solute carries an electric charge (as does, for example, the potassium ion), the electrical component of the chemical potential must also be considered. Suppose the membrane is permeable to K^+ and Cl^- rather than to sucrose. Because the ionic species (K^+ and Cl^-) diffuse independently, each has its own chemical potential. Thus, for inward K^+ diffusion,

$$\Delta\tilde{\mu}_K = \tilde{\mu}_K^i - \tilde{\mu}_K^o \quad (6.6)$$

Substituting the appropriate terms from Equation 6.1 into Equation 6.6, we get

$$\Delta\tilde{\mu}_s = (RT \ln [K^+]^i + zFE^i) - (RT \ln [K^+]^o + zFE^o) \quad (6.7)$$

and because the electrostatic charge of K^+ is +1, $z = +1$, and

$$\Delta\tilde{\mu}_K = RT \ln \frac{[K^+]^i}{[K^+]^o} + F(E^i - E^o) \quad (6.8)$$

The magnitude and sign of this expression will indicate the driving force and direction for K^+ diffusion across the membrane. A similar expression can be written for Cl^- (but remember that for Cl^-, $z = -1$).

Equation 6.8 shows that ions, such as K^+, diffuse in response to both their concentration gradients ($[K^+]^i/[K^+]^o$) and any electrical potential difference between the two compartments ($E^i - E^o$). One important implication of this equation is that ions can be driven passively against their concentration gradients if an appropriate voltage (electric field) is applied between the two compartments. Because of the importance of electric fields in the biological transport of any charged molecule, $\tilde{\mu}$ is often called the **electrochemical potential**, and $\Delta\tilde{\mu}$ is the difference in electrochemical potential between two compartments.

Transport of Ions across Membrane Barriers

If two ionic solutions are separated by a biological membrane, diffusion is complicated by the fact that the ions must move through the membrane as well as across the open solutions. The extent to which a membrane permits the movement of a substance is called **membrane permeability**. As we will discuss later, permeability depends on the composition of the membrane as well as on the chemical nature of the solute. In a loose sense, permeability can be expressed in terms of a diffusion coefficient for the solute through the membrane. However, permeability is influenced by several additional factors, such as the ability of a substance to enter the membrane, that are difficult to measure.

Despite its theoretical complexity, we can readily measure permeability by determining the rate at which a solute passes through a membrane under a specific set of conditions. Generally the membrane will hinder diffusion and thus reduce the speed with which equilibrium is reached. For any particular solute, however, the permeability or resistance of the membrane itself cannot alter the final equilibrium conditions. Equilibrium occurs when $\Delta\tilde{\mu}_j = 0$.

In the sections that follow we will discuss the factors that influence the distribution of ions across a membrane. These parameters can be used to predict the relationship between the electrical gradient and the concentration gradient of an ion.

Different diffusion rates for cations and anions produce diffusion potentials

When salts diffuse across a membrane, an electrical membrane potential (voltage) can develop. Consider the two KCl solutions separated by a membrane in **Figure 6.2**. The K^+ and Cl^- ions will permeate the membrane independently as they diffuse down their respective gradients of electrochemical potential. And unless the membrane is very porous, its permeability to the two ions will differ.

As a consequence of these different permeabilities, K^+ and Cl^- will initially diffuse across the membrane at different rates. The result is a slight separation of charge, which instantly creates an electrical potential across the membrane. In biological systems, membranes are usually more permeable to K^+ than to Cl^-. Therefore, K^+ will diffuse out of the cell (see compartment A in Figure 6.2) faster than Cl^-, causing the cell to develop a negative electric charge with respect to the extracellular medium. A potential that develops as a result of diffusion is called a **diffusion potential**.

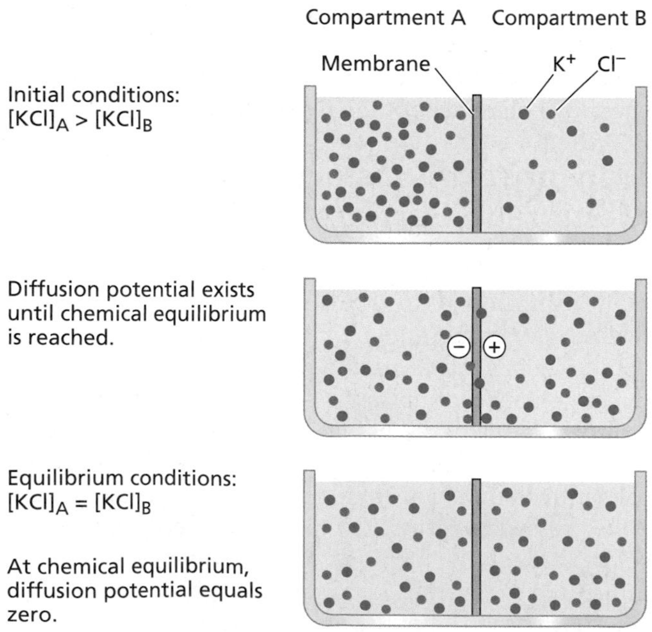

The principle of electrical neutrality must always be kept in mind when the movement of ions across membranes is considered: bulk solutions always contain equal numbers of anions and cations. The existence of a membrane potential implies that the distribution of charges across the membrane is uneven; however, the actual number of unbalanced ions is negligible in chemical terms. For example, a membrane potential of –100 millivolts (mV), like that found across the plasma membranes of many plant cells, results from the presence of only 1 extra anion out of every 100,000 within the cell—a concentration difference of only 0.001%! As Figure 6.2 shows, all of these extra anions are found immediately adjacent to the surface of the membrane; there is no charge imbalance throughout the bulk of the cell.

In our example of KCl diffusion across a membrane, electrical neutrality is preserved, because as K^+ moves ahead of Cl^- in the membrane, the resulting diffusion potential retards the movement of K^+ and speeds that of Cl^-. Ultimately, both ions diffuse at the same rate, but the diffusion potential persists and can be measured. As the system moves toward equilibrium and the concentration gradient collapses, the diffusion potential also collapses.

Figure 6.2 Development of a diffusion potential and a charge separation between two compartments separated by a membrane that is preferentially permeable to potassium ions. If the concentration of potassium chloride is higher in compartment A ($[KCl]_A > [KCl]_B$), potassium and chloride ions will diffuse into compartment B. If the membrane is more permeable to potassium ions than to chloride, potassium ions will diffuse faster than chloride ions, and a charge separation (+ and –) will develop, resulting in establishment of a diffusion potential.

How does membrane potential relate to ion distribution?

Because the membrane in the preceding example is permeable to both K^+ and Cl^- ions, equilibrium will not be reached for either ion until the concentration gradients decrease to zero. However, if the membrane were permeable only to K^+, diffusion of K^+ would carry charges across the membrane until the membrane potential balanced the concentration gradient. Because a change in potential requires very few ions, this balance would be reached instantly. Potassium ions would then be at equilibrium, even though the change in the concentration gradient for K^+ would be negligible.

When the distribution of any solute across a membrane reaches equilibrium, the passive flux, J (i.e., the amount of solute crossing a unit area of membrane per unit time), is the same in the two directions—outside to inside and inside to outside:

$$J_{o \to i} = J_{i \to o}$$

Fluxes are related to $\Delta\tilde{\mu}$ (for a discussion on fluxes and $\Delta\tilde{\mu}$, see **WEB APPENDIX 1**); thus, at equilibrium the electrochemical potentials will be the same:

$$\tilde{\mu}_j^o = \tilde{\mu}_j^i$$

and for any given ion (the ion is symbolized here by the subscript j),

$$\mu_j^* + RT \ln C_j^o + z_j F E^o = \mu_j^* + RT \ln C_j^i + z_j F E^i \qquad (6.9)$$

By rearranging Equation 6.9, we obtain the difference in electrical potential between the two compartments at equilibrium ($E^i - E^o$):

$$E^i - E^o = \frac{RT}{z_jF}\left(\ln\frac{C_j^o}{C_j^i}\right)$$

This electrical-potential difference is known as the **Nernst potential** (ΔE_j) for that ion,

$$\Delta E_j = E^i - E^o$$

and

$$\Delta E_j = \frac{RT}{z_jF}\left(\ln\frac{C_j^o}{C_j^i}\right) \quad (6.10)$$

or

$$\Delta E_j = \frac{2.3RT}{z_jF}\left(\log\frac{C_j^o}{C_j^i}\right)$$

This relationship, known as the *Nernst equation*, states that at equilibrium, the difference in concentration of an ion between two compartments is balanced by the voltage difference between the compartments. The Nernst equation can be further simplified for a univalent cation at 25°C:

$$\Delta E_j = 59\text{mV}\ \log\frac{C_j^o}{C_j^i} \quad (6.11)$$

Note that a tenfold difference in concentration corresponds to a Nernst potential of 59 mV ($C_o/C_i = 10/1$; log 10 = 1). That is, a membrane potential of 59 mV would maintain a tenfold concentration gradient of an ion whose movement across the membrane is driven by passive diffusion. Similarly, if a tenfold concentration gradient of an ion existed across the membrane, passive diffusion of that ion down its concentration gradient (if it were allowed to come to equilibrium) would result in a difference of 59 mV across the membrane.

All living cells exhibit a membrane potential that is due to the asymmetric ion distribution between the inside and outside of the cell. We can determine these membrane potentials by inserting a microelectrode into the cell and measuring the voltage difference between the inside of the cell and the extracellular medium (**Figure 6.3**).

The Nernst equation can be used at any time to determine whether a given ion is at equilibrium across a membrane. However, a distinction must be made between equilibrium and steady state. *Steady state* is the condition in which influx and efflux of a given solute are equal, and therefore the ion concentrations are constant over time. Steady state is not necessarily the same as equilibrium (see Figure 6.1); in steady state, the existence of active transport across the membrane prevents many diffusive fluxes from ever reaching equilibrium.

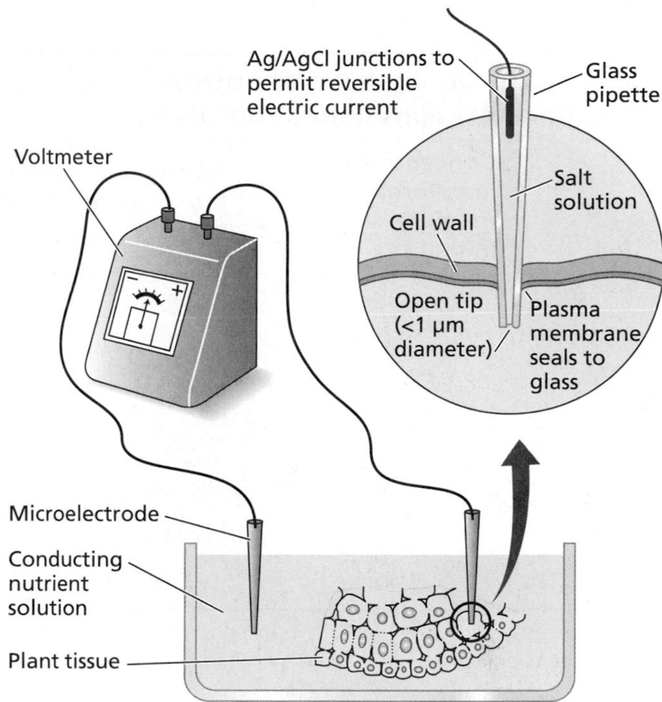

Figure 6.3 Diagram of a pair of microelectrodes used to measure membrane potentials across plasma membranes. One of the glass micropipette electrodes is inserted into the cell compartment under study (usually the vacuole or the cytoplasm), while the other is kept in an electrolytic solution that serves as a reference. The microelectrodes are connected to a voltmeter, which records the electrical-potential difference between the cell compartment and the solution. Typical membrane potentials across plant plasma membranes range from –60 to –240 mV. The insert shows how electrical contact with the interior of the cell is made through the open tip of the glass micropipette, which contains an electrically conducting salt solution.

The Nernst equation distinguishes between active and passive transport

Table 6.1 shows how experimental measurements of ion concentrations at steady state in pea root cells compare with predicted values calculated from the Nernst equation. In this example, the concentration of each ion in the external solution bathing the tissue and the measured membrane potential were substituted into the Nernst equation, and the internal concentration of each ion was predicted.

Prediction using the Nernst equation assumes passive ionic distribution, but notice that, of all the ions shown in Table 6.1, only K^+ is at or near equilibrium. The anions NO_3^-, Cl^-, $H_2PO_4^-$, and SO_4^{2-} all have higher internal concentrations than predicted, indicating that their uptake is active. The cations Na^+, Mg^{2+}, and Ca^{2+} have lower internal concentrations than predicted; therefore, these ions enter the cell by diffusion down their electrochemical-potential gradients and are then actively exported.

TABLE 6.1 Comparison of observed and predicted ion concentrations in pea root tissue

Ion	Concentration in external medium (mmol L^{-1})	Internal concentration[a] (mmol L^{-1})	
		Predicted	Observed
K^+	1	74	75
Na^+	1	74	8
Mg^{2+}	0.25	1340	3
Ca^{2+}	1	5360	2
NO_3^-	2	0.0272	28
Cl^-	1	0.0136	7
$H_2PO_4^-$	1	0.0136	21
SO_4^{2-}	0.25	0.00005	19

Source: Data from Higinbotham et al. 1967.

Note: The membrane potential was measured as –110 mV.

[a]Internal concentration values were derived from ion content of hot water extracts of 1–2 cm intact root segments.

The example shown in Table 6.1 is an oversimplification; plant cells have several internal compartments, each of which can differ in its ionic composition from the others. The cytosol and the vacuole are the most important intracellular compartments in determining the ionic relations of plant cells. In most mature plant cells, the central vacuole occupies 90% or more of the cell's volume, and the cytosol is restricted to a thin layer around the periphery of the cell.

Because of its small volume, the cytosol of most angiosperm cells is difficult to assay chemically. For this reason, much of the early work on the ionic relations of plants focused on certain green algae, such as *Chara* and *Nitella*, whose cells are several inches long and may contain an appreciable volume of cytosol. In brief:

- Potassium ions are accumulated passively by both the cytosol and the vacuole. When extracellular K^+ concentrations are very low, K^+ may be taken up actively.
- Sodium ions are pumped actively out of the cytosol into the extracellular space and vacuole.
- Excess protons, generated by intermediary metabolism, are also actively extruded from the cytosol. This process helps maintain the cytosolic pH near neutrality, while the vacuole and the extracellular medium are generally more acidic by one or two pH units.
- Anions are taken up actively into the cytosol.
- Calcium ions are actively transported out of the cytosol at both the plasma membrane and the vacuolar membrane, which is called the tonoplast.

Many different ions permeate the membranes of living cells simultaneously, but K^+ has the highest concentrations in plant cells, and it exhibits high permeabilities. A modified version of the Nernst equation, the **Goldman equation**, includes all permeant ions (all ions for which mechanisms of transmembrane movement exist) and therefore gives a more accurate value for the diffusion potential. When permeabilities and ion gradients are known, it is possible to calculate a diffusion potential across a biological membrane from the Goldman equation. The diffusion potential calculated from the Goldman equation is termed the *Goldman diffusion potential* (for a detailed discussion of the Goldman equation, see **WEB TOPIC 6.1**).

Proton transport is a major determinant of the membrane potential

In most eukaryotic cells, K^+ has both the greatest internal concentration and the highest membrane permeability, so the diffusion potential may approach E_K, the Nernst potential for K^+. In some cells of some organisms—particularly in some mammalian cells such as neurons—the normal resting potential of the cell may also be close to E_K. This is not the case with plants and fungi, however, which often show experimentally measured membrane potentials (often –200 to –100 mV) that are much more negative than those calculated from the Goldman equation, which are usually only –80 to –50 mV. Thus, in addition to the diffusion potential, the membrane potential must have a second component. The excess voltage is provided by the electrogenic plasma membrane H^+-ATPase.

Whenever an ion moves into or out of a cell without being balanced by countermovement of an ion of opposite charge, a voltage is created across the membrane. Any active transport mechanism that results in the movement of a net electric charge will tend to move the membrane potential away from the value predicted by the Goldman equation. Such transport mechanisms are called *electrogenic pumps* and are common in living cells.

The energy required for active transport is often provided by the hydrolysis of ATP. We can study the dependence of the plasma membrane potential on ATP by observing the effect of cyanide (CN^-) on the membrane potential (**Figure 6.4**). Cyanide rapidly poisons the mitochondria, and the cell's ATP consequently becomes depleted. As ATP synthesis is inhibited, the membrane potential falls to the level of the Goldman diffusion potential (see **WEB TOPIC 6.1**).

Thus, the membrane potentials of plant cells have two components: a diffusion potential and a component resulting from electrogenic ion transport (transport that results in the generation of a membrane potential). When cyanide inhibits electrogenic ion transport, the pH of the external medium increases while the cytosol becomes acidic because protons remain inside the cell. This observation

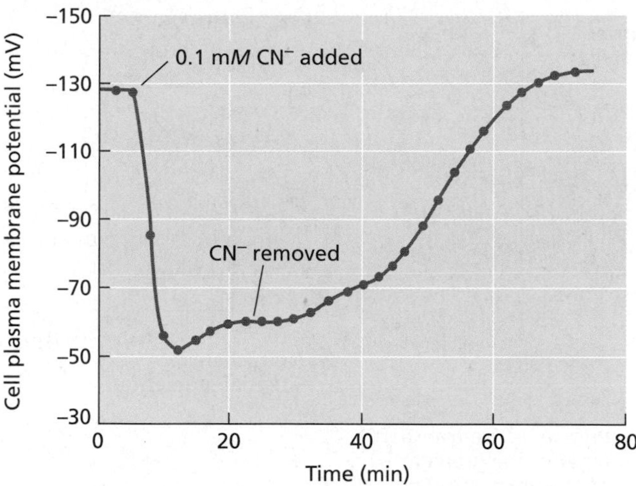

Figure 6.4 The plasma membrane potential of a pea cell collapses when cyanide (CN^-) is added to the bathing solution. Cyanide blocks ATP production in the cell by poisoning the mitochondria. The collapse of the membrane potential upon addition of cyanide indicates that an ATP supply is necessary for maintenance of the potential. Washing the cyanide out of the tissue results in a slow recovery of ATP production and restoration of the membrane potential. (After Higinbotham et al. 1970.)

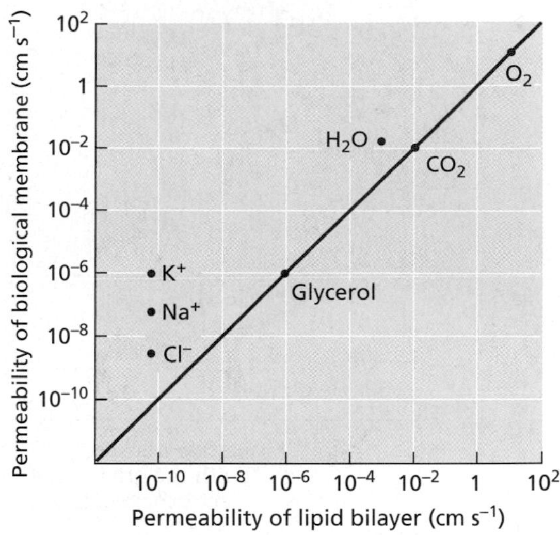

Figure 6.5 Typical values for the permeability of a biological membrane to various substances compared with those for an artificial phospholipid bilayer. For nonpolar molecules such as O_2 and CO_2, and for some small uncharged molecules such as glycerol, permeability values are similar in both systems. For ions and selected polar molecules, including water, the permeability of biological membranes is increased by one or more orders of magnitude because of the presence of transport proteins. Note the logarithmic scale.

is one piece of evidence that it is the active transport of protons out of the cell that is electrogenic.

A change in membrane potential caused by an electrogenic pump will change the driving forces for diffusion of all ions that cross the membrane. For example, the outward transport of H^+ can create an electrical driving force for the passive diffusion of K^+ into the cell. Protons are transported electrogenically across the plasma membrane not only in plants but also in bacteria, algae, fungi, and some animal cells, such as those of the kidney epithelia.

ATP synthesis in mitochondria and chloroplasts also depends on a H^+-ATPase. In these organelles, this transport protein is usually called an *ATP synthase* because it forms ATP rather than hydrolyzing it (see Chapter 12). We will discuss the structure and function of membrane proteins involved in active and passive transport in plant cells in detail later in this chapter.

Membrane Transport Processes

Artificial membranes made of pure phospholipids have been used extensively to study membrane permeability. When the permeability of artificial phospholipid bilayers to ions and molecules is compared with that of biological membranes, important similarities and differences become evident (**Figure 6.5**).

Biological and artificial membranes have similar permeabilities to nonpolar molecules and many small polar molecules. On the other hand, biological membranes are much more permeable to ions, to some large polar molecules, such as sugars, and to water than artificial bilayers are. The reason is that, unlike artificial bilayers, biological membranes contain **transport proteins** that facilitate the passage of selected ions and other molecules. The general term *transport proteins* encompasses three main categories of proteins: channels, carriers, and pumps (**Figure 6.6**), each of which we will describe in more detail later in this section.

Transport proteins exhibit specificity for the solutes they transport, so cells require a great diversity of transport proteins. The simple prokaryote *Haemophilus influenzae*, the first organism for which the complete genome was sequenced, has only 1743 genes, yet more than 200 of those genes (more than 10% of the genome) encode various proteins involved in membrane transport. In Arabidopsis, out of a predicted 33,602 protein-coding genes, as many as 1800 may encode proteins with transport functions.

Although a particular transport protein is usually highly specific for the kinds of substances it will transport, its specificity is often not absolute. In plants, for example, a K^+ transporter in the plasma membrane may transport K^+, Rb^+, and Na^+ with different preferences. In contrast, most K^+ transporters are completely ineffective in transporting anions such as Cl^- or uncharged solutes such as sucrose. Similarly, a protein involved in the transport of neutral

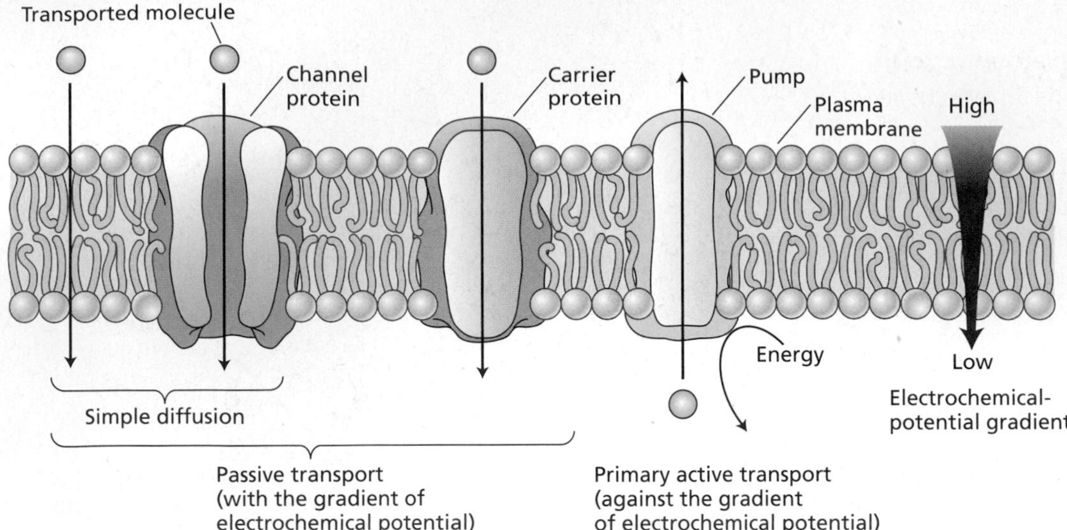

Figure 6.6 Three classes of membrane transport proteins: channels, carriers, and pumps. Channels and carriers can mediate the passive transport of a solute across a membrane (by simple diffusion or facilitated diffusion) down the solute's gradient of electrochemical potential. Channel proteins act as membrane pores, and their specificity is determined primarily by the biophysical properties of the channel. Carrier proteins bind the transported molecule on one side of the membrane and release it on the other side. (The different types of carrier proteins are described in more detail in Figure 6.10.) Primary active transport is carried out by pumps and uses energy directly, usually from ATP hydrolysis, to pump solutes against their gradient of electrochemical potential.

amino acids may move glycine, alanine, and valine with equal ease, but may not accept aspartic acid or lysine.

In the next several pages we will consider the structures, functions, and physiological roles of the various membrane transporters found in plant cells, especially in the plasma membrane and tonoplast. We will begin with a discussion of the role of certain transporters (channels and carriers) in promoting the diffusion of solutes across membranes. We will then distinguish between primary and secondary active transport and discuss the roles of the electrogenic H^+-ATPase and various symporters (proteins that transport two substances in the same direction simultaneously) in driving H^+-coupled secondary active transport.

Channels enhance diffusion across membranes

Channels are transmembrane proteins that function as selective pores through which ions, and in some cases neutral molecules, can diffuse across the membrane. The size of a pore and the density and nature of the surface charges on its interior lining determine its transport specificity. Transport through channels is always passive, and the specificity of transport depends on pore size and electric charge more than on selective binding (**Figure 6.7**).

As long as the channel pore is open, substances that can penetrate the pore diffuse through it extremely rapidly: about 10^8 ions per second through an ion channel. Channel pores are not open all the time, however. Channel proteins contain particular regions called **gates** that open and close the pore in response to signals. Signals that can regulate channel activity include membrane potential changes, ligands, hormones, light, and posttranslational modifications such as phosphorylation. For example, voltage-gated channels open or close in response to changes in the membrane potential (see Figure 6.7B). Another intriguing regulatory signal is mechanical force, which changes the conformation and thereby controls the gating of mechanosensitive ion channels in plants and other organisms.

Individual ion channels can be studied in detail by an electrophysiological technique called patch clamping (see **WEB TOPIC 6.2**), which can detect the electrical current carried by ions diffusing through a single open channel or a collection of channels. Patch clamp studies reveal that for a given ion, such as K^+, a membrane has a variety of different channels. These channels may open over different voltage ranges, or in response to different signals, which may include K^+ or Ca^{2+} concentrations, pH, reactive oxygen species, and so on. This specificity enables the transport of each ion to be fine-tuned to the prevailing conditions. Thus, the ion permeability of a membrane is a variable that depends on the mix of ion channels that are open at a particular time.

As we saw in the experiment presented in Table 6.1, the distribution of most ions is not close to equilibrium across the membrane. Therefore, we know that channels for most ions are usually closed. Plant cells generally accumulate more anions than could occur via a strictly passive mechanism. Therefore, when anion channels open, anions flow out of the cell, and active mechanisms are required for anion uptake. Calcium ion channels are tightly regulated and essentially open only during signal transduction.

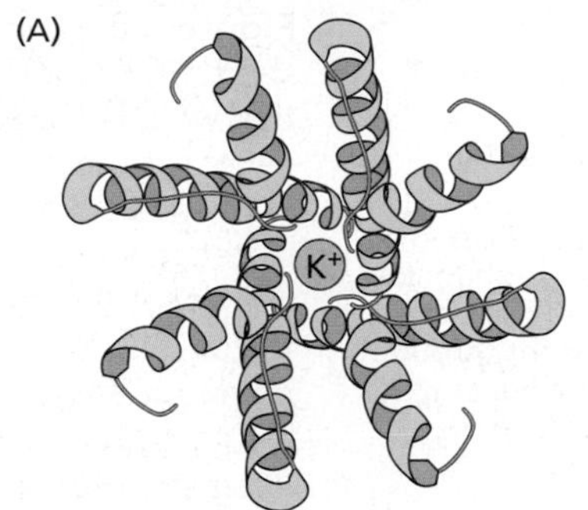

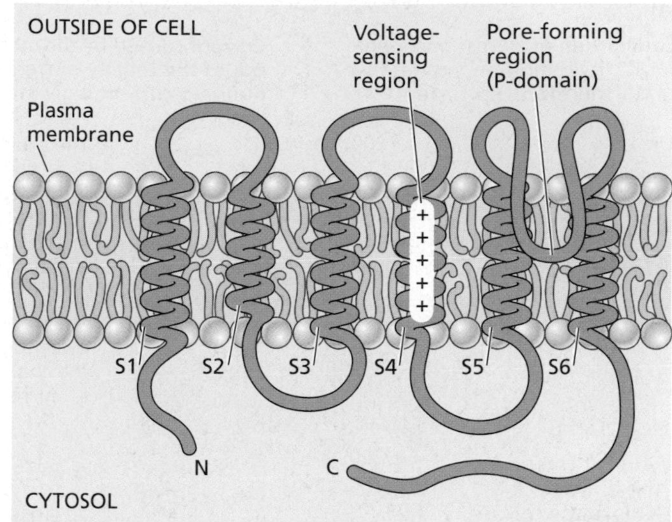

Figure 6.7 Models of K^+ channels in plants. (A) Top view of a channel, looking through the pore of the protein. Membrane-spanning helices of four subunits come together in an inverted teepee with the pore at the center. The pore-forming regions of the four subunits dip into the membrane, forming a K^+ selectivity finger region at the outer part of the pore (more details on the structure of this channel can be found in **WEB ESSAY 6.1**). (B) Side view of an inwardly rectifying K^+ channel, showing a polypeptide chain of one subunit, with six membrane-spanning helices (S1–S6). The fourth helix contains positively charged amino acids and acts as a voltage sensor. The pore-forming region (P-domain) is a loop between helices 5 and 6. (A after Leng et al. 2002; B after Buchanan et al. 2000.)

Calcium ion channels function only to allow Ca^{2+} flux into the cytosol, and Ca^{2+} must be expelled from the cytosol by active transport. In contrast, K^+ can diffuse either inward or outward through channels, depending on whether the membrane potential is more negative or more positive than E_K, the potassium ion equilibrium potential.

K^+ channels that open only at potentials more negative than the prevailing Nernst potential for K^+ are specialized for inward diffusion of K^+ and are known as **inwardly rectifying**, or simply *inward*, K^+ channels. Conversely, K^+ channels that open only at potentials more positive than the Nernst potential for K^+ are **outwardly rectifying**, or *outward*, K^+ channels (**Figure 6.8**) (see **WEB ESSAY 6.1**). Inward K^+ channels function in the accumulation of K^+ from the apoplast, as occurs, for example, during K^+ uptake by guard cells in the process of stomatal opening (see Figure 6.8). Various outward K^+ channels function in the closing of stomata and in the release of K^+ into the xylem or apoplast.

Carriers bind and transport specific substances

Unlike channels, carrier proteins do not have pores that extend completely across the membrane. In transport mediated by a carrier, the substance being transported is initially bound to a specific site on the carrier protein. This requirement for binding allows carriers to be highly selective for a particular substrate to be transported. **Carriers** therefore specialize in the transport of specific inorganic or organic ions as well as other organic metabolites. Binding causes a conformational change in the protein, which exposes the substance to the solution on the other side of the membrane. Transport is complete when the substance dissociates from the carrier's binding site.

Because a conformational change in the protein is required to transport an individual molecule or ion, the rate of transport by a carrier is many orders of magnitude slower than that through a channel. Typically, carriers may transport 100 to 1000 ions or molecules per second, while millions of ions can pass through an open ion channel in the same amount of time. The binding and release of molecules at a specific site on a carrier protein are similar to the binding and release of molecules by an enzyme in an enzyme-catalyzed reaction. As we will discuss later in this chapter, enzyme kinetics have been used to characterize transport carrier proteins.

Carrier-mediated transport (unlike transport through channels) can be either passive transport or secondary active transport (secondary active transport is discussed in a subsequent section). Passive transport via a carrier is sometimes called **facilitated diffusion**, although it resembles diffusion only in that it transports substances down their gradient of electrochemical potential, without an additional input of energy. (The term "facilitated diffusion" might seem more appropriately applied to transport through channels, but historically it has not been used in this way.)

Primary active transport requires energy

To carry out active transport, a carrier must couple the energetically uphill transport of a solute with another, energy-releasing event so that the overall free-energy change is negative. **Primary active transport** is coupled directly to a source of energy other than $\Delta\tilde{\mu}_j$, such as ATP hydrolysis, an oxidation–reduction reaction (as in the electron transport chain of mitochondria and chloroplasts), or the absorption

(A)

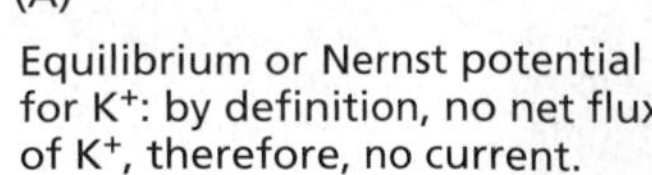

Equilibrium or Nernst potential for K+: by definition, no net flux of K+, therefore, no current.

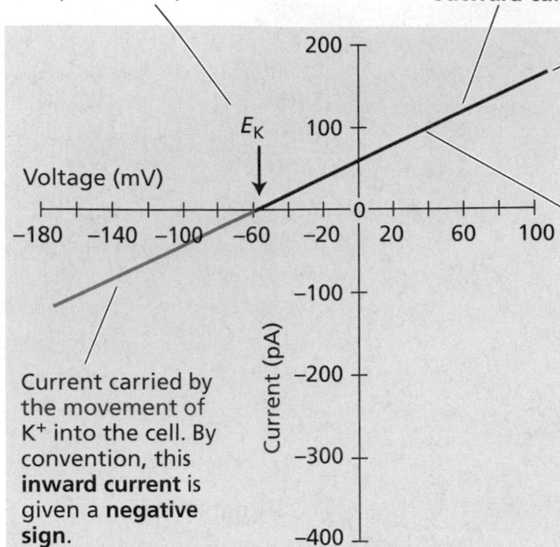

(B)

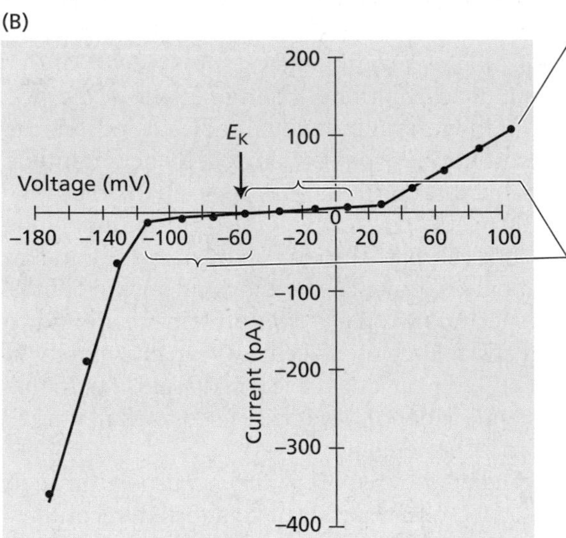

(C)

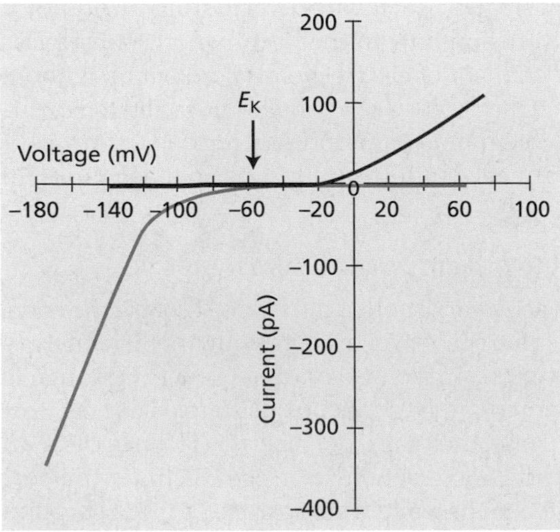

Figure 6.8 Current–voltage relationships. (A) Diagram showing the current that would result from K^+ flux through a set of hypothetical plasma membrane K^+ channels that were not voltage-regulated, given a K^+ concentration in the cytosol of 100 m*M* and an extracellular K^+ concentration of 10 m*M*. Note that the current would be linear, and that there would be zero current at the equilibrium (Nernst) potential for K^+ (E_K). (B) Actual K^+ current data from an Arabidopsis guard cell protoplast, with the same intracellular and extracellular K^+ concentrations as in (A). These currents result from the activities of voltage-regulated K^+ channels. Note that, again, there is zero net current at the equilibrium potential for K^+. However, there is also zero net current over a broader voltage range because the channels are closed over this voltage range in these conditions. When the channels are closed, no K^+ can flow through them, hence zero current is observed over this voltage range. (C) The current–voltage relationship in (B) actually results from the activities of two sets of channels—the inwardly rectifying K^+ channels and the outwardly rectifying K^+ channels—which together produce the current–voltage relationship. (B after L. Perfus-Barbeoch and S. M. Assmann, unpublished data.)

of light by the carrier protein (such as bacteriorhodopsin in halobacteria).

Membrane proteins that carry out primary active transport are called **pumps** (see Figure 6.6). Most pumps transport inorganic ions, such as H^+ or Ca^{2+}. However, as we will see later in this chapter, pumps belonging to the ATP-binding cassette (ABC) family of transporters can carry large organic molecules.

Ion pumps can be further characterized as either electrogenic or electroneutral. In general, **electrogenic transport** refers to ion transport involving the net movement of charge across the membrane. In contrast, **electroneutral transport**, as the name implies, involves no net movement of charge. For example, the Na^+/K^+-ATPase of animal cells pumps three Na^+ out for every two K^+ in, resulting in a net outward movement of one positive charge. The Na^+/K^+-ATPase is therefore an electrogenic ion pump. In contrast, the H^+/K^+-ATPase of the animal gastric mucosa pumps one H^+ out of the cell for every one K^+ in, so there is no net movement of charge across the membrane. Therefore, the H^+/K^+-ATPase is an electroneutral pump.

For the plasma membranes of plants, fungi, and bacteria, as well as for plant tonoplasts and other plant and animal endomembranes, H^+ is the principal ion that is electrogenically pumped across the membrane. The **plasma membrane H⁺-ATPase** generates the gradient of electrochemical potential of H^+ across the plasma membrane, while the **vacuolar H⁺-ATPase** (usually called the **V-ATPase**) and the **H⁺-pyrophosphatase** (H^+-PPase) electrogenically pump protons into the lumen of the vacuole and the Golgi cisternae.

Secondary active transport uses stored energy

In plant plasma membranes, the most prominent pumps are those for H^+ and Ca^{2+}, and the direction of pumping is outward from the cytosol to the extracellular space. Another mechanism is needed to drive the active uptake of mineral nutrients such as NO_3^-, SO_4^{2-}, and $H_2PO_4^-$; the uptake of amino acids, peptides, and sucrose; and the export of Na^+, which at high concentrations is toxic to plant cells. The other important way that solutes are actively transported across a membrane against their gradient of electrochemical potential is by coupling the uphill transport of one solute to the downhill transport of another. This type of carrier-mediated cotransport is termed **secondary active transport** (**Figure 6.9**).

Secondary active transport is driven indirectly by pumps. In plant cells, protons are extruded from the cytosol by electrogenic H^+-ATPases operating in the plasma membrane and at the vacuolar membrane. Consequently, a membrane potential and a pH gradient are created at the expense of ATP hydrolysis. This gradient of electrochemical potential for H^+, referred to as $\Delta\tilde{\mu}_{H^+}$, or (when expressed in other units) the **proton motive force** (**PMF**), represents stored free energy in the form of the H^+ gradient (see **WEB TOPIC 6.3**).

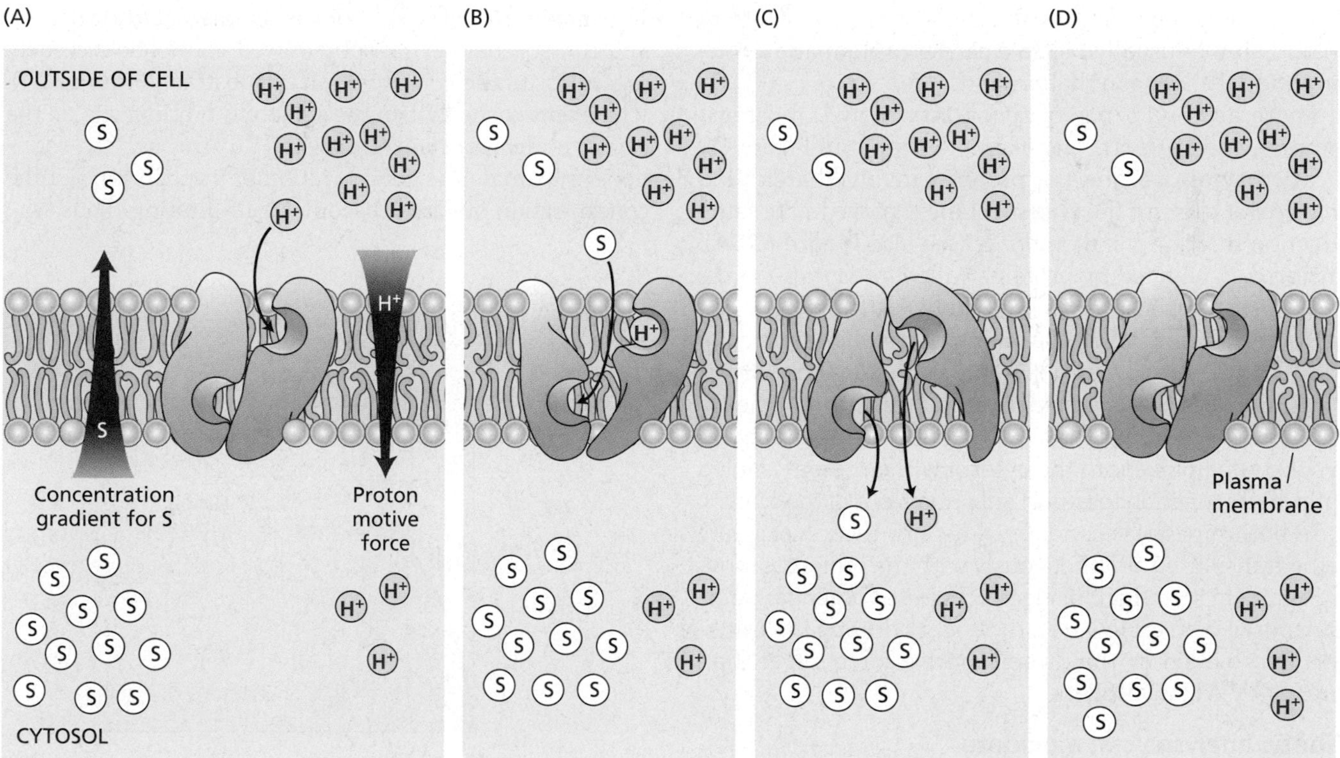

Figure 6.9 Hypothetical model of secondary active transport. In secondary active transport, the energetically uphill transport of one solute is driven by the energetically downhill transport of another solute. In the illustrated example, energy that was stored as proton motive force ($\Delta\tilde{\mu}_{H^+}$, symbolized by the red arrow on the right in [A]) is being used to take up a substrate (S) against its concentration gradient (red arrow on the left). (A) In the initial conformation, the binding sites on the protein are exposed to the outside environment and can bind a proton. (B) This binding results in a conformational change that permits a molecule of S to be bound. (C) The binding of S causes another conformational change that exposes the binding sites and their substrates to the inside of the cell. (D) Release of a proton and a molecule of S to the cell's interior restores the original conformation of the carrier and allows a new pumping cycle to begin.

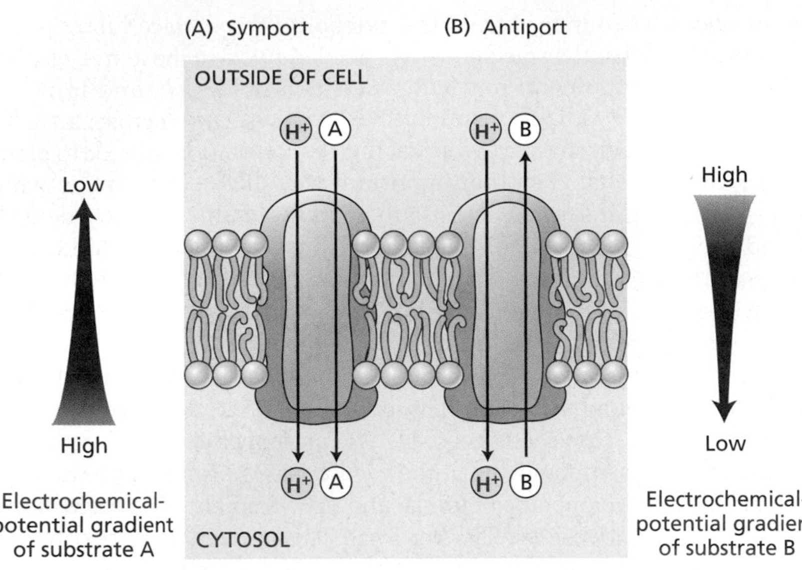

Figure 6.10 Two examples of secondary active transport coupled to a primary proton gradient. (A) In symport, the energy dissipated by a proton moving back into the cell is coupled to the uptake of one molecule of a substrate (e.g., a sugar) into the cell. (B) In antiport, the energy dissipated by a proton moving back into the cell is coupled to the active transport of a substrate (e.g., a sodium ion) out of the cell. In both cases, the substrate under consideration is moving against its gradient of electrochemical potential. Both neutral and charged substrates can be transported by such secondary active transport processes.

The proton motive force generated by electrogenic H^+ transport is used in secondary active transport to drive the transport of many other substances against their gradients of electrochemical potential. Figure 6.9 shows how secondary active transport may involve the binding of a substrate (S) and an ion (usually H^+) to a carrier protein and a conformational change in that protein.

There are two types of secondary active transport: symport and antiport. The example shown in Figure 6.9 is called **symport** (and the proteins involved are called *symporters*) because the two substances move in the same direction through the membrane (see also **Figure 6.10A**). **Antiport** (facilitated by proteins called *antiporters*) refers to coupled transport in which the energetically downhill movement of one solute drives the active (energetically uphill) transport of another solute in the opposite direction (**Figure 6.10B**). Considering the direction of the H^+ gradient, proton-coupled symporters generally function in substrate uptake into the cytosol while proton-coupled antiporters function to export substrates out of the cytosol.

In both types of secondary transport, the ion or solute being transported simultaneously with the protons is moving against its gradient of electrochemical potential, so its transport is active. However, the energy driving this transport is provided by the proton motive force rather than directly by ATP hydrolysis.

Kinetic analyses can elucidate transport mechanisms

Thus far we have described cellular transport in terms of its energetics. However, cellular transport can also be studied by use of enzyme kinetics because it involves the binding and dissociation of molecules at active sites on transport proteins (see **WEB TOPIC 6.4**). One advantage of the kinetic approach is that it gives new insights into the regulation of transport.

In kinetic experiments, the effects of external ion (or other solute) concentrations on transport rates are measured. The kinetic characteristics of the transport rates can then be used to distinguish between different transporters. The maximum rate (V_{max}) of carrier-mediated transport, and often of channel transport as well, cannot be exceeded, regardless of the concentration of substrate (**Figure 6.11**). V_{max} is approached when the substrate-binding site on the carrier is always occupied or when flux through the channel is maximal. The concentration of transporter, not the concentration of solute, becomes rate-limiting. Thus, V_{max}

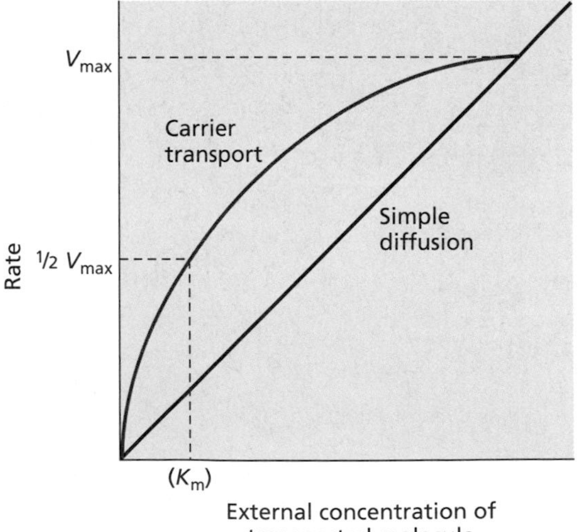

Figure 6.11 Carrier transport often shows enzyme kinetics, including saturation (V_{max}) (see **WEB APPENDIX 1**). In contrast, simple diffusion through open channels is ideally directly proportional to the concentration of the transported solute or, for an ion, to the difference in electrochemical potential across the membrane.

is an indicator of the number of molecules of the specific transport protein that are functioning in the membrane.

The constant K_m (which is numerically equal to the solute concentration that yields half the maximum rate of transport) tends to reflect the properties of the particular binding site. Low K_m values indicate high binding affinity of the transport site for the transported substance. Such values usually imply the operation of a carrier system. Higher values of K_m indicate a lower affinity of the transport site for the solute. The affinity is often so low that in practice V_{max} is never reached. In such cases, kinetics alone cannot distinguish between carriers and channels.

Cells or tissues often display complex kinetics for the transport of a solute. Complex kinetics usually indicate the presence of more than one type of transport mechanism—for example, both high- and low-affinity transporters. **Figure 6.12** shows the rate of sucrose uptake by soybean cotyledon protoplasts as a function of the external sucrose concentration. Uptake increases sharply with concentration and begins to saturate at about 10 m*M*. At concentrations above 10 m*M*, uptake becomes linear and nonsaturable within the concentration range tested. Inhibition of ATP synthesis with metabolic poisons blocks the saturable component, but not the linear one.

The interpretation of the pattern shown in Figure 6.12 is that sucrose uptake at low concentrations is an energy-dependent, carrier-mediated process (H^+–sucrose symport). At higher concentrations, sucrose enters the cells by diffusion down its concentration gradient and is therefore insensitive to metabolic poisons. Consistent with these data, both H^+–sucrose symporters and sucrose facilitators (i.e., transport proteins that mediate transmembrane sucrose flux down its free-energy gradient) have been identified at the molecular level.

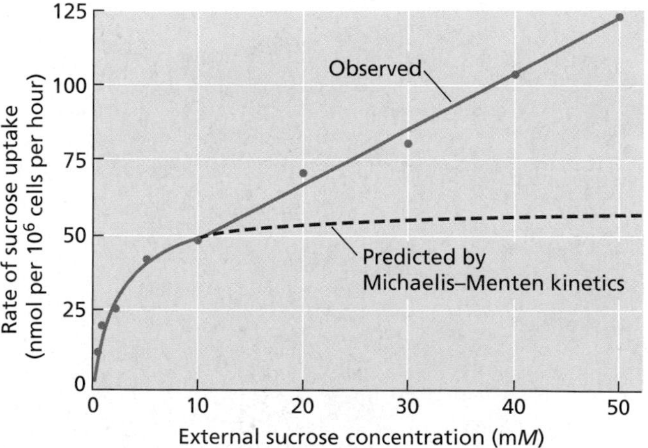

Figure 6.12 The transport properties of a solute can change with solute concentrations. For example, at low concentrations (1–10 m*M*), the rate of uptake of sucrose by soybean cells shows saturation kinetics typical of carriers. A curve fitted to these data is predicted to approach a maximum rate (V_{max}) of 57 nmol per 10^6 cells per hour. Instead, at higher sucrose concentrations, the uptake rate continues to increase linearly over a broad range of concentrations, consistent with the existence of additional facilitated transport mechanisms for sucrose uptake. (After Lin et al. 1984.)

Membrane Transport Proteins

Numerous representative transport proteins located in the plasma membrane and the tonoplast are illustrated in **Figure 6.13**. Typically, transport across a biological membrane is energized by one primary active transport system coupled to ATP hydrolysis. The transport of one ionic species—for example, H^+—generates an ion gradient and an electrochemical potential. Many other ions or neutral organic substrates can then be transported by a variety of secondary active transport proteins, which energize the transport of their substrates by simultaneously carrying one or two H^+ down their energy gradient. Thus, protons circulate across the membrane, outward through the primary active transport proteins, and back into the cell through the secondary active transport proteins. Most of the ion gradients across membranes of higher plants are generated and maintained by electrochemical-potential gradients of H^+, which are generated by electrogenic H^+ pumps.

Evidence suggests that in plants, Na^+ is transported out of the cell by a Na^+–H^+ antiporter and that Cl^-, NO_3^-, $H_2PO_4^-$, sucrose, amino acids, and other substances enter the cell via specific H^+ symporters. What about potassium ions? Potassium ions can be taken up from the soil or the apoplast by symport with H^+ (or under some conditions, Na^+). When the free-energy gradient favors passive K^+ uptake, K^+ can enter the cell by flux through specific K^+ channels. However, even influx through channels is driven by the H^+-ATPase, in the sense that K^+ diffusion is driven by the membrane potential, which is maintained at a value more negative than the K^+ equilibrium potential by the action of the electrogenic H^+ pump. Conversely, K^+ efflux requires the membrane potential to be maintained at a value more positive than E_K, which can be achieved by efflux of Cl^- or other anions through anion channels.

We have seen in preceding sections that some transmembrane proteins operate as channels for the controlled diffusion of ions. Other membrane proteins act as carriers for other substances (uncharged solutes and ions). Active transport uses carrier-type proteins that are energized either directly by ATP hydrolysis or indirectly as in the case of symporters and antiporters. The latter systems use the energy of ion gradients (often a H^+ gradient) to drive the energetically uphill transport of another ion or molecule. In the pages that follow we will examine in more detail the molecular properties, cellular locations, and genetic manipulations of some of the transport proteins that mediate the movement of organic and inorganic nutrients, as well as water, across plant plasma membranes.

Figure 6.13 Overview of the various transport proteins in the plasma membrane and tonoplast of plant cells.

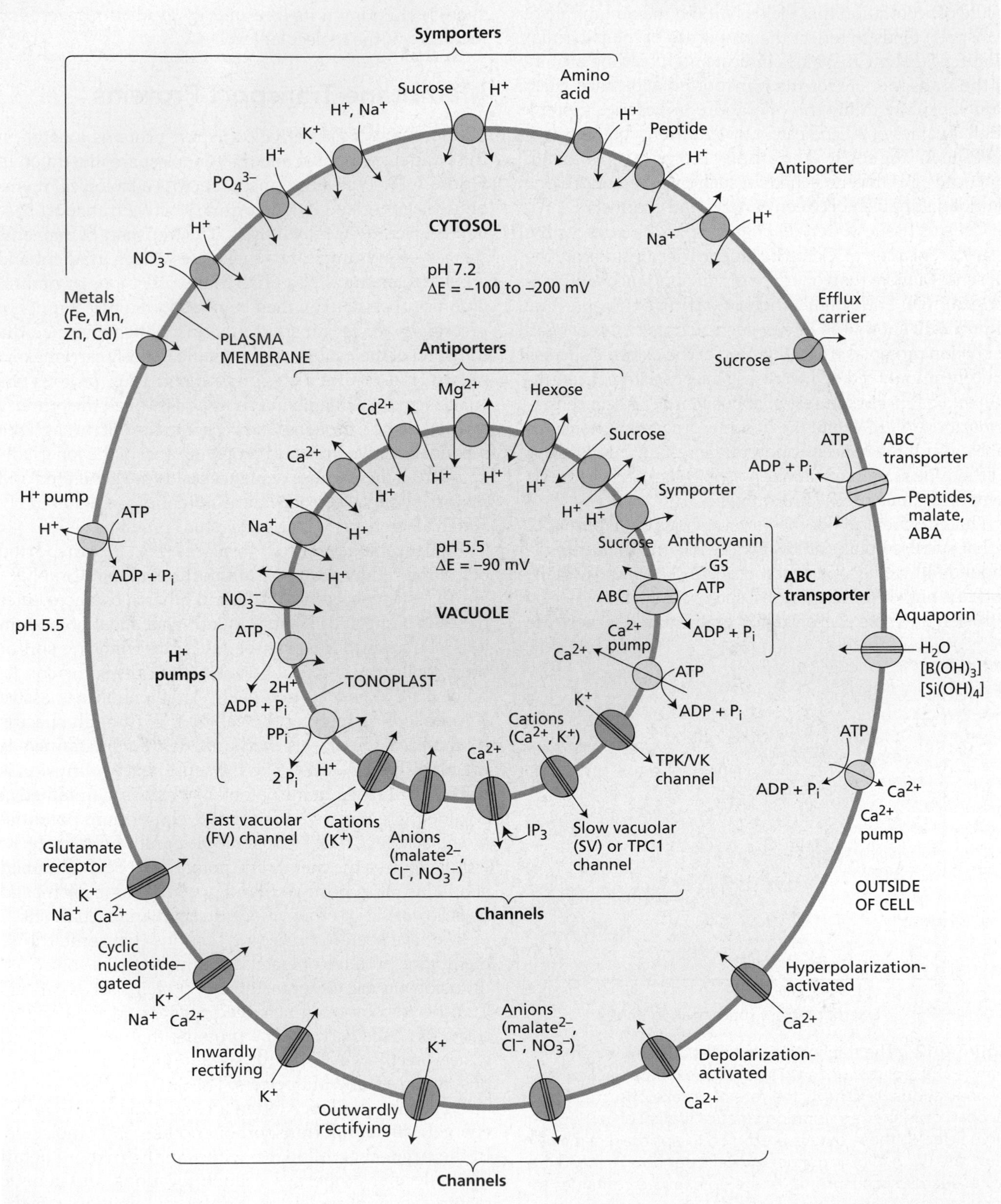

The genes for many transporters have been identified

Transporter gene identification has revolutionized the study of transporter proteins. One way to identify transporter genes is to screen plant complementary DNA (cDNA) libraries for genes that complement (i.e., compensate for) transport deficiencies in yeast. Many yeast transporter mutants have been used to identify corresponding plant genes by complementation. In the case of genes for ion channels, researchers have also studied the behavior of the channel proteins by expressing the genes in oocytes of the frog *Xenopus*, which, because of their large size, are convenient for electrophysiological studies. Genes for both inwardly and outwardly rectifying K^+ channels have been cloned and characterized in this way, and coexpression of ion channels and putative regulatory proteins such as protein kinases in oocytes has provided information on regulatory mechanisms of channel gating. As the number of sequenced genomes has increased, it has become common to identify putative transporter genes by phylogenetic analysis, in which sequence comparison with genes encoding transporters of known function in another organism allows one to predict function in the organism of interest. Based on such analyses, it has become evident that gene families, rather than individual genes, exist in plant genomes for most transport functions. Within a gene family, variations in transport kinetics, in modes of regulation, and in differential tissue expression give plants a remarkable plasticity to acclimate to and prosper under a broad range of environmental conditions. In the next sections we discuss the functions and diversity of transporters for the major categories of solutes found within the plant body (note that sucrose transport was discussed earlier in this chapter and is also discussed in Chapter 11).

Transporters exist for diverse nitrogen-containing compounds

Nitrogen, one of the macronutrients, can be present in the soil solution as nitrate (NO_3^-), ammonia (NH_3), or ammonium (NH_4^+). Plant NH_4^+ transporters are facilitators that promote NH_4^+ uptake down its free-energy gradient. Plant NO_3^- transporters are of particular interest because of their complexity. Kinetic analysis shows that NO_3^- transport, like the sucrose transport shown in Figure 6.12, has both high-affinity (low K_m) and low-affinity (high K_m) components. Both of these components are mediated by more than one gene product. In contrast to sucrose, NO_3^- is negatively charged, and such an electric charge imposes an energy requirement for uptake of nitrate. The energy is provided by symport with H^+. Nitrate transport is also strongly regulated according to NO_3^- availability: The enzymes required for NO_3^- transport, as well as for NO_3^- assimilation (see Chapter 13), are induced in the presence of NO_3^- in the environment, and uptake can be repressed if NO_3^- accumulates in the cells.

Mutants with defects in NO_3^- transport or NO_3^- reduction can be selected by growth in the presence of chlorate (ClO_3^-). Chlorate is a NO_3^- analog that is taken up and reduced in wild-type plants to the toxic product chlorite. If plants resistant to ClO_3^- are selected, they are likely to show mutations that block NO_3^- transport or reduction. Several such mutations have been identified in Arabidopsis. The first transporter gene identified in this way, named *CHL1*, encodes an inducible NO_3^-–H^+ symporter that functions as a dual-affinity carrier, with its mode of action (high affinity or low affinity) being switched by its phosphorylation status. Remarkably, this transporter also functions as a NO_3^- sensor that regulates NO_3^--induced gene expression.

Once nitrogen has been incorporated into organic molecules, there are a variety of mechanisms that distribute it throughout the plant. Peptide transporters provide one such mechanism. Peptide transporters are important for mobilizing nitrogen reserves during seed germination and senescence. In the carnivorous pitcher plant *Nepenthes alata*, high levels of expression of a peptide transporter are found in the pitcher, where the transporter presumably mediates uptake of peptides from digested insects into the internal tissues.

Some peptide transporters operate by coupling with the H^+ electrochemical gradient. Other peptide transporters are members of the ABC family of proteins, which directly use energy from ATP hydrolysis for transport; thus, this transport does not depend on a primary electrochemical gradient (see **WEB TOPIC 6.5**). The ABC family is an extremely large protein family, and its members transport diverse substrates, ranging from small inorganic ions to macromolecules. For example, large metabolites such as flavonoids, anthocyanins, and secondary products of metabolism are sequestered in the vacuole via the action of specific ABC transporters, while other ABC transporters mediate the transmembrane transport of the hormone abscisic acid.

Amino acids constitute another important category of nitrogen-containing compounds. The plasma membrane amino acid transporters of eukaryotes have been divided into five superfamilies, three of which rely on the proton gradient for coupled amino acid uptake and are present in plants. In general, amino acid transporters can provide high- or low-affinity transport, and they have overlapping substrate specificities. Many amino acid transporters show distinct tissue-specific expression patterns, suggesting specialized functions in different cell types. Amino acids constitute an important form in which nitrogen is distributed long distances in plants, so it is not surprising that the expression patterns of many amino acid transporter genes include expression in living vascular tissue.

Amino acid and peptide transporters have important roles in addition to their function as distributors of nitrogen resources. Because plant hormones are frequently found conjugated with amino acids and peptides, transporters for those molecules may also be involved in the distribution of hormone conjugates throughout the plant body. The hormone auxin is derived from tryptophan, and the genes encoding auxin transporters are related to those for some amino acid transporters. In another example, proline is an amino acid that accumulates under salt stress. This accumulation lowers the water potential of the cell, thereby promoting cellular water retention under stress conditions.

Cation transporters are diverse

Cations are transported by both cation channels and cation carriers. The relative contributions of each type of transport mechanism differ depending on the membrane, cell type, and prevailing conditions.

CATION CHANNELS On the order of 50 genes in the Arabidopsis genome encode channels mediating cation uptake across the plasma membrane or intracellular membranes such as the tonoplast. Some of these channels are highly selective for specific ionic species, such as potassium ions. Others allow passage of a variety of cations, sometimes including Na^+, even though this ion is toxic when over-accumulated. As described in **Figure 6.14**, cation channels are categorized into six types based on their deduced structures and cation selectivity.

Of the six types of plant cation channels, the Shaker channels have been the most thoroughly characterized. These channels are named after a *Drosophila* K^+ channel whose mutation causes the flies to shake or tremble. Plant Shaker channels are highly K^+-selective and can be either inwardly or outwardly rectifying or weakly rectifying. Some members of the Shaker family can:

- Mediate K^+ uptake or efflux across the guard cell plasma membrane.
- Provide a major conduit for K^+ uptake from the soil.
- Participate in K^+ release to dead xylem vessels from the living stelar cells.
- Play a role in K^+ uptake in pollen, a process that promotes water influx and pollen tube elongation.

Some Shaker channels, such as those in roots, can mediate high-affinity K^+ uptake, allowing passive K^+ uptake at micromolar external K^+ concentrations as long as the membrane potential is sufficiently hyperpolarized to drive this uptake.

Not all ion channels are as strongly regulated by membrane potential as the majority of Shaker channels are. Some ion channels, such as the TPK/VK channels (see Figure 6.13), are not voltage-regulated, and the voltage sensitivity of others, such as the KCO3 channel, has not yet been determined. Cyclic nucleotide–gated cation chan-

(A) K^+ channels

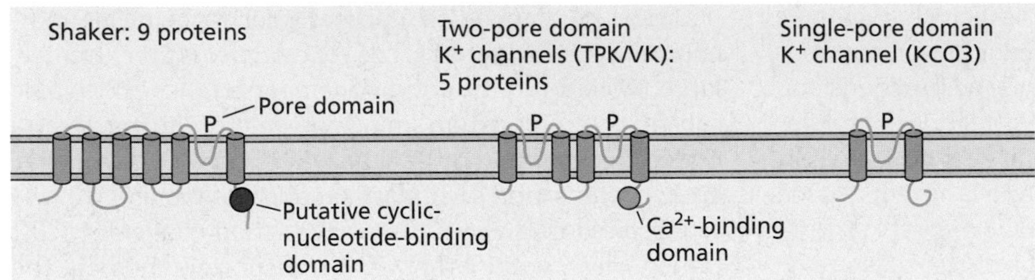

(B) Poorly selective cation channels (C) Ca^{2+}-permeable channels (D) Cation selective, Ca^{2+} permeable

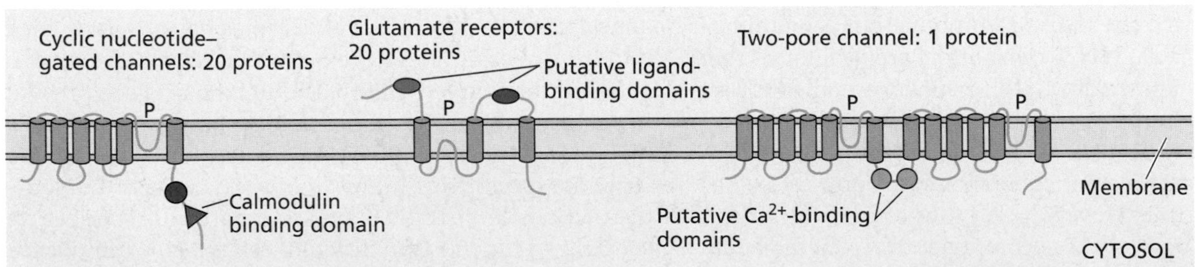

Figure 6.14 Six families of Arabidopsis cation channels. Some channels have been identified from sequence homology with channels of animals, while others have been experimentally verified. (A) K^+-selective channels. (B) Weakly selective cation channels with activity regulated by binding of cyclic nucleotides. (C) Putative glutamate receptors; based on measurements of cytosolic Ca^{2+} changes, these proteins probably function as Ca^{2+}-permeable channels. (D) Two-pore channel: one protein (TPC1) is the sole two-pore channel of this type encoded in the Arabidopsis genome. TPC1 is permeable to mono- and divalent cations, including Ca^{2+}. (After Very and Sentenac 2002; Lebaudy et al. 2007.)

nels are one example of a ligand-gated channel, with activity promoted by the binding of cyclic nucleotides such as cGMP. These channels exhibit weak selectivity with permeability to K^+, Na^+, and Ca^{2+}. Cyclic nucleotide–gated cation channels are involved in diverse physiological processes, including disease resistance, senescence, temperature sensing, and pollen tube growth and viability. Another interesting set of ligand-gated channels are the glutamate receptor channels. These channels are homologous to a class of glutamate receptors in the mammalian nervous system that function as glutamate-gated cation channels, and are activated in plants by glutamate and some other amino acids. Plant glutamate receptor channels are permeable to Ca^{2+}, K^+, and Na^+ to varying extents but have been particularly implicated in Ca^{2+} uptake and signaling in nutrient acquisition in roots and in guard cell and pollen tube physiology.

Ion fluxes must also occur in and out of the vacuole, and both cation- and anion-permeable channels have been characterized in the vacuolar membrane (see Figure 6.13). Plant vacuolar cation channels include the KCO3 K^+ channel (see Figure 6.14A), the Ca^{2+}-activated TPC1/SV cation channel (see Figures 6.13 and 6.14B), and most TPK/VK channels (see Figure 6.13), which are highly selective K^+ channels that are activated by Ca^{2+}. In addition, Ca^{2+} efflux from internal storage sites such as the vacuole plays an important signaling role. Ca^{2+} release from stores is triggered by several second messenger molecules, including cytosolic Ca^{2+} itself and inositol trisphosphate ($InsP_3$). For a more detailed description of these signal transduction pathways, see Chapter 15.

CATION CARRIERS A variety of carriers also move cations into plant cells. One family of transporters that specializes in K^+ transport across plant membranes is the HAK/KT/KUP family (which we will refer to here as the HAK family). The HAK family contains both high-affinity and low-affinity transporters, some of which also mediate Na^+ influx at high external Na^+ concentrations. High-affinity HAK transporters are thought to take up K^+ via H^+–K^+ symport, and these transporters are particularly important for K^+ uptake from the soil when soil K^+ concentrations are low. A second family, the cation–H^+ antiporters (CPAs), mediates electroneutral exchange of H^+ and other cations, including K^+ in some cases. A third family consists of the Trk/HKT transporters (which we will refer to here as the HKT transporters), which can operate as K^+–H^+ or K^+–Na^+ symporters, or as Na^+ channels under high external Na^+ concentrations. The importance of the HKT transporters for K^+ transport remains incompletely elucidated, but as described below, HKT transporters are central elements in plant tolerance of saline conditions.

Irrigation increases soil salinity, and salinization of croplands is an increasing problem worldwide. Although halophytic plants, such as those found in salt marshes, are adapted to a high-salt environment, such environments are deleterious to other, glycophytic, plant species, including the majority of crop species. Plants have evolved mechanisms to extrude Na^+ across the plasma membrane, to sequester salt in the vacuole, and to redistribute Na^+ within the plant body.

At the plasma membrane, a Na^+–H^+ antiporter was uncovered in a screen to identify Arabidopsis mutants that showed enhanced sensitivity to salt, hence this antiporter was named Salt Overly Sensitive, or SOS1. SOS1-type antiporters in the root extrude Na^+ from the plant, thereby lowering internal concentrations of this toxic ion.

Vacuolar Na^+ sequestration occurs by activity of Na^+–H^+ antiporters—a subset of CPA proteins—which couple the energetically downhill movement of H^+ into the cytosol with Na^+ uptake into the vacuole. When the Arabidopsis *AtNHX1* Na^+–H^+ antiporter gene is overexpressed, it confers greatly increased salt tolerance to both Arabidopsis and crop species such as maize (corn; *Zea mays*), wheat, and tomato.

Whereas SOS1 and NHX antiporters reduce cytosolic Na^{2+} concentrations, HKT1 transporters transport Na^+ from the apoplast into the cytosol. However, Na^+ uptake by HKT1 transporters in the plasma membrane of root xylem parenchyma cells is important in retrieving Na^+ from the transpiration stream, thereby reducing Na^+ concentrations and attendant toxicity in photosynthetic tissues. Presumably such Na^+ is then excluded from the root cytosol by the action of SOS1 and NHX transporters. Transgenic expression of a HKT1 transporter in a durum (pasta) variety of wheat greatly increases grain yield in wheat grown on saline soils.

Just as for Na^+, there is a large free-energy gradient for Ca^{2+} that favors its entry into the cytosol from both the apoplast and intracellular stores. This entry is mediated by Ca^{2+}-permeable channels, which we described above. Calcium ion concentrations in the cell wall and in the apoplast are usually in the millimolar range; in contrast, free cytosolic Ca^{2+} concentrations are kept in the hundreds of nanomolar (10^{-9} *M*) to 1 micromolar (10^{-6} *M*) range, against the large electrochemical-potential gradient for Ca^{2+} diffusion into the cell. Calcium ion efflux from the cytosol is achieved by Ca^{2+}-ATPases found at the plasma membrane and in some endomembranes such as the tonoplast (see Figure 6.13) and endoplasmic reticulum. Much of the Ca^{2+} inside the cell is stored in the central vacuole, where it is sequestered via Ca^{2+}-ATPases and via Ca^{2+}–H^+ antiporters, which use the electrochemical potential of the proton gradient to energize the vacuolar accumulation of Ca^{2+}.

Because small changes in cytosolic Ca^{2+} concentration drastically alter the activities of many enzymes, cytosolic Ca^{2+} concentration is tightly regulated. The Ca^{2+}-binding protein, calmodulin (CaM), participates in this regulation. Although CaM has no catalytic activity of its own, Ca^{2+}-bound CaM binds to many different classes of target proteins and regulates their activity (see **WEB ESSAY 6.2**).

Ca^{2+}-permeable cyclic nucleotide–gated channels are CaM-binding proteins, and there is evidence that this CaM binding results in downregulation of channel activity. One class of Ca^{2+}-ATPases also binds CaM. CaM binding releases these ATPases from autoinhibition, resulting in increased Ca^{2+} extrusion into the apoplast, endoplasmic reticulum, and vacuole. Together, these two regulatory effects of CaM provide a mechanism whereby increases in cytosolic Ca^{2+} concentration initiate a negative feedback loop, via activated CaM, that aids in restoration of resting cytosolic Ca^{2+} levels.

Anion transporters have been identified

Nitrate (NO_3^-), chloride (Cl^-), sulfate (SO_4^{2-}), and phosphate ($H_2PO_4^-$) are the major inorganic ions in plant cells, and malate^{2-} is a major organic anion. The free-energy gradient for all of these anions is in the direction of passive efflux. Several types of plant anion channels have been characterized by electrophysiological techniques, and most anion channels appear to be permeable to a variety of anions. In particular, several anion channels with differential voltage-dependence and anion permeabilities have been shown to be important for anion efflux from guard cells during stomatal closure.

In contrast to the relative lack of specificity of anion channels, anion carriers that mediate the energetically uphill transport of anions into plant cells exhibit selectivity for particular anions. In addition to the transporters for nitrate uptake described above, plants have transporters for various organic anions, such as malate and citrate. As we will discuss in Chapter 10, malate uptake is an important contributor to the increase in intracellular solute concentration that drives the water uptake into guard cells that leads to stomatal opening. One member of the ABC family, AtABCB14, has been assigned this malate import function.

Phosphate availability in the soil solution often limits plant growth. In Arabidopsis, a family of about nine plasma membrane phosphate transporters, some high affinity and some low affinity, mediates phosphate uptake in symport with protons. These transporters are expressed primarily in the root epidermis and root hairs, and their expression is induced upon phosphate starvation. Other phosphate–H^+ symporters have also been identified in plants and have been localized to membranes of intracellular organelles such as plastids and mitochondria. Another group of phosphate transporters, the phosphate translocators, are located in the inner plastid membrane, where they mediate exchange of inorganic phosphate with phosphorylated carbon compounds (see **WEB TOPIC 8.11**).

Transporters for metal and metalloid ions transport essential micronutrients

Several metals are essential nutrients for plants, although they are required in only trace amounts. One example is iron. Iron deficiency is the most common human nutritional disorder worldwide, so an increased understanding of how plants accumulate iron may also benefit efforts to improve the nutritional value of crops. Over 25 ZIP transporters mediate the uptake of iron, manganese, and zinc ions into plants, and other transporter families that mediate the uptake of copper and molybdenum ions have been identified. Metal ions are usually present at low concentrations in the soil solution, so these transporters are typically high-affinity transporters. Some metal ion transporters mediate the uptake of cadmium or lead ions, which are undesirable in crop species because cadmium and lead ions are toxic to humans. However, this property may prove useful in the detoxification of soils by uptake of contaminants into plants (phytoremediation), which can then be removed and properly discarded.

Once in the plant, metal ions, usually chelated with other molecules, must be transported into the xylem for distribution throughout the plant body via the transpiration stream, and metals must also be sent to their appropriate subcellular destinations. For example, most of the iron in plants is found in chloroplasts, where it is incorporated into chlorophyll and components of the electron transport chain (see Chapter 7). Overaccumulation of ionic forms of metals such as iron and copper can lead to production of toxic reactive oxygen species (ROS). Compounds that chelate metal ions guard against this threat, and transporters that mediate metal uptake into the vacuole are also important in maintaining metal concentrations at nontoxic levels.

Metalloids are elements that have properties of both metals and nonmetals. Boron and silicon are two metalloids that are used by plants. Both play important roles in cell wall structure—boron by cross-linking cell wall polysaccharides, and silicon by increasing structural rigidity. Boron (as boric acid [$B(OH)_3$; also written H_3BO_3]) and silicon (as silicic acid [$Si(OH)_4$; also written H_4SiO_4]) both enter cells via aquaporin-type channels (see below) and are exported via efflux transporters, probably by secondary active transport. Due to similarities in chemical structure, arsenite (a form of arsenic) can also enter plant roots via the silicon channel and be exported to the transpiration stream via the silicon transporter. Rice is particularly efficient at taking up arsenite, and as a result, arsenic poisoning from human consumption of rice is a significant problem in regions of southeast Asia.

Aquaporins have diverse functions

Aquaporins are a class of transporters that are relatively abundant in plant membranes and are also common in animal membranes (see Chapters 3 and 4). The Arabidopsis genome is predicted to encode approximately 35 aquaporins. As the name implies, many aquaporin proteins mediate the flux of water across membranes, and it has been hypothesized that aquaporins function as sensors of gradients in osmotic potential and turgor pressure. In addition, some aquaporin proteins mediate the influx of mineral nutrients (e.g., boric acid and silicic acid). There is also some evidence that aquaporins can act as conduits for carbon dioxide, ammonia (NH_3), and hydrogen peroxide (H_2O_2) movement across plant plasma membranes.

Aquaporin activity is regulated by phosphorylation as well as by pH, Ca^{2+} concentration, heteromerization, and reactive oxygen species. Such regulation may account for the ability of plant cells to quickly alter their water permeability in response to circadian rhythm and to stresses such as salt, chilling, drought, and flooding (anoxia). Regulation also occurs at the level of gene expression. Aquaporins are highly expressed in epidermal and endodermal cells and in the xylem parenchyma, which may be critical points for control of water movement.

Plasma membrane H^+-ATPases are highly regulated P-type ATPases

As we have seen, the outward active transport of protons across the plasma membrane creates gradients of pH and electrical potential that drive the transport of many other substances (ions and uncharged solutes) through the various secondary active transport proteins. H^+-ATPase activity is also important for the regulation of cytosolic pH and for the control of cell turgor, which drives organ (leaf and flower) movement, stomatal opening, and cell growth. **Figure 6.15** illustrates how a membrane H^+-ATPase might work.

Plant and fungal plasma membrane H^+-ATPases and Ca^{2+}-ATPases are members of a class known as P-type ATPases, which are phosphorylated as part of the catalytic cycle that hydrolyzes ATP. Plant plasma membrane H^+-ATPases are encoded by a family of about a dozen genes. The roles of each H^+-ATPase isoform are starting to be understood, based on information from gene expression patterns and functional analysis of Arabidopsis plants harboring null mutations in individual H^+-ATPase genes. Some H^+-ATPases exhibit cell-specific patterns of expression. For example, several H^+-ATPases are expressed in guard cells, where they energize the plasma membrane to drive solute uptake during stomatal opening (see Chapter 4).

In general, H^+-ATPase expression is high in cells with key functions in nutrient movement, including root endodermal cells and cells involved in nutrient uptake from the apoplast that surrounds the developing seed. In cells in which multiple H^+-ATPases are coexpressed, they may be differentially regulated or may function redundantly, perhaps providing a "fail-safe" mechanism to this all-important transport function.

Figure 6.16 shows a model of the functional domains of a yeast plasma membrane H^+-ATPase, which is similar to those of plants. The protein has ten membrane-spanning domains that cause it to loop back and forth across the membrane. Some of the membrane-spanning domains make up the pathway through which protons are pumped. The catalytic domain, which catalyzes ATP hydrolysis, including the aspartic acid residue that becomes phosphorylated during the catalytic cycle, is on the cytosolic face of the membrane.

Like other enzymes, the plasma membrane H^+-ATPase is regulated by the concentration of substrate (ATP), pH, temperature, and other factors. In addition, H^+-ATPase molecules can be reversibly activated or deactivated by specific signals, such as light, hormones, or pathogen attack. This type of regulation is mediated by a specialized auto-

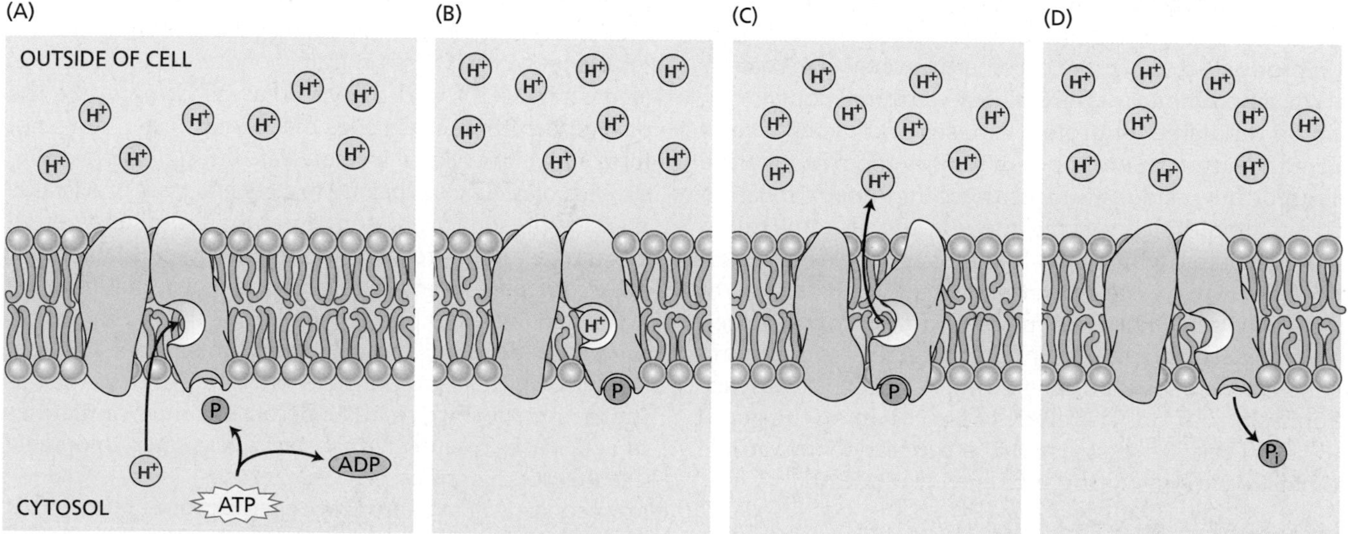

Figure 6.15 Hypothetical steps in the transport of a proton against its chemical gradient by H^+-ATPase. The pump, embedded in the membrane, (A) binds the proton on the inside of the cell and (B) is phosphorylated by ATP. (C) This phosphorylation leads to a conformational change that exposes the proton to the outside of the cell and makes it possible for the proton to diffuse away. (D) Release of the phosphate ion (P_i) from the pump into the cytosol restores the initial configuration of the H^+-ATPase and allows a new pumping cycle to begin.

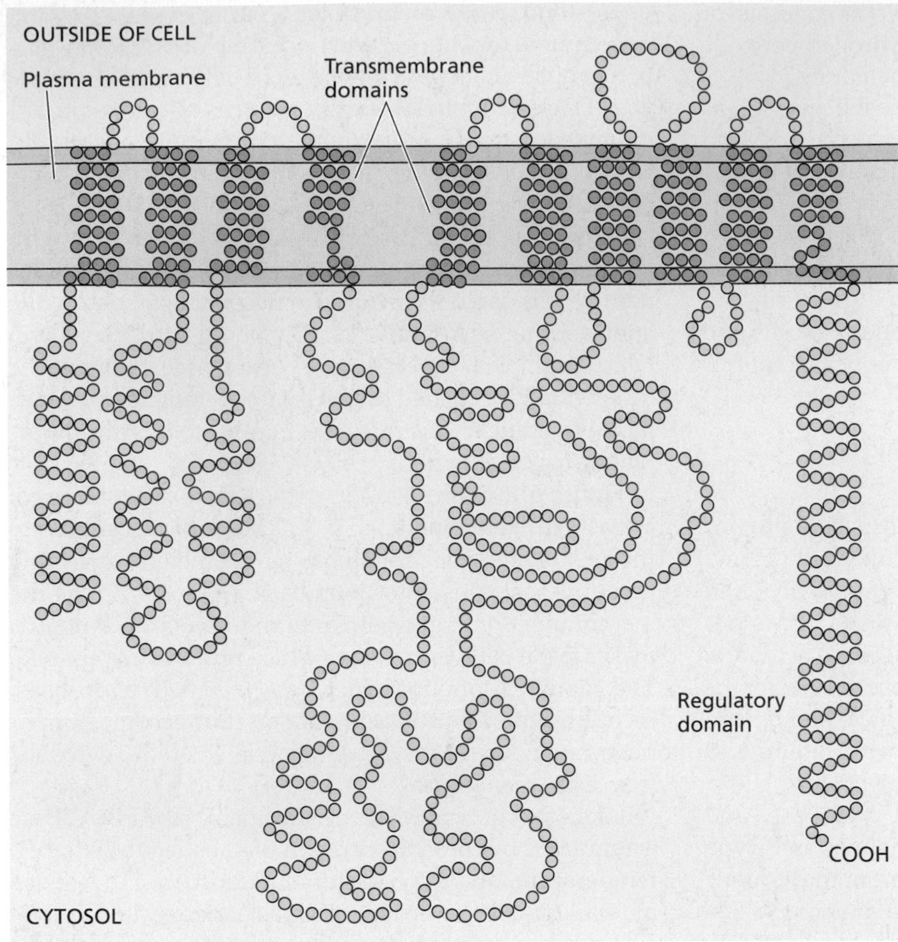

Figure 6.16 Two-dimensional representation of a plasma membrane H⁺-ATPase from yeast. Each small circle represents an amino acid. The H⁺-ATPase protein has ten transmembrane segments. The regulatory domain is an autoinhibitory domain. Posttranslational modifications that lead to displacement of the autoinhibitory domain result in H⁺-ATPase activation. (After Palmgren 2001.)

inhibitory domain at the C-terminal end of the polypeptide chain, which acts to regulate the activity of the H^+-ATPase (see Figure 6.16). If the autoinhibitory domain is removed by a protease, the enzyme becomes irreversibly activated.

The autoinhibitory effect of the C-terminal domain can also be regulated by protein kinases and phosphatases that add phosphate groups to or remove them from serine or threonine residues on this domain. Phosphorylation recruits ubiquitous enzyme-modulating proteins called 14-3-3 proteins, which bind to the phosphorylated region and thus displace the autoinhibitory domain, leading to H^+-ATPase activation. The fungal toxin **fusicoccin**, which is a strong activator of the H^+-ATPase, activates this pump by increasing 14-3-3 binding affinity even in the absence of phosphorylation. The effect of fusicoccin on the guard cell H^+-ATPases is so strong that it can lead to irreversible stomatal opening, wilting, and even plant death.

The tonoplast H^+-ATPase drives solute accumulation in vacuoles

Plant cells increase their size primarily by taking up water into a large central vacuole. Therefore, the osmotic pressure of the vacuole must be kept sufficiently high for water to enter from the cytosol. The tonoplast regulates the traffic of ions and metabolites between the cytosol and the vacuole, just as the plasma membrane regulates their uptake into the cell. Tonoplast transport became a vigorous area of research following the development of methods for the isolation of intact vacuoles and tonoplast vesicles (see WEB TOPIC 6.6). These studies elucidated a diversity of anion and cation channels in the tonoplast membrane (see Figure 6.13) and led to the discovery of a new type of proton-pumping ATPase, the vacuolar H^+-ATPase, which transports protons into the vacuole (see Figure 6.13).

The vacuolar H^+-ATPase differs both structurally and functionally from the plasma membrane H^+-ATPase. The vacuolar ATPase is more closely related to the F-ATPases of mitochondria and chloroplasts (see Chapter 12), and the vacuolar ATPase, unlike the plasma membrane ATPases discussed earlier, does not form a phosphorylated intermediate during ATP hydrolysis. Vacuolar ATPases belong to a general class of ATPases that are present on the endomembrane systems of all eukaryotes. They are large enzyme complexes, about 750 kDa, composed of multiple subunits. These subunits are organized into a peripheral complex, V_1, that is responsible for ATP hydrolysis, and an integral membrane channel complex, V_0, that is responsible for H^+ translocation across the membrane (**Figure 6.17**). Because of their similarities to F-ATPases, vacuolar ATPases are assumed to operate like tiny rotary motors (see Chapter 12).

Vacuolar ATPases are electrogenic proton pumps that transport protons from the cytosol to the vacuole and generate a proton motive force across the tonoplast. Electrogenic proton pumping accounts for the fact that the vacuole is typically 20 to 30 mV more positive than the cytosol, although it is still negative relative to the external medium. To allow maintenance of bulk electrical neutrality, anions

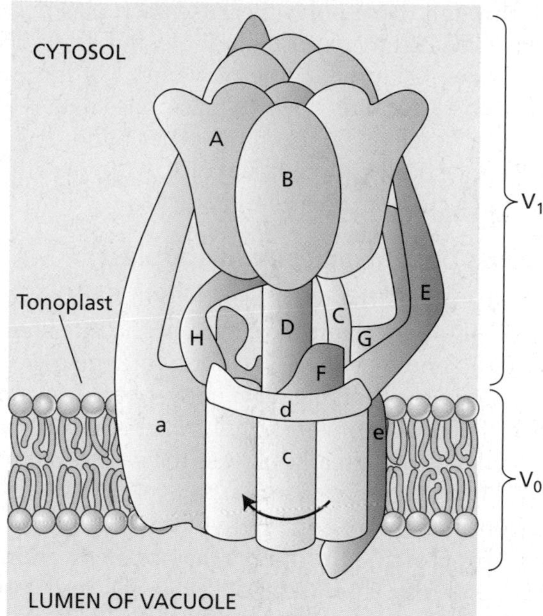

Figure 6.17 Model of the V-ATPase rotary motor. Many polypeptide subunits come together to make up this complex enzyme. The V_1 catalytic complex, which is easily dissociated from the membrane, contains the nucleotide-binding and catalytic sites. Components of V_1 are designated by uppercase letters. The integral membrane complex mediating H^+ transport is designated V_0, and its subunits are designated by lowercase letters. It is proposed that ATPase reactions catalyzed by each of the A subunits, acting in sequence, drive the rotation of the shaft (D) and the six c subunits. The rotation of the c subunits relative to the a subunit is thought to drive the transport of H^+ across the membrane. (After Kluge et al. 2003.)

such as Cl^- or malate^{2-} are transported from the cytosol into the vacuole through channels in the tonoplast. The conservation of bulk electrical neutrality by anion transport makes it possible for the vacuolar H^+-ATPase to generate a large concentration gradient of protons (pH gradient) across the tonoplast. This gradient accounts for the fact that the pH of the vacuolar sap is typically about 5.5, while the cytosolic pH is typically 7.0 to 7.5. Whereas the electrical component of the proton motive force drives the uptake of anions into the vacuole, the electrochemical-potential gradient for H^+ ($\tilde{\mu}_{H^+}$) is harnessed to drive the uptake of cations and sugars into the vacuole via secondary transport (antiporter) systems (see Figure 6.13).

Although the pH of most plant vacuoles is mildly acidic (about 5.5), the pH of the vacuoles of some species is much lower—a phenomenon termed *hyperacidification*. Vacuolar hyperacidification is the cause of the sour taste of certain fruits (lemons) and vegetables (rhubarb). Biochemical studies have suggested that the low pH of lemon fruit vacuoles (specifically, those of the juice sac cells) is due to a combination of factors:

- The low permeability of the vacuolar membrane to protons permits a steeper pH gradient to build up.
- A specialized vacuolar ATPase is able to pump protons more efficiently (with less wasted energy) than normal vacuolar ATPases can.
- Organic acids such as citric, malic, and oxalic acids accumulate in the vacuole and help maintain its low pH by acting as buffers.

H^+-pyrophosphatases also pump protons at the tonoplast

Another type of proton pump, a H^+-pyrophosphatase (H^+-PPase), works in parallel with the vacuolar ATPase to create a proton gradient across the tonoplast (see Figure 6.13). This enzyme consists of a single polypeptide that harnesses energy from the hydrolysis of inorganic pyrophosphate (PP_i) to drive H^+ transport.

The free energy released by PP_i hydrolysis is less than that from ATP hydrolysis. However, the H^+-PPase transports only one H^+ ion per PP_i molecule hydrolyzed, whereas the vacuolar ATPase appears to transport two H^+ ions per ATP hydrolyzed. Thus, the energy available per H^+ transported appears to be approximately the same, and the two enzymes seem to be able to generate comparable proton gradients. Interestingly, the plant H^+-PPase is not found in animals or in yeast, although similar enzymes are present in some bacteria and protists.

Both the V-ATPase and the H^+-PPase are found in other compartments of the endomembrane system in addition to the vacuole. Consistent with this distribution, evidence is emerging that these ATPases regulate not only H^+ gradients per se, but also vesicle trafficking and secretion. In addition, increased auxin transport and cell division in Arabidopsis plants overexpressing a H^+-PPase, and the opposite phenotypes in plants with reduced H^+-PPase activity, indicate connections between H^+-PPase activity and the synthesis, distribution, and regulation of auxin transporters.

Ion Transport in Roots

Mineral nutrients absorbed by the root are carried to the shoot by the transpiration stream moving through the xylem (see Chapter 4). Both the initial uptake of nutrients and water and the subsequent movement of these substances from the root surface across the cortex and into the xylem are highly specific, well-regulated processes.

Ion transport across the root obeys the same biophysical laws that govern cellular transport. However, as we have seen in the case of water movement (see Chapter 4), the anatomy of roots imposes some special constraints on the pathway of ion movement. In this section we will discuss the pathways and mechanisms involved in the radial movement of ions from the root surface to the tracheary elements of the xylem.

Solutes move through both apoplast and symplast

Thus far, our discussion of cellular ion transport has not included the cell wall. In terms of the transport of small molecules, the cell wall is a fluid-filled lattice of polysaccharides through which mineral nutrients diffuse readily. Because all plant cells are separated by cell walls, ions can diffuse across a tissue (or be carried passively by water flow) entirely through the cell wall space without ever entering a living cell. This continuum of cell walls is called the **extracellular space**, or **apoplast** (see Figure 4.4). Typically, 5 to 20% of the plant tissue volume is occupied by cell walls.

Just as the cell walls form a continuous phase, so do the cytoplasms of neighboring cells, collectively referred to as the **symplast**. Plant cells are interconnected by cytoplasmic bridges called plasmodesmata (see Chapter 1), cylindrical pores 20 to 60 nm in diameter (**Figure 6.18** and Figure 1.6). Each plasmodesma is lined with plasma membrane and contains a narrow tubule, the *desmotubule*, which is a continuation of the endoplasmic reticulum.

In tissues where significant amounts of intercellular transport occur, neighboring cells contain large numbers of plasmodesmata, up to 15 per square micrometer of cell surface. Specialized secretory cells, such as floral nectaries and leaf salt glands, have high densities of plasmodesmata.

By injecting dyes or by making electrical-resistance measurements on cells containing large numbers of plasmodesmata, investigators have shown that inorganic ions, water, and small organic molecules can move from cell to cell through these pores. Because each plasmodesma is partly occluded by the desmotubule and its associated proteins (see Chapter 1), the movement of large molecules such as proteins through plasmodesmata requires special mechanisms. Ions, on the other hand, appear to move symplastically through the plant by simple diffusion through plasmodesmata (see Chapter 4).

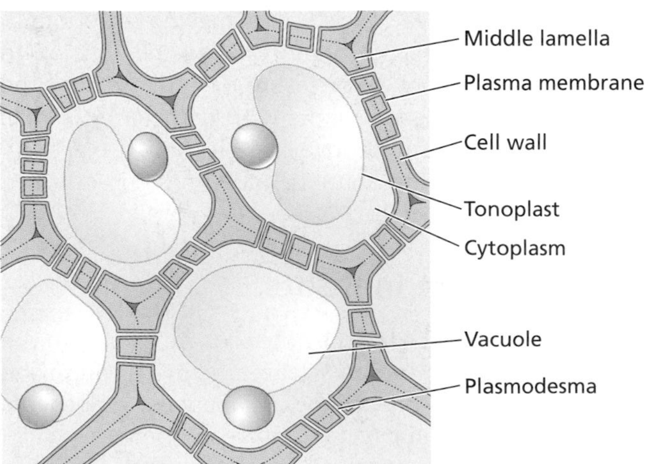

Figure 6.18 Plasmodesmata (singular is "plasmodesma") connect the cytoplasms of neighboring cells, thereby facilitating cell-to-cell communication.

Ions cross both symplast and apoplast

Ion absorption by the root (see Chapter 5) is more pronounced in the root hair zone than in the meristem and elongation zones. Cells in the root hair zone have completed their elongation but have not yet begun secondary growth. The root hairs are simply extensions of specific epidermal cells that greatly increase the surface area available for ion absorption.

An ion that enters a root may immediately enter the symplast by crossing the plasma membrane of an epidermal cell, or it may enter the apoplast and diffuse between the epidermal cells through the cell walls. From the apoplast of the cortex, an ion (or other solute) may either be transported across the plasma membrane of a cortical cell, thus entering the symplast, or diffuse radially all the way to the endodermis via the apoplast. The apoplast forms a continuous phase from the root surface through the cortex. However, in all cases, ions must enter the symplast before they can enter the stele, because of the presence of the Casparian strip. As discussed in Chapters 4 and 5, the Casparian strip is a lignified or suberized layer that forms as rings around the specialized cells of the endodermis (**Figure 6.19**) and effectively blocks the entry of water and solutes into the stele via the apoplast.

The stele consists of dead tracheary elements surrounded by living pericycle and xylem parenchyma cells. Once an ion has entered the stele through the symplastic connections across the endodermis, it continues to diffuse through the living cells. Finally, the ion is released into the apoplast and diffuses into the conducting cells of the xylem—because these cells are dead, their interiors are continuous with the apoplast. The Casparian strip allows nutrient uptake to be selective; it also prevents ions from diffusing back out of the root through the apoplast. Thus, the presence of the Casparian strip allows the plant to maintain a higher ion concentration in the xylem than exists in the soil water surrounding the roots.

Xylem parenchyma cells participate in xylem loading

The process whereby ions exit the symplast of a xylem parenchyma cell and enter the conducting cells of the xylem for translocation to the shoot is called **xylem loading**. Xylem loading is a highly regulated process. Xylem

(A)

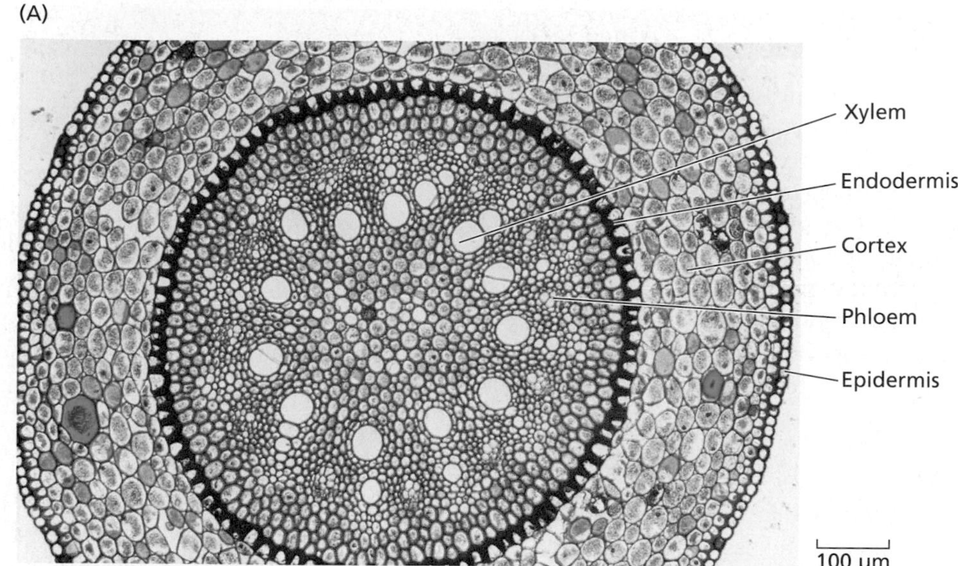

(B)

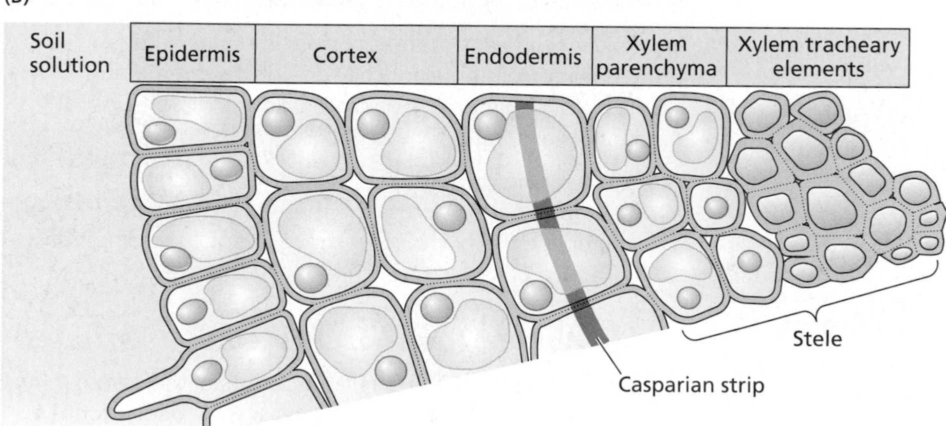

Figure 6.19 Tissue organization in roots. (A) Cross section through a root of carrion flower (genus *Smilax*), a monocot, showing the epidermis, cortex parenchyma, endodermis, xylem, and phloem. (B) Schematic diagram of a root cross section, illustrating the cell layers through which solutes pass from the soil solution to the xylem tracheary elements.

parenchyma cells, like other living plant cells, maintain plasma membrane H^+-ATPase activity and a negative membrane potential. Transporters that specifically function in the unloading of solutes to the tracheary elements have been identified by electrophysiological and genetic approaches. The plasma membranes of xylem parenchyma cells contain proton pumps, aquaporins, and a variety of ion channels and carriers specialized for influx or efflux.

In Arabidopsis xylem parenchyma, the stelar outwardly rectifying K^+ channel (SKOR) is expressed in cells of the pericycle and xylem parenchyma, where it functions as an efflux channel, transporting K^+ from the living cells out to the tracheary elements. In mutant Arabidopsis plants lacking the SKOR channel protein, or in plants in which SKOR has been pharmacologically inactivated, K^+ transport from the root to the shoot is severely reduced, confirming the function of this channel protein.

Several types of anion-selective channels have also been identified that participate in unloading of Cl^- and NO_3^- from the xylem parenchyma. Drought, abscisic acid (ABA) treatment, or elevation of cytosolic Ca^{2+} concentrations (which often occurs as a response to ABA), all reduce the activity of SKOR and anion channels of the root xylem parenchyma, a response that could help maintain cellular hydration in the root under desiccating conditions.

Other, less selective ion channels found in the plasma membrane of xylem parenchyma cells are permeable to K^+, Na^+, and anions. Other transport molecules have also been identified that mediate loading of boron (as boric acid $[B(OH)_3]$ or borate $[B(OH)_4$-]), Mg^{2+}, and $H_2PO_4^{2-}$. Thus, the flux of ions from the xylem parenchyma cells into the xylem tracheary elements is under tight metabolic control through the regulation of plasma membrane H^+-ATPases, ion efflux channels, and carriers.

SUMMARY

The biologically regulated movement of molecules and ions from one location to another is known as transport. Plants exchange solutes within their cells, with their local environment, and among their tissues and organs. Both local and long-distance transport processes in plants are controlled largely by cellular membranes. Ion transport in plants is vital to their mineral nutrition and stress tolerance, and modulation of plant transport components and properties has potential to improve the nutritive value, stress tolerance, and yield of crops.

Passive and Active Transport

- Concentration gradients and electrical-potential gradients, the main forces that drive transport across biological membranes, are integrated by a term called the electrochemical potential (**Equation 6.8**).
- Movement of solutes across membranes down their free-energy gradient is facilitated by passive transport mechanisms, whereas movement of solutes against their free-energy gradient is known as active transport and requires energy input (**Figure 6.1**).

Transport of Ions across Membrane Barriers

- The extent to which a membrane permits the movement of a substance is a property known as membrane permeability (**Figure 6.5**).
- Permeability depends on the lipid composition of the membrane, the chemical properties of the solutes, and particularly on the membrane proteins that facilitate the transport of specific substances.
- For each permeant ion, the distribution of that particular ionic species across a membrane that would occur at equilibrium is described by the Nernst equation (**Equation 6.10**).
- Transport of H^+ across the plant plasma membrane by H^+-ATPases is a major determinant of the membrane potential (**Figures 6.15, 6.16**).

Membrane Transport Processes

- Biological membranes contain specialized proteins—channels, carriers, and pumps—that facilitate solute transport (**Figure 6.6**).
- The net result of membrane transport processes is that most ions are maintained in disequilibrum with their surroundings.
- Channels are regulated protein pores that, when open, greatly enhance fluxes of ions and, in some cases, neutral molecules across membranes (**Figures 6.6, 6.7**).
- Organisms have a great diversity of ion channel types. Depending on the channel type, channels can be nonselective or highly selective for just one ionic species. Channels can be regulated by many parameters, including voltage, intracellular signaling molecules, ligands, hormones, and light (**Figures 6.8, 6.13, 6.14**).
- Carriers bind specific substances and transport them at a rate several orders of magnitude lower than that of channels (**Figures 6.6, 6.11**).
- Pumps require energy for transport. Active transport of H^+ and Ca^{2+} across plant plasma membranes is mediated by pumps (**Figure 6.6**).
- Secondary active transporters in plants harness energy from energetically downhill movement of protons to mediate energetically uphill transport of another solute (**Figure 6.9**).
- In symport, both transported solutes move in the same direction across the membrane, whereas in antiport, the two solutes move in opposite directions (**Figure 6.10**).

Membrane Transport Proteins

- Many channels, carriers, and pumps of the plant plasma membrane and tonoplast have been identified at the molecular level (**Figure 6.13**) and characterized using electrophysiological (**Figure 6.8**) and biochemical techniques.
- Transporters exist for diverse nitrogenous compounds, including NO_3^-, amino acids, and peptides.
- Plants have a great variety of cation channels that can be classified according to their ionic selectivity and regulatory mechanisms (**Figure 6.14**).
- Several different classes of cation carriers mediate K^+ uptake into the cytosol (**Figure 6.13**).
- Na^+–H^+ antiporters on the tonoplast and plasma membrane extrude Na^+ into the vacuole and apoplast, respectively, thereby opposing accumulation of toxic levels of Na^+ in the cytosol (**Figure 6.13**).
- Ca^{2+} is an important second messenger in signal transduction cascades, and its cytosolic concentration is tightly regulated. Ca^{2+} enters the cytosol pas-

sively, via Ca^{2+}-permeable channels, and is actively removed from the cytosol by Ca^+ pumps and Ca^{2+}–H^+ antiporters (**Figure 6.13**).

- Selective carriers that mediate NO_3^-, Cl^-, SO_4^-, and $H_2PO_4^-$ uptake into the cytosol and anion channels that nonselectively mediate anion efflux from the cytosol regulate cellular concentrations of these macronutrients (**Figure 6.13**).
- Both essential and toxic metal ions are transported by high-affinity ZIP transport proteins (**Figure 6.13**).
- Aquaporins facilitate flux of water and other specific molecules, including boric acid, silicic acid, and arsenite, across plant plasma membranes, and their regulation allows for rapid changes in water permeability in response to environmental stimuli.
- Plasma membrane H^+-ATPases are encoded by a multigene family, and their activity is reversibly controlled by an autoinhibitory domain (**Figure 6.16**).
- Like the plasma membrane, the tonoplast also contains both cation and anion channels, as well as a diversity of other transporters.
- Two types of H^+ pumps found in the vacuolar membrane, V-ATPases and H^+-pyrophosphatases, regulate the proton motive force across the tonoplast, which in turn drives the movement of other solutes across this membrane via antiport mechanisms (**Figures 6.13, 6.17**).

Ion Transport in Roots

- Solutes such as mineral nutrients move between cells either through the extracellular space (the apoplast) or from cytoplasm to cytoplasm (via the symplast). The cytoplasm of neighboring cells is connected by plasmodesmata, which facilitate symplastic transport (**Figure 6.18**).
- When a solute enters the root, it may be taken up into the cytosol of an epidermal cell, or it may diffuse through the apoplast into the root cortex and then enter the symplast through a cortical or endodermal cell.
- The presence of the Casparian strip prevents apoplastic diffusion of solutes into the stele. Solutes enter the stele via diffusion from endodermal cells to pericyle and xylem parenchyma cells.
- During xylem loading, solutes are released from xylem parenchyma cells to the conducting cells of the xylem, and then move to the shoot in the transpiration stream (**Figure 6.19**).

WEB MATERIAL

- **WEB TOPIC 6.1 Relating the Membrane Potential to the Distribution of Several Ions across the Membrane: The Goldman Equation** The Goldman equation is used to calculate the membrane permeability to more than one ion.
- **WEB TOPIC 6.2 Patch Clamp Studies in Plant Cells** Patch clamping is applied to plant cells for electrophysiological studies.
- **WEB TOPIC 6.3 Chemiosmosis in Action** The chemiosmotic theory explains how electrical and concentration gradients are used to perform cellular work.
- **WEB TOPIC 6.4 Kinetic Analysis of Multiple Transporter Systems** Application of principles of enzyme kinetics to transport systems provides an effective way to characterize different carriers.
- **WEB TOPIC 6.5 ABC Transporters in Plants** ATP-binding cassette (ABC) transporters are a large family of active transport proteins energized directly by ATP.
- **WEB TOPIC 6.6 Transport Studies with Isolated Vacuoles and Membrane Vesicles** Certain experimental techniques enable the isolation of tonoplast and plasma membrane vesicles for study.

- **WEB ESSAY 6.1 Potassium Channels** Several plant K^+ channels have been characterized.
- **WEB ESSAY 6.2 Calmodulin: A Simple but Multifaceted Signal Transducer** This essay describes how CaM interacts with a broad array of cellular proteins and how these protein–protein interactions act to transduce changes in Ca^{2+} concentration into a complex web of biochemical responses.

available at plantphys.net

Suggested Reading

Barbier-Brygoo, H., Vinauger, M., Colcombet, J., Ephritikhine, G., Frachisse, J., and Maurel, C. (2000) Anion channels in higher plants: Functional characterization, molecular structure and physiological role. *Biochim. Biophys. Acta* 1465: 199–218.

Buchanan, B. B., Gruissem, W., and Jones, R. L., eds. (2000) *Biochemistry and Molecular Biology of Plants.* American Society of Plant Physiologists, Rockville, MD.

Burch-Smith, T. M., and Zambryski, P. C. (2012) Plasmodesmata paradigm shift: Regulation from without versus within. *Annu. Rev. Plant. Biol.* 63: 239–260.

Harold, F. M. (1986) *The Vital Force: A Study of Bioenergetics.* W. H. Freeman, New York.

Jammes, F., Hu, H. C., Villiers, F., Bouten, R., and Kwak, J. M. (2011) Calcium-permeable channels in plant cells. *FEBS J.* 278: 4262–4276.

Li, G., Santoni, V., and Maurel, C. (2013) Plant aquaporins: Roles in plant physiology. *Biochim. Biophys. Acta* 1840: 1574–1582.

Marschner, H. (1995) *Mineral Nutrition of Higher Plants.* Academic Press, London.

Martinoia, E., Meyer, S., De Angeli, A., and Nagy, R. (2012) Vacuolar transporters in their physiological context. Annu. Rev. Plant Biol. 63: 183–213.

Munns, R., James, R. A., Xu, B., Athman, A., Conn, S. J., Jordans, C., Byrt, C. S., Hare, R. A., Tyerman, S. D., Tester, M., et al. (2012) Wheat grain yield on saline soils is improved by an ancestral Na^+ transporter gene. *Nat. Biotechnol.* 30: 360–364.

Nobel, P. (1991) *Physicochemical and Environmental Plant Physiology.* Academic Press, San Diego, CA.

Palmgren, M. G. (2001) Plant plasma membrane H^+-ATPases: Powerhouses for nutrient uptake. *Annu. Rev. Plant Physiol. Plant Mol. Biol.* 52: 817–845.

Roelfsema, M.R., and Hedrich, R. (2005) In the light of stomatal opening: New insights into "the Watergate." New Phytol. 167: 665–691.

Schroeder, J. I., Delhaize, E., Frommer, W. B., Guerinot, M. L., Harrison, M. J., Herrera-Estrella, L., Horie, T., Kochian, L. V., Munns, R., Nishizawa, N. K., et al. (2013) Using membrane transporters to improve crops for sustainable food production. *Nature* 497: 60–66.

Ward, J. M., Mäser, P., and Schroeder, J. I. (2009). Plant ion channels: Gene families, physiology, and functional genomics analyses. *Annu. Rev. Plant Biol.* 71: 59–82.

Yamaguchi, T., Hamamoto, S., and Uozumi, N. (2013) Sodium transport system in plant cells. *Front. Plant Sci.* 4: 410.

Yazaki, K., Shitan, N., Sugiyama, A., and Takanashi, K. (2009) Cell and molecular biology of ATP-binding cassette proteins in plants. *Int. Rev. Cell Mol. Biol.* 276: 263–299.

UNIT II

Biochemistry and Metabolism

UNIT II

Chapter 7
Photosynthesis: The Light Reactions

Chapter 8
Photosynthesis: The Carbon Reactions

Chapter 9
Photosynthesis: Physiological and Ecological Considerations

Chapter 10
Stomatal Biology

Chapter 11
Translocation in the Phloem

Chapter 12
Respiration and Lipid Metabolism

Chapter 13
Assimilation of Inorganic Nutrients

7 Photosynthesis: The Light Reactions

Life on Earth ultimately depends on energy derived from the sun. Photosynthesis is the only process of biological importance that can harvest this energy. A large fraction of the planet's energy resources results from photosynthetic activity in either recent or ancient times (fossil fuels). This chapter introduces the basic physical principles that underlie photosynthetic energy storage and the current understanding of the structure and function of the photosynthetic apparatus.

The term *photosynthesis* means literally "synthesis using light." As we will see in this chapter, photosynthetic organisms use solar energy to synthesize complex carbon compounds. More specifically, light energy drives the synthesis of carbohydrates and generation of oxygen from carbon dioxide and water:

$$\underset{\text{Carbon dioxide}}{6\,CO_2} + \underset{\text{Water}}{6\,H_2O} \rightarrow \underset{\text{Carbohydrate}}{C_6H_{12}O_6} + \underset{\text{Oxygen}}{6\,O_2}$$

Energy stored in these carbohydrate molecules can be used later to power cellular processes in the plant and can serve as the energy source for all forms of life.

This chapter deals with the role of light in photosynthesis, the structure of the photosynthetic apparatus, and the processes that begin with the excitation of chlorophyll by light and culminate in the synthesis of ATP and NADPH.

Photosynthesis in Higher Plants

The most active photosynthetic tissue in higher plants is the mesophyll of leaves. Mesophyll cells have many chloroplasts, which contain the specialized light-absorbing green pigments, the **chlorophylls**. In photosynthesis, the plant uses solar energy to oxidize water, thereby releasing oxygen, and to reduce carbon dioxide, thereby forming large carbon compounds, primarily sugars. The complex series of reactions that culminate in the reduction of CO_2 include the thylakoid reactions and the carbon fixation reactions.

The **thylakoid reactions** of photosynthesis take place in the specialized internal membranes of the chloroplast called thylakoids

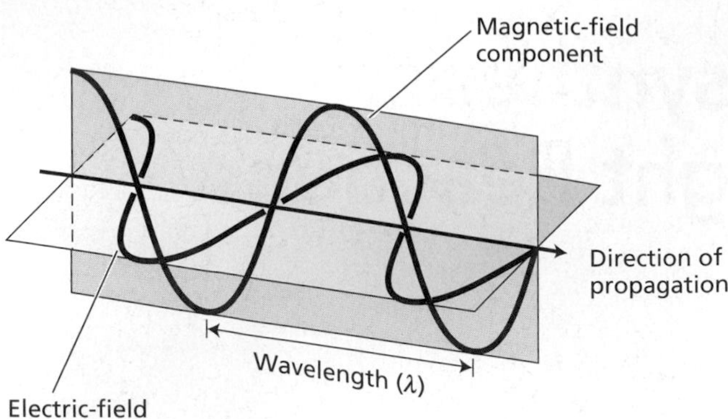

Figure 7.1 Light is a transverse electromagnetic wave, consisting of oscillating electric and magnetic fields that are perpendicular to each other and to the direction of propagation of the light. Light moves at a speed of 3.0×10^8 m s^{-1}. The wavelength (λ) is the distance between successive crests of the wave.

(see Chapter 1). The end products of these thylakoid reactions are the high-energy compounds ATP and NADPH, which are used for the synthesis of sugars in the **carbon fixation reactions**. These synthetic processes take place in the stroma of the chloroplast, the aqueous region that surrounds the thylakoids. The thylakoid reactions, also called the "light reactions" of photosynthesis, are the subject of this chapter; the carbon fixation reactions will be discussed in Chapter 8.

In the chloroplast, light energy is converted into chemical energy by two different functional units called *photosystems*. The absorbed light energy is used to power the transfer of electrons through a series of compounds that act as electron donors and electron acceptors. The majority of electrons are extracted from H_2O, which is oxidized to O_2, and ultimately reduce $NADP^+$ to NADPH. Light energy is also used to generate a proton motive force (see Chapter 6) across the thylakoid membrane; this proton motive force is used to synthesize ATP.

General Concepts

In this section we will explore the essential concepts that provide a foundation for an understanding of photosynthesis. These concepts include the nature of light, the properties of pigments, and the various roles of pigments.

Light has characteristics of both a particle and a wave

A triumph of physics in the early twentieth century was the realization that light has properties of both particles and waves. A wave (**Figure 7.1**) is characterized by a **wavelength**, denoted by the Greek letter lambda (λ), which is the distance between successive wave crests. The **frequency**, represented by the Greek letter nu (ν), is the number of wave crests that pass an observer in a given time. A simple equation relates the wavelength, the frequency, and the speed of any wave:

$$c = \lambda \nu \tag{7.1}$$

where c is the speed of the wave—in the present case, the speed of light (3.0×10^8 m s^{-1}). The light wave is a transverse (side-to-side) electromagnetic wave, in which both electric and magnetic fields oscillate perpendicularly to the direction of propagation of the wave and at 90° with respect to each other.

Light is also a particle, which we call a **photon**. Each photon contains an amount of energy that is called a **quantum** (plural *quanta*). The energy content of light is not continuous but rather is delivered in discrete packets, the quanta. The energy (E) of a photon depends on the frequency of the light according to a relation known as Planck's law:

$$E = h\nu \tag{7.2}$$

where h is Planck's constant (6.626×10^{-34} J s).

Sunlight is like a rain of photons of different frequencies. Our eyes are sensitive to only a small range of frequencies—the visible-light region of the electromagnetic spectrum (**Figure 7.2**). Light of slightly higher frequencies (or shorter wavelengths) is in the ultraviolet region of

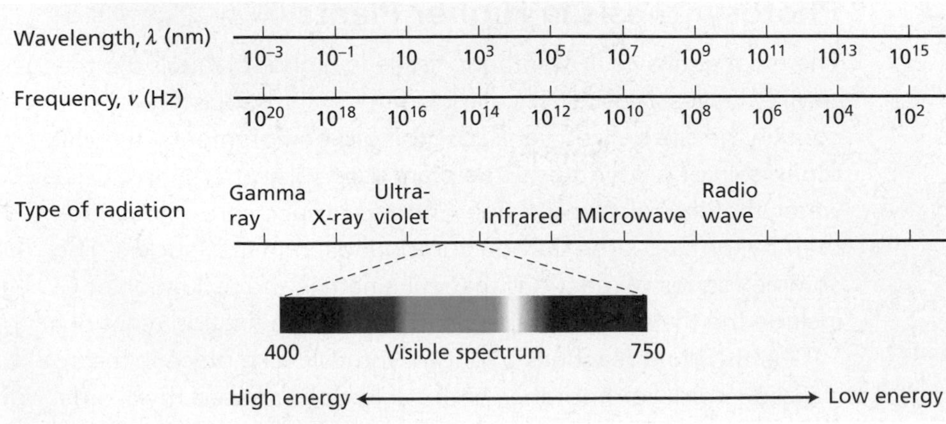

Figure 7.2 Electromagnetic spectrum. Wavelength (λ) and frequency (ν) are inversely related. Our eyes are sensitive to only a narrow range of wavelengths of radiation, the visible region, which extends from about 400 nm (violet) to about 700 nm (red). Short-wavelength (high-frequency) light has a high energy content; long-wavelength (low-frequency) light has a low energy content.

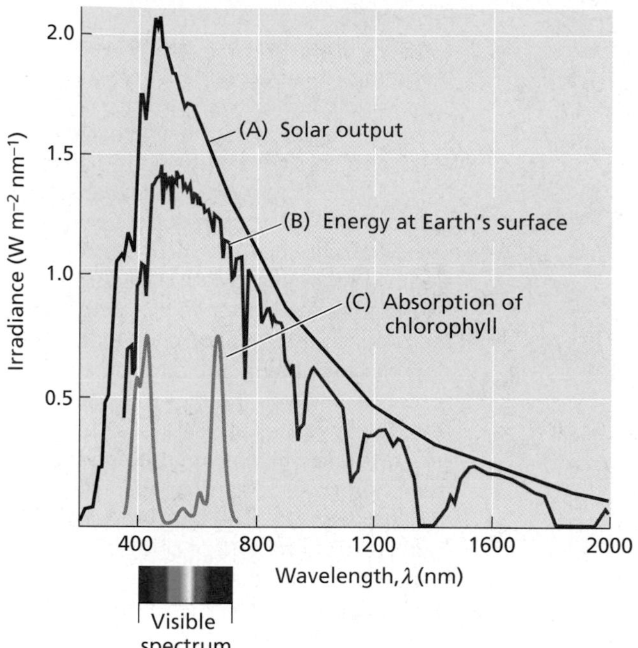

Figure 7.3 The solar spectrum and its relation to the absorption spectrum of chlorophyll. Curve A is the energy output of the sun as a function of wavelength. Curve B is the energy that strikes the surface of Earth. The sharp valleys in the infrared region beyond 700 nm represent the absorption of solar energy by molecules in the atmosphere, chiefly water vapor. Curve C is the absorption spectrum of chlorophyll, which absorbs strongly in the blue (about 430 nm) and the red (about 660 nm) portions of the spectrum. Because the green light in the middle of the visible spectrum is not efficiently absorbed, some of it is reflected into our eyes and gives plants their characteristic green color.

the spectrum, and light of slightly lower frequencies (or longer wavelengths) is in the infrared region. The output of the sun is shown in **Figure 7.3**, along with the energy density that strikes the surface of Earth. The **absorption spectrum** (plural *spectra*) of chlorophyll *a* (green curve in Figure 7.3) indicates the approximate portion of the solar output that is used by plants.

An absorption spectrum provides information about the amount of **light energy** taken up or absorbed by a molecule or substance as a function of the wavelength of the light. The absorption spectrum for a particular substance in a nonabsorbing solvent can be determined by a spectrophotometer, as illustrated in **Figure 7.4**. Spectrophotometry, the technique used to measure the absorption of light by a sample, is more completely discussed in WEB TOPIC 7.1.

When molecules absorb or emit light, they change their electronic state

Chlorophyll appears green to our eyes because it absorbs light mainly in the red and blue parts of the spectrum, so only some of the light enriched in green wavelengths (about 550 nm) is reflected into our eyes (see Figure 7.3).

The absorption of light is represented by Equation 7.3, in which chlorophyll (Chl) in its lowest energy, or ground, state absorbs a photon (represented by $h\nu$) and makes a transition to a higher energy, or excited, state (Chl*):

$$\text{Chl} + h\nu \rightarrow \text{Chl}^* \quad (7.3)$$

The distribution of electrons in the excited molecule is somewhat different from the distribution in the ground-state molecule (**Figure 7.5**). Absorption of blue light excites the chlorophyll to a higher energy state than absorption of red light, because the energy of photons is higher when their wavelength is shorter. In the higher excited state, chlorophyll is extremely unstable; it rapidly gives up some of its energy to the surroundings as heat, and enters the **lowest excited state**, where it can be stable for a maximum of several nanoseconds (10^{-9} s). Because of the inherent instability of the excited state, any process that captures its energy must be extremely rapid.

In the lowest excited state, the excited chlorophyll has four alternative pathways for disposing of its available energy:

1. Excited chlorophyll can re-emit a photon and thereby return to its ground state—a process known as **fluorescence**. When it does so, the wavelength of fluorescence is slightly longer (and of lower energy) than the wavelength of absorption, because a portion of the excitation energy is converted into heat before the fluorescent photon is emitted. Chlorophylls fluoresce in the red region of the spectrum.

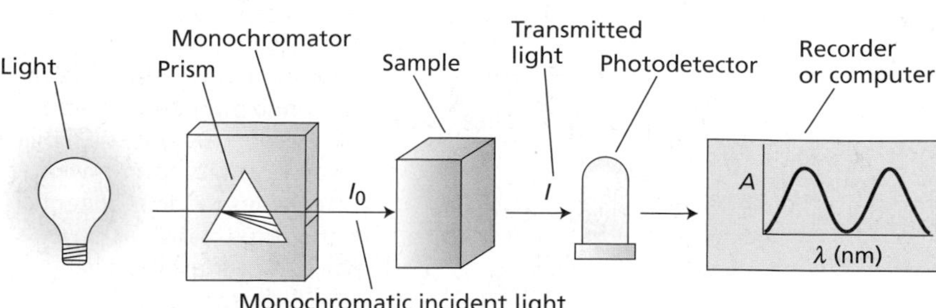

Figure 7.4 Schematic diagram of a spectrophotometer. The instrument consists of a light source, a monochromator that contains a wavelength selection device such as a prism, a sample holder, a photodetector, and a recorder or computer. The output wavelength of the monochromator can be changed by rotation of the prism; the graph of absorbance (*A*) versus wavelength (λ) is called a spectrum.

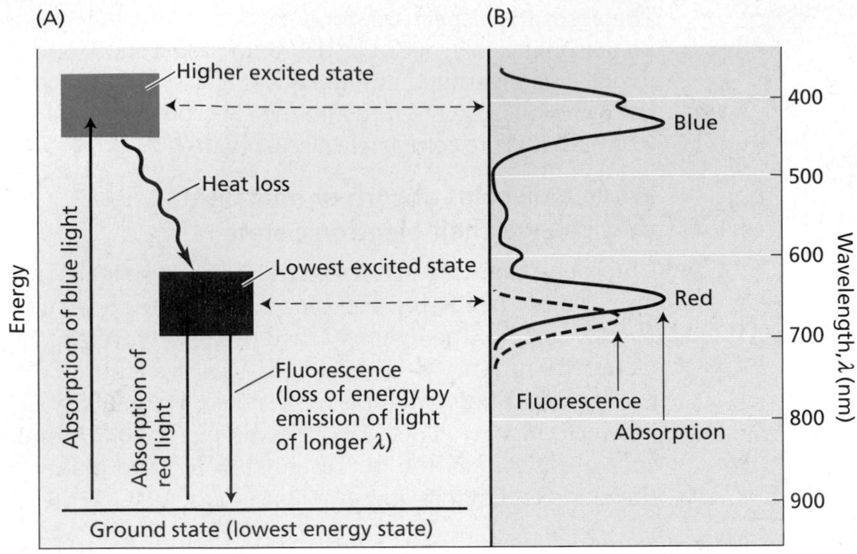

Figure 7.5 Light absorption and emission by chlorophyll. (A) Energy level diagram. Absorption or emission of light is indicated by vertical arrows that connect the ground state with excited electron states. The blue and red absorption bands of chlorophyll (which absorb blue and red photons, respectively) correspond to the upward vertical arrows, signifying that energy absorbed from light causes the molecule to change from the ground state to an excited state. The downward-pointing arrow indicates fluorescence, in which the molecule goes from the lowest excited state to the ground state while re-emitting energy as a photon. (B) Spectra of absorption and fluorescence. The long-wavelength (red) absorption band of chlorophyll corresponds to light that has the energy required to cause the transition from the ground state to the first excited state. The short-wavelength (blue) absorption band corresponds to a transition to a higher excited state.

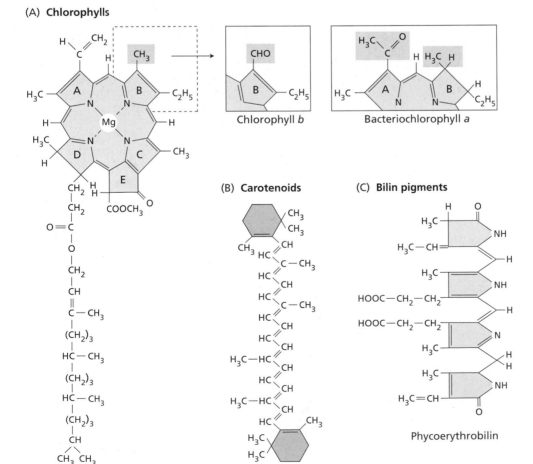

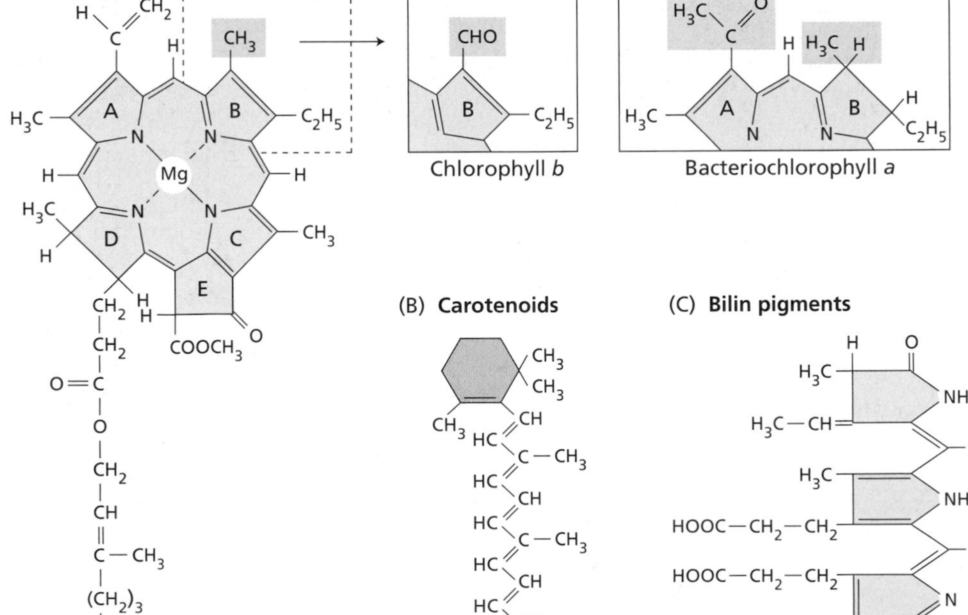

Figure 7.6 Molecular structure of some photosynthetic pigments. (A) The chlorophylls have a porphyrin-like ring structure with a magnesium ion (Mg) coordinated in the center and a long hydrophobic hydrocarbon tail that anchors them in the photosynthetic membrane. The porphyrin-like ring is the site of the electron rearrangements that occur when the chlorophyll is excited, and of the unpaired electrons when it is either oxidized or reduced. Various chlorophylls differ chiefly in the substituents around the rings and the pattern of double bonds. (B) Carotenoids are linear polyenes that serve as both antenna pigments and photoprotective agents. (C) Bilin pigments are open-chain tetrapyrroles found in antenna structures known as phycobilisomes that occur in cyanobacteria and red algae.

2. The excited chlorophyll can return to its ground state by directly converting its excitation energy into heat, with no emission of a photon.
3. Chlorophyll may participate in **energy transfer**, during which an excited chlorophyll transfers its energy to another molecule.
4. A fourth process is **photochemistry**, in which the energy of the excited state causes chemical reactions to occur. The photochemical reactions of photosynthesis are among the fastest known chemical reactions. This extreme speed is necessary for photochemistry to compete with the three other possible reactions of the excited state just described.

Photosynthetic pigments absorb the light that powers photosynthesis

The energy of sunlight is first absorbed by the pigments of the plant. All pigments active in photosynthesis are found in the chloroplast. Structures and absorption spectra of several photosynthetic pigments are shown in **Figures 7.6** and **Figure 7.7**, respectively. The chlorophylls and **bacteriochlorophylls** (pigments found in certain bacteria) are the typical pigments of photosynthetic organisms.

Chlorophylls *a* and *b* are abundant in green plants, and *c*, *d*, and *f* are found in some protists and cyanobacteria. A number of different types of bacteriochlorophyll have been found; type *a* is the most widely distributed. WEB TOPIC 7.2 shows the distribution of pigments in different types of photosynthetic organisms.

All chlorophylls have a complex ring structure that is chemically related to the porphyrin-like groups found in hemoglobin and cytochromes (see Figure 7.6A). A long hydrocarbon tail is almost always attached to the ring structure. The tail anchors the chlorophyll to the hydrophobic portion of its environment. The ring structure contains some loosely bound electrons and is the part of the molecule involved in electronic transitions and redox (reduction–oxidation) reactions.

The different types of **carotenoids** found in photosynthetic organisms are all linear molecules with multiple conjugated double bonds (see Figure 7.6B). Absorption bands in the 400 to 500 nm region give carotenoids their characteristic orange color. The color of carrots, for example, is due to the carotenoid β-carotene, whose structure and absorption spectrum are shown in Figures 7.6 and 7.7, respectively.

Carotenoids are found in all natural photosynthetic organisms. Carotenoids are integral constituents of the thylakoid membrane and are usually associated intimately with many of the proteins that make up the photosynthetic apparatus. The light energy absorbed by the carotenoids is transferred to chlorophyll for photosynthesis; because of this role they are called **accessory pigments**. Carotenoids also help protect the organism from damage caused by light (see p. 196 of this chapter and Chapter 9.)

Key Experiments in Understanding Photosynthesis

Establishing the overall chemical equation of photosynthesis required several hundred years and contributions by many scientists (literature references for historical developments can be found on the web site for this book). In 1771, Joseph Priestley observed that a sprig of mint growing in air in which a candle had burned out improved the air so that another candle could burn. He had discovered oxygen evolution by plants. A Dutch biologist, Jan Ingenhousz, documented the essential role of light in photosynthesis in 1779.

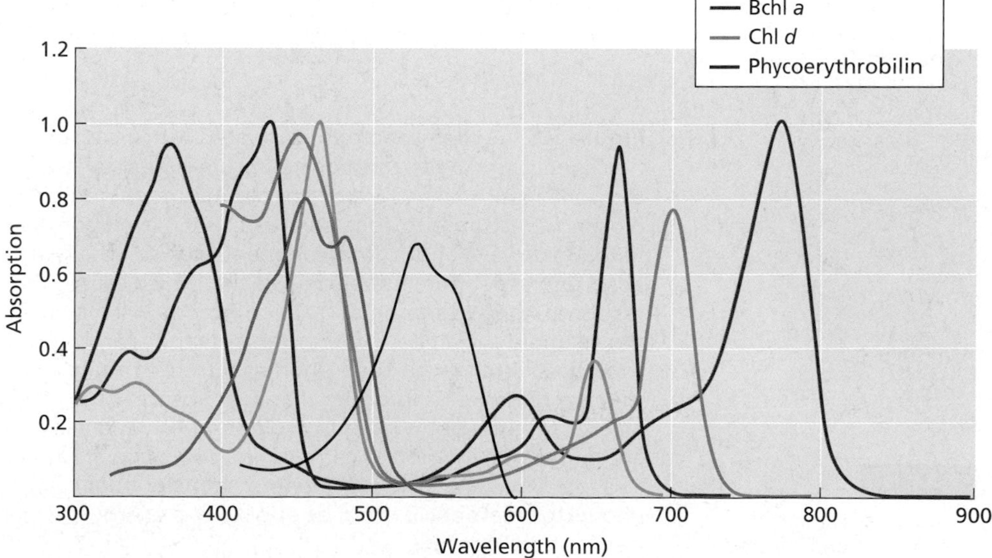

Figure 7.7 Absorption spectra of some photosynthetic pigments, including β-carotene, chlorophyll *a* (Chl *a*), chlorophyll *b* (Chl *b*), bacteriochlorophyll *a* (Bchl *a*), chlorophyll *d* (Chl *d*), and phycoerythrobilin. The absorption spectra shown are for pure pigments dissolved in nonpolar solvents, except phycoerythrin, a protein from cyanobacteria that contains a phycoerythrobilin chromophore covalently attached to the peptide chain. In many cases the spectra of photosynthetic pigments in vivo are substantially affected by the environment of the pigments in the photosynthetic membrane.

Other scientists established the roles of CO_2 and H_2O and showed that organic matter, specifically carbohydrate, is a product of photosynthesis along with oxygen. By the end of the nineteenth century, the balanced overall chemical reaction for photosynthesis could be written as follows:

$$6\ CO_2 + 6\ H_2O \xrightarrow{\text{Light, plant}} C_6H_{12}O_6 + 6\ O_2 \quad (7.4)$$

where $C_6H_{12}O_6$ represents a simple sugar such as glucose. As we will discuss in Chapter 8, glucose is not the actual product of the carbon fixation reactions, so this part of the equation should not be taken literally. However, the energetics for the actual reaction are approximately the same as represented here.

The chemical reactions of photosynthesis are complex. At least 50 intermediate reaction steps have now been identified, and additional steps undoubtedly will be discovered. An early clue to the chemical nature of the essential chemical process of photosynthesis came in the 1920s from investigations of photosynthetic bacteria that did not produce oxygen as an end product. From his studies on these bacteria, C. B. van Niel concluded that photosynthesis is a redox process. This conclusion has served as a fundamental concept on which all subsequent research on photosynthesis has been based.

We will now turn to the relationship between photosynthetic activity and the spectrum of absorbed light. We will discuss some of the critical experiments that have contributed to our present understanding of photosynthesis, and we will consider equations for the essential chemical reactions of photosynthesis.

Action spectra relate light absorption to photosynthetic activity

The use of action spectra has been central to the development of our current understanding of photosynthesis. An **action spectrum** depicts the magnitude of a response of a biological system to light as a function of wavelength. For example, an action spectrum for photosynthesis can be constructed from measurements of oxygen evolution at different wavelengths (**Figure 7.8**). Often an action spectrum can identify the *chromophore* (pigment) responsible for a particular light-induced phenomenon.

Some of the first action spectra were measured by T. W. Engelmann in the late 1800s (**Figure 7.9**). Engelmann used a prism to disperse sunlight into a rainbow that was allowed to fall on an aquatic algal filament. A population of O_2-seeking bacteria was introduced into the system. The bacteria congregated in the regions of the filaments that evolved the most O_2. These were the regions illuminated by blue light and red light, which are strongly absorbed by chlorophyll. Today, action spectra can be measured in room-sized spectrographs in which a huge monochromator bathes the experimental samples in monochromatic light. The technology is more sophisticated, but the principle is the same as that of Engelmann's experiments.

Action spectra were very important for the discovery of two distinct photosystems operating in O_2-evolving photosynthetic organisms. Before we introduce the two photosystems, however, we need to describe the light-gathering antennas and the energy needs of photosynthesis.

Photosynthesis takes place in complexes containing light-harvesting antennas and photochemical reaction centers

A portion of the light energy absorbed by chlorophylls and carotenoids is eventually stored as chemical energy via the formation of chemical bonds. This conversion of energy from one form to another is a complex process that depends on cooperation between many pigment molecules and a group of electron transfer proteins.

The majority of the pigments serve as an **antenna complex**, collecting light and transferring the energy to the **reaction center complex**, where the chemical oxidation and reduction reactions leading to long-term energy stor-

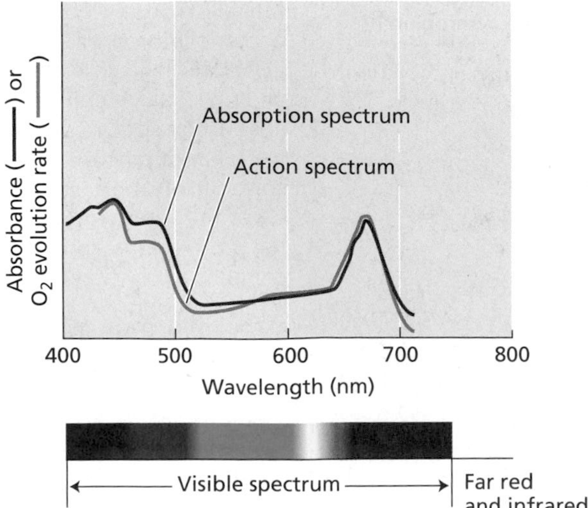

Figure 7.8 Action spectrum compared with an absorption spectrum. The absorption spectrum is measured as shown in Figure 7.4. An action spectrum is measured by plotting a response to light, such as oxygen evolution, as a function of wavelength. If the pigment used to obtain the absorption spectrum is the same as that which causes the response, the absorption and action spectra will match. In the example shown here, the action spectrum for oxygen evolution matches the absorption spectrum of intact chloroplasts quite well, indicating that light absorption by the chlorophylls mediates oxygen evolution. Discrepancies are found in the region of carotenoid absorption, from 450 to 550 nm, indicating that energy transfer from carotenoids to chlorophylls is not as effective as energy transfer between chlorophylls.

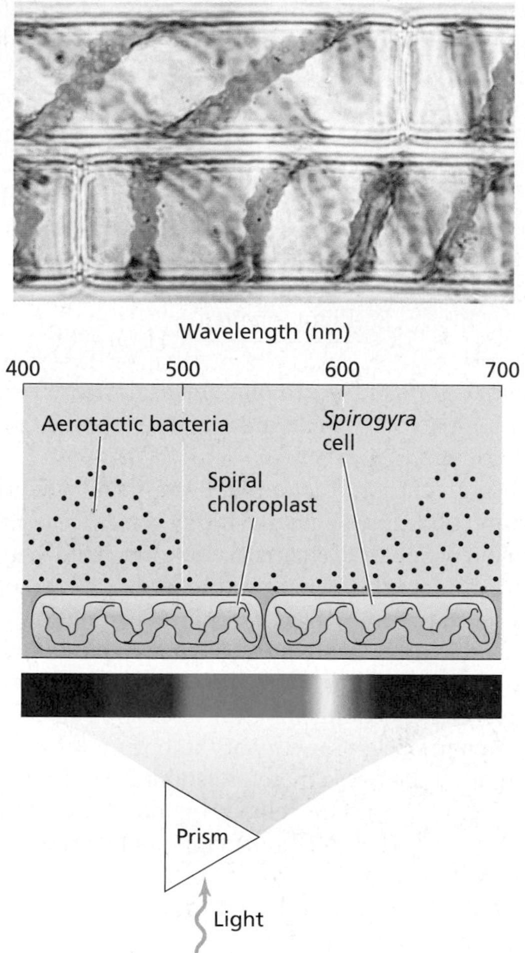

Figure 7.9 Schematic diagram of the action spectrum measurements by T. W. Engelmann. Engelmann projected a spectrum of light onto the spiral chloroplast of the filamentous green alga *Spirogyra* and observed that O_2-seeking bacteria introduced into the system collected in the region of the spectrum where chlorophyll pigments absorb. This action spectrum gave the first indication of the effectiveness of light absorbed by pigments in driving photosynthesis.

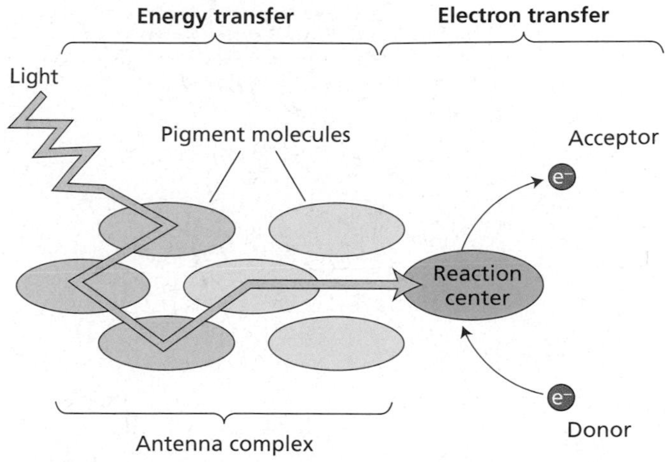

Figure 7.10 Basic concept of energy transfer during photosynthesis. Many pigments together serve as an antenna, collecting light and transferring its energy to the reaction center, where chemical reactions store some of the energy by transferring electrons from a chlorophyll pigment to an electron acceptor molecule. An electron donor then reduces the chlorophyll again. The transfer of energy in the antenna is a purely physical phenomenon and involves no chemical changes.

age take place (**Figure 7.10**). Molecular structures of some of the antenna and reaction center complexes will be discussed later in the chapter.

How does the plant benefit from this division of labor between antenna and reaction center pigments? Even in bright sunlight, a single chlorophyll molecule absorbs only a few photons each second. If there were a reaction center associated with each chlorophyll molecule, the reaction center enzymes would be idle most of the time, only occasionally being activated by photon absorption. However, if a reaction center receives energy from many pigments at once, the system is kept active a large fraction of the time.

In 1932, Robert Emerson and William Arnold performed a key experiment that provided the first evidence for the cooperation of many chlorophyll molecules in energy conversion during photosynthesis. They delivered very brief (10^{-5} s) flashes of light to a suspension of the green alga *Chlorella pyrenoidosa* and measured the amount of oxygen produced. The flashes were spaced about 0.1 s apart, a time that Emerson and Arnold had determined in earlier work was long enough for the enzymatic steps of the process to be completed before the arrival of the next flash. The investigators varied the energy of the flashes and found that at high energies the oxygen production did not increase when a more intense flash was given: The photosynthetic system was saturated with light (**Figure 7.11**).

In their measurement of the relationship of oxygen production to flash energy, Emerson and Arnold were surprised to find that under saturating conditions, only 1 molecule of oxygen was produced for each 2500 chlorophyll molecules in the sample. We know now that several hundred pigments are associated with each reaction center and that each reaction center must operate four times to produce 1 molecule of oxygen—hence the value of 2500 chlorophylls per O_2.

The reaction centers and most of the antenna complexes are integral components of the photosynthetic membrane. In eukaryotic photosynthetic organisms, these membranes are found within the chloroplast; in photosynthetic prokaryotes, the site of photosynthesis is the plasma membrane or membranes derived from it.

The graph shown in Figure 7.11 permits us to calculate another important parameter of the light reactions of photosynthesis, the quantum yield. The **quantum yield** of photochemistry (Φ) is defined as follows:

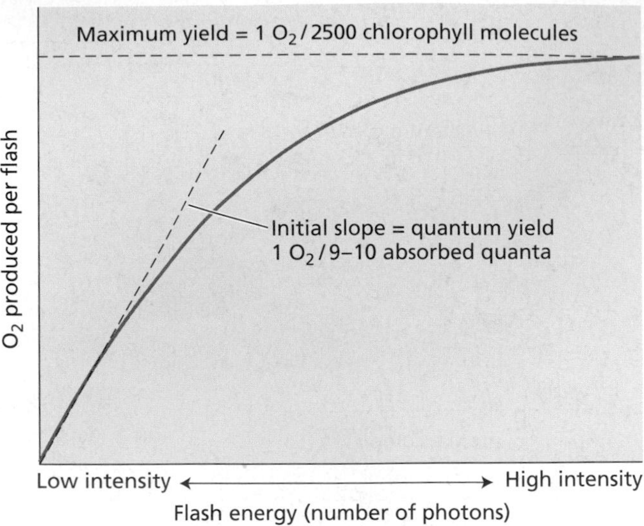

Figure 7.11 Relationship of oxygen production to flash energy, the first evidence for the interaction between the antenna pigments and the reaction center. At saturating energies, the maximum amount of O_2 produced is 1 molecule per 2500 chlorophyll molecules.

$$\Phi = \frac{\text{Number of photochemical products}}{\text{Total number of quanta absorbed}} \tag{7.5}$$

In the linear portion (low light intensity) of the curve, an increase in the number of photons stimulates a proportional increase in oxygen evolution. Thus, the slope of the curve measures the quantum yield for oxygen production. The quantum yield for a particular process can range from 0 (if that process does not respond to light) to 1.0 (if every photon absorbed contributes to the process by forming a product). A more detailed discussion of quantum yields can be found in **WEB TOPIC 7.3**.

In functional chloroplasts kept in dim light, the quantum yield of photochemistry is approximately 0.95, the quantum yield of fluorescence is 0.05 or lower, and the quantum yields of other processes are negligible. Thus, the most common result of chlorophyll excitation is photochemistry. Products of photosynthesis such as O_2 require more than a single photochemical event to be formed, and therefore have a lower quantum yield of formation than the photochemical quantum yield. It takes about ten photons to produce one molecule of O_2, so the quantum yield of O_2 production is about 0.1, even though the photochemical quantum yield for each step in the process is nearly 1.0.

The chemical reaction of photosynthesis is driven by light

It is important to realize that equilibrium for the chemical reaction shown in Equation 7.4 lies very far in the direction of the reactants. The equilibrium constant for Equation 7.4, calculated from tabulated free energies of formation for each of the compounds involved, is about 10^{-500}. This number is so close to zero that one can be quite confident that in the entire history of the universe no molecule of glucose has formed spontaneously from H_2O and CO_2 without external energy being provided. The energy needed to drive the photosynthetic reaction comes from light. Here's a simpler form of Equation 7.4:

$$CO_2 + H_2O \xrightarrow{\text{Light, plant}} (CH_2O) + O_2 \tag{7.6}$$

where (CH_2O) is one-sixth of a glucose molecule. About nine or ten photons of light are required to drive the reaction of Equation 7.6.

Although the photochemical quantum yield under optimum conditions is nearly 100%, the *efficiency* of the conversion of light into chemical energy is much less. If red light of wavelength 680 nm is absorbed, the total energy input (see Equation 7.2) is 1760 kJ per mole of oxygen formed. This amount of energy is more than enough to drive the reaction in Equation 7.6, which has a standard-state free-energy change of +467 kJ mol^{-1}. The efficiency of conversion of light energy at the optimal wavelength into chemical energy is therefore about 27%. Most of this stored energy is used for cellular maintenance processes; the amount diverted to the formation of biomass is much less (see Chapter 9).

There is no conflict in the fact that the photochemical quantum efficiency (quantum yield) is nearly 1.0 (100%), the energy conversion efficiency is only 27%, and the overall efficiency of conversion of solar energy is only a few percent. The *quantum efficiency* is a measure of the fraction of absorbed photons that engage in photochemistry; the *energy efficiency* is a measure of how much energy in the absorbed photons is stored as chemical products; and the *solar energy storage efficiency* is a measure of how much of the energy in the entire solar spectrum is converted to usable form. The numbers indicate that almost all the absorbed photons engage in photochemistry, but only about a fourth of the energy in each photon is stored, the remainder being converted to heat, and only about half of the solar spectrum is absorbed by the plant. The overall energy conversion efficiency into biomass, including all loss processes and considering the entire solar spectrum as energy source, is significantly lower still—approximately 4.3% for C_3 plants and 6% for C_4 plants.

Light drives the reduction of $NADP^+$ and the formation of ATP

The overall process of photosynthesis is a redox chemical reaction, in which electrons are removed from one chemical species, thereby oxidizing it, and added to another species, thereby reducing it. In 1937, Robert Hill found that in the light, isolated chloroplast thylakoids reduce a variety of compounds, such as iron salts. These compounds serve as oxidants in place of CO_2, as the following equation shows:

$$4\ Fe^{3+} + 2\ H_2O \rightarrow 4\ Fe^{2+} + O_2 + 4\ H^+ \qquad (7.7)$$

Many compounds have since been shown to act as artificial electron acceptors in what has come to be known as the Hill reaction. The use of artificial electron acceptors has been invaluable in elucidating the reactions that precede carbon reduction. The demonstration of oxygen evolution linked to the reduction of artificial electron acceptors provided the first evidence that oxygen evolution could occur in the absence of carbon dioxide and led to the now accepted and proven idea that the oxygen in photosynthesis originates from water, not from carbon dioxide.

We now know that during the normal functioning of the photosynthetic system, light reduces nicotinamide adenine dinucleotide phosphate ($NADP^+$), which in turn serves as the reducing agent for carbon fixation in the Calvin–Benson cycle (see Chapter 8). ATP is also formed during the electron flow from water to $NADP^+$, and it too is used in carbon reduction.

The chemical reactions in which water is oxidized to oxygen, $NADP^+$ is reduced to NADPH, and ATP is formed are known as the *thylakoid reactions* because almost all the reactions up to $NADP^+$ reduction take place in the thylakoids. The carbon fixation and reduction reactions are called the *stroma reactions* because the carbon reduction reactions take place in the aqueous region of the chloroplast, the stroma. Although this division is somewhat arbitrary, it is conceptually useful.

Oxygen-evolving organisms have two photosystems that operate in series

By the late 1950s, several experiments were puzzling the scientists who studied photosynthesis. One of these experiments, carried out by Emerson, measured the quantum yield of photosynthesis as a function of wavelength and revealed an effect known as the red drop (**Figure 7.12**).

If the quantum yield is measured for the wavelengths at which chlorophyll absorbs light, the values found throughout most of the range are fairly constant, indicating that any photon absorbed by chlorophyll or other pigments is as effective as any other photon in driving photosynthesis. However, the yield drops dramatically in the far-red region of chlorophyll absorption (greater than 680 nm).

This drop cannot be caused by a decrease in chlorophyll absorption, because the quantum yield measures only light that has actually been absorbed. Thus, light with a wavelength greater than 680 nm is much less efficient than light of shorter wavelengths.

Another puzzling experimental result was the **enhancement effect**, also discovered by Emerson. He measured the rate of photosynthesis separately with light of two different wavelengths and then used the two beams simultaneously. When red and far-red light were given together, the rate of photosynthesis was greater than the sum of the individual rates, a startling and surprising observation. These and others observations were eventually explained by experiments performed in the 1960s (see **WEB TOPIC 7.4**) that led to the discovery that two photochemical complexes, now known as **photosystems I** and **II** (**PSI** and **PSII**), operate in series to carry out the early energy storage reactions of photosynthesis.

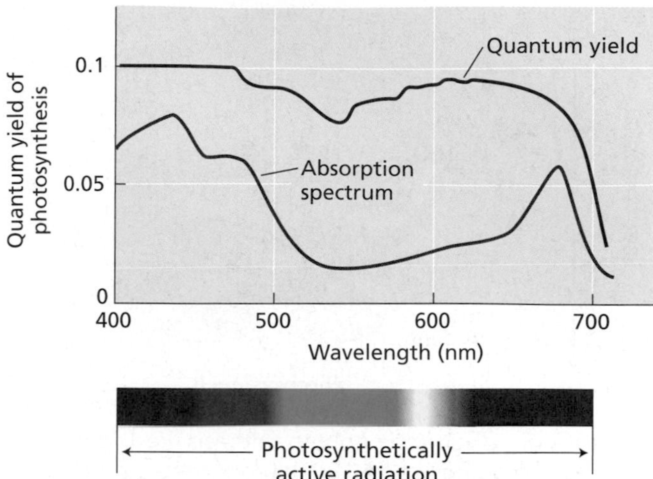

Figure 7.12 Red drop effect. The quantum yield of oxygen evolution (upper, black curve) falls off drastically for far-red light of wavelengths greater than 680 nm, indicating that far-red light alone is inefficient in driving photosynthesis. The slight dip near 500 nm reflects the somewhat lower efficiency of photosynthesis using light absorbed by accessory pigments, carotenoids.

PSI preferentially absorbs far-red light of wavelengths greater than 680 nm; PSII preferentially absorbs red light of 680 nm and is driven very poorly by far-red light. This wavelength dependence explains the enhancement effect and the red drop effect. Another difference between the photosystems is that:

- PSI produces a strong reductant, capable of reducing $NADP^+$, and a weak oxidant.
- PSII produces a very strong oxidant, capable of oxidizing water, and a weaker reductant than the one produced by PSI.

The reductant produced by PSII re-reduces the oxidant produced by PSI. These properties of the two photosystems are shown schematically in **Figure 7.13**.

The scheme of photosynthesis depicted in Figure 7.13, called the Z (for *zigzag*) *scheme*, has become the basis for understanding O_2-evolving (oxygenic) photosynthetic organisms. It accounts for the operation of two physically and chemically distinct photosystems (I and II), each with its own antenna pigments and photochemical reaction center. The two photosystems are linked by an electron transport chain.

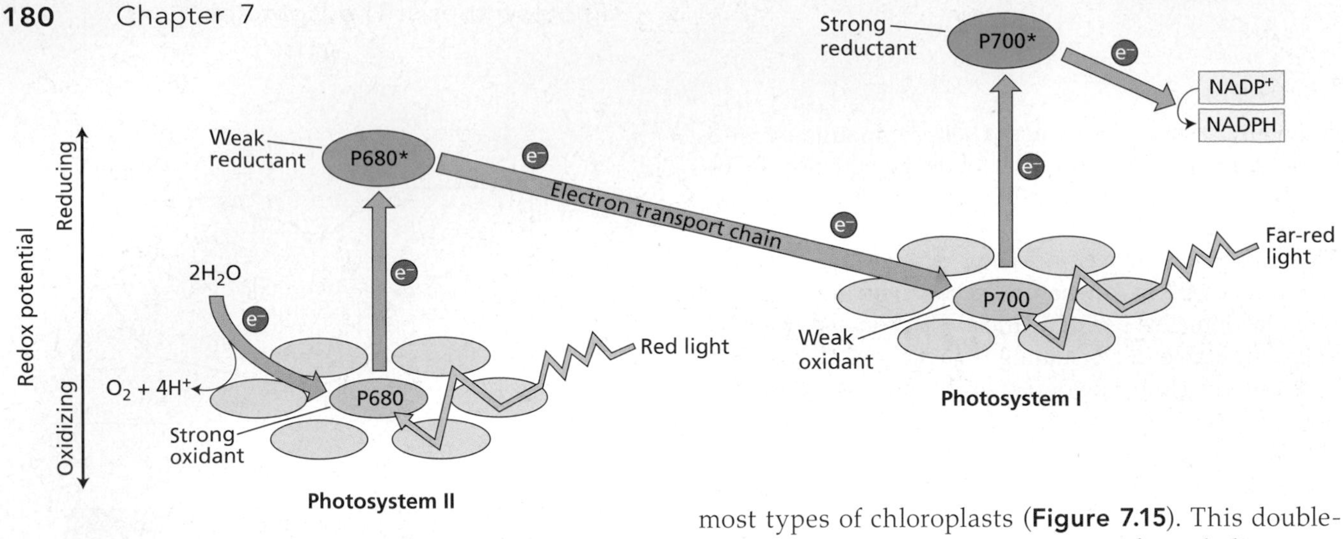

Figure 7.13 Z scheme of photosynthesis. Red light absorbed by photosystem II (PSII) produces a strong oxidant and a weak reductant. Far-red light absorbed by photosystem I (PSI) produces a weak oxidant and a strong reductant. The strong oxidant generated by PSII oxidizes water, while the strong reductant produced by PSI reduces $NADP^+$. This scheme is basic to an understanding of photosynthetic electron transport. P680 and P700 refer to the wavelengths of maximum absorption of the reaction center chlorophylls in PSII and PSI, respectively.

Organization of the Photosynthetic Apparatus

The previous section explained some of the physical principles underlying photosynthesis, some aspects of the functional roles of various pigments, and some of the chemical reactions carried out by photosynthetic organisms. We will now turn to the architecture of the photosynthetic apparatus and the structure of its components, and learn how the molecular structure of the system leads to its functional characteristics.

The chloroplast is the site of photosynthesis

In photosynthetic eukaryotes, photosynthesis takes place in the subcellular organelle known as the chloroplast. **Figure 7.14** shows a transmission electron micrograph of a thin section from a pea chloroplast. The most striking aspect of the structure of the chloroplast is the extensive system of internal membranes known as **thylakoids**. All the chlorophyll is contained within this membrane system, which is the site of the light reactions of photosynthesis.

The carbon reduction reactions, which are catalyzed by water-soluble enzymes, take place in the **stroma**, the region of the chloroplast outside the thylakoids. Most of the thylakoids appear to be very closely associated with each other. These stacked membranes are known as **grana lamellae** (singular *lamella;* each stack is called a *granum*), and the exposed membranes in which stacking is absent are known as **stroma lamellae**.

Two separate membranes, each composed of a lipid bilayer and together known as the **envelope**, surround most types of chloroplasts (**Figure 7.15**). This double-membrane system contains a variety of metabolite transport systems. The chloroplast also contains its own DNA, RNA, and ribosomes. Some of the chloroplast proteins are products of transcription and translation within the chloroplast itself, whereas most of the others are encoded by nuclear DNA, synthesized on cytoplasmic ribosomes, and then imported into the chloroplast. This remarkable division of labor, extending in many cases to different subunits of the same enzyme complex, will be discussed in more detail later in this chapter. For some dynamic structures of chloroplasts, see WEB ESSAY 7.1.

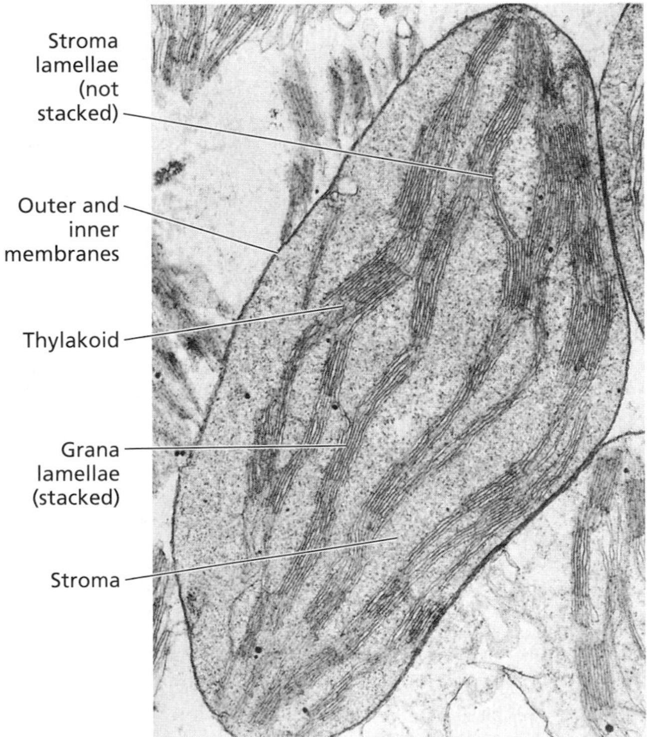

Figure 7.14 Transmission electron micrograph of a chloroplast from pea (*Pisum sativum*) fixed in glutaraldehyde and OsO_4, embedded in plastic resin, and thin-sectioned with an ultramicrotome. (14,500×) (Courtesy of J. Swafford.)

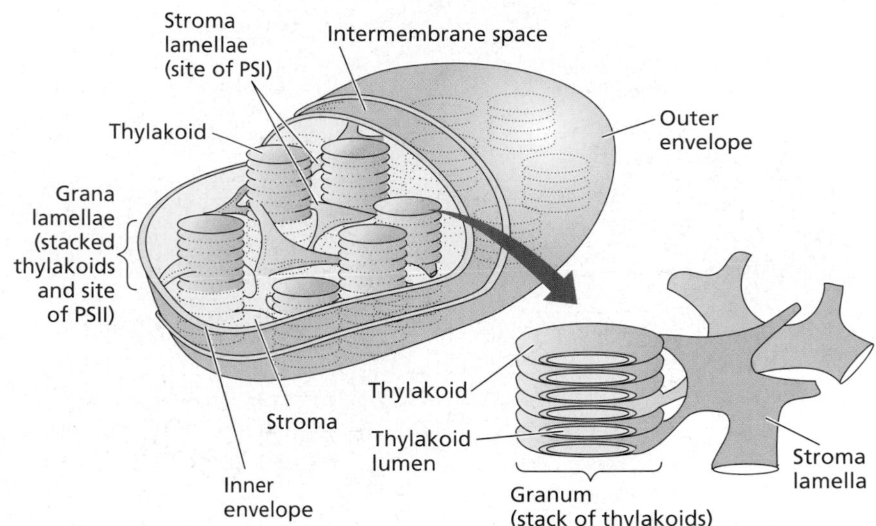

Figure 7.15 Schematic picture of the overall organization of the membranes in the chloroplast. The chloroplast of higher plants is surrounded by the inner and outer membranes (envelope). The region of the chloroplast that is inside the inner membrane and surrounds the thylakoid membranes is known as the stroma. It contains the enzymes that catalyze carbon fixation and other biosynthetic pathways. The thylakoid membranes are highly folded and appear in many pictures to be stacked like coins (the granum), although in reality they form one or a few large interconnected membrane systems, with a well-defined interior and exterior with respect to the stroma. (After Becker 1986.)

Thylakoids contain integral membrane proteins

A wide variety of proteins essential to photosynthesis are embedded in the thylakoid membranes. In many cases, portions of these proteins extend into the aqueous regions on both sides of the thylakoids. These **integral membrane proteins** contain a large proportion of hydrophobic amino acids and are therefore much more stable in a nonaqueous medium such as the hydrocarbon portion of the membrane (see Figure 1.7A).

The reaction centers, the antenna pigment–protein complexes, and most of the electron carrier proteins are all integral membrane proteins. In all known cases, integral membrane proteins of the chloroplast have a unique orientation within the membrane. Thylakoid membrane proteins have one region pointing toward the stromal side of the membrane and the other oriented toward the interior space of the thylakoid, known as the *lumen* (see Figure 7.15).

The chlorophylls and accessory light-gathering pigments in the thylakoid membrane are always associated in a noncovalent, but highly specific, way with proteins, thereby forming pigment–protein complexes. Both antenna and reaction center chlorophylls are associated with proteins that are organized within the membrane so as to optimize energy transfer in antenna complexes and electron transfer in reaction centers, while at the same time minimizing wasteful processes.

Photosystems I and II are spatially separated in the thylakoid membrane

The PSII reaction center, along with its antenna chlorophylls and associated electron transport proteins, is located predominantly in the grana lamellae (**Figure 7.16A**). The PSI reaction center and its associated antenna pigments and electron transfer proteins, as well as the ATP synthase enzyme that catalyzes the formation of ATP, are found almost exclusively in the stroma lamellae and at the edges of the grana lamellae. The cytochrome b_6f complex of the electron transport chain that connects the two photosystems is evenly distributed between stroma and grana lamellae. The structures of all these complexes are shown in **Figure 7.16B**.

Thus, the two photochemical events that take place in O_2-evolving photosynthesis are spatially separated. This separation implies that one or more of the electron carriers that function between the photosystems diffuses from the grana region of the membrane to the stroma region, where electrons are delivered to PSI. These diffusible carriers are the blue-colored copper protein plastocyanin (PC) and the organic redox cofactor plastoquinone (PQ). These carriers will be discussed in more detail later in this chapter.

In PSII, the oxidation of two water molecules produces four electrons, four protons, and a single O_2 (see the section *Mechanisms of Electron Transport* for details). The protons produced by this oxidation of water must also be able to diffuse to the stroma region, where ATP is synthesized. The functional role of this large separation (many tens of nanometers) between photosystems I and II is not entirely clear but is thought to improve the efficiency of energy distribution between the two photosystems.

The spatial separation between photosystems I and II indicates that a strict one-to-one stoichiometry between the two photosystems is not required. Instead, PSII reaction centers feed reducing equivalents into a common intermediate pool of lipid-soluble electron carriers (plastoquinone). The PSI reaction centers remove the reducing equivalents from the common pool, rather than from any specific PSII reaction center complex.

Most measurements of the relative quantities of photosystems I and II have shown that there is an excess of PSII in chloroplasts. Most commonly, the ratio of PSII to PSI is about 1.5:1, but it can change when plants are grown under different light conditions. In contrast to the situation in

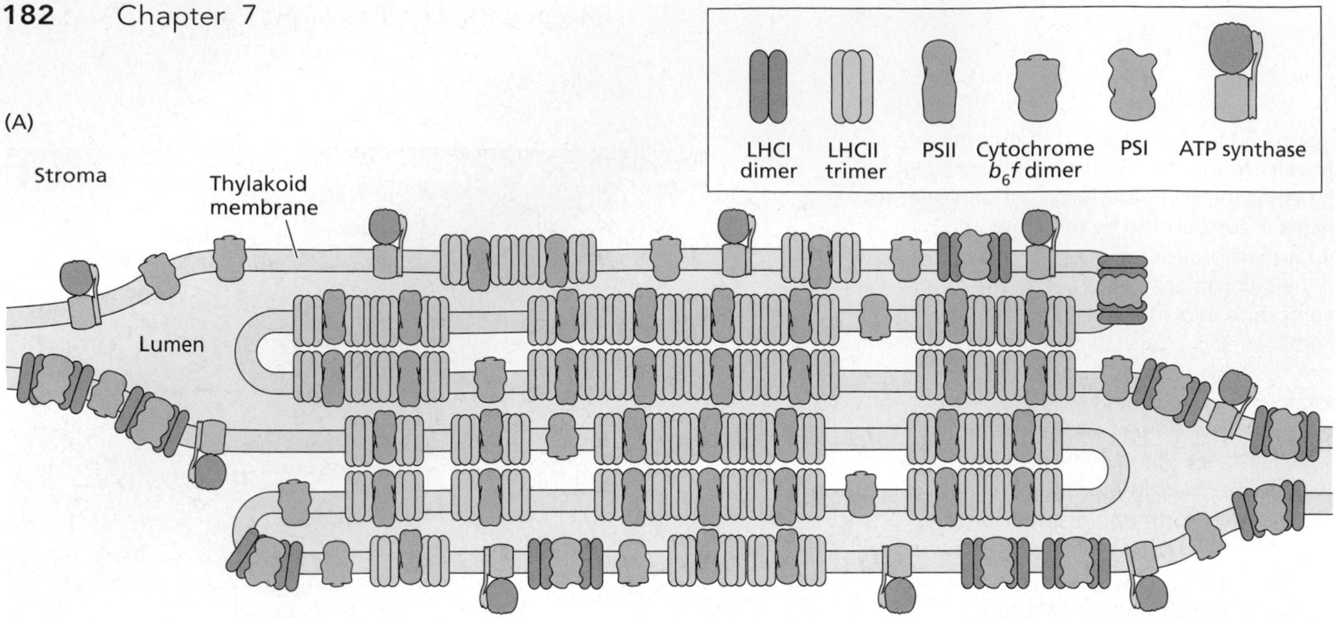

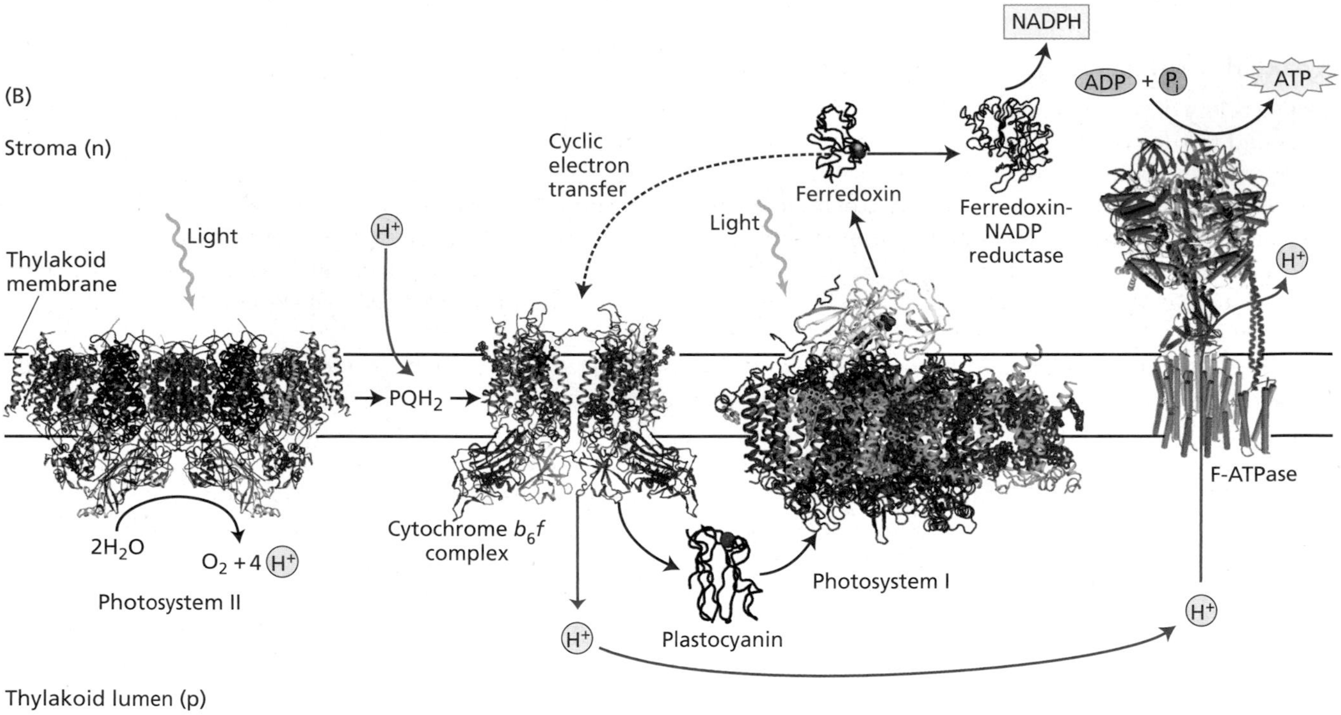

Figure 7.16 Organization and structure of the four major protein complexes of the thylakoid membrane. (A) PSII is located predominantly in the stacked regions of the thylakoid membrane; PSI and ATP synthase are found in the unstacked regions protruding into the stroma. Cytochrome b_6f complexes are evenly distributed. This lateral separation of the two photosystems requires that electrons and protons produced by PSII be transported a considerable distance before they can be acted on by PSI and the ATP-coupling enzyme. (B) Structures of the four main protein complexes of the thylakoid membrane. Shown also are the two diffusible electron carriers—plastocyanin, which is located in the thylakoid lumen, and plastohydroquinone (PQH_2) in the membrane. The lumen has a positive electrical charge (p) with respect to the stroma (n). (A after Allen and Forsberg 2001; B after Nelson and Ben-Shem 2004.)

chloroplasts of eukaryotic photosynthetic organisms, cyanobacteria usually have an excess of PSI over PSII.

Anoxygenic photosynthetic bacteria have a single reaction center

Non-O_2-evolving (anoxygenic) organisms contain only a single photosystem similar to either photosystem I or II. These simpler organisms have been very useful for detailed structural and functional studies that have contributed to a better understanding of oxygenic photosynthesis. In most cases, these anoxygenic photosystems carry out cyclic electron transfer with no net reduction or oxidation. Part of the energy of the photon is conserved as a proton motive force (see p. 153) and is used to make ATP.

Reaction centers from purple photosynthetic bacteria were the first integral membrane proteins to have structures determined to high resolution (see Figures 7.5.A and 7.5.B in **WEB TOPIC 7.5**). Detailed analysis of these structures, along with the characterization of numerous mutants, has revealed many of the principles involved in the energy storage processes carried out by all reaction centers.

The structure of the purple bacterial reaction center is thought to be similar in many ways to that found in PSII from oxygen-evolving organisms, especially in the electron acceptor portion of the chain. The proteins that make up the core of the bacterial reaction center are relatively similar in sequence to their PSII counterparts, implying an evolutionary relatedness. A similar situation is found with respect to the reaction centers from the anoxygenic green sulfur bacteria and the heliobacteria, compared with PSI. The evolutionary implications of this pattern will be discussed later in this chapter.

Organization of Light-Absorbing Antenna Systems

The antenna systems of different classes of photosynthetic organisms are remarkably varied, in contrast to the reaction centers, which appear to be similar in even distantly related organisms. The variety of antenna complexes reflects evolutionary adaptation to the diverse environments in which different organisms live, as well as the need in some organisms to balance energy input to the two photosystems. In this section we will learn how energy transfer processes absorb light and deliver energy to the reaction center.

Antenna systems contain chlorophyll and are membrane-associated

Antenna systems function to deliver energy efficiently to the reaction centers with which they are associated. The size of the antenna system varies considerably in different organisms, ranging from a low of 20 to 30 bacteriochlorophylls per reaction center in some photosynthetic bacteria, to generally 200 to 300 chlorophylls per reaction center in higher plants, to a few thousand pigments per reaction center in some types of algae and bacteria. The molecular structures of antenna pigments are also quite diverse, although all of them are associated in some way with the photosynthetic membrane. In almost all cases, the antenna pigments are associated with proteins to form pigment–protein complexes.

The physical mechanism by which excitation energy is conveyed from the chlorophyll that absorbs the light to the reaction center is thought to be **fluorescence resonance energy transfer**, often abbreviated as FRET. By this mechanism the excitation energy is transferred from one molecule to another by a nonradiative process.

A useful analogy for resonance transfer is the transfer of energy between two tuning forks. If one tuning fork is struck and properly placed near another, the second tuning fork receives some energy from the first and begins to vibrate. The efficiency of energy transfer between the two tuning forks depends on their distance from each other and their relative orientation, as well as on their vibrational frequencies, or pitches. Similar parameters affect the efficiency of energy transfer in antenna complexes, with energy substituted for pitch.

Energy transfer in antenna complexes is usually very efficient: Approximately 95 to 99% of the photons absorbed by the antenna pigments have their energy transferred to the reaction center, where it can be used for photochemistry. There is an important difference between energy transfer among pigments in the antenna and the electron transfer that occurs in the reaction center: Whereas energy transfer is a purely physical phenomenon, electron transfer involves chemical (redox) reactions.

The antenna funnels energy to the reaction center

The sequence of pigments within the antenna that funnel absorbed energy toward the reaction center has absorption maxima that are progressively shifted toward longer red wavelengths (**Figure 7.17**). This red shift in absorption maximum means that the energy of the excited state is somewhat lower nearer the reaction center than in the more peripheral portions of the antenna system.

As a result of this arrangement, when excitation is transferred, for example, from a chlorophyll *b* molecule absorbing maximally at 650 nm to a chlorophyll *a* molecule absorbing maximally at 670 nm, the difference in energy between these two excited chlorophylls is lost to the environment as heat.

For the excitation to be transferred back to the chlorophyll *b*, the energy lost as heat would have to be resupplied. The probability of reverse transfer is therefore smaller simply because thermal energy is not sufficient to make up the deficit between the lower-energy and higher-energy pigments. This effect gives the energy-trapping process a degree of directionality or irreversibility and makes the delivery of excitation to the reaction center more efficient. In essence, the system sacrifices some energy from each quantum so that nearly all of the quanta can be trapped by the reaction center.

Many antenna pigment–protein complexes have a common structural motif

In all eukaryotic photosynthetic organisms that contain both chlorophyll *a* and chlorophyll *b*, the most abundant antenna proteins are members of a large family of structurally related proteins. Some of these proteins are associated primarily with PSII and are called **light-harvesting complex II (LHCII)** proteins; others are associated with PSI and are called LHCI proteins. These antenna complexes are also known as **chlorophyll *a/b* antenna proteins**.

The structure of one of the LHCII proteins has been determined (**Figure 7.18**). The protein contains three α-helical regions and binds 14 chlorophyll *a* and *b* molecules, as

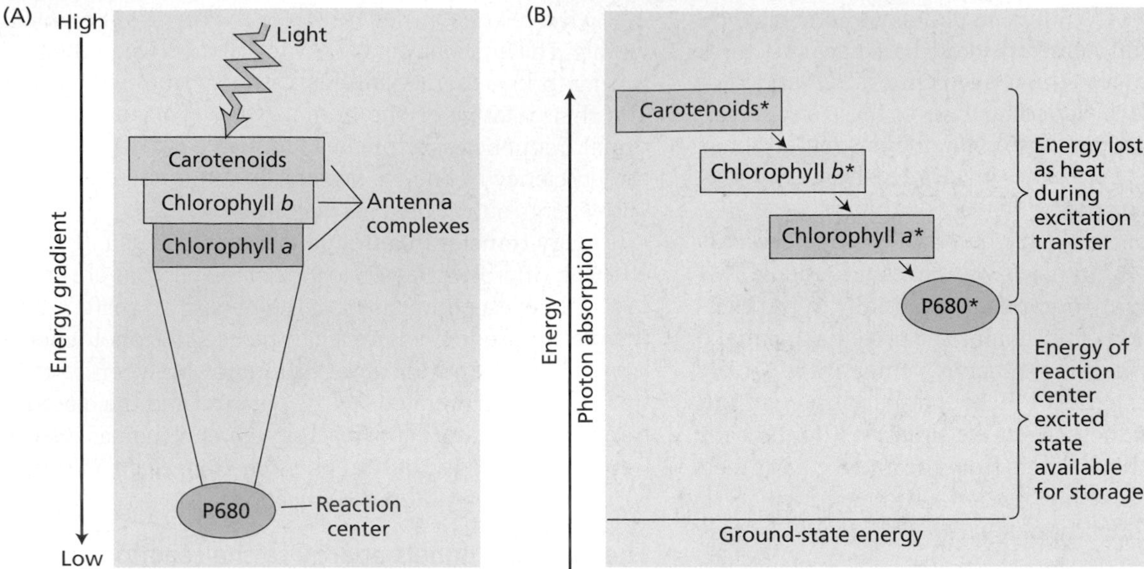

Figure 7.17 Funneling of excitation from the antenna system toward the reaction center. (A) The excited-state energy of pigments increases with distance from the reaction center; that is, pigments closer to the reaction center are lower in energy than those farther from the reaction center. This energy gradient ensures that excitation transfer toward the reaction center is energetically favorable and that excitation transfer back out to the peripheral portions of the antenna is energetically unfavorable. (B) Some energy is lost as heat to the environment by this process, but under optimal conditions almost all the excitation energy absorbed in the antenna complexes can be delivered to the reaction center. The asterisks denote excited states.

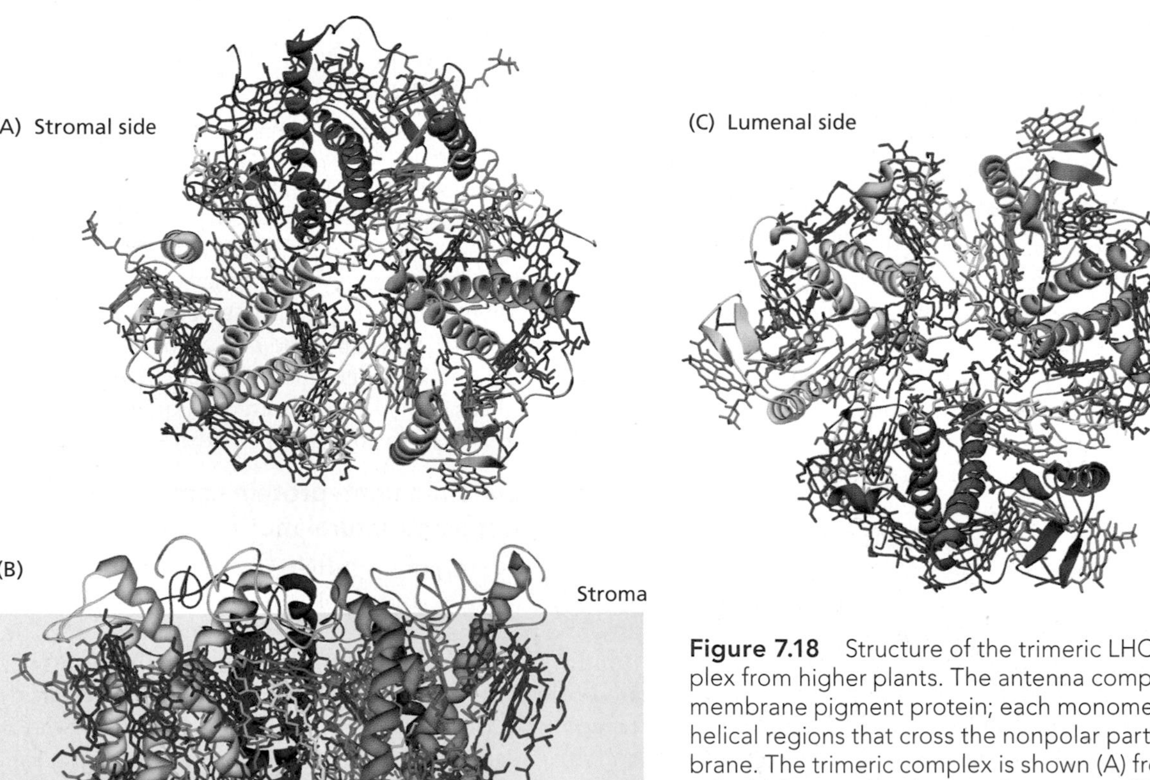

Figure 7.18 Structure of the trimeric LHCII antenna complex from higher plants. The antenna complex is a transmembrane pigment protein; each monomer contains three helical regions that cross the nonpolar part of the membrane. The trimeric complex is shown (A) from the stromal side, (B) from within the membrane, and (C) from the lumenal side. Gray, polypeptide; dark blue, Chl *a*; green, Chl *b*; dark orange, lutein; light orange, neoxanthin; yellow, violaxanthin; pink, lipids. (After Barros and Kühlbrandt 2009.)

well as four carotenoids. The structure of the LHCI proteins is generally similar to that of the LHCII proteins. All of these proteins have significant sequence similarity and are almost certainly descendants of a common ancestral protein.

Light absorbed by carotenoids or chlorophyll *b* in the LHC proteins is rapidly transferred to chlorophyll *a* and then to other antenna pigments that are intimately associated with the reaction center. The LHCII complex is also involved in regulatory processes, which we will discuss later in the chapter.

Mechanisms of Electron Transport

Some of the evidence that led to the idea of two photochemical reactions operating in series was discussed earlier in this chapter. In this section we will consider in detail the chemical reactions involved in electron transfer during photosynthesis. We will discuss the excitation of chlorophyll by light and the reduction of the first electron acceptor, the flow of electrons through photosystems II and I, the oxidation of water as the primary source of electrons, and the reduction of the final electron acceptor ($NADP^+$). The chemiosmotic mechanism that mediates ATP synthesis will be discussed in detail later in the chapter (see the section *Proton Transport and ATP Synthesis in the Chloroplast*).

Electrons from chlorophyll travel through the carriers organized in the Z scheme

Figure 7.19 shows a current version of the Z scheme, in which all the electron carriers known to function in electron flow from H_2O to $NADP^+$ are arranged vertically at their midpoint redox potentials (see **WEB TOPIC 7.6** for further detail). Components known to react with each other are connected by arrows, so the Z scheme is really

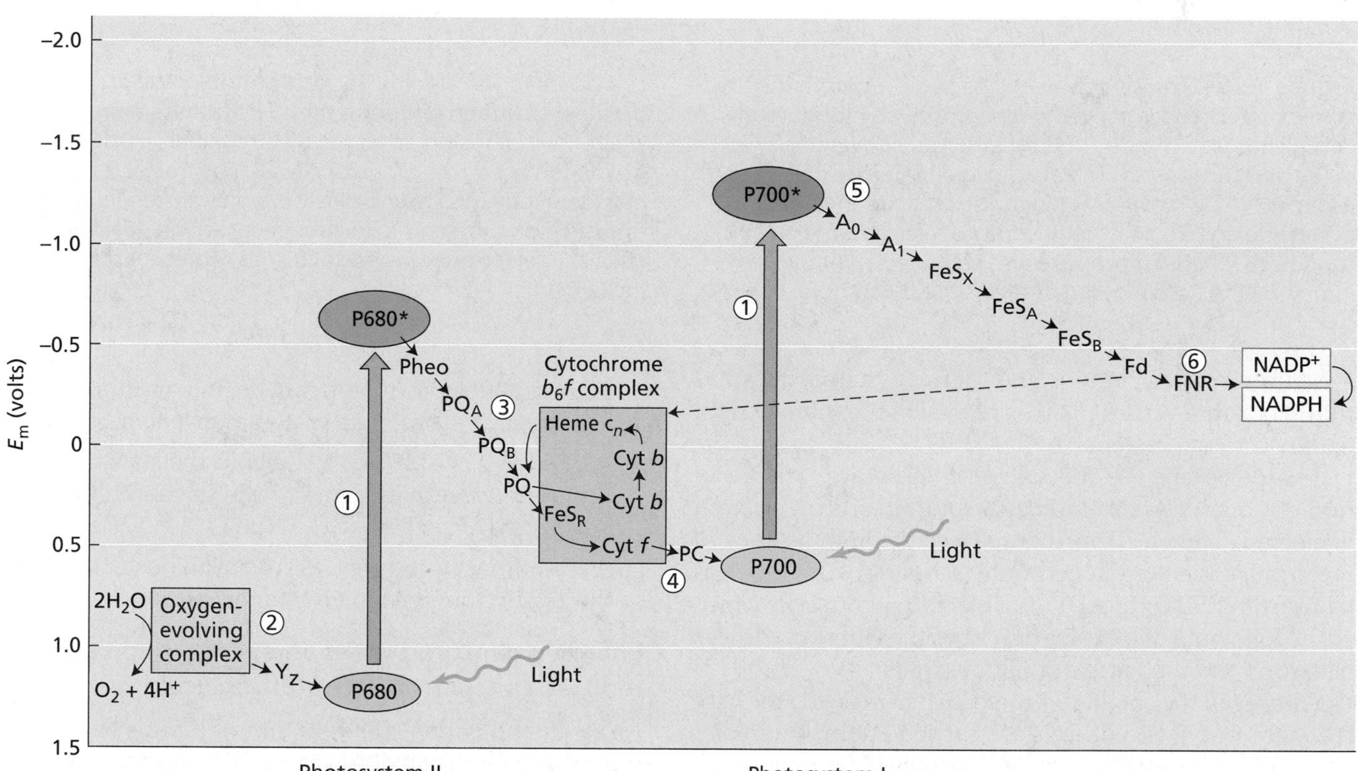

Figure 7.19 Detailed Z scheme for O_2-evolving photosynthetic organisms. The redox carriers are placed at their midpoint redox potentials (at pH 7). (1) The vertical arrows represent photon absorption by the reaction center chlorophylls: P680 for photosystem II (PSII) and P700 for photosystem I (PSI). The excited PSII reaction center chlorophyll, P680*, transfers an electron to pheophytin (Pheo). (2) On the oxidizing side of PSII (to the left of the arrow joining P680 with P680*), P680 oxidized by light is re-reduced by Y_z, which has received electrons from oxidation of water. (3) On the reducing side of PSII (to the right of the arrow joining P680 with P680*), pheophytin transfers electrons to the acceptors PQ_A and PQ_B, which are plastoquinones. (4) The cytochrome b_6f complex transfers electrons to plastocyanin (PC), a soluble protein, which in turn reduces $P700^+$ (oxidized P700). (5) The acceptor of electrons from P700* (A_0) is thought to be a chlorophyll, and the next acceptor (A_1) is a quinone. A series of membrane-bound iron–sulfur proteins (FeS_X, FeS_A, and FeS) transfers electrons to soluble ferredoxin (Fd). (6) The soluble flavoprotein ferredoxin–$NADP^+$ reductase (FNR) reduces $NADP^+$ to NADPH, which is used in the Calvin–Benson cycle to reduce CO_2 (see Chapter 8). The dashed line indicates cyclic electron flow around PSI. (After Blankenship and Prince 1985.)

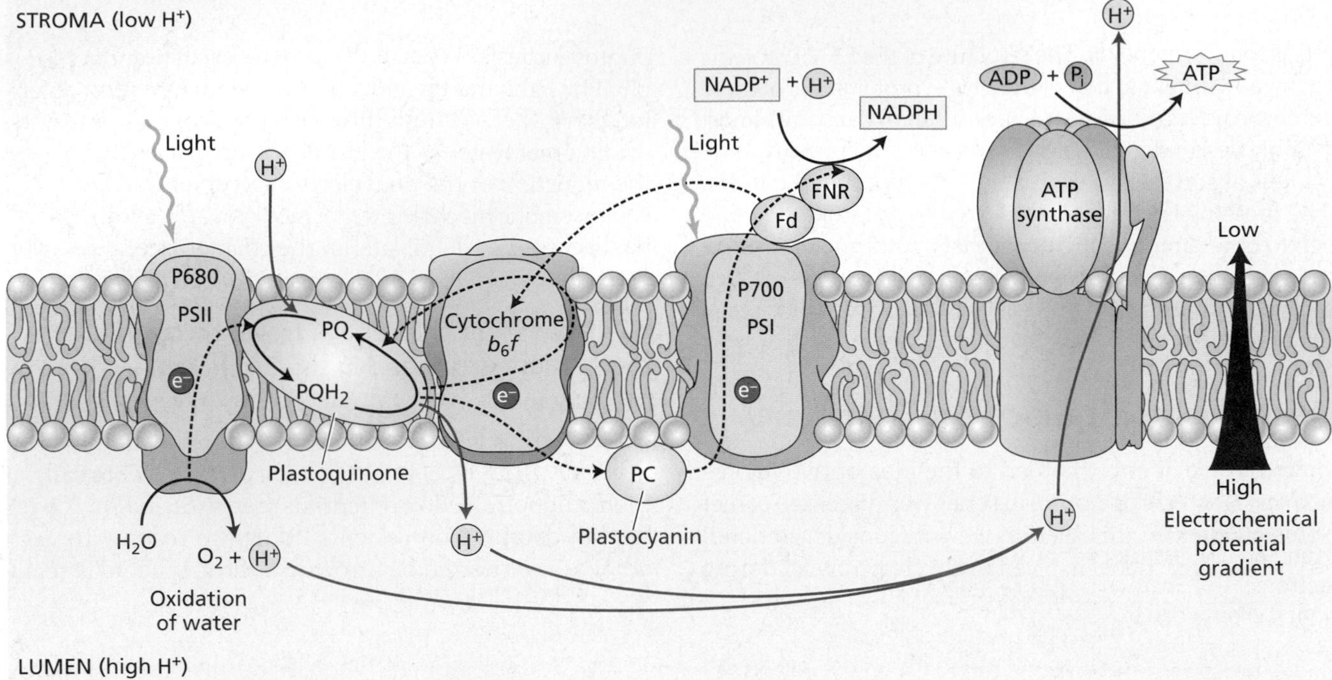

Figure 7.20 Transfer of electrons and protons in the thylakoid membrane is carried out vectorially by four protein complexes (see Figure 7.16B for structures). Water is oxidized and protons are released in the lumen by PSII. PSI reduces $NADP^+$ to NADPH in the stroma, via the action of ferredoxin (Fd) and the flavoprotein ferredoxin–$NADP^+$ reductase (FNR). Protons are also transported into the lumen by the action of the cytochrome b_6f complex and contribute to the electrochemical proton gradient. These protons must then diffuse to the ATP synthase enzyme, where their diffusion down the electrochemical potential gradient is used to synthesize ATP in the stroma. Reduced plastoquinone (PQH_2) and plastocyanin transfer electrons to cytochrome b_6f and to PSI, respectively. The dashed lines represent electron transfer; solid lines represent proton movement.

a synthesis of both kinetic and thermodynamic information. The large vertical arrows represent the input of light energy into the system.

Photons excite the specialized chlorophyll of the reaction centers (P680 for PSII; P700 for PSI), and an electron is ejected. The electron then passes through a series of electron carriers and eventually reduces P700 (for electrons from PSII) or $NADP^+$ (for electrons from PSI). Much of the following discussion describes the journeys of these electrons and the nature of their carriers.

Almost all the chemical processes that make up the light reactions of photosynthesis are carried out by four major protein complexes: PSII, the cytochrome b_6f complex, PSI, and the ATP synthase. These four integral membrane complexes are vectorially oriented in the thylakoid membrane to function as follows (**Figure 7.20**; also see Figure 7.16):

- PSII oxidizes water to O_2 in the thylakoid lumen and in the process releases protons into the lumen. The reduced product of photosystem II is plastohydroquinone (PQH_2).
- Cytochrome b_6f oxidizes PQH_2 molecules that were reduced by PSII and delivers electrons to PSI via the soluble copper protein plastocyanin. The oxidation of PQH_2 is coupled to proton transfer into the lumen from the stroma, generating a proton motive force.
- PSI reduces $NADP^+$ to NADPH in the stroma by the action of ferredoxin (Fd) and the flavoprotein ferredoxin–$NADP^+$ reductase (FNR).
- ATP synthase produces ATP as protons diffuse back through it from the lumen into the stroma.

Energy is captured when an excited chlorophyll reduces an electron acceptor molecule

As discussed earlier, the function of light is to excite a specialized chlorophyll in the reaction center, either by direct absorption or, more frequently, via energy transfer from an antenna pigment. This excitation process can be envisioned as the promotion of an electron from the highest-energy filled orbital of the chlorophyll to the lowest-energy unfilled orbital (**Figure 7.21**). The electron in the upper orbital is only loosely bound to the chlorophyll and is easily lost if a molecule that can accept the electron is nearby.

The first reaction that converts electron energy into chemical energy—that is, the primary photochemical event—is the transfer of an electron from the excited state of a chlorophyll in the reaction center to an acceptor

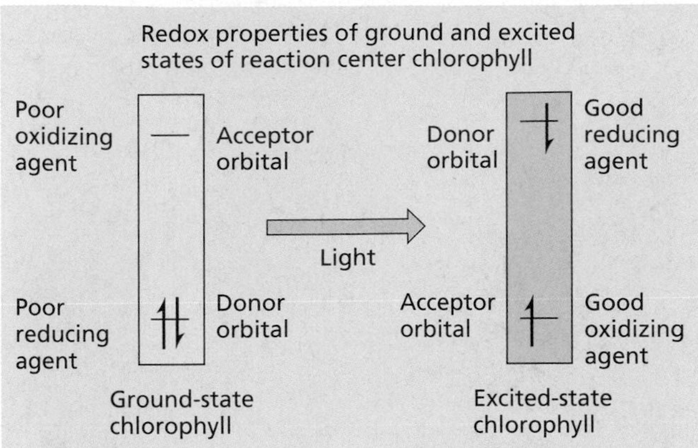

Figure 7.21 Orbital occupation diagram for the ground and excited states of reaction center chlorophyll. In the ground state the molecule is a poor reducing agent (loses electrons from a low-energy orbital) and a poor oxidizing agent (accepts electrons only into a high-energy orbital). In the excited state the situation is markedly different, and an electron can be lost from the high-energy orbital, making the molecule an extremely powerful reducing agent. This is the reason for the extremely negative excited-state redox potential shown by P680* and P700* in Figure 7.19. The excited state can also act as a strong oxidant by accepting an electron into the lower-energy orbital, although this pathway is not significant in reaction centers. (After Blankenship and Prince 1985.)

molecule. An equivalent way to view this process is that the absorbed photon causes an electron rearrangement in the reaction center chlorophyll, followed by an electron transfer process in which part of the energy in the photon is captured in the form of redox energy.

Immediately after the photochemical event, the reaction center chlorophyll is in an oxidized state (electron deficient, or positively charged), and the nearby electron acceptor molecule is reduced (electron rich, or negatively charged). The system is now at a critical juncture. The lower-energy orbital of the positively charged oxidized reaction center chlorophyll shown in Figure 7.21 has a vacancy and can accept an electron. If the acceptor molecule donates its electron back to the reaction center chlorophyll, the system will be returned to the state that existed before the light excitation, and all the absorbed energy will be converted into heat.

This wasteful *recombination* process, however, does not appear to occur to any substantial degree in functioning reaction centers. Instead, the acceptor transfers its extra electron to a secondary acceptor and so on down the electron transport chain. The oxidized reaction center of the chlorophyll that had donated an electron is re-reduced by a secondary donor, which in turn is reduced by a tertiary donor. In plants, the ultimate electron donor is H_2O, and the ultimate electron acceptor is $NADP^+$ (see Figure 7.19).

The essence of photosynthetic energy storage is thus the initial transfer of an electron from an excited chlorophyll to an acceptor molecule, followed by a very rapid series of secondary chemical reactions that separate the positive and negative charges. These secondary reactions separate the charges to opposite sides of the thylakoid membrane in approximately 200 picoseconds (1 picosecond = 10^{-12} s).

With the charges thus separated, the reversal reaction is many orders of magnitude slower, and the energy has been captured. Each of the secondary electron transfers is accompanied by a loss of some energy, thus making the process effectively irreversible. The quantum yield for the production of stable products in purified reaction centers from photosynthetic bacteria has been measured as 1.0; that is, every photon produces stable products, and no reversal reactions occur.

Measured quantum requirements for O_2 production in higher plants under optimal conditions (low-intensity light) indicate that the values for the primary photochemical events are also very close to 1.0. The structure of the reaction center appears to be extremely fine-tuned for maximum rates of productive reactions and minimum rates of energy-wasting reactions.

The reaction center chlorophylls of the two photosystems absorb at different wavelengths

As discussed earlier in the chapter, PSI and PSII have distinct absorption characteristics. Precise measurements of absorption maxima are made possible by optical changes in the reaction center chlorophylls in the reduced and oxidized states. The reaction center chlorophyll is transiently in an oxidized state after losing an electron and before being re-reduced by its electron donor.

In the oxidized state, chlorophylls lose their characteristic strong light absorbance in the red region of the spectrum; they become **bleached**. It is therefore possible to monitor the redox state of these chlorophylls by time-resolved optical absorbance measurements in which this bleaching is monitored directly (see **WEB TOPIC 7.1**).

Using such techniques, it was found that the reaction center chlorophyll of PSI absorbs maximally at 700 nm in its reduced state. Accordingly, this chlorophyll is named **P700** (the P stands for *pigment*). The analogous optical transient of PSII is at 680 nm, so its reaction center chlorophyll is known as **P680**. The reaction center bacteriochlorophyll from purple photosynthetic bacteria was similarly identified as **P870**.

The X-ray structure of the bacterial reaction center (see Figures 7.5.A and 7.5.B in **WEB TOPIC 7.5**) clearly indicates that P870 is a closely coupled pair or dimer of bacteriochlorophylls rather than a single molecule. The primary donor of PSI, P700, is also a dimer of chlorophyll *a* molecules. PSII also contains a dimer of chlorophylls, although the primary electron transfer event may not originate from

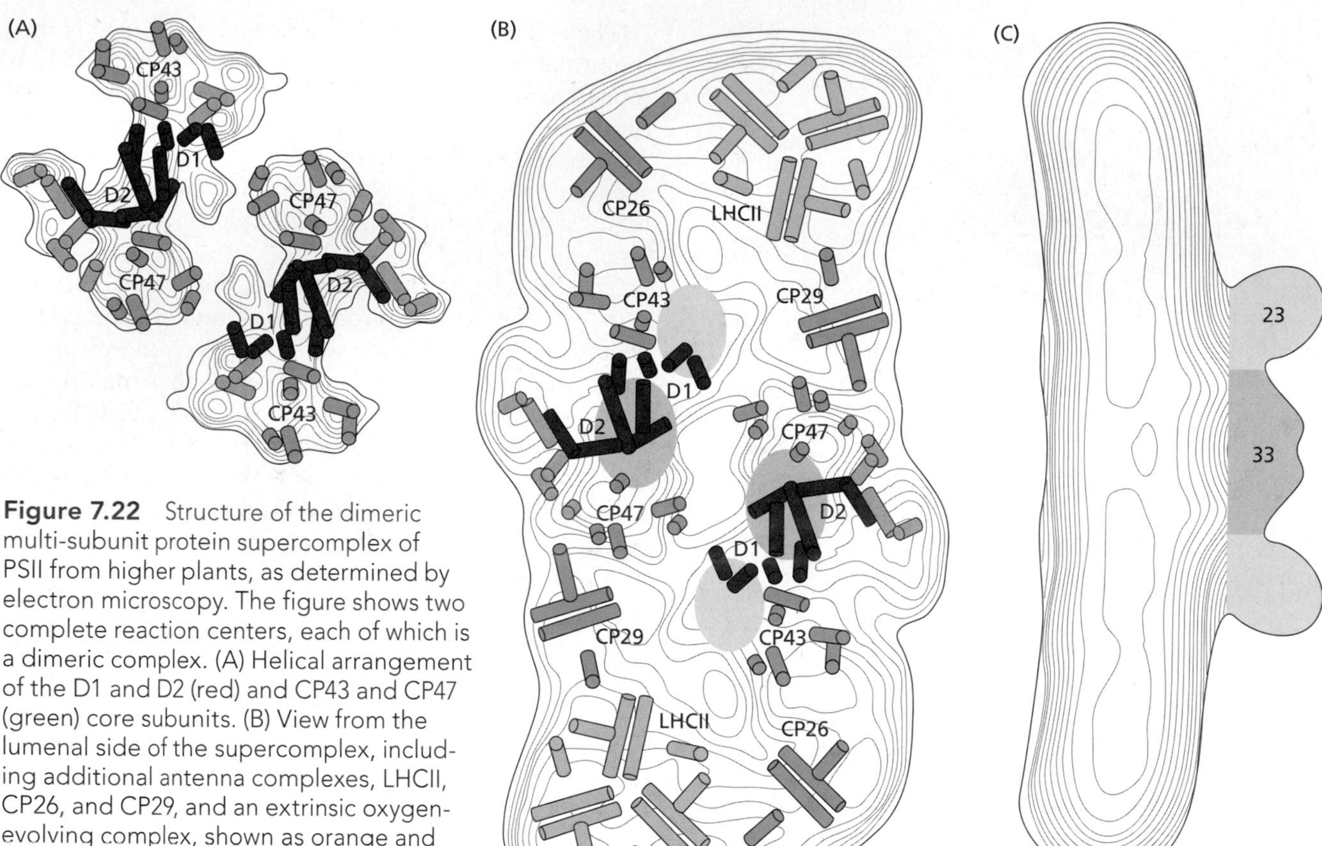

Figure 7.22 Structure of the dimeric multi-subunit protein supercomplex of PSII from higher plants, as determined by electron microscopy. The figure shows two complete reaction centers, each of which is a dimeric complex. (A) Helical arrangement of the D1 and D2 (red) and CP43 and CP47 (green) core subunits. (B) View from the lumenal side of the supercomplex, including additional antenna complexes, LHCII, CP26, and CP29, and an extrinsic oxygen-evolving complex, shown as orange and yellow circles. Other helices are shown in gray. (C) Side view of the complex illustrating the arrangement of the extrinsic proteins of the oxygen-evolving complex. (After Barber et al. 1999.)

these pigments. In the oxidized state, reaction center chlorophylls contain an unpaired electron.

Molecules with unpaired electrons can often be detected by a magnetic-resonance technique known as **electron spin resonance (ESR)**. ESR studies, along with the spectroscopic measurements already described, have led to the discovery of many intermediate electron carriers in the photosynthetic electron transport system.

The PSII reaction center is a multi-subunit pigment–protein complex

PSII is contained in a multi-subunit protein supercomplex (**Figure 7.22**). In higher plants, the multi-subunit protein supercomplex has two complete reaction centers and some antenna complexes. The core of the reaction center consists of two membrane proteins known as D1 and D2, as well as other proteins, as shown in **Figure 7.23** and WEB TOPIC 7.7.

The primary donor chlorophyll, additional chlorophylls, carotenoids, pheophytins, and plastoquinones (two electron acceptors described below) are bound to the membrane proteins D1 and D2. These proteins have some sequence similarity to the L and M peptides of purple bacteria. Other proteins serve as antenna complexes or are involved in oxygen evolution. Some, such as cytochrome b_{559}, have no known function but may be involved in a protective cycle around PSII.

Water is oxidized to oxygen by PSII

Water is oxidized according to the following chemical reaction:

$$2\,H_2O \rightarrow O_2 + 4\,H^+ + 4\,e^- \qquad (7.8)$$

This equation indicates that four electrons are removed from two water molecules, generating an oxygen molecule and four hydrogen ions. (For more on oxidation–reduction reactions, see WEB APPENDIX 1 and WEB TOPIC 7.6.)

Water is a very stable molecule. Oxidation of water to form molecular oxygen is very difficult: The photosynthetic oxygen-evolving complex is the only known biochemical system that carries out this reaction, and is the source of almost all the oxygen in Earth's atmosphere.

Many studies have provided a substantial amount of information about the process (see WEB TOPIC 7.7). The protons produced by water oxidation are released into the lumen of the thylakoid, not directly into the stromal compartment (see Figure 7.20). They are released into

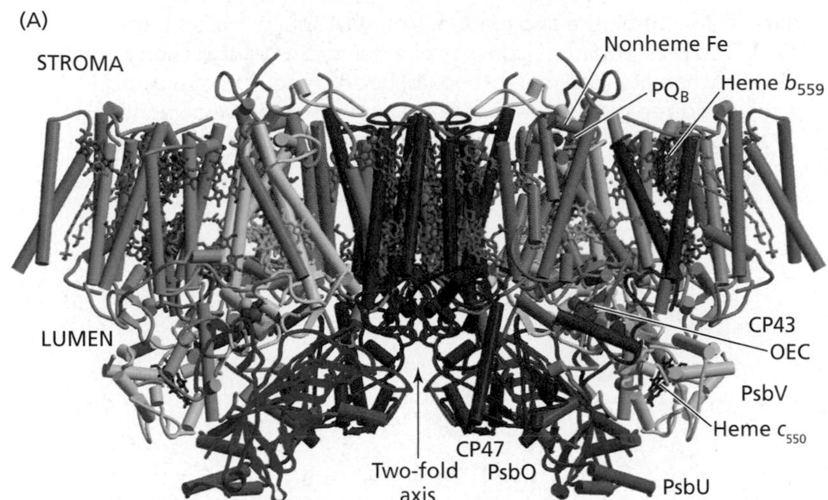

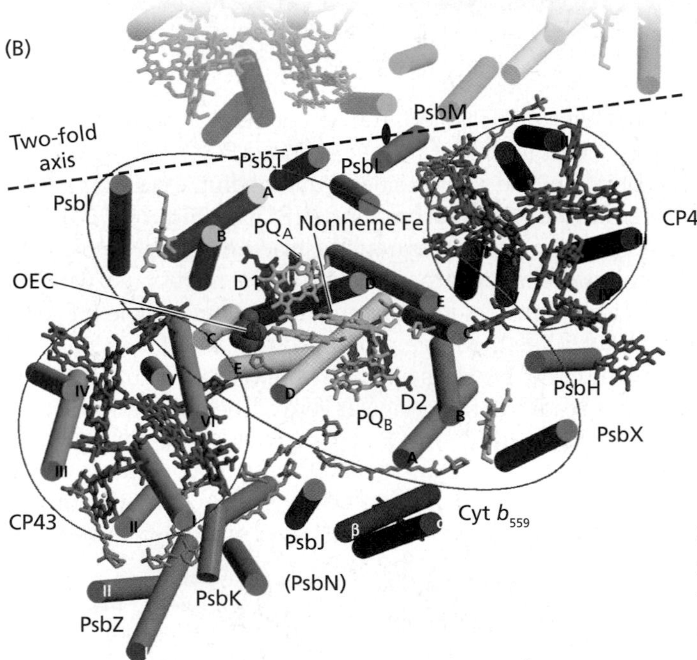

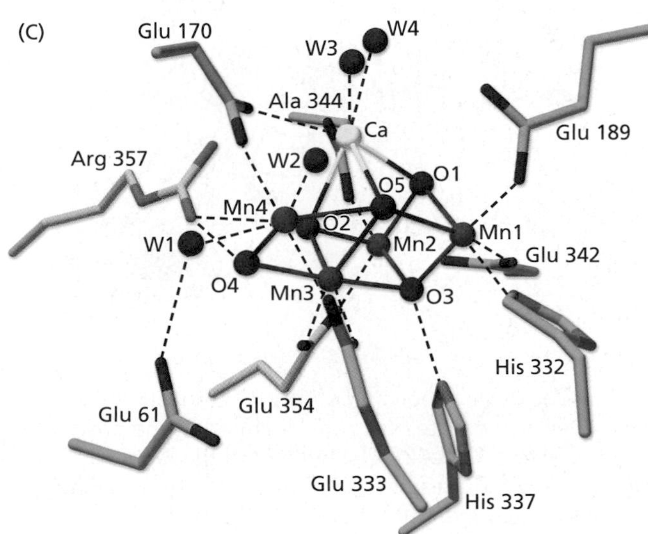

Figure 7.23 Structure of the PSII reaction center from the cyanobacterium *Thermosynechococcus elongatus*, resolved at 3.5 Å. The structure includes the D_1 (yellow) and D_2 (orange) core reaction center proteins, the CP43 (green) and CP47 (red) antenna proteins, cytochromes b_{559} and c_{550}, the extrinsic 33-kDa oxygen evolution protein PsbO (dark blue), and the pigments and other cofactors. (A) Side view parallel to the membrane plane. (B) View from the lumenal surface, perpendicular to the plane of the membrane. (C) Detail of the Mn-containing water-splitting complex. (A, B after Ferreira et al. 2004; C after Umena et al. 2011.)

the lumen because of the vectorial nature of the membrane and the fact that the oxygen-evolving complex is localized near the interior surface of the thylakoid membrane. These protons are eventually transferred from the lumen to the stroma by translocation through the ATP synthase. In this way, the protons released during water oxidation contribute to the electrochemical potential driving ATP formation.

It has been known for many years that manganese (Mn) is an essential cofactor in the water-oxidizing process (see Chapter 5), and a classic hypothesis in photosynthesis research postulates that Mn ions undergo a series of oxidations—known as *S states* and labeled S_0, S_1, S_2, S_3, and S_4 (see **WEB TOPIC 7.7**)—that are perhaps linked to H_2O oxidation and the generation of O_2. This hypothesis has received strong support from a variety of experiments, most notably X-ray absorption and ESR studies, both of which detect the manganese ions directly. Analytical experiments indicate that four Mn ions are associated with each oxygen-evolving complex. Other experiments have shown that Cl^- and Ca^{2+} ions are essential for O_2 evolution (see **WEB TOPIC 7.7**). The detailed chemical mechanism of the oxidation of water to O_2 is not yet well understood, but with structural information now available, rapid progress is being made in this area.

One electron carrier, generally identified as Y_z, functions between the oxygen-evolving complex and P680 (see Figure 7.19). To function in this region, Y_z needs to have a very strong tendency to retain its electrons. This species has been identified as a radical formed from a tyrosine residue in the D1 protein of the PSII reaction center.

Pheophytin and two quinones accept electrons from PSII

Spectral and ESR studies have revealed the structural arrangement of the carriers in the electron acceptor complex. **Pheophytin**, a chlorophyll in which the central mag-

(A)

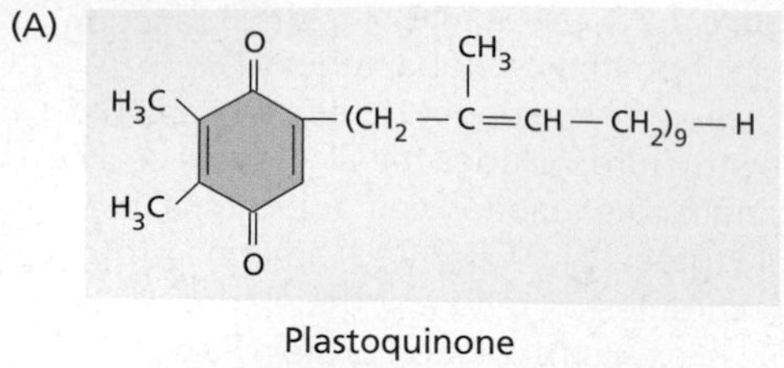

Plastoquinone

(B)

Plastoquinone (PQ)

Plastosemiquinone ($PQ^{\bar{\bullet}}$)

Plastohydroquinone (PQH_2)

Figure 7.24 Structure and reactions of plastoquinones that operate in PSII. (A) The plastoquinone consists of a quinoid head and a long nonpolar tail that anchors it in the membrane. (B) Redox reactions of plastoquinone. The fully oxidized plastoquinone (PQ), anionic plastosemiquinone ($PQ^{\bar{\bullet}}$), and reduced plastohydroquinone (PQH_2) forms are shown; R represents the side chain.

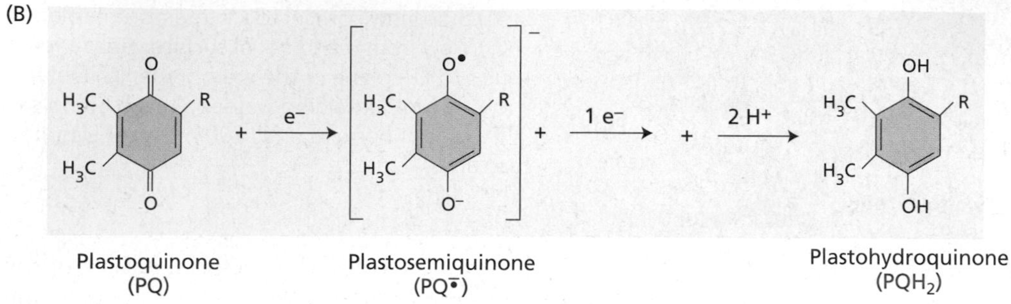

nesium ion has been replaced by two hydrogen ions, acts as an early acceptor in PSII. The structural change gives pheophytin chemical and spectral properties that are slightly different from those of Mg-based chlorophylls. Pheophytin passes electrons to a complex of two plastoquinones in close proximity to an iron ion. The processes are very similar to those found in the reaction center of purple bacteria (for details, see Figure 7.5.B in **WEB TOPIC 7.5**).

The two plastoquinones, PQ_A and PQ_B, are bound to the reaction center and receive electrons from pheophytin in a sequential fashion. Transfer of the two electrons to PQ_B reduces it to PQ_B^{2-}, and the reduced PQ_B^{2-} takes two protons from the stroma side of the medium, yielding a fully reduced **plastohydroquinone (PQH_2)** (**Figure 7.24**). The PQH_2 then dissociates from the reaction center complex and enters the hydrocarbon portion of the membrane, where it in turn transfers its electrons to the cytochrome b_6f complex. Unlike the large protein complexes of the thylakoid membrane, PQH_2 is a small, nonpolar molecule that diffuses readily in the nonpolar core of the membrane bilayer.

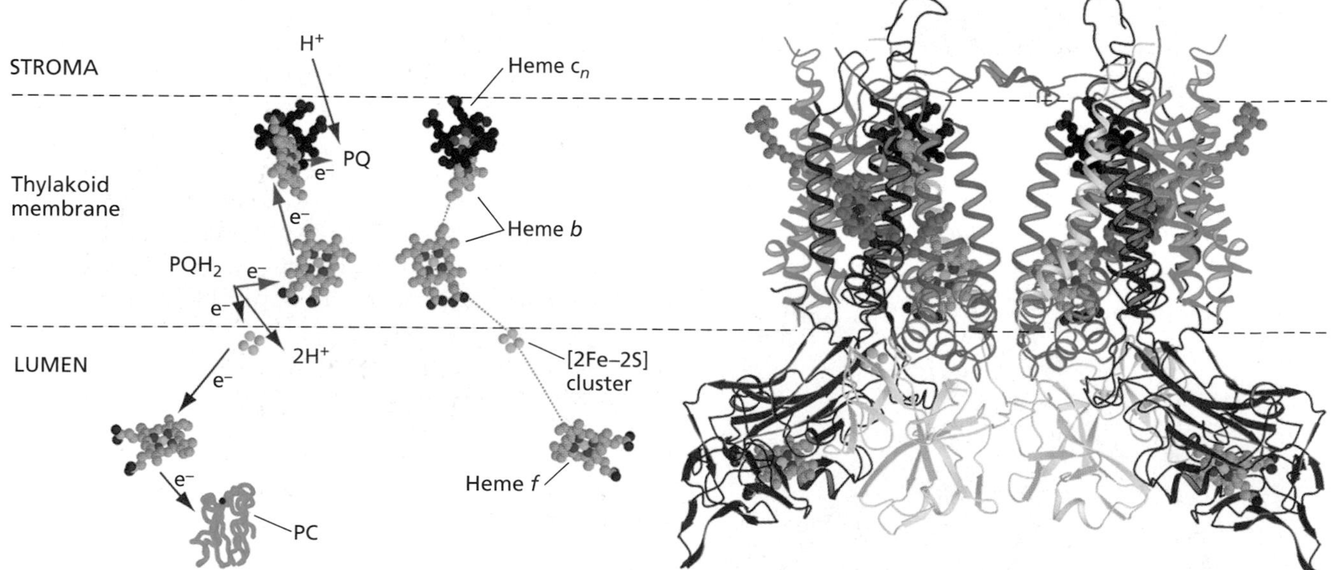

Figure 7.25 Structure of the cytochrome b_6f complex from cyanobacteria. The diagram on the right shows the arrangement of the proteins and cofactors in the complex. Cytochrome b_6 protein is shown in blue, cytochrome *f* protein in red, Rieske iron–sulfur protein in yellow, and other smaller subunits in green and purple. On the left, the proteins have been omitted to more clearly show the positions of the cofactors. [2 Fe–2S] cluster, part of the Rieske iron–sulfur protein; PC, plastocyanin; PQ, plastoquinone; PQH_2, plastohydroquinone. (After Kurisu et al. 2003.)

Electron flow through the cytochrome b_6f complex also transports protons

The **cytochrome b_6f complex** is a large multi-subunit protein with several prosthetic groups (**Figure 7.25**). It contains two *b*-type hemes and one *c*-type heme (**cytochrome *f***). In *c*-type cytochromes the heme is covalently attached to the peptide; in *b*-type cytochromes the chemically similar protoheme group is not covalently attached (see **WEB TOPIC 7.8**). In addition, the complex contains a **Rieske iron–sulfur protein** (named for the scientist who discovered it), in which two iron ions are bridged by two sulfide ions. The functional roles of all these cofactors are reasonably well understood, as described below. However, the cytochrome b_6f complex also contains additional cofactors—including an additional heme group (called heme c_n), a chlorophyll, and a carotenoid—whose functions are yet to be resolved.

The structures of the cytochrome b_6f complex and the related cytochrome bc_1 complex in the mitochondrial electron transport chain (see Chapter 12) suggest a mechanism for electron and proton flow. The precise way by which electrons and protons flow through the cytochrome b_6f complex is not yet fully understood, but a mechanism known as the **Q cycle** accounts for most of the observations. In this mechanism, plastohydroquinone (also called plastoquinol) (PQH_2) is oxidized, and one of the two electrons is passed along a linear electron transport chain toward PSI, while the other electron goes through a cyclic process that increases the number of protons pumped across the membrane (**Figure 7.26**).

In the linear electron transport chain, the oxidized Rieske protein (**FeS_R**) accepts an electron from PQH_2 and transfers it to cytochrome *f* (see Figure 7.26A). Cytochrome *f* then transfers an electron to the blue-colored copper protein plastocyanin (PC), which in turn reduces oxidized P700 of PSI. In the cyclic part of the process (see Figure 7.26B), the plastosemiquinone (see Figure 7.24) transfers its other electron to one of the *b*-type hemes, releasing both of its protons to the lumenal side of the membrane.

The first *b*-type heme transfers its electron through the second *b*-type heme to an oxidized plastoquinone molecule, reducing it to the semiquinone form near the stromal surface of the complex. Another similar sequence of electron flow fully reduces the plastoquinone, which picks up protons from the stromal side of the membrane and is released from the b_6f complex as plastohydroquinone.

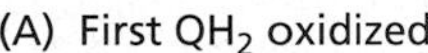

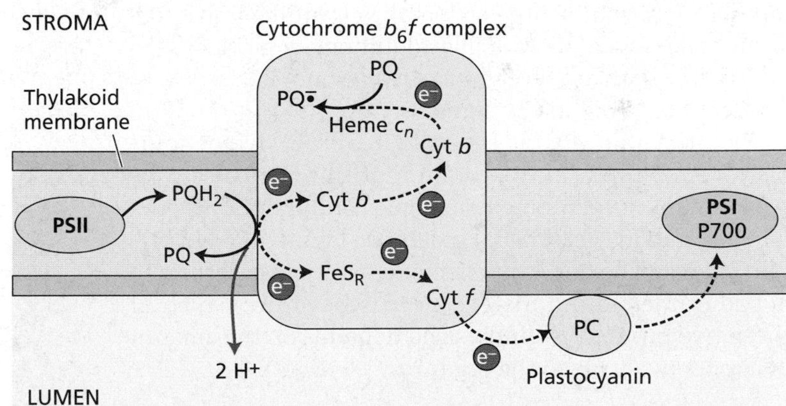

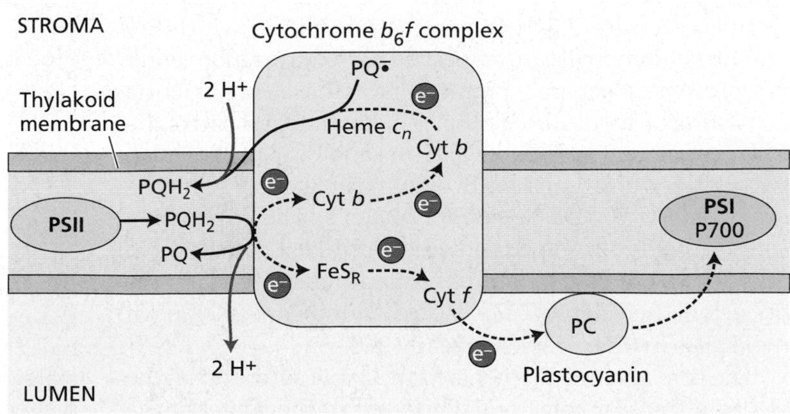

Figure 7.26 Mechanism of electron and proton transfer in the cytochrome b_6f complex. This complex contains two *b*-type cytochromes (Cyt *b*), a *c*-type cytochrome (Cyt *c*, historically called cytochrome *f*), a Rieske Fe–S protein (FeS_R), and two quinone oxidation–reduction sites. (A) The noncyclic or linear processes: A plastohydroquinone (PQH_2) molecule produced by the action of PSII (see Figure 7.24) is oxidized near the lumenal side of the complex, transferring its two electrons to the Rieske Fe–S protein and one of the *b*-type cytochromes and simultaneously expelling two protons to the lumen. The electron transferred to FeS_R is passed to cytochrome *f* (Cyt *f*) and then to plastocyanin (PC), which reduces P700 of PSI. The reduced *b*-type cytochrome transfers an electron to the other *b*-type cytochrome, which reduces a plastoquinone (PQ) to the plastosemiquinone (PQ•) state (see Figure 7.24). (B) The cyclic processes: A second PQH_2 is oxidized, with one electron going from FeS_R to PC and finally to P700. The second electron goes through the two *b*-type cytochromes and reduces the plastosemiquinone to the plastohydroquinone, at the same time picking up two protons from the stroma. Overall, four protons are transported across the membrane for every two electrons delivered to P700.

The overall result of two turnovers of the complex is that two electrons are transferred to P700, two plastohydroquinones are oxidized to the plastoquinone form, and one oxidized plastoquinone is reduced to the plastohydroquinone form. In the process of oxidizing the plastoquinones, four protons are transferred from the stromal to the lumenal side of the membrane.

By this mechanism, electron flow connecting the acceptor side of the PSII reaction center to the donor side of the PSI reaction center also gives rise to an electrochemical potential across the membrane, due in part to H^+ concentration differences on the two sides of the membrane. This electrochemical potential is used to power the synthesis of ATP. The cyclic electron flow through the cytochrome *b* and plastoquinone increases the number of protons pumped per electron beyond what could be achieved in a strictly linear sequence.

Plastoquinone and plastocyanin carry electrons between photosystems II and I

The location of the two photosystems at different sites on the thylakoid membranes (see Figure 7.16) requires that at least one component is capable of moving along or within the membrane in order to deliver electrons produced by PSII to PSI. The cytochrome b_6f complex is distributed equally between the grana and the stroma regions of the membranes, but its large size makes it unlikely that it is the mobile carrier. Instead, plastoquinone or plastocyanin or possibly both are thought to serve as mobile carriers to connect the two photosystems.

Plastocyanin (**PC**) is a small (10.5 kDa), water-soluble, copper-containing protein that transfers electrons between the cytochrome b_6f complex and P700. This protein is found in the lumenal space (see Figure 7.26). In certain green algae and cyanobacteria, a *c*-type cytochrome is sometimes found instead of plastocyanin; which of these two proteins is synthesized depends on the amount of copper available to the organism.

The PSI reaction center reduces NADP+

The PSI reaction center complex is a large multi-subunit complex (**Figure 7.27**). In contrast to PSII, in which the antenna chlorophylls are associated with the reaction center but present on separate pigment–proteins, a core antenna consisting of about 100 chlorophylls is an integral part of the PSI reaction center. The core antenna and P700 are bound to two proteins, PsaA and PsaB, with molecular masses in the range of 66 to 70 kDa (see **WEB TOPIC 7.8**). The PSI reaction center complex from pea contains four LHCI complexes in addition to the core structure similar to that found in cyanobacteria (see Figure 7.27). The total number of chlorophyll molecules in this complex is nearly 200.

The core antenna pigments form a bowl surrounding the electron transfer cofactors, which are in the center of the

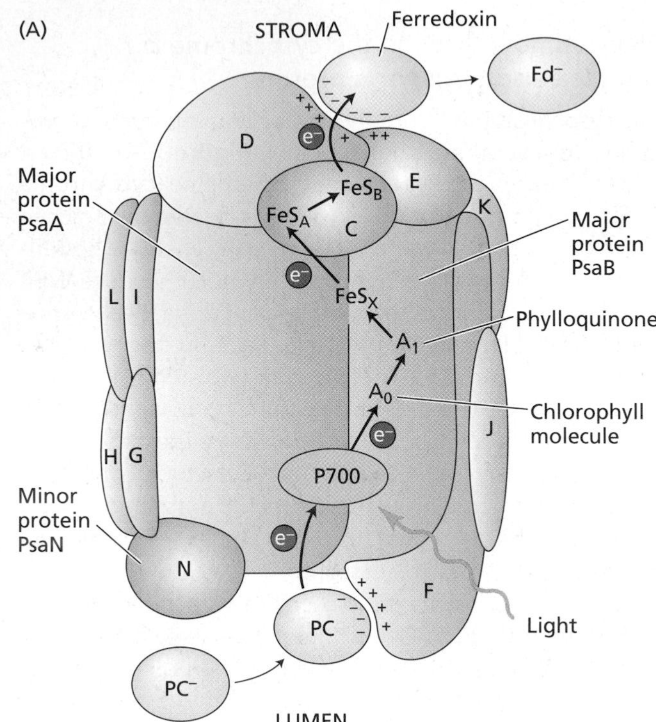

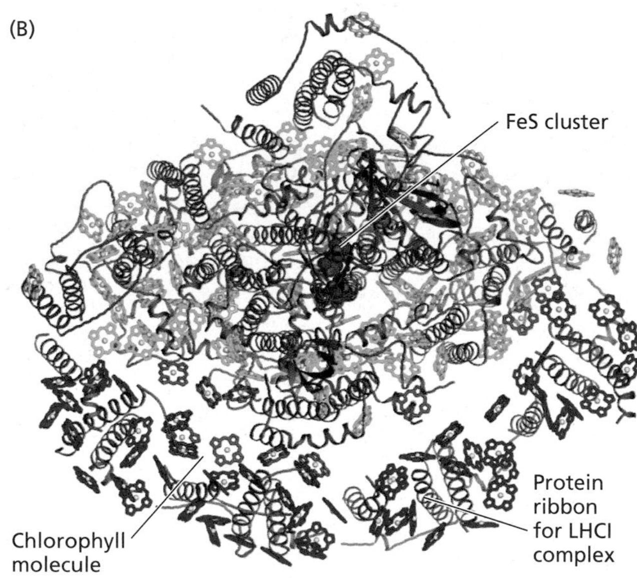

Figure 7.27 Structure of PSI. (A) Structural model of the PSI reaction center from higher plants. Components of the PSI reaction center are organized around two major core proteins, PsaA and PsaB. Minor proteins PsaC to PsaN are labeled C to N. Electrons are transferred from plastocyanin (PC) to P700 (see Figures 7.19 and 7.20) and then to a chlorophyll molecule (A_0), to phylloquinone (A_1), to the Fe–S centers FeS_X, FeS_A, and FeS_B, and finally to the soluble iron–sulfur protein ferredoxin (Fd). (B) Structure of the PSI reaction center complex from pea at 4.4 Å resolution, including the LHCI antenna complexes. This is viewed from the stromal side of the membrane. (A after Buchanan et al. 2000; B after Nelson and Ben-Shem 2004.)

complex. In their reduced form, the electron carriers that function in the acceptor region of PSI are all extremely strong reducing agents. These reduced species are very unstable and thus difficult to identify. Evidence indicates that one of these early acceptors is a chlorophyll molecule, and another is a quinone species, phylloquinone, also known as vitamin K_1.

Additional electron acceptors include a series of three membrane-associated iron–sulfur proteins, also known as **Fe–S centers**: **FeS_X**, **FeS_A**, and **FeS_B** (see Figure 7.27). Fe–S center X is part of the P700-binding protein; centers A and B reside on an 8-kDa protein that is part of the PSI reaction center complex. Electrons are transferred through centers A and B to **ferredoxin (Fd)**, a small, water-soluble iron–sulfur protein (see Figures 7.19 and 7.27). The membrane-associated flavoprotein **ferredoxin–NADP⁺ reductase (FNR)** reduces $NADP^+$ to NADPH, thus completing the sequence of noncyclic electron transport that begins with the oxidation of water.

In addition to the reduction of $NADP^+$, reduced ferredoxin produced by PSI has several other functions in the chloroplast, such as the supply of reductants to reduce nitrate and the regulation of some of the carbon-fixation enzymes (see Chapter 8).

Cyclic electron flow generates ATP but no NADPH

Some of the cytochrome b_6f complexes are found in the stroma region of the membrane, where PSI is located. Under certain conditions, **cyclic electron flow** is known to occur from the reducing side of PSI via plastohydroquinone and the b_6f complex and back to P700. This cyclic electron flow is coupled to proton pumping into the lumen, which can be used for ATP synthesis but does not oxidize water or reduce $NADP^+$ (see Figure 7.16B). Cyclic electron flow is especially important as an ATP source in the bundle sheath chloroplasts of some plants that carry out C_4 carbon fixation (see Chapter 8). The molecular mechanism of cyclic electron flow is not well understood. Some proteins involved in regulating the process are just being discovered, and this remains an active area of research.

Some herbicides block photosynthetic electron flow

The use of herbicides to kill unwanted plants is widespread in modern agriculture. Many different classes of herbicides have been developed. Some act by blocking amino acid, carotenoid, or lipid biosynthesis or by disrupting cell division. Other herbicides, such as dichlorophenyldimethylurea (DCMU, also known as diuron) and paraquat, block photosynthetic electron flow (**Figure 7.28**).

DCMU blocks electron flow at the quinone acceptors of PSII, by competing for the binding site of plastoquinone that is normally occupied by PQ_B. Paraquat accepts electrons from the early acceptors of PSI and then reacts with oxygen to form superoxide, O_2^-, a species that is very damaging to chloroplast components.

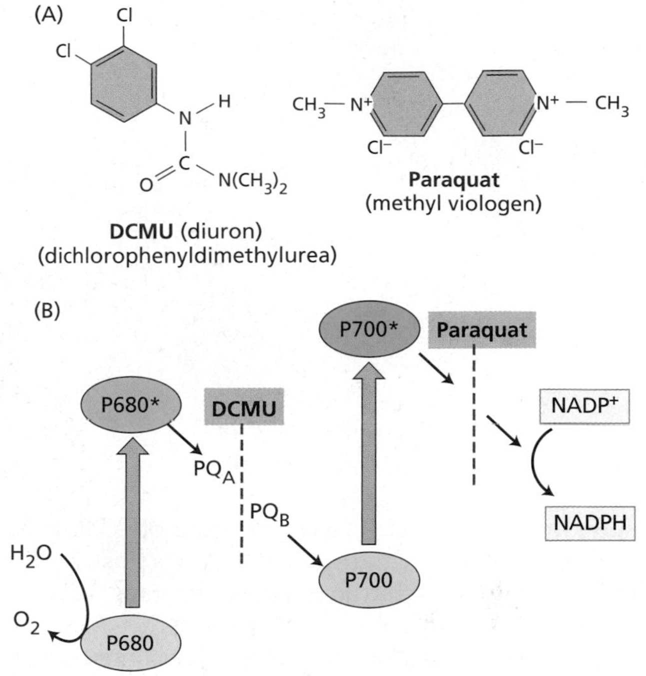

Figure 7.28 Chemical structure and mechanism of action of two important herbicides. (A) Chemical structure of dichlorophenyldimethylurea (DCMU) and methyl viologen (paraquat), two herbicides that block photosynthetic electron flow. DCMU is also known as diuron. (B) Sites of action of the two herbicides. DCMU blocks electron flow at the plastoquinone acceptors of PSII by competing for the binding site of plastoquinone. Paraquat acts by accepting electrons from the early acceptors of PSI.

Proton Transport and ATP Synthesis in the Chloroplast

In the preceding sections we learned how captured light energy is used to reduce $NADP^+$ to NADPH. Another fraction of the captured light energy is used for light-dependent ATP synthesis, which is known as **photophosphorylation**. This process was discovered by Daniel Arnon and his coworkers in the 1950s. Under normal cellular conditions, photophosphorylation requires electron flow, although under some conditions electron flow and photophosphorylation can take place independently of each other. Electron flow without accompanying phosphorylation is said to be **uncoupled**.

It is now widely accepted that photophosphorylation works via the chemiosmotic mechanism. This mechanism was first proposed in the 1960s by Peter Mitchell. The same general mechanism drives phosphorylation during aerobic respiration in bacteria and mitochondria (see Chapter 12), as well as the transfer of many ions and metabolites across membranes (see Chapter 6). Chemiosmosis appears to be a unifying aspect of membrane processes in all forms of life.

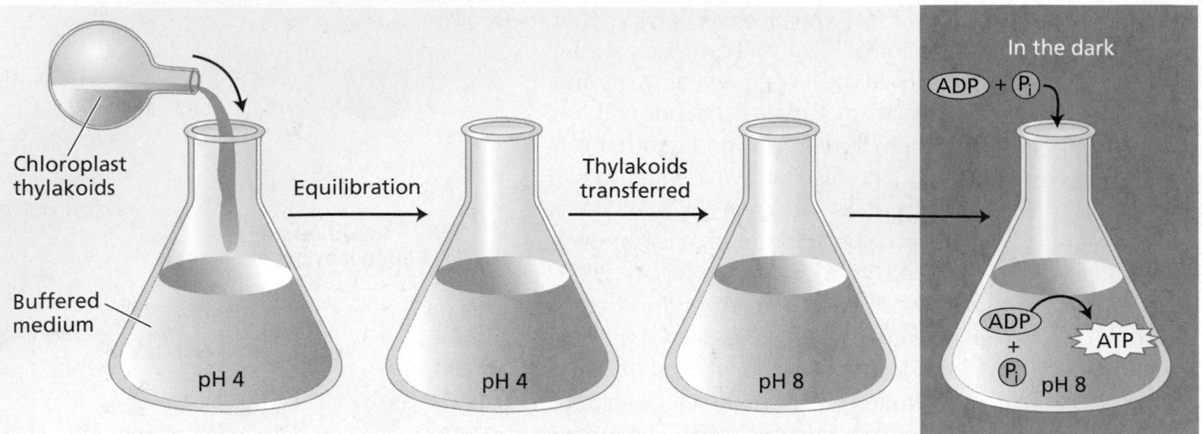

Figure 7.29 Summary of the experiment carried out by Jagendorf and coworkers. Isolated chloroplast thylakoids kept previously at pH 8 were equilibrated in an acid medium at pH 4. The thylakoids were then transferred to a buffer at pH 8 that contained ADP and P_i. The proton gradient generated by this manipulation provided a driving force for ATP synthesis in the absence of light. This experiment verified a prediction of the chemiosmotic theory stating that a chemical potential across a membrane can provide energy for ATP synthesis. (After Jagendorf 1967.)

In Chapter 6 we discussed the role of ATPases in chemiosmosis and ion transport at the cell's plasma membrane. The ATP used by the plasma membrane ATPase is synthesized by photophosphorylation in the chloroplast and oxidative phosphorylation in the mitochondrion. Here we are concerned with chemiosmosis and transmembrane proton concentration differences used to make ATP in the chloroplast.

The basic principle of chemiosmosis is that ion concentration differences and electrical potential differences across membranes are sources of free energy that can be used by the cell. As described by the second law of thermodynamics (see WEB APPENDIX 1 for a detailed discussion), any nonuniform distribution of matter or energy represents a source of energy. Differences in **chemical potential** of any molecular species whose concentrations are not the same on opposite sides of a membrane provide such a source of energy.

The asymmetric nature of the photosynthetic membrane and the fact that proton flow from one side of the membrane to the other accompanies electron flow were discussed earlier. The direction of proton translocation is such that the stroma becomes more alkaline (fewer H^+ ions) and the lumen becomes more acidic (more H^+ ions) as a result of electron transport (see Figures 7.20 and 7.26).

Some of the early evidence supporting a chemiosmotic mechanism of photosynthetic ATP formation was provided by an elegant experiment carried out by André Jagendorf and coworkers (**Figure 7.29**). They suspended chloroplast thylakoids in a pH 4 buffer, and the buffer diffused across the membrane, causing the interior, as well as the exterior, of the thylakoid to equilibrate at this acidic pH. They then rapidly transferred the thylakoids to a pH 8 buffer, thereby creating a pH difference of four units across the thylakoid membrane, with the inside acidic relative to the outside.

They found that large amounts of ATP were formed from ADP and P_i by this process, with no light input or electron transport. This result supports the predictions of the chemiosmotic hypothesis, described in the paragraphs that follow.

Mitchell proposed that the total energy available for ATP synthesis, which he called the **proton motive force** (Δp), is the sum of a proton chemical potential and a transmembrane electrical potential. These two components of the proton motive force from the outside of the membrane to the inside are given by the following equation:

$$\Delta p = \Delta E - 59(\text{pH}_i - \text{pH}_o) \tag{7.9}$$

where ΔE is the transmembrane electrical potential, and $\text{pH}_i - \text{pH}_o$ (or ΔpH) is the pH difference across the membrane. The constant of proportionality (at 25°C) is 59 mV per pH unit, so a transmembrane pH difference of one pH unit is equivalent to a membrane potential of 59 mV. Most evidence suggests that the steady-state electrical potential is relatively small in chloroplasts, so that most of the proton motive force is derived from the pH gradient.

In addition to the need for mobile electron carriers discussed earlier, the uneven distribution of photosystems II and I, and of ATP synthase at the thylakoid membrane (see Figure 7.16), poses some challenges for the formation of ATP. ATP synthase is found only in the stroma lamellae and at the edges of the grana stacks. Protons pumped across the membrane by the cytochrome b_6f complex or protons produced by water oxidation in the middle of the grana must move laterally up to several tens of nanometers to reach an ATP synthase.

The ATP is synthesized by an enzyme complex (mass ~400 kDa) known by several names: **ATP synthase**, **ATPase** (after the reverse reaction of ATP hydrolysis), and **CF_o–CF_1**.

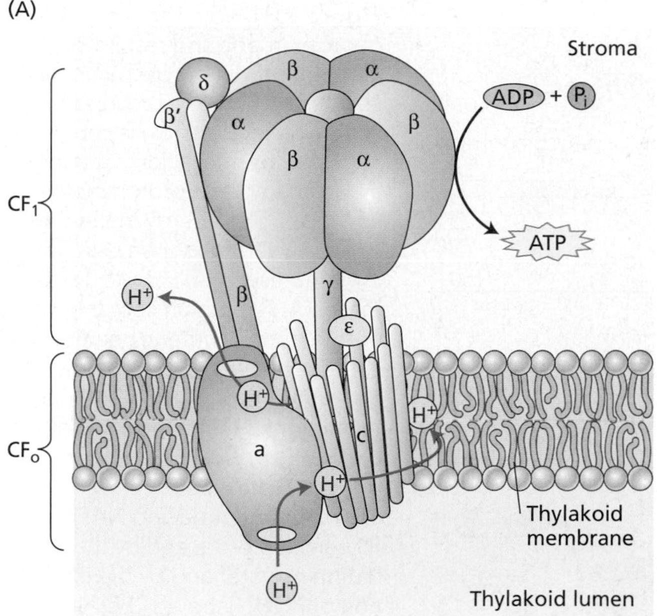

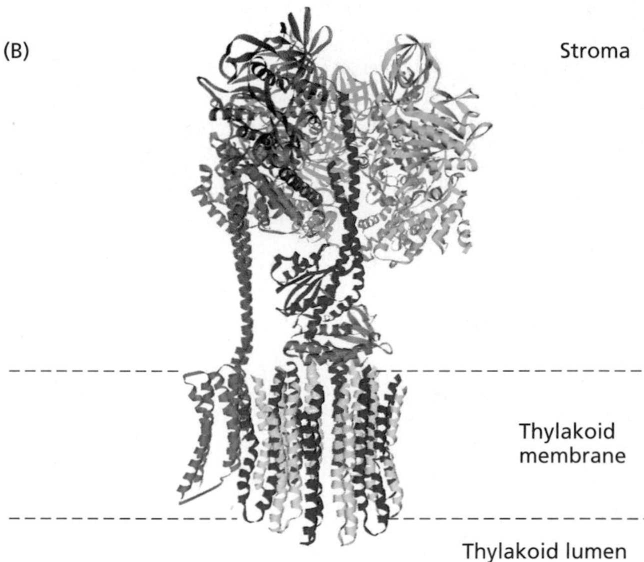

Figure 7.30 Subunit composition (A) and compiled crystal structure (B) of chloroplast F_1F_o ATP synthase. This enzyme consists of a large multi-subunit complex, CF_1, attached on the stromal side of the membrane to an integral membrane portion, known as CF_o. CF_1 consists of five different polypeptides, with stoichiometry of $\alpha_3\,\beta_3\,\gamma\,\delta\,\varepsilon$. CF_o contains probably four different polypeptides, with a stoichiometry of a b b′ c_{14}. Protons from the lumen are transported by the rotating *c* polypeptide and ejected on the stroma side. The structure closely resembles that of the mitochondrial F_1F_o ATP synthase (see Chapter 12) and the vacuolar V-type ATPase (see Chapter 6). (Figure courtesy of W. Frasch.)

This enzyme consists of two parts: a hydrophobic membrane-bound portion called CF_o and a portion that sticks out into the stroma called CF_1 (**Figure 7.30**). CF_o appears to form a channel across the membrane through which protons can pass. CF_1 is made up of several peptides, including three copies of each of the α and β peptides arranged alternately much like the sections of an orange. Whereas the catalytic sites are located largely on the β polypeptide, many of the other peptides are thought to have primarily regulatory functions. CF_1 is the portion of the complex that synthesizes ATP.

The molecular structure of the mitochondrial ATP synthase has been determined by X-ray crystallography. Although there are significant differences between the chloroplast and mitochondrial enzymes, they have the same overall architecture and probably nearly identical catalytic sites. In fact, there are remarkable similarities in the way electron flow is coupled to proton translocation in chloroplasts, mitochondria, and purple bacteria (**Figure 7.31**). Another remarkable aspect of the mechanism of the ATP synthase is that the internal stalk and probably much of the CF_o portion of the enzyme rotate during catalysis. The enzyme is actually a tiny molecular motor (see **WEB TOPICS 7.9 AND 12.4**). Three molecules of ATP are synthesized for each rotation of the enzyme.

Direct microscopic imaging of the CF_o part of the chloroplast ATP synthase indicates that it contains 14 copies of the integral membrane subunit. Each subunit can translocate one proton across the membrane each time the complex rotates. This suggests that the stoichiometry of protons translocated to ATP formed is 14/3, or 4.67. Measured values of this parameter are usually somewhat lower than this value, and the reasons for this discrepancy are not yet understood.

Repair and Regulation of the Photosynthetic Machinery

Photosynthetic systems face a special challenge. They are designed to absorb large amounts of light energy and process it into chemical energy. At the molecular level, the energy in a photon can be damaging, particularly under unfavorable conditions. In excess, light energy can lead to the production of toxic species such as superoxide, singlet oxygen, and peroxide, and damage can occur if the light energy is not dissipated safely. Photosynthetic organisms therefore contain complex regulatory and repair mechanisms.

Some of these mechanisms regulate energy flow in the antenna system, to avoid excess excitation of the reaction centers and ensure that the two photosystems are equally driven. Although very effective, these processes are not entirely fail-safe, and sometimes toxic compounds are produced. Additional mechanisms are needed to dissipate these compounds—in particular, toxic oxygen species. In this section we will examine how some of these processes work to protect the system against photodamage.

(A) Purple bacteria

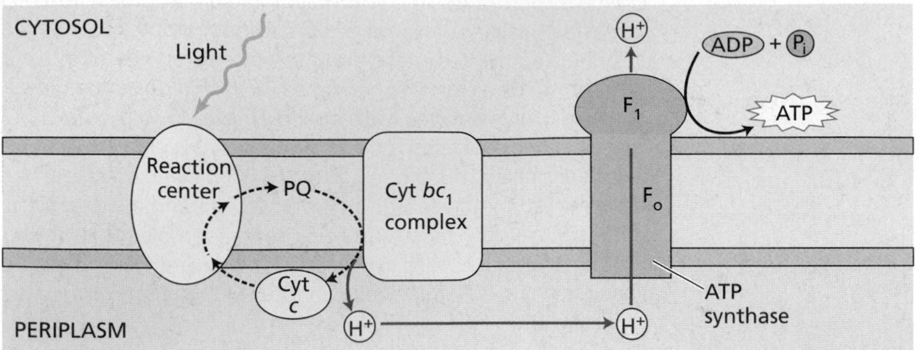

(B) Chloroplasts

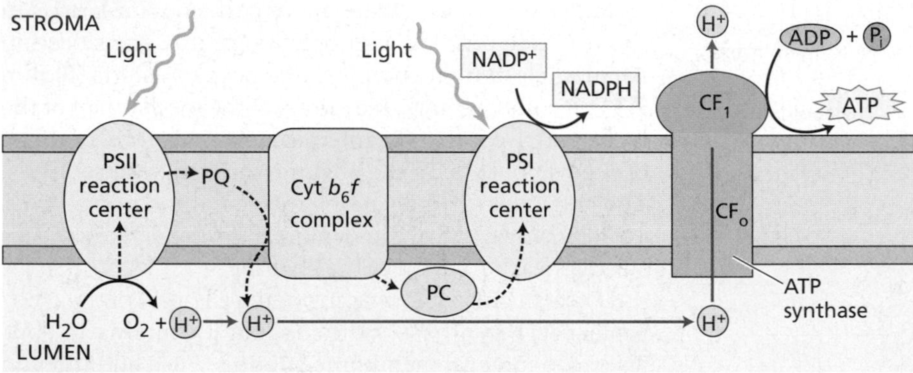

(C) Mitochondria

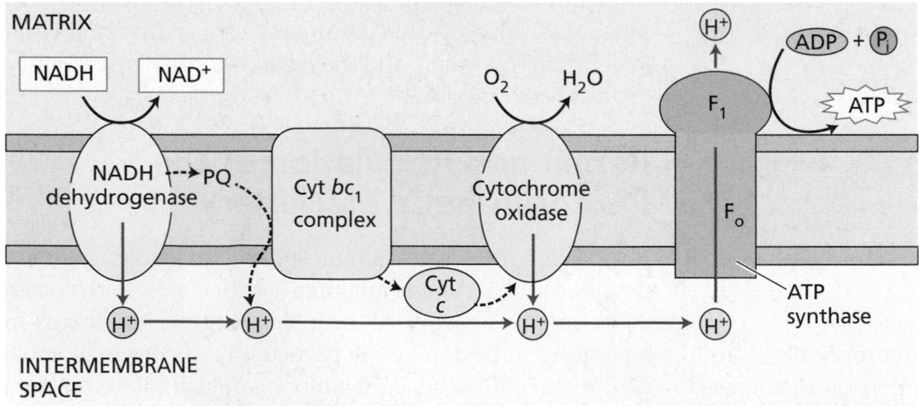

Figure 7.31 Similarities of photosynthetic and respiratory electron flow in purple bacteria, chloroplasts, and mitochondria. In all three, electron flow is coupled to proton translocation, creating a transmembrane proton motive force (Δp). The energy in the proton motive force is then used for the synthesis of ATP by ATP synthase. (A) A reaction center in purple photosynthetic bacteria carries out cyclic electron flow, generating a proton potential by the action of the cytochrome bc_1 complex. (B) Chloroplasts carry out noncyclic electron flow, oxidizing water and reducing $NADP^+$. Protons are produced by the oxidation of water and by the oxidation of PQH_2 (labeled "PQ" in the illustration) by the cytochrome b_6f complex. (C) Mitochondria oxidize NADH to NAD^+ and reduce oxygen to water. Protons are pumped by the enzyme NADH dehydrogenase, the cytochrome bc_1 complex, and cytochrome oxidase. The ATP synthases in the three systems are very similar in structure.

Despite these protective and scavenging mechanisms, damage can occur, and additional mechanisms are required to repair the system. **Figure 7.32** provides an overview of the several levels of the regulation and repair systems.

Carotenoids serve as photoprotective agents

In addition to their role as accessory pigments, carotenoids play an essential role in **photoprotection**. The photosynthetic membrane can easily be damaged by the large amounts of energy absorbed by the pigments if this energy cannot be stored by photochemistry; this is why a protection mechanism is needed. The photoprotection mechanism can be thought of as a safety valve, venting excess energy before it can damage the organism. When the energy stored in chlorophylls in the excited state is rapidly dissipated by excitation transfer or photochemistry, the excited state is said to be **quenched**.

If the excited state of chlorophyll is not rapidly quenched by excitation transfer or photochemistry, it can react with molecular oxygen to form an excited state of oxygen known

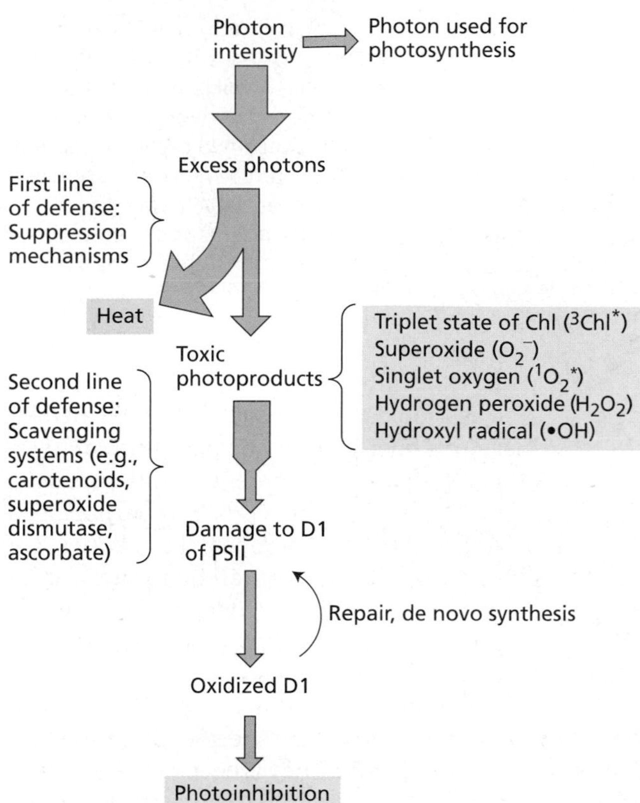

Figure 7.32 Overall picture of the regulation of photon capture and the protection and repair of photodamage. Protection against photodamage is a multilevel process. The first line of defense is suppression of damage by quenching of excess excitation as heat. If this defense is not sufficient and toxic photoproducts form, a variety of scavenging systems eliminate the reactive photoproducts. If this second line of defense also fails, the photoproducts can damage the D1 protein of PSII. This damage leads to photoinhibition. The D1 protein is then excised from the PSII reaction center and degraded. A newly synthesized D1 is reinserted into the PSII reaction center to form a functional unit. (After Asada 1999.)

as **singlet oxygen** ($^1O_2^*$). The extremely reactive singlet oxygen goes on to react with and damage many cellular components, especially lipids. Carotenoids exert their photoprotective action by rapidly quenching the excited state of chlorophyll. The excited state of carotenoids does not have sufficient energy to form singlet oxygen, so it decays back to its ground state while losing its energy as heat.

Mutant organisms that lack carotenoids cannot live in the presence of both light and molecular oxygen—a rather difficult situation for an O_2-evolving photosynthetic organism. Mutants of non-O_2-evolving photosynthetic bacteria that lack carotenoids can be maintained under laboratory conditions if oxygen is excluded from the growth medium.

Some xanthophylls also participate in energy dissipation

Nonphotochemical quenching, a major process regulating the delivery of excitation energy to the reaction center, can be thought of as a "volume knob" that adjusts the flow of excitations to the PSII reaction center to a manageable level, depending on the light intensity and other conditions. The process appears to be an essential part of the regulation of antenna systems in most algae and plants.

Nonphotochemical quenching is the quenching of chlorophyll fluorescence (see Figure 7.5) by processes other than photochemistry. As a result of nonphotochemical quenching, a large fraction of the excitations in the antenna system caused by intense illumination are quenched by conversion into heat. Nonphotochemical quenching is thought to be involved in protecting the photosynthetic machinery against overexcitation and subsequent damage.

The molecular mechanism of nonphotochemical quenching is not well understood, and evidence suggests that there are several distinct quenching processes that may have different underlying mechanisms. It is clear that the pH of the thylakoid lumen and the state of aggregation of the antenna complexes are important factors. Three carotenoids, called **xanthophylls**, are involved in nonphotochemical quenching: violaxanthin, antheraxanthin, and zeaxanthin (**Figure 7.33**).

In high light, violaxanthin is converted into zeaxanthin, via the intermediate antheraxanthin, by the enzyme violaxanthin de-epoxidase. When light intensity decreases, the process is reversed. Binding of protons and zeaxanthin to light-harvesting antenna proteins is thought to cause conformational changes that lead to quenching and heat dissipation.

Nonphotochemical quenching appears to be preferentially associated with a peripheral antenna complex of PSII, the PsbS protein. Recent evidence suggests that a transient electron transfer process may be an important part of the molecular quenching mechanism, although other molecular explanations have also been proposed. This remains an active and controversial research area.

The PSII reaction center is easily damaged

Another effect that appears to be a major factor in the stability of the photosynthetic apparatus is photoinhibition, which occurs when excess excitation arriving at the PSII reaction center leads to its inactivation and damage. **Photoinhibition** is a complex set of molecular processes defined as the inhibition of photosynthesis by excess light.

As we will discuss in detail in Chapter 9, photoinhibition is reversible in early stages. Prolonged inhibition, however, results in damage to the system such that the PSII reaction center must be disassembled and repaired. The main target of this damage is the D1 protein that makes up part of the PSII reaction center complex (see Figure 7.22). When D1 is damaged by excess light, it must be removed from

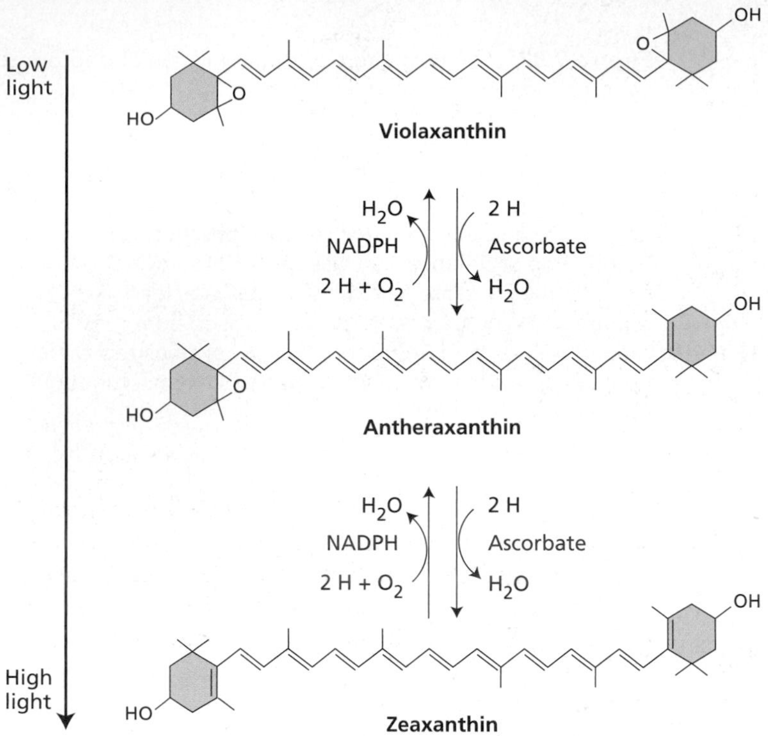

Figure 7.33 Chemical structure of violaxanthin, antheraxanthin, and zeaxanthin. The highly quenched state of PSII is associated with zeaxanthin, the unquenched state with violaxanthin. Enzymes interconvert these two carotenoids, with antheraxanthin as the intermediate, in response to changing conditions, especially changes in light intensity. Zeaxanthin formation uses ascorbate as a cofactor, and violaxanthin formation requires NADPH.

the membrane and replaced with a newly synthesized molecule. The other components of the PSII reaction center are not damaged by excess excitation and are thought to be recycled, so the D1 protein is the only component that needs to be synthesized (see Figure 7.32).

PSI is protected from active oxygen species

PSI is particularly vulnerable to damage from reactive oxygen species. The ferredoxin acceptor of PSI is a very strong reductant that can easily reduce molecular oxygen to form superoxide (O_2^-). This reduction competes with the normal channeling of electrons to the reduction of $NADP^+$ and other processes. Superoxide is one of a series of reactive oxygen species that can be very damaging to biological membranes, but when formed in this way it can be eliminated by the action of a series of enzymes, including superoxide dismutase and ascorbate peroxidase.

Thylakoid stacking permits energy partitioning between the photosystems

The fact that photosynthesis in higher plants is driven by two photosystems with different light-absorbing properties poses a special problem. If the rate of delivery of energy to PSI and PSII is not precisely matched and conditions are such that the rate of photosynthesis is limited by the available light (low light intensity), the rate of electron flow will be limited by the photosystem that is receiving less energy. In the most efficient situation, the input of energy would be the same to both photosystems. However, no single arrangement of pigments would satisfy this requirement because at different times of day the light intensity and spectral distribution tend to favor one photosystem or the other.

This problem can be solved by a mechanism that shifts energy from one photosystem to the other in response to different conditions. Such a regulating mechanism has been shown to operate under different experimental conditions. The observation that the overall quantum yield of photosynthesis is nearly independent of wavelength (see Figure 7.12) strongly suggests that such a mechanism exists.

Thylakoid membranes contain a protein kinase that can phosphorylate a specific threonine residue on the surface of LHCII, one of the membrane-bound antenna pigment proteins described earlier in the chapter (see Figure 7.18). When LHCII is not phosphorylated, it delivers more energy to PSII, and when it is phosphorylated, it delivers more energy to PSI.

The kinase is activated when plastoquinone, one of the electron carriers between PSI and PSII, accumulates in the reduced state. Reduced plastoquinone accumulates when PSII is being activated more frequently than PSI. The phosphorylated LHCII then migrates out of the stacked regions of the membrane into the unstacked regions (see Figure 7.16), probably because of repulsive interactions with negative charges on adjacent membranes.

Genetics, Assembly, and Evolution of Photosynthetic Systems

Chloroplasts have their own DNA, mRNA, and protein synthesis machinery, but most chloroplast proteins are encoded by nuclear genes and imported into the chloroplast. In this section we will consider the genetics, assembly, and evolution of the main chloroplast components.

Chloroplast genes exhibit non-Mendelian patterns of inheritance

Chloroplasts and mitochondria reproduce by division rather than by de novo synthesis. This mode of reproduction is not surprising, since these organelles contain genetic information that is not present in the nucleus. During cell division, chloroplasts are divided between the two daughter cells. In most sexual plants, however, only the maternal plant contributes chloroplasts to the zygote. In these plants the normal Mendelian pattern of inheritance does not apply to chloroplast-encoded genes, because the offspring receive chloroplasts from only one parent. The result is **non-Mendelian**, or **maternal**, **inheritance**. Numerous traits are inherited in this way; one example is the herbicide-resistance trait discussed in WEB TOPIC 7.10.

Most chloroplast proteins are imported from the cytoplasm

Chloroplast proteins can be encoded by either chloroplastic or nuclear DNA. The chloroplast-encoded proteins are synthesized on chloroplast ribosomes; the nuclear-encoded proteins are synthesized on cytoplasmic ribosomes and then transported into the chloroplast. Many nuclear genes contain introns—that is, base sequences that do not code for protein. The mRNA is processed to remove the introns, and the proteins are then synthesized in the cytosol.

The genes needed for chloroplast function are distributed in the nucleus and in the chloroplast genome with no evident pattern, but both sets are essential for the viability of the chloroplast. Some chloroplast genes are necessary for other cellular functions, such as heme and lipid synthesis. Control of the expression of the nuclear genes that code for chloroplast proteins is complex and dynamic, involving light-dependent regulation mediated by both phytochrome and blue light (see Chapter 16), as well as other factors.

The transport of chloroplast proteins that are synthesized in the cytosol is a tightly regulated process. For example, the enzyme rubisco (see Chapter 8), which functions in carbon fixation, has two types of subunits, a chloroplast-encoded large subunit and a nuclear-encoded small subunit. Small subunits of rubisco are synthesized in the cytosol and transported into the chloroplast, where the enzyme is assembled.

In this and other known cases, the nuclear-encoded chloroplast proteins are synthesized as precursor proteins containing an N-terminal amino acid sequence known as a **transit peptide**. This terminal sequence directs the precursor protein to the chloroplast, facilitates its passage through both the outer and the inner envelope membranes, and is then clipped off. The electron carrier plastocyanin is a water-soluble protein that is encoded in the nucleus but functions in the lumen of the chloroplast. It therefore must cross three membranes to reach its destination in the lumen. The transit peptide of plastocyanin is very large and is processed in more than one step as it directs the protein through two sequential translocations across the inner envelope membrane and the thylakoid membrane.

The biosynthesis and breakdown of chlorophyll are complex pathways

Chlorophylls are complex molecules exquisitely suited to the light absorption, energy transfer, and electron transfer functions that they carry out in photosynthesis (see Figure 7.6). Like all other biomolecules, chlorophylls are made by a biosynthetic pathway in which simple molecules are used as building blocks to assemble more complex molecules. Each step in the biosynthetic pathway is enzymatically catalyzed.

The chlorophyll biosynthetic pathway consists of more than a dozen steps (see WEB TOPIC 7.11). The process can be divided into several phases (**Figure 7.34**), each of which can be considered separately, but in the cell they are highly coordinated and regulated. This regulation is essential because free chlorophyll and many of the biosynthetic intermediates are damaging to cellular components. The damage results largely because chlorophylls absorb light efficiently, but in the absence of accompanying proteins, they lack a pathway for disposing of the energy, with the result that toxic singlet oxygen is formed.

The breakdown pathway of chlorophyll in senescing leaves is quite different from the biosynthetic pathway. The first step is removal of the phytol tail by an enzyme known as chlorophyllase, followed by removal of the magnesium ion by magnesium de-chelatase. Next the porphyrin structure is opened by an oxygen-dependent oxygenase enzyme to form an open-chain tetrapyrrole.

The tetrapyrrole is further modified to form water-soluble, colorless products. These colorless metabolites are then exported from the senescent chloroplast and transported to the vacuole, where they are stored. The chlorophyll metabolites are not further processed or recycled, although the proteins associated with them in the chloroplast are subsequently recycled into new proteins. The recycling of proteins is thought to be important for the nitrogen economy of the plant.

Complex photosynthetic organisms have evolved from simpler forms

The complicated photosynthetic apparatus found in plants and algae is the end product of a long evolutionary sequence. Much can be learned about this evolutionary process from analysis of simpler prokaryotic photosynthetic organisms, including the anoxygenic photosynthetic bacteria and the cyanobacteria.

The chloroplast is a semiautonomous cell organelle, with its own DNA and a complete protein synthesis apparatus. Many of the proteins that make up the photosynthetic apparatus, as well as all the chlorophylls and lipids, are synthesized in the chloroplast. Other proteins are imported from the cytoplasm and are encoded by nuclear genes. How did this curious division of labor come about? Most experts now agree that the chloroplast is the descendant of a symbiotic relationship between a cyanobacterium and a simple nonphotosynthetic eukaryotic cell. This type of relationship is called **endosymbiosis**.

Originally the cyanobacterium was capable of independent life, but over time much of its genetic information needed for normal cellular functions was lost, and

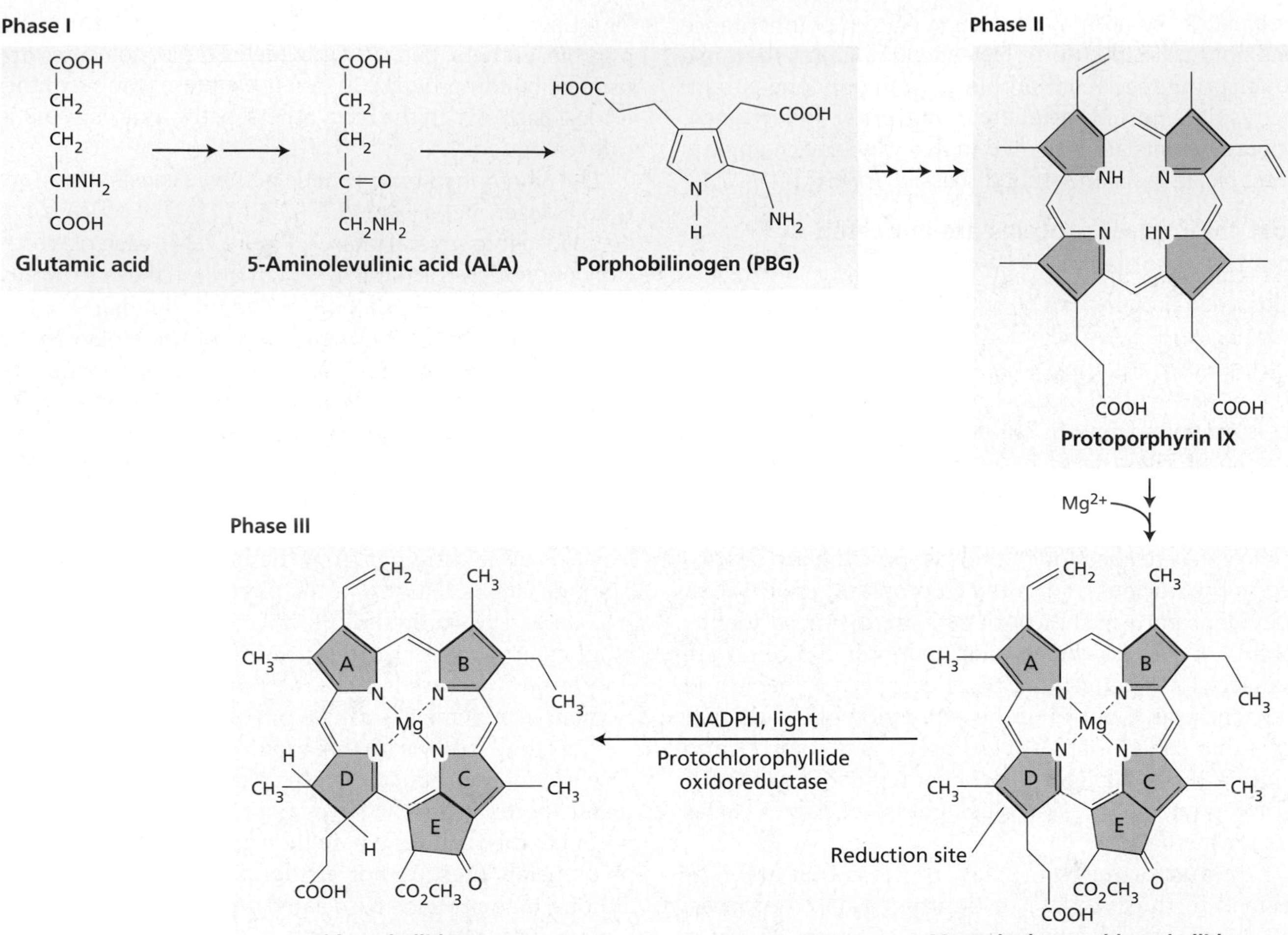

Figure 7.34 The biosynthetic pathway of chlorophyll. The pathway begins with glutamic acid, which is converted to 5-aminolevulinic acid (ALA). Two molecules of ALA are condensed to form porphobilinogen (PBG). Four PBG molecules are linked to form protoporphyrin IX. The magnesium ion is then inserted, and the light-dependent cyclization of ring E, the reduction of ring D, and the attachment of the phytol tail complete the process. Many steps in the process are omitted in this figure.

a substantial amount of information needed to synthesize the photosynthetic apparatus was transferred to the nucleus. So the chloroplast was no longer capable of life outside its host and eventually became an integral part of the cell.

In some types of algae, chloroplasts have arisen by endosymbiosis of eukaryotic photosynthetic organisms. In these organisms the chloroplast is surrounded by three (and in some cases four) membranes, which are thought to be remnants of the plasma membranes of the earlier organisms. Mitochondria are also thought to have originated by endosymbiosis in a separate event much earlier than chloroplast formation.

The answers to other questions related to the evolution of photosynthesis are less clear. These include the nature of the earliest photosynthetic systems, how the two photosystems became linked, and the evolutionary origin of the oxygen-evolving complex.

SUMMARY

Photosynthesis in plants uses light energy for the synthesis of carbohydrates and generation of oxygen from carbon dioxide and water. Energy stored in carbohydrates is used to power cellular processes in the plant and serves as energy resources for all forms of life.

Photosynthesis in Higher Plants

- Within chloroplasts, chlorophylls absorb light energy for the oxidation of water, releasing oxygen and generating NADPH and ATP (thylakoid reactions).
- NADPH and ATP are used to reduce carbon dioxide to form sugars (carbon fixation reactions).

General Concepts

- Light behaves as both a particle and a wave, delivering energy as photons, some of which are absorbed and used by plants (**Figures 7.1–7.3**).
- Light-energized chlorophyll may fluoresce, transfer energy to another molecule, or use its energy to drive chemical reactions (**Figures 7.5, 7.10**).
- All photosynthetic organisms contain a mixture of pigments with distinct structures and light-absorbing properties (**Figures 7.6, 7.7**).

Key Experiments in Understanding Photosynthesis

- An action spectrum for photosynthesis shows algal oxygen evolution at certain wavelengths (**Figures 7.8, 7.9**).
- Antenna pigment–protein complexes collect light energy and transfer it to the reaction center complexes (**Figure 7.10**).
- Light drives the reduction of $NADP^+$ and the formation of ATP. Oxygen-evolving organisms have two photosystems (PSI and PSII) that operate in series (**Figures 7.12, 7.13**).

Organization of the Photosynthetic Apparatus

- Within the chloroplast, thylakoid membranes contain the reaction centers, the light-harvesting antenna complexes, and most of the electron carrier proteins. PSI and PSII are spatially separated in thylakoids (**Figure 7.16**).

Organization of Light-Absorbing Antenna Systems

- The antenna system funnels energy to the reaction center (**Figure 7.17**).
- Light-harvesting proteins of both photosystems are structurally similar (**Figure 7.18**).

Mechanisms of Electron Transport

- The Z scheme identifies the flow of electrons through carriers in PSII and PSI from H_2O to $NADP^+$ (**Figures 7.13, 7.19**).
- Four large protein complexes transfer electrons: PSII, the cytochrome b_6f complex, PSI, and ATP synthase (**Figures 7.16, 7.20**).
- PSI reaction center chlorophyll has maximum absorption at 700 nm; PSII reaction center chlorophyll absorbs maximally at 680 nm.
- The PSII reaction center is a multi-subunit protein–pigment complex (**Figures 7.22, 7.23**).
- Manganese ions are required to oxidize water.
- Two hydrophobic plastoquinones accept electrons from PSII (**Figures 7.20, 7.24**).
- Protons are transported into the thylakoid lumen when electrons pass through the cytochrome b_6f complex (**Figures 7.20, 7.25**).
- Plastoquinone and plastocyanin carry electrons between PSII and PSI (**Figure 7.26**).
- $NADP^+$ is reduced by the PSI reaction center, using three Fe–S centers and ferredoxin as electron carriers (**Figure 7.27**).
- Cyclic electron flow generates ATP, but no NADPH, by proton pumping.
- Herbicides may block photosynthetic electron flow (**Figure 7.28**).

Proton Transport and ATP Synthesis in the Chloroplast

- In vitro transfer of pH 4–equilibrated chloroplast thylakoids to a pH 8 buffer resulted in the formation of ATP from ADP and P_i, supporting the chemiosmotic hypothesis (**Figure 7.29**).
- Protons move down an electrochemical gradient (proton motive force), passing through an ATP synthase and forming ATP (**Figure 7.30**).
- During catalysis, the CF_o portion of the ATP synthase rotates like a miniature motor.
- Proton translocation in chloroplasts, mitochondria, and purple bacteria shows significant similarities (**Figure 7.31**).

Repair and Regulation of the Photosynthetic Machinery

- Protection and repair of photodamage consists of quenching and heat dissipation, neutralizing toxic photoproducts, and synthetic repair of PSII (**Figure 7.32**).
- Xanthophylls (carotenoids) participate in nonphotochemical quenching (**Figure 7.33**).
- Kinase-mediated phosphorylation of LHCII causes its migration to stacked thylakoids and its delivery of energy to PSI. Upon dephosphorylation, LHCII migrates to unstacked thylakoids and delivers more energy to PSII.

Genetics, Assembly, and Evolution of Photosynthetic Systems

- Chloroplasts have their own DNA, mRNA, and protein synthesis system. They import most proteins into the chloroplast. These proteins are encoded by nuclear genes and synthesized in the cytosol.
- Chloroplasts show a non-Mendelian, maternal pattern of inheritance.
- Chlorophyll biosynthesis can be divided into four phases (**Figure 7.34**).
- The chloroplast is descended from a symbiotic relationship between a cyanobacterium and a simple nonphotosyntheic eukaryotic cell.

WEB MATERIAL

- **WEB TOPIC 7.1 Principles of Spectrophotometry** Spectroscopy is a key technique for the study of light reactions.
- **WEB TOPIC 7.2 The Distribution of Chlorophylls and Other Photosynthetic Pigments** The content of chlorophylls and other photosynthetic pigments varies among plant kingdoms.
- **WEB TOPIC 7.3 Quantum Yield** Quantum yields measure how effectively light drives a photobiological process.
- **WEB TOPIC 7.4 Antagonistic Effects of Light on Cytochrome Oxidation** Photosystems I and II were discovered in some ingenious experiments.
- **WEB TOPIC 7.5 Structures of Two Bacterial Reaction Centers** X-ray diffraction studies resolved the atomic structure of the reaction center of PSII.
- **WEB TOPIC 7.6 Midpoint Potentials and Redox Reactions** The measurement of midpoint potentials is useful for analyzing electron flow through PSII.
- **WEB TOPIC 7.7 Oxygen Evolution** The S state mechanism is a model for water splitting in PSII.
- **WEB TOPIC 7.8 Photosystem I** The PSI reaction is a multiprotein complex.
- **WEB TOPIC 7.9 ATP Synthase** The ATP synthase functions as a molecular motor.
- **WEB TOPIC 7.10 Mode of Action of Some Herbicides** Some herbicides kill plants by blocking photosynthetic electron flow.
- **WEB TOPIC 7.11 Chlorophyll Biosynthesis** Chlorophyll and heme share early steps of their biosynthetic pathways.

- **WEB ESSAY 7.1 A Novel View of Chloroplast Structure** Stromules extend the reach of the chloroplasts.

available at plantphys.net

Suggested Reading

Blankenship, R. E. (2014) *Molecular Mechanisms of Photosynthesis 2nd Ed.*, Wiley-Blackwell, Oxford, UK.

Blankenship, R. E., Madigan, M. T., and Bauer, C. E., eds. (1995) *Anoxygenic Photosynthetic Bacteria* (*Advances in Photosynthesis*, vol. 2). Kluwer, Dordrecht, Netherlands.

Cramer, W. A., and Knaff, D. B. (1990) *Energy Transduction in Biological Membranes: A Textbook of Bioenergetics.* Springer, New York.

Frank, H. A., Young, A. J., Britton, G., and Cogdell, R. J. (1999) *The Photochemistry of Carotenoids,* (*Advances in Photosynthesis*, vol. 8). Kluwer, Dordrecht, Netherlands.

Hohmann-Marriott, M. F. and Blankenship, R. E. (2011) Evolution of photosynthesis. *Annu. Rev. Plant Biol.* 62: 515–548.

Ke, B. (2001) *Photosynthesis Photobiochemistry and Photobiophysics* (*Advances in Photosynthesis*, vol. 10). Kluwer, Dordrecht, Netherlands.

Nicholls, D. G., and Ferguson, S. J. (2013) *Bioenergetics*, 4th ed. Academic Press, San Diego.

Ort, D. R., and Yocum, C. F., eds. (1996) *Oxygenic Photosynthesis: The Light Reactions* (*Advances in Photosynthesis*, vol. 4). Kluwer, Dordrecht, Netherlands.

Scheer, H. (1991) *Chlorophylls*. CRC Press, Boca Raton, FL.

Walker, D. (1992) *Energy, Plants and Man*, 2nd ed. Oxygraphics, Brighton, East Sussex, England.

Zhu, X.-G., Long, S. P. and Ort, D. R. (2010) Improving photosynthetic efficiency for greater yield. *Ann. Rev. Plant Biol.* 61: 235–261.

8 Photosynthesis: The Carbon Reactions

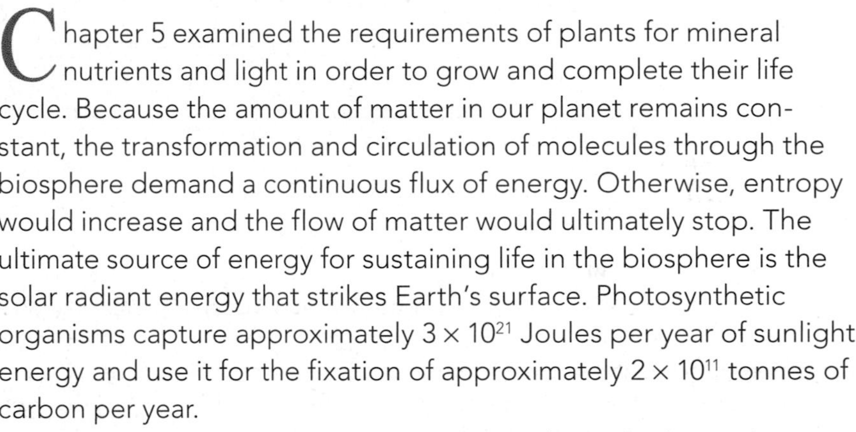

Chapter 5 examined the requirements of plants for mineral nutrients and light in order to grow and complete their life cycle. Because the amount of matter in our planet remains constant, the transformation and circulation of molecules through the biosphere demand a continuous flux of energy. Otherwise, entropy would increase and the flow of matter would ultimately stop. The ultimate source of energy for sustaining life in the biosphere is the solar radiant energy that strikes Earth's surface. Photosynthetic organisms capture approximately 3×10^{21} Joules per year of sunlight energy and use it for the fixation of approximately 2×10^{11} tonnes of carbon per year.

More than 1 billion years ago, heterotrophic cells dependent on abiotically produced organic molecules acquired the ability to convert sunlight into chemical energy through primary endosymbiosis with an ancient cyanobacterium. Comparisons of the amino acid sequences of proteins from plastids, cyanobacteria, and eukaryotes have led us to group the progeny of this ancient event under the denomination of Archaeplastidae, which comprises three major lineages: Chloroplastidae (Viridiplantae: green algae, land plants), Rhodophyceae (red algae), and Glaucophytae (unicellular algae containing cyanobacteria-like plastids called cyanelles). The genetic integration of the cyanobacterium with its host reduced some functions by loss of genes and established a complex mechanism in the outer and inner membrane for targeting (1) nucleus-encoded proteins to the endosymbiont and (2) plastid-encoded proteins to the host. The endosymbiotic events entailed the gain of new metabolic pathways. The ancestral endosymbiont conveyed the ability not only to carry out oxygenic photosynthesis, but also to synthesize novel compounds, such as *starch*.

In Chapter 7 we learned how the energy associated with the photochemical oxidation of water to molecular oxygen at thylakoid membranes generates ATP, reduced ferredoxin, and NADPH. Subsequently, the products of the light reactions, ATP and NADPH, flow from thylakoid membranes to the surrounding fluid phase (stroma) and drive the enzyme-catalyzed reduction of atmospheric

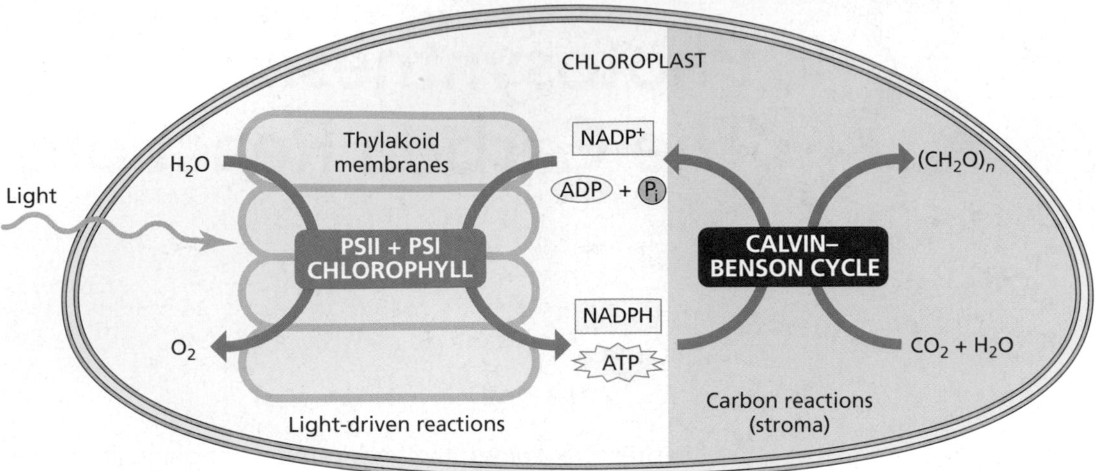

Figure 8.1 Light and carbon reactions of photosynthesis in chloroplasts of land plants. In thylakoid membranes, the excitation of chlorophyll in the photosynthetic electron transport system [photosystem II (PSII) + photosystem I (PSI)] by light elicits the formation of ATP and NADPH (see Chapter 7). In the stroma, both ATP and NADPH are consumed by the Calvin–Benson cycle in a series of enzyme-driven reactions that reduce the atmospheric CO_2 to carbohydrates (triose phosphates).

CO_2 to carbohydrates and other cell components (**Figure 8.1**). The latter reactions in the stroma of chloroplasts were long thought to be independent of light and, as a consequence, were for many years referred to as the *dark reactions*. However, these stroma-localized reactions are more properly referred to as the *carbon reactions of photosynthesis* because products of the photochemical processes not only provide substrates for enzymes but also control the catalytic rate.

In the beginning of this chapter we will analyze the metabolic cycle that incorporates atmospheric CO_2 into organic compounds appropriate for life: the **Calvin–Benson cycle**. Next we will consider how the unavoidable phenomenon of photorespiration releases part of the assimilated CO_2. Because a side reaction with molecular oxygen decreases the efficiency of photosynthetic CO_2 assimilation, we will also examine biochemical mechanisms for mitigating the loss of CO_2: CO_2 pumps (see **WEB TOPIC 8.1**), C_4 metabolism, and crassulacean acid metabolism. Finally, we will consider the formation of the two major products of the photosynthetic CO_2 fixation: **starch**, the reserve polysaccharide that accumulates transiently in chloroplasts; and **sucrose**, the disaccharide that is exported from leaves to developing and storage organs of the plant.

The Calvin–Benson Cycle

A requisite for maintaining life in the biosphere is the fixation of atmospheric CO_2 into skeletons of organic compounds that are compatible with the needs of the cell. These endergonic transformations are driven by the energy coming from physical and chemical sources. The predominant pathway of autotrophic CO_2 fixation is the Calvin–Benson cycle, which is found in many prokaryotes and in all photosynthetic eukaryotes, from the most primitive algae to the most advanced angiosperms. This pathway decreases the oxidation state of carbon from the highest value, found in CO_2 (+4), to levels found in sugars (e.g., +2 in keto groups —CO—; 0 in secondary alcohols —CHOH—). In view of its notable capacity for lowering the carbon oxidation state, the Calvin–Benson cycle is also aptly named the *reductive pentose phosphate cycle* and *photosynthetic carbon reduction cycle*. In this section we will examine how CO_2 is fixed via the Calvin–Benson cycle through the use of ATP and NADPH that are generated by the light reactions (see Figure 8.1), and how the cycle is regulated.

The Calvin–Benson cycle has three phases: carboxylation, reduction, and regeneration

In the 1950s, a series of ingenious experiments carried out by M. Calvin, A. Benson, J. A. Bassham, and their colleagues provided convincing evidence for the Calvin–Benson cycle (see **WEB TOPIC 8.2**). The Calvin–Benson cycle proceeds in three phases that are highly coordinated in the chloroplast (**Figure 8.2**):

1. *Carboxylation* of the CO_2 acceptor molecule. The first committed enzymatic step in the cycle is the reaction of CO_2 and water with a five-carbon acceptor molecule (ribulose 1,5-bisphosphate) to generate two molecules of a three-carbon intermediate (3-phosphoglycerate).
2. *Reduction* of 3-phosphoglycerate. The 3-phosphoglycerate is converted to three-carbon carbohydrates (triose phosphates) by enzymatic reactions driven by photochemically generated ATP and NADPH.

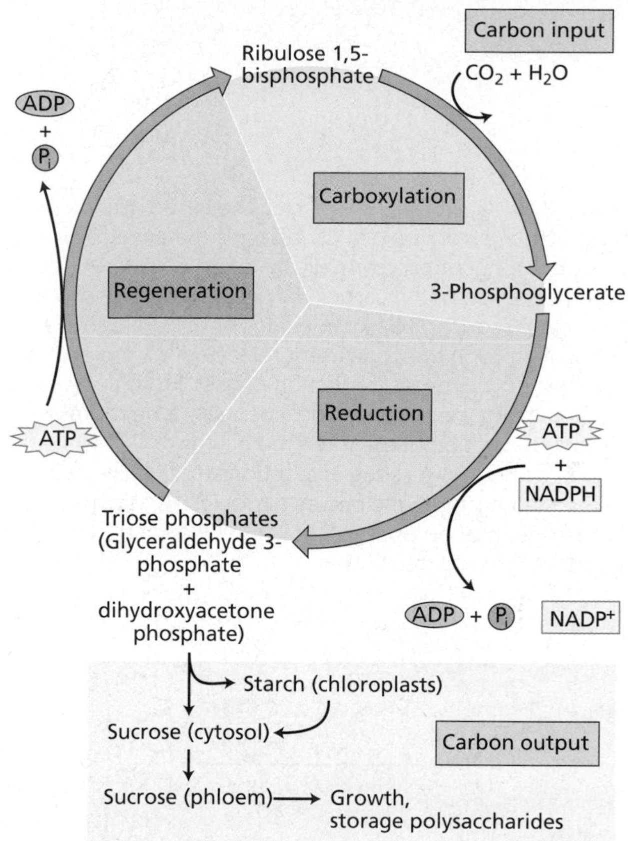

Figure 8.2 The Calvin–Benson cycle proceeds in three phases: (1) *carboxylation*, which covalently links atmospheric carbon (CO_2) to a carbon skeleton; (2) *reduction*, which forms a carbohydrate (triose phosphate) at the expense of photochemically generated ATP and reducing equivalents in the form of NADPH; and (3) *regeneration*, which restores the CO_2 acceptor ribulose 1,5-bisphosphate. At steady state, the input of CO_2 equals the output of triose phosphates. The latter either serve as precursors of starch biosynthesis in the chloroplast or flow to the cytosol for sucrose biosynthesis and other metabolic reactions. Sucrose is loaded into the phloem sap and used for growth or polysaccharide biosynthesis in other parts of the plant.

3. *Regeneration* of the CO_2 acceptor ribulose 1,5-bisphosphate. The cycle is completed by regeneration of ribulose 1,5-bisphosphate through a series of ten enzyme-catalyzed reactions, one requiring ATP.

The carbon output as triose phosphates balances the carbon input provided by atmospheric CO_2. Triose phosphates generated by the Calvin–Benson cycle are converted to starch in the chloroplast or exported to the cytosol for the formation of sucrose. Sucrose is transported in

TABLE 8.1 Reactions of the Calvin–Benson cycle

Enzyme	Reaction
1. Ribulose 1,5-bisphosphate carboxylase/oxygenase (rubisco)	Ribulose 1,5-bisphosphate + CO_2 + H_2O → 2 3-phosphoglycerate
2. 3-Phosphoglycerate kinase	3-Phosphoglycerate + ATP → 1,3-bisphosphoglycerate + ADP
3. NADP–glyceraldehyde-3-phosphate dehydrogenase	1,3-Bisphosphoglycerate + NADPH + H^+ → glyceraldehyde 3-phosphate + $NADP^+$ + P_i
4. Triose phosphate isomerase	Glyceraldehyde 3-phosphate → dihydroxyacetone phosphate
5. Aldolase	Glyceraldehyde 3-phosphate + dihydroxyacetone phosphate → fructose 1,6-bisphosphate
6. Fructose 1,6-bisphosphatase	Fructose 1,6-bisphosphate + H_2O → fructose 6-phosphate + P_i
7. Transketolase	Fructose 6-phosphate + glyceraldehyde 3-phosphate → erythrose 4-phosphate + xylulose 5-phosphate
8. Aldolase	Erythrose 4-phosphate + dihydroxyacetone phosphate → sedoheptulose 1,7-bisphosphate
9. Sedoheptulose 1,7-bisphosphatase	Sedoheptulose 1,7-bisphosphate + H_2O → sedoheptulose 7-phosphate + P_i
10. Transketolase	Sedoheptulose 7-phosphate + glyceraldehyde 3-phosphate → ribose 5-phosphate + xylulose 5-phosphate
11a. Ribulose 5-phosphate epimerase	Xylulose 5-phosphate → ribulose 5-phosphate
11b. Ribose 5-phosphate isomerase	Ribose 5-phosphate → ribulose 5-phosphate
12. Phosphoribulokinase (ribulose 5-phosphate kinase)	Ribulose 5-phosphate + ATP → ribulose 1,5-bisphosphate + ADP + H^+

Note: P_i stands for inorganic phosphate.

the phloem to heterotrophic plant organs for sustaining growth and the synthesis of storage products.

The fixation of CO_2 via carboxylation of ribulose 1,5-bisphosphate and the reduction of the product 3-phosphoglycerate yield triose phosphates

In the carboxylation step of the Calvin–Benson cycle, one molecule of CO_2 and one molecule of H_2O react with one molecule of ribulose 1,5-bisphosphate to yield two molecules of 3-phosphoglycerate (**Figure 8.3** and **Table 8.1**, reaction 1). This reaction is catalyzed by the chloroplast enzyme ribulose 1,5-bisphosphate carboxylase/oxygenase, referred to as **rubisco** (see WEB TOPIC 8.3). In the first partial reaction, a H^+ is extracted from carbon 3 of ribulose

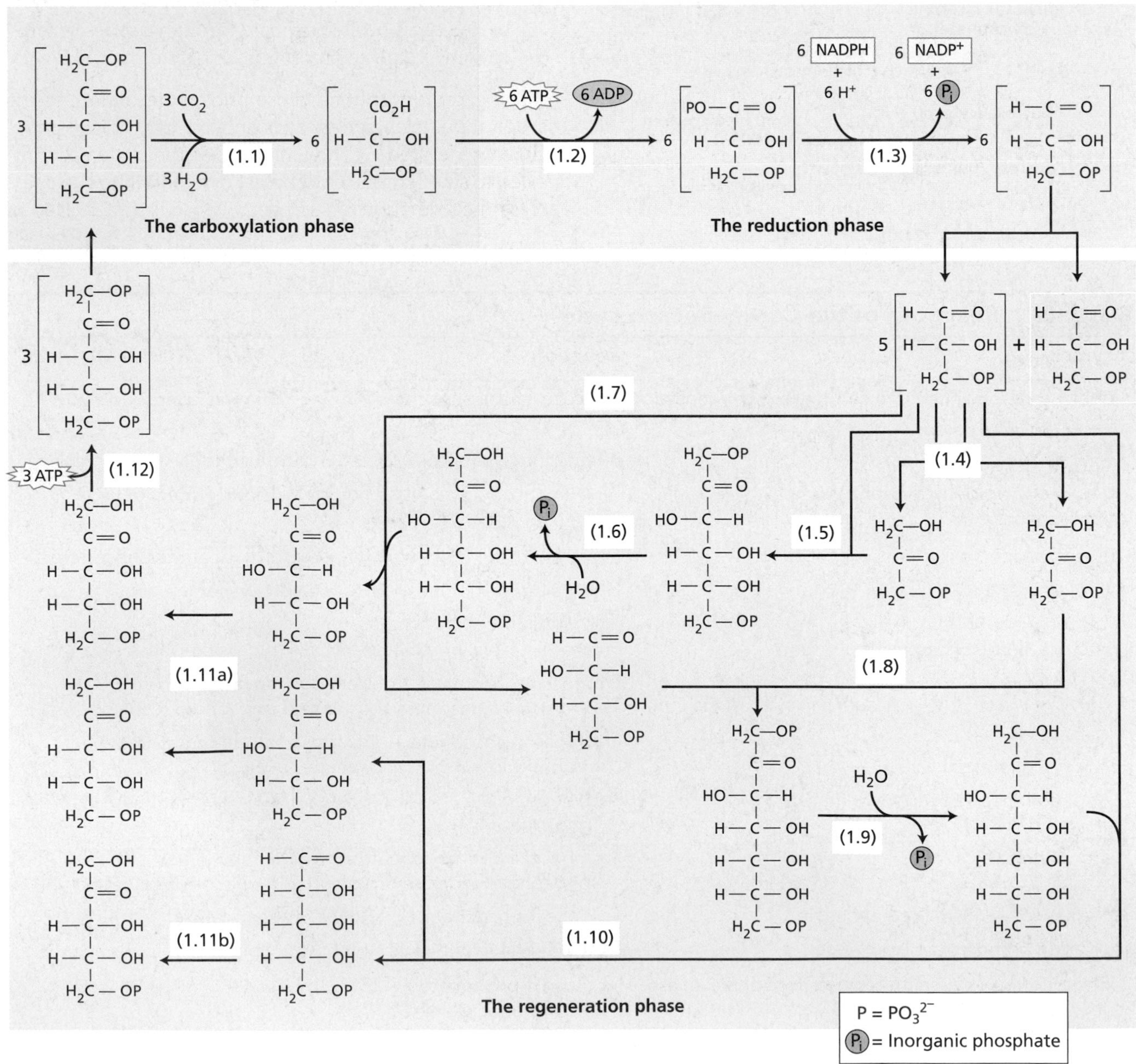

Figure 8.3 Calvin–Benson cycle. The carboxylation of three molecules of ribulose 1,5-bisphosphate yields six molecules of 3-phosphoglycerate (carboxylation phase). After phosphorylation of the carboxylic group, 1,3-bisphosphoglycerate is reduced to six molecules of glyceraldehyde-3-phosphate with the concurrent release of six molecules of inorganic phosphate (reduction phase). From the total six molecules of glyceraldehyde 3-phosphate, one represents the net assimilation of the three molecules of CO_2 while the other five undergo a series of reactions that finally regenerate the starting three molecules of ribulose 1,5-bisphosphate (regeneration phase). See Table 8.1 for a description of each numbered reaction.

1,5-bisphosphate (**Figure 8.4**). The addition of CO_2 to the unstable rubisco-bound enediol intermediate drives the second partial reaction to the irreversible formation of 2-carboxy-3-ketoarabinitol 1,5-bisphosphate. The hydration of 2-carboxy-3-ketoarabinitol 1,5-bisphosphate yields two molecules of 3-phosphoglycerate.

In the reduction phase of the Calvin–Benson cycle, two successive reactions reduce the carbon of the 3-phosphoglycerate produced by the carboxylation phase (see Figure 8.3 and Table 8.1, reactions 2 and 3):

1. First, ATP formed by the light reactions phosphorylates 3-phosphoglycerate at the carboxyl group, yielding a mixed anhydride, 1,3-bisphosphoglycerate, in a reaction catalyzed by 3-phosphoglycerate kinase.
2. Next, NADPH, also generated by the light reactions, reduces 1,3-bisphosphoglycerate to glyceraldehyde 3-phosphate, in a reaction catalyzed by the chloroplast enzyme NADP–glyceraldehyde-3-phosphate dehydrogenase.

The operation of three carboxylation and reduction phases yields six molecules of glyceraldehyde 3-phosphate (6 molecules × 3 carbons/molecule = 18 carbons total) when three molecules of ribulose 1,5-bisphosphate (3 molecules × 5 carbons/molecule = 15 carbons total) react with three molecules of CO_2 (3 carbons total) and the six molecules of 3-phosphoglycerate are reduced (see Figure 8.3).

The regeneration of ribulose 1,5-bisphosphate ensures the continuous assimilation of CO_2

In the regeneration phase, the Calvin–Benson cycle facilitates the continuous uptake of atmospheric CO_2 by restoring the CO_2 acceptor ribulose 1,5-bisphosphate. To this end, three molecules of ribulose 1,5-bisphosphate (3 mol-

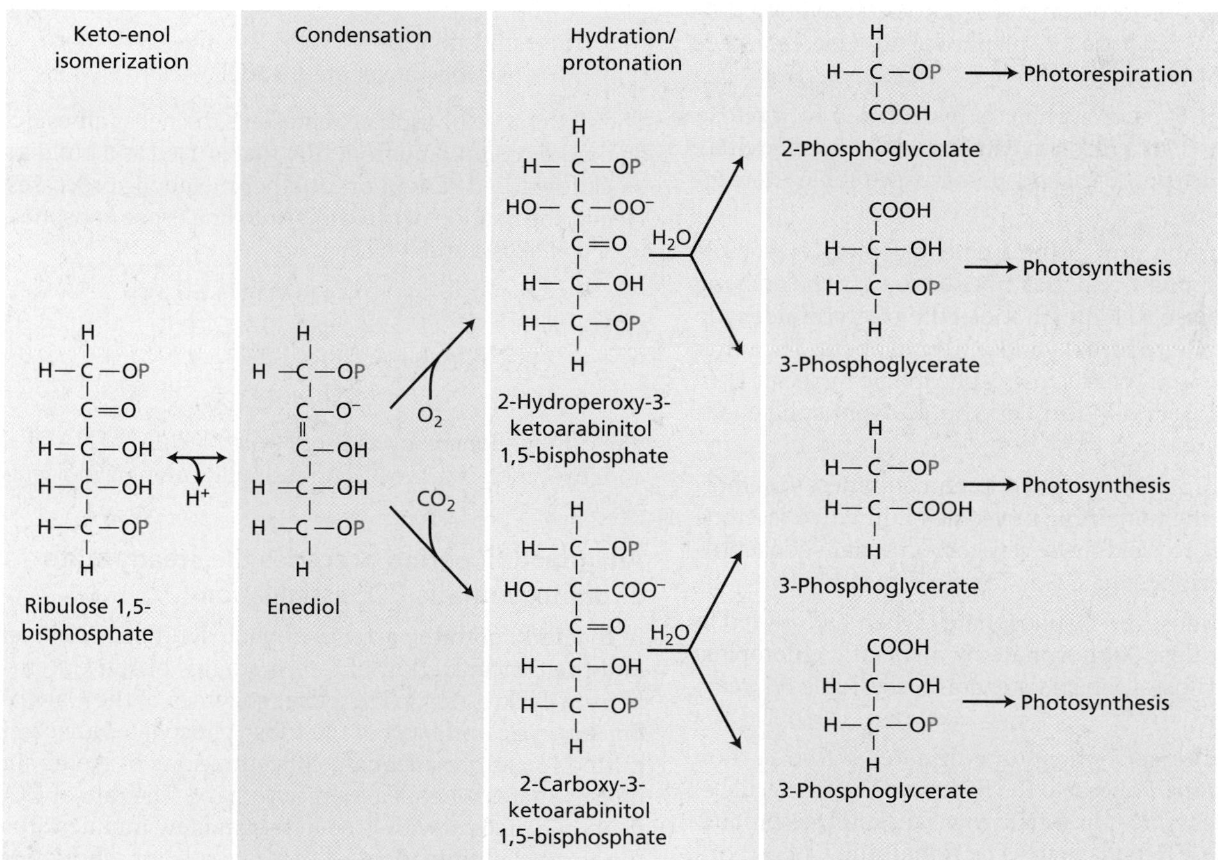

Figure 8.4 Carboxylation and oxygenation of ribulose 1,5-bisphosphate catalyzed by rubisco. The binding of ribulose 1,5-bisphosphate to rubisco facilitates the formation of an enzyme-bound enediol intermediate that can be attacked by CO_2 or O_2 at carbon 2. With CO_2, the product is a six-carbon intermediate (2-carboxy-3-ketoarabinitol 1,5-bisphosphate); with O_2, the product is a five-carbon reactive intermediate (2-hydroperoxy-3-ketoarabinitol 1,5-bisphosphate). The hydration of these intermediates at carbon 3 triggers the cleavage of the carbon–carbon bond between carbons 2 and 3, yielding two molecules of 3-phosphoglycerate (carboxylase activity) or one molecule of 2-phosphoglycolate and one molecule of 3-phosphoglyerate (oxygenase activity). The important physiological effect of the oxygenase activity is described in the section *The C_2 Oxidative Photosynthetic Carbon Cycle.*

ecules × 5 carbons/molecule = 15 carbons total) are formed by reactions that reshuffle the carbons from five molecules of glyceraldehyde 3-phosphate (5 molecules × 3 carbons/molecule = 15 carbons) (see Figure 8.3). The sixth molecule of glyceraldehyde 3-phosphate (1 molecule × 3 carbons/molecule = 3 carbons total) represents the net assimilation of three molecules of CO_2 and becomes available for the carbon metabolism of the plant. The reshuffling of the other five molecules of glyceraldehyde 3-phosphate to yield three molecules of ribulose 1,5-bisphosphate proceeds through reactions 4 to 12 in Table 8.1 and Figure 8.3:

- Two molecules of glyceraldehyde 3-phosphate are converted to dihydroxyacetone phosphate in the reaction catalyzed by triose phosphate isomerase (see Table 8.1, reaction 4). Glyceraldehyde 3-phosphate and dihydroxyacetone phosphate are collectively designated *triose phosphates*.
- One molecule of dihydroxyacetone phosphate undergoes aldol condensation with a third molecule of glyceraldehyde 3-phosphate, a reaction catalyzed by aldolase, to give fructose 1,6-bisphosphate (see Table 8.1, reaction 5).
- Fructose 1,6-bisphosphate is hydrolyzed to fructose 6-phosphate in a reaction catalyzed by a specific chloroplast fructose 1,6-bisphosphatase (see Table 8.1, reaction 6).
- A two-carbon unit of the fructose 6-phosphate molecule (carbons 1 and 2) is transferred via the enzyme transketolase to a fourth molecule of glyceraldehyde 3-phosphate to form xylulose 5-phosphate. The other four carbons of the fructose 6-phosphate molecule (carbons 3, 4, 5, and 6) form erythrose 4-phosphate (see Table 8.1, reaction 7).
- The erythrose 4-phosphate then combines, via aldolase, with the remaining molecule of dihydroxyacetone phosphate to yield the seven-carbon sugar sedoheptulose 1,7-bisphosphate (see Table 8.1, reaction 8).
- Sedoheptulose 1,7-bisphosphate is then hydrolyzed to sedoheptulose 7-phosphate by a specific chloroplast sedoheptulose 1,7-bisphosphatase (see Table 8.1, reaction 9).
- Sedoheptulose 7-phosphate donates a two-carbon unit (carbons 1 and 2) to the fifth (and last) molecule of glyceraldehyde 3-phosphate, via transketolase, producing xylulose 5-phosphate. The remaining five carbons (carbons 3–7) of sedoheptulose 7-phosphate molecule become ribose 5-phosphate (see Table 8.1, reaction 10).
- Two molecules of xylulose 5-phosphate are converted to two molecules of ribulose 5-phosphate by a ribulose 5-phosphate epimerase (see Table 8.1, reaction 11a), while a third molecule of ribulose 5-phosphate originates from ribose 5-phosphate by the action of ribose 5-phosphate isomerase (see Table 8.1, reaction 11b).
- Finally, phosphoribulokinase (also called ribulose 5-phosphate kinase) catalyzes the phosphorylation of three molecules of ribulose 5-phosphate with ATP, thus regenerating the three molecules of ribulose 1,5-bisphosphate needed for restarting the cycle (see Table 8.1, reaction 12).

In summary, triose phosphates are formed in the carboxylation and reduction phases of the Calvin–Benson cycle using energy (ATP) and reducing equivalents (NADPH) generated by the illuminated photosystems of chloroplast thylakoid membranes:

$$3\ CO_2 + 3\ \text{ribulose 1,5-bisphosphate} + 3\ H_2O + 6\ \text{NADPH} + 6\ H^+ + 6\ \text{ATP}$$
$$\downarrow$$
$$6\ \text{Triose phosphates} + 6\ \text{NADP}^+ + 6\ \text{ADP} + 6\ P_i$$

From these six triose phosphates, five are used in the regeneration phase that restores the CO_2 acceptor (ribulose 1,5-bisphosphate) for the continuous functioning of the Calvin–Benson cycle:

$$5\ \text{Triose phosphates} + 3\ \text{ATP} + 2\ H_2O \rightarrow 3\ \text{ribulose 1,5-bisphosphate} + 3\ \text{ADP} + 2\ P_i$$

The sixth triose phosphate represents the net synthesis of an organic compound from CO_2 that is used as a building block for stored carbon or for other metabolic processes. Hence, the fixation of three CO_2 into one triose phosphate uses 6 NADPH and 9 ATP:

$$3\ CO_2 + 5\ H_2O + 6\ \text{NADPH} + 9\ \text{ATP}$$
$$\downarrow$$
$$\text{Glyceraldehyde 3-phosphate} + 6\ \text{NADP}^+ + 9\ \text{ADP} + 8\ P_i$$

The Calvin–Benson cycle uses two molecules of NADPH and three molecules of ATP to assimilate a single molecule of CO_2.

An induction period precedes the steady state of photosynthetic CO_2 assimilation

In the dark, both the activity of photosynthetic enzymes and the concentration of intermediates of the Calvin–Benson cycle are low. Therefore, enzymes of the Calvin–Benson cycle and most of the triose phosphates are committed to restore adequate concentrations of metabolic intermediates when leaves receive light. The rate of CO_2 fixation increases with time in the first few minutes after the onset of illumination—a time lag called the **induction period**. The acceleration of the photosynthesis rate is due to both the activation of enzymes by light (discussed later in this chapter) and an increase in the concentration of intermediates of the Calvin–Benson cycle. In short, the six triose phosphates formed in the carboxylation and reduction phases of the Calvin–Benson cycle during the induction period are used mainly for the regeneration of the CO_2 acceptor ribulose 1,5-bisphosphate.

When photosynthesis reaches a steady state, five of the six triose phosphates formed contribute to the regeneration of the CO_2 acceptor ribulose 1,5-bisphosphate, while a sixth triose phosphate is used in the chloroplast for starch formation and in the cytosol for sucrose synthesis and other metabolic processes (see Figure 8.2). For a detailed analysis of the efficiency of the Calvin–Benson cycle in the use of energy, see **WEB TOPIC 8.4**.

Many mechanisms regulate the Calvin–Benson cycle

The efficient use of energy in the Calvin–Benson cycle requires the existence of specific regulatory mechanisms which ensure not only that all intermediates in the cycle are present at adequate concentrations in the light, but also that the cycle is turned off in the dark. To produce the necessary metabolites in response to environmental stimuli, chloroplasts achieve the appropriate rates of biochemical transformations through the modification of enzyme levels (μmoles of enzyme/chloroplast) and catalytic activities (μmoles of substrate converted/minute/μmole of enzyme).

Gene expression and protein biosynthesis determine the concentrations of enzymes in cell compartments. The amounts of enzymes present in the chloroplast stroma are regulated by the concerted expression of nuclear and chloroplast genomes. Nucleus-encoded enzymes are translated on 80S ribosomes in the cytosol and subsequently transported into the plastid. Plastid-encoded proteins are translated in the stroma on prokaryotic-like 70S ribosomes.

Light modulates the expression of stromal enzymes encoded by the nuclear genome via specific photoreceptors (e.g., phytochrome and blue-light receptors). However, nuclear gene expression needs to be synchronized with the expression of other components of the photosynthetic apparatus in the organelle. Most of the regulatory signaling between nucleus and plastids is anterograde—that is, the products of nuclear genes control the transcription and translation of plastid genes. Such is the case, for example, in the assembly of stromal rubisco from eight nucleus-encoded small subunits (S) and eight plastid-encoded large subunits (L). However, in some cases (e.g., the synthesis of proteins associated with chlorophyll), regulation can be retrograde—that is, the regulatory signal flows from the plastid to the nucleus.

In contrast to slow changes in catalytic rates caused by variations in the concentration of enzymes, posttranslational modifications rapidly change the specific activity of chloroplast enzymes (μmoles of substrate converted/minute/μmole of enzyme). Two general mechanisms accomplish the light-mediated modification of the kinetic properties of stromal enzymes:

1. Change in covalent bonds that result in a chemically modified enzyme, such as the carbamylation of amino groups [$Enz–NH_2 + CO_2 \leftrightarrow Enz–NH–CO_2^- + H^+$] or the reduction of disulfide bonds [$Enz–(S)_2 + Prot–(SH)_2 \leftrightarrow Enz–(SH)_2 + Prot–(S)_2$]
2. Modification of noncovalent interactions caused by changes in (1) ionic composition of the cellular milieu (e.g., pH, Mg^{2+}), (2) binding of enzyme effectors, (3) close association with regulatory proteins in supramolecular complexes, or (4) interaction with thylakoid membranes

In our further discussion of regulation, we will examine light-dependent mechanisms that regulate the specific activity of five pivotal enzymes within minutes of the light–dark transition:

- Rubisco
- Fructose 1,6-bisphosphatase
- Sedoheptulose 1,7-bisphosphatase
- Phosphoribulokinase
- NADP–glyceraldehyde-3-phosphate dehydrogenase

Rubisco-activase regulates the catalytic activity of rubisco

Most life forms in the biosphere depend on photosynthetic organisms that capture inorganic carbon from the environment through the Calvin–Benson cycle. That point withstanding, the maximum number of molecules of CO_2 that rubisco converts to products per catalytic site (*turnover rate*) is extremely slow (1–12 CO_2 fixed/s). George Lorimer and colleagues found that rubisco must be activated before acting as a catalyst. Chemical modifications, site-directed mutagenesis, molecular dynamics calculations, and high-resolution crystal structures showed that the CO_2 molecule plays a dual role in the activity of rubisco: CO_2 transforms the enzyme from an inactive to an active form (*activation*) and is also the substrate for the carboxylase reaction (*catalysis*).

The catalytic activities of rubisco—carboxylation and oxygenation—require the formation of a lysyl-carbamate (rubisco–NH_2-CO_2^-) by a molecule of CO_2 called *activator CO_2* ("see Rubisco activation" in **Figure 8.5**). The subsequent binding of Mg^{2+} to the carbamate stabilizes the carbamylated rubisco (rubisco–NH_2–CO_2^- • Mg^{2+}) and converts rubisco to a catalytically competent enzyme. Another molecule of CO_2—*substrate CO_2*—can then react with ribulose 1,5-bisphosphate at the active site of rubisco (see "Catalytic cycle" in Figure 8.5), releasing two molecules of 3-phosphoglycerate (see "Products" in Figure 8.5).

Sugar phosphates (such as xylulose 1,5-bisphosphate and the naturally occurring inhibitor 2-carboxyarabinitol 1-phosphate) and the substrate (ribulose 1,5-bisphosphate) prevent activation and inhibit catalysis by binding tightly to the uncarbamylated rubisco and to the carbamylated rubisco, respectively. Plants and green algae overcome this inhibition with the protein rubisco-activase,

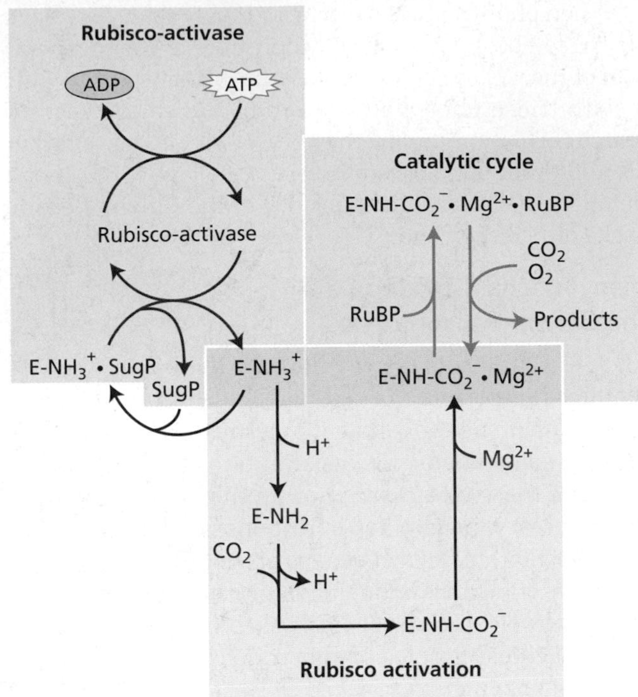

Figure 8.5 CO_2 functions both as an activator and as a substrate in the reaction catalyzed by rubisco. *Activation*: The reaction of activator CO_2 with rubisco (E) causes formation of the E–carbamate adduct (E–NH–CO_2^-), whose stabilization by Mg^{2+} yields the E–carbamate adduct (E–NH–CO_2^- • Mg^{2+}) at the enzyme's active site (Rubisco activation, bottom panel). In the stroma of illuminated chloroplasts, the increase in both pH (lower H^+ concentration) and concentration of Mg^{2+} facilitates the formation of the (E–NH–CO_2^- • Mg^{2+}) complex, which represents the catalytically active form of rubisco. The tight binding of sugar phosphates (SugP), such as ribulose 1,5-bisphosphate (RuBP), either impedes the production of the E–carbamate adduct or blocks the binding of substrates to the carbamylated enzyme. In the rubisco-activase–mediated cycle (Rubisco-activase, left panel), the hydrolysis of ATP by rubisco-activase elicits a conformational change of rubisco that reduces its binding affinity for sugar phosphates. *Catalysis*: Upon the formation of the complex (E–NH–CO_2^- • Mg^{2+}) at the active site of the enzyme, rubisco combines with ribulose 1,5-bisphosphate and subsequently with either the substrate CO_2 or O_2, initiating the carboxylase or oxygenase activities, respectively (see Figure 8.4) (Catalytic cycle, right panel). *Products*: The products of the catalytic cycle are either two molecules of 3-phosphoglycerate (carboxylase activity) or one molecule each of 3-phosphoglycerate and 2-phosphoglycolate (oxygenase activity).

which removes the sugar phosphates from the uncarbamylated and carbamylated rubisco, thus allowing rubisco to be activated through carbamylation and Mg^{2+} binding ("Rubisco-activase" in Figure 8.5; also see **WEB TOPIC 8.5**). Rubisco-activase requires the hydrolysis of ATP to release the tightly bound inhibitors.

RUBISCO-ACTIVASE In many plant species, the alternative splicing of a unique pre-mRNA yields two identical rubisco-activases that differ only at the carboxyl terminus: the long-form α (46 kDa) and the short-form β (42 kDa). The C-extension of the long-form α carries two cysteines that modulate the sensitivity of ATPase activity to the ATP:ADP ratio by thiol–disulphide exchange. In this way, regulation of rubisco-activase is linked to light via the ferredoxin–thioredoxin system described in the next section. However, other components still unknown may be involved because light also stimulates the activity of rubisco in species that naturally produce only the short-form β lacking the regulatory cysteines (e.g., tobacco).

Light regulates the Calvin–Benson cycle via the ferredoxin–thioredoxin system

Light regulates the catalytic activity of four enzymes of the Calvin–Benson cycle directly via the **ferredoxin–thioredoxin system**. This mechanism uses ferredoxin reduced by the photosynthetic electron transport chain together with two chloroplast proteins (ferredoxin–thioredoxin reductase and thioredoxin) to regulate fructose 1,6-bisphosphatase, sedoheptulose 1,7-bisphosphatase, phosphoribulokinase, and NADP–glyceraldehyde-3-phosphate dehydrogenase (**Figure 8.6**).

Light transfers electrons from water to ferredoxin via the photosynthetic electron transport system (see Chapter 7). Reduced ferredoxin converts the disulfide bond of the regulatory protein thioredoxin (—S—S—) to the reduced state (—SH HS—) with the iron–sulfur enzyme ferredoxin–thioredoxin reductase. Subsequently, reduced thioredoxin cleaves a specific disulfide bridge (oxidized cysteines) of the target enzyme, forming free (reduced) cysteines. The cleavage of the enzyme disulfide bonds causes a conformational change that increases catalytic activity (see Figure 8.6 and **WEB TOPIC 8.6**). Deactivation of thioredoxin-activated enzymes takes place when darkness relieves the "electron pressure" from photosynthetic electron transport. However, details of the deactivation process are unknown.

Advances in structural studies and bioinformatics led to recognition that enzymes regulated by thioredoxin do not exhibit a cysteine-containing consensus sequence. The target enzymes may carry the regulatory cysteines in the core of the polypeptide (fructose 1,6-bisphosphatase: —Cys155—Cys174—), at the C terminus (glyceraldehyde-3-phosphate dehydrogenase: —Cys349—Cys358—), or in the active site (phosphoribulokinase: —Cys16—Cys55—).

Proteomic studies have shown that the ferredoxin–thioredoxin system regulates enzymes functional in numerous chloroplast processes other than carbon fixa-

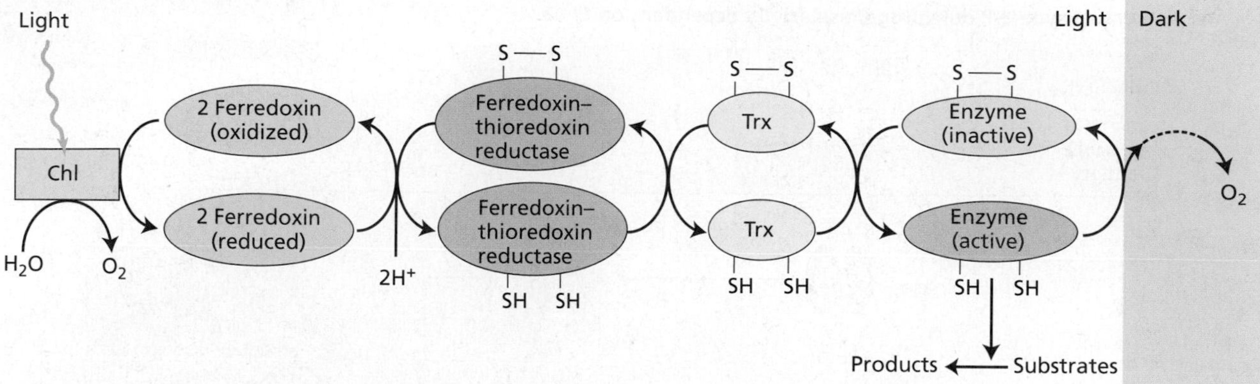

Figure 8.6 Ferredoxin–thioredoxin system. The ferredoxin–thioredoxin system links the light signal sensed by thylakoid membranes to the activity of enzymes in the chloroplast stroma. The activation of enzymes of the Calvin–Benson cycle starts in the light with the reduction of ferredoxin by the photosynthetic electron transport chain (Chl) (see Chapter 7). Reduced ferredoxin, together with two protons, is used to reduce a catalytically active disulfide bond (—S—S—) of the iron–sulfur enzyme ferredoxin–thioredoxin reductase, which in turn reduces the unique disulfide bond (—S—S—) of the regulatory protein thioredoxin (Trx) (see WEB TOPIC 8.6 for details). The reduced form of thioredoxin (—SH HS—) then reduces regulatory disulfide bonds of the target enzyme, triggering its conversion to the catalytically active state that catalyzes the transformation of substrates into products. Darkness halts the electron flow from ferredoxin to the enzyme, and thioredoxin becomes oxidized. Although the mechanism for the deactivation of thioredoxin-activated enzymes in the dark is not fully clear, it appears that O_2-actuated oxidations cause the formation of oxidized thioredoxin. Next, the unique disulfide bond (—S—S—) of thioredoxin brings the reduced form (—SH HS—) of the enzyme back to the oxidized form (—S—S—) with the concurrent loss of the catalytic capacity. In contrast to thioredoxin-activated enzymes, a chloroplast enzyme of the oxidative pentose phosphate cycle, glucose 6-phophate dehydrogenase, does not operate in the light but is functional in the dark because thioredoxin reduces the disulfide critical for the activity of the enzyme. The ability of thioredoxin to regulate functional enzymes in different pathways minimizes futile cycling.

tion. Thioredoxin also protects proteins against damage caused by reactive oxygen species, such as hydrogen peroxide (H_2O_2), the superoxide anion ($O_2{\bullet}^-$), and the hydroxyl radical (OH•).

Light-dependent ion movements modulate enzymes of the Calvin–Benson cycle

Upon illumination, the flow of protons from the stroma into the thylakoid lumen is coupled to the release of Mg^{2+} from the intrathylakoid space to the stroma. These light-actuated ion fluxes decrease the stromal concentration of protons (the pH increases from 7 to 8) and increase that of Mg^{2+} by 2 to 5 m*M*. The light-mediated increase of pH and the concentration of Mg^{2+} activate enzymes of the Calvin–Benson cycle that require Mg^{2+} for catalysis and are more active at pH 8 than at pH 7: rubisco, fructose 1,6-bisphosphatase, sedoheptulose 1,7-bisphosphatase, and phosphoribulokinase. The modifications of ionic composition of the chloroplast stroma are reversed rapidly upon darkening.

Light controls the assembly of chloroplast enzymes into supramolecular complexes

The formation of supramolecular complexes with regulatory proteins also has important effects on the catalytic activity of chloroplast enzymes. For example, glyceraldehyde-3-phosphate dehydrogenase binds noncovalently with phosphoribulokinase and CP12—a protein of approximately 8.5 kDa containing four conserved cysteines able to form two disulfide bridges (**Figure 8.7**). The three proteins form a ternary complex (CP12–phosphoribulokinase–glyceraldehyde 3-phosphate dehydrogenase) wherein glyceraldehyde-3-phosphate dehydrogenase and phosphoribulokinase are catalytically inactive. Light regulates the stability of the ternary complex through the ferredoxin–thioredoxin system. Reduced thioredoxin cleaves the disulfide bonds of both phosphoribulokinase and CP12, releasing glyceraldehyde-3-phosphate dehydrogenase and phosphoribulokinase in the catalytically active conformations.

The C_2 Oxidative Photosynthetic Carbon Cycle

Rubisco catalyzes both the carboxylation and the oxygenation of ribulose 1,5-bisphosphate (see Figure 8.4). Carboxylation yields two molecules of 3-phosphoglycerate, and oxygenation produces one molecule each of 3-phosphoglycerate and 2-phosphoglycolate. The oxygenase activity of rubisco causes partial loss of the carbon

Glyceraldehyde-3-P dehydrogenase acivity dependent on CP12

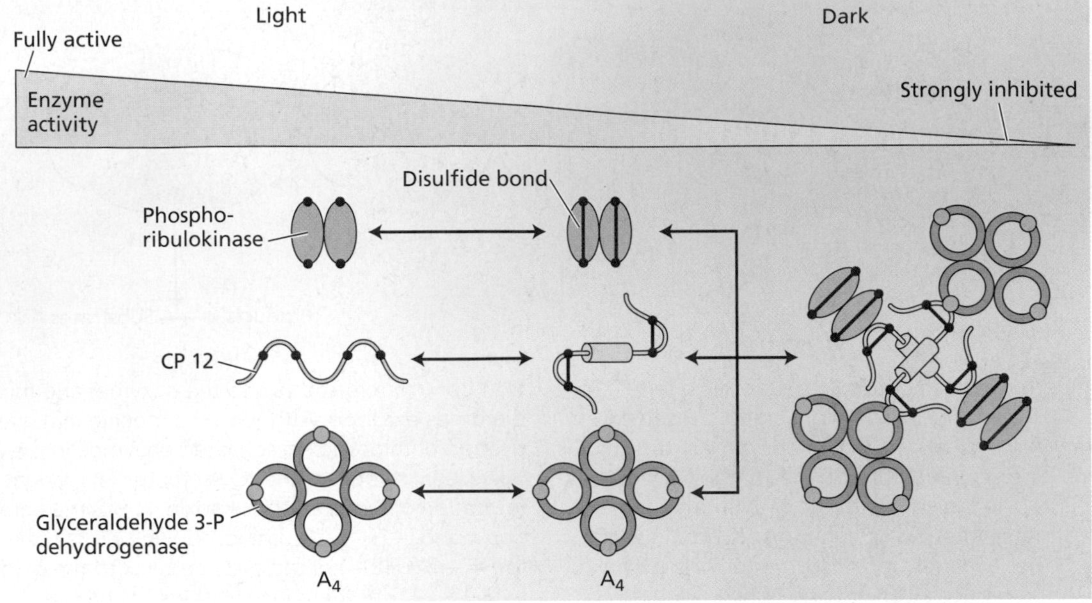

Glyceraldehyde-3-P dehydrogenase acivity dependent on C-terminal extension

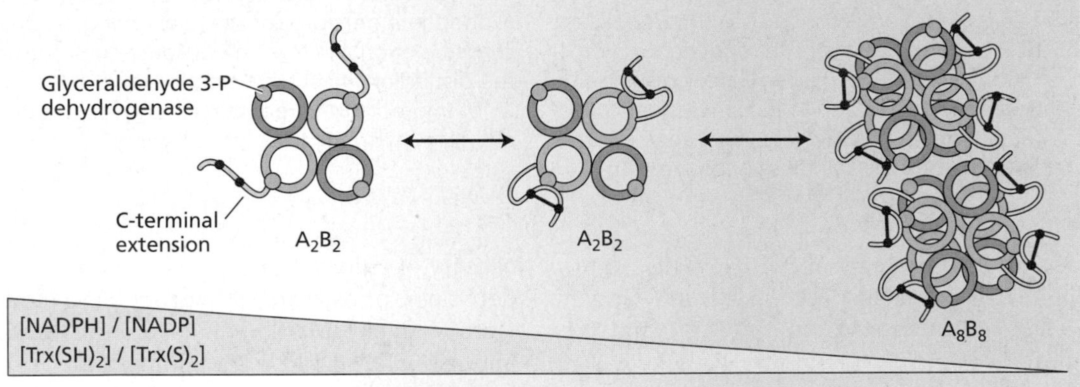

Figure 8.7 Regulation of chloroplast phosphoribulokinase and glyceraldehyde-3-phosphate dehydrogenase. Chloroplasts contain two isoforms of glyceraldehyde-3-phosphate dehydrogenases, named A_4 and A_2B_2. The A_4 isoform is a catalytically active tetramer. The A and B polypeptides of the A_2B_2 isoform are similar except that a C-terminal extension of the subunit B holds two cysteines able to form a disulfide bridge. Moreover, the A_2B_2 glyceraldehyde-3-phosphate dehydrogenase can form the A_8B_8 oligomer. Under "dark conditions," the interaction of oxidized phosphoribulokinase with the A_4-glyceraldehyde-3-phosphate dehydrogenase and oxidized CP12 stabilizes the complex [(A_4-glyceraldehyde-3-phosphate dehydrogenase)$_2$•(phosphoribulokinase)$_2$•(CP12)$_4$]. Both A_4-glyceraldehyde-3-phosphate dehydrogenase and phosphoribulokinase are catalytically inactive in the ternary complex. Under "light conditions," reduced thioredoxin cleaves the disulfide bonds of CP12 and phosphoribulokinase. The reduction of phosphoribulokinase and CP12 separates the components of the ternary complex releasing the active phosphoribulokinase and the active A_4B_4-glyceraldehyde 3-phosphate dehydrogenase. Reduced thioredoxin (Trx) cleaves the disulfide bond in subunit B of A_8B_8-glyceraldehyde-3-phosphate dehydrogenase. The reduction converts the inactive oligomer into the active A_2B_2-glyceraldehyde-3-phosphate dehydrogenase.

fixed by the Calvin–Benson cycle and yields 2-phosphoglycolate, an inhibitor of two chloroplast enzymes: triose phosphate isomerase and phosphofructokinase. To avoid both drain of carbon out of the Calvin–Benson cycle and enzyme inhibition, 2-phosphoglycolate is metabolized through the C_2 oxidative photosynthetic carbon cycle. This network of coordinated enzymatic reactions, also known as **photorespiration**, occurs in chloroplasts, leaf peroxisomes, and mitochondria (**Figure 8.8**, **Table 8.2**) (see WEB TOPIC 8.7).

TABLE 8.2 Reactions of the C_2 oxidative photosynthetic carbon cycle

Reaction[a]	Enzyme
1. 2 Ribulose 1,5-bisphosphate + 2 O_2 → 2 2-phosphoglycolate + 2 3-phosphoglycerate	Rubisco
2. 2 2-Phosphoglycolate + 2 H_2O → 2 glycolate + 2 P_i	Phosphoglycolate phosphatase
3. 2 Glycolate + 2 O_2 → 2 glyoxylate + 2 H_2O_2	Glycolate oxidase
4. 2 H_2O_2 → 2 H_2O + O_2	Catalase
5. 2 Glyoxylate + 2 glutamate → 2 glycine + 2 2-oxoglutarate	Glutamate:glyoxylate aminotransferase
6. Glycine + NAD^+ + [GDC] → CO_2 + NH_4^+ + NADH + [GDC-THF-CH_2]	Glycine decarboxylase complex (GDC)
7. [GDC-THF-CH_2] + glycine + H_2O → serine + [GDC]	Serine hydroxymethyl transferase
8. Serine + 2-oxoglutarate → hydroxypyruvate + glutamate	Serine:2-oxoglutarate aminotransferase
9. Hydroxypyruvate + NADH + H^+ → glycerate + NAD^+	Hydroxypyruvate reductase
10. Glycerate + ATP → 3-phosphoglycerate + ADP	Glycerate kinase
11. Glutamate + NH_4^+ + ATP → glutamine + ADP + Pi	Glutamine synthetase
12. 2-Oxoglutarate + glutamine + 2 Fd_{red} + 2 H^+ → 2 glutamate + 2 Fd_{oxid}	Ferredoxin-dependent glutamate synthase (GOGAT)

Net reaction of the C_2 oxidative photosynthetic carbon cycle

2 Ribulose 1,5-bisphosphate + 3 O_2 + H_2O + glutamate
↓ **(reactions 1 to 9)**
Glycerate + 2 3-phosphoglycerate + NH_4^+ + CO_2 + 2 P_i + 2-oxoglutarate

Two reactions in the chloroplasts restore the molecule of glutamate:

2-Oxoglutarate + NH_4^+ + [(2 Fd_{red} + 2 H^+), ATP]
↓ **(reactions 11 and 12)**
Glutamate + H_2O + [(2 Fd_{oxid}), ADP + P_i]

and the molecule of 3-phosphoglycerate:

Glycerate + ATP
↓ **(reaction 10)**
3-Phosphoglycerate + ADP

Hence, the consumption of three molecules of atmospheric oxygen in the C_2 oxidative photosynthetic carbon cycle (two in the oxygenase activity of rubisco and one in peroxisomal oxidations) elicits

- the release of one molecule of CO_2 and
- the consumption of two molecules of ATP and two molecules of reducing equivalents (2 Fd_{red} + 2 H^+)

for

- incorporating a three-carbon skeleton back into the Calvin–Benson Cycle, and
- restoring glutamate from NH_4^+ and 2-oxoglutarate.

[a]Locations: Chloroplasts; peroxisomes; mitochondria. Fd: ferredoxin.

Recent studies have shown that the C_2 oxidative photosynthetic carbon cycle is an ancillary component of photosynthesis that not only salvages part of the assimilated carbon but also links to other pathways of contemporary land plants. In this section we will begin with the relevant features of the C_2 oxidative photosynthetic carbon cycle in land plants and cyanobacteria. Next we will describe the integration of photorespiration into plant metabolism and then turn to different approaches to increase the yield of crop biomass through the modification of leaf photorespiration.

The oxygenation of ribulose 1,5-bisphosphate sets in motion the C_2 oxidative photosynthetic carbon cycle

In evolutionary terms, rubisco appears to have evolved from an ancient enolase in the archaea methionine-salvage pathway. Billions of years ago, oxygenation of

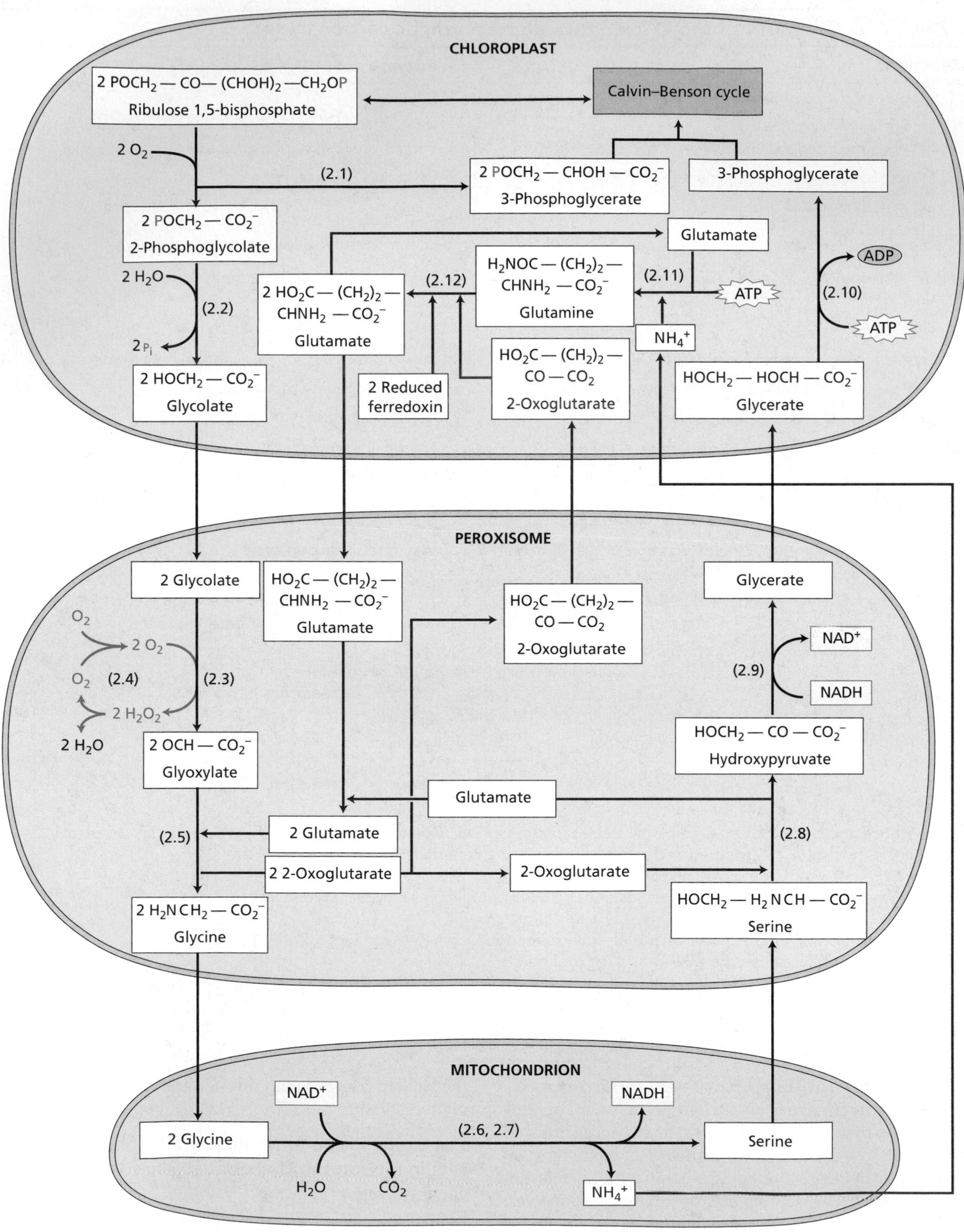
CHLOROPLAST
2 POCH2 — CO— (CHOH)2 —CH2OP
Ribulose 1,5-bisphosphate
Calvin–Benson cycle
2 O2
(2.1)
2 POCH2 — CHOH — CO2−
3-Phosphoglycerate
3-Phosphoglycerate
2 POCH2 — CO2−
2-Phosphoglycolate
Glutamate
ADP
2 H2O
(2.2)
2 Pi
2 HO2C — (CH2)2 —
CHNH2 — CO2−
Glutamate
(2.12)
H2NOC — (CH2)2 —
CHNH2 — CO2−
Glutamine
(2.11)
ATP
(2.10)
ATP
NH4+
2 HOCH2 — CO2−
Glycolate
2 Reduced
ferredoxin
HO2C — (CH2)2 —
CO — CO2
2-Oxoglutarate
HOCH2 — HOCH — CO2−
Glycerate
PEROXISOME
2 Glycolate
HO2C — (CH2)2 —
CHNH2 — CO2−
Glutamate
HO2C — (CH2)2 —
CO — CO2
2-Oxoglutarate
Glycerate
O2
2 O2
O2
(2.4)
(2.3)
2 H2O2
2 H2O
NAD+
(2.9)
NADH
2 OCH — CO2−
Glyoxylate
HOCH2 — CO — CO2−
Hydroxypyruvate
Glutamate
(2.5)
2 Glutamate
(2.8)
2 2-Oxoglutarate
2-Oxoglutarate
2 H2NCH2 — CO2−
Glycine
HOCH2 — H2NCH — CO2−
Serine
MITOCHONDRION
NAD+
NADH
2 Glycine
(2.6, 2.7)
Serine
H2O
CO2
NH4+

Figure 8.8 Operation of the C_2 oxidative photosynthetic cycle. Enzymatic reactions are distributed in three organelles: chloroplasts, peroxisomes, and mitochondria. *In chloroplasts*, the oxygenase activity of rubisco yields two molecules of 2-phosphoglycolate, which, upon the action of phosphoglycolate phosphatase, form two molecules of glycolate and two molecules of inorganic phosphate. Two molecules of glycolate (four carbons) flow concurrently with one molecule of glutamate from chloroplasts to peroxisomes. *In peroxisomes*, the glycolate is oxidized by O_2 to glyoxylate in a reaction catalyzed by the enzyme glycolate oxidase. The glutamate:glyoxylate aminotransferase catalyzes the conversion of glyoxylate and glutamate to glycine and 2-oxoglutarate. The amino acid glycine flows from peroxisomes to mitochondria. *In mitochondria*, two molecules of glycine (four carbons) yield a molecule of serine (three carbons) with the concurrent release of CO_2 (one carbon) and NH_4^+ by the successive action of the glycine decarboxylase complex and serine hydroxymethyltransferase. The amino acid serine is then transported back to the peroxisome and transformed to glycerate (three carbons) by the successive action of serine:2-oxoglutarate aminotransferase and hydroxypyruvate reductase. Glycerate and 2-oxoglutarate from peroxisomes, and NH_4^+ from mitochondria, return to chloroplasts in a process that recovers part of the carbon (three carbons) and all the nitrogen lost in photorespiration. Glycerate is phosphorylated to 3-phosphoglycerate and incorporated back into the Calvin–Benson cycle. The chloroplast stroma recover the nitrogen lost in the exported glutamate using the inorganic nitrogen (NH_4^+) and 2-oxoglutarate for the successive action of glutamine synthetase and ferredoxin-dependent glutamate synthase (GOGAT). See Table 8.2 for a description of each numbered reaction.

ribulose 1,5-bisphosphate was insignificant in non-oxygenic prokaryotes due to the lack of O_2 and high levels of CO_2 in the ancient atmosphere. The high concentrations of O_2 and low levels of CO_2 in the extant atmosphere boost the oxygenase activity of rubisco, making inevitable the formation of the toxic 2-phosphoglycolate. All rubiscos catalyze the incorporation of O_2 into ribulose 1,5-bisphosphate. Even homologs from anaerobic autotrophic bacteria exhibit the oxygenase activity, demonstrating that the oxygenase reaction is intrinsically linked to the active site of rubisco and not an adaptive response to the appearance of O_2 in the biosphere.

The oxygenation of the 2,3-enediol isomer of ribulose 1,5-bisphosphate with one molecule of O_2 yields an unstable intermediate that rapidly splits into one molecule each of 3-phosphoglycerate and 2-phosphoglycolate (see Figure 8.4, Figure 8.8, and Table 8.2, reaction 1). In chloroplasts of land plants, 2-phosphoglycolate phosphatase catalyzes the rapid hydrolysis of 2-phosphoglycolate to glycolate (see Figure 8.8 and Table 8.2, reaction 2). The subsequent transformations of glycolate take place in the peroxisomes and mitochondria (see Chapter 1). Glycolate leaves the chloroplasts through a specific transporter in the envelope inner membrane and diffuses to the peroxisomes (see Figure 8.8). In peroxisomes, glycolate oxidase catalyzes the oxidation of glycolate by O_2, producing H_2O_2 and glyoxylate (see Table 8.2, reaction 3). The peroxisomal catalase breaks down the H_2O_2, releasing O_2 and H_2O (see Figure 8.8 and Table 8.2, reaction 4). The glutamate:glyoxylate aminotransferase catalyzes the transamination of glyoxylate with glutamate, yielding the amino acid glycine (see Figure 8.8 and Table 8.2, reaction 5).

Glycine exits the peroxisomes and enters the mitochondria where a multienzyme complex of glycine decarboxylase (GDC) and serine hydroxymethyltransferase catalyzes the conversion of two molecules of glycine and one molecule of NAD^+ into one molecule each of serine, NADH, NH_4^+, and CO_2 (see Figure 8.8 and Table 8.2, reactions 6 and 7). First, GDC uses one molecule of NAD^+ for the oxidative decarboxylation of one molecule of glycine, yielding one molecule each of NADH, NH_4^+, and CO_2 and the activated one-carbon unit methylene tetrahydrofolate (THF) bound to GDC ($GDC\text{-}THF\text{-}CH_2$):

$$\text{Glycine} + NAD^+ + \text{GDC-THF} \rightarrow NADH + NH_4^+ + CO_2 + \text{GDC-THF-}CH_2$$

Next, serine hydroxymethyltransferase catalyzes the addition of the methylene unit to a second molecule of glycine, forming serine and regenerating THF to ensure high levels of glycine decarboxylase activity:

$$\text{Glycine} + \text{GDC-THF-}CH_2 \rightarrow \text{Serine} + \text{GDC-THF}$$

The oxidation of carbon atoms (two molecules of glycine [oxidation states C1: +3; C2: -1] → serine [oxidation states C1: +3; C2: 0; C3: –1] and CO_2 [oxidation state C: +4]) drives the reduction of oxidized pyridine nucleotide:

$$NAD^+ + H^+ + 2\ e^- \rightarrow NADH$$

The reaction products of the enzyme glycine decarboxylase are metabolized at different locations in leaf cells. NADH is oxidized to NAD^+ in the mitochondria. NH_4^+ and CO_2 are exported to chloroplasts, where they are assimilated to form glutamate (see below) and 3-phosphoglycerate, respectively.

The newly formed serine diffuses from the mitochondria back to the peroxisomes for donation of its amino group to 2-oxoglutarate via transamination catalyzed by serine:2-oxoglutarate aminotransferase, forming glutamate and hydroxypyruvate (see Figure 8.8 and Table 8.2, reaction 8). Next, an NADH-dependent reductase catalyzes the transformation of hydroxypyruvate into glycerate (see Figure 8.8 and Table 8.2, reaction 9). Finally, glycerate reenters the chloroplast, where it is phosphorylated by ATP to yield 3-phosphoglycerate and ADP (see Figure 8.8 and Table 8.2, reaction 10). Hence, the formation of 2-phosphoglycolate (via rubisco) and the phosphorylation of glycerate (via glycerate kinase) metabolically link the

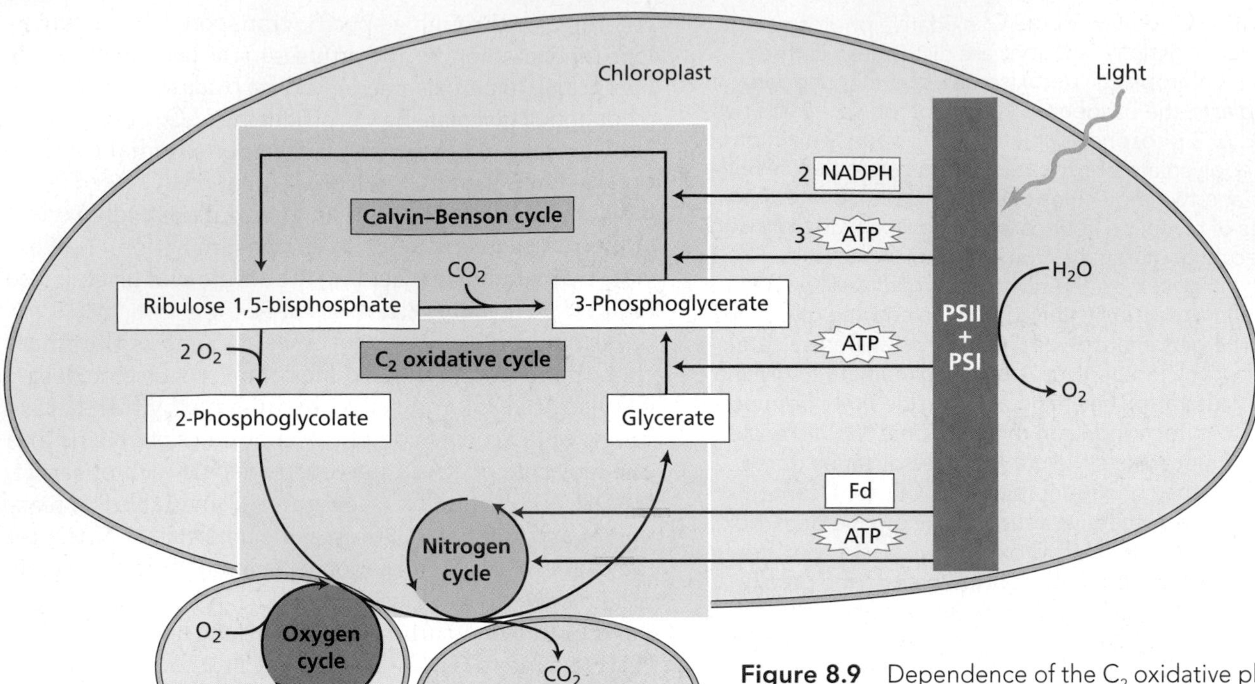

Figure 8.9 Dependence of the C_2 oxidative photosynthetic carbon cycle on chloroplast metabolism. The supply of ATP and reducing equivalents from light reactions in thylakoid membranes is needed for the functioning of the C_2 oxidative photosynthetic cycle in three compartments: chloroplasts, peroxisomes and mitochondria. The *carbon cycle* uses (1) NADPH and ATP to maintain the adequate level of ribulose 1,5-bisphosphate in the Calvin–Benson cycle and (2) ATP to convert glycerate to 3-phosphoglycerate in the C_2 oxidative photosynthetic carbon cycle. The *nitrogen cycle* employs ATP and reducing equivalents to restore glutamate from NH_4^+ and 2-oxoglutarate coming from the photorespiratory cycle. In the peroxisome, the *oxygen cycle* contributes to the removal of H_2O_2 formed in the oxidation of glycolate by O_2.

Calvin–Benson cycle to the C_2 oxidative photosynthetic carbon cycle.

The NH_4^+ released in the oxidation of glycine diffuses rapidly from the matrix of the mitochondria to the chloroplasts (see Figure 8.8). In the chloroplast stroma, glutamine synthetase catalyzes the ATP-dependent incorporation of NH_4^+ into glutamate, yielding glutamine, ADP, and inorganic phosphate (see Figure 8.8 and Table 8.2, reaction 11). Subsequently, glutamine and 2-oxoglutarate are substrates of ferredoxin-dependent glutamate synthase (GOGAT) for the production of two molecules of glutamate (see Table 8.2, reaction 12). The reassimilation of NH_4^+ into the photorespiratory cycle restores glutamate for the action of the peroxisomal glutamate:glyoxylate aminotransferase in the conversion of glyoxylate to glycine (see Table 8.2, reaction 5).

Carbon, nitrogen, and oxygen atoms circulate through photorespiration (**Figure 8.9**).

- In the *carbon* cycle, the chloroplasts transfer two molecules of glycolate (four carbon atoms) to peroxisomes and recovers one molecule of glycerate (three carbon atoms). The mitochondria release one molecule of CO_2 (one carbon atom).
- In the *nitrogen* cycle, the chloroplasts transfer one molecule of glutamate (one nitrogen atom) and recover one molecule of NH_4^+ (one nitrogen atom).
- In the *oxygen* cycle, rubisco and glycolate oxidase catalyze the incorporation of two molecules of O_2 each (eight oxygen atoms) when two molecules of ribulose 1,5-bisphosphate enter the C_2 oxidative photosynthetic carbon cycle (see Table 8.2, reactions 1 and 3). However, catalase releases one molecule of O_2 from two molecules of H_2O_2 (two oxygen atoms) (see Table 8.2, reaction 4). Hence, three molecules of O_2 (six oxygen atoms) are reduced in the photorespiratory cycle.

In vivo, three aspects regulate the distribution of metabolites between the Calvin–Benson cycle and the C_2 oxidative photosynthetic carbon cycle: one inherent to the plant (the kinetic properties of rubisco) and two linked to the environment (the concentration of atmospheric CO_2 and O_2, and temperature).

The specificity factor (Ω) estimates the preference of rubisco for CO_2 versus O_2:

$$\Omega = [V_C/K_C]/[V_O/K_O]$$

where V_C and V_O are the maximum velocities for carboxylation and oxygenation, respectively, and K_C and K_O are the Michaelis–Menten constants for CO_2 and O_2, respectively. Ω settles the ratio of the velocity of carboxylation (v_C) to that of oxygenation (v_O) at environmental concentrations of CO_2 and O_2:

$$\Omega = v_C/v_O \times [(O_2)/(CO_2)]$$

The specificity factor (Ω) estimates the relative capacity of rubisco for carboxylation and oxygenation [v_C/v_O] when the concentration of CO_2 around the active site is equal to that of O_2 [$(O_2)/(CO_2) = 1$]. Ω is a constant for every rubisco that denotes the relative efficiency with which O_2 competes with CO_2 at a given temperature. Rubiscos from different organisms exhibit variations in the value of Ω: the Ω of rubisco from cyanobacteria (Ω ~40) is lower than that from C_3 plants (Ω ~82–90) and from C_4 species (Ω ~70–82).

The ambient temperature exerts an important influence over Ω and the concentrations of CO_2 and O_2 around the active site of rubisco. Warmer environments have the effect of:

- increasing the oxygenase activity of rubisco more than the carboxylase activity. The greater increase of K_C for CO_2 than of K_O for O_2 diminishes the Ω of rubisco.
- lowering the solubility of CO_2 to a greater extent than O_2. The increase of $[O_2]/[CO_2]$ decreases the v_C/v_O ratio; that is, the oxygenase activity of rubisco prevails over the carboxylase activity (see **WEB TOPIC 8.8**).
- reducing the stomatal aperture to conserve water. The stomatal closure lowers the uptake of atmospheric CO_2, thereby decreasing CO_2 at the active site of rubisco.

Overall, warmer environments significantly limit the efficiency of photosynthetic carbon assimilation because the progressive increase in temperature tilts the balance away from photosynthesis (carboxylation) and toward photorespiration (oxygenation) (see Chapter 9).

Photorespiration is linked to the photosynthetic electron transport system

Photosynthetic carbon metabolism in intact leaves reflects the competition for ribulose 1,5-bisphosphate between two mutually opposing cycles, the Calvin–Benson cycle and the C_2 oxidative photosynthetic carbon cycle. These cycles are interlocked with the photosynthetic electron transport system for the supply of ATP and reducing equivalents (reduced ferredoxin and NADPH) (see Figure 8.9). For salvaging two molecules of 2-phosphoglycolate by conversion to one molecule of 3-phosphoglycerate, photophosphorlyation provides one molecule of ATP necessary for the transformation of glycerate to 3-phosphoglycerate (see Table 8.2, reaction 10), while the consumption of NADH by hydroxypyruvate reductase (see Table 8.2, reaction 9) is counterbalanced by its production by glycine decarboxylase (see Table 8.2, reaction 6).

In photorespiration, nitrogen:

- *enters* into the peroxisome through the transamination step catalyzed by the glutamate:glyoxylate aminotransferase (two nitrogen atoms) (see Table 8.2, reaction 5), and
- *leaves* the peroxisome (1) as NH_4^+ (one nitrogen atom), in the reaction catalyzed by the glycine decarboxylase–serine hydroxymethyltransferase complex (see Table 8.2, reactions 6 and 7) and (2) in the transamination step catalyzed by serine:2-oxoglutarate aminotransferase (one nitrogen atom) (see Table 8.2, reaction 8).

The photosynthetic electron transport system supplies one molecule of ATP and two molecules of reduced ferredoxin needed for salvaging one molecule of NH_4^+ through its incorporation into glutamate via glutamine synthetase (see Table 8.2, reaction 11) and ferredoxin-dependent glutamate synthase (GOGAT) (see Table 8.2, reaction 12).

In summary,

$$\text{2 Ribulose 1,5-bisphosphate} + 3\ O_2 + H_2O + \text{ATP} + [2\ \text{ferredoxin}_{red} + 2\ H^+ + \text{ATP}]$$
$$\downarrow$$
$$\text{3 3-Phosphoglycerate} + CO_2 + 2\ P_i + \text{ADP} + [2\ \text{ferredoxin}_{oxid} + \text{ADP} + P_i]$$

Because of the additional provision of ATP and reducing power for the operation of the photorespiratory cycle, the quantum requirement for CO_2 fixation under photorespiratory conditions (high $[O_2]$ and low $[CO_2]$) is higher than under nonphotorespiratory conditions (low $[O_2]$ and high $[CO_2]$).

Enzymes of the plant C_2 oxidative photosynthetic carbon cycle derive from different ancestors

The complete genomes of different organisms have demonstrated that all photorespiratory enzymes are present in plants and in red and green algae. Moreover, these phylogenetic studies suggest that the distribution of enzymes in plants correlates with the origin of compartments involved in the C_2 oxidative photosynthetic carbon cycle. Chloroplast enzymes evolved from a cyanobacterial endosymbiont, while mitochondrial enzymes have a proteobacterial ancestor. For example, chloroplast glycerate kinase is of cyanobacterial origin, and mitochondrial glycine decarboxylase comes from an ancient proteobacteria.

Cyanobacteria use a proteobacterial pathway for bringing carbon atoms of 2-phosphoglycolate back to the Calvin–Benson cycle

Cyanobacterial genomes code for all enzymes of the plant C_2 oxidative photosynthetic carbon cycle. The presence of photorespiration in the first O_2 producers indicates an ancient mechanism linked closely to oxygenic photosynthesis that emerged as an adaptation to cope with

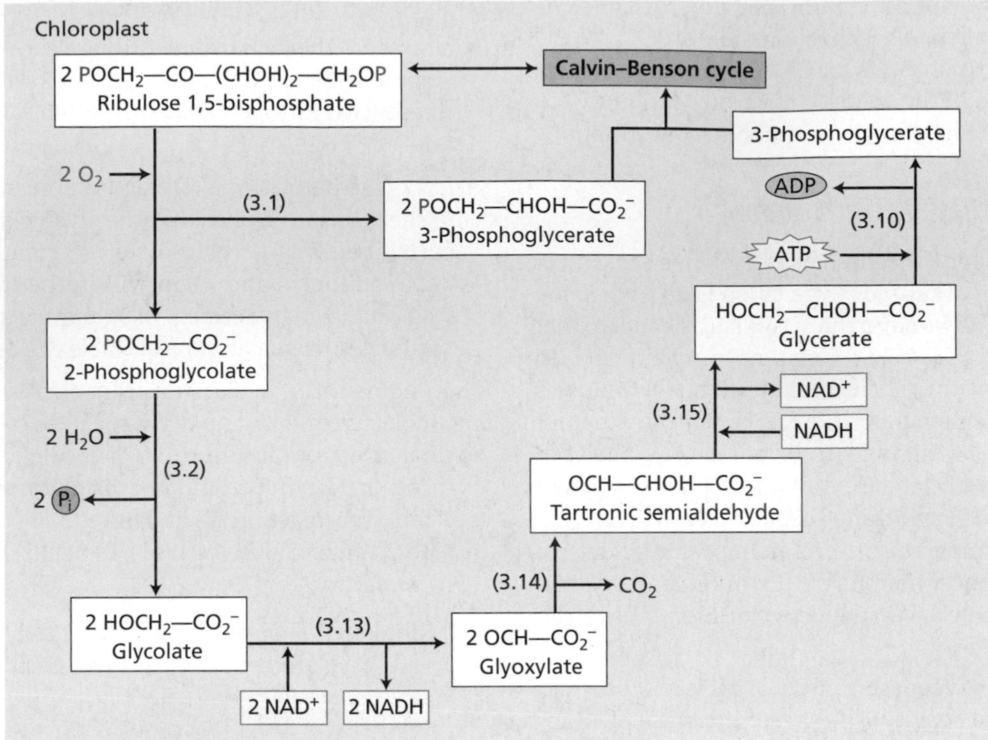

Figure 8.10 C_2 oxidative photosynthetic carbon cycle of cyanobacteria. Similarly to plants, the photorespiratory metabolism of cyanobacteria starts with the oxygenase activity of rubisco, followed by the hydrolytic activity of 2-phosphoglycolate phosphatase (reactions 3.1 and 3.2). At this stage, glycolate dehydrogenase couples the oxidation of glycolate to glyoxylate with the reduction of NAD^+ (reaction 3.13). Next, tartronic semialdehyde synthase catalyzes the conversion of two molecules of glyoxylate into tartronic semialdehyde and CO_2 (reaction 3.14). Finally, tartronic semialdehyde reductase catalyzes the reduction of tartronic semialdehyde to glycerate (reaction 3.15). The phosphorylation of glycerate catalyzed by glycerate kinase brings the 3-phosphoglycerate back to the Calvin–Benson cycle (reaction 3.10).

intracellular O_2. Although all "plantlike" photorespiratory enzymes are present, extant cyanobacteria use enzymes from proteobacterial ancestors for recovering carbon lost in the C_2 oxidative photosynthetic carbon cycle (**Figure 8.10** and **Table 8.3**, reactions 1 and 2).

Initially, the enzyme glycolate dehydrogenase (see Table 8.3, reaction 13) converts photorespiratory glycolate into glyoxylate [glycolate + NAD^+ → glyoxylate + NADH + H^+].

Next, two enzymes catalyze the conversion of glyoxylate into glycerate:

- Tartronic semialdehyde synthase [glyoxylate → tartronate semialdehyde + CO_2] (see Table 8.3, reaction 14)
- Tartronic semialdehyde reductase [tartronate semialdehyde + NADH + H^+ → glycerate + NAD^+] (see Table 8.3, reaction 15)

Finally, cyanobacterial glycerate kinase phosphorylates glycerate, yielding 3-phosphoglycerate that reenters the Calvin–Benson cycle [glycerate + ATP → 3-phosphoglycerate + ADP] (see Table 8.3, reaction 10).

As in land plants, the alternative photorespiratory cycle of cyanobacteria releases one carbon atom (see Table 8.3, reaction 14) and incorporates a three-carbon skeleton back into the Calvin–Benson Cycle (see Table 8.3, reaction 10). The requirements of ATP and reductants for this alternative pathway are different from those used by land plants in the C_2 oxidative photosynthetic carbon cycle because cyanobacteria circumvent the release and the refixation of NH_4^+ (compare *Net reaction of the C_2 oxidative photosynthetic carbon cycle* in Tables 8.2 and 8.3).

The C_2 oxidative photosynthetic carbon cycle interacts with many metabolic pathways

Early research suggested that the C_2 oxidative photosynthetic carbon cycle served to salvage carbon diverted by the oxygenase activity of rubisco and to protect plants from stressful conditions such as high light, drought, and salt stress. The negative impact of photorespiration on photosynthetic CO_2 assimilation originated from plant mutants that do not survive in air (21% O_2; 0.04% CO_2) but resume their normal growth in high-CO_2 environ-

Table 8.3 Reactions of the C_2 oxidative photosynthetic carbon cycle in cyanobacteria

Reaction[a]	Enzyme
1. 2 Ribulose 1,5-bisphosphate + 2 O_2 → 2 2-phosphoglycolate + 2 3-phosphoglycerate	Rubisco
2. 2 2-Phosphoglycolate + 2 H_2O → 2 glycolate + 2 P_i	Phosphoglycolate phosphatase
13. 2 Glycolate + 2 NAD^+ → 2 glyoxylate + 2 NADH + 2 H^+	Glycolate dehydrogenase
14. 2 Glyoxylate + H^+ → tartronic semialdehyde + CO_2	Tartronic semialdehyde synthase
15. Tartronic semialdehyde + NADH + H^+ → glycerate + NAD^+	Tartronic semialdehyde reductase
10. Glycerate + ATP → 3-phosphoglycerate + ADP	Glycerate kinase

Net reaction of the C_2 oxidative photosynthetic carbon cycle in cyanobacteria

2 Ribulose 1,5-bisphosphate + 2 O_2 + 2 H_2O + NAD^+
↓ **(reactions 1, 2, 13, 14, and 15)**
Glycerate + 2 3-phosphoglycerate + CO_2 + 2 P_i + NADH

The phosphorylation of glycerate catalyzed by glycerate kinase restores the molecule of 3-phosphoglycerate to the Calvin–Benson cycle:

Glycerate + ATP
↓ **(reaction 10)**
3-Phosphoglycerate + ADP

Hence, the consumption of two molecules of O_2 in the oxygenase activity of rubisco starts in cyanobacteria a bacterial-like glycerate pathway that releases one molecule of CO_2, forms one molecule of the reductant NADH, and consumes one molecule of ATP for recovering a three-carbon skeleton back into the Calvin–Benson cycle.

[a]Location: Chloroplasts.

ments (2% CO_2). This feature, called the *photorespiratory phenotype,* serves for the identification of unknown components of the C_2 oxidative cycle. For example, Arabidopsis mutants lacking glycerate kinase accumulate glycerate and concurrently are incapable of growing in normal air but are viable in atmospheres with elevated levels of CO_2.

However, the C_2 oxidative photosynthetic carbon cycle requires the participation of three organelles—chloroplasts, mitochondria, and peroxisomes—that are integrated into whole-cell metabolism. Recent studies have revealed a tight connection between photorespiration and other pathways of plant metabolism. The C_2 oxidative photosynthetic carbon cycle interacts with:

- *Nitrogen metabolism at multiple levels*: Photorespiration reassimilates NH_4^+ formed in the mitochondria, uses glutamate in peroxisomal transaminations and produces amino acids (serine, glycine) for other metabolic pathways.
- *Cell redox homeostasis*: H_2O_2 formed by the peroxisomal glycolate oxidase regulates the redox state of leaves. The formation of H_2O_2 induces suicide programs in catalase-deficient barley plants that exhibit the photorespiratory phenotype. Although H_2O_2 damages key cellular molecules such as DNA and lipids, the current view recognizes this reactive oxygen species as a signaling molecule linked to hormone and stress responses.
- *C_1 metabolism*: 5,10-methylene tetrahydrofolate is the cofactor required by glycine decarboxylase–serine hydroxymethyltransferase in the conversion of glycine to serine in the mitochondria. Reactions mediated by folates transfer one-carbon units in the synthesis of precursors for proteins, nucleic acids, lignin, and alkaloids.
- *Expression of transcription factors*: More than 200 transcription factors are differentially expressed when plants are transferred from atmospheres with high levels of CO_2 to normal air. Photorespiration enhances the expression of genes encoding the components of cyclic electron flow pathways, consistent with the additional demand of energy of the photorespiratory pathway. Photorespiration decreases transcripts encoding proteins involved in synthesis of starch and sucrose and in the metabolism of nitrogen and sulphur.

Production of biomass may be enhanced by engineering photorespiration

Solutions to the current food and energy shortages depend on the degree to which land plants can be adapted to higher CO_2 assimilation. When O_2 outcompetes CO_2, the oxygenase activity of rubisco lowers the amount of carbon that enters the Calvin–Benson cycle. Therefore, to understand how to engineer leaf cells for improving the photosynthetic efficiency, scientists are tackling various aspects of the C_2 oxidative photosynthetic carbon cycle, from modification of the active site of rubisco to the introduction of parallel photorespiratory pathways by genetic engineering. Despite considerable efforts, the modification of rubisco for alleviating photorespiration has not met with success.

Because the C_2 oxidative photosynthetic carbon cycle is essential for land plants, an attractive possibility is to incorporate different mechanisms for retrieving the carbon atoms of 2-phosphoglycolate. Two approaches lower the flux of photorespiratory metabolites through peroxisomes and mitochondria, releasing photorespired CO_2 in the chloroplast where it can be directly refixed. One approach introduces a bacterial (*E. coli*) glycolate catabolic pathway into chloroplasts of land plants (Arabidopsis) (see Figure 8.10). Chloroplasts of these transgenic plants have an entirely functional photorespiratory cycle while additionally accommodating the bacterial enzymes glycolate dehydrogenase, tartronic semialdehyde synthase, and tartronic semialdehyde reductase (see Table 8.3, reactions 13, 14, and 15). The engineered plants grow faster, have increased biomass, and contain higher levels of soluble sugars.

Alternatively, the overexpression of three enzymes in the chloroplast stroma of Arabidopsis—glycolate oxidase, catalase, and malate synthase—causes the release of CO_2 from glycolate. First, the oxidation of glycolate by the novel chloroplast glycolate oxidase yields glyoxylate and H_2O_2, and catalase catalyzes the subsequent decomposition of H_2O_2 [2 glycolate + 2 $O_2 \rightarrow$ 2 glyoxylate + 2 H_2O_2; 2 $H_2O_2 \rightarrow 2 H_2O + O_2$]. Next, the successive action of two enzymes converts two molecules of glyoxylate (two carbon atoms) into pyruvate (three carbon atoms) and CO_2 (one carbon atom):

- Malate synthase catalyzes the condensation of glyoxylate with acetyl-CoA [CoA-S~CO-CH_3], yielding malate [2 glyoxylate + CoA-S~CO-CH_3 → malate + CoA-SH].
- The chloroplastic NADP–malic enzyme catalyzes the decarboxylation of malate to pyruvate with the concurrent formation of NADPH [malate + $NADP^+$ → pyruvate + CO_2 + NADPH + H^+].

Finally, chloroplastic pyruvate dehydrogenase catalyzes the conversion of pyruvate into acetyl-CoA, yielding NADH and another molecule of CO_2 [pyruvate + CoA-SH + NAD^+ → CoA-S~CO-CH_3 + CO_2 + NADH + H^+]. As a result of this alternative cycle, one molecule of glycolate (two carbon atoms) is converted to two molecules of CO_2 (two carbon atoms). The oxidation of carbon atoms yields reducing power in the form of NADPH and NADH.

These novel pathways depart from plant photorespiration in sidestepping the mitochondrial and peroxisomal reactions. As a consequence, the shift of glycolate from plant photorespiration to the engineered pathways releases CO_2 in the immediate vicinity of rubisco, allowing rapid CO_2 fixation, and at the same time avoids the use of energy (ATP and reductant) required to recover NH_4^+.

Inorganic Carbon–Concentrating Mechanisms

Except for some photosynthetic bacteria, photoautotrophic organisms in the biosphere use the Calvin–Benson cycle to assimilate atmospheric CO_2. The pronounced reduction in CO_2 and rise in O_2 levels that commenced about 350 million years ago triggered a series of adaptations to handle an environment that promoted photorespiration in photosynthetic organisms. These adaptations include various strategies for active uptake of CO_2 and HCO_3^- from the surrounding environment and accumulation of inorganic carbon near rubisco. The immediate consequence of higher levels of CO_2 around rubisco is a decrease in the oxygenation reaction. CO_2 and HCO_3^- pumps at the plasma membrane have been studied extensively in prokaryotic cyanobacteria, eukaryotic algae, and aquatic plants (see **WEB TOPIC 8.1**).

In land plants, the diffusion of CO_2 from the atmosphere to the chloroplast plays a crucial role in net photosynthesis. To be incorporated into sugar compounds, inorganic carbon has to cross four barriers: the cell wall, plasma membrane, cytoplasm, and chloroplast envelope. Recent evidence has revealed that pore-forming membrane proteins (aquaporins) function as diffusion facilitators for various small molecules, decreasing the resistance of the mesophyll to the transport of CO_2.

Land plants evolved two carbon-concentrating mechanisms for increasing the concentration of CO_2 at the rubisco carboxylation site:

- C_4 photosynthetic carbon fixation (C_4)
- Crassulacean acid metabolism (CAM)

The uptake of atmospheric CO_2 by these carbon-concentrating mechanisms precedes CO_2 assimilation through the Calvin–Benson cycle.

Inorganic Carbon–Concentrating Mechanisms: The C_4 Carbon Cycle

C_4 photosynthesis has evolved as one of the major carbon-concentrating mechanisms used by land plants to compensate for limitations associated with low levels of atmospheric CO_2. Some of the most productive crops on the planet (e.g., corn; sugarcane, sorghum) use this mechanism to enhance the catalytic capacity of rubisco. In this section we will examine:

- The biochemical and anatomical attributes of C_4 photosynthesis that minimize the oxygenase activity of rubisco and the concurrent loss of carbon through the photorespiratory cycle
- The concerted action of different types of cells for the incorporation of inorganic carbon into carbon skeletons
- The light-mediated regulation of enzyme activities, and
- The importance of C_4 photosynthesis for sustaining plant growth in many tropical areas

Malate and aspartate are the primary carboxylation products of the C_4 cycle

In the late 1950s, H. P. Kortschack and Y. Karpilov observed that ^{14}C label appeared initially in the four-carbon acids malate and aspartate when $^{14}CO_2$ was provided to leaves of sugarcane and maize in the light. This finding was unexpected because a three-carbon acid, 3-phosphoglycerate, is the first labeled product in the Calvin–Benson cycle. M. D. Hatch and C. R. Slack explained that particular distribution of radioactive carbon by proposing an alternative mechanism to the Calvin–Benson cycle. This pathway is named the *C_4 photosynthetic carbon cycle* (also known as the Hatch–Slack cycle or the C_4 cycle).

Hatch and Slack found that (1) malate and aspartate are the first stable intermediates of photosynthesis and (2) that the carbon 4 of these four-carbon acids subsequently becomes carbon 1 of 3-phosphoglycerate. These transformations take place in two morphologically distinct cell types—the mesophyll and bundle sheath cells—that are separated by their respective walls and membranes ("Diffusion barrier" on **Figure 8.11**).

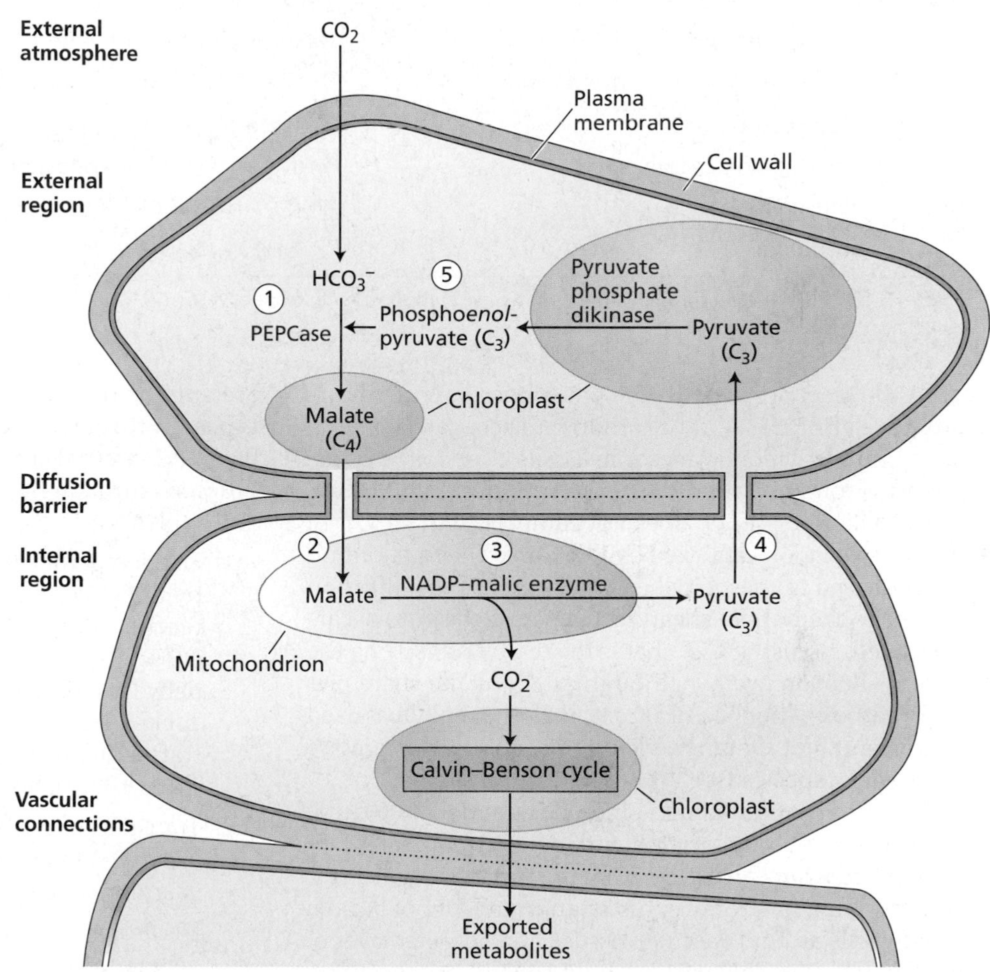

Figure 8.11 The C_4 photosynthetic carbon cycle involves five successive stages in two different compartments. (1) In the periphery of leaf cells (*external region*), the enzyme phospho*enol*pyruvate carboxylase (PEPCase) catalyzes the reaction of HCO_3^-, provided by the uptake of atmospheric CO_2, with phospho*enol*pyruvate, a three-carbon compound. Depending on the plant, the reaction product, oxaloacetate, a four-carbon compound, is further transformed to malate or aspartate by the action of the enzymes NADP-malate dehydrogenase or aspartate aminotransferase, respectively (see Table 8.4). For simplicity, malate is shown here (for differences among plant species in reactions sustaining the C_4 photosynthetic pathways, see **WEB TOPIC 8.9**). (2) The four-carbon acid flows across a *diffusion barrier* to the *internal region*, close to vascular connections. (3) The decarboxylating enzyme (e.g., NAD–malic enzyme) releases the CO_2 from the four-carbon acid, yielding a three-carbon acid (e.g., pyruvate). The uptake of the released CO_2 by the chloroplasts in the internal region builds up a large excess of CO_2 over O_2 around rubisco, thereby facilitating the assimilation of CO_2 through the Calvin–Benson cycle. (4) The residual three-carbon acid flows back to the external region. (5) Closing the C_4 cycle, the enzyme pyruvate–phosphate dikinase catalyzes the regeneration of phospho*enol*pyruvate, the acceptor of the HCO_3^-, for another turn of the cycle. The consumption of two molecules of ATP per molecule of fixed CO_2 (see Table 8.4, reactions 7 and 8) drives the C_4 cycle in the direction of the arrows, thus pumping CO_2 from the atmosphere to the Calvin–Benson cycle. The assimilated carbon leaves the chloroplast and, after transformation to sucrose in the cytoplasm, enters the phloem for translocation to other parts of the plant.

Table 8.4 Reactions of C_4 photosynthesis

Enzyme	Reaction
1. PEPCase	Phospho*enol*pyruvate + HCO_3^- → oxaloacetate + P_i
2. NADP–malate dehydrogenase	Oxaloacetate + NADPH + H^+ → malate + $NADP^+$
3. Aspartate aminotransferase	Oxaloacetate + glutamate → aspartate + 2-oxoglutarate
Decarboxylating enzymes	
4a. NADP–malic enzyme	Malate + $NADP^+$ → pyruvate + CO_2 + NADPH + H^+
4b. NAD–malic enzyme	Malate + NAD^+ → pyruvate + CO_2 + NADH + H^+
5. Phospho*enol*pyruvate carboxykinase	Oxaloacetate + ATP → phospho*enol*pyruvate + CO_2 + ADP
6. Alanine aminotransferase	Pyruvate + glutamate → alanine + 2-oxoglutarate
7. Pyruvate–phosphate dikinase	Pyruvate + P_i + ATP → phospho*enol*pyruvate + AMP + PP_i
8. Adenylate kinase	AMP + ATP → 2 ADP
9. Pyrophosphatase	PP_i + H_2O → 2 P_i

Note: P_i and PP_i stand for inorganic phosphate and pyrophosphate, respectively.

In the C_4 cycle, the enzyme phospho*enol*pyruvate carboxylase (PEPCase), rather than rubisco, catalyzes the initial carboxylation in mesophyll cells close to the external atmosphere (**Table 8.4**, reaction 1) (see **WEB ESSAY 8.1**). Unlike rubisco, O_2 does not compete with HCO_3^- in the carboxylation catalyzed by PEPCase. The four-carbon acids formed in mesophyll cells flow across the diffusion barrier to the bundle sheath cells, where they are decarboxylated, releasing CO_2 that is refixed by rubisco via the Calvin–Benson cycle. Although all C_4 plants share primary carboxylation via PEPCase, the other enzymes used to concentrate CO_2 in the vicinity of rubisco vary among different C_4 species (see **WEB TOPIC 8.9**).

Since the seminal studies of the 1950s and the 1960s, the C_4 cycle has been associated with a particular leaf structure, called **Kranz anatomy** (*Kranz*, German for "wreath"). Typical Kranz anatomy exhibits an inner ring of bundle sheath cells around vascular tissues and an outer layer of mesophyll cells. This particular leaf anatomy generates a diffusion barrier that (1) separates the uptake of atmospheric carbon in mesophyll cells from CO_2 assimilation by rubisco in bundle sheath cells and (2) limits the leakage of CO_2 from bundle sheath to mesophyll cells. However, there are now clear examples of single-cell C_4 photosynthesis in a number of green algae, diatoms, and aquatic and land plants (**Figure 8.12A**) (see **WEB TOPIC 8.10**). In summary, diffusion gradients—not only *between* but also *within* the cells—guide the shuttling of metabolites between the two compartments that operate the C_4 cycle.

The C_4 cycle assimilates CO_2 by the concerted action of two different types of cells

The key features of the C_4 cycle were initially described in leaves of plants such as maize whose vascular tissues are surrounded by two distinctive photosynthetic cell types. In this anatomical context, the transport of CO_2 from the external atmosphere to the bundle sheath cells proceeds through five successive stages (see Figure 8.11 and Table 8.4):

1. Fixation of the HCO_3^- in phospho*enol*pyruvate by PEPCase in the mesophyll cells (see Table 8.4, reaction 1). The reaction product, oxaloacetate, is subsequently reduced to malate by NADP–malate dehydrogenase in the mesophyll chloroplasts (see Table 8.4, reaction 2) or converted to aspartate by transamination with glutamate in the cytosol (see Table 8.4, reaction 3).
2. Transport of the four-carbon acids (malate or aspartate) to bundle sheath cells that surround the vascular bundles.
3. Decarboxylation of the four-carbon acids and generation of CO_2, which is then reduced to carbohydrate via the Calvin–Benson cycle. Prior to this reaction, an aspartate aminotransferase catalyzes the conversion of aspartate back to oxaloacetate in some C_4 plants (see Table 8.4, reaction 3). Different types of C_4 plants make use of different decarboxylases to release CO_2 for the effective suppression of the oxygenase reaction of rubisco (see Table 8.4, reactions 4a, 4b, and 5) (see **WEB TOPIC 8.9**).
4. Transport of the three-carbon backbone (pyruvate or alanine) formed by the decarboxylation step back to the mesophyll cells.
5. Regeneration of phospho*enol*pyruvate, the HCO_3^- acceptor. ATP and inorganic phosphate convert pyruvate to phospho*enol*pyruvate, releasing AMP and pyrophosphate (see Table 8.4, reaction 7). Two

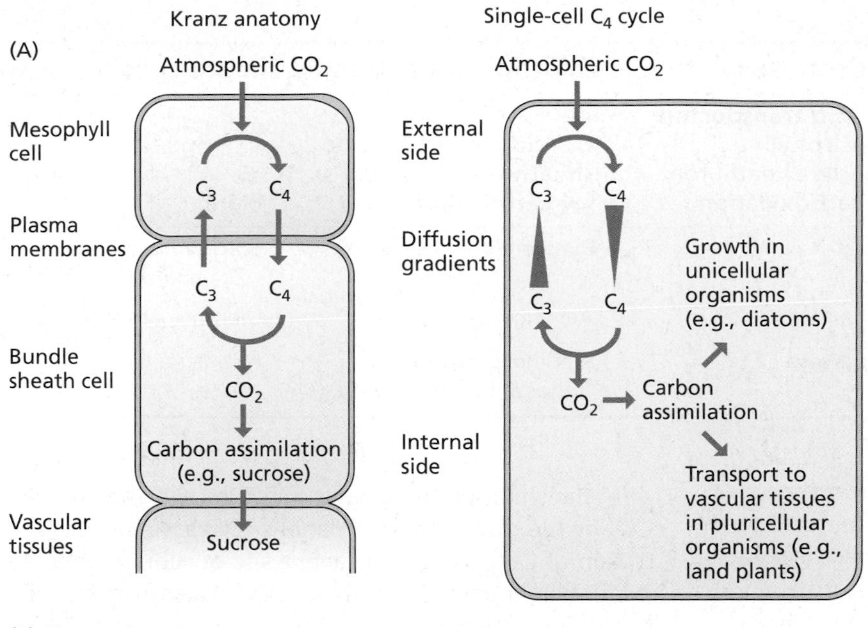

(B) Kranz anatomy

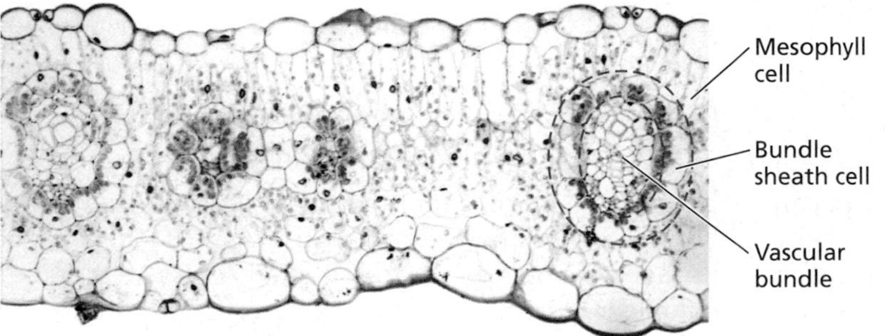

(C) Single-cell C_4 cycle

CO_2
Cytosol (external region)
C_3
C_4
Cytosol (diffusion barrier)
Cytosol (internal region)
C_3
C_4
$[CO_2]$
Cytosol (internal region)
Cytosol (external region)
CO_2
C_3
C_4
C_3
C_4
$[CO_2]$

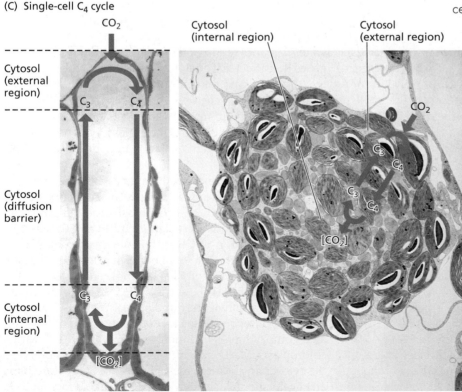

Figure 8.12 C_4 photosynthetic pathway in leaves of different plants. (A) In almost all known C_4 plants, photosynthetic CO_2 assimilation requires the development of Kranz anatomy (left panel). This anatomical feature compartmentalizes photosynthetic reactions in two distinct types of cells that are arranged concentrically around the veins: mesophyll and bundle sheath cells. Bundle sheath cells surround the vascular tissue, while an outer ring of mesophyll cells is peripheral to the bundle sheath and adjacent to intercellular spaces. Membranes that separate cells assigned to CO_2 fixation from cells destined to reduce carbon are essential for the efficient function of C_4 photosynthesis in land plants. Some unicellular organisms (e.g., diatoms) and a few land plants (typified by *Suaeda aralocaspica* [formerly *Borszczowia aralocaspica*] and two *Bienertia* species) contain the equivalents of the C_4 compartmentalization in a single cell (right panel). Studies on the key photosynthetic enzymes of these plants also indicate two dimorphic chloroplasts located in different cytoplasmic compartments having functions analogous to mesophyll and bundle sheath cells in Kranz anatomy. Products of CO_2 assimilation sustain growth in unicellular organisms and leave the cytosol for vascular tissues in multicellular organisms. (B) Kranz anatomy. Light micrograph of transverse section of leaf blade of *Flaveria australasica* (NAD–malic enzyme type C_4 photosynthesis). (C) Single-cell C_4 photosynthesis. Diagrams of the C_4 cycle are superimposed on electron micrographs of *Suaeda aralocaspica* (left) and *Bienertia cycloptera* (right). (B courtesy of Athena McKown; C from Edwards et al 2004.)

Table 8.5 Mechanisms of C_4 acid decarboxylation in chloroplasts of bundle sheath cells

Decarboxylating enzyme	C_4 acid transported [mesophyll → bundle sheath] for decarboxylation	C_3 acid moved [bundle sheath → mesophyll] for carboxylation	Plant
NADP–malic enzyme (NADP–ME)	Malate	Pyruvate	*Sorghum bicolor, Zea mays*
NAD–malic enzyme (NAD–ME)	Aspartate	Alanine	*Cleome, Atriplex*
PEP carboxykinase (PEPCK)	Aspartate	Alanine, pyruvate phospho*enol*pyruvate,	*Pannicum maximum*

molecules of ATP are consumed in the conversion of pyruvate to phospho*enol*pyruvate: one in the reaction catalyzed by pyruvate–phosphate dikinase (see Table 8.4, reaction 7) and another in the transformation of AMP to ADP catalyzed by adenylate kinase (see Table 8.4, reaction 8). When alanine is the three-carbon compound exported by the bundle sheath cells, the formation of pyruvate by alanine aminotransferase precedes phosphorylation by pyruvate–phosphate dikinase (see Table 8.4, reaction 6).

The compartmentalization of enzymes ensures that inorganic carbon from the surrounding atmosphere can be taken up initially by mesophyll cells, fixed subsequently by the Calvin–Benson cycle of bundle sheath cells, and finally exported to the phloem (see Figure 8.11).

The C_4 cycle uses different mechanisms for decarboxylation of four-carbon acids transported to bundle sheath cells

C_4 photosynthesis transports different four-carbon acids from mesophyll to bundle sheath cells, employs different mechanisms to decarboxylate the four-carbon acid in the bundle sheath cells, and recovers in mesophyll cells different three-carbon acids from bundle sheath cells (**Table 8.5**). Malate and aspartate produced in the chloroplasts and cytosol of mesophyll cells, respectively, are transported to bundle sheath cells.

In the NADP–malic enzyme (NADP–ME) type of C_4 photosynthesis, malate enters the chloroplast of bundle sheath cells where is decarboxylated by NADP–ME (see Table 8.4, reaction 4a).

In the NAD–malic enzyme (NAD–ME) and PEP carboxykinase (PEPCK) types of C_4 photosynthesis, cytosolic aspartate aminotransferase of the bundle sheath cells catalyzes the conversion of aspartate back to oxaloacetate [aspartate + pyruvate → oxaloacetate + alanine]. The decarboxylation of oxaloacetate in both cases takes place in the mitochondria of bundle sheath cells via NAD–ME (see Table 8.4, reaction 4b) and via PEPCK (see Table 8.4, reaction 5). The released CO_2 diffuses from the mitochondria to the chloroplasts of bundle sheath cells.

In the chloroplasts of bundle sheath cells, the released CO_2 by the three decarboxylations increases the concentration of CO_2 around the active site of rubisco, thereby minimizing the inhibition by O_2. Pyruvate (from NADP–ME type C_4 photosynthesis) and alanine (from NAD–ME and PEPCK types) are transported from bundle sheath cells to the mesophyll cells for the regeneration of phospho*enol*pyruvate.

Bundle sheath cells and mesophyll cells exhibit anatomical and biochemical differences

Originally described for tropical grasses and *Atriplex*, the C_4 cycle is now known to occur in at least 62 independent lineages of angiosperms distributed across 19 different families. C_4 plants evolved from C_3 ancestors around 30 million years ago in response to multiple environmental stimuli such as atmospheric changes (decline of CO_2, increase of O_2), modification of global weather, periods of drought, and intense solar radiation. The transition from C_3 to C_4 plants requires the coordinated modification of genes that affect leaf anatomy, cell ultrastructure, metabolite transport, and regulation of metabolic enzymes. The analyses of (i) specific genes and elements that control their expression; (ii) mRNAs and the deduced amino acid sequences; and (iii) C_3 and C_4 genomes and transcriptomes indicate that convergent evolution underlies the multiple origins of C_4 plants.

Except in three terrestrial plants (see below), the distinctive Kranz anatomy increases the concentration of CO_2 in bundle sheath cells to almost 10-fold higher than in the external atmosphere (**Figure 8.12B and C**). The efficient accumulation of CO_2 in the vicinity of chloroplast rubisco reduces the rate of photorespiration to 2 to 3% of photosynthesis. Mesophyll and bundle sheath cells show large biochemical differences. PEPCase and rubisco are located in mesophyll and bundle sheath cells, respectively, while the decarboxylases are found in different intracellular compartments of bundle sheath cells: NADP–ME in chloroplasts, NAD–ME in mitochondria, and PEPCK in the cytosol. In addition, mesophyll cells contain randomly arranged chloroplasts with stacked thylakoid membranes, while chloroplasts in bundle sheath cells are concentrically arranged and exhibit unstacked thylakoids. These chloroplasts correlate with the energy requirements of the type of C_4 photosynthesis. For example, C_4 species of the NADP–ME type, in which malate is shuttled from mesophyll chloroplasts to bundle sheath cells, exhibit functional photosystems II and I in mesophyll chloroplasts, whereas bundle sheath chloroplasts are deficient in photosystem II. NADP–ME species require NADPH in mesophyll chloroplasts for the reduction of oxaloacetate to malate.

The C_4 cycle also concentrates CO_2 in single cells

The finding of C_4 photosynthesis in organisms devoid of Kranz anatomy disclosed a much greater diversity in modes of C_4 carbon fixation than had previously been thought to exist (see **WEB TOPIC 8.10**). Three plants that grow in Asia, *Suaeda aralocaspica* (formerly *Borszczowia aralocaspica*) and two *Bienertia* species, perform complete C_4 photosynthesis within single chlorenchyma cells (see Figure 8.12A and C). The external region, proximal to the external atmosphere, carries out the initial carboxylation and regeneration of phospho*enol*pyruvate, whereas the internal region functions in the decarboxylation of four-carbon acids and the refixation of the liberated CO_2 via rubisco. The cytosol of these Chenopodiaceae species houses dimorphic chloroplasts with different subsets of enzymes.

Diatoms—photosynthetic eukaryotic algae found in marine and freshwater systems—also accomplish C_4 photosynthesis within a single cell. The importance of the C_4 pathway in carbon fixation was confirmed by using inhibitors specific for PEPCase and by identifying nucleotide sequences encoding enzymes essential for C_4 metabolism (PEPCase, PEPCK, and pyruvate–phosphate dikinase) in the genomes of two diatoms, *Thalassiosira pseudonana* and *Phaeodactylum tricornutum*. Although the discovery of these genes suggests that carbon is assimilated through the C_4 pathway, diatoms also possess bicarbonate transporters and carbonic anhydrases that may function to elevate the concentration of CO_2 at the active site of rubisco. Biochemical analyses of C_4-essential enzymes and HCO_3^- transporters will be required to assess the functional importance of the different concentrating mechanisms in diatoms.

Light regulates the activity of key C_4 enzymes

In addition to supplying ATP and NADPH for the operation of the C_4 cycle, light is essential for the regulation of several participating enzymes. Variations in photon flux density elicit changes in the activities of NADP–malate dehydrogenase, PEPCase, and pyruvate–phosphate dikinase by two different mechanisms: thiol–disulfide exchange [Enz–(Cys-S)$_2$ $\leftrightarrow$ Enz–(Cys-SH)$_2$] and phosphorylation–dephosphorylation of specific amino acid residues [e.g., serine, Enz–Ser-OH $\leftrightarrow$ Enz–Ser-OP].

NADP–malate dehydrogenase is regulated via the ferredoxin–thioredoxin system as in C_3 plants (see Figure 8.6). The enzyme is reduced (activated) by thioredoxin when leaves are illuminated, and is oxidized (deactivated) in the dark. The diurnal phosphorylation of PEPCase by a specific kinase, named PEPCase kinase, increases the uptake of ambient CO_2, and the nocturnal dephosphorylation by protein phosphatase 2A brings PEPCase back to low activity. A highly unusual enzyme regulates the dark–light activity of pyruvate–phosphate dikinase. Pyruvate–phosphate dikinase is modified by a bifunctional threonine kinase–phosphatase that catalyzes both ADP-dependent phosphorylation and P_i-dependent dephosphorylation of pyruvate–phosphate dikinase. Darkness promotes the phosphorylation of pyruvate–phosphate dikinase (PPDK) by the regulatory kinase–phosphatase [(PPDK)$_{active}$ + ADP $\rightarrow$ (PPDK-P)$_{inactive}$ + AMP], causing the loss of enzyme activity. The phosphorolytic cleavage of the phosphoryl group in the light by the same enzyme restores the catalytic capacity of PPDK [(PPDK-P)$_{inactive}$ + P_i $\rightarrow$ (PPDK)$_{active}$ + PP_i].

Photosynthetic assimilation of CO_2 in C_4 plants demands more transport processes than in C_3 plants

The chloroplasts export part of the fixed carbon to the cytosol during active photosynthesis while importing the phosphate released from biosynthetic processes to replenish ATP and other phosphorylated metabolites in the stroma. In C_3 plants, the major factors that modulate the partitioning of assimilated carbon between the chloroplast and cytosol are the relative concentrations of triose phosphates and inorganic phosphate. Triose phosphate isomerases rapidly interconvert dihydroxyacetone phosphate and glyceraldehyde 3-phosphate in the plastid and cytosol (**Table 8.6**, reaction 1). The triose phosphate translocator—a protein complex in the inner membrane of the chloroplast envelope—exchanges chloroplast triose phosphates for cytosol phosphate (see Table 8.6, reaction 2) (see **WEB TOPIC 8.11**). Thus, C_3 plants require one transport process across the chloroplast envelope to export triose phosphates (three molecules of CO_2 assimilated) from the chloroplasts to the cytosol.

In C_4 plants, the distribution of photosynthetic CO_2 assimilation over two different cells entails a massive flux of metabolites between mesophyll cells and bundle sheath cells. Moreover, three different pathways accomplish the assimilation of inorganic carbon in C_4 photosynthesis. In this context, different metabolites flow from the cytosol of leaf cells to chloroplasts, mitochondria, and conducting tissues. Therefore, the composition and the function of translocators in organelles and plasma membrane of C_4 plants depend on the pathway used for CO_2 assimilation. For example, mesophyll cells of NADP–ME type C_4 photosynthesis use four transport steps across the chlo-

Table 8.6 Reactions in the conversion of photosynthetically formed triose phosphates to sucrose

1. *Triose phosphate isomerase*
   Dihydroxyacetone phosphate → glyceraldehyde 3-phosphate

2. *Phosphate/triose phosphate translocator*
   Triose phosphate (*chloroplast*) + P_i (*cytosol*) → triose phosphate (*cytosol*) + P_i (*chloroplast*)

3. *Fructose 1,6-bisphosphate aldolase*
   Dihydroxyacetone phosphate + glyceraldehyde 3-phosphate → fructose 1,6-bisphosphate

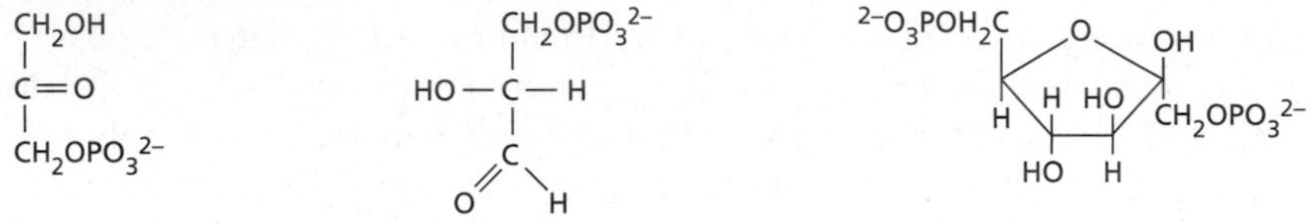

4. *Fructose 1,6-bisphosphatase*
   Fructose 1,6-bisphosphate + H_2O → fructose 6-phosphate + P_i

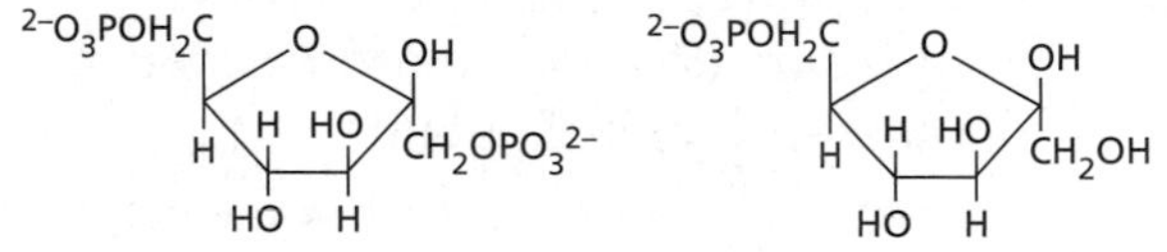

5a. *Fructose 6-phosphate 1-kinase (phosphofructokinase)*
   Fructose 6-phosphate + ATP → fructose 1,6-bisphosphate + ADP

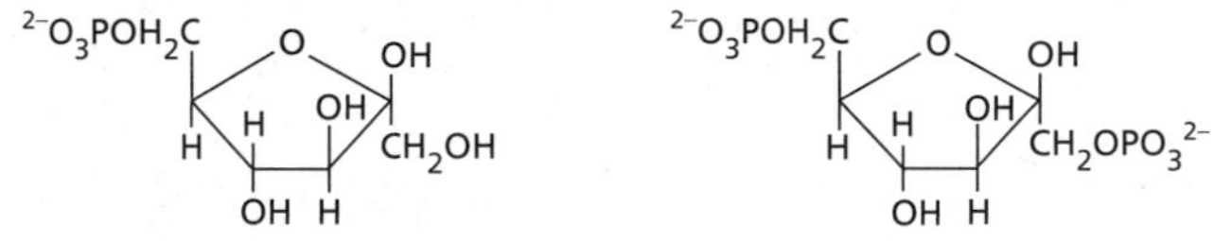

5b. *PP_i-linked phosphofructokinase*
   Fructose 6-phosphate + PP_i → fructose 1,6-bisphosphate + P_i

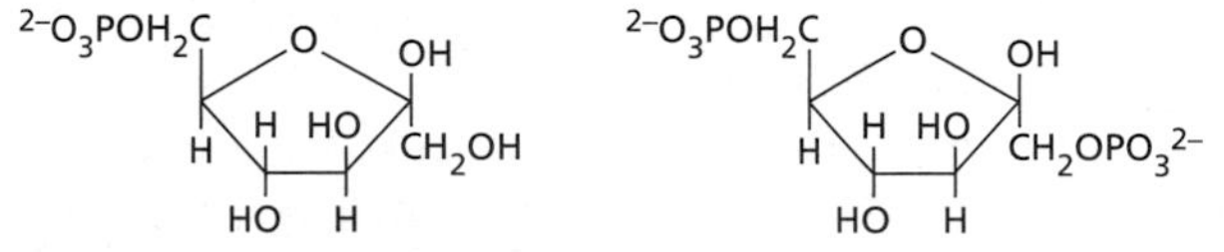

5c. *Fructose 6-phosphate 2-kinase*
   Fructose 6-phosphate + ATP → fructose 2,6-bisphosphate + ADP

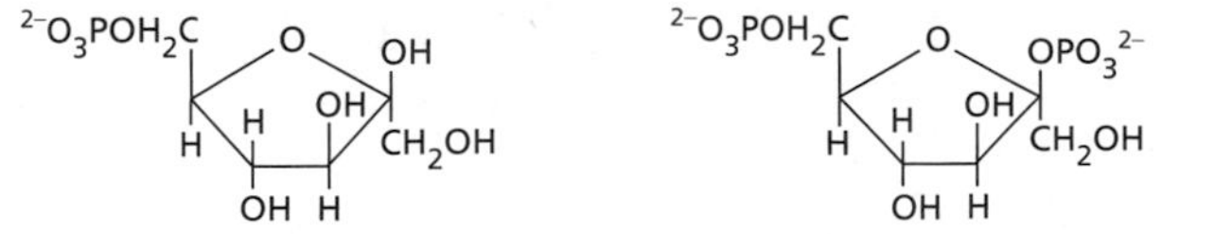

6. *Fructose 2,6-bisphosphatase*
   Fructose 2,6-bisphosphate + H_2O → fructose 6-phosphate + P_i

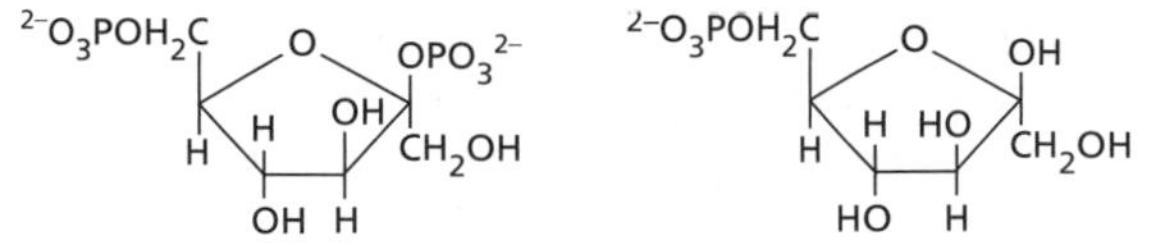

Table 8.6 *(continued)*

7. *Hexose phosphate isomerase*
   Fructose 6-phosphate → glucose 6-phosphate

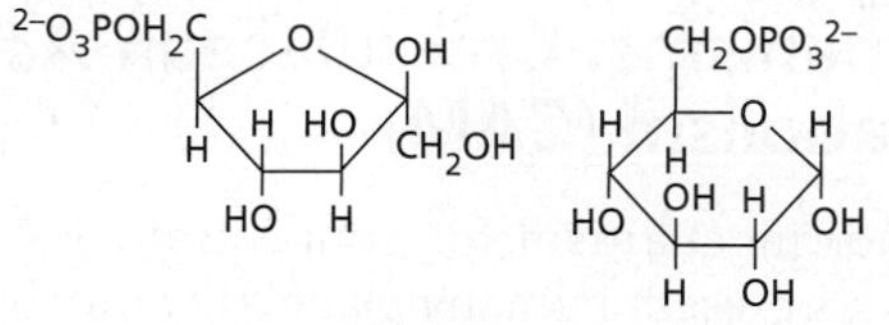

8. *Phosphoglucomutase*
   Glucose 6-phosphate → glucose 1-phosphate

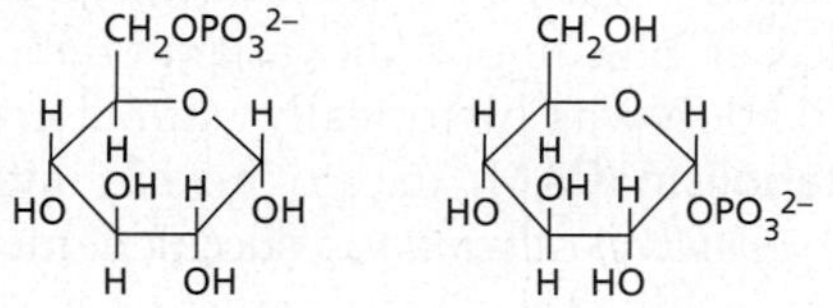

9. *UDP-glucose pyrophosporylase*
   Glucose 1-phosphate + UTP → UDP-glucose + PP_i

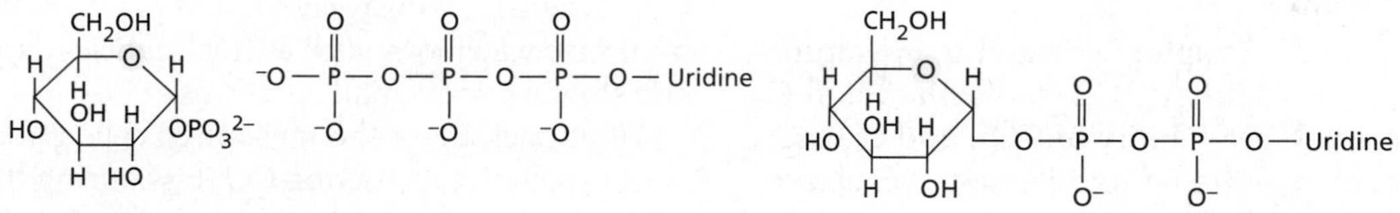

10. *Sucrose 6^F^-phosphate synthase*
    UDP-glucose + fructose 6-phosphate → UDP + sucrose 6^F-phosphate

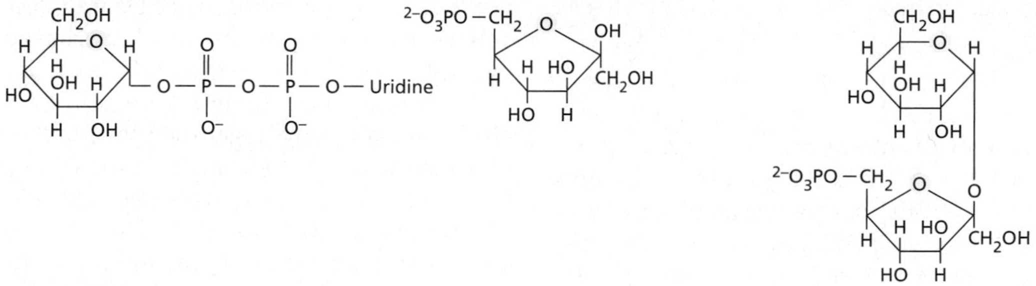

11. *Sucrose 6^F-phosphate phosphatase*
    Sucrose 6^F-phosphate + H_2O → sucrose + P_i

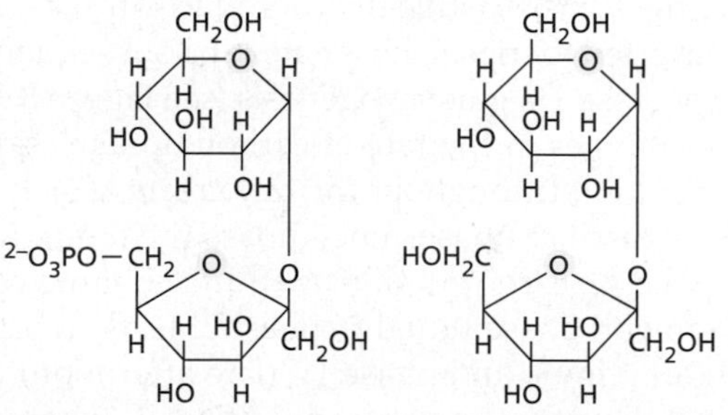

Note: The triose phosphate isomerase (reaction 1) catalyzes the equilibrium between dihydroxyacetone phosphate and glyceraldehyde 3-phosphate in the chloroplast stroma, while the P_i *translocator* (reaction 2) facilitates the exchange between triose phosphates and P_i across the chloroplast inner envelope membrane. All other enzymes catalyze cytosol reactions.

P_i and PP_i stand for inorganic phosphate and pyrophosphate, respectively.

roplast envelope to fix one molecule of atmospheric CO_2: (1) import of cytosolic pyruvate (unknown transporter); (2) export of stromal phospho*enol*pyruvate (phospho*enol*-pyruvate phosphate translocator); (3) import of cytosolic oxaloacetate (dicarboxylate transporter); and (4) export of stromal malate (dicarboxylate transporter).

The adaptation of chloroplast envelopes to the requirements of C_4 photosynthesis was revealed when chloroplast membranes from mesophyll cells of pea (a C_3 plant) and maize (a C_4 plant) were analyzed by liquid chromatography and tandem mass spectroscopy. Chloroplasts from mesophyll cells of C_3 and C_4 plants exhibit qualitatively similar but quantitatively different proteomes in their envelope membranes. In particular, translocators that participate in the transport of triose phosphates and phospho*enol*pyruvate are more abundant in the envelopes of C_4 plants than in those of C_3 plants. This higher abundance ensures that fluxes of metabolic intermediates across the chloroplast envelope in C_4 plants are higher than in C_3 plants.

In hot, dry climates, the C_4 cycle reduces photorespiration

As noted earlier on this chapter, elevated temperatures limit the rate of photosynthetic CO_2 assimilation in C_3 plants by decreasing the solubility of CO_2, and the ratio of the carboxylation to oygenation reactions of rubisco. Because of the decrease in the photosynthetic activity of rubisco, the energy demands associated with photorespiration increase in warmer areas of the world. In C_4 plants, two features contribute to overcome the deleterious effects of high temperature:

- First, atmospheric CO_2 enters the cytoplasm of mesophyll cells where carbonic anhydrase rapidly and reversibly converts CO_2 into bicarbonate [$CO_2 + H_2O \rightarrow HCO_3^- + H^+$] ($K_{eq} = 1.7 \times 10{-}^4$). Warm climates decrease the levels of CO_2, but the low concentrations of cytosolic HCO_3^- saturate PEPCase because the affinity of the enzyme for its substrate is sufficiently high. Thus, the high activity of PEPCase enables C_4 plants to reduce their stomatal aperture at high temperatures and thereby conserve water while fixing CO_2 at rates equal to or greater than those of C_3 plants.
- Second, the high concentration of CO_2 in bundle sheath chloroplasts minimizes the operation of the C_2 oxidative photosynthetic carbon cycle.

The response of net CO_2 assimilation to temperature controls the distribution of C_3 and C_4 species on Earth. The optimal photosynthetic efficiency of C_3 species generally occurs at temperatures lower than for C_4 species: approximately 20–25°C and 25–35°C, respectively. By enabling more efficient assimilation of CO_2 at higher temperatures, C_4 species become more abundant in the tropics and subtropics and less abundant when latitudes depart from the equator. Although C_4 photosynthesis is commonly dominant in warm environments, a group of perennial grasses (*Miscanthus*, *Spartina*) are chilling-tolerant C_4 crops that thrive in areas where the weather is moderately cold.

Inorganic Carbon–Concentrating Mechanisms: Crassulacean Acid Metabolism (CAM)

Another mechanism for concentrating CO_2 around rubisco is present in many plants that inhabit arid environments with seasonal water availability, including commercially important plants such as pineapple (*Ananas comosus*), agave (*Agave* spp.), cacti (Cactaceae), and orchids (Orchidaceae). This important variant of photosynthetic carbon fixation was historically named **crassulacean acid metabolism** (**CAM**), to recognize its initial observation in *Bryophyllum calycinum*, a succulent member of the Crassulaceae. Like the C_4 mechanism, CAM appears to have originated during the last 35 million years to conserve water in habitats where rainfall is insufficient for crop growth. The leaves of CAM plants have traits that minimize water loss, such as thick cuticles, large vacuoles, and stomata with small apertures.

Tight packing of the mesophyll cells enhances CAM performance by restricting CO_2 loss during the day. In all CAM plants, the initial capture of CO_2 into four-carbon acids takes place at night, and the posterior incorporation of CO_2 into carbon skeletons occurs during the day (**Figure 8.13**). At night, cytosolic PEPCase fixes atmospheric and respiratory CO_2 into oxaloacetate using phospho*enol*-pyruvate formed via the glycolytic breakdown of stored carbohydrates (see Table 8.4, reaction 1). A cytosolic NADP–malate dehydrogenase converts the oxaloacetate to malate, which is stored in the acidic solution of vacuoles for the remainder of the night (see Table 8.4, reaction 2). During the day, the stored malate exits the vacuole for decarboxylation by mechanisms similar to those in C_4 plants—that is, by a cytosolic NADP–ME or mitochondrial NAD–ME (see Table 8.4, reactions 4a and 4b). The released CO_2 is made available to chloroplasts for fixation via the rubisco, while the coproduced three-carbon acid is converted to triose phosphates and subsequently to starch or sucrose via gluconeogenesis (see Figure 8.13).

Changes in the rate of carbon uptake and in enzyme regulation throughout the day create a 24-h CAM cycle. Four distinct phases encompass the temporal control of C_4 and C_3 carboxylations within the same cellular environment: phase I (night), phase II (early morning), phase III (daytime), and phase IV (late afternoon) (**WEB TOPIC 8.12**). During the nocturnal phase I, when stomata are open, CO_2 is captured and stored as malate in the vacuole. CO_2 uptake by PEPCase dominates phase I. In the diurnal phase III, when stomata are closed and leaves

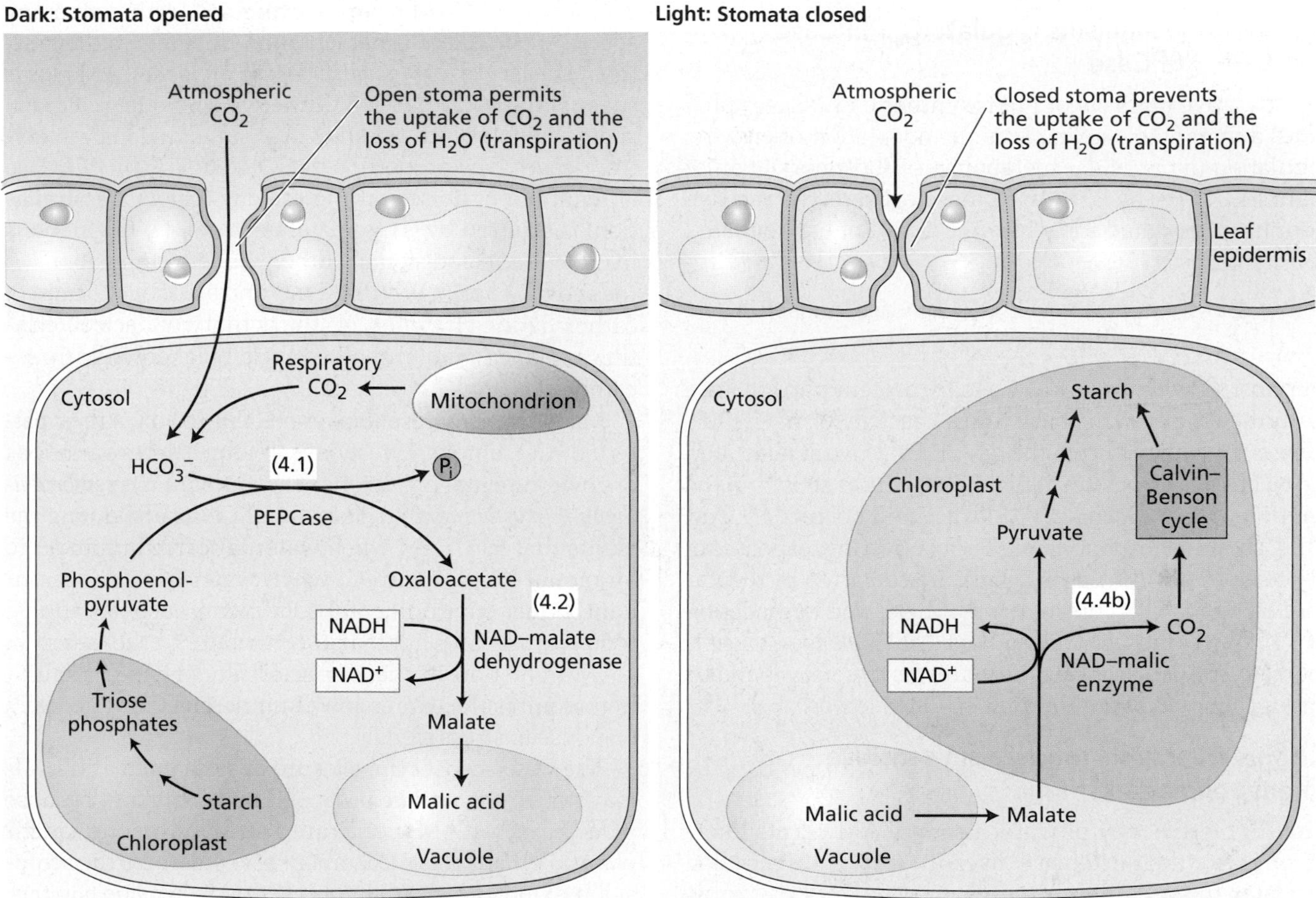

Figure 8.13 Crassulacean acid metabolism (CAM). In CAM metabolism, CO_2 uptake is separated temporally from refixation via the Calvin–Benson cycle. The uptake of atmospheric CO_2 takes place at night when stomata are open. At this stage, gaseous CO_2 in the cytosol, coming from the external atmosphere and the mitochondrial respiration, increases the levels of HCO_3^- [$CO_2 + H_2O \leftrightarrow HCO_3^- + H^+$]. Then cytosolic PEPCase catalyzes a reaction between HCO_3^- and phospho*enol*pyruvate provided by the nocturnal breakdown of chloroplast starch. The four-carbon acid formed, oxaloacetate, is reduced to malate, which in turn proceeds to the acid milieu of the vacuole. During the day, the malic acid stored in the vacuole flows back to the cytosol. The action of the NAD–malic enzyme transforms the malate, releasing CO_2, which is refixed into carbon skeletons by the Calvin–Benson cycle. In essence, the diurnal accumulation of starch in the chloroplast constitutes the net gain of the nocturnal uptake of inorganic carbon. The adaptive advantage of the stomatal closure during the day is that it prevents not only water loss by transpiration, but also the exchange of internal CO_2 with the external atmosphere. See Table 8.4 for a description of numbered reactions.

are photosynthesizing, the stored malate is decarboxylated. This results in high concentrations of CO_2 around the active site of rubisco, thereby alleviating the adverse effects of photorespiration. The transient phases II and IV shift the metabolism in preparation for phases III and I, respectively. In phase II, rubisco activity increases, but it decreases in phase IV. In contrast, the activity of PEPCase increases in phase IV but declines in phase II. The contribution of each phase to the overall carbon balance varies considerably among different CAM plants and is sensitive to environmental conditions. Constitutive CAM plants use the nocturnal uptake of CO_2 at all times, while their facultative counterparts resort to the CAM pathway only when induced by water or salt stress.

Whether the triose phosphates produced by the Calvin–Benson cycle are stored as starch in chloroplasts or used for the synthesis of sucrose depends on the plant species. However, these carbohydrates ultimately ensure not only plant growth, but also the supply of substrates for the next nocturnal carboxylation phase. To sum up, the temporal separation of nocturnal initial carboxylation from diurnal decarboxylation increases the concentration of CO_2 near rubisco and reduces the inevitable inefficiency of the oxygenase activity.

Different mechanisms regulate C_4 PEPCase and CAM PEPCase

Comparative analysis of photosynthetic PEPCases provides a remarkable example of the adaptation of enzyme regulation to particular metabolisms. Phosphorylation of plant PEPCases by PEPCase kinase converts the inactive nonphosphorylated form into the active phosphorylated counterpart:

$$\text{PEPCase}_{\text{inactive}} + \text{ATP} \rightarrow [\text{PEPCase kinase}] \rightarrow \text{PEPCase–P}_{\text{active}} + \text{ADP}$$

Dephosphorylation of PEPCase by protein phosphatase 2A brings the enzyme back to the inactive form. C_4 PEPCase is functional during the day and inactive at night, and CAM PEPCase operates at night and reduces activity in the daytime. Thus, diurnal C_4 PEPCase and nocturnal CAM PEPCase are phosphorylated. The contrasting responses of photosynthetic PEPCases to light are conferred by regulatory elements that control the synthesis and degradation of PEPCase kinases. The synthesis of PEPCase kinase is mediated by light-sensing mechanisms in C_4 leaves and by endogenous circadian rhythms in CAM leaves.

CAM is a versatile mechanism sensitive to environmental stimuli

The high efficiency of water use in CAM plants likely accounts for their extensive diversification and speciation in water-limited environments. CAM plants that grow in deserts, such as cacti, open their stomata during the cool nights and close them during the hot, dry days. The potential advantage of terrestrial CAM plants in arid environments is well illustrated by the unintentional introduction of the African prickly pear (*Opuntia stricta*) into the Australian ecosystem. From a few plants in 1840, the population of *O. stricta* progressively expanded to occupy 25 million ha in less than a century.

Closing the stomata during the day minimizes the loss of water in CAM plants, but because H_2O and CO_2 share the same diffusion pathway, CO_2 must then be taken up by the open stomata at night (see Figure 8.13). The availability of light mobilizes the reserves of vacuolar malic acid for the action of specific decarboxylating enzymes—NAD(P)–ME and PEPCK—and the assimilation of the resulting CO_2 via the Calvin–Benson cycle. CO_2 released by decarboxylation does not escape from the leaf because stomata are closed during the day. As a consequence, the internally generated CO_2 is fixed by rubisco and converted to carbohydrates by the Calvin–Benson cycle. Thus, stomatal closure not only helps conserve water, but also assists in the buildup of the elevated internal concentration of CO_2 that enhances the photosynthetic carboxylation of ribulose 1,5-bisphosphate.

Genotypic attributes and environmental factors modulate the extent to which the biochemical and physiological capacity of CAM plants is expressed. Although many species of succulent ornamental houseplants in the family Crassulaceae (e.g., *Kalanchoë*) are obligate CAM plants that exhibit circadian rhythmicity, others (e.g., *Clusia*) show C_3 photosynthesis and CAM simultaneously in distinct leaves. The proportion of CO_2 taken up by PEPCase at night or by rubisco during the day (net CO_2 assimilation) is adjusted by (1) stomatal behavior, (2) fluctuations in organic acid and storage carbohydrate accumulation, (3) the activity of primary (PEPCase) and secondary (rubisco) carboxylating enzymes, (4) the activity of decarboxylating enzymes, and (5) synthesis and breakdown of three-carbon skeletons.

Many CAM representatives are able to adjust their pattern of CO_2 uptake in response to longer-term variations of environmental conditions. The ice plant (*Mesembryanthemum crystallinum* L.), agave, and *Clusia* are among the plants that use CAM when water is scarce but undergo a gradual transition to C_3 when water becomes abundant. Other environmental conditions, such as salinity, temperature, and light, also contribute to the extent of CAM induction in these species. This form of regulation requires the expression of numerous CAM genes in response to stress signals.

The water-conserving closure of stomata in arid lands may not be the unique basis of CAM evolution, because paradoxically, CAM species are also found among aquatic plants. Perhaps this mechanism also enhances the acquisition of inorganic carbon (as HCO_3^-) in aquatic habitats, where high resistance to gas diffusion restricts the availability of CO_2.

Accumulation and Partitioning of Photosynthates—Starch and Sucrose

Metabolites accumulated in the light—photosynthates—become the ultimate source of energy for plant development. The photosynthetic assimilation of CO_2 by most leaves yields sucrose in the cytosol and starch in the chloroplasts. During the day, sucrose flows continuously from the leaf cytosol to heterotrophic sink tissues, while starch accumulates as dense granules in chloroplasts (**Figure 8.14**) (WEB TOPIC 8.13). The onset of darkness not only stops the assimilation of CO_2, but also starts the degradation of chloroplast starch. The content of starch in the chloroplasts falls through the night because breakdown products flow to the cytosol to sustain the export of sucrose to other organs. The large fluctuation of stromal starch in the light versus the dark is why the polysaccharide stored in chloroplasts is called *transitory starch*. Transitory starch functions as (1) an overflow mechanism that stores photosynthate when the synthesis and transport of sucrose are limited during the day and (2) an energy reserve to provide an adequate supply of carbohydrate at night when sugars are not formed by photosynthesis.

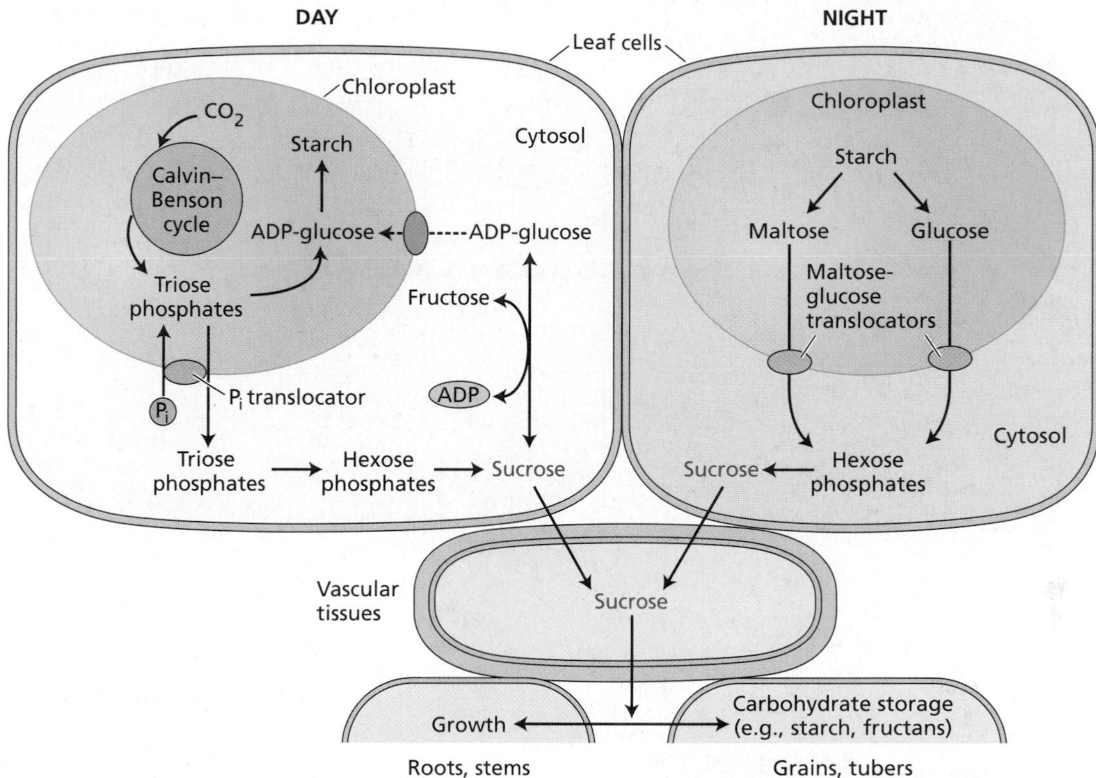

Figure 8.14 Carbon mobilization in land plants. During the day, carbon assimilated photosynthetically is used either for the formation of starch in the chloroplast or is exported to the cytosol for the synthesis of sucrose. External and internal stimuli control the partitioning between starch and sucrose. Triose phosphates from the Calvin–Benson cycle may be used for (1) the synthesis of chloroplast ADP-glucose (the glucosyl donor for starch synthesis) or (2) translocation to the cytosol for the synthesis of sucrose. At night, the cleavage of the glycosidic linkages of starch releases both maltose and glucose, which flow across the chloroplast envelope to supplement the hexose phosphate pool and contribute to sucrose synthesis. Transport across the chloroplast envelope, carried out by translocators for phosphate and maltose and glucose, conveys information between the two compartments. As a consequence of the diurnal synthesis and nocturnal breakdown, the levels of chloroplast starch are maximal during the day and minimal at night. This transitory starch serves as the nocturnal energy reserve that provides an adequate supply of carbohydrates to land plants, and also as a diurnal overflow that accepts the carbon excess when photosynthetic CO_2 assimilation proceeds faster than the synthesis of sucrose. Daily, sucrose links the assimilation of inorganic carbon (CO_2) in leaves to the utilization of organic carbon for growth and storage in nonphotosynthetic parts of the plant.

Plants vary widely in the extent to which they accumulate starch and sucrose in leaves (see Figure 8.14). In some species (e.g., soybean, sugar beet, Arabidopsis), the proportion of starch to sucrose in the leaf is almost constant throughout the day. In others (e.g., spinach, French beans), starch accumulates when sucrose exceeds the storage capacity of the leaf or the demand of sink tissues.

The carbon metabolism of leaves also responds to the requirements of sink tissues for energy and growth. Regulatory mechanisms ensure that physiological processes in the chloroplast are synchronized not only with the cytoplasm of the leaf cell but also with other parts of the plant during the day–night cycle. An abundance of sugars in leaves promotes plant growth and carbohydrate storage in reserve organs, while low levels of sugars in sink tissues stimulate the rate of photosynthesis. Sucrose transport links the availability of carbohydrates in source leaves to the use of energy and the formation of reserve polysaccharides in sink tissues (see Chapter 11).

Formation and Mobilization of Chloroplast Starch

Starch is the main storage carbohydrate in plants, being surpassed only by cellulose as the most abundant polysaccharide. In the light, chloroplasts store part of the assimilated carbon as insoluble starch granules that are degraded at night. The 24-h rhythm of starch turnover adjusts to the ambient situation. For example, Arabidopsis plants grown in short days (6-h day/18-h night) allocate into starch more photosynthate than plants grown in long days (18-h day/6-h night), but in both cases the

(A) Amylose

CH$_2$OH CH$_2$OH CH$_2$OH CH$_2$OH CH$_2$OH CH$_2$OH
O
α-1,4
OH
n
m

Amylopectin

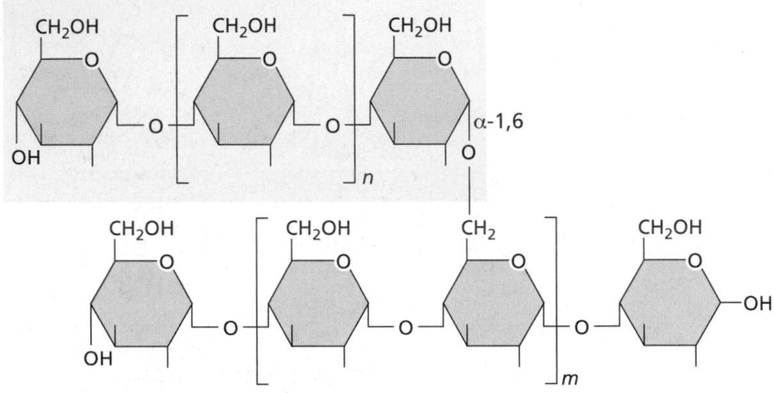

(B)

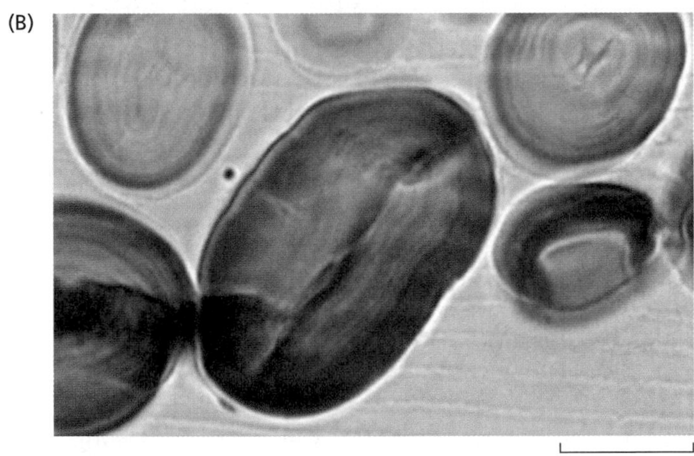

(C) Amylopectin cluster (0.1–1 nm)

Lamella (10 nm)

Crystalline

Amorphous

Blocklet (20–250 nm)

Normal blocklets

Defective blocklets

Starch granule (1000 nm)

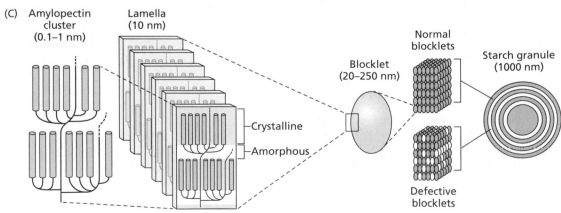

transitory starch is consumed by dawn. In the sections that follow, we will consider the chloroplast processes associated with the diurnal accumulation and nocturnal degradation of starch.

Chloroplast stroma accumulates starch as insoluble granules during the day

Starch, like glycogen, is a complex polysaccharide built up from a single monosaccharide—glucose—that consists of two major components, amylopectin and amylose (**Figure 8.15A**). The α-D-glucosyl units associate in long linear chains linked through α-D-1,4 glycosidic linkages wherein α-D-1,6 glycosidic linkages are formed as branch points. The contribution of α-D-1,6 glycosidic linkages to total bonds is lower in amylose (less than 1%) than in amylopectin (~5–6%); thus, the former is essentially linear and the latter branched. The molecular weight of amylose (500–20,000 glucose units) is lower than that of amylopectin (~10^6 glucose units). The structure, size, and proportions of amylose and amylopectin in the starch granule vary among plant species.

Chloroplasts store large amounts of reduced carbon without changing the osmotic balance of the cell by packing amylose and amylopectin into insoluble starch granules (**Figure 8.15B**) (WEB TOPIC 8.13). The amylose content and the ratio of long to short branch chains in amylopectin regulate the structure and size of the starch granule.

◀ **Figure 8.15** Composition and structure of the starch granule. (A) Starch is made up of amylose and amylopectin. Glucose units are linked almost exclusively through α-D-1,4 glucosidic bonds in amylose. Amylopectin also contains α-D-1,4-linked glucose chains (6 < n,m < 100 glucose residues), but they are interspersed with α-D-1,6-glycosidic bonds (branch points) that give a treelike structure to the macromolecule. (B) Concentric layers of the starch granule are revealed by light microscopy of iodine-stained sections of pea seed starch. Iodine reacts primarily with the amylose component. (C) Four levels of organization make up the starch granule: the cluster of amylopectin molecules (0.1–1 nm), the lamella (~10 nm), the blocklet (20–250 nm), and the whole granule (> 1000 nm). Amylopectin molecules are closely packed with other molecules of amylopectin, forming clusters of double helices. The crystalline lamella is created by the association of amylopectin double helices interspersed with amorphous regions. The blocklet is the ordered aggregation of several crystalline-amorphous lamellae into an asymmetrical structure with an axial ratio of 3:1 (named "normal blocklets"). Amylose and other materials (e.g., water, lipids) disturb the regular formation of blocklets, introducing "defects" (named "defective blocklets"). The ordered aggregation of normal and defective blocklets forms the concentric rings of hard (crystalline) and soft (semicrystalline) shells in the starch granule. (B from Ridout et al 2003.)

Moreover, the association of stromal components (phosphate monoesters, lipids, phospholipids, and proteins) with the granule also controls the molecular architecture (**Figure 8.15C**). As the accumulation of starch granules in the stroma exerts tension on the envelope, ion channels perceive the mechanical stimuli and rapidly adjust the volume and shape of chloroplasts. The fluctuation of transitory starch stems from changes in the size of a fixed number of starch granules.

The biosynthesis of amylose and amylopectin proceeds through successive steps: initiation, elongation, branching, and termination of the polysaccharide chain. Numerous studies have improved our understanding of elongation and branching, but knowledge of initiation and termination remains limited.

The sugar nucleotide ADP-glucose provides the glucosyl moiety for the biosynthesis of the α-D-1,4 glycosidic linkages of amylose. Although the origin of chloroplast ADP-glucose is controversial, the chloroplast enzyme ADP-glucose pyrophosphorylase (AGPase) catalyzes the synthesis of most of this starch precursor (**Figure 8.16A**, reaction 1). The elongation of amylose proceeds via the enzyme starch synthase, which catalyzes the transfer of the glucosyl moiety of ADP-glucose to the nonreducing end of a preexisting α-D-1,4-glucan primer. The glucose added to the glucan retains the α-configuration in the new glycosidic bond (see Figure 8.16A, reaction 2). Multiple isoforms of starch synthase are located in the soluble stroma and in association with the particulate starch granules (see below).

During the elongation process, starch-branching enzymes transfer a segment of an α-D-1,4-glucan chain to a carbon 6 of glucosyl moieties in the same glucan, forming a novel α-D-1,6 glycosidic linkages (see Figure 8.16A, reaction 3). Starch-branching enzymes are also present in various isoforms that differ not only in the length of the glucan chain transferred, but also in their location in the stroma and starch granules.

The randomly branched amylopectin generally does not integrate into the granule of starch. The isoamylases and the **disproportionating enzyme** (**D-enzyme**) process inappropriately positioned branches. Isoamylases trim the branches that impede formation of the crystalline regions of amylopectin, and the trimmed polysaccharide can be integrated into the starch granule (**Figure 8.16B**, reaction 4). The D-enzyme recycles the residual oligosaccharides back to the biosynthesis of starch through the glucan transferase reaction:

$$(\text{Glucose})_m + (\text{glucose})_n \rightarrow (\text{glucose})_{m+n-x} + (\text{glucose})_x$$

where m and $n \geq 3$ and $x \leq 4$ (see Figure 8.16B, reaction 5). The products of this reaction become substrates for the action of starch synthases and branching enzymes (see Figure 8.16B, reactions 2 and 3).

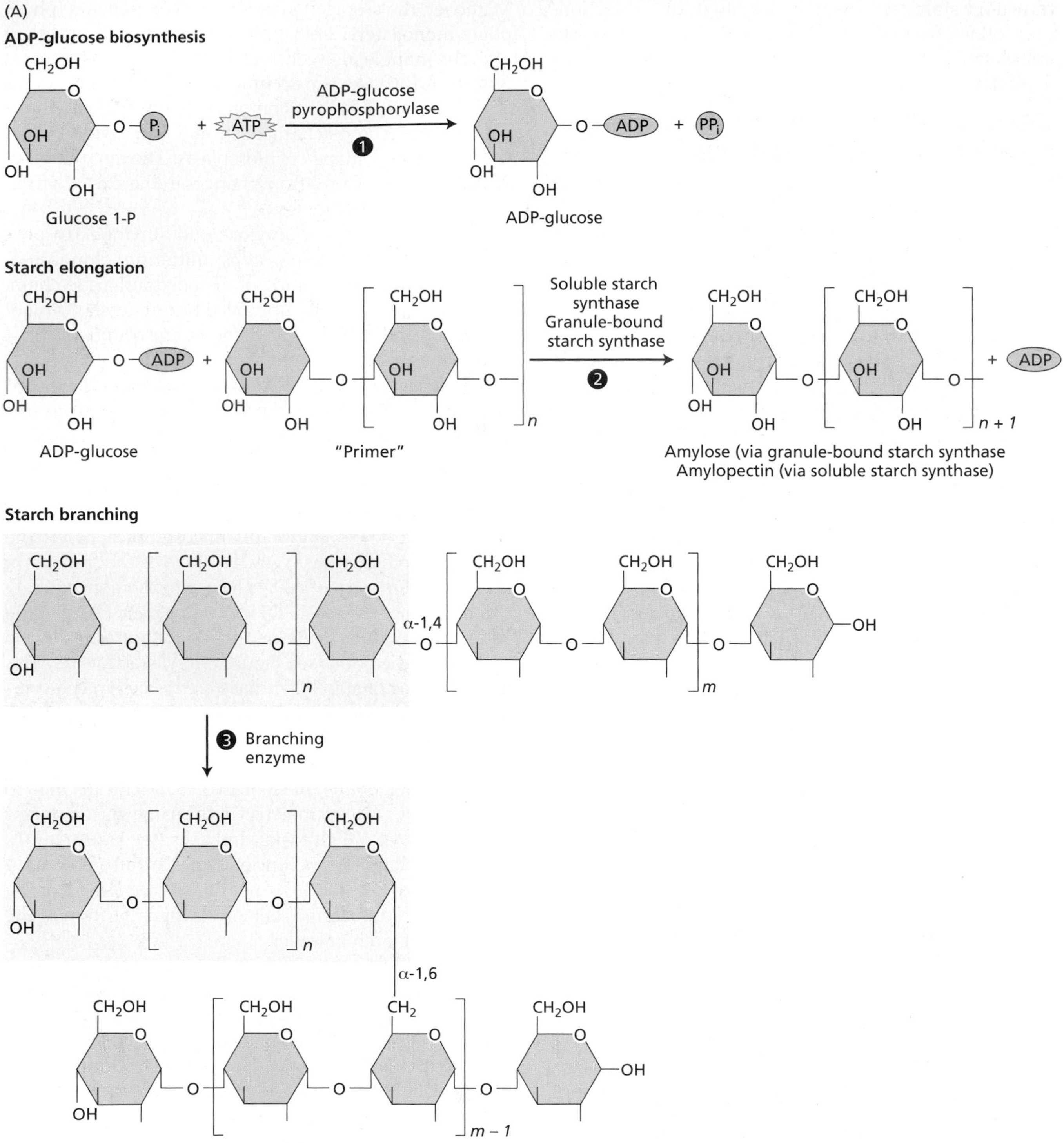

(A)

ADP-glucose biosynthesis

Glucose 1-P + ATP → (❶ ADP-glucose pyrophosphorylase) → ADP-glucose + PP_i

Starch elongation

ADP-glucose + "Primer" → (❷ Soluble starch synthase, Granule-bound starch synthase) → + ADP

Amylose (via granule-bound starch synthase
Amylopectin (via soluble starch synthase)

Starch branching

α-1,4

❸ Branching enzyme

α-1,6

Figure 8.16 Pathway of starch synthesis. Starch biosynthesis in plants is a complex process that includes biosynthesis of the sugar nucleotide ADP-glucose, formation of the "primer," elongation of the linear α-D-1,4-linked glucan, and branching of the amylose molecule for the biosynthesis of amylopectin. (A) Starch elongation and branching. (1) The first committed step of starch biosynthesis is the formation of ADP-glucose. The enzyme ADP-glucose pyrophosphorylase catalyzes the formation of ADP-glucose from ATP and glucose 1-phosphate with the concurrent release of pyrophosphate. (2) The next step in the formation of starch is the successive addition of glucosyl moieties through α-D-1,4 linkages that elongate the polysaccharide. Starch synthases transfer the glucosyl moiety of ADP-glucose to the nonreducing end of a preexisting α-D-1,4-glucan primer, retaining the anomeric configuration of the glucose in the glycosidic bond. The biosynthetic pathway for the formation of the primer remains elusive. The multiple isoforms of starch syn-

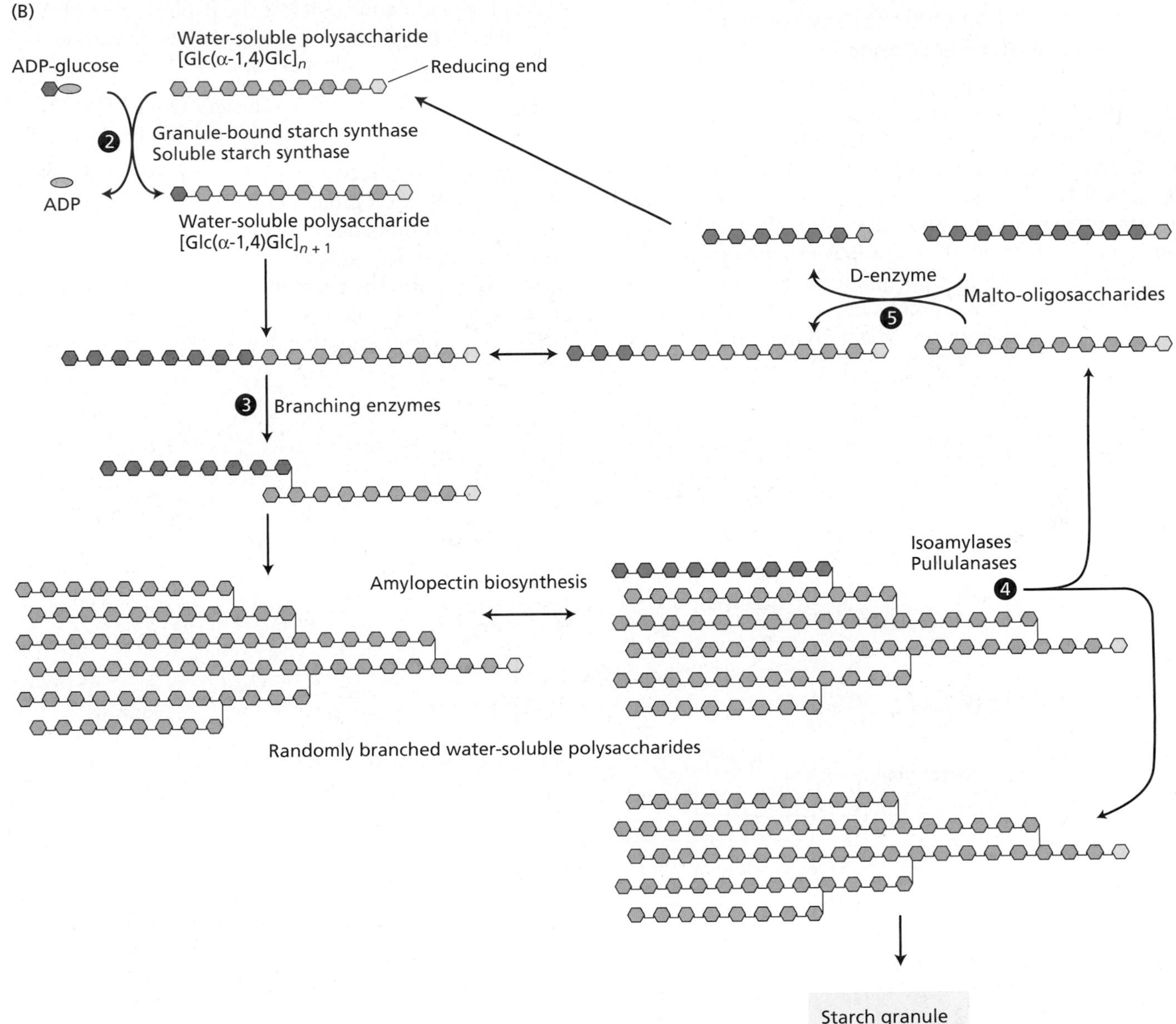

thases found in plant tissues are the granule-bound starch synthase, located essentially within the granule matrix, and the soluble starch synthases, which are partitioned between granular and stromal fractions according to species, tissues, and developmental stages. (3) Branching enzymes catalyze the formation of branch points within the glucan chains through cleavage of α-D-1,4 linkages and the transfer of the released oligosaccharide to a linear glucan, forming an α-D-1,6 linkage. (B) Biosynthesis of amylopectin. Reactions 2 and 3 are as in (A). (4) The yellow unit illustrates the reducing end of the polysaccharide, that is, the glucose moiety whose aldehyde groups do not form a glycosidic linkage. Debranching enzymes cleave α-D-1,6 linkages off randomly branched water-soluble polysaccharides, yielding small linear α-D-1,4-glucans (malto-oligosaccharides). Depending on their substrate requirements, these enzymes are either isoamylases or pullulanases. The former are active toward loosely spaced branches of amylopectin, while the latter exhibit high activity toward the tightly spaced branches of the glucan polymer. Released malto-oligosaccharides may in turn constitute adequate primers for granule-bound starch synthases or serve as substrate for the disproportionating enzyme (D-enzyme). (5) The D-enzyme alters (disproportionates) the chain length distribution of pools of malto-oligosaccharides [(α-1,4-glucan)$_m$ + (α-D-1,4-glucan)$_n$ → (α-D-1,4-glucan)$_{m-x}$ + (α-D-1,4-glucan)$_{n+x}$]. In essence, the D-enzyme catalyzes the cleavage and subsequent transfer of α-D-1,4-linked glucans moieties [x] from a malto-oligosaccharide donor [(α-D-1,4-glucan)$_m$] to a counterpart acceptor [(α-D-1,4-glucan$_n$]. At this stage, the shortened malto-oligosaccharide may serve as "primer" in the elongation (2), while the enlarged malto-oligosaccharide may serve as water-soluble polysaccharides in the branching process (3).

Starch degradation at night requires the phosphorylation of amylopectin

Creative molecular approaches for construction of transgenic plants, biochemical analyses, and genome sequence information have conceived a new picture of the pathway involved in the nocturnal degradation of transitory starch (**Figure 8.17**). At night, starch must be phosphorylated for the formation of maltose, the predominant form of carbon exported from the chloroplast to the cytosol. Glucan–water dikinase and phosphoglucan–water dikinase incorporate phosphoryl groups into the transitory starch. Unlike most kinases, glucan–water dikinase releases inorganic phosphate and transfers the β-phosphate of ATP (indicated by blue P in the equation below) to carbon 6 of glucosyl moieties of amylopectin:

$$\text{Adenosine-P-}\mathbf{P}\text{-P (ATP)} + \text{(glucan)–O–H} + H_2O \rightarrow \text{adenosine-P (AMP)} + \text{(glucan)–O–}\mathbf{P} + P_i$$

Although phosphoryl groups occur infrequently in leaf starch (1 phosphoryl group per 2000 glucosyl residues in Arabidopsis), diminished activities of glucan–water dikinase in transgenic plants decreases starch degradation. As a consequence, the content of starch in mature leaves of transgenic Arabidopsis lines (named *starch excess 1*, or *sex1*) is up to seven times that of wild-type leaves. Thiore-

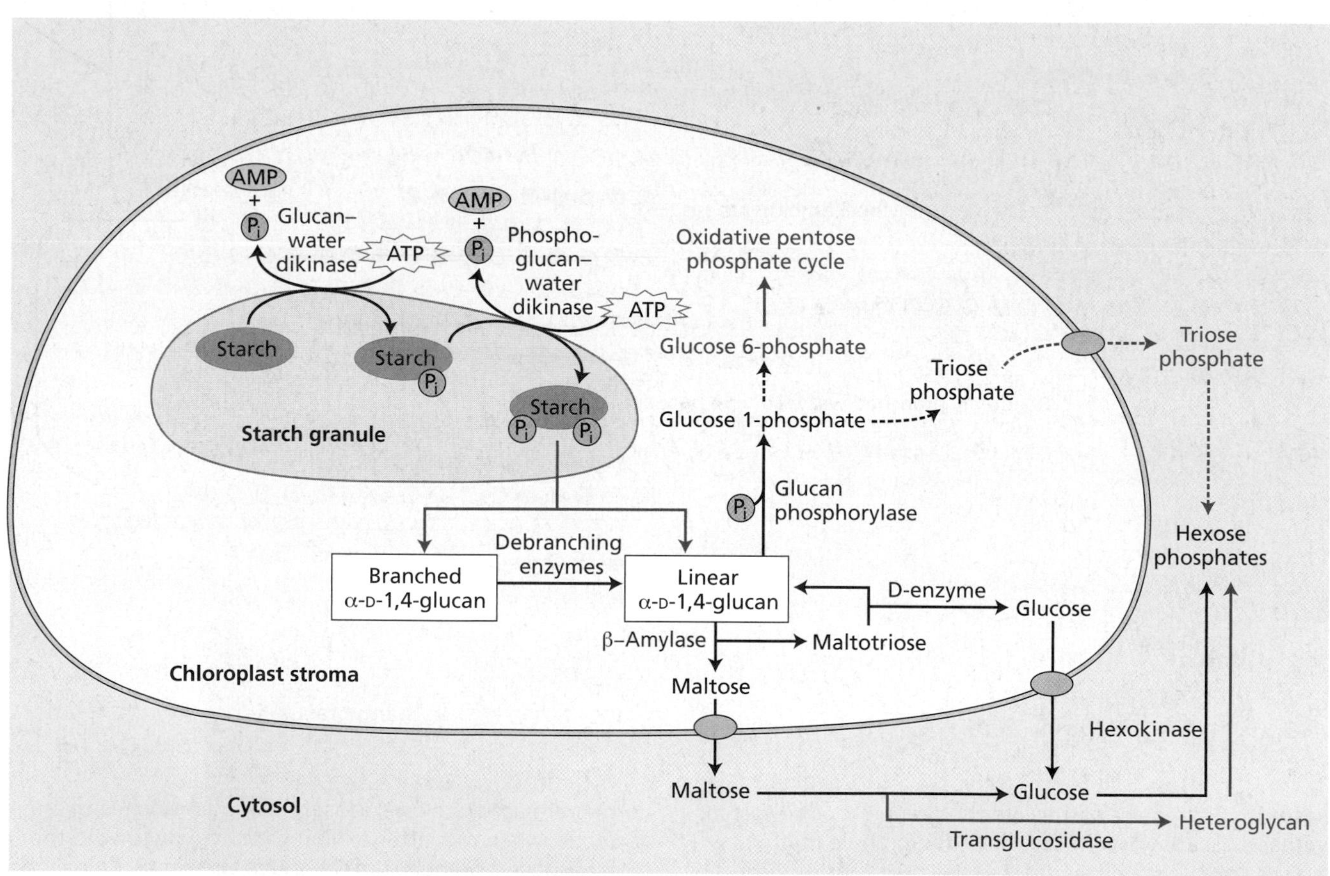

Figure 8.17 Nocturnal starch degradation in Arabidopsis leaves. The release of soluble glucans from the starch granule at night requires prior phosphorylation of the polysaccharide via glucan–water dikinase and phosphoglucan–water dikinase. At this stage, debranching enzymes transform the branched starch into linear glucans, which in turn can be converted into maltose via β-amylolysis catalyzed by chloroplast β-amylase. The residual maltotriose is transformed to maltopentaose and glucose via the D-enzyme. The maltopentaose product is adequate for hydrolysis via chloroplast β-amylase, while the glucose can be exported to the cytosol. Under stress conditions, the phosphorolytic cleavage of α-D-1,4-glucans catalyzed by chloroplast glucan phosphorylase yields glucose 1-phosphate, which can be cleaved to triose phosphate and exchanged for phosphate, or incorporated into the oxidative pentose phosphate cycle. Two transporters in the chloroplast envelope, one for maltose and another for glucose, facilitate the flow of the products of starch degradation to the cytosol. The use of maltose in the leaf cytosol proceeds via a transglucosidase that transfers a glucosyl moiety to a heteroglycan and concurrently releases a molecule of glucose. The cytosolic glucose can be phosphorylated via a hexokinase to glucose 6-phosphate for incorporation into the pool of hexose phosphates.

doxin-dependent processes regulate (1) the catalytic activity and (2) the distribution of the enzyme between the stroma and the starch granule.

Land plants contain a second enzyme, phosphoglucan–water dikinase, that catalyzes a reaction similar to glucan–water dikinase but requires strictly a phosphorylated glucan as substrate. Phosphoglucan–water dikinase adds the β-phosphate of ATP to carbon 3 of glucosyl moieties of amylopectin and releases inorganic phosphate (see Figure 8.17):

$$\text{Adenosine-P-}\mathbf{P}\text{-P (ATP)} + [\mathbf{P}\text{–glucan}]\text{–O–H} + H_2O$$
$$\rightarrow \text{adenosine-P (AMP)} + [\mathbf{P}\text{–glucan}]\text{–O–}\mathbf{P} + P_i$$

Mutants lacking phosphoglucan–water dikinase also contain increased levels of starch but, unlike *sex1* mutants, do not exhibit altered content of phosphorylated amylopectin.

The export of maltose prevails in the nocturnal breakdown of transitory starch

Two mechanisms accomplish the cleavage of the α-D-1,4 glycosidic bond of phosphorylated starch (see Figure 8.17).

1. Hydrolysis, catalyzed by amylases:

$$[\text{Glucose}]_n + H_2O \rightarrow [\text{glucose}]_{n-m} + [\text{glucose}]_m$$
$$[\alpha\text{-amylase}]$$

$$[\text{Glucose}]_n + H_2O \rightarrow \text{linear } [\text{glucose}]_{n-2} + \text{maltose}$$
$$[\beta\text{-amylase}]$$

2. Phosphorolysis, catalyzed by α-glucan phosphorylases:

$$[\text{Glucose}]_n + P_i \rightarrow [\text{glucose}]_{n-1} + \text{glucose 1-phosphate}$$

As maltose is the major product of starch breakdown that chloroplasts export to the cytoplasm at night, β-amylases form the disaccharide by acting on the starch granule or on oligosaccharides released from the granule by α-amylases. However, neither α-amylases nor β-amylases hydrolyze the α-D-1,6 glycosidic bond that constitutes 4 to 5% of the glycosidic linkages in amylopectin (see Figure 8.17). Two debranching enzymes, pullulanase (limit dextrinase) and isoamylase, are essential for the complete breakdown of starch granules to linear glucans (see Figure 8.17). The linear glucans provided by pullulanases and isoamylases are further degraded at night by chloroplast β-amylase.

The production of maltose leads unavoidably to the formation of low amounts of maltotriose, because the exhaustive action of β-amylase cannot further process the trisaccharide (see Figure 8.17). The D-enzyme catalyzes the following transformation:

$$2\,[\text{Glucose}]_3 \rightarrow [\text{glucose}]_5 + \text{glucose}$$
$$(\text{Maltotriose}) \qquad (\text{maltopentaose})$$

The formation of maltopentaose, which is processed via β-amylases, and the export of glucose to the cytosol, via the glucose transporter in the innter chloroplast membrane, prevent the accumulation of the maltotriose as starch breaks down during the night.

A protein of the inner chloroplast membrane, maltose transporter, selectively transports maltose across the envelope. The use of maltose in the leaf cytosol follows a biochemical pathway unsuspected before the advent of transgenic plants. Transgenic lines devoid of a cytosolic transglucosidase degrade starch poorly and accumulate maltose to levels much higher than in wild-type plants. The transglucosylation reaction catalyzed by this enzyme transfers a glucosyl moiety from maltose to cytosolic heteroglycans constituted of arabinose, galactose, and glucose [(heteroglycans) + maltose → (heteroglycans)-glucose + glucose]. Phosphorylation of the remaining glucose by hexokinase adds glucose 6-phosphate to the hexose phosphate pool for conversion to sucrose.

The synthesis and degradation of the starch granule are regulated by multiple mechanisms

Numerous mechanisms regulate the activity of enzymes involved in starch metabolism.

REDOX CONTROL The importance of reducing and oxidizing conditions in the control of starch breakdown comes from biochemical experiments (AGPase, glucan–water dikinase, phosphoglucan phosphatase, and β-amylase 1) and potential targets of thioredoxin in proteomics screens (β-amylase, α-glucan phosphorylase, ADP-glucose translocator, and starch branching enzyme IIa).

PROTEIN PHOSPHORYLATION The rapid response is the distinctive feature of signaling via protein phosphorylation. In the plastid, specific protein kinases catalyse the transfer of the γ-phosphate from ATP to specific amino acids (generally serine, threonine, and tyrosine) of enzymes related to starch metabolism (phosphoglucoisomerase, phosphoglucomutase, AGPase, glucan–water dikinases, transglucosidase [dpe2], α-amylase 3, β-amylases, limit dextrinase, starch branching enzymes, starch synthases, granule-bound starch synthase, α-glucan phosphorylase, glucose transporter, and maltose transporter). The physiological role of these phosphorylations is unknown.

FORMATION OF COMPLEXES WITH PROTEINS Many enzymes involved in the formation of the granule (soluble and granule-bound starch synthases, α-amylases, and glucan–water dikinases) bind to scaffold proteins that possess starch-binding domains. The formation of these heterocomplexes markedly changes the activity of enzymes.

ALLOSTERIC EFFECTORS (LOW MOLECULAR WEIGHT METABOLITES) Small molecules interact with enzyme sites distal to the active site and thereby perturb the catalytic activity over a distance—that is, they have an allosteric effect. Thus, low molecular weight metabolites participate actively in the starch synthesis. For example, the disaccharide trehalose [α-D-Gluc-(1→1)-α- D-Gluc] does not accumulate to any great extent in the vast majority of plants, but trehalose 6-phosphate significantly increases the reductive activation of ADP-glucose pyrophosphorylase.

Sucrose Biosynthesis and Signaling

The production of sucrose in the leaf cytosol, coupled to loading and translocation in the phloem, ensures a suitable supply of carbohydrates for the optimal development of the plant. Additionally, sucrose communicates the carbon and energy status of tissues sustaining autotrophic assimilation (leaves) to compartments performing heterotrophic consumption (e.g., roots, tubers, and grains). Thus, sucrose not only provides carbon skeletons for growth and polysaccharide biosynthesis, but is also a key signaling molecule that regulates carbon partitioning between source leaves and sink tissues. This section will describe mainly the mechanisms that allocate products of photosynthetic CO_2 assimilation to the cytosol for the synthesis of sucrose.

Triose phosphates from the Calvin–Benson cycle build up the cytosolic pool of three important hexose phosphates in the light

During active photosynthesis, the accumulation of dihydroxyacetone phosphate and glyceraldehyde 3-phosphate in the cytosol increases the formation of fructose 1,6-bisphosphate catalyzed by cytosolic aldolase ($\Delta G^{0\prime}$ = 24 kJ/mol) (**Figure 8.18**; also see Table 8.6, reaction 3).

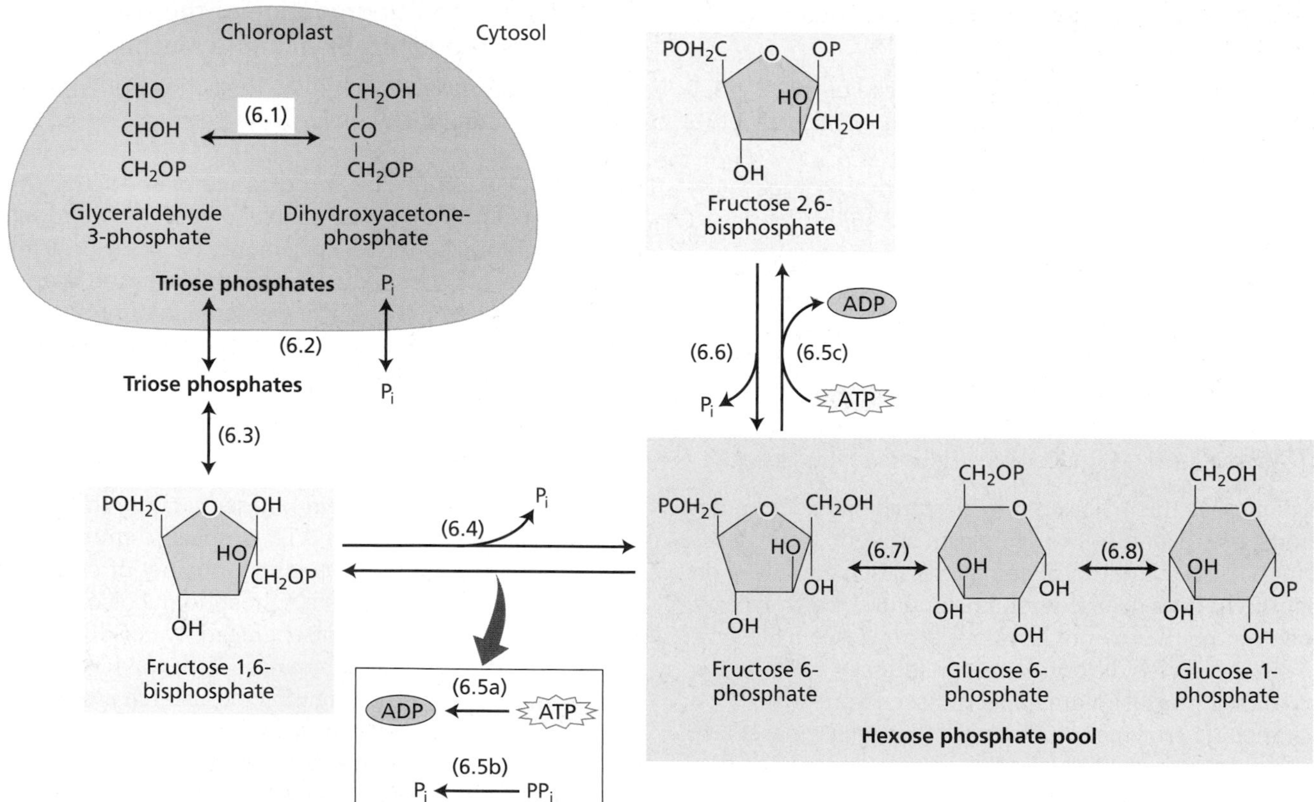

Figure 8.18 Interconversion of hexose phosphates. Fructose 1,6-bisphosphate, formed from triose phosphates by the action of aldolase, is cleaved at the carbon 1 position by cytosolic fructose-1,6-bisphosphatase, which differs structurally and functionally from the chloroplast counterpart. Fructose 6-phosphate constitutes the starting substrate for three transformations. *First*, land plants employ two different phosphorylation reactions of fructose 6-phosphate at the carbon 1 position of the furanose ring: the classic ATP-dependent phosphofructokinase (see glycolysis in Chapter 12), and a pyrophosphate-dependent phosphofructokinase that catalyzes the readily reversible phosphorylation of fructose 6-phosphate using pyrophosphate as substrate. *Second*, fructose 6-phosphate 2-kinase catalyzes the ATP-dependent phosphorylation of fructose 6-phosphate to fructose 2,6-bisphosphate and, in turn, fructose 2,6-bisphosphate phosphatase catalyzes the hydrolysis of fructose 2,6-bisphosphate, releasing the phosphoryl group and again yielding fructose 6-phosphate. *Third*, hexose phosphate isomerase and glucose 6-phosphate isomerase, respectively, favor the isomerization of fructose 6-phosphate to glucose 6-phosphate and of glucose 6-phosphate to glucose 1-phosphate. Collectively, fructose 6-phosphate, glucose 6-phosphate, and glucose 1-phosphate constitute the pool of hexose phosphates. See Table 8.6 for a description of numbered reactions.

Given that the cytosolic aldolase catalyzes the reaction of two triose phosphates, the K_{eq} for this reaction is:

K_{eq} = [dihydroxyacetone phosphate] × [glyceraldehyde 3-phosphate] / [fructose 1,6-bisphosphate]
= [triose phosphates]^2 / [fructose 1,6-bisphosphate],

implying that the concentration of fructose 1,6-bisphosphate varies exponentially in response to changes in the concentration of triose phosphates. Hence, a constant input of triose phosphates from photosynthetically active chloroplasts biases the aldolase reaction in the cytosol of leaf cells toward the formation of fructose 1,6-bisphosphate. The reverse reaction—the aldol cleavage of fructose 1,6-bisphosphate to dihydroxyacetone phosphate and glyceraldehyde 3-phosphate—takes place when the proportion of fructose 1,6-bisphosphate is high in relation to triose phosphates, for example in glycolysis.

Cytosolic fructose 1,6-bisphosphatase subsequently catalyzes the hydrolysis of fructose 1,6-bisphosphate at the carbon 1 position, yielding fructose 6-phosphate and phosphate ($\Delta G^{0\prime}$ = 16.7 kJ/mol) (see Figure 8.18 and Table 8.6, reaction 4).

Cytosolic fructose 6-phosphate can proceed to different destinations through:

1. Phosphorylation of carbon 1, which restores fructose 1,6-bisphosphate, catalyzed by two enzymes, phosphofructokinase and pyrophosphate-dependent phosphofructokinase (see Table 8.6, reactions 5a and b).
2. Phosphorylation of carbon 2, which yields fructose 2,6-bisphosphate, catalyzed by a unique bifunctional enzyme confined to the cytosol. Fructose 6-phosphate 2-kinase/fructose 2,6-bisphosphate phosphatase catalyzes both the incorporation and the hydrolysis of the phosphoryl group (see Table 8.6, reaction 5c and 6).
3. Isomerization, which produces glucose 6-phosphate, catalyzed by hexose phosphate isomerase (see Table 8.6, reaction 7).

The cytosolic concentration of fructose 6-phosphate is kept close to equilibrium with glucose 6-phosphate and glucose 1-phosphate by readily reversible reactions catalyzed by hexose phosphate isomerase ($\Delta G^{0\prime}$ = 8.7 kJ/mol) and phosphoglucomutase ($\Delta G^{0\prime}$ = 7.3 kJ/mol) (see Table 8.6, reactions 7 and 8). These three sugar phosphates are collectively named *hexose phosphates* (see Figure 8.18).

Fructose 2,6-bisphosphate regulates the hexose phosphate pool in the light

The regulatory metabolite cytosolic fructose 2,6-bisphosphate regulates the exchange of triose phosphates and phosphate for the formation of the hexose phosphate pool. A high ratio of triose phosphates to phosphate in the cytosol, typical of photosynthetically active leaves, suppresses the formation of fructose 2,6-bisphosphate, because triose phosphates strongly inhibit the kinase activity of the bifunctional enzyme fructose 6-phosphate 2-kinase/fructose 2,6-bisphosphate phosphatase. By contrast, a low ratio of triose phosphates to phosphate, typical of limited photosynthesis, promotes the synthesis of fructose 2,6-bisphosphate because phosphate stimulates the fructose 6-phosphate 2-kinase activity and inhibits the fructose 2,6-bisphosphatase activity. Higher concentrations of fructose 2,6-bisphosphate inhibit the activity of cytosolic fructose 1,6-bisphosphatase and, in so doing, deplete the level of hexose phosphates in the cytosol.

In turn, fructose 6-phosphate inhibits the bisphosphatase activity and activates the kinase activity of the bifunctional enzyme fructose 6-phosphate 2-kinase/fructose 2,6-bisphosphate phosphatase, and thereby increases the concentration of fructose 2,6-bisphosphate. As fructose 2,6-bisphosphate inhibits fructose 1,6-bisphosphatase, the concentration of fructose 6-phosphate decreases. Thus, fructose 2,6-bisphosphate modulates the pool of hexose phosphates in response not only to photosynthesis, but also to the demands of the cytosolic hexose phosphate pool itself.

Sucrose is continuously synthesized in the cytosol

Photosynthate produced in leaves is transported, primarily as sucrose, to meristems and developing organs such as growing leaves, roots, flowers, fruits, and seeds (see Figure 8.14). The concentration of sucrose in the cytosol of leaves is dependent on two processes:

1. Carbon import, which conveys diurnal triose phosphates and nocturnal maltose from chloroplasts to the leaf cytosol for sucrose synthesis.
2. Carbon export, which delivers sucrose from the leaf cytosol to other tissues for energy demands and polysaccharide synthesis

Cell fractionation, the physical separation of organelles for analysis of their intrinsic enzyme activities, has shown that sucrose is synthesized in the cytosol from the hexose phosphate pool as depicted in **Figure 8.19**, using the reactions described in Table 8.6.

The conversion of hexose phosphates to sugar nucleotides precedes the formation of sucrose. In the cytosol, glucose 1-phosphate reacts with UTP to yield UDP-glucose and pyrophosphate in a reaction catalyzed by UDP-glucose pyrophosphorylase (see Table 8.6, reaction 9). Two consecutive reactions complete the synthesis of sucrose from UDP-glucose. Sucrose 6^F-phosphate synthase (the superscript F indicates that sucrose is phosphorylated at carbon 6 of the fructose moiety) first catalyzes the formation of sucrose 6^F-phosphate from fructose 6-phosphate and UDP-glucose (see Table 8.6, reaction 10). Subsequently, sucrose 6^F-phosphate phosphatase releases inorganic phosphate from sucrose 6^F-phosphate, yielding sucrose (see Table 8.6, reaction 11).

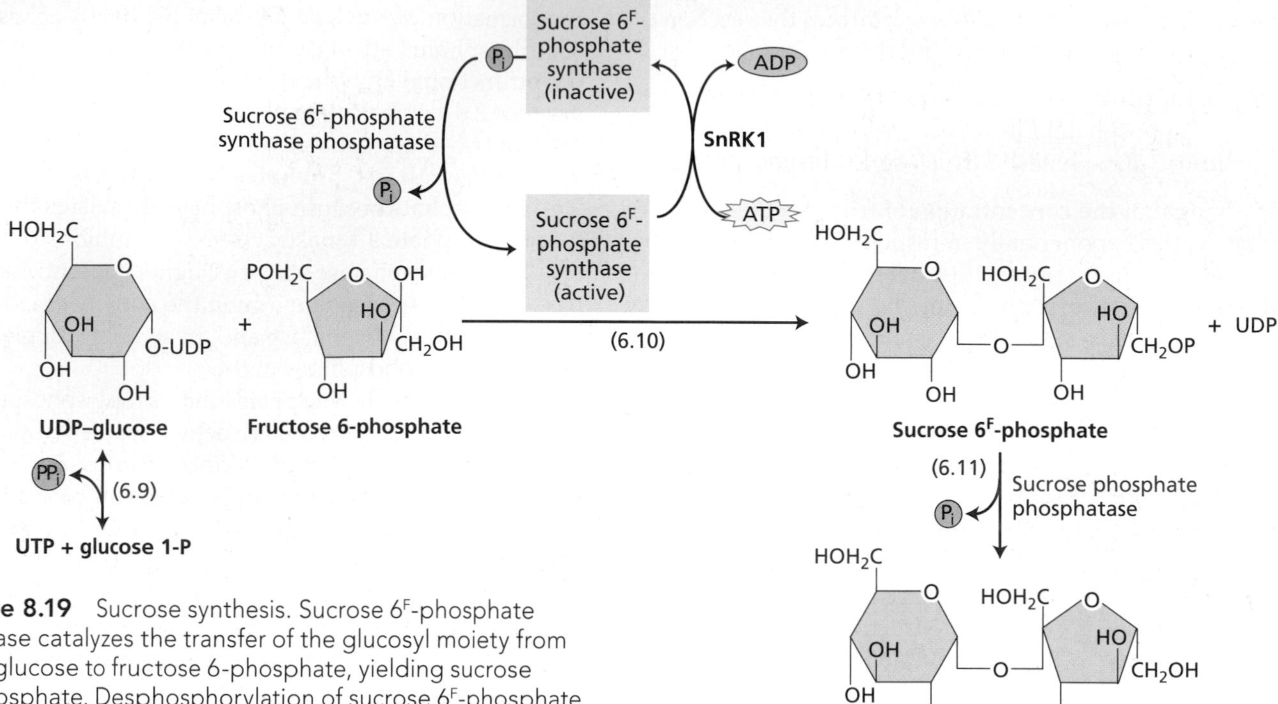

Figure 8.19 Sucrose synthesis. Sucrose 6^F-phosphate synthase catalyzes the transfer of the glucosyl moiety from UDP-glucose to fructose 6-phosphate, yielding sucrose 6^F-phosphate. Desphosphorylation of sucrose 6^F-phosphate by the enzyme sucrose 6^F-phosphate phosphatase releases the disaccharide sucrose. The posttranslational transformation of covalent bonds (via phosphorylation–dephosphorylation) and noncovalent interactions (via allosteric effectors) regulates the activity of sucrose 6^F-phosphate synthase. The phosphorylation of a specific serine residue on the enzyme by the concerted action of ATP and a specific kinase, SnRK1, yields an inactive enzyme. The release of phosphate from phosphorylated sucrose 6^F-phosphate synthase by a specific sucrose 6^F-phosphate synthase–phosphatase restores the basal activity. (The 6^F notation in sucrose 6^F-phosphate indicates that sucrose is phosphorylated at the carbon 6 of the fructose moiety.) See Table 8.6 for a description of numbered reactions.

The reversible formation of sucrose 6^F-phosphate ($\Delta G^{0\prime}$= –5.7 kJ/mol) followed by its irreversible hydrolysis ($\Delta G^{0\prime}$ = –16.5 kJ/mol) renders the synthesis of sucrose essentially irreversible in vivo. Moreover, the association of these enzymes into macromolecular complexes facilitates the direct transfer of sucrose 6^F-phosphate to sucrose 6^F-phosphate phosphatase without mixing with other metabolites.

Sucrose 6^F-phosphate synthase is regulated by posttranslational modifications (protein phosphorylation) and metabolites (allosteric control) (see Figure 8.19). In the dark, the phosphorylation of sucrose 6^F-phosphate synthase by a specific kinase lowers its catalytic activity. The kinase, SnRK1 (*S*ucrose *n*on-fermenting-*1*-*R*elated protein *K*inase), is a hub within a network of signaling pathways that phosphorylates and inactivates other enzymes (nitrate reductase, trehalose-phosphate synthase, and fructose 6-phosphate 2-kinase/fructose 2,6-bisphosphate phosphatase). In the light, the inactive sucrose 6^F-phosphate synthase is activated by dephosphorylation via a protein phosphatase. The phosphorylation of sucrose 6^F-phosphate synthase is also regulated by cytosolic metabolites: glucose 6-phosphate inhibits the kinase SnRK1, and phosphate inhibits the phosphatase (**Figure 8.20**). In addition to its regulation by phosphorylation–dephosphorylation, the active form of sucrose 6^F-phosphate synthase is stimulated by glucose 6-phosphate and inhibited by phosphate. Thus, the increased levels of hexose phosphates and decreased levels of phosphate in the cytosol, caused by high rates of photosynthesis, enhance the synthesis of sucrose. Conversely, sucrose 6^F-phosphate synthase is inefficient when increased levels of phosphate in the cytosol, caused by lower rates of photosynthesis, diminish hexose phosphates.

Sucrose synthesized in the cytosol of leaf cells is loaded into the phloem, transported to distant destinations, and unloaded in tissues such as developing leaves, apical meristems, and different organs (stems, tubers, grains). Specific membrane proteins, called sucrose transporters, drive the mass flow of sucrose to distant parts of the plant. Sucrose transport acts in concert with other signaling mechanisms—both tissue- and cell-specific—as a long-distance signal that promotes developmental responses through regulation of hormonal responses at the sink level. Thus, the loading and unloading of phloem vessels with sucrose conveys bidirectional information on nutrients and energy between the source leaves and the sink tissues.

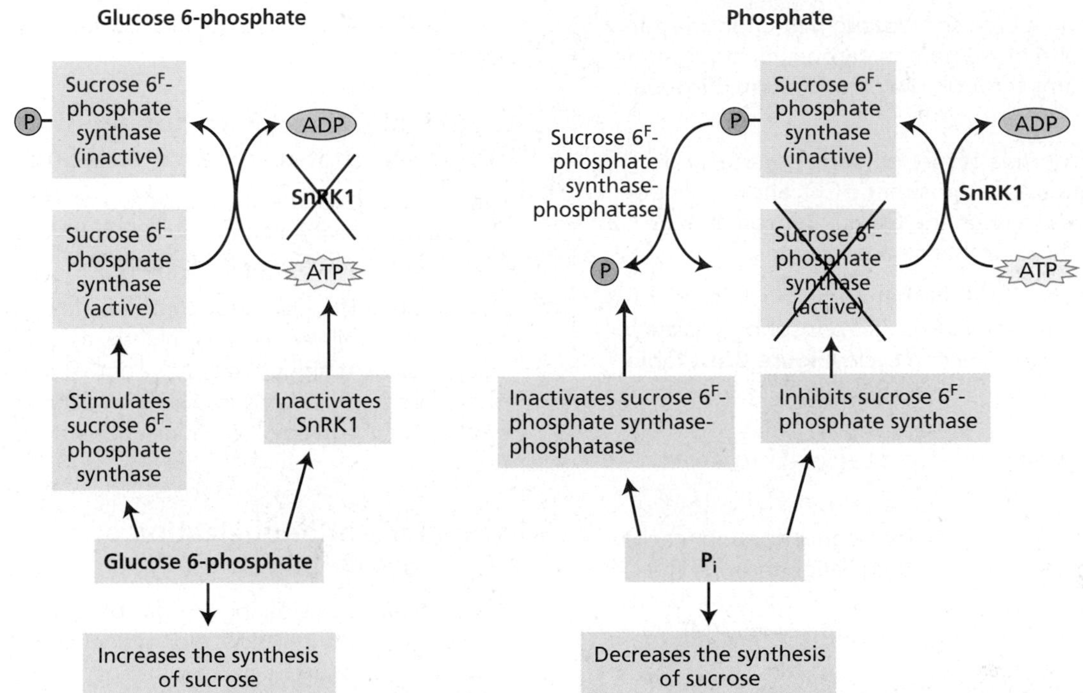

Figure 8.20 Glucose 6-phosphate and phosphate regulate the synthesis of sucrose. Glucose 6-phosphate increases the synthesis of sucrose by modulating the activity of two associated enzymes. Glucose 6-phosphate enhances the activity of sucrose 6^F-phosphate synthase itself, and also impedes formation of the inactive form of sucrose 6^F-phosphate synthase by inhibiting the SnRK1 kinase that phosphorylates and deactivates the enzyme. Phosphate lowers the synthesis of sucrose in a converse manner. Phosphate inhibits the activity of sucrose 6^F-phosphate synthase, and deactivates sucrose 6^F-phosphate synthase–phosphatase, the enzyme that converts sucrose 6^F-phosphate synthase to its active form. The transition of leaves from dark to light increases the concentration of glucose 6-phosphate and concurrently decreases the concentration of phosphate in the cytosol. Thus, the higher level of glucose 6-phosphate and the low level of phosphate jointly enhance sucrose synthesis in the light. The red Xs indicate inactive enzymes.

SUMMARY

Sunlight ultimately provides energy for the assimilation of inorganic carbon into organic material (autotrophy). The Calvin–Benson cycle is the predominant pathway for this conversion in many prokaryotes and in all plants.

The Calvin–Benson Cycle

- NADPH, and ATP generated by light in chloroplast thylakoids drive the endergonic fixation of atmospheric CO_2 through the Calvin–Benson cycle in the chloroplast stroma (**Figure 8.1**).
- The Calvin–Benson cycle has three phases: (1) carboxylation of ribulose 1,5-bisphosphate with CO_2 catalyzed by rubisco, yielding 3-phosphoglycerate; (2) reduction of the 3-phosphoglycerate to triose phosphates using ATP and NADPH; and (3) regeneration of the CO_2 acceptor molecule ribulose 1,5-bisphosphate (**Figures 8.2, 8.3**).
- CO_2 and O_2 compete in the carboxylation and oxygenation reactions catalyzed by rubisco (**Figure 8.4**).
- Rubisco-activase controls the activity of rubisco wherein CO_2 functions as both activator and substrate (**Figure 8.5**).
- Light regulates the activity of rubisco-activase and four enzymes of the Calvin–Benson cycle via the ferredoxin–thioredoxin system and changes in Mg^{2+} concentration and pH (**Figures 8.6, 8.7**).

The C_2 Oxidative Photosynthetic Carbon Cycle

- The C_2 oxidative photosynthetic carbon cycle (photorespiration) minimizes the loss of fixed CO_2 by the oxygenase activity of rubisco (**Table 8.2**).

- Chloroplasts, peroxisomes, and mitochondria participate in the movement of carbon, nitrogen, and oxygen atoms through photorespiration (**Figures 8.8, 8.9**).
- Kinetic properties of rubisco, temperature, and concentrations of atmospheric CO_2 and O_2 control the balance between the Calvin–Benson and the C2 oxidative photosynthetic carbon cycles.
- Cyanobacteria have alternative mechanisms for retrieving the carbon atoms of 2-phosphoglycolate for use in the Calvin–Benson cycle (**Figure 8.10; Table 8.3**).

Inorganic Carbon–Concentrating Mechanisms

- Land plants have two carbon-concentrating mechanisms that precede CO_2 assimilation through the Calvin–Benson cycle: C_4 photosynthetic carbon fixation (C_4) and crassulacean acid metabolism (CAM).

Inorganic Carbon–Concentrating Mechanisms: The C_4 Carbon Cycle

- The C_4 photosynthetic carbon cycle fixes atmospheric CO_2 via PEPCase into carbon skeletons in one compartment. The four-carbon acid products flow to another compartment where CO_2 is released and refixed via rubisco (**Figure 8.11; Table 8.4**).
- The C_4 cycle may be driven by diffusion gradients within a single cell as well as by gradients between mesophyll and bundle sheath cells (Kranz anatomy) (**Figure 8.12; Table 8.5**).
- Light regulates the activity of key C_4 cycle enzymes: NADP–malate dehydrogenase, PEPCase, and pyruvate–phosphate dikinase.
- The C_4 cycle reduces photorespiration and water loss in hot, dry climates.

Inorganic Carbon–Concentrating Mechanisms: Crassulacean Acid Metabolism (CAM)

- CAM photosynthesis captures atmospheric CO_2 and scavenges respiratory CO_2 in arid environments.
- CAM is generally associated with anatomical features that minimize water loss.
- In CAM plants, the initial capture of CO_2 and its final incorporation into carbon skeletons are temporally separated (**Figure 8.13**).
- Genetics and environmental factors determine CAM expression.

Accumulation and Partitioning of Photosynthates—Starch and Sucrose

- In most leaves, sucrose in the cytosol and starch in chloroplasts are the end products of photosynthetic CO_2 assimilation (**Figure 8.14; Table 8.6**).
- During the day, sucrose flows from the leaf cytosol to sink tissues, while starch accumulates as granules in chloroplasts. At night, the starch content of chloroplasts falls to provide carbon skeletons for sucrose synthesis in the cytosol to nourish heterotrophic tissues.

Formation and Mobilization of Chloroplast Starch

- Starch biosynthesis during the day proceeds through successive steps: initiation, elongation, branching, and termination of the polysaccharide chain (**Figures 8.15, 8.16**).
- Starch degradation at night requires prior phosphorylation of the polysaccharide. Glucan–water dikinase and phosphoglucan–water dikinase catalyze transfer of the β–phosphate of ATP to starch (**Figure 8.17**).
- The breakdown of linear glucans by chloroplast β-amylases yields maltose, which is exported to the cytosol for sucrose synthesis.

Sucrose Biosynthesis and Signaling

- During the day, the ratio of triose phosphates to inorganic phosphate modulates the partitioning of carbon between chloroplasts and cytosol. The accumulation of triose phosphates in the cytosol builds up the pool of hexose phosphates. Hexose phosphates are precursors in the cytosolic synthesis of sucrose catalyzed by sucrose 6^F-phosphate synthase and sucrose 6^F-phosphate phosphatase (**Figures 8.18, 8.19**).
- Phosphorylation and noncovalent interactions with metabolites regulate sucrose 6^F-phosphate synthase activity (**Figure 8.20**).
- In addition to providing carbon for growth and polysaccharide biosynthesis, sucrose acts as a signal in regulating genes that encode enzymes, transporters, and storage proteins.

WEB MATERIAL

- **WEB TOPIC 8.1 CO_2 Pumps** Cyanobacteria contain protein complexes (CO_2 pumps) and supramolecular complexes for the uptake and fixation of inorganic carbon.
- **WEB TOPIC 8.2 How the Calvin–Benson Cycle was Elucidated** Experiments carried out in the 1950s led to the discovery of the path of CO_2 fixation.
- **WEB TOPIC 8.3 Rubisco: A Model Enzyme for Studying Structure and Function** As the most abundant enzyme on Earth, rubisco was obtained in quantities sufficient for elucidating its structure and catalytic properties.
- **WEB TOPIC 8.4 Energy Demands for Photosynthesis in Land Plants** Evaluation of NADPH and ATP budget during the asimilation of CO_2.
- **WEB TOPIC 8.5 Rubisco Activase** Rubisco is unique among Calvin–Benson Cycle enzymes in its regulation by a specific protein, rubisco activase.
- **WEB TOPIC 8.6 Thioredoxins** First found to regulate chloroplast enzymes, thioredoxins are now known to play a regulatory role in all types of cells.
- **WEB TOPIC 8.7 Operation of the C_2 Oxidative Photosynthetic Carbon Cycle** The enzymes of the C_2 oxidative photosynthetic carbon cycle are localized in three different organelles.
- **WEB TOPIC 8.8 Carbon Dioxide: Some Important Physicochemical Properties** Plants have adapted to the properties of CO_2 by altering the reactions catalyzing its fixation.
- **WEB TOPIC 8.9 Three Variations of C_4 Metabolism** Certain reactions of the C_4 photosynthetic pathway differ among plant species.
- **WEB TOPIC 8.10 Single-Cell C_4 Photosynthesis** Some marine organisms and land plants accomplish C_4 photosynthesis in a single cell.
- **WEB TOPIC 8.11 Chloroplast Phosphate Translocators** Chloroplast phosphate translocators are antiporters that catalyze a strict 1:1 exchange of phosphate with other metabolites between the chloroplast and the cytosol.
- **WEB TOPIC 8.12 Photorespiration in CAM Plants** During the day, stomatal closing and photosynthesis in CAM leaves lead to very high intracellular concentrations of both oxygen and carbon dioxide. These unusual conditions pose interesting adaptive challenges to CAM leaves.
- **WEB TOPIC 8.13 Starch Architecture** The morphology and composition of the starch granule influence the synthesis and degradation of the polysaccharide.

- **WEB ESSAY 8.1 Modulation of Phosphoenolpyruvate Carboxylase in C_4 and CAM Plants** The CO_2-fixing enzyme phosphoenolpyruvate carboxylase is regulated differently in C_4 and CAM species.

available at plantphys.net

Suggested Reading

Balsera, M., Uberegui, E., Schürmann, P., and Buchanan, B. B. (2014) Evolutionary development of redox regulation in chloroplasts. *Antioxid. Redox Signal.* 21: 1327–1355.

Bordych, C., Eisenhut, M., Pick, T. R., Kuelahoglu, C., and Weber, A. P. M. (2013) Co-expression analysis as tool for the discovery of transport proteins in photorespiration. *Plant Biol.* 15: 686–693.

Christin, P. A., Arakaki, M., Osborne, C. P., Bräutigam, A., Sage, R. F., Hibberd, J. M., Kelly, S., Covshoff, S., Wong, G. S., Hancock, L. et al. (2014) Shared origins of a key enzyme during the evolution of C4 and CAM metabolism. *J. Exp. Bot.* 65: 3609–3621.

Denton, A. K., Simon, R., and Weber, A. P. M. (2013) C4 photosynthesis: From evolutionary analyses to strategies for synthetic reconstruction of the trait. Curr. Opin. *Plant Biol.* 16: 315–321.

Ducat, D. C., and Silver, P. A. (2012) Improving carbon fixation pathways. Curr. Opin. Chem. Biol. 16: 337–344.

Florian, A., Araújo, W. L., and Fernie, A. R. (2013) New insights into photorespiration obtained from metabolomics. *Plant Biol.* 15: 656–666.

Hagemann, M., Fernie, A. R., Espie, G. S., Kern, R. Eisenhut, M., Reumann, S., Bauwe, H., and Weber, A. P. M. (2013) Evolution of the biochemistry of the photorespiratory C2 cycle. *Plant Biol.* 15: 639–647.

Henderson, J. N., Kuriata, A. M., Fromme, R., Salvucci, M. E., and Wachter, R. M. (2011) Atomic resolution X-ray structure of the substrate recognition domain of higher plant ribulose-bisphosphate carboxylase/oxygenase (rubisco) activase. *J. Biol. Chem.* 286: 35683–35688.

Hibberd, J. M., and Covshoff, S. (2010) The regulation of gene expression required for C4 photosynthesis. *Annu. Rev. Plant Biol.* 61: 181–207.

Peterhansel, C., and Offermann, S. (2012) Re-engineering of carbon fixation in plants –Challenges for plant biotechnology to improve yields in a high-CO2 world. *Curr. Opin. Biotechnol.* 23: 204–208.

Sage, R. F., Christin, P. A., and Edwards, E. J. (2011) The C4 plant lineages of planet Earth. *J. Exp. Bot.* 62: 3155–3169.

Sage, R. F., Khoshravesh, R., and Sage, T. L. (2014) From proto-Kranz to C4 Kranz: Building the bridge to C4 photosynthesis. *J. Exp. Bot.* 65: 3341–3356.

Timm, S., and Bauwe, H. (2013) The variety of photorespiratory phenotypes – Employing the current status for future research directions on photorespiration. *Plant Biol.* 15: 737–747.

Erb, T. J., Evans, B. S., Cho, K., Warlick, B. P., Sriram, J., Wood, B. M., Imker, H. J., Sweedler, J. V., Tabita, F. R., and Gerlt, J. A. (2012) A rubisCO-like protein links SAM metabolism with isoprenoid biosynthesis. *Nat. Chem. Biol.* 8: 926–932.

9 Photosynthesis: Physiological and Ecological Considerations

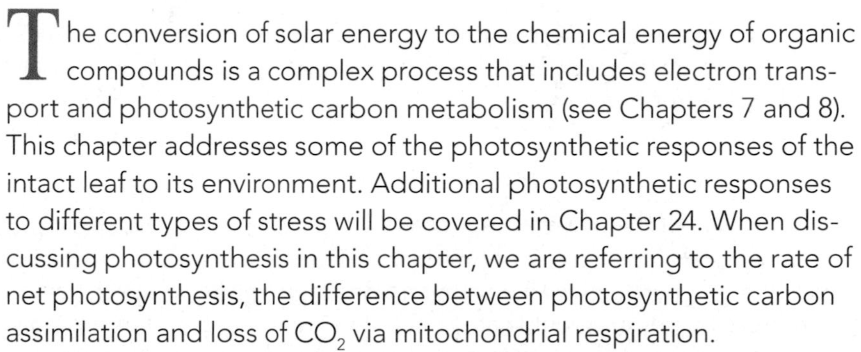

The conversion of solar energy to the chemical energy of organic compounds is a complex process that includes electron transport and photosynthetic carbon metabolism (see Chapters 7 and 8). This chapter addresses some of the photosynthetic responses of the intact leaf to its environment. Additional photosynthetic responses to different types of stress will be covered in Chapter 24. When discussing photosynthesis in this chapter, we are referring to the rate of net photosynthesis, the difference between photosynthetic carbon assimilation and loss of CO_2 via mitochondrial respiration.

The impact of the environment on photosynthesis is of broad interest, especially to physiologists, ecologists, evolutionary biologists, climate change scientists, and agronomists. From a physiological standpoint, we wish to understand the direct responses of photosynthesis to environmental factors such as light, ambient CO_2 concentrations, and temperature, as well as the indirect responses (mediated through the effects of stomatal control) to environmental factors such as humidity and soil moisture. The dependence of photosynthetic processes on environmental conditions is also important to agronomists because plant productivity, and hence crop yield, depends strongly on prevailing photosynthetic rates in a dynamic environment. To the ecologist, photosynthetic variation among different environments is of great interest in terms of adaptation and evolution.

In studying the environmental dependence of photosynthesis, a central question arises: How many environmental factors can limit photosynthesis at one time? The British plant physiologist F. F. Blackman hypothesized in 1905 that, under any particular conditions, the rate of photosynthesis is limited by the slowest step in the process, the so-called *limiting factor*. The implication of this hypothesis is that at any given time, photosynthesis can be limited either by light or by CO_2 concentration, for instance, but not by both factors. This hypothesis has had a marked influence on the approach used by plant physiologists to study photosynthesis—that is, varying one factor and keeping all other environmental

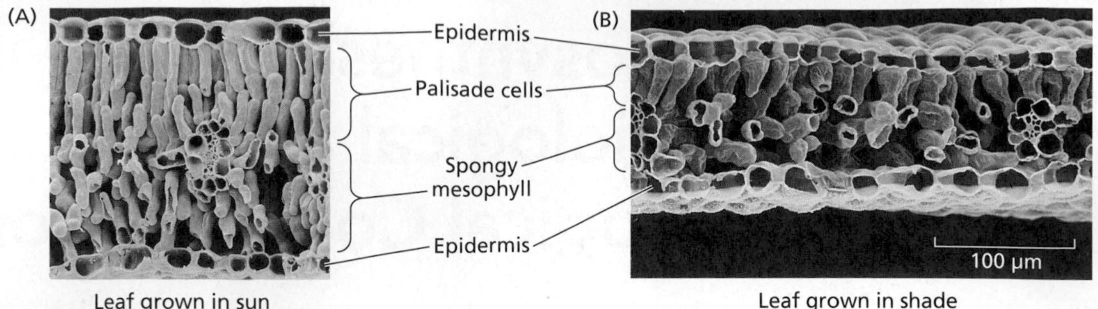

Figure 9.1 Scanning electron micrographs of the leaf anatomy of a legume (*Thermopsis montana*) grown in different light environments. Note that the sun leaf (A) is much thicker than the shade leaf (B) and that the palisade (columnlike) cells are much longer in the leaves grown in sunlight. Layers of spongy mesophyll cells can be seen below the palisade cells. (Courtesy of T. Vogelmann.)

conditions constant. In the intact leaf, three major metabolic processes have been identified as important for photosynthetic performance:

- Rubisco capacity
- Regeneration of ribulose bisphosphate (RuBP)
- Metabolism of the triose phosphates

Graham Farquhar and Tom Sharkey added a fundamentally new perspective to our understanding of photosynthesis by pointing out that we should think of the controls on the overall net photosynthetic rates of leaves in economic terms, considering "supply" and "demand" functions for carbon dioxide. The metabolic processes referred to above take place in the palisade cells and spongy mesophyll of the leaf (**Figure 9.1**). These biochemical activities describe the "demand" for CO_2 by photosynthetic metabolism in the cells. However, the rate of CO_2 "supply" to these cells is largely determined by diffusion limitations resulting from stomatal regulation and subsequent resistance in the mesophyll. The coordinated actions of "demand" by photosynthetic cells and "supply" by guard cells affect the leaf photosynthetic rate as measured by net CO_2 uptake.

In the following sections we will focus on how naturally occurring variation in light and temperature influences photosynthesis in leaves and how leaves in turn adjust or acclimate to such variation. We will also explore how atmospheric carbon dioxide influences photosynthesis, an especially important consideration in a world where CO_2 concentrations are rapidly increasing as humans continue to burn fossil fuels for energy production.

Photosynthesis Is Influenced by Leaf Properties

Scaling up from the chloroplast (the focus of Chapters 7 and 8) to the leaf adds new levels of complexity to photosynthesis. At the same time, the structural and functional properties of the leaf make possible other levels of regulation.

We will start by examining the capture of light, and how leaf anatomy and leaf orientation maximize light absorption for photosynthesis. Then we will describe how leaves acclimate to their light environment. We will see that the photosynthetic response of leaves grown under different light conditions also reflects the ability of a plant to grow under different light environments. However, there are limits to the extent to which photosynthesis in a species can acclimate to very different light environments. For example, in some situations photosynthesis is limited by an inadequate supply of light. In other situations, absorption of too much light would cause severe problems if special mechanisms did not protect the photosynthetic system from excessive light. While plants have multiple levels of control over photosynthesis that allow them to grow successfully in constantly changing environments, there are ultimately limits to what is possible.

Consider the many ways in which leaves are exposed to different spectra (qualities) and quantities of light that result in photosynthesis. Plants grown outdoors are exposed to sunlight, and the spectrum of that sunlight will depend on whether it is measured in full sunlight or under the shade of a canopy. Plants grown indoors may receive either incandescent or fluorescent lighting, each of which is spectrally different from sunlight. To account for these differences in spectral quality and quantity, we need uniformity in how we measure and express the light that affects photosynthesis.

The light reaching the plant is a flux, and that flux can be measured in either energy or photon flux units. **Irradiance** is the amount of energy that falls on a flat sensor of known area per unit time, expressed in watts per square meter (W m^{-2}). Recall that time (seconds) is contained within the term watt: 1 W = 1 joule (J) s^{-1}. Quantum flux, or photon flux density (PFD), is the number of incident **quanta** (singular *quantum*) striking the leaf, expressed in moles per square meter per second (mol m^{-2} s^{-1}), where *moles* refers to the number of photons (1 mol of light = 6.02 × 10^{23} photons, Avogadro's number). Quanta and energy

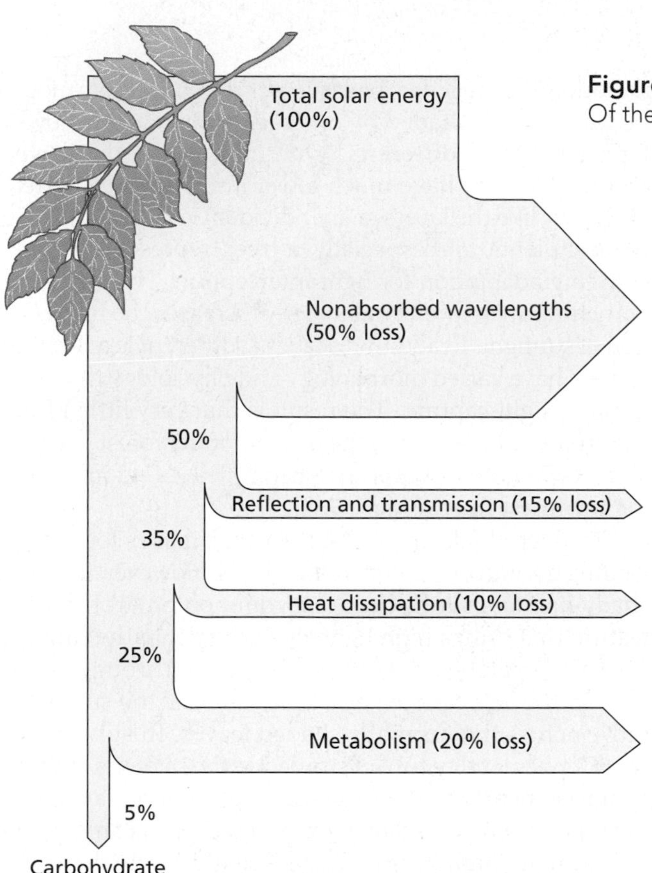

Figure 9.2 Conversion of solar energy into carbohydrates by a leaf. Of the total incident energy, only 5% is converted into carbohydrates.

units for sunlight can be interconverted relatively easily, provided that the wavelength of the light, λ, is known. The energy of a photon is related to its wavelength as follows:

$$E = \frac{hc}{\lambda}$$

where c is the speed of light (3×10^8 m s^{-1}), h is Planck's constant (6.63×10^{-34} J s), and λ is the wavelength of light, usually expressed in nanometers (1 nm = 10^{-9} m). From this equation it can be shown that a photon at 400 nm has twice the energy of a photon at 800 nm (see **WEB TOPIC 9.1**).

When considering photosynthesis and light, it is appropriate to express light as photosynthetic photon flux density (PPFD)—the flux of light (usually expressed as micromoles per square meter per second [µmol m^{-2} s^{-1}]) within the photosynthetically active range (400–700 nm). How much light is there on a sunny day? Under direct sunlight on a clear day, PPFD is about 2000 µmol m^{-2} s^{-1} at the top of a dense forest canopy, but may be only 10 µmol m^{-2} s^{-1} at the bottom of the canopy because of light absorption by the leaves overhead.

Leaf anatomy and canopy structure maximize light absorption

On average, about 340 W of radiant energy from the sun reach each square meter of Earth's surface. When this sunlight strikes the vegetation, only 5% of the energy is ultimately converted into carbohydrates by photosynthesis (**Figure 9.2**). The reason this percentage is so low is that a major percentage of the light is of a wavelength either too short or too long to be absorbed by the photosynthetic pigments (**Figure 9.3**). Furthermore, of the photosynthetically active radiation (400–700 nm) that is incident on a leaf, a small percentage is transmitted through the leaf and some is also reflected from its surface. Because chlorophyll absorbs strongly in the blue and red regions of the spectrum (see Figure 7.3), green wavelengths are the ones most dominant in the transmitted and reflected light (see Figure 9.3)—hence the green color of vegetation. Lastly, a percentage of the photosynthetically active radiation that is initially absorbed by the leaf is lost through metabolism and a smaller amount is lost as heat (see Chapter 7).

The anatomy of the leaf is highly specialized for light absorption. The outermost cell layer, the epidermis, is typically transparent to visible light, and the individual cells are often convex. Convex epidermal cells can act as lenses and focus light so that the amount reaching some of the chloroplasts can be many times greater than the amount of ambient light. Epidermal focusing is common among herbaceous plants and is especially prominent among tropical plants that grow in the forest understory, where light levels are very low.

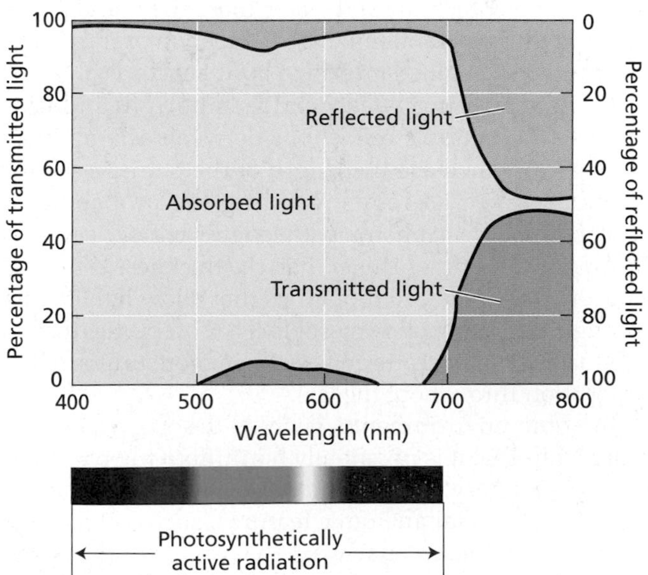

Figure 9.3 Optical properties of a bean leaf. Shown here are the percentages of light absorbed, reflected, and transmitted, as a function of wavelength. The transmitted and reflected green light in the wave band at 500–600 nm gives leaves their green color. Note that most of the light above 700 nm is not absorbed by the leaf. (After Smith 1986.)

Below the epidermis, the top layers of photosynthetic cells are called **palisade cells**; they are shaped like pillars that stand in parallel columns one to three layers deep (see Figure 9.1). Some leaves have several layers of columnar palisade cells, and we may wonder how efficient it is for a plant to invest energy in developing multiple cell layers when the high chlorophyll content of the first layer would appear to allow little transmission of incident light to the leaf interior. In fact, more light than might be expected penetrates the first layer of palisade cells because of the *sieve effect* and *light channeling.*

The **sieve effect** occurs because chlorophyll is not uniformly distributed throughout cells but instead is confined to the chloroplasts. This packaging of chlorophyll results in shading between the chlorophyll molecules and creates gaps between the chloroplasts where light is not absorbed—hence the reference to a sieve. Because of the sieve effect, the total absorption of light by a given amount of chlorophyll in chloroplasts of a palisade cell is less than the light that would be absorbed by the same amount of chlorophyll were it uniformly distributed in solution.

Light channeling occurs when some of the incident light is propagated through the central vacuoles of the palisade cells and through the air spaces between the cells, an arrangement that results in the transmission of light into the leaf interior. In the interior, below the palisade layers, is the **spongy mesophyll**, where the cells are very irregular in shape and are surrounded by large air spaces (see Figure 9.1). The large air spaces generate many interfaces between air and water that reflect and refract the light, thereby randomizing its direction of travel. This phenomenon is called **interface light scattering**.

Light scattering is especially important in leaves because the multiple refractions between cell–air interfaces greatly increase the length of the path over which photons travel, thereby increasing the probability of absorption. In fact, photon path lengths within leaves are commonly four times longer than the thickness of the leaf. Thus, the palisade cell properties that allow light to pass through, and the spongy mesophyll cell properties that are conducive to light scattering, result in more uniform light absorption throughout the leaf.

In some environments, such as deserts, there is so much light that it is potentially harmful to the photosynthetic machinery of leaves. In these environments leaves often have special anatomic features, such as hairs, salt glands, and epicuticular wax, that increase the reflection of light from the leaf surface, thereby reducing light absorption. Such adaptations can decrease light absorption by as much as 60%, thereby reducing overheating and other problems associated with the absorption of too much solar energy.

At the whole-plant level, leaves at the top of a canopy absorb most of the sunlight, and reduce the amount of radiation that reaches leaves lower down in the canopy. Leaves that are shaded by other leaves experience lower light levels and different light quality than the leaves above them and have much lower photosynthetic rates. However, like the layers of an individual leaf, the structure of most plants, and especially of trees, represents an outstanding adaptation for light interception. The elaborate branching structure of trees vastly increases the interception of sunlight. In addition, leaves at different levels of the canopy have varied morphology and physiology that help improve light capture. The result is that very little PPFD penetrates all the way to the bottom of the forest canopy; almost all of the PPFD is absorbed by leaves before reaching the forest floor (**Figure 9.4**).

The deep shade of a forest floor thus makes for a challenging growth environment for plants. However, in many shady habitats **sunflecks** are a common environmental feature that brings high light levels deep into the canopy. These are patches of sunlight that pass through small gaps in the leaf canopy; as the sun moves, the sunflecks move across the normally shaded leaves. In spite of the short, ephemeral nature of sunflecks, the photons in them comprise nearly 50% of the total light energy available during the day. In a dense forest, sunflecks can change the sunlight impinging on a shade leaf by more than tenfold within seconds. This critical energy is available for only a few minutes, and in a very high dose. Many deep-

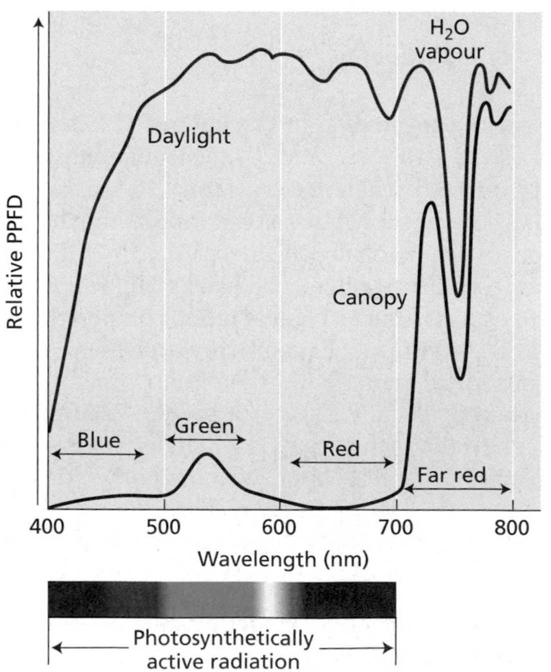

Figure 9.4 Relative spectral distributions of sunlight at the top of a canopy and in the shade under the canopy. Most photosynthetically active radiation is absorbed by leaves in the canopy. (After Smith 1994.)

shade species that experience sunflecks have physiological mechanisms for taking advantage of this burst of light when it occurs. Sunflecks also play a role in the carbon metabolism of densely planted crops, where the lower leaves are shaded by leaves higher up on the plant.

Leaf angle and leaf movement can control light absorption

The angle of the leaf relative to the sun determines the amount of sunlight incident on it. Incoming sunlight can strike a flat leaf surface at a variety of angles depending on the time of day and the orientation of the leaf. Maximum incident radiation occurs when sunlight strikes a leaf perpendicular to its surface. When the rays of light deviate from perpendicular, however, the incident sunlight on a leaf is proportional to the angle at which the rays hit the surface.

Under natural conditions, leaves exposed to full sunlight at the top of the canopy tend to have steep leaf angles so that less than the maximum amount of sunlight is incident on the leaf blade; this allows more sunlight to penetrate into the canopy. For this reason, it is common to see the angle of leaves within a canopy decrease (become more horizontal) with increasing depth in the canopy.

Some leaves maximize light absorption by **solar tracking**; that is, they continuously adjust the orientation of their laminae (blades) such that they remain perpendicular to the sun's rays (**Figure 9.5**). Many species, including alfalfa, cotton, soybean, bean, and lupine, have leaves capable of solar tracking.

Solar-tracking leaves present a nearly vertical position at sunrise, facing the eastern horizon. Individual leaf blades then begin to track the rising sun, following its movement across the sky with an accuracy of ±15° until sunset, when the laminae are nearly vertical, facing the west. During the night the leaves take a horizontal position and reorient just before dawn so that they face the eastern horizon, ready for another sunrise. Leaves track the sun only on clear days, and they stop moving when a cloud obscures the sun. In the case of intermittent cloud cover, some leaves can reorient as rapidly as 90° per hour and thus can catch up to the new solar position when the sun emerges from behind a cloud.

Solar tracking is a blue-light response (see Chapter 16), and the sensing of blue light in solar-tracking leaves occurs in specialized regions of the leaf or stem. In species of *Lavatera* (Malvaceae), the photosensitive region is located in or near the major leaf veins, but in many species, notably legumes, leaf orientation is controlled by a specialized organ called the **pulvinus** (plural *pulvini*), found at the junction between the blade and the petiole. In lupines (*Lupinus*, Fabaceae), for example, leaves consist of five or more leaflets, and the photosensitive region is in a pulvinus located at the basal part of each leaflet lamina (see Figure 9.5). The pulvinus contains motor cells that change their osmotic potential and generate mechanical forces that determine laminar orientation. In other plants, leaf orientation is controlled by small mechanical changes along the length of the petiole and by movements of the younger parts of the stem.

Heliotropism is another term used to describe leaf orientation by solar tracking. Leaves that maximize light interception by solar tracking are referred to as *diaheliotropic*. Some solar-tracking plants can also move their leaves so that they *avoid* full exposure to sunlight, thus minimizing heating and water loss. These sun-avoiding leaves are called *paraheliotropic*. Some plant species, such as soybean, have leaves that can display diaheliotropic movements when they are well watered and paraheliotropic movements when they experience water stress.

Leaves acclimate to sun and shade environments

Acclimation is a developmental process in which leaves express a set of biochemical and morphological adjustments that are suited to the particular environment in

(A)

(B)

Figure 9.5 Leaf movement in a sun-tracking plant. (A) Initial leaf orientation in the lupine *Lupinus succulentus*, with no direct sunlight. (B) Leaf orientation 4 hours after exposure to oblique light. Arrows indicate the direction of the light beam. Movement is generated by asymmetric swelling of a pulvinus, found at the junction between the lamina and the petiole. In natural conditions, the leaves track the sun's trajectory in the sky. (From Vogelmann and Björn 1983, courtesy of T. Vogelmann.)

which the leaves are exposed. Acclimation can occur in mature leaves and in newly developing leaves. **Plasticity** is the term we use to define how much adjustment can take place. Many plants have developmental plasticity to respond to a range of light regimes, growing as sun plants in sunny areas and as shade plants in shady habitats. The ability to acclimate is important, given that shady habitats can receive less than 20% of the PPFD available in an exposed habitat, and deep-shade habitats receive less than 1% of the PPFD incident at the top of the canopy.

In some plant species, individual leaves that develop under very sunny or deep shady environments are often unable to persist when transferred to the other type of habitat. In such cases, the mature leaf will abscise and a new leaf will develop that is better suited for the new environment. You may notice this if you take a plant that developed indoors and transfer it outdoors; after some time, if it is the right type of plant, a new set of leaves will develop that are better suited to high sunlight. However, some plant species are not able to acclimate when transferred from a sunny to a shady environment, or vice versa. The lack of acclimation indicates that these plants are specialized for either a sunny or a shady environment. When plants adapted to deep-shade conditions are transferred into full sunlight, the leaves experience chronic photoinhibition and leaf bleaching, and they eventually die. We will discuss photoinhibition later in this chapter.

Sun and shade leaves have contrasting biochemical and morphological characteristics:

- Shade leaves increase light capture by having more total chlorophyll per reaction center, a higher ratio of chlorophyll *b* to chlorophyll *a*, and usually thinner laminae than sun leaves.
- Sun leaves increase CO_2 assimilation by having more rubisco and can dissipate excess light energy by having a large pool of xanthophyll-cycle components (see Chapter 7). Morphologically they have thicker leaves and a larger palisade layer than shade leaves (see Figure 9.1).

These morphological and biochemical modifications are associated with specific acclimation responses to the *amount* of sunlight in a plant's habitat, but light *quality* can also influence such responses. For example, far-red light, which is absorbed primarily by photosystem I (PSI), is proportionally more abundant in shady habitats than in sunny ones (see Chapter 18). To better balance the flow of energy through PSII and PSI, the adaptive response of some shade plants is to produce a higher ratio of PSII to PSI reaction centers, compared with that found in sun plants. Other shade plants, rather than altering the ratio of PSII to PSI reaction centers, add more antenna chlorophyll to PSII to increase absorption by this photosystem. These changes appear to enhance light absorption and energy transfer in shady environments.

Effects of Light on Photosynthesis in the Intact Leaf

Light is a critical resource that limits plant growth, but at times leaves can be exposed to too much rather than too little light. In this section we will describe typical photosynthetic responses to light as measured by light-response curves. We will also consider how features of a light-response curve can help explain contrasting physiological properties between sun and shade plants, and between C_3 and C_4 species. The section will conclude with descriptions of how leaves respond to excess light.

Light-response curves reveal photosynthetic properties

Measuring net CO_2 fixation in intact leaves across varying PPFD levels generates light-response curves (**Figure 9.6**) In near darkness there is little photosynthetic carbon assimilation, but because mitochondrial respiration continues, CO_2 is given off by the plant (see Chapter 12). CO_2 uptake is negative in this part of the light-response curve.

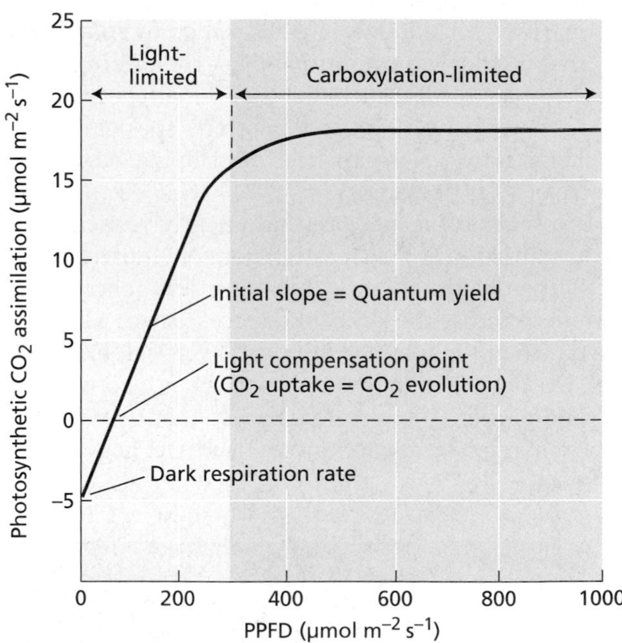

Figure 9.6 Response of photosynthesis to light in a C_3 plant. In darkness, respiration causes a net efflux of CO_2 from the plant. The light compensation point is reached when photosynthetic CO_2 assimilation equals the amount of CO_2 evolved by respiration. Increasing light above the light compensation point proportionally increases photosynthesis, indicating that photosynthesis is limited by the rate of electron transport, which in turn is limited by the amount of available light. This portion of the curve is referred to as light-limited. Further increases in photosynthesis are eventually limited by the carboxylation capacity of rubisco or the metabolism of triose phosphates. This part of the curve is referred to as carboxylation-limited.

At higher PPFD levels, photosynthetic CO_2 assimilation eventually reaches a point at which CO_2 uptake exactly balances CO_2 evolution. This is called the **light compensation point**. The PPFD at which different leaves reach the light compensation point can vary among species and developmental conditions. One of the more interesting differences is found between plants that normally grow in full sunlight and those that grow in the shade (**Figure 9.7**). Light compensation points of sun plants range from 10 to 20 μmol m^{-2} s^{-1}, whereas corresponding values for shade plants are 1 to 5 μmol m^{-2} s^{-1}.

Why are light compensation points lower for shade plants? For the most part, this is because respiration rates in shade plants are very low; therefore only a little photosynthesis is necessary to bring the net rates of CO_2 exchange to zero. Low respiratory rates allow shade plants to survive in light-limited environments through their ability to achieve positive CO_2 uptake rates at lower PPFD values than sun plants.

A linear relationship between PPFD and photosynthetic rate persists at light levels above the light compensation point (see Figure 9.6). Throughout this linear portion of the light-response curve, photosynthesis is light-limited; more light stimulates proportionately more

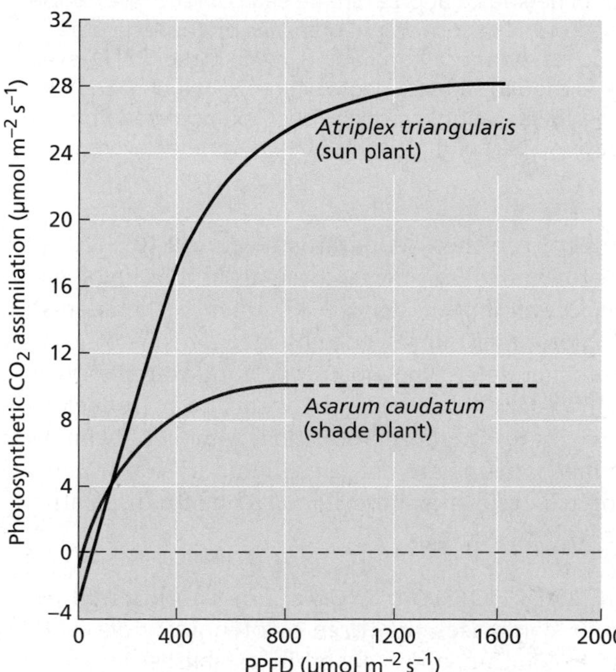

Figure 9.7 Light-response curves of photosynthetic carbon fixation in sun and shade plants. Triangle orache (*Atriplex triangularis*) is a sun plant, and wild ginger (*Asarum caudatum*) is a shade plant. Typically, shade plants have a low light compensation point and have lower maximum photosynthetic rates than sun plants. The dashed line has been extrapolated from the measured part of the curve. (After Harvey 1979.)

photosynthesis. When corrected for light absorption, the slope of this linear portion of the curve provides the **maximum quantum yield** of photosynthesis for the leaf. Leaves of sun and shade plants show very similar quantum yields despite their different growth habitats. This is because the basic biochemical processes that determine quantum yield are the same for these two types of plants. But quantum yield can vary among plants with different photosynthetic pathways.

Quantum yield is the ratio of a given light-dependent product to the number of absorbed photons (see Equation 7.5). Photosynthetic quantum yield can be expressed on either a CO_2 or an O_2 basis, and as explained in Chapter 7, the quantum yield of photochemistry is about 0.95. However, the maximum photosynthetic quantum yield of an integrated process such as photosynthesis is lower than the theoretical yield when measured in chloroplasts (organelles) or whole leaves. Based on the biochemistry discussed in Chapter 8, we expect the theoretical maximum quantum yield for photosynthesis to be 0.125 for C_3 plants (one CO_2 molecule fixed per eight photons absorbed). But under today's atmospheric conditions (400 ppm CO_2, 21% O_2), the quantum yields for CO_2 of C_3 and C_4 leaves vary between 0.04 and 0.07 mole of CO_2 per mole of photons.

In C_3 plants the reduction from the theoretical maximum is caused primarily by energy loss through photorespiration. In C_4 plants the reduction is caused by the additional energy requirements of the CO_2-concentrating mechanism and potential cost of refixing CO_2 that has diffused out from within the bundle sheath cells. If C_3 leaves are exposed to low O_2 concentrations, photorespiration is minimized and the maximum quantum yield increases to about 0.09 mole of CO_2 per mole of photons. In contrast, if C_4 leaves are exposed to low O_2 concentrations, the quantum yields for CO_2 fixation remain constant at about 0.05 to 0.6 mole of CO_2 per mole of photons. This is because the carbon-concentrating mechanism in C_4 photosynthesis eliminates nearly all CO_2 evolution via photorespiration.

At higher PPFD along the light-response curve, the photosynthetic response to light starts to level off (see Figures 9.6 and 9.7) and eventually approaches *saturation*. Beyond the light saturation point, net photosynthesis no longer increases, indicating that factors other than incident light, such as electron transport rate, rubisco activity, or the metabolism of triose phosphates, have become limiting to photosynthesis. Light saturation levels for shade plants are substantially lower than those for sun plants (see Figure 9.7). This is also true for leaves of the same plant when grown in sun versus shade (**Figure 9.8**). These levels usually reflect the maximum PPFD to which a leaf was exposed during growth.

The light-response curve of most leaves saturates between 500 and 1000 μmol m^{-2} s^{-1}, well below full sunlight (which is about 2000 μmol m^{-2} s^{-1}). An exception to

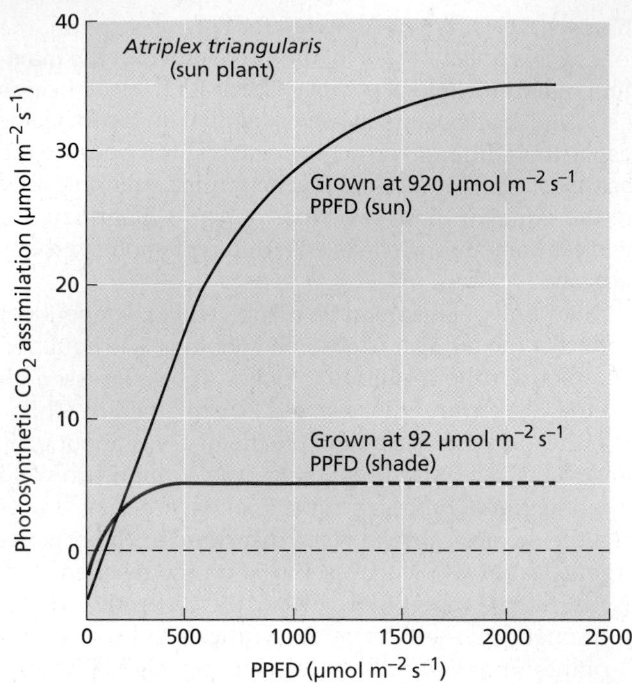

Figure 9.8 Light-response curve of photosynthesis of a sun plant grown under sun versus shade conditions. The upper curve represents an *A. triangularis* leaf grown at a PPFD level ten times higher than that of the lower curve. In the plant grown at the lower light levels, photosynthesis saturates at a substantially lower PPFD, indicating that the photosynthetic properties of a leaf depend on its growing conditions. The dashed red line has been extrapolated from the measured part of the curve. (After Björkman 1981.)

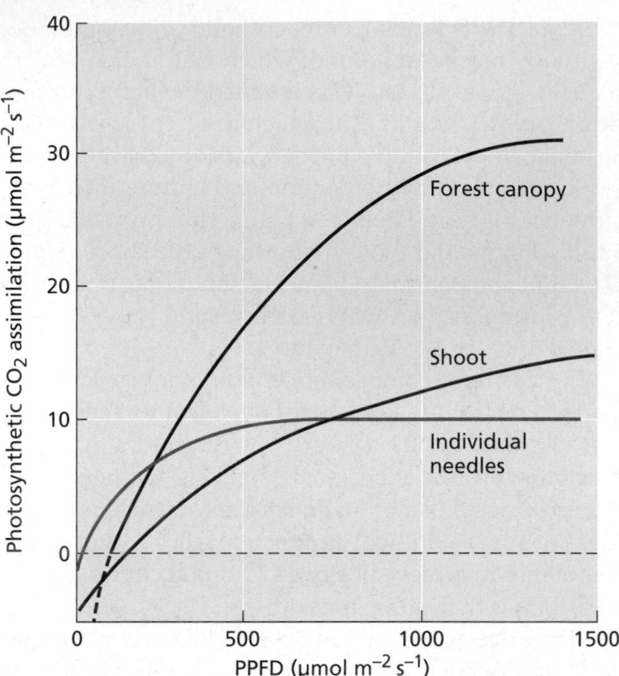

Figure 9.9 Changes in photosynthesis (expressed on a per-square-meter basis) in individual needles, a complex shoot, and a forest canopy of Sitka spruce (*Picea sitchensis*) as a function of PPFD. Complex shoots consist of groupings of needles that often shade each other, similar to the situation in a canopy where branches often shade other branches. As a result of shading, much higher PPFD levels are needed to saturate photosynthesis. The dashed portion of the forest canopy trace has been extrapolated from the measured part of the curve. (After Jarvis and Leverenz 1983.)

this is well-fertilized crop leaves, which often saturate above 1000 μmol $m^{-2} s^{-1}$. Although individual leaves are rarely able to use full sunlight, whole plants usually consist of many leaves that shade each other. Thus, at any given time of the day only a small proportion of the leaves are exposed to full sun, especially in plants with dense canopies. The rest of the leaves receive subsaturating photon fluxes that come from sunflecks that pass through gaps in the leaf canopy, diffuse light, and light transmitted through other leaves.

Because the photosynthetic response of the intact plant is the sum of the photosynthetic activity of all the leaves, only rarely is photosynthesis light-saturated at the level of the whole plant (**Figure 9.9**). It is for this reason that crop productivity is usually related to the total amount of light received during the growing season, rather than to single-leaf photosynthetic capacity. Given enough water and nutrients, the more light a crop receives, the higher the biomass produced.

Leaves must dissipate excess light energy

When exposed to excess light, leaves must dissipate the surplus absorbed light energy to prevent damage to the photosynthetic apparatus (**Figure 9.10**). There are several routes for energy dissipation that involve *nonphotochemical quenching* (see Chapter 7), the quenching of chlorophyll fluorescence by mechanisms other than photochemistry. The most important example involves the transfer of absorbed light energy away from electron transport toward heat production. Although the molecular mechanisms are not yet fully understood, the xanthophyll cycle is an important avenue for dissipation of excess light energy.

THE XANTHOPHYLL CYCLE The xanthophyll cycle, which comprises the three carotenoids violaxanthin, antheraxanthin, and zeaxanthin, establishes an ability to dissipate excess light energy in the leaf (see Figure 7.33). Under high light, violaxanthin is converted to antheraxanthin and then to zeaxanthin. Both of the aromatic rings of violaxanthin have a bound oxygen atom. In antheraxanthin only one of the two rings has a bound oxygen, and in zeaxanthin neither does. Zeaxanthin is the most effective of the three xanthophylls in heat dissipation, and antheraxanthin is only half as effective. Whereas the level

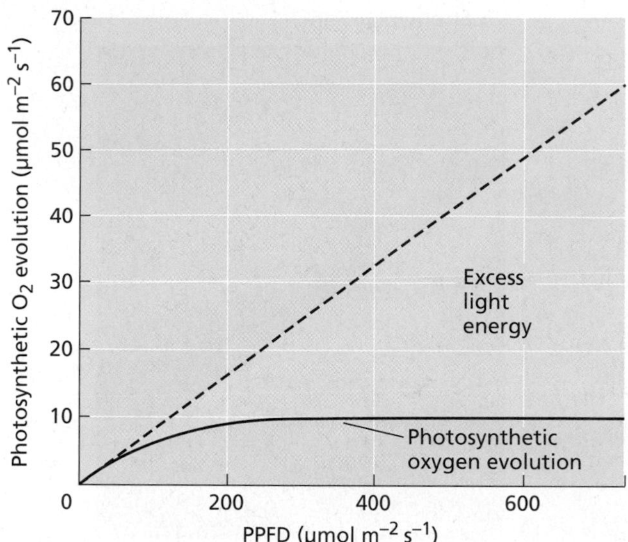

Figure 9.10 Excess light energy in relation to a light-response curve of photosynthetic oxygen evolution in a shade leaf. The broken line shows theoretical oxygen evolution in the absence of any rate limitation to photosynthesis. At PPFD levels up to 150 μmol m^{-2} s^{-1}, a shade plant is able to use the absorbed light. Above 150 μmol m^{-2} s^{-1}, however, photosynthesis saturates, and an increasingly larger amount of the absorbed light energy must be dissipated. At higher PPFD levels there is a large difference between the fraction of light used by photosynthesis versus that which must be dissipated (excess light energy). The differences are much greater in a shade plant than in a sun plant. (After Osmond 1994.)

of antheraxanthin remains relatively constant throughout the day, the zeaxanthin content increases at high PPFD and decreases at low PPFD.

In leaves growing under full sunlight, zeaxanthin and antheraxanthin can make up 40% of the total xanthophyll-cycle pool at maximum PPFD levels attained at midday (**Figure 9.11**). In these conditions a substantial amount of excess light energy absorbed by the thylakoid membranes can be dissipated as heat, thus preventing damage to the photosynthetic machinery of the chloroplast (see Chapter 7). Leaves that grow in full sunlight contain a substantially larger xanthophyll pool than do shade leaves, so they can dissipate higher amounts of excess light energy. Nevertheless, the xanthophyll cycle also operates in plants that grow in the low light of the forest understory, where they are occasionally exposed to sunflecks. Exposure to just one sunfleck results in the conversion of much of the violaxanthin in the leaf to zeaxanthin.

The xanthophyll cycle is also important in species that remain green during winter, when photosynthetic rates are very low yet light absorption remains high. Unlike in the diurnal cycling of the xanthophyll pool observed in the summer, zeaxanthin levels remain high all day during the winter. This mechanism maximizes dissipation of light energy, thereby protecting the leaves against photo-oxidation when winter cold prevents carbon assimilation.

CHLOROPLAST MOVEMENTS An alternative means of reducing excess light energy is to move the chloroplasts so that they are no longer exposed to high light. Chloroplast movement is widespread among algae, mosses, and leaves of higher plants. If chloroplast orientation and location are controlled, leaves can regulate how much incident light is absorbed. In darkness (**Figure 9.12A and B**), chloroplasts gather at the cell surfaces parallel to the plane of the leaf so that they are aligned perpendicularly to the incident light—a position that maximizes absorption of light.

Under high light (**Figure 9.12C**), the chloroplasts move to the cell surfaces that are parallel to the incident light, thus avoiding excess absorption of light. Such chloroplast rearrangement can decrease the amount of light absorbed by the leaf by about 15%. Chloroplast movement in leaves is a typical blue-light response (see Chapter 16). Blue light also controls chloroplast orientation in many of the lower plants, but in some algae, chloroplast movement is controlled by phytochrome. In leaves, chloroplasts move along actin microfilaments in the cytoplasm, and calcium regulates their movement.

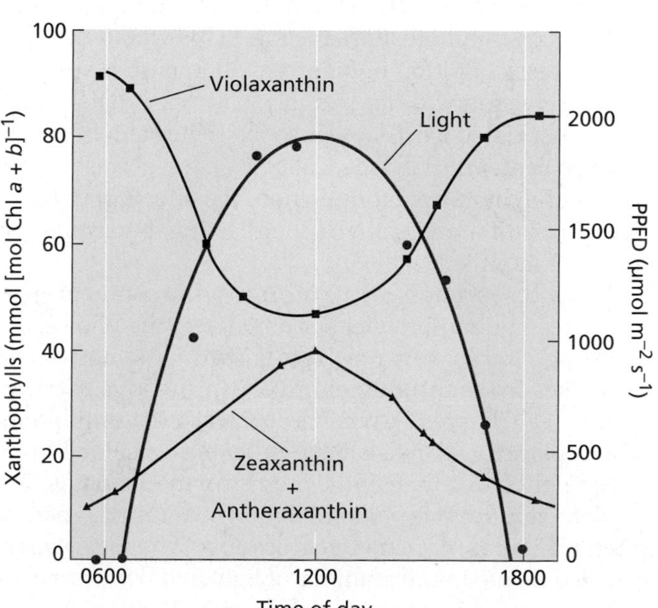

Figure 9.11 Diurnal changes in xanthophyll content as a function of PPFD in sunflower (*Helianthus annuus*). As the amount of light incident to a leaf increases, a greater proportion of violaxanthin is converted to antheraxanthin and zeaxanthin, thereby dissipating excess excitation energy and protecting the photosynthetic apparatus. (After Demmig-Adams and Adams 1996.)

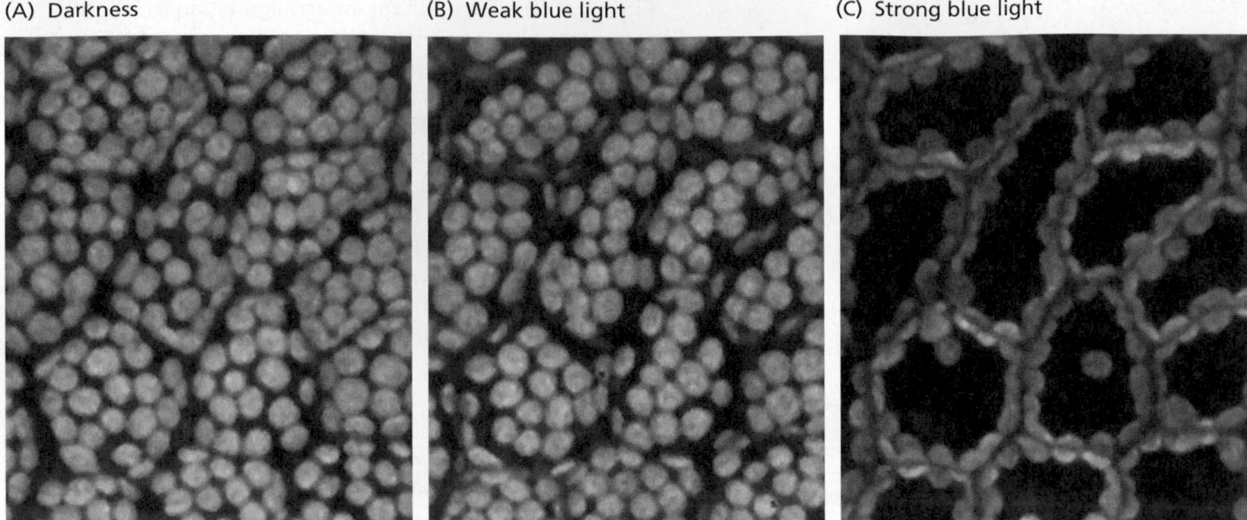

Figure 9.12 Chloroplast distribution in photosynthesizing cells of the duckweed *Lemna*. These surface views show the same cells under three conditions: (A) darkness, (B) weak blue light, and (C) strong blue light. In (A) and (B), chloroplasts are positioned near the upper surface of the cells, where they can absorb maximum amounts of light. When the cells are irradiated with strong blue light (C), the chloroplasts move to the side walls, where they shade each other, thus minimizing the absorption of excess light. (Courtesy of M. Tlalka and M. D. Fricker.)

LEAF MOVEMENTS Plants have also evolved responses that reduce the excess radiation load on whole leaves during high sunlight periods, especially when transpiration and its cooling effects are reduced because of water stress. These responses often involve changes in the leaf orientation relative to the incoming sunlight. For example, heliotropic leaves of both alfalfa and lupine track the sun but at the same time can reduce incident light levels by folding leaflets together so that the leaf laminae become nearly parallel to the sun's rays (paraheliotrophic). These movements are accomplished by changes in the turgor pressure of pulvinus cells at the tip of the petiole. Another common response is mild wilting, as seen in many sunflowers, whereby a leaf droops to a vertical orientation, again effectively reducing the incident heat load and reducing transpiration and incident light levels. Many grasses are able to effectively "twist" through loss of turgor in bulliform cells, resulting in reduced incident PPFD.

Absorption of too much light can lead to photoinhibition

When leaves are exposed to more light than they can use (see Figure 9.10), the reaction center of PSII is inactivated and often damaged in a phenomenon called **photoinhibition** (see Chapter 7). The characteristics of photoinhibition in the intact leaf depend on the amount of light to which the plant is exposed. The two types of photoinhibition are dynamic photoinhibition and chronic photoinhibition.

Under moderate excess light, **dynamic photoinhibition** is observed. Quantum yield decreases, but the maximum photosynthetic rate remains unchanged. Dynamic photoinhibition is caused by the diversion of absorbed light energy toward photoprotective heat dissipation—hence the decrease in quantum yield. This decrease is often temporary, and quantum yield can return to its initial higher value when PPFD decreases below saturation levels. **Figure 9.13** shows how photons from sunlight are allocated to photosynthetic reactions versus being thermally dissipated as excess energy over the course of a day under favorable and stressed environmental conditions.

Chronic photoinhibition results from exposure to high levels of excess light that damage the photosynthetic system and decrease both instantaneous quantum yield and maximum photosynthetic rate. This would happen if the stress condition in Figure 9.13B persisted because photoprotection was not possible. Chronic photoinhibition is associated with damage to the D1 protein from the reaction center of PSII (see Chapter 7). In contrast to the effects of dynamic photoinhibition, the effects of chronic photoinhibition are relatively long-lasting, persisting for weeks or months.

Early researchers of photoinhibition interpreted decreases in quantum yield as damage to the photosynthetic apparatus. It is now recognized that short-term decreases in quantum yield reflect protective mechanisms (see Chapter 7), whereas chronic photoinhibition represents actual damage to the chloroplast resulting from excess light or a failure of the protective mechanisms.

How significant is photoinhibition in nature? Dynamic photoinhibition appears to occur daily, when leaves are exposed to maximum amounts of light and there is a corresponding reduction in carbon fixation. Photoinhibition

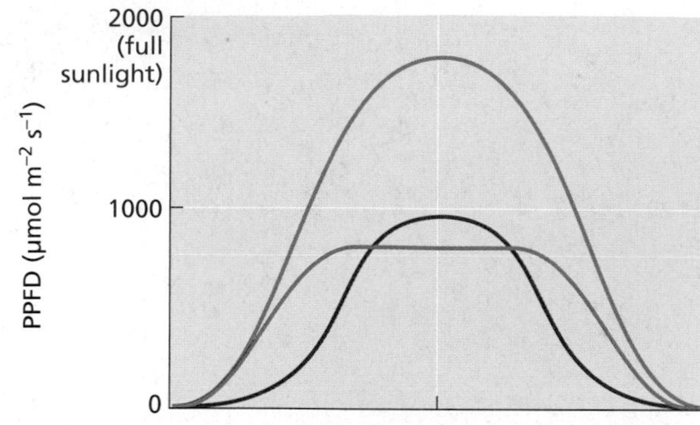

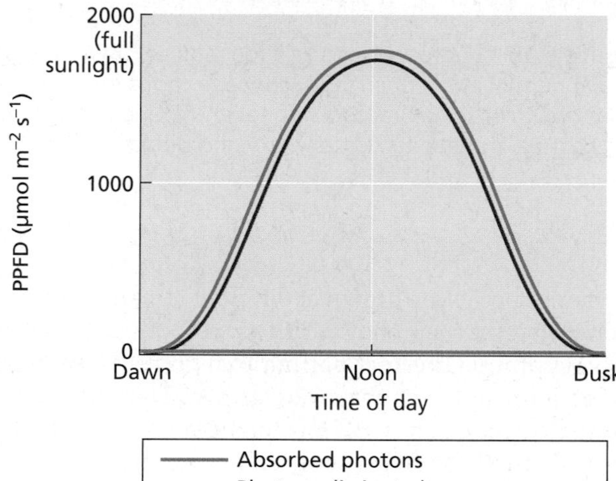

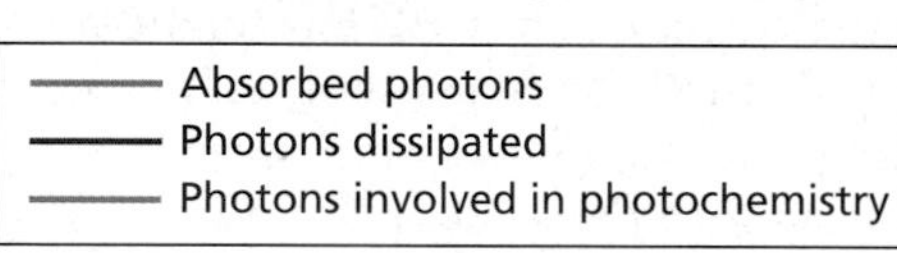

Figure 9.13 Changes over the course of a day in the allocation of photons absorbed by sunlight. Shown here are contrasts in how the photons striking a leaf are either involved in photochemistry or thermally dissipated as excess energy under favorable conditions (A) and stress conditions (B). (After Demmig-Adams and Adams 2000.)

is more pronounced at low temperatures, and becomes chronic under more extreme climatic conditions.

Effects of Temperature on Photosynthesis in the Intact Leaf

Photosynthesis (CO_2 uptake) and transpiration (H_2O loss) share a common pathway. That is, CO_2 diffuses into the leaf, and H_2O diffuses out, through the stomatal opening regulated by the guard cells. While these are independent processes, vast quantities of water are lost during photosynthetic periods, with the molar ratio of H_2O loss to CO_2 uptake often exceeding 250. This high water-loss rate also removes heat from leaves through evaporative cooling, keeping them relatively cool even under full sunlight conditions. Transpirational cooling is important, since photosynthesis is a temperature-dependent process, but the concurrent water loss means that cooling comes at a cost, especially in arid and semiarid ecosystems.

Leaves must dissipate vast quantities of heat

The heat load on a leaf exposed to full sunlight is very high. In fact, under normal sunny conditions with moderate air temperatures, a leaf would warm up to a dangerously high temperature if all incident solar energy were absorbed and none of the heat was dissipated. However, this does not occur because leaves absorb only about 50% of the total solar energy (300–3000 nm), with most of the absorption occurring in the visible portion of the spectrum (see Figures 9.2 and 9.3). This amount is still large. The typical heat load of a leaf is dissipated through three processes (**Figure 9.14**):

- Radiative heat loss: All objects emit long-wave radiation (at about 10,000 nm) in proportion to their temperature to the fourth power (Stephan Boltzman equation). However, the maximum emitted wavelength is inversely proportional to the leaf temperature, and leaf temperatures are low enough that the wavelengths emitted are not visible to the human eye.

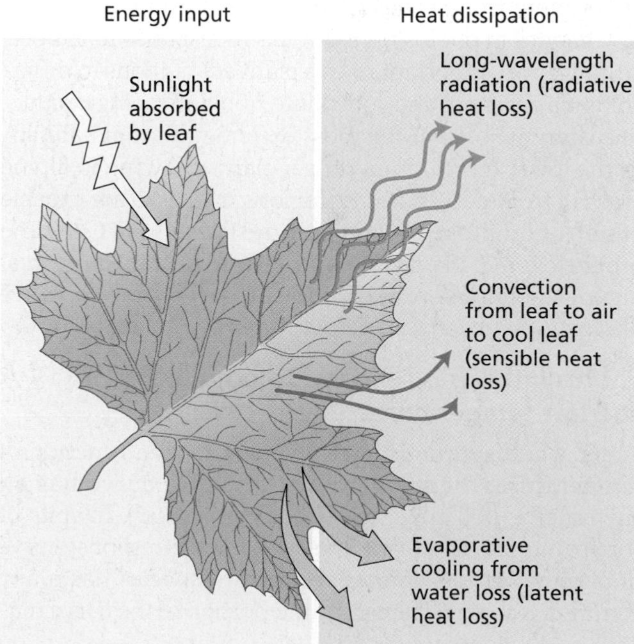

Figure 9.14 Absorption and dissipation of energy from sunlight by the leaf. The imposed heat load must be dissipated in order to avoid damage to the leaf. The heat load is dissipated by emission of long-wavelength radiation, by sensible heat loss to the air surrounding the leaf, and by the evaporative cooling caused by transpiration.

- Sensible heat loss: If the temperature of the leaf is higher than that of the air circulating around the leaf, the heat is convected (transferred) away from the leaf to the air. The size and shape of a leaf influence the amount of sensible heat loss.
- Latent heat loss: Because the evaporation of water requires energy, when water evaporates from a leaf (transpiration), it removes large amounts of heat from the leaf and thus cools it. The human body is cooled by the same principle, commonly known as perspiration.

Sensible heat loss and evaporative heat loss are the most important processes in the regulation of leaf temperature, and the ratio of the two fluxes is called the **Bowen ratio**:

$$\text{Bowen ratio} = \frac{\text{Sensible heat loss}}{\text{Evaporative heat loss}}$$

In well-watered crops, transpiration (see Chapter 4), and hence water evaporation from the leaf, is high, so the Bowen ratio is low (see WEB TOPIC 9.2). Conversely, when evaporative cooling is limited, the Bowen ratio is high. For example, in a water-stressed crop, partial stomatal closure reduces evaporative cooling and the Bowen ratio is increased. The amount of evaporative heat loss (and thus the Bowen ratio) is influenced by the degree to which stomata remain open.

Plants with very high Bowen ratios conserve water, but consequently may also experience high leaf temperatures. However, the temperature difference between the leaf and the air does increase the amount of sensible heat loss. Reduced growth is usually correlated with high Bowen ratios, because a high Bowen ratio is indicative of at least partial stomatal closure.

There is an optimal temperature for photosynthesis

Maintaining favorable leaf temperatures is crucial to plant growth because maximum photosynthesis occurs within a relatively narrow temperature range. The peak photosynthetic rate across a range of temperatures is the *photosynthetic thermal optimum*. When the optimal temperature for a given plant is exceeded, photosynthetic rates decrease. The photosynthetic thermal optimum reflects biochemical, genetic (adaptation), and environmental (acclimation) components.

Species adapted to different thermal regimes usually have an optimal temperature range for photosynthesis that reflects the temperatures of the environment in which they evolved. A contrast is especially clear between the C_3 plant *Atriplex glabriuscula*, which commonly grows in cool coastal environments, and the C_4 plant *Tidestromia oblongifolia*, from a hot desert environment (**Figure 9.15**). The ability to acclimate or biochemically adjust to temperature can also be found within species. When plants of

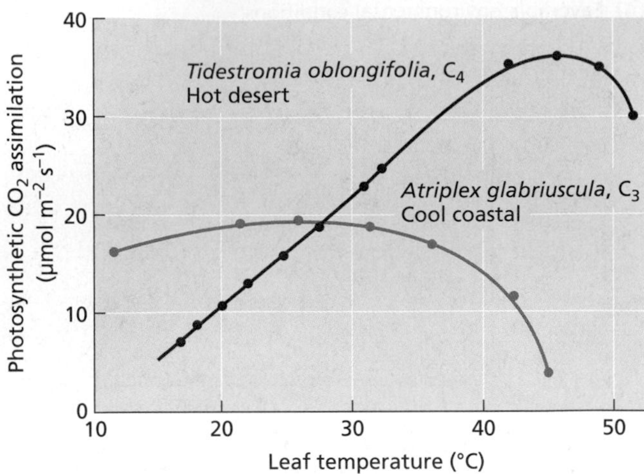

Figure 9.15 Photosynthesis as a function of leaf temperature at normal atmospheric CO_2 concentrations for a C_3 plant grown in its natural cool habitat and a C_4 plant growing in its natural hot habitat. (After Berry and Björkman 1980.)

the same species are grown at different temperatures and then tested for their photosynthetic response, they show photosynthetic thermal optima that correlate with the temperature at which they were grown. That is, plants of the same species grown at low temperatures have higher photosynthetic rates at low temperatures, whereas those same plants grown at high temperatures have higher photosynthetic rates at high temperatures. The ability to adjust morphologically, physiologically, or biochemically in response to changes in the environment is referred to as plasticity. Plants with a high thermal plasticity are capable of growing over a wide range of temperatures.

Changes in photosynthetic rates in response to temperature play an important role in plant adaptations to different environments and contribute to plants being productive even in some of the most extreme thermal habitats. In the lower temperature range, plants growing in alpine areas of Colorado and arctic regions in Alaska are capable of net CO_2 uptake at temperatures close to 0°C. At the other extreme, plants living in Death Valley, California, one of the hottest places on Earth, can achieve positive photosynthetic rates at temperatures approaching 50°C.

Photosynthesis is sensitive to both high and low temperatures

When photosynthetic rates are plotted as a function of temperature, the temperature-response curve has an asymmetric bell-type shape (see Figure 9.15). In spite of some differences in shape, the temperature-response curve of photosynthesis among and within species has many common features. The ascending portion of the curve represents a temperature-dependent stimulation of enzymatic activities; the flat top is the temperature range that is optimal for photosynthesis; and the descending portion of the curve is associated with temperature-sensitive deleterious effects, some of which are reversible while others are not.

What factors are associated with the decline in photosynthesis above the photosynthetic temperature optimum? Temperature affects all biochemical reactions of photosynthesis as well as membrane integrity in chloroplasts, so it is not surprising that the responses to temperature are complex. Cellular respiration rates increase as a function of temperature, but they are not the primary reason for the sharp decrease in net photosynthesis at high temperatures. A major impact of high temperature is on membrane-bound electron transport processes, which become uncoupled or unstable at high temperatures. This cuts off the supply of reducing power needed to fuel net photosynthesis and leads to a sharp overall decrease in photosynthesis.

Under ambient CO_2 concentrations and with favorable light and soil moisture conditions, the photosynthetic thermal optimum is often limited by the activity of rubisco. In leaves of C_3 plants, the response to increasing temperature reflects conflicting processes: an increase in carboxylation rate and a decrease in the affinity of rubisco for CO_2 with a corresponding increase in photorespiration (see Chapter 8). (There is also evidence that rubisco activity decreases because of negative heat effects on rubisco activase at higher [>35°C] temperatures; see Chapter 8.) The reduction in the affinity for CO_2 and the increase in photorespiration attenuate the potential temperature response of photosynthesis under ambient CO_2 concentrations. By contrast, in plants with C_4 photosynthesis, the leaf interior is CO_2-saturated, or nearly so (as we discussed in Chapter 8), and the negative effect of high temperature on rubisco affinity for CO_2 is not realized. This is one reason that leaves of C_4 plants tend to have a higher photosynthetic temperature optimum than do leaves of C_3 plants (see Figure 9.15).

At low temperatures, C_3 photosynthesis can also be limited by factors such as phosphate availability in the chloroplast. When triose phosphates are exported from the chloroplast to the cytosol, an equimolar amount of inorganic phosphate is taken up via translocators in the chloroplast membrane. If the rate of triose phosphate use in the cytosol decreases, phosphate uptake into the chloroplast is inhibited and photosynthesis becomes phosphate-limited. Starch synthesis and sucrose synthesis decrease rapidly with decreasing temperature, reducing the demand for triose phosphates and causing the phosphate limitation observed at low temperatures.

Photosynthetic efficiency is temperature-sensitive

Photorespiration (see Chapter 8) and the quantum yield (light-use efficiency) differ between C_3 and C_4 photosynthesis, with changes particularly noticeable as temperatures vary. **Figure 9.16** illustrates quantum yield for photosynthesis as a function of leaf temperature in C_3 plants and C_4 plants in today's atmosphere of 400 ppm CO_2. In the C_4 plants the quantum yield remains constant with temperature, reflecting low rates of photorespiration. In the C_3 plants the quantum yield decreases with temperature, reflecting a stimulation of photorespiration by temperature and an ensuing higher energy cost for net CO_2 fixation.

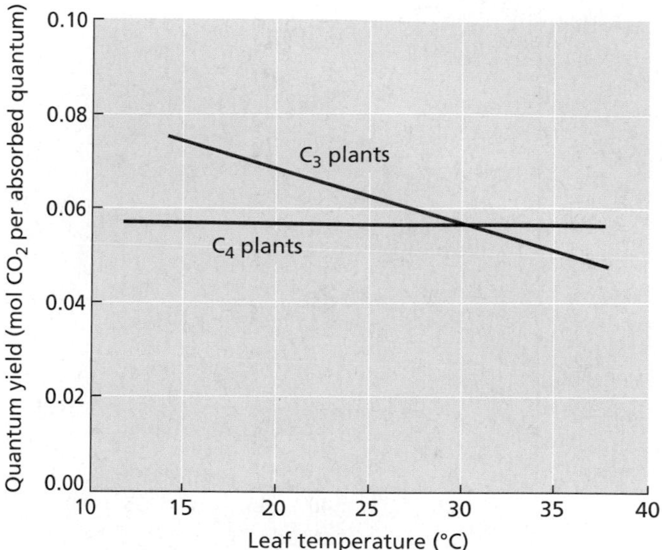

Figure 9.16 Quantum yield of photosynthetic carbon fixation in C_3 and C_4 plants as a function of leaf temperature. Photorespiration increases with temperature in C_3 plants, and the energy cost of net CO_2 fixation increases accordingly. This higher energy cost is reflected in lower quantum yields at higher temperatures. In contrast, photorespiration is very low in C_4 plants and the quantum yield does not show a temperature dependence. Note that at lower temperatures the quantum yield of C_3 plants is higher than that of C_4 plants, indicating that C_3 photosynthesis is more efficient at lower temperatures. (After Ehleringer et al. 1997.)

The combination of reduced quantum yield and increased photorespiration leads to expected differences in the photosynthetic capacities of C_3 and C_4 plants in habitats with different temperatures. The predicted relative rates of primary productivity of C_3 and C_4 grasses along a latitudinal transect in the Great Plains of North America from southern Texas in the United States to Manitoba in Canada are shown in **Figure 9.17**. This decline in C_4 relative to C_3 productivity moving northward closely parallels the shift in abundance of plants with these pathways in the Great Plains: C_4 species are more common below 40°N, and C_3 species dominate above 45°N (see **WEB TOPIC 9.3**).

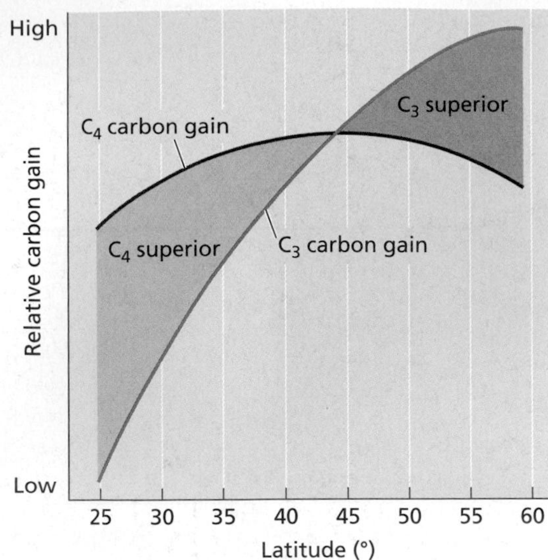

Figure 9.17 Relative rates of photosynthetic carbon gain predicted for identical C_3 and C_4 grass canopies as a function of latitude across the Great Plains of North America. (After Ehleringer 1978.)

Effects of Carbon Dioxide on Photosynthesis in the Intact Leaf

We have discussed how light and temperature influence leaf physiology and anatomy. Now we will turn our attention to how CO_2 concentration affects photosynthesis. CO_2 diffuses from the atmosphere into leaves—first through stomata, then through the intercellular air spaces, and ultimately into cells and chloroplasts. In the presence of adequate amounts of light, higher CO_2 concentrations support higher photosynthetic rates. The reverse is also true: low CO_2 concentrations can limit the amount of photosynthesis in C_3 plants.

In this section we will discuss the concentration of atmospheric CO_2 in recent history, and its availability for carbon-fixing processes. Then we will consider the limitations that CO_2 places on photosynthesis and the impact of the CO_2-concentrating mechanisms of C_4 plants.

Atmospheric CO_2 concentration keeps rising

Carbon dioxide presently accounts for about 0.040%, or 400 ppm, of air. The partial pressure of ambient CO_2 (c_a) varies with atmospheric pressure and is approximately 40 pascals (Pa) at sea level (see **WEB TOPIC 9.4**). Water vapor usually accounts for up to 2% of the atmosphere and O_2 for about 21%. The largest constituent in the atmosphere is diatomic nitrogen, at about 77%.

Today the atmospheric concentration of CO_2 is almost twice the concentration that prevailed over the last 400,000 years, as measured from air bubbles trapped in glacial ice in Antarctica (**Figure 9.18A and B**), and it is higher than any experienced on Earth in the last 2 million years. Most extant plant taxa are therefore thought to have evolved in a low-CO_2 world (~180–280 ppm CO_2). Only when one looks back about 35 million years does one find CO_2 concentrations of much higher levels (>1000 ppm). Thus, the geologic trend over these many millions of years was one of decreasing atmospheric CO_2 concentrations (see **WEB TOPIC 9.5**).

Currently, the CO_2 concentration of the atmosphere is increasing by about 1 to 3 ppm each year, primarily because of the burning of fossil fuels (e.g., coal, oil, and natural gas) and deforestation (**Figure 9.18C**). Since 1958, when C. David Keeling began systematic measurements of CO_2 in the clean air at Mauna Loa, Hawaii, atmospheric CO_2 concentrations have increased by more than 25%. By 2100 the atmospheric CO_2 concentration could reach 600 to 750 ppm unless fossil fuel emissions and deforestation are diminished (see **WEB TOPIC 9.6**).

CO_2 diffusion to the chloroplast is essential to photosynthesis

For photosynthesis to occur, CO_2 must diffuse from the atmosphere into the leaf and to the carboxylation site of rubisco. The diffusion rate depends on the CO_2 concentration gradient in the leaf (see Chapters 3 and 6) and resistances along the diffusion pathway. The cuticle that covers the leaf is nearly impermeable to CO_2, so the main port of entry of CO_2 into the leaf is the stomatal pore. (The same path is traveled in the reverse direction by H_2O.) CO_2 diffuses through the pore into the substomatal cavity and into the intercellular air spaces between the mesophyll cells. This portion of the diffusion path of CO_2 into the chloroplast is a gaseous phase. The remainder of the diffusion path to the chloroplast is a liquid phase, which begins at the water layer that wets the walls of the mesophyll cells and continues through the plasma membrane, the cytosol, and the chloroplast. (For the properties of CO_2 in solution, see **WEB TOPIC 8.8**.)

The sharing of the stomatal entry pathway by CO_2 and H_2O presents the plant with a functional dilemma. In air of high relative humidity, the diffusion gradient that drives water loss is about 50 times larger than the gradient that drives CO_2 uptake. In drier air, this difference can be much larger. Therefore, a decrease in stomatal resistance through the opening of stomata facilitates higher CO_2 uptake but is unavoidably accompanied by substantial water loss. Not surprisingly, many adaptive features help counteract this water loss in plants in arid and semiarid regions of the world.

Each portion of the CO_2 diffusion pathway imposes a resistance to CO_2 diffusion, so the supply of CO_2 for photosynthesis meets a series of different points of resistance. The gaseous phase of CO_2 diffusion into the leaf can be divided into three components—the boundary layer, the stomata, and the intercellular spaces of the leaf—each

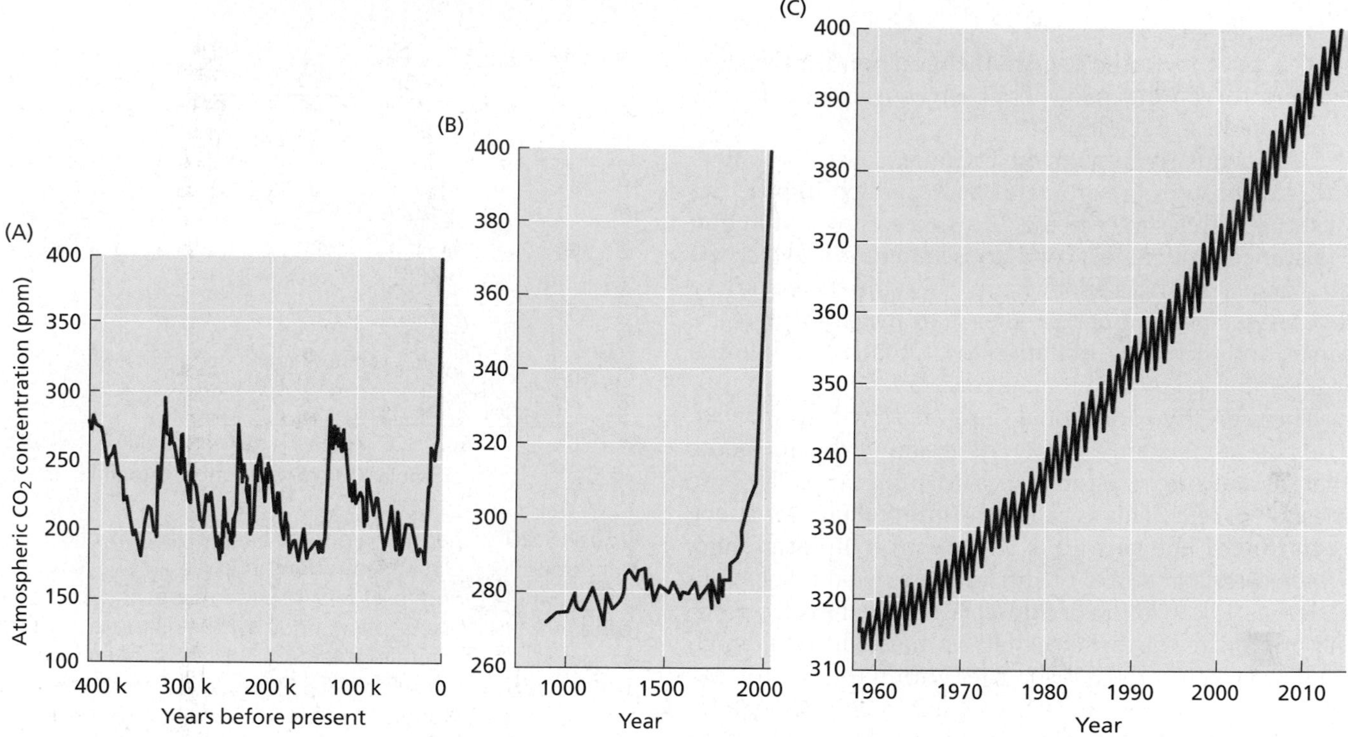

Figure 9.18 Concentration of atmospheric CO_2 from 420,000 years ago to the present. (A) Past atmospheric CO_2 concentrations, determined from bubbles trapped in glacial ice in Antarctica, were much lower than current levels. (B) In the last 1000 years, the rise in atmospheric CO_2 concentration coincides with the Industrial Revolution and the increased burning of fossil fuels. (C) Current atmospheric concentrations of CO_2 measured at Mauna Loa, Hawaii, continue to rise. The wavy nature of the trace is caused by change in atmospheric CO_2 concentrations associated with seasonal changes in relative balance between photosynthesis and respiration rates. Each year the highest CO_2 concentration is observed in May, just before the Northern Hemisphere growing season, and the lowest concentration is observed in October. (After Barnola et al. 1994, Keeling and Whorf 1994, Neftel et al. 1994, and Keeling et al. 1995; updated using data from http://www.esrl.noaa.gov/gmd/ccgg/trends/.)

of which imposes a resistance to CO_2 diffusion (**Figure 9.19**). An evaluation of the magnitude of each point of resistance is helpful for understanding CO_2 limitations to photosynthesis.

The boundary layer consists of relatively unstirred air near the leaf surface, and its resistance to diffusion is called the **boundary layer resistance**. The boundary layer resistance affects all diffusive processes, including water and CO_2 diffusion as well as sensible heat loss, discussed earlier. The boundary layer resistance decreases with smaller leaf size and greater wind speed. Smaller leaves thus have a lower resistance to CO_2 and water diffusion, and to sensible heat loss. Leaves of desert plants are usually small, facilitating sensible heat loss. In contrast, large leaves are often found in the humid tropics, especially in the shade. These leaves have large boundary layer resis-

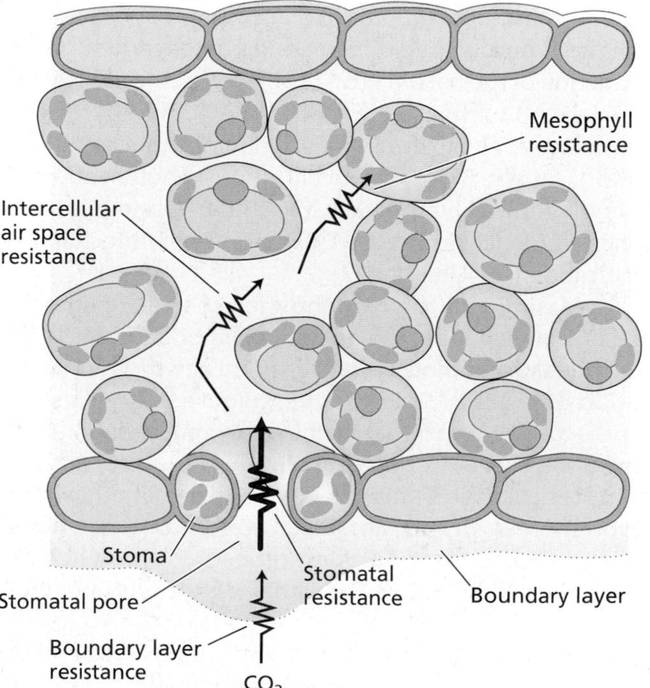

Figure 9.19 Points of resistance to the diffusion and fixation of CO_2 from outside the leaf to the chloroplasts. The stomatal aperture is the major point of resistance to CO_2 diffusion into the leaf.

tances, but they can dissipate the radiation heat load by evaporative cooling made possible by the abundant water supply in these habitats.

After diffusing through the boundary layer, CO_2 enters the leaf through the stomatal pores, which impose the next type of resistance in the diffusion pathway, **stomatal resistance**. Under most conditions in nature, in which the air around a leaf is seldom completely still, the boundary layer resistance is much smaller than the stomatal resistance, and the main limitation to CO_2 diffusion into the leaf is imposed by the stomatal resistance.

There are two additional resistances within the leaf. The first is resistance to CO_2 diffusion in the air spaces that separate the substomatal cavity from the walls of the mesophyll cells. This is called the **intercellular air space resistance**. The second is the **mesophyll resistance**, which is resistance to CO_2 diffusion in the liquid phase in C_3 leaves. Localization of chloroplasts near the cell periphery minimizes the distance that CO_2 must diffuse through liquid to reach carboxylation sites within the chloroplast. The mesophyll resistance to CO_2 diffusion is thought to be approximately 1.4 times the combined boundary layer resistance and stomatal resistance when the stomata are fully open. Because the stomatal guard cells can impose a variable and potentially large resistance to CO_2 influx and water loss in the diffusion pathway, regulating the stomatal aperature provides the plant with an effective way to control gas exchange between the leaf and the atmosphere (see **WEB TOPIC 9.4**).

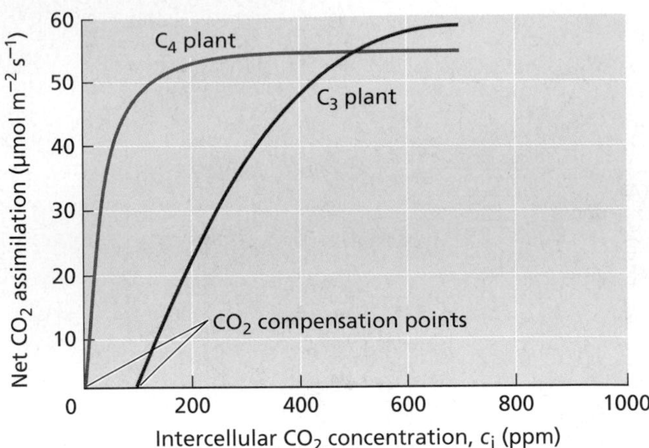

Figure 9.20 Changes in photosynthesis as a function of intercellular CO_2 concentrations in Arizona honeysweet (*Tidestromia oblongifolia*), a C_4 plant, and creosote bush (*Larrea tridentata*), a C_3 plant. Photosynthetic rate is plotted against calculated intercellular CO_2 concentration inside the leaf (see Equation 5 in **WEB TOPIC 9.4**). The intercellular CO_2 concentration at which net CO_2 assimilation is zero defines the CO_2 compensation point. (After Berry and Downton 1982.)

CO_2 imposes limitations on photosynthesis

For C_3 plants growing with adequate light, water, and nutrients, CO_2 enrichment above natural atmospheric concentrations results in increased photosynthesis and enhanced productivity. Expressing photosynthetic rate as a function of the partial pressure of CO_2 in the intercellular air space (c_i) within the leaf (see **WEB TOPIC 9.4**) makes it possible to evaluate limitations to photosynthesis imposed by CO_2 supply. At low c_i concentrations, photosynthesis is strongly limited by the low CO_2. In the absence of atmospheric CO_2, leaves release CO_2 because of mitochondrial respiration (see Chapter 12).

Increasing c_i to the concentration at which photosynthesis and respiration balance each other defines the **CO_2 compensation point**. This is the point at which net assimilation of CO_2 by the leaf is zero (**Figure 9.20**). This concept is analogous to that of the light compensation point discussed earlier in this chapter. The CO_2 compensation point reflects the balance between photosynthesis and respiration as a function of CO_2 concentration, whereas the light compensation point reflects that balance as a function of PPFD under constant CO_2 concentration.

C_3 VERSUS C_4 PLANTS In C_3 plants, increasing c_i above the compensation point increases photosynthesis over a wide concentration range (see Figure 9.20). At low to intermediate CO_2 subambient concentrations, photosynthesis is limited by the carboxylation capacity of rubisco. At higher c_i concentrations, photosynthesis begins to saturate as the net photosynthetic rate becomes limited by another factor (remember Blackman's concept of limiting factors). At these higher c_i levels, net photosynthesis becomes limited by the capacity of the light reactions to generate sufficient NADPH and ATP to regenerate the acceptor molecule ribulose-1,5-bisphosphate. Most leaves appear to regulate their c_i values by controlling stomatal opening, so that c_i remains at an intermediate subambient concentration between the limits imposed by carboxylation capacity and the capacity to regenerate ribulose-1,5-bisphosphate. In this way, both the light and dark reactions of photosynthesis are co-limiting. A plot of net CO_2 assimilation as a function of c_i tells us how photosynthesis is regulated by CO_2, independent of the functioning of stomata (see Figure 9.20).

Comparing such a plot for C_3 and C_4 plants reveals interesting differences between the two pathways of carbon metabolism:

- In C_4 plants, photosynthetic rates saturate at c_i values of about 100–200 ppm, reflecting the effective CO_2-concentrating mechanisms operating in these plants (see Chapter 8).
- In C_3 plants, increasing c_i levels continue to stimulate photosynthesis over a much broader CO_2 range than for C_4 plants.

- In C_4 plants, the CO_2 compensation point is zero or nearly zero, reflecting their very low levels of photorespiration (see Chapter 8).
- In C_3 plants, the CO_2 compensation point is about 50–100 ppm at 25°C, reflecting CO_2 production because of photorespiration (see Chapter 8).

These responses reveal that C_3 plants will be more likely than C_4 plants to benefit from ongoing increases in today's atmospheric CO_2 concentrations (see Figure 9.20). Because photosynthesis in C_4 plants is saturated at low CO_2 concentrations, C_4 plants do not benefit much from increases in atmospheric CO_2 concentrations.

From an evolutionary perspective, the ancestral photosynthetic pathway is C_3 photosynthesis, and C_4 photosynthesis is a derived pathway. During earlier geologic time periods when atmospheric CO_2 concentrations were much higher than they are today, CO_2 diffusion through stomata into leaves would have resulted in higher c_i values and therefore higher photosynthetic rates in C_3 plants, but not in C_4 plants. The evolution of C_4 photosynthesis is one biochemical adaptation to a CO_2-limited atmosphere. Our current understanding is that C_4 photosynthesis evolved recently in geological terms, more than 20 million years ago.

If the ancient Earth of more than 50 million years ago had atmospheric CO_2 concentrations that were well above those of today, under what atmospheric conditions might we expect C_4 photosynthesis to become a major photosynthetic pathway found in Earth's ecosystems? Jim Ehleringer's group suggests that C_4 photosynthesis first became a prominent component of terrestrial ecosystems in the warmest growing regions of Earth when global CO_2 concentrations decreased below some critical and as yet unknown threshold (**Figure 9.21**). Concurrently, the negative impacts of high photorespiration and CO_2 limitation on C_3 photosynthesis would have been greatest under these warm to hot growing conditions, and low atmospheric CO_2. C_4 plants would have been most favored during periods of Earth's history when CO_2 levels were lowest. There are now data to indicate that C_4 photosynthesis was more prominent during the glacial periods when atmospheric CO_2 levels were below 200 ppm (see Figure 9.18). Other factors may have contributed to the spread of C_4 plants, but certainly low atmospheric CO_2 was one important factor favoring their evolution and ultimately geographic expansion.

Because of the CO_2-concentrating mechanisms in C_4 plants, CO_2 concentration at the carboxylation sites within C_4 chloroplasts is typically close to saturating for rubisco activity. As a result, plants with C_4 metabolism need less rubisco than C_3 plants to achieve a given rate of photosynthesis, and thus require less nitrogen to grow. In addition, the CO_2-concentrating mechanism allows the leaf to maintain high photosynthetic rates at lower c_i values. This

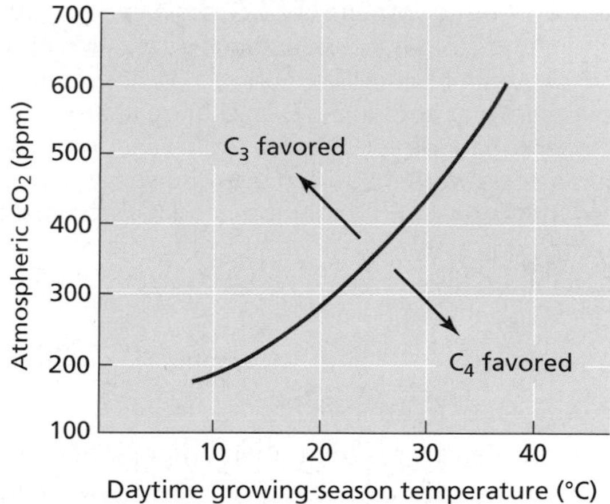

Figure 9.21 Combination of global atmospheric CO_2 levels and daytime growing-season temperatures that are predicted to favor C_3 versus C_4 grasses. At any point in time, Earth is at a single atmospheric CO_2 concentration, resulting in the expectation that C_4 plants would be most common in habitats with the warmest growing seasons. (After Ehleringer et al. 1997.)

permits stomates to remain relatively closed, resulting in less water loss for a given rate of photosynthesis. Thus, the CO_2-concentrating mechanism helps C_4 plants use water and nitrogen more efficiently than C_3 plants. However, the additional energy cost required by the CO_2-concentrating mechanism (see Chapter 8) reduces the light-use efficiency of C_4 photosynthesis. This is probably one reason that most shade-adapted plants in temperate regions are not C_4 plants.

CAM PLANTS Plants with crassulacean acid metabolism (CAM), including many cacti, orchids, bromeliads, and other succulents, have stomatal activity patterns that contrast with those found in C_3 and C_4 plants. CAM plants open their stomata at night and close them during the day, exactly the opposite of the pattern observed in leaves of C_3 and C_4 plants (**Figure 9.22**). At night, atmospheric CO_2 diffuses into CAM plants where it is combined with phosphoenolpyruvate and fixed into oxaloacetate, which is reduced to malate (see Chapter 8). Because stomata are open primarily at night, when lower temperatures and higher humidity reduce transpiration demand, the ratio of water loss to CO_2 uptake is much lower in CAM plants than it is in either C_3 or C_4 plants.

The main photosynthetic constraint on CAM metabolism is that the capacity to store malic acid is limited, and this limitation restricts the total amount of CO_2 uptake. However, the daily cycle of CAM photosynthesis can be very flexible. Some CAM plants are able to enhance total photosynthesis during wet conditions by fixing CO_2 via the Calvin–Benson cycle at the end of the day, when tem-

Figure 9.22 Photosynthetic net CO_2 assimilation, H_2O evaporation, and stomatal conductance of a CAM plant, the cactus *Opuntia ficus-indica*, during a 24-hour period. The whole plant was kept in a gas-exchange chamber in the laboratory. The dark period is indicated by shaded areas. Three parameters were measured over the study period: (A) photosynthetic rate, (B) water loss, and (C) stomatal conductance. In contrast to plants with C_3 or C_4 metabolism, CAM plants open their stomata and fix CO_2 at night. (After Gibson and Nobel 1986.)

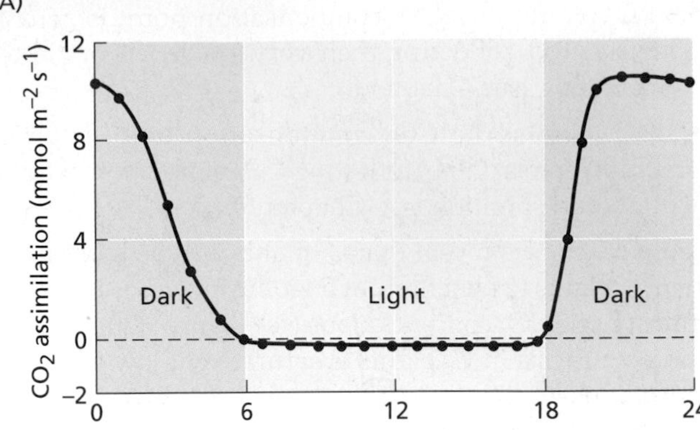

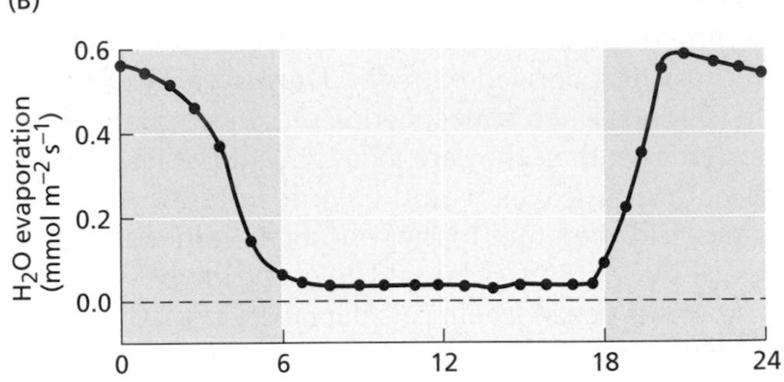

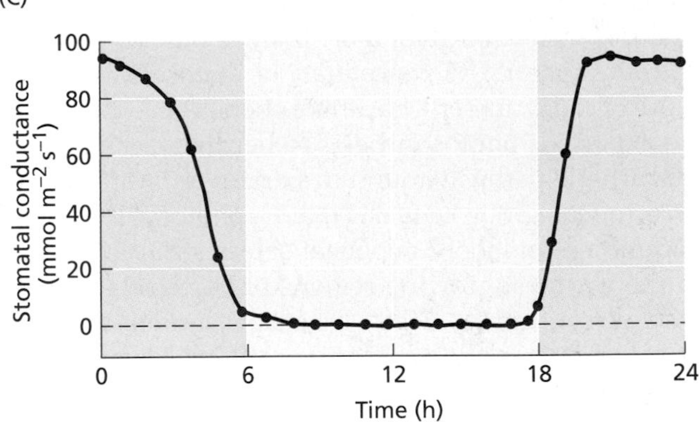

perature gradients are less extreme. Other plants may use CAM as a survival mechanism during severe water limitations. For example, cladodes (flattened stems) of cacti can survive after detachment from the plant for several months without water. Their stomata are closed all the time, and the CO_2 released by respiration is refixed into malate. This process, which has been called *CAM idling*, also allows the intact plant to survive for prolonged drought periods while losing remarkably little water.

How will photosynthesis and respiration change in the future under elevated CO_2 conditions?

The consequences of increasing global atmospheric CO_2 are under intense scrutiny by scientists and government agencies, particularly because of predictions that the greenhouse effect is altering the world's climate. The **greenhouse effect** refers to the warming of Earth's climate that is caused by the trapping of long-wavelength radiation by the atmosphere.

A greenhouse roof transmits visible light, which is absorbed by plants and other surfaces inside the greenhouse. Some of the absorbed light energy is converted to heat, and some of it is reemitted as long-wavelength radiation. Because glass transmits long-wavelength radiation very poorly, this radiation cannot leave the greenhouse through the glass roof, and the greenhouse heats up. Certain gases in the atmosphere, particularly CO_2 and methane, play a role similar to that of the glass roof in a greenhouse. The increased CO_2 concentrations, and elevated temperatures associated with the greenhouse effect, have multiple influences on photosynthesis and plant growth. At today's atmospheric CO_2 concentrations, photosynthesis in C_3 plants is CO_2-limited, but this situation will change as atmospheric CO_2 concentrations continue to rise.

A central question in plant physiology today is: How will photosynthesis and respiration differ by the year 2100 when global CO_2 levels have reached 500 ppm, 600 ppm, or even higher? This question is particularly relevant as humans continue to add CO_2 derived from fossil fuel combustion to Earth's atmosphere. Under well-watered and highly fertilized laboratory conditions, most C_3 plants grow about 30% faster when the CO_2 concentration reaches 600 to 750 ppm than they do today; above that atmospheric CO_2 concentration the growth rate becomes more limited by the nutrients available to the plant. To study this question in the field, scientists need to be able to create realistic simulations of future environments. A promising approach to the study of plant physiology and ecology in environments with elevated CO_2 levels has been the use of Free Air CO_2 Enrichment (FACE) experiments.

For FACE experiments, entire fields of plants or natural ecosystems are encircled by emitters that add CO_2 to the air to create the high-CO_2 environment we might expect to have 25 to 50 years from now. **Figure 9.23** shows FACE experiments in three different vegetation types.

FACE experiments have provided key new insights into how plants and ecosystems will respond to the CO_2 levels expected in the future. One key observation is that

Figure 9.23 Free Air CO_2 Enrichment (FACE) experiments are used to study how plants and ecosystems will respond to future CO_2 levels. Shown here are FACE experiments in a deciduous forest (A) and a crop canopy (B). (C) Under elevated CO_2 levels, leaf stomata are more closed, resulting in higher leaf temperatures, as shown by the infrared image of a crop canopy. (A courtesy of D. Karnosky; B courtesy of the USDA/ARS; C from Long et al. 2006.)

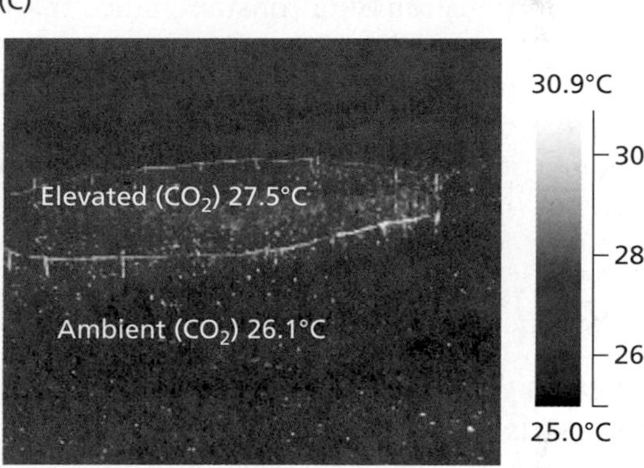

plants with the C_3 photosynthetic pathway are much more responsive than C_4 plants under well-watered conditions, with net photosynthetic rates increasing 20% or more in C_3 plants and little to not at all in C_4 plants. Photosynthesis increases in C_3 plants because c_i levels increase (see Figure 9.20). At the same time, there is a down-regulation of photosynthetic capacity manifested in reduced activity of the enzymes associated with the dark reactions of photosynthesis.

Elevated CO_2 levels will affect many plant processes. For instance, leaves tend to keep their stomata more closed under elevated CO_2 levels. As a direct consequence of reduced transpiration, leaf temperatures are higher (see Figure 9.23C), which may feed back on basic mitochondrial respiration. This is indeed an exciting and promising area of current research. From FACE studies, it has become increasingly clear that an acclimation process occurs under higher CO_2 levels in which respiration rates are different than they would be under today's atmospheric conditions, but not as high as would have been predicted without the down-regulation acclimation response.

While CO_2 is indeed important for photosynthesis and respiration, other factors are important for growth under elevated CO_2. For example, a common FACE observation is that plant growth under elevated CO_2 levels quickly becomes constrained by nutrient availability (remember Blackman's rule of limiting factors). A second, surprising, observation is that the presence of pollutant trace gases, such as ozone, can reduce the net photosynthetic response below the maximum values predicted from initial FACE and greenhouse studies of a decade ago.

As a result of increased atmospheric CO_2, warmer and drier conditions are also predicted to occur in the near future, and nutrient limitations are predicted to increase. Important progress is being made by studying how growth of irrigated and fertilized crops compares with that of plants in natural ecosystems in a world with elevated CO_2. Understanding these responses is crucial as society looks for increased agricultural outputs to support rising human populations and to provide raw materials for biofuels.

Stable Isotopes Record Photosynthetic Properties

We can learn more about the different photosynthetic pathways in plants by measuring the relative abundances of stable isotopes in plants. In particular, the stable isotopes of carbon atoms in a leaf contain useful information about photosynthesis.

Recall that isotopes are simply different forms of an element. In the different isotopes of an element, the number of protons remains constant, since that defines the element, but the number of neutrons varies. Radioactive isotopes of an element decay to form different elements over time. In contrast, stable isotopes of an element remain constant and unchanged over time. The two stable isotopes of carbon are ^{12}C and ^{13}C, differing in composition only by the addition of a neutron in ^{13}C. ^{11}C and ^{14}C are radioactive isotopes of carbon that are frequently used in biological tracer experiments.

How do we measure the stable carbon isotopes of plants?

Atmospheric CO_2 contains the naturally occurring stable carbon isotopes ^{12}C and ^{13}C in the proportions 98.9% and 1.1%, respectively. The chemical properties of $^{13}CO_2$ are identical to those of $^{12}CO_2$, but plants assimilate less $^{13}CO_2$ than $^{12}CO_2$. In other words, leaves discriminate against the heavier isotope of carbon during photosynthesis, and therefore they have smaller $^{13}C/^{12}C$ ratios than are found in atmospheric CO_2.

The $^{13}C/^{12}C$ isotope composition is measured by use of a mass spectrometer, which yields the following ratio:

$$R = \frac{^{13}C}{^{12}C} \quad (9.1)$$

The **carbon isotope ratio** of plants, $\delta^{13}C$, is quantified on a parts per thousand (per mil, ‰) basis:

$$\delta^{13}C‰ = \left(\frac{R_{sample}}{R_{standard}} - 1\right) \times 1000 \quad (9.2)$$

where the standard represents the carbon isotopes contained in a fossil belemnite from the Pee Dee limestone formation of South Carolina. The $\delta^{13}C$ of atmospheric CO_2 has a value of –8‰, meaning that there is less ^{13}C in atmospheric CO_2 than is found in the carbonate of the belemnite standard.

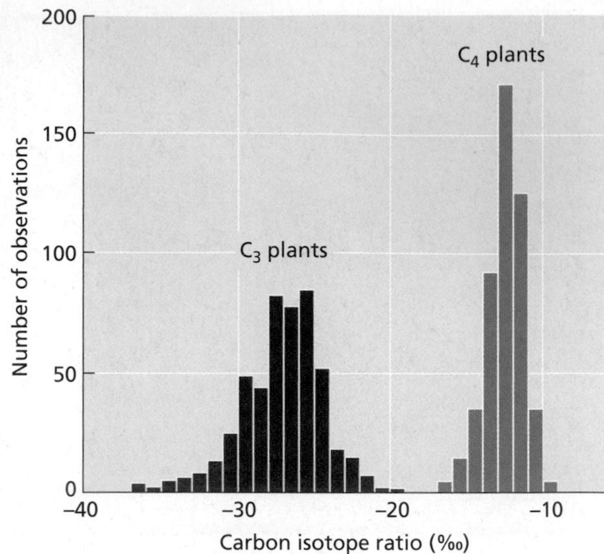

Figure 9.24 Frequency histograms for observed carbon isotope ratios in C_3 and C_4 plant taxa from around the world. (After Cerling et al. 1997.)

What are some typical values for carbon isotope ratios of plants? C_3 plants have a mean $\delta^{13}C$ value of about –28‰; C_4 plants have a mean value of about –14‰. Both C_3 and C_4 plants have less ^{13}C than does CO_2 in the atmosphere, which means that leaf tissues discriminate against ^{13}C during the photosynthetic process. Thure Cerling and colleagues provided $\delta^{13}C$ data for a large number of C_3 and C_4 plants from around the world (**Figure 9.24**). What becomes clear from Figure 9.24 is that there is a wide spread of $\delta^{13}C$ values in C_3 and C_4 plants, with averages of –28‰ and –14‰, respectively. These $\delta^{13}C$ variations actually reflect the consequences of small variations in physiology associated with changes in stomatal conductance in different environmental conditions. Thus, $\delta^{13}C$ values can be used both to distinguish between C_3 and C_4 photosynthesis and to further reveal details about stomatal conditions for plants grown in different environments, such as C_3 plants in the tropics versus those in deserts.

Differences in carbon isotope ratio are easily detectable with mass spectrometers that allow for precise measurements of the abundance of ^{12}C and ^{13}C. Many of our foods grown in temperate climates, such as wheat (*Triticum aestivum*), rice (*Oryza sativa*), potatoes (*Solanum tuberosum*), and beans (*Phaseoulus* spp.) are products of C_3 plants. Yet many of our most productive crops, especially those grown under warm summer conditions, are C_4 plants such as maize (corn; *Zea mays*), sugarcane (*Saccharum officinarum*), and sorghum (*Sorghum bicolor*). Starches and sugars extracted from all of these foods may be chemically identical, but these carbohydrates can be related back to their C_3 or C_4 plant source on the basis of their $\delta^{13}C$ values. For example, measuring the $\delta^{13}C$ values of table sugar (sucrose) makes it possible to determine if the sucrose came from sugar beet (*Beta vulgaris*; a C_3 plant) or sugarcane (a C_4 plant) (see **WEB TOPIC 9.7**).

Why are there carbon isotope ratio variations in plants?

What is the physiological basis for ^{13}C depletion in plants relative to CO_2 in the atmosphere? It turns out that both the diffusion of CO_2 into the leaf and the carboxylation selectivity for $^{12}CO_2$ play a role.

We can predict the carbon isotope ratio of a C_3 leaf as

$$\delta^{13}C_L = \delta^{13}C_A - a - (b - a)(c_i/c_a) \qquad (9.3)$$

where $\delta^{13}C_L$ and $\delta^{13}C_A$ are the carbon isotope ratios of the leaf and atmosphere, respectively; a is the diffusion fraction; b is the net carboxylase fraction in the leaf; and c_i/c_a is the ratio of intercellular to ambient CO_2 concentrations.

CO_2 diffuses from air outside of the leaf to the carboxylation sites within leaves in both C_3 and C_4 plants. We express this diffusion using the term a. Because $^{12}CO_2$ is lighter than $^{13}CO_2$, it diffuses slightly faster toward the carboxylation site, creating an effective diffusion fractionation factor of 4.4‰. Thus, we would expect leaves to have a more negative $\delta^{13}C$ value simply because of this diffusion effect. Yet this factor alone is not sufficient to explain the $\delta^{13}C$ values of C_3 plants as shown in Figure 9.24.

The initial carboxylation event is a determining factor in the carbon isotope ratio of plants. Rubisco represents the only carboxylation reaction in C_3 photosynthesis and has an intrinsic discrimination value against ^{13}C of 30‰. By contrast, PEP carboxylase, the primary CO_2 fixation enzyme of C_4 plants, has a much different isotope discrimination effect—about 2‰. Thus, the inherent difference between the two carboxylating enzymes contributes to the different isotope ratios observed in C_3 and C_4 plants. We use b to describe the net carboxylation effect.

Other physiological characteristics of plants affect their carbon isotope ratio. One primary factor is the partial pressure of CO_2 in the intercellular air spaces of leaves (c_i). In C_3 plants the potential isotope discrimination by rubisco of −30‰ is not fully expressed during photosynthesis because the availability of CO_2 at the carboxylation site becomes a limiting factor restricting the discrimination by rubisco. Greater discrimination against $^{13}CO_2$ occurs when c_i is high, as when stomata are open. Yet open stomata also facilitate water loss. Thus, lower ratios of photosynthesis to transpiration are correlated with greater discrimination against ^{13}C. When leaves are exposed to water stress, stomata tend to close, reducing c_i values. As a consequence, C_3 plants grown under water-stress conditions tend to have higher carbon isotope ratios (i.e., less discrimination against ^{13}C).

The application of carbon isotope ratios in plants has become very productive, because Equation 9.3 provides a strong link between the carbon isotope ratio measurement and the intercellular CO_2 value in a leaf. Intercellular CO_2 levels are then directly linked with aspects of photosynthesis and stomatal constraints. As stomata close in C_3 plants or as water stress increases, we find that the leaf carbon isotope ratio increases. The carbon isotope ratio measurement then becomes a direct proxy to estimate several aspects of short-term water stress. These applications include using carbon isotopes to measure plant performance in both agricultural and ecological studies.

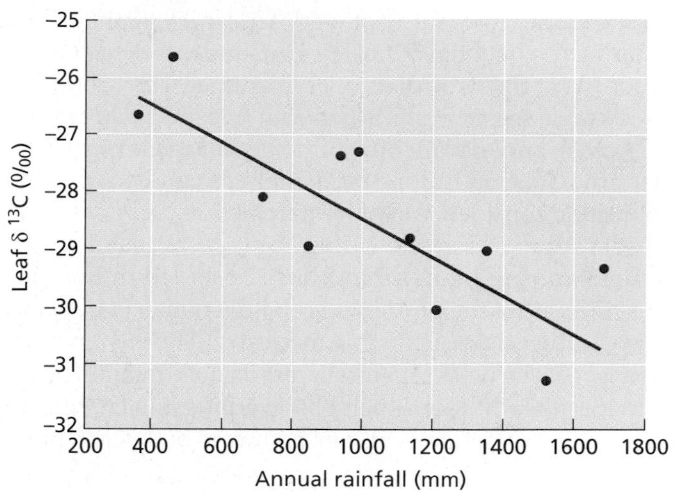

Figure 9.25 Vegetation changes occur along rainfall gradients in southern Queensland, Australia. Here we see that changes in carbon isotope ratios of vegetation appear to be strongly related to precipitation amounts in a region, suggesting that decreased moisture levels influence c_i values and therefore carbon isotope ratios in C_3 species along a geographical gradient in Australian taxa. (After Stewart et al. 1995.)

One emergent environmental pattern is that, on average, leaf carbon isotope ratios decrease as precipitation increases under natural conditions. **Figure 9.25** illustrates this pattern in a transect across Australia. Here we see that the $\delta^{13}C$ values are highest in the arid regions of Australia and become progressively lower along a precipitation gradient from desert to tropical rainforest ecosystems. Applying Equation 9.3 to interpret these $\delta^{13}C$ data, we conclude that intercellular CO_2 levels of leaves of desert plants are lower than what we typically see in leaves of rainforest plants. Because of the sequential nature of tree ring formation (see Chapter 19), $\delta^{13}C$ observations in tree rings can help identify the long-term effects of reduced water availability on plants (e.g., desert versus rainforest habitats) compared with short-term effects that would be recorded in leaves (e.g., seasonal drought cycles).

Carbon isotope ratio analyses are commonly used today to determine the dietary patterns of humans and other animals. The proportion of C_3 to C_4 foods in an animal's diet is recorded in its tissues—teeth, bones, muscles, and hair. Thure Cerling and colleagues described an interesting application of carbon isotope ratio analysis to the eating habits of a family of wild African elephants. They examined sequential $\delta^{13}C$ values in segments of tail hair to reconstruct the daily diets of each animal. They observed predictable seasonal shifts between trees (C_3) and grasses (C_4) as resource availability changed because of rainfall patterns. Carbon isotope ratio analyses can be expanded to include consideration of human diets. One broad-scale observation is that the carbon isotope ratios of North Americans are higher than those observed in Europeans, indicating the prominent role that maize (a C_4 plant) plays in the diets of North Americans. Another application is measuring $\delta^{13}C$ in fossil, carbonate-containing soils and fossil teeth. From such observations it is possible to reconstruct the photosynthetic pathways of the prevailing vegetation from the ancient past. These approaches have been used to determine that C_4 plants became prevalent in grasslands between 6 and 10 million years ago. They have also helped reconstruct the diets of ancient and modern animals (see **WEB TOPIC 9.8**).

CAM plants can have $\delta^{13}C$ values that are very close to those of C_4 plants. In CAM plants that fix CO_2 at night via PEP carboxylase, $\delta^{13}C$ is expected to be similar to that of C_4 plants. However, when some CAM plants are well watered, they can switch to C_3 mode by opening their stomata and fixing CO_2 during the day via rubisco. Under these conditions the isotope composition shifts toward that of C_3 plants. Thus, the $\delta^{13}C$ values of CAM plants reflect how much carbon is fixed via the C_3 pathway versus the C_4 pathway.

SUMMARY

In considering photosynthetic performance, both the limiting factor hypothesis and an "economic perspective" emphasizing CO_2 "supply" and "demand" have guided research.

Photosynthesis Is Influenced by Leaf Properties

- Leaf anatomy is highly specialized for light absorption (**Figure 9.1**).
- About 5% of the solar energy reaching Earth is converted into carbohydrates by photosynthesis. Much absorbed light is lost in reflection and transmission, in metabolism, and as heat (**Figures 9.2, 9.3**).
- In dense forests, almost all photosynthetically active radiation is absorbed by leaves (**Figure 9.4**).
- Leaves of some plants maximize light absorption by solar tracking (**Figure 9.5**).
- Some plants respond to a range of light regimes. However, sun and shade leaves have contrasting morphological and biochemical characteristics.
- To increase light absorption, some shade plants produce a higher ratio of PSII to PSI reaction centers, while others add antenna chlorophyll to PSII.

Effects of Light on Photosynthesis in the Intact Leaf

- Light-response curves show the PPFD where photosynthesis is limited by light or by carboxylation capacity. The slope of the linear portion of the light-response curve measures the maximum quantum yield (**Figure 9.6**).
- Light compensation points for shade plants are lower than for sun plants because respiration rates in shade plants are very low (**Figures 9.7, 9.8**).
- Beyond the saturation point, factors other than incident light, such as electron transport, rubisco activity, or triose phosphate metabolism, limit photosynthesis. Rarely is an entire plant light-saturated (**Figure 9.9**).
- The xanthophyll cycle dissipates excess absorbed light energy to avoid damaging the photosynthetic apparatus (**Figures 9.10, 9.11**). Chloroplast movements also limit excess light absorption (**Figure 9.12**).
- Dynamic photoinhibition temporarily diverts excess light absorption to heat but maintains maximum photosynthetic rate (**Figure 9.13**). Chronic photoinhibition is irreversible.

Effects of Temperature on Photosynthesis in the Intact Leaf

- Plants are remarkably plastic in their adaptations to temperature. Optimal photosynthetic temperatures have strong biochemical, genetic (adaptation), and environmental (acclimation) components.
- Leaf absorption of light energy generates a heat load that must be dissipated (**Figure 9.14**).
- Temperature-sensitivity curves identify (a) a temperature range where enzymatic events are stimulated, (b) a range for optimal photosynthesis, and (c) a range where deliterious events occur (**Figure 9.15**).
- Below 30°C the quantum yield of C_3 plants is higher than that of C_4 plants; above 30°C, the situation is reversed (**Figure 9.16**). Because of photorespiration,

the quantum yield is strongly dependent on temperature in C_3 plants but is nearly independent of temperature in C_4 plants.

- Reduced quantum yield and increased photorespiration due to temperature effects lead to differences in the photosynthetic capacities of C_3 and C_4 plants and result in a shift of species dominance across different latitudes (**Figure 9.17**).

Effects of Carbon Dioxide on Photosynthesis in the Intact Leaf

- Atmospheric CO_2 levels have been increasing since the Industrial Revolution because of human use of fossil fuels and deforestation (**Figure 9.18**).
- Concentration gradients drive the diffusion of CO_2 from the atmosphere to the carboxylation site in the leaf, using both gaseous and liquid routes. There are multiple resistances along the CO_2 diffusion pathway, but under most conditions stomatal resistance has the greatest effect on CO_2 diffusion into a leaf (**Figure 9.19**).
- Enrichment of CO_2 above natural atmospheric levels results in increased photosynthesis and productivity (**Figure 9.20**).
- C_4 photosynthesis may have become prominent in Earth's warmest regions when global atmospheric CO_2 concentrations fell below a threshold value (**Figure 9.21**).
- Opening at night and closing during the day, the stomatal activity of CAM plants contrasts with that found in C_3 and C_4 plants (**Figure 9.22**).
- Free Air CO_2 Enrichment (FACE) experiments show that C_3 plants are more responsive to elevated CO_2 than are C_4 plants (**Figure 9.23**).

Stable Isotopes Record Photosynthetic Properties

- The carbon isotope ratios of leaves can be used to distinguish photosynthetic pathway differences among different plant species.
- Both C_3 and C_4 plants have less ^{13}C than does CO_2 in the atmosphere, indicating that leaf tissues discriminate against ^{13}C during photosynthesis (**Figure 9.24**).
- Conditions that cause stomata to close in C_3 plants, such as water stress, cause the leaf carbon isotope ratio to increase. Thus, the carbon isotope ratio of a leaf can be used as a direct estimate of physiological responses to the environment (e.g., short-term water stress) (**Figure 9.25**).

WEB MATERIAL

- **WEB TOPIC 9.1 Working with Light** Amount, direction, and spectral quality are important parameters for the measurement of light.
- **WEB TOPIC 9.2 Heat Dissipation from Leaves: The Bowen Ratio** Sensible heat loss and evaporative heat loss are the most important processes in the regulation of leaf temperature.
- **WEB TOPIC 9.3 The Geographic Distributions of C_3 and C_4 Plants** The geographic distributions of C_3 and C_4 plants correspond closely with growing-season temperatures in today's world.
- **WEB TOPIC 9.4 Calculating Important Parameters in Leaf Gas Exchange** Gas exchange methods allow us to measure photosynthesis and stomatal conductance in the intact leaf.
- **WEB TOPIC 9.5 Prehistoric Changes in Atmospheric CO_2** Over the past 800,000 years, atmospheric CO_2 levels changed between 180 ppm (glacial periods) and 280 ppm (interglacial periods) as Earth moved between ice ages.
- **WEB TOPIC 9.6 Projected Future Increases in Atmospheric CO_2** Atmospheric CO_2 reached 400 ppm in 2014 and is expected to reach 500 ppm in this century.
- **WEB TOPIC 9.7 Using Carbon Isotopes to Detect Adulteration in Foods** Carbon isotopes are frequently used to detect the substitution of C_4 sugars into C_3 food products, such as the introduction of sugarcane into honey to increase yield.
- **WEB TOPIC 9.8 Reconstruction of the Expansion of C_4 Taxa** The $\delta^{13}C$ of animal teeth faithfully records the carbon isotope ratios of food sources and can be used to reconstruct the abundances of C_3 and C_4 plants eaten by mammalian grazers.

available at plantphys.net

Suggested Reading

Adams, W. W., Zarter, C. R., Ebbert, V., and Demmig-Adams, B. (2004) Photoprotective strategies of overwintering evergreens. *Bioscience* 54: 41–49.

Bjørn, L. O., and Vogelmann, T. C. (1994) Quantification of light. In *Photomorphogenesis in Plants*, 2nd ed., R. E. Kendrick and G. H. M. Kronenberg, eds., Kluwer, Dordrecht, Netherlands, pp. 17–25.

Bowes, G. (1993) Facing the inevitable: Plants and increasing atmospheric CO_2. *Annu. Rev. Plant Physiol. Plant Mol. Biol.* 44: 309–332.

Demmig-Adams, B., and Adams, W. (1996) The role of xanthophyll cycle carotenoids in the protection of photosynthesis. *Trends Plant Sci.* 1: 21–26.

Ehleringer, J. R., Cerling, T. E., and Helliker, B. R. (1997) C_4 photosynthesis, atmospheric CO_2, and climate. *Oecologia* 112: 285–299.

Evans, J. R., von Caemmerer, S., and Adams, W. W. (1988) *Ecology of Photosynthesis in Sun and Shade.* CSIRO, Melbourne.

Farquhar, G. D., Ehleringer, J. R., and Hubick, K. T. (1989) Carbon isotope discrimination and photosynthesis. *Annu. Rev. Plant Physiol. Plant Mol. Biol.* 40: 503–537.

Haupt, W., and Scheuerlein, R. (1990) Chloroplast movement. *Plant Cell Environ.* 13: 595–614.

Kirk, J. T. (1994) *Light and Photosynthesis in Aquatic Ecosystems.* Cambridge University Press, Cambridge.

Koller, D. (2000) Plants in search of sunlight. *Adv. Bot. Res.* 33: 35–131.

Laisk, A., and Oja, V. (1998) *Dynamics of leaf photosynthesis.* CSIRO, Melbourne.

Long, S. P., Ainsworth, E. A., Leakey, A. D., Nosberger, J., and Ort, D. R. (2006) Food for thought: Lower-than-expected crop stimulation with rising CO_2 concentrations. *Science* 312: 1918–1921.

Long, S. P., Ainsworth, E. A., Rogers, A., and Ort, D. R. (2004) Rising atmospheric carbon dioxide: Plants FACE the future. *Annu. Rev. Plant Biol.* 55: 591–628.

Long, S. P., Humphries, S., and Falkowski, P. G. (1994) Photoinhibition of photosynthesis in nature. *Annu. Rev. Plant Physiol. Plant Mol. Biol.* 45: 633–662.

Ort, D. R., and Yocum, C. F. (1996) *Oxygenic Photosynthesis: The Light Reactions.* Kluwer, Dordrecht, Netherlands.

Osmond, C. B. (1994) What is photoinhibition? Some insights from comparisons of shade and sun plants. In *Photoinhibition of Photosynthesis*, N. R. Baker and J. R. Bowyer, eds., BIOS Scientific, Oxford, pp. 1–24.

Sharkey, T. D. (1996) Emission of low molecular mass hydrocarbons from plants. *Trends Plant Sci.* 1: 78–82.

Terashima, I. (1992) Anatomy of non-uniform leaf photosynthesis. *Photosyn. Res.* 31: 195–212.

Terashima, I., and Hikosaka, K. (1995) Comparative ecophysiology of leaf and canopy photosynthesis. *Plant Cell Environ.* 18: 1111–1128.

Vogelmann, T. C. (1993) Plant tissue optics. *Annu. Rev. Plant Physiol. Plant Mol. Biol.* 44: 231–251.

von Caemmerer, S. (2000) *Biochemical models of leaf photosynthesis.* CSIRO, Melbourne.

Zhu, X. G., Long, S. P., and Ort, D. R. (2010) Improving photosynthetic efficiency for greater yield. *Annu. Rev. Plant Biol.* 61: 235–261.

10 Stomatal Biology

Stomata, from the Greek word for "mouth," are structural features of aerial organs of most plants. The term *stoma* (singular) denotes a microscopic pore or hole through the surface of the plant organ, which allows communication between the interior of the leaf and the external environment, and a pair of specialized cells—the guard cells—that surround the pore.

Guard cells respond to environmental signals by changing their dimensions, thereby regulating the size of the pore, or stomatal aperture. The botanist Hugo von Mohl proposed in 1856 that turgor changes in guard cells provide the mechanical force for changes in stomatal apertures (see Chapter 4). The guard cells are continuously swelling or contracting, and the ensuing wall deformations cause changes in pore dimensions. These dimension changes are the end result of the perception of environmental signals by the guard cells.

Visualize the external surface of a leaf from the perspective of a bee (see Figure 4.12C). Within a sea of epidermal cells, pairs of guard cells appear interspaced, with a pore at the center of each cell pair. In some species, the guard cells stand on their own; in others, they are flanked by specialized subsidiary cells that separate them from regular epidermal cells.

Inspection of the distribution of stomata in leaves growing in environments with different levels of available water gives a clue to the role of stomata in plant adaptations. Leaves from aquatic plants living under water are devoid of stomata. Leaves that float in water, commonly in ponds, have stomata in their upper surface growing in contact with air, but lack them in the surfaces in contact with water. Aerial leaves have stomata on both surfaces, although the frequency and distribution of the stomata vary drastically with phylogeny and environment.

Why are stomata required in leaf surfaces in contact with air? The requirement is a critical adaptation to avoid desiccation. When aquatic plants invaded terrestrial habitats, they developed an impermeable cuticle which prevents water loss. That adaptation, however, presented a different problem for plant survival: Any substance that effectively blocks the outward diffusion of water also acts as a barrier against the inward diffusion of CO_2, an essential substrate for photosynthesis (see Chapter 8).

Stomata offer a temporal solution to this problem. Stomata close at night because there is no photosynthesis without light, and they open during the day. In general, stomata close when water is limiting, thus preventing excessive, deleterious water loss. Stomata open under conditions favoring photosynthesis (see Chapter 9).

The driving force for stomatal movements is turgor pressure, discussed in detail in Chapter 4. Environmental stimuli associated with a high CO_2 demand inside the leaf are transduced into a higher turgor pressure, which leads to guard cell swelling and a widening of the stomatal pore. Stimuli associated with a need to reduce water use of the plant are transduced into a decrease in turgor, and stomatal closure. Guard cells, then, are turgor valves.

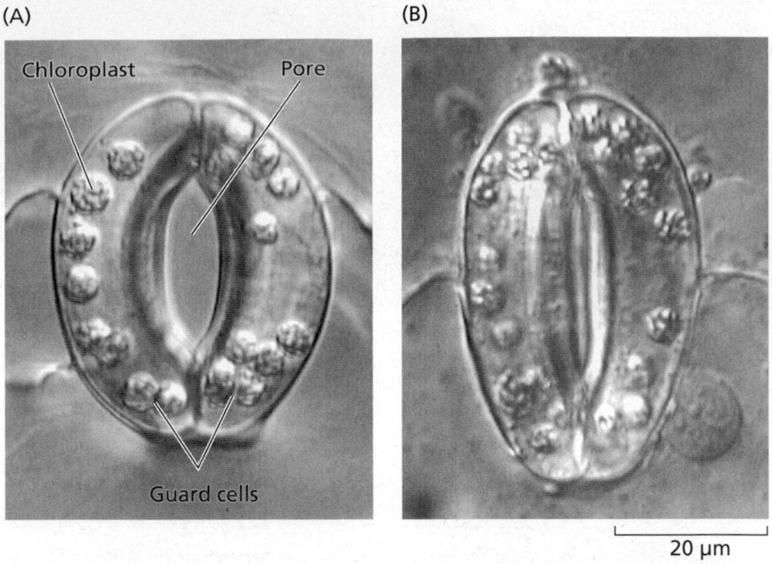

Figure 10.1 Light-stimulated stomatal opening in detached epidermis of *Vicia faba*. An open, light-treated stoma (A) is shown in the dark-treated, closed state in (B). Stomatal opening is quantified by microscopic measurement of the width of the stomatal pore. (Courtesy of E. Raveh.)

Light-dependent Stomatal Opening

Under temperate conditions, light is the dominant stimulus causing stomatal opening (see WEB TOPIC 10.1). The two major factors involved with light-dependent opening are (1) photosynthesis in the guard cell chloroplast and (2) a specific response to blue light. In addition, increases in mesophyll photosynthesis lower intercellular CO_2 concentrations, and low intercellular CO_2 opens stomata.

Guard cells respond to blue light

Several characteristics of stomatal movements that depend on blue light make guard cells a valuable experimental system for the study of blue-light responses:

- The stomatal response to blue light is rapid and reversible, and it is localized in a single cell type, the **guard cell** (**Figure 10.1**).
- The stomatal response to blue light regulates stomatal movements *throughout the life of the plant*. This is unlike phototropism and hypocotyl elongation, which are functionally important only at early stages of development (see Chapter 16).
- The signal transduction process that links the perception of blue light with the opening of stomata is understood in considerable detail.

Stomata open as light levels reaching the leaf surface increase, and close as light decreases. In greenhouse-grown leaves of broad bean (*Vicia faba*), stomatal movements closely correlate with incident solar radiation at the leaf surface (**Figure 10.2**). This light dependence of stomatal movements has been documented in many species and conditions.

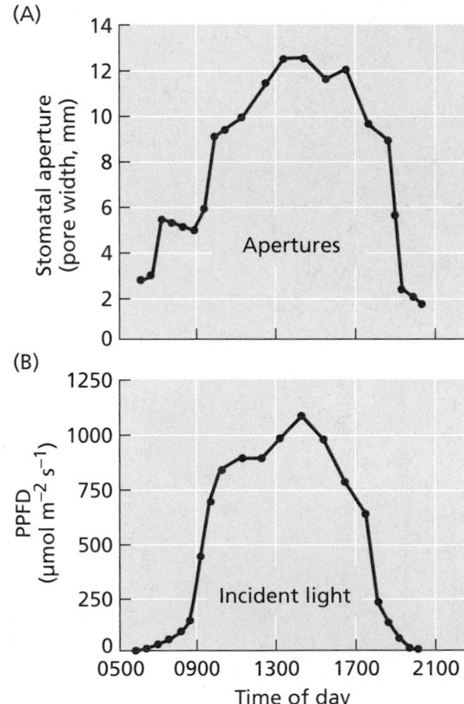

Figure 10.2 Stomatal opening tracks photosynthetically active radiation at the leaf surface. Stomatal opening in the lower surface of leaves of *V. faba* grown in a greenhouse, measured as the width of the stomatal pore (A), closely follows the levels of photosynthetically active radiation (400–700 nm) incident to the leaf (B), indicating that the response to light is the dominant response regulating stomatal opening. PPFD, Photosynthetic photon flux density. (After Srivastava and Zeiger 1995.)

Studies of the stomatal response to light have shown that dichlorophenyldimethylurea (DCMU), an inhibitor of photosynthetic electron transport (see Figure 7.28), causes a partial inhibition of light-stimulated stomatal opening. These results indicate that photosynthesis in the guard cell chloroplast plays a role in light-dependent stomatal opening, but why is the response only partial? This partial response to DCMU points to the involvement of a DCMU-insensitive, nonphotosynthetic component of the stomatal response to light. Detailed studies carried out under colored light have shown that light activates two distinct responses of guard cells: photosynthesis in the guard cell chloroplast (see **WEB ESSAY 10.1**) and a specific blue-light response.

Since blue light stimulates both the specific blue-light response of stomata and guard cell photosynthesis (see the action spectrum for photosynthesis in Figure 7.8, and **WEB ESSAY 10.1**), blue light alone cannot be used to study the specific stomatal response to blue light. To achieve a clear-cut separation between the two light responses, researchers use dual-beam experiments. First, high fluence rates of red light are used to *saturate* the photosynthetic response; such saturation prevents any further stomatal opening mediated by photosynthesis in response to further increases in red light. Then, low photon fluxes of blue light are added after the response to the saturating red light has been established (**Figure 10.3**). The addition of blue light causes substantial additional stomatal opening that, as just explained, cannot be due to further stimulation of guard cell photosynthesis, because the background red light has saturated photosynthesis.

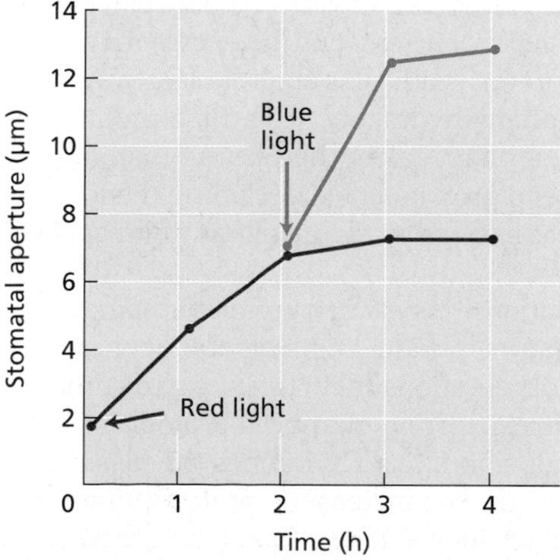

Figure 10.3 Response of stomata to blue light under a red-light background. Stomata from detached epidermis of common dayflower (*Commelina communis*) were treated with saturating photon fluxes of red light (red trace). In a parallel treatment, stomata illuminated with red light were also illuminated with blue light, as indicated by the arrow (blue trace). The increase in stomatal opening above the level reached in the presence of saturating red light indicates that a different photoreceptor system, stimulated by blue light, is mediating the additional increases in opening. Experiments carried out with detached epidermis eliminate CO_2 effects from the mesophyll. (From Schwartz and Zeiger 1984.)

An action spectrum for the stomatal response to blue light under saturating background red illumination shows a three-finger pattern (**Figure 10.4**). This action spectrum, typical of blue-light responses and distinctly different from the action spectrum for photosynthesis, further indicates that guard cells respond specifically to blue light.

When guard cells are treated with cellulolytic enzymes that digest the cell walls, **guard cell protoplasts** are released and can be used for experimentation. In the laboratory, guard cell protoplasts swell when illuminated with blue light (**Figure 10.5**), indicating that blue light is sensed within the guard cells proper. The swelling of guard cell protoplasts also illustrates how intact guard cells function. Light stimulates the uptake of ions, and the accumulation of organic solutes in the guard cell protoplasts decreases the cell's osmotic potential (increases the osmotic pressure). As a result, water flows in, and the guard cell protoplasts swell. In guard cells with intact cell walls, this increase in turgor leads to the deformation of the cell walls and an increase in the stomatal aperture (see Chapter 4).

Blue light activates a proton pump at the guard cell plasma membrane

When guard cell protoplasts from broad bean (*V. faba*) are irradiated with blue light under saturating background red-light illumination, the pH of the suspension medium becomes more acidic (**Figure 10.6**). This blue light–induced acidification is blocked by uncouplers that dissipate pH gradients, such as CCCP (discussed shortly), and by inhibitors of the proton-pumping H^+-ATPase (discussed in Chapter 6), such as orthovanadate (see Figure

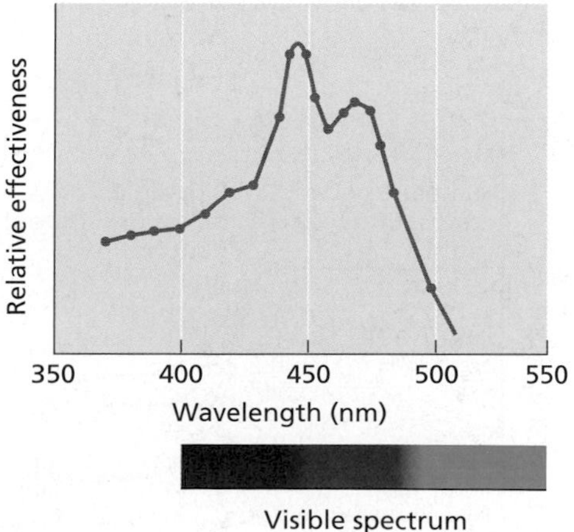

Figure 10.4 Action spectrum for blue light–stimulated stomatal opening (under a red-light background). (After Karlsson 1986.)

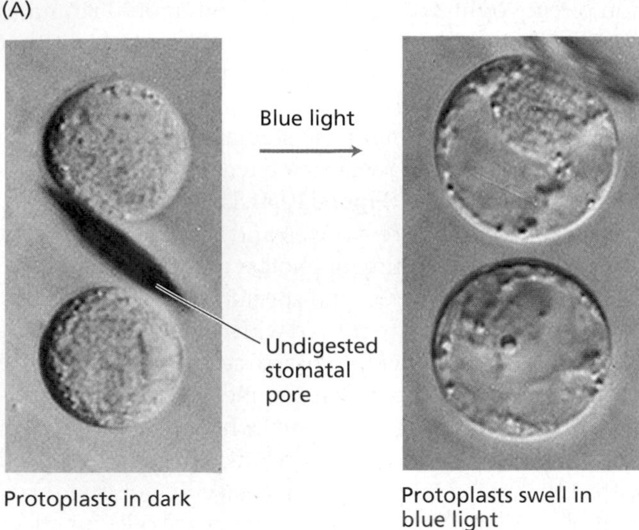

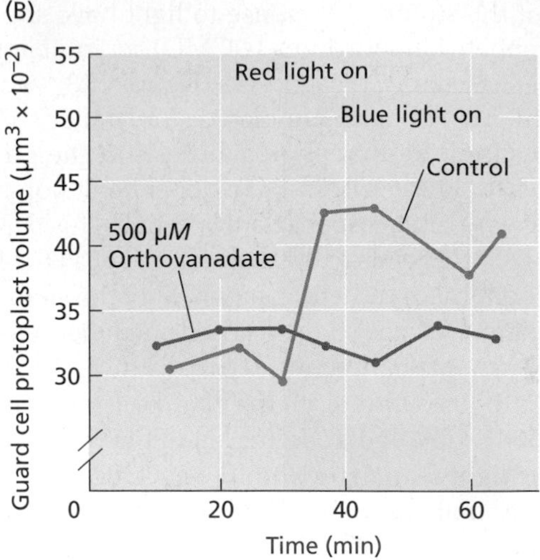

Figure 10.5 Blue light–stimulated swelling of guard cell protoplasts. (A) In the absence of a rigid cell wall, guard cell protoplasts of onion (*Allium cepa*) swell. (B) Blue light stimulates the swelling of guard cell protoplasts of broad bean (*V. faba*), and orthovanadate, an inhibitor of the H^+-ATPase, inhibits this swelling. Blue light stimulates ion and water uptake in the guard cell protoplasts, which in the intact guard cells provides a mechanical force working against the rigid cell wall that distorts the guard cells and drives increases in stomatal apertures. (A from Zeiger and Hepler 1977; B after Amodeo et al. 1992.)

10.5B). Such acidification studies have made it clear that *blue light activates a proton-pumping ATPase in the guard cell plasma membrane*.

In the intact leaf, this blue-light stimulation of proton pumping lowers the pH of the apoplastic space surrounding the guard cells, and generates the driving force needed for ion uptake and stomatal opening. The plasma membrane ATPase from guard cells has been isolated and extensively characterized.

The activation of electrogenic pumps such as the proton-pumping ATPase can be measured in patch clamp experiments as an outward electric current at the plasma membrane (see **WEB TOPIC 6.2** for a description of patch clamping). **Figure 10.7A** shows a patch clamp recording of a guard cell protoplast treated in the dark with the fungal toxin fusicoccin, a well-characterized activator of plasma membrane ATPases. Exposure to fusicoccin stimulates an outward electric current, which generates a proton gradient. This proton gradient is abolished by carbonyl cyanide *m*-chlorophenylhydrazone (CCCP), a proton ionophore (uncoupler) that makes the plasma membrane highly permeable to protons, thus precluding the formation of a proton gradient across the membrane and abolishing net proton efflux.

The relationship between proton pumping at the guard cell plasma membrane and stomatal opening is evident from the observations that (1) fusicoccin stimulates both proton extrusion from guard cell protoplasts and stomatal opening, and (2) CCCP inhibits the fusicoccin-stimulated opening. The increase in proton-pumping rates as a function of fluence rates of blue light further indicates that increasing the number of blue photons in the solar radiation reaching the leaf should cause a larger stomatal opening (see Figure 10.6). A pulse of blue light given under a saturating red-light background can also stimulate an

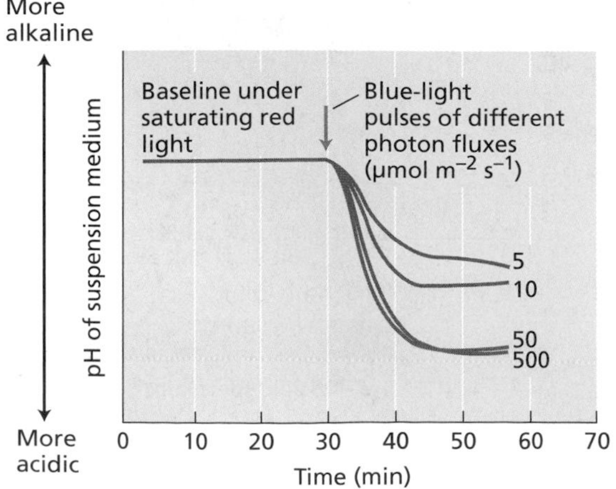

Figure 10.6 Acidification of a suspension medium of guard cell protoplasts of *V. faba* stimulated by a 30-s pulse of blue light. The acidification results from the stimulation of an H^+-ATPase at the plasma membrane by blue light, and it is associated with protoplast swelling (see Figure 10.5). (After Shimazaki et al. 1986.)

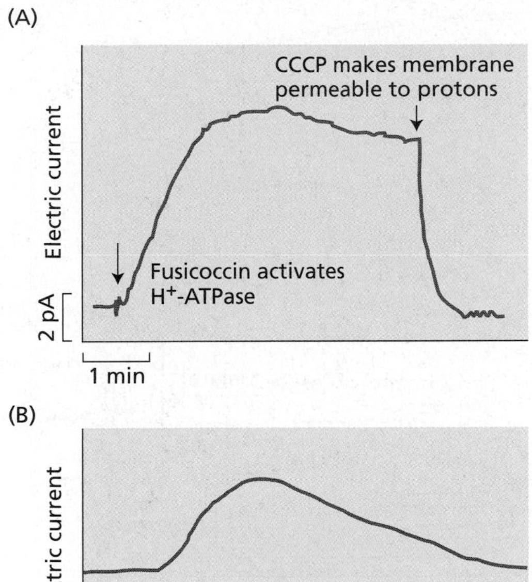

Figure 10.7 Activation of the H^+-ATPase at the plasma membrane of guard cell protoplasts by fusicoccin and blue light can be measured as electric current in patch clamp experiments. (A) Outward electric current (measured in picoamps, pA) at the plasma membrane of a guard cell protoplast stimulated by the fungal toxin fusicoccin, an activator of the H^+-ATPase. The current is abolished by the proton ionophore carbonyl cyanide *m*-chlorophenylhydrazone (CCCP). (B) Outward electric current at the plasma membrane of a guard cell protoplast stimulated by a blue-light pulse. These results indicate that blue light stimulates the H^+-ATPase. (A after Serrano et al. 1988; B after Assmann et al. 1985.)

outward electric current from guard cell protoplasts (**Figure 10.7B**). The relationship between these pulse-stimulated electrical currents and the acidification response to blue light pulses, shown in Figure 10.6, indicates that the measured electric current is carried by protons moving from the cell interior to the apoplast.

Blue-light responses have characteristic kinetics and lag times

The stomatal responses to blue-light pulses illustrate two important properties of blue-light responses: a persistence of the response after the light signal has been switched off, and a significant lag time separating the onset of the light signal and the beginning of the response. In contrast to typical photosynthetic responses, which are activated very quickly after a "light-on" signal and cease when the light goes off, blue-light responses proceed at maximum rates for several minutes after application of the pulse (see Figures 10.6 and 10.7B).

This persistence of the blue-light response after the "light-off" signal can be explained by a physiologically inactive form of the blue-light photoreceptor that is converted to an active form by blue light, with the active form reverting slowly to the physiologically inactive form after the blue light is switched off. How quickly a response to a blue-light pulse takes place thus depends on the time course of the reversion of the active form to the inactive one.

Another property of the response to blue-light pulses is a lag time, which lasts about 25 s in both the acidification response and the outward electric currents stimulated by blue light (see Figures 10.6 and 10.7). This time interval is probably required for the signal transduction process to proceed from the photoreceptor site to the proton-pumping ATPase and for the proton gradient to form. Similar lag times have been measured for blue light–dependent inhibition of hypocotyl elongation (see Chapter 16).

Blue light regulates the osmotic balance of guard cells

Blue light modulates guard cell osmoregulation by means of its activation of proton pumping, solute uptake, and stimulation of the synthesis of organic solutes (see **WEB TOPIC 10.2**). Before discussing these blue-light responses, let us briefly consider the major osmotically active solutes in guard cells.

The plant physiologist F. E. Lloyd hypothesized in 1908 that guard cell turgor is regulated by osmotic changes resulting from starch–sugar interconversions, a concept that led to a starch–sugar hypothesis of stomatal movements. The discovery of potassium ion fluxes in guard cells in Japan in the 1940s, and their rediscovery in the West in the 1960s, replaced the starch–sugar hypothesis with the modern theory of guard cell osmoregulation by potassium ions and their counterions, Cl^- and malate^{2-}.

The potassium ion concentration in guard cells increases severalfold when stomata open, from 100 mM in the closed state to 400 to 800 mM in the open state, depending on the species and the experimental conditions. In most species, these large concentration changes in K^+ are electrically balanced by varying amounts of the anions Cl^- and malate^{2-} (**Figure 10.8**; also see **WEB TOPIC 10.2**). However, in some species of the genus *Allium*, such as onion (*A. cepa*), K^+ is balanced solely by Cl^-.

Chloride anions are taken up into the guard cells from the apoplast during stomatal opening and extruded during stomatal closing. Malate anions, by contrast, are synthesized in the guard cell cytosol, in a metabolic pathway that uses carbon skeletons generated by starch hydrolysis (see Figure 10.8A). The malate content of guard cells decreases during stomatal closing, but it remains unclear

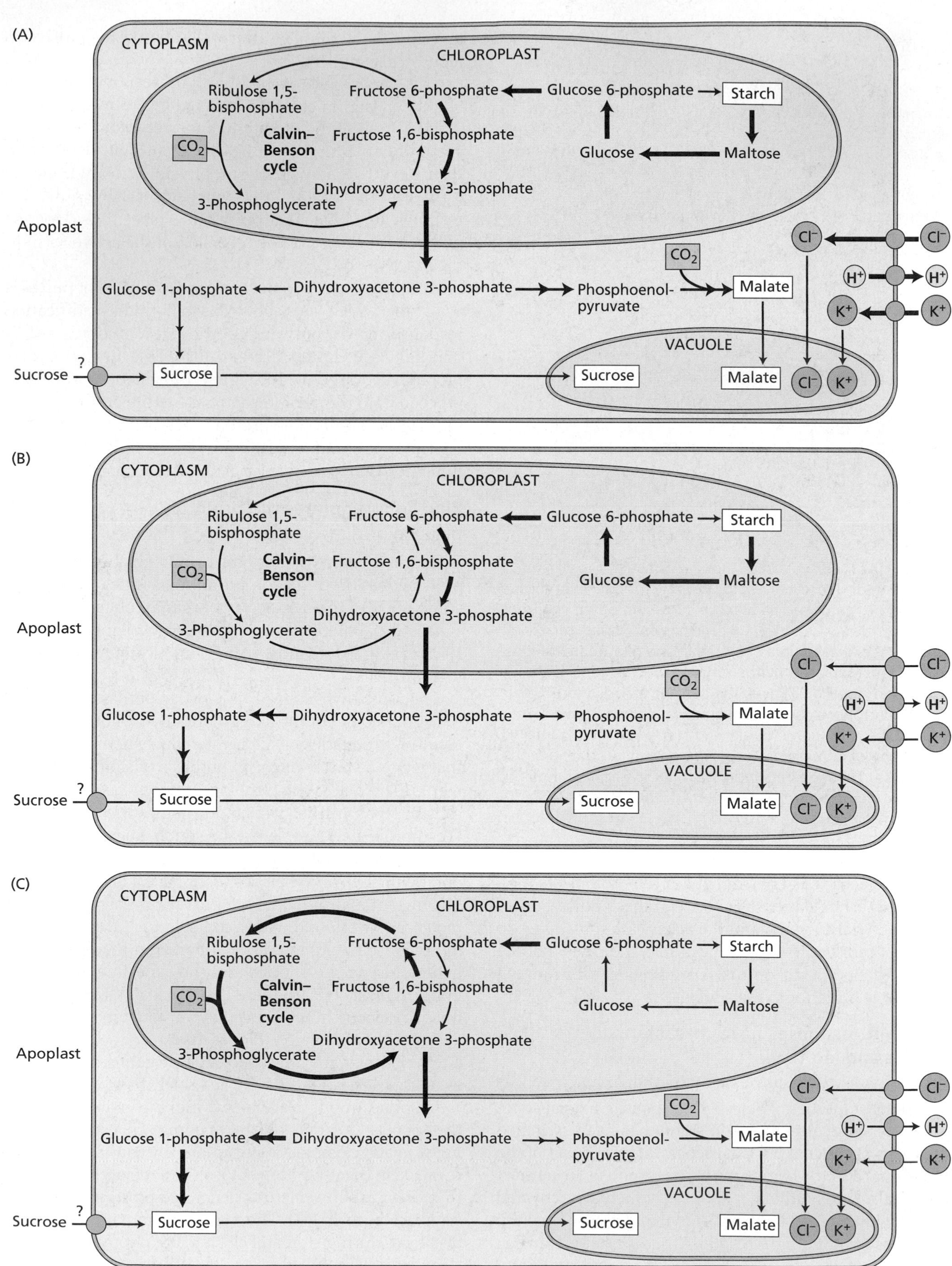
(A)
CYTOPLASM
CHLOROPLAST
Ribulose 1,5-bisphosphate
Fructose 6-phosphate
Glucose 6-phosphate
Starch
CO_2
Calvin–Benson cycle
Fructose 1,6-bisphosphate
Glucose
Maltose
Dihydroxyacetone 3-phosphate
3-Phosphoglycerate
Apoplast
Cl^-
CO_2
Glucose 1-phosphate
Dihydroxyacetone 3-phosphate
Phosphoenol-pyruvate
Malate
H^+
K^+
VACUOLE
Sucrose
?
Sucrose
Sucrose
Malate
(B)
CYTOPLASM
CHLOROPLAST
Ribulose 1,5-bisphosphate
Fructose 6-phosphate
Glucose 6-phosphate
Starch
CO_2
Calvin–Benson cycle
Fructose 1,6-bisphosphate
Glucose
Maltose
Dihydroxyacetone 3-phosphate
3-Phosphoglycerate
Apoplast
Cl^-
CO_2
Glucose 1-phosphate
Dihydroxyacetone 3-phosphate
Phosphoenol-pyruvate
Malate
H^+
K^+
VACUOLE
Sucrose
?
Sucrose
Sucrose
Malate
(C)
CYTOPLASM
CHLOROPLAST
Ribulose 1,5-bisphosphate
Fructose 6-phosphate
Glucose 6-phosphate
Starch
CO_2
Calvin–Benson cycle
Fructose 1,6-bisphosphate
Glucose
Maltose
Dihydroxyacetone 3-phosphate
3-Phosphoglycerate
Apoplast
Cl^-
CO_2
Glucose 1-phosphate
Dihydroxyacetone 3-phosphate
Phosphoenol-pyruvate
Malate
H^+
K^+
VACUOLE
Sucrose
?
Sucrose
Sucrose
Malate

◀ **Figure 10.8** Three distinct osmoregulatory pathways in guard cells. The thick, dark arrows identify the major metabolic steps of each pathway that lead to the accumulation of osmotically active solutes in the guard cells. (A) Potassium and its counterions. Potassium and chloride are taken up in secondary transport processes driven by a proton gradient; malate is formed from the hydrolysis of starch. (B) Accumulation of sucrose from starch hydrolysis. (C) Accumulation of sucrose from photosynthetic carbon fixation. The possible uptake of apoplastic sucrose is also indicated. (From Talbott and Zeiger 1998.)

whether malate is catabolized in mitochondrial respiration, extruded into the apoplast, or both.

Potassium ions and chloride are taken up into guard cells via secondary transport mechanisms driven by the electrochemical potential for H^+, $\Delta\mu_{H^+}$, generated by the proton pump discussed earlier in the chapter (see Figure 10.8; see also Chapter 6). Proton extrusion makes the electrical potential difference across the guard cell plasma membrane more negative; light-dependent hyperpolarizations as high as 64 mV have been measured. In addition, proton pumping generates a pH gradient of about 0.5 to 1 pH unit.

The electrical component of the proton gradient provides a driving force for the passive uptake of K^+ via voltage-regulated K^+ channels that were discussed in Chapter 6. Chloride is thought to be taken up through a proton–chloride symporter. Thus, both components of the electrochemical proton gradient generated by blue light–dependent proton pumping play key roles in the uptake of ions for stomatal opening.

Guard cell chloroplasts (see Figure 10.1) contain large starch grains. In contrast to what happens in mesophyll chloroplasts, starch content in guard cells decreases during stomatal opening in the morning and increases during closing at the end of the day. Starch, an insoluble, high molecular weight polymer of glucose, does not contribute to the cell's osmotic potential, but the hydrolysis of starch into glucose and fructose and the subsequent accumulation of sucrose (see Figure 10.8B) cause a decrease in the osmotic potential (or increase in osmotic pressure) of guard cells. In the reverse process, starch synthesis decreases the sugar concentration, resulting in an increase of the cell's osmotic potential, which the starch–sugar hypothesis predicted to be associated with stomatal closing.

With the discovery of the major role of potassium ions and their counterions in guard cell osmoregulation, the starch–sugar hypothesis was no longer considered important. However, we will now see that sucrose, a major osmoregulatory molecule in the starch–sugar hypothesis, does play an important role in guard cell osmoregulation (see **WEB ESSAY 10.2**).

Sucrose is an osmotically active solute in guard cells

Recent studies of daily courses of stomatal movements in intact leaves have shown that the K^+ content in guard cells increases in parallel with early-morning opening, but it decreases in the early afternoon under conditions in which apertures continue to increase. In contrast, sucrose content increases slowly in the morning, and upon K^+ efflux, sucrose becomes the dominant osmotically active solute in guard cells. Stomatal closing at the end of the day parallels a decrease in the sucrose content (**Figure 10.9**). These osmoregulatory patterns have been seen in both *V. faba* and onion guard cells grown in greenhouse or growth chamber conditions.

One implication of these osmoregulatory features is that stomatal opening is associated primarily with K^+ uptake, and closing is associated with a decrease in sucrose content (see Figure 10.9). The function of distinct K^+- and sucrose-dominated osmoregulatory phases is unclear; K^+ might be the solute of choice for the daily opening that occurs at sunrise. The sucrose phase might be associated with the coordination of stomatal movements in the epidermis with rates of photosynthesis in the mesophyll.

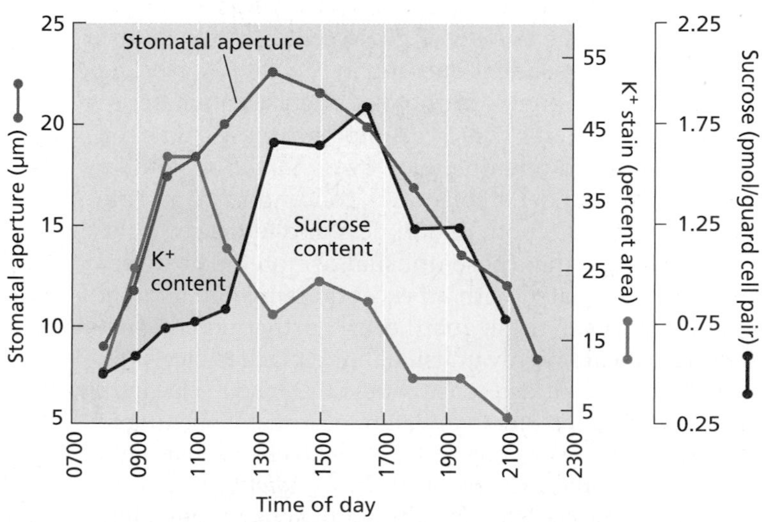

Figure 10.9 Daily course of changes in stomatal aperture, and in potassium and sucrose content, of guard cells from intact leaves of broad bean (*V. faba*). These results indicate that the changes in osmotic potential required for stomatal opening in the morning are mediated by potassium and its counterions, whereas the afternoon changes are mediated by sucrose. (After Talbott and Zeiger 1998.)

Where do osmotically active solutes originate? Three major distinct metabolic pathways that can supply osmotically active solutes to guard cells have been clearly characterized (see Figure 10.8):

1. The uptake of K^+ and Cl^- from the apoplast, coupled to the biosynthesis of malate^{2-} within the guard cells (see Figure 10.8A)
2. The production of sucrose in the guard cell cytoplasm from precursors originating from starch hydrolysis in the guard cell chloroplast (see Figure 10.8B)
3. The production of sucrose from precursors made in the photosynthetic carbon fixation pathway at the guard cell chloroplast (see Figure 10.8C)

Depending on conditions, one or more osmoregulatory pathways can be active. For instance, during red light–stimulated stomatal opening in detached epidermis of *V. faba* kept under ambient CO_2 concentrations, the dominant solute in guard cells is sucrose generated by the photosynthetic carbon fixation pathway in the guard cell chloroplast, with no detectable K^+ uptake or starch degradation (see Figure 10.8C). However, in CO_2-free air, photosynthetic carbon fixation is inhibited and the red light–stimulated opening is associated with K^+ accumulation (see Figure 10.8A; see also **WEB TOPIC 10.2**).

Some unusual osmoregulatory pathways are used in nature. Chloroplasts from guard cells of the orchid *Paphiopedilum* are devoid of chlorophyll. *Paphiopedilum* stomata open in response to blue light, and fail to show a typical red light–stimulated opening. In contrast, stomata from the fern *Adiantum* lack a specific blue-light response and open in response to red light. *Adiantum* guard cells have an unusually large number of guard cell chloroplasts, and the red light–dependent opening in *Adiantum* is blocked by DCMU, an inhibitor of photosynthetic electron transport. This implicates guard cell photosynthesis in the red light–stimulated opening. However, *Adiantum* stomata accumulate K^+ under ambient CO_2 concentrations and are insensitive to CO_2 both in darkness and under red light. It is intriguing that these unusual osmoregulatory features are associated with an exceptionally high number of chloroplasts in these guard cells. Furthermore, *Adiantum* stomata are highly unusual in their lack of blue-light sensitivity. In contrast, sensitivity to CO_2 and to blue light were shown to be linearly related to stomatal opening in *V. faba*. These results suggest that the lack of a blue-light response and the CO_2 insensitivity in *Adiantum* could be associated with a defective blue-light sensory transducing system.

The contrasting, unusual features of *Paphiopedilum* and *Adiantum* illustrate the remarkable functional plasticity of guard cells, also shown in other studies in the intact leaf. These plastic features include acclimations of the responses to blue light and CO_2, and daily changes in guard cell photosynthetic rates. They are described in detail in **WEB ESSAY 10.3**.

Mediation of Blue-light Photoreception in Guard Cells by Zeaxanthin

Stomata from the Arabidopsis mutant *npq1* (*n*on*p*hotochemical *q*uenching) lack a specific blue-light response. Such specificity is critical because, as stated earlier, guard cells have different mechanisms mediating responses to blue light. The *npq1* mutant has a lesion in the enzyme

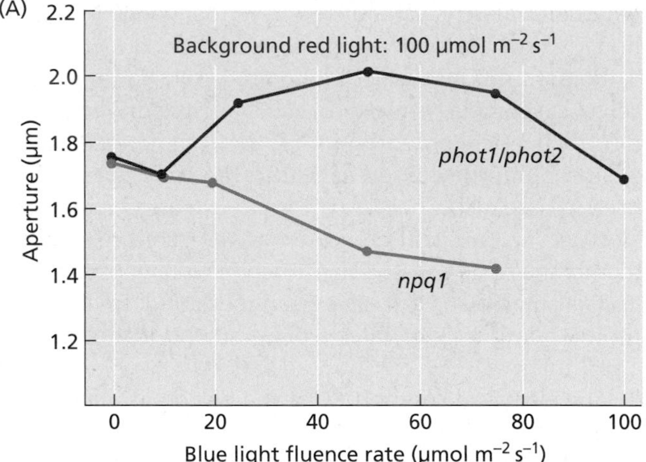

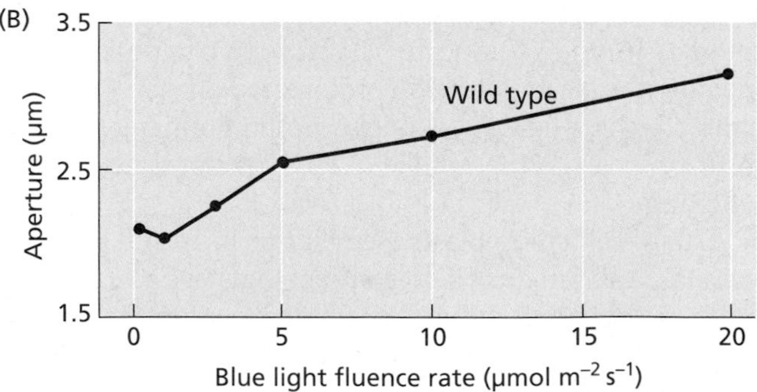

Figure 10.10 (A) Blue-light sensitivity of the zeaxanthin-less mutant *npq1* and of the phototropin-less double mutant *phot1/phot2*. The blue-light responses are assayed under 100 μmol m^{-2} s^{-1} red light to prevent stomatal opening resulting from the stimulation of photosynthesis by blue light. Darkness is shown as zero fluence rate. Neither mutant shows opening when illuminated with 10 μmol m^{-2} s^{-1} blue light. The *phot1/phot2* mutant opens at higher fluence rates of blue light, whereas the *npq1* mutant fails to show any blue light–stimulated opening. In fact, *npq1* stomata close, most likely because of an inhibitory effect of the additional blue light on photosynthesis-driven opening. (B) Blue light–stimulated opening in the wild type. Note the reduced scale of the y-axis, showing the reduced magnitude of the opening of *phot1/phot2* stomata, as compared with the wild type. (After Talbott et al. 2002.)

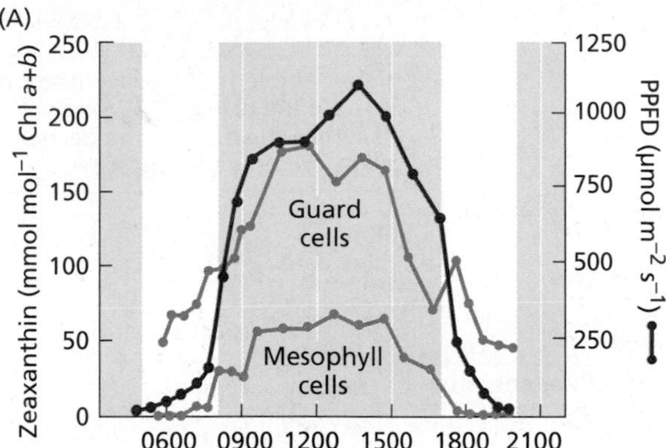

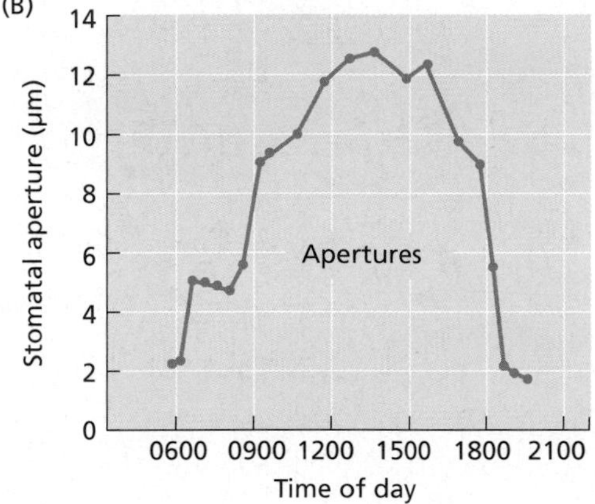

Figure 10.11 Zeaxanthin content of guard cells closely tracks photosynthetically active radiation and stomatal apertures. (A) Daily course of photosynthetically active radiation reaching the leaf surface (red trace), and of zeaxanthin content of guard cells (blue trace) and mesophyll cells (green trace) of *V. faba* leaves grown in a greenhouse. The white areas within the graph highlight the contrasting sensitivity of the xanthophyll cycle in mesophyll and guard cell chloroplasts under the low irradiances prevailing early and late in the day. (B) Stomatal apertures in the same leaves used to measure guard cell zeaxanthin content. PPFD, photosynthetic photon flux density. (After Srivastava and Zeiger 1995.)

that converts the carotenoid violaxanthin to zeaxanthin. Recall from Chapters 7 and 9 that zeaxanthin is a component of the xanthophyll cycle of chloroplasts (see Figure 7.33), which protects photosynthetic pigments from excess excitation energy. In addition, zeaxanthin functions as a blue-light photoreceptor in guard cells, mediating blue light–stimulated stomatal opening. Compelling evidence for this role of zeaxanthin ensues from the observation that, in the absence of zeaxanthin, guard cells from *npq1* lack a specific blue-light response (**Figure 10.10**).

Additional evidence further indicates that zeaxanthin is a blue-light photoreceptor in guard cells:

- In daily opening of stomata in intact leaves, incident radiation, zeaxanthin content of guard cells, and stomatal apertures are closely correlated (**Figure 10.11**).
- The absorption spectrum of zeaxanthin (**Figure 10.12**) closely matches the action spectrum for blue light–stimulated stomatal opening (see Figure 10.4).
- The sensitivity of guard cells to blue light increases as a function of their zeaxanthin concentration. The conversion of violaxanthin to zeaxanthin depends on the pH of the thylakoid lumen. Light-driven proton pumping at the thylakoid membrane acidifies the lumen compartment and increases the concentration of zeaxanthin (**Figure 10.13**). Because of this property of the xanthophyll cycle, guard cells illuminated with red light accumulate zeaxanthin. When guard cells from detached epidermis treated with increasing fluence rates of red light are exposed to a short blue-light pulse, the resulting blue light–stimulated stomatal opening is linearly related to the fluence rate of the red-light pretreatment, and to the zeaxanthin content of the guard cells at the time of application of the blue-light pulse.
- Blue light–stimulated stomatal opening is inhibited by 3 m*M* dithiothreitol (DTT), and the inhibition is concentration dependent. Zeaxanthin formation is blocked by DTT, a reducing agent that reduces S—S bonds to —SH groups and effectively inhibits the enzyme that converts violaxanthin into zeaxanthin. DTT does not block red light–stimulated opening.
- The facultative CAM species *Mesembryanthemum crystallinum* shifts its carbon metabolism from C_3 to CAM mode in response to salt stress. In the C_3 mode, stomata accumulate zeaxanthin and open in response to blue light. CAM induction inhibits both zeaxanthin accumulation and the ability of guard cells to open in response to blue light.

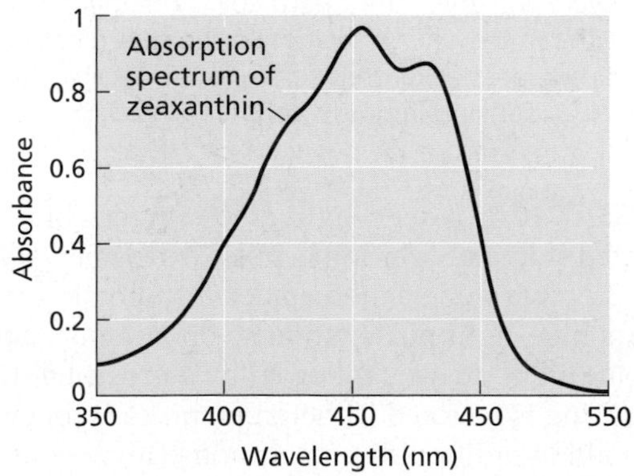

Figure 10.12 Absorption spectrum of zeaxanthin in ethanol. (Courtesy of Professor Wieslaw Gruszecki.)

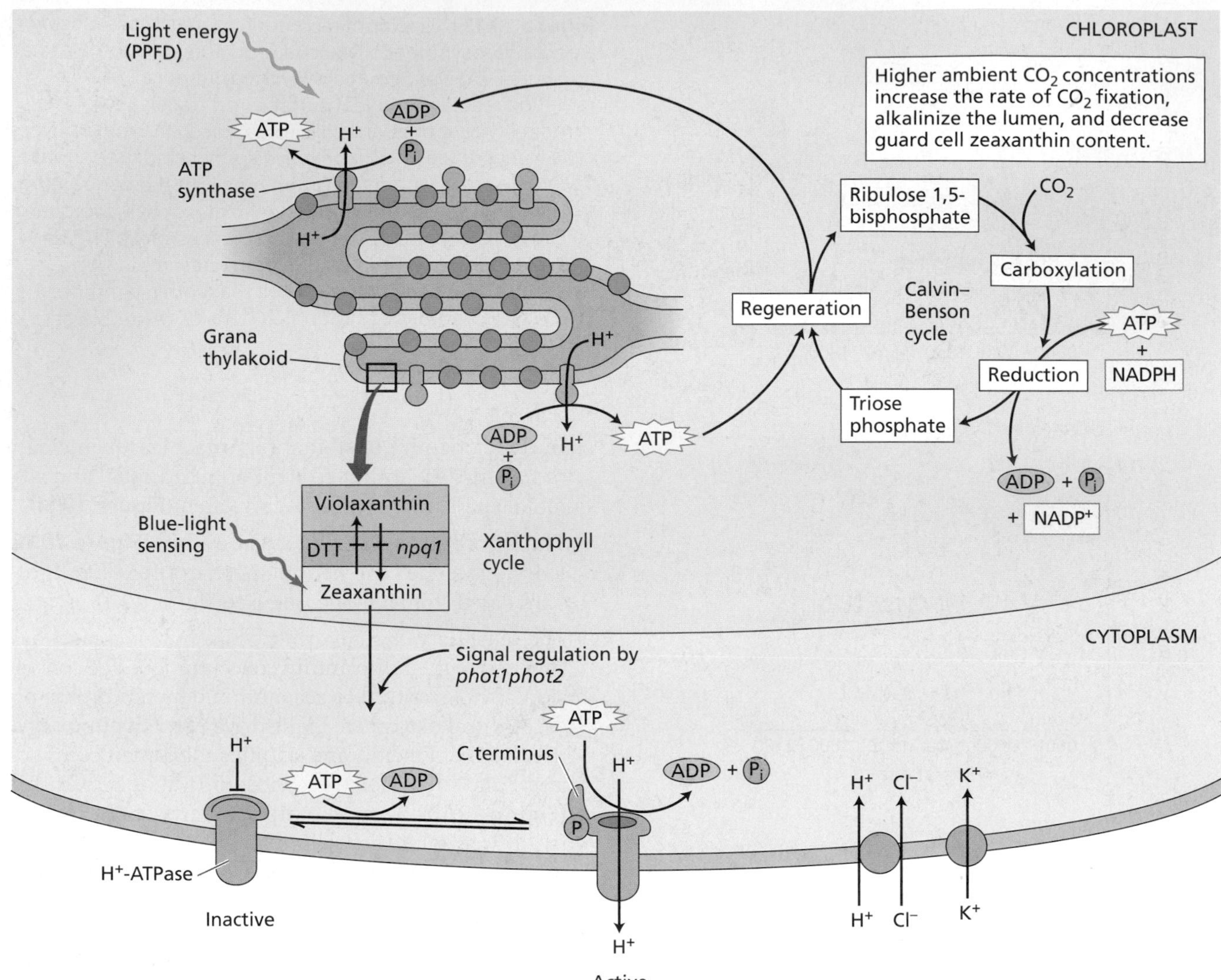

Figure 10.13 Role of zeaxanthin in blue-light sensing in guard cells. Zeaxanthin concentration in guard cells varies with the activity of the xanthophyll cycle. The enzyme that converts violaxanthin to zeaxanthin is an integral thylakoid protein showing a pH optimum of 5.2. Acidification of the lumen stimulates zeaxanthin formation, and alkalinization favors violaxanthin formation. Lumen pH depends on photosynthetic photon flux density (most effective at blue and red wavelengths; see Chapter 7), and on the rate of ATP synthesis, which consumes energy and dissipates the pH gradient across the thylakoid. Thus, photosynthetic activity in the guard cell chloroplast, lumen pH, zeaxanthin content, and blue-light sensitivity play an interactive role in the regulation of stomatal apertures. Compared with their mesophyll counterparts, guard cell chloroplasts are enriched in photosystem II, and they have unusually high rates of photosynthetic electron transport and low rates of photosynthetic carbon fixation. These properties favor lumen acidification at low photon fluxes, and they explain zeaxanthin formation in the guard cell chloroplast early in the day (see Figure 10.11). The regulation of zeaxanthin content by lumen pH, and the tight coupling between lumen pH and Calvin–Benson cycle activity in the guard cell chloroplast, further suggests that rates of CO_2 fixation in the guard cell chloroplast can regulate zeaxanthin concentrations and integrate light and CO_2 sensing in guard cells (see **WEB ESSAY 10.3**).

Reversal of Blue Light–Stimulated Opening by Green Light

Blue light–stimulated opening is abolished by green light in the 500 to 600 nm region of the spectrum. The blue-light response is suppressed when guard cells are simultaneously illuminated with both blue and green light (see **WEB ESSAY 10.4**). Green light also reverses blue light–stimulated stomatal opening in pulse experiments (**Figure 10.14**). Stomata in detached epidermis open in response to a 30-s blue-light pulse, and the opening is abolished if the blue-light pulse is followed by a green-light pulse. The opening is restored if the green pulse is followed by a second blue-light pulse, in a response analogous to the

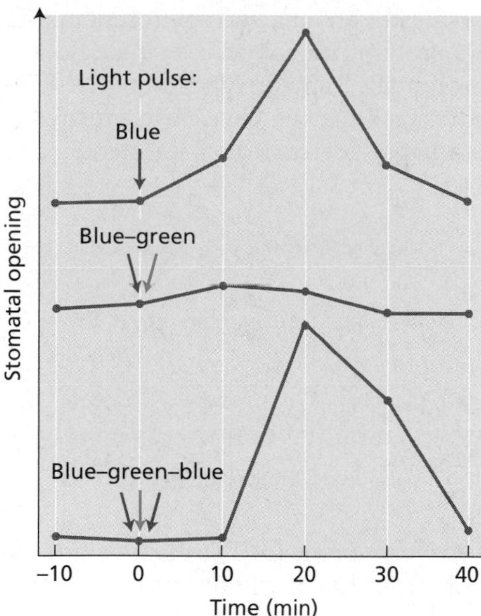

Figure 10.14 Blue/green reversibility of stomatal movements. Stomata open when given a 30-s blue-light pulse (1800 μmol m^{-2} s^{-1}) under a background of continuous red light (120 μmol m^{-2} s^{-1}). A green-light pulse (3600 μmol m^{-2} s^{-1}) applied after the blue-light pulse blocks the blue-light response. The opening is restored upon application of a second blue-light pulse given after the green-light pulse. (After Frechilla et al. 2000.)

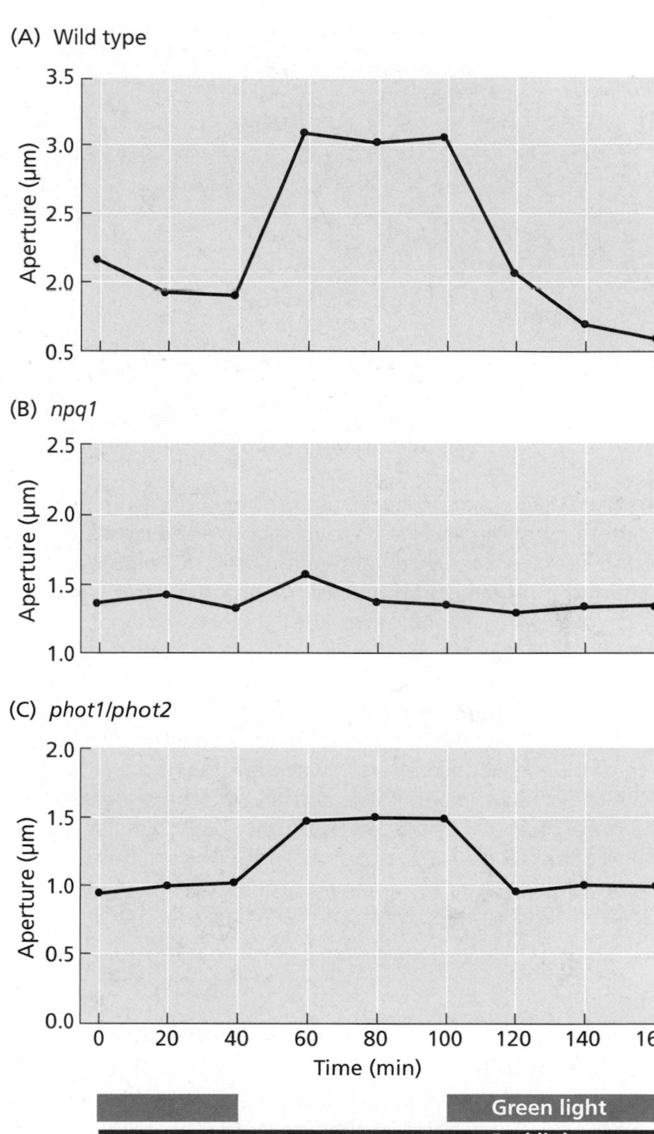

Figure 10.15 Green light regulates stomatal apertures in the intact leaf. Stomata from intact, attached leaves of Arabidopsis grown in a growth chamber under blue, red, and green light open when green light is removed and close when green light is restored. Blue light is required for the expression of this stomatal sensitivity to green light. Stomata from the zeaxanthin-less *npq1* mutant fail to respond to green light, whereas stomata from the *phot1/phot2* double mutant have a response similar to that of the wild type. (After Talbott et al. 2006.)

red/far-red reversibility of phytochrome responses. The blue/green reversibility response has been observed in stomata from detached epidermis of several species, and it is also seen in intact leaves (see **WEB ESSAY 10.4**).

Stomata from intact, attached leaves of Arabidopsis illuminated with blue, red, and green light in a growth chamber increase their aperture when the green light is turned off and close when the green light is turned on again (**Figure 10.15**). This green-light response cannot be mediated by photosynthesis in the mesophyll or in the guard cells, because stomata would be expected to close in response to lower photosynthetic rates ensuing from the green "light-off" signal. The green light–mediated opening is not observed if green light is turned off in experiments with leaves illuminated only with red and green light. Thus, the opening response to green light is seen only in the presence of blue light, as observed in experiments with detached epidermis peels. An important ecophysiological implication of these stomatal responses to green light in the intact leaf is that green photons from solar radiation are expected to down-regulate the stomatal response to blue light under natural conditions.

Stomata from the phototropin-less double mutant *phot1/phot2* respond to blue light and open further when green light is turned off, but stomata from the zeaxanthin-less mutant *npq1* do not (see Figure 10.15). These results indicate that the green reversal of the blue-light response requires zeaxanthin but not phototropin.

An action spectrum for the green reversal of blue light–stimulated opening shows a maximum at 540 nm and two minor peaks at 490 and 580 nm (**Figure 10.16**). Such an action spectrum rules out the involvement of phytochrome

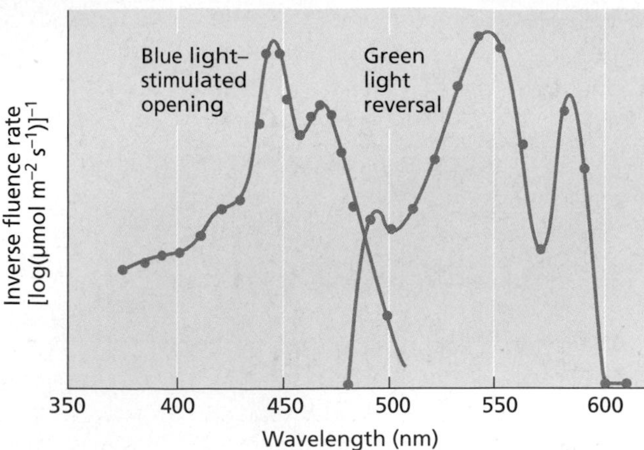

Figure 10.16 Action spectrum for blue light–stimulated stomatal opening and for its reversal by green light. The action spectrum for blue light–stimulated opening was obtained from measurements of transpiration as a function of wavelength in wheat leaves kept under a red-light background. The action spectrum for the green-light reversal of blue light–stimulated opening was calculated from measurements of aperture changes in stomata from detached epidermis of *Vicia faba* irradiated with a constant fluence rate of blue light and different wavelengths of green light. Note that the two spectra are similar, with the spectrum for the green reversal displaced by about 90 nm. Similar spectral red shifts have been observed upon the isomerization of carotenoids in a protein environment. (Left curve from Karlsson 1986; right curve from Frechilla et al. 2000.)

or chlorophylls. Rather, the action spectrum is remarkably similar to the action spectrum for blue light–stimulated stomatal opening, red-shifted (displaced toward the longer, red wave-band of the spectrum) by about 90 nm. Similar spectral red shifts have been observed upon the isomerization of carotenoids in a protein environment. As discussed earlier, the action spectrum for blue light–stimulated stomatal opening matches the absorption spectrum of zeaxanthin (see Figure 10.12). Spectroscopic studies have shown that green light is very effective in the isomerization of zeaxanthin. Isomerization of zeaxanthin changes the orientation of the molecule within the membrane, a transition that would be very effective as a transduction signal.

A carotenoid–protein complex senses light intensity

A carotenoid–protein complex that functions as a light-intensity sensor provides a model system for the blue–green photocycle in guard cells (see **WEB ESSAY 10.5**). The **orange carotenoid protein** (**OCP**) is a soluble protein associated with the phycobilisome antenna of photosystem II in cyanobacteria. Recall from Chapter 7 that cyanobacteria are photosynthetic bacteria common to freshwater and marine environments. The OCP is a 35 kDa protein that contains a single noncovalently bound carotenoid, 3′-hydroxyechinenone. Zeaxanthin and 3′-hydroxyechinenone have closely related chemical structures and both derive from β-carotene.

OH
HO
Zeaxanthin

OH
O
3′-Hydroxyechinenone

The active form of the OCP is essential for the induction of photoprotection in photosynthetic cyanobacteria. In addition, the photocycle resulting from the interconversion of the blue- and green-absorbing forms of the OCP functions as an effective light-intensity sensor.

These findings strongly suggest that the reversibility of stomatal opening in response to consecutive blue and green pulses results from the operation of a photocycle, most likely mediated by protein-bound zeaxanthin, converted by blue light into a physiologically active, green light–absorbing form and back-converted by green light into an inactive, blue light–absorbing form. It is also of interest that blue light–stimulated fluorescence quenching, likely to be associated with photoprotection, has been observed with guard cell and coleoptile chloroplasts of higher plants, paralleling the observation with the OCP in cyanobacteria.

In the 1940s, the discovery of carotenoids in the blue light–sensing coleoptile tip (see **WEB TOPIC 10.3**) implicated the blue light–absorbing carotenoids as possible blue-light photoreceptors, but the hypothesis was ruled out because of the very short half-life of the excited carotenoid molecule. The OCP represents the first clearly documented case of a protein-bound carotenoid that functions as a photoreceptor, and of a protein–carotenoid complex sensing light intensity. The striking similarity between some of the photobiological properties of the OCP of cyanobacteria and of blue-light sensing by zeaxanthin in the guard cell chloroplast should stimulate further research in both systems.

The Resolving Power of Photophysiology

Let us consider a mind experiment. You are assigned to research the stomatal response to blue light in a new plant species recently discovered in the Serengeti desert in Africa. Assume that, as in most plant species in real life, epidermal cells of this species have no chloroplasts. Since the species grows in a high-light environment, you first check the stomatal response to high light after a period of darkness. You use a blue-light filter that transmits maximally at around 450 nm, and expose a leaf to sunlight filtered through that filter. You observe that after a few minutes, stomata open. What can you tell about the stomatal response? It could be that the leaf mesophyll responded to blue light with high rates of photosynthesis, and intercellular CO_2 concentrations decreased. The stomata would then have opened in response to a decrease in intercellular CO_2. Alternatively, the stomata could have responded to blue light directly. You can distinguish between the two possibilities by removing epidermal strips from the leaf and incubating the isolated stomata under blue light. If the stomata open in isolation, the experiment shows that the stomata have a direct response to blue light.

Next, you want to ask questions about the nature of the guard cell photoreceptors. Photosynthesis in the guard cell chloroplast could have mediated the response. If that was the case, can you get opening by replacing the blue light with red light? A positive result with red light would implicate photosynthesis, and you could confirm the operation of guard cell photosynthesis by incubating the stomata under blue or red light in the presence of the photosynthetic inhibitor DCMU (see Figure 7.28).

What if red light gave no response? If so, photosynthesis can be ruled out, and you could test responses to blue light, or to red/far-red light. You could test the operation of a blue-light photoreceptor by checking whether green light reverses the blue light–stimulated opening. Alternatively, a typical blue-light photoreceptor would show increasing stomatal apertures under constant, low fluence rates of blue light, and increasing fluence rates of background red light.

A phytochrome response (such as those in **WEB TOPIC 10.4**) could be resolved by replacing the exciting blue light with red light, and then finding out whether far-red light closes the stomata and red light reopens them.

The principles used in this mind experiment can be applied to research results. Take, for instance, the remarkable results shown in Figure 10.15. These were growth-chamber experiments with intact leaves of Arabidopsis, showing that stomata from leaves illuminated with blue, red, and green light open when the green light is turned off and close when the green light is turned on again. Could we be dealing with a photosynthesis-driven response in this experiment? Unlikely, because the total light fluence decreased when the green light was turned off, yet stomatal opening increased. However, turning the green light on and off in the presence of red light alone did not show any changes in stomatal apertures, indicating that blue light is needed for the response to green light, and that we are most likely dealing with the blue–green cycling.

Photophysiological analysis can also be very useful for the interpretation of research results. For instance, the phototropin-less double mutant *phot1/phot2* responds to green light in the green-/blue-/red-light experiments with a small but fully reproducible opening. However, the zeaxanthin-less mutant, *npq*, fails to respond to green light. These results have important implications. Both *phot1/phot2* and *npq1* have lesions in their sensory transduction mechanisms associated with their responses to blue light and the blue–green cycle, yet *phot1/phot2* responds to green light while *npq1* does not. This implies that the genetic lesion in *npq1* has disabled the blue–green cycle, while the response in *phot1/phot2* appears unaffected. In a 2013 publication on *phot1/phot2*, Ken-Ichiro Shimazaki and his coworkers discuss how phototropins are associated with several blue-light responses without a common sensory transduction cascade. Is the *npq1* cascade yet a different one, sharing blue-light sensitivity but underscoring different components?

Photophysiological analysis can help us understand these questions. The specific blue-light response can be reversed by green light; the blue component of the photosynthetic response is blocked under saturating red light; and the phytochrome response is far-red reversible. Applications of these approaches have been illustrated earlier. For example, obtained results showed that blue light–stimulated stomatal opening observed in the Arabidopsis mutant *npq1* cannot be reversed by green light, but it is reversed by far-red light, indicating that the involved photoreceptor is phytochrome. In contrast, the blue light–stimulated opening observed with *phot1/phot2* stomata is green light–reversible, indicating that a specific blue-light photoreceptor is mediating the response.

These exciting results illustrate how the use of well-defined genetic mutants, combined with high-resolution physiological tools, can answer many unresolved questions of guard cell photobiology.

SUMMARY

Stomata are structural features of most plants. Each stoma consists of a microscopic pore that allows communication between the interior of the leaf and the external environment, and a pair of guard cells that surround the pore. The guard cells can be flanked by specialized subsidiary cells that separate them from regular epidermal cells. Guard cells respond to environmental signals by changing their dimensions, thereby regulating the size of the pore, or stomatal aperture. Stomata are a critical adaptation to avoid desiccation; they close when water is limiting and open under conditions favoring photosynthesis. The driving force for stomatal movements is turgor pressure.

Light-dependent Stomatal Opening

- Light is the dominant stimulus causing stomatal opening. The two major driving forces for light-dependent stomatal opening are photosynthesis in the guard cell chloroplast and a specific response to blue light.
- An inhibitor of photosynthetic electron transport, DCMU, also inhibits stomatal opening, indicating that the photosynthetic process plays a role in stomatal opening. The inhibition is only partial, however, meaning that other opening mechanisms must be active. A major second mechanism is a specific stomatal response to blue light (**Figure 10.1**).
- Researchers use dual-beam experiments to study the stomatal response to blue light. An action spectrum for the stomatal response to blue light obtained under saturated red light shows a characteristic three-peak pattern (**Figures 10.3, 10.4**).
- Light-stimulated stomatal movements are driven by changes in the osmoregulation of guard cells. Blue light stimulates an H^+-ATPase at the guard cell plasma membrane, generating an electrochemical-potential gradient that drives ion uptake (**Figures 10.5–10.7**).
- Blue light also stimulates starch degradation and malate biosynthesis. Accumulation of sucrose and K^+ and its counterions within guard cells leads to stomatal opening (**Figure 10.8**).
- Guard cell chloroplasts usually contain large starch grains. In contrast to what happens in mesophyll chloroplasts, starch content in guard cell chloroplasts decreases during stomatal opening in the morning and increases during closing at the end of the day.
- Light quality can change the activity of different osmoregulatory pathways that modulate stomatal movements. Stomatal opening is associated primarily with K^+ uptake. Stomatal closing is associated with a loss of K^+ and a decrease in sucrose content (**Figure 10.9**).

Mediation of Blue-light Photoreception in Guard Cells by Zeaxanthin

- The chloroplast carotenoid zeaxanthin has been implicated in blue-light photoreception in guard cells (**Figure 10.10**).
- Daily stomatal opening, incident radiation, zeaxanthin content of guard cells, and stomatal apertures are closely related (**Figure 10.11**).
- The absorption spectrum of zeaxanthin matches the action spectrum for blue light–stimulated stomatal opening (**Figures 10.4, 10.12**).
- Blue light–stimulated stomatal opening is blocked if zeaxanthin accumulation in guard cells is prevented. Manipulation of zeaxanthin content in guard cells permits regulation of their response to blue light (**Figure 10.13**).

Reversal of Blue Light–Stimulated Opening by Green Light

- The blue-light response of guard cells is reversed by green light (**Figure 10.14**).
- The reversibility of the stomatal response to blue light by green light can be observed in the intact leaf, indicating that this modulation of the stomatal response has functional implications under natural conditions (**Figure 10.15**).
- The zeaxanthin-less mutant *npq1* does not show blue/green reversibility, indicating that zeaxanthin is required for the response. Phototropin-less mutants show normal blue/green reversibility.
- The action spectrum of the green reversal resembles the action spectrum of blue light–stimulated opening and the absorption spectrum of zeaxanthin (**Figure 10.16**).
- A carotenoid–protein complex in cyanobateria, the orange carotenoid protein (OCP), shows blue/green reversibility and functions as a light sensor. The OCP provides a molecular model for blue-light sensing by zeaxanthin in guard cells.

The Resolving Power of Photophysiology

- Photophysiological principles add excellent diagnostic power to the analysis of research with guard cells. For example, the specific blue-light response can be reversed by green light, the blue component of the photosynthetic response is blocked under saturating red light, and the phytochrome response is far-red reversible. Blue light–stimulated stomatal opening observed in the Arabidopsis mutant *npq1* cannot be reversed by green light, but it is reversed by far-red light, indicating that the involved photoreceptor is phytochrome. In contrast, the blue light–stimulated opening observed with *phot1/phot2* stomata is green light–reversible, indicating that a specific blue-light photoreceptor is mediating the response.

WEB MATERIAL

- **WEB TOPIC 10.1 Blue-light Sensing and Light Gradients** Light gradients within organs might serve as sensing mechanisms.
- **WEB TOPIC 10.2 Guard Cell Osmoregulation and a Blue Light–Activated Metabolic Switch** Blue light controls major osmoregulatory pathways in guard cells and unicellular algae.
- **WEB TOPIC 10.3 The Coleoptile Chloroplast** Both the coleoptile and the guard cell chloroplast specialize in sensory transduction.
- **WEB TOPIC 10.4 Phytochrome-mediated Responses in Stomata** Studies in the orchid *Paphiopedilum* and the zeaxanthin-less mutant of Arabidopsis, *npq1*, show that phytochrome regulates stomatal movements.
- **WEB ESSAY 10.1 Guard Cell Photosynthesis** Photosynthesis in the guard cell chloroplast shows unique regulatory features.
- **WEB ESSAY 10.2 Sucrose Metabolism in Guard Cells** Sucrose is involved in stomatal function through its action as an osmolyte and a substrate.
- **WEB ESSAY 10.3 The Plasticity of Guard Cells** The remarkable functional plasticity of the guard cells shapes our knowledge about stomatal function.
- **WEB ESSAY 10.4 The Blue/Green Reversibility of the Blue-light Response of Stomata** The responses of guard cells to blue and green light drive a unique photocycle.
- **WEB ESSAY 10.5 The Orange Carotenoid Protein** A unique photoreceptor protein measures time and uses a carotenoid chromophore.

available at plantphys.net

Suggested Reading

Assmann, S. M. (2010) Hope for Humpty Dumpty. Systems biology of cellular signaling. *Plant Physiol.* 152: 470–449.

Frechilla, S., Zhu, J., Talbott, L. D., and Zeiger, E. (1999) Stomata from *npq1*, a zeaxanthin-less *Arabidopsis* mutant, lack a specific response to blue light. *Plant Cell Physiol.* 40: 949–954.

Frechilla, S., Talbott, L. D., Bogomolni, R. A., and Zeiger, E. (2000) Reversal of blue light-stimulated stomatal opening by green light. *Plant Cell Physiol.* 41: 171–176.

Iino, M., Ogawa, T., and Zeiger, E. (1985) Kinetic properties of the blue light response of stomata. *Proc. Natl. Acad. Sci. USA* 82: 8019–8023.

Karlsson, P. E. (1986) Blue light regulation of stomata in wheat seedlings. II. Action spectrum and search for action dichroism. *Physiol. Plant.* 66: 207–210.

Kirilovsky, D. and Kerfeld, C. A. (2013) The orange carotenoid protein: A blue-green light photoactive protein. *Photochem. Photobiol. Sci.* 12: 1135–1143.

Lawson, T. (2009) Guard cell photosynthesis and stomatal function. *New Phytol.* 181: 13–34.

Milanowska, J., and Gruszecki, W. I. (2005) Heat-induced and light-induced isomerization of the xanthophyll pigment zeaxanthin. *J. Photochem. Photobiol. B.* 80: 178–186.Punginelli, C., Wilson, A., Routaboul, J. M., and Kirilovsky, D. (2009) Influence of zeaxanthin and echinenone binding on the activity of the orange carotenoid protein. *Biochim. Biophys. Acta* 1787: 280–288.

Roelfsema, M. R. G., Steinmeyer, R., Staal, M., and Hedrich, R. (2001) Single guard cell recordings in intact plants: Light-induced hyperpolarization of the plasma membrane. *Plant J.* 26: 1–13.

Roelfsema, M. R. G. and Kollist, H. (2013) Tiny pores with a global impact. *New Phytol.* 197: 11–15.

Spalding, E. P. (2000) Ion channels and the transduction of light signals. *Plant Cell Environ.* 23: 665–674.

Srivastava, A., and Zeiger, E. (1995) Guard cell zeaxanthin tracks photosynthetic active radiation and stomatal apertures in *Vicia faba* leaves. *Plant Cell Environ.* 18: 813–817.

Talbott, L. D., Zhu, J., Han, S. W., and Zeiger, E. (2002) Phytochrome and blue light-mediated stomatal opening in the orchid, *Paphiopedilum*. *Plant Cell Physiol.* 43: 639–646.

Talbott, L. D., Shmayevich, I. J., Chung, Y., Hammad, J. W., and Zeiger, E. (2003) Blue light and phytochrome-mediated stomatal opening in the *npq1* and *phot1 phot2* mutants of *Arabidopsis*. *Plant Physiol.* 133: 1522–1529.

Talbott, L. D., Hammad, J. W., Harn, L. C, Nguyen, V., Patel, J., and Zeiger, E. (2006) Reversal by green light of blue light-stimulated stomatal opening in intact, attached leaves of *Arabidopsis* operates only in the potassium dependent, morning phase of movement. *Plant Cell Physiol.* 47: 333–339.

Zeiger, E., Talbott, L. D., Frechilla, S., Srivastava, A., and Zhu, J. X. (2002) The guard cell chloroplast: A perspective for the twenty-first century. *New Phytol.* 153: 415–424.

11 Translocation in the Phloem

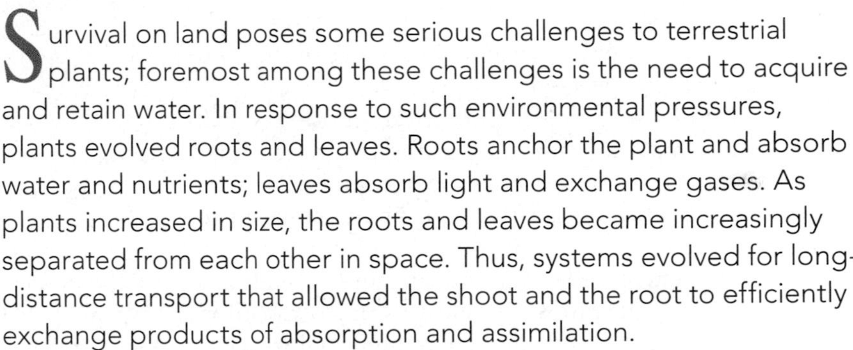

Survival on land poses some serious challenges to terrestrial plants; foremost among these challenges is the need to acquire and retain water. In response to such environmental pressures, plants evolved roots and leaves. Roots anchor the plant and absorb water and nutrients; leaves absorb light and exchange gases. As plants increased in size, the roots and leaves became increasingly separated from each other in space. Thus, systems evolved for long-distance transport that allowed the shoot and the root to efficiently exchange products of absorption and assimilation.

You will recall from Chapters 4 and 6 that the xylem is the tissue that transports water and minerals from the root system to the aerial portions of the plant. The **phloem** is the tissue that translocates the products of photosynthesis—particularly sugars—from mature leaves to areas of growth and storage, including the roots.

The phloem also transmits signals between sources and sinks in the form of regulatory molecules, and redistributes water and various compounds throughout the plant body. All of these molecules appear to move with the transported sugars. The compounds to be redistributed, some of which initially arrive in the mature leaves via the xylem, can be either transferred out of the leaves without modification or metabolized before redistribution.

The discussion that follows emphasizes translocation in the phloem of angiosperms, because most of the research has been conducted on that group of plants. We will compare gymnosperms briefly with angiosperms in terms of the anatomy of their conducting cells and implications for their mechanism of translocation.

First we will examine some aspects of translocation in the phloem that have been researched extensively and are thought to be relatively well understood, including the pathway and patterns of translocation, materials translocated in the phloem, and rates of movement. In the second part of the chapter we will explore aspects of translocation in the phloem that need further investigation. These include the mechanism of phloem transport, including the details of sieve element ultrastructure and the magnitude of pressure gradients between sources and sinks; phloem loading and unloading; and the allocation and partitioning of photosynthetic products. Finally, we will explore an area of intensive research at present: the phloem as a transport pathway for signaling molecules such as proteins and RNA.

Pathways of Translocation

The two long-distance transport pathways—the phloem and the xylem—extend throughout the plant body. The phloem is generally found on the outer side of both primary and secondary vascular tissues (**Figures 11.1 and 11.2**). In plants with secondary growth, the phloem constitutes the inner bark. Although phloem is commonly found in a position external to the xylem, it is *also* found on the inner side in many eudicot families. In these families the phloem in the two positions is called external and internal phloem, respectively.

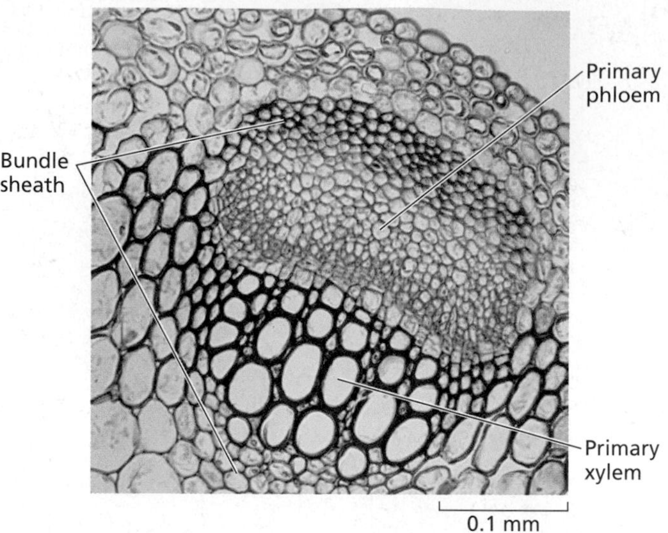

Figure 11.1 Transverse section of a vascular bundle of trefoil, a clover (*Trifolium*). The primary phloem is toward the outside of the stem. Both the primary phloem and the primary xylem are surrounded by a bundle sheath of thick-walled sclerenchyma cells, which isolate the vascular tissue from the ground tissue. Fibers and xylem vesssels are stained red.

The cells of the phloem that conduct sugars and other organic materials throughout the plant are called **sieve elements**. *Sieve element* is a comprehensive term that includes both the highly differentiated **sieve tube elements** typical of the angiosperms and the relatively unspecialized **sieve cells** of gymnosperms. In addition to sieve elements, the phloem tissue contains companion cells (discussed below) and parenchyma cells (which store and release food molecules). In some cases the phloem tissue also includes fibers and sclereids (for protection and strengthening of the tissue) and laticifers (latex-containing cells). However, only the sieve elements are directly involved in translocation.

The small veins of leaves and the primary vascular bundles of stems are often surrounded by a **bundle sheath** (see Figure 11.1), which consists of one or more layers of compactly arranged cells. (You will recall the bundle sheath cells involved in C_4 metabolism discussed in Chapter 8.) In the vascular tissue of leaves, the bundle sheath surrounds the small veins all the way to their ends, isolating the veins from the intercellular spaces of the leaf.

We will begin our discussion of translocation pathways with the experimental evidence demonstrating that the sieve elements are the conducting cells in the phloem. Then we will examine the structure and physiology of these unusual plant cells.

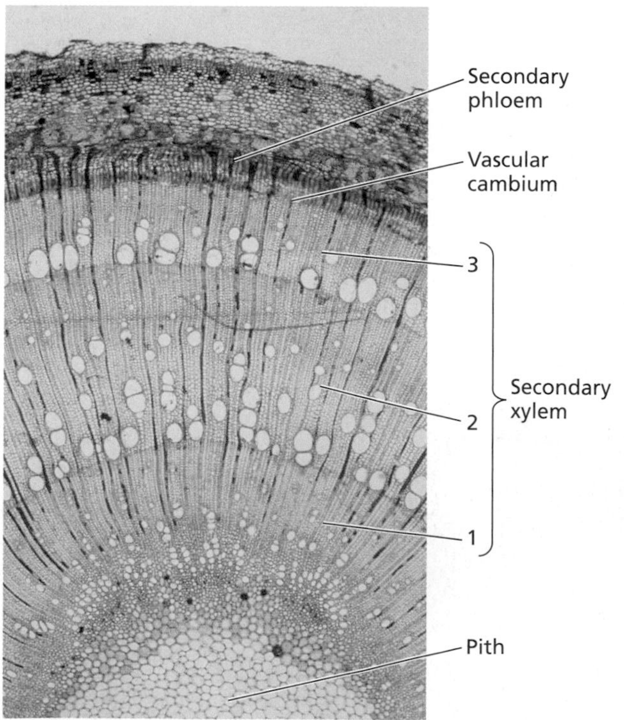

Figure 11.2 Transverse section of a 3-year-old stem of an ash (*Fraxinus excelsior*) tree. (27×) The numbers 1, 2, and 3 indicate growth rings in the secondary xylem. The old (outer) secondary phloem has been crushed by expansion of the xylem. Only the most recent (innermost) layer of secondary phloem is functional.

Sugar is translocated in phloem sieve elements

Early experiments on phloem transport date back to the nineteenth century, indicating the importance of long-distance transport in plants (see WEB TOPIC 11.1). These classic experiments demonstrated that removal of a ring of bark around the trunk of a tree, which removes the phloem, effectively stops sugar transport from the leaves to the roots without altering water transport through the xylem. When radioactive compounds became available, $^{14}CO_2$ was used to show that sugars made in the photosynthetic process are translocated through the phloem sieve elements (see WEB TOPIC 11.1).

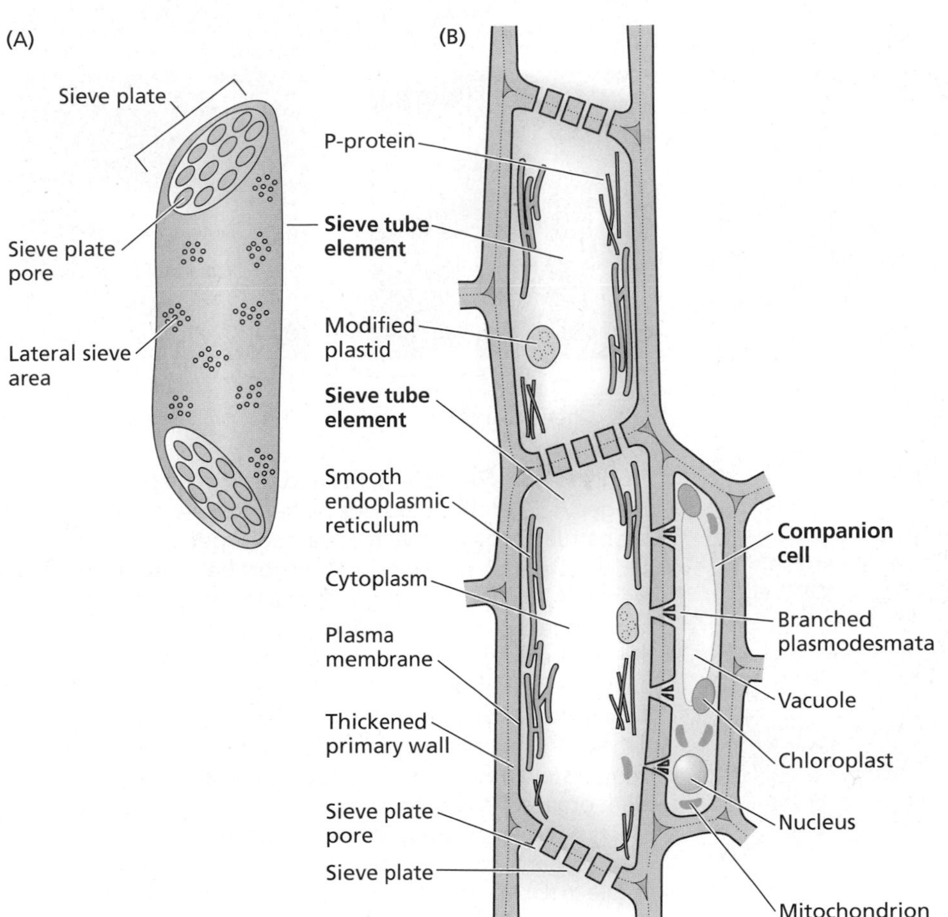

Figure 11.3 Schematic drawings of mature sieve elements (sieve tube elements) joined together to form a sieve tube. (A) External view, showing sieve plates and lateral sieve areas. (B) Longitudinal section, showing two sieve tube elements joined together to form a sieve tube. The pores in the sieve plates between the sieve tube elements are open channels for transport through the sieve tube. The plasma membrane of a sieve tube element is continuous with that of its neighboring sieve tube element. Each sieve tube element is associated with one or more companion cells, which take over some of the essential metabolic functions that are reduced or lost during differentiation of the sieve tube elements. Note that the companion cell has many cytoplasmic organelles, whereas the sieve tube element has relatively few organelles. An ordinary companion cell of a leaf is depicted here.

Mature sieve elements are living cells specialized for translocation

Detailed knowledge of the ultrastructure of sieve elements is critical to any discussion of the mechanism of translocation in the phloem. Mature sieve elements are unique among living plant cells (**Figures 11.3 and 11.4**). They lack many structures normally found in living cells, even in the undifferentiated cells from which they are formed. For example, sieve elements lose their nuclei and tonoplast (vacuolar membranes) during development. Microfilaments, microtubules, Golgi bodies, and ribosomes are also generally absent from the mature cells. In addition to the plasma membrane, organelles that are retained include somewhat modified mitochondria, plastids, and smooth endoplasmic reticulum. The walls are nonlignified, though they are secondarily thickened in some cases.

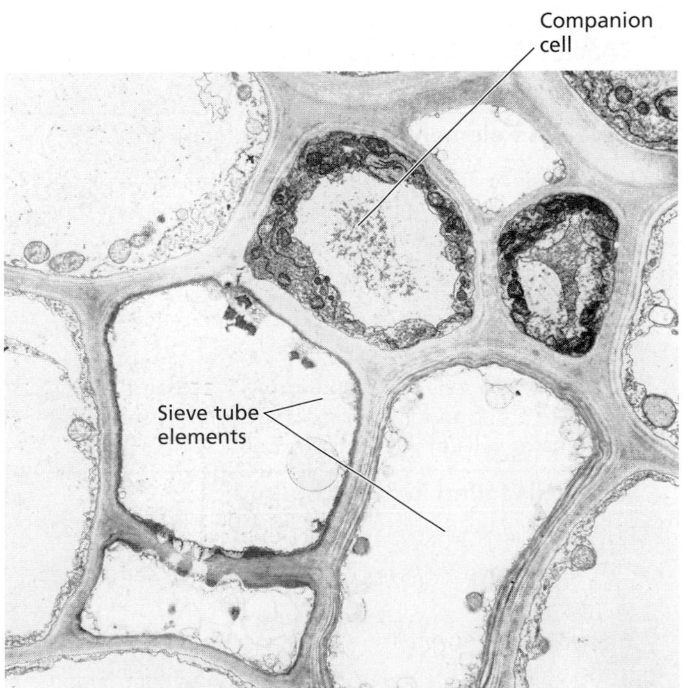

Figure 11.4 Electron micrograph of a transverse section of ordinary companion cells and mature sieve tube elements. The cellular components are distributed along the walls of the sieve tube elements, where they offer less resistance to mass flow. (From Warmbrodt 1985.)

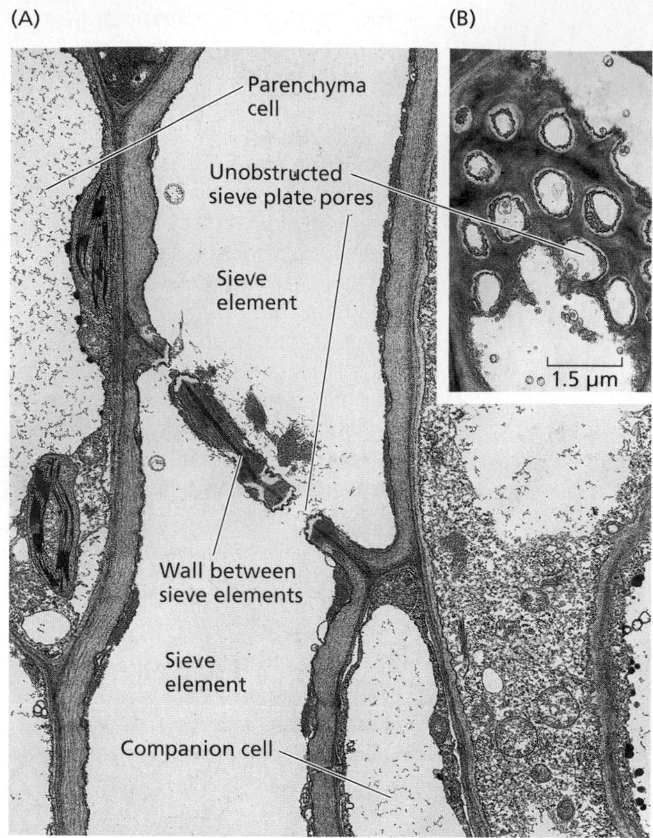

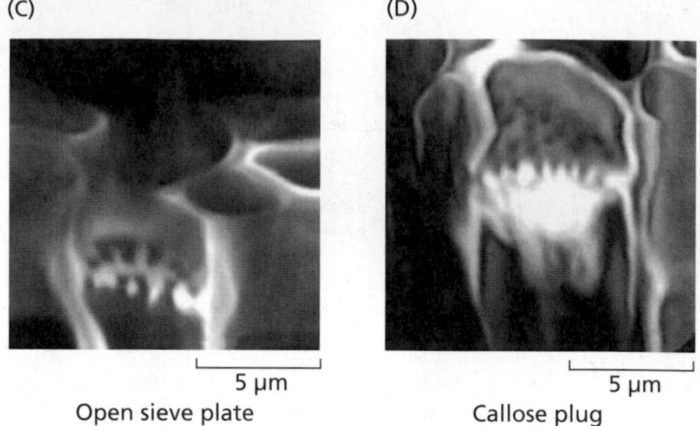

Figure 11.5 Sieve elements and sieve plate pores. In images A, B, and C, the sieve plate pores are open—that is, unobstructed by P-protein or callose. Open pores provide a low-resistance pathway for transport between sieve elements. (A) Electron micrograph of a longitudinal section of two mature sieve elements (sieve tube elements), showing the wall between the sieve elements (called a sieve plate) in the hypocotyl of winter squash (*Cucurbita maxima*). (B) The inset shows sieve plate pores in face view. (C and D) Three-dimensional reconstructions of Arabidopsis sieve plates using a staining technique that can be used to image entire plant organs with confocal laser scanning microscopy. Open sieve pores are visible in (C), while a callose plug, such as that formed in response to sieve tube damage, is visible in (D). (A and B from Evert 1982; C and D from Truernit et al. 2008.)

Thus, the cellular structure of sieve elements is different from that of tracheary elements of the xylem, which lack a plasma membrane, have lignified secondary walls, and are dead at maturity. As we will see, living cells are critical to the mechanism of translocation in the phloem.

Large pores in cell walls are the prominent feature of sieve elements

Sieve elements (sieve cells and sieve tube elements) have characteristic sieve areas in their cell walls, where pores interconnect the conducting cells (**Figure 11.5**). The sieve area pores range in diameter from less than 1 μm to approximately 15 μm. Unlike sieve areas of gymnosperms, the sieve areas of angiosperms can differentiate into **sieve plates** (see Figure 11.5 and **Table 11.1**). Sieve plates have larger pores than the other sieve areas in the cell and are generally found on the end walls of sieve tube elements, where the individual cells are joined together to form a longitudinal series called a **sieve tube** (see Figure 11.3).

The distribution of sieve tube contents, especially within the sieve plate pores, has been debated for many years and is a critical question when considering the mechanism of phloem transport. Early micrographs showed blocked or occluded pores, thought to be artifacts due to damage as the tissues were prepared for observation. (See next page, *Damaged sieve elements are sealed off.*) Later, less invasive

TABLE 11.1 Characteristics of the two types of sieve elements in seed plants

Sieve tube elements found in angiosperms
1. Some sieve areas are differentiated into sieve plates; individual sieve tube elements are joined together into a sieve tube.
2. Sieve plate pores are open channels.
3. P-protein is present in all eudicots and many monocots.
4. Companion cells are sources of ATP and perhaps other compounds. In some species, they serve as transfer cells or intermediary cells.
Sieve cells found in gymnosperms
1. There are no sieve plates; all sieve areas are similar.
2. Pores in sieve areas appear blocked with membranes.
3. There is no P-protein.
4. Albuminous cells sometimes function as companion cells.

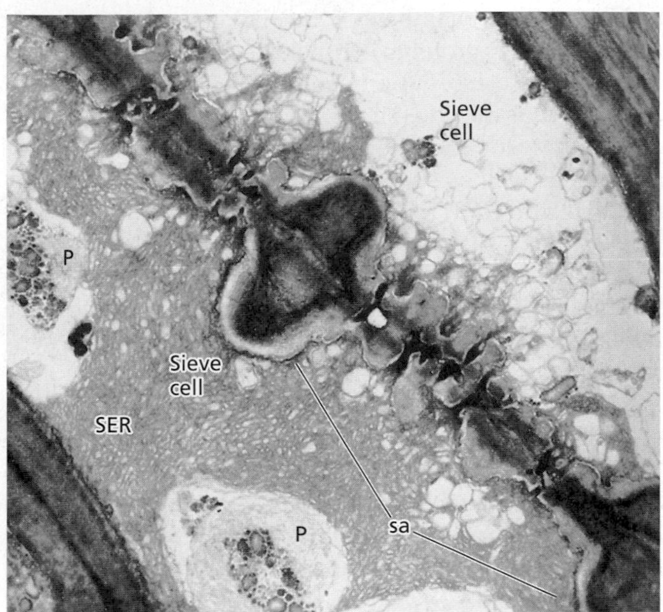

Figure 11.6 Electron micrograph showing a sieve area (sa) linking two sieve cells in the secondary phloem of a conifer (*Pinus resinosa*). Smooth endoplasmic reticulum (SER) covers the sieve area on both sides and is also found within the pores and the extended median cavity. Such obstructed pores would result in a high resistance to mass flow between sieve cells. P, plastid. (From Schulz 1990.)

techniques often showed the sieve plate pores of sieve tube elements to be open channels that allow unfettered transport between cells (see Figure 11.5A–C). A later section (*Sieve plate pores appear to be open channels*) will further consider the distribution of sieve element contents within the cells and within the sieve plate pores.

In contrast to pores in sieve tube elements of angiosperms, sieve areas in gymnosperms do not appear to be open channels. All of the sieve areas in gymnosperms such as conifers are structurally similar, though they can be more numerous on the overlapping end walls of sieve cells. The pores of gymnosperm sieve areas meet in large median cavities in the middle of the wall. Smooth endoplasmic reticulum (SER) covers the sieve areas (**Figure 11.6**) and is continuous through the sieve pores and median cavity, as indicated by ER-specific staining. Observation of living material with confocal laser scanning microscopy confirms that the observed distribution of SER is not an artifact of fixation.

Table 11.1 lists characteristics of sieve tube elements and sieve cells.

Damaged sieve elements are sealed off

Sieve element sap is rich in sugars and other organic molecules. (*Sap* is a general term used to refer to the fluid contents of plant cells.) These molecules represent an energy investment for the plant, and their loss must be prevented when sieve elements are damaged. Short-term sealing mechanisms involve sap proteins, while the principal long-term mechanism for preventing sap loss entails closing sieve plate pores with callose, a glucose polymer.

The main phloem proteins involved in sealing damaged sieve elements are structural proteins called **P-proteins** (see Figure 11.3B). (In earlier scientific literature, P-protein was called *slime*.) The sieve tube elements of most angiosperms, including those of all eudicots and many monocots, are rich in P-protein. However, P-protein is absent in gymnosperms. P-protein occurs in several different forms (tubular, fibrillar, granular, and crystalline), depending on the species and maturity of the cell. In immature cells, P-protein is most evident as discrete bodies in the cytosol known as **P-protein bodies**. P-protein bodies may be spheroidal, spindle-shaped, or twisted and coiled. They often disperse into tubular or fibrillar forms during cell maturation.

P-protein appears to function in sealing off damaged sieve elements by plugging sieve plate pores. Sieve tubes are under very high internal turgor pressure, and the sieve elements in a sieve tube are connected through sieve plate pores that appear to be open. When a sieve tube is cut or punctured, the release of pressure causes the contents of the sieve elements to surge toward the cut end, from which the plant could lose much sugar-rich phloem sap if there were no sealing mechanism. When surging occurs, however, P-protein is trapped on the sieve plate pores, helping to seal the sieve element and prevent further loss of sap. Direct support for the sealing function of P-protein has been found in both tobacco and Arabidopsis, in which mutants lacking P-protein lose significantly more transport sugar by sap exudation after wounding than do wild-type plants. (See *Phloem sap can be collected and analyzed* below for more information about exudation.) No visible phenotypic differences were observed between the mutant and wild-type plants.

Protein crystals released from ruptured plastids may play a similar sealing role in some monocots as P-protein does in eudicots. The sieve element organelles (mitochondria, plastids, and ER), however, appear to be anchored to each other or to the sieve element plasma membrane by minute protein "clamps." Which organelles are anchored depends on the species.

Another mechanism for blocking wounded sieve tubes with proteins occurs in plants in the legume family (Fabaceae). These plants contain large crystalloid P-protein that do not disperse during development. However, following damage or osmotic shock, the P-protein rapidly disperse and block the sieve tube. The process is reversible and controlled by calcium ions. These P-protein, known as **forisomes**, occur only in certain legumes and are encoded by members of the sieve element occlusion (SEO) gene family.

Homologous members of the SEO gene family encode conventional P-proteins in other species. These are called SEOR (sieve element occlusion related) genes. The term *P-protein* thus includes similar molecules that are involved in blocking wounded sieve elements in all eudicot angiosperms, as well as special P-proteins, such as forisomes and PP1 and PP2, found in *Cucurbita maxima* (see **WEB TOPIC 11.2**).

A longer-term solution to sieve tube damage is the production of the glucose polymer **callose** in the sieve pores (see Figure 11.5D). Callose, a β-1,3-glucan, is synthesized by an enzyme in the plasma membrane (callose synthase) and is deposited between the plasma membrane and the cell wall. Callose is synthesized in functioning sieve elements in response to damage and other stresses, such as mechanical stimulation and high temperatures, or in preparation for normal developmental events, such as dormancy. The deposition of **wound callose** in the sieve pores efficiently seals off damaged sieve elements from surrounding intact tissue, with complete occlusion occurring about 20 min after injury. In all cases, as sieve elements recover from damage or break dormancy, the callose disappears from the sieve pores; its dissolution is mediated by a callose-hydrolyzing enzyme. While Arabidopsis and tobacco mutants lacking P-protein show no visible phenotypic changes, Arabidopsis mutants lacking one callose synthase show reduced influorescence growth, apparently due to reduced assimilate transport to the influorescence.

Callose deposition is induced and callose synthase genes are up-regulated in rice (*Oryza sativa*) plants attacked by a phloem-feeding insect (brown planthopper); this occurs both in plants resistant to the insect and in susceptible plants. In the susceptible plants, however, feeding by the insects also activates genes for a callose-hydrolyzing enzyme. This unplugs the pores, allows continued feeding, and results in decreased sucrose and starch levels in the leaf sheath being attacked. Sealing off sieve elements that have been penetrated by insect mouthparts can thus play a key role in herbivore resistance.

Companion cells aid the highly specialized sieve elements

Each sieve tube element is usually associated with one or more **companion cells** (see Figures 11.3B, 11.4, and 11.5). The division of a single mother cell forms the sieve tube element and the companion cell. Numerous plasmodesmata (see Chapter 1) penetrate the walls between sieve tube elements and their companion cells; the plasmodesmata are often complex and branched on the companion cell side. The presence of abundant plasmodesmata suggests a close functional relationship between a sieve element and its companion cell, an association that is demonstrated by the rapid exchange of solutes, such as fluorescent dyes, between the two cells.

Companion cells play a role in the transport of photosynthetic products from producing cells in mature leaves to the sieve elements in the minor (small) veins of the leaf. They also take over some of the critical metabolic functions, such as protein synthesis, that are reduced or lost during differentiation of the sieve elements. In addition, the numerous mitochondria in companion cells may supply energy as ATP to the sieve elements.

There are at least three different types of companion cells in the minor veins of mature, exporting leaves: "ordinary" companion cells, transfer cells, and intermediary cells. All three cell types have dense cytoplasm and abundant mitochondria.

Ordinary companion cells (**Figure 11.7A**) have chloroplasts with well-developed thylakoids and a cell wall with a smooth inner surface. The number of plasmodesmata connecting ordinary companion cells to surrounding cells is quite variable and apparently reflects the pathway taken by sugars as they move from the mesophyll to the minor veins (discussed in the section *Phloem Loading*).

Transfer cells are similar to ordinary companion cells, except for the development of fingerlike wall ingrowths, particularly on the cell walls that face away from source sieve elements (**Figure 11.7B**). These wall ingrowths greatly increase the surface area of the plasma membrane, thus increasing the potential for solute transfer across the membrane. Relatively few plasmodesmata connect this type of companion cell to any of the surrounding cells except its own sieve element. As a result, the symplast of the sieve element and its transfer cell is relatively, if not entirely, symplastically isolated from that of surrounding cells. Xylem parenchyma cells can also be modified as transfer cells, serving to retrieve and reroute solutes moving in the xylem, which is also part of the apoplast. Transfer cells occur most frequently at nodes in path phloem, as well as in source phloem and in post–sieve element unloading pathways.

Although transfer cells (and some ordinary companion cells) are relatively isolated symplastically from surrounding cells, there are some plasmodesmata in the walls of these cells. The function of these plasmodesmata is not known. The fact that they are present indicates that they must have a function, and an important one, since the cost of having them is high: they are the avenues by which viruses become systemic in the plant. They are, however, difficult to study because they are so inaccessible.

In contrast to transfer cells, **intermediary cells** appear well suited for taking up solutes via cytoplasmic connections (**Figure 11.7C**). Intermediary cells have numerous plasmodesmata connecting them to bundle sheath cells. Although the presence of many plasmodesmatal connections to surrounding cells is their most characteristic feature, intermediary cells are also distinctive in having numerous small vacuoles, as well as poorly developed thylakoids and a lack of starch grains in the chloroplasts.

(A)

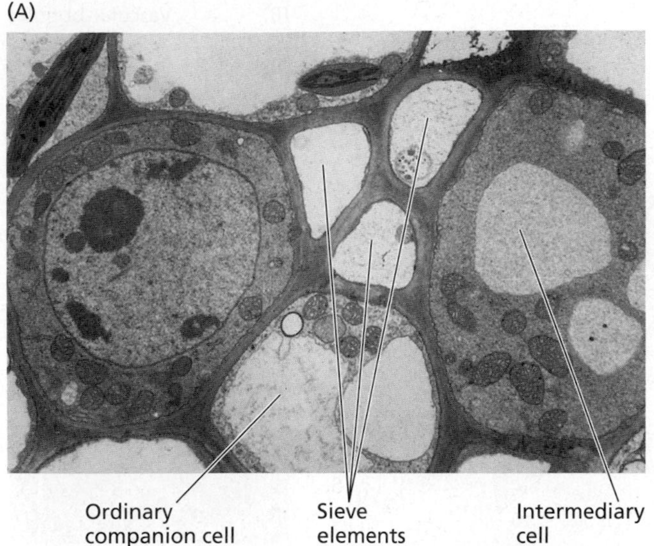

(B)

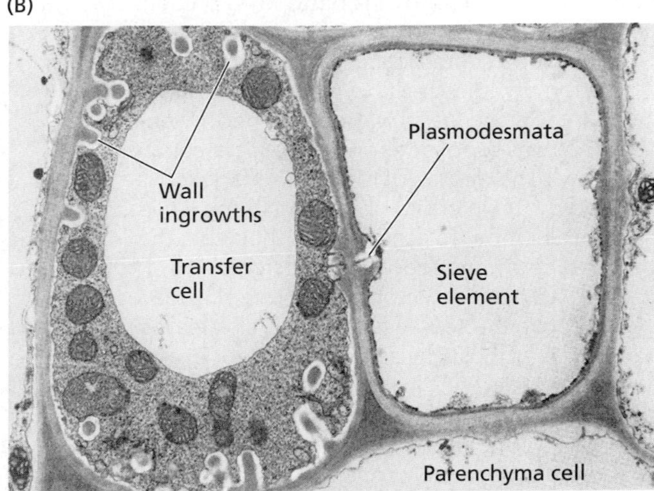

(C)

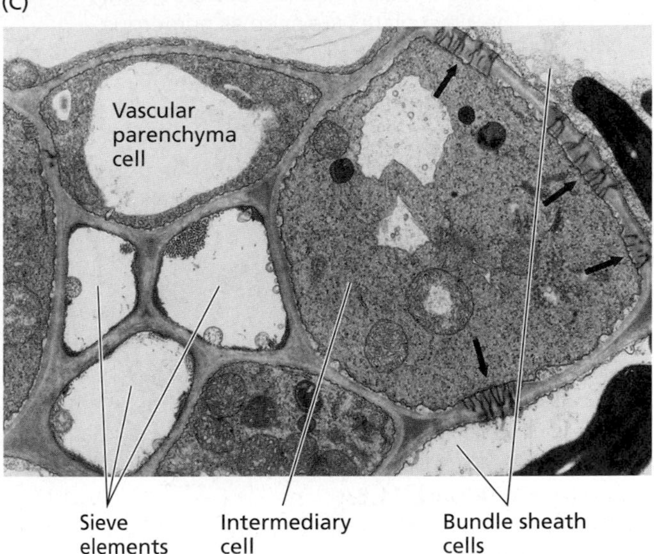

Figure 11.7 Electron micrographs of companion cells in minor veins of mature leaves. (A) Three sieve elements abut two intermediary cells and a less dense ordinary companion cell in a minor vein from scarlet monkey flower (*Mimulus cardinalis*). (6585×) (B) A sieve element adjacent to a transfer cell with numerous wall ingrowths in pea (*Pisum sativum*). (8020×) Such ingrowths greatly increase the surface area of the transfer cell's plasma membrane, thus increasing the transfer of materials from the mesophyll to the sieve elements. (C) A typical intermediary cell with numerous fields of plasmodesmata (arrows) connecting it to neighboring bundle sheath cells. These plasmodesmata are branched on both sides, but the branches are longer and narrower on the intermediary cell side. Minor-vein phloem was taken from heartleaf maskflower (*Alonsoa warscewiczii*). (4700×) (A and C from Turgeon et al. 1993, courtesy of R. Turgeon; B from Brentwood and Cronshaw 1978.)

In general, transfer cells are found in plants where transport sugars enter the apoplast during the movement of sugars from mesophyll cells to sieve elements. Transfer cells transport sugars from the apoplast to the symplast of the sieve elements and companion cells in the source. Intermediary cells, by contrast, function in symplastic transport of sugars from mesophyll cells to sieve elements. Ordinary companion cells can function in either symplastic or apoplastic short-distance transport in source leaves, depending in part on plasmodesmatal frequencies. (See the section *Phloem Loading*.)

Patterns of Translocation: Source to Sink

Sap in the phloem is not translocated exclusively in either an upward or a downward direction, and translocation in the phloem is not defined with respect to gravity. Rather, sap is translocated from areas of supply, called *sources*, to areas of metabolism or storage, called *sinks*. Because of their roles in sugar transport, the sieve elements of sources are often referred to as **collection phloem**, the sieve elements of the connecting pathway as *transport phloem*, and the sieve elements of sinks as **release phloem**.

Sources include exporting organs, typically mature leaves that are capable of producing photosynthate in excess of their own needs. The term **photosynthate** refers to products of photosynthesis. Another type of source is a storage organ during the exporting phase of its development. For example, the storage root of the biennial wild beet (*Beta maritima*) is a sink during the growing season of the first year, when it accumulates sugars received from the source leaves. During the second growing season, the same root becomes a source; the sugars are remobilized and used to produce a new shoot, which ultimately becomes reproductive.

Sinks include all nonphotosynthetic organs of the plant and organs that do not produce enough photosynthetic

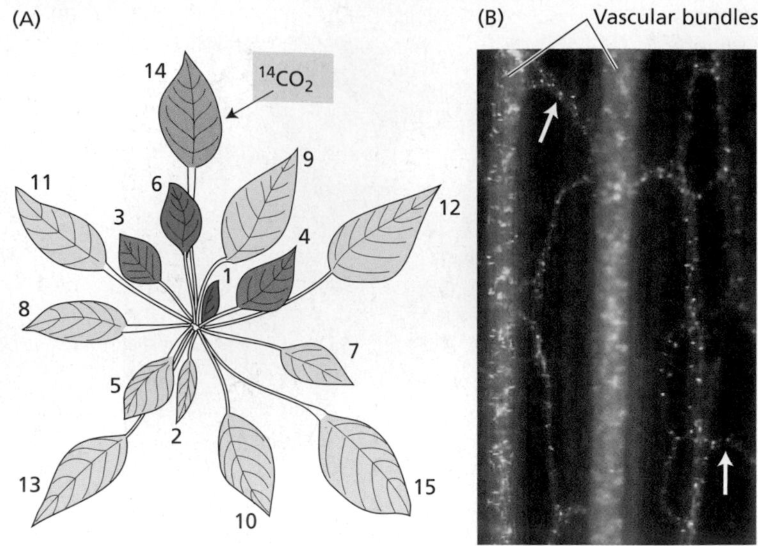

Figure 11.8 Source-to-sink patterns of phloem translocation. (A) Distribution of radioactivity from a single labeled source leaf in an intact plant. The distribution of radioactivity in leaves of a sugar beet plant (*Beta vulgaris*) was determined 1 week after $^{14}CO_2$ was supplied for 4 h to a single source leaf (arrow). The degree of radioactive labeling is indicated by the intensity of shading of the leaves. Leaves are numbered according to their age; the youngest, newly emerged leaf is designated 1. The ^{14}C label was translocated mainly to the sink leaves directly above the source leaf (that is, sink leaves on the same orthostichy as the source; for example, leaves 1 and 6 are sink leaves directly above source leaf 14). (B) Longitudinal view of a typical three-dimensional structure of the phloem in a thick section (from an internode of dahlia [*Dahlia pinnata*]), viewed here after clearing, staining with aniline blue, and observing under an epifluorescence microscope. The sieve plates are seen as numerous small dots because of the yellow staining of callose in the sieve areas. Two large longitudinal vascular bundles are prominent. This staining reveals the delicate sieve tubes forming the phloem network; two phloem anastomoses (vascular interconnections) are marked by arrows. (A based on data from Joy 1964; B courtesy of R. Aloni.)

products to support their own growth or storage needs. Roots, tubers, developing fruits, and immature leaves, which must import carbohydrate for normal development, are all examples of sink tissues. Both girdling and labeling studies support the source-to-sink pattern of translocation in the phloem (**Figure 11.8A**).

Although the overall pattern of transport in the phloem can be stated simply as source-to-sink movement, the specific pathways involved are often more complex, depending on proximity, development, vascular connections (**Figure 11.8B**), and modification of translocation pathways. Not all sources supply all sinks on a plant; rather, certain sources preferentially supply specific sinks (see WEB TOPIC 11.1).

Materials Translocated in the Phloem

Water is the most abundant substance in the phloem. Dissolved in the water are the translocated solutes, including carbohydrates, amino acids, hormones, some inorganic ions, RNAs and proteins, and some secondary compounds involved in defense and protection. Carbohydrates are the most significant and concentrated solutes in phloem sap (**Table 11.2**), with sucrose being the sugar most commonly transported in sieve elements. There is always some sucrose in sieve element sap, and it can reach concentrations of 0.3 to 0.9 *M*. Sugars, potassium ions, and amino acids and their amides are the principal molecules contributing to the osmotic potential of the phloem.

Complete identification of solutes that are mobile in the phloem and that have a significant function has been difficult; no one method of sampling phloem sap is free of artifacts or provides a complete picture of mobile solutes. We will begin this discussion with a brief examination of the available sampling methods, and then continue with a description of the solutes that are currently accepted as significant mobile substances in the phloem.

Phloem sap can be collected and analyzed

The collection of phloem sap is experimentally challenging because of the high turgor pressure in the sieve elements and the wound reactions described previously (see *Damaged sieve elements are sealed off* above and WEB TOPIC 11.3). Because of processes that plug sieve plate pores, only a few species exude phloem sap from wounds that sever sieve elements. Considerable challenges and problems

TABLE 11.2 The composition of phloem sap from castor bean (*Ricinus communis*), collected as an exudate from cuts in the phloem

Component	Concentration (mg mL^{-1})
Sugars	80.0–106.0
Amino acids	5.2
Organic acids	2.0–3.2
Protein	1.45–2.20
Potassium	2.3–4.4
Chloride	0.355–0.675
Phosphate	0.350–0.550
Magnesium	0.109–0.122

Source: Hall and Baker 1972.

present themselves when exuded sap is collected from cuts or wounds:

- The initial samples may be contaminated by the contents of surrounding damaged cells.
- In addition to plugging the sieve plate pores, sudden pressure release in sieve elements can disrupt cellular organelles and proteins and even pull substances from surrounding cells, especially the companion cells. Some materials, such as the ribulose bisphosphate carboxylase small subunit, are expected to be present only in tissues surrounding the phloem; failure to detect these materials in collected sap provides evidence that contamination from surrounding tissues has not occurred.
- The exudate is substantially diluted by the influx of water from the xylem and surrounding cells when the pressure/tension in the vascular tissue is released.
- Cucurbit sap has been used in many studies of translocated materials. Cucurbit species such as cucumber (*Cucumis sativa*) and pumpkin (*Cucurbita maxima*) have a complex phloem, including both internal and external sieve tubes (see the section *Pathways of Translocation* above), as well as sieve tubes outside the vascular bundles. In addition to the concerns listed above, the source of exudate in these species could be any of the sieve tubes present and could differ among species.

Exudation of sap from cut petioles or stems, enhanced by the inclusion of EDTA in the collection fluid, has also been used in several studies. Chelating agents such as EDTA bind calcium ions, thus inhibiting callose synthesis (which requires calcium ions) and allowing exudation to occur for extended periods. However, exudation into EDTA is subject to several additional technical problems, such as the leakage of solutes, including hexoses, from the affected tissues and is not a reliable method of obtaining phloem sap for analysis.

A preferable approach for collecting exuded sap is to use an aphid stylet as a "natural syringe." Aphids are small insects that feed by inserting their mouthparts, consisting of four tubular stylets, into a single sieve element of a leaf or stem. Sap can be collected from aphid stylets cut from the body of the insect, usually with a laser, after the aphid has been anesthetized with CO_2. The high turgor pressure in the sieve element forces the cell contents through the stylet to the cut end, where they can be collected. However, quantities of collected sap are small, and the method is technically difficult. Furthermore, exudation from severed stylets can continue for hours, suggesting that substances in aphid saliva prevent the plant's normal sealing mechanisms from operating and potentially altering the sap contents. Nonetheless, this method is thought to yield relatively pure sap from the sieve elements and companion cells and to provide a fairly accurate picture of the composition of phloem sap (see **WEB TOPIC 11.3**).

Sugars are translocated in a nonreducing form

Results from many analyses of collected sap indicate that the translocated carbohydrates are nonreducing sugars. Reducing sugars, such as the hexoses glucose and fructose, contain an exposed aldehyde or ketone group (**Figure 11.9A**). In a nonreducing sugar, such as sucrose, the ketone or aldehyde group is reduced to an alcohol or combined with a similar group on another sugar (**Figure 11.9B**).

Most researchers believe that the nonreducing sugars are the major compounds translocated in the phloem because they are less reactive than their reducing counterparts. In fact, reducing sugars such as hexoses are quite reactive and may be as much of a threat as reactive oxygen and nitrogen species. Animals can tolerate transporting glucose because it is present in fairly low concentrations in the blood, but hexoses cannot be tolerated in the phloem, in which very high sugar levels are maintained. Mechanistically, hexoses are sequestered in the vacuoles of plant cells and so have no direct access to the phloem.

Sucrose is the most commonly translocated sugar; many of the other mobile carbohydrates contain sucrose bound to varying numbers of galactose molecules. Raffinose consists of sucrose and one galactose molecule, stachyose consists of sucrose and two galactose molecules, and verbascose consists of sucrose and three galactose molecules (see Figure 11.9B). Translocated sugar alcohols include mannitol and sorbitol.

Other solutes are translocated in the phloem

Nitrogen is found in the phloem largely in amino acids—especially glutamate and aspartate—and their respective amides, glutamine and asparagine. Reported levels of amino acids and organic acids vary widely, even for the same species, but they are usually low compared with carbohydrates. (See **WEB TOPIC 11.4** for more information on nitrogen transport in the phloem.) A variety of proteins and RNAs occur in phloem sap in relatively low concentrations. RNAs found in the phloem include mRNAs, pathogenic RNAs, and small regulatory RNA molecules.

Almost all the endogenous plant hormones, including auxin, gibberellins, cytokinins, and abscisic acid have been found in sieve elements. The long-distance transport of hormones, especially auxin, is thought to occur at least partly in the sieve elements. Nucleotide phosphates have also been found in phloem sap.

Some inorganic ions move in the phloem, including potassium, magnesium, phosphate, and chloride (see Table 11.2). In contrast, nitrate, calcium, sulfur, and iron are relatively immobile in the phloem.

Proteins found in the phloem include structural P-proteins such as PP1 and PP2 (involved in the sealing of wounded sieve elements in cucurbit species), as well as a number of water-soluble proteins. The function of many of the proteins commonly found in phloem sap is related to

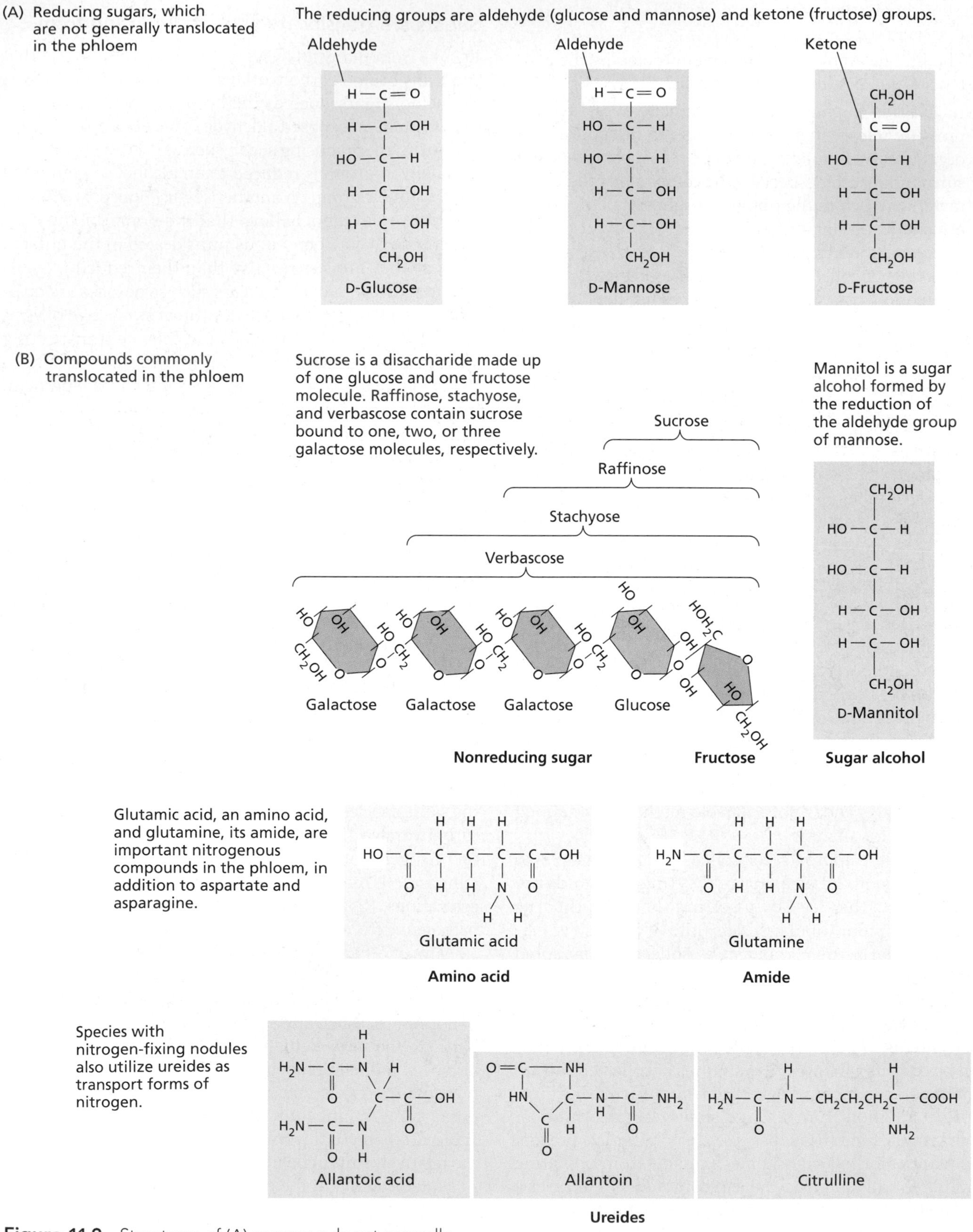

(A) Reducing sugars, which are not generally translocated in the phloem

The reducing groups are aldehyde (glucose and mannose) and ketone (fructose) groups.

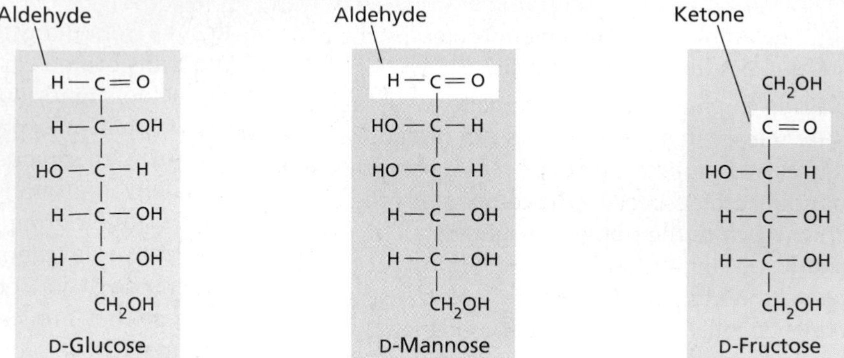

(B) Compounds commonly translocated in the phloem

Sucrose is a disaccharide made up of one glucose and one fructose molecule. Raffinose, stachyose, and verbascose contain sucrose bound to one, two, or three galactose molecules, respectively.

Sucrose

Raffinose

Stachyose

Verbascose

Galactose Galactose Galactose Glucose

Nonreducing sugar

Fructose

Mannitol is a sugar alcohol formed by the reduction of the aldehyde group of mannose.

D-Mannitol

Sugar alcohol

Glutamic acid, an amino acid, and glutamine, its amide, are important nitrogenous compounds in the phloem, in addition to aspartate and asparagine.

Glutamic acid

Amino acid

Glutamine

Amide

Species with nitrogen-fixing nodules also utilize ureides as transport forms of nitrogen.

Allantoic acid

Allantoin

Citrulline

Ureides

Figure 11.9 Structures of (A) compounds not normally translocated in the phloem and (B) compounds commonly translocated in the phloem.

stress and defense reactions (see the table in **WEB TOPIC 11.12**). The possible roles of RNAs and proteins as signal molecules will be further discussed at the end of the chapter.

Rates of Movement

It should be noted that in early publications reporting on rates of transport in the phloem, the units of velocity were centimeters per hour (cm h^{-1}), and the units of mass transfer were grams per hour per square centimeter (g h^{-1} cm^{-2}) of phloem or sieve elements. However, the currently preferred units (SI units) are meters (m) or millimeters (mm) for length, seconds (s) for time, and kilograms (kg) for mass. Rates from earlier works have been converted to SI units and are found in parentheses below.

The rate of movement of materials in the sieve elements can be expressed in two ways: as **velocity**, the linear distance traveled per unit time, or as **mass transfer rate**, the quantity of material passing through a given cross section of phloem or sieve elements per unit time. Mass transfer rates based on the cross-sectional area of the sieve elements are preferred because the sieve elements are the conducting cells of the phloem. Values for mass transfer rate range from 1 to 15 g h^{-1} cm^{-2} of sieve elements (in SI units, 2.8–41.7 µg s^{-1} mm^{-2}) (see **WEB TOPIC 11.5**).

Both velocities and mass transfer rates can be measured with radioactive tracers. (Methods of measuring mass transfer rates are described in **WEB TOPIC 11.5**.) In the simplest type of experiment for measuring velocity, ^{11}C- or ^{14}C-labeled CO_2 is applied for a brief period of time to a source leaf (pulse labeling), and the arrival of label at a sink tissue or at a particular point along the pathway is monitored with an appropriate detector.

In general, velocities measured by a variety of conventional techniques average about 1.0 m h^{-1} (0.28 mm sec^{-1}) and range from 0.3 to 1.5 m h^{-1} (in SI units, 0.08–0.42 mm s^{-1}). More recent measurements of velocity using NMR spectrometry and magnetic resonance imaging yielded an average velocity for castor bean of 0.25 mm s^{-1}, which is remarkably close to the average obtained using older methods. Transport velocities in the phloem are clearly quite high and exceed the rate of diffusion by many orders of magnitude. Any proposed mechanism of phloem translocation must account for these high velocities.

The Pressure-Flow Model, a Passive Mechanism for Phloem Transport

The most widely accepted mechanism of phloem translocation in angiosperms is the pressure-flow model. The pressure-flow model explains phloem translocation as a flow of solution (mass flow or bulk flow) driven by an osmotically generated pressure gradient between source and sink. This section will describe the pressure-flow model, predictions arising from mass flow, and data, both supporting and challenging. At the end of the section we will briefly explore the question of whether the model can possibly be applied to gymnosperms.

In early research on phloem translocation, both active and passive mechanisms were considered. All theories, both active and passive, assume an energy requirement in both sources and sinks. In sources, energy is necessary to synthesize the materials for transport and, in some cases, to move photosynthate into the sieve elements by active membrane transport. The movement of photosynthate into the sieve elements is called *phloem loading*, and we will discuss it in detail later in the chapter. In sinks, energy is essential for some aspects of movement from sieve elements to sink cells, which store or metabolize the sugar. This movement of photosynthate from sieve elements to sink cells is called *phloem unloading* and will also be discussed later.

The passive mechanisms of phloem transport further assume that energy is required in the sieve elements of the path between sources and sinks simply to maintain structures such as the cell plasma membrane and to recover sugars lost from the phloem by leakage. The pressure-flow model is an example of a passive mechanism. The active theories, by contrast, postulate an additional expenditure of energy by path sieve elements in order to drive translocation itself. While the active theories have largely been discounted, interest in certain aspects of these models may be revived, based on observations of pressures present in large plants, such as trees. (See the discussion below *Pressure gradients in the sieve elements may be modest*.)

An osmotically generated pressure gradient drives translocation in the pressure-flow model

Diffusion is far too slow to account for the velocities of solute movement observed in the phloem. Translocation velocities average 1 m h^{-1}; the rate of diffusion would be 1 m per 32 years! (See Chapter 3 for a discussion of diffusion velocities and the distances over which diffusion is an effective transport mechanism.)

The **pressure-flow model**, first proposed by Ernst Münch in 1930, states that a flow of solution in the sieve elements is driven by an osmotically generated *pressure gradient* between source and sink (Ψ_p). Phloem loading at the source and phloem unloading at the sink establish the pressure gradient.

As we will see later (see the section *Phloem loading can occur via the apoplast or the symplast*), three different mechanisms exist to generate high concentrations of sugars in the sieve elements of the source: photosynthetic metabolism in the mesophyll, conversion of photoassimilate to transport sugars in intermediary cells (polymer trapping), and active membrane transport. Recall from Chapter 3 (Equation 3.5) that $\Psi = \Psi_s + \Psi_p$; that is, $\Psi_p = \Psi - \Psi_s$. In source tissues, an accumulation of sugars in the sieve elements generates a low (negative) solute potential (Ψ_s) and causes a steep drop in the water potential (Ψ). In response to the water potential gradient, water enters the sieve elements and causes the turgor pressure (Ψ_p) to increase.

At the receiving end of the translocation pathway, **phloem unloading** leads to a lower sugar concentration in the sieve elements, generating a higher (less negative) solute potential in the sieve elements of sink tissues. As the water potential of the phloem rises above that of the xylem, water tends to leave the phloem in response to the water potential gradient, causing a decrease in turgor pressure in the sieve elements of the sink. **Figure 11.10** illustrates the pressure-flow hypothesis; the figure specifically shows the case in which active membrane transport from the apoplast generates a high sugar concentration in the source sieve elements.

The phloem sap moves by mass flow rather than by osmosis. That is, no membranes are crossed during transport from one sieve tube to another, and solutes move at the same rate as the water molecules. Since this is the case, mass flow can occur from a source organ with a lower water potential to a sink organ with a higher water potential, or vice versa, depending on the identities of the source and sink organs. In fact, Figure 11.10 illustrates an example in which the flow is against the water potential gradient. Such water movement does not violate the laws of thermodynamics, because it is an example of mass flow, which is driven by a pressure gradient, as opposed to osmosis, which is driven by a water potential gradient.

According to the pressure-flow model, movement in the translocation pathway is driven by transport of solutes and water into source sieve elements and out of sink sieve elements. The passive, pressure-driven, long-distance translocation in the sieve tubes ultimately depends on the mechanisms involved in phloem loading and unloading. These mechanisms are responsible for setting up the pressure gradient.

Some predictions of pressure flow have been confirmed, while others require further experimentation

Some important predictions emerge from the model of phloem translocation as mass flow described above:

- No true bidirectional transport (i.e., simultaneous transport in both directions) in a single sieve element can

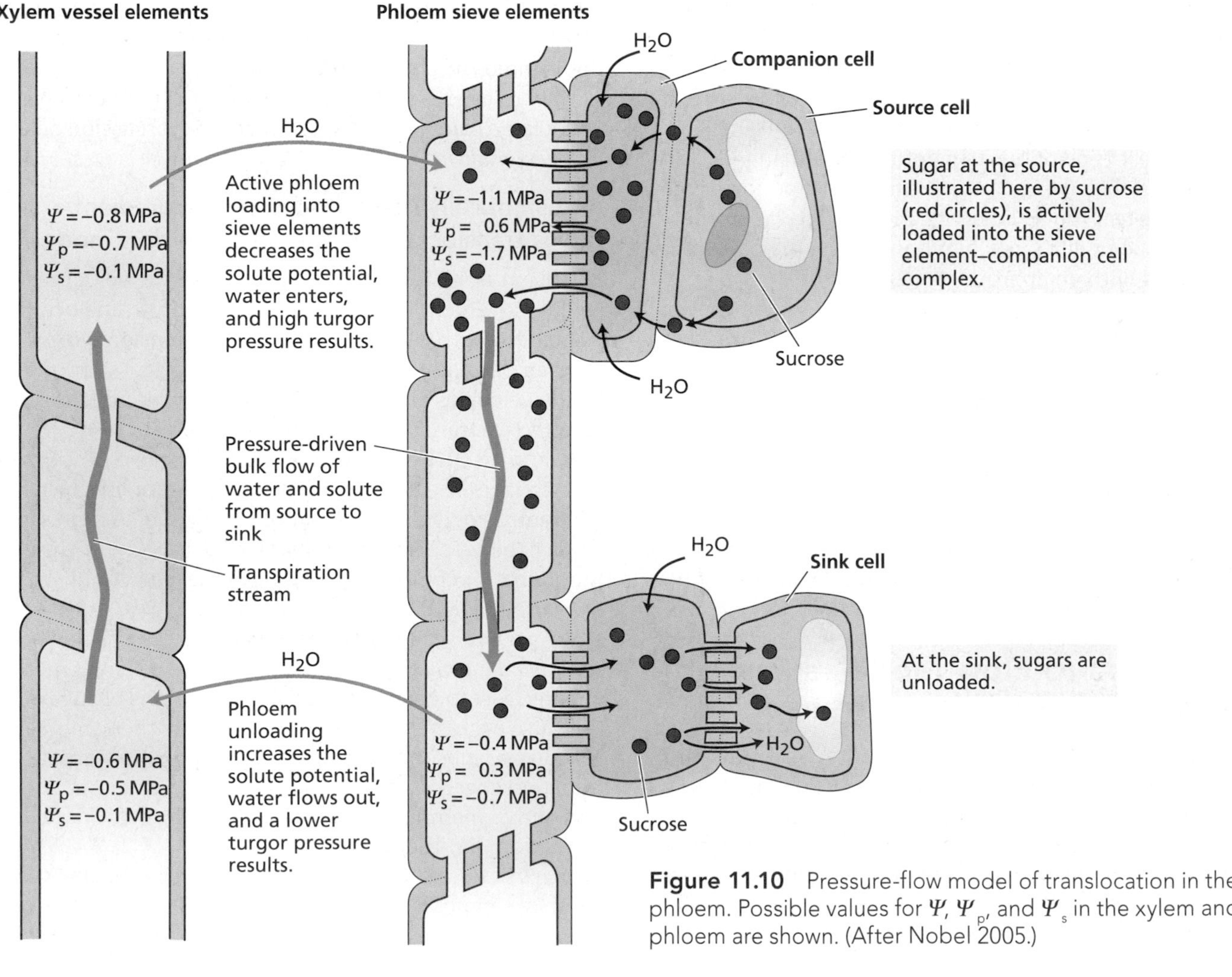
Xylem vessel elements
Phloem sieve elements
H2O
Companion cell
Source cell
Sugar at the source, illustrated here by sucrose (red circles), is actively loaded into the sieve element–companion cell complex.
Ψ = −0.8 MPa
Ψp = −0.7 MPa
Ψs = −0.1 MPa
H2O
Active phloem loading into sieve elements decreases the solute potential, water enters, and high turgor pressure results.
Ψ = −1.1 MPa
Ψp = 0.6 MPa
Ψs = −1.7 MPa
Sucrose
H2O
Pressure-driven bulk flow of water and solute from source to sink
Transpiration stream
H2O
Sink cell
H2O
At the sink, sugars are unloaded.
Phloem unloading increases the solute potential, water flows out, and a lower turgor pressure results.
Ψ = −0.6 MPa
Ψp = −0.5 MPa
Ψs = −0.1 MPa
Ψ = −0.4 MPa
Ψp = 0.3 MPa
Ψs = −0.7 MPa
H2O
Sucrose

Figure 11.10 Pressure-flow model of translocation in the phloem. Possible values for Ψ, Ψ_p, and Ψ_s in the xylem and phloem are shown. (After Nobel 2005.)

occur. A mass flow of solution precludes such bidirectional movement because a solution can flow in only one direction in a pipe at any one time. Solutes within the phloem can move bidirectionally but in different sieve elements or at different times. Furthermore, water and solutes must move at the same velocity in a flowing solution.

- No great expenditures of energy are required in order to drive translocation in the tissues along the path. Therefore, treatments that restrict the supply of ATP in the path, such as low temperature, anoxia, and metabolic inhibitors, should not stop translocation. However, energy *is* required to maintain the structure of the sieve elements, to reload and retrieve any sugars lost to the apoplast by leakage, and perhaps to reload sugars at the termination of sieve tubes.
- The sieve tube lumen and the sieve plate pores must be largely unobstructed. If P-protein or other materials block the pores, the resistance to flow of the sieve element sap might be too great.
- The pressure-flow hypothesis predicts the presence of a positive pressure gradient, with turgor pressure higher in sieve elements of sources than in those of sinks. According to the traditional picture of mass flow, the pressure difference must be large enough to overcome the resistance of the pathway and to maintain flow at the observed velocities. Thus, pressure gradients should be larger in long transport pathways, for example, in trees, than in short transport pathways, as in herbaceous plants.

The available evidence testing these predictions is presented below.

There is no bidirectional transport in single sieve elements, and solutes and water move at the same velocity

Researchers have investigated bidirectional transport by applying two different radiotracers to two source leaves, one above the other. Each leaf receives one of the tracers, and a point between the two sources is monitored for the presence of both tracers.

Transport in two directions has often been detected in sieve elements of different vascular bundles in stems. Transport in two directions has also been seen in adjacent sieve elements of the same bundle in petioles. Bidirectional transport in adjacent sieve elements can occur in the petiole of a leaf that is undergoing the transition from sink to source and simultaneously importing and exporting photosynthates through its petiole. However, simultaneous bidirectional transport in a single sieve element has never been demonstrated.

Measured velocities for transport in the phloem are remarkably similar, whether measured using carbon-labeled solutes or using NMR techniques, which detect water flow. Solutes and water move at the same velocity.

Both of these observations—the lack of bidirectional transport in a single sieve element and similar velocities for solutes and water—support the existence of mass flow in the sieve elements of the phloem.

The energy requirement for transport through the phloem pathway is small in herbaceous plants

In herbaceous plants that can survive periods of low temperature, such as sugar beet (*Beta vulgaris*), rapidly chilling a short segment of the petiole of a source leaf to approximately 1°C does not cause sustained inhibition of mass transport out of the leaf (**Figure 11.11**). Rather, there is a brief period of inhibition (minutes to a few hours), after which transport slowly returns to the control rate. Chilling reduces respiration rate and both the synthesis and the consumption of ATP in the petiole by about 90%, at a time when translocation has recovered and is proceeding normally. These experiments show that the energy requirement for actual transport through the pathway of these herbaceous plants is small, consistent with mass flow. Many of the effects of chilling treatments have, in fact, been attributed to loss and retrieval mechanisms along the path, rather than to the transport mechanism itself.

Chilling experiments in large plants, such as trees, generally extend over longer time periods (days to a few weeks). Chilling of the stem in such experiments often inhibits phloem transport over the treatment period. However, the methods used to evaluate transport, such as radial growth rates below the treatment zone or soil

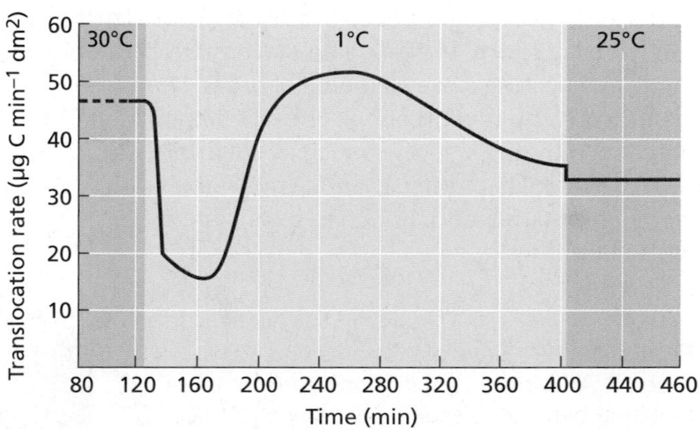

Figure 11.11 The energy requirement for translocation in the path is small in herbaceous plants. Loss of metabolic energy resulting from the chilling of a source leaf petiole partially reduces the rate of translocation in sugar beet. However, translocation rates recover with time despite the fact that ATP production and use are still largely inhibited by chilling. $^{14}CO_2$ was supplied to a source leaf, and a 2-cm portion of its petiole was chilled to 1°C. Translocation was monitored by the arrival of ^{14}C at a sink leaf. (1 dm [decimeter] = 0.1 m) (After Geiger and Sovonick 1975.)

CO_2 efflux, do not permit short-term, transient changes in transport to be observed.

It should be noted that extreme treatments that inhibit all energy metabolism do inhibit translocation even in herbaceous plants. For example, in bean, treating the petiole of a source leaf with a metabolic inhibitor (cyanide) inhibited translocation out of the leaf. However, examination of the treated tissue by electron microscopy revealed blockage of the sieve plate pores by cellular debris. Clearly, these results do not bear on the question of whether energy is required for translocation along the pathway.

Sieve plate pores appear to be open channels

Ultrastructural studies of sieve elements are challenging because of the high internal pressure in these cells. When the phloem is excised or killed slowly with chemical fixatives, the turgor pressure in the sieve elements is released. The contents of the cells, particularly P-protein, surge toward the point of pressure release and, in the case of sieve tube elements, accumulate on the sieve plates. This accumulation is probably the reason that many earlier electron micrographs show sieve plates that are obstructed.

Newer, rapid freezing and fixation techniques provide reliable pictures of undisturbed sieve elements. The use of confocal laser scanning microscopy, which allows for the direct observation of translocation through living sieve elements, addresses the additional question of whether the sieve plate pores and sieve element lumen are open in intact, tranlocating tissues.

When young Arabidopsis plants are rapidly frozen by plunging them in slush nitrogen, then freeze-substituted and fixed, sieve plate pores are often unobstructed in the tissue (**Figure 11.12A**). The sieve plate pores of living, translocating sieve elements of broad bean were also observed to be mostly open. The open condition of the pores seen in many species, such as cucurbits, sugar beet, bean (*Phaseolus vulgaris*), and Arabidopsis (see Figures 11.5 and 11.12A), is consistent with mass flow.

What about the distribution of P-protein in the sieve tube lumen? Electron micrographs of sieve tube members prepared by rapid freezing and fixation have often shown P-protein along the periphery of the sieve tube members or evenly distributed throughout the lumen of the cell. Furthermore, the sieve plate pores often contain P-protein in similar positions, lining the pore or in a loose network.

When a sieve element occlusion related protein (SEOR1 in Arabidopsis) was fused to yellow fluorescent protein (YFP) and observed with confocal microscopy, however, a somewhat different picture emerged. While a meshwork of protein filaments was often shown to extend throughout the lumen (**Figure 11.12B**), masses or agglomerates of protein were frequently observed to fill large portions of the sieve tube lumen at or close to the sieve plate. The structure of these masses was highly variable, but multiple large masses sometimes filled the entire lumen of the sieve tube (**Figure 11.12C**). These structures were observed in sieve elements of intact, living, translocating sieve elements. The researchers concluded that mass flow is still possible in Arabidopsis; however, knowledge of the porosity of the protein masses, as well as the degree of interaction of the protein with surrounding water molecules, will be needed to fully assess the impact of SEOR1 in Arabidopsis.

Pressure gradients in the sieve elements may be modest; pressures in herbaceous plants and trees appear to be similar

Mass flow or bulk flow is the combined movement of all the molecules in a solution, driven by a pressure gradient. What are the pressure values in sieve elements, and how can they be determined? Does a pressure gradient exist between sources and sinks, and if so, is the gradient modest or substantial? Do large plants, such as trees, have proportionally higher pressures in the phloem than small, herbaceous species?

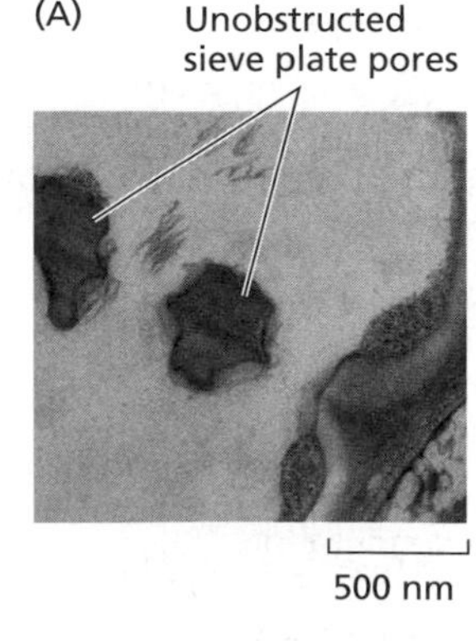

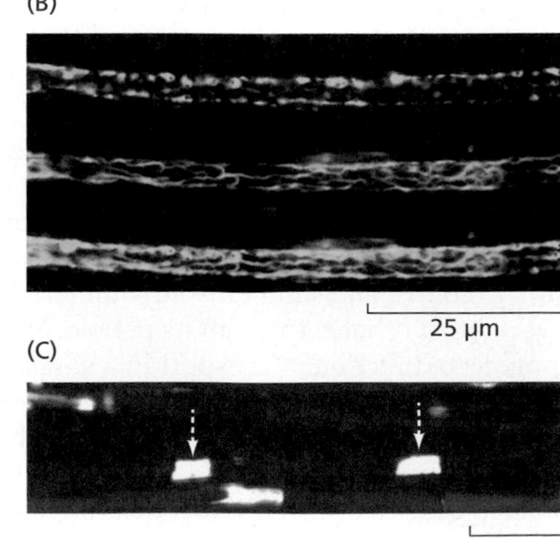

Figure 11.12 Sieve plate pores and sieve tubes in Arabidopis. (A) In plunge-frozen and freeze-substituted tissues, sieve plate pores are often unobstructed and do not contain any detectable callose. (B) Living root sieve tubes observed with confocal microscopy show the endoplasmic reticulum (green) surrounded by a fine SEOR1–yellow fluorescent protein (YFP) filament meshwork (cyan). (C) Masses or agglomerates of SEOR1–YFP protein sometimes fill the entire lumen of the sieve tube in confocal images; see dotted arrows. The sieve tubes in both (B) and (C) were living and functional. (From Froelich et al. 2011.)

Turgor pressure in sieve elements can either be calculated from the water potential and solute potential ($\Psi_p = \Psi - \Psi_s$) or measured directly. The most effective technique uses micromanometers or pressure transducers sealed over exuding aphid stylets (see **WEB TOPIC 11.3**). The data obtained are accurate because aphids pierce only a single sieve element, and the plasma membrane apparently seals well around the aphid stylet. Pressures measured using the aphid stylet technique range from approximately 0.7 to 1.5 MPa in both herbaceous plants and small trees.

Studies using calculated turgor pressures have detected pressure gradients sufficent to drive mass flow in a few herbaceous plants such as soybean. However, no systematic studies have been made of turgor gradients measured using aphid stylets in any plant. The data are critical to any evaluation of the pressure-flow hypothesis. Ideally, techniques that can measure turgor differences along the same continuous sieve tube, both in herbaceous plants and in large plants such as trees, must be developed. This will be an enormous technical challenge.

One observation is fairly certain, however, and that is that turgor pressures in trees are not proportionally higher than those in herbaceous plants. One study compared calculated turgor pressures (the technique often used in trees) and pressures measured using aphid stylets (a technique used in herbaceous plants) in small willow saplings. The two techniques yielded comparable values, averaging 0.6 MPa for the calculated pressures and 0.8 MPa for the measured pressures. Calculated pressures were as high as 2.0 MPa in large white ash trees. These values are not substantially different from those measured in herbaceous plants, as noted above. (Herbaceous plants and trees do often differ in their phloem loading strategies, in a way that is consistent with the relatively low pressures in trees; see the section *Phloem loading is passive in several tree species* below.)

Alternative models for translocation by mass flow have been suggested

We should not leave the topic of mechanisms for phloem transport without consideration of alternative models. One such model is the high-pressure manifold model, which is similar to the pressure-flow model but with several key differences. In the high-pressure manifold model:

- High pressures in sieve elements are generated primarily in the source, and phloem loading capacity often exceeds unloading capacity.
- The major resistance to mass flow occurs not in the sieve tubes or sieve plates of the path, but in the plasmodesmata between sieve element–companion cell complexes and sink tissues, particularly vascular parenchyma cells.
- Bulk flow thus would extend all the way from the sieve elements of sources to the sieve elements of sinks through the plasmodesmata linking the sieve elements of the sink to the vascular parenchyma. Since the highest resistance occurs in the plasmodesmata, small pressure gradients would occur between source and sink sieve elements, but the pressure differences between sink sieve elements and phloem parenchyma cells would be large.
- The resulting system could efficiently and rapidly transmit information on changes in the pressure or concentration of sap over long distances (see **WEB TOPIC 11.6**).

Another model, called the relay model, proposes that the phloem consists of functional units joined in series and that solutes are transported actively from one unit to the next, increasing the pressure available to drive transport over long distances such as those that exist in trees. While both of these models account for some of the observations on sieve tube turgor pressures noted above, the relay model also requires energy expenditure along the path, at least in trees. It is not known if the energy requirement along the path is small in trees, as it appears to be in herbaceous plants.

Mathematical models can also provide insight into the mechanism of phloem transport. Sequential enzymatic digestion of cellular contents has allowed more accurate measurements of sieve tube parameters, such as pore radius and the number of pores per plate. The sieve tube–specific conductivity (in μm^2) calculated from these measurements shows an inverse relationship with phloem sap velocity, measured with magnetic resonance imaging. This is an unexpected result if sieve tube conductivities regulate transport. In that case, a higher conductance tube would have a lower resistance, and a given pressure would be expected to result in a higher, not a lower, sap velocity. (See **WEB TOPIC 11.6** for more information on recent mathematical models of phloem transport.)

What can we conclude from the experiments and data described here? Some observations are consistent with the operation of mass flow and specifically the pressure-flow mechanism in angiosperm phloem: the movement of solutes and water at the same velocity; the lack of an energy requirement in the pathway of herbaceous plants; the presence of open sieve plate pores; and the failure to detect bidirectional transport. The significance of other observations to pressure flow is more problematic; in particular, the presence of protein masses blocking some sieve tube members and the similar pressures in the sieve elements of herbaceous plants and trees are puzzling indeed. More data are needed—clearly we don't have the whole picture.

Does translocation in gymnosperms involve a different mechanism?

Although mass flow explains translocation in angiosperms, it may not be sufficient for gymnosperms. Very little physiological information on gymnosperm phloem is available (but see *Phloem loading is passive in several tree species* below), and speculation about translocation in these species is based almost entirely on interpretations of electron micrographs. As discussed previously, the sieve cells of gymnosperms are similar in many respects to sieve tube elements of angiosperms, but the sieve areas of sieve cells are relatively unspecialized and do not appear to consist of open pores (see Figure 11.6).

The pores in gymnosperms are filled with numerous membranes that are continuous with the smooth endoplasmic reticulum adjacent to the sieve areas. Such pores seem to be inconsistent with the requirements of mass flow. Although these electron micrographs might be artifactual and fail to show conditions in the intact tissue, translocation in gymnosperms might involve a different mechanism—a possibility that requires further investigation.

Phloem Loading

Several transport steps are involved in the movement of photosynthate from the mesophyll chloroplasts to the sieve elements of mature leaves:

1. Triose phosphate formed by photosynthesis during the day (see Chapter 8) is transported from the chloroplast to the cytosol, where it is converted to sucrose. During the night, carbon from stored starch exits the chloroplast primarily in the form of maltose and is converted to sucrose. (Other transport sugars are later synthesized from sucrose in some species, while sugar alcohols are synthesized using hexose phosphate and in some cases hexose as the starting molecules.)
2. Sucrose moves from producing cells in the mesophyll to cells in the vicinity of the sieve elements in the smallest veins of the leaf (**Figure 11.13**). This **short-distance transport** pathway usually covers a distance of only a few cell diameters.
3. In the process called **phloem loading**, sugars are transported into the sieve elements and companion cells. Note that with respect to loading, the sieve elements and companion cells are often considered a functional unit, called the *sieve element–companion cell complex.* Once inside the sieve elements, sucrose and other solutes are translocated away from the source, a process known as **export**. Translocation through the vascular system to the sink is referred to as **long-distance transport**.

As discussed earlier, the processes of loading at the source and perhaps unloading at the sink provide the driving force for long-distance transport and are thus of considerable basic, as well as agricultural, importance. A thorough understanding of these mechanisms should provide the basis for technology aimed at enhancing crop productivity by increasing the accumulation of photosynthate in edible sink tissues, such as cereal grains.

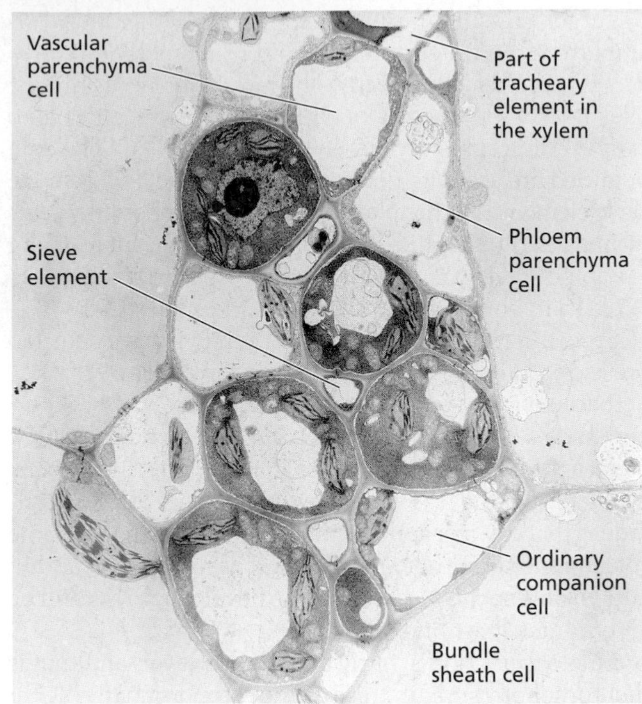

Figure 11.13 Electron micrograph showing the relationship between the various cell types of a small vein in a source leaf of sugar beet. (5000×) Photosynthetic cells (mesophyll cells) surround the compactly arranged cells of the bundle sheath layer. Photosynthate from the mesophyll must move a distance equivalent to only several cell diameters before being loaded into the sieve elements. Movement from the mesophyll to the sieve elements is thus known as short-distance transport. (From Evert and Mierzwa 1985, courtesy of R. Evert.)

Phloem loading can occur via the apoplast or symplast

We have seen that solutes (mainly sugars) in source leaves must move from the photosynthesizing cells in the mesophyll to the sieve elements. The initial short-distance pathway is probably symplastic (**Figure 11.14**). However, sugars might move entirely through the symplast (cytoplasm) to the sieve elements via the plasmodesmata (see Figure 11.14A), or they might enter the apoplast prior to phloem loading (see Figure 11.14B). (See Figure 4.4 for a general description of the symplast and apoplast.) One of the two routes, apoplastic or symplastic, is dominant in some species; many species, however, show evidence of being able to use more than one loading mechanism. For simplicity's sake, we will initially consider the pathways separately, then return to the subject of loading diversity.

Several mechanisms for phloem loading are now recognized: apoplastic loading, symplastic loading with

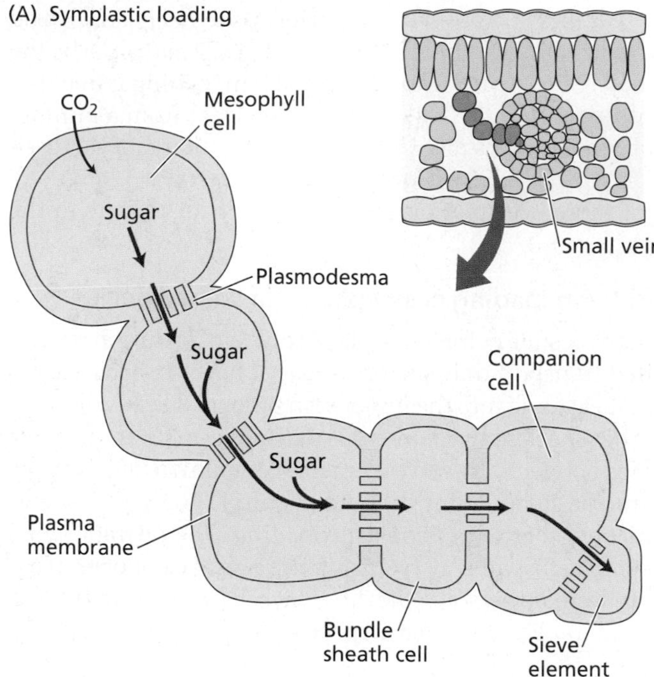

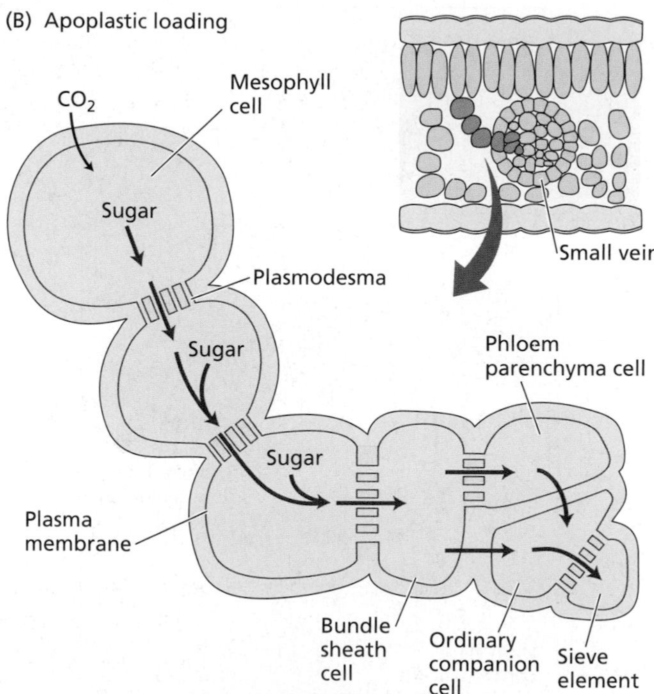

Figure 11.14 Schematic diagram of pathways of phloem loading in source leaves. (A) In the totally symplastic pathway, sugars move from one cell to another in the plasmodesmata, all the way from the mesophyll to the sieve elements. (B) In the partly apoplastic pathway, sugars initially move through the symplast but enter the apoplast just prior to loading into the companion cells and sieve elements. Sugars loaded into the companion cells are thought to move through plasmodesmata into the sieve elements.

polymer trapping, and passive symplastic loading. Early research on phloem loading focused on the apoplastic pathway, probably because it is very common in herbaceous plants and therefore crops. (In fact, much of our knowledge of plant physiology is probably slanted by its primary focus on herbaceous crops. As it turns out, the apoplastic pathway apparently is the most common mechanism.) In this section we will discuss apoplastic loading first, and then introduce the two types of symplastic loading (polymer trapping and passive symplastic loading) in the order in which their importance was recognized.

Abundant data support the existence of apoplastic loading in some species

In the case of apoplastic loading, the sugars enter the apoplast quite near the sieve element–companion cell complex. Sugars are then actively transported from the apoplast into the sieve elements and companion cells by an energy-driven, selective transporter located in the plasma membranes of these cells. Efflux into the apoplast is highly localized, probably into the walls of phloem parenchyma cells. The sucrose transporters that mediate the efflux of sucrose, most likely from the phloem parenchyma to the apoplast near the sieve element–companion cell complexes, have recently been identified in Arabidopsis and rice as a subfamily of the SWEET transporters.

Apoplastic phloem loading leads to three basic predictions:

1. Transported sugars should be found in the apoplast.
2. In experiments in which sugars are supplied to the apoplast, the exogenously supplied sugars should accumulate in sieve elements and companion cells.
3. Inhibition of sugar efflux from the phloem parenchyma or of uptake from the apoplast should result in inhibition of export from the leaf.

Many studies devoted to testing these predictions have provided solid evidence for apoplastic loading in several species (see **WEB TOPIC 11.7**).

Sucrose uptake in the apoplastic pathway requires metabolic energy

In many of the species initially studied, sugars become more concentrated in the sieve elements and companion cells than in the mesophyll. This difference in solute concentration can be demonstrated through measurement of the osmotic potential (Ψ_s) of the various cell types in the leaf (see Chapter 3).

In sugar beet, the osmotic potential of the mesophyll is approximately –1.3 MPa, and the osmotic potential of the sieve elements and companion cells is about –3.0 MPa. Most of this difference in osmotic potential is thought to result from accumulated sugar, specifically sucrose, because sucrose is the major transport sugar in this species. Experimental studies have also demonstrated that

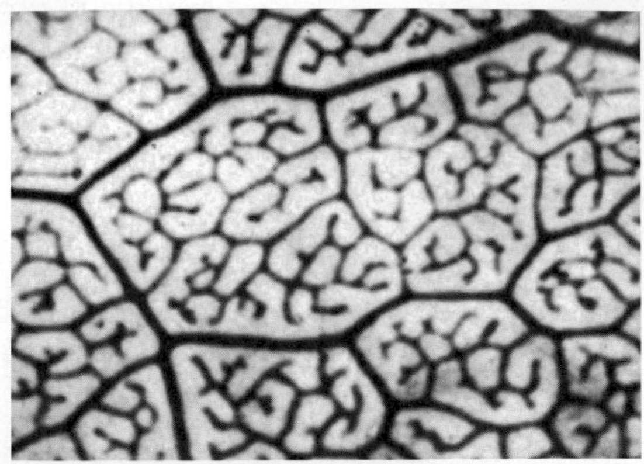

Figure 11.15 This autoradiograph shows that labeled sugar moves against its concentration gradient from the apoplast into sieve elements and companion cells of a sugar beet source leaf. A solution of ^{14}C-labeled sucrose was applied for 30 min to the upper surface of a sugar beet leaf that had previously been kept in darkness for 3 h. The leaf cuticle was removed to allow penetration of the solution to the interior of the leaf. The sieve elements and companion cells of the small veins in the source leaf contain high concentrations of labeled sugar, shown by the black accumulations, indicating that sucrose is actively transported against its concentration gradient (From Fondy 1975, courtesy of D. Geiger.)

both externally supplied sucrose and sucrose made from photosynthetic products accumulate in the sieve elements and companion cells of the minor veins of sugar beet source leaves (**Figure 11.15**) (see also **WEB TOPIC 11.7**).

The fact that sucrose is at a higher concentration in the sieve element–companion cell complex than in surrounding cells indicates that sucrose is actively transported against its chemical-potential gradient. The dependence of sucrose accumulation on active transport is supported by the fact that treating source tissue with respiratory inhibitors both decreases ATP concentration and inhibits loading of exogenous sugar.

Plants that load sugars apoplastically into the phloem may also load amino acids and sugar alcohols (sorbitol and mannitol) actively. In contrast, other metabolites, such as organic acids and hormones, may enter sieve elements passively. (See **WEB TOPIC 11.7** for a discussion of these topics.)

Phloem loading in the apoplastic pathway involves a sucrose–H⁺ symporter

A sucrose–H^+ symporter is thought to mediate the transport of sucrose from the apoplast into the sieve element–companion cell complex. Recall from Chapter 6 that symport is a secondary transport process that uses the energy generated by the proton pump (see Figure 6.10A). The energy dissipated by protons moving back into the cell is coupled to the uptake of a substrate, in this case sucrose (**Figure 11.16**).

Several sucrose–H^+ symporters have been cloned and localized in the phloem. SUT1 and SUC2 appear to be the major sucrose transporters in phloem loading into either companion cells or sieve elements. Data from a number of other studies also support the operation of a sucrose–H^+ symporter in phloem loading. (See **WEB TOPIC 11.7** for more information about sucrose transporters in the phloem.)

Phloem loading is symplastic in some species

Many results point to apoplastic phloem loading in species that transport only sucrose and that have few plasmodesmata leading into the minor vein phloem. However, many other species have numerous plasmodesmata at the interface between the sieve element–companion cell complex and the surrounding cells (see Figure 11.7C), which seems inconsistent with apoplastic loading. The operation of a symplastic pathway requiring the presence of open plasmodesmata between the different cells in the pathway has been implicated in these species.

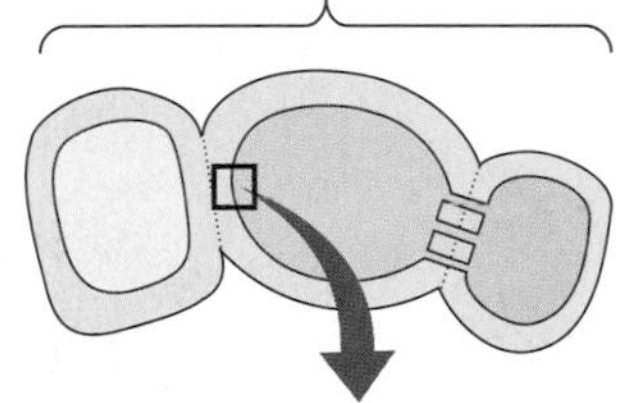

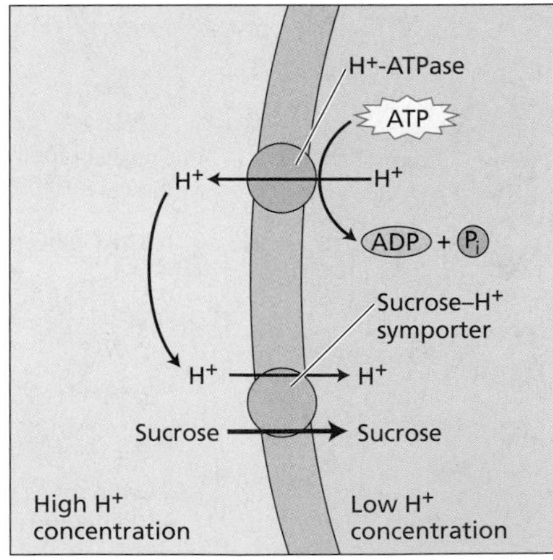

Figure 11.16 ATP-dependent sucrose transport in apoplastic sieve element loading. In the cotransport model of sucrose loading into the symplast of the sieve element–companion cell complex, the plasma membrane ATPase pumps protons out of the cell into the apoplast, establishing a higher proton concentration in the apoplast and a membrane potential of approximately –120 mV. The energy in the proton gradient is then used to drive the transport of sucrose into the symplast of the sieve element–companion cell complex through a sucrose–H^+ symporter.

The polymer-trapping model explains symplastic loading in plants with intermediary-type companion cells

A symplastic pathway has become evident in species that transport raffinose and stachyose, in addition to sucrose, in the phloem; that have intermediary cells in the minor veins; and that have abundant plasmodesmata leading into the minor veins. Some examples of such species are common coleus (*Coleus blumei*), pumpkin and squash (*Cucurbita pepo*), and melon (*Cucumis melo*). Remember that intermediary cells are specialized compantion cells; see *Companion cells aid the highly specialized sieve elements* above.

Two majors questions emerge concerning symplastic loading:

1. In many species the composition of sieve element sap is different from the solute composition in tissues surrounding the phloem. This difference indicates that certain sugars are specifically selected for transport in the source leaf. The involvement of symporters in apoplastic phloem loading provides a clear mechanism for selectivity, because symporters are specific for certain sugar molecules. Symplastic loading, in contrast, depends on the diffusion of sugars from the mesophyll to the sieve elements via the plasmodesmata. How can diffusion through plasmodesmata during symplastic loading be selective for certain sugars?
2. Data from several species showing symplastic loading indicate that sieve elements and companion cells have a higher osmolyte content (more negative osmotic potential) than the mesophyll. How can diffusion-dependent symplastic loading account for the observed selectivity for transported molecules and the accumulation of sugars against a concentration gradient?

The **polymer-trapping model** (**Figure 11.17**) has been developed to address these questions in species such as coleus and cucurbits. This model states that the sucrose synthesized in the mesophyll diffuses from the bundle sheath cells into the intermediary cells through the abundant plasmodesmata that connect the two cell types. In the intermediary cells, raffinose and stachyose (polymers made of three and four hexose sugars, respectively; see Figure 11.9B) are synthesized from the transported sucrose and from galactinol (a metabolite of galactose). Because of the anatomy of the tissue and the relatively large size of raffinose and stachyose, the polymers cannot diffuse back into the bundle sheath cells, but they can diffuse into the sieve element. Sugar concentrations in the sieve elements of these plants can reach levels equivalent to those in plants that load apoplastically. Sucrose can continue to diffuse into the intermediary cells, because its synthesis in the mesophyll and its utilization in the intermediary cells maintain the concentration gradient (see Figure 11.17).

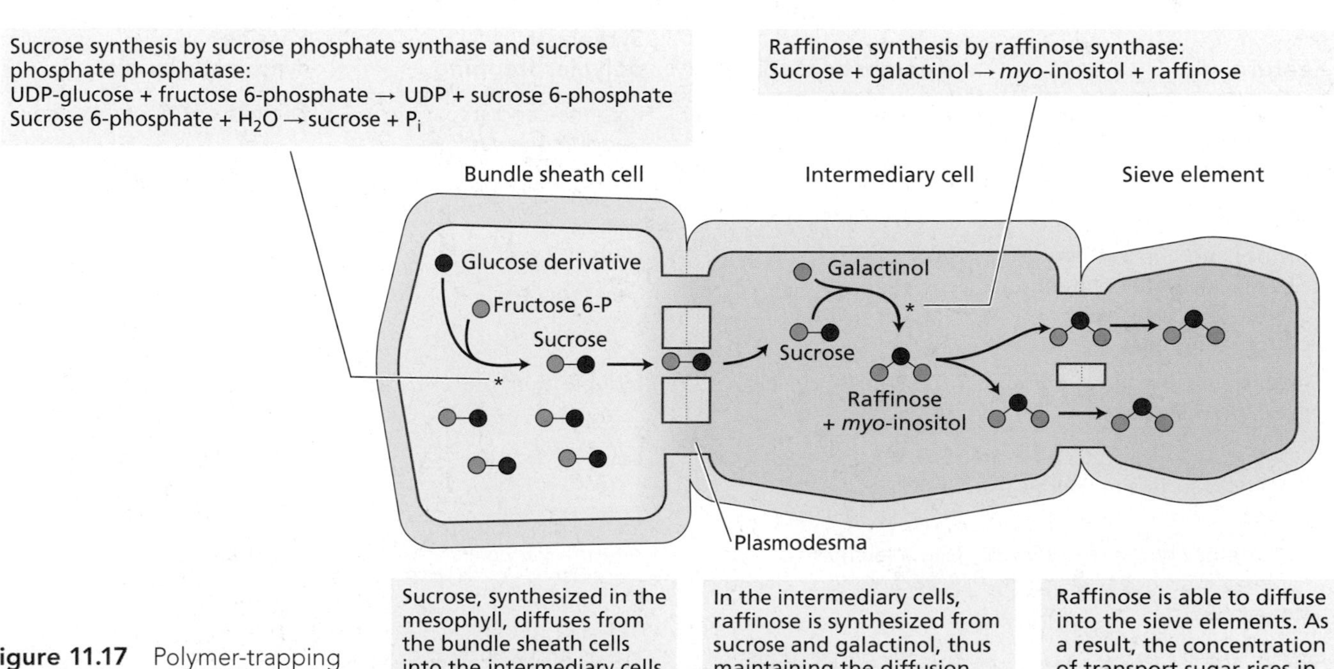
Sucrose synthesis by sucrose phosphate synthase and sucrose phosphate phosphatase:
UDP-glucose + fructose 6-phosphate → UDP + sucrose 6-phosphate
Sucrose 6-phosphate + H_2O → sucrose + P_i
Raffinose synthesis by raffinose synthase:
Sucrose + galactinol → *myo*-inositol + raffinose
Bundle sheath cell
Intermediary cell
Sieve element
Glucose derivative
Fructose 6-P
Sucrose
*
Galactinol
*
Sucrose
Raffinose
+ *myo*-inositol
Plasmodesma
Sucrose, synthesized in the mesophyll, diffuses from the bundle sheath cells into the intermediary cells through the abundant plasmodesmata.
In the intermediary cells, raffinose is synthesized from sucrose and galactinol, thus maintaining the diffusion gradient for sucrose. Because of its larger size, raffinose is not able to diffuse back into the mesophyll.
Raffinose is able to diffuse into the sieve elements. As a result, the concentration of transport sugar rises in the intermediary cells and the sieve elements. Note that stachyose is not shown here for clarity.

Figure 11.17 Polymer-trapping model of phloem loading. For simplicity, the trisaccharide stachyose is omitted. (After van Bel 1992.)

The polymer-trapping model makes three predictions:

1. Sucrose should be more concentrated in the mesophyll than in the intermediary cells.
2. The enzymes for raffinose and stachyose synthesis should be preferentially located in the intermediary cells.
3. The plasmodesmata linking the bundle sheath cells and the intermediary cells should exclude molecules larger than sucrose. Plasmodesmata between the intermediary cells and sieve elements must be wider to allow passage of raffinose and stachyose.

Several studies support the polymer-trapping model of symplastic loading in some species. However, recent modeling results suggest that additional, unkown factors must be present to enable plasmodesmata to block transport of oligosaccharides, such as raffinose and stachyose, back into the mesophyll, while permitting sufficient sucrose flux into the intermediary cells to maintain observed transport rates. (See **WEB TOPIC 11.7** for further discussion of these issues.)

Phloem loading is passive in several tree species

Passive symplastic phloem loading has recently been recognized as a mechanism that is widespread among plant species. While the data supporting this mechanism are relatively recent, passive symplastic loading was actually a part of Münch's original conception of pressure flow.

It has become apparent that several tree species possess abundant plasmodesmata between the sieve element–companion cell complex and surrounding cells but do not have intermediary-type companion cells and do not transport raffinose and stachyose. Willow (*Salix babylonica*) and apple (*Malus domestica*) trees are among the species that fall into this category, as does the gymnosperm *Pinus sylvestris*. These plants have no concentrating step in the pathway from the mesophyll into the sieve element–companion cell complex. Since a concentration gradient from the mesophyll into the phloem drives diffusion along this short-distance pathway, the absolute levels of sugars in the source leaves of these species must be high in order to maintain the required high solute concentrations and the resulting high turgor pressures in the sieve elements. Although there is wide variation (over 50-fold) and considerable overlap between groups of plants with different loading mechanisms, source leaf sugar concentrations are generally higher in the tree species that load passively.

The type of phloem loading is correlated with several significant characteristics

As discussed above, the operation of apoplastic and symplastic phloem-loading pathways is correlated with several defining characteristics, listed in **Table 11.3**.

TABLE 11.3 Patterns in apoplastic and symplastic loading

Feature	Apoplastic loading	Symplastic polymer trapping	Passive symplastic loading
Transport sugar	Sucrose	Raffinose and stachyose in addition to sucrose	Sucrose and sugar alcohols
Characteristic companion cells	Ordinary companion cells or transfer cells	Intermediary cells	Ordinary companion cells
Number and conductivity of plasmodesmata connecting the SE–CC complex to surrounding cells	Low	High	High
Dependence on active carriers in SE–CC complex	Transporter driven	Independent of transporters	Independent of transporters
Overall concentration of transport sugars in source leaves	Low	Low	High
Cell type in which driving force for long-distance transport is generated	Sieve element–companion cell complex	Intermediary cells	Mesophyll
Growth habit	Mainly herbaceous	Herbs and woody species	Mainly trees

Sources: Gamalei 1985; van Bel et al. 1992; Rennie and Turgeon 2009.

Note: Plants using all three mechanisms of phloem loading may also transport sugar alcohols. In addition, some species may load both apoplastically and symplastically, since different types of companion cells can be found within the veins of a single species. SE–CC complex, sieve element–companion complex.

- Species that have apoplastic phloem loading as their dominant loading strategy translocate sucrose almost exclusively and have either ordinary companion cells or transfer cells in the minor veins. These species usually possess few connections between the sieve element–companion cell complex and the surrounding cells. Active carriers in the sieve element–companion cell complex concentrate sucrose in the phloem cells and generate the driving force for long-distance transport.
- Species that use symplastic phloem loading with polymer trapping translocate oligosaccharides such as raffinose in addition to sucrose. They have intermediary-type companion cells in the minor veins, with abundant connections between the sieve element–companion cell complex and the surrounding cells. Polymer trapping probably concentrates transport sugars in the phloem cells and generates the driving force for long-distance transport.
- Species that have passive symplastic phloem loading translocate sucrose and sugar alcohols and have ordinary companion cells in the minor veins. These species also possess abundant connections between the sieve element–companion cell complex and the surrounding cells. Species with passive symplastic loading are characterized by high overall sugar concentration in the source leaves, which maintains a concentration gradient between the mesophyll and the sieve element–companion cell complex. The high sugar concentrations give rise to the high turgor pressures in the sieve elements of source leaves, generating the driving force for long-distance transport. Many of the species with passive symplastic loading are trees.

WEB TOPIC 11.7 discusses the relationships between loading characteristics (type of companion cell, transport sugars, and abundance of plasmodesmata) and the loading mechanisms in various species.

In the discussion above, apoplastic loading, symplastic loading with polymer trapping, and passive loading were considered separately. However, increasing evidence shows that many, if not all, plants are capable of using more than one loading mechanism, at least to some extent. For example, both structural and physiological data indicate that some polymer-trapping plants are also capable of apoplastic loading. Some of these plants, such as bear's breeches, or oyster plant (*Acanthus mollis*), possess both transfer cells and intermediary cells in their minor veins. Another polymer-trapping plant, *Alonsoa meridionalis*, expresses a stachyose synthase gene in intermediary cells, which is indicative of polymer trapping, but does not express it in ordinary companion cells; the same plants express a sucrose transporter in ordinary companion cells, indicative of apoplastic loading, but do not express it in intermediary cells. Other species, such as ash (*Fraxinus*), possibly use all three loading strategies. At the other end of the spectrum are species that load almost entirely by one mechanism. Even a mild reduction in sucrose–H^+ symporter activity (and thus the ability to load from the apoplast) resulted in significant loading inhibition in tobacco (*Nicotiana tabacum*), an "apoplastic loader," but even severe reductions in symporter activity had little effect on mullein (*Verbascum phoeniceum*), a "symplastic loader."

Plasmodesmatal frequencies suggest that the passive loading strategy is ancestral in the angiosperms, while apoplastic loading and polymer trapping evolved later. However, it is possible that the ability to load by multiple mechanisms may have been present even in the earliest angiosperms. Multiple loading mechanisms may allow plants to adapt quickly to abiotic stresses, such as low temperatures. Switching mechanisms may also reflect biotic stresses, such as viral infection. Certainly, the evolution of different loading types and the environmental pressures related to their evolution will continue to be important research areas in the future, as loading pathways are clarified in more species.

Phloem Unloading and Sink-to-Source Transition

Now that we have learned about the events leading up to the export of sugars from sources, let's take a look at **import** into sinks such as developing roots, tubers, and reproductive structures. In many ways the events in sink tissues are simply the reverse of the events in sources. The following steps are involved in the import of sugars into sink cells.

1. *Phloem unloading.* This is the process by which imported sugars leave the sieve elements of sink tissues.
2. *Short-distance transport.* After unloading, the sugars are transported to cells in the sink by means of a short-distance transport pathway. This pathway has also been called post–sieve element transport.
3. *Storage and metabolism.* In the final step, sugars are stored or metabolized in sink cells.

In this section we will discuss the following questions: Are phloem unloading and short-distance transport symplastic or apoplastic? Is sucrose hydrolyzed during the process? Do phloem unloading and subsequent steps require energy? Finally, we will examine the transition process by which a young, importing leaf becomes an exporting source leaf.

Phloem unloading and short-distance transport can occur via symplastic or apoplastic pathways

In sink organs, sugars move from the sieve elements to the cells that store or metabolize them. Sinks vary from growing vegetative organs (root tips and young leaves) to stor-

(A) Symplastic phloem unloading and short-distance transport

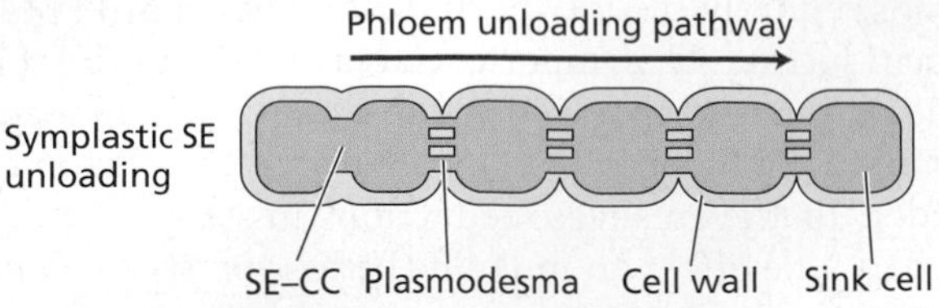

(B) Apoplastic phloem unloading and short-distance transport

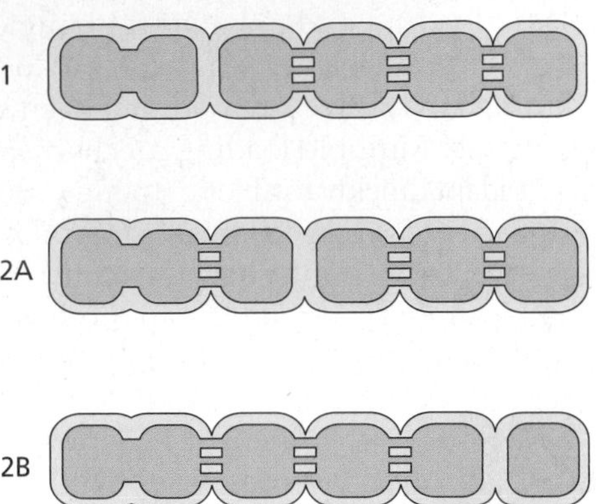

Type 1: This short-distance pathway is designated apoplastic because one step, phloem unloading from the sieve element–companion cell complex, occurs in the apoplast. Once the sugars are taken back up into the symplast of adjoining cells, transport is symplastic.

Type 2: These pathways also have an apoplastic step. However, phloem unloading from the sieve element–companion cell complex is symplastic. The apoplastic step occurs later in the pathways. The upper figure (type 2A) shows an apoplastic step close to the sieve element–companion cell complex; the lower figure (type 2B), an apoplastic step that is farther removed.

Figure 11.18 Pathways for phloem unloading and short-distance transport. The sieve element–companion cell complex (SE–CC) is considered a single functional unit. The presence of plasmodesmata is assumed to provide functional symplastic continuity. An absence of plasmodesmata between cells indicates an apoplastic transport step. (A) Symplastic phloem unloading and short-distance transport. All steps are symplastic. (B) Apoplastic phloem unloading and short-distance transport.

age tissues (roots and stems) to organs of reproduction and dispersal (fruits and seeds). Because sinks vary so greatly in structure and function, there is no single mechanism of phloem unloading and short-distance transport. Differences in import pathways due to differences in sink types are emphasized in this section; however, the pathway often depends on the stage of sink development as well.

As in sources, the sugars may move entirely through the symplast via the plasmodesmata in sinks, or they may enter the apoplast at some point. **Figure 11.18** diagrams the several possible pathways in sinks. Both unloading and the short-distance pathway appear to be completely symplastic in some young eudicot leaves, such as sugar beet and tobacco (see Figure 11.18A). Meristematic and elongating regions of primary root tips also appear to unload symplastically.

While symplastic import predominates in most sink tissues, part of the short-distance pathway is apoplastic in some sink organs at some stages of development—for example, in fruits, seeds, and other storage organs that accumulate high concentrations of sugars (see Figure 11.18B). The pathway can switch between symplastic and apoplastic in these sinks, with an apoplastic step being required when sink sugar concentrations are high. The apoplastic step could be located at the site of unloading itself (Type 1 in Figure 11.18B) or farther removed from the sieve elements (Type 2). This arrangement (Type 2), typical of developing seeds, appears to be the most common in apoplastic pathways.

An apoplastic step is required in developing seeds because there are no symplastic connections between the maternal tissues and the tissues of the embryo. Sugars exit the sieve elements (phloem unloading) via a symplastic pathway and are transferred from the symplast to the apoplast at some point removed from the sieve element–companion cell complex (Type 2 in Figure 11.18B). The apoplastic step permits membrane control over the substances that enter the embryo, because two membranes must be crossed in the process.

When an apoplastic step occurs in the import pathway, the transport sugar can be partly metabolized in the apoplast, or it can cross the apoplast unchanged (see WEB TOPIC 11.8). For example, sucrose can be hydrolyzed into glucose and fructose in the apoplast by invertase, a sucrose-splitting enzyme, and glucose and/or fructose would then enter the sink cells. Such sucrose-cleaving enzymes play a role in the control of phloem transport by sink tissues (see WEB TOPIC 11.10).

Transport into sink tissues requires metabolic energy

Inhibitor studies have shown that import into sink tissues is energy dependent. Growing leaves, roots, and storage sinks in which carbon is stored as starch or in protein appear to use symplastic phloem unloading and short-distance transport. Transport sugars are used as substrates for respiration and are metabolized into storage polymers and into compounds needed for growth. Sucrose metabolism thus results in a low sucrose concentration in the sink cells, maintaining a concentration gradient for sugar uptake. In this pathway, no membranes are crossed during sugar uptake into the sink cells, and transport is passive:

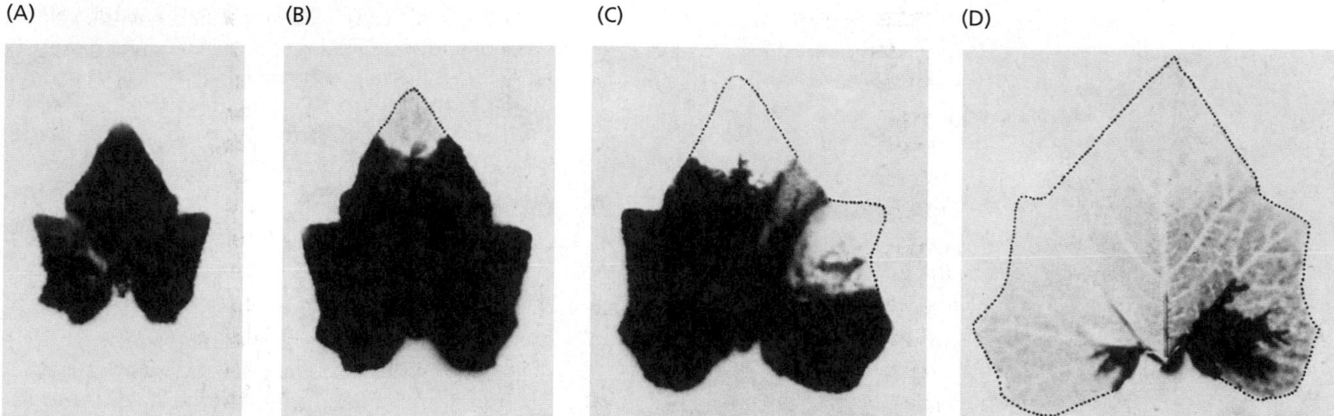

Figure 11.19 Autoradiographs of a leaf of summer squash (*Cucurbita pepo*), showing the transition of the leaf from sink to source status. In each case, the leaf imported ^{14}C from the source leaf on the plant for 2 h. Label is visible as black accumulations. (A) The entire leaf is a sink, importing sugar from the source leaf. (B–D) The base is still a sink. As the tip of the leaf loses the ability to unload and stops importing sugar (as shown by the loss of black accumulations), it gains the ability to load and to export sugar. (From Turgeon and Webb 1973.)

transport sugars move from a high concentration in the sieve elements to a low concentration in the sink cells. Metabolic energy is thus required in these sink organs mainly for respiration and for biosynthetic reactions.

In apoplastic import, sugars must cross at least two membranes: the plasma membrane of the cell that is releasing the sugar and the plasma membrane of the sink cell. When sugars are transported into the vacuole of the sink cell, they must also traverse the tonoplast. As discussed earlier, transport across membranes in an apoplastic pathway may be energy dependent. While some evidence indicates that both efflux and uptake of sucrose can be active (see **WEB TOPIC 11.8**), the transporters have yet to be completely characterized.

Since these transporters have been shown to be bidirectional in some studies, some of the same sucrose transporters described above for sucrose loading could also be involved in sucrose unloading; the direction of transport would depend on the sucrose gradient, the pH gradient, and the membrane potential. Furthermore, symporters important in phloem loading have been found in some sink tissues—for example, SUT1 in potato tubers. The symporter may function in sucrose retrieval from the apoplast, in import into sink cells, or in both. Monosaccharide transporters must be involved in uptake into sink cells when sucrose is hydrolyzed in the apoplast.

The transition of a leaf from sink to source is gradual

Leaves of eudicots such as tomato or bean begin their development as sink organs. A transition from sink to source status occurs later in development, when the leaf is approximately 25% expanded, and it is usually complete when the leaf is 40 to 50% expanded. Export from the leaf begins at the tip or apex of the blade and progresses toward the base until the whole leaf becomes a sugar exporter. During the transition period, the tip exports sugar, while the base imports it from the other source leaves (**Figure 11.19**).

The maturation of leaves is accompanied by a large number of functional and anatomic changes, resulting in a reversal of transport direction from importing to exporting. In general, the cessation of import and the initiation of export are independent events. In albino leaves of tobacco, which have no chlorophyll and therefore are incapable of photosynthesis, import stops at the same developmental stage as in green leaves, even though export is not possible. Therefore, some change besides the initiation of export must occur in developing leaves of tobacco that causes them to cease importing sugars.

Sugars are unloaded and loaded almost entirely via different veins in tobacco (**Figure 11.20**), contributing to the conclusion that import cessation and export initiation are two separate events. The minor veins that are eventually responsible for most of the loading in tobacco and other *Nicotiana* species do not mature until about the time import ceases and cannot play a role in unloading.

The change that stops import must thus involve blockage of unloading from the large veins at some point in the development of mature leaves. Factors that could account for the cessation of unloading include plasmodesmatal closure and a decrease in plasmodesmatal frequency. Experimental data have shown that both plasmodesmatal closure and elimination of plasmodesmata can occur.

Export of sugars begins when events have occurred that close the importing pathway and activate apoplastic loading and when loading has accumulated sufficient photosynthate in the sieve elements to drive translocation

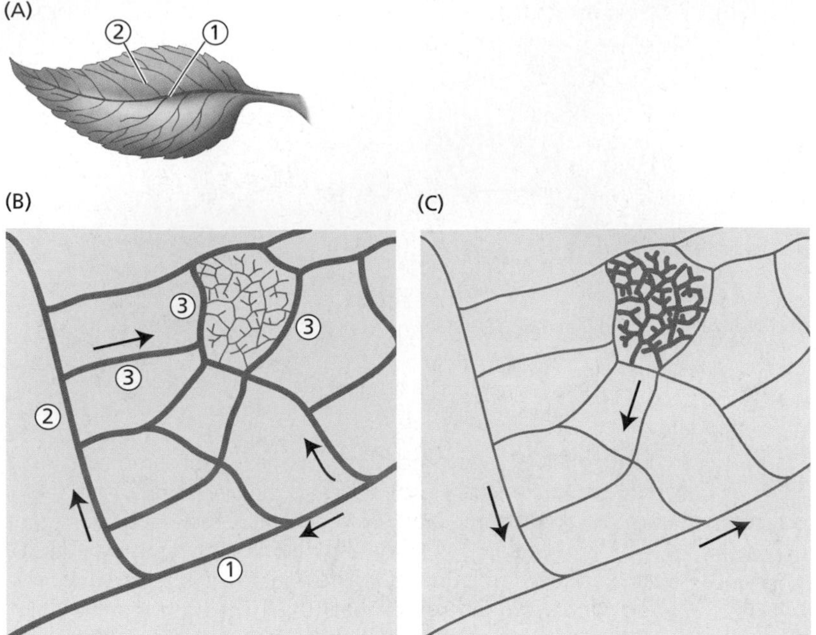

Figure 11.20 Division of labor in the veins of a tobacco leaf is shown in (A). When the leaf is immature and still in its sink phase (B), photosynthate is imported from mature leaves and distributed (arrows) throughout the blade (or lamina) via the larger, major veins (thicker lines). The major veins are numbered, with the midrib being the first-order vein. The imported photosynthate unloads from the same major veins into the mesophyll. The smallest, minor veins are shown within the areas enclosed by the third-order veins. The minor veins do not function in import and unloading because they are immature. In a source leaf (C), import has ceased, and export has begun. Photosynthate loads into the minor veins (thicker lines), while the larger veins serve only in export (arrows); they can no longer unload. Although (B) is drawn to scale from an autoradiograph, (C) is not to scale or in correct proportions, since the lamina grows considerably as the leaf matures. (After Turgeon 2006.)

out of the leaf. The following conditions are necessary for export to begin:

- The leaf is synthesizing photosynthate in sufficient quantity that some is available for export. The sucrose-synthesizing genes are being expressed.
- Minor veins responsible for loading have matured. A regulatory element (enhancer) has been identified in the DNA of Arabidopsis that acts as part of a cascade of events leading to minor vein maturation. The enhancer can activate a reporter gene fused to a companion cell–specific promoter and does so in the same tip-to-base pattern as in the sink-to-source transition.
- The sucrose–H^+ symporter is expressed and in place in the plasma membrane of the sieve element–companion cell complex. Regulation of these events is being investigated. For example, the promoter of the *SUC2* gene in Arabidopsis becomes active in companion cells in a pattern that corresponds to that in the sink-to-source transition (**Figure 11.21**). Binding sites for transcription factors have been identified within the *SUC2* promoter that mediate this source-specific and companion cell–specific gene expression.

In leaves of plants such as sugar beet and tobacco, the ability to accumulate exogenous sucrose in the sieve element–companion cell complex is acquired as the leaves undergo the sink-to-source transition, suggesting that the symporter required for loading has become functional. In developing leaves of Arabidopsis, expression of the symporter that is thought to transport sugars during loading begins in the tip and proceeds to the base during a sink-to-source transition. This is the same basipetal pattern that is seen in the development of export capacity.

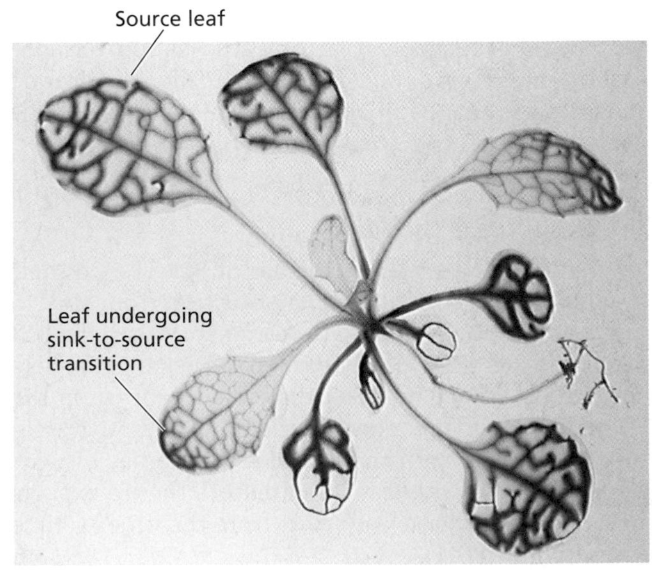

Figure 11.21 Export from source tissue depends on the placement and activity of active sucrose transporters. The figure shows an Arabidopsis rosette transformed with a construct consisting of a reporter gene under control of the *AtSUC2* promoter. SUC2, a sucrose–H^+ symporter, is one of the major sucrose transporters functioning in phloem loading. The reporter system used (GUS) forms a visible product (blue) where the promoter is active. Staining is visible only in the vascular tissue of source leaves and in the tips of leaves undergoing the sink-to-source transition. (From Schneidereit et al. 2008.)

Photosynthate Distribution: Allocation and Partitioning

The photosynthetic rate determines the total amount of fixed carbon available to the leaf. However, the amount of fixed carbon available for translocation depends on subsequent metabolic events. The regulation of the distribution of fixed carbon into various metabolic pathways is termed **allocation** in this chapter.

The vascular bundles in a plant form a system of "pipes" that can direct the flow of photosynthates to various sinks: young leaves, stems, roots, fruits, or seeds. However, the vascular system is highly interconnected, forming an open network that allows source leaves to communicate with multiple sinks. Under these conditions, what determines the volume of flow to any given sink? The differential distribution of photosynthates within the plant is termed **partitioning** in this chapter. (The terms *allocation* and *partitioning* are sometimes used interchangeably in current publications.)

After giving an overview of allocation and partitioning, we will examine the coordination of starch and sucrose synthesis. Throughout this section, keep in mind that a limited number of species have been studied, mainly those that load sucrose actively from the apoplast. It is likely that the mechanism of phloem loading affects the regulation of allocation, and so studies of allocation will have to be extended to a wider range of species. We will conclude by discussing how sinks compete, how sink demand might regulate photosynthetic rate in the source leaf, and how sources and sinks communicate with each other.

Allocation includes storage, utilization, and transport

The carbon fixed in a source cell can be used for storage, metabolism, and transport:

- *Synthesis of storage compounds.* Starch is synthesized and stored within chloroplasts and, in most species, is the primary storage form that is mobilized for translocation during the night. Plants that store carbon primarily as starch are called *starch storers*.
- *Metabolic utilization.* Fixed carbon can be utilized within various compartments of the photosynthesizing cell to meet the energy needs of the cell or to provide carbon skeletons for the synthesis of other compounds required by the cell.
- *Synthesis of transport compounds.* Fixed carbon can be incorporated into transport sugars for export to various sink tissues. A portion of the transport sugar can also be stored temporarily in the vacuole.

Allocation is also a key process in sink tissues. Once the transport sugars have been unloaded and enter the sink cells, they can remain as such or can be transformed into various other compounds. In storage sinks, fixed carbon can be accumulated as sucrose or hexose in vacuoles or as starch in amyloplasts. In growing sinks, sugars can be used for respiration and for the synthesis of other molecules required for growth.

Various sinks partition transport sugars

Sinks compete for the photosynthate being exported by the sources. Such competition determines the partitioning of transport sugars among the various sink tissues of the plant, at least in the short term. The allocation of sugar within a sink (storage or metabolism) affects its ability to compete for available sugars. In this way, the processes of partitioning and allocation interact.

Of course, events in sources and sinks must be synchronized. Partitioning determines the patterns of growth, and growth must be balanced between shoot growth (photosynthetic productivity) and root growth (water and mineral uptake) in such a way that the plant can respond to the challenges of a variable environment. The goal is *not* a constant root-to-shoot ratio but one that secures a supply of carbon and mineral nutrients appropriate to the needs of the plant.

So an additional level of control lies in the interaction between areas of supply and demand. Turgor pressure in the sieve elements could be an important means of communication between sources and sinks, acting to coordinate rates of loading and unloading. Chemical messengers are also important in communicating the status of one organ to the other organs in the plant. Such chemical messengers include plant hormones and nutrients, such as potassium and phosphate ions, and even the transport sugars themselves. Recent findings suggest that macromolecules (RNA and protein) may also play a role in photosynthate partitioning, perhaps by influencing transport through plasmodesmata.

Attainment of higher yields of crop plants is one goal of research on photosynthate allocation and partitioning. Whereas grains and fruits are examples of edible yields, total yield includes inedible portions of the shoot. Harvest index, the ratio of economical yield (edible grain) to total aboveground biomass, has increased over the years largely due to the efforts of plant breeders. One goal of modern plant physiology is to further increase yield based on a fundamental understanding of metabolism, development, and in the present context, partitioning.

However, allocation and partitioning in the whole plant must be coordinated such that increased transport to edible tissues does not occur at the expense of other essential processes and structures. Crop yield may also be improved if photosynthates that are normally "lost" by the plant are retained. For example, losses due to nonessential respiration or exudation from roots could be reduced. In the latter case, care must be taken not to disrupt essential processes outside the plant, such as growth of beneficial microbial species in the vicinity of the root that obtain nutrients from the root exudate.

Source leaves regulate allocation

Increases in the rate of photosynthesis in a source leaf generally result in an increase in the rate of translocation from the source. Control points for the allocation of photosynthate (**Figure 11.22**) include the distribution of triose phosphates to the following processes:

- Regeneration of intermediates in the C_3 photosynthetic carbon reduction cycle (the Calvin–Benson cycle; see Chapter 8)
- Starch synthesis
- Sucrose synthesis, as well as distribution of sucrose between transport and temporary storage pools

Various enzymes operate in the pathways that process the photosynthate, and the control of these steps is complex. The research described below focuses on species that load sucrose actively from the apoplast, specifically during the daylight hours. Further studies will be needed to extend our knowledge to plants using other loading strategies and to the regulation of allocation in those species.

During the day the rate of starch synthesis in the chloroplast must be coordinated with sucrose synthesis in the cytosol. Triose phosphates (glyceraldehyde-3-phosphate and dihydroxyacetone phosphate) produced in the chloroplast by the Calvin–Benson cycle (see Chapter 8) can be used for either starch or sucrose synthesis or in respiration. Sucrose synthesis in the cytosol diverts triose phosphate away from starch synthesis and storage. For example, it has been shown that when the demand for sucrose by other parts of a soybean plant is high, less carbon is stored as starch by the source leaves. The key enzymes involved in the regulation of sucrose synthesis in the cytosol and of starch synthesis in the chloroplast are sucrose phosphate synthase and fructose-1,6-bisphosphatase in the cytosol and ADP-glucose pyrophosphorylase in the chloroplast (see Figure 11.22 and Chapter 8).

However, there is a limit to the amount of carbon that normally can be diverted from starch synthesis in species that store carbon primarily as starch. Studies of allocation between starch and sucrose under different conditions suggest that a fairly steady rate of translocation throughout the 24-h period is a priority for most plants. See **WEB TOPIC 11.9** for further discussion of the balance between starch and sucrose synthesis in source leaves.

Sink tissues compete for available translocated photosynthate

As discussed earlier, translocation to sink tissues depends on the position of the sink in relation to the source and on the vascular connections between source and sink. Another factor determining the pattern of transport is competition between sinks, for example, between terminal sinks or between terminal sinks and axial sinks along the transport pathway. For example, young leaves might compete with roots for photosynthates in the translocation stream. Competition has been shown by numerous experiments in which removal of a sink tissue from a plant generally results in increased translocation to alternative, and hence competing, sinks. Conversely, increased sink size, for example, increased fruit load, decreases translocation to other sinks, especially the roots.

In the reverse type of experiment, the source supply can be altered while the sink tissues are left intact. When the supply of photosynthates from sources to competing sinks is suddenly and drastically reduced by shading of all the source leaves but one, the sink tissues become dependent on a single source. In sugar beet and

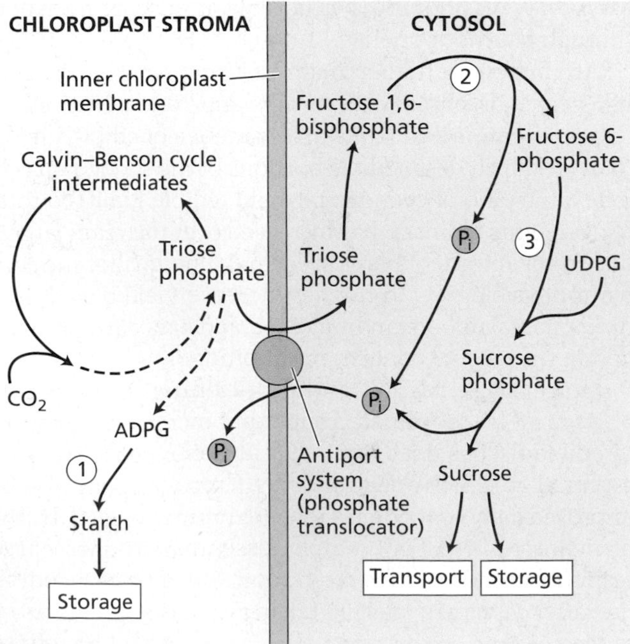

Figure 11.22 A simplified scheme for starch and sucrose synthesis during the day. Triose phosphate, formed in the Calvin–Benson cycle, can either be used in starch formation in the chloroplast or transported into the cytosol in exchange for inorganic phosphate (P_i) via the phosphate translocator in the inner chloroplast membrane. The outer chloroplast membrane (omitted here for clarity) is permeable to small molecules. In the cytosol, triose phosphate can be converted to sucrose for storage in the vacuole or transport or can be degraded via glycolysis. Key enzymes involved are starch synthetase (1), fructose-1,6-bisphosphatase (2), and sucrose phosphate synthase (3). The second and third enzymes, along with ADP-glucose pyrophosphorylase, which forms adenosine diphosphate glucose (ADPG), are regulated enzymes in sucrose and starch synthesis (see Chapter 8). UDPG, uridine diphosphate glucose. (After Preiss 1982.)

bean plants, the rates of photosynthesis and export from the single remaining source leaf usually do not change over the short term (approximately 8 h). However, the roots receive less sugar from the single source, while the young leaves receive relatively more. Shading in general decreases partitioning to roots. Presumably, the young leaves can deplete the sugar content of the sieve elements more readily and thus increase the pressure gradient and the rate of translocation toward themselves.

Treatments such as making the sink water potential more negative increase the pressure gradient and enhance transport to the sink. Treatment of the root tips of pea (*Pisum sativum*) seedlings with mannitol solutions increased the import of sucrose over the short term by more than 300%, possibly because of a turgor decrease in the sink cells. Longer-term experiments show the same trend. Moderate water stress induced with polyethylene glycol treatment of the roots increased the proportion of assimilates transported to the roots of apple plants over a period of 15 days but decreased the proportion transported to the shoot apex. This contrasts with shading treatments (above) in which source limitation diverts more sugar to the young leaves.

Sink strength depends on sink size and activity

The ability of a sink to mobilize photosynthate toward itself is often described as **sink strength**. Sink strength depends on two factors—sink size and sink activity—as follows:

$$\text{Sink strength} = \text{sink size} \times \text{sink activity}$$

Sink size is the total biomass of the sink tissue, and **sink activity** is the rate of uptake of photosynthates per unit biomass of sink tissue. Altering either the size or the activity of the sink results in changes in translocation patterns. For example, the ability of a pea pod to import carbon depends on the dry weight of that pod as a proportion of the total number of pods.

Changes in sink activity can be complex, because various activities in sink tissues can potentially limit the rate of uptake by the sink. These activities include unloading from the sieve elements, metabolism in the cell wall, uptake from the apoplast, and metabolic processes that use the photosynthate for either growth or storage.

Experimental treatments to manipulate sink strength have often been unspecific. For example, cooling a sink tissue, which would be expected to inhibit all activities that require metabolic energy, often results in a decrease in the velocity of transport toward the sink. More recent experiments take advantage of our ability to specifically over- or underexpress enzymes related to sink activity—for example, those involved in sucrose metabolism in the sink. The two major enzymes that split sucrose are acid invertase and sucrose synthase, both of which can catalyze the first step in sucrose utilization. WEB TOPIC 11.10 discusses evidence for a correlation between the activity of sucrose-splitting enzymes, particularly invertase, and sink demand.

The source adjusts over the long term to changes in the source-to-sink ratio

If all but one of the source leaves of a soybean plant are shaded for an extended period (e.g., 8 days), many changes occur in the single remaining source leaf. These changes include a decrease in starch concentration and increases in photosynthetic rate, rubisco activity, sucrose concentration, transport from the source, and orthophosphate concentration. Thus, in addition to the observed short-term changes in the distribution of photosynthate among different sinks, there are adjustments in the source leaf's metabolism in response to altered conditions over a longer term.

Photosynthetic rate (the net amount of carbon fixed per unit leaf area per unit time) often increases over several days when sink demand increases, and it decreases when sink demand decreases. An accumulation of photosynthate (sucrose or hexoses) in the source leaf can account for the linkage between sink demand and photosynthetic rate in starch-storing plants (see WEB TOPIC 11.11). Sugars act as signaling molecules that regulate many metabolic and developmental processes in plants. In general, carbohydrate depletion enhances the expression of genes for photosynthesis, reserve mobilization, and export processes, while abundant carbon resources favor genes for storage and utilization.

Sucrose or hexoses that would accumulate as a result of decreased sink demand are well known to repress photosynthetic genes. Interestingly, the genes for invertase and sucrose synthase, both of which can catalyze the first step in sucrose utilization, and genes for sucrose–H^+ symporters, which play a key role in apoplastic loading, are also among those regulated by carbohydrate supply.

Such regulation of photosynthesis by sink demand suggests that sustained increases in photosynthesis in response to elevated CO_2 in the atmosphere may depend on increasing sink strength (increasing the strength of existing sinks or developing new sinks). See Chapter 9 for a discussion of the results of increased CO_2 levels in the atmosphere on photosynthesis and growth of plants.

Transport of Signaling Molecules

Besides its major function in the long-distance transport of photosynthate, the phloem is also a conduit for the transport of signaling molecules from one part of the organism to another. Such long-distance signals coordinate the activities of sources and sinks and regulate plant growth and development. As indicated earlier, the signals between sources and sinks might be physical or chemical. Physical signals such as turgor change are transmitted rapidly via the interconnecting system of sieve elements.

Molecules traditionally considered to be chemical signals, such as proteins and plant hormones, are found in the phloem sap, as are mRNAs and small RNAs, which have more recently been added to the list of signal molecules. The translocated carbohydrates themselves may also act as signals.

Turgor pressure and chemical signals coordinate source and sink activities

Turgor pressure may play a role in coordinating the activities of sources and sinks. For example, if phloem unloading were rapid under conditions of rapid sugar utilization at the sink tissue, turgor pressures in the sieve elements of sinks would be reduced, and this reduction would be transmitted to the sources. If loading were controlled in part by sieve element turgor, loading would increase in response to this signal from the sinks. The opposite response would be seen when unloading was slow in the sinks. Loading of sugars from storage within cells along the axial pathway also responds to changes in solute demand. Some data suggest that cell turgor can modify the activity of the proton-pumping ATPase at the plasma membrane and therefore alter membrane transport rates.

Shoots produce growth regulators such as auxin, which can be rapidly transported to the roots via the phloem, and roots produce cytokinins, which move to the shoots through the xylem. Gibberellins (GA) and abscisic acid (ABA) are also transported throughout the plant in the vascular system. Plant hormones play a role in regulating source–sink relationships. They affect photosynthate partitioning in part by controlling sink growth, leaf senescence, and other developmental processes. Plant defense responses against herbivores and pathogens can also change allocation and partitioning of photoassimilates, with plant defense hormones such as jasmonic acid mediating the responses.

Loading of sucrose has been shown to be stimulated by exogenous auxin but inhibited by ABA in some source tissues, while exogenous ABA enhances, and auxin inhibits, sucrose uptake by some sink tissues. Hormones might regulate apoplastic loading and unloading by influencing the levels of active transporters in plasma membranes. Other potential sites of hormone regulation of unloading include tonoplast transporters, enzymes for metabolism of incoming sucrose, wall extensibility, and plasmodesmatal permeability in the case of symplastic unloading (see the next section).

As indicated earlier, carbohydrate levels can influence the expression of genes encoding photosynthesis components, as well as genes involved in sucrose hydrolysis. Many genes have been shown to be responsive to sugar depletion and abundance. Thus, not only is sucrose transported in the phloem, but sucrose or its metabolites can also act as signals that modify the activities of sources and sinks. For example, sucrose–H^+ symporter mRNA declines in sugar beet source leaves fed exogenous sucrose through the xylem. The decline in symporter mRNA is accompanied by a loss of symporter activity in plasma membrane vesicles isolated from the leaves. A working model includes the following steps:

1. Decreased sink demand leads to high sucrose levels in the vascular tissue.
2. High sucrose levels lead to down-regulation of the symporter in the source.
3. Decreased loading results in increased sucrose concentrations in the source.

Increased sucrose concentrations in the source can result in a lower photosynthetic rate (see **WEB TOPIC 11.11**). An increase of starch accumulation in source leaves of plants transformed with antisense DNA to the sucrose–H^+ symporter SUT1 also supports this model .

Sugars and other metabolites have been shown to interact with hormonal signals to control and integrate many plant processes. Gene expression in some source–sink systems responds to both sugar and hormonal signals.

Proteins and RNAs function as signal molecules in the phloem to regulate growth and development

It has long been known that viruses can move in the phloem, traveling as complexes of proteins and nucleic acids or as intact virus particles. More recently, endogenous RNA molecules and proteins have been found in phloem sap, and at least some of these can function as signal molecules or generate phloem-mobile signals.

To be assigned a signaling role in plants, a macromolecule must meet a number of significant criteria:

- The macromolecule must move from source to sink in the phloem.
- The macromolecule must be able to leave the sieve element–companion cell complex in sink tissues. Alternatively, the macromolecule might trigger the formation of a second signal that transmits information to the sink tissues surrounding the phloem; that is, it might initiate a signal cascade.
- Perhaps most important, the macromolecule must be able to modify the functions of specific cells in the sink.

How well do various macromolecules in the phloem meet these criteria?

At least some proteins synthesized in companion cells can clearly enter the sieve elements through the plasmodesmata that connect the two cell types and can move with the translocation stream to sink tissues. For example, passive movement of proteins from companion cells to sieve elements has been demonstrated in Arabidopsis and tobacco plants. These plants were transformed with the gene for green fluorescent protein (GFP) from jellyfish, under control of the *SUC2* promoter from Arabidopsis. The

SUC2 sucrose–H^+ symporter is synthesized within the companion cells, so proteins expressed under the control of its promoter, including GFP, are also synthesized in the companion cells. GFP, which is localized by its fluorescence after excitation with blue light, moves through plasmodesmata from companion cells into sieve elements of source leaves (**Figure 11.23A**) and migrates within the phloem to sink tissues, as do larger GFP-fusion proteins. However, only free GFP is able to move symplastically into sink tissues of the root (**Figure 11.23B**). Limited evidence exists for movement of proteins from cells outside the sieve element–companion cell complex into source phloem or for movement of proteins from the phloem into sink tissues outside the sieve element–companion cell complex. However, phloem transport of proteins that modify cellular functions has been demonstrated, implying that some signal, either the protein itself or some other signal molecule, moves between the sieve element–companion cell complexes and surrounding tissues of sources and sinks. A classic example is the FLOWERING LOCUS T (FT) protein, which appears to be a significant component of the floral stimulus that moves from source leaf to apex and that induces flowering at the apex in response to inductive conditions (see Chapter 20). The FT protein has been shown to move from companion cells of source leaves, where it is expressed, into the sieve elements of sources, probably by diffusion through the plasmodesmata. Movement of FT protein into the apical tissues has also been demonstrated and is thought to occur by a selective pathway. (See next section.)

RNAs transported in the phloem consist of endogenous mRNAs, pathogenic RNAs, and small RNAs associated with gene silencing (see Chapter 2). Most of these RNAs appear to travel in the phloem as complexes of RNA and protein (ribonucleoproteins [RNPs]). As with proteins in the phloem, direct evidence for movement of RNAs between sieve element–companion cell complexes and surrounding tissues is somewhat limited. However, some mRNAs transported in the phloem have been shown to cause visible changes in sinks after being unloaded into target tissues. For example, mRNA for a regulator of gibberellic acid responses (called GAI) was localized to sieve elements and companion cells of pumpkin (*Cucurbita pepo*) and was found in pumpkin phloem sap. Transgenic tomato plants expressing a mutant version of the regulator gene were dwarf and dark green. The mRNA for the mutant regulator was localized to sieve elements, was able to be transported across graft unions into wild-type scions, and was unloaded into apical tissues. As a result, the mutant phenotype developed in new growth on the wild-type scion.

Only a few specific mRNAs appear to be transported over long distances in the phloem. Motifs in coding sequences and in untranslated regions of the RNA both play important roles in the long-distance movement of GAI RNA. Similar results have been obtained for mRNA for the transcription factor BEL5 in potato (*Solanum tuberosum*). *BEL5* transcripts formed in the leaves move in the phloem across graft unions to stolon tips, the site of tuber induction, and the movement is correlated with enhanced tuber production. Preferential accumulation of the mRNA occurs when untranslated regions are present in addition to coding regions.

See WEB TOPIC 11.12 for further discussion of these topics.

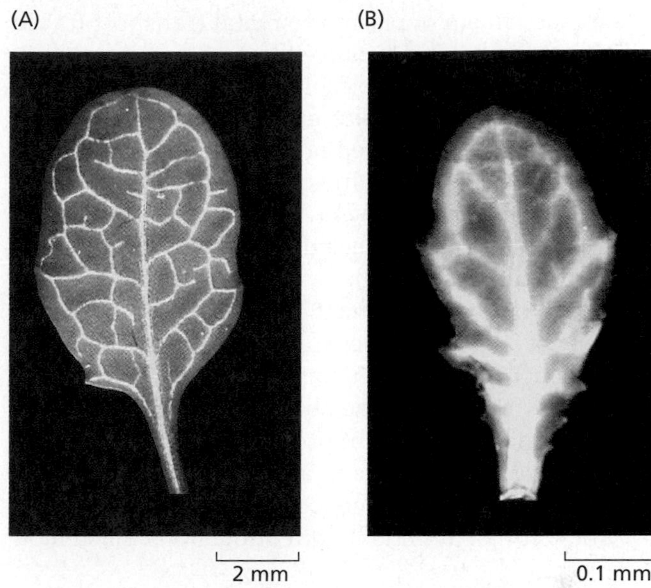

Figure 11.23 GFP fluorescence in source and sink leaves from transgenic Arabidopsis plants expressing GFP under control of the *SUC2* promoter indicates that GFP moves through plasmodesmata from companion cells into sieve elements of source leaves and from sieve elements into the surrounding mesophyll of sink leaves. (A) GFP is synthesized in companion cells and moves into the sieve elements of the source, as indicated by the bright fluorescence in the veins. (B) Free GFP is imported into sink leaves and moves into the surrounding mesophyll. Because GFP has moved into surrounding tissues, the veins are no longer distinctly delineated, and GFP fluorescence is much more diffuse. Even though the source leaf in (A) appears to be the same size as the sink leaf in (B), the source leaf is actually much larger. Note that the scales in (A) and (B) are different. (From Stadler et al. 2005.)

Plasmodesmata function in phloem signaling

Plasmodesmata have been implicated in nearly every aspect of phloem translocation, from loading to long-distance transport (pores in sieve areas and sieve plates are modified plasmodesmata) to allocation and partitioning. What role might plasmodesmata play in macromolecular signaling in the phloem?

The mechanism of plasmodesmatal transport (called trafficking) can be either passive (nontargeted) or selective and regulated. When a molecule moves passively, its size must be smaller than the **size exclusion limit (SEL)** of the plasmodesmata. As indicated earlier, GFP moves passively through plasmodesmata. In contrast, when a molecule moves in a selective fashion, it must possess a trafficking signal or be targeted in some other way to the plasmodesmata. The transport of some developmental transcription factors and of viral movement proteins appears to occur by means of a selective mechanism. Viral movement proteins interact directly with plasmodesmata to allow the passage of viral nucleic acids between cells. Once at the plasmodesmata, movement proteins act to increase the SEL of the plasmodesmata to allow the viral genome to move between cells. Endogenous proteins are thought to carry out similar functions for endogenous macromolecules such as FT protein and some P-proteins (see **WEB TOPIC 11.12**). Interaction with components at or within the plasmodesmata, such as chaperones, is also required.

It is fitting to end this chapter with research topics that will continue to engage plant physiologists of the future: regulation of growth and development via the transport of endogenous RNA and protein signals, the nature of the proteins that facilitate the transport of signals through plasmodesmata, and the possibility of targeting signals to specific sinks in contrast to mass flow. Many other potential areas of inquiry have been indicated in this chapter as well, such as the mechanism of phloem transport in gymnosperms, the nature and role of proteins in the lumen of the sieve elements, and the magnitude of pressure gradients in the sieve elements, especially in trees. As always in science, an answer to one question generates more questions!

SUMMARY

Phloem translocation moves the products of photosynthesis from mature leaves to areas of growth and storage. It also transmits chemical signals and redistributes ions and other substances throughout the plant body.

Pathways of Translocation

- Sieve elements of the phloem conduct sugars and other organic materials throughout the plant (**Figures 11.1–11.3**).
- During development, sieve elements lose many organelles, retaining only the plasma membrane and modified mitochondria, plastids, and smooth endoplasmic reticulum (**Figures 11.3, 11.4**).
- Sieve elements are interconnected through pores in their cell walls (**Figure 11.5**).
- In gymnosperms, smooth ER covers the sieve areas and is continuous through the sieve pores and median cavity (**Figure 11.6, Table 11.1**).
- P-proteins and callose seal off damaged phloem to limit loss of sap.
- Companion cells aid transport of photosynthetic products to the sieve elements. They also supply proteins and ATP to the sieve elements (**Figures 11.3–11.5; 11.7**).

Patterns of Translocation: Source to Sink

- Phloem translocation is not defined by gravity. Sap is translocated from sources to sinks, and the pathways involved are often complex (**Figure 11.8**).

Materials Translocated in the Phloem

- The composition of sap has been determined; nonreducing sugars are the main transported molecules (**Table 11.2; Figure 11.9**).
- Sap includes proteins, many of which may have functions related to stress and defense reactions.

Rates of Movement

- Transport velocities in the phloem are high and exceed the rate of diffusion over long distances by orders of magnitude.

The Pressure-Flow Model, a Passive Mechanism for Phloem Transport

- The pressure-flow model explains phloem translocation as a bulk flow of solution driven by an osmotically generated pressure gradient between source and sink.
- Phloem loading at the source and phloem unloading at the sink establish the pressure gradient for passive, long-distance bulk flow (**Figure 11.10**).
- Pressure gradients in the phloem sieve elements may be modest; pressures in herbaceous plants and trees appear to be similar. Alternative models for translocation by mass flow are being developed.

Phloem Loading

- The export of sugars from sources involves allocation of photosynthate to transport, short-distance transport, and phloem loading.

- Phloem loading can occur by way of the symplast or apoplast (**Figure 11.14**).
- Sucrose is actively transported into the sieve element–companion cell complex in the apoplastic pathway (**Figures 11.15, 11.16**).
- The polymer-trapping model holds that polymers are synthesized from sucrose in the intermediary cells; the larger oligosaccharides can only diffuse into the sieve elements (**Figure 11.17**).
- Apoplastic and symplastic phloem-loading pathways have defining characteristics (**Table 11.3**).

Phloem Unloading and Sink-to-Source Transition

- The import of sugars into sink cells involves phloem unloading, short-distance transport, and storage or metabolism.
- Phloem unloading and short-distance transport may operate by symplastic or apoplastic pathways in different sinks (**Figure 11.18**).
- Transport into sink tissues is energy dependent.
- Import cessation and export initiation are separate events, and there is a gradual transition from sink to source (**Figures 11.19, 11.20**).
- The transition from sink to source requires a number of conditions, including the expression and localization of the sucrose–H^+ symporter (**Figure 11.21**).

Photosynthate Distribution: Allocation and Partitioning

- Allocation in source leaves includes synthesis of storage compounds, metabolic utilization, and synthesis of transport compounds.
- The regulation of allocation must thus control the distribution of fixed carbon to the Calvin–Benson cycle, starch synthesis, sucrose synthesis, and respiration (**Figure 11.22**).
- A variety of chemical and physical signals are involved in partitioning resources among the various sinks.
- In competing for photosynthate, sink strength depends on sink size and sink activity.
- In response to altered conditions, short-term changes alter the distribution of photosynthate among different sinks, while long-term changes take place in source metabolism and alter the amount of photosynthate available for transport.

Transport of Signaling Molecules

- Turgor pressure, cytokinins, gibberellins, and abscisic acid have signaling roles in coordinating source and sink activities.
- Some proteins can move from companion cells into sieve elements of source leaves, and through the phloem to sink leaves (**Figure 11.23**).
- Proteins and RNAs transported in the phloem can alter cellular functions.
- Changes in the size exclusion limit (SEL) may control what passes through plasmodesmata.

WEB MATERIAL

- **WEB TOPIC 11.1 Sieve Elements as the Transport Cells between Sources and Sinks** Various methods demonstrate that sugar is transported in the sieve elements of the phloem; anatomical and developmental factors affect the basic source-to-sink pattern of transport.
- **WEB TOPIC 11.2 An Additional Mechanism for Blocking Wounded Sieve Elements in the Legume Family** P-protein bodies rapidly disperse and block legume sieve tubes following wounding.
- **WEB TOPIC 11.3 Sampling Phloem Sap** Exudation from wounds and from severed aphid stylets yields sufficient phloem sap for analysis.
- **WEB TOPIC 11.4 Nitrogen Transport in the Phloem** Soybean is an economically important species widely studied in terms of nitrogen transport in the phloem.
- **WEB TOPIC 11.5 Monitoring Traffic on the Sugar Freeway: Sugar Transport Rates in the Phloem** A variety of techniques measure mass transfer rate in the phloem, the dry weight moving through a cross-sectional area of sieve elements per unit time.
- **WEB TOPIC 11.6 Alternative Models for Translocation by Mass Flow** Some mathematical models suggest that the pressure gradient in the sieve elements of angiosperms is small.
- **WEB TOPIC 11.7 Experiments on Phloem Loading** Evidence exists for apoplastic loading of sieve

elements in some species and for symplastic loading (polymer trapping) in others. While active carriers have been identified and characterized for some substances entering the phloem, other substances may enter sieve elements passively.

- **WEB TOPIC 11.8 Experiments on Phloem Unloading** Apoplastic unloading varies in its energy requirements and in the role of the cell-wall invertase.
- **WEB TOPIC 11.9 Allocation in Source Leaves: The Balance between Starch and Sucrose Synthesis** Experiments with mutants and transgenic plants reveal flexibility in the regulation of starch and sucrose synthesis in source leaves.
- **WEB TOPIC 11.10 Partitioning: The Role of Sucrose-metabolizing Enzymes in Sinks** Increases in cell-wall invertase activity can enhance transport to a sink, while decreases in activity can inhibit transport to the sink.
- **WEB TOPIC 11.11 Possible Mechanisms Linking Sink Demand and Photosynthetic Rate in Starch Storers** Photosynthate accumulation decreases the photosynthetic rate.
- **WEB TOPIC 11.12 Proteins and RNAs: Signal Molecules in the Phloem** Some proteins and RNAs are transported between companion cells and sieve elements, travel in sieve elements between sources and sinks, and can modify cellular functions in the sinks. Little evidence exists for a movement of proteins outside the companion cells.

available at plantphys.net

Suggested Reading

Andriunas, F. A., Zhang, H.-M., Xia, X., Patrick, J. W., and Offler, C. E. (2013) Intersection of transfer cells with phloem biology—Broad evolutionary trends, function, and induction. *Front. Plant Sci.* 4: 221. [DOI: 10.3389/fpls.2013.00221]

Holbrook, N. M., and Zwieniecki, M. A., eds. (2005) *Vascular Transport in Plants.* Elsevier Academic Press, Burlington, MA.

Jekat, S. B., Ernst, A. M., von Bohl, A., Zielonka, S., Twyman, R. M., Noll, G. A., and Prufer, D. (2013) P-proteins in *Arabidopsis* are heteromeric structures involved in rapid sieve tube sealing. *Front. Plant Sci.* 4: 225. [DOI: 10.3389/fpls.2013.00225]

Knoblauch, M. and Oparka, K. (2012) The structure of the phloem – Still more questions than answers. *Plant J.* 70: 147–156.

Liesche, J., and Schulz, A. (2013) Modeling the parameters for plasmodesmatal sugar filtering in active symplasmic phloem loaders. *Front. Plant Sci.* 4: 207. [DOI: 10.3389/fpls.2013.00207]

Mullendore, D. L., Windt, C. W., Van As, H., and Knoblauch, M. (2010) Sieve tube geometry in relation to phloem flow. *Plant Cell* 22: 579–593.

Patrick, J. W. (2013) Does Don Fisher's high-pressure manifold model account for phloem transport and resource partitioning? *Front. Plant Sci.* 4: 184.

Slewinski, T. L., Zhang, C., and Turgeon, R. (2013) Structural and functional heterogeneity in phloem loading and transport. *Front. Plant Sci.* 4: 244. [DOI: 10.3389/fpls.2013.00244]

Thompson, G. A. and van Bel, A. J. E., eds. (2013) *Phloem: Molecular Cell Biology, Systemic Communication, Biotic Interactions.* Wiley-Blackwell, Ames, IA.

Turgeon, R. (2010) The puzzle of phloem pressure. *Plant Physiol.* 154: 578–581.

Yoo, S.-C., Chen, C., Rojas, M., Daimon, Y., Ham, B.-K., Araki, T., and Lucas, W. J. (2013) Phloem long-distance delivery of FLOWERING LOCUS T (FT) to the apex. *Plant J.* 75: 456–468.

Zhang, C., Yu, X., Ayre, B. G., and Turgeon, R. (2012) The origin and composition of cucurbit "phloem" exudate. *Plant Physiol.* 158: 1873–1882.

12 Respiration and Lipid Metabolism

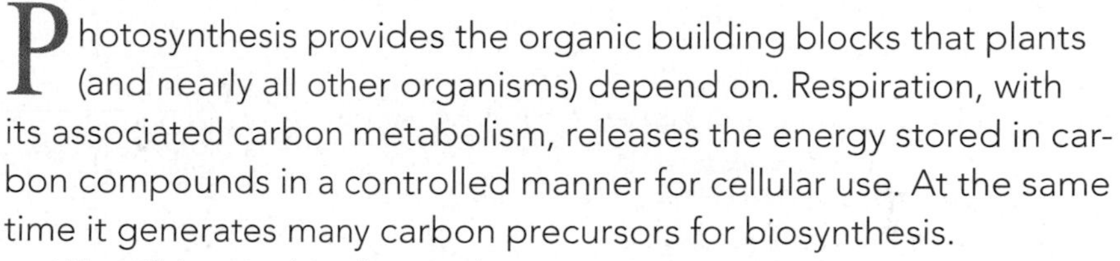

Photosynthesis provides the organic building blocks that plants (and nearly all other organisms) depend on. Respiration, with its associated carbon metabolism, releases the energy stored in carbon compounds in a controlled manner for cellular use. At the same time it generates many carbon precursors for biosynthesis.

We will begin this chapter by reviewing respiration in its metabolic context, emphasizing the interconnections among the processes involved and the special features that are peculiar to plants. We will also relate respiration to recent developments in our understanding of the biochemistry and molecular biology of plant mitochondria and respiratory fluxes in intact plant tissues. Then we will describe the pathways of lipid biosynthesis that lead to the accumulation of fats and oils, which many plant species use for energy and carbon storage. We will also examine lipid synthesis and the influence of lipids on membrane properties. Finally, we will discuss the catabolic pathways involved in the breakdown of lipids and the conversion of their degradation products into sugars that occurs during the germination of fat-storing seeds.

Overview of Plant Respiration

Aerobic (oxygen-requiring) respiration is common to nearly all eukaryotic organisms, and in its broad outlines the respiratory process in plants is similar to that found in animals and other aerobic eukaryotes. However, some specific aspects of plant respiration distinguish it from its animal counterpart. **Aerobic respiration** is the biological process by which reduced organic compounds are oxidized in a controlled manner. During respiration, energy is released and transiently stored in a compound, **adenosine triphosphate** (**ATP**), which is used by the cellular reactions for maintenance and development.

Glucose is usually cited as the substrate for respiration. In most plant cell types, however, reduced carbon is derived from sources such as the disaccharide sucrose, other sugars, organic acids, triose phosphates from photosynthesis, and metabolites from lipid and protein degradation (**Figure 12.1**).

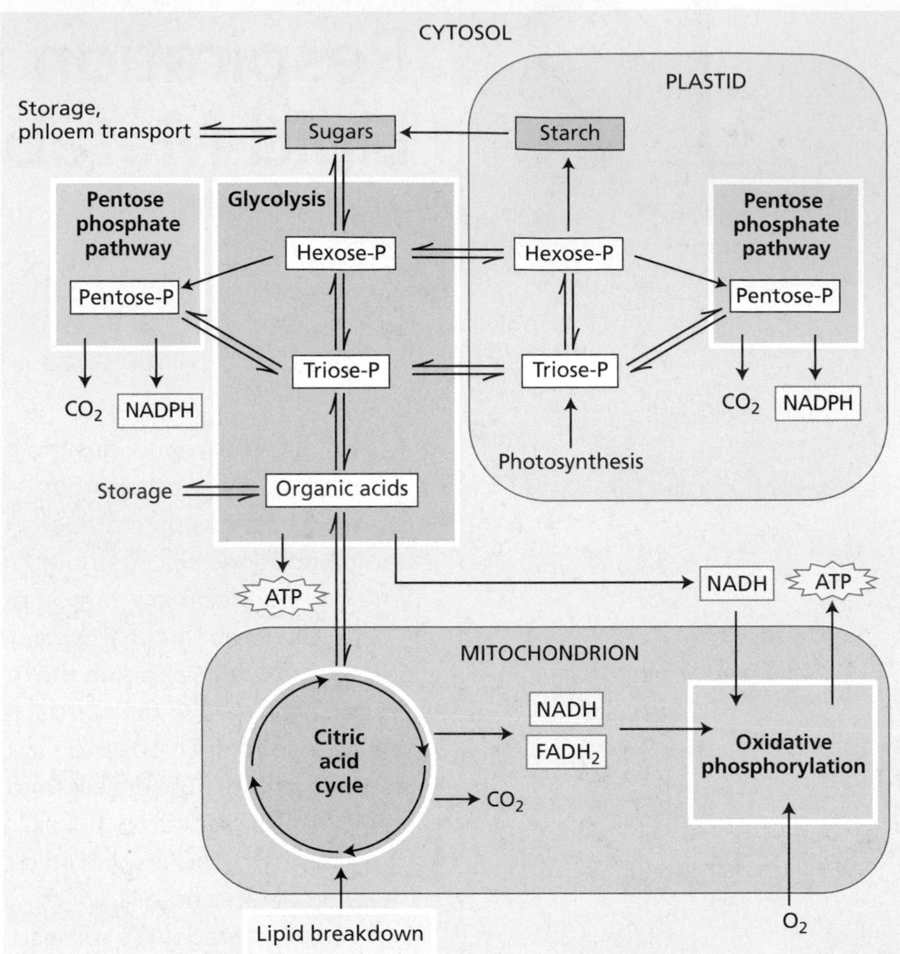

Figure 12.1 Overview of respiration. Substrates for respiration are generated by other cellular processes and enter the respiratory pathways. Glycolysis and the oxidative pentose phosphate pathways in the cytosol and plastids convert sugars into organic acids such as pyruvate, via hexose phosphates and triose phosphates, generating NADH or NADPH, and ATP. The organic acids are oxidized in the mitochondrial citric acid cycle, and the NADH and $FADH_2$ produced provide the energy for ATP synthesis by the electron transport chain and ATP synthase in oxidative phosphorylation. In gluconeogenesis, carbon from lipid breakdown is broken down in the glyoxysomes, metabolized in the citric acid cycle, and then used to synthesize sugars in the cytosol by reverse glycolysis.

From a chemical standpoint, plant respiration can be expressed as the oxidation of the 12-carbon molecule sucrose and the reduction of 12 molecules of O_2:

$$C_{12}H_{22}O_{11} + 13\ H_2O \rightarrow 12\ CO_2 + 48\ H^+ + 48\ e^-$$

$$12\ O_2 + 48\ H^+ + 48\ e^- \rightarrow 24\ H_2O$$

giving the following net reaction:

$$C_{12}H_{22}O_{11} + 12\ O_2 \rightarrow 12\ CO_2 + 11\ H_2O$$

This reaction is the reversal of the photosynthetic process; it represents a coupled redox reaction in which sucrose is completely oxidized to CO_2 while oxygen serves as the ultimate electron acceptor and is reduced to water in the process. The change in standard **Gibbs free energy** ($\Delta G^{0\prime}$) for the net reaction is –5760 kJ per mole (342 g) of sucrose oxidized. This large negative value means that the equilibrium point is strongly shifted to the right, and much energy is therefore released by sucrose degradation. The controlled release of this free energy, along with its coupling to the synthesis of ATP, is the primary, although by no means only, role of respiratory metabolism.

To prevent damage by heating of cellular structures, the cell oxidizes sucrose in a series of step-by-step reactions. These reactions can be grouped into four major processes: glycolysis, the oxidative pentose phosphate pathway, the citric acid cycle, and oxidative phosphorylation. These pathways do not function in isolation, but exchange metabolites at several levels. The substrates of respiration enter the respiratory process at different points in the pathways, as summarized in Figure 12.1:

- **Glycolysis** involves a series of reactions catalyzed by enzymes located in both the cytosol and the plastids. A sugar—for example, sucrose—is partly oxidized via six-carbon sugar phosphates (hexose phosphates) and three-carbon sugar phosphates (triose phosphates) to produce an organic acid—mainly pyruvate. The process yields a small amount of energy as ATP and reducing power in the form of a reduced nicotinamide nucleotide, NADH.
- In the **oxidative pentose phosphate pathway**, also located in both the cytosol and the plastids, the six-carbon glucose 6-phosphate is initially oxidized to the five-carbon ribulose 5-phosphate. Carbon is lost as CO_2, and reducing power is conserved in the form of another reduced nicotinamide nucleotide, NADPH. In subsequent near-equilibrium reactions of the pentose phosphate pathway, ribulose 5-phosphate is converted

into sugar phosphates containing three to seven carbon atoms. These intermediates can be used in biosynthetic pathways or reenter glycolysis.

- In the **citric acid cycle**, pyruvate is oxidized completely to CO_2, via stepwise oxidations of organic acids in the innermost compartment of the mitochondrion—the matrix. This process mobilizes the major amount of reducing power (16 NADH + 4 $FADH_2$ per sucrose) and a small amount of energy (ATP) from the breakdown of sucrose.
- In **oxidative phosphorylation**, electrons are transferred along an electron transport chain consisting of a series of protein complexes embedded in the inner of the two mitochondrial membranes. This system transfers electrons from NADH (and related species)—produced by glycolysis, the oxidative pentose phosphate pathway, and the citric acid cycle—to oxygen. This electron transfer releases a large amount of free energy, much of which is conserved through the synthesis of ATP from ADP and P_i (inorganic phosphate), catalyzed by the enzyme ATP synthase. Collectively, the redox reactions of the electron transport chain and the synthesis of ATP are called oxidative phosphorylation.

Nicotinamide adenine dinucleotide (NAD^+/NADH) is an organic cofactor (coenzyme) associated with many enzymes that catalyze cellular redox reactions. NAD^+ is the oxidized form that undergoes a reversible two-electron reduction to yield NADH (**Figure 12.2**). The standard reduction potential for the NAD^+/NADH redox couple is about –320 mV. This tells us that NADH is a relatively strong reductant (i.e., electron donor), which can conserve the free energy carried by the electrons released during the stepwise oxidations of glycolysis and the citric acid cycle. A related compound, nicotinamide adenine dinucleotide phosphate ($NADP^+$/NADPH), has a similar function in photosynthesis (see Chapters 7 and 8) and the oxida-

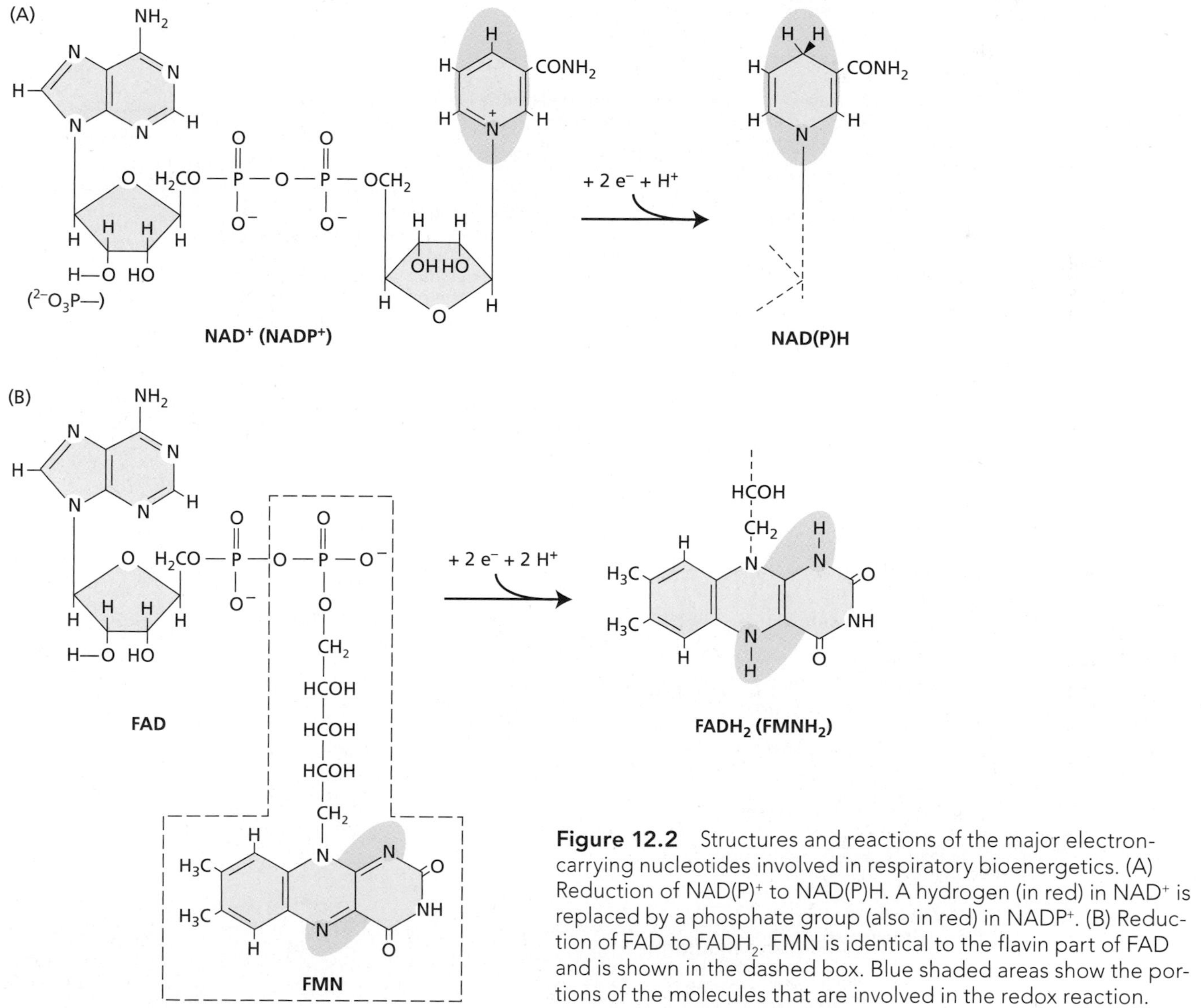

Figure 12.2 Structures and reactions of the major electron-carrying nucleotides involved in respiratory bioenergetics. (A) Reduction of $NAD(P)^+$ to NAD(P)H. A hydrogen (in red) in NAD^+ is replaced by a phosphate group (also in red) in $NADP^+$. (B) Reduction of FAD to $FADH_2$. FMN is identical to the flavin part of FAD and is shown in the dashed box. Blue shaded areas show the portions of the molecules that are involved in the redox reaction.

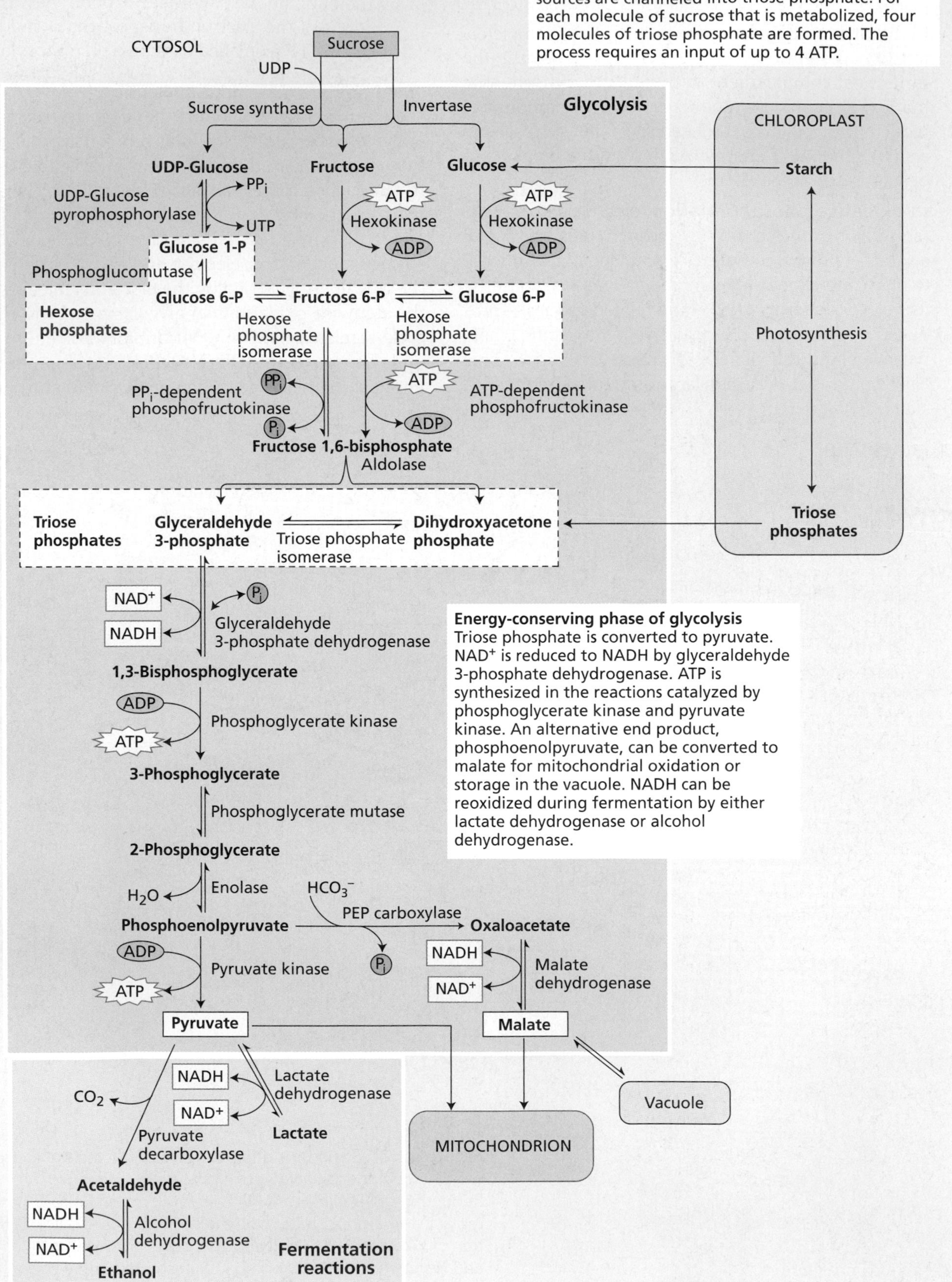

(A)
Initial phase of glycolysis Substrates from different sources are channeled into triose phosphate. For each molecule of sucrose that is metabolized, four molecules of triose phosphate are formed. The process requires an input of up to 4 ATP.
CYTOSOL
Sucrose
UDP
Sucrose synthase
Invertase
Glycolysis
CHLOROPLAST
UDP-Glucose
Fructose
Glucose
Starch
PP_i
UDP-Glucose pyrophosphorylase
UTP
ATP
Hexokinase
ADP
ATP
Hexokinase
ADP
Glucose 1-P
Phosphoglucomutase
Glucose 6-P
Fructose 6-P
Glucose 6-P
Hexose phosphates
Hexose phosphate isomerase
Hexose phosphate isomerase
Photosynthesis
PP_i
PP_i-dependent phosphofructokinase
P_i
ATP
ATP-dependent phosphofructokinase
ADP
Fructose 1,6-bisphosphate
Aldolase
Triose phosphates
Glyceraldehyde 3-phosphate
Triose phosphate isomerase
Dihydroxyacetone phosphate
Triose phosphates
NAD^+
NADH
P_i
Glyceraldehyde 3-phosphate dehydrogenase
Energy-conserving phase of glycolysis Triose phosphate is converted to pyruvate. NAD^+ is reduced to NADH by glyceraldehyde 3-phosphate dehydrogenase. ATP is synthesized in the reactions catalyzed by phosphoglycerate kinase and pyruvate kinase. An alternative end product, phosphoenolpyruvate, can be converted to malate for mitochondrial oxidation or storage in the vacuole. NADH can be reoxidized during fermentation by either lactate dehydrogenase or alcohol dehydrogenase.
1,3-Bisphosphoglycerate
ADP
Phosphoglycerate kinase
ATP
3-Phosphoglycerate
Phosphoglycerate mutase
2-Phosphoglycerate
H_2O
Enolase
HCO_3^-
PEP carboxylase
Phosphoenolpyruvate
Oxaloacetate
ADP
Pyruvate kinase
P_i
NADH
Malate dehydrogenase
NAD^+
ATP
Pyruvate
Malate
CO_2
NADH
Lactate dehydrogenase
NAD^+
Lactate
Pyruvate decarboxylase
Vacuole
MITOCHONDRION
Acetaldehyde
NADH
Alcohol dehydrogenase
NAD^+
Ethanol
Fermentation reactions

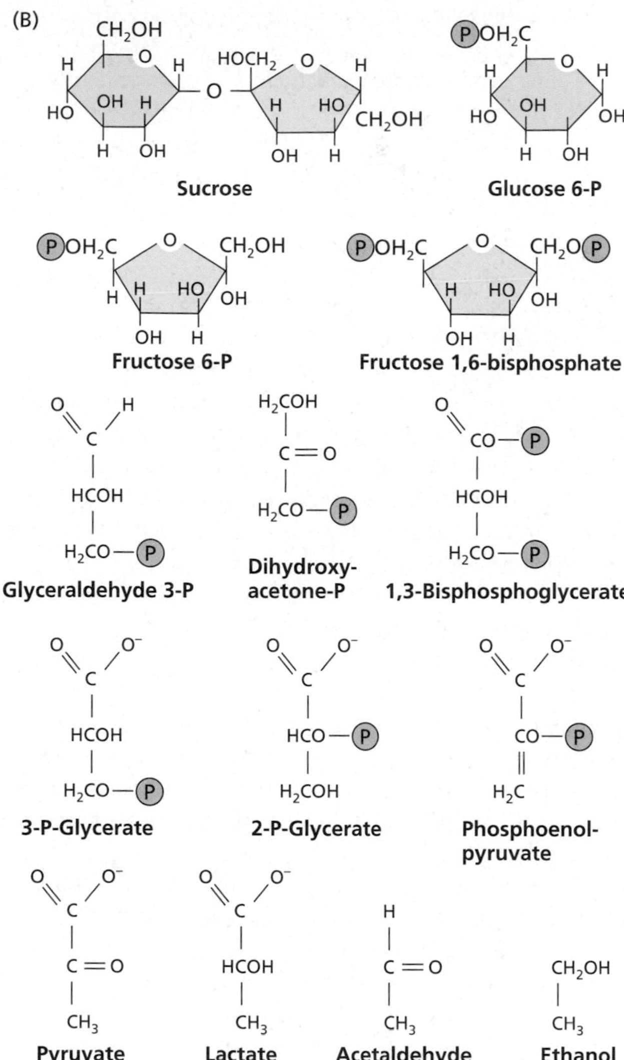

Figure 12.3 Reactions of plant glycolysis and fermentation. (A) In the main glycolytic pathway, sucrose is oxidized via hexose phosphates and triose phosphates to the organic acid pyruvate, but plants also carry out alternative reactions. All the enzymes included in this figure have been measured at levels sufficient to support the respiration rates observed in intact plant tissues, and fluxes through the pathway have been observed in vivo. The double arrows denote reversible reactions; the single arrows, essentially irreversible reactions. (B) The structures of the carbon intermediates. P, phosphate group.

tive pentose phosphate pathway, and also takes part in mitochondrial metabolism. These roles will be discussed later in the chapter.

The oxidation of NADH by oxygen via the electron transport chain releases free energy (220 kJ mol^{-1}) that drives the synthesis of approximately 60 ATP (as we will see later). We can now formulate a more complete picture of respiration as related to its role in cellular energy metabolism by coupling the following two reactions:

$$C_{12}H_{22}O_{11} + 12\ O_2 \rightarrow 12\ CO_2 + 11\ H_2O$$

$$60\ ADP + 60\ P_i \rightarrow 60\ ATP + 60\ H_2O$$

Keep in mind that not all the carbon that enters the respiratory pathway ends up as CO_2. Many respiratory carbon intermediates are the starting points for pathways that synthesize amino acids, nucleotides, lipids, and many other compounds.

Glycolysis

In the early steps of glycolysis (from the Greek words *glykos*, "sugar," and *lysis*, "splitting"), carbohydrates are converted into hexose phosphates, each of which is then split into two triose phosphates. In a subsequent energy-conserving phase, each triose phosphate is oxidized and rearranged to yield one molecule of pyruvate, an organic acid. Besides preparing the substrate for oxidation in the citric acid cycle, glycolysis yields a small amount of chemical energy in the form of ATP and NADH.

When molecular oxygen is unavailable—for example, in plant roots in flooded soils—glycolysis can be the main source of energy for cells. For this to work, the *fermentative pathways*, which are carried out in the cytosol, must reduce pyruvate to recycle the NADH produced by glycolysis. In this section we will describe the basic glycolytic and fermentative pathways, emphasizing features that are specific to plant cells. In the next section we will discuss the pentose phosphate pathway, another pathway for sugar oxidation in plants.

Glycolysis metabolizes carbohydrates from several sources

Glycolysis occurs in all living organisms (prokaryotes and eukaryotes). The principal reactions associated with the classic glycolytic pathway in plants are almost identical to those in animal cells (**Figure 12.3**). However, plant glycolysis has unique regulatory features, alternative enzymatic routes for several steps, and a parallel partial glycolytic pathway in plastids.

In animals, the substrate of glycolysis is glucose, and the end product is pyruvate. Because sucrose is the major translocated sugar in most plants, and is therefore the form of carbon that most nonphotosynthetic tissues import, sucrose (not glucose) can be argued to be the true sugar substrate for plant glycolysis. The end products of plant glycolysis include another organic acid, malate.

In the early steps of glycolysis, sucrose is split into its two monosaccharide units—glucose and fructose—which can readily enter the glycolytic pathway. Two pathways for the splitting of sucrose are known in plants, both of which take part in the use of sucrose from phloem unloading (see Chapter 11): the invertase pathway and the sucrose synthase pathway.

Invertases hydrolyze sucrose in the cell wall, vacuole, or cytosol into its two component hexoses (glucose and fructose). The hexoses are then phosphorylated in the cytosol by a hexokinase that uses ATP to form **hexose phosphates**. Alternatively, *sucrose synthase* combines sucrose with UDP to produce fructose and UDP-glucose in the cytosol. UDP-glucose pyrophosphorylase then converts UDP-glucose and pyrophosphate (PP_i) into UTP and glucose 6-phosphate (see Figure 12.3). While the sucrose synthase reaction is close to equilibrium, the invertase reaction is essentially irreversible, driving the flux in the forward direction.

Through studies of transgenic plants lacking specific invertases or sucrose synthase, each enzyme has been found to be essential for specific life processes, but differences are observed among plant tissues and species. For example, sucrose synthase and cell wall invertase are needed for normal fruit development in several crop species, whereas the invertase degrading cytosolic sucrose is necessary for optimum root cell wall integrity and leaf respiration in *Arabidopsis thaliana*. Both sucrose synthase and invertases can degrade sucrose for glycolysis, and if one of the enzymes is absent, for example in a mutant, the other enzyme(s) can still maintain respiration. The existence of different pathways that serve a similar function and can replace each other without a clear loss in function is called **metabolic redundancy**; it is a common feature in plant metabolism. In plastids, a partial glycolysis occurs that produces metabolites for plastidial biosynthetic reactions, for example synthesis of fatty acids, tetrapyrroles, and aromatic amino acids. Starch is both synthesized and catabolized only in plastids, and carbon obtained from starch degradation (for example, in a chloroplast at night) enters the glycolytic pathway in the cytosol primarily as glucose (see Chapter 8). In the light, photosynthetic products can enter the glycolytic pathway directly as triose phosphate. In overview, glycolysis works like a funnel with an initial phase collecting carbon from different carbohydrate sources, depending on the physiological situation.

In the initial phase of glycolysis, each hexose unit is phosphorylated twice and then split, producing two molecules of **triose phosphate**. This series of reactions consumes two to four molecules of ATP per sucrose unit, depending on whether the sucrose is split by sucrose synthase or invertase. These reactions also include two of the three essentially irreversible reactions of the glycolytic pathway, which are catalyzed by hexokinase and phosphofructokinase (see Figure 12.3). As we will see later, the phosphofructokinase reaction is one of the control points of glycolysis in both plants and animals.

The energy-conserving phase of glycolysis extracts usable energy

The reactions discussed thus far convert carbon from the various substrate pools to triose phosphates. Once *glyceraldehyde 3-phosphate* is formed, the glycolytic pathway can begin to extract usable energy in the energy-conserving phase. The enzyme *glyceraldehyde 3-phosphate dehydrogenase* catalyzes the oxidation of the aldehyde to a carboxylic acid, reducing NAD^+ to NADH. This reaction releases sufficient free energy to allow the phosphorylation (using inorganic phosphate) of glyceraldehyde 3-phosphate to produce 1,3-bisphosphoglycerate. The phosphorylated carboxylic acid on carbon 1 of 1,3-bisphosphoglycerate (see Figure 12.3) has a large standard free-energy change ($\Delta G^{0'}$) of hydrolysis (–49.3 kJ mol^{-1}). Thus 1,3-bisphosphoglycerate is a strong donor of phosphate groups.

In the next step of glycolysis, catalyzed by *phosphoglycerate kinase*, the phosphate on carbon 1 is transferred to a molecule of ADP, yielding ATP and 3-phosphoglycerate. For each sucrose entering the pathway, four ATPs are generated by this reaction—one for each molecule of 1,3-bisphosphoglycerate.

This type of ATP synthesis, traditionally referred to as **substrate-level phosphorylation**, involves the direct transfer of a phosphate group from a substrate molecule to ADP to form ATP. ATP synthesis by substrate-level phosphorylation is mechanistically distinct from ATP synthesis by the ATP synthases involved in oxidative phosphorylation in mitochondria (which we will describe later in this chapter) or in photophosphorylation in chloroplasts (see Chapter 7).

In the subsequent two reactions, the phosphate on 3-phosphoglycerate is transferred to carbon 2, and then a molecule of water is removed, yielding the compound *phosphoenolpyruvate* (*PEP*). The phosphate group on PEP has a high $\Delta G^{0'}$ of hydrolysis (–61.9 kJ mol^{-1}), which makes PEP an extremely good phosphate donor for ATP formation. Using PEP as substrate, the enzyme *pyruvate kinase* catalyzes a second substrate-level phosphorylation to yield ATP and pyruvate. This final step, which is the third essentially irreversible step in glycolysis, yields four additional molecules of ATP for each sucrose molecule that enters the pathway.

Plants have alternative glycolytic reactions

The glycolytic degradation of sugars to pyruvate occurs in most organisms, but many organisms can also operate a similar pathway in the opposite direction. This process, to synthesize sugars from organic acids, is known as **gluconeogenesis**.

Gluconeogenesis is particularly important in plants (such as the castor oil plant *Ricinus communis* and sunflower) that store carbon in the form of oils (triacylglycerols) in the seeds. When such a seed germinates, the oil is converted by gluconeogenesis into sucrose, which is transported to the growing cells in the seedling. In the initial phase of glycolysis, gluconeogenesis overlaps with the pathway for synthesis of sucrose from photosynthetic triose phosphate described in Chapter 8, which is typical of leaf cells.

Because the glycolytic reaction catalyzed by *ATP-dependent phosphofructokinase* is essentially irreversible (see Figure 12.3), an additional enzyme, *fructose-1,6-bisphosphate phosphatase,* converts fructose 1,6-bisphosphate irreversibly into fructose 6-phosphate and P_i during gluconeogenesis. ATP-dependent phosphofructokinase and fructose-1,6-bisphosphate phosphatase represent a major control point of carbon flux through the glycolytic/gluconeogenic pathways of both plants and animals as well as in sucrose synthesis in plants (see Chapter 8).

In plants, the interconversion of fructose 6-phosphate and fructose 1,6-bisphosphate is made more complex by the presence of an additional (cytosolic) enzyme, *PP_i-dependent phosphofructokinase* (pyrophosphate:fructose 6-phosphate 1-phosphotransferase), which catalyzes the following reversible reaction (see Figure 12.3):

$$\text{Fructose 6-P} + PP_i \rightarrow \text{fructose 1,6-bisphosphate} + P_i$$

where -P represents bound phosphate. PP_i-dependent phosphofructokinase is found in the cytosol of most plant tissues at levels that are considerably higher than those of ATP-dependent phosphofructokinase. The reaction catalyzed by PP_i-dependent phosphofructokinase is readily reversible, but it is unlikely to operate in sucrose synthesis. Suppression of PP_i-dependent phosphofructokinase in transgenic plants has shown that it contributes to glycolytic conversion of hexose phosphates to triose phosphates, but that it is not essential for plant survival, indicating that the ATP-dependent phosphofructokinase can take over its function. The three enzymes that interconvert fructose 6-phosphate and fructose 1,6-bisphosphate are all regulated to match the plant demands for both respiration and synthesis of sucrose and polysaccharides. As a consequence, the operation of the glycolytic pathway in plants has several unique characteristics (see **WEB ESSAY 12.1**).

At the end of the glycolytic process, plants have alternative pathways for metabolizing PEP. In one pathway PEP is carboxylated by the ubiquitous cytosolic enzyme **PEP carboxylase** to form the organic acid oxaloacetate. The oxaloacetate is then reduced to malate by the action of *malate dehydrogenase,* which uses NADH as a source of electrons (see Figure 12.3). The resulting malate can be stored by export to the vacuole or transported to the mitochondrion, where it can be used in the citric acid cycle (discussed later). Thus, the action of pyruvate kinase and PEP carboxylase can produce pyruvate or malate for mitochondrial respiration, although pyruvate dominates in most tissues.

In the absence of oxygen, fermentation regenerates the NAD^+ needed for glycolysis

Oxidative phosphorylation does not function in the absence of oxygen. Glycolysis then cannot continue because the cell's supply of NAD^+ is limited and once the NAD^+ becomes tied up in the reduced state (NADH), the glyceraldehyde-3-phosphate dehydrogenase comes to a halt. To overcome this limitation, plants and other organisms can further metabolize pyruvate by carrying out one or more forms of **fermentation** (see Figure 12.3).

Alcoholic fermentation is common in plants, although more widely known from brewer's yeast. Two enzymes, pyruvate decarboxylase and alcohol dehydrogenase, act on pyruvate, ultimately producing ethanol and CO_2 and oxidizing NADH in the process. In lactic acid fermentation (common in mammalian muscle, but also found in plants), the enzyme lactate dehydrogenase uses NADH to reduce pyruvate to lactate, thus regenerating NAD^+.

Plant tissues may be subjected to low (hypoxic) or zero (anoxic) concentrations of ambient oxygen. The best-studied example involves flooded or waterlogged soils in which the diffusion of oxygen is sufficiently reduced for root tissues to become hypoxic. Such conditions force the tissues to carry out fermentative metabolism. In maize (corn; *Zea mays*), the initial metabolic response to low oxygen concentrations is lactic acid fermentation, but the subsequent response is alcoholic fermentation. Ethanol is thought to be a less toxic end product of fermentation because it can diffuse out of the cell, whereas lactate accumulates and promotes acidification of the cytosol. In numerous other cases, plants or plant parts function under near-anoxic conditions by carrying out some form of fermentation.

It is important to consider the efficiency of fermentation. *Efficiency* is defined here as the energy conserved as ATP relative to the energy potentially available in a molecule of sucrose. The standard free-energy change ($\Delta G^{0\prime}$) for the complete oxidation of sucrose to CO_2 is –5760 kJ mol^{-1}. The $\Delta G^{0\prime}$ for the synthesis of ATP is 32 kJ mol^{-1}. However, under the nonstandard conditions that normally exist in both mammalian and plant cells, the synthesis of ATP requires an input of free energy of approximately 50 kJ mol^{-1}.

Normal glycolysis leads to a net synthesis of four ATP molecules for each sucrose molecule converted into pyruvate. With ethanol or lactate as the final product, the efficiency of fermentation is only about 4%. Most of the energy available in sucrose remains in the ethanol or lactate. Changes in the glycolytic pathway under oxygen deficiency can increase the ATP yield. This is the case when sucrose is degraded via sucrose synthase instead of invertase, avoiding ATP consumption by the hexokinase in the initial phase of glycolysis. Such modifications emphasize the importance of energetic efficiency for plant survival in the absence of oxygen (see **WEB ESSAY 12.1**).

Because of the low energy recovery of fermentation, an increased rate of carbohydrate breakdown is needed to sustain the ATP production necessary for cell survival. The increased glycolytic rate is called the *Pasteur effect* after the French microbiologist Louis Pasteur, who first noted it when yeast switched between aerobic respiration and fermentation. Glycolysis is up-regulated by changes in metabolite levels and by the induction of genes encoding the enzymes of glycolysis and fermentation. Low-oxygen-induced genes are regulated by oxygen-dependent degradation of the gene regulatory factors.

In contrast to the products of fermentation, the pyruvate produced by glycolysis during aerobic respiration is further oxidized by mitochondria, resulting in a much more efficient use of the free energy available in sucrose.

Plant glycolysis is controlled by its products

In vivo, glycolysis appears to be regulated at the level of fructose 6-phosphate phosphorylation and PEP turnover. Unlike in animals, AMP and ATP are not major effectors of plant phosphofructokinase and pyruvate kinase. A more important regulator of plant glycolysis is the cytosolic concentration of PEP, which is a potent inhibitor of the plant ATP-dependent phosphofructokinase.

This inhibitory effect of PEP on phosphofructokinase is strongly decreased by inorganic phosphate, making the cytosolic ratio of PEP to P_i a critical factor in the control of plant glycolytic activity. Pyruvate kinase and PEP carboxylase, the enzymes that metabolize PEP in the last steps of glycolysis (see Figure 12.3), are in turn sensitive to feedback inhibition by citric acid cycle intermediates and their derivatives, including malate, citrate, 2-oxoglutarate, and glutamate.

In plants, therefore, the control of glycolysis comes from the "bottom up" (as we will discuss later in the chapter), with primary regulation at the level of PEP metabolism by pyruvate kinase and PEP carboxylase. Secondary regulation is exerted by PEP at the conversion of fructose 6-phosphate into fructose 1,6-bisphosphate (see Figure 12.3). In contrast, regulation in animals operates from the "top down," with primary activation occurring at the phosphofructokinase and secondary activation at the pyruvate kinase.

One possible benefit of bottom-up control of glycolysis is that it permits plants to regulate net glycolytic flux to pyruvate independently of related metabolic processes such as the Calvin–Benson cycle and sucrose–triose phosphate–starch interconversion. Another benefit of this control mechanism is that glycolysis can adjust to the demand for biosynthetic precursors.

A consequence of bottom-up control of glycolysis is that its rate can influence cellular concentrations of sugars, in combination with sugar-supplying processes such as phloem transport. Glucose and sucrose are potent signaling molecules that make the plant adjust its growth and development to its carbohydrate status. For example, hexokinase not only functions as a glycolytic enzyme but also as a glucose receptor, which induces sugar-dependent gene expression.

The presence of more than one enzyme metabolizing PEP in plant cells—pyruvate kinase and PEP carboxylase—may have consequences for the control of glycolysis. Although the two enzymes are inhibited by similar metabolites, PEP carboxylase can, under some conditions, catalyze a reaction that bypasses pyruvate kinase. The resulting malate can then enter the mitochondrial citric acid cycle.

Experimental support for multiple pathways of PEP metabolism comes from the study of transgenic tobacco plants with less than 5% of the normal level of cytosolic pyruvate kinase in their leaves. In these plants, neither rates of leaf respiration nor rates of photosynthesis differed from those in controls with wild-type levels of pyruvate kinase. However, reduced root growth in the transgenic plants indicated that the pyruvate kinase reaction could not be circumvented without some detrimental effects.

Fructose 2,6-bisphosphate also affects the phosphofructokinase reaction, but unlike PEP, it affects the reaction in both the forward and reverse direction (see Chapter 8 for a detailed discussion). Therefore, fructose 2,6-bisphosphate mediates control of the partitioning of sugars between respiration and biosynthesis.

Another level of regulation may ensue from changes in the location of the glycolytic enzymes. These enzymes were believed to be soluble in the cytosol; however, it is now clear that under high respiratory demand, there is a substantial pool of glycolytic enzymes bound to the mitochondrial outer surface. This positioning allows direct movement of intermediates from one enzyme to the next (called *substrate channeling*), which separates mitochondrially bound glycolysis from glycolysis in the cytosol. The latter can then contribute carbon intermediates to other processes without interfering with pyruvate production for respiration.

Understanding of the regulation of glycolysis requires the study of temporal changes in metabolite levels. Rapid extraction, separation, and analysis of many metabolites can be achieved by an approach called *metabolic profiling* (see **WEB ESSAY 12.2**).

The Oxidative Pentose Phosphate Pathway

The glycolytic pathway is not the only route available for the oxidation of sugars in plant cells. The oxidative pentose phosphate pathway (also known as the *hexose monophosphate shunt*) can also accomplish this task (**Figure 12.4**). The reactions are carried out by soluble enzymes present in the cytosol and in plastids. Under most conditions, the pathway in plastids predominates over that in the cytosol.

The first two reactions of this pathway involve the oxidative events that convert the six-carbon molecule glucose 6-phosphate into the five-carbon unit **ribulose 5-phosphate**, with loss of a CO_2 molecule and generation of two molecules of NADPH (not NADH). The remaining reac-

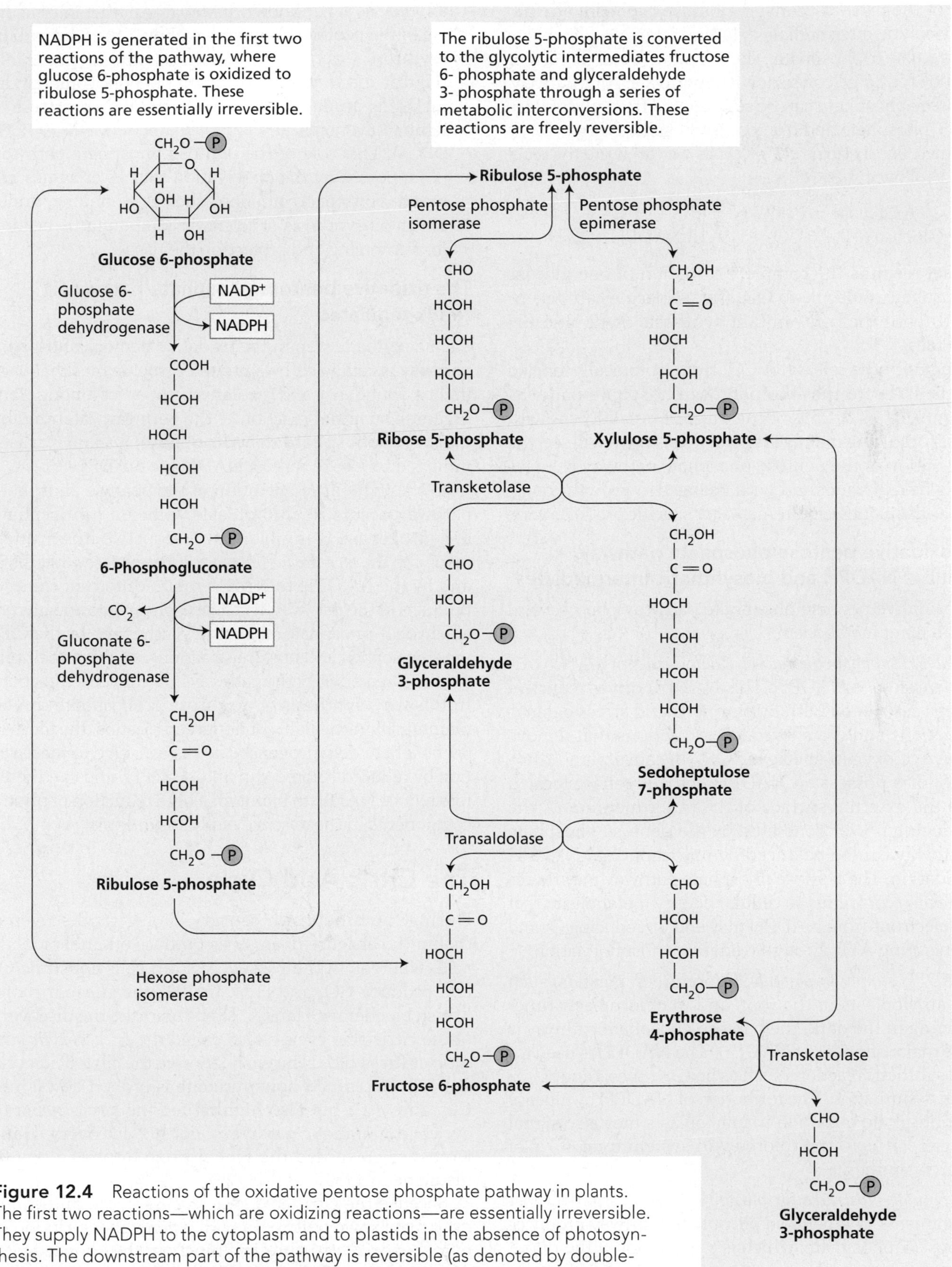

Figure 12.4 Reactions of the oxidative pentose phosphate pathway in plants. The first two reactions—which are oxidizing reactions—are essentially irreversible. They supply NADPH to the cytoplasm and to plastids in the absence of photosynthesis. The downstream part of the pathway is reversible (as denoted by double-headed arrows), so it can supply five-carbon substrates for biosynthesis even when the oxidizing reactions are inhibited—for example, in chloroplasts in the light.

tions of the pathway convert ribulose 5-phosphate into the glycolytic intermediates glyceraldehyde 3-phosphate and fructose 6-phosphate. These products can be further metabolized by glycolysis to yield pyruvate. Alternatively, glucose 6-phosphate can be regenerated from glyceraldehyde 3-phosphate and fructose 6-phosphate by glycolytic enzymes. For six turns of this cycle, we can write the reaction as follows:

$$6 \text{ Glucose 6-P} + 12 \text{ NADP}^+ + 7 \text{ H}_2\text{O} \rightarrow$$
$$5 \text{ Glucose 6-P} + 6 \text{ CO}_2 + \text{P}_i + 12 \text{ NADPH} + 12 \text{ H}^+$$

The net result is the complete oxidation of one glucose 6-phosphate molecule to CO_2 (five molecules are regenerated) with the concomitant synthesis of 12 NADPH molecules.

Studies of the release of CO_2 from isotopically labeled glucose indicate that the pentose phosphate pathway accounts for 10 to 25% of the glucose breakdown, with the rest occurring mainly via glycolysis. As we will see, the contribution of the pentose phosphate pathway changes during development and with changes in growth conditions as the plant's requirements for specific products vary.

The oxidative pentose phosphate pathway produces NADPH and biosynthetic intermediates

The oxidative pentose phosphate pathway plays several roles in plant metabolism:

- *NADPH supply in the cytosol.* The product of the two oxidative steps is NADPH. This NADPH drives reductive steps associated with biosynthetic and stress defense reactions and is a substrate for reactions that remove reactive oxygen species (ROS). Because plant mitochondria possess an NADPH dehydrogenase located on the external surface of the inner membrane, the reducing power generated by the pentose phosphate pathway can be balanced by mitochondrial NADPH oxidation. The pentose phosphate pathway may therefore also contribute to cellular energy metabolism; that is, electrons from NADPH may end up reducing O_2 and generating ATP through oxidative phosphorylation.
- *NADPH supply in plastids.* In nongreen plastids, such as amyloplasts in the root, and in chloroplasts functioning in the dark, the pentose phosphate pathway is the major supplier of NADPH. The NADPH is used for biosynthetic reactions such as lipid synthesis and nitrogen assimilation. The formation of NADPH by glucose 6-phosphate oxidation in amyloplasts may also signal sugar status to the thioredoxin system for control of starch synthesis.
- *Supply of substrates for biosynthetic processes.* In most organisms, the pentose phosphate pathway produces ribose 5-phosphate, which is a precursor of the ribose and deoxyribose needed in the synthesis of nucleic acids. In plants, however, ribose appears to be synthesized by another, as yet unknown, pathway. Another intermediate in the pentose phosphate pathway, the four-carbon erythrose 4-phosphate, combines with PEP in the initial reaction that produces plant phenolic compounds, including aromatic amino acids and the precursors of lignin, flavonoids, and phytoalexins (see **WEB APPENDIX 4**). This role of the pentose phosphate pathway is supported by the observation that its enzymes are induced by stress conditions such as wounding, under which biosynthesis of aromatic compounds is needed for reinforcing and protecting the tissue.

The oxidative pentose phosphate pathway is redox-regulated

Each enzymatic step in the oxidative pentose phosphate pathway is catalyzed by a group of isoenzymes that vary in their abundance and regulatory properties among plant organs. The initial reaction of the pathway, catalyzed by **glucose 6-phosphate dehydrogenase**, is in many cases inhibited by a high ratio of NADPH to $NADP^+$.

In the light, little operation of the pentose phosphate pathway occurs in chloroplasts. Glucose 6-phosphate dehydrogenase is inhibited by a reductive inactivation involving the *ferredoxin–thioredoxin system* (see Chapter 8) and by the NADPH to $NADP^+$ ratio. Moreover, the end products of the pathway, fructose 6-phosphate and glyceraldehyde 3-phosphate, are being synthesized by the Calvin–Benson cycle. Thus, mass action will drive the nonoxidative reactions of the pathway in the reverse direction. In this way, synthesis of erythrose 4-phosphate can be maintained in the light. In nongreen plastids, the glucose 6-phosphate dehydrogenase is less sensitive to inactivation by reduced thioredoxin and NADPH, and can therefore reduce $NADP^+$ to maintain a high reduction of plastid components in the absence of photosynthesis.

The Citric Acid Cycle

During the nineteenth century, biologists discovered that in the absence of air, cells produce ethanol or lactic acid, whereas in the presence of air, cells consume O_2 and produce CO_2 and H_2O. In 1937 the German-born British biochemist Hans A. Krebs reported the discovery of the citric acid cycle—also called the *tricarboxylic acid cycle* or *Krebs cycle*. The elucidation of the citric acid cycle not only explained how pyruvate is broken down into CO_2 and H_2O, but also highlighted the key concept of cycles in metabolic pathways. For his discovery, Hans Krebs was awarded the Nobel Prize in physiology or medicine in 1953.

Because the citric acid cycle occurs in the mitochondrial matrix, we will begin with a general description of mitochondrial structure and function, the knowledge of which was obtained mainly through experiments on isolated mitochondria (see **WEB TOPIC 12.1**). We will then

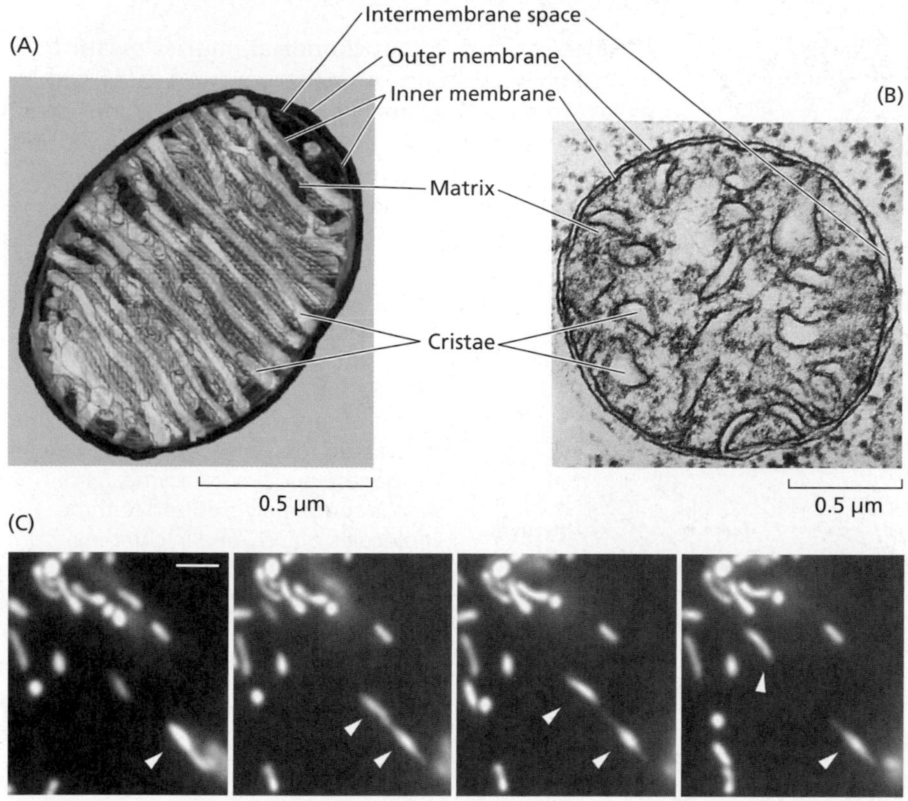

Figure 12.5 Structure of mitochondria from animals and plants. (A) Three-dimensional tomography picture of a chicken brain mitochondrion, showing the invaginations of the inner membrane, called cristae, as well as the locations of the matrix and intermembrane space (see also Figure 12.10). (B) Electron micrograph of a mitochondrion in a mesophyll cell of broad bean (*Vicia faba*). Typically, individual mitochondria are 1 to 3 μm long in plant cells, which means they are substantially smaller than nuclei and plastids. (C) Time-lapse pictures showing a dividing mitochondrion in an Arabidopsis epidermal cell (arrowheads). All the visible organelles are mitochondria labeled with green fluorescent protein. The pictures shown were taken 2 s apart. Scale bar = 1 μm. See **WEB ESSAY 12.3** for the complete video. (A from Perkins et al. 1997; B from Gunning and Steer 1996; C courtesy of David C. Logan.)

review the steps of the citric acid cycle, emphasizing the features that are specific to plants and how they affect respiratory function.

Mitochondria are semiautonomous organelles

The breakdown of sucrose into pyruvate releases less than 25% of the total energy in sucrose; the remaining energy is stored in the four molecules of pyruvate. The next two stages of respiration (the citric acid cycle and oxidative phosphorylation) take place within an organelle enclosed by a double membrane, the **mitochondrion** (plural *mitochondria*).

Plant mitochondria are usually spherical or rodlike and range from 0.5 to 1.0 μm in diameter and up to 3 μm in length (**Figure 12.5**). Like chloroplasts, they are semiautonomous organelles, because they contain ribosomes, RNA, and DNA, which encodes a limited number of mitochondrial proteins. Plant mitochondria are thus able to carry out the various steps of protein synthesis and to transmit their genetic information. The number and sizes of mitochondria in a cell can vary dynamically due to mitochondrial division and fusion (see Figure 12.5C and WEB ESSAY 12.3) while keeping up with cell division. Metabolically active tissues usually contain more mitochondria than less active tissues, reflecting the mitochondrial role in energy metabolism. Guard cells, for example, are unusually rich in mitochondria.

The ultrastructural features of plant mitochondria are similar to those of mitochondria in other organisms (see Figure 12.5A and B). Plant mitochondria have two membranes: a smooth **outer mitochondrial membrane** completely surrounds a highly invaginated **inner mitochondrial membrane**. The invaginations of the inner membrane are known as **cristae** (singular *crista*). As a consequence of its greatly enlarged surface area, the inner membrane can contain more than 50% of the total mitochondrial protein. The region between the two mitochondrial membranes is known as the **intermembrane space**. The compartment enclosed by the inner membrane is referred to as the mitochondrial **matrix.** It has a very high content of macromolecules, approximately 50% by weight. Because there is little water in the matrix, mobility is restricted, and it is likely that matrix proteins are organized into multienzyme complexes (so-called metabolons) to facilitate substrate channeling.

Intact mitochondria are osmotically active; that is, they take up water and swell when placed in a hypoosmotic medium. Ions and polar molecules are generally unable to diffuse freely through the inner membrane, which functions as the osmotic barrier. The outer membrane is permeable to solutes that have a molecular mass of less than approximately 10,000 Da—that is, most cellular metabolites and ions, but not proteins. The lipid fraction of both membranes is primarily made up of phospholipids, 80% of which are either phosphatidylcholine or phosphatidylethanolamine. About 15% is diphosphatidylglycerol (also called cardiolipin), which occurs in cells only in the inner mitochondrial membrane.

Pyruvate enters the mitochondrion and is oxidized via the citric acid cycle

The citric acid cycle is also known as the tricarboxylic acid cycle because of the importance of the tricarboxylic acids citric acid (citrate) and isocitric acid (isocitrate) as early intermediates (**Figure 12.6**). This cycle constitutes the second stage in respiration and takes place in the mitochondrial matrix. Its operation requires that the pyruvate generated in the cytosol during glycolysis be transported through the impermeable inner mitochondrial membrane via a specific transport protein (as we will describe shortly).

Once inside the mitochondrial matrix, pyruvate is decarboxylated in an oxidation reaction catalyzed by **pyruvate dehydrogenase**, a large complex consisting of several enzymes. The products are NADH, CO_2, and acetyl-CoA, in which the acetyl group derived from pyruvate is linked by a thioester bond to a cofactor, coenzyme A (CoA) (see Figure 12.6).

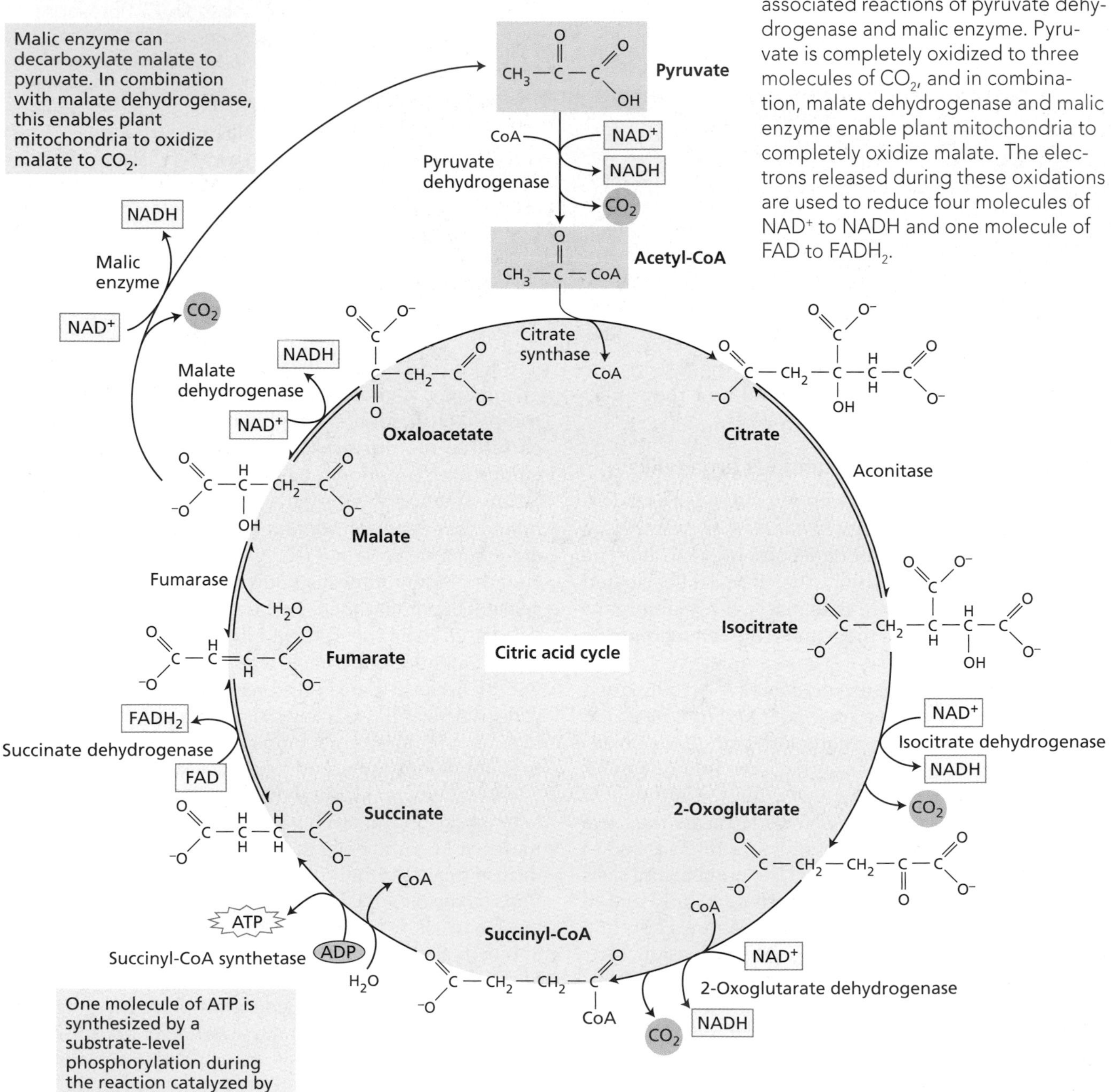

Figure 12.6 The plant citric acid cycle and associated reactions. The reactions and enzymes of the citric acid cycle are displayed, along with the associated reactions of pyruvate dehydrogenase and malic enzyme. Pyruvate is completely oxidized to three molecules of CO_2, and in combination, malate dehydrogenase and malic enzyme enable plant mitochondria to completely oxidize malate. The electrons released during these oxidations are used to reduce four molecules of NAD^+ to NADH and one molecule of FAD to $FADH_2$.

In the next reaction, the enzyme citrate synthase, formally the first enzyme in the citric acid cycle, combines the acetyl group of acetyl-CoA with a four-carbon dicarboxylic acid (*oxaloacetate*) to give a six-carbon tricarboxylic acid (citrate). Citrate is then isomerized to isocitrate by the enzyme aconitase.

The following two reactions are successive oxidative decarboxylations, each of which produces one NADH and releases one molecule of CO_2, yielding a four-carbon product bound to CoA, succinyl-CoA. At this point, three molecules of CO_2 have been produced for each pyruvate that entered the mitochondrion, or 12 CO_2 for each molecule of sucrose oxidized.

In the remainder of the citric acid cycle, succinyl-CoA is oxidized to oxaloacetate, allowing the continued operation of the cycle. Initially the large amount of free energy available in the thioester bond of succinyl-CoA is conserved through the synthesis of ATP from ADP and P_i via a substrate-level phosphorylation catalyzed by *succinyl-CoA synthetase*. (Recall that the free energy available in the thioester bond of acetyl-CoA was used to form a carbon–carbon bond in the step catalyzed by citrate synthase.) The resulting succinate is oxidized to fumarate by *succinate dehydrogenase*, which is the only membrane-associated enzyme of the citric acid cycle and also part of the electron transport chain.

The electrons and protons removed from succinate end up not on NAD^+, but on another cofactor involved in redox reactions: **flavin adenine dinucleotide** (**FAD**). FAD is covalently bound to the active site of succinate dehydrogenase and undergoes a reversible two-electron reduction to produce $FADH_2$ (see Figure 12.2B).

In the final two reactions of the citric acid cycle, fumarate is hydrated to produce malate, which is subsequently oxidized by *malate dehydrogenase* to regenerate oxaloacetate and produce another molecule of NADH. The oxaloacetate produced is now able to react with another acetyl-CoA and continue the cycling.

The stepwise oxidation of one molecule of pyruvate in the mitochondrion gives rise to three molecules of CO_2, and much of the free energy released during these oxidations is conserved in the form of four NADH and one $FADH_2$. In addition, one molecule of ATP is produced by a substrate-level phosphorylation.

The citric acid cycle of plants has unique features

The citric acid cycle reactions outlined in Figure 12.6 are not all identical to those carried out by animal mitochondria. For example, the step catalyzed by succinyl-CoA synthetase produces ATP in plants and GTP in animals. These nucleotides are energetically equivalent.

A feature of the plant citric acid cycle that is absent in many other organisms is the presence of **malic enzyme** in the mitochondrial matrix of plants. This enzyme catalyzes the oxidative decarboxylation of malate:

$$\text{Malate} + NAD^+ \rightarrow \text{pyruvate} + CO_2 + NADH$$

The activity of malic enzyme enables plant mitochondria to operate alternative pathways for the metabolism of PEP derived from glycolysis (see **WEB ESSAY 12.1**). As already described, malate can be synthesized from PEP in the cytosol via the enzymes PEP carboxylase and malate dehydrogenase (see Figure 12.3). For degradation, malate is transported into the mitochondrial matrix, where malic enzyme can oxidize it to pyruvate. This reaction makes possible the complete net oxidation of citric acid cycle intermediates such as malate (**Figure 12.7A**) or citrate (**Figure 12.7B**). Many plant tissues, not only those that carry out crassulacean acid metabolism (see Chapter 8), store significant amounts of malate or other organic acids in their vacuoles. Degradation of malate via mitochondrial malic enzyme is important for regulating levels of organic acids in cells—for example, during fruit ripening.

Instead of being degraded, the malate produced via PEP carboxylase can replace citric acid cycle intermediates used in biosynthesis. Reactions that replenish intermediates in a metabolic cycle are known as *anaplerotic*. For example, export of 2-oxoglutarate for nitrogen assimilation in the chloroplast causes a shortage of malate for the citrate synthase reaction. This malate can be replaced through the PEP carboxylase pathway (**Figure 12.7C**).

Gamma-aminobutyric acid (GABA) is an amino acid that accumulates under several biotic and abiotic stress conditions in plants. GABA is synthesized from 2-oxoglutarate and degraded into succinate by the so-called **GABA shunt**, which bypasses the citric acid cycle enzymes. The functional relationship between GABA accumulation and stress remains poorly understood.

Mitochondrial Electron Transport and ATP Synthesis

ATP is the energy carrier used by cells to drive life processes, so chemical energy conserved during the citric acid cycle in the form of NADH and $FADH_2$ must be converted into ATP to perform useful work in the cell. This O_2-dependent process, called oxidative phosphorylation, occurs in the inner mitochondrial membrane.

In this section we will describe the process by which the energy level of the electrons from NADH and $FADH_2$ is lowered in a stepwise fashion and conserved in the form of an electrochemical proton gradient across the inner mitochondrial membrane. Although fundamentally similar in all aerobic cells, the electron transport chain of plants (and many fungi and protists) contains multiple NAD(P)H dehydrogenases and an alternative oxidase, none of which are found in mammalian mitochondria.

We will also examine the enzyme that uses the energy of the proton gradient to synthesize ATP: the F_oF_1-ATP

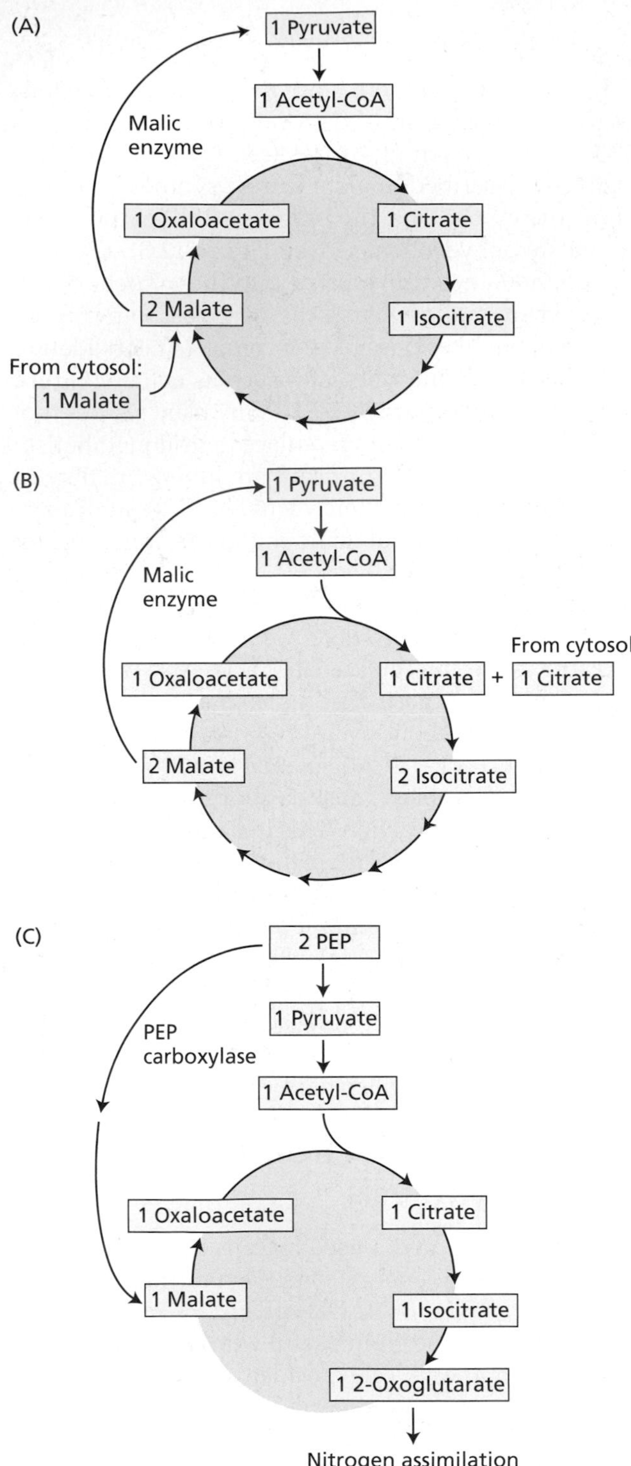

Figure 12.7 Malic enzyme and PEP carboxylase provide plants with metabolic flexibility for the metabolism of PEP and pyruvate. Malic enzyme converts malate into pyruvate and thus allows plant mitochondria to oxidize both (A) malate and (B) citrate to CO_2 without involving pyruvate delivered by glycolysis. (C) With the added action of PEP carboxylase to the standard pathway, glycolytic PEP is converted into 2-oxoglutarate, which is used for nitrogen assimilation.

synthase. After examining the various stages in the production of ATP, we will summarize the energy conservation steps at each stage, as well as the regulatory mechanisms that coordinate the different pathways.

The electron transport chain catalyzes a flow of electrons from NADH to O_2

For each molecule of sucrose oxidized through glycolysis and the citric acid cycle, 4 molecules of NADH are generated in the cytosol, and 16 molecules of NADH plus 4 molecules of $FADH_2$ (associated with succinate dehydrogenase) are generated in the mitochondrial matrix. These reduced compounds must be reoxidized, or the entire respiratory process will come to a halt.

The electron transport chain catalyzes a transfer of two electrons from NADH (or $FADH_2$) to oxygen, the final electron acceptor of the respiratory process. For the oxidation of NADH, the reaction can be written as

$$\text{NADH} + \text{H}^+ + \tfrac{1}{2}\,\text{O}_2 \rightarrow \text{NAD}^+ + \text{H}_2\text{O}$$

From the reduction potentials for the NADH–NAD^+ pair (–320 mV) and the H_2O–½ O_2 pair (+810 mV), it can be calculated that the standard free energy released during this overall reaction ($-nF\Delta E^{0\prime}$) is about 220 kJ per mole of NADH. Because the succinate–fumarate reduction potential is higher (+30 mV), only 152 kJ per mole of succinate is released. The role of the electron transport chain is to bring about the oxidation of NADH (and $FADH_2$) and, in the process, use some of the free energy released to generate an electrochemical proton gradieµnt, $\Delta\tilde{\mu}_{H^+}$, across the inner mitochondrial membrane.

The electron transport chain of plants contains the same set of electron carriers found in the mitochondria of other organisms (**Figure 12.8**). The individual electron transport proteins are organized into four transmembrane multiprotein complexes (identified by roman numerals I through IV), all of which are localized in the inner mitochondrial membrane. Three of these complexes are engaged in proton pumping (I, III, and IV).

COMPLEX I (NADH DEHYDROGENASE) Electrons from NADH generated by the citric acid cycle in the mitochondrial matrix are oxidized by complex I (an **NADH dehydrogenase**). The electron carriers in complex I include a tightly bound cofactor (**flavin mononucleotide**, or **FMN**, which is chemically similar to FAD; see Figure 12.2B) and several iron–sulfur centers. Complex I then transfers these electrons to ubiquinone. Four protons are pumped from the matrix into the intermembrane space for every electron pair passing through the complex.

Ubiquinone, a small lipid-soluble electron and proton carrier, is localized within the inner membrane. It is not tightly associated with any protein, and it can diffuse within the hydrophobic core of the membrane bilayer.

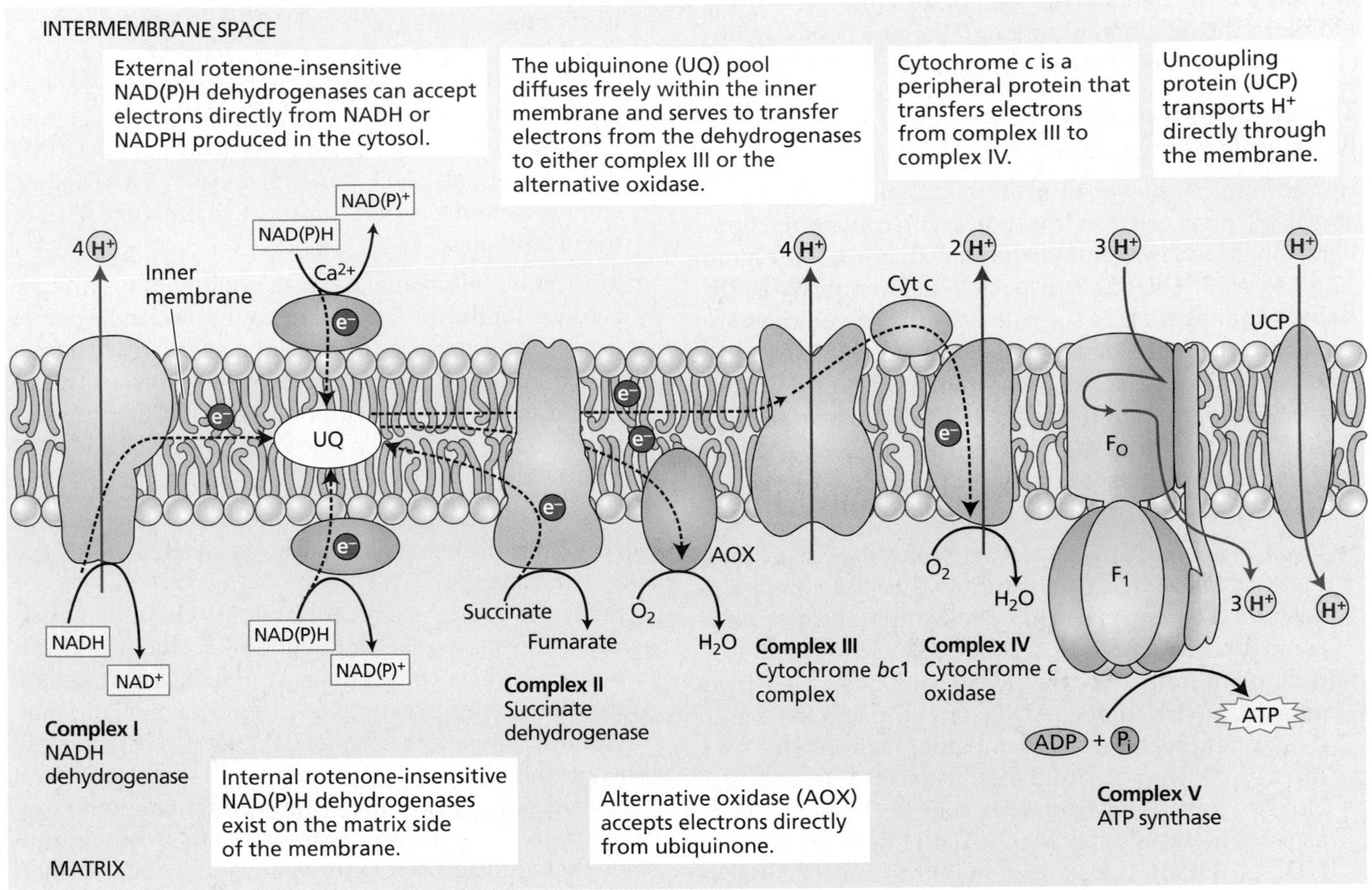
INTERMEMBRANE SPACE
External rotenone-insensitive NAD(P)H dehydrogenases can accept electrons directly from NADH or NADPH produced in the cytosol.
The ubiquinone (UQ) pool diffuses freely within the inner membrane and serves to transfer electrons from the dehydrogenases to either complex III or the alternative oxidase.
Cytochrome c is a peripheral protein that transfers electrons from complex III to complex IV.
Uncoupling protein (UCP) transports H+ directly through the membrane.
NAD(P)+
NAD(P)H
4 H+
Inner membrane
Ca2+
e−
4 H+
2 H+
3 H+
H+
Cyt c
UCP
UQ
Fo
AOX
O2
H2O
F1
3 H+
H+
NADH
NAD+
NAD(P)H
NAD(P)+
Succinate
Fumarate
O2
H2O
Complex III
Cytochrome bc1 complex
Complex IV
Cytochrome c oxidase
ATP
ADP + Pi
Complex I
NADH dehydrogenase
Complex II
Succinate dehydrogenase
Internal rotenone-insensitive NAD(P)H dehydrogenases exist on the matrix side of the membrane.
Alternative oxidase (AOX) accepts electrons directly from ubiquinone.
Complex V
ATP synthase
MATRIX

Figure 12.8 Organization of the electron transport chain and ATP synthesis in the inner membrane of the plant mitochondrion. Mitochondria from nearly all eukaryotes contain the four standard protein complexes: I, II, III, and IV. The structures of all complexes have been determined, but they are shown here as simplified shapes. The electron transport chain of the plant mitochondrion contains additional enzymes (depicted in green) that do not pump protons. Additionally, uncoupling proteins directly bypass the ATP synthase by allowing passive proton influx. This multiplicity of bypasses in plants, whereas mammals have only the uncoupling protein, gives a greater flexibility to plant energy coupling.

COMPLEX II (SUCCINATE DEHYDROGENASE) Oxidation of succinate in the citric acid cycle is catalyzed by this complex, and the reducing equivalents are transferred via $FADH_2$ and a group of iron–sulfur centers to ubiquinone. Complex II does not pump protons.

COMPLEX III (CYTOCHROME bc_1 COMPLEX) Complex III oxidizes reduced ubiquinone (ubiquinol) and transfers the electrons via an iron–sulfur center, two *b*-type cytochromes (b_{565} and b_{560}), and a membrane-bound cytochrome c_1 to cytochrome *c*. Four protons per electron pair are pumped out of the matrix by complex III using a mechanism called the **Q cycle** (see WEB TOPIC 12.2).

Cytochrome c is a small protein loosely attached to the outer surface of the inner membrane and serves as a mobile carrier to transfer electrons between complexes III and IV.

COMPLEX IV (CYTOCHROME *c* OXIDASE) Complex IV contains two copper centers (Cu_A and Cu_B) and cytochromes a and a_3. This complex is the terminal oxidase and brings about the four-electron reduction of O_2 to two molecules of H_2O. Two protons are pumped out of the matrix per electron pair (see Figure 12.8).

Both structurally and functionally, ubiquinone and the cytochrome bc_1 complex are very similar to plastoquinone and the cytochrome b_6f complex, respectively, in the photosynthetic electron transport chain (see Chapter 7).

Reality may be more complex than the description above implies. Plant respiratory complexes contain a number of plant-specific subunits whose function is still unknown. Several of the complexes contain subunits that participate in functions other than electron transport, such as protein import. Finally, several of the complexes appear to be present in supercomplexes, instead of freely mobile in the membrane, although the functional significance of these supercomplexes is not clear.

The electron transport chain has supplementary branches

In addition to the set of protein complexes described above, the plant electron transport chain contains components not found in mammalian mitochondria (see Figure 12.8 and **WEB TOPIC 12.3**). Especially, additional **NAD(P)H dehydrogenases** and a so-called **alternative oxidase** are bound to the inner membrane. They do not pump protons, so the energy released from oxidation of NADH is not conserved as ATP but instead is turned into heat (**Figure 12.9**). These enzymes therefore are often called *nonphosphorylating,* in contrast to the proton-pumping complexes I, III, and IV.

- Plant mitochondria have two pathways for oxidizing matrix NADH. Electron flow through complex I, described above, is sensitive to inhibition by several compounds, including rotenone and piericidin. In addition, plant mitochondria have a rotenone-insensitive dehydrogenase, ND_{in}(NADH), on the matrix surface of the inner mitochondrial membrane. This enzyme oxidizes NADH derived from the citric acid cycle, and may also be a bypass engaged when complex I is overloaded, as we will see shortly. An NADPH dehydrogenase, ND_{in}(NADPH), is also present on the matrix surface, but very little is known about this enzyme.
- Rotenone-insensitive NAD(P)H dehydrogenases, mostly Ca^{2+}-dependent, are also attached to the outer surface of the inner membrane facing the intermembrane space. They oxidize either NADH or NADPH from the cytosol. Electrons from these external NAD(P)H dehydrogenases—ND_{ex}(NADH) and ND_{ex}(NADPH)—enter the main electron transport chain at the level of the ubiquinone pool.
- Most, if not all, plants have an additional respiratory pathway for the oxidation of ubiquinol and reduction of oxygen. This pathway involves the alternative oxidase, which, unlike cytochrome *c* oxidase, is insensitive to inhibition by cyanide, carbon monoxide, and the signal molecule nitric oxide (see **WEB TOPIC 12.3** and **WEB ESSAY 12.4**).

The physiological significance of these supplementary electron transport enzymes will be considered more fully later in the chapter.

Some additional electron transport chain dehydrogenases present in plant mitochondria directly perform important carbon conversions. A *proline dehydrogenase* oxidizes the amino acid proline. Proline accumulates during osmotic stress (see Chapter 24), and it is degraded by this mitochondrial pathway when water status returns to normal. An electron transfer flavoprotein:quinone oxidoreductase mediates the degradation of several amino acids that are used by plants as a reserve under carbon starvation conditions induced by light deprivation. Finally,

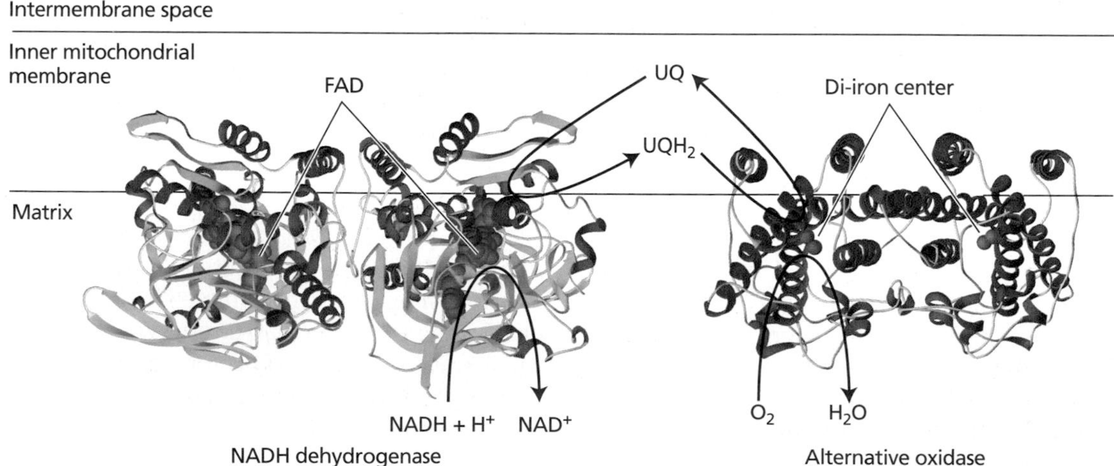

Figure 12.9 Nonphosphorylating electron transport. The internal rotenone-insensitive NADH dehydrogenase and the alternative oxidase are both homodimers, and the reactions are drawn for one monomer each. The enzymes are partly embedded in the inner leaflet of the inner mitochondrial membrane. Therefore, transfer of electrons between the hydrophilic redox couples NADH/NAD^+ and H_2O/O_2, and the hydrophobic ubiquinone (UQH_2/UQ), via single internal redox centers (FAD or di-iron groups), cannot involve proton pumping; the energy released by the reaction is instead given off as heat. Proton pumping across the inner mitochondrial membrane requires large transmembrane protein complexes. (Model of NADH dehydrogenase [from brewer's yeast, *Saccharomyces cerevisiae*] based on data from Iwata et al. 2012; model of alternative oxidase [from the sleeping sickness parasite, *Trypanosoma brucei*] based on data from Shiba et al. 2013.)

a galactono-gamma-lactone dehydrogenase, specific to plants, performs the last step in the major pathway for synthesis of the antioxidant *ascorbic acid* (also known as vitamin C). The enzyme uses cytochrome *c* as its electron acceptor, in competition with normal respiration.

ATP synthesis in the mitochondrion is coupled to electron transport

In oxidative phosphorylation, the transfer of electrons to oxygen via complexes I, III, and IV is coupled to the synthesis of ATP from ADP and P_i via the F_oF_1-ATP synthase (complex V). The number of ATPs synthesized depends on the nature of the electron donor.

In experiments conducted on isolated mitochondria, electrons donated to complex I (e.g., generated by malate oxidation) give ADP:O ratios (the number of ATPs synthesized per two electrons transferred to oxygen) of 2.4 to 2.7 (**Table 12.1**). Electrons donated to complex II (from succinate) and to the external NADH dehydrogenase give values in the range of 1.6 to 1.8, while electrons donated directly to cytochrome *c* oxidase (complex IV) via artificial electron carriers give values of 0.8 to 0.9. Results such as these (for both plant and animal mitochondria) have led to the general concept that there are three sites of energy conservation along the electron transport chain, at complexes I, III, and IV.

The experimental ADP:O ratios agree quite well with the values calculated on the basis of the number of H^+ pumped by complexes I, III, and IV and the cost of 4 H^+ for producing one ATP (see next section and Table 12.1). For instance, electrons from external NADH pass only complexes III and IV, so a total of 6 H^+ are pumped, giving 1.5 ATP (when the alternative oxidase pathway is not used).

TABLE 12.1 Theoretical and experimental ADP:O ratios in isolated plant mitochondria

Electrons feeding into	ADP:O ratio	
	Theoretical[a]	Experimental
Complex I	2.5	2.4–2.7
Complex II	1.5	1.6–1.8
External NADH dehydrogenase	1.5	1.6–1.8
Complex IV	1.0[b]	0.8–0.9

[a]It is assumed that complexes I, III, and IV pump 4, 4, and 2 H^+ per 2 electrons, respectively; that the cost of synthesizing 1 ATP and exporting it to the cytosol is 4 H^+; and that the nonphosphorylating pathways are not active.

[b]Cytochrome *c* oxidase (complex IV) pumps only 2 protons. However, 2 electrons move from the outer surface of the inner membrane (where the electrons are donated) across the inner membrane to the inner, matrix side. As a result, 2 H^+ are consumed on the matrix side. This means that the net movement of H^+ and charges is equivalent to the movement of a total of 4 H^+, giving an ADP:O ratio of 1.0.

The mechanism of mitochondrial ATP synthesis is based on the **chemiosmotic hypothesis**, described in Chapter 7, which was first proposed in 1961 by Nobel laureate Peter Mitchell as a general mechanism of energy conservation across biological membranes. According to the chemiosmotic hypothesis, the orientation of electron carriers within the inner mitochondrial membrane allows for the transfer of protons across the inner membrane during electron flow (see Figure 12.8).

Because the inner mitochondrial membrane is highly impermeable to protons, an **electrochemical proton gradient** can build up. As discussed in Chapters 6 and 7, the free energy associated with the formation of an electrochemical proton gradient ($\Delta\tilde{\mu}_{H^+}$, also referred to as a *proton motive force*, Δp, when expressed in units of volts) is made up of an electrical transmembrane potential component (ΔE) and a chemical-potential component (ΔpH) according to the following approximate equation:

$$\Delta p = \Delta E - 59\Delta \text{pH (at 25°C)}$$

where

$$\Delta E = E_{\text{inside}} - E_{\text{outside}}$$

and

$$\Delta \text{pH} = \text{pH}_{\text{inside}} - \text{pH}_{\text{outside}}$$

ΔE results from the asymmetric distribution of a charged species (H^+ and other ions) across the membrane, and ΔpH is due to the H^+ concentration difference across the membrane. Because protons are translocated from the mitochondrial matrix to the intermembrane space, the resulting ΔE across the inner mitochondrial membrane has a negative value. Under normal conditions, the ΔpH is approximately 0.5 and the ΔE approximately 0.2 V. Because the membrane is only 7 to 8 nm thick, this ΔE corresponds to an electric field of at least 25 million V/m (or 10 times the field generating a lightning flash in a thunderstorm), emphasizing the enormous forces involved in electron transport.

As this equation shows, both ΔE and ΔpH contribute to the proton motive force in plant mitochondria, although ΔpH constitutes the smaller part, probably because of the large buffering capacity of both cytosol and matrix, which prevents large pH changes. This situation contrasts with that in the chloroplast, where almost all of the proton motive force across the thylakoid membrane is due to ΔpH (see Chapter 7).

The free-energy input required to generate $\Delta\tilde{\mu}_{H^+}$ comes from the free energy released during electron transport. How electron transport is coupled to proton translocation is not completely understood in all cases. Because of the low permeability (conductance) of the inner membrane to protons, the proton electrochemical gradient can be consumed to carry out chemical work (ATP synthesis). The $\Delta\tilde{\mu}_{H^+}$ is coupled to the synthesis of ATP by an additional protein complex associated with the inner membrane, the F_oF_1-ATP synthase.

The **F_oF_1-ATP synthase** (also called *complex V*) consists of two major components, F_o and F_1 (see Figure 12.8). **F_o** (subscript "o" for oligomycin-sensitive) is an integral membrane protein complex of at least three different polypeptides. They form the channel through which protons cross the inner membrane. The other component, **F_1**, is a peripheral membrane protein complex that is composed of at least five different subunits and contains catalytic sites for converting ADP and P_i into ATP. This complex is attached to the matrix side of F_o.

The passage of protons through the channel is coupled to the catalytic cycle of the F_1 component of the ATP synthase, allowing the ongoing synthesis of ATP and the simultaneous use of the $\Delta\tilde{\mu}_{H^+}$. For each ATP synthesized, 3 H^+ pass through the F_o component from the intermembrane space to the matrix, down the electrochemical proton gradient.

A high-resolution structure for the F_1 component of the mammalian ATP synthase provided evidence for a model in which a part of F_o rotates relative to F_1 to couple H^+ transport to ATP synthesis (see **WEB TOPIC 12.4**). The structure and function of the CF_0CF_1-ATP synthase in chloroplasts are similar to those of the mitochondrial ATP synthase (see Chapter 7).

The operation of a chemiosmotic mechanism of ATP synthesis has several implications. First, the true site of ATP formation on the inner mitochondrial membrane is the ATP synthase, not complex I, III, or IV. These complexes serve as sites of energy conservation whereby electron transport is coupled to the generation of a $\Delta\tilde{\mu}_{H^+}$. The synthesis of ATP decreases the $\Delta\tilde{\mu}_{H^+}$ and, as a consequence, its restriction on the electron transport complexes. Electron transport is therefore stimulated by a large supply of ADP.

The chemiosmotic hypothesis also explains the action mechanism of **uncouplers**. These are a wide range of chemically unrelated, artificial compounds (including 2,4-dinitrophenol and *p*-trifluoromethoxycarbonylcyanide phenylhydrazone [FCCP]) that decrease mitochondrial ATP synthesis but stimulate the rate of electron transport (see **WEB TOPIC 12.5**). All of these uncoupling compounds make the inner membrane leaky to protons, which prevents the buildup of a sufficiently large $\Delta\tilde{\mu}_{H^+}$ to drive ATP synthesis or restrict electron transport.

Transporters exchange substrates and products

The electrochemical proton gradient also plays a role in the movement of the organic acids of the citric acid cycle, and of the substrates and products of ATP synthesis, into and out of mitochondria (**Figure 12.10**). Although ATP is synthesized in the mitochondrial matrix, most of it is used outside the mitochondrion, so an efficient mechanism is needed for moving ADP into and ATP out of the organelle.

The ADP/ATP (adenine nucleotide) transporter performs the active exchange of ADP and ATP across the inner membrane. The movement of the more negatively charged ATP^{4-} out of the mitochondrion in exchange for ADP^{3-}—that is, one net negative charge out—is driven by the electrical-potential gradient (ΔE, positive outside) generated by proton pumping.

The uptake of inorganic phosphate (P_i) involves an active phosphate transporter protein that uses the chemical-potential component (ΔpH) of the proton motive force to drive the electroneutral exchange of P_i^- (in) for OH^- (out). As long as a ΔpH is maintained across the inner membrane, the P_i content within the matrix remains high. Similar reasoning applies to the uptake of pyruvate, which is driven by the electroneutral exchange of pyruvate for OH^-, leading to continued uptake of pyruvate from the cytosol (see Figure 12.10).

The total energetic cost of taking up one phosphate and one ADP into the matrix and exporting one ATP is the movement of one H^+ from the intermembrane space into the matrix:

- Moving one OH^- out in exchange for P_i^- is equivalent to one H^+ in, so this electroneutral exchange consumes the ΔpH, but not the ΔE.
- Moving one negative charge out (ADP^{3-} entering the matrix in exchange for ATP^{4-} leaving) is the same as moving one positive charge in, so this transport lowers only the ΔE.

This proton, which drives the exchange of ATP for ADP and P_i, should also be included in our calculation of the cost of synthesizing one ATP. Thus, the total cost is 3 H^+ used by the ATP synthase plus 1 H^+ for the exchange across the membrane, or a total of 4 H^+.

The inner membrane also contains transporters for dicarboxylic acids (malate or succinate) exchanged for P_i^{2-} and for the tricarboxylic acids (citrate, aconitate, or isocitrate) exchanged for dicarboxylic acids (see Figure 12.10 and **WEB TOPIC 12.5**).

Aerobic respiration yields about 60 molecules of ATP per molecule of sucrose

The complete oxidation of a sucrose molecule leads to the net formation of

- Eight molecules of ATP by substrate-level phosphorylation (four from glycolysis and four from the citric acid cycle)
- Four molecules of NADH in the cytosol
- Sixteen molecules of NADH plus four molecules of $FADH_2$ (via succinate dehydrogenase) in the mitochondrial matrix

On the basis of theoretical ADP:O values (see Table 12.1), we can estimate that 52 molecules of ATP will be generated per molecule of sucrose by oxidative phosphoryla-

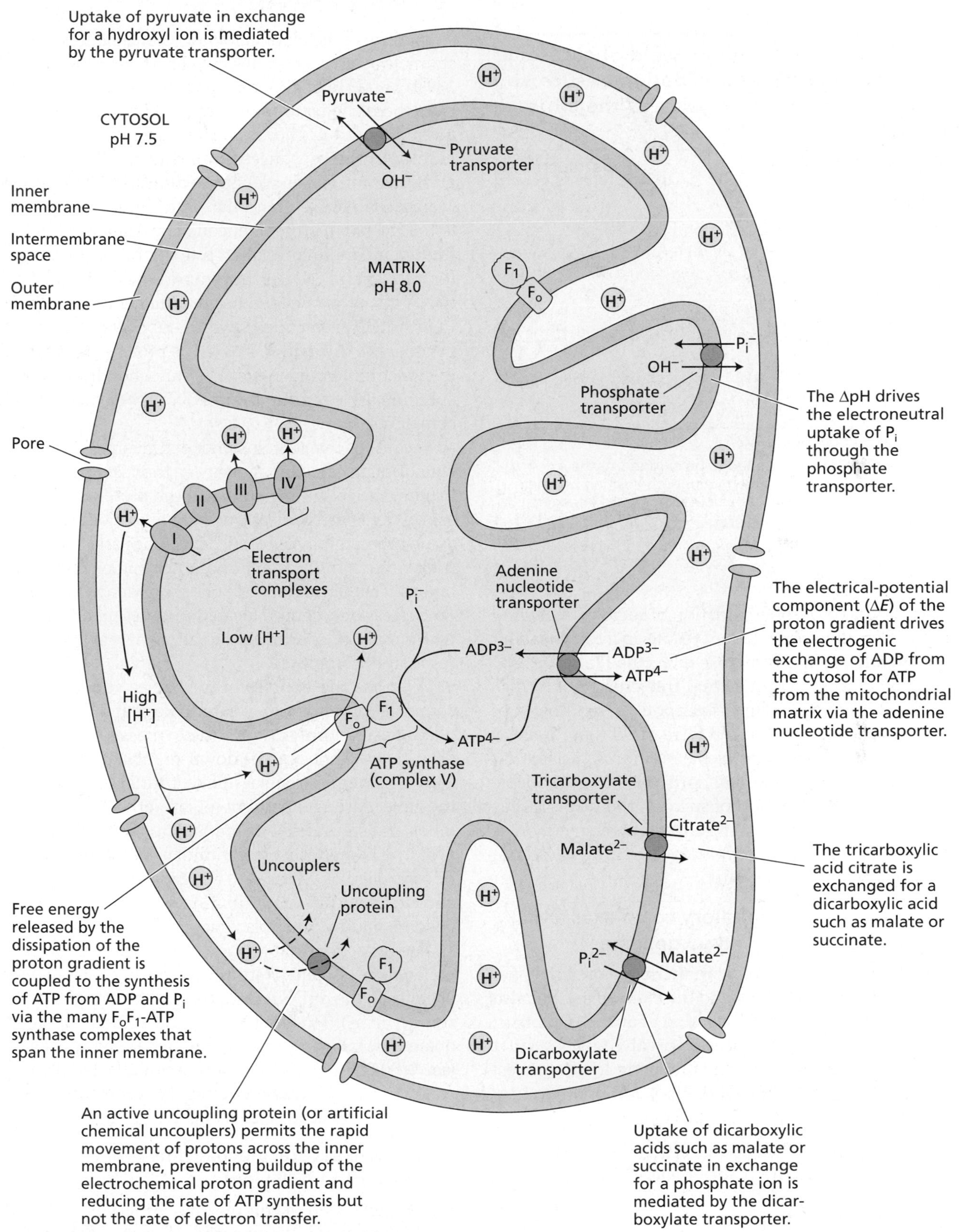

Uptake of pyruvate in exchange for a hydroxyl ion is mediated by the pyruvate transporter.
CYTOSOL
pH 7.5
Pyruvate⁻
OH⁻
Pyruvate transporter
Inner membrane
Intermembrane space
Outer membrane
MATRIX
pH 8.0
F1
Fo
Pi⁻
OH⁻
Phosphate transporter
The ΔpH drives the electroneutral uptake of Pi through the phosphate transporter.
Pore
I
II
III
IV
Electron transport complexes
Adenine nucleotide transporter
The electrical-potential component (ΔE) of the proton gradient drives the electrogenic exchange of ADP from the cytosol for ATP from the mitochondrial matrix via the adenine nucleotide transporter.
Low [H⁺]
High [H⁺]
ADP3–
ATP4–
ATP synthase (complex V)
Tricarboxylate transporter
Citrate2–
Malate2–
The tricarboxylic acid citrate is exchanged for a dicarboxylic acid such as malate or succinate.
Uncouplers
Uncoupling protein
Pi2–
Malate2–
Free energy released by the dissipation of the proton gradient is coupled to the synthesis of ATP from ADP and Pi via the many FoF1-ATP synthase complexes that span the inner membrane.
Dicarboxylate transporter
An active uncoupling protein (or artificial chemical uncouplers) permits the rapid movement of protons across the inner membrane, preventing buildup of the electrochemical proton gradient and reducing the rate of ATP synthesis but not the rate of electron transfer.
Uptake of dicarboxylic acids such as malate or succinate in exchange for a phosphate ion is mediated by the dicarboxylate transporter.

Figure 12.10 Transmembrane transport in plant mitochondria. An electrochemical proton gradientμ, $\Delta\tilde{\mu}_{H^+}$, consisting of an electrical-potential component (ΔE, –200 mV, negative inside) and a chemical-potential component (ΔpH, alkaline inside), is established across the inner mitochondrial membrane during electron transport. The $\Delta\tilde{\mu}_{H^+}$ is used by specific transporters that move metabolites across the inner membrane. (After Douce 1985.)

TABLE 12.2 The maximum yield of cytosolic ATP from the complete oxidation of sucrose to CO_2 via aerobic glycolysis and the citric acid cycle

Part reaction	ATP per sucrose[a]
Glycolysis	
4 substrate-level phosphorylations	4
4 NADH	4 × 1.5 = 6
Citric acid cycle	
4 substrate-level phosphorylations	4
4 $FADH_2$	4 × 1.5 = 6
16 NADH	16 × 2.5 = 40
Total	**60**

Source: Adapted from Brand 1994.

Note: Cytosolic NADH is assumed to be oxidized by the external NADH dehydrogenase. The other nonphorphorylating pathways (e.g., the alternative oxidase) are assumed not to be engaged.

[a]Calculated using the theoretical ADP:O values from Table 12.1.

tion. The complete aerobic oxidation of sucrose (including substrate-level phosphorylation) results in a total of about 60 ATPs synthesized per sucrose molecule (**Table 12.2**).

Using 50 kJ mol^{-1} as the actual free energy of formation of ATP in vivo, we find that about 3010 kJ mol^{-1} of free energy is conserved in the form of ATP per mole of sucrose oxidized during aerobic respiration. This amount represents about 52% of the standard free energy available from the complete oxidation of sucrose; the rest is lost as heat. It also represents a vast improvement over fermentative metabolism, in which only 4% of the energy available in sucrose is converted into ATP.

Several subunits of respiratory complexes are encoded by the mitochondrial genome

The genetic system of the plant mitochondrion differs not only from that of the nucleus and the chloroplast, but also from those found in the mitochondria of animals, protists, and fungi. Most notably, processes involving RNA differ between plant mitochondria and mitochondria from most other organisms (see **WEB TOPIC 12.6**). Major differences are found in

- RNA splicing (for example, special introns are present)
- RNA editing (in which the nucleotide sequence is changed)
- Signals regulating RNA stability
- Translation (plant mitochondria use the universal genetic code, whereas mitochondria in other eukaryotes have deviant codons)

The size of the plant mitochondrial genome varies substantially even among closely related plant species, but at 180 to 11,000 kilobase pairs (kbp), it is always much larger than the compact and uniform 16 kbp genome found in mammalian mitochondria. The size differences are due mainly to the presence of noncoding DNA, including numerous introns, in plant **mitochondrial DNA (mtDNA)**. Mammalian mtDNA encodes only 13 proteins, in contrast to the 35 known proteins encoded by Arabidopsis mtDNA. Both plant and mammalian mitochondria contain genes for rRNAs and tRNAs, but importantly, several nuclear tRNA genes are needed to give the complete tRNA set. Plant mtDNA encodes several subunits of respiratory complexes I through V, as well as proteins that take part in cytochrome biogenesis. The mitochondrially encoded subunits are essential for the activity of the respiratory complexes.

Except for the proteins encoded by mtDNA, all mitochondrial proteins (possibly more than 2000) are encoded by nuclear DNA—including all the proteins in the citric acid cycle. These nuclear-encoded mitochondrial proteins are synthesized by cytosolic ribosomes and imported via translocators in the outer and inner mitochondrial membranes. Therefore, oxidative phosphorylation is dependent on expression of genes located in two separate genomes, which must be coordinated to allow new synthesis of the respiratory complexes.

Whereas the expression of nuclear genes for mitochondrial proteins is regulated like that of other nuclear genes, less is known about the expression of mitochondrial genes. Genes can be down-regulated by a decreased copy number for the segment of mtDNA that contains the gene. Also, gene promoters in mtDNA are of several kinds and show different transcriptional activities. However, the biogenesis of respiratory complexes appears to be controlled by changes in the expression of the nuclear-encoded subunits; coordination with the mitochondrial genome mainly takes place posttranslationally.

The mitochondrial genome is especially important for pollen development. Naturally occurring rearrangements of genes in the mtDNA lead to so-called cytoplasmic male sterility (cms). This trait leads to perturbed pollen development by inducing a premature **programmed cell death** (see **WEB ESSAY 12.5**) on otherwise unaffected plants. The cms traits are used in breeding of several crop plants for making hybrid seed stocks.

Plants have several mechanisms that lower the ATP yield

As we have seen, a complex machinery is required for conserving energy in oxidative phosphorylation. So it is perhaps surprising that plant mitochondria have several functional proteins that reduce this efficiency (see **WEB TOPIC 12.3**). Plants are probably less limited by energy supply (sunlight) than by other factors in the environment (e.g., access to water and nutrients). As a consequence, metabolic flexibility may be more important to them than energetic efficiency.

In the following subsections we will discuss the role of three nonphosphorylating mechanisms and their possible usefulness in the life of the plant: the alternative oxidase, the uncoupling protein, and the rotenone-insensitive NAD(P)H dehydrogenases.

THE ALTERNATIVE OXIDASE Most plants display a capacity for *cyanide-resistant respiration* that is comparable to the capacity of the cyanide-sensitive cytochrome *c* oxidase pathway. The cyanide-resistant oxygen uptake is catalyzed by the alternative oxidase (see Figure 12.9 and WEB TOPIC 12.3).

Electrons feed off the main electron transport chain into this alternative pathway at the level of the ubiquinone pool (see Figure 12.8). The alternative oxidase, the only component of the alternative pathway, catalyzes a four-electron reduction of oxygen to water and is specifically inhibited by several compounds, most notably salicylhydroxamic acid. When electrons pass to the alternative pathway from the ubiquinone pool, two sites of proton pumping (at complexes III and IV) are bypassed. Because there is no energy conservation site in the alternative pathway between ubiquinone and oxygen, the free energy that would normally be conserved as ATP is lost as heat when electrons are shunted through this pathway.

How can a process as seemingly energetically wasteful as the alternative pathway contribute to plant metabolism? One example of the functional usefulness of the alternative oxidase is its activity in so-called thermogenic flowers of several plant families —for example, the voodoo lily (*Sauromatum guttatum*) (see WEB ESSAY 12.6). Just before pollination, parts of the inflorescence exhibit a dramatic increase in the rate of respiration caused by a greatly increased expression of alternative oxidase or uncoupling protein (depending on the species). As a result, the temperature of the upper appendix increases by as much as 25°C over the ambient temperature. During this extraordinary burst of heat production, certain amines, indoles, and terpenes are volatilized, and the plant therefore gives off a putrid odor that attracts insect pollinators. Salicylic acid has been identified as the signal initiating this thermogenic event in the voodoo lily and was later found also to be involved in plant pathogen defense (see Chapter 23).

In most plants, the respiratory rates are too low to generate sufficient heat to raise the temperature significantly. What other role(s) does the alternative pathway play? To answer that question, we need to consider the regulation of the alternative oxidase: Its transcription is often specifically induced, for example, by various types of abiotic and biotic stress. The activity of the alternative oxidase, which functions as a dimer, is regulated by reversible oxidation–reduction of an intermolecular sulfhydryl bridge, by the reduction level of the ubiquinone pool, and by pyruvate. The first two factors ensure that the enzyme is most active under reducing conditions, while the last factor ensures that the enzyme has high activity when there is plenty of substrate for the citric acid cycle (see WEB TOPIC 12.3).

If the respiration rate exceeds the cell's demand for ATP (i.e., if ADP levels are very low), the reduction level in the mitochondrion will be high, and the alternative oxidase will be activated. Thus, the alternative oxidase makes it possible for the mitochondrion to adjust the relative rates of ATP production and synthesis of carbon skeletons for use in biosynthetic reactions.

Another possible function of the alternative pathway is in the response of plants to a variety of stresses (phosphate deficiency, chilling, drought, osmotic stress, and so on), many of which can inhibit mitochondrial respiration (see Chapter 24). In response to stress, the electron transport chain leads to increased formation of reactive oxygen species (ROS), initially superoxide but also hydrogen peroxide and the hydroxyl radical, which act as a signal for the activation of alternative oxidase expression. By draining off electrons from the ubiquinone pool (see Figure 12.8), the alternative pathway prevents overreduction, thus limiting the production of ROS and minimizing the detrimental effects of stress on respiration (see WEB ESSAY 12.7). The up-regulation of alternative oxidase is an example of *retrograde regulation,* in which nuclear gene expression responds to changes in organellar status (**Figure 12.11**).

THE UNCOUPLING PROTEIN A protein found in the inner membrane of mammalian mitochondria, the **uncoupling protein**, can dramatically increase the proton permeability of the membrane and thus act as an uncoupler. As a result, less ATP and more heat are generated. Heat production appears to be one of the uncoupling protein's main functions in mammalian cells.

It had long been thought that the alternative oxidase in plants and the uncoupling protein in mammals were simply two different means of achieving the same end. It was therefore surprising when a protein similar to the uncoupling protein was discovered in plant mitochondria. This protein is induced by stress and stimulated by ROS. In knockout mutants, photosynthetic carbon assimilation and growth were decreased consistent with the interpretation that the uncoupling protein, like the alternative oxidase, functions to prevent overreduction of the electron transport chain and formation of ROS (see WEB TOPIC 12.3 and WEB ESSAY 12.7).

ROTENONE-INSENSITIVE NADH DEHYDROGENASES Multiple rotenone-insensitive dehydrogenases oxidizing NADH or NADPH are found in plant mitochondria (see

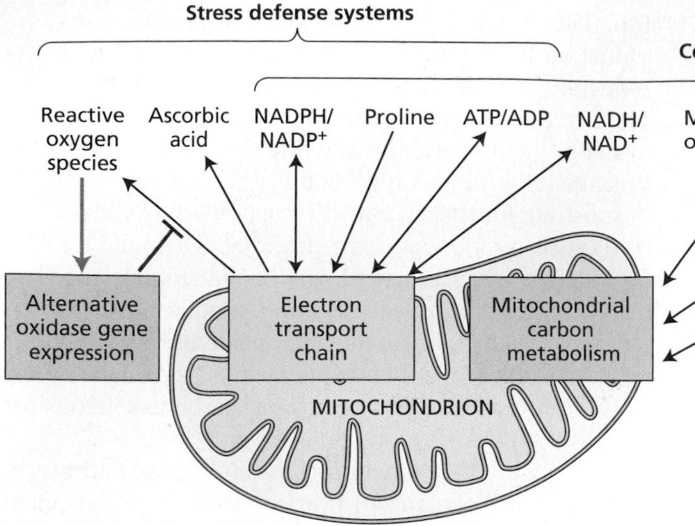

Figure 12.11 Metabolic interactions between mitochondria and cytosol. Mitochondrial activities can influence the cytosolic levels of redox and energy molecules involved in stress defense and in central carbon metabolism (such as growth processes and photosynthesis). An exact distinction between stress defense and carbon metabolism cannot be made because they have components in common. Arrows denote influences caused by changes in mitochondrial synthesis (e.g., reactive oxygen species [ROS], ATP, or ascorbic acid) or degradation (e.g., NAD[P]H, proline, or glycine). The ROS-mediated activation of expression of nuclear genes for the alternative oxidase is an example of retrograde regulation.

Figure 12.9 and **WEB TOPIC 12.3**). The internal, rotenone-insensitive NADH dehydrogenase (ND_{in}[NADH]) may work as a non-proton-pumping bypass when complex I is overloaded. Complex I has a higher affinity (ten times lower K_m) for NADH than ND_{in}(NADH). At lower NADH levels in the matrix, typically when ADP is available, complex I dominates, whereas when ADP is rate-limiting, NADH levels increase and ND_{in}(NADH) is more active. ND_{in}(NADH) and the alternative oxidase probably recycle the NADH into NAD^+ to maintain pathway activity. Since reducing power can be shuttled from the matrix to the cytosol by the exchange of different organic acids, external NADH dehydrogenases can have bypass functions similar to those of ND_{in}(NADH). Taken together, these NADH dehydrogenases and the NADPH dehydrogenases are likely to make plant respiration more flexible and allow control of specific redox homeostasis of NADH and NADPH in mitochondria and cytosol (see Figure 12.11).

Short-term control of mitochondrial respiration occurs at different levels

The substrates of ATP synthesis—ADP and P_i—appear to be key short-term regulators of the rates of glycolysis in the cytosol and of the citric acid cycle and oxidative phosphorylation in the mitochondria. Control points exist at all three stages of respiration; here we will give just a brief overview of some major features of respiratory control.

The best-characterized site of posttranslational regulation of mitochondrial respiratory metabolism is the pyruvate dehydrogenase complex, which is phosphorylated by a *regulatory protein kinase* and dephosphorylated by a *protein phosphatase*. Pyruvate dehydrogenase is inactive in the phosphorylated state, and the regulatory protein kinase is inhibited by pyruvate, allowing the enzyme to be active when substrate is available (**Figure 12.12**). Pyruvate dehy-

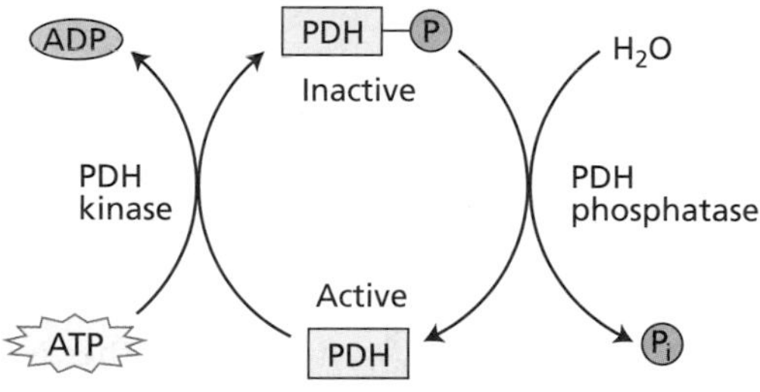

Pyruvate + CoA + NAD^+ ⟶ Acetyl-CoA + CO_2 + NADH + H^+

Effect on PDH activity	Mechanism
Activating	
Pyruvate	Inhibits kinase
ADP	Inhibits kinase
Mg^{2+} (or Mn^{2+})	Stimulates phosphatase
Inactivating	
NADH	Inhibits PDH Stimulates kinase
Acetyl-CoA	Inhibits PDH Stimulates kinase
NH_4^+	Inhibits PDH Stimulates kinase

Figure 12.12 Metabolic regulation of pyruvate dehydrogenase (PDH) activity, directly and by reversible phosphorylation. Upstream and downstream metabolites regulate PDH activity by direct actions on the enzyme itself or by regulating its protein kinase or protein phosphatase.

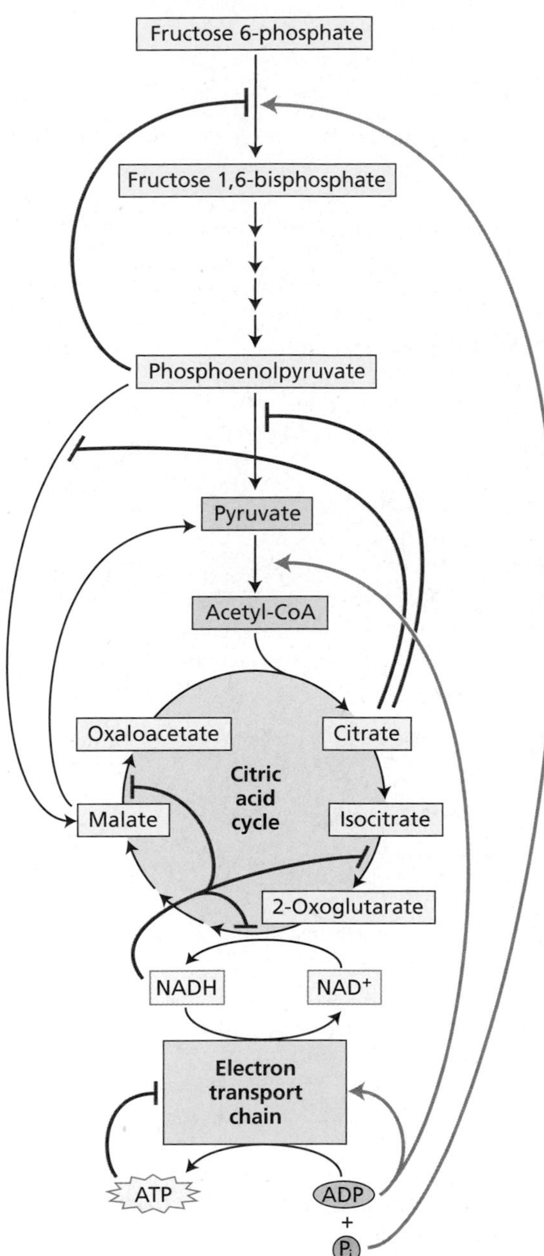

Figure 12.13 Model of bottom-up regulation of plant respiration. Several substrates for respiration (e.g., ADP) stimulate enzymes in early steps of the pathways (green arrows). In contrast, accumulation of products (e.g., ATP) inhibits upstream reactions (red lines) in a stepwise fashion. For instance, ATP inhibits the electron transport chain, leading to an accumulation of NADH. NADH inhibits citric acid cycle enzymes such as isocitrate dehydrogenase and 2-oxoglutarate dehydrogenase. Citric acid cycle intermediates such as citrate inhibit the PEP-metabolizing enzymes in the cytosol. Finally, PEP inhibits the conversion of fructose 6-phosphate into fructose 1,6-bisphosphate and restricts carbon flow into glycolysis. In this way, respiration can be up- or down-regulated in response to changing demands for either of its products: ATP and organic acids.

drogenase forms the entry point to the citric acid cycle, so this regulation adjusts the activity of the cycle to the cellular demand.

Thioredoxins control many enzymes by reversible redox dimerization of cysteine residues (see Chapter 8). Numerous mitochondrial enzymes, representing virtually all pathways, are potentially modified by thioredoxins. Although the detailed mechanisms have not been worked out yet, it is likely that mitochondrial redox status exerts an important control on respiratory processes.

The citric acid cycle oxidations, and subsequently respiration, are dynamically controlled by the cellular level of adenine nucleotides. As the cell's demand for ATP in the cytosol decreases relative to the rate of synthesis of ATP in the mitochondria, less ADP is available, and the electron transport chain operates at a reduced rate (see Figure 12.10). This slowdown could be signaled to citric acid cycle enzymes through an increase in matrix NADH, which inhibits the activity of several citric acid cycle dehydrogenases.

The buildup of citric acid cycle intermediates (such as citrate) and their derivates (such as glutamate) inhibits the action of cytosolic pyruvate kinase, increasing the cytosolic PEP concentration, which in turn reduces the rate of conversion of fructose 6-phosphate into fructose 1,6-bisphosphate, thus inhibiting glycolysis.

In summary, plant respiratory rates are *allosterically controlled* from the "bottom up" by the cellular level of ADP (**Figure 12.13**). ADP initially regulates the rate of electron transfer and ATP synthesis, which in turn regulates citric acid cycle activity, which, finally, regulates the rates of the glycolytic reactions. This bottom-up control allows the respiratory carbon pathways to adjust to the demand for biosynthetic building blocks, thereby increasing respiratory flexibility.

Respiration is tightly coupled to other pathways

Glycolysis, the oxidative pentose phosphate pathway, and the citric acid cycle are linked to several other important metabolic pathways, some of which we will cover in greater detail in **WEB APPENDIX 4**. The respiratory pathways produce the central building blocks for synthesis of a wide variety of plant metabolites, including amino acids, lipids and related compounds, isoprenoids, and porphyrins (**Figure 12.14**). Indeed, much of the reduced carbon that is metabolized by glycolysis and the citric acid cycle is diverted to biosynthetic purposes and not oxidized to CO_2.

Mitochondria are also integrated into the cellular redox network. Variations in consumption or production of redox and energy-carrying compounds such as NAD(P)H and organic acids are likely to affect metabolic pathways in the cytosol and in plastids. Of special importance is the synthesis of ascorbic acid, a central redox and stress

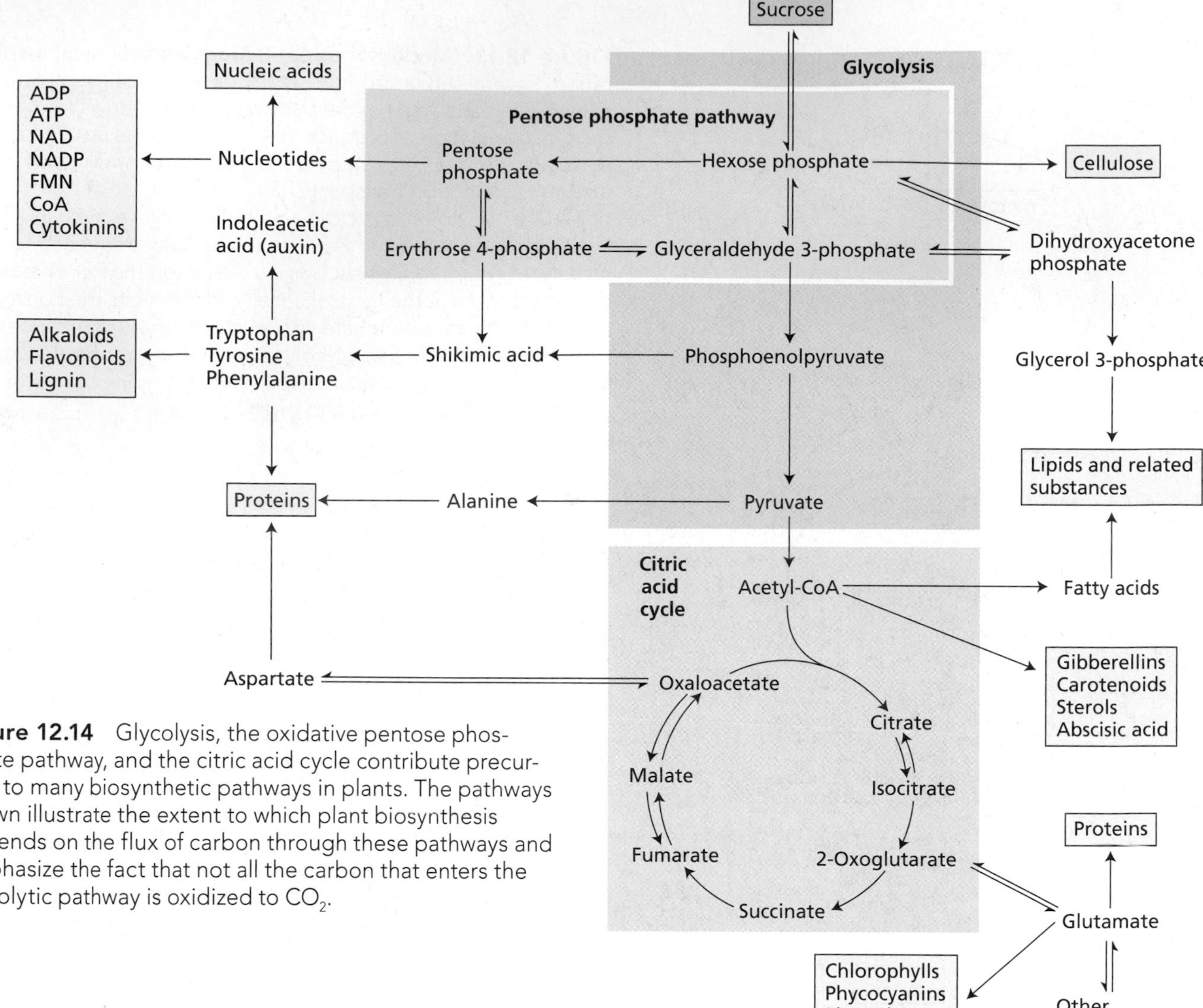

Figure 12.14 Glycolysis, the oxidative pentose phosphate pathway, and the citric acid cycle contribute precursors to many biosynthetic pathways in plants. The pathways shown illustrate the extent to which plant biosynthesis depends on the flux of carbon through these pathways and emphasize the fact that not all the carbon that enters the glycolytic pathway is oxidized to CO_2.

defense molecule in plants, by the electron transport chain (see Figure 12.11). Mitochondria also carry out steps in the biosynthesis of coenzymes necessary for many metabolic enzymes in other cell compartments (see **WEB ESSAY 12.8**).

Respiration in Intact Plants and Tissues

Many rewarding studies of plant respiration and its regulation have been carried out on isolated organelles and on cell-free extracts of plant tissues. But how does this knowledge relate to the function of the whole plant in a natural or agricultural setting?

In this section we will examine respiration and mitochondrial function in the context of the whole plant under a variety of conditions. First we will explore what happens when green organs are exposed to light: Respiration and photosynthesis operate simultaneously and are functionally integrated in the cell. Next we will discuss rates of respiration in different tissues, which may be under developmental control. Finally, we will look at the influence of various environmental factors on respiration rates.

Plants respire roughly half of the daily photosynthetic yield

Many factors can affect the respiration rate of an intact plant or of its individual organs. Relevant factors include the species and growth habit of the plant, the type and age of the specific organ, and environmental variables such as light, external O_2 and CO_2 concentrations, temperature, and nutrient and water supply (see Chapter 24). By measuring different oxygen isotopes, it is possible to measure the in vivo activities of the alternative oxidase and cytochrome *c* oxidase simultaneously. Therefore, we know that a significant part of respiration in most tissues takes place via the "energy-wasting" alternative pathway (see **WEB ESSAY 12.9**).

Whole-plant respiration rates, particularly when considered on a fresh-weight basis, are generally lower than respiration rates reported for animal tissues. This difference is mainly due to the presence in plant cells of a large vacuole and a cell wall, neither of which contains mitochondria. Nonetheless, respiration rates in some plant tissues are as high as those observed in actively respiring animal tissues, so the respiratory process in plants is not inherently slower than in animals. In fact, isolated plant mitochondria respire as fast as or faster than mammalian mitochondria.

The contribution of respiration to the overall carbon economy of the plant can be substantial. Whereas only green tissues photosynthesize, all tissues respire, and they do so 24 hours a day. Even in photosynthetically active tissues, respiration, if integrated over the entire day, uses a substantial fraction of gross photosynthesis. A survey of several herbaceous species indicated that 30 to 60% of the daily gain in photosynthetic carbon is lost to respiration, although these values tend to decrease in older plants. Trees respire a similar fraction of their photosynthetic production, but their respiratory loss increases with age as the ratio of photosynthetic to nonphotosynthetic tissue decreases. In general, unfavourable growth conditions will increase respiration relative to photosynthesis, and thus lower the overall carbon yield of the plant.

Respiration operates during photosynthesis

Mitochondria are involved in the metabolism of photosynthesizing leaves in several ways. The glycine generated by photorespiration is oxidized to serine in the mitochondrion in a reaction involving mitochondrial oxygen consumption (see Chapter 8). At the same time, mitochondria in photosynthesizing tissue carry out normal mitochondrial respiration (i.e., via the citric acid cycle). Relative to the maximum rate of photosynthesis, rates of mitochondrial respiration measured in green tissues in the light are far slower, generally by a factor of 6- to 20-fold. Given that rates of photorespiration can often reach 20 to 40% of the gross photosynthetic rate, daytime photorespiration is a larger provider of NADH for the respiratory chain than the normal respiratory pathways.

The activity of pyruvate dehydrogenase, one of the ports of entry into the citric acid cycle, decreases in the light to 25% of its activity in darkness. Consistently, the overall rate of mitochondrial respiration decreases in the light, but the extent of the decrease remains uncertain at present. It is clear, however, that the mitochondrion is a major supplier of ATP to the cytosol (e.g., for driving biosynthetic pathways) even in illuminated leaves.

Another role of the respiratory pathways during photosynthesis is to supply precursors for biosynthetic reactions, such as the 2-oxoglutarate needed for nitrogen assimilation (see Figures 12.7C and 12.14). The formation of 2-oxoglutarate also produces NADH in the matrix, linking the process to oxidative phosphorylation or to nonphosphorylating respiratory chain activities.

Additional evidence for the involvement of mitochondrial respiration in photosynthesis has been obtained in studies with mitochondrial mutants defective in respiratory complexes. Compared with the wild type, these plants have slower leaf development and photosynthesis because changes in levels of redox-active metabolites are communicated between mitochondria and chloroplasts, negatively affecting photosynthetic function.

Different tissues and organs respire at different rates

Respiration is often considered to have two components of comparable magnitude. **Maintenance respiration** is needed to support the function and turnover of the tissues already present. **Growth respiration** provides the energy needed for converting sugars into the building blocks that make up new tissues. A useful rule of thumb is that the greater the overall metabolic activity of a given tissue, the higher its respiration rate. Developing buds usually show very high rates of respiration, and respiration rates of vegetative organs usually decrease from the point of growth (e.g., the leaf tip in eudicots and the leaf base in monocots) to more differentiated regions. A well-studied example is the growing barley leaf.

In mature vegetative organs, stems generally have the lowest respiration rates, whereas leaf and root respiration varies with the plant species and the conditions under which the plants are growing. Low availability of soil nutrients, for example, increases the demand for respiratory ATP production in the root. This increase reflects increased energy costs for active ion uptake and root growth in search of nutrients. (See **WEB TOPIC 12.7** for a discussion of how crop yield is affected by changes in respiration rates.)

When a plant organ has reached maturity, its respiration rate either remains roughly constant or decreases slowly as the tissue ages and ultimately senesces. An exception to this pattern is the marked rise in respiration, known as the *climacteric,* that accompanies the onset of ripening in many fruits (e.g., avocado, apple, and banana) and senescence in detached leaves and flowers. During fruit ripening, massive conversion of, for example, starch (banana) or organic acids (tomato and apple) into sugars occurs, accompanied by a rise in the hormone ethylene (see Chapter 21) and the activity of the cyanide-resistant alternative pathway.

Different tissues can use different substrates for respiration. Sugars dominate overall, but in specific organs other compounds, such as organic acids in maturing apples or lemons and lipids in germinating sunflower or canola seedlings, may provide the carbon for respiration. These compounds are built with different ratios of carbon to oxygen atoms. Therefore, the ratio of CO_2 release to O_2 consumption, which is called the **respiratory quotient**, or **RQ**, varies with the substrate oxidized. Lipids, sugars, and organic acids represent a series of rising RQ because lipids contain little oxygen per carbon, and organic acids much. Alcoholic fermentation releases CO_2 without consuming O_2, so a high RQ is also a marker for fermentation. Since RQ can be determined in the field, it is an important parameter in analyses of carbon metabolism on a larger scale.

Environmental factors alter respiration rates

Many environmental factors can alter the operation of metabolic pathways and change respiratory rates. Here we will examine the roles of environmental oxygen (O_2), temperature, and carbon dioxide (CO_2).

OXYGEN Oxygen can affect plant respiration because of its role as a substrate in the overall respiratory process. At 25°C, the equilibrium concentration of O_2 in an air-saturated (21% O_2) aqueous solution is about 250 μ*M*. The K_m value for oxygen in the reaction catalyzed by cytochrome *c* oxidase is well below 1 μ*M*, so there should be no apparent dependence of the respiration rate on external O_2 concentrations. However, respiration rates decrease if the atmospheric oxygen concentration is below 5% for whole organs or below 2 to 3% for tissue slices. These findings show that oxygen supply can impose a limitation on plant respiration.

Oxygen diffuses slowly in aqueous solutions. Compact organs such as seeds and potato tubers have a noticeable O_2 concentration gradient from the surface to the center, which restricts the ATP/ADP ratio. Diffusion limitation is even more significant in seeds with a thick seed coat or in plant organs submerged in water. When plants are grown hydroponically, the solutions must be aerated to keep oxygen levels high in the vicinity of the roots (see Chapter 5). The problem of oxygen supply is particularly important in plants growing in very wet or flooded soils (see also Chapter 24).

Some plants, particularly trees, have a restricted geographic distribution because of the need to maintain a supply of oxygen to their roots. For instance, the dogwood *Cornus florida* and tulip tree poplar (*Liriodendron tulipifera*) can survive only in well-drained, aerated soils. On the other hand, many plant species are adapted to grow in flooded soils. For example, rice and sunflower rely on a network of intercellular air spaces (called **aerenchyma**) running from the leaves to the roots to provide a continuous gaseous pathway for the movement of oxygen to the flooded roots. If this gaseous diffusion pathway throughout the plant did not exist, the respiration rates of many plants would be limited by an insufficient oxygen supply.

Limitation in oxygen supply can be more severe for trees with very deep roots that grow in wet soils. Such roots must survive on anaerobic (fermentative) metabolism or develop structures that facilitate the movement of oxygen to the roots. Examples of such structures are outgrowths of the roots, called *pneumatophores*, that protrude out of the water and provide a gaseous pathway for oxygen diffusion into the roots. Pneumatophores are found in *Avicennia* and *Rhizophora*, both trees that grow in mangrove swamps under continuously flooded conditions.

TEMPERATURE Respiration operates over a wide temperature range (see WEB ESSAYS 12.6 and 12.9). It typically increases with temperatures between 0 and 30°C and reaches a plateau at 40 to 50°C. At higher temperatures, it again decreases because of inactivation of the respiratory machinery. The increase in respiration rate for every 10°C increase in temperature is commonly called the **temperature coefficient, Q_{10}**. This coefficient describes how respiration responds to short-term temperature changes, and it varies with plant development and external factors. On a longer time scale, plants acclimate to low temperatures by increasing their respiratory capacity so that ATP production can be continued.

Low temperatures are used to retard postharvest respiration during the storage of fruits and vegetables, but those temperatures must be adjusted with care. For instance, when potato tubers are stored at temperatures above 10°C, respiration and ancillary metabolic activities are sufficient to allow sprouting. Below 5°C, respiration rates and sprouting are reduced, but the breakdown of stored starch and its conversion into sucrose impart an unwanted sweetness to the tubers. Therefore, potatoes are best stored at 7 to 9°C, which prevents the breakdown of starch while minimizing respiration and germination (see also WEB ESSAY 12.4).

CARBON DIOXIDE It is common practice in commercial storage of fruits to take advantage of the effects of oxygen concentration and temperature on respiration by storing fruits at low temperatures under 2 to 3% O_2 and 3 to 5% CO_2 concentrations. The reduced temperature lowers the respiration rate, as does the reduced O_2 level. Low levels of oxygen, instead of anoxic conditions, are used to avoid lowering tissue oxygen tensions to the point at which fermentative metabolism sets in. Carbon dioxide has a limited direct inhibitory effect on respiration at the artificially high concentration of 3 to 5%.

The atmospheric CO_2 concentration is currently (2014) around 400 ppm, but it is increasing as a result of human activities, and it is projected to increase to 700 ppm before the end of the twenty-first century (see Chapter 9). The flux of CO_2 between plants and the atmosphere by photosynthesis and respiration is much larger than the flux of CO_2 to the atmosphere caused by the burning of fossil fuels. Therefore, the effects of elevated CO_2 concentrations on plant respiration will strongly influence future global atmospheric changes. Laboratory studies have shown that 700 ppm CO_2 does not directly inhibit plant respiration, but measurements on whole ecosystems indicate that respiration per biomass unit may decrease with increased CO_2 concentrations. The mechanism behind the latter

Glycerol	Triacylglycerol (the major stored lipid)	Polar glycerolipid
CH_2OH	$H_2C-O-C(=O)-(CH_2)_n-CH_3$	$H_2C-O-C(=O)-(CH_2)_n-CH_3$
$CHOH$	$HC-O-C(=O)-(CH_2)_n-CH_3$	$HC-O-C(=O)-(CH_2)_n-CH_3$
CH_2OH	$H_2C-O-C(=O)-(CH_2)_n-CH_3$	H_2C-O-X

X	
X = H	Diacylglycerol (DAG)
X = HPO_3^-	Phosphatidic acid
X = $PO_3^- -CH_2-CH_2-\overset{+}{N}(CH_3)_3$	Phosphatidylcholine
X = $PO_3^- -CH_2-CH_2-NH_2$	Phosphatidylethanolamine
X = galactose	Galactolipids

Figure 12.15 Structural features of triacylglycerols and polar glycerolipids in higher plants. The carbon chain lengths of the fatty acids, which always have an even number of carbons, range from 12 to 20 but are typically 16 or 18. Thus, the value of n is usually 14 or 16.

effect is not yet clear, and it is at present not possible to fully predict the potential importance of plants as a sink for anthropogenic CO_2.

Lipid Metabolism

Whereas animals use fats for energy storage, plants use them for both energy and carbon storage. Fats and oils are important storage forms of reduced carbon in many seeds, including those of agriculturally important species such as soybean, sunflower, canola, peanut, and cotton. Oils serve a major storage function in many nondomesticated plants that produce small seeds. Some fruits, such as olives and avocados, also store fats and oils.

In this final part of the chapter we will describe the biosynthesis of two types of glycerolipids: the *triacylglycerols* (the fats and oils stored in seeds) and the *polar glycerolipids* (which form the lipid bilayers of cellular membranes) (**Figure 12.15**). We will see that the biosynthesis of triacylglycerols and polar glycerolipids requires the cooperation of two organelles: the plastids and the endoplasmic reticulum. We will also examine the complex process by which germinating seeds obtain carbon skeletons and metabolic energy from the oxidation of fats and oils.

Fats and oils store large amounts of energy

Fats and oils belong to the general class termed *lipids*, a structurally diverse group of hydrophobic compounds that are soluble in organic solvents and highly insoluble in water. Lipids represent a more reduced form of carbon than carbohydrates, so the complete oxidation of 1 g of fat or oil (which contains about 40 kJ of energy) can produce considerably more ATP than the oxidation of 1 g of starch (about 15.9 kJ). Conversely, the biosynthesis of lipids requires a correspondingly large investment of metabolic energy.

Other lipids are important for plant structure and function but are not used for energy storage. These lipids include the phospholipids and galactolipids that make up plant membranes, as well as sphingolipids, which are also important membrane components; waxes, which make up the protective cuticle that reduces water loss from exposed plant tissues; and terpenoids (also known as isoprenoids), which include carotenoids involved in photosynthesis and sterols present in many plant membranes.

Triacylglycerols are stored in oil bodies

Fats and oils exist mainly in the form of **triacylglycerols** (*acyl* refers to the fatty acid portion), in which fatty acid molecules are linked by ester bonds to the three hydroxyl groups of glycerol (see Figure 12.15).

The fatty acids in plants are usually straight-chain carboxylic acids having an even number of carbon atoms. The carbon chains can be as short as 12 units and as long as 30 or more, but most commonly are 16 or 18 carbons long. *Oils* are liquid at room temperature, primarily because of the presence of carbon–carbon double bonds (unsaturation) in their component fatty acids; *fats*, which have a higher proportion of saturated fatty acids, are solid at room temperature. The major fatty acids in plant lipids are shown in **Table 12.3**.

The proportions of fatty acids in plant lipids vary with the plant species. For example, peanut oil is about 9% palmitic acid, 59% oleic acid, and 21% linoleic acid, and cottonseed oil is 25% palmitic acid, 15% oleic acid, and 55% linoleic acid. We will discuss the biosynthesis of these fatty acids shortly.

In most seeds, triacylglycerols are stored in the cytoplasm of either cotyledon or endosperm cells in organelles known as **oil bodies** (also called *spherosomes* or *oleosomes*) (see Chapter 1). The oil-body membrane is a single layer of phospholipids (i.e., a half-bilayer) with the hydrophilic ends of the phospholipids exposed to the cytosol and the hydrophobic acyl hydrocarbon chains facing the triacylglycerol interior (see Chapter 1). The oil body is stabilized by the presence of specific proteins, called oleosins, that coat its outer surface and prevent the phospholipids of adjacent oil bodies from coming in contact and fusing with it.

TABLE 12.3 Common fatty acids in higher plant tissues

Name[a]	Structure
Saturated fatty acids	
Lauric acid (12:0)	$CH_3(CH_2)_{10}CO_2H$
Myristic acid (14:0)	$CH_3(CH_2)_{12}CO_2H$
Palmitic acid (16:0)	$CH_3(CH_2)_{14}CO_2H$
Stearic acid (18:0)	$CH_3(CH_2)_{16}CO_2H$
Unsaturated fatty acids	
Oleic acid (18:1)	$CH_3(CH_2)_7CH{=}CH(CH_2)_7CO_2H$
Linoleic acid (18:2)	$CH_3(CH_2)_4CH{=}CH{-}CH_2{-}CH{=}CH(CH_2)_7CO_2H$
Linolenic acid (18:3)	$CH_3CH_2CH{=}CH{-}CH_2{-}CH{=}CH{-}CH_2{-}CH{=}CH{-}(CH_2)_7CO_2H$

[a]Each fatty acid has a numerical abbreviation. The number before the colon represents the total number of carbons; the number after the colon is the number of double bonds.

The unique membrane structure of oil bodies results from the pattern of triacylglycerol biosynthesis. Triacylglycerol synthesis is completed by enzymes located in the membranes of the endoplasmic reticulum (ER), and the resulting fats accumulate between the two monolayers of the ER membrane bilayer. The bilayer swells apart as more fats are added to the growing structure, and ultimately a mature oil body buds off from the ER.

Polar glycerolipids are the main structural lipids in membranes

As outlined in Chapter 1, each membrane in the cell is a bilayer of *amphipathic* (i.e., having both hydrophilic and hydrophobic regions) lipid molecules in which a polar head group interacts with the aqueous environment while hydrophobic fatty acid chains form the core of the membrane. This hydrophobic core prevents unregulated diffusion of solutes between cell compartments and thereby allows the biochemistry of the cell to be organized.

The main structural lipids in membranes are the **polar glycerolipids** (see Figure 12.15), in which the hydrophobic portion consists of two 16-carbon or 18-carbon fatty acid chains esterified to positions 1 and 2 of a glycerol backbone. The polar head group is attached to position 3 of the glycerol. There are two categories of polar glycerolipids:

1. **Glyceroglycolipids**, in which sugars form the head group (**Figure 12.16A**)
2. **Glycerophospholipids**, in which the head group contains phosphate (**Figure 12.16B**)

Plant membranes have additional structural lipids, including sphingolipids and sterols (see Chapter 15), but these are minor components. Other lipids perform specific roles in photosynthesis and other processes. Included among these lipids are chlorophylls, plastoquinone, carotenoids, and tocopherols, which together account for about one-third of the lipids in plant leaves.

Figure 12.16 shows the nine major polar glycerolipid classes in plants, each of which can be associated with many different fatty acid combinations. The structures shown in Figure 12.16 illustrate some of the more common molecular species.

Chloroplast membranes, which account for 70% of the membrane lipids in photosynthetic tissues, are dominated by glyceroglycolipids; other membranes of the cell contain glycerophospholipids (**Table 12.4**). In nonphotosynthetic tissues, glycerophospholipids are the major membrane glycerolipids.

Fatty acid biosynthesis consists of cycles of two-carbon addition

Fatty acid biosynthesis involves the cyclic condensation of two-carbon units derived from acetyl-CoA. In plants, fatty acids are synthesized primarily in the plastids, while in animals they are synthesized primarily in the cytosol.

The enzymes of the biosynthesis pathway are thought to be held together in a complex that is collectively referred to as *fatty acid synthase*. The complex probably allows the series of reactions to occur more efficiently than it would if the enzymes were physically separated from one another. In addition, the growing acyl chains are covalently bound to a low-molecular-weight acidic protein called the **acyl carrier protein** (**ACP**). When conjugated to the acyl carrier protein, an acyl chain is referred to as **acyl-ACP**.

Figure 12.16 Major polar glycerolipid classes found in ▶ plant membranes: (A) glyceroglycolipids and a sphingolipid and (B) glycerophospholipids. Two of at least six different fatty acids may be attached to the glycerol backbone. One of the more common molecular species is shown for each glycerolipid class. The numbers given below each name refer to the number of carbons (number before the colon) and the number of double bonds (number after the colon).

(A) Glyceroglycolipids

Monogalactosyldiacylglycerol
(18:3 | 16:3)

Glucosylceramide

Sulfolipid (sulfoquinovosyldiacylglycerol)
(18:3 | 16:0)

Digalactosyldiacylglycerol
(16:0 | 18:3)

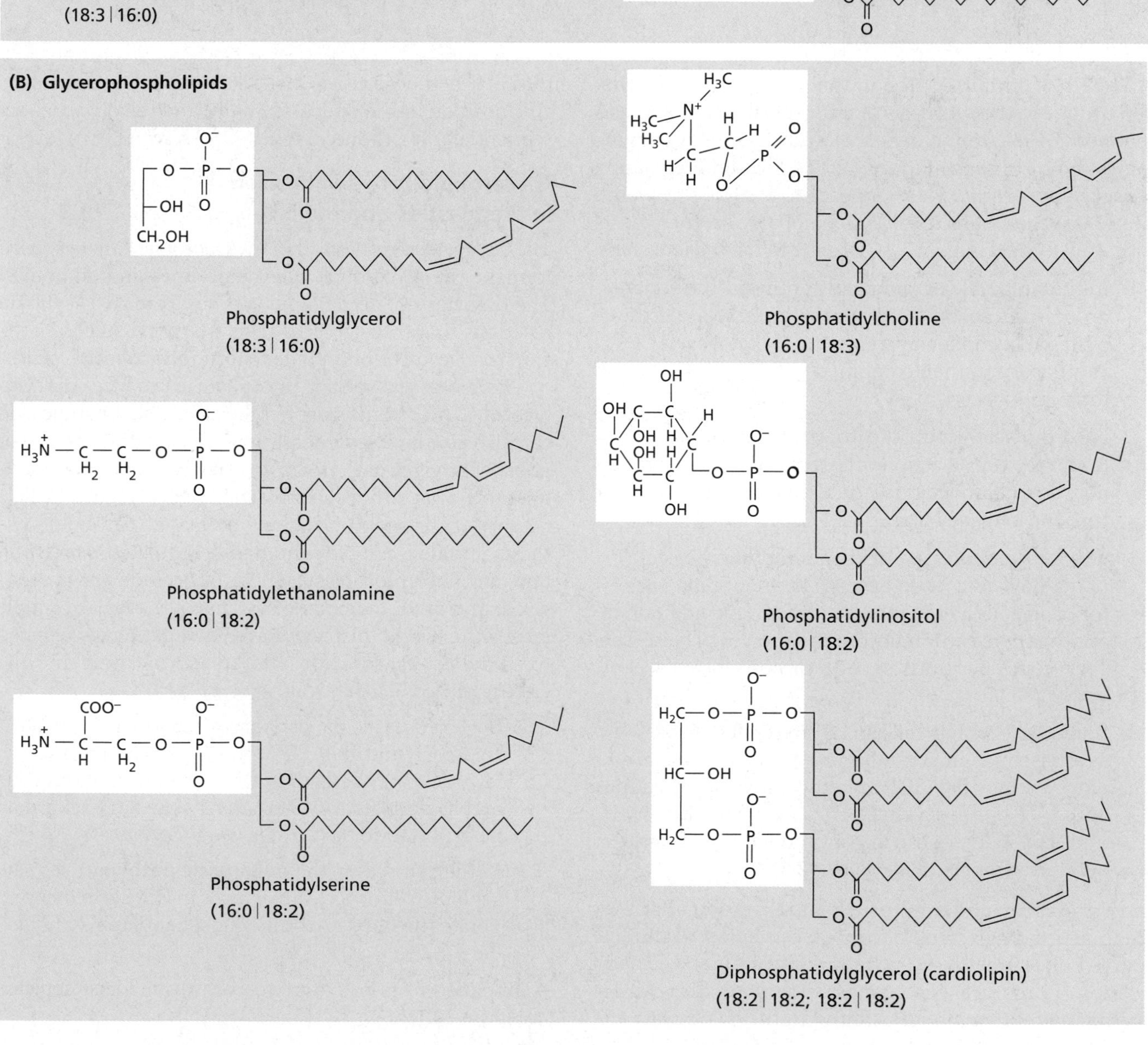

TABLE 12.4 Glycerolipid components of cellular membranes

Lipid	Lipid composition (percentage of total)		
	Chloroplast	Endoplasmic reticulum	Mitochondrion
Phosphatidylcholine	4	47	43
Phosphatidylethanolamine	—	34	35
Phosphatidylinositol	1	17	6
Phosphatidylglycerol	7	2	3
Diphosphatidylglycerol	—	—	13
Monogalactosyldiacylglycerol	55	—	—
Digalactosyldiacylglycerol	24	—	—
Sulfolipid	8	—	—

The first committed step in the pathway (i.e., the first step unique to the synthesis of fatty acids) is the synthesis of malonyl-CoA from acetyl-CoA and CO_2 by the enzyme *acetyl-CoA carboxylase* (**Figure 12.17**). The tight regulation of acetyl-CoA carboxylase appears to control the overall rate of fatty acid synthesis. The malonyl-CoA then reacts with ACP to yield malonyl-ACP in the following four steps:

1. In the first cycle of fatty acid synthesis, the acetate group from acetyl-CoA is transferred to a specific cysteine of condensing enzyme (3-ketoacyl-ACP synthase) and then combined with malonyl-ACP to form acetoacetyl-ACP.
2. Next the keto group at carbon 3 is removed (reduced) by the action of three enzymes to form a new acyl chain (butyryl-ACP), which is now four carbons long (see Figure 12.17).
3. The four-carbon fatty acid and another molecule of malonyl-ACP then become the new substrates for condensing enzyme, resulting in the addition of another two-carbon unit to the growing chain. The cycle continues until 16 or 18 carbons have been added.
4. Some 16:0-ACP is released from the fatty acid synthase machinery, but most molecules that are elongated to 18:0-ACP are efficiently converted into 18:1-ACP by a desaturase enzyme. Thus, 16:0-ACP and 18:1-ACP are the major products of fatty acid synthesis in plastids (**Figure 12.18**).

Fatty acids may undergo further modification after they are linked with glycerol to form glycerolipids. Additional double bonds are placed in the 16:0 and 18:1 fatty acids by a series of desaturase isozymes. *Desaturase isozymes* are integral membrane proteins found in the chloroplast and the ER. Each desaturase inserts a double bond at a specific position in the fatty acid chain, and the enzymes act sequentially to produce the final 18:3 and 16:3 products.

Glycerolipids are synthesized in the plastids and the ER

The fatty acids synthesized in the chloroplast are next used to make the glycerolipids of membranes and oil bodies. The first steps of glycerolipid synthesis are two acylation reactions that transfer fatty acids from acyl-ACP or acyl-CoA to glycerol-3-phosphate to form **phosphatidic acid**.

The action of a specific phosphatase produces **diacylglycerol** (**DAG**) from phosphatidic acid. Phosphatidic acid can also be converted directly into phosphatidylinositol or phosphatidylglycerol; DAG can give rise to phosphatidylethanolamine or phosphatidylcholine (see Figure 12.18).

The localization of the enzymes of glycerolipid synthesis reveals a complex and highly regulated interaction between the chloroplast, where fatty acids are synthesized, and other membrane systems of the cell. In simple terms, the biochemistry involves two pathways referred to as the prokaryotic (or chloroplast) pathway and the eukaryotic (or ER) pathway:

1. In chloroplasts, the **prokaryotic pathway** uses the 16:0-ACP and 18:1-ACP products of chloroplast fatty acid synthesis to synthesize phosphatidic acid and its derivatives. Alternatively, the fatty acids may be exported to the cytoplasm as CoA esters.
2. In the cytoplasm, the **eukaryotic pathway** uses a separate set of acyltransferases in the ER to incorporate the fatty acids into phosphatidic acid and its derivatives.

A simplified version of this two-pathway model is depicted in Figure 12.18.

$CH_3-C(=O)-SCoA$

Acetyl-CoA

1. This is the first committed step in fatty acid biosynthetic pathway.

ATP

CO_2

ADP + P_i

Acetyl-CoA carboxylase

$^{-}OOC-CH_2-C(=O)-SCoA$

Malonyl-CoA

2. Malonyl group is transferred to acyl carrier protein (ACP).

ACP

$^{-}OOC-CH_2-C(=O)-SACP$

Malonyl-ACP

3. The first cycle of fatty acid synthesis begins here.

Condensing enzyme

Decarboxylation step

CO_2

$CH_3-C(=O)-CH_2-C(=O)-SACP$

Acetoacetyl-ACP

5. The second cycle of fatty acid synthesis begins here.

Decarboxylation step

ACP, CO_2

Condensing enzyme

(Continues to 16- to 18-carbon chain length)

ACP

Completed fatty acid

7. ACP is removed from completed fatty acid in a transferase reaction.

$CH_3-CH_2-CH_2-C(=O)-SACP$

Butyryl-ACP (acyl-ACP)

6. The cycle continues multiple times adding acetate (2-carbon) units from malonyl-ACP.

2 NADPH

2 $NADP^+$

4. The keto group at carbon 3 is removed in three steps.

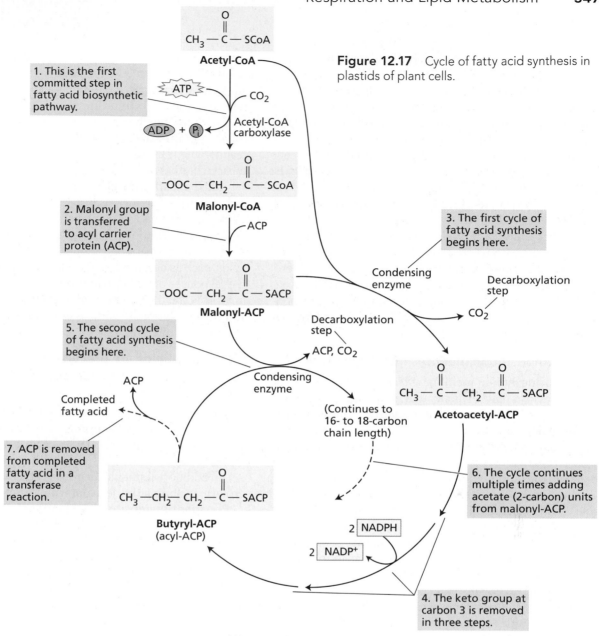

Figure 12.17 Cycle of fatty acid synthesis in plastids of plant cells.

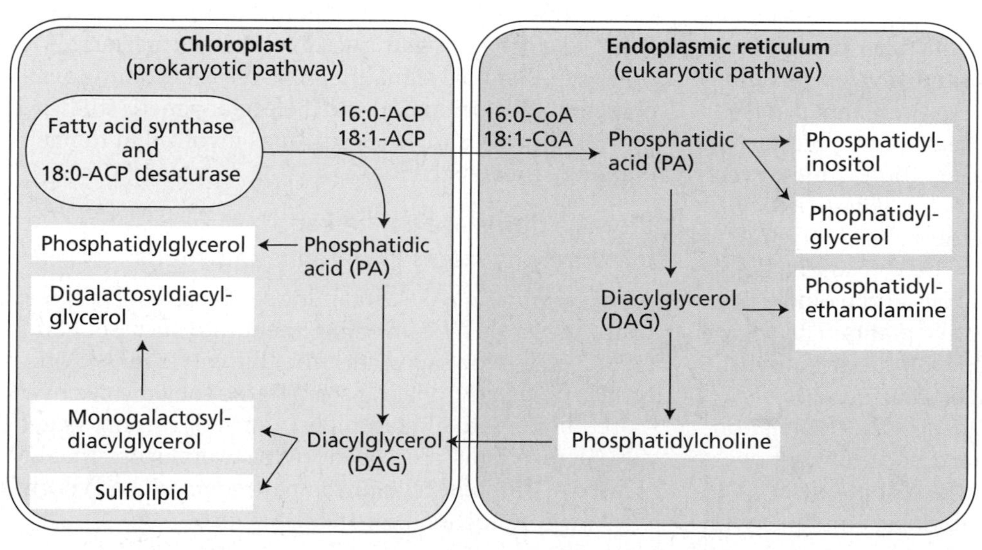

Figure 12.18 The two pathways for glycerolipid synthesis in the chloroplast and endoplasmic reticulum of Arabidopsis leaf cells. The major membrane components are shown in boxes. Glycerolipid desaturases in the chloroplast, and enzymes in the ER, convert 16:0 and 18:1 fatty acids into the more highly unsaturated fatty acids shown in Figure 12.16.

In some higher plants, including Arabidopsis and spinach, the two pathways contribute almost equally to chloroplast lipid synthesis. In many other angiosperms, however, phosphatidylglycerol is the only product of the prokaryotic pathway, and the remaining chloroplast lipids are synthesized entirely by the eukaryotic pathway.

The biochemistry of triacylglycerol synthesis in oilseeds is generally the same as described for the glycerolipids: 16:0-ACP and 18:1-ACP are synthesized in the plastids of the cell and exported as CoA thioesters for incorporation into DAG in the ER (see Figure 12.18).

The key enzymes in oilseed metabolism (not shown in Figure 12.18) are *acyl-CoA:DAG acyltransferase* and *PC:DAG acyltransferase,* which catalyze triacylglycerol synthesis. As noted earlier, triacylglycerol molecules accumulate in specialized subcellular structures—the oil bodies—from which they can be mobilized during germination and converted into sugars.

Lipid composition influences membrane function

A central question in membrane biology is the functional reason behind lipid diversity. Each membrane system of the cell has a characteristic and distinct complement of lipid types, and within a single membrane, each class of lipids has a distinct fatty acid composition (see Table 12.4).

A simple view of a membrane is one in which lipids make up the fluid, semipermeable bilayer that is the matrix for the functional membrane proteins. Since this bulk lipid role could be satisfied by a single unsaturated species of phosphatidylcholine, such a simple model is obviously unsatisfactory. Why is lipid diversity needed? One aspect of membrane biology that might offer answers to this central question is the relationship between lipid composition and the ability of organisms to adjust to temperature changes. For example, chill-sensitive plants experience sharp reductions in growth rate and development at temperatures between 0 and 12°C (see Chapter 24). Many economically important crops, such as cotton, soybean, maize, rice, and many tropical and subtropical fruits, are classified as chill-sensitive. In contrast, most plants that originate from temperate regions are able to grow and develop at chilling temperatures and are classified as chill-resistant plants.

It has been suggested that, because of the decrease in lipid fluidity at lower temperatures, the primary event of chilling injury is *a transition from a liquid-crystalline phase to a gel phase* in cellular membranes. According to this hypothesis, such a transition would result in alterations in the metabolism of chilled cells and would lead to injury and death of the chill-sensitive plants. The degree of unsaturation of the fatty acids would determine the temperature at which such damage occurred.

Recent research, however, suggests that the relationship between membrane unsaturation and plant responses to temperature is more subtle and complex (see WEB TOPIC 12.8). The responses of Arabidopsis mutants with increased saturation of fatty acids to low temperatures are not what is predicted by the chill-sensitivity hypothesis, suggesting that normal chilling injury may not be strictly related to the level of unsaturation of membrane lipids.

On the other hand, experiments with transgenic tobacco plants that are chill-sensitive show opposite results. The transgenic expression of exogenous genes in tobacco has been used specifically to decrease the level of saturated phosphatidylglycerol or to bring about a general increase in membrane unsaturation. In each case, damage caused by chilling was alleviated to some extent.

These new findings make it clear that either the extent of membrane unsaturation or the presence of particular lipids, such as desaturated phosphatidylglycerol, can affect the responses of plants to low temperatures. As discussed in WEB TOPIC 12.8, more work is required to fully understand the relationship between lipid composition and membrane function.

Membrane lipids are precursors of important signaling compounds

Plants, animals, and microbes all use membrane lipids as precursors for compounds that are used for intracellular or long-range signaling. For example, jasmonate hormone—derived from linolenic acid (18:3)—activates plant defenses against insects and many fungal pathogens (see Chapter 23). In addition, jasmonate regulates other aspects of plant growth, including the development of anthers and pollen.

Phosphatidylinositol 4,5-bisphosphate (PIP_2) is the most important of several phosphorylated derivatives of phosphatidylinositol known as *phosphoinositides*. In animals, receptor-mediated activation of phospholipase C leads to the hydrolysis of PIP_2 into inositol trisphosphate ($InsP_3$) and diacylglycerol, both of which act as intracellular secondary messengers.

The action of $InsP_3$ in releasing Ca^{2+} into the cytoplasm (through Ca^{2+}-sensitive channels in the tonoplast and other membranes), and thereby regulating cellular processes, has been demonstrated in several plant systems, including the stomatal guard cells. Information about other types of lipid signaling in plants is becoming available through biochemical and molecular genetic studies of phospholipases and other enzymes involved in the generation of these signals.

Storage lipids are converted into carbohydrates in germinating seeds

After germinating, oil-containing seeds metabolize stored triacylglycerols by converting them into sucrose. Plants are not able to transport fats from the cotyledons to other tissues of the germinating seedling, so they must convert stored lipids into a more mobile form of carbon, generally sucrose. This process involves several steps that are located in different cellular compartments: oil bodies, glyoxysomes, mitochondria, and the cytosol.

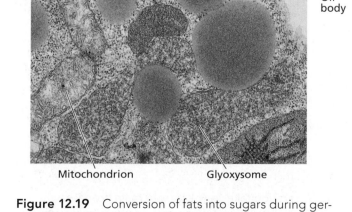

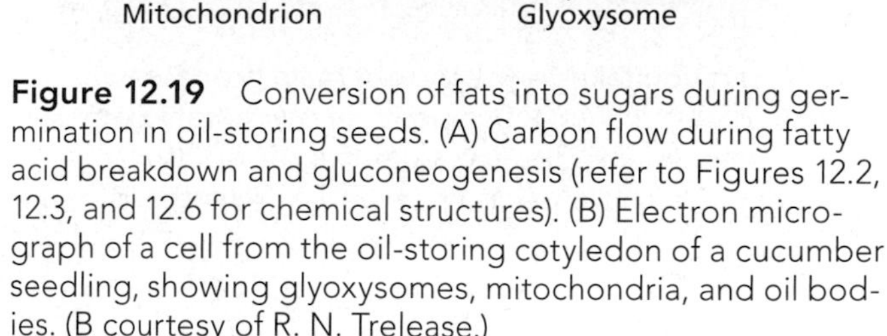

Figure 12.19 Conversion of fats into sugars during germination in oil-storing seeds. (A) Carbon flow during fatty acid breakdown and gluconeogenesis (refer to Figures 12.2, 12.3, and 12.6 for chemical structures). (B) Electron micrograph of a cell from the oil-storing cotyledon of a cucumber seedling, showing glyoxysomes, mitochondria, and oil bodies. (B courtesy of R. N. Trelease.)

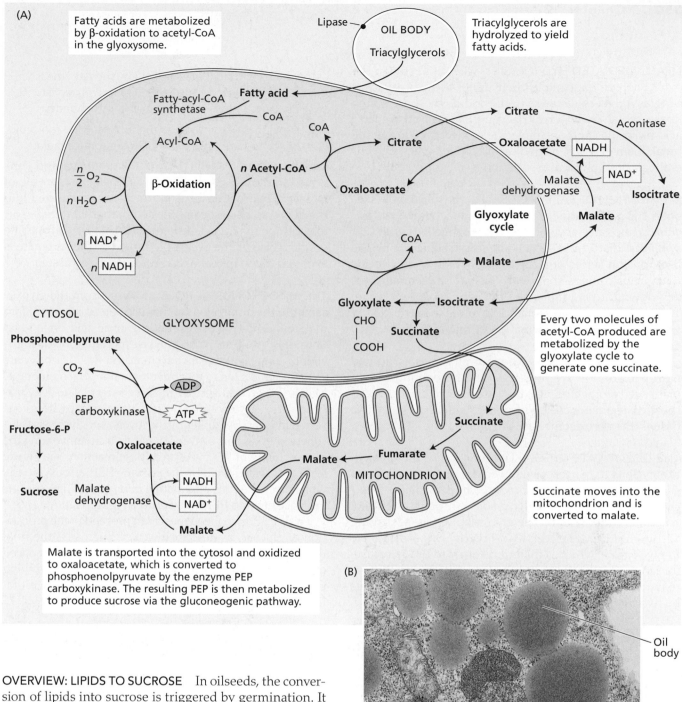

OVERVIEW: LIPIDS TO SUCROSE In oilseeds, the conversion of lipids into sucrose is triggered by germination. It begins with the hydrolysis of triacylglycerols stored in oil bodies into free fatty acids, followed by oxidation of those fatty acids to produce acetyl-CoA (**Figure 12.19**). The fatty acids are oxidized in a type of peroxisome called a **glyoxysome**, an organelle enclosed by a single membrane bilayer that is found in the oil-rich storage tissues of seeds. Acetyl-CoA is metabolized in the glyoxysome and cytoplasm (see Figure 12.19A) to produce succinate, which is transported from the glyoxysome to the mitochondrion, where it is converted first into fumarate and then into malate. The process ends in the cytosol with the conversion of malate into glucose via gluconeogenesis, and then into sucrose. In most oilseeds, approximately 30% of the acetyl-CoA is used for energy production via respiration, and the rest is converted into sucrose.

LIPASE-MEDIATED HYDROLYSIS The initial step in the conversion of lipids into carbohydrates is the breakdown of triacylglycerols stored in oil bodies by the enzyme lipase, which hydrolyzes triacylglycerols into three fatty acid molecules and one molecule of glycerol. During the breakdown of lipids, oil bodies and glyoxysomes are generally in close physical association (see Figure 12.19B).

β-OXIDATION OF FATTY ACIDS The fatty acid molecules enter the glyoxysome, where they are activated by conversion into fatty-acyl-CoA by the enzyme *fatty-acyl-CoA synthetase*. Fatty-acyl-CoA is the initial substrate for the **β-oxidation** series of reactions, in which C_n fatty acids (fatty acids composed of *n* carbons) are sequentially broken down into *n*/2 molecules of acetyl-CoA (see Figure 12.19A). This reaction sequence involves the reduction of ½ O_2 to H_2O and the formation of one NADH for each acetyl-CoA produced.

In mammalian tissues, the four enzymes associated with β-oxidation are present in the mitochondrion. In plant seed storage tissues, they are located exclusively in the glyoxysome or the equivalent organelle in vegetative tissues, the peroxisome (see Chapter 1).

THE GLYOXYLATE CYCLE The function of the **glyoxylate cycle** is to convert two molecules of acetyl-CoA into succinate. The acetyl-CoA produced by β-oxidation is further metabolized in the glyoxysome through a series of reactions that make up the glyoxylate cycle (see Figure 12.19A). Initially, the acetyl-CoA reacts with oxaloacetate to give citrate, which is then transferred to the cytoplasm for isomerization to isocitrate by aconitase. Isocitrate is reimported into the glyoxysome and converted into malate by two reactions that are unique to the glyoxylate cycle:

1. First, isocitrate (C_6) is cleaved by the enzyme isocitrate lyase to give succinate (C_4) and glyoxylate (C_2). The succinate is exported to the mitochondria.
2. Next, malate synthase combines a second molecule of acetyl-CoA with glyoxylate to produce malate.

Malate is then transferred to the cytoplasm and converted into oxaloacetate by the cytoplasmic isozyme of malate dehydrogenase. Oxaloacetate is reimported into the glyoxysome and combines with another acetyl-CoA to continue the cycle (see Figure 12.19A). The glyoxylate produced keeps the cycle operating, but the succinate is exported to the mitochondria for further processing.

THE MITOCHONDRIAL ROLE Moving from the glyoxysomes to the mitochondria, the succinate is converted into malate by the two corresponding citric acid cycle reactions. The resulting malate can be exported from the mitochondria in exchange for succinate via the dicarboxylate transporter located in the inner mitochondrial membrane. Malate is then oxidized to oxaloacetate by malate dehydrogenase in the cytosol, and the resulting oxaloacetate is converted into carbohydrates by the reversal of glycolysis (gluconeogenesis). This conversion requires circumventing the irreversibility of the pyruvate kinase reaction (see Figure 12.3) and is facilitated by the enzyme PEP carboxykinase, which uses the phosphorylating ability of ATP to convert oxaloacetate into PEP and CO_2 (see Figure 12.19A).

From PEP, gluconeogenesis can proceed to the production of glucose, as described earlier. Sucrose is the final product of this process, and is the primary form of reduced carbon translocated from the cotyledons to the growing seedling tissues. Not all seeds quantitatively convert fat into sugar, however (see **WEB TOPIC 12.9**).

SUMMARY

Using the building blocks provided by photosynthesis, respiration releases the energy stored in carbon compounds in a controlled manner for cellular use. At the same time it generates many carbon precursors for biosynthesis.

Overview of Plant Respiration

- In plant respiration, reduced cellular carbon generated by photosynthesis is oxidized to CO_2 and water, and this oxidation is coupled to the synthesis of ATP.
- Respiration takes place by four main processes: glycolysis, the oxidative pentose phosphate pathway, the citric acid cycle, and oxidative phosphorylation (the electron transport chain and ATP synthesis) (**Figure 12.1**).

Glycolysis

- In glycolysis, carbohydrates are converted into pyruvate in the cytosol, and a small amount of ATP is synthesized via substrate-level phosphorylation. NADH is also produced (**Figure 12.3**).
- Plant glycolysis has alternative enzymes for several steps. These allow differences in substrates used, products made, and the direction of the pathway.
- When insufficient O_2 is available, fermentation regenerates NAD^+ for glycolysis. Only a minor fraction of the energy available in sugars is conserved by fermentation (**Figure 12.3**).
- Plant glycolysis is regulated from the "bottom up" by its products.

The Oxidative Pentose Phosphate Pathway

- Carbohydrates can be oxidized via the oxidative pentose phosphate pathway, which provides building blocks for biosynthesis and reducing power as NADPH (**Figure 12.4**).

The Citric Acid Cycle

- Pyruvate is oxidized to CO_2 within the mitochondrial matrix through the citric acid cycle, generating a large number of reducing equivalents in the form of NADH and $FADH_2$ (**Figures 12.5, 12.6**).
- In plants, the citric acid cycle is involved in alternative pathways that allow oxidation of malate or citrate and export of intermediates for biosynthesis (**Figures 12.6, 12.7**).

Mitochondrial Electron Transport and ATP Synthesis

- Electron transport from NADH and $FADH_2$ to oxygen is coupled by enzyme complexes to proton transport across the inner mitochondrial membrane. This generates an electrochemical proton gradient used for powering synthesis and export of ATP (**Figures 12.8–12.10**).
- During aerobic respiration, up to 60 molecules of ATP are produced per molecule of sucrose (**Table 12.2**).
- Typical for plant respiration is the presence of several proteins (alternative oxidase, NAD[P]H dehydrogenases, and uncoupling protein) that lower the energy recovery (**Figures 12.8, 12.9**).
- The main products of the respiratory process are ATP and metabolic intermediates used in biosynthesis. The cellular demand for these compounds regulates respiration via control points in the electron transport chain, the citric acid cycle, and glycolysis (**Figures 12.11–12.14**).

Respiration in Intact Plants and Tissues

- More than 50% of the daily photosynthetic yield may be respired by a plant.
- Many factors can affect the respiration rate observed at the whole-plant level. These factors include the nature and age of the plant tissue and environmental factors such as light, temperature, nutrient and water supply, and O_2 and CO_2 concentrations.

Lipid Metabolism

- Triacylglycerols (fats and oils) are an efficient form for storage of reduced carbon, particularly in seeds. Polar glycerolipids are the primary structural components of membranes (**Figures 12.15, 12.16; Tables 12.3, 12.4**).
- Triacylglycerols are synthesized in the ER and accumulate within the phospholipid bilayer, forming oil bodies.
- Fatty acids are synthesized in plastids using acetyl-CoA, in cycles of two-carbon addition. Fatty acids from the plastids can be transported to the ER, where they are further modified (**Figures 12.17, 12.18**).
- The function of a membrane may be influenced by its lipid composition. The degree of unsaturation of the fatty acids influences the sensitivity of plants to cold, but does not seem to be involved in normal chilling injury.
- Some lipid derivatives, such as jasmonate, are important plant hormones.
- During germination in oil-storing seeds, the stored lipids are metabolized to carbohydrates in a series of reactions that include the glyoxylate cycle. The glyoxylate cycle takes place in glyoxysomes, and subsequent steps occur in the mitochondria (**Figure 12.19**).
- The reduced carbon generated during lipid breakdown in the glyoxysomes is ultimately converted into carbohydrates in the cytosol by gluconeogenesis (**Figure 12.19**).

WEB MATERIAL

- **WEB TOPIC 12.1 Isolation of Mitochondria** Intact, functional mitochondria can be purified for analysis in vitro.
- **WEB TOPIC 12.2 The Q Cycle Explains How Complex III Pumps Protons across the Inner Mitochondrial Membrane** A cyclic process allows for a higher proton-to-electron stoichiometry.
- **WEB TOPIC 12.3 Multiple Energy Conservation Bypasses in Oxidative Phosphorylation of Plant Mitochondria** The enigmatic "energy-wasting" nonphosphorylating pathways of respiration are important for metabolic flexibility.

- **WEB TOPIC 12.4 F_oF_1-ATP Synthases: The World's Smallest Rotary Motors** Rotation of the γ subunit brings about the conformational changes that couple proton flux to ATP synthesis.
- **WEB TOPIC 12.5 Transport Into and Out of Plant Mitochondria** Plant mitochondria transport metabolites, coenzymes, and macromolecules.
- **WEB TOPIC 12.6 The Genetic System in Plant Mitochondria Has Several Special Features** The mitochondrial genome encodes about 40 mitochondrial proteins.
- **WEB TOPIC 12.7 Does Respiration Reduce Crop Yields?** Crop yield is correlated with low respiration rates in a way that is not fully understood.
- **WEB TOPIC 12.8 The Lipid Composition of Membranes Affects the Cell Biology and Physiology of Plants** Lipid mutants are expanding our understanding of the ability of organisms to adapt to temperature changes.
- **WEB TOPIC 12.9 Utilization of Oil Reserves in Cotyledons** In some species, only part of the stored lipid in the cotyledons is exported as carbohydrate.

- **WEB ESSAY 12.1 Metabolic Flexibility Helps Plants to Survive Stress** The ability of plants to carry out a metabolic step in different ways increases plant survival under stress.
- **WEB ESSAY 12.2 Metabolic Profiling of Plant Cells** Metabolic profiling complements genomics and proteomics.
- **WEB ESSAY 12.3 Mitochondrial Dynamics: When Form Meets Function** Fluorescence microscopy has shown that mitochondria dynamically change shape, size, number, and distribution in vivo.
- **WEB ESSAY 12.4 Seed Mitochondria and Stress Tolerance** Seeds experience a wide range of stresses and are dependent on respiration for germination.
- **WEB ESSAY 12.5 Balancing Life and Death: The Role of the Mitochondrion in Programmed Cell Death** Programmed cell death is an integral part of the life cycle of plants, often directly involving mitochondria.
- **WEB ESSAY 12.6 Respiration by Thermogenic Flowers** The temperature of thermogenic flowers, such as the *Arum* lilies, can increase up to 35°C above their surroundings.
- **WEB ESSAY 12.7 Reactive Oxygen Species (ROS) and Plant Respiration** The production of reactive oxygen species is an unavoidable consequence of aerobic respiration.
- **WEB ESSAY 12.8 Coenzyme Synthesis in Plant Mitochondria** Pathways for synthesis of coenzymes are often split between organelles.
- **WEB ESSAY 12.9 In Vivo Measurement of Plant Respiration** The activities of the alternative oxidase and cytochrome *c* oxidase can be simultaneously measured.

available at plantphys.net

Suggested Reading

Atkin, O. K., and Tjoelker, M. G. (2003) Thermal acclimation and the dynamic response of plant respiration to temperature. *Trends Plant Sci.* 8: 343–351.

Bates P. D., Stymne, S., and Ohlrogge, J. (2013) Biochemical pathways in seed oil synthesis. *Curr. Opin. Plant Biol.* 16: 358–364.

Gonzalez-Meler, M. A., Taneva, L., and Trueman, R. J. (2004) Plant respiration and elevated atmospheric CO_2 concentration: Cellular responses and global significance. *Ann. Bot.* 94: 647–656.

Markham, J. E., Lynch, D. V., Napier, J. A., Dunn, T. M., and Cahoon, E. B. (2013) Plant sphingolipids: function follows form. *Curr. Opin. Plant Biol.* 16: 350–357.

Millar, A. H., Whelan, J., Soole, K. L. and Day, D. A. (2011) Organization and regulation of mitochondrial respiration in plants. *Annu. Rev. Plant Biol.* 62: 79–104.

Møller, I. M. (2001) Plant mitochondria and oxidative stress. Electron transport, NADPH turnover and metabolism of reactive oxygen species. *Annu. Rev. Plant Physiol. Plant Mol. Biol.* 52: 561–591.

Nicholls, D. G., and Ferguson, S. J. (2013) *Bioenergetics 4*, 4th ed. Academic Press, San Diego, CA.

Plaxton, W. C. and Podestá, F. E. (2006) The functional organization and control of plant respiration. *Crit. Rev. Plant Sci.* 25: 159–198.

Rasmusson, A. G., Geisler, D. A., and Møller, I. M. (2008) The multiplicity of dehydrogenases in the electron transport chain of plant mitochondria. *Mitochondrion* 8: 47–60.

Sweetlove, L. J., Beard, K. F. M., Nunes-Nesi, A., Fernie, A. R. and Ratcliffe, R. G. (2010) Not just a circle: Flux modes in the plant TCA cycle. *Trends Plant Sci.* 15: 462–470.

Vanlerberghe, G. C. (2013) Alternative oxidase: A mitochondrial respiratory pathway to maintain metabolic and signaling homeostasis during abiotic and biotic stress in plants. *Int. J. Mol. Sci.* 14: 6805–6847.

Wallis, J. G., and Browse, J. (2010) Lipid biochemists salute the genome. *Plant J.* 61: 1092–1106.

13 Assimilation of Inorganic Nutrients

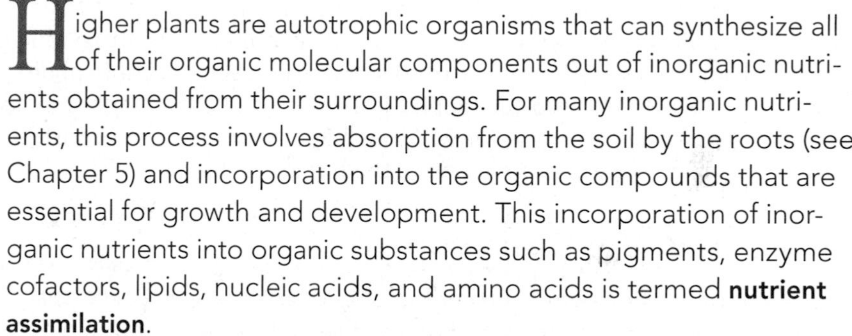

Higher plants are autotrophic organisms that can synthesize all of their organic molecular components out of inorganic nutrients obtained from their surroundings. For many inorganic nutrients, this process involves absorption from the soil by the roots (see Chapter 5) and incorporation into the organic compounds that are essential for growth and development. This incorporation of inorganic nutrients into organic substances such as pigments, enzyme cofactors, lipids, nucleic acids, and amino acids is termed **nutrient assimilation**.

Assimilation of some nutrients—particularly nitrogen and sulfur—involves a complex series of biochemical reactions that are among the most energy-consuming reactions in living organisms.

- In nitrate (NO_3^-) assimilation, the nitrogen in NO_3^- is converted to a higher-energy (more reduced) form in nitrite (NO_2^-), then to a yet-higher-energy (even more reduced) form in ammonium (NH_4^+), and finally into the amide nitrogen of the amino acid glutamine. This process consumes the equivalent of 12 ATPs per amide nitrogen.
- Plants such as legumes form symbiotic relationships with nitrogen-fixing bacteria to convert molecular nitrogen (N_2) into ammonia (NH_3). Ammonia (NH_3) is the first stable product of natural fixation; at physiological pH, however, ammonia is protonated to form the ammonium ion (NH_4^+). The process of biological nitrogen fixation, together with the subsequent assimilation of NH_3 into an amino acid, consumes the equivalent of about 16 ATPs per amide nitrogen.
- The assimilation of sulfate (SO_4^{2-}) into the amino acid cysteine via the two pathways found in plants consumes about 14 ATPs.

For some perspective on the enormous energies involved, consider that if these reactions run rapidly in reverse—say, from NH_4NO_3 (ammonium nitrate) to N_2—they become explosive, liberating vast amounts of energy as motion, heat, and light. Nearly all explosives, including nitroglycerin, TNT (trinitrotoluene), and gunpowder, are based on the rapid oxidation of nitrogen or sulfur compounds.

Assimilation of other nutrients, especially the macronutrient and micronutrient cations (see Chapter 5), involves the formation of complexes with organic compounds. For example, Mg^{2+} associates with chlorophyll pigments, Ca^{2+} associates with pectates within the cell wall, and Mo^{6+} associates with enzymes such as nitrate reductase and nitrogenase. These complexes are highly stable, and removal of the nutrient from the complex may result in total loss of function.

This chapter outlines the primary reactions through which the major nutrients (nitrogen, sulfur, phosphate, cations such as Mg^{2+} and K^+, and oxygen) are assimilated and discusses the organic products of these reactions. We will emphasize the physiological implications of the required energy expenditures and introduce the topic of symbiotic nitrogen fixation. Plants serve as the major conduit through which nutrients pass from slower geophysical domains into faster biological ones; this chapter will thus highlight the vital role of plant nutrient assimilation in the human diet.

Nitrogen in the Environment

Many prominent biochemical compounds in plant cells contain nitrogen (see Chapter 5). For example, nitrogen is found in the nucleotides and amino acids that form the building blocks of nucleic acids and proteins, respectively. Only the elements oxygen, carbon, and hydrogen are more abundant in plants than nitrogen. Most natural and agricultural ecosystems show dramatic gains in productivity after fertilization with inorganic nitrogen, attesting to the importance of this element and to the fact that it is present in suboptimal amounts.

In this section we will discuss the biogeochemical cycle of nitrogen, the crucial role of nitrogen fixation in the conversion of molecular nitrogen into ammonium and nitrate, and the fate of ammonium and nitrate in plant tissues.

Nitrogen passes through several forms in a biogeochemical cycle

Nitrogen is present in many forms in the biosphere. The atmosphere contains vast quantities (about 78% by volume) of molecular nitrogen (N_2) (see Chapter 9). For the most part, this large reservoir of nitrogen is not directly available to living organisms. Acquisition of nitrogen from the atmosphere requires the breaking of an exceptionally stable triple covalent bond between two nitrogen atoms ($N \equiv N$) to produce ammonia (NH_3) or nitrate (NO_3^-). These reactions, known as **nitrogen fixation**, occur through both industrial and natural processes.

TABLE 13.1 The major processes of the biogeochemical nitrogen cycle

Process	Definition	Rate (10^{13} g y^{-1})[a]
Industrial fixation	Industrial conversion of molecular nitrogen to ammonia	10
Atmospheric fixation	Lightning and photochemical conversion of molecular nitrogen to nitrate	1.9
Biological fixation	Prokaryotic conversion of molecular nitrogen to ammonia	17
Plant acquisition	Plant absorption and assimilation of ammonium or nitrate	120
Immobilization	Microbial absorption and assimilation of ammonium or nitrate	N/C
Ammonification	Bacterial and fungal catabolism of soil organic matter to ammonium	N/C
Anammox	Anaerobic ammonium oxidation: bacterial conversion of ammonium and nitrite to molecular nitrogen	N/C
Nitrification	Bacterial (*Nitrosomonas* sp.) oxidation of ammonium to nitrite and subsequent bacterial (*Nitrobacter* sp.) oxidation of nitrite to nitrate	N/C
Mineralization	Bacterial and fungal catabolism of soil organic matter to mineral nitrogen through ammonification or nitrification	N/C
Volatilization	Physical loss of gaseous ammonia to the atmosphere	10
Ammonium fixation	Physical embedding of ammonium into soil particles	1
Denitrification	Bacterial conversion of nitrate to nitrous oxide and molecular nitrogen	21
Nitrate leaching	Physical flow of nitrate dissolved in groundwater out of the topsoil and eventually into the oceans	3.6

Note: Terrestrial organisms, the soil, and the oceans contain about 5.2×10^{15} g, 95×10^{15} g, and 6.5×10^{15} g, respectively, of organic nitrogen that is active in the cycle. Assuming that the amount of atmospheric N_2 remains constant (inputs = outputs), the *mean residence time* (the average time that a nitrogen molecule remains in organic forms) is about 370 years [(pool size)/(fixation input) = $(5.2 \times 10^{15}$ g $+ 95 \times 10^{15}$ g$)/(8 \times 10^{13}$ g $y^{-1} + 1.9 \times 10^{13}$ g $y^{-1} + 17 \times 10^{13}$ g $y^{-1})$] (Schlesinger 1997).

[a]N/C, not calculated.

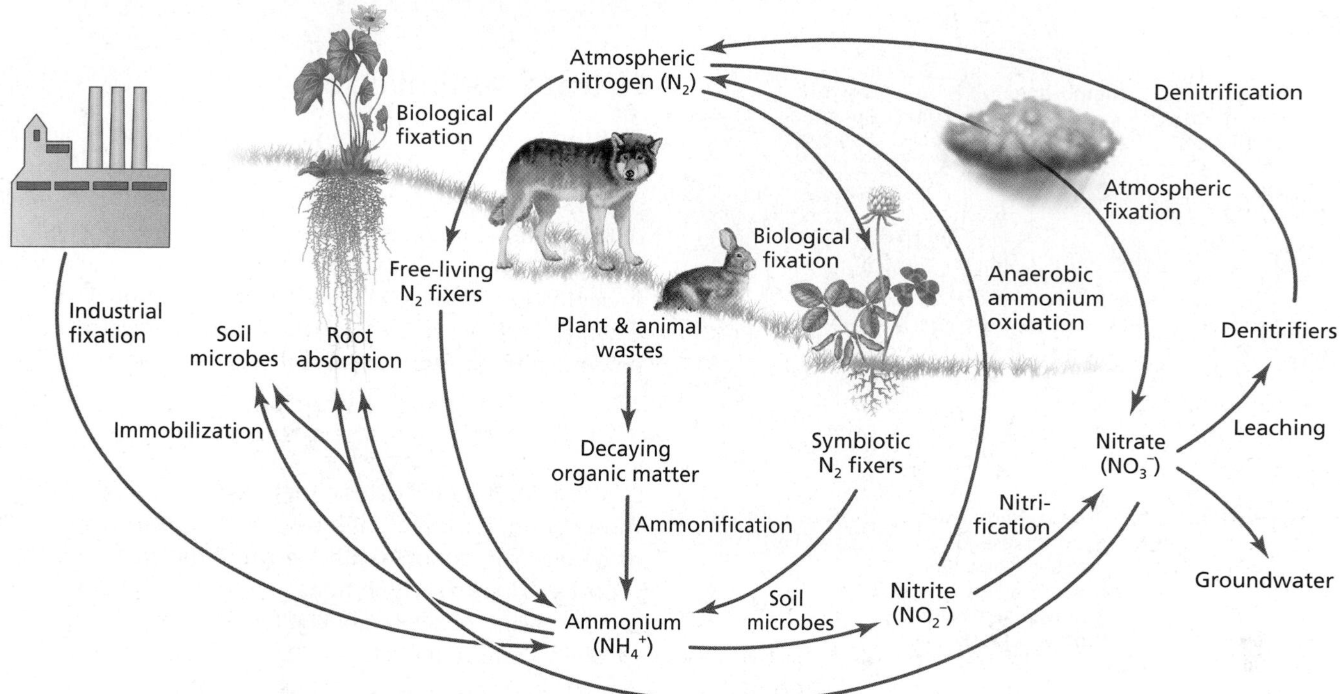

Figure 13.1 Nitrogen cycles through the atmosphere, changing from a gas to soluble ions before it is incorporated into organic compounds in living organisms. Some of the steps involved in the nitrogen cycle are shown.

N_2 combines with hydrogen to form ammonia under elevated temperature (about 200°C) and high pressure (about 200 atmospheres) and in the presence of a metal catalyst (usually iron). The extreme conditions are required to overcome the high activation energy of the reaction. This nitrogen fixation reaction, called the *Haber–Bosch process*, is a starting point for the manufacture of many industrial and agricultural products, including nitrogen fertilizers. Worldwide industrial production of nitrogen fertilizers amounts to more than 110 million metric tons per year (11×10^{13} g y^{-1}).

The following natural processes fix about 190 million metric tons per year of nitrogen (**Table 13.1**):

- *Lightning.* Lightning is responsible for about 8% of the nitrogen fixed by natural processes. Lightning converts water vapor and oxygen into highly reactive hydroxyl free radicals, free hydrogen atoms, and free oxygen atoms that attack molecular nitrogen (N_2) to form nitric acid (HNO_3). This nitric acid subsequently falls to Earth with rain.
- *Photochemical reactions.* Approximately 2% of the nitrogen fixed derives from photochemical reactions between gaseous nitric oxide (NO) and ozone (O_3) that produce nitric acid (HNO_3).
- *Biological nitrogen fixation.* The remaining 90% results from biological nitrogen fixation, in which bacteria or cyanobacteria (blue-green algae) fix N_2 into ammonia (NH_3). This ammonia dissolves in water to form ammonium (NH_4^+):

$$NH_3 + H_2O \rightarrow NH_4^+ + OH^- \quad (13.1)$$

From an agricultural standpoint, biological nitrogen fixation is critical, because industrially produced nitrogen fertilizers are economically and environmentally costly and not affordable by many poor farmers.

Once fixed into ammonia or nitrate, nitrogen enters a biogeochemical cycle and passes through several organic or inorganic forms before it eventually returns to molecular nitrogen (**Figure 13.1**; see also Table 13.1). The ammonium (NH_4^+) and nitrate (NO_3^-) ions in the soil solution that are generated through fixation or released through decomposition of soil organic matter become the object of intense competition among plants and microorganisms. To be competitive, plants have evolved mechanisms for scavenging these ions rapidly from the soil solution (see Chapter 5). Under the elevated soil concentrations that occur after fertilization, the absorption of ammonium and nitrate by the roots may exceed the capacity of a plant to assimilate these ions, leading to their accumulation within the plant's tissues.

Unassimilated ammonium or nitrate may be dangerous

Ammonium, if it accumulates to high levels in living tissues, is toxic to both plants and animals. Ammonium dissipates transmembrane proton gradients (**Figure 13.2**) that are required for photosynthetic and respiratory electron transport (see Chapters 7 and 12), for sequestering metabolites in the vacuole (see Chapter 6), and for transporting nutrients across biological membranes (see Chapter 6). Because high levels of ammonium are dangerous, animals have developed a strong aversion to its

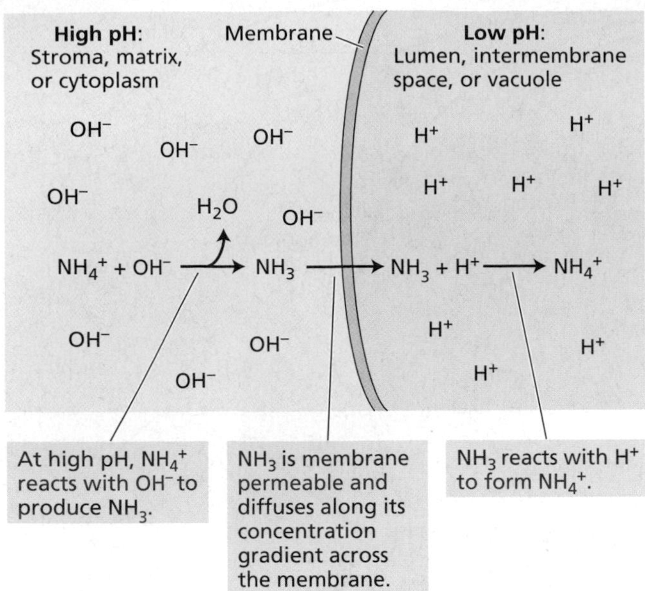

Figure 13.2 NH_4^+ toxicity derives from dissipation of pH gradients. The left side represents the stroma, matrix, or cytoplasm, where the pH is high; the right side represents the lumen, intermembrane space, or vacuole, where the pH is low; and the membrane represents the thylakoid, inner mitochondrial, or tonoplast membrane for a chloroplast, mitochondrion, or root cell, respectively. The net result of the reaction shown is that both the OH^- concentration on the left side and the H^+ concentration on the right side have been diminished; that is, the pH gradient has been dissipated. (After Bloom 1997.)

smell. The active ingredient in smelling salts, a medicinal vapor released under the nose to revive a person who has fainted, is ammonium carbonate. Plants assimilate ammonium near the site of absorption or generation and rapidly store any excess in their vacuoles, thus avoiding toxic effects on membranes and the cytosol.

Unlike the case with ammonium, plants can store high levels of nitrate, and they can translocate it from tissue to tissue without deleterious effect. Yet if livestock or humans consume plant material that is high in nitrate, they may suffer methemoglobinemia, a disease in which the liver reduces nitrate to nitrite, which combines with hemoglobin and renders hemoglobin unable to bind oxygen. Humans and other animals may also convert nitrate into nitrosamines, which are potent carcinogens, or into nitric oxide, a potent signaling molecule involved in many physiological processes such as widening of blood vessels. Some countries limit the nitrate content in plant materials sold for human consumption.

In the next sections we will discuss the process by which plants assimilate nitrate into organic compounds via the enzymatic reduction of nitrate first into nitrite, next into ammonium, and then into amino acids.

Nitrate Assimilation

Plant roots actively absorb nitrate from the soil solution via several low- and high-affinity nitrate–proton cotransporters (see Chapter 6). Plants eventually assimilate most of this nitrate into organic nitrogen compounds. The first step of this process is the conversion of nitrate to nitrite in the cytosol, a reduction reaction (for redox properties, see Chapter 12) that involves the transfer of two electrons. The enzyme **nitrate reductase** catalyzes this reaction:

$$NO_3^- + NAD(P)H + H^+ \rightarrow NO_2^- + NAD(P)^+ + H_2O \qquad (13.2)$$

where NAD(P)H indicates either NADH or NADPH. The most common form of nitrate reductase uses only NADH as an electron donor; another form of the enzyme that is found predominantly in nongreen tissues such as roots can use either NADH or NADPH.

The nitrate reductases of higher plants are composed of two identical subunits, each containing three prosthetic groups: flavin adenine dinucleotide (FAD), heme, and a molybdenum ion complexed to an organic molecule called a *pterin*.

A pterin (fully oxidized)

Nitrate reductase is the main molybdenum-containing protein in vegetative tissues; one symptom of molybdenum deficiency is the accumulation of nitrate that results from diminished nitrate reductase activity.

X-ray crystallography and comparison of the amino acid sequences for nitrate reductase from several species with the sequences of other well-characterized proteins that bind FAD, heme, or molybdenum ions have led to a multiple-domain model for nitrate reductase; a simplified three-domain model is shown in **Figure 13.3**. The FAD-binding domain accepts two electrons from NADH or NADPH. The electrons then pass through the heme domain to the molybdenum complex, where they are transferred to nitrate.

Many factors regulate nitrate reductase

Nitrate, light, and carbohydrates influence nitrate reductase at the transcription and translation levels. In barley seedlings, nitrate reductase mRNA was detected approximately 40 minutes after addition of nitrate, and maximum levels were attained within 3 hours (**Figure 13.4**). In contrast to the rapid mRNA accumulation, there was a gradual linear increase in nitrate reductase activity, reflecting that the synthesis of the nitrate reductase protein requires the presence of the nitrate reductase mRNA.

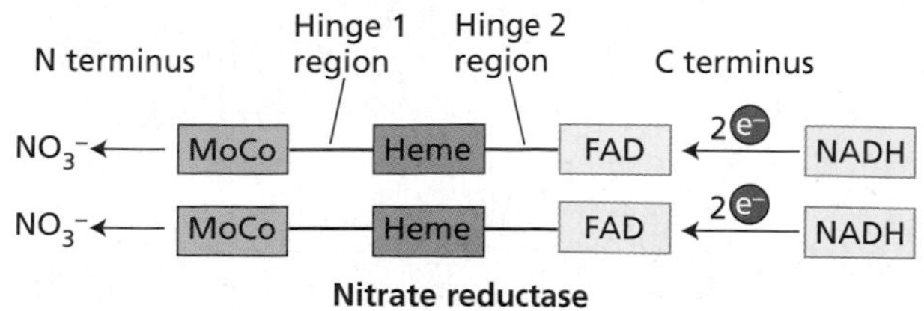

Figure 13.3 A model of the nitrate reductase dimer, illustrating the three binding domains whose polypeptide sequences are similar in eukaryotes: molybdenum complex (MoCo), heme, and FAD. The NADH binds at the FAD-binding region of each subunit and initiates a two-electron transfer from the carboxyl (C) terminus, through each of the electron transfer components, to the amino (N) terminus. Nitrate is reduced at the molybdenum complex near the amino terminus. The polypeptide sequences of the hinge regions are highly variable among species.

In addition, the protein is subject to posttranslational modification (involving a reversible phosphorylation) that is analogous to the regulation of sucrose phosphate synthase (see Chapters 8 and 11). Light, carbohydrate levels, and other environmental factors stimulate a protein phosphatase that dephosphorylates a key serine residue in the hinge 1 region of nitrate reductase (between the molybdenum complex and heme-binding domains; see Figure 13.3) and thereby activates the enzyme.

Operating in the reverse direction, darkness and Mg^{2+} stimulate a protein kinase that phosphorylates the same serine residues, which then interact with a 14-3-3 inhibitor protein, and thereby inactivate nitrate reductase. *Regulation of nitrate reductase activity through phosphorylation and dephosphorylation provides more rapid control than can be achieved through synthesis or degradation of the enzyme (minutes versus hours).*

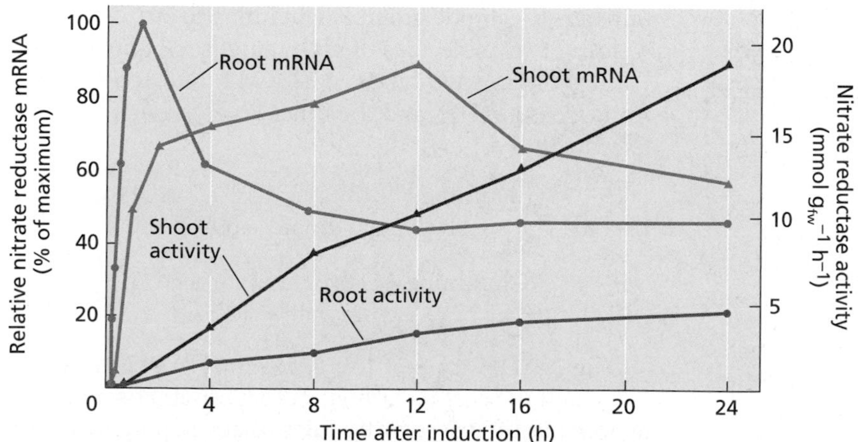

Figure 13.4 Stimulation of nitrate reductase activity follows the induction of nitrate reductase mRNA in shoots and roots of barley; g_{fw}, grams fresh weight. (After Kleinhofs et al. 1989.)

Nitrite reductase converts nitrite to ammonium

Nitrite (NO_2^-) is a highly reactive, potentially toxic ion. Plant cells immediately transport the nitrite generated by nitrate reduction (see Equation 13.2) from the cytosol into chloroplasts in leaves and plastids in roots. In these organelles, the enzyme nitrite reductase reduces nitrite to ammonium, a reaction that involves the transfer of six electrons, according to the following overall reaction:

$$NO_2^- + 6\ Fd_{red} + 8\ H^+ \rightarrow NH_4^+ + 6\ Fd_{ox} + 2\ H_2O \quad (13.3)$$

where Fd is ferredoxin and the subscripts *red* and *ox* stand for *reduced* and *oxidized*, respectively. Reduced ferredoxin is derived from photosynthetic electron transport in the chloroplasts (see Chapter 7) and from NADPH generated by the oxidative pentose phosphate pathway in nongreen tissues (see Chapter 12).

Chloroplasts and root plastids contain different forms of the enzyme, but both forms consist of a single polypeptide containing two prosthetic groups: an iron–sulfur cluster (Fe_4S_4) and a specialized heme. These groups act together to bind nitrite and reduce it to ammonium. Although no nitrogen compounds of intermediate redox states accumulate, a small percentage (0.02–0.2%) of the nitrite reduced is released as nitrous oxide (N_2O), a greenhouse gas. The electron flow through ferredoxin, Fe_4S_4, and heme can be represented as in **Figure 13.5**.

Nitrite reductase is encoded in the nucleus and synthesized in the cytoplasm with an N-terminal transit peptide that targets it to the plastids. Elevated concentrations of NO_3^- or exposure to light induce the transcription of nitrite reductase mRNA. Accumulation of the end products in the process—asparagine and glutamine—represses this induction.

Both roots and shoots assimilate nitrate

In many plants, when the roots receive small amounts of nitrate, nitrate is reduced primarily in the roots. As the supply of nitrate increases, a greater proportion of the absorbed nitrate is translocated to the shoot and assimilated there. Even under similar conditions of nitrate supply, the balance between root and shoot nitrate metabolism—as indicated by the proportion of nitrate reductase activity in each of the two organs or by the relative concentrations of nitrate and reduced nitrogen in the xylem sap—varies from species to species.

In plants such as cocklebur (*Xanthium strumarium*), nitrate metabolism is restricted to the shoot; in other

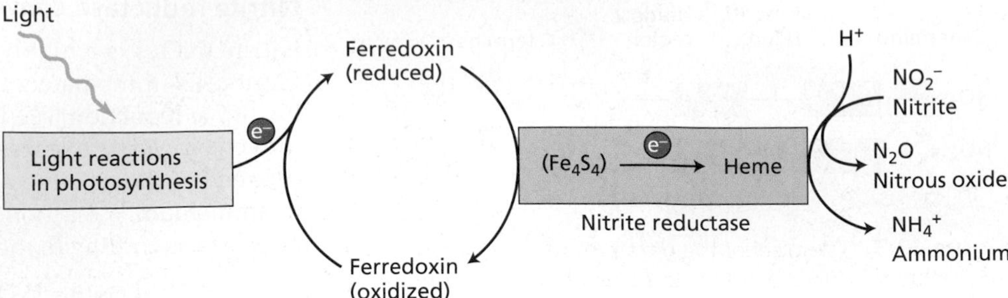

Figure 13.5 Model for coupling of photosynthetic electron flow, via ferredoxin, to the reduction of nitrite by nitrite reductase. The enzyme contains two prosthetic groups, Fe_4S_4 and heme, which participate in the reduction of nitrite to ammonium.

plants, such as white lupine (*Lupinus albus*), most nitrate is metabolized in the roots (**Figure 13.6**). Generally, species native to temperate regions rely more heavily on nitrate assimilation by the roots than do species of tropical or subtropical origins.

Ammonium Assimilation

Plant cells avoid ammonium toxicity by rapidly converting the ammonium generated from nitrate assimilation or photorespiration (see Chapter 8) into amino acids. The primary pathway for this conversion involves the sequential actions of glutamine synthetase and glutamate synthase. In this section we will discuss the enzymatic processes

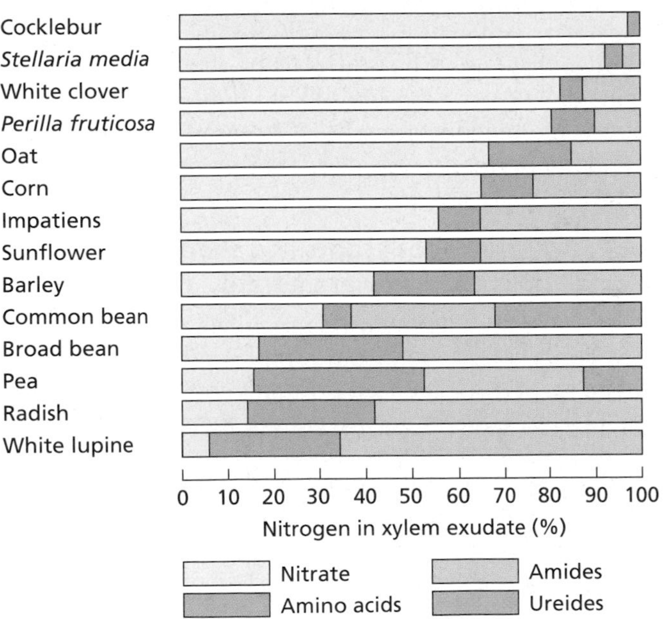

Figure 13.6 Relative amounts of nitrate and other nitrogen compounds in the xylem sap of various plant species. The plants were grown with their roots exposed to nitrate solutions, and xylem sap was collected by severing the stem. Note the presence of ureides in common bean and pea; only legumes of tropical origin export nitrogen in such compounds. (After Pate 1983.)

that mediate the assimilation of ammonium into essential amino acids, and the role of amides in the regulation of nitrogen and carbon metabolism.

Converting ammonium to amino acids requires two enzymes

Glutamine synthetase (**GS**) combines ammonium with glutamate to form glutamine (**Figure 13.7A**):

$$\text{Glutamate} + NH_4^+ + \text{ATP} \rightarrow \text{glutamine} + \text{ADP} + P_i \qquad (13.4)$$

This reaction requires the hydrolysis of one ATP and involves a divalent cation such as Mg^{2+}, Mn^{2+}, or Co^{2+} as a cofactor. Plants contain two classes of GS, one in the cytosol and the other in root plastids or shoot chloroplasts. The cytosolic forms are expressed in germinating seeds or in the vascular bundles of roots and shoots and produce glutamine for intercellular nitrogen transport. The GS in root plastids generates amide nitrogen for local consumption; the GS in shoot chloroplasts reassimilates photorespiratory NH_4^+. Light and carbohydrate levels alter the expression of the plastid forms of the enzyme, but they have little effect on the cytosolic forms.

Elevated plastid levels of glutamine stimulate the activity of **glutamate synthase** (also known as *glutamine:2-oxoglutarate aminotransferase*, or **GOGAT**). This enzyme transfers the amide group of glutamine to 2-oxoglutarate, yielding two molecules of glutamate (see Figure 13.7A). Plants contain two types of GOGAT; one accepts electrons from NADH, and the other accepts electrons from ferredoxin (Fd):

$$\text{Glutamine} + \text{2-oxoglutarate} + \text{NADH} + H^+ \rightarrow 2\ \text{glutamate} + \text{NAD}^+ \qquad (13.5)$$

$$\text{Glutamine} + \text{2-oxoglutarate} + \text{Fd}_{red} \rightarrow 2\ \text{glutamate} + \text{Fd}_{ox} \qquad (13.6)$$

The NADH type of the enzyme (NADH-GOGAT) is located in plastids of nonphotosynthetic tissues such as roots or the vascular bundles of developing leaves. In roots, NADH-GOGAT is involved in the assimilation of NH_4^+ absorbed from the rhizosphere (the soil near the

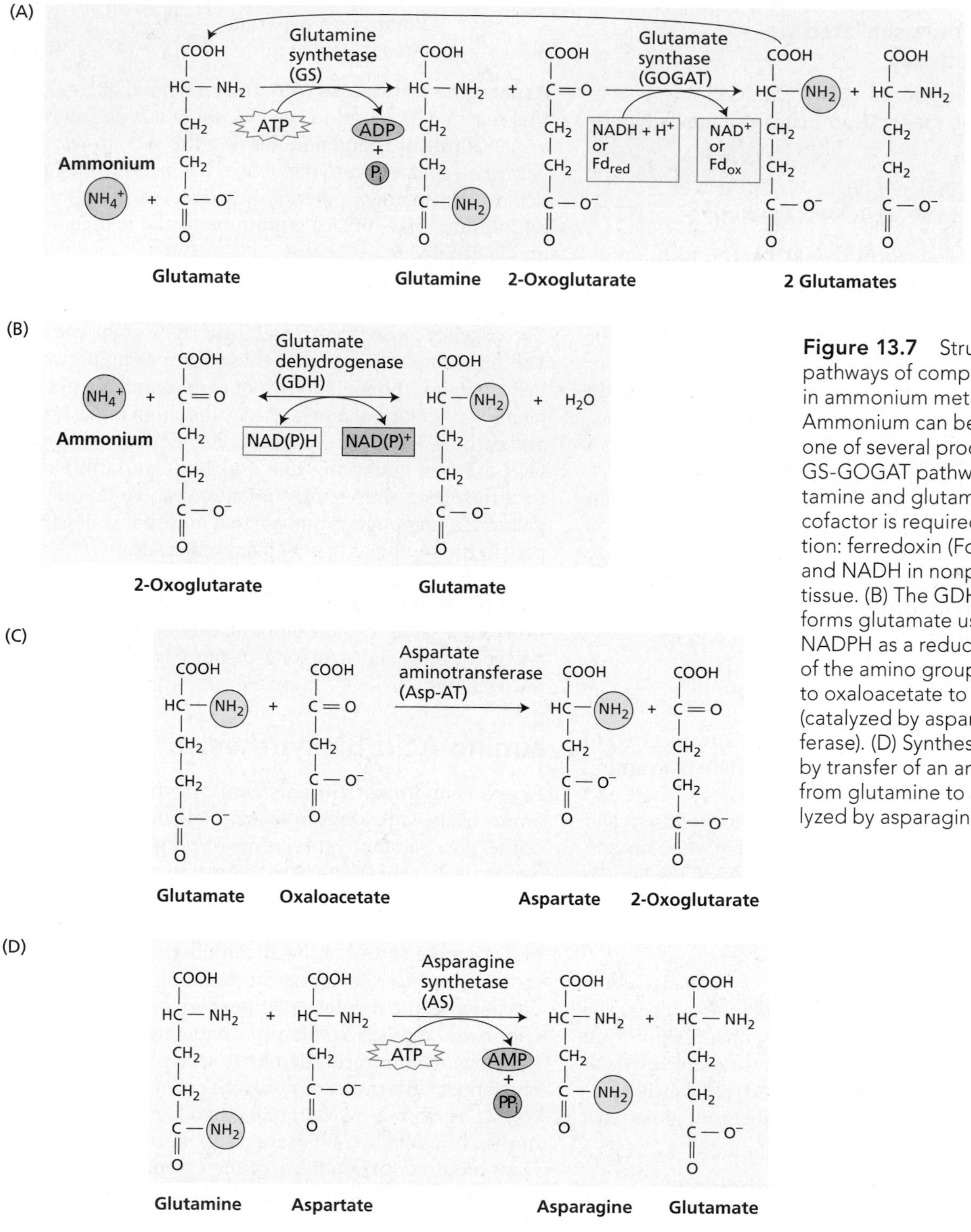

Figure 13.7 Structure and pathways of compounds involved in ammonium metabolism. Ammonium can be assimilated by one of several processes. (A) The GS-GOGAT pathway that forms glutamine and glutamate. A reduced cofactor is required for the reaction: ferredoxin (Fd) in green leaves and NADH in nonphotosynthetic tissue. (B) The GDH pathway that forms glutamate using NADH or NADPH as a reductant. (C) Transfer of the amino group from glutamate to oxaloacetate to form aspartate (catalyzed by aspartate aminotransferase). (D) Synthesis of asparagine by transfer of an amino acid group from glutamine to aspartate (catalyzed by asparagine synthetase).

surface of the roots); in vascular bundles of developing leaves, NADH-GOGAT assimilates glutamine translocated from roots or senescing leaves.

The ferredoxin-dependent type of glutamate synthase (Fd-GOGAT) is found in chloroplasts and serves in photorespiratory nitrogen metabolism. Both the amount of protein and its activity increase with light levels. Roots, particularly those under nitrate nutrition, have Fd-GOGAT in plastids. Fd-GOGAT in the roots presumably functions to incorporate the glutamine generated during nitrate assimilation. Electrons to reduce Fd in roots are generated by the oxidative pentose phosphate pathway (see Chapter 12).

Ammonium can be assimilated via an alternative pathway

Glutamate dehydrogenase (**GDH**) catalyzes a reversible reaction that synthesizes or deaminates glutamate (**Figure 13.7B**):

$$\text{2-Oxoglutarate} + NH_4^+ + NAD(P)H \leftrightarrow \text{glutamate} + H_2O + NAD(P)^+ \qquad (13.7)$$

An NADH-dependent form of GDH is found in mitochondria, and an NADPH-dependent form is localized in the chloroplasts of photosynthetic organs. Although both forms are relatively abundant, they cannot substitute for the GS–GOGAT pathway for assimilation of ammonium, and their primary function is in deaminating glutamate during the reallocation of nitrogen (see Figure 13.7B).

Transamination reactions transfer nitrogen

Once assimilated into glutamine and glutamate, nitrogen is incorporated into other amino acids via transamination reactions. The enzymes that catalyze these reactions are known as aminotransferases. An example is **aspartate aminotransferase** (**Asp-AT**), which catalyzes the following reaction (**Figure 13.7C**):

$$\text{Glutamate} + \text{oxaloacetate} \rightarrow \text{2-oxoglutarate} + \text{aspartate} \qquad (13.8)$$

in which the amino group of glutamate is transferred to the carboxyl group of oxaloacetate. Aspartate is an amino acid that participates in the malate–aspartate shuttle to transfer reducing equivalents from the mitochondrion and chloroplast into the cytosol (see **WEB TOPIC 12.5**) and in the transport of carbon from the mesophyll to the bundle sheath for C_4 carbon fixation (see Chapter 8). All transamination reactions require pyridoxal phosphate (vitamin B6) as a cofactor.

Aminotransferases are found in the cytoplasm, chloroplasts, mitochondria, glyoxysomes, and peroxisomes. The aminotransferases in chloroplasts may have a significant role in amino acid biosynthesis, because plant leaves or isolated chloroplasts exposed to radioactively labeled carbon dioxide rapidly incorporate the label into glutamate, aspartate, alanine, serine, and glycine.

Asparagine and glutamine link carbon and nitrogen metabolism

Asparagine, isolated from asparagus as early as 1806, was the first amide to be identified. It serves not only as a component of proteins, but also as a key compound for nitrogen transport and storage because of its stability and high nitrogen-to-carbon ratio (2 N to 4 C for asparagine compared with 2 N to 5 C for glutamine and 1 N to 5 C for glutamate).

The major pathway for asparagine synthesis involves the transfer of the amide nitrogen from glutamine to aspartate (**Figure 13.7D**):

$$\text{Glutamine} + \text{aspartate} + ATP \rightarrow \text{glutamate} + \text{asparagine} + AMP + PP_i \qquad (13.9)$$

Asparagine synthetase (**AS**), the enzyme that catalyzes this reaction, is found in the cytosol of leaves and roots and in nitrogen-fixing nodules (see the section *Bacterial Nitrogen Fixation*). In maize (corn; *Zea mays*) roots, particularly those under potentially toxic levels of ammonia, ammonium may replace glutamine as the source of the amide group.

High levels of light and carbohydrate—conditions that stimulate plastid GS and Fd-GOGAT—inhibit the expression of genes coding for AS and the activity of the enzyme. The opposing regulation of these competing pathways helps balance the metabolism of carbon and nitrogen in plants. Conditions of ample energy (i.e., high levels of light and carbohydrates) stimulate GS (see Equation 13.4) and GOGAT (see Equations 13.5 and 13.6), and inhibit AS; thus they favor nitrogen assimilation into glutamine and glutamate, compounds that are rich in carbon and participate in the synthesis of new plant materials.

In contrast, energy-limited conditions inhibit GS and GOGAT, stimulate AS, and thus favor nitrogen assimilation into asparagine, a compound that is rich in nitrogen and sufficiently stable for long-distance transport or long-term storage.

Amino Acid Biosynthesis

Humans and most animals cannot synthesize certain amino acids—histidine, isoleucine, leucine, lysine, methionine, phenylalanine, threonine, tryptophan, and valine, and in the case of young humans, arginine (adult humans can synthesize arginine)—and thus must obtain these so-called essential amino acids from their diet. In contrast, plants synthesize all of the 20 amino acids that are common in proteins. The nitrogen-containing amino group, as discussed in the previous section, derives from transamination reactions with glutamine or glutamate. The carbon skeleton for amino acids derives from 3-phosphoglycerate, phosphoenolpyruvate, or pyruvate generated during glycolysis, or from 2-oxoglutarate or oxaloacetate generated in the citric acid cycle (**Figure 13.8**). Parts of these pathways required for synthesis of the essential amino acids are appropriate targets for herbicides (such as Roundup; see Chapter 2), because they are missing from animals, so substances that block these pathways are lethal to plants but in low concentrations do not injure animals.

Biological Nitrogen Fixation

Biological nitrogen fixation accounts for most of the conversion of atmospheric N_2 into ammonium, and thus serves as the key entry point of molecular nitrogen into the biogeochemical cycle of nitrogen (see Figure 13.1). In

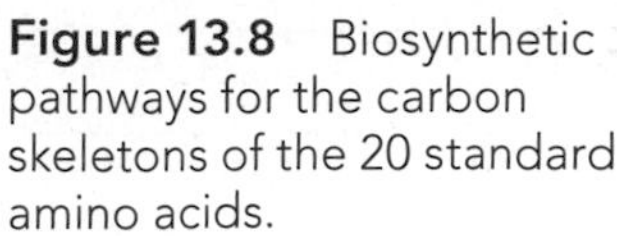

Figure 13.8 Biosynthetic pathways for the carbon skeletons of the 20 standard amino acids.

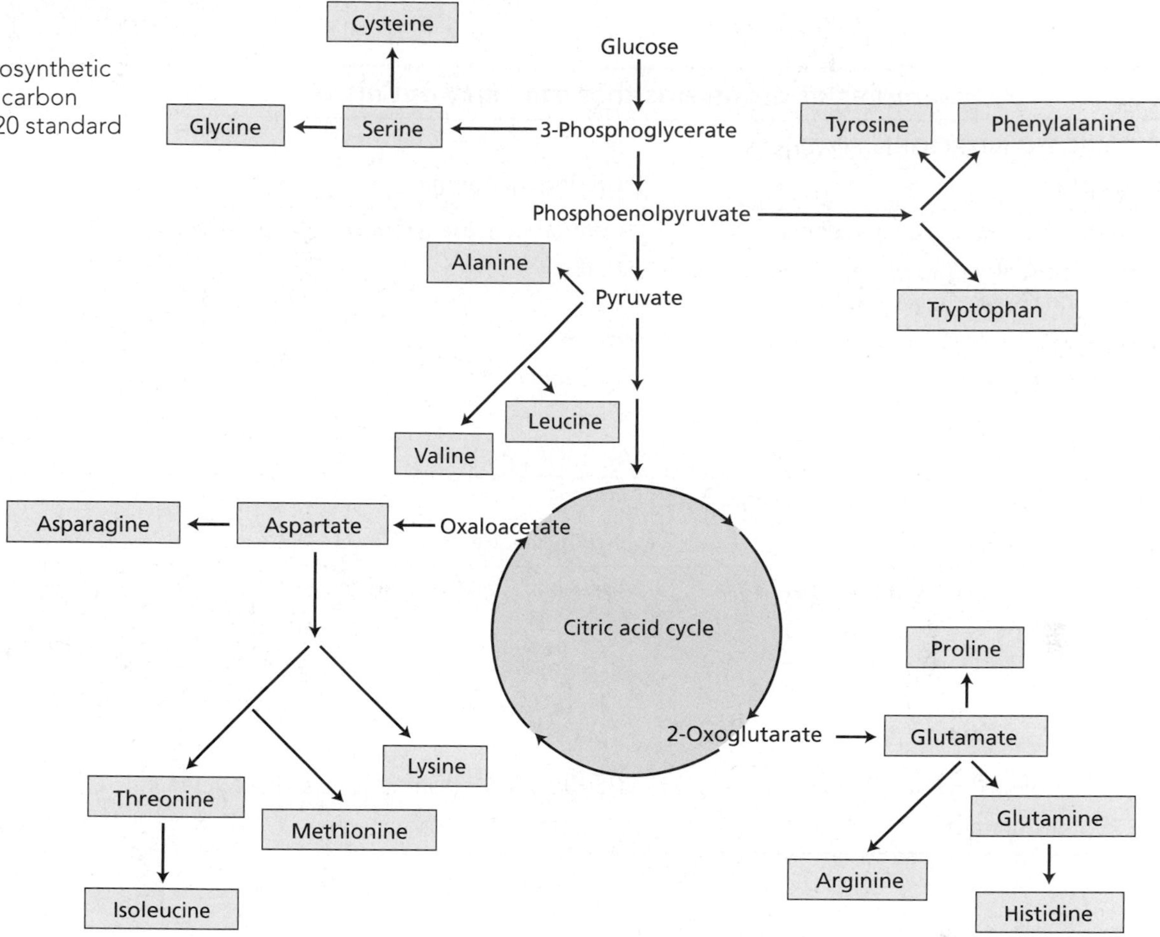

this section we will describe the symbiotic relationship between nitrogen-fixing organisms and higher plants; nodules, the specialized structures that form in roots when infected by nitrogen-fixing bacteria; the genetic and signaling interactions that regulate nitrogen fixation by symbiotic prokaryotes and their hosts; and the properties of the nitrogenase enzymes that fix nitrogen.

Free-living and symbiotic bacteria fix nitrogen

Some bacteria, as stated earlier, can convert atmospheric nitrogen into ammonia (**Table 13.2**). Most of these nitrogen-fixing prokaryotes live in the soil, generally independent of other organisms. Several form symbiotic associations with higher plants in which the prokaryote directly provides the host plant with fixed nitrogen in exchange for other nutrients and carbohydrates (see Table 13.2, top). Such symbioses occur in nodules that form on the roots of the plant and contain the nitrogen-fixing bacteria.

The most common type of symbiosis occurs between members of the plant family Fabaceae (Leguminosae) and soil bacteria of the genera *Azorhizobium, Bradyrhizobium, Mesorhizobium, Rhizobium,* and *Sinorhizobium* (collectively called **rhizobia**; **Table 13.3** and **Figure 13.9**). Another common type of symbiosis occurs between several woody plant species, such as alder trees, and soil bacteria of the genus

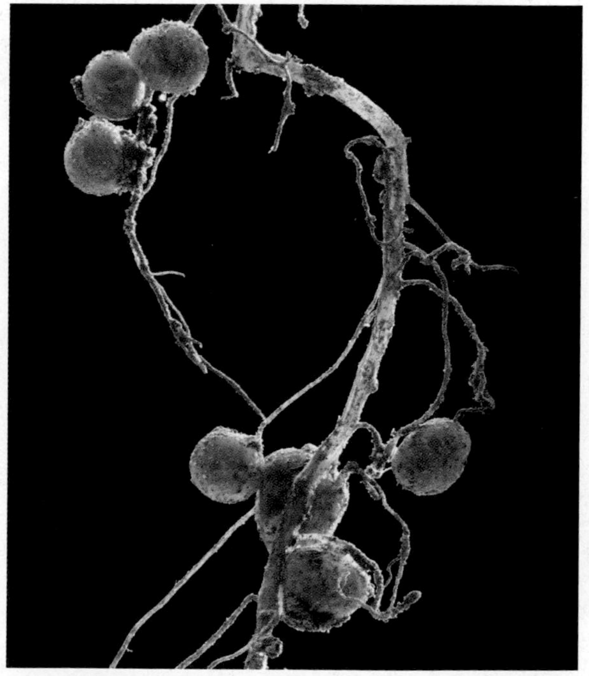

Figure 13.9 Root nodules on a common bean (*Phaseolus vulgaris*). The nodules, the spherical structures, are a result of infection by *Rhizobium* sp.

TABLE 13.2 Examples of organisms that can carry out nitrogen fixation

SYMBIOTIC NITROGEN FIXATION	
Host plant	**N-fixing symbionts**
Leguminous: legumes, *Parasponia*	*Azorhizobium, Bradyrhizobium, Mesorhizobium, Rhizobium, Sinorhizobium*
Actinorhizal: alder (tree), *Ceanothus* (shrub), *Casuarina* (tree), *Datisca* (shrub)	*Frankia*
Gunnera	*Nostoc*
Azolla (water fern)	*Anabaena*
Sugarcane	*Acetobacter*
Miscanthus	*Azospirillum*
FREE-LIVING NITROGEN FIXATION	
Type	**N-fixing genera**
Cyanobacteria (blue-green algae)	*Anabaena, Calothrix, Nostoc*
Other bacteria	
Aerobic	*Azospirillum, Azotobacter, Beijerinckia, Derxia*
Facultative	*Bacillus, Klebsiella*
Anaerobic	
Nonphotosynthetic	*Clostridium, Methanococcus* (archaebacterium)
Photosynthetic	*Chromatium, Rhodospirillum*

Frankia; these plants are known as **actinorhizal** plants. Still other types of nitrogen-fixing symbioses involve the South American herb *Gunnera* and the tiny water fern *Azolla,* which form associations with the cyanobacteria *Nostoc* and *Anabaena,* respectively (**Figure 13.10**; also see Table 13.2). Finally, several types of nitrogen-fixing bacteria are associated with C_4 grasses such as sugarcane and *Miscanthus.*

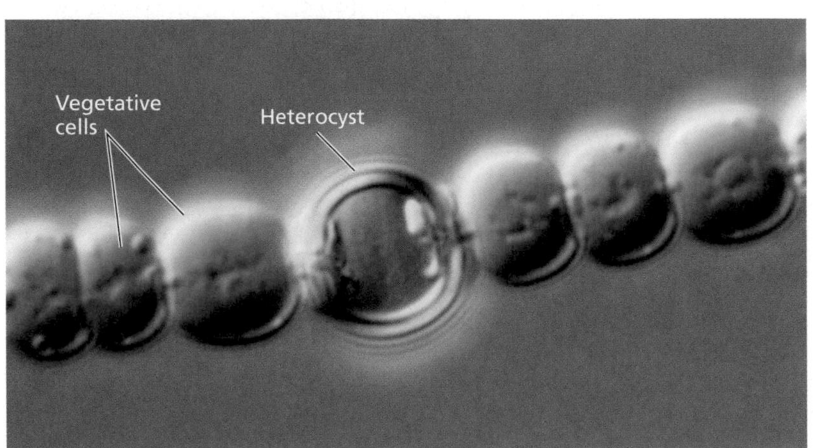

Figure 13.10 A heterocyst in a filament of the nitrogen-fixing cyanobacterium *Anabaena,* which forms associations with the water fern *Azolla.* The thick-walled heterocysts, interspersed among vegetative cells, have an anaerobic inner environment that allows cyanobacteria to fix nitrogen in aerobic conditions.

Nitrogen fixation requires microanaerobic or anaerobic conditions

Because nitrogen fixation involves the expenditure of large amounts of energy, the nitrogenase enzymes that catalyze these reactions have sites that facilitate the high-energy exchange of electrons. Oxygen, being a strong electron acceptor, can damage these sites and irreversibly inactivate nitrogenase, so nitrogen must be fixed under anaerobic conditions. Each of the nitrogen-fixing organisms listed in Table 13.2 either functions under natural anaerobic conditions or creates an internal, local anaerobic environment (microanaerobic) separated from the oxygen in the atmosphere that surrounds it.

In cyanobacteria, anaerobic conditions are created in specialized cells called *heterocysts* (see Figure 13.10). Heterocysts are thick-walled cells that differentiate when filamentous cyanobacteria are deprived of NH_4^+. These cells lack photosystem II, the oxygen-producing photosystem of chloroplasts (see Chapter 7), so they do not generate oxygen. Heterocysts appear to represent an adaptation for nitrogen fixation, in that they are widespread among aerobic cyanobacteria that fix nitrogen.

TABLE 13.3 Associations between host plants and rhizobia

Plant host	Rhizobial symbiont
Parasponia (a nonlegume, formerly called *Trema*)	*Bradyrhizobium* spp.
Soybean (*Glycine max*)	*Bradyrhizobium japonicum* (slow-growing type); *Sinorhizobium fredii* (fast-growing type)
Alfalfa (*Medicago sativa*)	*Sinorhizobium meliloti*
Sesbania (aquatic)	*Azorhizobium* (forms both root and stem nodules; the stems have adventitious roots)
Bean (*Phaseolus*)	*Rhizobium leguminosarum* bv. *phaseoli*; *R. tropicii*; *R. etli*
Clover (*Trifolium*)	*Rhizobium leguminosarum* bv. *trifolii*
Pea (*Pisum sativum*)	*Rhizobium leguminosarum* bv. *viciae*
Aeschynomene (aquatic)	Photosynthetic *Bradyrhizobium* clade (photosynthetically active rhizobia that form stem nodules, probably associated with adventitious roots)

Cyanobacteria can fix nitrogen under anaerobic conditions such as those that occur in flooded fields. In Asian countries, nitrogen-fixing cyanobacteria of both the heterocyst and nonheterocyst types are a major means for maintaining an adequate nitrogen supply in the soil of rice fields. These microorganisms fix nitrogen when the fields are flooded and die as the fields dry, releasing the fixed nitrogen to the soil. Another important source of available nitrogen in flooded rice fields is the water fern *Azolla*, which associates with the cyanobacterium *Anabaena*. The *Azolla–Anabaena* association can fix as much as 0.5 kg of atmospheric nitrogen per hectare per day, a rate of fertilization that is sufficient to attain moderate rice yields.

Free-living bacteria that are capable of fixing nitrogen are aerobic, facultative, or anaerobic (see Table 13.2, bottom):

- *Aerobic* nitrogen-fixing bacteria such as *Azotobacter* are thought to maintain a low oxygen concentration (microaerobic conditions) through their high levels of respiration. Others, such as *Gloeothece*, evolve O_2 photosynthetically during the day and fix nitrogen during the night when respiration lowers oxygen levels.
- *Facultative* organisms, which are able to grow under both aerobic and anaerobic conditions, generally fix nitrogen only under anaerobic conditions.
- Obligate *anaerobic* nitrogen-fixing bacteria that grow in environments devoid of oxygen can be either photosynthetic (e.g., *Rhodospirillum*) or nonphotosynthetic (e.g., *Clostridium*).

Symbiotic nitrogen fixation occurs in specialized structures

Some symbiotic nitrogen-fixing prokaryotes dwell within **nodules**, the special organs of the plant host that enclose the nitrogen-fixing bacteria (see Figure 13.9). In the case of *Gunnera*, these organs are preexisting stem glands that develop independently of the symbiont. In the case of legumes and actinorhizal plants, the nitrogen-fixing bacteria induce the plant to form root nodules.

Grasses can also develop symbiotic relationships with nitrogen-fixing organisms, but in these associations root nodules are not produced. Instead, the nitrogen-fixing bacteria anchor to the root surfaces, mainly around the elongation zone and the root hairs, or live as endophytes, colonizing plant tissues without causing disease. For example, the nitrogen-fixing bacteria *Acetobacter diazotrophicus* and *Herbaspirillum* spp. live in the apoplast of stem tissues in sugarcane and may provide their host with about 30% of its nitrogen, which lessens the need for fertilization. The potential for associative and endophytic nitrogen-fixing bacteria to supplement the nitrogen nutrition of maize, rice, and other grains has been explored, but the diversity of bacterial species found on roots and in tissues, and the variety of plant responses to these bacteria, have impeded progress.

Legumes and actinorhizal plants regulate gas permeability in their nodules, maintaining oxygen concentrations of 20 to 40 nanomolar (n*M*) within the nodule (about 10,000 times lower than equilibrium concentrations in water). These levels can support respiration but are sufficiently low to avoid inactivation of the nitrogenase. Gas permeability increases in the light and decreases under drought or upon exposure to nitrate. The mechanism for regulating gas permeability is not yet known, but it may involve potassium ion fluxes into and out of infected cells.

Nodules contain oxygen-binding heme proteins called **leghemoglobins**. Leghemoglobins are the most abundant proteins in nodules, giving them a heme-pink color, and are crucial for symbiotic nitrogen fixation. Leghemoglobins have a high affinity for oxygen (a K_m of about 10 n*M*), about ten times higher than the β chain of human hemoglobin.

Although leghemoglobins were once thought to provide a buffer for nodule oxygen, more recent studies indicate that they store only enough oxygen to support nodule respiration for a few seconds. Their function is to increase the rate of oxygen transport to the respiring symbiotic bacterial cells, which decrease substantially the steady-state level of oxygen in infected cells. To continue aerobic respiration under such conditions, the bacteroid uses a specialized electron transport chain (see Chapter 12) in which the terminal oxidase has an affinity for oxygen even higher than that of leghemoglobins, a K_m of about 7 n*M*.

Establishing symbiosis requires an exchange of signals

The symbiosis between legumes and rhizobia is not obligatory. Legume seedlings germinate without any association with rhizobia, and they may remain unassociated throughout their life cycle. Rhizobia also occur as free-living organisms in the soil. Under nitrogen-limited conditions, however, the symbionts seek each other out through an elaborate exchange of signals. This signaling, the subsequent infection process, and the development of nitrogen-fixing nodules involve specific genes in both the host and the symbionts.

Plant genes specific to nodules are called **nodulin** genes; rhizobial genes that participate in nodule formation are called **nodulation** (*nod*) **genes**. The *nod* genes are classified as common *nod* genes or host-specific *nod* genes. The common *nod* genes—*nodA, nodB,* and *nodC*—are found in all rhizobial strains; the host-specific *nod* genes—such as *nodP, nodQ,* and *nodH;* or *nodF, nodE,* and *nodL*—differ among rhizobial species and determine the host range (the plants that can be infected). Only one of the *nod* genes, the regulatory *nodD,* is constitutively expressed, and as we will explain in detail, its protein product (NodD) regulates the transcription of the other *nod* genes.

The first stage in the formation of the symbiotic relationship between the nitrogen-fixing bacteria and their host is migration of the bacteria toward the roots of the host plant. This migration is a chemotactic response mediated by chemical attractants, especially (iso)flavonoids and betaines, secreted by the roots. These attractants activate the rhizobial NodD protein, which then induces transcription of the other *nod* genes. The promoter region of all *nod* operons, except that of *nodD,* contains a highly conserved sequence called the *nod* box. Binding of the activated NodD to the *nod* box induces transcription of the other *nod* genes.

Nod factors produced by bacteria act as signals for symbiosis

The *nod* genes, which NodD activates, code for nodulation proteins, most of which are involved in the biosynthesis of Nod factors. **Nod factors** are lipochitin oligosaccharide signal molecules, all of which have a chitin β-1→4-linked *N*-acetyl-D-glucosamine backbone (varying in length from three to six sugar units) and a fatty acid chain on the C-2 position of the nonreducing sugar (**Figure 13.11**).

Three of the *nod* genes (*nodA, nodB,* and *nodC*) encode enzymes (NodA, NodB, and NodC, respectively) that are required for synthesizing this basic structure:

1. NodA is an *N*-acyltransferase that catalyzes the addition of a fatty acyl chain.
2. NodB is a chitin-oligosaccharide deacetylase that removes the acetyl group from the terminal nonreducing sugar.
3. NodC is a chitin-oligosaccharide synthase that links *N*-acetyl-D-glucosamine monomers.

Host-specific *nod* genes that vary among rhizobial species are involved in the modification of the fatty acyl chain or the addition of groups important in determining host specificity:

- NodE and NodF determine the length and degree of saturation of the fatty acyl chain; those of *Rhizobium leguminosarum* bv. *viciae* and *R. meliloti* result in the synthesis of an 18:4 and a 16:2 fatty acyl group, respectively. (Recall from Chapter 12 that the number before the colon gives the total number of carbons in the fatty acyl chain, and the number after the colon gives the number of double bonds.)
- Other enzymes, such as NodL, influence the host specificity of Nod factors through the addition of specific substitutions at the reducing or nonreducing sugar moieties of the chitin backbone.

A particular legume host responds to a specific Nod factor. The legume receptors for Nod factors are protein kinases with extracellular sugar-binding LysM domains (for lysin motif, a widespread protein module originally identified in enzymes that degrade bacterial cell walls but also present in many other proteins) in the root hairs. Nod factors activate these domains, inducing oscillations in the concentrations of free calcium ions in the nuclear regions of root epidermal cells. Recognition of the calcium ion oscillations requires a calcium ion/calmodulin-dependent protein kinase (CaMK) that is associated with a protein of unknown function named CYCLOPS. Once the plant epidermal cell recog-

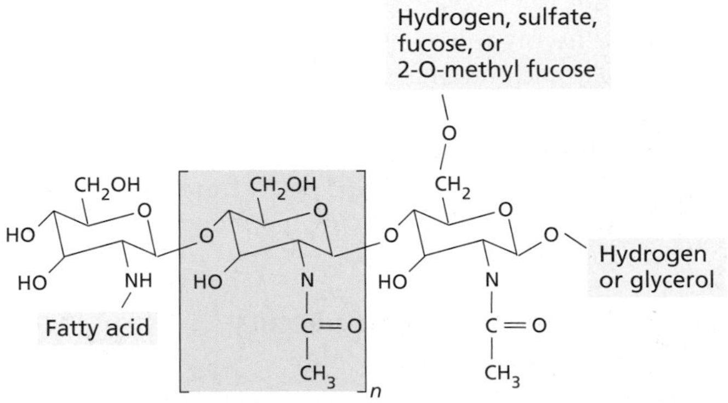

Figure 13.11 Nod factors are lipochitin oligosaccharides. The fatty acid chain typically has 16 to 18 carbons. The number of repeated middle sections (*n*) is usually two or three. (After Stokkermans et al. 1995.)

nizes ongoing calcium ion oscillations, Nod factor–responsive transcriptional regulators directly associate with the promoters of Nod factor–inducible genes. The overall process links Nod factor perception at the plasma membrane to gene expression changes in the nucleus and is called the symbiotic pathway because it shares elements with the process through which arbuscular mycorrhizal fungi initially interact with their hosts (see Chapters 5 and 23).

Nodule formation involves phytohormones

Two processes—infection and nodule organogenesis—occur simultaneously during root nodule formation. Rhizobia usually infect root hairs by first releasing Nod factors that induce a pronounced curling of the root hair cells (**Figure 13.12A and B**). The rhizobia become enclosed in the small compartment formed by the curling. The cell wall

Figure 13.12 The infection process during nodule organogenesis. (A) Rhizobia bind to an emerging root hair in response to chemical attractants sent by the plant. (B) In response to factors produced by the bacteria, the root hair exhibits abnormal curling growth, and rhizobia cells proliferate within the coils. (C) Localized degradation of the root hair wall leads to infection and formation of the infection thread from Golgi secretory vesicles of root cells. (D) The infection thread reaches the end of the cell, and its membrane fuses with the plasma membrane of the root hair cell. (E) Rhizobia are released into the apoplast and penetrate the compound middle lamella to the subepidermal cell plasma membrane, leading to the initiation of a new infection thread, which forms an open channel with the first. (F) The infection thread extends and branches until it reaches target cells, where vesicles composed of plant membrane that enclose bacterial cells are released into the cytosol.

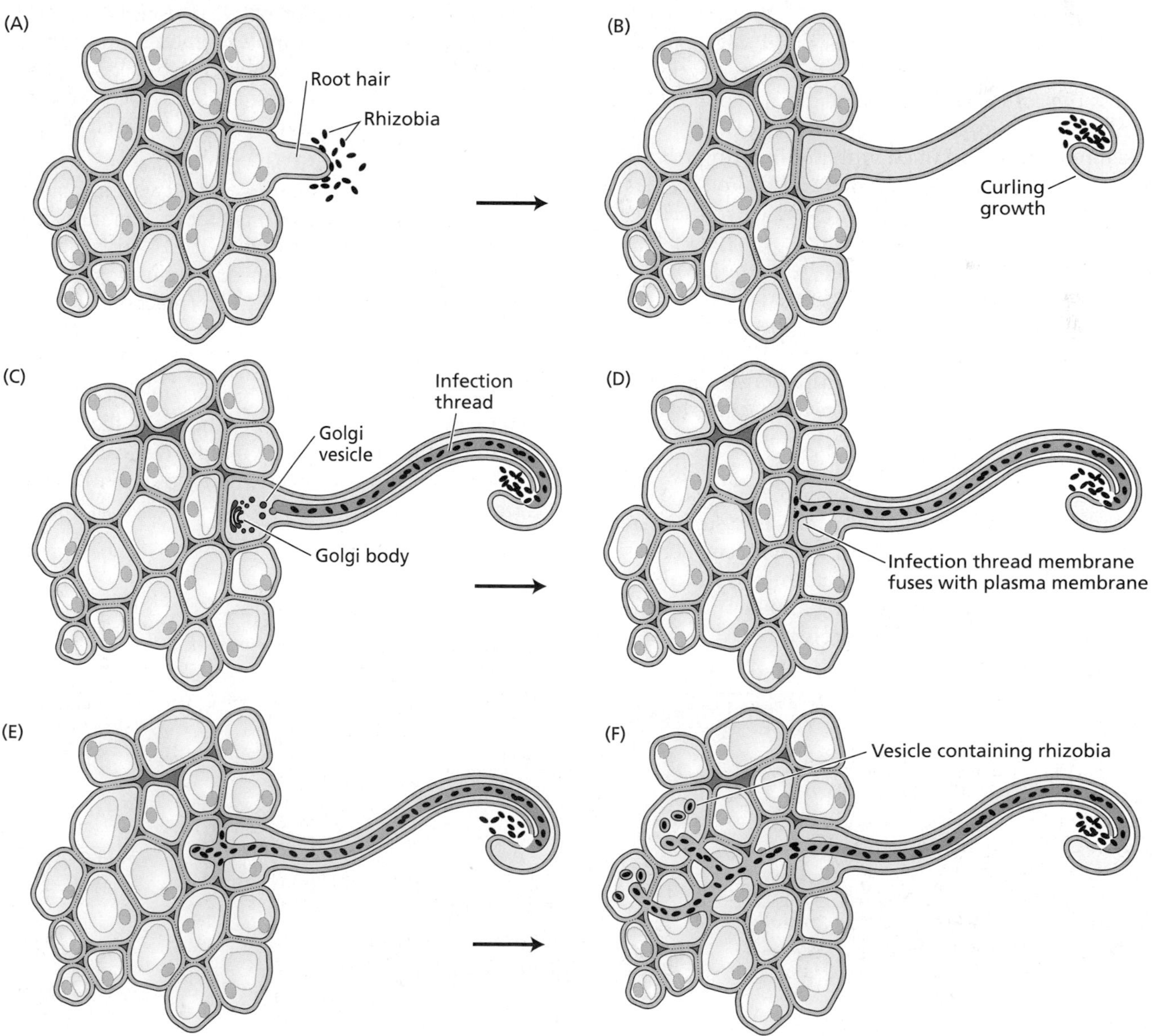

of the root hair degrades in these regions, also in response to Nod factors, allowing the bacterial cells direct access to the outer surface of the plant plasma membrane.

The next step is formation of the **infection thread** (**Figure 13.12C**), an internal tubular extension of the plasma membrane that is produced by the fusion of Golgi-derived membrane vesicles at the site of infection. The thread grows at its tip by the fusion of secretory vesicles to the end of the tube. Deeper into the root cortex, near the xylem, cortical cells dedifferentiate and start dividing, forming a distinct area within the cortex, called a *nodule primordium,* from which the nodule will develop. The nodule primordia form opposite the protoxylem poles of the root vascular bundle (see WEB TOPIC 13.1).

Different signaling compounds, acting either positively or negatively, control the development of nodule primordia. Nod factors activate localized cytokinin signaling in the root cortex and pericycle, leading to the localized suppression of polar auxin transport, which in turn induces nodule morphogenesis and stimulates cell division. Ethylene is synthesized in the region of the pericycle, diffuses into the cortex, and blocks cell division opposite the phloem poles of the root.

The infection thread filled with proliferating rhizobia elongates through the root hair and cortical cell layers, in the direction of the nodule primordium. When the infection thread reaches the nodule primordium, its tip fuses with the plasma membrane of a host cell and penetrates into the cytoplasm (**Figure 13.12D**). Bacterial cells are subsequently released into the cytoplasm, surrounded by the host plasma membrane, resulting in the formation of an organelle called the *symbiosome*. Branching of the infection thread inside the nodule enables the bacteria to infect many cells (**Figure 13.12E and F**).

At first the bacteria within symbiosomes continue to divide, and the surrounding symbiosome membrane (also called the *peribacteroid membrane*) increases in surface area to accommodate this growth by fusing with smaller vesicles. Soon thereafter, upon an undetermined signal from the plant, the bacteria stop dividing and begin to differentiate into nitrogen-fixing **bacteroids**.

The nodule as a whole develops such features as a vascular system (which facilitates the exchange of fixed nitrogen produced by the bacteroids for nutrients contributed by the plant) and a layer of cells to exclude O_2 from the root nodule interior. In some temperate legumes (e.g., pea), the nodules are elongated and cylindrical because of the presence of a *nodule meristem*. The nodules of tropical legumes, such as soybean and peanut, lack a persistent meristem and are spherical.

The nitrogenase enzyme complex fixes N_2

Biological nitrogen fixation, like industrial nitrogen fixation, produces ammonia from molecular nitrogen. The overall reaction is

$$N_2 + 8\,e^- + 8\,H^+ + 16\,ATP \rightarrow 2\,NH_3 + H_2 + 16\,ADP + 16\,P_i \qquad (13.10)$$

Note that the reduction of N_2 to 2 NH_3, a six-electron transfer, is coupled to the reduction of two protons to evolve H_2. The **nitrogenase enzyme complex** catalyzes this reaction.

The nitrogenase enzyme complex can be separated into two components—the Fe protein and the MoFe protein—neither of which has catalytic activity by itself (**Figure 13.13**):

- The Fe protein is the smaller of the two components and has two identical subunits that vary in mass from 30 to 72 kDa each, depending on the bacterial species. Each subunit contains an iron–sulfur cluster (4 Fe and 4 S^{2-}), which participates in the redox reactions that convert N_2 to NH_3. The Fe protein is irreversibly inactivated by O_2 with typical half-decay times of 30 to 45 seconds.
- The MoFe protein has four subunits, with a total molecular mass of 180 to 235 kDa, depending on the bacterial species. Each subunit has two Mo–Fe–S clusters. The MoFe protein is also inactivated by O_2, with a half-decay time in air of 10 minutes.

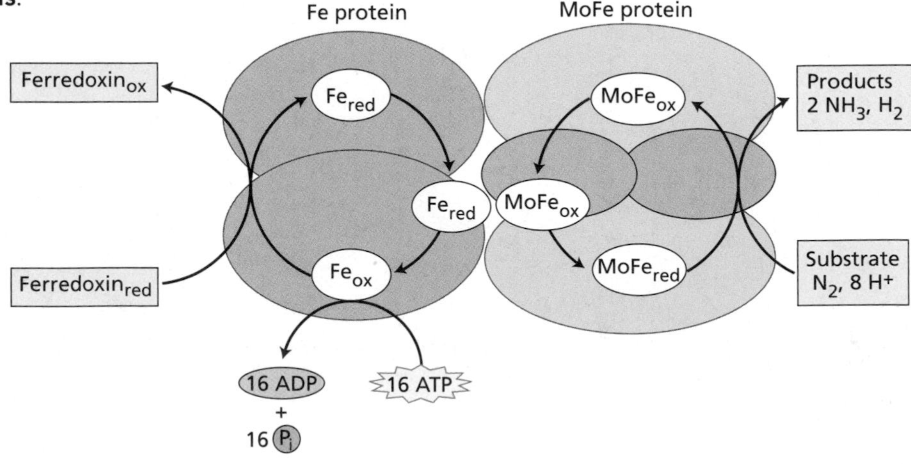

Figure 13.13 The reaction catalyzed by nitrogenase. Ferredoxin reduces the Fe protein. Binding and hydrolysis of ATP to the Fe protein are thought to cause a conformational change of the Fe protein that facilitates the redox reactions. The Fe protein reduces the MoFe protein, and the MoFe protein reduces the N_2. (After Dixon and Wheeler 1986; Buchanan et al. 2000.)

TABLE 13.4 Reactions catalyzed by nitrogenase

$N_2 \rightarrow NH_3$	Molecular nitrogen fixation
$N_2O \rightarrow N_2 + H_2O$	Nitrous oxide reduction
$N_3^- \rightarrow N_2 + NH_3$	Azide reduction
$C_2H_2 \rightarrow C_2H_4$	Acetylene reduction
$2\ H^+ \rightarrow H_2$	H_2 production
$ATP \rightarrow ADP + P_i$	ATP hydrolytic activity

In the overall nitrogen reduction reaction (see Figure 13.13), ferredoxin serves as an electron donor to the Fe protein, which in turn hydrolyzes ATP and reduces the MoFe protein. The MoFe protein can then reduce numerous substrates (**Table 13.4**), although under natural conditions it reacts only with N_2 and H^+. One of the reactions catalyzed by nitrogenase, the reduction of acetylene to ethylene, is used in estimating nitrogenase activity (see WEB TOPIC 13.2).

The energetics of nitrogen fixation are complex. The production of NH_3 from N_2 and H_2 is an exergonic reaction (see WEB APPENDIX 1 for a discussion of exergonic reactions), with a $\Delta G^{0\prime}$ (change in free energy) of –27 kJ mol^{-1}. However, industrial production of NH_3 from N_2 and H_2 is *endergonic*, requiring a very large energy input because of the activation energy needed to break the triple bond in N_2. For the same reason, the enzymatic reduction of N_2 by nitrogenase also requires a large investment of energy (see Equation 13.10), although the exact changes in free energy are not yet known.

Calculations based on the carbohydrate metabolism of legumes show that a plant respires 9.3 moles of CO_2 per mole of N_2 fixed. On the basis of Equation 13.10, the $\Delta G^{0\prime}$ for the overall reaction of biological nitrogen fixation is about –200 kJ mol^{-1}. Because the overall reaction is highly exergonic, ammonium production is limited by the slow operation (number of N_2 molecules reduced per unit time is about 5 s^{-1}) of the nitrogenase complex. To compensate for this slow turnover rate, the bacteroid synthesizes large amounts of nitrogenase (up to 20% of the total protein in the cell).

Under natural conditions, substantial amounts of H^+ are reduced to H_2 gas, and this process can compete with N_2 reduction for electrons from nitrogenase. In rhizobia, 30 to 60% of the energy supplied to nitrogenase may be lost as H_2, diminishing the efficiency of nitrogen fixation. Some rhizobia, however, contain hydrogenase, an enzyme that can split the H_2 formed and generate electrons for N_2 reduction, thus improving the efficiency of nitrogen fixation.

Amides and ureides are the transported forms of nitrogen

The symbiotic nitrogen-fixing prokaryotes release ammonia that, to avoid toxicity, must be rapidly converted into organic forms in the root nodules before being transported to the shoot via the xylem. Nitrogen-fixing legumes can be classified as amide exporters or ureide exporters, depending on the composition of the xylem sap. Amides (principally the amino acids asparagine or glutamine) are exported by temperate-region legumes, such as pea (*Pisum*), clover (*Trifolium*), broad bean (*Vicia*), and lentil (*Lens*).

Ureides are exported by legumes of tropical origin, such as soybean (*Glycine*), common bean (*Phaseolus*), peanut (*Arachis*), and southern pea (*Vigna*). The three major ureides are allantoin, allantoic acid, and citrulline (**Figure 13.14**). Allantoin is synthesized in peroxisomes from uric acid, and allantoic acid is synthesized from allantoin in the endoplasmic reticulum. The site of citrulline synthesis from the amino acid ornithine has not yet been determined. All three compounds are ultimately released into the xylem and transported to the shoot, where they are rapidly catabolized to ammonium. This ammonium enters the assimilation pathway described earlier.

Sulfur Assimilation

Sulfur is among the most versatile elements in living organisms. Disulfide bridges in proteins play structural and regulatory roles (see Chapter 8). Sulfur participates in electron transport through iron–sulfur clusters (see Chapters 7 and 12). The catalytic sites for several enzymes and coenzymes, such as urease and coenzyme A, contain sulfur. Secondary metabolites (compounds that are not involved in primary pathways of growth and development) that contain sulfur range from the rhizobial Nod factors

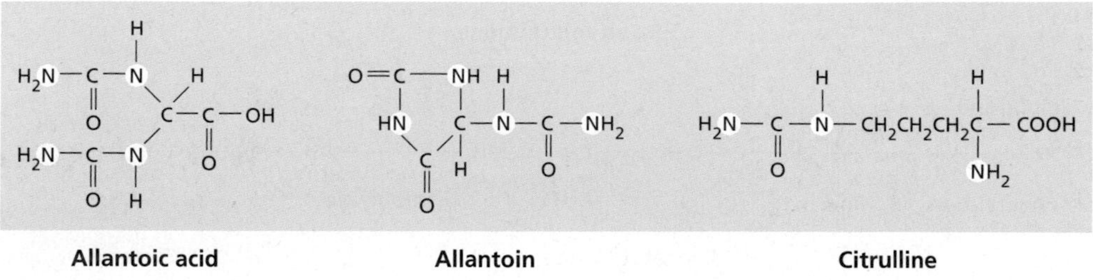

Allantoic acid

Allantoin

Citrulline

$H_2N-C(=O)-N(H)-CH_2CH_2CH_2C(H)(NH_2)-COOH$

Figure 13.14 The major ureide compounds used to transport nitrogen from sites of fixation to sites where their deamination will provide nitrogen for amino acid and nucleoside synthesis.

discussed in the previous section to the antiseptic alliin in garlic and the anticarcinogen sulforaphane in broccoli.

The versatility of sulfur derives in part from the property that it shares with nitrogen: *multiple stable oxidation states*. In this section we will discuss the enzymatic steps that mediate sulfur assimilation, and the biochemical reactions that catalyze the reduction of sulfate into the two sulfur-containing amino acids, cysteine and methionine.

Sulfate is the form of sulfur transported into plants

Most of the sulfur in higher-plant cells derives from sulfate (SO_4^{2-}) transported via an H^+–SO_4^{2-} symporter (see Chapter 6) from the soil solution. Sulfate in the soil comes predominantly from the weathering of parent rock material. Industrialization, however, adds an additional source of sulfate: atmospheric pollution. The burning of fossil fuels releases several gaseous forms of sulfur, including sulfur dioxide (SO_2) and hydrogen sulfide (H_2S), which find their way to the soil in rain.

In the gaseous phase, sulfur dioxide reacts with a hydroxyl radical and oxygen to form sulfur trioxide (SO_3). SO_3 dissolves in water to become sulfuric acid (H_2SO_4), a strong acid, which is the major source of acid rain. Plants can metabolize sulfur dioxide taken up in the gaseous form through their stomata. Nonetheless, prolonged exposure (more than 8 hours) to high atmospheric concentrations (greater than 0.3 ppm) of SO_2 causes extensive tissue damage because of the formation of sulfuric acid.

Sulfate assimilation requires the reduction of sulfate to cysteine

The first steps in the synthesis of sulfur-containing organic compounds involve the reduction of sulfate and synthesis of the amino acid cysteine (**Figure 13.15**). Sulfate is very stable and thus needs to be activated before any subsequent reactions may proceed. Activation begins with the reaction between sulfate and ATP to form adenosine-

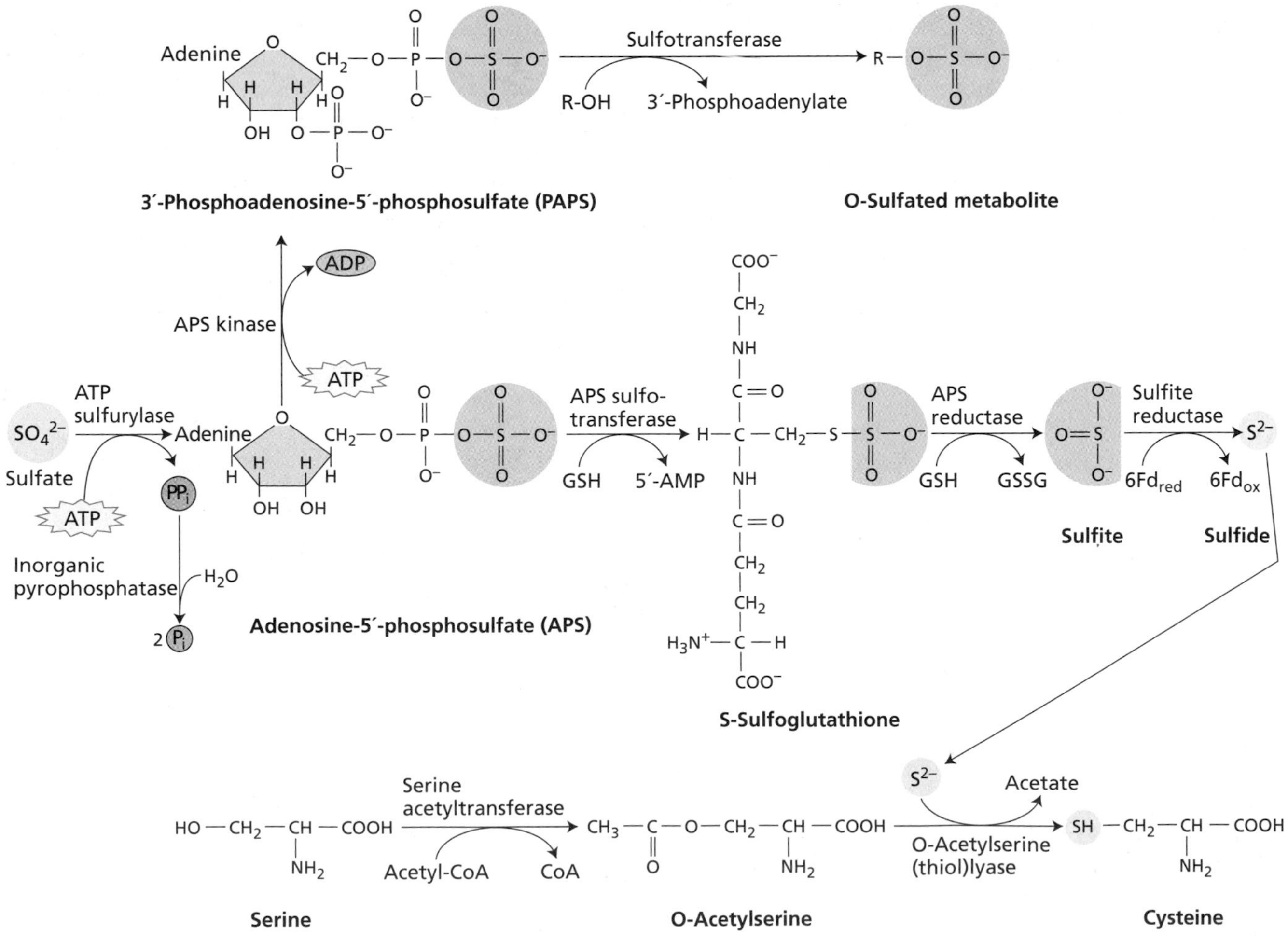

Figure 13.15 Structure and pathways of compounds involved in sulfur assimilation. The enzyme ATP sulfurylase cleaves pyrophosphate from ATP and replaces it with sulfate. Sulfide is produced from APS through reactions involving reduction by glutathione and ferredoxin. The sulfide reacts with *O*-acetylserine to form cysteine. Fd, ferredoxin; GSH, glutathione, reduced; GSSG, glutathione, oxidized.

5′-phosphosulfate (APS) and pyrophosphate (PP_i) (see Figure 13.15):

$$SO_4^{2-} + ATP \rightarrow APS + PP_i \qquad (13.11)$$

The enzyme that catalyzes this reaction, ATP sulfurylase, has two forms: The major one is found in plastids, and a minor one is found in the cytoplasm. The activation reaction is energetically unfavorable. To drive this reaction forward, the products APS and PP_i must be converted immediately to other compounds. PP_i is hydrolyzed to inorganic phosphate (P_i) by inorganic pyrophosphatase according to the following reaction:

$$PP_i + H_2O \rightarrow 2\ P_i \qquad (13.12)$$

The other product, APS, is rapidly reduced or phosphorylated. Reduction is the dominant pathway.

The reduction of APS is a multistep process that occurs exclusively in the plastids. First, APS reductase transfers two electrons, apparently from reduced glutathione (GSH), to produce sulfite (SO_3^{2-}):

$$\begin{aligned} APS + 2\ GSH &\rightarrow SO_3^{2-} \\ &+ 2\ H^+ + GSSG + AMP \end{aligned} \qquad (13.13)$$

where GSSG stands for oxidized glutathione. (The *SH* in GSH and the *SS* in GSSG stand for S—H and S—S bonds, respectively.)

Second, sulfite reductase transfers six electrons from ferredoxin (Fd_{red}) to produce sulfide (S^{2-}):

$$SO_3^{2-} + 6\ Fd_{red} \rightarrow S^{2-} + 6\ Fd_{ox} \qquad (13.14)$$

The resultant sulfide then reacts with *O*-acetylserine (OAS) to form cysteine and acetate. The *O*-acetylserine that reacts with S^{2-} is formed primarily in the mitochondria from a reaction catalyzed by serine acetyltransferase:

$$\text{Serine} + \text{acetyl-CoA} \rightarrow \text{OAS} + \text{CoA} \qquad (13.15)$$

The cytoplasm produces most of a cell's cysteine through a reaction catalyzed by OAS (thiol)lyase:

$$\text{OAS} + S^{2-} \rightarrow \text{cysteine} + \text{acetate} \qquad (13.16)$$

The phosphorylation of APS, localized in plastids and the cytosol, is the alternative pathway. First, APS kinase catalyzes a reaction of APS with ATP to form 3′-phosphoadenosine-5′-phosphosulfate (PAPS):

$$APS + ATP \rightarrow PAPS + ADP \qquad (13.17)$$

Sulfotransferases in the cytosol then may transfer the sulfate group from PAPS to various compounds, including choline, brassinosteroids, flavonol, gallic acid glucoside, glucosinolates, peptides, and polysaccharides.

Sulfate assimilation occurs mostly in leaves

The reduction of sulfate to cysteine changes the oxidation number of sulfur from +6 to –2, thus involving the transfer of eight electrons. Glutathione, ferredoxin, NAD(P)H, or *O*-acetylserine may serve as electron donors at various steps of the pathway (see Figure 13.15). In Arabidopsis, all the enzymes of sulfate assimilation—with the exception of sulfite reductase and the enzymes catalyzing the synthesis of reduced glutathione—are encoded by small multigene families. It is not yet clear whether this is a functional redundancy or if all genes have a specific function or location.

Leaves are generally much more active than roots in sulfur assimilation, presumably because photosynthesis provides reduced ferredoxin, and photorespiration generates serine, that may stimulate the production of *O*-acetylserine (see Chapter 8). Sulfur assimilated in leaves is exported via the phloem to sites of protein synthesis (shoot and root apices, and fruits) mainly as glutathione:

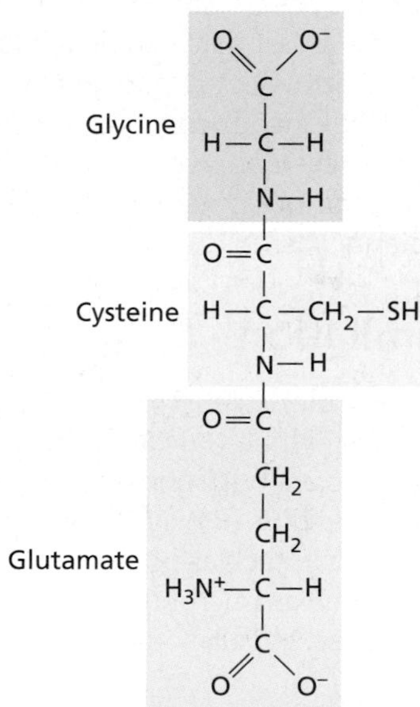

Reduced glutathione

Glutathione also acts as a signal that coordinates the transport of sulfate into roots and the assimilation of sulfate by the shoot.

Methionine is synthesized from cysteine

Methionine, the other sulfur-containing amino acid found in proteins, is synthesized in plastids from cysteine (see **WEB TOPIC 13.3** for further detail). After cysteine and methionine are synthesized, sulfur can be incorporated into proteins and a variety of other compounds, such as acetyl-CoA and *S*-adenosylmethionine. The latter compound is important in the synthesis of ethylene (see Chapter 15) and in reactions involving the transfer of methyl groups, as in lignin synthesis (see Chapter 23).

Phosphate Assimilation

Phosphate (HPO_4^{2-}) in the soil solution is readily transported into plant roots via an H^+–HPO_4^{2-} symporter (see Chapter 6) and incorporated into a variety of organic compounds, including sugar phosphates, phospholipids, and nucleotides. The main entry point of phosphate into assimilatory pathways occurs during the formation of ATP, the energy "currency" of the cell. In the overall reaction for this process, inorganic phosphate is added to the second phosphate group in adenosine diphosphate to form a phosphate ester bond.

In mitochondria, the energy for ATP synthesis derives from the oxidation of NADH or succinate by oxidative phosphorylation (see Chapter 12). ATP synthesis is also driven by light-dependent photophosphorylation in the chloroplasts (see Chapter 7). In addition to these reactions in mitochondria and chloroplasts, reactions in the cytosol such as glycolysis also assimilate phosphate.

Glycolysis incorporates inorganic phosphate into 1,3-bisphosphoglyceric acid, forming a high-energy acyl phosphate group. This phosphate can be donated to ADP to form ATP in a substrate-level phosphorylation reaction (see Chapter 12). Once incorporated into ATP, the phosphate group may be transferred via many different reactions to form the various phosphorylated compounds found in higher-plant cells.

Cation Assimilation

Cations taken up by plant cells form complexes with organic compounds in which the cation becomes bound to the complex by noncovalent bonds (for a discussion of noncovalent bonds, see **WEB APPENDIX 1**). Plants assimilate macronutrient cations such as potassium, magnesium, and calcium ions, as well as micronutrient cations such as copper, iron, manganese, cobalt, sodium, and zinc ions, in this manner. In this section we will describe coordination bonds and electrostatic bonds, which mediate the assimilation of several cations that plants require as nutrients, and the special requirements for the absorption of iron by roots and subsequent assimilation of iron within plants.

Cations form noncovalent bonds with carbon compounds

The noncovalent bonds formed between cations and carbon compounds are of two types: coordination bonds and electrostatic bonds. In the formation of a coordination complex, several oxygen or nitrogen atoms of a carbon compound donate unshared electrons to form a bond with the cation nutrient. As a result, the positive charge on the cation is neutralized.

Coordination bonds typically form between polyvalent cations and carbon compounds—for example, complexes between copper ions and tartaric acid (**Figure 13.16A**)

Figure 13.16 Examples of coordination complexes. Coordination complexes form when oxygen or nitrogen atoms of a carbon compound donate unshared electron pairs (represented by dots) to form a bond with a cation. (A) Copper ions share electrons with the hydroxyl oxygens of tartaric acid. (B) Magnesium ions share electrons with nitrogen atoms in chlorophyll *a*. Dashed lines represent a coordination bond between unshared electrons from the nitrogen atoms and the magnesium cation. (C) The "egg box" model of the interaction of polygalacturonic acid, a major constituent of pectins in cell walls, and calcium ions. At right is an enlargement of a single calcium ion forming a coordination complex with the hydroxyl oxygens of the galacturonic acid residues. (After Rees 1977.)

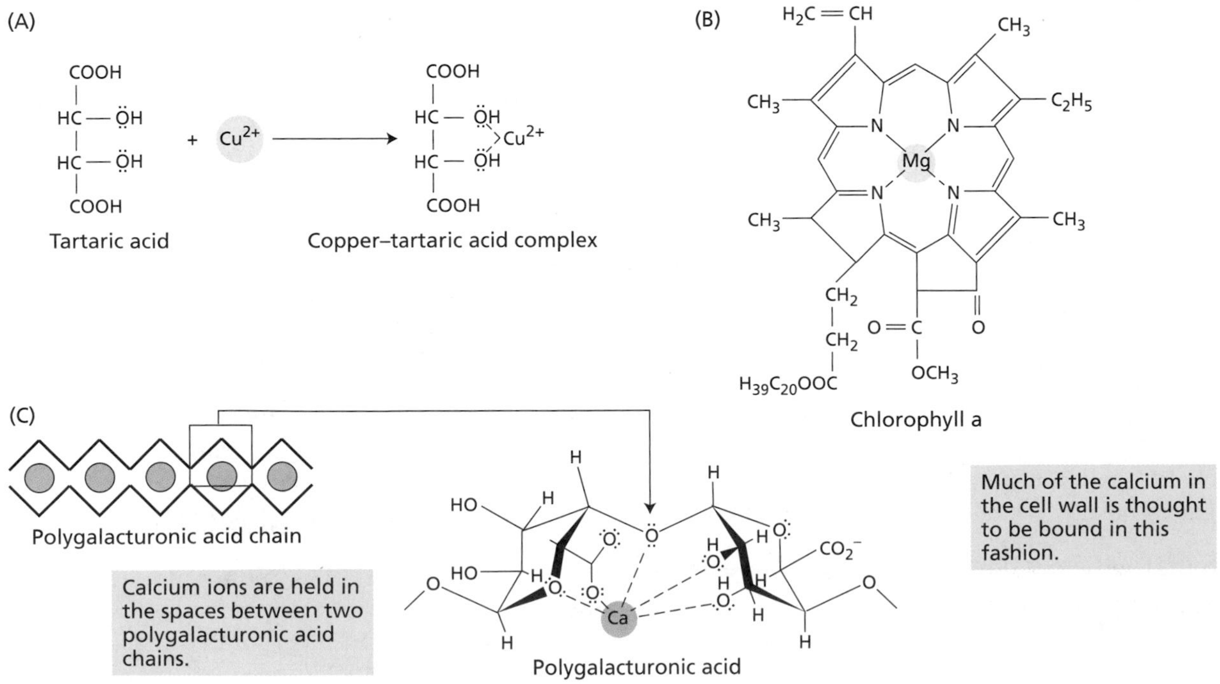

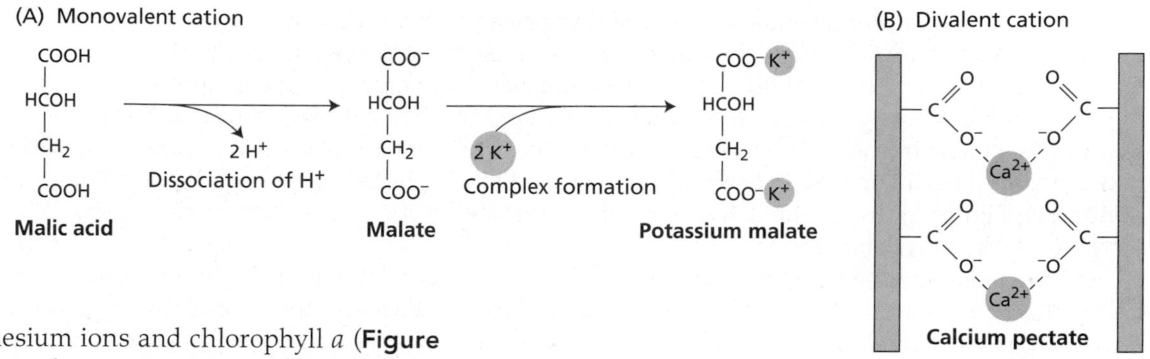

Figure 13.17 Examples of electrostatic (ionic) complexes. (A) The monovalent cation K^+ and malate form the complex potassium malate. (B) The divalent cation Ca^{2+} and pectate form the complex calcium pectate. Divalent cations can form cross-links between parallel strands that contain negatively charged carboxyl groups. Calcium cross-links play a structural role in the cell walls.

or between magnesium ions and chlorophyll *a* (**Figure 13.16B**). The nutrients that are assimilated as coordination complexes include copper, zinc, iron, and magnesium ions. Calcium ions can also form coordination complexes with the polygalacturonic acid of cell walls (**Figure 13.16C**).

Electrostatic bonds form because of the attraction of a positively charged cation for a negatively charged group such as carboxylate (—COO^-) on a carbon compound. Unlike the situation in coordination bonds, the cation in an electrostatic bond retains its positive charge. Monovalent cations such as the potassium ion can form electrostatic bonds with the carboxylic groups of many organic acids (**Figure 13.17A**). Nonetheless, many of the potassium ions that are accumulated by plant cells and function in osmotic regulation and enzyme activation remain in the cytosol and the vacuole as free ions. Divalent ions such as the calcium ion form electrostatic bonds with pectates (**Figure 13.17B**) and the carboxylic groups of polygalacturonic acid (see Chapter 14).

In general, cations such as magnesium and calcium ions are assimilated by the formation of both coordination complexes and electrostatic bonds with amino acids, phospholipids, and other negatively charged molecules.

Roots modify the rhizosphere to acquire iron

Iron is important in iron–sulfur proteins (see Chapter 7) and as a catalyst in enzyme-mediated redox reactions (see Chapter 5), such as those of nitrogen metabolism discussed earlier. Plants obtain iron from the soil, where it is present primarily as ferric iron (Fe^{3+}) in oxides such as $Fe(OH)^{2+}$, $Fe(OH)_3$, and $Fe(OH)_4^-$. At neutral pH, ferric iron is highly insoluble. To absorb sufficient amounts of iron from the soil solution, roots have developed several mechanisms that increase iron solubility and thus its availability (**Figure 13.18**). These mechanisms include:

- Soil acidification, which increases the solubility of ferric iron, followed by the reduction of ferric iron to the more soluble ferrous form (Fe^{2+}).
- Release of compounds that form stable, soluble complexes with iron. Recall from Chapter 5 that such compounds are called iron chelators (see Figure 5.3).

Roots generally acidify the soil around them. They export protons during the import and assimilation of cations, particularly ammonium, and release organic acids, such as

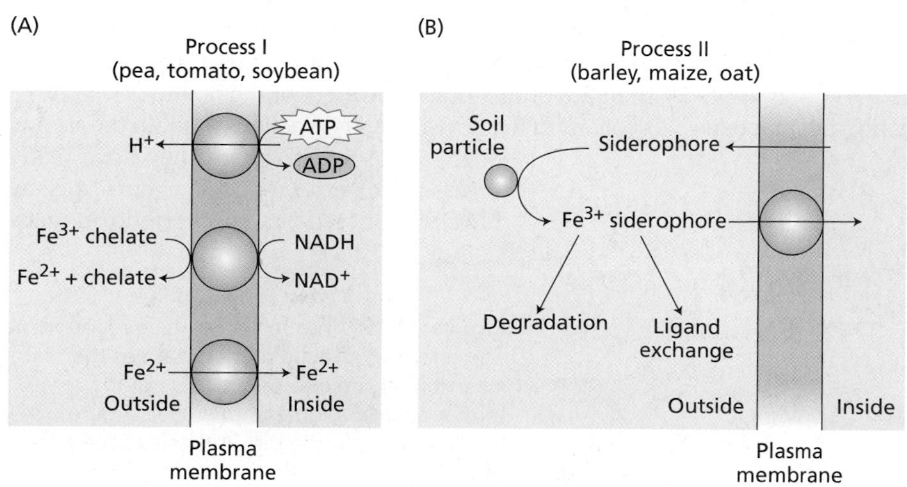

Figure 13.18 Two processes through which plants roots absorb iron. (A) A process common to dicots such as pea, tomato, and soybean. The chelates include organic compounds such as malic acid, citric acid, phenolics, and piscidic acid. (B) A process common to grasses such as barley, maize, and oat. After the grass excretes the siderophore and it removes iron from soil particles, the complex may degrade and release the iron to the soil, exchange iron for another ligand, or be transported into the root. (After Guerinot and Yi 1994.)

malic acid and citric acid, that enhance iron and phosphate availability (see Figure 5.5). Iron deficiencies stimulate the export of protons by roots. In addition, plasma membranes in roots contain an enzyme, called *iron-chelate reductase,* that reduces ferric iron (Fe^{3+}) to the ferrous (Fe^{2+}) form, with cytosolic NADH or NADPH serving as the electron donor (see Figure 13.18A). The activity of this enzyme increases under iron deprivation.

Several compounds secreted by roots form stable chelates with iron. Examples include malic acid, citric acid, phenolics, and piscidic acid. Grasses produce a special class of iron chelators called *siderophores*. Siderophores are made of amino acids that are not found in proteins, such as mugineic acid, and form highly stable complexes with Fe^{3+}. Root cells of grasses have Fe^{3+}–siderophore transport systems in their plasma membranes that bring the chelate into the cytoplasm. Under iron deficiency, grass roots release more siderophores into the soil and increase the capacity of their Fe^{3+}–siderophore transport system (see Figure 13.18B).

Iron cations form complexes with carbon and phosphate

After the roots absorb iron cations or an iron chelate, they oxidize it to a ferric form and translocate much of it to the leaves as an electrostatic complex with citrate or nicotianamine.

Once in the leaves, an iron cation undergoes an important assimilatory reaction through which it is inserted into the porphyrin precursor of heme groups found in the cytochromes located in chloroplasts and mitochondria (see Chapter 7). This reaction is catalyzed by the enzyme ferrochelatase (**Figure 13.19**). Most of the iron in the plant is found in heme groups. In addition, iron–sulfur proteins of the electron transport chain (see Chapter 7) contain nonheme iron covalently bound to the sulfur atoms of cysteine residues in the apoprotein. Iron is also found in Fe_2S_2 centers, which contain two irons (each complexed with the sulfur atoms of cysteine residues) and two inorganic sulfides.

Free iron (iron that is not complexed with carbon compounds) may interact with oxygen to form highly damaging hydroxyl radicals, OH•. Plant cells may limit such damage by storing surplus iron in an iron–protein complex called **ferritin**. Mutants of Arabidopsis show that, although ferritins are essential for protection against oxidative damage, they do not serve as a major iron pool for either seedling development or proper functioning of the photosynthetic apparatus. Ferritin consists of a protein shell with 24 identical subunits forming a hollow sphere that has a molecular mass of about 480 kDa. Within this sphere is a core of 5400 to 6200 iron atoms present as a ferric oxide–phosphate complex.

How iron is released from ferritin is uncertain, but breakdown of the protein shell appears to be involved. The level of free iron in plant cells regulates the de novo biosynthesis of ferritin. Interest in ferritin is high because iron in this protein-bound form may be highly available to humans, and foods rich in ferritin, such as soybean, may address dietary anemia problems.

Oxygen Assimilation

Respiration accounts for the bulk (about 90%) of the O_2 assimilated by plant cells (see Chapter 12). Another major pathway for the assimilation of O_2 into organic compounds involves the incorporation of O_2 from water (see reaction 1 in Table 8.1). A small proportion of oxygen can be directly assimilated into organic compounds in the process of *oxygen fixation* via enzymes known as *oxygenases*. The most prominent oxygenase in plants is ribulose-1,5-bisphosphate carboxylase/oxygenase (rubisco) which during photorespiration incorporates oxygen into an organic compound and releases energy (see Chapter 8). Other oxygenases are discussed in **WEB TOPIC 13.4**.

The Energetics of Nutrient Assimilation

Nutrient assimilation generally requires large amounts of energy to convert stable, low-energy, highly oxidized inorganic compounds into high-energy, highly reduced organic compounds. For example, the reduction of nitrate to nitrite and then to ammonium requires the transfer of about eight electrons and accounts for about 25% of the total energy expenditures in both roots and shoots. Consequently, a plant may use one-fourth of its energy to assimilate nitrogen, a constituent that accounts for less than 2% of the total dry weight of the plant.

Many of these assimilatory reactions occur in the stroma of the chloroplast, where they have ready access to powerful reducing agents, such as NADPH, thioredoxin, and ferredoxin, gen-

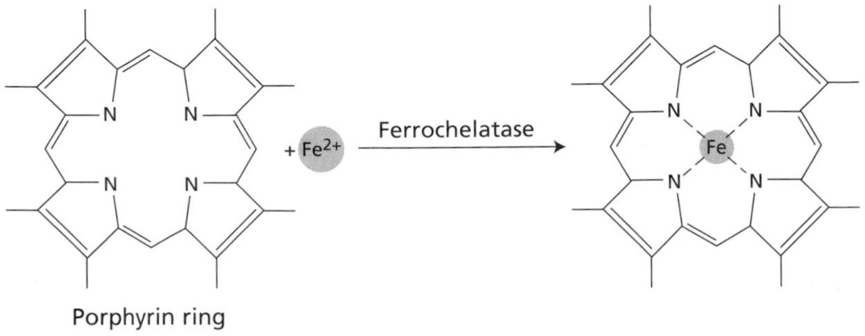

Porphyrin ring

+ Fe^{2+} Ferrochelatase

Figure 13.19 The ferrochelatase reaction. The enzyme ferrochelatase catalyzes the insertion of iron into the porphyrin ring to form a coordination complex. See Figure 7.34 for illustration of the biosynthesis of the porphyrin ring.

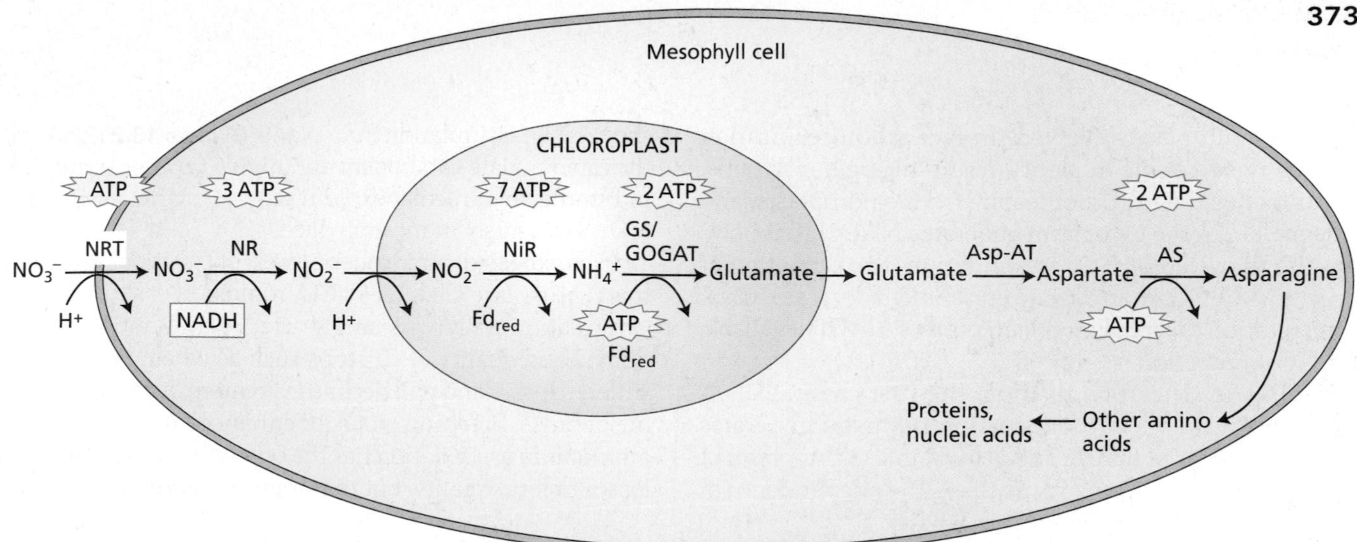

Figure 13.20 Summary of the processes involved in the assimilation of mineral nitrogen in the leaf. Nitrate translocated from the roots through the xylem is absorbed by a mesophyll cell via one of the nitrate–proton cotransporters (NRT) into the cytoplasm. There it is reduced to nitrite via nitrate reductase (NR). Nitrite is translocated into the stroma of the chloroplast along with a proton. In the stroma, nitrite is reduced to ammonium via nitrite reductase (NiR) and this ammonium is converted into glutamate via the sequential action of glutamine synthetase (GS) and glutamate synthase (GOGAT). Once again in the cytoplasm, the glutamate is transaminated to aspartate via aspartate aminotransferase (Asp-AT). Finally, asparagine synthetase (AS) converts aspartate into asparagine. The approximate amounts of ATP equivalents are given above each reaction.

erated during photosynthetic electron transport. This process—coupling nutrient assimilation to photosynthetic electron transport—is called **photoassimilation** (**Figure 13.20**).

Photoassimilation and the C_3 carbon fixation cycle occur in the same compartment. However, only when photosynthetic electron transport generates reductant in excess of the needs of the C_3 carbon fixation cycle—for example, under conditions of high light and low CO_2—does photoassimilation proceed. High levels of CO_2 inhibit nitrate assimilation in the shoots of C_3 plants (**Figure 13.21A**) (see WEB ESSAY 13.1).

One physiological mechanism responsible for this phenomenon involves photorespiration (see Chapter 8). Photorespiration has been erroneously portrayed as a wasteful process, a vestige of the high CO_2 atmospheres

Figure 13.21 Shoot NO_3^- assimilation as a function of shoot internal CO_2 concentration (C_i) for (A) nine C_3 species and (B) three C_4 species. Shoot NO_3^- assimilation is assessed by ΔAQ (decrease in the ratio of shoot CO_2 consumption to O_2 evolution with a shift from NH_4^+ to NO_3^- nutrition). (After Searles and Bloom 2003; Bloom et al. 2012.)

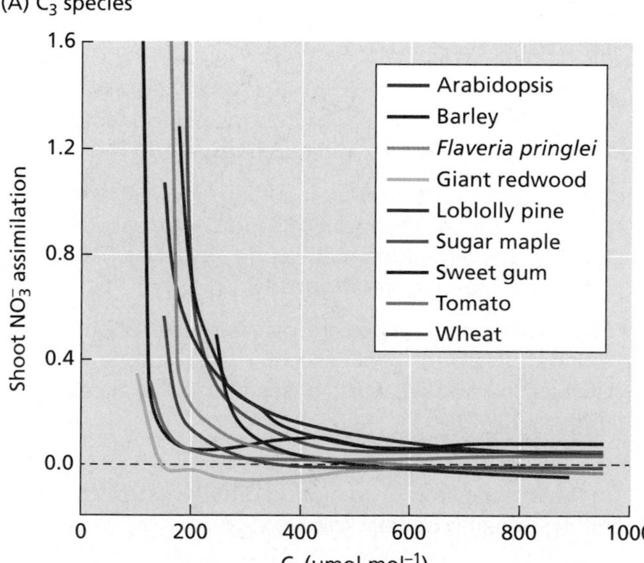

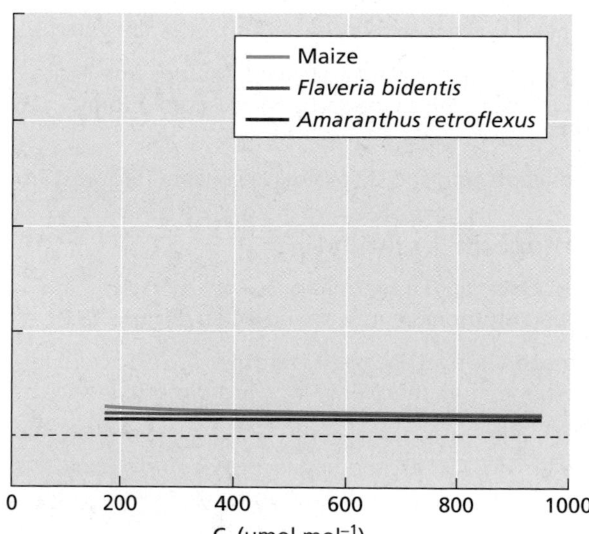

under which plants evolved. In fact, photorespiration plays a positive role in plant carbon–nitrogen relations. It stimulates the export of malate from chloroplasts, and this malate in the cytoplasm generates NADH that powers the first step of NO_3^- assimilation, the reduction of NO_3^- to NO_2^-. Carbon dioxide enrichment decreases photorespiration, decreasing the amount of NADH available for NO_3^- reduction.

Unlike in C_3 carbon fixation, the first carboxylation reaction in the C_4 carbon fixation pathway generates ample amounts of malate and NADH in the cytoplasm of mesophyll cells. This explains the CO_2-independence of shoot NO_3^- assimilation in C_4 plants (**Figure 13.21B**). Also, the rapid malate catabolism and high CO_2 concentrations in bundle sheath cells explain why C_4 plants assimilate NO_3^- exclusively in mesophyll cells.

If, as expected, atmospheric levels of CO_2 double during this century (see Chapter 9), CO_2 inhibition of shoot nitrate assimilation will increasingly affect plant–nutrient relations. Food quality of C_3 crops such as wheat has already suffered losses and will decline even more during the coming decades. Breeding crops for enhanced root nitrate and ammonium assimilation has the potential to mitigate such losses in food quality, but this approach is yet untapped.

SUMMARY

Nutrient assimilation is the often energy-requiring process by which plants incorporate inorganic nutrients into the carbon constituents necessary for growth and development.

Nitrogen in the Environment

- When nitrogen is fixed into ammonia (NH_3) or nitrate (NO_3^-), it passes through several organic or inorganic forms before it eventually returns to molecular nitrogen (N_2) (**Figure 13.1**).
- At high concentrations, ammonium (NH_4^+) is toxic to living tissues, but nitrate can be safely stored and translocated in plant tissues (**Figure 13.2**).

Nitrate Assimilation

- Plant roots actively absorb nitrate, then reduce it to nitrite (NO_2^-) in the cytosol (**Figure 13.3**).
- Nitrate, light, and carbohydrates affect the transcription and translation of nitrate reductase (**Figure 13.4**).
- Darkness and Mg^{2+} can inactivate nitrate reductase. Such inactivation is faster than regulation by decreased synthesis or degradation of the enzyme.
- In chloroplasts and root plastids, the enzyme nitrite reductase reduces nitrite to ammonium (**Figure 13.5**).
- Both roots and shoots assimilate nitrate (**Figure 13.6**).

Ammonium Assimilation

- Plant cells avoid ammonium toxicity by rapidly converting ammonium into amino acids (**Figure 13.7**).
- Nitrogen is incorporated into other amino acids via transamination reactions involving glutamine and glutamate.
- The amino acid asparagine is a key compound for nitrogen transport and storage.

Amino Acid Biosynthesis

- The carbon skeleton for amino acids derives from intermediates of glycolysis and the citric acid cycle (**Figure 13.8**).

Biological Nitrogen Fixation

- Biological nitrogen fixation accounts for most of the ammonia formed from atmospheric N_2 (**Figure 13.1; Tables 13.1, 13.2**).
- Several types of nitrogen-fixing bacteria form symbiotic associations with higher plants (**Figures 13.9, 13.10; Table 13.3**).
- Nitrogen fixation requires anaerobic or microanaerobic conditions.
- Symbiotic nitrogen-fixing prokaryotes function within specialized structures formed by the plant host (**Figure 13.9**).
- The symbiotic relationship is initiated by the migration of the nitrogen-fixing bacteria toward the roots of the host plant, which is mediated by chemical attractants secreted by the roots.
- Attractants activate the rhizobial NodD protein, which then induces the biosynthesis of Nod factors that act as signals for symbiosis (**Figure 13.11**).
- Nod factors induce root hair curling, sequestration of rhizobia, cell wall degradation, and bacterial access to the root hair plasma membrane, from which an infection thread forms (**Figure 13.12**).
- Filled with proliferating rhizobia, the infection thread elongates through root tissue in the direction of the developing nodule, which arises from cortical cells (**Figure 13.12**).
- In response to a signal from the plant, the bacteria in the nodule stop dividing and differentiate into nitrogen-fixing bacteroids.

- The reduction of N_2 to NH_3 is catalyzed by the nitrogenase enzyme complex (**Figure 13.13**).
- Fixed nitrogen is transported as amides or ureides (**Figure 13.14**).

Sulfur Assimilation

- Most assimilated sulfur derives from sulfate (SO_4^{2-}) absorbed from the soil solution, but plants can also metabolize gaseous sulfur dioxide (SO_2) entering via stomata.
- Synthesis of sulfur-containing organic compounds begins with the reduction of sulfate to the amino acid cysteine (**Figure 13.15**).
- Sulfate is assimilated in leaves and exported as glutathione via the phloem to growing sites.

Phosphate Assimilation

- Roots absorb phosphate (HPO_4^{2-}) from the soil solution, and its assimilation occurs with the formation of ATP.
- From ATP, the phosphate group may be transferred to many different carbon compounds in plant cells.

Cation Assimilation

- Polyvalent cations form coordination bonds with carbon compounds (**Figure 13.16**).
- Monovalent cations form electrostatic bonds with the carboxylate groups (**Figure 13.17**).
- Roots use several mechanisms to absorb sufficient amounts of insoluble ferric iron (Fe^{3+}) from the soil solution (**Figure 13.18**).
- Once in the leaves, iron undergoes an important assimilatory reaction (**Figure 13.19**).
- To limit the free radical damage that free iron can cause, plant cells may store surplus iron as ferritin.

Oxygen Assimilation

- Respiration and the oxygenase activity of rubisco account for most of the O_2 assimilated by plant cells, but direct oxygen fixation also is catalyzed by other oxygenases.

The Energetics of Nutrient Assimilation

- Energy-requiring nutrient assimilation is coupled to photosynthetic electron transport, which generates powerful reducing agents (**Figure 13.20**).
- Photoassimilation operates only when photosynthetic electron transport generates reductant in excess of the needs of the C_3 carbon fixation cycle.
- Rising levels of atmospheric CO_2 inhibit nitrate assimilation in the shoots of C_3 plants (**Figure 13.21**).

WEB MATERIAL

- **WEB TOPIC 13.1 Development of a Root Nodule** Nodule primordia form opposite the protoxylem poles of the root vascular bundles.
- **WEB TOPIC 13.2 Measurement of Nitrogen Fixation** Acetylene reduction is used as an indirect measurement of nitrogen reduction.
- **WEB TOPIC 13.3 The Synthesis of Methionine** Methionine is synthesized in plastids from cysteine.
- **WEB TOPIC 13.4 Oxygenases** Oxygenases are enzymes that catalyze oxygen assimilation.

- **WEB ESSAY 13.1 Elevated CO_2 and Nitrogen Photoassimilation** In leaves grown under high CO_2 concentrations, CO_2 inhibits nitrogen photoassimilation because it competes for reductant and inhibits photorespiration and nitrite transport.

available at **plantphys.net**

Suggested Reading

Andrews, M. (1986) The partioning of nitrate assimilation between roots and shoot of higher plants. *Plant Cell Environ.* 9: 511–519.

Appleby, C. A. (1984) Leghemoglobin and *Rhizobium* respiration. *Annu. Rev. Plant Physiol.* 35: 443–478.

Beevers, L. (1976) *Nitrogen Metabolism in Plants.* Elsevier, London.

Bloom, A. J., Burger, M., Asensio, J. S. R., and Cousins, A. B. (2010) Carbon dioxide enrichment inhibits nitrate assimilation in wheat and *Arabidopsis*. *Science* 328: 899–903.

Bloom, A. J., Rubio-Asensio, J. S., Randall, L., Rachmilevitch, S., Cousins, A. B., and Carlisle, E. A. (2012) CO_2 enrichment inhibits shoot nitrate assimilation in C_3

but not C_4 plants and slows growth under nitrate in C_3 plants. *Ecology* 93: 355–367.

Brady, N. C. (1979) *Nitrogen and Rice*. International Rice Research Institute, Manila.

Crawford, N. M., and Forde, B. J. (2002) Molecular and developmental biology of inorganic nitrogen nutrition. In: *The Arabidopsis Book*, C. Somerville and E. Meyerowitz, eds., American Society of Plant Physiologists, Rockville, MD. DOI: 10.1199/tab.0011, http://www.aspb.org/publications/arabidopsis/

Dixon, R. O. D., and Wheeler, C. T. (1986) *Nitrogen Fixation in Plants*. Chapman and Hall, New York.

Epstein, E., and Bloom, A. J. (2005) *Mineral Nutrition of Plants: Principles and Perspectives*, 2nd ed. Sinauer Associates, Sunderland, MA.

Foyer, C. H., Bloom, A. J., Queval, G., and Noctor, G. (2009) Photorespiratory metabolism: Genes, mutants, energetics, and redox signaling. *Annu. Rev. Plant Biol.* 60: 455–484.

George, E., Marschner, H., and Jakobsen, I. (1995) Role of arbuscular mycorrhizal fungi in uptake of phosphorus and nitrogen from soil. *Crit. Rev. Biotechnol.* 15: 257–270.

Geurts, R., Lillo, A., and Bisseling, T. (2012) Exploiting an ancient signalling machinery to enjoy a nitrogen fixing symbiosis. *Curr. Opin. Plant Biol.* 15: 438–443.

Guerinot, M. L., and Yi, Y. (1994) Iron: Nutritious, noxious, and not readily available. *Plant Physiol.* 104: 815–820.

Herridge, D. F., Peoples, M. B., and Boddey, R. M. (2008) Global inputs of biological nitrogen fixation in agricultural systems. *Plant Soil* 311: 1–18.

Jeong, J., and Guerinot, M. L. (2009) Homing in on iron homeostasis in plants. *Trends Plant Sci.* 14: 280–285.

Kaiser, W. M., Weiner, H., and Huber, S. C. (1999) Nitrate reductase in higher plants: A case study for transduction of environmental stimuli into control of catalytic activity. *Physiol. Plant.* 105: 385–390.

Lam, H.-M., Coschigano, K. T., Oliveira, I. C., Melo-Oliveira, R., and Coruzzi, G. M. (1996) The molecular-genetics of nitrogen assimilation into amino acids in higher plants. *Annu. Rev. Plant Physiol. Plant Mol. Biol.* 47: 569–593.

Leustek, T., Martin, M. N., Bick, J.-A., and Davies, J. P. (2000) Pathways and regulation of sulfur metabolism revealed through molecular and genetic studies. *Annu. Rev. Plant Physiol. Plant Mol. Biol.* 51: 141–165.

Long, S. R. (1996) *Rhizobium* symbiosis: Nod factors in perspective. *Plant Cell* 8: 1885–1898.

Maillet, F., Poinsot, V., Andre, O., Puech-Pages, V., Haouy, A., Gueunier, M., Cromer, L., Giraudet, D., Formey, D., Niebel, A., et al. (2011) Fungal lipochitooligosaccharide symbiotic signals in arbuscular mycorrhiza. *Nature* 469: 58–63.

Marschner, H., and Marschner, P. (2012) *Marschner's Mineral Nutrition of Higher Plants*, 3rd ed. Elsevier/Academic Press, London; Waltham, MA.

Mathews, C. K., and Van Holde, K. E. (1996) *Biochemistry, 2nd ed.* Benjamin/Cummings, Menlo Park, CA.

Mendel, R. R. (2005) Molybdenum: Biological activity and metabolism. *Dalton Trans.* 2005: 3404–3409.

Metzler, D. E. (1977) *Biochemistry*. Academic Press, New York.

Miller, A. J., Fan, X. R., Orsel, M., Smith, S. J., and Wells, D. M. (2007) Nitrate transport and signalling. *J. Exp. Bot.* 58: 2297–2306.

Oldroyd, G. E., Murray, J. D., Poole, P. S., and Downie, J. A. (2011) The rules of engagement in the legume-rhizobial symbiosis. *Annu. Rev. Gen.* 45: 119–144.

Olsen, R. A., Clark, R. B., and Bennet, I. H. (1981) The enhancement of soil fertility by plant roots. *Am. Sci.* 69: 378–384.

Quesada, A., Gomez-Garcia, I., and Fernandez, E. (2000) Involvement of chloroplast and mitochondria redox valves in nitrate assimilation. *Trends Plant Sci.* 5: 463–464.

Reis, V. M., Baldani, J. I., Baldani, V. L. D., and Dobereiner, J. (2000) Biological dinitrogen fixation in Gramineae and palm trees. *Crit. Rev. Plant Sci.* 19: 227–247.

Roberts, G. R., Keys, A. I., and Whittingham, C. P. (1970) The transport of photosynthetic products from the chloroplast of tobacco leaves. *J. Exp. Bot.* 21: 683–692.

Rolfe, B. G., and Gresshoff, P. M. (1988) Genetic analysis of legume nodule initiation. *Annu. Rev. Plant Physiol. Plant Mol. Biol.* 39: 297–319.

Santi, C., Bogusz, D., and Franche, C. (2013) Biological nitrogen fixation in non-legume plants. *Ann. Bot.* 111: 743–767.

Schlesinger, W. H. (1997) *Biogeochemistry: An analysis of global change*. 2nd ed. Academic Press, San Diego, Calif.

Schubert, K. R., Lennigs, N. T., and Evans, H. I. (1978) Hydrogen reactions of nodulated leguminous plants. *Plant Physiol.* 61: 398–401.

Sivasankar, S., and Oaks, A. (1996) Nitrate assimilation in higher plants—The effect of metabolites and light. *Plant Physiol. Biochem.* 34: 609–620.

Smart, D. R., and Bloom, A. J. (2001) Wheat leaves emit nitrous oxide during nitrate assimilation. *Proc. Natl. Acad. Sci. USA* 98: 7875–7878.

Smith, R. I., Bouton, L. H., Schank, S. C., Quesenberry, K. H., Tyler, M. E., Milam, T. R., Gaskins, M. H., and Littell, R. C. (1976) Nitrogen fixation in grasses inoculated with Spirillum lipoferum. *Science* 193: 1003–1005.

Thomine, S. and Vert, G. (2013) Iron transport in plants: Better be safe than sorry. *Curr. Opin. Plant Biol.* 16: 322–327.

Trelstad, R. L., Lawley, K. R., and Holmes, L. B. (1981) Nonenzymatic hydroxlations of proline and lysine by reduced oxygen derivatives. *Nature* 289: 310–312.

Vande Broek, A., and Vanderleyden, J. (1995) Review: Genetics of the Azospirillum-plant root association. *Crit. Rev. Plant Sci.* 14: 445–466.

Welch, R. M., and Graham, R. D. (2004) Breeding for micronutrients in staple food crops from a human nutrition perspective. *J. Exp. Bot.* 55: 353–364.

UNIT III

Growth and Development

UNIT III

Chapter 14
Cell Walls: Structure, Formation, and Expansion

Chapter 15
Signals and Signal Transduction

Chapter 16
Signals from Sunlight

Chapter 17
Embryogenesis

Chapter 18
Seed Dormancy, Germination, and Seedling Establishment

Chapter 19
Vegetative Growth and Organogenesis

Chapter 20
The Control of Flowering and Floral Development

Chapter 21
Gametophytes, Pollination, Seeds, and Fruits

Chapter 22
Plant Senescence and Cell Death

Chapter 23
Biotic Interactions

Chapter 24
Abiotic Stress

14 Cell Walls: Structure, Formation, and Expansion

Plant cells, unlike animal cells, are surrounded by a mechanically strong cell wall. This thin layer consists of a scaffold of cellulose fibrils embedded in a matrix of polysaccharides, proteins, and other polymers produced by the cell. The matrix polymers and cellulose fibrils assemble into a strong network linked by a mixture of covalent and noncovalent bonds. The matrix may also contain enzymes and other materials that modify the wall's physical and chemical characteristics. Additionally, the hydration state of the cell wall greatly influences its physical properties and mechanical strength.

The cell walls of prokaryotes, fungi, algae, and plants differ from each other in chemical composition and molecular structure, yet they all serve three common functions: regulating cell volume, determining cell shape, and mechanically protecting the delicate protoplast against biochemical and physical assaults. Plant cell walls have acquired additional functions not evident in the cell walls of other organisms, and these diverse functions are reflected in the cell wall's structural complexity and diversity in composition and form.

In addition to these biological functions, the plant cell wall is the foundation and raw material for many products important to human society. Plant cell walls are used in the production of paper, textiles (such as cotton, flax, and linen), and lumber as well as other wood-based products. Plant cell walls are also used to make synthetic fibers (such as rayon), plastics, films, coatings, adhesives, gels, and thickeners. Today, major efforts are underway worldwide to develop cost-effective methods for converting "cellulosic biomass" into biofuels to replace petroleum-based transportation fuels such as gasoline. According to some scenarios, a billion tons of cellulosic biomass will need to be harvested each year in the United States to replace nearly a third of the petroleum currently used for transportation. As the most abundant reservoir of organic carbon in nature and the major sink for photosynthetically captured carbon, the plant cell wall also takes part in the processes of carbon flow through ecosystems.

We will begin this chapter with a description of the general functions and composition of cell walls and the mechanisms of their biosynthesis and assembly. We'll then turn to the role of the primary cell wall in cell expansion. The mechanisms of tip growth, which occurs in a few specialized cell types, will be contrasted with those of diffuse growth, particularly with respect to the establishment of cell polarity and the control of the rate of cell expansion. Many cells—most notably those of the xylem that are involved in long-distance water transport and structural support of the stem—produce a thickened and lignified cell wall inside the primary wall. We will describe current thinking of the structure of this wall and the process of lignification.

Overview of Plant Cell Wall Functions and Structures

Without their cell walls, plants would be very different organisms from what we know. Instead of stately trees, we would find formless blobs of single amoeba-like cells. Indeed, the cell wall is critical for many essential processes in plant growth, development, and daily functions:

- Cell walls determine the mechanical strength of plant structures, allowing plants to grow to great heights.
- Cells are glued together by their walls, preventing cell sliding, slippage, and motility. This constraint on cell movement contrasts markedly with the situation in animal cells, and it dictates the way in which plants develop. Control of cell adhesion—and selective release of adhesion—is important for the development of intercellular air spaces for gas exchange and for cell separation during leaf abscission and other developmentally controlled tissue separations.
- Plant morphogenesis ultimately depends on the control of cell wall properties, because the physical enlargement of plant cells is limited principally by the ability of the cell wall to expand.
- As a mechanically strong layer encapsulating the cell, the wall acts as a cellular "exoskeleton" that controls cell shape and allows high turgor pressures to develop. Without a cell wall to resist the forces generated by turgor pressure, plant water relations would be profoundly different (see Chapter 3).
- Transpirational water flow in the xylem requires a mechanically strong wall that resists collapse in response to the negative pressure in the xylem. Defects in cell wall formation often result in a collapsed xylem phenotype.
- The cell wall acts as a diffusion barrier that limits the size and kinds of molecules that can reach the plasma membrane, both through sieving effects and through ionic and hydrophobic interactions. Fixed negative charges in the walls profoundly influence the distribution of ions and charged macromolecules.
- Numerous sensory proteins are partly anchored in the cell wall and form a bridge to the plasma membrane, providing a mechanism for sensing cell integrity.
- Cell walls present a major structural and chemical barrier to invasion and spread of pathogens and parasites and to tissue harvesting by herbivores. Moreover, oligosaccharides released from the cell wall by the action of lytic enzymes from invading microbes act as important signaling molecules that elicit defense responses against pathogens and symbionts.
- The cuticle, which is a complex hydrophobic layer integrated into the outer epidermal cell walls of the shoot, serves as a major barrier to water loss and pathogen invasion.

Much of the carbon assimilated in photosynthesis is channeled into polysaccharides that make up the wall. During specific phases of development or periods of sugar starvation, some of these polymers may be hydrolyzed into their constituent sugars to be scavenged by the cell and used to meet the cell's needs. This role is most notable in seeds with large storage reserves in the thick cell walls of endosperm or cotyledons. Easily digested polysaccharides are packed into these cell walls during seed development and are rapidly mobilized during germination to feed the growing embryo.

The functional diversity and varying roles of the cell wall require diverse cell wall structures. We'll begin this section with a brief description of the morphology and basic architecture of plant cell walls. Then we'll discuss the organization, composition, and synthesis of the wall in some of its diverse forms.

Plant cell walls vary in structure and function

In stained sections of plant organs, the most obvious visual objects under a microscope are cell walls, which may vary greatly in appearance and composition in different cell types (**Figure 14.1**). For example, cell walls of parenchyma in the pith and cortex are generally thin (~100 nm) and have few distinguishing features. In contrast, epidermal cells, collenchyma, xylem vessels and tracheids, phloem fibers, and other forms of sclerenchyma cells have thicker walls (~1,000 nm or more, often multilayered). These walls may be intricately sculpted and impregnated with substances such as lignin, cutin, suberin, waxes, silicate polymers, or structural proteins, which alter the wall's chemical and physical properties.

The walls on the different sides of a cell may vary in thickness, in amount and type of impregnating substances, in sculpting, and in frequency of pitting and plasmodesmata—tiny membrane-lined channels that allow passive transport of small molecules and active transport of proteins and nucleic acids between the cytoplasm of

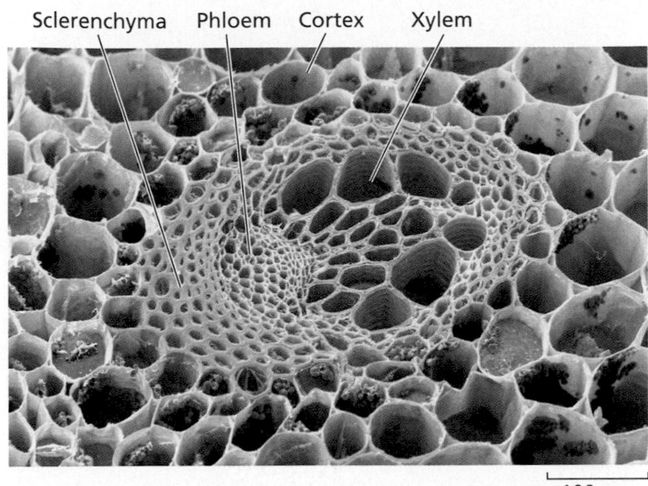

Figure 14.1 Cross section of a stem of a buttercup (*Ranunculus repens*), showing cells with varying wall morphology in different tissue types (see labels). Note the highly thickened walls of the sclerenchyma fiber cells and the pitted walls of the xylem vessels.

adjacent cells (see Figure 1.4). For example, the outer wall of the epidermis lacks plasmodesmata, is much thicker than the other walls of the cell, and is coated externally with cutin and waxes. Its polysaccharide composition may also differ from that of other walls, and in grasses the epidermal wall may contain a layer of polymerized silicate. In guard cells, the wall adjacent to the stomatal pore is much thicker than the walls on the other sides of the cell. Such variations in wall architecture for a single cell reflect the cell's polarity and differentiated functions and arise from targeted secretion of wall components to the cell surface.

Despite this morphological diversity, cell walls commonly are classified into two major types: primary walls and secondary walls. This classification is based not on structural or biochemical differences, but rather on the developmental state of the cell that is producing the cell wall. **Primary walls** are defined as walls formed during cell growth. Usually they are thin and architecturally simple (**Figure 14.2A** and **Figure 14.3A**), but some primary walls may be thick and multilayered, such as those found in collenchyma or in the epidermis (**Figure 14.2B and C**).

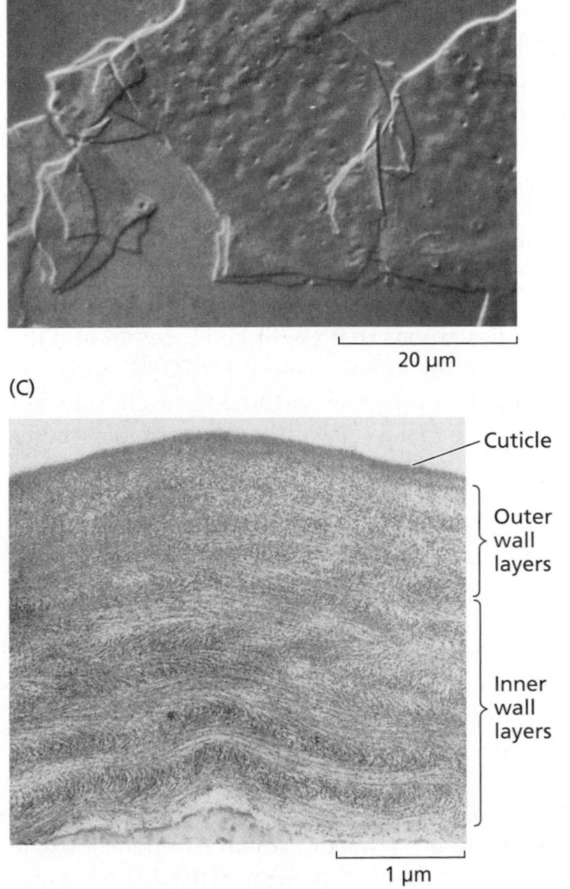

Figure 14.2 Three views of primary cell walls. (A) This surface view of cell wall fragments from onion parenchyma was taken with a light microscope using Nomarski optics. Note that at this scale the wall looks like a very thin sheet with small surface depressions; these depressions may be pit fields, places where plasmodesmatal connections between cells are concentrated. (B) The inner surface of an unextracted, never-dried wall of onion scale epidermis, imaged under water by atomic force microscopy. Note the fibrous texture of the wall and the presence of multiple lamellae with fibrils in different orientations. The narrowest fibrils are approximately 3 nm in diameter. They aggregate to form larger bundles. (C) Electron micrograph of the outer epidermal cell wall (cross section) from the growing region of a bean hypocotyl. Multiple layers are visible within the wall. The inner layers are thicker and more defined than the outer layers, because the outer layers are the older regions of the wall and have been stretched and thinned by cell expansion. (A from McCann et al. 1990; B from Zhang et al. 2014; C from Roland et al. 1982.)

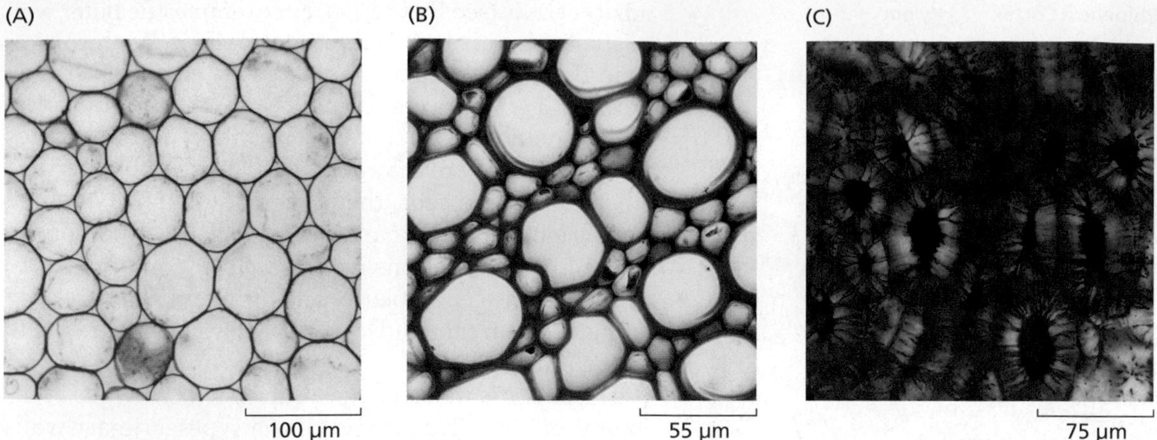

Figure 14.3 Diversity of cell wall structure. The thin walls of buttercup (*Ranunculus occidentalis*) stem parenchyma (A) contrast with the thickened secondary cell walls of tracheids in a vascular bundle of a sunflower (*Helianthus* sp.) stem (B) and sclereids from a cherry pit (*Prunus* sp.).

Secondary walls are formed after cell enlargement stops. They are deposited between the plasma membrane and the primary cell wall. Secondary walls may be highly specialized in structure and composition, reflecting the differentiated state of the cell (**Figure 14.3B and C**). In the water-conducting tissue (xylem), fiber cells, tracheids, and vessels are notable for possessing thickened, multilayered secondary walls that are strengthened and waterproofed by **lignin**. However, not all secondary walls are lignified or thickened. **Pits** and **pit fields** are thin areas where the primary wall is not covered by a secondary wall and are populated with plasmodesmata.

A thin layer called the **middle lamella** is found at the interface where the walls of neighboring cells come into contact. The middle lamella is typically enriched with acidic polysaccharides (pectins), which may be complexed with hydroxyproline-rich glycoproteins (HRGPs). While the origin of the middle lamella can be traced to the cell plate formed during cell division, additional materials must be recruited to this layer as the cells expand. One of its key functions is to serve as a flexible adhesive layer between cells.

Components differ for primary and secondary cell walls

Cell walls contain several types of polysaccharides that are named after the principal sugars they contain (**Figure 14.4** and WEB TOPIC 14.1). For example, a **glucan** is a polymer of glucose units linked end to end, a **galactan** is a polymer of galactose, a **xylan** is a polymer of xylose, a **mannan** is a polymer of mannose, and so on. **Glycan** is the general term for a polymer made up of sugars and is synonymous with polysaccharide.

Polysaccharides may be linear unbranched chains of sugar residues (units), or they may contain side branches attached to the backbone. For branched polysaccharides, the backbone of the polysaccharide is usually indicated by the last part of the name. For example, **xyloglucan** has a glucan backbone (a linear chain of glucose residues) with xylose sugars attached as side chains. **Arabinoxylan** has a xylan backbone (a chain of xylose residues) with arabinose side chains. The names can get lengthy. For example, **glucuronoarabinoxylan** (**GAX**) is an arabinoxylan decorated with a low frequency of glucuronic acid units. However, a compound name does not necessarily imply a branched structure. For example, **rhamnogalacturonan I** is the name given to a polymer containing both rhamnose and galacturonic acid in its backbone (it also has galactan and arabinan side chains that are not included in the name). Thus, the name is based on the major sugars in the polymer but does not indicate its structural details.

The specific linkages between sugar rings, including the specific carbons that are linked together and the configuration of the linkage (see WEB TOPIC 14.1), are important for the properties of the polysaccharide. For instance, amylose (not a cell wall component, but a component of starch in the plastid) is an α(1,4)-linked glucan (C-1 and C-4 carbons of adjacent glucose rings are linked through an O-glycosidic bond in an α configuration), whereas cellulose is a glucan made of β(1,4)-linkages, and callose is a wound-inducible glucan composed primarily of β(1,3)-linkages. These differences in linkages make huge differences in the physical properties, enzymatic digestibility, and functional roles of these three polymers of glucose. This illustrates the diversity and versatility of polysaccharides that can be made from the same sugar building block.

Cell wall polysaccharides are classified into three groups. **Cellulose** is the major fibrillar component of the cell wall and is composed of an array of β(1,4)-linked glu-

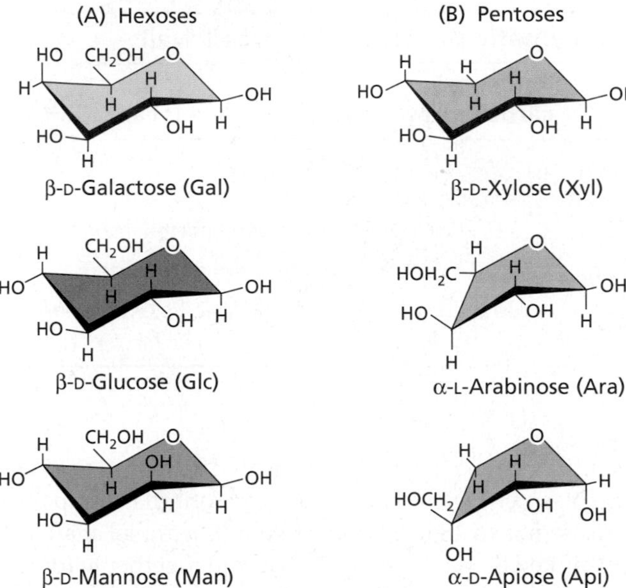

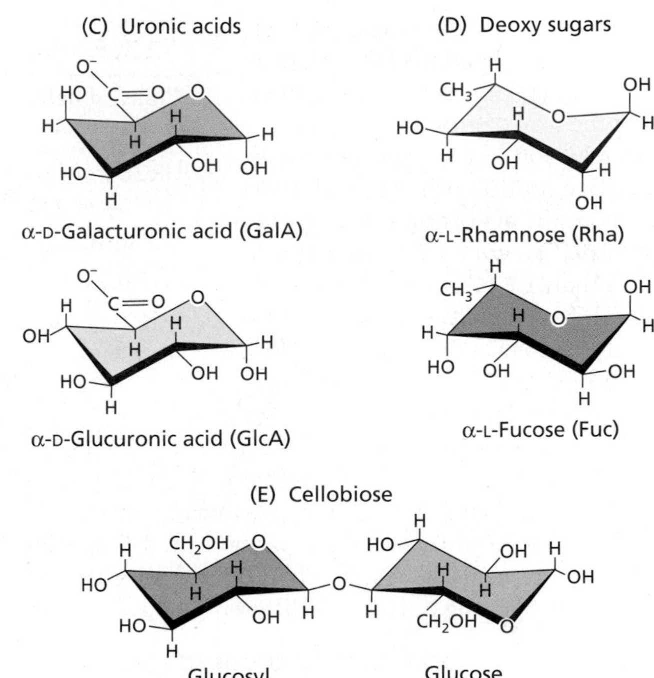

Figure 14.4 Conformational structures of sugars commonly found in plant cell walls. (A) Hexoses (six-carbon sugars). (B) Pentoses (five-carbon sugars). (C) Uronic acids (acidic sugars). (D) Deoxy sugars. (E) Cellobiose, showing the (1,4)-β-D-linkage between two glucose residues in inverted orientation. All sugars are shown in their pyranose (six-membered ring) forms except arabinose and apiose, which are shown in the furanose (five-membered ring) form.

cans coalesced to form a microfibril with well-ordered and less-ordered regions, which is insoluble in water and has high tensile strength (see details in the next section). **Pectin** is the name we give to a complex and diverse group of hydrophilic, gel-forming polysaccharides rich in acidic sugar residues. Many pectins are readily solubilized from the wall with hot water or with calcium chelators. The third group of wall polysaccharides are collectively called **hemicelluloses**. They usually require a strong extractant, such as 1–4 *M* NaOH, to be solubilized from the cell wall. Chemically hemicelluloses have been defined as polysaccharides with β(1,4)-linked backbones linked in an equatorial configuration (meaning the link between residues is in line with the plane of the ring). Pectins and hemicelluloses are also called **matrix polysaccharides**.

As detailed below, plant cell walls are constructed of a scaffold of cellulose microfibrils embedded in a polymeric matrix that varies by plant species, cell type, and region of the cell wall (**Table 14.1**). Typical primary cell walls of eudicots are rich in pectins, with smaller amounts of cellulose and hemicelluloses, whereas secondary cell walls are high in cellulose and a different form of hemicellulose, with varying amounts of lignin, an aromatic polymer that we will describe later in the chapter. As a result of the high pectin content, primary walls have a relatively high water content, which is important for maintaining the ability of the wall to expand during cell enlargement. In contrast, the cellulose–hemicellulose–lignin structure of secondary cell walls is densely packed and contains less water—a structure well designed for strength and compression resistance.

Primary cell walls may also contain 2 to 10% **nonenzymatic proteins** whose exact functions are uncertain. Such proteins may be localized in the walls of specific cell types or more widespread (**Table 14.2**) and usually are identified by short motifs or repeating sequences of amino acids or a high degree of glycosylation. A variety of functions

TABLE 14.1 Structural components of plant cell walls

Class	Examples
Cellulose	Microfibrils of (1,4)-β-D-glucan
Pectins	Homogalacturonan Rhamnogalacturonan I with arabinan, galactan, and arabinogalactan side chains Rhamnogalacturonan II
Hemicelluloses	Xyloglucan Glucuronarabinoxylan variants include glucuronoxylan and arabinoxylan Glucomannan Mixed linkage (1,3;1,4)-β-D-glucan
Nonenzymic proteins	(See Table 14.2)
Lignin	(See Figure 14.22)

have been suggested for these proteins, including consolidation of the cell plate following cytokinesis and strengthening of the wall of growing root hairs.

In addition to these proteins with repetitive motifs, primary cell walls also contain **arabinogalactan proteins (AGPs)**, which usually amount to less than 1% of the dry mass of the wall. These water-soluble proteins are heavily glycosylated. More than 90% of the mass of AGPs may be sugar residues—primarily galactose and arabinose (**Figure 14.5**). Multiple AGP forms are found in plant tissues, either in the wall or associated with the external face of the plasma membrane (via a GPI anchor), and they display tissue- and cell-specific expression patterns. AGPs may function in cell adhesion and in cell signaling during cell differentiation.

TABLE 14.2 Nonenzymatic proteins of the cell wall

Class of cell wall protein	Percentage carbohydrate	Notable tissue localization
HRGP (hydroxyproline-rich glycoprotein)	~55	Cambium, vascular parenchyma
PRP (proline-rich protein)	0–20	Xylem, fibers, cortex, root hairs
GRP (glycine-rich protein)	0	Primary xylem and phloem
AGP (arabinogalactan protein)	up to 90	Varied cell-specific expression

Cellulose microfibrils have an ordered structure and are synthesized at the plasma membrane

The simplest **cellulose microfibrils** are narrow structures, approximately 3 nm wide (1 nm = 10^{-9} meter), that strengthen the cell wall, sometimes reinforcing more in one direction than in another, depending on how the microfibrils are deposited in the wall (i.e., they give structural bias; see Figure 14.2B). Each microfibril is composed of an estimated 18 to 24 (most likely 18) parallel chains of (1,4)-linked β-D-glucose tightly packed together to form a highly ordered (crystalline) core with extensive hydrogen bonding within and between glucan chains (**Figure 14.6**). The chains surrounding the core are more flexible, and their positions are influenced by interactions with water and matrix polysaccharides at the surface. Moreover, there is evidence of periodic disorder along the microfibril, that is, short segments where crystalline order is interrupted at intervals of 150 to 300 nm.

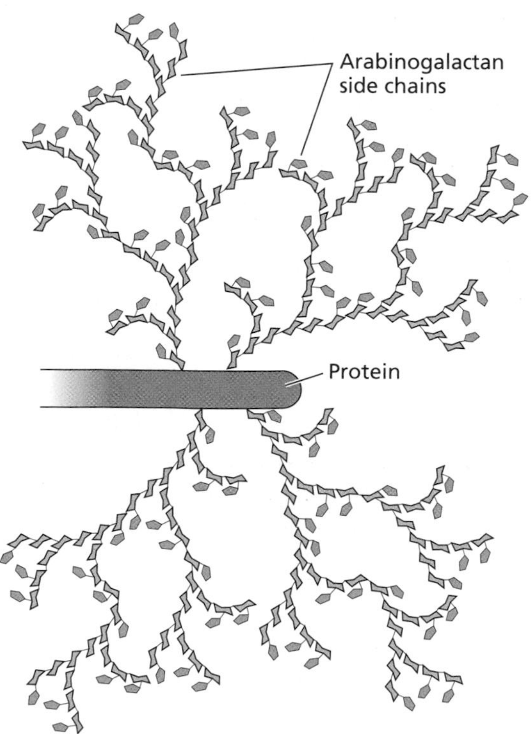

Figure 14.5 An arabinogalactan protein molecule showing the highly branched arabinogalactan side chains attached to the protein backbone. AGPs have a wide diversity of structures. (After Carpita and McCann 2000.)

Native cellulose in plants is found in two variant crystalline forms, called allomorphs Iα and Iβ, which differ slightly in the way the parallel glucan chains are packed. Cellulose Iβ is the more dominant allomorph in land plants. The biological significance of these two crystalline forms is unclear at present. Microfibrils have hydrophilic surfaces, populated by polar –OH groups extending from the sides of stacked glucose chains, and hydrophobic surfaces, populated by nonpolar C–H groups populating the plane of the sugar rings (see Figure 14.6E). These surfaces bind water and matrix polymers differently, and as a result the microfibril shape is an important factor for wall construction. It is also important for enzymatic attack by microbial cellulases, which dock onto the hydrophobic surface and remove one glucan chain at a time. A major barrier to enzymatic attack of cellulose is the energetic cost of stripping an individual glucan from this crystalline microfibril.

Cellulose microfibrils in nature vary considerably in width and degree of order, depending on their biological source. For instance, cellulose microfibrils in primary cell walls of land plants are approximately 3 nm wide, whereas those formed by some algae may be up to 20 nm wide and may be more highly ordered (more crystalline) than those found in land plants. This variation corresponds to the number of chains that make up the cross section of a microfibril. Individual microfibrils can also bundle together to form larger **macrofibrils**; this is most common in the cell walls of woody tissues, where cellulose has a higher degree of order (crystallinity) than in primary cell

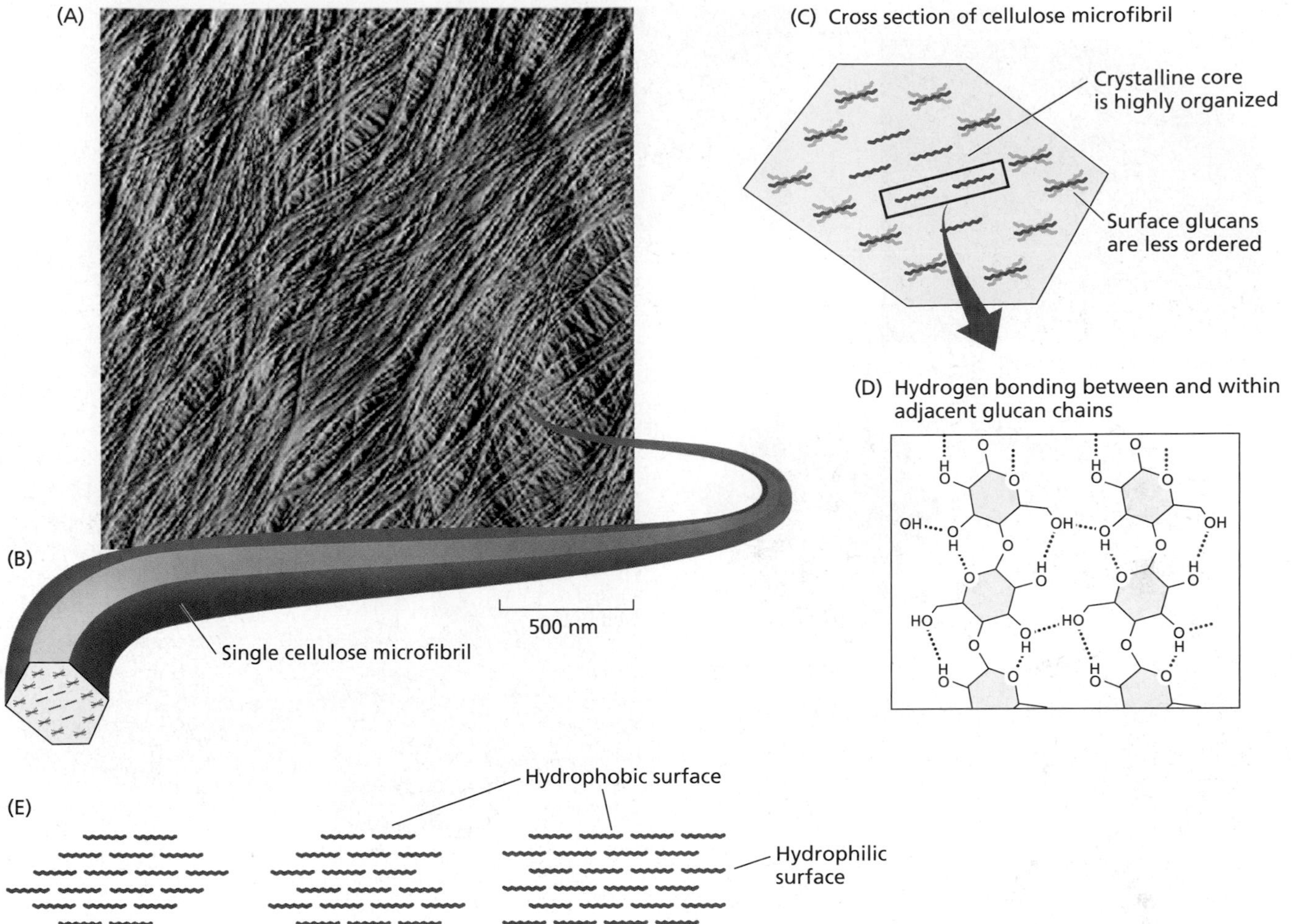

Figure 14.6 Structure of a cellulose microfibril. (A) Atomic force image of the primary cell wall from onion epidermis. Note its fibrillar texture, which arises from layers of cellulose microfibrils. (B) A single cellulose microfibril composed of (1,4)-β-D-glucan chains tightly bonded to each other to form a crystalline microfibril. (C) Cross section of a cellulose microfibril, illustrating one model of cellulose structure, with a crystalline core of highly ordered (1,4)-β-D-glucans surrounded by a less organized layer. (D) The crystalline regions of cellulose have precise alignment of glucans, with hydrogen bonding within, but not between, layers of (1,4)-β-D-glucans. (E) Potential cross section shapes of microfibrils. Note that the hydrophobic surface area varies greatly with shape. (After Matthews et al. 2006; micrograph from Zhang et al. 2014.)

walls. Cellulose chain length (or DP, degree of polymerization) ranges from approximately 2,000 to more than 25,000 glucose residues, corresponding to a fully extended length of 1 to 13 μm. The microfibril may be longer than the individual glucans because of overlap and staggering of the glucans in the microfibril. It is very difficult to obtain accurate measures of the length of microfibrils in the cell wall, but our best estimates are in the 1 to 13 μm range.

Evidence from electron microscopy indicates that cellulose microfibrils are synthesized by large, ordered protein complexes, called cellulose synthase complexes, which are embedded in the plasma membrane (**Figure 14.7**). These rosette-like structures are made up of six subunits, each of which is believed to contain three to six units of **cellulose synthase**, the enzyme that synthesizes the individual glucans that make up the microfibril. The cellulose synthase complexes probably contain additional proteins, but these have not yet been identified.

Cellulose synthases in plants are encoded by a gene family named ***CESA*** (*Cellulose Synthase A*), which is a multigene family found in all land plants. Genetic evidence indicates that three different *CESA* family members are involved in cellulose synthesis in primary cell walls and that a different set of three is used to make cellulose in secondary walls of woody tissues. Experimentally, CESA units have been swapped between the cellulose synthase complexes of primary and secondary walls and the complexes have still synthesized cellulose microfibrils.

The *CESA* gene family is part of a larger superfamily (cellulose synthase superfamily) that contains closely

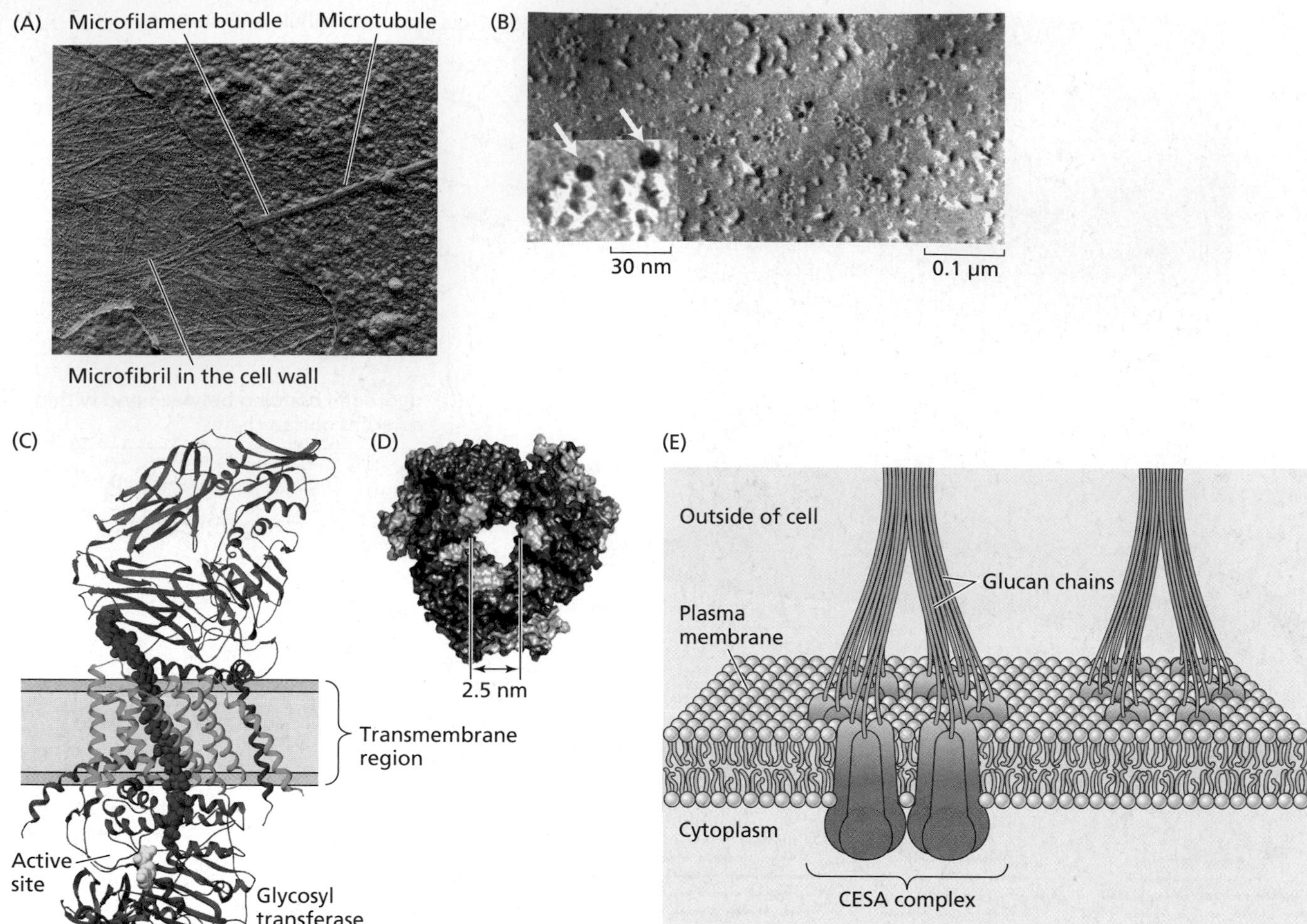

Figure 14.7 Cellulose microfibrils are synthesized at the cell surface by membrane-bound complexes containing cellulose synthase (CESA) proteins. (A) Electron micrograph showing newly synthesized cellulose microfibrils immediately exterior to the plasma membrane. (B) Freeze-fracture replicas showing binding of nanogold antibody against cellulose synthase to rosette structures in the membrane. The inset shows an enlarged view of two selected particle rosettes with immunogold labeling, indicating that the rosette structures contain CESA. The nanogold particles are the dark circles indicated with arrows. (C) Structure of a bacterial cellulose synthase. The brown region indicates the catalytic domain of the glycosyl transferase (GT) region where the catalytic site is located; this is the business end of the protein that transfers a glucose from uridine diphosphate glucose (UDP-glucose) to the glucan (blue). The green region indicates the transmembrane (TM) region, which forms a tunnel for the glucan to cross the membrane. The purple region is a domain not present in plant CESAs. (D) One possible oligomeric form of CESA in which three CESAs form a trimeric complex corresponding to one of the particles in the rosette structure seen in (B). (E) Computational model of a CESA complex extruding glucan chains that coalesce to form a microfibril. (A from Gunning and Steer 1996; B from Kimura et al. 1999; C data from Morgan et al. 2013; D from Sethaphong et al. 2013; E after image courtesy of Yara Yingling.)

related *CSL* (*cellulose-synthase like*) gene families named *CSLA, CSLB, . . .to CSLH. . . .*etc. Some authors refer to this superfamily as the *CESA/CSL* superfamily. ***CSLA*** genes encode synthases for (1,4)-β-D-mannan, ***CSLF*** and ***CSLH*** genes encode synthases for so-called mixed-linkage (1,3;1,4)-β-D-glucan, and ***CSLC*** genes probably encode synthases for the (1,4)-β-D-glucan backbone of xyloglucan. There is evidence that ***CSLD*** participates in mannan synthesis and in cellulose synthesis in root hairs. The other *CSL* families probably encode enzymes that synthesize the backbones of other hemicelluloses. However, the xylan backbone may be synthesized by a distinctly different group of synthases, including synthases named **GT43** (glycosyl transferase family 43). All of these synthases are **sugar–nucleotide polysaccharide glycosyltransferases**, which transfer monosaccharides from sugar nucleotides to the growing end of the polysaccharide chain.

The catalytic domain of cellulose synthase, which is located on the cytoplasmic side of the plasma membrane, transfers a glucose residue from a sugar nucleotide donor, uridine diphosphate glucose (UDP-glucose), to the growing glucan chain. Recent studies of the structure of a bacterial cellulose synthase have yielded insights into the details of the formation of glucan and its transport across the membrane through a tunnel in the synthase (see Figure 14.7C). Computational modeling indicates that a similar catalytic mechanism operates in plant CESAs. The modeling also leads to hypotheses of how multiple synthases could be grouped within the cellulose synthesis complex to produce multiple parallel glucan chains that coalesce to form a microfibril immediately after synthesis (see Figure 14.7D and E). There is some evidence that hemicelluloses may get entrapped in the microfibril as it forms; this may create disorder in the crystalline microfibril and also anchor the microfibril to the matrix.

Other proteins are implicated in cellulose microfibril formation, but their detailed functions are not yet certain. Defects in a class of membrane-associated (1,4)-β-D-endoglucanases, named KORRIGAN, result in reduced synthesis and crystallinity of cellulose, suggesting they may function in cellulose crystallization Likewise, members of the COBRA family, which contain a cellulose-binding domain, have been implicated in assembly of crystalline microfibrils. It is quite likely that proper formation of cellulose microfibrils requires a number of other proteins as well.

Matrix polymers are synthesized in the Golgi apparatus and secreted via vesicles

The matrix is a hydrated, polymeric phase between the crystalline cellulose microfibrils. Matrix polysaccharides are synthesized by membrane-bound glycosyltransferases in the Golgi apparatus and are delivered to the cell wall via exocytosis of tiny vesicles (**Figure 14.8** and **WEB TOPIC 14.2**). As described above, genes in the *CSL* families encode glycosyltransferases for synthesis of the backbone of some of these matrix polysaccharides. Additional sugar residues may be added as branches to the polysaccharide

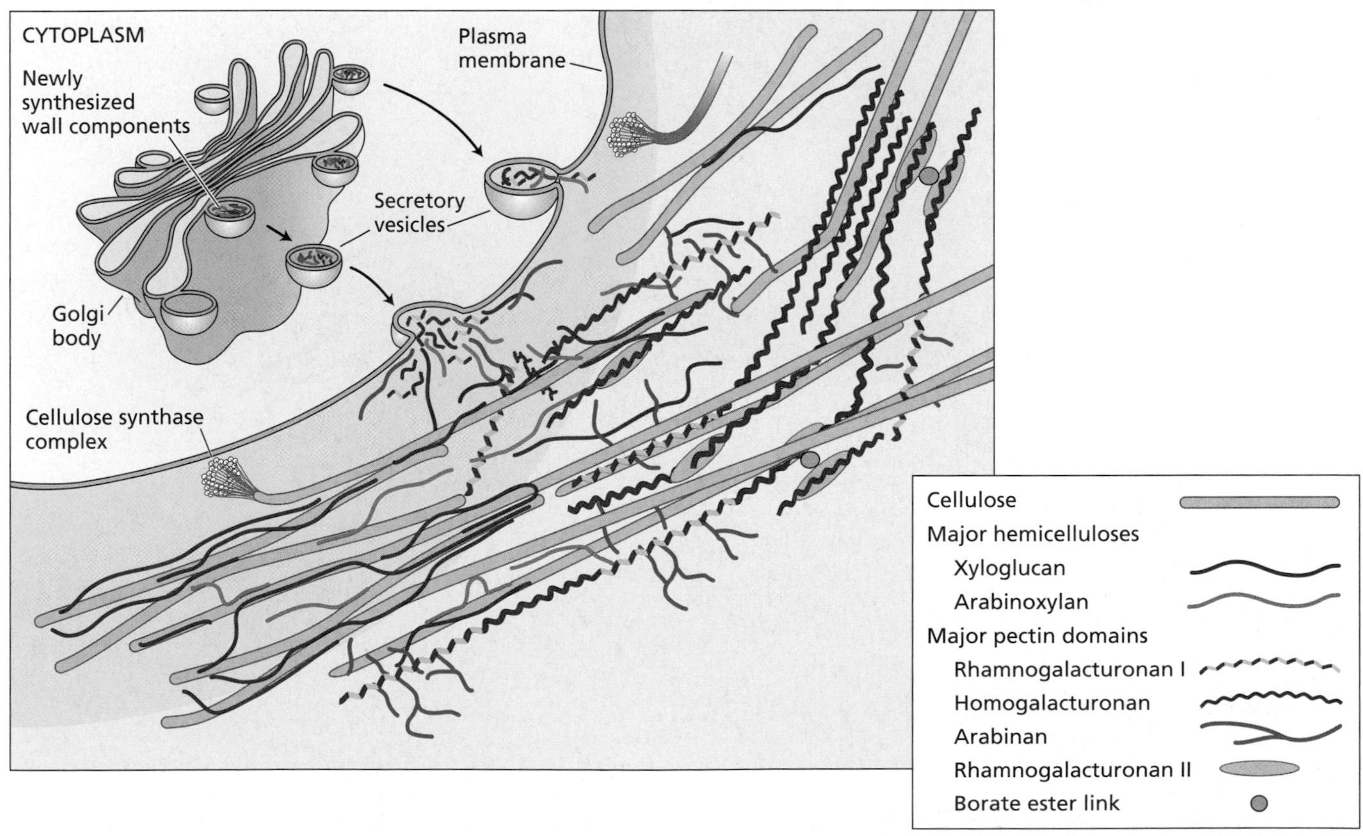

Figure 14.8 Schematic diagram of the major structural components of the primary cell wall and their possible arrangement. Cellulose microfibrils (gray rods) are synthesized at the cell surface and are partially coated with hemicelluloses (blue and purple strands), which may separate microfibrils from one another. Pectins (red, yellow, and green strands) form an interlocking matrix that controls microfibril spacing and wall porosity. Pectins and hemicelluloses are synthesized in the Golgi apparatus and delivered to the wall via vesicles that fuse with the plasma membrane and thus deposit these polymers to the cell surface. For clarity, the hemicellulose–cellulose network is emphasized on the left, and the pectin network is emphasized on the right. (After Cosgrove 2005.)

backbone by other sets of glycosyltransferases, probably acting coordinately in membrane-bound complexes.

Unlike cellulose, which forms a crystalline microfibril, the matrix polysaccharides are much less ordered and are often described as amorphous. This noncrystalline character is a consequence of the structure of these polysaccharides—their branching and their nonlinear conformation. Nevertheless, studies using various physical techniques, including infrared spectroscopy and nuclear magnetic resonance (NMR), indicate partial order in the orientation of hemicelluloses and pectins in the cell wall, probably as a result of a physical tendency for these polymers to become aligned along the long axis of cellulose. Such realignment of pectins after deposition to the cell wall has been visualized by confocal microscopy combined with metabolic labeling with a fucose molecule coupled to a fluorescence dye.

Pectins are hydrophilic gel-forming components of the primary cell wall

Pectins comprise the most abundant component of most primary cell walls, forming a hydrated gel phase in which cellulose and hemicelluloses are embedded. They act as hydrophilic filler to prevent aggregation and collapse of the cellulose network, and they also determine the porosity of the cell wall to macromolecules. They are particularly concentrated in the middle lamella, notably at tricellular junctions, and are important for cell adhesion. Release of oligosaccharides from pectins during fungal invasion of plant tissue elicits defense responses that limit pathogen invasion (see Chapter 23).

Pectins constitute a heterogeneous group of polysaccharides, characteristically containing galacturonic acid

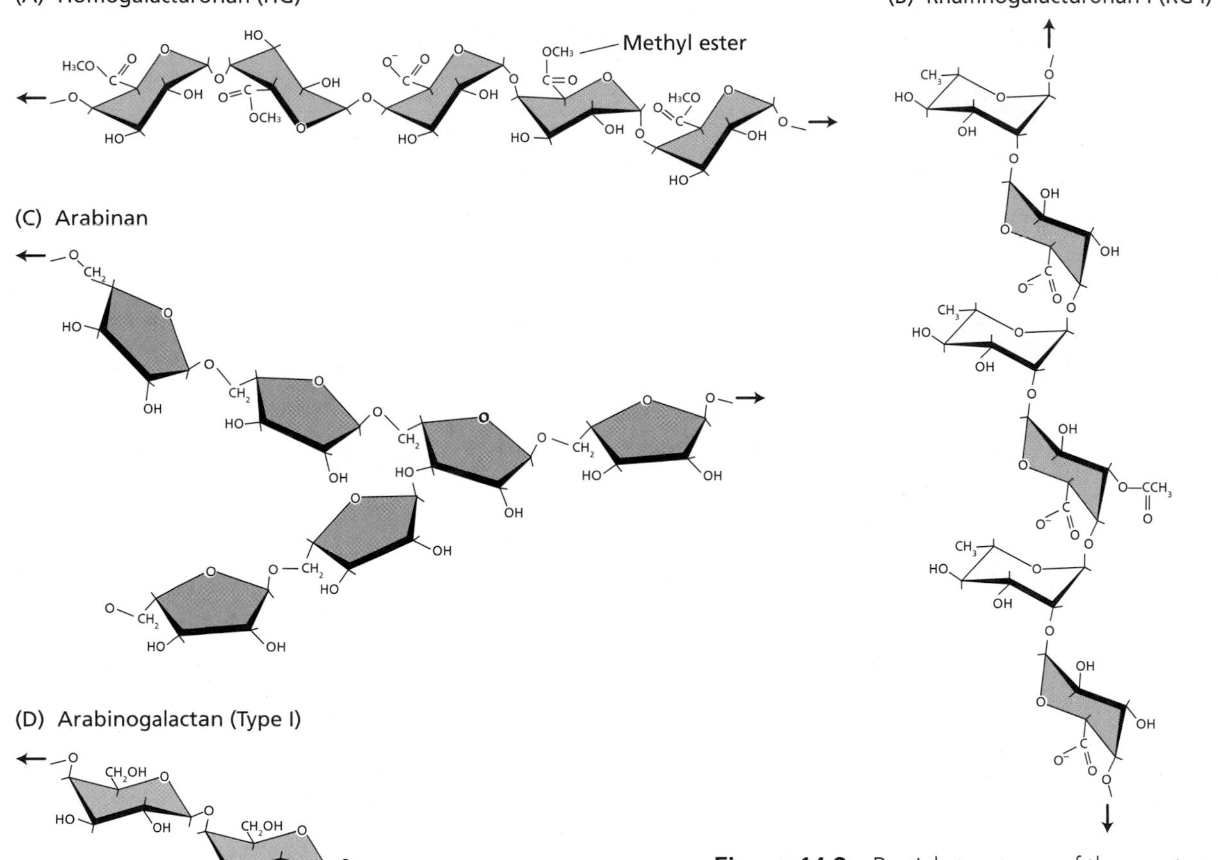

Figure 14.9 Partial structures of the most common pectins. (A) Homogalacturonan, also known as polygalacturonic acid or pectic acid, is made up of (1,4)-linked α-D-galacturonic acid (GalA). The carboxyl residues are often methyl esterified. (B) Rhamnogalacturonan I (RG I) is a very large pectin domain, with a backbone of alternating GalA and (1,2)-α-D-rhamnose (Rha). Side chains are attached to rhamnose and are composed principally of arabinans (C), galactans, and arabinogalactans (D). These side chains may be short or quite long. The galacturonic acid residues are often methyl esterified. (After Carpita and McCann 2000.)

and neutral sugars such as rhamnose, galactose, and arabinose. These different polysaccharides are often, but not always, covalently linked to one another, forming large macromolecular structures (~10^6 Da). NMR studies indicate that pectins make contact with cellulose surfaces in the wall, and binding studies have shown that the neutral side chains of pectins can bind to cellulose surfaces, although more weakly than do hemicelluloses. NMR results also indicate that pectins make intimate contact with xyloglucan. There also is evidence for covalent linkages between pectins and hemicelluloses, and a recent study identified a covalent complex containing an arabinogalactan protein, pectin, and xylan, but the extent and significance of such cross-linking for primary wall function is not yet certain.

The three major pectic polysaccharides, sometimes called pectin domains, are **homogalacturonan** (**HG**), rhamnogalacturonan I (RG I), and **rhamnogalacturonan II** (**RG II**) (**Figure 14.9**). HG is a linear chain of (1,4)-linked α-D-galacturonic acid residues, some of which are methyl esterified. It is the most abundant pectin in primary cell walls. RG I has a long backbone of alternating rhamnose and galacturonic acid residues and carries long side-chains of **arabinans**, galactans, and so-called **type-1 arabinogalactans**, collectively known as neutral pectic polysaccharides. RG II, which is the least abundant of these pectin domains, contains an HG backbone decorated with side chains made of at least ten different sugars in a complicated pattern of linkages. Although RG I and RG II have similar names, they have very different structures.

It has been proposed that in the wall these pectin domains are covalently linked end to end. **Figure 14.10**

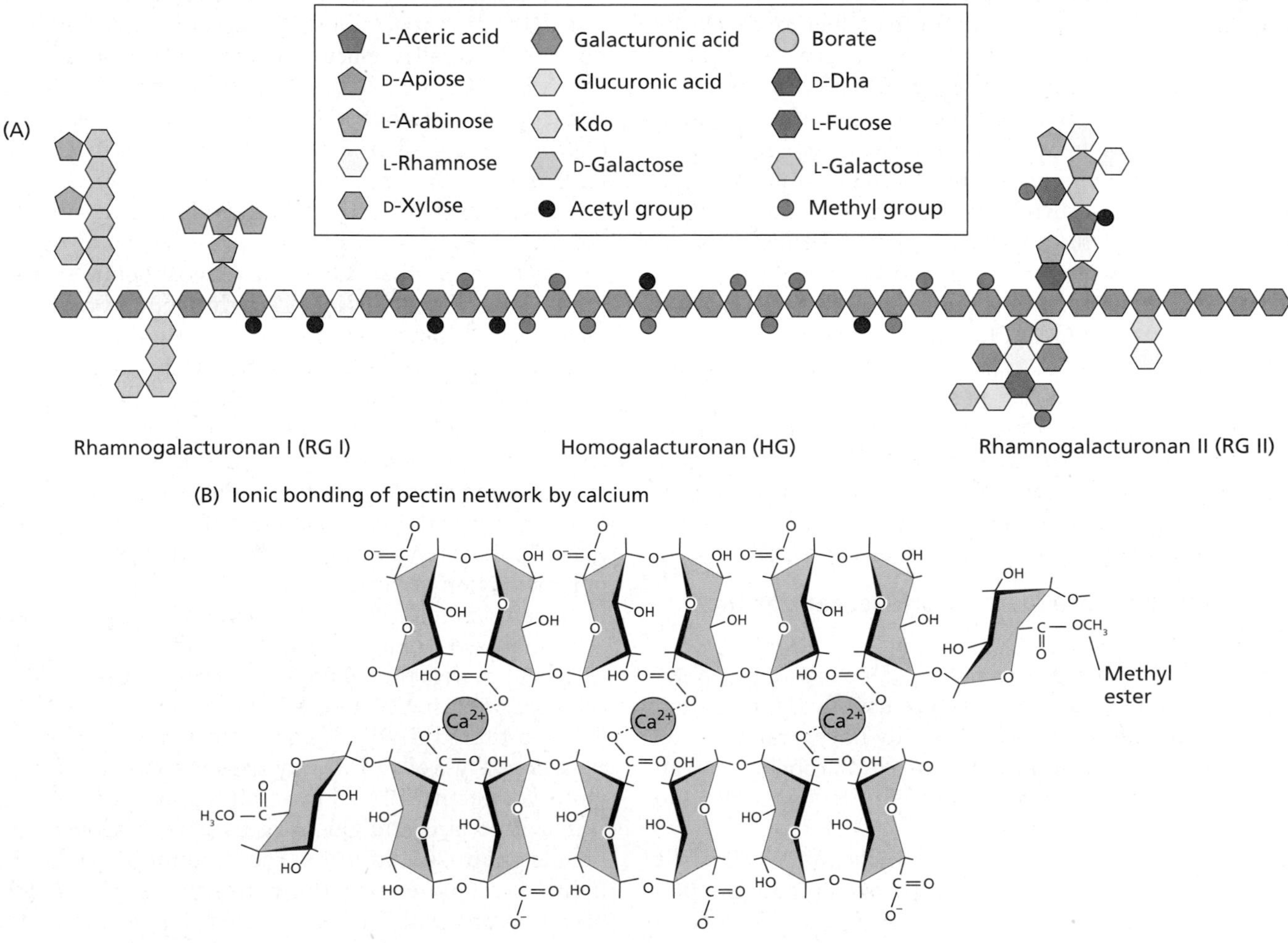

Figure 14.10 (A) Schematic model illustrating the linear arrangement of the various pectin domains to each other, including rhamnogalacturonan I (RG I), homogalacturonan (HG), and rhamnogalacturonan II (RG II). The structure is not quantitatively accurate: HG should be about ten times more abundant and RG I about two times more abundant. Kdo = 3-Deoxy-D-manno-2-octulosonic acid; D-Dha = dihydroxyacetone. (B) Formation of a pectin network involves ionic bridging of the nonesterified carboxyl groups (COO^-) by calcium ions. When blocked by methyl-esterified groups, the carboxyl groups cannot participate in this type of interchain network formation. Likewise, the presence of side chains on the backbone interferes with network formation. (A after Mohnen 2008; B after Carpita and McCann 2000.)

illustrates a hypothetical scheme for linkage of HG, RG I, and RG II. Not all the pectic polysaccharides are attached to such large structures, however. For instance, most arabinans and galactans were not linked to acidic polysaccharides in cell walls from growing pea stems and HG was solubilized without other pectic components from maize cell walls by gentle, nonenzymatic methods.

Additional cross-linking of pectic polysaccharides occurs via borate diesters between two RG II domains. Such cross-linking is important for wall structure and the mechanical strength of tissues. In some plant groups such as Amaranthaceae, which includes spinach and sugar beet, pectic arabinans and galactans are esterified with ferulic acid which can undergo oxidative reactions to form diferulate cross-links.

When HG is initially synthesized, many of the acidic carboxyl groups are methyl esterified, giving a less charged polysaccharide. Removal of methyl esters in the cell wall by pectin methyl esterase enzymes facilitates ionic cross-linking of HG and gel formation. Extensive block-wise de-esterification of HG restores the charged carboxyl group and enables calcium ions to form ionic bridges between adjacent chains, resulting in a relatively stiff gel. The solubilization of pectins by calcium chelators is based on the removal of these calcium bridges. Ionic gel formation by HG is important for the adhesion of cells by the middle lamella and makes the primary cell wall less extensible. De-esterification of HG also plays a role in leaf primordial initiation at the shoot apical meristem and in pollen tube growth. By creating free carboxyl groups, de-esterification also increases the electric-charge density in the wall, which in turn may influence the concentration of ions in the wall, the activities of wall enzymes, and possibly the distribution of charged signaling molecules.

HG is synthesized in the Golgi apparatus by a glycosyl transferase named GAUT1, which transfers galacturonic acid from a UDP donor to an HG acceptor. GAUT1 is part of a protein complex that is anchored to the inner face of the Golgi apparatus membrane by a related but enzymatically inactive protein, GAUT7. The Golgi apparatus is thought to contain numerous other enzymes that participate in the synthesis of other wall polysaccharides, but these enzymes have not yet been well characterized.

Hemicelluloses are matrix polysaccharides that bind to cellulose

Hemicelluloses comprise a heterogeneous group of polysaccharides (**Figure 14.11**) that are usually tightly bound in the wall. Hemicelluloses typically have a pronounced ability to bind to cellulose in vitro and probably play an important role in the assembly of cellulose microfibrils to form a coherent cell wall in vivo.

The dominant hemicellulose in primary cell walls of most land plants is xyloglucan, which consists of a (1,4)-β-D-glucan decorated with (1,6)-linked α-D-xylosyl residues (see Figure 14.11A). Xyloglucan structure shows some variability among species. In most eudicots, 30 to 40% of the xylose residues are appended with a galactose residue, which in turn may bear a terminal fucose residue. A concise nomenclature has been developed to refer to the branching pattern of xyloglucan (see Figure 14.11B): for example, G is used for an unsubstituted glucose residue; X means glucose is substituted solely with xylose; L is used for a xylose–galactose side chain; and F denotes a xylose–galactose–fucose side chain.

Xyloglucan has a repeating substructure in which one of every four glucose residues in the backbone is unsubstituted (does not carry a sugar side chain). Endoglucanase digestion of xyloglucan from most eudicot sources yields three major oligosaccharides with four glucose residues in the backbone, designated XXXG, XXFG, and XLFG. In contrast, xyloglucan in grass cell walls is mostly made up of XXGG, XXGGG, and XXGGGG repeat units. Plants in the Solanaceae, such as tomato, use an arabinose residue in place of galactose, which appears to be functionally equivalent for cell wall mechanics. Glycosidases are able to remove the side chain sugars, resulting in xyloglucans with a lower degree of substitution, which bind more tightly to cellulose.

Unlike in most land plants, the dominant hemicellulose in the primary cell wall of grasses (Poaceae) is arabinoxylan (also called glucuronoarabinoxylan or GAX; see Figure 14.11C). Xyloglucans and pectins are also present in grass cell walls, but are greatly reduced in abundance. GAX has a (1,4)-β-D-xylan backbone substituted with (1,3)-α-L-arabinose residues; approximately 1 residue in 50 is substituted with (1,2)-α-D-glucuronic acid. The degree of arabinose substitution varies widely, from more than 80% to less than 10%. Unlike most hemicelluloses, highly substituted GAX is not tightly bound to the cell wall, does not bind cellulose in vitro, and is readily solubilized from the cell wall under mild conditions used to extract pectin. Some of the arabinose residues bear ferulate groups attached by an ester linkage. Oxidative coupling of ferulate groups results in cross-links between GAX; such cross-linking reduces the digestibility of grasses (i.e., for feeding cows and sheep) and may reduce cell wall extensibility. Ferulates also function as nucleation sites for lignin polymerization in grass walls.

In addition to GAX, primary cell walls of grasses also contain mixed-linkage **(1,3;1,4)-β-D-glucan**. The mixed-linkage glucan is thought to bind tightly to the surface of cellulose, reducing cellulose–cellulose interactions, while less-substituted GAX may serve a cross-linking function.

Secondary walls of woody tissues contain little xyloglucan or pectin; instead the matrix polysaccharides are primarily xylans and glucomannans with a low degree of side chain substitution. These hemicelluloses bind tightly to cellulose and require strong alkali to be solubilized from the wall. The major hemicellulose of secondary

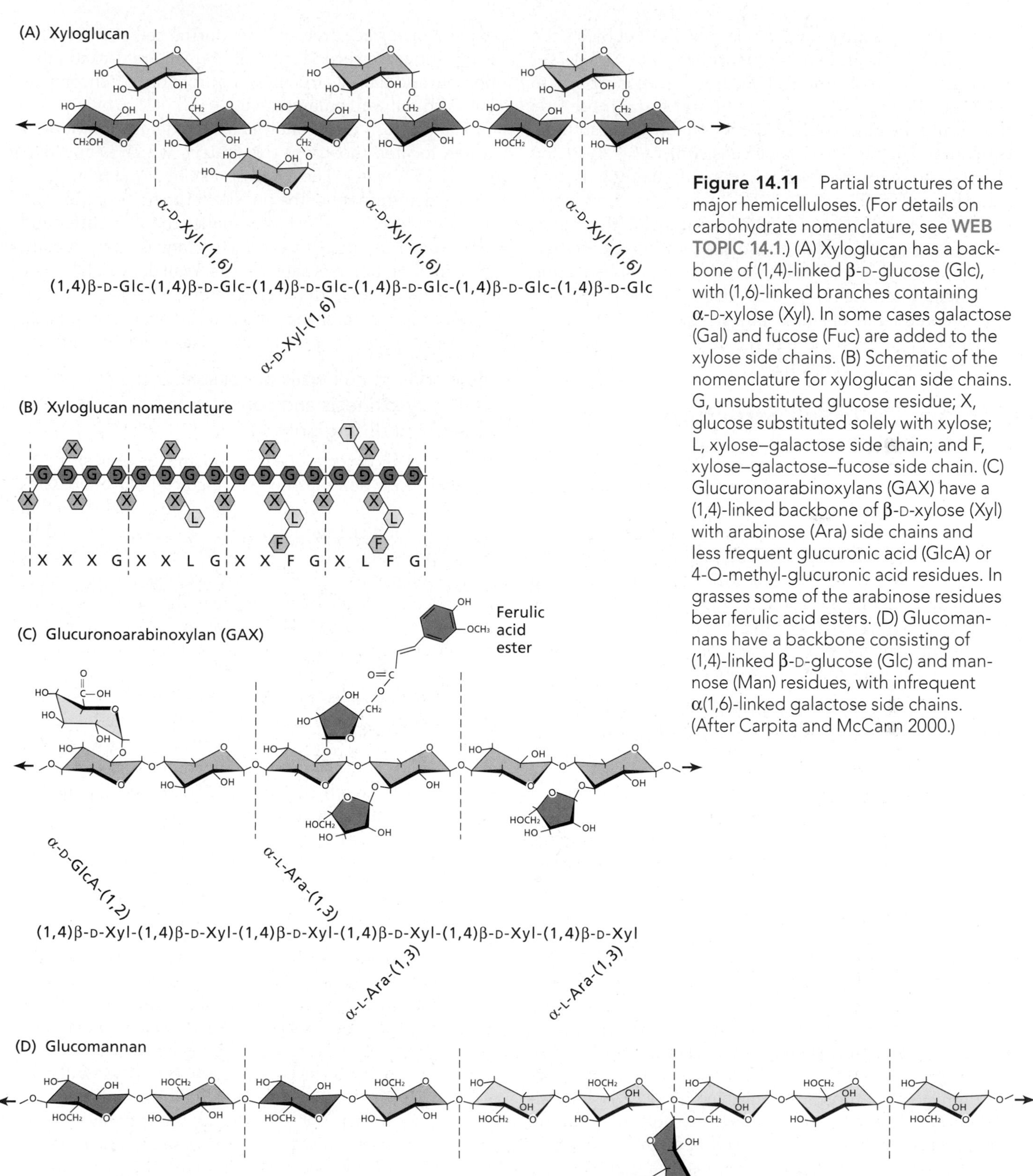

Figure 14.11 Partial structures of the major hemicelluloses. (For details on carbohydrate nomenclature, see **WEB TOPIC 14.1**.) (A) Xyloglucan has a backbone of (1,4)-linked β-D-glucose (Glc), with (1,6)-linked branches containing α-D-xylose (Xyl). In some cases galactose (Gal) and fucose (Fuc) are added to the xylose side chains. (B) Schematic of the nomenclature for xyloglucan side chains. G, unsubstituted glucose residue; X, glucose substituted solely with xylose; L, xylose–galactose side chain; and F, xylose–galactose–fucose side chain. (C) Glucuronoarabinoxylans (GAX) have a (1,4)-linked backbone of β-D-xylose (Xyl) with arabinose (Ara) side chains and less frequent glucuronic acid (GlcA) or 4-O-methyl-glucuronic acid residues. In grasses some of the arabinose residues bear ferulic acid esters. (D) Glucomannans have a backbone consisting of (1,4)-linked β-D-glucose (Glc) and mannose (Man) residues, with infrequent α(1,6)-linked galactose side chains. (After Carpita and McCann 2000.)

walls varies by source: in secondary walls of eudicots the dominant hemicellulose is **glucuronoxylan**, with lesser amounts of glucomannans. Glucuronoxylan is similar to GAX (see Figure 14.11C) but without the arabinose side chains and the glucuronic acid is 4-O-methyl substituted. **Glucomannan** has a backbone consisting of β(1,4)-linked glucose and mannose residues, with infrequent galactose side chains (see Figure 14.11D). In gymnosperm wood the major hemicellulose is glucomannan, with smaller amounts of arabinoxylan substituted with 4-O-methyl-glucuronyl residues. GAX of low degree of substitution is the major hemicellulose of secondary walls in grasses. The low frequency of side chains in these hemicelluloses enables them to bind more tightly to cellulose and to pack more tightly in the cell wall.

Primary Cell Wall Structure and Function

Early in their life, plant cells form a pliant cell wall that is extensible and able to incorporate new structural material as the wall extends. The general wall structure consists of thin layers made of long cellulose microfibrils embedded in a hydrated matrix of noncellulosic polysaccharides and a small amount of nonenzymatic proteins (see Figure 14.8, Table 14.1). This structure imparts an ideal combination of flexibility and strength to the growing cell wall, which must be both extensible and strong at the same time.

The primary cell wall is composed of cellulose microfibrils embedded in a matrix of pectins and hemicelluloses

By dry mass, primary cell walls typically contain approximately 40% pectin, 25% cellulose, and 20% hemicellulose, with perhaps 5% protein and the remaining percentage composed of diverse other materials. However, large deviations from these typical values may be found among species. For example, the walls of grass coleoptiles consist of 60 to 70% hemicellulose (GAX), 20 to 25% cellulose, and only about 10% pectin. Cereal endosperm walls may contain as little as 2% cellulose, with hemicellulose making up most of the wall. Parenchyma cell walls of celery and sugar beets contain mostly cellulose and pectin, with as little as 4% hemicellulose. The wall at the tip of pollen tubes appears to be mostly pectin, with small amounts of cellulose to reinforce the tip structure. Wall compositions and polysaccharide structures are not static, but can change developmentally as a result of altered patterns of synthesis and the action of enzymes that can trim side chains and digest pectins and hemicelluloses. Thus, it may be misleading to talk about "typical" primary cell walls, as they may be quite diverse.

What primary walls have in common is that they are formed by growing cells, contain a highly hydrated matrix between the cellulose microfibrils, and have the ability to expand in surface area, at least during cell growth. This is in contrast to secondary walls, which are packed more densely and have a structural, reinforcing role incompatible with cell wall enlargement.

The primary wall contains a considerable amount of water, located mostly in the matrix, which is approximately 75% water. The hydration state of the matrix is a critical determinant of the physical properties of the wall; for example, removal of water makes the wall stiffer and less extensible, and this is a factor contributing to plant growth inhibition by water deficits. Wall dehydration may also be important in the strengthening of cell walls during *lignification*, a process that drives water out of the cell wall and results in a stiffer wall that resists enzymatic attack.

New primary cell walls are assembled during cytokinesis and continue to be assembled during growth

Primary walls originate de novo during the final stages of cell division, when the newly formed **cell plate** separates the two daughter cells and solidifies into a stable wall that is capable of bearing the physical stresses generated by turgor pressure.

The cell plate forms when Golgi vesicles and endoplasmic reticulum cisternae aggregate in the spindle midzone area of a dividing cell. This aggregation is organized by the **phragmoplast**, a complex assembly of microtubules, membranes, and vesicles that forms during late anaphase or early telophase (see Chapter 1). The membranes of the vesicles fuse with each other and with the lateral plasma membrane to become the new plasma membrane separating the daughter cells. The contents of the vesicles are the precursors from which the new middle lamella and the primary wall are assembled.

The "life" of an individual polymer may be outlined as follows:

Synthesis → deposition → assembly → modification

At any given moment, wall polymers may populate any or all of these stages. The synthesis and deposition of the major wall polymers were described earlier. Modifications may alter interactions between wall components or may be part of the processes of polysaccharide turnover and wall disassembly. Here we will consider wall polymer assembly into a cohesive network, and later we will consider modifications that affect cell enlargement.

After their secretion into the apoplast, the wall polymers must be assembled into a cohesive structure; that is, the individual polymers must attain the physical arrangement and bonding relationships that are characteristic of the primary (growing) cell wall and that confer on it both tensile strength and extensibility. Although the details of wall assembly are not fully understood, the prime candidates for this process are self-assembly and enzyme-mediated assembly.

SELF-ASSEMBLY Self-assembly is an attractive concept because it is mechanistically simple. Many polysaccharides possess a marked tendency to aggregate spontaneously into organized structures. Aggregation can make the separation of hemicelluloses into distinct polymers technically difficult. In contrast, pectins are more soluble and tend to form dispersed, isotropic (randomly arranged) networks (gels). Self-assembly may not be the entire story because when hemicelluloses are bound in cellulose in vitro, their binding is much weaker than is the case in real cell walls. This discrepancy hints at the involvement of other processes needed to make strong networks in the wall.

ENZYME-MEDIATED ASSEMBLY In addition to self-assembly, enzymes may facilitate wall assembly. A prime candidate for enzyme-mediated wall assembly is **xyloglucan endotransglucosylase** (**XET**). This enzyme, which belongs to a large family of enzymes named **xyloglucan endotransglucosylase/hydrolases** (**XTHs**), has the ability to cut the backbone of a xyloglucan and to join one end of the cut xyloglucan with the free end of an acceptor xyloglucan (**Figure 14.12**). Such a transfer reaction integrates newly synthesized xyloglucans into the wall, potentially strengthening the cell wall. Transglycosylases with other substrate specificities have recently been detected in plant cell walls, but their biological functions have not yet been assessed.

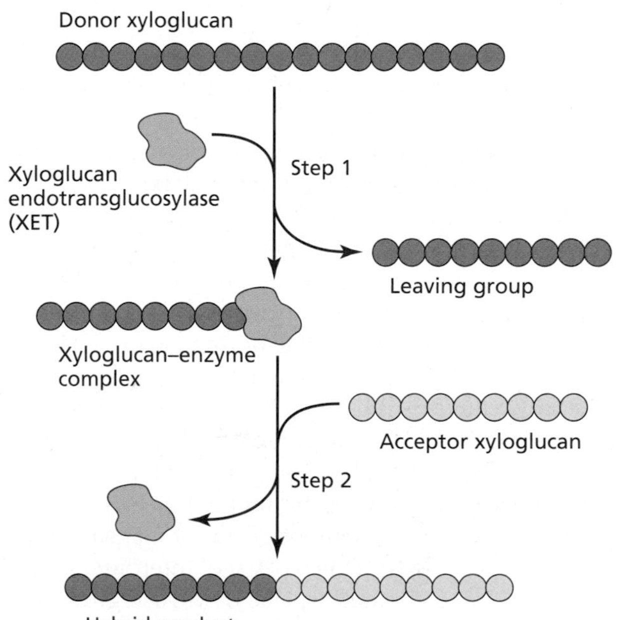

Figure 14.12 Action of xyloglucan endotransglucosylase (XET) to cut and join xyloglucan polymers into new configurations. Step 1: The enzyme cuts a xyloglucan molecule (the donor xyloglucan), forming a long-lived complex where the xyloglucan is covalently attached to the enzyme. Step 2: Subsequently the enzyme transfers the xyloglucan chain to the nonreducing end of a second xyloglucan (the acceptor xyloglucan), resulting in a hybrid product. (After Fry 2004.)

Other wall enzymes that might aid in assembly of the wall include glycosidases, pectin methyl esterase, and various oxidases. Some glycosidases remove the side chains of hemicelluloses, increasing the tendency of hemicelluloses to adhere to each other and to the surface of cellulose microfibrils. As we described in the previous section, pectin methyl esterase removes methyl esters that block acidic groups of HG, thereby enhancing HG's ability to form a Ca^{2+}-bridged gel network. Oxidases such as peroxidase catalyze cross-links between phenolic groups (tyrosine, phenylalanine, ferulic acid) in wall proteins, pectins, and other wall polymers. Such oxidative cross-linking is also the basis of lignin formation, which we'll discuss later in the chapter.

Mechanisms of Cell Expansion

During plant cell enlargement, new wall polymers are continuously synthesized and secreted at the same time that the preexisting wall is expanding. Wall expansion may be highly localized (as in the case of **tip growth**) or more dispersed over the wall surface (**diffuse growth**) (**Figure 14.13**). Tip growth is characteristic of root hairs and pollen tubes; it is closely linked to cytoskeletal processes, especially those of actin microfilaments (see **WEB ESSAY 14.1**). Most of the other cells in the plant body exhibit diffuse growth, which is linked to the activities of both microtubules and actin microfilaments. Cells such as fibers, some

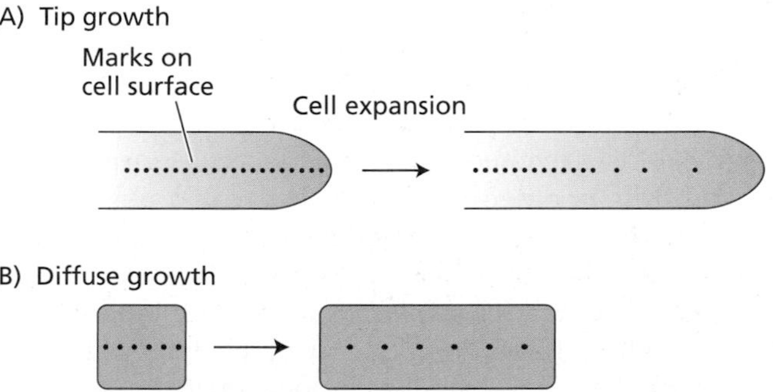

Figure 14.13 The cell surface expands differently during tip growth and diffuse growth. (A) Expansion of a tip-growing cell is confined to an apical dome at one end of the cell. If marks are placed on the cell surface and the cell is allowed to continue to grow, only the marks that were initially within the apical dome grow farther apart. Root hairs and pollen tubes are examples of plant cells that exhibit tip growth. (B) If marks are placed on the surface of a diffuse-growing cell, the distance between all the marks increases as the cell grows. Most cells in multicellular plants grow by diffuse growth.

sclereids, and trichomes grow in a pattern that is intermediate between that of diffuse growth and tip growth.

Even in cells with diffuse growth, however, different parts of the wall may enlarge at different rates or in different directions. For example, in cortical cells of the stem, the end walls grow much less than side walls. This difference may be due to structural or enzymatic variations in specific walls or to variations in the stresses borne by different walls. As a consequence of this uneven pattern of wall expansion, plant cells may assume irregular forms.

Microfibril orientation influences growth directionality of cells with diffuse growth

During growth, the loosened cell wall is extended by physical forces generated from cell turgor pressure. Turgor pressure creates an outward-directed force, equal in all directions. The directionality of growth is determined in large part by the structure of the cell wall—specifically, the orientation of cellulose microfibrils.

When cells first form in the meristem, they are isodiametric; that is, they have equal diameters in all directions. If the orientation of cellulose microfibrils in the primary cell wall is randomly arranged, the cells grow isotropically (equally in all directions), expanding radially to generate a sphere (**Figure 14.14A**). In most plant cell walls, however, cellulose microfibrils are aligned in a preferential direction, resulting in **anisotropic growth** (e.g., in the stem, the cells increase in length much more than in width).

In the lateral walls of elongating cells such as cortical and vascular cells of stems and roots, or the giant internode cells of the filamentous green alga *Nitella*, cellulose microfibrils are deposited circumferentially (transversely), at right angles to the long axis of the cell. The circumferential arrangement of cellulose microfibrils restricts growth in girth and promotes growth in length (**Figure 14.14B**).

Cell wall deposition continues as cells enlarge. According to the **multinet growth hypothesis**, each successive wall layer is stretched and thinned as cells grow, so the microfibrils in older cell wall layers would be expected to become passively reoriented in the longitudinal direction as cells elongate. Evidence for passive reorientation has been reported for growing cells in Arabidopsis roots stained with a fluorescent dye to enable imaging of cellulose microfibril bundles by confocal microscopy.

Other observations cast doubt on the universality of multinet growth. In a study to test the ability of cell wall microfibrils to passively reorient in response to wall tension, isolated walls from growing hypocotyls were allowed to undergo slow extension under conditions that mimicked normal growth, and the effect of this extension on the orientation of the cellulose microfibrils on the inner wall surface was examined by electron microscopy. Allowing the wall to extend slowly by 20 to 30% failed to alter the transverse angle of the microfibrils at the inner wall surface, suggesting that the microfibrils separated

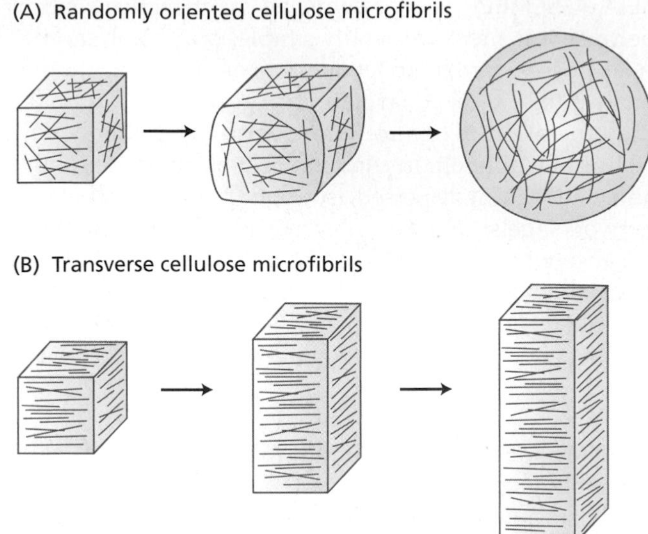

Figure 14.14 The orientation of newly deposited cellulose microfibrils determines the direction of cell expansion. (A) If the cell wall is reinforced by randomly oriented cellulose microfibrils, the cell will expand equally in all directions, forming a sphere. (B) When most of the reinforcing cellulose microfibrils have the same orientation, the cell expands at right angles to the microfibril orientation and is constrained in the direction of the reinforcement. Here the microfibril orientation is transverse, so cell expansion is longitudinal.

from each other in a coordinated fashion. These and other results suggest that wall expansion involves selective loosening of junctions that hold microfibrils together, rather than a generalized loosening of the matrix.

Other experiments suggest that the older layers of the cell wall (i.e., the outer half) may be so fragmented as a result of their history of enlargement that they may contribute little to the control of growth. By this hypothesis, the inner one-fourth of the wall dominates control of cell expansion (see WEB TOPIC 14.3).

So far we have considered only a simple pattern of diffuse growth. So-called pavement cells in the epidermis of many eudicot leaves, however, present a more complicated situation. These cells are highly lobed, creating an interlocking pattern resembling that of a jigsaw puzzle (**Figure 14.15A and B**). This pattern of interdigitating cell wall expansion combines aspects of diffuse growth and tip growth and requires the action of small, GTP-binding proteins called **ROP (*Rho*-like from plants) GTPases** and their activating proteins called RICs (*R*OP-*i*nteracting CRIB motif-containing proteins) (**Figure 14.15C**). These proteins organize the cytoskeleton (actin microfilaments and tubulin microtubules), which delivers materials and catalysts for local control of cell wall growth. As we will describe next, the cytoskeleton plays a central role in regulating cell wall growth.

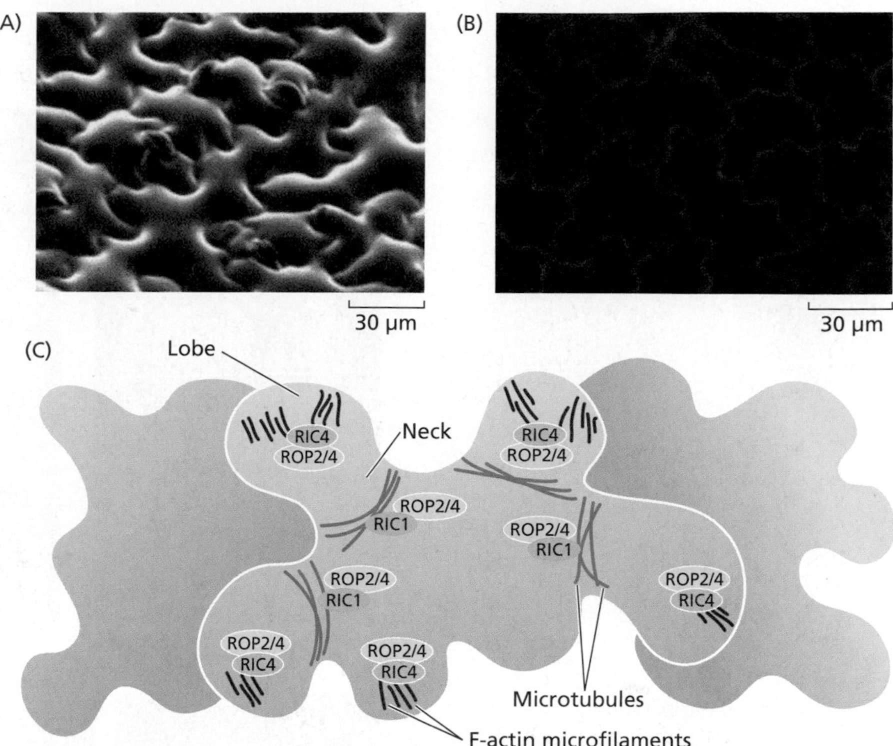

Figure 14.15 Interdigitating cell growth of leaf pavement cells and its regulation by ROP GTPases. (A) Scanning electron micrograph of pavement cells from an Arabidopsis leaf. Note the jigsaw puzzle–like appearance. (B) Immunofluorescence image of pavement cells shows more clearly the lobes and indentations formed by interdigitated cells. (C) A model to explain the role of ROP GTPases and their effectors (RICs) in leaf morphogenesis. ROP2/4 GTPases, when activated by RIC4, promote actin microfilament formation in regions of growing lobes, whereas when activated by RIC1 they promote microtubule bundling at neck regions. These cytoskeletal changes somehow act as signals to direct the direction of wall growth. (A courtesy of Daniel Szymanski; B from Settleman 2005, courtesy of J. Settleman; C after Fu et al. 2005.)

Cortical microtubules influence the orientation of newly deposited microfibrils

Newly deposited cellulose microfibrils usually are coaligned with microtubule arrays in the cytoplasm, close to the plasma membrane (**Figure 14.16**). A striking example occurs in xylem vessel elements, where bands of cortical microtubules mark the sites of secondary wall thickenings and also the sites of CESA localization. Moreover, experimental disruptions of microtubule organization with drugs or by genetic defects often leads to disorganized wall structure and disorganized growth. For example, several drugs bind to tubulin, the subunit protein of microtubules, causing them to depolymerize. When growing roots are treated with a microtubule-depolymerizing drug, such as oryzalin,

Figure 14.16 The orientation of microtubules in the cortical cytoplasm mirrors the orientation of newly deposited cellulose microfibrils in the walls of cells that are elongating. (A) The arrangement of microtubules can be revealed with fluorescently labeled antibodies to the microtubule protein tubulin. In this differentiating tracheary element from a *Zinnia* cell suspension culture, the pattern of microtubules (green) mirrors the orientation of the cellulose microfibrils in the wall, as shown by calcofluor staining (blue). (B) The alignment of cellulose microfibrils in the cell wall can sometimes be seen in grazing sections prepared for electron microscopy, as in this micrograph of a developing sieve tube element in a root of *Azolla* (a water fern). The longitudinal axis of the root and the sieve tube element runs vertically. Both the wall microfibrils (double-headed arrows) and the cortical microtubules (single-headed arrows) are aligned transversely. (A courtesy of Robert W. Seagull; B courtesy of A. Hardham.)

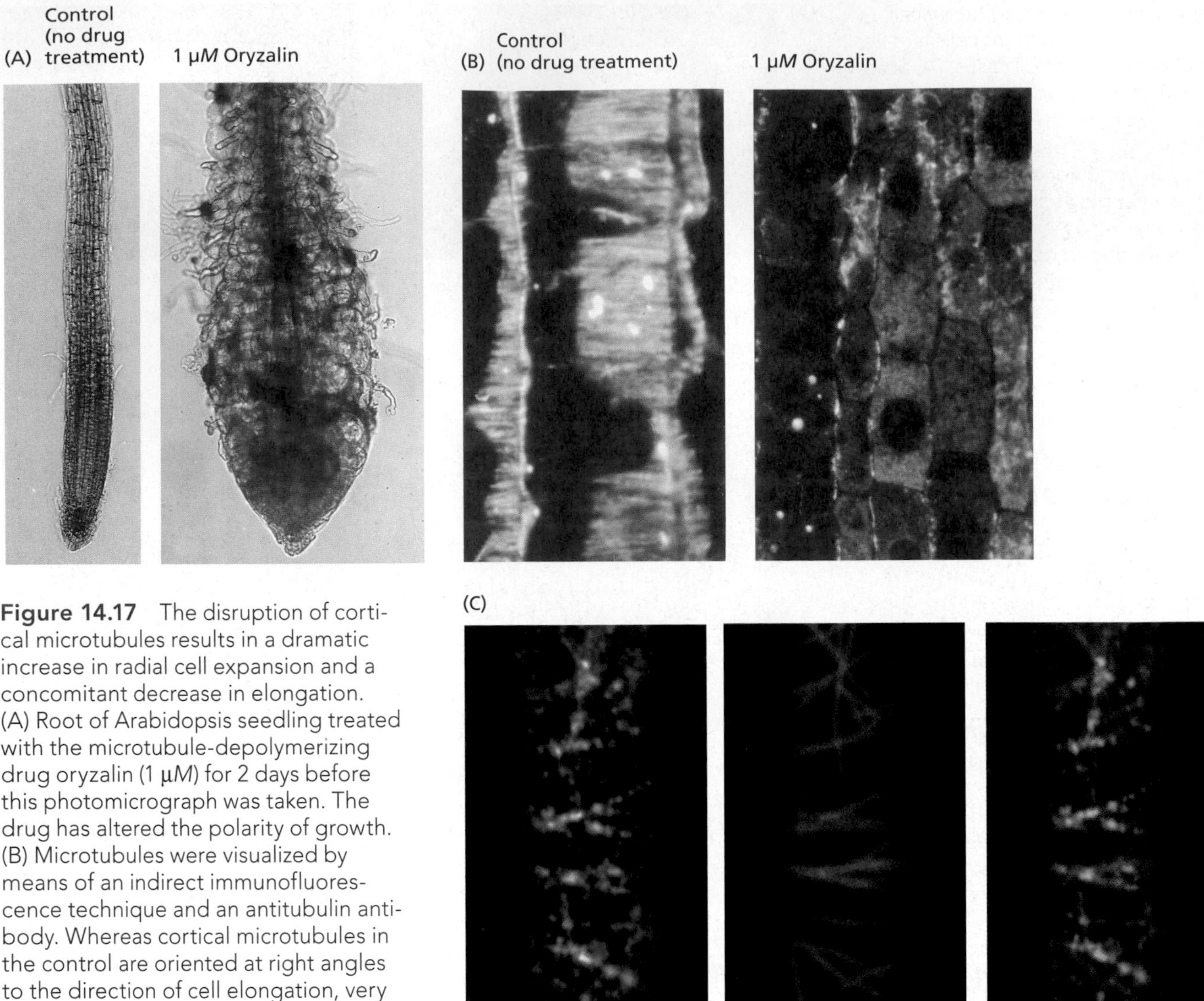

Figure 14.17 The disruption of cortical microtubules results in a dramatic increase in radial cell expansion and a concomitant decrease in elongation. (A) Root of Arabidopsis seedling treated with the microtubule-depolymerizing drug oryzalin (1 μ*M*) for 2 days before this photomicrograph was taken. The drug has altered the polarity of growth. (B) Microtubules were visualized by means of an indirect immunofluorescence technique and an antitubulin antibody. Whereas cortical microtubules in the control are oriented at right angles to the direction of cell elongation, very few microtubules remain in roots treated with 1 μ*M* oryzalin. (C) Images of fluorescently tagged CESA proteins (left panel) and microtubules (middle panel) indicate that microtubules guide the trajectories of CESA movement in the plasma membrane, thus guiding the orientation of cellulose microfibrils. The right panel shows the superposition of the two images. (A and B from Baskin et al. 1994, courtesy of T. Baskin; C from Gutierrez et al. 2009.)

the region of elongation expands laterally, becoming bulbous and tumorlike (**Figure 14.17A and B**). This disrupted growth is due to the isotropic expansion of the cells; that is, they enlarge like a sphere instead of elongating. The drug-induced destruction of microtubules in the growing cells interferes with the transverse deposition of cellulose. Cellulose microfibrils continue to be synthesized in the absence of microtubules, but they are deposited randomly and consequently the cells expand equally in all directions.

These and related observations have led to the suggestion that microtubules serve as tracks that guide or direct the movement of CESA complexes as they synthesize microfibrils (see **WEB ESSAY 14.2**). The movement of CESA in living cells was made visible by expressing a fusion of CESA with a fluorescent protein. The CESA units were observed to move within the plasma membrane along microtubule tracks (**Figure 14.17C**); they were also seen to be inserted into the plasma membrane from the Golgi apparatus at microtubule-tethered compartments. A molecular linker between CESA and microtubules was recently identified as CSI1 (CESA interactive protein 1), providing a link between the cytoskeleton and cellulose orientation. These results, obtained by confocal microscopy and genetics, reveal new details of how the cytoskeleton directs cell wall organization.

The Extent and Rate of Cell Growth

Plant cells typically expand ten to a thousandfold in volume before reaching maturity. In extreme cases, cells may enlarge more than ten thousandfold in volume compared with their meristematic initials (e.g., xylem vessel elements). The cell wall undergoes this massive expansion without losing its mechanical integrity and without becoming thinner. Thus, newly synthesized polymers are integrated into the wall without destabilizing it. Exactly how this integration is accomplished is uncertain, although self-assembly and xyloglucan endotransglucosylase (XET) probably play important roles, as we described earlier in the chapter.

This integrating process may be particularly critical for rapidly growing root hairs, pollen tubes, and other specialized cells that exhibit tip growth, in which the region of wall deposition and surface expansion is localized to the hemispherical dome at the apex of the tubelike cell, and where cell expansion and wall deposition must be closely coordinated.

In rapidly growing cells with tip growth, the wall doubles its surface area and is displaced to the nonexpanding part of the cell within minutes. This is a much greater rate of wall expansion than is typically found in cells with diffuse growth, where growth rates are approximately 1 to 10% per hour. Because of their fast expansion rates, tip-growing cells are highly susceptible to wall thinning and bursting. Mechanical and cytological models of tip growth in pollen tubes have given hints about how expansion and additions of wall components need to be coordinated for stable tip growth. Although diffuse growth and tip growth appear to be different growth patterns, both types of wall expansion must have analogous, if not identical, processes of polymer integration, wall stress relaxation, and wall polymer creep.

Many factors influence the rate of cell wall expansion. Cell type and age are important developmental factors. So, too, are hormones such as auxin and gibberellin. Environmental conditions such as light and water availability may likewise modulate cell expansion. These internal and external factors are most likely to modify cell expansion by altering the way in which the cell wall is loosened, so that it yields (stretches irreversibly) differently. In this context we speak of the **yielding properties of the cell wall**.

Stress relaxation of the cell wall drives water uptake and cell expansion

Because the cell wall is the major mechanical restraint that limits cell expansion, much attention has been given to its physical properties. As a hydrated polymeric material, the plant cell wall has physical properties that are intermediate between those of a solid and those of a liquid. We call these **viscoelastic**, or **rheological** (flow), **properties**. Walls of cells that are growing are generally less rigid than those of mature cells, and under appropriate conditions they exhibit a long-term irreversible stretching, or **yielding**, that is absent or nearly absent in mature walls.

Stress relaxation is a crucial concept for understanding how cell walls enlarge. The term "stress" is used here in the mechanical sense, as force per unit area. Wall stresses arise as an inevitable consequence of cell turgor. The turgor pressure in growing plant cells is typically between 0.3 and 1.0 megapascals (MPa). Turgor pressure stretches the cell wall and generates a counterbalancing physical stress or tension in the wall. Because of cell geometry (a large pressurized volume contained by a thin wall), this wall tension is equivalent to 10 to 100 MPa of tensile stress—a very large stress indeed.

This simple fact has important consequences for the mechanics of cell enlargement. Whereas animal cells can change shape in response to cytoskeleton-generated forces, such forces are negligible compared with the turgor-generated forces that are resisted by the plant cell wall. To change shape, plant cells must thus control the direction and rate of wall expansion, which they do by depositing cellulose in a biased orientation (this determines the directionality of cell wall expansion) and by selectively loosening the bonding between microfibrils. This biochemical loosening enables movement or slippage of cellulose microfibrils and their associated matrix polysaccharides, thereby increasing the wall surface area. At the same time, such loosening reduces the physical stress in the wall.

Wall stress relaxation is crucial because it allows growing plant cells to reduce their turgor and water potentials, which enables them to absorb water and to expand. Without stress relaxation, wall synthesis would only thicken the wall, not expand it; indeed, wall deposition and wall expansion are not closely linked in many cases. During secondary-wall deposition in nongrowing cells, for example, stress relaxation does not occur and consequently polysaccharide deposition results in a thickened cell wall.

When plant cells undergo expansive growth, the increase in volume is generated mostly by water uptake. This water ends up mainly in the vacuole, which takes up an ever larger proportion of the cell volume as the cell enlarges. WEB ESSAY 14.3 describes how growing cells regulate their water uptake and how this uptake is coordinated with wall yielding.

Acid-induced growth and wall stress relaxation are mediated by expansins

A common characteristic of growing cell walls is that they extend much faster at acidic pH than at neutral pH. This phenomenon is called **acid growth**. In living cells, acid growth is evident when growing cells are treated with acid buffers or with the drug fusicoccin, which induces acidification of the cell wall solution by activating an H^+-ATPase in the plasma membrane.

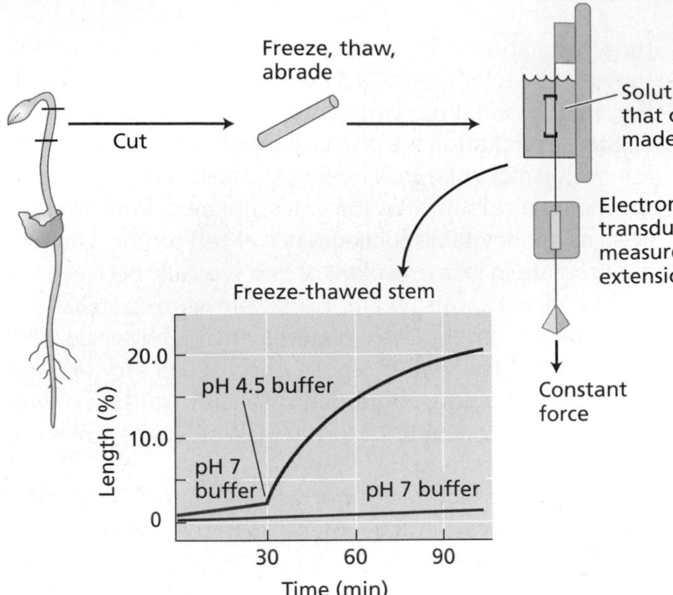

Figure 14.18 Acid-induced extension of isolated cell walls, measured in an extensometer. The wall sample from killed cells is clamped and put under tension in an extensometer that measures the length with an electronic transducer attached to a clamp. When the solution surrounding the wall is replaced with an acidic buffer (e.g., pH 4.5), the wall extends irreversibly in a time-dependent fashion (it creeps). (After Durachko and Cosgrove 2009.)

An example of acid-induced growth can be found in the initiation of the root hair, where the local wall pH drops to a value of 4.5 at the time when the epidermal cell begins to bulge outward. Auxin-induced growth is also associated with wall acidification, but it is probably not sufficient to account for the entire growth induction by this hormone (see Chapter 19), and other wall-loosening processes may additionally be involved. Nevertheless, this pH-dependent mechanism of wall extension appears to be an evolutionarily conserved process common to all land plants and is involved in a variety of growth processes.

Acid growth may also be observed in isolated cell walls, which lack normal cellular, metabolic, and synthetic processes. Such an observation entails using an extensometer to place the walls in tension and to measure long-term wall extension or "creep" (**Figure 14.18**).

The term **creep** refers to a time-dependent irreversible extension, typically the result of slippage of wall polymers relative to one another. When growing walls are incubated in neutral buffer (pH 7) and clamped in an extensometer, the walls extend briefly when tension is applied, but extension soon ceases. When transferred to an acidic buffer (pH 5 or less), the wall begins to extend rapidly, in some instances continuing for many hours.

This acid-induced creep is characteristic of walls in growing cells, but it is not observed in mature (nongrowing) walls. When walls are pretreated with heat, proteases, or other agents that denature proteins, they lose their acid growth ability. Such results indicate that acid growth is not due simply to the physical chemistry of the wall (e.g., a weakening of the pectin gel), but is catalyzed by one or more wall proteins.

The idea that proteins are required for acid growth was confirmed in reconstitution experiments in which heat-inactivated walls were restored to nearly full acid-growth responsiveness by the addition of proteins extracted from growing walls (**Figure 14.19**). The active components proved to be a group of proteins that were named **expansins**. Expansins catalyze the pH-dependent extension and stress relaxation of cell walls. They are effective in catalytic amounts (about 1 part protein per 5000 parts wall, by dry

Figure 14.19 Scheme for the reconstitution of extensibility of isolated cell walls. (A) Cell walls are prepared as in Figure 14.18 and briefly heated to inactivate the endogenous acid extension response. To restore this response, proteins are extracted from growing walls and added to the solution surrounding the wall. (B) Addition of proteins containing expansins restores the acid extension properties of the wall. (After Cosgrove 1997.)

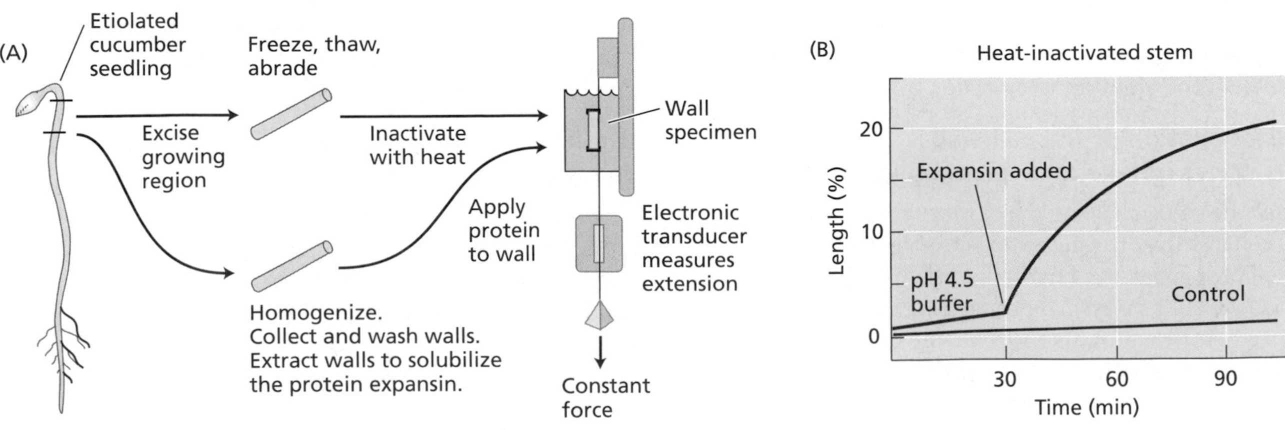

weight), but they do not exhibit lytic or other enzymatic activities.

With the complete sequencing of several plant genomes, we now know that expansins belong to a large superfamily of proteins, divided into two major expansin families, **α-expansins** (**EXPAs**) and **β-expansins** (**EXPBs**), plus two smaller families of unknown function. In extension assays with isolated cell walls, EXPAs are more active on eudicot cell walls whereas EXPBs are more active on grass cell walls. Current evidence indicates that EXPAs loosen cellulose–cellulose junctions containing xyloglucan, whereas EXPBs loosen wall complexes containing GAX.

Expansins have also been discovered in a small set of bacteria and fungi, where they facilitate colonization of plant tissues. Evolutionary analysis indicates that the bacterial expansins probably arose from one or more horizontal gene transfers from a plant to a bacterium, followed by additional horizontal gene transfers among various bacterial species that colonize the vascular system of plants.

The molecular basis for expansin action on wall rheology is still uncertain, but most evidence indicates that expansins cause wall creep by loosening noncovalent adhesion between wall polysaccharides. Studies of protein structure and binding suggest that expansins act at sites in the cell wall where cellulose microfibrils stick to one another.

Cell wall models are hypotheses about how molecular components fit together to make a functional wall

To understand how plant cells grow, it is essential to understand how the cell wall polymers are connected to produce a structure with enough tensile strength to resist turgor pressure, yet at the same time flexible enough allow for irreversible expansion of the wall fabric and incorporation of new polymers to strengthen the wall.

The earliest molecular model of primary cell wall architecture envisioned a covalently linked complex of xyloglucans, pectins, and structural proteins that was noncovalently bonded to cellulose microfibrils. This model was later replaced by an alternative concept in which xyloglucans fully coated the surfaces of cellulose microfibrils and directly tethered them into a load-bearing network, with pectins and glycoproteins forming an independent, interpenetrating matrix (**Figure 14.20A**).

Serious doubts about this "tethered network" model have emerged in recent years. The idea that xyloglucans cover most cellulose surfaces has been undermined by NMR data indicating that only about 10% of cellulose microfibril surfaces are coated by xyloglucan and that pectins directly contact cellulose surfaces. *Arabidopsis thaliana* mutants have been generated that completely lack xyloglucan, yet they have a relatively minor growth phenotype—a surprising result indeed, demonstrating that xyloglucan is not essential for at least some aspects of primary cell wall structure and function. These results demonstrate that plants are extremely adaptive with regard to their wall composition.

Biomechanical analyses of cell walls digested with substrate-specific endoglucanases show that the majority of xyloglucan does not contribute to wall mechanics and that cellulose is not directly linked by extensive xyloglucan

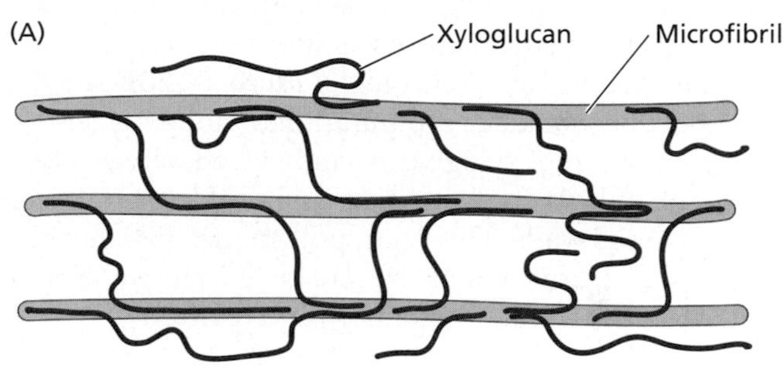

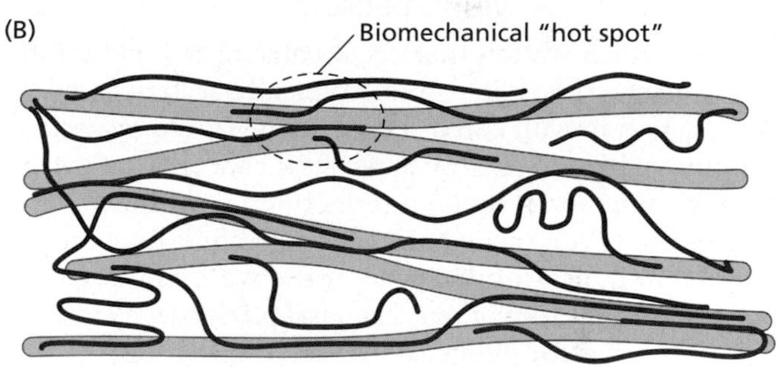

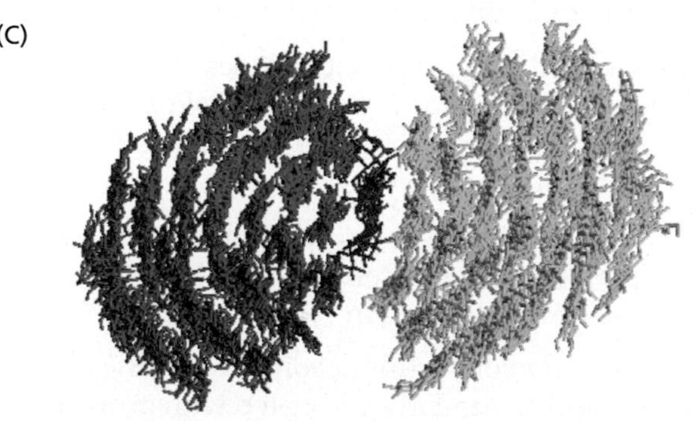

Figure 14.20 Alternative concepts of the structural role of xyloglucan. (A) The tethered-network model proposes that xyloglucans bind extensively to cellulose surfaces and form cross-bridges that tether microfibrils together. (B) The biomechanical "hot spot" model proposes that most of the xyloglucan is not load-bearing and that wall extension and mechanics are controlled at limited regions where cellulose microfibrils come in close contact, aided by entrapped xyloglucan. (C) A computational simulation of two cellulose microfibrils (blue and green, shown in cross section) bonded together by a xyloglucan chain (red). (B after Park and Cosgrove 2012; C from Zhao et al. 2013.)

tethers. Instead, the results suggest that a quantitatively minor component of xyloglucan intertwines with cellulose to form structurally important junctions that control wall creep and mechanical extensibility. Moreover, expansin has been found to target a site with similar properties, that is, one containing both xyloglucan and cellulose with altered crystalline structure.

From these studies a new concept is emerging about the functional architecture of growing cell walls. The revised view posits a microscale network containing biomechanical "hot spots," which are limited junctions of bundled cellulose microfibrils where wall extensibility and mechanics are controlled (**Figure 14.20B and C**). In support of this idea, computational modeling shows that a monolayer of xyloglucan sandwiched between cellulose microfibrils could provide appreciable mechanical strength to cell walls. This "hot spot" model, like the previous models before it, must be considered a hypothesis in need of further testing, validation, and revision.

Many structural changes accompany the cessation of wall expansion

The growth cessation that occurs during cell maturation is generally irreversible and is typically accompanied by a reduction in wall extensibility, as measured by various biophysical methods. These physical changes in the wall might come about by (a) a reduction in wall-loosening processes, (b) an increase in wall cross-linking, or (c) an alteration in the composition of the wall, making for a more rigid structure or one less susceptible to wall loosening. There is some evidence for each of these ideas.

Several modifications of the maturing wall may contribute to wall rigidification:

- Newly secreted matrix polysaccharides may be altered in structure so as to form tighter complexes with cellulose or other wall polymers, or they may be resistant to wall-loosening activities.
- Removal of (1,3;1,4)-β-D-glucan in grass cell walls is coincident with growth cessation in these walls and may cause wall rigidification.
- De-esterification of pectins, leading to pectin gels that are more rigid, is similarly associated with growth cessation in both grasses and eudicots.
- Cross-linking of phenolic groups in the wall (such as tyrosine residues in HRGPs, ferulic acid residues attached to matrix polysaccharides, and lignin) generally coincides with wall maturation and is believed to be mediated by peroxidase, a putative wall rigidification enzyme.

Thus, many structural changes in the wall occur during and after cessation of growth, and it has not yet been possible to identify the significance of individual processes for cessation of wall expansion.

Secondary Cell Wall Structure and Function

The secondary cell wall (SCW) is a hierarchical structure assembled by some living cells inside the primary cell wall after wall expansion has ceased (**Figure 14.21A and B**). The best-studied SCWs are from cells that are highly lignified and dead at maturity, such as tracheids, xylem vessels, and fibers in woody tissues, but other notable examples include phloem and interfascicular fibers, stone cells, and epidermal cells such as cotton fibers, which are not lignified.

SCWs generally serve a structural, reinforcing role. In contrast to primary cell walls, which can extend dynamically, incorporate new materials, and withstand tensile forces generated by cell turgor, SCWs are structurally designed to resist compressive and tensile forces generated by gravity, external forces causing organ bending, and negative hydrostatic pressures arising during transpiration. One striking consequence of defects in SCW cellulose synthesis is xylem vessel collapse. The mechanical properties of SCWs are stable, enduring even after cell death, and are determined by wall architecture and physical interactions among cell wall polymers. Although SCW structure and function differ from those of more dynamic primary cell walls, the two probably share some principles of formation.

As mentioned previously, SCW cellulose in woody tissues is synthesized by a set of three CESAs different from the three CESAs used for primary cell wall cellulose. The significance of this fact for cellulose structure is unclear at this time, but possibly it has an impact on how cellulose synthase complexes function, either separately or in clusters to make macrofibrils. Another important distinction of SCWs is that their hemicelluloses have xylan and (gluco-) mannan backbones with low degrees of substitution (few side chains), whereas the hemicelluloses of primary cell walls are highly substituted. This difference has a major impact on hemicellulose properties, such as conformation, solubility, and binding to cellulose and probably has a substantial effect on packing of cellulose microfibrils in the cell wall.

Secondary cell walls are rich in cellulose and hemicellulose and often have a hierarchical organization

The best-studied SCWs consist of sequentially formed concentric layers named S_1, S_2, S_3, and so on, but the number of layers varies with cell type (see Figure 14.21B). SCWs with two or three layers are common in wood and fibers. Cellulose orientation is distinct for each layer, with the first-deposited layer (S_1) oriented in a shallow, nearly transverse helix, whereas cellulose in the thicker S_2 layer is oriented more longitudinally.

MACROFIBRIL FORMATION, STRUCTURE, AND ADHESION

SCW lamellae contain highly aligned cellulose microfibrils that are assembled into tightly packed macrofibrils,

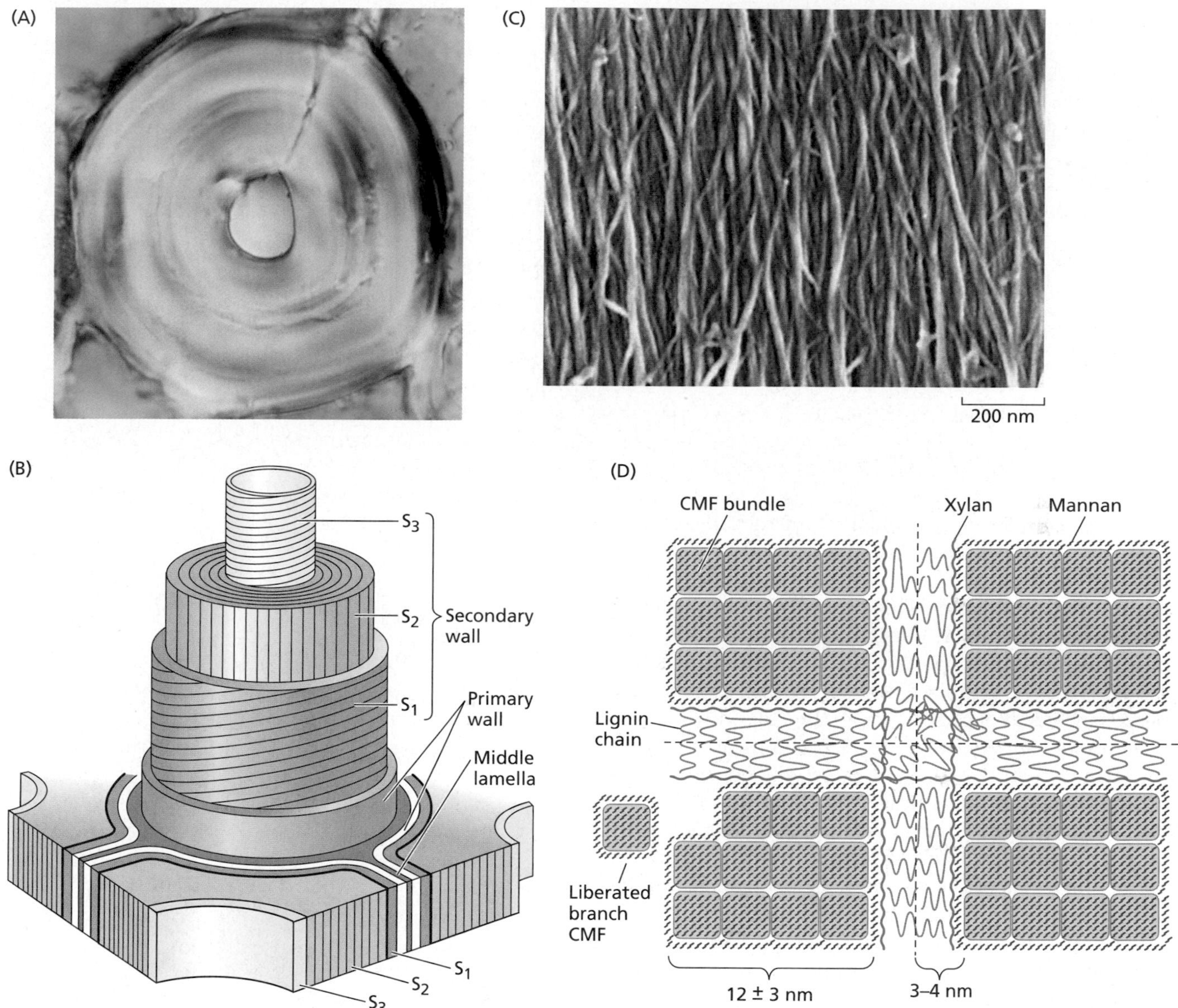

Figure 14.21 (A) Cross section of a *Podocarpus* sclereid, in which multiple layers in the secondary wall are visible. (B) Diagram of the cell wall organization often found in tracheids and other cells with thick secondary walls. Three distinct layers (S_1, S_2, and S_3) are formed interior to the primary wall. (C) Macrofibrils visible on the inner surface of a *Ginko* tracheid cell wall, as seen by field emission scanning electron microscopy. (D) A model of macrofibril structure and packing. Here the macrofibril is shown as a three-by-four array of elementary cellulose microfibrils (CMFs) that are tightly packed and coated with glucomannan. Binding the microfibrils together is a lignin–xylan layer. (A © David Webb; C from Terashima et al. 2004; D after Terashima et al. 2009.)

which in turn coalign with each other and are separated by hemicellulose and lignin (**Figure 14.21C**). Although molecular models have been central to primary cell wall studies for more than 40 years, molecular representations of SCWs have only recently been formulated and as yet have undergone relatively little testing and refinement. One model by Terashima and coworkers (**Figure 14.21D**) illustrates some basic concepts of macrofibril construction at the nanoscale, but details of the model are hypothetical, and alternative relationships among cellulose, hemicellulose, and lignin have been proposed.

The concept of the macrofibril as an aggregate of numerous individual microfibrils is based largely on high-resolution electron microscopy of de-lignified and partially deconstructed cell walls. The appearance of macrofibrils in the wall suggests that their formation is well organized and begins at the earliest stages of cellulose microfibril formation. One possibility is that clusters of cellulose synthesis

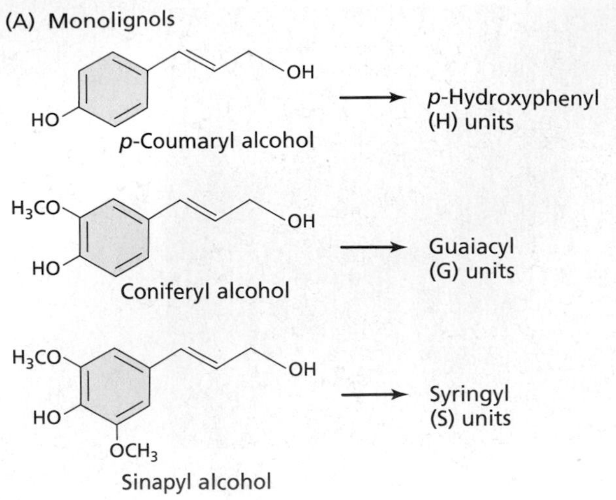

(A) Monolignols

p-Coumaryl alcohol → *p*-Hydroxyphenyl (H) units

Coniferyl alcohol → Guaiacyl (G) units

Sinapyl alcohol → Syringyl (S) units

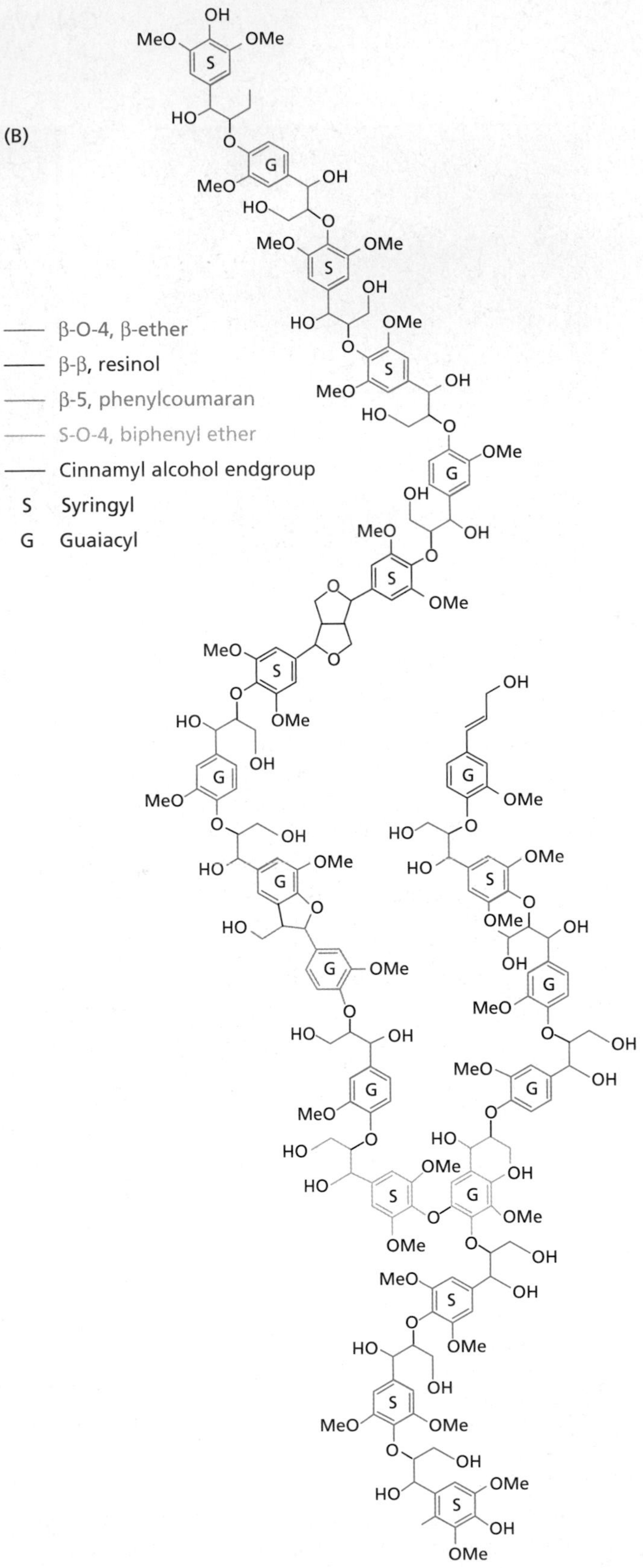

(B)

β-O-4, β-ether
β-β, resinol
β-5, phenylcoumaran
5-O-4, biphenyl ether
Cinnamyl alcohol endgroup
S Syringyl
G Guaiacyl

Figure 14.22 (A) Monolignols, which become the H, G, and S units of the lignin polymer, differ in the number of methoxy substituents on the phenolic ring. (B) Current model of the structure of poplar lignin, composed of S and G monolignol units that are cross-linked by free radicals generated by peroxidase and laccase. Note that this is one of billions of possible isomers. (B after Ralph et al. 2007.)

complexes—one for each elementary microfibril—coordinately produce microfibrils that align and coalesce immediately to form a macrofibril, with hemicellulose interaction occurring thereafter. This process may be mediated by auxiliary proteins. Although specific candidates have not yet been identified, proteins in the COBRA and KORRIGAN families might be involved, as their mutant phenotypes include reduced cell wall organization.

The model shown in Figure 14.21D indicates a structured arrangement of matrix polymers, with glucomannans coating the macrofibril surface, xylans positioned in the next layer, and lignin linking xylans and filling the remaining space between macrofibrils. Other authors suggest that lignin is intercalated and intertwined among hemicellulose chains. The details of macrofibril structure may differ for species with different hemicellulose composition.

Physical and computational studies indicate that macrofibrils are prevented from fusing into one massive cellulose crystal by water trapped between the constituent microfibrils. Misalignment and twisting of individual microfibrils may also help prevent such crystallization. In another study, macrofibril diameter was seen to vary according to cell type and lamella, correlating with lignin content, but nothing is known about the underlying controls for this process.

Lignification transforms the SCW into a hydrophobic structure resistant to deconstruction

SCWs are often lignified, a process that starts after SCW formation is well underway and can even continue after cell death, evidently in some cell types by metabolic contributions from neighboring, living cells. The major building blocks of lignin, called monolignols, are sinapyl and coniferyl alcohols, with minor amounts of *p*-coumaryl alcohol (**Figure 14.22A**). Monolignols are synthesized in the cell from phenylalanine via the phenylpropanoid pathway (see **WEB APPENDIX 4**).

Monolignols are exported across the plasma membrane, possibly by ABC transporters, to the cell wall where they undergo oxidative coupling, resulting in syringyl (S), guaiacyl (G), and *p*-hydroxphenyl (H) lignin units. The S unit is unbranched, whereas the G and H units are capable of forming branched structures. Lignin in most species is a mixture of all three units, but this can vary spatially and developmentally, as well as among species. Angiosperm lignin is mostly composed of G and S units, whereas gymnosperm lignin contains mostly G units. Grasses have slightly elevated levels of H units. Recent studies show that lignin polymerization is very flexible and can incorporate a variety of phenolic subunits.

Lignin formation involves the oxidative radical-mediated coupling of monolignols in the wall, catalyzed by peroxidases and laccases to form a random combinatorial polymer (**Figure 14.22B**). A great deal of work has characterized lignin structure, the monolignol biosynthesis pathway, and strategies for modifying this pathway to manipulate lignification.

In woody tissues, lignin polymerization usually starts at cell corners in the primary cell wall (including the middle lamella), then spreads progressively to the SCW lamellae. The basis for this pattern of lignification is not well understood, but it is generally speculated that nucleation sites exist in the pectin-rich middle lamella where lignification starts and that physical characteristics of the wall matrix may influence radical-based monolignol polymerization and cross-linking to wall polysaccharides.

A special case of lignification occurs in a narrow region of the root endodermal wall called the Casparian strip, which forms a hydrophobic barrier between the stele and the cortex. For many years the Casparian strip was thought to be composed of suberin, but recent advances have shown that it contains lignin that is polymerized in a very restricted part of the cell wall. Key factors controlling its synthesis include the CASP1 protein, which organizes membrane proteins at the Casparian strip, an NADPH oxidase that generates hydrogen peroxide, and a peroxidase that generates the monolignol radical intermediate (**Figure 14.23**). In addition, the wall

(A)

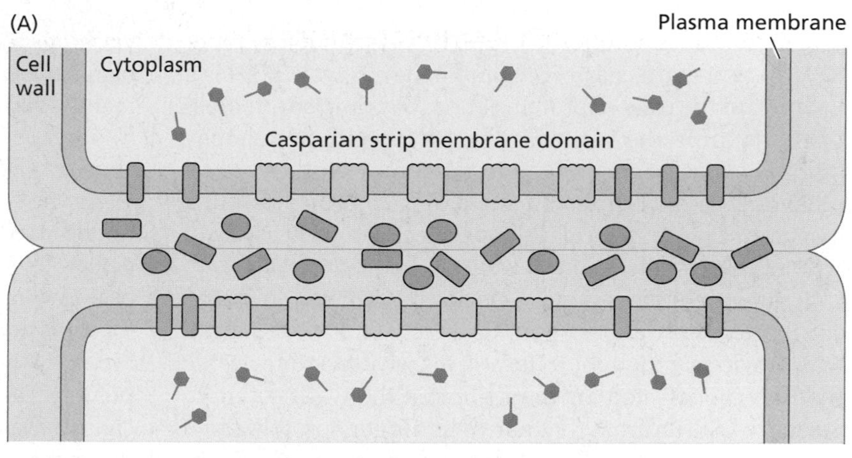

(B)

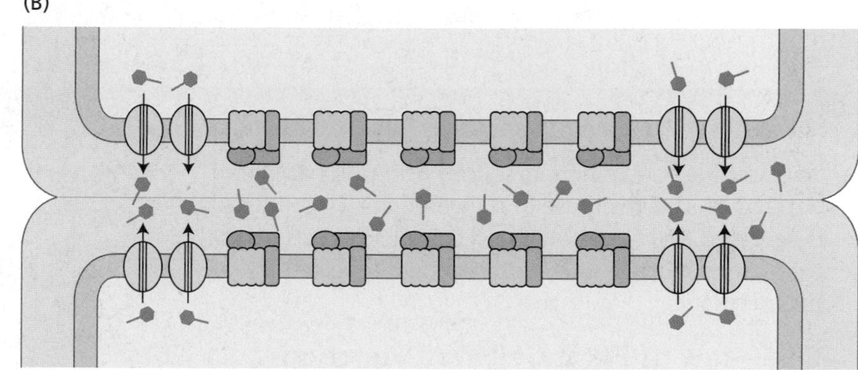

(C)

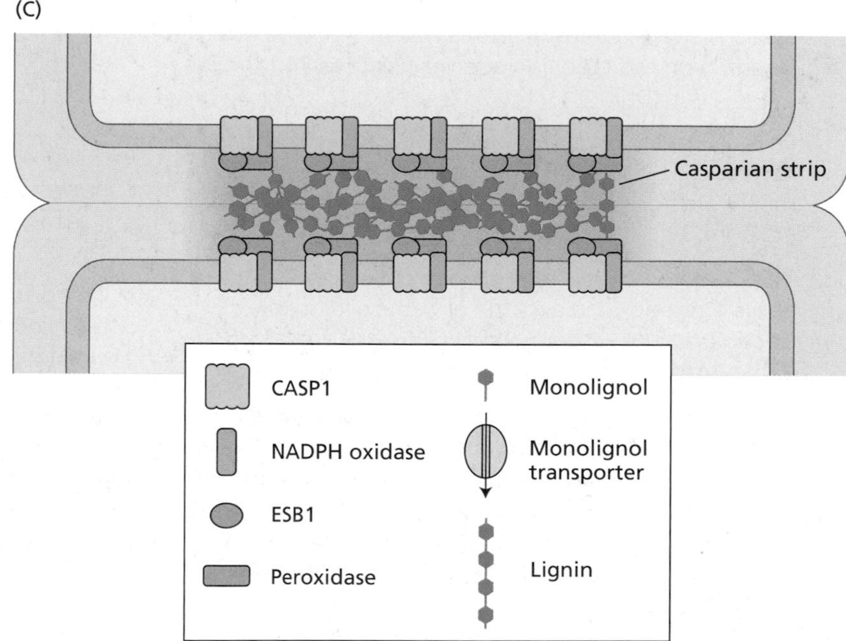

Figure 14.23 Schematic representation of Casparian strip deposition. (A) CASPs are initially distributed uniformly around the plasma membrane but soon aggregate at a central domain, referred to as the Casparian strip membrane domain (CSD). (B) NADPH oxidase and peroxidase are recruited to the CSD, and monolignols are exported in a nontargeted process to the apoplast. (C) Lignin polymerization occurs exclusively in the cell wall adjacent to the CSD because the enzymes are localized there. (After Roppolo and Geldner 2012.)

protein **Enhanced Suberin 1** (**ESB1**) is essential for proper lignification in this narrow region of the cell wall. ESB1 is a member of the class of proteins known as **dirigent-domain proteins** (from the Latin *dirigere*, "to direct"), which may guide the stereochemistry of a compound synthesized by other enzymes. The exact function of ESB1 is unclear, but it may nucleate lignin formation specifically in the Casparian strip of the endodermal cell wall.

Although lignification is associated with wall strengthening, the physical basis for this effect is not clear. Lignin was previously thought to form a massive macromolecule that interpenetrated and cross-linked the wall, but more recent results indicate that native lignin (or "protolignin") is smaller than generally believed. Notably, the S-rich lignin of transgenic poplar overexpressing ferulate 5-hydroxylase has a degree of polymerization of only 10, yet the plants appear phenotypically normal. Technical obstacles make it difficult to assess the extent of lignin cross-linking in other cell walls, but extensive cross-linking does not appear to be essential for wood formation in poplar.

As the secondary cell wall becomes lignified, water is removed and is replaced by the hydrophobic lignin molecules. This tends to enhance noncovalant interactions between lignin and polysaccharides, perhaps accounting for some of the wall strengthening. There is also evidence for extensive covalent linking between lignin and wall polysaccharides, but these linkages have been difficult to characterize in detail. In grass cell walls, lignin–carbohydrate linkages are largely via ferulate groups attached to arabinose residues in GAX (see Figure 14.11C).

SUMMARY

The architecture, mechanics, and function of plants depend on the structure of the cell wall. The wall is secreted and assembled as a complex structure that varies in form and composition as the cell differentiates.

Overview of Plant Cell Wall Functions and Structures

- Cell walls vary greatly in form and composition, depending on cell type and species (**Figures 14.1–14.3**).
- Primary cell walls are synthesized in actively growing cells, whereas secondary cell walls are deposited in certain cells, such as xylem vessel elements and sclerenchyma fibers, after cell expansion ceases (**Figures 14.2**, **14.3**).
- The primary cell wall is a network of cellulose microfibrils embedded in a matrix of hemicelluloses, pectins, and structural proteins (**Figures 14.4**, **14.5**; **Table 14.1**).
- Cellulose microfibrils are highly ordered arrays of glucan chains synthesized at the surface of the cell by protein complexes called cellulose synthase complexes. These resette-like structures contain three to six units of cellulose synthase that associate with each other to form a hexameric subunit (**Figures 14.6**, **14.7**).
- Matrix polysaccharides are synthesized in the Golgi apparatus and secreted via vesicles (**Figure 14.8**).
- Pectins form hydrophilic gels that can become cross-linked by calcium ions, and hemicelluloses bind microfibrils together (**Figures 14.9–14.11**).
- Secondary walls in woody tissues typically contain xylans and glucomannans instead of xyloglucan and pectin.

Primary Cell Wall Structure and Function

- Wall assembly occurs partly by spontaneous self-assembly but may also be mediated by enzymes. Xyloglucan endotransglucosylase has the ability to carry out transglycosylation reactions that integrate newly synthesized xyloglucans into the wall (**Figure 14.12**).

Mechanisms of Cell Expansion

- Wall expansion may be highly localized (tip growth) or more dispersed over the wall surface (diffuse growth) (**Figure 14.13**).
- In diffuse-growing cells, the orientation of cell growth is determined by the orientation of cellulose microfibrils, which is determined by the orientation of microtubules in the cytoplasm (**Figures 14.14**, **14.16**).
- Complicated cell growth patterns, such as those like the "jigsaw" pattern present in the leaf epidermis of eudicots, involve GTP-binding proteins that organize the cytoskeletal elements, thereby directing the local pattern of wall growth (**Figure 14.15**).

The Extent and Rate of Cell Growth

- Biochemical loosening of the cell wall leads to wall stress relaxation, which dynamically links water uptake with cell wall expansion in the growing cell.
- The actions of hormones (such as auxin and gibberellin) and environmental conditions (such as light and water availability) modulate cell expansion by altering wall extensibility or the yield properties of the wall.
- Acid-induced cell wall extension is characteristic of walls in growing cells and is mediated by the protein expansin, which loosens the noncovalent adhesions between wall polysaccharides (**Figures 14.18**, **14.19**).

- Cessation of cell growth during cell maturation involves multiple mechanisms of cell wall cross-linking and rigidification.

Secondary Cell Wall Structure and Function

- Secondary cell walls are typically thick layers deposited between the plasma membrane and the primary cell wall. They add strength and compression resistance to stems and other organs.
- The secondary cell walls of woody tissues are comprised of two or more layers containing cellulose, hemicellulose and lignin.
- Lignin is formed within the wall by oxidative coupling of monolignols into a random polymer of phenolic subunits. It locks the secondary cell wall into a hydrophobic material that is resistant to enzymatic deconstruction (**Figures 14.21**, **14.22**).

WEB MATERIAL

- **WEB TOPIC 14.1 Terminology for Polysaccharide Chemistry** A brief review of terms used to describe the structures, bonds, and polymers in polysaccharide chemistry is provided.
- **WEB TOPIC 14.2 Matrix Components of the Cell Wall** The secretion of xyloglucan and glycosylated proteins by the Golgi apparatus can be demonstrated at the ultrastructural level.
- **WEB TOPIC 14.3 The Mechanical Properties of Cell Walls: Studies with *Nitella*** Experiments have demonstrated that the inner 25% of the cell wall determines the directionality of cell expansion.
- **WEB ESSAY 14.1 Calcium Gradients and Oscillations in the Growing Pollen Tube** Calcium plays a role in regulating pollen tube tip growth.
- **WEB ESSAY 14.2 Microtubules, Microfibrils, and Growth Anisotropy** The orientations of microtubules or microfibrils are not always correlated with the directionality of growth.
- **WEB ESSAY 14.3 Biophysical Coordination of Water Uptake and Cell Wall Enlargement** A physical model provides a quantitative framework for relating the physics of water absorption to wall extension and for assessing the limiting physical factors for cell growth.

available at **plantphys.net**

Suggested Reading

Albersheim, P., Darvill, A., Roberts, K., Sederoff, R., and Staehelin, A. (2011) *Plant Cell Walls.* Garland Science, New York.

Baskin, T. I. (2005) Anisotropic expansion of the plant cell wall. *Annu. Rev. Cell Dev. Biol.* 21: 203–222.

Boerjan, W., Ralph, J., and Baucher, M. (2003) Lignin biosynthesis. *Annu. Rev. Plant Biol.* 54: 519–546.

Cosgrove, D. J. (2005) Growth of the plant cell wall. *Nat. Rev. Mol. Cell Biol.* 6: 850–861.

Cosgrove, D. J., and Jarvis, M. C. (2012) Comparative structure and biomechanics of plant primary and secondary cell walls. *Front. Plant Sci.* 3: 204.

Lu, F., and Ralph, J. (2010) Lignin. In *Cereal Straw as a Resource for Sustainable Biomaterials and Biofuels*, R. C. Sun ed., Elsevier, Amsterdam, pp. 169–207.

Mohnen, D. (2008) Pectin structure and biosynthesis. *Curr. Opin. Plant Biol.* 11: 266–277.

Paredez, A. R., Somerville, C. R., and Ehrhardt, D. W. (2006) Visualization of cellulose synthase demonstrates functional association with microtubules. *Science* 312: 1491–1495.

Plomion, C., Leprovost, G., and Stokes, A. (2001). Wood Formation in Trees. *Plant Physiol.* 127: 1513–1523.

Sampedro, J., and Cosgrove, D. J. (2005) The expansin superfamily. *Genome Biol.* 6: 242.

Waldron, K. W., and Brett, C. T. (2007) The role of polymer cross-linking in intercellular adhesion. In *Plant Cell Separation and Adhesion,* J. Roberts and Z. Gonzalez-Carranza eds., Blackwell, Oxford, pp. 183–204.

Zhong, R., and Ye, Z. H. (2007) Regulation of cell wall biosynthesis. *Curr. Opin. Plant Biol.* 10: 564–572.

15 Signals and Signal Transduction

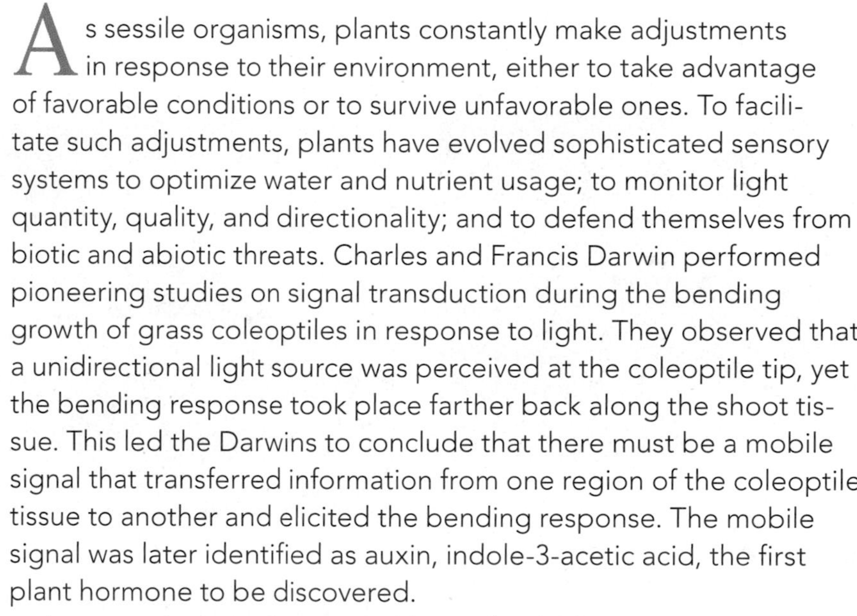

As sessile organisms, plants constantly make adjustments in response to their environment, either to take advantage of favorable conditions or to survive unfavorable ones. To facilitate such adjustments, plants have evolved sophisticated sensory systems to optimize water and nutrient usage; to monitor light quantity, quality, and directionality; and to defend themselves from biotic and abiotic threats. Charles and Francis Darwin performed pioneering studies on signal transduction during the bending growth of grass coleoptiles in response to light. They observed that a unidirectional light source was perceived at the coleoptile tip, yet the bending response took place farther back along the shoot tissue. This led the Darwins to conclude that there must be a mobile signal that transferred information from one region of the coleoptile tissue to another and elicited the bending response. The mobile signal was later identified as auxin, indole-3-acetic acid, the first plant hormone to be discovered.

In general, an environmental input that initiates one or more plant responses is referred to as a signal, and the physical component that biochemically responds to that signal is designated a receptor. Receptors are either proteins or, in the case of light receptors, pigments associated with proteins. Once receptors sense their specific signal, they must *transduce* the signal (i.e., convert it from one form to another) in order to amplify the signal and trigger the cellular response. Receptors often do this by modifying the activity of other proteins or by employing intracellular signaling molecules called **second messengers**; these molecules then alter cellular processes such as gene transcription. Hence, all signal transduction pathways typically involve the following chain of events:

Signal → receptor → signal transduction → response

In many cases the initial response is the production of secondary signals, such as hormones, which are then transported to the site of action to evoke the main physiological response. Many of the specific events and intermediate steps involved in plant signal transduction have now been identified, and these intermediates constitute the **signal transduction pathways**.

We will begin this chapter by providing a brief overview of the types of external cues that direct plant growth. Next we will discuss how plants employ signal transduction pathways to regulate gene expression and post-translational responses. A surprising discovery has been that, in the majority of cases, plant signal transduction pathways function by inactivating, degrading, or removing repressor proteins that modulate transcription. Signal amplification via second messengers is required, as well as mechanisms for signal transmission to coordinate responses throughout the plant. Finally, we will examine how individual stimulus-response cascades are often integrated with other signaling pathways, termed *cross regulation*, to shape plant responses to their environment in time and space.

Temporal and Spatial Aspects of Signaling

Plant signal transduction mechanisms may be relatively rapid or extremely slow (**Figure 15.1**). When some carnivorous plants, most notably Venus flytrap (*Dionaea muscipula*), catch insects, they use modified leaf traps that close within milliseconds after touch stimulation. Similarly, the sensitive plant (*Mimosa pudica*) folds its leaflets rapidly upon being touched. Young seedlings reorient them-

Figure 15.1 Timing of plant responses to the environment ranges from very rapid to extremely slow. (A) Insect movements on modified leaves of a Venus flytrap (*Dionaea muscipula*) activate trigger hairs, inducing rapid closure of the leaf lobes. (B) The leaves of sundew plants (*Drosera anglica*) capture insects in a sticky fluid produced by stalked glands called tentacles, then roll up to secure the prey and begin digestion. (C) Hawthorn tree (*Crataegus* spp.) subjected to prevailing onshore winds responds slowly by growing away from the wind. (D) Tree trunks and branches can respond slowly to mechanical stress by producing reaction wood. In this case the tree is an angiosperm, which produces *tension wood* on the upper surface. Gymnosperms produce *compression wood* on the lower surface. (E) Cross section through a gymnosperm tree branch with compression wood, creating an asymmetrical ring structure.

selves with respect to gravity minutes after being placed horizontally. In general, such rapid response mechanisms involve electrochemical responses to transduce signals, since gene transcription and protein translation mechanisms are too slow. In contrast, plants attacked by insect herbivores may emit volatiles to attract insect predators within a few hours. Processes occurring on this timescale often involve new transcription and translation activity (see Chapter 2).

Longer-term environmental responses modify developmental programs to shape plant architecture over the entire life of the plant. Examples of long-term responses include modulation of root branching in response to nutrient availability, growth of sun or shade leaves to adjust for light conditions, and activation of lateral bud outgrowth when the shoot apex is damaged by grazing herbivores. Long-term plant responses can operate over timescales of months or years. For example, a long period of low temperature, termed *vernalization*, is required by many plant species for flowering to occur (see Chapter 20). Chromatin remodeling is often involved in such long-term responses (see Chapter 2).

Plant responses to environmental signals also differ spatially. In a **cell autonomous response** to an environmental signal, both signal reception and response occur in the same cell. In contrast, a **non–cell autonomous response** is one in which signal reception occurs in one cell and the response occurs in distal cells, tissues, or organs. An example of cell autonomous signaling is the opening of guard cells, where blue light activates membrane ion transporters to swell guard cells via the phototropin blue-light photoreceptors (see Chapters 10 and 16). An example of non–cell autonomous signaling in the same organs would be the formation of additional stomata when mature leaves are exposed to high light intensity in a process that requires transmission of information from one organ to another (see Chapter 19).

Signal Perception and Amplification

Although highly varied in nature and makeup, all signal transduction pathways share common features: an initial stimulus is perceived by a receptor and transmitted via intermediate processes to sites where physiological responses are initiated (**Figure 15.2**). The stimulus may derive from developmental programming or from the external environment. When the response mechanism reaches an optimal point, feedback mechanisms attenuate the processes and reset the sensor mechanism.

Receptors are located throughout the cell and are conserved across kingdoms

Receptors can be located at the plasma membrane, cytosol, endomembrane system, or nucleus, as exemplified by hormone and touch receptors (**Figure 15.3**). In some cases, receptors move from one compartment to another. Many plant receptors resemble those found in bacterial systems. For example, homologs of the bacterial mechanosensitive ion channel, **MscS** (***M**echanosensitive **c**hannel of **S**mall* **conductance**), are found both in the plasma membrane and chloroplast envelope (probably the inner membrane) of plant cells. Mechanosensitive channels act as stretch receptors and help cells and plastids adjust to osmotically induced swelling. Plant receptors that perceive the presence of the hormones cytokinin and ethylene, described later in this chapter, are derived from bacterial "two-com-

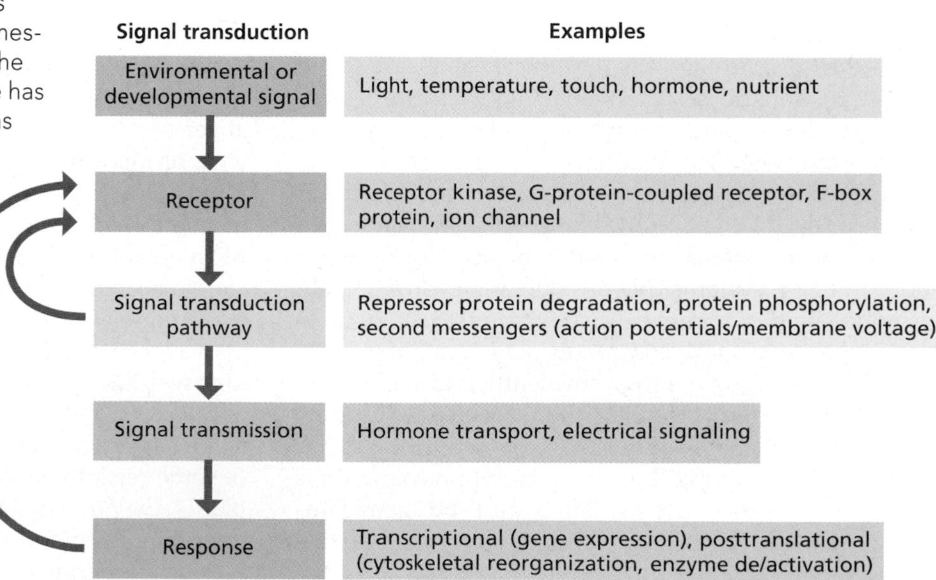

Figure 15.2 General scheme for signal transduction. Environmental or developmental signals are perceived by specialized receptors. A signaling cascade is then activated that involves second messengers and leads to a response by the plant cell. When an optimal response has been achieved, feedback mechanisms attenuate the signal.

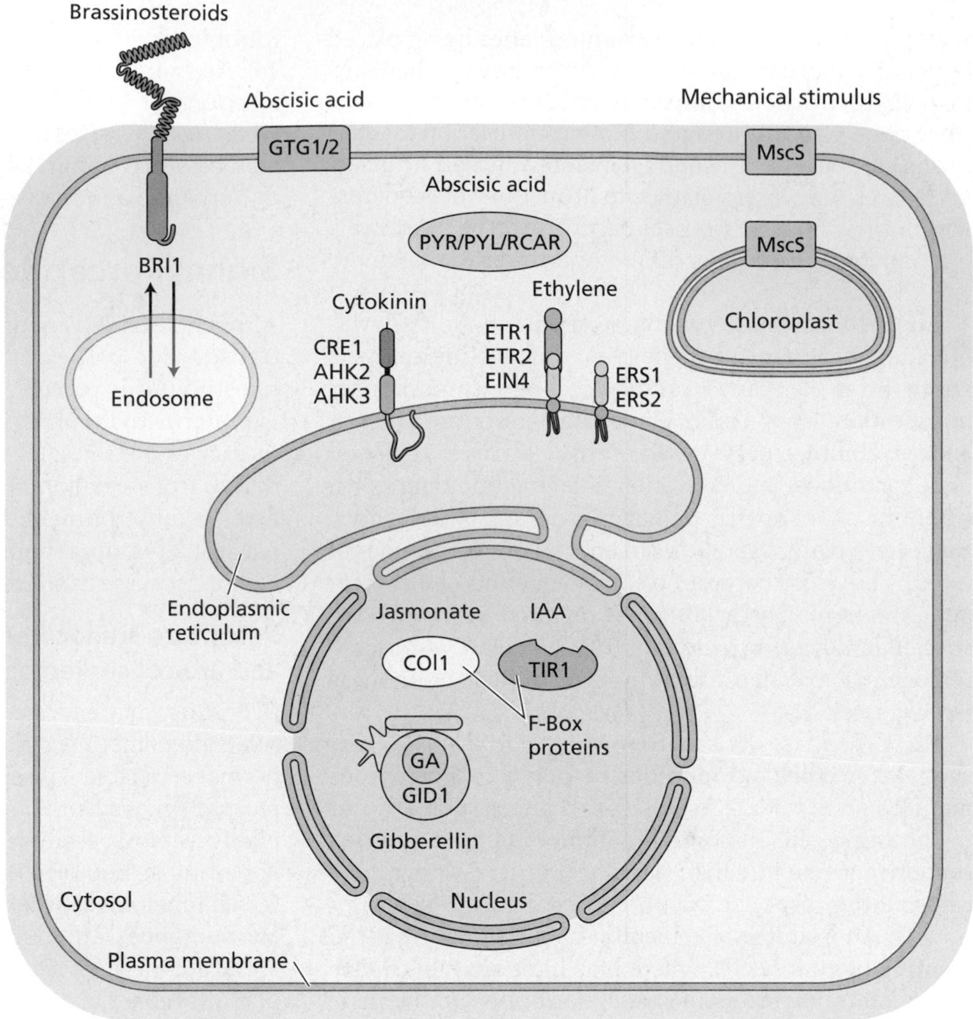

Figure 15.3 Primary locations of plant hormone receptors and mechanosensitive receptors (MscS) in the cell. The individual receptors are discussed later in the chapter. (After Santer and Estelle 2009.)

ponent" systems. Several plant photoreceptors diverged from similar proteins in bacteria and have taken on new functions. For example, bacterial members of the cryptochrome/photolyase superfamily are flavoproteins that repair pyrimidine dimers produced in DNA by UV light. In plants, cryptochromes lack the critical residues required for DNA repair, and instead mediate light control of stem elongation, leaf expansion, photoperiodic flowering, and the circadian clock (see Chapter 16).

Other plant receptors are more similar to those found in animals and fungi, but often have additional or modified components. Examples are found in plant F-box receptor/ubiquitin ligase systems that are integral to several plant hormone receptor complexes (see Figure 15.3). Eukaryotic E3 ubiquitin ligase complexes, which are present in both the cytosol and nucleus, covalently attach ubiquitin to substrate proteins, tagging them for degradation by the 26S proteasome. In the SCF (*S*kp, *C*ullin, and *F*-box protein) subfamily of E3 ligases, substrate recognition is mediated by **F-box proteins**. The plant F-box gene family has greatly expanded in plants to accommodate this expansion in function.

A **kinase** is an enzyme that catalyzes phosphorylation—that is, the addition of a phosphate group from ATP to a substrate, such as a protein, thus modifying its properties. When a protein functions as a receptor and transduces that signal by phosphorylating another molecule, it is called a **receptor kinase**. Depending on the type of receptor kinase, a target protein can be phosphorylated at various amino acid residues (serine, threonine, tyrosine, or histidine) to alter its biological activity. Receptor kinases, which function in diverse animal signaling mechanisms, have a limited, but important, role in plants. Most notable of these is the receptor system for brassinosteroid hormones, wherein the BRI1 receptor kinase plays a central role in development (see Figure 15.3). There are also a large number of **receptor-like serine/threonine kinases (RLKs)** in plants compared to other kingdoms, and RLKs play a prominent role in plant–pathogen interactions (see Chapter 23). However, although components of some receptor systems found in animals are found in plants, they may not participate in analogous functions. For example, animals systems contain a large number of plasma membrane G protein–coupled receptors (GPCRs) that detect a diverse array of extracellular signals, ranging from hormones to odors and flavors, and signal via a large family of heterotrimeric G proteins. Although plants possess a small number of heterotrimeric G proteins, no analogous GPCR function has been clearly demonstrated in plants to date.

Signals must be amplified intracellularly to regulate their target molecules

If a receptor is considered the gateway through which a signal enters a signaling network, the receptor location to some extent prescribes the length of the subsequent signaling pathway; such pathways can consist either of a few signaling steps or an elaborate cascade of signaling events. Perception of signals at the plasma membrane often activates transduction pathways with many intermediates. In the case of signaling pathways that must eventually reach the nucleus to regulate gene expression, signal strength along the pathway will dissipate unless it is reinforced by signal amplification events. In the absence of amplification, any activated signaling intermediate that must traverse the cytosol to translocate to the nucleus will become diluted due to diffusion and deactivation (for example, by dephosphorylation, degradation, or sequestration). Furthermore, many chemical signals are present at very low concentrations, and receptors can similarly occur at very low density, such that the initial signal may be quite weak. Signal amplification cascades serve to maintain or even enhance signal strength over larger distances. To elevate weak initial signaling events above the threshold of detection or to propagate them across the cytoplasm, cells employ amplification mechanisms such as phosphorylation cascades and second messengers.

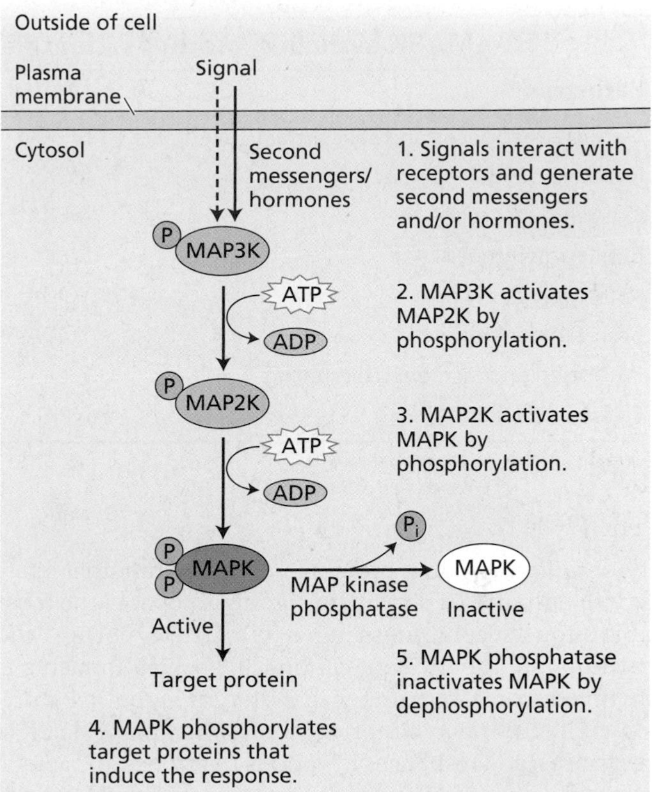

Figure 15.4 Mitogen-activated protein kinase (MAPK) pathways amplify signals to achieve a rapid and massive response to an environmental or developmental stimulus. The continuous and dotted arrows leading from the signal indicate direct and indirect activation, respectively. See Table 15.1 for specific MAP kinase intermediates in plants.

The MAP kinase signal amplification cascade is present in all eukaryotes

The **MAP (*m*itogen-*a*ctivated *p*rotein) kinase cascade** plays an important role in signal amplification in plants and other eukaryotes. The MAP kinase cascade owes its name to a series of protein kinases (signaling modules) that phosphorylate each other in a specific sequence, much like runners passing a baton in a relay race. MAP kinase cascades are phylogenetically ancient and conserved signaling modules that are involved in many important signaling pathways, including those regulating hormone, abiotic stress, and defense responses. Often, elements of a MAPK cascade represent points of convergence for several different signaling pathways. The first kinase in the sequence is a MAP kinase kinase kinase (MAP3K). When the MAP3K is activated by a receptor, it phosphorylates MAP kinase kinase (MAP2K), which phosphorylates MAP kinase (MAPK). MAPK, the "anchor" of the relay team, phosphorylates specific transcription factors and regulatory proteins, which causes changes to gene expression (**Figure 15.4**). Several MAPK signaling modules have been identified in plants, many of them related to stress responses (**Table 15.1**). The role of MAP kinases in plant responses to abiotic stress will be discussed in Chapter 24.

In the MAP kinase cascade, each kinase that is phosphorylated may modify the activity of many more of its own target proteins. A signaling cascade composed of several kinases is therefore theoretically able to alter the phosphorylation status (and hence activity) of thousands of target proteins in response to relatively few ligand molecules originally binding the receptor at the plasma membrane. However, the number of such interactions is likely to be much more limited if MAPKs are assembled into complexes by scaffolding proteins. Such MAPK modules have been identified in animal MAPK signaling pathways, where they are thought to promote response specificity.

Ca^{2+} is the most ubiquitous second messenger in plants and other eukaryotes

Second messengers—small molecules and ions that are rapidly produced or mobilized at relatively high levels after signal perception, and which can modify the activ-

Table 15.1 MAPK signaling modules identified in plants

Pathways	MAP3K	MAP2K	MAPK
Defense responses and salicylic acid synthesis	MEKK1	MKK1/2	MPK4
Reactive oxygen species homeostasis	MEKK1	MKK1/2	MPK4
Cold and salt stress	MEKK1	MKK2	MPK4/6
Ethylene synthesis	MEKK	MKK4/5	MPK3/6
Pathogen signaling	YODA	MKK4/5	MPK3/6
Stomata development	YODA	MKK4/5	MPK3/6
Pathogen and jasmonate signaling	?	MKK3	MPK1/2/7/14
Cytokinesis	NPK1	NtMEK2	Ntf6

Source: Suarez Rodriguez et al. 2010.

ity of target signaling proteins—represent another strategy to enhance or propagate signals. Probably the most ubiquitous second messenger in all eukaryotes is the calcium ion, the divalent cation Ca^{2+}, which in plants is involved in a vast number of different signaling pathways, including symbiotic interactions, plant defense responses, and responses to various hormones and abiotic stresses. Cytosolic Ca^{2+} levels increase rapidly when Ca^{2+}-permeable ion channels open to allow passive Ca^{2+} influx from Ca^{2+} stores into the cytosol (**Figure 15.5**). Channel activity must be tightly regulated to maintain precise control over the timing and duration of cytosolic Ca^{2+} elevation. Generally, ion channels are gated, meaning the channel pores are opened or closed by changes in transmembrane electrical potential, membrane tension, posttranslational modification, or binding of a ligand. Several families of Ca^{2+}-permeable channels have been identified in plants; these include plasma membrane–localized glutamate-like receptors (GLRs) and cyclic nucleotide-gated channels (CNGCs). Electrophysiological and other evidence supports the presence of Ca^{2+}-permeable channels at the tonoplast and the endoplasmic reticulum (ER).

Once receptor-mediated signaling activates Ca^{2+}-permeable channels, Ca^{2+} sensor proteins play a pivotal role as signaling intermediaries, linking Ca^{2+} signals to changes in cellular activities. Most plant genomes contain four major multigene families of Ca^{2+} sensors: the calmodulin (CaM) and calmodulin-like proteins, the Ca^{2+}-dependent protein kinases (CDPKs), Ca^{2+}/calmodulin-dependent protein kinases (CCaMKs), and calcineurin-B like proteins (CBLs), which function in concert with CBL-interacting protein kinases (CIPKs). Members of these sensor families modulate the activity of target proteins either by binding to (CaM) or by phosphorylating (CDPK, CCaMK, CBL/CIPK) the target protein in a Ca^{2+}-dependent manner (see Figure 15.5). Target proteins include transcription factors, various protein kinases, Ca^{2+}-ATPases, enzymes producing reactive oxygen species (ROS), and ion channels. Finally, Ca^{2+} pumps and Ca^{2+} exchangers in organelle and plasma membranes actively remove Ca^{2+} from the cytosol to terminate Ca^{2+} signaling (see Figure 15.5).

Changes in the cytosolic or cell wall pH can serve as second messengers for hormonal and stress responses

Plant cells use the **proton motive force** (i.e., the electrochemical proton gradient) across cellular membranes to drive ATP synthesis (see Chapters 7 and 12) and to energize secondary active transport (see Chapter 6). In addition to having such "housekeeping" activity, protons also appear to have signaling activity and function as second messengers. In a resting cell, cytosolic pH is typically kept constant at approximately pH 7.5, whereas the cell wall is acidified to pH 5.5 or lower. Extracellular pH can change rapidly in response to a variety of different endogenous and environmental signals, while intracellular pH changes occur more slowly because of cellular buffering capacity. In growing hypocotyls, for example, the plant hormone auxin triggers activation of the plasma membrane H^+-ATPase through phosphorylation of its C-terminus. This causes the cell wall to become more acidic, which is thought to promote cell expansion by activating cell wall–loosening enzymes such as expansins (see Chapter 14). In roots, however, where auxin inhibits cell expansion, auxin triggers the rapid alkalinization of the cell wall, a process that has been shown to be Ca^{2+}-dependent. Similar Ca^{2+}-dependent pH changes are observed in many environmental stress responses of plants (see Chapter 24).

Which transporters are activated or deactivated by Ca^{2+} to facilitate extra- and intracellular pH changes is currently unknown, as are, for the most part, the downstream targets of these pH changes. Certainly, the pH of the cell wall will affect the protonation status of weakly acidic small molecules such as plant hormones and consequently affect their ability to enter cells by diffusion.

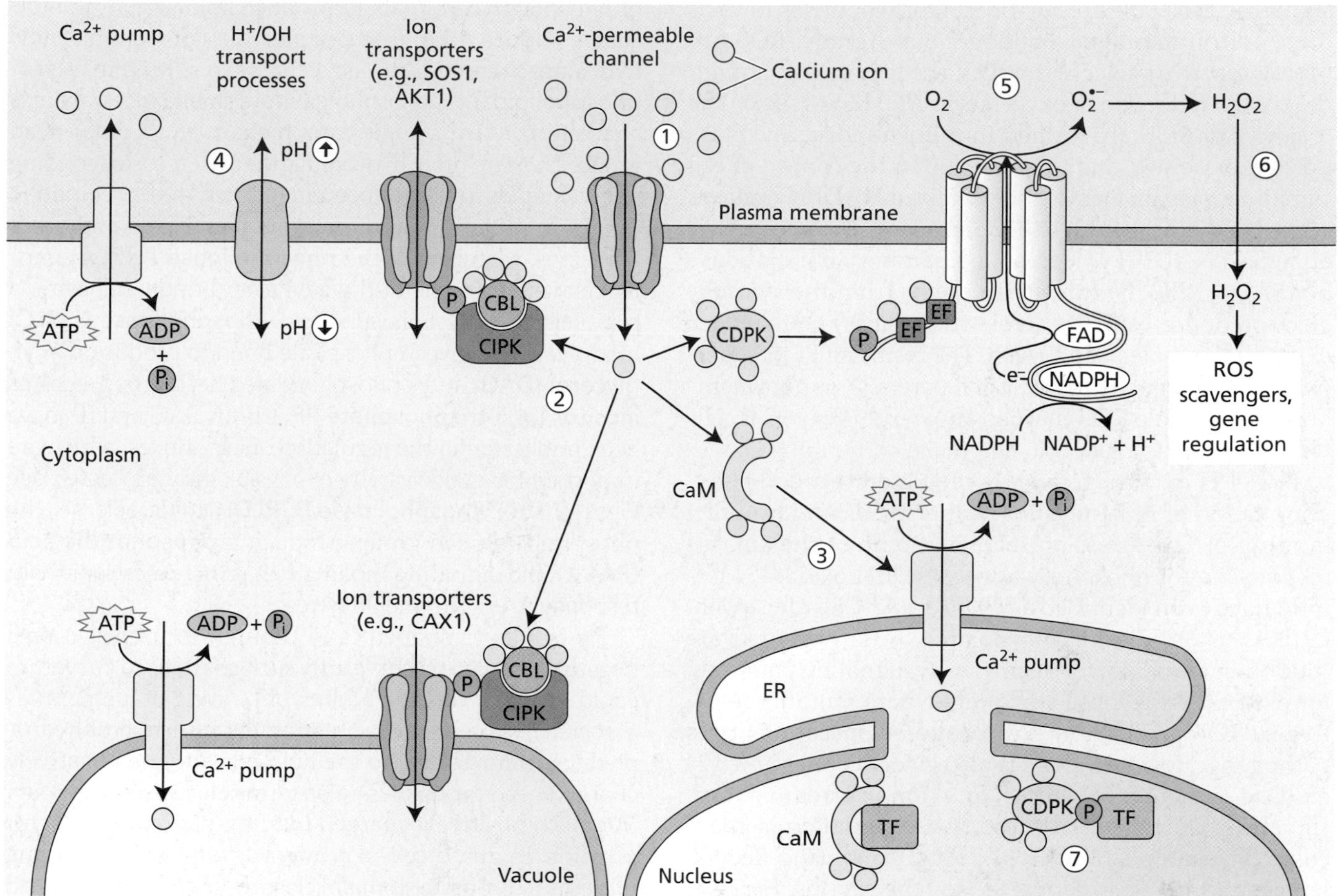

1. Signal-induced activation of Ca^{2+}-permeable ion channels leads to an increase in the concentration of free cytosolic Ca^{2+}.

2. Activated CBL/CIPK Ca^{2+} sensor proteins interact with ion transporters on the plasma membrane and the vacuolar membrane.

3. Activated calmodulin (CaM) stimulates Ca^{2+} pumps on the ER.

4. Ca^{2+} activation of membrane transporters triggers changes in the intracellular and extracellular pH.

5. Plasma membrane NADPH oxidases are activated by direct binding of Ca^{2+} to N-terminal EF-hand Ca^{2+} binding domains or by N-terminal phosphorylation by CDPK, promoting the formation of superoxide. Superoxide can dismutate via superoxide dismutase (SOD) to the ROS H_2O_2.

6. Diffusion of H_2O_2 into the cytosol alters the cellular redox status, which may regulate the activity of transcription factors and change gene expression.

7. Activated CDPK can regulate transcription factors in the nucleus.

Figure 15.5 Calcium ions, pH, and ROS function as second messengers that amplify signals and regulate the activity of target signaling proteins to trigger physiological responses. An increase in the $[Ca^{2+}]_{cyt}$ activates calcium sensor proteins (calmodulins [CaMs], Ca^{2+}-dependent protein kinases [CDPKs], and calcineurin-B like proteins/CBL-interacting protein kinases [CBL/CIPKs]), which are located at different subcellular sites.

Cell wall pH regulation may thus represent a mechanism to fine-tune hormone uptake and signaling. Evidence also exists for pH-dependent gating of potassium ion channels and aquaporins. What has made identification of pH signaling targets difficult is that the presence of acidic and basic amino acids makes all proteins sensitive to pH. Whether this sensitivity is physiologically relevant depends on the pK_a values (dissociation constants) of these amino acids and on how critical their protonation status is to the protein's ability to interact with other proteins, substrates, or ligands.

Reactive oxygen species act as second messengers mediating both environmental and developmental signals

In recent years, **reactive oxygen species** (**ROS**) have emerged not just as cytotoxic by-products of metabolic processes such as respiration and photosynthesis, but as signaling molecules regulating plant responses to various environmental and endogenous signals. ROS are highly reactive molecules which are generated through the partial reduction of oxygen (see **WEB ESSAY 12.7**). The majority of ROS are formed in mitochondria and plastids, peroxisomes, and the cell wall. In the context of cell signaling, plasma membrane–localized NADPH oxidases make up the best-understood family of ROS-producing enzymes. NADPH oxidases (or respiratory burst oxidase homologs, RBOHs) transfer electrons from the cytosolic electron donor NADPH across the membrane to reduce extracellular molecular oxygen. The resulting ROS, superoxide, can dismutate to hydrogen peroxide, a more membrane-permeable ROS that can apparently also enter cells through specific aquaporin channels.

NADPH oxidase activity is regulated through phosphorylation of its N-terminal amino acids and by direct binding of Ca^{2+} (see Figure 15.5). Some of the kinases responsible for phosphorylating NADPH oxidase N termini have been identified as CDPKs and CBL-dependent CIPKs. NADPH oxidase–mediated oxidative bursts are thus often found downstream of Ca^{2+} signaling pathways, for example in defense signaling, where mutants defective in ROS production exhibit altered susceptibility to pathogens. However, there is also evidence that NADPH oxidase–generated ROS can function upstream of Ca^{2+} signaling. During abscisic acid (ABA) signaling in guard cells, for example, a Sucrose Non-Fermenting Related Kinase2 (SnRK2) is activated (see the section *Hormone Signaling Pathways*) and subsequently phosphorylates and activates the NADPH oxidase RBOHF. The resulting production of ROS appears to signal the influx of Ca^{2+} through Ca^{2+}-permeable plasma membrane ion channels.

Targets of ROS signaling are just beginning to be identified. The thiol side chain of cysteine amino acid residues, in particular, can be oxidatively modified to form intramolecular (within the polypeptide/protein) or intermolecular (oxidative cross-linking of different [poly]peptides/proteins) disulfide bonds. Direct redox regulation has been shown to alter the DNA-binding activity or cellular localization of several transcription factors and transcriptional activators. In the cell wall, tyrosine residues of structural proteins, feruloyl (ferulic acid) conjugates of polysaccharides, and monolignols are all potential targets of ROS that may be oxidatively cross-linked to modify the strength or the barrier properties of the cell wall.

Lipid signaling molecules act as second messengers that regulate a variety of cellular processes

Phosphoglycerolipids and sphingolipids are primary lipid components of plant plasma membranes and are important determinants of their physical properties (e.g., membrane surface charge, fluidity, local membrane curvature). Several phospholipase enzymes hydrolyze specific bonds of phosphoglycerolipids to produce lipid signaling molecules (**Figure 15.6**) (see Chapter 12). For example, **acyl hydrolases** remove fatty acyl chains, resulting in a **lysophospholipid**. Lysophospholipids are small bioactive lipids characterized by a single carbon chain and a polar head group. They are more hydrophilic than their corresponding phospholipids and have been implicated in the regulation of proton pumping on the plasma membrane and other processes. Members of the **phospholipase A** (**PLA**) family cleave either one of the acyl ester bonds, releasing a fatty acid and a lysophospholipid. **Phospholipase C** (**PLC**) hydrolyzes the glycerophosphate bond to produce **diacylglycerol** (**DAG**) and a phosphorylated head group, such as **inositol 1,4,5-trisphosphate** (**IP$_3$**). Both DAG and IP_3 have been implicated in the regulation of Ca^{2+} fluxes, which are important for a wide variety of physiological processes (see Figure 15.6B). **Phospholipase D** (**PLD**) activity releases the phospholipid head group, producing **phosphatidic acid** (**PA**), a lipid signaling molecule that increases rapidly in response to environmental stress.

PA is a cone-shaped lipid thought to increase local negative membrane curvature; such changes in curvature could promote vesicle budding or binding of membrane-associated proteins by facilitating the insertion of hydrophobic amino acids into the lipid bilayer. The negatively charged head group of PA also forms electrostatic interaction with positively charged binding pockets of effector proteins. In guard cells, PA interacts with ABA signaling proteins to promote stomatal closure (see Chapters 16 and 24). PA also appears to modulate the dynamics of both microtubule and actin cytoskeleton. It enhances actin filament formation by binding to, and thereby negatively regulating, the activity of the actin capping protein, a protein that binds in a Ca^{2+}-independent manner to the growing ends of actin filaments, blocking subunit exchange (see Chapter 1).

Hormones and Plant Development

The form and function of multicellular organisms would not be possible without efficient communication among cells, tissues, and organs. In higher plants, regulation and coordination of metabolism, growth, and morphogenesis often depend on chemical signals from one part of the plant to another. This idea originated in the nineteenth century with the German botanist Julius von Sachs (1832–1897).

Sachs proposed that chemical messengers are responsible for the formation and growth of different plant organs. He also suggested that external factors such as gravity could affect the distribution of these substances within a plant. Indeed, it has since become apparent that a majority of signaling networks that translate environmental cues into growth and developmental responses regulate the metabolism or redistribution of these endogenous

(A)

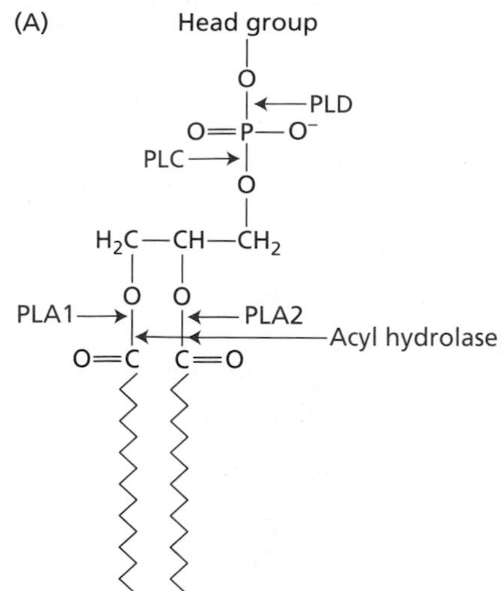

Head group	Lipid	
Choline	PtdCholine	PC
Ethanolamine	PtdEtn	PE
Glycerol	PtdGlycerol	PG
Serine	PtdSerine	PS
Inositol	PtdInositol	PI
Inositolmonophosphate	PtdInsP	PIP
Inositolbisphosphate	PtdInP$_2$	PIP$_2$
OH	Phosphatidic acid	PA

(B)

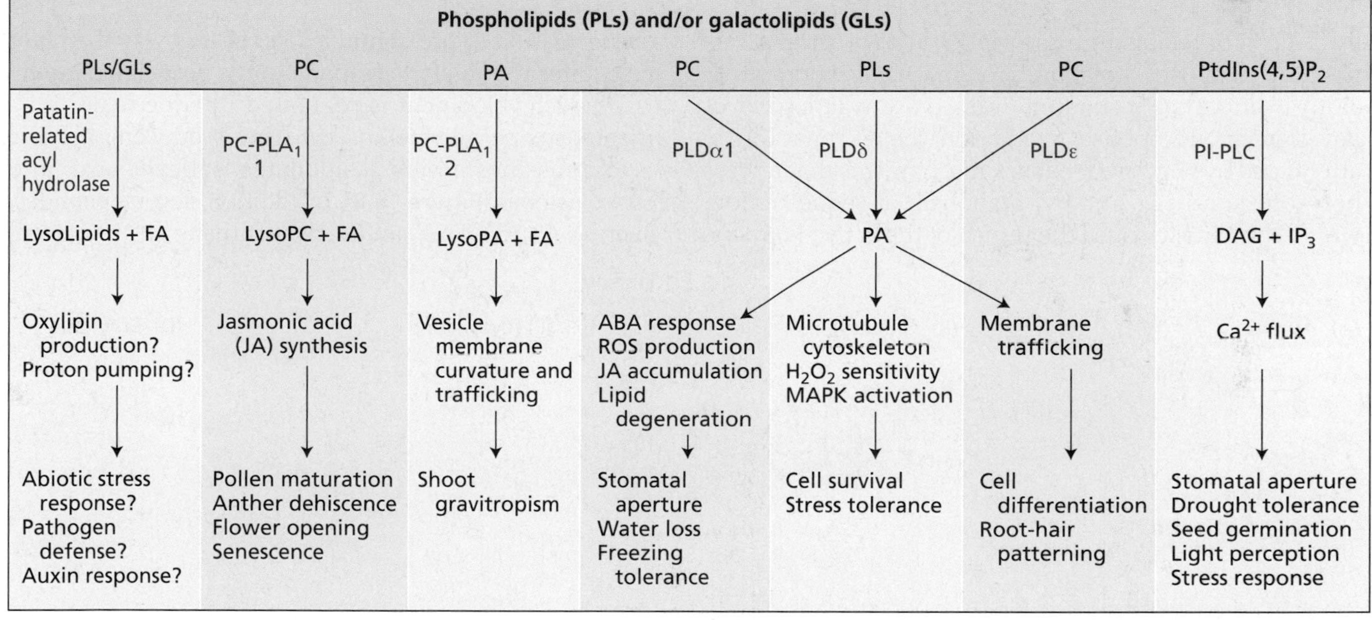

Figure 15.6 Lipid-modifying enzymes remodel cellular membranes and produce lipid signaling molecules. (A) Structure, hydrolysis, names, and abbreviations of the common phospholipids. (Left) The general structure of a phospholipid is shown, consisting of two fatty acyl chains esterified to a glycerol backbone, a phosphate (creating the "phosphatidyl" moiety [Ptd), and a variable head group. The positions subject to phospholipase action (PLA1, PLA2, PLC, and PLD) are indicated by the dashed red arrows. (Right) A table of possible head groups along with their abbreviations. (B) Membrane lipid substrates and messengers produced by different phospholipid or galactolipid-hydrolyzing enzymes, and their downstream cellular and physiological effects. (After Wang 2004.)

chemical messengers. Although Sachs did not know the identity of these messengers, his ideas led to their eventual discovery.

Hormones are chemical messengers that are produced in one cell and modulate cellular processes in another cell by interacting with specific proteins that function as receptors linked to cellular signal transduction pathways. As is the case with animal hormones, most plant hormones are capable of activating responses in target cells at vanishingly low concentrations. Although the details of hormonal control of development are quite diverse, all of the basic hormonal pathways share common features (**Figure 15.7**). For example, both signal perception and developmental programming often result in increases or decreases in hormone biosynthesis. The hormone is then transported to a site of action. Perception of the hormone

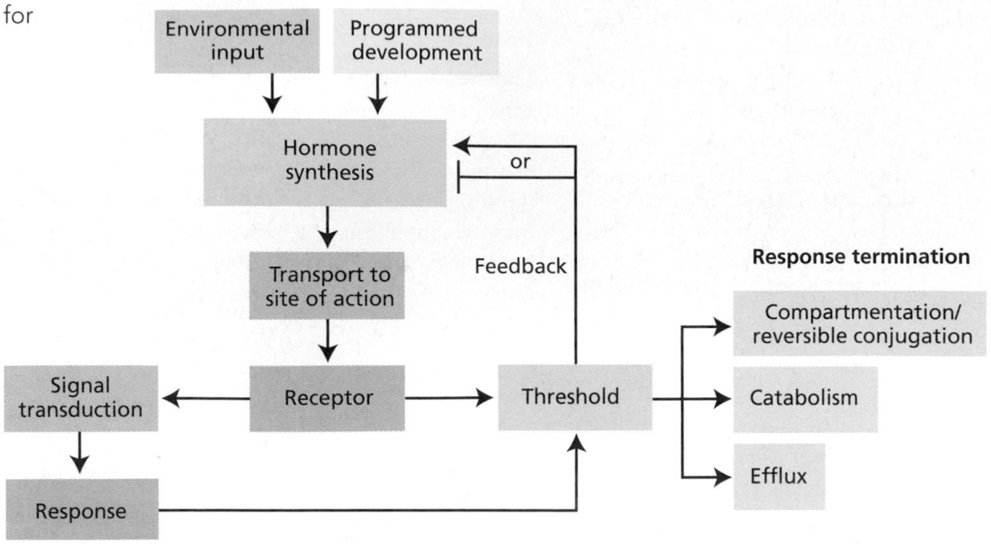

Figure 15.7 Common scheme for hormonal regulation.

by a receptor results in transcriptional or posttranscriptional (e.g., phosphorylation, protein turnover, ion extrusion) events that ultimately induce a physiological or developmental response. In addition, the response can be attenuated by negative feedback mechanisms that repress hormone synthesis and by catabolism or sequestration, which combine to cause the return of the active hormone concentration to pre-signal levels. In this way the plant reacquires the ability to respond to the next signal input.

Plant development is regulated by nine major hormones: auxins, gibberellins, cytokinins, ethylene, abscisic acid, brassinosteroids, jasmonates, salicylic acid, and strigolactones (**Figure 15.8**). In addition, several peptides, such as CLAVATA3, act over short distances to control

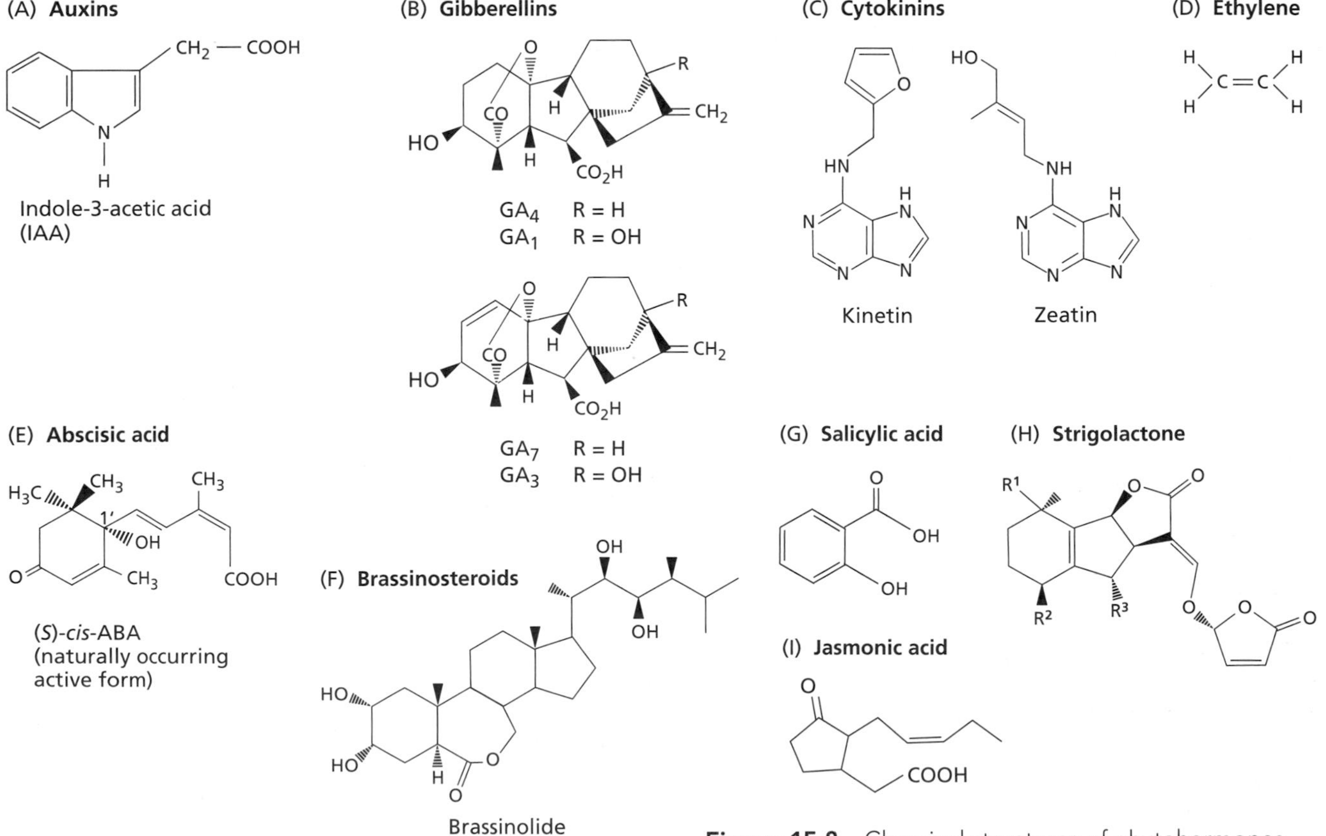

Figure 15.8 Chemical structures of phytohormones.

embryo development and apical meristem patterning. Indeed, the list of signaling molecules and growth regulators is likely to continue to expand in the coming years. Here we briefly introduce auxins, gibberellins, cytokinins, ethylene, abscisic acid, and brassinosteroids; the roles of jasmonates and salicylic acid during biotic interactions will be discussed in Chapter 23.

Auxin was discovered in early studies of coleoptile bending during phototropism

Auxin is essential to plant growth, and auxin signaling functions in virtually every aspect of plant development. Auxin was the first growth hormone to be studied in plants, and was discovered after the prediction of its existence by Charles and Francis Darwin in *The Power of Movement in Plants* (1881). The Darwins studied the bending of seedling sheath leaves (coleoptiles) of canary grass (*Phalaris canariensis*) and seedling hypocotyls of other species in response to unidirectional light, and concluded that a signal produced at the apex travels downward and causes lower cells on the shaded side to grow faster than on the illuminated side. Subsequently it was shown that the signal was a chemical that could diffuse through gelatin blocks (**Figure 15.9**). Plant physiologists named the chemical signal auxin from the Greek *auxein*, meaning "to increase" or "to grow," and identified indole-3-acetic acid (IAA) as the primary plant auxin. In some species, 4-chloro-IAA and phenylacetic acid function as natural auxins, but IAA is by far the most abundant and physiologically important form (see Figure 15.8A). Because the structure of IAA is relatively simple, researchers were quickly able to synthesize a wide array of molecules with auxin activity. Some of these compounds, such as 1-naphthalene acetic acid (NAA), 2,4-dichlorophenoxyacetic acid (2,4-D), and 2-methoxy-3,6-dichlorobenzoic acid (dicamba), are now used widely as growth regulators and herbicides in horticulture and agriculture.

Gibberellins promote stem growth and were discovered in relation to the "foolish seedling disease" of rice

A second group of plant hormones are the **gibberellins** (abbreviated to GA and numbered in the chronological sequence of their discovery). This group comprises a large number of compounds, all of which are tetracyclic (four-ringed) diterpenoid acids, but only a few of which, primarily GA_1, GA_3, GA_4, and GA_7, have intrinsic biological activity (see Figure 15.8B). One of the most striking effects of biologically active gibberellins, achieved through their role in promoting cell elongation, is the induction of internode elongation in dwarf seedlings. Gibberellins have other diverse roles during the plant life cycle: for example, they can promote seed germination (see Chapter 18), the transition to flowering (see Chapter 20), pollen development and pollen tube growth (see Chapter 21), and fruit development (see Chapter 21).

Gibberellins were first recognized by Eichi Kurosawa in 1926 and were isolated by Teijiro Yabuta and Yusuke Sumuki in the 1930s as natural products in the fungus *Gibberella fujikuroi* (now renamed *Fusarium fujikuroi*), from which the hormones derive their unusual name. Rice plants infected with *F. fujikuroi* grow abnormally tall, which leads to lodging (falling over) and reduced yield; hence the name *bakanae*, or "foolish seedling disease." Such overgrowth

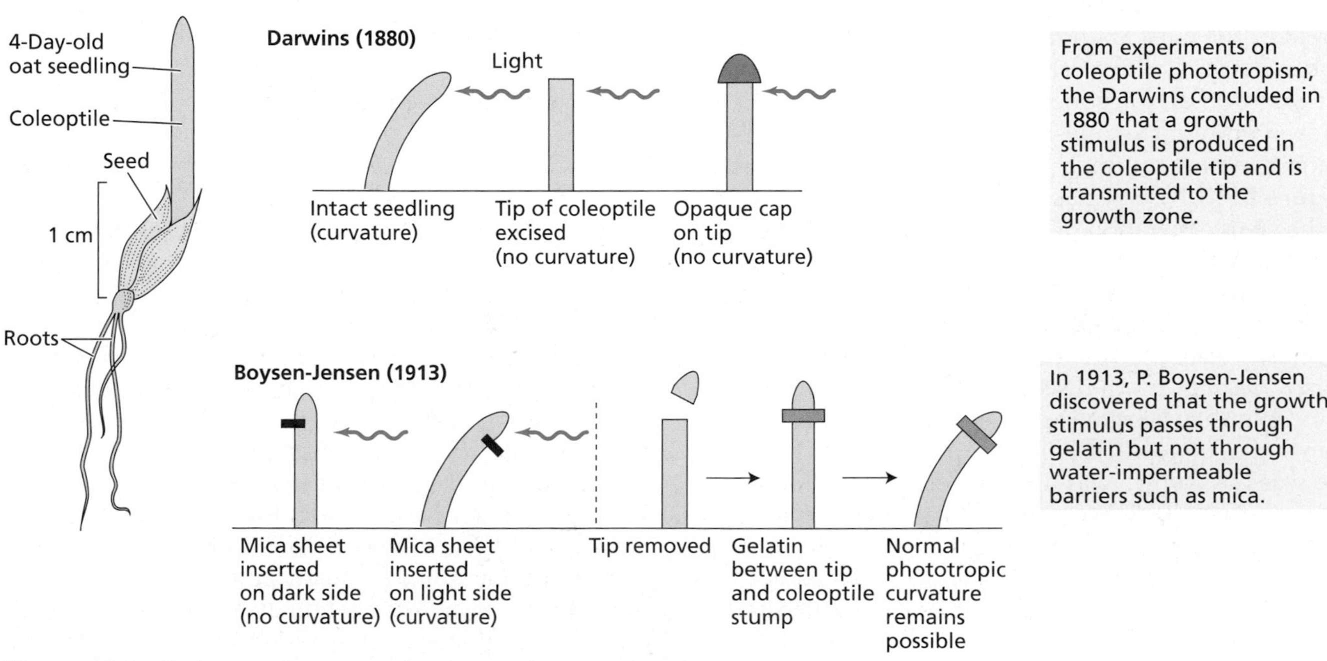

Figure 15.9 Early experiments on the chemical nature of auxin.

(A)

(B)

(C)

Figure 15.10 (A) Gibberellin induces growth in "Thompson Seedless" grapes. Untreated grapes normally remain small because of natural seed abortion. The bunch on the left is untreated. The bunch on the right was sprayed with GA_3 during fruit development, leading to larger fruits and elongation of the pedicels (fruit stalks). (B) The effect of exogenous GA_1 on wild-type (labeled as "normal" in the photograph) and dwarf mutant (*d1*) maize. Gibberellin stimulates dramatic stem elongation in the dwarf mutant but has little or no effect on the tall, wild-type plant. (C) Cabbage, a long-day plant, remains a low-growing rosette in short days, but it can be induced to bolt (grow long internodes) and flower by applications of GA_3. In the case illustrated, giant flowering stalks were produced. (B courtesy of B. Phinney.)

can be reproduced by applying gibberellins to uninfected rice seedlings. *F. fujikuroi* produces several different gibberellins, the most abundant of which is GA_3, also called gibberellic acid, which can be obtained commercially for horticultural and agronomic use. For example, GA_3 is sprayed onto grape vines to produce the large, seedless grapes we now routinely purchase in the grocery store (**Figure 15.10A**). Spectacular responses were obtained in the stem elongation of dwarf and rosette plants, particularly in genetically dwarf peas (*Pisum sativum*), dwarf maize (corn; *Zea mays*) (**Figure 15.10B**), and many rosette plants (**Figure 15.10C**).

Shortly after the first characterization of gibberellins from *F. fujikuroi*, it was discovered that plants also contain gibberellin-like substances but in much lower abundance than in the fungus. The first plant gibberellin to be identified was GA_1, which was discovered in extracts of runner bean seeds in 1958. We now know that gibberellins are ubiquitous in plants and are also present in several fungi in addition to *F. fujikuroi*. The majority of plants studied to date contain GA_1 and/or GA_4, so these are the gibberellins to which we assign "hormonal" function. In addition to GA_1 and GA_4, plants contain many inactive gibberellins which represent the precursors or deactivation products of the bioactive gibberellins.

Cytokinins were discovered as cell division–promoting factors in tissue culture experiments

Cytokinins were discovered in a search for factors that stimulated plant cells to divide (i.e., undergo cytokinesis) in conjunction with the phytohormone auxin. A small molecule was identified that could, in the presence of auxin, stimulate tobacco pith parenchyma tissue to proliferate in culture (**Figure 15.11A**). The cytokinesis-inducing molecule was named *kinetin*. While kinetin is a synthetic cytokinin, its structure is similar to that of naturally occurring cytokinins (see Figure 15.8C).

As we will see in later chapters, cytokinins have been shown to have effects on many physiological and developmental processes, including leaf senescence (see Chapter 22), apical dominance (see Chapter 18), formation and activity of apical meristems (see Chapter 17), gametophytic development (see Chapter 21), promotion of sink activity, vascular development, and breaking of bud dormancy (see Chapter 19). In addition, cytokinins play important roles in the interaction of plants with both biotic and abiotic factors, including salinity and drought stresses, macronutrients (including nitrate, phosphorus, iron, and sulfate), and symbiotic nitrogen-fixing bacteria and arbuscular mycorrhizal fungi, as well as pathogenic bacteria, fungi, nematodes, and viruses (**Figure 15.11B**) (also see Chapters 23 and 24).

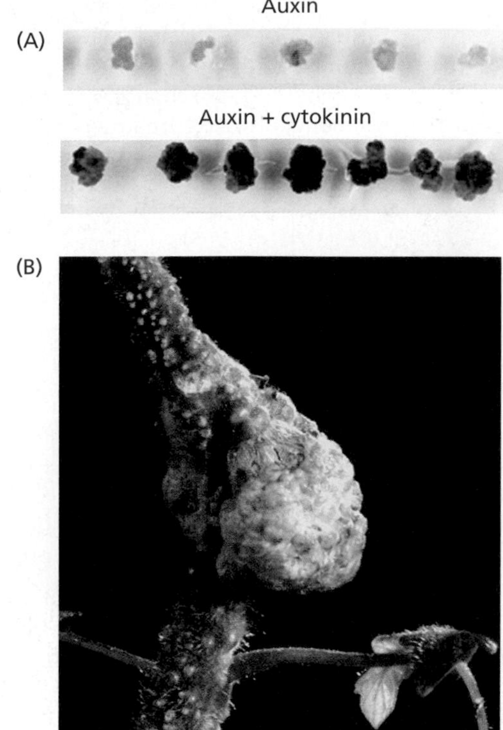

Figure 15.11 Cytokinin enhances cell division and greening. (A) Wild-type Arabidopsis leaf explants were induced to form callus (undifferentiated cells) by culturing in the presence of auxin alone (top) or auxin plus cytokinin (bottom). Cytokinin was required for callus growth and greening in the presence of light. (B) Tumor that formed on a tomato stem infected with the crown gall bacterium, *Agrobacterium tumefaciens*. Two months before this photo was taken, the stem was wounded and inoculated with a virulent strain of the crown gall bacterium. (A from Riou-Khamlichi et al. 1999; B from Aloni et al. 1998, courtesy of R. Aloni.)

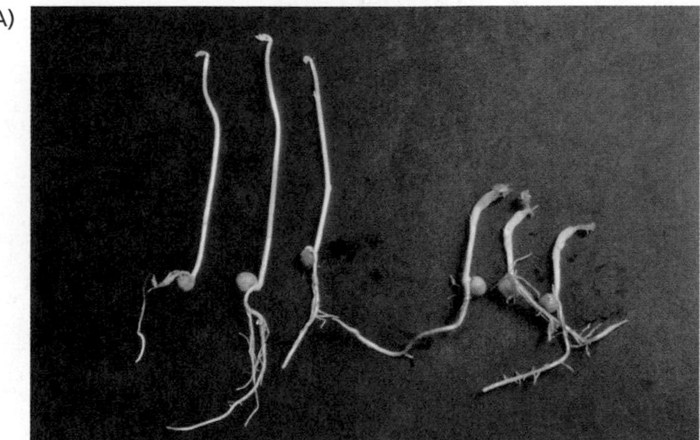

Figure 15.12 Ethylene responses. (A) Triple response of etiolated pea seedlings. Six-day-old pea seedlings were grown in the dark in the presence of 10 ppm (parts per million) ethylene (right) or left untreated (left). The treated seedlings show radial swelling, inhibition of elongation of the epicotyl, and horizontal growth of the epicotyl (diagravitropism). (B) Leaf epinasty in tomato. Epinasty, or downward bending of the tomato leaves (right), is caused by ethylene treatment. An untreated tomato is on the left. Epinasty results when the cells on the upper side of the petiole grow faster than those on the bottom. (Courtesy of S. Gepstein.)

Ethylene is a gaseous hormone that promotes fruit ripening and other developmental processes

Ethylene is a gas with a simple chemical structure (see Figure 15.8D) and was first identified as a plant growth regulator in 1901 by Dimitry Neljubov when he demonstrated its ability to alter the growth of etiolated pea seedlings in the laboratory (**Figure 15.12A**). Subsequently, ethylene was identified as a natural product synthesized by plant tissues.

Ethylene regulates a wide range of responses in plants, including seed germination and seedling growth, cell expansion and differentiation, leaf and flower senescence and abscission (see Chapters 18 and 22), and responses to biotic and abiotic stresses (see Chapters 23 and 24), including epinasty (**Figure 15.12B**).

Abscisic acid regulates seed maturation and stomatal closure in response to water stress

Abscisic Acid (ABA) is a ubiquitous hormone in vascular plants and has also been found in mosses, some phytopathogenic fungi, and a wide range of metazoans. ABA is a 15-carbon terpenoid (see Figure 15.8E) that was identified in the 1960s as a growth-inhibiting compound associated with the onset of bud dormancy and promotion of cotton fruit abscission. However, later work showed that ABA promotes senescence, the process preceding abscission, rather than abscission itself. Since then, ABA has also been shown to be a hormone that regulates salinity, dehydration, and temperature stress responses, including stomatal closure (**Figure 15.13**) (see Chapter 24). ABA also promotes seed maturation and dormancy (see Chapter 18) and regulates the growth of roots and shoots, heterophylly (the production of different leaf types on an individual plant), flowering, and some responses to pathogens (see Chapter 23).

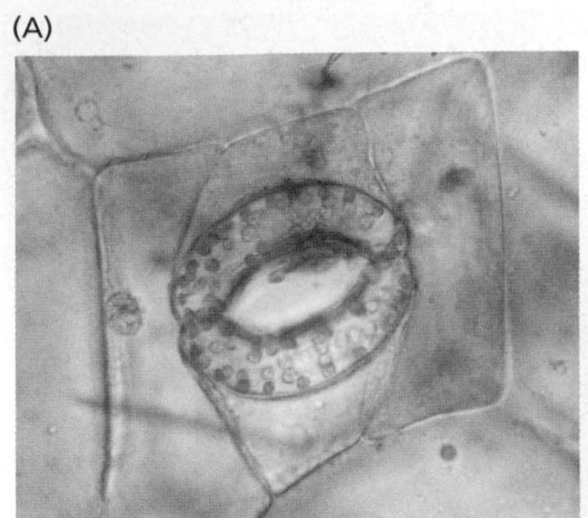

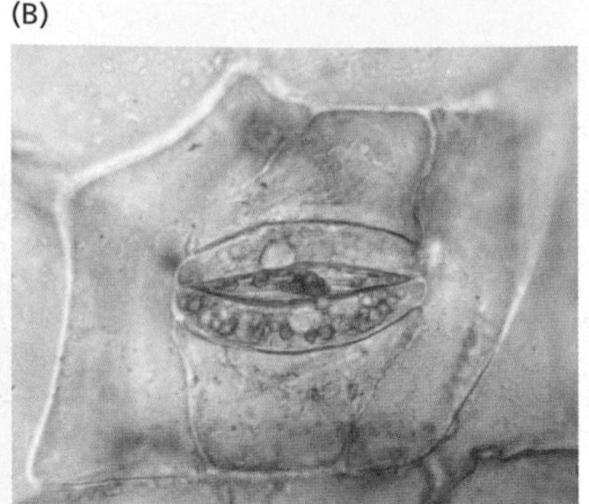

Figure 15.13 Stomatal closure in response to ABA. Stomata are open in the light for gas exchange with the environment (left). ABA treatment closes stomata in the light (right). This reduces water loss during the day under drought stress conditions.

Brassinosteroids regulate photomorphogenesis, germination, and other developmental processes

Brassinosteroids, initially named brassins, were first discovered as growth-promoting substances present in the pollen of *Brassica napus* (rape plant). Subsequent X-ray analysis showed that the most bioactive brassin in eudicots, which was named **brassinolide**, is a polyhydroxylated steroid similar to animal steroid hormones (see Figure 15.8F).

Many brassinosteroids have been identified, primarily intermediates of the brassinolide biosynthetic or catabolic pathways. Of these, the two known active brassinosteroid forms are brassinolide and its immediate precursor castasterone, although one form is predominant, depending on the plant species and tissue type. Brassinosteroids are ubiquitous plant hormones that, like auxins and gibberellins, appear to predate the evolution of land plants. In the angiosperms, brassinosteroids are found at low levels

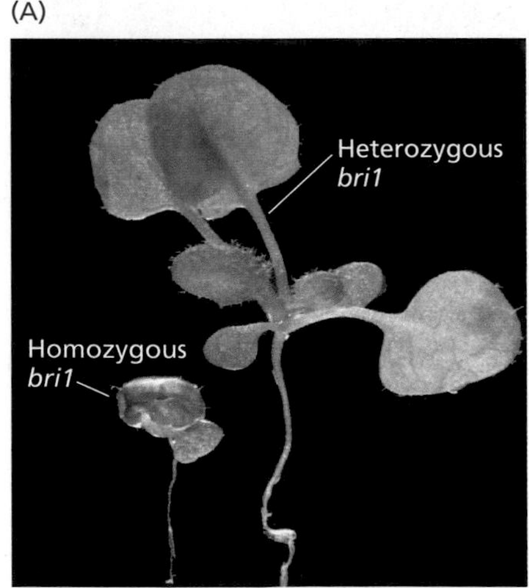

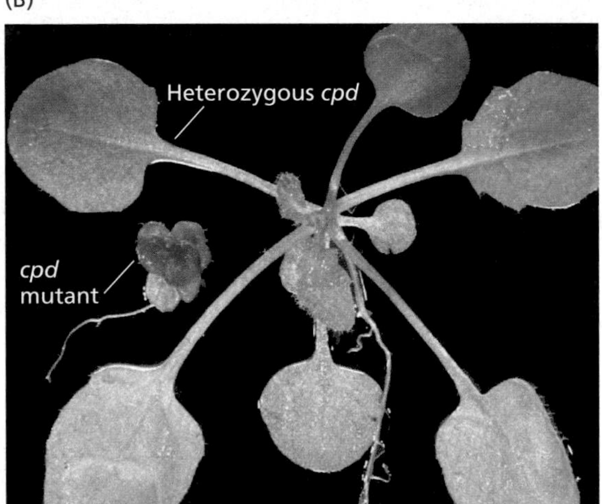

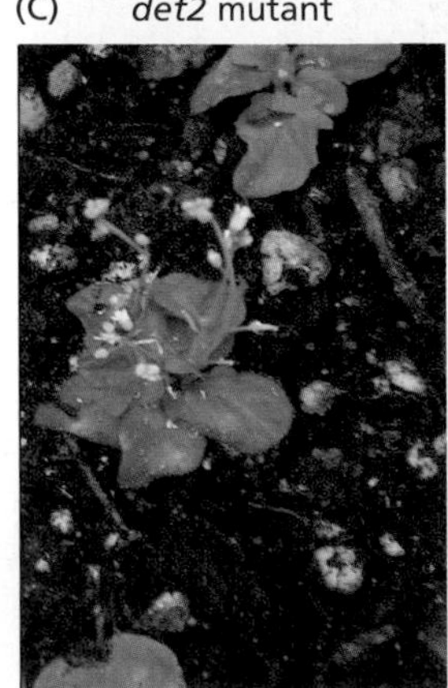

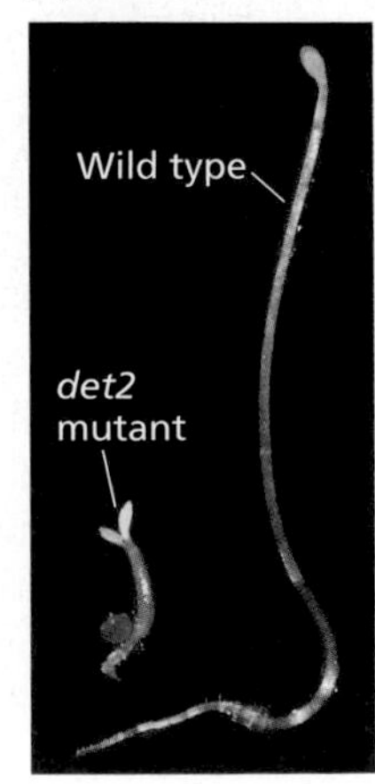

Figure 15.14 Phenotypes of Arabidopsis brassinosteroid mutants. (A) The 3-week-old light-grown homozygous *bri1* mutant (left) is a severe dwarf compared with the heterozygous *bri1* mutant (right), which exhibits wild-type morphology. (B) The 3-week-old light-grown homozygous *cpd* (constitutive photomorphogenisis and dwarfism) mutant (left) also exhibits a dwarf phenotype; the heterozygous mutant with a wild-type phenotype is on the right. (C) The light-grown adult *det2* mutant is dwarfed compared with the wild-type plant. (D) The dark-grown *det2* mutant on the left has short, thick hypocotyls and expanded cotyledons; the dark-grown wild type is on the right. (Courtesy of S. Savaldi-Goldstein.)

in various organs (e.g., flowers, leaves, roots) and at higher relative levels in pollen, immature seeds, and fruits.

Brassinosteroids play pivotal roles in a wide range of developmental phenomena in plants, including cell division, cell elongation, cell differentiation, photomorphogenesis, reproductive development, germination, leaf senescence, and stress responses. Mutants deficient in brassinosteroid synthesis, such as *det2* and *cpd*, show growth and developmental abnormalities, including dwarfism (**Figure 15.14**) and reduced apical dominance (see Chapter 19). Brassinosteroid-deficient mutants in Arabidopsis also exhibit de-etiolated growth when grown in the dark (see Figure 15.14D), and in maize show feminized male flowers.

Strigolactones suppress branching and promote rhizosphere interactions

Strigolactones, which occur in about 80% of plant species, are a group of terpenoid lactones (see Figure 15.8H) that were originally discovered as host-derived germination stimulants for root parasitic plants, such as the witchweeds (*Striga* spp.) and broomrapes (*Orobanche* and *Phelipanche* spp.) (**Figure 15.15**). They also promote symbiotic interactions with arbuscular mycorrhizal fungi, facilitating phosphate uptake from the soil. In addition, strigolactones suppress shoot branching and stimulate cambial activity and secondary growth (see Chapter 19). Strigolactones have analogous functions in roots, where they reduce adventitious and lateral root formation and promote root hair growth.

Figure 15.15 Rice plants colonized by root parasitic plants. (A) Flowering purple witchweed (*Striga hermonthica*) parasitizing a rice plant. (B) Etiolated *Striga hermonthica* seedling invading a rice root. (Photos courtesy of Ken Shirasu.)

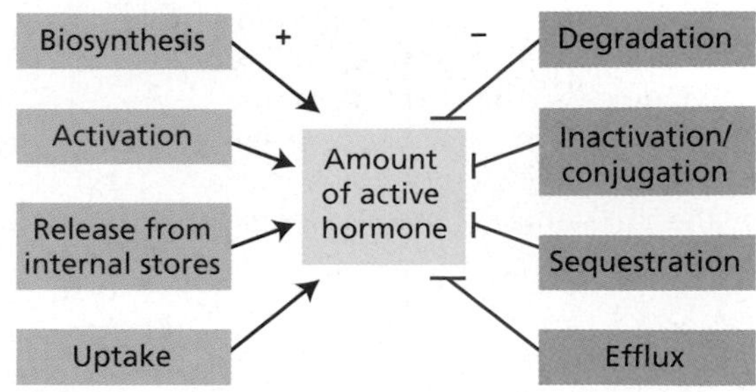

Figure 15.16 Homeostatic regulatory mechanisms that influence hormone concentration. Both positive and negative factors work in concert to maintain hormone homeostasis.

Phytohormone Metabolism and Homeostasis

To be effective signals, the concentrations of plant hormones must be tightly regulated in a cell type–specific and tissue-specific manner. In the simplest terms, a hormone's concentration in any given tissue or cell is determined by the balance between the rate of increase in its concentration (e.g., by local synthesis/activation or by import from elsewhere in the plant) and the rate of decrease in its concentration (e.g., by inactivation, degradation, sequestration, or efflux) (**Figure 15.16**). However, the regulation of hormone levels is complicated by many factors. First, primary hormone biosynthetic pathways may be augmented by secondary biosynthetic mechanisms. Second, there may be multiple, structural variants of a hormone, which vary widely in biological activity. Finally, as we will see later, there may be multiple mechanisms for removing active hormone from a system.

In this section we will discuss mechanisms for modulating hormone concentrations locally (within a cell or tissue). We will cover hormone transport between different parts of a plant in the next section.

Indole-3-pyruvate is the primary intermediate in auxin biosynthesis

IAA is structurally related to the amino acid tryptophan, and is primarily synthesized in a two-step process using indole-3-pyruvate (IPyA) as the intermediate (**Figure 15.17**). The second step in the pathway is carried out by the *YUCCA* gene product, a tryptophan aminotransferase. As we will see in later chapters, *YUCCA* genes play many important roles in plant development. They were first identified in Arabidopsis as a dominant mutation that causes elevated levels of free auxin. The mutant's name was derived from the phenotype of the adult plant, which exhibits increased apical dominance, tall inflorescences, and narrow, epinastic leaves, reminiscent of a yucca plant. A similar phenotype can be seen in an Arabidopsis mutant overexpressing the *YUC6* gene (**Figure 15.18**).

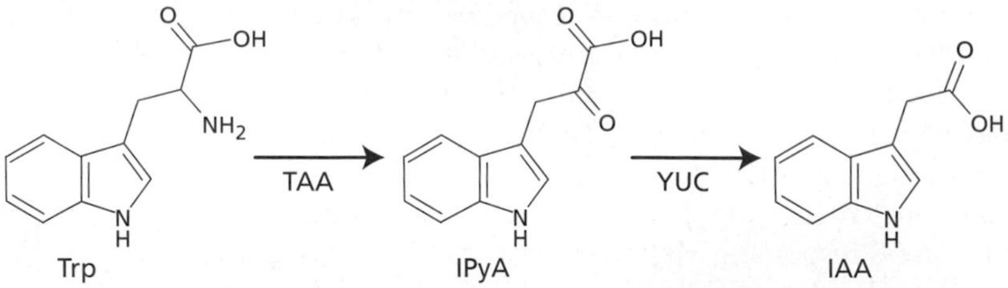

Figure 15.17 Auxin biosynthesis from tryptophan (Trp). In the first step, Trp is converted to indole-3-pyruvate (IPyA) by the TAA family of tryptophan amino transferases. Subsequently, IAA is produced from IPyA by the YUC family of flavin monooxygenases.

IAA biosynthesis is associated with rapidly dividing and growing tissues, especially in shoots. Although virtually all plant tissues appear to be capable of producing low levels of IAA, shoot apical meristems, young leaves, and young fruits are the primary sites of auxin synthesis. In plants that produce indole glucosinolate defense compounds (see Chapter 23), IAA can also be synthesized from tryptophan via a pathway with indole acetonitrile as an intermediate (see **WEB APPENDIX 3**). In maize kernels, IAA also appears to be synthesized by a tryptophan-independent pathway.

Figure 15.18 Arabidopsis mutant overexpressing the *YUC6* gene. The dominant *yuc6-1D* activation mutant (right) contains elevated levels of free IAA relative to the wild type (left) due to the overexpression of *YUCCA6*. Note the taller growth, reduced branching and delayed senescence of the mutant. (Courtesy of Dr. Jeong Im Kim.)

Auxin is toxic at high cellular concentrations, and without homeostatic controls the hormone could easily build up to toxic levels. Auxin catabolism by conjugation to hexose sugars and oxidative degradation ensures the permanent removal of active hormone when the concentration exceeds the optimal level or when the response to the hormone is complete. Covalent conjugation of amino acids to IAA can also result in permanent inactivation. However, most amino acyl conjugates serve as storage forms from which IAA can be rapidly released by enzymatic processes. Indole-3-butyric acid (IBA) is a compound that is routinely used in horticulture to promote rooting of cuttings and is rapidly converted by β-oxidation in the peroxisome to IAA. Both free and conjugated IBA are thought to occur naturally in plants and to serve as auxin sources for specific developmental processes. Auxin has also been shown to be conjugated to peptides, complex glycans (multiple sugar units), or glycoproteins in some plant species, but the precise physiological role of these conjugates is still unknown. A diagram of the storage and catabolic fates of auxin is shown in **Figure 15.19**.

Sequestration of auxin in endomembrane compartments, primarily the ER, also appears to regulate auxin levels available for signaling. Proteins that mediate IAA movement across the ER membrane have been identified, and a large store of the extracellular auxin receptor AUXIN BINDING PROTEIN1 (ABP1) (discussed later in the chapter) is primarily found in the ER lumen.

The well-documented toxicity of exogenously applied auxin, especially on eudicot species, forms the basis for a family of synthetic auxins, such as 2,4-dichlorophenoxyacetic acid (2,4-D), that have long been used as herbicides. Mutations causing auxin overexpression (see Figure 15.18) would tend to be lethal if it were not for the homeostatic control of auxin levels. The reason that synthetic auxins are more effective as herbicides than natural auxins is that the synthetic auxins are much less subject to homeostatic control—degradation, conjugation, transport, and sequestration—than natural auxins are.

Gibberellins are synthesized by oxidation of the diterpene *ent*-kaurene

Gibberellins are synthesized in several parts of a plant, including developing seeds, germinating seeds, developing leaves, and elongating internodes. The biosynthetic

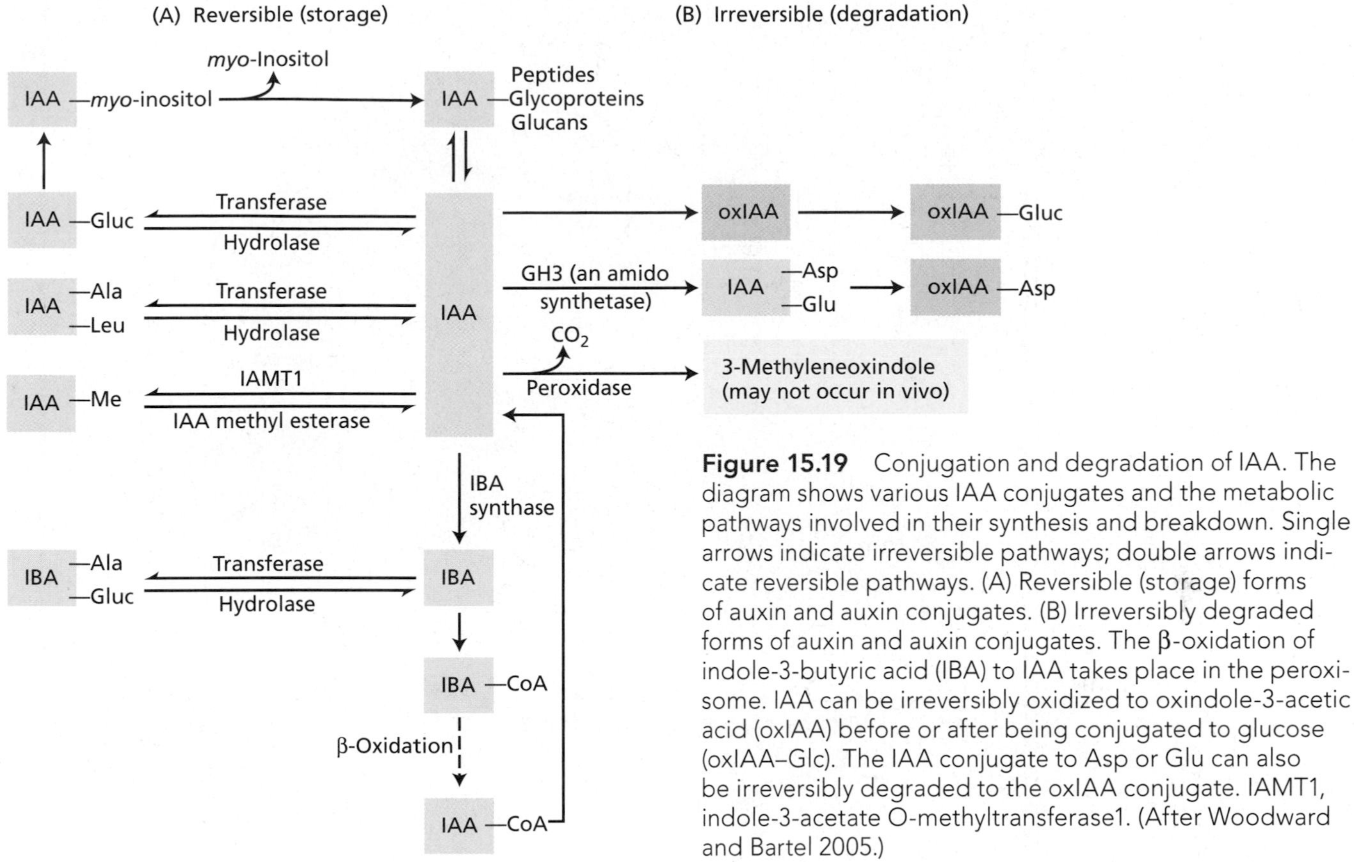

Figure 15.19 Conjugation and degradation of IAA. The diagram shows various IAA conjugates and the metabolic pathways involved in their synthesis and breakdown. Single arrows indicate irreversible pathways; double arrows indicate reversible pathways. (A) Reversible (storage) forms of auxin and auxin conjugates. (B) Irreversibly degraded forms of auxin and auxin conjugates. The β-oxidation of indole-3-butyric acid (IBA) to IAA takes place in the peroxisome. IAA can be irreversibly oxidized to oxindole-3-acetic acid (oxIAA) before or after being conjugated to glucose (oxIAA–Glc). The IAA conjugate to Asp or Glu can also be irreversibly degraded to the oxIAA conjugate. IAMT1, indole-3-acetate O-methyltransferase1. (After Woodward and Bartel 2005.)

pathway, which starts in plastids, leads to the production of a linear (straight chain) precursor molecule containing 20 carbon atoms, geranylgeranyl diphosphate or GGPP, which gets converted into *ent*-kaurene. This compound is oxidized sequentially by enzymes associated with the ER, leading to GA_{12}, the first gibberellin formed in all plants studied to date. Dioxygenase enzymes in the cytosol are able to oxidize GA_{12} to all other gibberellins in pathways that may be interconnected in such a way as to form a complex metabolic grid. A summary of gibberellin synthetic pathways is shown in **Figure 15.20**.

The pathways involved in gibberellin biosynthesis and catabolism are under tight genetic control. Several mechanisms have been described to date. These include gibberellin inactivation via a family of enzymes termed GA 2-oxidases, methylation via a methyl transferase, and conjugation to sugars. Genetic modulation of these pathways plays an important role in plant development. For example, as we will see in Chapter 19, *KNOXI* gene expression in the shoot apical meristem, which is crucial for proper function of the meristem, reduces gibberellin levels by inhibiting gibberellin biosynthesis and promoting gibberellin inactivation. Gibberellin biosynthesis is also regulated by feedback inhibition when cellular gibberellin exceeds threshold levels. The application of exogenous gibberellin causes downregulation of the *GA20ox* and *GA3ox* genes, whose products catalyze the two final two steps in the formation of bioactive gibberellins (GA_1 and GA_4).

Cytokinins are adenine derivatives with isoprene side chains

Cytokinins are adenine derivatives, and the most common class of cytokinins has isoprenoid side chains, including isopentenyl adenine (iP), dihydrozeatin (DHZ), and the most abundant cytokinin in higher plants, zeatin. Cytokinins are made from ADP/ATP and dimethylallyl diphosphate (DMAPP), primarily in plastids. A simplified schematic of the cytokinin biosynthetic pathway is shown in **Figure 15.21**.

In addition to the free bases, which are the only active forms, cytokinins are also present in the plant as ribosides (in which a ribose sugar is attached to the 9 nitrogen of the purine ring), ribotides (in which the ribose sugar moiety contains a phosphate group), or **glycosides** (in which a sugar molecule is attached to the 3, 7, or 9 nitrogen of the purine ring, or to the oxygen of the zeatin or dihydrozeatin side chain). In addition to this glycosylation-mediated inactivation, active cytokinin levels are also decreased catabolically through irreversible cleavage by cytokinin oxidases.

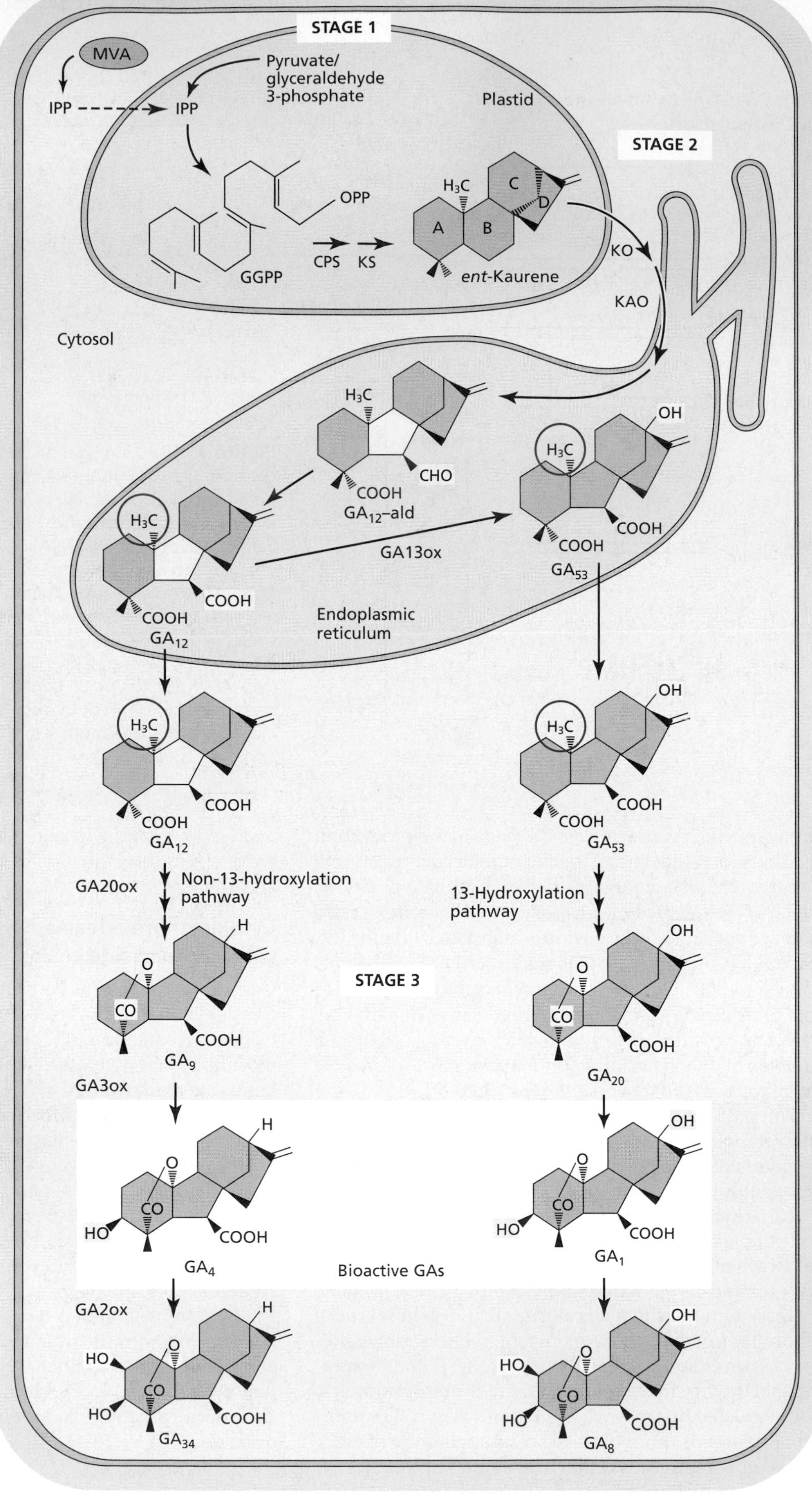
STAGE 1
MVA
Pyruvate/
glyceraldehyde
3-phosphate
IPP
IPP
Plastid
STAGE 2
OPP
GGPP
CPS
KS
H3C
A
B
C
D
ent-Kaurene
KO
KAO
Cytosol
CHO
COOH
GA12–ald
OH
H3C
COOH
COOH
GA53
GA13ox
H3C
COOH
COOH
GA12
Endoplasmic
reticulum
GA12
GA53
GA20ox
Non-13-hydroxylation
pathway
13-Hydroxylation
pathway
STAGE 3
GA9
GA20
GA3ox
GA4
Bioactive GAs
GA1
GA2ox
GA34
GA8

◀ **Figure 15.20** The three stages of gibberellin biosynthesis. Yellow highlighting indicates the part of the molecule that has been modified in the previous reaction. In stage 1 in the plastid, geranylgeranyl diphosphate (GGPP) is converted to *ent*-kaurene. In stage 2 in the endoplasmic reticulum, *ent*-kaurene is converted to GA_{12}-aldehyde and GA_{12}. GA_{12} is converted to GA_{53} by hydroxylation at C-13. In stage 3 in the cytosol, GA_{12} and GA_{53} are converted, via parallel pathways, to other gibberellins. This conversion proceeds with a series of oxidations at C-20 (red circles), resulting in the eventual loss of C-20 and the formation of C_{19}- gibberellins. 3-β-Hydroxylation then produces GA_4 and GA_1 as the bioactive gibberellins in each pathway. Hydroxylation at C-2 then converts GA_4 and GA_1 to the inactive forms GA_{34} and GA_8, respectively. In most plants the 13-hydroxylation pathway predominates, although in Arabidopsis and some others, the non-13-hydroxylation pathway is the main pathway. MVA, mevalonic acid; IPP, isopentenyl diphosphate; CPS, *ent*-copalyl diphosphate synthase; KS, *ent*-kaurene synthase; KO, *ent*-kaurene oxidase; KAO, *ent*-kaurenoic acid oxidase; GA20ox, GA 20-oxidase; GA3ox, GA 3-oxidase; GA2ox, GA 2-oxidase; GA13ox, GA 13-oxidase.

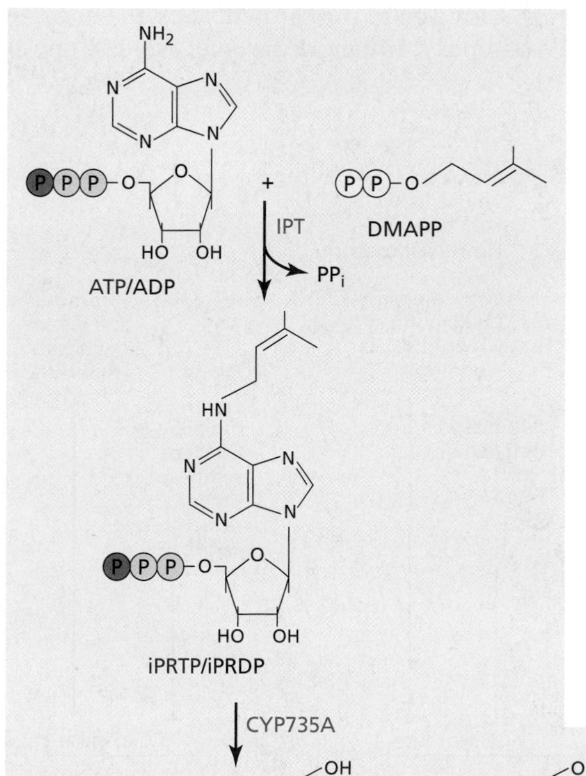

Figure 15.21 Simplified biosynthetic pathway for cytokinin biosynthesis. The first committed step in cytokinin biosynthesis, catalyzed by isopentenyl transferase (IPT), is the addition of the isopentenyl side chain from DMAPP (dimethylallyl diphosphate) to an adenosine moiety (ATP or ADP). iPRTP or iPRDP is converted to ZTP or ZDP respectively by cytochrome P450 monooxygenase (CYP735A) and eventually is converted to zeatin. Dihydrozeatin (DHZ) cytokinins are made from the various forms of *trans*-zeatin by an unknown enzyme (not shown). The ribotide and riboside forms of *trans*-zeatin can be interconverted, and free *trans*-zeatin can be formed from the ribotide by the LONELY GUY (LOG) family of cytokinin nucleoside 5′ monophosphate phosphohydrolase enzymes. iPRDP, isopentenyladenine riboside 5′-diphosphate; iPRTP, isopentenyladenine riboside 5′-triphosphate; ZTP, *trans*-Zeatin riboside 5′ triphosphate; ZDP, *trans*-Zeatin riboside 5′ diphosphate.

Consistent with its role in promoting cell division, cytokinin is required for proper functioning of the shoot apical meristem and is therefore tightly regulated (see Chapter 19). While inhibiting gibberellin levels, *KNOX* gene expression increases cytokinin levels in the shoot apical meristem by upregulating the cytokinin biosynthetic gene *ISOPENTENYL TRANSFERASE7* (*IPT7*) (see Figure 15.21).

Ethylene is synthesized from methionine via the intermediate ACC

Ethylene can be produced by almost all parts of higher plants, although the rate of production depends on the type of tissue, the stage of development, and environmental inputs. For example, certain mature fruits undergo a respiratory burst in response to ethylene, and ethylene levels increase in these fruits at the time of ripening (see Chapter 21). Ethylene is derived from the amino acid methionine and the intermediate *S*-adenosylmethionine, which is generated in the Yang cycle (**Figure 15.22**). The first committed and generally rate-limiting step in the biosynthesis is the conversion of *S*-adenosylmethionine to 1-aminocyclopropane-1-carboxylic acid (ACC) by the enzyme ACC synthase. ACC is then converted to ethylene by enzymes called **ACC oxidases**. As ethylene is a gaseous hormone, there is no evidence of ethylene catabolism in plants, and ethylene rapidly diffuses out of plant tissues when biosynthesis is pharmacologically interrupted.

Abscisic acid is synthesized from a carotenoid intermediate

ABA is synthesized in almost all cells that contain chloroplasts or amyloplasts and has been detected in every major organ and tissue. ABA is a 15-carbon terpenoid, or sesquiterpenoid, which is synthesized in plants by an indirect pathway via 40-carbon carotenoid intermediates (**Figure 15.23**). Early steps of this pathway occur in plastids. Cleavage of the carotenoid by the enzyme NCED (9-*cis*-epoxycarotenoid dioxygenase) is a rate-limiting,

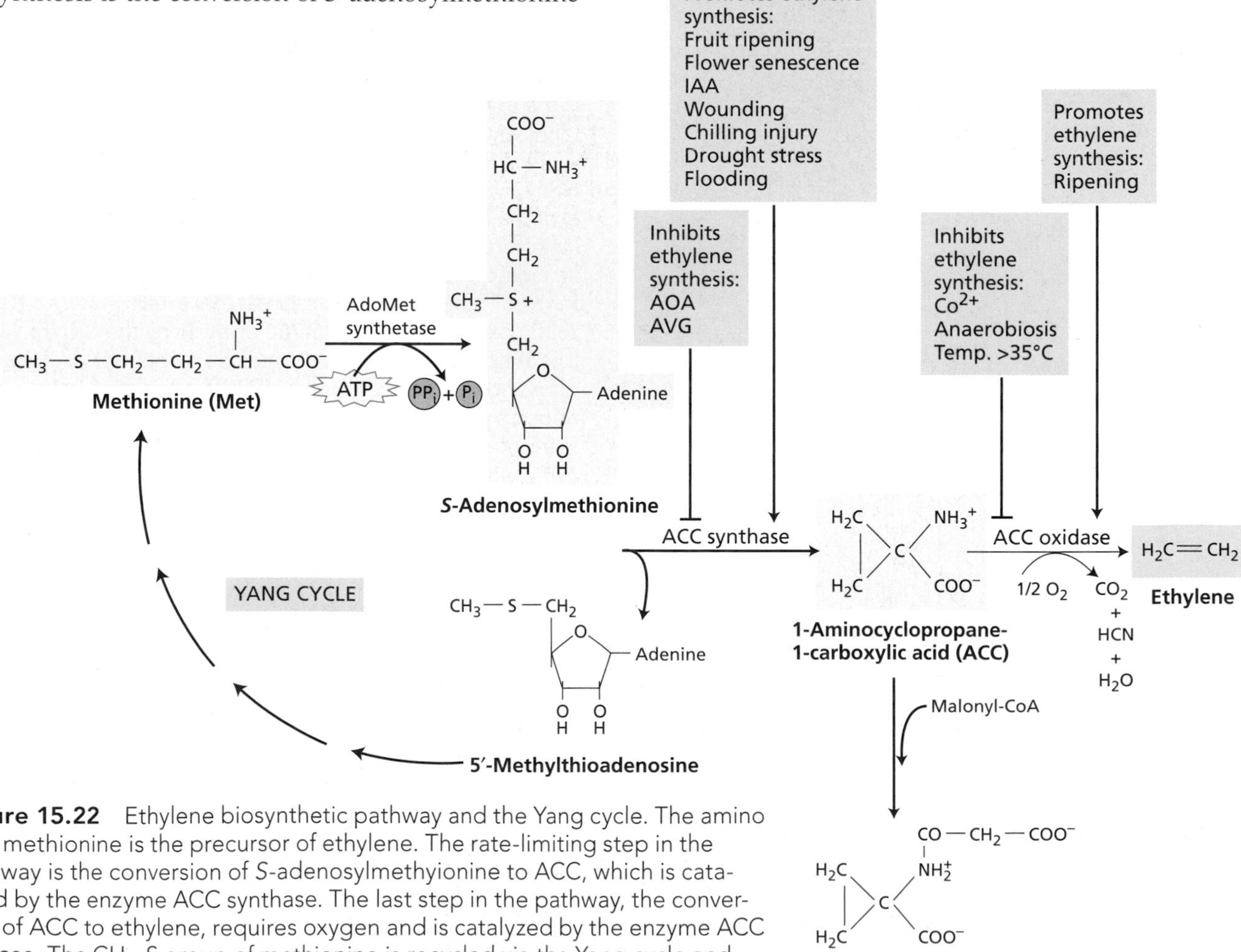

Figure 15.22 Ethylene biosynthetic pathway and the Yang cycle. The amino acid methionine is the precursor of ethylene. The rate-limiting step in the pathway is the conversion of *S*-adenosylmethyionine to ACC, which is catalyzed by the enzyme ACC synthase. The last step in the pathway, the conversion of ACC to ethylene, requires oxygen and is catalyzed by the enzyme ACC oxidase. The CH_3–S group of methionine is recycled via the Yang cycle and thus conserved for continued synthesis. In addition to being converted to ethylene, ACC can be conjugated to *N*-malonyl ACC. AOA, aminooxyacetic acid; AVG, aminoethoxy-vinylglycine. (After McKeon et al. 1995.)

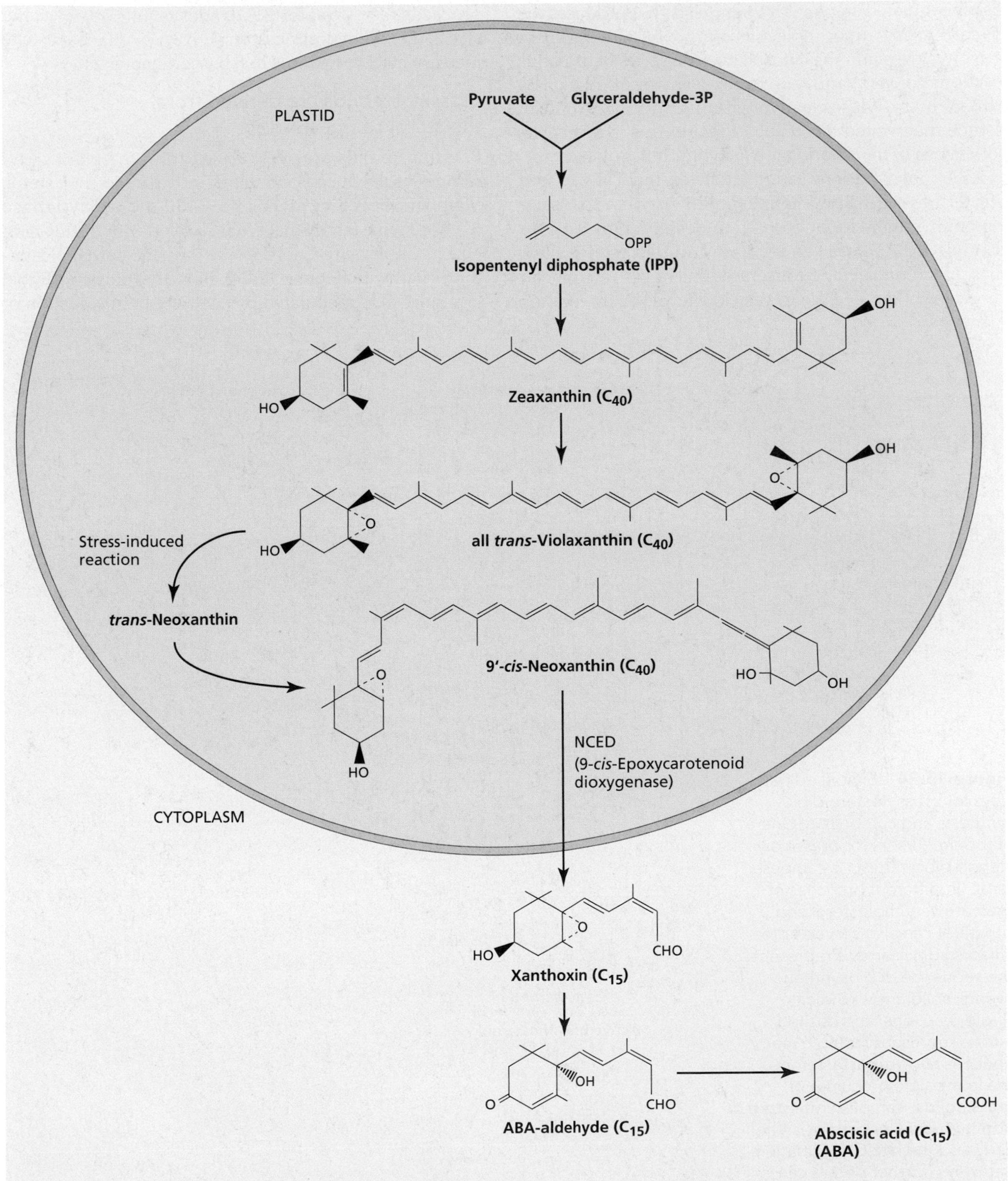

Figure 15.23 Simplified diagram of ABA biosynthesis via the terpenoid pathway. The initial stages occur in plastids, where isopentenyl diphosphate (IPP) is converted to the C_{40} xanthophyll zeaxanthin. Zeaxanthin is further modified to 9-*cis*-neoxanthin, which is cleaved by the enzyme NCED (9-*cis*-epoxycarotenoid dioxygenase) to form the C_{15} inhibitor, xanthoxin. Xanthoxin then is converted to ABA in the cytosol. ABA-deficient mutants that have been helpful in elucidating the pathway are shown in **WEB APPENDIX 3**.

highly regulated step in ABA synthesis that produces the 15-carbon precursor molecule xanthoxin, which subsequently moves into the cytosol, where a series of oxidative reactions convert xanthoxin to ABA. Further oxidation by ABA-8′-hydroxylases lead to ABA inactivation. ABA can also be inactivated by conjugation, but this is reversible. Both types of inactivation are also tightly regulated.

ABA concentrations can fluctuate dramatically in specific tissues during development or in response to changing environmental conditions. In developing seeds, for example, ABA levels can increase 100-fold within a few days, reaching average concentrations in the micromolar range, and then decline to very low levels as maturation proceeds (see Chapter 21). Under conditions of water stress (i.e., dehydration stress), ABA in the leaves can increase 50-fold within 4 to 8 h (see Chapter 24).

Brassinosteroids are derived from the sterol campesterol

Brassinosteroids are synthesized from the plant sterol campesterol, which is similar in structure to cholesterol. Members of the **cytochrome P450 monooxygenase** (**CYP**) enzyme family that are associated with the ER catalyze most of the reactions in the brassinosteroid biosynthesis pathway (**Figure 15.24**). Bioactive brassinosteroid levels are also modulated by a variety of inactivation or

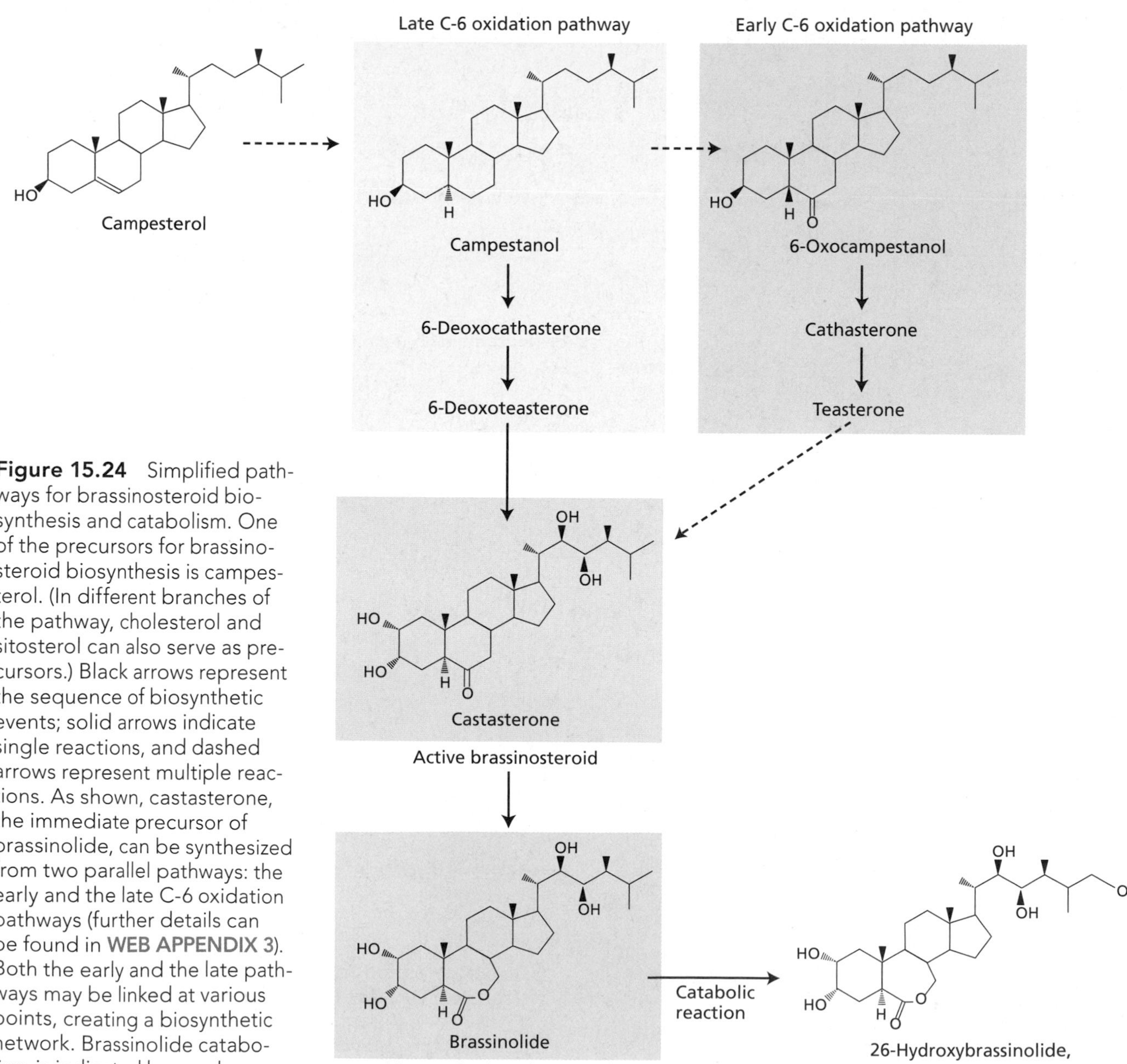

Figure 15.24 Simplified pathways for brassinosteroid biosynthesis and catabolism. One of the precursors for brassinosteroid biosynthesis is campesterol. (In different branches of the pathway, cholesterol and sitosterol can also serve as precursors.) Black arrows represent the sequence of biosynthetic events; solid arrows indicate single reactions, and dashed arrows represent multiple reactions. As shown, castasterone, the immediate precursor of brassinolide, can be synthesized from two parallel pathways: the early and the late C-6 oxidation pathways (further details can be found in WEB APPENDIX 3). Both the early and the late pathways may be linked at various points, creating a biosynthetic network. Brassinolide catabolism is indicated by a red arrow.

catabolic reactions, including epimerization, oxidation, hydroxylation, sulfonation, and conjugation to glucose or lipids. However, only few of the enzymes responsible for brassinosteroid catabolism and inactivation have been identified so far.

The levels of active brassinosteroids are also regulated by brassinosteroid-dependent negative feedback mechanisms in which hormone concentrations above a certain threshold cause a decrease in brassinosteroid biosynthesis. This attenuation is brought about by the downregulation of brassinosteroid biosynthetic genes and the upregulation of genes involved in brassinosteroid catabolism. Thus, mutants impaired in their ability to respond to brassinolide accumulate high levels of the active brassinosteroids compared with wild-type plants.

Strigolactones are synthesized from β-carotene

Like ABA, strigolactones are derived from carotenoid precursors in plastids in a pathway that is conserved up to synthesis of the intermediate carlactone, beyond which strigolactone biosynthesis diverges in a species-specific manner (**Figure 15.25**). This divergence is attributed to the functional diversity of MAX1 cytochrome P450 isoforms, which act on carlactone.

The strigolactone signaling pathway will be discussed in Chapter 19.

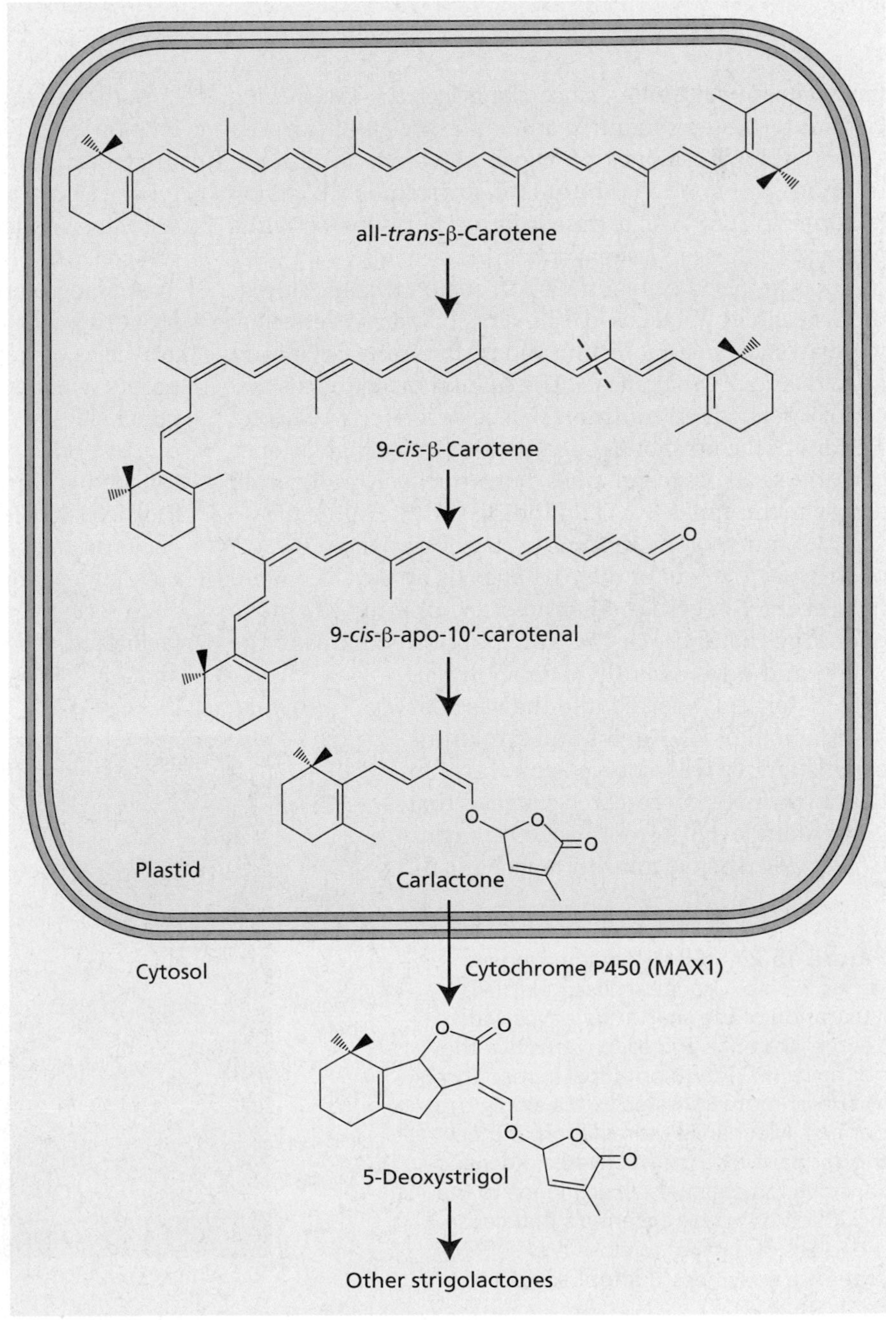

Figure 15.25 Strigolactone biosynthetic pathway and signaling proteins. All-*trans*-β-carotene is isomerized to 9-*cis*-β-carotene by a β-carotene isomerase. 9-*cis*-β-Carotene is cleaved (at the red, dotted line) by carotenoid cleavage dioxygenases, to produce carlactone. The final stages of strigolactone synthesis and signaling occur in the cytosol.

Signal Transmission and Cell–Cell Communication

Hormonal signaling typically involves the transmission of the hormone from its site of synthesis to its site of action. In general, hormones that are transported to sites of action in tissues distant from their site of synthesis are referred to as *endocrine* hormones, while those that act on cells adjacent to the source of synthesis are referred to as *paracrine* hormones (**Figure 15.26**). Hormones can also function in the same cells in which they are synthesized, in which case they are referred to as *autocrine effectors*. Most plant hormones

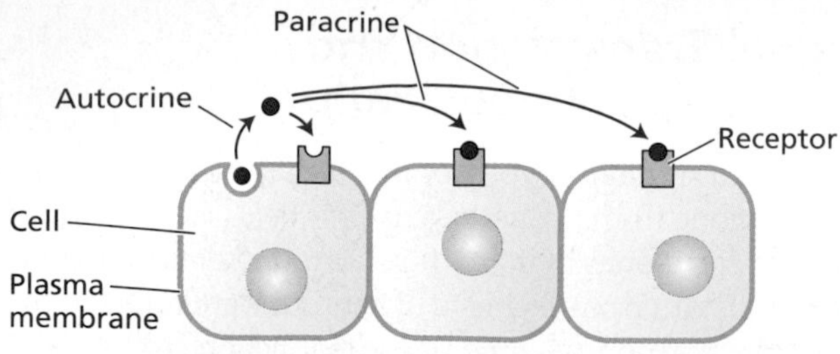

Figure 15.26 Autocrine versus paracrine signaling. Autocrine signals bind to receptors on the same cell in which they are synthesized. In contrast, paracrine signals bind to receptors on cells located a short distance away from the site of synthesis. Signaling that involves transport over greater distances is called endocrine signaling.

have paracrine activities, since plants lack the fast-moving circulatory systems found in animals associated with classic endocrine hormones. However, slower, long-distance hormone transport via the vascular system is a common feature in plants, despite the absence of hormone-secreting glands like those in animal endocrine systems.

For example, the polar transport of auxin via highly regulated cellular uptake and efflux mechanisms is essential to auxin's role in establishing and maintaining polar plant growth and organogenesis. The cellular mechanisms that control polar auxin transport will be described in Chapter 17. Lipophilic hormones such as ABA and strigolactones can diffuse across membranes, but are actively transported across membranes by ATP Binding Casette subfamily G (ABCG) transporters in some tissues. Polarized transport of strigolactone out of the root apex by an ABCG protein has recently been demonstrated. Cytokinins can move over long distances in the transpiration streams of the xylem and have recently been shown to be activiely transported into the vascular system in the root. Auxins and cytokinins can also move with source–sink fluxes in the phloem. Recent research suggests that gibberellin levels in root tissues are controlled via an active transport mechanism, resulting in accumulation of this growth hormone in expanding endodermal cells that control root elongation. As a gaseous compound, ethylene is more soluble in lipid bilayers than in the aqueous phase and can freely pass through the plasma membrane. In contrast, its precursor, ACC, is water soluble and is thought to be transported via the xylem to shoot tissues. It is currently unknown whether brassinosteroids have endocrine or paracrine activity. Brassinosteroids do not seem to undergo root-to-shoot and shoot-to-root translocation, as experiments in pea and tomato indicate that reciprocal stock/scion grafting of wild-type to brassinosteroid-deficient mutants do not rescue the phenotype of the latter. Instead, components of the brassinosteroid signaling and biosynthesis pathway are expressed throughout the plant, especially in young growing tissues.

Although plants lack nervous systems, like animals they employ long-distance electrical signaling to communicate between distant parts of the plant body. The most common type of electrical signaling in plants is the **action potential**, the transient depolarization of the plasma membrane of a cell generated by voltage-gated ion channels (see Chapter 6). Action potentials have been shown to mediate touch-induced leaflet closure in sensitive plant (*Mimosa pudica*), as well as the rapid closure (~0.1 s) of Venus flytrap, which occurs when an insect touches the sensitive hairs on the upper sides of the traplike leaf lobes (**Figure 15.27A**). For the response to be activated, either two hairs must be touched within 20 s of each other, or one hair must be touched twice in rapid succession. Since each displacement elicits an action potential (**Figure 15.27B**), the leaf must have a mechanism for counting action potentials.

In recent years, electrical signaling has been shown to facilitate rapid communication between remote parts of

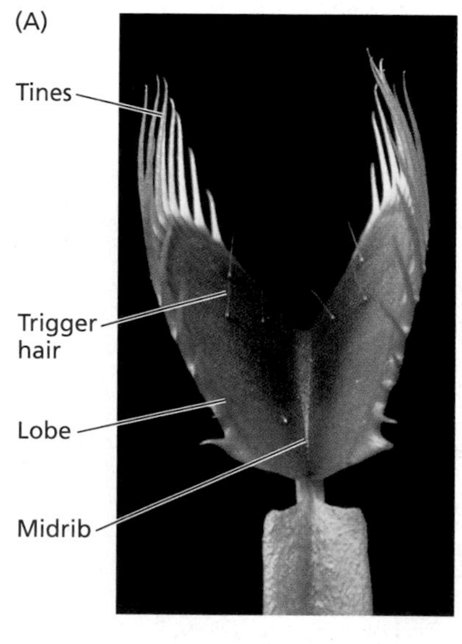

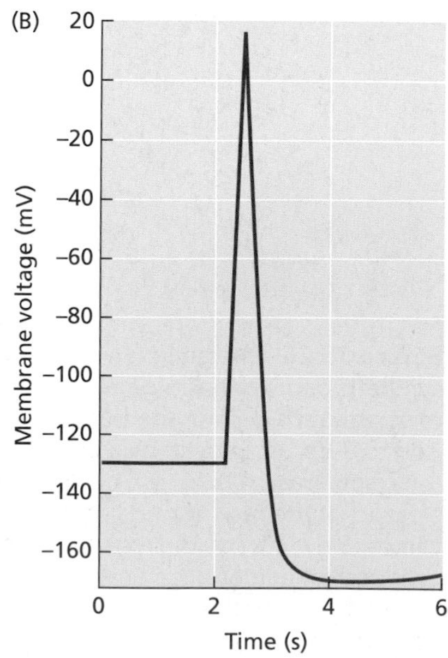

Figure 15.27 Electrical signaling in Venus flytrap (*Dionaea muscipula*). (A) Illustration of the snap-trap leaves with needle-like tines and touch-sensitive trigger hairs. (B) Action potential in response to two or more contacts with a single trigger hair. Mechanical stimulation of the trigger hairs by a prey activates mechanosensitive ion channels, which leads to the induction of action potentials that cause the lobes of the leaf to close and secrete digestive enzymes. (B after Escalante-Pérez et al. 2011.)

plants in response to various types of stress, indicating that electrical signaling is a general physiological feature of plants. As we will discuss in Chapter 23, electrical signals can be propagated throughout the plant via the vasculature in response to damage caused by chewing insects. However, unlike in animal nervous systems, plants lack synapses that transmit electrical signals from neuron to neuron via the secretion of neurotransmitters. The mechanism of the much slower electrical signal transmission along the vascular systems of plants is still poorly understood.

Hormonal Signaling Pathways

The sites of action of hormones are cells that possess specific receptors that can bind the hormone and initiate a signal transduction cascade. Plants employ large numbers of receptor kinases and signal transduction kinases to bring about the physiological responses of hormone target cells. In the following sections we will examine the types of receptors and signal transduction pathways that are associated with each of the main plant hormones.

The cytokinin and ethylene signal transduction pathways are derived from the bacterial two-component regulatory system

In bacteria, **two-component regulatory systems** are important signaling systems that mediate a wide range of responses to environmental stimuli. The two components of this signaling system consist of a membrane-bound histidine kinase **sensor protein** and a soluble **response regulator** protein (**Figure 15.28A**). Sensor proteins receive the input signal, undergo autophosphorylation on a histidine residue, and pass the signal on to response regulators by transferring the phosphoryl group to a conserved aspartate residue on the response regulator. The phosphorylation-activated response regulators, many of

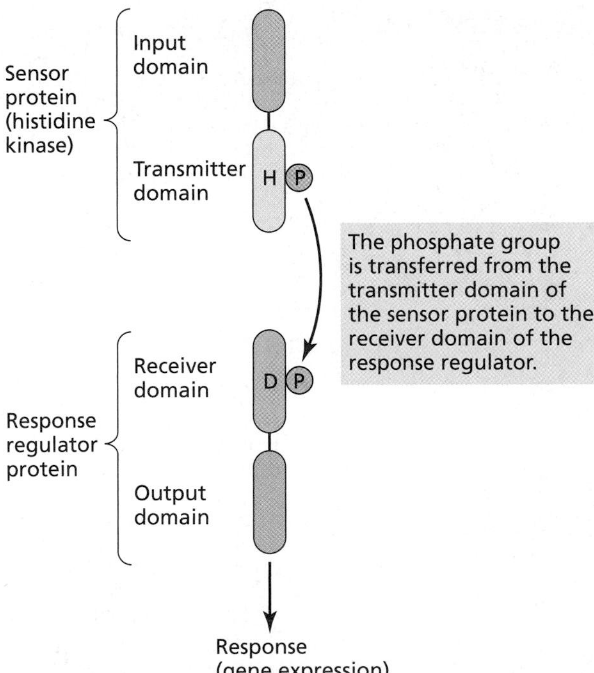

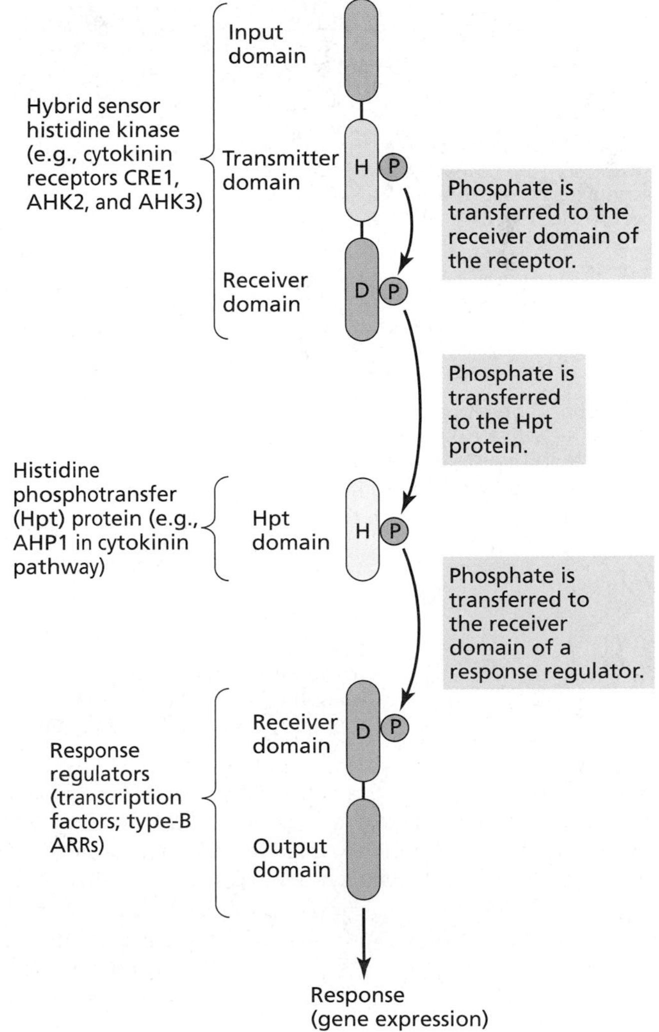

Figure 15.28 Two-component signaling systems of bacteria and plants. (A) The bacterial two-component system, consisting of a sensor protein and a response regulator protein, is found only in prokaryotes. (B) A derived multistep version of the two-component system, involving a phosphotransfer protein intermediate, is found in both prokaryotes and eukaryotes. The plant two-component receptor protein includes a receiver domain fused to the transmitter domain. A separate histidine phosphotransfer protein transfers phosphates from the receiver domain of the receptor to the receiver domain of the response regulator. H, histidine residue; D, aspartate residue.

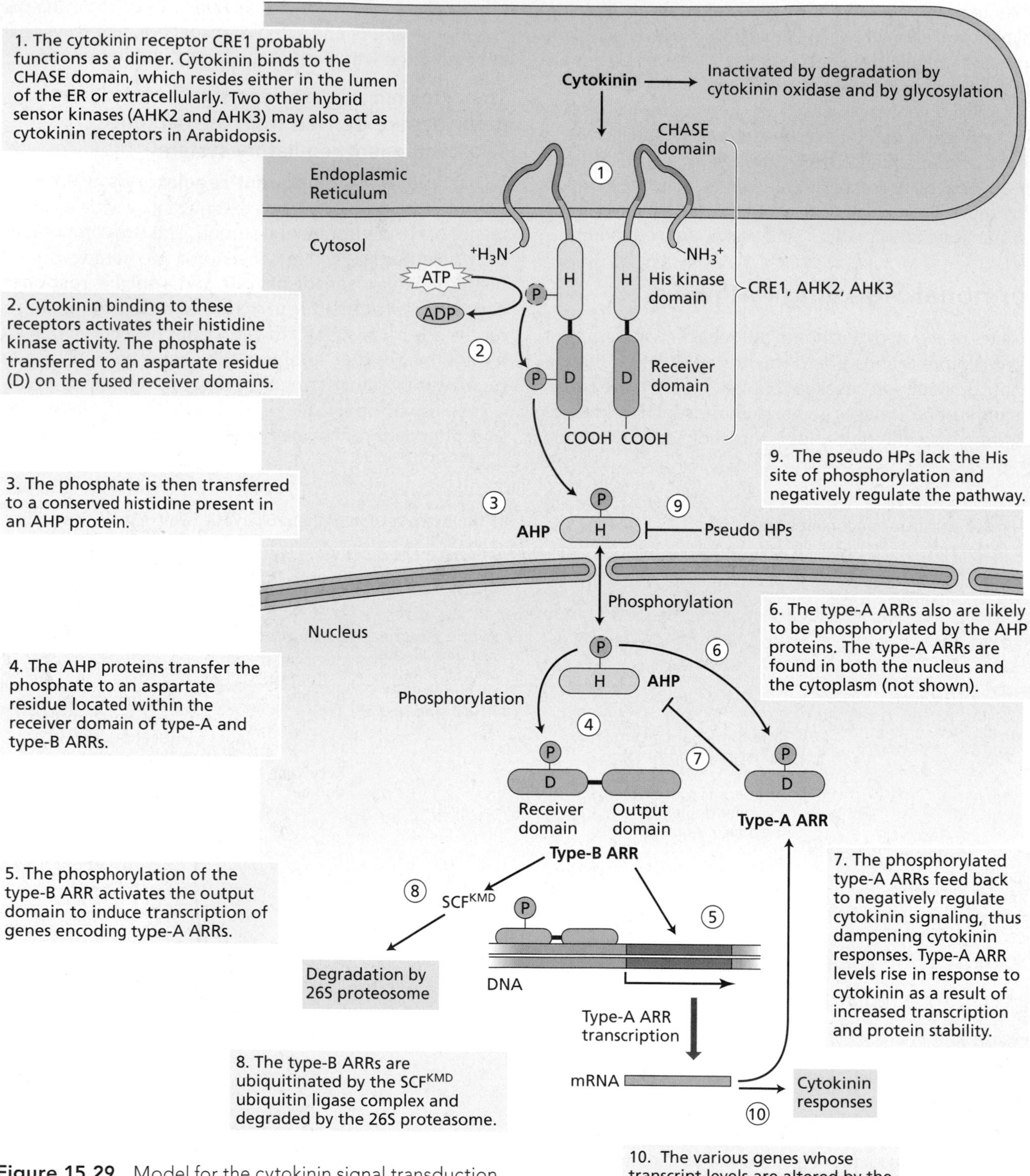

1. The cytokinin receptor CRE1 probably functions as a dimer. Cytokinin binds to the CHASE domain, which resides either in the lumen of the ER or extracellularly. Two other hybrid sensor kinases (AHK2 and AHK3) may also act as cytokinin receptors in Arabidopsis.
Cytokinin
Inactivated by degradation by cytokinin oxidase and by glycosylation
CHASE domain
Endoplasmic Reticulum
1
Cytosol
+H3N
NH3+
ATP
ADP
P
H
H
His kinase domain
CRE1, AHK2, AHK3
2. Cytokinin binding to these receptors activates their histidine kinase activity. The phosphate is transferred to an aspartate residue (D) on the fused receiver domains.
2
P
D
D
Receiver domain
COOH COOH
9. The pseudo HPs lack the His site of phosphorylation and negatively regulate the pathway.
3. The phosphate is then transferred to a conserved histidine present in an AHP protein.
3
P
9
AHP
H
Pseudo HPs
Phosphorylation
6. The type-A ARRs also are likely to be phosphorylated by the AHP proteins. The type-A ARRs are found in both the nucleus and the cytoplasm (not shown).
Nucleus
P
6
H
AHP
4. The AHP proteins transfer the phosphate to an aspartate residue located within the receiver domain of type-A and type-B ARRs.
Phosphorylation
4
P
7
P
D
D
Receiver domain
Output domain
Type-A ARR
Type-B ARR
5. The phosphorylation of the type-B ARR activates the output domain to induce transcription of genes encoding type-A ARRs.
8
SCFKMD
P
5
7. The phosphorylated type-A ARRs feed back to negatively regulate cytokinin signaling, thus dampening cytokinin responses. Type-A ARR levels rise in response to cytokinin as a result of increased transcription and protein stability.
Degradation by 26S proteosome
DNA
Type-A ARR transcription
8. The type-B ARRs are ubiquitinated by the SCFKMD ubiquitin ligase complex and degraded by the 26S proteasome.
mRNA
Cytokinin responses
10
10. The various genes whose transcript levels are altered by the type-B ARRs, as well as their downstream targets, mediate the response of the cell to cytokinin.

Figure 15.29 Model for the cytokinin signal transduction pathway. Cytokinin binds to the dimerized CRE1 receptor on the endoplasmic reticulum, which initiates a phosphorylation cascade leading to the cytokinin response. KMDs, KISS ME DEADLY proteins; AHP, Arabidopsis histidine phosophotransfer protein; ARR, Arabidopsis response regulator; pseudo HP, inhibits cytokinin signaling by competing with AHP1-5 for phosphotransfer.

which function as transcription factors, then bring about the cellular response. Sensor proteins have two domains, an *input domain,* which receives the environmental signal, and a *transmitter domain,* which transmits the signal to the response regulator. The response regulator proteins also have two domains, a *receiver domain,* which receives the signal from the transmitter domain of the sensor protein, and an *output domain,* which mediates the response.

Modifications of this simple bacterial two-component system are found in the signal transduction pathways activated by the plant hormones cytokinin and ethylene. Cytokinin signaling is mediated by a phosphorylation relay system that consists of a transmembrane cytokinin receptor, a phosphotransfer protein, and a nuclear response regulator (**Figure 15.28B**). The cytokinin receptors, designated CRE1, AHK2, and AHK3 in Arabidopsis, are related in amino acid sequence to the histidine kinases in two-component systems. However, these cytokinin receptors are described as *hybrid sensor histidine kinases* since they contain both the bacterial sensor input and histidine kinase (transmitter) domains as well as the receiver domain of a bacterial response regulator protein.

It was originally assumed that cytokinin receptors are located in the plasma membrane, reflected in the name of the ligand binding domain, CYCLASE HISTIDINE KINASE ASSOCIATED SENSORY EXTRACELLULAR (CHASE). However, the majority of the cytokinin receptors of Arabidopsis and maize actually reside in the ER. Cytokinin binding to the CHASE domain of its receptor triggers autophosphorylation of a histidine residue on the transmitter domain, followed by transfer of this same phosphate to the aspartate residue on the receiver domain (**Figure 15.29**). The phosphate is then transferred to **ARABIDOPSIS HISTIDINE PHOSPHOTRANSFER** (**AHP**) proteins. The newly phosphorylated AHPs function as signaling intermediates that transmit membrane-perceived cytokinin signals to nuclear-localized response regulators (termed **ARABIDOPSIS RESPONSE REGULATOR** or **ARR**) by transferring the phosphate group to an aspartate on the receiver domain of the ARR (see Figure 15.29). This phosphorylation of the ARRs alters their activity, which brings about the cellular response.

The ARR response regulators are encoded by multigene families. They fall into two basic classes: the **type-A ARR** genes, the products of which are made up solely of a receiver domain, and the **type-B ARR** genes, which also include an output domain containing DNA-binding transcription activation sites (see Figure 15.29). Type-A ARRs negatively regulate cytokinin signaling by interacting with other proteins in a manner dependent on the phosphorylation state of the type-A ARR. Type-B ARRs are activated by phosphorylation, which enables them to regulate the transcription of a set of target genes, including those encoding the type-A ARRs, which gives rise to the cellular changes involved in the cytokinin response. A family of F-box proteins called KISS ME DEADLY (KMD) proteins negatively regulates the cytokinin response by targeting type-B ARR proteins for degradation via the E3 ubiquitin ligase complex, SCF^{KMD}.

Ethylene receptors are encoded by a multigene family (in Arabidopsis, ETR1, ETR2, ERS1, ERS2, and EIN4) that is also evolutionarily related to bacterial two-component histidine kinases. However, only two of the ethylene receptors in Arabidopsis (ETR1 and ERS1) have intrinsic histidine kinase activity, and this activity does not appear to play an essential role in signaling. In contrast to cytokinin signaling, the ethylene-signaling pathway thus does not involve a phosphorylation relay system. Ethylene receptors are localized to the ER membrane and interact with two downstream signaling proteins, CTR1 (CONSTITUTIVE TRIPLE RESPONSE1) and EIN2 (ETHYLENE-INSENSITIVE2) (**Figure 15.30**). CTR1 is a soluble serine/threonine kinase that is always physically associated with ethylene receptors. EIN2 is an ER transmembrane protein with a cytosolic C-terminal domain that is a target for CTR1 kinase activity. EIN2 is required to stabilize transcription factors of the EIN3 (ETHYLENE-INSENSITIVE3) family, which activate the transcription of ethylene-responsive genes.

Ethylene receptors function as *negative regulators* that actively repress the hormone response in the absence of the hormone. In the absence of ethylene (when receptors are turned on), the ethylene receptors activate the CTR1 kinase, which then directly phosphorylates and thereby inactivates EIN2 (see Figure 15.30). Active CTR1 is thus also a negative regulator of the ethylene response pathway.

When ethylene binds to the N-terminal transmembrane domain of the ethylene receptors, the receptors are inactivated and CTR1 is "switched off." This leads to the dephosphorylation of EIN2 by an as-yet-unidentified phosphatase and the subsequent proteolytic cleavage of its cytosolic C terminus by an unidentified protease. Interaction of CTR1 with EIN2 and similar proteins also regulates the stability of the receptor to ensure that ethylene response mechanisms can rapidly reset. The released EIN2 C-terminal domain then migrates into the nucleus where it activates EIN3, either directly or indirectly. The activated EIN3 family of transcription factors regulates the transcription of most genes that are rapidly induced by ethylene, including the transcription factor **ERF1** (**ETHYLENE RESPONSE FACTOR1**; see Figure 15.30). The activation of EIN3 and the ERFs serves to alter the expression of a large number of genes to bring about the numerous changes in the function of plant cells in response to ethylene.

Receptor-like kinases mediate brassinosteroid and certain auxin signaling pathways

The largest class of plant receptor kinases consists of the receptor-like serine/threonine kinases (RLKs). Many RLKs localize to the plasma membrane as transmembrane

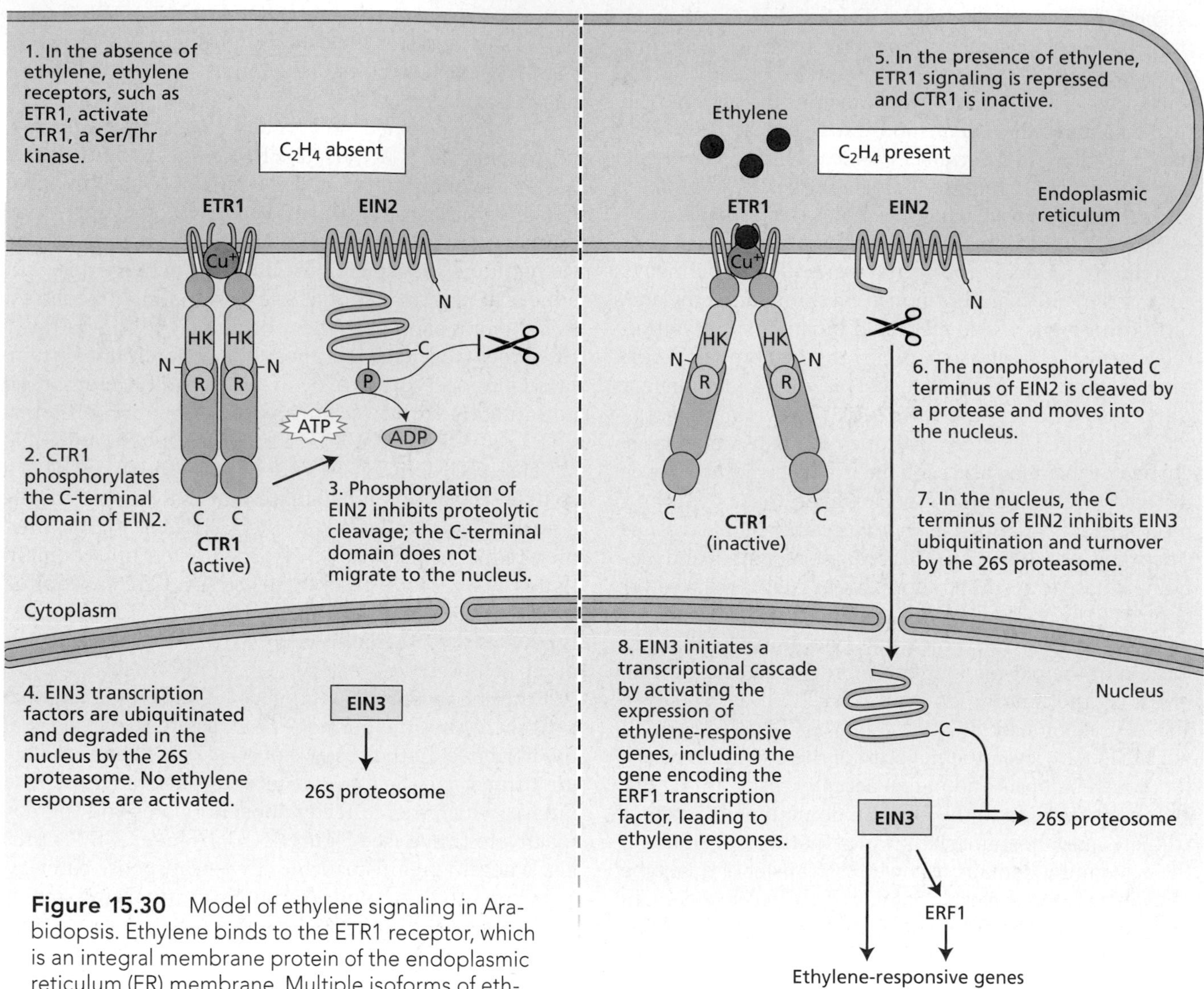

1. In the absence of ethylene, ethylene receptors, such as ETR1, activate CTR1, a Ser/Thr kinase.
C_2H_4 absent
ETR1
EIN2
Cu+
HK
R
N
C
P
ATP
ADP
2. CTR1 phosphorylates the C-terminal domain of EIN2.
CTR1 (active)
3. Phosphorylation of EIN2 inhibits proteolytic cleavage; the C-terminal domain does not migrate to the nucleus.
Cytoplasm
4. EIN3 transcription factors are ubiquitinated and degraded in the nucleus by the 26S proteasome. No ethylene responses are activated.
EIN3
26S proteosome
5. In the presence of ethylene, ETR1 signaling is repressed and CTR1 is inactive.
Ethylene
C_2H_4 present
Endoplasmic reticulum
CTR1 (inactive)
6. The nonphosphorylated C terminus of EIN2 is cleaved by a protease and moves into the nucleus.
7. In the nucleus, the C terminus of EIN2 inhibits EIN3 ubiquitination and turnover by the 26S proteasome.
8. EIN3 initiates a transcriptional cascade by activating the expression of ethylene-responsive genes, including the gene encoding the ERF1 transcription factor, leading to ethylene responses.
Nucleus
26S proteosome
ERF1
Ethylene-responsive genes

Figure 15.30 Model of ethylene signaling in Arabidopsis. Ethylene binds to the ETR1 receptor, which is an integral membrane protein of the endoplasmic reticulum (ER) membrane. Multiple isoforms of ethylene receptors may be present in a cell; only ETR1 is shown for simplicity. The receptor is a dimer, held together by disulfide bonds. Ethylene binds within the transmembrane domain, through a copper cofactor, which is assembled into the ethylene receptors. (After Ju et al. 2012.)

proteins that harbor extracellular ligand-binding domains and cytoplasmic kinase domains, which relay information to the cell interior via phosphorylation of serine or threonine residues of target proteins. Some RLKs have also been shown to phosphorylate tyrosine residues. The ligands for several RLKs have been identified and include chemical signals produced by biotic interactors as well as endogenous plant hormones such as brassinosteroids, auxin, and peptide hormones.

The RLK-mediated brassinosteroid signaling pathway combines signal amplification and repressor inactivation strategies to transduce an extracellular brassinosteroid hormone signal into a transcriptional response. In brief, brassinolide binding to the brassinosteroid receptor kinase BRASSINOSTEROID-INSENSITIVE1 (BRI1) on the plasma membrane triggers a phosphorylation cascade that causes a repressor protein, BRASSINOSTEROID-INSENSITIVE2 (BIN2), to become inactivated. This results in activation of the transcription factors BRI1-EMS SUPPRESSOR1 (BES1) and BRASSINAZOLE-RESISTANT1 (BZR1) and subsequent gene expression (**Figure 15.31**).

The BRI1 receptor belongs to the plasma membrane leucine rich repeat (LRR) subfamily of RLKs and contains an N-terminal extracellular domain that binds brassinolide, a single transmembrane domain, and a cytoplasmic kinase domain with specificity toward tyrosine, serine, or threonine residues (see Figure 15.31). Upon binding brassinolide, homodimers of BRI1 are activated and heterooligomerize with the RLK BRI1-ASSOCIATED RECEPTOR KINASE1 (BAK1) (see Figure 15.31); both RLKs undergo auto- and

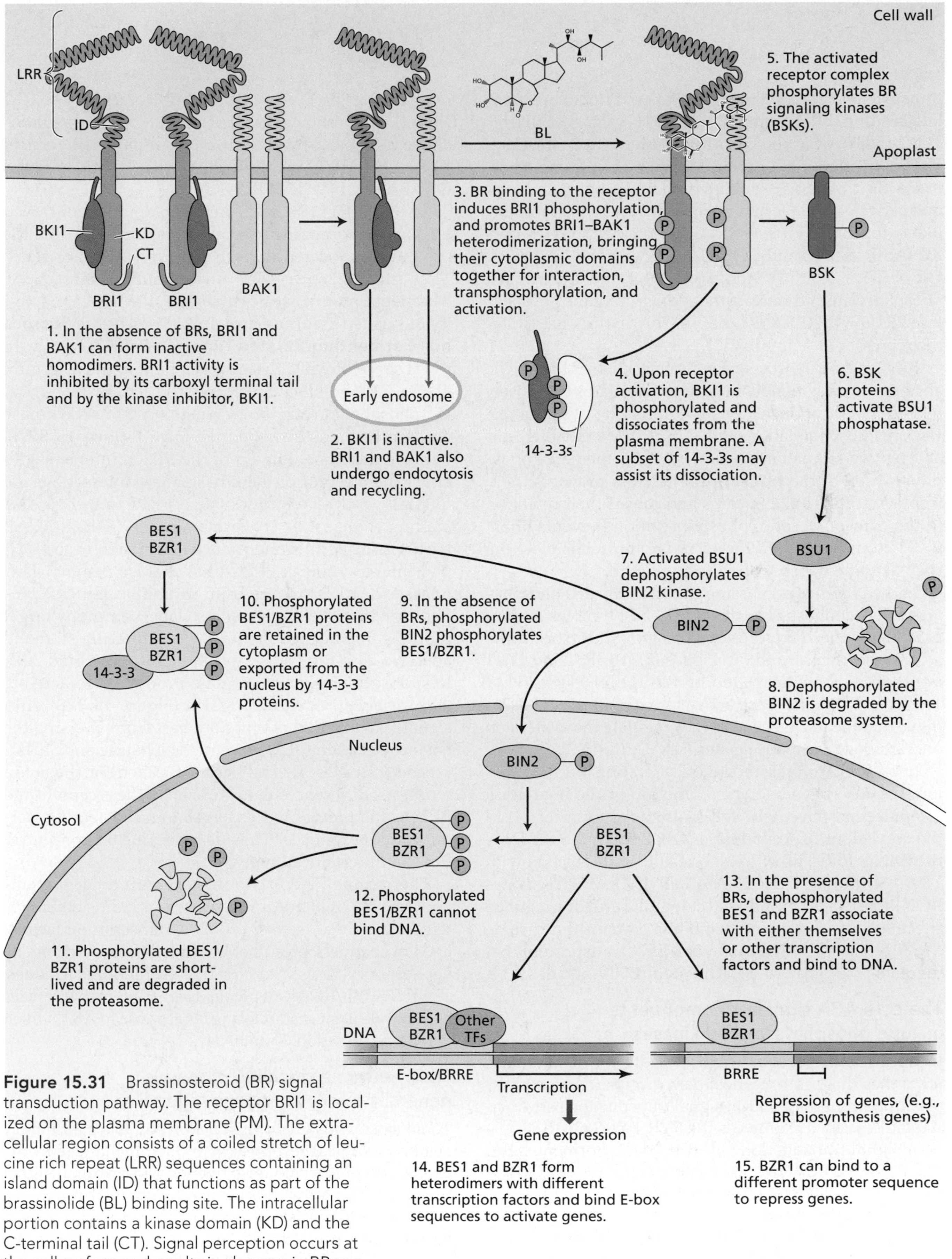

Cell wall
LRR
ID
BKI1
KD
CT
BRI1
BRI1
BAK1
BL
Apoplast
5. The activated receptor complex phosphorylates BR signaling kinases (BSKs).
3. BR binding to the receptor induces BRI1 phosphorylation, and promotes BRI1–BAK1 heterodimerization, bringing their cytoplasmic domains together for interaction, transphosphorylation, and activation.
BSK
1. In the absence of BRs, BRI1 and BAK1 can form inactive homodimers. BRI1 activity is inhibited by its carboxyl terminal tail and by the kinase inhibitor, BKI1.
Early endosome
2. BKI1 is inactive. BRI1 and BAK1 also undergo endocytosis and recycling.
14-3-3s
4. Upon receptor activation, BKI1 is phosphorylated and dissociates from the plasma membrane. A subset of 14-3-3s may assist its dissociation.
6. BSK proteins activate BSU1 phosphatase.
BES1
BZR1
BSU1
7. Activated BSU1 dephosphorylates BIN2 kinase.
10. Phosphorylated BES1/BZR1 proteins are retained in the cytoplasm or exported from the nucleus by 14-3-3 proteins.
9. In the absence of BRs, phosphorylated BIN2 phosphorylates BES1/BZR1.
BIN2
14-3-3
8. Dephosphorylated BIN2 is degraded by the proteasome system.
Nucleus
Cytosol
12. Phosphorylated BES1/BZR1 cannot bind DNA.
13. In the presence of BRs, dephosphorylated BES1 and BZR1 associate with either themselves or other transcription factors and bind to DNA.
11. Phosphorylated BES1/BZR1 proteins are short-lived and are degraded in the proteasome.
Other TFs
DNA
E-box/BRRE
Transcription
Gene expression
BRRE
Repression of genes, (e.g., BR biosynthesis genes)
14. BES1 and BZR1 form heterodimers with different transcription factors and bind E-box sequences to activate genes.
15. BZR1 can bind to a different promoter sequence to repress genes.

Figure 15.31 Brassinosteroid (BR) signal transduction pathway. The receptor BRI1 is localized on the plasma membrane (PM). The extracellular region consists of a coiled stretch of leucine rich repeat (LRR) sequences containing an island domain (ID) that functions as part of the brassinolide (BL) binding site. The intracellular portion contains a kinase domain (KD) and the C-terminal tail (CT). Signal perception occurs at the cell surface and results in changes in BR-regulated gene expression. BRRE, brassinoisteroid response element. (After Jiang et al. 2013.)

transphosphorylation during activation. Prior to brassinolide binding, BRI1 interacts with BRI1-KINASE INHIBITOR1 (BKI1), which prevents association with BAK1. Upon BRI1 activation, BKI1 is released from the plasma membrane, BRI1 and BAK1 dimerize, and BRI1 phosphorylates and activates two plasma membrane–anchored receptor-like cytoplasmic kinases (RLCKs), the BR-SIGNALING KINASE1 (BSK1) and CONSTITUTIVE DIFFERENTIAL GROWTH1 (CDG1). Activated BSK1 and CDG1 then phosphorylate and activate the serine/threonine phosphatase BRI1 SUPPRESSOR1 (BSU1). This, in turn, inactivates the repressor protein BIN2.

BIN2 is a serine/threonine protein kinase that, in the absence of brassinolide, negatively regulates the closely related transcription factors BES1 and BZR1 by phosphorylation. Phosphorylation of BES1/BZR1 by active BIN2 has at least two regulatory roles. First, BIN2-mediated phosphorylation of the transcription factors prevents them from shuttling to the nucleus and causes their retention in the cytosol. Second, phosphorylation prevents BES1/BZR1 from binding to target promoters, thus blocking their activity as transcriptional regulators.

In the presence of brassinolide, the activated phosphatase BSU1 dephosphorylates BIN2 and promotes its degradation by the 26S proteasome system, thus blocking its activity (see steps 6 and 7 in Figure 15.31). BES1 and BZR1 are then dephosphorylated by PROTEIN PHOSPHATASE2A (PP2A), and the active forms of BES1 and BZR1 move into the nucleus where they regulate the expression of brassinolide response genes (see Figure 15.31).

In addition to brassinosteroid signaling, an RLK system has also been shown to function in the regulation of epidermal pavement cell lobing (see Chapter 14) in the cotyledons of Arabidopsis. Auxin binding to AUXIN BINDING PROTEIN1 (ABP1) results in interaction with TRANSMEMBRANE KINASE1 (TMK1), which activates ROP kinases. ROPs and associated RIC proteins regulate subcellular trafficking of membrane proteins that include the PINFORMED (PIN) auxin efflux transporters that regulate polar growth (see Chapters 17, 19, and 21).

The core ABA signaling components include phosphatases and kinases

In addition to protein kinases, protein phosphatases (enzymes that remove phosphate groups from proteins) play important roles within signal transduction pathways. A well-described example is the PYR/PYL/RCAR-dependent signal transduction pathway of the hormone ABA. The **PYR/PYL/RCAR** members of the START (*ST*EROIDOGENIC *A*CUTE *R*EGULATORY PROTEIN-RELATED LIPID-*T*RANSFER) domain protein superfamily, which contains a predicted hydrophobic ligand-binding pocket, constitute the initial step of the core ABA signal transduction pathway. Fourteen members of this subfamily have been identified in Arabidopsis. Their nomenclature reflects their discoveries: PYRABACTIN RESISTANCE1 (PYR1) which shows resistance to the synthetic sulfonamide compound pyrabactin, which mimics ABA action; PYR1-LIKE (PYL) and REGULATORY COMPONENTS OF ABA RECEPTORS (RCARs).

The PYR/PYL/RCAR protein subfamily is conserved among plants ranging from eudicots to mosses, and the proteins are located in both the cytosol and nucleus. They interact with PP2C phosphatases in an ABA-dependent manner to regulate the downstream activity of serine/threonine protein kinases of the **Sucrose non-Fermenting Related Kinase2** (**SnRK2**) family. In the absence of ABA, these PP2Cs bind to the C termini of SnRK2s and block SnRK2 kinase activity by removing phosphate groups from a region within the kinase domain termed the *activation loop* (**Figure 15.32A**). Because the same domain of the PP2Cs interacts with either the receptor or the kinase, these interactions are mutually exclusive for individual PP2C isoforms. ABA binding changes the conformation of PYR/PYL/RCAR receptors to permit or enhance interaction with PP2C and thereby repress PP2C phosphatase activity. This releases SnRK2 kinases from inhibition. SnRK2 proteins are then free to phosphorylate their many target proteins, including ion channels regulating stomatal aperture and the transcription factors that bind ABA Response Elements to in gene promoters to activate ABA-responsive gene expression (**Figure 15.32B**). ABA signal transduction is therefore based on reversing the balance between PP2C protein phosphatase and SnRK2 kinase activities. As has been described for the auxin receptors, differences in expression of the receptors and PP2Cs, and their affinities for ABA and each other, permit varied responses to a wide range of ABA concentrations in different cell types.

These same PP2Cs also interact with other proteins implicated in cellular ABA responses, including other protein kinases, Ca^{2+} sensor proteins, transcription factors, and ion channels, presumably regulating their activity by dephosphorylating specific serine or threonine residues. The PYR/PYL/RCAR-dependent signaling pathway plays a major role in stomatal closing in response to ABA, which will be discussed in Chapter 24.

Plant hormone signaling pathways generally employ negative regulation

The majority of signal transduction pathways ultimately elicit a biological response by inducing changes in the expression of selected target genes. Most animal signal transduction pathways induce a response through the activation of a cascade of positive regulators. In contrast, *the majority of plant transduction pathways induce a response by inactivating repressor proteins*. For example, ethylene binding to ETR1 results in dissociation from the repressor CTR1 and activation of the transcription factor EIN3

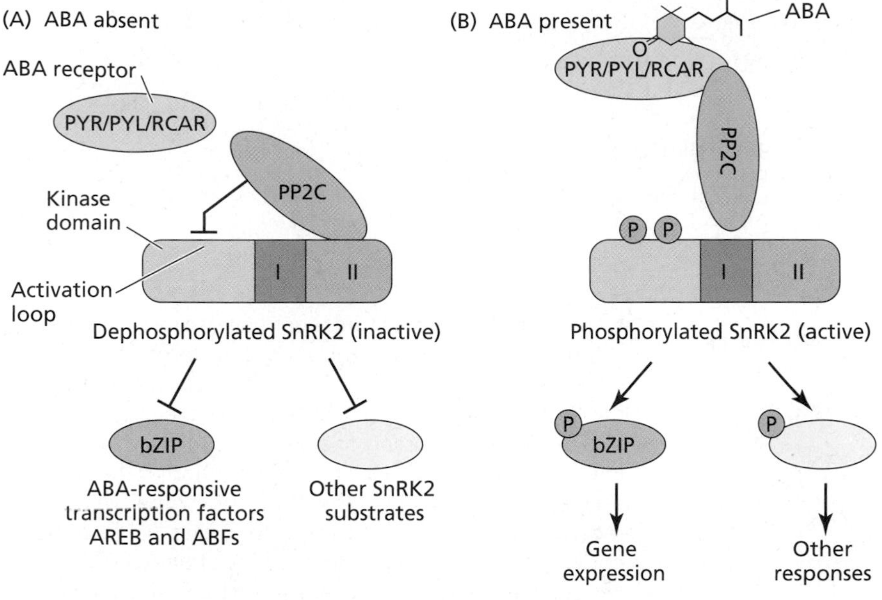
(A) ABA absent
ABA receptor
PYR/PYL/RCAR
PP2C
Kinase domain
Activation loop
I
II
Dephosphorylated SnRK2 (inactive)
bZIP
ABA-responsive transcription factors AREB and ABFs
Other SnRK2 substrates
In the absence of ABA, the protein phosphatase PP2C keeps the protein kinase SnRK2 dephosphorylated and thereby inactivated.
(B) ABA present
ABA
PYR/PYL/RCAR
PP2C
P
Phosphorylated SnRK2 (active)
bZIP
Gene expression
Other responses
When ABA is present, its receptor prevents dephosphorylation of SnRK2 by PP2C. Phosphorylated (active) SnRK2 phosphorylates downstream substrates, thereby inducing ABA responses.

Figure 15.32 Abscisic acid (ABA) signaling involves kinase and phosphatase activities. (A) In the absence of ABA, the protein phosphatase PP2C dephosphorylates and inactivates the SnRK2 kinase. (B) In the presence of ABA, the ABA receptor protein PYR/PYL/RCAR interacts with PP2C, blocking phosphatase action and releasing SnRK2 from negative regulation. The activated SnRK2 phosphorylates ABA-responsive transcription factors (bZIP) and other unknown substrates to induce an ABA response. SnRK2, SNF1-related protein kinase 2; PP2C, protein phosphatase 2C; AREB, ABA-responsive element binding protein; ABF, ABA-responsive element-binding factor.

(see Figure 15.30). Similarly, brassinosteroid binding to the receptor kinase BRI1 causes the repressor protein BIN2 to become inactivated, resulting in activation of the transcription factors BES1 and BZR1 (see Figure 15.31).

Why have plant cells evolved signaling pathways based on negative regulation rather than positive regulation, as occurs in animal cells? Mathematical modeling of signal transduction pathways employing negative regulators suggests that negative regulators result in faster induction of downstream response genes. The speed of a response, particularly to an environmental stress such as drought, may be crucial to the survival of the sessile plant. Hence, the adoption by plants of negative regulatory signaling pathways in the majority of cases is likely to have conferred a selective advantage during evolution.

Several different molecular mechanisms have been described in plant cells to inactivate repressor proteins: these include dephosphorylation to modulate repressor activity, retargeting of the repressor to another cellular compartment, and degradation of the repressor protein. As noted above, protein dephosphorylation is employed by the brassinosteroid pathway to inactivate the repressor protein BIN2 (see Figure 15.31).

Several plant hormone receptors encode components of the ubiquitination machinery and mediate signaling via protein degradation

Protein degradation as a mechanism to inactivate repressor proteins was first described as part of the auxin signaling pathway. Since then, the **ubiquitin–proteasome pathway** has been shown to be central to most, if not all, hormone signaling pathways. In brief, a small protein called ubiquitin is first activated by an enzyme termed an *E1 ubiquitin-activating enzyme* in an ATP-dependent manner (**Figure 15.33A**; see also Figure 2.18). The ubiquitin tag is transferred to a second enzyme called an E2 ubiquitin-conjugating enzyme. This enzyme then associates with one of a family of large protein complexes called **S-PHASE KINASE-ASSOCIATED PROTEIN1 (Skp1)/Cullin/F-box (SCF) complexes**, which function as E3 ubiquitin ligases. A superscript term is applied to an E3 ligase name (e.g., SCF^{TIR1}) to indicate which F-box protein the complex contains. F-box proteins typically recruit target proteins to the SCF complex so that they can be tagged with multiple copies of ubiquitin by the E3 ligase (see Figure 15.33A). Such polyubiquitination acts as a marker targeting the protein for degradation by the 26S proteasome, a large multiprotein complex that degrades ubiquitin-tagged proteins. In plants, the F-box gene family has been greatly expanded to many hundreds of genes, which presumably degrade a similar number of distinct targets. For instance, the KMD proteins described previously function as part of an SCF^{KMD} E3 ubiquitin ligase complex and directly interact with type-B ARR proteins to negatively regulate the cytokinin signaling pathway.

Several of these F-box proteins function as hormone receptor complexes (**Figure 15.33B and C**). In many

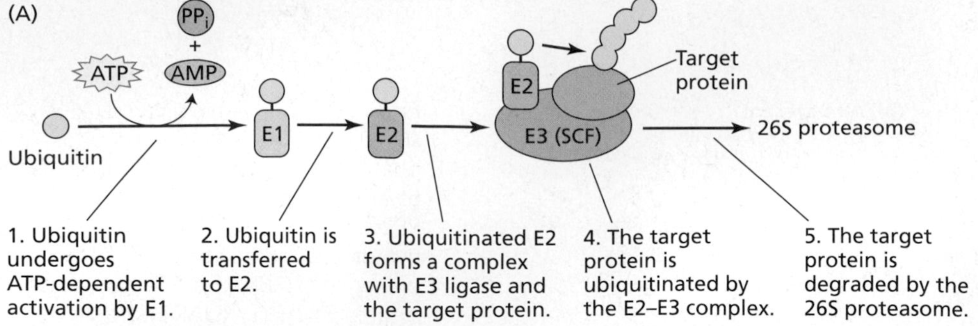

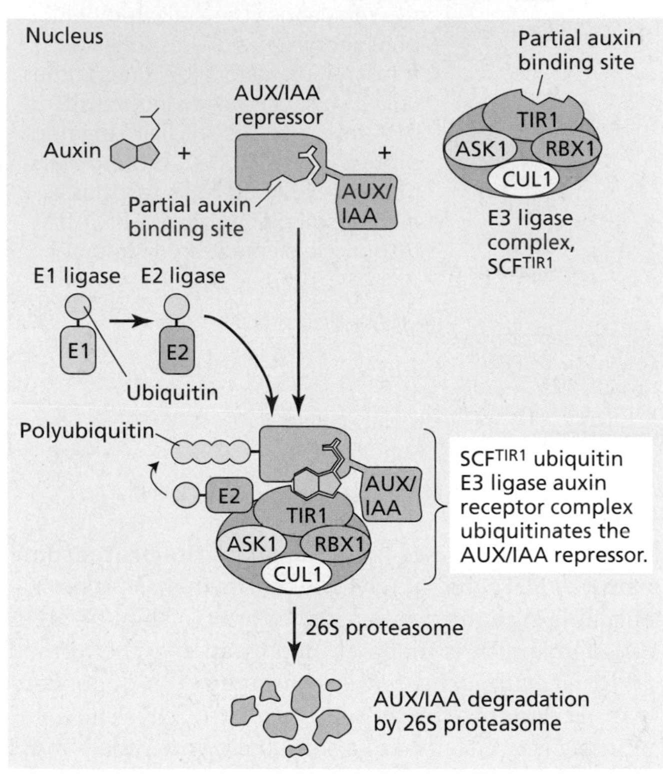

Figure 15.33 Signal transduction pathways in plants often function by inactivating repressor proteins. (A) Schematic diagram of the ubiquitin–proteasome degradation pathway, which occurs in both the cytosol and nucleus. (B) The binding of auxin to its receptor complex initiates ubiquitin-dependent degradation of the AUX/IAA repressor protein by the 26S proteasome. The auxin receptor is composed of two proteins: the SCF complex component TIR1 and the repressor protein AUX/IAA. Ubiquitin moieties are first activated by E1 ligase and added to target proteins by E2 ligase. TIR1 recruits AUX/IAA proteins to the SCF^{TIR1} complex in an auxin-dependent manner. Once recruited by auxin, AUX/IAA proteins are ubiquitinated by the E3 ligase activity of the SCF^{TIR1} complex, which marks the protein for destruction by the 26S proteasome. (C) The binding of gibberellin (GA) to its receptor leads to the degradation of the DELLA repressor by the 26S proteasome. (Top) In the nucleus, gibberellin binds to the GID1 receptor and induces a conformational change in the N-terminal domain of the receptor, allowing the receptor to interact with the TVHYNP and DELLA domains of the DELLA repressor. (Bottom) The formation of the GID1-repressor complex promotes the interaction between the repressor and the E3 ubiquitin ligase SCF^{SLY}, leading to the ubiquitination and degradation of the DELLA repressor by the 26S proteasome.

(C)

Formation of the GA-GID1-DELLA complex

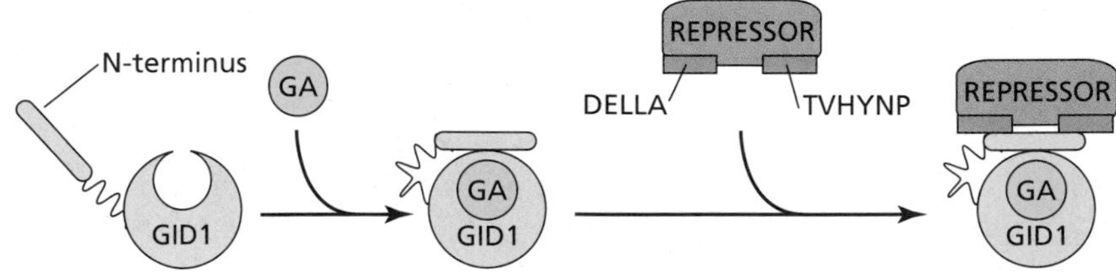

Proteasome-dependent degradation of DELLAs

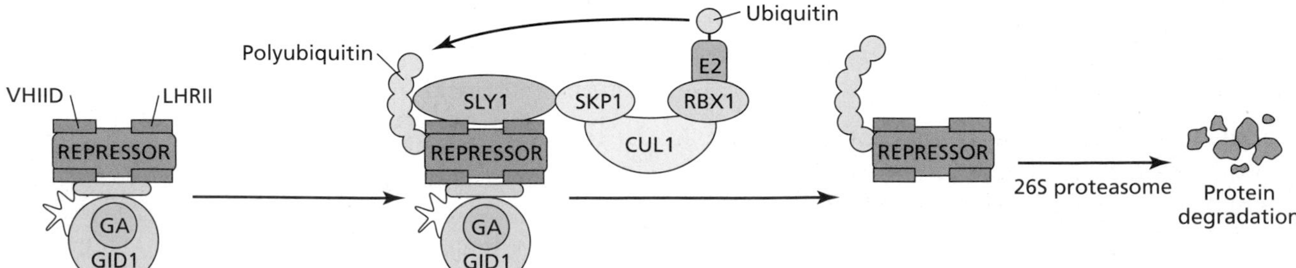

hormone signaling pathways, the proteins targeted for degradation are transcriptional repressors.

In the auxin signaling pathway, genes in the auxin receptor gene family, *TIR1/AFB1-5*, encode F-box components of the SCF complex that functions in the degradation of **AUXIN/INDOLE-3-ACETIC ACID (AUX/IAA)** repressors of auxin-responsive gene transcription (**Figure 15.34A**). Auxin-responsive genes typically have **auxin response element (AuxRE)** binding sites located in their promoter regions. **Auxin response factors (ARFs)** are transcription factors that bind to these AuxRE motifs to stimulate or repress transcription (see Figure 15.34A). To activate transcription, ARFs form homodimers via conserved interaction domains and recruit chromatin remodeling factors. When auxin concentrations are low, AUX/IAA repressor proteins containing interaction domains similar to those found in the ARFs form heterodimers with the ARFs and thus repress transcriptional activation. In the presence of auxin, AUX/IAA repressors are recruited to the TIR1/AFB receptor complex and are tagged with ubiquitin for degradation by the 26S proteasome (see Figure 15.33B). This allows ARFs to dimerize or even oligomerize and activate gene transcription. TIR1 (and AFB1-5) therefore function as auxin co-receptors, with auxin acting as a "molecular glue" without a requirement for receptor phosphorylation, unlike what is observed in analogous E3 ligase systems. Among the many target genes are genes encoding auxin-metabolizing enzymes and AUX/IAA repressors, which eventually serve to reduce active auxin levels and terminate ARF-dependent signaling.

The plant hormones jasmonate and gibberellin also promote the interaction between an F-box protein of a SCF ubiquitin E3 ligase and its transcriptional repressor target proteins (**Figure 15.34B and C**). The F-box protein CORONATINE-INSENSITIVE1 (COI1) functions as a jasmonate receptor. Like auxin, jasmonate (conjugated to the amino acid isoleucine) promotes the interaction between COI1 and repressors of jasmonate-induced gene expression called **JASMONATE ZIM-DOMAIN (JAZ)** proteins (see Figure 15.34B), thereby targeting JAZ proteins for degradation (see Figure 23.19). Analogous to AUX/IAA proteins, JAZ repressor proteins suppress transcription of jasmonate-responsive genes by binding to basic helix-loop-helix (bHLH) MYC transcription factors. Jasmonate-induced ubiquitin-dependent degradation of JAZ repressor proteins results in the release and activation of MYC transcription factors, triggering the induction of jasmonate-responsive gene expression.

Gibberellin signaling also involves components of the SCF complex (see Figure 15.34C). However, the gibberellin receptor **GIBBERELLIN INSENSITIVE DWARF 1 (GID1)** does not itself function as an F-box protein. Instead, when GID1 binds gibberellin, the receptor undergoes a conformational change that promotes the binding of DELLA repressor proteins. This in turn induces a conformational change in the DELLA protein and facilitates the interaction of GID1-bound DELLA to SCF^{SLY1}, an E3 ubiquitin ligase in Arabidopsis that contains the F-box protein SLY1 (see Figure 15.33C). In effect, the binding of gibberellin receptor GID1 to DELLA repressor proteins triggers ubiquitination of the latter via the F-box protein SLY1 and subsequent degradation of DELLA proteins by the 26S proteasome. For example, DELLA degradation results in the release and activation of **phytochrome interacting factor (PIF)** transcription factors, such as PIF3 and PIF4, as well as other bHLH transcription factors, thus triggering changes in gene expression.

As the above discussion indicates, auxin, jasmonate, and gibberellins signal by *directly* targeting the stability of nuclear-localized repressor proteins and thereby inducing a transcriptional response. Such a short signal transduction pathway provides the means for a very rapid change in nuclear gene expression. However, there is no opportunity for signal amplification as in the case of a signaling pathway involving a kinase cascade or second messengers. Instead, any resulting transcriptional response is directly related to the abundance of the signal molecule, since this determines the number of repressor molecules that are degraded. This important feature in the organization of signal transduction pathways may help explain why comparatively high concentrations of signals such as auxin and gibberellin are required to elicit a biological response.

Plants have evolved mechanisms for switching off or attenuating signaling responses

Arguably, the ability to switch off a response to a signal is just as important as the ability to initiate it. Plants terminate signaling through a variety of mechanisms.

As previously discussed, chemical signals such as plant hormones can be degraded or rendered inactive by oxidation or conjugation to sugars or amino acids. They may also be sequestered into other cellular compartments to spatially separate them from receptors.

Receptors and signaling intermediates that are activated by phosphorylation can be inactivated by phosphatase-mediated dephosphorylation. Activated components of the MAP kinase pathway, for example, are inactivated by MAP kinase phosphatases, ensuring tight cellular control over the duration and intensity of MAP kinase–mediated signaling (see Figure 15.4). Similarly, ion transporters and cellular scavengers can quickly lower elevated concentrations of second messengers to switch off signal amplification (see Figure 15.5). As we have seen, protein degradation provides another mechanism for the plant cell to regulate the abundance of key components of the signal transduction pathway, such as the receptor or a transcription factor.

Feedback regulation represents another key mechanism employed to attenuate a response. For example, the *AUX/IAA* genes, which encode the AUX/IAA auxin

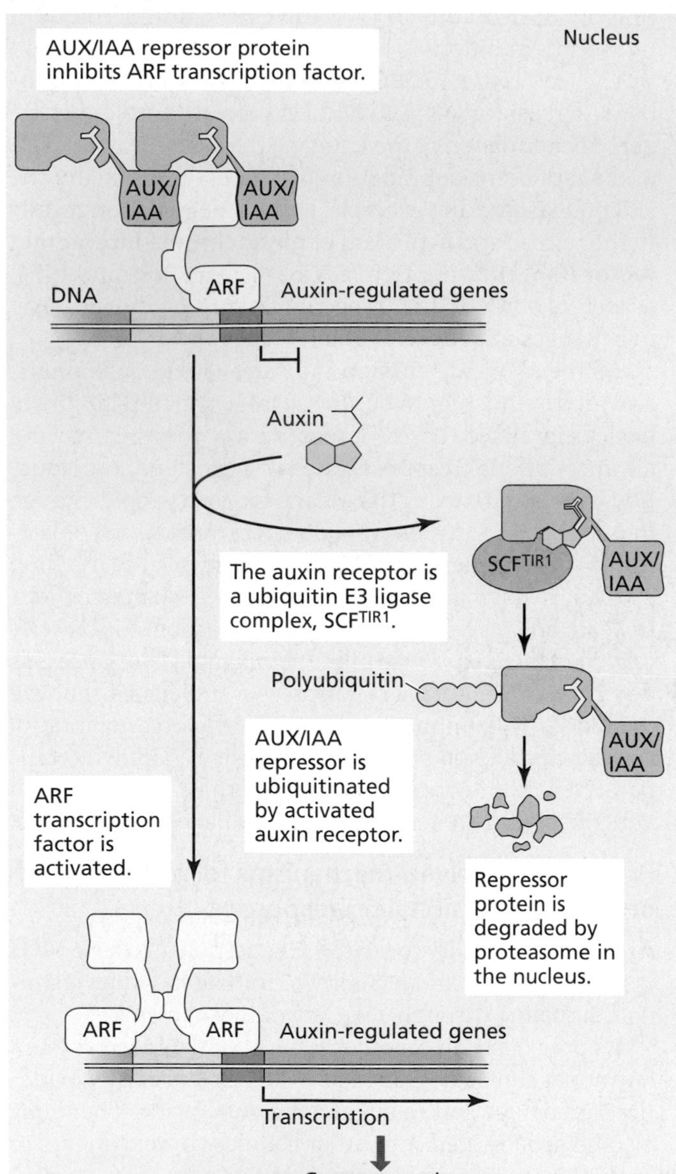

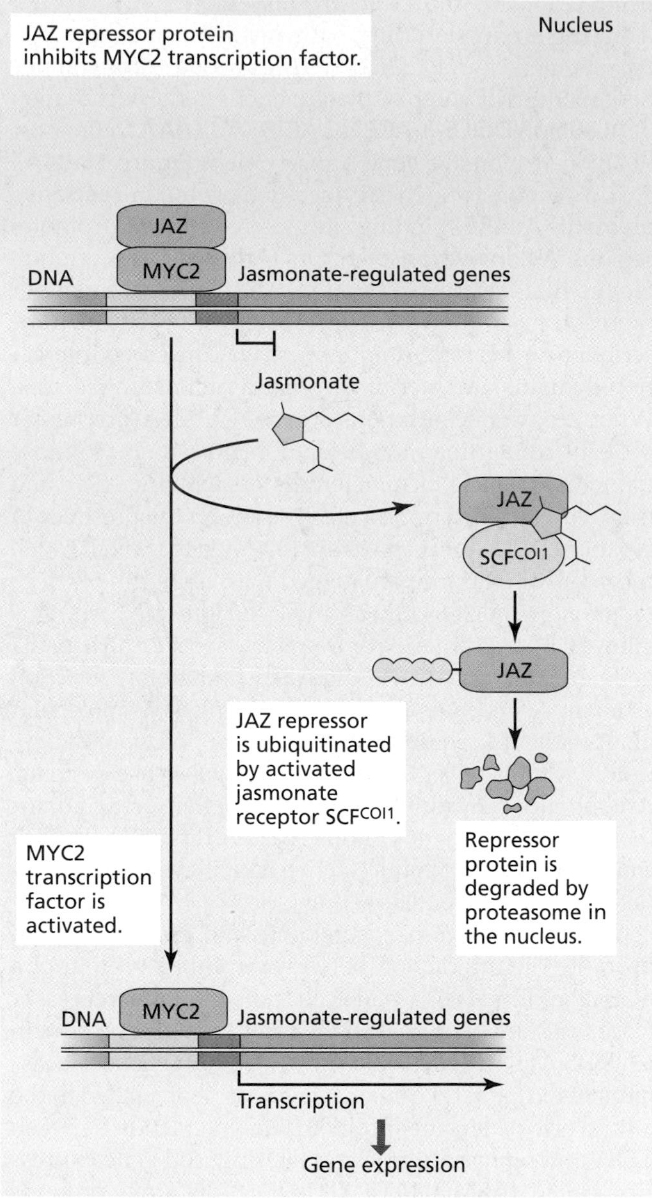

Figure 15.34 Several plant hormone receptors are part of SCF ubiquitination complexes. Auxin, jasmonate (JA), and gibberellin (GA) signal by promoting interaction between components of the SCF ubiquitination machinery and repressor proteins operating in each hormone's signal transduction pathway. Auxin (A) and JA (B) directly promote interaction between the SCF^{TIR1} and SCF^{COI1} complexes and the AUX/IAA and JAZ repressors, respectively. The structural characteristics of the ARF and AUX/IAA proteins that function in auxin signaling have been determined by X-ray crystallography and are reflected in the figure. The structural characteristics of the JAZ repressor protein have not yet been determined. (C) In contrast, gibberellin additionally requires a receptor protein, GID1, to form the complex between SCF^{SLY1} and DELLA proteins. Addition of multiple ubiquitins (polyubiquitin) marks these repressor proteins for degradation. This triggers the activation of ARF, MYC2, and PIF3/4 transcription factors, resulting in auxin-, jasmonate-, and gibberellin-induced changes in gene expression.

repressor proteins, have auxin response element binding sites located in their promoter regions. Thus, AUX/IAA proteins can bind to the promoters of their own genes and repress their own expression. When auxin signaling triggers the degradation of AUX/IAA repressors, the ensuing transcription of auxin response genes leads to the replacement of AUX/IAA proteins and thus to response attenuation or termination (see Figure 15.34A).

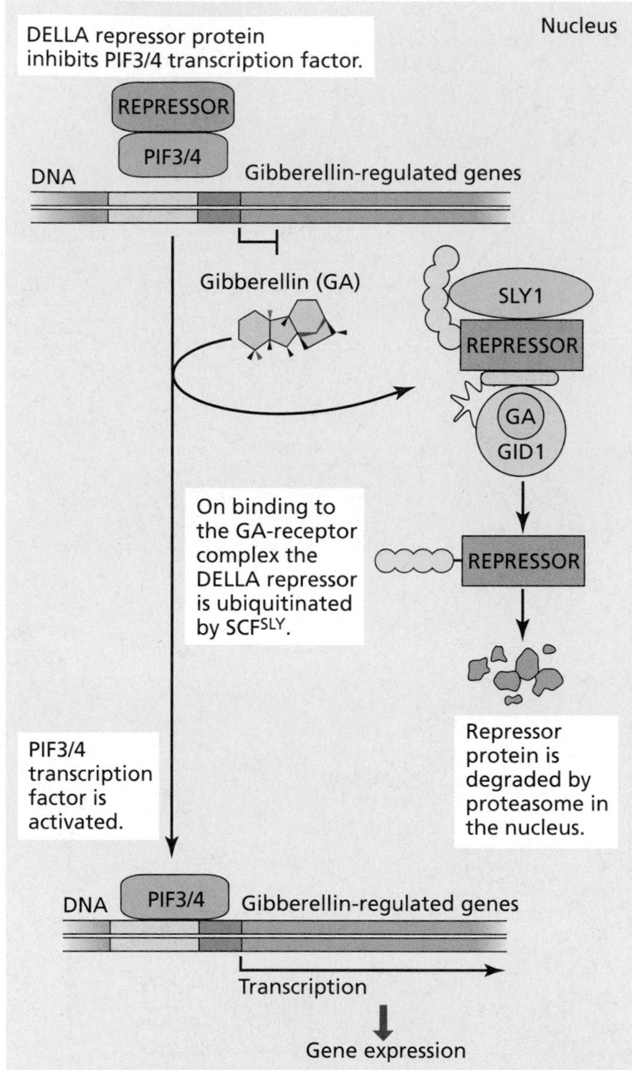

Hormone signaling pathways are often subject to several loops of negative feedback regulation. This is nicely illustrated by the gibberellin pathway (**Figure 15.35**). Bioactive gibberellin (GA_4 in this example) is synthesized through a complex biosynthetic pathway that involves multiple enzyme-catalyzed reactions. The last two enzymes in this pathway are encoded by members of the *GA20ox* and *GA3ox* gene families. As shown in Figure 15.35, in the absence of gibberellin, DELLA transcriptional regulators promote the expression of the genes coding for the GA20ox and GA3ox enzymes, which leads to increased gibberellin biosynthesis. At the same time, DELLA inhibits the expression of genes encoding the gibberellin catabolism enzyme GA2ox, which leads to decreased gibberellin degradation. As of result of these two effects of DELLA, gibberellin concentrations increase. In the presence of gibberellin, DELLA proteins are degraded by the proteasomal pathway. As a result, gibberellin biosynthesis decreases and gibberellin catabolism is increased. Thus, gibberellin negatively regulates its own concentration in the cell. These positive and negative feedback loops help ensure that the appropriate gibberellin levels and responses are maintained during plant development.

The cellular response output to a signal is often tissue-specific

Many environmental and endogenous signals can trigger a variety of often highly diverse plant responses. Typically, particular tissue or cell types do not display the entire range of potential responses when exposed to a signal, but exhibit distinct response specificity. The plant hormone auxin, for example, promotes cell expansion in growing aerial tissues while inhibiting cell expansion in roots. It elicits lateral root initiation in a subset of cells of the root pericycle while inducing leaf primorida at the shoot apical meristem and controls vascular differentiation in developing plant organs. How can the developmental context of tissues and cells determine such diverse responses to one simple signal? As discussed, auxin signal transduction involves the auxin-dependent interaction of TIR1/AFB receptors and AUX/IAA repressor proteins, leading to AUX/IAA degradation and release of AUX/IAA-mediated repression of ARF transcription factor activity (see Figure 15.34A). All of these signaling components are encoded by multigene families (in Arabidopsis there are 6 *TIR1/AFB* genes, 29 *AUX/IAA*s, and 23 *ARF*s) and have distinct expression patterns, biochemical properties, and biological functions. Where within the plant these components are expressed, how highly they are expressed, the strength of their binding affinity, and the cellular auxin levels they experience all influence the shape of the final auxin response. For example, while it appears that all TIR1/AFBs can potentially interact with many different AUX/IAAs in an auxin-dependent manner, not all of these proteins are expressed in all cell types. Furthermore, the dosage at which auxin promotes these interactions varies significantly with different receptor/repressor combinations, such that some TIR1/AFB–AUX/IAA complexes form at very low auxin concentrations whereas others require substantially higher auxin levels to stably interact. Differential sensitivity and expression may also be mechanisms to achieve tissue specificity in other hormone signal transduction pathways where receptors or other signaling components are encoded by multigene families.

Cross-regulation allows signal transduction pathways to be integrated

Within plant cells, signal transduction pathways never function in isolation, but operate as part of a complex web of signaling interactions. These interactions account for the fact that plant hormones often exhibit *agonistic* (additive or positive) or *antagonistic* (inhibitory or negative) interactions with other signals. Classic examples include

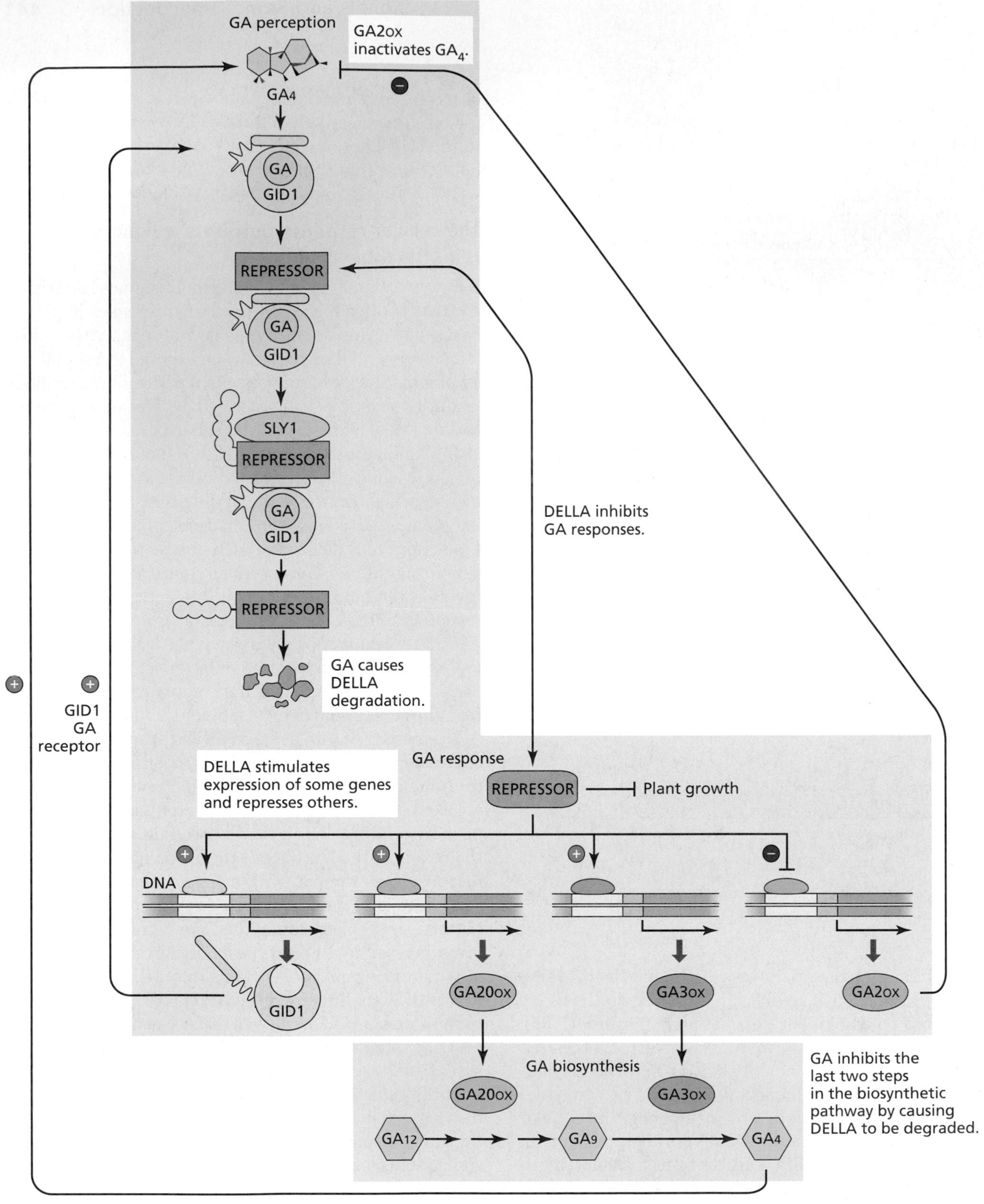

Figure 15.35 The gibberellin (GA) response is regulated by a series of feedback mechanisms involving components of both gibberellin signal transduction and gibberellin biosynthesis. *GA20ox* and *GA3ox* genes encode enzymes that catalyze the last steps of the gibberellin biosynthetic pathway, while GA2ox catalyzes the inactivation of the bioactive gibberellin, GA_4. *GID1* encodes the gibberellin receptor that, following ligand binding, recruits DELLA repressor proteins to the SCF^{SLY1} complex for ubiquitination, triggering their degradation. In the absence of gibberellin, the DELLA proteins positively regulate *GID1*, *GA20ox*, and *GA3ox* (plus signs), and negatively regulate *GA2ox* (minus sign). Conversely, both bioactive gibberellin and the GID1 receptor enhance DELLA repressor degradation (plus signs), while GA2ox blocks DELLA repressor degradation (minus signs).

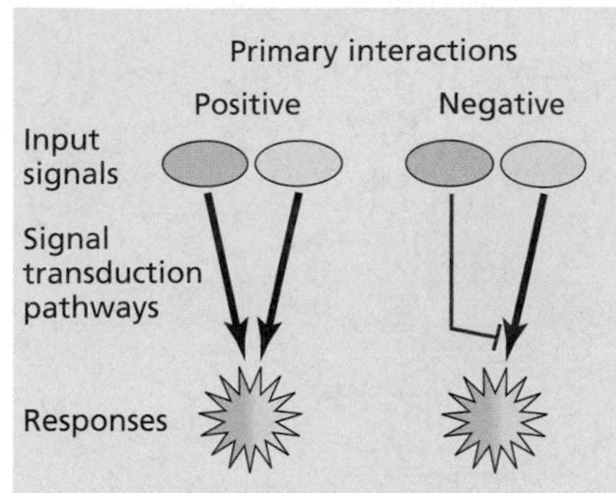

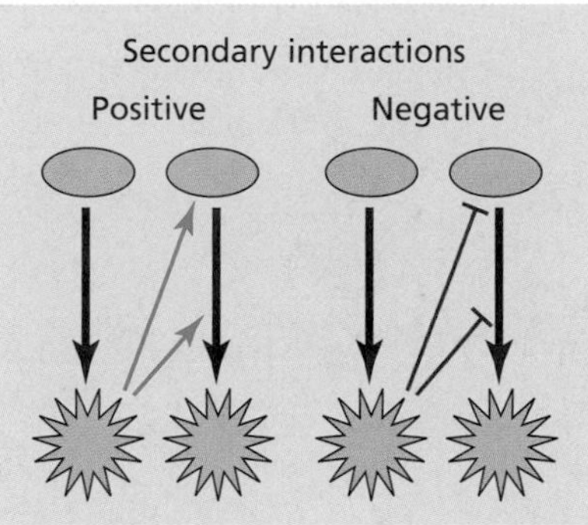

Tertiary interactions
Positive
Negative

Two input pathways regulate a single shared protein or multiple shared proteins controlling a response. Both pathways have the same effect on the response.

Two input pathways converge on shared protein(s), but one of the pathways inhibits the effect of the other.

Two input pathways regulate separate responses. In addition, one pathway enhances the input levels or perception of the other pathway.

As in the positive interaction, except that one pathway represses the input levels or perception of the other pathway.

The response of one of the signaling pathways promotes the response of the other pathway.

The response of one of the signaling pathways inhibits the response of the other pathway.

Figure 15.36 Signal transduction pathways operate as part of a complex web of signaling interactions. Three types of cross-regulation have been proposed: primary, secondary, and tertiary. Input signals are shown here as ovals, signal transduction pathways are indicated by heavy arrows, and responses (pathway outputs) are shown as stars. Green (positive) or red (negative) colored lines indicate where one pathway influences the other pathway. The three types of cross-regulation can be either positive or negative.

the antagonistic interaction between gibberellin and ABA in the control of seed germination (see Chapter 18).

The interaction between signaling pathways has been termed **cross-regulation**, and three categories have been proposed (**Figure 15.36**):

1. **Primary cross-regulation** involves distinct signaling pathways regulating a shared transduction component in a positive or a negative manner.
2. **Secondary cross-regulation** involves the output of one signaling pathway regulating the abundance or perception of a second signal.
3. **Tertiary cross-regulation** involves the outputs of two distinct pathways exerting influences on one another.

The integration of growth and stress response mechanisms in Arabidopsis roots by the ABA INSENSITIVE4 (ABI4) transcription factor involves all three types of cross-regulation. As shown in **Figure 15.37A**, both cytokinin and ABA exhibit primary cross regulation when they induce transcription of *ABI4* to, in turn, regulate transcription of a variety of biosynthetic and stress response genes. However, cytokinin and ABA regulation of *ABI4* transcription also provides an example of secondary cross-regulation, as induction of *ABI4* transcription acts on auxin signaling by reducing rootward auxin flows within root vascular tissues that initiate lateral root formation (**Figure 15.37B**). Finally, ABI4 also functions in tertiary cross-regulation. ABA and cytokinin also reduce lateral root elongation, which is positively regulated by shootward transport of auxin from the root apex via AUX1, PIN2, and ABCB4. ABI4 positively regulates the endosomal ASPARTYL PROTEASE2 (APA2), which degrades the ABCB4 auxin transporter that regulates auxin flows in root epidermal cells (**Figure 15.37C**). ABCB4 abundance on the plasma membrane is rapidly reduced after ABA or cytokinin treatment in wild type plants, but is unaffected in *abi4* or *apa2* mutants.

Thus, plant signaling is not based on a simple linear sequence of transduction events, but involves cross-regulation among many pathways. Understanding how such complex signaling pathways operate will demand a new scientific approach. This approach is often referred to as **systems biology** and employs mathematical and computational models to simulate these nonlinear biological networks and better predict their outputs.

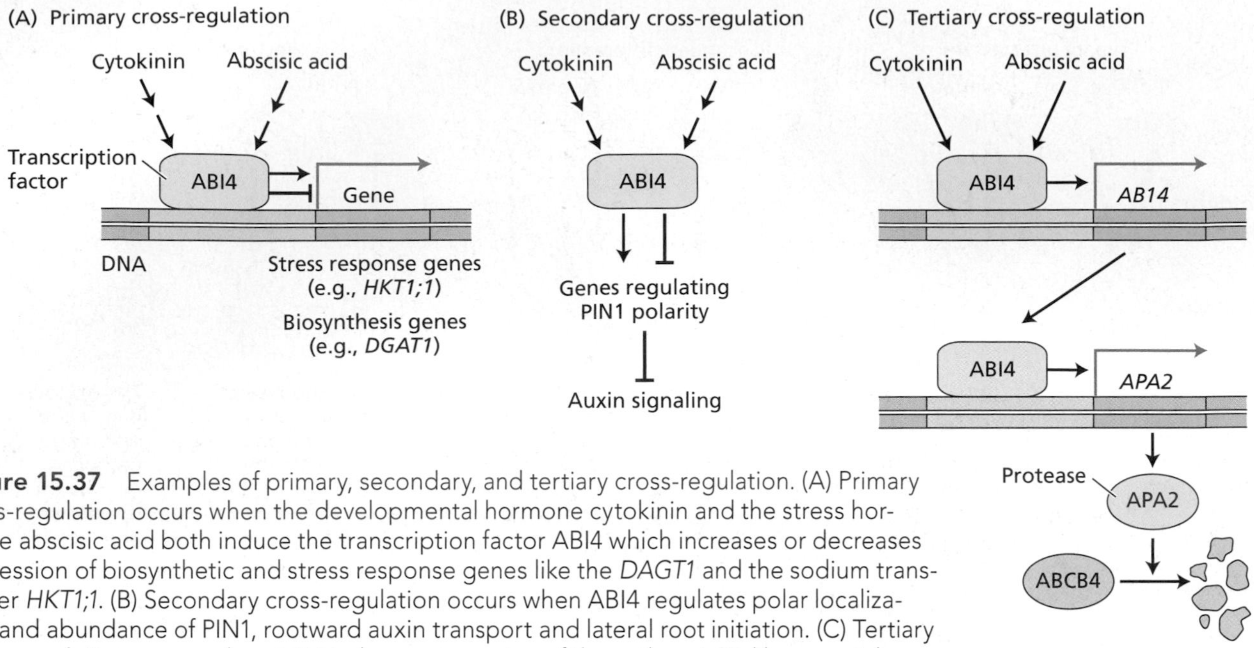

(A) Primary cross-regulation

Cytokinin Abscisic acid

Transcription factor ABI4 Gene

DNA Stress response genes (e.g., *HKT1;1*) Biosynthesis genes (e.g., *DGAT1*)

(B) Secondary cross-regulation

Cytokinin Abscisic acid

ABI4

Genes regulating PIN1 polarity

Auxin signaling

(C) Tertiary cross-regulation

Cytokinin Abscisic acid

ABI4 *AB14*

ABI4 *APA2*

Protease APA2

ABCB4

Figure 15.37 Examples of primary, secondary, and tertiary cross-regulation. (A) Primary cross-regulation occurs when the developmental hormone cytokinin and the stress hormone abscisic acid both induce the transcription factor ABI4 which increases or decreases expression of biosynthetic and stress response genes like the *DAGT1* and the sodium transporter *HKT1;1*. (B) Secondary cross-regulation occurs when ABI4 regulates polar localization and abundance of PIN1, rootward auxin transport and lateral root initiation. (C) Tertiary cross-regulation occurs when ABI4 induces expression of the cathepsin D-like aspartyl protease APA2 to induce degradation of ABCB4. ABCB4 contributes to shootward auxin transport that regulates lateral root elongation.

SUMMARY

Both short- and long-term physiological responses to external and internal signals arise from the transformation (transduction) of signals into mechanistic pathways. In order to activate areas that may be distal from the initial signaling location, signaling intermediates are amplified before dissemination (transmission). Once in play, signaling pathways often overlap into complex signaling networks, a phenomenon termed cross-regulation, to coordinate integrated physiological responses.

Temporal and Spatial Aspects of Signaling

- Plants use signal transduction to coordinate both rapid and slow responses to stimuli (**Figures 15.1, 15.2**).

Signal Perception and Amplification

- Receptors are present throughout cells and are conserved amoung bacteria, plant, animal, and fungal kingdoms (**Figure 15.3**).
- Signaling intermediates must be amplified to prevent dilution of the signaling cascade; the MAPK amplification pathway is conserved among eukaryotes (**Figure 15.4**).
- Signals can also be amplified by second messengers such as Ca^{2+}, H^+, reactive oxygen species (ROS), and modified lipids (lipid signaling molecules), although it can be challenging to parse their signaling targets (**Figures 15.5, 15.6**).

Hormones and Plant Development

- Hormones are conserved chemical messengers that can, at very low concentrations, transmit signals between cells and initiate physiological responses (**Figures 15.7, 15.8**).
- The first growth hormone to be identified was auxin, during studies of coleoptile bending due to phototropism (**Figure 15.9**).
- Studies on the "foolish seedling disease" of rice led to the discovery of the gibberellin group of growth hormones (**Figure 15.10**).
- Tissue-culture experiments revealved the role of cytokinins as cell division–promoting factors (**Figure 15.11**).
- Ethylene is a gaseous hormone that promotes fruit ripening and other developmental processes (**Figure 15.12**).

- Abscisic acid regulates seed maturation and stomatal closure in response to water stress (**Figure 15.13**).
- Brassinosteroids are lipid-soluble hormones that regulate many processes, including photomorphogenesis and germination (**Figure 15.14**).
- Strigolactones reduce shoot branching and promote rhizosphere interactions (**Figure 15.15**).

Phytohormone Metabolism and Homeostasis

- The concetration of hormones is tightly regulated so that signals produce timely responses without compromising sensitivity to the same signal in the future (**Figure 15.16**).
- Indole-3-pyruvate (IPyA) is the primary auxin intermediate, and its concentration is also tightly regulated (**Figures 15.17–15.19**).
- Gibberellins (GAs) are all derived from GA_{12}, which is oxidized in the cytosol (**Figure 15.20**).
- Cytokinins are adenine derivatives. *KNOX* genes promote cytokinin concentrations in the shoot apical meristem while inhibiting gibberellin levels (**Figure 15.21**).
- Ethylene is synthesized from methionine and diffuses rapidly out of plants as a gas; there is no evidence of ethylene catabolism (**Figure 15.22**).
- Abscisic acid is synthesized from 40-carbon carotenoids; its concentrations can fluctuate dramatically during developmental processes (**Figure 15.23**).
- Brassinosteroids arise from campesterol, which is similar in structure to cholesterol (**Figure 15.24**).
- Strigolactones are synthesized from careteniods, similarly to abscisic acid (**Figure 15.25**).

Signal Transmission and Cell–Cell Communication

- Hormones can signal cells within, nearby, or far away from their site of synthesis (**Figure 15.26**).
- Plants can also employ fast-acting, long-distance electrical signaling using action potentials, though the transmission of such signals is poorly understood (**Figure 15.27**).

Hormonal Signaling Pathways

- Cytokinin and ethylene pathways use derived two-component regulatory systems, which involve membrane-bound sensor proteins and soluble response regulator proteins (**Figures 15.28–15.30**).
- Brassinosteriod and certain auxin pathways use transmembrane receptor-like kinases (RLKs) to phosphorylate serine or threonine regions of target proteins (**Figure 15.31**).
- Abscisic acid pathways use phosphatases as well as kinases (**Figure 15.32**).
- In contrast to animal hormone pathways, plant hormone pathways generally employ negative regulators (inactivating repressors), allowing faster activation of downstream response genes (**Figures 15.33, 15.34**).
- Switching off signaling pathways is accomplished by degradation or sequestration of chemical signals via feedback mechanisms (**Figure 15.35**).
- Although hormones can effect a wide variety of responses, tissues exhibit response specificity.
- Integration of signal transduction pathways is accomplished through cross-regulation (**Figures 15.36, 15.37**).

Suggested Reading

Davière, J.-M., and Achard, P. (2013) Gibberellin signaling in plants. *Development* 140: 1147–1151.

Hwang, I., Sheen, J., and Müller, B. (2012) Cytokinin signaling networks. *Annu. Rev. Plant Biol.* 63: 353–380.

Jiang, J., Zhang, C., and Wang, X. (2013) Ligand perception, activation, and early signaling of plant steroid receptor brassinosteroid insensitive 1. J. *Integr. Plant Biol.* 55: 1198–1211.

Ju, C., and Chang, C. (2012) Advances in ethylene signalling: Protein complexes at the endoplasmic reticulum membrane. *AoB Plants 2012: pls031. DOI: 10.1093/aobpla/pls031*

Santner, A., and Estelle, M. (2009) *Recent advances and emerging trends in plant hormone signaling.* Nature (Lond.) 459: 1071–1078.

Suarez-Rodriguez, M. C., Petersen, M., and Mundy, J. (2010)Mitogen-activated protein kinase signaling in plants. *Annu. Rev. Plant Biol.* 61: 621–649.

Xuemin, W. (2004) Lipid signaling. *Curr. Opin. Plant Biol.* 7: 329–336.

16 Signals from Sunlight

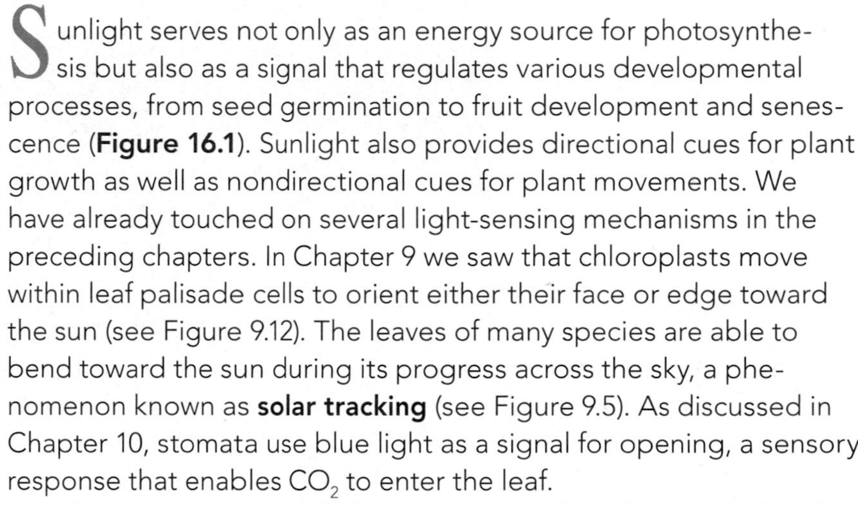

Sunlight serves not only as an energy source for photosynthesis but also as a signal that regulates various developmental processes, from seed germination to fruit development and senescence (**Figure 16.1**). Sunlight also provides directional cues for plant growth as well as nondirectional cues for plant movements. We have already touched on several light-sensing mechanisms in the preceding chapters. In Chapter 9 we saw that chloroplasts move within leaf palisade cells to orient either their face or edge toward the sun (see Figure 9.12). The leaves of many species are able to bend toward the sun during its progress across the sky, a phenomenon known as **solar tracking** (see Figure 9.5). As discussed in Chapter 10, stomata use blue light as a signal for opening, a sensory response that enables CO_2 to enter the leaf.

In later chapters we will encounter examples of light-regulated plant development. For example, many seeds require light to germinate, a process called **photoblasty**. Sunlight inhibits stem growth and stimulates leaf expansion in growing seedlings, two of several light-induced phenotypic changes collectively referred to as **photomorphogenesis** (**Figure 16.2**; also see Chapter 18). Most of us are familiar with the observation that the branches of houseplants placed near a window grow toward the incoming light. This phenomenon, called **phototropism**, is an example of how plants alter their growth patterns in response to the direction of incident radiation (**Figure 16.3**; also see Chapter 18). In some species the leaves fold up at night (**nyctinasty**) and open at dawn (**photonasty**). Photonastic movements are plant movements in response to nondirectional light. As we will discuss in Chapter 20, many plants flower at specific times of the year in response to changing day length, a phenomenon called **photoperiodism**.

In addition to visible light (**Figure 16.4**), sunlight also contains ultraviolet (UV) radiation, which can damage membranes, DNA, and proteins (see Chapter 24). Many plants can sense the presence of UV radiation and protect themselves against cellular damage by synthesizing simple phenolics and flavonoids that act as sunscreens and remove damaging oxidants and free radicals that are induced by the high-energy UV photons.

Figure 16.1 Sunlight exerts multiple influences on plants. Plants expose their leaves to sunlight to transform solar energy into chemical energy, and they also use sunlight for a wide range of developmental signals that optimize photosynthesis and detect seasonal changes.

All of the photoresponses noted above, including the responses to UV radiation, involve receptors that detect specific wavelengths of light and induce developmental or physiological changes. As we saw in Chapter 15, hormone signal transduction involves a chain of reactions beginning with a hormone receptor and ending with a physiological response. The receptor molecules that plants use to detect sunlight are termed **photoreceptors**. Like hormone receptors, photoreceptors respond to a signal, in this case light, by initiating signaling reactions that typically involve second messengers and phosphorylation cascades (see Figure 15.2).

In this chapter we will discuss the signaling mechanisms involved in light-regulated growth and development, focusing primarily on the red-light (620–700 nm), far-red light (710–850 nm), blue light (350–500 nm), and UV-B radiation (290-320 nm) receptors.

Plant Photoreceptors

Pigments, such as chlorophyll and the accessory pigments of photosynthesis, are molecules that absorb visible light at specific wavelengths, and reflect or transmit the nonabsorbed wavelengths, which are perceived as colors. Unlike the photosynthetic pigments, photoreceptors absorb a photon of a given wavelength and use this energy as a signal that initiates a photoresponse. With the exception of UVR8 (discussed at the end of this chapter), all the known photoreceptors consist of a protein plus a light-absorbing prosthetic group (a nonprotein molecule attached to the photoreceptor protein) called a **chromo-**

Figure 16.2 Comparison of seedlings grown in the light versus the dark. (Left) Cress seedlings grown in the light. (Right) Cress seedlings grown in the dark. The dark-grown seedlings exhibit etiolation, characterized by elongated hypocotyls and a lack of chlorophyll.

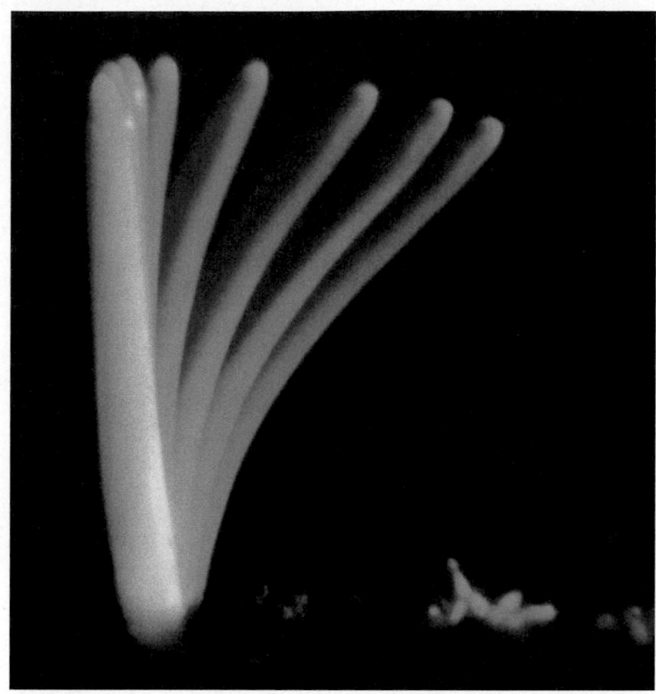

Figure 16.3 Time-lapse photograph of a maize (corn; *Zea mays*) coleoptile growing toward unilateral blue light given from the right. In the first image on the left, the coleoptile is about 3 cm long. The consecutive exposures were made 30 min apart. Note the increasing angle of curvature as the coleoptile bends. (Courtesy of M. A. Quiñones.)

phore. As we will see later, the protein structures of the different photoreceptors vary. Other common aspects of photoreceptors include sensitivity to light quantity (number of photons), light quality (wavelength dependency and associated action spectra), light intensity, and the duration of the exposure to light. In each case, perception of light by specific photoreceptors initiates cellular signals that ultimately regulate specific photoresponses.

Among the photoreceptors that can promote photomorphogenesis in plants, the most important are those that absorb red and blue light. **Phytochromes** are photoreceptors that absorb red and far-red light most strongly (600–750 nm), but they also absorb blue light (350-500 nm) and UV-A radiation (320–400 nm). Phytochromes mediate many aspects of vegetative and reproductive development, as will be described in the chapters that follow. Three main classes of photoreceptors mediate the effects of UV-A/blue light: the cryptochromes, the phototropins, and the **ZEITLUPE** (**ZTL**, German for "slow motion") family. Cryptochromes, like the phytochromes, play a major role in plant photomorphogenesis, while **phototropins** primarily regulate phototropism, chloroplast movements, and stomatal opening. The ZTL family of photoreceptors plays roles in day length perception and circadian rhythms. As in the case of hormone signaling, light signaling typically involves interactions between multiple photoreceptors and their signaling intermediates.

By convention, photoreceptors are designated in lower case (e.g., phy, cry, phot) when the holoprotein (protein plus chromophore) is described, and in upper case (PHY, CRY, PHOT) when the apoprotein (protein minus chromophore) is described. To be consistent with genetic conventions, we will use upper case italics (*PHY, CRY, PHOT*) for the genes encoding the photoreceptor apoproteins.

Recently, a unique photoreceptor system has been isolated from Arabidopsis that is specific for the perception of ultraviolet radiation (**UV RESISTANCE LOCUS 8**, or **UVR8**) and is responsible for several UV-B-induced photomorphogenic responses. UVR8 will be discussed at the end of the chapter.

Photoresponses are driven by light quality or spectral properties of the energy absorbed

As in the case of hormone receptors (see Chapter 15), the different photoreceptor systems in plants can interact with each other, and it can be difficult to separate specific responses within the full solar spectrum since many photoreceptors may be absorbing energy at the same time.

Figure 16.4 Plants can use visible light, UV-A, and UV-B radiation as developmental signals (all wavelengths in nm).

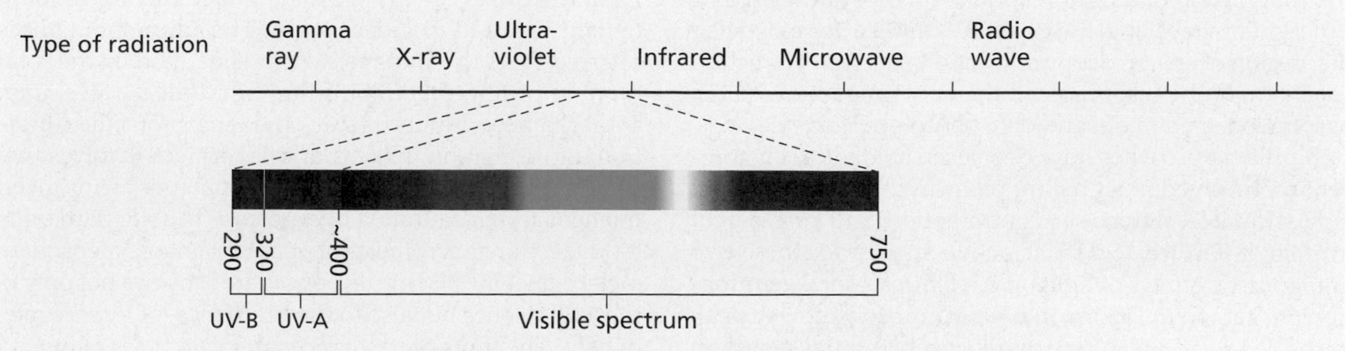

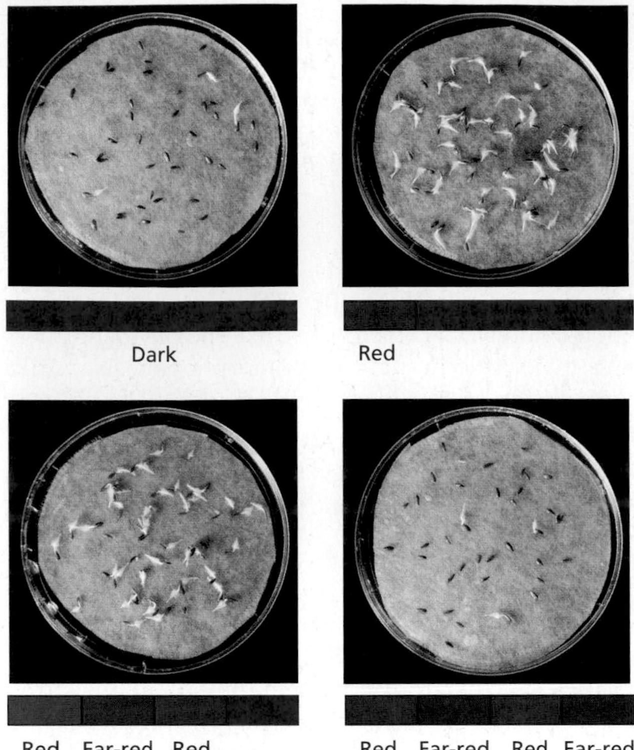

Figure 16.5 Lettuce seed germination is a typical photoreversible response controlled by phytochrome. Red light promotes lettuce seed germination, but this effect is reversed by far-red light. Imbibed (hydrated) seeds were given alternating treatments of red followed by far-red light. The effect of the light treatment depended on the last treatment given. Very few seeds germinated following the last far-red treatment.

For example, the process of **de-etiolation**, characterized by the production of chlorophyll in dark-grown (etiolated) seedlings when exposed to light, results from the coaction of phytochrome absorbing red light and cryptochrome absorbing blue light from sunlight. How, then, can we functionally distinguish intrinsic responses to individual photoreceptors? In many cases, a contribution from photosynthesis cannot be excluded since photosynthetic pigments also absorb red light and blue light.

To determine which wavelengths of light are necessary to bring about a particular plant response, photobiologists typically produce what is known as an action spectrum. Action spectra describe the wavelength specificity of a biological response to sunlight. Each photoreceptor differs in its atomic composition and arrangement and thus exhibits different absorption characteristics. As we saw in Chapter 7, a photosynthetic action spectrum is a graph that plots the magnitude of a light response (photosynthesis) as a function of wavelength (see WEB TOPIC 7.1 for a detailed discussion of spectroscopy and action spectra). The action spectrum of the response can then be compared with the absorption spectra of candidate photoreceptors.

Similar approaches have been used to identify photoreceptors involved in signaling pathways. For example, red light stimulates lettuce seed germination, and far-red light inhibits it (**Figure 16.5**). The action spectra for these two antagonistic effects of light on Arabidopsis seed germination are shown in **Figure 16.6A**. Stimulation shows a peak in the red region (660 nm), while inhibition has a peak in the far-red region (720 nm). When the absorption spectra of each of the two forms of phytochrome (Pr and Pfr) are measured separately in a spectrophotometer designed to study photoreversible molecules, they correspond closely to the action spectra for the stimulation and inhibition of seed germination, respectively (**Figure 16.6B**). As discussed next, the close correspondence between the action and absorption spectra of phytochrome not only confirmed its identity as the photoreceptor involved in regulating seed germination, it also established that the red/far-red reversibility of seed germination was due to the photoreversibility of phytochrome itself.

Similarly, action spectra for blue light–stimulated phototropism, stomatal movements, and other key blue-light responses all exhibit a peak in the UV-A region (at 370 nm) and a peak in the blue region (~410–500 nm) that has a characteristic "three-finger" fine structure (**Figure 16.7A**), suggesting a common photoreceptor. The absorption spectrum for the LOV2 domain of phototropin, which contains the chromophore flavin mononucleotide (FMN), is identical to the action spectrum for phototropism (**Figure 16.7B**), consistent with phototropin acting as the photoreceptor for these responses. The mechanism of phototropin action will be discussed later in the chapter.

Plants responses to light can be distinguished by the amount of light required

Light responses can also be distinguished by the amount of light required to induce them. The amount of light is referred to as the **fluence**, which is defined as the total number of photons impinging on a unit surface area. Total fluence = fluence rate × the length of time (duration) of irradiation. Note that this formula involves two components: the number of incident photons at any given moment and the duration of exposure. The standard units for fluence are micromoles of quanta (photons) per square meter (μmol m^{-2}). Some responses are sensitive not only to the total fluence but also to the **irradiance**, or fluence rate, of light. The units of irradiance are micromoles of quanta

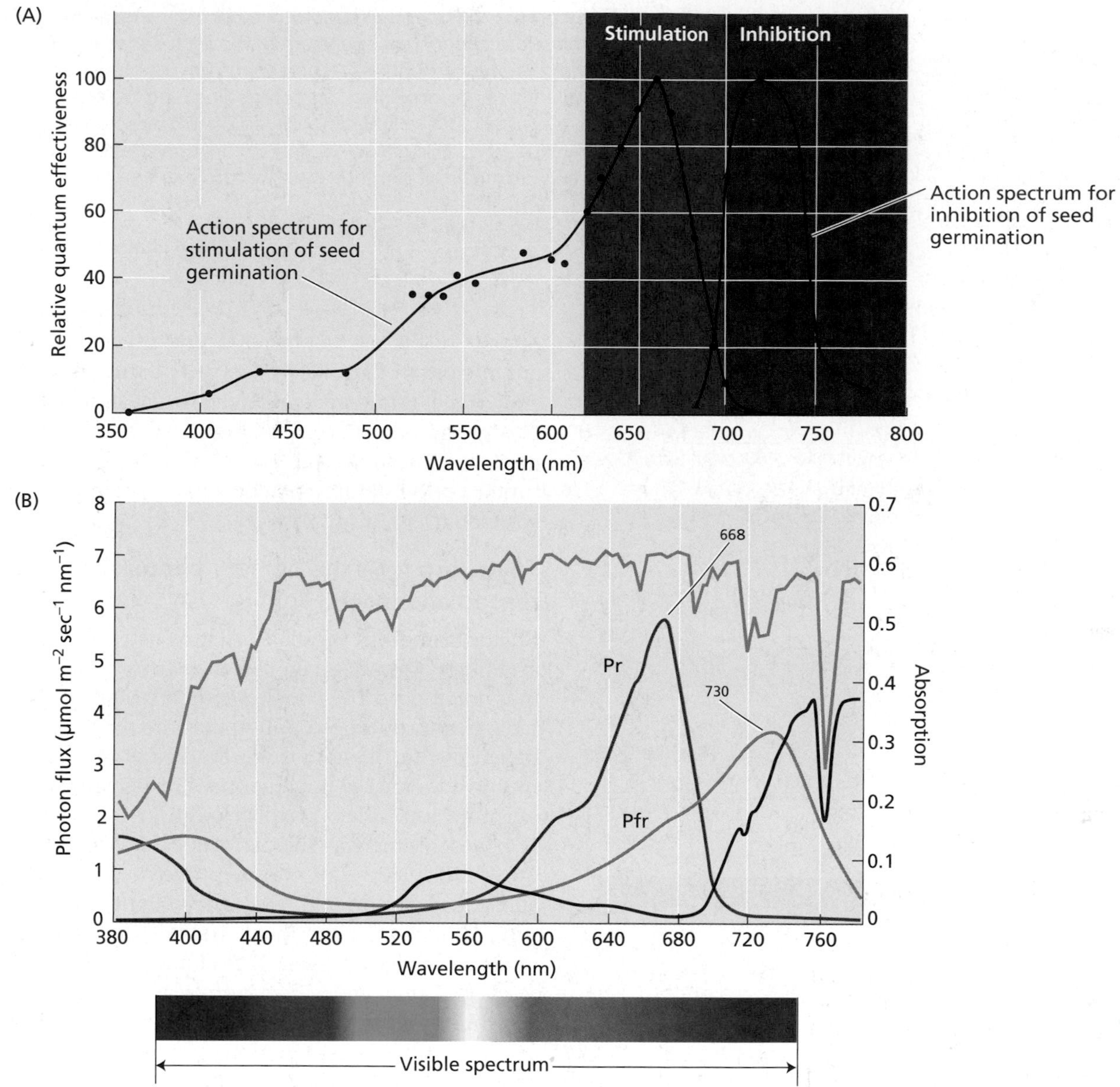

Figure 16.6 The action spectrum of phytochrome function matches its absorption spectrum. (A) Action spectra for the photoreversible stimulation and inhibition of seed germination in Arabidopsis. (B) Absorption spectra of purified oat phytochrome in the Pr (red line) and Pfr (green line) forms overlap. At the top of the canopy, there is a relatively uniform distribution of visible-spectrum light (blue line), but under a dense canopy much of the red light is absorbed by plant pigments, resulting in transmittance of mostly far-red light. The black line shows the spectral properties of light that is filtered through a leaf. Thus, the relative proportions of Pr and Pfr are determined by the degree of vegetative shading in the canopy. (A after Shropshire et al. 1961; B after Kelly and Lagarias 1985, courtesy of Patrice Dubois.)

per square meter per second (µmol m^{-2} s^{-1}). (For definitions of these and other terms used in light measurement, see Chapter 9 and **WEB TOPIC 9.1**).

Because photochemical responses are stimulated only when a photon is absorbed by its photoreceptor, there can be a difference between incident irradiation and absorption. For example, in photosynthesis the apparent quantum efficiency is assessed as the electron transport rate or total carbon assimilation as a function of the incident photosynthetically active radiation (PAR). However, this measure underestimates the *actual* quantum efficiency because not all the incident photons are absorbed. This caveat is also important in assessing the dose response of the photomorphogenic responses of green plants to red or blue light, because much of the light is absorbed by chlorophyll. The same principle applies to the responses to UV radiation, since the epidermis may absorb just under 100% of the incident UV radiation. Thus, the amount of radiation required

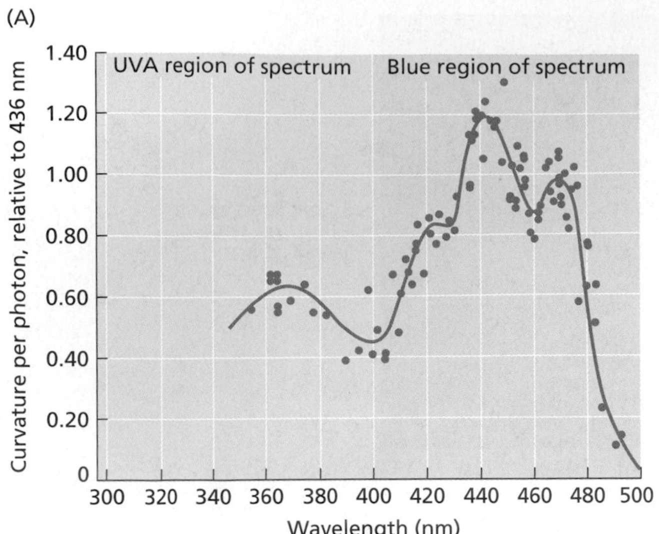

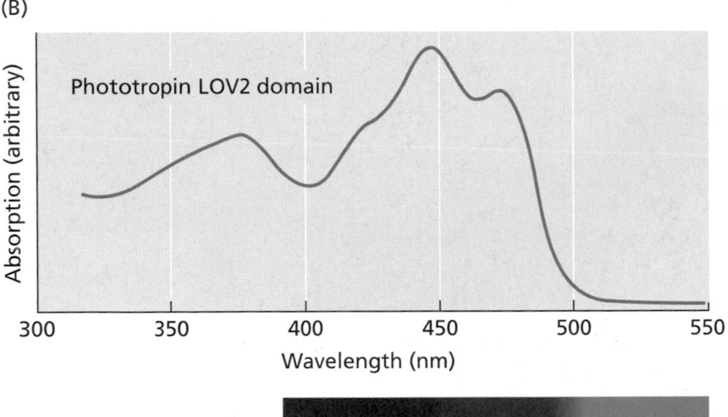

Figure 16.7 The action spectrum of phototropism matches the absorption spectrum of the light-sensing LOV domain of phototropin. (A) Action spectrum for blue light–stimulated phototropism in oat coleoptiles. The "three-finger" pattern in the 400–500 nm region is characteristic of many blue-light responses. (B) The absorption spectrum of the LOV2 domain of phototropin. (A after Thimann and Curry 1960; B after Swartz et al. 2001.)

to induce a photoresponse may be quite high based on the amount of incident irradiation required and quite low based on actual photon absorption by the photoreceptor.

Phytochromes

Phytochromes were first identified in flowering plants as the photoreceptors responsible for photomorphogenesis in response to red and far-red light. However, they are members of a gene family present in all land plants, and have also been found in streptophyte algae, cyanobacteria, other bacteria, fungi, and diatoms. For example, **bacterial phytochrome-like proteins** (**BphPs**) regulate the biosynthesis of the photosynthetic apparatus in *Rhodopseudomonas palustris,* and of pigments in *Deinococcus radiodurans* and *Rhodospirillum centenum*. The phytochrome from the filamentous fungus *Aspergillus nidulans* appears to play a role in sexual development. These functions of bacterial and fungal phytochromes are thus conceptually analogous to photomorphogenesis in flowering plants.

Because neither red nor far-red light penetrate water to depths greater than a few meters, phytochrome would appear to be less useful as a photoreceptor for aquatic organisms. However, recent studies have shown that different algal phytochromes can sense orange, green, or even blue light, suggesting that phytochromes have the potential to be spectrally tuned during natural selection to absorb different wavelengths.

Phytochrome is the primary photoreceptor for red and far-red light

Phytochrome is a cyan-blue (midway between green and blue) or cyan-green protein with a molecular mass of about 125 kilodaltons (kDa). Many of the biological properties of phytochrome were established in the 1930s through studies of red light–induced morphogenic responses, especially seed germination. A key breakthrough in the history of phytochrome was the discovery that the effects of red light (620–700 nm) could be reversed by a subsequent irradiation with far-red light (710–850 nm). This phenomenon was first demonstrated in germinating lettuce seeds (see Figure 16.5) but was also observed in stem and leaf growth, as well as in floral induction and other developmental phenomena (**Table 16.1**). The reversibility of the red and far-red responses ultimately led to the discovery that a single photoreversible photoreceptor, phytochrome, was responsible for both activities. It was subsequently demonstrated that the two forms of phytochrome could be distinguished spectroscopically (see Figure 16.6B).

Phytochrome can interconvert between Pr and Pfr forms

In dark-grown, or etiolated, seedlings, phytochrome is present in a red light–absorbing form, referred to as **Pr**. This cyan-blue-colored inactive form is converted by red light to a far-red light–absorbing form called **Pfr**, which is pale cyan-green in color and is considered to be the active form of phytochrome. Pfr can revert back to inactive Pr in darkness, but this is a relatively slow process. However, Pfr can be rapidly converted to Pr by irradiation with far-red light. This conversion–reconversion property, termed **photoreversibility** (also referred to as *photochromism*), is the most distinctive feature of phytochrome and can be

TABLE 16.1 Typical photoreversible responses induced by phytochrome in a variety of higher and lower plants

Group	Genus	Stage of development	Effect of red light
Angiosperms	*Lactuca* (lettuce)	Seed	Promotes germination
	Avena (oat)	Seedling (etiolated)	Promotes de-etiolation (e.g., leaf unrolling)
	Sinapis (mustard)	Seedling	Promotes formation of leaf primordia, development of primary leaves, and production of anthocyanin
	Pisum (pea)	Adult	Inhibits internode elongation
	Xanthium (cocklebur)	Adult	Inhibits flowering (photoperiodic response)
Gymnosperms	*Pinus* (pine)	Seedling	Enhances rate of chlorophyll accumulation
Pteridophytes	*Onoclea* (sensitive fern)	Young gametophyte	Promotes growth
Bryophytes	*Polytrichum* (moss)	Germling	Promotes replication of plastids
Chlorophytes	*Mougeotia* (alga)	Mature gametophyte	Promotes orientation of chloroplasts to directional dim light

measured in vivo or in vitro with almost identical results. It is often diagrammed as follows:

$$\mathrm{Pr} \underset{\text{Far-red light}}{\overset{\text{Red light}}{\rightleftarrows}} \mathrm{Pfr}$$

Photoreversibility is thus a defining feature of phytochromes. Even algal phytochromes with peak absorbances in the orange, green, or blue regions of the spectrum exhibit photoreversibility at a different wavelength.

It is important to note that the phytochrome pool is never fully converted to the Pfr or Pr forms following red or far-red irradiation, because the absorption spectra of the Pfr and Pr forms overlap. Thus, when Pr molecules are exposed to red light, most of them absorb the photons and are converted to Pfr, but some of the Pfr made also absorbs the red light and is converted back to Pr (see Figure 16.6B). The proportion of phytochrome in the Pfr form after saturating irradiation by red light is only about 88%. Similarly, the very small amount of far-red light absorbed by Pr makes it impossible to convert Pfr entirely to Pr by broad-spectrum far-red light. Instead, an equilibrium of 98% Pr and 2% Pfr is achieved. This equilibrium is termed the **photostationary state**.

Pfr is the physiologically active form of phytochrome

Because phytochrome responses are induced by red light, they could, in theory, result from either the appearance of Pfr or the disappearance of Pr. In most cases studied, a quantitative relationship holds between the magnitude of the physiological response and the amount of Pfr generated by light, but no such relationship holds between the physiological response and the loss of Pr. Evidence such as this has led to the conclusion that Pfr is the physiologically active form of phytochrome.

The use of narrow waveband red (R) and far-red (FR) light was central to the discovery and eventual isolation of phytochrome. However, unlike the plants used in laboratory-based photobiological experiments, a plant growing outdoors is never exposed to pure "red" or "far-red" light. In natural settings plants are exposed to a much broader spectrum of light, and it is under these conditions that phytochrome must work to regulate developmental responses to changes in the light environment. Indeed, as shown in Figure 16.6B, the plant canopy itself can have a dramatic effect on both the quantity and quality of incident light reaching individual plants. Of particular importance is the R:FR ratio, which is strongly affected by the presence of a canopy because chlorophyll absorbs red but not far-red light. Thus, as we will discuss in Chapter 18, plants growing beneath a canopy use phytochrome to sense the R:FR ratio in regulating such processes as shade avoidance, competitive interactions, and seed germination.

The phytochrome chromophore and protein both undergo conformational changes in response to red light

Phytochrome in the functionally active dimeric form is a soluble protein with a molecular mass of about 250 kDa. The evolutionary origin of phytochrome is very ancient, predating the appearance of eukaryotes. Bacterial phytochromes are light-dependent histidine kinases that function as **sensor proteins** that phosphorylate corresponding **response regulator** proteins (see Chapter 15). However, as discussed below, phytochromes appear to lack a functional histidine kinase domain, which is characteristic of bacterial two-component systems.

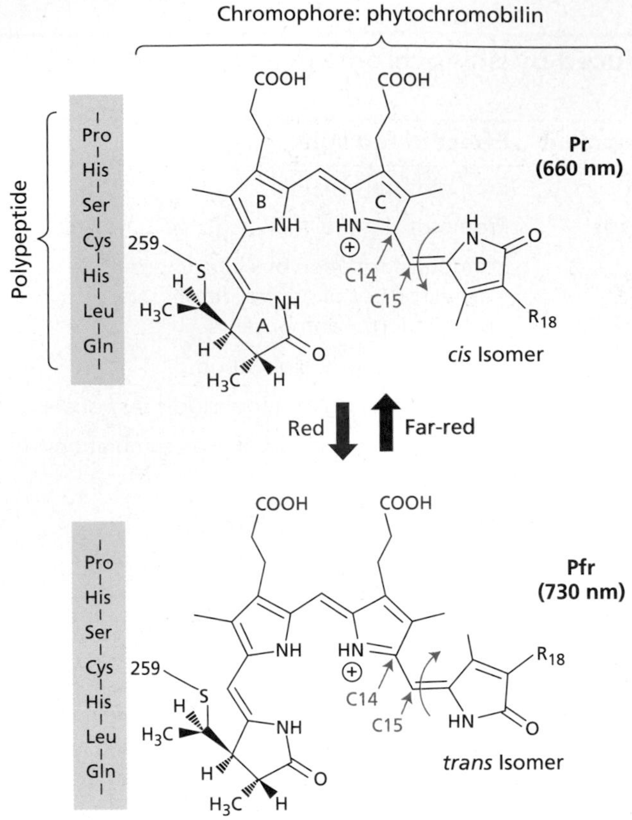

Chromophore: phytochromobilin

Polypeptide

Pro His Ser Cys His Leu Gln

259

COOH COOH

B C NH HN ⊕ C14 C15 H N D O R18 A NH O H3C H H3C H S H

Pr (660 nm)

cis Isomer

Red Far-red

Pfr (730 nm)

trans Isomer

Figure 16.8 Structure of the Pr and Pfr forms of the chromophore (phytochromobilin) and the peptide region bound to the chromophore through a thioether linkage. The chromophore undergoes a *cis–trans* isomerization at carbon 15 in response to red and far-red light. (Courtesy of Clark Lagarias.)

In higher plants the phytochrome chromophore is a linear tetrapyrrole called **phytochromobilin** (**Figure 16.8**). Phytochromobilin is synthesized inside plastids and is derived from heme via a pathway that branches from the chlorophyll biosynthetic pathway. Phytochromobilin is exported from the plastid to the cytosol where it autocatalytically attaches to the PHY apoprotein through a thioether linkage to a cysteine residue. (Thioether linkages are ethers in which the oxygen is replaced by a sulfur: R^1—S—R^2.) There are five phytochrome isoforms in angiosperms (phyA–E), with each isoform encoded by a separate gene and each playing a unique role in development. In Arabidopsis, all five are present, whereas only three are present in rice and only two in poplar.

Figure 16.9A illustrates several of the structural domains in phytochrome. The N-terminal half of phytochrome contains a **PAS** domain, a **GAF** domain with bilin-lyase activity, which binds the chromophore, and the **PHY** domain, which stabilizes phytochrome in the Pfr form. The **PAS-GAF-PHY** domains comprise the chromophore-binding, photosensory region of phytochrome. A "hinge" region separates the N-terminal and C-terminal halves of the molecule.

Downstream of the hinge region are two **PAS-related domain** (**PRD**) repeats that mediate phytochrome dimerization. The PRD domain has been implicated in targeting the Pfr form of phyB to the nucleus, although it lacks a canonical nuclear localization signal (NLS). The C-terminal region of plant phytochromes contains a histidine kinase–related domain (HKRD). However, as noted earlier, higher plant phytochromes, unlike bacterial phytochromes, lack a functional histidine kinase domain.

A comparison of the domain structures of plant phytochrome with the prokaryotic phytochromes Cph1 (*c*yanobacterial *ph*ytochrome *1*) and BphPs (bacterial phytochrome-like proteins) highlights several differences between plant and prokaryotic phytochromes, including the absence of the two PRD domains and the presence of the HKRD domain in place of a functional histidine kinase domain of prokaryotes (see Figure 16.10A). Although all phytochromes contain tetrapyrrole chromophores, phytochromobilin differs from the prokaryotic chromophores in the chemical groups attached to the tetrapyrrole rings (see **WEB ESSAY 16.1**).

Exposure of the Pr form of phytochrome to red light causes atomic-scale structural changes in the chromophore phytochromobilin: the Pr chromophore undergoes a *cis–trans* isomerization between carbons 15 and 16 and rotation of the C14—C15 single bond (see Figure 16.8). The change in the chromophore leads to the rearrangement of crucial secondary-structure elements in the protein.

The crystal structure of the light-sensing N-terminal half of Arabidopsis phyB is shown in **Figure 16.9B**. Two structural elements thought to be important for photoconversion from Pr to Pfr are the **β-hairpin region** and the helical spine. Based on studies with the bacterial phytochrome of *Deinococcus radiodurans* and Arabidopsis phytochrome, a **toggle model** has been proposed for phytochrome interconversion, as illustrated in **Figure 16.9C**. According to the model, the structure of the β-hairpin region of Pr is changed to an α helix during the conversion of Pr to Pfr, which initiates other conformational changes in the protein. However, because of the significant differences between the chromophores and domain structures of plant and bacterial phytochromes further studies are needed to confirm the toggle model.

Pfr is partitioned between the cytosol and the nucleus

In the cytosol, phytochrome holoproteins dimerize in the inactive Pr state (**Figure 16.10**). The conversion of Pr to Pfr by red light is associated with a conformational change in the dimer that is still unresolved. Both phyA and phyB move from the cytosol into the nucleus in a light-dependent fashion (**Figure 16.11**), but they do so by different mechanisms. Neither phyA nor phyB contains a canoni-

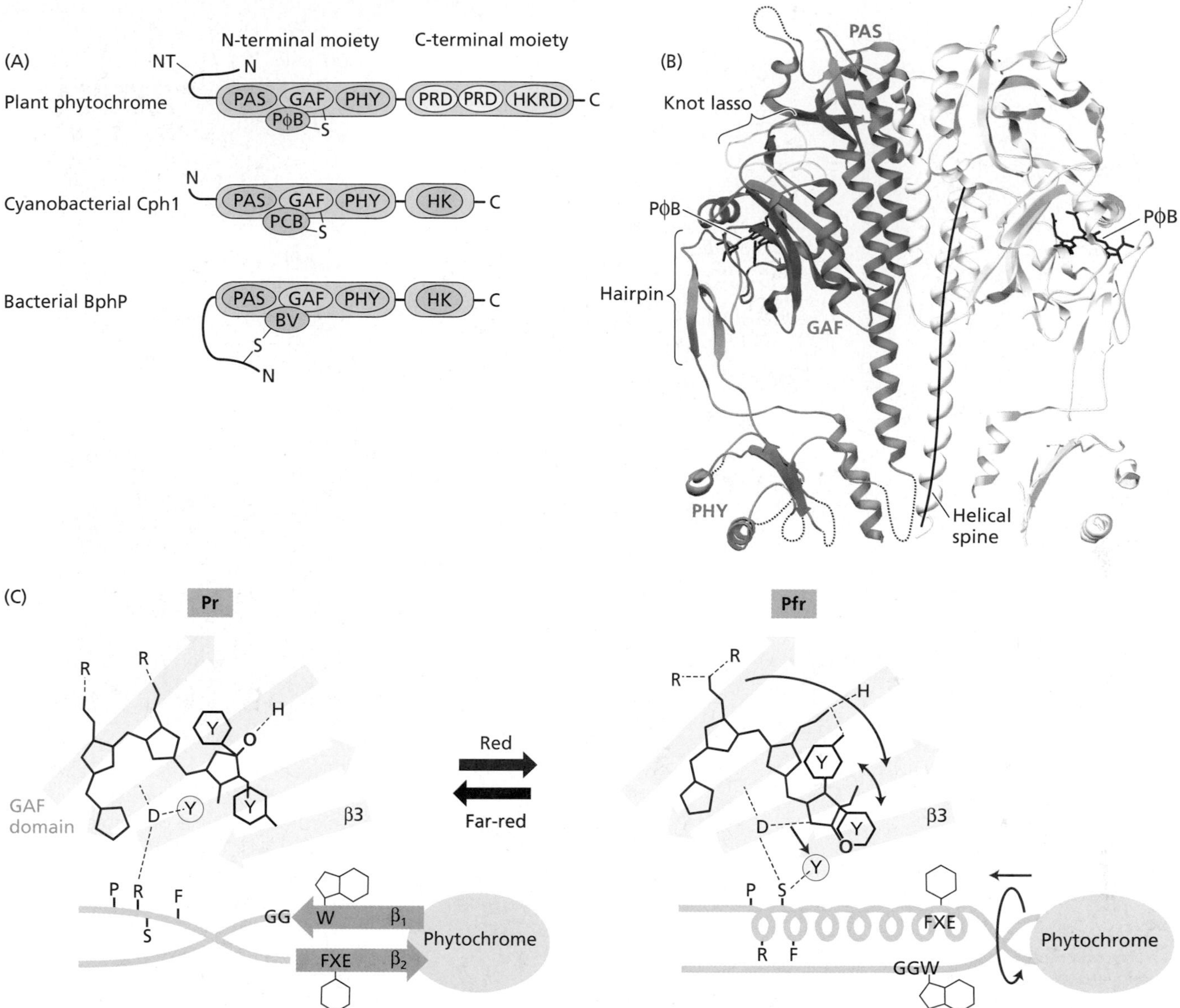

Figure 16.9 Phytochrome domains and their functions. (A) Schematic representation of plant phytochrome (PHY), prokaryotic Cph1 (cyanobacterial phytochrome 1), and BphP (bacterial phytochrome-like protein). The chromophore is attached to cysteine residues in the proteins via a thioether link (–S–). Note that the cysteine residue that forms the linkage is located in the GAF domain in canonical phytochromes such as PHY and Cph1, whereas it is located in the N-terminal extension in BphP-type bacterial phytochromes. NT, N-terminal extension; HK, histidine kinase domain; HKRD histidine kinase–related domain. (B) Ribbon diagram of the N-terminal, light-sensing half of Pr form of the Arabidopsis PhyB dimer. The three domains are colored as follows: PAS, blue; GAF, green; and PHY, orange. The chromophore phytochromobilin (PΦB) is indicated in light blue. (C) Toggle model for the light-induced conformational change of Pr to Pfr. The red light-induced rotation of the D ring of the chromophore causes the β-hairpin to become helical and exert a tug on the helical spine. β2 in the β-hairpin changes conformation to an α-helix after rotation. The chromophore is red; the tyrosines (Y) in the hexagons and histidine (H) near the chromophore rotate in opposite directions from the D ring during the conformational change. The letters in part (C) refer to amino acids. For example, "FXE" stands for "phenylalanine-any residue-glutamate." (B and C after Burgie et al. 2014.)

cal nuclear localization signal (NLS). The PRD domain of phyB can potentially serve as an NLS, but it is apparently masked in the Pr form. Conversion of Pr to Pfr by red light may expose the functional NLS of the PRD domain of phyB, facilitating phyB's import into the nucleus. In contrast, the PRD domain of phyA cannot act as an NLS, and it is therefore dependent on other proteins, such as **FAR-RED ELONGATED HYPOCOTYL1** (**FHY1**) and its homolog FHY1-LIKE (FHL), to transport it into the nucleus (see Figure 16.10).

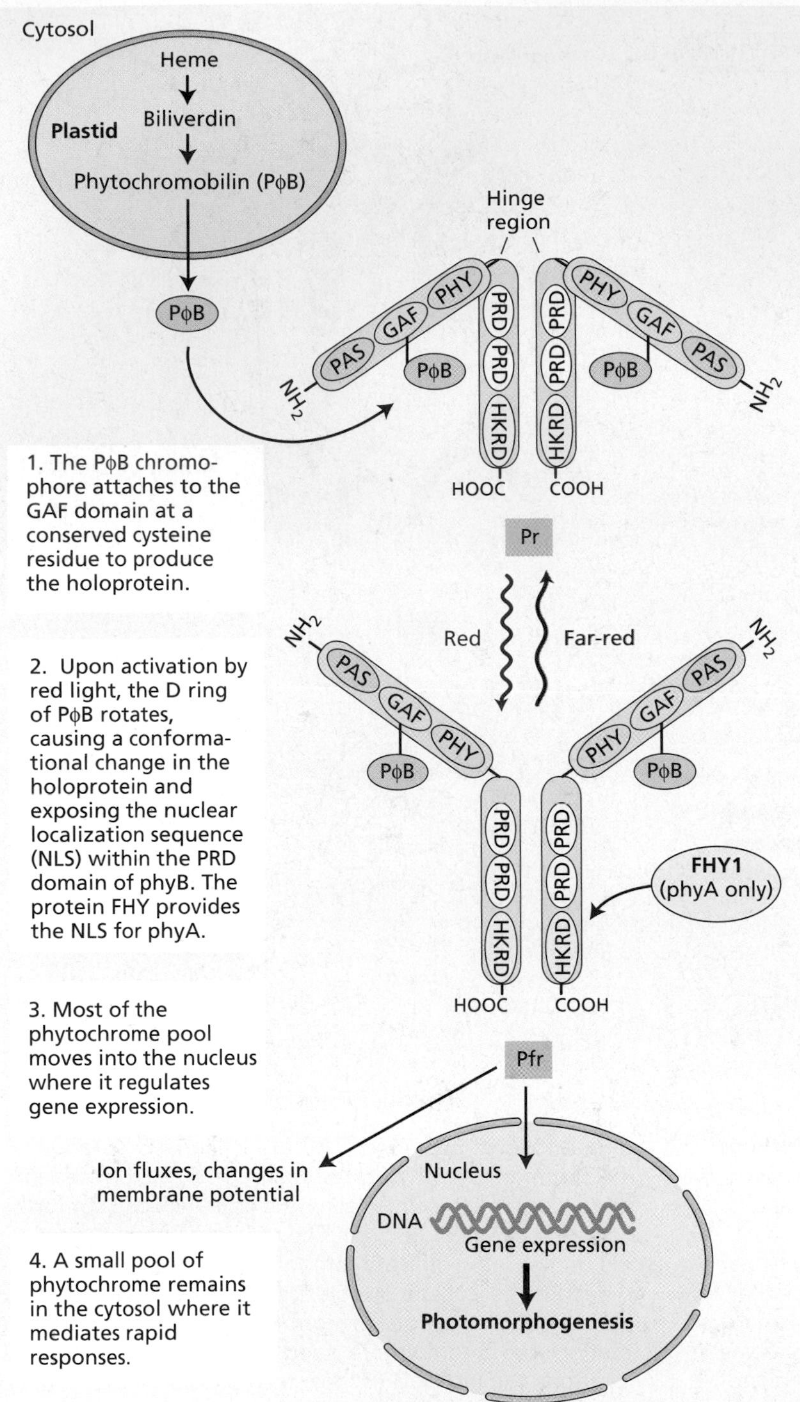

Cytosol
Heme
Plastid
Biliverdin
Phytochromobilin (PϕB)
PϕB
Hinge region
PAS
GAF
PHY
PRD
HKRD
NH2
HOOC
COOH
Pr
Red
Far-red
Pfr
FHY1 (phyA only)
1. The PϕB chromophore attaches to the GAF domain at a conserved cysteine residue to produce the holoprotein.
2. Upon activation by red light, the D ring of PϕB rotates, causing a conformational change in the holoprotein and exposing the nuclear localization sequence (NLS) within the PRD domain of phyB. The protein FHY provides the NLS for phyA.
3. Most of the phytochrome pool moves into the nucleus where it regulates gene expression.
Ion fluxes, changes in membrane potential
4. A small pool of phytochrome remains in the cytosol where it mediates rapid responses.
Nucleus
DNA
Gene expression
Photomorphogenesis

Figure 16.10 After synthesis of phytochromobilin in the plastid and assembly with the apoprotein (1), phytochrome is activated by red light (2) and moves into the nucleus (3) to modulate gene expression. A small pool of phytochrome remains in the cytosol, where it may regulate rapid biochemical changes (4). Whereas phyB has its own nuclear localization signal, phyA requires the protein FHY1 to enter the nucleus. Several conserved domains of phytochrome are shown: PAS, GAF (contains bilin-lyase domain), PHY, PRD (PAS-related domain), and HKRD (histidine kinase–related domain). PϕB, phytochromobilin. (After Montgomery and Lagarias 2002.)

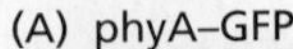

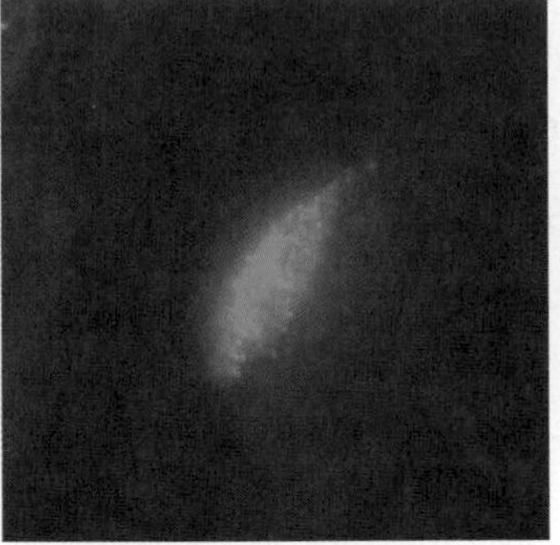

(B) phyB–GFP

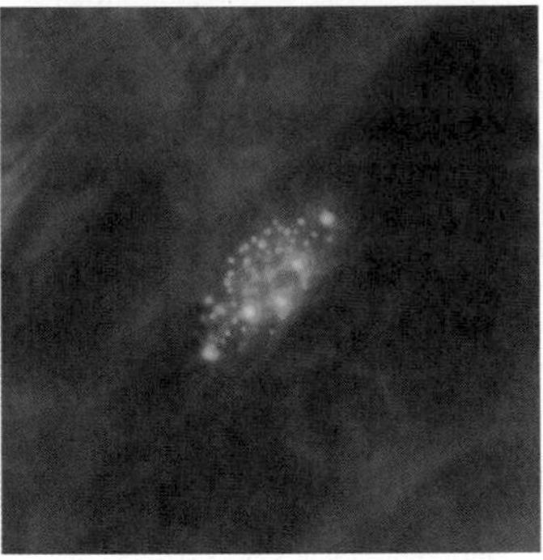

Figure 16.11 Nuclear localization of phy–GFP fusion proteins in epidermal cells of Arabidopsis hypocotyls. Transgenic Arabidopsis plants expressing phyA–GFP (A) or phyB–GFP (B) were exposed to either continuous far-red light (A) or white light (B) and observed under a fluorescence microscope. Only nuclei are visible, demonstrating that the light treatments induced nuclear accumulation of the phy–GFP fusion proteins. In darkness, phy is absent from the nucleus. These results indicate a role for nuclear–cytoplasmic partitioning in controlling phytochrome signaling. The smaller bright green dots inside the nucleus in (B) are called "speckles." The number and size of these speckles have been correlated with light responsiveness. (From Yamaguchi et al. 1999, courtesy of A. Nagatani.)

Once in the nucleus, phytochromes interact with transcriptional regulators to mediate changes in gene transcription. Thus, one important function of phytochrome is to serve as a light-activated switch to bring about global changes in gene expression. However, as discussed below, several phytochrome responses, such as the inhibition of stem elongation, occur extremely rapidly, within minutes or even seconds following exposure to red or far-red light. Thus, phytochromes can also play important roles in the cytosol, regulating membrane potentials and ion fluxes in response to red and far-red light (see Figure 16.10).

Phytochrome Responses

The variety of different phytochrome responses in intact plants is extensive, in terms of both the kinds of responses (see Table 16.1) and the quantity of light needed to induce the responses. A survey of this variety will show how diversely the effects of a single photoevent—the absorption of light by Pr—are manifested throughout the plant. For ease of discussion, phytochrome-induced responses may be logically grouped into two types:

- Rapid biochemical events
- Slower morphological changes, including movements and growth

Some of the early biochemical reactions affect later developmental responses. The nature of these early biochemical events, which comprise signal transduction pathways, will be treated in detail later in the chapter. Here we will focus on the effects of phytochrome on whole-plant responses. As we will see, such responses can be classified into various types depending on the amount and duration of light required and on their action spectra.

Phytochrome responses vary in lag time and escape time

Morphological responses to the photoactivation of phytochrome are often observed visually after a *lag time*—the time between stimulation and the observed response. The lag time may be as brief as a few minutes or as long as several weeks. These differences in response times result from the multiple signal transduction pathways that function downstream of phytochrome signaling as well as interactions with other developmental mechanisms. The more rapid of these responses are usually reversible movements of organelles (see WEB TOPIC 16.1) or reversible volume changes (swelling, shrinking) in cells, but even some growth responses are remarkably fast. For instance, red-light inhibition of the stem elongation rate of light-grown pigweed (*Chenopodium album*) and Arabidopsis is observed within minutes after the proportion of Pfr to Pr in the stem is increased. However, lag times of several weeks are observed for the induction of flowering in Arabidopsis and other species.

Variety in phytochrome responses can also be seen in the phenomenon called **escape from photoreversibility**. Red light–induced events are reversible by far-red light for only a limited period of time, after which the response is said to have "escaped" from reversal control by light. This escape phenomenon can be explained by a model based on the assumption that phytochrome-controlled morphological responses are the end result of a multistep sequence of linked biochemical reactions in the responding cells. Early stages in the sequence may be fully reversible by removing Pfr, but at some point in the sequence a point of no return is reached, beyond which the reactions proceed irreversibly toward the response. The escape time therefore represents the amount of time it takes before the overall sequence of reactions becomes irreversible—essentially, the time it takes for Pfr to complete its primary action. The escape time for different phytochrome responses ranges remarkably, from less than a minute to hours.

Phytochrome responses fall into three main categories based on the amount of light required

As **Figure 16.12** shows, phytochrome responses fall into three major categories based on the amount of light they require: very low fluence responses (VLFRs), low-fluence responses (LFRs), and high-irradiance responses (HIRs). VLFRs and LFRs have a characteristic range of light flu-

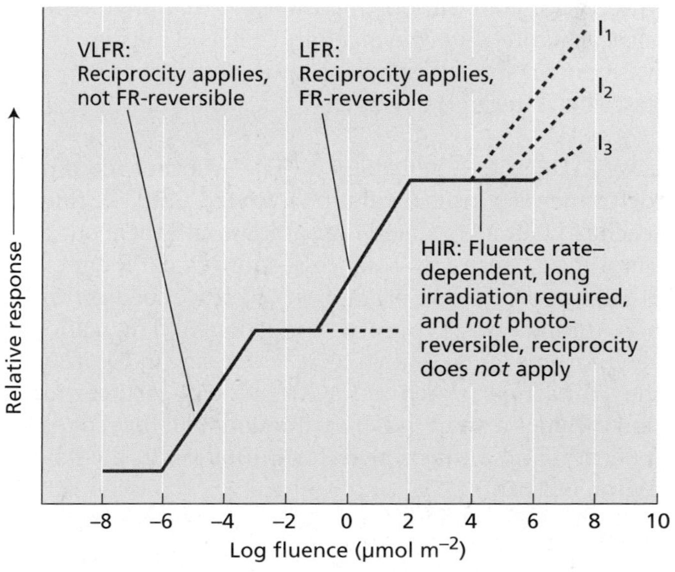

Figure 16.12 Three types of phytochrome responses, based on their sensitivities to fluence. The relative magnitudes of representative responses are plotted against increasing fluences of red light. Short light pulses activate very low fluence responses (VLFRs) and low-fluence responses (LFRs). Because high-irradiance responses (HIRs) are proportional to irradiance as well as to fluence, the effects of three different irradiances given continuously are illustrated ($I_1 > I_2 > I_3$). (After Briggs et al. 1984.)

ences within which the magnitude of the response is proportional to the fluence. HIRs, on the other hand, are proportional to the irradiance.

VERY LOW FLUENCE RESPONSES (VLFRs) Some phytochrome responses can be initiated by fluences as low as 0.0001 μmol m^{-2} (a few seconds of starlight, or one-tenth of the amount of light emitted by a firefly in a single flash), and they become saturated (i.e., reach a maximum) at about 0.05 μmol m^{-2}. For example, Arabidopsis seeds can be induced to germinate with red light in the range of 0.001 to 0.1 μmol m^{-2}. In dark-grown oat (*Avena* spp.) seedlings, red light can stimulate the growth of the coleoptile and inhibit the growth of the mesocotyl (the elongated axis between the coleoptile and the root) at similarly low fluences. (The ecological implications of the VLFR in seed germination are discussed in **WEB ESSAY 16.1**.)

VLFRs are non-photoreversible. The minute amount of light needed to induce VLFRs converts less than 0.02% of the total phytochrome to Pfr. Because the far-red light that would normally reverse a red-light effect converts only 98% of the Pfr to Pr (as discussed earlier), about 2% of the phytochrome remains as Pfr—significantly more than the 0.02% needed to induce VLFRs. In other words, far-red light cannot lower the Pfr concentration below 0.02%, so it is unable to inhibit VLFRs. Although VLFRs are non-photoreversible, the action spectrum for VFLR responses (e.g. seed germination) are similar to those of LFR responses (discussed next), supporting the view that phytochrome is the photoreceptor involved in VFLRs. This hypothesis was confirmed using phytochrome-deficient mutants, as described later in the chapter.

LOW-FLUENCE RESPONSES (LFRs) Another set of phytochrome responses cannot be initiated until the fluence reaches 1.0 μmol m^{-2}, and they are saturated at about 1000 μmol m^{-2}. These low-fluence responses (LFRs) include processes such as the promotion of lettuce seed germination, inhibition of hypocotyl elongation, and regulation of leaf movements (see Table 16.1). As we saw in Figure 16.6, the LFR action spectrum for Arabidopsis seed germination includes a main peak for stimulation in the red region (660 nm) and a major peak for inhibition in the far-red region (720 nm).

Both VLFRs and LFRs can be induced by brief pulses of light, provided that the total amount of light energy adds up to the required fluence. The total fluence is a function of two factors: the fluence rate (μmol m^{-2} s^{-1}) and the time of irradiation. Thus, a brief pulse of red light will induce a response, provided that the light is sufficiently bright, and conversely, very dim light will work if the irradiation time is long enough. This reciprocal relationship between fluence rate and time is known as the **law of reciprocity**. VLFRs and LFRs both obey the law of reciprocity; that is, the magnitude of the response (e.g., percent germination or degree of inhibition of hypocotyl elongation) is dependent on the product of the fluence rate and the time of irradiation.

However, reciprocity holds true only when photon absorption by the photoreceptor studied is the rate-limiting step in the response being studied. Reciprocity is confounded when any step between photoreceptor activation and the response measured (e.g., hypocotyl elongation) becomes limiting. Thus, the concept of reciprocity is difficult to demonstrate for many responses.

HIGH-IRRADIANCE RESPONSES (HIRs) Phytochrome responses of the third type are termed high-irradiance responses (HIRs), several of which are listed in **Table 16.2**. HIRs require prolonged or continuous exposure to light of relatively high irradiance. The response is proportional to the irradiance until the response saturates and additional light has no further effect (see **WEB TOPIC 16.2**). The reason these responses are called high-irradiance responses rather than high-fluence responses is that they are proportional to fluence rate—the number of photons striking the plant tissue per second—rather than to fluence—the total number of photons striking the plant in a given period of illumination. HIRs saturate at much higher fluences than LFRs—at least 100 times higher. Because neither continuous exposure to dim light nor transient exposure to bright light can induce HIRs, these responses do not to obey the law of reciprocity.

Many of the LFRs listed in Table 16.1, particularly those involved in de-etiolation, also qualify as HIRs. For example, at low fluences the action spectrum for anthocyanin production in seedlings of white mustard (*Sinapis alba*) is indicative of phytochrome and shows a single peak in the red region of the spectrum. The effect is reversible with far-red light (a photochemical property unique to phytochrome), and the response obeys the law of reciprocity. However, if the dark-grown seedlings are instead exposed to high-irradiance light for several hours, the action spectrum contains peaks in the far-red and blue regions, the effect is no longer photoreversible, and the response becomes proportional to the irradiance. Thus, the same effect can be either an LFR or an HIR, depending on the

TABLE 16.2 Some plant photomorphogenic responses induced by high irradiances

Synthesis of flavonoids, including anthocyanins, in various dicot seedlings and in apple skin segments
Inhibition of hypocotyl elongation in mustard, lettuce, and petunia seedlings
Induction of flowering in henbane (Hyoscyamus)
Plumular hook opening in lettuce
Enlargement of cotyledons in mustard
Production of ethylene in sorghum

history of a seedling's exposure to light. As we will discuss below, different phytochrome molecules are responsible for these various types of responses.

Phytochrome A mediates responses to continuous far-red light

As noted earlier, Arabidopsis contains five genes encoding phytochromes, *PHYA–PHYE*. Four of the five phytochromes, phyB–phyE, appear mostly light-stable in the plant and function primarily in the regulation of LFRs and of shade avoidance involving changes in the R:FR ratio. In contrast, phyA is rapidly degraded as Pfr and controls plant responses to VLFRs and the far-red HIRs. Recent studies suggest that phyB also is degraded in the nucleus along with its PIF targets during signaling. Thus, Pfr turnover appears to be a conserved property of plant phytochromes.

In early mutant screens of Arabidopsis, mutations in phyB were identified in mutants with altered hypocotyl elongation under continuous white light, collectively termed *hy* mutants. Continuous white light is detected by the light-stable phytochromes, phyB–phyE. Since far-red HIRs were known to require light-labile phytochrome, it was suspected that phyA must be the photoreceptor involved in the perception of continuous far-red light. Screens of mutants that fail to respond to continuous far-red light and instead grow tall and spindly led to the identification of phyA mutants as well as additional mutants that were defective in chromophore formation, indicating that phyA mediates the response to continuous far-red light.

Mutants lacking phyA also failed to germinate in response to millisecond pulses of light, but showed a normal response to red light in the low-fluence range. This result demonstrates that phyA also functions as the primary photoreceptor for this VLFR. When *phyA/phyB* double mutants are grown under high-fluence red light (>100 µmol m^{-2} s^{-1}), they are even more elongated than the *phyB* single mutant. PhyA has also been shown to play a role in the photoperiod control of flowering in Arabidopsis and rice.

Phytochrome B mediates responses to continuous red or white light

The characterization of the *hy3* mutant revealed an important role for phyB in de-etiolation, since mutant seedlings grown in continuous white light had long hypocotyls. The *phyB* mutant is deficient in chlorophyll and in some mRNAs that encode chloroplast proteins, and it is impaired in its ability to respond to plant hormones.

In addition to regulating white and red light–mediated HIRs, phyB also appears to regulate LFRs, such as photoreversible seed germination, the phenomenon that originally led to the discovery of phytochrome. Wild-type Arabidopsis seeds require light for germination, and the response shows red/far-red reversibility in the low-fluence range (see Figure 16.6A). Mutants that lack phyA respond normally to red light, whereas mutants deficient in phyB are unable to respond to low-fluence red light. This experimental evidence strongly suggests that phyB mediates photoreversible seed germination.

PhyB also plays an important role in regulating plant responses to shade treatments. Plants that are deficient in phyB often look like wild-type plants that are grown under dense vegetative canopies. In fact, mediating responses to vegetative shade such as accelerated flowering and increased elongation growth may be one the most ecologically important roles of phytochromes (see Chapter 18).

Roles for phytochromes C, D, and E are emerging

Although phyA and phyB are the predominant forms of phytochrome in Arabidopsis, phyC, phyD, and phyE play unique roles in regulating responses to red and far-red light. The creation of double and triple mutants has made it possible to assess the relative role of each phytochrome in a given response. phyD and phyE are structurally similar to phyB but are not functionally redundant. Responses mediated by phyD and phyE include petiole and internode elongation and the control of flowering time (see Chapter 20). The characterization of *phyC* mutants in Arabidopsis suggests a complex interplay between phyC, phyA, and phyB response pathways. This specialization in phytochrome gene function is likely to be important in fine-tuning phytochrome responses to daily and seasonal changes in light regimes.

Phytochrome Signaling Pathways

All phytochrome-regulated changes in plants begin with absorption of light by the photoreceptor. After light absorption, the molecular properties of phytochrome are altered, affecting the interaction of the phytochrome protein with other cellular components that ultimately bring about changes in the growth, development, or position of an organ (see Tables 16.1 and 16.2).

Molecular and biochemical techniques are helping unravel the early steps in phytochrome action and the signal transduction pathways that lead to physiological or developmental responses. These responses fall into two general categories:

- Ion fluxes, which cause relatively rapid turgor responses
- Altered gene expression, which typically results in slower, longer-term responses

In this section we will examine the effects of phytochrome on both membrane permeability and gene expression, as well as the possible chain of events constituting the signal transduction pathways that bring about these effects.

Phytochrome regulates membrane potentials and ion fluxes

Phytochrome can rapidly alter the properties of membranes, within seconds of a light pulse. Such rapid modulation has been measured in individual cells and has been inferred from the effects of red and far-red light on the surface potential of roots and oat coleoptiles, in which the lag time between the production of Pfr and the onset of measurable hyperpolarization (membrane potential changes) occurs within seconds. Changes in the electrical potential of cells imply changes in the flux of ions across the plasma membrane and suggest that some of the cytosolic responses of phytochrome are initiated at or near the plasma membrane (see **WEB TOPIC 16.3**).

One long-standing conundrum has been how the filamentous green alga *Mougeotia* uses red light to stimulate rapid chloroplast movement (see **WEB TOPIC 16.1**). In many species, including Arabidopsis, chloroplast movements are mediated by blue light through the action of phototropin photoreceptor proteins. In *Mougeotia*, the photoreceptors regulating chloroplast movement consist of a fusion between phytochrome and a phototropin known as **neochrome**, and show typical bilin binding as well as red/far-red reversibility. Thus, *Mougeotia* appears to have evolved the ability to exploit red light as a signal to induce a response (chloroplast movement) that is typically mediated by blue light.

Phytochrome regulates gene expression

As the term *photomorphogenesis* implies, plant development is profoundly influenced by light. Elongated stems, folded cotyledons, and the absence of chlorophyll characterize the development of dark-grown, etiolated seedlings. Complete reversal of these symptoms by light involves major long-term alterations in metabolism that can only be brought about by changes in gene expression. Light-regulated plant promoters are similar to those of other eukaryotic genes: a collection of modular elements, the number, position, flanking sequences, and binding activities of which can lead to a wide range of transcriptional patterns. No single DNA sequence or binding protein is common to all phytochrome-regulated genes.

At first it may appear paradoxical that light-regulated genes have such a range of regulatory elements, any combination of which can confer light-regulated expression. However, this array of sequences allows for the differential light- and tissue-specific regulation of many genes through the action of multiple photoreceptors.

The stimulation and repression of transcription by light can be very rapid, with lag times as short as 5 min. Using DNA microarray analysis, global patterns of gene expression in response to changes in light can be monitored. (For a discussion of methods for transcriptional analysis see **WEB TOPIC 2.4**. These studies have indicated that nuclear import triggers a transcriptional cascade involving thousands of genes that are involved in photomorphogenic development. By monitoring gene expression profiles over time following a shift of plants from darkness to light, both early and late targets of *PHY* gene action have been identified.

The nuclear import of phyA and phyB is highly correlated with the light quality that stimulates their activities. That is, nuclear import of phyA is activated by either red or far-red light, or low-fluence broad-spectrum light, whereas phyB import is driven by red-light exposure and is reversible by far-red light. Nuclear import of the phytochrome proteins represents a major control point in phytochrome signaling.

Some of the early gene products that are rapidly upregulated following a shift from darkness to light are themselves transcription factors that activate the expression of other genes. The genes encoding these rapidly upregulated proteins are called **primary response genes**. Expression of the primary response genes depends on signal transduction pathways (discussed next) and is independent of protein synthesis. In contrast, the expression of the late genes, or **secondary response genes**, requires the synthesis of new proteins.

Phytochrome interacting factors (PIFs) act early in signaling

Phytochrome interacting factors (**PIFs**) are a family of proteins that act primarily as negative regulators of photomorphogenic responses. A quadruple mutant that disrupts the functions of multiple PIF family members displays constitutive photomorphogenic development when plants are grown in the dark. PIFs regulate various aspects of phytochrome-mediated photomorphogenesis, including seed germination, chlorophyll biosynthesis, shade avoidance, and hypocotyl elongation. PIFs promote etiolated development in the dark (**skotomorphogenesis**) primarily by serving as transcriptional activators of dark-induced genes (**Figure 16.13A**), and also by repressing some light-induced genes (**Figure 16.13B**). In both cases, red light-induced Pfr formation initiates the degradation of PIF proteins through phosphorylation, followed by degradation via the proteasome complex (see Chapters 2 and 15). The rapid degradation of PIFs may provide a mechanism for modulating light responses that is tightly coupled to the activities of phy proteins.

PIFs that interact with *either* phyA or phyB define branch points in the phy signaling networks, whereas proteins that interact with *both* phyA and phyB are likely to represent points of convergence. One of the most extensively characterized of these factors is **PIF3**, a basic helix-loop-helix (bHLH) transcription factor that interacts with *both* phyA and phyB. PIF3 and several related PIF or **PIF-like proteins** (**PILs**) are particularly notable because at least five members of this gene family selectively interact with phytochromes in their active Pfr conformations. The fact that these proteins are localized to the nucleus and can bind to DNA suggests an intimate association between phytochrome and gene transcription.

(A) PIFs as constitutive transcriptional activators in the dark

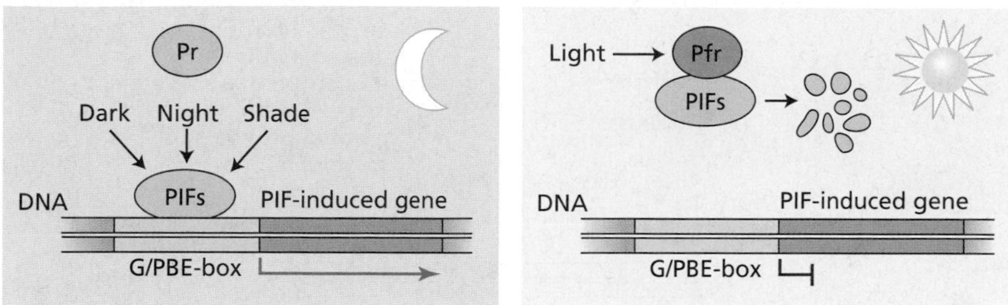

(B) PIFs as constitutive transcriptional repressors in the dark

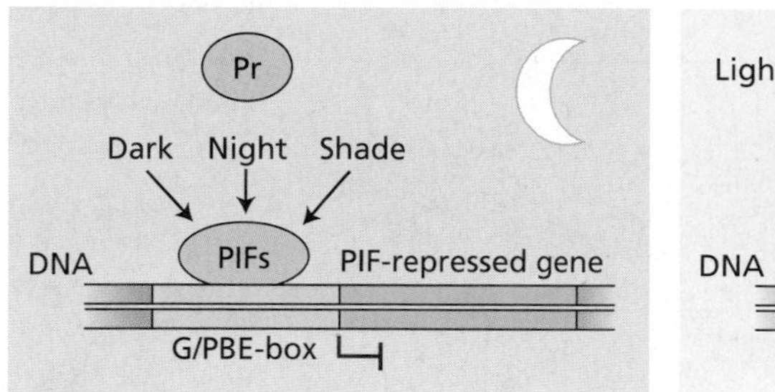

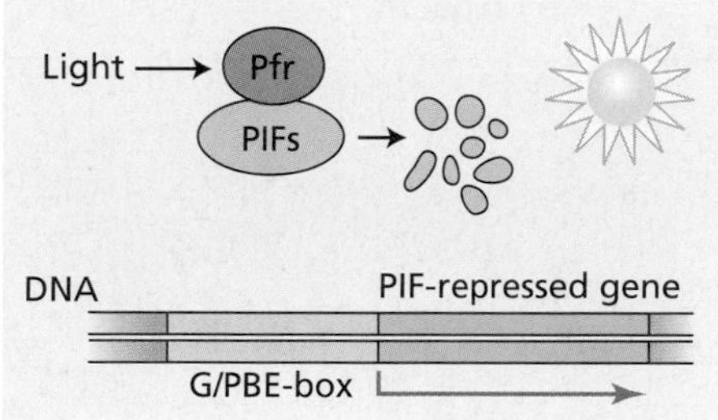

Figure 16.13 Phytochrome interacting factors (PIFs) act as negative regulators of photomorphogenesis. (A) Most PIFs are constitutive activators of genes that are expressed in the dark or in response to shade. In the light, Pfr promotes the degradation of the PIFs, blocking the transcription of skotomorphogenesis genes. (B) During de-etiolation, PIFs also can act as constitutive repressors of some light-induced genes. Pfr causes the turnover these PIFs, allowing the expression of photomorphogenesis genes. (After Leivar and Monte 2014.)

Phytochrome signaling involves protein phosphorylation and dephosphorylation

A group of membrane-associated **phytochrome kinase substrate** (**PKS**) proteins appears to modify phytochrome activity via phosphorylation either directly or via interactions with other kinases. PKS1 interacts with phyA and phyB in both the active Pfr and inactive Pr forms. Molecular and genetic analyses suggest that these proteins act selectively to promote phyA-mediated VLFR. A number of phosphatases have also been shown to interact with phy and regulate its phosphorylation state.

Phytochrome-induced photomorphogenesis involves protein degradation

As discussed in Chapter 15, the majority of plant signal transduction pathways involve the inactivation, degradation, or removal of repressor proteins. The phytochrome signaling pathway is consistent with this general principle. For example, phyA is rapidly degraded following its activation by light. Thus, protein degradation, in addition to phosphorylation, is emerging as a ubiquitous mechanism regulating many cellular processes, including light and hormone signaling, circadian rhythms, and flowering time (for examples, see Chapters 15 and 20).

Genetic screens conducted independently by several groups identified mutants that exhibited light-grown phenotypes when grown in the dark. The genes identified in these screens were called ***CONSTITUTIVE PHOTOMORPHOGENESIS1*** (***COP1***), *DE-ETIOLATED* (*DET*), and *FUSCA* (*FUS*) (for the dark red color of the anthocyanins that accumulate in light-grown seedlings). Many of these genes are allelic or encode proteins that are part of the same complexes, and they are collectively known as COP/DET/FUS. These genes encode proteins of the **COP1–SUPPRESSOR OF PHYA** (**COP1–SPA**) **complex**, the **COP9 signalosome** (**CSN**) **complex**, and other complexes that are involved in the ubiquitination and proteasomal degradation of photomorphogenesis-promoting proteins.

COP1, another negative regulator of photomorphogenesis, is a component of the E3 ubiquitin ligase complex that targets photomorphogenesis-promoting proteins for degradation such as phyA, phyB, and several transcription factors. COP1 is found in the nucleus in the dark and in the cytoplasm in the light. COP1 movement into the nucleus in the dark requires the COP9 signalosome (CSN) complex, although it is not yet clear how CSN targets COP1 to the nucleus (**Figure 16.14**). In the nucleus, COP1 interacts directly with SPA1, which promotes PHYA destruction. The COP1–SPA1-E3 ligase complex is also responsible for the ubiquitination and proteasomal degradation of photomorphogenesis-promoting proteins, such as the bZIP transcription factor HY5 (see Figure 16.14). As a result, skotomorphogenesis becomes the default pathway of development.

In the presence of light, COP1 activity is repressed, although the complete mechanism underlying COP1 inactivation in the light is unknown. Light-dependent export of COP1 to the cytoplasm is a slow process, requires long exposure to light (more than 24 h), and is likely a mechanism to suppress COP1 activation under extended light conditions. Together, the repression of COP1 activity and export to the cytoplasm allow transcription factors to bind to promoter elements in genes that mediate photomorphogenic development.

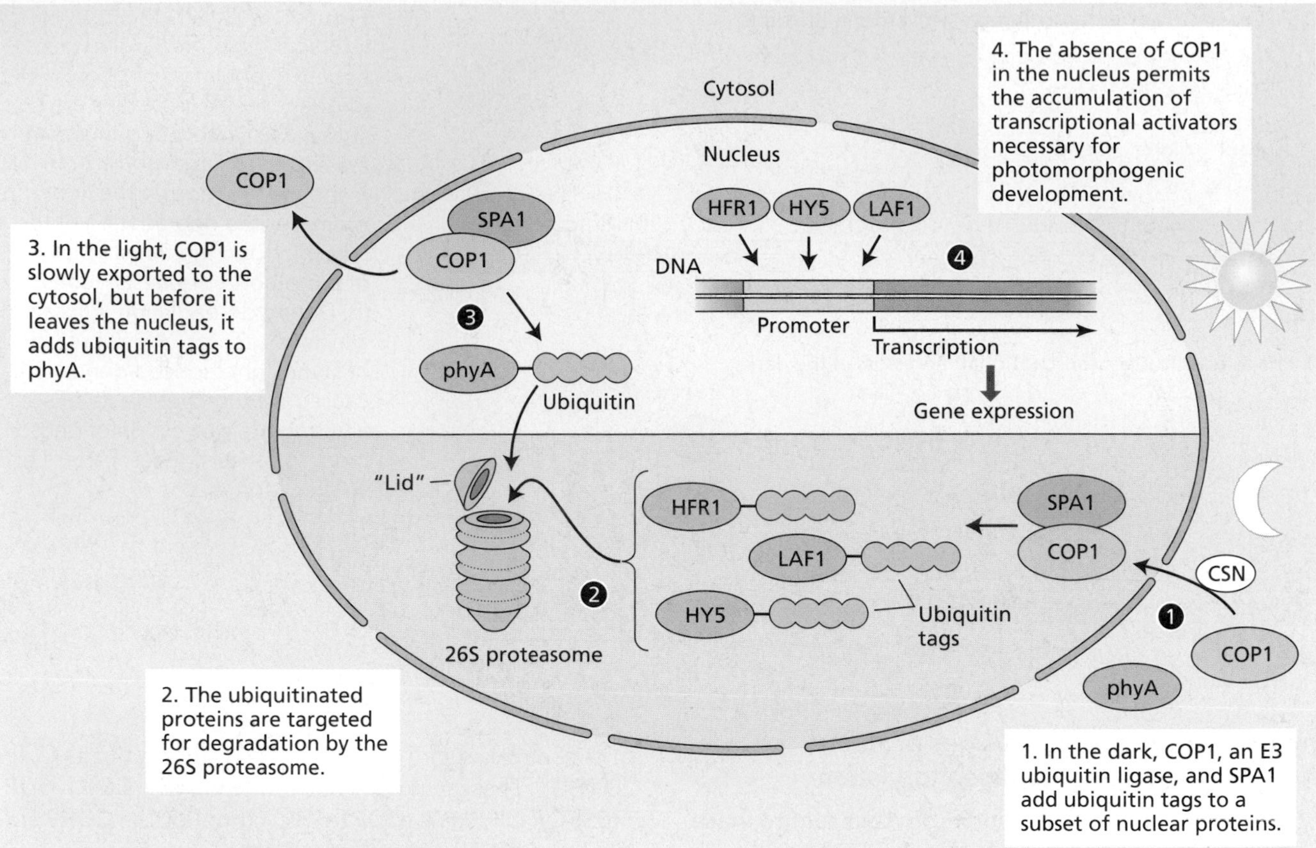

Figure 16.14 COP proteins regulate the turnover of proteins required for photomorphogenic development. During the night, COP1 enters the nucleus with the help of the COP9 signalosome complex (CSN). COP1 forms a complex with SPA1, and the COP1–SPA1 complex adds ubiquitin to a subset of transcriptional activators that promote photomorphogenesis. The transcription factors are then degraded by the proteasome complex. During the day, COP1 exits the nucleus, allowing the transcriptional activators to accumulate.

As will be discussed in Chapter 20, COP1 is also responsible for the degradation of the flowering regulators CONSTANS (CO) and GIGANTEA (GI).

Blue-Light Responses and Photoreceptors

Blue-light responses have been reported in higher plants, algae, ferns, fungi, and prokaryotes. In addition to phototropism, these responses include anion uptake in algae, inhibition of seedling hypocotyl (stem) elongation, stimulation of chlorophyll and carotenoid synthesis, activation of gene expression, and enhancement of respiration. Among motile unicellular organisms such as certain algae and bacteria, blue light mediates *phototaxis,* the movement of motile unicellular organisms toward or away from light. Blue light also stimulates the infection process in bacteria, such as the animal pathogen *Brucella abortus.* Some blue-light responses were introduced in relation to photosynthesis in Chapters 9 and 10, including chloroplast movements, solar tracking, and stomatal opening. In Chapter 18, several key blue-light responses—photoblasty, phototropism, and photomorphogenesis—will be discussed in the context of seed germination and seedling establishment.

Three distinct classes of photoreceptors mediate the effects of UV-A/blue light (320–500 nm): the cryptochromes, phototropins and the ZEITLUPE (ZTL) family of proteins. Cryptochromes (cry), like the phytochromes, play a major regulatory role in plant photomorphogenesis. Phototropins (phots), by contrast, are involved in directing organ, chloroplast, and nuclear movements, solar tracking, and stomatal opening, all of which are light-dependent processes that optimize the photosynthetic efficiency of plants. The ZTL family, has been shown to participate in the control of circadian clocks and flowering.

Blue-light responses have characteristic kinetics and lag times

The inhibition of stem elongation and the stimulation of stomatal opening by blue light illustrate two important temporal properties of blue-light responses:

1. A significant lag time separating the light signal and the maximum response rate
2. Persistence of the response after the light signal has been switched off

Blue-light responses can be relatively rapid compared with most photomorphogenic changes. However, in contrast to typical photosynthetic responses, which are fully activated almost instantaneously after a "light on" signal, and which cease as soon as the light goes off, blue-light responses exhibit a lag time of variable duration and proceed at maximum rates for several minutes after application of a light pulse.

For example, blue light induces a decrease in growth rate and a transient membrane depolarization in etiolated cucumber seedlings only after a lag time of about 25 s (**Figure 16.15**). The persistence of blue-light responses in the absence of blue light has been studied using blue-light pulses. For example, blue light–induced activation of the H^+-ATPase in guard cells decays following a pulse of blue light, but only after several minutes have elapsed (see Figures 10.6 and 10.7). This persistence of the blue-light response after the pulse can be explained by a photochemical cycle in which the physiologically active form of the photoreceptor, which has been converted from the inactive form by blue light, slowly reverts to the inactive form after the blue light is switched off. As we will discuss shortly, in the case of phototropins this cycle appears to involve four main processes: receptor dephosphorylation by a protein phosphatase, breaking the covalent carbon–sulfur bond, dissociation of the receptor from its target molecules, and the dark reversion of light-driven conformational changes. The rate of decay of the response to a blue-light pulse would thus depend on the time course of the reversion of the active form of the photoreceptor back to the inactive form.

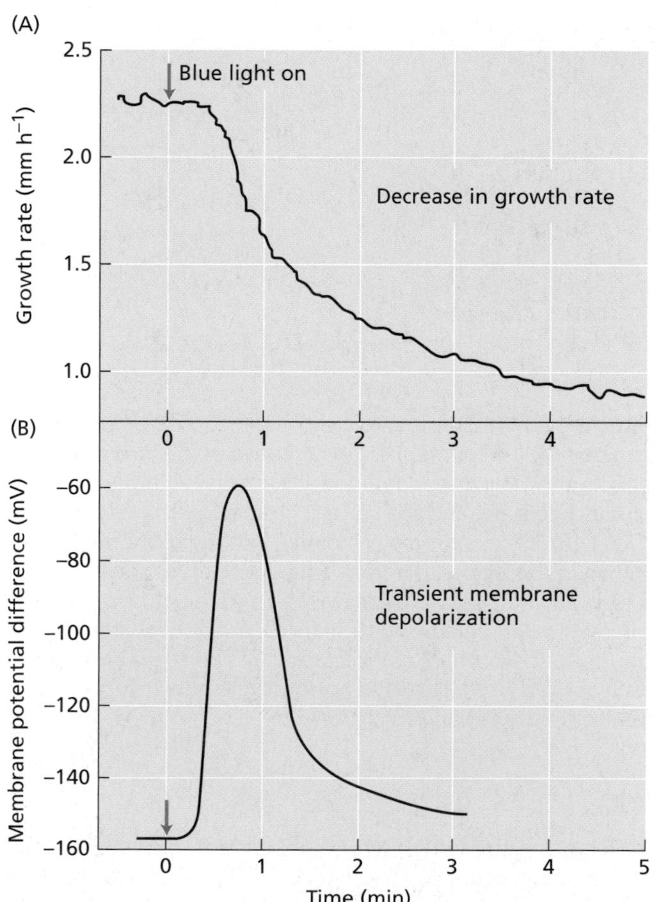

Figure 16.15 (A) Blue light–induced changes in elongation rates of etiolated cucumber hypcotyls. (B) Blue light–induced transient membrane depolarization of hypocotyl cells. (After Spalding and Cosgrove 1989.)

Cryptochromes

Cryptochromes are blue-light photoreceptors that mediate several blue-light responses, including suppression of hypocotyl elongation, promotion of cotyledon expansion, membrane depolarization, inhibition of petiole elongation, anthocyanin production, and circadian clock entrainment. ***CRYPTOCHROME1*** (***CRY1***), was originally identified in Arabidopsis using genetic screens for mutants whose hypocotyls were elongated when grown in white light because they lacked the light-stimulated inhibition of hypocotyl elongation described above. Further analyses showed that the long hypocotyl phenotype of one of the mutants, *hy4*, was specific to blue light-inhibition of hypocotyl elongation. That is, hypocotyl elongation was still inhibited by red light in the *hy4* mutant. As will be discussed later in the chapter, cryptochromes are responsible for the long-term blue light-induced inhibition of hypocotyl elongation, while phototropins mediate the rapid inhibitory response.

The *HY4* gene encodes a 75-kDa protein with significant sequence homology to microbial **photolyase**, a blue light–activated enzyme that repairs pyrimidine dimers in DNA caused by exposure to ultraviolet radiation. In view of this sequence similarity, the HY4 protein, later named cry1, was proposed to be a blue-light photoreceptor mediating stem elongation. Cryptochromes, however, show no photolyase activity. Cryptochrome proteins were later discovered in many organisms, including cyanobacteria, ferns, algae, fruit flies, mice, and humans. Arabidopsis contains three cryptochrome genes: *CRY1, CRY,2* and *CRY3.*

The activated FAD chromophore of cryptochrome causes a conformational change in the protein

The domain structure of Arabidopsis cryptochromes is shown in **Figure 16.16A**. Similar to a major class of photolyases, cryptochromes bind a **flavin adenine dinucleotide** (**FAD**) and the pterin 5,10-methyltetrahydrofolate (MTHF) as chromophores (**Figure 16.16B and C**). Pterins

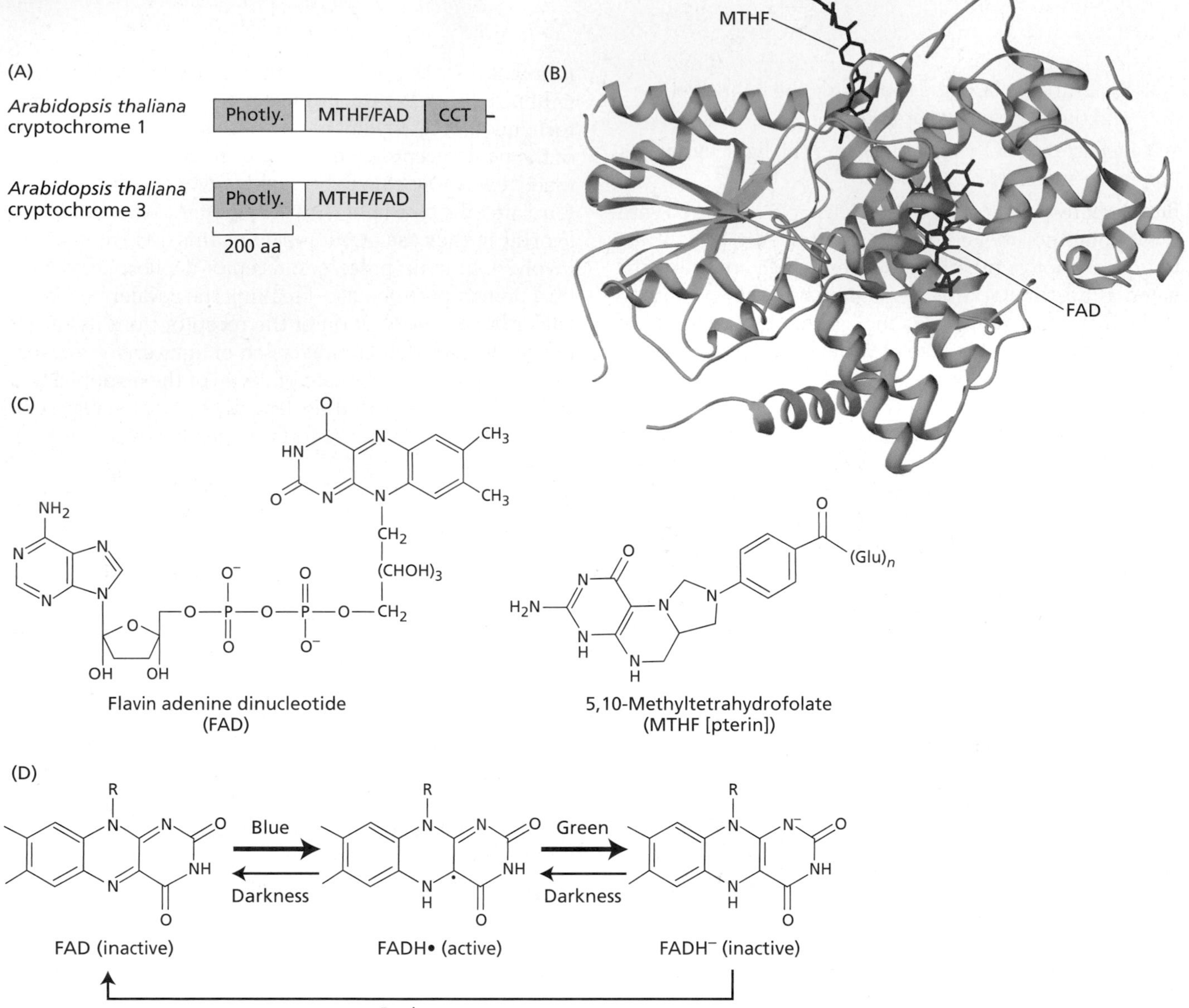

Figure 16.16 Cryptochrome domain and chromophore structure. (A) Alignment of two cryptochromes from Arabidopsis showing the photolyase-like domain (Photly.), FAD-binding domain, and the cryptochrome C-terminal domain (CCT). (B) Cryptochrome is a dimer, but the monomer is shown in this ribbon diagram. The light-harvesting cofactor 5,10-methyltetrahydrofolate (MTHF) and the catalytic cofactor flavin adenine dinucleotide (FAD) are noncovalently bound to the protein, as indicated. (C) The structures of FAD and MTHF. (D) The FAD photocycle of cryptochrome. (B after Huang et al. 2006.)

are light-absorbing pteridine derivatives often found in pigmented cells of insects, fishes, and birds. In photolyases, blue light is absorbed by the pterin, and the excitation energy is then transferred to FAD. A similar mechanism may operate in cryptochrome, but definitive evidence is lacking. However, it is clear that FAD is the primary chromophore regulating cryptochrome activity.

Blue-light absorption alters the redox status of the bound FAD chromophore, and it is this primary event that triggers photoreceptor activation (**Figure 16.16D**). As occurs in phytochromes and phototropins, this activation mechanism involves protein conformational changes. In the case of cryptochromes, light absorption by the N-terminal photolyase region is thought to alter the conformation of a C-terminal extension, which is necessary for signaling. This C-terminal extension is absent from photolyase enzymes but is clearly essential for cryptochrome signaling. We can therefore view plant cryptochrome as a molecular light switch where absorption of blue photons at the N-terminal photosensory region results in protein conformational changes at the C terminus, which, in turn, initiates signaling by binding to specific partner proteins. As in phytochromes, dimerization of cryptochromes, mediated by the photolyase-like domain, may be important for their signaling.

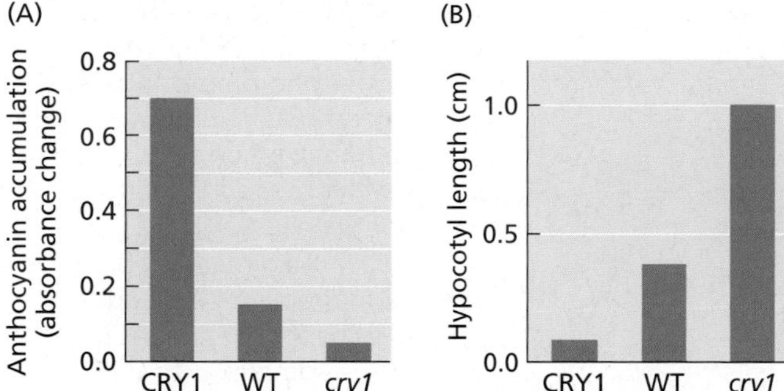

Figure 16.17 Blue light stimulates the accumulation of anthocyanin (A) and the inhibition of stem elongation (B) in transgenic and mutant seedlings of Arabidopsis. These bar graphs show the phenotypes of a transgenic plant overexpressing the gene that encodes CRY1 (CRY1 OE), the wild type (WT), and *cry1* mutants. The enhanced blue-light response of the CRY1 overexpressor demonstrates the important role of this gene product in stimulating anthocyanin biosynthesis and inhibiting stem elongation. (After Ahmad et al. 1998.)

cry1 and cry2 have different developmental effects

Overexpression of the CRY1 apoprotein in transgenic tobacco or Arabidopsis plants results in a stronger blue light–induced inhibition of hypocotyl elongation, as well as increased production of anthocyanin (**Figure 16.17**). A second cryptochrome, named cry2, was later isolated from Arabidopsis. Both cry1 and cry2 appear to be ubiquitous throughout the plant kingdom. A major difference between them is that the cry2 protein is preferentially degraded under blue light, whereas cry1 is much more stable. Transgenic plants overexpressing the *CRY2* gene show only a small enhancement of the inhibition of hypocotyl elongation found in the wild type, indicating that unlike cry1, cry2 does not play a primary role in inhibiting stem elongation. However, transgenic plants overexpressing *CRY2* show a large increase in blue light–stimulated cotyledon expansion. In addition, cry1, and to a lesser extent cry2, is involved in the setting of the circadian clock in Arabidopsis, whereas cry2 plays a major role in the induction of flowering (see Chapter 20). Cryptochrome homologs have also been shown to function in regulating the circadian clock in flies, mice, and humans.

It is also worth noting that in Arabidopsis the nuclear and cytoplasmic pools of cry1 have been shown to have distinct biological functions. Contrary to expectations, nuclear, rather than cytoplasmic, cry1 molecules were found to mediate blue light–mediated changes in membrane depolarization. This response, showing a time course of several seconds, is one of the fastest cry1-mediated responses to blue light. The mechanism involved in this blue light–dependent, anion-channel activation is not yet known.

While cry1 and cry2 are generally found in the nucleus, cry3 is localized to chloroplasts and mitochondria. The function of cry3 is not yet known, although it has been shown to have photolyase activity specific to single-stranded DNA lesions. Furthermore, the mechanism of cry3 signaling is obviously different from that of cry1 and cry2 since it lacks a prominent C-terminal extension.

Nuclear cryptochromes inhibit COP1-induced protein degradation

Both cry1 and cry2 are present in the nucleus and the cytoplasm, and there is no evidence that cryptochrome moves into the nucleus in response to light. **Figure 16.18** shows that in darkness, COP1, in concert with SPA1 and other factors, acts to degrade transcription factors such as HY5, which induce the expression of genes required for photomorphogenesis (see also Figure 16.14). Upon activation by blue light, cry1 in the nucleus forms a complex with SPA1 and COP1 that prevents them from acting, thereby preventing the degradation of HY5 and other transcription factors that promote photomorphogenesis. As in the case of phytochrome signaling, the increased levels of HY5 and other transcription factors promote photomorphogenic development.

It is the C terminus of cryptochrome that binds to SPA1 and prevents SPA1/COP1 action. Arabidopsis plants overexpressing only the C-terminal region of cryptochrome (CCT) show phenotypes similar to those of *cop* mutants, which resemble light-grown seedlings when grown in darkness. The model shown in Figure 16.18 can explain the phenotype of CCT-overexpressing plants. Without the photosensory N-terminal domain, the CCT region can adopt an active conformation that sequesters the activity of COP1 and SPA1 even in the absence of light, thereby promoting an increase in HY5 protein levels and transcription of key photomorphogenic genes.

Blue light-induced phosphorylation of cryptochrome also appears to be important in modulating its activity and, in the case of cry2, promoting its degradation. The protein kinases involved are not fully understood, but phosphorylation may be important in maintaining the C terminus of cry1 in an active conformation (see Figure 16.18).

Cryptochrome can also bind to transcriptional regulators directly

In addition to controlling the levels of transcription factors, cryptochrome can also directly bind to and regulate the activity of specific DNA-binding proteins. In the case of flowering, cry2 has been shown to bind directly to bHLH transcription factors such as Cry-Interacting bHLH1 (CIB1). CIB1 regulates floral initiation by binding to the promoter of *FLOWERING LOCUS T (FT)*. FT is the mobile transcriptional regulator that migrates from leaves to the apical meristem and activates transcription of floral meristem identity genes (see Chapter 20). Plants overexpressing CIB1 flower earlier than wild-type plants.

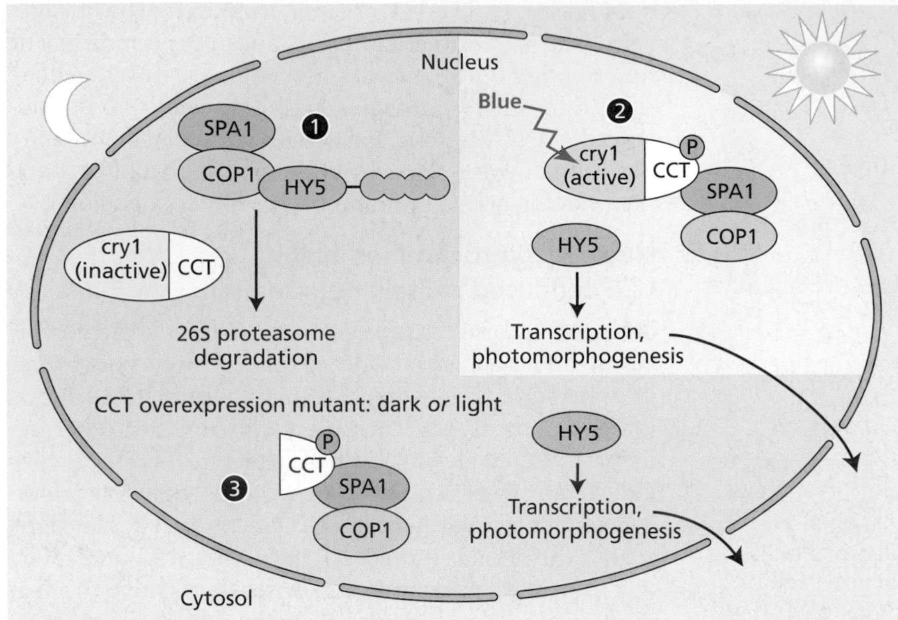

Figure 16.18 Model of cry1 interactions with COP1/SPA1 in the regulation of photomorphogenesis. (1) In the dark, COP1/SPA1 acts to degrade transcription factors such as HY5, which is required for photomorphogenesis. (2) In the light, cry1 is activated directly by blue light and indirectly by blue light-induced phosphorylation. Activated cry1 forms a complex with COP1 and SPA1 via the C-terminal domain, preventing them from degrading protein targets such as HY5. (3) In the absence of the photosensory N terminus as in the truncation mutant diagrammed on the bottom, the CCT region can adopt an active conformation that sequesters COP1/SPA1 in the absence of light, thereby promoting an increase in HY5 protein levels and transcription of key morphogenic genes.

The Coaction of Cryptochrome, Phytochrome, and Phototropins

Coaction between crytochrome and phytochrome was long suspected because several developmental processes, such as photomorphogenesis and flowering, were known to be under phytochrome control, yet mutations in the *CRY2* gene led to alterations in these responses. We now understand that coaction exists between several of the plant photoreceptors. Many of the developmental processes affected by coaction can be grouped into three general categories: stem or hypocotyl elongation, flowering, and regulation of circadian rhythms.

Stem elongation is inhibited by both red and blue photoreceptors

As noted above, the stems of seedlings growing in the dark elongate very rapidly, and the inhibition of stem elongation by light is a key photomorphogenic response of the seedling emerging from the soil surface (see Chapter 18). Although phytochrome is involved in this response, the action spectrum for the decrease in elongation rate also shows strong activity in the blue region, which cannot be explained by the absorption properties of phytochrome. In fact, the 400- to 500-nm blue region of the action spectrum for the inhibition of stem elongation closely resembles that of phototropism.

Experimentally, it is possible to separate a reduction in elongation rates mediated by phytochrome from a reduction mediated by a specific blue-light response. If lettuce seedlings are given low fluence rates of blue light under a strong background of yellow light, their hypocotyl elongation rate is reduced by more than 50%. The background yellow light establishes a well-defined Pr:Pfr ratio. Adding blue light at low fluence rates does not significantly change this ratio, ruling out a phytochrome effect on the reduction in elongation rate observed upon the addition of blue light. These results indicate that the elongation rate of the hypocotyl is controlled by a specific blue-light response that is independent of the phytochrome-mediated response.

A specific blue light–mediated hypocotyl response can also be distinguished from one mediated by phytochrome by their contrasting time courses. Whereas phytochrome-mediated changes in elongation rates can be detected within approximately 10 to 90 min, depending on the species, blue-light responses show lag times of less than 1 min. High-resolution analysis of the changes in growth rate mediating the inhibition of hypocotyl elongation by blue light has provided valuable information about the interactions among phototropin, cry1, cry2, and phyA. After a lag of 30 s, blue light–treated, wild-type Arabidopsis seedlings show a rapid decrease in elongation rates during the first 30 min, and then they grow very slowly for several days.

Another rapid response elicited by blue light is a depolarization of the membrane of hypocotyl cells that precedes the inhibition of growth rate (see Figure 16.15B). This membrane depolarization is caused by the activation of anion channels (see Chapter 6), which facilitates the efflux of anions such as chloride. Application of an anion channel blocker, NPPB (5-nitro-2-[4-phenylbutylamino]-benzoate), prevents the blue light–dependent membrane depolarization and decreases the inhibitory effect of blue light on hypocotyl elongation.

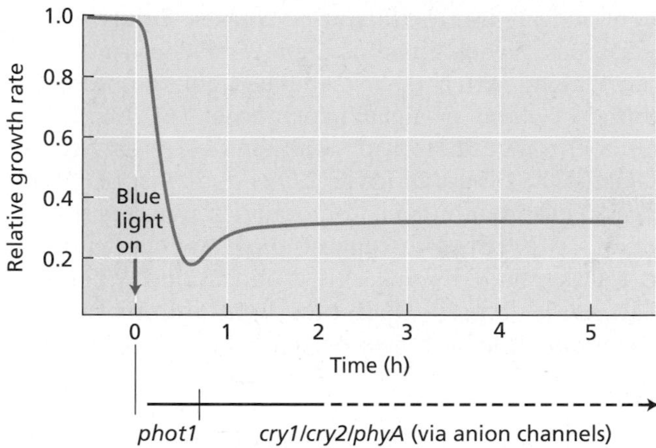

Figure 16.19 Sensory transduction process of blue light–stimulated inhibition of stem elongation in Arabidopsis. Elongation rates in the dark (0.25 mm h^{-1}) were normalized to 1. Within 30 s of the onset of blue-light irradiation, growth rates decreased; they approached zero within 30 min, then continued at very reduced rates for several days. If blue light was applied to a *phot1* mutant, dark-growth rates remained unchanged for the first 30 min, indicating that the inhibition of elongation in the first 30 min is under phototropin control. Similar experiments with *cry1*, *cry2*, and *phyA* mutants indicated that the respective gene products control elongation rates at later times. (After Parks et al. 2001.)

Analysis of the same response in *phot1*, *cry1*, *cry2*, and *phyA* mutants has shown that suppression of stem elongation by blue light during seedling de-etiolation is initiated by phot1, with cry1, and to a limited extent cry2, modulating the response after 30 min (**Figure 16.19**). The slow growth rate of stems in blue light–treated seedlings is primarily a result of the persistent action of cry1, and this is the reason that *cry1* mutants of Arabidopsis show a long hypocotyl, compared with the short hypocotyl of the wild type. PhyA appears to play a role in at least the early stages of blue light–regulated growth, because growth inhibition does not progress normally in *phyA* mutants.

Phytochrome interacts with cryptochrome to regulate flowering

In Arabidopsis, continuous blue or far-red light promotes flowering, and red light inhibits flowering. Far-red light acts through phyA, and the antagonistic effect of red light is produced by phyB. One might expect the *cry2* mutant to be delayed in flowering, since blue light promotes flowering. However, *cry2* mutants flower at the same time as the wild type under either continuous blue or continuous red light. A delay is observed only if both blue light and red light are given together. Therefore, cry2 probably promotes flowering in blue light by repressing phyB function. Cry2 apparently inhibits phyB function by suppressing the activity of EARLY FLOWERING 3 (ELF3), which interacts with phytochrome, indicating that these signaling pathways converge.

The circadian clock is regulated by multiple aspects of light

As noted previously in this chapter, a number of plant processes show oscillations of activity that roughly correspond to a 24-h, or circadian, cycle. This endogenous rhythm uses an oscillator that must be **entrained** (synchronized) to the daily light–dark cycles of the external environment. In experiments designed to characterize the role of photoreceptors in this process, phytochrome-deficient mutants were crossed with lines carrying the luciferase reporter gene that is regulated by the circadian clock. The pace of the oscillator was slowed (i.e., the period length increased) when *phyA* mutants were grown under dim red light, but not under high-irradiance red light. However, *phyB* mutants showed timing defects only under high-irradiance red light. The cryptochromes cry1 and cry2 were required for blue light–mediated entrainment of the circadian clock. These studies indicate that both phytochromes and cryptochromes entrain the circadian clock in Arabidopsis. This light input appears to be modulated by the genes *EARLY FLOWERING 3* (*ELF3*) and *TIME FOR COFFEE* (*TIC*). Mutations in *ELF3* stop the oscillations of the clock at dusk, whereas mutations in *TIC* stop the clock at dawn. The *elf3/tic* double mutant is completely arrhythmic, suggesting that *TIC* and *ELF* interact with different components of the clock at different phases in the rhythm.

Phototropins

Early attempts to identify blue-light photoreceptor mutants of Arabidopsis with defective phototropic responses were later extended by Winslow Briggs and colleagues and resulted in the isolation of several nonphototropic hypocotyl (*nph*) mutants, which showed impaired phototropic responses to low intensities of blue light. Subsequent cloning of the *NPH1* locus resulted in the identification of the photoreceptor for phototropism. The encoded protein was named phototropin after its role in mediating phototropic responses, but these receptors also control several blue-light responses that collectively function to optimize photosynthetic efficiency and promote plant growth, particularly under low light conditions.

Angiosperms contain two phototropin genes, *PHOT1* and *PHOT2*. phot1 is the primary phototropic receptor in Arabidopsis and mediates phototropism in response to low and high fluence rates of blue light. phot2 mediates phototropism in response to high light intensities. Similar overlaps in the functions of the phot1 and phot2 photoreceptors are observed for other blue-light responses in Arabidopsis, including chloroplast movements, stomatal opening, leaf movements, and leaf expansion. Together with phototropism, these processes integrate efficient light capture and CO_2 uptake for photosynthesis. Consequently, growth of phototropin-deficient mutants is severely compromised, particularly under low light intensities.

Blue light induces changes in FMN absorption maxima associated with conformation changes

In contrast to cryptochromes, which are predominantly localized in the nucleus, phototropin receptors are associated with the plasma membrane, where they function as light-activated serine/threonine kinases. **Figure 16.20A** illustrates the domain structure of Arabidopsis phototropin 1, along with three related blue-light photoreceptors found in plants or algae: neochrome, ZEITLUPE, and aureochrome. Phototropin contains two light-sensing **LIGHT-OXYGEN-VOLTAGE (LOV)** domains, LOV1 and LOV2, each binding a chromophore flavin mononucleotide (FMN). Spectroscopic studies have shown that in the dark, one FMN molecule is noncovalently bound to each LOV domain. Upon blue-light illumination, the FMN molecule becomes covalently bound to a cysteine

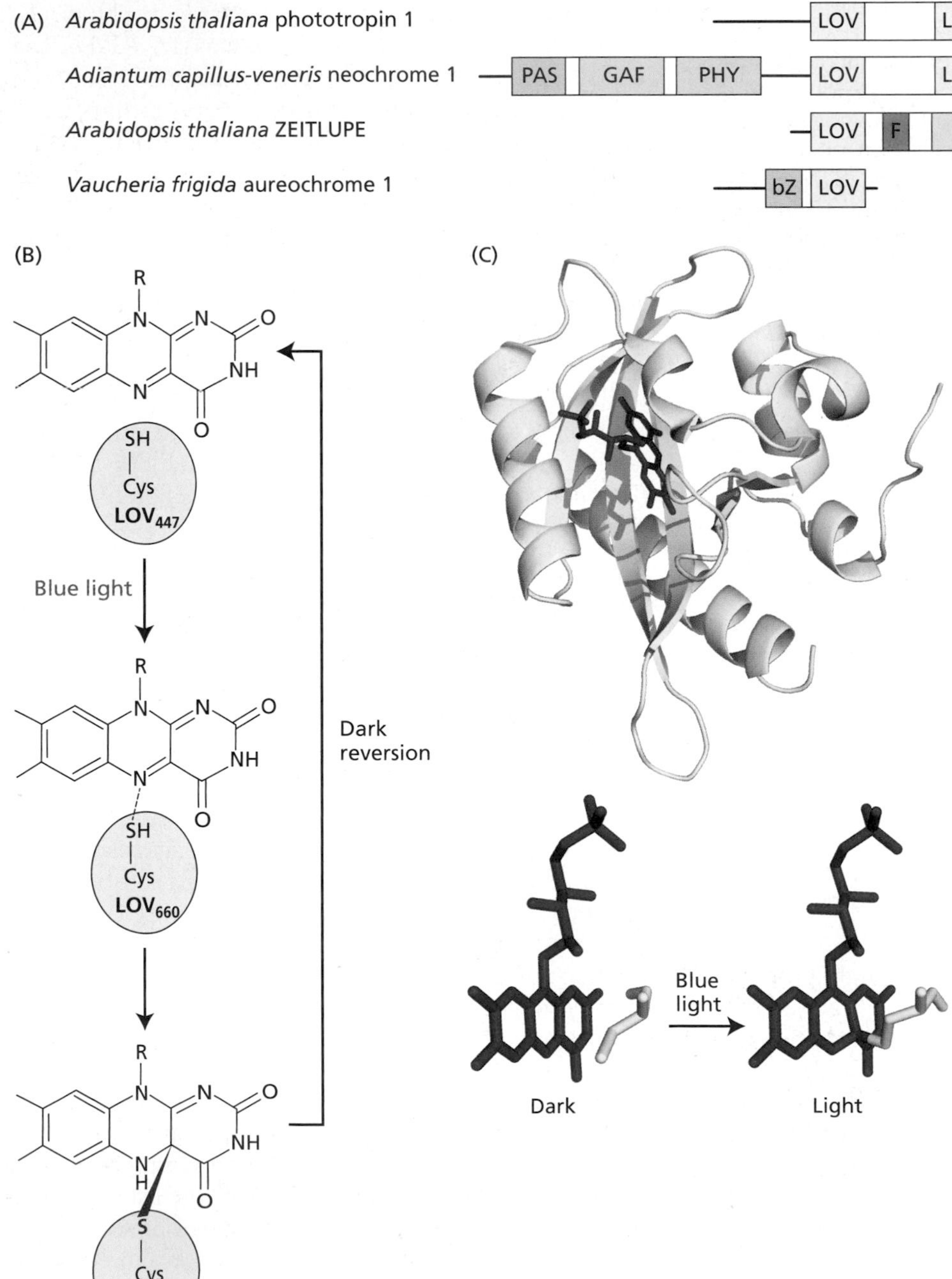

Figure 16.20 Phototropin domain composition, photocycle, and LOV domain structure. (A) The domain compositions of phototropin and related LOV-domain photoreceptors. (B) The FMN photocycle of phototropin. In the dark, the maximum absorption of the FMN chromophore is around 450 nm. Blue light induces the formation of a covalent bond between FMN and a cysteine residue, shifting the absorption maximum to 390 nm via a LOV_{660} intermediate form. The reaction is reversible in darkness. (C) Crystal structure of the oat phot1 LOV2 domain in darkness (intact phototropin has not yet been crystallized). The protein is in yellow and the FMN cofactor is blue. The J-helix is on the left of the LOV2 core. The two diagrams below show just the flavin and the formation of the cysteine adduct after irradiation with blue light. (After Christie 2007.)

residue in the phototropin molecule, forming a cysteine-flavin covalent adduct (**Figure 16.20B**). As discussed below, this reaction induces a major protein conformational change, which can be reversed by a dark treatment. The three-dimensional structure of the LOV2 domain resembles a closed molecular hand holding the FMN tightly by noncovalent interactions inside its core (**Figure 16.20C**). The same figure also shows the formation of the covalent bond between the flavin cofactor and a cysteine residue in response to blue light.

The LOV2 domain is primarily responsible for kinase activation in response to blue light

As demonstrated in mutagenesis experiments, the LOV2 domain, in particular, is essential for blue light–induced kinase activation and autophosphorylation of the phototropin photoreceptor. Mutating the conserved cysteine in the LOV1 domain of phot1 does not affect phototropic responsiveness (**Figure 16.21A and B**), whereas the equivalent mutation in LOV2 abolishes the response (**Figure 16.21C**). These and other studies have demonstrated the importance of LOV2 in controlling phototropin function. This is due in part to the position of LOV2 within the phototropin molecule, where it is coupled to a protein region known as the Jα-helix that is important for propagating light-driven changes within LOV2 to the kinase domain. The function of LOV1 is still poorly understood, but the domain is thought to play a role in receptor dimerization.

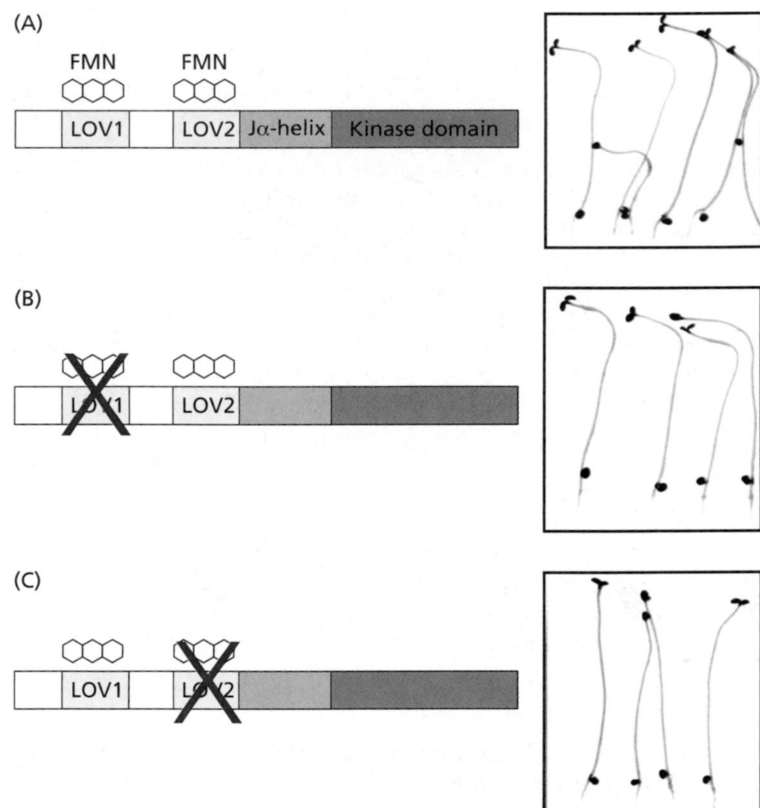

Figure 16.21 Phototropism in Arabidopsis seedlings can be used as the bioassay for phototropin activity. (A) Wild-type, with both LOV1 and LOV2 domains. (B) Mutating the cysteine in the LOV1 domain of phot1 does not affect phototropic responsiveness (the seedlings bend toward blue light). (C) The equivalent mutation in the LOV2 domain abolishes the response, demonstrating that only the LOV2 domain is required for phototropism. (Courtesy of John Christie.)

Blue light induces a conformational change that "uncages" the kinase domain of phototropin and leads to autophosphorylation

Although a three-dimensional structure for the entire phototropin molecule is still lacking, many genetic, biochemical, and biophysical studies have provided a good understanding of how the phototropin light switch works. As with cryptochrome and phytochrome, the N-terminal photosensory region of the phototropins controls the activity of the C-terminal half of the protein, which contains a serine/threonine kinase domain (see Figure 16.21A). In the dark, the N-terminal region, including the LOV domains, "cages" and inhibits the activity of the kinase domain (**Figure 16.22**). Absorption of blue photons by the LOV domains results in primary photochemical changes that lead to the uncaging of the kinase domain and its activation by the unfolding of the Jα-helix. The activation of the C-terminal kinase domain then leads to receptor autophosphorylation on multiple serine residues. Autophosphorylation of the kinase domain is required for all phototropin-mediated responses in Arabidopsis. A type 2A protein phosphatase mediates dephosphorylation and inactivation of phototropin in the dark (see Figure 16.22).

Phototropism requires changes in auxin mobilization

Activation of phototropin kinases triggers signal transduction events that establish a variety of different responses. One of these responses is phototropism, which occurs in both mature plants and seedlings. As mentioned in Chapter 15, observations of this phenomenon by Charles and Francis Darwin initiated a series of experiments that culminated in the discovery of the hormone auxin. The interactions of phototropins and auxin in the control of phototropism will be described in Chapter 18 on seedling establishment.

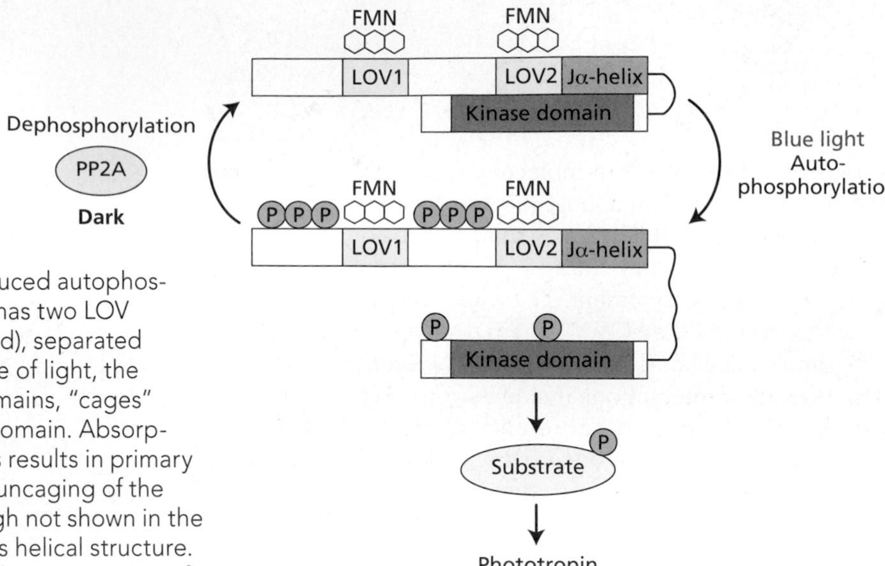

Figure 16.22 Model for blue light–induced autophosphorylation of phototropin. Phototropin has two LOV domains (yellow) and a kinase domain (red), separated by an α-helical region (Jα). In the absence of light, the N-terminal region, including the LOV domains, "cages" and represses the activity of the kinase domain. Absorption of blue photons by the LOV domains results in primary photochemical changes that lead to the uncaging of the kinase domain and its activation. Although not shown in the diagram, the Jα-helix completely loses its helical structure. Photoexcitation of the LOV domains results in activation of the C-terminal kinase domain, which leads to receptor autophosphorylation on multiple serine residues. Autophosphorylation within the kinase domain is essential for initiating all phototropin-mediated responses in Arabidopsis. Dephosphorylation resulting in inactivation occurs in the dark. (After Inoue et al. 2010.)

Phototropins regulate chloroplast movements via F-actin filament assembly

Leaves can alter the intracellular distribution of their chloroplasts in response to changing light conditions. As discussed in Chapter 9, this feature is adaptive, as the redistribution of chloroplasts within the cells modulates light absorption and prevents photodamage (see Figure 9.12). Under weak illumination, chloroplasts gather near the upper and lower walls of the leaf palisade cells (accumulation), thus maximizing light absorption (**Figure 16.23**). Under strong illumination, the chloroplasts move to the lateral walls that are parallel to the incident light (avoidance), thus minimizing light absorption and avoiding photodamage. In the dark, the chloroplasts move to the bottom of the cell, although the physiological function of this position is unclear. The action spectrum for the redistribution response shows the typical three-finger fine structure typical of specific blue-light responses (see Figure 16.7).

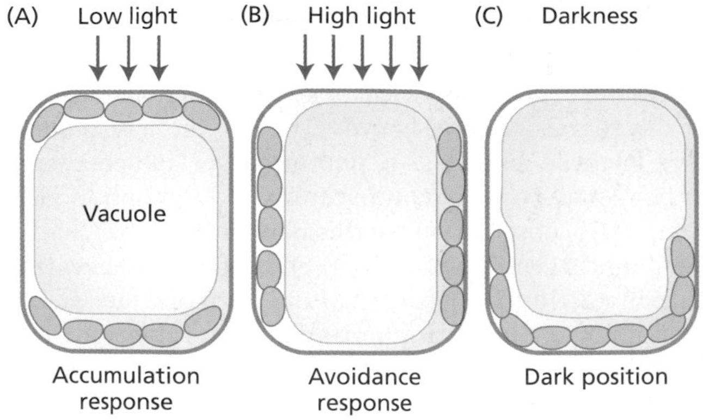

(A) Low light (B) High light (C) Darkness

Vacuole

Accumulation response — Avoidance response — Dark position

Figure 16.23 Schematic diagram of chloroplast distribution patterns in Arabidopsis palisade cells in response to different light intensities. (A) Under low light conditions, chloroplasts optimize light absorption by accumulating at the upper and lower sides of palisade cells. (B) Under high light conditions, chloroplasts avoid sunlight by migrating to the side walls of palisade cells. (C) Chloroplasts move to the bottom of the cell in darkness. (After Wada 2013.)

Arabidopsis *phot1* mutants have a normal avoidance response and a poor accumulation response. In contrast, *phot2* mutants lack the avoidance response but retain a fairly normal accumulation response. Cells from the *phot1/phot2* double mutant lack both the avoidance and accumulation responses. These results indicate that phot2 plays a key role in the avoidance response, and that both phot1 and phot2 contribute to the accumulation response. Studies have shown that phot2 mutants in fact do not survive in the field under conditions of full sunlight due to photooxidative damage.

The isolation of Arabidopsis mutants impaired in the chloroplast avoidance response led to the identification of a novel F-actin-binding protein, CHLOROPLAST UNUSUAL POSITIONING1 (CHUP1), consistent with earlier work showing that chloroplast movements occur through changes in the cytoskeleton. CHUP1 localizes to the chloroplast envelope and functions in chloroplast positioning and movement. A model of chloroplast movement in Arabidopsis is shown in **Figure 16.24**. Both phot1 and phot2 mediate the accumulation response and are localized at the plasma membrane. phot2, which mediates the avoidance response, is also localized on the chloroplast envelope. In the presence of full sunlight, CHUP1, which appears to anchor to the plasma membrane via protein interactions, binds to the chloroplast envelope. CHUP1 recruits G-actin and actin-polymerizing proteins to extend an existing F-actin filament (see Figure

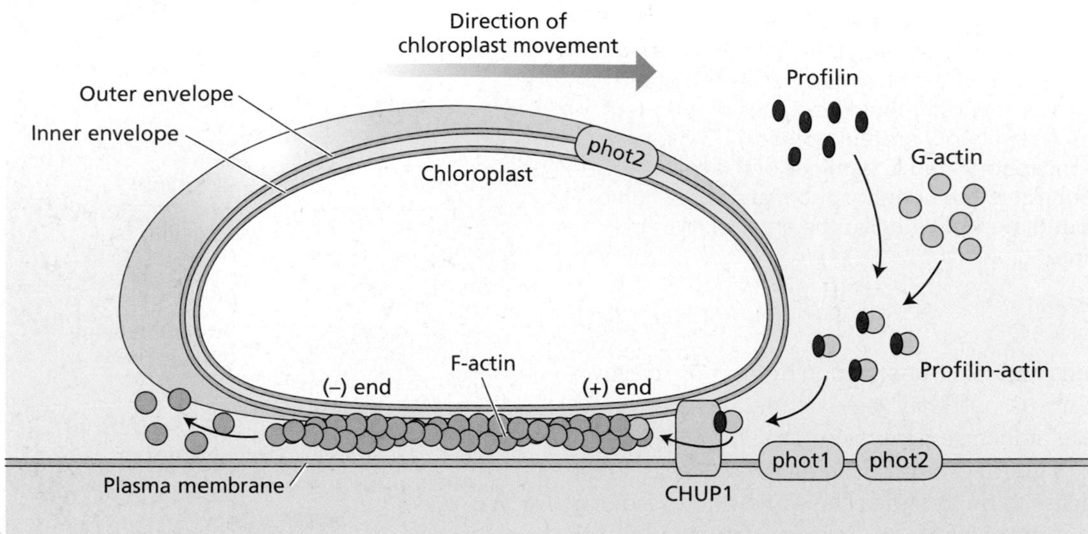

Figure 16.24 Model for phototropin-mediated chloroplast movement in *Arabidopsis thaliana*. Both phot1 and phot2 mediate the accumulation response and are localized at the plasma membrane. phot2 is also localized on the chloroplast envelope and probably mediates the avoidance response. CHUP1 binds to the chloroplast envelope via its N terminus and may also be anchored to the plasma membrane. CHUP1 initiates actin polymerization to extend an existing F-actin filament. As a result, the actin filament lengthens and CHUP1 and the chloroplast are pushed forward. The actin filaments are depolymerized at their minus ends. The green arrow shows the direction of chloroplast movement. See Chapter 1. (After Wada 2013.)

1.25). CHUP1 and the chloroplast are then pushed by the inserted G-actin, generating the motive force for chloroplast movement.

Stomatal opening is regulated by blue light, which activates the plasma membrane H^+-ATPase

Stomatal photophysiology and sensory transduction in relation to water and photosynthesis were discussed in Chapters 4 and 9, and will be discussed again in Chapter 18. Unlike all other blue-light responses, stomatal opening is stimulated by blue light and inhibited by green light. Based on studies with *phot1/phot2* double mutants the primary blue-light photoreceptor for stomatal opening has been definitively identified as phototropin. The identity of the green-light photoreceptor of guard cells is still unresolved. Cryptochrome is generally considered to be the most likely candidate, but there is evidence implicating the carotenoid zeaxanthin as well (see Chapter 10).

An extensive body of work has been carried out on the mechanism of blue light–induced stomatal opening. As a result, phototropin-mediated stomatal opening is arguably the best-understood signaling pathway of all the phototropin responses. Several key steps in the sensory transduction process of phototropin-stimulated stomatal opening have been identified. In particular, the guard cell proton-pumping H^+-ATPase plays a central role in the regulation of stomatal movements (**Figure 16.25**; see also Figures 10.6 and 10.7). The activated H^+-ATPase transports H^+ across the membrane and increases the inside-negative electrical potential, driving the K^+ uptake through the voltage-gated inward-rectifying K^+ channels. The accumulation of K^+ facilitates the influx of water into the guard cells, leading to an increase in turgor pressure and stomatal opening. The C terminus of the H^+-ATPase has an autoinhibitory domain that regulates the activity of the enzyme. If this autoinhibitory domain is experimentally removed by a protease, the H^+-ATPase becomes irreversibly activated. The autoinhibitory domain of the C terminus is thought to lower the activity of the enzyme by blocking its catalytic site. Conversely, the fungal toxin fusicoccin appears to activate the enzyme by displacing the autoinhibitory domain away from the catalytic site.

Upon blue-light irradiation, the H^+-ATPase shows a lower K_m for ATP and a higher V_{max}, indicating that blue light activates the H^+-ATPase. Activation of the enzyme involves the phosphorylation of serine and threonine residues of the C-terminal domain of the H^+-ATPase. Inhibitors of protein kinases, which might block phosphorylation of the H^+-ATPase, prevent blue light–stimulated proton pumping and stomatal opening. As with fusicoccin, phosphorylation of the C-terminal domain appears also to displace the autoinhibitory domain of the C terminus from the catalytic site of the enzyme.

A regulatory protein termed 14-3-3 protein has been found to bind to the phosphorylated C terminus of the guard cell H^+-ATPase, but not to the nonphosphorylated one (see Figure 16.25). The 14-3-3 proteins are ubiquitous regulatory proteins in eukaryotic organisms. In plants,

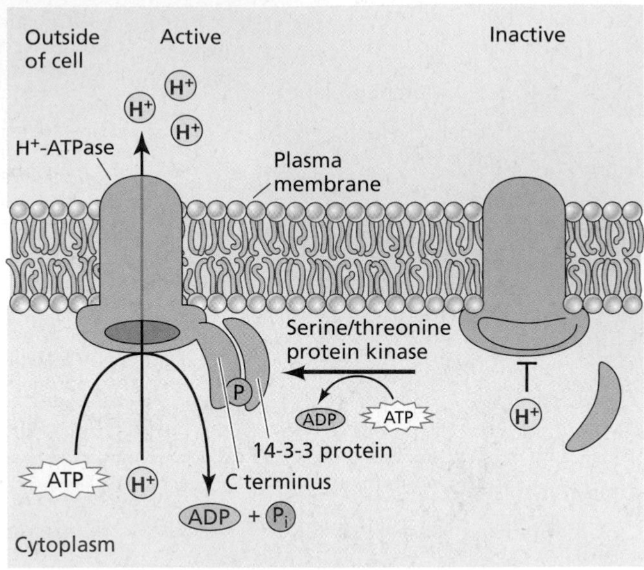

Figure 16.25 Role of the proton-pumping ATPase in the regulation of stomatal movement. Blue light activates the H⁺-ATPase. Activation of the enzyme involves the phosphorylation of serine and threonine residues of its C-terminal domain. A regulatory protein termed 14-3-3 protein binds to the phosphorylated C terminus of the guard cell H⁺-ATPase, but not to the nonphosphorylated one. Pumping protons out of the cell requires the entry of K⁺ to balance the charge.

14-3-3 proteins regulate transcription by binding to activators in the nucleus, and they regulate metabolic enzymes such as nitrate reductase. Only one of the four 14-3-3 isoforms found in guard cells binds to the H⁺-ATPase, so the binding appears to be specific. The same 14-3-3 isoform binds to the guard cell H⁺-ATPase in response to both fusicoccin and blue-light treatments. The 14-3-3 protein dissociates from the H⁺-ATPase upon dephosphorylation of the C-terminal domain.

The main signal transduction events of phototropin-mediated stomatal opening have been identified

Phototropins do not phosphorylate the H⁺-ATPase directly. The kinase involved in phosphorylating the H⁺-ATPase has not yet been identified. However, early signal transduction events following phototropin excitation at the guard cell plasma membrane have been identified and are illustrated in **Figure 16.26**. The membrane-associated, guard cell–specific protein kinase called BLUE LIGHT SIGNALING1 (BLUS1) is phosphorylated by phot1 and phot2 redundantly. Arabidopsis mutants lacking BLUS1 show no blue light–induced stomatal opening but are not impaired in other phototropin responses, including phototropism and chloroplast relocation. This phosphorylation event is essential in initiating the early transduction events that ultimately lead to phosphorylation and activation of the H⁺-ATPase.

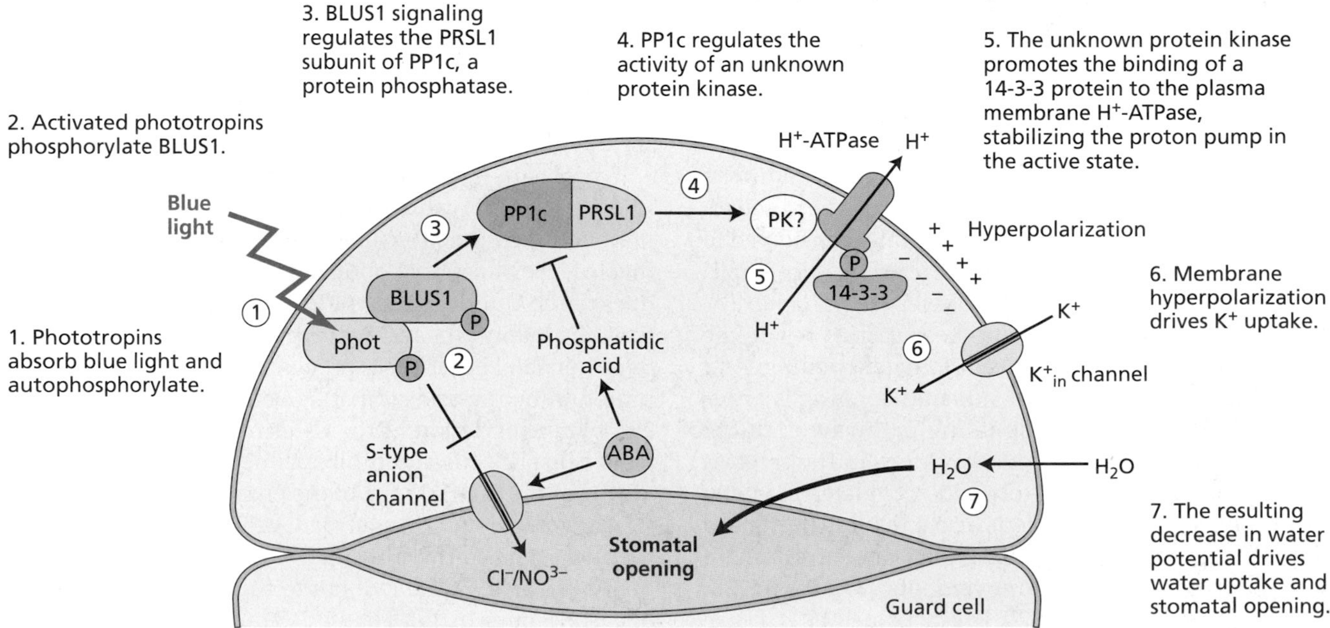

Figure 16.26 Phototropin signal transduction leading to stomatal opening. ABA antagonizes phototropin through phosphatidic acid, which interacts with PP1c in the phototropin pathway. (Courtesy of Ken-ichiro Shimazaki.)

The signals from BLUS1 converge on **PROTEIN PHOSPHATASE1 (PP1)**, a signaling intermediate that leads to activation of the H^+-ATPase. PP1 is a serine/threonine protein phosphatase composed of a catalytic subunit (PP1c) and a regulatory subunit, PRSL1 (PP1 REGULATORY SUBUNIT2-LIKE PROTEIN1), that modulates catalytic activity, subcellular localization, and substrate specificity. PP1c positively regulates stomatal opening through blue-light signaling between phototropins and the plasma membrane H^+-ATPase in guard cells.

As we will discuss in detail in Chapter 23 in relation to drought stress, abscisic acid (ABA) causes stomatal closure in the light. ABA induces the production of phosphatidic acid, a lipid signaling molecule (see Chapter 15). As shown in Figure 16.26 phosphatidic acid blocks PP1 activity, one of the steps in the phototropin pathway. ABA also activates S-type anion channels (see Chapter 6), which are inhibited by blue light.

Responses to Ultraviolet Radiation

In addition to its cytotoxic effects, UV-B radiation can elicit a wide range of photomorphogenic responses, some of which are listed in **Table 16.3**. The photoreceptor responsible for UV-B-induced developmental responses, UVR8, is a seven-bladed β-propeller protein, which forms functionally inactive homodimers in the absence of UV-B (**Figure 16.27**). Unlike phytochrome, cryptochrome, and phototropin, UVR8 lacks a prosthetic chromophore. The two identical subunits of UVR8 are linked in the dimer by a network of salt bridges formed between tryptophan residues, which serve as the primary UV-B sensors, and nearby arginine residues.

TABLE 16.3 Photomorphogenic responses to UV-B

Response
Gene regulation
UV-B tolerance
Flavonoid biosynthesis
Hypocotyl growth suppression
Leaf/epidermal cell expansion
Endoreduplication in epidermal cells
Stomatal density
Entrainment of circadian clock
Increased photosynthetic efficiency

Source: Jenkins 2014.

Upon absorbing UV-B photons, the tryptophans undergo structural changes that break the salt bridges, leading to the dissociation of the two functionally active monomers. The monomers then interact with COP1–SPA complexes to activate gene expression, as illustrated in **Figure 16.28**. Thus, although the COP1-SPA acts as negative regulator that targets transcription factors for degradation during phytochrome and cryptochrome responses (see Figures 16.14 and 16.16), COP1-SPA acts as a positive regulator during UV-B signaling by interacting with the C-terminal region of UVR8 in the nucleus. The UVR8-COP1-SPA complex then activates the transcription of the major transcription factor HY5, which controls the expression of many of the genes induced by UV-B.

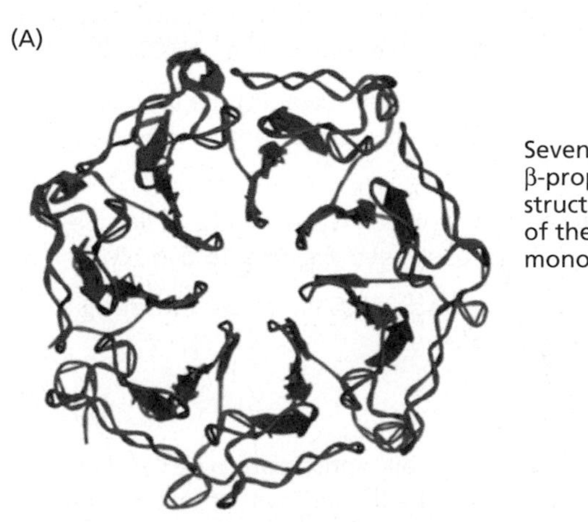

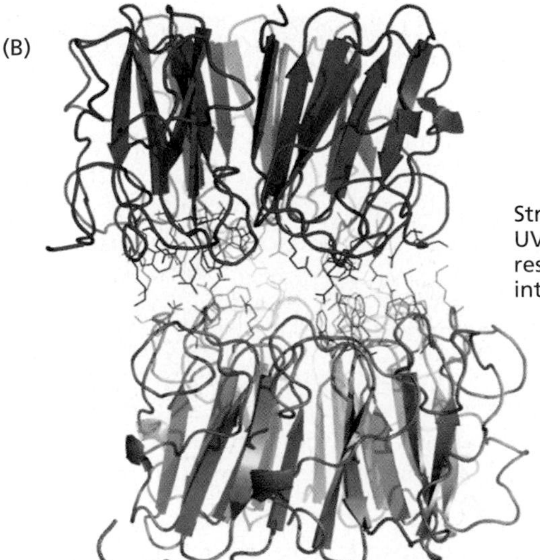

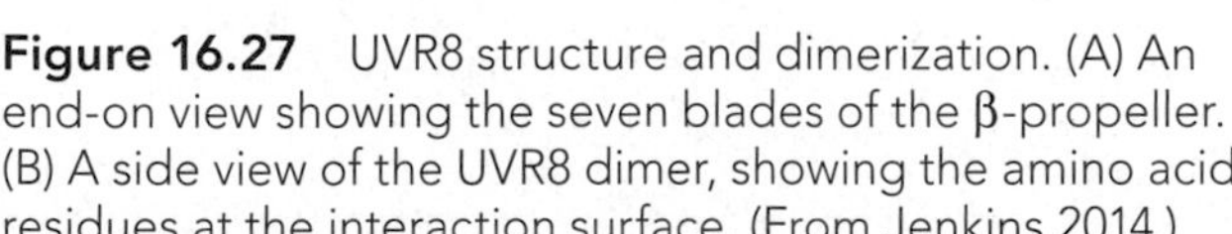

Figure 16.27 UVR8 structure and dimerization. (A) An end-on view showing the seven blades of the β-propeller. (B) A side view of the UVR8 dimer, showing the amino acid residues at the interaction surface. (From Jenkins 2014.)

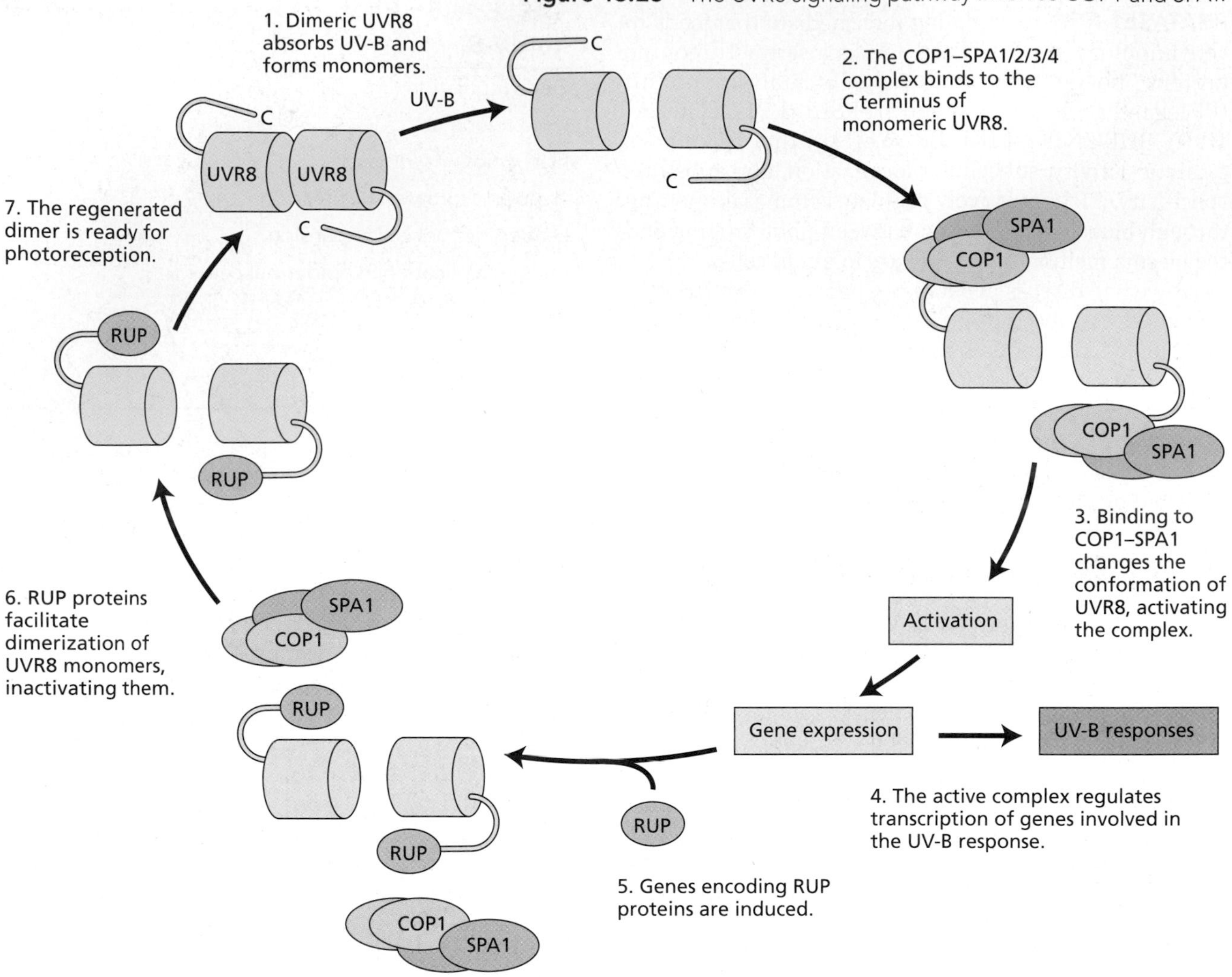
1. Dimeric UVR8 absorbs UV-B and forms monomers.
UV-B
C
UVR8
UVR8
C
C
C
2. The COP1–SPA1/2/3/4 complex binds to the C terminus of monomeric UVR8.
SPA1
COP1
COP1
SPA1
3. Binding to COP1–SPA1 changes the conformation of UVR8, activating the complex.
Activation
Gene expression
UV-B responses
4. The active complex regulates transcription of genes involved in the UV-B response.
RUP
5. Genes encoding RUP proteins are induced.
RUP
RUP
COP1
SPA1
6. RUP proteins facilitate dimerization of UVR8 monomers, inactivating them.
SPA1
COP1
RUP
RUP
7. The regenerated dimer is ready for photoreception.

Figure 16.28 The UVR8 signaling pathway involves COP1 and SPA1.

SUMMARY

Photoreceptors, including phytochromes, cryptochromes, and phototropins, help plants regulate developmental processes over their lifetimes by sensitizing plants to incident light. Photoreceptors also initiate protective processes in response to harmful radiation.

Plant Photoreceptors

- Sunlight regulates developmental processes over the life of the plant and provides directional and nondirectional cues for growth and movement. Sunlight also contains UV radiation that can harm plant tissues (**Figures 16.1–16.4**).
- Phytochromes (which absorb red and far-red light) and phototropins and cryptochromes (which absorb blue light and UV-A) are photoreceptors that are sensitive to light quantity, quality, and duration.
- Action spectra and absorption spectra help researchers determine which wavelengths of light lead to specific photoresponses (**Figures 16.5–16.7**).
- Light fluence and irradiance also govern whether a photoresponse occurs.

Phytochromes

- Phytochrome is generally sensitive to red and far-red light, and it exhibits the ability to interconvert between Pr and Pfr forms.
- The physiologically active form of phytochrome is Pfr.

- Red light triggers conformational changes in both the phytochrome chromophore and protein (**Figures 16.8–16.10**).
- Pfr movement from the cytosol to the nucleus enables phytochrome-regulated transcription in the nucleus (**Figure 16.11**).

Phytochrome Responses

- Photoresponses exhibit various lag times (between exposure to light and the subsequent response) and escape times (wherein the response is only reversible for a certain amount of time).
- Phytochrome-initiated responses fall into one of three main categories: very low fluence responses (VLFRs), low-fluence responses (LFRs), or high-irradiance responses (HIRs) (**Figure 16.12**).
- Phytochrome A mediates responses to continuous far-red light.
- Phytochrome B mediates responses to continuous red or white light.

Phytochrome Signaling Pathways

- Phytochrome can rapidly change membrane potentials and ion fluxes.
- Phytochrome regulates gene expression through a wide range of modular elements.
- Phytochrome itself can be phosphorylated and dephosphorylated.
- Phytochrome-induced photomorphogenesis involves protein degradation (**Figure 16.13**).

Blue-Light Responses and Photoreceptors

- In contrast to red- and far-red-light responses, blue-light responses generally exhibit longer lag times and more persistance after the disappearance of the light signal (**Figures 16.14, 16.15**).

Cryptochromes

- Activation of the flavin adenine dinucleotide (FAD) chromophore causes a conformational change in cryptochrome, enabling cryptochrome to bind to other protein partners.
- Cryptochrome homologs 1, 2, and 3 have different developmental effects, and are localized differently than phytochromes (**Figure 16.16**).
- Whereas phytochrome promotes protein degradation via COP1, nuclear cryptochromes inhibit COP1-induced protein degradation, leading to photomorphogenesis (compare **Figures 16.13 and 16.17**).

The Coaction of Cryptochrome, Phytochrome, and Phototropins

- Both phytochrome and cryptochrome inhibit stem elongation (**Figure 16.18**).
- Phytochrome interacts with cryptochrome to regulate flowering, and both types of photoreceptors are necessary to maintain circadian cycles.

Phototropins

- Similarly to cryptochromes, phototropins mediate photoresponses to blue light; phototropin 1 and 2 are sensitive to different and overlapping intensities of blue light.
- Phototropins are located in the plasma membrane, and each has two flavin mononucleotide (FMN) chromophores that can induce conformational changes (**Figures 16.19, 16.20**).
- When phototropins are activated by blue light, their kinase domain is "uncaged," causing autophosphorylation (**Figure 16.21**).
- Phototropins mediate chloroplast accumulation and avoidance responses to weak and strong light via F-actin filament assembly (**Figures 16.22, 16.23**).
- Blue light, sensed by phototropins, causes activation of plasma membrane H^+-ATPases and ultimately regulates stomatal opening. However, the kinase that activates H^+-ATPases has not been identified (**Figures 16.24, 16.25**).

Responses to Ultraviolet Radiation

- The photoreceptor involved responses to UV-B irradiation is UVR8.
- Unlike other phytochromes, cryptochromes, phototropins, UVR8 lacks a prosthetic chromophore.
- UVR8 interacts with the COP1-SPA complex to activate the transcription of UV-B-induced genes.

WEB MATERIAL

- **WEB TOPIC 16.1 *Mougeotia*: A Chloroplast with a Twist** Microbeam irradiation experiments have been used to localize phytochrome in this filamentous green alga.
- **WEB TOPIC 16.2 Phytochrome and High-Irradiance Responses** Dual-wavelength experiments helped demonstrate the role of phytochrome in HIRs.
- **WEB TOPIC 16.3 Phytochrome Effects on Ion Fluxes** Phytochrome regulates ion fluxes across membranes by altering the activities of ion channels and the plasma membrane proton pump.

- **WEB ESSAY 16.1 Awakened by a Flash of Sunlight** When placed in the proper soil environment, seeds acquire extraordinary sensitivity to light, such that germination can be stimulated by less than 1 second of exposure to sunlight during soil cultivation.

available at plantphys.net

Suggested Reading

Burgie, E. S., Bussell, A. N., Walker, J. M., Dubiel, K., and Vierstra, R. D. (2014) Crystal structure of the photosensing module from a red/far-red light-absorbing plant phytochrome. *Proc. Natl. Acad. Sci. USA* 111: 10179–10184.

Christie, J. M., and Murphy, A. S. (2013) Shoot phototropism in higher plants: New light through old concepts. *Am. J. Bot.* 100: 35–46.

Christie, J. M., Kaiserli, E., and Sullivan, S. (2011) Light sensing at the plasma membrane. In *Plant Cell Monographs,* Vol. 19: *The Plant Plasma Membrane,* A. S. Murphy, W. Peer, and B. Schulz, eds., Springer-Verlag, Berlin, Heidelberg, pp. 423–443.

Inoue, S.-I., Takemiya, A., and Shimazaki, K.-I. (2010) Phototropin signaling and stomatal opening as a model case. *Curr. Opin. Plant Biol.* 13: 587–593.

Leivar, P., and Monte, E. (2014) PIFs: Systems integrators in plant development. *Plant Cell* 26: 56–78.

Liscum, E., Askinosie, S. K., Leuchtman, D. L., Morrow, J., Willenburg, K. T., and Coats, D. R. (2014) Phototropism: Growing towards an understanding of plant movement. *Plant Cell* 26: 38–55.

Rizzini, L., Favory, J.-J., Cloix, C., Faggionato, D., O'Hara, A., Kaiserli, E., Baumeister, R., Schäfer, E., Nagy, F., Jenkins, G. I., et al. (2011) Perception of UV-B by the *Arabidopsis* UVR8 protein. *Science* 332: 103–106.

Rockwell, R. C., Duanmu, D., Martin, S. S., Bachy, C., Price, D. C., Bhattachary, D., Worden, A. Z., and Lagariasa, J. K. (2014) Eukaryotic algal phytochromes span the visible spectrum. *Proc. Natl. Acad. Sci. USA* 111: 3871–3876.

Swartz, T. E., Corchnoy, S. B., Christie, J. M., Lewis, J. W., Szundi, I., Briggs, W. R. and Bogomolni, R. A. (2001) The photocycle of a flavin-binding domain of the blue light photoreceptor phototropin. *J. Biol. Chem.* 276: 36493–36500.

Takala, H., Bjorling, A., Berntsson, O., Lehtivuori1, H., Niebling, S., Hoernke, M., Kosheleva, I., Henning, R., Menzel, A., Janne, A., et al. (2014) Signal amplification and transduction in phytochrome photosensors. *Nature* 509: 245–249.

Takemiya, A., Sugiyama, N., Fujimoto, H., Tsutsumi, T., Yamauchi, S., Hiyama, A., Tadao, Y., Christie, J. M., and Shimazaki, K.-I. (2013) Phosphorylation of BLUS1 kinase by phototropins is a primary step in stomatal opening. *Nat. Commun.* 4: 2094. DOI: 10.1038/ncomms3094

Takemiya, A., Yamauchi, S., Yano, T., Ariyoshi, C., and Shimazaki, K.-I. (2013) Identification of a regulatory subunit of protein phosphatase 1, which mediates blue light signaling for stomatal opening. *Plant Cell Physiol.* 54: 24–35.

Wada, M. (2013) Chloroplast movement. *Plant Sci.* 210: 177–182.

17 Embryogenesis

Plants offer intriguing developmental contrasts to animals, not only with respect to their diverse forms, but also in how those forms arise. A sequoia tree, for example, may grow for thousands of years before reaching a size big enough for an automobile to drive through its trunk. In contrast, an Arabidopsis plant can complete its life cycle in little more than a month, making hardly more than a handful of leaves (**Figure 17.1**). Dissimilar as they may be, both species employ growth mechanisms common to all multicellular plants, in which form is elaborated gradually through adaptive postembryonic growth processes. Animals, by contrast, typically have a more predictable pattern of development in which the basic body plan is largely determined during embryogenesis.

These differences between plants and animals can be understood partly in terms of contrasting survival strategies. Being photosynthetic, plants rely on flexible patterns of growth that allow them to adapt to fixed locations where conditions may be less than ideal, especially with respect to sunlight, and may vary over time. Animals, being heterotrophic, evolved mechanisms for mobility instead. In this chapter we will consider the essential characteristics of plant development and the nature of the mechanisms that guide these flexible patterns of plant growth.

Biologists who wish to understand plant development are faced with two general issues. The first is the challenge of formulating clear and relevant descriptions of changes that occur over time. As an organism grows, are there corresponding increases in its complexity, and if so, how can this complexity be most simply described? To what extent is growth coupled to cell division, cell expansion, and specific differentiation processes? How do environmental factors influence growth processes?

With a detailed description of growth in place, biologists can begin to address a second set of questions that relates to the nature of the underlying mechanisms: How can characteristic patterns of growth be explained by genetically determined processes? How are these intrinsic programs of development coupled to external influences such as nutrient levels, energy inputs, and stress?

(A)

(B)

Figure 17.1 Two contrasting examples of plant form arising from indeterminate growth processes. (A) The Chandelier Tree, a famous *Sequoia sempervirens* that has adapted to many challenges during its roughly 2400-year existence. (B) The compact form and rapid life cycle of the much smaller *Arabidopsis thaliana* have made it a useful model for understanding mechanisms that guide plant growth and development.

What types of mechanisms mediate this coupling? What physical components are involved, how are they organized at the cellular and tissue levels, and how are their dynamic behaviors regulated in time and space?

To address these issues, this chapter will begin with a brief overview of essential aspects of the organization and life cycle of plants and how they relate to basic growth processes. As background to this discussion, various approaches that can be used to provide a detailed and quantitative description of growth and development are reviewed in WEB APPENDIX 2. Building on this foundation, we will then consider how physiological, molecular, and genetic approaches can provide valuable insights into how these processes are regulated.

Overview of Plant Growth and Development

An essential aspect of almost all land plants is their sedentary lifestyle. By virtue of their ability to photosynthesize, favorably positioned plants can readily obtain both the energy and the nutrients that they need to grow and survive. Relieved of the need to move, plants have never evolved the sort of anatomical complexity that enables mobility in animals. In its place, one finds a relatively rigid anatomy adapted to the capture of light energy and nutrients. As a consequence, plant cells, unlike animal cells, are firmly attached to their neighbors in a relatively inflexible, often woody, matrix. This rigid anatomy imposes constraints on how the plant grows. Cells are added progressively to the body through the activity of localized structures called meristems. By contrast, many aspects of animal development, including the formation of primary tissue layers, are characterized by the migration of cells to new locations.

While the sedentary habit of plants allows a relatively simple organization, this lack of mobility presents significant challenges. Because plants are unable to relocate to optimal habitats, they must instead adapt to their local environments. While this adaptation can occur on a physiological level, it may also be achieved through the flexible patterns of development that characterize vegetative growth. A key element of this adaptive growth is the presence of meristematic tissues, which contain reservoirs of cells whose fate remains undetermined. Through the regulated proliferation and differentiation of these cells, plants are able to produce a variety of complex forms adapted to the local environment.

Sporophytic development can be divided into three major stages

The development of the seed plant sporophyte can be broken down into three major stages (**Figure 17.2**): embryogenesis, vegetative development, and reproductive development.

EMBRYOGENESIS The term *embryogenesis* describes the process by which a single cell is transformed into a multicellular entity having a characteristic, but typically rudimentary, organization. In most seed plants, embryogenesis takes place within the confines of the ovule, a specialized structure formed within the carpels of the flower. The overall sequence of embryonic development is highly predictable, perhaps reflecting the need for the embryo to be effectively packaged within the maternally derived integuments to form the seed. With this consistency, embryogenesis affords some of the clearest examples of basic patterning processes in plants.

Among these processes are those responsible for establishing polarity, thus providing a framework in which cells differentiate according to their positions in the embryo. Within this framework, groups of cells become functionally specialized to form epidermal, cortical, and vascular tissues. Certain groups of cells, known as apical meristems, are established at the growing points of the shoot and root and enable the elaboration of additional tissues and organs during subsequent vegetative growth. At the conclusion of embryogenesis, a number of physiological changes occur to enable the embryo to withstand long periods of dormancy and harsh environmental conditions (see **WEB TOPIC 17.1**).

VEGETATIVE DEVELOPMENT With germination, the embryo breaks its dormant state and, by mobilizing stored reserves, commences a period of vegetative growth. Depending on the species, germination occurs in response to a combination of factors, which may include time, moisture, and extended cold, heat, and light (see **WEB TOPIC 17.1** and Chapter 18). Drawing initially on reserves stored in its cotyledons (e.g., beans) or in endosperm (e.g., grasses), the seedling builds on its rudimentary form through the activity of the root and shoot apical meristems. Through photomorphogenesis (see Chapter 16) and further development of the shoot, the seedling becomes photosynthetically competent, thus enabling further vegetative growth.

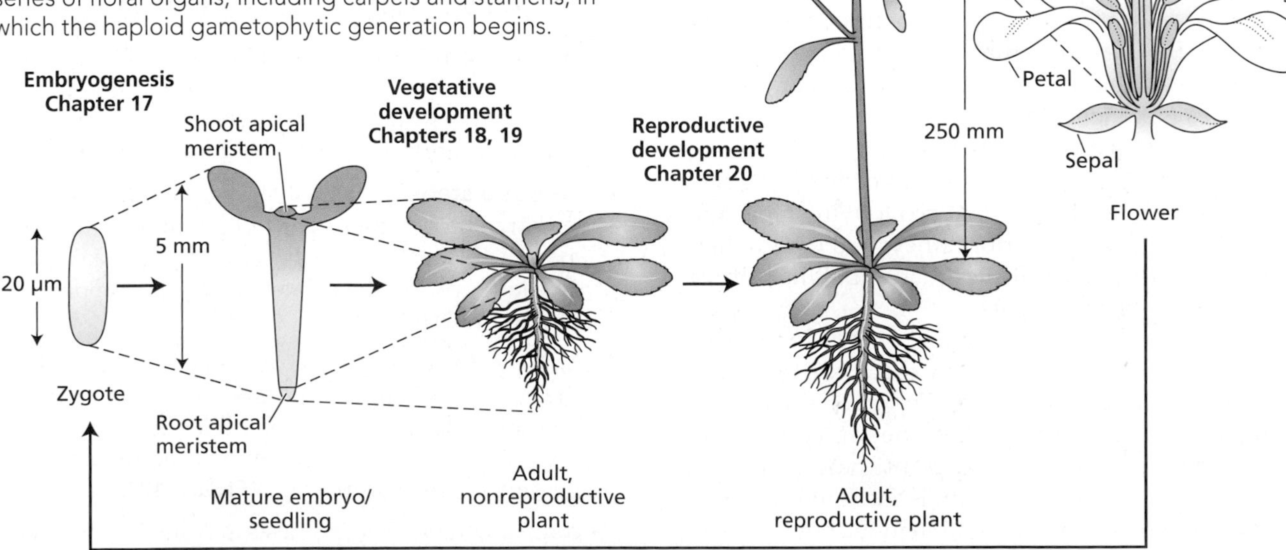

Figure 17.2 Major phases of sporophyte development. During embryogenesis, the single-celled zygote elaborates a rudimentary but polar organization that features groups of undetermined cells contained in the shoot and root apical meristems. During vegetative development, indeterminate patterns of growth, which reflect inputs from both intrinsic programs and environmental factors, yield a variable shoot and root architecture. During reproductive development, vegetative shoot apical meristems (SAMs) are reprogrammed to produce a characteristic series of floral organs, including carpels and stamens, in which the haploid gametophytic generation begins.

Unlike the growth of animals, vegetative growth is typically indeterminate—not predetermined, but subject to variation with no definite end point. This indeterminate growth is characterized by reiterated programs of lateral organ development that allow the plant to elaborate an architecture best suited to the local environment.

REPRODUCTIVE DEVELOPMENT After a period of vegetative growth, plants respond to a combination of internal and external cues, including size, temperature, and photoperiod, to undergo the transition to reproductive development. In flowering plants, this transition involves the formation of specialized floral meristems that give rise to flowers. The processes by which floral meristems are specified and then develop to produce a stereotyped sequence of organ formation have provided some of the best-studied examples of plant development and are described in detail in Chapter 20.

In the following sections we will examine several fundamental examples of plant development and consider how molecular and genetic methods have contributed to our understanding of how regional differences in growth are achieved.

Embryogenesis: The Origins of Polarity

In seed plants, embryogenesis transforms a single-celled zygote into the considerably more complex individual contained in the mature seed. As such, embryogenesis provides many examples of developmental processes by which the basic architecture of the plant is established, including the elaboration of forms (**morphogenesis**), the associated formation of functionally organized structures (**organogenesis**), and the **differentiation** of cells to produce anatomically and functionally distinct tissues (**histogenesis**). An essential feature of this basic architecture is the presence of apical meristems at the tips of the shoot and root axes (see Figure 17.2), which are key to sustaining indeterminate patterns of vegetative growth. Finally, the development of the embryo features complex changes in physiology that enable the embryo to withstand prolonged periods of inactivity (**dormancy**) and to recognize and interpret environmental cues that signal the plant to resume growth (**germination**).

In the sections that follow, we will examine from several perspectives how the complexity of the embryo arises. We will begin with a detailed description of Arabidopsis embryogenesis, highlighting similarities and differences to embryogenesis in other higher plant species. Next we will consider the nature of the signals that guide complex patterns of growth and differentiation in the embryo, with several lines of evidence highlighting the importance of position-dependent cues. Finally, we will explore examples that illustrate how molecular and genetic approaches provide insight into the mechanisms that translate these cues into organized patterns of growth.

Embryogenesis differs between eudicots and monocots, but also features common fundamental processes

Anatomical comparisons highlight differences in the patterns of embryogenesis seen among different seed plant groups, such as those between monocots and eudicots. Arabidopsis (a eudicot) and rice (a monocot) provide two examples of embryogenesis that differ in detail but that share certain fundamental features relating to the establishment of major growth axes. Here we will describe Arabidopsis embryogenesis in detail. An account of the somewhat distinct pattern of embryogenesis in monocots as exemplified by rice is provided in WEB TOPIC 17.2.

ARABIDOPSIS EMBRYOGENESIS By virtue of the relatively small size of the Arabidopsis embryo, the patterns of cell division by which it arises are relatively simple and easily followed. Five stages, each of which is linked to the shape of the embryo, are widely recognized:

1. **Zygotic stage**. The first stage of the diploid life cycle commences with the fusion of the haploid egg and sperm to form the single-celled zygote. Polarized growth of this cell, followed by an asymmetric transverse division, gives rise to a small apical cell and an elongated basal cell (**Figure 17.3A**).
2. **Globular stage**. The apical cell undergoes a series of divisions (**Figure 17.3B–D**) to generate a spherical, eight-cell (**octant**) globular embryo exhibiting radial symmetry (see Figure 17.3C). Additional cell divisions increase the number of cells in the globular embryo (see Figure 17.3D) and create the outer layer, the *protoderm*, which later becomes the epidermis.
3. **Heart stage**. Focused cell division in two regions occurs on either side of the future shoot apical meristem to form the two cotyledons, giving the embryo bilateral symmetry (**Figure 17.3E and F**).
4. **Torpedo stage**. Cell elongation and cellular differentiation processes occur throughout the embryonic axis. Visible distinctions between the adaxial and abaxial tissues of the cotyledons become apparent (**Figure 17.3G**).
5. **Mature stage**. Toward the end of embryogenesis, the embryo and seed lose water and become metabolically inactive as they enter dormancy (discussed in Chapter 18). Storage compounds accumulate in the cells at the mature stage (**Figure 17.3H**).

A comparison of embryogenesis in Arabidopsis, a eudicot, with that of rice, a monocot, as well as many other plants illustrates differences in embryo size, shape, cell number, and division patterns. Despite these differences,

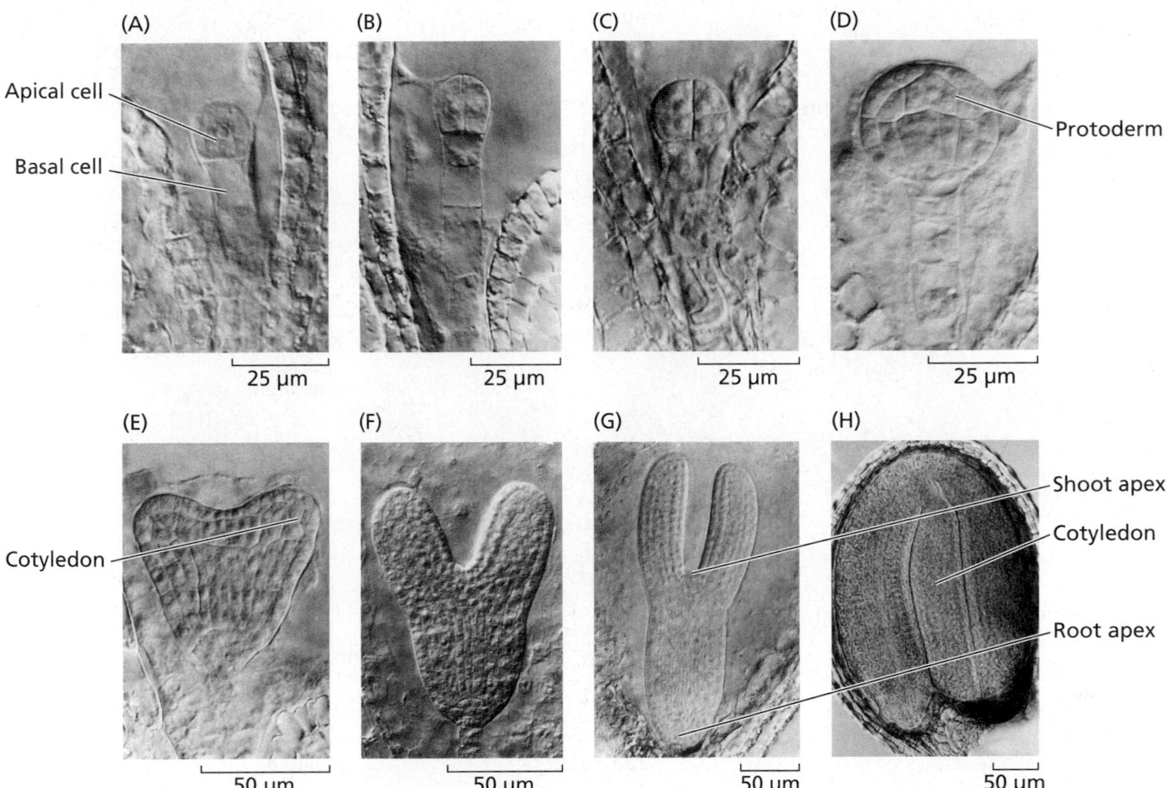

Figure 17.3 The stages of Arabidopsis embryogenesis are characterized by precise patterns of cell division. (A) One-cell embryo after the first division of the zygote, which forms the apical and basal cells. (B) Two-cell embryo. (C) Eight-cell embryo. (D) Mid-globular stage, which has developed a distinct protoderm (surface layer). (E) Early heart stage. (F) Late heart stage. (G) Torpedo stage. (H) Mature embryo. (From West and Harada 1993; photographs taken by K. Matsudaira Yee; courtesy of John Harada, © American Society of Plant Biologists, reprinted with permission.)

several common themes emerge that can be generalized to all seed plants. Perhaps the most fundamental of these relates to **polarity**. Beginning with the single-celled zygote, embryos become progressively more polarized throughout their development along two axes: an **apical–basal axis**, which runs between the tips of the embryonic shoot and root, and a **radial axis**, perpendicular to the apical–basal axis, which extends from the center of the plant outward (**Figure 17.4**).

In the following section we will consider how these axes are established and discuss how specific molecular processes guide their development. Much of our discussion will focus on Arabidopsis, which is not only a powerful model for molecular and genetic studies, but also displays simple and highly stereotyped cell divisions during the early stages of its embryonic development. By observing changes in this simple pattern, we can more easily recognize both physiological and genetic factors that influence embryonic development. A graphic diagrammatic depiction of the earliest cell divisions in Arabidopsis, provided in **Figure 17.5**, offers a convenient guide for the discussion that follows. (For a discussion of the establishment of polarity in a simpler, algal zygote, see **WEB TOPIC 17.3**.)

Apical–basal polarity is established early in embryogenesis

A characteristic feature of seed plants is a polarity in which tissues and organs are arrayed in a stereotyped order along an axis that extends from the shoot apical meri-

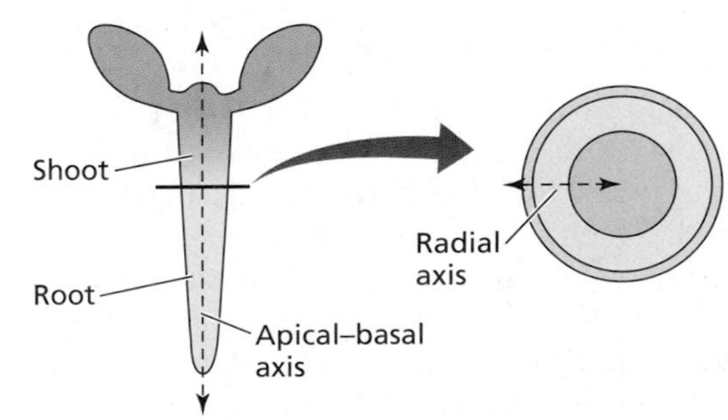

Figure 17.4 In longitudinal cross section (left), the apical-basal axis extends between the tips of the embryonic root and shoot. In transverse cross section (right), the radial axis extends from the center to the surface across vascular, ground and epidermal tissues.

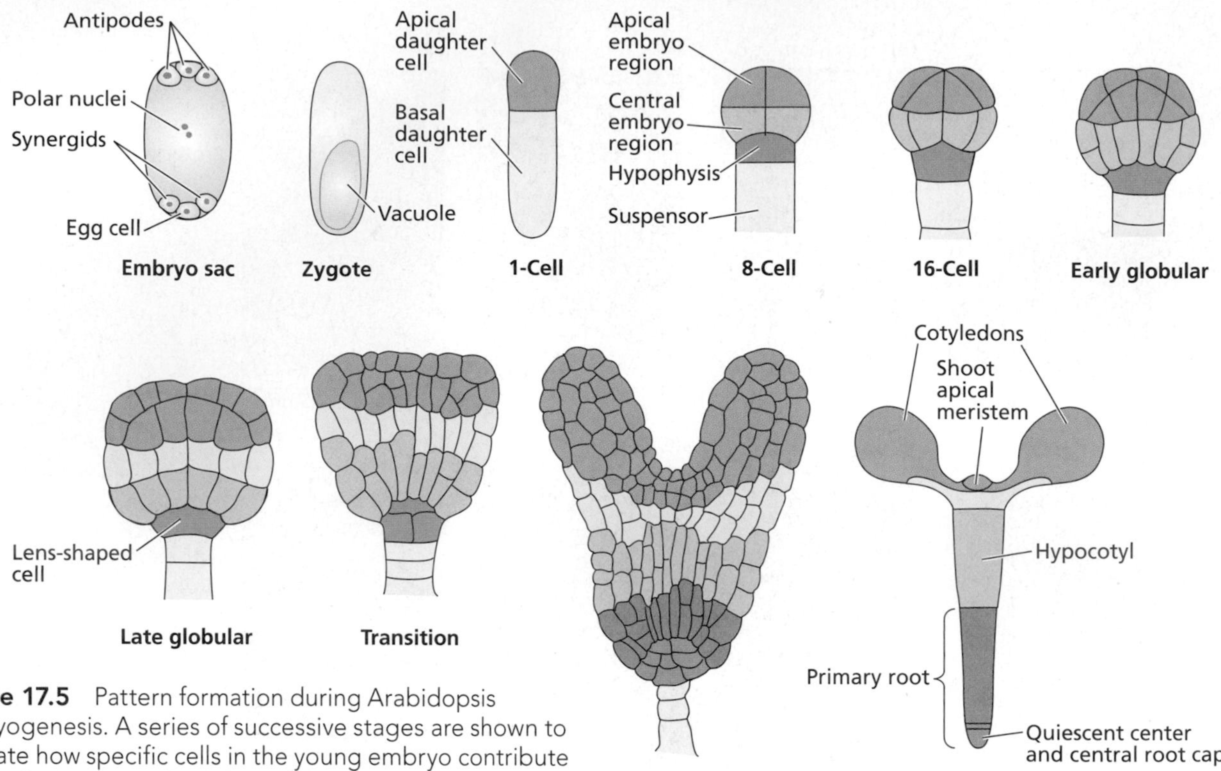

Figure 17.5 Pattern formation during Arabidopsis embryogenesis. A series of successive stages are shown to illustrate how specific cells in the young embryo contribute to specific anatomically defined features of the seedling. Clonally related groups of cells (cells that can be traced back to a common progenitor) are indicated by distinct colors. Following the asymmetric division of the zygote, the smaller, apical daughter cell divides to form an eight-cell embryo consisting of two tiers of four cells each. The upper tier gives rise to the shoot apical meristem and most of the flanking cotyledon primordia. The lower tier produces the hypocotyl and some of the cotyledons, the embryonic root, and the upper cells of the root apical meristem. The basal daughter cell produces a single file of cells that make up the suspensor. The uppermost cell of the suspensor becomes the hypophysis (blue), which is part of the embryo. The hypophysis divides to form the quiescent center and the stem cells (initials) that form the root cap. (After Laux et al. 2004.)

stem to the root apical meristem. An early manifestation of this apical–basal axis is seen in the zygote itself, which elongates approximately threefold and becomes polarized with respect to its intracellular composition. The apical end of the zygote is densely cytoplasmic, in contrast to the basal end, which contains a large central vacuole. These differences in cytoplasmic density are captured when the zygote divides asymmetrically to give a short, cytoplasmically dense **apical cell** and a longer, vacuolated **basal cell** (see Figures 17.3A and 17.5).

The two cells produced by the division of the zygote are also distinguished by their subsequent developmental fates. Nearly the entire embryo, and ultimately the mature plant, is derived from the smaller apical cell, which first undergoes two longitudinal divisions, then a set of transverse divisions (producing new cell walls at right angles to the apical–basal axis) to generate the eight-cell (octant) globular embryo (see Figures 17.3C and 17.5).

The basal cell has a more limited developmental potential. A series of transverse divisions produces the filamentous **suspensor**, which attaches the embryo to the vascular system of the parent plant. Only the uppermost of the division products, known as the **hypophysis**, becomes incorporated into the mature embryo. Through further cell division, the hypophysis contributes to essential parts of the root apical meristem, including the columella and associated root cap tissues, and the quiescent center (see Figure 17.5), which we will discuss later in the chapter.

In the cells that make up the octant globular embryo, there is little, apart from position, to distinguish the appearance of the upper and lower tiers of cells. All eight cells then divide **periclinally** (new cell walls form parallel to the tissue surface) (**Figure 17.6**) to form a new cell layer called the **protoderm**, which ultimately forms the epidermis. As the embryo increases in volume, cells of the protoderm divide **anticlinally** (new cell walls form perpendicular to the tissue surface) to increase the area of this one-cell-thick tissue. By the early globular stage, broad distinctions between the fates of cells from the upper and lower tiers begin to emerge:

- The apical region, derived from the apical quartet of cells, gives rise to the cotyledons and the shoot apical meristem.

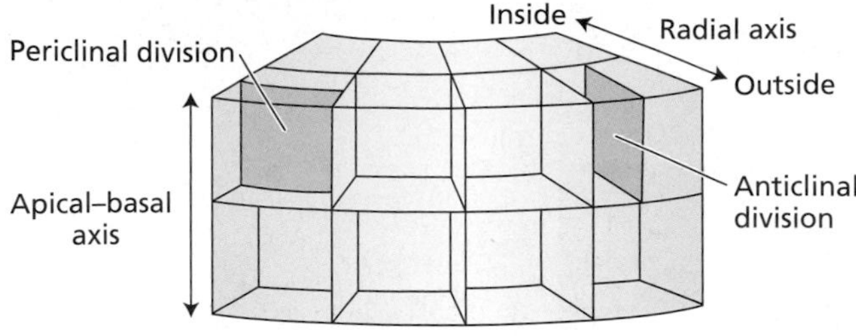

Figure 17.6 Periclinal and anticlinal cell division. Periclinal divisions produce new cell walls parallel to the tissue surface, and thus contribute to the establishment of a new layer. Anticlinal divisions produce new cell walls perpendicular to the tissue surface, and thus increase the number of cells within a layer.

- The middle region, derived from the basal quartet of cells, gives rise to the hypocotyl (embryonic stem), the root, and the apical regions of the root meristem.
- The hypophysis, derived from the uppermost cell of the suspensor, gives rise to the rest of the root meristem.

Position-dependent mechanisms guide embryogenesis

The reproducible patterns of cell division during early embryogenesis in Arabidopsis might suggest that a fixed sequence of cell division is essential to this phase of development. This consistency would be expected if the fates of individual cells within the embryo became fixed, or determined, early; once their fates were established, these cells would be committed to fixed programs of development. Such a *lineage-dependent* mechanism can be likened to assembling a structure from a standard set of parts according to self-contained instructions.

Although many examples of lineage-dependent mechanisms have been documented in animal development, this type of model by itself does not easily explain several general features of plant embryogenesis. First, such lineage-dependent mechanisms are difficult to reconcile with the more variable patterns of cell division typically seen during embryogenesis in many other plant species, including rice and even close relatives of Arabidopsis. Second, even for Arabidopsis, some limited variation in cell division behavior during normal embryogenesis can be seen by following the fates of individual cells with sensitive fate-mapping techniques (**Figure 17.7**). Finally, one

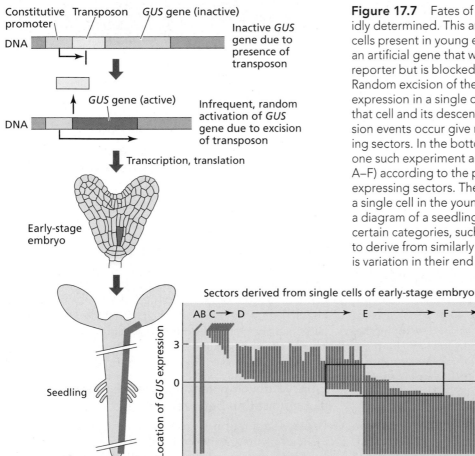

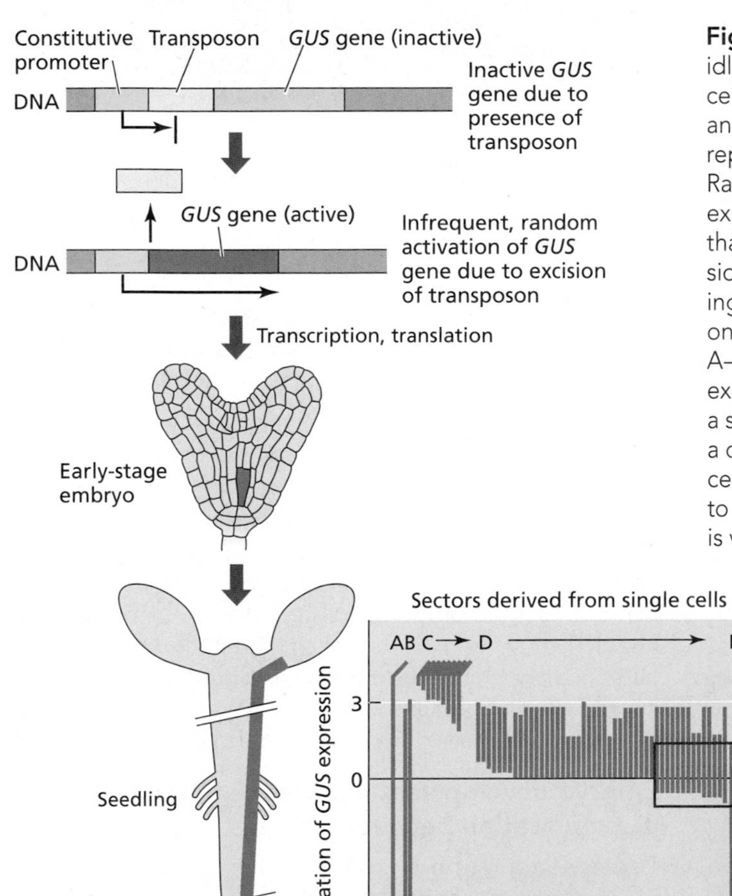

Figure 17.7 Fates of specific embryonic cells are not rigidly determined. This analysis tracks the fates of individual cells present in young embryos. The top diagram shows an artificial gene that would constitutively express a *GUS* reporter but is blocked by the presence of a transposon. Random excision of the transposon activates *GUS* gene expression in a single cell, providing a heritable marker for that cell and its descendants. Embryos in which these excision events occur give rise to seedlings with *GUS*-expressing sectors. In the bottom diagram, the seedlings from one such experiment are sorted into categories (labeled A–F) according to the positions and extents of their *GUS*-expressing sectors. These sectors, each of which arose from a single cell in the young embryo, are shown aligned with a diagram of a seedling to the left. Although sectors within certain categories, such as E and F, are similar and are likely to derive from similarly positioned cells in the embryo, there is variation in their end points. For example, the top ends of sectors in category E overlap with the bottom ends of some sectors in category D. Similar variability can be seen in the end points for other classes of sectors. This variability is inconsistent with a strictly lineage dependent mechanism for cell fate determination, but is more easily explained by mechanisms that respond to feedback from position dependent cues. (After Scheres et al. 1994.)

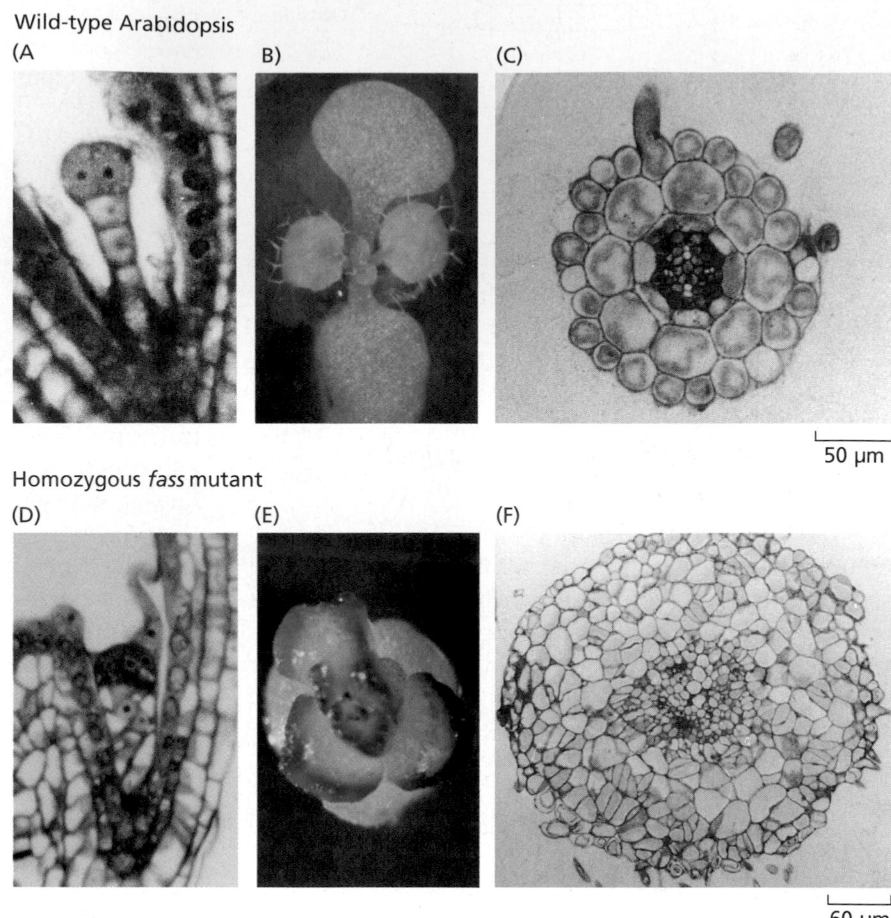

Figure 17.8 Extra cell divisions do not block the establishment of basic radial pattern elements. Arabidopsis plants with mutations in the *FASS* (alternatively, *TON2*) gene are unable to form a preprophase band of microtubules in cells at any stage of division. Plants carrying this mutation are highly irregular in their cell division and expansion planes, and as a result are severely deformed. However, they continue to produce recognizable tissues and organs in their correct positions. Although the organs and tissues produced by these mutant plants are highly abnormal, a radially oriented tissue pattern is still evident. (Top) Wild-type Arabidopsis: (A) early globular stage embryo; (B) seedling seen from the top; (C) cross section of a root. (Bottom) Comparable stages of Arabidopsis homozygous for the *fass* mutation: (D) early embryogenesis; (E) mutant seedling seen from the top; (F) cross section of a mutant root, showing the random orientation of the cells but a nearly wild-type tissue order: an outer epidermal layer covers a multicellular cortex, which in turn surrounds the vascular cylinder. (From Traas et al. 1995.)

can consider the extreme examples provided by certain Arabidopsis mutants that have markedly different patterns of cell division, but still retain the ability to form basic embryonic features (**Figure 17.8**). From this perspective, it seems that the relatively predictable pattern of cell division seen in Arabidopsis may simply reflect the small size of its embryo, which places physical limits on the polarity and probable positions of early cell divisions. Therefore, embryogenesis would seem to involve a variety of mechanisms, including those that do not solely rely on a fixed sequence of cell divisions.

Intercellular signaling processes play key roles in guiding position-dependent development

Given that the morphogenesis of the embryo can accommodate variable patterns of cell division, developmental processes that rely on **position-dependent** mechanisms that determine cell fate seem likely to play significant roles. Such mechanisms would operate by modulating the behavior of cells in a manner that reflects their position in the developing embryo, rather than their lineage. This type of mechanism would explain how equivalent forms can arise through different patterns of cell division. Such position-dependent determination processes could be expected to feature three general kinds of functional elements:

1. There must be cues that signify unique positions within the developing structure.
2. Individual cells must have the means to assess their location in relation to the positional cues.
3. Cells must have the ability to respond in an appropriate way to the positional cues.

These basic requirements focus attention on the cellular context in which signaling processes operate. How is the propagation of signals across space and time affected by the physical makeup of the cell and its relationship to surrounding tissue? Do physical features such as membranes and cell walls merely represent obstacles to intercellular communication, or are they integral to mechanisms that enable signaling outputs to be regulated in response to additional inputs? In the following section we will consider several examples that illustrate how genetically defined signaling processes contribute to embryo development.

Embryo development features regulated communication between cells

Perhaps in a manner analogous to individuals within a social group, individual cells within the developing embryo display a range of facilities that may serve to enable, limit, and transform information during communication. One remarkable aspect of early-stage embryos is the relatively small effect cell walls have on the intercellular movement of certain classes of large molecules. Studies in intact plants show that large artificial dyes and fluorescently tagged protein molecules can move from cell to cell throughout the embryo (**Figure 17.9**), probably via cytoplasmic bridges provided by plasmodesmata. As development progresses, movement of these molecules becomes more size-restricted and spatially limited, suggesting that plasmodesmata-regulated flow of information becomes more important for later stages of development, perhaps to enable regionalized patterns of histogenesis. Paradoxically, during these same early stages of development, the movement of certain classes of relatively small molecules, including the wide-ranging plant hormone auxin, appears more restricted. As we will see, this regulated intercellular movement of molecules plays an essential role in a variety of developmental processes, including the establishment of the axial architecture of the embryo.

The analysis of mutants identifies genes for signaling processes that are essential for embryo organization

Various types of mutants have been analyzed to gain insight into the processes that help establish the basic polarity of the embryo. Many of these processes affect proteins that are likely to contribute to some aspect of signal transduction. To isolate mutations that specifically affect embryonic patterning processes, rather than some essential but more general metabolic activity, screens

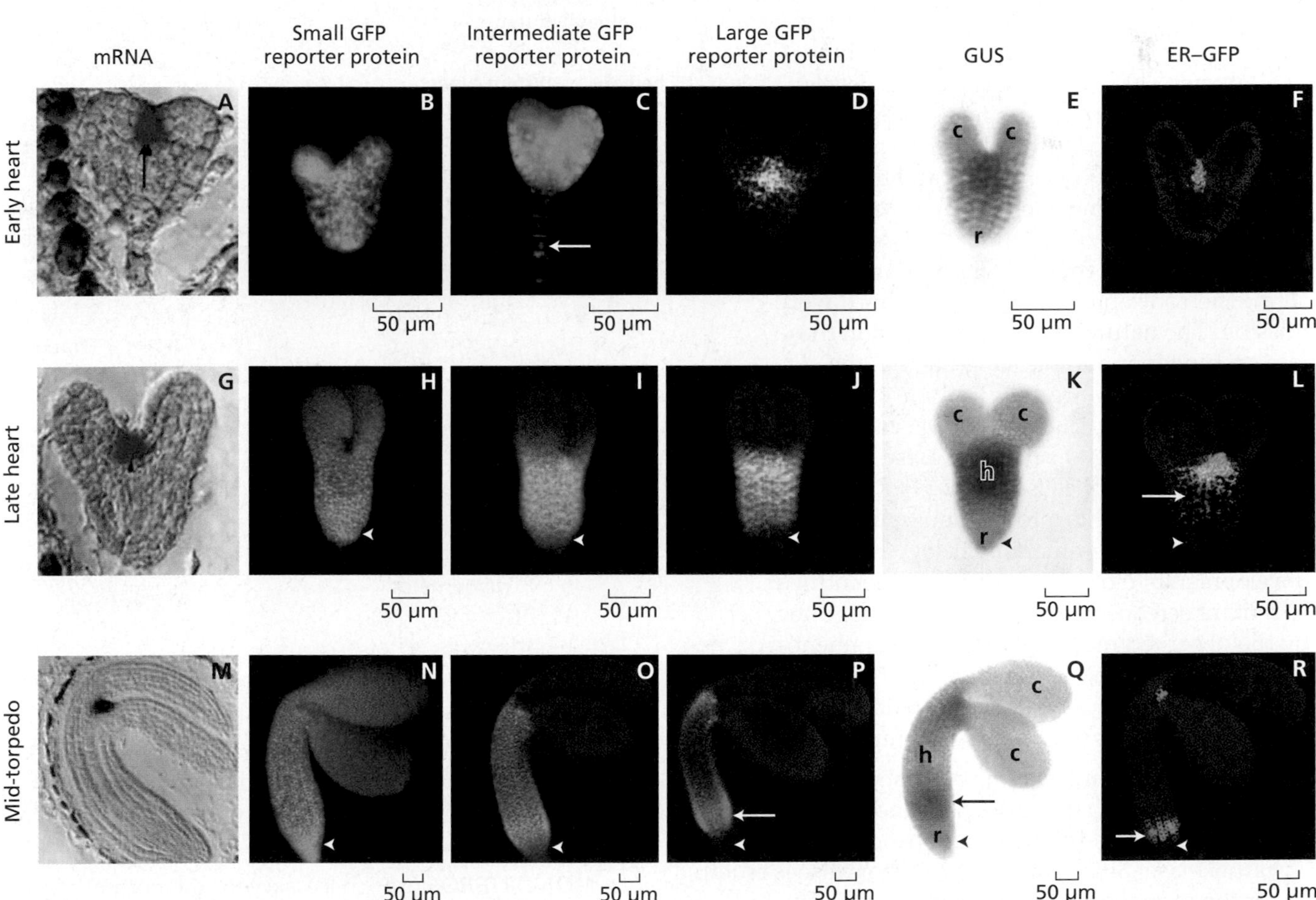

Figure 17.9 The potential for intercellular protein movement changes during development. Images show the distribution of small (B, H, N), intermediate (C, I, O), and large (D, J, P) GFP reporter proteins in embryos of different ages (early heart, A–F; late heart, G–L; mid-torpedo, M–R). All constructs are transcribed from an *STM* promoter, which produces transcripts in relatively small regions of the embryos, as shown by in situ hybridization (A, G, M) or by fusion to the nondiffusible GUS (E, K, Q) or to ER–GFP reporters (F, L, R). Small proteins appear to move readily in all stages of embryogenesis (B, H, N), but the mobility of larger proteins is lower and becomes more restricted in older embryos (C and D, I and J, O and P). Arrows indicate the nucleus in suspensor cells (C) and ectopic expression of the *STM* promoter in hypocotyls (L, P–R). Arrowheads indicate the root. Abbreviations: c, cotyledons; h, hypocotyl; r, root. (From Kim et al. 2005.)

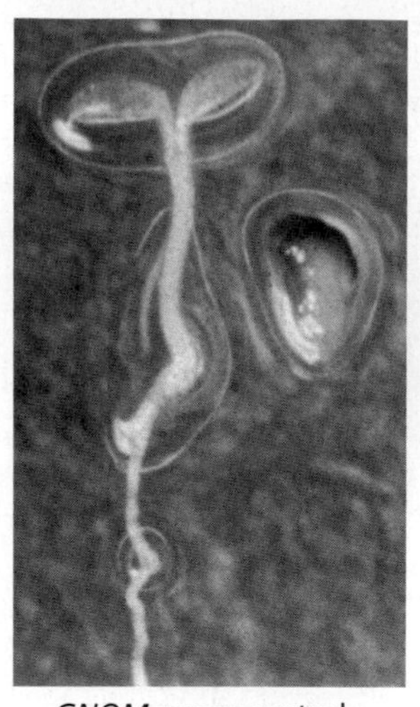

GNOM genes control apicalbasal polarity

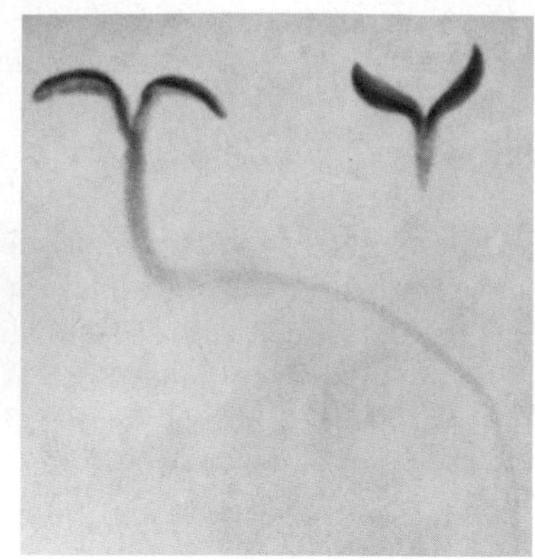

MONOPTEROS genes control formationof the primary root

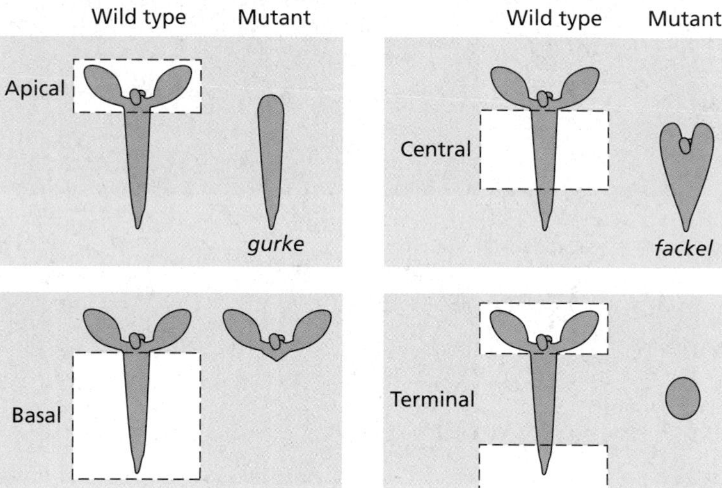

Figure 17.10 Genes essential for Arabidopsis embryogenesis have been identified from their mutant phenotypes. The development of mutant seedlings is contrasted here with that of wild-type seedlings at the same stage of development. (A) The *GNOM* gene helps establish apical–basal polarity. A plant homozygous for the *gnom* mutation is shown on the right. (B) The *MONOPTEROS* gene is necessary for basal patterning and formation of the primary root. A plant homozygous for the *monopteros* mutation (on the right) has a hypocotyl, a normal shoot apical meristem, and cotyledons, but lacks the primary root. (C) Schematic of four deletion mutant types. In each pair, the boxed regions of the wild-type plant on the left are missing from the mutant on the right. (A from Mayer et al. 1993; B from Berleth and Jürgens 1993; C from Mayer et al. 1991.)

were performed for *seedling defective* mutants. The mutants obtained were capable of developing into mature seeds, suggesting a relatively intact metabolism, but displayed an abnormal organization when germinated and examined as seedlings. Among such mutants were those in which the normal apical–basal morphology was disrupted so that the shoot apical meristem, the root apical meristem, or both were missing. The nature of the defects seen in these loss-of-function mutants suggests that the corresponding genes are required for establishing the normal apical–basal pattern (**Figure 17.10**). The cloning of several of these genes by map-based techniques (see **WEB TOPIC 2.2**) has offered some insights into their molecular functions, which are summarized below. As a note, developmental biologists name genes identified in mutant screens with clever names suggested by the observed mutant morphological phenotypes. For instance, the werewolf (*wer*) mutant, was so named for the disordered epidermal cell files and root hairs observed in the mutant.

- ***GURKE*** (***GK***), named for the cucumber-like shape of the mutant, in which the cotyledons and shoot apical meristem are reduced or missing, encodes an acetyl-CoA carboxylase. Since acetyl-CoA carboxylase is required for the proper synthesis of very-long-chain fatty acids (VLCFA) and sphingolipids, these molecules or their derivatives appear to be crucial for proper patterning of the apical portion of the embryo.
- ***FACKEL*** (***FK***) was originally interpreted to be required for hypocotyl formation. Mutants exhibit complex pattern formation defects that include malformed cotyledons, short hypocotyl and root, and often multiple shoot and root meristems. *FK* encodes a sterol C-14 reductase, suggesting that sterols are critical for pattern formation during embryogenesis.
- ***GNOM*** (***GN***) encodes a guanine nucleotide exchange factor (GEF), which enables the directional transport of auxin by establishing a polar distribution of PIN auxin efflux carriers.
- ***MONOPTEROS*** (***MP***), necessary for the normal formation of basal elements such as the root and hypocotyl, encodes an auxin response transcription factor (ARF).

This small collection of mutants highlights the potential significance of specific signaling processes to embryogenesis. Though it is not well understood how mutations to *GK* and *FK* lead to characteristic embryonic pattern defects, the predicted biochemical activities of the proteins encoded by both genes are consistent with disruption of some form of lipid-mediated signaling. Similarly, both

GN and *MP* can be linked to signaling processes, both of which feature auxin. Given the wealth of background information on auxin-dependent responses, we will next consider the importance of polarized auxin transport and the specific roles of *GN* and *MP* in more detail, including how they contribute to an auxin-dependent establishment of the apical–basal axis of the developing embryo.

Auxin functions as a mobile chemical signal during embryogenesis

As is seen in some aspects of animal development, substances termed **morphogens** play key roles in providing positional cues. Through combinations of synthesis, transport, and turnover, morphogen molecules attain a graded distribution within tissues, which in turn evokes a range of concentration-dependent responses. The varied levels and mobility of certain plant hormones and the range of physiological responses they evoke suggest the potential of these molecules to act as morphogens. Although auxin, cytokinins, and abscisic acid (ABA) have been shown to move in xylem transpiration and phloem source–sink streams (see Chapter 11), auxin is the only plant hormone that is polarly transported from cell to cell in an energy-dependent manner. Auxin (indole-3-acetic acid, or IAA) and its synthetic analogs are known as morphogens, as they can be used to induce the formation of embryos from somatic cells and can elicit specific concentration-dependent responses in target tissues. These responses correlate with discrete gradients that occur during embryonic development and which are created by a combination of localized auxin synthesis and intercellular processes that are collectively described as polar auxin transport.

Plant polarity is maintained by polar auxin streams

Polar auxin transport is found in almost all plants, including bryophytes and ferns. Early studies of this phenomenon focused on auxin movement in apical and epidermal tissues during seedling phototropic responses (see Chapter 18). Long-distance polar auxin transport through the vascular parenchyma from sites of synthesis in apical tissues and young leaves to the root tip was shown to regulate stem elongation, apical dominance, and lateral branching (see Chapter 19). Auxin flows redirected at the root apex into the root epidermis were shown to be necessary for root gravitropic responses (see Chapter 18).

Polar auxin transport has been verified by radiolabeled auxin tracer assays and mass spectroscopic analyses of auxin content in discrete tissues. More recently, the use of auxin reporters to report relative auxin concentrations in individual cells and tissues has become a preferred means of visualizing auxin levels in intact plants. The most commonly used reporters are based on DR5, an artificial auxin-responsive promoter that is fused to a reporter gene (whose activity is easily visualized). DR5 fusions to β-glucuronidase (GUS), which produces a blue color when incubated with chromogenic substrates such as p-nitrophenyl β-D-glucuronide, and green fluorescent protein (GFP) or similar fluorescent proteins, are widely used (for example, see Figure 1.30). However, DR5-based reporters require gene transcription to function, which delays the response to auxin. A more dynamic auxin reporter, DII-Venus, is based on a fusion of a yellow fluorescent protein variant to a portion of the AUX/IAA auxin co-receptor protein, which is rapidly degraded in the presence of auxin (see Chapter 15). DII-Venus degrades (disappears) rapidly when auxin is present.

By convention, auxin transport from the shoot and root apices to the root–shoot transition zone is referred to as a *basipetal* flow, whereas downward auxin flow in the root is referred to as *acropetal* transport. As this terminology can be confusing, a newer terminology assigns the term *rootward* transport to all auxin flows toward the root apex and the term *shootward* transport to any directional flow away from the root apex. Both shootward and rootward polar auxin transport are primary mechanisms for effecting programmed and plastic directional growth.

Polar transport proceeds in a cell-to-cell fashion, rather than via the symplast; that is, auxin exits a cell through the plasma membrane, diffuses across the cell wall, and enters the next cell through its plasma membrane (**Figure 17.11**). The overall process requires metabolic energy, as evidenced by the sensitivity of polar transport to O_2 deprivation, sucrose depletion, and metabolic inhibitors. The velocity of polar auxin transport can exceed 10 mm h^{-1} in some tissues, which is faster than diffusion but much slower than phloem translocation rates (see Chapter 11). Polar transport is specific for all natural and some synthetic auxins; other weak organic acids, inactive auxin analogs, and IAA conjugates are poorly transported. Although polar auxin concentration gradients in the embryo appear to be initially established by localized auxin synthesis, they are amplified and extended by specific transporter proteins on the plasma membrane.

AUXIN UPTAKE IAA is a weak acid (pK_a = 4.75). In the apoplast, where plasma membrane H^+-ATPases normally maintain a cell wall solution of pH 5 to 5.5, 15 to 25% of the auxin is present in a lipophilic, undissociated form (IAAH) that diffuses passively across the plasma membrane down a concentration gradient. Auxin uptake is accelerated by secondary active transport of the amphipathic, anionic IAA^- present in the apoplast via AUXIN1/LIKE AUXIN1 (AUX1/LAX) symporters that cotransport two protons along with the auxin anion. This secondary active transport of auxin allows for greater auxin accumulation than does simple diffusion because anionic auxin is driven across the membrane by the proton motive force (i.e., the high proton concentration in the apoplastic solution). Although polarized localization of AUX1 on the plasma membrane occurs in some cells, such as the pro-

(A)

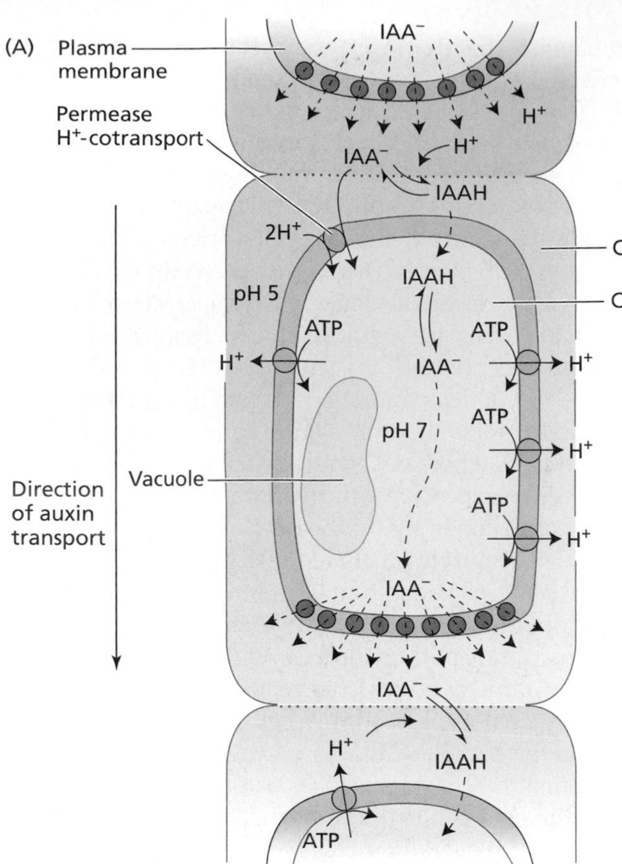

1. IAA enters the cell either passively in the undissociated form (IAAH) or by secondary active cotransport in the anionic form (IAA⁻).

2. The cell wall is maintained at an acidic pH by the activity of the plasma membrane H⁺-ATPase.

3. In the cytosol, which has a neutral pH, the anionic form (IAA⁻) predominates.

4. The anions exit the cell via auxin anion efflux carriers that are concentrated at the basal ends of each cell in the longitudinal pathway.

Figure 17.11 (A) Simplified chemiosmotic model of polar auxin transport. Shown here is one elongated cell in a column of auxin-transporting cells. Additional export mechanisms contribute to transport by preventing reuptake of IAA at sites of export and in adjoining cell files. (B) Model for polar auxin transport in small cells with significant back-diffusion of auxin due to a high surface-to-volume ratio. ABCB proteins are thought to maintain polar streams by preventing reuptake of auxin exported at carrier sites. In larger cells, ABCB transporters appear to exclude movement of auxin out of polar streams into adjoining cell files.

(B)

1. The plasma membrane H⁺-ATPase (purple) pumps protons into the apoplast. The acidity of apoplast affects the rate of auxin transport by altering the ratio of IAAH and IAA⁻ present in the apoplast.

2. IAAH can enter the cell via proton symporters such as AUX1 (blue) or diffusion (dashed arrows). Once inside the cytosol, IAA is an anion, and may only exit the cell via active transport.

3. ABCB proteins are localized (red) nonpolarly on the plasma membrane and can drive active (ATP-dependent) auxin efflux.

4. Synergistically enhanced active polar transport occurs when polarly localized PIN proteins (brown) associate with ABCB proteins, overcoming the effects of back-diffusion.

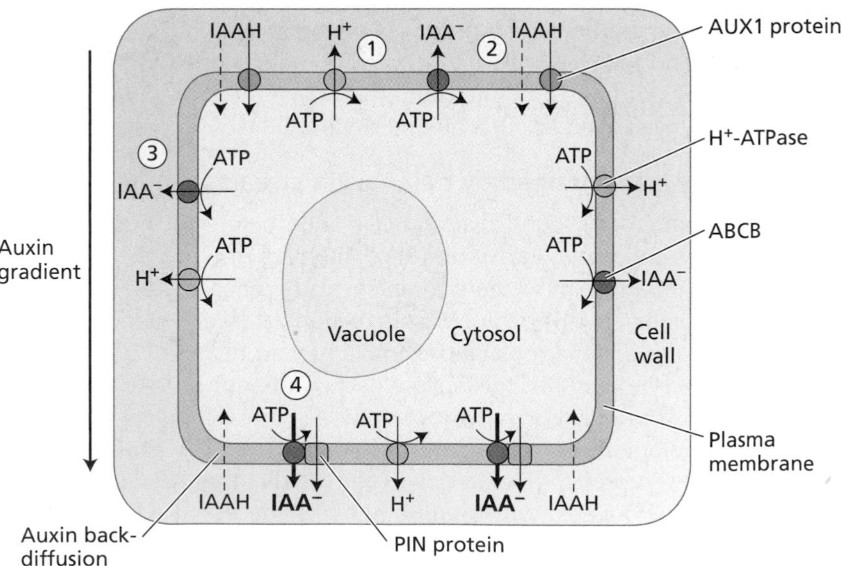

tophloem, the most important contribution of AUX1 is its role in creating cellular sinks that drive polar auxin transport streams. Shootward auxin flows in the *aux1* mutant of Arabidopsis are completely disrupted resulting in agravitropic root growth, but expression of *AUX1* under the control of a promoter asociated with the lateral root cap completely restores gravitropic growth. The compound 1-naphthoxyacetic acid is often used as an inhibitor of the auxin uptake activity of AUX1/LAX proteins.

AUXIN EFFLUX In the neutral pH of the cytosol, the anionic form of auxin, IAA⁻, predominates.Transport of IAA⁻ out of the cell is driven by the negative membrane potential inside the cell. However, because the lipid bilayer

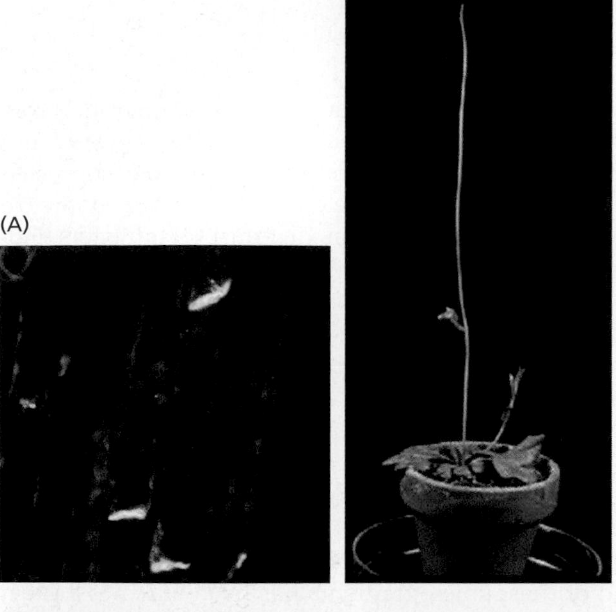

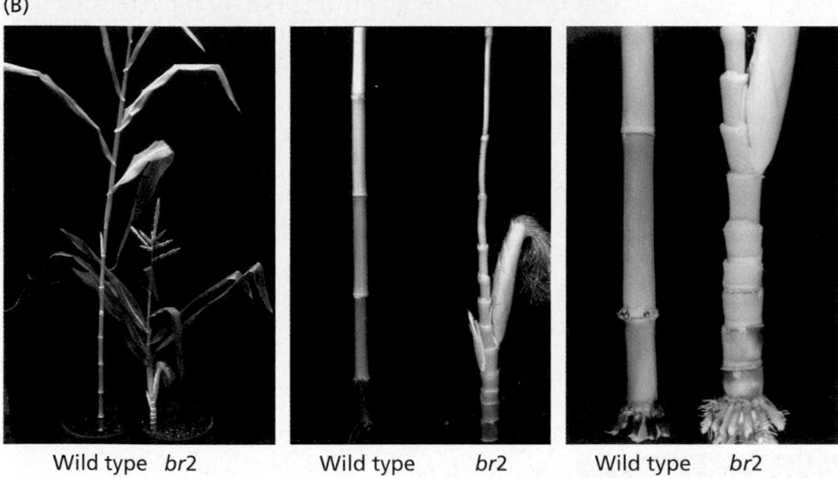

Figure 17.12 (A) PIN1 in Arabidopsis. (Left) Localization of the PIN1 protein at the basal ends of conducting cells in Arabidopsis inflorescences as seen by immunofluorescence microscopy. (Right) The *pin1* mutant of Arabidopsis. A normal wild-type Arabidopsis plant can be seen in Figure 17.1B. (B) The *BR2* (*Brachytic 2*) gene encodes an ABCB required for normal auxin transport in maize, and *br2* mutants have short internodes. The mutant was created by insertional mutagenesis with the Mutator transposon. Unknown to the investigators, the Mu8 transposon contained a fragment of the *BR2* gene. Expression of the *BR2* gene fragment produced interfering RNA (RNAi), which silenced *BR2* expression (see Chapter 2). The *br2* mutants have compact lower stalks (middle and right) but normal tassels and ear (left and middle). (A courtesy of L. Gälweiler and K. Palme; B from Multani et al. 2003.)

of the membrane is impermeable to the anion, auxin export out of the cell must occur via transport proteins on the plasma membrane. Where **PIN auxin efflux carrier proteins** are polarly localized—that is, present on the plasma membrane at only one end of a cell—auxin uptake into the cell and subsequent efflux via PIN give rise to a net polar transport (see Figure 17.11B). (The PIN family of proteins is named after the pin-shaped inflorescences formed by the *pin1* mutant of Arabidopsis; **Figure 17.12A**.) Different PIN family members mediate auxin efflux in each tissue, and *pin* mutants exhibit phenotypes consistent with function in these tissues. Of the PIN proteins, PIN1 is the most studied, as it is essential to virtually every aspect of polar development and organogenesis in plant shoots.

A subset of ATP-dependent transporters from the large superfamily of ATP-binding cassette (ABC) integral membrane transporters amplifies efflux and prevents reuptake of exported auxin, especially in small cells where auxin concentrations are high. Defective *ABCB* (ABC "B" class) genes in Arabidopsis, maize (corn; *Zea mays*), and sorghum result in dwarf mutations of varying severity and in altered gravitropism and reduced auxin efflux (**Figure 17.12B**).

In general, ABCBs are uniformly, rather than polarly, distributed on the plasma membranes of cells in shoot and root apices (see Figure 17.11B). However, when specific ABCB and PIN proteins co-occur in the same location in the cell, the specificity of auxin transport is enhanced; PINs function synergistically with ABCBs to stimulate directional auxin transport. The compound *N*-1-naphthylphthalamic acid (NPA) binds to ABCB auxin transport proteins and their regulators and is used as an inhibitor of auxin efflux activity.

Auxin transport is regulated by multiple mechanisms

As would be expected for such an important function, auxin transport is regulated by both transcriptional and posttranscriptional mechanisms. Genes encoding enzymes that function in auxin metabolism (see WEB APPENDIX 2), signaling (see Chapter 15), and transport are regulated by developmental programs and environmental cues. Almost all known plant hormones have an effect on auxin transport or auxin-dependent gene expression. Auxin itself regulates expression of the genes encoding auxin transporters to increase or decrease their abundance and, thus, to regulate auxin levels.

As is common with many signal transduction pathways, phosphorylation of auxin transporters is a key regulatory mechanism. For instance, the kinase D6PK activates the auxin transport activity of a subset of PIN proteins, and the phototropin 1 photoreceptor kinase inactivates the efflux activity of ABCB19 in phototropic responses (see Chapter 18). Membrane composition and cell wall structure also regulate transporter activity, as both PIN1 and ABCB19 localization on the plasma membrane is dependent on structural sterols or sphingolipids, and PIN1 polar localization is abolished in cellulose synthase-deficient mutants of Arabidopsis. Furthermore, some natural compounds, primarily flavonoids, function as auxin efflux inhibitors. Flavonoids act as reactive oxygen species (ROS) scavengers and are inhibitors of some metalloenzymes, kinases, and phosphatases. Their effects on auxin transport appear to result primarily from these activities.

Regulation of the cellular trafficking of auxin transport proteins to and from the plasma membrane plays a particularly important role in plant development. Specific chaperone proteins are required for successful direction of auxin transporters to the plasma membrane. For instance, the AXR4 protein regulates the trafficking of AUX1, and the immunophilin-like protein TWISTED DWARF 1 (named for the phenotype of the *twd1* mutant in Arabidopsis) regulates the folding and trafficking to the plasma membrane of multiple ABCB auxin transporters. But the most important cellular trafficking processes that regulate polar auxin transport in embryo development are those directing the polar localization of the PIN1 efflux transporter.

One of the great breakthroughs in plant developmental biology has been the combined use of *DR5::GFP* and DII-Venus with GFP fusions of auxin transport proteins to visualize the processes by which microscopic auxin concentration gradients *canalize* (create a channel for) directional auxin transport streams directed by PIN1 as the embryo develops. In other words, small directional flows of auxin are amplified and stabilized by establishment of transport proteins and vascular tissue in configurations that maintain directional flows of auxin to growing tissues. The connection between PIN1 polarity and polar development in the embryo was initially suggested in studies where immature embryos propagated in vitro were treated with auxin or auxin transport inhibitors (**Figure 17.13A and B**). The cup-shaped apical regions induced by artificially perturbing auxin levels were subsequently recognized to be similar to those of *pin1-1* mutants, in which localized polar auxin gradients are disrupted (**Figure 17.13C and D**).

By applying various measures of auxin (summarized in **Table 17.1**), provisional maps have been developed that suggest how auxin synthesis and directed transport combine to generate a patterned distribution of auxin across the developing embryo (**Figure 17.14**). Visualization of these microgradients has been aided to a great extent by the use of auxin-responsive reporters such as DR5 and DII-Venus (see Chapter 15) in combination with fusions of fluorescent proteins to PIN1, the primary efflux component that reinforces and extends embryonic polar auxin flows. Auxin efflux mediated by PIN1 is regulated by kinases that activate the transport activity of the protein, as well as by PINOID-dependent phosphorylation of the central "loop" region of PIN1 that regulates its polar localization.

Polar localization of the PIN auxin efflux proteins is thought to involve three processes:

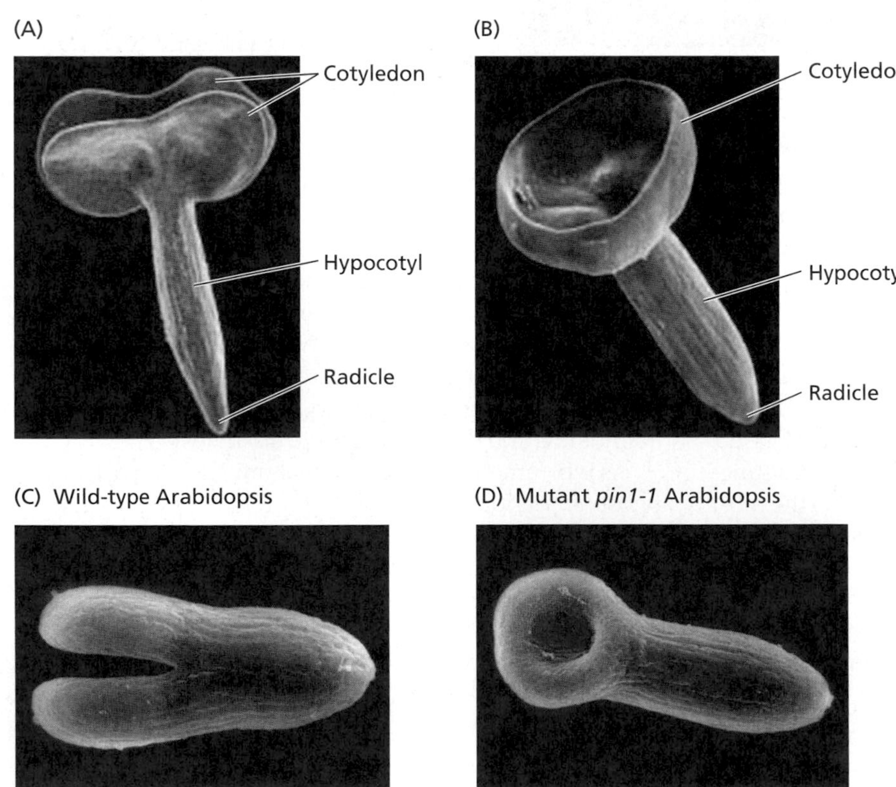

Figure 17.13 Evidence of a role for auxin in embryonic development. (A) A *Brassica juncea* embryo cultured in vitro and showing normal morphology. The radicle is the embryonic root. (B) Altered morphology of a *Brassica juncea* embryo, caused by culturing it for 10 days in the presence of the auxin transport inhibitor *N*-1-naphthylphthalamic acid (NPA). Scale bars in A and B = 250 μm. (C) Wild-type Arabidopsis embryo. (D) A *pin1-1* mutant Arabidopsis embryo. Note the similar failure in cotyledon separation caused by chemical inhibition of auxin transport in vitro and by disruption of auxin transport by mutations in the *PIN* gene. (A and B from Hadfi et al. 1998; C and D from Liu et al. 1993.)

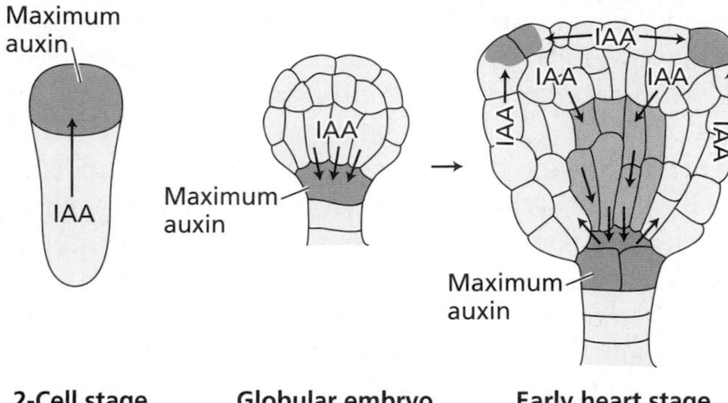

Figure 17.14 PIN1-dependent movement of auxin (IAA) during early stages of embryogenesis. Auxin movement, as inferred from the asymmetric distribution of the PIN1 protein and the activity of a DR5 auxin-responsive reporter, is depicted by arrows. Blue areas denote cells with maximum auxin concentrations. Auxin maximums resulting from synthesis of the hormone create gradients that are then reinforced by the polar orientation of PIN1.

- Initial isotropic (nondirectional) trafficking to the plasma membrane. Multiple experimental approaches show that trafficking of PINs to the plasma membrane involves conserved secretory processes (see Chapter 1).
- Transcytosis and concentration in polarized plasma membrane domains. This process is not well characterized but has been observed with PIN2 in root cells. However, PIN2 polar localization is much less dynamic and auxin-sensitive compared with PIN1 and is thought to be determined primarily by developmental programming. Polar alignments of PIN1 and PIN7 with auxin gradients observed during embryogenesis are presumed to result from transcytosis. However, to date, transcytosis has been documented at the sub cellular level only with PIN2 in epidermal cells of the mature root.
- Stabilization via interactions with the cell wall. Genetic or pharmacological disruption of cell wall biosynthesis results in a complete loss of PIN1 polarity in Arabidopsis.

The GNOM protein establishes a polar distribution of PIN auxin efflux proteins

Central to the dynamic nature of PIN1 and PIN2 localization is their trafficking through a subcellular compartment characterized by the presence of the GNOM protein. The GNOM protein establishes a polar distribution of auxin efflux proteins, and *gnom* mutants have severe developmental defects (see Figure 17.10A and C). When the *GNOM* gene was initially cloned, the similarity of its predicted protein with guanine nucleotide exchange factors (GEFs) did not immediately suggest how the gene contributes to the formation of the apical and basal regions of the embryo. It had been noted, however, that many aspects of the *gnom* mutant phenotype can be mimicked, or **phenocopied**, by application of auxin transport inhibitors, suggesting that GNOM activity might be necessary for normal auxin transport.

An explanation for how GNOM could enable auxin transport emerged through experiments demonstrating that the GEF activity of GNOM is required for the polarized localization of PIN proteins. GNOM, like other related GEF proteins, promotes the intracellular movement of vesicles that deliver specific proteins to targeted sites within the cell. Mutation of *GNOM* disrupts the normal polarized distribution of PIN proteins, though this did not necessarily prove that decreased GEF activity was the cause. However, further experiments demonstrated that the GEF activity of GNOM is crucial for PIN localization. Disruption of PIN localization is observed in cells treated with brefeldin A, an inhibitor of GEF activity, but not in cells that contain an altered form of GNOM to which brefeldin A is unable to bind. The notion that the altered pattern of embryonic development in *gnom*

TABLE 17.1 Methods used to determine auxin levels in plants

Method	Sensitivity	Specificity	Resolution	Comments
Mass spectroscopy	Medium	High	Tissue or organ level	Can discriminate between different forms of auxin
Immunodetection	High*	Medium	Cellular	*Depends on the accessibility of auxin to antibody binding and specificity of antibody
Reporters	High	High	Cellular	Indicates location of auxin-dependent responses, but reporter activity may in some cases be limited by other factors; these may be artificial promoters (DR5, DII-Venus) or fusions with auxin-responsive gene promoters
PIN localization	Medium	Medium	Cellular	Polarized distribution of PIN1 and PIN2 auxin transporters is used to infer directional auxin flows

mutants reflects a disruption of PIN activity is supported by the similar developmental defects that result from directly disrupting genes that encode PIN proteins.

These results suggest that apical–basal patterning of the embryo relies on differences in the distribution of auxin across the embryo, which are created, at least in part, through PIN-directed movement of auxin. In support of this model, the distribution of auxin inferred from auxin reporters at various stages of embryonic development is consistent with that inferred from the polarized distribution of PIN proteins (see Figure 17.14). At the two-cell stage, the preferential accumulation of PIN proteins in the apical wall of the basal cell can be linked to the higher auxin levels in the apical cell. Later in the development of the embryo, the distribution of PIN proteins is reversed, with higher levels along the basal faces of apical cells, which in turn leads to higher auxin levels in basal regions (see Figure 17.14, globular stage). During the early heart stage that follows, the distribution of PIN proteins becomes more complex, resulting in a downward internal flow of auxin that is balanced by an upward flow through superficial cell layers (see Figure 17.14, early heart stage).

MONOPTEROS encodes a transcription factor that is activated by auxin

The cloning of the *MONOPTEROS* (*MP*) gene (see Figure 17.10B and C) revealed that it encodes a member of a family of proteins called **auxin response factors** (**ARFs**), implicating it in auxin-dependent processes. In the presence of auxin, ARFs regulate the transcription of specific genes involved in auxin responses. In the absence of auxin, the activity of these proteins is inhibited through their physical association with specific repressors, termed IAA/AUX proteins. Auxin-dependent responses occur when auxin triggers the targeted degradation of these repressors, allowing ARFs to interact with their target genes (see Chapter 15).

Several lines of evidence support the view that *MP* mediates at least a subset of auxin responses. *mp* mutants not only lack the basal region of the embryo (see Figure 17.10B and C) but also have defects in vascular patterning similar to those observed when auxin levels or movements are artificially disturbed, suggesting that MP is likely to regulate genes that guide auxin-dependent vascular development. Separate genetic studies have confirmed that MP activity is regulated by auxin. These studies focused on a mutant termed *bodenlos* (*bdl*), which, like *mp* mutants, lacks the basal region of the embryo. This similarity suggested that the two genes might be functionally related. Molecular cloning of *BDL* showed that it encodes one of many IAA/AUX repressor proteins. The normal form of BDL associates with MP to repress MP activity, but this repression can be relieved by auxin-induced degradation of BDL. Biochemical studies demonstrated that the mutant form of BDL is resistant to auxin-induced degradation and would thus remain bound to MP, repressing its activity and producing a phenotype similar to that of *mp*.

Taken together, GNOM and MP can be seen to form part of a more complex mechanism by which the movement of auxin and the responses this elicits help guide the establishment of the apical–basal axis. While it is tempting to attribute auxin-dependent phenomena to concentration-dependent responses, it is important to appreciate other potential models, including those that involve the polarization of cells and tissues that result from directional auxin flows, rather than to a response tied to some absolute level of auxin. More detailed genetic analyses, as well as more refined methods for measuring auxin levels and responses, should help discriminate between these alternatives.

Radial patterning guides formation of tissue layers

In addition to the distinctions among cells and tissues positioned along the apical–basal axis of the developing embryo, differences can also be seen along a radial axis that runs perpendicular to the apical–basal axis, extending from the interior to the surface. In Arabidopsis, differentiation of tissues along the radial axis is first observed in the globular embryo (**Figure 17.15**), where periclinal divisions separate the embryo into three radially defined regions. The outermost cells form a one-cell-thick surface layer known as the protoderm, which eventually differentiates into the epidermis. Below this layer lie cells that will later become the **ground tissue**, which in turn differentiates into cortex (the ground tissue between the vascular system and the epidermis) and, in the root and hypocotyl, the endodermis (the layer of suberized cells that restricts water and ion movements into and out of the stele via the apoplast; see Chapter 4). In the most central domain lies the **procambium**, which generates the vascular tissues, including the pericycle of the root.

As was seen for apical–basal patterning of the embryo, a precisely defined sequence of cell division does not appear essential for the establishment of basic radial pattern elements. Significant variability in the patterns of cell divisions associated with the formation of radial patterns can be seen among related species, and basic pattern elements can still be established in mutants with disturbed patterns of cell division, suggesting a prominent role for position-dependent mechanisms. In the following sections we will discuss experiments that address the nature of these mechanisms, providing further examples of the utility of molecular genetic analyses. Further discussion of the physical aspects of cell division can be found in **WEB ESSAY 17.1**.

The origin of epidermis: a boundary and interface at the edge of the radial axis

One obvious and singular aspect of the radial axis of the embryo is provided by the protoderm, a tissue that can be uniquely defined by its superficial position and which

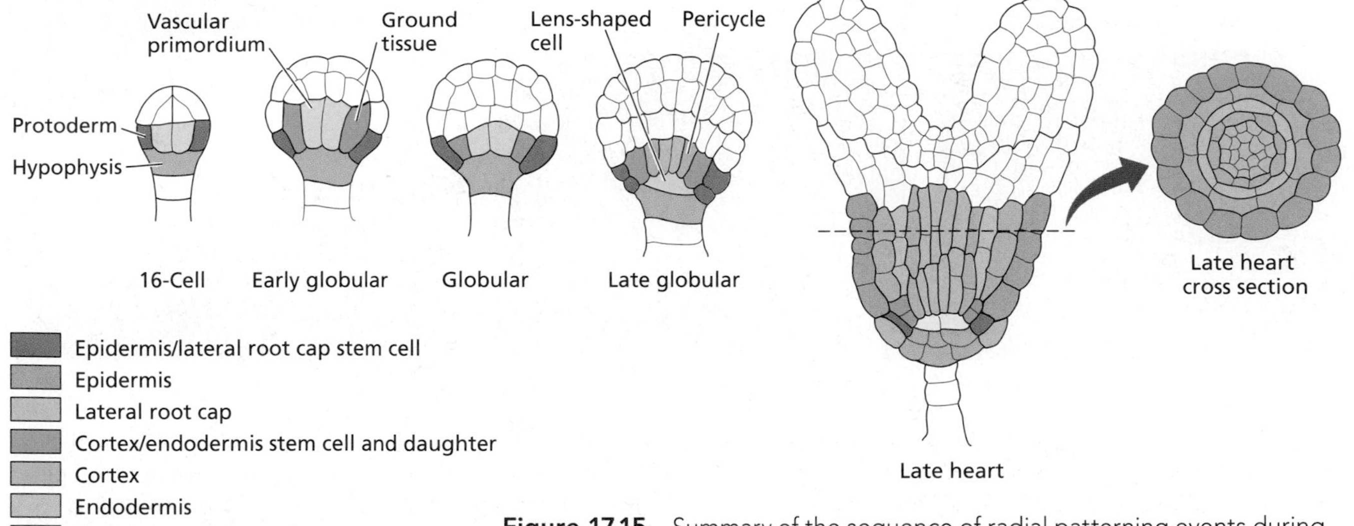

Figure 17.15 Summary of the sequence of radial patterning events during Arabidopsis embryogenesis. The five successive embryonic stages shown in longitudinal section illustrate the origin of distinct tissues, beginning with the delineation of the protoderm (left) and ending with the formation of the vascular tissues (right). Note how the number of tissues increases through the action of stem cells. A cross-sectional view of the basal portion of the late heart stage embryo is shown at the far right (the level of the cross section is shown by the line in the longitudinal section to its left).

eventually produces the epidermis, a critical tissue that mediates communication between the plant and the outside world. Originating early in embryogenesis, protodermal cells have a set of exposed walls that could, in theory, facilitate the exchange of signals with the external environment or, alternatively, act as a boundary as signals move from cell to cell within the embryo. In either case, the protoderm would exhibit unique properties distinguishing it from the internal cell layers, and thus provide potential cues for radial patterning. For example, studies in *Citrus* have shown the presence of a cuticle layer on the surface of the embryo from the earliest zygotic stages through maturity, suggesting that the walls of protodermal cells form a communication boundary. Some studies also suggest that the epidermis can act as a physical constraint to the growth of more internal layers.

Genetic studies have helped us understand the processes that contribute to the unique character of the epidermis. For example, two genes, *Arabidopsis thaliana MERISTEM LAYER1* (*ATML1*) and *PROTODERMAL FACTOR2* (*PDF2*), have been identified as having essential roles in promoting the epidermal identity of superficially positioned cells. Both genes encode homeodomain transcription factors and are expressed from early stages of embryogenesis in the outer cells of the embryo proper. This expression appears necessary for the establishment of normal epidermal identity, since loss-of-function mutant plants have an abnormal epidermis in which cells display characteristics normally associated with mesophyll cells (**Figure 17.16A and B**). Conversely, ectopic (in the wrong location) expression of *ATML1* in internal tissues has been shown to induce abnormal epidermal characteristics. Together, these results suggest that *ATML1* and the related *PDF2* are likely to function by promoting the activity of downstream genes that mediate the development of epidermal characters. Molecular analysis supports and refines this model, showing that the protein products of both genes bind to specific eight-base-pair recognition sequences shared by the promoters of genes that are transcribed at higher levels in the epidermis (**Figure 17.16C**). The *ATML1* and *PDF2* genes themselves contain this same recognition sequence, suggesting that their expression is maintained by a positive feedback loop. However, the nature of the signals that confine the expression of these two genes to the epidermis remains unclear.

Procambial precusors for the vascular stele lie at the center of the radial axis

It is easy to imagine that the unique geometric properties at the center of the developing embryo would afford further potential positional cues for patterning tissues along the radial axis, with vascular tissues of the stele eventually occupying the most central positions. Genetic and developmental analyses suggest that this process is progressive, with periclinal divisions first producing additional layers of cells along the radial axis that then become patterned to particular fates by the activity of specific gene networks. For example, Arabidopsis

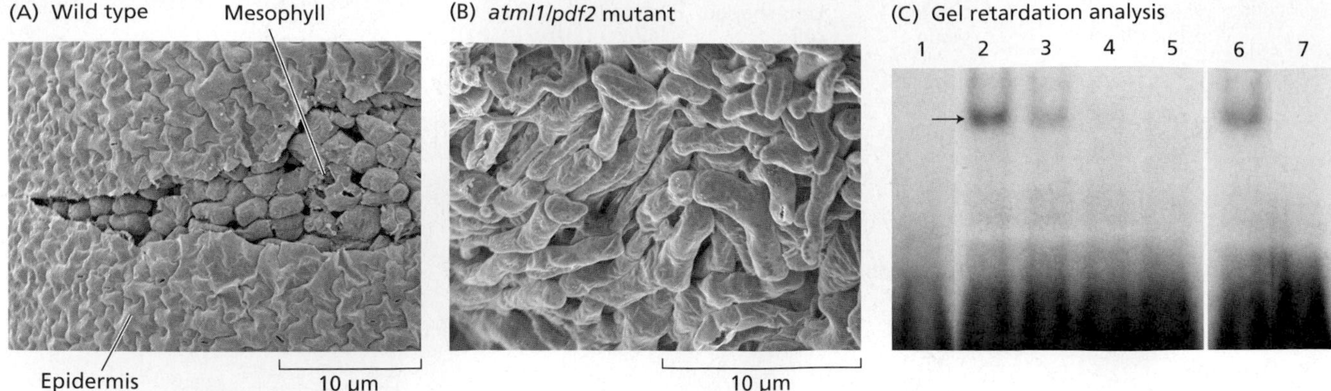

Figure 17.16 *ATML1* and *PDF2* are required for the establishment of a normal epidermis. Comparison of (A) a wild-type plant and (B) a double *atml1/pdf2* mutant shows the resemblance between the superficial layers of the mutant with the mesophyll of the wild-type plant (partially exposed in A). (C) Gel retardation analysis shows that the PDF2 protein binds specifically to a defined sequence found in promoters of genes regulated by PDF2, such as PDF1. A labeled 21-nucleotide probe (L1) with the same sequence as the L1 box of the PDF1 promoter was mixed with maltose-binding protein fused to PDF2 (MBP–PDF2). The DNA probe bound to the protein, producing a labeled complex that can be seen as a band in the gel (lane 2, arrow). No complex was produced if L1 was mixed with maltose-binding protein alone (lane 1) or if MBP–PDF2 was mixed with a mutated L1 probe (lane 7). Labeling of the complex diminished when unlabeled L1 probe (competitor) was added in increasing amounts (100-, 300-, or 1000-fold excess; lanes 3, 4, and 5). (From Abe et al. 2003.)

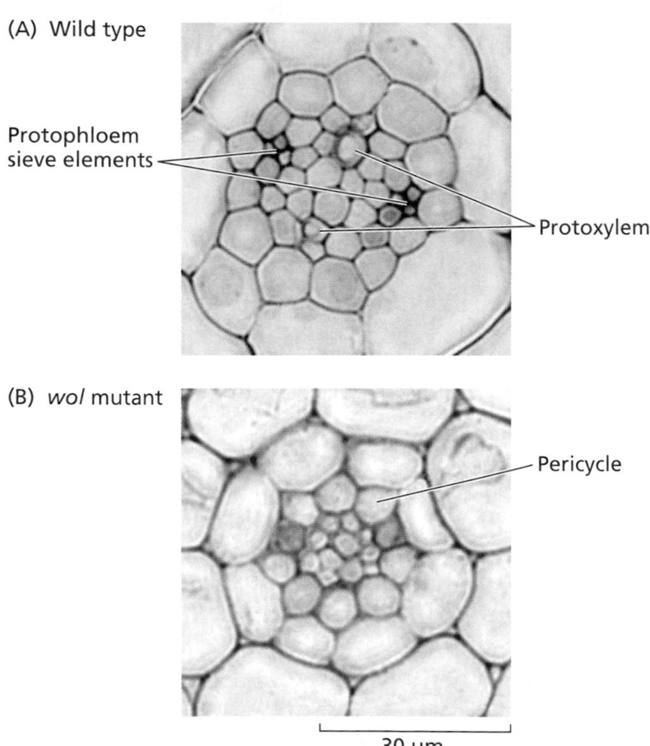

Figure 17.17 The cytokinin receptor encoded by the Arabidopsis *WOODEN LEG* (*WOL*) gene (see Chapter 15) is required for normal phloem development. Comparison of (A) wild-type and (B) *wol* mutant roots shows an absence of phloem elements in *wol* that is accompanied by an apparent decrease in the number of cell layers. (From Mähönen et al. 2000.)

mutants that are deficient for the *WOODEN LEG* (*WOL*) gene fail to undergo a critical round of cell division that normally produces precursors for xylem and phloem (**Figure 17.17**). This defect leads to the development of a vascular system that contains xylem, but not phloem. *WOL* (also known as *CYTOKININ RESPONSE1* [*CRE1*]) encodes one of several related receptors for cytokinin, implicating this hormone in the establishment of radial pattern elements (see Chapter 15). However, these defects can be **rescued** (reversing a phenotype by altering a second factor) by *fass* (i.e., by making a *wol/fass* double mutant), which causes extra rounds of cell division. Thus, it appears that the absence of phloem in *wol* may simply reflect the absence of an appropriately positioned precursor cell layer rather than the inability to specify phloem cell identity.

The differentiation of cortical and endodermal cells involves the intercellular movement of a transcription factor

The development of endodermal and cortical tissues provides a classic example of how the radial patterning process can be regulated by gene activity communicated between adjacent layers. Two Arabidopsis genes, *SCARECROW* (*SCR*) and *SHORT-ROOT* (*SHR*), are both essential for the normal formation of cortical and endodermal cell layers. The similar protein sequences encoded by these two genes place them in the *GRAS* family of transcription factors, whose name derives from the first known members, *GIBBERELLIN-INSENSITIVE* (*GAI*), *REPRESSOR OF GA1–3* (*RGA*), and *SCR*.

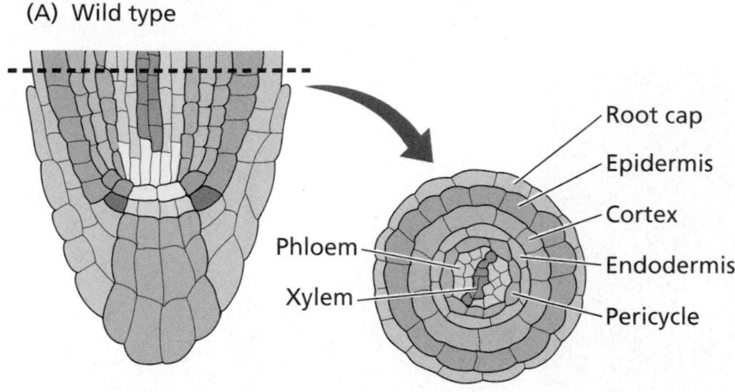

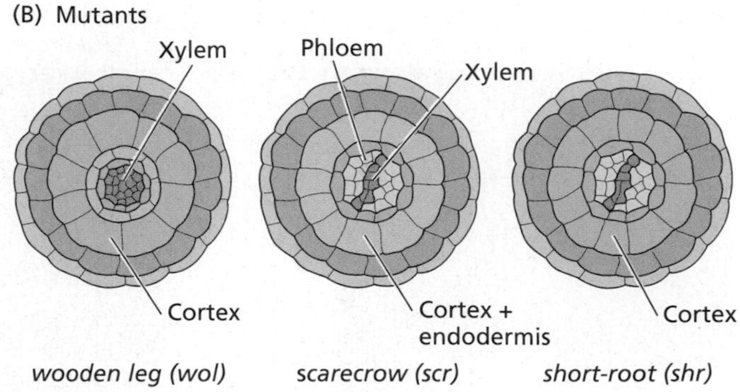

Figure 17.18 A comparison of normal and mutant radial root patterns shows the spatially defined functions of specific genes. (A) Wild-type root. (B) Defective radial root patterns of three Arabidopsis mutants: *wooden leg* (*wol*), *scarecrow* (*scr*), and *short-root* (*shr*). (After Nakajima and Benfey 2002.)

Mutants in which either *SCR* or *SHR* activity is reduced fail to undergo a round of cell division that produces the two layers that later differentiate as separate cortex and endodermis. Mutations in either gene block the round of cell division that creates these separate layers (**Figure 17.18**). In *scr* mutants, the single layer that remains exhibits characteristics of both endodermis and cortex, suggesting that the mutant is still able to express these characteristics but is unable to separate them into discrete layers. This interpretation is supported by the ability of *fass* to restore more normal growth patterns. Much as it rescues *wol*, *fass* appears to compensate for the division defect of *scr*, and thus provides separate layers in which distinct endodermal and cortical traits can be expressed.

The mutant *shr* not only exhibits a cell division defect similar to that of *scr*, but is also unable to elaborate cellular characteristics typical of the endodermis. The single undivided layer in *shr* lacks endodermal traits, such as the Casparian strip, and instead displays gene activities that are normally limited to the cortex. This apparent requirement for *SHR* gene activity to specify endodermal traits is puzzling, since the expression of *SHR* mRNA is normally restricted to more internal, provascular tissues.

More detailed analyses involving the use of fluorescently tagged proteins have addressed this paradox, showing that although *SHR* mRNA is confined to the vascular cylinder, its translation product is not. The SHR protein is able to move into the adjacent, more external layer via plasmodesmata, where it has several activities, including promoting enhanced transcription of *SCR*. Following the translation of the *SCR* mRNA, SHR forms a heterodimer with the SCR protein to enhance the transcription of genes associated with endodermal programs of development (**Figure 17.19**). The contribution of the SHR protein to the differentiation of cortical and endodermal cells provides a clear example of how the functions of specific transcription factors may depend on their movement between cell layers.

Meristematic Tissues: Foundations for Indeterminate Growth

The development of plants shows a remarkable degree of plasticity, which to a large extent can be attributed to specialized tissues called **meristems**. A meristem can be broadly defined as a group of cells that retain the ability to proliferate and whose ultimate fate is not rigidly determined, but is subject to modification by external factors, thus enabling the plant to best exploit the prevailing environment. Several types of meristems that contribute to the vegetative development of plants can be distinguished based on their position in the plant.

The **root apical meristem** (**RAM**) and **shoot apical meristem** (**SAM**) are found at the tips of the root and shoot, respectively. **Intercalary meristems**, such as the **vascular cambium**, represent proliferative tissues that are flanked by differentiated tissues. **Marginal meristems** function in a similar manner at the edges of developing organs. Small, superficial clusters of cells, known as **meristemoids**, give rise to structures such as trichomes or stomata (see WEB ESSAY 17.2 for a historical overview of plant meristems). In the following sections we will consider the basic features of the root and shoot apical meristems, as well as the vascular cambium, that make them useful models for understanding the mechanisms that control the division of cells and determination of their fates.

The root and shoot apical meristems use similar strategies to enable indeterminate growth

Although it might seem difficult to imagine two parts of a plant more different than a shoot and a root, certain features of the RAM and SAM and the roles they play in enabling indeterminate patterns of growth invite comparisons. Each of these structures features a spatially defined cluster of cells, termed **initials**, that are distinguished by their slow rate of division and undetermined fate. As the descendants of initials are displaced away by polarized

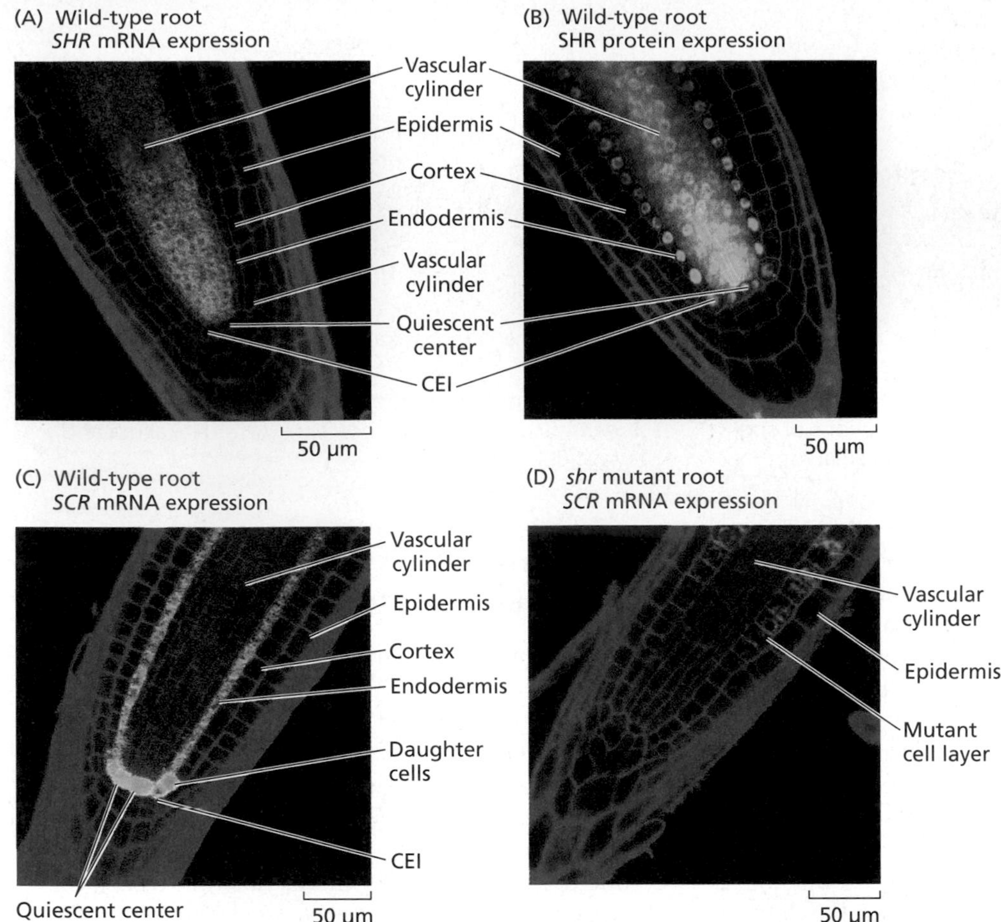

Figure 17.19 The *SHORT-ROOT* (*SHR*) and *SCARECROW* (*SCR*) genes in Arabidopsis control tissue patterning during root development. Here, the mRNAs or proteins for *SHR* and *SCR* have been localized by confocal laser scanning microscopy. (A and B) *SHR* expression. (A) During early root development, *SHR* promoter activity is restricted to the stele (as visualized using an *SHR* promoter–green fluorescent protein [GFP] fusion). (B) The SHR protein shows a distinct pattern of localization, which includes the central stele and also the nuclei of adjacent endodermis (as visualized using an *SHR* promoter + coding region + GFP fusion). (C and D) *SCR* expression (monitored using an *SCR* promoter–GFP fusion). (C) In wild-type roots, the *SCR* is transcribed in the quiescent center (QC), endodermis, and cortical–endodermal stem cell (CEI). It is not present in the cortex, vascular cylinder, or epidermis. (D) The expression of *SCR* is markedly reduced in the *shr* mutant root, and appears only in the mutant cell layer that has characteristics of both endodermis and cortex. (From Helariutta et al. 2000.)

patterns of cell division, they take on various differentiated fates that contribute to the radial and longitudinal organization of the root or shoot and to the development of lateral organs.

From this perspective, it is clear that both the RAM and the SAM must have mechanisms that balance the production of new cells with the ongoing recruitment of cells into differentiated tissues. Is it possible that common aspects of RAM and SAM behavior can be traced to similar underlying mechanisms? How are these mechanisms regulated to maintain the characteristic organizations of the shoot and root and to enable adaptive growth responses to a range of environments? Do the distinct patterns of growth and organogenesis in root and shoot impose special requirements on RAM and SAM function? To address these questions, we will discuss basic features of the RAM and the SAM as well as examples of genetically defined signaling pathways that contribute to their establishment and maintenance.

The Root Apical Meristem

Many aspects of root growth reflect adaptations to a demanding environment. Roots, which anchor the plant and absorb water and mineral nutrients from the soil, display complex patterns of growth and tropisms that allow them to explore and exploit a heterogeneous environment laden with obstacles (see Chapter 18). Although cells produced by the RAM divide, differentiate, and elongate as they are displaced away from the tip, much like their counterparts in the shoot do, lateral outgrowths such as root hairs or lateral branches emerge farther away from the root tip in regions where cell elongation is complete. This spatial separation, which helps prevent damage to lateral organs from shearing forces, affords a useful opportunity to focus on just those processes at the root tip that serve to maintain a set of initials and regulate their division activity. In the section that follows, we will consider the generation of root organization at the apex in more detail, discussing regional differences in cellular behavior that contribute to the growth and functionality of the root. We will then review experimental evidence suggesting that the coordinated growth of the root depends on a combination of auxin-dependent and cytokinin-dependent programs of gene activity that are coordinated by specific classes of transcription factors and response regulators.

The root tip has four developmental zones

The basic features of root development can best be described by first distinguishing zones within the root with distinct cellular behaviors. Although it is impossible to define their boundaries with absolute precision, the division of the root into the following zones provides a useful spatial framework that is relevant to our discussion of the underlying mechanisms (**Figure 17.20**).

- The **root cap** occupies the most distal part of the root. It represents a unique set of initial derivatives that are displaced distally away from the meristematic zone. The differentiated products of these divisions cover the apical meristem and protect it from mechanical injury as the root tip is pushed through the soil. Other functions of the root cap include perception of gravity, to enable gravitropism, and the secretion of compounds that help the root penetrate the soil and mobilize mineral nutrients.
- The **meristematic zone** lies just under the root cap. It contains a cluster of cells that act as initials, dividing with characteristic polarities to produce cells that divide further and differentiate into the various mature tissues that make up the root. Cells surrounding these initials have small vacuoles and expand and divide rapidly.
- The **elongation zone** is the site of rapid and extensive cell elongation. Although some cells continue to divide while they elongate within this zone, the rate of division decreases progressively to zero with increasing distance from the meristem.
- The **maturation zone** is the region in which cells acquire their differentiated characteristics. Cells enter the maturation zone after division and elongation have ceased, and in this region lateral organs such as lateral roots and root hairs may begin to form. Differentiation may begin much earlier, but cells do not achieve their mature state until they reach this zone.

In Arabidopsis these four developmental zones occupy little more than the first millimeter of the root tip. In many other species these zones extend over a longer distance, but growth is still confined to the distal regions of the root.

The origin of different root tissues can be traced to specific initial cells

Given the progressive and linear development of the tissues that make up the root, it is relatively simple to trace

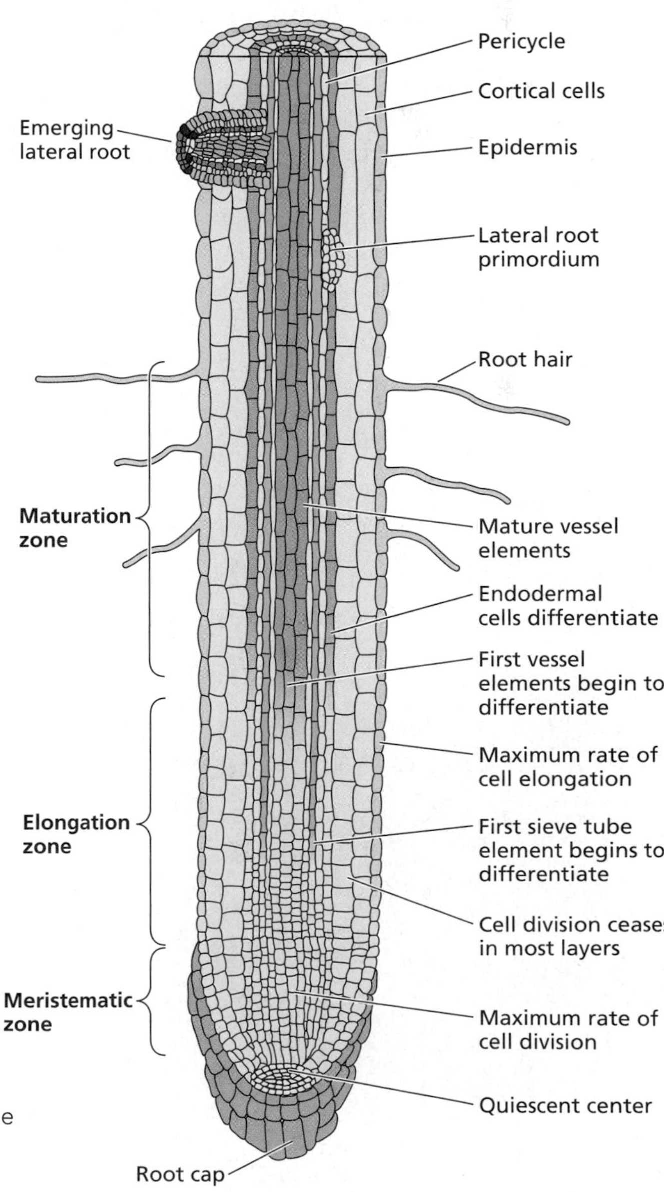

Figure 17.20 Simplified diagram of a primary root showing the root cap, the meristematic zone, the elongation zone, and the maturation zone.

their origin back to specific initial cells in the subapical region. In most plant roots, a medial longitudinal cross section reveals long cell files that converge in the subapical region of the root (**Figure 17.21A**). At the center of this convergence zone is the so-called **quiescent center** (**QC**), named for its relatively low rate of cell division compared with those of surrounding tissues.

The close physical association between the initials that give rise to various tissues and the cells that make up the adjacent QC suggests a close functional interdependence between these cell types. Some have argued that the distinction between the QC and adjacent meristematic cells is somewhat artificial because, in the roots of many higher plants, the cells that make up the QC occasionally divide to replace adjacent initials. In a similar line of reasoning, attention can be drawn to other plant species in which the relationship between the QC and initials is different. In some of these, the QC may include dozens or hundreds of cells, and this number may change during the plant's life cycle. By contrast, in some lower vascular plants, such as the water fern *Azolla*, a single, centrally positioned apical cell appears to fulfill the roles of both QC and initials by retaining low but consistent mitotic activity throughout vegetative development (see WEB TOPIC 17.4).

Like the patterns of cell division associated with embryogenesis, the behavior of the QC and the surrounding initials varies among plant species, suggesting that position-dependent mechanisms play an important role in specifying these cell types. As was the case for embryogenesis, considerable insight into the underlying mechanisms is afforded by models such as Arabidopsis, in which the behavior of individual cells can be easily monitored. The roots of Arabidopsis are well suited to this approach, given their small size and relatively transparent nature. Observations are also simplified by the relatively small number of cells of the Arabidopsis root and its accessibility, which enable real-time microscopic monitoring of developmental processes.

The QC of Arabidopsis consists of only four cells, and because division of these cells during postembryonic development is rare, factors that perturb the activity of the QC or the surrounding initials are easily recognized. In Arabidopsis, four distinct sets of initials, all of which are adjacent to the QC, can be defined in terms of their position and the tissues they produce (**Figure 17.21B**):

1. **Columella initials**. Located directly below (distal to) the QC, these initials give rise to the central portion (columella) of the root cap.
2. **Epidermal–lateral root cap initials**. Located to the side of the QC, these initials first divide anticlinally to set off daughter cells, which then divide periclinally to form two files of cells that mature into the lateral root cap and epidermis.
3. **Cortical–endodermal initials**. Located interior and adjacent to the epidermal–lateral root cap initials, the

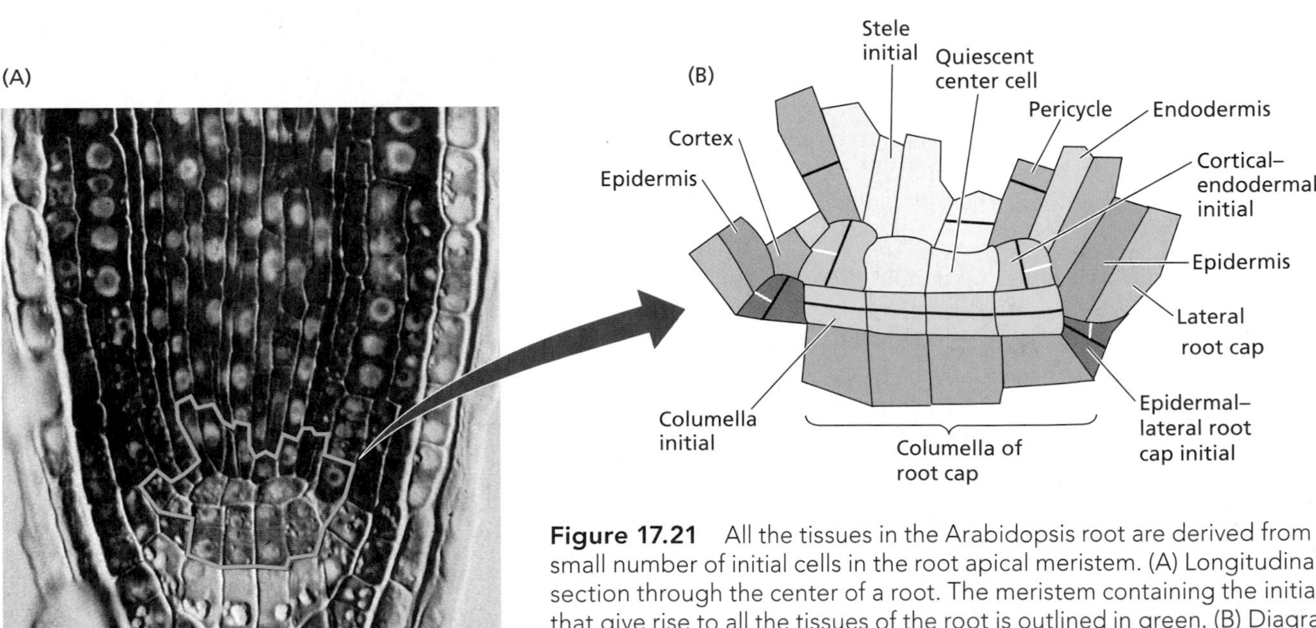

Figure 17.21 All the tissues in the Arabidopsis root are derived from a small number of initial cells in the root apical meristem. (A) Longitudinal section through the center of a root. The meristem containing the initials that give rise to all the tissues of the root is outlined in green. (B) Diagram of the region outlined in A. Only two of the four quiescent center cells are depicted in this section. The heavier black lines indicate the cell division planes that occur in the initials. The white lines indicate the secondary cell divisions that occur in the cortical–endodermal and epidermal–lateral root cap initials. (From Schiefelbein et al. 1997, courtesy of J. Schiefelbein, © American Society of Plant Biologists, reprinted with permission.)

cortical–endodermal initials divide anticlinally to set off daughter cells, which then divide periclinally to form the cortical and endodermal cell layers.

4. **Stele initials**. Located directly above (proximal to) the QC, these initials give rise to the vascular system, including the pericycle.

Cell ablation experiments implicate directional signaling processes in determination of cell identity

To test and refine the hypothesis that the behavior of the QC and surrounding initials is influenced by position-dependent signaling processes, a series of experiments was performed to assess the contributions of specific cells to the determination process. To evaluate these contributions, the highly stereotyped patterns of cell division in the RAM of normal Arabidopsis were compared with those in plants in which one or more specific cells had been destroyed (or ablated) using microscopically focused laser beams.

Ablation of the QC led to abnormal division and precocious differentiation of the adjacent initials (see Figure 17.21B), suggesting that the QC produces a mobile signal that acts on the adjacent initials to prevent their differentiation and thus maintain their ability to divide. In a related experiment, ablation of differentiated cells that were adjacent to initials caused the initials to take on abnormal identities, as revealed by the cell types they produced. These results suggest that the specification of the particular identities of initials relies on signals that emanate from more differentiated tissues.

Auxin contributes to the formation and maintenance of the RAM

Just as auxin seems to play a role in establishing apical–basal polarity in the embryo, a convincing case can be made for the involvement of auxin in positioning the RAM and guiding its complex behavior. In normal roots, the position of the QC coincides with an auxin concentration maximum. When the position of this maximum is shifted by chemical treatments, the position of the QC shows corresponding changes. By contrast, treatments that abolish this maximum lead to the loss of the QC.

Responses to auxin are mediated by several distinct families of transcription factors

Even with an understanding of how a graded distribution of auxin across the root can be achieved, some explanation is still required for how these concentration differences evoke a variety of downstream responses, including in the localized zones of cell division, elongation, and differentiation (see Figure 17.20). One part of the explanation involves auxin response factors (whose regulation by auxin is described in more detail in Chapter 15). Above some threshold concentration, auxin triggers the breakdown of IAA/AUX repressors, which would otherwise bind to ARFs such as MONOPTEROS and thus block their ability to regulate transcription. As in the establishment of the root during embryogenesis, MP and other ARFs play auxin-dependent roles to maintain the root during vegetative growth.

Genetic approaches have revealed additional types of transcription factors that act downstream of ARFs to coordinate specific aspects of root growth (**Figure 17.22**). Two of these transcription factors, belonging to the AP2/Ethylene

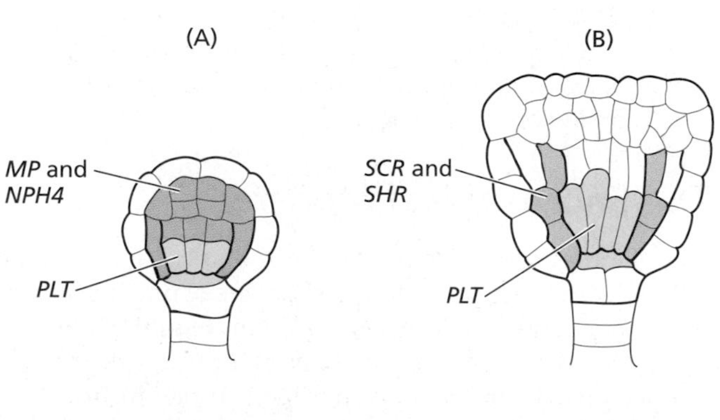

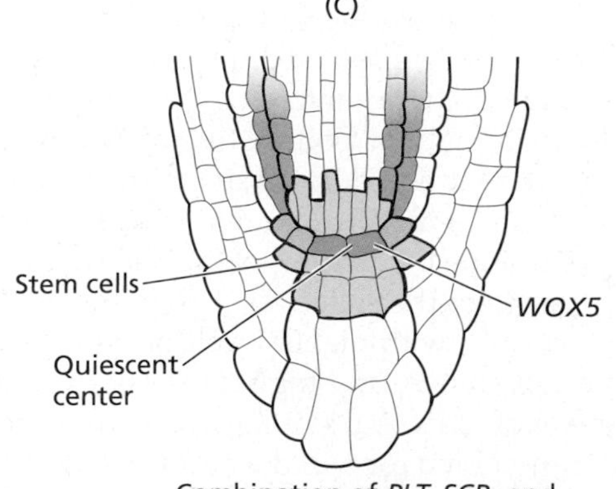

Figure 17.22 Model for the specification of cell identity in the root. (A) Early auxin-dependent expression of the *MONOPTEROS* (*MP*) and *NONPHOTOTROPIC HYPOCOTYL 4* (*NPH4*) genes. *MP* and *NPH4* promote *PLETHORA* (*PLT*) expression in a basal domain. (B) *PLT* promotes the expression of *SCARECROW* (*SCR*) and *SHORT-ROOT* (*SHR*). (C) The combination of *PLT*, *SCR*, and *SHR* gene expression directs centrally positioned cells to become the quiescent center (QC), and also induces the expression of *WOX5*, which contributes to the maintenance of surrounding initial cells. The area outlined in red contains stem cells. (From Aida et al. 2004.)

Responsive Factor class, are encoded by the *PLETHORA 1* (*PLT1*) and *PLETHORA 2* (*PLT2*) genes. In the zone of high auxin concentration that includes the QC, the expression of these two *PLT* genes is activated. Mutants in which *PLT* genes have been disrupted are unable to form or maintain a functional QC, suggesting that these genes normally regulate programs of transcription that are essential for these processes. Conversely, the artificial expression of *PLT* genes in more proximal regions of the root leads to the formation of an ectopic QC. Together, these experiments support models in which auxin provides positional cues that lead to specific programs of transcription, which in turn mediate specific cellular behaviors that contribute to the formation and maintenance of the RAM.

WOX genes (for ***W****USCHEL homeob****ox***) encode a third family of transcription factors that play key roles not only in the RAM, but also, as we will see, in the SAM and vascular cambium. Genes belonging to this family contain a distinctive form of the homeobox DNA binding motif that was first described in *WUS*, a gene that is essential for both the formation and maintenance of the SAM. Similar to that of *PLT* genes, the expression of several root-specific *WOX* genes appears sensitive to auxin, as shown by changes in the distribution of *WOX* gene transcripts in *mp* or *bdl* mutants (which lack certain root-related auxin-dependent activities). One of these *WOX* genes, *WOX5*, is expressed in a small group of cells in the root tip that includes the QC and the surrounding initials. The very focused pattern of *WOX5* expression is determined by a combination of *PLT, SCR*, and *SHR* activities. Similar to *WUS*, which functions in the SAM to maintain a population of undifferentiated initials (as we will discuss later in this chapter), *WOX5* appears to play an analogous role in the root, where its expression in the QC helps maintain adjacent initial cells by preventing their premature differentiation (see further discussion below on comparable mechanisms in the SAM and vascular cambium).

Cytokinin is required for normal root development

Although much of our discussion of the growth and development of the root has focused on auxin, recent studies have also drawn attention to cross talk between cytokinin and auxin signaling. Contrasting activities of these hormones were first noted in physiological studies with undifferentiated cell cultures known as *callus* (see Figure 15.11A). Application of mixtures of these hormones in different proportions to the callus resulted in the generation of shoots or roots, with higher levels of cytokinin favoring formation of shoots, and higher levels of auxin promoting formation of roots. Auxin is largely synthesized in the shoot and transported rootward via ABCB and PIN auxin transporters, while cytokinin synthesized in the root moves shootward in the xylem. These observations resulted in the concept of antagonistic auxin–cytokinin regulation of shoot and root development. The widespread use of shoot decapitation to remove auxin supply and application of additional auxin to manipulate apical dominance in horticulture further supported this model. However, more recently, strigolactone has been shown to be another major hormone that interacts with auxin to regulate shoot architecture. The factors controlling apical dominance and shoot branching are described in more detail in Chapter 19.

Although the known elements that make up the cytokinin and auxin signal transduction pathways are largely distinct, similar experimental approaches have proved useful for the analysis of both pathways. Approaches analogous to the development of DR5 and DII-Venus reporter fusions, which provide a measure of auxin activity (see Chapter 15), have been developed for cytokinin, in which promoter sequences that are activated by cytokinin are fused to GUS or GFP reporters. The results of this reporter-based approach suggest that cytokinin signaling begins early in root development in the hypophysis of the globular embryo. Upon division of the hypophysis, cytokinin expression is lost in the basal cell but is retained in the apical cell, which divides further to form the QC. At the same time, DR5-based reporters for auxin show an inverse pattern of expression, suggesting that auxin and cytokinin have opposing activities (**Figure 17.23**).

Further molecular and genetic analyses suggest that the loss of cytokinin activity in the basal cell causes changes in the organization of the RAM and is a direct consequence of high auxin activity. Two genes that suppress cytokinin responses, *ARR7* and *ARR15*, have *a*uxin *r*esponse *e*lements (AuxRE) in their promoters, suggesting that, like DR5 reporters, they are regulated by auxin. Artificial deletion of these elements lowers *ARR7* and *ARR15* expression in the basal cell, leading to ectopic cytokinin activity. Perturbing the expression of *ARR7* and *ARR15* results in abnormal phenotypes, suggesting that suppression of cytokinin signaling in the basal cell is essential for normal development. More recent work has reinforced the view that cytokinin-based signaling and its antagonism with auxin signaling and transport play a significant role in the RAM, enabling infrequent cell divisions that are occassionally observed in the QC.

The Shoot Apical Meristem

Like the root apical meristem, the shoot apical meristem is faced with the task of maintaining sets of undetermined cells that enable indeterminate growth (**Figure 17.24**). As previously discussed, however, there are significant differences between the two meristem types in how the descendants of those cells become incorporated into organs. Whereas lateral root initiation occurs well back of the root tip (see Chapter 18), leaves and associated axillary branches form in close proximity to apical initials in the

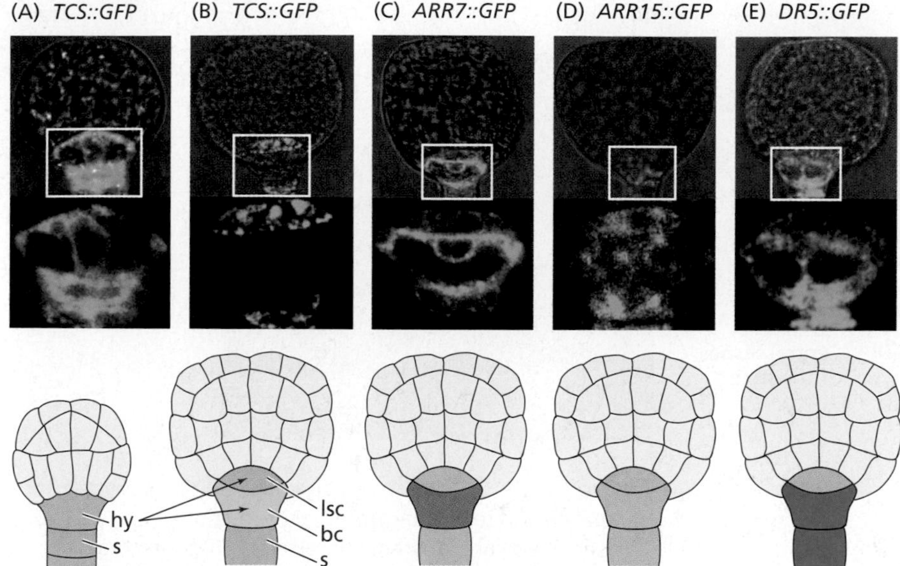

Figure 17.23 Inverse correlation between cytokinin and auxin signaling in the embryo. (A) Expression of *TCS::GFP* (a reporter for cytokinin) in the hypophysis at the early globular stage. (B) Down-regulation of *TCS::GFP* expression in the basal cell lineage at the late globular stage. (C) Expression of *ARR7::GFP* is highest in the basal cell lineage. (D) Expression pattern of *ARR15::GFP*. (*ARR7* and *ARR15* are genes that suppress cytokinin responses.) (E) Expression of *DR5::GFP* (an auxin-responsive reporter) is highest in the basal cell lineage. The boxed sections in the upper panels are magnified underneath; schematic interpretations are shown at the bottom. Abbreviations: hy, hypophysis; bc, basal cell; lsc, lens-shaped cell; s, suspensor. (From Müller and Sheen 2008.)

shoot. In place of the root cap that protects the apical initials of the root, young leaf primordia overlap and enclose the shoot tip.

Given the concentrated set of activities in the shoot tip, specific anatomical terminology has proved useful in their description. In this context, the term *shoot apical meristem* refers specifically to the initial cells and their undifferentiated derivatives, but excludes adjacent regions of the apex that contain cells that are fully committed to particular developmental fates. The more inclusive term **shoot apex** (plural *apices*) refers to the apical meristem plus the most recently formed leaf primordia.

As in the previously considered examples involving embryos and roots, the size, shape, and organization of the SAM vary according to a number of parameters, including plant species, developmental stage, and growth conditions. Cycads have the largest SAM among vascular plants, measuring over 3 mm in diameter; at the other extreme, the SAM of Arabidopsis is less than 50 μm in diameter and contains only a few dozen cells. Within a given species, significant variations in SAM size can also occur over time, and SAM shapes may range from flat to mounded. Some of these variations are associated with successive rounds of leaf initiation, in which groups of cells on the flanks of the SAM become committed to a determinate fate. Further variations may be related to seasonal differences in growth rate, including the onset of dormancy or flowering.

In the following sections we will first consider the basic organization of the SAM, discussing in detail the regional differences in cellular behavior that contribute to its function. We will then discuss evidence suggesting that, like the RAM, the SAM relies on localized differences in hormone and transcription factor activities for its formation and maintenance.

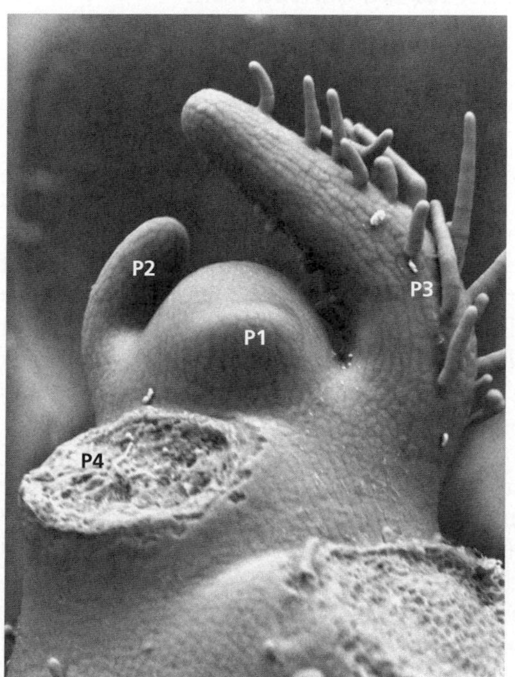

Figure 17.24 Shoot apex of a tomato plant. This SEM micrograph shows the basic features of the shoot apex, including a central dome-shaped region, which maintains uncommitted initials, and a series of leaf primordia (P1, P2, P3), which have successively emerged at lateral positions on the flanks of the shoot apex. P4 indicates the base of an older leaf primordium that was removed to expose the younger primordia. (From Kuhlemeier and Reinhardt 2001; courtesy of D. Reinhardt.)

(A)

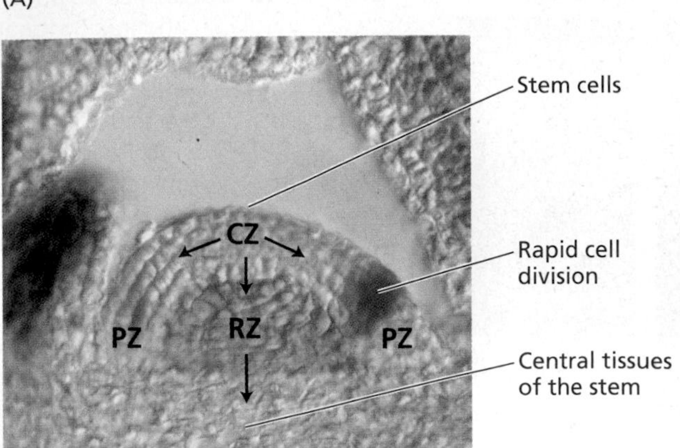

(B)

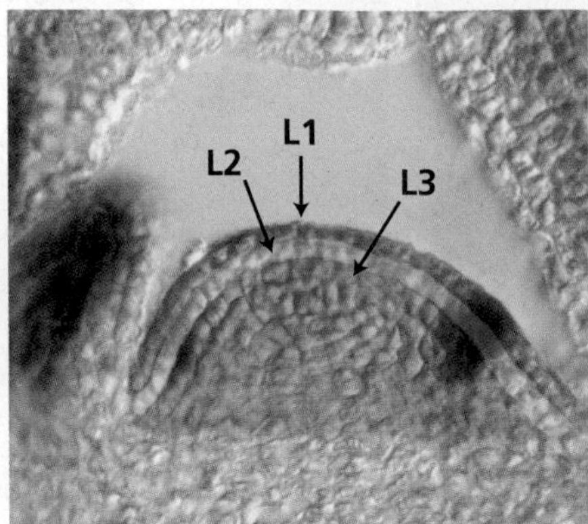

Figure 17.25 The Arabidopsis shoot apical meristem can be analyzed in terms of cytological zones or cell layers. (A) The shoot apical meristem has cytological zones that represent regions with different identities and functions. The central zone (CZ) contains meristematic cells that divide slowly but are the ultimate source of the tissues that make up the plant body. The peripheral zone (PZ), in which cells divide rapidly, surrounds the central zone and produces the leaf primordia. A rib zone (RZ) lies interior to the central zone and generates the central tissues of the stem. (B) The shoot apical meristem also has cell layers that contribute to specific tissues of the shoot. Most cell divisions are anticlinal in the outer, L1 and L2, layers; the planes of cell divisions are more randomly oriented in the L3 layer. The outermost (L1) layer generates the shoot epidermis; the L2 and L3 layers generate internal tissues. (From Bowman and Eshed 2000.)

The shoot apical meristem has distinct zones and layers

A discussion of the cellular organization of the SAM provides a useful framework for a more detailed description of its growth and development. Its organization is best appreciated by microscopic examination of shoot apices. Longitudinal sections of shoot apices reveal **zonation**, a term originally developed to describe regional cytological differences in the organization of the SAM of gymnosperms, but which has been extended to other seed plants to describe regional differences in cell division (**Figure 17.25**).

At the center of an active SAM lies the **central zone** (**CZ**), containing a cluster of infrequently dividing cells that can be compared to similar cells that make up the QC of roots. A flanking region, known as the **peripheral zone** (**PZ**), consists of cytoplasmically dense cells that divide more frequently to produce cells that later become incorporated into lateral organs such as leaves. A centrally positioned **rib zone** (**RZ**) more proximal to the CZ contains dividing cells that give rise to the internal tissues of the stem (see Figure 17.25A).

In addition to these regional differences in the frequency of division, distinct patterns in the polarity of cell division are also observed. In most angiosperm species these differences are reflected in the layered organization of superficial cells, which are sometimes collectively referred to as the **tunica**. One or more adjacent layers that make up the tunica are defined by a consistent pattern of anticlinal cell division, which has the effect of producing a tissue of uniform thickness that can be easily recognized in cross section. By contrast, the cells lying interior to the tunica, known as the **corpus**, display more variable division polarities, which lead to increases in tissue volume. In the next section we will discuss elegant methods to track patterns of cell division that provide insights into processes that maintain the characteristic organization of the SAM.

Shoot tissues are derived from several discrete sets of apical initials

Studies of cell lineage relationships indicate that, like root tissues, shoot tissues are derived from a small number of apical initials. In classic studies, the chemical colchicine was applied to shoot apices to induce the occasional formation of polyploid cells. These cells grow relatively normally but can be easily recognized by their increased nuclear volume and cell size (**Figure 17.26**). Examination of sectioned shoot apices of plants that had been treated and allowed to grow for a time revealed large sectors of polyploid cells that were confined to specific layers and which extended into apical regions. The size and shape of each sector could be explained by supposing that it originated from one of a small number of apical initials.

Analyses of a large number of such marked sectors, from both superficial layers and deeper tissues, indicate that several discrete sets of initials are typically maintained in the SAM. One set of superficial initials gives rise to a clonally distinct epidermal layer, termed **L1**, while more internal sets of initials give rise to the subepidermal **L2** layer and a centrally positioned **L3** layer (see Figures

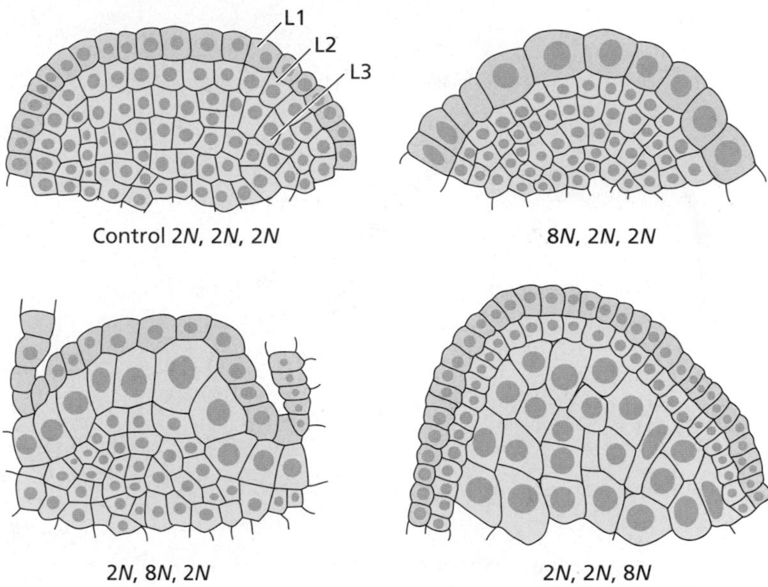

Figure 17.26 In shoot apices treated with colchicine, one of the cell layers contains enlarged, polyploid (8*N*) nuclei, demonstrating the presence of clonally distinct layers in the shoot apical meristem. (From Steeves and Sussex 1989.)

17.25B and 17.26). In many cases the marked sectors encompass only a portion of the shoot's circumference, suggesting that each layer derives from a small number of initials.

Analyses of cell lineages show that the identities of initial cells are determined by position-dependent mechanisms. Marked sectors that extend to apical regions of the shoot may exhibit abrupt changes in width or thickness over time. These changes can be explained by occasional divisions that lead to a marked initial displacing, or being displaced by, adjacent initials. This dynamic behavior indicates that the identities of apical initials, including their characteristic division patterns, reflect their relative position near the tip of the shoot apex, rather than a rigidly programmed identity. Similarly, the identities of cells derived from these initials also appear to be largely determined by position-dependent mechanisms. If a rare periclinal division leads to the derivative of an L2 cell adopting a superficial position, that cell will typically adopt an epidermal identity that reflects its new location.

Factors involved in auxin movement and responses influence SAM formation

The establishment of the SAM, like that of the RAM, is linked to complex patterns of intercellular auxin transport. During early stages of embryogenesis, the polar distribution of PIN proteins, especially PIN1, leads to the accumulation of auxin in apical regions, but by the early heart stage, a reversal in the distribution of PIN proteins leads to a basally directed redistribution of auxin. The factors that determine these changes are not completely understood, but changes in the phosphorylation state of PINs mediated by the kinase PINOID and the phosphatase PP2 can have significant effects on PIN localization (see Chapter 19). Additional, but less direct, inputs into PIN localization are suggested by the phenotypes that result from disrupting genes that encode several distinct classes of transcription factors, including members of the KANADI, DORNRÖSCHEN, and HD ZIP III families. The altered patterns of embryonic development associated with these mutants have been interpreted to be caused by changes in the distribution of PIN proteins, which precede any overt changes in growth or cell division.

One consequence of the complex pattern of auxin movement in the embryo is the formation of a central apical region where auxin-dependent activities are low relative to those in flanking regions (**Figure 17.27**). Transport of auxin away from this region converges with upward superficial flows along the flanks of the embryo to create auxin maximums at the tips of the developing cotyledons. These pools of auxin feed into downward flows that converge in the hypocotyl, then continue to form the previously discussed auxin maximum in the QC. ARFs such as MP and the closely related NONPHOTOTROPIC HYPOCOTYL 4 (NPH4) are activated by auxin to promote vascular development, further reinforcing this pattern of directional transport. Mutants that lack both MP and NPH4 are not only deficient in the basal structures, such as the root, but also lack cotyledons. The similarities of these phenotypes to those associated with mutations affecting PIN-mediated auxin transport are consistent with models in which MP and NPH4 enable auxin-dependent responses.

Embryonic SAM formation requires the coordinated expression of transcription factors

Although many types of genes are likely to be important in the formation and maintenance of the SAM, screens for mutants that block the formation of the SAM highlight the significance of three additional classes of transcription factors. One of these is encoded by *WUS* and is notable for belonging to the same family of homeodomain transcription factors that includes WOX5, which was previously described as playing an important role in the RAM. *WUS* is expressed in subapical regions as early as the 16-cell embryonic stage (**Figure 17.28**) and, as we will discuss in more detail, plays an important role in specifying and maintaining the identities of the apical initials of the SAM. Later, during the transition stage, transcription factors of the NAC class, encoded by *CUP-SHAPED COTYLEDON* (*CUC*) *1* and *2*, are expressed in an apically positioned

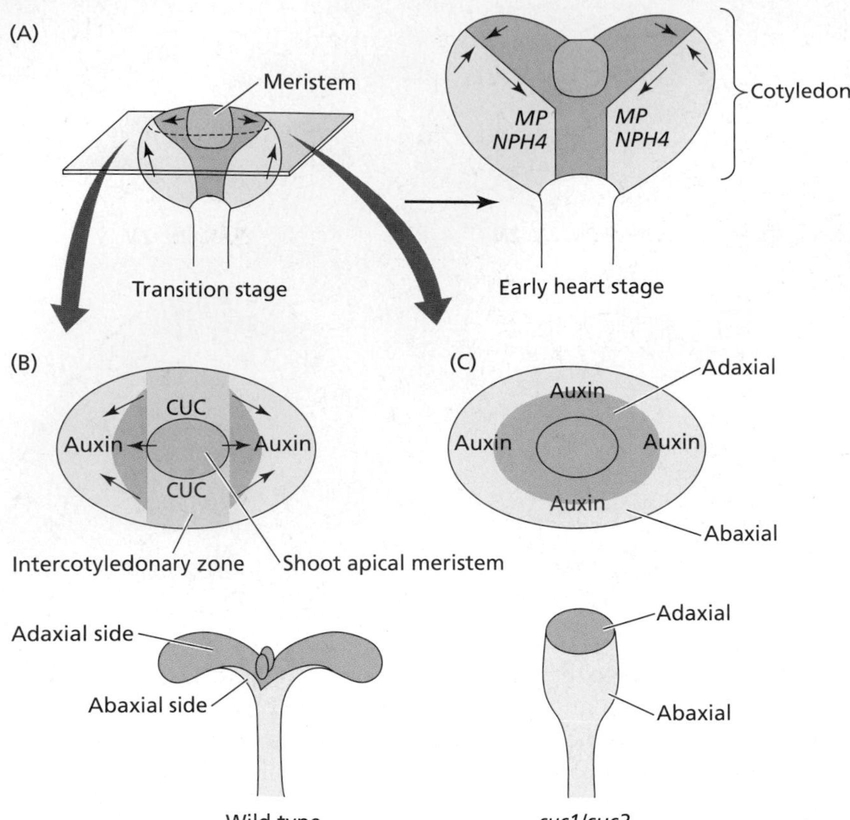

Figure 17.27 Model for auxin-dependent patterning of the shoot apex. (A) The direction of auxin transport (black arrows) during transition stage and early heart stage Arabidopsis embryos. (B and C) Cross sections (as shown in A) through the apical region of a wild-type embryo (B) and a *CUP-SHAPED COTYLEDON* (*CUC*) gene double mutant (*cuc1/cuc2*) (C), showing the region in the embryo that will develop into the shoot apical meristem, the intercotyledonary zones, and the adaxial and abaxial domains of the cotyledon. In the wild-type embryo, the SAM and intercotyledonary zones have low auxin levels and, consequently, high CUC levels, whereas the opposite pattern is seen in the flanking cotyledon primordia. In the *cuc1/cuc2* mutant, the cotyledons fail to separate, thus preventing the formation of a shoot apical meristem. (After Jenik and Barton 2005.)

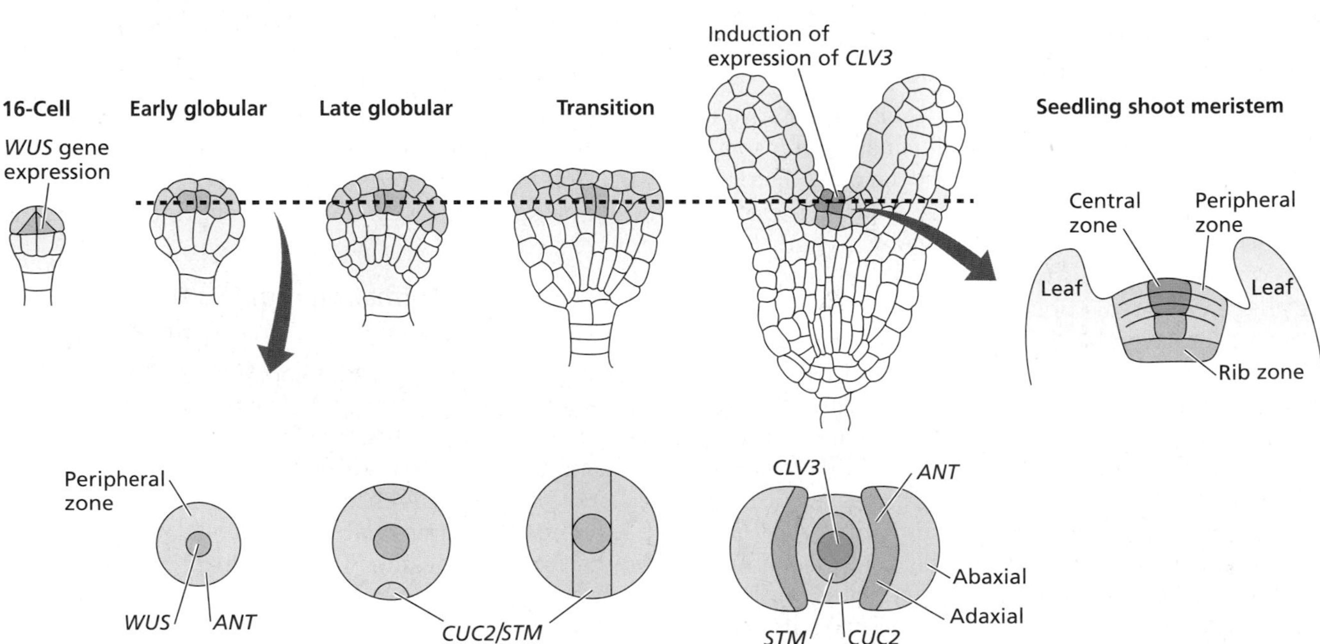

Figure 17.28 Formation of the apical region involves a defined sequence of gene expression. The top row illustrates the early onset of *WUS* expression in an internal layer, which induces the expression of *CLAVATA3* (*CLV3*) in adjacent external cell layers. Expression of *AINTEGUMENTA* (*ANT*) provides a marker for leaf or cotyledon identity. The bottom row shows cross sections at the level indicated by the dashed line and emphasizes the gene expression patterns that demarcate the emerging cotyledonary and shoot apical regions. (After Laux et al. 2004.)

stripe between the two developing cotyledons (also see Figure 17.27B). Finally, during the heart stage, this sequence of gene activation concludes as another class of homeodomain transcription factors, encoded by *SHOOT MERISTEMLESS* (*STM*), becomes expressed in a circular domain contained within the *CUC* expression domain. Together, *WUS* and *STM* appear to help maintain cells in a state in which they can proliferate and thus ensure that the growth and differentiation of shoot tissues are balanced with the production of new undetermined cells.

The localized expression of *CUC* genes and the subsequent appearance of STM appear to reflect the relatively low level of auxin-dependent activities in central apical regions compared with that in flanking tissues (see Figure 17.27B and C). For example, blocking auxin signaling in the flanking cotyledonary regions, with mutations of *MP* and *NPH4*, leads to ectopic expression of *CUC* genes in these regions. A role for auxin signaling is further supported by the observation that normal embryos treated with auxin transport inhibitors (see Figure 17.13) exhibit defects (cup-shaped cotyledons) similar to those observed among *cuc* mutants. The expression of *CUC*-like genes in the central apical region of the embryo provides an environment that enables further patterning processes, including the localized expression of the *STM* gene, which initially coincides with the stripelike CUC expression domain but later becomes focused in a central circular domain. This expression depends on *CUC* gene activities, since *STM* expression does not occur in *cuc* mutant embryos. The final stages of SAM establishment sees the expression of *CLAVATA3* (*CLV3*), which as we will discuss shortly, plays a key role in limiting the number of cells that function as apical initials.

A combination of positive and negative interactions determines apical meristem size

Given the ongoing recruitment of cells into various tissues and organs of the shoot, we would expect a finely tuned mechanism to adjust the rate at which new cells are produced, in order to maintain a consistently sized SAM. The activity of the *WUS* gene appears essential for maintaining the identity of apical initials. Loss of *WUS* activity in mutants leads to differentiation of apical initials in the central zone, blocking their ability to divide and replace cells that are recruited into differentiating tissues in the peripheral zone.

To learn more about how WUS might alter transcription to maintain the identity of apical initials, researchers transiently overexpressed *WUS* using an inducible system, then monitored global changes in transcript levels using microarrays. Transcript levels for several type-A *ARR* cytokinin response regulators, which act to repress cytokinin responses, decreased significantly when *WUS* was overexpressed. Further experiments using qRT-PCR to monitor *ARR* gene expression showed that these genes were repressed within 4 hours and that the same repression occurred even when protein synthesis was inhibited by cycloheximide. These results suggest that WUS represses *ARR* genes directly, rather than relying on the synthesis of an intermediate transcription factor. The direct interaction of WUS with the promoters of *ARR* genes is also supported by experiments in which complexes between the WUS protein and *ARR7* promoter sequences were detected with antibodies to WUS. Complementary genetic approaches, in which artificial overexpression of *ARR7* resulted in a *wus*-like phenotype, provided further evidence for a role of cytokinin in maintaining the SAM (**Figure 17.29**).

At the same time that WUS is acting to promote the activity of apical initials, a distinct set of genes known as *CLAVATAs* (from the Latin for "club-shaped") acts in an opposing way to limit apical initial activity. Three distinct *CLV* genes, *CLV1*, *CLV2*, and *CLV3*, were first described in Arabidopsis in terms of their mutant phenotypes, in which the SAM becomes grossly enlarged. The enlarged meristem phenotypes shared by mutants suggest that the proteins encoded by these genes act interdependently. Molecular and biochemical analyses support this idea,

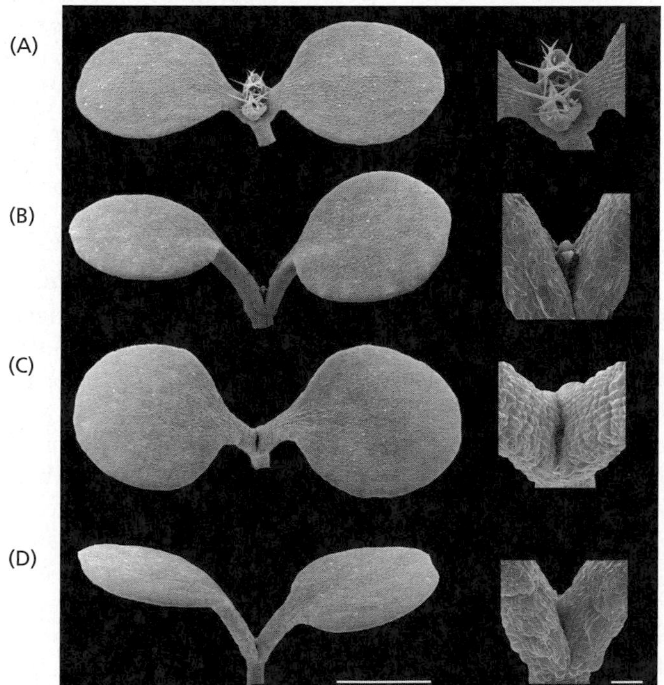

Figure 17.29 Overexpression of a type-A ARR cytokinin response regulator (*ARR7*) in Arabidopsis phenocopies *wus*. (A) A weakly expressing line has a wild-type morphology. (B) A more highly expressing line has an intermediate phenotype. (C) A strongly expressing line has a phenotype very similar to that of *wus*. (D) A *wus* mutant seedling. Scale bars = 1 mm for seedlings and 100 μm for meristem insets. (From Leibfried et al. 2005.)

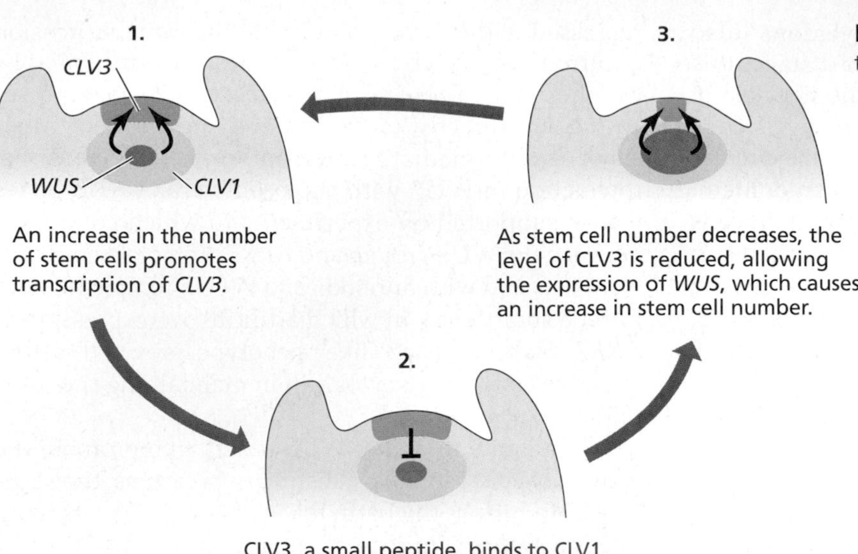

1.
CLV3
WUS
CLV1
An increase in the number of stem cells promotes transcription of CLV3.
2.
CLV3, a small peptide, binds to CLV1 and suppresses the expression of WUS. WUS is required for the maintenance of stem cell number.
3.
As stem cell number decreases, the level of CLV3 is reduced, allowing the expression of WUS, which causes an increase in stem cell number.

Figure 17.30 Model of the feedback loop that maintains initial cells in the SAM.

providing evidence that the CLV proteins physically interact with each other to function as a protein kinase signaling relay, whose outputs act to limit meristem size. *CLV1* encodes a leucine-rich repeat receptor kinase (LRRK), a type of transmembrane protein whose intracellular kinase activity is activated by binding of specific ligands to the extracellular leucine rich receptor domain. (See Chapter 15 for further information about receptor kinase signaling pathways.) *CLV2* is very similar to *CLV1* but lacks the intracellular kinase domain, and thus appears to depend on interaction with other intracellular proteins for its signaling outputs.

A key aspect of the meristem-limiting signaling outputs of CLV1 and CLV2 is the requirement that these proteins bind a small 11-kD peptide encoded by *CLV3*, which normally appears to be the limiting factor in determining signaling output levels. By virtue of this dependency, meristem size can be efficiently controlled by regulating the amount of CLV3 protein, which then activates a CLV1- and CLV2-dependent signaling cascade to somehow repress meristem growth. Not surprisingly, the meristem-promoting gene *WUS* has been implicated by genetic evidence to be a major target of *CLV*-mediated repression. For example, in *clv* mutants, transcription of *WUS* is increased, which leads to the increase in the size of the SAM. Conversely, when *CLV3* is overexpressed, *WUS* transcription is repressed, leading to a phenocopy of the *wus* mutant phenotype in which the meristem is lost.

It seems plausible that a variety of physiological and growth parameters could influence meristem size via control of CLV3 levels. One instructive example of such control is seen in the regulation of *CLV3* by *WUS*, whose expression promotes the transcription of the *CLV3* gene. Although this interaction might seem somewhat counterintuitive, given that *WUS* itself is a target of *CLV*-mediated repression, the combination of the activation of *CLV* signaling by *WUS*, alongside inhibition of *WUS* by *CLV* signaling, provides a mechanism to stabilize WUS levels, and hence meristem size (**Figure 17.30**). By this model, increases in WUS levels would promote expression of genes that promote apical initial identity in cells that lie within the central zone. At the same time, increased WUS levels would also activate transcription of *CLV3*, thereby increasing levels of the CLV3 peptide and leading to *CLV*-mediated repression of WUS. This example in which *WUS* activity is self-limiting via *CLV*-mediated feedback inhibition provides one of the classic examples of homeostatic regulation in plant development.

KNOX class homeodomain genes help maintain the proliferative ability of the SAM through regulation of cytokinin and GA levels

Although maintenance of stem cells in the SAM and RAM may rely on similar mechanisms, both involving WOX family transcription factors (WUS and WOX5), certain aspects of these mechanisms in the SAM are unique. For example, the production of lateral organs in close proximity to the more pluripotent cells of the meristem appears to require an additional level of regulation, which is mediated by members of another family of homeodomain transcription factors. The first functional analyses of these genes focused on how their ectopic expression promoted the formation of contorted, knotlike protuberances on leaves. It soon became clear, however, that this class of genes, termed *KNOX* after the original maize mutant

Figure 17.31 Model for how expression of the KNOX transcription factor STM elevates cytokinin levels while repressing GA in the SAM. P4 is a developing leaf, and P0 is the site where the next leaf primordium will form. (After Hudson 2005.)

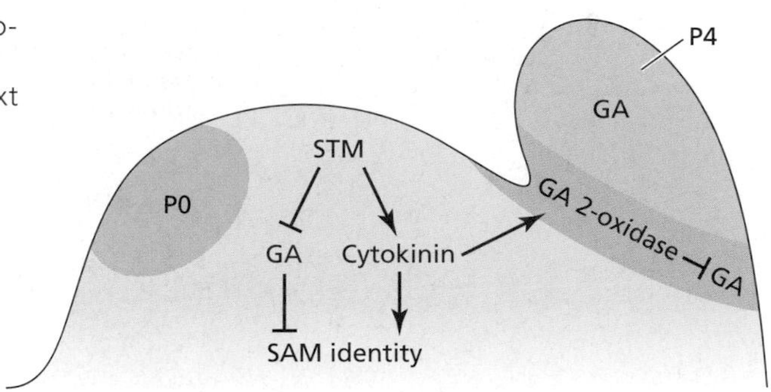

KNOTTED1 and the homeobox protein it encodes, has prominent roles in the maintenance of the SAM. A typical example involves the previously discussed Arabidopsis *STM* gene. This *KNOX* gene is expressed throughout almost the entire body of the meristem, but not in the groups of cells in flanking positions that are committed to become leaf primordia (P0 in **Figure 17.31**). In mutants where *STM* expression is deficient, the SAM either fails to form or is not maintained during vegetative growth.

Insights into how KNOX proteins function have been gained in experiments that show that the meristem instability associated with loss of *STM* activity can be rescued by exogenous application of cytokinin. Furthermore, *STM* expression in the SAM activates the transcription of genes encoding isopentenyl transferases that are involved in the biosynthesis of cytokinin. These observations support the idea that in the *stm* mutants, exogenously applied cytokinin rescues the SAM by compensating for reduced biosynthesis of the hormone and that cytokinin acts to stabilize meristems (see Figure 17.31; also see Figure 17.29).

A second key function of *KNOX* class genes is to suppress the accumulation of GA in the SAM. In a variety of species, *KNOX* genes have been shown to directly repress transcription of *GA 20-OXIDASE1*, which encodes an enzyme for a rate-limiting step in biosynthesis of the active form of GA (see Figure 17.31). KNOX proteins also suppress GA activity in the meristem indirectly via cytokinin, which stimulates expression of *GA 2-OXIDASE* at the boundaries between emerging leaves and the SAM. *GA 2-OXIDASE* encodes an enzyme that breaks down biologically active GA, and this additional mechanism is thought to prevent movement of active GA into the SAM from nearby developing leaves (e.g., P4 in Figure 17.31). Genetic experiments have shown that artificial activation of GA signaling in the SAM destabilizes the meristem, demonstrating that the restriction of GA levels in the SAM is likely to be a key mechanism by which *KNOX* genes contribute to meristem stability.

Localized zones of auxin accumulation promote leaf initiation

A long-standing question in plant biology is how the characteristic arrangement of leaves on the shoot, or **phyllotaxy**, is achieved. Three basic phyllotactic patterns, termed alternate, decussate (opposite), and spiral, can be directly linked to the pattern of initiation of leaf primordia on the shoot apical meristem (**Figure 17.32**). These patterns depend on a number of factors, including intrinsic factors that tend to produce a phyllotaxy that is characteristic of a species. However, environmental factors or mutations (e.g., the maize *abphyll* mutant, or the Arabidopsis *clavata* mutants) that lead to changes in meristem size or shape can also affect phyllotaxy, suggesting that position-dependent mechanisms play important roles. Position dependency is also supported by classic experiments in which surgical cuts to the shoot apex were shown to perturb the positioning of nearby leaf primordia.

Studies involving experimental manipulation of the shoot apex in Arabidopsis and tomato have shown that auxin can influence the positions of leaves. For example, leaf primordia can be induced to form at abnormal positions on the shoot apex by applying small quantities of auxin directly to the shoot apical meristem, suggesting that auxin is a key factor in determining the position of leaf initiation (**Figure 17.33**). Further support for this hypothesis is provided by the changes in patterns of leaf initiation that result from experimental applications of auxin transport inhibitors.

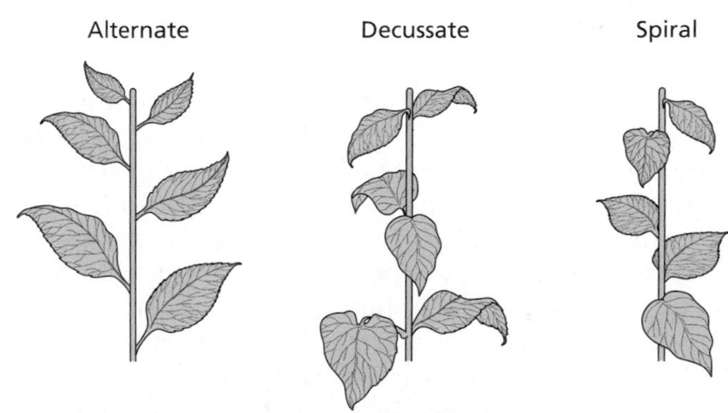

Figure 17.32 Three types of leaf arrangements (phyllotactic patterns) along the shoot axis. The same terms are also used for inflorescences and flowers.

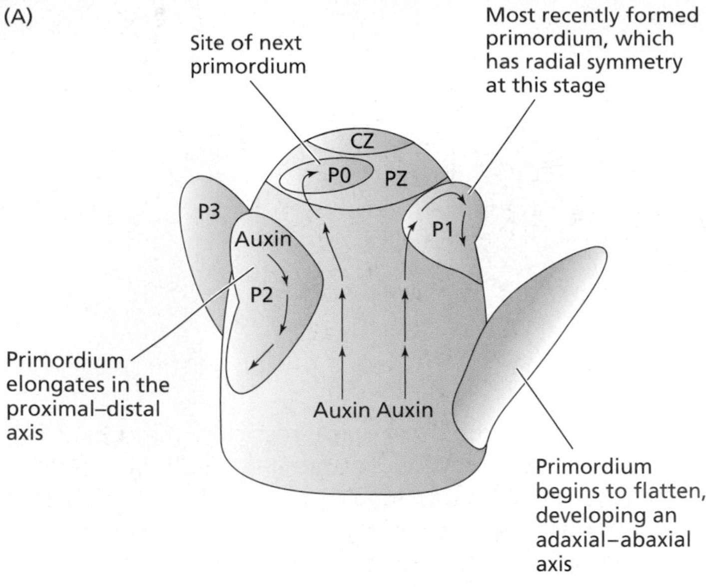

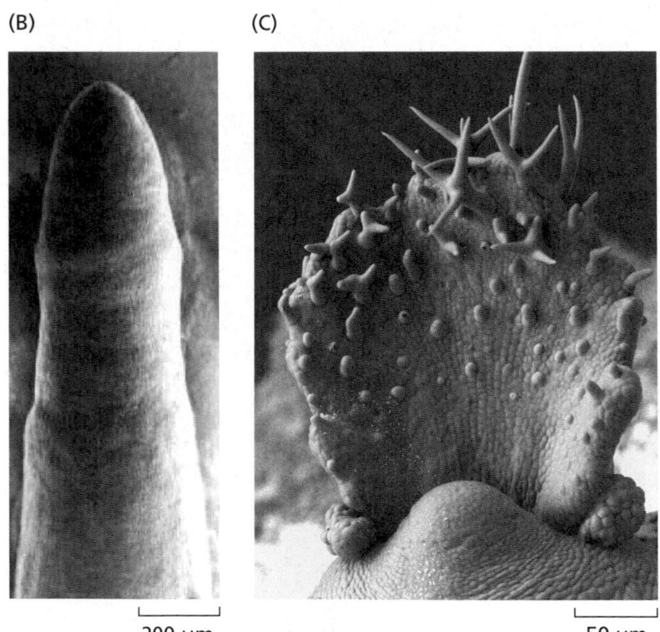

Figure 17.33 (A) Sites of leaf formation are related to patterns of polar auxin transport. Patterns of auxin movement (arrows) can be inferred from asymmetric localization of PIN proteins. P0, P1, P2, and P3 refer to the ages of leaf primordia; P0 corresponds to the stage at which the leaf begins its overt development, and P1, P2, and P3 represent increasingly older leaves. Leaf primordia are initiated where auxin accumulates. Acropetal (toward tip) movement of auxin is blocked at the boundary separating the central and peripheral zones (CZ and PZ, respectively), leading to increased auxin levels at this position and the initiation of a leaf (P0). The newly formed leaf primordium (P1) acts as an auxin sink, thus preventing initiation of new leaves directly above it. The displacement of a more mature leaf (P2) away from the PZ allows acropetal auxin movements to become reestablished, thus enabling the initiation of another leaf. (B) Scanning electron micrograph of a *pin1* inflorescence meristem that fails to produce leaf primordia. See Figure 17.12A for a photo of a *pin1* mutant plant. (C) Leaf primordium induced on the inflorescence meristem of a *pin1* mutant by placing a microdrop of IAA in lanolin paste on the side of the meristem. (A after Reinhardt et al. 2003; B from Vernoux et al. 2000; C from Reinhart et al. 2003.)

Several complementary approaches have now provided compelling evidence that sites of leaf initiation correspond to localized zones of auxin accumulation. Although it is difficult to measure auxin levels directly in such small regions, localized concentration maximums can be inferred from DR5 reporters, whose activity shows a close correspondence to leaf initiation sites. The formation of these maximums can be explained by the asymmetric distribution of PIN proteins in cells, which would mediate the convergence of superficial auxin flows from basal parts of the shoot with downward and lateral flows from the shoot apex (see Figure 17.33A). The developmental processes that regulate leaf growth are described in detail in Chapter 19.

The Vascular Cambium

In contrast to the terminally positioned SAM and RAM, the vascular cambium presents a very distinct meristem organization that extends along nearly the entire length of the apical–basal axis of the plant, and which functions to produce vascular tissues along the radial axis. Although the precise organization of the cambium varies considerably among seed plants, a consistent feature is seen in the positioning of one or more layers of cells that act as initials for vascular tissues. Typically, derivatives of initials that are displaced inward develop as xylem, while those that are displaced outward develop as phloem. By maintaining these initials, the cambium provides a means to increase the capacity of the stem or root for vascular transport as the plant grows.

The maintenance of undetermined initials in various meristem types depends on similar mechanisms

Despite some significant distinctions in the organization of cambium compared with the SAM and RAM, certain common aspects, such as the need to maintain sets of stable initials, suggest that these different meristem types could rely on similar mechanisms. Several lines of analysis have been used to explore this idea. On a descriptive level, the cambium shows elevated expression of genes that are known to be important in other meristem types, including

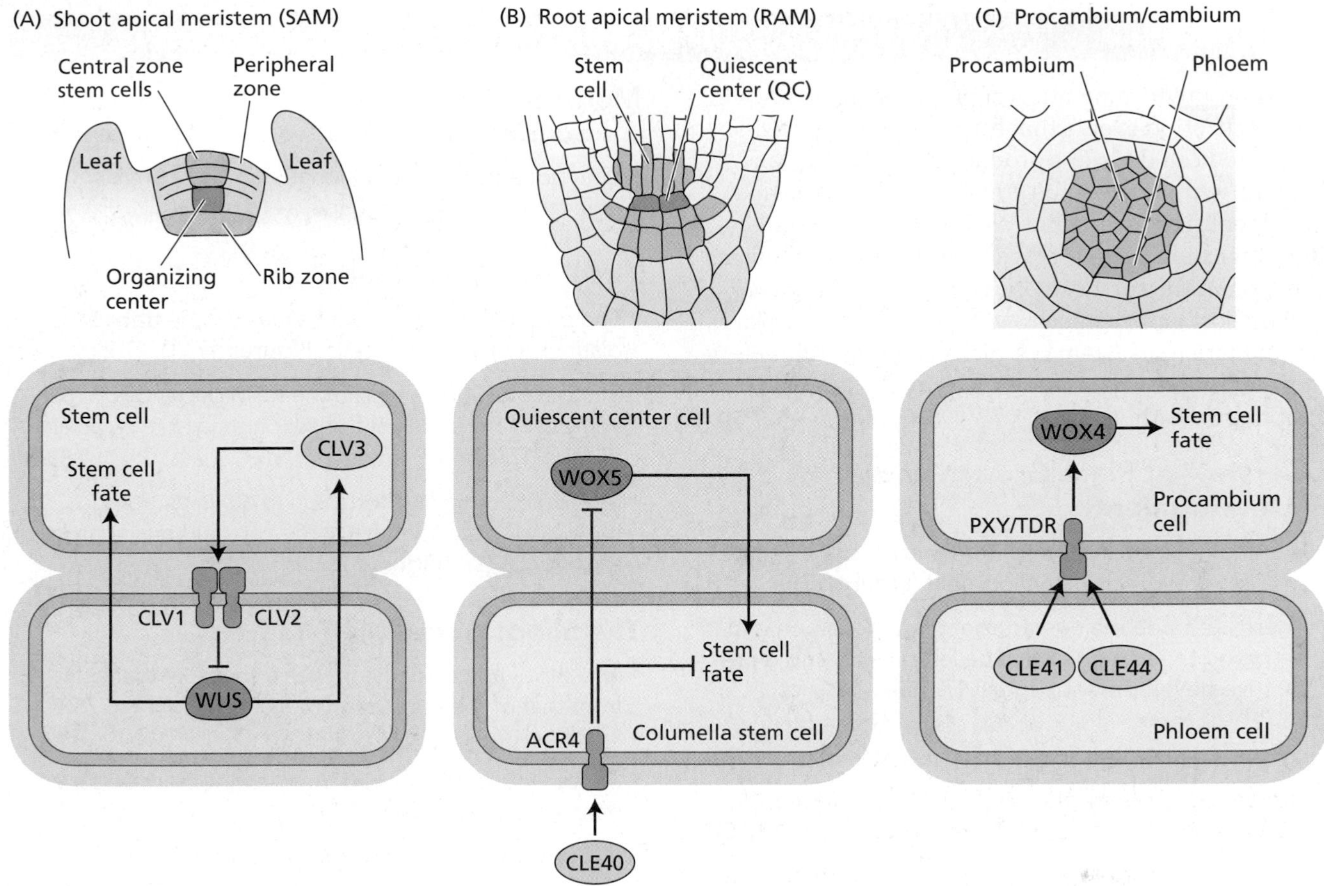

Figure 17.34 Comparison of three distinct patterning processes that exploit regulatory interactions between small peptides and WOX class transcription factors. (A) WUS promotion of apical initials in the SAM. (B) WOX5 promotion of initials in the RAM. (C) WOX4 promotion of initials in the vascular cambium. (After Miyashima et al. 2013.)

KNOX and *WOX* genes. In recent years, genetic analyses have provided examples where the expression of these genes is seen to contribute to cambium function, providing further detail on the functional relationship between meristem types.

An instructive comparison that highlights the similar mechanisms between meristem types is seen in the roles played by WOX class transcription factors. As noted in our earlier discussion, the activities of initials in both the SAM and the RAM depend on the activity of WOX class transcription factors: *WUS* promotes the activity of initials in the SAM (**Figure 17.34A**), while *WOX5* contributes to the function of initials in the RAM (**Figure 17.34B**). Remarkably, experiments involving artificial expression of *WOX5* and *WUS* have demonstrated that these genes are functionally interchangeable if expressed in the appropriate tissue. Further parallels are seen in the manner in which the activities of *WOX5* and *WUS* are suppressed by small related peptides, CLE40 and CLV3, respectively. Moreover, in both cases, the small peptides appear to repress *WOX* activity via their interaction with LRRKs.

More recent work has suggested that the vascular cambium relies on a similar mechanism, where the maintenance of initials depends on the activity of *WOX4*, which like *WUS* and *WOX5*, is regulated via the interaction of small peptides with an LRRK, in this case PHLOEM INTERCALATED WITH XYLEM(PXY)/TDIF RECEPTOR (TDR) (**Figure 17.34C**). However, unlike in the apical meristems, in which the expression of the small peptides leads to reduced transcription of *WOX5* and *WUS*, thereby limiting apical initial activity, the expression of CLE41 and CLE44 peptides instead promotes *WOX4* activity. Despite this difference, the common theme of peptide-mediated regulation of transcription factors does suggest a form of control that might be especially well suited to maintaining initial cells in an undetermined state, and which can operate over relatively short distances. Further analyses should help clarify whether the functionally related aspects of these regulatory modules reflect their derivation from a common ancestral mechanism versus convergent evolution that has exploited elements well suited to maintaining cells in an indeterminate state.

SUMMARY

The sporophyte generation of plants begins with the fertilization events that initiate embryogenesis. Regulated cell divisions produce the polar axis and bilateral symmetry of the embryo. Both mobile and positional signals function as morphogenic regulators. An extended set of these regulatory mechanisms function in the further elaboration of plant organs during post embryonic growth. Postembryonic plants retain meristems (stem cell niches) that are sites of undifferentiated cell division to provide for plastic or adaptive growth.

Overview of Plant Growth and Development

- Meristematic cells are undetermined and are central to plant growth and development (**Figure 17.1**).
- There are three major stages of plant development: embryogenesis, vegetative development, and reproductive development (**Figure 17.2**).

Embryogenesis: The Origins of Polarity

- Among seed plants, the apical–basal polarity is established early in embryogenesis (**Figures 17.3–17.5**).
- Position-dependent mechanisms for determining cell fate guide embryogenesis (**Figure 17.7**). Arabidopsis mutants demonstrate that some process other than a fixed sequence of cell divisions must guide radial pattern formation (**Figure 17.8**).
- The potential for intercellular protein movement changes during development (**Figure 17.9**).
- Screens for seedling defective mutants reveal genes and that are essential for normal Arabidopsis embryogenesis (**Figure 17.10**).
- Auxin (indole-3-acetic acid) may function as a mobile chemical signal during embryogenesis (**Figures 17.11–17.14; Table 17.1**).
- Radial patterning guides formation of tissue layers (**Figure 17.15**).
- Two Arabidopsis genes establish normal epidermal identity (**Figure 17.16**).
- Different genes establish internal tissues, including the vasculature and cortex (**Figures 17.17–17.19**).

Meristematic Tissues: Foundations for Indeterminate Growth

- The root and shoot apical meristems use similar strategies to enable indeterminate growth.

The Root Apical Meristem

- The origin of different root tissues can be traced to distinct types of initial cells (**Figures 17.20, 17.21**).
- The behaviour of initials in the RAM depends on the activation of a series of transcription factors by auxin (**Figure 17.22**).
- Cytokinin acting in opposition to auxin establishes the apical–basal identity of the two cells arising from the hypophysis (**Figure 17.23**).

The Shoot Apical Meristem

- The shoot apical meristem has a structure distinct from that of the root apical meristem (**Figures 17.24, 17.25**).
- Shoot tissues are derived from several discrete sets of apical initials (**Figure 17.26**).
- PIN proteins determine auxin levels across the SAM, leading to auxin flows away from initials and triggering formation of leaf primordia (**Figure 17.27**).
- Embryonic SAM formation requires the coordinated expression of specific transcription factors to establish a set of undetermined cells with potential for further proliferation (**Figure 17.28**).
- The WUS transcription factor maintains the identity of apical initials in the SAM by up-regulating cytokinin signaling in the SAM (**Figure 17.29**).
- WUS activity is self-limiting via CLV-mediated feedback (**Figure 17.30**).
- Expression of KNOX transcription factors promotes cytokinin production in the SAM, while limiting GA levels (**Figure 17.31**).
- Phyllotactic patterns are directly linked to the pattern of leaf formation (**Figure 17.32**).
- Leaf initiation sites are determined at sites of localized auxin accumulation (**Figure 17.33**).

The Vascular Cambium

- WOX transcription factors and small peptide signaling modules are used in several contexts to promote stem cell identity (**Figure 17.34**).

WEB MATERIAL

- **WEB TOPIC 17.1 Embryonic Dormancy** The ability of seeds to lie dormant for long periods and then germinate under favorable conditions reflects the activity of complex physiological programs.
- **WEB TOPIC 17.2 Rice Embryogenesis** Embryogenesis in rice is typical of that found in most monocots, and is distinct from that of Arabidopsis.
- **WEB TOPIC 17.3 Polarity of *Fucus* Zygotes** A wide variety of external gradients can polarize the growth of cells that are initially apolar.
- **WEB TOPIC 17.4 *Azolla* Root Development** Anatomical studies of the root of the aquatic fern *Azolla* have provided insights into cell fate during root development.

- **WEB ESSAY 17.1 Division Plane Determination in Plant Cells** Plant cells appear to utilize mechanisms different from those used by other eukaryotes to control their division planes.
- **WEB ESSAY 17.2 Plant Meristems: A Historical Overview** Scientists have used many approaches to unraveling the secrets of plant meristems.

available at plantphys.net

Suggested Reading

Aichinger, E., Kornet, N., Friedrich, T. and Laux, T. (2012) Plant stem cell niches. *Annu. Rev. Plant Biol.* 63: 615–636.

Aloni, R. (1995) The induction of vascular tissue by auxin and cytokinin. In *Plant Hormones and their Role in Plant growth Development*, 2nd ed., P. J. Davies ed., Kluwer, Dordrecht, Netherlands, pp. 531–546.

Barlow, P. W. (1994) Evolution of structural initial cells in apical meristems of plants. *J. Theor. Biol.* 169: 163–177.

Esau, K. (1965) *Plant Anatomy*, 2nd ed. Wiley, New York.

Hudson, A. (2005) Plant meristems: Mobile mediators of cell fate. *Curr. Biol.* 15: R803–R805.

Jenik, P. D., and Barton, M. K. (2005) Surge and destroy: The role of auxin in plant embryogenesis. *Development* 132: 3577–3585.

Laux, T., Wurschum, T., and Breuninger, H. (2004) Genetic regulation of embryonic pattern formation. *Plant Cell* 16 (Suppl): S190–S202.

Maule, A. J., Benitez-Alfonso, Y., and Faulkner, C. (2011) Plasmodesmata - Membrane tunnels with attitude. *Curr. Opin. Plant Biol.* 14: 683–690.

Meyerowitz, E. M. (1996) Plant development: Local control, global patterning. *Curr. Opin. Genet. Dev.* 6: 475–479.

Miyashima, S., Sebastian, J., Lee, J.-Y. and Helariutta, Y. (2013) Stem cell function during plant vascular development. *EMBO J.* 32: 178–193.

Poethig, R. S. (1997) Leaf morphogenesis in flowering plants. *Plant Cell* 9: 1077–1087.

Raghavan, V. (1986) *Embryogenesis in Angiosperms.* Cambridge University Press, Cambridge, UK.

Reinhardt, D., Pesce, E. R., Stieger, P., Mandel, T., Baltensperger, K., Bennett, M., Traas, J., Friml, J. and Kuhlemeier, C. (2003) Regulation of phyllotaxis by polar auxin transport. *Nature* 426: 255–260.

Sachs, T. (1991) Cell polarity and tissue patterning in plants. *Development* 113 (Supplement 1): 83–93.

Scheres, B., Wolkenfelt, H., Willemsen, V., Terlouw, M., Lawson, E., Dean, C., and Weisbeek, P. (1994) Embryonic origin of the *Arabidopsis* primary root and root meristem initials. *Development* 120: 2475–2487.

Scheres, B. (2013) Rooting plant development. *Development,* 140: 939–941.

Silk, W. K. (1984) Quantitative descriptions of development. *Ann. Rev. Plant Physiol.* 35: 479–518.

Sparks, E., Wachsman, G. and Benfey, P. N. (2013) Spatiotemporal signalling in plant development. *Nat. Rev. Genet.* 14: 631–644.

Steeves, T. A., and Sussex, I. M. (1989) *Patterns in Plant Development.* Cambridge University Press, Cambridge, UK.

18 Seed Dormancy, Germination, and Seedling Establishment

"It's not dead, it's resting."
–Monty Python

In Chapter 17 we discussed the early stages of embryogenesis that occur in the developing seeds of angiosperms. Seeds are specialized dispersal units unique to the Spermatophyta, or seed plants. In both angiosperms and gymnosperms, seeds develop from ovules, which contain the female gametophyte, which we will discuss in Chapter 21. After fertilization, the resulting zygote develops into the embryo. The packaging of the embryo into a self-contained seed was one of many adaptations that freed plant reproduction from a dependence on water. The evolution of seed plants thus represents an important milestone in the adaptation of plants to dry land.

In this chapter we will continue our discussion of the developmental sequence by describing the processes of seed germination and seedling establishment—by which we mean the production of the first photosynthetic leaves and a minimal root system. Between embryogenesis and germination there is typically a period of *seed maturation* followed by *quiescence*, during which seed dissemination occurs. Germination is thus delayed until the water, oxygen, and temperature conditions are favorable for seedling growth. Some seeds require an additional treatment, such as light or physical abrasion, before they can germinate, a condition known as *dormancy*.

In addition to serving as a protective barrier during embryogenesis, seeds also provide nourishment during both embryogenesis and early seedling development. The food reserves of seeds are stored in several types of tissues. Because the process of germination is tightly coupled to the mobilization of stored food reserves, we will begin with a description of seed structure and composition. Next we will consider various types of seed dormancy, which, in some cases, must be overcome for germination to take place. We will then discuss the mobilization of stored food reserves in different types of seeds, and we will explore the role of hormones in coordinating the processes of seedling growth and food mobilization.

During seedling establishment, plant responses to light, gravity, and touch help orient plant roots and shoots within their respective environments. In addition, vascular tissue differentiation provides the crucial link between root and shoot for the movement of water and minerals. Finally, we will describe the process of root branching, a critical stage in seedling establishment.

Seed Structure

In this chapter we will focus on the seeds of angiosperms because of their extraordinary diversity and importance to agriculture, but it is important to appreciate the basic differences between angiosperms and gymnosperms. Some of the major anatomical changes associated with the evolution of seeds are discussed in WEB TOPIC 18.1. All seeds are surrounded by a protective outer layer of dead cells called the **testa**, or **seed coat**. However, the testa may sometimes be fused to the **pericarp,** or fruit wall, derived from the ovary wall. In this case the "seed" is actually a fruit. **Table 18.1** lists some familiar examples of true seeds versus fruits that resemble seeds.

Seed anatomy varies widely among different plant groups

The angiosperm embryo is a relatively simple structure consisting of the embryonic axis and one or two cotyledons. The axis is composed of the **radicle**, or embryonic root, the **hypocotyl**, to which the cotyledons are attached, and the **shoot apex** bearing the **plumule**, or first true leaf primordia. Despite the relative simplicity of the embryo, and the limited number of tissues surrounding it, seed anatomy exhibits considerable diversity among the different plant groups. Seeds come in all shapes and sizes, ranging from the smallest orchid seed, weighing a microgram (10^{-6} g), to the huge seed of the double coconut palm, tipping the scales at 30 kg (66 lbs)!

Some representative examples of seeds from eudicots and monocots are shown in **Figure 18.1**. Seeds can be categorized broadly as endospermic or non-endospermic, depending on the presence or absence of a well-formed triploid endosperm at maturity. For example, beet seeds are non-endospermic because the triploid endosperm is largely used up during embryo development. Instead, the perisperm and storage cotyledons serve as the main sources of nutrients during germination (see Figure 18.1). The **perisperm** is derived from the **nucellus**, the maternal tissue that gives rise to the ovule (see Chapter 21). Garden bean seeds (*Phaseolus vulgaris*) and legume seeds in general are also non-endospermic, relying on their large storage cotyledons, which make up most of the bulk of the seed, for their food reserves. In contrast, Castor bean (*Ricinus communis*), onion (*Allium cepa*), wheat (*Triticum* spp.), and maize (corn; *Zea mays*) seeds are all endospermic.

In keeping with its role as a food storage tissue, the endosperm is typically rich in starch, oils, and protein. Some endosperm tissue has thick cell walls that break down during germination, releasing a variety of sugars. The outermost layer of the endosperm in some species differentiates into a specialized secretory tissue with thickened primary walls called the *aleurone layer,* so called because it is composed of cells filled with **protein storage vacuoles**, originally called *aleurone grains.* As we will see later in the chapter, the aleurone layer plays an important role in regulating dormancy in certain eudicot seeds. In wheat seeds and those of other members of the Poaceae (grass family), secretory aleurone layers are also responsible for the mobilization of stored food reserves during germination.

The embryos of cereal grains are highly specialized and merit closer examination both because of their agricultural importance and because they have been widely used as model systems to study the hormonal regulation of food mobilization during germination. Specialized

Table 18.1 Seeds or fruits?

Seed	Fruit (and type)
Brassica species (e.g., rapeseed, mustard, cabbage)	Ash, maple, elm (samara)
Brazil nut	Buckwheat, anemone, avens (achene)
Castor bean	Cereals (caryopsis)
Coffee bean	Hazel, walnut, oak (nut)
Cotton	Lettuce, sunflower, and other Compositae (cypsela)
Legumes (e.g., peas, beans)	
Squashes (e.g., cucumber, pumpkin)	
Tomato	

Source: Bewley et al. 2013, p. 3.

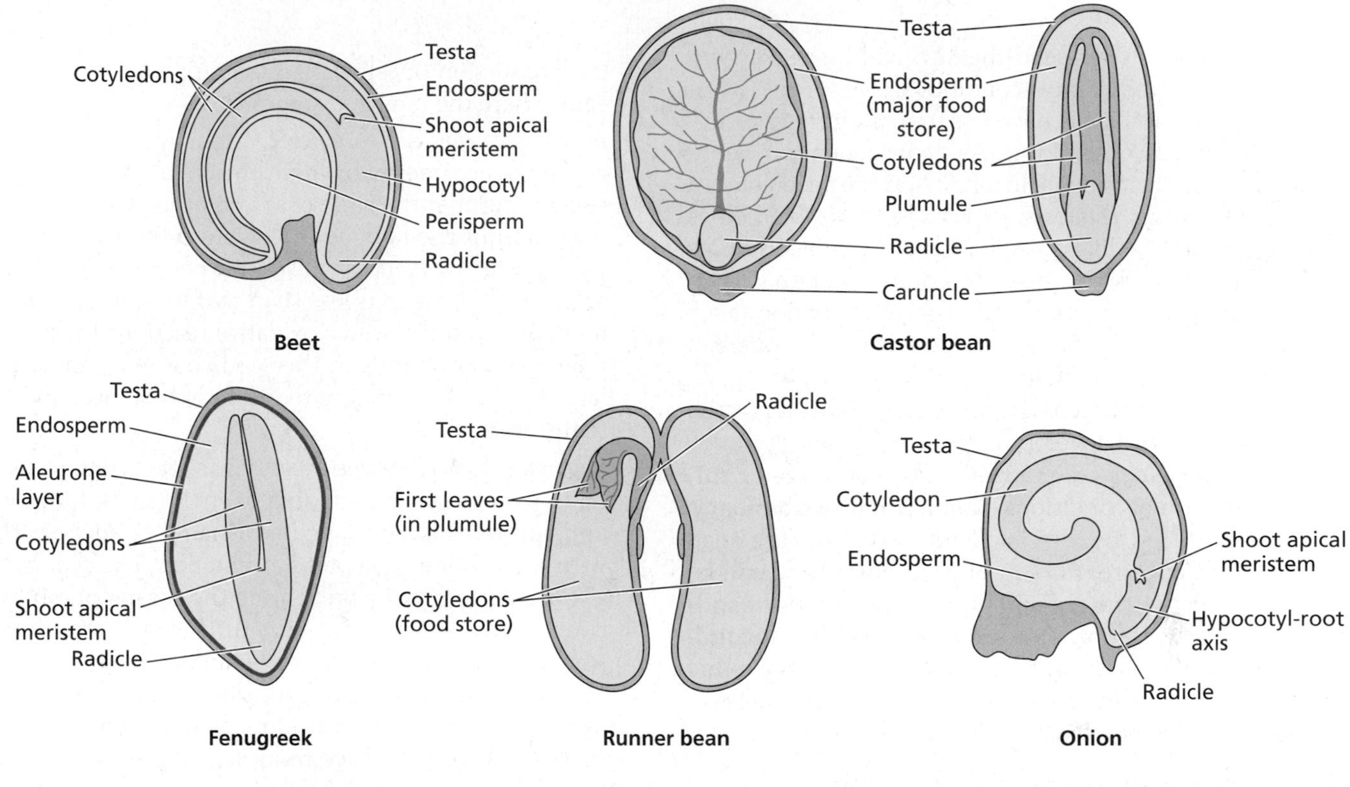

Figure 18.1 Seed structure of selected eudicots and monocots.

embryonic structures peculiar to the grass family include the following (see Figure 18.1):

- The single cotyledon has been modified by evolution to form an absorptive organ, the **scutellum**, which forms the interface between the embryo and the starchy endosperm tissue.
- The basal sheath of the scutellum has elongated to form a **coleoptile** that covers and protects the first leaves while buried beneath the soil.
- The base of the hypocotyl has elongated to form a protective sheath around the radicle called the **coleorhiza**.
- In some species, such as maize, the upper hypocotyl has been modified to form a **mesocotyl**. During seedling development, the growth of the mesocotyl helps raise the leaves to the soil surface, especially in the case of deeply planted seeds (see WEB TOPIC 18.2).

Seed Dormancy

During seed maturation, the embryo dehydrates and enters a quiescent phase. Seed germination requires rehydration and can be defined as the resumption of growth of the embryo in the mature seed. However, germination encompasses all the events that take place between the start of imbibition of the dry seed and the emergence of the embryo, usually the radicle, from the structures that surround it. Successful completion of germination depends on the same environmental conditions as vegetative growth (see Chapter 19): water and oxygen must be available, and the temperature must be suitable. However, a viable (living) seed may not germinate even if the appropriate environmental requirements are satisfied, a phenomenon known as **seed dormancy**. Seed dormancy is an intrinsic temporal block to the completion of germination that provides additional time for seed dispersal over greater geographic distances, or for seasonal dormancy cycling in the soil seed bank (described later in the chapter). It also maximizes seedling survival by preventing germination in unfavorable conditions.

Mature seeds typically have less than 0.1 g water g^{-1} dry weight at the time of shedding. As a consequence of dehydration, metabolism comes to a halt and the seed enters a quiescent ("resting") state. In some cases the seed becomes dormant as well. Unlike quiescent seeds, which germinate upon rehydration, dormant seeds require additional treatments or signals for germination to occur. After dormancy is broken, the seed is able to germinate over the range of conditions permissible for the particular genotype.

Different types of seed dormancy can be distinguished on the basis of the developmental timing of dormancy onset. Newly dispersed, mature seeds that fail to germinate under normal conditions exhibit **primary dormancy**, typically induced by abscisic acid (ABA) during seed maturation. (ABA regulation of seed dormancy will be discussed later in the chapter.) Once primary dormancy has been lost, non-dormant seeds may acquire **secondary dormancy** if exposed to unfavorable conditions that inhibit germination over a period of time. For examples of secondary dormancy see WEB TOPIC 18.3.

Dormancy can be imposed on the embryo by the surrounding tissues

Seed dormancy may result from embryo dormancy, from the inhibitory effects of tissues surrounding the embryo, or from both. Physiological dormancy imposed on the embryo by the seed coat and other enclosing tissues, such as endosperm, pericarp, or extrafloral organs, is known as **coat-imposed dormancy**. The embryos of such seeds germinate readily in the presence of water and oxygen once the seed coat and other surrounding tissues have either been removed or damaged. There are several mechanisms by which seed coats can impose dormancy on the embryo:

- *Water impermeability*. This type of coat-imposed dormancy is common in plants found in arid and semiarid regions, especially among legumes, such as clover (*Trifolium* spp.) and alfalfa (*Medicago* spp.). The classic example is Indian lotus (*Nelumbo nucifera*) seeds, which have survived up to 1200 years because of their impermeable coats. Waxy cuticles, suberized layers, and palisade layers of lignified sclereids all combine to restrict the penetration of water into the seed. This type of dormancy can be broken by mechanical or chemical scarification. In the wild, passage through the digestive tracts of animals can cause chemical scarification.
- *Mechanical constraint*. The first visible sign of germination is typically the radicle (embryonic root) breaking through its surrounding structures, such as the endosperm, if present, and seed coat. In some cases, however, the thick-walled endosperm may be too rigid for the radicle to penetrate, as in Arabidopsis, tomato, coffee, and tobacco. For such seeds to complete germination, the endosperm cell walls must be weakened by the production of cell wall–degrading enzymes, typically where the radicle emerges.
- *Interference with gas exchange*. Dormancy in some seeds can be overcome by oxygen-enriched atmospheres, suggesting that the seed coat and other surrounding tissues limit the supply of oxygen to the embryo. In wild mustard (*Sinapis arvensis*), the permeability of the seed coat to oxygen is less than that to water by a factor of 10^4. In other seeds, oxidative reactions involving phenolic compounds in the seed coat may consume large amounts of oxygen, reducing oxygen availability to the embryo.
- *Retention of inhibitors*. Dormant seeds often contain secondary metabolites, including phenolic acids, tannins, and coumarins, and repeated rinsing of such seeds with water often promotes germination. The coat may impose dormancy by preventing the escape of inhibitors from the seed, or they may diffuse into the embryo from the coat and prevent germination. The *transparent testa* (*tt*) mutant of Arabidopsis, which contains reduced amounts of proanthocyanidins (condensed tannins) in its seed coat, displays reduced dormancy but also reduced longevity.

Embryo dormancy may be caused by physiological or morphological factors

Seed dormancy that is intrinsic to the embryo and is not due to any influence of the seed coat or other surrounding tissues is called **embryo dormancy**. In some cases, embryo dormancy can be relieved by removal of the cotyledons. Species in which the cotyledons exert an inhibitory effect include European hazel (*Corylus avellana*) and European ash (*Fraxinus excelsior*).

Seeds also may fail to germinate because the embryos have not reached their full size or maturity. These embryos simply require additional time to enlarge under appropriate conditions before they can emerge from the seed. Familiar examples of dormancy caused by undersized embryos are celery (*Apium graveolens*) and carrot (*Daucus carota*) (**Figure 18.2**). Seeds with undifferentiated embryos are usually tiny and include the parasitic plant broomrapes (*Orobanche* and *Phelipanche* spp.) and orchids.

Non-dormant seeds can exhibit vivipary and precocious germination

In some estuarine species the mature seeds not only lack dormancy, but also germinate while still on the mother plant, a phenomenon known as **vivipary**. True vivipary, the germination of immature seeds on the mother plant, is extremely rare in angiosperms and is largely restricted to mangroves and other plants growing in estuarine or riparian ecosystems in the tropics and subtropics. A well-known example of a viviparous species is the red mangrove (*Rhizophora mangle*) (**Figure 18.3**). Seeds of this

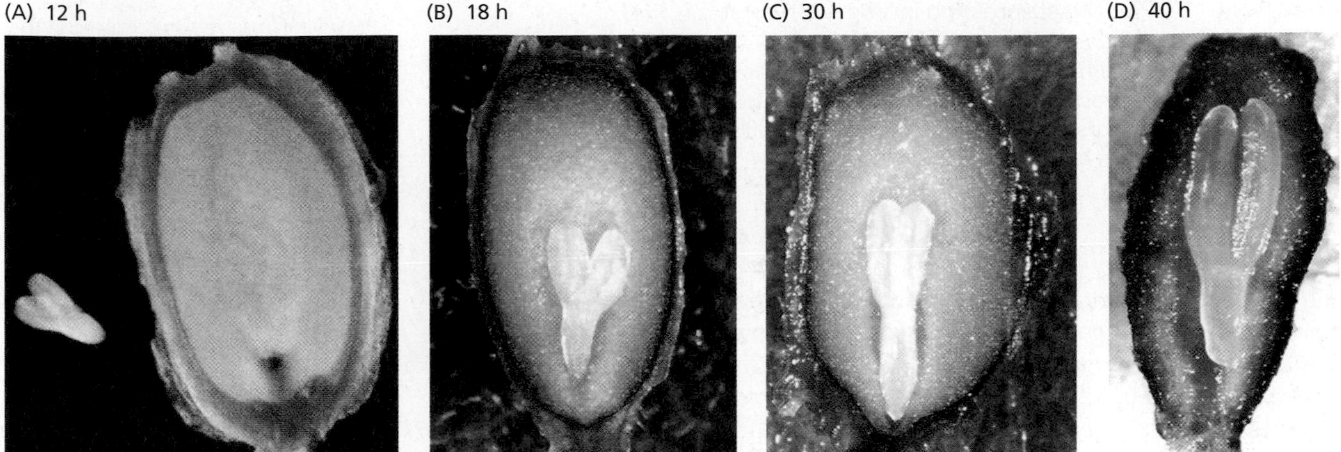

Figure 18.2 Growth of the undersized carrot embryo during inbibition of seeds for 12 (A), 18 (B), 30 (C), and 40 h (D). The tiny embryo on the left, removed from the seed for better visibility, is embedded in a cavity in the endosperm formed by the release of cell wall–degrading enzymes. Germination begins with the emergence of the radicle from the seed 2 to 4 days after imbibition. (From Homrichhausen et al. 2003.)

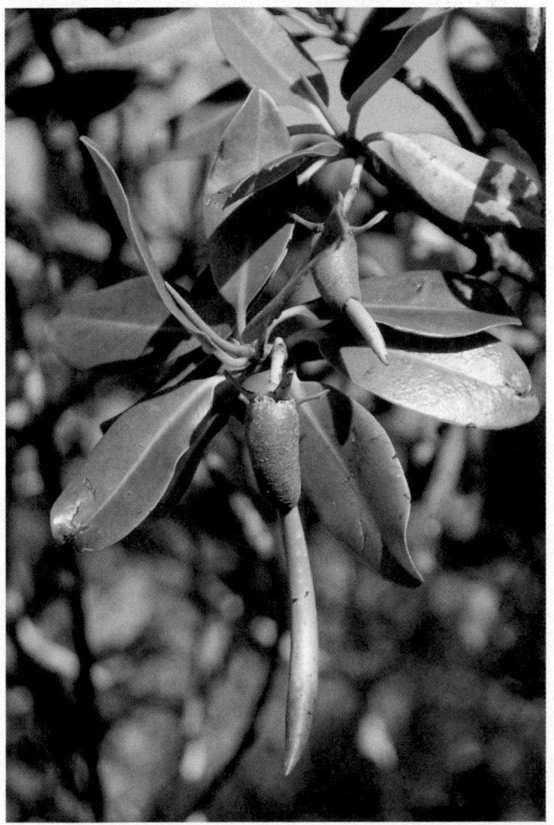

Figure 18.3 Viviparous seeds of red mangrove (*Rhizophora mangle*).

species germinate while still inside the attached fruit and produce an elongated, dartlike propagule that can drop from the tree and root itself in the surrounding soft mud.

The germination of physiologically mature seeds on the mother plant is known as **preharvest sprouting** and is characteristic of some grain crops when they mature in wet weather (**Figure 18.4A**). Preharvest sprouting in cereals (e.g., wheat, barley, rice, and sorghum) reduces grain quality and causes serious economic losses. In maize, *viviparous* (*vp*) mutants have been selected in which the embryos germinate directly on the cob while still attached to the parent plant, referred to as **precocious germination** (**Figure 18.4B**). Several of these mutants are ABA deficient (*vp2*, *vp5*, *vp7*, *vp9*, and *vp14*); one is ABA insensitive (*vp1*). Vivipary in the ABA-deficient mutants can be partially prevented by treatment with exogenous ABA. Vivipary in maize also requires synthesis of gibberellin (GA) early in embryogenesis as a positive signal; double mutants deficient in gibberellin and ABA do not exhibit this phenomenon. This shows that the ABA:GA ratio is what regulates germination, not the actual amount of ABA.

The ABA:GA ratio is the primary determinant of seed dormancy

It has long been known that ABA exerts an inhibitory effect on seed germination, while gibberellin exerts a positive influence. According to the **hormone balance theory**, the ratio of these two hormones serves as the primary determinant of seed dormancy and germination. The relative hormonal activities of ABA and gibberellin in the seed depend on two main factors: the amounts of each hormone present within the target tissues, and the ability of the target tissues to detect and respond to each of the hormones. Hormonal sensitivity, in turn, is a function of the hormonal signaling pathways in the target tissues.

The amounts of the two hormones are regulated by their rates of synthesis versus deactivation (see Chapter 15). For ABA, the reaction involving **NCED**, the enzyme

Figure 18.4 (A) Preharvest sprouting in a head of wheat (*Triticum aestivum*). (B) Precocious germination in the ABA-deficient *vivipary14* (*vp14*) mutant of maize. The VP14 protein catalyzes the cleavage of 9-*cis*-epoxycarotenoids to form xanthoxin, a precursor of ABA. (A from Li et al. 2009; B courtesy of Bao Cai Tan and Don McCarty.)

that cleaves 9-*cis* xanthophylls to xanthoxin, appears to be the rate-limiting step in its biosynthetic pathway. ABA deactivation is mainly carried out by the ABA-8′-hydroxylase, CYP707A2.

The rate-limiting step in the gibberellin biosynthetic pathways is the final reaction that gives rise to the active form of the hormone (GA_9 to GA_4 in Arabidopsis; GA_{20} to GA_1 in lettuce), catalyzed by the enzyme **GA 3-oxidase** (**GA3ox**). The major enzyme deactivating gibberellin is **GA 2-oxidase** (**GA2ox**), which negatively regulates germination by reducing the gibberellin content of the seed.

The balance between the two biosynthetic and deactivation pathways is regulated at the gene level by the action of transcription factors. Gibberellin promotion of germination requires the destruction of DELLA family proteins that repress germination, in part by increasing expression of proteins that promote ABA biosynthesis. The increased ABA then promotes the expression of ABI-class protein phosphatase-regulated transcription factor(s) (see Chapter 15) and the germination-inhibiting DELLA protein (discussed later in this chapter), creating a positive feedback loop.

According to a recent model, the balance between the activities of ABA and gibberellin in seeds is under both developmental and environmental control (**Figure 18.5**). During the early stages of seed development, ABA sensitivity is high and gibberellin sensitivity is low, which favors dormancy over germination. Later in seed development, ABA sensitivity declines and gibberellin sensitivity increases, favoring germination. At the same time, the seed becomes progressively more sensitive to envi-

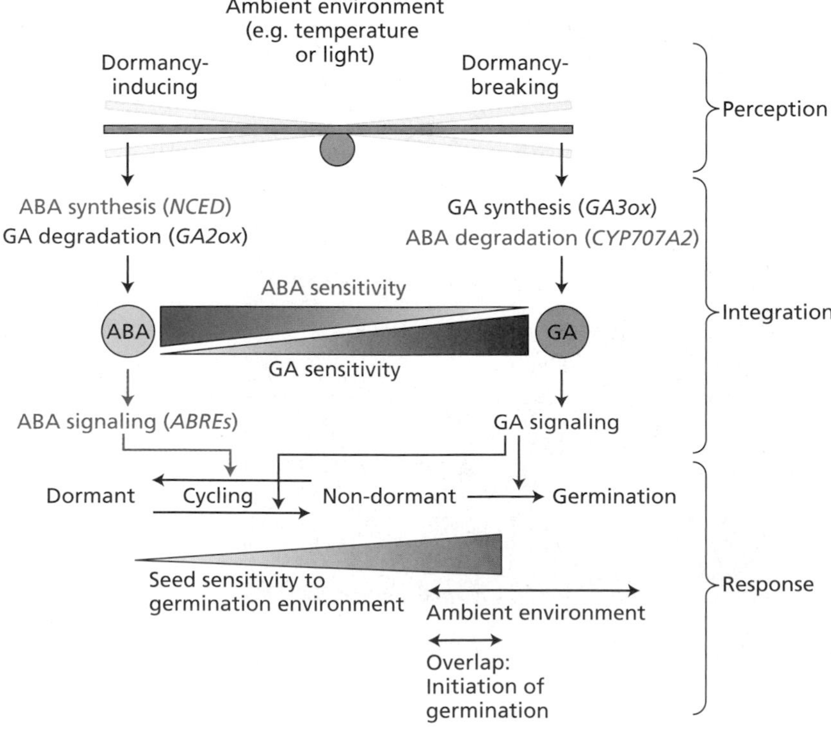

Figure 18.5 Model for ABA and gibberellin (GA) regulation of dormancy and germination in response to environmental factors. Environmental factors such as temperature affect the ABA:GA ratios and the responsiveness of the embryo to ABA and gibberellin. In dormancy, gibberellin is catabolized and ABA synthesis and signaling predominate. In the transition to germination, ABA is catabolized and gibberellin synthesis and signaling predominate. The complex interplay between ABA and gibberellin synthesis, degradation, and sensitivity in response to ambient environmental conditions can result in cycling between dormant and non-dormant states (dormancy cycling). Germination can proceed to completion when there is an overlap between favorable environmental conditions and non-dormancy. Key target genes of ABA and gibberellin are in parentheses. (After Finch-Savage and Leubner-Metzger 2006.)

ronmental cues, such as temperature and light, that can either stimulate or inhibit germination.

However, ABA and gibberellin are by no means the only hormones regulating seed dormancy. Both ethylene and brassinosteroids reduce the ability of ABA to inhibit germination, apparently by negatively regulating the ABA signal transduction pathway. ABA also inhibits ethylene biosynthesis, while brassinosteroids enhance it. Thus, hormonal networks are likely involved in regulating seed dormancy, as they are in regulating most developmental phenomena.

Release from Dormancy

Breaking dormancy involves a metabolic state change in the seed that allows the embryo to reinitiate growth. Because germination is an irreversible process that commits the seed to grow into a seedling, many species have developed sophisticated mechanisms for sensing the optimal environmental conditions for this to occur. Often there are seasonal components to the ultimate "decision" of a seed to germinate, as in the examples of secondary dormancy noted earlier in the chapter. In this section we will discuss some of the environmental cues that bring about the release from dormancy. Although each external signal will be discussed separately, seeds in nature must integrate their responses with multiple environmental factors, perceived either simultaneously or in succession.

Because the ABA:GA ratio plays such a decisive role in maintaining seed dormancy, environmental conditions that break dormancy ultimately are thought to operate at the level of gene networks that affect the balance between the responses to ABA and gibberellin. This hypothesis is consistent with the fact that treatment of seeds with gibberellin often can substitute for a positive environmental signal in breaking dormancy.

Light is an important signal that breaks dormancy in small seeds

Many seeds have a light requirement for germination (termed *photoblasty*), which may involve only a brief exposure, as in the case of the 'Grand Rapids' cultivar of lettuce (*Lactuca sativa*); an intermittent treatment (e.g., succulents of the genus *Kalanchoë*); or even a specific photoperiod involving short or long days. For example, birch (*Betula* spp.) seeds require long days to germinate, while seeds of the conifer eastern hemlock (*Tsuga canadensis*) require short days. Phytochrome, which senses red (R) and far-red (FR) wavelengths of light (see Chapter 16), is the primary sensor for light-regulated seed germination. All light-requiring seeds exhibit coat-imposed dormancy, and removal of the outer tissues—specifically, the endosperm—allows the embryo to germinate in the absence of light. The effect that light has on the embryo is thus to enable the radicle (the embryonic root) to penetrate the endosperm, a process facilitated in some species by enzymatic weakening of the cell walls in the micropylar region, next to the radicle.

Light is required by the small seeds of numerous herbaceous and grassland species, many of which remain dormant if they are buried below the depth to which light penetrates. Even when such seeds are on or near the soil surface, the amount of shading by the vegetation canopy (i.e., the R:FR light ratio the seeds receive) is likely to affect their germination. We will return to the effects of the R:FR ratio later in the chapter in relation to the phenomenon of shade avoidance.

Some seeds require either chilling or after-ripening to break dormancy

Many seeds require a period of low temperature (0–10°C) to germinate. In temperate-zone species, this requirement is of obvious survival value, since such seeds will not germinate in the fall, but only in the following spring. Chilling seeds to break their dormancy is referred to as **stratification**, named for agricultural practice of overwintering dormant seeds in layered mounds of soil or moist sand. Today seeds are simply stored moist in a refrigerator. Stratification has the added benefit that it synchronizes germination, which ensures that plants will mature at the same time. **Figure 18.6A** shows the effects of chilling on apple seed germination. Intact seeds require 80 days of chilling for maximum germination; in contrast, isolated embryos achieve this at approximately 50 days. Thus, the presence of the seed coat and endosperm increases the chilling requirement of the embryo by about 30 days.

Some seeds may require a period of **after-ripening**, meaning dry storage at room temperature, before they can germinate. The duration of the after-ripening requirement may be as short as a few weeks (e.g., barley, *Hordeum vulgare*) or as long as 5 years (e.g. curly dock, *Rumex crispus*). In the field, after-ripening may occur in winter annuals in which dormancy is broken by high summer temperatures, allowing the seeds to germinate in the fall. In contrast, moist chilling during the cold winter months is effective in many summer annuals. After-ripening of horticultural and agricultural crop seeds is usually performed in special drying ovens that maintain the appropriate temperature, aeration, and low moisture conditions.

The effect of the duration of after-ripening on the germination of seeds of *Nicotiana plumbaginifolia* is shown in **Figure 18.6B**. Seeds after-ripened for only 14 days began to germinate after about 10 days of subsequent wetting, while seeds after-ripened for 10 months began germinating after only 3 days. The mechanism by which after-ripening brings about a release from dormancy is poorly understood. Seeds are considered "dry" when their water content drops below 20%. In several species, ABA decreases during after-ripening, and even a small decline

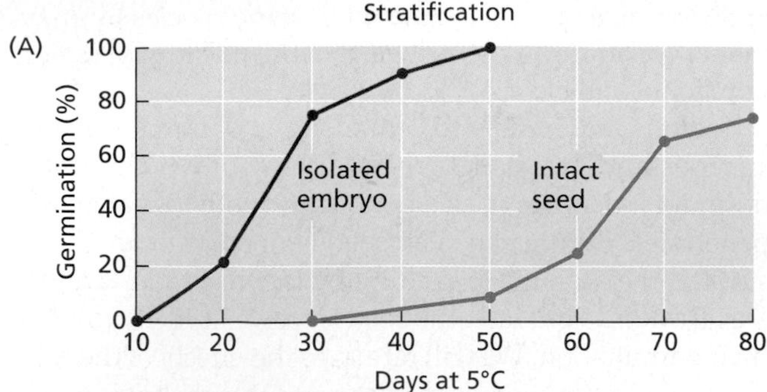

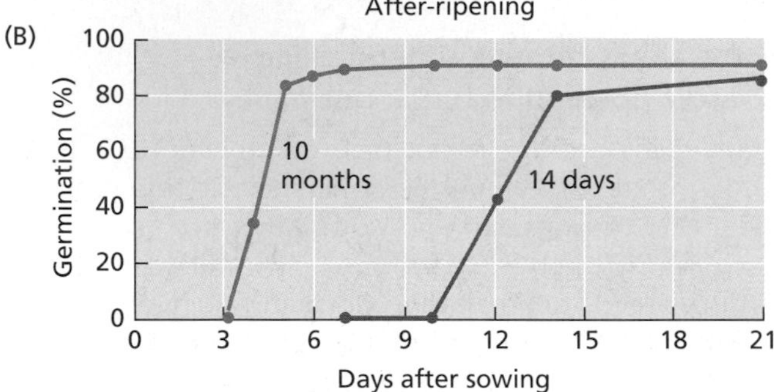

Figure 18.6 Seed dormancy can be overcome by stratification and after-ripening. (A) Release of apple seeds from dormancy by stratification, or moist chilling. Imbibed seeds were stored at 5°C and periodically removed to test the seeds or isolated embryos for germination. The germination of intact seeds was significantly delayed compared with that of isolated embryos. (B) The effect of after-ripening (dry storage at room temperature) on the germination of seeds of *Nicotiana plumbaginifolia*. After-ripening for 10 months or longer greatly accelerated germination compared with after-ripening of only 14 days. (A after Visser 1956; B after Grappin et al. 2000.)

may be sufficient to break dormancy. For example, in *N. plumbaginifolia* seeds, the ABA content decreases by about 40% during after-ripening. However, if seeds become too dry (5% water content or less), the effectiveness of after-ripening is diminished.

Seed dormancy can by broken by various chemical compounds

Numerous chemicals, such as respiratory inhibitors, sulfhydryl compounds, oxidants, and nitrogenous compounds, have been shown to break seed dormancy in specific species. However, only a few of these occur naturally in the environment. Of these, nitrate, often in combination with light, is probably the most important. Some plants, such as hedge mustard (*Sysymbrium officinale*), have an absolute requirement for nitrate and light for seed germination. Another chemical agent that can break dormancy is nitric oxide (NO), a signaling molecule found in both plants and animals (see Chapters 23 and 24). Arabidopsis mutants unable to synthesize NO exhibit reduced germination, and the effect can be reversed by treating the seeds with exogenous NO. Another strong chemical stimulant of seed germination in many species under natural conditions is smoke, which is produced during forest fires. Smoke is likely to contain multiple germination stimulants, but one of the most active of these is **karrikinolide**, a member of the class karrikins, which structurally resemble strigolactones (see Chapters 15 and 17).

In all three of the above examples, the chemical stimulants appear to break dormancy by the same basic mechanism: by down-regulating ABA synthesis or signaling, and up-regulating gibberellin synthesis or signaling, thus altering the ABA:GA ratio.

Seed Germination

Germination is the process that begins with water uptake by the dry seed and ends with the emergence of the embryonic axis, usually the radicle, from its surrounding tissues. Strictly speaking, germination does not include seedling growth after radicle emergence, which is referred to as **seedling establishment**. Similarly, the rapid mobilization of stored food reserves that fuels the initial growth of the seedling is considered a postgermination process.

Germination requires permissive ranges of water, temperature, and oxygen, and often light and nitrate as well. Of these, water is the most essential factor. The water content of mature, air-dried seeds is in the range of 5 to 15%, well below the threshold required for fully active metabolism. In addition, water uptake is needed to generate the turgor pressure that powers cell expansion, the basis of vegetative growth and development. As we discussed in Chapter 3, water uptake is driven by the gradient in water potential (ψ) from the soil to the seed. For example, incubating tomato seeds at high ambient water potential (ψ = 0 MPa) allows 100% germination, whereas incubation at low water potential (ψ = –1.0 MPa), which nullifies the gradient in water potential, completely suppresses germination (**Figure 18.7**).

Germination can be divided into three phases corresponding to the phases of water uptake

Under normal conditions, water uptake by the seed is triphasic (**Figure 18.8**):

- *Phase I.* The dry seed takes up water rapidly by the process of imbibition.
- *Phase II.* Water uptake by imbibition declines and metabolic processes, including transcription and translation, are reinitiated. The embryo expands, and the radicle emerges from the seed coat.
- *Phase III.* Water uptake resumes due to a decrease in ψ as the seedling grows, and the stored food reserves of the seed are fully mobilized.

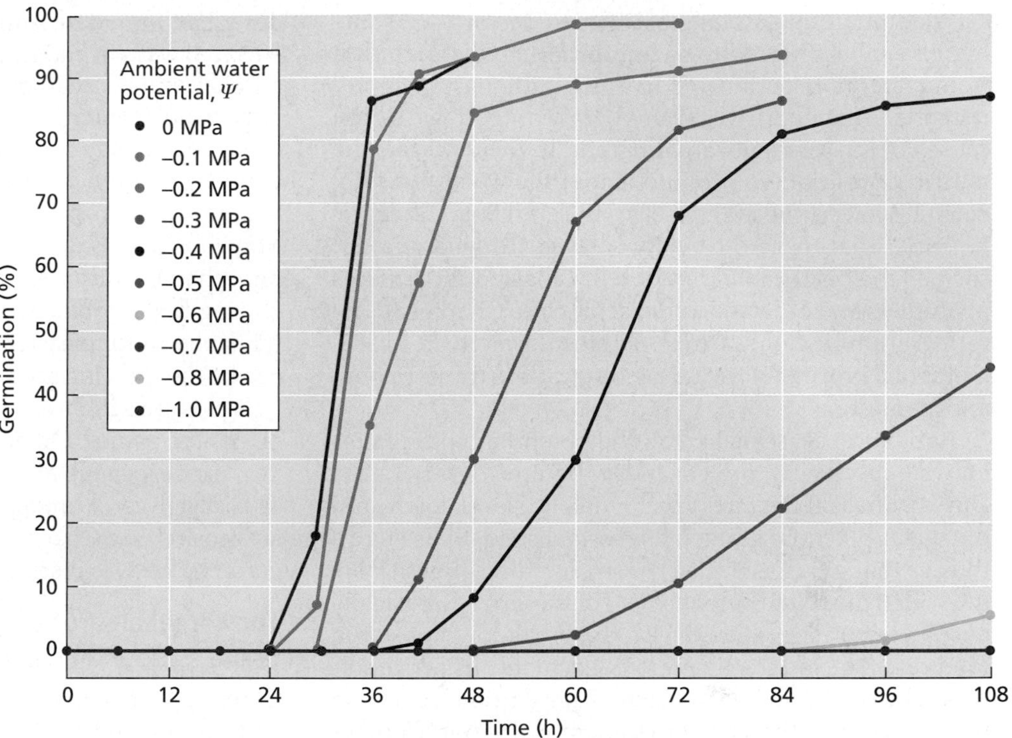

Figure 18.7 Time course of tomato seed germination at different ambient water potentials. (After G. Leubner [http://www.seedbiology.de] using data from Liptay and Schopfer 1983.)

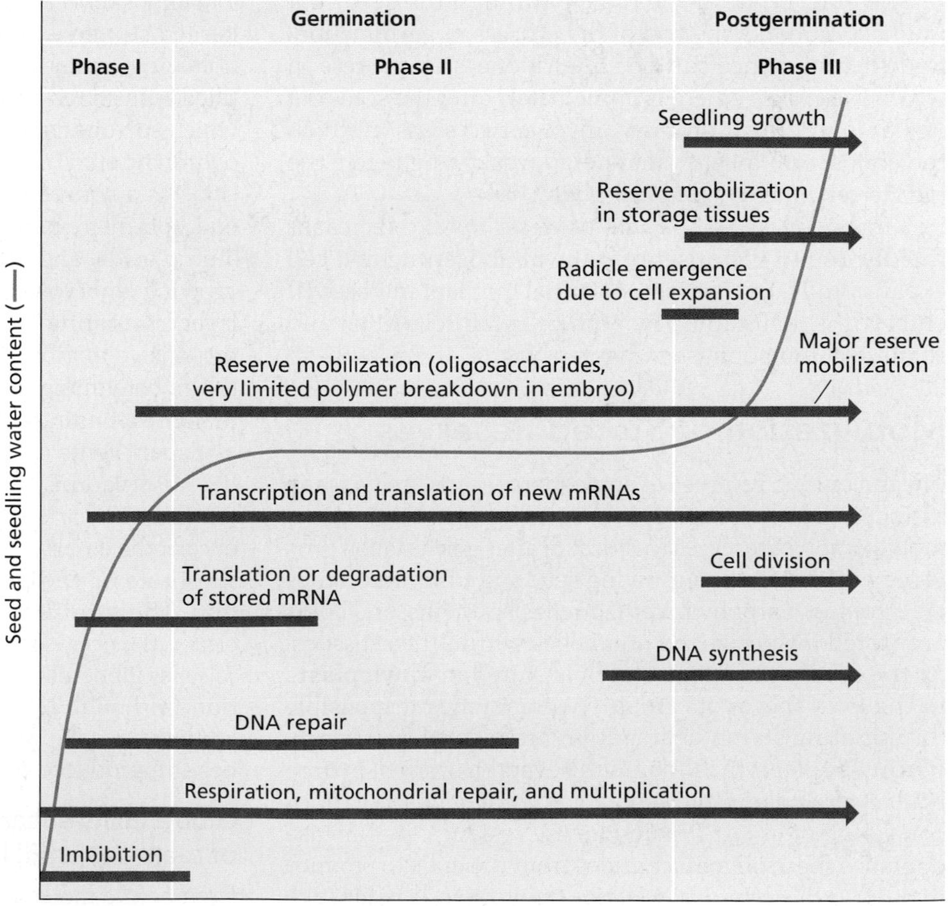

Figure 18.8 Phases of seed imbibition. In Phase I, dry seeds imbibe, or uptake, water rapidly. Since water flows from the higher to the lower water potential, the water uptake by the seed stops when the difference in water potential between the seed and the environment becomes zero. During Phase II the cells expand and the radicle emerges from the seed. Metabolic activity increases and cell wall loosening occurs. In Phase III, water uptake resumes as the seedling becomes established. (After Nonogaki et al. 2010.)

The initial rapid uptake of water by the dry seed during Phase I is referred to as **imbibition**, to distinguish it from water uptake during Phase III. Although the water potential gradient drives water uptake in both cases, the causes of the gradients are different. In the dry seed, the **matric potential** (ψ_m) component of the water potential equation lowers the ψ and creates the gradient. The matric potential arises from the binding of water to solid surfaces, such as the microcapillaries of cell walls and the surfaces of proteins and other macromolecules (see Chapter 3). The rehydration of cellular macromolecules activates basal metabolic processes, including respiration, transcription, and translation.

Imbibition ceases when all the potential binding sites for water become saturated, and ψ_m becomes less negative. During Phase II the rate of water uptake slows down until the water potential gradient is reestablished. Phase II can thus be thought of as the lag phase preceding growth, during which the solute potential (ψ_s) of the embryo gradually becomes more negative due to the breakdown of stored food reserves and the liberation of osmotically active solutes. The seed volume may increase as a result, rupturing the seed coat. At the same time, additional metabolic functions come online, such as the re-formation of the cytoskeleton and the activation of DNA repair mechanisms.

The emergence of the radicle through the seed coat in Phase II marks the end of the process of germination. Radicle emergence can be either a one-step process in which the radicle emerges immediately after the seed coat (testa) is ruptured, or it may involve two steps in which the endosperm must first undergo weakening before the radicle can emerge (see WEB TOPIC 18.4).

During Phase III the rate of water uptake increases rapidly due to the onset of cell wall loosening and cell expansion. Thus, the water potential gradient in Phase III embryos is maintained by both cell wall relaxation and solute accumulation (see Chapter 14).

Mobilization of Stored Reserves

The major food reserves of angiosperm seeds are typically stored in the cotyledons or in the endosperm. The massive mobilization of reserves that occurs after germination provides nutrients to the growing seedling until it becomes autotrophic. Carbohydrates (starches), proteins, and lipids are stored in specialized organelles within these tissues. At the subcellular level, starch is stored in **amyloplasts** in the endosperm of cereals. Two enzymes responsible for initiating starch degradation are α- and β-amylase. α-Amylase (of which there are several isoforms) hydrolyzes starch chains internally to produce oligosaccharides consisting of α(1,4)-linked glucose residues. β-Amylase degrades these oligosaccharides from the ends to produce maltose, a disaccharide. Maltase then converts maltose to glucose. The hormonal regulation of these enzymes is described in more detail in the next section. The thickened cell walls of endosperm tissue in some seeds provide another source of carbohydrates for the growing seedling during mobilization.

Protein storage vacuoles are the primary source of amino acids for new protein synthesis in the seedling. In addition, protein storage vacuoles contain **phytin**, the K^+, Mg^{2+}, and Ca^{2+} salt of phytic acid a (*myo*-inositol hexaphosphate), a major storage form of phosphate in seeds. During food mobilization, the enzyme **phytase** hydrolyzes phytin to release phosphate and the other ions for use by the growing seedling.

Lipids are a high-energy carbon source that is stored in oil or lipid bodies. Oil bodies from rape, mustard, cotton, flax, maize, peanut, and sesame seeds contain lipids, such as triacylglycerols and phospholipids, and proteins, such as oleosins (see Chapter 1). Lipid catabolism during seed germination was discussed in Chapter 12.

The cereal aleurone layer is a specialized digestive tissue surrounding the starchy endosperm

Cereal grains consist of three parts: the embryo, the endosperm, and the fused testa–pericarp (**Figure 18.9**). The embryo, which will grow into the new seedling, has a specialized absorptive organ, the scutellum. The triploid endosperm is composed of two tissues: the centrally located **starchy endosperm** and the **aleurone layer**. The nonliving starchy endosperm consists of thin-walled cells filled with starch grains. Living cells of the aleurone layer, which surrounds the endosperm, synthesize and release hydrolytic enzymes into the endosperm during germination. As a consequence, the stored food reserves of the endosperm are broken down, and the solubilized sugars, amino acids, and other products are transported to the growing embryo via the scutellum. The isolated aleurone layer, consisting of a homogeneous population of cells responsive to gibberellin, has been widely used to study the gibberellin signal transduction pathway in the absence of nonresponding cell types.

Experiments carried out in the 1960s confirmed earlier observations that the secretion of starch-degrading enzymes by barley aleurone layers depends on the presence of the embryo. It was soon discovered that GA_3 could substitute for the embryo in stimulating starch degradation. The significance of the gibberellin effect became clear when it was shown that the embryo synthesizes and releases gibberellins into the endosperm during germination. Although aleurone layers respond to GA_3, genetic studies have shown that GA_1 is the only bioactive gibberellin produced by cereals.

Gibberellins enhance the transcription of *α-amylase* mRNA

Even before molecular biological approaches were developed, there was already physiological and biochemical

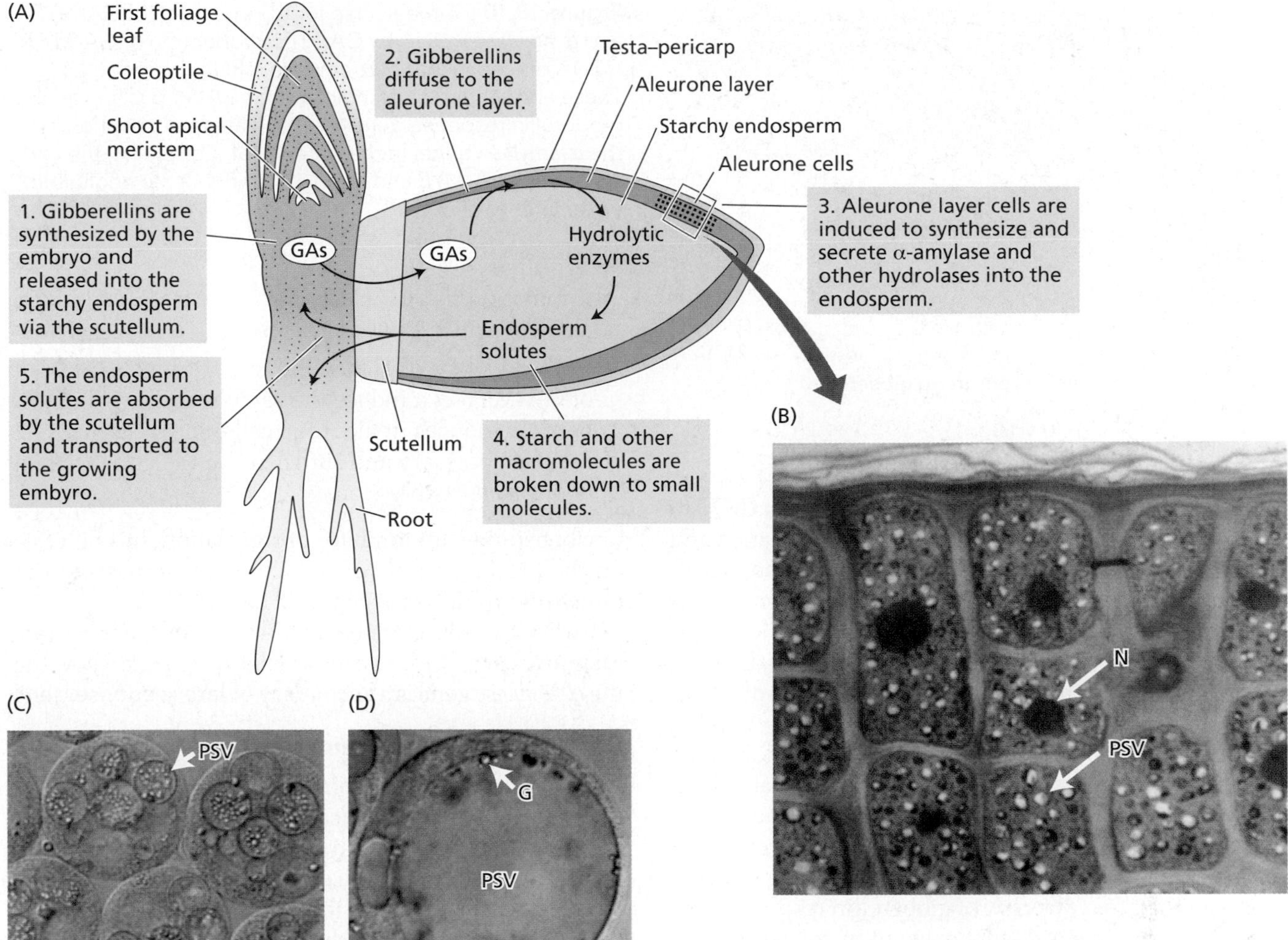

Figure 18.9 Structure of a barley grain and the functions of various tissues during germination. (A) Diagram of germination-initiated interactions. (B–D) Micrographs of the barley aleurone layer (B) and barley aleurone protoplasts at an early (C) and a late (D) stage of amylase production. Multiple protein storage vesicles (PSVs) in (C) coalesce to form a large vesicle in (D), which will provide amino acids for α-amylase synthesis. G, phytin globoid that sequesters minerals; N, nucleus. (B–D from Bethke et al. 1997, courtesy of P. Bethke.)

evidence that gibberellin enhanced α-amylase production at the level of gene transcription. The two main lines of evidence were:

- GA_3-stimulated α-amylase production was shown to be blocked by inhibitors of transcription and translation.
- Isotope-labeling studies demonstrated that the stimulation of α-amylase activity by bioactive gibberellin involved de novo synthesis of the enzyme from amino acids, rather than activation of preexisting enzyme.

Cereal grains can be cut in two, and "half-seeds" that lack the embryo (the source of bioactive gibberellin in intact grain) make a convenient experimental system for studying the action of applied gibberellin. Microarray studies have confirmed the up-regulation of genes encoding several α-amylase isoforms in rice half-seeds that have been treated for 8 h with gibberellin. In these half-seeds, the only living cells—and the only cells in which gibberellin signaling occurs—are in the aleurone layer. Of all genes in the microarray analyses, those encoding α-amylase isoforms show the highest degree of up-regulation after gibberellin treatment, followed closely by proteases and other hydrolases.

The gibberellin receptor, GID1, promotes the degradation of negative regulators of the gibberellin response

As discussed in Chapter 15, the gibberellin receptor GIBBERELLIN INSENSITIVE DWARF 1 (GID1) undergoes a conformational change when it binds to gibberellin, which promotes the binding of DELLA repressor proteins. The DELLA protein also undergoes a conformational change, facilitating interaction with the E3 ubiquitin ligase SCF^{SLY1}.

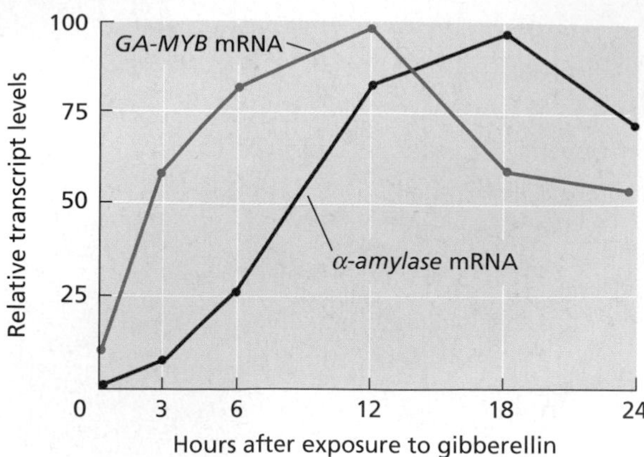

Figure 18.10 Time course for the induction of *GA-MYB* and *α-amylase* mRNA by GA_3. The production of *GA-MYB* mRNA precedes that of *α-amylase* mRNA by about 3 h. These and other results indicate that *GA-MYB* is an early gibberellin response gene that regulates transcription of the *α-amylase* gene. In the absence of gibberellin, the levels of both *GA-MYB* and *α-amylase* mRNAs are negligible. (After Gubler et al. 1995.)

As a result, the binding of gibberellin receptor GID1 to the DELLA repressor proteins triggers ubiquitination and subsequent degradation by the 26S proteasome, which allows the gibberellin response to proceed (see Figures 15.33 and 15.34).

The aleurone layers of rice *gid1* mutants with a defective gibberellin receptor are unable to synthesize α-amylase, clearly implicating the soluble gibberellin receptor GID1 in this classic gibberellin response. Other evidence obtained before the characterization of GID1 suggested that gibberellin may also bind to a protein in the plasma membrane of aleurone cells. Evidence for more than one receptor has been obtained for auxin and abscisic acid (see Chapter 15). Given the great diversity of gibberellin responses, the existence of multiple receptors may not be too surprising, though at the present there is no definitive identification of a plasma membrane–localized gibberellin receptor. Within the aleurone cells there are both Ca^{2+}-independent and Ca^{2+}-dependent gibberellin signaling pathways. The former leads to the production of α-amylase, while the latter regulates its secretion.

GA-MYB is a positive regulator of α-amylase transcription

The sequence of the **gibberellic acid response element** (**GARE**) in the *α-amylase* gene promoter (TAACAAA) is similar to the DNA sequence to which **MYB proteins** bind. MYB proteins are a class of transcription factors in all eukaryotes including plants. In barley, rice, and Arabidopsis, a subset of MYBs have been implicated in GA signaling.

In barley, a member of this subfamily, **GA-MYB**, has been implicated in giberellin signaling. Evidence that GA-MYB activates *α-amylase* gene expression (i.e., that GA-MYB is a *positive regulator* of *α-amylase*) includes the following:

- Synthesis of *GA-MYB* mRNA begins to increase as early as 1 h after gibberellin treatment, preceding the increase in *α-amylase* mRNA by several hours (**Figure 18.10**).
- A mutation in the GARE that prevents MYB binding also prevents *α-amylase* expression.
- In the absence of gibberellin, constitutive expression of GA-MYB can induce the same responses that gibberellin induces in aleurone cells, showing that GA-MYB is necessary and sufficient for the enhancement of *α-amylase* expression.

Cycloheximide, an inhibitor of translation, has no effect on the production of *GA-MYB* mRNA, indicating that protein synthesis is not required for *GA-MYB* expression. *GA-MYB* can therefore be defined as a **primary** or **early response gene**. In contrast, similar experiments show that the *α-amylase* gene is a **secondary** or **late response gene**.

DELLA repressor proteins are rapidly degraded

Drawing together our information for the cereal aleurone system (**Figure 18.11**), we can hypothesize that the binding of bioactive gibberellin to GID1 leads to degradation of the DELLA protein. As a consequence of DELLA degradation, and via some intermediary steps that have not yet been defined, the expression of *GA-MYB* is up-regulated. Finally, the GA-MYB protein binds to a highly conserved GARE in the promoter of the gene for α-amylase, activating its transcription. α-Amylase is secreted from aleurone cells by a pathway that requires Ca^{2+} accumulation. Starch breakdown occurs in cells of the starchy endosperm by the action of α-amylase and other hydrolases, and the resultant sugars are exported to the growing embryo.

Some of the genes encoding other hydrolytic enzymes whose synthesis is promoted by gibberellin also have GA-MYB–binding motifs in their promoters, indicating that this is a common pathway for gibberellin responses in aleurone layers.

ABA inhibits gibberellin-induced enzyme production

In addition to the ABA–gibberellin antagonism affecting seed dormancy, ABA inhibits the gibberellin-induced

Figure 18.11 Composite model for the induction of α-amylase synthesis in barley aleurone layers by gibberellin. A Ca^{2+}-independent pathway induces *α-amylase* gene transcription; a calcium-dependent pathway is involved in α-amylase secretion. ▶

1. GA_1 from the embryo enters an aleurone cell.

2. Once inside the cell, GA_1 may initiate a calcium-calmodulin–dependent pathway necessary for α-amylase secretion.

3. GA_1 binds to GID1 in the nucleus.

4. Upon binding GA_1, the GID1 receptor undergoes an allosteric change that facilitates its binding to a DELLA repressor.

5. Once the DELLA protein has bound the GA_1–GID complex, an F-box protein (part of an SCF complex) poly- ubiquitinates the GRAS domain of the DELLA protein.

6. The polyubiquitinated DELLA protein is degraded by the 26S proteasome.

7. Once the DELLA protein has been degraded, transcription of an early gene is activated. (*GA-MYB* is shown, in this model, as an early gene, although there is evidence that transcriptional regulation of other early genes may occur first.) The mRNA for *GA-MYB* is translated in the cytosol.

8. The newly synthesized GA-MYB transcription factor enters the nucleus and binds the promoters of α-amylase and genes encoding other hydrolytic enzymes.

9. Transcription of these genes is activated.

10. α-Amylase and other hydrolases are synthesized on the rough ER, processed, and packaged into secretion vesicles by the Golgi body.

11. Proteins are secreted by exocytosis.

12. The secretory pathway requires GA stimulation of the calcium-calmodulin–dependent pathway.

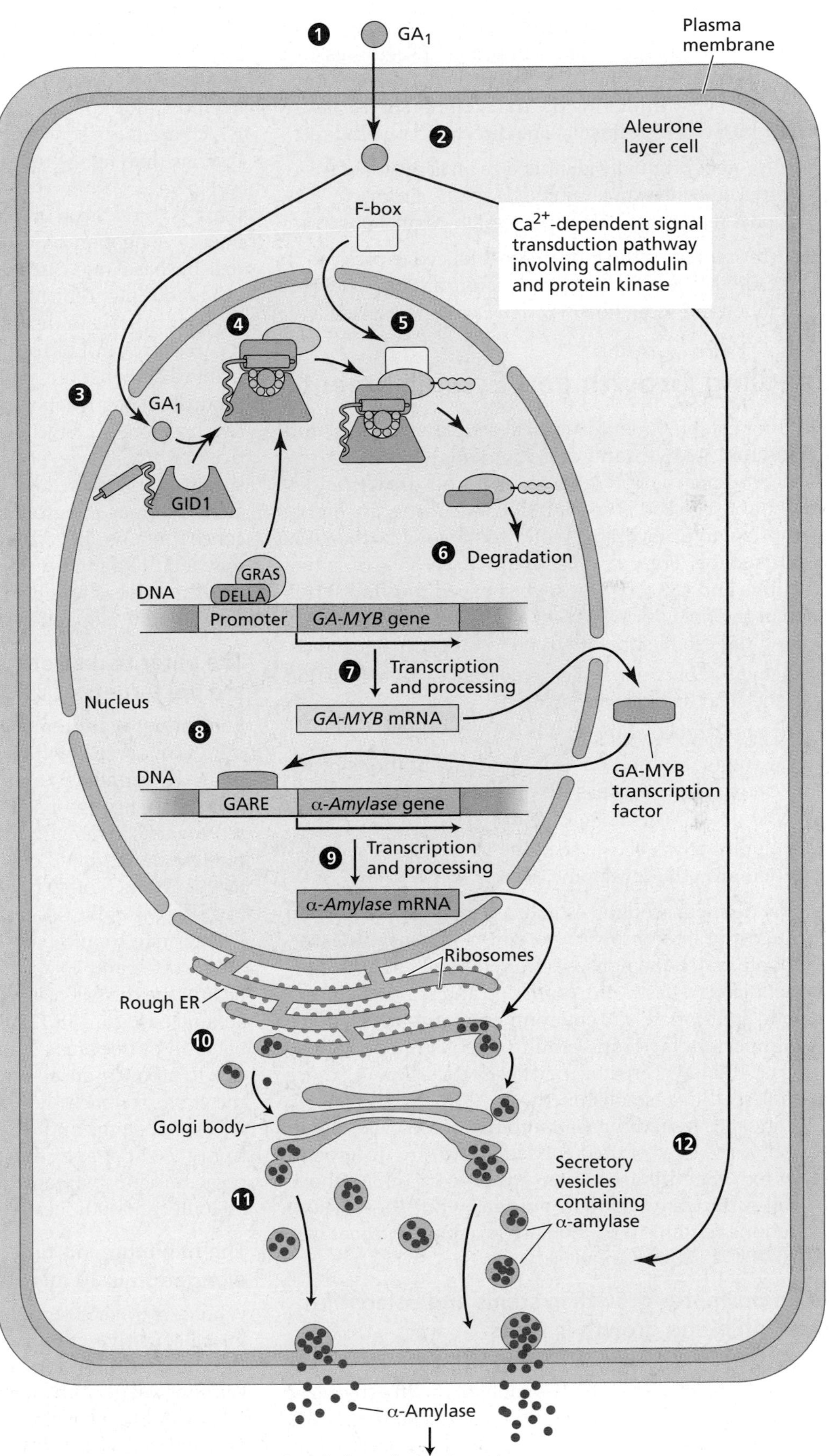

synthesis of hydrolytic enzymes that are essential for the breakdown of storage reserves in growing seedlings (see Figure 18.9). ABA inhibits gibberellin-dependent α-amylase synthesis by inhibiting the transcription of *α-amylase* mRNA by two mechanisms, one direct and one indirect:

1. A protein originally identified as an activator of ABA-induced gene expression, VP1, acts as a transcriptional repressor of some gibberellin-regulated genes.
2. ABA represses the gibberellin-induced expression of GA-MYB, a transcription factor that mediates the gibberellin induction of *α-amylase* gene expression.

Seedling Growth and Establishment

Seedling establishment is critical for plant survival and subsequent growth and development. This transition between germination (emergence) and growth independent of the seed is crucial, since seedlings are highly susceptible to unfavorable biotic and abiotic factors during this stage. For example, about 10 to 55% of maize seedlings and 48 to 70% of soybean seedlings fail at this stage in the field.

Seedling establishment has been variously defined as:

- The period between radicle emergence and exhaustion of the seed reserve (physiology)
- The appearance of the first leaf (agronomy)
- The stage at which environmental conditions begin to exert selective pressure on seedling survivorship (ecology)
- The point at which the seedling is capable of self-sustained growth (development)

Broadly defined, seedling establishment is the stage when the seedling becomes competent to photosynthesize, assimilate water and nutrients from the soil, undergo normal cellular and tissue differentiation and maturation, and respond appropriately to environmental stimuli. Seed size is an important factor in seedling establishment because larger seeds have greater food reserves, allowing more time for seedling establishment.

Angiosperm seedlings fall into two main classes with respect to the fates of their cotyledons during the growth of the axis. Seedlings that raise their cotyledons above the soil surface are said to be **epigeal**, while those whose cotyledons remain in the soil are termed **hypogeal** (see WEB TOPIC 18.2 for examples).

Auxin promotes growth in stems and coleoptiles, while inhibiting growth in roots

Auxin synthesized in the shoot apex is transported toward the tissues below. The steady supply of auxin arriving at the subapical region of the stem or coleoptile is required for the continued elongation of these cells. Because the level of endogenous auxin in the elongation region of a normal healthy plant is nearly optimal for growth, spraying the plant with exogenous auxin causes only a modest and short-lived stimulation in growth. Such spraying may even be inhibitory in the case of dark-grown seedlings, which are more sensitive to supraoptimal auxin concentrations than light-grown plants are.

However, when the endogenous source of auxin is removed by excision of stem or coleoptile sections containing the elongation zone, the growth rate rapidly decreases to a low basal rate. Such excised sections often respond to exogenous auxin by rapidly increasing their growth rate back to the level in the intact plant (**Figure 18.12**).

Auxin control of root elongation has been more difficult to demonstrate, perhaps because auxin induces the production of ethylene, which inhibits root growth. These two hormones interact differentially in root tissue to control growth. However, even if ethylene biosynthesis is specifically blocked, low concentrations (10^{-10} to 10^{-9} *M*) of auxin promote the growth of intact roots, whereas higher concentrations (10^{-6} *M*) inhibit growth. Thus, while roots may require a minimum concentration of auxin to grow, root growth is strongly inhibited by auxin concentrations that promote elongation in stems and coleoptiles.

The outer tissues of eudicot stems are the targets of auxin action

Eudicot stems are composed of many types of tissues and cells, only some of which may limit the growth rate. This point is illustrated by a simple experiment. When sections from growing regions of an etiolated eudicot stem, such as pea, are split lengthwise and incubated in buffer alone, the two halves bend outward. This result indicates that in the absence of auxin, the central tissues—including the pith, vascular tissues, and inner cortex—elongate at a faster rate than the outer tissues, which consist of the outer cortex and epidermis. Thus, the outer tissues must be limiting the extension rate of the stem in the absence of auxin (see Figure 18.12). When similar sections are incubated in buffer plus auxin, the two halves bend inward, due to auxin-induced elongation of the outer tissues of the stem. To reach these outer tissues of the elongating regions of stems and stemlike structures, auxin derived from the shoot apex must be diverted laterally from the polar transport stream in vascular parenchyma cells to the outer shoot tissues.

The minimum lag time for auxin-induced elongation is 10 minutes

When a stem or coleoptile section is excised and inserted into a sensitive growth-measuring device, the growth response to auxin can be monitored at high resolution. Without auxin in the medium, the growth rate declines rapidly. Addition of auxin markedly stimulates the growth rates of oat (*Avena sativa*) coleoptile and soybean (*Glycine max*) hypocotyl sections after a lag period of only 10 to 12

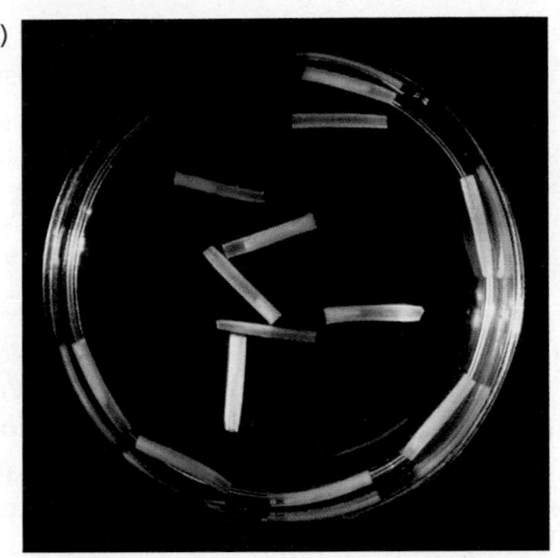

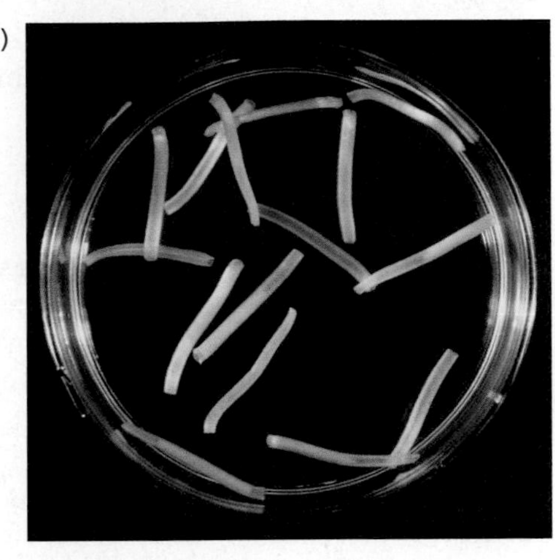

Figure 18.12 Auxin stimulates the elongation of oat coleoptile sections that have been depleted of endogenous auxin. These coleoptile sections were incubated for 18 h in either water (A) or auxin (B). The yellow inside the translucent coleoptile represents primary leaf tissue. (Photos © M. B. Wilkins.)

min (**Figure 18.13A**). The maximum growth rate, which represents a five- to tenfold increase over the basal rate, is reached after 30 to 60 min of auxin treatment. As is shown in **Figure 18.13B**, a threshold concentration of auxin must be reached to initiate this response. Beyond the optimum concentration, auxin becomes inhibitory.

The stimulation of growth by auxin requires energy, and metabolic inhibitors inhibit the response within minutes. Auxin-induced growth is also sensitive to inhibitors of protein synthesis such as cycloheximide, suggesting

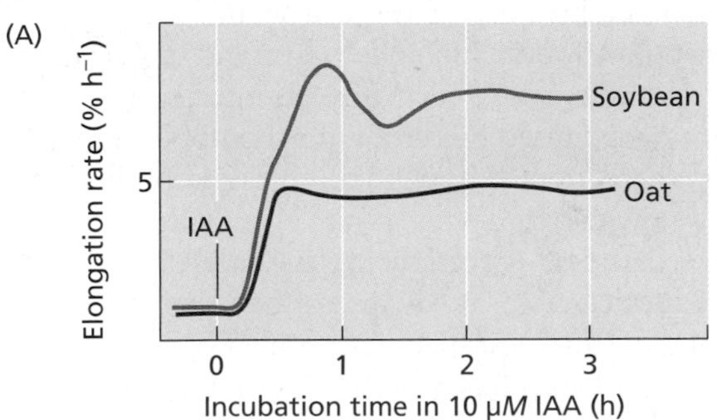

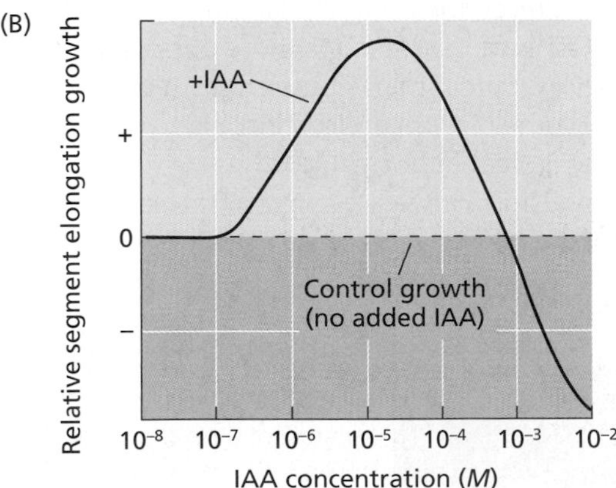

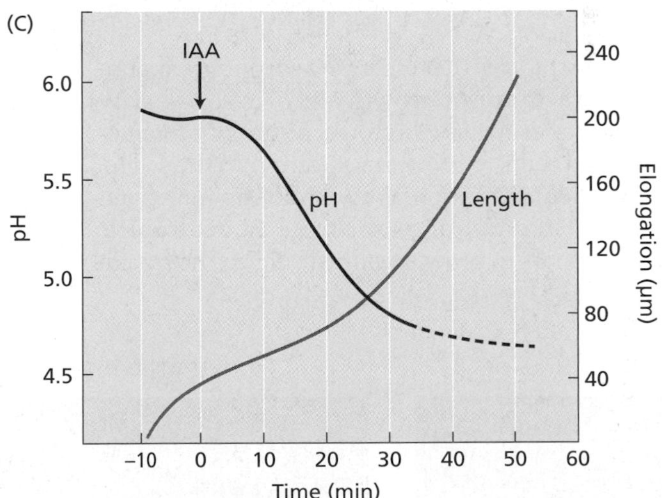

Figure 18.13 Time course and dose-response to auxin. (A) Comparison of the growth kinetics of oat coleoptile and soybean hypocotyl sections incubated with 10 μ*M* IAA (indole-3-acetic acid) and 2% sucrose. Growth is plotted as the rate of elongation, rather than as absolute length, at each time point. The growth rate of the soybean hypocotyl oscillates after 1 h, whereas that of the oat coleoptile is constant. (B) Typical dose-response curve for IAA-induced growth in pea stem or oat coleoptile sections. Elongation growth of excised sections of coleoptiles or young stems is plotted versus increasing concentrations of exogenous IAA. At concentrations above 10^{-5} *M*, IAA becomes less and less effective. Above about 10^{-4} *M* it becomes inhibitory, as shown by the fact that the stimulation decreases and the curve eventually falls below the dashed line, which represents growth in the absence of added IAA. (C) Kinetics of auxin-induced elongation and cell wall acidification in maize coleoptiles. The pH of the cell wall was measured with a pH microelectrode. Note the similar lag times (10–15 min) for both cell wall acidification and the increase in the rate of elongation. (A after Cleland 1995; C after Jacobs and Ray 1976.)

that protein synthesis is required for the response. Inhibitors of RNA synthesis also inhibit auxin-induced growth after a slightly longer delay.

Auxin-induced proton extrusion induces cell wall creep and cell elongation

According to the **acid growth hypothesis**, hydrogen ions act as an intermediate between auxin and cell wall loosening (see Chapter 14). The source of the hydrogen ions is the plasma membrane H^+-ATPase, whose activity is thought to increase in response to auxin. Auxin stimulates proton extrusion into the cell wall after 10 to 15 min of lag time, consistent with the growth kinetics, as shown in **Figure 18.13C**. As discussed in Chapter 14, cell wall–loosening proteins called **expansins** loosen the cell walls by weakening the hydrogen bonds between the polysaccharide components of the wall when the pH is acidic.

Tropisms: Growth in Response to Directional Stimuli

Plants respond to external stimuli by altering their growth and development patterns. During seedling establishment, abiotic factors such as gravity, touch, and light influence the initial growth habit of the young plant. **Tropisms** are directional growth responses in relation to environmental stimuli caused by the asymmetric growth of the plant axis (stem or root). Tropisms may be positive (growth toward the stimulus) or negative (growth away from the stimulus).

One of the first forces that emerging seedlings encounter is gravity. **Gravitropism**, growth in response to gravity, enables shoots to grow upward toward sunlight for photosynthesis, and roots to grow downward into the soil for water and nutrients. As soon as the shoot tip penetrates the soil surface, it encounters sunlight. **Phototropism** enables leafy shoots to grow toward sunlight, thus maximizing photosynthesis, while some roots grow away from sunlight. **Thigmotropism**, differential growth in response to touch, helps roots grow around obstacles and twining vines and tendrils to wrap around other structures for support.

Gravitropism involves the lateral redistribution of auxin

When dark-grown *Avena* seedlings are oriented horizontally, the coleoptiles bend upward in response to gravity. According to the **Cholodny–Went hypothesis**, a general model that applies to all tropism responses, auxin in a horizontally oriented coleoptile tip is transported laterally to the lower side, causing the lower side of the coleoptile to grow faster than the upper side. Early experimental evidence indicated that the tip of the coleoptile could perceive gravity and redistribute auxin to the lower side (see Chapter 16). For example, if coleoptile tips are oriented horizontally, a greater amount of auxin diffuses into the agar block from the lower half than from the upper half, as demonstrated by bioassay (**Figure 18.14**).

Tissues below the tip are able to respond to gravity as well. For example, when vertically oriented maize coleoptiles are decapitated by removing the upper 2 mm of the tip and then oriented horizontally, gravitropic bending occurs at a slow rate for several hours even without the tip. Application of indole-3-acetic acid (IAA), the major auxin, to the cut surface restores the rate of bending to normal levels. This finding indicates that both the perception of the gravitational stimulus and the asymmetric diversion of auxin can occur in the tissues below the tip, although the tip is still required for auxin production.

Lateral redistribution of auxin is more difficult to demonstrate in shoot apical meristems than in coleoptiles because of the presence of auxin recirculation in developing leaf and apical shoot primordia, similar to that observed in root tips. However, some of the same differ-

(A)

(B)

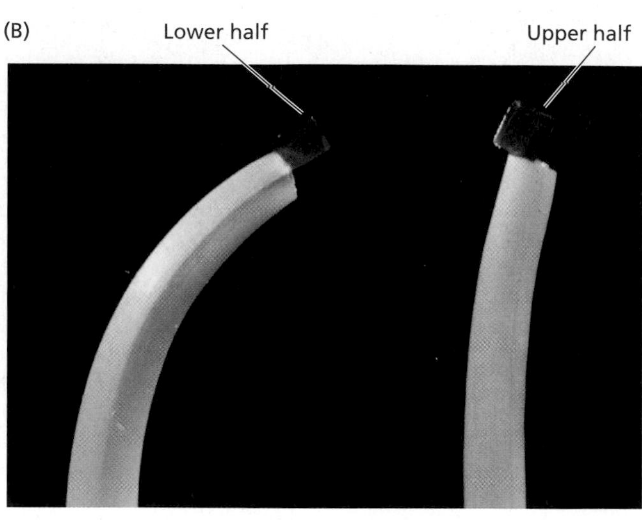

Figure 18.14 Auxin is transported to the lower side of a horizontally oriented oat coleoptile tip. (A) Auxin from the upper and lower halves of a horizontal tip is allowed to diffuse into two agar blocks. (B) The agar block from the lower half (left) induces greater curvature in a decapitated coleoptile than the agar block from the upper half (right) does. (Photos © M. B. Wilkins.)

ential auxin transport mechanisms seen in phototropic bending are also involved in shoot gravitropic bending.

Polar auxin transport requires energy and is gravity independent

The polarity of auxin transport in the developing embryo was discussed in Chapter 17. **Figure 18.15A** illustrates the use of the terms **basipetal** (toward the base) and **acropetal** (toward the apex) when discussing the direction of auxin movement. For simplicity we will sometimes use the terms *rootward* and *shootward* to refer to the downward and upward movements of auxin, respectively.

Early studies of polar auxin transport were carried out using the *donor–receiver agar block method* (**Figure 18.15B**). An agar block containing radioisotope-labeled auxin (donor block) is placed on one end of a tissue segment, and a receiver block is placed on the other end. The movement of auxin through the tissue into the receiver block can be determined over time by measuring the radioactivity in the receiver block. This method has been refined to allow for the deposition of much smaller droplets of radiolabeled auxin onto discrete surfaces of plants, improving the accuracy of transport studies over short distances.

From such studies, the general properties of polar auxin transport have emerged. Tissues differ in the degree of polarity of auxin transport. In coleoptiles, vegetative stems, leaf petioles, and the root epidermis, transport from the apices predominates, whereas in the stelar tissues of the root, auxin is transported toward the root apex. Polar auxin transport is not affected by the orientation of the tissue (at least over short periods of time), so it is independent of gravity.

A demonstration of the lack of gravity effects on polar auxin transport is shown in **Figure 18.16**. In this experiment, grape hardwood cuttings are placed in a moist chamber, allowing the formation of adventitious roots at the basal ends of the cuttings, while adventitious shoots form at the apical ends. The same polarity of root and shoot formation occurs even when the cuttings are inverted. Roots form at the base because root differentiation is stimulated by auxin accumulation due to polar transport. Shoots tend to form at the apical end where the auxin concentration is lowest.

Polar auxin transport proceeds in a cell-to-cell fashion, rather than via the symplast; that is, auxin exits the cell through the plasma membrane, diffuses across the compound middle lamella, and enters the next cell through its plasma membrane. The export of auxin from cells is termed *auxin efflux*; the entry of auxin into cells is called *auxin uptake* or *influx*. The overall process requires metabolic energy, as

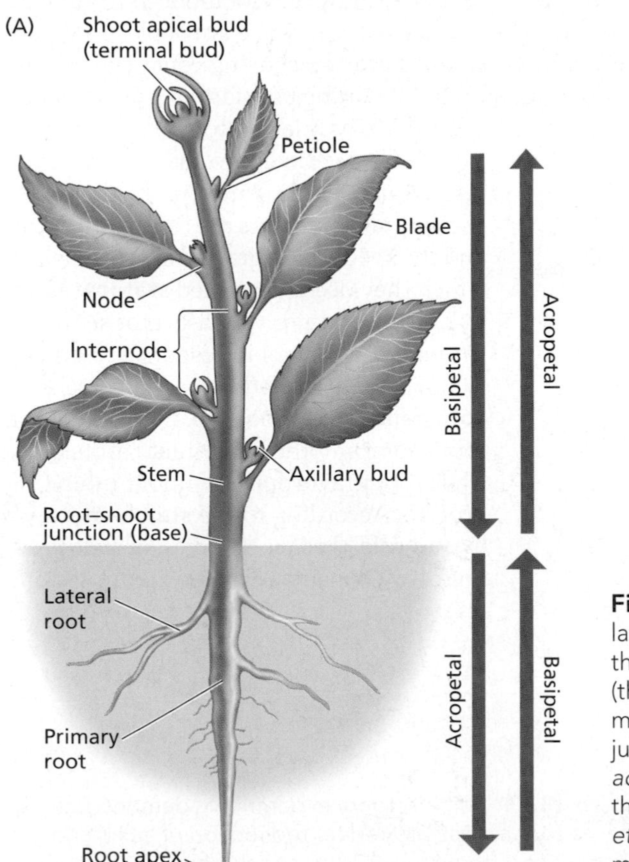

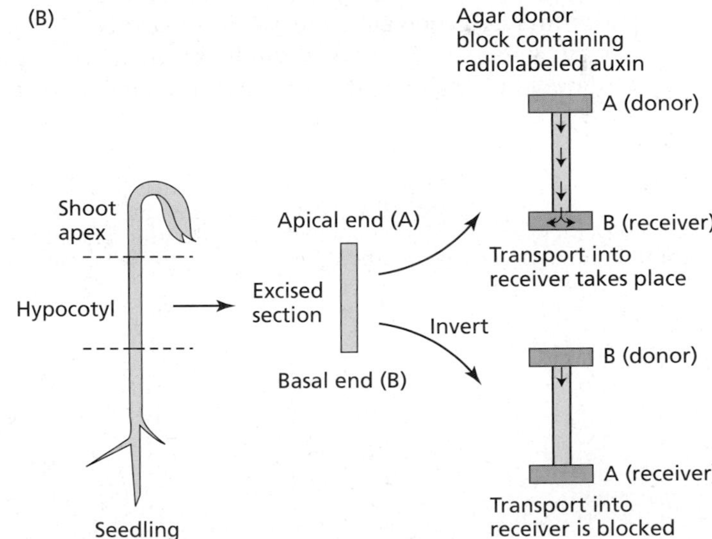

Figure 18.15 Demonstration of polar auxin transport with radiolabeled auxin. (A) Polar auxin transport is described in terms of the direction of its movement in relation to the base of the plant (the root–shoot junction). Auxin moving downward from the shoot moves *basipetally* (toward the base) until it reaches the root–shoot junction. From that point, downward movement is described as *acropetal* (toward the apex). Movement of auxin from the apex of the root toward the root–shoot junction is also described as *basipetal* (toward the base). (B) Donor–receiver agar block method for measuring polar auxin transport. The polarity of transport is independent of the orientation of the plant tissue with respect to gravity.

Adventitious roots
Adventitious shoot
Adventitious shoot

Figure 18.16 Adventitious roots grow from the basal ends of grape hardwood cuttings, and adventitious shoots grow from the apical ends, whether the cuttings are maintained in the inverted orientation (the two cuttings on the left) or the upright orientation (the cuttings on the right). The roots always form at the basal ends because polar auxin transport is independent of gravity. (From Hartmann and Kester 1983.)

evidenced by the sensitivity of polar auxin transport to O_2 deprivation, sucrose depletion, and metabolic inhibitors.

The velocity of polar auxin transport can exceed 3 mm h^{-1} in some tissues, which is faster than diffusion but slower than phloem translocation rates (see Chapter 11). Higher rates of polar auxin transport are observed in tissues immediately adjacent to the shoot and root apical meristems. Polar transport is specific for active auxins, both natural and synthetic; other weak organic acids, inactive auxin analogs, and IAA conjugates are poorly transported. The specificity of polar auxin transport indicates that it is mediated by protein carriers on the plasma membrane.

According to the starch–statolith hypothesis, specialized amyloplasts serve as gravity sensors in root caps

In addition to protecting the sensitive cells of the apical meristem as the tip penetrates the soil, the root cap is the site of gravity perception. Because the cap is some distance away from the elongation zone where bending occurs, graviresponsive signaling events initiated in the root cap must induce production of a chemical messenger that modulates growth in the elongation zone. Microsurgery experiments in which half of the cap was removed showed that the cap supplies a root growth inhibitor, later identified as auxin, to the lower side of the root during gravitropic bending (**Figure 18.17**).

The primary mechanism by which gravity can be detected by cells is via the motion of a falling or sedimenting body. Obvious candidates for intracellular gravity sensors in plants are the large, dense amyloplasts that are present in specialized gravity-sensing cells. These large amyloplasts (starch-containing plastids) are of sufficiently high density relative to the cytosol that they readily sediment to the bottom of the cell (**Figure 18.18**). Amyloplasts that function as gravity sensors are called **statoliths**, and the specialized gravity-sensing cells in which they occur are called **statocytes**.

Large, dense amyloplasts that sediment through the cytosol in response to gravity (statoliths) are located in the central cells, or **columella**, of the root cap. Removal of the root cap from otherwise intact roots abolishes root gravitropism without inhibiting growth. According to the **starch–statolith hypothesis**, these cells represent statocytes, or gravity-sensing cells (see Figure 18.18).

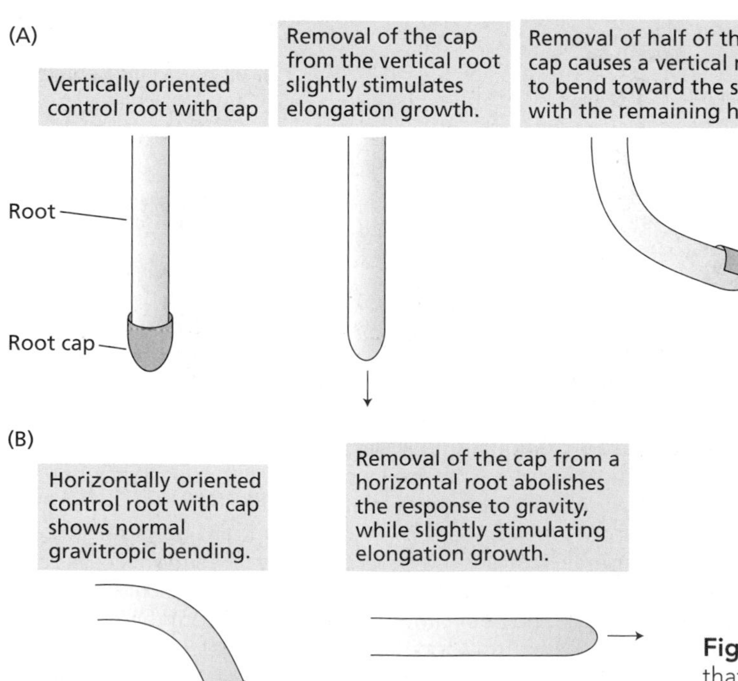
(A)
Vertically oriented control root with cap
Removal of the cap from the vertical root slightly stimulates elongation growth.
Removal of half of the cap causes a vertical root to bend toward the side with the remaining half-cap.
Root
Root cap
(B)
Horizontally oriented control root with cap shows normal gravitropic bending.
Removal of the cap from a horizontal root abolishes the response to gravity, while slightly stimulating elongation growth.

Figure 18.17 Microsurgery experiments demonstrate that the root cap is required for redirection of auxin and subsequent differential inhibition of elongation in root gravitropic bending. (After Shaw and Wilkins 1973.)

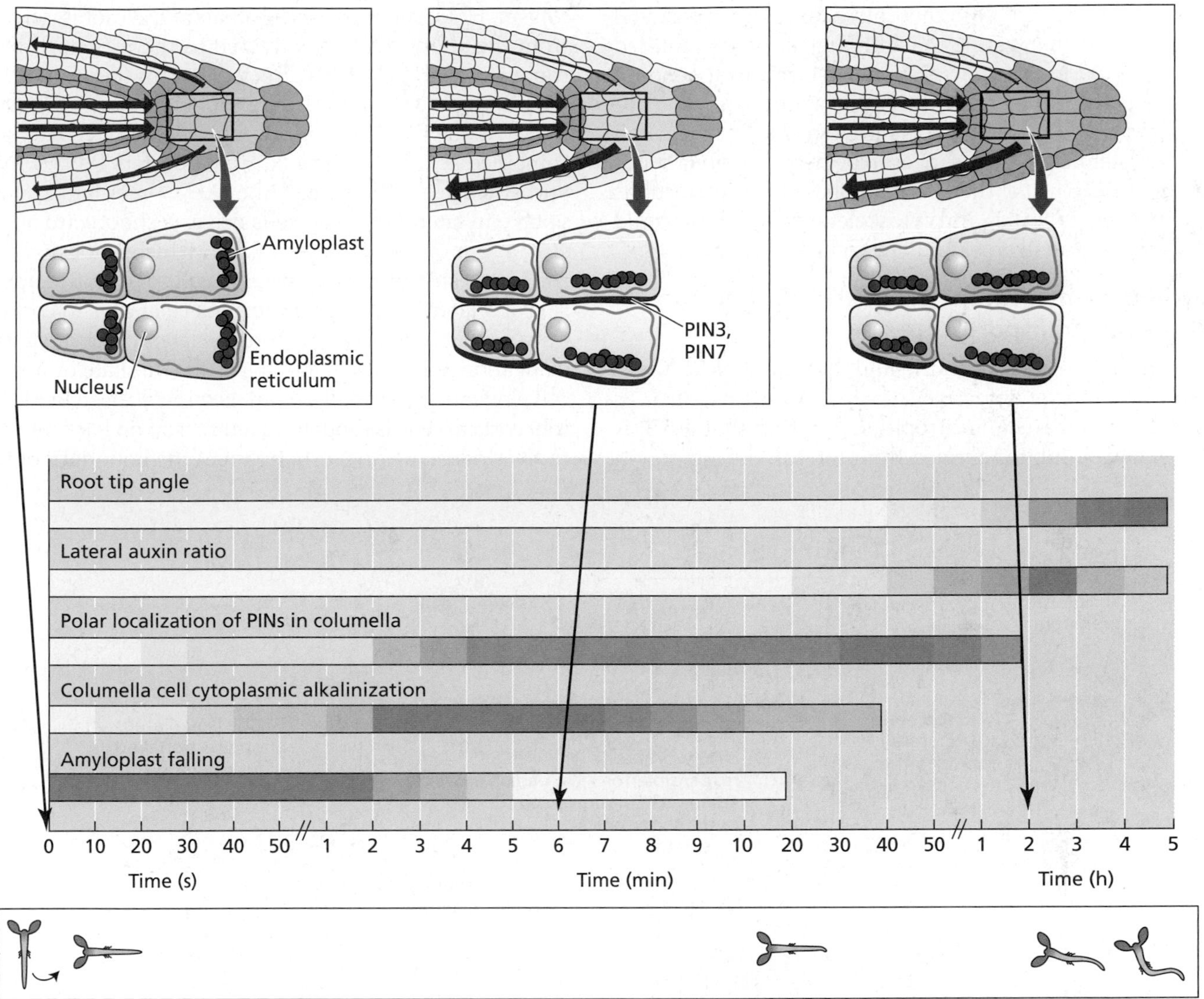

Figure 18.18 Sequence of events following gravistimulation of an Arabidopsis root. The time scale on the bottom is nonlinear. The shaded horizontal bars in the graph indicate the timing of various events in gravistimulated roots, with the darkest shading corresponding to the greatest change. The growth of the seedling at different stages of the response is illustrated below the time scale. Three stages of statolith sedimentation are shown on the top. The left figure shows time zero, when the seedling is first rotated 90°. The second and third stages shown are at about 6 min and 2 h after rotation. The red arrows indicate auxin flow, with thicker arrows indicating more flow. Cells with relatively high auxin concentrations are shown in orange. Columella cells of the root tip are shown in green at time zero; the color changes to blue and then to blue-green at later stages to indicate the degree of alkalinization of the cytoplasm. The distribution of PIN3 is diagrammed as a purple outline on the plasma membrane of the columella cells. (After Baldwin et al. 2013.)

Precisely how the statocytes sense their falling statoliths is still poorly understood. According to one hypothesis, contact or pressure resulting from the amyloplasts resting on the ER on the lower side of the cell triggers the response (see Figure 18.18). The predominant form of ER in columella cells is the tubular type, but an unusual form of ER, called "nodal ER," is also present and may play a role in the gravity response.

The starch–statolith hypothesis of gravity perception in roots is supported by several lines of evidence. Amyloplasts are the only organelles that consistently sediment in the columella cells of different plant species, and the rate of sedimentation correlates closely with the time required to perceive the gravitational stimulus (see Figure 18.18). The gravitropic responses of starch-deficient mutants are generally much slower than those of wild-type plants.

Nevertheless, starchless mutants exhibit some residual gravitropism, suggesting that although starch is required for a normal gravitropic response, starch-independent gravity perception mechanisms may also exist.

Other organelles, such as nuclei, may be dense enough to act as statoliths. It may not even be necessary for a statolith to come to rest at the bottom of the cell, as interactions with endomembranes and cytoskeletal components could transduce a gravitropic signal in an unknown manner.

Auxin movements in the root are regulated by specific transporters

Although root caps contain small amounts of IAA, the roots of mutants defective in auxin transport, such as *aux1* and *pin2,* are agravitropic, suggesting that auxin is the growth inhibitor derived from the cap during gravitropism. However, most of the auxin in the root is derived from the shoot. IAA is delivered to the root apex by a rootward PIN1/ABCB19–directed stream (**Figure 18.19**). IAA is also synthesized in the root meristem. However, the hormone is excluded from root cap apical cells by the combined activity of the auxin transporters PIN3, PIN4, and ABCB1. At the same time, AUX1-mediated auxin uptake in lateral root cap cells drives a shootward auxin stream out of the root apex. PIN2, which is localized at the upper side of root epidermal cells and at the upper side and lateral side facing the epidermal cells in cortical cells, conducts auxin away from the lateral root cap to the elongation zone, where auxin acts to stimulate or inhibit cell elongation. In addition, an *auxin reflux loop* model in root cortical cells is thought to redirect auxin back into the rootward stelar transport stream at the boundary of the

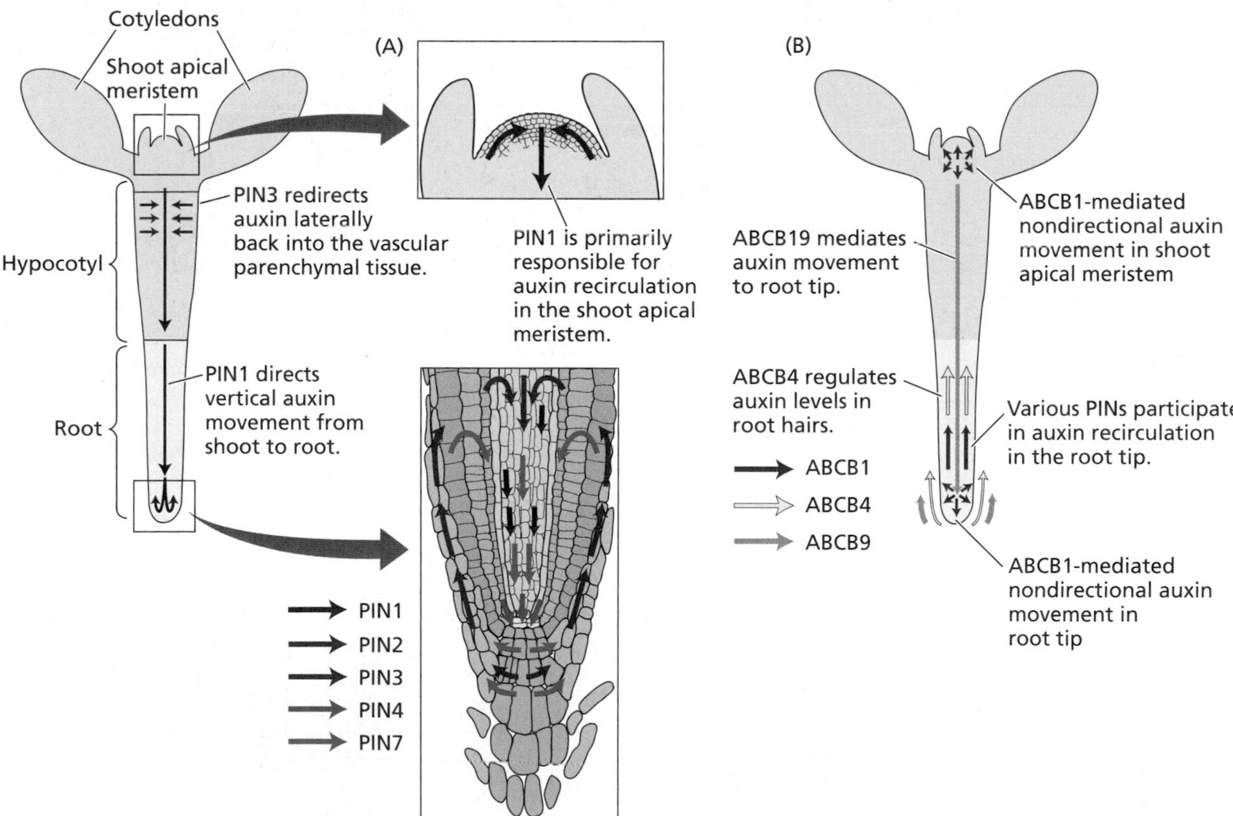

Figure 18.19 In Arabidopsis, PIN and ABCB transport proteins direct the auxin efflux component of polar auxin transport throughout the plant. (A) PIN proteins determine the basal direction of auxin movement. Directional auxin movement is associated with the tissue-specific distribution of PIN efflux carrier proteins. PIN1 mediates vertical transport of IAA from the shoot to the root along the embryonic apical–basal axis (see Figure 17.4) and creates an auxin sink that drives basipetal auxin transport upward from the root apex via PIN2 efflux carriers. Since some lateral diffusion of auxin may occur, PIN7 and PIN3 are thought to redirect auxin back into vascular parenchymal tissue, where polar transport takes place. The two inserts show PIN1-mediated auxin movement in the shoot apical meristem (upper) and PIN-regulated auxin circulation in the root tip (lower). (B) Auxin flow associated with ATP-dependent ABCB transport proteins. The multidirectional arrows at the shoot and root apices indicate nondirectional auxin transport. However, when combined with polarly localized PIN proteins, directional transport occurs. ABCB4 regulates auxin levels in elongating root hairs. (A, root model after Blilou et al. 2005.)

elongation zone (see Figure 18.19). Auxin circulation at the growing tip may allow root growth to continue for a time independent of auxin from the shoot, as well as move auxin synthesized from the root tip into the reflux stream.

The gravitropic stimulus perturbs the symmetric movement of auxin from the root tip

According to the current model for gravitropism, shootward auxin transport in a vertically oriented root is equal on all sides. When the root is oriented horizontally, however, the signals from the cap redirect most of the auxin to the lower side, thus inhibiting the growth of that lower side (see Figure 18.18). Consistent with this model, the transport of [^{3}H]IAA across a horizontally oriented root cap is polar, with a preferential downward movement. The downward movement of auxin across a horizontal root cap has been confirmed using a reporter gene construct, *DR5:GFP,* consisting of green fluorescent protein (GFP) expressed under the control of the auxin-sensitive *DR5* promoter.

One of the members of the PIN protein family, PIN3, is thought to participate in the redirection of auxin in roots that are displaced from the vertical orientation. In a vertically oriented root, PIN3 is uniformly distributed around the columella cells, but when the root is placed on its side, PIN3 is preferentially targeted to the lower side of these cells (see Figure 18.18). This redistribution of PIN3 is thought to accelerate auxin transport to the lower side of the cap. However, as *pin3* mutants are not completely agravitropic, other asymmetric events might act along with PIN3 localization to alter auxin flows. The most likely event would be an asymmetric change in apoplastic acidification, which would impose an asymmetric chemiosmotic potential to redirect auxin flow. This would cause PIN3 redistribution, which would amplify the flow of auxin in the new direction (canalization; see Chapter 19).

Gravity perception in eudicot stems and stemlike organs occurs in the starch sheath

In eudicot stems and stemlike organs, the statoliths involved in gravity perception are located in the **starch sheath**, the innermost layer of cortical cells that surrounds the ring of vascular bundles of the shoot (**Figure 18.20**). The starch sheath is continuous with the endodermis of the root, but unlike in the endodermis, its cells contain amyloplasts that are redistributed when the gravity vector is changed.

Genetic studies have confirmed the central role of the starch sheath in shoot gravitropism. Arabidopsis mutants lacking amyloplasts in the starch sheath display agravitropic shoot growth but normal gravitropic root growth. As noted in Chapter 17 in the *scarecrow* (*scr*) mutant of Arabidopsis the cell layer from which the endodermis and the starch sheath are derived remains undifferentiated. As a result, the hypocotyl and inflorescence of the *scr* mutant

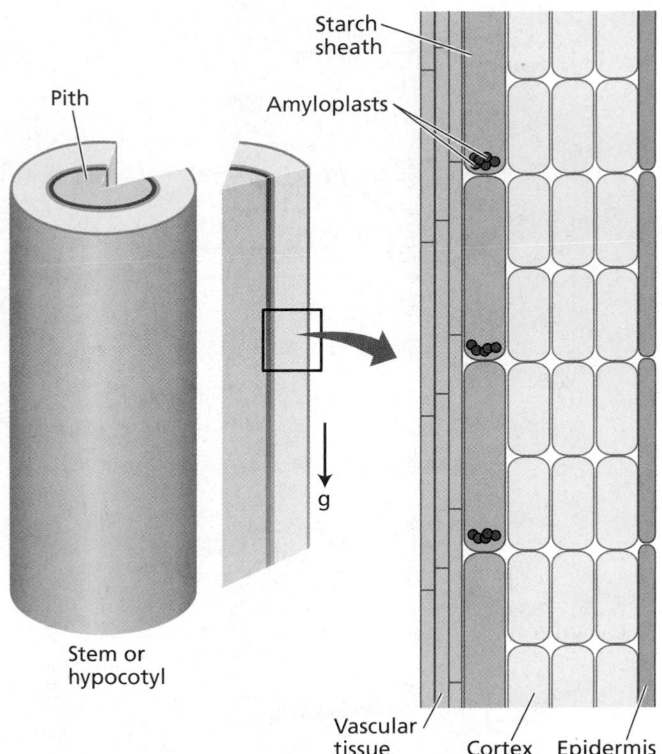

Figure 18.20 Diagram of the starch sheath located outside the ring of vascular tissue. The cutaway view shows the amyloplasts at the bottom of the cells. (After Volkmann et al. 1979.)

are agravitropic, although the root exhibits a normal gravitropic response.

As in the case of root gravitropism, the site of gravity perception (starch sheath) is located at some distance from the site of the auxin-mediated gravity response (outer cortex and epidermis). Auxin transporters play a pivotal role in directing auxin to its target tissues. The cells of the starch sheath contain ABCB19 and PIN3, which function coordinately to restrict auxin streams to the vascular system (**Figure 18.21**). Selective regulation of the downward auxin transport stream conducted by PIN1 inside the vascular cylinder and selective restriction of lateral auxin movement into the starch sheath cells by ABCB19 and PIN3 appear to play a fundamental role in tropic bending.

Gravity sensing may involve pH and calcium ions (Ca^{2+}) as second messengers

A variety of experiments suggest that localized changes in pH and Ca^{2+} gradients are part of the signaling that occurs during gravitropism. Changes in intracellular pH can be detected early in root columella cells responding to gravity (see Figure 18.18). When pH-sensitive dyes were used to monitor both intracellular and extracellular pH in Arabidopsis roots, rapid changes were observed after roots were

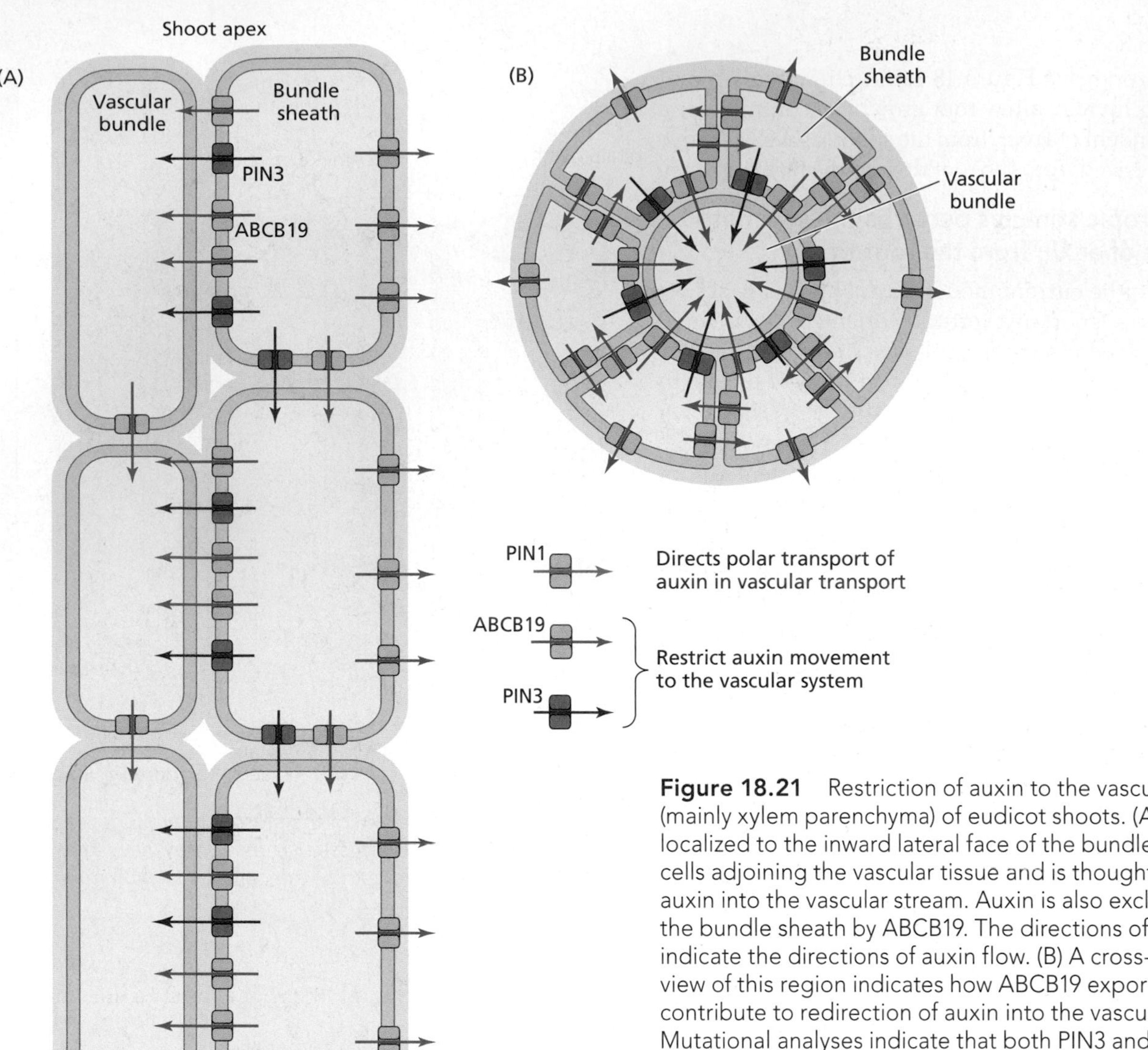

Figure 18.21 Restriction of auxin to the vascular tissue (mainly xylem parenchyma) of eudicot shoots. (A) PIN3 is localized to the inward lateral face of the bundle sheath cells adjoining the vascular tissue and is thought to redirect auxin into the vascular stream. Auxin is also excluded from the bundle sheath by ABCB19. The directions of the arrows indicate the directions of auxin flow. (B) A cross-sectional view of this region indicates how ABCB19 export would contribute to redirection of auxin into the vascular cylinder. Mutational analyses indicate that both PIN3 and ABCB19 function in lateral redistribution of auxin in tropic bending.

rotated to a horizontal position. Within 2 min of gravistimulation, the cytoplasmic pH of the columella cells of the root cap increased from 7.2 to 7.5 (**Figure 18.22**), while the apoplastic pH declined from 5.5 to 4.5. These changes preceded any detectable tropic curvature by about 10 min.

The alkalinization of the cytosol combined with the acidification of the apoplast suggests that activation of the plasma membrane H^+-ATPase is one of the initial events that mediates root gravity perception or signal transduction. The chemiosmotic model of polar auxin transport (see Figure 17.11) predicts that differential acidification of the apoplast and alkalinization of the cytosol would result in increased directional uptake and efflux of IAA from the affected cells.

Early physiological studies suggested that Ca^{2+} release from storage pools might be involved in root gravitropic signal transduction. For example, treatment of maize roots with EGTA [ethylene glycol-bis(β-aminoethyl ether)-N,N,N′,N′-tetraacetic acid], a compound that can chelate (form a complex with) Ca^{2+}, prevents Ca^{2+} uptake by cells and inhibits root gravitropism. As in the case of [^{3}H]IAA, $^{45}Ca^{2+}$ is polarly transported to the lower half of a root cap that is stimulated by gravity. Auxin-dependent Ca^{2+} and pH signaling thus appears to regulate root gravitropic bending through the propagation of a Ca^{2+}-dependent signaling pathway. Changes in extracellular pH may also be an important signaling element that could modulate auxin responses by altering the chemisosmotic proton gradient.

For a discussion of how root gravitropism interacts with **circumnutation**, the endogenously regulated spiraling growth pattern of the root tip, and thigmotropism, see WEB TOPIC 18.5.

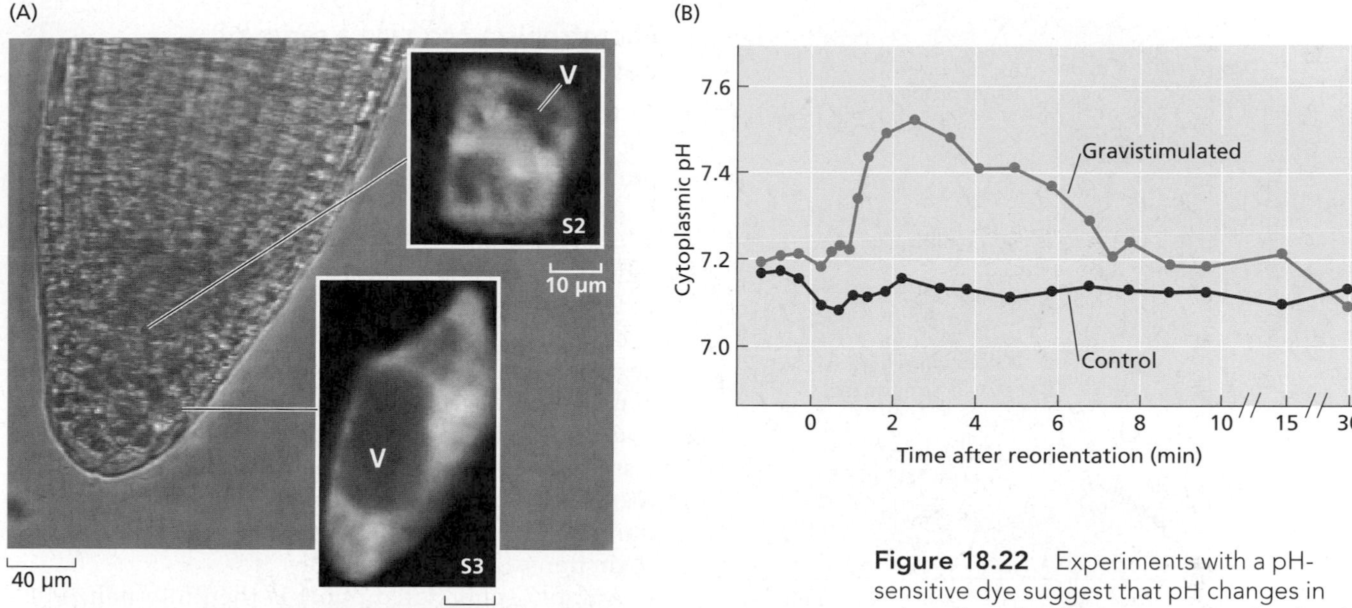

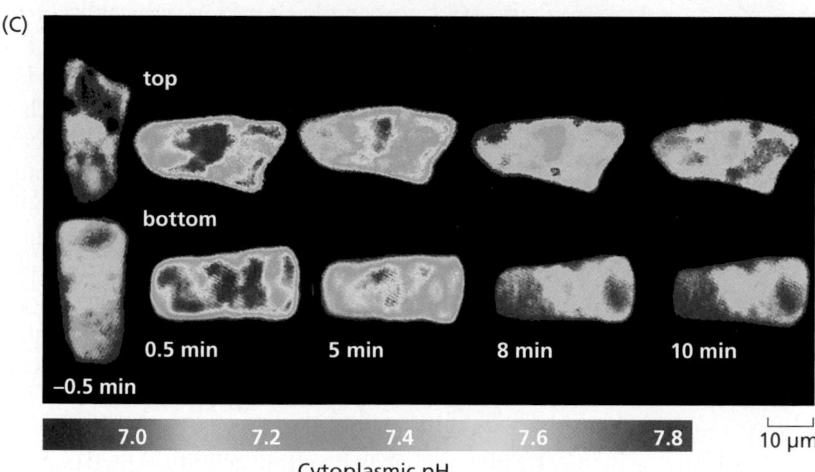

Figure 18.22 Experiments with a pH-sensitive dye suggest that pH changes in columella cells of the root cap are involved in gravitropic signal transduction. (A) Micrograph showing magnification of the root tip and two columella cells at different levels (stories) of the root cap, labeled S2 (story 2) and S3 (story 3) (insets). The cytosols of the two columella cells are fluorescing because the cells have been microinjected with a pH-sensitive fluorescent dye. The vacuoles (labeled V) contain no dye and therefore appear dark. (B) Cytoplasmic pH increases in less than 1 min after gravistimulation. (C) Imaging of pH-sensitive dyes in the response of the two columella cells in (A) to gravitropic stimulus. The color scale below was used to generate the data in (B). (From Fasano et al. 2001.)

Phototropism

Whatever the angle of sunlight, an emergent seedling is able to bend toward it to optimize light absorption, a phenomenon known as phototropism. As we saw in Chapter 16, blue light is particularly effective in inducing phototropism, and two flavoproteins, **phototropins 1** and **2**, are the photoreceptors for phototropic bending. Phototropism requires downstream signaling events that are posttranslational and which occur rapidly to cause bending growth. As in the case of gravitropism, the bending response to directional blue light can be explained by the Cholodny–Went model of lateral auxin redistribution.

Phototropism is mediated by the lateral redistribution of auxin

Charles and Francis Darwin provided the first clue concerning the mechanism of phototropism on coleoptiles by demonstrating that while light is perceived at the tip, bending occurs in the region below the tip. The Darwins proposed that some "influence" was transported from the tip to the growing region, thus causing the observed asymmetric growth response. This influence was later shown to be IAA.

When a shoot is growing vertically, auxin is transported polarly from the growing tip to the elongation zone. The polarity of auxin transport from shoot to root is independent of gravity. However, auxin can also be transported laterally, and this lateral diversion of auxin lies at the heart of the Cholodny–Went model for tropisms. In gravitropic bending, auxin from the root tip that is redirected to the lower side of the root inhibits cell elongation, causing the root to bend downward. In phototropic bending, the auxin for the shoot tip that is redirected to the shaded side of the axis stimulates cell

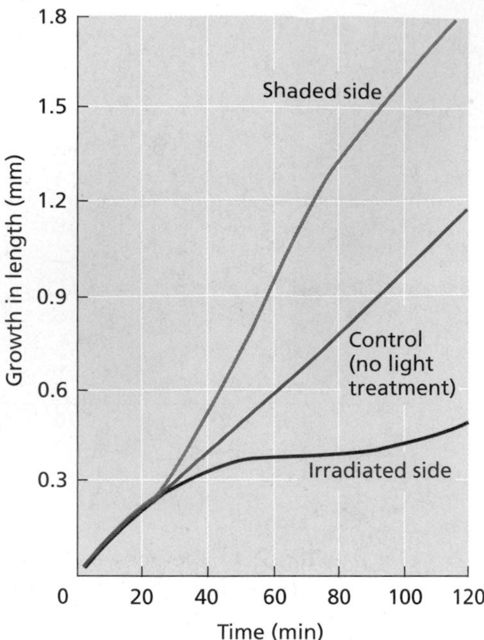

Figure 18.23 Time course of growth on the irradiated and shaded sides of a coleoptile responding to a 30-s pulse of unidirectional blue light. Control coleoptiles were not given a light treatment. (After Iino and Briggs 1984.)

elongation. The resulting differential growth causes the shoot to bend toward the light (**Figure 18.23**).

Although phototropic mechanisms appear to be highly conserved across plant species, the precise sites of auxin production, light perception, and lateral transport have been difficult to define. In maize coleoptiles, auxin accumulates in the upper 1 to 2 mm of the tip. The zones of photosensing and lateral transport extend farther, within the upper 5 mm of the tip. The response is also strongly dependent on the light fluence (the number of photons per unit area). Similar zones of auxin synthesis/accumulation, light perception, and lateral transport are seen in the true shoots of all monocots and eudicots examined to date.

Acidification of the apoplast appears to play a role in phototropic growth: The apoplastic pH on the shaded side of phototropically bending stems or coleoptiles is more acidic than on the irradiated side. Decreased pH increases auxin transport by increasing both the rate of IAA entry into the cell and the chemiosmotic proton potential–driven auxin efflux mechanisms. According to the acid growth hypothesis, this acidification would also be expected to enhance cellular elongation. Both processes—enhanced auxin transport and increased cell elongation on the shaded side—would be expected to contribute to bending toward light.

For a description of negative phototropism in roots, see WEB TOPIC 18.6.

Phototropism occurs in a series of posttranslational events

As we mentioned earlier, the events in phototropic bending occur rapidly. Although phototropins are hydrophilic proteins, they are associated with the plasma membrane. In Arabidopsis, low-fluence blue light is perceived by the cells on the irradiated side of the hypocotyl and a series of signal transduction events is initiated.

During the first minute, new longitudinally oriented microtubules are formed and preexisting microtubules are degraded (see WEB TOPIC 18.7). After approximately 3 min of unilateral blue-light irradiation, phototropin 1 (phot1) undergoes autophosphorylation and some of the proteins dissociate from the plasma membrane. Next, the activated phot1 on the plasma membrane phosphorylates the auxin transporter, ABCB19, inhibiting its activity and blocking auxin transport (see below).

Autophosphorylated phot1 is then internalized by clathrin-mediated endocytosis. The function of internalization is unclear, but may play a role in either phototropin signaling or receptor desensitization. For example, the protein NON-PHOTOTROPIC HYPOCOTYL 3 (NPH3), originally identified as a nonbending mutant (*nph3*), is a substrate adapter for a ubiquitin ligase. NPH3 is localized on the plasma membrane and is dephosphorylated following exposure to blue light. The dephosphorylated NPH3 interacts with phot1, and phot1 is then targeted for degradation by the 26S proteasome. Counterintuitively, NPH3-mediated ubiquitination of phot1 appears to be required for phototropic bending. There is circumstantial evidence that ubiquitination may enhance the endosomal transport of phot1 to other parts of the cell.

As noted above, blue light–induced phosphorylation of ABCB19 by phot1 inhibits its efflux activity. ABCB19 plays an important role in transporting auxin out of the shoot apex, and it also maintains long-distance auxin transport streams by preventing cellular reuptake and diffusion into neighboring tissues. ABCB19 thus functions with PIN1 to facilitate polar auxin transport from the apical tissues to the roots. As a result of ABCB19 inhibition, auxin accumulates above the cotyledonary node and less auxin is delivered to the elongation zone, causing hypocotyl elongation to cease (**Figure 18.24**). After the pause in elongation, PIN3-mediated basipetal auxin transport preferentially resumes on the shaded side of the seedling. Auxin accumulation on the shaded side of the upper hypocotyl can be detected after approximately 15 min of exposure to unilateral blue light. In addition, there is an increase in the auxin concentration in the vascular cylinder of the hypocotyl at and below the elongation zone (see Figure 18.24). Bending toward the blue-light source begins after approximately 2 h.

Although phototropins are the primary photoreceptors for phototropism, phytochromes and cryptochromes can also contribute to the response (see WEB TOPIC 18.8).

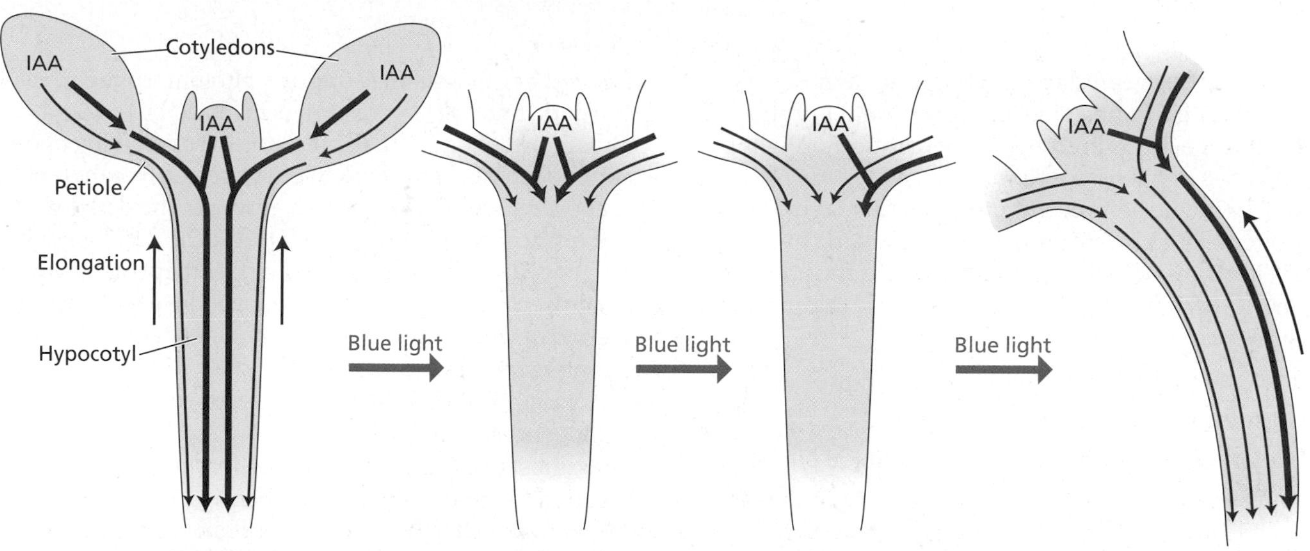
Cotyledons
IAA
IAA
IAA
Petiole
Elongation
Hypocotyl
Blue light
IAA
Blue light
IAA
Blue light
IAA
1. In the dark, auxin primarily moves from the shoot to the root through the vascular tissues in the petioles and hypocotyl, and through the epidermis.
2. After exposure to unidirectional blue light, auxin movement briefly stops at the cotyledonary node and the seedling stops growing vertically.
3. Auxin is redistributed to the shaded side and polar transport resumes.
4. The cells on the shaded side of the hypocotyl elongate, resulting in differential growth, and the seedling bends toward the light source.

Figure 18.24 Model of basipetal auxin movement (red lines) associated with phototropism in dark-acclimated seedlings of Arabidopsis. (After Christie et al. 2011.)

Photomorphogenesis

The shoots of dark-grown seedlings are **etiolated**—that is, they have long, spindly hypocotyls, an apical hook, closed cotyledons, and nonphotosynthetic proplastids, which causes the unexpanded leaves to have a pale yellow color. In contrast, light-grown seedlings have shorter, thicker hypocotyls, open cotyledons, and expanded leaves with photosynthetically active chloroplasts (**Figure 18.25**). Development in the dark is termed **skotomorphogenesis**, while development in the presence of light is called **photomorphogenesis**. When dark-grown seedlings are transferred to the light, photomorphogenesis takes over and the seedlings are said to be **de-etiolated**.

(A) Light-grown maize

(B) Dark-grown maize

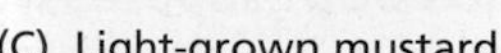

(C) Light-grown mustard

(D) Dark-grown mustard

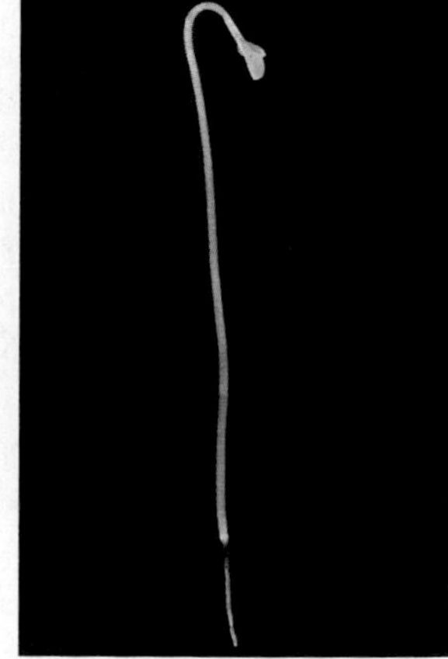

Figure 18.25 Light- and dark-grown monocot and eudicot seedlings. (A and B) Maize (corn; *Zea mays*) and (C and D) mustard (*Eruca* sp.) seedlings grown either in the light (A and C) or the dark (B and D). Symptoms of etiolation in maize, a monocot, include the absence of greening, reduction in leaf width, failure of leaves to unroll, and elongation of the coleoptile and mesocotyl. In mustard, a eudicot, etiolation symptoms include absence of greening, reduced leaf size, hypocotyl elongation, and maintenance of the apical hook. (A and B, photos courtesy of Patrice Dubois; C and D, photos by David McIntyre.)

The switch between dark- and light-grown development involves genome-wide transcriptional and translational changes triggered by the perception of light by several classes of photoreceptors (see Chapter 16). Despite the complexity of the process, the transition from skotomorphogenesis to photomorphogenesis is surprisingly rapid. Within minutes of applying a single flash of light to a dark-grown bean seedling, several developmental changes occur:

- A decrease in the rate of stem elongation
- The beginning of apical hook opening
- Initiation of the synthesis of photosynthetic pigments

Light thus acts as a signal to induce a change in the form of the seedling, from one that facilitates growth beneath the soil to one that will enable the plant to efficiently harvest light energy and convert it into the essential sugars, proteins, and lipids necessary for growth.

Among the different photoreceptors that can promote photomorphogenic responses in plants, the most important are those that absorb red and blue light. Phytochrome is a protein-pigment photoreceptor that absorbs red and far-red light most strongly, but it also absorbs blue light. Phytochrome mediates several aspects of vegetative and reproductive development, including germination, photomorphogenesis, and flowering (see Chapter 20). Cryptochromes are flavoproteins that mediate many blue-light responses involved in photomorphogenesis, including the inhibition of hypocotyl elongation, cotyledon expansion, and petiole elongation.

As we saw in Chapter 16, photomorphogenesis is negatively regulated. In the dark, many of the transcription factors that regulate photomorphogenesis are degraded in the nucleus via COP1-mediated ubiquitination and the 26S proteasome, while in the light this process is prevented, allowing photomorphogenesis to proceed. Plant hormones act to coordinate these changes throughout the plant.

Gibberellins and brassinosteroids both suppress photomorphogenesis in the dark

In the dark, the level of phytochrome in the Pfr (far red–absorbing) form is low. Since Pfr inhibits hypocotyl sensitivity to gibberellins, endogenous gibberellins promote hypocotyl cell elongation to a greater extent in the dark than in the light, causing the spindly appearance of dark-grown seedlings. In the light, Pr (the red-absorbing form of phytochrome) is converted to Pfr, which causes the hypocotyl to become less sensitive to gibberellins. As a result, hypocotyl elongation is greatly reduced and the seedling undergoes de-etiolation. For this reason, gibberellin-deficient mutant peas grown in the dark appear de-etiolated, although they lack chlorophyll, which requires light for synthesis (see Chapter 7). Taken together, these results indicate that gibberellins suppress photomorphogenesis in the dark, and the suppression is reversed by red light.

Brassinosteroids play a parallel role in suppressing photomorphogenesis in the dark. Genetic screens for mutants that appeared de-etiolated when grown in the dark led to the identification of the *DE-ETIOLATED2* (*DET2*) gene, which encodes a brassinosteroid biosynthetic gene. *det2* loss-of-function mutants have reduced levels of brassinosteroids, resulting in a de-etiolated appearance of the seedling even when grown in the dark (**Figure 18.26**). Thus, brassinosteroids, like gibberellins, suppress photomorphogenesis in the dark.

Brassinosteroids are also required for promotion of cell elongation by gibberellins, and gibberellin-induced degradation of the DELLA repressor enhances the brassinosteroid response. Finally, the signal transduction pathways of these two hormones interact with the phytochrome pathway through their regulation of phytochrome interacting factors (PIFs) (see Chapter 16).

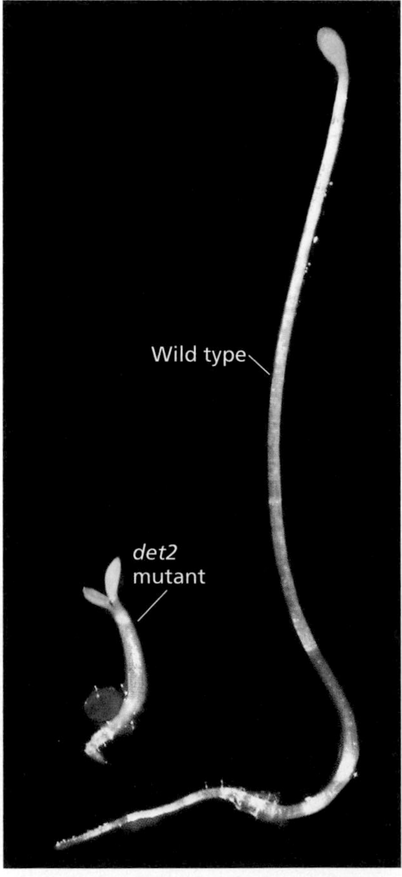

Figure 18.26 The dark-grown *det2* mutant seedling of Arabidopsis on the left has a short, thick hypocotyl and open cotyledons. The dark-grown wild type is on the right. (Courtesy of S. Savaldi-Goldstein.)

Hook opening is regulated by phytochrome and auxin

Etiolated eudicot seedlings are usually characterized by a hook region located just behind the shoot apex. Hook formation and maintenance in the dark result from ethylene-induced asymmetric growth (**Figure 18.27A**). The closed shape of the hook is a consequence of the more rapid elongation of the outer side of the stem compared with the inner side. When the hook is exposed to white light it opens, because the elongation rate of the inner side increases, equalizing the growth rates on both sides (see WEB APPENDIX 2).

Red light induces hook opening, and far-red light reverses the effect of red, indicating that phytochrome is the photoreceptor involved in this process. A close interaction between phytochrome and ethylene controls hook opening. As long as ethylene is produced by the hook tissue in the dark, elongation of the cells on the inner side is inhibited. Red light inhibits ethylene formation, promoting growth on the inner side, thereby causing the hook to open.

The auxin-insensitive mutant *axr1* does not develop an apical hook; and treatment of wild-type Arabidopsis seedlings with NPA (*N*-1-naphthylphthalamic acid), an inhibitor of polar auxin transport, blocks apical hook formation. These and other results indicate a role for auxin in maintaining hook structure. The more rapid growth of the outer tissues relative to the inner tissues could reflect an ethylene-dependent lateral redistribution of auxin, analogous to the lateral auxin gradient that develops during phototropic curvature.

Ethylene induces lateral cell expansion

At concentrations above 0.1 $\mu L\ L^{-1}$, ethylene changes the growth pattern of eudicot seedlings by reducing the rate of elongation and increasing lateral expansion, leading to swelling of the hypocotyl or the epicotyl. As discussed in Chapter 14, the directionality of plant cell expansion is determined by the orientation of the cellulose microfibrils in the cell wall. Transverse microfibrils reinforce the cell wall in the lateral direction, so that turgor pressure is channeled into cell elongation. The orientation of the microfibrils is in turn determined by the orientation of the cortical array of microtubules in the cortical (peripheral) cytoplasm. In typical elongating plant cells, the cortical microtubules are arranged transversely, giving rise to transversely arranged cellulose microfibrils.

During the seedling response to ethylene, the transverse pattern of microtubule alignment in the cells of the hypocotyl is disrupted, and the microtubules switch over to a longitudinal orientation (**Figure 18.27B**). This 90 degree shift in microtubule orientation leads to a parallel shift in cellulose microfibril deposition. The newly deposited wall is reinforced in the longitudinal direction rather than the transverse direction, which promotes lateral expansion instead of elongation (**Figure 18.28**).

(A)

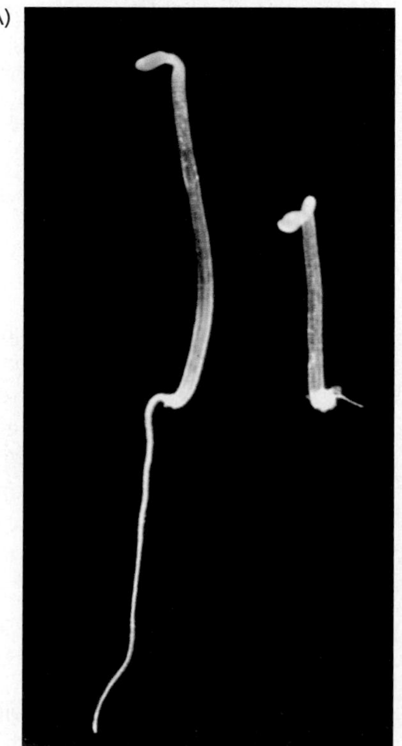

(B)

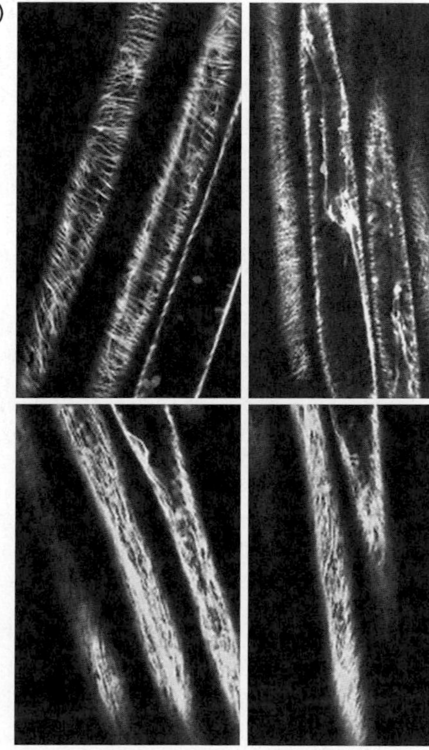

Figure 18.27 Effects of ethylene on growth and microtubule orientation in Arabidopsis seedlings. (A) The ethylene triple response in Arabidopsis. Three-day-old etiolated seedlings grown in the presence (right) or absence (left) of 10 ppm ethylene. Note the shortened hypocotyl, reduced root elongation, and exaggeration of the curvature of the apical hook that result from the presence of ethylene. (B) Ethylene affects microtubule orientation. Microtubule orientation is horizontal in hypocotyls of control dark-grown transgenic Arabidopsis seedlings expressing a tubulin gene tagged with green fluorescent protein (see upper panel). Microtubule orientation is longitudinal in hypocotyl cells from seedlings treated with the ethylene precursor, ACC, which increases ethylene production (see lower panel). (A courtesy of Joe Kieber; B from Le et al. 2005.)

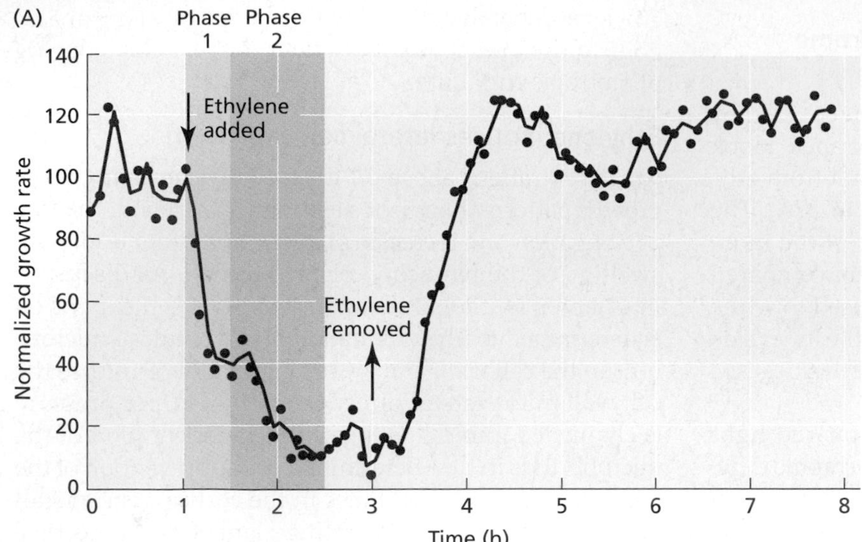

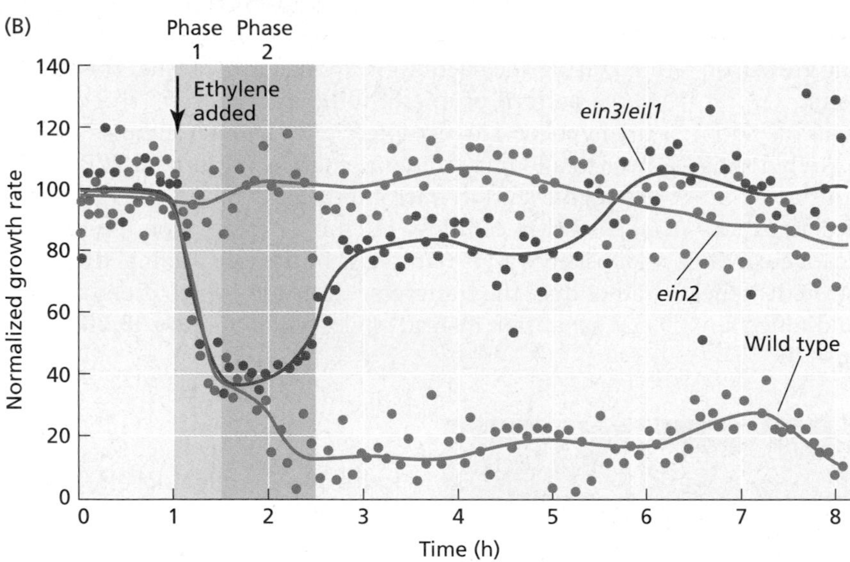

Figure 18.28 Kinetics of the effects of ethylene on hypocotyl elongation in dark-grown Arabidopsis seedlings. (A) Growth rate of etiolated wild-type Arabidopsis after exposure to ethylene and subsequent removal of ethylene at the times indicated by the arrows. Note that the reduction in the growth rate following exposure to ethylene occurs in two distinct phases. (B) Growth rate of etiolated wild-type seedlings and *ein2*, and *ein3/eil1* mutant seedlings following exposure to ethylene at the time indicated by the arrow. Note that the phase 1 response of the *ein3/eil1* mutant seedlings with a defective ethylene signaling pathway (see Chapter 15) is identical to that of the wild type, but that the phase 2 response is absent. (After Binder et al. 2004a, b.)

Shade Avoidance

Seedlings that germinate beneath other plants must immediately compete for light resources that are required for seedling establishment. Shade avoidance is the enhanced stem elongation that occurs in certain plants in response to shading by leaves. The response is specific for shade produced by green leaves, which act as filters for red and blue light, and is not induced by other types of shade.

In this section we will discuss the central role of phytochrome in both shade sensing and shade avoidance. Other regulatory systems that contribute to shade avoidance include the blue-light photoreceptor cryptochrome and the plant hormone brassinosteroid.

Phytochrome enables plants to adapt to changes in light quality

The presence of a red/far-red reversible pigment in all green plants, from algae to eudicots, suggests that these wavelengths of light provide information that helps plants adjust to their environment. What environmental conditions change the relative levels of these two wavelengths in natural radiation?

The ratio of red light (R) to far-red light (FR) varies remarkably in different environments. This ratio can be defined as follows:

$$\text{R/FR} = \frac{\text{Photon fluence rate in 10 nm band centered on 660 nm}}{\text{Photon fluence rate in 10 nm band centered on 730 nm}}$$

Table 18.2 compares the total fluence rate (related to light intensity) in photons (400–800 nm) and the R:FR values in eight natural conditions and environments. Compared with daylight, there is proportionally more far-red light during sunset, under 5 mm of soil, or under the canopy of other plants (as on the floor of a forest). The canopy phenomenon results from the fact that green leaves absorb red light because of their high chlorophyll content, but are relatively transparent to far-red light.

Decreasing the R:FR ratio causes elongation in sun plants

An important function of phytochrome is that it enables plants to sense shading by other plants. Plants that increase stem extension in response to shading are said to

Table 18.2 Ecologically important light parameters

	Fluence rate (μmol m^{-2} s^{-1})	R:FR[a]
Daylight	1900	1.19
Sunset	26.5	0.96
Moonlight	0.005	0.94
Ivy canopy	17.7	0.13
Soil, at a depth of 5 mm	8.6	0.88
Lakes, at a depth of 1 m		
Black Loch	680	17.2
Loch Leven	300	3.1
Loch Borralie	1200	1.2

Source: Smith 1982, p. 493.

Note: The light intensity factor (400–800 nm) is given as the photon flux density, and phytochrome-active light is given as the R:FR ratio.

[a]Absolute values taken from spectroradiometer scans; the values should be taken to indicate the relationships between the various natural conditions and not as actual environmental means.

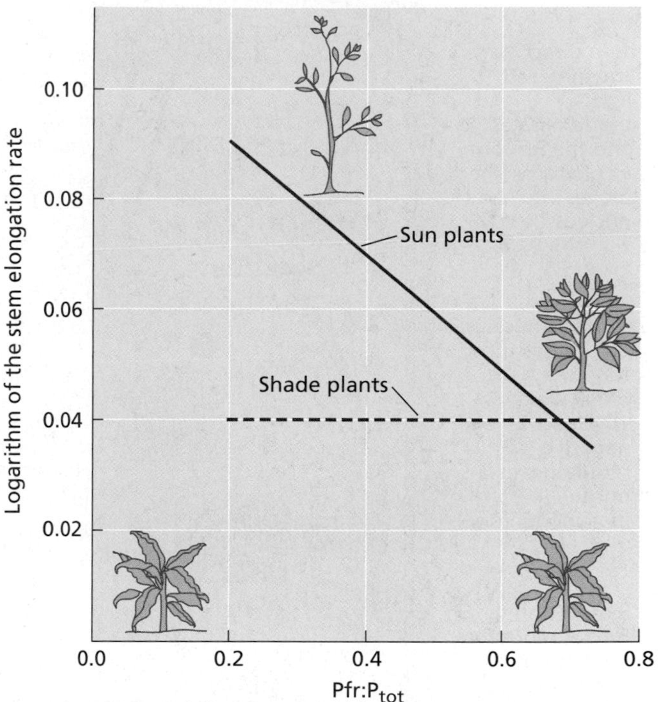

Figure 18.29 Phytochrome appears to play a predominant role in controlling stem elongation rate in sun plants (solid line) but not in shade plants (dashed line). (After Morgan and Smith 1979.)

exhibit a shade avoidance response. As shading increases, the R:FR ratio decreases (see Figure 16.6B). A higher proportion of far-red light converts more Pfr to Pr, and the ratio of Pfr to total phytochrome (Pfr/P_{total}) decreases.

When "sun plants" (plants adapted to an open-field habitat) were grown in natural light under a system of shades so that R:FR was controlled, stem extension rates increased in response to a higher far-red content (i.e., a lower Pfr:P_{total} ratio) (**Figure 18.29**). In other words, simulated canopy shading (high levels of far-red light, low Pfr:P_{total} ratio) induced these plants to allocate more of their resources to growing taller. This correlation was not as strong for "shade plants," which normally grow under a leaf canopy. Shade plants showed less reduction in their stem extension rate than did sun plants when they were exposed to higher R:FR values (see Figure 18.29). Thus, there appears to be a systematic relationship between phytochrome-controlled growth and species habitat. Such results are taken as an indication of the involvement of phytochrome in shade perception.

For a "sun plant" or "shade-avoiding plant," there is a clear adaptive value in allocating its resources toward more rapid extension growth when it is shaded by another plant. In this way it can enhance its chances of growing above the canopy and acquiring a greater share of unfiltered, photosynthetically active light. The price for increased internode elongation is usually reduced leaf area and reduced branching, but at least in the short run, this adaptation to canopy shade increases plant fitness. When the plant grows above the canopy or a canopy gap occurs when a tree falls in the forest, then the plant is released from shade avoidance and competition for light.

Genetic analyses of Arabidopsis have indicated that, of the five phytochrome isoforms (phyA–E) found in angiosperms, phyB plays the predominant role in mediating many shade avoidance responses, but phyD and phyE also contribute. phyA also plays a role by antagonizing the responses mediated by phyB, D, and E.

When plants are grown under high R:FR, as in an open canopy, phy proteins become nuclear localized and inactivate PIF proteins, which act as negative regulators of the phytochrome photomorphogenic response (**Figure 18.30**). Under low R:FR, a pool of phytochrome is excluded from the nucleus, enabling the accumulation of PIF proteins that promote elongation responses (see Figure 16.13). In addition to interactions with phy, PIF proteins also are subject to negative regulation by DELLA proteins, which are components of the gibberellin signaling pathway. Thus, PIF proteins appear to integrate numerous light signals in the transition from skotomorphogenesis to photomorphogenesis (e.g., chlorophyll biosynthesis) as well as fine-tune responses to changes in light quality (e.g., shade avoidance).

Other photoreceptors and hormones, such as cryptochromes, auxin, and brassinosteroids, also participate in photomorphogenesis. For a discussion see **WEB TOPIC 18.9**.

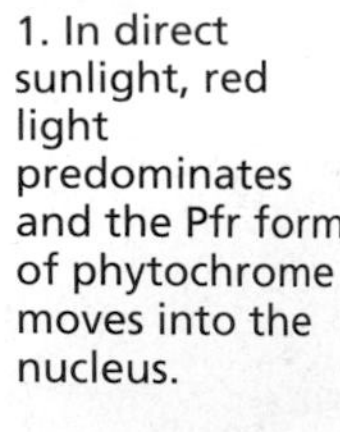

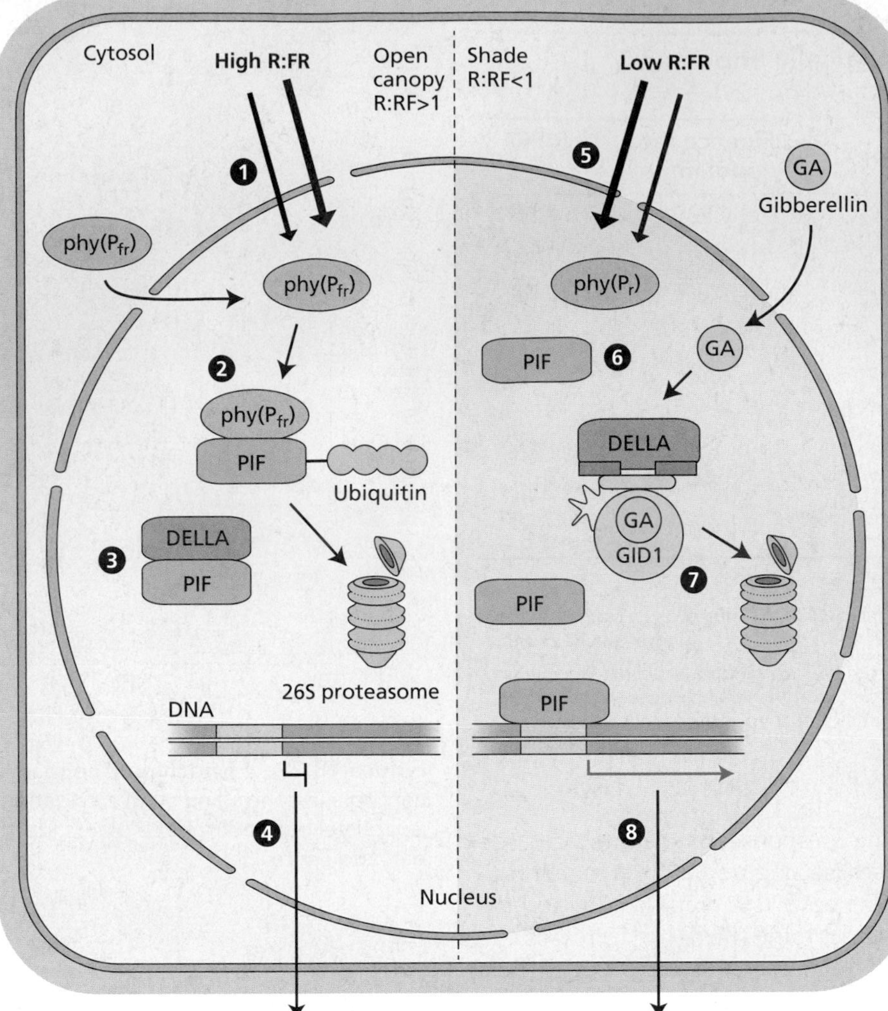

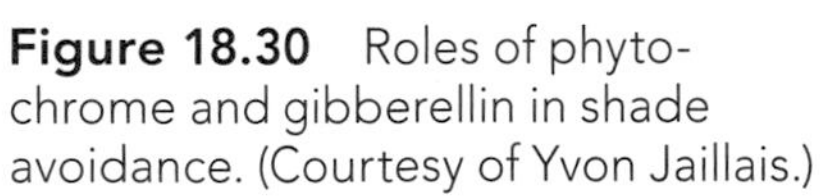

Figure 18.30 Roles of phytochrome and gibberellin in shade avoidance. (Courtesy of Yvon Jaillais.)

Reducing shade avoidance responses can improve crop yields

Shade avoidance responses may be highly adaptive in a natural setting to help plants outcompete neighboring vegetation. (For a discussion of ecotypic variation in the phytochrome responses, see **WEB TOPIC 18.10**.) But for many crop species, a reallocation of resources from reproductive to vegetative growth can reduce crop yield. In recent years, yield gains in crops such as maize have come largely through the breeding of new varieties with a higher tolerance to crowding (which induces shade avoidance responses) rather than through increases in basic yield per plant. As a consequence, today's maize crops can be grown at higher densities than older varieties without suffering decreases in plant yield (**Figure 18.31**).

Vascular Tissue Differentiation

During embryogenesis within the seed, symplastic and apoplastic transport are sufficient to distribute water, nutrients, and signals throughout the embryo by the process of diffusion. Following germination, however, the emergent seedling requires a continuous vascular system to distribute materials quickly and efficiently throughout the plant. The vascular system of the embryo consists only

Figure 18.31 High-density planting and crop yield. Modern maize varieties are planted at high density. Traditionally, Native Americans grew maize on small hills or mounds, which were separated by more than a meter. The plants were short and often produced multiple small ears. In contrast, modern hybrids are machine-planted in dense rows with little space between them (typically 74,000 to 94,000 plants per hectare). Although yield per plant has not increased dramatically for many years in commercial hybrids, overall yields have continued to increase, largely because of better performance of plants at high planting density. As shown in this image from upstate New York, modern varieties of corn have upright leaves that help the plants capture sunlight energy under crowded conditions. (Courtesy of T. Brutnell.)

of procambrial strands—immature vascular tissue. During seedling emergence, the first protoxylem and protophloem cells appear, followed by the larger metaxylem and metaphloem cells (**Figure 18.32**). Protophloem and metaphloem cells can differentiate into sieve elements, companion cells, fibers, or parenchyma cells. Protoxylem and metaxylem cells can become xylem vessels and tracheids, fibers, or parenchyma.

Auxin and cytokinin are required for normal vascular development

Auxin and cytokinin interactions are important for directing vascular development. For example, in Chapter 17 we saw that the *WOODEN LEG* (*WOL*) gene encodes a cytokinin receptor and is required for vascular development. In *wol* mutants, which are defective in cytokinin signaling, protophloem and protoxylem cells fail to develop, suggesting that cytokinin signaling is required for the specification of cell types by procambial cells.

Similarly, *AXR3* genes are members of the AUX/IAA gene family of transcriptional regulators that are rapidly induced by auxin. AXR3 is required for auxin signaling, and protoxylem development is blocked in *axr3* mutants. Thus, auxin signaling is required for protoxylem development.

Because of the difficulty of studying xylem differentiation in tissues with multiple cell types, much of our current understanding of the process comes from the study of xylogenesis in zinnia suspension-cultured cells.

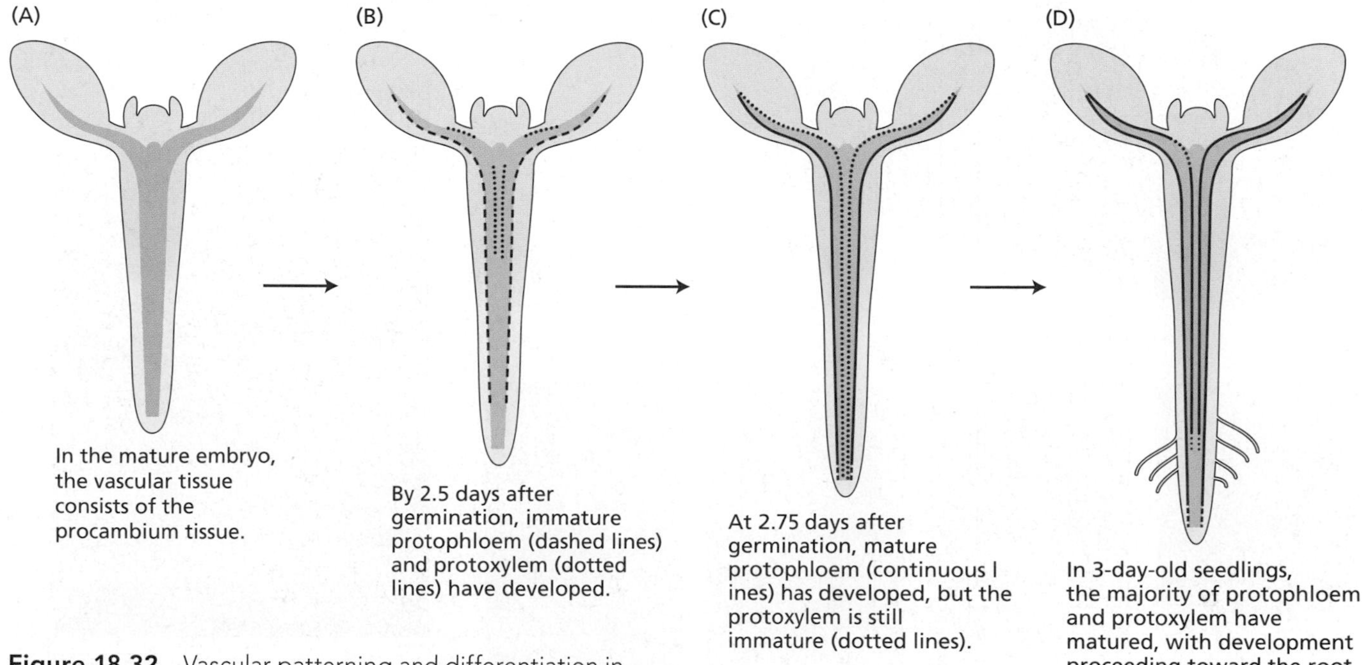

Figure 18.32 Vascular patterning and differentiation in Arabidopsis embryos and seedlings. (After Busse and Evert 1999.)

Zinnia suspension-cultured cells can be induced to undergo xylogenesis

Zinnia (*Zinnia elegans*) suspension cell cultures, derived from leaf mesophyll cells, can be induced to differentiate directly from mature parenchyma cells to xylem tracheary elements. Three developmental stages of zinnia xylogenesis have been identified, each associated with specific physiological states, morphological changes, and gene expression patterns:

- Stage I consists of the de-differentiation of mesophyll cells and the acquisition of competence to re-differentiate into tracheary elements. De-differentiation involves many of the same genes involved in the wound response of plants.
- During Stage II, the synthesis, patterning, and deposition of secondary wall material begin. Autophagy (see Chapter 22) becomes active at this time, contributing to cell autolysis.
- In Stage III, the deposition of secondary wall thickenings is completed and lignification takes place. Simultaneously, the vacuolar membrane breaks down, which leads to complete protoplast autolysis, including the plasma membrane. At the end of the process, only a hollow tube consisting of an outer granular layer, a middle layer of primary wall, and an inner layer with lignified secondary wall thickenings remains (**Figure 18.33**).

Xylogenesis involves chemical signaling between neighboring cells

Auxin and cytokinin are both required for the initiation of xylogenesis in zinnia cell cultures, and brassinosteroid acts at the later stages to promote lignification and programmed cell death. Other signaling agents are undoubtedly involved. For example, vessel elements formed in zinnia suspension-cultured cells have closed end walls (see Figure 18.33A–E), in contrast to vessel elements formed in vivo, which have open end walls. This discrepancy suggests that normal development of tracheary elements involves signaling between the upper and lower cells of a file. A known example of such signaling between developing tracheary elements is the proteoglycan-like factor **xylogen**, which mediates xylem differentiation in zinnia suspension cell cultures. Xylogen normally accumulates in the meristem, procambium, and xylem of zinnia seedlings, and it is concentrated at the apical ends of the cell walls of differentiating tracheary elements (Figure 18.33F). The polar distribution of xylogen suggests a role in cell-to-

(A)

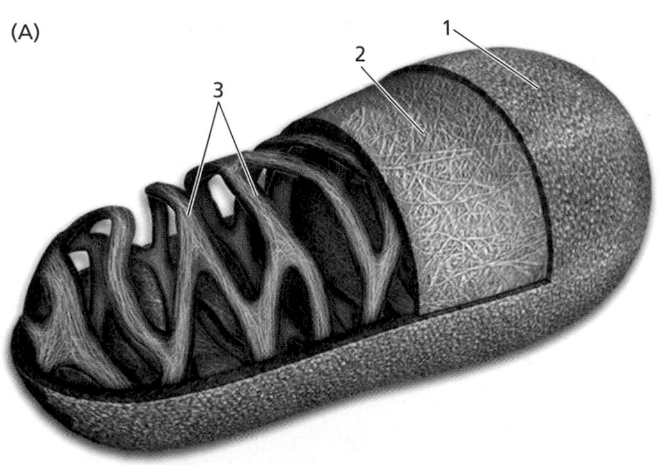

(B)

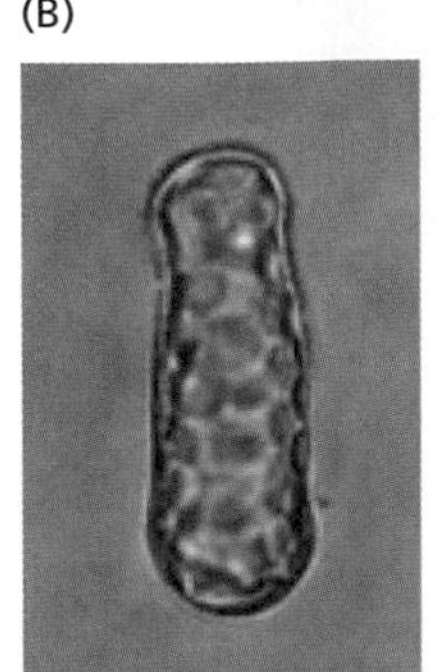

(C)

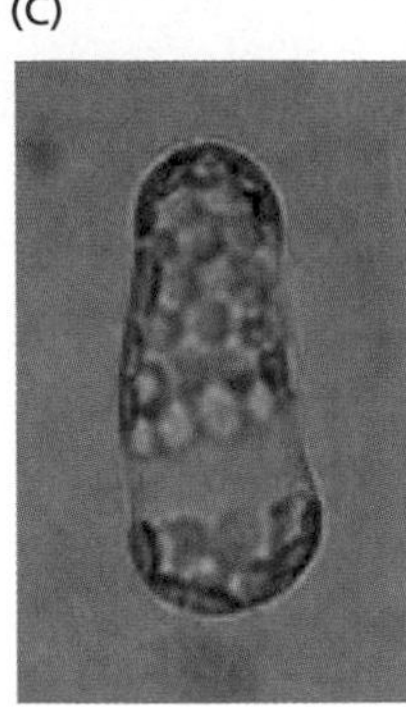

(D)

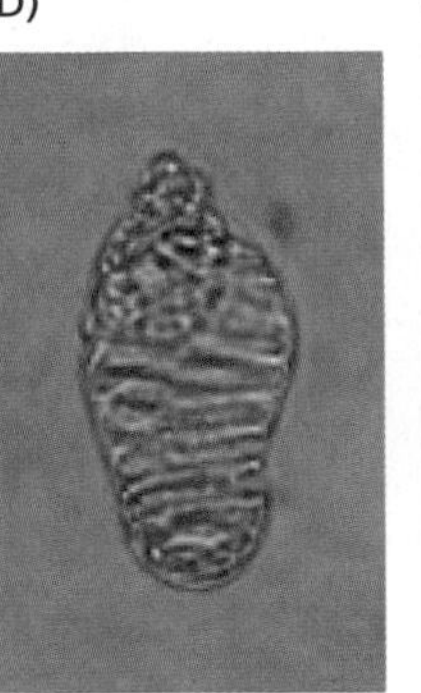

(E)

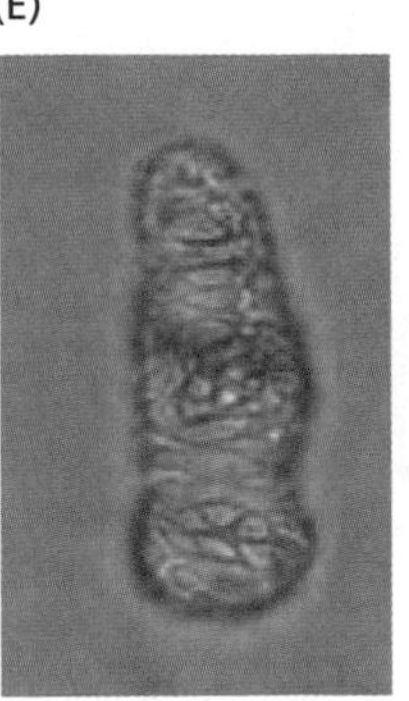

(F)

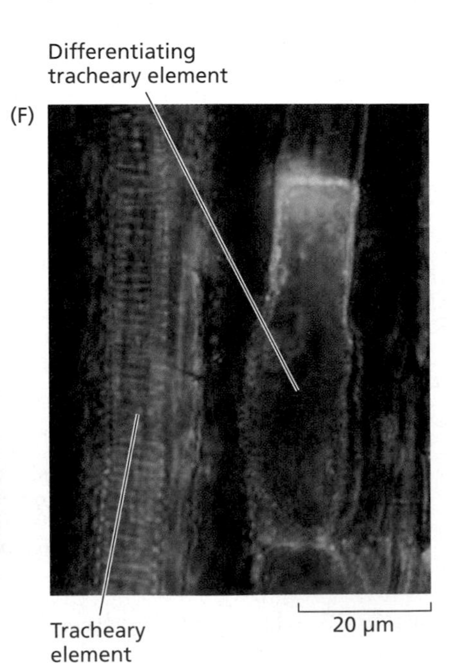

Figure 18.33 Xylem formation in zinnia in the plant and in cell culture. (A) Zinnia xylem cell walls have three main layers: (1) an outer granular matrix, (2) a primary cell wall, and (3) a secondary cell wall. (B–E) Sequence showing zinnia tracheary element differentiation in mesophyll cell suspension culture. (F) Xylogen localization in a 14-day-old zinnia seedling. (A from Lacayo et al. 2010, art by Sabrina Fletcher; B–E from Novo-Uzal et al. 2013; F from Motose et al. 2004.)

cell communication during the formation of vessels that is specific for the end walls.

Root Growth and Differentiation

The earliest land plants lacked roots and instead used shallow, rhizome-like structures (underground stems) for anchorage and absorption. Roots evolved independently at least twice, and are present in all extant groups of vascular land plants. The developing root axis can be divided into three fundamental zones: the **meristematic zone**, the **elongation zone**, and the **differentiation zone** (**Figure 18.34**). Specialized cells, tissues, and organs, including root hairs, endodermis, xylem and phloem conducting elements, and lateral root primordia, all reach maturity in the differentiation zone. The formation of the Casparian strip in the endodermis was described in Chapter 14.

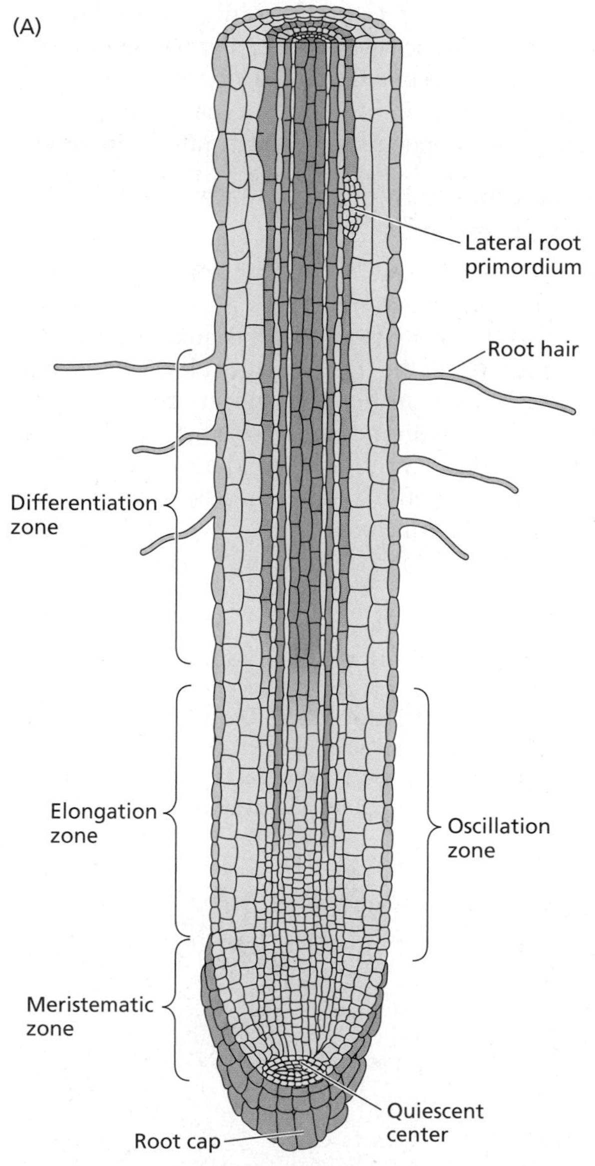

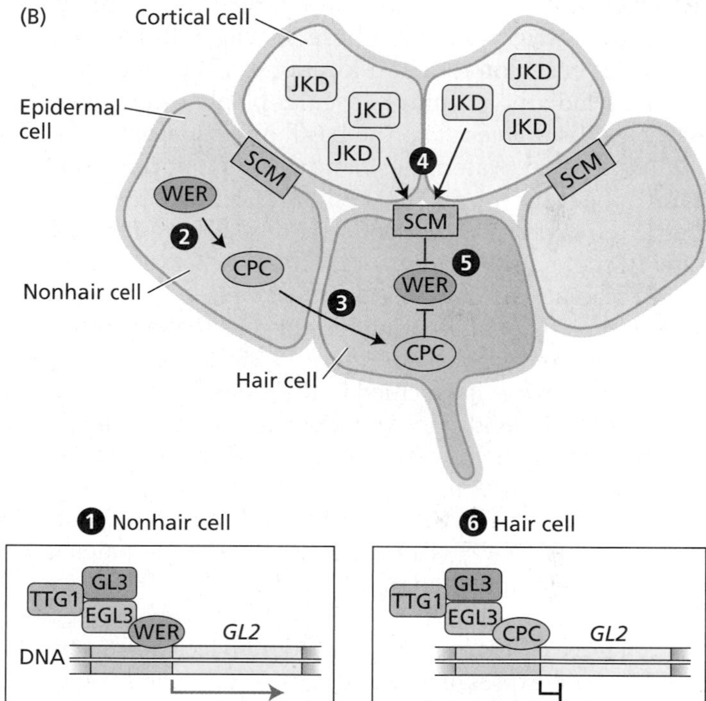

1. In nonhair cells, WER forms a transcriptional complex with TTG1, GL3, and EGL3 to activate the gene GL2, which results in nonhair cell fate.

2. The WER transcriptional complex induces the expression of the *CPC* gene.

3. The CPC protein moves into the presumptive hair cell and prevents WER from forming the transcription complex.

4. Cells in the cortex release a signal dependent on the *JKD* gene, which activates the SCM protein of the presumptive hair cell.

5. Activated SCM further represses WER.

6. In the absence of WER activity *GL2* is not expressed, which leads to hair cell specification.

Figure 18.34 Sites of lateral root and root hair initiation in Arabidopsis roots. (A) Longitudinal section through the root showing overlapping developmental zones. Cell division occurs in the meristematic zone, and cell expansion and elongation occur in the elongation zone. The transition region between the meristematic and elongation zones (oscillation zone) is also indicated. Cell differentiation occurs in the differentiation zone, marked by the formation of root hairs by trichoblasts. (B) Root hair specification in Arabidopsis involves multiple genes. *TTG1, TRANSPARENT TESTA GLABRA1; GL3, GLABRA3; GL2, GLABRA2; EGL3, ENHANCER OF GLABRA3; WER, WEREWOLF; CPC, CAPRICE; JKD, JACKDAW; SCM, SCRAMBLED.*

Root epidermal development follows three basic patterns

Root hairs are important for water and nutrient uptake. They also serve a mechanical role to help anchor plants in soil. The majority of plant species (including most ferns and dicots, as well as many monocots) exhibit Type I root hair development, in which every root epidermal cell can potentially differentiate into a root hair (**Figure 18.35A**). In the remaining species the epidermis consists of a mixture of cells, some having the potential to form root hairs (**trichoblasts**), and others that are incapable of forming root hairs (**atrichoblasts**). These species fall into two categories based on root hair location. In Type II plants, which include the primitive vascular plants *Lycopodium*, *Selaginella* and *Equisetum*, the basal angiosperm family Nymphaeaceae (water lilies), and some monocots, root hairs arise from the smaller cell produced by an asymmetric cell division in the root meristem (**Figure 18.35B**). Type III root hair development is found exclusively in the Brassicacaceae. In Arabidopsis, for example, the root epidermis consists of alternating files of cells that are either atrichoblasts or trichoblasts (**Figure 18.35C**). Thus Type III trichoblast cell fate is specified in the meristem.

Trichoblast identity in Arabidopsis root meristems is determined by the interaction of transcription factors. The transcription factors WEREWOLF (WER), TRANSPARENT TESTA GLABRA1 (TTG1), GLABRA3 (GL3), and ENHANCER OF GLABRA3 (EGL3) form a complex that promotes expression of *GL2* and *CAPRICE* (*CPC*) in the atrichoblast (see Figure 18.34B). In the trichoblasts, CPC displaces WER from the transcription factor complex, so *GL2* is not expressed. In addition, *WER* expression is suppressed in trichoblast cells via signaling through the receptor-like kinase SCRAMBLED (SCM). Trichoblast identity is also promoted by JACKDAW (JKD), a zinc finger transcription factor, and by small signaling peptides called **root meristem growth factors**. According to one hypothesis, a root epidermal cell develops into a trichoblast because it has more surface area in contact with two cortical cells, so there is more signaling peptide to bind to the SCM receptor (see Figure 18.34B).

Each root hair cell has a long, fingerlike extension that usually grows from the basal end of the epidermal cell (see Figure 18.35). The cells extend by tip growth and are associated with calcium gradients similar to those of growing pollen tubes (see Chapter 21). The bulge that forms the nascent root hair at the base of the epidermal cell is correlated with cell wall acidification and loosening. However, since exogenous acidification does not change the position of the bulge, other endogenous factors must be involved.

Auxin and other hormones regulate root hair development

An auxin transporter, ABCB4 in Arabidopsis, plays a role in root hair emergence by maintaining intracellular auxin concentrations. ABCB4 is a reversible auxin transporter localized throughout the trichoblasts. At low intracellular auxin levels, ABCB4 functions as an influx carrier, importing auxin and promoting root hair growth. Once the intracellular auxin concentration rises to a threshold level, ABCB4 switches to efflux mode, thereby slowing auxin import and moderating root hair growth. ABCB4 thus promotes and regulates root hair length. The PIN2 auxin

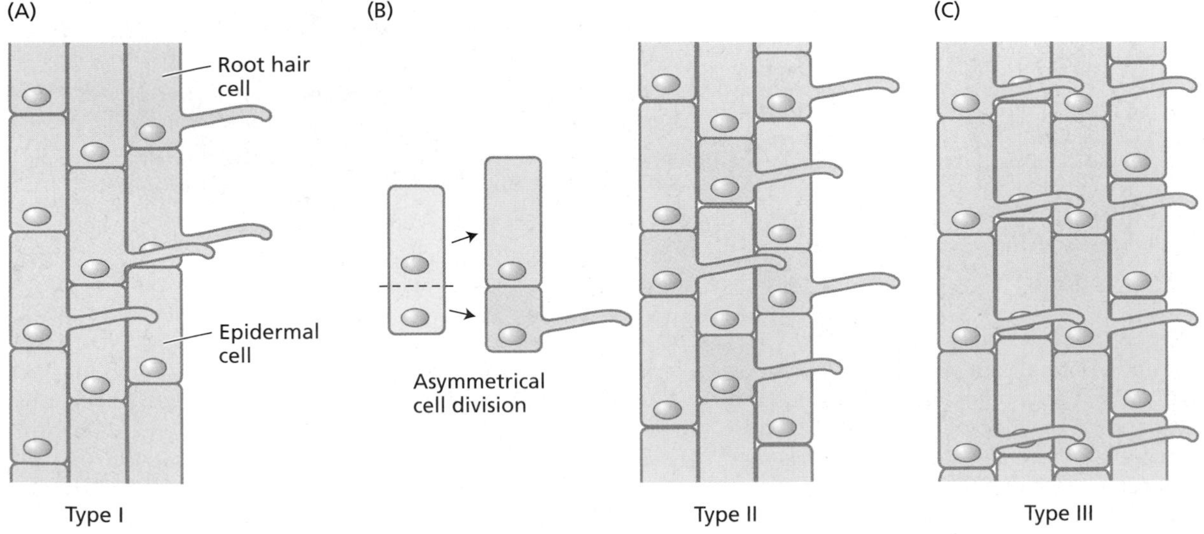

Figure 18.35 Three patterns of trichoblast differentiation. (A) Type I, in which all root epidermal cells have the potential to become trichoblasts. (B) Type II, in which the trichoblast results from an asymmetric cell division. (C) Type III, in which trichoblasts and atrichoblasts occur in alternating cell files. (After Bibikova and Gilroy 2003.)

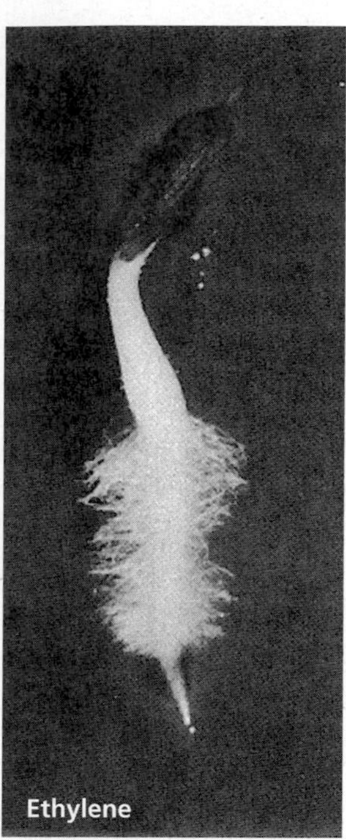

Figure 18.36 Promotion of root hair formation by ethylene in lettuce seedlings. Two-day-old seedlings were treated with air (left) or 10 ppm ethylene (right) for 24 h before the photo was taken. Note the profusion of root hairs on the ethylene-treated seedling. (From Abeles et al. 1992, courtesy of F. Abeles.)

efflux carrier is localized in both trichoblasts and atrichoblasts. In parallel, the auxin symporter AUX1 is localized in atrichoblasts, which have higher auxin concentrations than trichoblasts. *aux1* and *pin2* mutants have shorter root hairs and *abcb4* mutants have longer root hairs compared with wild-type plants.

In ethylene-treated roots, cells not overlying a cortical cell junction differentiate into hair cells and produce root hairs in abnormal locations (**Figure 18.36**). Seedlings grown in the presence of ethylene inhibitors (such as the silver ion, Ag^+), as well as ethylene-insensitive mutants, display a reduction in root hair formation. These observations suggest that ethylene acts as a positive regulator in the differentiation of root hairs. Jasmonic acid has also been shown to enhance root hair growth, but brassinosteroids inhibit root hair growth, possibly by inhibiting auxin responses by enhancing AUX/IAA expression.

Lateral root formation and emergence depend on endogenous and exogenous signals

In gymnosperms and most eudicots, lateral root primordia are initiated in the pericycle cells adjacent to the xylem poles. However, in grasses lateral root primordia form in the pericycle and endodermal cells adjacent to the phloem poles. In the majority of plants, anticlinal divisions in the pericycle cells precede periclinal divisions. These lateral root primordia cells continue cell division and cell expansion until the new lateral root emerges through the cortical and epidermal cell layers (**Figure 18.37**).

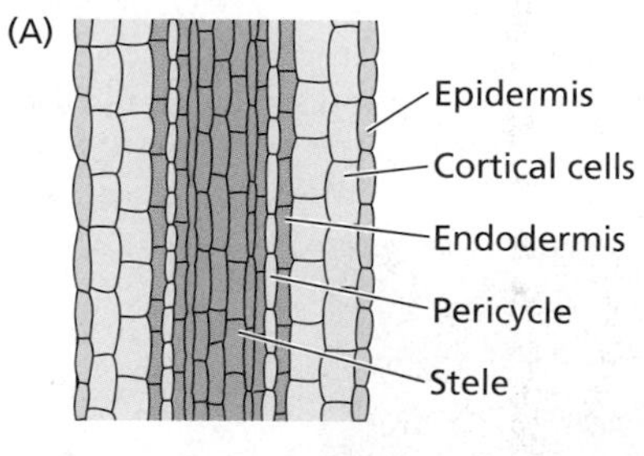

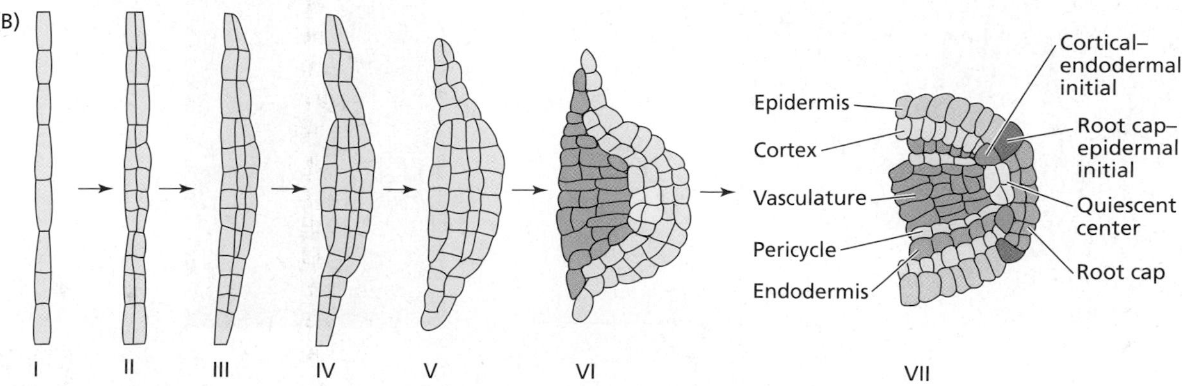

Figure 18.37 Lateral root development. (A) Longitudinal section of a root. Anticlinal cell divisions in the pericycle initiate lateral root formation. (B) Stages of lateral root development. Stage I consists of a single layer of pericycle. During stage II, the pericycle cells divide periclinally to form inner and outer layers. In stages III and IV, the lateral root primordium forms a dome shape and periclinal and anticlinal divisions continue. In stage V, the cortical cells loosen so that the lateral root primordium can expand between the cells of the primary root. In stage VI, the lateral root primordium recapitulates the tissues of the primary root: epidermal, cortical, and endodermal cell layers. In stage VII, the stele differentiates, the epidermal cells separate, and the lateral root primordium emerges. (After Petricka et al. 2012.)

The lateral root contains all of the cell types of the primary root, and the vascular tissue of the lateral root is continuous with that of the primary root. Lateral roots initiate in the zone of differentiation of the primary root (see Figure 18.34A). Whereas the apical meristems of primary roots are usually determinate due to a combination of genetic and environmental factors (see **WEB TOPIC 18.11**), those of lateral roots are indeterminate and lateral roots may form branches as well, greatly increasing the total surface area of the root system.

Regions of lateral root emergence correspond with regions of auxin maxima

The newest/youngest lateral root primordia are usually located closest to the apical meristem. However, the site of lateral root emergence can vary according to endogenous and exogenous factors. The signal(s) that determines the site of future lateral root primordia and inititates the first anticlinal divisions has not yet been determined. Oscillations of auxin, pH, Ca^{2+}, and other signals occur in the elongation zone of the primary root (see Figure 18.34A). These oscillations, and their downstream signals, could potentially explain the regular pattern of lateral roots normally observed based on periodic lateral root primordia initiation. The sites of lateral root emergence have been correlated with regions of high auxin activity (**Figure 18.38 A and B**).

Genetic studies have identified several genes that form part of the regulatory network for lateral root initiation. For example, mutations in the gene *SHATTERPROOF* alter periodic root branching, and mutations in genes involved in cell-to-cell communication alter lateral root spacing. SOLITARY-ROOT (SLR)/IAA14 is a transcriptional repressor of auxin-responsive genes that is important for periclinal cell divisions in the lateral root primordia and is required for lateral root emergence. Auxin signaling mutants—such as *tir1/afb*, *axr1*, and many *aux/iaa* mutants—affect lateral root initiation and therefore have either a reduced number of lateral roots or none at all. Shoot-derived auxin is important for lateral root initiation, whereas root-derived auxin is required for lateral root emergence.

Figure 18.38 Arabidopsis lateral root primordia develop at sites where the root bends. Prebranch sites form in the oscillation zone immediately below the zone of differentiation. (A) Expression of the luciferase gene driven by the auxin-sensitive DR5 promoter in an Arabidopsis root. Auxin levels fluctuate over time, as indicated by the dashed arrow. The asterisk indicates a prebranch site. OZ, oscillation zone. (B) Auxin levels measured by luminescence in two Arabidopsis roots oscillate with a period of about 6 h in this region. ADU, analog digital units. (C) Auxin maxima along the root (light blue areas) correspond to sites of bending, lateral root formation, and lateral root emergence. Sites of lateral root formation are indicated by arrows. (From Van Norman et al. 2013.)

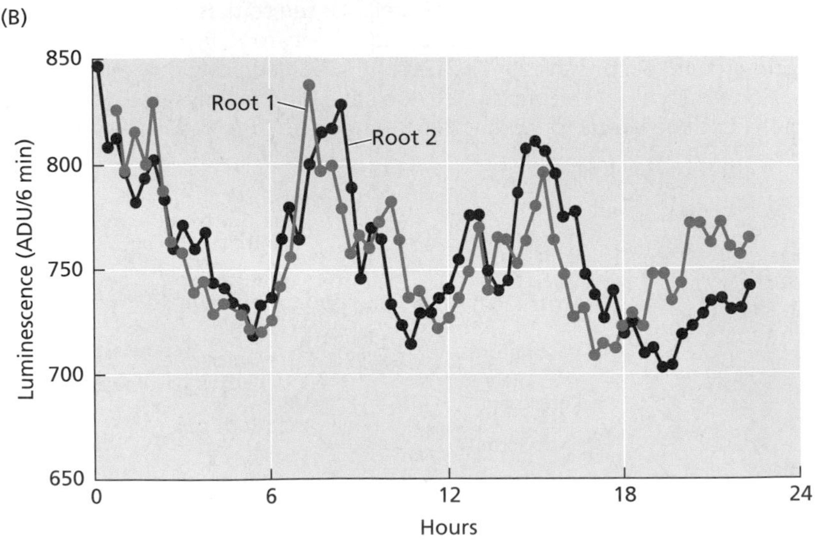

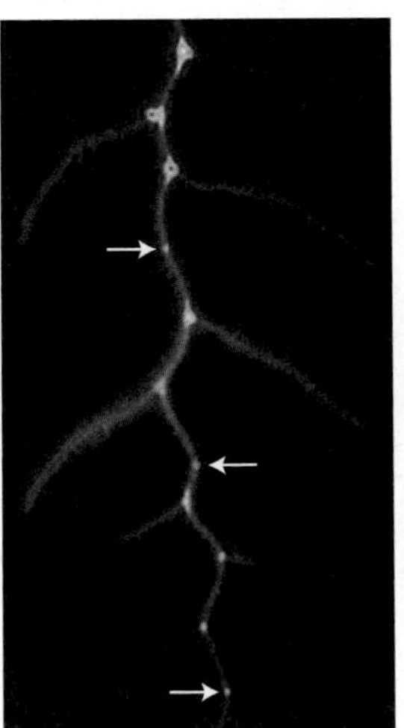

Figure 18.39 The set point angle of tree branches creates different patterns of tree architecture.

One of the predictable places where lateral roots emerge that is not based on periodic patterning is at the point where the root bends. Following gravitropic stimuli (or mechanical bending in the lab), lateral root primordia form and emerge on the outside of the curve at the site of the bending (**Figure 18.38C**). The bending stimulus induces a Ca^{2+} spike and results in a localized increase in auxin that leads to lateral root initiation. In contrast to auxin, ethylene inhibits lateral root development.

Lateral roots and shoots have gravitropic setpoint angles

The angle at which gravitropic organs are maintained with respect to gravity is known as the **gravitropic setpoint angle**. By convention, a primary root growing vertically downward has a gravitropic setpoint angle of 0 degrees, while that of a primary shoot growing vertically upward is 180 degrees. As soon as the lateral root has developed an elongation zone, it becomes competent to respond to gravity. However, its gravitropic setpoint angle is different from that of the primary root.

It is common for graviresponsive lateral roots to grow at nonvertical angles (between 0 and 180°; see Figure 18.38). In general, if a graviresponsive branch (root or shoot) is mechanically displaced, upward or downward, from its gravitropic setpoint angle, it will undergo tropic growth to shift it back toward that gravitropic setpoint angle. This means that lateral roots with nonvertical gravitropic setpoint angles can be negatively gravitropic—that is, grow against the gravity vector. Conversely, nonvertical shoot branches can be positively gravitropic and grow downward with the gravity vector locator (**Figure 18.39**). This observation provides a simple demonstration that the mechanistic basis for the maintenance of graviresponsive nonvertical growth cannot lie solely in differences in gravitropic competence between primary and lateral organs. There must be another, unidentified mechanism that can drive upward growth in lateral roots and downward growth in lateral shoots.

SUMMARY

Seeds require rehydration, and sometimes additional treatments, to germinate. During germination and establishment, food reserves maintain the seedling until it is autotrophic, tropisms help orient roots and shoots, vascular tissue and root hairs differentiate, and lateral roots are formed.

Seed Structure

- Seed anatomy varies widely in the types and distributions of stored food resources and the nature of the seed coat (**Figure 18.1**).

Seed Dormancy

- Seed dormancy may arise from the embryo itself or from the surrounding tissues, such as the endosperm and seed coat (**Figure 18.2**).
- Seeds that do not become dormant may exhibit vivipary and precocious germination (**Figures 18.3, 18.4**).
- The primary hormones regulating seed dormancy are abscisic acid and gibberellins (**Figure 18.5**).

Release from Dormancy

- Light breaks dormancy in many small seeds.
- Some seeds require chilling or after-ripening to break dormancy (**Figure 18.6**).
- Nitrate, nitric oxide, and smoke can break dormancy.

Seed Germination

- Germination takes place in three phases relating to water uptake (**Figures 18.7, 18.8**).

Mobilization of Stored Reserves

- The cereal aleurone layer responds to gibberellins by secreting hydrolytic enzymes (including α-amylase) into the surrounding endosperm, making starches available to the embryo (**Figure 18.9**).
- Gibberellins secreted by the embryo also enhance the transcription of *α-amylase* mRNA, which initiates starch degradation.
- The gibberellin receptor GID1 promotes the degradation of negative regulators of α-amylase production, including DELLA proteins, thereby up-regulating GA-MYB proteins and α-amylase transcription (**Figures 18.10, 18.11**).
- Abscisic acid inhibits α-amylase transcription.

Seedling Growth and Establishment

- At optimum concentrations, auxin promotes stem and coleoptile growth and inhibits root growth. However, above optimum concentrations, auxin can inhibit stem and coleoptile growth (**Figures 18.12, 18.13**).

Tropisms: Growth in Response to Directional Stimuli

- Lateral redistribution of auxin allows plants to exhibit gravitropism (**Figure 18.14**).
- Polarization of auxin requires energy and is gravity independent (**Figures 18.15, 18.16**).
- Statoliths in statocytes serve as gravity sensors in root caps (**Figures 18.17, 18.18**).
- Most of the auxin in the root is derived from the shoot. (**Figure 18.19**).
- A displaced, horizontal root redirects auxin to the lower side, inhibiting growth there.
- Gravitropism is enabled in eudicot stems and stem-like organs by statoliths in the starch sheath (**Figures 18.20, 18.21**).
- pH and calcium ions (Ca^{2+}) act as second messengers in the signaling that occurs during gravitropism (**Figure 18.22**).

Phototropism

- Like gravitropism, phototropism involves lateral redistribution of auxin (**Figure 18.23**).
- The first step in phototropic bending occurs within minutes of irradiation when phototropin 1 phosphorylates the ABCB19 auxin transporter to inhibit rootward auxin transport (**Figure 18.24**).
- Lateral auxin redirection at the shoot apex begins within 30 min, and bending begins after approximately 2 h (**Figure 18.24**).

Photomorphogenesis

- Seedlings transition from skotomorphogenesis (development in the dark; i.e., underground) to photomorphogenesis (development in the presence of light) at the first instance of light (**Figure 18.25**).
- In etiolated shoots, gibberellins and brassinosteroids suppress photomorphogenesis (**Figure 18.26**).
- Phytochrome, auxin, and ethylene regulate hook opening and lateral cell expansion (**Figures 18.27, 18.28**).

Shade Avoidance

- Phytochrome, a red/far-red reversible pigment, sensitizes plants to changes in light quality and mediates growth toward optimal light conditions (**Figures 18.29, 18.30**).
- Increasing tolerance to crowding (i.e., reducing shade avoidance responses) has increased crop yields in maize (**Figure 18.31**).

Vascular Tissue Differentiation

- Auxin and cytokinin mediate the development of the vascular system (**Figure 18.32**).
- Zinnia suspension cell cultures have been used to study the regulation of tracheary element differentiation in vitro. (**Figure 18.33**).

Root Growth and Differentiation

- Root hairs are specialized epidermal cells that reach maturity in the differentiation zone of the root axis (**Figures 18.34–18.36**).
- Lateral roots are initiated in the pericycle and emerge through cortical and epidermal cells (**Figure 18.37**).
- Lateral roots emerge in regions of high auxin and carotenoid activity (**Figures 18.38, 18.39**).
- Lateral roots and shoots can grow both with and against the gravity vector in accordance with their gravitropic set point angle.

WEB MATERIAL

- **WEB TOPIC 18.1 The Evolution of Seeds** The changes in seed anatomy from extinct seed ferns to the angiosperms are described.
- **WEB TOPIC 18.2 Seedling Growth Can Be Divided into Two Types: Epigeal and Hypogeal** Examples of epigeal versus hypogeal seedling establishment are illustrated.
- **WEB TOPIC 18.3 Seeds Exhibit Both Primary and Secondary Dormancy** Secondary dormancy can be observed in nature in the seed dormancy cycles of annual dicot weed species.
- **WEB TOPIC 18.4 Phase III of Germination Can Be Either a One-Step or a Two-Step Process** Phase III of germination can either be a one-step process in which the radicle emerges immediately after the seed coat is ruptured, or it may involve two steps in which the endosperm undergoes weakening before the radicle can emerge.
- **WEB TOPIC 18.5 Thigmotropism, Gravitropism, and Circumnutation are Integrated Signals** The "wavy root" syndrome, first demonstrated by Charles Darwin in *The Power of Movement in Plants*, is a good example of the integration of gravitropism, thigmotropism, and circumnutation signaling.
- **WEB TOPIC 18.6 Roots Exhibit Negative Phototropism** Blue light has been shown to mediate negative phototropism in roots.
- **WEB TOPIC 18.7 Blue Light Causes Cortical Microtubules to Reorient in the Longitudinal Direction** After exposure of hypocotyls to unilateral blue light, new microtubules are formed in less than one minute that are longitudinally oriented and parallel to the hypocotyl axis.
- **WEB TOPIC 18.8 Phytochrome and Cryptochromes Contribute to Phototropism** The molecular link between PHYA and PHOT1 is PHYTOCHROME KINASE SUBSTRATE 1 (PKS1).
- **WEB TOPIC 18.9 Shade Avoidance Is Regulated by Cryptochromes, Auxin, and Brassinosteroids** In addition to cryptochrome, both auxin and brassinosteroid are required for hypocotyl elongation in seedlings under low blue light conditions.
- **WEB TOPIC 18.10 Phytochrome Responses Show Ecotypic Variation** Surveys of the light responses in Arabidopsis and maize have revealed tremendous ecotypic variation, both in the physiology of their light responses and in their phytochrome gene families.
- **WEB TOPIC 18.11 Maintenance of Meristem Activity Is Crucial for Seedling Establishment** CLAVATA small signal peptides *CLE19* and *CLE40* are expressed in roots and promote meristem maintenance.

available at plantphys.net

Suggested Reading

Baldwin, K. L., Strohm, A. K., and Masson, P. H. (2013) Gravity sensing and signal transduction in vascular plant primary roots. *Am. J. Bot.* 100: 126–142.

Bewley, J. D., Bradford, K. J., Hilhorst, H. W. M., and Nonogaki, H. (2013) *Seeds: Physiology of Development, Germination and Dormancy*, 3rd ed. Springer, New York.

Casal, J. J. (2013) Photoreceptor signaling networks in plant responses to shade. *Annu. Rev. Plant Biol.* 64: 403–427.

Finch-Savage, W. E., and Leubner-Metzger, G. (2006) Seed dormancy and the control of germination. *New Phytol.* 171: 501–523

Graeber, K., Kakabayashi, K., Miatton, E., Leubner-Metzger, G., and Soppe, W. J. J. (2012) Molecular mechanisms of seed dormancy. *Plant Cell Environ.* 35: 1769–1786.

Lacayo, C. I., Malkin, A. J., Holman, H.-Y. N., Chen, L., Ding, S.-Y., Hwang, M. S., and Thelen, M. P. (2010) Imaging cell wall architecture in single *Zinnia elegans* tracheary elements. *Plant Physiol.* 154: 121–133.

Lia, Y.-C., Rena, J.-P., Cho, M.-J., Zhou, S.-M., Kim, Y.-B., Guo, H.-X., Wong, J. H., Niu, H.-B., Kim, H.-K., Morigasaki, S., et al. (2009) The level of expression of thioredoxin is linked to fundamental properties and applications of wheat seeds. *Mol. Plant* 2: 430–441.

Migliaccio, F., Tassone, P., and Fortunati, A. (2013) Circumnutation as an autonomous root movement in plants. *Am. J. Bot.* 100: 4–13.

Novo-Uzal, E., Fernández-Pérez, F., Herrero, J., Gutiérrez, J., Gómez-Ros, L. V., Ángeles Bernal, M., Díaz, J., Cuello, J., Pomar, F., and Ángeles Pedreño, M. (2013) From *Zinnia* to *Arabidopsis*: Approaching the involvement of peroxidases in lignification. *J. Exp. Bot.* 64: 3499–3518.

Palmieri, M., and Kiss, J. Z. (2007) The role of plastids in gravitropism. In *The Structure and Function of Plastids*, R. R. Wise, and J. K. Hoober, eds., Springer, Berlin, pp. 507–525.

Petricka, J. J., Winter, C. M., and Benfey, P. N. (2012) Control of *Arabidopsis* root development. *Annu. Rev. Plant Biol.* 63: 563–590.

Sawchuk, M. G., Edgar, A., and Scarpella, E. (2013) Patterning of leaf vein networks by convergent auxin transport pathways. *PLOS Genet.* 9: 1–13.

Van Norman, J. M., Xuan, W., Beeckman, T., and Benfey, P. N. (2014) Periodic root branching in *Arabidopsis* requires synthesis of an uncharacterized carotenoid derivative. *Proc. Natl. Acad. Sci USA* 111(13): E1300–E1309. DOI: 10.1073/pnas.1403016111

Van Norman, J. M., Zhang, J., Cazzonelli, C. I., Pogson, B. J., Harrison, P. J., Bugg, T. D. H., Chan, K. X., Thompson, A. J., and Benfey, P. N. (2013) To branch or not to branch: The role of pre-patterning in lateral root formation. *Development* 140: 4301–4310.

19 Vegetative Growth and Organogenesis

Although embryogenesis and seedling establishment play critical roles in establishing the basic polarity and growth axes of the plant, many other aspects of plant form reflect developmental processes that occur after seedling establishment. For most plants, shoot architecture depends critically on the regulated production of determinate lateral organs, such as leaves, as well as the regulated formation and outgrowth of indeterminate branch systems. Root systems, though typically hidden from view, have comparable levels of complexity that result from the regulated formation and outgrowth of indeterminate lateral roots (see Chapter 18). In addition, secondary growth is the defining feature of the vegetative growth of woody perennials, providing the structural support that enables trees to attain great heights. In this chapter we will consider the molecular mechanisms that underpin these growth patterns. Like embryogenesis, vegetative organogenesis and secondary growth rely on local differences in the interactions and regulatory feedback among hormones, which trigger complex programs of gene expression that drive specific aspects of organ development.

Leaf Development

Morphologically, the leaf is the most variable of all the plant organs. The collective term for any type of leaf on a plant, including structures that evolved from leaves, is **phyllome**. Phyllomes include the photosynthetic **foliage leaves** (what we usually mean by "leaves"), protective **bud scales**, **bracts** (leaves associated with inflorescences, or flowers), and **floral organs**. In angiosperms, the main part of the foliage leaf is expanded into a flattened structure, the **blade**, or **lamina**. The appearance of a flat lamina in seed plants in the middle to late Devonian was a key event in leaf evolution. A flat lamina maximizes light capture and also creates two distinct leaf domains: **adaxial** (upper surface) and **abaxial** (lower surface) (**Figure 19.1**). Several types of leaves have evolved based on their adaxial–abaxial leaf structure (see WEB TOPIC 19.1).

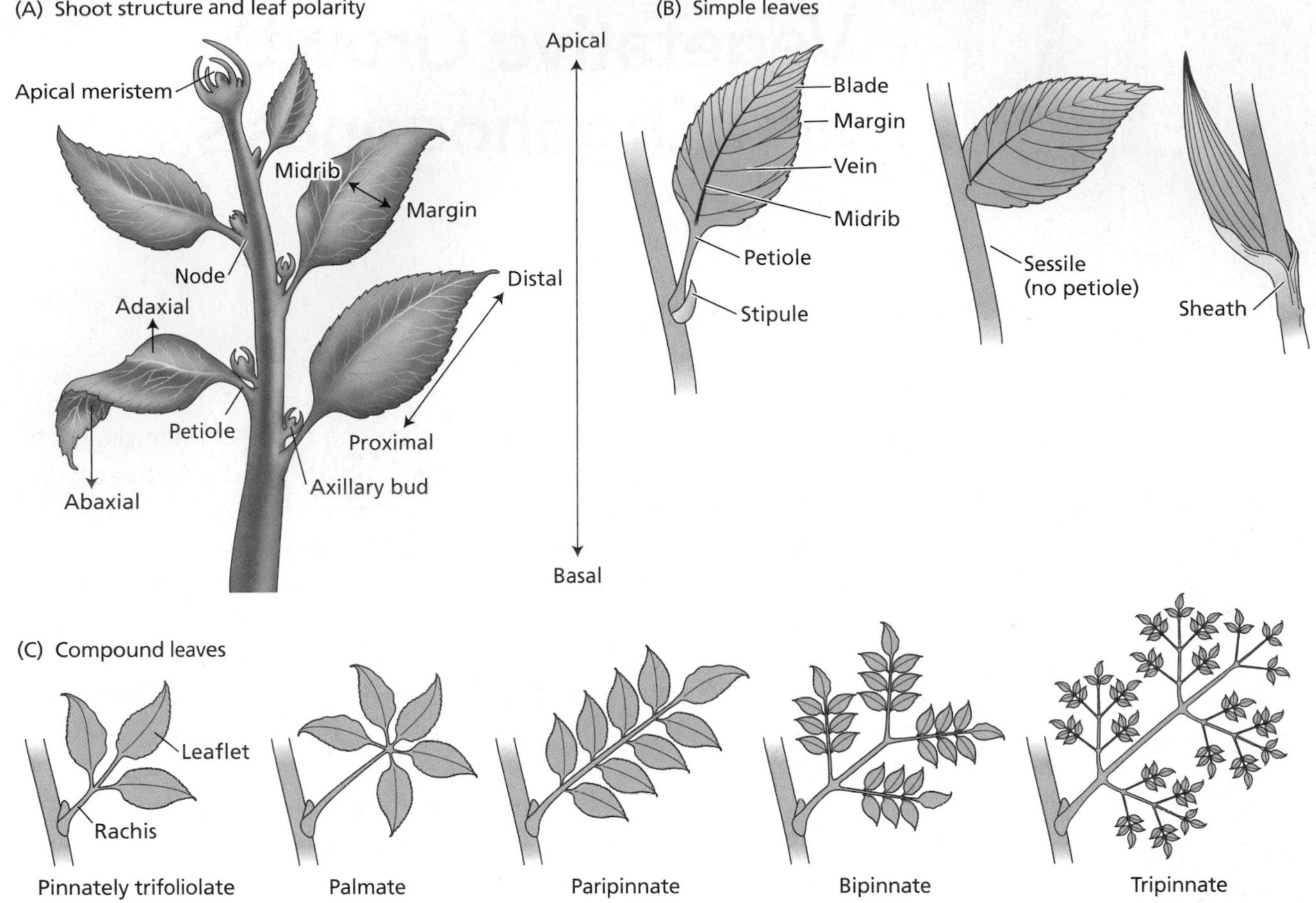

Figure 19.1 Overview of leaf structure. (A) Shoot structure, showing three types of leaf polarity: adaxial–abaxial, distal–proximal, and midrib–margin. (B) Examples of simple leaves. Variations in lower leaf structure include the presence or absence of stipules and petioles, and leaf sheaths. (C) Examples of compound leaves.

In the majority of plants, the leaf blade is attached to the stem by a stalk called the **petiole**. However, some plants have **sessile leaves**, with the leaf blade attached directly to the stem (see Figure 19.1B). In most monocots and certain eudicots the base of the leaf is expanded into a sheath around the stem. Many eudicots have **stipules**, small outgrowths of the leaf primordia, located on the abaxial side of the leaf base. Stipules protect the young developing foliage leaves and are sites of auxin synthesis during early leaf development.

Leaves may be **simple** or **compound** (see Figures 19.1B and C). A simple leaf has one blade, whereas a compound leaf has two or more blades, the **leaflets**, attached to a common axis, or **rachis**. Some leaves, like the adult leaves of some *Acacia* species, lack a blade and instead have a flattened petiole simulating the blade, the **phyllode**. In some plants the stems themselves are flattened like blades and are called **cladodes**, as in the cactus *Opuntia*.

We will begin our discussion of foliage leaf development with the production of leaf primordia. Next we will examine blade formation in simple leaves, which involves the marginal expansion of leaf tissues, differentiation into adaxial and abaxial domains, and morphogenesis along the proximal-distal axis. Compound leaves are produced by variations on these developmental pathways. Finally, we will discuss the gene networks and hormonal signals that control the development of the specialized cells of the epidermis and vascular tissue.

The Establishment of Leaf Polarity

All leaves and modified leaves begin as small protuberances, called primordia, on the flanks of the shoot apical meristem (SAM) (see Chapter 17). All SAMs in higher plants share a common structure: a central, often dome-shaped domain surrounded by several emerging primordia, which may be leaf primordia or, in the case of an inflorescence meristem, flower primordia. Cells in the apical meristem are considered undifferentiated and pluripotent. Nevertheless, as we saw in Chapter 17, the cells of the SAM are organized into three more or less stable tissue layers—L1, L2, and L3—although some plants,

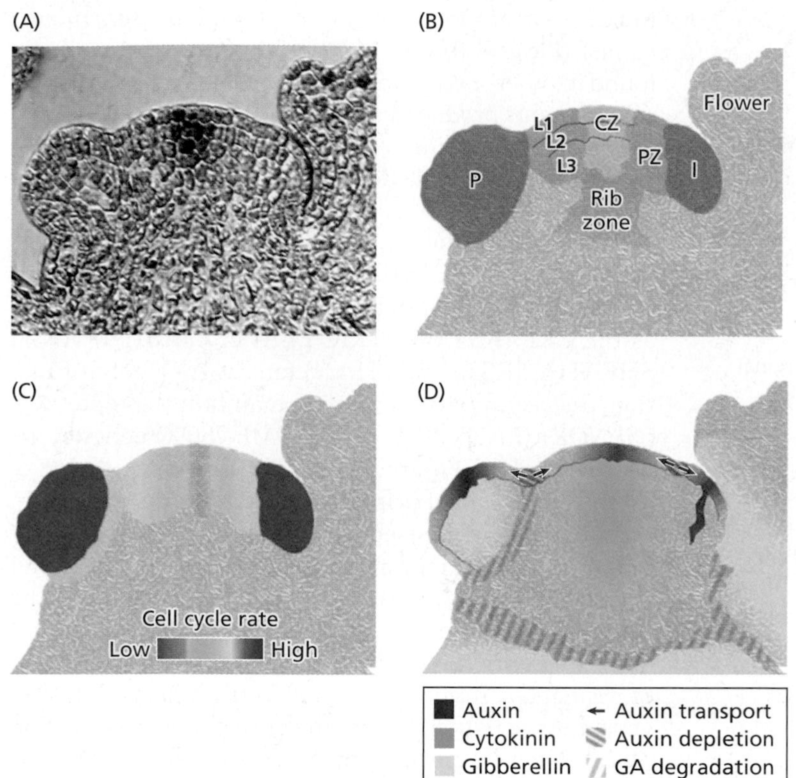

Figure 19.2 Longitudinal section of the Arabidopsis inflorescence meristem and diagrams showing its functional organization. (A) Light micrograph of the Arabidopsis inflorescence meristem showing the localization of *CLAVATA3* (*CLV3*) gene expression (brown stain) (see Chapter 17). (B) Anatomical zonation of the inflorescence meristem showing the central zone (CZ), peripheral zone (PZ), flower primordium (P), flower primoridium initial (I), L1–L3 layers, flower, and rib zone. (C) Spatial variations in the rate of cell division (indicated by color bar), showing the highest rates in the flower primordia. (D) Proposed distribution of the three main hormones, auxin, cytokinin, and gibberellins, as well as sites of auxin transport, auxin depletion, and gibberellin (GA) degradation. (From Besnard et al. 2011.)

such as maize (corn; *Zea mays*), lack L3. These tissue layers can be further demarcated into three histological zones: the central zone (CZ), peripheral zone (PZ), and rib zone (RZ) (**Figure 19.2A and B**).

To a large extent, positional information dictates the fate of cells in the SAM. For example, the highest rates of cell division in the inflorescence meristem of Arabidopsis are found in the primordium (P) and primordium initial (I), followed by the PZ and the CZ (**Figure 19.2C**). Position also determines the patterns of intracellular and intercellular signaling. As we discussed in Chapter 17, the different histological zones exhibit distinctive patterns of gene expression that maintain the growing SAM as a stable structure.

Hormonal signals play key roles in regulating leaf primordia emergence

As we discussed in Chapter 17, polar auxin transport in the L1 layer of the SAM is essential for leaf primordia emergence, and is responsible for leaf phyllotaxy (the pattern of leaf emergence from the stem; see Figure 17.32). When shoot apices are cultured in the presence of auxin transport inhibitors they fail to form primordia, and application of auxin to the SAM results in the induction of primordia at the application site. Auxin synthesis via the YUCCA pathway (see Chapter 15) generates auxin concentration gradients which, in turn, regulate expression and asymmetric distribution of PIN auxin efflux carriers to enhance or *canalize* localized polar auxin transport streams. (We will discuss auxin canalization in more detail later in this chapter.) Other hormones, such as cytokinin, gibberellins, and brassinosteroids, also play critical roles in maintaining SAM structure and activity. The distributions of auxin, cytokinin, and gibberellins in the SAM are shown in **Figure 19.2D**.

The initiation of leaf primordia was recently shown to be light-dependent in a manner that is independent of photosynthesis; tomato and Arabidopsis apices cease producing new leaf primordia when the plants are grown in the dark. This cessation is correlated with decreased auxin synthesis and loss of the polar localization of PIN1 in the SAM. However, because organ initiation in dark-grown apices can be restored only after application of both auxin and cytokinin, it appears that cytokinin may be involved in a light-dependent leaf initiation pathway. Phytochrome B has been implicated as the photoreceptor involved in the response to light (see Chapter 16), as it regulates auxin synthesis and overall auxin levels in the plant.

In addition to hormonal signals, mechanical stress in the SAM has been shown to alter microtubule arrangements as well as PIN1 distribution, which can affect leaf primordia initiation (see WEB TOPIC 19.2).

A signal from the SAM initiates adaxial–abaxial polarity

Since leaf primordia develop from a group of cells on the flank of the SAM, leaves possess inherent positional relationships with the SAM: the adaxial side of a leaf primordium is derived from cells adjacent to the SAM, while the abaxial side is derived from cells farther away. Microsurgical studies in the 1950s demonstrated that some type of communication between the SAM and the leaf primordium is required for the establishment of leaf adaxial–abaxial polarity. For example, a transverse incision divid-

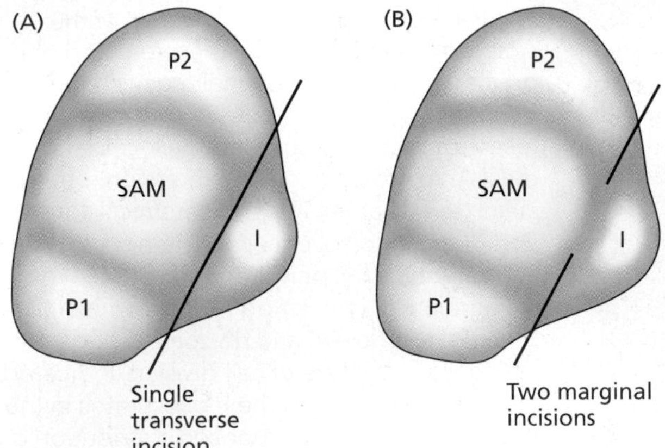

Figure 19.3 Microsurgical experiment demonstrating the influence of the SAM on leaf primordium (P) adaxial–abaxial development in potato (*Solanum tuberosum*). (A) A primordium initial (I) isolated from the SAM by a transverse incision grows radially and contains only abaxial tissues. (B) A primordium initial (I) that has not been completely isolated from the SAM shows normal adaxial–abaxial symmetry. (After Sussex 1951.)

ing the SAM from the primordium initial (I) caused the initial to develop radially without forming any adaxial tissue (**Figure 19.3A**). The resulting "leaf" was cylindrical and contained only abaxial tissues (it was *abaxialized*). However, two marginal incisions that allowed unimpeded communication between the SAM and the primordium initial led to the development of normal adaxial–abaxial symmetry (**Figure 19.3B**). Later refinements of these surgical experiments using laser ablation and microdissection techniques yielded similar results, suggesting that a signal from the SAM is required for specification or maintenance of adaxial identity. However, the nature of this signal remains a mystery.

ARP genes promote adaxial identity and repress the *KNOX1* gene

Insights into the molecular basis of adaxial and abaxial identity came from analysis of the loss-of-function mutants *phantastica* (*phan*) in snapdragon (*Antirrhinum majus*) (**Figure 19.4A**). *phan* mutants have since been found in other species, including Arabidopsis and tobacco. *phan* mutants produce leaves with altered adaxial–abaxial symmetry, ranging from abaxialized needlelike leaves that fail to produce lamina, to leaves with blades exhibiting a mosaic of adaxial and abaxial characters (**Figure 19.4B**).

The *PHAN* gene of *Antirrhinum,* and its orthologs, such as *ASYMMETRIC LEAVES1* (*AS1*) in Arabidopsis, encode MYB class transcription factors referred to as the ARP (ASYMMETRIC LEAVES1 [AS1], ROUGH SHEATH2 [RS2], and PHAN) family. ARP genes function, at least in part, by helping maintain the repression of *KNOX1* (*KNOTTED1-LIKE HOMEOBOX*) genes in the developing leaf (**Figure 19.5A**). The down-regulation of *KNOX* genes in leaf primordia, which occurs initially in response to the focused accumulation of auxin at leaf initiation sites, is required for adaxial development and is essential for normal adaxial–abaxial patterning of the leaf in many, but not all, species. The importance of this down-regulation is exemplified by the abnormal leaves of *phan* and *as1* mutants, as well as of plants with *KNOX* gene mutations that prevent the normal down-regulation of *KNOX* gene expression in leaves. In Arabidopsis, however, mutations in *AS1* alone do not affect abaxial–adaxial polarity, and thus other factors seem to be involved. Since ARP genes are expressed uniformly in the leaf primordia, it is assumed that their role in adaxial fate specification depends on the presence of interacting protein partners.

A large part of KNOX1 protein function appears to be mediated by its inhibitory effects on gibberellin levels in the SAM (see Chapter 17; Figure 17.31). While acting to inhibit gibberellin biosynthesis and to promote gibberellin inactivation, KNOX transcription factors also activate the cytokinin biosynthetic gene *ISOPENTENYL TRANSFERASE7* (*IPT7*), which increases cytokinin levels.

Adaxial leaf development requires HD-ZIP III transcription factors

Adaxial development also depends critically on a group of transcription factors known as HD-ZIP III proteins, so named because of the presence of both a DNA-binding

(A)

(B)

Figure 19.4 Effects of *phan* mutations on leaf morphology in *Antirrhinum majus.* (A) The vegetative shoot of a wild-type plant with normal leaves. (B) The vegetative shoot of a *phan* mutant with narrow (n), needlelike (n-l), and mosaic (m) leaves. (From Waites and Hudson 1995.)

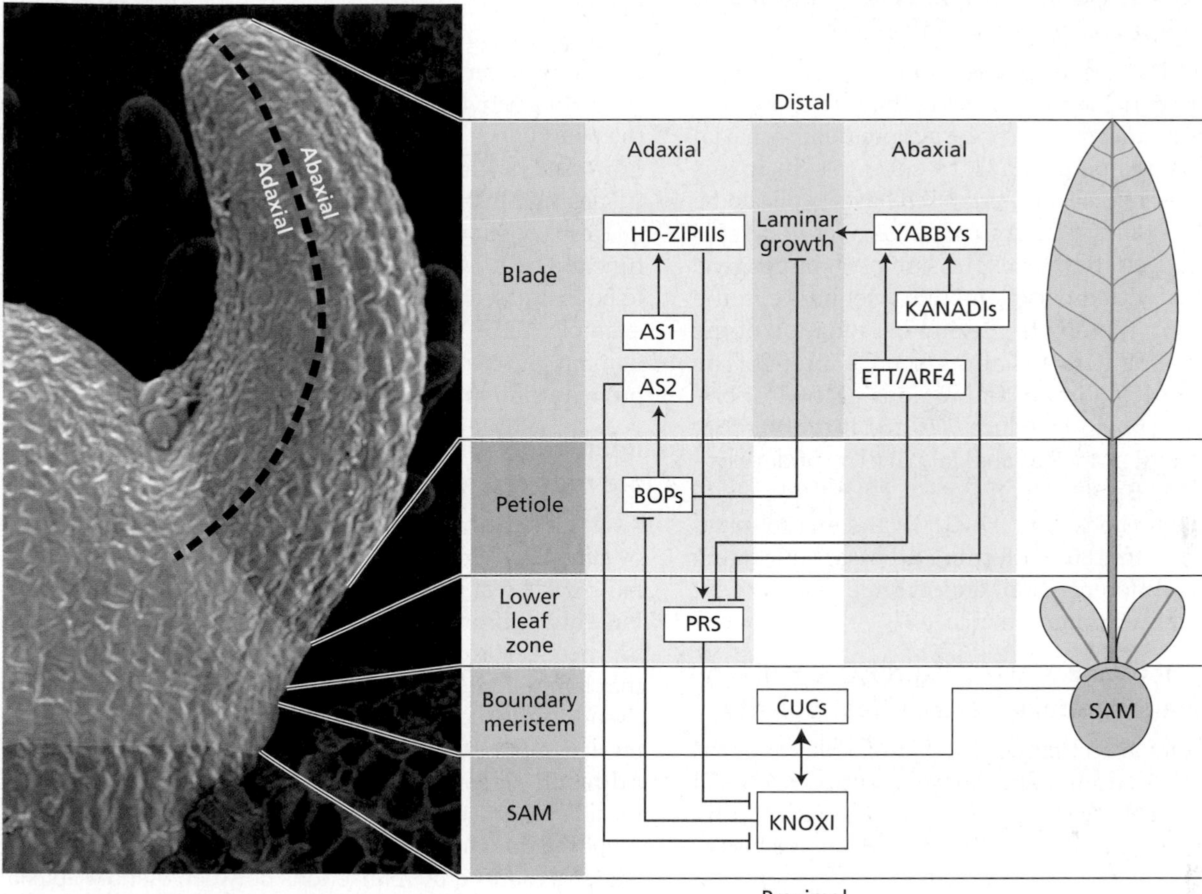

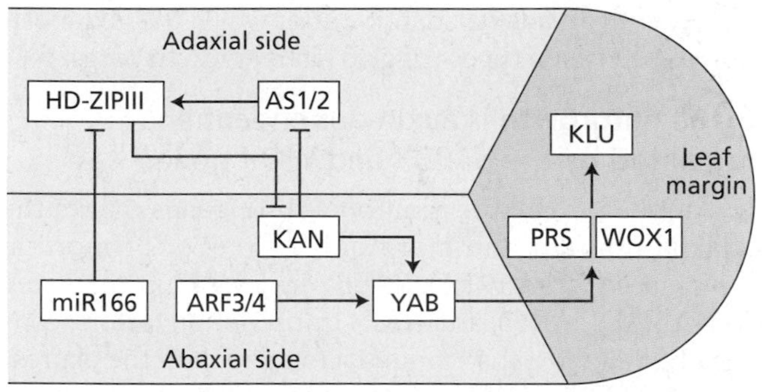

Figure 19.5 Gene networks regulating leaf polarity. (A) Regulation of proximal–distal polarity. Various genes involved in proximal–distal patterning interact with specific genes in the abaxial–adaxial gene network. (B) Gene networks involved in leaf margin growth and adaxial–abaxial polarity. SAM, shoot apical meristem. (See text for discussion.) (A after Townsley and Sinha 2012; B after Fukushima and Hasebe 2013.)

homeodomain and a leucine zipper dimerization domain. HD-ZIP III transcription factors are also distinguished by a putative lipid/sterol-binding domain, suggesting that their activity could be regulated by types of signaling molecules that are currently unknown in plants. These transcription factors also feature a conserved sequence motif that mediates protein–protein interactions, providing additional scope for the regulation of their activity.

Expression of the HD-ZIP III genes, such as *PHABULOSA* (*PHB*) and *PHAVOLUTA* (*PHV*), is normally limited to the adaxial domains of the leaf primordia (see Figure 19.5A). When these genes are abnormally expressed throughout the leaf, as occurs in some *phb* and *phv* mutants, abaxial tissues take on adaxial characteristics. For example, in mutants in which *PHB* is ectopically expressed in abaxial domains of the leaf, axillary buds, normally limited to the adaxial side of the leaf base, now form on both sides. Conversely, mutations that block the function of *PHB* and *PHV* genes in their normal, adaxial domains of expression lead to loss of adaxial characters, but only if the activity of both genes is blocked. Together, these results suggest that *PHB* and *PHV* act redundantly to promote adaxial identities in tissues where they are expressed.

The expression of HD-ZIP III genes is antagonized by miR166 in abaxial regions of the leaf

Because HD-ZIP III genes promote the acquisition of an adaxial identity in those tissues where they are expressed, their expression must somehow be suppressed in abaxial regions of the developing leaf. In an effort to explain this restricted expression, several analyses have implicated a class of small regulatory RNAs known as microRNAs (or miRs). A microRNA inhibits expression of its target gene by base pairing with a complementary sequence in the gene's transcript, thereby triggering degradation of the mRNA or blocking its translation (see Chapter 2). The expression of miR166 in abaxial regions of the leaf primordia has been shown to reduce *PHB* and *PHV* transcript levels, thus enabling normal abaxial patterns of development (**Figure 19.5B**).

The antagonism between HD-ZIP III and miR166 plays multiple roles in different patterning processes, including vascular tissue differentiation, endodermis development in the root, and SAM maintenance.

Antagonism between KANADI and HD-ZIP III is a key determinant of adaxial–abaxial leaf polarity

Transcription factors in the KANADI family play a central role in the specification of abaxial cell identity. KANADI genes appear to have overlapping functions with YABBY genes (discussed below), with the most dramatic loss of abaxial identity being observed when loss-of-function mutations of the two types of genes are combined. Conversely, abnormal formation of abaxial tissues is observed when KANADI genes are overexpressed. Although it is not entirely clear how KANADI transcription factors promote abaxial identity, young embryos that are deficient in KANADI activity exhibit changes in the polar distribution of PIN auxin efflux carriers that precede any overt changes in development. The suggestion that abaxial development is closely coupled to polar transport of auxin is reinforced by the observation that members of the AUXIN RESPONSE FACTOR gene family, *ARF3* and *ARF4*, are required for the normal establishment of abaxial fate (see Figure 19.5B and Chapter 15). KANADI genes and HD-ZIP III genes play antagonistic roles in adaxial–abaxial patterning in both leaves and vasculature (see Figure 19.5B).

The YABBY gene family of transcription factors, named after the Australian freshwater crayfish, appears to act redundantly with the KANADI genes. YABBY gene mutants were among the earliest leaf polarity mutants discovered in Arabidopsis. The first member of this gene family identified, *CRABS CLAW* (*CRC*), was defined by the phenotype of its Arabidopsis loss-of-function mutant, in which the organization of the carpels (parts of the flower) is disturbed. The more general activity of Arabidopsis YABBY genes is revealed when mutations affecting several members of this family are combined. These multiple mutants have defective floral and vegetative leaflike organs in which abaxial characters have been replaced by adaxial characters, suggesting that there is functional redundancy among members of the YABBY gene family. The abaxial-promoting activity of YABBY genes is further supported by the phenotypes of plants in which YABBY genes are overexpressed. Such plants show ectopic formation of abaxial tissues, and in some circumstances loss of the SAM.

Despite their redundant action with KANADI genes, the function of YABBY genes is more enigmatic and appears to be associated predominantly with growth. In maize, for example, YABBY genes are expressed in the adaxial leaf domain and hence their role in maize is thought to be to promote lamina outgrowth rather than abaxialization.

Interactions between adaxial and abaxial tissues are required for blade outgrowth

As described above, the abaxialized primordia produced by surgically isolating primordia from the apical meristem fail to form leaf blades (see Figure 19.3). Similarly, in *phan* mutants, leaf primordia with no adaxial tissues develop into needlelike leaves. Together these observations suggest that lamina outgrowth requires both adaxial and abaxial tissues. Indeed, the mosaic leaves sometimes produced by *phan* mutants have bladelike outgrowths called **lamina ridges** that are formed specifically at the boundaries of the adaxial and abaxial domains (**Figure 19.6**). It has been proposed that the normal lateral growth of the lamina (leaf blade) is induced by interactions between distinct abaxial and adaxial tissue types. According to this model, the primary function of *PHAN* is to enable the development of tissues with an adaxial identity, after which the juxtaposition of two tissue types triggers lateral growth programs.

Blade outgrowth is auxin dependent and regulated by the YABBY and WOX genes

In Arabidopsis, expression of YABBY genes marks the abaxial domain and marginal regions of primordial leaves (see Figure 19.5A). YABBY genes are up-regulated by KANADI, ARF3, and ARF4 transcription factors; conversely, YABBY transcription factors promote the expression of *KAN1* and *ARF4* genes, forming positive feedback loops. In the absence of all YABBY gene activity, leaf primordia establish adaxial–abaxial polarity but fail to initiate lamina outgrowth. These findings indicate that YABBY genes mediate the induction of growth activity related to adaxial–abaxial polarity.

YABBY transcription factors positively regulate a member of the WOX gene family, *PRS* (*PRESSED FLOWER*), which is expressed in the leaf margin and promotes blade outgrowth (see Figure 19.5B). PRS and WOX1 transcription factors function cooperatively, and the *prs/wox1* double mutant exhibits a narrow-leaf phenotype in Arabidopsis, similar to the leaf phenotype of *phan* mutants. PRS- and WOX1-dependent blade outgrowth is, in part,

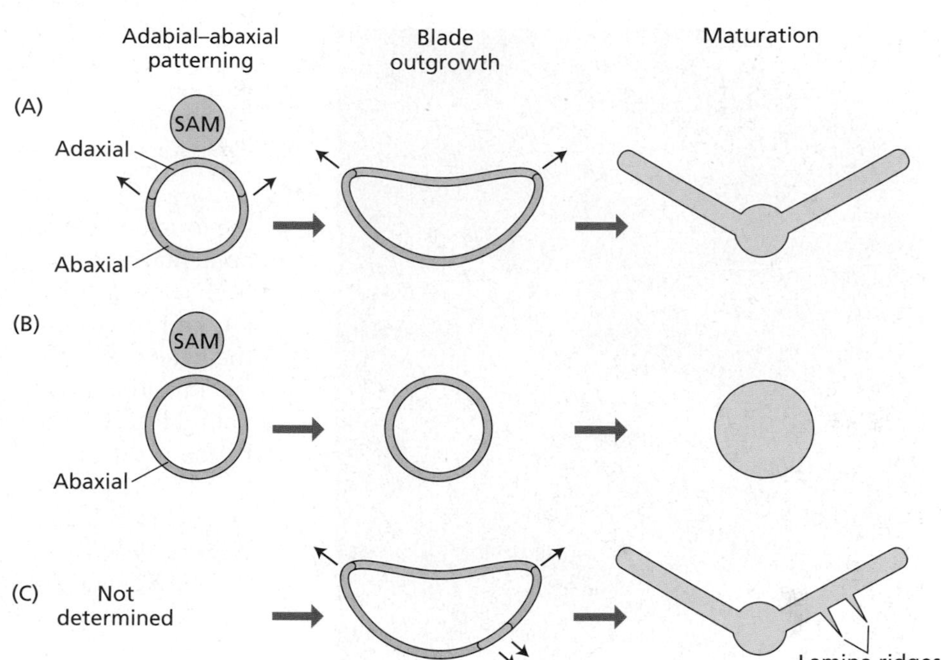

Figure 19.6 Leaf development in relation to the adaxial–abaxial boundaries in different types of leaves. The diagrams shows cross-sectional outlines of leaf primordia at the establishment of adaxial–abaxial patterning (left), at an early stage in blade outgrowth (middle), and at leaf maturity (right). (A) Conventional bifacial leaf, as in wild-type Arabidopsis. (B) *phan* mutant of snapdragon (*Antirrhinum majus*) and *PHAN* ortholog mutants of tobacco (*Nicotiana sylvestris*). (C) Maize *milkweed pod1* mutant. Note the outgrowths on the surfaces where the adaxial and abaxial tissues come in contact. (From Fukushima and Hasebe 2013.)

mediated by an as yet unidentified mobile signal(s) processed by KLU, a cytochrome P450 monooxygenase (see Figure 19.5B). KLU promotes cell division activity in aerial organs, including leaves, and a loss-of-function mutant of the *KLU* gene produces smaller organs. Auxin appears to be another signal acting in blade formation, independent of KLU. Multiple loss-of-function mutants of the *YUCCA* (*YUC*) auxin biosynthetic genes exhibit defective blade outgrowth, raising the possibility that auxin participates in the regulatory network for directed growth of the leaf.

Leaf proximal–distal polarity also depends on specific gene expression

In addition to adaxial–abaxial polarity, leaf development also exhibits polarity along its length, called **proximal–distal polarity.** Developing leaf primordia can be divided lengthwise into four main zones extending from the meristem: boundary meristem, lower-leaf zone, petiole, and blade (see Figure 19.5A).

Proximal–distal polarity becomes evident as the primordium begins to grow out and away from the SAM. The **boundary meristem**, although not considered part of the leaf, is important for normal leaf initiation. The initiation of leaves from the peripheral zone requires the creation of meristem-to-organ boundaries, buffer zones that separate these two cell groups with distinct gene expression programs and morphologies. The boundary meristem itself expresses a unique set of transcription factors that participate in the local repression of cell proliferation, a prerequisite for the development of physically separate organs. *CUC* (*CUP-SHAPED COTYLEDON*) 1 and 2 genes in Arabidopsis encode plant-specific NAC (NAM; ATAF1,2; CUC2) transcription factors that regulate cotyledon formation (see Chapter 17; Figure 17.27). Later in development, these *CUC* genes also control the specification of organ boundaries during leaf initiation. As typically happens in the case of genes regulating boundary functions, *cuc1/cuc2* double mutants display organ fusions and growth arrest. As in cotyledon development during embryogenesis, there is interdependence between *CUC* gene expression and auxin-dependent leaf primordium initiation.

The **lower-leaf zone** (**LLZ**) plays an important role in leaves that develop stipules or form leaf sheaths (see Figure 19.1). In these cases, the founder cells (which give rise to the leaf primordium) recruit additional cells into the primordium through a mechanism that in Arabidopsis is dependent on the expression of orthologs of the WOX gene *PRS* (*PRESSED FLOWER*). Cells that are recruited to become stipules or sheath are taken from the flanks of the primordium.

The region of the leaf primordium destined to become the petiole is characterized by the expression of *BOP* (*Blade on Petiole*) genes, which encode transcriptional activators that are required to establish petiole identity in the proximal portion of the leaf in Arabidopsis (see Figure 19.5A). The double mutant *bop1/bop2* lacks the proper distinction between leaf blade and petiole, and both single mutants show laminar development on what would be the petiole. *BOP1* and *BOP2* are both expressed in the adaxial domain, where they act redundantly to suppress laminar outgrowth in the petiole region.

In compound leaves, de-repression of the *KNOX1* gene promotes leaflet formation

Compound leaves have evolved independently many times from simple leaf forms. Despite wide variations in

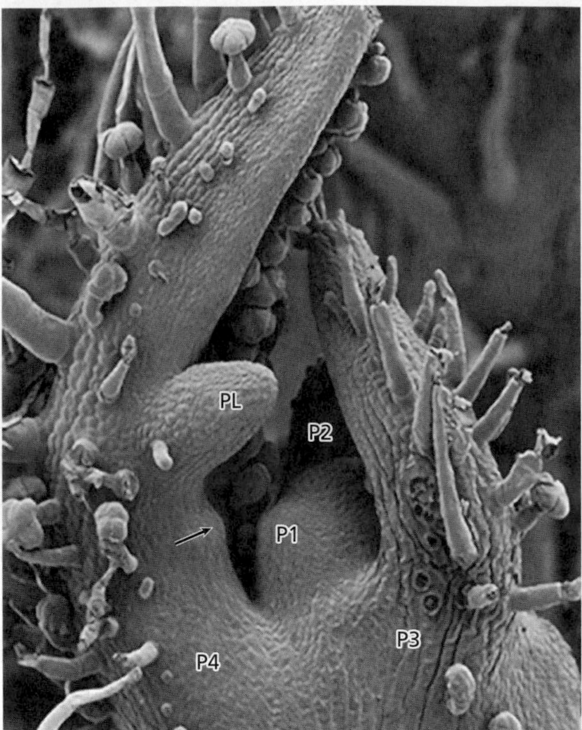
PL
P2
P1
P3
P4
100 μm

Figure 19.7 Scanning electron micrograph of tomato shoot tip showing developing compound leaf. Primordia 1 through 4 (P1–P4) are shown. The first primary leaflet pair (PL) and the second leaflet pair (arrow) are visible on P4. (From Kang and Sinha 2010.)

the form and complexity of compound leaves, the developmental mechanisms that lead to their formation have been converged on repeatedly. By delaying the differentiation process, individual leaf primordia can redeploy the gene regulatory networks used by the SAM during leaf initiation to form leaflet primordia, resulting in compound leaf development (**Figure 19.7**). Similar to what happens during the initiation of leaf primordia on the SAM, PIN1 proteins focus auxin flow, leading to formation of local auxin maxima on the flanks of the primordia (**Figure 19.8**).

KNOX1 genes are important components of the regulatory network involved in compound leaf development (see Figure 19.8). *CUC* genes are required for the de-repression of *KNOX* genes. Cytokinins act downstream of KNOX proteins in promoting leaflet development. For example, overexpression of the cytokinin biosynthetic gene, *IPT7*, in tomato leaf primordia causes an increase in the number of leaflets. Conversely, overexpression of the cytokinin degradation gene, *CKX3*, results in a decrease in the number of leaflets. A parallel role of *KNOX* and *CUC* genes in the formation of leaf serrations is discussed in **WEB TOPIC 19.3**.

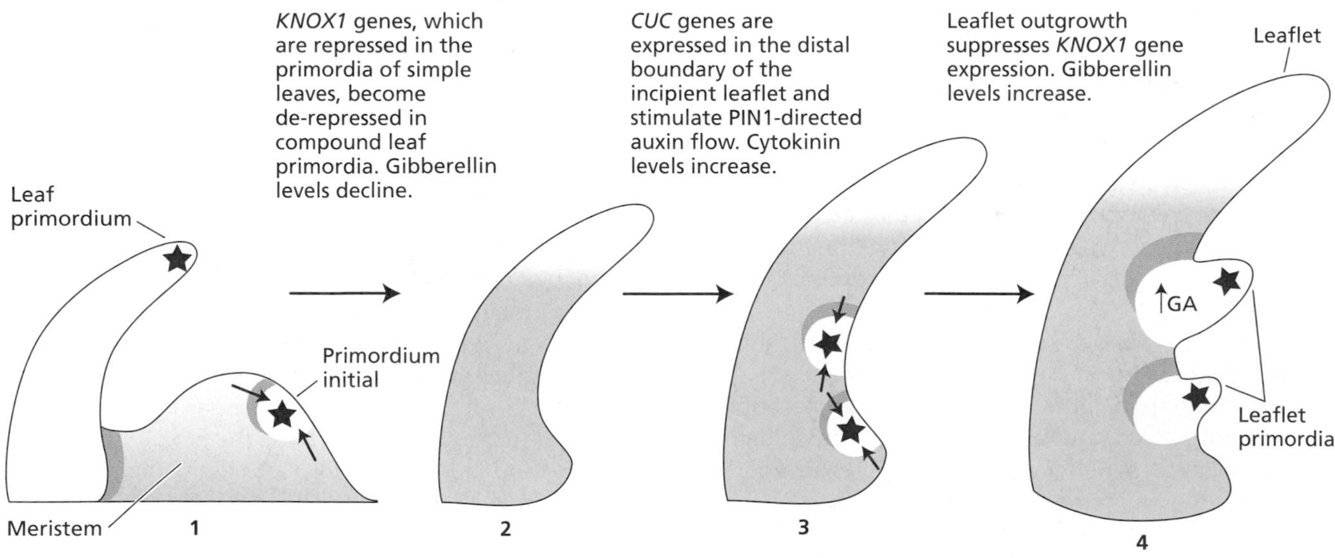
KNOX1 genes, which are repressed in the primordia of simple leaves, become de-repressed in compound leaf primordia. Gibberellin levels decline.
CUC genes are expressed in the distal boundary of the incipient leaflet and stimulate PIN1-directed auxin flow. Cytokinin levels increase.
Leaflet outgrowth suppresses KNOX1 gene expression. Gibberellin levels increase.
Leaflet
Leaf primordium
Primordium initial
Meristem
1
2
3
4
GA
Leaflet primordia
KNOX1 expression
CUC expression
Auxin flow
Peak of auxin response
GA Gibberellin

Figure 19.8 Development of compound leaves. The initial stages of simple and compound leaf development are similar. *KNOX1* genes are repressed in the primordium initial (1) and are subsequently reactivated (2), thereby maintaining the primordium in an undifferentiated state. Leaflet primordia are then initiated in a process resembling the initiation of leaf primordia involving PIN1-mediated auxin flow (3 and 4). (From Hasson et al. 2010.)

Differentiation of Epidermal Cell Types

In addition to the palisade parenchyma and spongy mesophyll, which are specialized for photosynthesis and gas exchange, the leaf epidermis also plays vital roles in leaf function. The epidermis is the outermost layer of cells on the primary plant body, including both vegetative and reproductive structures. The epidermis usually consists of a single layer of cells derived from the L1 layer, or **protoderm**. In some plants, such as members of the Moraceae and certain species of the Begoniaceae and Piperaceae, the epidermis has two to several cell layers derived from periclinal divisions of the protoderm.

There are three main types of epidermal cells found in all angiosperms: pavement cells, trichomes, and guard cells. **Pavement cells** are relatively unspecialized epidermal cells that can be regarded as the default developmental fate of the protoderm. **Trichomes** are unicellular or multicellular extensions of the shoot epidermis that take on diverse forms, structures, and functions, including protection against insect and pathogen attack, reduction of water loss, and increased tolerance of abiotic stress conditions. **Guard cells** are pairs of cells that surround the **stomata**, or pores, which are present in the photosynthetic parts of the shoot. Guard cells regulate gas exchange between the leaf and the atmosphere by undergoing tightly regulated turgor changes in response to light and other factors (see Chapter 10). Other specialized epidermal cells, such as **lithocysts**, **bulliform cells**, **silica cells**, and cork cells (**Figure 19.9**), are found only in certain groups of plants and are not as well studied.

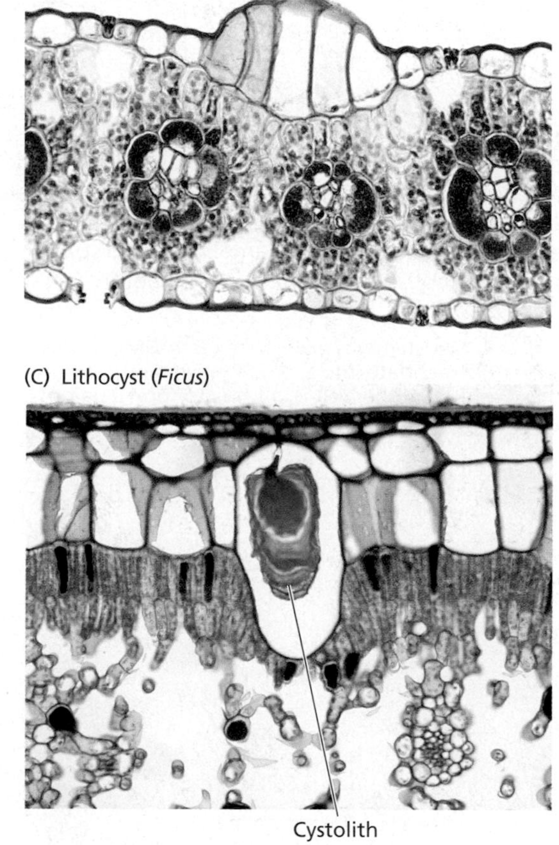

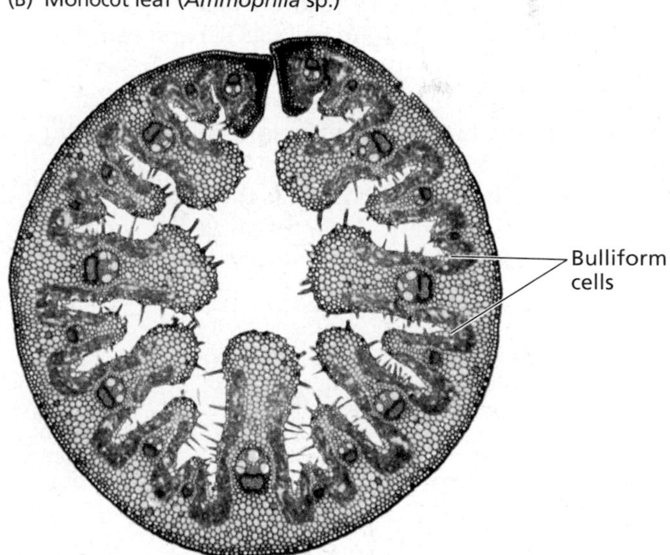

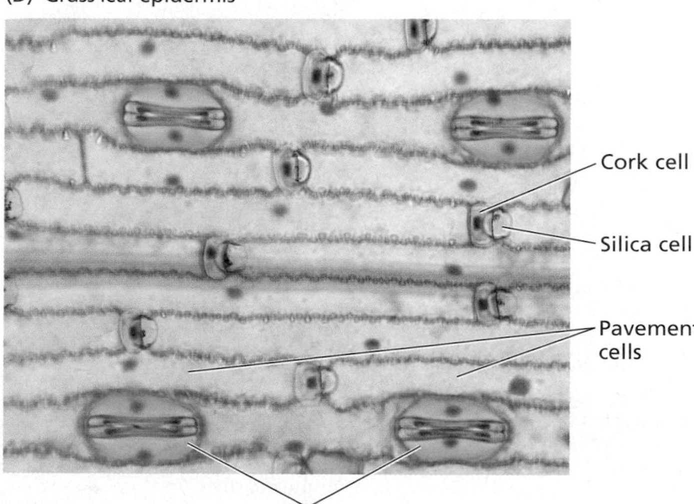

Figure 19.9 Examples of specialized epidermal cells. (A) Bulliform cells of maize. (B) Rolled leaf of marram grass (*Ammophila* sp.). Rolling and unrolling in grass leaves is driven by turgor changes in the bulliform cells. (C) Lithocyst cell in a *Ficus* leaf containing a cystolith, composed of calcium carbonate deposited on a cellulosic stalk attached to the upper cell wall. (D) Wheat (*Triticum aestivum*) leaf epidermis with pairs of silica and cork cells interspersed among the pavement cells.

The formation of pavement cells, the default pathway for epidermal cell development, was discussed in Chapter 14 (see Figure 14.15). Here we will describe the development of two types of specialized epidermal cells, guard cells and trichomes, which have been studied intensively as model systems for pattern formation and cytodifferentiation.

Guard cell fate is ultimately determined by a specialized epidermal lineage

Developing plant leaves exhibit a tip-to-base developmental gradient, with cell division prevalent at the base of the leaf and differentiation occurring near the tip. In Arabidopsis, guard cell differentiation also follows this trend, but is ultimately governed by the stomatal cell lineage (**Figure 19.10**). In the developing protoderm (which will give rise to the leaf epidermis), a population of **meristemoid mother cells (MMCs)** is established. Each MMC divides asymmetrically (the so-called **entry division**) to give rise to two morphologically distinct daughter cells—a larger **stomatal lineage ground cell (SLGC)** and a smaller **meristemoid** (see Figure 19.10). An SLGC can either differentiate into a pavement cell or become an MMC and found secondary or satellite lineages. The meristemoid can undergo a variable number of asymmetric **amplifying divisions** giving rise to as many as three SLGCs, with the meristemoid ultimately differentiating into a **guard mother cell (GMC)**, which is recognizable because of its rounded morphology. The GMC then undergoes one symmetrical division, forming a pair of guard cells surrounding a pore—the stomate. Although this lineage is called the "stomatal lineage," the ability of meristemoids and SLGCs to undergo repeated divisions means that this lineage is actually responsible for generating the majority of the epidermal cells in the leaves.

Following amplifying divisions of the meristemoid, the resulting SLGCs can differentiate into pavement cells, which are the most abundant cell type in the epidermis of a mature leaf, or they can divide asymmetrically (**spacing divisions**) to give rise to a secondary meristemoid. The orientation of division in asymmetrically dividing SLGCs is important for enforcing the "one-cell-spacing rule," according to which stomata must be situated at least one cell length apart to maximize gas exchange between the leaf and the atmosphere. Incorrect stomatal patterning results when genes controlling critical stages in the lineage are mutated.

Figure 19.10 Stomatal development in Arabidopsis. Three related transcription factors, SPCH, MUTE, and FAMA, form heterodimers with SCRM and are required for the production of meristemoids, GMCs, and guard cells. They are also required for the amplifying and spacing pathways as well (not shown). (After Lau and Bergmann 2012.)

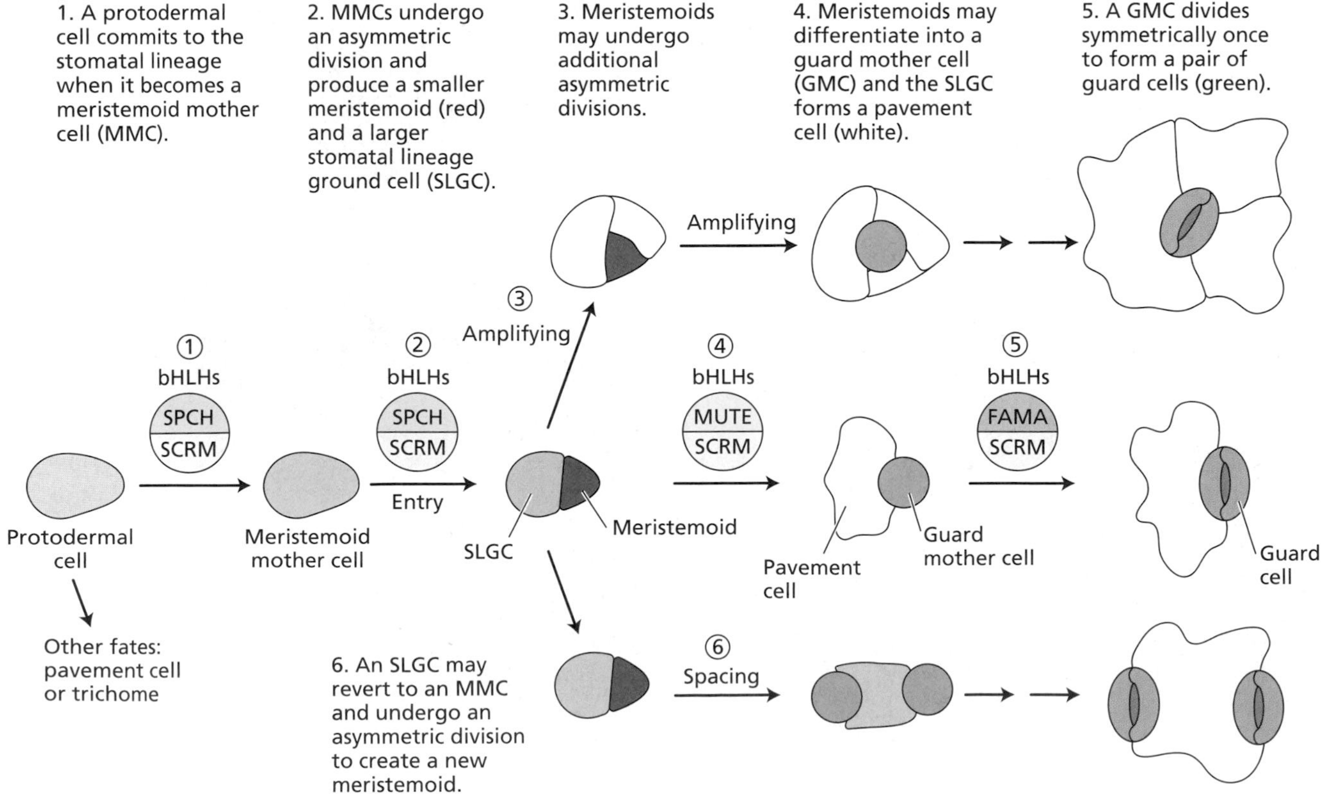

1. A protodermal cell commits to the stomatal lineage when it becomes a meristemoid mother cell (MMC).
2. MMCs undergo an asymmetric division and produce a smaller meristemoid (red) and a larger stomatal lineage ground cell (SLGC).
3. Meristemoids may undergo additional asymmetric divisions.
4. Meristemoids may differentiate into a guard mother cell (GMC) and the SLGC forms a pavement cell (white).
5. A GMC divides symmetrically once to form a pair of guard cells (green).
Amplifying
③
Amplifying
①
bHLHs
SPCH
SCRM
②
bHLHs
SPCH
SCRM
Entry
④
bHLHs
MUTE
SCRM
⑤
bHLHs
FAMA
SCRM
Protodermal cell
Meristemoid mother cell
SLGC
Meristemoid
Pavement cell
Guard mother cell
Guard cell
Other fates: pavement cell or trichome
6. An SLGC may revert to an MMC and undergo an asymmetric division to create a new meristemoid.
⑥
Spacing

Two groups of bHLH transcription factors govern stomatal cell fate transitions

The various stages in stomatal development highlight three specific cell-state transitions: (1) MMC to meristemoid, (2) meristemoid to GMC, and (3) GMC to mature guard cells. Each of these transitions is associated with, and requires the specific expression of, one of three basic helix-loop-helix (bHLH) transcription factors: SPEECHLESS (SPCH), MUTE, and FAMA (named after the Roman goddess of rumor) (see Figure 19.10). SPCH drives MMC formation and the asymmetric entry division of these cells, as well as the subsequent asymmetric amplifying and spacing divisions. MUTE terminates stem cell behavior by promoting the differentiation of meristemoids into GMCs, and FAMA promotes the terminal cell division and differentiation of GMCs into guard cells. In addition, two related bHLH leucine zipper (bHLH-LZ) proteins, SCREAM (SCRM) and SCRM2, have been identified as the partners of SPCH, MUTE, and FAMA.

Peptide signals regulate stomatal patterning by interacting with cell surface receptors

Leucine-rich repeat receptor-like kinases (LRR-RLKs) are single-pass transmembrane proteins with an extracellular ligand-binding domain and an intracellular kinase domain for downstream signaling. The ERECTA family (ERf) of receptor-like kinases (RLKs) has three members—ERECTA, ERL1, and ERL2—all of which control the proper patterning and differentiation of stomata. For example, ERECTA, which is expressed strongly in the protodermal cells but is undetectable thereafter, restricts asymmetric entry division in MMCs (**Figure 19.11**).

A receptor-like protein, TOO MANY MOUTHS (TMM), is also required for stomatal patterning. TMM is expressed within the stomatal lineage and appears to provide specificity to the more widely expressed ERECTA gene family (see Figure 19.11). Receptor-like proteins lack a C-terminal kinase domain and thus are thought to be incapable of transducing signals on their own. Like the ERf, the receptor-like protein TMM inhibits stomatal lineage proliferation and guides spacing divisions in leaves.

The EPIDERMAL PATTERNING FACTOR-LIKE (EPFL) protein family is a recently identified group of 11 small, secreted cysteine-rich peptides that have been shown to regulate stomatal development. Two founding members of the family, EPF1 and EPF2, are stomatal lineage–specific factors, and they repress stomatal development at specific stages when ERECTA genes are being expressed. According to current models, EPF2 and EPF1 are secreted by MMCs/meristemoids and GMCs, respectively, and are perceived by ERECTA family receptors in surrounding cells. As a result, the ERECTA receptor inhibits stomatal development (see Figure 19.11). In this way the EPF2–ERECTA pair regulates the number and density of stomata. Different pairings between EPFL peptides and ERECTA family receptors regulate different aspects of stomatal patterning, while TMM apparently modulates the signaling pathway.

An unexpected wrinkle in the above scenario is the discovery that the mesophyll also contributes to stomatal patterning. One of the EPFL peptides, **STOMAGEN**, a positive regulator of stomatal density, is produced by the underlying mesophyll and released to the epidermis. Experiments have shown that the depletion of STOMAGEN results in a decrease in stomatal numbers, indicating that its function is important for normal stomatal development. The stomata-inducing phenotype of STOMAGEN overexpression or its exogenous application requires TMM, leading to the proposal that TMM may act as a receptor for STOMAGEN. However, the mechanism by which STOMAGEN stimulates stomatal development is still unknown.

Genetic screens have led to the identification of positive and negative regulators of trichome initiation

Trichome development has been most thoroughly studied in the rosette leaves of Arabidopsis. Arabidopsis trichomes are unicellular and branched, with a distinctive tricorn (three-horned) structure (**Figure 19.12**).

Arabidopsis trichomes develop from single protodermal cells. The first recognizable change from a proto-

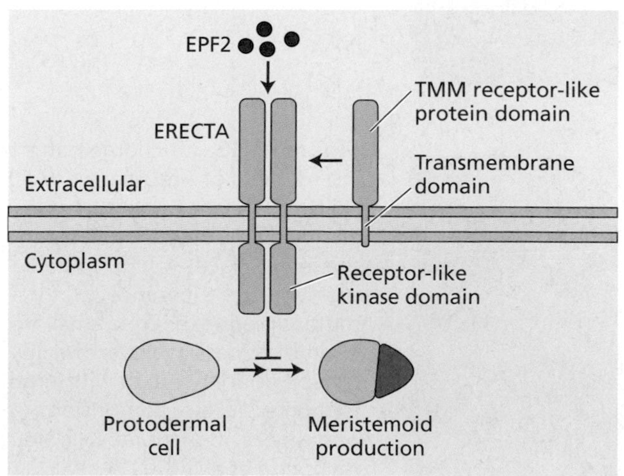

Figure 19.11 EPF2 peptide signaling negatively regulates stomatal density and patterning. EPF2 is synthesized and secreted by meristemoid mother cells and early meristemoids. The presence of extracellular EPF2 is detected by the receptor-like kinase ERECTA of protodermal cells. Together with the receptor-like protein TMM, the EPF2–ERECTA complex activates an intracellular signaling cascade that represses the production of new meristemoids. (After Lau and Bergmann 2012.)

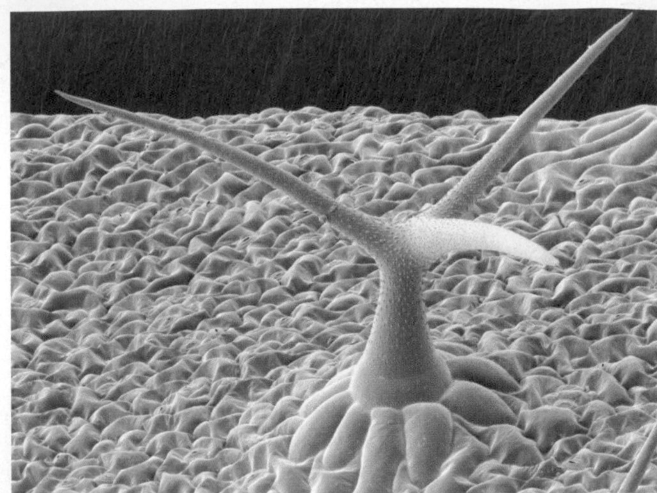

Figure 19.12 Arabidopsis trichome showing the typical tricorn branching pattern.

dermal cell to an incipient trichome cell is an increase in nuclear size due to the initiation of **endoreduplication**, replication of the nuclear genome in the absence of nuclear or cell divisions (see Chapter 2). Trichome cell morphogenesis is characterized by an initial outgrowth, followed by two successive branching events resulting in the tricorn morphology.

Trichomes are initiated at the base of the developing leaf, where they are typically separated by three or four protodermal cells that do not develop into trichomes. This regular spacing suggests the existence of developmental fields between neighboring trichomes that inhibit trichome initiation in the intervening protodermal cells. As the leaf expands, new trichomes are initiated at the leaf base, and the previously formed trichomes are further separated by cell divisions of the intervening epidermal cells.

Genetic screens for mutants affecting trichome development have led to the discovery of genes regulating trichome patterning—especially trichome density and spacing (**Figure 19.13**). The mutants generally fall into two classes. One class shows fewer or no trichomes, indicative of the absence of proteins that are positive regulators of trichome formation (see Figure 19.13B). These genes include *TRANSPARENT TESTA GLABRA1* (*TTG1*), *GLABRA1* (*GL1*), and *GLABRA3* (*GL3*). *TTG1* encodes a protein with WD40 repeat domains (a 40 amino acid motif with conserved tryptophan [W] and aspartate [D] residues) that generally function as protein–protein interaction domains. *GL1* encodes a MYB-related transcription factor, and *GL3* encodes a bHLH-like transcription factor. GL1, GL3, and TTG1 function together as a GL1–GL3–TTG1 protein complex that regulates the expression of other genes.

The second class of trichome patterning mutants has either more trichomes or unevenly spaced trichomes (trichome clusters), and the corresponding genes therefore

Figure 19.13 Trichome patterning mutants of Arabidopsis. (A) Wild-type plant with more or less evenly distributed trichomes on the leaf surfaces. (B) *gl1* mutant plant lacking trichomes. (C) *try* mutant plant exhibiting small trichome clusters (white arrow). (D) *try/cpc* double mutant with large trichome clusters comprising as many as 40 trichomes. (From Balkunde et al. 2010.)

encode proteins that act as negative regulators of trichome development (see Figure 19.13C and D). These negative regulators include *TRYPTICON* (*TRY*), which encodes a MYB protein lacking a transcriptional activation domain. TRY is expressed in developing trichomes and moves to the surrounding cells, where it inactivates the GL1–GL3–TTG1 complex by displacing GL1 (**Figure 19.14**). Inactivation of the GL1–GL3–TTG1 complex prevents trichome formation in the surrounding cells and thus establishes the regular spacing of trichomes on the leaf epidermis.

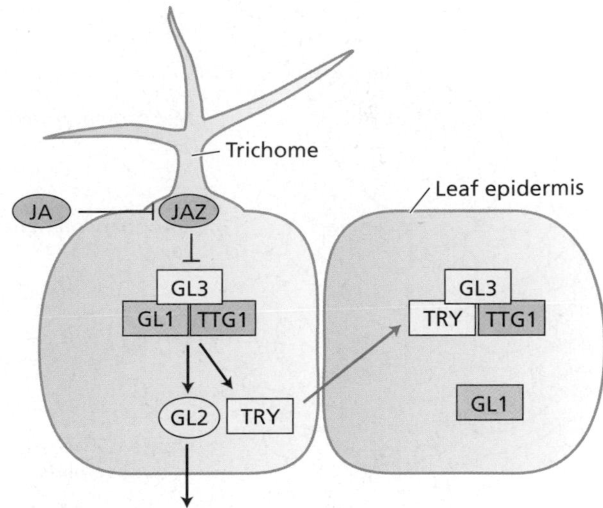

Figure 19.14 Role of *GLABRA2* (*GL2*) in leaf trichome formation. Cells that will form trichomes strongly express the *GL2* and *TRY* genes (black arrows). GL2 protein acts as a positive regulator of trichome cell differentiation. TRY protein moves to neighboring epidermal cells (blue arrow), where it inhibits trichome formation. (After Qing and Aoyama 2012.)

GLABRA2 acts downstream of the GL1–GL3–TTG1 complex to promote trichome formation

GLABRA2 (*GL2*) was originally identified as a gene that, when mutated, caused aborted trichomes with aberrant cell expansion. *GL2*, which is activated in trichome cells by the GL1–GL3–TTG1 complex, encodes a homeodomain leucine zipper transcription factor (see Figure 19.14). *GL2* expression is thought to represent the rate-limiting step in trichome formation. In wild-type plants, high levels of *GL2* promoter activity have been observed in the entire leaf at early leaf development stages; however, later on this activity is limited to developing trichomes and cells surrounding early-stage trichomes. Extensive analyses of gene expression patterns indicate that a large number of genes are regulated downstream of *GL2* during trichome differentiation.

Whereas *GL2* promotes trichome formation in the leaf epidermis, the gene has the opposite effect in roots. *gl2* mutants form ectopic root hairs, indicating that the gene product acts as a suppressor of root hair development.

Jasmonic acid regulates Arabidopsis leaf trichome development

Jasmonic acid (JA) and its derivative compounds function as key signaling molecules in trichome formation, and addition of exogenous jasmonic acid causes an increase in the number of leaf trichomes in Arabidopsis. In Arabidopsis, jasmonate ZIM-domain (JAZ) proteins repress trichome formation by binding to GL3 and GL1, key partners of the activation complex. Also, jasmonic acid participates in trichome initiation by degrading JAZ proteins, thereby abolishing the interactions of JAZ proteins with bHLH and MYB factors, which activate the transcription of trichome activators (see Figure 19.14).

Venation Patterns in Leaves

The leaf vascular system is a complex network of interconnecting veins consisting of two main conducting tissue types, xylem and phloem, as well as nonconducting cells, such as parenchyma, sclerenchyma, and fibers. The spatial organization of the leaf vascular system—its **venation pattern**—is both species- and organ-specific. Venation patterns fall into two broad categories: *reticulate venation*, found in most eudicots, and *parallel venation*, typical of many monocots (**Figure 19.15**).

(A)

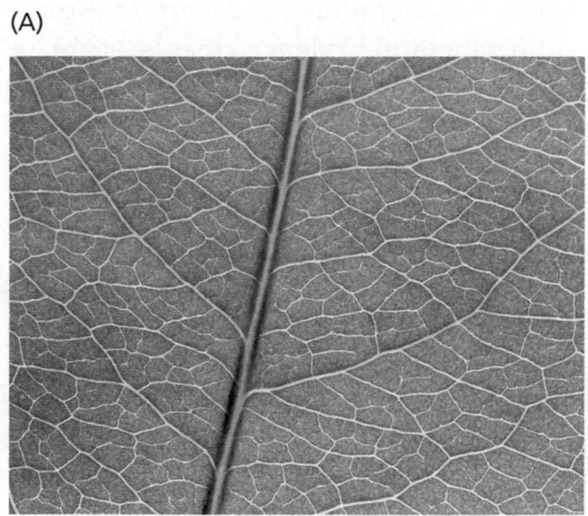

(B)

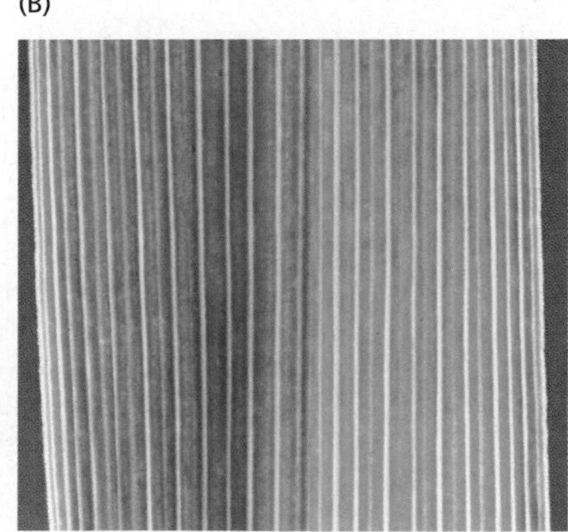

Figure 19.15 Two basic patterns of leaf venation in angiosperms. (A) Reticulate venation in *Prunus serotina*, a eudicot. (B) Parallel venation in *Iris sibirica*, a monocot.

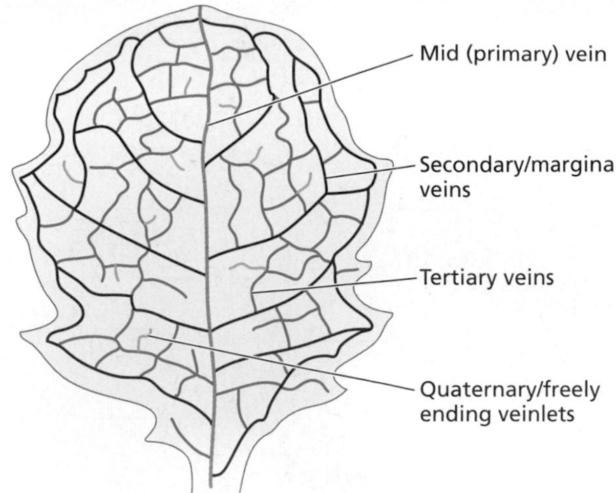

Figure 19.16 Hierarchy of venation in the mature Arabidopsis leaf based on the diameter of the veins at the site of attachment to the parent vein. (After Lucas et al. 2013.)

Figure 19.17 Development of the shoot vascular system. ▶ (A) Longitudinal section through the shoot tip of perennial flax (*Linum perenne*), showing the early stage in the differentiation of the leaf trace procambium at the site of a future leaf primordium. The leaf primordia and leaves are numbered, beginning with the youngest initial. (B) Early vascular development in a shoot with decussate phyllotaxy. Dense stippling in the tip indicates SAM, young leaf primordia, and procambial strands. Leaf traces develop basipetally to the mature vascular system and form a sympodium. The region where the leaf trace diverges from the continuous vascular bundle is called the leaf gap. Numbers correspond to the leaf order, starting with the primordia (not all leaves are shown). (After Esau 1953.)

Despite the diversity of leaf venation patterns, they all share a hierarchical organization. Veins are organized into distinct size classes—primary, secondary, tertiary, and so on—based on their width at the point of attachment to the parent vein (**Figure 19.16**). The smallest veinlets end blindly in the mesophyll. The hierarchical structure of the leaf vascular system reflects the hierarchical functions of different-sized veins, with larger diameter veins functioning in the **bulk transport** of water, minerals, sugars, and other metabolites, and smaller diameter veins functioning in **phloem loading** (see Chapter 11).

The question of how leaf venation patterns develop has long intrigued plant biologists. For the leaf vascular system to carry out its long-distance transport functions effectively, its many cell types must be arranged properly within the radial and longitudinal dimensions of the vascular bundle. It is not surprising, then, that the differentiation of vascular tissues is under strict developmental control. In this section we will first describe the development of a leaf's vascular connection to the rest of the plant. Then we will discuss how the higher-order venation pattern of a leaf is established.

The primary leaf vein is initiated discontinuously from the preexisting vascular system

In the mid-nineteenth century, the Swiss plant anatomist Carl Wilhelm von Nägeli made a surprising discovery while tracing the source of vascular bundles in the primary shoot. In the mature part of the stem of seed plants, the longitudinal vascular bundles form a continuous conducting system that begins at the root–shoot juncture and ends near the growing tips. Nägeli had assumed that the vascular system must grow upward (acropetally) from the preexisting vascular system to the growing tips of the shoot. Instead, he discovered that the leaf vascular bundles, arising from vascular precursor cells called the **procambium**, were initiated discontinuously in association with the emerging leaf primordia in the SAM (**Figure 19.17A**). From there the vascular bundles differentiated downward (basipetally) toward the node directly below the leaf and formed a connection to the older vascular bundle. The portion of the vascular bundle that enters the leaf was later called the **leaf trace** (**Figure 19.17B**).

What Nägeli had discovered was that the continuous longitudinal vascular bundles in the stem are actually composed of individual leaf traces. Species may differ in the exact course of leaf trace development, but the basic interpretation of the seed plant shoot primary vascular system as a sympodium of leaf traces appears to be universal.

Auxin canalization initiates development of the leaf trace

Several lines of evidence indicate that auxin stimulates formation of vascular tissues. An example is the role of auxin in regeneration of vascular tissue after wounding (**Figure 19.18A**). Vascular regeneration is prevented by removal of the leaf and shoot above the wound but can be restored by the application of auxin to the cut petiole above the wound, suggesting that auxin from the leaf is required for vascular regeneration. As shown in **Figure 19.18B**, the files of regenerating xylem elements originate at the source of auxin at the upper cut end of the vascular bundle, and progress basipetally until they reconnect with the cut end of the vascular bundle below, matching the

Figure 19.18 Auxin-induced xylem regeneration around ▶ a wound in cucumber (*Cucumis sativus*) stem tissue. (A) Method for carrying out the wound regeneration experiment. (B) Fluorescence micrograph showing regenerating vascular tissue around the wound. The arrow indicates the wound site where auxin accumulates and xylem differentiation begins. (B courtesy of R. Aloni.)

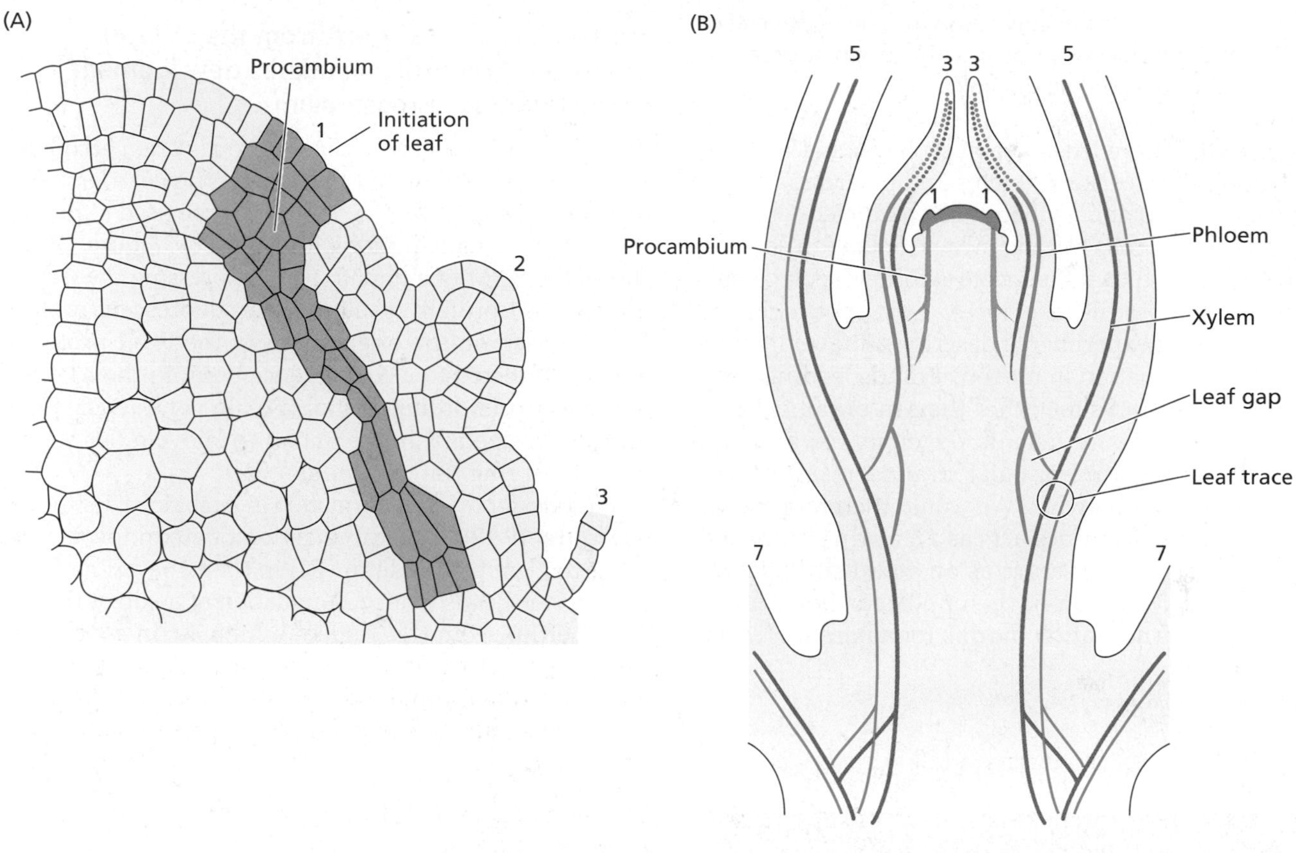
(A)
Procambium
1
Initiation of leaf
2
3
(B)
5
3 3
5
1 1
Procambium
Phloem
Xylem
Leaf gap
Leaf trace
7
7

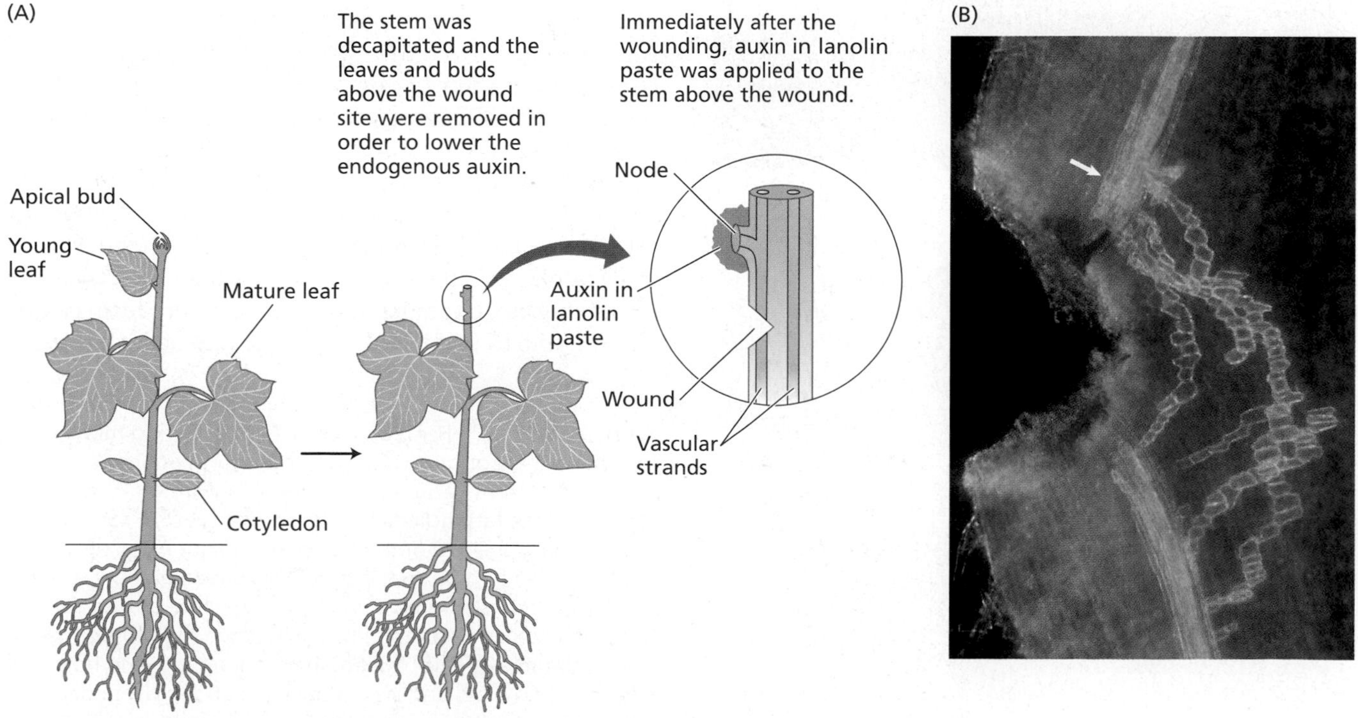
(A)
The stem was decapitated and the leaves and buds above the wound site were removed in order to lower the endogenous auxin.
Immediately after the wounding, auxin in lanolin paste was applied to the stem above the wound.
(B)
Apical bud
Young leaf
Mature leaf
Cotyledon
Node
Auxin in lanolin paste
Wound
Vascular strands
Intact cucumber plant
Decapitated and wounded cucumber plant

presumed direction of the flow of auxin. The upper end of the cut vascular bundle thus acts as the **auxin source** and the lower cut end as the **auxin sink**.

These and similar observations in other systems, such as bud grafting, have led to the hypothesis that as auxin flows through tissues it stimulates and polarizes its own transport, which gradually becomes channeled–or canalized–into files of cells leading away from auxin sources; these cell files can then differentiate to form vascular tissue.

Consistent with this idea, local auxin application (as in the wounding experiments described above) induces vascular differentiation in narrow strands leading away from the application site, rather than in broad fields of cells. New vasculature usually develops toward, and unites with, preexisting vascular strands, resulting in a connected vascular network. We would therefore predict that a developing leaf trace acts as an auxin source and the existing stem vasculature as an auxin sink. Recent studies on leaf venation have supported this source–sink model, or **canalization model**, for auxin flow at the molecular level.

(A)

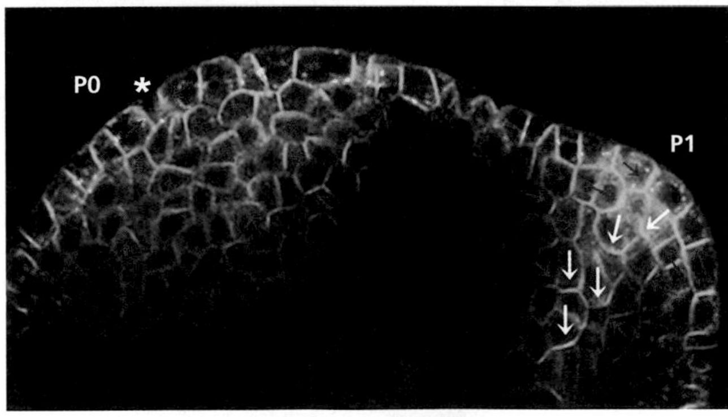

(B)

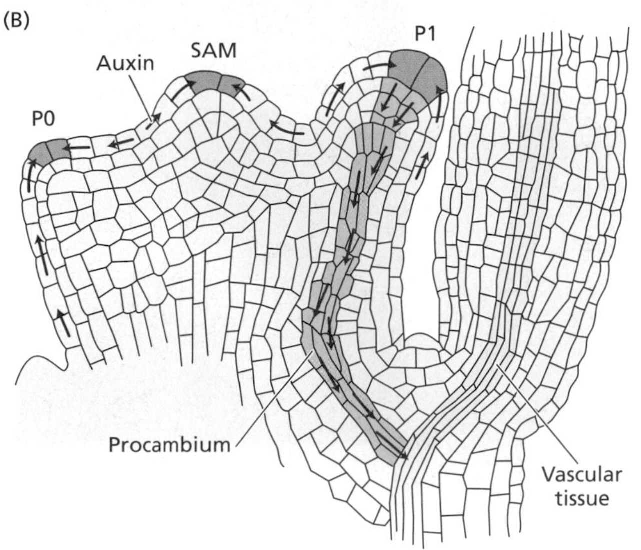

Basipetal auxin transport from the L1 layer of the leaf primordium initiates development of the leaf trace procambium

As we saw in Chapter 18, canalization is often accompanied by redistribution of PIN1 auxin efflux carriers. Furthermore, the distribution of PIN1 can be used to predict the direction of auxin flow within a tissue. **Figure 19.19A** shows the SAM of a tomato plant expressing the Arabidopsis PIN1 protein fused to green fluorescent protein (GFP). Based on the orientations of the PIN1 proteins, auxin is directed to a convergence point in the L1 layer of the leaf primordium initial (P0). In contrast, auxin is directed basipetally in the initiating midvein (leaf trace) of the emerging leaf primordium (P1).

A model for midvein formation in Arabidopsis is shown in **Figure 19.19B**. The canalization of auxin toward the tip of the leaf primordium (P1) in the L1 layer via PIN1 transporters leads to an accumulation of auxin at the tip. Auxin efflux from this region of high auxin concentration becomes canalized via PIN1 proteins in the basipetal direction toward the older leaf trace directly below it. This induces the differentiation of the procambium in the basipetal direction.

The existing vasculature guides the growth of the leaf trace

Microsurgical experiments have shown that the existing vascular bundle in the stem is required for the directional development of the leaf trace procambium. **Figure 19.20A** shows PIN1 distribution in the apex of a tomato plant expressing Arabidopsis PIN1 fused to GFP. The leaf trace emerging from the leaf primordium initial (P0) has connected to the existing leaf trace of the leaf primordium below it (P3), as shown diagrammatically in **Figure 19.20C**. However, if P3 is surgically removed, the leaf trace from P0 connects instead to the vascular bundle of the leaf primordium on the other side of the stem (P2) (**Figure 19.20B and D**). These results suggest that either the existing vascular bundle is serving as an auxin sink and thus facilitating auxin canalization, or that it is producing a different signal that guides the development of the leaf trace.

Figure 19.19 PIN1-mediated auxin flow during midvein formation. (A) Longitudinal section through a tomato vegetative meristem expressing AtPIN1:GFP (green). The red arrows on the left indicate the direction of auxin movement toward the site of the leaf primoridium initial (I1, white star). The red arrows on the right indicate auxin flow toward the emerging leaf primordium (P1). The white arrows show basipetal auxin movement, which initiates the differentiation of the midvein. (B) Schematic diagram of auxin flow through the L1, L2, and L3 tissue layers and midvein differentiation during the formation of leaf primordia. Primordium initial (P0), primordium (P1). (A from Bayer et al. 2009.)

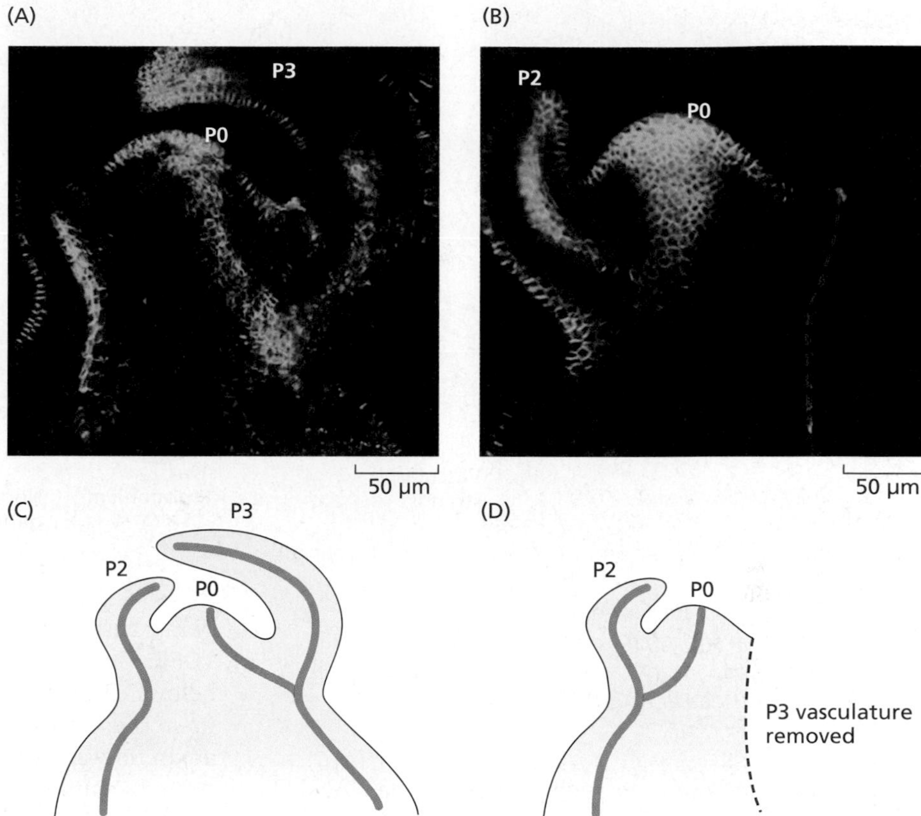

Figure 19.20 Preexisting vascular bundle guides basipetal development of the leaf trace. (A and C) In the control Arabidopsis meristem expressing AtPIN1:GFP (green), the newly initiated leaf trace (I1) grows toward, and connects to, the leaf trace associated with P3 directly below. (B and D) When the P3 vasculature is surgically removed (dashed red line), the P0 leaf trace connects instead to the P2 leaf trace on the other side of the stem. (From Bayer et al 2009.)

Primary phloem is the first vascular tissue to form from procambial cells, and its differentiation begins at the vascular bundle below and proceeds acropetally into the leaf primordium. In contrast, primary xylem differentiation lags behind primary phloem, is discontinuous, and proceeds both acropetally into the leaf primordium and basipetally toward the vascular bundle below.

Higher-order leaf veins differentiate in a predictable hierarchical order

The hierarchical order of leaf vascularization has been best studied in Arabidopsis. In general, vein development and patterning progress in the basipetal direction (**Figure 19.21A**, black arrow). In other words, venation is generally at a more advanced stage of development at the tip of a developing leaf than it is at the base.

During vein formation, **ground meristem cells** differentiate into **pre-procambium** cells—a stable intermediate state between ground cells and procambium cells that is characterized, in Arabidopsis, by the expression of the transcription factor ATHB8. Pre-procambial cells are isodiametric in shape (approximately cube-shaped) and are anatomically indistinguishable from ground meristem cells. Cell divisions of the pre-procambium are parallel to the direction of growth of the vascular strand, resulting in the elongated cells characteristic of the procambium (**Figure 19.21B**).

The pattern of vein formation follows a stereotypical course in Arabidopsis. The first procambium that forms in the leaf primordium—the leaf trace—represents the future **primary vein** or **midvein**. Secondary pre-procambium of the first pair of looped, secondary veins (orange arrows in Figure 19.21B) develops out from the midvein. The pre-procambium of the second pair of secondary vein loops progresses either basipetally or acropetally. Third and higher secondary vein loop pairs progress out from the midvein toward the leaf margin and reconnect with other extending strands (black arrows in Figure 19.21A).

The procambium differentiates from the pre-procambium simultaneously along the procambial strand (green lines in Figure 19.21A). Xylem differentiation occurs approximately 4 days later and can develop either continuously, or as discontinuous islands, along the vascular strand (magenta arrows in Figure 19.21A).

Proper differentiation of the vascular tissues within the veins depends on normal adaxial–abaxial polarity of the leaf. The four circles shown in **Figure 19.21C** represent vascular differentiation in the presence and absence of adaxial–abaxial polarity. The green circle on the left represents the undifferentiated procambial strand. Under conditions of normal adaxial–abaxial polarity, xylem develops on the adaxial side and phloem on the abaxial side. However, if the leaf has become adaxialized, as is *phan* mutants, the xylem cells surround the phloem, whereas

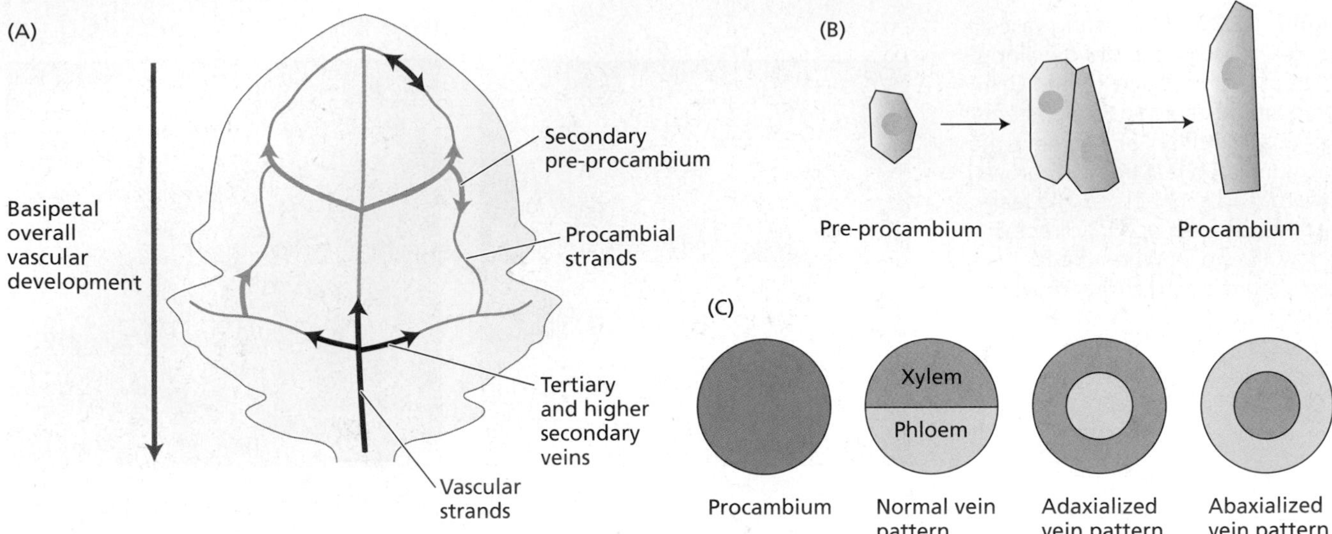

Figure 19.21 (A) Development of vein pattern in young leaves. (B) Formation of procambial cells from a pre-procambium cell. (C) Radial vein pattern in leaves. (Left to right): Procambial strand; normal vein pattern; vein pattern in adaxialized mutants; and vein pattern in abaxialized mutants. (From Lucas et al. 2013.)

in the abaxialized mutants, such as those of the KANADI gene family, phloem cells surround the xylem cells.

Auxin canalization regulates higher-order vein formation

As it does during leaf trace development, PIN1 is also thought to regulate auxin canalization during the formation of higher-order leaf veins. PIN1 in the epidermal layer of the developing leaf directs auxin to **convergence points** along the leaf margin (**Figure 19.22A**). These convergence points correspond to locations where serrations (see **WEB TOPIC 19.3**) and water pores called *hydathodes* (discussed below) can develop. As the auxin concentration builds up in these regions, auxin efflux induces PIN1-mediated auxin flow away from the convergence points toward the primary vein, which in turn causes the differentiation of pre-procambium along the path of auxin flow, eventually forming a secondary leaf vein. In Arabidopsis leaves, tertiary vein formation can result in loops that connect the primary and secondary veins. Again, this tertiary vein formation is guided by canalization mediated by PIN1 proteins (**Figure 19.22B**).

Despite the abundant evidence correlating PIN1 distribution in the leaf with auxin canalization and vein formation, *pin1* mutants have surprisingly mild phenotypes (**Fig-**

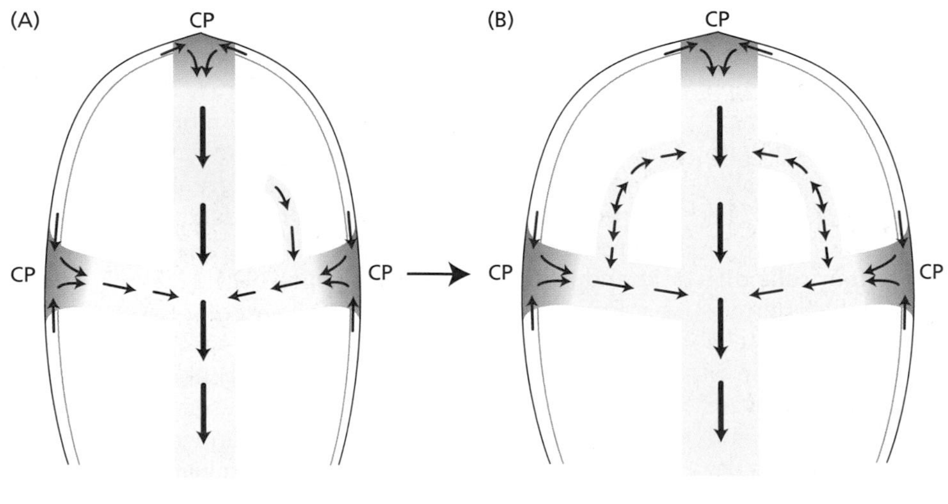

Figure 19.22 Model for higher-order leaf vein formation in Arabidopsis. (A) Auxin accumulates at convergence points (CPs) on the leaf margins, where PIN1 proteins direct auxin transport. Canalization of polar auxin transport leads to the differentiation of the procambium of secondary veins. (B) Tertiary veins can form when auxin becomes diverted by PIN1 proteins associated with the midvein. Such tertiary veins may form loops that connect to the secondary veins. The red arrows indicate the direction of PIN1-mediated auxin flow. (After Petrášek and Friml 2009.)

(A)

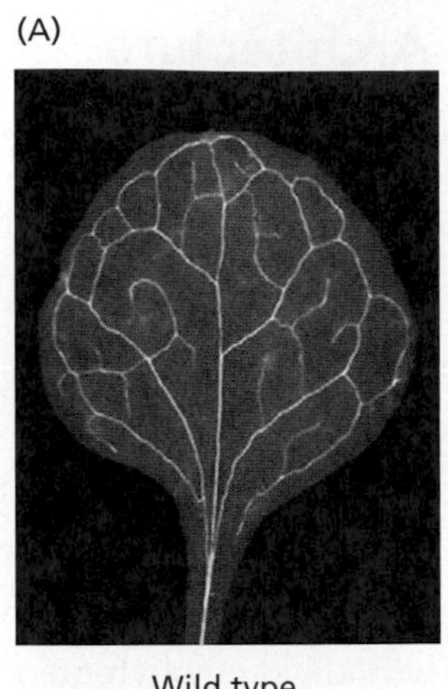

Wild type

(B)

pin1/pin6

(C)

yuc1/yuc2/yuc4/yuc6

Figure 19.23 Mutations that affect auxin transport or auxin biosynthesis alter leaf venation patterns. (A) Wild-type (WT) leaf. (B) *pin1/pin6* double mutant. Although the venation pattern of the mutant is defective, it retains the normal hierarchy of veins. (C) *yuc1/yuc2/yuc4/yuc6* quadruple mutant. In the absence of significant auxin biosynthesis, the venation pattern is highly reduced. (A and B from Sawchuk et al. 2013; C from Cheng et al. 2006.)

ure 19.23). For example, the *pin1/pin6* double mutant leaf shown in Figure 19.23B has an altered shape and a defective venation pattern, but the basic hierarchical structure of the veins is still intact, indicating that other factors also contribute to auxin canalization. For example, other auxin transporters, such as ABCB19, which helps to narrow canalized auxin streams by excluding auxin from neighboring cells, and AUX1/LAX permeases, which create uptake sinks that increase auxin flow (see Chapter 17), may be able to maintain canalization in the absence of PIN1.

Localized auxin biosynthesis is critical for higher-order venation patterns

An additional cause of auxin accumulation on the leaf margin, in addition to canalization by PIN1, is based on localized auxin biosynthesis. As discussed earlier in the chapter, the adaxial–abaxial interface triggers the expression of the *YUCCA* (*YUC*) genes. Auxin production at the leaf margins is thought to stimulate the expansion of the lamina. Auxin accumulation is concentrated in the hydathode regions along the leaf margin, where *YUCCA* genes are known to be expressed (**Figure 19.24A**). **Hydathodes** are specialized pores associated with vein endings at the leaf margin, from which xylem sap may exude in the presence of root pressure (see Chapter 4). **Figure 19.24B** stunningly illustrates the canalization of auxin from its site of synthesis in the hydathode region to its sink—a developing vein.

The importance of auxin synthesis for leaf venation is dramatically demonstrated by the phenotypes of *YUCCA* gene mutants in Arabidopsis. In contrast to what is seen in the mild phenotype of the *pin1/pin6* double mutant (see Figure 19.23B), the normal venation pattern is almost entirely eliminated in *yuc1/yuc2/yuc4/yuc6* quadruple mutants, in which auxin biosynthesis is substantially reduced (see Figure 19.23C). The few remaining veins suggest that either residual auxin is being synthesized by a different biosynthetic pathway or that an auxin-independent pathway can direct a limited amount of vein formation.

(A)

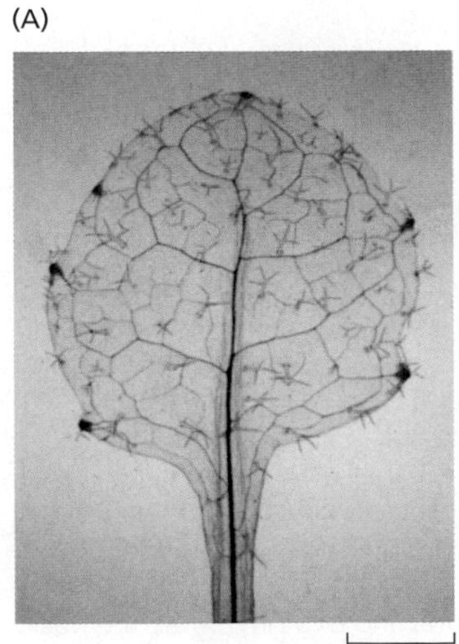

1 mm

(B)

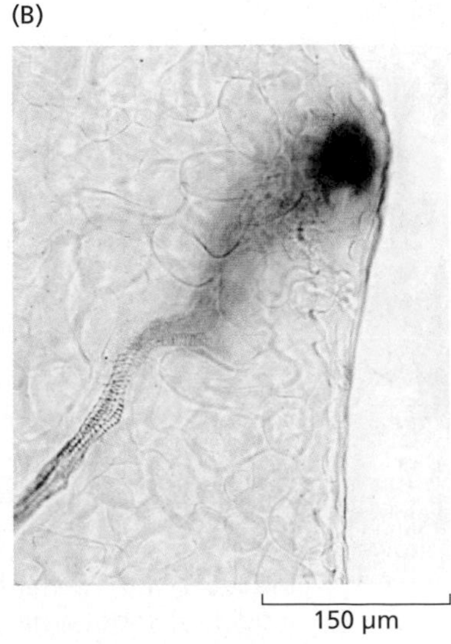

150 μm

Figure 19.24 Auxin biosynthesis at the hydathodes of Arabidopsis leaves, as indicated by the expression of the *GUS* reporter gene driven by the auxin-responsive DR5 promoter. (A) An Arabidopsis leaf that has been cleared to reveal the blue stain. (B) Auxin flow and canalization from the hydathode toward a developing leaf vein. (From Aloni et al. 2003.)

Based on the abundance of evidence from other studies, we can reconstruct the process of vein formation as follows:

1. Auxin is synthesized by YUCCA proteins and accumulates in the hydathode regions.
2. Auxin efflux from the margin induces PIN1 formation and polar orientation in nearby cells, promoting auxin flux away from the site of auxin synthesis.
3. ABCB exporters enhance canalization by excluding auxin from all but a narrow zone that leads directly to the developing leaf vein, while AUX1/LAX uptake transporters create sinks that enhance auxin flows.
4. Auxin is taken up by the developing cells of the vein, which maintains auxin flux until the vein is fully differentiated.

(A)

(B)

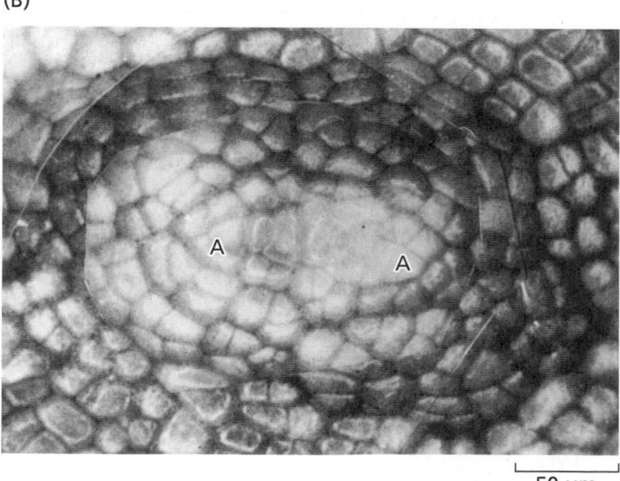

Figure 19.25 Dichotomous branching in the primitive vascular plant *Psilotum nudum* (image shows sporangia). (A) Shoot showing dichotomous branching. (B) Shoot tip showing the formation of two SAMs during branch formation. A, shoot apical meristem. (B from Takiguchi et al. 1996.)

Shoot Branching and Architecture

The shoot and inflorescence architecture of flowering plants is determined to a large extent by the branching patterns established during postembryonic development. The earliest vascular plants branched **dichotomously** at the SAM, producing two equal shoots. This condition exists today in some lower vascular plants (**Figure 19.25**) and a few angiosperms, such as certain cacti.

In contrast, shoot architecture in seed plants is characterized by multiple repetitions of a basic module called the **phytomer**, which consists of an internode, a node, a leaf, and an **axillary meristem** (**Figure 19.26**). Modification of the position, size, and shape of the individual phytomer, and variations in the regulation of axillary bud outgrowth, provided the morphological basis for the remarkable diversity of shoot architecture among seed plants. Vegetative and inflorescence branches, as well as the floral primordia produced by inflorescences, are derived from axillary meristems initiated in the axils of leaves. During vegetative development, axillary meristems, like the apical meristem, initiate the formation of leaf primordia, resulting in axillary buds. These buds either become dormant or develop into lateral shoots depending on their position

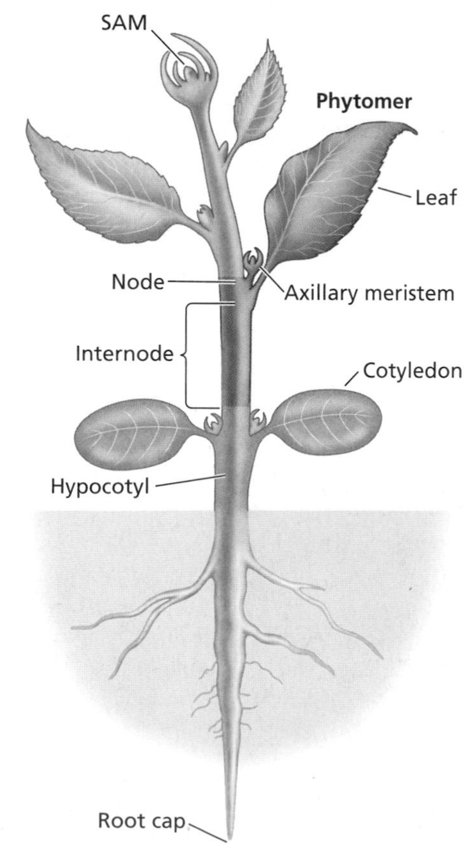

Figure 19.26 Schematic drawing of a phytomer, the basic module of shoot organization in seed plants.

along the shoot axis, the developmental stage of the plant, and environmental factors. During reproductive development, axillary meristems initiate formation of inflorescence branches and flowers. Hence, the growth habit of a plant depends not only on the patterns of axillary meristem formation, but also on meristem identity and its subsequent growth characteristics.

Figure 19.27 The tomato mutant *lateral suppressor* (*ls*) shows defects in axillary bud formation. (A) A wild-type plant. (B) The *ls* mutant. Axillary buds fail to form in most of the leaf axils. (Courtesy of Klaus Theres.)

Axillary meristem initiation involves many of the same genes as leaf initiation and lamina outgrowth

Auxin biosynthesis, transport, and signaling are all required for the initiation of axillary meristems, as demonstrated by the fact that mutants defective in these pathways fail to form new axillary meristems. Axillary meristem initiation involves three main steps: correct positioning of initial cells, delineation of the meristem boundaries, and establishment of the meristem proper. As discussed earlier in the chapter, PIN1-mediated auxin transport helps determine the sites of leaf primordia, and it is also important for axillary meristem formation.

Not surprisingly, genetic evidence indicates considerable overlap in the gene networks involved in the initiation of leaf primordia, leaf margin serrations, and axillary meristems. For example, mutations in the *LATERAL SUPPRESSOR* (*LAS*) genes from tomato (*Solanum lycopersicum*) (**Figure 19.27**) and Arabidopsis cause a complete block in axillary bud formation during the vegetative phase of development, and similar results have been observed in rice (*Oryza sativa*). Consistent with this finding, *LAS* mRNA has been shown to accumulate in the axils of leaf primordia, where new axillary meristems develop (**Figure 19.28**). *LAS* expression patterns are similar to those for *CUC* genes, which, as discussed earlier, regulate embryonic shoot meristem formation and specify lateral organ boundaries. Two other genes that are known to be required for normal axillary bud formation in Arabidopsis are the bHLH protein gene *REGULATOR OF AXILLARY MERISTEM FORMATION* (*ROX*) and the MYB transcription factor gene *REGULATOR OF AXILLARY MERISTEMS* (*RAX*).

Auxin, cytokinins, and strigolactones regulate axillary bud outgrowth

Once the axillary meristems are formed, they may enter a phase of highly restricted growth (dormancy), or they may be released to grow into axillary branches. The "go or no go" decision is determined by developmental programming and environmental responses mediated by plant hormones that act as local and long-distance signals. Interactions of hormonal signaling pathways coordinate the relative growth rates of different branches and the shoot tip, which ultimately determine shoot architecture. The main hormones involved are auxin, cytokinins, and strigolactones (see Chapter 15). All three hormones are produced in varying quantities in the root and shoot, but translocation of the hormones allows them to exert effects far from their site of synthesis (**Figure 19.29**).

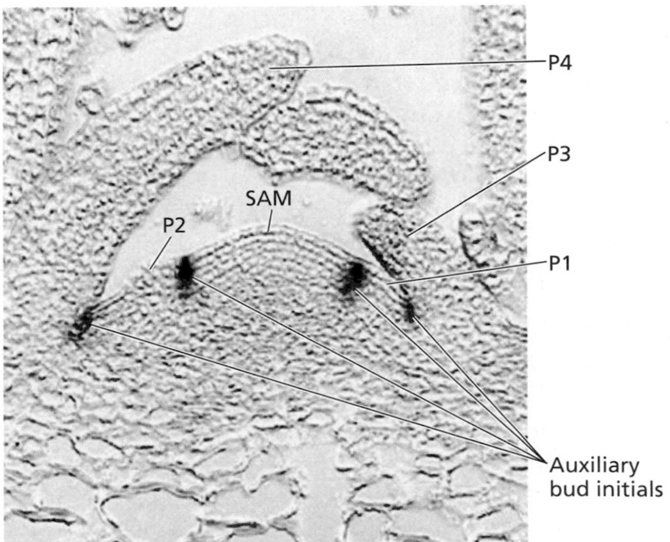

Figure 19.28 Accumulation of *LATERAL SUPPRESSOR* mRNA in the axillary bud regions of an Arabidopsis shoot tip. P1–P4 = leaf primordia. (From Greb et al. 2003.)

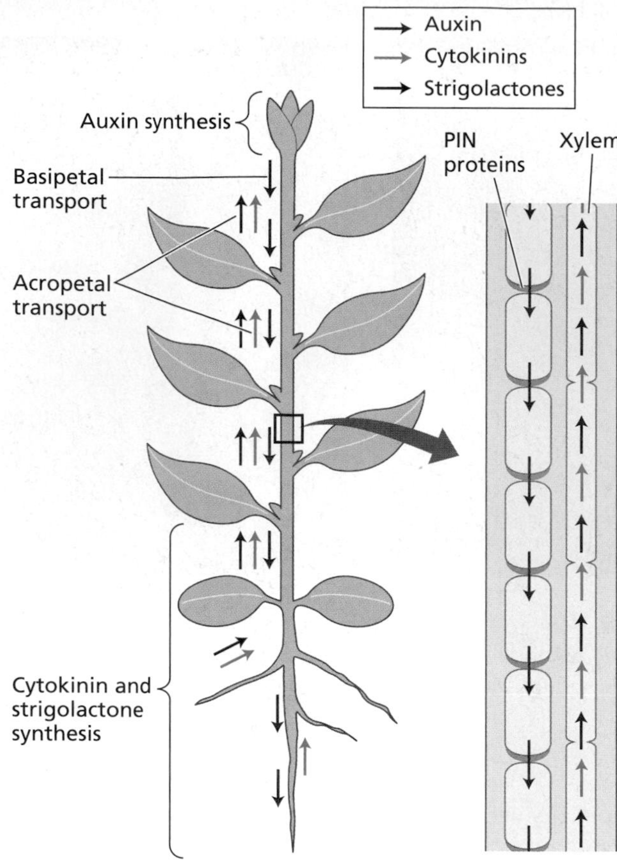

Figure 19.29 Long-distance transport of three hormones that regulate shoot branching: auxins, cytokinins, and strigolactones. Auxin is produced primarily in young expanding leaves and is transported basipetally by PIN1-mediated polar auxin transport. Strigolactones and cytokinins are synthesized mainly in the root and can be translocated acropetally to the shoot in the xylem. These two hormones can also be synthesized in shoot tissues adjacent to axillary buds. (After Domagalska and Leyser 2011.)

Auxin is synthesized predominantly in young leaves and the shoot apex and is transported rootward in a specialized polar auxin transport stream via ABCB and PIN proteins in the vascular cylinder and starch sheath or endodermis (see Chapter 18). Auxin can also be transported in the phloem, where it moves by mass flow from source to sink (see Chapter 11).

Strigolactones are transported out of sites of synthesis via plasma membrane ABC transporters and have been shown to move in the xylem from root to shoot. Cytokinins, too, are transported from sites of synthesis to the xylem by an ABC transporter, and they can also move in the phloem. Consequently, there is considerable scope for long-distance communication via these hormones.

Auxin from the shoot tip maintains apical dominance

The role of auxin in regulating axillary bud growth is most easily demonstrated in experiments on apical dominance. **Apical dominance** is the control exerted by the shoot tip over axillary buds and branches below. Plants with strong apical dominance are typically weakly branched and show a strong branching response to decapitation (removal of the growing or expanding leaves and shoot tip). Plants with weak apical dominance are typically highly branched and show little, if any, response to decapitation.

More than a century of experimental evidence suggests that in plants with strong apical dominance, auxin produced in the shoot tip inhibits axillary bud outgrowth. In plants with strong apical dominance, mutants with decreased rootward auxin transport exhibit increased branching, and treatment of the shoot apex with auxin transport inhibitors results in increased branching. Addition of auxin to the shoot at the point of apical excision inhibits outgrowth, while application of auxin transport inhibitors to the stem releases the axillary buds below from apical dominance (**Figure 19.30A**). Gardeners take advantage of this phenomenon when they "pinch back" chrysanthemums with strong apical dominance to create dense, dome-shaped bushes of blossoms.

Strigolactones act locally to repress axillary bud growth

Strigolactones are thought to act in conjuction with auxin during apical dominance. Arabidopsis mutants defective either in strigolactone biosynthesis (*max1* [*more axillary growth1*], *max3*, or *max4*) or signaling (*max2*) show increased branching without decapitation (**Figure 19.30B**).

Figure 19.30 Axillary bud outgrowth is inhibited by auxin ▶ and strigolactones. (A) Classic physiological experiments demonstrating the role of auxin in apical dominance. In detipped shoots the axillary bud is released from apical dominance. Replacing the missing shoot tip with auxin prevents bud outgrowth. Applying an inhibitor of polar auxin transport to the stem causes bud outgrowth below the application site. (B) Grafting experiments carried out with mutants defective in strigolactone biosynthesis or signaling that have increased branching. Grafting the shoots of the strigolactone biosynthesis mutants (*max1*, *max3*, or *max4*) to wild-type roots restored shoot branching of the mutant to wild-type levels. Grafting the roots of the strigolactone signaling mutant *max2* onto the shoots of the wild type and synthesis mutants *max1*, *max3*, or *max4* also prevented bud outgrowth, demonstrating that *max2* can produce the signal in the roots, even though it cannot respond to it. The branch-inhibiting hormone can also be produced in the shoot, as grafting of the wild-type shoot to the strigolactone-deficient roots (*max1*, *max3*, or *max4*) did not increase the number of branches. (After Domagalska and Leyser 2011.)

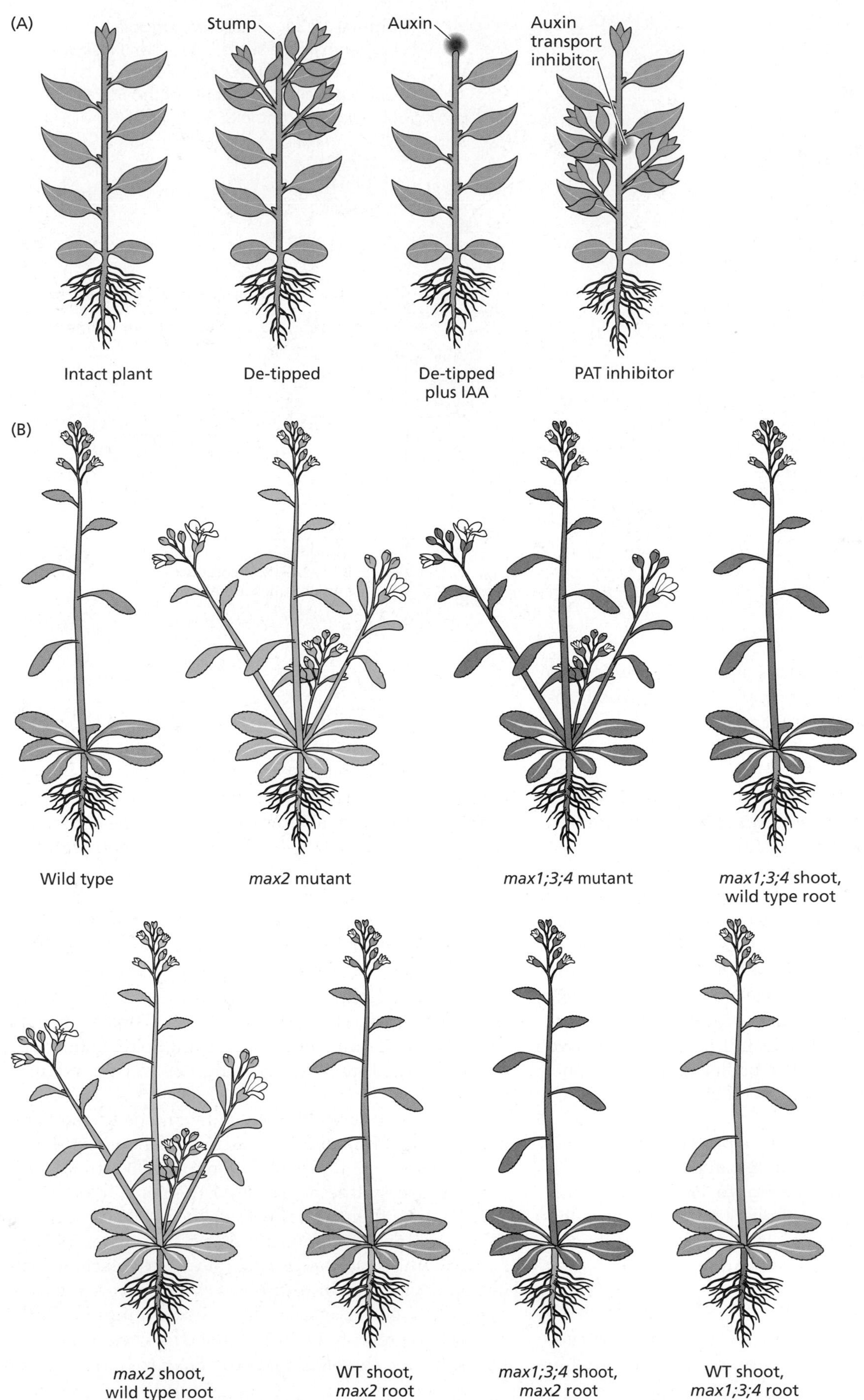
(A)
Stump
Auxin
Auxin transport inhibitor
Intact plant
De-tipped
De-tipped plus IAA
PAT inhibitor
(B)
Wild type
max2 mutant
max1;3;4 mutant
max1;3;4 shoot, wild type root
max2 shoot, wild type root
WT shoot, *max2* root
max1;3;4 shoot, *max2* root
WT shoot, *max1;3;4* root

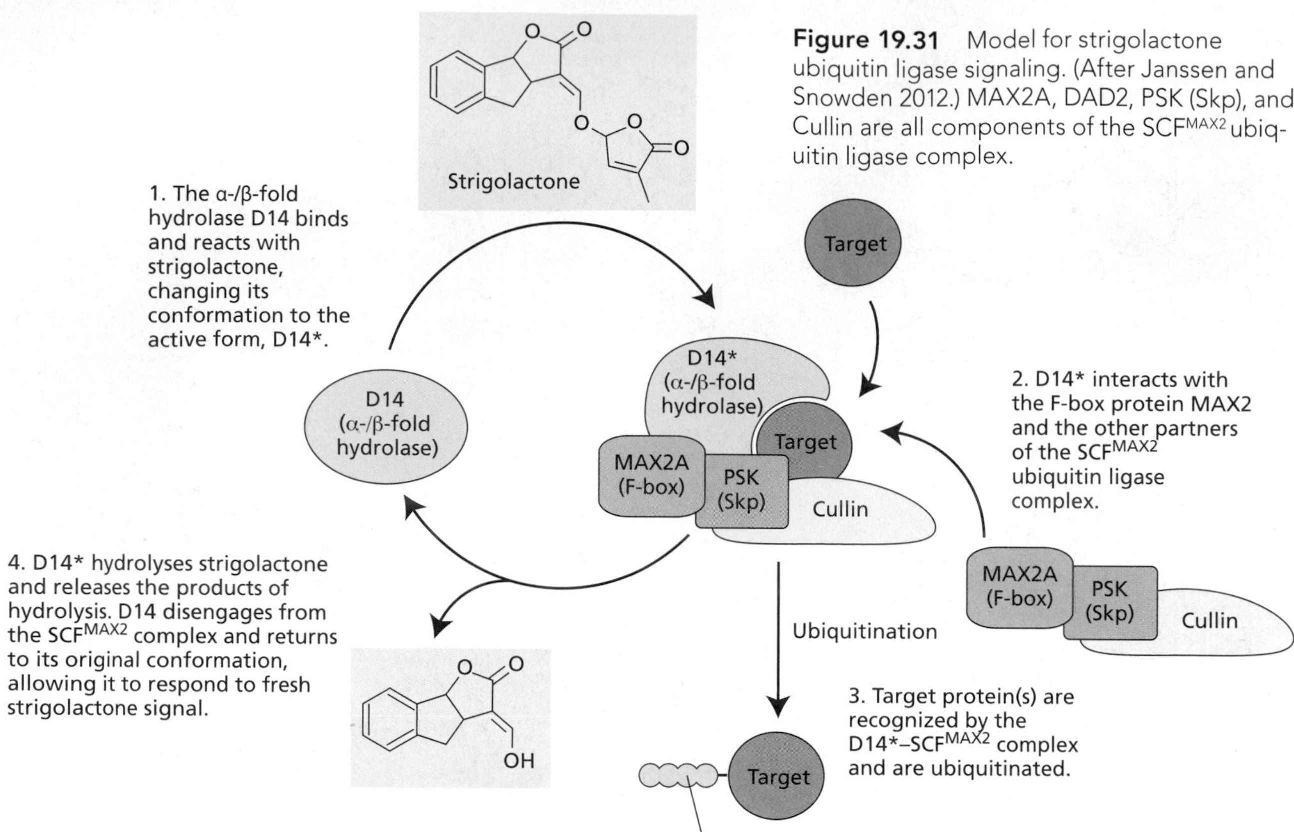
Strigolactone
1. The α-/β-fold hydrolase D14 binds and reacts with strigolactone, changing its conformation to the active form, D14*.
D14 (α-/β-fold hydrolase)
D14* (α-/β-fold hydrolase)
Target
MAX2A (F-box)
PSK (Skp)
Cullin
2. D14* interacts with the F-box protein MAX2 and the other partners of the SCFMAX2 ubiquitin ligase complex.
4. D14* hydrolyses strigolactone and releases the products of hydrolysis. D14 disengages from the SCFMAX2 complex and returns to its original conformation, allowing it to respond to fresh strigolactone signal.
OH
Ubiquitination
3. Target protein(s) are recognized by the D14*–SCFMAX2 complex and are ubiquitinated.
Ubiquitin

Figure 19.31 Model for strigolactone ubiquitin ligase signaling. (After Janssen and Snowden 2012.) MAX2A, DAD2, PSK (Skp), and Cullin are all components of the SCF^{MAX2} ubiquitin ligase complex.

Grafting the biosynthesis mutant shoot onto a wild-type root restores apical dominance, indicating that strigolactone can move from the root to the shoot. However, root-derived strigolactone is not required for bud repression, since wild-type shoots grafted onto strigolactone-deficient roots have normal apical dominance. This result suggests that the strigolactones that repress bud growth normally come from within the shoot.

Genes regulating strigolactone biosynthesis and reception are conserved among higher plants. Strigolactones are produced in the plastids from β-carotene by three sequentially acting plastid enzymes that must be co-located in the same cell: D27, a carotenoid isomerase, and CCD7 and CCD8, which are carotenoid cleavage dioxygenases (see Chapter 15). The product, an apocarotenoid named carlactone, can move between cells but must undergo two oxygenation steps to produce a bioactive strigolactone. The oxygenation steps are catalyzed by a cytosolic cytochrome P450.

Strigolactones are perceived by a protein complex containing an α-/β-fold hydrolase protein and an F-box protein (D14 and MAX2, respectively) (**Figure 19.31**). The signaling mechanism appears to be similar to that for gibberellin signaling (see Figures 15.33 and 15.34) and involves targeting proteins for degradation by ubiquitination.

Cytokinins antagonize the effects of strigolactones

Direct application of cytokinin to axillary buds stimulates their growth, suggesting that cytokinins are involved in breaking apical dominance. Consistent with this hypothesis, the expression of two cytokinin biosynthetic genes (*IPT1* and *IPT2*) increases in the second nodal stem of peas following decapitation, suggesting that auxin from the shoot apex normally represses these genes. This was confirmed by incubating excised stem segments with and without auxin; *IPT1* and *IPT2* expression persisted only in segments incubated without auxin. In addition, application of the auxin transport inhibitor 2,3,5-triiodobenzoic acid (TIBA) around the internode led to increased *IPT1* and *IPT2* expression below the site of application, demonstrating that these genes are normally repressed by auxin transported down from the shoot apex. Thus, it appears that the cytokinins involved in breaking apical dominance are synthesized locally at the node, not transported from the root.

A simplified model for the antagonistic interactions between cytokinin and strigolactone is shown in **Figure 19.32**. Auxin maintains apical dominance by stimulating strigolactone synthesis via the *MAX4* gene. In eudicots, strigolactone then activates the gene for BRANCHED1 (BRC1), a transcription factor which suppresses axillary bud growth. Besides activating *BRC1*, strigolactone also inhibits cytokinin biosynthesis by negatively regulating the IPT genes. In contrast, cytokinin inhibits *BRC1* action, and prevents auxin-induced strigolactone biosynthesis. In rice, the *BRC1* homolog *FINE CULM1* (*FC1*) is

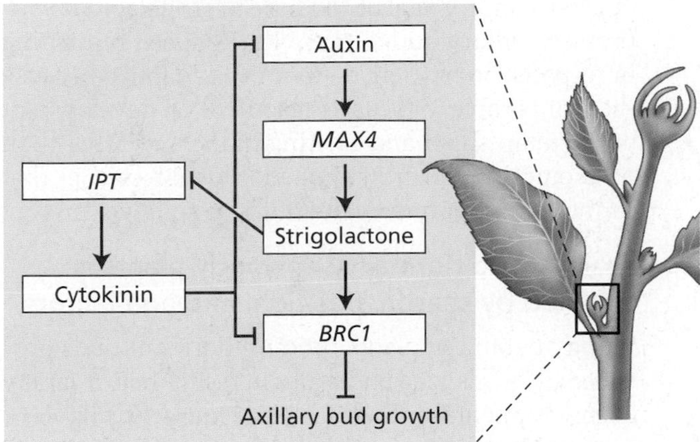

Figure 19.32 Hormonal network regulating apical dominance. Auxin from the shoot apex promotes strigolactone synthesis in the nodal area via the *MAX4* gene. In eudicots, strigolactone up-regulates the *BRANCHED1* (*BRC1*) gene and down-regulates *IPT* genes. BRC1 inhibits axillary bud growth. Strigolactone also inhibits cytokinin biosynthesis, which otherwise would prevent BRC1 production. (From El-Showk et al. 2013.)

the target of strigolactone signaling, while in maize the *TEOSINTE BRANCHED1* (*TB1*) is the primary gene regulating branching. This gene is responsible for a major trait involved in the domestication of maize, transforming the highly branched maize progenitor teosinte into the more desirable reduced branching phenotype of modern maize (**Figure 19.33**).

The initial signal for axillary bud growth may be an increase in sucrose availability to the bud

Recent evidence indicates that sucrose itself may serve as the initial signal in controlling bud outgrowth (**Figure 19.34**). In pea plants, axillary bud growth is initiated approximately 2.5 h after decapitation. This is 24 h prior to any detectable decline in the auxin level in the stem adjacent to the axillary bud, which suggests that a decrease in auxin from the tip occurs much too slowly to initiate bud outgrowth.

In contrast, studies using [^{14}C]sucrose demonstrated that the concentration of leaf-derived sucrose in the stem adjacent to the bud begins to decline in as little as 2 h after decapitation. This decline is due to the uptake of sugars by the axillary bud. Thus, following decapitation, bud outgrowth on the lower stem is initiated *prior to* auxin depletion but *after* sucrose depletion in the stem adjacent to the bud. As a result of decapitation, the endogenous carbon supply to the axillary buds increases within the timeframe sufficient to induce bud release. Apical dominance is thus regulated by the strong sink activity of the growing tip, which limits sugar availability to the axillary buds. However, sustained bud outgrowth requires the depletion of auxin in the stem adjacent to the bud as well.

Integration of environmental and hormonal branching signals is required for plant fitness

In some cases, a plant can adjust its default shoot branching pattern in response to environmental conditions. Two classic examples are the shade avoidance response and

(A)

Teosinte (*Zea mays* ssp. *parviglumis*)

(B)

Maize (*Zea mays* ssp. *mays*)

Figure 19.33 Comparison of teosinte (*Zea mays* ssp. *parviglumis*) and modern maize (*Zea mays* ssp. *mays*). (Left photo courtesy of Paul Gepts.)

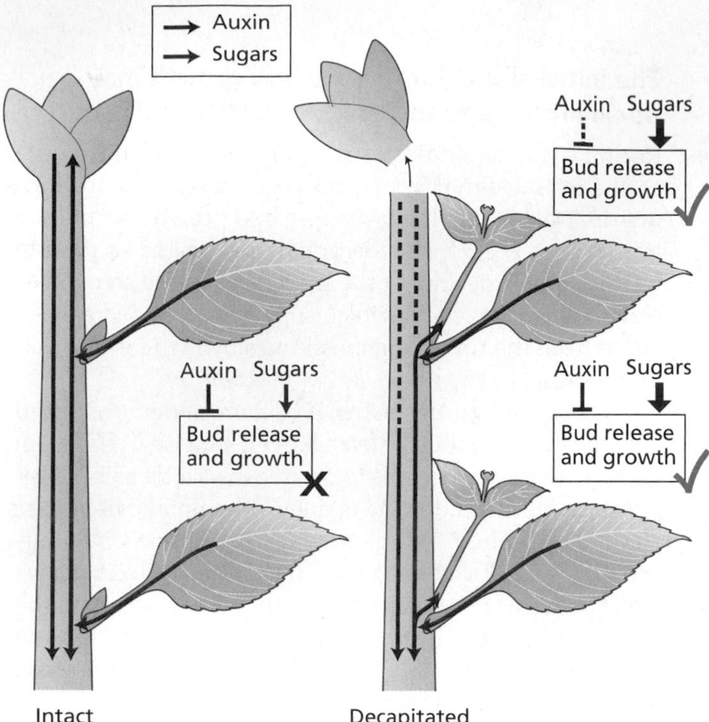

Figure 19.34 Apical dominance is regulated by sugar availability. After decapitation, sugars, which normally flow toward the shoot tip via the phloem, rapidly accumulate in axillary buds, stimulating bud outgrowth. At the same time, the loss of the apical supply of auxin results in a depletion of auxin in the stem. However, auxin depletion is relatively slow and therefore the growing buds in the upper shoot are affected before those lower on the stem. In this model, auxin is involved predominately in the later stages of branch growth. (After Mason et al. 2014.)

the nutrient deficiency response. Both of these responses involve the regulatory pathways described above.

Plants avoid shade through enhanced shoot elongation and suppressed branching. Shade avoidance involves phytochrome B signaling in response to the decreased R:FR light ratio that results when sunlight is filtered through green leaves containing chlorophyll (see Chapter 18). Genetic studies in Arabidopsis have shown that phytochrome B requires both auxin and strigolactone signaling pathways, as well as the budspecific *BRC1* and *BRC2* genes to inhibit axillary bud outgrowth under shading conditions.

The response to nutrient deficiency is mediated by strigolactones. Well-nourished plants are bushy, whereas plants growing under poor nutrient conditions tend to be weakly branched. The involvement of strigolactones in this branching response presumably relates, evolutionarily, to the role of these hormones in enhancing nutrient acquisition. Mycorrhizal plant species secrete strigolactones into the rhizosphere to promote the mycorrhizal symbiosis and enhance nutrient uptake. The details vary among different species, but even in non-mycorrhizal plant species, strigolactone levels in the shoot are elevated under low nutrient conditions. The increase in strigolactone suppresses axillary bud outgrowth. Reduced branching in response to nutrient deficiency is adaptive because the plant is able to focus its resources on development of the main shoot and existing branches, rather than on promoting the growth of additional branches that cannot be supported by the nutrient supply.

Axillary bud dormancy in woody plants is affected by season, position, and age factors

Woody perennial plants produce dormant buds protected by specialized bud scales in response to a variety of environmental and age factors (**Figure 19.35**). Major environmental factors that influence bud dormancy include temperature, light, photoperiod, water, and nutrients. Bud position and plant age are also important factors. The circadian clock and flowering genes such as *FT, CO,* and *TFL1,* together with phytochrome A, are involved in controlling dormancy in deciduous trees in relation to photoperiod and chilling requirements. In aspen trees, for example, an established target of this regulatory system is the cell cycle in buds. Even in herbaceous plants, pathways that regulate flowering in response to photoperiod interact with pathways regulating axillary bud outgrowth. For example, strigolactone branching mutants of garden pea (*Pisum sativum*) show dramatic shifts in the position and number of axillary branches when grown under different photo-

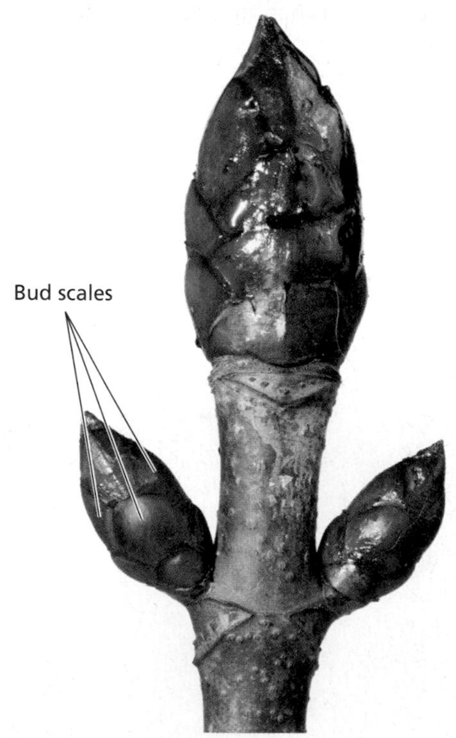

Figure 19.35 Dormant axillary buds of horse chestnut (*Aesculus hippocastanum*) surrounded by bud scales.

periods (even prior to flower opening), and flowering genes affect branching at cauline nodes in Arabidopsis.

Root System Architecture

Plant root systems are the critical link between the growing shoot and the rhizosphere, providing both vital nutrients and water to sustain growth. In addition, roots anchor and stabilize the plant, enabling the growth of aboveground vegetative and reproductive organs. Because roots function in heterogeneous and often changing soil conditions, roots must be capable of adapting to ensure a steady flow of water and nutrients to the shoot under a variety of conditions. Recent research on the structure of root systems has been driven by advances in root system phenotyping (see **WEB TOPIC 19.4**). These and other studies have shown that plants have evolved complex control mechanisms that regulate root system architecture.

Plants can modify their root system architecture to optimize water and nutrient uptake

Root system architecture is the spatial configuration of the entire root system in the soil. More specifically, root system architecture refers to the geometric arrangement of individual roots within the plant's root system in the three-dimensional soil space. Root systems are composed of different root types, and plants are able to modify and control the types of roots they produce, the root angles, the rates of root growth, and the degree of branching. Variations in root system architecture within and across species have been linked to resource acquisition and growth. As illustrated in **Figure 19.36**, root system architecture varies widely among species, even those living in the same habitat.

Figure 19.36 Diversity of root systems in prairie plants.

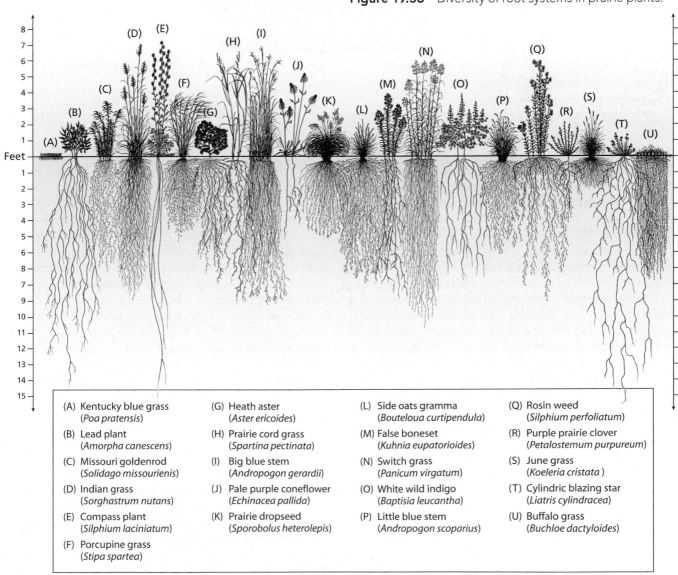

(A) Kentucky blue grass (*Poa pratensis*)
(B) Lead plant (*Amorpha canescens*)
(C) Missouri goldenrod (*Solidago missourienis*)
(D) Indian grass (*Sorghastrum nutans*)
(E) Compass plant (*Silphium laciniatum*)
(F) Porcupine grass (*Stipa spartea*)
(G) Heath aster (*Aster ericoides*)
(H) Prairie cord grass (*Spartina pectinata*)
(I) Big blue stem (*Andropogon gerardii*)
(J) Pale purple coneflower (*Echinacea pallida*)
(K) Prairie dropseed (*Sporobolus heterolepis*)
(L) Side oats gramma (*Bouteloua curtipendula*)
(M) False boneset (*Kuhnia eupatorioides*)
(N) Switch grass (*Panicum virgatum*)
(O) White wild indigo (*Baptisia leucantha*)
(P) Little blue stem (*Andropogon scoparius*)
(Q) Rosin weed (*Silphium perfoliatum*)
(R) Purple prairie clover (*Petalostemum purpureum*)
(S) June grass (*Koeleria cristata*)
(T) Cylindric blazing star (*Liatris cylindracea*)
(U) Buffalo grass (*Buchloe dactyloides*)

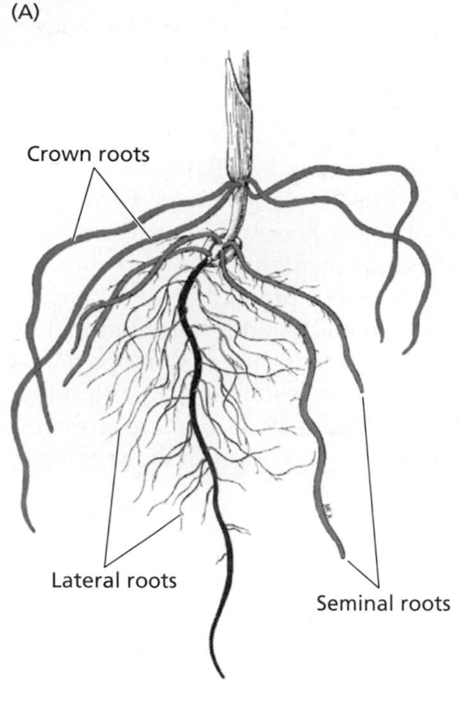

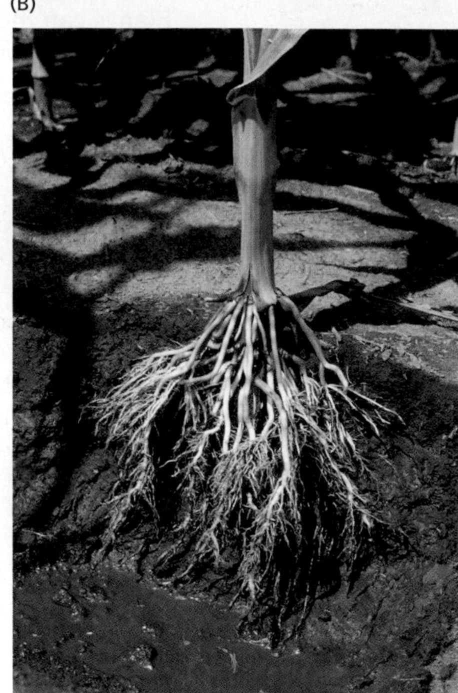

Figure 19.37 Root system of a 14-day-old maize seedling composed of primary root derived from the embryonic radicle, the seminal roots derived from the scutellar node, postembryonically formed crown roots that arise at nodes above the mesocotyl, and lateral roots. (B) Mature maize root system. (A from Hochholdinger and Tuberosa 2009.)

Monocots and eudicots differ in their root system architecture

Before examining the complexities of root system architecture, it is important to understand how monocot and eudicot root systems are organized and how they differ. Both monocot and eudicot root systems are roughly similar in structure, consisting of an embryonically derived primary root (the radicle), lateral roots, and adventitious roots. However, there are significant differences in their root systems. Monocot root systems are generally more fibrous and complex than the root systems of eudicots, especially in the cereals. For example, the maize seedling root system consists of a **primary root** that develops from the radicle, **seminal roots** (adventitious roots that branch from the scutellar node), and postembryonically derived **crown roots** (**Figure 19.37**). The primary and seminal roots are highly branched and fibrous. The crown roots, also called "prop roots," are adventitious roots derived from the lowermost nodes of the stem. Although crown roots are relatively unimportant in seedlings, in contrast to the primary and seminal roots, crown roots continue to form, develop, and branch throughout vegetative growth. Thus the crown root system makes up the vast majority of the root system in adult maize plants.

The root system of a young eudicot consists of the primary (or tap) root and its branch roots. As the root system matures, basal roots arise from the base of the tap root. In addition, adventitious roots can arise from subterranean stems or from the hypocotyl, and can be considered to be loosely analogous to the adventitious crown roots in the cereals. The root system of a soybean plant as a representative eudicot is depicted in **Figure 19.38** where the tap, branch, basal, and adventitious roots can be seen.

Root system architecture changes in response to phosphorous deficiencies

Phosphorus is, along with nitrogen, the most limiting mineral nutrient for crop production (see Chapters 5 and 13). Phosphorus limitation is a particular problem in tropical soils, where the highly weathered acidic soils tend to bind phosphorus tightly, rendering much of it unavailable for acquisition by roots. Plant root systems undergo well-documented morphological alterations in response to phosphorus deficiency. These responses can vary somewhat from species to species, but in general they include a reduction in primary root elongation, an increase in lateral root proliferation and elongation, and an increase in the number of root hairs.

Phosphorus, always in the form of the phosphate anion, is immobile in soil because it binds tightly to iron and aluminum oxides on the surfaces of clay particles or is fixed as biological phosphorus within soil microorganisms. Hence, the majority of the phosphorus, especially in low-phosphorus soils, is trapped in the surface horizons (layers) of the soil. Phosphorus deficiency can trigger "topsoil foraging" by plants. Some bean genotypes, for example, respond to phosphorus deficiency by producing more adventitious lateral roots, decreasing the growth angle of these roots (relative to the shoot) so that they are more shallow, increasing the number of lateral roots emerging from the tap root, and increasing root hair den-

Figure 19.38 Soybean root system showing primary (tap) root, branch roots, basal roots, and adventitious roots. (Courtesy of Leon Kochian.)

sity and length (**Figure 19.39**). These changes in root system architecture combine to locate more of the roots in the topsoil where the majority of the phosphorus resides. These and similar discoveries of "phosphorus-efficient" genotypes have enabled researchers to breed for root system architecture traits in bean and soybean that better adapt these crops to low-phosphorus soils.

Once roots are placed in soil with adequate supplies of phosphorus, they still must solubilize and absorb the phosphate. These tasks are facilitated by a range of biochemical processes, such as the release of organic acids into the soil to solubilize phosphate from aluminum and iron phosphates, the release of phosphatases to solubilize organic phosphorus, acidification of the rhizosphere, and

Figure 19.39 Topsoil foraging of phosphorus by phosphorus-efficient bean genotypes. (After Lynch 2007.)

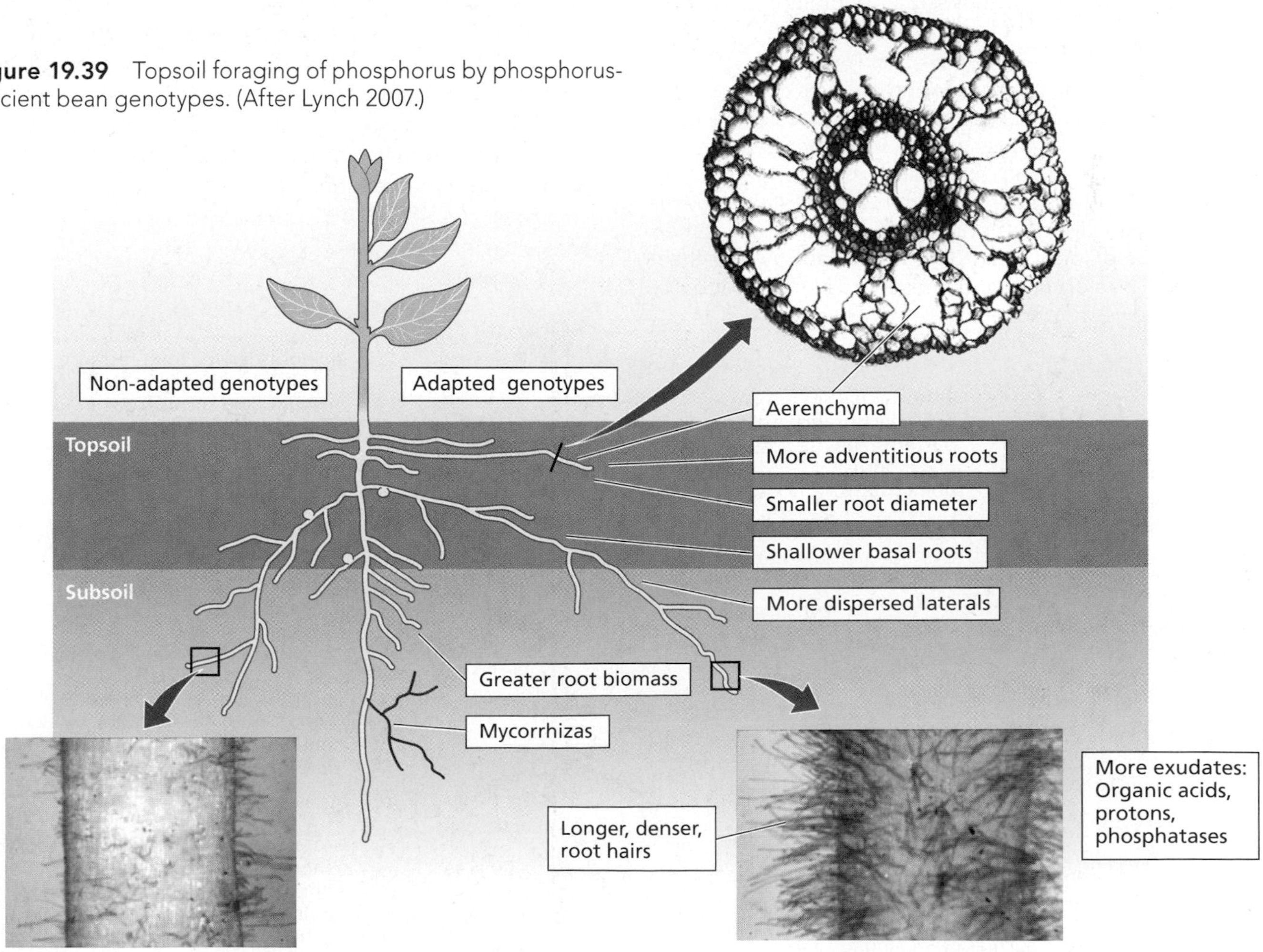

an increased abundance of phosphate transporters on the plasma membrane (see Chapters 5 and 13). Phosphorus-efficient plants are enhanced in these biochemical adaptations, maximizing their ability to extract phosphorus from low-phosphorus soils.

Root system architecture responses to phosphorus deficiency involve both local and systemic regulatory networks

Both local and systemic regulatory networks are involved in the adaptation of root system architecture to phosphorous deficiency. Individual roots are able to respond locally to phosphorus-deficient patches in the rhizosphere; hormones play important roles in reprogramming root development locally to facilitate more efficient phosphorus capture by that part of the root system. However, if a plant experiences prolonged phosphorus deficiency, systemic signaling comes into play.

The systemic regulation of phosphorus-deficiency responses is summarized in **Figure 19.40**. This initially involves long-distance transport of signals from the root to the shoot via the xylem. The root-to-shoot signals may include the phosphate ion itself, as well as sugars, cytokinins, strigolactones, and possibly other signals that have not yet been identified. The arrival of these stress signals at target cells in source leaves triggers additional signaling events. Subsequently, long-distance signals from the shoot—including siRNAs and miRNAs, mRNAs, proteins, sucrose, and other unidentified signals—are transported via the phloem to various sinks, where they regulate plant growth and phosphorus homeostasis. These sinks include both root and shoot apical meristems. For example, the miRNA miR399 has been shown to be induced and transported via the phloem to the root under phosphorus-stress conditions, where it suppresses expression of a putative ubiquitin E2 conjugase, which is involved in the degradation of root phosphate transporters. The miR399-mediated suppression of ubiquitination in response to phosphorus deficiency results in the promotion of root phosphate transport. In addition, genes acting downstream of the miR399 signaling pathway regulate phosphate loading into the xylem and encode a plasma membrane phosphate transporter.

Thus far it appears that all of the phosphorus-deficiency miRNAs that are transported to the root via the phloem are involved in the regulation of root phosphate transport processes rather than root development. However, it is likely that other, as yet unidentified, phloem-mobile signals play a role in altering root system architecture. As the model in Figure 19.40 indicates, phosphorus deficiency increases the abundance of *IAA18* and *IAA28* mRNAs, and it has been shown that these mRNAs are transported to the root in tomato plants, altering root auxin sensitivity and lateral root formation.

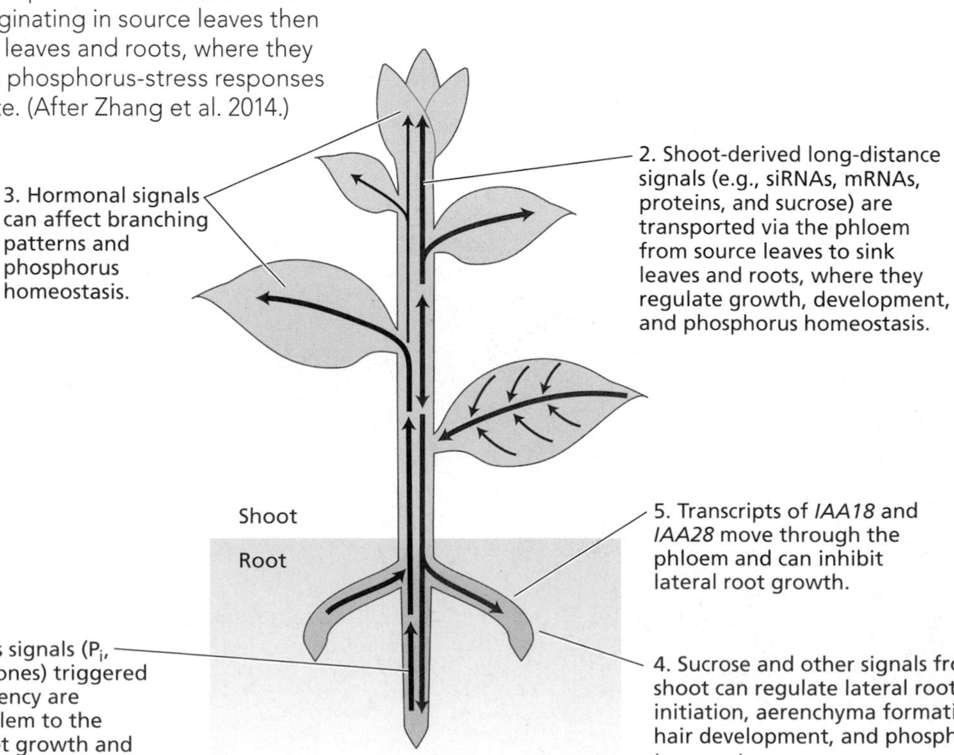

2. Shoot-derived long-distance signals (e.g., siRNAs, mRNAs, proteins, and sucrose) are transported via the phloem from source leaves to sink leaves and roots, where they regulate growth, development, and phosphorus homeostasis.
3. Hormonal signals can affect branching patterns and phosphorus homeostasis.
Shoot
Root
5. Transcripts of *IAA18* and *IAA28* move through the phloem and can inhibit lateral root growth.
1. Root-derived stress signals (P_i, cytokinins, strigolactones) triggered by phosphorus deficiency are transported in the xylem to the shoot, affecting shoot growth and architecture, such as branching.
4. Sucrose and other signals from the shoot can regulate lateral root initiation, aerenchyma formation, root hair development, and phosphate transport.

Figure 19.40 Phosphate sensing involves communication between the root and shoot. Phosphate deficiency in the soil results in the movement of various stress signals in the xylem to the shoot (black arrows), where they alter development and trigger phosphorus homeostatic mechanisms. Additional signals originating in source leaves then move via the phloem to sink leaves and roots, where they can affect development and phosphorus-stress responses (purple arrows). P_i, phosphate. (After Zhang et al. 2014.)

Mycorrhizal networks augment root system architecture in all major terrestrial ecosystems

As we discussed earlier in Chapter 5, fungal mycorrhizas are nearly ubiquitous in nature and play an important role in the mineral nutrition of individual plants. In addition, recent studies have shown that entire communities of plants are typically linked by mycorrhizal associations, which form nutritional networks.

A **mycorrhizal network** is defined as a common mycorrhizal mycelium linking the roots of two or more plants. For decades, scientists have been fascinated by evidence that mycorrhizal networks can transfer organic and inorganic nutrients, especially phosphate, between the root systems of otherwise separate individuals. Long-distance nutrient transfer through direct hyphal pathways appears to occur by mass flow driven by source–sink gradients generated by nutrient differences between plants. Through their effects on plant nutrition, mycorrhizal networks have been shown to facilitate seedling establishment, promote vegetative growth, and enhance plant responses to biotic and abiotic stress in a wide range of ecosystems. At the ecosystem level, mycorrhizal networks play an important role in carbon, nutrient, and water cycling.

In recent years the use of molecular techniques has shed light on the nature and extent of mycorrhizal networks among forest trees and between overstory and understory plants. For example, **microsatellite DNA sequences** (short tandem repeats) have been employed as molecular markers to study the spatial topology of **genets** (clonal colonies) of ectomycorrhizal fungi in the genus *Rhizopogon*. In a forest of Douglas-fir (*Pseudotsuga menziesii*), the majority of trees in a 30 × 30 m plot were found to be interconnected by a complex mycorrhizal network of *Rhizopogon vesiculosus* and *R. vinicolor*. The most highly connected tree was linked to 47 other trees through eight *R. vesiculosus* genets and three *R. vinicolor* genets. The interconnectivity of the trees in this forest illustrates the surprising complexity of nutrient flow in forest ecosystems, which will need to be taken into account in future studies of the effects of climate change on forest productivity.

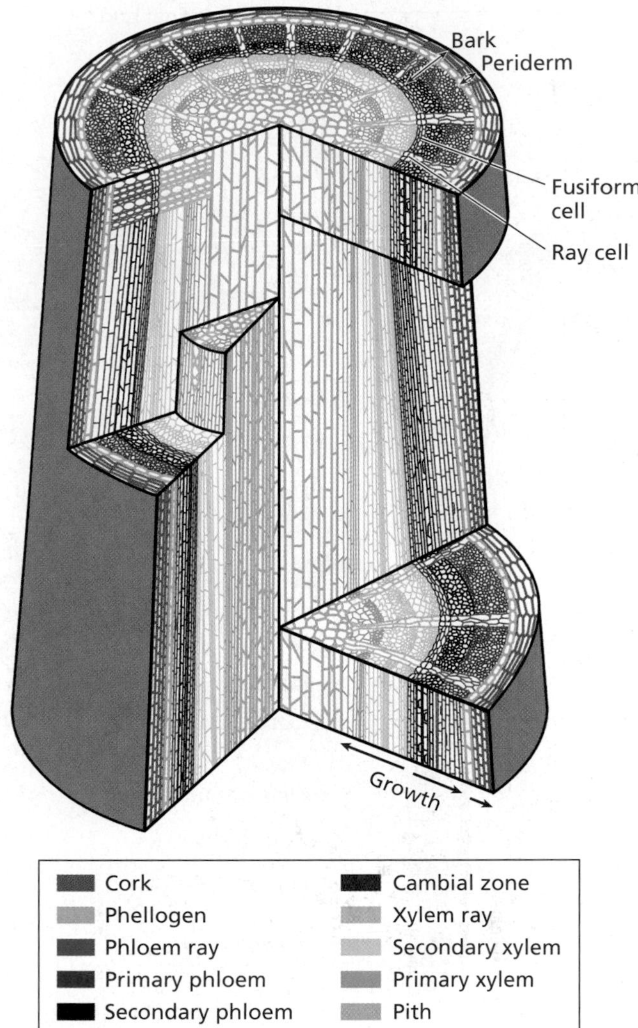

Figure 19.41 Internal anatomy of a woody stem. The *zone* of the vascular cambium (red region) consists of a single layer of cambial stem cells and its immediate derivatives on either side, and is surrounded by an outer layer of secondary phloem cells (black) and an inner layer of secondary xylem cells (light green). The primary phloem (dark blue), primary xylem (dark green), and pith (light blue) are also shown. The periderm includes the phellogen (tan cell layer) and phellem (cork cells, in brown). The bark layer includes all tissues external to the vascular cambium. Most angiosperm and gymnosperm tree species also contain radial files of ray cells that play a role in nutrient transport and storage. (From Risopatron et al. 2010.)

Secondary Growth

All gymnosperms and most eudicots—including woody shrubs and trees, as well as large herbaceous species—develop lateral meristems that cause radial growth (growth in width) in stems and roots. Growth that results from lateral meristems is called **secondary growth** (**Figure 19.41**; see also Figure 1.5). Two types of lateral meristems are involved in secondary growth: the **vascular cambium**, which produces secondary vascular tissues, and the **cork cambium**, or **phellogen**, which produces the outer protective layers of the secondary plant body called the **periderm**. Secondary growth via the vascular cambium has arisen repeatedly during the evolution of vascular plants, and many extinct groups exhibit conspicuous secondary vascular tissues. Monocots as a group lack a vascular cambium and therefore do not exhibit secondary growth. Even tree-form members of the Palmae lack a vascular cambium and increase their width solely by means of a **primary thickening meristem**, located in the "meristem cap" just below the leaf primordia, which produces additional

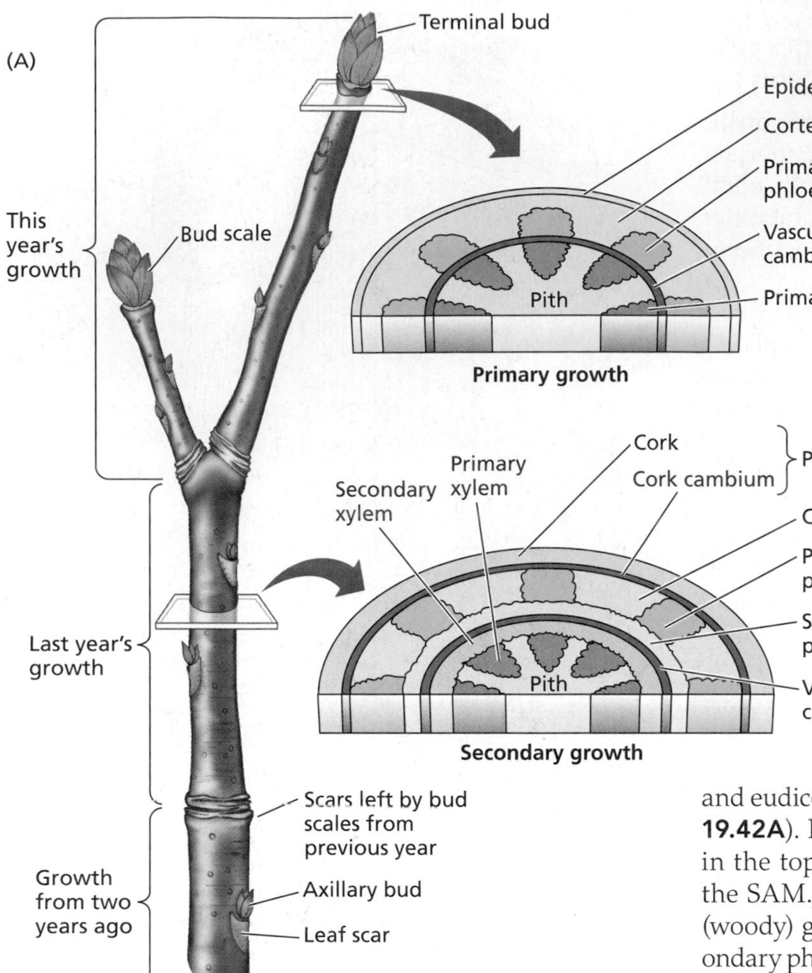

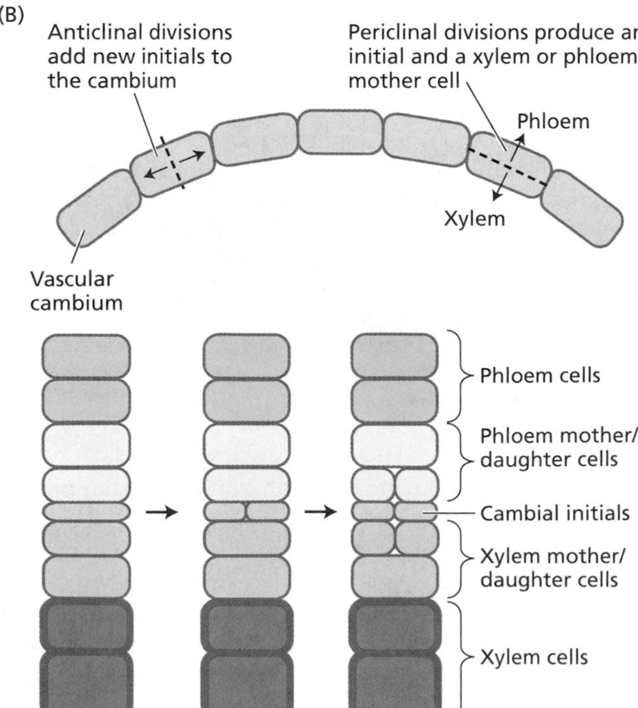

Figure 19.42 The development of secondary vascular tissue. (A) Primary growth in woody stems occurs in the spring, followed by secondary growth. (B) Orientations of the cell division planes in the cambial zone maintain the proper balance between growth in diameter versus circumference. Cambial cells initially divide anticlinally to produce new initials and increase the circumference of the cambium. The same initials also divide periclinally to produce xylem and phloem mother cells, always leaving behind another initial.

primary tissues, including vascular bundles. The primary thickening meristem persists in some species, producing additional parenchyma tissue and vascular bundles.

The transition from primary to secondary growth in gymnosperms and eudicots is readily visible along the shoot axis (**Figure 19.42A**). In poplar, for example, primary growth occurs in the top eight internodes, approximately 15 cm from the SAM. Primary growth then gives way to secondary (woody) growth that produces secondary xylem and secondary phloem. The primary and secondary growth zones are separated both spatially and temporally, are easily discernible, and develop rapidly (within 1–2 months) in fast-growing species such as poplar.

The vascular cambium and cork cambium are the secondary meristems where secondary growth originates

Secondary growth originates in the vascular cambium—a lateral meristem that displays perennial growth patterns in woody species. Many herbaceous species also have a vascular cambium, but its formation is usually conditional (e.g., in response to stress) or is very short-lived.

The vascular cambium consists of meristem cells (cambial initials) organized in radial files that form a continuous cylinder around the stem. Cambial initials divide to produce xylem, phloem, and ray mother cells, which in turn undergo several rounds of divisions to form a zone of relatively undifferentiated cells that usually comprise six to eight cell files and are known as the cambium zone. These cells then differentiate into several cell types. The vascular cambium of all modern extant seed plants is bifacial—that is, it produces xylem inward and phloem outward (see Figure 1.5).

In addition to secondary growth involved in secondary phloem and xylem, most woody eudicots and gymnosperms develop a secondary cambium known as cork cambium or phellogen that gives rise to the periderm (see Figure 19.41). Collectively the periderm consists of phellogen, phellem, and phelloderm. **Phellem**, or **cork**, is the multilayered protective tissue of dead cells with suberized walls, which is formed outward by the phellogen. The **phelloderm** is living parenchyma tissue that is formed inward. The term **bark**, often applied incorrectly to the periderm alone, actually consists of all the tissues outside the vascular cambium, including functional secondary phloem, crushed nonfunctional secondary phloem, crushed primary phloem, and the periderm (phellogen, phellem, and phelloderm). Bark tends to peel easily from a tree because the vascular cambium, with its dividing cell layers, is much more fragile than the secondary tissues on either side.

The degree of phellogen activity giving rise to phellem varies among tree species, with cork oak (*Quercus suber*) representing an extreme example that contains a permanent phellogen layer producing cork or phellem indefinitely. The thick, corky layer probably protects the main trunk from dehydration in the hot and dry Mediterranean climate.

Secondary growth evolved early in the evolution of land plants

Fossil vestiges of primitive secondary growth activity can be found very early in plant evolution, possibly predating modern seed plants. Secondary growth likely predates the evolution of gymnosperm plants and is speculated to be the ancestral life form of all modern angiosperms. For example, the woody perennial *Amborella* genus is the most basal lineage in the angiosperm clade, thought to be the ancestor of all modern angiosperms. The woody perennial habit has been lost and reacquired during the evolution of seed plants, with some lineages showing intermediary forms, known as insular woodiness. Secondary woody growth has evolved into various forms and shapes that are likely adaptive in nature. For example, many lianas display flat stems resulting from differential proliferation of xylem tissues in particular parts of the stem circumference. Alternatively, stems of some lianas (most notably from the Bignonieae family) maintain a cylindrical shape but produce internally wedge-shaped sectors of parenchyma tissues. These changes, seen predominantly in lianas, are thought to facilitate stem flexibility, wound healing associated with twisting, and recovery from loss of xylem conductivity associated with severe twisting.

Secondary growth from the vascular cambium gives rise to secondary xylem and phloem

Vascular cambium displays two main division patterns—**anticlinal** (perpendicular to the stem surface) and **periclinal** (parallel to the stem surface) (**Figure 19.42B**). Anticlinal divisions add more cells to the cambium to accommodate the increasing girth of the stem and are thought to indicate the position of the cambium initials, which are otherwise morphologically indistinguishable from the other cells in the cambium zone. The peak of anticlinal division is typically within the first to second cell file proximal to the phloem and is usually used to pinpoint the approximate position of the vascular cambium.

In a typical bifacial cambium, the periclinal divisions produce phloem outward and xylem inward in the woody stem. The proliferation of xylem is disproportionately geater and in many ways more complex since it encompasses the full life cycle of the tracheary cells in a matter of days. In addition to phloem and xylem cells, the vascular cambium produces ray cells, parenchyma cells that serve as conduits for lateral transport in the stem and for storage during unfavorable conditions such as winter dormancy. Ray cells can be arranged in one (uniseriate) or multiple (multiseriate) files to form a tissue known as *rays* that traverses the phloem, cambium, and xylem (see Figure 19.41).

Phytohormones have important roles in regulating vascular cambium activity and differentiation of secondary xylem and phloem

As with many other processes in plants, hormones play important roles in the regulation of secondary growth. Several hormones provide positional cues and signals for growth and differentiation of different cell types and tissues (**Figure 19.43**). Here we focus on four hormones because a significant amount of experimental evidence supports their role in regulation of secondary growth. However, this does not imply that they have more significant roles than other hormones.

Although auxin movements in trees has not been extensively studied, it is assumed that auxin is produced in leaves and apical meristems and transported via polar auxin transport to the stem and vascular cambium. Measurements of auxin concentrations from vascular cambium into differentiating xylem and phloem in both angiosperm and gymnosperm trees have shown that the peak of the gradient is located in the cambium initials, and tapers toward the differentiating xylem and phloem. The decrease is steeper toward the phloem and much more gradual toward the xylem. This concentration gradient across the cambial zone has lead to the speculation that the role of auxin in xylem and phloem differentiation is based on a radial morphogen gradient.

The critical role of auxin is also supported by exogenous treatments showing that auxin application in decapitated trees, in which the vascular cambium has become inactive, leads to reactivation of the cambium. More recently, direct manipulation of the auxin response in transgenic poplar trees has shown that auxin sensitivity is critical for both periclinal and anticlinal divisions in the cambium and affects the growth and differentiation of xylem cells.

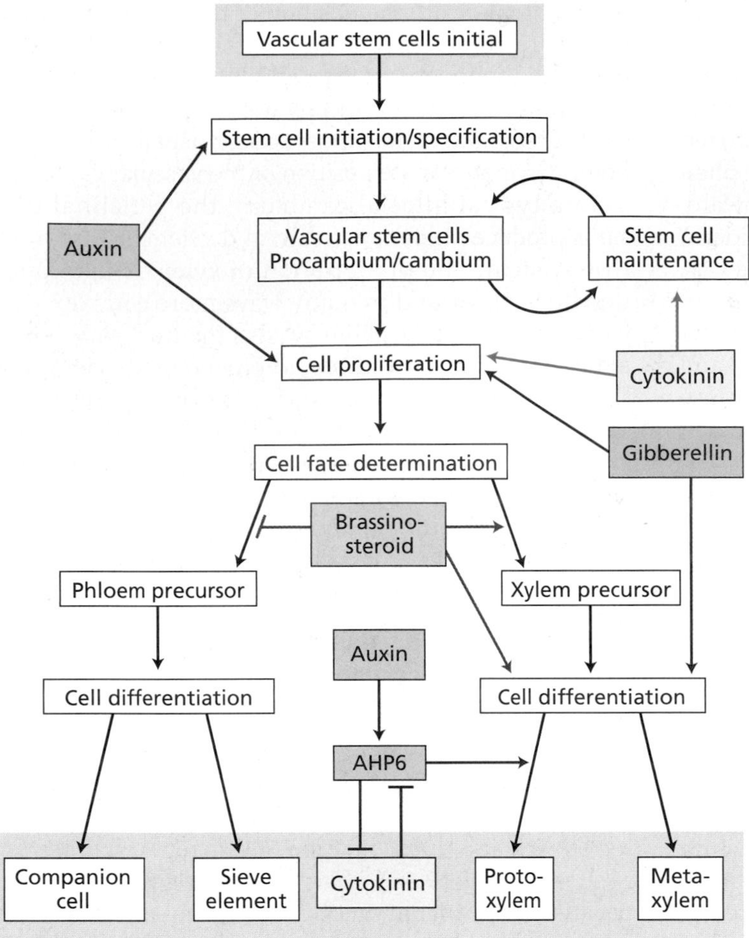

Figure 19.43 Hormones are involved in regulating key stages of secondary vascular tissue development. ARABIDOPSIS HISTIDINE PHOSPHOTRANSFER PROTEIN (AHP6) acts as a cytokinin signaling inhibitor that restricts the domain of cytokinin activity, thus allowing protoxylem differentiation in a spatially specific manner.

Gibberellins also play major and distinct role in secondary growth. Like auxins, bioactive gibberellins exhibit a concentration gradient across the wood-forming zone, but unlike in auxin, the peak is shifted toward the developing xylem. Exogenous treatment of decapitated seedlings lacking auxin with gibberellins resulted in the activation of cambium cell divisions. However the dividing cells lost their typical shape and failed to differentiate into xylem. Simultaneous application of auxin and gibberellins prevented the abnormalities observed in the gibberellin treatment alone, and stimulated cambium division to an extent not seen in the gibberellin and auxin treatment alone, suggesting that the two hormones act synergistically (see Figure 19.43).

Metabolic profiling and expression of several genes from the gibberellin biosynthetic pathway indicate that gibberellin metabolism in the wood-forming tissues also involves transport of gibberellin precursors from the phloem laterally through the rays into the differentiating xylem where they are then converted to bioactive forms (see Figure 19.43). Both exogenous treatments and transgenic manipulations indicate that gibberellins have a positive effect on elongation of fiber cells, suggesting a role for gibberellins in xylem cell differentiation and growth.

Cytokinin has also been implicated in regulation of secondary growth (see Figure 19.43). A specific decrease in the cytokinin concentration in the cambium zone of transgenic poplar trees led to significant impairment of radial growth and cell division in the cambium. This result was correlated with the expression in the cambium zone of a gene encoding the cytokinin receptor and primary response regulator involved in cytokinin signaling. This suggests that cytokinin is an important regulator of cell proliferation in the cambium.

Ethylene is yet another hormone that has been strongly implicated as playing a regulatory role in secondary growth. The concentration of the ethylene precursor, 1-aminopropane-1-carboxylic acid (ACC), was found to be high at the cambium zone, but unlike with auxin and gibberellin, no gradient was detected. Ethylene treatment and ACC feeding experiments demonstrated that ethylene is a positive regulator of cambial activity, radial growth, and secondary xylem formation. These results are also consistent with the results of transgenic manipulation of ethylene biosynthesis and response in poplar. Ethylene plays an important role in the formation of **tension wood**, a specialized type of reaction wood in angiosperms formed in response to stem bending or tilting. Expression of both ethylene biosynthesis and signaling genes is elevated in the tension wood–forming zone, and ethylene insensitive transgenic poplars fail to produce tension wood.

Genes involved in stem cell maintenance, proliferation, and differentiation regulate secondary growth

Like other developmental processes in plants, secondary growth involves several key steps. It can be divided into three developmental stages:

- Maintenance of the initial cell microenvironment, or stem cell niche
- Proliferation and growth of cells derived from the stem cells
- Differentiation of the dividing cell into different cell types, tissues, and organs

It is therefore not surprising that the developmental and growth patterns during secondary growth are governed by processes and genes that are similar to those that regulate the development of the SAM. This similarity has aided the molecular dissection of the mechanisms of secondary growth.

Perhaps the best studied of these processes is the maintenance of the stem cell niche. In the SAM of Arabidopsis, KNOX1 transcription factors, such as SHOOT MERISTEMLESS (STM) and BREVIPEDICELLUS (BP), are involved in maintaining stem cell identity. Orthologs of the *STM* and *BP* genes in poplar, known as *ARBORKNOX 1* and *2*, play similar roles in the cambium. Overexpression of the two genes in transgenic plants leads to delayed differentiation and enlargement of the cambium zone.

Proliferation of the dividing cells in the SAM is typically regulated by genes such as *AINTEGUMENTA*. AINTEGUMENTA is an AP2-type transcription factor involved in regulation of organ size in Arabidopsis through the activation of cell proliferation. In aspen, an ortholog of the *AINTEGUMENTA* gene was found to be highly expressed in the cells that display high cell proliferation in the cambium zone.

The best example of similarity between the regulation of the SAM and the cambium is arguably at the differentiation step. The HD-ZIP III and KANADI transcription factors play important roles in defining the adaxial and abaxial polarity in the emerging leaf. This polarity results in the differentiation of xylem on the adaxial side of the leaf vein, and phloem on the abaxial side (see Figure 19.5). The HD-ZIP III and KANADI genes also regulate the patterning of the primary vascular bundles, as well as the differentiation of secondary vascular tissues later in development, probably through their effects on polar auxin transport.

The three developmental stages considered above also require spatial separation. This is achieved through transcription factors that define the developmental boundaries. One such class of transcription factors involved in the regulation of the SAM is LATERAL ORGAN BOUNDARIES (LBD). LBD genes help establish the boundary meristem (discussed earlier in the chapter), which separates the undifferentiated cells in the SAM from the differentiating tissues of the leaf primordium. Members of the LBD family were found to play a similar role in secondary growth by separating the cambium zone from the differentiating secondary phloem and xylem.

Environmental factors influence vascular cambium activity and wood properties

Plants are sessile and require robust responses to unfavorable conditions for survival (see Chapter 24). This is particularly important for woody perennial plants such as trees that can occupy a site for hundreds and even thousands of years. One distinct challenge that trees face is climate seasonality, which poses risks to trees' survival during prolonged (seasonal) unfavorable or lethal conditions, such as these encountered during winter months in the temperate and boreal regions. To endure the dehydration and freezing stress during the winter months, trees alternate between periods of active growth and dormancy. The annual transition from active growth to dormancy in the cambium results in the formation of tree rings that record the amount of lateral growth of the tree each year. The molecular mechanisms that control cambium growth during growth–dormancy cycles are poorly understood. Phytohormones such as auxin and gibberellins are thought to play major roles in the reactivation and cessation of growth in the cambium.

Growth seasonality also imposes a significant challenge with respect to nutrient use, storage, and recycling. Nitrogen is the most abundant macronutrient in plants. Although all plant species have mechanisms to recycle, store, and remobilize nitrogen during the growing season, the seasonal cycling of nitrogen is a hallmark of the perennial life habit. For example, nitrogen from the senescing leaves is stored in the form of **bark storage proteins** (**BSPs**) in small vacuoles of the phloem parenchyma (inner bark). These proteins are synthesized early in the autumn but are rapidly mobilized during the spring as growth is reinitiated. The signaling mechanisms involved are still unclear but may involve the transport of hormonal signals from the SAM.

Physiologically, wood serves transport, storage, and mechanical functions, and thus the response to various environmental factors reflects changes that best accommodate these functions. These three functions are also reflected in the main cell types that are encountered in the xylem. For example, in a typical angiosperm the transport, mechanical, and storage functions are carried out by vessels, fibers, and parenchyma cells, respectively. The proportion of these cell types changes dramatically in response to different stress factors and reflects compensatory shifts that reinforce one function or another.

The mechanical function of wood is strongly reinforced during reaction wood formation (see Figure 15.1D and E). Reaction wood forms when stems are displaced from their vertical position. In angiosperm trees, reaction wood develops on the top part of the stem and is known as tension wood. Tension wood is distinct from wood developed under vertical orientation in that it contains more fibers (e.g., the cells that serve the mechanical function), and its cell walls are enriched with highly crystalline cellulose that reinforces the mechanical function. In contrast, water deficit or significant osmotic stress results in changes that support the transport function and are in a way the opposite of those in tension wood. Wood developed under drought conditions typically shows increased vessel density and cell walls that produce more lignin than cellulose. These changes improve the water transport and retention function of the xylem.

SUMMARY

After embryogenesis and germination, vegetative growth is controlled by developmental processes involving molecular interactions and regulatory feedback. These mechanisms create root and shoot polarity, allowing plants to produce lateral organs (e.g., leaves and branching systems), which form an overall vegetative architecture.

Leaf Development

- The development of flat laminas in spermatophytes was a key evolutionary event; since then, phyllome morphology has diversified dramatically (**Figure 19.1**).

The Establishment of Leaf Polarity

- In addition to positional information, hormone distribution also affects leaf primordia emergence (**Figure 19.2**).
- Adaxial–abaxial polarity in a leaf primordium is established by a signal from the SAM (**Figure 19.3**).
- ARP transcription factors interact with protein partners to promote adaxial identity and repress the *KNOX1* gene (**Figures 19.4, 19.5**).
- Adaxial identity is also supported by HD-ZIP III transcription factors, which are abaxially suppressed by the microRNA miR166.
- Specification of abaxial identity is promoted by KANADI and YABBY gene families and is antagonized by HD-ZIP III.
- Normal blade growth depends on juxaposition of adaxial and abaxial tissues and is regulated by auxin and by YABBY and WOX genes (**Figure 19.5**).
- Leaf primordia also differentiate proximally–distally into a boundary meristem, lower-leaf zone, petiole, and blade (**Figure 19.5**).
- Similar genes and transcription factors govern compound leaf formation (**Figures 19.7, 19.8**).

Differentiation of Epidermal Cell Types

- The epidermis is derived from the protoderm (L1) and has three main cell types: pavement cells, trichomes, and stomatal guard cells, as well as other cell types (**Figure 19.9**).
- Not only guard cells, but the majority of leaf epidermal cells, arise from specialized meristemoid mother cells (MMCs), stomatal lineage ground cells (SLGCs), meristemoids, and guard mother cells (GMCs) (**Figure 19.10**).
- Basic helix-loop-helix (bHLH) transcription factors govern the cell-state transitions from MMCs to meristemoids, meristemoids to GMCs, and GMCs to mature guard cells (**Figure 19.10**).
- Stomatal-lineage cells and mesophyll cells excrete peptide signals that interact with transmembrane receptors to regulate stomatal patterning (**Figure 19.11**).
- Genes in protoderm cells regulate trichome differentiation and distribution (**Figures 19.12–19.14**).
- The GL2 transcription factor is the rate-limiting element in trichome formation.
- Jasmonic acid regulates development of leaf trichomes in Arabidopsis.

Venation Patterns in Leaves

- Leaf venation patterns indicate the spatial organization of the vasculature (**Figures 19.15, 19.16**).
- Triggered by auxin that is transported downward, leaf veins are initiated separately from established vasculature and grow *down* to rejoin it, directed by the vascular bundle in the stem (**Figures 19.17–19.20**).
- Similarly to initial development of veins, development of higher-order veins proceeds from tip to base and is regulated by auxin canalization. However, auxin transport is less dependent on PIN1 (**Figures 19.21–19.23**).
- Localized auxin biosynthesis allows for development of higher-order veins (**Figure 19.24**).

Shoot Branching and Architecture

- Shoot architecture can be based on continuously branching, equal shoots, or on repeating units of hierarchical shoots, leading to axillary branches (**Figures 19.25, 19.26**).
- Branch initiation involves some of the same genes and hormones as leaf initiation and outgrowth (**Figures 19.27–19.29**).
- There is strong experimental and empirical evidence that auxin and strigolactones from the shoot tip maintain apical dominance (**Figures 19.30, 19.31**).
- Cytokinins break apical dominance and promote axillary dominance (**Figures 19.32, 19.33**).
- Sucrose also serves as an initial signal for axillary bud growth (**Figure 19.34**).
- Environmental signals can override default hormonal signals to shape vegetative architecture. For example, woody perennial plants produce dormant buds in response to temperature and to availability of water, nutrients, and light (**Figure 19.35**).

Root System Architecture

- Species-specific root system architecture optimizes water and nutrient uptake (**Figure 19.36**).
- Monocot roots systems are composed largely of seminal and crown roots, while eudicot root systems are derived largely from the primary (tap) root (**Figures 19.37, 19.38**).
- Phosphorus availability can alter root system architecture both locally and systemically (**Figures 19.39, 19.40**).
- Mycorrhizal relationships with root systems are terrestrially ubiquitous.

Secondary Growth

- Growth in width is accomplished by the vascular cambium and cork cambium, which are secondary meristems that give rise to secondary vascular tissue (**Figures 19.41, 19.42**).
- Secondary growth predates the appearance of gynosperms.
- Auxin, gibberellins, cytokinins, and ethylene regulate vascular cambium activity and differentiation of secondary vascular tissues (**Figure 19.43**).
- Genes regulate the cellular microenvironment for stem cell maintenance, proliferation, and differentiation.
- Vascular cambium activity is sensitive to environmental factors that ultimately influence wood properties.

WEB MATERIAL

- **WEB TOPIC 19.1 Bifacial, Unifacial, and Equifacial Leaves** Bifacial, unifacial, and equifacial leaves can be distinguished based on their anatomical and morphological differences.
- **WEB TOPIC 19.2 Mechanical Stress Alters Microtubule Orientation and PIN1 Distribution in the SAM** The apical meristem can be thought of as a giant cell whose shape generates stress patterns that can influence PIN1 localization.
- **WEB TOPIC 19.3 Leaf Serrations Are Coordinated by the Action of a CUC2–Auxin Feedback Loop** While marginal serrations are modified by many genes, the key components are auxin and *CUC2*.
- **WEB TOPIC 19.4 Advances in Root System Phenotyping** Modern root system image capture methods include two-dimensional (2D) and three-dimensional (3D) imaging techniques.

available at plantphys.net

Suggested Reading

Balkunde, R., Pesch, M., and Hülskamp, M. (2010) Trichome patterning in *Arabidopsis thaliana*: From genetic to molecular models. *Curr. Top. Dev. Biol.* 91: 299–321.

Bayer, I., Smith, R. S., Mandel, T., Nakayama, N., Sauer, M., Prusinkiewicz, P., and Kuhlemeier, C. (2009) Integration of transport-based models for phyllotaxis and midvein formation. *Genes Dev.* 23: 373–384.

Besnard, F., Vernoux, T., and Hamant, O. (2011) Organogenesis from stem cells in planta: Multiple feedback loops integrating molecular and mechanical signals. *Cell. Mol. Life Sci.* 68: 2885–2906.

Byrne, M. E. (2012) Making leaves. *Curr. Opin. Plant Biol.* 15: 24–30.

Caño-Delgado, A., Lee, J. Y., and Demura, T. (2010) Regulatory mechanisms for specification and patterning of plant vascular tissues. *Annu. Rev. Cell Dev. Biol.* 26: 605–637.

Domagalska, M.A., and Leyser, O. (2011) Signal integration in the control of shoot branching. *Nat. Rev. Mol. Cell Biol.* 12: 211–221.

Fukushima, K., and M. Hasebe1 (2014) Adaxial–abaxial polarity: The developmental basis of leaf shape diversity. *Genesis* 52: 1–18.

Greb, T., Clarenz, O., Schafer, E., Muller, D., Herrero, R., Schmitz, G., and Theres, K. (2003) Molecular analysis of the *LATERAL SUPPRESSOR* gene in *Arabidopsis* reveals a conserved control mechanism for axillary meristem formation. *Genes Dev.* 17: 1175–1187.

Hay, A., and Tsiantis, M. (2010) KNOX genes: Versatile regulators of plant development and diversity. *Development* 137: 3153–3165.

Heisler, M. G., Hamant, O., Krupinski, P., Uyttewaal, M., Ohno, C., Jönsson, H., Traas, J., and Meyerowitz, E. M. (2010) Alignment between PIN1 polarity and microtubule orientation in the shoot apical meristem reveals a tight coupling between morphogenesis and auxin transport. *PLOS Biol.* 8(10): e1000516. DOI:10.1371/journal.pbio.100051

Lau, S. and Bergmann, D. C. (2012) Stomatal development: A plant's perspective on cell polarity, cell fate transitions and intercellular communication. *Development* 139: 3683–3692.

Lucas, W.J., Groover, A., Lichtenberger, R., Furuta, K., Yadav, S. R., Helariutta, Y., He, X. Q., Fukuda, H., Kang, J., Brady, S. M., et al. (2013) The plant vascular system: Evolution, development and functions. *J. Integr. Plant Biol.* 55: 294–388.

Mason, M. G., Ross, J. J., Babst, B. A., Wienclaw, B. N., and Beveridge, C. A. (2014) Sugar demand, not auxin, is the initial regulator of apical dominance. *Proc. Natl. Acad. Sci. USA* 111: 6092–6097.

Qing, L., and Aoyama, T. (2012) Pathways for epidermal cell differentiation via the homeobox gene *GLABRA2*: Update on the roles of the classic regulator. *J. Integr. Plant Biol.* 54: 729–737.

Risopatron, J. P. M., Sun, Y., and Jones, B. J. (2012) The vascular cambium: Molecular control of cellular structure. *Protoplasma* 247:145–161.

Townsley, B. T., and Sinha, N. R. (2012) A new development: Evolving concepts in leaf ontogeny. *Annu. Rev. Plant Biol.* 63: 535–562.

Yang, F., Wang, Q., Schmitz, G., Müller, D., and Theres, K. (2012) The bHLH protein ROX acts in concert with RAX1 and LAS to modulate axillary meristem formation in Arabidopsis. *Plant J.* 71: 61–70.

Zhang, Z., Liao, H., and Lucas, W. J. (2014) Molecular mechanisms underlying phosphate sensing, signaling, and adaptation in plants. *J. Integr. Plant Biol.* 56: 192–220.

20 The Control of Flowering and Floral Development

Most people look forward to the spring season and the profusion of flowers it brings. Many vacationers carefully time their travels to coincide with specific blooming seasons: *Citrus* along Blossom Trail in southern California, tulips in Holland. In Washington, DC, and throughout Japan, the cherry blossoms are received with spirited ceremonies. As spring progresses into summer, summer into fall, and fall into winter, wildflowers bloom at their appointed times. Flowering at the correct time of year is crucial for the reproductive fitness of the plant; plants that are cross-pollinated must flower in synchrony with other individuals of their species as well as with their pollinators at the time of year that is optimal for seed set.

Although the strong correlation between flowering and seasons is common knowledge, the phenomenon poses fundamental questions that will be addressed in this chapter:

- How do plants keep track of the seasons of the year and the time of day?
- Which environmental signals influence flowering, and how are those signals perceived?
- How are environmental signals transduced to bring about the developmental changes associated with flowering?

In Chapter 19 we discussed the role of the root and shoot apical meristems in vegetative growth and development. The transition to flowering involves major changes in the pattern of morphogenesis and cell differentiation at the shoot apical meristem. Ultimately, as we will see, this process leads to the production of the floral organs—sepals, petals, stamens, and carpels.

Floral Evocation: Integrating Environmental Cues

A particularly important developmental decision during the plant life cycle is when to flower. The process by which the shoot apical meristem becomes committed to forming flowers is termed **floral evocation**. Delaying this commitment to flower will increase the carbohydrate reserves that will be available for mobilization, allowing more and better-provisioned seeds to mature. Delaying flowering, however, also potentially increases the danger that the plant will be eaten, killed by abiotic stress, or outcompeted by other plants before it reproduces. Reflecting this, plants have evolved an extraordinary range of reproductive adaptations—for example, annual versus perennial life cycles.

Annual plants such as groundsel (*Senecio vulgaris*) may flower within a few weeks after germinating. But trees may grow for 20 or more years before they begin to produce flowers. Across the plant kingdom, different species flower at a wide range of ages, indicating that the age, or perhaps the size, of the plant is an internal factor controlling the switch to reproductive development.

The case in which flowering occurs strictly in response to internal developmental factors, independently of any particular environmental condition, is referred to as *autonomous regulation*. In species that exhibit an absolute requirement for a specific set of environmental cues in order to flower, flowering is considered to be an *obligate* or *qualitative* response. If flowering is promoted by certain environmental cues but will eventually occur in the absence of such cues, the flowering response is *facultative* or *quantitative*. A species with a facultative flowering response, such as Arabidopsis, relies on both environmental and autonomous signals to promote reproductive growth.

Photoperiodism and vernalization are two of the most important mechanisms underlying seasonal responses. Photoperiodism (see Chapter 16) is a response to the length of day or night; vernalization is the promotion of flowering by prolonged cold temperature. Other signals, such as light quality, ambient temperature, and abiotic stress, are also important external cues for plant development.

The evolution of both internal (autonomous) and external (environment-sensing) control systems enables plants to precisely regulate flowering so that it occurs at the optimal time for reproductive success. For example, in many populations of a particular species, flowering is synchronized, which favors crossbreeding. Flowering in response to environmental cues also helps ensure that seeds are produced under favorable conditions, particularly with respect to water and temperature. However, this makes plants especially vulnerable to rapid climate change, such as global warming, which can alter the regulatory networks that govern floral timing (**WEB TOPIC 20.1**).

The Shoot Apex and Phase Changes

All multicellular organisms pass through a series of more or less defined developmental stages, each with its characteristic features. In humans, infancy, childhood, adolescence, and adulthood represent four general stages of development, with puberty as the dividing line between the nonreproductive and the reproductive phases. Similarly, plants pass through distinct developmental phases. The timing of these transitions often depends on environmental conditions, allowing plants to adapt to a changing environment. This is possible because plants continuously produce new organs from the shoot apical meristem.

The transitions between different phases are tightly regulated developmentally, since the plant must integrate information from the environment as well as autonomous signals to maximize its reproductive fitness. The following sections will describe the major pathways that control these decisions.

Plant development has three phases

Postembryonic development in plants can be divided into three phases:

1. The juvenile phase
2. The adult vegetative phase
3. The adult reproductive phase

The transition from one phase to another is called **phase change**.

The primary distinction between the juvenile and the adult vegetative phases is that the latter has the ability to form reproductive structures: flowers in angiosperms, cones in gymnosperms. However, flowering, which represents the expression of the reproductive competence of the adult phase, often depends on specific environmental and developmental signals. Thus, the absence of flowering itself is not a reliable indicator of juvenility.

The transition from juvenile to adult is frequently accompanied by changes in vegetative characteristics, such as leaf morphology, phyllotaxy (the arrangement of leaves on the stem), thorniness, rooting capacity, and leaf retention in deciduous plants such as English ivy (*Hedera helix*) (**Figure 20.1**; see also **WEB TOPIC 20.2**). Such changes are most evident in woody perennials, but they are apparent in many herbaceous species as well. Unlike the abrupt transition from the adult vegetative phase to the reproductive phase, the transition from juvenile to adult vegetative is usually gradual, involving intermediate forms.

Juvenile tissues are produced first and are located at the base of the shoot

The time sequence of the three developmental phases results in a spatial gradient of juvenility along the shoot axis. Because growth in height is restricted to the apical meristem, the juvenile tissues and organs, which form

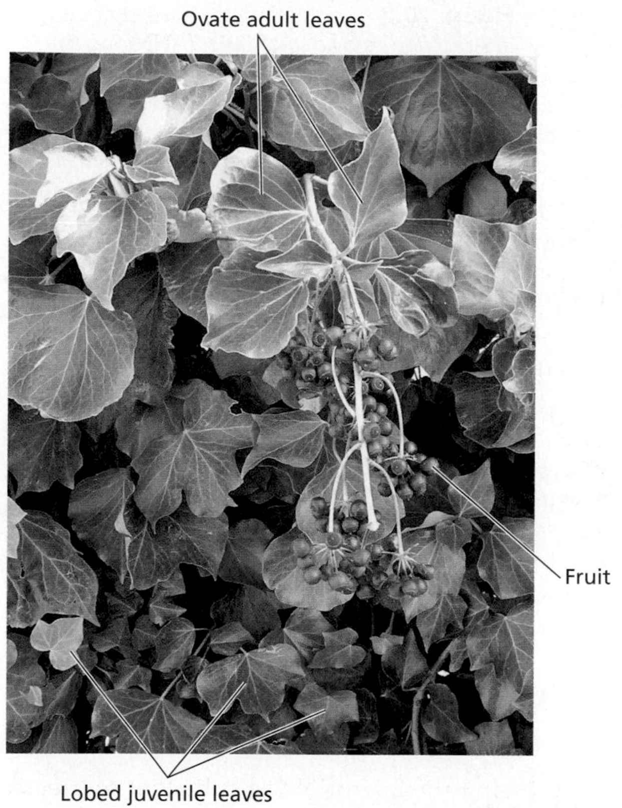

Figure 20.1 Juvenile and adult forms of English ivy (*Hedera helix*). The juvenile form has lobed palmate leaves arranged alternately, a climbing growth habit, and no flowers. The adult form (projecting out to the right) has entire ovate leaves arranged in spirals, an upright growth habit, and flowers that develop into fruits. (Courtesy of L. Rignanese.)

first, are located at the base of the shoot. In rapidly flowering herbaceous species, the juvenile phase may last only a few days, and few juvenile structures are produced. In contrast, woody species have a more prolonged juvenile phase, in some cases lasting 30 to 40 years (**Table 20.1**). In these cases the juvenile structures can account for a significant portion of the mature plant.

Once the meristem has switched to the adult phase, only adult vegetative structures are produced, culminating in flowering. The adult and reproductive phases are therefore located in the upper and peripheral regions of the shoot.

Attainment of a sufficiently large size appears to be more important than the plant's chronological age in determining the transition to the adult phase. Conditions that retard growth, such as mineral deficiencies, low light, water stress, defoliation, and low temperature, tend to prolong the juvenile phase or even cause reversion to juvenility of adult shoots. In contrast, conditions that promote vigorous growth accelerate the transition to the adult phase. When growth is accelerated, exposure to the correct flower-inducing treatment can result in flowering.

TABLE 20.1 Length of juvenile period in some woody plants

Species	Length of juvenile period
Rose (*Rosa* [hybrid tea])	20–30 days
Grape (*Vitis* spp.)	1 year
Apple (*Malus* spp.)	4–8 years
Citrus spp.	5–8 years
English ivy (*Hedera helix*)	5–10 years
Redwood (*Sequoia sempervirens*)	5–15 years
Sycamore maple (*Acer pseudoplatanus*)	15–20 years
English oak (*Quercus robur*)	25–30 years
European beech (*Fagus sylvatica*)	30–40 years

Source: Clark 1983.

Although plant size seems to be the most important factor, it is not always clear which specific component associated with size is critical. In some *Nicotiana* species, it appears that plants must produce a certain number of leaves to transmit a sufficient amount of the floral stimulus to the apex.

Once the adult phase has been attained, it is relatively stable and is maintained during vegetative propagation or grafting. For example, cuttings taken from the basal region of mature plants of English ivy develop into juvenile plants, while those taken from the tip develop into adult plants. When scions were taken from the base of a flowering silver birch (*Betula verrucosa*) and grafted onto seedling rootstocks, there were no flowers on the grafts for the first 2 years. In contrast, grafts taken from the top of the mature tree flowered freely.

The term *juvenility* has different meanings for herbaceous and woody species. Whereas juvenile herbaceous meristems flower readily when grafted onto flowering adult plants (see **WEB TOPIC 20.3**), juvenile woody meristems generally do not. Juvenile woody meristems are thus said to lack the competence to flower (see **WEB TOPIC 20.4**).

Phase changes can be influenced by nutrients, gibberellins, and other signals

The transition at the shoot apex from the juvenile to the adult phase can be affected by transmissible factors from the rest of the plant. In many plants, exposure to low-light conditions prolongs juvenility or causes reversion to juvenility. A major consequence of a low-light regime is a reduction in the supply of carbohydrates to the apex; thus carbohydrate supply, especially sucrose, may play a role in the transition between juvenility and maturity. Carbohydrate supply as a source of energy and raw material

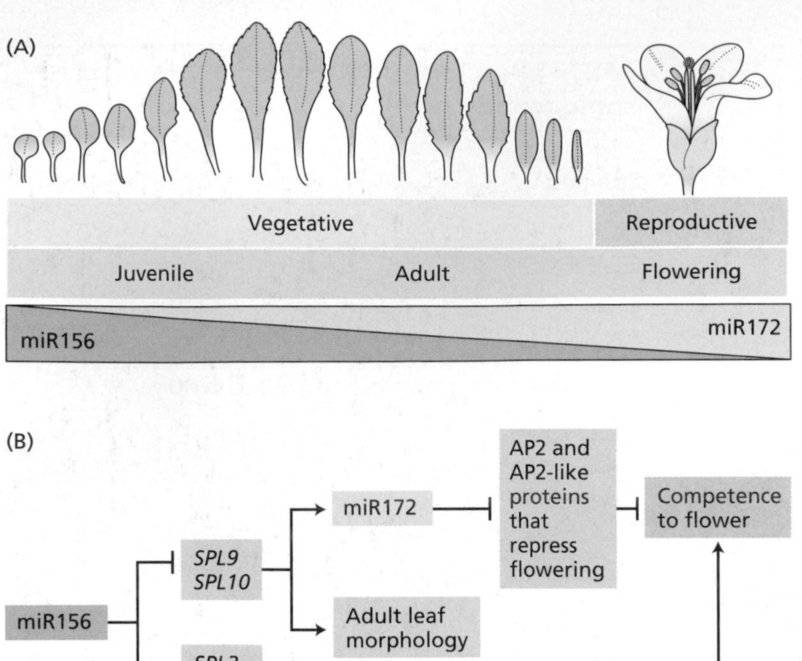

Figure 20.2 Regulation of phase change in Arabidopsis by microRNAs. (A) During the earliest stages of development, the level of miR156 is very high and the level of miR172 is very low, promoting the juvenile vegetative growth phase. Juvenile leaves are small and round, and exhibit trichomes only on their adaxial side. Over time, the level of miR156 declines, and the level of miR172 increases, promoting the transition to the adult vegetative phase. Adult vegetative leaves are larger and more elongated, with abaxial trichomes. (B) The decline in the level of miR156 allows the expression of the *SPL9* and *SPL10* genes, which up-regulate miR172. miR172 down-regulates six AP2-like transcription factors that repress flowering. Release from repression, combined with the up-regulation of the flower-promoting genes *SPL3–5*, makes the plant competent to flower, allowing the transition to flowering. The decline in adult leaf size reflects a gradual shift in the allocation of sugars from the leaves to the developing reproductive structures.

can affect the size of the apex. For example, in the florist's chrysanthemum (*Chrysanthemum morifolium*), flower primordia are not initiated until a minimum apex size has been reached. In Arabidopsis, carbohydrate status in the plant is transmitted by the small signaling molecule trehalose 6-phosphate, a disaccharide. Plants lacking trehalose 6-phosphate are very late flowering, even under inductive conditions, and trehalose 6-phosphate activates flowering pathways in both the leaves and shoot apex.

The apex receives a variety of hormonal and other factors from the rest of the plant in addition to carbohydrates and other nutrients. Experimental evidence shows that the application of gibberellins causes reproductive structures to form in young, juvenile plants of several conifer families. The involvement of endogenous gibberellins in the control of reproduction is also indicated by the fact that other treatments that accelerate cone production in pines (e.g., root removal, water stress, and nitrogen starvation) often also result in a buildup of gibberellins in the plant.

A major class of conserved molecules that control phase transitions in plants is the microRNAs. MicroRNAs are small noncoding RNA molecules that target the mRNA transcripts of other genes through short regions of sequence homology, thus interfering with their function (see Chapter 2). In Arabidopsis and many other plants, including trees, the microRNA miR156 is key for controlling the juvenile-to-adult transition (**Figure 20.2**). Some of the target genes of miR156 promote the transition to flowering. The level of miR156 decreases over time, and once it falls below a threshold the targeted genes are expressed and phase change becomes possible. Overexpressing the microRNA is sufficient to greatly delay phase changes in Arabidopsis and poplar trees.

In addition to miR156, the microRNA miR172 has been implicated in phase transitions in Arabidopsis. miR172 levels increase during development as miR156 levels decline. In contrast to miR156, the abundance of which is controlled by plant age, miR172 expression appears to be under photoperiodic control (discussed later in this chapter). The targets of miR172 include several transcripts that encode transcription factors involved in the repression of flowering. Thus, miR172 promotes the phase change from adult vegetative growth to reproductive growth.

Circadian Rhythms: The Clock Within

Organisms are normally subjected to daily cycles of light and darkness, and both plants and animals often exhibit rhythmic behavior in association with these changes. Examples of such rhythms include leaf and petal movements (day and night positions), stomatal opening and closing, growth and sporulation patterns in fungi (e.g., *Pilobolus* and *Neurospora*), time of day of pupal emergence (the fruit fly *Drosophila*), and activity cycles in rodents, as well as daily changes in the rates of metabolic processes such as photosynthesis and respiration.

When organisms are transferred from daily light–dark cycles to continuous darkness or continuous light, many of these rhythms continue to be expressed, at least for several days. Under such uniform conditions the period of the rhythm is close to 24 h, and consequently the term *circadian rhythm* (from the Latin *circa*, "about," and *diem*, "day") is applied (see Chapter 16). Because they continue in a constant light or dark environment, these circadian rhythms cannot be direct responses to the presence or absence of light but must be based on an internal pacemaker, often called an endogenous oscillator. A molecular model for a plant endogenous oscillator was described in Chapter 16.

The endogenous oscillator is coupled to a variety of physiological processes, such as leaf movement or photosynthesis, and it maintains the rhythm. For this reason the endogenous oscillator can be considered the clock mechanism, and the physiological functions that are being regulated, such as leaf movements or photosynthesis, are sometimes referred to as the hands of the clock.

Circadian rhythms exhibit characteristic features

Circadian rhythms arise from cyclic phenomena that are defined by three parameters:

1. **Period** is the time between comparable points in the repeating cycle. Typically the period is measured as the time between consecutive maxima (peaks) or minima (troughs) (**Figure 20.3A**).
2. **Phase*** is any point in the cycle that is recognizable by its relationship to the rest of the cycle. The most obvious phase points are the peak and trough positions.

*The term *phase* in this context should not be confused with the term *phase change* in meristem development discussed earlier.

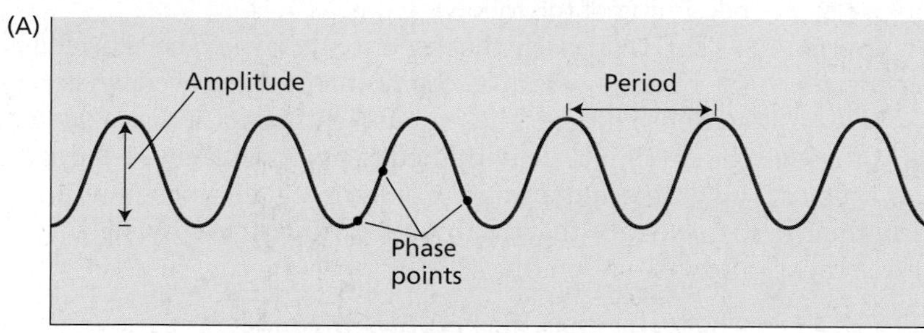

A typical circadian rhythm. The period is the time between comparable points in the repeating cycle; the phase is any point in the repeating cycle recognizable by its relationship with the rest of the cycle; the amplitude is the distance between peak and trough.

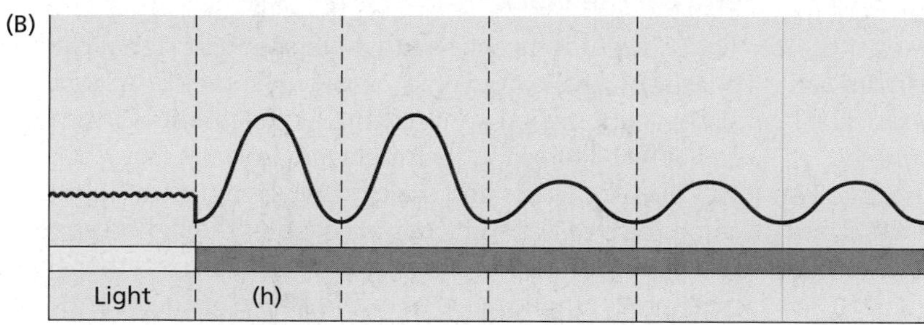

Suspension of a circadian rhythm in continuous bright light and the release or restarting of the rhythm following transfer to darkness.

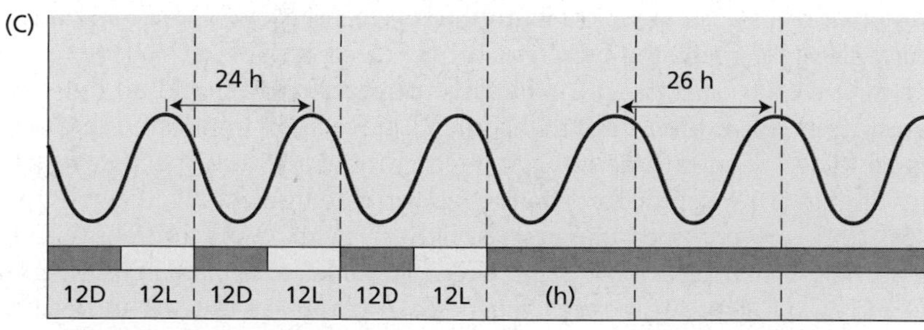

A circadian rhythm entrained to a 24-h light–dark (L–D) cycle and its reversion to the free-running period (26 h in this example) following transfer to continuous darkness.

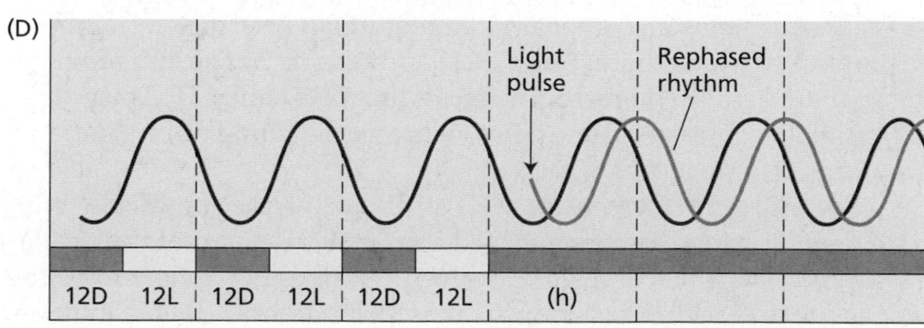

Typical phase-shifting response to a light pulse given shortly after transfer to darkness. The rhythm is rephased (delayed) without its period being changed.

Figure 20.3 Some characteristics of circadian rhythms.

3. **Amplitude** is usually considered to be the distance between peak and trough. The amplitude of a biological rhythm can often vary while the period remains unchanged (as, for example, in **Figure 20.3B**).

In constant light or darkness, rhythms depart from an exact 24-h period. The rhythms then drift in relation to solar time, either gaining or losing time depending on whether the period is shorter or longer than 24 h. Under natural conditions, the endogenous oscillator is **entrained** (synchronized) to a true 24-h period by environmental signals, the most important of which are the light-to-dark transition at dusk and the dark-to-light transition at dawn (**Figure 20.3C**).

Such environmental signals are termed **zeitgebers** (German for "time givers"). When such signals are removed—for example, by transfer to continuous darkness—the rhythm is said to be **free-running**, and it reverts to the circadian period that is characteristic of the particular organism (see Figure 20.3B).

Although the rhythms are generated internally, they normally require an environmental signal, such as exposure to light or a change in temperature, to initiate their expression. In addition, many rhythms damp out (i.e., the amplitude decreases) when the organism is subjected to a constant environment for several cycles. When this occurs, an environmental zeitgeber, such as a transfer from light to dark or a change in temperature, is required to restart the rhythm (see Figure 20.3C). Note that *the clock itself does not damp out; only the coupling between the molecular clock (endogenous oscillator) and the physiological function is affected.*

The circadian clock would be of no value to the organism if it could not keep accurate time under the fluctuating temperatures experienced in natural conditions. Indeed, temperature has little or no effect on the period of the free-running rhythm. The feature that enables the clock to keep time at different temperatures is called **temperature compensation**. Although all of the biochemical steps in the pathway are temperature sensitive, their temperature responses probably cancel each other. For example, changes in the rates of synthesis of intermediates could be compensated for by parallel changes in their rates of degradation. In this way, the steady-state levels of clock regulators would remain constant at different temperatures.

Phase shifting adjusts circadian rhythms to different day–night cycles

In circadian rhythms, physiological responses are coupled to a specific time point of the endogenous oscillator so that the response occurs at a particular time of day. A single oscillator can be coupled to multiple circadian rhythms, which may even be out of phase with each other.

How do such responses remain on time when the daily durations of light and darkness change with the seasons? Investigators typically test the response of the endogenous oscillator by placing the organism in continuous darkness and examining the response to a short pulse of light (usually less than 1 h) given at different phase points in the free-running rhythm. When an organism is entrained to a cycle of 12 h light and 12 h dark and then allowed to free-run in constant light or darkness, the phase of the rhythm that coincides with the light period of the previous entraining cycle is called the **subjective day**, and the phase that coincides with the dark period is called the **subjective night**.

If a light pulse is given during the first few hours of the subjective night, the rhythm is delayed; the organism interprets the light pulse as the end of the previous day (**Figure 20.3D**). In contrast, a light pulse given toward the end of the subjective night advances the phase of the rhythm; now the organism interprets the light pulse as the beginning of the following day.

This is precisely the response that would be expected if the rhythm were able to stay on local time even when the seasons change. These phase-shifting responses enable the rhythm to be entrained to approximately 24-h cycles with different durations of light and darkness, and they demonstrate that the rhythm can adjust to seasonal variations in day length.

Phytochromes and cryptochromes entrain the clock

The molecular mechanism whereby a light signal causes phase shifting is not yet known, but studies in Arabidopsis have identified some of the key elements of the circadian oscillator and its inputs and outputs (see Chapter 16). The low levels and specific wavelengths of light that can induce phase shifting indicate that the light response must be mediated by specific photoreceptors rather than by photosynthetic rate. For example, the red-light entrainment of rhythmic day/night leaf movements in *Samanea*, a semitropical leguminous tree, is a low-fluence response mediated by phytochrome (see Chapter 16).

Arabidopsis has five **phytochromes**, and all but one of them (phytochrome C) have been implicated in clock entrainment. Each phytochrome acts as a specific photoreceptor for red, far-red, or blue light. As well as phytochromes, plants sense light through cryptochromes (CRY), and the CRY1 and CRY2 proteins participate in blue-light entrainment of the clock in plants, as they do in insects and mammals (see Chapter 18). Surprisingly, CRY proteins also appear to be required for normal entrainment by red light. Since these proteins do not absorb red light, this requirement suggests that CRY1 and CRY2 may act as intermediates in phytochrome signaling during entrainment of the clock.

In *Drosophila*, CRY proteins interact physically with clock components and thus constitute part of the oscillator mechanism. However, this does not appear to be the case in Arabidopsis, in which *cry1/cry2* double mutants are impaired in entrainment but otherwise have normal circadian rhythms. In plants it has been shown that photoactivated CRY2 is able to activate flowering in response to blue light by directly up-regulating expression of a key flowering gene, *FLOWERING LOCUS T* (*FT*) (which we will discuss in detail later in this chapter).

Photoperiodism: Monitoring Day Length

As we have seen, the circadian clock enables organisms to repeat particular molecular or biochemical events at specific times of day or night. **Photoperiodism**, or the ability of an organism to detect day length, makes it possible for an event to occur at a particular time of year, thus allowing for a seasonal response. Circadian rhythms and photoperiodism have the common property of responding to cycles of light and darkness.

Precisely at the equator, day length and night length are equal and constant throughout the year. As one moves away from the equator toward the poles, the days become longer in summer and shorter in winter (**Figure 20.4**). Plant species have evolved the ability to detect these seasonal changes in day length, and their specific photoperiodic responses are strongly influenced by the latitude in which they originated.

Photoperiodic phenomena are found in both animals and plants. In the animal kingdom, day length controls such seasonal activities as hibernation, development of summer and winter coats, and reproductive activity. Plant responses controlled by day length are numerous; they include the initiation of flowering, asexual reproduction, the formation of storage organs, and the onset of dormancy.

Plants can be classified according to their photoperiodic responses

Numerous plant species flower during the long days of summer, and for many years plant physiologists believed that the correlation between long days and flowering was a consequence of the accumulation of photosynthetic products synthesized during long days.

This hypothesis was shown to be incorrect by the work of Wightman Garner and Henry Allard, conducted in the 1920s at the U.S. Department of Agriculture laboratories in Beltsville, Maryland. Garner and Allard found that a mutant variety of tobacco, 'Maryland Mammoth', grew profusely to about 5 m in height but failed to flower in the prevailing conditions of summer (**Figure 20.5**). However, the plants flowered in the greenhouse during the winter under natural light conditions.

These results ultimately led Garner and Allard to test the effect of artificially shortened days by covering plants grown during the long days of summer with a light-tight tent from late in the afternoon until the following morning. These artificial short days also caused the plants to flower. Garner and Allard concluded that day length, rather than the accumulation of photosynthate, was the determining factor in flowering. They were able to confirm their hypothesis in many different species and conditions. This work laid the foundations for the extensive subsequent research on photoperiodic responses.

(A)

Hours of daylight: 20, 19, 18, 17, 16, 15, 14, 13, 12, 11, 10, 9, 8, 7, 6

60°, 50°, 40°, 30°, 20°, 10°, 0°

March equinox — June solstice — September equinox

J F M A M J J A S O N D

Month of year

(B)

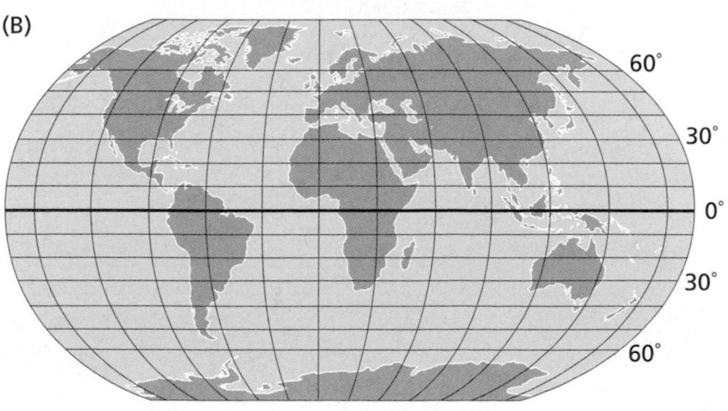

Figure 20.4 (A) Effect of latitude on day length at different times of the year in the northern hemisphere. Day length was measured on day 20 of each month. (B) Global map showing longitudes and latitudes.

Figure 20.5 'Maryland Mammoth' mutant of tobacco (right) compared with wild-type tobacco (left). Both plants were grown during summer in the greenhouse. (University of Wisconsin graduate students used for scale.) (Courtesy of R. Amasino.)

Although many other aspects of plants' development may also be affected by day length, flowering is the response that has been studied the most. Flowering species tend to fall into one of two main photoperiodic response categories: short-day plants and long-day plants.

- **Short-day plants (SDPs)** flower only in short days (*qualitative* SDPs), or their flowering is accelerated by short days (*quantitative* SDPs).
- **Long-day plants (LDPs)** flower only in long days (*qualitative* LDPs), or their flowering is accelerated by long days (*quantitative* LDPs).

The essential distinction between long-day and short-day plants is that flowering in LDPs is promoted only when the day length *exceeds* a certain duration, called the **critical day length**, in every 24-h cycle, whereas promotion of flowering in SDPs requires a day length that is *less than* the critical day length. The absolute value of the critical day length varies widely among species, and only when flowering is examined for a range of day lengths can the correct photoperiodic classification be established (**Figure 20.6**).

LDPs can effectively measure the lengthening days of spring or early summer and delay flowering until the critical day length is reached. Many varieties of wheat (*Triticum aestivum*) behave in this way. SDPs often flower in the fall when the days shorten below the critical day length, as in many varieties of *Chrysanthemum morifolium*. However, day length alone is an ambiguous signal, because it cannot distinguish between spring and fall.

Plants exhibit several adaptations for avoiding the ambiguity of the day-length signal. One is the presence of a juvenile phase that prevents the plant from responding to day length during the spring. Another mechanism for avoiding the ambiguity of day length is the coupling of a temperature requirement to a photoperiodic response. Certain plant species, such as winter wheat, do not respond to photoperiod until after a cold period (vernalization or overwintering) has occurred. (We will discuss vernalization later in this chapter.)

Other plants avoid seasonal ambiguity by distinguishing between *shortening* and *lengthening* days. Such "dual–day length plants" fall into two categories:

- **Long–short-day plants (LSDPs)** flower only after a sequence of long days followed by short days. LSDPs, such as *Bryophyllum*, *Kalanchoe*, and night-blooming jasmine (*Cestrum nocturnum*), flower in the late summer and fall, when the days are shortening.

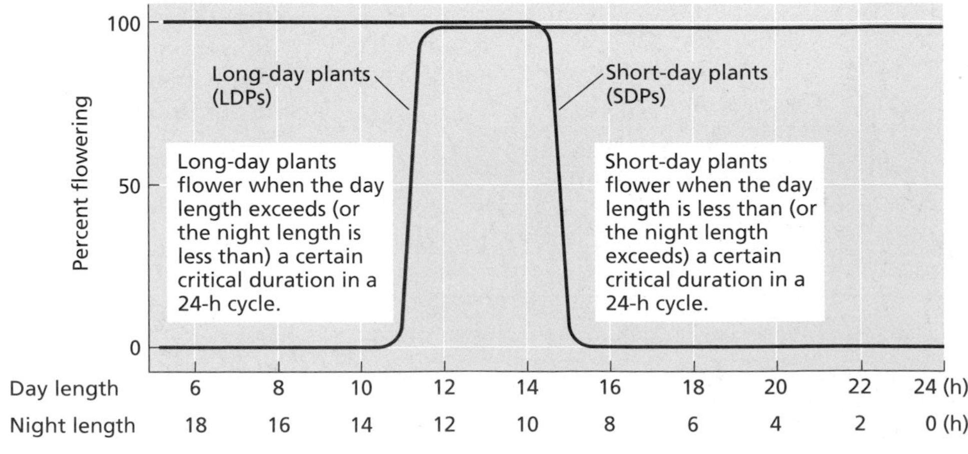

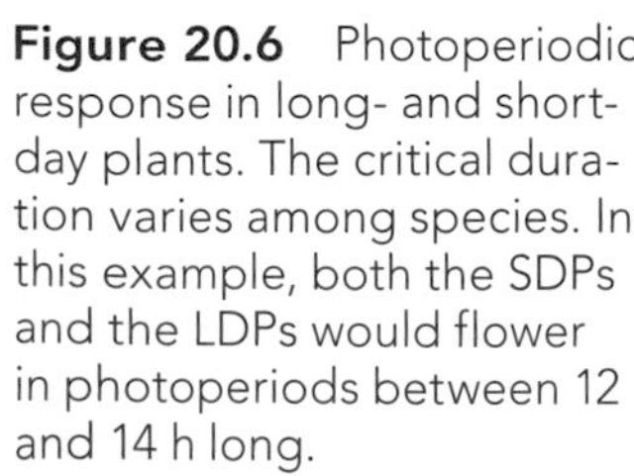

Figure 20.6 Photoperiodic response in long- and short-day plants. The critical duration varies among species. In this example, both the SDPs and the LDPs would flower in photoperiods between 12 and 14 h long.

- **Short–long-day plants (SLDPs)** flower only after a sequence of short days followed by long days. SLDPs, such as white clover (*Trifolium repens*), Canterbury bells (*Campanula medium*), and echeveria (*Echeveria harmsii*), flower in the early spring in response to lengthening days.

Finally, species that flower under any photoperiodic condition are referred to as day-neutral plants. **Day-neutral plants (DNPs)** are insensitive to day length. Flowering in DNPs is typically under autonomous regulation—that is, internal developmental control. Some day-neutral species, such as kidney bean (*Phaseolus vulgaris*), evolved near the equator where the day length is constant throughout the year. Many desert annuals, such as desert paintbrush (*Castilleja chromosa*) and desert sand verbena (*Abronia villosa*), evolved to germinate, grow, and flower quickly whenever sufficient water is available. These are also DNPs.

The leaf is the site of perception of the photoperiodic signal

The photoperiodic stimulus in both LDPs and SDPs is perceived by the leaves. For example, treatment of a single leaf of the SDP *Xanthium* with short photoperiods is sufficient to cause the formation of flowers, even when the rest of the plant is exposed to long days. Thus, in response to photoperiod the leaf transmits a signal that regulates the transition to flowering at the shoot apex. The photoperiod-regulated processes that occur in the leaves resulting in the transmission of a floral stimulus to the shoot apex are referred to collectively as **photoperiodic induction**.

Photoperiodic induction can take place in a leaf that has been separated from the plant. For example, in the SDP *Perilla crispa* (a member of the mint family), an excised leaf exposed to short days can cause flowering when subsequently grafted to a noninduced plant maintained in long days. This result indicates that photoperiodic induction depends on events that take place exclusively in the leaf.

Plants monitor day length by measuring the length of the night

Under natural conditions, day and night lengths configure a 24-h cycle of light and darkness. In principle, a plant could perceive a critical day length by measuring the duration of either light or darkness. Much experimental work in the early studies of photoperiodism was devoted to establishing which part of the light–dark cycle is the controlling factor in flowering. Results showed that flowering of SDPs is determined primarily by the duration of darkness (**Figure 20.7A**). It was possible to induce flowering in SDPs with light periods longer than the critical value, provided that these were followed by sufficiently long nights (**Figure 20.7B**). Similarly, SDPs did not flower when short days were followed by short nights.

More detailed experiments demonstrated that photoperiodic timekeeping in SDPs is a matter of measuring the duration of darkness. For example, flowering occurred only when the dark period exceeded 8.5 h in cocklebur (*Xanthium strumarium*) or 10 h in soybean (*Glycine max*). The duration of darkness was also shown to be important in LDPs (see Figure 20.7). These plants were found to flower in short days, provided that the accompanying night length was also short; however, a regime of long days followed by long nights was ineffective.

Night breaks can cancel the effect of the dark period

A feature that underscores the importance of the dark period is that it can be made ineffective by interruption with a short exposure to light, called a **night break** (see Figure 20.7A). In contrast, interrupting a long day with a brief dark period does not cancel the effect of the long day (see Figure 20.7B). Night-break treatments of only a few minutes are effective in *preventing* flowering in many SDPs, including *Xanthium* and *Pharbitis*, but much longer exposures are often required to *promote* flowering in LDPs.

In addition, the effect of a night break varies greatly according to the time when it is given. For both LDPs and SDPs, a night break was found to be most effective when given near the middle of a dark period of 16 h (**Figure 20.8**).

The discovery of the night-break effect, and its time dependence, had several important consequences. It established the central role of the dark period and provided a valuable probe for studying photoperiodic timekeeping. Because only small amounts of light are needed, it became possible to study the action and identity of the photoreceptor without the interfering effects of photosynthesis and other nonphotoperiodic phenomena. This discovery has also led to the development of commercial methods for regulating the time of flowering in horticultural species, such as *Kalanchoë*, chrysanthemum, and poinsettia (*Euphorbia pulcherrima*).

Photoperiodic timekeeping during the night depends on a circadian clock

The decisive effect of night length on flowering indicates that measuring the passage of time in darkness is central to photoperiodic timekeeping. Most of the available evidence favors a mechanism based on a circadian rhythm. According to the **clock hypothesis**, photoperiodic timekeeping depends on an endogenous circadian oscillator of the type discussed earlier in the chapter (see also Chapter 16). The central oscillator is coupled to various physiological processes that involve gene expression, including flowering in photoperiodic species.

Measurements of the effect of a night break on flowering can be used to investigate the role of circadian rhythms in photoperiodic timekeeping. For example,

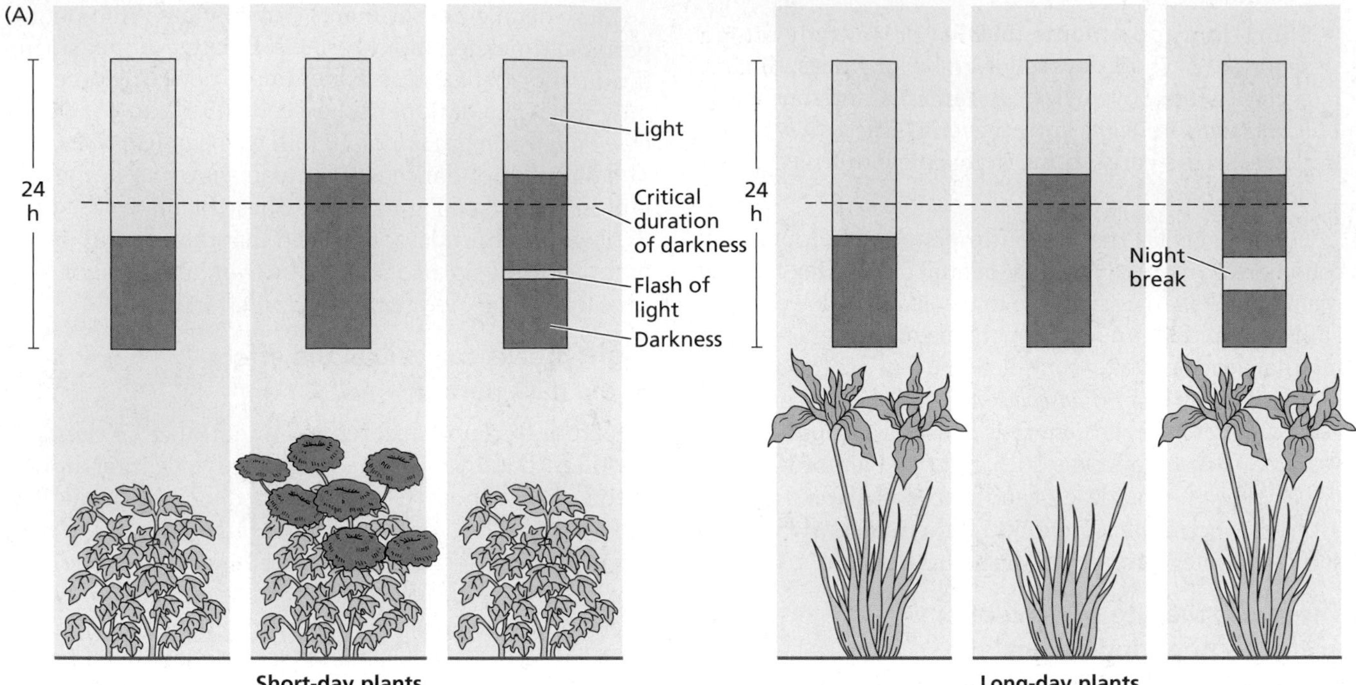

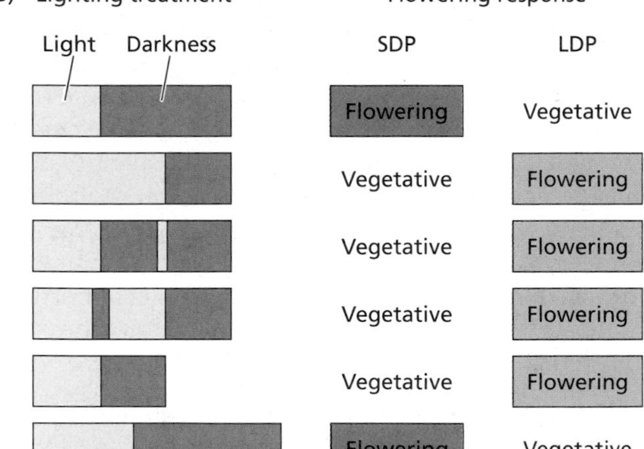

Figure 20.7 Photoperiodic regulation of flowering. (A) Effects on SDPs and LDPs. (B) Effects of the duration of the dark period on flowering. Treating SDPs and LDPs with different photoperiods clearly shows that the critical variable is the length of the dark period.

when soybean plants, which are SDPs, are transferred from an 8-h light period to an extended 64-h dark period, the flowering response to night breaks shows a circadian rhythm (**Figure 20.9**).

This type of experiment provides strong support for the clock hypothesis. If this SDP were simply measuring the length of night by the accumulation of a particular intermediate in the dark, any dark period greater than the critical night length should cause flowering. Yet long dark periods are not inductive for flowering if the light break is given at a time that does not properly coincide with a certain phase of the endogenous circadian oscillator. This finding demonstrates that flowering in SDPs requires both a dark period of sufficient duration and a dawn signal at an appropriate time in the circadian cycle (see Figure 20.3).

Further evidence for the role of a circadian oscillator in photoperiod measurement is the observation that the photoperiodic response can be phase-shifted by light treatments (see **WEB TOPIC 20.5**).

The coincidence model is based on oscillating light sensitivity

How does an oscillation with a 24-h period measure a critical duration of darkness of, say, 8 to 9 h, as in the SDP *Xanthium*? In 1936 Erwin Bünning proposed that the control of flowering by photoperiodism is achieved by an oscillation of phases with different sensitivities to light. This proposal has evolved into the **coincidence model**, in which the circadian oscillator controls the timing of light-sensitive and light-insensitive phases.

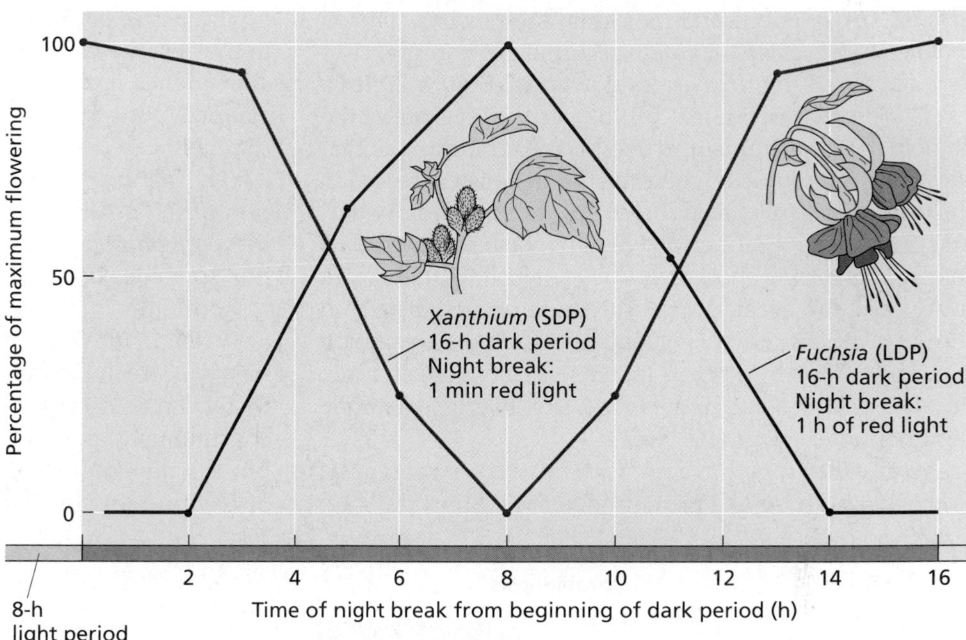

Figure 20.8 The time at which a night break is given determines the flowering response. When given during a long dark period, a night break promotes flowering in LDPs and inhibits flowering in SDPs. In both cases, the greatest effect on flowering occurs when the night break is given near the middle of the 16-h dark period. The LDP *Fuchsia* was given a 1-h exposure to red light in a 16-h dark period. *Xanthium* was exposed to red light for 1 min in a 16-h dark period. (Data for *Fuchsia* from Vince-Prue 1975; data for *Xanthium* from Salisbury 1963 and Papenfuss and Salisbury 1967.)

The ability of light either to promote or to inhibit flowering depends on the phase in which the light is given. When a light signal is administered during the light-sensitive phase of the rhythm, the effect is either to *promote* flowering in LDPs or to *prevent* flowering in SDPs. As shown in Figure 20.9, the phases of sensitivity and insensitivity to light continue to oscillate in darkness. Flowering in SDPs is induced only when exposure to light from a night break or from dawn occurs after completion of the light-sensitive phase of the rhythm.

If a similar experiment is performed with an LDP, flowering is induced only when the night break occurs *during* the light-sensitive phase of the rhythm. In other words, *flowering in both SDPs and LDPs is induced when the light exposure is coincident with the appropriate phase of the rhythm*. This continued oscillation of sensitive and insensitive phases in the absence of dawn and dusk light signals is characteristic of a variety of processes controlled by the circadian oscillator.

The coincidence of *CONSTANS* expression and light promotes flowering in LDPs

According to the coincidence model, plant flowering responses are sensitive to light only at certain times of the day–night cycle. A key component of a regulatory pathway that promotes flowering of Arabidopsis in long days is a gene called ***CONSTANS*** **(*CO*)**, which encodes a zinc finger protein that regulates the transcription of other genes. *CO* was first identified in an Arabidopsis mutant, *co,* that was incapable of a photoperiodic flowering response. The expression of *CO* is controlled by the circadian clock, with the peak of activity occurring 12 h after dawn (**Fig-**

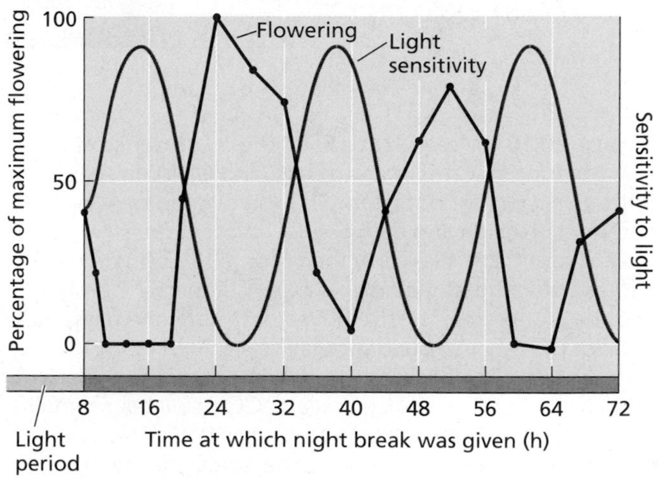

Figure 20.9 Rhythmic flowering in response to night breaks. In this experiment, the SDP soybean (*Glycine max*) received cycles of an 8-h light period followed by a 64-h dark period. A 4-h night break was given at various times during the long inductive dark period. The flowering response, plotted as the percentage of the maximum, was then plotted for each night break given. Note that a night break given at 26 h induced maximum flowering, while no flowering was obtained when the night break was given at 40 h. Moreover, this experiment demonstrates that the sensitivity to the effect of the night break shows a circadian rhythm. These data support a model in which flowering in SDPs is induced only when dawn (or a night break) occurs after the completion of the light-sensitive phase. In LDPs the light break must coincide with the light-sensitive phase for flowering to occur. (Data from Coulter and Hamner 1964.)

ure 20.10A). Genetic and molecular studies have shown that in Arabidopsis CO protein accumulates in response to long days, and this accelerates flowering (**Figure 20.10B**).

As indicated in Figure 20.10B, a critical feature of the coincidence mechanism in the LDP Arabidopsis is that flowering is promoted when the *CO* gene is expressed in the leaf (the site of perception of the photoperiodic stimulus) during the light period. The increase in *CO* mRNA that occurs during short days does not lead to an increase in CO protein, because *CO* expression occurs entirely in the dark. In contrast, during long days *CO* gene expression is accompanied by a sharp increase in CO protein level, because at least some of the expression overlaps with the light period (see Figure 20.10B).

As a result, long days are inductive for Arabidopsis flowering because CO protein increases. Short days are noninductive because CO protein level does not increase in the absence of light. Thus, an important feature of the coincidence model is that there must be overlap (coincidence) between *CO* mRNA synthesis and daylight so that light can permit active CO protein to accumulate to a level that promotes flowering. The circadian oscillation of *CO* mRNA provides an explanation for the link between photoperiod perception and the circadian clock. But how does daylight bring about the accumulation of CO protein?

A clue to the function of light was provided by experiments in which *CO* was expressed from a constitutive promoter. Under these conditions, *CO* mRNA was expressed continuously, and its level remained constant throughout the day–night cycle. Nevertheless, the abundance of CO protein continued to cycle, suggesting that CO

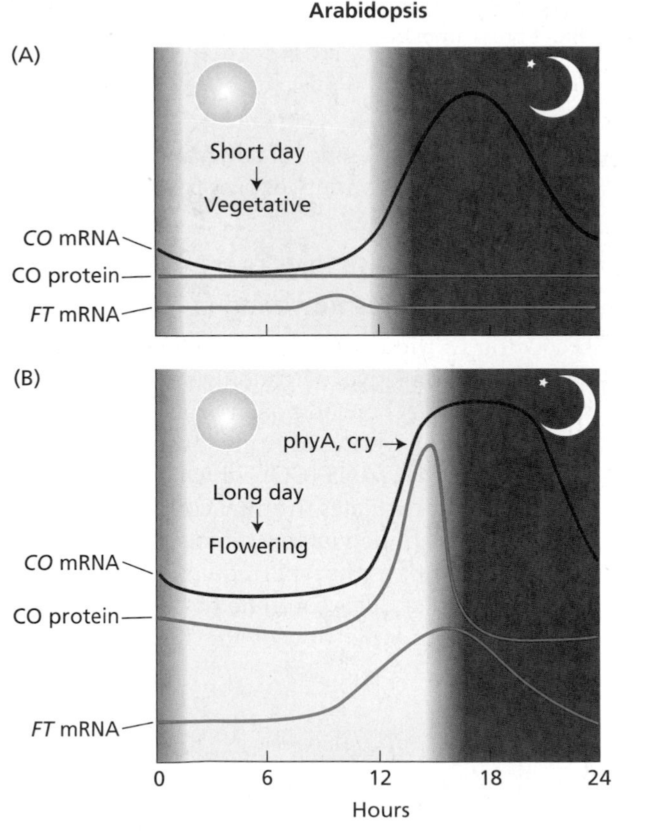

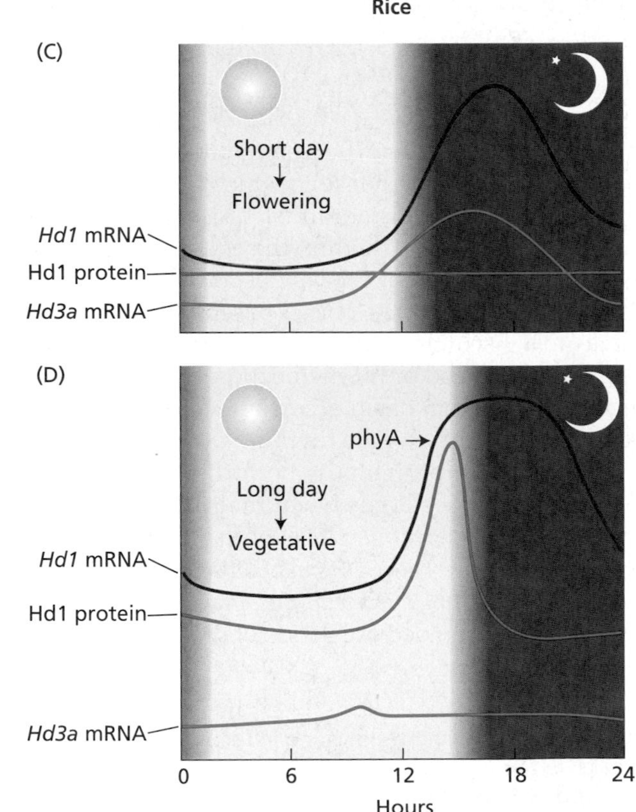

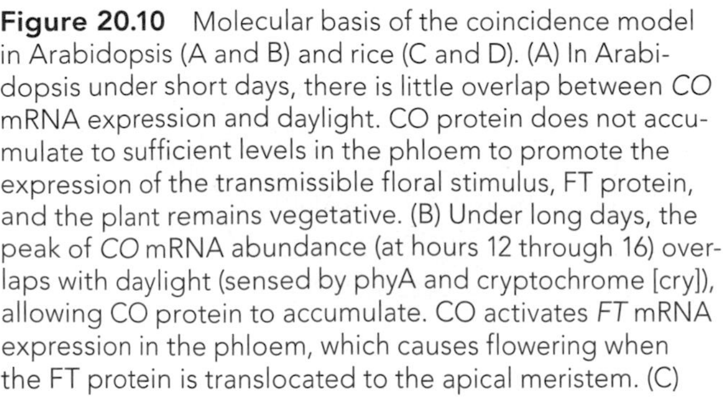

Figure 20.10 Molecular basis of the coincidence model in Arabidopsis (A and B) and rice (C and D). (A) In Arabidopsis under short days, there is little overlap between *CO* mRNA expression and daylight. CO protein does not accumulate to sufficient levels in the phloem to promote the expression of the transmissible floral stimulus, FT protein, and the plant remains vegetative. (B) Under long days, the peak of *CO* mRNA abundance (at hours 12 through 16) overlaps with daylight (sensed by phyA and cryptochrome [cry]), allowing CO protein to accumulate. CO activates *FT* mRNA expression in the phloem, which causes flowering when the FT protein is translocated to the apical meristem. (C) In rice under short days, the lack of coincidence between *Hd1* mRNA expression and daylight prevents the accumulation of the Hd1 protein, which acts as a repressor of the gene encoding the rice transmissible floral stimulus and FT relative, Hd3a. In the absence of the Hd1 protein repressor, *Hd3a* mRNA is expressed and the protein it encodes is translocated to the apical meristem where it causes flowering. (D) Under long days (sensed by phytochrome), the peak of *Hd1* mRNA expression overlaps with the day, allowing the accumulation of the Hd1 repressor protein. As a result, *Hd3a* mRNA is not expressed, and the plant remains vegetative. (After Hayama and Coupland 2004.)

protein abundance is regulated by a posttranscriptional mechanism.

The posttranscriptional mechanism is based in part on differences in the rates of CO degradation in the light versus the dark. During the dark, CO is tagged with ubiquitin and rapidly degraded by the 26S proteasome (see Chapter 2). Light appears to enhance the *stability* of the CO protein, allowing it to accumulate during the day. This explains why *CO* promotes flowering only when its mRNA expression coincides with the light period. In the dark, CO protein does not accumulate, because it is rapidly degraded.

However, the situation is more complicated than a simple light–dark switch regulating CO turnover. The effect of light on CO stability depends on the photoreceptor involved. Different photoreceptors not only contribute to setting the phase of the circadian rhythm, but more directly, they also affect CO protein accumulation and flowering. In the morning, phyB signaling appears to enhance CO degradation, whereas in the evening (when CO protein accumulates in long days), cryptochromes and phyA antagonize this degradation and allow the CO protein to build up (see Figure 20.10; WEB TOPIC 20.6).

How does the CO protein stimulate flowering in long day plants? CO, a transcriptional regulator, promotes flowering by stimulating the expression of a key floral signal, *FLOWERING LOCUS T* (*FT*). As we will describe later in the chapter, there is now evidence that FT protein is the phloem-mobile signal that stimulates flower evocation in the meristem. A similar pathway is used to promote flowering in SDPs, as we'll discuss next.

SDPs use a coincidence mechanism to inhibit flowering in long days

Studies of flowering in the SDP rice have shown that the basic coincidence mechanism for photoperiod-sensing is conserved in rice and Arabidopsis. In the long history of rice cultivation, breeders have identified variant alleles of several genes that modify flowering behavior. The rice genes ***Heading-date1*** (***Hd1***) and ***Heading-date3a*** (***Hd3a***) encode proteins homologous to Arabidopsis CO and FT, respectively. In transgenic plants, overexpression of *FT* in Arabidopsis, and of *Hd3a* in rice, results in rapid flowering regardless of photoperiod, demonstrating that both *FT* and *Hd3a* are strong promoters of flowering. Moreover, expression of both the native *FT* and *Hd3a* genes is substantially elevated during inductive photoperiods (long days in Arabidopsis and short days in rice) (**Figure 20.10C**). In addition, rice *Hd1* and Arabidopsis *CO* exhibit similar patterns of circadian mRNA accumulation.

The difference between rice and Arabidopsis is that in the SDP rice, Hd1 acts as an *inhibitor* of *Hd3a* expression. That is, in rice the coincidence of *Hd1* expression and light signaling through phytochrome suppresses flowering by inhibiting the expression of *Hd3a* (**Figure 20.10D**). In contrast, CO *promotes* the expression of its downstream gene, *FT*, in the LDP Arabidopsis. Flowering in the SDP rice thus occurs only when *Hd1* is expressed exclusively in the dark. Remarkably, the different responses to photoperiod of SDPs versus LDPs are due in part to the opposite effects of this one component, CO/Hd1, of the photoperiodic sensing system.

However, it is important to note that photoperiodism is highly complex, and other regulatory mechanisms that fine-tune the responses of SDPs and LDPs to changing day length are certain to be present.

Phytochrome is the primary photoreceptor in photoperiodism

Night-break experiments are well suited for studying the nature of the photoreceptors (see Chapter 16) involved in the reception of light signals during the photoperiodic response. The inhibition of flowering in SDPs by night breaks was one of the first physiological processes shown to be under the control of phytochrome (**Figure 20.11**).

In many SDPs, a night break becomes effective only when the supplied dose of light is sufficient to saturate the photoconversion of **Pr** (phytochrome that absorbs red light) to **Pfr** (phytochrome that absorbs far-red light) (see Chapter 16). A subsequent exposure to far-red light, which photoconverts the pigment back to the physiologically inactive Pr form, restores the flowering response.

Action spectra for the inhibition and restoration of the flowering response in SDPs are shown in **Figure 20.12**. A peak at 660 nm, the absorption maximum of Pr, is obtained when dark-grown *Pharbitis* seedlings are used to avoid interference from chlorophyll. In contrast, the spectra for *Xanthium* provide an example of the response in green plants, in which the presence of chlorophyll can cause some discrepancy between the action spectrum and the absorption spectrum of Pr. These action spectra plus the red/far-red reversibility of the night break responses confirm the role of phytochrome as the photoreceptor that is involved in photoperiod measurement in SDPs.

Another demonstration of the critical role of phytochrome in photoperiodism in SDPs comes from genetic analyses. In rice, the gene *PHOTOPERIOD SENSITIVITY5* (*Se5*) encodes a protein similar to Arabidopsis HY1. Se5 and HY1 are enzymes that catalyze a step in the biosynthesis of phytochrome chromophore. Mutations in *Se5* cause rice to flower extremely rapidly regardless of the day length.

Night break experiments with LDPs have also implicated phytochrome. Thus, in some LDPs a night break of red light promotes flowering, and a subsequent exposure to far-red light prevents this response (see Figure 20.11).

A circadian rhythm in the promotion of flowering by far-red light has been observed in the LDPs barley (*Hordeum vulgare*), darnel ryegrass (*Lolium temulentum*), and

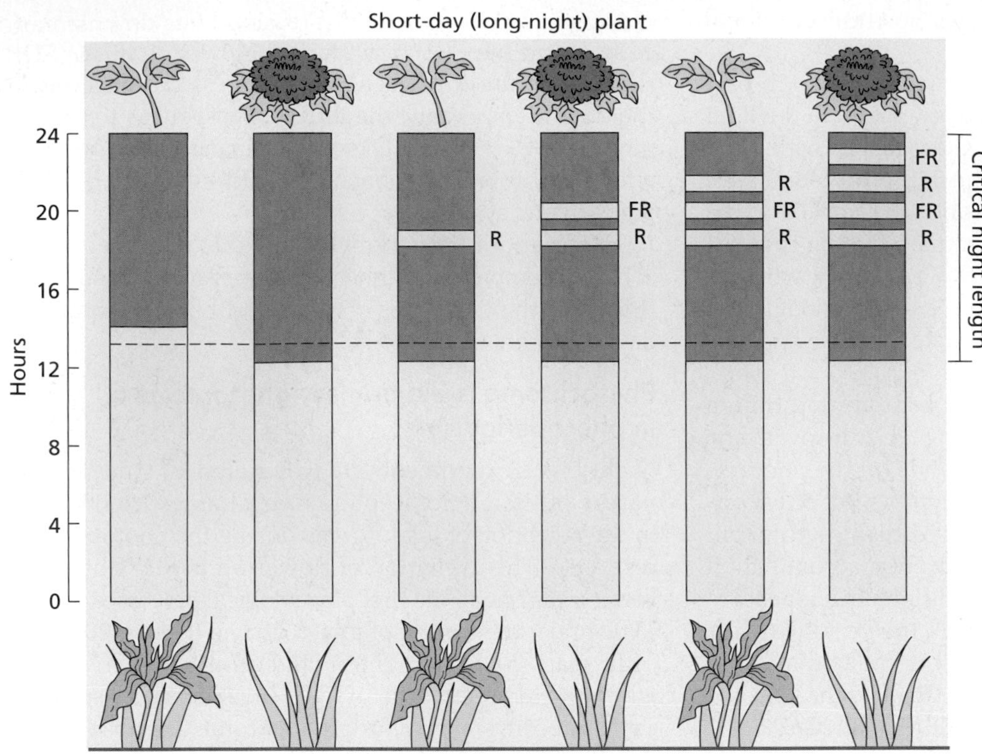

Figure 20.11 Phytochrome control of flowering by red (R) and far-red (FR) light. A flash of red light during the dark period induces flowering in an LDP, and the effect is reversed by a flash of far-red light. This response indicates the involvement of phytochrome. In SDPs, a flash of red light prevents flowering, and the effect is reversed by a flash of far-red light.

Arabidopsis (**Figure 20.13**). The response is proportional to the irradiance and duration of far-red light and is therefore a high-irradiance response (HIR; see Chapter 16). As in other HIRs, phyA is the phytochrome that mediates the response to far-red light. Consistent with a role of phyA in the flowering of LDPs, mutations in the *PHYA* gene delay flowering in Arabidopsis. However, in some LDPs the role of phytochrome is more complex than in SDPs, because a blue-light photoreceptor also participates in the response.

A blue-light photoreceptor regulates flowering in some LDPs

In some LDPs, such as Arabidopsis, blue light can promote flowering, suggesting the possible participation of a blue-light photoreceptor in the control of flowering. As we discussed in Chapter 18, the cryptochromes, encoded by the *CRY1* and *CRY2* genes, are blue-light photoreceptors that control seedling growth in Arabidopsis.

As noted earlier, the CRY protein has also been implicated in the entrainment of the circadian oscillator. The role of blue light in flowering and its relationship to circadian rhythms have been investigated by use of the luciferase reporter gene construct mentioned in **WEB TOPIC**

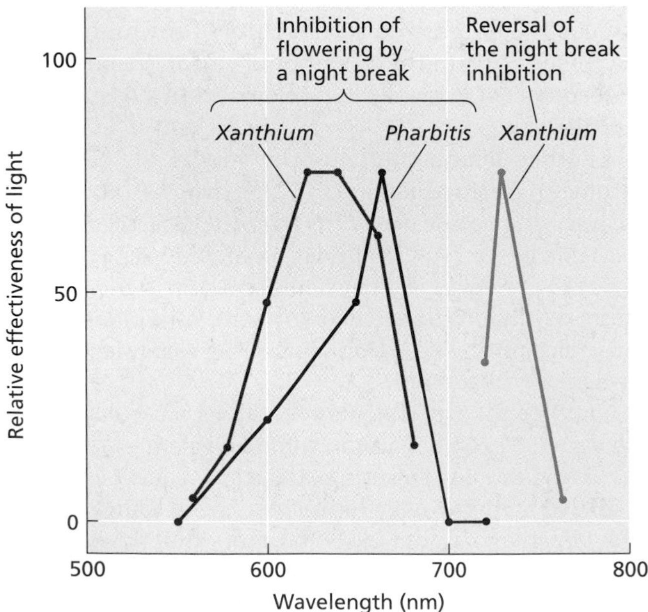

Figure 20.12 Action spectra for the control of flowering by night breaks implicate phytochrome. Flowering in SDPs is inhibited by a short light treatment (night break) given in an otherwise inductive period. In the SDP *Xanthium strumarium*, red-light night breaks of 620 to 640 nm are the most effective. Reversal of the red-light effect is maximal at 725 nm. In the dark-grown SDP *Pharbitis nil*, which is devoid of chlorophyll and its interference with light absorption, night breaks of 660 nm are the most effective. This 660 nm maximum coincides with the absorption maximum of phytochrome. (Data for *Xanthium* from Hendricks and Siegelman 1967; data for *Pharbitis* from Saji et al. 1983.)

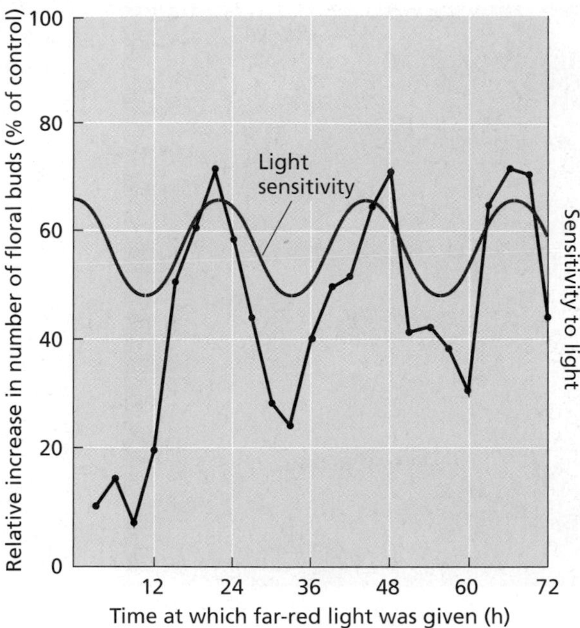

Figure 20.13 Effect of far-red light on floral induction in Arabidopsis. At the indicated times during a continuous 72-h daylight period, 4 h of far-red light were added. Data points in the graph are plotted at the centers of the 6-h treatments. The data show a circadian rhythm of sensitivity to the far-red promotion of flowering (red line). This supports a model in which flowering in LDPs is promoted when the light treatment (in this case far-red light) coincides with the peak of light sensitivity. (After Deitzer 1984.)

20.7. In continuous white light, the cyclic luminescence has a period of 24.7 h, but in constant darkness the period lengthens to between 30 and 36 h. Either red or blue light, given individually, shortens the period to 25 h.

To distinguish between the effects of phytochrome and a blue-light photoreceptor, researchers transformed phytochrome-deficient *hy1* mutants, which are defective in chromophore synthesis and are therefore deficient in *all* phytochromes, with the luciferase construct to determine the effect of the mutation on the period length. Under continuous white light, the *hy1* plants had a period similar to that of the wild type, indicating that little or no phytochrome is required for white light to affect the period. Furthermore, under continuous red light, which would be perceived only by phyB, the period of *hy1* was significantly lengthened (i.e., it became more like constant darkness), whereas the period was not lengthened by continuous blue light. These results indicate that both phytochrome and a blue-light photoreceptor are involved in period control.

The role of blue light in regulating both circadian rhythmicity and flowering is also supported by studies with an Arabidopsis flowering-time mutant, *elf3* (*e*arly *f*lowering *3*) (see **WEB TOPICS 20.7, 20.8**). Confirmation that a blue-light photoreceptor is involved in sensing inductive photoperiods in Arabidopsis was provided by experiments demonstrating that mutations in one of the cryptochrome genes, *CRY2* (see Chapter 18), caused a delay in flowering and an inability to perceive inductive photoperiods.

In contrast, plants carrying a gain-of-function allele of *CRY2* flowered much earlier than the wild type. In addition, the *cry1/cry2* double mutants flowered slightly later than *cry2* in long days, indicating some functional redundancy of CRY1 and CRY2 in promoting flowering time in Arabidopsis.

In addition to their role in entraining the circadian clock, it is likely that the cryptochromes, like phyA, also regulate flowering directly by stabilizing the CO protein, allowing it to accumulate under long day conditions. As noted above, the CO protein acts as a promoter of flowering in LDPs.

Vernalization: Promoting Flowering with Cold

Vernalization is the process whereby repression of flowering is alleviated by a cold treatment given to a hydrated seed (i.e., a seed that has imbibed water) or to a growing plant (dry seeds do not respond to the cold treatment because vernalization is an active metabolic process). Without the cold treatment, plants that require vernalization show delayed flowering or remain vegetative, and they are not competent to respond to floral signals such as inductive photoperiods. In many cases these plants grow as rosettes with no elongation of the stem (**Figure 20.14**).

In this section we will examine some of the characteristics of the cold requirement for flowering, including the range and duration of the inductive temperatures, the sites of perception, the relationship to photoperiodism, and a possible molecular mechanism.

Vernalization results in competence to flower at the shoot apical meristem

Plants differ considerably in the age at which they become sensitive to vernalization. Winter annuals, such as the winter forms of cereals (which are sown in the fall and flower in the following summer), respond to low temperature very early in their life cycle. In fact, many winter annuals can be vernalized before germination (i.e., radicle emergence from the seed) if the seeds have imbibed water and become metabolically active. Other plants, including most biennials (which grow as rosettes during the first season after sowing and flower in the following summer), must reach a minimal size before they become sensitive to low temperature for vernalization.

The effective temperature range for vernalization is from just below freezing to about 10°C, with a broad optimum usually between about 1 and 7°C. The effect of cold increases with the duration of the cold treatment until the

Figure 20.14 Vernalization induces flowering in the winter-annual types of *Arabidopsis thaliana*. The plant on the left is a winter-annual type that has not been exposed to cold. The plant on the right is a genetically identical winter-annual type that was exposed to 40 days of temperatures slightly above freezing (4°C) as a seedling. It flowered 3 weeks after the end of the cold treatment with about nine leaves on the primary stem. (Courtesy of Colleen Bizzell.)

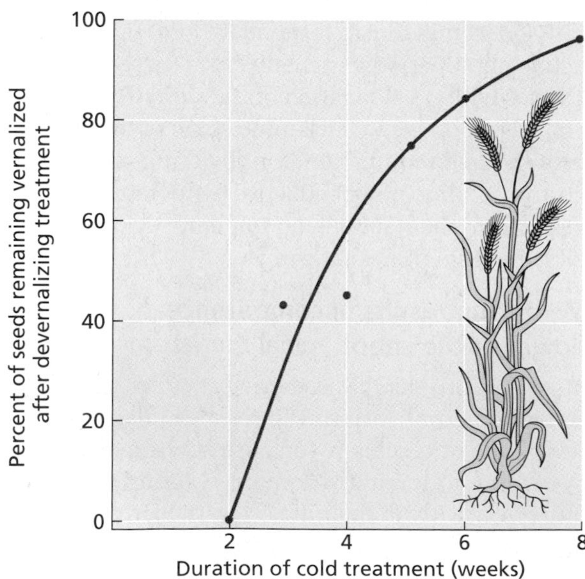

Figure 20.15 Duration of exposure to low temperature increases the stability of the vernalization effect. The longer that winter rye (*Secale cereale*) is exposed to a cold treatment, the greater the number of plants that remain vernalized when the cold treatment is followed by a devernalizing treatment. In this experiment, seeds of rye that had imbibed water were exposed to 5°C for different lengths of time, then immediately given a devernalizing treatment of 3 days at 35°C. (Data from Purvis and Gregory 1952.)

response is saturated. The response usually requires several weeks of exposure to low temperature, but the precise duration varies widely with species and variety.

Vernalization can be lost as a result of exposure to devernalizing conditions, such as high temperature (**Figure 20.15**), but the longer the exposure to low temperature, the more permanent the vernalization effect.

Vernalization appears to take place primarily in the shoot apical meristem. Localized cooling causes flowering when only the stem apex is chilled, and this effect appears to be largely independent of the temperature experienced by the rest of the plant. Excised shoot tips have been successfully vernalized, and where seed vernalization is possible, fragments of embryos consisting essentially of the shoot tip are sensitive to low temperature.

In developmental terms, vernalization results in the acquisition of competence of the meristem to undergo the floral transition. Yet, as discussed earlier in the chapter, competence to flower does not guarantee that flowering will occur. A vernalization requirement is often linked to a requirement for a particular photoperiod. The most common combination is a requirement for cold treatment *followed* by a requirement for long days—a combination that leads to flowering in early summer at high latitudes (see WEB TOPIC 20.9).

Vernalization can involve epigenetic changes in gene expression

For vernalization to occur, active metabolism is required during the cold treatment. Sources of energy (sugars) and oxygen are required, and temperatures below freezing at which metabolic activity is suppressed are not effective for vernalization. Furthermore, cell division and DNA replication also appear to be required. In some plant species, vernalization causes a stable change in the competence of the meristem to form an inflorescence.

One model for how vernalization stably affects competence is that there are changes in the pattern of gene expression in the meristem after cold treatment that persist into the spring and throughout the remainder of the life cycle. Stable changes in gene expression that do not involve alterations in the DNA sequence and which can be passed on to descendant cells through mitosis or meiosis are known as epigenetic changes. As such, epigenetic changes in gene expression are stable even after the signal (in this case cold) that induced the change is no longer present. Epigenetic changes of gene expression occur in many organisms, from yeast to mammals, and often require cell division and DNA replication, as is the case for vernalization.

The involvement of epigenetic regulation of a specific target gene in the vernalization process has been confirmed in the LDP Arabidopsis. In winter-annual types of Arabidopsis that require both vernalization and long days for flowering to be accelerated, a gene that acts as a repressor of flowering has been identified: *FLOWERING LOCUS C* (*FLC*). *FLC* is highly expressed in nonvernalized shoot apical regions. After vernalization, this gene is epigenetically switched off for the remainder of the plant's life cycle, permitting flowering in response to long days to occur (**Figure 20.16**). In the next generation, however, the gene is switched on again, restoring the requirement for cold. Thus, in Arabidopsis the state of expression of the *FLC* gene represents a major determinant of meristem competence. In Arabidopsis it has been shown that FLC works by directly repressing the expression of the key floral signal *FT* in the leaves as well as the transcription factors SOC1 and FD at the shoot apical meristem.

The epigenetic regulation of *FLC* involves stable changes in chromatin structure resulting from **chromatin remodeling** (see Chapter 2). Vernalization causes the chromatin of the *FLC* gene to lose histone modifications characteristic of euchromatin (transcriptionally active DNA) and to acquire modifications, such as the methylation of specific lysine residues, characteristic of heterochromatin (transcriptionally inactive DNA). The cold-induced conversion of *FLC* from euchromatin to heterochromatin effectively silences the gene.

A range of vernalization pathways may have evolved

Many vernalization-requiring plants germinate in the fall, taking advantage of the cool and moist conditions optimal for their growth. The vernalization requirement of such plants ensures that flowering does not occur until spring, allowing the plants to survive winter vegetatively (flowers are especially sensitive to frost). A vernalizing plant must not only sense cold exposure but also must have a mechanism to measure the duration of cold exposure. For example, if a plant is exposed to a short period of cold early in the fall followed by a return of warm temperatures later that fall, it is important for the plant not to perceive the brief exposure to cold as winter and the following warm weather as spring. Accordingly, vernalization occurs only after exposure to a duration of cold sufficient to indicate that a complete winter season has passed.

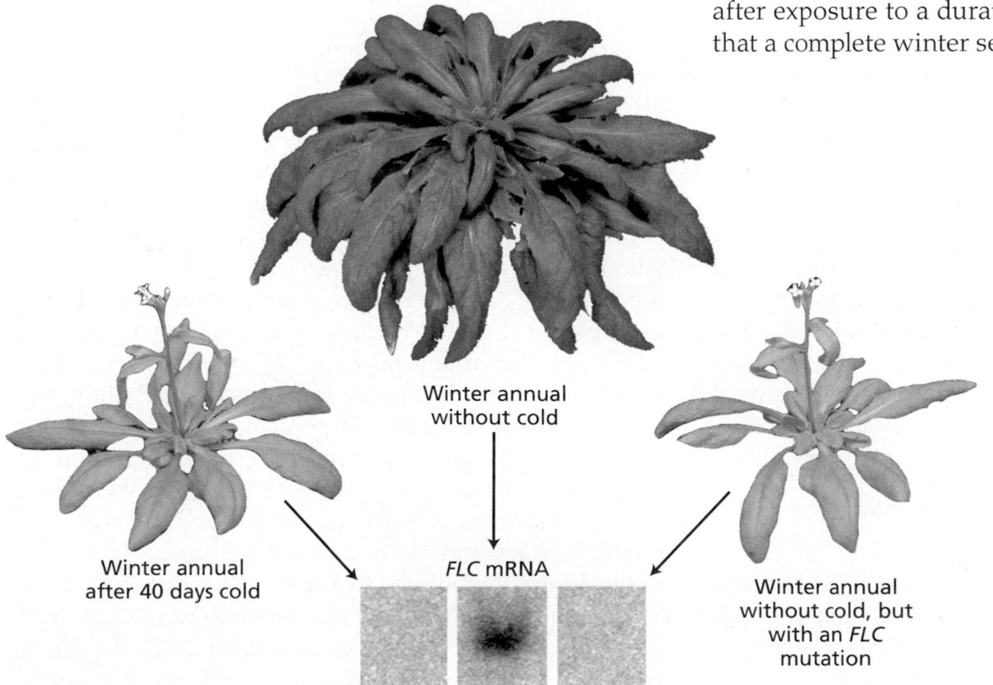

Figure 20.16 Plants with a vernalization requirement are either quite delayed in flowering or do not flower unless they experience a period of prolonged cold. (Left) Vernalization blocks the expression of the gene *FLOWERING LOCUS C* (*FLC*) in cold-requiring winter-annual ecotypes of Arabidopsis. (Right) A winter annual with an *FLC* mutation exhibits rapid flowering without cold treatment. (Photos courtesy of R. Amasino.)

A similar system of measuring the duration of cold before buds can be released from dormancy operates in many perennials that grow in temperate climates. The mechanism that plants have evolved to measure the duration of cold is not known, but in Arabidopsis there are genes that are induced only after exposure to a long period of cold, and these genes are critical to the vernalization process.

There does not appear to be a particular vernalization pathway that is conserved among all flowering plants. As discussed above, *FLC* is the flowering repressor that is responsible for the vernalization requirement in Arabidopsis. *FLC* encodes a MADS box protein that is related to regulatory proteins that will be discussed later in this chapter, such as DEFICIENS and AGAMOUS, that are involved in floral development. In cereals, a gene encoding a different type of protein, a zinc finger–containing protein called VRN2 (*vernalization 2*), acts as the flowering repressor that creates a vernalization requirement.

It appears that the major groups of flowering plants evolved in warm climates and therefore did not evolve a mechanism to measure the duration of winter. Over geologic time, regions of Earth gradually developed a temperate climate due to continental drift and other factors. Members of many groups of plants adapted to these new temperate niches with the development of responses such as vernalization and bud dormancy, and thus these responses are likely to have evolved independently in different groups of plants.

Figure 20.17 Demonstration by grafting of a leaf-generated floral stimulus in the SDP *Perilla crispa*. (Left) Grafting an induced leaf from a plant grown under short days onto a noninduced shoot causes the axillary shoots to produce flowers. The donor leaf has been trimmed to facilitate grafting, and the upper leaves have been removed from the stock to promote phloem translocation from the scion to the receptor shoots. (Right) Grafting a noninduced leaf from a plant grown under long days results in the formation of vegetative branches only. (Courtesy of J. A. D. Zeevaart.)

Long-distance Signaling Involved in Flowering

Although floral evocation occurs at the apical meristems of shoots, in photoperiodic plants inductive photoperiods are sensed by the leaves. This suggests that a long-range signal must be transmitted from the leaves to the apex, which has been shown experimentally through extensive grafting experiments in many different plant species. The biochemical nature of this signal had long baffled physiologists. The problem was finally solved using molecular genetic approaches, and the floral stimulus was identified as a protein. In this section we will review the background for the discovery of the floral stimulus, referred to as **florigen**, which serves as the long-distance signal during flowering. We will also describe various other biochemical signals that can serve either as activators or as inhibitors of flowering.

Grafting studies provided the first evidence for a transmissible floral stimulus

The production in photoperiodically induced leaves of a biochemical signal that is transported to a distant target tissue (the shoot apex) where it stimulates a response (flowering) satisfies an important criterion for a hormonal effect. In the 1930s, Mikhail Chailakhyan, working in Russia, postulated the existence of a universal flowering hormone, which he named florigen.

The evidence in support of florigen comes mainly from experiments in which noninduced receptor plants were stimulated to flower by having a leaf or shoot from a photoperiodically induced donor plant grafted to them. For example, in the SDP *Perilla crispa*, grafting a leaf from a plant grown under inductive short days onto a plant grown under noninductive long days causes the latter to flower (**Figure 20.17**). Moreover, the floral stimulus seems to be the same in plants with different photoperiodic requirements. Thus, grafting an induced shoot from the LDP *Nicotiana sylvestris*, grown under long days, onto the SDP 'Maryland Mammoth' tobacco caused the latter to flower under noninductive (long-day) conditions.

The leaves of DNPs have also been shown to produce a graft-transmissible floral stimulus (**Table 20.2**). For example, grafting a single leaf of a day-neutral variety of soy-

TABLE 20.2 Transmission of the flowering signal occurs through a graft junction

Donor plants maintained under flower-inducing conditions	Photoperiod type[a,b]	Vegetative receptor plant induced to flower	Photoperiod type[a,b]
Helianthus annus	DNP in LD	*H. tuberosus*	SDP in LD
Nicotiana tabacum 'Delcrest'	DNP in SD	*N. sylvestris*	LDP in SD
Nicotiana sylvestris	LDP in LD	*N. tabacum* 'Maryland Mammoth'	SDP in LD
Nicotiana tabacum 'Maryland Mammoth'	SDP in SD	*N. sylvestris*	LDP in SD

Note: The successful transfer of a flowering induction signal by grafting between plants of different photoperiodic response groups shows the existence of a transmissible floral hormone that is effective.

[a]LDPs = Long-day plants; SDPs = Short-day plants; DNPs = Day-neutral plants.

[b]LD, Long days; SD, short days.

bean, 'Agate', onto the short-day variety, 'Biloxi', caused flowering in 'Biloxi' even when the latter was maintained in noninductive long days. Similarly, a shoot from a day-neutral variety of tobacco (*Nicotiana tabacum*, cv. Trapezond) grafted onto the LDP *Nicotiana sylvestris* induced the latter to flower under noninductive short days.

Grafting studies also showed that in some species, such as *Xanthium* (SDP), *Bryophyllum* (SLDP), and *Silene* (LDP), not only can flowering be induced by grafting, but the induced state itself appears to be self-propagating (see **WEB TOPIC 20.10**). In a few cases, flowering has even been induced by grafts between different genera. The SDP *Xanthium strumarium* flowered under long-day conditions when shoots of flowering *Calendula officinalis* were grafted onto a vegetative *Xanthium* stock. Similarly, grafting a shoot from the LDP *Petunia hybrida* onto a stock of the cold-requiring biennial henbane (*Hyoscyamus niger*) caused the latter to flower under long days, even though it was nonvernalized (**Figure 20.18**).

In *Perilla crispa* (see Figure 20.17), the movement of the floral stimulus from a donor leaf to the stock across the graft union correlated closely with the translocation of ^{14}C-labeled assimilates from the donor, and this movement was dependent on the establishment of vascular continuity across the graft union. These results confirmed earlier girdling studies showing that the floral stimulus is translocated along with photoassimilates in the phloem.

Florigen is translocated in the phloem

The leaf-derived photoperiodic floral stimulus is translocated via the phloem to the shoot apical meristem, where it promotes floral evocation. Treatments that block phloem translocation, such as girdling or localized heat-killing, block flowering by preventing the movement of the floral stimulus out of the leaf.

It is possible to measure rates of movement of florigen by removing a leaf at different times after induction, and comparing the time it takes for the signal to reach two buds located at different distances from the induced leaf. The rationale for this type of measurement is that a threshold amount of the signaling compound has reached

Figure 20.18 Successful transfer of the floral stimulus between different genera: The scion (right branch) is the LDP *Petunia hybrida*, and the stock is nonvernalized henbane (*Hyoscyamus niger*). The graft combination was maintained under long days. (Courtesy of J. A. D. Zeevaart.)

the bud when flowering takes place, despite the removal of the leaf. In this way, the time for a sufficient amount of signal to exit the leaf can be determined. Furthermore, comparing the times of induction for two differently positioned buds provides a measure of the rate of movement of the signal along the stem.

Studies using this method have shown that the rate of movement of the flowering signal is comparable to, or somewhat slower than, the rate of translocation of sugars in the phloem (see Chapter 11). For example, export of the floral stimulus from adult leaves of the SDP *Chenopodium* is complete within 22.5 h from the beginning of the long night period. In the LDP *Sinapis*, movement of the floral stimulus out of the leaf is complete as early as 16 h after the start of the long-day treatment.

Because the floral stimulus is translocated along with sugars in the phloem, it is subject to source–sink relations. An induced leaf positioned close to the shoot apex is more likely to cause flowering than an induced leaf at the base of a stem, which normally feeds the roots. Similarly, noninduced leaves positioned between the induced leaf and the apical bud will tend to inhibit flowering by serving as the preferred source leaves for the bud, thus preventing the floral stimulus from the more distal induced leaf from reaching its target.

The Identification of Florigen

Pioneering grafting experiments of the kind described above established the importance of a long-range signal from the leaf to the apical meristem to stimulate flowering. Since the 1930s, there have been many unsuccessful attempts to isolate and characterize florigen. A major breakthrough was the identification of *FLOWERING LOCUS T* (*FT*) in Arabidopsis through genetic screens.

The Arabidopsis protein FLOWERING LOCUS T (FT) is florigen

According to the coincidence model, flowering in LDPs such as Arabidopsis occurs when the *CONSTANS* gene is expressed during the light period. *CO* gene expression appears to be highest in the companion cells of the phloem of leaves and stems. The downstream target gene of *CO*, ***FLOWERING LOCUS T*** (***FT***), is also specifically expressed in the companion cells.

Consistent with a phloem localization of CO, *co* mutants with a defective photoperiodic response could be rescued by expressing *CO* specifically in the phloem of the minor veins of mature leaves using a promoter construct specific for companion cells. In contrast, expressing *CO* in the apical meristems of *co* mutants did not restore the photoperiodic response. Thus, *CO* seems to act specifically in the phloem of leaves to stimulate flowering in response to long days. In addition, flowering could be induced in the *co* mutant by grafting transgenic shoots expressing *CO* in the phloem of their leaves onto the mutant. This observation suggests that *CO* expression gives rise to a graft-transmissible floral stimulus that can cause flowering at the apical meristem.

The signaling output of CO activity is mediated by the expression of *FT*. In Arabidopsis, *CO* expression during long days results in an increase in *FT* mRNA. However, unlike *CO*, *FT* stimulates flowering when expressed in either the companion cells or the apical meristem.

Biochemically, FT is a small globular protein that is related to a family of regulatory proteins conserved between budding yeast and vertebrates. Expression of the *FT* gene (or its relatives, such as *Hd3a* in rice, discussed earlier) is induced in a range of species during their floral-inductive photoperiods. When the *FT* gene is introduced into a range of plant species whose flowering is not influenced by photoperiod, it causes photoperiod-independent flowering. In addition, FT protein can move from the leaves to the apical meristem and thus exhibits all the properties that would be expected of florigen.

According to the current model, FT protein moves via the phloem from the leaves to the meristem under inductive photoperiods. There are two critical steps in this process: the export of FT from companion cells to the sieve tube elements, and the activation of FT target genes at the shoot apex, which triggers flower development. The endoplasmic reticulum (ER) is one of the major routes for the transport of proteins from the companion cells to the sieve tube elements. The ER-localized protein, FT INTERACTING PROTEIN1 (FTIP1), is required for FT movement into the phloem translocation stream, which takes it to the meristem (**Figure 20.19**). Once in the floral meristem, the FT protein enters the nucleus and forms a complex with **FLOWERING D** (**FD**), a basic leucine zipper (bZIP) transcription factor that is expressed in the meristem. The complex of FT and FD then activates floral identity genes such as *APETALA1* (*AP1*).

In Arabidopsis, these events set in motion positive feedback loops that keep the meristem in a flowering state. After being activated by the FT protein, FD triggers the expression of *SOC1* and *AP1*. Both of these targets activate *LEAFY* (*LFY*; a floral identity gene we will discuss later in this chapter), and *LFY* directly activates the expression of *AP1* and *FD*, forming two positive feedback loops (see Figure 20.19). Because of the action of these positive feedback loops, floral initiation in Arabidopsis is irreversible. However, the meristems of some species lack such positive feedback loops and, as a consequence, revert to producing leaves in the absence of a continuous inductive photoperiod.

Gibberellins and ethylene can induce flowering

Among the naturally occurring growth hormones, gibberellins (see Chapter 15) can have a strong influence on flowering (see **WEB TOPIC 20.11**). Exogenous gibberellin

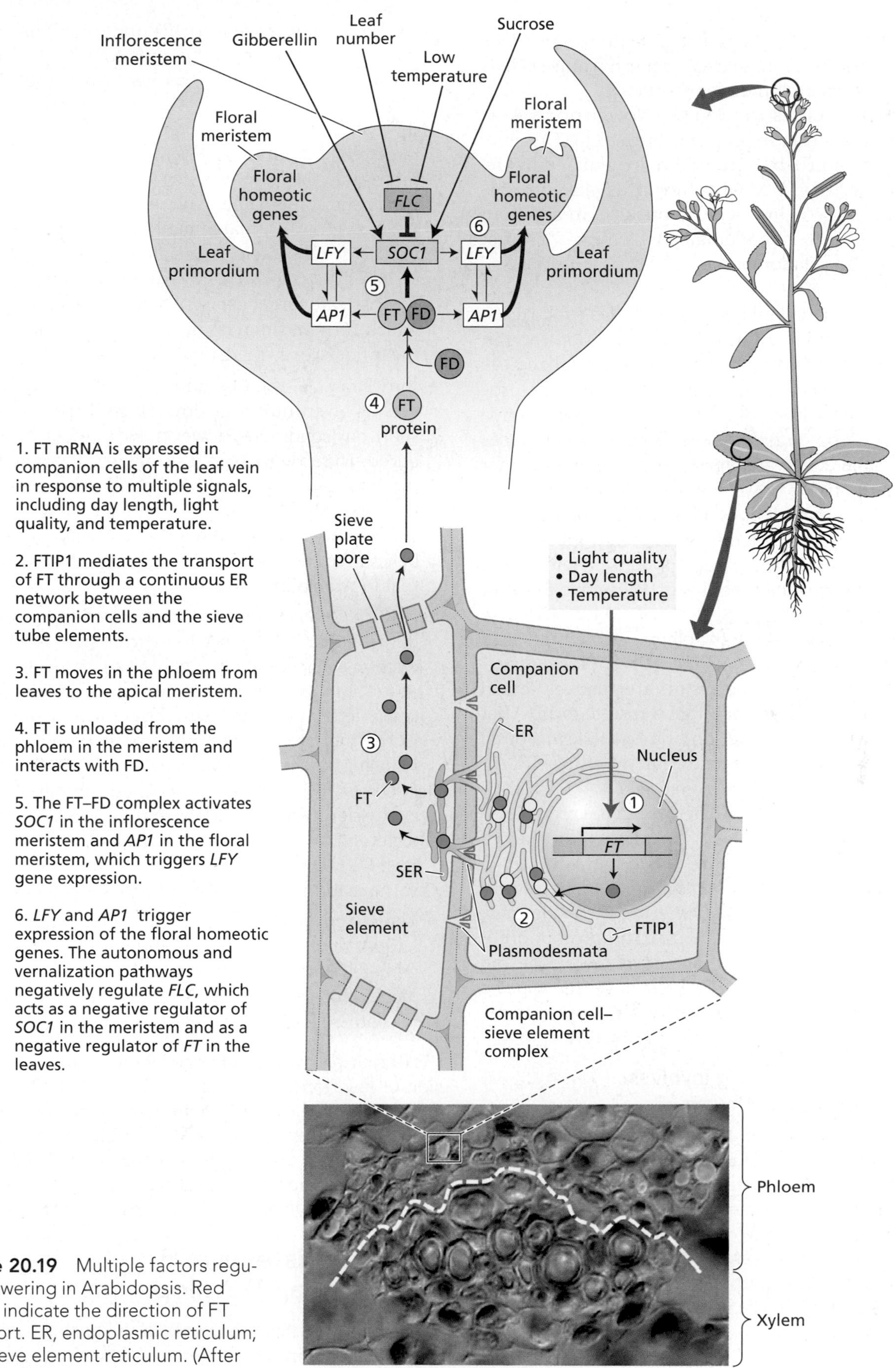
Inflorescence meristem
Gibberellin
Leaf number
Low temperature
Sucrose
Floral meristem
Floral meristem
Floral homeotic genes
Floral homeotic genes
FLC
SOC1
LFY
LFY
AP1
AP1
FT
FD
FD
Leaf primordium
Leaf primordium
⑤
⑥
④ FT protein
1. FT mRNA is expressed in companion cells of the leaf vein in response to multiple signals, including day length, light quality, and temperature.
2. FTIP1 mediates the transport of FT through a continuous ER network between the companion cells and the sieve tube elements.
3. FT moves in the phloem from leaves to the apical meristem.
4. FT is unloaded from the phloem in the meristem and interacts with FD.
5. The FT–FD complex activates *SOC1* in the inflorescence meristem and *AP1* in the floral meristem, which triggers *LFY* gene expression.
6. *LFY* and *AP1* trigger expression of the floral homeotic genes. The autonomous and vernalization pathways negatively regulate *FLC*, which acts as a negative regulator of *SOC1* in the meristem and as a negative regulator of *FT* in the leaves.
Sieve plate pore
• Light quality
• Day length
• Temperature
Companion cell
ER
Nucleus
③
FT
①
FT
SER
Sieve element
②
FTIP1
Plasmodesmata
Companion cell–sieve element complex
Phloem
Xylem
Leaf vein (transverse view)

Figure 20.19 Multiple factors regulate flowering in Arabidopsis. Red arrows indicate the direction of FT transport. ER, endoplasmic reticulum; SER, sieve element reticulum. (After Liu et al. 2013.)

can evoke flowering when applied either to rosette LDPs such as Arabidopsis, or to dual–day length plants such as *Bryophyllum*, when grown under short days.

Gibberellin appears to promote flowering in Arabidopsis by activating expression of the *LEAFY* gene. The activation of *LFY* by gibberellin is mediated by the transcription factor GA-MYB, which is negatively regulated by DELLA proteins (see Chapter 18). In addition, GA-MYB levels are also modulated by a microRNA that promotes the degradation of the GA-MYB transcript (see Chapter 18).

Exogenously applied gibberellins can also evoke flowering in a few SDPs in noninductive conditions and in cold-requiring plants that have not been vernalized. As previously discussed, cone formation can also be promoted in juvenile plants of several gymnosperm families by the addition of gibberellins. Thus, in some plants exogenous gibberellins can bypass the endogenous trigger of age in autonomous flowering, as well as the primary environmental signals of day length and temperature. As discussed in Chapter 18, plants contain many gibberellin-like compounds. Most of these compounds are either precursors to, or inactive metabolites of, the active forms of gibberellin.

Gibberellin metabolism in the plant is strongly affected by day length. For example, in the LDP spinach (*Spinacia oleracea*), the levels of gibberellins are relatively low in short days, and the plants maintain a rosette form. After the plants are transferred to long days, the levels of all the gibberellins of the 13-hydroxylated pathway ($GA_{53} \rightarrow GA_{44} \rightarrow GA_{19} \rightarrow GA_{20} \rightarrow GA_1$; see **WEB APPENDIX 3**) increase. However, the fivefold increase in the physiologically active gibberellin, GA_1, is what causes the marked stem elongation that accompanies flowering.

In addition to gibberellins, other growth hormones can either inhibit or promote flowering. One commercially important example is the striking promotion of flowering in pineapple (*Ananas comosus*) by ethylene and ethylene-releasing compounds—a response that appears to be restricted to members of the pineapple family (Bromeliaceae).

The transition to flowering involves multiple factors and pathways

It is clear that the transition to flowering involves a complex system of interacting factors. Leaf-generated transmissible signals are required for determination of the shoot apex in both autonomously regulated and photoperiodic species.

Genetic studies have established that there are four distinct developmental pathways that control flowering in the LDP Arabidopsis (see Figure 20.19):

- The *photoperiodic pathway* begins in the leaf and involves phytochromes and cryptochromes. (Note that phyA and phyB have contrasting effects on flowering; see **WEB TOPIC 20.6**) In LDPs under long-day conditions, the interaction of these photoreceptors with the circadian clock initiates a pathway that results in the expression of *CO* in the phloem companion cells of the leaf. *CO* activates the expression of its downstream target gene, *FT*, in the phloem. FT protein ("florigen") moves into the sieve tube elements and is translocated to the apical meristem, where it stimulates flowering. As shown in the enlargement of the meristem in Figure 20.19, FT protein forms a complex with the transcription factor FD. The FD–FT complex then activates downstream target genes such as *SOC1*, *AP1*, and *LFY*, which turn on floral homeotic genes on the flanks of the inflorescence meristem.
- In rice, a SDP, the *CO* homolog *Heading-date 1* (*Hd1*) acts as an inhibitor of flowering. During inductive short-day conditions, however, Hd1 protein is not produced. The absence of Hd1 stimulates the expression of the *Hd3a* gene in the phloem companion cells (Hd3a is a relative of FT). Hd3a protein is then translocated via the sieve tubes to the apical meristem, where it is believed to stimulate flowering by a pathway similar to that in Arabidopsis.
- In the *autonomous* and *vernalization pathways*, flowering occurs either in response to internal signals—the production of a fixed number of leaves—or to low temperatures. In the autonomous pathway of Arabidopsis, all of the genes associated with the pathway are expressed in the meristem. The autonomous pathway acts by reducing the expression of the flowering repressor gene *FLOWERING LOCUS C* (*FLC*), an inhibitor of *SOC1* expression. Vernalization also represses *FLC*, but by a different mechanism (an epigenetic switch). Because the *FLC* gene is a common target, the autonomous and vernalization pathways are grouped together.
- The *gibberellin pathway* is required for early flowering and for flowering under noninductive short days. The gibberellin pathway involves GA-MYB as an intermediary, which promotes *LFY*; gibberellin may also interact with *SOC1* by a separate pathway.

All four pathways converge by increasing the expression of key floral regulators: *FT* in the vasculature and *SOC1*, *LFY*, and *AP1* in the meristem (see Figure 20.19). Expression of genes such as *SOC1*, *LFY*, and *AP1* in turn activates downstream genes required for floral organ development such as *AP3*, *PISTILLATA* (*PI*), and *AGAMOUS* (*AG*), as we will see in the next part of this chapter.

Floral Meristems and Floral Organ Development

Once flowering has been evoked, the business of building flowers begins. The shapes of flowers are extremely diverse, reflecting adaptations to protect developing game-

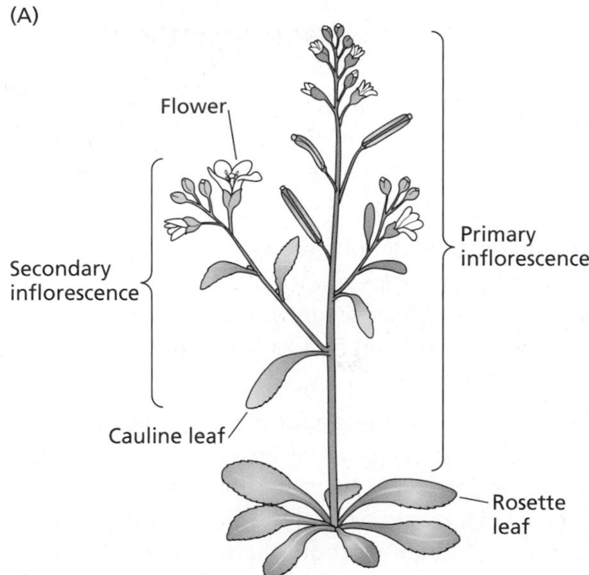

Figure 20.20 (A) The shoot apical meristem in *Arabidopsis thaliana* generates different organs at different stages of development. Early in development the shoot apical meristem forms a rosette of basal leaves. When the plant makes the transition to flowering, the shoot apical meristem is transformed into a primary inflorescence meristem that ultimately produces an elongated stem bearing flowers. Leaf primordia initiated prior to the floral transition develop on the stem (cauline leaves), and secondary inflorescences develop in the axils of these stem-borne leaves. (B) Photograph of a flowering Arabidopsis plant. (Courtesy of Richard Amasino.)

tophytes, attract pollinators, promote self-pollination or cross-pollination as appropriate, and produce and disperse fruits and seeds. Despite this diversity, genetic and molecular studies have now identified a network of genes that control floral morphogenesis in flowers as different as those of Arabidopsis and snapdragon (*Antirrhinum majus*). Variations on this regulatory network now seem to account for floral morphogenesis in other species as well.

In this section we will focus on floral development in Arabidopsis, which has been studied extensively. First we will outline the basic morphological changes that occur during the transition from the vegetative to the reproductive phase. Next we will consider the arrangement of the floral organs in four whorls on the meristem, and the types of genes that govern the normal pattern of floral development.

The shoot apical meristem in Arabidopsis changes with development

Floral meristems can usually be distinguished from vegetative meristems by their larger size. In the vegetative meristem, the cells of the central zone complete their division cycles slowly. The transition from vegetative to reproductive development is marked by an increase in the frequency of cell divisions within the central zone of the shoot apical meristem (see Chapter 17). The increase in the size of the meristem is largely a result of the increased division rate of these central cells.

During the vegetative phase of growth, the Arabidopsis apical meristem produces leaves with very short internodes, resulting in a basal rosette of leaves (**Figure 20.20**). When reproductive development is initiated, the vegetative meristem is transformed into the primary inflorescence meristem. The **primary inflorescence meristem** produces an elongated inflorescence axis bearing two types of lateral organs: stem-borne (or inflorescence) leaves and flowers. The axillary buds of the stem-borne leaves develop into **secondary inflorescence meristems**, and their activity repeats the pattern of development of the primary inflorescence meristem. The Arabidopsis inflorescence meristem has the potential to grow indefinitely and thus exhibits *indeterminate* growth. Flowers arise from **floral meristems** that form on the flanks of the inflorescence meristem (**Figure 20.21**). In contrast to the inflorescence meristem, the floral meristem is determinate.

The four different types of floral organs are initiated as separate whorls

Floral meristems initiate four different types of floral organs: sepals, petals, stamens, and carpels. These sets of organs are initiated in concentric rings, called **whorls**, around the flanks of the meristem (**Figure 20.22**). The initiation of the innermost organs, the carpels, consumes all of the meristematic cells in the apical dome, and only the floral organ primordia (localized regions of cell division) are present as the floral bud develops. In Arabidopsis, the whorls are arranged as follows:

- The first (outermost) whorl consists of four sepals, which are green at maturity.
- The second whorl is composed of four petals, which are white at maturity.
- The third whorl contains six stamens (the male reproductive structures), two of which are shorter than the other four.
- The fourth (innermost) whorl is a single complex organ, the gynoecium or pistil (the female reproductive structure), which is composed of an ovary with two fused carpels, each containing numerous ovules, and a short style capped with a stigma.

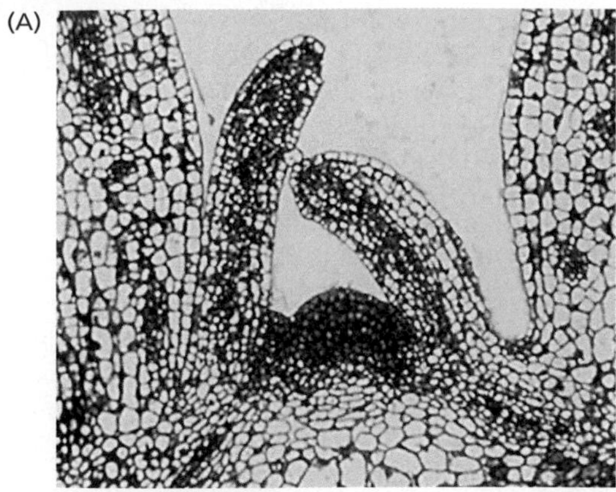

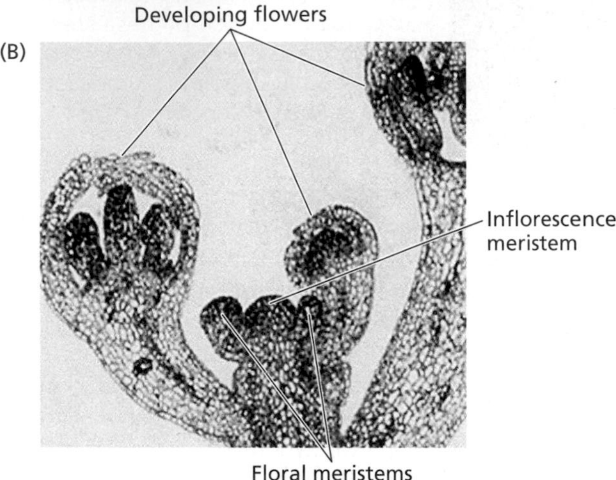

Figure 20.21 Longitudinal sections through a vegetative (A) and a reproductive (B) shoot apical region of Arabidopsis. (Courtesy of V. Grbic and M. Nelson.)

Two major categories of genes regulate floral development

Studies of mutations have enabled identification of two key categories of genes that regulate floral development: floral meristem identity genes and floral organ identity genes.

1. Floral meristem identity genes encode transcription factors that are necessary for the initial induction of floral organ identity genes. They are the positive regulators of floral organ identity in the developing floral meristem.
2. **Floral organ identity genes** directly control floral organ identity. The proteins encoded by these genes are transcription factors that interact with other protein cofactors to control the expression of downstream genes whose products are involved in the formation or function of floral organs.

While certain genes fit neatly within these two general categories, it is important to keep in mind that floral development involves complex, nonlinear gene networks. In these networks, individual genes often play multiple roles. For example, evolution has recruited the same transcription factor, APETALA2, to first regulate floral meristem identity and then floral organ identity (**Table 20.3**).

Floral meristem identity genes regulate meristem function

Floral meristem identity genes must be active for the immature primordia formed at the flanks of the shoot apical meristem or inflorescence meristem to become floral meristems. (Recall that an apical meristem that is forming floral meristems on its flanks is known as an

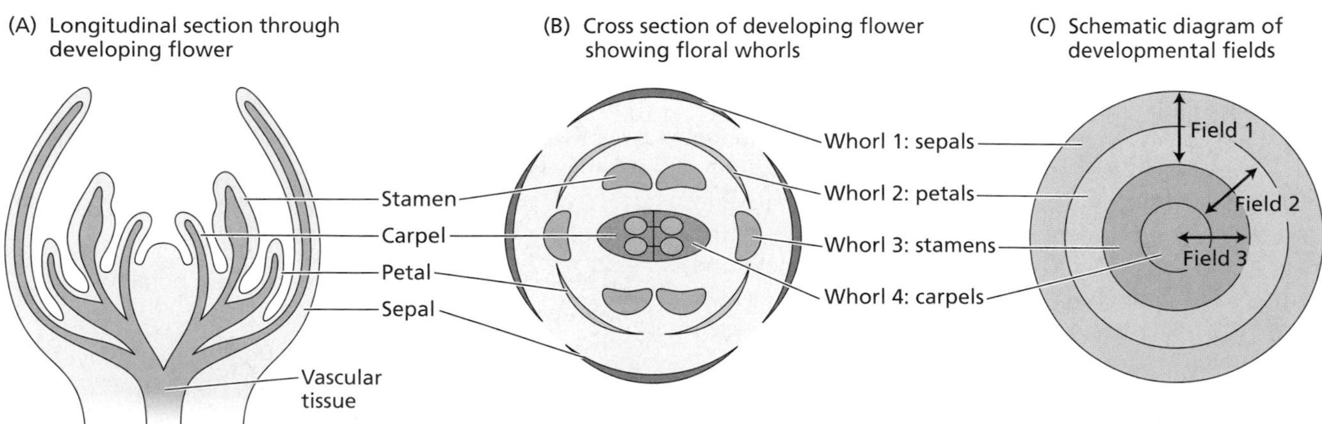

Figure 20.22 Floral organs are initiated sequentially by the floral meristem of Arabidopsis. (A, B) The floral organs are produced as successive whorls (concentric circles), starting with the sepals and progressing inward. (C) According to the combinatorial model, the functions of each whorl are determined by three overlapping developmental fields. These fields correspond to the expression patterns of specific floral organ identity genes. (After Bewley et al. 2000.)

TABLE 20.3 Genes that regulate flowering

Gene	Transcription Factor Family	Functions	Expression Domain	Orthologs
CONSTANS (*CO*)	Zinc finger	Activates flowering in response to long photoperiods	In leaves under long photoperiods	*AtCO* (potato); *Hd1* (rice)
FLOWERING D (*FD*)	bZIP	Receptor for florigen, activates flowering via AP1	In shoot apex	*OsFD1* (rice)
SUPPRESSOR OF OVEREXPRESSION OF CONSTANS1 (*SOC1*)	MADS	Activates flowering downstream of florigen	Leaves and apex	-
PHYTOCHROME INTERACTING FACTOR4 (*PIF4*)	bHLH	Activates florigen in response to high temperature	Leaves and apex	-
FLOWERING LOCUS C	MADS	Floral repressor	Leaves and apex	-
SHORT VEGETATIVE PHASE (*SVP*)	MADS	Represses flowering at low temperature	Leaves and apex	-
FLOWERING LOCUS M (*FLM*)	MADS	Represses flowering	Leaves and apex	-
LEAFY (*LFY*)	LFY	Floral meristem identity gene	Shoot apex	*RLF* (rice); *FLORICAULA* (*Antirrhinum*)
APETALA1 (*AP1*)	MADS	Class A homeotic gene, meristem identity	Floral meristems, whorl 1	*SQUAMOSA* (*Antirrhinum*); *ZAP1*, *GLOSSY15* (maize; corn [*Zea mays*])
APETALA2 (*AP2*)	AP2/EREBP	Class A homeotic gene, floral meristem identity	Floral meristems, whorl 1	*BRANCHED FLORETLESS1* (maize)
PISTILLATA (*PI*)	MADS	Class B homeotic gene	Whorls 2 and 3	*GLOBOSA* (*Antirrhinum*)
AGAMOUS (*AG*)	MADS	Class C homeotic gene	Whorls 3 and 4	*PLENA* and *FARINELLI* (*Antirrhinum*); *ZAG1* and *ZMM2* (maize)
SEPALLATA (*SEP*) *1, 2, 3, 4*	MADS	Class E homeotic genes	Whorls 1–4	*DEFH49, DEFH200, DEFH72, AmSEP3B* (*Antirrhinum*); *ZMM3, 8, 14* (maize)
CAULIFLOWER (*CAL*)	MADS	Meristem identity	Floral meristem	-
FRUITFULL (*FUL*)	MADS	Floral meristem identity	Floral meristem and cauline leaves	-

inflorescence meristem; see Figure 20.21.) For example, mutants of snapdragon (*Antirrhinum*) that have a defect in the floral meristem identity gene *FLORICAULA* (*FLO*) develop an inflorescence that does not produce flowers. Instead of developing floral meristems in the axils of the bracts, *flo* mutants develop additional inflorescence meristems in the bract axils. Thus, the wild-type *FLO* gene controls the determination step that establishes floral meristem identity.

In Arabidopsis, *LEAFY* (*LFY*), *FLOWERING D* (*FD*), *SUPPRESSOR OF OVEREXPRESSION OF CONSTANS1* (*SOC1*), and *APETALA1* (*AP1*) are among the critical genes in the genetic pathway that must be activated to establish floral meristem identity (see Table 20.3). *LFY* is the Arabidopsis version of the snapdragon *FLO* gene. As we saw earlier in the chapter, *LFY*, *FD*, and *SOC1* play central roles in floral evocation by integrating signals from several different pathways involving both environmental and internal cues. *lfy* and *fd* double mutants fail to form flowers, highlighting the roles of *LFY* and *FD* as floral meristem identity genes that serve as master regulators for the initiation of floral development.

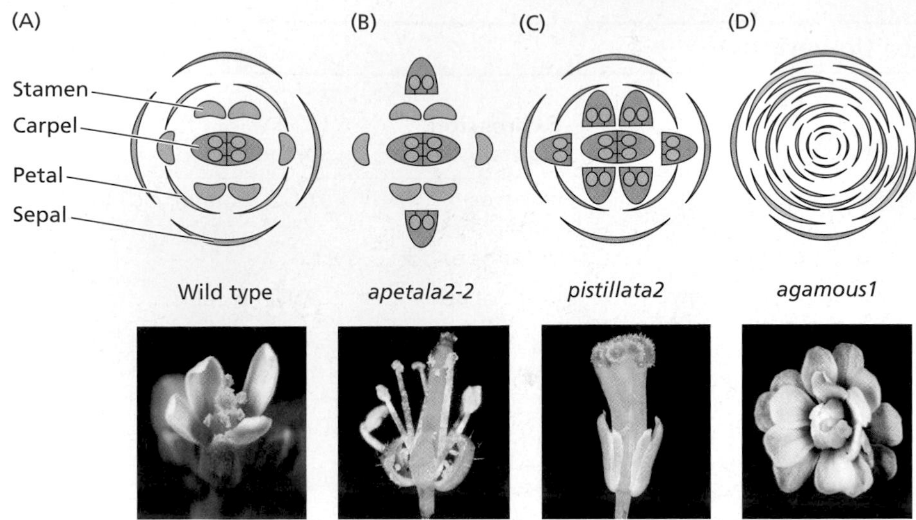

Figure 20.23 Mutations in the floral organ identity genes dramatically alter the structure of the flower. (A) Wild-type Arabidopsis shows normal structure in all four floral components. (B) *apetala2-2* mutants lack sepals and petals. (C) *pistillata2* mutants lack petals and stamens. (D) *agamous1* mutants lack both stamens and carpels. (Photos from Meyerowitz 2002; courtesy of J. L. Riechmann.)

Homeotic mutations led to the identification of floral organ identity genes

The genes that determine floral organ identity were discovered as floral homeotic mutants. Mutations in the fruit fly, *Drosophila*, led to the identification of a set of homeotic genes encoding transcription factors that determine the locations at which specific structures develop. Homeotic genes act as major developmental switches that activate the entire genetic program for a particular structure. The expression of homeotic genes thus gives organs their identity.

The floral organ identity genes were first identified as single-gene homeotic mutations that altered floral organ identity, causing some of the floral organs to appear in the wrong places. Five key genes initially were identified in Arabidopsis that specify floral organ identity: *APETALA1* (*AP1*), *APETALA2* (*AP2*), *APETALA3* (*AP3*), *PISTILLATA* (*PI*), and *AGAMOUS* (*AG*). Mutations in these genes dramatically altered the structure, and thus the identity, of the floral organs produced in two adjacent whorls (**Figure 20.23**). For example, plants with the *ap2* mutation lacked sepals and petals (see Figure 20.23B). Plants bearing *ap3* or *pi* mutations produced sepals instead of petals in the second whorl, and carpels instead of stamens in the third whorl (see Figure 20.23C). Plants homozygous for the *ag* mutation lacked both stamens and carpels (see Figure 20.23D). Because mutations in these genes change floral organ identity without affecting the initiation of flowers, they are, by definition, homeotic genes.

The role of organ identity genes in floral development is dramatically illustrated by experiments in which two or three activities are eliminated by loss-of-function mutations. In quadruple-mutant Arabidopsis plants (*ap1, ap2, ap3/pi*, and *ag*) floral meristems no longer produce floral organs but rather produce green leaflike structures; these leaflike organs are produced with a whorled phyllotaxy typical of normal flowers (**Figure 20.24**). This experimental result shows that leaves are the "ground state" of organs produced by shoot meristems, and that the activity of additional genes such as *AP1* and *AP2* are required to convert the leaflike, "ground-state" organs into petals, sepals, stamens, and pistils. This result supports the idea of the eighteenth-century German poet and natural scientist Johann Wolfgang von Goethe (1749–1832), who speculated that floral organs are highly modified leaves.

The ABC model partially explains the determination of floral organ identity

The five floral organ identity genes described above fall into three classes—A, B, and C—defining three differ-

Figure 20.24 A quadruple mutant of Arabidopsis (*ap1, ap2, ap3/pi, ag*) produces leaflike structures in place of floral organs. (Courtesy of John Bowman.)

Figure 20.25 Interpretation of the phenotypes of floral homeotic mutants based on the ABC model. (A) All three activity classes are functional in the wild type. (B) Loss of Class C activity results in expansion of Class A activity throughout the floral meristem. (C) Loss of Class A activity results in the spread of Class C activity throughout the meristem. (D) Loss of Class B activity results in the expression of only Class A and C activities.

(A) Wild type

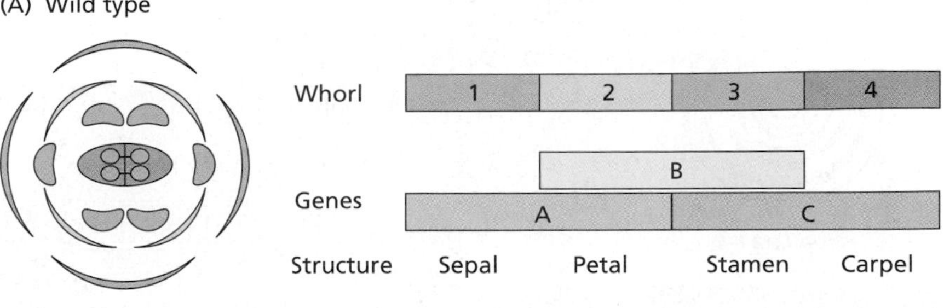

(B) Loss of Class C activity

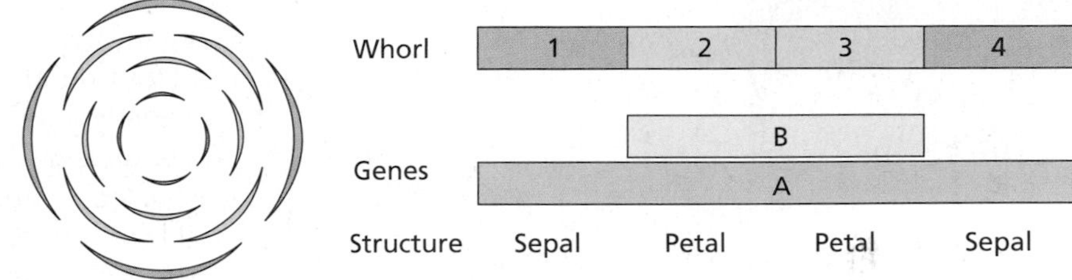

(C) Loss of Class A activity

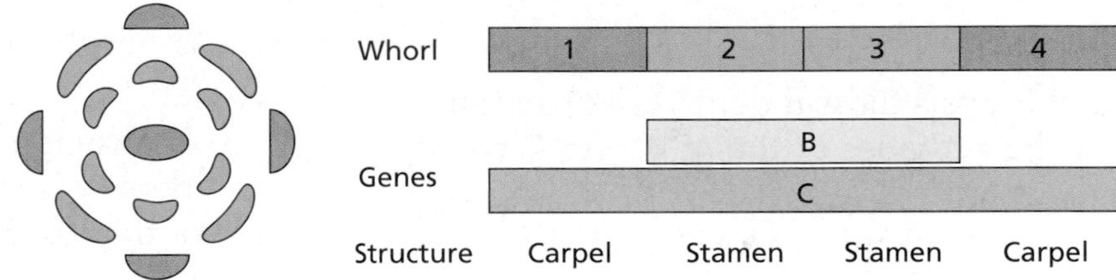

(D) Loss of Class B activity

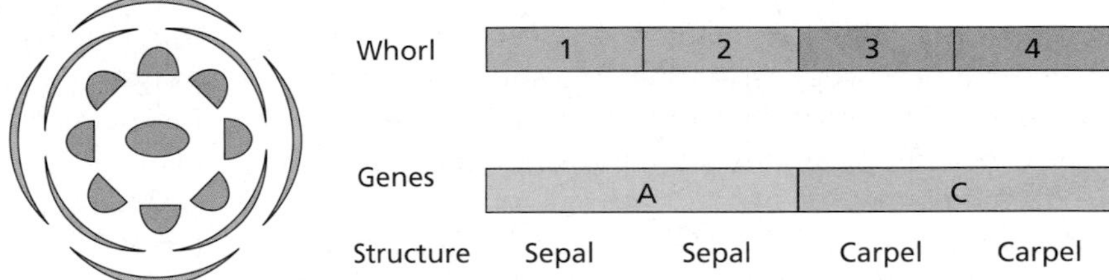

ent kinds of activities encoded by three distinct types of genes (**Figure 20.25**):

- Class A activity, encoded by *AP1* and *AP2*, controls organ identity in the first and second whorls. Loss of Class A activity results in the formation of carpels instead of sepals in the first whorl, and of stamens instead of petals in the second whorl.
- Class B activity, encoded by *AP3* and *PI*, controls organ determination in the second and third whorls. Loss of Class B activity results in the formation of sepals instead of petals in the second whorl, and of carpels instead of stamens in the third whorl.
- Class C activity, encoded by *AG*, controls events in the third and fourth whorls. Loss of Class C activity results in the formation of petals instead of stamens in the third whorl. Moreover, in the absence of Class C activity, the fourth whorl (normally a carpel) is replaced by a *new flower*. As a result, the fourth whorl of an *ag* mutant flower is occupied by sepals. The floral meristem is no longer determinate. Flowers continue to form *within* flowers, and the pattern of organs (from outside to inside) is: sepal, petal, petal; sepal, petal, petal; and so on.

The **ABC model** accounts for many observations in two distantly related eudicot species (snapdragon and Arabidopsis), and provides a way of understanding how relatively few key regulators can combinatorially provide a complex outcome. The ABC model postulates that organ identity in each whorl is determined by a unique combination of the three organ identity gene activities (see Figure 20.25):

- Class A activity alone specifies sepals.
- Class A and B activities are required for the formation of petals.
- Class B and C activities form stamens.
- Class C activity alone specifies carpels.

The model further proposes that Class A and C activities mutually repress each other; that is, both A- and C-class genes exclude each other from their expression domains, in addition to their function in determining organ identity.

Although the patterns of organ formation in wild-type flowers and most of the mutants are predicted and explained by this model, not all observations can be accounted for by the ABC genes alone. For example, expression of the ABC genes throughout the plant does not transform vegetative leaves into floral organs. Thus, the ABC genes, while necessary, are not sufficient to impose

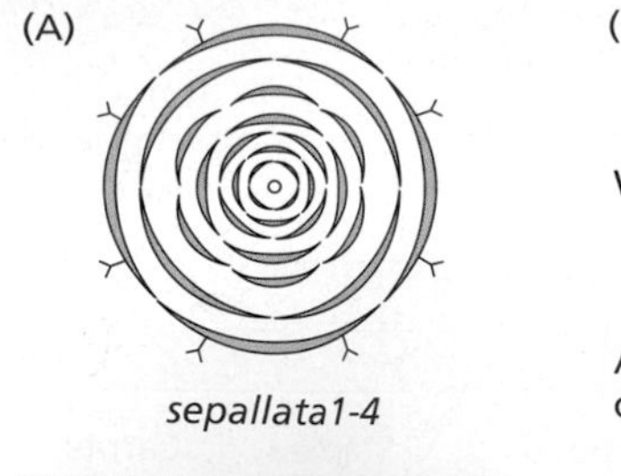

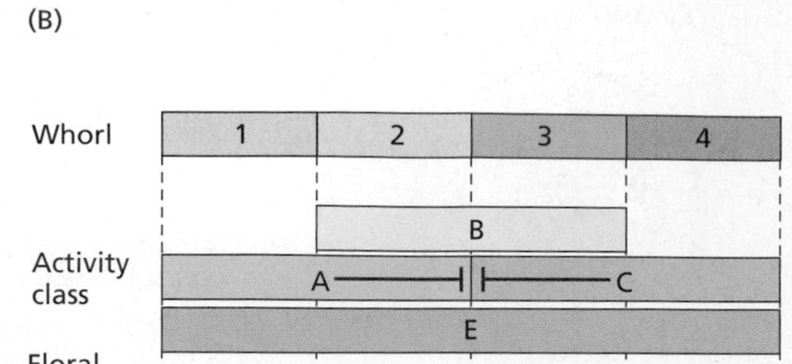

Figure 20.26 ABCE model for floral development. (A) In *sepallata1-4* mutants, all of the floral organs resemble vegetative leaves, suggesting that *SEP* genes are required for floral meristem identity. (B) ABCE model for floral organ determination in which *SEP*s act as Class E genes required for floral organ identity. (From Krizek and Fletcher 2005.)

floral organ identity onto a leaf developmental program. As we will discuss next, transcription factors encoded by floral meristem identity genes are also required for the formation of petals, stamens, and carpels.

Arabidopsis Class E genes are required for the activities of the A, B, and C genes

Since the A, B, and C genes were identified, another class of floral homeotic genes, the Class E genes, has been discovered. Mutations in three of the other genes identified in mutant screens for floral homeotic mutants, *AGAMOUS-LIKE1-3* (*AGL1-3*), produced only subtle phenotypes when mutated individually. However, the flowers of the *agl1/agl2/agl3* triple mutants consisted of sepal-like structures only, suggesting that the subtle phenotypes previously observed in the three individually mutated *AGL* genes were due to functional redundancy. Because of the sepal-rich phenotype of the triple mutant, the three *AGL* genes were renamed *SEPALLATA1-3* (*SEP1-3*) and were added to the ABC model as Class E genes (**Figure 20.26**). (Class D genes are required for ovule formation and are described later.)

Another *SEPALLATA* gene, *SEP4*, is required redundantly with the other three *SEP* genes to confer sepal identity, and contributes to the development of the other three organ types. *sep* quadruple mutants show a conversion of all four floral organ types into leaflike structures, similar to the *ap1, ap2, ap3/pi,* and *ag* quadruple mutant (see Figures 20.24 and 20.26). Remarkably, by expressing Class E genes in combination with Class A and B genes, it is possible to convert both cotyledons and vegetative leaves into petals (**Figure 20.27**).

The ABCE model was formulated based on genetic experiments in Arabidopsis and *Antirrhinum*. Flowers from different species have evolved diverse structures by modifying the regulatory networks described by the ABCE model (see WEB TOPIC 20.12).

According to the Quartet Model, floral organ identity is regulated by tetrameric complexes of the ABCE proteins

All homeotic genes that have been identified so far, in both plants and animals, encode transcription factors. However, unlike animal homeotic genes, which contain homeobox sequences, most plant homeotic genes belong to a class of related sequences known as **MADS box genes**. The acronym MADS is based on the four founding members (*MCM1, AGAMOUS, DEFICIENS,* and *SRF*) of a large gene family.

Many of the genes that determine floral organ identity are MADS box genes, including the *DEFICIENS* gene of snapdragon and the *AGAMOUS* (*AG*), *PISTILLATA* (*PI*), and *APETALA3* (*AP3*) genes of Arabidopsis (see Table 20.3). The MADS box genes share a characteristic, conserved nucleotide sequence known as a MADS box, which encodes a protein structure known as the MADS domain (**Figure 20.28A**). Adjacent to the MADS domain is an intervening region followed by the K domain, which is a coiled-coil region primarily involved in protein–protein interactions. The MADS box gene transcription factors form tetramers that bind to $CC(A/T)_6GG$ sequences, the

Figure 20.27 Conversion of cotyledons and vegetative leaves into petals by ectopic expression of Class E genes in combination with Class A and B genes. Arabidopsis plants overexpressing *SEP3/AP1/AP3/PI* (A) or *AP1/AP3/PI/SEP2/SEP3* (B) transgenes. (From Pelaz et al. 2001.)

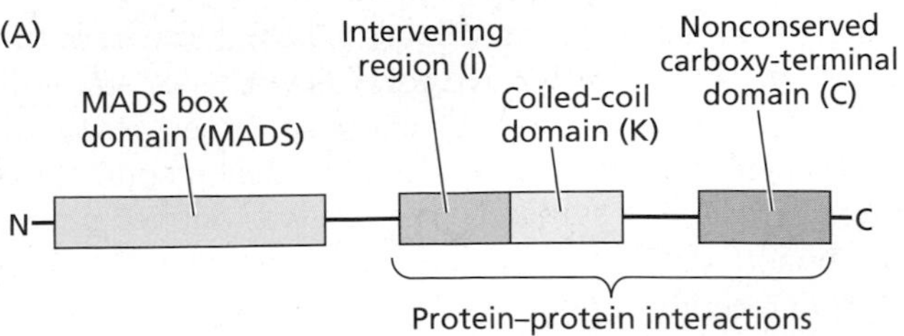

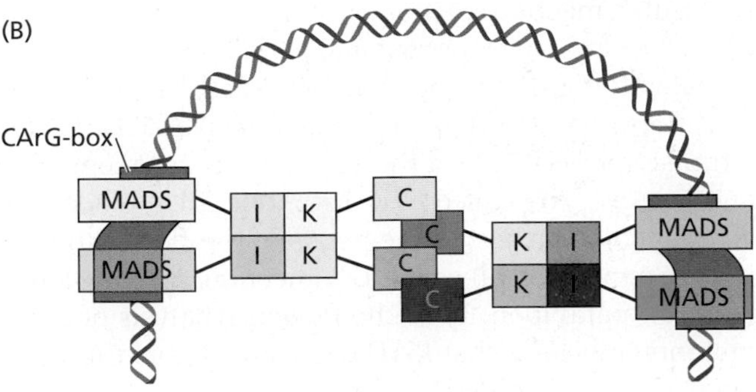

Figure 20.28 Model of MADS box domain interaction with target genes. (A) Domain structure of MADS box transcription factors. (B) Tetramers of MADS box transcription factors bind to a pair of CArG-box motifs in the regulatory regions of their target genes, which causes DNA bending. DNA bending may either activate or repress the target genes.

so-called CArG-box, in the regulatory regions of their target genes. When the tetramers bind two different CArG-boxes on the same target gene, the boxes are brought into close proximity, causing DNA bending (**Figure 20.28B**).

Not all homeotic genes are MADS box genes, and not all genes containing the MADS box domain are homeotic genes. For example, the homeotic gene *AP2* is a member of the AP2/ERF (ethylene-responsive element-binding factor) family of transcription factors, and the floral meristem identity gene *SOC1* is a MADS box gene.

To gain a more mechanistic understanding of the ABCE model, a biochemical interaction model, called the **Quartet Model**, has been proposed (**Figure 20.29**). In the Quartet Model, tetramers of combinations of the ABCE genes directly bind DNA and specify floral organs. The model is based on the observation that MADS box genes dimerize, and two dimers can come together, forming a tetramer. These tetramers are hypothesized to bind CArG-boxes on target genes and modify their expression (see Figure 20.28B). Although all MADS box proteins can form higher-order complexes, not all of these are able to bind DNA. For example, Class B factors (AP3 and PI) bind DNA only as heterodimers, whereas both homodimers and heterodimers of Classes A, C, and E can bind DNA. According to the model, tetramers composed of different homodimers and heterodimers of MADS domain proteins can exert combinatorial control over floral organ identity. For example, the AP3–PI heterodimer interacts *directly* with AP1 and SEP3 to promote petal formation, and *indirectly* with AG with the help of SEP3 acting as a scaffold. In general, the SEP proteins seem to act as cofactors that provide flower-specific activity to the ABC genes by making complexes of their products.

Class D genes are required for ovule formation

According to the ABCE model, carpel formation requires the activities of the Class C and E genes. However, it appears that a third group of MADS box genes closely related to the Class C genes is required for ovule forma-

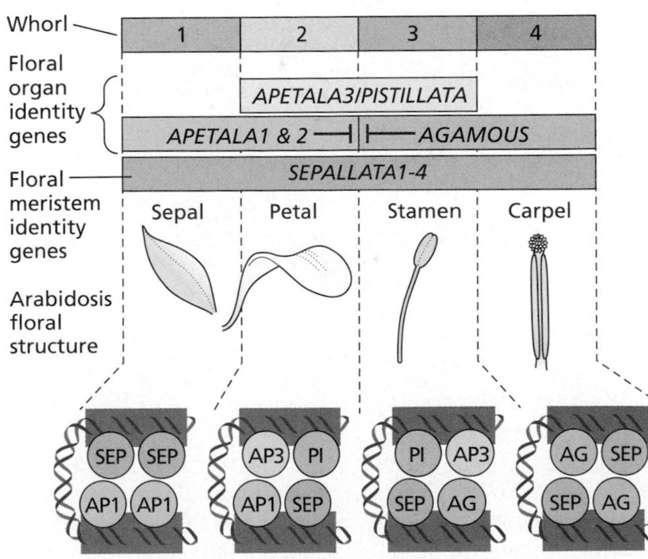

Figure 20.29 Quartet model of floral organ specification in *Arabidopsis*. In whorl 1, expression of Class A (*AP1* and *AP2*) and E (*SEP*) genes results in the formation of sepals. In whorl 2, expression of Class A (*AP1*, *AP2*), B (*AP3*, *PI*), and E (*SEP*) genes results in the formation of petals. In whorl 3, the expression of Class B (*AP3*, *PI*), C (*AG*), and E (*SEP*) genes causes the formation of stamens. In whorl 4, Class C (*AG*) and E (*SEP*) genes specify carpels. In addition, Class A activity (*AP1* and *AP2*) represses Class C activity (*AG*) in whorls 1 and 2, while Class C activity represses Class A activity in whorls 3 and 4. According to the Quartet Model, the identity of each of the floral organs is determined by four combinations of the floral homeotic proteins known as the MADS box proteins. Two dimers of each tetramer recognize two different DNA sites (termed CArG-boxes, shown here in yellow) on the same strand of DNA, which are brought into close proximity by DNA bending. Note that SEPALLATA proteins are present in all four complexes, serving to recruit the other proteins to the complex. The exact structures of the multimeric complexes are hypothetical.

tion. These ovule-specific genes have been called Class D genes. Since the ovule is a structure within the carpel, Class D genes are not, strictly speaking, "organ identity genes," although they function in much the same way in specifying ovules. Class D activities were first discovered in petunia. Silencing two MADS box genes known to be involved in floral development in petunia, *FLORAL-BINDING PROTEIN7/11* (*FBP7/11*), resulted in the growth of styles and stigmas in the locations normally occupied by ovules. When the *FBP11* was overexpressed in petunia, ovule primoridia formed on the sepals and petals.

In Arabidopsis, the ectopic expression of either *SHATTERPROOF1* or *SHATTERPROOF2* (*SHP1, SHP2*) or *SEEDSTICK* (*STK*) is sufficient to induce the transformation of sepals into carpeloid organs bearing ovules. Moreover, *stk/shp1/shp2* triple mutants lack normal ovules. Thus, in addition to the Class C and E genes, Class D genes are required for normal ovule development.

Floral asymmetry in flowers is regulated by gene expression

While many flowers, such as those of Arabidopsis, are radially symmetrical, many plants have evolved flowers with bilateral symmetry, which allowed them to form specialized structures to attract pollinators. For example, *Antirrhinum* flowers show distinct differences in the shapes of the upper (dorsal) petals compared with the lower (ventral) petals (**Figure 20.30**). How has this occurred? Again, as with the ABCE model, genetics has provided the answer. Mutations that disrupt the development of zygomorphic flowers have been known since the eighteenth century. Carl Linnaeus was the first to describe a naturally occurring mutant of toadflax (*Linaria vulgaris*) that converted the bilaterally symmetrical flower to a radially symmetrical form (**Figure 20.31**). Flowers of the genus *Linaria* normally have corollas with four stamens and a single nectar spur. The bizarre specimen described by Linnaeus had five stamens and five nectar spurs. This abnormal, radially symmetrical state was dubbed *peloria* by Linnaeus, from the Greek word for "monster."

More recently, analogous ("peloric") mutants in *Antirrhinum majus* have allowed a genetic dissection of the molecular mechanisms of floral symmetry specification. The cloning of the mutated gene *RADIALIS* (*RAD*) revealed a regulatory mechanism by which *RAD* controls floral asymmetry (see Figure 20.30). *RAD* encodes a transcription factor of the MYB family that represses another key gene called *DIVARICATA* (*DIV*). When *DIV* is mutated, it causes all the petals of the flower to look like the upper (dorsal) petals. *DIV* therefore specifies lower (ventral) petal identity in the flower. Analysis of other mutants indicated that *RAD* determines upper (dorsal) petal identity. The RAD transcription factor is activated by two other genes, *CYCLOIDEA* and *DICHOTOMA*, that are expressed in the dorsal petals. The expression of *RAD* thus allows *DIV* to be repressed in the dorsal part of the flower. Where *RAD* is not expressed in the bottom of the flower, *DIV* is expressed and specifies ventral fate.

Thus far, our understanding of floral development has been based primarily on two model species, Arabidopsis and *Antirrhinum*. One of the challenges of the future will be to explore the variations in the gene networks that regulate floral development across a wider spectrum of flowering plants. A second challenge will be to try to understand how floral developmental pathways evolved from nonflowering ancestors. Such studies may one day lead to the solution of Darwin's "abominable mystery"—the evolution of angiosperms.

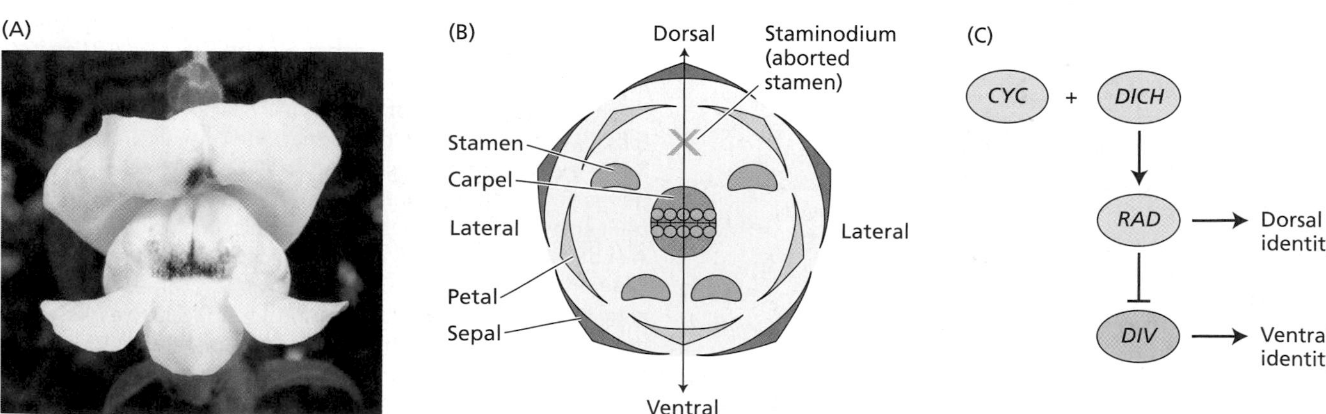

Figure 20.30 Floral asymmetry in *Antirrhinum*. (A, B) *Antirrhinum* flowers exhibit bilateral symmetry. (C) The *DIVARICATA* (*DIV*) gene encodes a MYB transcription factor that promotes ventral identity throughout the flower. *CYCLOIDEA* (*CYC*) and *DICHOTOMA* (*DICH*) encode related transcription factors that activate the *RADIALIS* (*RAD*) gene. The RAD protein antagonizes DIV in the dorsal part of the flower and limits its activity to the lateral and ventral domains. (B from Krizek and Fletcher 2005.)

Figure 20.31 The *peloria* mutant of toadflax (*Linaria vulgaris*). The normal flower of *Linaria* with bilateral symmetry is shown on the left, and the radially symmetrical *peloria* mutant is shown on the right. The peloric toadflax flower is now known to be caused by inactivation of the *Linaria CYCLOIDEA* gene through DNA methylation. (From Busch and Zachgo 2009.)

SUMMARY

Formation of the floral organs (sepals, petals, stamens, and carpels) occurs at the shoot apical meristem and is linked to both internal (autonomous) and external (environmental) signals. A network of genes that control floral morphogenesis has been identified in several species.

Floral Evocation: Integrating Environmental Cues

- Internal (autonomous) and external (environment-sensing) control systems enable plants to precisely regulate and time flowering for reproductive success.
- Two of the most important seasonal responses that affect floral development are photoperiodism (response to changes in day length) and vernalization (response to prolonged cold).
- Synchronized flowering favors crossbreeding and helps ensure seed production under favorable conditions.

The Shoot Apex and Phase Changes

- In plants, the transition from juvenile to adult is usually accompanied by changes in vegetative characteristics (**Figure 20.1**).

Circadian Rhythms: The Clock Within

- Circadian rhythms are based on an endogenous oscillator, not the presence or absence of light; they are defined by three parameters: period, phase, and amplitude (**Figure 20.3**).
- Temperature compensation prevents temperature changes from affecting the period of the circadian clock.
- Phytochromes and cryptochromes entrain the circadian clock.

Photoperiodism: Monitoring Day Length

- Plants can detect seasonal changes in day length at latitudes away from the equator (**Figure 20.4**).
- Flowering in LDPs requires that the day length exceed a certain duration, called the critical day length. Flowering in SDPs requires a day length that is less than the critical day length (**Figure 20.6**).
- Leaves perceive the photoperiodic stimulus in both LDPs and SDPs.
- Plants monitor day length by measuring the length of the night; flowering in both SDPs and LDPs is determined primarily by the duration of darkness (**Figure 20.7**).
- For both LDPs and SDPs, the dark period can be made ineffective by interruption with a short exposure to light (a night break) (**Figure 20.8**).
- The flowering response to night breaks shows a circadian rhythm, supporting the clock hypothesis (**Figure 20.9**).
- In the coincidence model, flowering is induced in both SDPs and LDPs when light exposure is coincident with the appropriate phase of the oscillator.
- CO (in Arabidopsis) and Hd1 (in rice) regulate flowering by controlling the transcription of floral stimulus genes (**Figure 20.10**).
- CO protein is degraded at different rates in the light versus the dark. Light enhances the *stability* of CO, allowing it to accumulate during the day; in the dark it is rapidly degraded.
- The effects of red and far-red night breaks implicate phytochrome in the control of flowering in SDPs and LDPs (**Figures 20.11**, **20.12**).

- Flowering in LDPs is promoted when the inductive light treatment coincides with a peak in light sensitivity, which follows a circadian rhythm (**Figure 20.13**).

Vernalization: Promoting Flowering with Cold

- In sensitive plants, a cold treatment is required for plants to respond to floral signals such as inductive photoperiods (**Figures 20.14**, **20.15**).
- For vernalization to occur, active metabolism is required during the cold treatment.
- After vernalization, the *FLC* gene is epigenetically switched off for the remainder of the plant's life cycle, permitting flowering in response to long days to occur in Arabidopsis (**Figure 20.16**).
- The epigenetic regulation of *FLC* involves stable changes in chromatin structure.
- A variety of vernalization pathways have evolved in flowering plants.

Long-distance Signaling Involved in Flowering

- In photoperiodic plants, a long-range signal is transmitted in the phloem from the leaves to the apex, permitting floral evocation (**Figures 20.17**, **20.18**).

The Identification of Florigen

- FT is a small, globular protein that exhibits the properties that would be expected of florigen.
- FT protein moves via the phloem from the leaves to the shoot apical meristem under inductive photoperiods. In the meristem, FT forms a complex with the transcription factor FD to activate floral identity genes (**Figure 20.19**).
- The four distinct pathways that control flowering converge to increase the expression of key floral regulators: *FT* in the vasculature and *SOC1*, *LFY*, and *AP1* in the meristem (**Figure 20.19**).

Floral Meristems and Floral Organ Development

- The four different types of floral organs are initiated sequentially in separate, concentric whorls (**Figure 20.22**).
- Formation of floral meristems requires active floral meristem identity genes such as *SOC1*, *AP1*, and *LFY* in Arabidopsis.
- Mutations in homeotic floral identity genes alter the types of organs produced in each of the whorls (**Figures 20.23, 20.24**).
- The ABC model suggests that organ identity in each whorl is determined by the combined activity of three organ identity genes (**Figure 20.25**).
- Expression of Class E floral meristem identity genes (e.g., *SEPALLATA*) is required for expression of Class A, B, and C genes (**Figure 20.26**).
- Many floral organ identity genes encode MADS domain–containing transcription factors that function as heterotetramers (**Figure 20.28, Table 20.3**). The Quartet Model describes how these transcription factors might act together to specify floral organs (**Figure 20.29**).
- Variations on the ABCE model can explain the diversity of angiosperm flower structures (**Figures 20.30, 20.31**).

WEB MATERIAL

- **WEB TOPIC 20.1 Climate Change has Caused Measurable Changes in Flowering Time of Wild Plants** Plants are able to sense as little as a 1°C difference in temperature, and increasing ambient temperature accelerates flowering in many species.
- **WEB TOPIC 20.2 Contrasting the Characteristics of Juvenile and Adult Phases of English Ivy (*Hedera helix*) and Maize (*Zea mays*)** A table of juvenile versus adult morphological characteristics is presented.
- **WEB TOPIC 20.3 Flowering of Juvenile Meristems Grafted to Adult Plants** The competence of juvenile meristems to flower can be tested in grafting experiments.
- **WEB TOPIC 20.4 Competence and Determination are Two Stages in Floral Evocation** Experiments have been carried out to define competence and determination during floral evocation.
- **WEB TOPIC 20.5 Characteristics of the Phase-Shifting Response in Circadian Rhythms** Petal movements in *Kalanchoë* have been used to study circadian rhythms.
- **WEB TOPIC 20.6 The Contrasting Effects of Phytochromes A and B on Flowering** PhyA and phyB affect flowering in Arabidopsis and other species.

- **WEB TOPIC 20.7 Support for the Role of Blue-Light Regulation of Circadian Rhythms** ELF3 plays a role in mediating the effects of blue light on flowering time.
- **WEB TOPIC 20.8 Genes That Control Flowering Time** A discussion of genes that control different aspects of flowering time is presented.
- **WEB TOPIC 20.9 Regulation of Flowering in Canterbury Bells by Both Photoperiod and Vernalization** Short days acting on the leaf can substitute for vernalization at the shoot apex in Canterbury bells.
- **WEB TOPIC 20.10 The Self-propagating Nature of the Floral Stimulus** In certain species, the induced state can be transferred by grafting almost indefinitely.
- **WEB TOPIC 20.11 Examples of Floral Induction by Gibberellins in Plants with Different Environmental Requirements for Flowering** A table of the effects of gibberellins on plants with different photoperiodic requirements is presented.
- **WEB TOPIC 20.12 Variations of the ABCE Model are Found in Other Species** Variations in the ABCE model are associated with contrasting floral morphology in different monocots and eudicots.

available at plantphys.net

Suggested Reading

Amasino R. (2010) Seasonal and developmental timing of flowering. *Plant J.* 61: 1001–1013. DOI: 10.1111/j.1365-313X.2010.04148.x.

Andrés, F., and Coupland, G. (2012) The genetic basis of flowering responses to seasonal cues. *Nat. Rev. Genet.* 13: 627–639. DOI: 10.1038/nrg3291

Busch, A., and Zachgo, S. (2009) Flower symmetry evolution: Towards understanding the abominable mystery of angiosperm radiation. *BioEssays* 31: 1181–1190.

Causiera, B., Schwarz-Sommerb, Z., and Davies, B. (2010) Floral organ identity: 20 years of ABCs. *Semin. Cell Dev. Biol.* 21: 73–79.

Huijser, P., and Schmid, M. (2011) The control of developmental phase transitions in plants. *Development* 138: 4117–4129. DOI:10.1242/dev.063511

Jaeger, E., Pullen, N., Lamzin, S., Morris, R. J., and Wigge, P. A. (2013) Interlocking feedback loops govern the dynamic behavior of the floral transition in *Arabidopsis*. *Plant Cell* 25: 820–833.

Krizek, B. A., and Fletcher, J. C. (2005) Molecular mechanisms of flower development: An armchair guide. *Nat. Rev. Genet.* 6: 688–698.

Lee, J., and Lee, I. (2010) Regulation and function of SOC1, a flowering pathway integrator. *J. Exp. Bot.* 61: 2247–2254.

Liu, L., Liu, C., Hou, X., Xi, W., Shen, L., Tao, Z., Wang, Y., and Yu, H. (2012) FTIP1 is an essential regulator required for florigen transport. *PLOS Biol.* 10(4): e1001313. DOI:10.1371/ journal.pbio.1001313

Liu, L., Zhu, Y., Shen, L., and Yu, H. (2013) Emerging insights into florigen transport. *Curr. Opin. Plant Biol.* 16: 607–613.

Rijpkemaa, A. S., Vandenbusscheb, M., Koesc, R., Heijmansd, K., and Gerats, T. (2010) Variations on a theme: Changes in the floral ABCs in angiosperms. *Semin. Cell Dev. Biol.* 21: 100–107.

Song, Y. H., Ito, S., and Imaizumi, T. (2013) Flowering time regulation: Photoperiod- and temperature-sensing in leaves. *Trends Plant Sci.* 18: 575–583.

Taoka, K.-I., Ohki, I., Tsuji, H., Kojima, C., and Shimamoto, K. (2013) Structure and function of florigen and the receptor complex. *Trends Plant Sci.* 18: 287–294.

21 Gametophytes, Pollination, Seeds, and Fruits

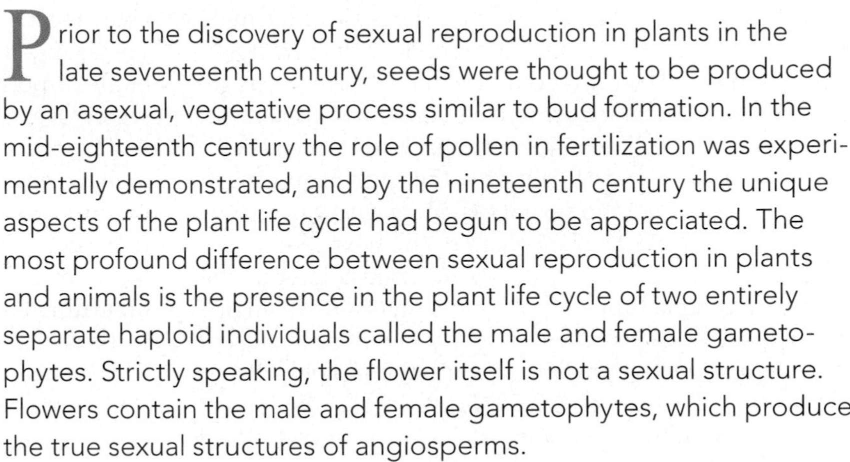

Prior to the discovery of sexual reproduction in plants in the late seventeenth century, seeds were thought to be produced by an asexual, vegetative process similar to bud formation. In the mid-eighteenth century the role of pollen in fertilization was experimentally demonstrated, and by the nineteenth century the unique aspects of the plant life cycle had begun to be appreciated. The most profound difference between sexual reproduction in plants and animals is the presence in the plant life cycle of two entirely separate haploid individuals called the male and female gametophytes. Strictly speaking, the flower itself is not a sexual structure. Flowers contain the male and female gametophytes, which produce the true sexual structures of angiosperms.

We will begin our discussion with an overview of the plant life cycle and how it has evolved from that of simpler algal forms to that of flowering plants. Next we will discuss the development of the male and female gametophytes, which produce gametes. As sessile organisms, plants depend on vectors such as wind or insects to bring about pollination and fertilization. As we will see, plants are not entirely passive in this process: they have evolved complex mechanisms, both anatomical and biochemical, that encourage outcrossing. The final step in the process is the development of the seed and fruit—the structures that protect and nourish the embryo, and deliver it to a suitable substrate on which to germinate and establish itself as a new seedling.

Development of the Male and Female Gametophyte Generations

The plant life cycle differs fundamentally from that of animals in that it encompasses two separate multicellular generations, a diploid (2*N*) *sporophyte generation* and a haploid (1*N*) *gametophyte generation* (see Chapter 1). The presence of two genetically distinct multicellular stages in the plant life cycle is called **alternation of generations**. Alternation of generations takes place in the male and female reproductive organs of the flower—the stamens (androecium) and the carpels (gynoecium).

Because of alternation of generations, there is a crucial difference between plant and animal life cycles in the fates of the products of meiosis. In animals, the haploid cells produced by meiosis differentiate directly into the gametes—sperm or egg. In contrast, the haploid cells produced by meiosis in plants differentiate into spores—microspores (male) or megaspores (female) (**Figure 21.1**) (see Chapter 2 for a review of meiosis). The microspores and megaspores undergo mitotic divisions to produce haploid individuals called *male gametophytes* (or *microgametophytes*) and *female gametophytes* (or *megagametophytes*). The male gametophytes form in the anther of the stamen, while the female gametophyte develops inside the ovule. At maturity, specialized cells within the male and female gametophytes divide mitotically to produce the gametes—sperm and egg. The presence of the haploid gametophyte generation in the plant life cycle means that gametes in plants are produced by mitosis rather than by meiosis.

In the final step—fertilization—the egg cell and one of the sperm cells undergo sexual fusion, or *syngamy*, to produce the 2*N zygote*, the first stage of the next sporophyte generation. In addition, as we will discuss later in the chapter, a unique type of gamete fusion occurs in angiosperms: a second sperm cell fuses with a diploid *central cell* of the female gametophyte to produce the triploid *primary endosperm cell*, which goes on to form the nutritive endosperm tissue of the seed. The participation of two sperm cells during fertilization, unique to flowering plants, is called *double fertilization*.

Based on reconstructed phylogenies of land plants, the plant life cycle has evolved from a condition with a dominant, free-living, haploid gametophyte to one with a dominant, free-living, diploid sporophyte. For a discussion of the evolution of diploidy in plants see **WEB TOPIC 21.1**.

Formation of Male Gametophytes in the Stamen

The male gametophyte is formed in the stamen of the flower. Typically, the stamen is made up of a delicate filament attached to an anther composed of four microsporangia arranged in opposite pairs (**Figure 21.2A**). Each pair of microsporangia is separated from the other by a central region of sterile tissue surrounding a vascular bundle.

The precise sequence of microsporangium development varies from species to species. In Arabidopsis, the mature anther contains **archesporial cells**, the cells that ultimately undergo meiosis, surrounded by four somatic layers: the epidermis, endothecium, middle layer, and tapetum. These layers are originally derived from the three layers of the floral meristem (L1, L2, and L3). The L1 layer becomes the epidermis, and the L2 layer gives rise

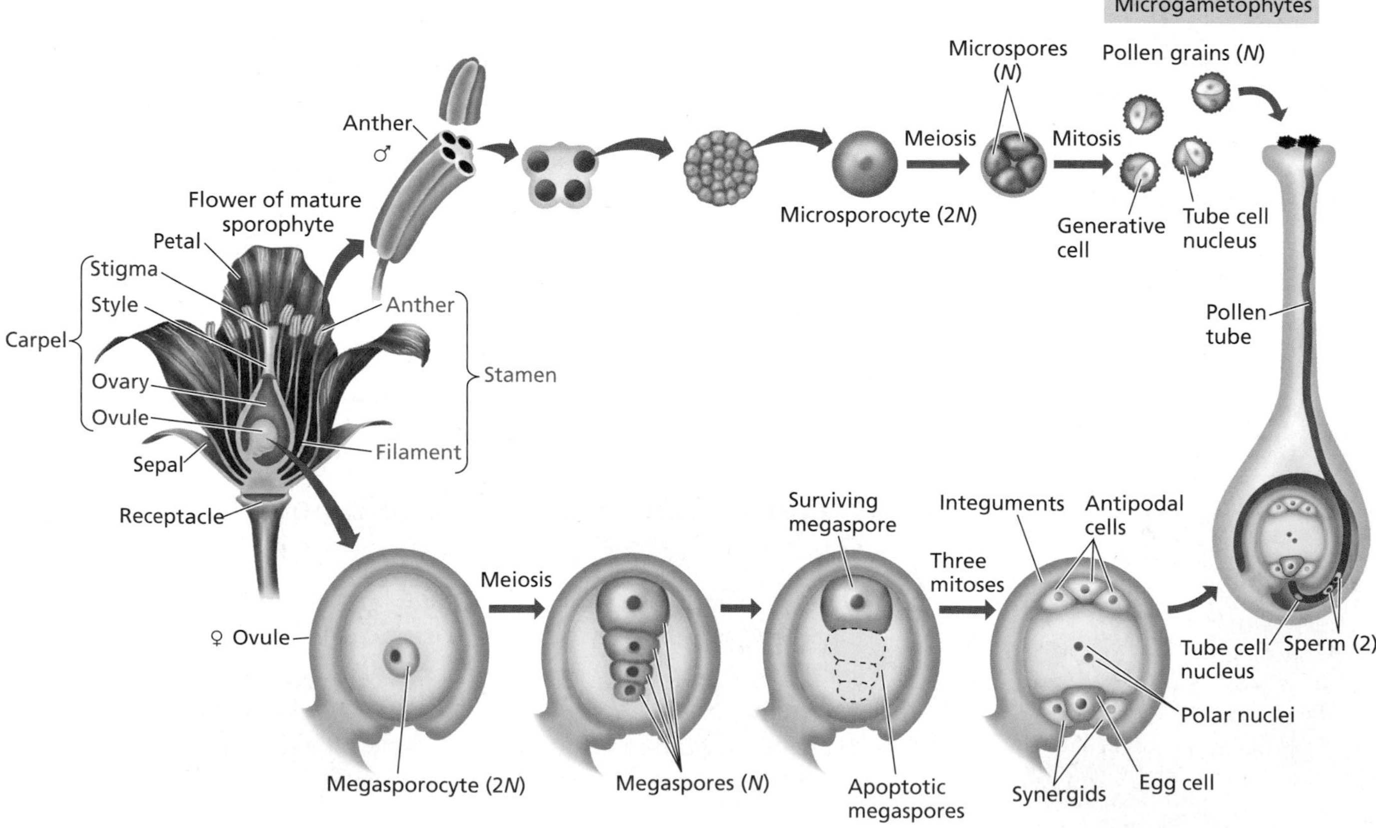

Figure 21.1 Angiosperm life cycle.

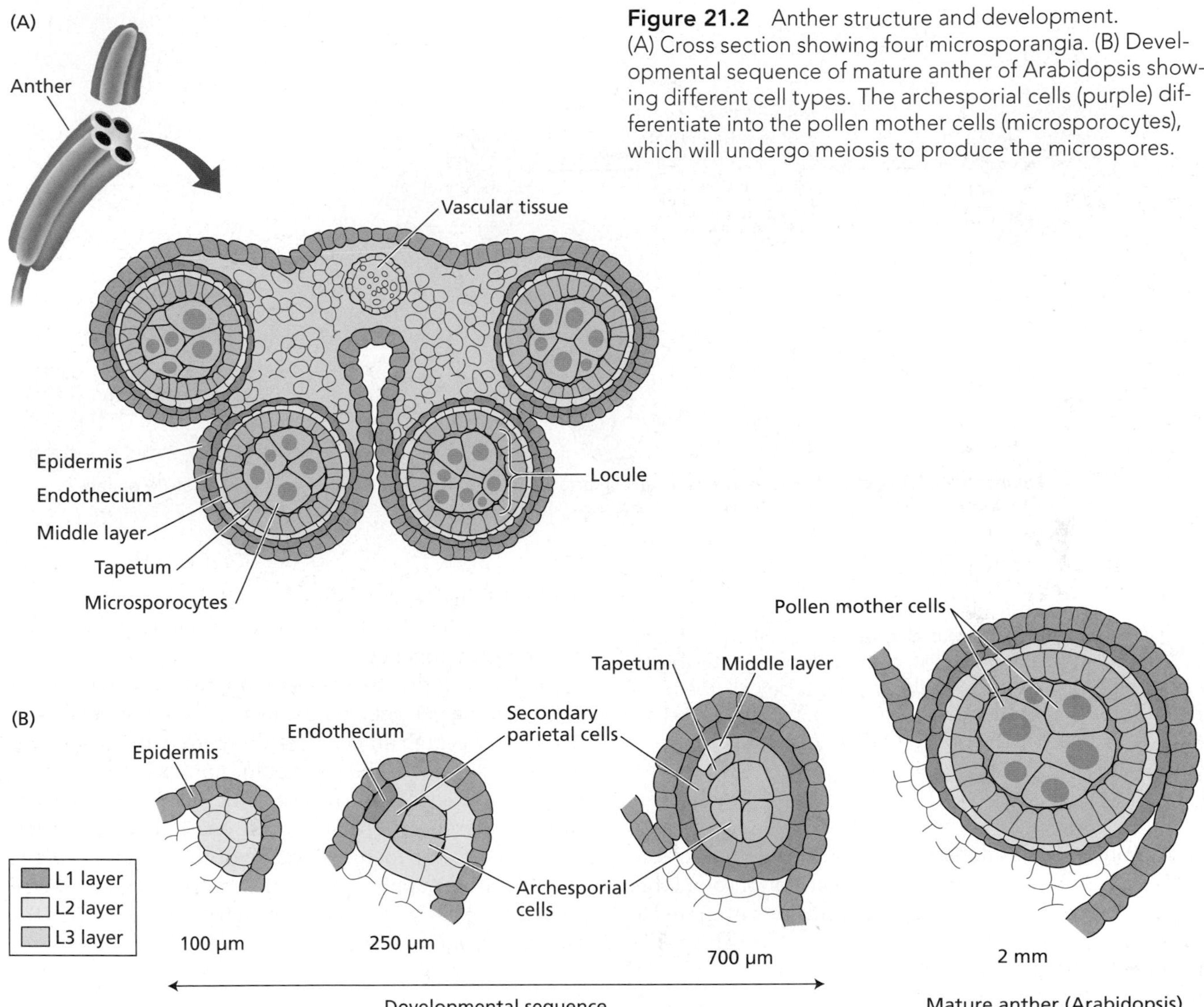

Figure 21.2 Anther structure and development. (A) Cross section showing four microsporangia. (B) Developmental sequence of mature anther of Arabidopsis showing different cell types. The archesporial cells (purple) differentiate into the pollen mother cells (microsporocytes), which will undergo meiosis to produce the microspores.

to the archesporial cells as well as the inner surrounding layers, as shown in **Figure 21.2B**. The central region containing the archesporial cells is called the **locule**.

Pollen grain formation occurs in two successive stages

The development of the male gametophyte, or pollen grain, is temporally divided into two phases: microsporogenesis and microgametogenesis. During **microsporogenesis**, archesporial cells within the locules differentiate into **microsporocytes**, or **pollen mother cells**—diploid cells capable of undergoing meiosis to produce the microspores (**Figure 21.3A**). The microsporocytes undergo meiosis, resulting in a tetrad of haploid **microspores** joined together at their cell walls, which are composed largely of the polysaccharide callose, a (1,3)-β-glucan. The **tapetum**, a layer of secretory cells surrounding the locule, secretes the hydrolytic enzyme *callase* and other cell wall–degrading enzymes into the locule; this partially digests the cell walls and separates the tetrad into individual microspores (see Figure 21.3A). In some insect-pollinated species, pollen is normally shed as tetrads, as in the common heather (*Calluna vulgaris*), or in even larger assemblages termed **polyads**, as in *Acacia*. Although wild-type Arabidopsis produces individual microspores, in *quartet* (*qrt*) mutants tetrad dissolution is blocked. Nevertheless, the pollen grains of *qrt* mutants develop normally and are fertile.

Once the microspores have formed inside the anther locules—either separated or as tetrads or polyads—the microsporogenesis phase of microgametophyte development is completed. The second stage is **microgametogenesis**, the formation of male gametes.

During microgametogenesis, the haploid microspore develops mitotically into the mature male gametophyte, composed of the **vegetative** (or **tube**) **cell** and two sperm cells (**Figure 21.3B**). Prior to the first mitotic division, the

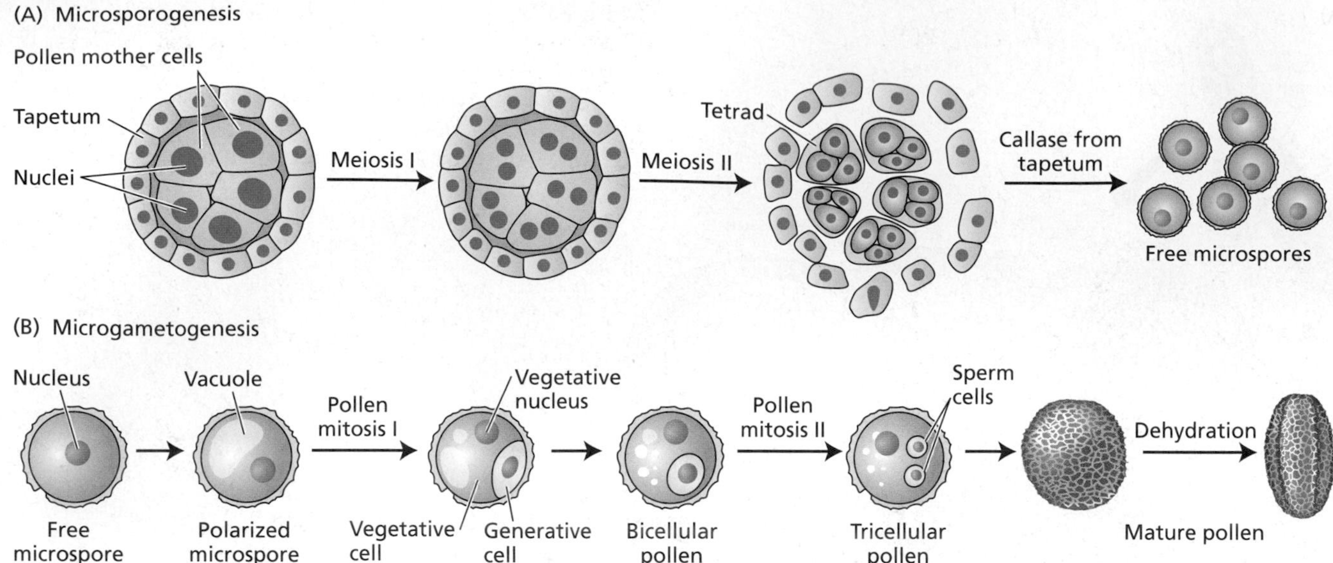

Figure 21.3 Male gametophyte development. (A) Microsporogenesis. Pollen mother cells undergo meiosis to produce a tetrad of microspores. (B) Microgametogenesis. The haploid nucleus divides mitotically to produce the tube cell (vegetative cell) and the generative cell (bicellular stage). After being engulfed by the tube cell, the generative cell divides mitotically to produce two sperm cells (tricellular stage). As the pollen grain matures, it forms a specialized cell wall.

microspore expands substantially, a process associated with cell wall biosynthesis and the formation of a large vacuole. In parallel, the microspore nucleus migrates to the cell wall, producing a *polarized microspore*. The polarized microspore then undergoes a highly asymmetric cell division (pollen mitosis I), giving rise to a large vegetative cell and a small *generative cell* (or male germ cell). At first the generative cell remains attached to the microspore cell wall and is enclosed by a hemispherical wall of callose, which also serves to separate the generative cell from the vegetative cell. This callose layer breaks down and the generative cell is engulfed by the vegetative cell, resulting in a unique anatomical structure: a cell within a cell (bicellular stage). The engulfed generative cell subsequently takes on an elongated or spindlelike shape, which may assist in passage of the generative cell through the dynamic protoplasm of the rapidly growing pollen tube. During maturation, pollen grains accumulate carbohydrate or lipid reserves to support the active metabolism required for rapid germination and pollen tube growth. At this stage the pollen is usually released from the anther by dehiscence (opening) of the anther wall, and the generative cell divides to produce the two sperm cells (pollen mitosis II) only after the pollen grain has landed on a stigma and a pollen tube has formed. In many plants, however, the generative cell undergoes pollen mitosis II while still inside the anther (tricellular stage). In either case, the production of the two sperm cells signals the end of microgametogenesis.

Depending on the species, tapetal cells may either remain at the periphery of the locule (as they do in Arabidopsis) or become amoeboid and migrate into the locule, intermingling with the developing microspores. In both cases tapetal cells perform a secretory function and eventually undergo programmed cell death, releasing their contents into the locule. Because of the essential role of tapetal cells in supplying enzymes, nutrients, and cell wall constituents to the developing pollen grains, defects in the tapetum usually cause abnormal pollen development and decreased fertility.

The multilayered pollen cell wall is surprisingly complex

The outer surfaces of pollen grain cell walls exhibit a remarkable variety of sculptural features that play important ecological roles in the transfer of pollen from flower to flower (**Figure 21.4A**). Equally complex, however, are the multiple subsurface wall layers that provide a labyrinth of internal spaces where lipids and proteins may be deposited (**Figure 21.4B**).

Initiation of pollen cell wall formation begins in the microspores immediately following meiosis. An ephemeral callose wall is the first of several layers to be deposited by the microspore onto the cell surface. This is followed by the *primexine* (a precursor of the *sexine*), the *nexine*, and finally the *intine*. (Note that because the microspore is the source of these layers, the innermost layer is the last to be deposited.)

The primexine, composed largely of polysaccharide, acts as a template that guides the accumulation of *sporopollenin*, the main structural component of the *exine*, or

(A)

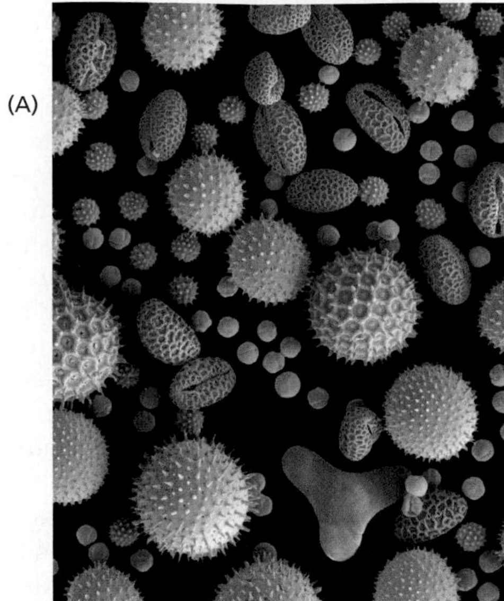

(B)

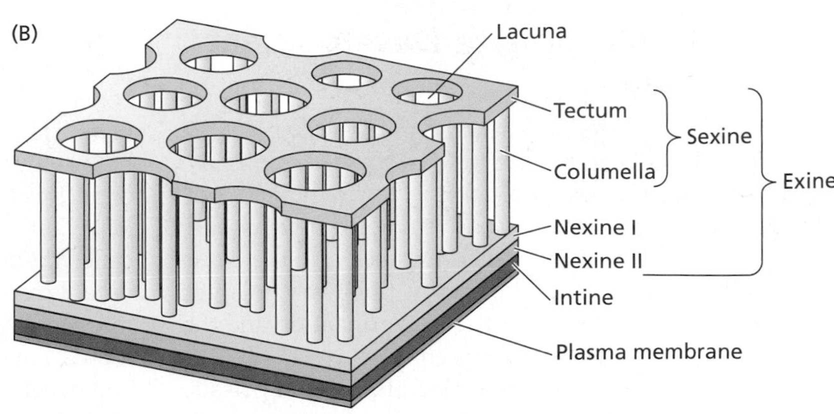

Figure 21.4 Pollen grain cell wall structure. (A) Scanning electron microscope image of pollen grains from different species exhibiting distinct ornamentation. (B) Architecture of a typical pollen cell wall, showing the inner and outer layers and sculptural elements. The sexine may be tectate (with a tectum), semi-tectate (with a partial tectum), or intectate (without a tectum). The diagram shows a pollen wall with a tectum, which creates a smooth surface.

outer layer, which includes both the nexine and sexine. While the microspores are still in a tetrad, the early exine is assembled from sporopollenin precursors synthesized and secreted by the microspores themselves. However, once the outermost, callose walls are dissolved and the microspores are released from the tetrad, most of the sporopollenin precursors are supplied by the tapetum. The *intine,* or inner layer, consists primarily of cellulose and pectins.

Recent studies in Arabidopsis suggest that the sporopollenin polymer consists of phenolic and fatty acid–derived constituents that are covalently coupled, similar to lignin and suberin. In addition, most pollen grain walls include elongated zones called *apertures* where the exine is either thin or missing (**Figure 21.5**). Pollen tubes emerge through the apertures when the pollen grain germinates on a compatible stigma.

The number of apertures and pattern of exine sculpturing are characteristic of an angiosperm family, genus, and often species. Smooth pollen is associated with wind pollination, as in oaks (*Quercus*) and grasses (maize, or corn [*Zea mays*]), while plants pollinated by insects, birds, and mammals tend to have highly sculptured patterns consisting of spines, hooks, or sticky threadlike projections, which enable the pollen to adhere to foraging pollinators. Because sporopollenin is resistant to decay, pollen is well represented in the fossil record, and the distinctive patterns of the exine are useful for identifying which species were present, as well as suggesting the conditions of early climates. In species with dry stigmas (discussed later in the chapter), such as Arabidopsis, the tapetum also coats the pollen grains with **tryphine**, a sticky, adhesive layer that covers the exine layer. Tryphine is rich in proteins, fatty acids, waxes, and other hydrocarbons.

Figure 21.5 Scanning electron microscope images of Arabidopsis pollen grain. (A) Arabidopsis pollen showing two of its three apertures, which are elongated furrows where the wall is weaker and thinner. (B) Higher magnification of the tectate exine of an Arabidopsis pollen grain. (Courtesy of D. Twell and S. Hyman.)

(A)

(B)

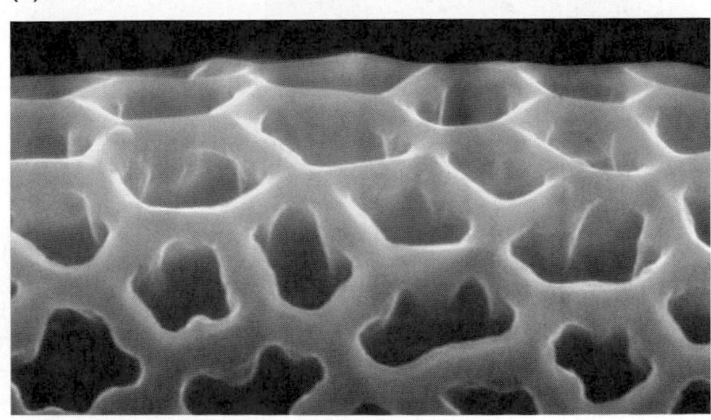

Female Gametophyte Development in the Ovule

In angiosperms, *ovules* are located within the *ovary* of the *gynoecium*, the collective term for the carpels. Ovules are the sites of both megasporogenesis and megagametogenesis. Upon fertilization of the female gamete, or egg, by a sperm cell, embryogenesis is initiated and the ovule develops into a seed. Simultaneously, the ovary enlarges and becomes a fruit. We will discuss fertilization and the development of fruits later in this chapter.

Ovule primordia arise in a specialized ovary tissue called the *placenta*. The locations of placental tissue vary among different plant groups, and include the marginal, parietal, axile, basal, and free-central types of placentation (see WEB TOPIC 21.2). The type of placentation within the ovary determines the positions and arrangement of the seeds within the fruit.

The Arabidopsis gynoecium is an important model system for studying ovule development

The gynoecium of Arabidopsis, as in many members of the Brassicaceae (mustard family), consists of two fused carpels, referred to as *valves*, separated by a medial partition called the *septum* (**Figure 21.6**). The edges of valves and the septum are joined at a strip of tissue called the *replum*, which plays an important role in the dehiscence of the dry fruit. In each carpel there are two strips of placental tissue associated with the septum on either side of the gynoecium.

Ovule primordia first appear along the placenta as conical projections with rounded tips (**Figure 21.7**). Three zones can already be distinguished at an early stage of primoridum development: the proximal region at the base, which gives rise to the stalklike *funiculus*; the distal or *micropylar* region at the tip, which produces the *nucellus*, where meiosis takes place; and the central region, called the *chalaza*, which gives rise to the *integuments*, the outer layers of the ovule. The cell that will differentiate into the *megaspore mother cell* is clearly visible in the primordial nucellus because of its large size, large nucleus, and dense cytoplasm.

Typically there are two integument layers: inner and outer. The inner integument forms a ridge some distance behind the apex of the nucellus, followed by the outer integument layer (see Figure 21.7). The two integument layers continue to grow over the nucellus until they reach the micropyle. At the same time, the funiculus curves slightly, causing the ovule to bend inward toward the septum. In this way the micropyle is brought closer to the *transmitting tract*, a specialized region within the septum, through which the pollen tube grows during pollination.

The vast majority of angiosperms exhibit *Polygonum*-type embryo sac development

The development of the female gametophyte, or *embryo sac*, is more complex and more diverse than that of the male gametophyte. According to one classification scheme, there are more than 15 different patterns of embryo sac development in angiosperms. The most common pattern was first described in the genus *Polygonum* ("knotweed") and is therefore called the *Polygonum* type of embryo sac. We will discuss this type of embryo sac development here; deviations from *Polygonum*-type development are described in WEB TOPIC 21.3.

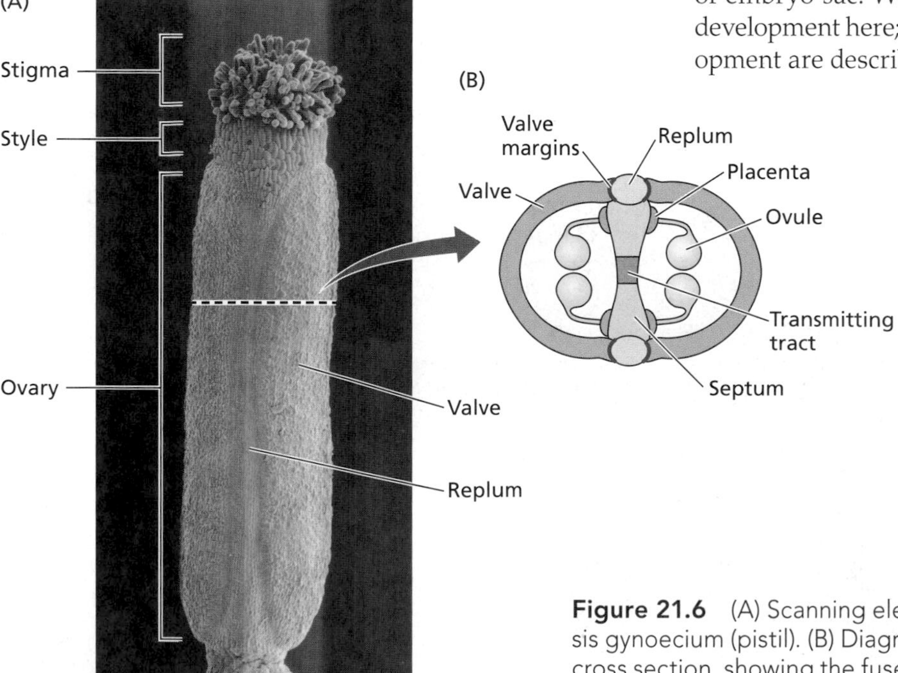

Figure 21.6 (A) Scanning electron microscope image of Arabidopsis gynoecium (pistil). (B) Diagram of the Arabidopsis ovary as seen in cross section, showing the fused carpel structure. Each valve represents a separate carpel. (A from Gasser and Robinson-Beers 1993.)

Figure 21.7 Ovule morphogenesis in Arabidopsis showing several stages of development and tissue types. The L1 layer (pink) gives rise to the epidermis, the L2 layer (yellow) gives rise to the bulk of the integuments and chalaza, and the L3 layer (blue) gives rise to the funiculus.

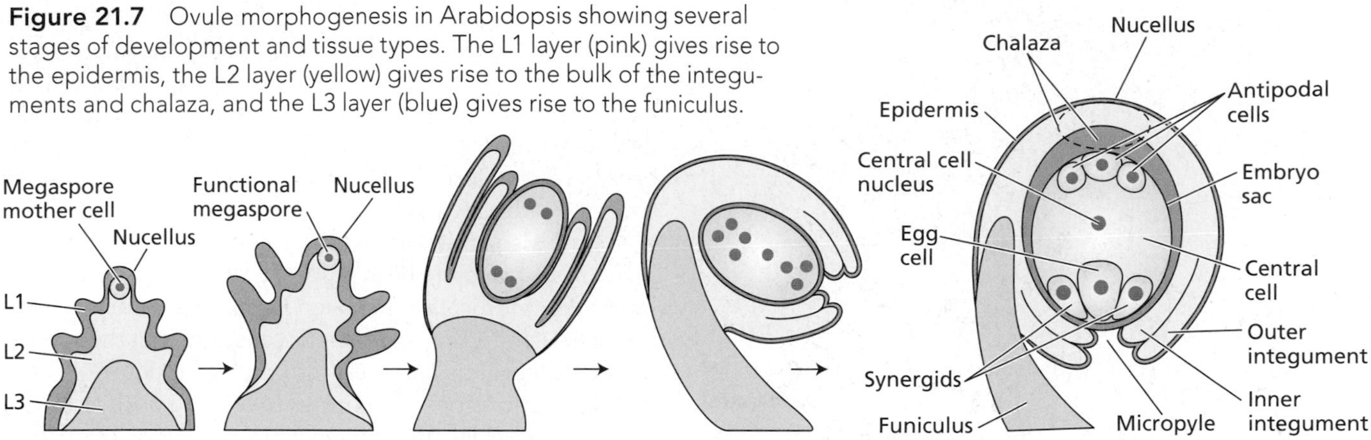

Functional megaspores undergo a series of free nuclear mitotic divisions followed by cellularization

The archesporial cell within the nucellus differentiates into the **megaspore mother cell**, the cell that undergoes meiosis. In the *Polygonum* type of embryo sac, meiosis of the diploid megaspore mother cell produces four haploid megaspores (**Figure 21.8**). Three of the megaspores, usually those at the micropylar end of the nucellus, subsequently undergo programmed cell death, leaving only one functional megaspore. Functional megaspores then undergo three rounds of free nuclear mitotic divisions (mitoses without cytokinesis) to produce a *syncytium*—a multinucleate cell formed by nuclear divisions. The result is an eight-nucleate, immature embryo sac. Four of the nuclei then migrate to the chalazal pole, and the other four migrate to the micropylar pole. Three of the nuclei at each pole undergo cellularization, while the remaining two nuclei, called **polar nuclei**, migrate toward the central region of the embryo sac, which also contains a large vacuole. The cytoplasm and the two polar nuclei develop their own plasma membrane and cell wall, giving rise to a large binucleate cell. The fully cellularized embryo sac represents the mature female gametophyte or embryo sac. At maturity, the *Polygonum*-type embryo sac consists of seven cells and eight nuclei.

The three cells at the chalazal end of the embryo sac are termed the **antipodal cells**. Ultrastructural studies have shown that the antipodal cells contain large membrane invaginations, perhaps indicative of a role in nutritional exchange or hormonal signaling. However, antipodal cells are absent in the order Nymphaeales, which includes the water lilies, as well as in members of the evening primrose family (Onagraceae). As a result, these two plant groups have only four-nucleate embryo sacs at maturity. In many other species, including Arabidopsis, the antipodal cells degenerate prior to fertilization, which suggests they do not play an essential role in fertilization. In contrast, in members of the grass family (Poaceae) the antipodal cells proliferate, so they may play a role in fertilization in the grasses.

The **egg cell** (the female gamete that combines with a sperm cell to form the zygote) and the two **synergid cells** are located at the micropylar end of the embryo sac and are collectively referred to as the **egg apparatus** (**Figure 21.9**). An additional feature is the presence of a **filiform apparatus** at the extreme micropylar end of each synergid. The filiform apparatus consists of a convoluted, thickened cell wall that increases the surface area of the plasma membrane. As we will discuss later in the chapter, synergid cells are involved in the final stages of pollen

Figure 21.8 Developmental stages in the *Polygonum*-type of embryo sac of Arabidopsis. The stages are as described in the text. The beige areas represent cytoplasm, the white areas represent vacuoles, and the purple circles represent nuclei. The chalazal pole is depicted up and the micropylar pole down. The central cell nucleus is formed by fusion of the polar nuclei.

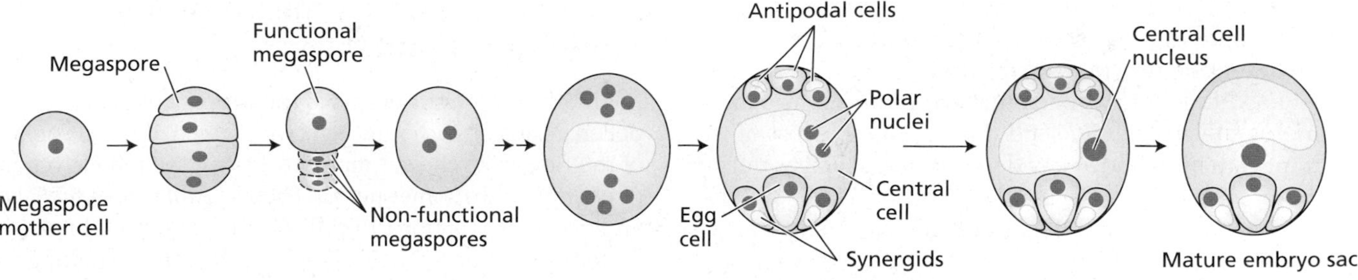

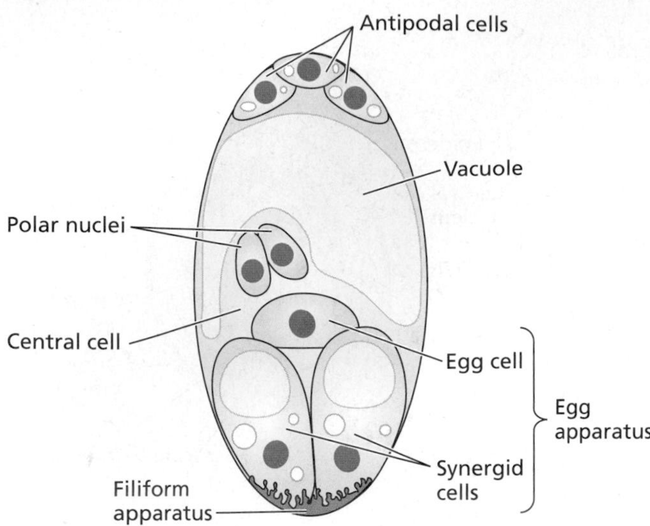

Figure 21.9 Diagram of the egg apparatus and filiform apparatus of the *Polygonum*-type embryo sac.

tube attraction, the discharge of pollen tube contents into the embryo sac, and gamete fusion.

The large binucleate cell in the middle of the embryo sac is called the **central cell**. Although its developmental fate is quite different from that of the egg, the central cell is also regarded as a gamete because it fuses with one of the sperm cells during double fertilization. In Arabidopsis, the two polar nuclei of the central cell fuse to form a single diploid nucleus prior to fusion with the sperm cell. Depending on the type of embryo sac, the number of polar nuclei may range from one in *Oenothera,* to eight or more in *Peperomia*. During double fertilization in the *Polygonum*-type embryo sac, one sperm cell fuses with the egg to produce the zygote, and the other fuses with the central cell to produce the triploid **primary endosperm cell**, which divides mitotically to give rise to the nutritive endosperm of the seed. Because different types of embryo sacs contain different numbers of polar nuclei, the ploidy level of the endosperm ranges from 2*N* in *Oenothera* to 15*N* in *Peperomia*.

Embryo sac development involves hormonal signaling between sporophytic and gametophytic generations

Out of a total of approximately 28,000 genes in Arabidopsis, it is believed that only a few thousand are specifically involved in female gametophyte development. Several hundred mutants have been analyzed that affect female gametogenesis in Arabidopsis, and these mutants have been used to identify genes that are required for female gametogenesis or early seed development. Most of the mutants are impaired at key developmental stages during gametogenesis, with a greater proportion either arrested before the first haploid mitosis or defective in steps occurring after cellularization.

The ovules of all gametophytic mutants found thus far have normal sporophytic (2*N*) cells—that is, normal nucellus, integuments, funiculus, and megaspore mother cell. In fact, the genetic screens designed to identify gametophytic defects would not be able to reveal sporophytic defects because the maternal plants are heterozygous. In contrast, several mutants with defects in the sporophytic tissues of the ovule also show anomalies in gametophyte development. It has therefore been proposed that a hierarchy exists in the communication between the female gametophyte and the surrounding sporophytic cells, and that the sporophytic maternal tissues exert the greater influence.

Three hormones—auxin, cytokinin, and brassinosteroids—have been implicated in the regulation of various stages of female gametophyte development in Arabidopsis. For example, two *YUCCA* genes, which encode flavin monooxygenases involved in local auxin biosynthesis, are expressed in the ovule, and the auxin efflux carrier PIN1 is expressed in the nucellus. Mutations in the latter have been shown to cause defects in the development of the female gametophyte, causing it to arrest at the one- or two-nucleate stage. These observations are consistent with the role of auxin as a cell fate determinant in female gametophytes.

Cytokinins synthesized in the chalazal region of the nucellus have been implicated in megasporogenesis. Triple mutants lacking functional AHK receptors, which are required for the cytokinin response (see Chapter 15), fail to develop functional megaspores. Brassinosteroids have been shown to be required for the initiation of mitotic divisions by the megaspore. The female gametophytes of Arabidopsis mutants with a defective *CYP85A1* gene, which codes for an enzyme that regulates brassinosteroid biosynthesis in the embryo sac, are arrested before the first nuclear mitotic division of the haploid functional megaspore. In other words, brassinosteroid biosynthesis inside the embryo sac is required for the initiation of megagametophyte development. However, brassinosteroid biosynthesis in the embryo sac appears to be controlled by a sporophytically expressed gene, *SPOROCYTELESS* (*SPL*). In *spl* mutants, archesporial cells are formed in both anther and ovule primordia, but they fail to develop further. Since the *CYP85A1* gene is abundantly expressed in embryo sacs of wild-type ovules, but not in those of *spl* ovules, the sporophytic *SPL* gene appears to regulate brassinosteroid biosynthesis in the gametophytic embryo sac.

Pollination and Fertilization in Flowering Plants

Pollination in angiosperms is the process of transferring pollen grains from the anther of the stamen, the male organ of the flower, to the stigma of the pistil, the female organ of the flower. In some species, such as *Arabidopsis thaliana* and rice, reproduction typically occurs via self-pollination, or selfing—that is, the pollen and the stigma belong to the same individual sporophyte. In other species, **cross-pollination**, or **outcrossing**, is the norm—the male parent and the female parent are separate sporophytic individuals. Many species can reproduce by either self- or cross-pollination; other species, as we will discuss later, have various mechanisms for promoting cross-pollination, and may even be incapable of reproducing by self-pollination.

In the case of cross-pollination, pollen might travel great distances before landing on a suitable stigma. Produced in excess, pollen grains are dispersed by wind, insects, birds, and mammals, which carry the nonmotile male gametes of angiosperms much farther than the motile sperm of lower plants could ever swim.

Successful pollination depends on several factors, including ambient temperature, timing, and the receptivity of the stigma of a compatible flower. Many pollen grains can tolerate desiccation and high temperatures during their journey to the stigma. However, some pollen grains, such as those of tomato, are damaged by heat. Understanding how some pollen grains tolerate periods of high temperature will help ensure our food supply as the global climate changes.

Delivery of sperm cells to the female gametophyte by the pollen tube occurs in six phases

The female gametes are well protected from the environment by ovary tissues. Consequently, to reach an unfertilized egg, sperm cells must be delivered by a pollen tube that grows from the stigma to the ovule. In the Arabidopsis gynoecium, which is similar to those of other angiosperms, this process has been divided into six phases (**Figure 21.10**).

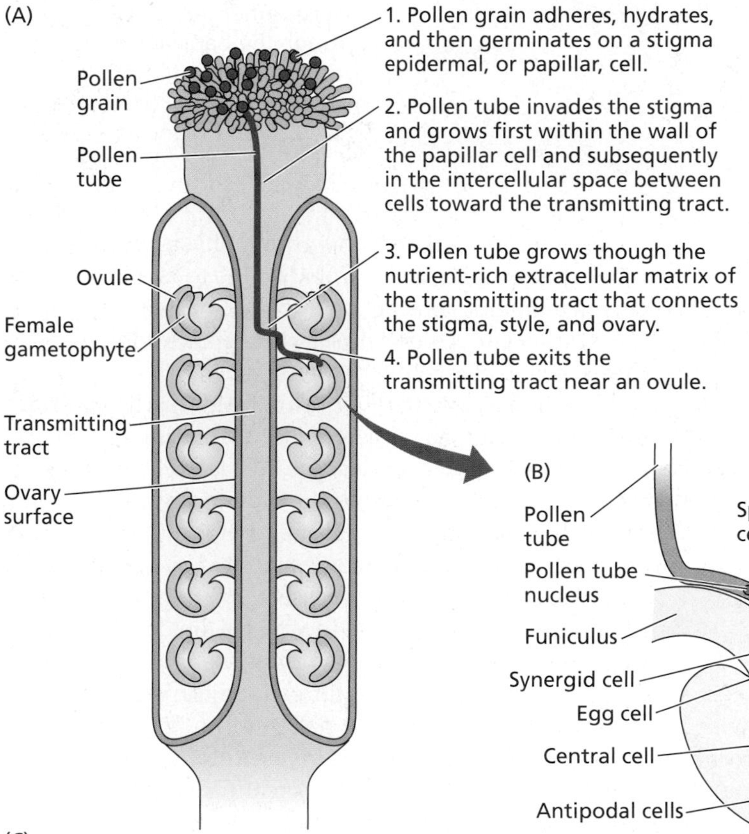

(A)
Pollen grain
Pollen tube
Ovule
Female gametophyte
Transmitting tract
Ovary surface
1. Pollen grain adheres, hydrates, and then germinates on a stigma epidermal, or papillar, cell.
2. Pollen tube invades the stigma and grows first within the wall of the papillar cell and subsequently in the intercellular space between cells toward the transmitting tract.
3. Pollen tube grows though the nutrient-rich extracellular matrix of the transmitting tract that connects the stigma, style, and ovary.
4. Pollen tube exits the transmitting tract near an ovule.

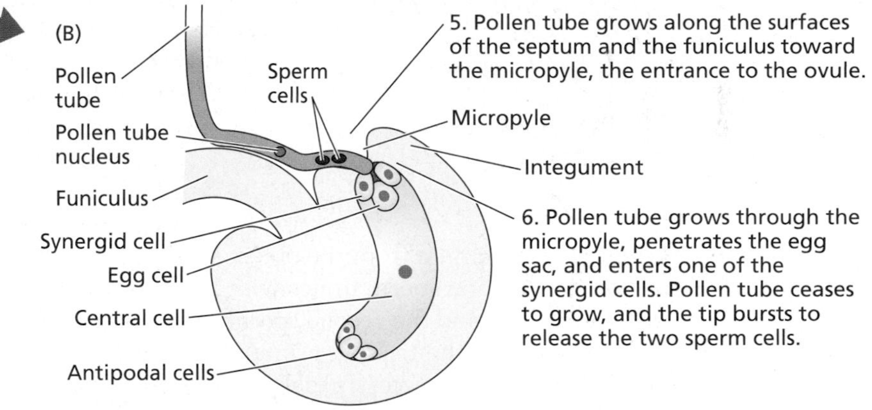

(B)
Pollen tube
Pollen tube nucleus
Funiculus
Synergid cell
Egg cell
Central cell
Antipodal cells
Sperm cells
5. Pollen tube grows along the surfaces of the septum and the funiculus toward the micropyle, the entrance to the ovule.
Micropyle
Integument
6. Pollen tube grows through the micropyle, penetrates the egg sac, and enters one of the synergid cells. Pollen tube ceases to grow, and the tip bursts to release the two sperm cells.

(C)
Monocots
Eudicots
Early divergents
Number of taxa
0
10
20
30
40
50
60
70
80
10
20
50
100
200
500
1000
2000
5000
10,000
20,000
>20,000
Pollen tube growth rate (μm/h)

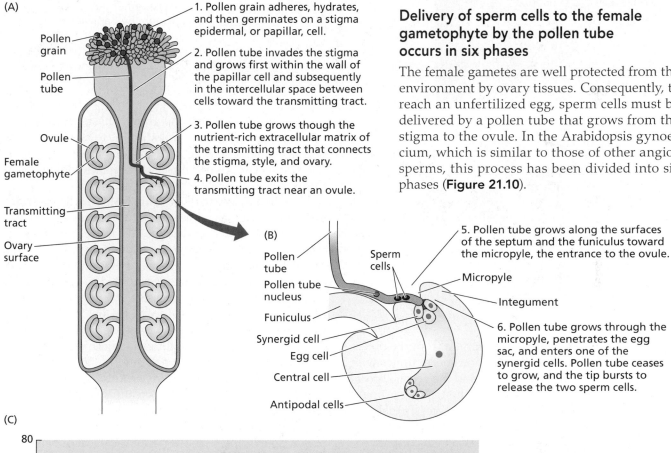

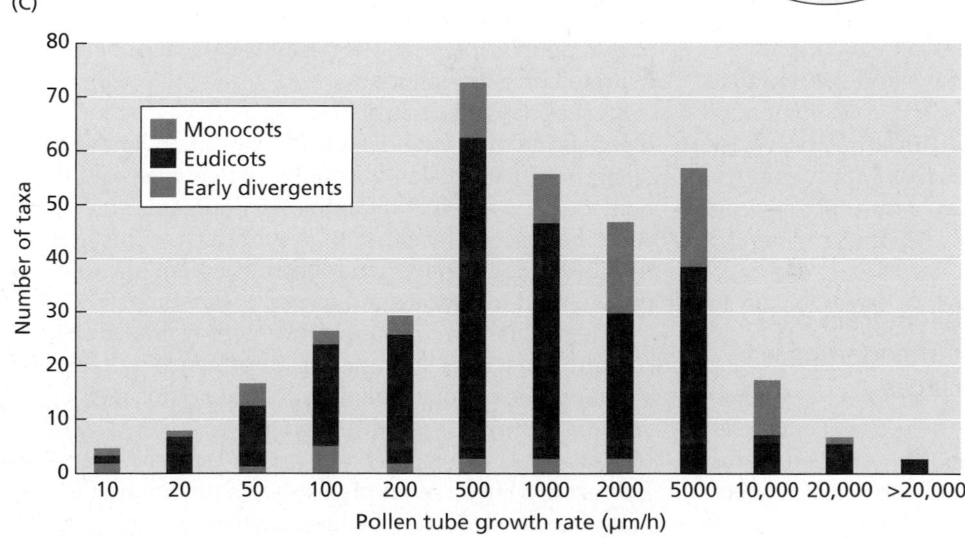

Figure 21.10 Pollination in Arabidopsis. (A) The six phases of pollen tube growth and guidance. (B) Pollen tube arrives at micropylar opening of a single ovule. (C) In vivo pollen tube growth rates of angiosperms measured in 352 species. Growth rates were calculated from the linear distance traveled by the longest tube divided by the actual period of active pollen tube growth. (A and B after Johnson and Lord 2006; C after Williams 2012.)

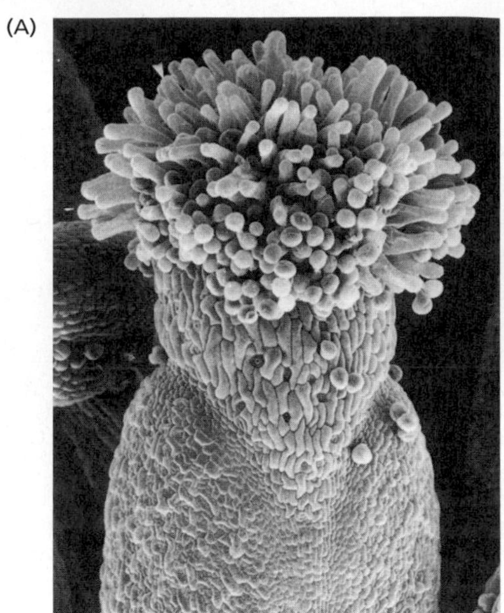

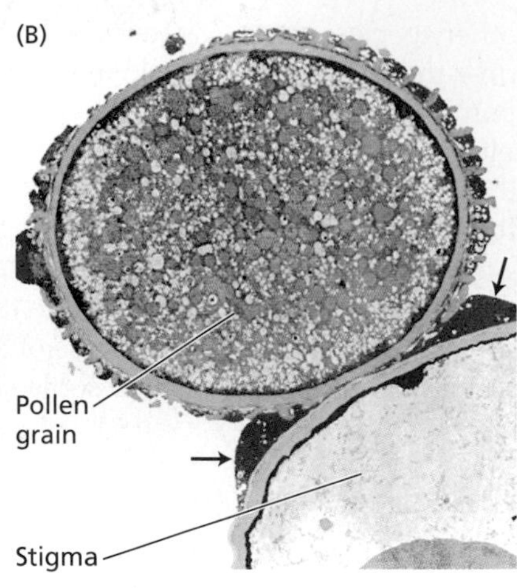

Figure 21.11 Adhesion and hydration of pollen grains on the stigmas of Arabidopsis flowers. (A) Scanning electron microscope image of stigmatic papillae. (B) Transmission electron microscope image showing contact between a pollen grain and a stigma papillus. A "foot" of lipid-rich material (arrows) collects between the two surfaces. (C) The four stages of pollen tube adhesion, hydration and foot formation, emergence, and growth through the papillar cell wall toward the style. (A from Bowman 1994; B from Edlund et al. 2004; C after Edlund et al. 2004.)

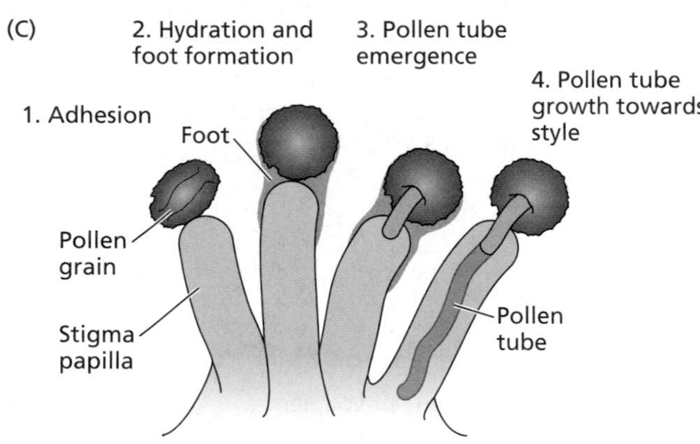

After the sperm cells are discharged from the pollen tube, double fertilization occurs: one sperm unites with the egg cell to produce the zygote, and the second sperm fuses with the central cell to form the triploid primary endosperm cell. As we'll discuss next, successful delivery of the two sperm cells to the two female gametes (egg and central cell) throughout the six phases of the process depends on extensive interactions and communication between the pollen tube, the pistil, and the female gametophyte. As shown in Figure 21.10C, the pollen tube growth rate of angiosperms ranges from approximately 10 μm per h to more than 20,000 μm (2 mm) per h, about 100 times faster than the growth rate of gymnosperm pollen tubes.

Adhesion and hydration of a pollen grain on a compatible flower depend on recognition between pollen and stigma surfaces

Angiosperm reproduction is highly selective. Female tissues are able to discriminate among diverse pollen grains, accepting those from the appropriate species and rejecting others from unrelated species. When pollen lands on a compatible stigma, the grains physically adhere to the stigma papillar cells, probably due to biophysical and chemical interactions between pollen proteins and lipids and stigma surface proteins. Pollen grains adhere poorly to stigmas of plants of other families.

Flowers have either wet or dry stigmas. The surface cells of wet stigmas release a viscous mixture of proteins, lipids, and polysaccharides; the surface cells of dry stigmas, such as those found in the Brassicaceae, are covered by a cell wall, cuticle, and protein pellicle (**Figure 21.11**). Whereas pollen grains become hydrated by default on wet stigmas, the hydration process on dry stigmas is highly regulated. After landing on the stigma, lipids and proteins from the pollen coat flow out onto the stigma and mingle with materials from papillar cells to form the "foot," a structure that attaches the pollen grain firmly to the tip of the papillar cell. During this process, the lipids in the foot are thought to reorganize, creating a capillary system through which water and ions can flow from the stigma to the pollen grain. This mechanism apparently allows the pollen grain to perform the paradoxical feat of becoming hydrated on a dry stigma.

In support of the role of lipids in pollen hydration, Arabidopsis mutants with defects in long-chain lipid metabolism produced pollen without a pollen coat, and these pollen grains failed to hydrate on the stigma. This defect could be rescued by either high humidity or the application of lipids to the stigma, both of which allowed the pollen grain to hydrate and form a pollen tube.

The mechanism of water movement from the papillar cell into the foot is still unclear. In principle, water could either diffuse out of the papillar cell via plasma membrane aquaporin channels (see Chapter 3) or be secreted by vesicular exocytosis. In favor of a secretory mechanism, pollen grains fail to hydrate on pistils with a mutation in a gene that is required for the normal exocytosis of Golgi vesicles.

Ca^{2+}-triggered polarization of the pollen grain precedes tube formation

During hydration, the pollen grain becomes physiologically activated. Calcium ion influx into the vegetative cell triggers reorganization of the cytoskeleton and causes the cell to become physiologically and ultrastructurally polarized. The source of the Ca^{2+} is unknown but may be either the cytoplasm or the cell wall of the papillar cell. Live imaging of free Ca^{2+} in Arabidopsis pollen grains has shown that the cytosolic Ca^{2+} concentration increases at the future germination site soon after hydration, and remains elevated until tube emergence. Both actin microfilaments and secretory vesicles accumulate below the germination pore, or aperture, and the vegetative nucleus migrates to a position that will allow it to enter the germinating pollen tube ahead of the sperm cells. In addition to water and Ca^{2+}, the stigma may supply a variety of other factors that promote pollen germination as well, but thus far these appear to be species-specific.

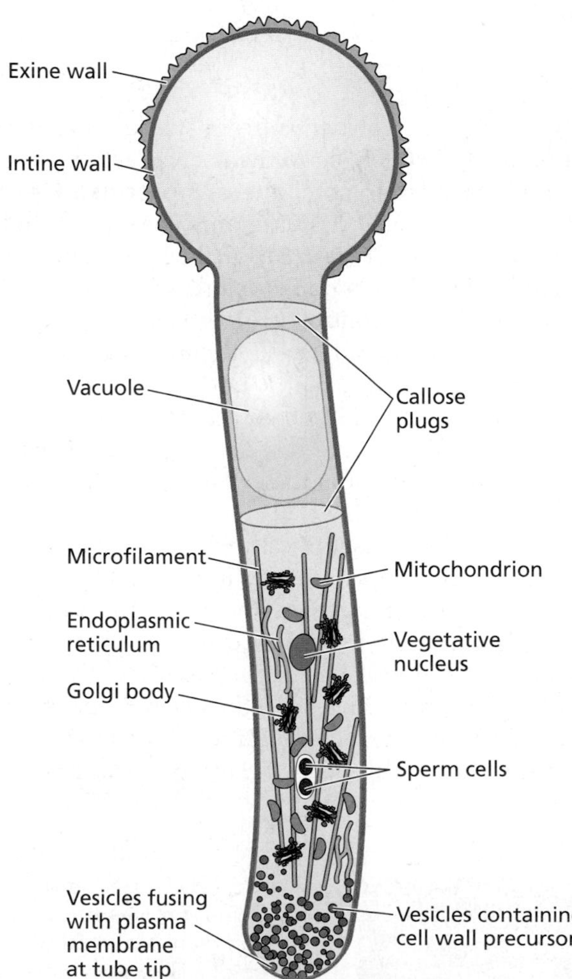

Figure 21.12 Pollen tube elongating by tip growth. The cytoplasm is concentrated in the growing region of the tube by large vacuoles and callose partitions. (After Konrad et al. 2011.)

Pollen tubes grow by tip growth

Following germination, the pollen tube begins to grow by *tip growth* (see Chapter 14). As noted earlier (see Figure 21.10C), the speed of pollen tube elongation in some flowering plants is extremely rapid, reaching rates of more than 5 μm per s in vivo, compared with 10 to 40 nm per s for the tip growth of root hairs. Moreover, tube length can reach up to 40 cm, as it does to travel the length of a maize silk (the style of maize carpels). After penetrating through gaps in the waxy cuticle of the papillar cell, the pollen tube enters the papillar cell wall (see Figure 21.11C).

Growing pollen tubes restrict the cytoplasm, the two sperm nuclei, and the vegetative nucleus to the growing apical region by forming large vacuoles and callose partitions to seal off the rear portion of the tube (**Figure 21.12**). At the apical end of the pollen tube is a region known as the *clear zone* (**Figure 21.13A**). Small secretory vesicles are found in the clear zone, but large organelles—such as nuclei, endoplasmic reticulum, and mitochondria—are excluded. The molecular basis for the clear zone seems to be related to the disruption or reorganization of the actin cables that drive cytoplasmic streaming, because streaming is observed in the region behind the clear zone, but not within it (**Figure 21.13B**). The cytoplasm is packed with small secretory vesicles delivering wall materials and new membranes to the growing tip.

How pollen tubes and other tip-growing cells regulate their polarity is a fundamental question in plant development. One hypothesis is that ion gradients at the growing tip are involved. For example, the tip of a growing pollen tube is polarized due to local Ca^{2+} and pH gradients (**Figure 21.14**). The cytosolic Ca^{2+} concentration is high at the extreme tip (3–10 μ*M*) and drops to basal levels (0.2–0.3 μ*M*) within 20 μm from the apex. In addition, the cytosolic pH is slightly acidic (pH 6.8) at the extreme tip of the clear zone and alkaline (pH 7.5) at the base of the clear zone. Both the Ca^{2+} concentration and the cytosolic pH oscillate in the clear zone with a periodicity that correlates with oscillations in the pollen tube growth rate, suggesting a link between the two. Electrical as well as chemical changes due to Ca^{2+} concentrations and pH are known to play roles in cell signaling, cytoskeletal dynamics, membrane trafficking, and exocytosis, all of which are involved in maintaining pollen tube polarity.

Receptor-like kinases are thought to regulate the ROP1 GTPase switch, a master regulator of tip growth

Plant cells use a conserved mechanism based on small GTPases (enzymes that hydrolyze GTP to GDP) to regulate polarity in a variety of cells. These regulatory GTPases

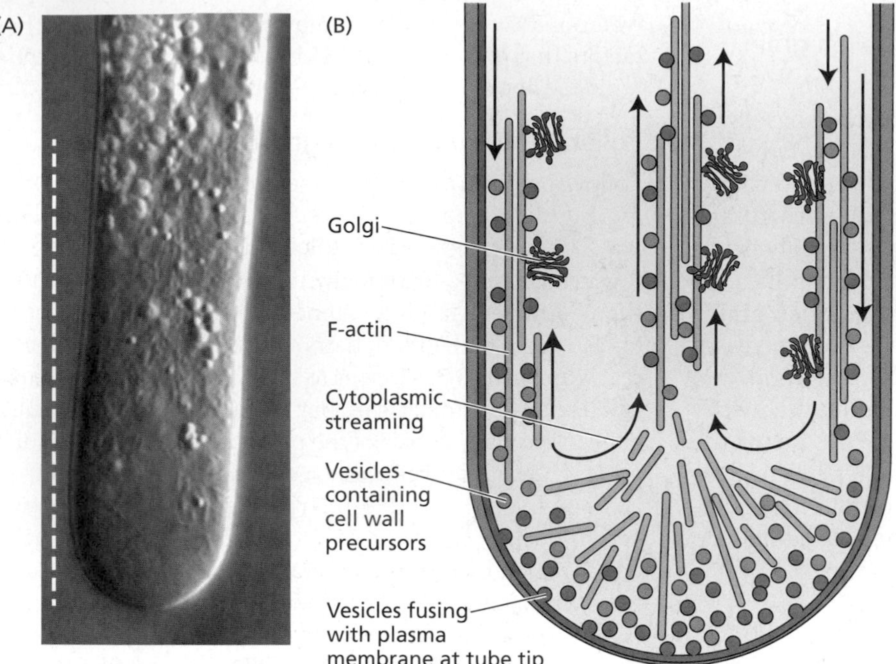

Figure 21.13 Clear zone of a growing pollen tube. (A) Micrograph of the clear zone (white dashed line) of elongating lily pollen. (B) Diagram of ultrastructural components of the clear zone. As indicated in the illustration, the "clear zone" is not really clear, but contains Golgi bodies, F-actin filaments, and numerous small vesicles. Arrows indicate the circularity of cytoplasmic streaming on either side of the central axis. (A courtesy of J. Feijo; B after Cheung et al. 2010.)

are molecular switches that can cycle between an active, GTP-bound form and an inactive, GDP-bound form (**Figure 21.15A**). When a regulatory GTPase is in its active form, it triggers downstream signal transduction pathways (see Chapter 15). Conversion from the active form back to the inactive form is catalyzed by the GTPase itself, which hydrolyzes the bound GTP to GDP. The GTPase switch is operated by other proteins that affect the rate of either GTP hydrolysis or GDP release. **Guanine nucleotide exchange factors** (**GEFs**) activate inactive GTPases by replacing GDP with GTP, while **GTPase-activating proteins** (**GAPs**) inactivate GTPases by promoting GTP hydrolysis.

In plants, tip growth and polar cell expansion are regulated by a unique family of small GTPases called ROPs (for Rho-like GTPase). Arabidopsis has 11 different *ROP* genes, 7 of which are either abundant or preferentially expressed in mature pollen grains and tubes.

ROP1 GTPase resides on the plasma membrane at the tips of growing pollen tubes and is a regulator of tip growth. As with other regulatory GTPases, ROP1 activity can be switched on and off by GEFs and GAPs, respectively. There is also evidence (from studies of root hair development) that GEFs are themselves activated by a signaling mechanism involving **receptor-like kinases** (**RLKs**), which are encoded by a large gene family in the Arabidopsis genome. **Figure 21.15B** illustrates a proposed mechanism by which a pollen-expressed RLK interacts directly with GEF to control tip growth. Upon being activated by an unidentified ligand, the RLK activates GEF, which in turn activates ROP1. Locally activated ROP1 then stimulates NADPH oxidase activity, resulting in the production of reactive oxygen species (ROS). ROS, in turn, promotes Ca^{2+} influx from the extracellular space, which enhances tip growth.

ROP1 also interacts specifically with a group of proteins called **ROP-interactive CRIB motif–containing proteins** (**RICs**). When overexpressed in Arabidopsis, RIC3 and RIC4 alter pollen tube polarity and exocytosis, suggesting they act downstream of ROP1. Further studies have shown that the RIC4 pathway promotes F-actin assembly and induces the accumulation of exocytic vesicles at the tip. Precisely how RIC3 and RIC4 alter tube polarity is still unresolved.

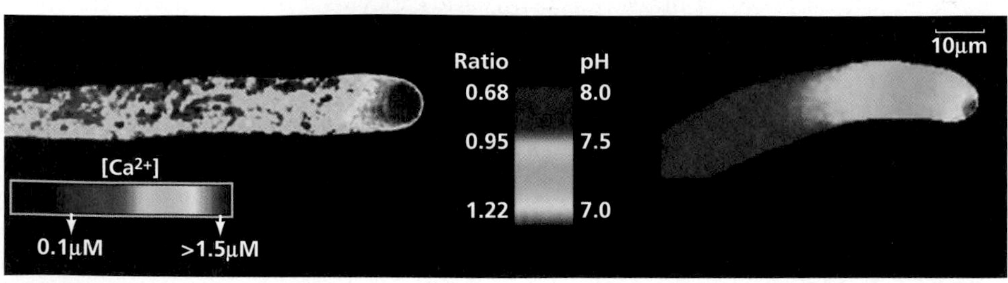

Figure 21.14 Ca^{2+} and pH gradients in pollen tubes. (Left) Ca^{2+} gradient at the growing tip of a lily pollen tube injected with a Ca^{2+}-sensitive dye. (Right) pH gradient from tip to base in tobacco pollen expressing a pH-sensitive indicator. (Courtesy of J. Feijo.)

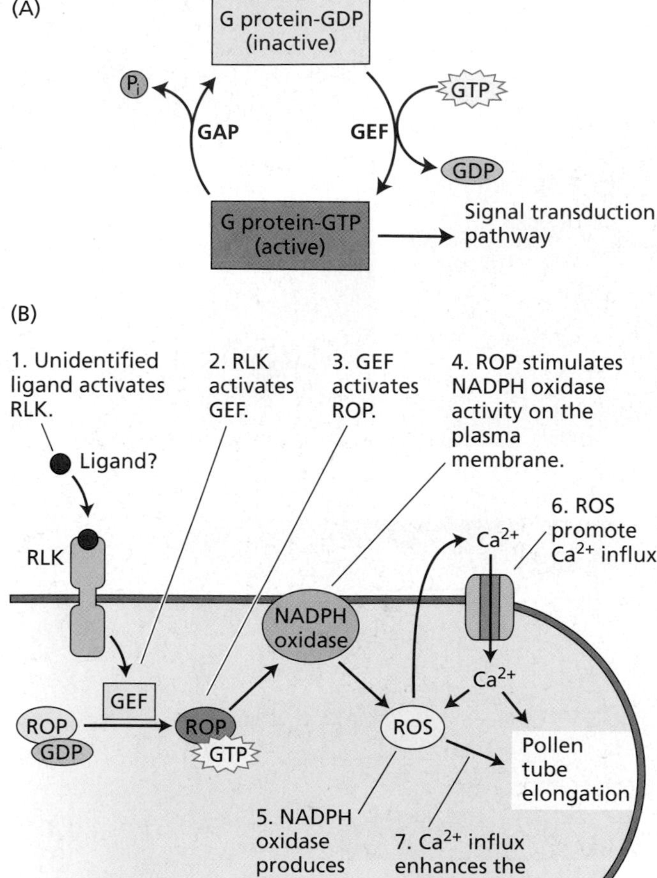

Figure 21.15 (A) Guanine nucleotide exchange factors (GEFs) and GTPase-activating proteins (GAPs) regulate the activities of small GTPases (ROPS), which act as molecular swtiches in pollen tubes. (B) Model for the regulation of pollen tube growth by receptor-like kinases (RLKs) and ROP GTPases. Recent studies suggest that ROP1 is concentrated in the apical cap region at the tip of the pollen tube.

Pollen tube tip growth in the pistil is directed by both physical and chemical cues

For successful fertilization to occur, the pollen tube must find its way to the micropyle of a ovule. In fact, there is often competition among pollen tubes to arrive at the micropyle first and thus be the one to fertilize the egg. The surrounding maternal tissues may even influence the outcome of this "race"—a type of female mate selection. What factors stimulate pollen tube growth and guide pollen tubes to an ovule?

Two major models have been proposed to explain the growth of pollen tubes toward an ovule: the mechanical hypothesis and the chemotropic hypothesis. In the **mechanical hypothesis**, the pistil architecture dictates the path of the tube, which follows a narrow **transmitting tract** leading to the ovule (see Figure 21.10A). During growth toward the ovule, the pollen tubes are in intimate contact with the components of the extracellular matrix of the transmitting tract. The transmitting tract **extracellular matrix** is a complex mixture of cell wall proteins, including arabinogalactan proteins, proline-rich glycoproteins, and hydroxyproline-rich glycoproteins (see Chapter 14). According to the mechanical hypothesis, these proteins provide adhesive molecules that keep the tube in place and provide traction for growth through the style. The extracellular matrix also supplies nutrients that could support the tube's metabolic activity.

According to the **chemotropic hypothesis**, a hierarchy of molecular cues directs the pollen tube to its destination by stimulating the tip to grow toward the ovule. A few pistil-expressed molecules have been identified that function in pollen tube guidance. In lily, a small, secreted protein—**stigma/style cysteine-rich adhesin** (**SCA**), a lipid transfer protein—is secreted by the transmitting tract epidermis that lines the hollow style, and is involved in growth and adhesion of the tube along the tract. Another small, secreted protein, **chemocyanin**, a member of the phytocyanin family of blue copper proteins, acts as a directional cue.

While the early stages of pollen tube growth are regulated by sporophytic cells in the transmitting tract, genetic analyses of Arabidopsis and in vitro guidance experiments in *Torenia fournieri* (discussed below) support the idea that chemical signals from the female gametophyte also play critical roles in directing pollen tubes to the ovule.

Style tissue conditions the pollen tube to respond to attractants produced by the synergids of the embryo sac

To get from the stigma to the ovary, a pollen tube passes through the style. Besides serving as a conduit for the pollen tube to reach the transmitting tract and ovary, stylar tissues also enable the pollen tube to become competent to perceive guidance signals from the female gametophyte. As described below, *Torenia fournieri* (a member of the Lamiales, which includes lavender and lilac) provides a convenient model system to study the production of pollen tube attractants by the female gametophyte, and has also been used to uncover the role of the style in priming the response of the pollen tube to the attractants released by the embryo sac.

In the vast majority of angiosperms, the sporophytic tissues of the ovule (integuments) cannot be readily removed from the embryo sac. However, in *T. fournieri* and several other species, the embryo sac grows out through the micropyle toward the funiculus (**Figure 21.16**). The egg cell, the two synergid cells, and approximately half of the central cell are thus located outside the ovule in this species. When ovules of *T. fournieri* are excised from the placenta, the naked embryo sacs are directly exposed to the medium. Experiments have demonstrated that when

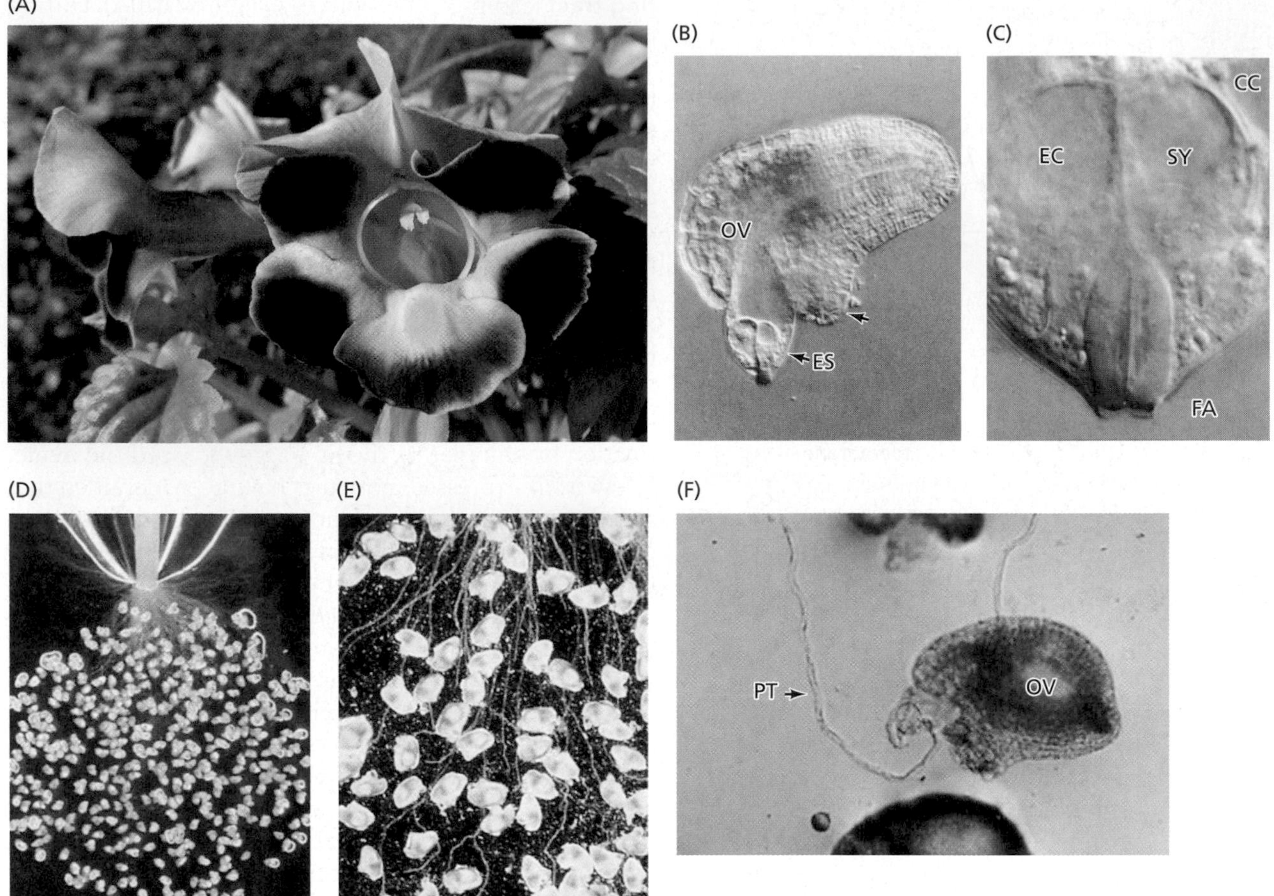

Figure 21.16 Use of excised *Torenia fournieri* ovules to study the influence of the style on the directed growth of the pollen tube. (A) Flower of *T. fournieri*. (B) *T. fournieri* embryo sac (ES) extending from the micropylar region of the excised ovule (OV). (C) Magnified view of the naked embryo sac showing the central cell (CC), the egg cell (EC), and one of the two synergids (SY) with its filiform apparatus (FA). (D) Ovules placed near a pollinated style. (E) Dark-field image showing pollen tubes growing toward ovules. (F) Micrograph of a pollen tube (PT) that has reached the micropylar end of a naked embryo sac in an ovule (OV). (B–F from Higashiyama et al. 1998.)

excised *T. fournieri* ovules are co-cultivated with pollen tubes that have been germinated in vitro (on a nutrient medium), the pollen tubes do not grow toward the ovule. However, if the *T. fournieri* pollen grains are first germinated on a living stigma and allowed to emerge from the cut end of the style, they do grow toward the micropylar end of the embryo sac (see Figure 21.16F). This experiment demonstrates that the pollen tube interacts with the female sporophyte and becomes conditioned in a way that enables it to respond to cues from the female gametophyte and grow toward the micropyle. Indeed, in Arabidopsis, comparisons of the transcriptomes of pollen tubes grown either in vitro or through pistil tissues have shown that significant changes in gene expression are induced by growth through the pistil tissues.

The cellular source of the pollen tube attractant in *T. fournieri* was identified by laser ablation of specific cells of the embryo sac. Pollen tubes failed to grow toward the ovule only if the synergid cells—but not the egg or central cell—had been killed. The pollen chemoattractants of *Torenia* have now been identified as cysteine-rich polypeptides called **LUREs**. LUREs are related to *defensins*, a group of antimicrobial proteins found in animals and plants. The various LUREs of *Torenia* apparently act in a species-specific manner. LURE-like proteins have also been identified in Arabidopsis, and *T. fournieri* ovules expressing an Arabidopsis LURE attract Arabidopsis pollen tubes preferentially.

Double fertilization occurs in three distinct stages

When the pollen tube senses chemical attractants secreted by the synergids, the tube grows through the micropyle, penetrates the embryo sac, and enters one of the synergid cells. Once inside the synergid, the pollen tube stops growing and the tip bursts, releasing the two sperm cells.

Based on live imaging of fluorescently tagged sperm cells, sperm cell behavior in Arabidopsis can be divided into three stages (**Figure 21.17**). First, the pollen tube bursts within a few seconds of entering the synergid, either during or just prior to the breakdown of the receptive synergid cell. Second, the two sperm cells remain stationary at a boundary region between the egg cell and the central cell for approximately 7 min. Third, one sperm fuses with the egg and the other fuses with the central cell, completing double fertilization.

Many questions about double fertilization remain. For example, how is the bursting of the pollen tube regulated? According to one model, a receptor-like kinase on the synergid plasma membrane becomes activated and stimulates ROS production and Ca^{2+} uptake. Since exogenously applied hydroxyl radicals are known to cause pollen tubes to burst in a Ca^{2+}-dependent manner, it is possible that a combination of hydroxyl radicals and high Ca^{2+} may cause the pollen tube to burst upon entering the synergid.

Another question is what determines the behavior of sperm cells after they are released from the pollen tube. It is likely that discharged sperm cells exchange additional signals with the female gametes to prepare for fusion. In Arabidopsis, for example, a cysteine-rich protein is released from the egg cell upon sperm arrival. The sperm responds by secreting a specific membrane protein onto its surface. This sperm surface protein apparently facilitates the fusion of the male and female gametes. Consistent with this hypothesis, mutant sperm cells lacking the surface protein are unable to fertilize either the egg or central cell.

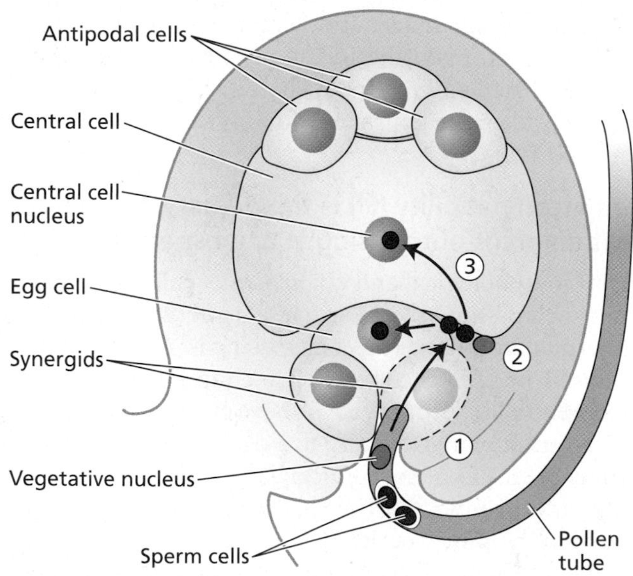

Antipodal cells
Central cell
Central cell nucleus
Egg cell
Synergids
Vegetative nucleus
Sperm cells
Pollen tube
1
2
3
1. The pollen tube bursts and discharges. Sperm cells are delivered rapidly from the pollen tube into the female gametophyte. The receptive synergid cell is likely to break down just after the start of pollen tube discharge.
2. Two sperm cells remain at the boundary region between the egg cell and the central cell for several minutes.
3. One sperm cell fuses with the egg cell and the other fuses with the central cell, and their nuclei move toward the target gamete nuclei.

Figure 21.17 Sperm cell behavior during double fertilization in Arabidopsis can be divided into three stages.

Selfing versus Outcrossing

Many plants have evolved mechanisms to prevent selfing and promote outcrossing, which increases both genetic diversity and the ability to adapt to different environmental conditions. The primary mechanism used by flowering plants to prevent selfing is pollen self-incompatibility, which we will discuss later in the section. Certain features of floral morphology or developmental timing can promote outcrossing, as when the stamens and pistils of a bisexual flower or monoecious plants mature at different times. Finally, the production of male-sterile (functionally female) individuals also functions to prevent selfing and promote outcrossing in plants.

Hermaphroditic and monoecious species have evolved floral features to ensure outcrossing

Because the majority of flowering plants—over 85%—are hermaphroditic, early botanists assumed they must be self-pollinating. It thus came as an surprise at the end of the eighteenth century when Christian Konrad Sprengel demonstrated that in the majority of angiosperms floral morphology seems to be optimized for attracting insect pollinators, and that these pollinators facilitate cross-pollination rather than self-pollination. Temporal and spatial features of flower morphology were identified that prevented self-pollination in both hermaphroditic and monoecious species. In **dichogamy**, the stamens and pistils mature at different times. There are two types of dichogamy: *protandry* and *protogyny*. In protandrous flowers the stamens mature before the pistils, while in protogynous flowers the pistils mature before the stamens (**Figure 21.18A**). Since individuals in a wild population are at different developmental stages at any given time, there will always be pollen available for every pistil, and vice versa.

Another flower feature that promotes outcrossing is **heterostyly**. In heterostylous species, two or three morphological types of flowers, termed *morphs*, exist in the same population. The flower morphs differ in the lengths of their pistil and stamens. In one morph the stamens are short and the pistils are long, while in the second morph it is the reverse (**Figure 21.18B**). The lengths of the stamens and pistils in the two morphs are adapted for pollination

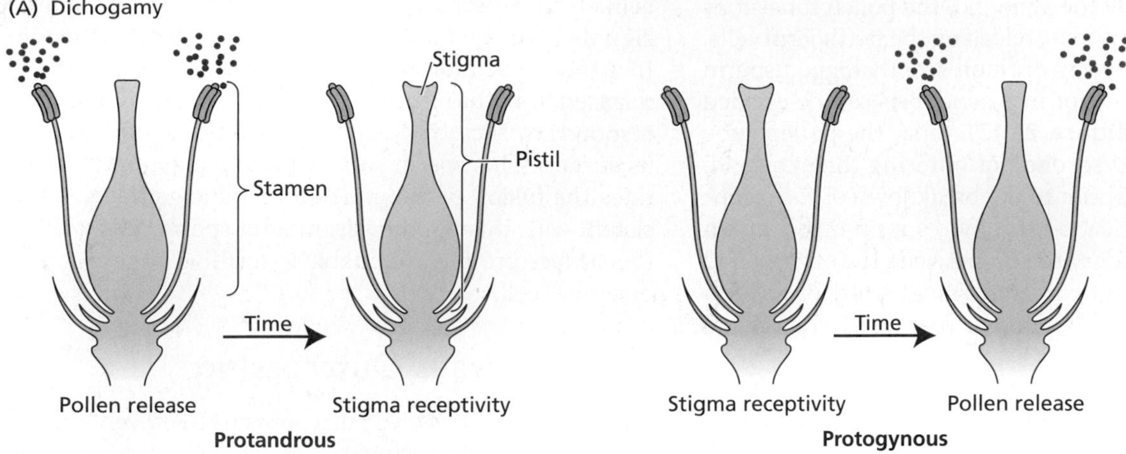

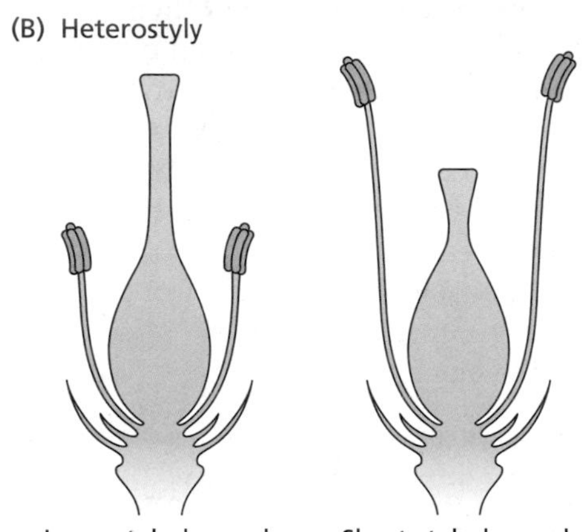

Figure 21.18 Morphological adaptations in flowers that promote outcrossing. (A) Dichogamy. In protandrous flowers, pollen release from the anthers occurs prior to stigma receptivity (indicated by the open stigma). In protogynous flowers, stigma receptivity precedes pollen release. (B) Heterostyly. Two anatomically different types of flowers are produced: long-styled morphs and short-styled morphs. Due to incompatibility reactions, the two types can pollinate each other but not themselves.

by different pollinators or by different body parts of the same pollinator, thus promoting outcrossing.

Cytoplasmic male sterility (CMS) occurs in the wild and is of great utility in agriculture

Male sterility—the inability to produce functional pollen—is widespread among plants and effectively prevents self-pollination. Male sterility is often maternally inherited, caused by gain-of-function mutations of the mitochondrial genome, and is therefore called *cytoplasmic male sterility* (*CMS*). CMS has been studied extensively in a wide variety of crops, where it has been exploited in breeding programs.

Most types of CMS mutations are caused by mitochondrial chromosomal rearrangements that produce chimeric genes with new functions. Plant mitochondrial genomes are large, variable in size, and tend to undergo recombination in specific regions (see Chapter 2 and **WEB TOPIC 12.6**). Mitochondrial genome rearrangements can result in fusions between two distinct mitochondrial sequences, sometimes producing a novel, functional gene. Although no two CMS mutations isolated so far have been the same, all seem to inhibit mitochondrial function when expressed in the anther, resulting in the production of ROS and ROS-mediated programmed cell death. For a discussion of the molecular mechanism of CMS in rice, and its reversal, see **WEB TOPIC 21.4**.

Self-incompatibility (SI) is the primary mechanism that enforces outcrossing in angiosperms

Floral morphological and cytoplasmic male sterility promote outcrossing in some species, but in the vast majority of hermaphroditic species outcrossing is strictly enforced by a self/nonself recognition mechanism termed **self-incompatibility** (**SI**). SI systems have evolved several times in flowering plants, leading to a diverse array of mechanisms. SI creates a biochemical barrier that prevents self-pollination, while allowing pollination by another individual in the same species.

The ability to discriminate between self and nonself is a ubiquitous and essential function of both multicellular and microbial species. In vertebrates, for example, recognition of nonself depends on the major histocompatibility complex (MHC), in which allelic variability, or polymorphisms, at MHC loci facilitates self/nonself discrimination. In plants, self/nonself recognition during sexual reproduction is mediated by the self-incompatibility locus, *S*, which directs the recognition and rejection of self-pollen. The *S* locus consists of multiple genes, or **determinants**, that are expressed either in the anther and pollen grain (male) or in the pistil (female). The male and female determinant genes are inherited as a single segregating unit and have many alleles. The allelic variants of this gene complex are called ***S* haplotypes**. A haplotype is any combination of alleles at adjacent loci on a chromosome that are inherited together.

During pollination, the proteins expressed by the alleles of the determinant genes "determine" whether the pollen will be perceived as self or nonself by the stigma. If the pollen grain and stigmatic cells carry alleles from the same *S* haplotype, an incompatible reaction occurs and the pollen is rejected. If, however, the *S* haplotypes of the pollen and stigma carry *different* alleles, pollination and fertilization are allowed to proceed.

There are two main categories of SI systems in plants, both of which are defined by the incompatibility phenotype of the pollen grain (**Figure 21.19**). In **sporophytic self-incompatibility** (**SSI**), the pollen grain's incompatibility phenotype is determined by the diploid genome of the pollen parent—specifically, the tapetum of the anther. If either of the *S* haplotypes of the pollen parent matches either of the *S* haplotypes in the pistil, then rejection will occur. SSI incompatibility reactions typically block pollen growth prior to hydration and germination. However, if the ungerminated SSI pollen is removed from the incompatible stigma and placed on a compatible stigma, it will recover.

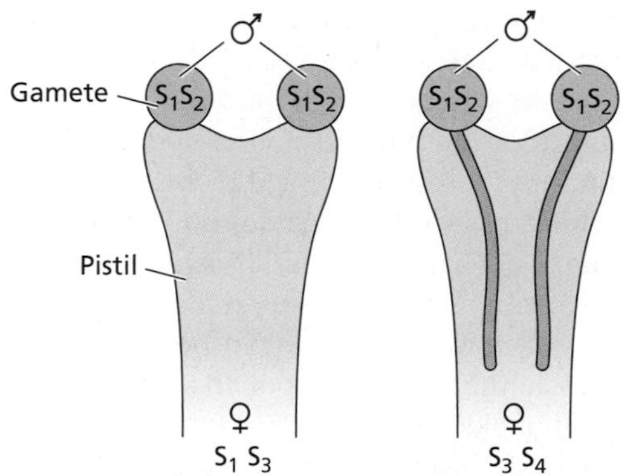

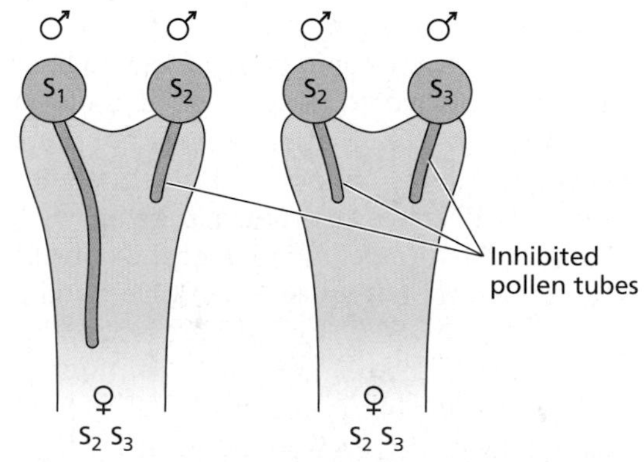

Figure 21.19 Comparison of gametophytic and sporophytic self-incompatibility. (A) Sporophytic self-incompatibility (SSI). Pollen tube growth proceeds only if the parent diploid genotype does not match the female parent. (B) Gametophytic self-incompatibility (GSI). Pollen tube growth proceeds only if the haploid genotype does not match the female *S* locus.

In **gametophytic self-incompatibility** (**GSI**), the incompatibility phenotype of the pollen is determined by the pollen's own (haploid) genotype. In this case, rejection occurs if the single *S* haplotype of the pollen grain matches either of the *S* haplotypes in the pistil. GSI incompatibility reactions typically arrest pollen tube development after the tube has grown partway through the style. In contrast to SSI, GSI incompatibility reactions typically kill the pollen tube.

Correlations have been made between the type of SI system and other reproductive features of the flower. For example, SSI is often associated with a dry stigma, while GSI has been correlated with a wet stigma. Thus, SSI pollen must obtain water from the stigma before the pollen tube can emerge, whereas GSI pollen becomes hydrated and metabolically active as soon as it lands on the stigma, allowing it to germinate relatively rapidly.

The Brassicaceae sporophytic SI system requires two *S*-locus genes

The only sporophytic SI system that has been characterized in any detail thus far is that of the Brassicaceae. In the Brassicaceae, two highly polymorphic genes of the *S* locus are involved in the SI response (**Figure 21.20**). The male *S*-determinant is a cysteine-rich protein located in the pollen coat and is called the ***S*-locus Cysteine-Rich protein** (**SCR**). Although SCRs are expressed in both the diploid tapetum and the haploid pollen grain, only the SCRs produced by the tapetum are essential for the self-incompatibility reaction. For this reason, the SI system in Brassicaceae is regarded as sporophytic. The female *S*-determinant is a serine/threonine receptor kinase, called the ***S*-locus receptor kinase** (**SRK**), which is located in the plasma membrane of stigma cells. The SRK has an extracellular domain that is highly variable among different *S* haplotypes, as expected for a protein involved in self-recognition.

During microgametogenesis, the diploid tapetum releases various proteins, including two types of SCRs—one from each *S* haplotype—which are incorporated into the exine layer of the pollen grain cell wall. After pollination, the SCRs diffuse onto the stigma surface and penetrate the papilla cell wall until they reach the plasma membrane. Because the stigma is diploid, the papilla cell

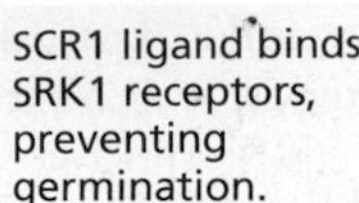

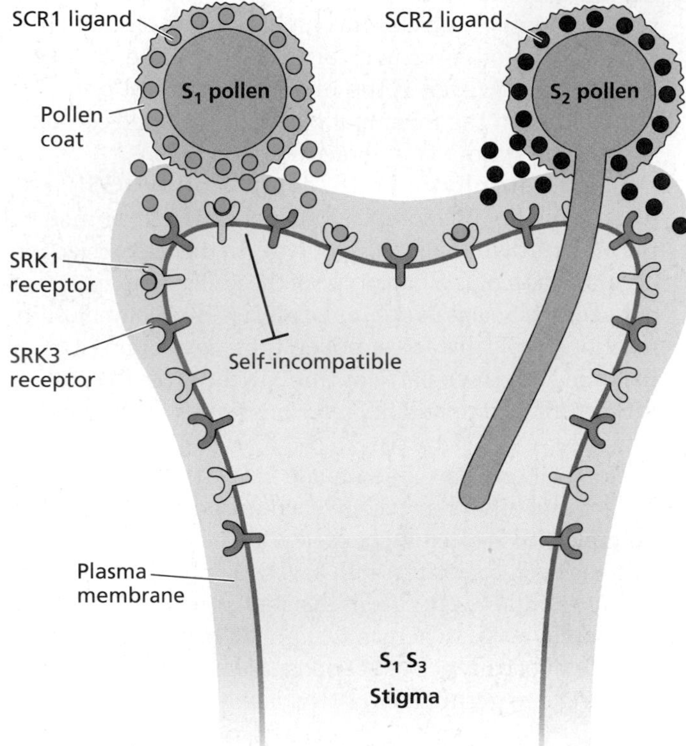

Figure 21.20 Receptor–ligand interactions and recognition of "self" pollen at the stigma epidermal surface. The diagram shows two pollen grains with different haplotypes (S_2 and S_1) on the stigma of a self-incompatible S_1S_3 heterozygote. The *S*-locus Cysteine-Rich protein (SCR) ligand of each pollen grain is located in the pollen cell wall and is delivered to the epidermal surface when the grain lands on the stigma. The SCR1 ligand from pollen grains expressing the S_1 haplotype binds and activates the SRK1 receptor on the surface of the S_1S_3 stigma cell, triggering a signaling cascade that leads to inhibition of hydration, germination, and tube growth. In contrast, a pollen grain derived from a plant expressing neither the S_1 nor S_3 haplotypes (e.g., the S_2 haplotype) produces an SCR2 ligand that fails to bind and activate SRK receptors, allowing pollen tube growth to proceed.

plasma membrane contains two types of SRKs, one for each *S* haplotype. Each SRK recognizes and binds only to its cognate SCR on the same *S*-locus haplotype. If this happens, binding of SCR to SRK causes autophosphorylation of the receptor. Phosphorylation of the SRK receptor initiates a signaling cascade that rapidly inhibits functions that would normally facilitate pollen hydration and germination. The SSI reaction occurs even if only one of the two *S* haplotypes represented in the pollen coat (as SCRs) is present in the genome of the stigma.

Gametophytic self-incompatibility (GSI) is mediated by cytotoxic S-RNases and F-box proteins

Gametophytic self-incompatibility (GSI) is the predominant form of self-incompatibility among flowering plants. GSI is controlled by a single multiallelic locus (*S* locus) containing two tightly linked genes, one encoding the pollen-expressed male determinant and the other encoding the pistil-expressed female determinant. In the Solanaceae, Scrophulariaceae, and Rosaceae families, the pollen determinant is specified by a gene encoding an F-box protein, SLF/SFB, which is involved in targeting proteins for degradation via the ubiquitination pathway (see Chapter 2). The pistil determinant is specified by a cytotoxic S-ribonuclease (S-RNase) gene, which is specifically expressed in the transmitting tract of the style. Pollen tube rejection occurs whenever there is a match between the *S*-determinant of the haploid pollen and one of the two *S*-determinants expressed in the diploid style.

The molecular basis for the interaction between the male and female *S*-determinants in GSI is poorly understood. A key advance was the discovery that the S-RNases produced in the transmitting tract can be taken up by the pollen tube, whether or not the pollen determinant is allelic to the pistil haplotype. In other words, the recognition between the S-RNase and the pollen *S*-determinant occurs inside the pollen tube, where only self S-RNases should be cytotoxic. This observation was consistent with the identification of the pollen determinant, SLF/SFB, as an F-box protein, a component of the E3 ligase complex SCF, which is involved in protein degradation through the ubiquitin–26S proteasome-dependent pathway. This suggested a simple model in which recognition of a nonself S-RNase by the pollen tube SCF^{SLF} leads to the ubiquitination and degradation of the nonself S-RNases in the pollen tube (**Figure 21.21**). Degradation of nonself S-RNase by SCF^{SLF} would prevent RNase cytotoxicity and allow the pollen tube to continue growing. In the case of self-pollen, however, SCF^{SLF} fails to bind to the S-RNase taken up from the transmitting tract. As a result, the S-RNase digests the RNA of the pollen tube vegetative cell, leading to cell death. This simple model accounts for some aspects of GSI, but not all. For example, sequestration of S-RNase in the tube cell vacuole appears to play an important role in protecting against cytotoxicity. During incompatible reactions, the breakdown of the vacuolar membrane may trigger programmed cell death of the tube cell.

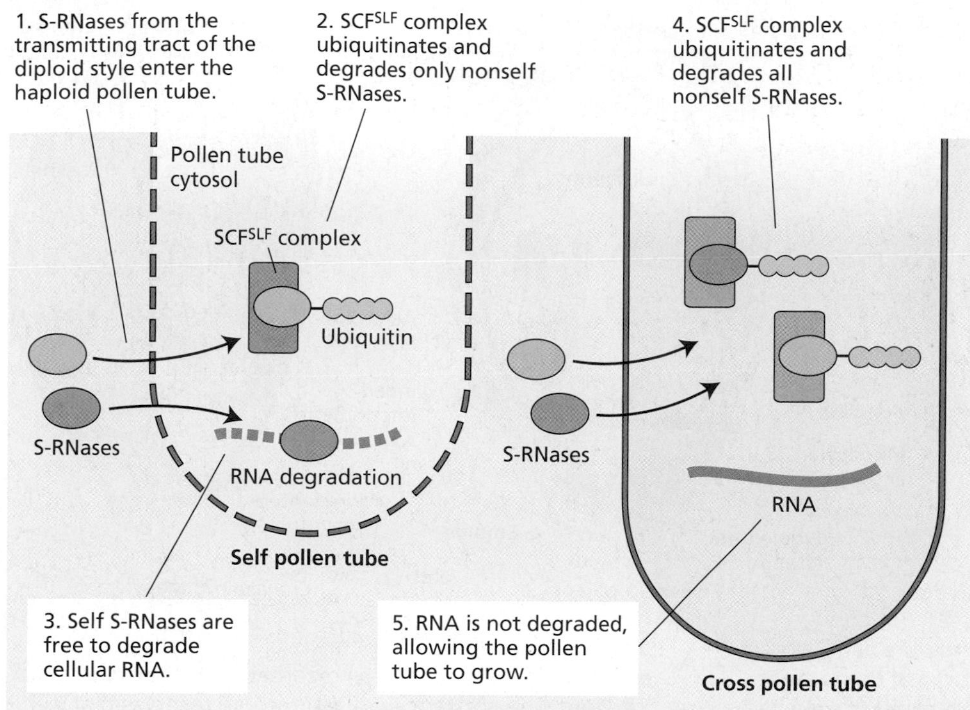
1. S-RNases from the transmitting tract of the diploid style enter the haploid pollen tube.
2. SCF^SLF complex ubiquitinates and degrades only nonself S-RNases.
4. SCF^SLF complex ubiquitinates and degrades all nonself S-RNases.
Pollen tube cytosol
SCF^SLF complex
Ubiquitin
S-RNases
RNA degradation
S-RNases
RNA
Self pollen tube
3. Self S-RNases are free to degrade cellular RNA.
5. RNA is not degraded, allowing the pollen tube to grow.
Cross pollen tube

Figure 21.21 RNase degradation model for gametophytic self-incompatibility (GSI). (Left) Self pollen tube. Because the pollen is haploid, its SCF^{SLF} complex recognizes and degrades only the nonself S-RNAse produced by the diploid transmitting tract. As a result, the remaining self S-RNase is free to degrade cellular RNA. (Right) Cross pollen tube. During cross-pollination, the pollen tube SCF^{SLF} complex recognizes and degrades both of the nonself S-RNases, which eliminates toxicity and allows pollen tube growth to proceed.

Apomixis: Asexual Reproduction by Seed

In some species the embryo is not produced as a result of meiosis and fertilization, but from a chromosomally unreduced (diploid) cell in the ovule that differentiates directly into a zygote and is therefore genetically identical to the female parent. This type of asexual or clonal reproduction by seed is known as *apomixis,* and the plants produced in this way are known as *apomicts*. Apomixis is found in approximately 0.1% of angiosperms, including both monocots and eudicots in over 40 angiosperm families. Common examples include citrus species, mango, dandelion, blackberry, crabapple, and the fodder grass *Panicum*. The various types of apomixis are described in **WEB TOPIC 21.5**.

Apomixis is not an evolutionary dead end

Because of its clonal nature, apomixis was once regarded as an evolutionary blind alley, genetically distinct from sexual reproduction. This hypothesis was based on the assumption that apomixis represented an irreversible phylogenetic branch point that would inevitably lead to the extinction of the lineage. This view has now been superseded by phylogenetic analyses that have shown that apomixis is not only widely distributed in both early- and late-branching lineages, but that it is reversible as well. That is, lineages that were once apomictic sometimes revert to obligate sexual reproduction.

The genetic control of apomixis is based on the altered expression of the same genes that control the normal development of the nucellus and megagametophyte. Because the vast majority of apomictic genotypes are polyploid, it has been suggested that the evolution of apomixis may have contributed to the fitness of polyploid species.

Elucidating the mechanism of apomixis could potentially provide plant breeders with an important new tool for improving crops. Many of our most productive crops, such as maize, are hybrids that have been developed to take advantage of the phenomenon of *heterosis,* or hybrid vigor (see Chapter 2). Because hybrid plants do not breed true, and therefore cannot be propagated by seed, hybrid seeds must be generated anew each season by repeating the original cross. But if apomixis were to be introduced into the F_1 hybrid, the hybrid would be able to produce seeds clonally, thus avoiding the problem of the loss of heterosis in the F_2 generation. Given the potential of such techniques to accelerate the progress of crop improvement, research on the mechanisms of sexual development in plants has intensified in recent years.

Endosperm Development

From both an ecological and agricultural perspective, the plant life cycle begins and ends with a seed. We now pick up the story of the angiosperm ovule immediately after double fertilization, and follow its transformation into a mature seed.

The endosperm develops from mitotic divisions of the primary endosperm nucleus resulting from double fertilization. There are three types of endosperm development in angiosperms: *nuclear, cellular,* and *helobial*. Of these, the nuclear type is the most common and has been extensively

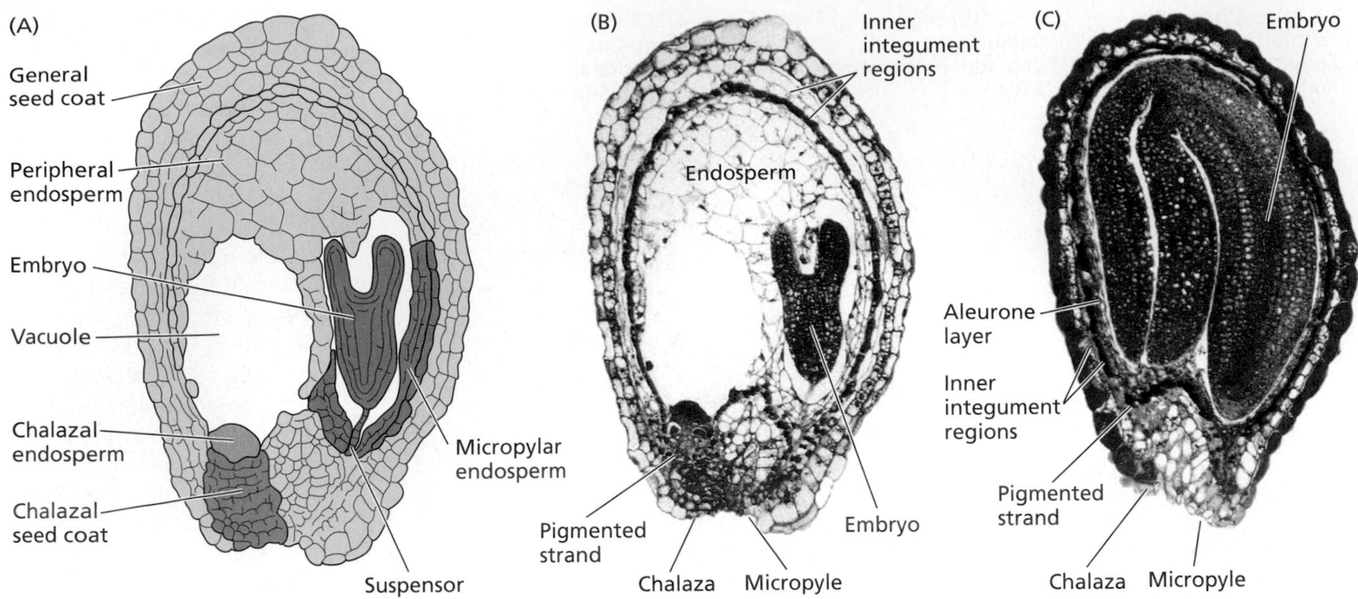

Figure 21.22 Arabidopsis seed structure. (A) Diagram of an Arabidopsis seed with the embryo at the torpedo stage of development. (B) Light micrograph of a stained section of an Arabidopsis seed at the same stage as in (A). The embryo is embedded in mature endosperm tissue. The seed is covered by a seed coat derived from the inner and outer integument tissues of the ovule. (C) Mature seed. The endosperm has mostly been resorbed, and the embryo fills the seed. The cotyledons contain stored reserves that will support early seedling growth following germination. (From Debeaujon et al. 2003.)

studied in seeds of cereals and Arabidopsis, as we will discuss in the following sections. (For a description of the other endosperm types, see WEB TOPIC 21.6.)

During seed morphogenesis, the endosperm provides nutrition to the developing embryo. In some species there is sufficient endosperm remaining to provide nutrition to the germinated seedling as well. In Arabidopsis and many other species, the endosperm is almost entirely resorbed (solubilized and absorbed) during embryogenesis, and the reserves that will support early seedling growth are stored in the embryo's cotyledons (**Figure 21.22**). The fleshy cotyledons of legumes are highly specialized for food storage (see Chapter 18). In cereals and other grasses, the endosperm persists during seed development and becomes the major site for the storage of starch and protein (**Figure 21.23**). Mobilization of these reserves

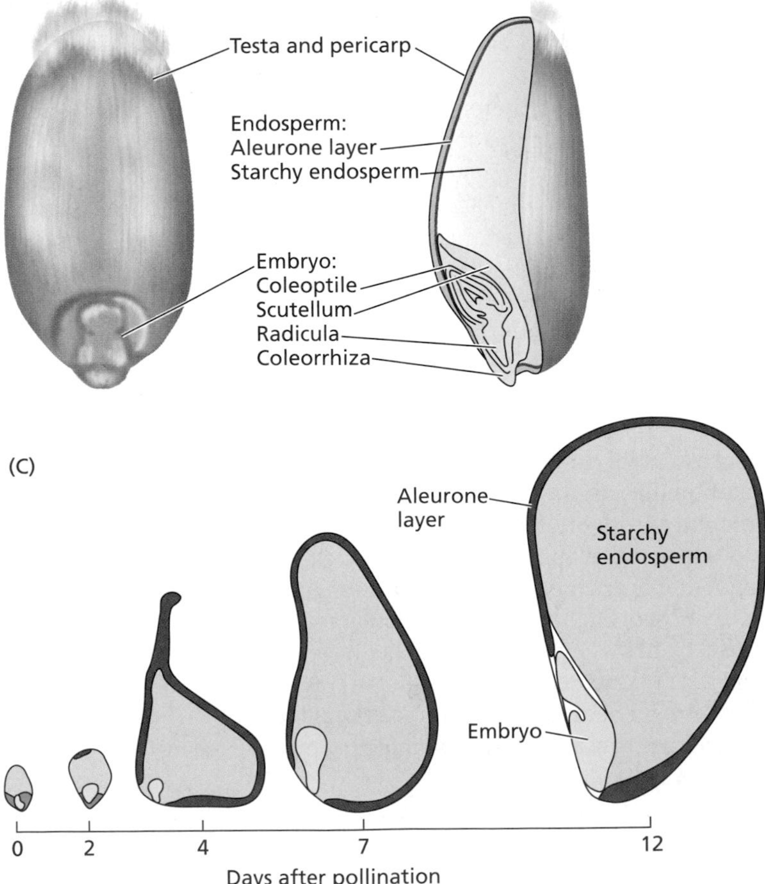

Figure 21.23 Cereal seed structure, as illustrated by wheat (*Triticum aestivum*). (A) Surface view of the seed, showing the location of the embryo in relation to the endosperm. (B) Longitudinal section through a seed. (C) Development of the embryo. (C after Cosségal et al. 2007.)

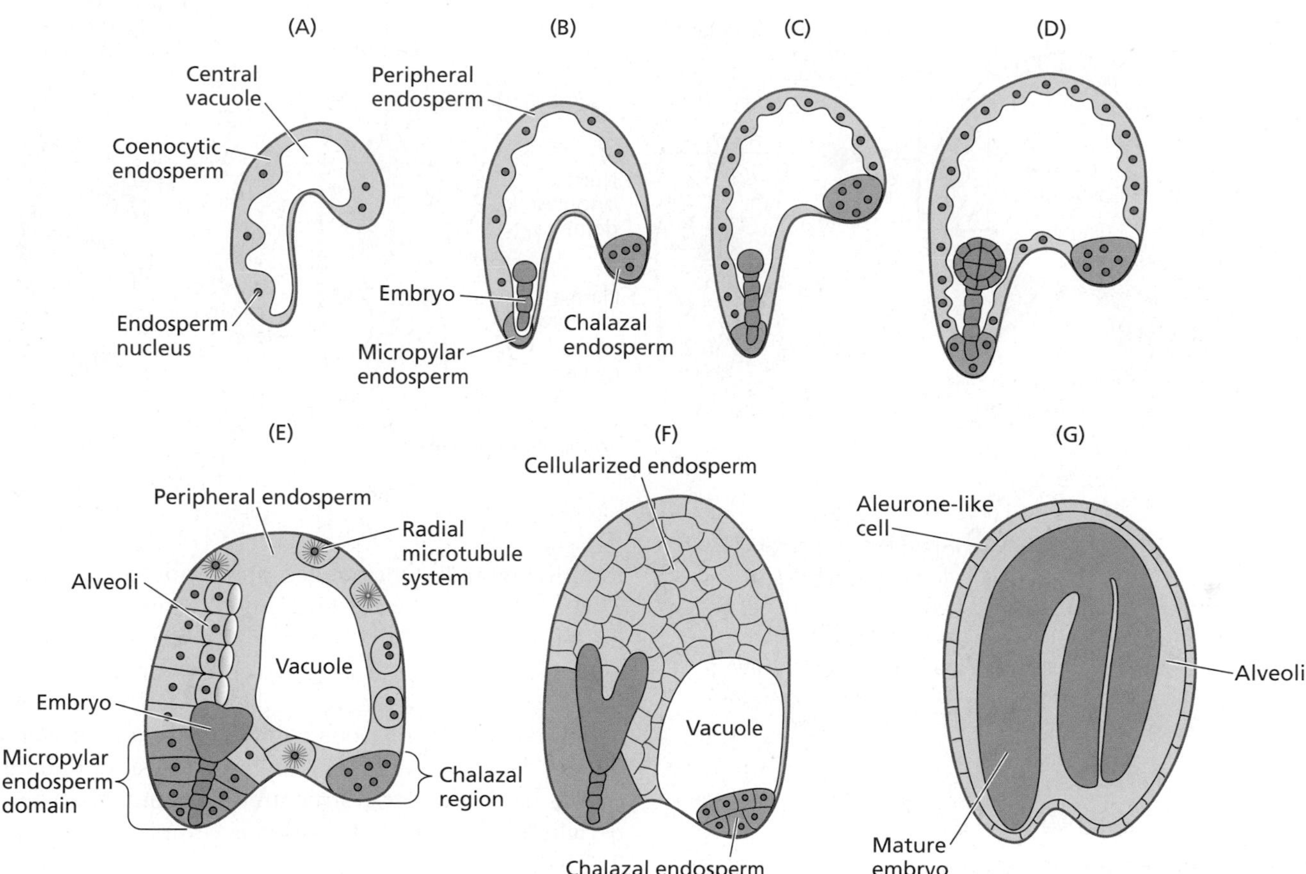

Figure 21.24 Development of the endosperm coenocyte of Arabidopsis. (A–D) The primary endosperm nucleus undergoes free nuclear divisions, and the resulting nuclei migrate to the periphery of the coenocytic central cell. (E–G) Cellularization of the endosperm coenocyte begins at the micropylar endosperm region and progresses to the chalazal region. All but a thin layer of endosperm at the periphery (the aleurone layer) is reabsorbed by the growing embryo during development. (After Olsen 2004.)

for transport to the embryo is the final function of the endosperm before it undergoes programmed cell death as the seedling becomes established.

In seeds with a nuclear-type endosperm, development proceeds in two phases: a *coenocytic phase* and a *cellular phase*. Just after double fertilization, the endosperm nucleus undergoes several rounds of mitosis without cytokinesis, forming a multinucleate coenocyte. At a specific time that varies with species, the coenocyte deposits cell walls around each nucleus as it gradually undergoes cellularization.

Cellularization of coenocytic endosperm in Arabidopsis progresses from the micropylar to the chalazal region

Several stages in the development of the endosperm coenocyte of Arabidopsis are shown in **Figure 21.24**. The primary endosperm nucleus undergoes a series of eight mitotic divisions without cytokinesis, leading to the production of about 200 nuclei, mostly located at the periphery of the large central cell. By the globular embryo stage, the endosperm coenocyte of Arabidopsis has three regions that become distinct as the seed grows: the **micropylar endosperm** that surrounds the embryo, the peripheral endosperm in the central chamber, and the chalazal endosperm.

Cellularization of the coenocytic endosperm in Arabidopsis begins in the micropylar endosperm region and progresses to the chalazal region (see Figure 21.24, parts E–G). The process is inititated during the globular stage of embryogenesis, at which time the coenocyte is organized into evenly spaced nuclear cytoplasmic domains defined by radial systems of microtubules (**Figure 21.25**, see parts A and B). Mini-phragmoplasts (see Chapter 1) assemble at the boundaries of adjacent nuclear cytoplasmic domains and vesicle fusion at the division plane (see Figure 21.25C). A coherent cell plate then develops by the fusion of tubular membrane into porous sheets. The last stage is the fusion of one side of the cell plate with the parental plasma membrane (see Figure 21.25D). Following the wall formation

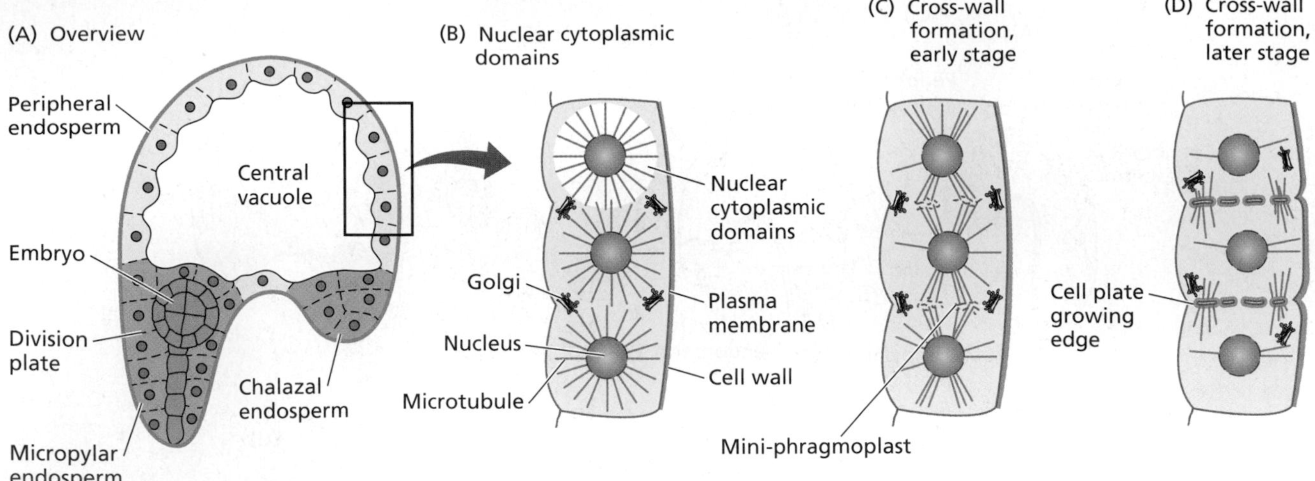

Figure 21.25 Cross wall formation in the peripheral endosperm of Arabidopsis. (A) Cellularization begins during the globular stage of embryogenesis. (B) The coenocyte is organized into nuclear cytoplasmic domains by radial microtubules. (C) Mini-phragmoplasts form at the boundaries between adjacent domains. (D) Vesicles fuse to form cross walls. (After Otegui 2007.)

between adjacent nuclear cytoplasmic domains, the cells are referred to as **alveolar cells** because of their tubelike nature: the end of the cell facing the central vacuole lacks a cross wall and is open to the cytoplasm of the central cell. Subsequent divisions of the alveolar cells inward lead to cross wall formation in the peripheral cell layers. Eventually the entire endosperm is cellularized.

The cellular endosperm of Arabidopsis is largely consumed as the embryo grows. At maturity, a massive embryo fills the seed and only a single layer of endosperm remains in the mature seed (see Figures 21.22C and 21.24G). As discussed in Chapter 18, the persistent endosperm layer, sometimes referred to as the *aleurone layer* by analogy to cereal grains, contributes to coat-imposed dormancy in Arabidopsis and other small-seeded species, and breakdown of its cell wall is required for the completion of germination.

Cellularization of the coenocytic endosperm of cereals progresses centripetally

In cereals, the endosperm is not consumed during embryogenesis, and as a result it takes up a much larger volume of the mature seed (see Figure 21.23).

During cereal endosperm development, the triploid primary endosperm nucleus undergoes a series of mitotic divisions without cytokinesis, and the nuclei migrate to the periphery of the central cell, which also contains a large central vacuole (**Figure 21.26**, see parts A–D). As in the Arabidopsis coenocyte, each of the nuclei is surrounded by radially arranged microtubules (see Figure 21.26E). Anticlinal walls form initially between adjacent nuclei, resulting in the tubelike alveolar cells, with the open end pointing toward the central vacuole (see Figure 21.26F). The alveolar nuclei then undergo one or more periclinal mitotic divisions followed by cytokinesis, producing daughter cells. The innermost layer of daughter cells remains alveolar in structure, and continues to divide periclinally until cellularization is complete (see Figure 21.26G and H).

The most important source of starchy endosperm cells is the interior cells of the cell files that are present at the completion of endosperm cellularization (see Figure 21.26H). Soon after this the cells undergo further divisions, with the division planes now oriented randomly so that the cell file pattern is soon lost. The second source of starchy endosperm cells is the inner daughter cells of the aleurone layer that divide periclinally. These cells redifferentiate to become the outer layers of the starchy endosperm.

Endosperm development and embryogenesis can occur autonomously

Although embryogenesis and endosperm formation occur concurrently and in close proximity, the two developmental programs are experimentally separable. For example, the ability to generate somatic (asexual) embryos in tissue culture, a routine procedure in many biotechnology laboratories, demonstrates that embryogenesis can occur in the absence of the surrounding seed tissues. In this case, the nutrient medium, which includes hormones, substitutes for the presence of a nutritive endosperm.

Conversely, in Arabidopsis, mutations in any one of three *FERTILIZATION-INDEPENDENT SEED* (*FIS*) genes (*FIS1*, *FIS2*, and *FIS3*) triggers autonomous endosperm development in the absence of fertilization and embryo formation. The mutant endosperm is diploid rather than triploid but is otherwise normal. Since the seed coat (testa)

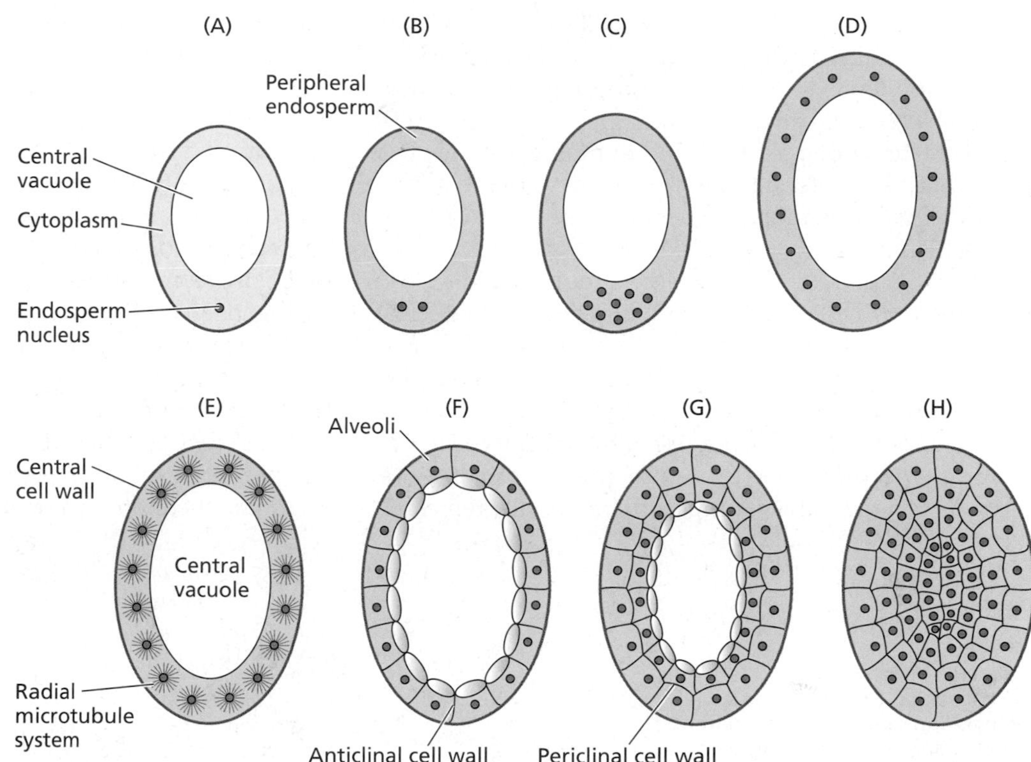

Figure 21.26 Development of the endosperm coenocyte of cereals. (A–D) The triploid endosperm nucleus is located in the basal cytoplasm of the central cell. After a series of free nuclear divisions, the nuclei migrate to the periphery of the large coenocytic cell. (E–H) Cellularization of the coenocytic endosperm of cereals. (After Olsen 2004.)

and fruit (silique) also are formed in the *fis* mutants, the development of the endosperm, testa, and ovary wall appears to be coordinated.

Many of the genes that control endosperm development are maternally expressed genes

Animal breeders have known about parent-of-origin effects for thousands of years. For example, crossing donkeys with horses produces hinnies when the horse is the male parent and mules when the horse is the female parent. In corn, certain alleles of the *R* and *B* genes that regulate anthocyanin accumulation produce pigmented kernels when inherited from one parent but not when inherited from the other parent. **Parent-of-origin effects** are defined as phenotypes that depend on the sex of the parent from which the trait is inherited.

A subset of parent-of-origin effects is caused by **imprinted gene expression**. Imprinted genes are expressed predominantly from either the maternal or the paternal allele, in contrast to nonimprinted genes in which alleles from both parents are expressed equally. Imprinted gene expression is considered epigenetic because alleles that have identical or nearly identical DNA sequences are expressed differentially. The differences in expression result from covalent modification of the DNA or its associated proteins (see Chapter 2).

Genes that are maternally expressed and paternally silent are referred to as **maternally expressed genes** (**MEGs**), and those that are paternally expressed and maternally silent are referred to as **paternally expressed genes** (**PEGs**). In flowering plants, imprinted gene expression is almost entirely confined to endosperm tissue, and in two different studies in Arabidopsis, the vast majority of them are MEGs (100–165 MEGs vs. 10–43 PEGs). The evolutionary significance of the role of MEGs in the endosperm is that the female parent controls the nutrition of the developing embryo. No imprinted genes have been identified in the embryo itself.

The FIS proteins are members of a Polycomb repressive complex (PRC2) that represses endosperm development

An important feature of Arabidopsis *fis* mutants was revealed following reciprocal crosses between the mutants and wild types. These crosses showed that defective endosperm development and embryo abortion were observed only after maternal inheritance of a mutant *fis* allele. This parent-of-origin effect is due to differential expression of paternal and maternal alleles caused by parental genomic imprinting. Such parent-of-origin effects are regulated in part by Polycomb group proteins.

Polycomb group proteins are evolutionarily conserved regulators that repress the transcription of their target genes, which often play essential roles in cell proliferation and differentiation. They mediate epigenetic changes through chromatin remodeling during both plant and animal development. Polychrome group protein complexes include multiple forms of Polycomb repressive complex 2 (PRC2), which catalyzes methylation of histones, which are components of nucleosomes whose methylation tends to inhibit transcription of the associated DNA (see Figure 2.14). Plants have multiple PRC2 complexes encoded by multiple homologs of subunit genes that have different roles in plant development. The FIS–PRC2 (Fertilization Independent Seed–PRC2) complex, which normally controls arrest of the Arabidopsis female gametophyte, is composed of four proteins: FIS1, also called MEDEA (MEA), FIS2, FIS3 (also called FERTILIZATION-INDEPENDENT ENDOSPERM [FIE]), and MUSASHI HOMOLOG 1 (MSI1). As noted earlier, loss-of-function mutations of any of the *FIS* genes causes the spontaneous onset of mitosis in the central cell in the absence of fertilization. Therefore, the FIS proteins are subunits of a regulatory complex that normally represses endosperm development in the absence of fertilization. Presumably, the normal function of the complex is to methylate histones associated with genes that promote endosperm development. In the presence of fertilization, the endosperm of *fis* mutants overproliferates and remains uncellularized. The effects of *msi1* mutations are pleiotropic. In addition to autonomous endosperm development, loss-of-function *msi1* mutations cause autonomous divisions of the egg cell, leading to a nonviable, parthenogenic embryo.

Gene imprinting in the endosperm also involves methylation and demethylation of DNA, carried out, respectively, by MET1 (a DNA methyltransferase) and DME (DEMETER), a DNA glycosylase. (DNA glycosylase removes 5-methylcytosine residues from DNA sequences, which are then replaced with unmethylated cytosines.)

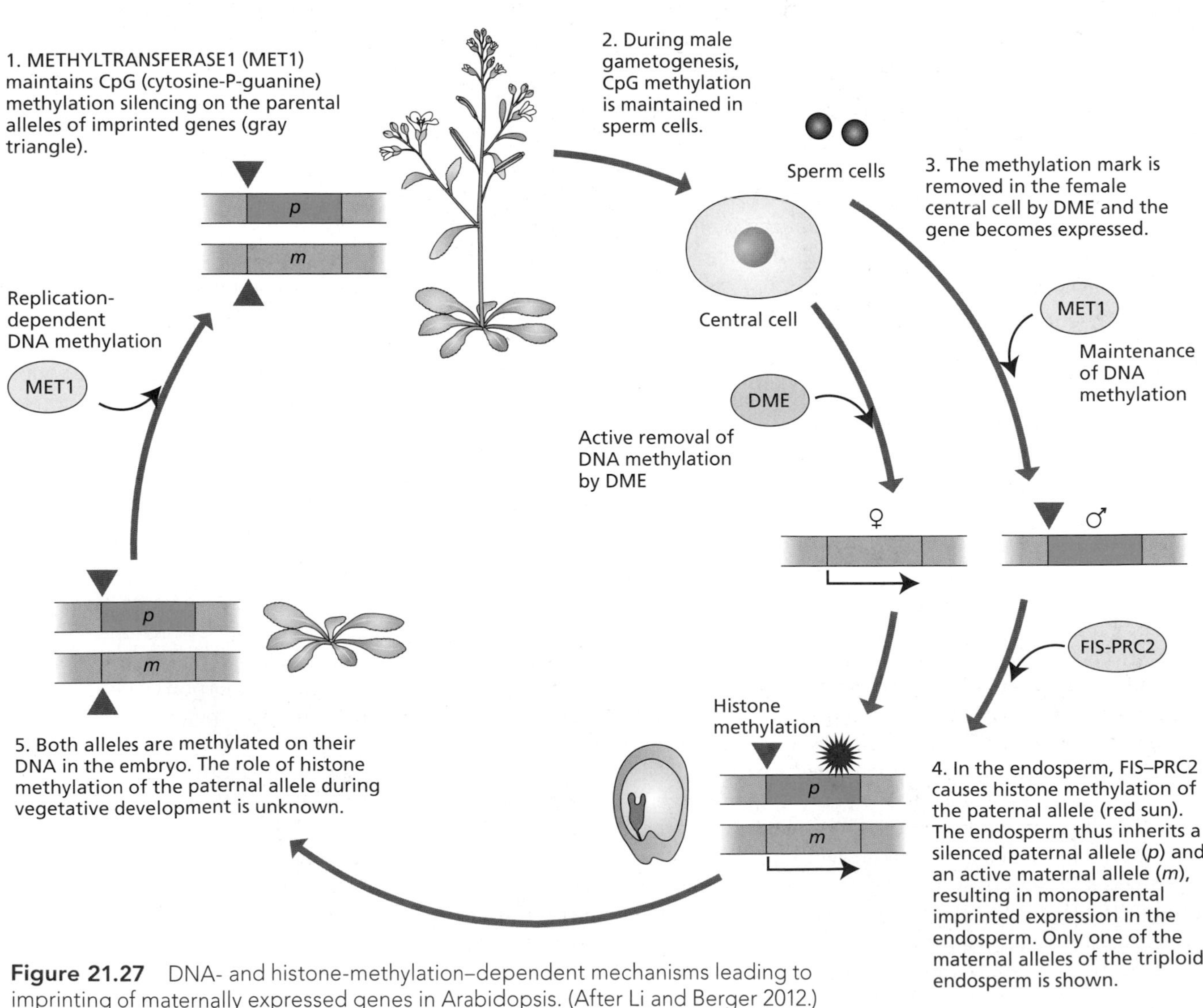

Figure 21.27 DNA- and histone-methylation–dependent mechanisms leading to imprinting of maternally expressed genes in Arabidopsis. (After Li and Berger 2012.)

A model showing how DNA demethylation, DNA methylation, and FIS–PRC2–mediated histone methylation may regulate MEG expression in the endosperm is shown in **Figure 21.27**. The imprinted alleles start out being DNA-methylated in both the haploid sperm and the diploid central cell nucleus, leading to partial gene inactivation. The presence of DME in the central cell demethylates the maternal alleles, restoring their full activity. After double fertilization, histone methylation by FIS–PRC2 completes the inactivation of the paternal allele, while the two maternal alleles from the central cell remain active.

As noted earlier, the maternal effect of the *FIS* gene mutations indicates that the early events of seed formation are under the control of the female parent. The role of the *FIS* genes in endosperm development indicates that maternal control is exerted over nutrient allocation to the embryo. It is therefore intriguing that the human homologs of the *FIS* genes are involved in the control of placental development, a tissue that also provides nutritional support for the growing embryo.

Cells of the starchy endosperm and aleurone layer follow divergent developmental pathways

Whereas seeds of many species store reserves as proteins and oils, the endosperm of cereals stores large quantities of starch. The **starchy endosperm** is a unique tissue, accounting for the bulk of the endosperm in cereal grains (see Figure 21.23). The major metabolic pathway in the starchy endosperm is, as the name implies, starch biosynthesis: the precursor molecule, ADP-glucose, is synthesized in the cytosol and then imported into the amyloplast, where it is enzymatically polymerized into amylose and amylopectin (see Chapter 8). The starchy endosperm of cereals also contains storage proteins, which are deposited in protein storage vacuoles.

Endoreduplication resulting in extremely high amounts of DNA appears to play a critical role in starchy endosperm development. In maize, for example, the DNA content may reach 96C—that is, 96 times the amount present in the haploid nucleus. Endoreduplication begins during reserve deposition, and the accumulation of DNA prevents further nuclear or cell division.

The starchy endosperm of cereals is dead at maturity due to programmed cell death, an event linked to the ethylene signaling pathway. In the maize mutant *shrunken2*, which overproduces ethylene, endosperm cell death is accelerated.

As discussed in Chapter 18, the aleurone layer (the outermost layer[s] of the endosperm) functions during early seedling growth by mobilizing starch and storage protein reserves in the starchy endosperm through the production of a α-amylase, protease, and other hydrolases in response to gibberellins produced by the embryo. Maize and wheat have one layer of aleurone cells, rice has one to several layers, and barley has three layers. In cereal grains, the aleurone layer is the only part of the endosperm that may become pigmented.

Aleurone cells become morphologically distinct in the barley endosperm at 8 days after pollination, comparable to the other cereals. Cytological evidence suggests that aleurone cell fate is specified earlier, after the first periclinal division of the alveolar nuclei. The basis for this conclusion is that at this stage, the aleurone precursor cells exhibit a hooplike cortical microtubule array that distinguishes them from cells that will become the starchy endosperm.

Two genes, *DEK1* and *CR4*, have been implicated in aleurone layer differentiation

The differentiation of the aleurone layer is under the control of several regulatory genes. For example, a loss-of-function mutation in the maize gene *DEFECTIVE KERNEL1* (*DEK1*) results in the production of seeds without aleurone layers (compare **Figure 21.28A and B**). When the *VIVIPAROUS1* (*Vp1*) gene promoter, which is expressed specifically in the aleurone layer, was fused to the *GUS* gene and used as a reporter for aleurone cells (*Vp1:GUS* transgene), the wild-type grain containing the transgene showed the blue color reaction indicating the presence of the aleurone layer (**Figure 21.28C**), whereas the *dek1* mutant seed containing the same transgene did not (**Figure 21.28D**). Similar effects of *dek1* mutations have been reported in Arabidopsis and rice seeds. The *DEK1* gene encodes a large, complex integral membrane protein that localizes to the plasma membrane. An extracellular loop in its structure suggests the DEK1 protein has the potential to interact with extracellular molecules, including signaling ligands.

The CRINKLY4 (CR4) protein is a receptor-like kinase that also functions as a positive regulator of aleurone cell fate. Mutants homozygous for the recessive *cr4* mutation show sporadic patches that lack an aleurone layer. The phenotypes of *cr4* mutants resemble those of a weak allele of *DEK1*. The evidence thus far suggests that CR4 acts downstream of DEK1, and immunolocalization studies have shown that DEK1 and CR4 proteins are located together in the plasma membrane.

There appears to be a functional connection between the aleurone layer of the endosperm and the epidermis of leaves. This first became evident in the maize *cr4* mutant, which disrupts aleurone layer specification and also perturbs the leaf epidermis in various ways: cells are often irregularly shaped with poorly developed cuticles, and the epidermis sometimes contains multiple cell layers. Similarly, weak alleles of *dek1* have a pronounced effect on the leaf epidermis of maize, rice, and Arabidopsis.

Although significant progress has been made in identifying genes implicated in aleurone cell development, the signaling pathways involved in its differentiation have not been worked out.

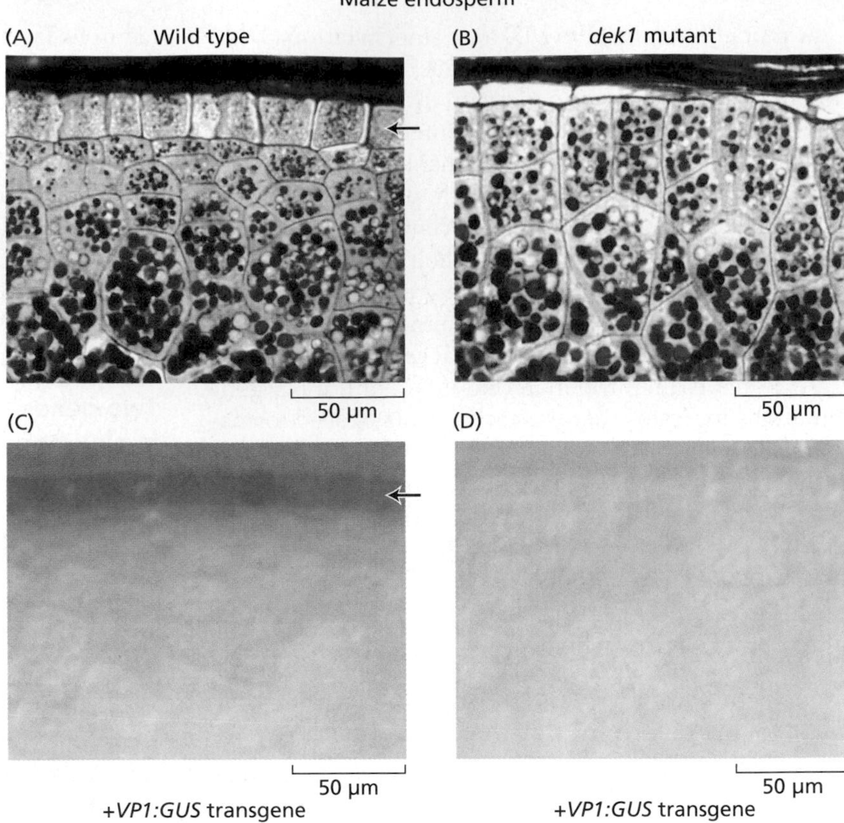

Figure 21.28 A loss-of-function mutation in the maize gene *DEFECTIVE KERNEL1* (*DEK1*) results in the production of seeds without aleurone layers. The wild type (A and C) and the *dek1* (B and D) mutant. (A and B) Starchy endosperm cells are filled with starch grains, which are stained pink. The arrow in (A) highlights the aleurone layer with dense granular cytoplasm and cuboidal cells. The aleurone layer is absent in the *dek1* mutant, and its surface cells have starchy endosperm identity. (C and D) A *VP1:GUS* transgene is a marker for aleurone cells. Wild-type endosperm shows aleurone-specific β-glucuronidase (GUS) activity (blue), while the marker is not expressed in the *dek1* mutant. (From Becraft and Yi 2011.)

Seed Coat Development

In response to fertilization, the Arabidopsis seed coat differentiates from the cells of the maternally derived ovule integuments over 2 to 3 weeks (**Figure 21.29**). Cells in both layers of the outer integument and all three layers of the inner integument undergo a dramatic period of growth in the first few days after fertilization through both cell division and expansion (see Figure 21.31B). The resulting five cell layers undergo one of four distinct fates. Cells of the innermost layer, derived from the ovule **endothelium**, synthesize **proanthocyanidin** flavonoid compounds, also known as **condensed tannins** (see WEB APPENDIX 4), which accumulate in the central vacuole of the endothelial cells during the first week after fertilization and later become oxidized, imparting a brown color to the differentiated cells (known as the pigmented cell layer) and the whole seed coat. By contrast, cells of the other two inner integument layers do not appear to differentiate further, undergo early programmed cell death, and are crushed as the seed develops (see Figure 21.29D and E).

Cells of both outer integument layers accumulate starch in amyloplasts during the initial growth phase (see Figure 21.29B) before their fates diverge. The subepidermal layer (layer 2), which differentiates into **palisade cells**, produces a thickened wall on the inner tangential side of the cells (see Figure 21.29C–E). The cells of the epidermal layer (layer 1) synthesize and secrete a large quantity of mucilage (a specialized secondary cell wall that contains some pectin) into the apoplast specifically at the junction of the outer tangential and radial cell walls (see Figure 21.29C). Hydrated mucilage appears to provide a moist environment for seed germination and protection from chemicals that may be present in an animal gut. In addition, the secondary cell walls of the outer two layers provide protection for the embryo, and tannins can be toxic to intruders.

Following mucilage synthesis, a cellulosic secondary cell wall is deposited that completely fills the space occupied by the cytoplasmic column, forming the **columella** (see Figure 21.29D and E). During the later stages of seed development, the cells of all remaining seed coat layers die. The structure of the epidermal cells is preserved by the columella, and the remaining layers are crushed by the end of seed maturation. Proanthocyanidins are apparently released from the endothelial cells and impregnate the inner three cell layers during this period (see Figure 21.29E).

Seed coat development appears to be regulated by the endosperm

Seed coat growth and differentiation are initiated by fertilization and normally proceed coordinately with the development of the embryo and endosperm. Because the seed coat surrounds the seed, its growth in surface area must be coordinated with the growth of the embryo and endosperm for the seed to reach its mature size. If the seed coat fails to expand, seed size is reduced.

Figure 21.29 Development of the ovule integuments into the seed coat of Arabidopsis following fertilization. Several stages (A–E) of seed development for the whole seed (left) and a detail of the developing seed coat (right) are shown. (A) Prior to fertilization. (B) Five days after fertilization. The two cell layers of the ovule outer integument (1 and 2) and three cell layers of the inner integument (3–5) have grown. (C–D) Ten days after fertilization. Cells of individual layers have almost completed differentiation into specialized cell types, including endothelium (5), palisade (2), and epidermis (1). (E) Fifteen days (seed maturity). The cells of all five layers are dead and have been crushed together, except for the epidermis, the shape of which is maintained by the thick secondary cell wall of the columella. Red arrows indicate starch-containing plastids (in B), mucilage in the apoplast (in C), and secondary cell wall forming in the epidermis (in D). The green arrows indicate the secondary cell wall of the palisade (in C and D). Al, endosperm aleurone layer; Em, embryo; En, endosperm; Es, embryo sac; Ii, inner integument; Oi, outer integument. (From Haughn and Chaudhury 2005.)

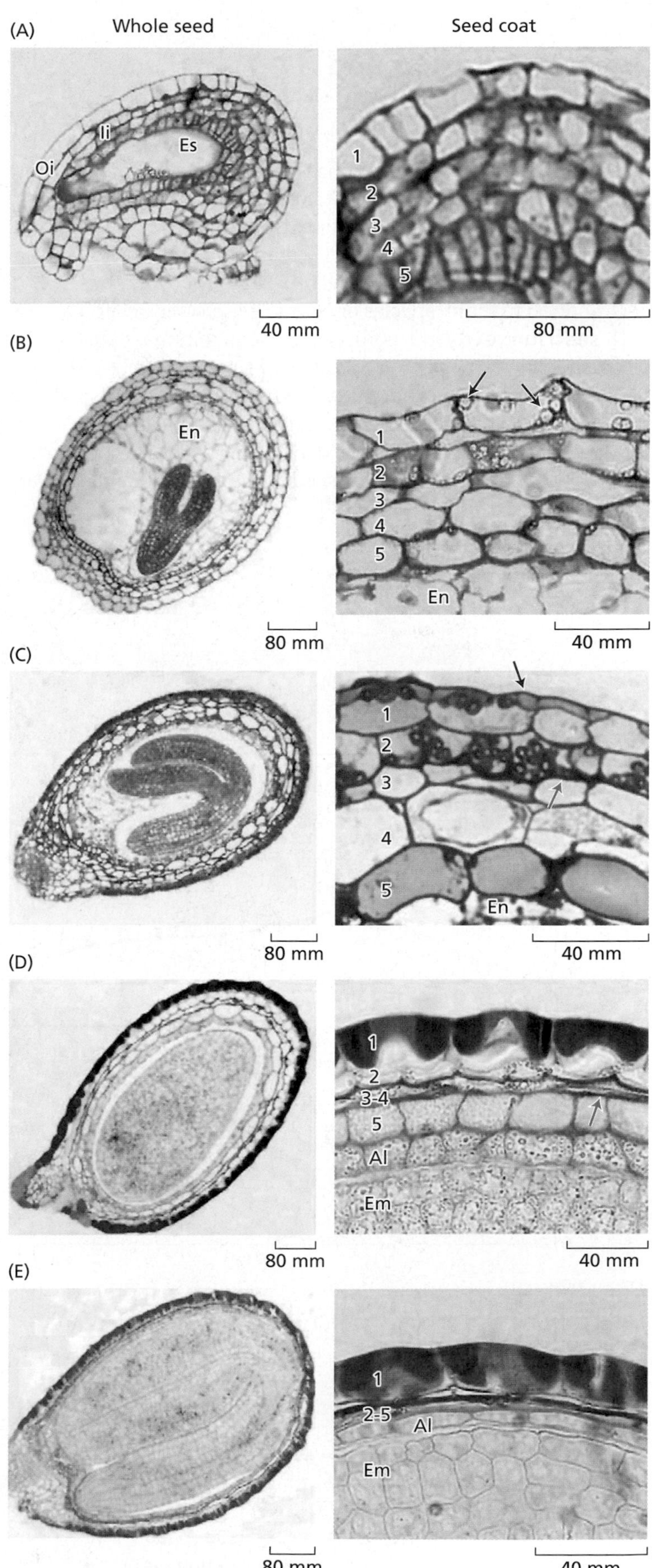

For example, the *TRANSPARENT TESTA GLABRA2* (*TTG2*) gene positively regulates proanthocyanidin biosynthesis as well as seed coat expansion. As a result, loss-of-function *ttg2* mutants have smaller seeds, presumably because the embryo and endosperm are mechanically constrained by the seed coat during development.

Conversely, mutations in the *HAIKU* gene result in limited growth of the coenocytic endosperm. This defect in endosperm growth also affects the growth of the developing seed coat, such that cell elongation in the expanding seed coat is restricted. This suggests that the growing endosperm regulates the extent to which the ovule integument cells elongate following the initiation of seed coat development.

As noted earlier, embryogenesis is blocked in the *fis* mutants, yet endosperm and seed coat development progress more or less normally. Thus, a signal from the coenocytic endosperm appears to be sufficient to initiate seed coat development in cells of the integument. Consistent with this idea, no significant growth of the seed coat occurs in seeds in which only the egg cell is fertilized, and seed coat development is strongly inhibited in seeds in which the endosperm has been experimentally destroyed.

Seed Maturation and Desiccation Tolerance

Thus far we have discussed seed histodifferentiation and reserve deposition. The final phase of seed development is termed maturation. For many species, maturation also includes the acquisition of **desiccation tolerance**. This involves the evaporative loss of water to produce a dry seed, a prerequisite for the quiescent state that precedes germination in many species of plants. It is also correlated with **seed longevity**, the ability of the seed to remain viable in the dry state over long periods of time.

The term **orthodox seed** has been used to denote those seeds that can tolerate desiccation and are storable in a dry state for variable periods of time, depending on the species. The world-champion orthodox seed is the 2,000-year-old Judaean date palm (*Phoenix dactylifera*) seed that was successfully germinated in 2005. In contrast, **recalcitrant seeds** are those that are released from the plant with a relatively high water content and active metabolism. Unlike orthodox seeds, recalcitrant seeds deteriorate upon dehydration and do not survive storage. Mango and avocado are examples of plants with recalcitrant seeds.

Seed filling and desiccation tolerance phases overlap in most species

The timing of desiccation tolerance and seed longevity, in relation to the attainment of mature size and seed dispersal, varies across species. For most species, the acquisition of desiccation tolerance occurs during seed filling. Subsequently, during late maturation, seeds progressively acquire longevity, the ability to remain alive in the dry state for prolonged periods of time.

For example, four stages of seed growth and development (embryogenesis, seed filling, late maturation, and pod abscission) for seeds of barrel clover (*Medicago truncatula*) are shown in **Figure 21.30A**. Embryogenesis (histodifferentiation) proceeds during the first 10 days after pollination, after which seed filling begins, as indicated by the increase in seed dry weight. Simultaneously, the water content of the seed declines (**Figure 21.30B**). The acquisition of desiccation tolerance begins approximately 24 days after pollina-

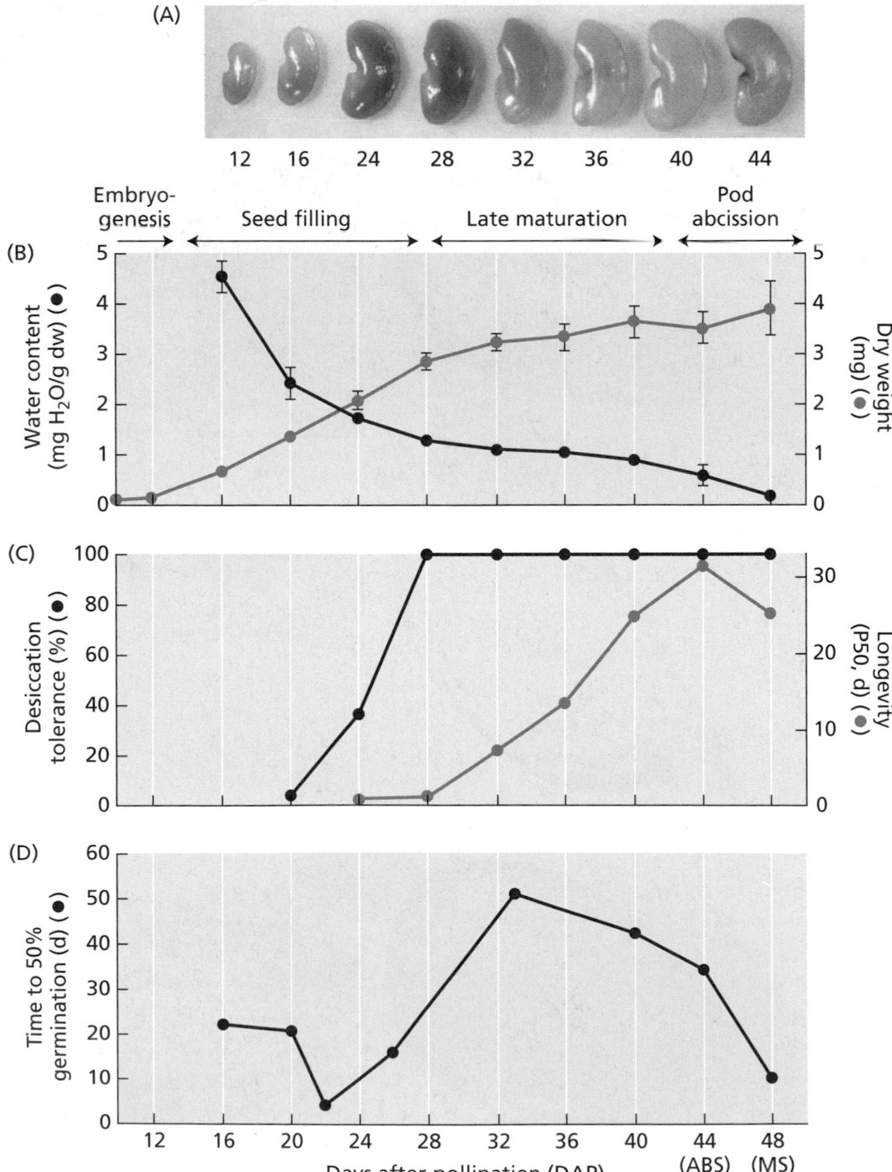

Figure 21.30 Metabolic and physiological changes during seed maturation of *Medicago truncatula*. Seed development is divided into four major phases: embryogenesis, seed filling, late maturation, and pod abscission. (A) Time course of seed development. (B) Water content and dry weight (dw) changes. (C) Acquisition of desiccation tolerance, measured as the percentage germination after rapid drying to 43% relative humidity, and longevity, determined as the time to reduce viability to 50% under storage at 75% relative humidity and 35°C. (D) Changes in germination speed or dormancy, determined as the time required for 50% of seeds to complete germination at 20°C. ABS, abscission; MS, mature seed. (After Verdier et al. 2013.)

tion, and overlaps with the seed-filling stage and dehydration phases. From 28 days after pollination onward, seeds gradually acquire longevity (storability) (**Figure 21.30C**). Freshly harvested seeds acquire the ability to germinate at approximately 16 days after pollination, and the ability to complete germination increases to 50% from 22 to 32 days after pollination, after which it declines to 10% due to the onset of dormancy (**Figure 21.30D**) (seed dormancy is discussed in Chapter 18). However, this dormancy can be overcome by dry storage for 6 months (after-ripening), after which the fully mature seeds germinate within 24 h.

The acquisition of desiccation tolerance involves many metabolic pathways

For orthodox seeds, seed desiccation involves more than just the physical drying of the seed. It is associated with distinct patterns of gene expression and metabolism that affect multiple physiological processes, including dormancy, after-ripening, and germination. During mid- to late embryogenesis of orthodox seeds, when seed abscisic acid (ABA) content is highest (see Chapter 18), multiple metabolic processes are activated that contribute to the acquisition of desiccation tolerance. In Arabidopsis, the expression patterns of more than 6,900 genes, approximately one-third of the genome, change during this period. The main metabolic processes that are activated as a consequence include:

- Accumulation of disaccharides and oligosaccharides
- Synthesis of storage proteins
- Synthesis of late-embryogenesis-abundant (LEA) proteins
- Synthesis of small heat shock proteins (smHSPs)
- Activation of antioxidative defenses
- Changes in the physical structure of cells
- Gradual increase in cell density

During the acquisition of desiccation tolerance, the cells of the embryo acquire a glassy state

Desiccation can severely damage membranes and other cellular constituents (see Chapter 24). Mature seeds have as little as 0.1 g water g^{-1} dry weight, with water potentials between –350 and –50 MPa. As seeds begin to dehydrate, embryos accumulate sugars and a specific set of proteins. These groups of molecules are thought to interact to produce a *glassy state*. In general, a glass is defined as an amorphous, metastable state that resembles a solid, brittle material but retains the disorder and physical properties of the liquid state. Biological glasses are highly viscous liquids with very slow molecular diffusion rates, and therefore can participate only in limited chemical reactions. Because nonreducing sugars such as sucrose, raffinose, and stachyose accumulate during the late stages of seed maturation, it was initially assumed that they were primarily responsible for cellular glass formation. However, the physical properties of sugar glasses are significantly different from those of desiccated embryos, which led to the hypothesis that proteins, specifically LEA proteins (see next section), are required for glass formation in seeds.

LEA proteins and nonreducing sugars have been implicated in seed desiccation tolerance

Late-embryogenesis-abundant (LEA) proteins are small, hydrophilic, largely disordered, and thermostable proteins that are synthesized in orthodox seeds during mid- to late maturation and in vegetative tissues in response to osmotic stress. They are thought to have a range of protective functions against desiccation with different efficiencies, including ion binding, antioxidant activity, hydration buffering, and membrane and protein stabilization. Since LEA proteins were first described in the early 1980s in cotton seeds, related proteins have been identified in seeds and pollen grains of other plant species, as well as in bacteria, cyanobacteria, and some invertebrates, and LEA proteins have been shown to increase osmotolerance in transgenic yeast. The ability of "resurrection plants" (e.g., *Craterostigma plantagineum*) to survive extreme desiccation has been linked to the accumulation of LEA proteins. In addition, LEA proteins may play a role in the response to freezing and salinity stress, both of which involve cellular dehydration (see Chapter 24).

Most LEA proteins show a biased amino acid composition, resulting in high hydrophilicity, and they are related to a group of proteins called the **dehydrins**. However, a distinguishing feature of dehydrins is their high glycine content, and since not all LEA proteins have this property, dehydrins and related proteins are considered to be subsets of the LEA protein family, which comprises nine groups. A key feature of LEA proteins is their ability to form hydrogen bonds with sucrose. Since sugars accumulate during seed maturation, it is thought that LEA proteins interact with sucrose and other disaccharides and oligosaccharides to form the glassy state required for desiccation tolerance.

Specific LEA proteins have been implicated in desiccation tolerance in *Medicago truncatula*

Considering that LEA proteins are a heterogeneous class of proteins, the question of whether there are specific LEA proteins involved in determining the formation of the glassy state in dehydrated cells remains to be addressed. Using a proteomic approach, researchers have identified a subset of LEA proteins in seeds of *Medicago truncatula* that correlates with survival in the dry state, making them possible candidates for the stabilization of the glassy state. Of the 38 LEA polypeptides detected in mature seeds of *M. truncatula*, a small subset of them accumulates specifically during the acquisition of desiccation tolerance, while a different subset accumulates during the acquisition of seed longevity.

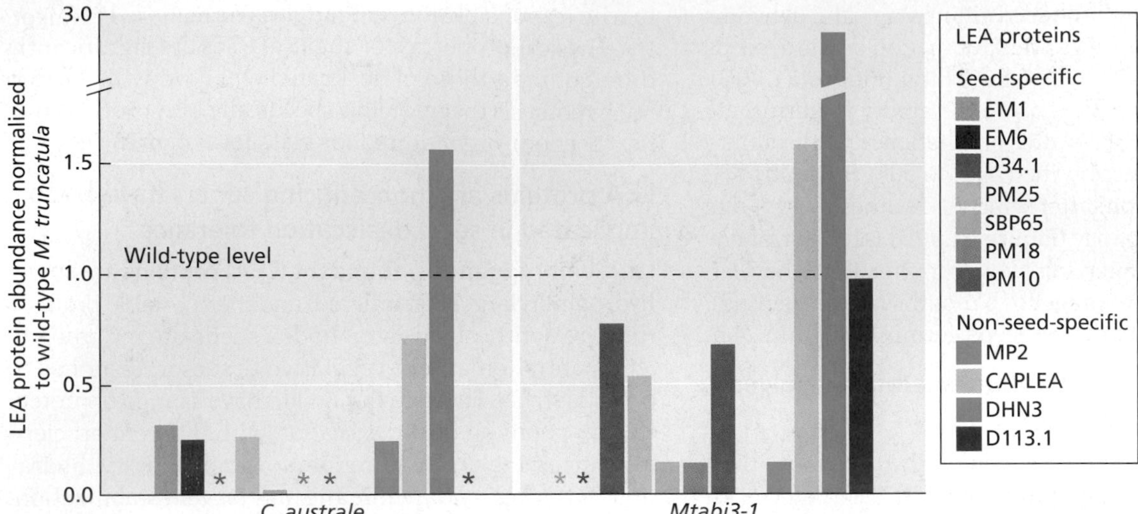

Figure 21.31 The LEA protein profile in desiccation-sensitive cotyledons of *Castanospermum australe* and seeds of *Mtabi3-1* mutants compared with desiccation-tolerant *Medicago truncatula* wild-type seeds—a value of 1 corresponds to wild-type values (*C. australe* and *Mtabi3*-1). The red, yellow, and green bars represent different LEA proteins specifically expressed only in the seed, while the blue bars are LEA proteins that are expressed throughout the plant. Polypeptides not detected are indicated by asterisks. (After Delahaie et al. 2013.)

In another study, the LEA proteins of recalcitrant and orthodox seeds were compared to determine whether any of those expressed in orthodox seeds were absent from recalcitrant seeds (**Figure 21.31**). Wild-type *M. truncatula* seeds were used as representative orthodox seeds. Two types of recalcitrant seeds were used: black bean (*Castanospermum australe*), a close relative of *M. truncatula*, and the ABA-insensitive *Mtabi3-1* mutant of *M. truncatula*, which fails to develop desiccation tolerance during seed maturation. All of the seed-specific LEA proteins, and some of the non-seed-specific LEA proteins, were found in much lower amounts in both types of recalcitrant seeds than in the orthodox seed, strongly implicating these LEA proteins in the acquisition of desiccation tolerance.

Abscisic acid plays a key role in seed maturation

As we saw in the previous section, the seeds of ABA-insensitive mutants of *M. truncatula* fail to develop desiccation tolerance and are therefore recalcitrant. The synthesis of LEA proteins, storage proteins, and lipids is promoted by ABA, as shown by physiological and genetic studies of cultured embryos of many species. ABA-deficient mutants fail to accumulate these proteins. Furthermore, synthesis of some LEA proteins, or of related family members, can be induced in vegetative tissues by ABA treatment. These results suggest that the synthesis of many LEA proteins is under ABA control during seed maturation.

As we discussed in Chapter 15, ABA induces changes in cellular metabolism by activating, either directly or indirectly, a network of transcription factors. In particular, ABI3 induces the synthesis of storage proteins and LEA proteins through interactions with bZIP transcription factors such as ABI5. An analysis of the gene regulatory network in *M. truncatula* seeds has shown that *ABI5* genes occupy a central position in the regulatory network and are highly connected to *LEA* and desiccation tolerance genes. Thus, *ABI3* and *ABI5*, along with several other genes, are the core components of the seed-specific ABA signaling pathway that regulates survival in the dry state.

Coat-imposed dormancy is correlated with long-term seed-viability

The seeds of many herbs and garden vegetables, such as onion, okra, and soybean, can remain viable under storage for only 1 to 2 years. Others, such as those of cucumbers and celery, remain viable for up to 5 years. In 1879, W. J. Beal initiated the longest-running experiment on seed longevity by burying the seeds of 21 different species in unstoppered bottles in a sandy hilltop near the Michigan Agricultural College in East Lansing. After 120 years (in the year 2000), only one species, moth mullein (*Verbascum blattaria*), remained viable. However, this is by no means the maximum longevity for seeds. For example, seeds of canna lily (*Canna compacta*) apparently can live for at least 600 years, while the oldest authenticated surviving seeds are those of sacred, Indian, or Asian lotus (*Nelumbo nucifera*) at nearly 1300 years, and date palm (*Phoenix dactylifera*), which was found buried at Masada in Israel, at a remarkable 2000 years. The two species with the greatest seed longevities (Indian lotus and date palm) have highly impermeable seed coats, suggesting that coat-imposed

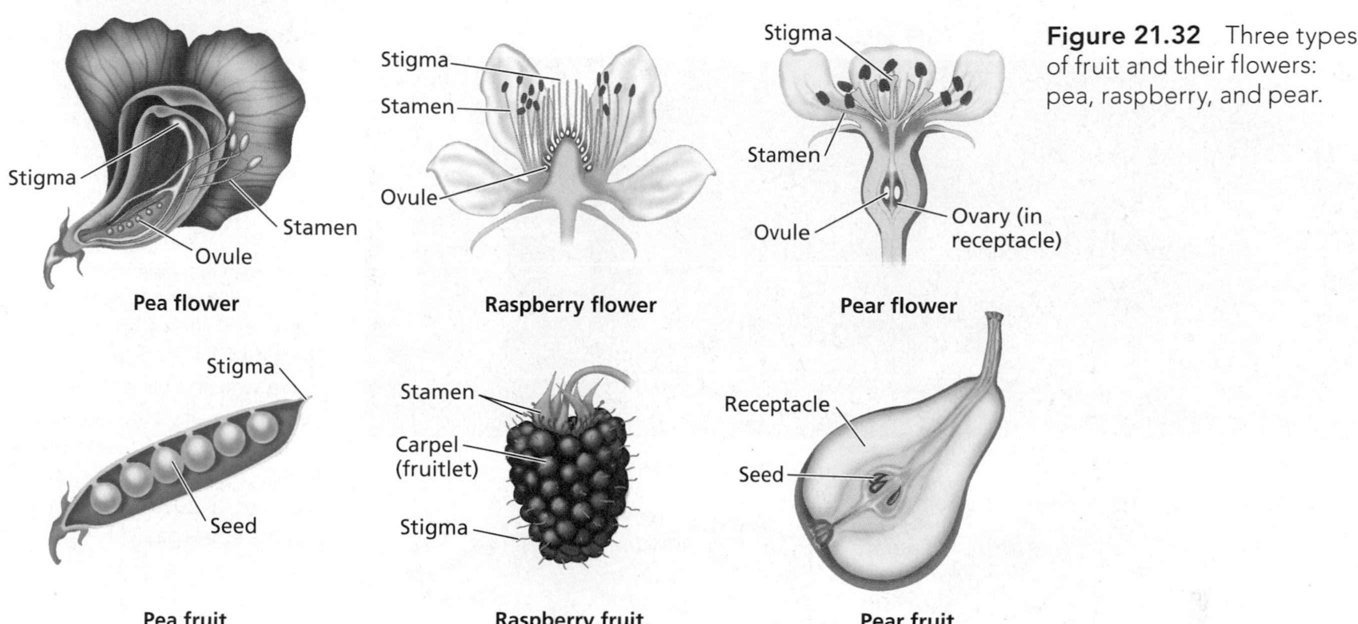

Figure 21.32 Three types of fruit and their flowers: pea, raspberry, and pear.

dormancy is associated with long-term viability. However, many orthodox seeds can be stored for a long time under seed-bank conditions at low temperature.

Fruit Development and Ripening

True fruits are found only in the flowering plants. In fact, fruits are a defining feature of the angiosperms, since *angio* means "vessel" or "container" in Greek and *sperm* means "seed." Diverse fruit types are represented in early Cretaceous fossils, including nuts and fleshy drupes and berries. Fruits are typically derived from a mature ovary containing seeds, but they can also include a variety of other tissues. For instance, the fleshy part of the strawberry is actually the receptacle, while the true fruits are the dry achenes embedded in this tissue.

Fruits are seed-dispersal units and they can be grouped according to several features (see WEB TOPIC 21.7). Based on their composition and moisture content, they can be either dry or fleshy. If the fruit splits to release its seeds, it is termed **dehiscent**. The fleshy fruits that we are most familiar with are **indehiscent** and occur in a variety of forms. Tomatoes, bananas, and grapes are defined botanically as **berries**, in which the seeds are embedded in a fleshy mass, while peaches, plums, apricots, and almonds are classified as **drupes**, in which the seed is enclosed in a stony endocarp. Apples and pears are **pome** fruits, in which the edible tissue is derived from accessory structures such as floral parts or the receptacle. Fruits can also be defined as *simple*, with a mature single or compound ovary, as in hazelnut, Arabidopsis, and tomato. Alternatively, they may be *aggregate*, where flowers have multiple carpels that are not joined together, as in raspberry. Finally, they may be *multiple*, where the fruit is formed from a cluster of flowers and each flower produces a fruit, as in pineapple. Some examples of fleshy and dry fruit types are illustrated in **Figure 21.32**.

The developmental switch that turns a pistil into a growing fruit depends on fertilization of the ovules. In most angiosperms, the gynoecium senesces and dies if unfertilized.

Arabidopsis and tomato are model systems for the study of fruit development

Arabidopsis has been a key model plant for the study of dry, dehiscent fruits. The gynoecium in Arabidopsis arises from the fusion of two carpels, referred to collectively as the pistil, and forms in the center of the flower. In Arabidopsis and many other members of Brassicaceae, several fruit tissues develop, including the carpel walls or pericarp (known also as the valves), a central replum, a false septum, and valve margins that form at the valve and replum borders (**Figure 21.33**). In Arabidopsis the valve margins differentiate into zones that will be involved in dehiscence; the margins are where the fruit will open. Dry fruits have relatively few cell layers in the carpel walls, and some of these may be lignified, especially in areas associated with fruit dehiscence.

Much of what we know about the development of fleshy, indehiscent fruits has come from work on tomato (*Solanum lycopersicum*), a member of the nightshade family (Solanaceae) (**Figure 21.34A**). In tomato, as in Arabidopsis, the fruit is derived from the fusion of carpels. The carpel walls are called the pericarp (equivalent to the valves in Arabidopsis), and the seeds are attached to the placenta. Unlike Arabidopsis fruits, tomato fruits are indehiscent and the carpels remain completely fused. In fleshy fruits, cell division is usually followed by massive cell expansion

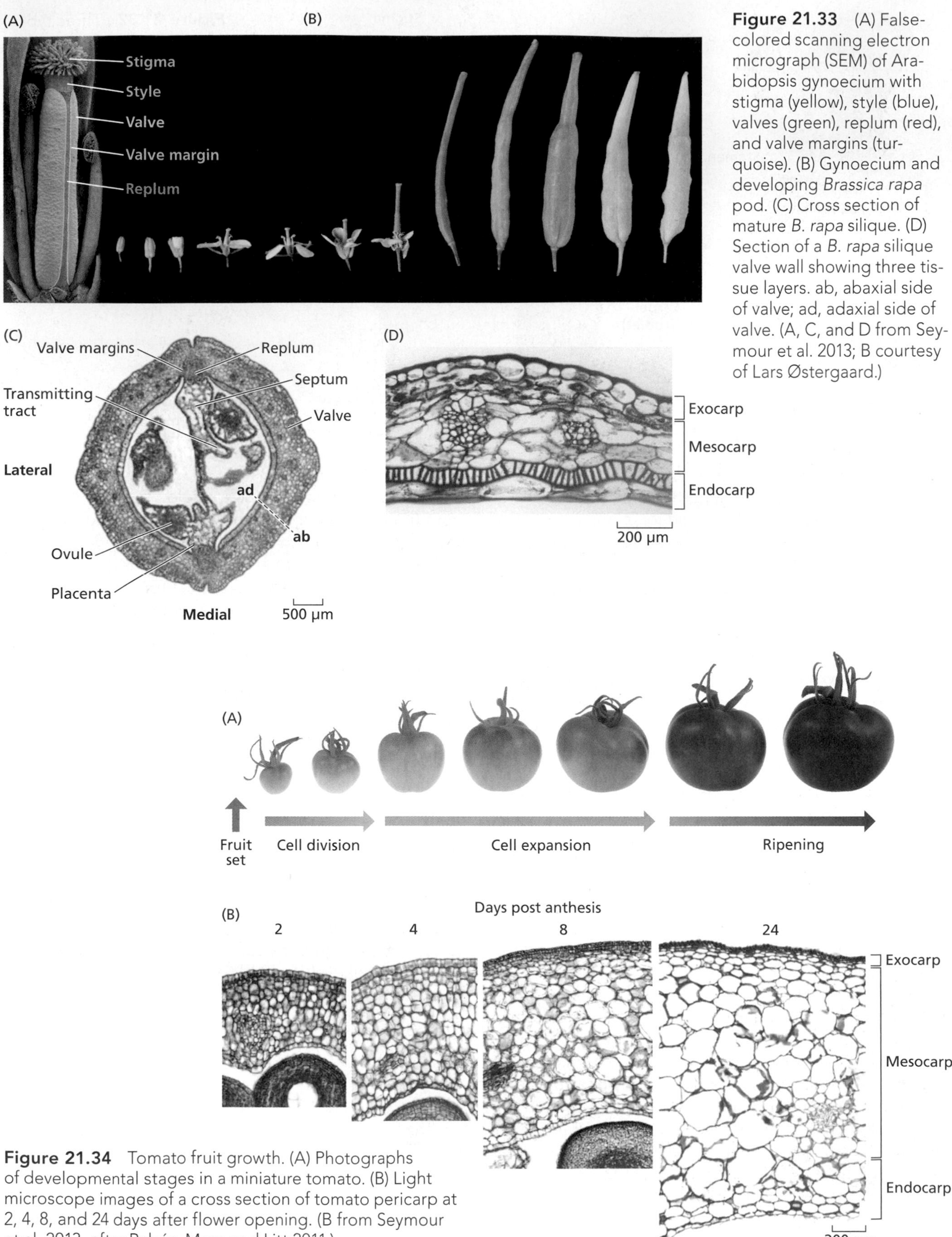

Figure 21.33 (A) False-colored scanning electron micrograph (SEM) of Arabidopsis gynoecium with stigma (yellow), style (blue), valves (green), replum (red), and valve margins (turquoise). (B) Gynoecium and developing *Brassica rapa* pod. (C) Cross section of mature *B. rapa* silique. (D) Section of a *B. rapa* silique valve wall showing three tissue layers. ab, abaxial side of valve; ad, adaxial side of valve. (A, C, and D from Seymour et al. 2013; B courtesy of Lars Østergaard.)

Figure 21.34 Tomato fruit growth. (A) Photographs of developmental stages in a miniature tomato. (B) Light microscope images of a cross section of tomato pericarp at 2, 4, 8, and 24 days after flower opening. (B from Seymour et al. 2013, after Pabón-Mora and Litt 2011.)

Figure 21.35 Phytoene synthase plays a role in lycopene production in tomato pericarp. The tomato on the left is a wild-type, red ripe fruit. The tomato on the right has reduced levels of phytoene synthase gene expression and therefore fails to accumulate the red pigment lycopene. (Images courtesy of R. G. Fray; also see Fray and Grierson 1993.)

(**Figure 21.34B**). In some varieties of tomato, for example, pericarp cell diameters may reach 0.5 mm. About 30 genetic loci, called quantitative trait loci (QTLs), have been shown to control fruit size in tomato, and several of the genes composing these QTLs have been cloned. One locus (*Fw2.2*) encodes a plant-specific and fruit-specific protein that regulates cell division in the fruit and therefore affects fruit size. Some fleshy fruits also have lignified cell layers, such as the hard endocarp, the "stone" or "pit" in drupes (e.g., peach).

Fleshy fruits undergo ripening

Ripening in fleshy fruits refers to the changes that make them attractive (to humans and other animals) and ready to eat. Such changes typically include color development, softening, starch hydrolysis, sugar accumulation, production of aroma compounds, and the disappearance of organic acids and phenolic compounds, including tannins. Dry fruits do not undergo a true ripening process, but as we will discuss later, many of the same families of genes that control dehiscence in dry fruits seem to have been recruited to new functions in ripening in fleshy fruits. Because of the importance of fruits in agriculture and their health benefits, the vast majority of studies on fruit ripening have focused on edible fruits. Tomato is the established model for studying fruit ripening, as it has proved highly amenable to biochemical, molecular, and genetic studies on the mechanism of ripening.

Ripening involves changes in the color of fruit

Fruits ripen from green to a range of colors, including red, orange, yellow, purple, and blue. The pigments involved not only affect the visual appeal of the fruit but also taste and aroma, and are known to have health benefits for humans. Fruits typically contain a mixture of pigments, including the green chlorophylls; yellow, orange, and red carotenoids; red, blue, and violet anthocyanins; and yellow flavonoids. The loss of the green pigment at the onset of ripening is caused by the degradation of chlorophyll and the conversion of chloroplasts to chromoplasts, which act as the site for the accumulation of carotenoids (see Chapter 1).

Carotenoids are responsible for the red color of tomato fruits. During ripening in tomato, the concentration of carotenoids increases between 10- and 14-fold, mainly due to the accumulation of the deep red pigment lycopene. Fruit ripening involves the active biosynthesis of carotenoids, the chemical precursors for which are synthesized in the plastids. The first committed step is the formation of the colorless molecule phytoene by the enzyme phytoene synthase. In tomato, phytoene is then converted to the red pigment lycopene through a series of further reactions. Experiments with transgenic tomatoes have demonstrated that silencing the gene for phytoene synthase prevents the formation of lycopene (**Figure 21.35**).

Anthocyanins are the pigments responsible for the blue and purple color in some berries (**Figure 21.36**). Anthocyanins are made through the phenylpropanoid pathway; that is, they are derived from the amino acid phenylalanine. Phenylpropanoids constitute some of the most important sets of secondary metabolites in plants. They contribute not only to the characteristic color and flavor of fruits but also to unfavorable traits, such as browning of fruit tissues via enzymatic oxidation of phenolic compounds by polyphenol oxidases. The genetic basis for anthocyanin biosynthesis is relatively well understood. At the molecular level, anthocyanin biosynthesis is regulated via coordinated transcrip-

Figure 21.36 Blueberries accumulate more than a dozen different anthocyanins during ripening, including malvidin-, delphinidin-, petunidin-, cyanidin-, and peonidin-glycosides, which give them a deep purple color.

tional control of the enzymes in the biosynthetic pathway by a range of transcription factors.

Fruit softening involves the coordinated action of many cell wall–degrading enzymes

Softening of fruit involves changes to the fruit's cell walls. In most fleshy fruits, the cell walls consist of a semirigid composite of cellulose microfibrils—thought to be tethered by a xyloglucan network—which is embedded in a gel-like pectin matrix. In tomato, more than 50 genes related to cell wall structure show changes in expression during ripening, indicating a highly complex set of events connected with cell wall remodeling during the ripening process.

Experiments in transgenic plants have demonstrated that no single cell wall–degrading enzyme can account for all aspects of softening in tomato or other fruits. It appears that texture changes result from the synergistic action of a range of cell wall–degrading enzymes and that suites of texture-related genes give different fruits their unique melting, crisp, or mealy textures. However, even in tomato, the precise contribution of each type of enzyme to fruit texture is still poorly understood. Changes in the cuticle of the fruit that affect water loss also affect texture and shelf life.

Taste and flavor reflect changes in acids, sugars, and aroma compounds

Fruits have evolved to act as vehicles for the dispersal of seeds, and most fleshy fruits that are consumed by humans undergo changes that make them especially palatable to eat when they are ripe. These chemical changes include alterations in sugars and acids and the release of aroma compounds. In many fruits, starch is converted at the onset of ripening to glucose and fructose, and citric and malic acids are also abundant. However, although sugars and acids are vital for taste, volatiles are what really determine the unique flavor of fruits such as tomato.

Flavor volatiles arise from a wide range of compounds. Some of the most detailed studies have been undertaken in tomato. They show that of the 400 or so volatiles produced by tomato, only a small number have a positive effect on flavor. The most important flavor volatiles in tomato are derived from the catabolism of fatty acids such as linoleic acid (hexanal) and linolenic acid (*cis*-3-hexenal, *cis*-3-hexenol, *trans*-2-hexenal) via lipoxygenase activity. Other important volatiles, including 2- and 3-methylbutanal, 3-methylbutanol, phenylacetaldehyde, 2-phenylethanol, and methyl salicylate, are derived from the essential amino acids leucine, isoleucine, and phenylalanine. A third class of volatiles are the apocarotenoids, which are derived via the oxidative cleavage of carotenoids. Apocarotenoids such as β-damascenone are important in tomato, apple, and grape.

Volatile production is intimately connected with the ripening process, but the regulation of these events is not well understood. It is probably controlled by some of the diverse transcription factors that show altered expression during ripening.

The causal link between ethylene and ripening was demonstrated in transgenic and mutant tomatoes

Ethylene has long been recognized as the hormone that can accelerate the ripening of many edible fruits. However, the definitive demonstration that ethylene is required for fruit ripening was provided by experiments in which ethylene biosynthesis was blocked by inhibiting expression of either ACC synthase (ACS) or ACC oxidase (ACO). ACS and ACO are the enzymes involved in the second-to-last and last steps in ethylene synthesis, respectively (see Figure 15.22). These two steps in the pathway are normally tightly regulated. Silencing the genes encoding either of these enzymes using antisense RNA constructs inhibits ripening in transgenic tomatoes (**Figure 21.37**). Exogenous ethylene restores normal ripening in the transgenic tomato fruits.

Further demonstration of the requirement for ethylene in fruit ripening came from analysis of the *Never-ripe* mutation in tomato. As the name implies, this mutation completely blocks the ripening of tomato fruit. Molecular analysis has revealed that the *Never-ripe* phenotype is caused by a mutation in an ethylene receptor that renders the receptor unable to bind ethylene. These results, together with the demonstration that inhibiting ethylene biosynthesis blocks ripening, provided unequivocal proof of the role of ethylene in fruit ripening.

The role of plant hormones other than ethylene in controlling ripening is much less well understood, although auxin, ABA, and gibberellins are known to have an effect on this important developmental process.

Climacteric and non-climacteric fruit differ in their ethylene responses

Fleshy fruits have traditionally been placed in two groups as defined by the presence or absence of a characteristic respiratory rise called a **climacteric** at the onset of ripening. Climacteric fruits show this respiratory increase, and also a spike of ethylene production immediately before, or coincident with, the respiratory rise (**Figure 21.38**). Apple, banana, avocado, and tomato are examples of climacteric fruits. In contrast, fruits such as citrus fruits and grape do not exhibit such large changes in respiration and ethylene production and are termed **non-climacteric** fruits.

In plants with climacteric fruits, two systems of ethylene production operate, depending on the stage of development:

- In System 1, which acts in immature climacteric fruit, ethylene inhibits its own biosynthesis by negative feedback.
- In System 2, which occurs in mature climacteric fruit and in senescing petals in some species, ethylene stimulates its own biosynthesis—that is, it is autocatalytic.

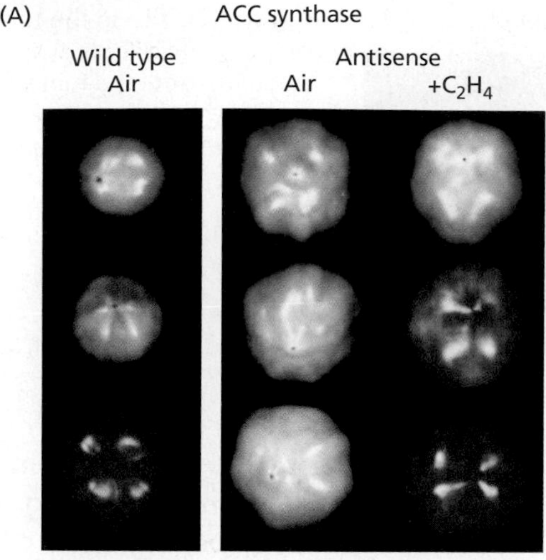

Figure 21.37 Antisense silencing of ACC synthase (A) and ACC oxidase (B) inhibits ripening and senescence (C). (A) Fruit expressing an *ACS2* (*ACC SYNTHASE2*) antisense gene, together with controls (wild type). Note that in air the antisense fruit did not ripen, but did senesce after 70 days (yellow); ripening could be restored by adding external ethylene (C_2H_4). (B) The *ACO1* (*ACC OXIDASE*) antisense gene only inhibited ethylene synthesis by approximately 95%: fruit ripened, but overripening and deterioration were greatly reduced. (C) In addition, leaf senescence was delayed in the *ACO1* antisense plant. (From Oeller et al. 1991, re-printed in Grierson 2013.)

The positive feedback loop for ethylene biosynthesis in System 2 ensures that the entire fruit ripens evenly once ripening has commenced.

When mature climacteric fruits are treated with ethylene, the onset of the climacteric rise and the changes associated with ripening are hastened. In contrast, when immature climacteric fruits are treated with ethylene, the respiration rate increases gradually as a function of the ethylene concentration, but the treatment does not trigger production of endogenous ethylene or induce ripening. Ethylene treatment of non-climacteric fruits, such as citrus, strawberry, and grape, does not cause an increase in respiration and is not required for ripening. However, it can alter ripening characteristics in some species, such as the enhancement of color in citrus.

Although the distinction between climacteric and non-climacteric fruits is a useful generalization, some non-climacteric fruits may also respond to ethylene; for example, citrus fruits de-green in response to exogenous ethylene. Indeed, the distinction between climacteric and non-climac-

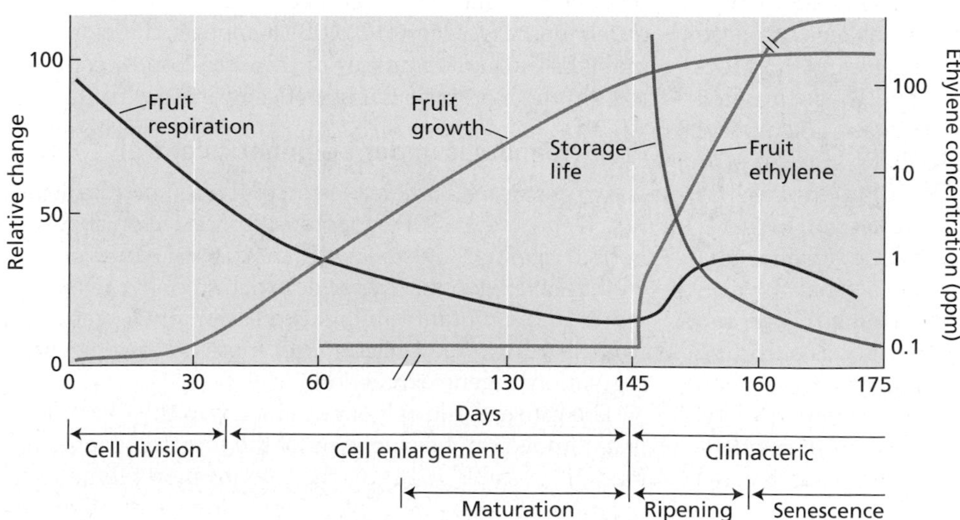

Figure 21.38 Growth and development of apple and pear fruits in relation to the effects of ethylene and ripening. Climacteric fruits show a characteristic respiratory increase, and a spike of ethylene production immediately before, and coincident with, the respiratory rise, which signals the onset of ripening. (After Dilley 1981.)

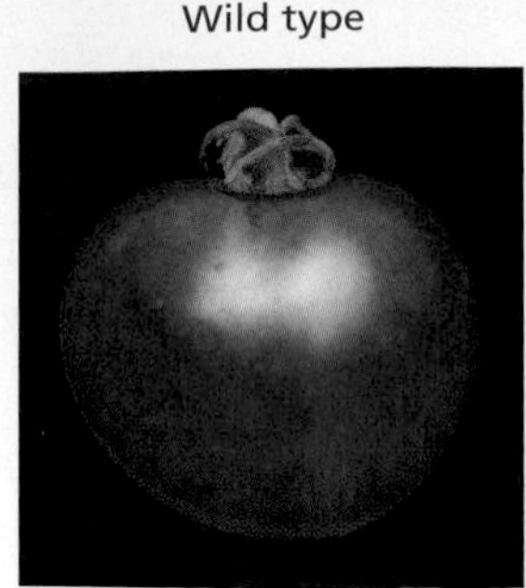

Figure 21.39 In the tomato mutants *ripening inhibitor* (*rin*) and *Colorless non-ripening* (*Cnr*), the mutation prevents normal ripening. (Courtesy of G. B. Seymour, University of Nottingham.)

teric fruits may be less dramatic than previously thought, with some species showing contrasting behavior depending on the cultivar. For example, melon (*Cucumis melo*) can be climacteric or non-climacteric depending on the variety.

The ripening process is transcriptionally regulated

Several rare, spontaneous, monogenic mutants in tomato show abnormal ripening or the complete abolition of this process. They include *ripening inhibitor* (*rin*) and *Colorless non-ripening* (*Cnr*) (**Figure 21.39**). The *rin* locus encodes a MADS box transcription factor termed MADS-RIN, which is induced at the onset of ripening, and the *Cnr* locus encodes the transcription factor CNR.

The *MADS-RIN* gene is a member of the *SEPALLATA* gene family, which also includes genes required for floral organ identity and floral meristem determinacy (see Chapter 20, Figure 20.29). Suppression of *MADS-RIN* expression in transgenic tomatoes yields nonripening fruits, and complementation of the *rin* mutant with the *MADS-RIN* gene rescues the *rin* mutant, demonstrating that MADS-RIN is necessary for normal ripening.

MADS-RIN interacts with the promoters of ACC synthase genes, suggesting that MADS-RIN regulates ethylene biosynthesis (see Chapter 15). MADS-RIN also binds to the regulatory regions of numerous ripening-related genes to directly control their expression (**Figure 21.40A**). These include genes that code for proteins involved in cell wall metabolism, such as polygalacturonase, galactanase and expansins; proteins that are involved in carotenoid formation, such as phytoene synthase ; and those involved in aroma biosynthesis, such as lipoxygenase and alcohol dehydrogenase. The binding of MADS-RIN to the promoters of ACC synthase genes and other target genes has been shown to depend on the CNR transcription factor cited above. It seems likely that both ethylene signaling and MADS-RIN act synergistically to promote normal ripening.

Since the cloning of the genes underlying the *rin* and *Cnr* mutations, a large number of other genes encoding transcription factors required for ripening have been described. These ripening-regulatory genes are involved in a network with downstream effectors to promote ethylene biosynthesis and the biochemical changes associated with ripening.

Angiosperms share a range of common molecular mechanisms controlling fruit development and ripening

MADS box genes are involved in the control of ripening in a wide variety of fleshy fruits, not just tomato. These include banana, strawberry, and bilberry. MADS box genes are also important in the development and maturation of dry fruits, and control the dehiscence process (**Figure 21.40B**). Indeed, the Arabidopsis *SHATTERPROOF* (*SHP*) and *FRUITFULL* (*FUL*) genes are probably orthologs of tomato *TAGL1* and *TDR4*.

There are two *SHP* genes in Arabidopsis, and silencing both of them leads to indehiscent fruits. The *FUL* gene is required for maintaining the identity of the silique valves. In *ful* mutants, the *SHP* genes are ectopically expressed in the valve tissue. FUL thus specifies valve cell fate at least in part by repressing the expression of valve margin identity genes in valve tissue. The *SHP* genes positively regulate the expression of another transcription factor known as *INDEHISCENT* (*IND*). Increases in *IND* gene expression are linked to alterations in auxin levels in the dehiscence zone and upregulation of cell wall–degrading enzymes such as polygalacturonase. Replum tissue identity on the median side of the valve margins is maintained by the expression of *REPLUMLESS* (*RPL*), and the floral homeotic gene *AP2* has been demonstrated to repress replum development (see Figure 21.40B). It appears, therefore, that many of the same families of transcription factors play roles in the maturation of both fleshy and dry fruits.

Fruit ripening is under epigenetic control

As we discussed earlier, a lesion in the tomato *Cnr* locus, which encodes an SBP-type transcription factor, abolishes normal ripening. Unexpectedly, the *Cnr* lesion was discovered to be epigenetic: hypermethylation of the *CNR* promoter in the mutant inhibits expression of the gene and fruit ripening. (See Chapter 2 for more on the epigenetic regulation of gene expression.)

The cause of the epigenetic change in the *Cnr* mutant is not known, but sequencing of the tomato genome has made it possible to study the tomato methylome—the position and type of DNA methylation associated with

(A)

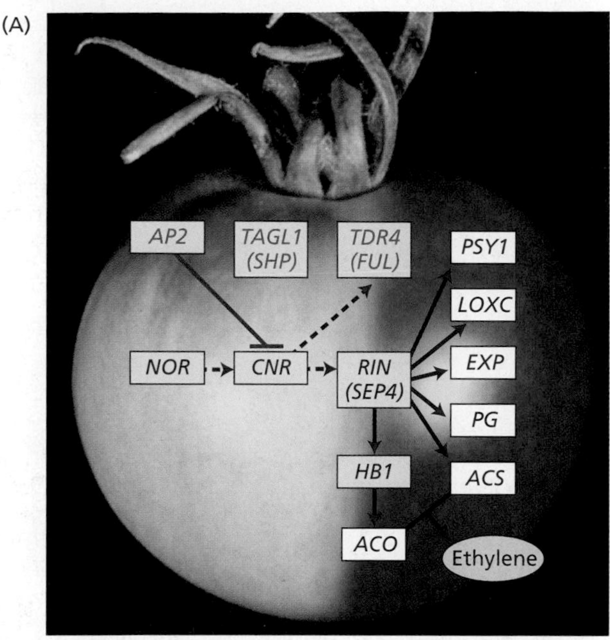

(B)

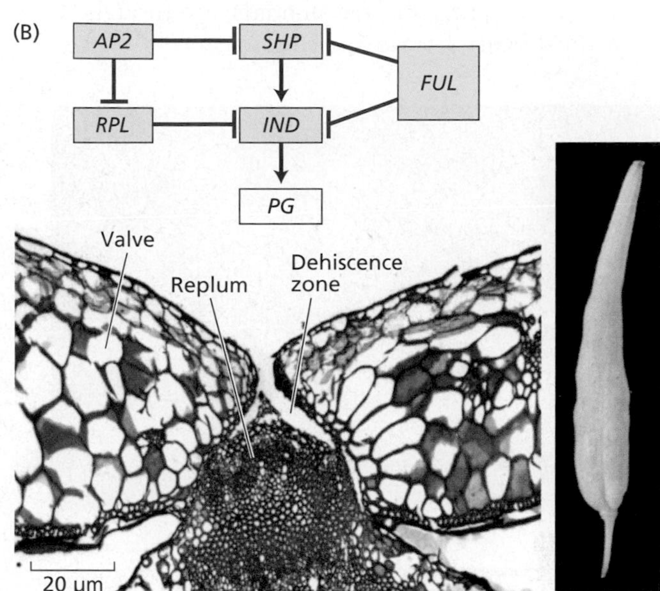

Figure 21.40 Ripening network in fleshy fruits and comparison with events in the dehiscence of dry fruits. (A) Major known regulators of ripening in tomato. The blue rectangles are transcription factors; the red labels are genes where orthologs are also found in dry dehiscent fruits. Downstream effectors are shown in white boxes. Solid lines between *RIN* and other genes indicates activation, while dashed lines indicate possible activation. The red line between *AP2* and *CNR* indicates repression. (B) *Brassica rapa* silique (right) and dehiscence zone (left), illustrating a dry-fruit gene network. (Image of tomato fruit with ripening inhibited by silver thiosulfate on the left side courtesy of Don Grierson and Kevin Davies; after Seymour et al. 2103.)

the genome sequences—during fruit ripening. This work has revealed that tomato ripening is associated with a reduction in the levels of DNA methylation in the promoters of ripening-related genes, which would be expected to increase the expression of these genes. It seems likely that this constitutes a new and hitherto unexplored layer of regulation governing the ripening process.

A mechanistic understanding of the ripening process has commercial applications

Fruits did not evolve solely for the benefit of humans but are nevertheless an important part of our diet. Fruits provide ready sources of vitamins A, C, E, and K; minerals such as potassium and iron; and secondary metabolites that have health-promoting properties, such as the red pigment lycopene. Fresh fruits are also economically valuable products, but they often have a short shelf life. Understanding fruit development and ripening is therefore important for agriculture to increase yield, nutritional value, and quality while maintaining postharvest life.

The control of fruit ripening is of substantial commercial significance. Elucidation of the role of ethylene in the ripening of climacteric fruits has resulted in many practical applications aimed at either uniform ripening or the delay of ripening. For example, banana bunches are picked unripe, when they are still green and hard, which helps them survive the journey from the fields of Central and South America to their final destinations all over the world. The clusters of unripe fruits—called *hands* (a single fruit is a *finger*)—are cut from the bunch, treated with fungicide, packed in cartons, and shipped overseas. On arrival at their destination, the bananas are placed in temperature-controlled rooms and treated with small amounts of ethylene gas to initiate ripening. This mirrors the natural ripening process but ensures that fruits at different stages of maturity will initiate ripening at the same time, making them easier to market.

In the case of fruits such as apples, ripening can be delayed using controlled atmosphere storage and refrigeration, thus extending the marketable period for the crop. In elite varieties of tomato, the *rin* mutation is widely used in the heterozygous form to slow the rate of ripening and extend shelf life. A major disadvantage of using the "*rin* gene" is that it delays aspects of ripening so that the fruits are often deficient in optimal levels of flavor, aroma, and other quality-associated compounds. A more effective approach would be to target individual ripening processes, for instance by extending the shelf life of fruits by slowing softening in the absence of detrimental effects on color and flavor. Access to the tomato genome sequence makes this goal a reality by allowing scientists to identify the genes underlying complex traits controlling individual aspects of fruit quality.

It is also possible to manipulate the quality of fruit. For example, anthocyanins, like carotenoids, are thought to protect against heart disease and certain cancers, as they are strong antioxidants that can scavenge excessive damaging free radicals. Anthocyanin levels in fruits can be manipulated by transgenic approaches, even to the extent of introducing high levels of these compounds in tomato flesh, where they do not normally occur (**Figure 21.41**). A greater understanding of the molecular determinants of other aspects of fruit development, such as the production of volatiles, will presumably offer other opportunities to improve fruit quality.

Figure 21.41 Anthocyanin production can be induced in tomato by overexpressing transcription factors that control the biosynthesis of these compounds in snapdragon (*Antirrhinum*).

SUMMARY

Plants exhibit alternation of generations, where diploids tend to be dominant but haploids produce the gametes. Genetic diversity is encouraged through outcrossing, which is enabled by vectors such as wind or insects, while inbreeding is minimized by active preventional mechanisms in the plant. The new diploid generation develops in the seed or fruit, which ripens and becomes attractive to vectors that disperse seeds.

Development of the Male and Female Gametophyte Generations

- Plants undergo both a diploid and a haploid generation in order to make gametes and reproduce (**Figure 21.1**).
- Diploidy allows individuals to mask deleterious recessive alleles and allows populations to exhibit greater genetic diversity.

Formation of Male Gametophytes in the Stamen

- Pollen forms in two stages: first microsporogenesis, then microgametogenesis (**Figures 21.2, 21.3**).
- Pollen cell walls are complex, with multiple layers for nutrient storage and for pollen dispersal (**Figures 21.4, 21.5**).

Female Gametophyte Development in the Ovule

- Eggs are formed in the female gametophyte (embryo sac) first by megasporogenesis and then by megagametogenesis (**Figures 21.6, 21.7**).
- Most angiosperms exhibit *Polygonum*-type megagametophyte development, wherein meiosis of a diploid mother cell produces four immature haploid megagametophytes, only one of which undergoes megagametogenesis.
- Megagametogenesis begins with three mitotic divisions without cytokinesis, followed by cellularization (**Figures 21.8, 21.9**).

Pollination and Fertilization in Flowering Plants

- Once pollen has been delivered to the stigma, sperm cells travel to the female gametophyte through a newly grown pollen tube (**Figure 21.10**).
- A pollen tube will not form unless there is recognition between pollen and stigma (**Figure 21.11**).
- Pollen tubes grow by tip growth (**Figures 21.12–21.14**).
- Pollen-expressed receptor-linked kinases (RLKs) may regulate a GTPase switch, enabling polar cell expansion of the pollen tube (**Figure 21.15**).
- The path of pollen tube growth is determined by physical and chemical cues from the pistil and the megagametophyte (**Figure 21.16**).
- Once the pollen tube has reached the ovule, two sperm are released to fertilize the egg and the central cell (**Figure 21.17**).

Selfing versus Outcrossing

- Outcrossing is ensured in hermaphroditic and monoecious species by dichogamy and heterostyly (**Figure 21.18**).
- Self-pollination is reduced by cytoplasmic male sterility; CMS can be reversed by a class of Restorer of Fertility (Rf) genes.

- Self-incompatibility (SI) biochemically prevents self-pollination in angiosperms (**Figure 21.19**).
- Sporophytic SI reactions require expression of two highly variable *S*-locus genes, while gametophytic SI is mediated by cytotoxic S-RNases and F-box proteins (**Figures 21.20, 21.21**).

Apomixis: Asexual Reproduction by Seed

- Apomixis, or clonal reproduction by a diploid cell, may contribute to the fitness of polyploid species.
- Being able to induce apomixis would reduce loss of hybrid vigor in agricultural crops.

Endosperm Development

- After fertilization, the diploid endosperm, which will provide nutrition to the embryo, becomes multinucleate, making it a coenocyte (**Figures 21.22, 21.23**).
- Cellularization of the coenocytic endosperm in Arabidopsis proceeds from the micropylar to the chalazal region, while cellularization of cereal endosperms proceeds centripetally (**Figures 21.24–21.26**).
- Endosperm development is controlled primarily by maternally expressed genes (MEGs), not by the embryo.
- Endosperm development is repressed until after fertilization by FIS proteins, which methylate and demethylate DNA and histones in the endosperm (**Figure 21.27**).
- The aleurone layer is differentiated from starchy endosperm cells, and while two genes, *DEK1* and *CR4*, have been implicated, the overall mechanism is unclear (**Figure 21.28**).

Seed Coat Development

- The seed coat arises from maternal integuments, but its development is regulated by the endosperm (**Figure 21.29**).

Seed Maturation and Desiccation Tolerance

- Seed filling and the acquisition of dessication tolerance overlap in most species (**Figure 21.30**).
- The acquisition of dessication tolerance is aided by LEA proteins, which form hydrogen bonds with nonreducing sugars, allowing embryo cells to acquire a glassy state that makes them more stable than cells that are simply dehydrated (**Figure 21.31**).
- Synthesis of LEA proteins is controlled by abscisic acid.
- Impermeable seed coats and low temperatures can increase seed longevity, which otherwise is highly variable across species.

Fruit Development and Ripening

- Fruits are seed-disperal units that arise from the pistil and contain the seed(s) (**Figures 21.32–21.34**).
- Fleshy fruits undergo ripening, which involves color changes, highly coordinated fruit softening, and other changes (**Figures 21.35, 21.36**).
- Acids, sugars, and volatiles determine the flavor of ripe and unripe fleshy fruits.
- Ethylene accelerates ripening, particularly in climacteric fruits (**Figures 21.37, 21.38**).
- Many molecular mechanisms directing fruit ripening are conserved among angiosperms (**Figure 21.40**).
- A mechanistic understanding of the ripening process has commercial applications (**Figure 21.41**).

WEB MATERIAL

- **WEB TOPIC 21.1 Evolution has Favored Diploidy in Plant Life Cycles** The possible selective advantages of diploidy over haploidy are discussed.
- **WEB TOPIC 21.2 Types of Placentation in Fruits** A diagram of the various types of placentation in fruits is presented.
- **WEB TOPIC 21.3 Variations in Gametophyte Development** Deviations from Polygonum-type placental development. The characteristics of monosporic, bisporic, and tetrasporic embryo sacs are described.
- **WEB TOPIC 21.4 The Molecular Mechanism of Cytoplasmic Sterility in Rice** The molecular mechanism of CYTOPLASMIC MALE STERILITY (CMS) has been elucidated in the "wild abortive" or CMS-WA rice system.
- **WEB TOPIC 21.5 Various Types of Apomixis** The mechanisms of sporophytic versus gametophytic apomixes are described.
- **WEB TOPIC 21.6 Three Types of Endosperm Development** Endosperm development falls into three basic categories: nuclear, cellular, and helobial.
- **WEB TOPIC 21.7 Fruit Types and Examples** A table of the commonly encountered fruit types and examples is presented.

available at plantphys.net

Suggested Reading

Angelovici, R., Galili, G., Fernie, A. R., and Fait, A. (2010) Seed desiccation: a bridge between maturation and germination. *Trends Plant Sci.* 15: 211–218.

Burg, S. P., and Burg. E, A. (1965) Ethylene action and ripening of fruits. *Science* 148: 1190–1196.

Craddock, C., Lavagi, I., and Yang, Z. (2012) New insights into Rho signaling from plant ROP/Rac GTPases. *Trends Cell Biol.* 22: 492–501.

Dinneny, J. R., and Yanofsky M. F. (2005) Drawing lines and borders: How the dehiscent fruit of *Arabidopsis* is patterned. *Bioessays* 27: 42–49.

Dresselhaus, T., and Franklin-Tong, N. (2013) Male-female crosstalk during pollen germination, tube growth and guidance, and double fertilization. *Mol. Plant* 6: 1018–1036.

Gehring, M. (2013) Genomic imprinting: Insights from plants. *Annu. Rev. Genet.* 47: 187–208.

Klee, H. J., and Giovannoni, J. J. (2011) Genetics and control of tomato fruit ripening and quality attributes. *Annu Rev Genet.* 45: 41–59. DOI: 10.1146/annurev-genet-110410-132507.

Knapp, S. (2002) Tobacco to tomatoes: A phylogenetic perspective on fruit diversity in the Solanaceae. *J. Exp. Bot.* 53: 2001–2022.

Knapp, S., and Litt, A. (2013) Fruit—An angiosperm innovation. *In The Molecular Biology and Biochemistry of Fruit Ripening*, G. B. Seymour, G. A. Tucker, M. Poole, and J. J. Giovannoni, eds., Wiley-Blackwell, Oxford, UK, p. 216.

Li, J., and Berger, F. (2012) Endosperm: Food for humankind and fodder for scientific discoveries. *New Phytol.* 195: 290–305.

Manning, K., Tor, M., Poole, M., Hong, Y., Thompson, A. J., King, G. J., Giovannoni, J. J., and Seymour, G. B. (2006) A naturally occurring epigenetic mutation in a gene encoding an SBP-box transcription factor inhibits tomato fruit ripening. *Nat. Genet.* 38: 948–952.

McCann, M., and Rose, J. (2010) Blueprints for building plant cell walls. *Plant Physiol.* 153: 365.

Nasrallah, J. B. (2011) Self-incompatibility in the Brassicaceae. *In Plant Genetics and Genomics: Crops and Models*, Vol. 9: *Genetics and Genomics of the Brassicaceae*, R. Schmidt, and I. Bancroft, eds., Springer, Berlin, pp. 389–412. DOI: 10.1007/978-1-4419-7118-0_14

Okuda, S., Tsutsui, H., Shiina, K., Sprunck, S., Takeuchi, H., Yui, R., Kasahara, R. D., Hamamura, Y., Mizukami, A., Susaki, D., et al. (2009) Defensin-like polypeptide LUREs are pollen tube attractants secreted from synergid cells. *Nature* 458: 357–361.

Rodrigues, J. C. M., Luo, M., Berger, F., and Koltunow, A. M. G. (2010) Polycomb group gene function in sexual and asexual seed development in angiosperms. *Sex. Plant Reprod.* 23: 123–133.

Seymour, G. B., Østergaard, L., Chapman, N. H., Knapp, S., and Martin, C. (2013) Fruit development and ripening. *Annu Rev Plant Biol.* 64: 219–241. DOI: 10.1146/annurev-arplant-050312-120057.

Spence, J., Vercher, Y., Gates, P., and Harris, N. (1996) "Pod shatter" in *Arabidopsis thaliana*, *Brassica napus* and *B. juncea*. *J. Microsc.* 181: 195–203.

Tomato Genome Consortium. (2012) The tomato genome sequence provides insights into fleshy fruit evolution. *Nature* 485: 635–641

Twell, D. (2010) Male gametophyte development. In: *Plant Developmental Biology—Biotechnological Perspectives*, Vol. 1, E. C. Pua and M. R. Davey, eds., Springer-Verlag, Berlin, pp. 225–244.

Vrebalov, J., Ruezinsky, D., Padmanabhan, V., White, R., Medrano, D., Drake, R., Schuch, W., and Giovannoni, J. (2002) A MADS-box gene necessary for fruit ripening at the tomato ripening-inhibitor (rin) locus. *Science* 296: 343–346.

Wilkinson, J. Q., Lanahan, M. B., Yen, H.-C., Giovannoni, J. J. and Klee, H. J. (1995) An ethylene-inducible component of signal transduction encoded by Never-ripe. *Science* 270: 1807–1809.

Yang, W.-C., Shi, D.-Q., and Chen, Y.-H. (2010) Female gametophyte development in flowering plants *Annu. Rev. Plant Biol.* 61: 89–108.

Zhong, S., Fei, Z., Chen, Y.-R., Zheng,Y., Huang, M., Vrebalov, J., McQuinn, R., Gapper, N., Liu, B., Xiang, J., et al. (2013) Single-base resolution methylomes of tomato fruit development reveal epigenome modifications associated with ripening. *Nat. Biotechnol.* 31: 154–159.

22 Plant Senescence and Cell Death

Every autumn, people living in temperate climates enjoy the spectacular color changes that can precede the loss of leaves from deciduous trees (**Figure 22.1**). Traditionally, poets have used the coloration and falling of autumn leaves as poignant reminders of old age and impending death, as in the opening lines from Shakespeare's sonnet 73:

> *That time of year thou mayst in me behold,*
> *When yellow leaves, or none, or few, do hang*
> *Upon those boughs which shake against the cold,*
> *Bare ruined choirs, where late the sweet birds sang.*

Autumn leaves turn yellow, orange, or red and fall from their branches in response to shorter day lengths and cooler temperatures, which trigger two related developmental processes: senescence and abscission. Although senescence ultimately leads to the death of targeted tissues, it is distinct from the related term *necrosis*. **Senescence** is an energy-dependent, autolytic (self-digesting) process that is controlled by the interaction of environmental factors with genetically regulated developmental programs. Although it has some overlap with senescence, **necrosis** is usually defined as death that is directly caused by physical damage, poisons (such as herbicides), or other external agents. **Abscission** refers to the separation of cell layers that occurs at the bases of leaves, floral parts, and fruits, which allows them to be shed easily without damaging the plant.

There are three types of plant senescence, based on the level of structural organization of the senescing unit: *programmed cell death*, *organ senescence*, and *whole plant senescence*. **Programmed cell death** (**PCD**) is a general term referring to the genetically regulated death of individual cells. During PCD, the protoplasm, and sometimes the cell wall, undergoes autolysis. In the case of the development of xylem tracheary elements and fibers, however, secondary wall layers are deposited prior to cell death. PCD is an essential aspect of normal plant development (**Figure 22.2**), but it can also be induced in response to both abiotic and biotic stress. Organ senescence (the senescence of whole leaves, branches, flowers, or fruits) occurs at various stages of vegetative and repro-

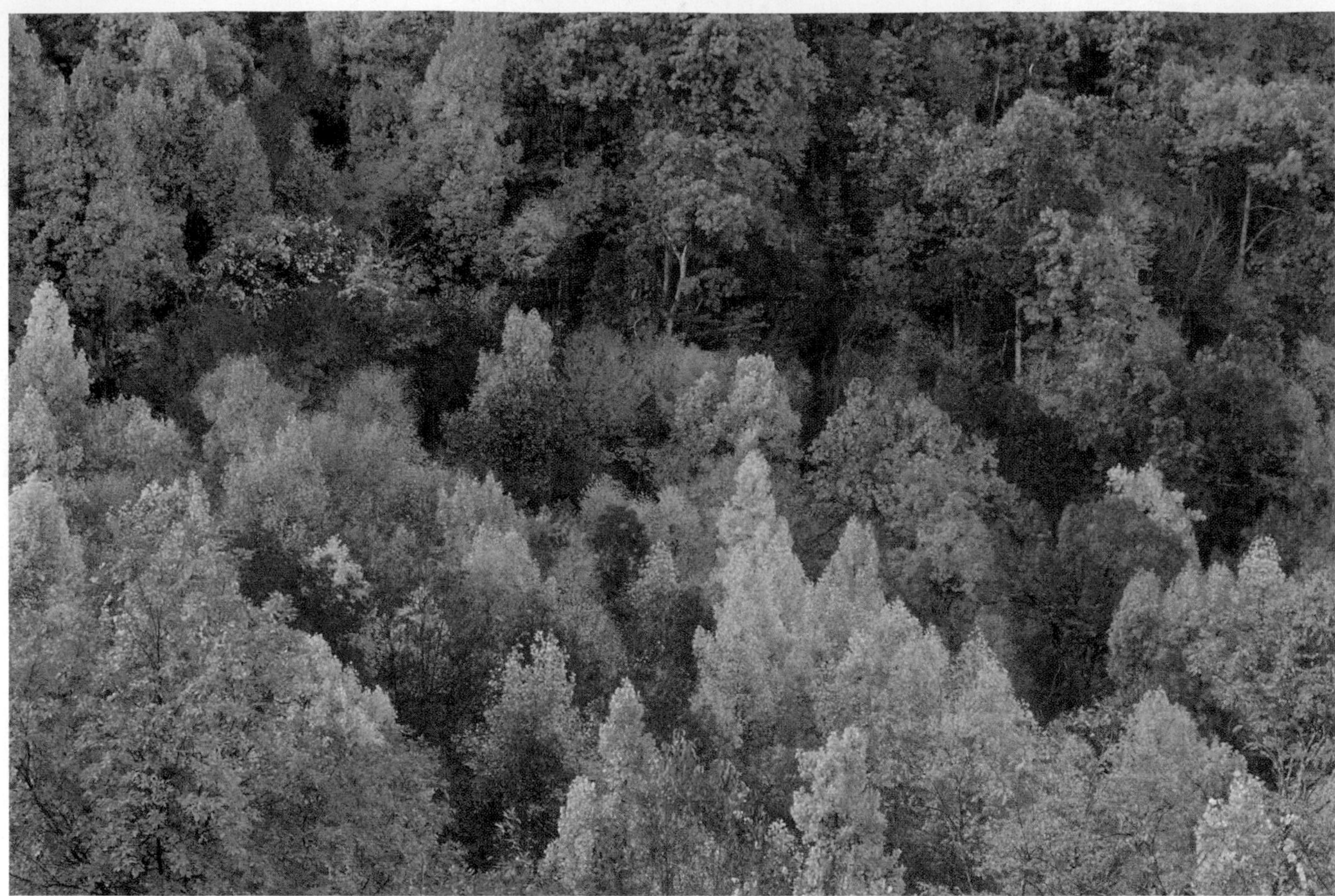

Figure 22.1 Fall colors along Blue Ridge Parkway in Virginia. The combination of several deciduous tree species produces a wide range of hues.

ductive development and typically includes abscission of the senescing organ. As previously noted, leaf senescence is strongly influenced by photoperiod and temperature. Finally, **whole plant senescence** involves the death of the entire plant. Whole plant senescence differs from aging in animals and is much more variable. For example, individual plant life spans may range from a few weeks for some desert annuals to as long as 4600 years for bristlecone pines. Clonal perennial plants can live even longer. Because of the presence of continuously dividing apical meristems, plants potentially could live forever, yet all apical meristems eventually fail and the plant dies. Why does this happen? As we will see, whole plant senescence is a complex function of the plant's genetic program, nutrient and water availability, age, and other factors.

The processes of PCD, organ senescence, and whole plant senescence differ with respect to the size, cell number, and complexity of their senescing units. They also differ with respect to the developmental and environmental cues that trigger them. However, it is important to note that, at the cellular level, PCD, organ senescence, and whole plant senescence all use the same or similar genetic pathways for cell autolysis. In other words, PCD is a common feature of all three types of senescence.

We will begin with a brief overview of the primary enzymatic mechanisms responsible for cell autolysis in plants and animals. We will then examine the various cytological changes that accompany PCD, including *autophagy* and the genetic pathways that regulate autophagy. Next we will turn to senescence at the organ level, focusing on leaf senescence. Finally, we will discuss the factors that govern the two different types of whole plant senescence: monocarpic and polycarpic senescence.

Programmed Cell Death and Autolysis

All eukaryotic organisms, including plants, animals, and fungi, have evolved mechanisms of cellular suicide that are collectively known as programmed cell death. In multicellular plants and animals, the organized destruction of cells is required for normal growth and development and for the removal of unwanted, damaged, or infected cells. PCD can be initiated by specific developmental signals or by potentially lethal events, such as pathogen attack or errors in DNA replication during cell division. It involves

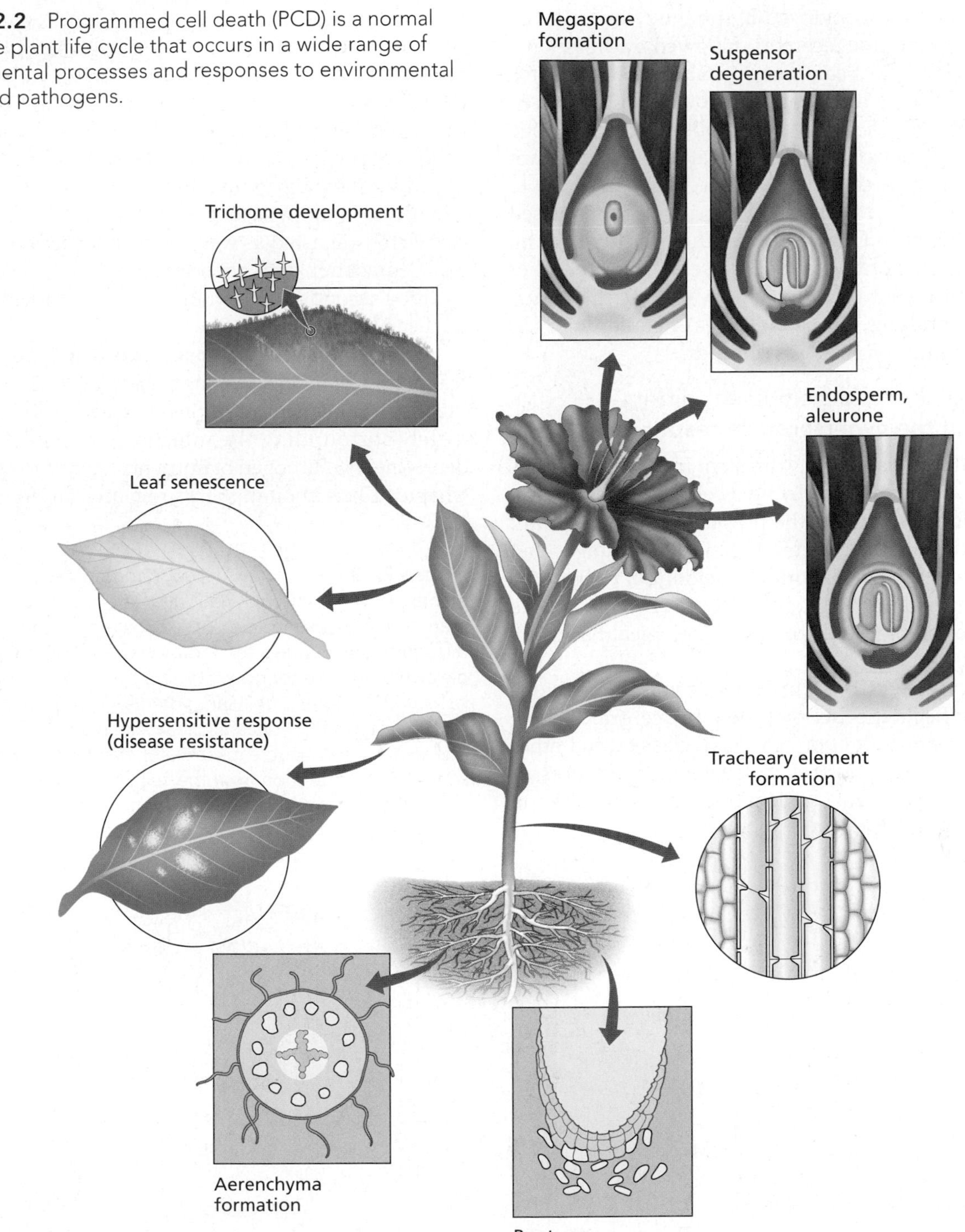

Figure 22.2 Programmed cell death (PCD) is a normal part of the plant life cycle that occurs in a wide range of developmental processes and responses to environmental signals and pathogens.

the expression of a characteristic set of genes that orchestrate the dismantling of cellular components, ultimately causing cell death.

PCD in animals is usually associated with a distinct set of morphological and biochemical changes called **apoptosis** (Greek for "falling off," as in autumn leaves). During apoptosis the cell nucleus condenses, and the chromosomes fragment as a result of endonuclease digestion of the DNA between specific nucleosomes; this process produces an ordered oligonucleotide "ladder" when the DNA is size-separated by gel electrophoresis. In addition to nucleases, *caspases* (cysteine-dependent *asp*artate-specific prote*ases*) target particular proteins by introducing single breaks after specific aspartate residues. Directed digestion of target proteins by caspases leads to the controlled death of the cell. During this process the plasma membrane forms irregular bulges, or blebs, and the cell fragments into numerous vesicles called *apoptotic bodies*, which are then phagocytosed by phagocytes with the appropriate membrane receptors, followed by digestion.

Autolysis in plants bears some resemblance to apoptosis in animals but is more variable. For example, ordered oligonucleotide ladders have been observed during the hypersensitive response to pathogen attack in some plant species, but in most cases DNA degradation gives rise to a DNA smear. (The hypersensitive response is discussed below, and again in Chapter 23.) Although there are no true caspases in plants, plant cells use a variety of other proteases during autolysis, including "caspase-like" cysteine endopeptidases, serine proteases, metalloproteases, and the ubiquitin–proteasome complex.

PCD during normal development differs from that of the hypersensitive response

Plant cells are fundamentally different from animals cells in that they are surrounded by rigid cell walls that prevent cell migration. Because of the presence of the cell wall, and the absence of phagocytes, the kinds of changes that occur during apoptosis in animals seldom, if ever, occur in plants. Instead, ultrastructural studies have led to the characterization of two distinctive cytological pathways of PCD in plants.

Vacuolar-type PCD occurs during normal development, and reflects the fact that the large central vacuole is the main repository of proteases, nucleases, and other lytic enzymes. Examples of such developmental processes include the development of xylem tracheary elements and fibers, leaf shaping during morphogenesis, leaf senescence, and megasporogenesis (see Figure 22.2). The cytological changes associated with tracheary element diffentiation are illustrated in **Figure 22.3A**. During vacuolar-type PCD, the vacuole swells and either permeabilizes or ruptures, releasing hydrolases into the cytosol and causing large-scale degradation. The cytosol and all its organelles, including the plasma membrane, are completely broken down, and in many cases the cell wall is either partially or completely digested as well, as in endosperm tissue. Cell wall degradation does not occur in cells that have acquired lignified secondary cell walls during the process, such as tracheary elements and fibers.

Hypersensitive response-type (HR-type) PCD is a plant defense mechanism against microbial attack. During the hypersensitive response in leaves, the cells immediately surrounding the infection site commit suicide, depriving the pathogen of nutrients needed to spread (see Chapter 23). Although there are many variations of HR-

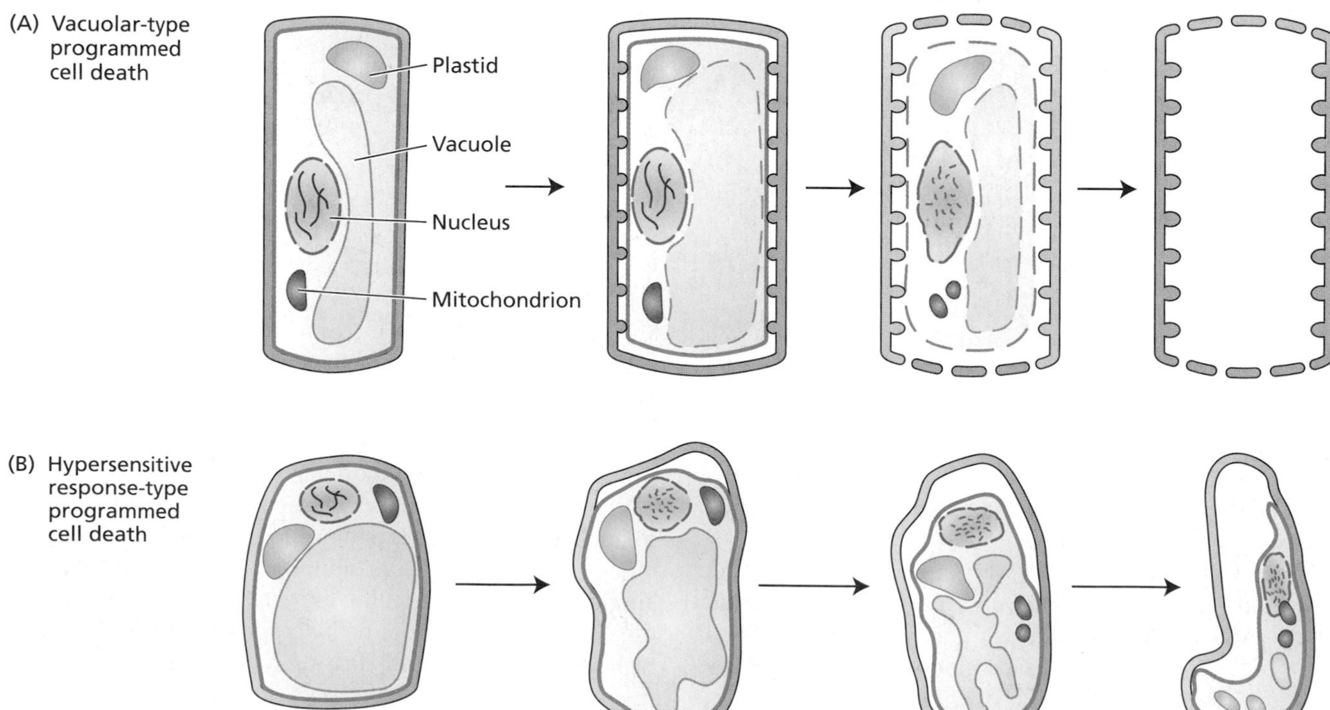

Figure 22.3 Two types of programmed cell death in plants. (A) Vacuolar-type PCD, also referred to as developmental PCD, is exemplified here by xylem tracheary element differentiation. During secondary wall deposition the vacuole swells and the tonoplast breaks down, releasing hydrolases that digest the cellular contents. (B) Hypersensitive response-type PCD occurs in leaves in response to microbial attack. The vacuole loses water, resulting in marked cell shrinkage, contraction from the cell wall, and nuclear DNA degradation. Continued water loss from the cytosol leads to the breakdown of the plasma membrane and the release of residual cellular contents into the apoplast.

type PCD, the one feature they have in common is that autolysis is *not* initiated by vacuolar swelling and leakage. Rather, as shown in **Figure 22.3B**, vacuolar water loss and cell shrinkage are the earliest events of HR-type PCD, followed by nuclear DNA degradation. The cell continues to contract due to water loss, and the cellular organelles and plasma membrane break down, releasing their contents into the apoplast.

The autophagy pathway captures and degrades cellular constituents within lytic compartments

Cells, like complex machines, experience wear and tear over time, and parts need to be replaced on an ongoing basis to extend their life spans. **Autophagy** (from the Greek words meaning "self-eating") was first characterized in animal cells as the catabolic mechanism that delivers cellular components to lysosomes, where they are degraded. Autophagy protects the cell from the harmful or lethal effects of damaged or unnecessary proteins and organelles. During starvation, the autophagic breakdown and recycling of cellular components also ensure cellular survival by maintaining cellular energy levels.

Two types of autophagy have been identified in animals and yeast that also occur in plants: macroautophagy and microautophagy, although the evidence for plant microautophagy is more controversial. In **macroautophagy**, the best-studied type of plant autophagy, specialized organelles called autophagosomes enclose the cytoplasmic components and fuse with the vacuole. **Microautophagy** involves the invagination of the tonoplast membrane and the formation of small intravacuolar vesicles called autophagic bodies, which are rapidly degraded by lytic enzymes inside the vacuole. Here we will focus on macroautophagy, which we will refer to simply as autophagy.

In autophagy, the endoplasmic reticulum (ER) first gives rise to a cup-shaped, membranous cisterna called the **phagophore** (**Figure 22.4**). In animals, the phagophore has been shown to form at a specialized site on the ER (discussed below). The early phagophore then acquires additional membrane lipids, expands, and pinches off from the ER. The expansion and fusion of the phagophore enable it to engulf cytoplasmic components targeted for destruction, including misfolded proteins, ribosomes, ER, and mitochondria. The phagophore becomes spherical, and the inner and outer phospholipid bilayers fuse to form the completed **autophagosome**, surrounded by a double membrane. In plants, the outer membrane of the autophagosome fuses with the vacuolar membrane, or tonoplast. In the process, a single-membrane vesicle called the **autophagic body** enters the vacuole and is degraded (see Figure 22.4). The monomers (amino acids, sugars, nucleosides, etc.) generated by hydrolytic breakdown of the autophagic body are returned to the cytosol for reuse, either as an energy source or as building blocks for new cellular structures. Electron micrographs of plant autophagosomes and autophagic bodies are shown in **Figure 22.5**.

A subset of the autophagy-related genes controls the formation of the autophagosome

The genes that regulate autophagy were first identified in yeast and are called **autophagy-related genes** or ***ATG* genes**. Many of the *ATG* genes are conserved in evolution, and homologs to the yeast genes have been found in both plants and mammals. In yeast, *ATG* genes have been shown to regulate starvation-induced autophagy, the cytoplasm-to-vacuole targeting pathway, and the selective autophagy of organelles. The "core autophagy machinery" controls the initiation and growth of the autophagosome and has been divided into three main protein groups:

- ATG9 and its cycling system, which includes the ATG1/ATG13 kinase complex
- The phosphatidylinositol 3-OH kinase (PI[3]K) complex
- The ubiquitin-like protein system, which includes the ATG12 complex and ATG8

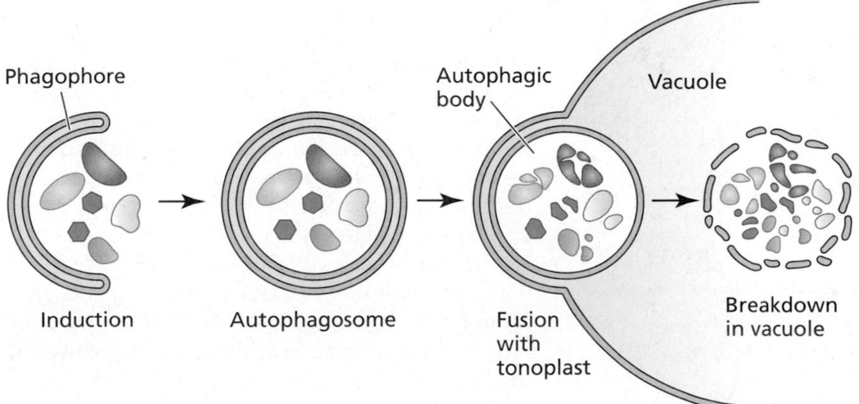

Figure 22.4 Autophagosome formation in eukaryotes. Autophagosome biogenesis begins with the formation of a cup-shaped double-membrane structure called a phagophore. The edges of the phagophore grow and engulf cargo (macromolecules and organelles). The edges then fuse, forming a double-membrane vesicle called the autophagosome. Some digestion occurs within the autophagosome during its transit toward the vacuole. Upon reaching the vacuole, the outer membrane of the autophagosome fuses with the tonoplast, and the remaining cargo enters the vacuole within a single-membrane vesicle (autophagic body), which can then be degraded by lytic enzymes.

(A)

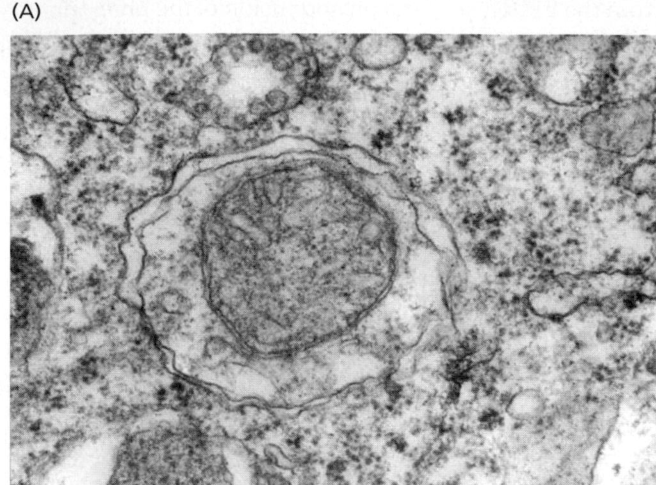

(B)

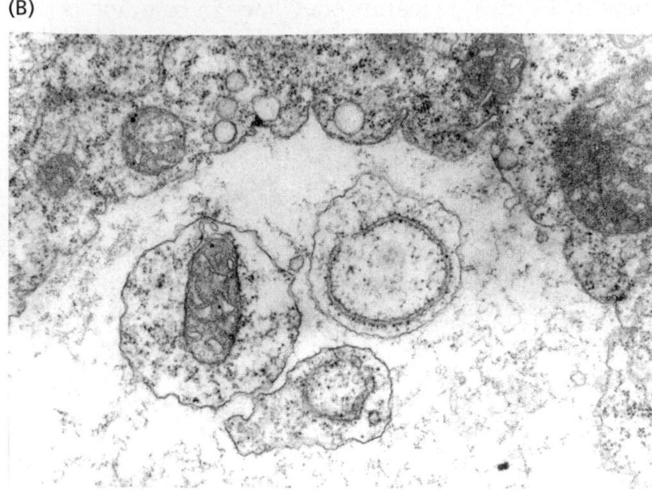

Figure 22.5 Autophagosome and autophagic body produced by sucrose-starved tobacco (*Nicotiana tabacum*) cells treated with an inhibitor that prevents the breakdown of autophagic bodies in the central vacuole. (A) A double-membrane autophagosome that has engulfed a mitochondrion in the cytosol. (B) Three single-membrane autophagic bodies in the vacuole, each containing organelles destined for autophagic turnover. (Courtesy of David G. Robinson.)

All three groups of proteins are localized on the **phagophore assembly site** of the ER (**Figure 22.6**). ATG9 plays a crucial role by shuttling back and forth between the phagophore assembly site and the *trans* Golgi network and other sites, supplying the expanding phagophore with membrane components. Other ATG proteins, such as the ATG1 kinase complex, are required for the efficient functioning of this membrane shuttling system. For example, the ATG1/ATG13 complex is required for movement of ATG9 from the phagophore assembly site to the peripheral site, where it obtains fresh membrane. Inhibition of this transport process blocks autophagy. The PI(3) K complex also serves to regulate the ATG9 membrane shuttle system.

An important milestone in our understanding of autophagy was the identification of TOR (target of rapamycin), a serine/threonine protein kinase, as a master switch controlling the *ATG* genes. The TOR pathway is a major metabolic and developmental switch in eukaryotes that integrates nutrient and energy signaling to promote cell proliferation and growth. Based on studies in yeast and mammals, TOR is thought to act as a negative regulator of autophagy in plants by phosphorylating the ATG1/ATG13 complex, which prevents it from binding to the phagophore assembly site (PAS) (see Figure 22.6). Without the ATG1/ATG13 complex on the PAS, ATG9 is

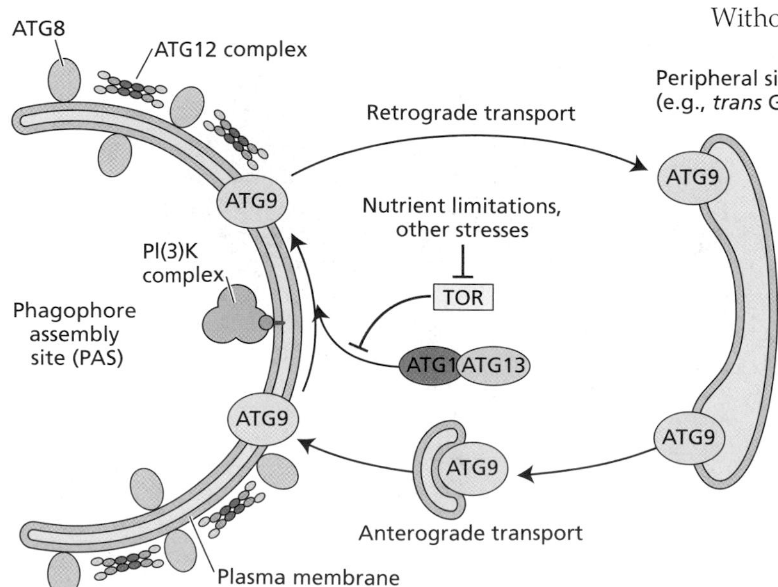

Figure 22.6 Simplified scheme for the "core autophagy machinery" localized on the phagophore assembly site. ATG9 drives the growth of the cup-shaped phagophore by shuttling between the phagophore assembly site and peripheral membrane sites. ATG9 requires the participation of the ATG1/ATG13 complex, as well as the phosphatidylinositol 3-OH kinase (PI[3]K) complex, the ATG12 complex, ATG8, and other proteins (not shown). The TOR (target of rapamycin) kinase complex acts as a negative regulator of autophagy by phosphorylating the ATG1/ATG13 complex.

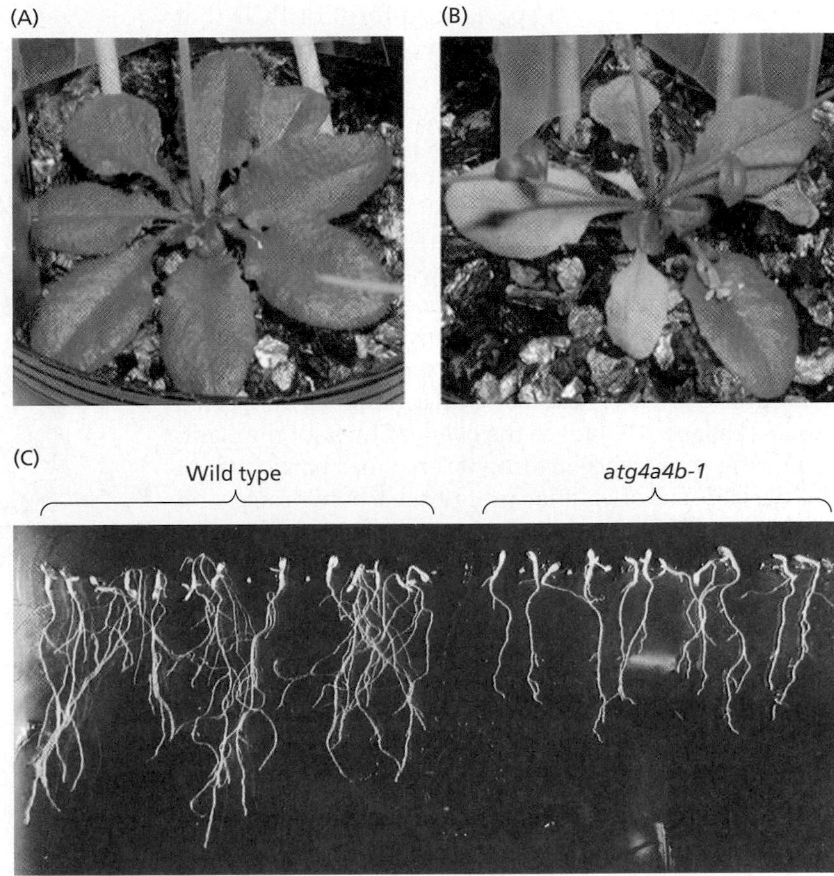

Figure 22.7 Phenotype of the autophagy-defective mutant *atg4a4b-1*. (A) Wild-type rosette leaves. (B) Autophagy-defective mutant rosette leaves exhibiting accelerated senescence. (C) Comparison of wild-type and mutant seedlings growing on agar under nitrogen-starved conditions. Root growth of the autophagy-defective seedlings is strongly inhibited. (A and B from Bassham et al. 2006; C from Yoshimoto et al. 2004.)

unable to cycle back to its peripheral site to obtain fresh membrane lipids for phagophore expansion, and autophagy comes to a halt. The ATG complex is also involved in recruiting other ATG proteins to the phagophore assembly site. TOR activity is, in turn, negatively regulated by nutrient limitation and other stresses. In this way, stresses of various kinds can stimulate autophagy via inhibition of TOR.

The autophagy pathway plays a dual role in plant development

In nonsenescing tissues, autophagy serves as a homeostatic mechanism that maintains the metabolic and structural integrity of the cell. The positive effect of autophagy on plant growth can be demonstrated by knocking out specific genes for autophagy in Arabidopsis. As shown in **Figure 22.7**, plants with defective autophagy exhibit accelerated senescence and reduced root growth compared with the controls. Autophagy can also have a negative effect on homeostasis, as occurs during the hypersensitive response (see Chapter 23). For example, transgenic Arabidopsis plants overexpressing the *RabG3b* (*Ras-related in brainG3b*) gene, which codes for a GTP-binding protein that activates autophagy, exhibit accelerated and unrestricted HR-type PCD over most of the leaf surface during pathogen infection. In contrast, PCD in control plants is restricted to much smaller regions.

The Leaf Senescence Syndrome

All leaves, including those of evergreens, undergo senescence, whether in response to age-dependent factors, environmental cues, biotic stress, or abiotic stress. Even the two permanent leaves of the unique South African species *Welwitschia mirabilis*, a distant relative of pine, die back continually from the tip, in equilibrium with the production of new leaf blade by the basal meristem (**Figure 22.8**).

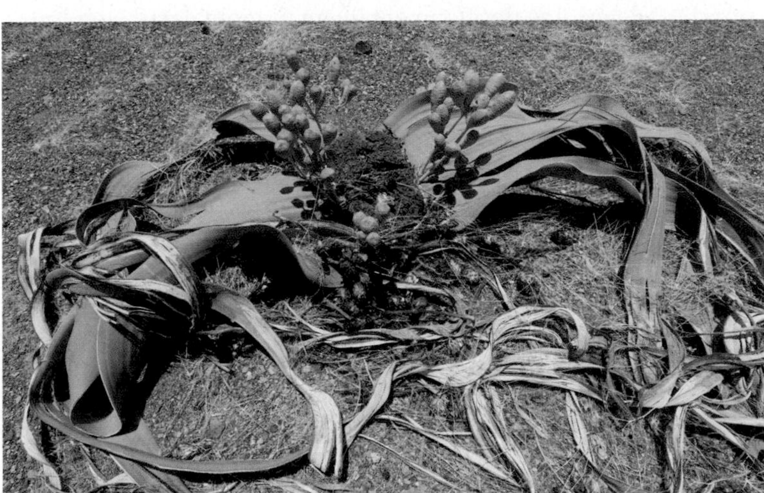

Figure 22.8 The South African gymnosperm *Welwitschia mirabilis* produces two permanent leaves that grow from a basal meristem. Senescence occurs at the tips. Over time the two leaves split longitudinally.

Leaf senescence is a specialized form of PCD that allows the efficient remobilization of nutrients from the source leaf to growing vegetative or reproductive sinks via the phloem. During senescence, leaf cells undergo genetically programmed changes in cell structure and metabolism. The earliest structural change is the breakdown of the chloroplast, which contains up to 70% of the leaf protein. Carbon assimilation is replaced by the breakdown and conversion of chlorophyll, proteins, and other macromolecules to exportable nutrients that can be translocated to growing vegetative organs or developing seeds and fruits. As is the case for other examples of PCD shown in Figure 22.2, leaf senescence is an evolutionarily selected process that contributes to the overall fitness of the plant.

During senescence, hydrolytic enzymes participate in the breakdown of cellular proteins, carbohydrates, and nucleic acids. The component sugars, nucleosides, and amino acids are then transported back into the main body of the plant via the phloem, where they will be reused for biosynthesis. Many minerals are also transported out of senescing organs back into the plant. Since senescence redistributes nutrients to growing parts of the plant, it can serve as a survival mechanism during environmentally adverse conditions, such as drought or temperature stress (see Chapter 24). However, leaf senescence occurs even under optimal growth conditions and is therefore part of the plant's normal developmental program. As new leaves are initiated at the shoot apical meristem, older leaves below may become shaded and lose the ability to function efficiently in photosynthesis, triggering senescence of the older leaves. In eudicots, senescence is usually followed by abscission, the process that enables plants to shed senescent leaves from the plants. Together, the coupled programs of leaf senescence and abscission help optimize the photosynthetic and nutrient efficiency of the plant.

The developmental age of a leaf may differ from its chronological age

Both internal and external cues influence the developmental age of leaf tissue, which may or may not correspond to the leaf's chronological age. The distinction between developmental and chronological age is nicely illustrated by a simple experiment carried out by the German plant physiologist Ernst Stahl in 1909. Stahl excised a small disc from a green leaf of mock orange (*Philadelphus grandiflora*), a deciduous shrub. He then incubated the disc on a simple nutrient solution in the laboratory until the fall, by which time the leaf attached to the plant had turned yellow. The drawing in **Figure 22.9** shows the disc superimposed on the intact leaf from which it was removed at the end of the experiment. Although the chronological ages of the leaf and the disc are the same, the leaf is now developmentally much older than the disc tissue. The intact leaf was subjected to a variety of internal signals coming from the surrounding leaf tissue and other parts of the plant, while the disc was literally cut off from these influences. Furthermore, the leaf attached to the plant remained outdoors exposed to the changing seasons, while the disc was cultured indoors under more or less constant conditions. Shielded from both internal and external cues, the leaf disc remained at the same developmental age as at the start of the experiment, while the attached leaf became developmentally older. We will discuss the factors that determine developmental age in more detail later in the chapter.

Figure 22.9 Early leaf senescence experiment showing the delayed senescence of a leaf disc cultured in the laboratory compared with the intact leaf of mock orange (*Philadelphus grandiflora*) from which the disc was excised. (From Stahl 1909.)

Leaf senescence may be sequential, seasonal, or stress-induced

Leaf senescence under normal growth conditions is governed by the developmental age of the leaf, which is a function of hormones and other regulatory factors. Under these circumstances, there is usually a senescence gradient from the youngest leaves located near to the growing tip to the oldest leaves near the base of the shoot—a pattern known as **sequential leaf senescence** (**Figure 22.10**). In contrast, the leaves of deciduous trees in temperate climates senesce all at once in response to the shorter days and cooler temperatures of autumn, a pattern known as **seasonal leaf senescence** (**Figure 22.11**). Both sequential

Figure 22.10 Sequential leaf senescence of wheat stems showing a gradient of senescence from the older leaves at the base to the younger leaves near the tip. (Courtesy of Andreas M. Fischer.)

and seasonal leaf senescence are variations of developmental senescence, since they occur under normal growing conditions. At the cellular level, both sequential and seasonal leaf senescence involve the vacuolar-type PCD pathway (see Figure 22.3A).

Leaf senescence can also occur prematurely under unfavorable, stressful environmental conditions. Among the abiotic stresses known to promote leaf senescence are drought, mineral deficiency, UV-B radiation, ozone, temperature extremes, high light, and darkness (see Chapter 24). Biotic stress, such as herbivory and pathogen infection, can also cause premature leaf senescence (see Chapter 23).

The morphological changes associated with stress-induced leaf senescence differ from those of developmental leaf senescence. In developmentally senescing leaves, senescence is coordinated at the whole leaf level, starting at the leaf tips or margins and spreading toward the leaf base (**Figure 22.12**). Environmental stress, by contrast, can be targeted to specific sites on a leaf. When localized stress occurs, the stressed tissue senesces earlier than the unstressed tissue. Mineral nutrient stress can also alter sequential leaf senescence (see Chapter 5).

Developmental leaf senescence consists of three distinct phases

Developmental leaf senescence can be divided into three distinct phases: the initiation phase, the degenerative phase, and the terminal phase (see Figure 22.12). During the *initiation phase,* the leaf receives developmental and environmental signals that initiate a decline in photosynthesis and a transition from being a nitrogen sink to a nitrogen source. Most of the autolysis of cellular organelles and macromolecules occurs during the *degenerative phase* of leaf senescence. The solubilized mineral and organic nutrients are then remobilized via the

(A) September 8 (B) September 13 (C) September 18

(D) September 25 (E) October 3 (F) October 8

Figure 22.11 Seasonal leaf senescence in an aspen tree (*Populus tremula*). All of the leaves begin to senesce in late September and undergo abscission in early October. (From Keskitalo et al. 2005.)

Figure 22.12 The three stages of leaf senescence.

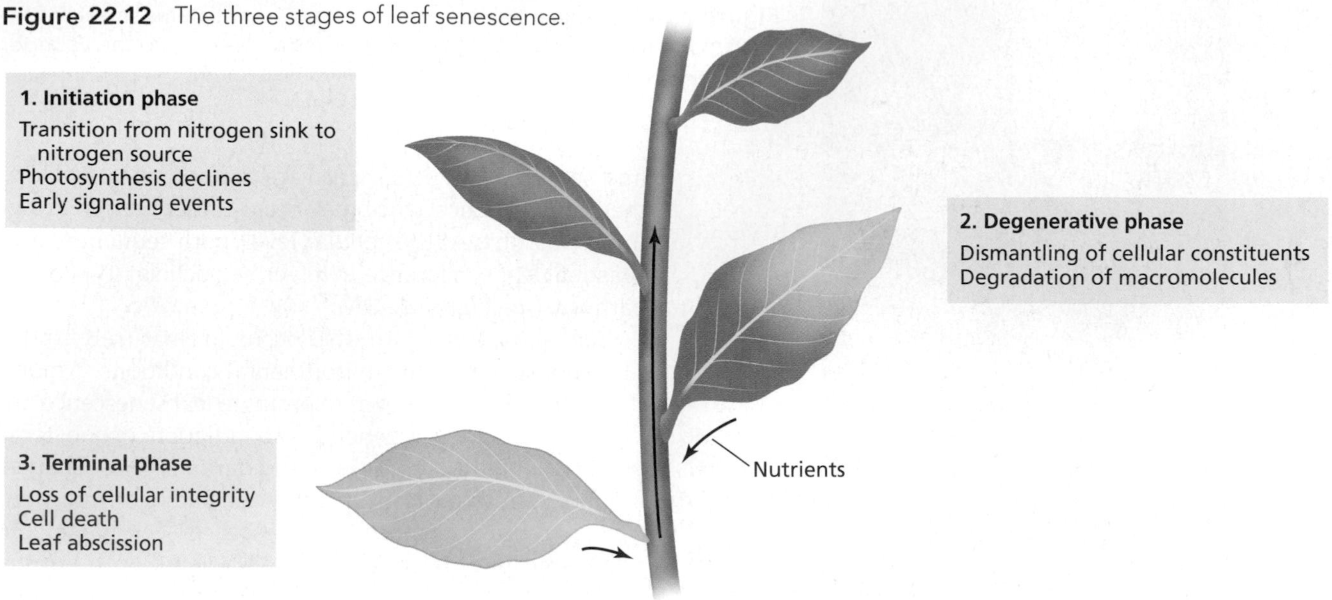

1. Initiation phase
Transition from nitrogen sink to nitrogen source
Photosynthesis declines
Early signaling events

2. Degenerative phase
Dismantling of cellular constituents
Degradation of macromolecules

3. Terminal phase
Loss of cellular integrity
Cell death
Leaf abscission

Nutrients

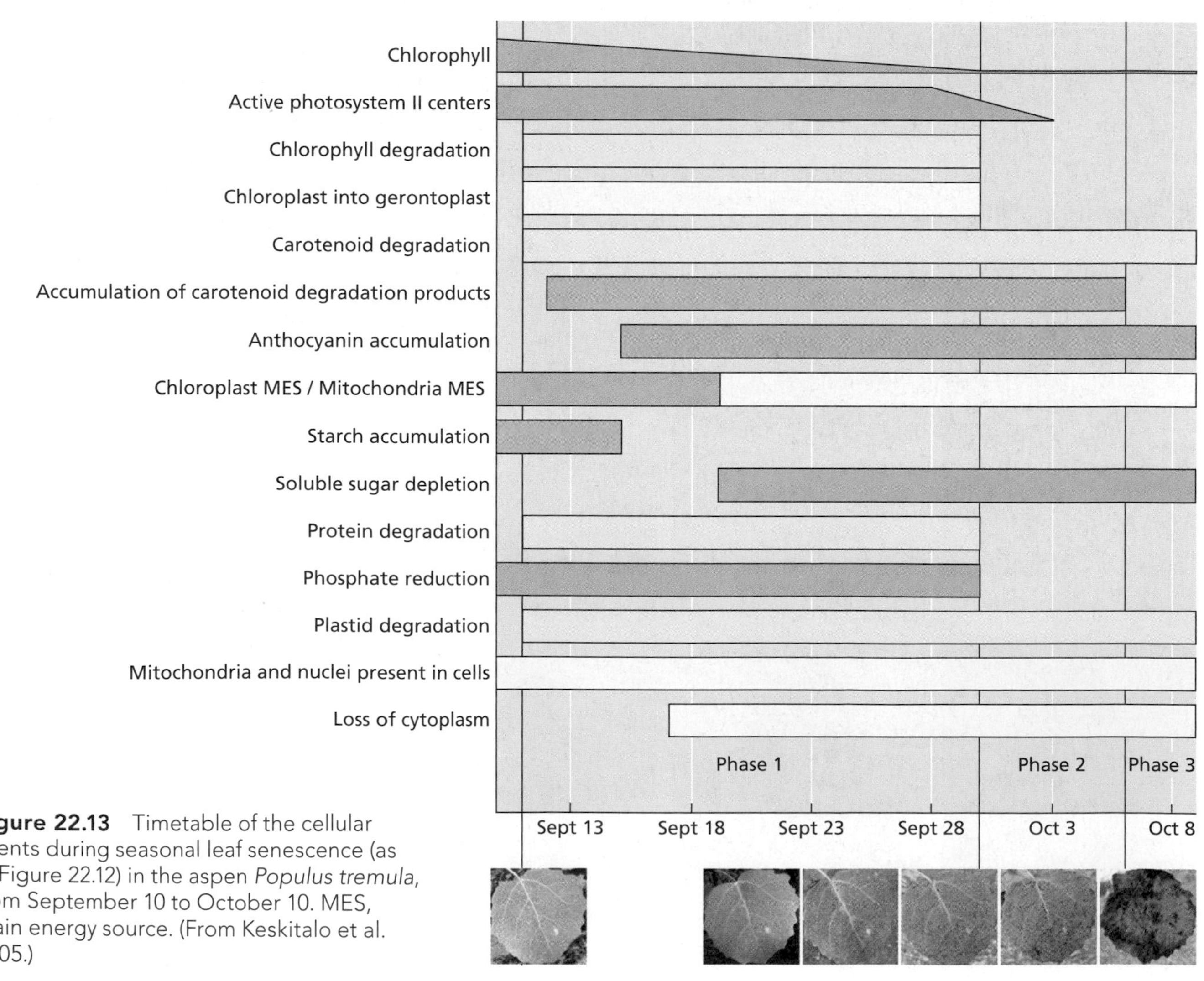

Figure 22.13 Timetable of the cellular events during seasonal leaf senescence (as in Figure 22.12) in the aspen *Populus tremula*, from September 10 to October 10. MES, main energy source. (From Keskitalo et al. 2005.)

phloem to growing sinks, such as young leaves, underground storage organs, or reproductive structures. The abscission layer forms during the degenerative phase of leaf senescence. During the *terminal phase*, autolysis is completed and cell separation takes place at the abscission layer, resulting in leaf abscission.

The earliest cellular changes during leaf senescence occur in the chloroplast

Chloroplasts contain about 70% of the total leaf protein, most of which consists of **ribulose-1,5-bisphosphate carboxylase/oxygenase** (**rubisco**) localized in the stroma, and **light-harvesting chlorophyll-binding protein II** (**LHCP II**) associated with the thylakoid membranes (see Chapters 7 and 8). Catabolism and remobilization of chloroplast proteins thus provides the primary source of amino acids and nitrogen for sink organs, and represents the earliest change that occurs during leaf senescence. This is illustrated in the cellular timetable of autumn leaf senescence in the aspen tree *Populus tremula*, shown in **Figure 22.13**. Chlorophyll loss begins around September 11, a week before the loss of other cytoplasmic constituents begins. The degenerative phase is largely complete by September 30, and the cell separation process in the abscission layer (discussed below) has begun, sealing off the phloem from further export of nutrients.

Chlorophyll and its primary degradation products are extremely photoreactive and potentially lethal to the cell. To prevent premature necrosis, the dismantling and degradation of chlorophyll-containing grana stacks and individual thylakoids must be carried out in a way that allows the safe removal and disposal of these potentially toxic compounds. During catabolism, chloroplasts are transformed into **gerontoplasts**, which resemble chromoplasts (see Figure 1.22). Ultrastructurally, gerontoplast formation involves the progressive unstacking of grana, the loss of thylakoid membranes, and a massive accumulation of **plastoglobuli** composed of lipids (**Figure 22.14**). The structural dismantling of the grana is accompanied by a decline in the primary photochemical reactions and in the efficiency of the Calvin–Benson cycle enzymes, including rubisco. Unlike chromoplasts, gerontoplasts retain the ability to divide, and their development is reversible up to a certain threshold when reversibility is lost and cells enter the terminal phase of senescence leading to cell death.

In contrast to chloroplasts, the nucleus and mitochondria, which are required for gene expression and energy production, remain intact until the later stages of senescence. However, not all chloroplasts senesce at the same rate. For example, the chloroplasts of guard cells are the last in a leaf to degrade, which suggests that they may continue to function even after mesophyll chloroplasts have become gerontoplasts. In the final stages of leaf senescence, typical symptoms of vacuolar-type PCD, such as tonoplast breakdown, nuclear condensation, and general autolysis, start at the leaf tip and spread downward toward the base.

(A)

(B)

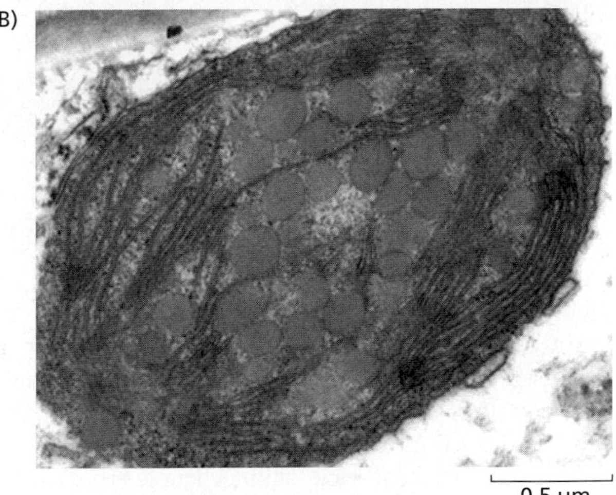

Figure 22.14 Ultrastructure of chloroplasts and gerontoplasts in mesophyll cells of barley leaf. (A) Chloroplasts before senescence and (B) gerontoplasts from leaves in which about 50% of the chlorophyll has been lost. (From Krupinska et al. 2012.)

The autolysis of chloroplast proteins occurs in multiple compartments

The degradation of chloroplast proteins during senescence involves both plastid-localized enzymes, including proteases, as well as other proteolytic systems outside the chloroplasts. For example, breakdown of rubisco and other stromal proteins occurs mainly outside the chloroplast via two kinds of autophagic structures, the **rubisco-containing bodies** and **senescence-associated vacuoles**. A major difference between these is that rubisco-containing bodies use the autophagy machinery, whereas senescence-associated vacuoles do not. Rubisco-containing bodies are surrounded by a double membrane and are thought to be formed when vesicles bud off from the senescing chloroplast, thus shrinking

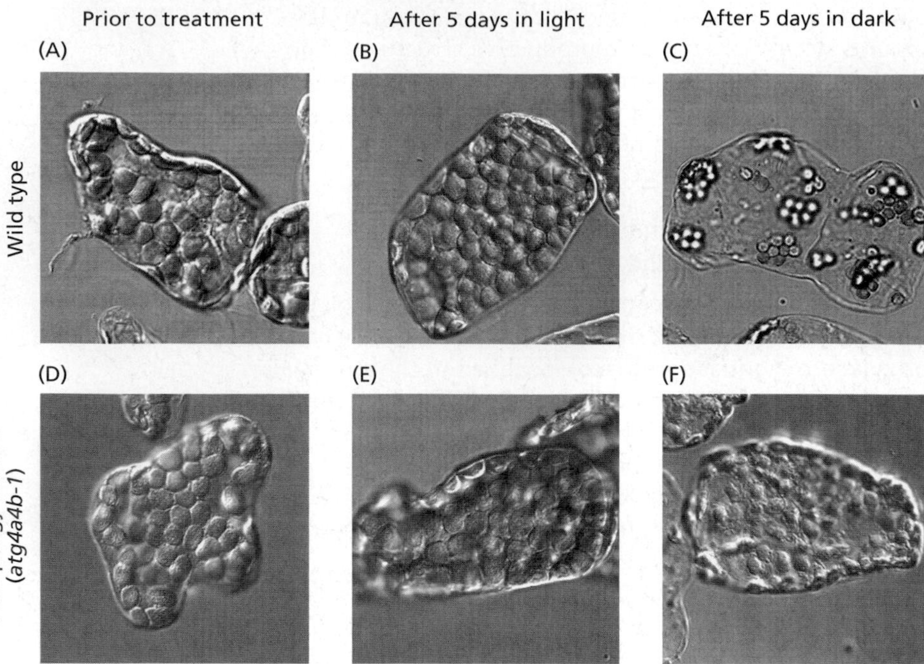

Figure 22.15 The autophagy pathway is required for chloroplast degradation in dark-induced leaf senescence in Arabidopsis. (A–C) Wild-type mesophyll cells. (D–F) Autophagy mutant (*atg4a4b-1*) mesophyll cells. (A, D) Cells from leaves prior to dark treatment. (B, E) Cells from leaves exposed to light for 5 days. (C, F) Cells from individually darkened leaves after 5 days in the dark. Chloroplast breakdown occurs in the dark in wild-type leaves (C) but not in autophagy mutant leaves (F). (From Wada et al. 2009.)

its size. Newly formed rubisco-containing bodies, which contain only rubisco and other stromal enzymes, are enclosed by autophagosomes that deliver their contents to the vacuole for subsequent degradation. In contrast to rubisco-containing bodies, senescence-associated vacuoles are small, protease-rich, acidic vacuoles that increase in number during senescence in the leaf mesophyll and guard cells, but not in the nongreen epidermal cells. Like rubisco-containing bodies, senescence-associated vacuoles contain rubisco and other stromal enzymes, and are capable of degrading them directly, although they may also fuse with the central vacuole.

Rubisco-containing bodies and senescence-associated vacuoles may reduce the size of the senescing chloroplast and degrade stromal proteins, but they are not involved in the breakdown of whole chloroplasts and their membranes. Studies have demonstrated that the autophagy pathway is required for whole chloroplast breakdown during dark-induced leaf senescence. As shown in **Figure 22.15**, the chloroplasts of an individually darkened leaf of wild-type Arabidopsis (see Figure 22.15C) are almost completely degraded compared with those of a control leaf kept in the light (see Figure 22.15B). However, the chloroplasts of the *atg4a4b-1* mutant are not broken down in the dark (see Figure 22.15F), suggesting that the autophagy pathway is involved in the degradation of whole chloroplasts. During this process whole chloroplasts can be engulfed by the vacuole.

It is likely that the early stages of chloroplast protein autolysis occur inside the chloroplast. Chloroplasts contain a number of ATP-dependent proteases of the *Clp* (*Caseinolytic protease*) and *FtsH* (*Filamentation temperature-sensitive H*) gene families, which are required for chloroplast development. Some of these proteases are specifically up-regulated during leaf senescence, although their precise roles in senescence remain unknown. Isolated chloroplasts can partially, but not completely, degrade rubisco in vitro. This suggests that chloroplast proteases participate in the early stages of the leaf senescence.

The STAY-GREEN (SGR) protein is required for both LHCP II protein recycling and chlorophyll catabolism

As discussed in Chapter 8, chlorophyll is tightly bound in complexes with proteins. During senescence these chlorophyll–protein complexes must be dismantled to allow the apoproteins to be recycled. STAY-GREEN (SGR) is a chloroplast protein that appears to act by destabilizing chlorophyll–protein complexes, and is thought to be required for the proteolysis of LHCP II within the chloroplast. Mutants of SGR stay green during senescence because chlorophyll cannot be catabolized when it is complexed to protein. The *green cotyledon* phenotype in Gregor Mendel's classic crossing experiments with peas (*Pisum sativum*) was caused by a mutation in the *SGR* gene. Despite their ability to retain their chlorophyll, *sgr* mutants exhibit the same decline in photosynthetic efficiency during senescence as do wild-type plants, because the turnover of the soluble proteins of the stroma is unaffected by the mutation.

Destabilization of chlorophyll–protein complexes by SGR, perhaps aided by partial proteolytic cleavage, releases LHCP II proteins for autolysis. The liberated chlorophyll molecules are then partially catabolized in

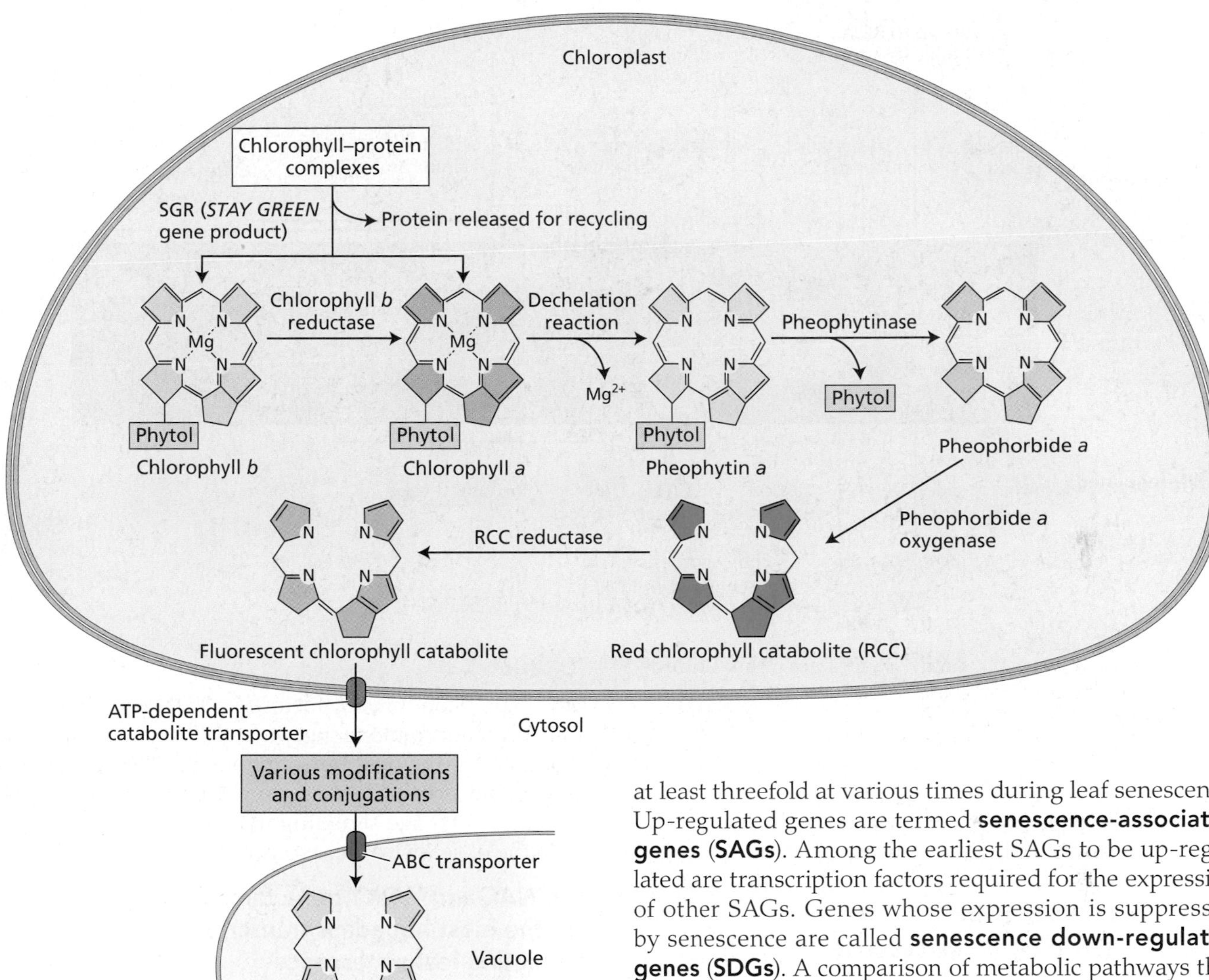

Figure 22.16 Chlorophyll catabolism pathway and compartmentation during leaf senescence.

the plastid and exported to the cytosol for further modification, before being permanently stored in the vacuole (**Figure 22.16**).

Leaf senescence is preceded by a massive reprogramming of gene expression

The transition from a mature, photosynthetically active leaf to a senescing leaf is a major phase change that requires the massive reprogramming of gene expression. A global analysis of gene expression in Arabidopsis has identified 827 genes whose transcript levels are increased at least threefold at various times during leaf senescence. Up-regulated genes are termed **senescence-associated genes** (**SAGs**). Among the earliest SAGs to be up-regulated are transcription factors required for the expression of other SAGs. Genes whose expression is suppressed by senescence are called **senescence down-regulated genes** (**SDGs**). A comparison of metabolic pathways that are either stimulated (by SAGs) or repressed (by SDGs) during sequential leaf senescence in Arabidopsis is shown in **Figure 22.17**. SAGs include many genes associated with abiotic and biotic stress, such as autophagy, the response to reactive oxygen species (ROS), metal ion binding, pectinesterase (cell wall breakdown), lipid breakdown, and genes involved in abscisic acid, jasmonic acid, and ethylene hormonal signaling (see Chapter 15).

Since senescence can have both internal and external causes, the question arises whether stress-related leaf senescence involves the same metabolic pathways and genetic programs as developmental leaf senescence. Comparisons have been made between the gene expression patterns of Arabidopsis leaves treated with a variety of abiotic stresses and those of naturally senescing leaves. At the earliest stages of treatment, the gene expression patterns of the stressed leaves were distinct from those of naturally senescing leaves. By the time the leaves had begun to yellow, however, the two sets of data converged. These findings suggest that abiotic stress initially involves specific stress-related signal transduction pathways, but

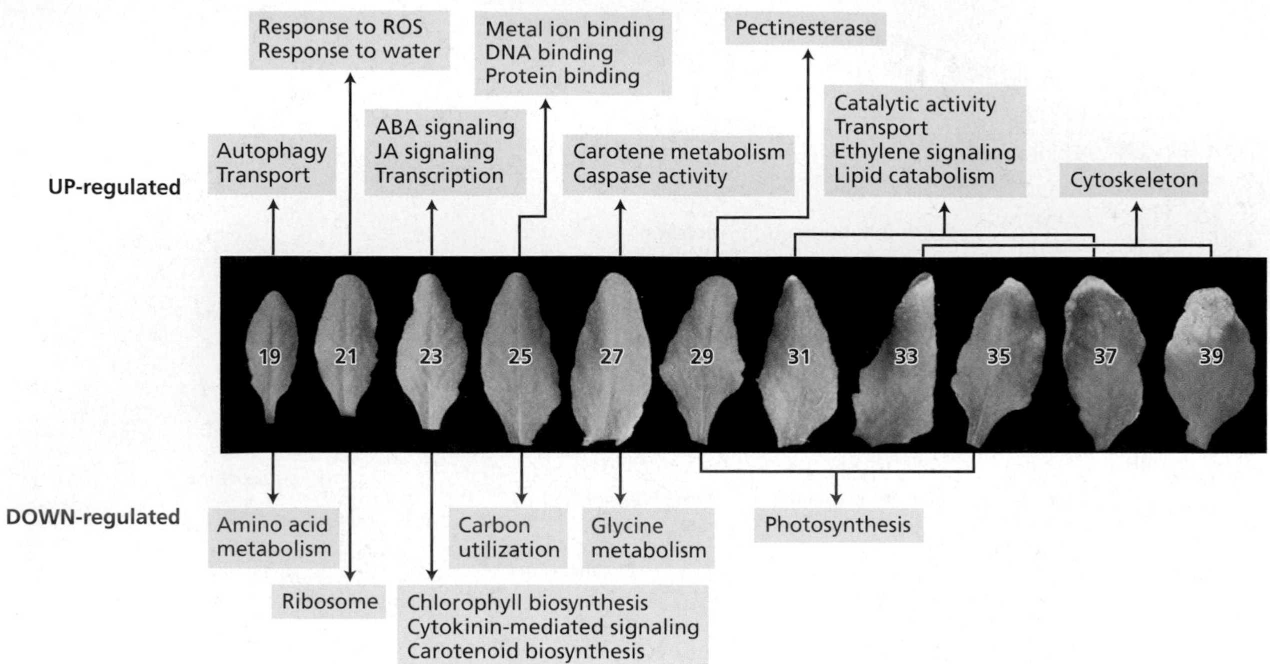

Figure 22.17 Metabolic pathways that are either up-regulated or down-regulated during senescence in Arabidopsis. (After Breeze et al. 2011.)

once PCD begins, the stress-induced pathways overlap with the developmental senescence pathway.

Leaf Senescence: The Regulatory Network

Much has been learned in recent years about the metabolic and gene regulatory pathways involved in senescence initiation, developmental age, and the leaf senescence program. An overview of the signaling pathways and regulatory networks that comprise the initiation phase of leaf senescence is shown in **Figure 22.18**. Among the important internal factors are plant hormones and other signaling molecules, such as salicylic acid. The developmental age of a leaf is also strongly affected by phase transitions, such as from the juvenile to the adult vegetative phase, and from the adult vegetative to the reproductive phase. External factors include seasonal changes, as well as biotic and abiotic stresses that subject the plant to extreme conditions outside its normal physiological range.

A network of overlapping signaling pathways integrates the input from the internal and external factors. These pathways include ROS-based signaling, the ubiquitin–proteasome pathway, protein kinases and phosphatases, mitogen-activated protein kinase (MAPK) signaling cascades (see Chapter 15), and hormonal signaling, all of which can alter gene expression by activating or repressing transcription factors. Epigenetic mechanisms also alter gene expression through histone and DNA modification, and chromatin remodeling. Small RNAs modulate gene expression at the posttranscriptional level. **Senescence-associated proteins** represent the end products of the developmental age signaling network that directly promote the onset of leaf senescence.

The *NAC* and *WRKY* gene families are the most abundant transcription factors regulating leaf senescence

The *NAC* and *WRKY* genes are the two most abundant families of differentially regulated transcription factors during senescence. NAC transcription factors (named after the related *NAM*, *ATAF*, and *CUC* gene families in different species) contain a highly conserved N-terminal DNA-binding domain and a variable regulatory C-terminal domain. NAC domain proteins comprise one of the largest groups of plant-specific transcription factors and are encoded by approximately 105 genes in Arabidopsis, approximately 140 genes in rice, and approximately 101 genes in soybean. They have been implicated in the regulation of a wide range of developmental processes.

NAC genes were first discovered in relation to leaf senescence in cereals. The presence of a functional *NAC* allele (called *NAM-B1*) causes earlier leaf senescence and nutrient retranslocation (nitrogen, iron, and zinc) to the developing grains of wild emmer wheat (*Triticum turgidum* ssp. *dicoccoides*), the ancestor of domesticated wheat varieties, allowing the grains to obtain the full benefit of the reclaimed nutrients from the leaves. In domesticated wheat varieties, such as the tetraploid pasta wheat (*Triticum turgidum* ssp. *durum*) and the hexaploid bread wheat

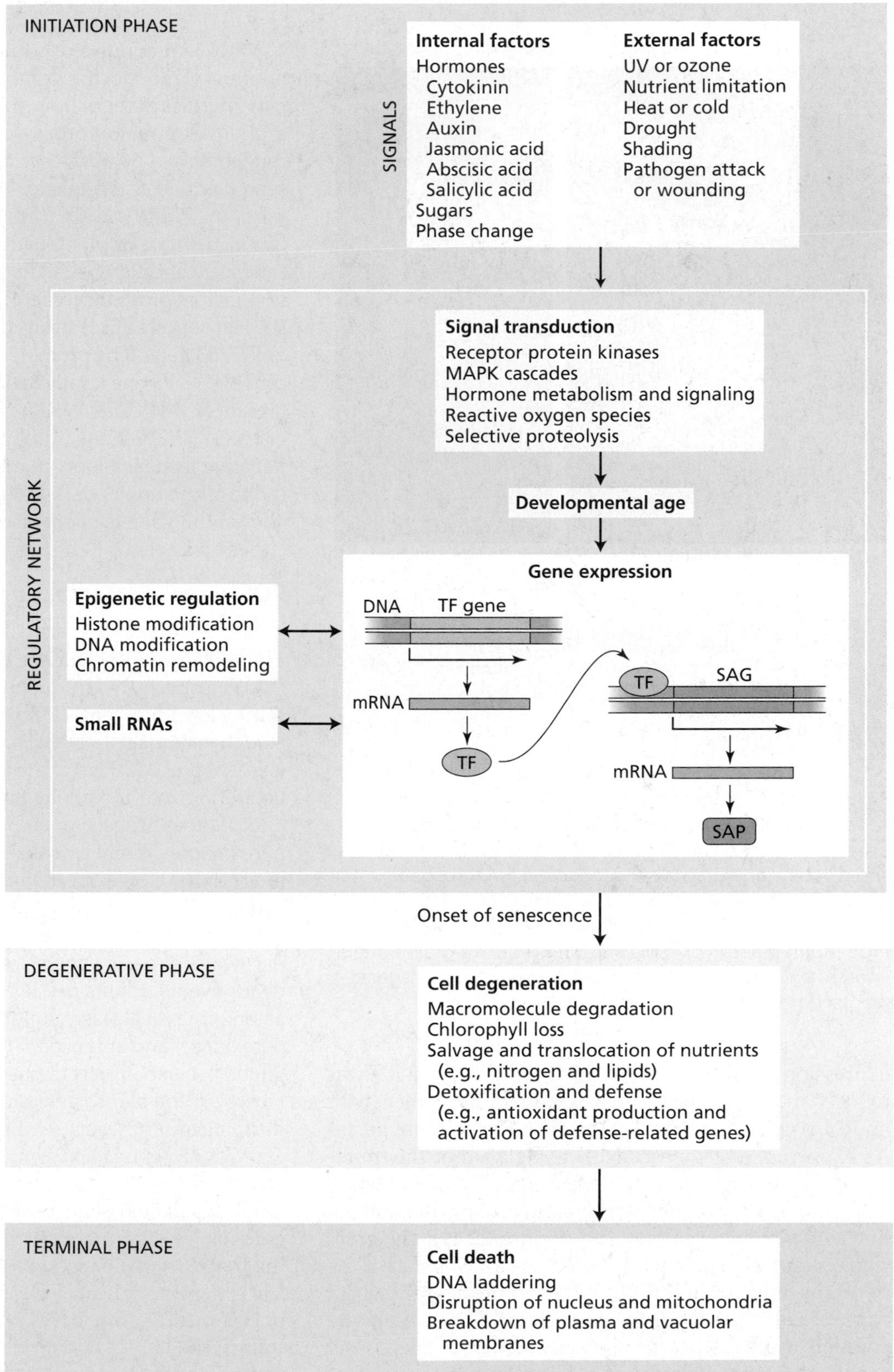

Figure 22.18 Overview of the signaling pathways and regulatory networks involved in the three main stages of leaf senescence (see text for discussion). MAPK, mitogen-activated protein kinase; TF, transcription factor; SAG, senescence-associated gene; SAP, senescence-associated protein.

(*Triticum turgidum* ssp. *aestivum*), a frameshift mutation results in a nonfunctional *NAM-B1* allele, which delays leaf senescence. This mutation seems to have been inadvertantly selected for during the early domestication of wheat. However, domesticated wheat varieties also contain two other closely related *NAC* genes, *NAM-A1* and *NAM-B2*, which lack the frameshift mutation and are therefore functional as senescence accelerators. To test the hypothesis that delayed leaf senescence results in reduced grain protein and mineral nutrient content, the

(C)

	Grain protein content (%)	Zn (ppm)	Fe (ppm)
Transgenic	13.27	52.45	37.40
Wild type	19.08	82.50	60.83

Figure 22.19 Repressing *NAC* gene expression delays senescence and reduces protein and mineral content in bread wheat (*Triticum turgidum* ssp. *aestivum*). (A) Whole shoots showing delayed senescence in the transgenic plants (left) compared with the wild type (right). (B) Comparisons of the spikes (ears) of transgenic (left) and wild-type (right) plants. (C) Table showing reduced grain protein content and lower levels of zinc and iron in the transgenic plants. (Photos and data from Uauy et al. 2006.)

expression of all three *NAM* alleles (*A1*, *B1*, and *B2*) was reduced in bread wheat plants by transforming them with an RNAi construct that specifically targeted these genes. As expected, leaf senescence was delayed in the transgenic plants compared with nontransgenic wild-type controls, and as a consequence, the grain protein and mineral nutrient contents were reduced (**Figure 22.19**). The grain size of the transgenic plants was the same as that of the nontransgenic controls, indicating that the delay in senescence did not translate into larger grains. These results are counterintuitive, since we normally associate delayed leaf senescence with higher yields. The fact that early senescence improves grain nutritional quality illustrates the crucial role of nutrient remobilization during leaf senescence for normal grain development. The importance of NAC transcription factors as regulators of leaf senescence has since been demonstrated in other species, including Arabidopsis and kidney bean (*Phaseolus vulgaris*).

WRKY (pronounced "worky") transcription factors are another plant-specific group of transcription factors that play important regulatory roles in many plant metabolic and developmental processes. WRKY transcription factors contain a 60 amino acid region named for the conserved amino acid sequence WRKYGQK in its N-terminal domain. Plant WRKY transcription factors are important regulators of plant–pathogen interactions as well as senescence. Like *NAC* gene products, WRKY transcription factors promote earlier leaf senescence. In Arabidopsis, leaf senescence is delayed in knockout mutants of the *WRKY53* gene. The promoters of several SAGs and many other *WRKY* gene family members are known to be direct targets of WRKY53. WRKY53 also binds to the promoter of the *WRKY53* gene, inhibiting its own expression in a negative feedback loop. In addition, *WRKY22* is involved in the regulation of dark-induced leaf senescence. Expression of the gene is suppressed by light and promoted by either darkness or ROS.

ROS serve as internal signaling agents in leaf senescence

There is growing evidence that reactive oxygen species (ROS), especially H_2O_2, play important roles as signals during leaf senescence. ROS are toxic chemicals that cause oxidative damage to DNA, proteins, and membrane lipids (see Chapter 24). They are produced primarily as by-products of normal metabolic processes, such as respiration and photosynthesis, in chloroplasts, mitochondria, and peroxisomes. They can also be produced on the plasma membrane. However, ROS do not trigger senescence by causing physicochemical damage to the cell, but rather they act as signals that activate genetically programmed pathways of gene expression that lead to regulated cell death events. Plants use ROS-scavenging systems, such as enzymes (catalase, superoxide dismutase, ascorbate peroxidase) and antioxidant molecules (e.g., ascorbate and glutathione) to protect themselves from oxidative damage. However, the plant's antioxidant concentrations decrease during leaf senescence, while ROS levels increase.

WRKY53 gene expression seems to act as a regulatory switch, controlling the expression of many SAGs during Arabidopsis leaf senescence. In Arabidopsis, *WRKY53* gene expression increases in leaves at the time of bolting (rapid stem elongation associated with flowering and leaf senescence). Leaf H_2O_2 levels also increase at the time of bolting. Treating leaves with H_2O_2 has been shown to induce the expression of *WRKY53*. Thus, there is good circumstantial evidence that H_2O_2 acts as a signal that triggers leaf senescence in Arabidopsis.

ROS signaling during leaf senescence is coupled to the activity of the MAPK signal transduction pathway. As described in Chapters 15 and 24, MAP kinases (MAPKs) are serine- or threonine-specific protein kinases that are involved in directing cellular responses to a diverse array of stimuli, including hormones and various types of stress. There is evidence that MAPK signaling acts upstream of *WRKY53* gene expression during leaf senescence in Arabidopsis.

Sugars accumulate during leaf senescence and may serve as a signal

Besides serving as an energy source and as building blocks for macromolecules, sugars may also act as signaling molecules, regulating metabolic pathways as well as developmental events. For example, as we saw in Chapter 8, trehalose 6-phosphate may serve as a signal that couples starch biosynthesis to the carbon status of the cytosol in leaves. Studies have shown that high concentrations of sugars lower photosynthetic activity, and may even trigger leaf senescence when the sugar exceeds a certain threshold. Sugar-induced senescence is especially important under conditions of low nitrogen availability. Recently it has been shown that both trehalose 6-phosphate and sugars accumulate in senescing Arabidopsis leaves, which suggests that trehalose 6-phosphate may play a role in the onset of leaf senescence, at least under conditions of high carbon availability.

Plant hormones interact in the regulation of leaf senescence

Leaf senescence is an evolutionarily selected, genetically regulated process that ensures efficient remobilization of nutrients to vegetative or reproductive sink organs. No mutation, treatment, or environmental condition has yet been found that abolishes the process completely, suggesting that leaf senescence is ultimately governed either by developmental or chronological age. However, both the timing and progression of senescence are flexible, and hormones are key developmental signals that accelerate or delay the timing of leaf senescence. Some hormones act as positive regulators of senescence, while others act as negative regulators. However, the same hormone can act either as a positive or negative regulator of the senescence process depending on the age of the leaf. In other words, leaves must reach a stage of maturity before they develop the *competency to senesce*. Only after competency is attained can the leaf respond to positive regulators of the senescence response. Hormones also mediate the responses to environmental cues, enabling the plant to maximize remobilization under varying environmental conditions.

In the discussion that follows we will discuss hormones individually, but it is important to keep in mind that hormonal pathways overlap and interact both cooperatively and antagonistically in the regulation of leaf senescence, consistent with a network-type control mechanism. In general, the hormones that regulate senescence can be divided into two basic categories based on their most commonly observed effects: positive (promoting) senescence regulators and negative (repressing) senescence regulators.

POSITIVE SENESCENCE REGULATORS

ETHYLENE Ethylene plays an important role in plant growth and development. Ethylene signaling regulates stress-related genes that are important for plant survival and growth. Ethylene is also regarded as a senescence-promoting hormone because ethylene treatment accelerates leaf and flower senescence, and inhibitors of ethylene synthesis and action can delay senescence. As we will discuss later in the chapter, ethylene plays an important role in abscission as well. The importance of ethylene signaling during senescence can also be inferred from the delayed senescence phenotype of ethylene-insensitive mutants in Arabidopsis, such as *etr1-1*. However, ethylene is not essential to the onset and progression of senescence. The senescence-accelerating effect of ethylene is elevated with increasing leaf age, and exposure of young leaves to ethylene has no effect on their senescence. Many of the transcripts for ethylene synthesis and signaling genes increase in abundance around the time that chlorophyll begins to decline. These observations suggest that ethylene signaling regulates the later stages of leaf senescence.

ABSCISIC ACID (ABA) ABA levels increase in senescing leaves, and exogenous application of ABA rapidly promotes the senescence syndrome and expression of several SAGs, which is consistent with ABA's effects on leaf senescence. However, like ethylene, ABA is considered an enhancer rather than a triggering factor of leaf senescence. During leaf senescence, genes associated with ABA synthesis and signaling are up-regulated and the endogenous ABA level increases. ABA levels are also significantly elevated under environmental stress conditions, which often induce leaf senescence (see Chapter 24). The NAC transcription factor VNI2 (VND-INTERACTING 2), which is up-regulated during leaf senescence, is also induced by either ABA or salt stress. Thus, there is a close interaction between ABA-induced stress signaling and leaf senescence signaling pathways.

ABA and water stress are coupled during leaf senescence. Senescing leaves dehydrate more rapidly than nonsenescing leaves because ABA-induced stomatal closure no longer functions. Stomata stay open because in senescing leaves ABA induces *SAG113*, a gene that encodes protein phosphatase 2C a negative regulatory component in the ABA signaling pathway (see Chapter 15). Protein phosphatase 2C inhibits stomatal closure specifically in senescing leaves. Knockout mutations of *SAG113* delay leaf senescence, whereas its overexpression accelerates the process. Prior to the onset of leaf senescence, ABA signaling induces stress tolerance processes, such as stomatal closure, which reduce water loss and delay senescence. But as the leaf ages, ABA signaling changes to induce transcripts such as *SAG113*, which inhibit ABA-induced stomatal closure, increasing water loss and accelerating senescence.

JASMONIC ACID (JA) Exogenous application of JA stimulates leaf senescence and controls the expression of a series of senescence-related genes. The Arabidopsis JA receptor, COI1 (CORONATINE-INSENSITIVE1), an F-box protein, is a key component of the JA signaling pathway (see Chapter 15). JA treatment accelerates leaf senescence in wild-type Arabidopsis plants, but not in the *coi1* mutants. In addition, the transcript abundance of genes involved in JA synthesis increases during developmental leaf senescence. Jasmonate content also increases in leaves as they senesce developmentally: 10-week-old Arabidopsis leaves have been shown to have 50-fold more JA than 6-week-old leaves. Despite the accumulation of JA during both natural- and dark-induced leaf senescence, the hormone is not essential for either the initiation or progression of these senescence processes. Thus, the *coi1* mutants do not exhibit delayed leaf senescence in Arabidopsis, although floral abscission is delayed. Thus, JA may play a more important role in flower senescence than in leaf senescence, at least in Arabidopsis. As in the cases of ethylene and ABA, the senescence-accelerating effect of JA is age-dependent. In Arabidopsis, for example, older leaves senesce much more rapidly in response to JA than do younger leaves.

BRASSINOSTEROIDS (BRs) Brassinosteroids appear to be positive regulators of senescence, since BR application accelerates senescence and BR-deficient mutants exhibit delayed senescence. The BR-insensitive Arabidopsis mutant *bri1*, in which the BR response has been inactivated, has a prolonged life span compared with wild-type plants, and also shows a reduction in the transcript levels of several SAGs. Conversely, a mutation that suppresses the *bri1* mutation exhibits accelerated senescence due to a constitutively active BR response pathway. However, the delayed senescence of BR mutants is associated with other phenotypic alterations, and thus it is possible that the delayed senescence is a secondary effect of the altered development. The results seem to suggest that BRs act as global regulators of leaf development, rather than as specific regulators of leaf senescence.

SALICYLIC ACID (SA) Salicylic acid is a phenolic phytohormone that regulates many aspects of plant growth and development, as well as various biotic and abiotic stress responses. Salicylic acid also positively regulates developmental leaf senescence. For example, Arabidopsis mutants defective in either SA biosynthesis or SA signaling exhibit delayed senescence compared with wild-type plants. Furthermore, the SA content of Arabidopsis leaves increases at the time chlorophyll concentrations begin to decline. Transcriptome analysis has confirmed that many of the genes involved in SA biosynthesis are up-regulated in senescing leaves, and about 20% of SAGs are up-regulated by the SA signaling pathway. SA treatment induces the expression of many SAGs, including *WRKY53*, which (as discussed above) acts as a master switch regulating other *WRKY* genes associated with leaf senescence. This suggests that SA plays a role in the onset of senescence, as well as in its progression.

NEGATIVE SENESCENCE REGULATORS

CYTOKININS The senescence-repressing role of cytokinins appears to be universal in plants and has been demonstrated in many types of studies. Although applied cytokinins do not prevent senescence completely, their effects can be dramatic, particularly when the cytokinin is sprayed directly on an intact plant. If only one leaf is treated, it remains green after other leaves of similar age have yellowed and dropped off the plant. If a small spot on a leaf is treated with cytokinin, that spot will remain green after the surrounding tissues on the same leaf begin to senesce. This "green island" effect can also be observed in leaves infected by some fungal pathogens, as well as in those hosting galls produced by insects. Such green islands have higher levels of cytokinins than surrounding leaf tissue. Unlike young leaves, mature leaves produce little, if any, cytokinin. During senescence, the transcript abundance of genes involved in cytokinin biosynthesis declines, whereas transcripts of genes involved in degrading cytokinins, such as cytokinin oxidase, increase during senescence. Mature leaves may therefore depend on root-derived cytokinins to postpone their senescence.

To test the role of cytokinin in regulating the onset of leaf senescence, tobacco plants were transformed with a chimeric gene in which a SAG-specific promoter was used to drive the expression of the *Agrobacterium tumefaciens* gene *ipt*, which encodes the enzyme that synthesizes cytokinin (**WEB APPENDIX 3**). The transformed plants had wild-type levels of cytokinins and developed normally until the onset of leaf senescence. As the leaves aged, however, the senescence-specific promoter was activated, triggering the expression of the *ipt* gene within leaf cells just as senescence would have been initiated. The resulting elevated cytokinin levels not only blocked senescence, but also limited further expression of the *ipt* gene, preventing cytokinin overproduction (**Figure 22.20**). This result suggests that cytokinins are natural regulators of leaf senescence. The AHK3 receptor appears to be the primary cytokinin receptor regulating leaf senescence in Arabidopsis (see Chapter 15). Elevated AHK3 function results in a significant delay in leaf senescence. Conversely, disruption of *AHK3*, but not other cytokinin receptor genes, results in premature leaf senescence.

Thus far, the molecular mechanism of cytokinin action in delaying leaf senescence remains unclear. According to a long-standing hypothesis, cytokinin represses leaf senescence by regulating nutrient mobilization and source–sink relations. This phenomenon can be observed when nutrients (sugars, amino acids, and so on) radiolabeled with ^{14}C or ^{3}H are fed to plants after one leaf or

Plant expressing *ipt* gene remains green and photosynthetic.
Age-matched control shows advanced senescence.

Figure 22.20 Leaf senescence is delayed in a transgenic tobacco plant containing a cytokinin biosynthesis gene, *ipt*, from *Agrobacterium tumefaciens* fused to a senescence-induced promoter. The *ipt* gene is expressed in response to signals that induce senescence. (From Gan and Amasino 1995, courtesy of R. Amasino.)

part of a leaf is treated with a cytokinin (**Figure 22.21**). Subsequent autoradiography of the whole plant reveals the pattern of movement and the sites at which the labeled nutrients accumulate. Experiments of this nature have demonstrated that nutrients are preferentially transported to and accumulated in the cytokinin-treated tissues. It has been postulated that the hormone causes nutrient mobilization by creating a new source–sink relationship. As discussed in Chapter 11, nutrients translocated in the phloem move from a site of production or storage (the source) to a site of utilization (the sink). The hormone may stimulate the metabolism of the treated area so that nutrients move toward it. For example, cytokinin-induced increases in extracellular invertase could regulate source–sink relations by hydrolyzing sucrose to hexoses, which are then transported into the cell. However, it is not necessary for the nutrient itself to be metabolized by the sink cells because even nonmetabolizable substrate analogs are mobilized by cytokinins (see Figure 22.21).

Cytokinin levels change in response to the concentration of nutrients to which plants are exposed. For example, application of nitrate to nitrogen-depleted maize (corn; *Zea mays*) seedlings results in a rapid rise in cytokinin levels in the roots, followed by mobilization of the cytokinins to the shoots via the xylem. This increase is due, at least in part, to an induction of expression of one member of the *IPT* gene family, *IPT3*. Cytokinin levels are also influenced by the concentration of phosphate in the environment, and cytokinins alter the expression of phosphate- and sulfate-responsive genes, suggesting an interaction between these response pathways.

AUXIN Elucidating the role of auxin in the regulation of leaf senescence has been complicated because auxin has been shown to

Figure 22.21 Effect of cytokinin on the movement of an amino acid in cucumber seedlings. A radioactively labeled amino acid that cannot be metabolized, such as aminoisobutyric acid, was applied as a discrete spot on the right cotyledon of each of these seedlings. The black stippling indicates the distribution of radioactivity. (Drawn from data in Mothes and Schütte 1961.)

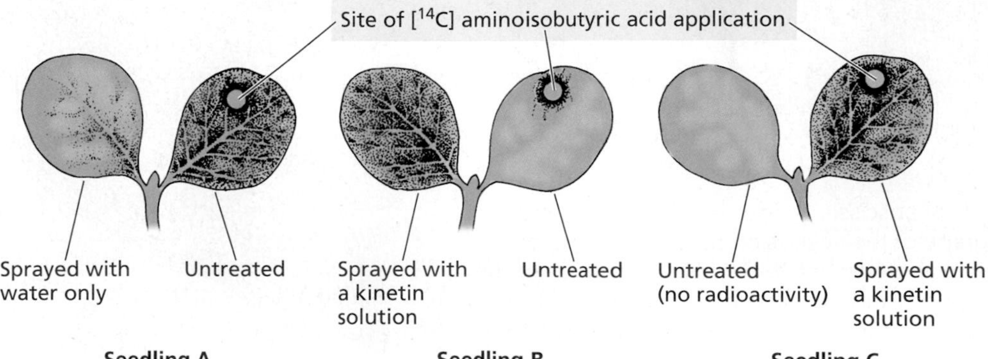
In seedling A, the left cotyledon was sprayed with water as a control. The left cotyledon of seedling B and the right cotyledon of seedling C were each sprayed with a solution containing 50 m*M* kinetin.
The dark stippling represents the distribution of the radioactive amino acid as revealed by autoradiography.
The results show that the cytokinin-treated cotyledon has become a nutrient sink. However, radioactivity is retained in the cotyledon to which the amino acid was applied when the labeled cotyledon is treated with kinetin (seedling C).
Site of [^{14}C] aminoisobutyric acid application
Sprayed with water only
Untreated
Sprayed with a kinetin solution
Untreated
Untreated (no radioactivity)
Sprayed with a kinetin solution
Seedling A
Seedling B
Seedling C

play a central role in so many aspects of plant growth and development. Adding to the complexity, high auxin concentrations stimulate ethylene production, which promotes senescence in mature leaves. However, much of the evidence obtained thus far points to a role for auxin as a negative regulator of leaf senescence. For example, application of exogenous auxin in Arabidopsis leads to a decrease in the expression of many SAGs. Overexpression of YUCCA6, the flavin-containing monooxygenase that catalyzes the rate-limiting step in auxin biosynthesis, delays leaf senescence and decreases SAG expression. In addition, the delayed-senescence Arabidopsis mutant *arf2* has a mutation in the gene *ARF2* (*AUXIN RESPONSE FACTOR2*), which is a repressor of auxin-response genes. By inactivating the *ARF2* repressor, the *arf2* mutation causes a constitutive auxin response, which delays leaf senescence.

GIBBERELLINS (GAs) Gibberellins are senescence-repressing hormones whose active forms decline in leaves as they age. For example, the senescence of excised leaf discs of *Taraxacum* and *Rumex* is delayed by treatment with gibberellic acid. Furthermore, expression of the gene encoding GA 2-oxidase, which is involved in the inactivation of GA, increased 18-fold during senescence, indicating that biologically active GA is removed during developmental leaf senescence. Furthermore, GA concentrations in leaves of romaine lettuce declined with the progression of senescence because of conversion of GA to an inactive GA-glucoside. Leaf senescence is inhibited by the availability of unconjugated, biologically active GA (GA_4 and GA_7).

Leaf Abscission

The shedding of leaves, fruits, flowers, and other plant organs is termed abscission (see WEB TOPIC 22.1). Abscission takes place within specific layers of cells called the **abscission zone**, located near the base of the petiole (**Figure 22.22**). The abscission zone becomes morphologically and biochemically differentiated during organ development, many months before organ separation actually takes place. Often the abscission

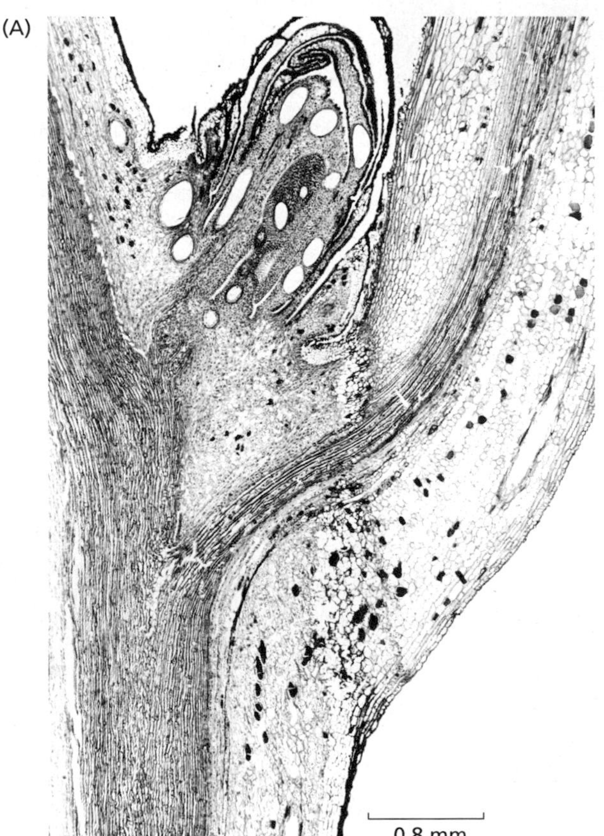

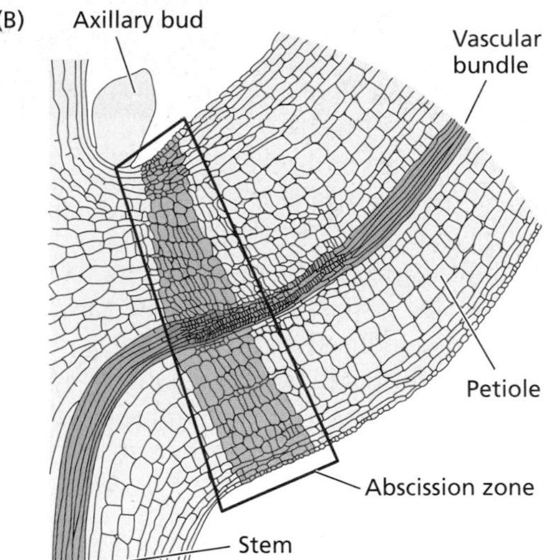

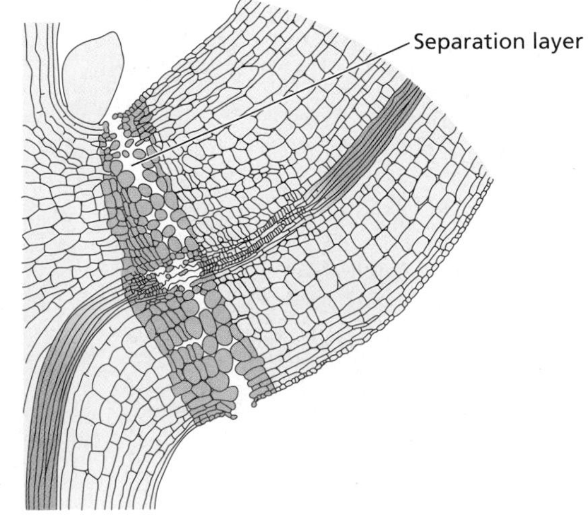

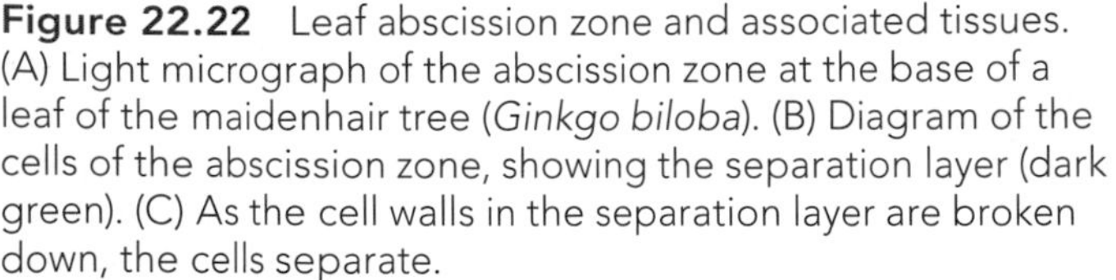

Figure 22.22 Leaf abscission zone and associated tissues. (A) Light micrograph of the abscission zone at the base of a leaf of the maidenhair tree (*Ginkgo biloba*). (B) Diagram of the cells of the abscission zone, showing the separation layer (dark green). (C) As the cell walls in the separation layer are broken down, the cells separate.

zone can be morphologically identified as one or more layers of isodiametrically flattened cells (see Figure 22.22).

Two mutants of tomato, *jointless* and *lateral suppressor*, fail to develop an abscission zone on the flower pedicel, and the mutated genes responsible for these phenotypes have been identified. The wild-type *JOINTLESS* gene encodes a MADS-box protein, one of a group of transcription factors that control many aspects of development, including floral organ identity. Prior to abscission, a **separation layer** forms within the abscission zone (see Figure 22.22). The dissolution of the walls between the cells of the separation layer, which does not result in cell death, subsequently results in the leaf being shed from the plant. The *LATERAL SUPPRESSOR* gene also regulates axillary bud development (see Chapter 19).

Figure 22.23 Effect of ethylene on abscission in the birch *Betula pendula*. The tree on the left is the wild type; the tree on the right was transformed with a mutated version of the Arabidopsis ethylene receptor gene, *ETR1*. The expression of this gene was under the transcriptional control of its own promoter. One of the characteristics of these mutant trees is that they do not drop their leaves when fumigated for 3 days with 50 ppm ethylene. (From Vahala et al. 2003.)

The timing of leaf abscission is regulated by the interaction of ethylene and auxin

Ethylene plays a key role in the activation of the events leading to cell separation within the abscission zone. The ability of ethylene gas to cause defoliation in birch trees is shown in **Figure 22.23**. The wild-type tree on the left has lost most of its leaves. Only the younger leaves at the top fail to abscise. The tree on the right has been transformed with a copy of the gene for the Arabidopsis ethylene receptor, *ETR1*, carrying the dominant *etr1* mutation (discussed earlier). This tree is unable to respond to ethylene and therefore does not shed any of its leaves after ethylene treatment.

The process of leaf abscission can be divided into three distinct developmental phases during which the cells of the abscission zone become competent to respond to ethylene (**Figure 22.24**).

1. *Leaf maintenance phase*. Prior to the perception of any signal (internal or external) that initiates the abscission process, the leaf remains healthy and fully functional. A gradient of auxin from the leaf blade to the stem maintains the abscission zone in a nonsensitive state.

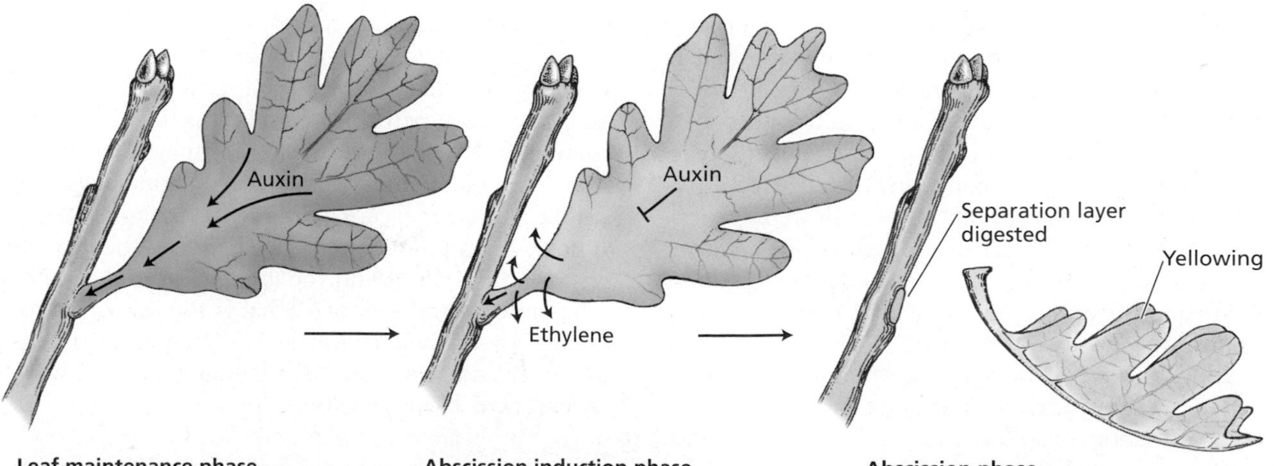
Auxin
Auxin
Ethylene
Separation layer digested
Yellowing
Leaf maintenance phase
High auxin from leaf reduces ethylene sensitivity of abscission zone and prevents leaf abscission.
Abscission induction phase
A reduction in auxin from the leaf increases ethylene sensitivity in the abscission zone, which triggers the abscission phase.
Abscission phase
Synthesis of enzymes that hydrolyze the cell wall polysaccharides results in cell separation and leaf abscission.

Figure 22.24 Schematic view of the roles of auxin and ethylene during leaf abscission. In the abscission induction phase, the level of auxin decreases, and the level of ethylene increases. These changes in the hormonal balance increase the sensitivity of the target cells to ethylene. (After Morgan 1984.)

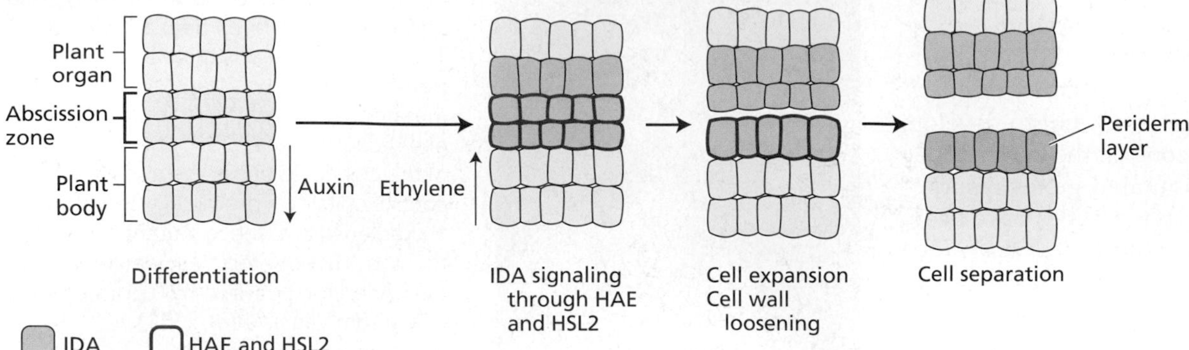

Figure 22.25 Model for peptide signaling during abscission. During abscission, specialized cells in the abscission zone capable of undergoing programmed cell separation respond to reduced levels of auxin from the leaf blade and augmented levels of ethylene and become competent to respond to abscission signals. The peptide signal IDA, indicated in purple, is expressed over a broader region than its receptors, HAE and HSL2 (dark blue outlines). Activation of the receptors by IDA leads to the transcription of cell wall remodeling genes, which causes cell expansion and cell separation, followed by the formation of a protective outer covering, or periderm, to block infection at the site. (After Aalen et al. 2013.)

2. *Abscission induction phase.* A reduction or reversal in the auxin gradient from the leaf blade, normally associated with leaf senescence, causes the abscission zone to become sensitive to ethylene. Treatments that enhance leaf senescence may promote abscission by interfering with auxin synthesis or transport in the leaf.
3. *Abscission phase.* The sensitized cells of the abscission zone respond to low concentrations of endogenous ethylene by synthesizing and secreting cell wall degrading enzymes and cell wall remodeling proteins, including β-1,4-glucanase (cellulase), polygalacturonase, xyloglucan endotransglucosylase/hydrolase, and expansin, resulting in cell separation and leaf abscission.

Early in the leaf maintenance phase, auxin from the leaf prevents abscission by maintaining the cells of the abscission zone in an ethylene-insensitive state. It has long been known that removal of the leaf blade (the site of auxin production) promotes petiole abscission. Application of exogenous auxin to petioles from which the leaf blade has been removed delays the abscission process.

In the abscission induction phase, typically associated with leaf senescence, the amount of auxin from the leaf blade decreases and the ethylene level rises. Ethylene appears to decrease the activity of auxin both by reducing its synthesis and transport and by increasing its destruction. The reduction in the concentration of free auxin increases the response of specific target cells in the abscission zone to ethylene. The abscission phase is characterized by the induction of abscission-related genes encoding specific hydrolytic and remodeling enzymes that loosen the cell walls in the abscission layer.

Mutations inhibiting flower abscission in Arabidopsis have led to the identification of several genes that regulate abscission initiation, including the small, secreted peptide INFLORESCENCE DEFICIENT IN ABSCISSION (IDA) and its probable receptors, leucine-rich repeat receptor-like kinases HAESA (HAE) and HAESA-LIKE2 (HSL2). Upon binding to IDA, the HAE/HSL2 receptor complex is thought to trigger the MAPK cascade, which leads to the transcriptional activation of genes encoding cell wall loosening enzymes, cell expansion, and cell separation (**Figure 22.25**).

Whole Plant Senescence

The programmed deaths of individual plant cells and organs are adaptations that benefit the plant as a whole by increasing its evolutionary fitness. The death of the whole plant, however, cannot easily be rationalized in evolutionary terms, even though the life spans of individual plants are to a large extent genetically determined and vary widely across species. In this final section of the chapter we will address some of the major questions that have been studied regarding whole plant senescence: Is whole plant senescence similar to aging in animals? What is the relationship, if any, between the life span of an individual plant and the longevity of its component cells, tissues, and organs? What is the role of reproduction in whole plant senescence? Why do meristems stop dividing, and does meristem failure lead to whole plant senescence? How does the senescence of single plants differ from that of clonal plants? And what role does the size of the plant play in determining its life span? As we will see, the regulation of source–sink relations figures prominently in all models for whole plant senescence that have been advanced thus far.

Figure 22.26 Monocarpic senescence in soybean (*Glycine max*). The entire plant on the left underwent senescence after flowering and producing fruit (pods). The plant on the right remained green and vegetative because its flowers were continually removed. (Courtesy of L. Noodén.)

Angiosperm life cycles may be annual, biennial, or perennial

Individual plant life spans vary from a few weeks in the case of desert ephemerals, which grow and reproduce rapidly in response to brief episodes of rain, to about 4,600 years in the case of bristlecone pine. In general, **annual plants** grow, reproduce, senesce, and die in a single season. **Biennial plants** devote their first year to vegetative growth and food storage, and their second year to reproduction, senescence, and death. Because annual and biennial plants undergo whole plant senescence following fruit and seed production, both are termed **monocarpic** because they reproduce only once (**Figure 22.26**).

Perennial plants live for 3 years or longer and may be herbaceous or woody. The range of maximum life spans for perennial plants is given in **Table 22.1**. Perennial plants are usually **polycarpic**, producing fruits and seeds over multiple seasons. However, there are also examples of monocarpic perennials, such as the century plant (*Agave americana*) (**Figure 22.27**) and Japanese timber bamboo (*Phyllostachys bambusoides*). The century plant grows vegetatively for 10 to 30 years before flowering, fruiting, and senescing, while Japanese timber bamboo can grow vegetatively for 60 to 120 years before reproduction and death. Remarkably, all clones from the same bamboo stock flower and senesce simultaneously, regardless of geographic location or climatic condition, which suggests the presence of a long-term biological clock of some kind.

Many perennial plants that form clones by asexual reproduction can proliferate into community-sized interconnected "individuals" that achieve astounding ages, such as King's lomatia (*Lomatia tasmanica*), a Tasmanian shrub in the Proteaceae family that can be over 43,000 years old. Each individual lomatia plant lives only about 300 years, but because it does not transfer any senescence signal to its clones, the clonal community apparently grows and proliferates indefinitely.

TABLE 22.1 Longevity of various individual and clonal plants

Species	Age (yr)
Individual plants	
Bristlecone pine (*Pinus longaeva*)	4600
Giant sequoia (*Sequoiadendron giganteum*)	3200
Stone pine (*Pinus cembra*)	1200
European beech (*Fagus sylvatica*)	930
Blackgum (*Nyssa sylvatica*)	679
Scots pine (*Pinus silvestris*)	500
Chestnut oak (*Quercus montana*)	427
Red oak (*Quercus rubra*)	326
European ash (*Fraxinus excelsior*)	250
English ivy (*Hedera helix*)	200
Flowering dogwood (*Cornus florida*)	125
Bigtooth aspen (*Populus grandidentata*)	113
Scots heather (*Calluna vulgaris*)	42
Spring heath (*Erica carnea*)	21
Scandinavian thyme (*Thymus chamaedrys*)	14
Clonal plants	
King's lomatia (*Lomatia tasmanica*)	43,000+
Creosote (*Larrea tridentata*)	11,000+
Bracken (*Pteridium aquilinum*)	1400
Sheep fescue (*Festuca ovina*)	1000+
Ground pine (*Lycopodium complanatum*)	850
Reed grass (*Calamagrostis epigeios*)	400+
Wood sage (*Teucrium scorodonia*)	10

Source: Thomas 2013.

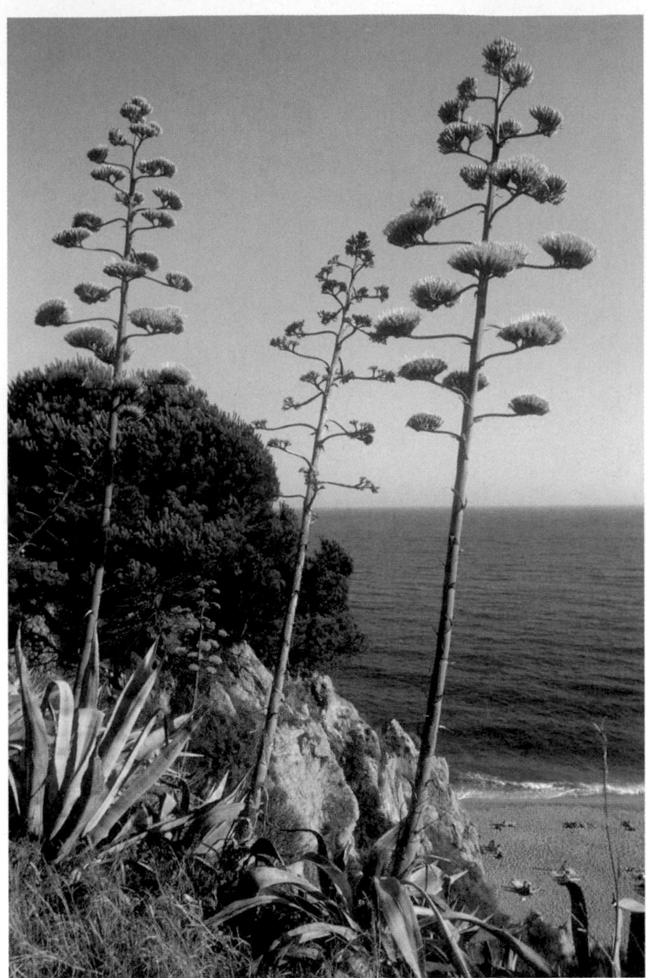

Figure 22.27 The century plant (*Agave americana*) flowers after 10–30 years of vegetative growth, after which it undergoes monocarpic senescence.

Whole plant senescence differs from aging in animals

Aging in animals is generally associated with gradual deterioration, the cumulative effect of wear and tear on the organism. According to some models, whole plant senescence is simply an accelerated form of aging. Senescing organs, tissues, and cells are programmed either to fail quickly or to be deficient in the mechanisms that would otherwise protect against physiological decline. According to this theory, the abilities of long-lived perennials to maintain the integrity of their meristems for thousands of years derives from developmental programs that successfully ward off the deteriorative effects of time.

One type of time-based cellular damage that has been investigated in plants is **mutational load**. Even the highest-fidelity cell replication mechanisms would be expected to propagate a significant number of errors over thousands of years. The mutation rate might even increase over time due to the build-up of reactive oxygen species (ROS). However, in bristlecone pine trees, no statistically significant relationship was found between the age of the individual and the frequency of mutations in pollen, seed, and seedlings. On the other hand, a significant decline in the average number of viable pollen grains per catkin per ramet was found in *Populus tremuloides* with increasing clone age. However, while decreasing pollen viability is consistent with mutational load, it would play no direct role in determining the longevity of clonal communities.

Age-dependent increases in somatic mutations leading to the production of chimeras and sports (plant parts that differ phenotypically from the parent plant) have also been observed in many perennial species. However, the evidence that such mutations contribute to whole plant senescence is very weak. Indeed, plants seem to have a high tolerance for **genetic mosaicism**, and they possess robust mechanisms for purging deleterious mutant cells.

Another type of time-based damage to cells that potentially could contribute to whole plant senescence is **telomere shortening**. Telomeres are regions of repetitive DNA that form the chromosome ends and protect them from degradation (see Chapter 1). Normal chromosome replication results in telomere shortening, and without some mechanism for telomere repair, they would eventually disappear after successive rounds of cell division. **Telomerase**—a ribonucleoprotein enzyme complex—extends the ends of telomeres after replication through the activity of telomerase reverse transcriptase. Although animals with dysfunctional telomerase age prematurely, Arabidopsis mutants lacking telomerase activity grow and reproduce for up to ten generations. In addition, observations on bristlecone pine and *Ginkgo biloba* plants failed to demonstrate progressive shortening of telomeres with increasing age. The cause of the differences between plant and animal telomeres with respect to aging is unclear.

The determinacy of shoot apical meristems is developmentally regulated

Plants are often described as having indeterminate growth due to the activities of the apical meristems, but the determinacy of the apical meristems is under strict developmental control. For example, shoot apical meristems may be continuously meristematic (indeterminate), or may cease activity (determinate), either by differentiating into a terminal organ such as a flower, or by undergoing growth arrest or senescence. Indeed, the growth habits, life cycles, and senescence profiles of different plant species are intimately connected to their patterns of apical meristem determinacy.

In monocarpic species, all indeterminate vegetative shoot apices become determinate floral apices, and the entire plant senesces and dies after seed dispersal. In contrast, polycarpic perennials retain a population of indeterminate shoot apices as well as those apices that become reproductive and determinate.

Monocarpic senescence typically involves three coordinated events: (1) senescence of somatic organs and tissues such as leaves; (2) growth arrest and senescence of shoot apical meristems; and (3) suppression of axillary buds. In peas, shoot apical meristem senescence has been shown to be regulated by both photoperiod and gibberellins. As discussed in Chapter 19, auxin from actively growing terminal buds suppresses the growth of axillary buds, a phenomenon known as apical dominance. Strigolactone and cytokinin play antagonistic roles during apical dominance, with strigolactone repressing axillary bud growth and cytokinin promoting axillar bud growth (see Chapter 19). Removal or death of the terminal bud reduces auxin transport and favors cytokinin signaling in the lateral buds, promoting branch formation. However, shoot apical meristem arrest during monocarpic senescence does not lead to activation of the axillary buds.

The molecular mechanism of axillary bud suppression during monocarpic senescence has been investigated in Arabidopsis. Expression of the gene for the transcription factor AtMYB2 in the basal internode is associated with the suppression of both cytokinin biosynthesis and branch formation during monocarpic senescence. T-DNA insertion mutants lacking a functional AtMYB2 protein are bushy as a result of increased cytokinin production. Significantly, senescence is delayed in the bushy mutant, which indicates that cytokinin acts as a negative regulator of whole plant senescence.

Nutrient or hormonal redistribution may trigger senescence in monocarpic plants

A diagnostic feature of monocarpic senescence is the ability to delay senescence well beyond the plant's normal life span by the removal of the reproductive structures. For example, repeated depodding enables soybean plants to remain vegetative for many years under favorable growing conditions, leading to a treelike appearance. What is the relationship between fruit development and whole plant senescence? One of the earliest explanations for monocarpic senescence was based on the redistribution of vital nutrients via the phloem from vegetative sources to reproductive sinks. This explanation still fits much of the currently available evidence. The alternative explanation, that developing fruits produce a hypothetical "death hormone," has never been convincingly demonstrated.

Many studies have shown that alterations in the source–sink relations of vegetative and reproductive tissues can affect the course of senescence. As discussed earlier in relation to leaf senescence, cytokinins increase sink strength in leaves and also delay leaf senescence. During monocarpic senescence in peas, high endogenous GA levels in the vegetative buds are correlated with high sink strength, vigorous vegetative growth, and delayed whole plant senescence. In contrast, high auxin levels in floral buds are correlated with high sink strength of the reproductive structures and rapid reproductive development followed by whole plant senescence.

If developing seeds and fruits are such strong sinks that they can trigger the senescence of the rest of the plant, why do male plants of dioecious species such as spinach (*Spinacea oleracea*), which produce neither seeds nor fruits, senesce at the same time as female plants, which produce copious seeds and fruits? Experiments carried out in the late 1950s showed that removing the tiny pollen-producing flowers from the male plants delayed senescence to the same extent that removing the female flowers did. This result seemed to contradict the resource redistribution model, since it was assumed that the use of carbohydrate resources by staminate spinach flowers would be negligible compared with the carbohydrate use of pistillate flowers. However, more recent studies have shown that, contrary to earlier assumptions, the nutritional demand of the staminate flowers actually *exceeds* the nutritional demand of the pistillate flowers, especially early in flower development, and thus could be a determining factor in triggering monocarpic senescence, even in male plants.

Although resource redistribution may well trigger monocarpic senescence, the critical compound is not carbohydrate, because many studies have now shown that the carbohydrate content of leaves actually *increases* during senescence. This observation is consistent with the ability of exogenous sugars to trigger leaf senescence. Rather than carbohydrate loss, alterations in source–sink relations caused by floral development may induce a global shift in the hormonal or nutrient balance of the vegetative organs. A loss of nitrogen coupled with a simultaneous accumulation of carbohydrate would cause an increase in the C:N ratio, which has been associated with vacuolar-type PCD in senescing leaves.

The rate of carbon accumulation in trees increases continuously with tree size

All individual trees die eventually, and it has long been assumed that the growth rate of trees declines with increasing tree size and mass. In fact, it is well established that as trees grow taller, their rate of growth in height decreases (**Figure 22.28**). To explain this decline in the rate of elongation growth over time, it has been reasoned that at some point the height of a tree will begin to approach the limits of the vascular system to deliver adequate supplies of water, minerals, and sugars to the growing tips of the extensive shoot and root systems. As water and other resources become limiting, declines in photosynthetic productivity should occur. The decline in leaf photosynthetic efficiency with increasing age of the tree is well documented. Age-related declining tree growth has also been viewed as an inevitable consequence of increasing resource allocation to reproduction.

Although the results of a few single-species studies have been consistent with decreases in growth rates

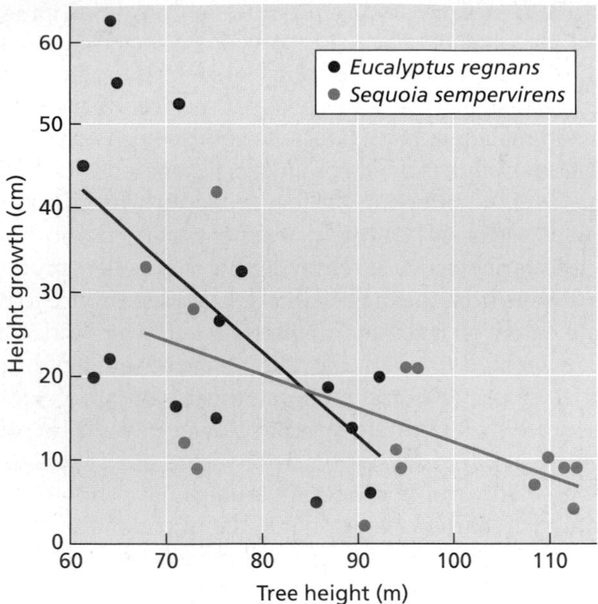

Figure 22.28 Annual growth in height of eucalyptus and sequoia trees as a function of the beginning height of the tree in the year 2006. In both cases, the growth in height declined with tree height. (After Sillett et al. 2010.)

as trees increase in height, most of the evidence cited in support of declining tree growth has not been based on measurements of individual tree mass, but on age-related declines in either the *net primary productivity* of forest stands in which the trees are all of a similar age, or in the *rate of mass gain per unit leaf area,* with the implicit assumption that productivity declines at the individual leaf level can be extrapolated to the entire tree.

Recently, however, an analysis was undertaken of the mass growth rates of 673,046 trees belonging to 403 tropical, subtropical, and temperate tree species on every forested continent. In every continent, the aboveground tree mass growth rates for most species increased continuously with the $\log_{10}$ of the tree mass. The results for North America are shown in **Figure 22.29**. In the case of the largest trees, 97% of the species exhibited this trend. In absolute terms, different species of trees with 100 cm trunk diameters typically added from 10 to 200 kg of aboveground dry mass each year, averaging 103 kg per year. This is nearly three times the rate for trees of the same species with 50 cm trunk diameters. In the case of the most massive tree species, such as *Eucalyptus regnans* and *Sequoia sempervirens,* individual trees can add as much as 600 kg to the aboveground mass every year.

The above findings demonstrate that although **growth efficiency** (tree mass growth per unit leaf area or leaf mass) often declines with increasing tree size, total tree leaf mass increases as the square of trunk diameter. A typical tree that experiences a 10-fold increase in diameter will therefore undergo a roughly 100-fold increase in total leaf mass and a 50- to 100-fold increase in total leaf area. Increases in the total leaf area are therefore sufficient to overcome the decline in growth efficiency and cause the whole-tree carbon accumulation rate to increase as the tree increases in size. However, at some point a limit is indeed reached and senescence occurs when the mass growth rate is at its peak. These findings suggest that whole tree senescence is caused by massive organ failure over a relatively short period of time, rather than by a slow decline due to aging. The extent to which such rapid tree senescence is caused by internal versus external factors, such as fire, nutrient depletion, water stress, or pathogen attack, is still poorly understood.

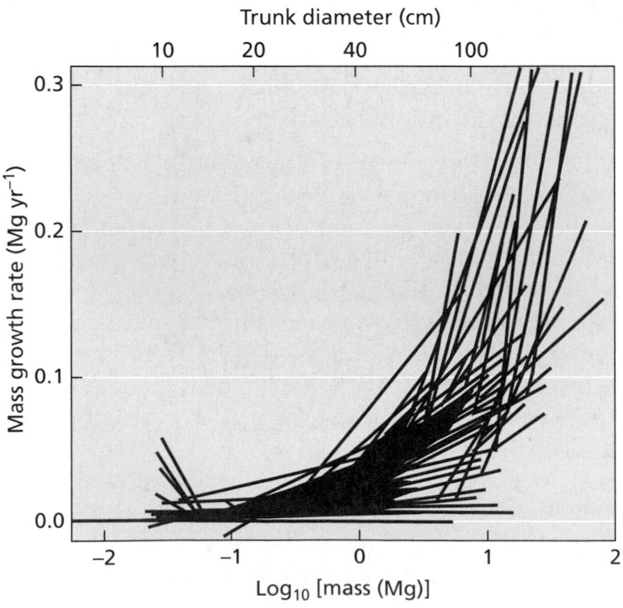

Figure 22.29 Aboveground mass growth rates for 110,153 trees belonging to 89 species in North America (USA). Trunk diameters (cm) are shown on the upper horizontal axis; the aboveground tree mass, expressed as $\log_{10}$ (mass in Mg [megagrams]), is shown on the lower horizontal axis; the mass growth rate (Mg yr^{-1}) is shown on the vertical axis. The mass growth rate increases with the aboveground tree mass. Similar results were obtained for 562,893 trees belonging to 314 species growing in five other continents. (After Stephenson et al. 2014.)

SUMMARY

Plant cells, organs, and organisms experience wear both from the effects of aging and from external stress. To break down old or damaged tissues, or to further some developmental pathways, plants undergo senescence, or genetically programmed cell death (**Figure 22.2**). Senescence differs from necrosis, which is the unexpected death of tissues caused by physical or chemical damage or other external agents.

Programmed Cell Death and Autolysis

- PCD during normal development occurs via vacuolar swelling and cell rupturing and is called vacuolar-type PCD, whereas PCD during the hypersensitive response occurs via vacuolar water loss and cell shrinkage and is called hypersensitive response-type PCD (**Figure 22.3**).
- Autophagosomes capture damaged cellular constituents and release their contents into the central vacuole to be degraded into reusable monomers (**Figures 22.4, 22.5**).
- A subset of the autophagy-related genes and proteins regulates the formation of autophagosomes (**Figure 22.6**).
- In addition to its role in senescence, the autophagy pathway is a homeostatic mechanism that maintains the metabolic and structural integrity of the cell (**Figure 22.7**).

The Leaf Senescence Syndrome

- Leaf senescence involves the breakdown of cellular proteins, carbohydrates, and nucleic acids and the redistribution of their components back into the main body of the plant, to actively growing areas. Minerals are also transported out of senescing leaves back into the plant.
- PCD can be manipulated to induce tissues to remain in less mature stages of development (**Figure 22.9**).
- Leaf senescence may exhibit a sequential or seasonal pattern or, if it is stress-induced, may be targeted to specific sites on a leaf (**Figures 22.10, 22.11**).
- Developmental leaf senescence consists of three phases: initiation, degeneration, and termination (**Figures 22.12, 22.13**).
- The earliest cellular changes during leaf senescence occur in the chloroplast. The transformation of chloroplasts into gerontoplasts allows for the safe removal and disposal of potentially toxic compounds produced by the degradation of chlorophyll (**Figure 22.14**).
- The autolysis of chloroplast proteins occurs in multiple compartments.
- The STAY-GREEN (SGR) protein is required for both LHCP II protein recycling and chlorophyll catabolism (**Figure 22.16**).
- Leaf senescence is preceded by a massive reprogramming of gene expression (**Figure 22.17**).

Leaf Senescence: The Regulatory Network

- A network of overlapping signaling pathways integrates internal and external input to regulate senescence through gene expression (**Figure 22.18**).
- The highly conserved *NAC* and *WRKY* gene families are the most abundant transcription factors regulating senescence.
- There is growing evidence that reactive oxygen species (ROS), especially H_2O_2, can serve as internal signals to promote senescence.
- High concentrations of sugars may also serve to signal leaf senescence, especially under conditions of low nitrogen availability.
- Plant hormones interact to regulate leaf senescence, though they are only effective at promoting senescence once the leaf reaches a certain stage of maturity.
- Senescence-promoting regulators include ethylene, abscisic acid, jasmonic acid, brassinosteroids, and salicylic acid.
- Senescence-repressing regulators include cytokinins, auxins, and gibberellins (**Figures 22.20, 22.21**).

Leaf Abscission

- Abscission is the shedding of leaves, fruits, flowers, or other plant organs, and takes place within specific layers of cells called the abscission zone (**Figure 22.22**).
- High levels of auxin keep leaf tissue in an ethylene-insensitive state, but as auxin levels drop, the abscission-promoting and auxin-repressing effects of ethylene become stronger (**Figures 22.23–22.25**).

Whole Plant Senescence

- In general, annuals and biennials reproduce only once before senescing, while perennials can reproduce multiple times before senescing.
- According to some models, whole plant senescence represents an accelerated form of aging where tissues are programmed to fail quickly once certain thresholds are reached.

- Nutrient or hormonal redistribution from vegetative structures to reproductive sinks may trigger whole plant senescence in monocarpic plants.
- While growth efficiency in trees declines with increasing tree size, leaf mass increases as the square of trunk diameter and can overcome this loss in efficiency, until internal or external factors initiate whole tree senescence (**Figures 22.28, 22.9**).

WEB MATERIAL

- **WEB TOPIC 22.1 Abscission and the Dawn of Agriculture** A short essay on the domestication of modern cereals based on artificial selection for nonshattering rachises.

available at plantphys.net

Suggested Reading

Aalen, R. B., Wildhagen, M., Stø, I. M., and Butenko, M. A. (2013) IDA: A peptide ligand regulating cell separation processes in *Arabidopsis*. *J. Exp. Bot.* 64: 5253–5261.

Breeze, E., Harrison, E., McHattie, S., Hughes, L., Hickman, R., Hill, C., Kiddle, S., Kim, Y.-S., Penfold, C. A., Jenkins, D., et al. (2011) High-resolution temporal profiling of transcripts during arabidopsis leaf senescence reveals a distinct chronology of processes and regulation. *Plant Cell* 23: 873–894.

Davies, P. J. and Gan, S. (2012) Towards an integrated view of monocarpic plant senescence. *Russ. J. Plant Physiol.* 59: 467–478.

Fischer, A. M. (2012) The complex regulation of senescence. *Crit. Rev. Plant Sci.* 31: 124–147.

Humbeck, K. (2013) Epigenetic and small RNA regulation of senescence. *Plant Mol. Biol.* 82: 529–537.

Lui, Y. and Bassham, D. C. (2012) Autophagy: Pathways for self-eating in plant cells. *Annu. Rev. Plant Biol.* 63: 215–237.

Luo, P. G., Deng, K. J., Hu, X. Y., Li, L. Q., Li, X., Chen, J. B., Zhang, H. Y., Tang, Z. X., Zhang, Y., Sun, Q. X., et al. (2013) Chloroplast ultrastructure regeneration with protection of photosystem II is responsible for the functional 'stay-green' trait in wheat. *Plant Cell Environ.* 36: 683–696.

Nakano, T. and Yasuhiro, I. (2013) Molecular mechanisms controlling plant organ abscission. *Plant Biotechnol.* 30: 209–216.

Noodén, L. D. (2013) Defining senescence and death in photosynthetic tissues. In: *Advances in Photosynthesis and Respiration*, Vol. 36: *Plastid Development in Leaves during Growth and Senescence*, B. Biswal, K. Krupinska, and U. C. Biswal, eds., Springer, pp. 283–306.

Ono, Y. (2013) Evidence for contribution of autophagy to rubisco degradation during leaf senescence in *Arabidopsis thaliana*. *Plant Cell Environ.* 36: 1147–1159.

Thomas, H. (2013) Senescence, ageing and death of the whole plant. *New Phytol.* 197: 696–711.

Wang, Y., Lin, A., Loake, G. J., and Chu, C. (2013) H_2O_2-induced leaf cell death and the crosstalk of reactive nitric/oxygen species. *J. Integr. Plant Biol.* 55: 202–208.

23 Biotic Interactions

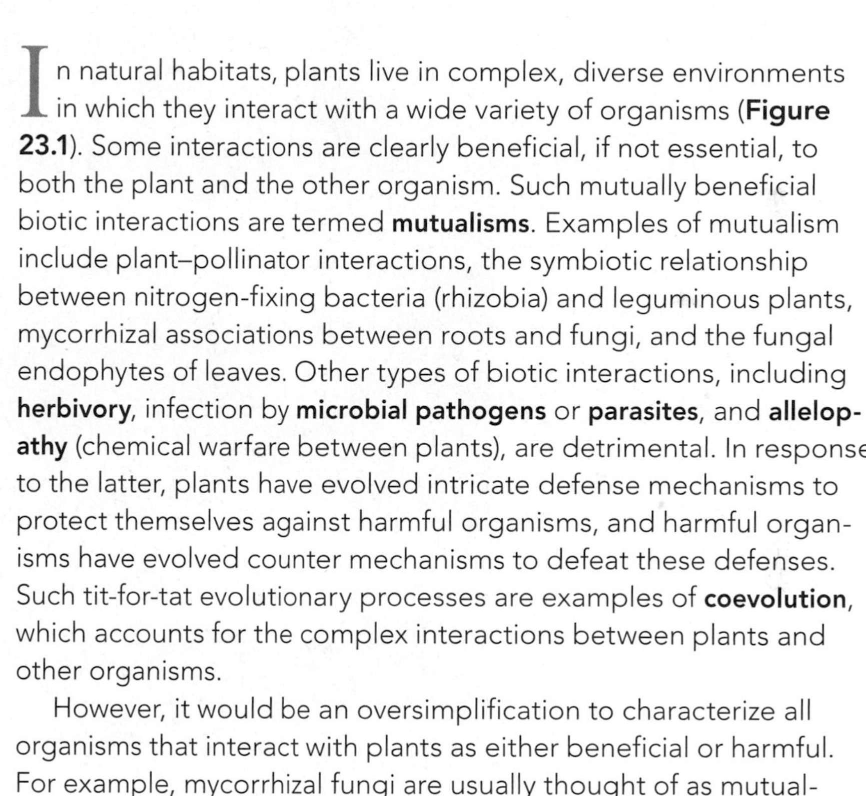

In natural habitats, plants live in complex, diverse environments in which they interact with a wide variety of organisms (**Figure 23.1**). Some interactions are clearly beneficial, if not essential, to both the plant and the other organism. Such mutually beneficial biotic interactions are termed **mutualisms**. Examples of mutualism include plant–pollinator interactions, the symbiotic relationship between nitrogen-fixing bacteria (rhizobia) and leguminous plants, mycorrhizal associations between roots and fungi, and the fungal endophytes of leaves. Other types of biotic interactions, including **herbivory**, infection by **microbial pathogens** or **parasites**, and **allelopathy** (chemical warfare between plants), are detrimental. In response to the latter, plants have evolved intricate defense mechanisms to protect themselves against harmful organisms, and harmful organisms have evolved counter mechanisms to defeat these defenses. Such tit-for-tat evolutionary processes are examples of **coevolution**, which accounts for the complex interactions between plants and other organisms.

However, it would be an oversimplification to characterize all organisms that interact with plants as either beneficial or harmful. For example, mycorrhizal fungi are usually thought of as mutualists that increase plant fitness (see Chapters 5 and 13), yet mycorrhizal plants can occupy various positions along the continuum from parasitism to mutualism. Similarly, the grazing of flowers by mammals decreases fitness in some plant species, but in others it can lead to an increase in the number of flowering stalks, thus increasing fitness. There are also organisms that benefit from their interaction with the plant without causing any harmful effects. Such neutral interactions (from the plant's perspective) are termed **commensalism**. Commensal organisms can become beneficial if they protect the plant from a second, harmful organism. For example, nonpathogenic soil rhizobacteria and fungi, which cause no harm to the plant, may nevertheless stimulate the plant's innate immune system (discussed later in the chapter) and thus protect the plant from pathogenic microorganisms.

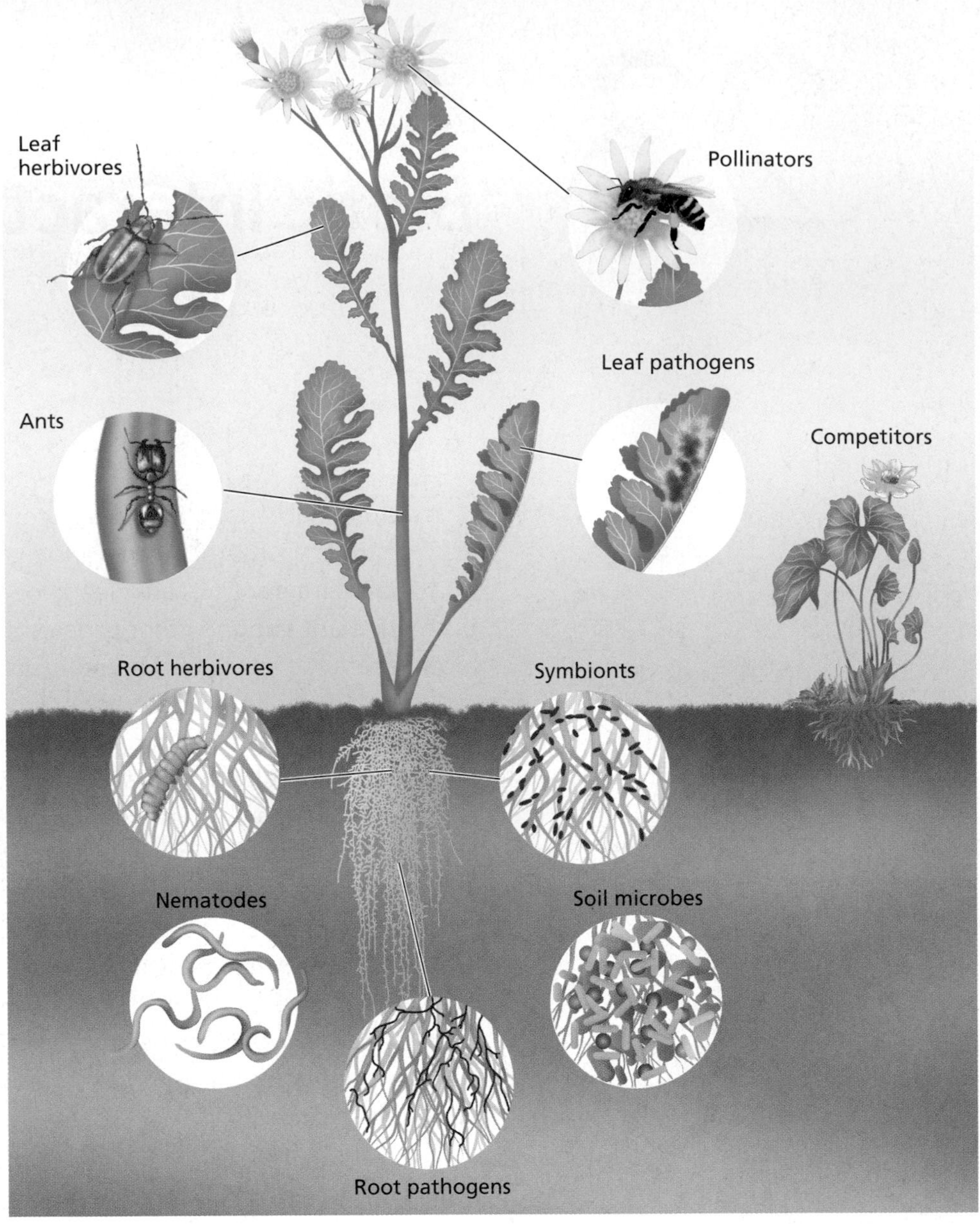

Figure 23.1 Virtually every part of the plant is adapted to coexist with the organisms in its immediate environment. (After van Dam 2009.)

The first line of defense against potentially harmful organisms is the plant surface. The cuticle (a waxy outer layer), the periderm, and other mechanical barriers help block bacterial, fungal, and insect entry (see **WEB TOPIC 23.1**). The second line of defense typically involves biochemical mechanisms that may be either constitutive or inducible. **Constitutive defenses** are always present, whereas **inducible defenses** are triggered in response to attack. Unlike constitutive defenses, inducible defenses require specific detection systems and signal transduction pathways that can sense the presence of an herbivore or pathogen and alter gene expression and metabolism accordingly.

We will begin our discussion of biotic interactions with examples of beneficial associations between plants and microorganisms. Next we will consider various types of harmful interactions between plants, herbivores, and pathogens. Mechanical barriers and toxic secondary metabolites are two main types of constitutive plant defenses against insects and other herbivores.

We will then survey the wide range of inducible defenses that plants have evolved to fend off insect herbivores, and the signaling molecules and signal transduction pathways that regulate them. We will also note the important roles that volatile organic compounds play in repelling herbivores, attracting insect predators, and acting as distress signals between different parts of plants and between neighboring plants.

Our description of inducible responses to herbivory will be followed by a discussion of inducible responses to microbial pathogens. Although plants lack animal-type immune systems, several plant-specific responses to biotic stress can confer both local and systemic resistance to pathogens. Finally, we will discuss the mechanisms of two other types of plant pathogens, nematodes and parasitic plants, and the ecological role that toxic root exudates play during plant–plant competition.

Beneficial Interactions between Plants and Microorganisms

It is likely that symbiotic associations between algae and fungi predate the emergence of the first land plants about 450–500 million years ago (MYA). For example, the first lichens, which are obligate associations between a fungus and either a green alga or a cyanobacterium, appear in the fossil record about 400 MYA—around the time that the first mycorrhizal associations with land plants made their appearance. This suggests that the invasion of land by green plants may have been aided by symbiotic associations with fungi. In the natural environment, land plants are colonized by a wide variety of beneficial microorganisms: endophytic and mycorrhizal fungi, bacteria in the form of biofilms on the surfaces of roots and leaves, endophytic bacteria, and nitrogen-fixing bacteria contained in the root or stem nodules.

In this section we will focus on the signaling mechanisms involved in beneficial interactions of plants with three types of microorganisms: nitrogen-fixing bacteria, mycorrhizal fungi, and rhizobacteria. In Chapters 5 and 13 we discussed these biotic interactions from an anatomical and physiological perspective. Here we will examine the molecular signaling mechanisms that control the formation of the associations.

The arbuscular mycorrhizal symbiosis is extremely ancient and, as noted above, appeared more than 400 MYA. In contrast, rhizobium–legume symbiosis is thought to have appeared about 60 MYA. The fungal partner in these arbuscular mycorrhizal associations belongs to the ancient phylum Glomeromycota, which has lost the ability to complete its life cycle outside the plant. Because of the difficulty of performing genetic analyses on an obligate symbiont, progress on arbuscular mycorrhizal signaling has proceeded slowly. As a result, the basic outlines of the symbiotic signaling pathway were first worked out for the rhizobium–legume association, which is thought to have evolved from the arbuscular mycorrhizal pathway.

Nod factors are recognized by the Nod factor receptor (NFR) in legumes

As described in Chapter 13, symbiotic nitrogen-fixing rhizobia release **nodulation** (**Nod**) **factors** as signaling agents as they near the surface of a legume root. The interaction between specific nod factors and their corresponding receptors is the basis for host–symbiont specificity. Nod factors are **lipochitin oligosaccharides** that bind to a specific class of **receptor-like kinases** (**RLKs**) that contain *N*-acetylglucosamine–binding **lysin motifs** (**LysM**) in the extracellular domain. (The term *lysin* refers to an enzyme that hydrolyzes bacterial peptidoglycan walls.) The first LysM receptor kinases were identified in the Japanese legume *Lotus japonicus*, which has two such receptors: Nod factor receptor 1 and 5 (NFR1 and NFR5). Both receptors have three extracellular LysM domains that recognize Nod factors in a species-specific manner (**Figure 23.2**). NFR1 and NFR5 also contain intracellular domains with similarity to plant serine/threonine protein kinases, but only NFR1 has kinase activity. Nevertheless, the two proteins are thought to bind Nod factors in the heterodimeric state (see Figure 23.2).

Upon binding to Nod factors, the NFR heterodimer initiates two separate processes. The first involves a signaling pathway that facilitates the infection process itself. The second involves the activation of a set of genes that regulate the formation of root nodules. A second type of receptor, called the **symbiosis receptor-like kinase** (**SYMRK**), is thought to participate in both processes. SYMRK possesses an extracellular domain that includes a leucine rich repeat (LRR) region and an intracellular protein kinase domain. Because the SYMRK receptor is required not only for the interaction of legumes with rhizobia, but also for the interactions of the actinobacterium *Frankia* with the roots of actinorhizal plants, such as the tree *Casuarina glauca* and the cucurbit *Datisca glomerata*, it is part of a **common symbiotic pathway** that is activated in both rhizobial and actinorhizal associations as well as in arbuscular mycorrhizal symbioses. Upon binding of Nod factors, SYMRK is thought to activate subsequent steps shared by these associations, including fluctuations in the calcium concentration, called calcium spiking, inside and around the nucleus of the infected epidermal cell, which causes the activation of the **core symbiotic genes** (**Figure 23.3**). The final steps leading to nodulation involve cytokinin signaling (see Chapter 15).

Arbuscular mycorrhizal associations and nitrogen-fixing symbioses involve related signaling pathways

In legumes, SYMRK and several other core symbiotic genes are required both for legume nodulation and for arbuscular mycorrhiza formation. This suggests that the interaction between legumes and nitrogen-fixing rhizobia evolved from the more ancient interaction between plants and mycorrhizal fungi. As noted above, the key symbiotic signals produced by rhizobia are lipochitin oligosaccharides called Nod factors. Similarly, the arbuscular mycorrhizal fungus *Glomus intraradices* releases lipochitin oligosaccharides called Myc factors that stimulate the formation of mycorrhizas in a wide variety of plants.

Nitrogen-fixing symbioses and arbuscular mycorrhizas may also involve related receptors. *Parasponia andersonii* is a nonlegume tropical tree that forms nitrogen-fixing symbioses with rhizobia. Even though this plant is only distantly related to the legumes, it has a LysM receptor that is required for both mycorrhiza formation *and* rhizobium-

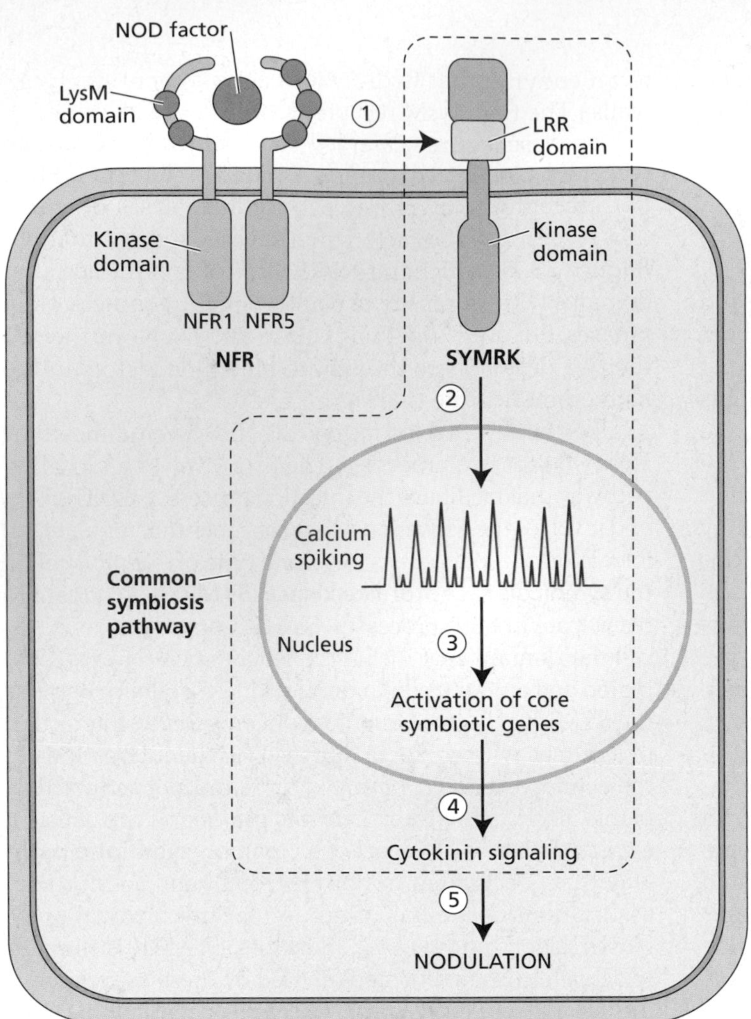

Figure 23.2 Model for Nod factor signaling in the root epidermis. (After Gough and Cullimore 2011 and Markmann and Parniske 2009.)

1. The binding of Nod factors to the Nod factor receptors (NFRs), which contain extracellular LysM motifs, initiates an interaction with the conserved SYMRK receptor-like kinases containing a leucine-rich repeat (LRR) domain.

2. The interaction between NFR and SYMRK initiates calcium spiking in the nucleus, presumably via a second messenger molecule.

3. Core symbiotic genes are activated.

4. Cytokinin signaling is initiated.

5. Cytokinin signaling brings about the morphological changes associated with nodulation.

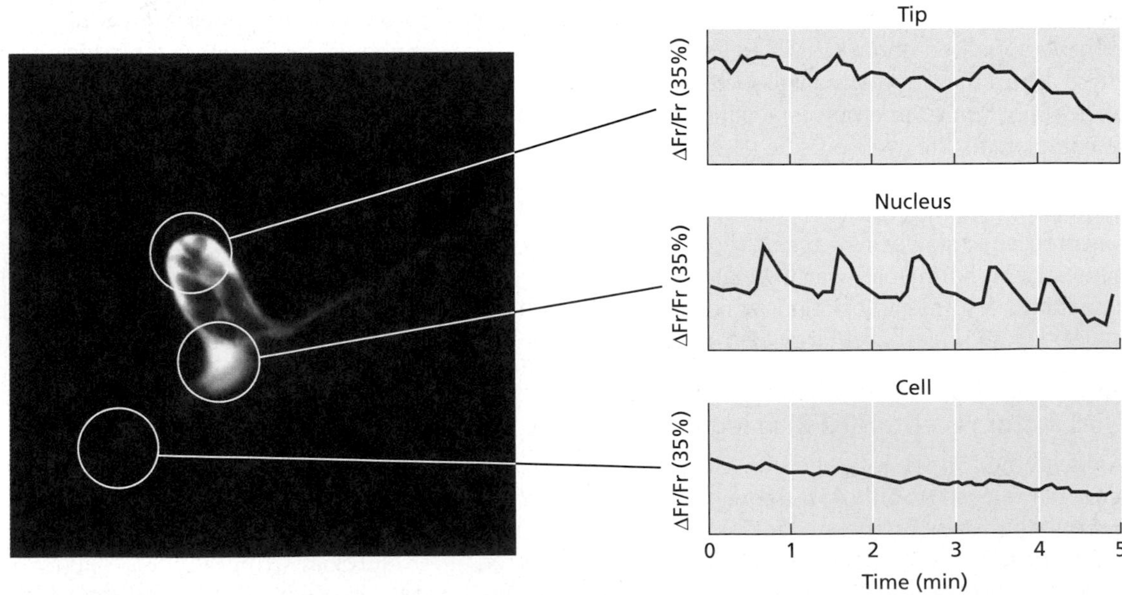

Figure 23.3 Calcium oscillations in an emerging root hair of pea (*Pisum sativum*) responding to the addition of Nod factor. The cell previously had been injected with a calcium-sensitive fluorescent dye. Prominent calcium spiking occurs around the nucleus, with weaker fluctuations apparent at the tip. No calcium fluorescence occurs in the main part of the cell. (From Oldroyd and Downie 2004, courtesy of S. Walker, John Innes Centre.)

induced root nodulation. The Nod factor receptors are also closely related to two receptors identified in the non-nodulating angiosperms Arabidopsis and rice (*Oryza sativa*). These receptors are required for the defense-related perception of chitin oligomers, compounds that are a chemical signature of fungi and are structurally related to Nod factors and Myc factors. This suggests that during evolution a LysM plant receptor involved in defense signaling has been recruited to activate genes involved in symbiotic associations.

Rhizobacteria can increase nutrient availability, stimulate root branching, and protect against pathogens

Plant roots provide a nutrient-rich habitat for the proliferation of soil bacteria that thrive on root exudates and lysates, which can represent as much as 40% of the total carbon fixed during photosynthesis. Population densities of bacteria in the rhizosphere can be up to 100-fold higher than in bulk soil, and up to 15% of the root surface may be covered by microcolonies of a variety of bacterial strains. While these bacteria use the nutrients that are released from the host plant, they also secrete metabolites into the rhizosphere.

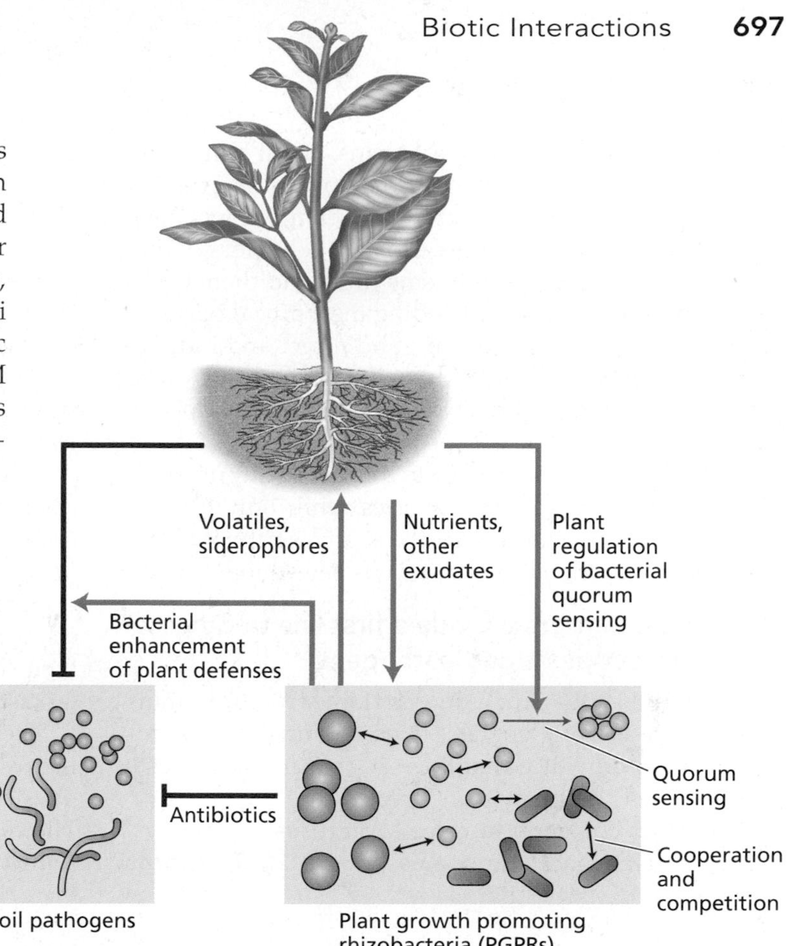

Figure 23.4 Diagram of the interactions between plants and plant growth promoting rhizobacteria, such as *Pseudomonas aeruginosa*, which is thought to release antibiotics or siderophores into the soil that alleviate plant abiotic or biotic stress. The plant exerts control over the bacterial population by regulating bacterial quorum sensing signaling pathways through the release of exudates by the roots. (After Goh et al. 2013.)

A loosely defined group of **plant growth promoting rhizobacteria** (**PGPR**) provides several beneficial services to growing plants (**Figure 23.4**). For example, volatiles produced by the bacterium *Bacillus subtilis* enhance proton release by Arabidopsis roots in iron-deficient growth media, thereby facilitating increased iron uptake. The resulting increase in iron content of the plants treated with *B. subtilis* volatiles is correlated with higher chlorophyll content, higher photosynthetic efficiency, and increased size. *B. subtilis* volatiles also alter root architecture by changing root length and lateral root density.

PGPR can also control the buildup of harmful soil organisms, as in the case of the suppression of the pathogenic fungus *Gaeumannomyces graminis* by a *Pseudomonas* species that synthesizes the antifungal compound 2,4-diacetylphloroglucinol. PGPR microbes can also provide cross-protection against pathogenic organisms by activating the *induced systemic resistance* pathway, which we will discuss later in the chapter. In addition, several studies have suggested that *Pseudomonas aeruginosa* can alleviate the symptoms of biotic and abiotic stress by the release of antibiotics or iron-scavening **siderophores** (see Chapter 13). The amounts of the compounds released by *P. aeruginosa* are controlled by **quorum sensing** signaling pathways, which are activated when the population density of the bacterium reaches a certain level. Plants can influence the amount of antibiotics or siderophores released by rhizobacteria through the production of root exudates that regulate bacterial quorum sensing pathways.

Harmful Interactions between Plants, Pathogens, and Herbivores

Plant pathology is the study of plant diseases. Microorganisms that cause infectious diseases in plants include fungi, oomycetes, bacteria, and viruses. The majority of fungal pathogens belong to the Ascomycetes, which bear meiospores inside a saclike *ascus*, and Basidiomycetes, which produce meiospores outside club-shaped cells called *basidia*. Oomycetes are funguslike organisms that include some of the most destructive plant pathogens in history, including the genus *Phytophthora*, cause of the disastrous potato late blight of the Great Irish Famine (1845–1849). Plant pathogenic bacteria also cause many serious diseases of plants, but they are fewer and less devastating than those caused by fungi and viruses.

In addition to microbial pathogens, about half of the nearly 1 million insect species feed on plants. In over 350 million years of plant–insect coevolution, insects have developed diverse feeding styles and behaviors, while plants have evolved mechanisms to defend themselves against insect herbivory, including mechanical barriers, constitutive chemical defenses, and direct and indirect inducible defenses. These defense mechanisms have apparently been quite effective, since most plant species are resistant to most insect species. Indeed, about 90% of insect herbivores are restricted to a single family of plants or a few closely related plant species, while only 10% are generalists. This suggests that the vast majority of plant–herbivore interactions have involved coevolution.

Mechanical barriers provide a first line of defense against insect pests and pathogens

Mechanical barriers, including surface structures, mineral crystals, and thigmonastic (touch-induced) leaf movements, often provide a first line of defense against pests and pathogens for many plant species.

The most common surface structures are thorns, spines, prickles, and trichomes (**Figure 23.5**). **Thorns** are modified branches, as in citrus and acacia; **spines** are modified leaves, as in cacti; and **prickles** are derived mainly from the epidermis and thus can be easily snapped off the stem, as in roses. These structures possess sharp, pointed tips that physically protect plants from larger herbivores, such as mammals, although they are less effective against smaller, insect herbivores, which can easily penetrate such defenses and reach the edible parts of the shoot. **Trichomes**, or hairs, provide a much more effective defense against insect pests, based on their physical and chemical deterrence mechanisms. Trichomes occur in a variety of forms, either as simple hairs or as glandular trichomes. Glandular trichomes store species-specific secondary metabolites (discussed in the next section) such as terpenoids and phenolics in a pocket formed between the cell wall and the cuticle. These pockets burst and release their contents upon contact, and the strong smell and bitter taste of these compounds repel insect herbivores.

The leaves of the stinging nettle (*Urtica dioica*) possess highly specialized "stinging hairs" that form an effective physical and chemical barrier against larger herbivores. The cell walls of these hollow, needlelike trichomes are reinforced with glass (silicates) and filled with a nasty

Figure 23.5 Examples of mechanical barriers developed by plants. (A) Thorns on a lemon tree (*Citrus* sp.) are modified branches, as can be seen by their position in the axil of a leaf. (B) Spines, which are characteristic of cacti (*Opuntia* spp.) in the New World, are modified leaves. (C) Prickles can be found on the stem and petioles of roses (*Rosa* spp.) and are formed by the epidermis. (D) Trichomes on shoot and leaves of tomato (*Solanum lycopersicum*) are also derived from epidermal cells. (Photos © J. Engelberth.)

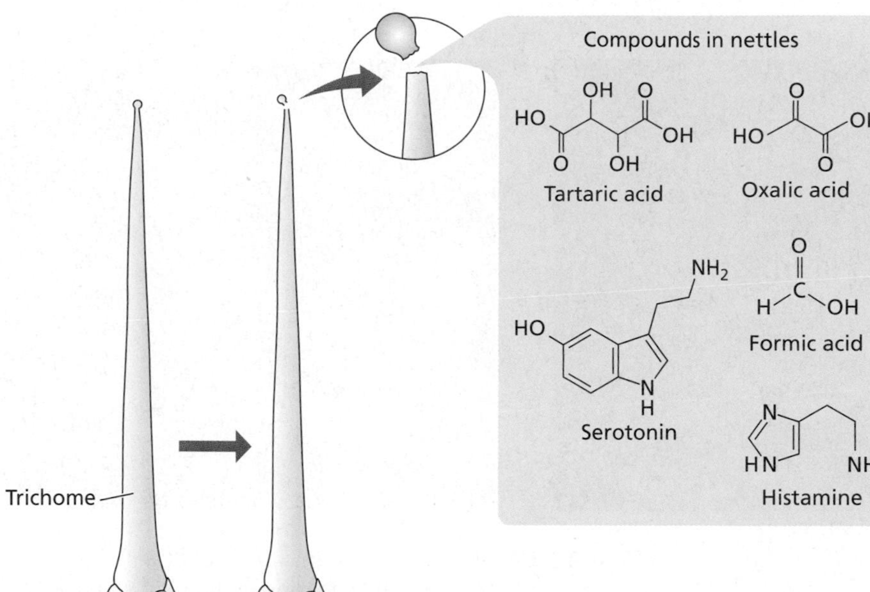

Figure 23.6 Trichomes of nettles (*Urtica dioica*) have a multicellular base with a single stinging cell protruding. The cell wall of this single cell is reinforced by silicates and breaks easily upon contact, releasing a "cocktail" of secondary metabolites that can cause severe irritation of the skin of animals.

"cocktail" of histamine, oxalic acid, tartaric acid, formic acid, and serotonin (**Figure 23.6**), which can cause severe irritation and inflammation. Prior to contact, the tip of the stinging hair is covered by a tiny glass bulb that readily snaps off when touched by an herbivore (or an unlucky human who happens to brush against it), leaving an extremely sharp point at the end. The pressure of contact pushes the needlelike trichome down onto the spongy tissue at the base, which acts like the plunger of a syringe to inject the cocktail into the skin.

Besides serving as barriers to insect herbivory, trichomes—when bent or damaged—may also act as herbivore sensors by sending electrical or chemical signals to surrounding cells. Such signals can trigger the formation of inducible defense compounds in the leaf mesophyll.

A different type of mechanical obstacle to herbivory is provided by mineral crystals that are present in many plant species. For example, silica crystals, called **phytoliths**, form in the epidermal cell walls, and sometimes in the vacuoles, of the Poaceae. Phytoliths add toughness to the cell walls and make the leaves of grasses difficult for insect herbivores to chew. The cell walls of the horsetail *Equisetum hyemale* contain so much silica that Native Americans and Mexicans used the stems to scour cooking pots.

Calcium oxalate crystals are present in the vacuoles of many species, and may be distributed evenly throughout the leaf or restricted to specialized cells called **idioblasts** (see Figure 1.35C). Some calcium oxalate crystals form bunches of needlelike structures called **raphides** (**Figure 23.7**), which can be harmful to larger herbivores. More than 200 families of plants contain these crystals, including species in the genera *Vitis*, *Agave*, and *Medicago*. Raphides have extremely sharp tips that can penetrate the soft tissue in the throat and esophagus of an herbivore. The raphide-rich *Dieffenbachia*, a popular tropical houseplant, is called "dumb cane" because chewing the leaves reputedly renders the victim speechless due to inflammation. Besides causing mechanical damage, raphides may also allow other toxic compounds produced by the plant to penetrate through the wounds they produce. Even prismatic calcium oxalate crystals have abrasive effects on the mouthparts of insect herbivores, especially the mandibles,

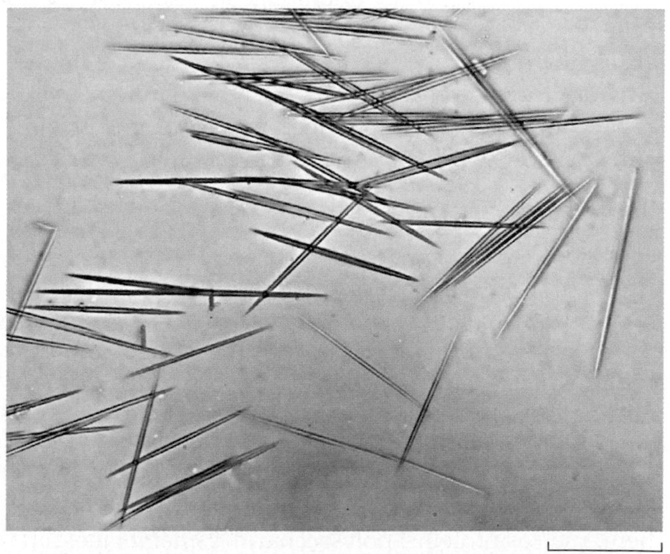

Figure 23.7 Calcium oxalate crystals (raphides) from the leaf of an agave (*Agave weberi*). These raphides are tightly packed in specialized cells called idioblasts and are released upon damage. Note the size and pointed ends of these structures.

(A)

(B)

Figure 23.8 Leaves of the sensitive plant (*Mimosa* spp.) respond rapidly to touch by folding in their individual leaflets within seconds. This rapid movement may deter potential insect herbivores. (A) Untouched (control) leaves. (B) Leaves 5 sec after touch. (Photos © J. Engelberth.)

thus serving as a mechanical deterrent for insects, mollusks, and other herbivores.

A very different means of avoiding herbivory is employed by the sensitive plant (*Mimosa* spp.). *Mimosa* leaves are composite leaves consisting of many individual leaflets that are connected to the midrib by a joint-like structure called the pulvinus. This pulvinus acts as a turgor-driven hinge, which allows each pair of leaflets to fold in response to various stimuli, including touch, damage, heat, diurnal cycles (in what is called *nyctinasty*, or sleep movements), and in response to water stress. If an insect herbivore attempts to nibble a *Mimosa* leaflet, the damaged leaflet immediately folds, and the response soon spreads to the other undamaged leaflets of the leaf. If the stress signal is strong enough, the entire leaf collapses downward due to the action of another pulvinus located at the base of the petiole. Such rapid movements of leaflets and leaves may deter feeding insects and grazing herbivores by startling them (**Figure 23.8**).

Plant secondary metabolites can deter insect herbivores

Chemical defense mechanisms comprise a second line of defense against plant pests and pathogens. Plants produce a wide array of chemicals that can be categorized into primary metabolites and secondary metabolites. **Primary metabolites** are those compounds that all plants produce and that are directly involved in growth and development. This includes sugars, amino acids, fatty acids and lipids, and nucleotides, as well as larger units that are made of them, such as proteins, polysaccharides, membranes, and DNA and RNA. In contrast, **secondary metabolites** (or *specialized metabolites*) are highly species-specific and usually belong to one of the three major classes of chemicals: terpenoids, phenolics, or alkaloids (**Figure 23.9**; also see WEB ESSAY 23.1). An important exception to this rule is the group of five plant hormones—cytokinins, gibberellins, brassinosteroids, abscisic acid, and strigolactones—which are all derivatives of one of these pathways, but which are considered primary metabolites because all plants require them for their growth and development and thus possess the biochemical machinery to synthesize them. Since the precursors of the hormones auxin and ethylene are amino acids, these hormones are synthesized via primary metabolism.

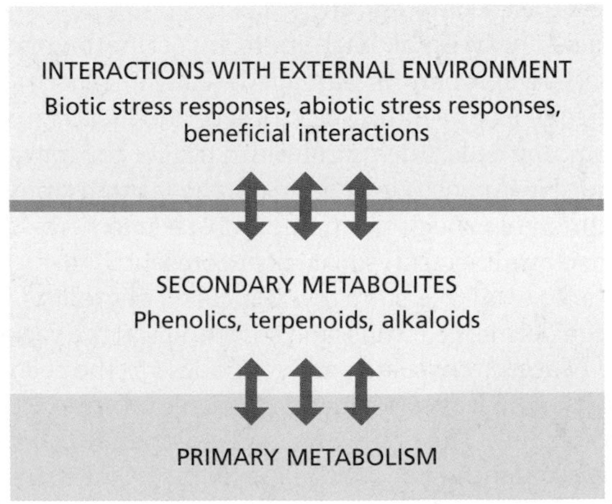

Figure 23.9 Secondary metabolites are located at the interface between primary metabolism and the interaction of organisms with their environment. As such, they play important roles in the plant defense response against pests and pathogens; the regulation of beneficial interactions, including the attraction of pollinators; and as modulators in response to abiotic stresses.

(A) *Cicuta* sp.

Cicutoxin

(B) *Digitalis* sp.

Digitoxin

Figure 23.10 Constitutive chemical defenses are effective against many different herbivores, including insects and mammals. Hemlock (*Cicuta* sp.) produces cicutoxin, a polyacetylene that prolongs the repolarization of neuronal action potentials. The active principle in foxglove (*Digitalis* sp.) is digitoxin, a cardiac glycoside that inhibits ATPase activity and can increase myocardial contraction. (Photos © J. Engelberth.)

Plants store constitutive toxic compounds in specialized structures

Plants can synthesize a broad range of secondary metabolites that have negative effects on the growth and development of other organisms and that can therefore be regarded as toxic. Classic examples of plants that are toxic to humans are hemlock (*Cicuta* spp.) and foxglove (*Digitalis* spp.) (**Figure 23.10**). The metabolites that cause symptoms in humans are well known and demonstrate the potential of secondary metabolites as defensive agents against mammalian herbivores. In some cases these compounds have proven useful for medicinal purposes. For example, the polyacetylene cicutoxin from hemlock prolongs the repolarization phase of neuronal action potentials, presumably through blockage of voltage-dependent K^+ channels. The active principle in foxglove, digitoxin, is a **cardenolide**, one of two groups of steroidal cardiac glycosides produced by plants. **Cardiac glycosides** are drugs used to treat congestive heart failure and cardiac arrhythmia. Digitoxin inhibits the Na^+/K^+ ATPase pump in the membranes of heart cells, leading to increased myocardial contraction.

Constitutively produced secondary metabolites that accumulate in cells potentially could have toxic effects on the plant itself. To avoid toxicity, these compounds must be safely stored in leak-proof cellular compartments, and they must also be relatively isolated from susceptible tissues in the event of leakage due to cellular damage. Plants therefore tend to accumulate toxic secondary metabolites in storage organelles such as vacuoles, or in specialized anatomical structures such as *resin ducts, laticifers* (latex-producing cells), or glandular trichomes. Following an attack by herbivores or pathogens, the toxins are released and become active at the site of damage, without adversely affecting vital growing areas. **Conifer resin ducts**, which are found in the cortex and phloem, contain a mixture of diverse terpenoids, including bicyclic monoterpenes such as α-pinene and β-pinene, monocyclic terpenes like limonene and terpinolene, and tricyclic sesquiterpenes, including longifolene, caryophyllene, and δ-cadinene, as well as resin acids, which are released immediately upon damage by herbivores (**Figure 23.11**). Once released, they can either be toxic to an attacking insect herbivore or act as an adhesive that may glue together the mouthparts of an herbivore. In extreme cases the resin may even engulf the whole insect or pathogen, effectively killing it.

Most of the resin ducts in coniferous plants are considered constitutive defenses, but they can also be induced upon herbivore damage. The formation of these adventitious resin ducts, sometimes referred to as *traumatic resin ducts*, as well as the biosynthesis of their resin, are regulated by the hormone methyl jasmonate, a derivative of jasmonic acid (discussed later in the chapter).

Laticifers are composed of cells that produce a milky fluid of emulsified components that coagulate upon exposure to air. This liquid is also often referred to as **latex**. Compared with resins, latex is usually much more complex and may also contain proteins and sugars besides toxic or repelling secondary metabolites. Laticifers may consist of

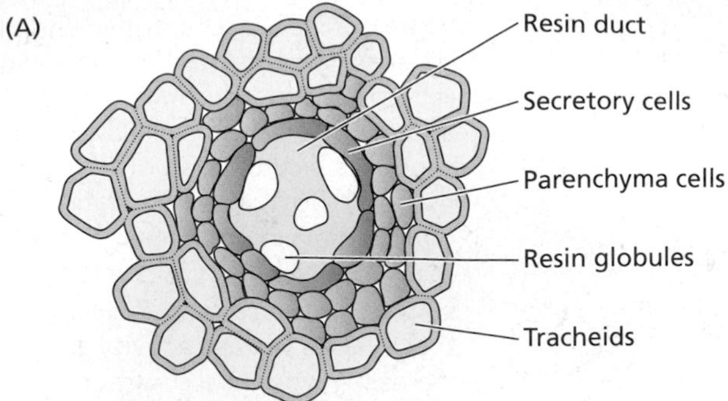

Figure 23.11 (A) Resin duct in the wood of a pine (*Araucaria* sp.). Note that the resin duct is surrounded by secretory cells that release resin components into the duct system. (B) Upon wounding, resin is released at the damage site. There it seals off the damage and serves as a repellent against further herbivory. (C) Common components of resin, in particular terpenoids. (Photo © J. Engelberth.)

a series of fused cells (articulated laticifers) or one long syncytial cell (non-articulated laticifers) (**Figure 23.12**). Most notable among the latex-producing plants is the rubber plant (*Hevea brasiliensis*), which has been grown commercially as a source of natural rubber. Upon wounding, this plant releases huge amounts of latex, which is collected and later converted into rubber. This rubber consists of a polymer of isoprene (*cis*-1,4-polyisoprene) and can have a molecular weight of up to 1 million Da. Under natural conditions, the rubber released by wounded trees

Figure 23.12 Laticifers are made of individual cells and can occur either as articulated systems (single cells connected by a small tube) or non-articulated systems (one large syncytial cell). Latex in the laticifers is released upon damage and often contains cardiac glycosides which fend off herbivores. While mulberry (*Morus* sp.) produces a milky latex in its articulated laticifers, oleander (*Nerium oleander*) releases a clear latex from non-articulated laticifers. (Photos © J. Engelberth.)

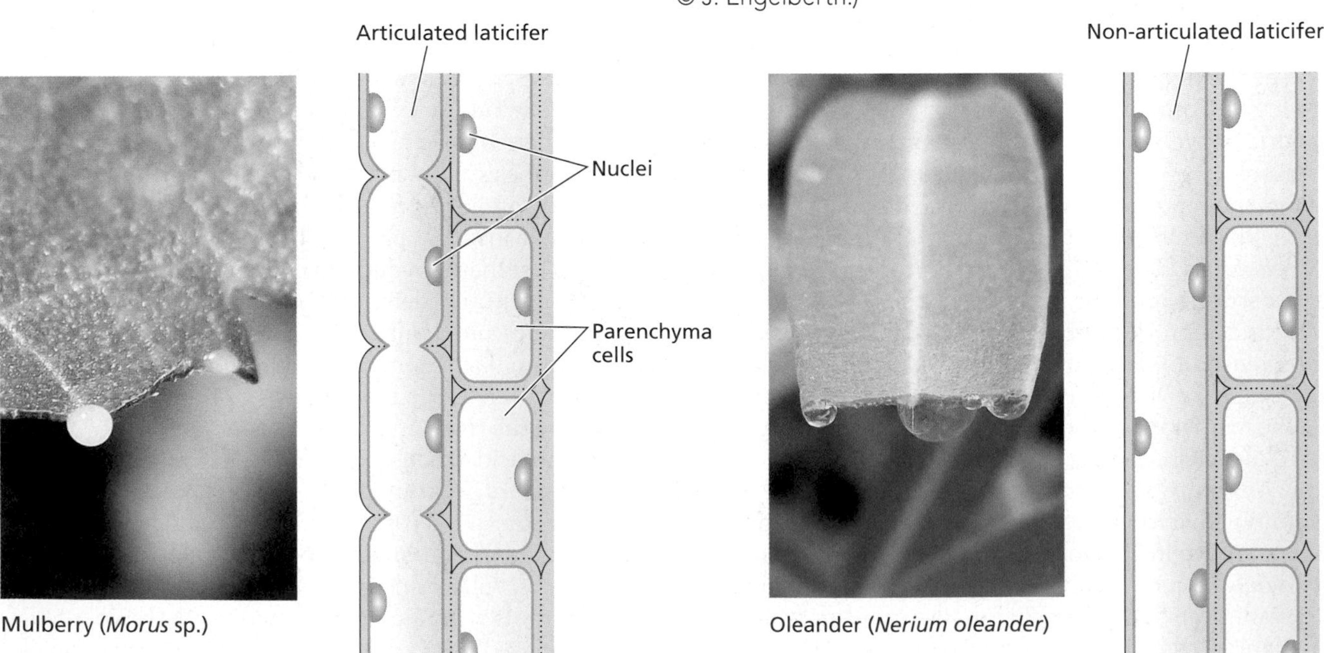

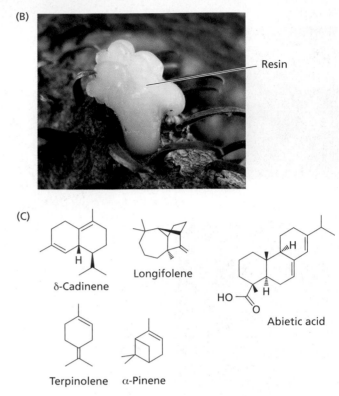

defends the plant against herbivores and pathogens, either by repelling or engulfing them.

Another commercially important plant that produces latex is the opium poppy (*Papaver somniferum*). The latex of this plant contains a high concentration of opiates, in particular morphine and codeine. When consumed, these compounds bind to opiate receptors in the nervous systems of herbivores and exert analgesic effects.

The milkweed (*Asclepias curassavica*) and related genera, such as oleander (*Nerium oleander*), produce latex that contains significant amounts of cardenolides, which are present at high concentrations in laticifers. The activity of these poisonous steroids is similar to that of digitoxin (see above) and at high concentrations may lead to heart arrest. Cardenolides also activate nerve centers in the vertebrate brain that induce vomiting. Generalist insect herbivores respond to these compounds either by being repelled or by suffering spasms that lead to death. In contrast, the specialist caterpillars of the monarch butterfly (*Danaus plexippus*) are insensitive to the toxins. They feed on milkweed leaves and retain the cardenolides. As a result, most insectivorous birds quickly learn to avoid eating monarch caterpillars and adult monarch butterflies. The bright and distinctive coloration of the caterpillars and butterflies serves to warn birds off. The large milkweed bug (*Oncopeltus fasciatus*) and the milkweed aphid (*Aphis nerii*) can also incorporate cardenolides into their bodies and become toxic themselves (**Figure 23.13**). Although all of these insects feed preferentially on milkweeds, the milkweed aphid and the large milkweed bug can also feed on oleander, which produces oleandrin as its major cardenolide.

In a further embellishment to the milkweed saga, the parasitic fly *Zenillia adamsoni* can obtain its cardenolide secondhand from the monarch caterpillar. When the female fly is ready to lay eggs, she seeks out a monarch caterpillar and deposits her eggs on its surface. Upon hatching, the fly larvae bore into the caterpillar and consume it from within. Besides using the caterpillar for nourishment, the fly larvae are able to store the caterpillar's toxic cardenolide and retain it throughout the adult stage.

Plants often store defensive chemicals as nontoxic water-soluble sugar conjugates in the vacuole

A common mechanism for storing toxic secondary metabolites is to conjugate them with a sugar, which also makes them more water soluble. As described above, most cardenolides and other related and toxic steroids are highly abundant as glycosides in the latex, and also in other compartments of the plant cell such as the vacuole. To become active, the glycosidic linkages often need to be hydrolyzed. Uncontrolled activation is prevented by the spatial separation of the activating hydrolases and their respective toxic substrates.

A good example of this spatial separation is found in the order Brassicales. Members of the Brassicales produce glucosinolates—sulfur-containing organic compounds derived from glucose and an amino acid—as their major defensive secondary metabolites (see **WEB APPENDIX 4**). The hydrolyzing enzyme, myrosinase (a thioglucosidase), is stored in different cells than its substrates. While myrosinase-containing cells are mostly glucosinolate-free, the so-called sulfur-rich cells (or S-cells) contain high concentrations of glucosinolates. When the tissue is damaged,

(A) Monarch butterfly (*Danaus plexippus*)

(B) Milkweed bug (*Oncopeltus fasciatus*)

(C) Milkweed aphid (*Aphis nerii*)

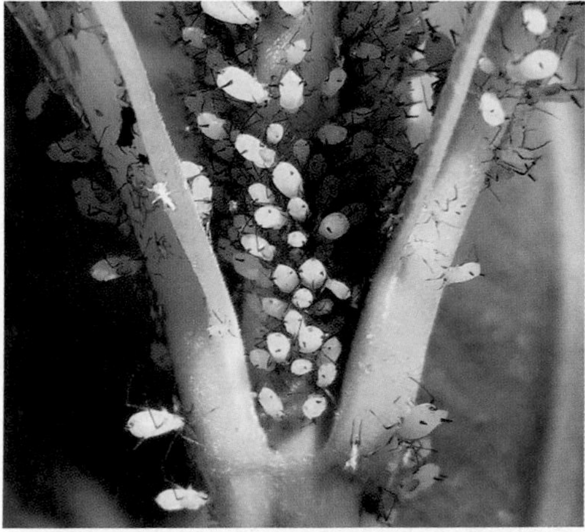

Figure 23.13 While most herbivores are very susceptible to the toxic metabolites in the latex of milkweed and oleander plants, some insect herbivores incorporate these compounds into their bodies and show this to potential predators by presenting bright colors. Shown here are three specialist insect herbivores that feed on these latex-producing plants: the caterpillar of the monarch butterfly (A), the milkweed bug (B), and the milkweed aphid (C). Of these, the milkweed bug and the milkweed aphid will use oleander as a food source if milkweed plants are not available. (Photos © J. Engelberth.)

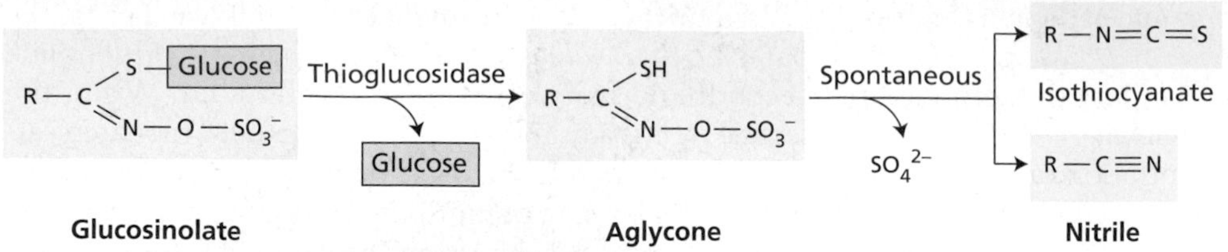

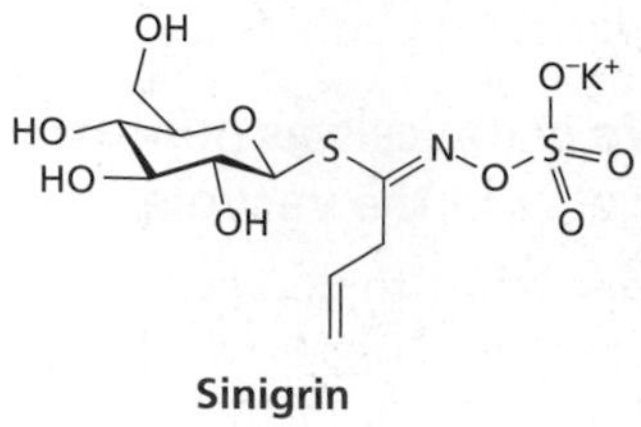

Figure 23.14 Hydrolysis of glucosinolates into mustard-smelling volatiles. R represents various alkyl or aryl substituents. For example, if R is $CH_2{=}CH{-}CH_2{-}$, the compound is sinigrin, a major glucosinolate of black mustard seeds and horseradish roots.

the released myrosinase and glucosinolates mix, resulting in the irreversible production of an unstable aglycone, which then rearranges into a variety of biologically active compounds, mostly nitriles and isothiocyanates (**Figure 23.14**). These "mustard oil bombs," in particular the isothiocyanates, are very effective against most generalist insect herbivores. The flavors of mustard, wasabi, radish, brussels sprouts, and other related vegetables are due to isothiocyanates.

Members of the Poaceae family, which includes all of the cereal crops, use benzoxazinoids (alkaloids derived from tryptophan) as constitutively produced defensive secondary metabolites. Benzoxazinoids, especially 2,4-dihydroxy-1,4-benzoxazin-3-one (DIBOA) and its derivative 2,4-dihydroxy-7-methoxy-1,4-benzoxazin-3-one (DIMBOA), are usually stored in the vacuole as glycosides coupled to D-glucose (Glc; **Figure 23.15**). Upon damage, the inactive glycosides are hydrolyzed, generating aglycones, which are toxic not only to insect herbivores but also against pathogens. Since the reaction is reversible, DIBOA and DIMBOA can be detoxified through the reformation of the glycosides, a reaction that is catalyzed by glycosyltransferases. Benzoxazinoids released into the soil by the roots of maize (corn; *Zea mays*) and other members of the Poaceae family are also toxic to nearby plants, but can be detoxified by these plants by hydroxylation and N-glycosylation reactions.

Cyanogenic glycosides represent a particularly toxic class of N-containing secondary metabolites. Upon tissue damage, these glycosides break down and release hydrogen cyanide (HCN). Cyanide inhibits cytochrome *c* oxidase in the mitochondria, which blocks the electron transport chain. As a consequence, electron transport and ATP synthesis come to a halt and the cell ultimately dies. Several plant species of economic and nutritional importance, including sorghum (*Sorghum bicolor*) and cassava (*Manihot esculenta*), produce different types of cyanogenic glycosides. The major cyanogenic glycoside in sorghum is dhurrin, which is derived from tyrosine. Dhurrin is stored as a glycoside, but when consumed by herbivores, the glycoside is quickly hydrolyzed to sugar and an aglycone, which is very unstable and releases HCN (**Figure 23.16**).

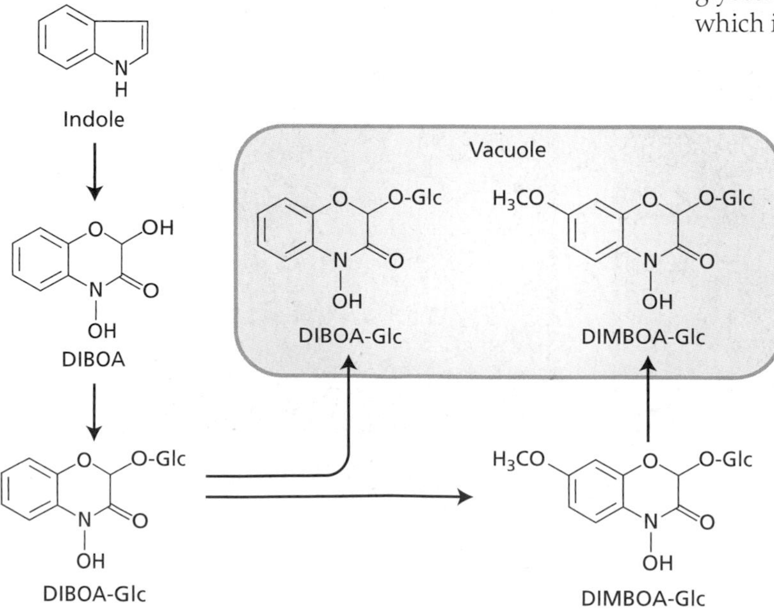

Figure 23.15 In members of the Poaceae family, benzoxazinoids, alkaloids derived from the tryptophan pathway, are major defense secondary metabolites. The compound 2,4-dihydroxy-1,4-benzoxazin-3-one (DIBOA) and its derivative 2,4-dihydroxy-7-methoxy-1,4-benzoxazin-3-one (DIMBOA) are stored in the vacuole as glycosides (coupled to D-glucose, Glc). After damage, the glycosides are hydrolyzed and release the toxic aglycones.

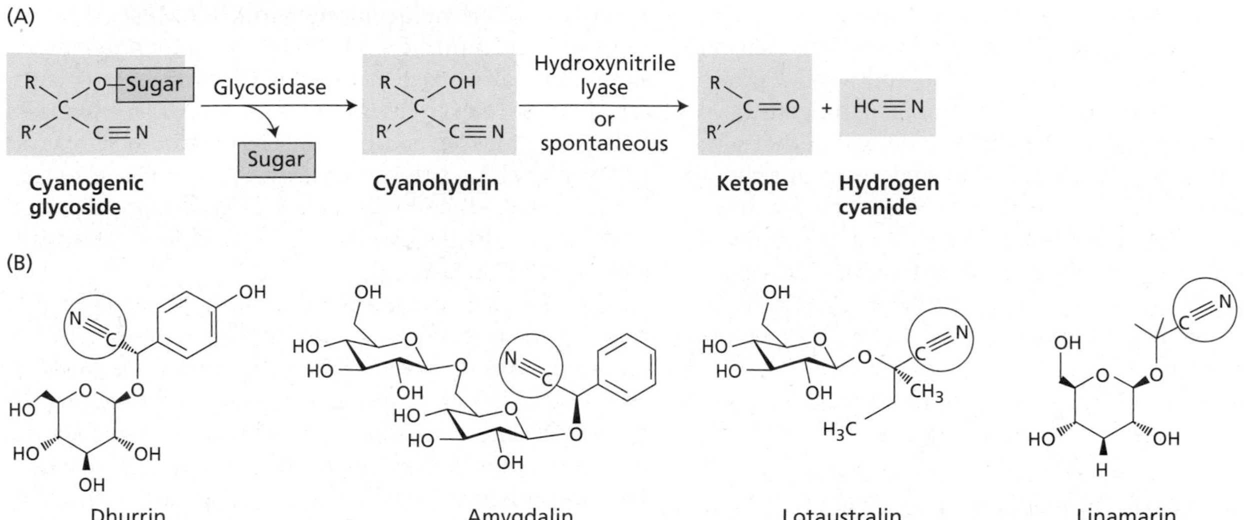

Figure 23.16 (A) Enzyme-catalyzed hydrolysis of cyanogenic glycosides to release hydrogen cyanide. (B) R and R′ in part (A) represent various alkyl or aryl substituents. For example, if R is phenyl, R′ is hydrogen, and the sugar is the disaccharide β-gentiobiose, then the compound is amygdalin (the common cyanogenic glycoside found in the seeds of almonds, apricots, cherries, and peaches). Other compounds that release toxic cyanide are dhurrin from sorghum as well as linamarin and lotaustralin from cassava. The cyanide group is marked by a circle.

Cassava accumulates linamarin and lotaustralin as its major cyanogenic glycosides (see Figure 23.16). Cassava roots are a major source of protein in tropical regions, but they must be carefully prepared to avoid cyanide toxicity. The continuous consumption of improperly processed cassava, even at low endogenous concentrations of cyanogenic glycosides, may lead to paralysis as well as liver and kidney damage.

There are many more plants that produce constitutive secondary metabolites and store them in specific cells or compartments, where they can be released upon damage by herbivores and pathogens. Despite the presence of defensive chemicals, humans often prize these plants, or parts of them, for their medicinal properties or culinary flavors. No spice rack would be complete without the dried leaves of basil, sage, thyme, rosemary, and oregano, even though the sole reason for the high contents of secondary metabolites in these plants is to protect them from damage by pests and pathogens.

Constitutive levels of secondary compounds are higher in young developing leaves than in older tissues

The indeterminate, modular nature of plant vegetative growth means that there will always be an age gradient from the mature leaves to the apical bud. Most plant defense mechanisms are not uniformly distributed across this age gradient, but are continuously adjusted by developmental and environmental cues. According to the **Optimal Defense Hypothesis**, the plant's limited supply of defense compounds is concentrated where it is most needed to maximize fitness. Since mature leaves undergo senescence sooner than younger leaves, and therefore have less value, the hypothesis predicts that secondary metabolite concentrations should be higher in younger leaves. Indeed, an overwhelming number of studies have shown that young, developing leaves have higher constitutive levels of secondary compounds than older leaves. The growing list of secondary metabolites that exhibit this particular behavior includes phenolic compounds, glucosinolates, alkaloids, cyanide, furanocoumarins, volatile organic compounds, and defensive proteins. Similar distribution pattern have been observed in belowground tissues as well.

Consistent with the Optimal Defense Hypothesis, young tissues not only contain the highest levels of constitutive secondary compounds, they are also more responsive to herbivory. Thus, younger leaves exhibit more robust inducible defenses to herbivores than older leaves.

Inducible Defense Responses to Insect Herbivores

While constitutive chemical defenses provide plants with basic protection against many pests and pathogens, and are common among plants in nature, there are disadvantages to this type of defense strategy. First, constitutive defenses are costly to the plant. The production of secondary metabolites requires a significant energy investment derived from primary metabolism, which is then unavailable for use in growth and reproduction. This trade-off is most obvious in agricultural crops, in which yield is increased, in part, by reducing the capacity of the plant to defend itself. Second, pests and pathogens can adapt to the plant's constitutive chemical defenses, as we saw in the case of the monarch caterpillar and milkweed. Certain species of insect herbivores and microbial pathogens have evolved physiological mechanisms to detoxify otherwise lethal secondary metabolites, and may even use these compounds to defend themselves against their own predators or parasites. Accordingly, most plants have evolved inducible defense systems in addition to whatever constitutive defenses they may have. Inducible defense systems enable plants to respond more flexibly to the full panoply of threats presented by pests and pathogens.

Based on their feeding behaviors, three major types of insect herbivores can be distinguished:

1. *Phloem feeders,* such as aphids and whiteflies, cause little damage to the epidermis and mesophyll cells. Phloem feeders insert their narrow *stylet,* which is an elongated mouthpart, into the phloem sieve tubes of leaves and stems. The plant defense response to phloem feeders more closely resembles the response to pathogens rather than to herbivores. Although the amount of direct injury to the plant is low, when these insects serve as vectors for plant viruses they can cause great damage.
2. *Cell-content feeders,* such as mites and thrips, are piercing-and-sucking insects that cause an intermediate amount of physical damage to plant cells.
3. *Chewing insects,* such as caterpillars (the larvae of moths and butterflies), grasshoppers, and beetles, cause the most significant damage to plants. In the discussion that follows, our definition of "insect herbivory" will mostly relate to this type of insect damage.

In the following paragraphs we will discuss some of the mechanisms by which plants recognize insect herbivores and how they signal defenses that include not only the de novo synthesis of toxic secondary metabolites and proteins, but also the recruitment of natural enemies of the attacker. We will also describe how plants send out signals to nearby plants to prepare against impending herbivory.

Plants can recognize specific components of insect saliva

To mount an effective inducible defense against pests or pathogens, the host plant must be able to distinguish between mechanical causes, such as wind or hail, and an actual biotic attack. Most plant responses to insect herbivores involve both a wound response and the recognition of certain compounds abundant in the insect's saliva or regurgitant. These compounds belong to a broad group of chemicals called **elicitors**, which trigger defense responses in plants to a wide variety of herbivores and pathogens. A recently coined term for insect-derived elicitors is **herbivore-associated molecular patterns** (**HAMPs**). Although repeated mechanical wounding can induce responses similar to those caused by insect herbivory in some plants, certain molecules in insect saliva can serve as enhancers of this stimulus. In addition, insect-derived elicitors can trigger signaling pathways *systemically*—that is, throughout the plant—thereby initiating defense responses that can minimize further damage in distal regions of the plant (also see **WEB ESSAY 23.1**).

The first elicitors identified in insect saliva were *fatty acid–amino acid conjugates* (or *fatty acid amides*) in the oral secretions of larvae of the beet armyworm (*Spodoptera exigua*). These compounds have been shown to elicit a response closely resembling the response to chewing insects, as opposed to the response to wounding alone. The biosynthesis of these conjugates depends on the plant as the source of the fatty acids linolenic acid (18:3) and linoleic acid (18:2).* After the insect ingests plant tissue containing these fatty acids, an enzyme in the gut conjugates the plant-derived fatty acid to an insect-derived amino acid, typically glutamine. In some caterpillars the resulting conjugate of linolenic acid and glutamine is further processed by the introduction of a hydroxyl group at position 17 of linolenic acid (**Figure 23.17A**). This compound, *N*-(17-hydroxylinolenoyl)-L-glutamine, was named **volicitin** for its potential to induce volatile secondary metabolites in maize plants.

Since the discovery of volicitin, a variety of fatty acid amides have been identified not only in lepidopteran species but also in crickets and fruit flies, and most of them were found to exhibit elicitor-like activities when applied to plants. While fatty acid amides exhibit a broad range of activity among various plant species, little is known about the immediate signaling events elicited by these compounds. Volicitin binds rapidly to plasma membranes isolated from maize leaves in a typical receptor–ligand fashion. While this implies the existence of a specific fatty acid amide receptor on the cell surface, no such protein has been identified to date.

Modified fatty acids secreted by grasshoppers act as elicitors of jasmonic acid accumulation and ethylene emission

A novel class of insect-derived elicitors has been isolated and characterized from the oral secretions of a grasshopper (*Schistocerca americana*). Thus far these elicitors have only been found in the suborder Caelifera, so they were named **caeliferins** (**Figure 23.17B**). Caeliferins are also fatty acid–based compounds with a chain length between 15 and 19 carbons and are usually saturated or monounsaturated. For caeliferins in the A group, hydroxyls in the α and ω

*Recall that the nomenclature for fatty acids is *X*:*Y*, where *X* is the number of carbon atoms and *Y* is the number of *cis* double bonds.

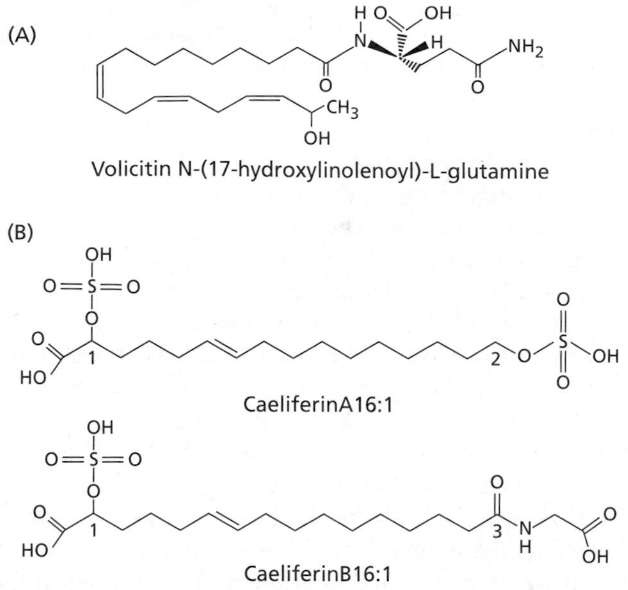

(A)

Volicitin N-(17-hydroxylinolenoyl)-L-glutamine

(B)

CaeliferinA16:1

CaeliferinB16:1

Figure 23.17 Structures of major insect-derived elicitors. (A) Linolenic acid–amino acid conjugates such as volicitin have been found to induce the release of volatile secondary metabolites in maize seedlings. These compounds and their linoleic acid analogs have been found in the regurgitant of the larvae of numerous lepidopteran species, and more recently in crickets and *Drosophila* larvae. (B) Caeliferins were isolated and identified from regurgitant of *Schistocerca americana*. Caeliferins in the A group with hydroxyls in the α (1) and ω (2) position are sulfated. Caeliferins in the B group are diacids with a sulfate in the α position (1) and a glycine conjugated to the ω carboxyl (3). Little is known to date about the biological activity of B-type caeliferins.

(omega) positions are sulfated (see Figure 23.17B). Caeliferins in the B group are diacids with a sulfate in the α position and a glycine conjugated to the ω carboxyl. In a volatile-based bioassay with maize seedlings, caeliferinA16:1 was found to be the most active compound among this group of elicitors, while caeliferinA16:0 was active in Arabidopsis. Application of caeliferin A to a wound site in Arabidopsis induced a transient spike of ethylene production and significantly higher jasmonic acid accumulation compared with mechanical wounding alone. Thus far, caeliferinA16:0 is the only insect-derived elicitor with biological activity in Arabidopsis.

The biological activity of caeliferins appears to be very species-specific. Neither legumes nor solanaceous plants respond to this elicitor with increased defense signaling. Unlike the fatty acid amides, the caeliferins are not derived from the plant. Not only do they display irregular chain lengths, but they are also characterized by a *trans*-configurated double bond. Neither feature is found in plants, strongly suggesting that they are of grasshopper origin.

Phloem feeders activate defense signaling pathways similar to those activated by pathogen infections

Although phloem feeders, such as aphids, cause little mechanical damage to plants, they are nonetheless serious agricultural pests that can significantly reduce crop yields. Plants in nature have evolved mechanisms to recognize and defend against phloem feeders. In contrast to chewing and piercing-and-sucking insects, which inflict severe tissue damage resulting in the activation of the jasmonic acid signaling pathway (discussed below), phloem feeders activate the **salicylic acid** signaling pathway, which is usually associated with pathogen infections. Because the defense response to phloem feeders involves receptor–ligand complexes that are closely related to those involved in the response to pathogens, we will describe the signaling mechanisms of this class of herbivores later in the chapter when we discuss microbial infections.

Calcium signaling and activation of the MAP kinase pathway are early events associated with insect herbivory

When plants recognize elicitors from insect saliva, a complex signal transduction network is activated. An increase in the cytosolic Ca^{2+} concentration ($[Ca^{2+}]_{cyt}$) is an early signal that mediates insect elicitor–induced responses. Ca^{2+} is a ubiquitous second messenger in multiple cellular responses of all eukaryotic systems (see Chapter 15). Under normal conditions, $[Ca^{2+}]_{cyt}$ is very low (~100 n*M*). Following stimulation by an elicitor, Ca^{2+} ions are rapidly released into the cytosol from storage compartments, such as mitochondria, endoplasmic reticulum, vacuole, and the cell wall. Increased Ca^{2+} levels in the cytosol then activate an array of target proteins, such as calmodulin and other Ca^{2+}-binding proteins, as well as Ca^{2+}-dependent protein kinases, which then activate downstream targets of the signaling pathway. These downstream targets typically include protein phosphorylation and transcriptional activation of stimulus-specific responses.

Although little is known about the detailed role of Ca^{2+} in the signaling of herbivore defenses, a picture is emerging that strongly suggests that Ca^{2+} plays an important role. In lima bean (*Phaseolus lunatus*), for example, the most significant increases in cytosolic Ca^{2+} concentration in response to insect herbivory occur in cell layers closest to the damage site, but are also detectable at lower levels in more distant tissues.

In Arabidopsis, a calmodulin-binding transcriptional regulator called IQD1 was identified as an important mediator of defense responses against insect herbivory. IQD1 binds calmodulin, a major Ca^{2+}-binding protein, in a Ca^{2+}-dependent manner and subsequently activates genes involved in glucosinolate biosynthesis. Accordingly, overexpression of IQD1 in Arabidopsis inhibits herbivore activity. These findings are consistent with an important role for Ca^{2+} in the regulation of antiherbivore defense mechanisms in plants.

In contrast to the defense-activating role of Ca^{2+}, a more recent study demonstrated that this signaling compound is also involved in the down-regulation of defense signaling, in particular the jasmonic acid pathway. Through virus-induced gene silencing of two calcium-dependent protein kinases (CDPKs) in a wild tobacco (*Nicotiana attenuata*), it was demonstrated that the accumulation of jasmonic acid after herbivory continued over a much longer period than it did in wild-type plants. Consequently, silenced plants also produced more defense metabolites and significantly slowed the growth of a specialist herbivore, the tobacco hornworm (*Manduca sexta*).

Insect herbivore–induced defense signaling also involves several types of mitogen-activated protein kinases (MAPKs). In tobacco, virus-induced gene silencing of the genes for the **wound-induced protein kinase** (**WIPK**) and the **salicylic acid–induced protein kinase** (**SIPK**), both members of the MAPK family, revealed that both are involved in the regulation of antiherbivore defenses. Both genes are significantly induced after insect herbivory and treatment with fatty acid amide elicitors. SIPK and WIPK also appear to be essential for different aspects of the jasmonic acid pathway (discussed later in this chapter). SIPK silencing mostly affects the early steps in the biosynthetic pathway for jasmonic acid, while WIPK-silenced plants are impaired in later steps of this pathway. In tomato, at least three different MAPKs are required to fully activate defenses against tobacco hornworm caterpillars. There, jasmonic acid accumulation and defense metabolite production were also shown to be significantly reduced in plants silenced for these three genes. These examples demonstrate the importance of a diverse array of MAP kinases in the regulation of plant defenses against insect herbivores.

Jasmonic acid activates defense responses against insect herbivores

A major signaling pathway involved in most plant defenses against insect herbivores is the *octadecanoid pathway*, which leads to the production of the hormone jasmonic acid (JA) (see Chapter 15). Together with other oxygenated fatty acid-derived products, octadecanoids belong to the family of **oxylipins**. JA levels rise steeply in response to insect herbivore damage and trigger the production of many proteins involved in plant defenses. Direct demonstration of the role of JA in insect resistance has come from research with JA-deficient mutant lines of Arabidopsis, tomato, and maize. Such mutants are easily killed by insect pests that normally cannot damage wild-type plants. Application of exogenous JA restores resistance nearly to the levels of the wild-type plant.

The structure and biosynthesis of JA have intrigued plant biologists because of the parallels to oxylipins that are central to inflammatory responses and other physiological processes in mammals. In plants, JA is synthesized from linolenic acid (18:3), which is released from membrane lipids and then converted to JA, as outlined in **Figure 23.18**. Two organelles participate in jasmonate biosynthesis, chloroplasts and peroxisomes. In the chloroplast, an intermediate derived from

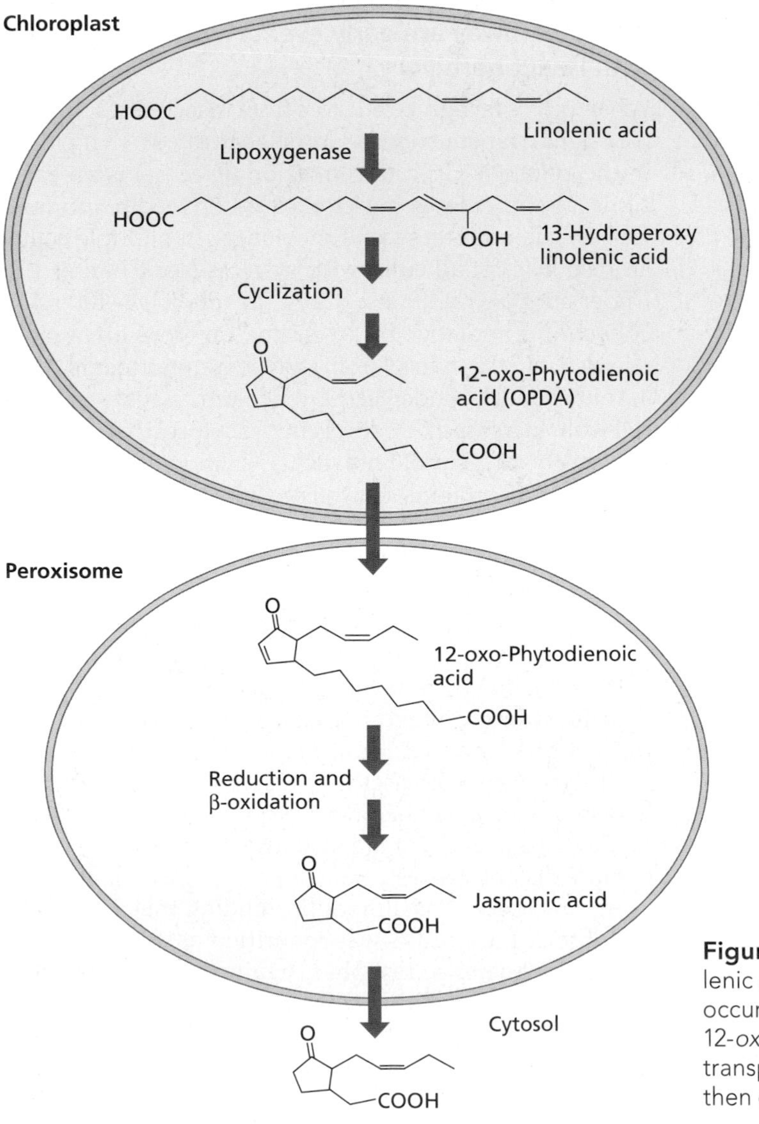

Figure 23.18 Steps in the pathway for conversion of linolenic acid (18:3) into jasmonic acid. The first enzymatic steps occur in the chloroplast, resulting in the cyclized product 12-oxo-phytodienoic acid (OPDA). This intermediate is transported to the peroxisome, where it is first reduced and then converted into jasmonic acid by β-oxidation.

linolenic acid is cyclized and then transported to the peroxisome, where enzymes of the β-oxidation pathway (see Chapter 12) complete the conversion to JA (see WEB ESSAY 23.3). JA induces the transcription of multiple genes that encode key enzymes in all major pathways for secondary metabolites.

Jasmonic acid acts through a conserved ubiquitin ligase signaling mechanism

Jasmonic acid not only activates defense-related genes, it also shuts down growth. JA-induced growth suppression allows the reallocation of resources to metabolic pathways involved in defense. JA acts through a conserved ubiquitin ligase–based signaling mechanism that bears close resemblance to those described for auxin and gibberellin (**Figure 23.19**) (see Chapter 15). Although unconjugated JA is hormonally active, many JA responses require activation of the hormone for optimal activity by conversion into an amino acid conjugate, such as jasmonic acid–isoleucine (JA–Ile). This conjugation is performed by enzymes referred to as **jasmonic acid resistance (JAR) proteins**, which belong to a family of carboxylic acid–conjugating enzymes. JAR1, for example, exhibits a high substrate specificity for JA and isoleucine and appears to be of particular importance for JA-dependent defense signaling.

When levels of bioactive JA are low, the expression of jasmonate-responsive genes is repressed by members of the **JAZ (JASMONATE ZIM-DOMAIN) protein family**, which are key regulators of the JA response. JAZ repressors act by binding to the **MYC2 transcription factor**, a major switch in the activation of JA-dependent genes. JAZ repressors also maintain the chromatin in a "closed" state that prevents JA-responsive transcription factors from binding to their targets.

To maintain chromatin in the inactive state, JAZ proteins bind to the F-box protein COI1, which is an essential component of the SCF protein complex SCF^{COI1}, a multiprotein E3 ubiquitin ligase. Two additional proteins and two histone deacetylase enzymes (HDA6 and HDA19) act as co-repressors along with the JAZ–COI1 complex and are instrumental in maintaining the chromatin in an inactive state (see Figure 23.19). The binding of JA–Ile to the JAZ–COI1 co-receptors leads to the ubiquitination of JAZ by the $SCF^{COI1\text{–}JA\text{–}Ile}$ complex, followed by JAZ degradation via the 26S proteasome (see Figure 23.19). Destruction of JAZ liberates the MYC2 transcription factor, which then recruits various other chromatin remodeling proteins and transcription factors that bring about the expression of the early JA-responsive genes.

Hormonal interactions contribute to plant–insect herbivore interactions

Several other signaling agents—including ethylene, salicylic acid, and methyl salicylate—are often induced by insect herbivory. In particular, ethylene appears to play an important role in this context. When applied alone to plants, ethylene has little effect on defense-related gene activation. However, when applied together with JA it seems to enhance JA responses. Similarly, when plants

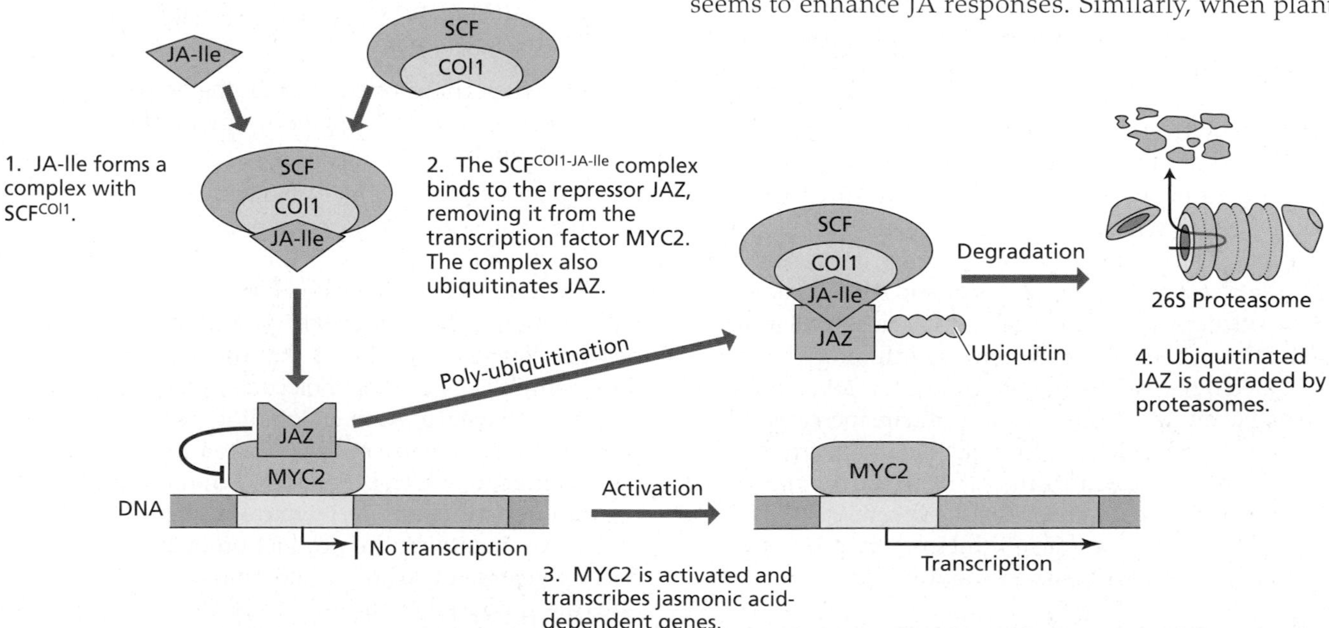

Figure 23.19 Jasmonic acid signaling. Jasmonic acid needs to be conjugated first to an amino acid (here: isoleucine) to bind to COI1 as part of a SCF^{COI1} protein complex. This complex targets JAZ, a repressor of transcription, leading to the degradation of this protein in a proteasome. Transcription factors such as MYC2 then initiate transcription of JA-dependent genes, including those for defense.

are treated with elicitors such as fatty acid amides (which by themselves do not induce the production of significant amounts of ethylene) in combination with ethylene, defense responses are significantly increased. Results such as these demonstrate that a concerted action of these signaling compounds is required for the full activation of induced defense responses. Multifactorial control allows plants to integrate numerous environmental signals in modulating the defense response.

JA initiates the production of defense proteins that inhibit herbivore digestion

Besides activating pathways for the production of toxic or repelling secondary metabolites, JA also initiates the biosynthesis of defense proteins. Most of these proteins interfere with the herbivore digestive system. For example, some legumes synthesize **α-amylase inhibitors**, which block the action of the starch-digesting enzyme α-amylase. Other plant species produce **lectins**, defensive proteins that bind to carbohydrates or carbohydrate-containing proteins. After ingestion by an herbivore, lectins bind to the epithelial cells lining the digestive tract and interfere with nutrient absorption.

A more direct attack on the insect herbivore's digestive system is performed by some plants through the production of a specific cysteine protease, which disrupt the peritrophic membrane that protects the gut epithelium of many insects. While none of these genes are essential for the vegetative growth of the plant, they have likely evolved from normal "housekeeping" genes during the coevolution of plants and their insect herbivores.

The best-known antidigestive proteins in plants are the **proteinase inhibitors**. Found in legumes, tomato, and other plants, these substances block the action of herbivore proteolytic enzymes. After entering the herbivore's digestive tract, they hinder protein digestion by binding tightly and specifically to the active site of protein-hydrolyzing enzymes such as trypsin and chymotrypsin. Insects that feed on plants containing proteinase inhibitors suffer reduced rates of growth and development that can be offset by supplemental amino acids in their diet.

The defensive role of proteinase inhibitors has been confirmed by experiments with transgenic tobacco. Plants that had been transformed to accumulate increased levels of proteinase inhibitors suffered less damage from insect herbivores than did untransformed control plants. As with glucosinolates, some insect herbivores have become adapted to plant proteinase inhibitors by production of digestive proteinases resistant to inhibition.

Herbivore damage induces systemic defenses

During herbivore attack, mechanical damage releases lytic enzymes from the plant that can potentially compromise the structural barriers of plant tissues. Some of the products generated by these enzymes can function as endogenous elicitors, called **damage associated molecular patterns (DAMPs)**. As we will discuss later in the chapter, DAMPs are recognized by pattern recognition receptors (PRRs) located on the cell surface. DAMPs usually appear in the apoplast and can induce protection against a broad range of organisms, a response known as *innate immunity*. For example, oligogalacturonides released from the cell wall can act as endogenous elicitors, although the perception system remains elusive.

In tomato, insect feeding leads to the rapid accumulation of proteinase inhibitors throughout the plant, even in undamaged areas far from the initial feeding site. The systemic production of proteinase inhibitors in young tomato plants is triggered by a complex sequence of events (**Figure 23.20**):

1. Wounded tomato leaves synthesize prosystemin, a large (200 amino acids) precursor protein.
2. Prosystemin is proteolytically processed to produce the short (18 amino acids) polypeptide DAMP called systemin.
3. Systemin is released from damaged cells into the apoplast.
4. In adjacent intact tissue (phloem parenchyma), systemin binds to a pattern recognition receptor on the plasma membrane (see the section *Plant Defenses against Pathogens* below and **WEB ESSAY 23.4**).
5. The activated systemin receptor becomes phosphorylated and activates a phospholipase A2 (PLA2).
6. The activated PLA2 generates the signal that initiates JA biosynthesis.
7. JA is then transported through the phloem to systemic parts of the plant by an unknown mechanism.
8. JA is taken up by target tissues and activates the expression of genes that encode proteinase inhibitors.

Although peptide signals such as systemin were originally thought to be restricted to solanaceous plants, it has become clear in recent years that other plants also produce peptides as signaling molecules in response to insect herbivory. Recently, such a signaling peptide, ZmPep3, was identified in maize in response to insect elictor treatment. ZmPep3, which is derived from the ZmPROPEP3 precursor, was found to elicit typical antiherbivore defense responses, including the production of benzoxazinoids and the release of volatile compounds. Since these peptides are produced in response to elicitor treatment, they appear to serve as enhancers of the defense response, as in the case of systemin. Orthologs of the ZmPROPEP3 propeptide are also found in other plant families, including Fabaceae (legumes), and may represent the functional analogs of systemin outside the Solanaceae.

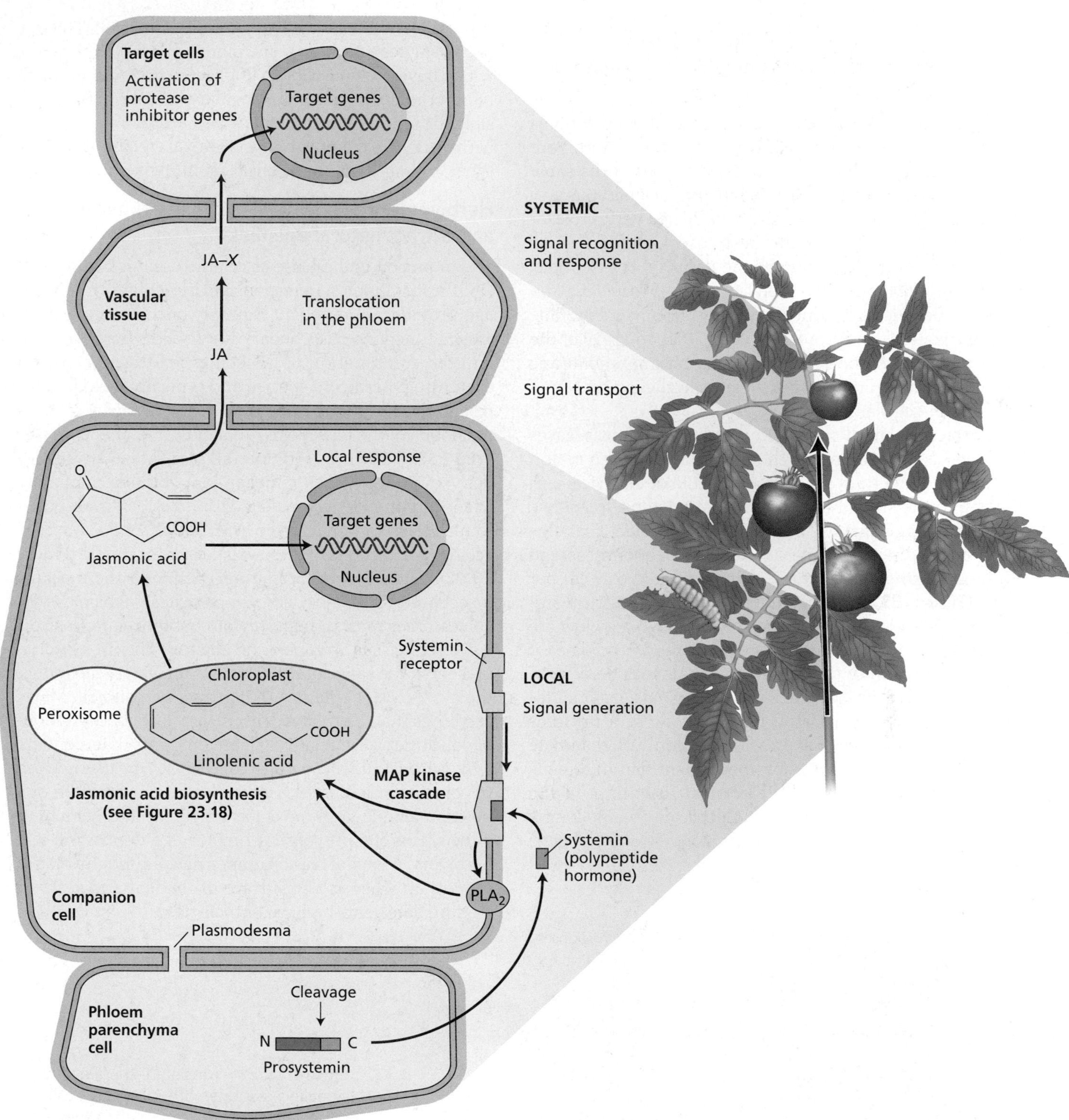

Figure 23.20 Proposed systemin signaling pathway for the rapid induction of protease inhibitor biosynthesis in a wounded tomato plant. Wounded leaves (bottom left of plant) synthesize prosystemin in phloem parenchyma cells, and the prosystemin is proteolytically processed to systemin. Systemin is released from phloem parenchyma cells and binds to receptors on the plasma membrane of adjacent companion cells. This binding activates a signaling cascade involving phospholipase A2 (PLA2) and mitogen-activated protein (MAP) kinases, which results in the biosynthesis of jasmonic acid (JA). JA is then transported via sieve elements, possibly in a conjugated form (JA–X), to unwounded leaves. There, JA initiates a signaling pathway in target mesophyll cells, resulting in the expression of genes that encode protease inhibitors. Plasmodesmata facilitate the spread of the signal at various steps in the pathway.

Glutamate receptor-like (GLR) genes are required for long-distance electrical signaling during herbivory

In response to herbivory, jasmonic acid accumulates within minutes—both locally, at the site of herbivore damage, and distally, in undamaged tissues of the same leaf and in other leaves. Although plants lack nervous systems, several lines of evidence are consistent with a role for electrical signaling in defense responses that occur some distance from the site of herbivore damage. For example, the feeding of Egyptian cotton leafworm (*Spodoptera littoralis*) on bean leaves induces a wave of plasma membrane depolarizations that spreads to undamaged areas of the leaf. In addition, ionophore-induced plasma membrane depolarization in tomato cells results in the expression of jasmonate-regulated genes.

Surface potential measurements of Arabidopsis leaves in response to feeding by the larvae of Egyptian cotton leafworm have now confirmed the role of electrical signaling in the spread of the jasmonate defense response to undamaged leaves. During feeding, electrical signals induced near the site of attack subsequently spread to neighboring leaves at a maximum speed of 9 cm per min (**Figure 23.21**). Since the relay of the electrical signal is most efficient for leaves directly above or below the wounded leaf, the vascular system is a good candidate for the transmission of the electrical signals to other leaves. At all sites that receive the electrical signals, jasmonate-mediated gene expression is turned on and initiates defense-response gene expression. A family of glutamate receptor-like (GLR) genes has been identified in screens for mutants with defective electrical signaling. In the *glr3.3/glr3.6* double mutant, the electrical wave no longer propagates after wounding, and jasmonate-response gene expression in leaves distal to wounds is reduced. The evidence suggests that, in some plants at least, GLR genes, which have previously been implicated in the recognition of other microbe-related molecular patterns, are responsible for long-distance defense signaling in response to herbivory. The relationship of electrical signaling to other types of long-distance defense signaling is unclear.

Herbivore-induced volatiles can repel herbivores and attract natural enemies

The induction and release of volatile organic compounds (VOCs), or volatiles, in response to insect herbivore damage provides an excellent example of the complex ecological functions of secondary metabolites in nature. The emitted combination of molecules is often specific for each insect herbivore species and typically includes representatives from the three major pathways of secondary metabolism: the terpenoids, alkaloids, and phenolics (see WEB ESSAY 23.1). In addition, all plants also emit lipid-derived products, such as **green-leaf volatiles** (a mixture of six-carbon aldehydes, alcohols, and esters) in response to mechanical damage (see WEB ESSAY 23.5, 23.6). The ecological functions of these volatiles are manifold (**Figure 23.22**). Often, they attract natural enemies of the attacking insect herbivores—predators or parasites—which use the volatile cues to find their prey or host for their offspring. As noted earlier, in maize the elicitor volicitin, which is present in the saliva of beet armyworm larvae, can induce the synthesis of volatiles that attract parasitoids. Maize seedlings that are treated with in very low concentrations of volicitin release relatively large amounts of terpenoids, which attract the tiny parasitoid wasp *Microplitis croceipes*. In contrast, volatiles released by leaves during moth oviposition (egg-laying) can act as repellents to other female moths, thereby preventing further egg deposition and herbivory. Many of these compounds, although volatile, remain attached to the surface of the leaf and serve as feeding deterrents because of their taste.

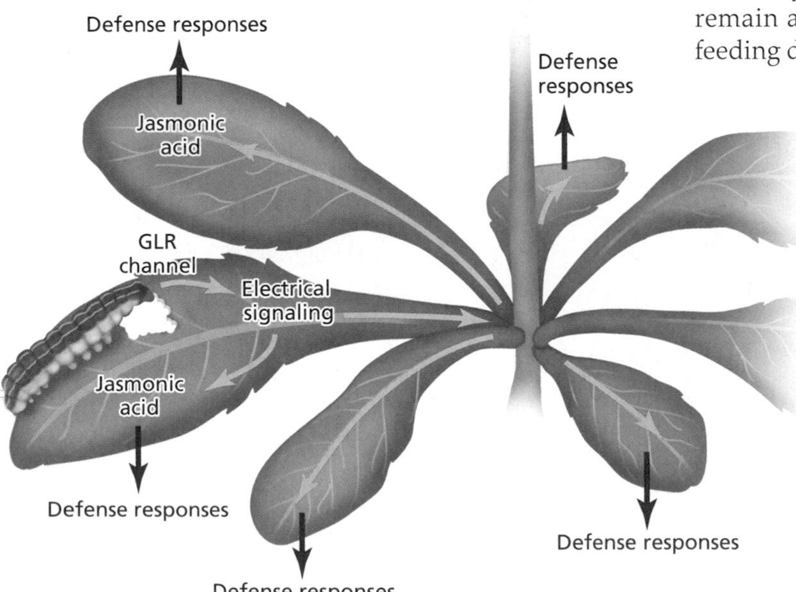

Figure 23.21 Model for the electrical signaling response of Arabidopsis to herbivore attack. Injury to the leaf caused by herbivory activates glutamate receptor-like (GLR) ion channels in the vascular system. The electrical signals are thought to travel through the vascular system and stimulate jasmonic acid (JA) production both locally and in other leaves. JA production then initiates defense responses that discourage further herbivory. (After Christmann and Grill 2013.)

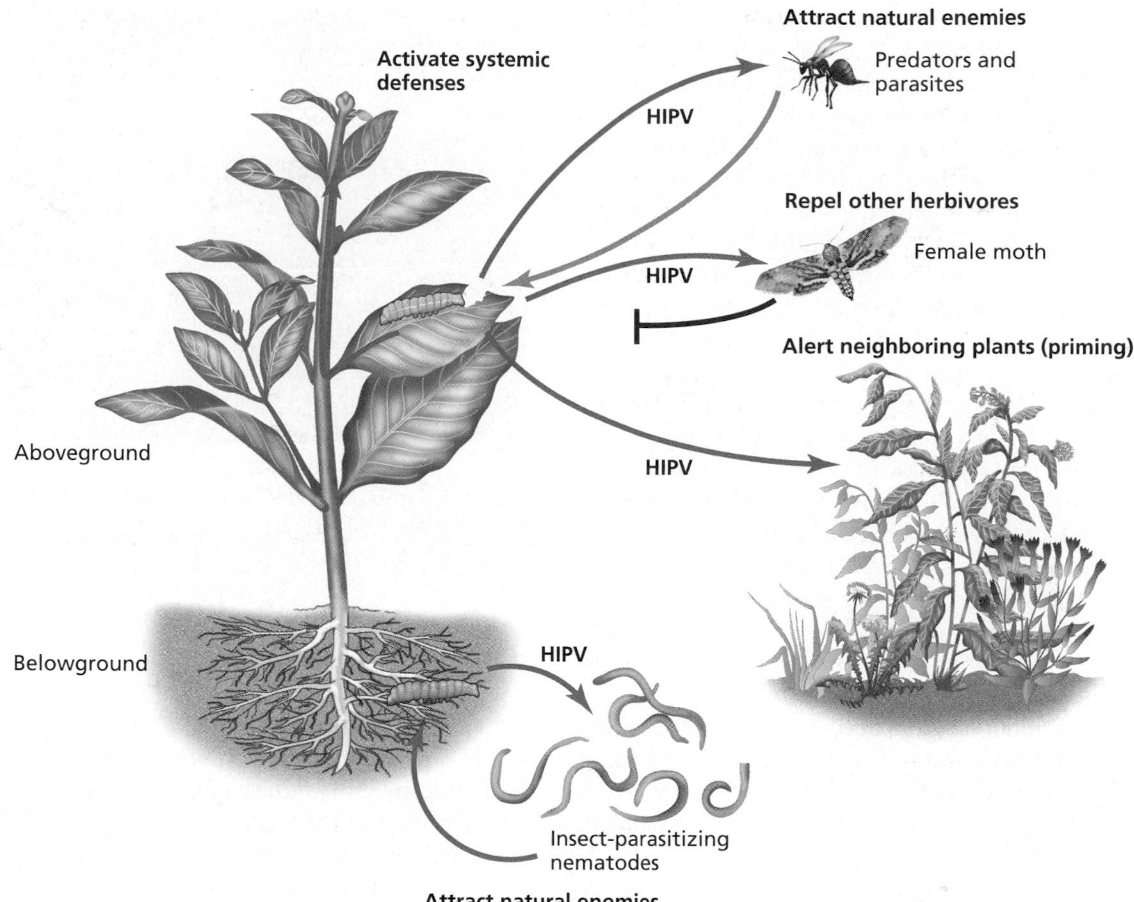

Figure 23.22 Ecological functions of insect herbivore–induced plant volatiles (HIPV). Many plants release a specific bouquet of volatile organic compounds when attacked by insect herbivores. These volatiles can consist of compounds from all major pathways for secondary metabolites including terpenoids (mono- and sesquiterpenes), alkaloids (indole), and phenylpropanes (methyl salicylate), as well as green-leaf volatiles. These volatiles can act as cues for natural enemies of the insect herbivore, for example parasitic wasps. Belowground parts of plants can also release volatiles when attacked by herbivores. It has been shown that these volatiles attract insect-parasitizing nematodes, which then attack the herbivore. Volatiles may also serve as a repellant for female moths, thereby avoiding further egg deposition. Most recently, volatiles have been found to act as a systemic defense signal in highly sectorial plants with interrupted vascular connections, and also between plants over short distances. There, these volatile signals prepare the receiving plant against impending herbivory by priming (preparing) defense responses, resulting in a stronger and faster response when the receiving plant is actually attacked.

Plants have the ability to distinguish among various insect herbivore species and to respond differentially. For example, following herbivory, *Nicotiana attenuata,* a wild tobacco that grows in the deserts of the Great Basin in the western United States, typically produces higher levels of nicotine, which poisons the insect central nervous system. However, when wild tobacco plants are attacked by nicotine-tolerant caterpillars, the plants show no increase in nicotine. Instead, they release volatile terpenes that attract insect predators of the caterpillars (see **WEB ESSAY 23.1**). Clearly, wild tobacco and other plants must have ways of determining what type of insect herbivore is damaging their foliage. Herbivores might signal their presence by the type of damage they inflict or the distinctive chemical compounds they release in their oral secretions.

Herbivore-induced volatiles can serve as long-distance signals between plants

The role of herbivore-induced plant volatiles is not limited to the mediation of ecological interactions between plants and insects. Certain volatiles emitted by infested plants can also serve as signals for neighboring plants to initiate expression of defense-related genes (see Figure 23.22). In addition to several terpenoids, green-leaf volatiles act as potent signals in this process. Green-leaf volatiles, which are like JA oxylipins produced from linolenic acid, are the major components of the familiar scent of freshly cut grasses (see **WEB ESSAY 23.5**). The biosynthetic pathway starts with 13-hydroperoxy-linolenic acid and is catalyzed by the enzyme hydroperoxide lyase (HPL). Major products of this pathway are Z-3-hexenal, Z-3-hexenol, and Z-3-hexenyl acetate and their respective E-2-enantiomers. Additionally, this pathway produces 12-oxo-Z-9-decenoic acid, the natural precursor of traumatin, the first wound hormone described for plants. Although the HPL pathway was first characterized 100 years ago, it has only recently gained significance, when it was shown that the volatile products of this pathway serve as potent signals in inter- and intraplant signaling. When maize plants were exposed to green-leaf volatiles, JA and JA-related gene expression were rapidly induced. More important, however, was the finding that exposure to green-leaf volatiles primed maize plant defenses to respond more strongly to subsequent attacks by insect herbivores. Green-leaf volatiles have been shown to prime or sensitize the defensive mechanisms of a variety of other plant species, including lima bean (*Phaseolus lunatus*), sagebrush (*Artemisia tridentata*), mouse ear cress (*Arabidopsis thaliana*), poplar (*Populus tremula*), and blueberry (*Vaccinium* spp.). Furthermore, they activate the production of phytoalexins and other antimicrobial compounds (discussed in the next section; also see **WEB ESSAY 23.6**) and appear to play an important role in the overall defense strategies of plants.

Herbivore-induced volatiles can also act as systemic signals within a plant

Besides providing a signal for neighboring plants, infested plants may well send a volatile signal to other parts of themselves (see Figure 23.22). From an evolutionary point of view, this may be the original function of those volatiles. Volatiles have been shown to act as inducers of herbivore resistance between different branches of sagebrush. It was found that airflow was essential for the induction of the induced resistance. Sagebrush, like other desert plants, is highly *sectorial*, meaning that the vascular system of the plant is not well integrated by interconnections. Although many plants are capable of responding systematically to herbivores by means of chemical signals that move internally through vascular interconnections, sagebrush and many other desert plants are unable to do so. Instead, volatiles are used to overcome these constraints and provide systemic signaling. A similar effect of volatiles was observed in lima bean, which uses extrafloral nectaries located at the base of leaf blades to attract predacious and parasitoid arthropods to protect them against various types of herbivores (**Figure 23.23**). For example, when leaf beetles attack lima bean, volatiles, in particular green-leaf volatiles, are released immediately from the damage site and signal other parts of the same plant to activate their defenses, including the production of **extrafloral nectar**.

Figure 23.23 Extrafloral nectaries of lima bean (*Phaseolus lunatus*).

Defense responses to herbivores and pathogens are regulated by circadian rhythms

Many aspects of plant metabolism and development are regulated by circadian rhythms (see Chapter 20). It has been estimated that about one-third of all plant genes exhibit circadian regulation in their expression. The list of genes with diurnally regulated transcription includes not only the predictable genes involved in photosynthesis, carbon metabolism, and water uptake, but numerous genes involved in plant defense. This observation has led to the proposal that resistance to insect herbivory may be under circadian control.

This hypothesis was recently confirmed by a study of the interactions between Arabidopsis and cabbage looper (*Trichoplusia ni*), a generalist lepidopteran herbivore (**Figure 23.24A**). Both cabbage looper herbivory and jasmonate-mediated plant defenses follow circadian rhythms that peak during the day. This suggests that the timing of the jasmonate-mediated defense response may be an adaptation that maximizes protection against herbivory. To test whether the plant circadian clock enhances defense against insect pests, herbivory was compared in Arabidopsis plants whose jasmonate-mediated defense responses were either in phase (**Figure 23.24B**) or out of phase (**Figure 23.24C**) with the circadian rhythm of cabbage looper feeding activity. After allowing cabbage loopers to feed freely on the plants for 72 h, plants whose defense responses were in phase with the loopers had visibly less tissue damage than plants whose circadian rhythm was out of phase with that of the insects (**Figure 23.24D**). As a result, cabbage loopers that fed on the phase-shifted Arabidopsis plants gained three times as much weight over the same period as the synchronized control plants did (**Figure 23.24E**).

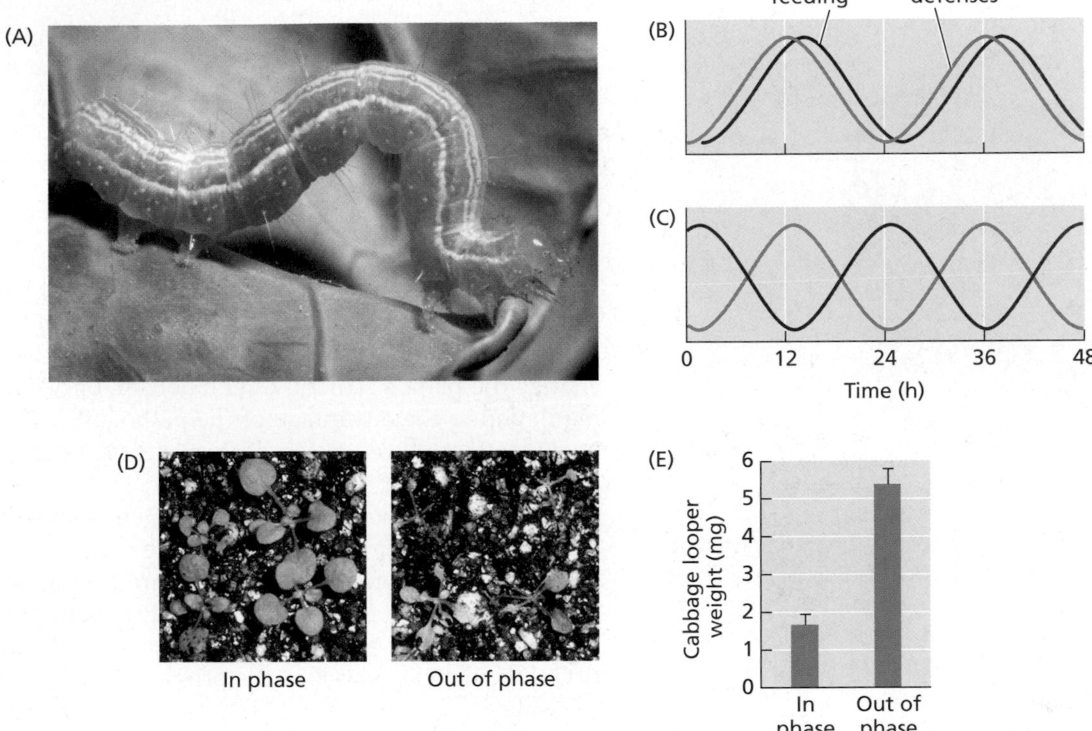

Figure 23.24 Example of circadian rhythms influencing plant defense against herbivory. (A) *Trichoplusia ni* (cabbage looper) caterpillar feeding on Arabidopsis. (B) Normally, the circadian clocks of the caterpillars and the plants are synchronized and both caterpillar feeding activity (red curve) and jasmonate-mediated plant defenses (green curve) peak during the day. This optimizes plant defenses and reduces caterpillar growth rate. (C) If the circadian rhythm of Arabidopsis is shifted by 12 h, the plant's defense response (green curve) is at a minimum when the caterpillar's feeding activity (red curve) is at a maximum, and the caterpillar grows more rapidly. (D) Out-of-phase Arabidopsis plants (right) suffered more damage than in-phase plants (left). (E) Comparison of weights of cabbage loopers growing on in-phase or out-of-phase Arabidopsis plants. (B–E after Goodspeed et al. 2012.)

Salicylic acid, which mediates defense responses to pathogens, showed the opposite accumulation phasing relative to jasmonates, with peaks occurring in the middle of the night. This diurnal accumulation of salicylates may contribute to the enhanced resistance of Arabidopsis to pathogenic bacteria when infection occurs in the early morning as opposed to the evening.

Insects have evolved mechanisms to defeat plant defenses

In spite of all the chemical mechanisms plants have evolved to protect themselves, herbivorous insects have evolved mechanisms for circumventing or overcoming these plant defenses by the process of *reciprocal evolutionary change between plant and insect*, a type of coevolution. These adaptations, like plant defense responses, can be either constitutive or induced. Constitutive adaptations are more widely distributed among specialist herbivorous insects, which can feed on only a few plant species, whereas induced adaptations are more likely to be found among insects that are dietary generalists. Although it is not always obvious, in most natural environments plant–insect interactions have led to a standoff in which each can develop and survive under suboptimal conditions.

Plant Defenses against Pathogens

Despite lacking an immune system comparable to that of animals, plants are surprisingly resistant to diseases caused by the fungi, bacteria, viruses, and nematodes that are ever present in the environment. In this section we will examine the diverse array of mechanisms that plants have evolved to resist infection locally, including microbe-associated molecular pattern (MAMP)-triggered immunity, effector-triggered immunity, the production of antimicrobial agents, and a type of programmed cell death called the hypersensitive response. We will also discuss two types of systemic plant immunity, referred to as *systemic acquired resistance* (*SAR*) and *induced systemic resistance* (*ISR*).

Microbial pathogens have evolved various strategies to invade host plants

Throughout their lives, plants are continuously exposed to a diverse array of pathogens. Successful pathogens have

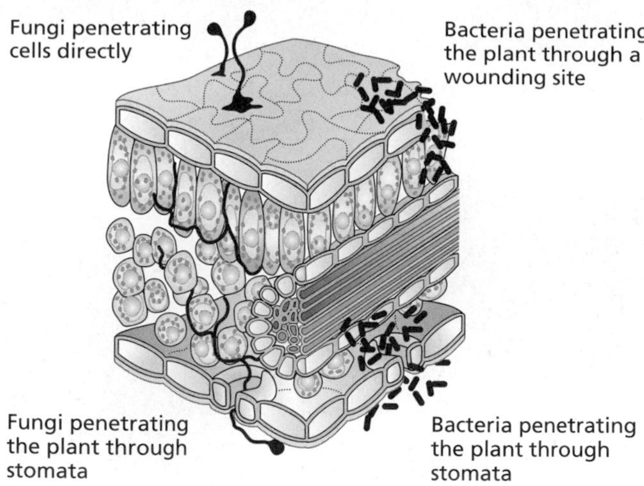

Figure 23.25 Plant pathogens such as bacteria and fungi have developed various methods for invading plants. Some fungi have mechanisms that allow them to directly penetrate the cuticle and cell wall of the plant. Other fungi, and also pathogenic bacteria, enter through natural openings such as stomata or through existing wounds caused by herbivores.

evolved various mechanisms to invade their host plant and cause disease (**Figure 23.25**). Some penetrate the cuticle and cell wall directly by secreting lytic enzymes, which digest these mechanical barriers. Others enter the plant through natural openings such as stomata, hydathodes, and lenticels. A third category invade the plant through wound sites, for example those caused by insect herbivores. Additionally, many viruses, as well as other types of pathogens, are transferred by insect herbivores, which serve as vectors, and invade the plant from the insect feeding site. Phloem feeders such as whiteflies and aphids deposit these pathogens directly into the vascular system, from which they can easily spread throughout the plant.

Once inside the plant, pathogens generally employ one of three main attack strategies to use the host plant as a substrate for their own proliferation. **Necrotrophic pathogens** attack their host by secreting cell wall–degrading enzymes or toxins, which eventually kill the affected plant cells, leading to extensive tissue maceration (softening of the tissues after death by autolysis). This dead tissue is then colonized by the pathogens and serves as a food source. A different strategy is used by **biotrophic pathogens**; after infection, most of the plant tissue remains alive and only minimal cell damage can be observed, as the pathogens feed on substrates provided by their host. **Hemibiotrophic pathogens** are characterized by an initial biotrophic stage, in which the host cells are kept alive as described for biotrophic pathogens. This phase is followed by a necrotrophic stage, in which the pathogens can cause extensive tissue damage.

Although these invasion and infection strategies are individually successful, plant disease epidemics are rare in natural ecosystems. This is because plants have evolved effective defense strategies against this diverse array of pathogens.

Pathogens produce effector molecules that aid in the colonization of their plant host cells

Plant pathogens can produce a wide array of effectors that support their ability to successfully colonize their host and gain nutritional benefits. **Effectors** are molecules that change the plant's structure, metabolism, or hormonal regulation to the advantage of the pathogen. They can be divided into three major classes: *enzymes*, *toxins*, and *growth regulators*. Because invasion of a suitable host is often the most difficult step for a pathogen, many pathogens produce enzymes that can degrade the plant cuticle and cell wall. Among those enzymes are cutinases, cellulases, xylanases, pectinases, and polygalacturonases. These enzymes have the ability to compromise the integrity of the cuticle as well as the primary and secondary cell walls.

Many pathogens also produce a wide array of toxins that act by targeting specific proteins of the plant (**Figure 23.26**). For example, the **HC-toxin** from the fungus *Cochliobolus carbonum*, which causes northern leaf blight disease, inhibits specific histone deacetylases in maize. In general, decreased deacetylation of histones, which are essential in the organization of the chromatin, tends to increase the expression of associated genes (see Figure 2.13). However, it is not yet known whether this is how HC-toxin causes disease in maize.

Fusicoccin (see Figure 23.26) is a nonspecific toxin from the fungus *Fusicoccum amygdali*. **Fusicoccin** constitutively activates the plant plasma membrane H^+-ATPase by first binding to a specific protein of the 14-3-3 group of regulators. This complex then binds to the C-terminal region of the H^+-ATPase and activates it irreversibly, leading to cell wall overacidification and plasma membrane hyperpolarization. These effects of fusicoccin are of particular importance for stomatal guard cells (see Chapter 10). Fusicoccin-induced plasma membrane hyperpolarization in guard cells causes massive K^+ uptake and permanent stomatal opening, which leads to wilting and ultimately the death of the plant. It is not yet clear if and how the pathogen benefits from the excessive wilting of its host.

Some pathogens produce effector molecules that significantly interfere with the hormonal balance of the plant host. The fungus *Gibberella fujikuroi*, which causes infected rice shoots to grow much faster relative to uninfected plants, produces gibberellic acid (GA_3) and other gibberellins. Gibberellins are thus responsible for the "foolish seedling disease" of rice. It is thought that fungal spores released from the taller, infected plants are more likely to spread to sur-

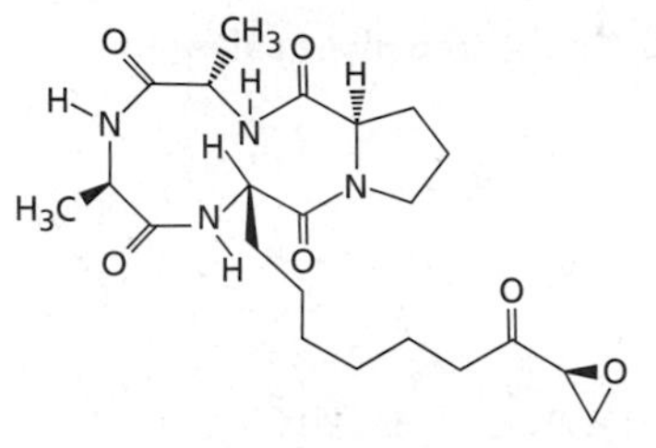

HC-toxin

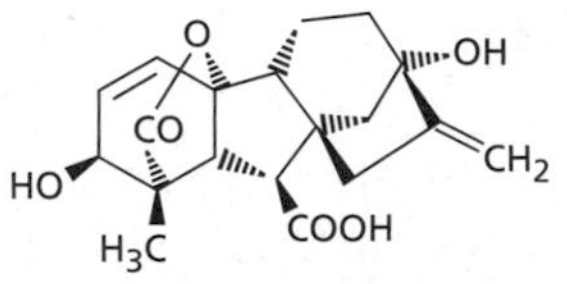

Gibberellic acid GA_3

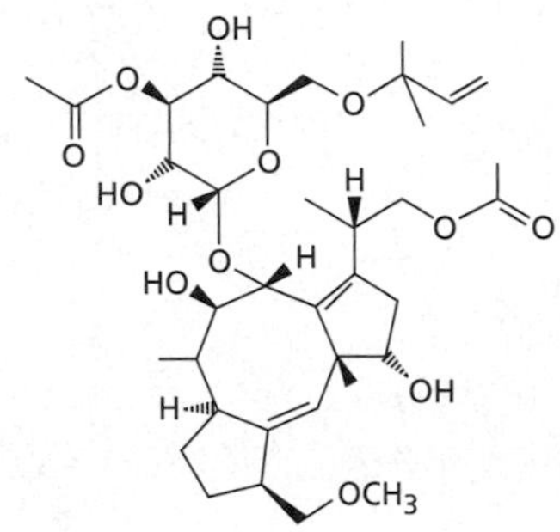

Fusicoccin

Figure 23.26 Effector molecules produced by pathogens help invade plants. Some pathogens produce specific effector molecules that significantly alter the physiology of the plant. The HC-toxin, a cyclic peptide, acts on the enzyme histone deacetylase in the nucleus, and may have a compromising effect on the expression of genes involved in defense. Fusicoccin binds to plant plasma membrane H^+-ATPases, in particular those in stomata, and activates them irreversibly. Gibberellins, produced by the fungus *Gibberella fujikuroi*, accelerate growth, resulting in bigger plants when compared with uninfested plants. The gibberellins produced by the fungi are identical to those produced endogenously by the plant.

rounding plants because of their height advantage. It was subsequently demonstrated that gibberellins are naturally occurring plant hormones (see Chapter 15).

The effectors of some pathogenic bacteria, such as *Xanthomonas*, are proteins that target the plant cell nucleus and cause marked changes in gene expression. These so-called transcription activator-like (TAL) effectors bind to the host plant DNA and activate the expression of genes beneficial to the pathogen's growth and dissemination.

Pathogen infection can give rise to molecular "danger signals" that are perceived by cell surface pattern recognition receptors (PRRs)

To distinguish between "self" and "nonself" during pathogen infection, plants possess **pattern recognition receptors** (**PRRs**) that perceive **microbe-associated molecular patterns** (**MAMPs**) which are conserved among a specific class of microorganisms (such as chitin for fungi, flagella for bacteria) but are absent in the host.

Receptor-like kinases (RLKs, which we introduced earlier in connection with beneficial plant–microbe interactions) and receptor-like proteins (RLPs) are key PRRs for microbe- and plant-derived molecular signals associated with pathogen infection (**Figure 23.27**). RLKs typically contain an extracellular domain such as a leucine-rich repeat (LRR) or Lysin Motif (LysM) domain, a transmembrane domain, and an intracellular kinase domain. RLPs contain an extracellular domain and a transmembrane domain but lack an intracellular kinase domain.

PRRs exist in protein complexes that are maintained in a resting state prior to ligand binding. The RLPs, which lack the typical cytoplasmic kinase domains, are thought to interact with RLKs to enable signals to be transduced to the cytoplasm. Upon binding to their molecular ligands, many cellular events are activated (see below), culminating in the transcriptional activation of a large number of defense response genes.

As we mentioned earlier, molecular alarm signals can also arise from the plant itself, either from damage caused by microbes or as the result of damage inflicted by chewing insects. Such plant-derived signals are collectively referred to as damage-associated molecular patterns (DAMPs) (see Figure 23.27).

Systemin, as discussed earlier, is an example of a plant-derived DAMP found in tomato that is produced in response to wounding associated with herbivore activity. Among the well-studied microorganism-derived MAMPs are Pep13, a 13 amino acid peptide from a cell wall–localized transglutaminase of the oomycete *Phytophthora*, the cause of the infamous potato blight in Ireland; flg22, a 22 amino acid peptide derived from the bacterial flagellin protein; and elf18, an 18 amino acid fragment of the bacterial elongation factor Tu. Since these molecules are common to many if not all species within groups of microorganisms, their recognition allows the plant to recognize entire classes of potentially pathogenic organisms, such as gram-positive versus gram-negative bacteria.

Perception of MAMPs or DAMPs by cell surface PRRs initiates a localized basal defense response called MAMP-triggered immunity which inhibits the growth and activity of nonadapted pathogens or pests. For example, control over stomatal aperture, a common site of pathogen invasion, serves as the first line of defense against pathogen invasion. When an Arabidopsis leaf is exposed to bacteria on the leaf surface, or to the MAMP flg22, the stomatal apertures decrease, thereby retarding pathogen invasion. As discussed in Chapter 10, stomatal opening is facilitated by the inward K^+ channels of guard cells that mediate K^+ uptake. The MAMP flg22 appears to induce partial stomatal closure by inhibiting K^+ uptake by guard cells. The flg22-elicited response is dependent on the presence of the LRR receptor-like kinase FLS2, as well as a heterotrimeric G protein.

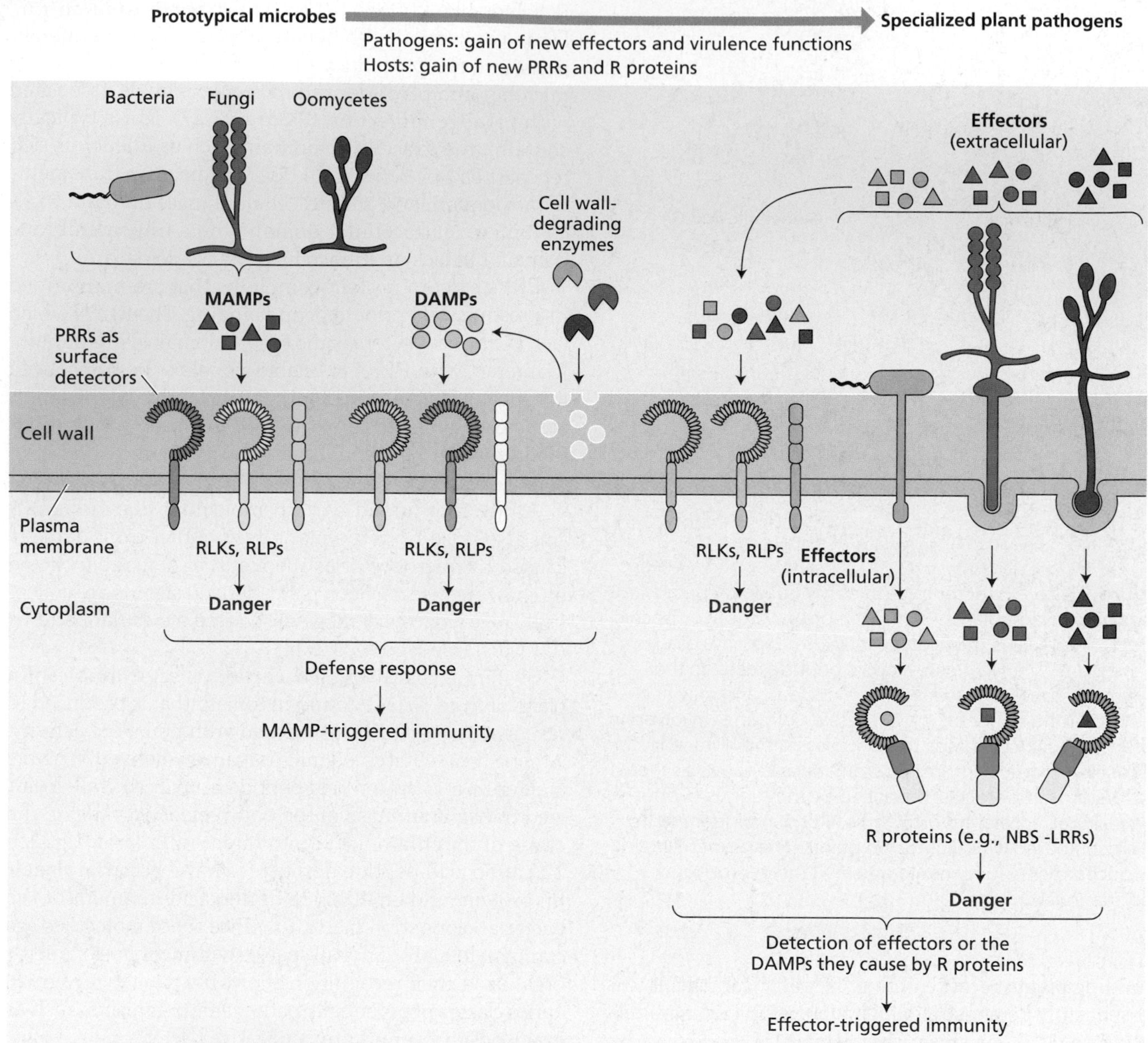

Figure 23.27 Plants have evolved defense responses to a variety of danger signals of biotic origin. These danger signals include microbe-associated molecular patterns (MAMPs), damage-associated molecular patterns (DAMPs), and effectors. Extracellular MAMPs produced by microbes, and DAMPs released by microbial enzymes, bind to pattern recognition receptors (PRRs) on the cell surface. As plants coevolved with pathogens, the pathogens acquired effectors as virulence factors, and plants evolved new PRRs to perceive extracellular effectors, and new resistance (R) proteins to perceive intracellular effectors. When MAMPs, DAMPs, and effectors bind to their PRRs and R proteins, two types of defense responses are induced: MAMP-triggered immunity and effector-triggered immunity. RLK, receptor-like kinase; RLP, receptor-like protein; NBS–LRR, nucleotide binding site–leucine rich repeat. (After Boller and Felix 2009.)

R genes provide resistance to individual pathogens by recognizing strain-specific effectors

Well-adapted microbial pathogens are able to subvert MAMP-triggered immunity by introducing a wide variety of effectors directly into the cytoplasm of the host cell. For example, pathogenic gram-negative bacteria with a Type III secretion system have evolved a syringe-shaped structure called the **injectisome** that spans the inner and outer bacterial membranes and includes a needlelike extracellular projection. Fungi and oomycetes have evolved other methods of transporting effectors directly into plant cells. Once inside the cells, these effectors can no longer be detected by the membrane-bound PRRs, and without a backup system the plant would be defenseless against the attack.

This microbial innovation placed plants under tremendous evolutionary pressure. For example, the bacterial toxin coronatine, produced by several pathogenic strains of *Pseudomonas syringae*, reverses the inhibitory effects of flg22 on K^+ uptake and stomatal opening. Plants, in turn, evolved a second line of defense based on a class of specialized **resistance (*R*) genes** that recognize these intracellular effectors and trigger defense responses to render them harmless. As a result, plants possess a second type of immunity called **effector-triggered immunity**, mediated by a set of highly specific intracellular receptors.*

There are several types of *R* gene products based on the arrangement of their functional domains. Of these, the most abundant group is the **nucleotide binding site–leucine rich repeat (NBS–LRR)** receptors. A subset of NBS–LRR receptors shuttles between the cytoplasm and the nucleus, where the receptors regulate gene expression, whereas others are tethered to the plasma membrane, where they can rapidly encounter an entering effector and trigger signal transduction pathways.

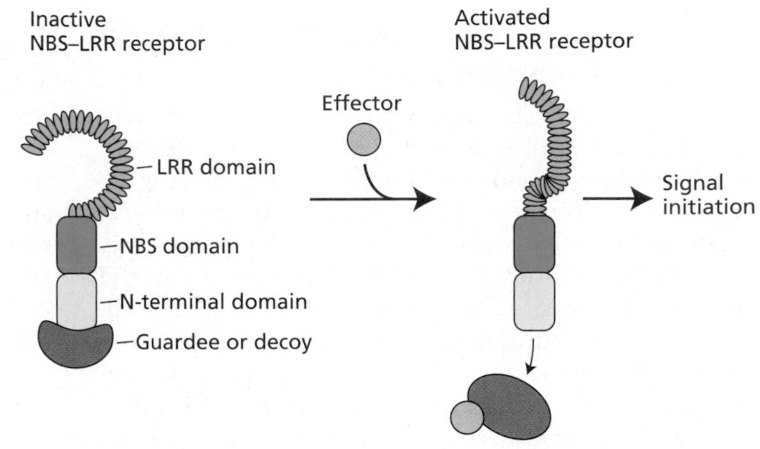

Figure 23.28 "Guard" hypothesis for *R* gene signaling during plant defense against pathogens. Soluble nucleotide binding site–leucine rich repeat (NBS–LRR) receptors bind to cellular proteins and use them as decoys ("guardees"). The NBS–LRR receptor is inactive as long as it is bound to its guardee. When effectors introduced into the cytoplasm by pathogens bind to the guardee, the guardee dissociates from the receptor, triggering the defense response.

Some NBS–LRRs become activated by binding directly to a pathogen effector, but most NBS–LRRs recognize and bind to plant proteins that are the targets of the pathogen effectors. According to the **guard hypothesis**, *R* gene products "guard" cellular proteins, which are called guardees or decoys (**Figure 23.28**). Interaction between the guardees and the pathogen effectors is required for successful infection by the pathogen. The NBS–LRR receptor remains inactive as long as it is bound to the guardee. However, when the effector interacts with the guardee, either altering its conformation or chemically modifying it, the NBS–LRR receptor becomes activated, triggering a signaling cascade that leads to the defense response.

Exposure to elicitors induces a signal transduction cascade

Within a few minutes after elicitors (effectors or MAMPs) have been recognized by an *R* gene product or a PRR, complex signaling pathways are set in motion that lead eventually to defense responses (see Figure 23.27). A common early element of these cascades is a transient change in the ion permeability of the plasma membrane. *R* gene product activation stimulates an influx of Ca^{2+} and H^+ ions into the cell and an efflux of K^+ and Cl^- ions out of the cell. The influx of Ca^{2+} activates the oxidative burst that may act directly in defense (as already described), as well as inducing other defense responses. Other components of pathogen-stimulated signal transduction pathways include nitric oxide, MAP kinases, calcium-dependent protein kinases, jasmonic acid, and salicylic acid.

*In the past, the microbial effector genes were confusingly called **avirulence (avr) genes**, based on the observation that they rendered the pathogen avirulent due to their "unintended" function in activating effector-triggered immunity, whereas the genes themselves coded for virulence effectors.

Effectors released by phloem-feeding insects also activate NBS–LRR receptors

Evidence from several plant species, such as rice, melon, and tomato, suggests that resistance genes recognize effectors from phloem-feeding insects and activate appropriate defenses. For example, the tomato *R* gene *Mi-1* confers resistance to aphids and whiteflies, the rice *R* gene *Bph14* confers resistance to the brown plant hopper, and the melon *R* gene *Vat* confers resistance to the cotton aphid. All of these *R* genes encode NBS–LRR receptors.

To date, several aphid salivary proteins have been identified that share functional features with the effectors of plant pathogens, including Mp10 and Mp42 in the aphid *Myzus persicae*. Overexpression of these two proteins together in tobacco (*Nicotiana benthamiana*) reduced the fecundity of aphids feeding on the transgenic plants. Interestingly, overexpression of Mp10 alone activated both jasmonic acid and salicylic acid signaling pathways in tobacco, and conferred partial resistance to the pathogenic oomycete *Phytophthora capsici*. Therefore, Mp10 and Mp42 appear to be effectors that trigger defense responses to both aphids and pathogens, even though the proteins are produced only by the aphids.

The hypersensitive response is a common defense against pathogens

A common physiological phenotype associated with effector-triggered immunity is the hypersensitive response, in which cells immediately surrounding the infection site die rapidly, depriving the pathogen of nutrients and preventing its spread. After a successful hypersensitive response, a small region of dead tissue is left at the site of the attempted invasion, but the rest of the plant is unaffected.

The hypersensitive response is often preceded by the rapid accumulation of reactive oxygen species (ROS) and nitric oxide (NO). Cells in the vicinity of the infection synthesize a burst of toxic compounds formed by the reduction of molecular oxygen, including the superoxide anion ($O_2\bullet^-$), hydrogen peroxide (H_2O_2), and the hydroxyl radical ($\bullet$OH). An NADPH-dependent oxidase located at the plasma membrane (**Figure 23.29**) is thought to produce $O_2\bullet^-$, which in turn is converted into $\bullet$OH and H_2O_2.

The hydroxyl radical is the strongest oxidant of these reactive oxygen species and can initiate radical chain reactions with a range of organic molecules, leading to lipid peroxidation, enzyme inactivation, and nucleic acid degradation. Reactive oxygen species may contribute to cell death as part of the hypersensitive response or act to kill the pathogen directly.

A rapid spike of NO production accompanies the oxidative burst in infected leaves. NO, which acts as a second messenger in many signaling pathways in animals and plants (see Chapter 15), is synthesized from the amino acid arginine by the enzyme nitric oxide (NO) synthase. An increase in the cytosolic calcium concentration appears to be required for the activation of NO synthase during the response. An increase in *both* NO and reactive oxygen species is required for the activation of the hypersensitive response: Increasing only one of these signals has little effect on the induction of cell death.

Many species react to fungal or bacterial invasion by synthesizing lignin or callose. These polymers are thought to serve as barriers, walling off the pathogen from the rest of the plant and physically blocking its spread. A related response is the modification of cell wall proteins. Certain proline-rich proteins of the wall become oxidatively cross-linked after pathogen attack in an H_2O_2-mediated reaction (see Figure 23.29). This process strengthens the walls of the cells in the vicinity of the infection site, thereby increasing their resistance to microbial digestion.

Another defense response to infection is the formation of hydrolytic enzymes that attack the cell wall of the pathogen. An assortment of glucanases, chitinases, and other hydrolases are induced by fungal invasion. Chitin, a polymer of *N*-acetylglucosamine residues, is a principal component of fungal cell walls. These hydrolytic enzymes belong to a group of **antimicrobial peptides** that are often induced during pathogen infection.

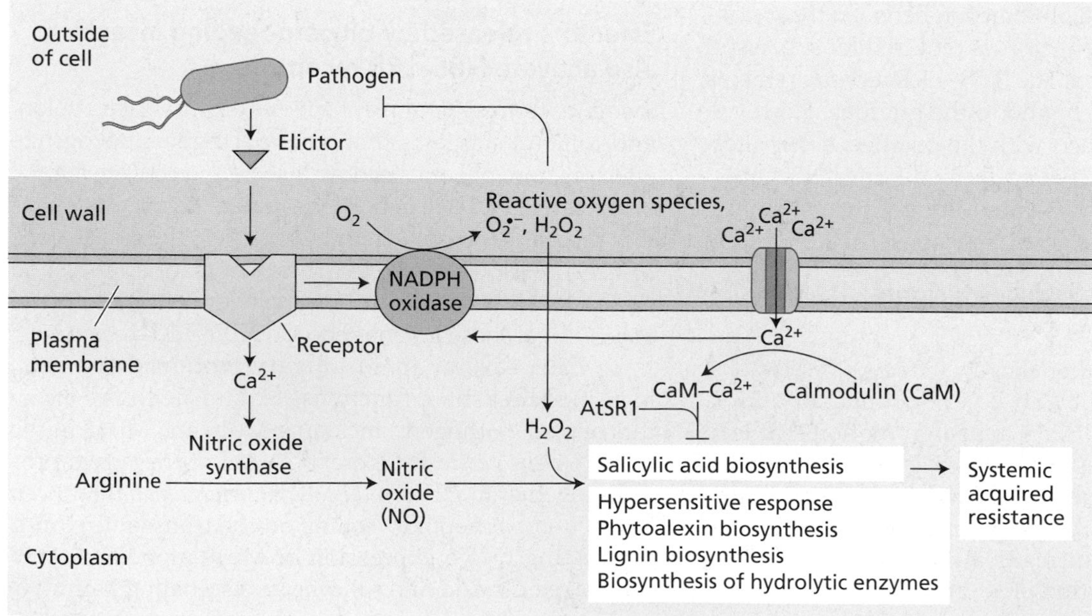

Figure 23.29 Many types of antipathogen defenses are induced by infection. Fragments of pathogen molecules called elicitors initiate a complex signaling pathway leading to the activation of defense responses. A burst of oxidation activity and nitric oxide production stimulates the hypersensitive response and other defense mechanisms. Note that Ca^{2+} is necessary for the activation of some defenses, while it is also a negative regulator of salicylic acid biosynthesis (see text for further details).

Phytoalexins with antimicrobial activity accumulate after pathogen attack

Phytoalexins are a chemically diverse group of secondary metabolites with strong antimicrobial activities that accumulate around the infection site. Phytoalexin production appears to be a common mechanism of resistance to pathogenic microbes in a wide range of plants. However, different plant families employ different types of secondary products as phytoalexins. For example, in leguminous plants, such as alfalfa and soybean, isoflavonoids are common phytoalexins, whereas in solanaceous plants, such as potato, tobacco, and tomato, various sesquiterpenes are produced as phytoalexins (**Figure 23.30**). (For discussions of the biosynthesis of these compounds, see WEB APPENDIX 4.)

Phytoalexins are generally undetectable in the plant prior to infection, but they are synthesized rapidly after microbial attack. The point of control is usually the expression of genes encoding enzymes for phytoalexin biosynthesis. Plants do not appear to store any of the enzymatic machinery required for phytoalexin synthesis. Instead, soon after microbial invasion they begin transcribing and translating the appropriate mRNAs and synthesizing the enzymes de novo.

Although phytoalexins accumulate in concentrations that have been shown to be toxic to pathogens in bioassays, the defensive significance of these compounds in the intact plant is not fully known. Experiments on genetically modified plants and pathogens have provided the first direct proof of phytoalexin function in vivo. For example, tobacco plants transformed with a gene for an enzyme catalyzing the biosynthesis of the phenylpropanoid phytoalexin resveratrol become much more resistant to a fungal pathogen than nontransformed control plants. Similarly, resistance of Arabidopsis to a fungal pathogen depends on the tryptophan-derived phytoalexin camalexin, because mutants deficient in camalexin production are more susceptible to the pathogen than wild-type Arabidopsis is. In other experiments, pathogens transformed with genes encoding phytoalexin-degrading enzymes were able to infect plants normally resistant to them.

A single encounter with a pathogen may increase resistance to future attacks

In addition to triggering defense responses locally, microbial pathogens also induce the production of signals such as salicylic acid, methyl salicylate, and other compounds that lead to systemic expression of the antimicrobial **pathogenesis-related** (**PR**) **genes**. PR genes comprise a small multigene family that encodes low-molecular-weight proteins (6–43 kD) composed of a diverse group of hydrolytic enzymes, wall-modifying enzymes, antifungal agents, and components of signaling pathways. PR proteins are localized either in vacuoles or in the apoplast and are most abundant in leaves, where they are presumed to confer protection against secondary infections. This phenomenon of local pathogen challenge enhancing resistance to secondary infection, called **systemic acquired resistance** (**SAR**), normally develops over a period of several days. SAR appears to result from increased levels of certain defense compounds that we have already mentioned, including chitinases and other hydrolytic enzymes.

Although the mechanism of SAR induction is still unknown, one of the endogenous signals is salicylic acid. The level of this benzoic acid derivative rises dramatically in the zone of infection after initial attack, and it is thought to establish SAR in other parts of the plant. However, grafting experiments in tobacco showed that infected, salicylic acid–deficient rootstocks could trigger SAR in wild-type scions. These results indicate that salicylic acid is neither the initial trigger at the infection site nor the mobile signal that induces SAR throughout the plant. Although free salicylic acid does not trigger the SAR response, there is evidence that methyl salicylate may be the mobile signal for SAR. Experiments in tobacco in which either salicylic acid methylation was blocked in the infected leaf, or methyl salicylate demethylation was inhibited in the systemic leaf, effectively prevented the SAR

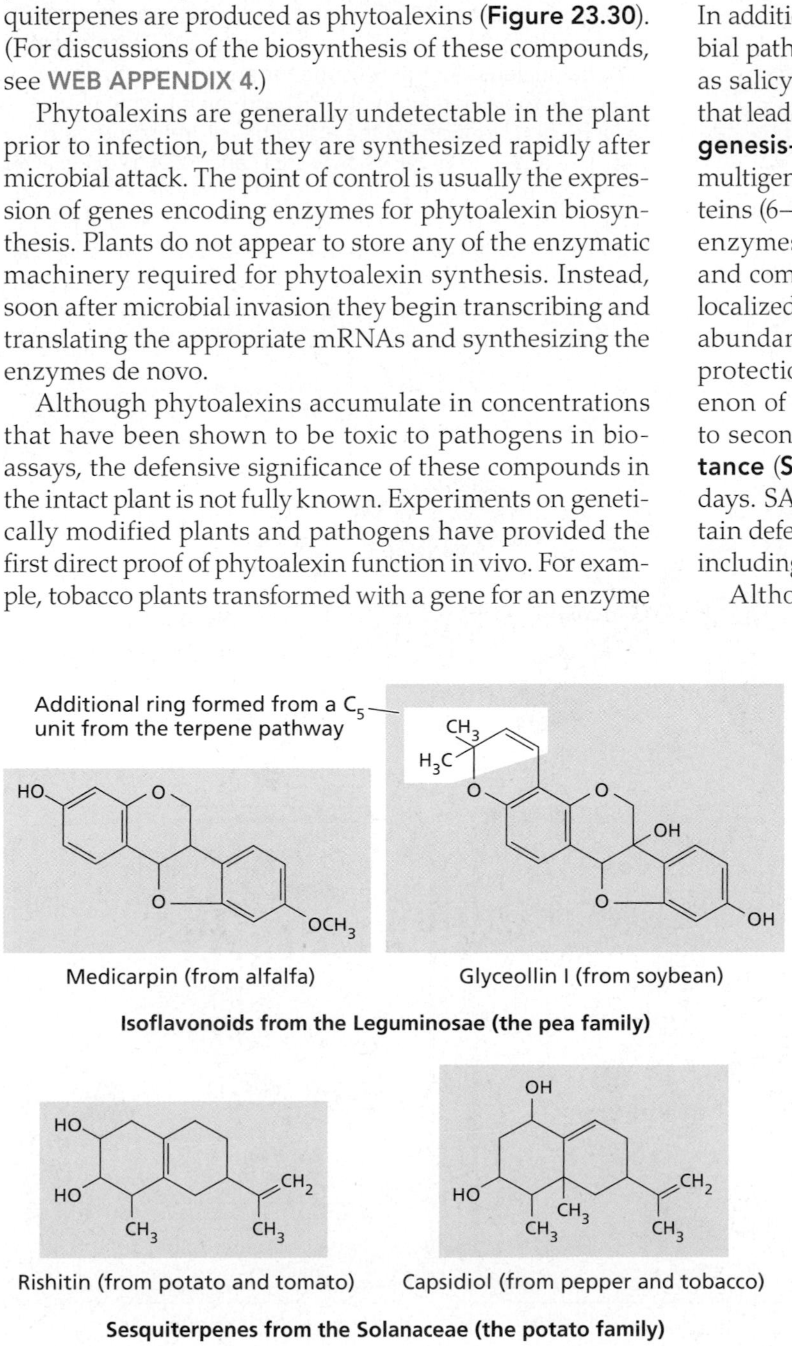

Additional ring formed from a C_5 unit from the terpene pathway

Medicarpin (from alfalfa) — Glyceollin I (from soybean)

Isoflavonoids from the Leguminosae (the pea family)

Rishitin (from potato and tomato) — Capsidiol (from pepper and tobacco)

Sesquiterpenes from the Solanaceae (the potato family)

Figure 23.30 Structure of some phytoalexins found in two different plant families.

response. Although methyl salicylate is volatile, it appears to be transported via the vascular system in tobacco.

Measurements of the rate of SAR transmission from the site of attack to the rest of the plant indicate that movement is too rapid (3 cm per h) for simple diffusion and further supports the hypothesis that the mobile signal must be transported through the vascular system. Most of the evidence points to the phloem as the primary pathway of translocation of the SAR signal. In Arabidopsis, mutations in the *DIR1* (Defective in Induced Resistance *1*) gene block the SAR response, and the *DIR1* gene is specifically expressed in the phloem. The *DIR1* gene encodes a lipid transfer protein, suggesting that the long-distance signal may involve a lipid.

In recent years several other signaling compounds that are potentially involved in the mediation of the SAR signal have been identified in plants. For example, azelaic acid, a nine-carbon dicarboxylic acid whose biosynthetic pathway is poorly understood, was found to have an essential role in the translocation of the SAR signal. Dehydroabietinal, a diterpenoid, was found to be rapidly translocated from the infection site throughout the plant and to activate SAR. Similarly, glycerol-3-phosphate, which is synthesized in the plastid, has also been implicated in long-distance SAR signaling.

While it is not clear if and how these diverse signals interact, it appears that they are all required to induce the full strength of SAR after pathogen infections.

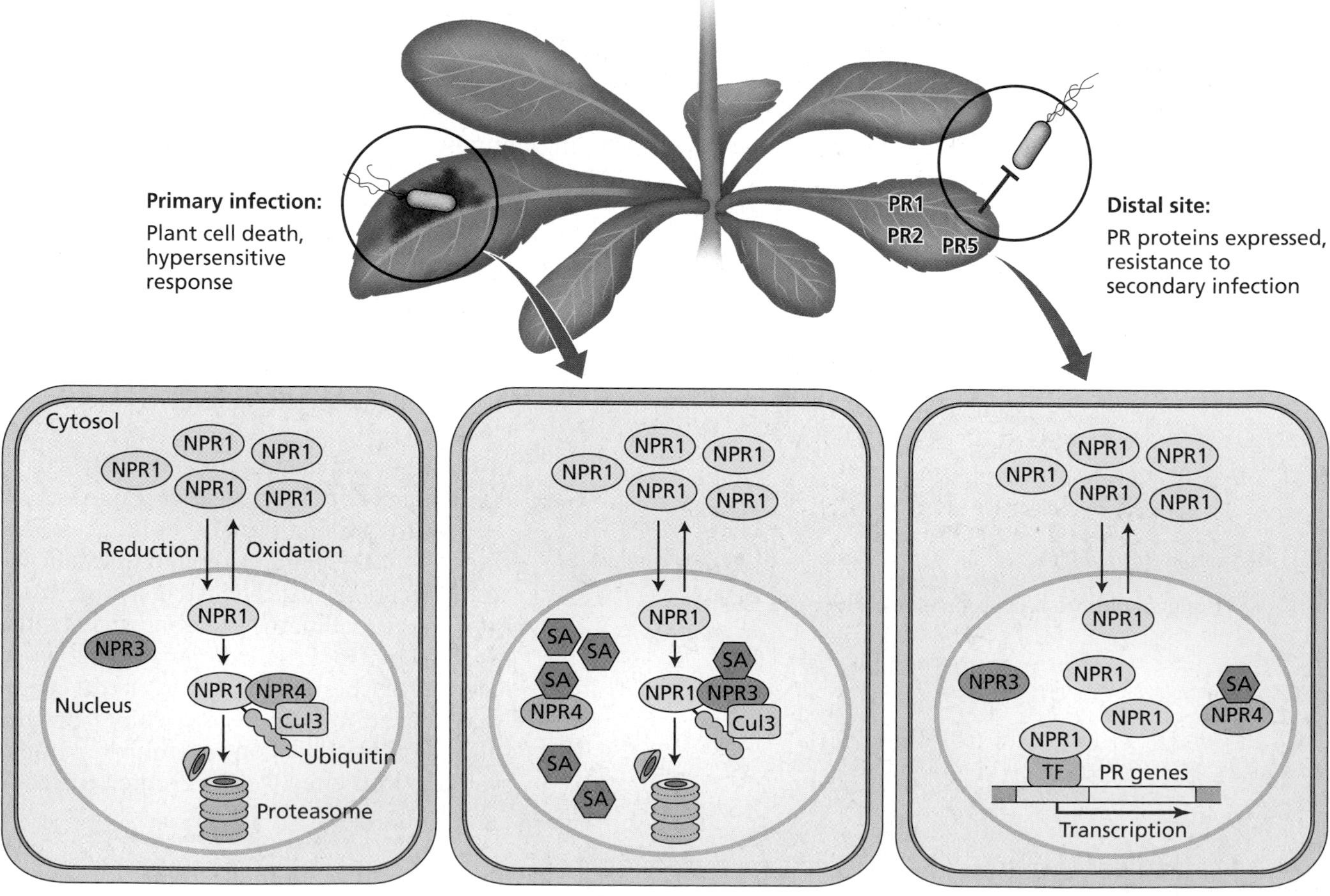

Figure 23.31 SAR signaling during the response to pathogens. A bacterial infection can induce effector-triggered immunity (ETI) and the hypersensitive response locally, as well as an increase in salicylic acid (SA) levels. The accumulation of SA causes the cellular redox state to oscillate, which releases NPR1 monomers from oligomers in the cytosol. The NPR1 monomers are then rapidly translocated into the nucleus. A high SA concentration in the nucleus promotes the association of NPR1 with NPR3, which leads to NPR1 degradation via the ubiquitin–proteasome pathway. The absence of NPR1 allows ETI and PCD to occur. The SA concentration of neighboring cells is lower. NPR1 degradation through binding to NPR3 does not occur, and NPR1 accumulates. NPR1 interacts with transcription factors (TFs) and activates gene expression involved in defenses against secondary infections. Ub, ubiquitin; Cul3, Cullin 3, a protein that functions as a scaffold for E3 ligases.

The main components of the salicylic acid signaling pathway for SAR have been identified

To identify the components of the salicylic acid signaling pathway during SAR, genetic screens were carried out to search for salicylic acid–insensitive mutants unable to synthesize PR proteins in response to salicylic acid. Multiple screens all identified a single genetic locus, *NPR1* (*nonexpressor of PR genes 1*). Subsequently, two paralogs (that is, related genes derived from gene duplication) of *NPR1* were discovered, *NPR3* and *NPR4*. Although NPR1 protein does not bind salicylic acid, NPR3 and NPR4 do, suggesting they might act as salicylic acid receptors. Structurally, the three proteins resemble adaptor proteins for the Cullin 3 E3 ubiquitin ligase pathway, suggesting that, similar to the auxin, gibberellin, and jasmonic acid receptors, they are involved in targeted protein degradation via the ubiquitin–proteasome pathway.

Figure 23.31 illustrates a model for salicylic acid regulation of both the hypersensitive response and effector-triggered immunity at the primary infection site, and SAR in distal tissues. According to the model, the function of NPR1 is to activate salicylic acid–responsive genes involved in defense, perhaps by promoting the degradation of repressor proteins. NPR1 exists in both an oligomeric form and a monomeric form. Oxidizing conditions promote oligomer formation in the cytoplasm, and reducing conditions favor the formation of monomers, which quickly enter the nucleus. Prior to infection there is little or no salicylic acid in the cell. Under these conditions, NPR1 associates with NPR4 and is rapidly degraded via the 26S proteasome pathway. This prevents defense responses from being activated unnecessarily. Upon infection, the intracellular concentration of salicylic acid rises sharply. Salicylic acid binds to NPR3, which facilitates the turnover of NPR1 via ubiquitination. The rapid destruction of NPR1 prevents the cells at the infection site from activating defense genes, and cell death ensues (the hypersensitive response).

In contrast, the concentration of salicylic acid is much lower in distal tissues, too low to bind to NPR3 but high enough to bind to NPR4 and prevent it from interacting with NPR1. Under these conditions, NPR1 accumulates and activates the massive transcriptional reprogramming involved in the SAR response. Proteins associated with the endomembrane system are also up-regulated, which allows the newly synthesisized PR proteins to be secreted into the apoplast. At the same time, epigenetic changes in chromatin structure contribute to the overall SAR syndrome.

Interactions of plants with nonpathogenic bacteria can trigger systemic resistance through a process called induced systemic resistance (ISR)

In contrast to SAR, which occurs as a consequence of actual pathogen infection, **induced systemic resistance (ISR)** is activated by nonpathogenic microbes (**Figure 23.32**). Rhizobacterium-mediated ISR is a broad-spectrum resistance response that is activated by selected strains of saprophytic rhizosphere bacteria. Beneficial rhizobacteria trigger ISR by priming the plant for potentiated activation of various cellular defense responses, which are subsequently induced upon pathogen attack. The potentiated responses include the oxidative burst, cell-wall reinforcement, accumulation of defense-related enzymes, and production of secondary metabolites.

The first evidence that potentiation of plant defense responses is involved in ISR came from experiments with carnation (*Dianthus caryophyllus*). Carnation plants develop an enhanced defensive capacity against the root rot fungus *Fusarium oxysporum* after colonization of the roots by a nonpathogenic strain of the bacterium *Pseu-*

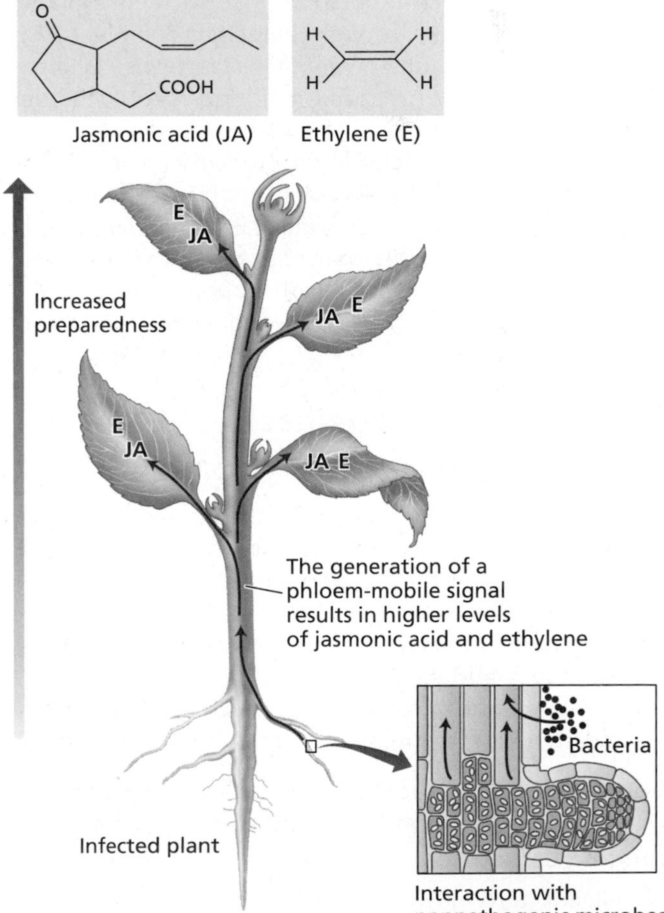

Figure 23.32 Exposure to nonpathogenic microorganisms may increase resistance to future pathogen attack through development of induced systemic resistance (ISR). Nonpathogenic microorganisms such as rhizobacteria activate signaling pathways involving jasmonic acid and ethylene that trigger ISR throughout the plant. Rather than activating immediate defensive measures, ISR is characterized by an increased level of preparedness against pathogen attack.

domonas fluorescens. Before challenge inoculation, no increase in phytoalexin levels could be detected in induced and uninduced plants, but upon subsequent inoculation with *F. oxysporum*, phytoalexin levels in ISR-expressing plants (i.e., those that had prior exposure to *P. fluorescens*) rose significantly faster than in uninduced plants.

Evidence for rhizobacterium-induced potentiation of host cell-wall strengthening has been described as well. In bean (*Phaseolus vulgaris*), a nonpathogenic strain of *Bacillus pumilus* induces ISR against *F. oxysporum*. By itself, colonization of the roots by the rhizobacteria did not induce morphological alterations of root tissue. However, upon challenge with *F. oxysporum*, root cell walls of ISR-expressing plants were rapidly strengthened at sites of attempted fungal penetration by large amounts of callose and phenolic materials, thereby effectively preventing fungal entry.

Nitrogen-fixing rhizobia can also influence plant–herbivore interactions through their effects on plant volatiles. Studies using lima bean (*Phaseolus lunatus*) showed that the presence of rhizobium-induced root nodules altered the composition of volatiles produced by the plant in response to the specialist herbivore Mexican bean beetle (*Epilachna varivestis*). Plants colonized by nitrogen-fixing rhizobia released higher amounts of indole compounds in response to jasmonic acid than they did in the absence of the rhizobia. This change in volatiles affected the behavior of beetles, which preferred the noncolonized plants over the colonized ones.

Plant Defenses against Other Organisms

While herbivorous insects and pathogenic microorganisms represent the greatest threat to plants, other organisms, including nematodes and parasitic plants, can also cause significant damage. However, relatively little is known about the factors that regulate the interactions of nematodes and parasitic plants with their respective hosts. There is, however, emerging evidence that secondary metabolites play an important role in this process.

Some plant parasitic nematodes form specific associations through the formation of distinct feeding structures

Nematodes, or roundworms, are water and soil inhabitants that often outnumber all other animals in their respective environments. Many nematodes exist as parasites relying on other living organisms, including plants, to complete their life cycle. Nematodes can cause severe losses of agricultural crops and ornamental plants. Plant parasitic nematodes can infect all parts of a plant, from roots to leaves, and may even live in the bark of forest trees. Nematodes feed through a hollow stylet that can easily penetrate plant cell walls. In the soil, nematodes can move from plant to plant, thereby causing extensive damage. Arguably the best studied among the plant parasitic nematodes are the **cyst nematodes** and those causing gall formation on infected roots, the so-called **root knot nematodes**. Both are endoparasites that depend on a living plant as host to complete their life cycles, and are therefore categorized as biotrophs. The life cycles of parasitic nematodes begins when dormant eggs recognize specific compounds secreted by the plant root (**Figure 23.33**). Once hatched, the juvenile nematodes swim to the root and penetrate it. There they migrate to the vascular tissue where they begin feeding on the cells of the vascular system.

At the permanent feeding site, usually in the root cortex, a cyst nematode larva pierces a cell with its stylet and injects saliva. As a result, the cell walls break down and neighboring cells are incorporated into a **syncytium** (see Figure 23.33A). The syncytium is a large, metabolically active feeding site that becomes multinucleate as neighboring plant cells are incorporated into it by cell-wall dissolution and cell fusion. The syncytium continues to spread centripetally toward the vascular tissue, incorporating pericycle cells and xylem parenchyma. The outer walls of the syncytium adjacent to the conducting elements form protuberances resembling those of transfer cells (see Chapter 11), indicating the syncytium now functions as a nutrient sink.

The cyst nematode, after establishing itself in such a feeding structure, grows and undergoes three molting stages while becoming a vermiform (wormlike) adult. At maturity, the female produces eggs internally, swells, and protrudes from the root surface. The mature male nematodes are released from the root into the soil and are attracted by pheromones to protruding females on the root surface. After fertilization, the female dies, forming a cyst containing the fertilized eggs.

Roots infected by root knot nematodes form large cells, resulting in the establishment of the characteristic knot or gall, which also remains in close contact with the vasculature and provides the nematode with nutrients (see Figure 23.33B).

As mentioned above, plant parasitic nematodes secrete a large number of effector molecules that affect the morphology and physiology of the plant. Among those effector molecules are also some that are specifically recognized by plants and activate defense responses through recognition by *R* gene products, as described above for plant–pathogen interactions. For example, the potato *R* gene *H1* binds specifically to nematode-derived elicitors of those strains that have a corresponding effector gene, and thus activates defense responses. Several of these plant *R* genes have been identified to date, and interestingly, all have been shown to participate in plant resistance to microbial pathogens as well.

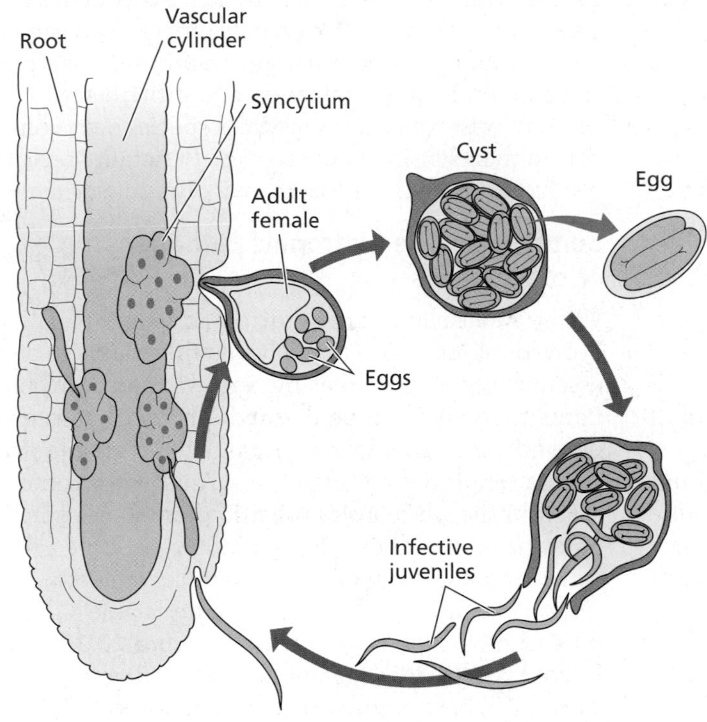

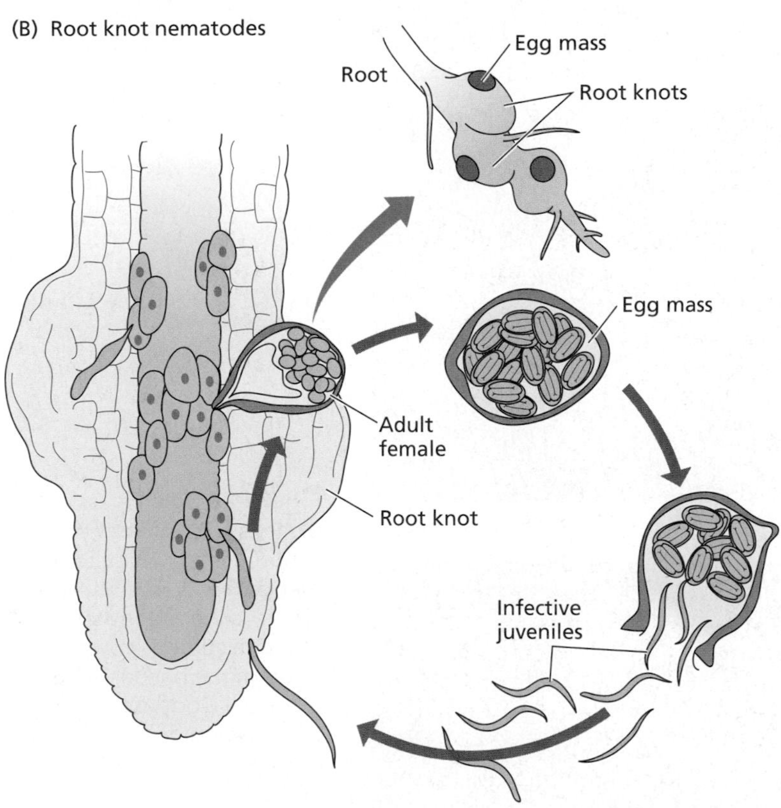

Figure 23.33 Nematodes can cause significant damage to plants. Most plant pathogenic nematodes attack the roots of plants. Free-living juvenile nematodes are attracted to secretions by the roots. After penetration, the nematode starts feeding on cells in the vasculature. (A) Cyst nematodes cause the formation of a specific feeding structure (syncytium) in the vasculature but do not cause other morphological changes. After fertilization the female cyst nematode dies thereby forming a cyst containing the fertilized eggs, from which a new generation of infective juveniles hatch. (B) Infection by root knot nematodes causes the formation of giant cells, resulting in the typical root knots. Upon maturation, the female nematode releases an egg mass from which new infective juveniles hatch and cause further infestations of plants.

Plants compete with other plants by secreting allelopathic secondary metabolites into the soil

Plants release compounds (**root exudates**) into their environment that change soil chemistry, thus increasing nutrient uptake or protecting against metal toxicity. Plants also secrete chemical signals that are essential for mediating interactions between plant roots and nonpathogenic soil bacteria, including nitrogen-fixing bacterial symbionts. However, microbes are not the only organisms that are influenced by secondary metabolites released by plant roots. Some of these chemicals are also involved in direct communication between plants. Plants release secondary metabolites to the soil to inhibit the roots of other plants, a phenomenon known as allelopathy.

Interest in allelopathy has increased in recent years because of the problem of invasive species that outcompete native species and take over natural habitats. A devastating example is the spotted knapweed (*Centaurea maculosa*), an invasive exotic weed introduced to North America that releases phytotoxic secondary metabolites into the soil. Spotted knapweed, a member of the aster family (Asteraceae), is native to Europe, where it is not a dominant or problematic species. However, in the northwestern United States it has become one of the worst invasive weeds, infesting over 1.8 million ha (~4.4 million acres) in Montana alone. Spotted knapweed often colonizes disturbed areas in North America, but it also invades rangelands,

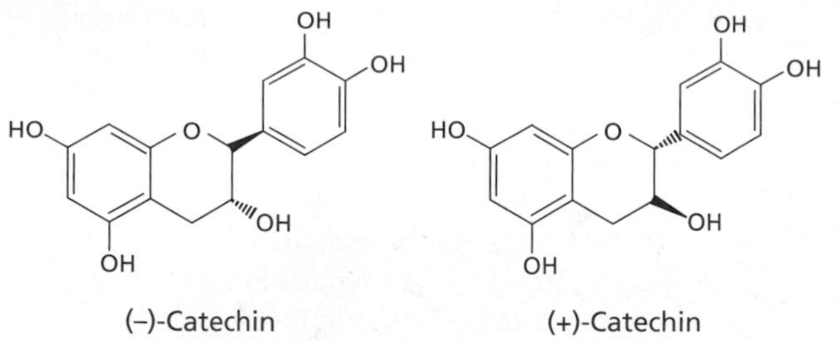

(–)-Catechin (+)-Catechin

Figure 23.34 Phytotoxic allelopathic compounds produced by spotted knapweed (*Centaurea maculosa*).

pastures, and prairies, where it displaces native species and establishes dense monocultures (see **WEB ESSAY 23.7**).

The phytotoxic secondary metabolites that spotted knapweed roots release into the soil have been identified as a racemic mixture of (±)-catechin (hereafter catechin) (**Figure 23.34**). The mechanism by which catechin acts as a phytotoxin has been elucidated. In susceptible species such as Arabidopsis, catechin triggers a wave of reactive oxygen species (ROS) initiated at the root meristem, which leads to a Ca^{2+} signaling cascade that triggers genome-wide changes in gene expression. In Arabidopsis, catechin doubled the expression of about 1,000 genes within 1 h of treatment. By 12 h many of these same genes were repressed, which may reflect the onset of cell death. Laboratory experiments examining the effects of catechin on plant germination and growth showed that native North American grassland species vary considerably in their sensitivity to catechin. Resistant species may produce root exudates that detoxify this allelochemical.

Some plants are biotrophic pathogens of other plants

While most plants are autotrophic, some plants have evolved into parasites that rely on other plants to provide essential nutrients for their own growth and development. Parasitic plants can be divided into two main groups depending on the degree of parasitism. **Hemiparasitic plants** retain the ability to perform at least some photosynthesis, while **holoparasitic plants** are completely parasitic on their host plants and have lost the ability to carry out photosynthesis. For example, mistletoe (genus *Viscum*), which has green leaves and is able to perform photosynthesis, is a hemiparasite (**Figure 23.35A and B**). In contrast, dodder (genus *Cuscuta*), which has lost the ability to photosynthesize and depends entirely on the host for sugars, is a holoparasite (**Figure 23.35C and D**).

Figure 23.35 Parasitic plants. (A) Mistletoe (*Viscum* sp.) on a mesquite tree (genus *Prosopis*). (B) Clearly visible is the green stem of the mistletoe growing through the bark of the host plant. (C) Dodder (*Cuscuta* sp.) growing on a patch of sand verbenas (*Abronia umbellata*) on dunes at the Pacific coast in California. (D) Close-up showing the high density of infestation of dodder on its host plant. (Photos © J. Engelberth.)

Parasitic plants have developed a specialized structure, the **haustorium**, which is a modified root (**Figure 23.36**). After establishing contact with its host plant, the haustorium penetrates the epidermis or bark and then the parenchyma to grow into the vascular tissue and absorb nutrients from the host. To reach the host plant, seeds of parasitic plants are either directly deposited by birds or are more randomly distributed by wind or other means. After germination, the seedlings must rely for a time on their seeds for their food supply, until they can find a suitable host. Recent research has shown that low amounts of species-specific plant volatiles may serve as cues for dodder seedlings and direct their growth toward the host. Alternatively, in the case of root parasites, such as *Striga*, compounds secreted by the host root guide the growth of the seedling roots toward the host. Upon contacting the host root, the *Striga* seedling root develops into a haustorium. The haustorium then penetrates the host root and grows directly into the vascular system through the pits of xylem vessels, where it absorbs the necessary nutrients through tubelike protoplasmic structures not covered by a cell wall.

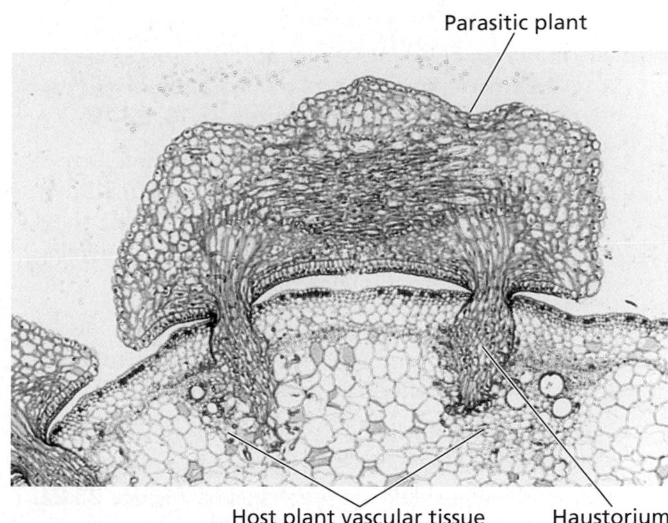

Figure 23.36 Micrograph showing the haustorium of dodder penetrating the tissues of its host plant.

The mechanisms of these interactions between parasitic plants and their hosts have been studied mostly at the morphological level, and little is known about the signaling mechanisms involved. It is clear that metabolites that are secreted or emitted as volatiles by the host plant provide important cues for the parasite. However, other factors, such as light, may also play a significant role in this process. There is also little known about the defense mechanisms of the host plant. It is likely that common defense signaling pathways, including jasmonic acid, salicylic acid, and ethylene, may play an important role in the defense against parasitic plants, but much more research is needed.

SUMMARY

Plants have evolved many strategies to cope with the threats by pests and pathogens. Strategies include sophisticated detection mechanisms and the production of toxic and repelling secondary metabolites. While some of these responses are constitutive, others are inducible. Overall, these strategies have led to a standoff in the coevolutionary race between plants and their pests.

Beneficial Interactions between Plants and Microorganisms

- Symbiotic nitrogen-fixing bacteria release Nod factors, which set off a series of reactions leading to infection and the formation of nodules (**Figures 23.2, 23.3**).
- Myc factors are released by mycorrhizal bacteria, leading to the formation of mycorrhizas.
- Rhizobacteria can release metabolites that assist plant growth by increasing nutrient availability and pathogen protection (**Figure 23.4**).

Harmful Interactions between Plants, Pathogens, and Herbivores

- Mechanical barriers that provide a first line of defense against pests and pathogens include thorns, spines, prickles, trichomes, and raphides (**Figures 23.5–23.8**).
- Secondary metabolites that serve defensive functions are stored in specialized structures that release their contents only upon damage (**Figures 23.10–23.12**).
- Some secondary metabolites are stored in the vacuole as water-soluble sugar conjugates that are spatially separated from their activating enzymes (**Figures 23.14–23.16**).

Inducible Defense Responses to Insect Herbivores

- Rather than producing defensive secondary metabolites continuously, plants can save energy by producing defensive compounds only when induced by mechanical damage or specific components of insect saliva (elicitors) (**Figure 23.17**).

- Jasmonic acid (JA) increases rapidly in response to insect damage and induces transcription of genes involved in plant defense (**Figures 23.18, 23.19**).
- Herbivore damage can induce systemic defenses by causing the synthesis of polypeptide signals. For example, systemin is released to the apoplast and binds to receptors in undamaged tissues, activating JA synthesis there (**Figure 23.20**).
- In addition to polypeptide signals, plants can also project electrical signals to initiate defense responses in as-yet undamaged tissues (**Figure 23.21**).
- Plants may release volatile compounds to attract natural enemies of herbivores, or to signal neighboring plants to initiate defense mechanisms (**Figure 23.22**).

Plant Defenses against Pathogens

- Pathogens can invade plants through cell walls by secreting lytic enzymes, through natural openings such as stomata and lenticels, and through wounds. Insect herbivores may also be pathogen vectors (**Figure 23.25**).
- Pathogens generally use one of three attack strategies: necrotrophism, biotrophism, or hemibiotrophism.
- Pathogens often produce effector molecules (enzymes, toxins, or growth regulators) that aid in initial infection (**Figure 23.26**).
- All plants have pattern recognition receptors (PRRs) that set off defense responses when activated by evolutionarily conserved microbe-associated molecular patterns (MAMPs; e.g., flagella, chitin) (**Figure 23.27**).
- Plant resistance (*R*) genes encode cytosolic receptors that recognize pathogen-derived effector gene products in the cytosol. Binding of an effector gene product to its receptor initiates antipathogen signaling pathways (**Figure 23.28**).
- Another antipathogen defense is the hypersensitive response, in which cells surrounding the infected site die rapidly, thereby limiting the spread of infection. The hypersensitive response is often preceded by rapid production of ROS and NO, which may kill the pathogen directly or aid in cell death (**Figure 23.29**).
- In response to infection, many plants produce phytoalexins, secondary metabolites with strong antimicrobial activity (**Figure 23.30**).
- A plant that survives local pathogen infection often develops increased resistance to subsequent attack, a phenomenon called systemic acquired resistance (SAR) (**Figure 23.31**).
- Interactions with nonpathogenic bacteria can trigger induced systemic resistance (ISR) (**Figure 23.32**).

Plant Defenses against Other Organisms

- Nematodes (roundworms) are parasites that can move between hosts and that induce formation of feeding structures and galls from vascular plant tissues. In response, plants use defensive signaling pathways similar to those used for pathogen infection (**Figure 23.33**).
- Some plants produce allelopathic secondary metabolites that enable them to outcompete nearby plant species.
- Some plants are parasitic on other plants. Parasitic plants can be divided into two main groups (hemiparasites and holoparasites) depending on their ability to perform some photosynthesis (**Figure 23.35**).
- Parasitic plants use a specialized structure, the haustorium, to penetrate their host, grow into the vasculature, and absorb nutrients (**Figure 23.36**).
- Some parasitic plants detect their host by the specific volatile profile that is constitutively released.

WEB MATERIAL

- **WEB TOPIC 23.1 Cutin, Waxes, and Suberin** Plant surfaces are covered with layers of lipid material protecting them against water losses and blocking the entry of pathogenic microorganisms.

- **WEB ESSAY 23.1 Unraveling the Function of Secondary Metabolites** Wild tobacco plants use alkaloids and terpenoids to defend themselves against herbivores.
- **WEB ESSAY 23.2 Early Signaling Events in the Plant Wound Response** A complex signaling network, which includes reactive oxygen species and rapid ion fluxes, is rapidly activated in wounded plants.
- **WEB ESSAY 23.3 Jasmonates and Other Fatty Acid–Derived Signaling Pathways in the Plant Defense Response** The importance of fatty acid–derived signaling pathways as regulators of diverse

plant defense strategies is becoming increasingly recognized. The complexity of the individual pathways and their mutual interactions are discussed in the context of direct and indirect defense strategies.

- **WEB ESSAY 23.4 The Systemin Receptor** The systemin receptor from tomato is an LRR-receptor kinase.
- **WEB ESSAY 23.5 The Plant Volatilome** The release of volatile organic compounds by plants provides an example of the diversity of secondary metabolites and the ecological implications thereof.
- **WEB ESSAY 23.6 Smelling the Danger and Getting Prepared: Volatile Signals as Priming Agents in the Defense Response** By releasing volatiles, herbivore-damaged plants not only attract natural enemies of the attacking insect herbivore, but also signal this event to neighboring plants, allowing them to prepare their defenses against impending herbivory.
- **WEB ESSAY 23.7 Secondary Metabolites and Allelopathy in Plant Invasions: A Case Study of *Centaurea maculosa*** The invasive weed *Centaurea maculosa*, which is rapidly taking over pastureland in the western United States, secretes the polyphenol catechin into the rhizosphere, which suppresses the growth and germination of neighboring plants.

available at plantphys.net

Suggested Reading

Belkhadir, Y., Yang, L., Hetzel, J., Dangl, J. L., and Chory, J. (In press 2014) The growth-defense pivot: Crisis management in plants mediated by LRR-RK surface receptors. *Trends Biochem Sci.* DOI: 10.1016/j.tibs.2014.06.006.

Elzinga, D. A., and Jander, G. (2013) The role of protein effectors in plant-aphid interactions. *Curr. Opin. Plant Biol.* 16: 451–456. DOI: 10.1016/j.pbi.2013.06.018.

Gleadow, R. M., and Møller, B. L. (2014) Cyanogenic glycosides: Synthesis, physiology, and phenotypic plasticity. *Annu. Rev. Plant Biol.* 65: 155–185. DOI: 10.1146/annurev-arplant-050213-040027.

Holeski, L. M., Jander, G., and Agrawal, A. A. (2012) Transgenerational defense induction and epigenetic inheritance in plants. *Trends Ecol. Evol.* 27: 618–626. DOI: 10.1016/j.tree.2012.07.011.

Jung, S. C., Martinez-Medina, A., Lopez-Raez, J. A., and Pozo, M. J. (2012) Mycorrhiza-induced resistance and priming of plant defenses. *J. Chem. Ecol.* 38: 651–664. DOI: 10.1007/s10886-012-0134-6.

Kachroo, A., and Robin, G. P. (2013) Systemic signaling during plant defense. *Curr. Opin. Plant Biol.* 16: 527–533. DOI: 10.1016/j.pbi.2013.06.019.

Kandoth, P. K., and Mitchum, M. G. (2013) War of the worms: How plants fight underground attacks. *Curr. Opin. Plant Biol.* 16: 457–463. DOI: 10.1016/j.pbi.2013.07.001.

Kazan, K., and Lyons, R. (2014) Intervention of phytohormone pathways by pathogen effectors. Plant Cell 26: 2285–2309.

Romeis, T., and Herde, M. (2014) From local to global: CDPKs in systemic defense signaling upon microbial and herbivore attack. *Curr. Opin. Plant Biol.* 20: 1–10. DOI: 10.1016/j.pbi.2014.03.002.

Selosse, M. A., Bessis, A., and Pozo, M. J. (In press 2014) Microbial priming of plant and animal immunity: Symbionts as developmental signals. *Trends Microbiol.* DOI: 10.1016/j.tim.2014.07.003.

Yan, S., and Dong, X. (2014) Perception of the plant immune signal salicylic acid. *Curr Opin Plant Biol.* 20: 64–68. DOI: 10.1016/j.pbi.2014.04.006

24

Abiotic Stress

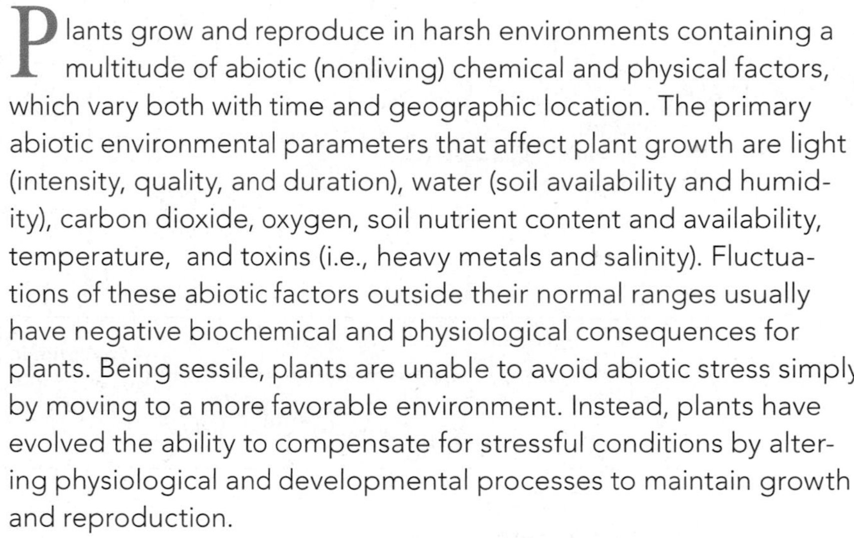

Plants grow and reproduce in harsh environments containing a multitude of abiotic (nonliving) chemical and physical factors, which vary both with time and geographic location. The primary abiotic environmental parameters that affect plant growth are light (intensity, quality, and duration), water (soil availability and humidity), carbon dioxide, oxygen, soil nutrient content and availability, temperature, and toxins (i.e., heavy metals and salinity). Fluctuations of these abiotic factors outside their normal ranges usually have negative biochemical and physiological consequences for plants. Being sessile, plants are unable to avoid abiotic stress simply by moving to a more favorable environment. Instead, plants have evolved the ability to compensate for stressful conditions by altering physiological and developmental processes to maintain growth and reproduction.

In this chapter we will provide an integrated view of how plants adapt and respond to abiotic stresses in the environment. Like all living organisms, plants are complex biological systems comprising thousands of different genes, proteins, regulatory molecules, signaling agents, and chemical compounds that form hundreds of interlinked pathways and networks. Under normal growing conditions, the different biochemical pathways and signaling networks must act in a coordinated manner to balance environmental inputs with the plant's genetic imperative to grow and reproduce. When exposed to unfavorable environmental conditions, this complex interactive system adjusts *homeostatically* to minimize the negative impacts of stress and maintain metabolic equilibrium (**Figure 24.1**).

We will begin by distinguishing between adaptation and acclimation in relation to abiotic stress. Next we will describe the various abiotic factors in the environment that can negatively affect plant growth and development. In the remainder of the chapter we will discuss plant stress-sensing mechanisms and the processes that transform sensory signals into physiological responses. Finally we will describe the specific metabolic, physiological, and anatomical changes that result from these signaling pathways and that enable plants to adapt or acclimate to abiotic stress.

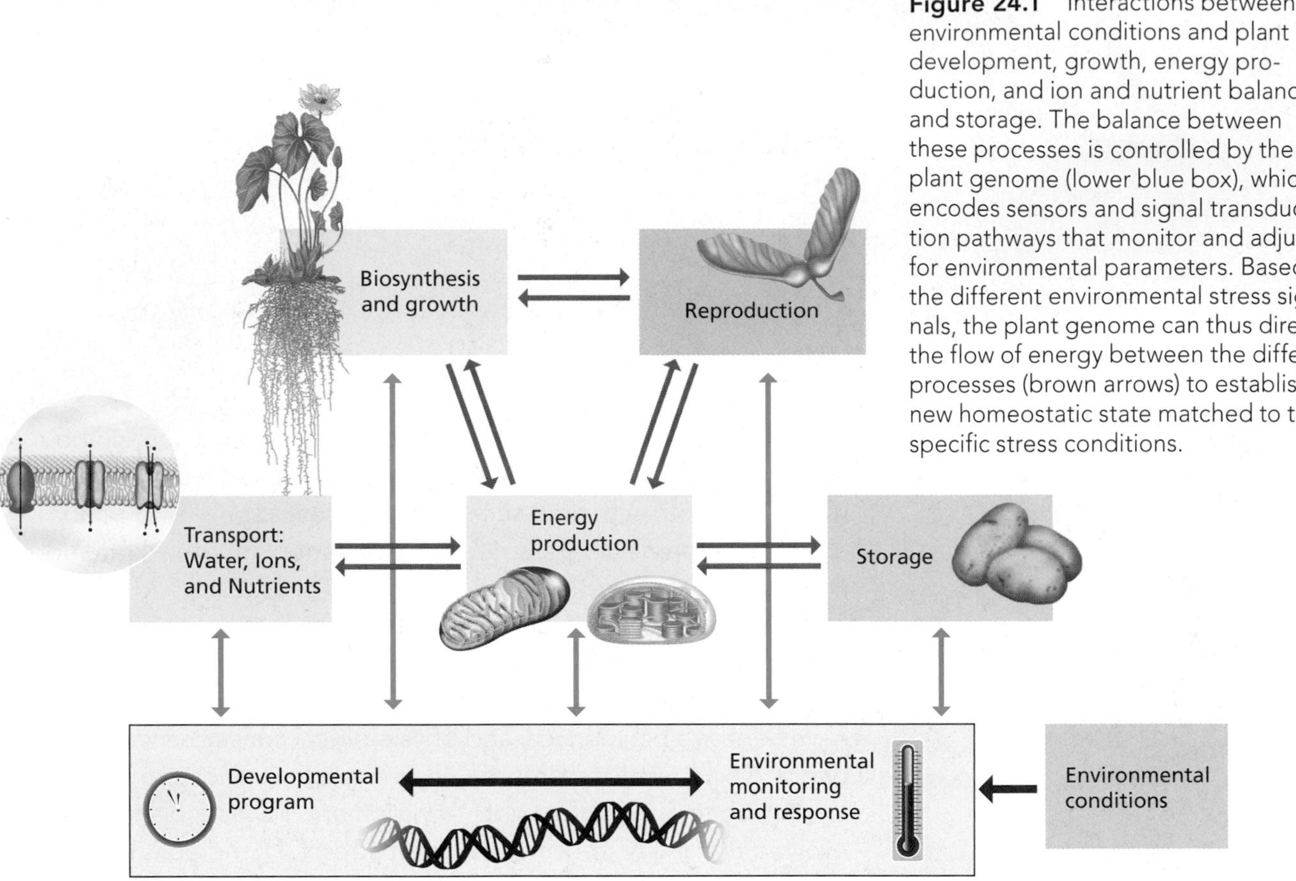

Figure 24.1 Interactions between environmental conditions and plant development, growth, energy production, and ion and nutrient balance and storage. The balance between these processes is controlled by the plant genome (lower blue box), which encodes sensors and signal transduction pathways that monitor and adjust for environmental parameters. Based on the different environmental stress signals, the plant genome can thus direct the flow of energy between the different processes (brown arrows) to establish a new homeostatic state matched to the specific stress conditions.

Defining Plant Stress

The ideal growth conditions for a given plant can be defined as the conditions that allow the plant to achieve its maximum growth and reproductive potential as measured by plant weight, height, and seed number, which together comprise the *total biomass* of the plant. **Stress** can be defined as any environmental condition that prevents the plant from achieving its full genetic potential. For example, a decrease in light intensity would cause a reduction in photosynthetic activity with a concomitant decrease in the energy supply to the plant. Under these conditions, the plant could compensate either by slowing down biosynthesis, thus reducing its growth rate, or by drawing on its stored food reserves in the form of starch (see Figure 24.1).

Similarly, a decrease in water availability would also have a deleterious effect on growth. One way that plants compensate for a decrease in water potential is by closing their stomata, which reduces water loss by transpiration. However, stomatal closure also decreases CO_2 uptake by the leaf, thereby reducing photosynthesis and suppressing growth. An example of the effects of two different drought treatments (moderate and severe) on the growth of rice plants is shown in **Figure 24.2**. Rice is able to tolerate moderate drought without any measurable effect on growth, but severe drought strongly inhibits vegetative growth.

Physiological adjustment to abiotic stress involves trade-offs between vegetative and reproductive development

How do changes in environmental conditions affect seed production? Under optimal growing conditions, the competition for resources among the different plant organs or developmental phases is minimal. The transition to reproductive growth occurs only after the vegetative adult phase completes its genetically determined developmental program (see Chapter 20). Under stress conditions, however, the vegetative growth program may terminate prematurely, and the plant may go immediately to the reproductive phase. In this case, the plant undergoes a transition to flowering, fertilization, and seed set before the plant has reached its full size, resulting in a smaller plant (see Figure 24.2). With fewer leaves to provide photosynthate, plants growing under suboptimal conditions may also produce fewer and smaller seeds.

The particular developmental pathway that a plant uses to maximize its reproductive potential under abiotic stress depends to a large extent on the plant's life cycle. For example, *annual plants* complete their life cycle in a

Figure 24.2 Comparison of control (non-drought) and drought-stressed rice plants. Whereas a moderate level of drought does not have a significant effect on plant growth, severe drought reduces the growth of rice plants. (Courtesy of Eduardo Blumwald.)

single season. It is thus advantageous for annual plants to adjust their metabolism and developmental programs so as to produce the maximum number of viable seeds under whatever environmental conditions are encountered during the season. By contrast, *perennial plants*, which have multiple seasons in which to produce seeds, tend to adjust their metabolism and developmental programs to ensure the optimal storage of food resources that will enable the plants to survive to the next season, even at the expense of seed production.

Acclimation and Adaptation

Individual plants respond to changes in the environment by directly altering their physiology or morphology to enhance survival and reproduction. Such responses require no new genetic modifications. If the response of the individual plant improves with repeated exposure to the environmental stress, then the response is termed **acclimation**. Acclimation represents a nonpermanent change in the physiology or morphology of the individual that can be reversed if the prevailing environmental conditions change. Epigenetic mechanisms that alter the expression of genes without changing the genetic code of an organism can extend the duration of acclimation responses and make them heritable. When genetic changes in an entire plant population have been fixed over many generations by selective environmental pressure, those changes are referred to as **adaptation**.

Adaptation to stress involves genetic modification over many generations

A remarkable example of adaptation to an extreme abiotic environment is the growth of plants in serpentine soils. Serpentine soils are characterized by low moisture, low concentrations of macronutrients, and elevated levels of heavy metals. These conditions would result in severe stress conditions for most plants. However, it is not unusual to find populations of plants that have become genetically adapted to serpentine soils growing not far from closely related nonadapted plants growing on "normal" soils. Simple transplant experiments have shown that only the adapted populations can grow and reproduce on the serpentine soil, and genetic crosses reveal the stable genetic basis of this adaptation.

The evolution of adaptive mechanisms in plants to a particular set of environmental conditions generally involves processes that allow *avoidance* of the potentially damaging effects of these conditions. For example, populations of the weed Yorkshire fog grass (*Holcus lanatus*) that are adapted to growth on arsenic-contaminated mine sites in southwestern England contain a specific genetic modification that reduces the uptake of arsenate, allowing the plants to avoid arsenic toxicity and thrive on contaminated mine sites. In contrast, populations growing on uncontaminated soils are less likely to contain this genetic modification.

Acclimation allows plants to respond to environmental fluctuations

In addition to genetic changes in entire populations, individual plants may acclimate to periodic changes in the environment by directly altering their morphology or physiology. The physiological changes associated with acclimation require no genetic modifications, and many are reversible. One example of acclimation from gardening is a process known as *hardening off*. To speed up the growth of plants, gardeners often start by growing them indoors in pots under optimal growth conditions. The gardeners then move the plants outdoors for part of the day over a period long enough to acclimate, or "harden," the plants to outdoor weather before moving them outdoors permanently.

Both genetic adaptation and acclimation can contribute to the plants' overall tolerance of extremes in their abiotic environment. In the example above, genetic adaptation in the arsenic-tolerant Yorkshire fog grass population only *reduces* arsenate uptake—it does not stop it. To mitigate the toxic effects of the arsenate that does accumulate, the adapted plants use the same biochemical mechanism that nonadapted plants use to respond to the toxic effects of arsenate accumulation in tissues. This mechanism involves the biosynthesis of low-molecular-weight, metal-binding molecules called *phytochelatins* (discussed in more detail later in the chapter), which reduce arsenic toxicity. Thus, the ability of Yorkshire fog grass to thrive on arsenic-contaminated mine waste depends on both a genetic adaptation specific to the tolerant population (arsenate exclusion; see *Exclusion and internal tolerance mechanisms allow plants to cope with toxic ions* later in the chapter) and on acclimation, which is common to all plants that respond to arsenic by producing phytochelatins.

Another example of acclimation is the response of salt-sensitive plants, termed *glycophytic* plants, to salinity. Although glycophytic plants are not genetically adapted to growth in saline environments, when exposed to elevated salinity they can activate several stress responses that allow the plants to cope with the physiological perturbations imposed by elevated salinity in their environment. For example, the SOS pathway (a signaling pathway discovered in *salt overly sensitive* mutants) leads to enhanced efflux of Na^+ from cells and a reduction in salinity-induced toxicity.

Environmental Factors and Their Biological Impacts on Plants

In this section we will briefly describe the ways in which various environmental stresses can disrupt plant metabolism. As with every biological system, plant survival and growth depends on complex networks of coupled anabolic and catabolic pathways that direct the flow of energy and resources within and between cells. Disruption of these networks by environmental factors can result in uncoupling of these pathways. For example, metabolic enzymes can, and often do, have different temperature optima. An increase or decrease in ambient temperature can inhibit a subset of enzymes without affecting other enzymes in the same or connected pathways. Such functional uncoupling of metabolic pathways could result in the accumulation of intermediate compounds that could be converted to toxic byproducts.

One of the most common group of toxic intermediates produced by stress are **Reactive oxygen species (ROS)** that are highly reactive forms of oxygen possessing at least one unpaired electron in their orbitals. They are capable of rapidly reacting with, and oxidizing, a wide variety of cellular constituents, including proteins, DNA, RNA, and lipids. The most common forms of ROS in plant cells are superoxide ($O_2{\bullet}^-$), singlet oxygen ($^1O_2$), hydrogen peroxide

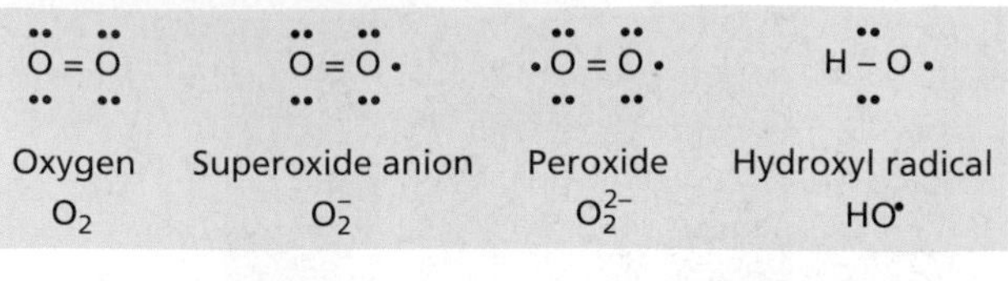

Figure 24.3 Chemistry of reactive oxygen species (ROS). Molecular oxygen does not have any available unpaired electrons in its orbitals, but different forms of ROS have at least one available unpaired electron and can accept electrons (e^-) from various cellular molecules, causing their oxidation.

(H_2O_2), and hydroxyl radicals (OH•) (**Figure 24.3**). ROS can also trigger an autocatalytic process of membrane oxidation that can result in the degradation of organelles and the plasma membrane, and cell death. Despite their mechanistic differences, most abiotic stresses result in the production of ROS (**Figure 24.4**).

Environmental stress can also disrupt compartmentalization of metabolic processes that isolates them from other cellular components. The same temperature extremes that can inhibit enzyme activity can also affect membrane fluidity: high temperature causes increased fluidity, and low temperature causes decreased fluidity.

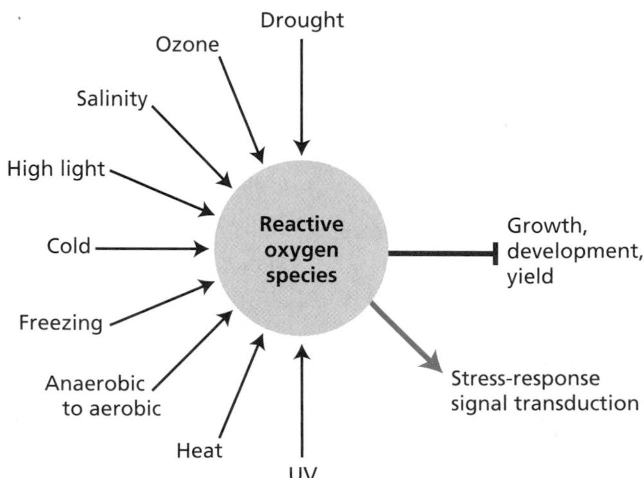

Figure 24.4 Dual role of reactive oxygen species (ROS) during abiotic stress. A variety of abiotic stresses result in the accumulation of ROS in cells. On the one hand, ROS have a negative effect on plant growth, development, and yield. On the other hand, ROS accumulation has a positive effect on cells by activating signal transduction pathways that induce acclimation mechanisms. These, in turn, counteract the negative effects of stress (including ROS accumulation).

Changes in membrane fluidity can disrupt the coupling between different protein complexes in chloroplast or mitochondrial membranes, resulting in the uncontrolled transfer of electrons to oxygen and the formation of reactive oxygen species.

Water deficit decreases turgor pressure, increases ion toxicity, and inhibits photosynthesis

As in most other organisms, water makes up the largest proportion of the cellular volume in plants and is the most limiting resource. About 97% of water taken up by plants is lost to the atmosphere (mostly by transpiration). About 2% is used for volume increase or cell expansion, and 1% for metabolic processes, predominantly photosynthesis (see Chapters 3 and 4). Water deficit (insufficient water availability) occurs in most natural and agricultural habitats and is caused mainly by intermittent to continuous periods without precipitation. *Drought* is the meteorological term for a period of insufficient precipitation that results in plant water deficit. However, this definition is somewhat misleading, since a crop can absorb water from the soil under conditions without rainfall, depending on the soil's water-holding capacity and the depth of the water table.

Water deficit can affect plants differently during vegetative versus reproductive growth. When plant cells experience water deficit, cell dehydration occurs. Cell dehydration adversely affects many basic physiological processes (**Table 24.1**). For example, during water deficit the water potential (Ψ) of the apoplast becomes more negative than that of the symplast, causing reductions in pressure potential (turgor) (Ψ_p) and volume. A secondary effect of cell dehydration is that ions become more con-

TABLE 24.1 Physiological and biochemical perturbations in plants caused by fluctuations in the abiotic environment

Environmental factor	Primary effects	Secondary effects
Water deficit	Water potential (Ψ) reduction Cell dehydration Hydraulic resistance	Reduced cell/leaf expansion Reduced cellular and metabolic activities Stomatal closure Photosynthetic inhibition Leaf abscission Altered carbon partitioning Cytorrhysis Cavitation Membrane and protein destabilization ROS production Ion cytotoxicity Cell death
Salinity	Water potential (Ψ) reduction Cell dehydration Ion cytotoxicity	Same as for water deficit (see above)
Flooding and soil compaction	Hypoxia Anoxia	Reduced respiration Fermentative metabolism Inadequate ATP production Production of toxins by anaerobic microbes ROS production Stomatal closure
High temperature	Membrane and protein destabilization	Photosynthetic and respiratory inhibition ROS production Cell death
Chilling	Membrane destabilization	Membrane dysfunction
Freezing	Water potential (Ψ) reduction Cell dehydration Symplastic ice crystal formation	Same as for water deficit (see above) Physical destruction
Trace element toxicity	Disturbed cofactor binding to proteins and DNA ROS production	Disruption of metabolism
High light intensity	Photoinhibition ROS production	Inhibition of PSII repair Reduced CO_2 fixation

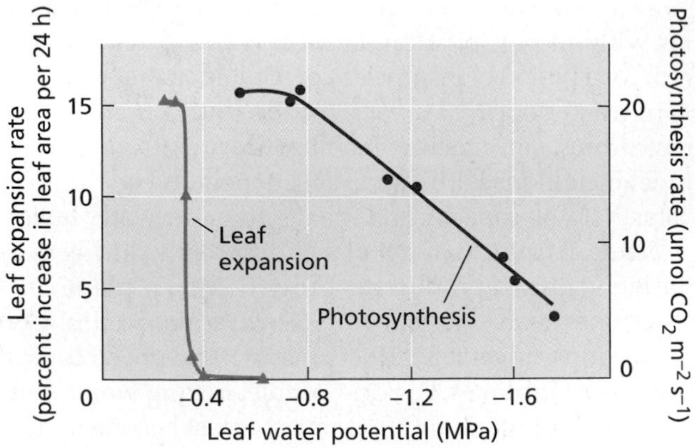

Figure 24.5 Effects of water stress on photosynthesis and leaf expansion of sunflower (*Helianthus annuus*). In this species, leaf expansion is completely inhibited under mild stress levels that barely affect photosynthetic rates. (After Boyer 1970.)

centrated and may become cytotoxic. Water deficit also induces the accumulation of abscisic acid (ABA), which promotes stomatal closure, reducing gas exchange and inhibiting photosynthesis (**Figure 24.5**). As a result of dehydration-induced uncoupling of the photosystems, free electrons produced by the reaction centers are not transferred to $NADP^+$, leading to the generation of ROS. Excess ROS damage DNA, inhibit protein synthesis, oxidize photosynthetic pigments, and cause the peroxidation of membrane lipids.

Salinity stress has both osmotic and cytotoxic effects

Excess soil salinity brought about by a combination of over-irrigation and poor soil drainage affects large regions of the world's land mass and has a severe impact on agriculture. It is estimated that 20% of all irrigated land is currently affected by salinity stress. Salinity stress has two components: nonspecific **osmotic stress** that causes water deficits, and specific ion effects resulting from the accumulation of toxic ions, which interfere with nutrient uptake and cause cytotoxicity. Salt-tolerant plants genetically adapted to salinity are termed **halophytes** (from the Greek word *halo*, "salty"), while less salt-tolerant plants that are not adapted to salinity are termed **glycophytes** (from the Greek word *glyco*, "sweet"). Under nonsaline conditions, the cytosol of higher plant cells contains about 100 m*M* K^+ and less than 10 m*M* Na^+, an ionic environment in which enzymes are optimally functional. In saline environments, cytosolic Na^+ and Cl^- increase to more than 100 m*M*, and these ions become cytotoxic. High concentrations of salt cause protein denaturation and membrane destabilization by reducing the hydration of these macromolecules. However, Na^+ is a more potent denaturant than K^+. At high concentrations, apoplastic Na^+ also competes for sites on transport proteins that are necessary for high-affinity uptake of K^+ (see Chapter 6), an essential macronutrient (see Chapter 5).

The effects of high salinity in plants occur through a two-phase process: a fast response to the high osmotic pressure at the root–soil interface and a slower response caused by the accumulation of Na^+ (and Cl^-) in the leaves. In the osmotic phase there is a reduction in shoot growth, with reduced leaf expansion and inhibition of lateral bud formation. The second phase starts with the accumulation of toxic amounts of Na^+ in the leaves, leading to the inhibition of photosynthesis and biosynthetic processes. Although in most species Na^+ reaches toxic concentrations before Cl^-, some plant species, such as citrus, grapevine, and soybean, are highly sensitive to excess Cl^-.

Light stress can occur when shade-adapted or shade-acclimated plants are subjected to full sunlight

Light stress can occur when excess high-intensity light absorbed by the plant overwhelms the capacity of the photosynthetic machinery to convert this light into sugars, as in the case of a shade-adapted or shade-acclimated plant suddenly subjected to full sunlight. In response to shade, most land plants either add more light-harvesting chlorophyll units (LHCII) to PSII, augmenting antenna size, or increase the number of PSII reaction centers relative to PSI to enhance light capture and energy transfer (see Chapter 7). If the shade-adapted or shade-acclimated plants are suddenly subjected to full sunlight, the excess light energy absorbed by the enlarged antenna complexes and transferred to the reaction centers can overwhelm the dark reaction's ability to convert the energy into sugars. Instead, the electrons feeding into the reaction centers are diverted to atmospheric oxygen, generating ROS, which can, in turn, cause cellular damage.

Temperature stress affects a broad spectrum of physiological processes

Temperature stress disrupts plant metabolism because of its differential effect on protein stability and enzymatic reactions, causing the uncoupling of different reactions and the accumulation of toxic intermediates and ROS. Heat stress increases membrane fluidity, whereas cold stress decreases membrane fluidity, causing the uncoupling of different multiprotein complexes, disruption of electron flow and energetic reactions, and disruption of ion homeostasis and regulation. Heat and cold stress can also destabilize and melt, or overstabilize and harden, RNA and DNA secondary structures, respectively, causing the disruption of transcription, translation, or RNA processing and turnover. In addition, temperature stress can block protein degradation, causing the buildup of protein aggregates. Such protein clumps disrupt normal cellular functions by interfering with the function of the cytoskeleton and associated organelles.

Flooding results in anaerobic stress to the root

When a field is flooded, the O_2 levels at the root surface decrease dramatically because most of the air in the soil is displaced by water, and the O_2 concentration of water is significantly lower than that of air: the atmosphere contains about 20% O_2 or 200,000 ppm, compared with less than 10 ppm dissolved O_2 in flooded soil. Under these conditions, respiration in roots is suppressed and fermentation is enhanced. This metabolic shift can cause energy depletion, acidification of the cytosol, and toxicity from ethanol accumulation. As a consequence of energy depletion, many processes such as protein synthesis are suppressed. Anaerobic stress can cause cell death within hours or days, depending on the degree of genetic adaptation of the species.

Even if the O_2-deprived plant is returned to normal O_2 levels, the recovery process itself can pose a hazard. While the roots are under anaerobic stress, the absence of O_2 prevents the formation of ROS. But if the O_2 level in the soil is rapidly increased, much of it is used to form ROS, causing oxidative damage to the root cells.

During freezing stress, extracellular ice crystal formation causes cell dehydration

Plants subjected to freezing temperatures must contend with the formation of ice crystals, either extracellularly or intracellulary. Intracellular ice crystal formation nearly always proves lethal to the cell. However, the water in the apoplast is relatively dilute and therefore has a higher freezing point than that of the more concentrated symplast. As a result, ice crystals tend to form in the apoplast and in the xylem tracheids and vessels, along which the ice can quickly propagate. The formation of ice crystals lowers the apoplastic water potential (Ψ), which becomes more negative than that of the symplast. Unfrozen water within the cell moves down this gradient out of the cell toward the ice crystals in the intercellular spaces. As water leaves the cell, the plasma membrane contracts and pulls away from the cell wall. During this process the plasma membrane, rigidified by the low temperature, may become damaged. The colder the temperatures, the more water travels down the gradient toward the frozen water. For example, at 14°F (–10°C), the symplast loses about 90% of its osmotically active water to the apoplast. In this respect, freezing stress has much in common with drought stress. As with drought stress, cells that are already dehydrated, like those in seeds and pollen, are less likely to undergo further dehydration by extracellular ice crystal formation.

Heavy metals can both mimic essential mineral nutrients and generate ROS

The uptake of heavy metals such as cadmium (Cd), arsenic (As), and aluminum (Al) by the plant cell can lead to the accumulation of ROS, inhibition of photosynthesis, disruption of membrane structure and ion homeostasis, inhibition of enzymatic reactions, and activation of programmed cell death (PCD). One of the reasons that heavy metals are so toxic is that they can mimic other essential metals (e.g., Ca^{2+} and Mg^{2+}), take their place in essential reactions, and disrupt these reactions. Cadmium, for example, can replace magnesium in chlorophyll or calcium in the calcium signaling protein calmodulin, disrupting both photosynthesis and signal transduction. The mimicking of essential elements can also explain the uptake of cadmium and other heavy metals into cells via channels that evolved to transport essential elements. Heavy metals can also bind to and inhibit different enzymes, as well as directly interact with oxygen to form ROS.

Mineral nutrient deficiencies are a cause of stress

As we discussed in Chapter 5, deficiencies in one or more essential mineral nutrients cause a range of plant metabolic disorders. Such deficiencies can occur even in the presence of an adequate nutrient supply, if the soil pH shifts the equilibrium of the nutrient into an insoluble form, making it unavailable for uptake. Most mineral nutrients are available between pH levels of 4.5 and 6.5 and become insoluble below or above this range (see Figure 5.5). Nutrient or pH stress almost always results in the suppression of plant growth and reproduction. The reason for this suppression is that mineral nutrients are components of essential enzymes and building blocks of the cell. An insufficient supply of iron or magnesium, for example, results in a decrease in heme content, which is required for chlorophyll and cytochrome biosynthesis. Without chlorophyll and cytochromes to conduct electron transfer, energy production in the cell ceases.

Ozone and ultraviolet light generate ROS that cause lesions and induce PCD

Ozone enters the plant through open stomata and is converted to different forms of ROS. These ROS cause lipid peroxidation and the oxidation of proteins, RNA, and DNA. These toxic effects induce the formation of lesions in leaves that are characteristic of the activation of programmed cell death (PCD). In general, the type of lesions (leaf chlorosis and tissue necrosis) and the severity of the injuries are dependent on the extent of exposure to ozone and can vary among different plant species. The thinning of the ozone layer in Earth's upper atmosphere reduces the filtering of ultraviolet (UV) radiation, resulting in an increase in UV radiation reaching Earth's surface. In addition to its effects on photosynthesis, UV radiation also induces the formation of ROS that can cause mutations during DNA replication. The UV-induced accumulation of ROS induces the activation of PCD and the formation of lesions. Both ozone and UV stress cause suppression of plant growth and reduction of agronomic yields.

Combinations of abiotic stresses can induce unique signaling and metabolic pathways

In the field, plants are often subjected to a combination of different abiotic stresses simultaneously. Drought and heat stress are examples of two different abiotic stresses that almost always occur together in the environment, with devastating results. Between 1980 and 2004 in the United States, the cost of crop damage due to drought plus heat was six times greater than the cost due to drought alone (**Figure 24.6A**).

The physiological acclimation of plants to a combination of different abiotic stresses is different from the acclimation of plants to different abiotic stresses applied individually. **Figure 24.6B** shows the effects of heat and drought, applied separately, on four physiological parameters of Arabidopsis: photosynthesis, respiration, stomatal conductance, and leaf temperature. The physiological profiles under the two stresses applied individually were quite different. Heat alone caused an elevation of leaf temperature and a large increase in stomatal conductance. Drought, however, was more inhibitory to photosynthesis and stomatal opening. The main effect of the combination of drought plus heat was a significant elevation in leaf temperature that could be deadly to the plant.

The combination of heat plus drought also induced different patterns of gene expression and metabolite biosynthesis than either stress alone. As shown in **Figure 24.6C**, drought plus heat caused the accumulation of 772 unique transcripts (yellow) and 5 unique metabolites (yellow), demonstrating that the acclimation of plants to the combination of drought and heat is different in many aspects from the acclimation of plants to drought or heat stress applied individually. Differences in physiological parameters, transcript accumulation, and metabolites could be a result of conflicting physiological responses to the two stresses. For example, during heat stress, plants *increase* their stomatal conductance, which cools their leaves by transpiration. However, if heat stress occurs simultaneously with drought, the stomata are closed, causing leaf temperature to be 2 to 5°C higher.

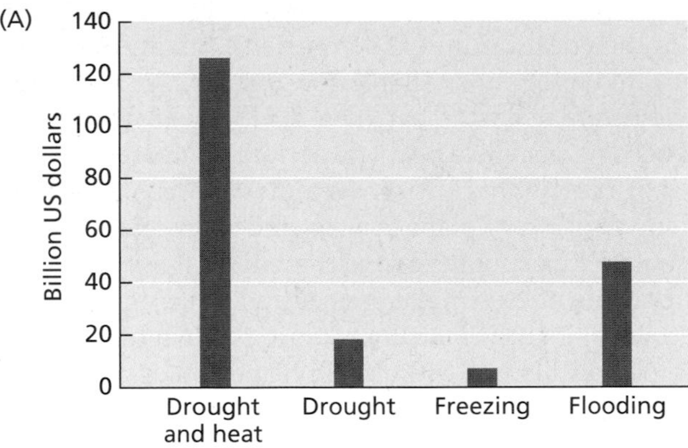

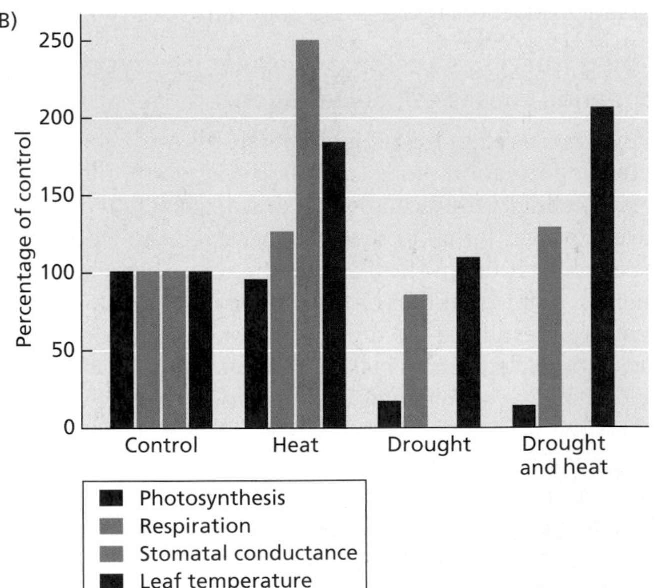

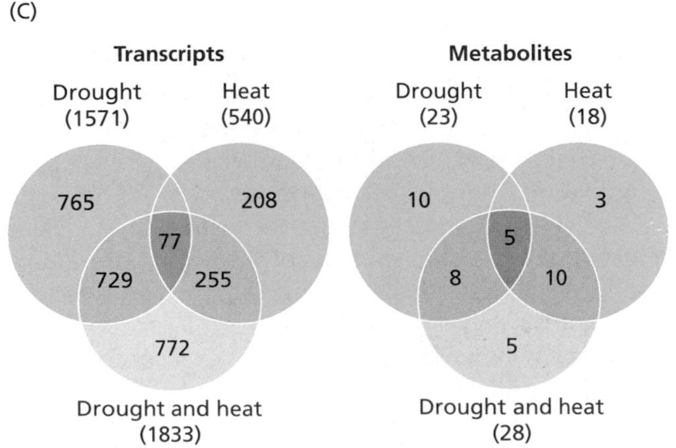

Figure 24.6 Effect of combined abiotic stresses on plant productivity, physiology, and molecular responses. (A) Losses to U.S. agriculture resulting from a combination of drought and heat stress were much higher than losses caused by drought, freezing, or flooding alone between 1980 and 2004. (B) The effect of combined drought and heat on plant physiology. Note the complete closure of plant stomata, which results in a higher leaf temperature. (C) Venn diagrams showing the effect of combined drought and heat on the transcriptome (left) and metabolome (right) of plants. (After Mittler 2006.)

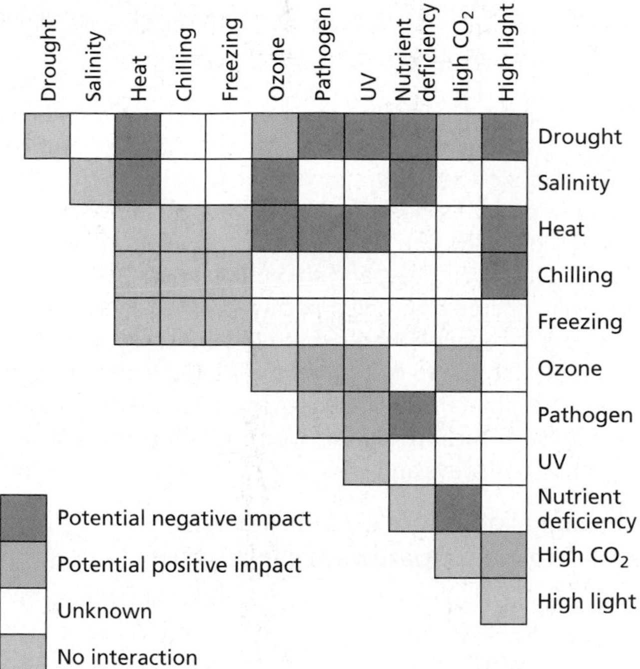

Figure 24.7 Stress matrix. Different combinations of potential environmental stresses can have different effects on field crops. The matrix is color-coded to indicate stress combinations that were studied with different crops and their overall effect on plant growth and yield. (After Mittler and Blumwald 2010.)

Salinity or heavy metal stress could pose a similar problem when combined with heat stress because enhanced transpiration could result in enhanced uptake of salt or heavy metals. In contrast, some stress combinations could have beneficial effects on plants compared with the individual stresses applied separately. For example, drought, which causes stomatal closure, could potentially enhance tolerance to ozone. The "stress matrix" shown in **Figure 24.7** summarizes different combinations of environmental conditions that could have a significant impact on agricultural production. Among the several stress interactions that could have a deleterious effect on crop productivity are drought and heat, salinity and heat, nutrient stress and drought, and nutrient stress and salinity. Interactions that could have a beneficial impact include drought and ozone, ozone and UV, and high CO_2 combined with drought, ozone, or high light.

Perhaps the most studied stress interactions are those of different abiotic stresses with biotic stresses, such as pests or pathogens. In most cases, prolonged exposure to abiotic stress conditions, such as drought or salinity, results in the weakening of plant defenses and enhanced susceptibility to pests or pathogens.

Sequential exposure to different abiotic stresses sometimes confers cross-protection

Several studies have reported that the application of a particular abiotic stress condition can enhance the tolerance of plants to a subsequent exposure to a different type of abiotic stress. This phenomenon is called **cross-protection**. It occurs because many stresses result in the accumulation of the same general stress-response proteins and metabolites—for example, ROS-scavenging enzymes, molecular chaperones, and osmoprotectants—and they persist in plants for some time even after the stress conditions have subsided. The application of a second stress to the same plants that experienced the initial stress may therefore have a decreased effect because the plants are already primed and ready to deal with several different aspects of the new stress condition. We will discuss the example of heat shock proteins later in the chapter.

Stress-Sensing Mechanisms in Plants

Plants use a variety of mechanisms to sense abiotic stress. As discussed above, environmental stress disrupts or alters many physiological processes in the plant by affecting protein or RNA stability, ion transport, the coupling of reactions, or other cellular functions. Any of these primary disruptions could be signaling the plant that a change in environmental conditions has occurred and that it's time to respond by altering existing pathways or by activating stress-response pathways. At least five different types of stress-sensing mechanisms can be distinguished:

- *Physical sensing* refers to the mechanical effects of stress on the plant or cell structure, for example, the contraction of the plasma membrane from the cell wall during drought stress.
- *Biophysical sensing* might involve changes in protein structure or enzymatic activity, such as the inhibition of different enzymes during heat stress.
- *Metabolic sensing* usually results from the detection of by-products that accumulate in cells due to the uncoupling of enzymatic or electron transfer reactions, such as the accumulation of ROS during stress caused by too much light.
- *Biochemical sensing* often involves the presence of specialized proteins that have evolved to sense a particular stress; for example, calcium channels that can sense changes in temperature and alter Ca^{2+} homeostasis.
- *Epigenetic sensing* refers to modifications of DNA or RNA structure that do not alter genetic sequences, such as the changes in chromatin that occur during temperature stress.

Each of these stress-sensing mechanisms can act individually or in combination to activate downstream signal transduction pathways.

Early-acting stress sensors provide the initial signal for the stress response

A diagram of the possible early events in the sensing of abiotic stress and the signal transduction and acclimation pathways activated by these events is presented in **Figure 24.8**. Thus far, several possible examples of stress-sensing mechanisms that act early in the pathway have been identified. They include:

- A calcium channel (cyclic nucleotide-gated calcium channel) identified in Arabidopsis that senses changes in temperature and is required for acclimation to heat stress
- A kinase (SnRK1, a SNF1-related kinase 1) that senses energy depletion during stress and activates hundreds of stress-response transcripts
- A plasma membrane histidine kinase (ATHK1) that senses osmotic stress and activates ABA-dependent and ABA-independent responses
- A leucine rich repeat receptor kinase (LRRK) protein (Srlk) that functions as an upstream regulator of salinity responses in Arabidopsis
- An endoplasmic reticulum transmembrane sensor inositol-requiring enzyme 1 (IRE1) that is a key protein required for heat tolerance in plants
- A member of the hypoxia-associated ethylene response factor group VII transcription factors, which senses oxygen levels during hypoxia stress

In recent years scientists have begun to use advanced tools such as transcriptomics, proteomics, and metabolomics to simultaneously study thousands of transcripts, proteins, and chemical compounds that are altered in plants in response to abiotic stress. These large-scale genomic analyses have enabled researchers to identify important stress-response pathways and networks involved in plant acclimation. In the following section we will discuss some of the major signaling pathways used by plants to transduce stress-specific signals and acclimate to new stress conditions.

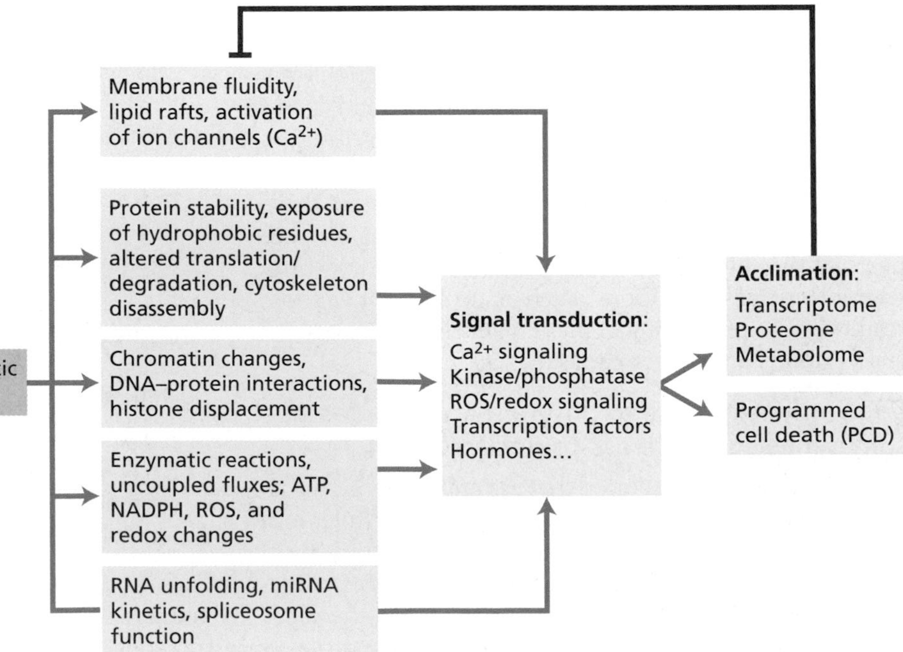

Figure 24.8 Early events in the sensing of abiotic stress by plants (blue boxes) and the signal transduction and acclimation pathways activated by these events (yellow boxes). (After Mittler et al. 2012.)

Signaling Pathways Activated in Response to Abiotic Stress

The initial stress-sensing mechanisms described above trigger a downstream response that comprises multiple signal transduction pathways. These pathways involve calcium, protein kinases, protein phosphatases, ROS signaling, activation of transcriptional regulators, accumulation of plant hormones, and so on. The stress-specific signals that emerge from these pathways, in turn, activate or suppress various networks that may either allow growth and reproduction to continue under stress conditions, or enable the plant to survive the stress until more favorable conditions return. In this section we will consider these signaling pathways and their interactions in greater detail.

The signaling intermediates of many stress-response pathways can interact

Stress-induced increases in the concentrations of cytosolic calcium and ROS are important early signaling events in many acclimation pathways. Cellular calcium levels are controlled by calcium channels, Ca^{2+}–H^+ antiporters, and Ca^{2+}-ATPases, which mediate the mobilization of calcium from storage compartments such as the vacuole, the endoplasmic reticulum, and the cell wall (see Chapter 6). Calcium regulates transcription factors by a variety of mechanisms. As shown in **Figure 24.9**, calcium can activate gene expression by binding directly to certain transcription factors. Alternatively, calcium forms Ca^{2+}–CaM complexes, which can activate transcription either directly or indirectly, by binding to a transcription factor. Calcium also activates various protein kinases and phosphatases that

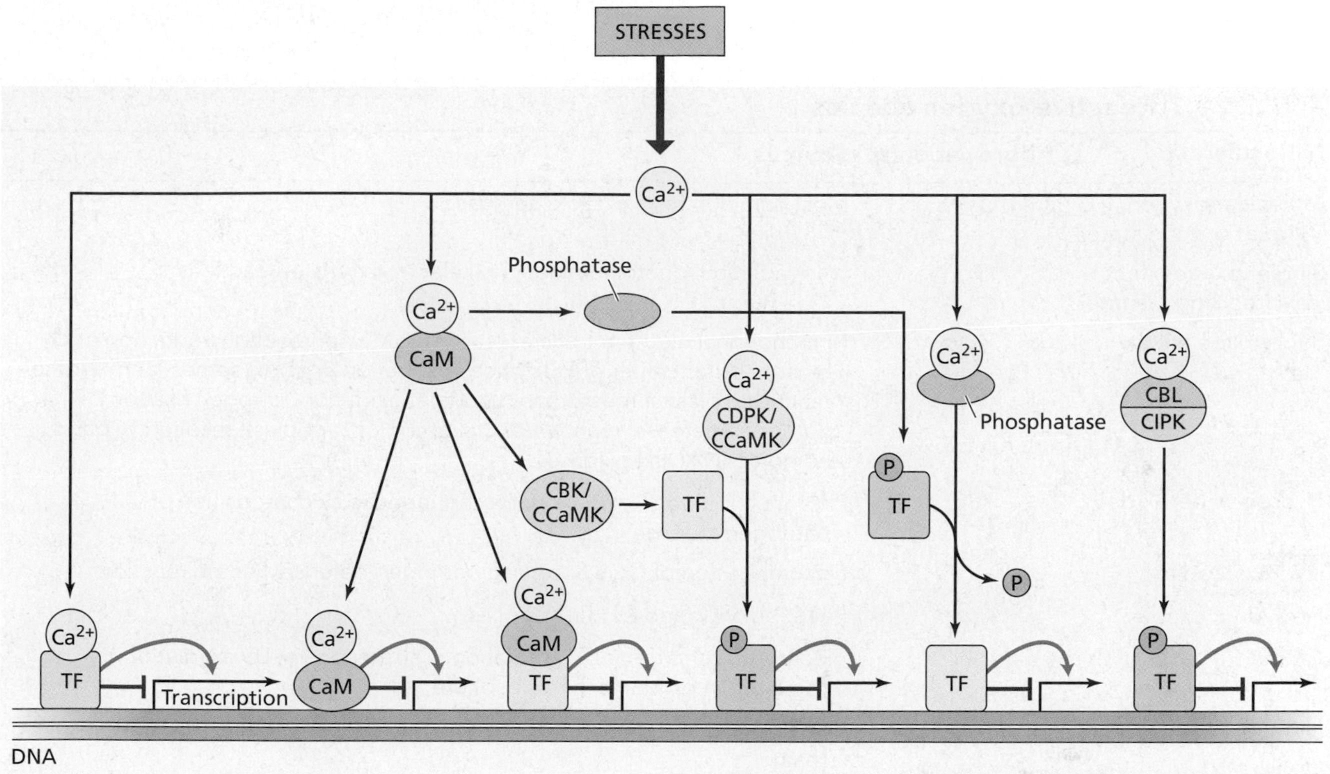

Figure 24.9 Stress-induced elevation of cellular calcium can regulate transcription by various mechanisms. Elevated calcium levels can result in calcium binding to different proteins, including transcription factors (TF), various calmodulins (CaM), kinases (e.g., calcium-dependent protein kinases [CDPK]) or kinase-binding proteins (e.g., CBLs [Calcineurin B-like proteins] that bind CIPKs [CBL-interacting protein kinases]), and phosphatases that directly or indirectly activate or suppress transcription, causing the activation of acclimation pathways. CCaMK, Ca^{2+} and calmodulin-dependent protein kinase; CBK, calmodulin-binding protein kinase. (After Reddy et al. 2011.)

regulate gene expression, either by phosphorylating (activating) or dephosphorylating (inhibiting) transcription factors. The vast cellular networks of protein kinases and phosphatases therefore play an essential role in integrating stress-response pathways.

The steady state level of ROS in the cell is governed by the balance between ROS- generating reactions and ROS-scavenging reactions (**Figure 24.10**). ROS generation occurs in several cellular compartments and as a result of the activities of specialized oxidases, such as NADPH oxidases, amine oxidases, and cell wall–bound peroxidases (**Table 24.2**). ROS scavenging is carried out by antioxidant molecules such as ascorbate, glutathione, vitamin E, and carotenoids, and by antioxidant enzymes such as superoxide dismutase, ascorbate peroxidase, and catalase. Many types of biotic and abiotic stresses trigger the production of ROS (see Figure 24.4). Because ROS can trigger the opening of calcium channels, and increases in cytosolic Ca^{2+} concentrations can activate calcium-dependent protein kinases (CDPKs), which activate NADPH oxidase, the calcium and ROS path-

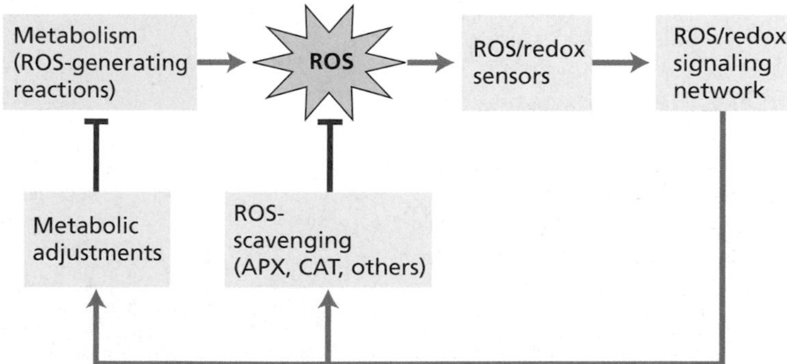

Figure 24.10 Basic ROS cycle. Normal metabolic reactions in cells, such as photosynthesis, respiration, photorespiration, and lipid oxidation, generate ROS. Various sensors monitor ROS levels in cells; an increase in ROS activates a signal transduction network that in turn activates ROS-scavenging mechanisms such as ascorbate peroxidase (APX), catalase (CAT), and superoxide dismutase (SOD). The signaling network also modulates various metabolic reactions and suppresses, when needed, some of the ROS-producing pathways. The overall result of the cycle is controlled maintenance of ROS levels in cells.

TABLE 24.2 Reactive oxygen species

Molecule	Abbreviation(s)	Sources
Molecular oxygen (triplet ground state)	O_2; $^3\Sigma$	Most common form of dioxygen gas
Singlet oxygen (first excited singlet state)	$^1O_2$; $^1\Delta$	UV irradiation, photoinhibition, PSII electron transfer reactions
Superoxide anion	$O_2^{\bullet-}$	Mitochondrial electron transfer reactions, Mehler reaction (reduction of O_2 by iron–sulfur center of PSI), photorespiration in glyoxysomes, peroxisome reactions, plasma membrane, paraquat oxidation, nitrogen fixation, pathogen defense, reaction of O_3 and OH^- in apoplast, respiratory burst homolog (NADPH oxidase)
Hydrogen peroxide	H_2O_2	Photorespiration, β-oxidation, proton-induced decomposition of $O_2^{\bullet-}$, pathogen defense
Hydroxyl radical	$HO^{\bullet-}$	Decomposition of O_3 in apoplast, pathogen defense, Fenton reaction
Perhydroxyl radical	$HO_2^{\bullet-}$	Reaction of O_3 and OH^- in apoplast
Ozone	O_3	Electrical discharge or UV irradiation in stratosphere, UV irradiation of combustion products in troposphere
Nitric oxide	NO	Nitrate reductase, nitrite reduction by the mitochondrial electron transport chain

Source: Jones et al. 2013.

ways interact in a positive feedback cycle (**Figure 24.11**). The elevation of calcium and ROS levels during the early stages of the stress response activates protein kinases and phosphatases that phosphorylate or dephosphorylate different transcription factors (see Figure 24.9). Activation or inhibition of transcription factors during abiotic stress can also result from changes in the redox status of the cell that are directly sensed by certain transcriptional regulators.

When plants are subjected to multiple stresses, cross talk can occur between the hormones, secondary messengers, and protein kinases or phosphatases involved in each of the stress pathways. For example, mitogen-activated protein

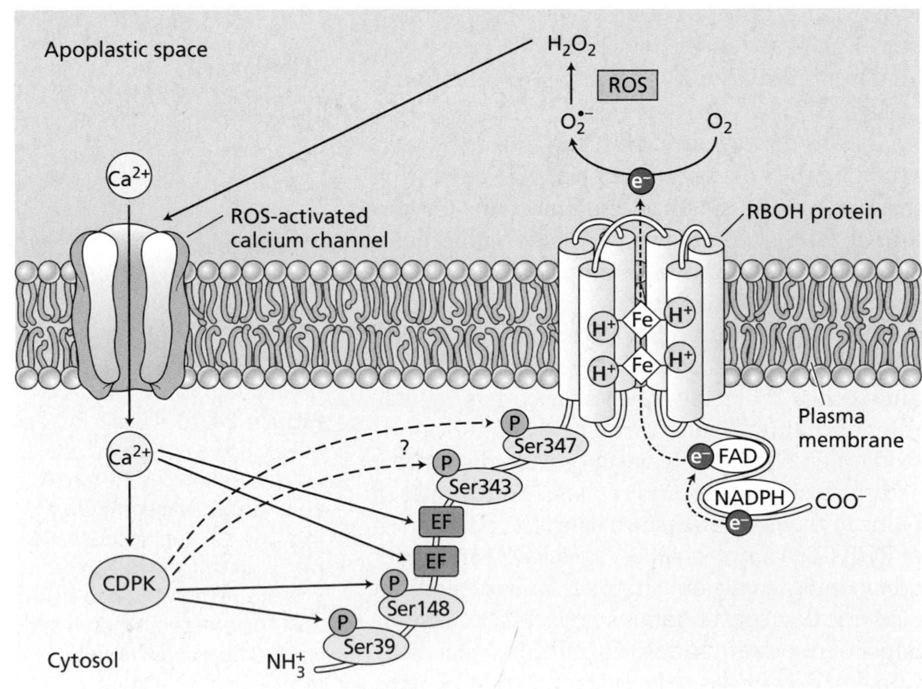

Figure 24.11 Interaction between ROS and calcium signaling mediated by respiratory burst oxidase homolog (RBOH) proteins (NADPH oxidases), calcium-dependent protein kinases (CDPKs), and ROS-activated calcium channels. ROS is shown to activate calcium channels on the plasma membrane (left). The elevated calcium levels in the cytosol then activate CDPKs (bottom) that phosphorylate and activate RBOH proteins (right) which generate more ROS. RBOH proteins have six transmembrane domains. The amino-terminal cytoplasmic domain of RBOH proteins contains four serines (Ser) that can be phosphorylated by CDPKs and two EF hands that can bind calcium directly. Similar ROS-activated calcium channels are found on the vacuole membrane (not shown).

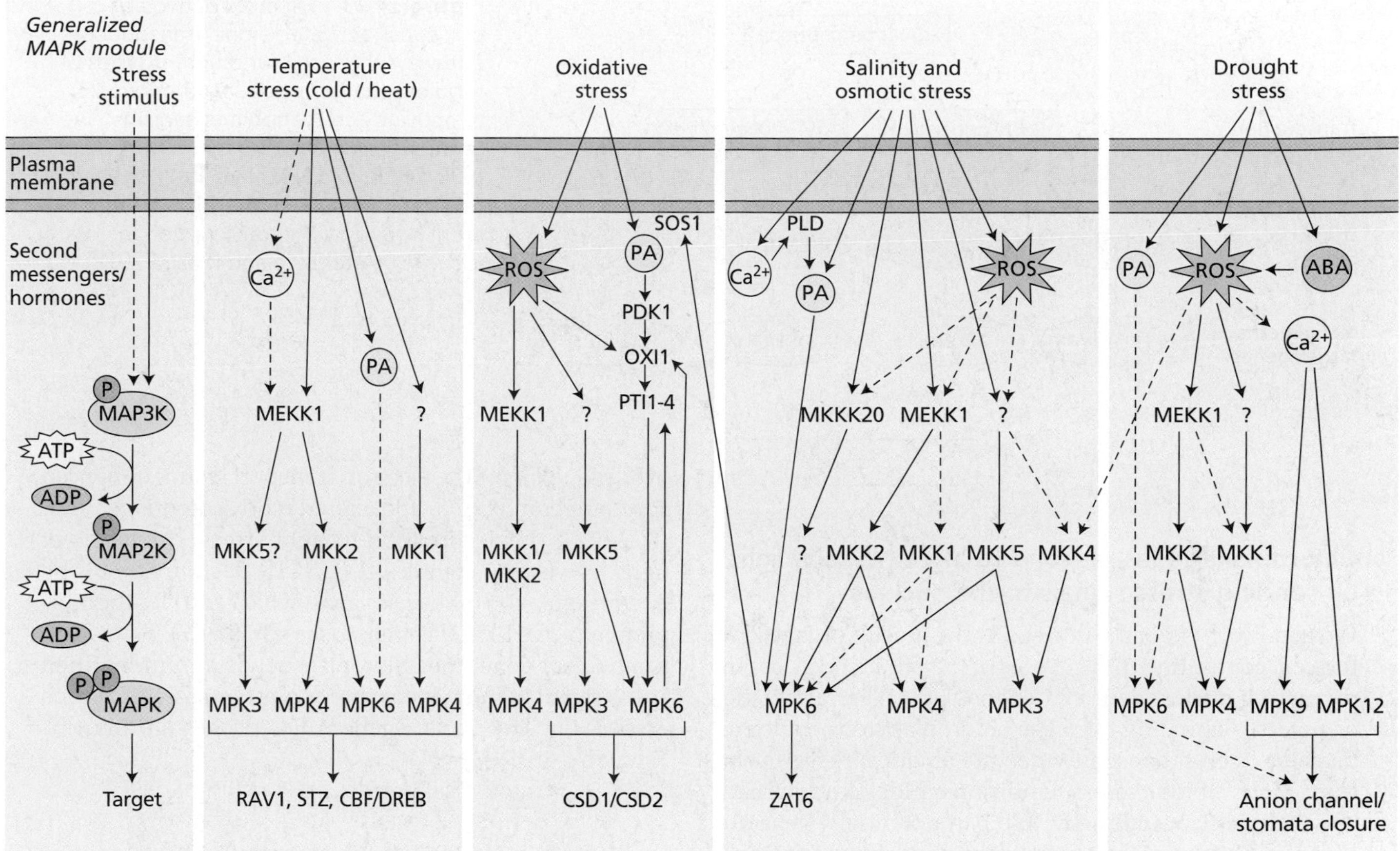

Figure 24.12 Schematic depiction of the cross talk between different secondary messengers, hormones, and MAPK modules that regulate abiotic stress responses in plants. The general progression of a MAP3K/MAP2K/MAPK cascade (i.e., a complete MAPK module) is shown on the left, and the pathways activated in response to different stresses are shown on the right. (After Smékalová et al. 2013.)

kinases (MAPKs) regulate several abiotic stress responses in Arabidopsis. MAPKs participate in MAP3K/MAP2K/MAPK cascades, collectively known as **MAPK modules**. As shown in **Figure 24.12**, responses to temperature stress, oxidative stress, salinity and osmotic stress, and drought stress are all regulated by modules of the same basic MAPK pathway. These four stress responses also share upstream signaling intermediates, such as calcium, phosphatidic acid, and ROS. Therefore, the production of signaling intermediates in any one of the stress responses can affect the other three responses.

Acclimation to stress involves transcriptional regulatory networks called *regulons*

Transcriptional regulators or transcription factors are proteins that bind specific DNA sequences and activate or suppress the expression of different genes. A particular transcription factor can bind to the promoters of hundreds of different genes and affect their expression simultaneously. A transcription factor can also bind to the promoter of a gene encoding another transcription factor, thereby activating or suppressing its expression. In this way a cascade of transcriptional regulation of gene expression can occur.

Combinations of different transcription factors can generate a gene network responding to a particular abiotic stimulus, with some genes being activated and some suppressed. Such transcriptional regulatory networks responding to abiotic stress have been termed **stress-response regulons**. An example of a stress-response regulon is shown in **Figure 24.13**. The advantage of using regulons to control the response of plants to a particular abiotic stress is that they simultaneously activate specific stress-response pathways while suppressing other pathways that are not needed or could even damage the plant during stress. For example, in response to high light conditions, certain genes that encode the photosynthetic antenna proteins might need to be suppressed, while other genes that encode for ROS scavenging might need to be activated.

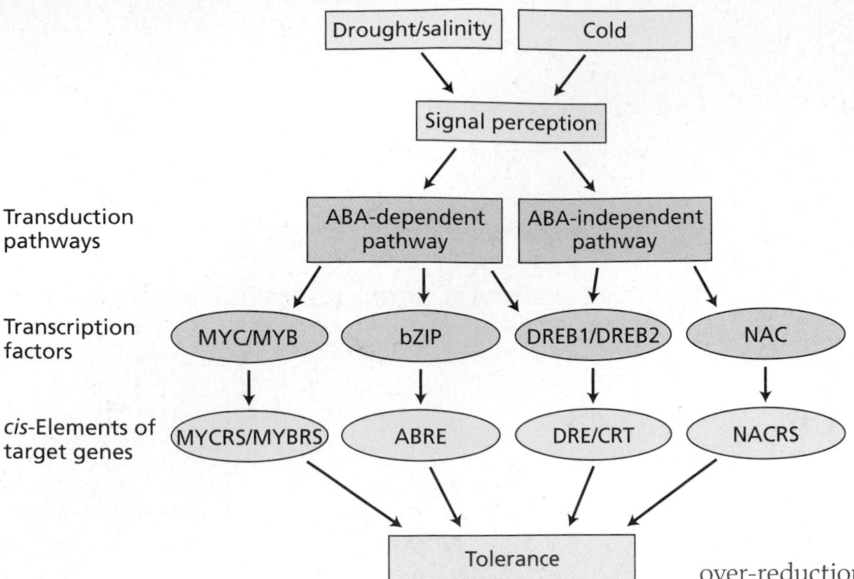

Figure 24.13 Example of two abiotic stress-activated signal transduction pathways that use four different types of regulons (transcription factor networks) to activate acclimation mechanisms. The regulons shown belong to the MYC/MYB, bZIP, DREB, and NAC families. For each regulon, the name of the DNA *cis*-element bound by the transcription factors is shown. (After Lata and Prasad 2011.)

Chloroplast genes respond to high-intensity light by sending stress signals to the nucleus

We usually think of the nucleus as the master organelle of the cell, controlling the activities of the other organelles by regulating nuclear gene expression. However, retrograde, or reverse, signaling from the chloroplast to the nucleus has also been proposed to mediate abiotic stress perception. Many abiotic stress conditions affect chloroplasts, either directly or indirectly, and can potentially generate signals that can influence nuclear gene expression and acclimation responses. Light stress, for example, can cause over-reduction of the electron transport chain, enhanced accumulation of ROS, and altered redox potential.

During the acclimation to light stress, the levels of light-harvesting complex II (LHCII) decline due to the down-regulation of the *Lhcb* gene, which encodes the apoprotein of the LHCII complex (see Chapter 7). Since *Lhcb* is a nuclear gene, the chloroplast sends an unidentified stress signal to the nucleus that down-regulates *Lhcb* gene expression. The nuclear gene *ABI4* encodes a transcrip-

Figure 24.14 Rapid systemic signaling in response to the physical sensing of a wound (arrow). (A) Time-lapse imaging of a rapid systemic signal initiated by a wound using a luciferase reporter fused to the promoter of the ROS-responsive *ZAT12* gene. Light is emitted from tissues where luciferase is expressed. (B) Schematic model of the ROS wave that is required to mediate rapid systemic signaling in response to abiotic stress. The ROS wave is generated by an active, self-propagating wave of ROS production (not diffusion) that starts at the initial tissue subjected to stress and spreads to the entire plant. Each cell along the path of the signal activates its RBOH proteins (NADPH oxidase) and generates ROS. When the signal reaches its systemic target, it activates acclimation mechanisms in the entire plant. The ROS wave is accompanied by a calcium wave and electrical signals. (After Mittler et al. 2011 and Suzuki et al. 2013.)

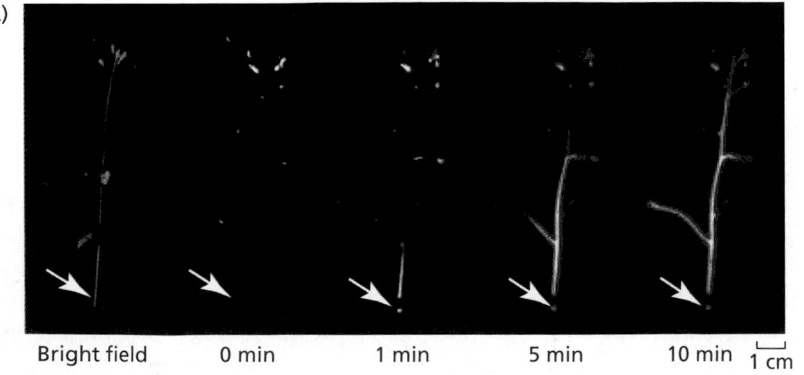

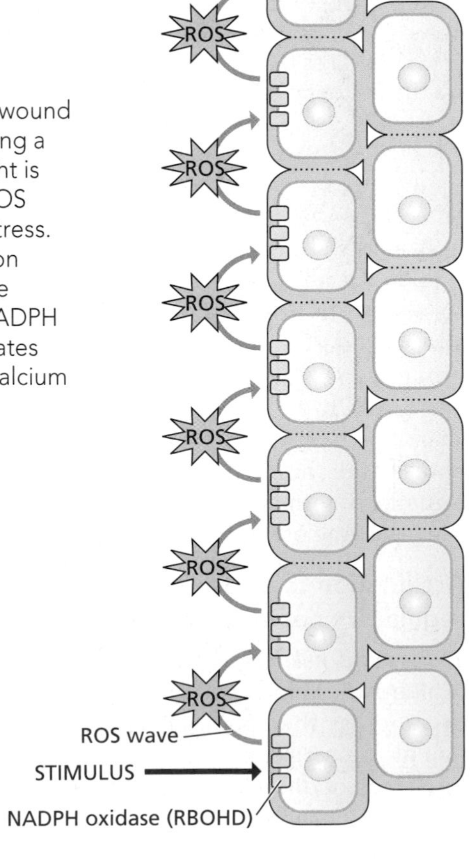

tion factor that suppresses the expression of *Lhcb* genes. In Arabidopsis there is evidence that the chloroplast gene *GUN1* acts upstream of *ABI4* during the acclimation to light stress. In other words, the GUN1 protein perceives the original stress signal in the chloroplast and either generates or transmits a second signal to the nucleus, which causes ABI4 to bind to the promoter of the *Lhcb* gene and block transcription.

A self-propagating wave of ROS mediates systemic acquired acclimation

As in systemic acquired resistance (SAR) during biotic stress (see Chapter 23), abiotic stress applied to one part of the plant can generate signals that can be transported to the rest of the plant, initiating acclimation even in parts of the plant that have not been subjected to the stress. This process is called **systemic acquired acclimation** (**SAA**). Rapid SAA responses to different abiotic stress conditions, including heat, cold, salinity, and high light intensity, have been demonstrated to be mediated by a self-propagating wave of ROS production, which travels at a rate of approximately 8.4 cm min^{-1} and is dependent on the presence of a specific NADPH oxidase, **respiratory burst oxidase homolog D** (**RBOHD**), which is located on the plasma membrane (**Figure 24.14**). The rapid rates of abiotic stress–systemic signals detected with luciferase imaging in these experiments suggest that many of the responses to abiotic stresses may occur at a rate much faster than previously thought.

Epigenetic mechanisms and small RNAs provide additional protection against stress

Thus far we have discussed responses to abiotic stress in terms of signaling cascades and altered gene expression—acclimation processes that can be reversed when more favorable conditions arise. Recently attention has focused on epigenetic changes, which potentially can provide long-term adaptation to abiotic stress. Because some chromatin modifications are mitotically and meiotically heritable, stress-induced epigenetic changes might even have evolutionary implications. Chromatin immunoprecipitation of DNA cross-linked to modified histones, coupled with modern sequencing technologies, has opened the door to genome-wide analyses of changes in the **epigenome**. Stable or heritable DNA methylation and histone modifications have now been linked with specific abiotic stresses (**Figure 24.15**).

The role of epigenetic regulation of flowering time has been studied in Arabidopsis in relation to genes known to be involved in abiotic stress. Mutations in some of the genes involved in epigenetic processes during stress cause changes in flowering times. For example, late flowering of the freezing-sensitive mutant *hos15* was shown to result from the deacetylation of the flowering genes *SOC* and *FT* (see Chapter 20). Normally the flowering repressor FLC (a MADS-box protein) is epigenetically repressed during vernalization, allowing acquisition of the competence to flower after exposure to prolonged low temperatures. This process was shown to involve several different proteins that could alter chromatin remodeling.

The involvement of small RNAs in abiotic stress responses has also received increased attention in recent years. Small RNAs belong to at least two different groups: microRNAs (miRNAs) and endogeneous short interfering RNAs (siRNAs). MicroRNAs and siRNAs can cause post-transcriptional gene silencing via RISC (RNA-induced silencing complex)-mediated degradation of mRNA in the cytosol (see Chapter 2). In addition, siRNA can suppress gene expression by altering chromatin properties in the nuclei via **RNA-induced transcriptional silencing** (**RITS**). The involvement of small RNAs in suppressing protein translation during stress has also been proposed. Both miRNAs and siRNAs have been shown to control gene expression during different abiotic stresses, including cold, nutrient deficiency, dehydration, salinity, and oxidative stresses.

Hormonal interactions regulate normal development and abiotic stress responses

Plant hormones mediate a wide range of adaptive responses and are essential for the ability of plants to adapt to abiotic stresses. Abscisic acid (ABA) biosynthesis is among the most rapid responses of plants to abiotic stress. ABA concentrations in leaves can increase up to 50-fold under drought conditions—the most dramatic change in concentration reported for any hormone in response to an environmental signal. Redistribution or biosynthesis of ABA is very effective in causing stomatal closure, and ABA accumulation in stressed leaves plays an important role in the reduction of water loss by transpiration under water-stress conditions (see Figure 24.24). Increases in humidity reduce ABA levels by increasing ABA breakdown, thereby permitting stomata to reopen. ABA-biosynthesis or response mutants are unable to close their stomata under drought conditions and are called *wilty* mutants. Many genes associated with ABA biosynthesis and genes encoding ABA receptors and downstream signaling components have been identified (see Chapter 15 and WEB APPENDIX 3). ABA also plays important roles in the adaptation of plants to cold temperatures and salinity stress. Cold stress induces the synthesis of ABA, and the exogenous application of ABA improves the cold tolerance of plants.

Another plant hormone that plays a key role in acclimation to various abiotic stresses is cytokinin. Cytokinin and ABA have antagonistic effects on stomatal opening, transpiration, and photosynthesis. Drought results in decreased cytokinin levels and increased ABA levels. Although ABA is normally required for stomatal closure, preventing excessive water loss, drought stress conditions can also inhibit photosynthesis and cause premature leaf

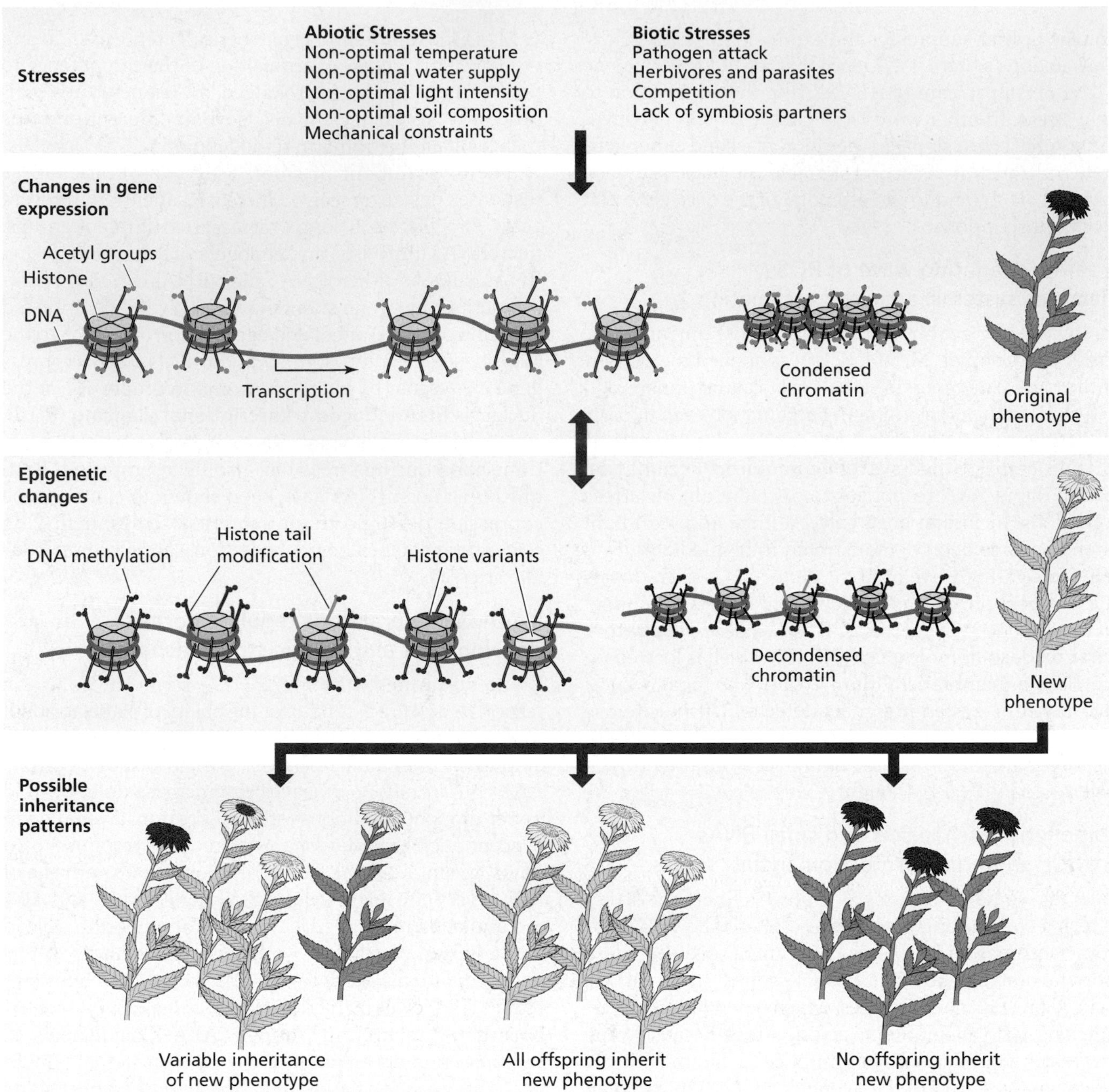

Figure 24.15 Stress-induced changes in gene expression can be mediated by protein / lipid / nucleic acid modification, second messengers, or hormones (e.g., abscisic acid, salicylic acid, jasmonic acid, and ethylene). Transcription changes or stress factors can affect chromatin via DNA methylation, histone tail modifications, histone variant replacements, or nucleosome loss and chromatin de-condensation. These changes are reversible and can modify plant metabolism or morphology under stress conditions. Usually, the new phenotypes are not transmitted to progeny, however, chromatin-associated changes have the potential to be heritable and could result in uniform maintenance of new features and epigenetic diversity. (After Gutzat and Mittelsten-Scheid 2012.)

senescence. Cytokinins appear to be able to ameliorate the effects of drought. As shown in **Figure 24.16**, transgenic plants that overexpress *IPT,* the gene encoding the enzyme isopentenyl transferase (the enzyme that catalyzes the rate-limiting step in cytokinin synthesis), exhibit enhanced drought tolerance compared with wild-type plants. Thus, cytokinins are able to protect biochemical processes associated with photosynthesis and delay senescence during drought stress.

In addition to ABA and cytokinin, gibberellic acid, auxin, salicylic acid, ethylene, jasmonic acid, and brassinosteroids also play important roles in the response of plants

(A) Wild type

(B) P_{SARK}::IPT

Figure 24.16 Effects of drought on wild-type and transgenic tobacco plants expressing isopentenyl transferase (a key enzyme in the production of cytokinin) under the control of P_{SARK} (promoter region of *S*enescence-*A*ssociated *R*eceptor *K*inase), a maturation and stress-induced promoter. Shown are wild-type (A) and transgenic (B) plants after 15 days of drought followed by 7 days of rewatering. (Courtesy of E. Blumwald.)

to abiotic stress, and the extensive overlapping among the different sets of hormone-regulated genes supports the existence of complex networks with significant cross talk between the different hormone signaling pathways. The synergistic or antagonistic hormone action and the coordination and mutual regulation of hormone biosynthetic pathways are of great importance for the ability of plants to acclimate to abiotic stress conditions.

Auxin, for example, can play crucial roles in the acclimation of plants to drought conditions. The gene *TLD1*, which encodes an indole-3-acetic acid (IAA)–amido synthetase, has been shown to induce the expression of genes encoding LEA (*L*ate *E*mbryogenesis *A*bundant) proteins that accumulate during seed maturation (see Chapter 21) and are also correlated with enhanced drought tolerance in rice. ABA produced by drought or salt stress activates proteases that degrade the ABCB4 auxin transporter that regulates root hair elongation. The expression of several genes linked with auxin synthesis, auxin transporters (*PIN1, PIN2, PIN4, AUX1*), and auxin-responsive transcription factors (*ARF2, ARF19*) was shown to be regulated by ethylene. Conversely, cellular auxin levels greatly influence ethylene biosynthesis. Several genes encoding ACC (1-aminocyclopropane-1-carboxylic acid) synthase, the rate-limiting step in ethylene biosynthesis, are regulated by auxin.

Gibberellic acid and brassinosteroids are two growth-promoting hormones that regulate many of the same physiological processes and could link growth regulation to abiotic stress responses. Furthermore, deficiencies of, or insensitivity to, either hormone result in similar phenotypes, such as dwarfism, reduced seed germination, and delayed flowering. Although gibberellic acid and brassinosteroids are known to act via different mechanisms, numerous genes regulated by both hormones have been identified, suggesting considerable overlap in their signaling pathways. In rice, for example, the protein OsGSR1, a member of the gibberellic acid–stimulated gene family, acts as a positive regulator of both the gibberellic acid response and brassinosteroid biosynthesis. OsGSR1 thus seems to serve as a regulatory link between the two hormonal pathways, mediating their interactions.

Gibberellic acid has also been shown to interact with salicylic acid. The application of gibberellic acid to Arabidopsis plants causes an increase in the expression of genes involved in both salicylic acid synthesis and action. Interactions between cytokinin and brassinosteroids occur as well. In transgenic plants expressing *IPT* under the control of a drought-specific promoter, the induction of cytokinin biosynthesis by drought resulted in the up-regulation of genes associated with the synthesis and regulation of brassinosteroids.

Developmental and Physiological Mechanisms That Protect Plants against Abiotic Stress

Thus far in the chapter we have discussed the various types of abiotic stress, the mechanisms by which plants sense abiotic stress, the signal transduction pathways that convert stress signals into altered gene expression, and the role of hormonal interactions in regulating networks of genetic pathways. In this section we will discuss the fruits of the labors of all these genetic networks—the metabolic, physiological, and anatomical changes that are produced to counter the effects of abiotic stress. Terrestrial plants first emerged onto land over 500 million years ago. Thus, they have had ample time to evolve mechanisms to cope with various types of abiotic stress. These mechanisms include the abilities to accumulate protective metabolites and proteins and to regulate growth, morphogenesis, photosynthesis, membrane transport, stomatal apertures, and resource allocation. The effects of these and other changes are to attain cellular homeostasis so that the plant's life cycle can be completed under the new environmental regime. Below we discuss some of the major physiological mechanisms of acclimation.

Plants adjust osmotically to drying soil by accumulating solutes

Water can only move through the soil–plant–atmosphere continuum if water potential decreases along that path (see Chapters 3 and 4). Recall from Chapter 3 that $\Psi = \Psi_s + \Psi_p$, where Ψ = water potential, Ψ_s = osmotic potential, and Ψ_p = pressure potential (turgor). When the water potential of the rhizosphere (the microenvironment surrounding the root) decreases due to water deficit or salinity, plants can continue to absorb water only as long as Ψ is lower (more negative) than it is in the soil water. **Osmotic adjustment** is the capacity of plant cells to accumulate solutes and use them to lower Ψ during periods of osmotic stress. The adjustment involves a net increase in solute content per cell that is independent of the volume changes that result from loss of water (**Figure 24.17**). The decrease in Ψ_s is typically limited to about 0.2 to 0.8 MPa, except in plants adapted to extremely dry conditions.

There are two main ways by which osmotic adjustment can take place, one involving the vacuole and the other the cytosol. A plant may take up ions from the soil, or transport ions from other plant organs to the root, so that the solute concentration of the root cells increases. For example, increased uptake and accumulation of potassium will lead to decreases in Ψ_s, due to the effect of the potassium ions on the osmotic pressure within the cell. This response is common in plants growing in saline soils, where ions such as potassium and calcium are readily available to the plant. The uptake of K^+ and other cations must be electrically balanced either by the uptake of inorganic anions, such as Cl^-, or by the production and vacuolar accumulation of organic acids such as malate or citrate.

There is a potential problem, however, when ions are used to decrease Ψ_s. Some ions, such as sodium or chloride, are essential to plant growth in low concentrations but in higher concentrations can have a detrimental effect on cellular metabolism. Other ions, such as potassium, are required in larger quantities but at high concentrations can still have a detrimental effect on the plant, mostly through disruption of plasma membranes or proteins. The accumulation of ions during osmotic adjustment is predominantly restricted to the vacuoles, where the ions are kept out of contact with cytosolic enzymes or organelles. For example, many halophytes (plants adapted to salinity) use vacuolar compartmentalization of Na^+ and Cl^- to facilitate osmotic adjustment that sustains or enhances growth in saline environments.

When the ion concentration in the vacuole increases, other solutes must accumulate in the cytosol to maintain water potential equilibrium between the two compart-

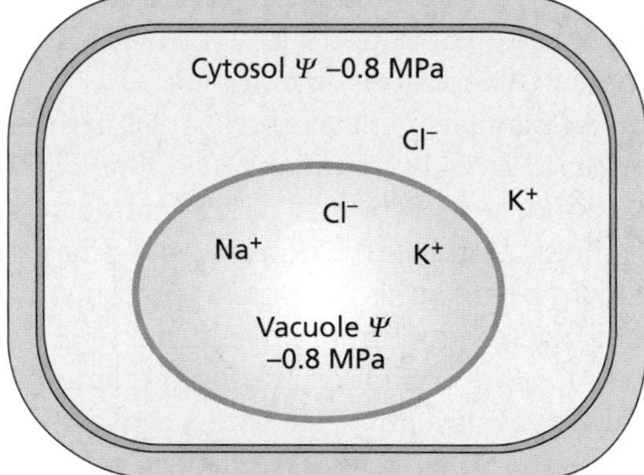

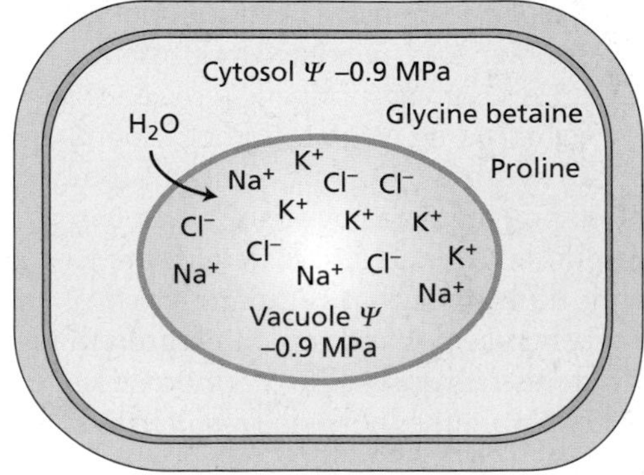

Figure 24.17 Solute adjustments during osmotic stress. The water potential of the cytosol and vacuole of cells must be slightly lower than that of the surrounding environment to maintain a water potential gradient that allows for water uptake. (A) Cell with an external water potential of –0.6 MPa. Equilibrium is maintained inside the cell by ion accumulation in the vacuole and cytosol. (B) Cell with an external water potential of –0.8 MPa because of salinity, drought, or other dehydration stresses. The cell can adjust osmotically by increasing the cellular concentration of solutes in the vacuole and cytosol. If inorganic ions are used for osmotic adjustment, they are typically stored in the vacuole, where they cannot affect metabolic processes in the cytosol. Equilibrium in the cytosol is maintained with compatible solutes (typically uncharged) such as proline and glycine betaine.

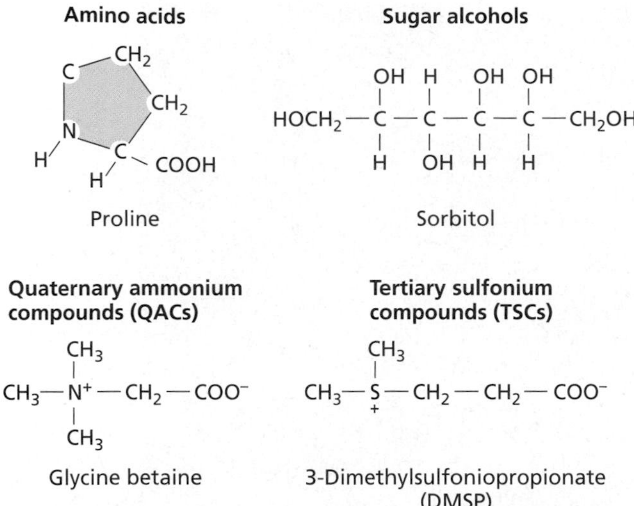

Figure 24.18 Four groups of molecules frequently serve as compatible solutes: amino acids, sugar alcohols, quaternary ammonium compounds, and tertiary sulfonium compounds. Note that these compounds are small and have no net charge.

ments. These solutes are called compatible solutes (or compatible osmolytes). **Compatible solutes** are organic compounds that are osmotically active in the cell but do not destabilize the membrane or interfere with enzyme function at high concentrations, as ions do. Plant cells can tolerate high concentrations of these compounds without detrimental effects on metabolism. Common compatible solutes include amino acids such as **proline**, sugar alcohols such as **sorbitol**, and quaternary ammonium compounds such as **glycine betaine** (**Figure 24.18**). Some of these solutes, such as proline, also seem to have an osmoprotectant function whereby they protect plants from toxic byproducts produced during periods of water shortage, and provide a source of carbon and nitrogen to the cell when conditions return to normal. Each plant family tends to use one or two compatible solutes in preference to others. Because the synthesis of compatible solutes is an active metabolic process, energy is required. The amount of carbon used for the synthesis of these organic solutes can be rather large, and for this reason the synthesis of these compounds tends to reduce crop yield.

Submerged organs develop aerenchyma tissue in response to hypoxia

We now turn to the mechanisms used by plants to cope with too much water. In most wetland plants, such as rice, and in many plants that acclimate well to wet conditions, the stem and roots develop longitudinally interconnected, gas-filled channels that provide a low-resistance pathway for the movement of O_2 and other gases. The gases (air) enter through stomata, or through lenticels (porous regions of the periderm that allow gas exchange) on woody stems and roots, and travel by molecular diffusion or by convection driven by small pressure gradients. In many plants adapted to wetland growth, root cells are separated by prominent, gas-filled spaces that form a tissue called aerenchyma. These cells develop in the roots of wetland plants independently of environmental stimuli. In some nonwetland monocots and eudicots, however, O_2 deficiency induces the formation of aerenchyma in the stem base and newly developing roots.

An example of induced aerenchyma occurs in maize (corn; *Zea mays*) (**Figure 24.19**). Hypoxia stimulates the activity of ACC synthase and ACC oxidase in the root tips of maize and causes ACC and ethylene to be produced faster. Ethylene triggers programmed cell death and disintegration of cells in the root cortex. The spaces formerly occupied by these cells provide gas-filled voids that facilitate movement of O_2. Ethylene-triggered cell death is highly selective; only some cells have the potential to initiate the developmental program that creates the aerenchyma.

When aerenchyma formation is induced, a rise in cytosolic Ca^{2+} concentration is thought to be part of the ethylene signal transduction pathway leading to cell death. Signals that elevate cytosolic Ca^{2+} concentration can promote cell death in the absence of hypoxia. Conversely, signals that lower cytosolic Ca^{2+} concentration block cell death in hypoxic roots that would normally form aerenchyma.

Some tissues can tolerate anaerobic conditions in flooded soils for an extended period (weeks or months) before developing aerenchyma. These include the embryo and coleoptile of rice (*Oryza sativa*) and rice grass (*Echinochloa crus-galli* var. *oryzicola*) and the rhizomes (underground horizontal stems) of giant bulrush (*Schoenoplectus lacustris*), salt marsh bulrush (*Scirpus maritimus*), and narrow-leafed cattail (*Typha angustifolia*). These rhizomes can survive for several months and expand their leaves under anaerobic conditions.

In nature, these rhizomes overwinter in anaerobic mud at the edges of lakes. In spring, once the leaves have expanded above the mud or water surface, O_2 diffuses down through aerenchyma into the rhizome. Metabolism then switches from an anaerobic (fermentative) to an aerobic mode, and roots begin to grow using the available O_2. Likewise, during germination of paddy (wetland) rice and of rice grass, the coleoptile breaks through the water surface and becomes a diffusion pathway for O_2 into submerged parts of the plant, including the roots. Although rice is a wetland species, its roots are as intolerant of anoxia as maize roots are. As the root extends into O_2-deficient soil, the continuous formation of aerenchyma just behind the tip allows O_2 movement within the root to supply the apical zone.

In roots of rice and other typical wetland plants, structural barriers composed of suberized and lignified cell walls prevent O_2 diffusion outward to the soil. The O_2 thus

(A)

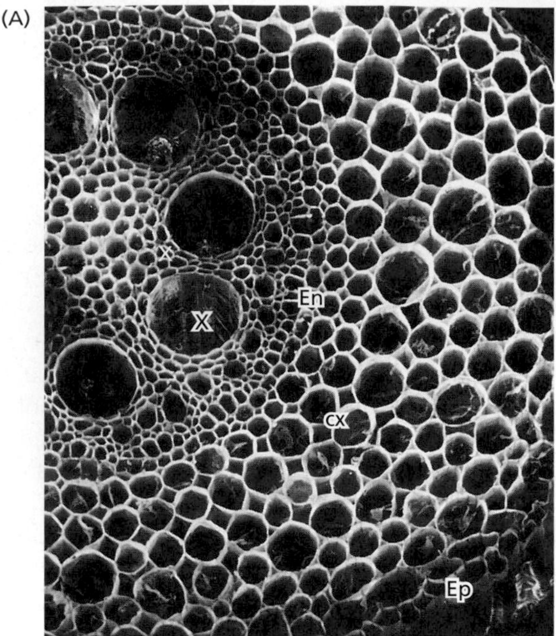

(B)

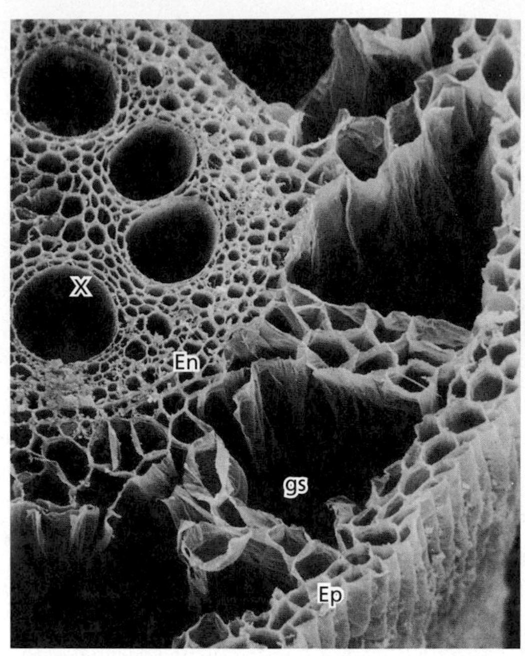

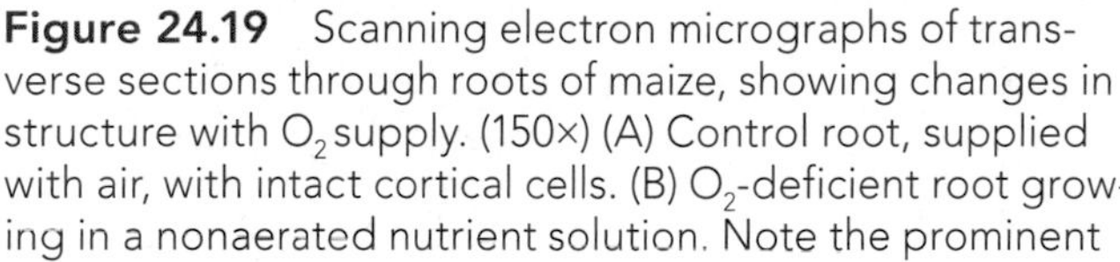

Figure 24.19 Scanning electron micrographs of transverse sections through roots of maize, showing changes in structure with O_2 supply. (150×) (A) Control root, supplied with air, with intact cortical cells. (B) O_2-deficient root growing in a nonaerated nutrient solution. Note the prominent gas-filled spaces (gs) in the cortex (cx), formed by degeneration of cells. The stele (all cells interior to the endodermis, En) and the epidermis (Ep) remain intact. X, xylem. (Courtesy of J. L. Basq and M. C. Drew.)

retained supplies the apical meristem and allows growth to extend 50 cm or more into anaerobic soil. In contrast, roots of nonwetland species, such as maize, leak O_2. Internal O_2 becomes insufficient for aerobic respiration in the root apex of these nonwetland plants, and this lack of O_2 severely limits the depth to which such roots can extend into anaerobic soil.

Antioxidants and ROS-scavenging pathways protect cells from oxidative stress

Reactive oxygen species accumulate in cells during many different types of environmental stresses and are detoxified by specialized enzymes and antioxidants, a process referred to as **ROS scavenging**. Biological antioxidants are small organic compounds or small peptides that can accept electrons from ROS such as superoxide or H_2O_2 and neutralize them. Common antioxidants in plants include the water-soluble ascorbate (vitamin C) and reduced tripeptide glutathione (GSH in reduced form, GSSG in oxidized form), and the lipid-soluble α-tocopherol (vitamin E) and β-carotene (vitamin A). To maintain an adequate supply of these compounds in the reduced state, cells rely on various reductases, such as glutathione reductase, dehydroascorbate reductase, and monodehydroascorbate reductase, which use the reducing power of NADH or NADPH produced by respiration or photosynthesis.

Some ROS may react spontaneously with cellular antioxidants, and some are unstable and decay before they cause cellular damage. However, plants have evolved several different **antioxidative enzymes** that dramatically increase the efficiency of these processes. For example, **superoxide dismutase** is an enzyme that simultaneously oxidizes and reduces the superoxide anion to produce hydrogen peroxide and oxygen according to the reaction: $O_2^{\bullet-} + 2\,H^+ \rightarrow O_2 + H_2O_2$. Variants of superoxide dismutase are found in chloroplasts, peroxisomes, mitochondria, cytosol, and apoplast. Different forms of **ascorbate peroxidase** are present in the same cellular compartments as superoxide dismutase. Ascorbate peroxidase catalyzes the destruction of hydrogen peroxide using ascorbic acid as a reducing agent in the following reaction: 2 L-ascorbate + H_2O_2 + 2 H^+ $\rightarrow$ 2 monodehydroascorbate + 2 H_2O. **Catalase** catalyzes the detoxification of hydrogen peroxide into water and oxygen in peroxisomes according to the reaction: $2\,H_2O_2 \rightarrow 2\,H_2O + O_2$. Reduced forms of **peroxiredoxins** (**Prx**) reduce hydrogen peroxide, and are themselves re-reduced by thioredoxin (Trx), according to the coupled reactions: Prx(reduced) + H_2O_2 $\rightarrow$ Prx(oxidized) + 2 H_2O, and Prx(oxidized) + Trx(reduced) $\rightarrow$ Prx(reduced) + Trx(oxidized). Finally, **glutathione peroxidase** catalyzes the detoxification of hydrogen peroxide using reduced glutathione (GSH) as a reducing agent: H_2O_2 + 2 GSH $\rightarrow$ GSSG + 2 H_2O.

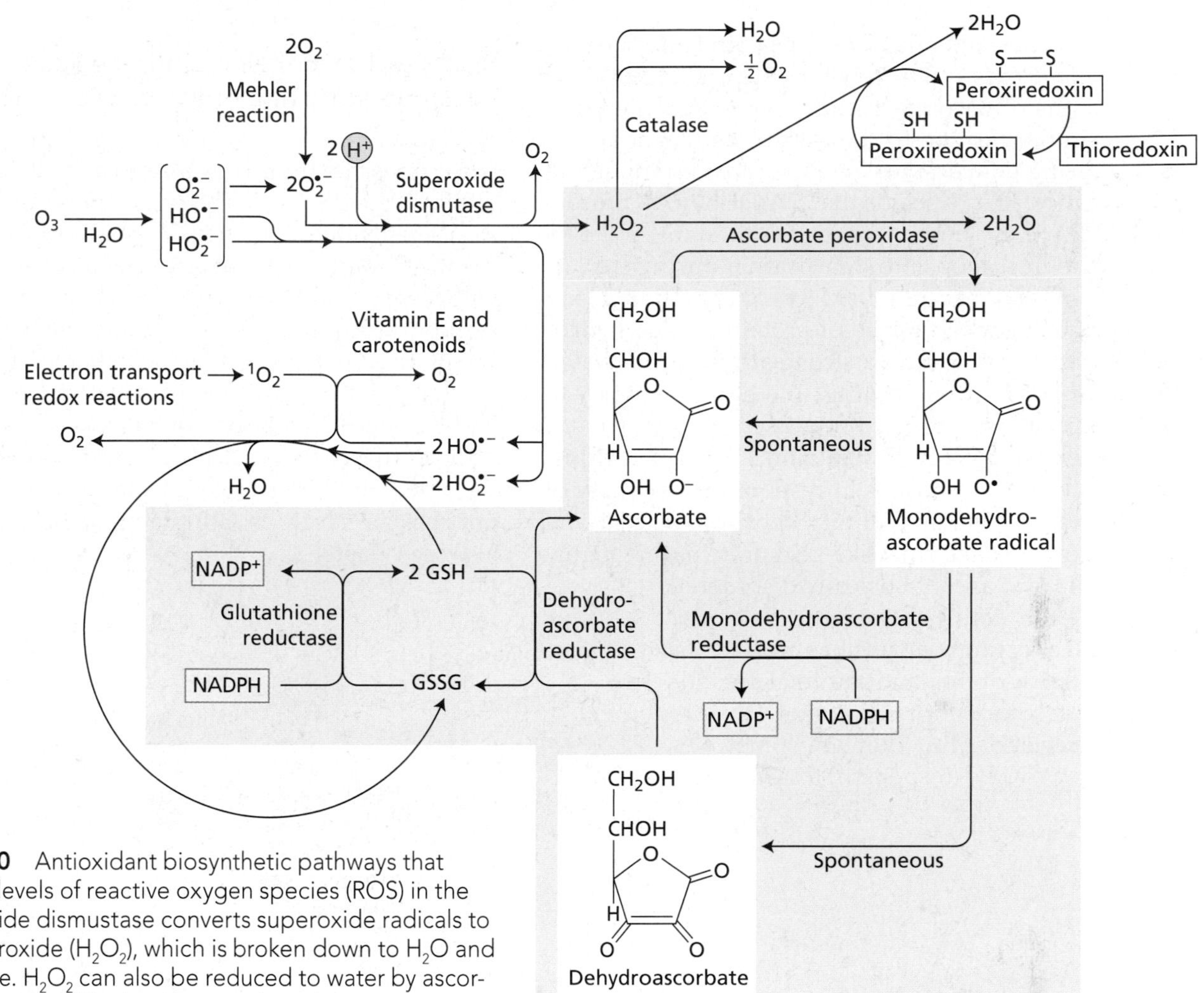

Figure 24.20 Antioxidant biosynthetic pathways that regulate the levels of reactive oxygen species (ROS) in the cell. Superoxide dismustase converts superoxide radicals to hydrogen peroxide (H_2O_2), which is broken down to H_2O and O_2 by catalase. H_2O_2 can also be reduced to water by ascorbate peroxidase, and hydroxyl ion and singlet oxygen can be consumed by the glutathione pathway, indicated by the colored background. (After Jones et al. 2013.)

ROS detoxification enzymes and antioxidants function in cells as a network supported by various antioxidant recycling systems that replenish the level of reduced antioxidants (**Figure 24.20**). This ROS-scavenging network maintains a safe level of ROS in cells, while allowing the cell to use ROS for signal transduction reactions.

Molecular chaperones and molecular shields protect proteins and membranes during abiotic stress

Protein structure is sensitive to disruption by changes in temperature, pH, or ionic strength associated with different types of abiotic stress. Plants have several mechanisms to limit or avoid such problems, including osmotic adjustment for maintenance of hydration, proton pumps to maintain pH homeostasis, and **molecular chaperone proteins** that physically interact with other proteins to facilitate protein folding, reduce misfolding, stabilize tertiary structure, and prevent aggregation or disaggregation. A unique set of chaperones, called **heat shock proteins (HSPs)**, are synthesized in response to a variety of environmental stresses. Cells that synthesize HSPs in response to heat stress show improved thermal tolerance and can tolerate subsequent exposures to higher temperatures that would otherwise be lethal. HSPs are induced by widely different environmental conditions, including water deficit, wounding, low temperature, and salinity. In this way, cells that have previously experienced one stress may gain cross-protection against another stress.

HSPs were discovered in the fruit fly (*Drosophila melanogaster*) and appear to be ubiquitous in plants, animals, fungi, and microorganisms. The heat shock response appears to be mediated by one or more signal transduction pathways, one of which involves a specific set of transcription factors, called **heat shock factors**, that regulate the transcription of HSP mRNAs. There are several different classes of HSPs, including HSP70s that bind and release misfolded proteins, HSP60s that produce huge barrel-like complexes that are used as chambers for protein folding, HSP101s that mediate the disaggregation of protein aggregates, and sHSPs and other HSPs that bind and stabilize different complexes and membranes (**Figure 24.21**).

Numerous other proteins have been identified that act in a similar fashion to stabilize proteins and membranes during dehydration, temperature extremes, and ion imbalance. These include the **LEA/DHN/RAB protein family**. LEA (Late Embryogenesis Abundant) proteins accumulate in response to dehydration during the latter stages of seed maturation. Most LEA proteins belong to a more widespread group of proteins called **hydrophilins**. Hydrophilins have a strong attraction for water, fold into α helices upon drying, and have the ability to reduce aggregation of dehydration-sensitive proteins, a property termed **molecular shielding**. DHNs (dehydrins) accumulate in plant tissues in response to a variety of abiotic stresses, including salinity, dehydration, cold, and freezing stress. Dehydrins, like LEA proteins, are highly hydrophilic and are intrinsically disordered proteins. Their ability to serve as molecular shields and as cryoprotectants has been attributed to their flexibility and minimal secondary structure. Since both LEAs and DHNs are often induced by ABA, they are sometimes referred to as RABs (responsive to ABA).

Plants can alter their membrane lipids in response to temperature and other abiotic stresses

As temperatures drop, membranes may go through a phase transition from a flexible liquid-crystalline structure to a solid gel structure. The phase transition temperature varies depending on the lipid composition of the membranes. Chilling-resistant plants tend to have membranes with more unsaturated fatty acids that increase their fluidity, whereas chilling-sensitive plants have a high percentage of saturated fatty acid chains that tend to solidify at low temperatures. In general, saturated fatty acids that have no double bonds solidify at higher temperatures than do lipids that contain polyunsaturated fatty acids, because the latter have kinks in their hydrocarbon chains and do not pack as closely as saturated fatty acids (**Table 24.3**; also see Chapter 1).

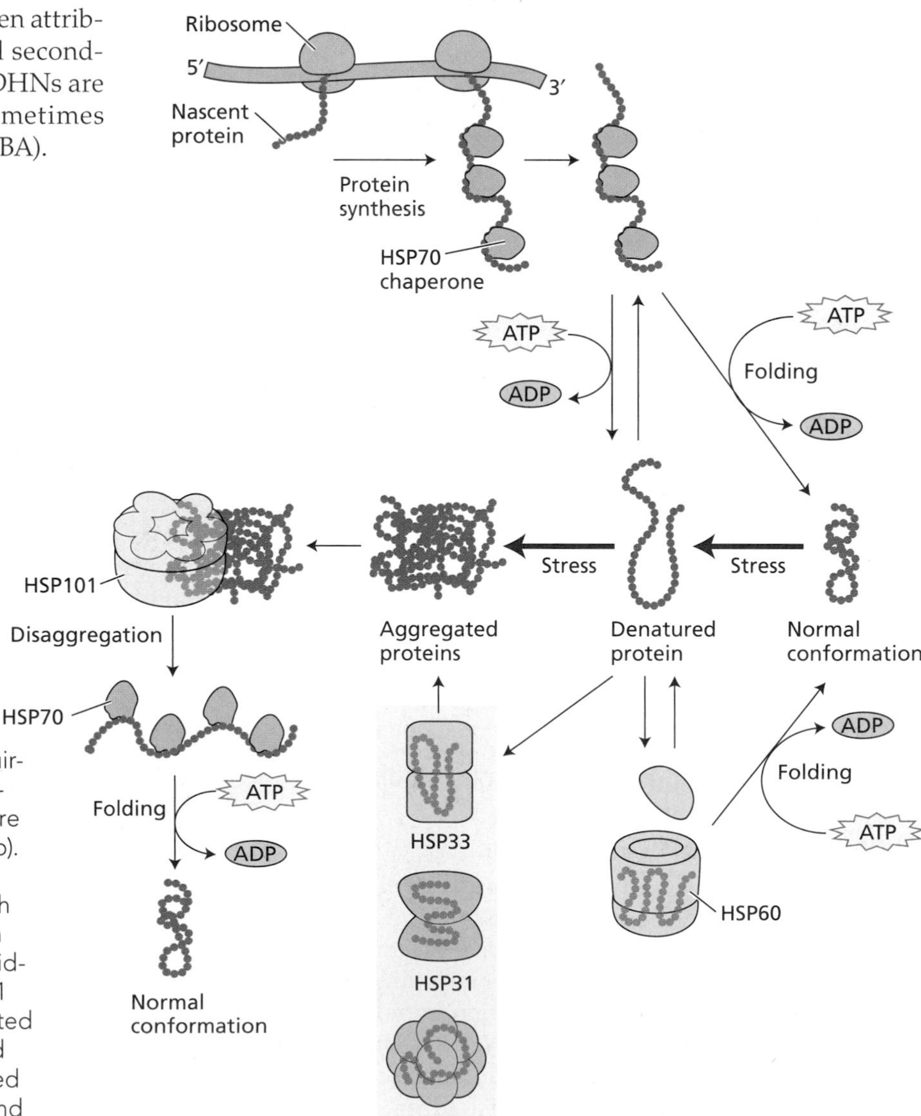

Figure 24.21 Molecular chaperone network in cells. Nascent proteins requiring the assistance of molecular chaperones to reach a proper conformation are associated with HSP70 chaperones (top). Native proteins that undergo denaturation during stress (right) associate with HSP70 and HSP60 chaperones (bottom right). If aggregates are formed (left middle), they are disaggregated by HSP101 and HSP70 (left). Additional stress-related chaperones such as HSP31, HSP33, and sHSPs can also associate with denatured proteins during stress. (After Baneyx and Mujacic 2004.)

TABLE 24.3 Fatty acid composition of mitochondria isolated from chilling-resistant and chilling-sensitive species

	Percent weight of total fatty acid content					
	Chilling-resistant species			Chilling-sensitive species		
Major fatty acids[a]	Cauliflower bud	Turnip root	Pea shoot	Bean shoot	Sweet potato	Maize shoot
Palmitic (16:0)	21.3	19.0	17.8	24.0	24.9	28.3
Stearic (18:0)	1.9	1.1	2.9	2.2	2.6	1.6
Oleic (18:1)	7.0	12.2	3.1	3.8	0.6	4.6
Linoleic (18:2)	16.1	20.6	61.9	43.6	50.8	54.6
Linolenic (18:3)	49.4	44.9	13.2	24.3	10.6	6.8
Ratio of unsaturated to saturated fatty acids	3.2	3.9	3.8	2.8	1.7	2.1

Source: After Lyons et al. 1964.

[a]Shown in parentheses are the number of carbon atoms in the fatty acid chain and the number of double bonds.

Prolonged exposure to extreme temperatures may result in an altered composition of membrane lipids, a form of acclimation. Certain transmembrane enzymes can alter lipid saturation, by introducing one or more double bonds into fatty acids. For example, during acclimation to cold temperatures, the activities of **desaturase enzymes** increase and the proportion of unsaturated lipids rises. This modification lowers the temperature at which the membrane lipids begin a gradual phase change from fluid to semicrystalline form and allows membranes to remain fluid at lower temperatures, thus protecting the plant against damage from chilling. Conversely, a greater degree of saturation of the fatty acids in membrane lipids makes the membranes less fluid. Certain mutants of Arabidopsis have reduced activity of omega-3 fatty acid desaturases. These mutants show increased thermotolerance of photosynthesis, presumably because the degree of saturation of chloroplast lipids is increased.

Exclusion and internal tolerance mechanisms allow plants to cope with toxic ions

Two basic mechanisms are employed by plants to tolerate the presence of high concentrations of toxic ions in the environment, including sodium (Na), arsenic (As), cadmium (Cd), copper (Cu), nickel (Ni), zinc (Zn), and selenium (Se): exclusion and internal tolerance. Exclusion refers to the ability to block the uptake of toxic ions into the cell, thereby preventing the concentrations of these ions from reaching a toxic threshold level. **Internal tolerance** typically involves biochemical adaptations that enable the plant to tolerate, compartmentalize, or chelate elevated concentrations of potentially toxic ions.

Glycophytes are salt-sensitive plants that generally rely on exclusion mechanisms to protect themselves from moderate levels of salinity in the soil. Glycophytes are able to tolerate moderate levels of salinity because of root mechanisms that either reduce the uptake of potentially harmful ions or actively pump these ions back into the soil. Calcium ions play a key role in minimizing the uptake of Na^+ ions from the external medium. As a charged ion, Na^+ has a very low permeability through the lipid bilayer, but it can be transported across the plasma membrane by both low- and high-affinity transport systems, many of which normally transport K^+ into root cells. External Ca^{2+} at millimolar concentrations (the normal physiological concentration of Ca^{2+} in the apoplast) increases the selectivity of the K^+ transporters and minimizes Na^+ uptake. Different Na^+/H^+ antiporters in the plasma membrane and tonoplast can also lower the cytosolic level of sodium by actively pumping it back into the apoplast or into the vacuole. The energy used to drive these processes is supplied by different H^+-pumping ATPases on these membranes (**Figure 24.22**).

In contrast to glycophytes, halophytes can tolerate high levels of Na^+ in the shoot because they have a greater capacity for vacuolar sequestration of ions in their leaf cells. Furthermore, halophytes seem to have a greater ability to restrict net Na^+ uptake into leaf cells. As a result of this increased vacuolar compartmentalization and reduced cellular uptake of Na^+ in the shoots, halophytes have an enhanced capacity to sustain an increased flux of Na^+ from roots in the transpirational stream.

An extreme example of internal tolerance to toxic ions is the **hyperaccumulation** of certain trace elements, which occurs in a limited number of species. Hyperaccumulating

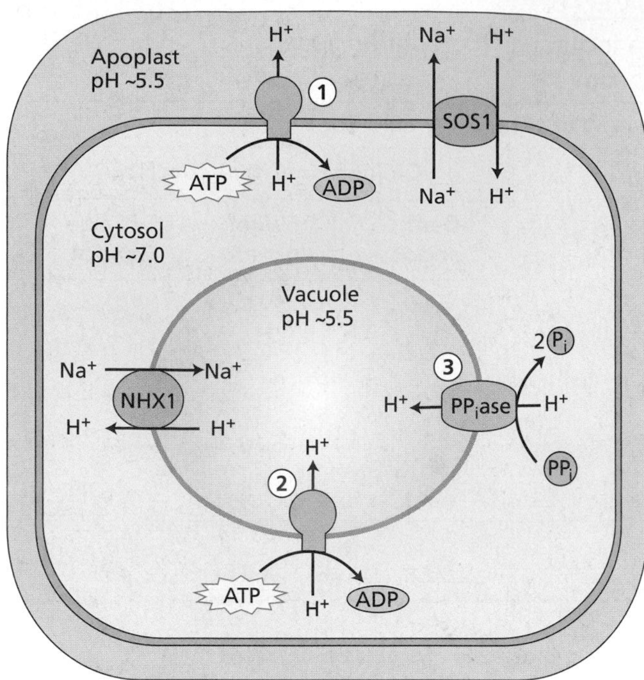

Figure 24.22 Primary and secondary active transport. The plasma membrane–localized H^+-pumping ATPase (P-type ATPase) (1) and the tonoplast-localized H^+-pumping ATPase (V-type ATPase) (2), and pyrophosphatase (PP_iase) (3) are primary active transport systems that energize the plasma membrane and the tonoplast, respectively. By coupling the energy released by hydrolysis of ATP or pyrophosphate, these H^+ pumps are able to transport H^+ across the plasma membrane and the tonoplast against an electrochemical gradient. The H^+–Na^+ antiporters SOS1 and NHX1 are secondary active transport systems that couple the transport of Na^+ against its electrochemical gradient with that of H^+ down its electrochemical gradient. SOS1 transports Na^+ out of the cell, whereas NHX1 transports Na^+ into the vacuole.

plants can tolerate foliar concentrations of various trace elements, such as arsenic, cadmium, nickel, zinc, and selenium, of up to 1% of their shoot dry weight (10 mg per gram dry weight). Hyperaccumulation is a relatively rare plant adaptation to potentially toxic ions that requires heritable genetic changes that enhance the expression of the ion transporters involved in the uptake and vacuolar compartmentalization of these ions.

Phytochelatins and other chelators contribute to internal tolerance of toxic metal ions

Chelation is the binding of an ion with at least two ligating atoms within a chelating molecule. Chelating molecules can have different atoms available for ligation, such as sulfur (S), nitrogen (N), or oxygen (O), and these different atoms have different affinities for the ions they chelate. By wrapping itself around the ion it binds to form a complex, the chelating molecule renders the ion less chemically active, thereby reducing its potential toxicity. The complex is then usually translocated to other parts of the plant, or stored away from the cytoplasm (typically in the vacuole). Long-distance transport of chelated ions from roots to shoots is also a critical process for hyperaccumulation of metals in shoot tissues. Both the iron chelator nicotianamine and the free amino acid histidine have been implicated in chelation of metals during this transport process. In addition, plants also synthesize other ligands for ion chelation, such as phytochelatins.

Phytochelatins are low-molecular-weight thiols consisting of the amino acids glutamate, cysteine, and glycine, with the general form of (γ-Glu-Cys)$_n$Gly. Phytochelatins are synthesized by the enzyme phytochelatin synthase. The thiol groups act as ligands for ions of trace elements such as cadmium and arsenic (**Figure 24.23**). Once formed, the phytochelatin–metal complex is transported into the vacuole for storage. Synthesis of phytochelatins has been shown to be necessary for resistance to cadmium and arsenic. In addition to chelation, active transport of metal ions into the vacuole and out of the cell also contributes to internal metal tolerance.

Plants use cryoprotectant molecules and antifreeze proteins to prevent ice crystal formation

During rapid freezing, the protoplast, including the vacuole, may **supercool**; that is, the cellular water may remain liquid even at temperatures several degrees below its theoretical freezing point. Supercooling is common in many species of the hardwood forests of southeastern Canada and the eastern United States. Cells can supercool only to about −40°C, the temperature at which ice forms spontaneously. Spontaneous ice formation sets the *low-temperature limit* at which many alpine and subarctic species that undergo deep supercooling can survive. It may also

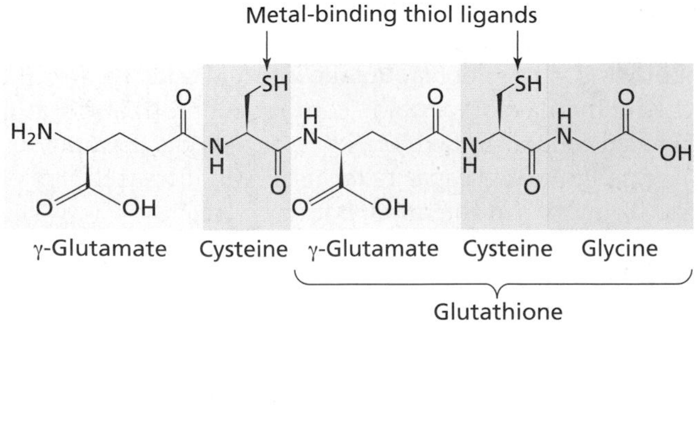

Figure 24.23 Molecular structure of the metal chelate phytochelatin. Phytochelatin uses the sulfur in cysteine to bind metals such as cadmium, zinc, and arsenic.

explain why the altitude of the timberline in mountain ranges is at or near the –40°C minimum isotherm.

Several specialized plant proteins, termed **antifreeze proteins**, limit the growth of ice crystals through a mechanism independent of lowering of the freezing point of water. Synthesis of these antifreeze proteins is induced by cold temperatures. The proteins bind to the surfaces of ice crystals to prevent or slow further crystal growth. Sugars, polysaccharides, osmoprotectant solutes, dehydrins, and other cold-induced proteins also have cryoprotective effects.

ABA signaling during water stress causes the massive efflux of K⁺ and anions from guard cells

As discussed earlier, hormones play an important signaling role in a variety of plant stress responses. During water stress, ABA increases dramatically in leaves, which leads to stomatal closure (**Figure 24.24**). Physiologically, stomatal closure is brought about by a reduction in turgor pressure that follows the massive efflux of K^+ and anions from guard cells. Activation of specialized ion efflux channels on the plasma membrane is required for such a large-scale loss of K^+ and anions from guard cells to occur. How does ABA bring this about?

Plasma membrane K^+ efflux channels are *voltage-gated* (see Chapter 6); that is, they open only if the plasma membrane becomes **depolarized**. ABA causes membrane depolarization by elevating cytosolic calcium in two ways: (1) by triggering a transient influx of Ca^{2+} ions and (2) by promoting the release of Ca^{2+} from internal stores, such as the endoplasmic reticulum and the vacuole. As a result, the cytosolic calcium concentration rises from 50 to 350 n*M* to as high as 1100 n*M* (1.1 μ*M*) (**Figure 24.25**). This increase in cytosolic calcium then opens calcium-activated anion channels on the plasma membrane.

The prolonged opening of anion channels permits large quantities of Cl^- and malate^{2-} ions to escape from the cell, moving down their electrochemical gradients. The outward flow of negatively charged Cl^- and malate^{2-} ions depolarizes the membrane, triggering the opening of the voltage-gated K^+ efflux channels. The elevated levels of cytosolic calcium also cause K^+ influx channels to close, reinforcing the depolarization effect.

In addition to increasing cytosolic calcium, ABA causes alkalinization of the cytosol from about pH 7.7 to pH 7.9. The increase in cytosolic pH has been shown to further stimulate the opening of the K^+ efflux channels. ABA also inhibits the activity of the plasma membrane H^+-ATPase, resulting in additional membrane depolarization. ABA inhibition of

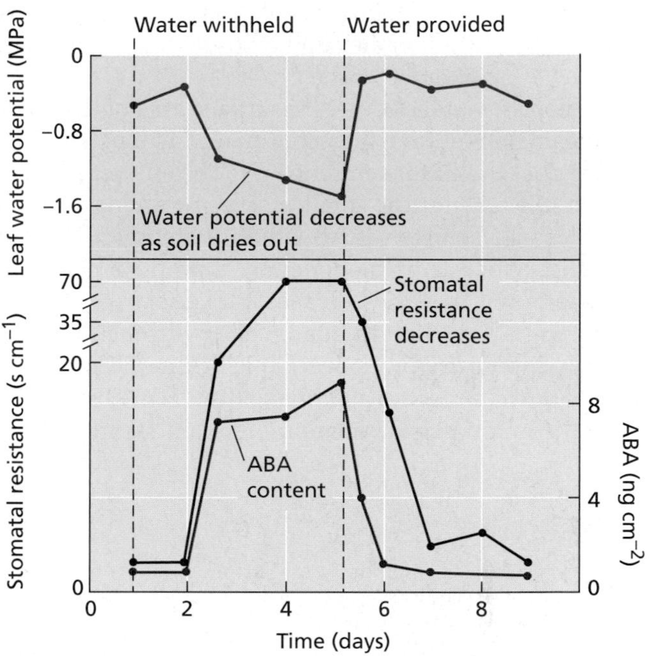

Figure 24.24 Changes in leaf water potential, stomatal resistance (the inverse of stomatal conductance), and ABA content in maize in response to water stress. As the soil dries out, the water potential of the leaf decreases, and the ABA content and stomatal resistance increase. Re-watering reverses the process. (After Beardsell and Cohen 1975.)

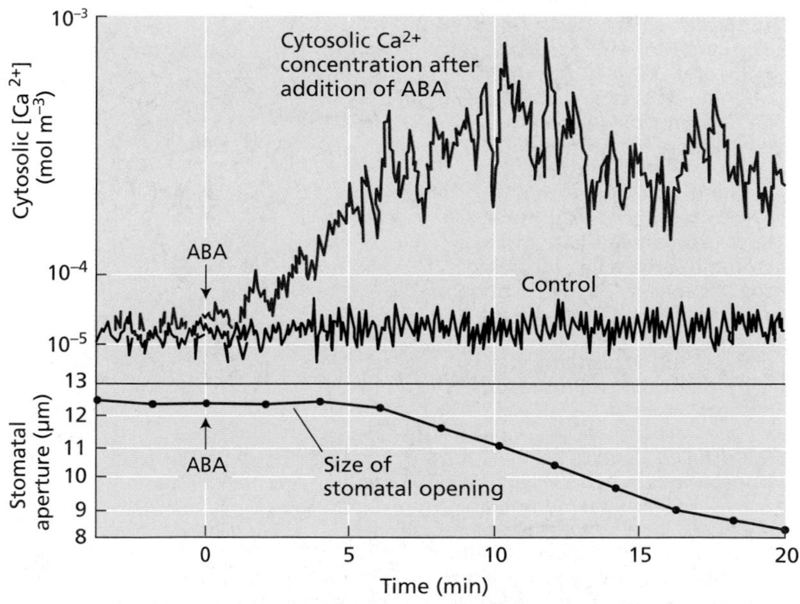

Figure 24.25 Time course of the ABA-induced increase in guard cell cytosolic Ca^{2+} concentration (upper panel) and ABA-induced stomatal aperture (lower panel). The rise in Ca^{2+} begins within approximately 3 min of the addition of ABA, followed by a steady decrease in the size of the stomatal aperture within an additional 5 min. (After McAinsh et al. 1990.)

the plasma membrane proton pump is apparently caused by the combination of elevated cytosolic Ca^{2+} concentration and alkalinization of the cytosol.

During stomatal closure, the surface area of the guard cell plasma membrane may contract by as much as 50%. Where does the extra membrane go? The answer seems to be that it is taken up as small vesicles by endocytosis—a process that also involves ABA-induced reorganization of the actin cytoskeleton mediated by a family of plant Rho GTPases, or ROPs (*Rho* GTPases in *P*lants).

Signal transduction in guard cells, with their multiple sensory inputs, involves protein kinases and phosphatases. For example, the activities of the H^+-ATPases that drive the guard plasma membrane potential are reduced by several protein kinases. Protein phosphatases have also been implicated in modifying specific H^+-ATPase activities, leading to changes in the anion channel activities. In view of these results, it appears that protein phosphorylation and dephosphorylation play important roles in the ABA signal transduction pathway in guard cells. A simpli-

Figure 24.26 Simplified model for ABA signaling in stomatal guard cells. The net effect is the loss of potassium (K^+) and its anion (Cl^- or $malate^{2-}$) from the cell. ROS, reactive oxygen species; CPK, Ca^{2+}-dependent protein kinase; OST1, OPEN STOMATAL1 protein kinase; PP2C, protein phosphatase 2C; RBOH, respiratory burst oxidase homolog, an NADPH oxidase. (After Benjamin Brandt and J. Schroeder, unpublished.)

1. Abscisic acid (ABA) binds cytosolic ABA receptors (RCAR/PYR/PYL) (see Chapter 15).

2. The ABA-bound receptors form a complex with type 2C protein phosphatases (PP2Cs), inhibiting their activity. PP2Cs represent a major negative regulator within this signaling network.

3. Inhibition of PP2C activity releases the positive regulator kinase OST1 from inhibition, resulting in both phosphorylation and activation of NADPH oxidases (RBOHs). RBOHs catalyze the formation of apoplastic reactive oxygen species (ROS) such as H_2O_2 and $O_2^{\bullet-}$, which triggers opening of plasma membrane Ca^{2+}-permeable ion channels. Ca^{2+} enters the cell.

4. The resulting elevation of cytosolic Ca^{2+}, augmented by Ca^{2+}-promoted release of Ca^{2+} stored in organelles, including the vacuole, leads to activation of calcium-dependent protein kinases (CPKs). CPKs can also activate RBOH proteins, further promoting Ca^{2+} influx into the cytosol.

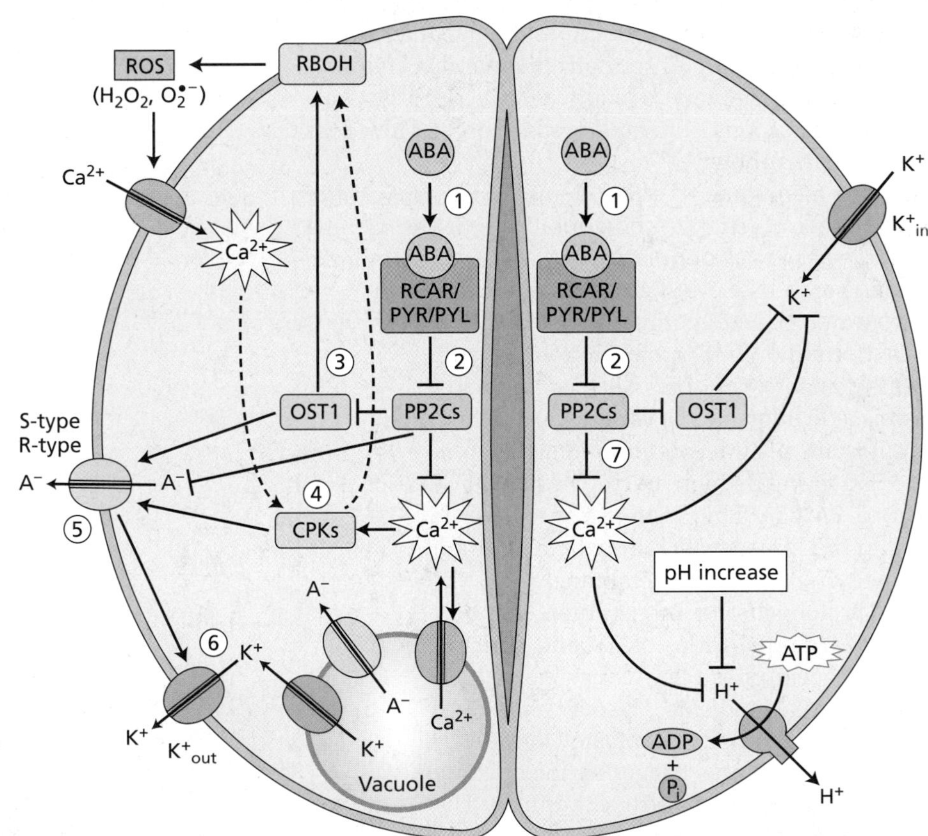

5. Both OST1 and CPKs phosphorylate and thereby activate plasma membrane anion channels, leading to efflux of anions (A^-). This process can be inhibited directly by PP2Cs in the absence of ABA. Two types of anion channels are activated: slow (S-type) and rapid (R-type).

6. Anion efflux leads to depolarization of the plasma membrane, which drives K^+ efflux via outward-rectifying potassium (K^+_{out}) channels. The majority of A^- and K^+ in a plant cell are stored in the vacuole and are released into the cytosol via Ca^{2+}-activated K^+ channels and anion release transporters present in the tonoplast. The efflux of ions (A^- and K^+) reduces the turgor pressure of guard cells, resulting in stomatal closure.

7. Inhibition of PP2C by ABA and RCAR/PYR/PYL also leads to inhibition of plasma membrane channels that mediate accumulation of ions during stomatal opening, such as H^+-ATPases and inward rectifying K^+ (K^+_{in}) channels. These channels would otherwise counteract the closure-promoting effects of ion efflux.

fied general model for ABA action in stomatal guard cells is shown in **Figure 24.26**.

Plants can alter their morphology in response to abiotic stress

In response to abiotic stress, plants can activate developmental programs that alter their phenotype, a phenomenon known as **phenotypic plasticity**. Phenotypic plasticity can result in adaptive anatomical changes that enable plants to avoid some of the harmful effects of abiotic stress.

An important example of phenotypic plasticity is the ability to alter leaf shape. As biological solar collectors, leaves must be exposed to sunlight and air, which makes them vulnerable to environmental extremes. Plants have thus evolved the ability to modify leaf morphology in ways that enable them to avoid or mitigate the effects of abiotic extremes. Such mechanisms include changes in leaf area, leaf orientation, leaf rolling, trichomes, and waxy cuticles, as outlined below.

LEAF AREA Large, flat leaves provide optimal surfaces for the production of photosynthate, but they can be detrimental to crop growth and survival under stressful conditions because they provide a large surface area for evaporation of water, which can lead to quick depletion of soil water or excessive, damaging absorption of solar energy. Plants can reduce their leaf area by reducing leaf cell division and expansion, by altering leaf shapes (**Figure 24.27A**), and by initiating senescence and abscission of leaves (**Figure 24.27B**). This phenomenon can lead to

Figure 24.27 Morphological changes in response to abiotic stress. (A) Altered leaf shape can occur in response to environmental changes. The oak (*Quercus* sp.) leaf on the left is from the outside of a tree canopy, where temperatures are higher than they are inside the canopy. The leaf on the right is from the inside of the canopy. The deeper sinuses of the leaf on the left result in a lower boundary layer, which allows for better evaporative cooling. (B) The leaves of young cotton (*Gossypium hirsutum*) plants abscise in response to water stress. The plants at left were watered throughout the experiment; those in the middle and at right were subjected to moderate stress and severe stress, respectively, before being watered again. Only a tuft of leaves at the top of the stem is left on the severely stressed plants. (C) Leaf movements in soybean in response to osmotic stress. Leaflet orientation of field-grown soybean (*Glycine max*) plants in the well-watered, unstressed position; during mild water stress; and during severe water stress. The large leaf movements induced by mild stress are quite different from wilting, which occurs during severe stress. Note that during mild stress the terminal leaflet has been raised, whereas the two lateral leaflets have been lowered; each is almost vertical. (A, photograph by David McIntyre; B courtesy of B. L. McMichael; C courtesy of D. M. Oosterhuis.)

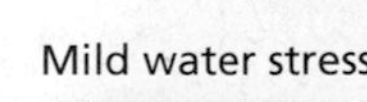

certain types of heterophylly, as in aquatic plants (WEB ESSAY 24.1.

LEAF ORIENTATION For protection against overheating during water deficit, the leaves of some plants may orient themselves away from the sun; such leaves are said to be **paraheliotropic**. Leaves that gain energy by orienting themselves perpendicular to the sunlight are referred to as **diaheliotropic**. Other factors that can alter the interception of radiation include wilting and leaf rolling. Wilting changes the angle of the leaf, and leaf rolling minimizes the profile of tissue exposed to the sun (**Figure 24.27C**).

TRICHOMES Many leaves and stems have hairlike epidermal cells known as trichomes or hairs. Trichomes can either be ephemeral or persist throughout the life of the organ. Some persisting trichomes remain alive, while others undergo programmed cell death, leaving only their cell walls. Densely packed trichomes on a leaf surface keep leaves cooler by reflecting radiation. Leaves of some plants have a silvery white appearance because the densely packed trichomes reflect a large amount of light. However, pubescent leaves are a disadvantage in the cooler spring months because the trichomes also reflect the visible light needed for photosynthesis.

CUTICLE The cuticle is a multilayered structure of waxes and related hydrocarbons deposited on the outer cell walls of the leaf epidermis. The cuticle, like trichomes, can reflect light, thereby reducing heat load. The cuticle appears to also restrict the diffusion of water and gases, as well as the entrance of pathogens. A developmental response to water deficit in some plants is the production of a thicker cuticle, which decreases transpiration.

ROOT:SHOOT RATIO The **root-to-shoot biomass ratio** is another important example of phenotypic plasticity. The root:shoot ratio appears to be governed by a functional balance between water uptake by the root and photosynthesis by the shoot. Within the limits set by the plant's genetic potential, a shoot tends to grow until water uptake by the roots becomes limiting to further growth; conversely, roots tend to grow until their demand for photosynthate from the shoot exceeds the supply. This functional balance is shifted if the water supply decreases. When water to the shoot becomes limiting, leaf expansion is reduced before photosynthetic activity is affected (see Figure 24.5). Inhibition of leaf expansion reduces the consumption of carbon and energy, and a greater proportion of the plant's assimilates can be allocated to the root system, where they can support further root growth (**Figure 24.28**). This root growth is sensitive to the water status of the soil microenvironment; the root apices in dry soil lose turgor, while roots in the soil zones that remain moist continue to grow.

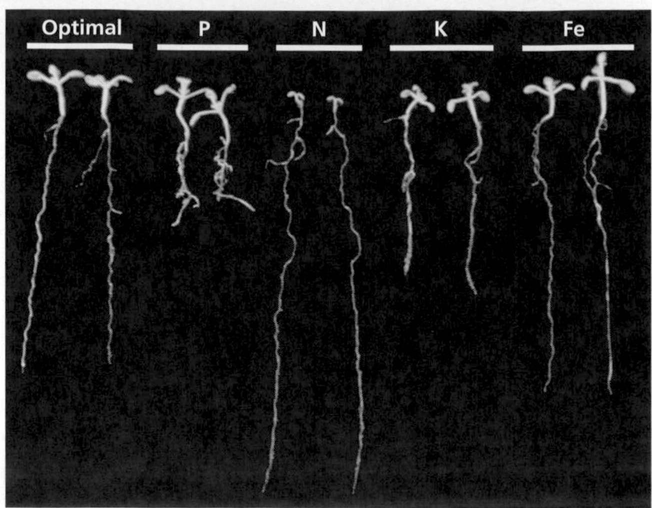

Figure 24.28 Effect of salt stress on the root-to-shoot ratio of tomato. (From Sánchez-Calderón et al. 2014.)

ABA plays an important role in regulating the root:shoot ratio during water stress. As shown in **Figure 24.29**, under water-stress conditions the root-to-shoot biomass ratio increases, allowing the roots to grow at the expense of the leaves. However, ABA-deficient mutants are unable to change their root:shoot ratio in response to water stress. Thus, ABA is required for the change in the root:shoot ratio to occur.

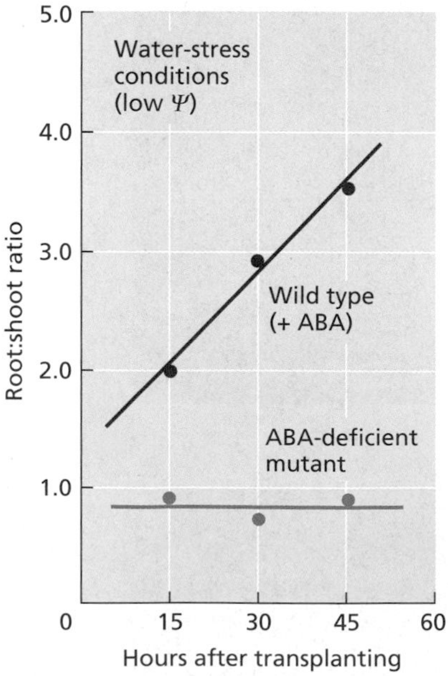

Figure 24.29 Under water-stress conditions (low Ψ, defined slightly differently for shoot and root), the ratio of root growth to shoot growth is much higher when ABA is present (i.e., in the wild type) than when it is absent (in the mutant). (After Saab et al. 1990.)

Metabolic shifts enable plants to cope with a variety of abiotic stresses

Changes in the environment may stimulate shifts in metabolic pathways that decrease the effect of stress on plant metabolism. For example, during anaerobic stress caused by flooding, roots ferment pyruvate to lactate through the action of lactate dehydrogenase (see Figure 12.3). Production of lactate (lactic acid) lowers the intracellular pH, inhibiting lactate dehydrogenase and activating pyruvate decarboxylase. These changes in enzyme activity quickly lead to a switch from lactate to ethanol production (see Figure 12.3). The net yield of ATP in fermentation is only 2 moles of ATP per mole of hexose sugar catabolized (compared with 36 moles of ATP per mole of hexose respired in aerobic respiration) and is therefore inadequate to support normal root growth. However, it is sufficient to keep root cells alive during temporary flooding, or until aerenchyma tissue forms in plants adapted to prolonged flooding.

Some plants possess metabolic adaptations, such as C_4 photosynthesis and crassulacean acid metabolism, that allow them to exploit more arid environments (see Chapters 8 and 9). **Crassulacean acid metabolism (CAM)** is an adaptation in which stomata open at night and close during the day. The leaf-to-air vapor pressure difference that drives transpiration during the day is greatly reduced at night, when both leaf and air are cool. As a result, the water-use efficiencies of CAM plants are among the highest measured. A CAM plant may gain 1 g of dry matter for only 125 g of water used—a ratio that is three to five times greater than the ratio for a typical C_3 plant. CAM is prevalent in succulent plants such as cacti. Some succulent species display **facultative CAM**, switching to CAM when subjected to water deficits or saline conditions. This switch in metabolism—which involves the synthesis of the enzymes phosphoenolpyruvate (PEP) carboxylase, pyruvate–orthophosphate dikinase (pyruvate–phosphate dikinase), and NADP–malic enzyme, among others—is a remarkable adaptation that allows the plant to acclimate to these conditions. As discussed in Chapters 8 and 9, CAM metabolism involves many structural, physiological, and biochemical features, including changes in carboxylation and decarboxylation patterns, transport of large quantities of malate into and out of the vacuoles, and reversal of the periodicity of stomatal movements.

The process of recovery from stress can be dangerous to the plant and requires a coordinated adjustment of plant metabolism and physiology

Once a plant acclimates to a specific set of environmental stress conditions, it achieves a state of metabolic homeostasis that enables it to grow optimally under these conditions. However, when the stress is removed, for example by rewatering in the case of a drought-stressed plant, the plant must shift its metabolism back to the new set of (nonstressed) conditions. In some instances, for example when the plant needs to shift from a highly reduced anaerobic environment to oxidized aerobic conditions during a relief from flooding stress, the metabolic shift may be highly dangerous to the plant because high levels of ROS could form and damage cells. The recovery process from stress is therefore as much a synchronized process as the acclimation to stress is. The plant must remove and recycle all the unneeded mRNAs, proteins, and protective chemicals and compounds that accumulated in cells during the acclimation process and throughout the stress period. In addition, the plant must shift its metabolic energy flow to prepare for and adjust to the new conditions. Reactivation of certain processes such as photosynthesis, respiration, and lipid biosynthesis may require a delicate and synchronized process because these pathways can produce high levels of ROS, and the ROS-scavenging pathways that protect the plant should be in place before the pathways are fully functional. Although from an energy standpoint it is best for the plant to remove and recycle all stress-response mechanisms when the stress subsides, some plants maintain a readiness to counter the reoccurrence of stress by keeping certain stress-response regulatory mechanisms, hormones, or epigenetic modifications active even after the stress subsides. This process is often referred to metaphorically as *memory* because the plant seems to "remember" the stress even long after it is over and will respond much faster to its reoccurrence compared with the first time it encountered it.

Developing crops with enhanced tolerance to abiotic stress conditions is a major goal of agricultural research

A major goal of studying stress responses in plants is to develop crops with enhanced tolerance to both biotic and abiotic stress conditions. Such crops would decrease the yield penalty associated with environmental stress and prevent annual losses of billions of dollars to agricultural production.

The central dogma of abiotic stress research in plants is to study how plants sense and acclimate to abiotic stress conditions, and then use this knowledge to develop plants and crops with enhanced tolerance to abiotic stresses. Strategies for the use of selected genes to improve tolerance to abiotic stresses in crops include gain- and loss-of-function approaches that target single genes at different levels. These genes could encode enzymes or regulatory proteins such as transcription factors or mitogen-activated protein kinases. Tissue-specific, constitutive, or stress-inducible promoters have been used to express the selected genes in order to achieve maximum efficiency in stress protection with as few as possible negative effects on growth and productivity. As described above, the availability of transcriptomic technology led to the identification of several different stress-response transcriptional networks. These studies identified several transcription factors that could activate multiple plant acclimation pathways and provide protection to plants against abiotic stress conditions in the lab. Nevertheless, only a few of these genes have thus far been demonstrated to work under field conditions. Quantitative trait loci (QTL) analysis and traditional breeding have also proven to be very useful for the identification of genes responsible for abiotic stress tolerance in crops. In combination with genetic engineering, these methods have resulted in the identification and successful use of genes responsible for tolerance of salinity in wheat and rice, boron and aluminum stress in wheat, sorghum, and barley, and anaerobic stress in rice.

SUMMARY

Plants sense changes in their environment and respond to them by means of dedicated stress response pathways. These pathways involve gene networks, regulatory proteins, and signaling intermediates, as well as proteins, enzymes and molecules that act to protect cells from the toxic effects of abiotic stress. Together these anti-stress mechanisms enable plants to acclimate or adapt to stresses such as drought, heat, cold, salinity and their possible combinations. A major research goal is to utilize some of these acclimation or adaptation mechanisms for the protection of crops under the adverse climatic conditions that are expected to result from global warming.

Defining Plant Stress

- Stress can be defined generally as any environmental condition that prevents the plant from achieving its full genetic potential under ideal growth conditions.
- Plant responses to abiotic stress involve trade-offs between vegetative and reproductive growth, which may differ depending on whether the plant is an annual or a perennial.

Acclimation and Adaptation

- Adaptation is characterized by genetic changes in an entire population that have been fixed by natural selection over many generations.
- Acclimation is the process whereby individual plants respond to periodic changes in the environment by directly altering their morphology or physiology. The physiological changes associated with acclimation require no genetic modifications, and many are reversible.
- Stress can be defined as any condition that prevents the plant from achieving its maximum growth and reproductive potential (**Figures 24.1, 24.2**).
- Trade-offs between vegetative and reproductive development occur during physiological adjustment to abiotic stress.

Environmental Factors and Their Biological Impacts on Plants

- Environmental stresses can disrupt plant metabolism through a variety of mechanisms, most of which result in the accumulation of reactive oxygen species (ROS) (**Figures 24.3, 24.4**).
- Cell dehydration leads to decreased turgor pressure, increased ion toxicity, and inhibition of photosynthesis (**Figure 24.5, Table 24.1**).
- Salinity stress causes protein denaturation and membrane destabilization, which reduce aboveground plant growth and inhibit photosynthesis (**Table 24.1**).
- Light stress occurs when plants receive more sunlight than they can use photosynthetically (**Table 24.1**).
- Temperature stress affects protein stability, enzymatic reactions, membrane fluidity, and the secondary structures of RNA and DNA (**Table 24.1**).
- Flooded soil is depleted of oxygen, leading to anaerobic stress to the root (**Table 24.1**).
- Freezing stress, like drought stress, causes cell dehydration (**Table 24.1**).
- Heavy metals can replace other essential metals and disrupt key reactions (**Table 24.1**).
- A lack of available mineral nutrients supresses plant growth and reproduction.
- Ozone and ultraviolet light induce the formation of ROS, which in turn induce the formation of leaf lesions and programmed cell death.
- Combinations of abiotic stresses can have effects on plant physiology and productivity that are different than those of individual stresses (**Figures 24.6, 24.7**).
- Plants can gain cross-protection when sequentially exposed to different abiotic stresses.

Stress-Sensing Mechanisms in Plants

- Plants use physical, biophysical, metabolic, biochemical, and epigenetic mechanisms to detect stresses and activate response pathways (**Figure 24.8**).

Signaling Pathways Activated in Response to Abiotic Stress

- Many stress-response pathways share signaling intermediates, allowing these pathways to be integrated (**Figures 24.9–24.12**).
- Regulons simultaneously activate specific stress-response pathways and suppress other pathways that are not needed or could even damage the plant during stress (**Figure 24.13**).
- Chloroplasts can send distress signals to the nucleus.
- A self-propagating wave of ROS production alerts as-yet unstressed parts of the plant of the need for a response (**Figure 24.14**).
- Epigenetic stress-response mechanisms may lead to heritable protection (**Figure 24.15**).
- Hormones act separately and together to regulate abiotic stress responses.

Developmental and Physiological Mechanisms That Protect Plants against Abiotic Stress

- Plants lower root Ψ to continue to absorb water in drying soil (**Figures 24.17, 24.18**).
- Aerenchyma tissue allows O_2 to diffuse down to submerged organs (**Figure 24.19**).
- ROS can be detoxified through scavenging pathways that reduce oxidative stress (**Figure 24.20**).
- Molecular chaperone proteins protect sensitive proteins and membranes during abiotic stress (**Figure 24.21**).
- Prolonged exposure to extreme temperatures can alter the composition of membrane lipids, thereby allowing plants to maintain membrane fluidity (**Table 24.3**).
- Plants cope with toxic ions through exclusion and internal tolerance mechanisms (**Figure 24.22**).
- Plants generate antifreeze proteins to prevent ice crystal formation.
- Stomatal closure is prompted by ABA-induced efflux of K^+ and anions from guard cells (**Figures 24.24–24.26**).
- Plants can alter their leaf morphology and their root-to-shoot biomass ratio to avoid or mitigate abiotic stress (**Figures 24.27–24.29**).
- Metabolic shifts enable plants to outlast ephemeral stresses such as flooding or environmental changes from day to night.
- Reversal of stress-response pathways must occur in a synchronized manner to avoid production of ROS.
- Agricultural researchers study how plants sense and acclimate to stressful conditions, and then try to develop crops with enhanced tolerance.

WEB MATERIAL

- **WEB ESSAY 24.1 Heterophylly in Aquatic Plants** Abscisic acid induces aerial-type leaf morphology in many aquatic plants.

available at plantphys.net

Suggested Reading

Ahuja, I., de Vos, R. C., Bones, A. M., and Hall, R. D. (2010) Plant molecular stress responses face climate change. *Trends Plant Sci.* 15: 664–674.

Atkinson, N. J., and Urwin, P. E. (2012) The interaction of plant biotic and abiotic stresses: From genes to the field. *J. Exp. Bot.* 63: 3523–3543.

Chinnusamy, V., and Zhu, J. K. (2009) Epigenetic regulation of stress responses in plants. *Curr. Opin. Plant Biol.* 12: 133–139.

Lobell, D. B., Schlenker, W., and Costa-Roberts, J. (2011) Climate trends and global crop production since 1980. *Science* 333: 616–620.

Mittler, R. (2002) Oxidative stress, antioxidants and stress tolerance. *Trends Plant Sci.* 7: 405–410.

Mittler, R., and Blumwald, E. (2010) Genetic engineering for modern agriculture: Challenges and perspectives. *Annu. Rev. Plant Biol.* 61: 443–462.

Mittler, R., Vanderauwera, S., Gollery, M., and Van Breusegem, F. (2004) Reactive oxygen gene network of plants. *Trends Plant Sci.* 9: 490–498.

Peleg, Z., and Blumwald, E. (2011) Hormone homeostasis and abiotic stress tolerance in crop plants. *Curr. Opin. Plant Biol.* 14: 1–6.

Peleg, Z., Apse, M. P., and Blumwald, E. (2011) Engineering salinity and water stress tolerance in crop plants: Getting closer to the field. *Adv. Bot. Res.* 57: 405–443.

Glossary

A

abaxial Refers to the lower surface of the leaf.

ABC model Proposal for the way in which floral homeotic genes control organ formation in flowers. According to the model, organ identity in each whorl is determined by a unique combination of the three organ identity gene activities.

abscission zone A region that contains the abscission layer and is located near the base of the petiole of leaves.

abscission The shedding of leaves, flowers, and fruits from a living plant. The process whereby specific cells in the leaf petiole (stalk) differentiate to form an abscission layer, allowing a dying/dead organ to separate from the plant.

absorption spectrum A graphic representation of the amount of light energy absorbed by a substance plotted against the wavelength of the light.

ACC oxidase Catalyzes the conversion of ACC to ethylene, the last step in ethylene biosynthesis.

accessory pigments Light-absorbing molecules in photosynthetic organisms that work with chlorophyll *a* in the absorption of light used for photosynthesis. They include carotenoids, other chlorophylls, and phycobiliproteins.

acclimation (hardening) The increase in plant stress tolerance due to exposure to prior stress. May involve gene expression. Contrast with adaptation.

acetylation The catalyzed chemical addition of an acetate group to another molecule.

acid growth hypothesis The hypothesis that cell wall acidification resulting from proton extrusion across the plasma membrane causes cell wall stress relaxation and extension.

acid growth A characteristic of growing cell walls in which they extend more rapidly at acidic pH than at neutral pH.

acropetal From the base to the tip of an organ, such as a stem, root, or leaf.

actin depolymerizing factor (ADF) One of a family of small proteins that bind actin filaments and promote their severing and depolymerization.

actin A major ATP-binding cytoskeletal protein. The monomer, globular actin, or G-actin, can bind either ADP or ATP. ATP-charged G-actin can self-associate to form long polar filaments of F-actin. In the F-actin form, the ATP is slowly hydrolyzed. The filaments grow by adding new monomers to the plus end (also called the barbed end) and shrink by releasing ADP-bound actin monomers from the minus end (also called the pointed end).

actinorhizal Pertaining to several woody plant species, such as alder trees, in which symbiosis occurs with soil bacteria of the nitrogen-fixing genus *Frankia*.

action potential A transient event in which the membrane potential difference rapidly rises (hyperpolarizes) and abruptly falls (depolarizes). Action potentials, which are triggered by the opening of ion channels, can be self-propagating along linear files of cells, especially in the vascular systems of plants.

action spectrum A graphic representation of the magnitude of a biological response to light as a function of wavelength.

activators In the control of transcription, positively acting transcription factors that bind to distal regulatory sequences usually located within 1000 bp of the transcription initiation site.

active transport The use of energy to move a solute across a membrane against a concentration gradient, a potential gradient, or both (electrochemical potential). Uphill transport.

acyl carrier protein (ACP) A low-molecular-weight, acidic protein to which growing acyl chains are covalently bonded on fatty acid synthetase.

acyl hydrolases Acyl hydrolases are enzymes that remove acyl groups (consisting of a carbonyl group and an alkyl group) from other functional groups.

acyl-ACP A fatty acid chain bonded to the acyl carrier protein.

adaptation An inherited level of stress resistance acquired by a process of selection over many generations. Contrast with acclimation.

adaxial Refers to the upper surface of a leaf.

adenosine triphosphate (ATP) The major carrier of chemical energy in the cell, which by hydrolysis is converted to adenosine diphosphate (ADP) or adenosine monophosphate (AMP).

adequate zone Range of mineral nutrient concentrations beyond which further addition of nutrients no longer increases growth or yield.

adhesion The attraction of water to a solid phase such as a cell wall or glass surface, due primarily to the formation of hydrogen bonds.

aerenchyma Anatomical feature of roots found in hypoxic conditions, showing large, gas-filled intercellular spaces in the root cortex.

aerobic respiration The complete oxidation of carbon compounds to CO_2 and H_2O, using oxygen as the final electron acceptor. Energy is released and conserved as ATP.

aeroponics The technique by which plants are grown without soil with their roots suspended in air while being sprayed continuously with a nutrient solution.

after-ripening Technique for breaking seed dormancy by storage at room temperature under dry conditions, usually for several months.

albuminous cells Sieve element-associated cells in the phloem of gymnosperms. Although similar to companion cells in angiosperms, they have a different developmental origin. Also called Strasburger cells.

aleurone layer Layer of aleurone cells surrounding and distinct from starchy endosperm of cereal grains.

allelopathy Release by plants of substances into the environment that have harmful effects on neighboring plants.

allocation The regulated diversion of photosynthate into storage, utilization, and/or transport.

allopolyploids Polyploids with multiple complete genomes derived from two separate species.

α-amylase inhibitors Substances synthesized by some legumes that interfere with herbivore digestion by blocking the action of the starch-digesting enzyme α-amylase.

α-expansins (EXPA) One of two major families of expansin proteins that catalyze the pH-dependent extension and stress relaxation of cell walls.

α-tubulin Along with β-tubulin, a component of the heterodimer monomer that polymerizes to form microtubules.

alternation of generations The presence of two genetically distinct multicellular stages, one haploid and one diploid, in the plant life cycle. The haploid gametophyte generate begins with meiosis, while the diploid sporophyte generation begins with the fusion of sperm and egg.

alternative oxidase An enzyme in the mitochondrial electron transport chain that reduces oxygen and oxidizes ubihydroquinone.

alveolar cells A layer of cells surrounded by tube-like cell walls, formed during the cellularization of coenocytic endosperm.

amplifying divisions In a differentiating leaf epidermis, a mechanism for amplifying the number of stomata in which themeristemoid undergoes a variable number of asymmetric divisions giving rise to as many as three stomatal lineage ground cells.

amplitude In a biological rhythm, the distance between peak and trough; it can often vary while the period remains unchanged.

amyloplast A starch-storing plastid found abundantly in storage tissues of shoots and roots, and in seeds. Specialized amyloplasts in the root cap also serve as gravity sensors.

anaphase A Early anaphase, during which the sister chromatids separate and begin to move toward opposite poles.

anaphase B Late anaphase, during which the polar microtubules slide past each other and elongate to push the spindle poles farther apart. Simultaneously, the sister chromosomes are pushed to their respective poles.

anaphase promoting complex During mitosis, this protein complex controls the proteasomal destruction of cyclin proteins, allowing the metaphase-aligned chromatids to segregate to their respective poles.

anaphase The stage of mitosis during which the two chromatids of each replicated chromosome are separated and move toward opposite poles.

anchored proteins Proteins that are bound to the membrane surface via lipid molecules, to which they are covalently attached.

aneuploidy A condition in which genomes contain additional or fewer individual chromosomes (not entire chromosome sets) than normal.

angiosperms The flowering plants. With their innovative reproductive organ, the flower, they are a more advanced type of seed plant and dominate the landscape. Distinguished from gymnosperms by the presence of a carpel that encloses the seeds.

anisotropic growth Enlargement that is greater in one direction over another; for example, elongating cells in the stem or root axis grow more in length than in width.

annual plant A plant that completes its life cycle from seed to seed, senesces, and dies within one year.

antenna complex A group of pigment molecules that cooperate to absorb light energy and transfer it to a reaction center complex.

antheridium The organ of the male that produces the sperm in the gametophyte generation of lower plants.

anticlinal Pertaining to the orientation of the cell plate at right angles to the longitudinal axis during cytokinesis.

antifreeze proteins Proteins that confer to aqueous solutions the property of thermal hysteresis. When induced by cold temperatures, these plant proteins bind to the surfaces of ice crystals to prevent or slow further crystal growth, thereby limiting or preventing freeze damage. Some antifreeze proteins may be identical to pathogenesis-related proteins.

antimicrobial peptides Small glycine/cysteine–enriched peptides produced by plants that inhibit bacterial growth.

antioxidative enzymes Proteins that detoxify reactive oxygen species.

antipodal cells Cells located at the challazal end of the embryo sac in a mature female gametophyte.

antiport A type of secondary active transport in which the passive (downhill) movement of protons or other ions drives the active (uphill) transport of a solute in the opposite direction.

apical cell In ferns and other primitive vascular plants, the single initial or stem cell of roots and shoots that gives rise to all the other cells of the organ. In angiosperm embryogenesis, the smaller, cytoplasm-rich cell formed by the first division of the zygote.

apical dominance In most higher plants, the growing apical bud's inhibition of the growth of lateral buds (axillary buds).

apical meristems Localized regions made up of undifferentiated cells undergoing cell division without differentiation at the tips of shoots and roots.

apical–basal axis An axis that extends from the shoot apical meristem to the root apical meristem.

apoplast The mostly continuous system of cell walls, intercellular air spaces, and xylem vessels in a plant.

apoplastic transport Movement of molecules through the cell wall continuum that is called the apoplast. Molecules may move through the linked cell walls of adjacent cells, and in that way move throughout the plant without crossing the plasma membrane.

apoptosis A type of programmed cell death found in animals showing characteristic morphological and biochemical changes, including fragmentation of nuclear DNA between the nucleosomes. Apoptosis-like changes also occur in some senescing plant tissues, differentiating xylem tracheary elements, and in the hypersensitive response against pathogens.

aquaporins Integral membrane proteins that form water-selective channels across the membrane. Such channels facilitate water movement across the membrane.

ARABIDOPSIS HISTIDINE PHOSPHOTRANSFER (AHP) Gene involved in cytokinin signal propagation from the plasma membrane receptor to the nucleus.

ARABIDOPSIS RESPONSE REGULATOR (ARR) Arabidopsis genes that are similar to bacterial two-component signaling proteins called response regulators. There are two classes: Type-A ARRs, whose transcription is up-regulated by cytokinin; type-B ARRs, whose expression is not affected by cytokinin.

arabinans Neutral polysaccharides with a backbone of 1,5-linked arabinose residues decorated with single or short side chains made of arabinose. Arabinans may be separate polymers or may be domains attached to the backbone of rhamnogalacturonan I.

arabinogalactan proteins (AGPs) A family of extensively glycosylated (mostly galactose and arabinose), water-soluble cell wall proteins that usually amount to less than 1% of the wall dry mass. Some may associate with the plasma membrane via a glycosylphosphatidylinositol anchor. They often display tissue- and cell-specific expression.

arabinoxylan A branched cell wall polysaccharide consisting of a backbone of xylose residues with arabinose side chains.

arbuscular mycorrhizas Symbioses between a newly described phylum of fungi, the Glomeromycota, and the roots of a broad range of angiosperms, gymnosperms, ferns, and liverworts. Facilitates the uptake of mineral nutrients by roots.

arbuscules Branched structures of mycorrhizal fungi that form within penetrated plant cells; the sites of nutrient transfer between the fungus and the host plant.

archegonium The organ of the female that produces the egg in the gametophyte generation of lower plants.

archesporial cells Cells that give rise either to the male pollen mother cell (microspore mother cell) or the female megaspore mother cell.

ARGONAUTE (AGO) A catalytic protein that is part of the RNA-induced silencing complex.

Arp 2/3 Actin related proteins 2 and 3 which bind to the side of a pre-existing actin filament and form a complex with actin to initiate growth of an actin filament branch.

ascorbate peroxidase An enzyme that converts peroxide and ascorbate to dehydroascorbate and water.

asparagine synthetase (AS) Enzyme that transfers nitrogen as an amino group from glutamine to aspartate, forming asparagine.

aspartate aminotransferase (Asp-AT) An aminotransferase that transfers the amino group from glutamate to the carboxyl atom of oxaloacetate to form aspartate.

ATP synthase The multi-subunit protein complex that synthesizes ATP from ADP and phosphate (P). F_oF_1 and CF_0-CF_1 types are present in mitochondria and chloroplasts, respectively. Also called Complex V.

atrichoblasts Root epidermal cells that are incapable of forming root hairs.

autonomous transposon A transposon that can move on its own without the need for other transposable elements.

autophagic body The single membrane-bound organelle, derived from the autophagosome, that enters the vacuole and releases its contents for degradation.

autophagosome A double membrane-bound organelle that delivers cellular components to the vacuole for degradation.

autophagy-related genes (ATG genes) A group of genes encoding proteins that are required for autophagy.

autophagy A catabolic mechanism that conveys cellular macromolecules and organelles via autophagosomes to lytic vacuoles where they are degraded and recycled.

autopolyploids Polyploids containing multiple complete genomes of a single species.

auxin response element (AuxRE) A DNA promoter sequence that modulates gene expression when bound by auxin-responsive transcription factors.

auxin response factors (ARFs) A family of proteins that regulate the transcription of specific genes involved in auxin responses; they are inhibited by association with specific Aux/IAA repressor proteins, which are degraded in the presence of auxin.

auxin sink A cell or tissue that takes up auxin from a nearby auxin source. Participates in auxin canalization during vascular differentiation.

auxin source A cell or tissue that exports auxin to other cells or tissues by polar transport.

AUXIN/INDOLE-3-ACETIC ACIDS (AUX/IAA) A family of short-lived small proteins that combine with the TIR1/AFB proteins to form the primary auxin receptor. This family of short-lived small proteins in Arabidopsis regulates auxin-induced gene expression by binding to ARF protein that is bound to DNA. If the specific ARF is a transcriptional activator, the Aux/IAA binding represses transcription.

avirulence genes (*avr* genes) Genes that encode specific elicitors of plant defense responses.

axillary buds Secondary meristems that are formed in the axils of leaves. If they are also vegetative meristems, they will have a structure and developmental potential similar to that of the vegetative apical meristem. Axillary buds can also form flowers, as in inflorescences.

axillary meristem Meristematic tissue in the axils of leaves that gives rise to axillary buds.

B

***Bacillus thuringiensis* (Bt)** A soil bacterium that is the source for a commonly used transgene encoding an insecticidal toxin.

bacterial phytochrome-like proteins (BphPs) Members of a widespread family of photosensors that include plant phytochromes (Phy family), cyanobacteria (Cph1 and Cph2), and purple and other nonphotosynthetic bacteria (BphP), and even fungi (Fph).

bacteriochlorophylls Light-absorbing pigments active in photosynthesis in anoxygenic photosynthetic organisms.

bacteroids Nitrogen-fixing organelles that develop from endosymbiotic bacteria upon a signal from the host plant.

bark storage proteins (BSPs) Storage proteins that accumulate in the phloem parenchyma (inner bark) of woody species at the end of the growing season in temperate climates. These proteins are mobilized to support growth in the spring.

bark Collective term for all the tissues outside the cambium of a woody stem or root, and composed of phloem and periderm.

basal cell In embryogenesis, the larger, vacuolated cell formed by the first division of the zygote. Gives rise to the suspensor.

basipetal From the growing tip of a shoot or root toward the base (junction of the root and shoot).

berry A simple, fleshy fruit produced from a single ovary and consisting of a an outer pigmented exocarp, a fleshy, juicy mesocarp, and a membranous inner endocarp.

β-expansins (EXPB) One of two major families of expansins; the number of EXPB genes is particularly numerous in grasses, where a subset are abundantly expressed in pollen and facilitate pollen tube penetration of the stigma.

β-hairpin region A basic protein structural motif consisting of two beta strands in an anti-parallel position resembling a hairpin. Also called a β-ribbon.

β-oxidation Oxidation of fatty acids into fatty acyl-CoA, and the sequential breakdown of the fatty acids into acetyl-CoA units. NADH is also produced.

β-tubulin Along with α-tubulin, a component of the heterodimer monomer that polymerizes to form microtubules.

biennial plant A plant that requires two growing seasons to flower and produce seed.

biolistics A procedure, also called the "gene gun" technique, in which tiny gold particles coated with the genes of interest are mechanically shot into cultured cells. Some of the DNA is randomly incorporated into the genome of the targeted cells.

biosphere Parts of the surface and atmosphere of the Earth that support life as well as the organisms living there.

biotrophic pathogens Pathogens that leave infected tissue alive and only minimally damaged while the pathogen continues to feed on host resources.

blade *See* lamina.

bleaching The loss of chlorophyll's characteristic absorbance due to its conversion into another structural state, often by oxidation.

boundary layer resistance (r_b) The resistance to the diffusion of water vapor due to the layer of unstirred air next to the leaf surface. A component of diffusional resistance.

boundary meristem A transitional zone that separates the leaf primordium from the shoot apical meristem.

Bowen ratio The ratio of sensible heat loss to evaporative heat loss, the two most important processes in the regulation of leaf temperature.

bract Small leaf-like structure with underdeveloped blade.

branched plasmodesmata Plasmodesmata that are connected together in the cell wall and, in turn, connect adjacent cells to each other. *See* plasmodesmata.

brassinolide A plant steroidal hormone with growth-promoting activity, first isolated from *Brassica napus* pollen. One of a group of plant steroidal hormones with similar activities called brassinosteroids.

brassinosteroids A group of plant steroid hormones that play important roles in many developmental processes, including cell division and cell elongation in stems and roots, photomorphogenesis, reproductive development, leaf senescence, and stress responses.

bryophyte *See* nonvascular plants.

bud scales Small, scale-like leaves that form a protective sheath around a dormant bud.

bulk transport Translocation of water and solutes by mass flow down a pressure gradient, as in the xylem or phloem.

bulliform cells Large, bubble-shaped cells usually grouped in clusters on the upper epidermis of grass leaves. The contraction and expansion of buliform cells in response to turgor changes regulates leaf rolling and unrolling.

bundle sheath One or more layers of closely packed cells surrounding the small veins of leaves and the primary vascular bundles of stems.

C

C_4 photosynthesis The photosynthetic carbon metabolism of certain plants in which the initial fixation of CO_2 and its subsequent reduction take place in different cells, the mesophyll and bundle sheath cells, respectively. The initial carboxylation is catalyzed by phosphoenylpyruvate carboxylase (not by rubisco as in

C_3 plants), producing a four-carbon compound (oxaloacetate), which is immediately converted to malate or aspartate.

caeliferins A family of sulfated α-hydroxy fatty acids that elicit plant volatile production and immune responses.

caleosins Calcium-binding proteins in the outer leaflet of oil body membranes, which, like the oleosins, have a large hydrophobic sequence that penetrates into the triglyceride storage lipid of the oil body.

callose A β-1,3-glucan synthesized in the plasma membrane and deposited between the plasma membrane and the cell wall. Synthesized by sieve elements in response to damage, stress, or as part of a normal developmental process.

calmodulin A conserved Ca^{2+}-binding protein found in all eukaryotes that regulates many Ca^{2+}-driven, metabolic reactions.

Calvin–Benson cycle The biochemical pathway for the reduction of CO_2 to carbohydrate. The cycle involves three phases: the carboxylation of ribulose 1,5-bisphosphate with atmospheric CO_2, catalyzed by rubisco, the reduction of the formed 3-phosphoglycerate to trioses phosphate by 3-phosphoglycerate kinase and NADP-glyceraldehyde-3-phosphate dehydrogenase, and the regeneration of ribulose 1,5-bisphospate through the concerted action of ten enzymatic reactions.

CAM *See* Crassulacean acid metabolism.

cambium Layer of meristematic cells between the xylem and phloem that produces cells of these tissues and results in the lateral (secondary) growth of the stem or root.

canalization model The hypothesis that asauxin flows through tissues it stimulates and polarizes its own transport, which gradually becomes channeled—or canalized—into files of cells leading away from auxin sources; these cell files can then differentiate to form vascular tissue.

capillarity The movement of water for small distances up a glass capillary tube or within the cell wall, due to water's cohesion, adhesion, and surface tension.

carbon fixation reactions The synthetic reactions occurring in the stroma of the chloroplast that use the high-energy compounds ATP and NADPH for the incorporation of CO_2 into carbon compounds.

carbon isotope ratio The ratio $^{13}C/^{12}C$ isotope composition of carbon compounds as measured by use of a mass spectrometer.

cardenolides Steroidal glycosides that taste bitter and are extremely toxic to higher animals through their action on Na^+K^+-activated ATPases. Extracted from foxglove (*Digitalis*) for treatment of human heart disorders.

cardiac glycosides Glycosylated organic plant defense compounds similar to oleandrin from Oleander bushes that are poisonous to animals and inhibit sodium/potassium channels to cause contractions in cardiac muscles.

carotenoids Linear polyenes arranged as a planar zigzag chain, with conjugated double bonds. These orange pigments serve both as antenna pigments and photoprotective agents.

carriers Membrane transport proteins that bind to a solute, undergo conformational change, and release the solute on the other side of the membrane.

Casparian strip A band in the cell walls of the endodermis that is impregnated with the waxlike, hydrophobic substance suberin. Prevents water and solutes from entering the xylem by moving between the endodermal cells.

catalase An enzyme that breaks hydrogen peroxide down into water. When it is abundant in peroxisomes it may form crystalline arrays.

cation exchange The replacement of mineral cations adsorbed to the surface of soil particles by other cations.

cavitation The collapse of tension in a column of water resulting from the indefinite expansion of a tiny gas bubble.

CCAAT box A sequence of nucleotides involved in the initiation of transcription in eukaryotes.

cell autonomous response A response to an environmental stimulus or genetic mutation that is localized to a particular cell.

cell plate Wall-like structure that separates newly divided cells. Formed by the phragmoplast and later becomes the cell wall.

cell wall The rigid cell surface structure external to the plasma membrane that supports, binds, and protects the cell. Composed of cellulose and other polysaccharides and proteins. *See also* primary cell walls and secondary cell walls.

cellulose microfibril Thin, ribbon-like structure of indeterminate length and variable width composed of 1→4-linked β-D-glucan chains tightly packed in crystalline arrays alternating with less organized amorphous regions. Provides structural integrity to the cell walls of plants and determines the directionality of cell expansion.

cellulose synthase Enzyme that catalyzes the synthesis of individual 1→4-linked β-D-glucans that make up cellulose microfibrils.

cellulose A linear chain of 1→4-linked β-D-glucose. The repeating unit is cellobiose.

central cell The cell in the embryo sac that fuses with the second sperm cell, giving rise to the primary endosperm cell.

central zone (CZ) A central cluster of relatively large, highly vacuolate, slow-dividing cells in shoot apical meristems, comparable to the quiescent center of root meristems.

centromere The constricted region on the mitotic chromosome where the kinetochore forms and to which spindle fibers attach.

***CESA* (*Cellulose synthase A*)** A multigene family of cellulose synthases found in all land plants.

channels Transmembrane proteins that function as selective pores for passive transport of ions or water across the membrane.

checkpoint A key regulatory point early in G_1 of the cell cycle that determines if the cell is committed to the initiation of DNA synthesis.

chelator A carbon compound that can form a noncovalent complex with certain cations facilitating their uptake (e.g., malic acid, citric acid).

chemical fertilizers Fertilizers that provide nutrients in inorganic forms.

chemical potential The free energy associated with a substance that is available to perform work.

chemiosmotic hypothesis The mechanism whereby the electrochemical gradient of protons established across a membrane by an electron transport process is used to drive energy-requiring ATP synthesis. It operates in mitochondria and chloroplasts.

chemocyanin A small, secreted protein in the style that acts as a directional cue during pollen tube growth.

chemotropic hypothesis The hypothesis that a hierarchy of molecular cues directs the pollen tube to its destination by stimulating the tip to grow toward the ovule.

chlorophyll *a*/*b* antenna proteins Chlorophyll-containing proteins associated with one or the other of the two photosystems in eukaryotic organisms. Also known as light-harvesting complex proteins (LHC proteins).

chlorophylls A group of light-absorbing green pigments active in photosynthesis.

chloroplast The organelle that is the site of photosynthesis in eukaryotic photosynthetic organisms.

chlorosis The yellowing of plant leaves such as occurs as a result of mineral deficiency. The leaves affected and the chlorosis localization on the leaf can be diagnostic for the type of deficiency.

Cholodny–Went hypothesis Early mechanism proposed for tropisms involving stimulation of the bending of the plant axis by lateral transport of auxin in response to a stimulus, such as light, gravity, or touch. The original model has been supported and expanded by recent experimental evidence.

chromatin The DNA–protein complex found in the interphase nucleus. Condensation of chromatin forms the mitotic and meiotic chromosomes.

30 nm chromatin fiber The irregular helical structure formed by nucleosomes wrapped with DNA.

chromatin remodeling Stable changes in chromatin structure accomplished by epigenetic factors.

chromophore A light-absorbing pigment molecule that is usually bound to a protein (an apoprotein).

chromoplasts Plastids that contain high concentrations of carotenoid pigments, rather than chlorophyll. Chromoplasts are responsible for the yellow, orange, or red colors of many fruits, flowers, and autumn leaves.

chromosome territory The specific region inside a nucleus that a chromosome occupies.

chromosomes The condensed form of chromatin that forms early in mitosis and meiosis.

chronic photoinhibition Photoinhibition of photosynthentic activity in which both quantum efficiency and the maximum rate of photosynthesis are decreased. Occurs under high levels of excess light.

circumnutation The tendency of a shoot or root tip to oscillate in a spiral pattern during growth.

***cis*-acting sequences** DNA sequences that bind transcription factors and are adjacent (*cis*) to the transcription units they regulate. Not to be confused with *cis*-elements.

***cis*-elements** Certain nucleotide sequences within the mRNA molecule by which mRNA stability is regulated. Not to be confused with *cis*-acting sequences in DNA that influence transcriptional activity.

cisgenics Genetic engineering techniques, in which genes are transferred between plants that could also otherwise be crossed sexually.

cisternae (singular *cisterna*) A network of flattened saccules and tubules that compose the endoplasmic reticulum, the ER.

citric acid cycle A cycle of reactions localized in the mitochondrial matrix that catalyzes the oxidation of pyruvate to CO_2. ATP and NADH are generated in the process.

cladodes Flattened, photosynthetic stems that perform the functions of leaves, as in the pads of prickly pear cacti (*Opuntia*).

clathrins Proteins that have a unique *triskelion* structure that spontaneously assemble into 100-nm cages that coat vesicles associated with endocytosis at the plasma membrane and other cellular trafficking events.

climacteric Marked rise in respiration at the onset of ripening that occurs in all fruits that ripen in response to ethylene, and in the senescence process of detached leaves and flowers.

clock hypothesis Currently accepted hypothesis of how plants measure night length. Proposes that photoperiodic timekeeping depends on the endogenous oscillator of circadian rhythms.

co-suppression Decreased expression of a gene when extra copies are introduced.

co-translational insertion The mechanism of inserting a protein into a membrane as it is being made, or translated, from mRNA. Most endomembrane proteins are co-translationally inserted first into the endoplasmic reticulum and then transported to their destination.

CO_2 compensation point The CO_2 concentration at which the rate of respiration balances the photosynthetic rate.

coat proteins Specific proteins on the surfaces of vesicles that determine the delivery of vesicle membrane and contents to the Golgi or the ER. COP1, COP2, and clathrins are coat proteins.

coat-imposed dormancy Dormancy imposed on the embryo by the seed coat and other enclosing tissues, such as endosperm, pericarp, or extrafloral organs.

coevolution Linked genetic adaptations of two or more organisms.

cohesion The mutual attraction between water molecules due to extensive hydrogen bonding.

coincidence model A model for flowering in photoperiodic plants in which the circadian oscillator controls the timing of light-sensitive and light-insensitive phases during the twenty-four hour cycle.

coleoptile A modified ensheathing leaf that covers and protects the young primary leaves of a grass seedling as it grows through the soil. Unilateral light perception, especially blue light, by the tip results in asymmetric growth and bending due to unequal auxin distribution in the lighted and shaded sides.

coleorhiza A protective sheath surrounding the embryonic radicle in members of the grass family.

collection phloem Sieve elements of sources.

collenchyma A specialized parenchyma with irregularly thickened, pectin-rich, primary cell walls that function in support in growing parts of a stem or leaf.

columella initials Located directly below (distal to) the quiescent center, these cells give rise to the central portion of the root cap.

columella The central cylinder of the root cap.

commensalism A relationship between two organisms where one organism benefits without negatively affecting the other.

common symbiotic pathway A sequence of common cellular events in plant roots that occurs in both mycorrhizal formation and root nodulation.

companion cells In angiosperms, metabolically active cells that are connected to their sieve element by large, branched plasmodesmata and take over many of the metabolic activities of the sieve element. In source leaves, they function in the transport of photosynthate into the sieve elements.

compatible solutes Organic compounds that are accumulated in the cytosol during osmotic adjustment. Compatible solutes do not inhibit cytosolic enzymes as do high concentrations of ions. Examples of compatible solutes include proline, sorbitol, mannitol, and glycine betaine.

complementation Genetic procedure by which two recessive mutations are introduced into the same cell to discover whether they effect the same genetic function and are therefore alleles. If the *trans* configuration (m +/+ m_1) exhibits a mutant phenotype, the mutations are allelic, but if they show a wild-type phenotype, they are nonallelic.

complexes I–V Also known as the respiratory complexes, they are NADH dehydrogenase (complex I), succinate dehydrogenase (complex II), the cytochrome *bc1* complex (complex III), cytochrome *c* oxidase (complex IV), and the F_oF_1-ATP synthase (complex V). Only complex V is not involved in electron transport.

complex V *See* ATP synthase.

compound leaf A leaf subdivided into leaflets.

condensed tannins Tannins that are polymers of flavonoid units. Require use of strong acid for hydrolysis.

conifer resin ducts Ducts or channels in conifer leaves and woody tissue that conduct terpenoid defense compounds. They may be constitutive or their formation may be induced by wounding/defense responses.

conifers Cone-bearing trees.

***CONSTANS* (*CO*)** Gene for a key component of a regulatory pathway that promotes flowering of Arabidopsis in long days; it encodes a protein that regulates the transcription of other genes.

constitutive defenses Plant defenses that are always immediately available or operational; that is, defenses that are not induced.

CONSTITUTIVE PHOTOMORPHOGENESIS 1 (COP1) A constitutive repressor of photomorphogenesis that interacts with photomorphogenesis-promoting factors such as HY5 to promote their degradation via the ubiquitin–proteasome pathway.

contact angle A quantitative measure of the degree to which a water molecule is attracted to a solid phase versus to itself.

convergence points Regions of maximum auxin concentration in the L1 layer of leaf primordia.

COP1-SUPPRESSOR OF PHYA (COP1-SPA) complex A protein that forms a complex with COP1 and represses photomorphogenesis.

COP9 signalosome (CSN) complex A protein complex that appears to facilitate COP1 entry into the nucleus.

COPI A vesicle-coating protein that directs vesicles involved in the retrograde movement within the Golgi and from the Golgi to the ER.

COPII A vesicle-coating protein that directs delivery of vesicle membrane and contents to the Golgi from the ER.

core promoter (minimum promoter) One part of the two-part eukaryotic promoter consisting of the minimum upstream sequence required for gene expression.

core promoter element The minimal portion of a promoter that is required to correctly initiate transcription.

core symbiotic genes Genes encoding components of the common symbiotic pathway.

cork cambium A layer of lateral meristem that develops within mature cells of the cortex and the secondary phloem. Produces the secondary protective layer, the periderm. Also called phellogen.

cork *See* phellem.

corpus The internal region of the shoot apical meristem in which the planes of cell division are not strongly polarized, leading to increases in the volume of the shoot.

cortex The outer layer of the root delimited on the outside by the epidermis and on the inside by the endodermis.

cortical ER The network of ER that lies just under the plasma membrane and is associated with the plasma at specific contact points. Distinct from internal ER which is found deeper in the cytoplasm and in trans-vacuolar strands.

cortical–endodermal initials A ring of stem cells that surround the quiescent center and generate the cortical and endodermal layers in roots.

Crassulacean acid metabolism (CAM) A biochemical process for concentrating CO_2 at the carboxylation site of rubisco. Found in the family Crassulaceae (*Crassula, Kalanchoë, Sedum*) and numerous other families of angiosperms. In CAM, CO_2 uptake and fixation take place at night, and decarboxylation and reduction of the internally released CO_2 occur during the day.

creep pH-dependent cell wall extension. Contributes to cell wall expansion along with polymer integration and wall stress relaxation.

cristae Folds in the inner mitochondrial membrane that project into the mitochondrial matrix.

critical concentration (of a nutrient) The minimum tissue content of a mineral nutrient that is correlated with maximal growth or yield.

critical day length The minimum length of the day required for flowering of a long-day plant; the maximum length of day that will allow short-day plants to flower. However, studies have shown that it is the length of the night, not the length of the day, that is important.

cross-pollination Pollination of a flower by pollen from the flower of a different plant.

cross-protection A plant response to one environmental stress that confers resistance to another stress.

cross-regulation The interaction of two or more signaling pathways.

crown gall A tumor-forming plant disease resulting from wound infection of the stem or trunk by the soil-dwelling bacterium *Agrobacterium tumefaciens.* A tumor resulting from the disease.

crown roots Adventitious roots that emerge from the lowermost nodes of a stem.

CRYPTOCHROME1 (CRY1) A flavoprotein implicated in many blue light responses that has homology with photolyase. Formerly HY4.

CSLA A family of cellulose synthase-like genes that encode synthases for (1,4)-β-D-mannan.

CSLC A family of cellulose synthase-like genes that encode synthases for the (1,4)-β-D-glucan backbone of xyloglucan.

CSLD A family of cellulose synthase-like genes, some of which are implicated in formation of crystalline cellulose in root hairs and other cells, whereas others are involved in mannan synthesis.

CSLF A family of cellulose synthase-like genes that encode synthases for 'mixed-linkage' (1,3;1,4)-β-D-glucans.

CSLH A family of cellulose synthase-like genes that encode synthases for 'mixed-linkage' (1,3;1,4)-β-D-glucans.

cyanogenic glycosides Nonalkaloid, nitrogenous protective compounds that break down to give off the poisonous gas hydrogen cyanide when the plant is crushed.

cyclic electron flow In photosystem I, flow of electrons from the electron acceptors through the cytochrome b_6f complex and back to P700, coupled to proton pumping into the lumen. This electron flow energizes ATP synthesis but does not oxidize water or reduce $NADP^+$.

cyclin-dependent kinases (CDKs) Protein kinases that regulate the transitions from G_1 to S, and from G_2 to mitosis, during the cell cycle.

cyclins Regulatory proteins associated with cyclin-dependent kinases that play a crucial role in regulating the cell cycle.

cyst nematodes Parasitic nematodes that invade roots and transform into a non-motile cyst. The soybean cyst nematode *Heterodera glycines* is a major threat to soybean production.

cytochrome b_6f complex A large multi-subunit protein complex containing two *b*-type hemes, one *c*-type heme (cytochrome *f*), and a Rieske iron–sulfur protein. A immobile protein distributed equally between the grana and the stroma regions of the membranes.

cytochrome bc_1 complex (complex III) A multi-subunit protein complex in the mitochondrial electron transport chain that catalyzes oxidation of reduced ubiquinone (ubiquinol) and reduction of cytochrome *c* linked to the pumping of protons from the matrix to the intermembrane space.

cytochrome *c* oxidase (complex IV) A multi-subunit protein complex in the mitochondrial electron transport chain that catalyzes oxidation of reduced cytochrome *c* and reduction of O_2 to H_2O linked to the pumping of protons from the matrix to the intermembrane space.

cytochrome *c* A peripheral, mobile component of the mitochondrial electron transport chain that oxidizes complex III and reduces complex IV.

cytochrome *f* A subunit in the cytochrome b_6f complex that plays a role in electron transport between Photosystem I and II.

cytochrome P450 monooxygenase (CYP) A generic term for a large number of related, but distinct, mixed-function oxidative enzymes localized on the endoplasmic reticulum. CYPs participate in a variety of oxidative processes, including steps in the biosynthesis of gibberellins and brassinosteroids.

cytokinesis In plant cells, following nuclear division, the separation of daughter nuclei by the formation of new cell wall.

cytoplasm The cellular matter enclosed by the plasma membrane exclusive of the nucleus, that contains the cytosol, ribosomes, and the cytoskeleton which, in eukaryotes, surrounds intracellular and membrane-limited organelles (chloroplasts, mitochondria, endoplasmic reticulum, etc.).

cytoplasmic sleeve The region of cytoplasm between the plasma membrane and the central, ER-derived desmotubule in a plasmodesma.

cytoplasmic streaming The coordinated movement of particles and organelles through the cytosol.

cytoskeleton Composed of polarized microfilaments of actin or microtubules of tubulin, the cytoskeleton helps control the organization and polarity of organelles and cells during growth.

cytosol The colloidal-aqueous phase of the cytoplasm containing dissolved solutes but excluding supramolecular structures, such as ribosomes and components of the cytoskeleton.

D

damage associated molecular patterns (DAMPs) Molecules originating from non-pathogenic sources that can initiate immune responses.

day-neutral plant (DNP) A plant whose flowering is not regulated by day length.

de-etiolation Rapid developmental changes associated with loss of the etiolated form due to the action of light. *See* photomorphogenesis.

deficiency zone Concentrations of a mineral nutrient in plant tissue below the critical concentration that reduces plant growth.

dehiscence The spontaneous opening of a mature anther or fruit, releasing its contents.

dehydrins Hydrophilic plant proteins that accumulate in response to drought stress and cold temperatures.

depolarized Refers to a decrease in the usually negative membrane potential difference across the plasma membrane of plant cells. May be caused by the activation of anion channels and loss of anions, such as chloride, from the cell interior, which is negative with respect to the outside.

dermal tissue The system that covers the outside of the plant body; the epidermis or periderm.

desaturase enzymes Enzymes that remove hydrogens in a carbon chain to create a double bond between carbons or to add an ethyl group to elongate a carbon chain.

desiccation tolerance Plant's ability to function while dehydrated.

desmotubule A narrow tubule of the ER that passes through plasmodesmata and connects the ER in adjacent cells.

determinants The two protein-coding regions of the S-locus: one expressed in the pistil (female determinant) and the other expressed in the anther (male determinant).

diacylglycerol (DAG) A molecule consisting of the three-carbon glycerol molecule to which two fatty acids are covalently attached by ester linkages.

diaheliotropic Refers to leaf movements that maximize light interception by solar tracking and minimize overexposure to light.

DICER-LIKE 1 (DCL1) One of the plant nuclear proteins that convert pri-miRNAs into miRNAs.

dichogamy The production of stamens and pistils at different times in hermaphroditic flowers—an adaptation that promotes cross-pollination.

dichotomous branching Branching that occurs by the division of the shoot apical meristem to produce two equal shoots.

difference in water vapor concentration Referring to the difference between the water vapor concentration of the air spaces inside the leaf and that of the air outside the leaf. One of the two major factors that drive transpiration from the leaf.

differentiation zone Developmental region of the root tip above the elongation zone in which cell differentiation occurs, including root hair formation and vascular tissue differentiation.

differentiation Process by which a cell acquires metabolic, structural, and functional properties that are distinct from those of its progenitor cell. In plants, differentiation is frequently reversible, when excised differentiated cells are placed in tissue culture.

diffuse growth A type of cell growth in plants in which expansion occurs more or less uniformly over the entire surface. Contrast with tip growth.

diffusion coefficient (D_s) The proportionality constant that measures how easily a specific substance *s* moves through a particular medium. The diffusion coefficient is a characteristic of the substance and depends on the medium.

diffusion potential The potential (voltage) difference that develops across a semipermeable membrane as a result of the differential permeability of solutes with opposite charges (for example K^+ and Cl^-).

diffusion The movement of substances due to random thermal agitation from regions of high free energy to regions of low free energy (e.g. from high to low concentration).

diffusional resistance Restriction posed by the boundary layer and the stomata to the free diffusion of gases from and into the leaf.

dioecious Refers to plants in which male and female flowers are found on different individuals, such as spinach (*Spinacia*) and hemp (*Cannabis sativa*). Contrast with monoecious.

diploid, 2*N* Having two of each chromosome; the 2*N* chromosome number characterizes the sporophyte generation.

directed organelle movement Movement of an organelle in a particular direction, which can be driven by the interaction with molecular motors associated with the cytoskeleton.

dirigent-domain proteins Homologs of a protein that positions two coniferyl alcohol substrates for oxidative radical dimerization in a specific stereospecific conformation for the formation of (+)-pinoresinol. Dirigent proteins have also been invoked in a disputed hypothesis about ordered lignin formation.

dispersed repeat A type of repeated sequence that is not restricted to a single location in the genome. May occur as microsatellites or transposons.

disproportionating enzyme (D-enzyme) One of two debranching enzymes that process inappropriately positioned oligosaccharide branches in the construction of starch granules. Catalyzes the transfer of a segment of a α-D-1,4-glucan to a new position in an acceptor, which may be glucose or a 1,4-linked α-D-glucan.

distal regulatory promoter sequences Located upstream of the proximal promoter sequences, these *cis*-acting sequences can exert either positive or negative control over eukaryotic promoters.

DNA transposons Dominant group of dispersed repeats found in heterochromatin that can move or be copied from one location to another within the genome of the same cell.

dolichol diphosphate Embedded in the ER membrane, this lipid is the assembly site for a branched oligosaccharide (*N*-acetylglucosamine, mannose, and glucose) that will be transferred to the free amino group of one or more asparagine residues of a protein in the ER that is destined for secretion.

domains (1) Regions (nucleotide sequences) within the gene that are similar to regions found in other genes. (2) Regions of a protein (amino acid sequence) with a particular structure or function. (3) The three major taxonomic groups of living organisms.

dormancy A living condition in which growth does not occur under conditions that are normally favorable to growth.

double fertilization A unique feature of all angiosperms whereby, along with the fusion of a sperm with the egg to create a zygote (with a diploid number of chromosomes), a second male gamete fuses with the polar nuclei in the embryo sac to generate the endosperm tissue (with a triploid or higher number of chromosomes).

downstream promoter element (DPE) A distinct type of core promoter element located approximately 30 nucleotides downstream from the transcription start site.

drupe A structure similar to a berry, but with a hardened, shell-like endocarp (pit or stone) that contains a seed.

dynamic instability The sequence of rapid catastrophic depolymerization and slower polymerizing rescue of the growing end of microtubules that occurs when the growing end is not stabilized with some microtubule associated proteins (MAPs).

dynamic photoinhibition Photoinhibition of photosynthesis in which quantum efficiency decreases but the maximum photosynthetic rate remains unchanged. Occurs under moderate, not high, excess light.

dynamin A large GTPase that is involved in the formation of many vesicles and organelles, including the cell plate.

E

early endosomes The small (100 nm) vesicles first formed in endocytosis. Initially they are coated with clathrin, but it is quickly lost. Part of the endomembrane system.

early response genes *See* primary response genes.

ectomycorrhizas Symbioses where the fungus typically forms a thick sheath, or mantle, of mycelium around roots. The root cells themselves are not penetrated by the fungal hyphae, but instead are surrounded by a network of hyphae called the Hartig net, which provides a large area of contact between the symbionts that is involved in nutrient transfers.

effector-triggered immunity Immune responses that are mediated by intracellular nucleotide binding leucine rich repeat proteins (NLRs) that are encoded by *R* genes.

effector A molecule that binds a protein to change its activity. Bacterial effectors are secreted by pathogens to act on proteins within a host cell.

egg apparatus The three cells at the micropylar end of the embryo sac consisting of the egg cell and two synergid cells.

egg cell The female gamete.

electrochemical potential The chemical potential of an electrically charged solute.

electrochemical proton gradient The sum of the electrical charge gradient and the pH gradient across the membrane, resulting from a concentration gradient of protons.

electrogenic transport Active ion transport involving the net movement of charge across a membrane.

electron spin resonance (ESR) A magnetic resonance technique that detects unpaired electrons in molecules. Instrumental measurements that identify intermediate electron carriers in the photosynthetic electron transport system.

electron transport chain A series of protein complexes in the inner mitochondrial membrane linked by the mobile electron carriers ubiquinone and cytochrome *c*, that catalyze the transfer of electrons from NADH to O_2. In the process a large amount of free energy is released. Some of that energy is conserved as an electrochemical proton gradient.

electronegative Having the capacity to attract electrons and thus producing a slightly negative electric charge.

electroneutral transport Active ion transport that involves no net movement of charge across a membrane.

elicitors Specific pathogen molecules or cell wall fragments that bind to plant proteins and thereby signal for plant defense against a pathogen. *See* avirulence genes.

elongation zone The region of rapid and extensive root cell elongation showing few, if any, cell divisions.

embryo dormancy Seed dormancy that is caused directly by the embryo and is not due to any influence of the seed coat or other surrounding tissues.

embryophyte *See* plants.

endocytosis The formation of small vesicles from the plasma membrane, which detach and move into the cytosol, where they fuse with elements of the endomembrane system.

endodermis A specialized layer of cells with a Casparian strip surrounding the vascular tissue in roots and some stems.

endopolyploidy Polyploidy caused by replication of chromosomes without division of the nucleus.

endoreduplication Cycles of nuclear DNA replication without mitosis resulting in polyploidization.

endosomes Early in endocytosis, vesicles that have lost their clathrin coats and moved away from the plasma membrane into the cell interior.

endosymbiosis A theory that explains the evolutionary origin of the chloroplast and mitochondrion through formation of a symbiotic relationship between a prokaryotic cell and a simple nonphotosynthetic eukaryotic cell, followed by extensive gene transfer to the nucleus.

endosymbiotic theory *See* endosymbiosis.

endothelium A cell layer derived from the innermost layer of the integument that surrounds the embryo sac and supplies it with nutrients, similar to the tapetal layer in anthers.

energy transfer In the light reactions of photosynthesis, the direct transfer of energy from an excited molecule, such as carotene, to another molecule, such as chlorophyll. Energy transfer can also take place between chemically identical molecules, such as chlorophyll-to-chlorophyll transfer.

enhanced suberin 1 (ESB1) A protein involved in the restricted formation of lignin in the narrow strip of cell wall that forms the Casparian strip in the root endodermis. Mutants in the *ESB1* gene are characterized by a spread of lignification beyond the Casparian strip as well as elevated levels of suberin in the root.

enhancement effect The synergistic (higher) effect of red and far-red light on the rate of photosynthesis, as compared with the sum of the rates when the two different wavelengths are delivered separately.

enhancers Positive regulatory sequences located tens of thousands of base pairs away from a gene's start site. Enhancers may be located either upstream or downstream from the promoter.

entrainment The synchronization of the period of biological rhythms by external controlling factors, such as light and darkness.

entry division During guard cell formation, the asymmetric cell division of the meristemoid mother cell to give rise to two morphologically distinct daughter cells—a larger stomatal lineage ground cell and a smaller meristemoid.

envelope The double-membrane system surrounding the chloroplast or the nucleus. The outer membrane of the nuclear envelope is continuous with the endoplasmic reticulum.

epidermal–lateral root cap initials Cells located to the side of the quiescent center. In Arabidopsis, these initials first divide anticlinally to set off daughter cells, which then divide periclinally to form two files of cells that will mature into the lateral root cap and epidermis.

epidermis The outermost layer of plant cells, typically one cell thick.

epigeal A type of seedling growth that results in the cotyledons being raised above soil level.

epigenetic modifications Chemical modifications to DNA and histones that cause heritable changes in gene activity without altering the underlying DNA sequence.

epigenome Heritable chemical modifications to DNA and chromatin including DNA methylation, histone methylation and acylation, and DNA sequences that generate noncoding RNA sequences that interfere with gene expression.

epigenomics The study of all epigenetic modifications of a genome.

ER exit sites (ERES) On the ER, specialized sites characterized by the coat protein COPII from which delivery of vesicles to the Golgi occurs.

***ERF1* (*ETHYLENE RESPONSE FACTOR1*)** A gene that encodes a protein belonging to the ERE-binding protein family of transcription factors.

escape from photoreversibility The loss of photoreversibility by far-red light of phytochrome-mediated red light–induced events after a short period of time.

essential element A chemical element that is part of a molecule that is an intrinsic component of the structure or metabolism of a plant. When the element is in limited supply, a plant suffers abnormal growth, development, or production.

ethylmethanesulfonate (EMS) A chemical mutagen that causes the addition of an ethyl group to a nucleotide, and results in a permanent mutation from G/C to A/T at that site.

etiolation Effects of seedling growth in the dark, in which the hypocotyl and stem are more elongated, cotyledons and leaves do not expand, and chloroplasts do not mature.

etioplast Photosynthetically inactive form of chloroplast found in etiolated seedlings. Does not synthesize chlorophyll or most of the enzymes and structural proteins required for the formation of thylakoids and operation of photosynthesis. Contains an elaborate system of interconnected membrane tubules called the prolamellar body.

euchromatin The dispersed, transcriptionally active form of chromatin. *See also* heterochromatin.

eudicots Abbreviated name for the eudicotyledons, one of two major classes of the angiosperms, and refers to the fact that plants in this class have two seed leaves (cotyledons).

eukaryotic pathway In the cytoplasm, the series of reactions for the synthesis of glycerolipids. *See also* prokaryotic pathway.

expansins Class of wall-loosening proteins that accelerate wall stress relaxation and cell expansion, typically with an optimum at acidic pH. Appear to mediate acid growth.

export The movement of photosynthate in sieve elements away from the source tissue.

extracellular matrix A general term that in plants usually refers to the cell wall.

extracellular space In plants, the space continuum outside the plasma membrane made up of interconnecting cell walls through which water and mineral nutrients readily diffuse. *See* apoplast.

extrafloral nectar Nectar produced outside of the flower and not involved in pollination events.

F

F_1 The ATP-binding matrix-facing part of the F_oF_1-ATP synthase.

F-actin Filamentous actin, the form of actin in the polymerized protofilament, which is formed from G-actin monomers.

F-box proteins Components of ubiquitin E3 ligase complexes.

facilitated diffusion Passive transport across a membrane using a carrier.

***FACKEL* (*FK*)** Gene that encodes a sterol C-14 reductase that seems to be critical for pattern formation during embryogenesis. Mutants exhibit pattern formation defects: malformed cotyledons, short hypocotyl and root, and often multiple shoot and root meristems.

facultative CAM Found in some plant species that switch between C_3 or C_4 metabolism and CAM metabolism under drought stress conditions.

far upstream element (FUE) A conserved genetic sequence located upstream of the polyA site in eukaryotic genes.

FAR-RED ELONGATED HYPOCOTYL1 (FHY1) A protein that facilitates the entry of phyA into the nucleus in response to light.

Fe–S centers Prosthetic groups consisting of inorganic iron and sulfur that are abundant in proteins in respiratory and photosynthetic electron transport.

fermentation The metabolism of pyruvate in the absence of oxygen, leading to the oxidation of the NADH generated in glycolysis to NAD^+. Allows glycolytic ATP production to function in the absence of oxygen.

ferredoxin A small, water-soluble iron–sulfur protein involved in electron transport in photosystem I.

ferredoxin–NADP$^+$ reductase (FNR) Membrane-associated flavoprotein that receives electrons from Photosystem I and reduces NADP$^+$ to NADPH.

ferredoxin–thioredoxin system Three chloroplast proteins (ferredoxin, ferredoxin-thioredoxin reductase, thioredoxin). The concerted action of the three proteins uses reducing power from the photosynthetic electron transport system to reduce protein disulfide bonds by a cascade of thiol/disulfide exchanges. As a result, light controls the activity of several enzymes of the Calvin–Benson cycle.

ferritin Protein acting in cellular iron storage in several compartments, including the mitochondria.

fertilization The formation of a diploid (2*N*) zygote from the cellular and nuclear fusion of two haploid (1*N*) gametes, the egg and the sperm.

FeSA Membrane-bound iron–sulfur protein that transfers electrons between Photosystem I and ferredoxin.

FeSB Membrane-bound iron–sulfur protein that transfers electrons between Photosystem I and ferredoxin.

FeSR An iron- and sulfur-containing subunit of the cytochrome b_6f complex, involved in electron and proton transfer. *See also* Rieske iron–sulfur protein.

FeSX Membrane-bound iron–sulfur protein that transfers electrons between Photosystem I and ferredoxin.

fiber An elongated, tapered sclerenchyma cell that provides support in vascular plants.

filiform apparatus A convoluted, thickened cell wall that increases the surface area of the plasma membrane at the extreme micropylar end of a synergid cell.

fimbrin An actin-binding protein that bundles F-actin filaments together into larger filamentous bundles.

fission The process by which portions of a membrane separate from the remaining membrane, forming vesicles.

flavin adenine dinucleotide (FAD) A riboflavin-containing cofactor that undergoes a reversible two-electron reduction to produce $FADH_2$.

flavin mononucleotide (FMN) A riboflavin-containing cofactor that undergoes a reversible one- or two-electron reduction to produce FMNH or $FMNH_2$.

flippases Enzymes that "flip" newly synthesized phospholipids across the bilayer from the outer (cytoplasmic) face of the membrane to the inner leaflet, thereby assuring symmetrical lipid composition of the membrane.

floral evocation The events occurring in the shoot apex that specifically commit the apical meristem to produce flowers.

floral meristem Forms floral (reproductive) organs: sepals, petals, stamens, and carpels. May form directly from vegetative meristems or indirectly via an inflorescence meristem.

floral organ identity genes Three types of genes that control the specific locations of floral organs in the flower.

floral organs Angiosperm organs involved directly or indirectly in sexual reproduction; sepals, petals, stamens, and carpels.

florigen The hypothetical, universal flowering hormone synthesized by leaves and translocated to the shoot apical meristem via the phloem. So far, it has not been isolated or characterized.

***FLOWERING LOCUS C* (*FLC*)** In Arabidopsis, a gene that represses flowering.

***FLOWERING LOCUS T* (*FT*)** The gene coding for the protein that acts as a florigen in Arabidopsis and other species.

flowering plants *See* angiosperms.

fluence The number of photons absorbed per unit surface area.

fluid-mosaic model The common molecular lipid–protein structure for all biological membranes. A double layer (bilayer) of polar lipids (phospholipids or, in chloroplasts, glycosylglycerides) has a hydrophobic, fluid-like interior. Membrane proteins are embedded in the bilayer and may move laterally due to its fluid-like properties.

fluorescence resonance energy transfer The physical mechanism by which excitation energy is conveyed from the pigment that absorbs the light to the reaction center.

fluorescence Following light absorption, the emission of light at a slightly longer wavelength (lower energy) than the wavelength of the absorbed light.

flux density (J_s) The rate of transport of a substance *s* across a unit area per unit time. J_s may have units of moles per square meter per second (mol m^{-2} s^{-1}).

F_o The integral membrane part of the F_oF_1-ATP synthase.

F_oF_1-ATP synthase A multi-subunit protein complex associated with the inner mitochondrial membrane that couples the passage of protons across the membrane to the synthesis of ATP from ADP and phosphate. The subscript 'o' in F_o refers to the binding of the inhibitor oligomycin. Similar to CF_0–CF_1 ATP synthase in photophosphorylation.

foliage leaves The principal lateral appendages of stems that carry out photosynthesis.

foliar application The application, and subsequent absorption, of some mineral nutrients to leaves as sprays.

forisomes Protein bodies that rapidly disperse and block the sieve tube. Occur only in certain legumes.

formins Proteins that bind actin and actin-profilin complexes to initiate polymerization of the actin filament.

free-running Designation of the biological rhythm that is characteristic for a particular organism when environmental signals are removed, as in total darkness. *See* zeitgeber.

frequency (ν) A unit of measurement that characterizes waves, in particular light energy. The number of wave crests that pass an observer in a given time.

fruit In angiosperms, one or more mature ovaries containing seeds and sometimes adjacent attached parts.

fusicoccin A fungal toxin that induces acidification of plant cell walls by activating an H^+-ATPase in the plasma membrane. Fusicoccin stimulates rapid acid growth in stem and coleoptile sections. It also stimulates stomatal opening by stimulating proton pumping at the guard cell plasma membrane.

fusion The coming together of the membranes from separate vesicles or organelles, usually resulting in the movement or mixing of content contained within the vesicle or organelle that results.

G

G_1 The phase of the cell cycle preceding the synthesis of DNA.

G_2 The phase of the cell cycle following the synthesis of DNA.

G protein GTP-binding protein involved in signal transduction.

G-actin The globular, monomeric form of actin from which F-actin is formed.

GA 2-oxidase (GA2ox) An enzyme that deactivates gibberellins.

GA 3-oxidase (GA3ox) An enzyme in stage 3 of the gibberellin biosynthetic pathway.

GA-MYB A MYB eukaryotic transcription factor implicated in GA signaling. Barley GAMYB is similar to that of three MYB proteins in Arabidopsis.

GABA shunt A pathway supplementing the citric acid cycle with the ability to form and degrade GABA.

GABA Gamma-aminobuteric acid.

GAF The chromophore-binding domain of phytochrome.

galactan A cell wall polysaccharide composed of galactose residues.

gamete A haploid (1*N*) reproductive cell.

gametophyte generation The stage, or generation, in the life cycle of plants that produces gametes. It alternates with the sporophyte generation, in a process called alternation of generations.

gametophyte The haploid (1*N*) multicellular structure that produces haploid gametes by mitosis and differentiation.

gametophytic self-incompatibility (GSI) The type of self-incompatibility in which the incompatibility phenotype of the pollen is determined by the pollen's own (haploid) genotype.

gate A structural domain of the channel protein that opens or closes the channel in response to external signals such as voltage changes, hormone binding, or light.

GC box A sequence of nucleotides involved in the initiation of transcription in eukaryotes.

gene fusion An artificial construct that links a promoter for one gene with the coding sequence of another gene. Often includes a reporter gene, such as green fluorescent protein gene, *GFP*, that produces a readily detected protein.

gene gun *See* biolistics.

general transcription factors Proteins that are required by RNA polymerases of eukaryotes for proper positioning at the transcription start site.

genet A group of genetically identical individuals, such as plants, fungi, or bacteria, clonally derived from the same ancestor and growing in the same general location.

genetic marker A DNA sequence that occurs at a known location on a chromosome that can be used to identify species or individuals.

genetic mosaicism The presence of two or more populations of cells with different genotypes caused by somatic mutations in a plant that has developed from a single fertilized egg.

genome Refers to all the genes in a haploid complement of eukaryotic chromosomes, in an organelle, a microbe, or the DNA or RNA content of a virus.

germination The events that take place between the start of imbibition of the dry seed and the emergence of the embryo, usually the radicle, from the structures that surround it. May also be applied to other quiescent structures, such as pollen grains or spores.

gerontoplast In senescing leaves, a modified chloroplast that has undergone progressive unstacking of grana, the loss of thylakoid membranes, and a massive accumulation of plastoglobuli composed of lipids.

GIBBERELLIN INSENSITIVE DWARF1 (GID1) Gibberellin receptor protein in rice.

gibberellin response element (GARE) Promoter sequence conferring GA responsiveness, located 200 to 300 base pairs upstream of the transcription start site.

gibberellins A large group of chemically related plant hormones synthesized by a branch of the terpenoid pathway and associated with the promotion of stem growth (especially in dwarf and rosette plants), seed germination, and many other functions.

Gibbs free energy The energy that is available to do work; in biological systems the work of synthesis, transport, and movement.

globular stage The first stage of embryogenesis. A radially symmetrical, but not developmentally uniform, sphere of cells produced by initially synchronous cell divisions of the zygote. *See* heart stage, torpedo stage.

glucan A polysaccharide made from glucose units.

(1,3;1,4)-β-D-glucan Mixed-linkage glucan found in cell walls of grasses. It may bind tightly to cellulose surface, producing a less sticky network.

glucomannan A polysaccharide made from both glucose and mannose units.

gluconeogenesis The synthesis of carbohydrates through the reversal of glycolysis.

glucose 6-phosphate dehydrogenase A cytosolic and plastidic enzyme that catalyzes the initial reaction of the oxidative pentose phosphate pathway.

glucuronoarabinoxylan A hemicellulose with a (1,4)-linked backbone of β-D-xylose (Xyl) and side chains containing arabinose (Ara) and 4-O-methyl-glucuronic acid (4-O-Me-α-D-GlcA).

glucuronoxylan A major hemicellulose in some secondary cell walls, consisting of a backbone of 1→4-linked β-D-xylose residues with occasional glucuronic acid side chains.

glutamate dehydrogenase (GDH) Catalyzes a reversible reaction that synthesizes or deaminates glutamate as part of the nitrogen assimilation process.

glutamate synthase (GOGAT) Enzyme that transfers the amide group of glutamine to 2-oxoglutarate, yielding two molecules of glutamate. Also known as glutamine:2-oxoglutarate aminotransferase.

glutamine synthetase (GS) Catalyzes the condensation of ammonium and glutamate to form glutamine. Reaction is critical for the assimilation of ammonium into essential amino acids. Two forms of GS exist—one in the cytosol and one in chloroplasts.

glutathione peroxidases A family of enzymes that reduce peroxide to water and lipid hydroperoxides to alcohols.

glycan A general term for a polymer made up of sugar units; it is synonymous with polysaccharide.

glyceroglycolipids Glycerolipids in which sugars form the polar head group. Glyceroglycolipids are the most abundant glycerolipids in chloroplast membranes.

glycerophospholipids Polar glycerolipids in which the hydrophobic portion consists of two 16-carbon or 18-carbon fatty acid chains esterified to positions 1 and 2 of a glycerol backbone. The phosphate-containing polar head group is attached to position 3 of the glycerol.

glycine betaine *N,N,N*-trimethylglycine, which functions in drought stress protection, and was originally identified in sugar beet (*Beta vulgaris*).

glycolysis A series of reactions in which a sugar is oxidized to produce two molecules of pyruvate. A small amount of ATP and NADH is produced.

glycophytes Plants that are not able to resist salts to the same degree as halophytes. Show growth inhibition, leaf discoloration, and loss of dry weight at soil salt concentrations above a threshold. Contrast with halophytes.

glycoproteins Proteins that have covalently attached sugar oligomers or polymers.

glycosides Compounds containing an attached sugar or sugars.

glycosylglycerides Polar lipid molecules that are found in the chloroplast membrane. In glycosylglycerides, there is no phosphate group, and the polar head group is galactose, digalactose, or a sulfated galactose.

glyoxylate cycle The sequence of reactions that convert two molecules of acetyl-CoA to succinate in the glyoxysome.

glyoxylate A two-carbon acid aldehyde that is an intermediate of the glyoxylate cycle.

glyoxysome An organelle found in the oil-rich storage tissues of seeds in which fatty acids are oxidized. A type of microbody.

glyphosate resistance Genetic capacity to survive a field application of the commercial herbicide Roundup, which kills weeds but does not harm resistant crop plants.

***GNOM* (*GN*)** Arabidopsis gene for the development of roots and cotyledons. Homozygous *GNOM* mutant produces seedlings lacking both roots and cotyledons.

GOGAT *See* glutamate synthase.

Goldman equation An equation that predicts the diffusion potential across a membrane, as a function of the concentrations and permeabilities of all ions (e.g., K^+, Na^+ and Cl^-) that permeate the membrane.

grana lamellae Stacked thylakoid membranes within the chloroplast. Each stack is called a granum, while the exposed membranes in which stacking is absent are known as stroma lamellae.

granum (plural *grana*) In the chloroplast, a stack of thylakoids.

gravitational potential The part of the chemical potential caused by gravity. It is only of a significant size when considering water transport into trees.

gravitropic response The growth initiated by the root cap's perception of gravity and the signal that directs the roots to grow downward.

gravitropic setpoint angle The angle at which gravitropic organs are maintained with respect to gravity.

gravitropism Plant growth in response to gravity, enabling roots to grow downward into the soil and shoots to grow upward.

green-leaf volatiles A mixture of lipid-derived six-carbon aldehydes, alcohols, and esters released by plants in response to mechanical damage.

greenhouse effect The warming of Earth's climate, caused by the trapping of long-wavelength radiation by CO_2 and other gases in the atmosphere. Term derived from the heating of a greenhouse resulting from the penetration of long-wavelength radiation through the glass roof, the conversion of the long-wave radiation to heat, and the blocking of the heat escape by the glass roof.

ground meristem cells In the plant meristems, cells that will give rise to the cortical and pith tissues and, in the root and hypocotyl, will produce the endodermis.

ground tissue The internal tissues of the plant, other than the vascular, transport tissues.

growth efficiency The growth in mass of a tree per unit leaf area or leaf mass.

growth respiration The respiration that provides the energy needed for converting sugars into the building blocks that make up new tissue. Contrast with maintenance respiration.

GT43 (glycosyl transferase family 43) Group of synthases that synthesize the backbone of xylan polysaccharides.

GTPase-activating proteins (GAPs) Proteins that inactivate GTPases by promoting GTP hydrolysis.

guanine nucleotide exchange factors (GEFs) Proteins that activate inactive GTPases by replacing GDP with GTP.

guard cell protoplasts Protoplasts prepared from guard cells by removing their walls through application of enzymes that degrade cell wall components.

guard cells A pair of specialized epidermal cells that surround the stomatal pore and regulate its opening and closing.

guard hypothesis An hypothesis wherein R proteins interact with pathogen effectors to prevent effector interactions with targets in plant cells.

guard mother cell (GMC) The cell that gives rise to a pair of guard cells to form a stoma.

GURKE (GK) Gene involved in pattern formation. Encodes an acetyl-CoA carboxylase that is required for the proper synthesis of very-long-chain fatty acids and sphingolipids, which are involved in the proper patterning of the apical portion of the embryo.

guttation An exudation of liquid from the leaves due to root pressure.

gymnosperms An early group of seed plants. Distinguished from angiosperms by having seeds borne unprotected (naked) in cones.

H

H^+-pyrophosphatase An electrogenic pump that moves protons into the vacuole, energized by the hydrolysis of pyrophosphate.

halophytes Plants that are native to saline soils and complete their life cycles in that environment. Contrast with glycophytes.

haploid, 1*N* Having a single set of chromosomes, in contrast to having a paired, diploid, set.

Hartig net A fungal network of hyphae that surround but do not penetrate the cortical cells of roots.

haustorium The hyphal tip of a fungus or root tip of a parasitic plant that penetrates host plant tissue.

HC-toxin A cell permeant cyclic tetrapeptide produced by the maize pathogen *Cochliobolus carbonum* that inhibits histone deacetylases.

Heading-date1 (Hd1) A gene for a *CO* homolog that acts as an inhibitor of flowering in rice.

Heading-date3a (Hd3a) The gene for the FT-like protein in rice that is translocated via the sieve tubes to the apical meristem, where it stimulates flowering.

heart stage The second stage of embryogenesis. A bilaterally symmetrical structure produced by rapid cell divisions in two regions on either side of the future shoot apex. *See* globular stage, torpedo stage.

heat shock factors Transcription factors that regulate expression of heat shock proteins.

heat shock proteins (HSPs) A specific set of proteins that are induced by a rapid rise in temperature, and by other factors that lead to protein denaturation. Most act as molecular chaperones.

heliotropism Movements of leaves toward or away from the sun.

hemibiotrophic pathogens Plant pathogens that show an initial biotrophic stage, which is followed by a necrotrophic stage, in which the pathogen causes extensive tissue damage.

hemicelluloses Heterogeneous group of polysaccharides that bind to the surface of cellulose, linking cellulose microfibrils together into a network. Typically solubilized by strong alkali solutions.

hemiparasitic plants Photosynthetic plants that are also parasites.

herbivore-associated molecular patterns (HAMPs) Plant immune responses initiated by interactions with herbivores.

herbivory Consumption of plants or parts of plants as a food source.

heterochromatin Chromatin that is densely packed, darkly staining, and transcriptionally inactive; it accounts for about 10% of the nuclear DNA.

heterochromatization The condensation of euchromatin into heterochromatin, resulting in gene silencing.

heterostyly The condition of having two or three different flower "morphs" in which the stamens and pistils are different lengths. In longistylous flowers the stamens are shorter than the pistils. In brevistylous flowers the stamens are longer than the pistils.

hexose phosphates Six-carbon sugars with phosphate groups attached.

histogenesis The differentiation of cells to produce various tissues.

histones A family of proteins that interact with DNA and around which DNA is wound to form a nucleosome.

Hoagland solution A nutrient solution for plant growth, originally formulated by Dennis R. Hoagland.

holoparasitic plants Non-photosynthetic plants that are obligate parasites.

homogalacturonan (HG) This pectin polysaccharide is a (1,4)-linked polymer of β-D-galacturonic acid residues; also called polygalacturonic acid.

hormone balance theory The hypothesis that seed dormancy and germination are regulated by the balance of ABA and gibberellin.

hybrid vigor (heterosis) The increased vigor often observed in the offspring of crosses between two inbred varieties of the same plant species.

hydathodes Specialized pores associated with vein endings at the leaf margin from which xylem sap may exude when there is positive hydrostatic pressure in the xylem. Also a site of auxin synthesis in immature leaves of Arabidopsis.

hydraulic conductivity Describes how readily water can move across a membrane; it is expressed in terms of volume of water per unit area of membrane per unit time per unit driving force (i.e., $m^3\ m^{-2}\ s^{-1}\ MPa^{-1}$).

hydrogen bonds Weak chemical bonds formed between a hydrogen atom and an oxygen or nitrogen atom.

hydrophilins Small proteins that function in seed dehydration/dormancy and drought stress responses.

hydroponics A technique for growing plants with their roots immersed in nutrient solution without soil.

hydrostatic pressure Pressure generated by compression of water into a confined space. Measured in units called pascals (Pa) or, more conveniently, megapascals (MPa).

hyperaccumulation Accumulation of metals in a healthy plant to levels much higher than those found in the soil and that are generally toxic to non-accumulators.

hypersensitive response-type (HR-type) PCD A common plant defense following microbial infection, in which cells immediately surrounding the infection site die rapidly, depriving the pathogen of nutrients and preventing its spread.

hyphal coils Branched structures of mycorrhizal fungi that form within penetrated plant cells; the sites of nutrient transfer between the fungus and the host plant. Also called arbuscules.

hypocotyl The region of the seedling stem below the cotyledons and above the root.

hypogeal A type of seedling growth in which the cotyledons remain below the soil surface.

hypophysis In seed plant embryogenesis, the apical-most progeny of the basal cell which contributes to the embryo and will form part of the root apical meristem.

I

idioblast A "special" cell that differs markedly in form, content, or size from the other cells in the same tissue.

imbibition The initial phase of water uptake in dry seeds which is driven by the matric potential component of the water potential, that is, by the binding of water to surfaces, such as the cell wall and cellular macromolecules.

import The movement of photosynthate in sieve elements into sink organs.

imprinted gene expression Imprinted genes are expressed predominantly from either the maternal or the paternal allele, in contrast to nonimprinted genes in which alleles from both parents are expressed equally.

indehiscent Lacking spontaneous opening of a mature anther or fruit.

induced systemic resistance (ISR) Plant defenses that are activated by non-pathogenic microbes such as rhizobacteria. A defense response elicited by a local infection is mediated by JA and ethylene and that leads to systemic and long-lasting disease resistance, effective against fungi, bacteria, and viruses.

inducible defenses Defense responses that exist at low levels until a biotic or abiotic stress is encountered.

induction period The period of time (time lag) elapsed between the perception of a signal and the activation of the response. In the Calvin–Benson cycle, the time elapsed between the onset of illumination and the full activitation of the cycle.

infection thread An internal tubular extension of the plasma membrane of root hairs through which rhizobia enter root cortical cells.

initials In the root and shoot meristems, a cluster of slowly dividing and undetermined cells. Their descendants are displaced away by polarized patterns of cell division and take on various differentiated fates, contributing to the radial and longitudinal organization of the root or shoot and to the development of lateral organs.

initiator element (INR) A conserved DNA sequence found in the core promoter region of eukaryotic genes.

injectisome A name for the type III secretion system appendage of some pathogenic bacteria.

inner mitochondrial membrane The inner of the two mitochondrial membranes containing the electron transport chain, the F_oF_1-ATP synthase, and numerous transporters.

inositol 1,4,5-triphosphate (IP_3) One of several second messengers that trigger the release of calcium from intracellular stores.

integral membrane proteins Proteins that are embedded in the lipid bilayer. Most span the entire width of the bilayer, so one part of the protein interacts with one side of the membrane, another part interacts with the hydrophobic core of the membrane, and a third part interacts with the other side of the membrane.

intercalary meristem Meristem located near the base, rather than the tip, of a stem or leaf, as in grasses.

intercellular air space resistance The resistance or hindrance that slows down the diffusion of CO_2 inside a leaf, from the substomatal cavity to the walls of the mesophyll cells.

interface light scattering The randomization of the direction of photon movement within plant tissues due to the reflecting and refracting of light from the many air–water interfaces. Greatly increases the probability of photon absorption within a leaf.

intermediary cell A type of companion cell with numerous plasmodesmatal connections to surrounding cells, particularly to the bundle sheath cells.

intermembrane space The fluid-filled space between the two mitochondrial membranes or between the two chloroplast envelope membranes.

internal tolerance Tolerance mechanisms that function in the symplast (as opposed to exclusion mechanisms).

internode Portion of a stem between nodes.

interphase Collectively the G_1, S, and G_2 phases of the cell cycle.

inverse PCR A type of PCR that can be used to amplify DNA when the sequence of only one strand is known.

inwardly rectifying Refers to ion channels that open only at potentials more negative than the prevailing Nernst potential for a cation, or more positive than the prevailing Nernst potential for an anion, and thus mediate inward current.

irradiance The amount of energy that falls on a flat sensor of known area per unit time. Expressed as watts per square meter (W m^{-2}). Note, time (seconds) is contained within the term watt: 1 W = 1 joule (J) s^{-1}, or as moles of quanta per square meter per second (mol m^{-2} s^{-1}), also referred to as fluence rate.

J

JASMONATE-ZIM DOMAIN (JAZ) A transcriptional repressor that serves as a switch for jasmonate signaling. In the presence of JA, JAZ is degraded, allowing positive transcriptional regulators to activate JA-induced genes.

jasmonic acid resistance (JAR) proteins Defense proteins that are induced by jasmonic acid.

JAZ (Jasmonate ZIM-Domain) protein family Transcriptional repressor proteins that are proteolytically degraded after jasmonate-induced tagging by an E3 ubiquitin ligase complex.

K

karrikinolide Component of smoke that stimulates seed germination; similar in structure to strigolactones.

katanin A microtubule severing protein named after a katana, a Samurai sword.

kinases Enzymes that have the capacity to transfer phosphate groups from ATP to other molecules.

kinesins Microtubule-binding motor proteins that bind ATP and interact with microtubules, as well as binding "cargo" molecules. They are responsible for movement of the cargo (with ATP hydrolysis) along microtubules or tethering of cargo, such as organelles or condensed chromosomes, to microtubules. Kinesin cargo can be other microtubules, and so kinesins drive spindle dynamics during cell division.

kinetochore The site of spindle fiber attachment to the chromosome in anaphase. A layered structure associated with the centromere that contains microtubule-binding proteins and kinesins that help depolymerize and shorten the kinetochore microtubules.

KNOLLE A target recognition protein involved in vesicle fusion during cell plate formation. Belongs to the SNARE family of proteins.

Kranz anatomy (G: *kranz*: wreath or halo.) The wreathlike arrangement of mesophyll cells around a layer of large bundle-sheath cells. The two concentric layers of photosynthetic tissue surround the vascular bundle. This anatomical feature is typical of leaves of many C_4 plants.

L

L1 A clonally distinct epidermal layer derived from one set of initials in the shoot apical meristem.

L2 A subepidermal layer of cells derived from an internal set of initials in the shoot apical meristem.

L3 A centrally positioned layer of cells derived from an internal set of initials in the shoot apical meristem.

lamina ridges In the mosaic leaves sometimes produced by *phan* mutants, blade-like outgrowths that arise at the boundaries of the adaxial and abaxial domains.

lamina The blade of a leaf.

late response genes *See* secondary response genes.

late-embryogenesis-abundant (LEA) proteins Proteins involved in desiccation tolerance. They interact to form a highly viscous liquid with very slow diffusion and therefore limited chemical reactions. Encoded by a group of genes that are regulated by osmotic stress, first characterized in desiccating embryos during seed maturation.

latent heat of vaporization The energy needed to separate molecules from the liquid phase and move them into the gas phase at constant temperature.

lateral roots Arise from the pericycle in mature regions of the root through establishment of secondary meristems that grow out through the cortex and epidermis, establishing a new growth axis.

latex A complex, often milky solution exuded from cut surfaces of some plants that represents the cytoplasm of laticifers and may contain defensive substances.

laticifers In many plants, an elongated, often interconnected network of separately differentiated cells that contain latex (hence the term laticifer), rubber, and other secondary metabolites.

law of reciprocity The reciprocal relationship between fluence rate (mol $m^{-2} s^{-1}$) and duration of light exposure characteristic of many photochemical reactions as well as some developmental responses of plants to light. Total fluence depends on two factors: the fluence rate and the irradiation time. A brief light exposure can be effective with bright light; conversely, dim light requires a long exposure time. Also referred to as Bunsen–Roscoe Law.

leaf blade The broad, expanded area of the leaf; also called the lamina.

leaf stomatal resistance Resistance to CO_2 diffusion imposed by the stomatal pores.

leaf trace The portion of the shoot primary vascular system that diverges into a leaf.

leaflet A subdivision of a compound leaf.

leaves The main lateral appendages radiating out from stems and branches. Green leaves are usually the major photosynthetic organs of the plant.

lectins Defensive plant proteins that bind to carbohydrates; or carbohydrate-containing proteins inhibiting their digestion by a herbivore.

leghemoglobin An oxygen-binding heme protein found in the cytoplasm of infected nodule cells that facilitates the diffusion of oxygen to the respiring symbiotic bacteria.

leucoplasts Nonpigmented plastids, the most important of which is the amyloplast.

light channeling In photosynthetic cells, the propagation of some of the incident light through the central vacuole of the palisade cells and through the air spaces between the cells.

light compensation point The amount of light reaching a photosynthesizing leaf at which photosynthetic CO_2 uptake exactly balances respiratory CO_2 release.

light energy The energy associated with photons.

light-harvesting chlorophyll-binding protein II (LHCP II) The light-harvesting or antenna complex for photosystem II.

light-harvesting complex II (LHCII) The most abundant antenna protein complex, associated primarily with photosystem II.

LIGHT-OXYGEN-VOLTAGE (LOV) domains Domains that are sites of binding of the FMN chromophore to phototropins and are thus the part of the protein that senses light.

lignin Highly branched phenolic polymer with a complex structure made up of phenylpropanoid alcohols that may be associated with celluloses and proteins. Deposited in secondary walls, it adds strength allowing upward growth and permitting conduction through the xylem under negative pressure. Lignin has significant defensive functions.

lipochitin oligosaccharides Bacterial signal molecules that mediate signaling between rhizobial bacteria and leguminous plants.

lithocyst An enlarged surface cell of certain leaves containing a cystolith, a concretion of calcium carbonate deposited on a cellulosic extension that hangs from the upper cell wall.

locules The pollen-containing cavities inside anthers. The term is also applied to the chambers within the ovary in which seeds develop.

long-day plant (LDP) A plant that flowers only in long days (qualitative LDP) or whose flowering is accelerated by long days (quantitative LDP).

long-distance transport Translocation through the phloem to the sink.

long–short-day plant (LSDP) A plant that flowers in response to a shift from long days to short days.

lower-leaf zone (LLZ) A zone at the base of leaf primordiathat in some species gives rise to stipules or leaf sheaths.

lowest excited state Excited state with the lowest energy attained when a chlorophyll molecule in a higher energy state gives up some of its energy to the surroundings as heat.

LUREs The pollen chemoattractants of *Torenia fournieri*, consisting of cysteine-rich polypeptides.

lysin motifs (LysM) N-acetylglucosamine (NAG)-containing molecules derived from fungal chitin that elicit plant responses.

lysophospholipid A phospholipid from which one or both fatty acid groups have been removed.

lytic vacuoles Analogous to lysosomes in animal cells, they release hydrolytic enzymes that degrade cellular constituents during senescence and autophagocytosis.

M

macroautophagy The main type of autophagy in plants, in which specialized organelles called autophagosomes enclose the cytoplasmic components and fuse with the vacuole.

macrofibrils Structures found in secondary cell walls of tracheids and fibers, consisting of ~10–20 aggregated cellulose microfibrils.

MADS box genes Genes encoding a family of transcription factors containing a conserved sequence called the MADS box. It is the family that includes most floral homeotic genes and some of the genes involved in regulating flowering time.

maintenance respiration The respiration needed to support the function and turnover of existing tissue. Contrast with growth respiration.

malic enzyme An enzyme that catalyzes the oxidation of malate to pyruvate, permitting plant mitochondria to oxidize malate or citrate to CO_2 without involving pyruvate generated by glycolysis.

mannan A hemicellulose with a backbone made of 1→4-linked β-D-mannose.

MAP (mitogen-activated protein) kinase cascade The binding of a ligand signal that results in the phosphorylation and activation of a series of kinase enzymes.

map-based cloning A technique that uses genetic analysis of the offspring of crosses between a mutant and a wild-type plant to narrow the location of the mutation to a short segment of the chromosome, which can then be sequenced.

marginal meristems Proliferative tissues that are flanked by differentiated tissues at the edges of developing organs.

margo A porous and relatively flexible region of the pit membranes in tracheids of conifer xylem surrounding a central thickening, the torus.

mass transfer rate The quantity of material passing through a given cross section of phloem or sieve elements per unit time.

maternal (non-Mendelian) inheritance A non-Mendelian pattern of inheritance in which offspring receive genes from only the female parent.

maternally expressed genes (MEGs) Genes for which only the maternal alleles are expressed.

matric potential (Ψ_m) The sum of osmotic potential (Ψ_s) + hydrostatic pressure (Ψ_p). Useful in situations (dry soils, seeds, and cell walls) where the separate measurement of Ψ_s and Ψ_p is difficult or impossible.

matrix polysaccharides Polysaccharides comprising the matrix of plant cell walls. In primary cell walls they consist of pectins, hemicelluloses, and proteins.

matrix The colloidal-aqueous phase contained within the inner membrane of a mitochondrion.

matrixules Protrusions of the outer and inner membrane in mitochondria.

maturation zone The region of the root that has completed its differentiation and shows root hairs for the absorption of water and solutes; and competent vascular tissue.

maximum quantum yield Ratio between photosynthetic product and the number of photons absorbed by a photosynthetic tissue. In a graphic plot of photon flux and photosynthetic rate, the quantum yield is given by the slope of the linear portion of the curve.

mechanical hypothesis A type of pollen tube growth that is dictated by pistil architecture.

megaspore mother cell The cell inside the ovule that gives rise to the megaspores by meiosis.

megaspore The haploid (1*N*) spore that develops into the female gametophyte.

megastrobili The strobili or cones that contain the female gametophytic tissue.

meiosis The "reduction division" whereby two successive cell divisions produce four haploid (1*N*) cells from one diploid (2*N*) cell. In plants with alternation of generations, spores are produced by meiosis. In animals, which don't have alternation of generations, gametes are produced by meiosis.

membrane permeability The extent to which a membrane permits or restricts the movement of a substance.

meristematic zone A region at the tip of the root containing the meristem that generates the body of the root. Located just above the root cap.

meristemoid mother cells (MMCs) The cells of the leaf protoderm that divide asymmetrically (the so-called entry division) to give rise to the meristemoid, a guard cell precursor.

meristemoids Small, superficial clusters of dividing cells that give rise to structures such as trichomes or stomata.

meristems Localized regions of ongoing cell division that enable growth during post-embryonic development.

mesocotyl In members of the grass family, the part of the elongating axis between the scutellum and the coleoptile.

mesophyll resistance The resistance to CO_2 diffusion imposed by the liquid phase inside leaves. The liquid phase includes diffusion from the intercellular leaf spaces to the carboxylation sites in the chloroplast.

mesophyll Leaf tissue found between the upper and lower epidermal layers, consisting of palisade parenchyma and spongy mesophyll.

metabolic redundancy A common feature of plant metabolism in which different pathways serve a similar function. They can therefore replace each other without apparent loss in function.

metabolomics The study of all of the metabolites in a cell, tissue, organ, or organism that are the products of cellular metabolism.

metaphase A stage of mitosis during which the nuclear envelope breaks down and the condensed chromosomes align in the middle of the cell.

methylation The chemical addition of methyl groups to alter structure or function. A common modification of cytosine residues in DNA.

microarray analysis A technique that uses a solid support onto which are spotted thousands of DNA sequences that are representative of single genes of a given species. The genes on an array can be investigated all in a single experiment, increasing the throughput of gene analysis manyfold over the classical methods.

microautophagy A less understood type of autophagy in plants that involves the invagination of the tonoplast membrane and the formation of small intravacuolar vesicles called autophagic bodies, which undergo lysis inside the vacuole.

microbial pathogens Bacterial or fungal organisms that cause disease in a host plant.

microbodies A class of spherical organelles surrounded by a single membrane and specialized for one of several metabolic functions, such as the β-oxidation of fatty acids and the metabolism of glyoxylate (in peroxisomes and glyoxysomes, respectively).

microfilament A component of the cell cytoskeleton made of actin; it is involved in organelle motility within cells.

microgametogenesis The process in the pollen grain that gives rise to the male gametes, the sperm cells.

micropyle The small opening at the distal end of the ovule, through which the pollen tube passes prior to fertilization.

microRNAs (miRNAs) Short (21–24 nt) RNAs that have double-stranded stem-loop structures and mediate RNA interference.

microsatellite DNA sequences One group of heterochromatic dispersed repeats that consist of sequences as short as two nucleotides repeated hundreds or even thousands of times. Also known as simple sequence repeats.

microspores The haploid (1*N*) cell that develops into the pollen tube or male gametophyte.

microsporocytes Cells that divides meiotically to produce microspores.

microsporogenesis The process in which microspores are formed by the microsporocyte.

microstrobili The strobili or cones that contain the male gametophytic tissue.

microtubule Component of the cell cytoskeleton made of tubulin, a component of the mitotic spindle, and a player in the orientation of cellulose microfibrils in the cell wall.

middle lamella A thin layer of pectin-rich material at the junction where the primary walls of neighboring cells come into contact. Originates as the cell plate during cell division.

midvein Also known as primary vein, the first formed vascular bundle that runs down the middle of the leaf blade in dicot leaves.

mineral nutrition The study of how plants obtain and use mineral nutrients.

mineralization Process of breaking down organic compounds by soil microorganisms that releases mineral nutrients in forms that can be assimilated by plants.

minus end The overall slow-growing, stationary, or depolymerizing end of a cytoskeletal polymer. In microtubules at the cell cortex, the minus end is the site of depolymerization, while the plus end is the site of polymerization, resulting in the phenomenon of treadmilling. The minus end is also called the "pointed end" in actin filaments.

mitochondrial DNA (mtDNA) The DNA found in mitochondria. Plant mtDNA consists of between approximately 200 and 2000 kb and is much larger than mitochondrial genomes of animals or fungi. Mitochondrial genes encode a variety of proteins necessary for cellular respiration.

mitochondrial genome *See* mitochondrial DNA.

mitochondrion (plural *mitochondria*) The organelle that is the site for most reactions in the respiratory process in eukaryotes.

mitosis The ordered cellular process by which replicated chromosomes are distributed to daughter cells formed by cytokinesis.

mitotic spindle The mitotic structure involved in chromosome movement. Polymerized from α- and β-tubulin monomers formed by the disassembly of the preprophase band in early metaphase.

model organisms Organisms that are particularly accessible and convenient for research and which provide information for hypothesis testing in other organisms.

molecular chaperone proteins Proteins that maintain and/or restore the active three dimensional structures of other macromolecules.

monocarpic Refers to plants, typically annuals, that produce fruits only once and then die.

monocot One of the two classes of flowering plants, characterized by a single seed leaf (cotyledon) in the embryo.

monoecious Refers to plants in which male and female flowers are found on the same individuals, such as cucumber (*Cucumis sativus*) and maize (corn; *Zea mays*). Contrast with dioecious.

***MONOPTEROS* (*MP*)** Gene involved in embryonic patterning. Encodes an auxin response factor that is necessary for the normal formation of basal elements such as the root and hypocotyls.

monosomy A type of aneuploidy in which only one chromosome of a given kind is present.

morphogenesis The developmental processes that give rise to biological form.

morphogens In animals, substances that play key roles in providing positional cues in certain types of position-dependent development.

movement proteins Non-structural proteins encoded by the virus genome that facilitate viral movement through the symplast.

MscS (Mechanosensitive channel of Small conductance) A mechanically gated ion channel that senses osmotically driven changes in cell volume or physical contact with an object, herbivore, or pathogen.

multinet growth hypothesis Concerns cell wall deposition during cell expansion. Holds that each successive wall layer is stretched and thinned during cell expansion, so the microfibrils would be expected to be passively reoriented in the direction of growth.

multivesicular body Part of the prevacuolar sorting compartment that functions in the degradation of vacuoles and their membranes.

mutant An individual that contains specific changes in its DNA sequence and may show an altered phenotype.

mutational load The total number of deleterious genes that have accumulated in the genome of an individual or a population, which may cause disease.

mutualism A symbiotic relationship in which both organisms benefit.

MYB proteins A class of transcription factors in eukaryotes. In plants, one subgroup of a large MYB family that has been implicated in GA signaling (GAMYB).

MYC2 transcription factor A basic helix-loop-helix leucine zipper motif protein that binds an extended G-box promoter. Its transcription is induced by dehydration stress and ABA. MYC2 regulates JA-dependent functions and some light responses.

mycorrhiza (plural *mycorrhizas* or *mycorrhizae*) The symbiotic (mutualistic) association of certain fungi and plant roots. Facilitates the uptake of mineral nutrients by roots.

mycorrhizal fungi Fungi that can form mycorrhizal symbioses with plants.

mycorrhizal network A common mycorrhizal mycelium linking the roots of two or more plants.

N

N-linked glycoprotein Glycan linked via a nitrogen atom to a protein. Formed by transfer of a 14-sugar glycan from the ER membrane-embedded dolichol diphosphate to the nascent polypeptide as it enters the lumen of the ER.

NAD(P)H dehydrogenases A collective term for membrane-bound enzymes that oxidize NADH or NADPH, or both, and reduce quinone. Several are present in the electron transport chain of mitochondria; for example, the proton-pumping complex I, but also simpler non-proton-pumping enzymes.

NADH dehydrogenase (complex I) A multi-subunit protein complex in the mitochondrial electron transport chain that catalyzes oxidation of NADH and reduction of ubiquinone linked to the pumping of protons from the matrix to the intermembrane space.

NCED An enzyme (9-*cis*-epoxycarotenoid dioxygenase) catalyzing the first committed step for ABA biosynthesis, forming an intermediate that is a neutral growth inhibitor and has physiological properties similar to those of ABA.

necrosis Death that is directly caused by physical damage, toxins, or other external agents.

necrotic spots Small spots of dead leaf tissue. For example, a characteristic of phosphorus deficiency.

necrotrophic Refers to cell and tissue killing by pathogens that attack their host plant first by secreting cell wall-degrading enzymes and/or toxins, which will lead to massive tissue laceration and plant death.

neochrome A photoreceptor in the alga *Mougeotia* that consists of a fusion between phytochrome and a phototropin.

Nernst potential The electrical potential described by the Nernst equation.

night break An interruption of the dark period with a short exposure to light that makes the entire dark period ineffective.

nitrate reductase Enzyme located in the cytosol that reduces nitrate (NO_3^-) to nitrite (NO_2^-). Catalyzes the first step by which nitrate absorbed by roots is assimilated into organic form.

nitrogen fixation The natural or industrial processes by which atmospheric nitrogen N_2 is converted to ammonia (NH_3) or nitrate (NO_3^-).

nitrogenase enzyme complex The two-component protein complex that conducts biological nitrogen fixation in which ammonia is produced from molecular nitrogen.

Nod factors Lipochitin oligosaccharide signal molecules active in regulating gene expression during nitrogen-fixing nodule formation. All Nod factors have a chitin β-1→4-linked *N*-acetyl-D-glucosamine backbone (varying in length from three to six sugar units) and a fatty acid chain on the C-2 position of the nonreducing sugar.

nodal roots Adventitious roots that form after the emergence of primary roots.

node Position on the stem where leaves are attached.

nodulation factors *See* nod factors.

nodulation (*nod*) genes Rhizobial genes, the products of which participate in nodule formation.

nodules Specialized organs of a plant host containing symbiotic nitrogen-fixing bacteria.

nodulin genes Plant genes specific to nodule formation.

non-autonomous transposons Transposons that require other transposable elements to move.

non-cell autonomous response A cellular response to an environmental stimulus or genetic mutation that is induced by other cells.

non-climacteric Refers to a type of fruit that does not undergo a climacteric, or respiratory burst, during ripening.

non-phosphorylating respiratory pathways Mitochondrial respiratory chain components not linked to proton pumping, (e.g., the alternative oxidase).

non-protein-coding RNAs (ncRNAs) RNAs that do not encode proteins, but may instead be involved in gene regulation or active in the RNA interference (RNAi) pathway.

non-seed plants Plant families that do not produce a seed.

nonenzymatic proteins Proteins without enzymatic activity, including arabinogalactan proteins, hydroproline-rich glycoproteins, various structural and signaling proteins. Expansins are included in this category.

nonphotochemical quenching The quenching of chlorophyll fluorescence by processes other than photochemistry—the converting of excess excitation into heat.

nonvascular plants Plants that do not have vascular tissues, such as xylem and phloem.

nucellus The maternal tissue of the ovary that gives rise to the ovule.

nuclear envelope The double membrane surrounding the nucleus.

nuclear genome The entire complement of DNA found in the nucleus.

nuclear localization signal A specific amino acid sequence required for a protein to gain entry into the nucleus.

nuclear pore complex (NPC) An elaborate structure, 120 nm wide, composed of more than a hundred different nucleoporin proteins arranged octagonally. The NPC forms a large, protein-lined pore in the nuclear membrane.

nuclear pores Sites where the two membranes of the nuclear envelope join, forming a partial opening between the interior of nucleus and the cytosol. Contains an elaborate structure of more than a hundred different nucleoporin proteins that form a nuclear pore complex (NPC).

nucleoids Organellar and prokaryotic genomes that are not enclosed in a nuclear envelope.

nucleolar organizer region (NOR) Associated with the nucleolus in the interphase nucleus. Site where portions of one or more chromosomes containing genes, repeated in tandem, coding for ribosomal RNA are clustered and are transcribed.

nucleolus (plural *nucleoli*) A densely granular region in the nucleus that is the site of ribosome synthesis.

nucleoporins Proteins that form the nuclear pore complex in the nuclear envelope.

nucleosome A structure consisting of eight histone proteins around which DNA is coiled.

nucleus (plural *nuclei*) The organelle that contains the genetic information primarily responsible for regulating cellular metabolism, growth, and differentiation.

nutrient assimilation The incorporation of mineral nutrients into carbon compounds such as pigments, enzyme cofactors, lipids, nucleic acids, or amino acids.

nutrient depletion zone The region surrounding the root surface showing diminished nutrient concentrations due to uptake into the roots and slow replacement by diffusion.

nutrient film growth system A form of hydroponic culture in which the plant roots lie on the surface of a trough, and the nutrient solution flows over the roots in a thin layer along the trough.

nutrient solution A solution containing only inorganic salts that supports the growth of plants in sunlight without soil or organic matter.

nyctinasty Sleep movements of leaves. Leaves extend horizontally to face the light during the day and fold together vertically at night.

O

O-linked oligosaccharides Oligosaccharides that are linked to proteins via the OH groups of hydroxyproline, serine, threonine, and tyrosine residues.

octant The spherical, eight-cell, globular embryo exhibiting radial symmetry.

oil bodies Also known as oleosomes or spherosomes, these organelles accumulate and store triacylglycerols. They are bounded by a single phospholipid leaflet ("half–unit membrane" or "phospholipid monolayer") derived from the ER.

oleosin A specific protein that coats oil bodies.

oligogalacturonans Pectin fragments (10 to 13 residues) resulting from plant cell wall degradation that elicit multiple defense responses. May also function during the normal control of cell growth and differentiation.

oligosaccharins Fragments resulting from plant cell wall degradation that affect plant defense and growth.

Optimal Defense Hypothesis A hypothesis that proposes that plants optimize survival and reproduction via a range of evolutionary adaptions that reduce herbivory.

orange carotenoid protein (OCP) A soluble protein associated with the phycobilisome antenna of photosystem II in cyanobacteria.

ordinary companion cell A type of companion cell with relatively few plasmodesmata connecting it to any of the surrounding cells other than its associated sieve element.

organelle tethering Attachment of an organelle to a cytoskeletal or membrane structure. Some classes of myosin or kinesin motors can attach organelles to F-actin or microtubules, respectively.

organic fertilizer A fertilizer that contains nutrient elements derived from natural sources without any synthetic additions.

organogenesis The formation of functionally organized structures during embryogenesis.

orthodox seed A seed that can tolerate desiccation and remain viable after storage in a dry state.

osmolarity A unit of concentration expressed as moles of total dissolved solutes per liter of solution (mol L^{-1}). In biology, the solvent is usually water.

osmosis The movement of water across a selectively permeable membrane toward the region of more negative water potential, Ψ (lower concentration of water).

osmotic adjustment The ability of the cell to accumulate compatible solutes and lower water potential during periods of osmotic stress.

osmotic potential (Ψ_s) The effect of dissolved solutes on water potential. Also called solute potential.

osmotic stress Stress imposed on cells or whole plants when the osmotic potential of external solutions is more negative than that of the solution inside the plant.

outcrossing The mating of two plants with different genotypes by cross-pollination.

outer mitochondrial membrane The outer of the two mitochondrial membranes, which appears to be freely permeable to all small molecules.

outwardly rectifying Refers to ion channels that open only at potentials more positive than the prevailing Nernst potential for a cation, or more negative than the prevailing Nernst potential for an anion, and thus mediate outward current.

oxidative pentose phosphate pathway A cytosolic and plastidic pathway that oxididizes glucose and produces NADPH and a number of sugar phosphates.

oxidative phosphorylation Transfer of electrons to oxygen in the mitochondrial electron transport chain that is coupled to ATP synthesis from ADP and phosphate by the ATP synthase.

oxylipins Oxygenated fatty acid derivatives that function in stress and pathogen responses in plants and animals.

P

P680 The chlorophyll of the photosystem II reaction center that absorbs maximally at 680 nm in its neutral state. The P stands for pigment.

P700 The chlorophyll of the photosystem I reaction center that absorbs maximally at 700 nm in its neutral state. The P stands for pigment.

P870 The reaction center bacteriochlorophyll from purple photosynthetic bacteria that absorbs maximally at 870 nm in its neutral state. The P stands for pigment.

P-protein bodies Discrete spheroidal, spindle-shaped, or twisted/coiled structures of P-proteins present in the cytosol of immature sieve tube elements of the phloem. Generally disperse into tubular or fibrillar forms during cell maturation.

P-proteins Phloem proteins that act to seal damaged sieve elements by plugging the sieve-element pores. Abundant in the sieve elements of most angiosperms, but absent from gymnosperms. Formerly called "slime."

paleopolyploids Species that show signs of ancient genome duplications followed by DNA loss.

palisade cells Below the leaf upper epidermis, the top one to three layers of pillar-shaped photosynthetic cells.

paraheliotropic Refers to movement of leaves away from incident sunlight.

parasite An organism that lives on or in an organism of another species, known as the host, from the body of which it obtains nutriment.

parenchyma Metabolically active plant tissue consisting of thin-walled cells, with air-filled spaces at the cell corners.

parent-of-origin effect A phenotypic difference in the progeny that depends on whether it was transmitted by the maternal or paternal parent.

partitioning The differential distribution of photosynthate to multiple sinks within the plant.

PAS-GAF-PHY The N-terminal half of phytochrome containing the photosensory domain.

PAS-related domain (PRD) On the phytochrome protein, two domains that mediate phytochrome dimerization.

PAS A domain of phytochrome, which is necessary for chromophore attachment to the protein.

passive transport Diffusion across a membrane. The spontaneous movement of a solute across a membrane in the direction of a gradient of (electro)chemical potential (from higher to lower potential). Downhill transport.

paternally expressed genes (PEGs) Genes for which only the paternal alleles are expressed.

pathogen-associated molecular patterns (PAMPS) Molecules originating from pathogenic sources that can initiate immune responses. These are a subset of microbe-associated molecular patterns (MAMPs).

pathogenesis-related (PR) genes Genes that encode small proteins that function either as antimicrobials or in initiating systemic defense responses.

pattern recognition receptors (PRRs) Primitive innate immune system proteins which are associated with PAMPs and DAMPs.

pavement cells The predominant type of leaf epidermal cells, which secrete a waxy cuticle and serve to protect the plant from dehydration and damage from ultraviolet radiation.

pectins A heterogeneous group of complex cell wall polysaccharides that form a gel in which the cellulose–hemicellulose network is embedded. Typically contain acidic sugars such as galacturonic acid and neutral sugars such as rhamnose, galactose, and arabinose. Often include calcium as a structural component, allowing extractions from the wall with chelators or dilute acids.

PEP carboxylase A cytosolic enzyme that forms oxaloacetate by the carboxylation of phosphoenolpyruvate.

perennial plants Plants that live for more than two years.

perforation plate The perforated end wall of a vessel element in the xylem.

pericarp The fruit wall surrounding a fruit, derived from the ovary wall.

periclinal Pertaining to the orientation of cell division such that the new cell walls form parallel to the tissue surface.

pericycle Meristematic cells forming the outermost layer of the vascular cylinder in the stem or root, interior to the endodermis. An internal meristematic tissue from which lateral roots arise.

periderm Tissue produced by the cork cambium that contributes to the outer bark of stems and roots during secondary growth of woody plants, replacing the epidermis. Also forms over wounds and abscission layers after the shedding of plant parts.

period In cyclic (rhythmic) phenomena, the time between comparable points in the repeating cycle, such as peaks or troughs.

peripheral proteins Proteins that are bound to the membrane surface by noncovalent bonds, such as ionic bonds or hydrogen bonds.

peripheral zone (PZ) A doughnut-shaped region surrounding the central zone in shoot apical meristems consisting of small, actively dividing cells with inconspicuous vacuoles. Leaf primordia are formed in the peripheral zone.

perisperm Storage tissue derived from the nucellus, often consumed during embryogenesis.

peroxiredoxins (Prx) A family of antioxidant enzymes that inactivate peroxides.

peroxisome Organelle in which organic substrates are oxidized by O_2. These reactions generate H_2O_2 that is broken down to water by the peroxysomal enzyme catalase.

peroxules Tubular protrusions from peroxisomes.

petiole The leaf stalk that joins the leaf blade to the stem.

Pfr The far-red light–absorbing form of phytochrome converted from Pr by the action of red light. The blue-green colored Pfr is converted back to Pr by far-red light. Pfr is the physiologically active form of phytochrome.

phagophore assembly site The cellular site where the phagophore is assembled during autophagy.

phagophore A double membrane that encloses and isolates cytoplasmic components during macroautophagy.

phase change The phenomenon in which the fates of the meristematic cells become altered in ways that cause them to produce new types of structures.

phase In cyclic (rhythmic) phenomena, any point in the cycle recognizable by its relationship to the rest of the cycle, for example, the maximum and minimum positions.

phellem A part of the secondary dermal system, or periderm, of woody plants, consisting of dead cells with secondary cell walls rich in suberin and lignin. Also called cork.

phelloderm In some plants, one or more layers of parenchyma-like tissue derived from the phellogen.

phellogen *See* cork cambium.

phenocopy A plant with growth characteristics produced by genetic or chemical interference that mimic those found in another genotype.

phenotypic plasticity Physiological or developmental responses of a plant to its environment that do not involve genetic changes.

pheophytin A chlorophyll in which the central magnesium atom has been replaced by two hydrogen atoms.

phloem fibers Elongated, tapering sclerenchyma cells associated with the other cells in the phloem.

phloem loading The movement of photosynthetic products from the mesophyll chloroplasts to the sieve elements of mature leaves. Includes short-distance transport steps and sieve-element loading. *See also* phloem unloading.

phloem unloading The movement of photosynthate from the sieve elements to the sink cells that store or metabolize them. Includes sieve-element unloading and short-distance transport. *See also* phloem loading.

phloem The tissue that transports the products of photosynthesis from mature leaves to areas of growth and storage, including the roots.

phosphate transporter Protein in the plasma membrane specific for the uptake of phosphate by the cell.

phosphatidic acid (PA) A diacylglycerol that has a phosphate on the third carbon of the glycerol backbone.

phosphatidylinositol bisphosphate (PIP_2) A group of phosphorylated derivatives of phosphatidylinositol.

phospholipase A (PLA) An enzyme that removes one of the fatty acid chains from a phospholipid.

phospholipase C (PLC) An enzyme whose action on phosphoinositides releases inositol triphosphate ($InsP_3$), along with diacylglycerol (DAG).

phospholipase D (PLD) An enzyme active in ABA signaling; it releases phosphatidic acid from phosphatidylcholine.

photoassimilation The coupling of nutrient assimilation to photosynthetic electron transport.

photoblasty Light-induced seed germination.

photochemistry Very rapid chemical reactions in which light energy absorbed by a molecule causes a chemical reaction to occur.

photoinhibition The inhibition of photosynthesis by excess light.

photolyase A blue light–activated enzyme that repairs pyrimidine dimers in DNA that have been damaged by ultraviolet radiation. Contains an FAD and a pterin.

photomorphogenesis The influence and specific roles of light on plant development. In the seedling, light-induced changes in gene expression to support aboveground growth in the light rather than belowground growth in the dark.

photon A discrete physical unit of radiant energy.

photonasty Plant movements in response to non-directional light.

photoperiodic induction The photoperiod-regulated processes that occur in leaves resulting in the transmission of a floral stimulus to the shoot apex.

photoperiodism A biological response to the length and timing of day and night, making it possible for an event to occur at a particular time of year.

photophosphorylation The formation of ATP from ADP and inorganic phosphate (P_i), catalyzed by the CF_0F_1-ATP synthase and using light energy stored in the proton gradient across the thylakoid membrane.

photoprotection A carotenoid-based system for dissipating excess energy absorbed by chlorophyll in order to avoid forming singlet oxygen and damaging pigments. Involves quenching.

photoreceptors Proteins that sense the presence of light and initiate a response via a signaling pathway.

photorespiration Uptake of atmospheric O_2 with a concomitant release of CO_2 by illuminated leaves. Molecular oxygen serves as substrate for rubisco and the formed 2-phosphoglycolate enters the photorespiratory carbon oxidation cycle. The activity of the cycle recovers some of the carbon found in 2-phosphoglycolate, but some is lost to the atmosphere.

photoreversibility The interconversion of the Pr and Pfr forms of phytochrome.

photostationary state Relating to phytochrome under natural light conditions, the equilibrium of 97% Pr and 3% Pfr.

photosynthate The carbon-containing products of photosynthesis.

photosystem I (PSI) A system of photoreactions that absorbs maximally far-red light (700 nm), oxidizes plastocyanin and reduces ferredoxin.

photosystem II (PSII) A system of photoreactions that absorbs maximally red light (680 nm), oxidizes water and reduces plastoquinone. Operates very poorly under far-red light.

phototropins 1 and 2 Two flavoproteins that are the photoreceptors for the blue-light signaling pathway that induces phototropic bending in Arabidopsis hypocotyls and in oat coleoptiles. They also mediate chloroplast movements and participate in stomatal opening in response to blue light. Phototropins are autophosphorylating protein kinases whose activity is stimulated by blue light.

phototropins Blue light photoreceptors that primarily regulate phototropism, chloroplast movements, and stomatal opening.

phototropism The alteration of plant growth patterns in response to the direction of incident radiation, especially blue light.

phragmoplast An assembly of microtubules, membranes, and vesicles that forms during late anaphase or early telophase and precedes fusion of vesicles to form cell plate.

PHY The designation for the phytochrome apoprotein (without the chromophore).

phyllode An expanded petiole resembling and having the function of a leaf, but without a true blade.

phyllome The collective term for all the leaves of a plant including structures that evolved from leaves, such as floral organs.

phyllotaxy The arrangement of leaves on the stem.

phytase Enzyme that breaks down the phosphate-rich seed storage compound phytin during seedling growth.

phytin The K^+, Mg^{2+}, and Ca^{2+} salt of phytic acid a (*myo*-inositol hexaphosphate), a major storage form of phosphate in seeds.

phytoalexins Chemically diverse group of secondary metabolites with strong antimicrobial activity that are synthesized following infection and accumulate at the site of infection.

phytochelatins Low-molecular-weight peptides synthesized from glutathione by the enzyme phytochelatin synthase. These peptides can bind a variety of metal(oids) and play an important role in tolerance of plants to As, Cd, and Zn.

phytochrome interacting factors (PIFs) Families of phytochrome-interacting proteins that may activate and repress gene transcription; some are targets for phytochrome-mediated degradation.

phytochrome kinase substrates (PKS) Proteins that participate in the regulation of phytochrome via direct phosphorylation or via phosphorylation by other kinases.

phytochrome A plant growth-regulating photoreceptor protein that absorbs primarily red light and far-red light, but also absorbs blue light. The holoprotein that contains the chromophore phytochromobilin.

phytochromobilin The linear tetrapyrrole chromophore of phytochrome.

phytoliths Discrete cells that accumulate silica in leaves or roots.

phytomer A developmental unit consisting of one or more leaves, the node to which the leaves are attached, the internode below the node, and one or more axillary buds.

PIF-like proteins (PILs) Nuclear, DNA-binding proteins that selectively interact with phytochromes in their active Pfr conformations.

PIF3 A basic helix-loop-helix transcription factor that interacts with both phyA and phyB.

PIN auxin efflux carrier proteins Plasma membrane transport proteins that amplify localized, directional auxin streams associated with embryonic development, organogenesis, and tropic growth.

pit fields Depressions in the primary cell walls where numerous plasmodesmata make connections with adjacent cells. When present, secondary walls are not deposited in the locations of pit fields, giving rise to pits.

pit membrane The porous layer in the xylem between pit pairs, consisting of two thinned primary walls and a middle lamella.

pit pair The adjacent pits of adjoined tracheid cells in the xylem. A low-resistance path for water movement between tracheids.

pit A microscopic region where the secondary wall of a tracheary element is absent and the primary wall is thin and porous, facilitating sap movement between one tracheid and the adjacent one.

pith The ground tissue in the center of the stem or root.

plant growth promoting rhizobacteria (PGPR) Soil bacteria associated with root surfaces that promote plant growth by producing plant growth regulators and/or fixing nitrogen.

plant tissue analysis In the context of mineral nutrition, the analysis of the concentrations of mineral nutrients in a plant sample.

plants All the families of plants inclusive of the non-seed, nonvascular plants.

plasma membrane A fluid mosaic structure composed of a bilayer of polar lipids (phospholipids or glycosylglycerides) and embedded proteins that together confer selective permeability on the structure. Also called plasmalemma.

plasma membrane H^+-ATPase An ATPase that pumps H^+ across the plasma membrane energized by ATP hydrolysis.

plasmodesmata (singular *plasmodesma*) Microscopic membrane-lined channel connecting adjacent cells through the cell wall and filled with cytoplasm and a central rod derived from the ER called the desmotubule. Allows the movement of molecules from cell to cell through the symplast. The pore size can apparently be regulated by globular proteins lining the channel inner surface and the desmotubule to allow particles as large as viruses to pass through.

plasticity The ability to adjust morphologically, physiologically, and biochemically in response to changes in the environment.

plastid genome The genome contained in chloroplasts and other plastids. Plastid genomes carry a subset of genes for plastid function, such as some of those involved in photosynthesis.

plastids Cellular organelles found in eukaryotes, bounded by a double membrane, and sometimes containing extensive membrane systems. They perform many different functions: photosynthesis, starch storage, pigment storage, and energy transformations.

plastocyanin (PC) A small (10.5 kDa), water-soluble, copper-containing protein that transfers electrons between the cytochrome b_6f complex and P700. This protein is found in the lumenal space.

plastoglobuli Lipid bodies that accumulate inside gerontoplasts during leaf senescence.

plastohydroquinone (PQH_2) The fully reduced form of plastoquinone.

plumule First true leaf of a growing seedling.

plus end The overall fast-growing, or polymerizing end of a cytoskeletal polymer. In microtubules at the cell cortex, the plus end is the site of polymerization, while the minus end is the site of depolymerization, resulting in the phenomenon of treadmilling. The plus end is also called the "barbed end" in actin filaments. The plus end generally has high-energy nucleotide triphosphates bound to it (GTP in microtubules, ATP in F-actin), while the minus end has lower-energy nucleotide diphosphates bound.

polar auxin transport Directional auxin streaming that functions in programmed development and plastic growth responses. Long-distance polar auxin transport maintains the overall polarity of the plant axis and supplies auxin for direction into localized streams.

polar glycerolipids The main structural lipids in membranes, in which the hydrophobic portion consists of two 16-carbon or 18-carbon fatty acid chains esterified to positions 1 and 2 of a glycerol.

polar nuclei The two haploid nuclei at the center of the embryo sac that normally fuse to form the diploid nucleus of the central cell.

polarity (1) Property of some molecules, such as water, in which differences in the electronegativity of some atoms results in a partial negative charge at one end of the molecule and a partial positive charge at the other end. (2) Refers to the distinct ends and intermediate regions along an axis. Beginning with the single-celled zygote, the progressive development of distinctions along two axes: an apical–basal axis and a radial axis.

pollen mother cell The microsporocyte that divides meiotically to produce the microspores in the anther.

pollen Small structures (microspores) produced by anthers of seed plants. Contain haploid male nuclei that will fertilize egg in ovule.

polyads Large assemblages of pollen that facilitate bulk transfer of multiple grains during insect-mediated pollination.

polycarpic Refers to perennial plants that produce fruit many times.

Polycomb group proteins A family of proteins, first discovered in *Drosophila*, that mediate chromatin remodeling, which typically leads to epigenetic gene silencing.

polymer-trapping model A model that explains the specific accumulation of sugars in the sieve elements of symplastically loading species.

polymorphic sequence A sequence that varies among members of a population or species.

polyploidy The condition of being polyploid, that is, having one or more extra sets of chromosomes.

polyribosomes Ribosomes which are "strung together" with messenger RNA and are in the process of translating protein from that mRNA.

pome A fruit, such as an apple, composed of one or more carpels surrounded by accessory tissue derived from the receptacle.

position-dependent Refers to mechanisms that operate by modulating the behavior of cells in a manner that depends on the position of the cells within the developing embryo.

posttranscriptional regulation Following transcription, the control of gene expression by altering mRNA stability or translation efficiency.

posttranslational Referring to events or modifications that occur after the synthesis, or translation, of a protein from its mRNA.

Pr Red light–absorbing phytochrome form. This is the form in which phytochrome is assembled. The blue-colored Pr is converted by red light to the far-red light–absorbing form, Pfr.

pre-procambium A stable intermediate state between ground cells and procambium cells that can be detected by the expression of specific transcription factors.

precocious germination Germination of viviparous mutant seeds while still attached to the mother plant.

preharvest sprouting Germination of physiologically mature wild-type seeds on the mother plant caused by wet weather.

preprophase band A circular array of microtubules and microfilaments formed in the cortical cytoplasm just prior to cell division that encircles the nucleus and predicts the plane of cytokinesis following mitosis.

preprophase In mitosis, the stage just before prophase during which the G_2 microtubules are completely reorganized into a preprophase band.

pressure potential (Ψ_p) The hydrostatic pressure of a solution in excess of ambient atmospheric pressure.

pressure-flow model A widely accepted model of phloem translocation in angiosperms. It states that transport in the sieve elements is driven by a pressure gradient between source and sink. The pressure gradient is osmotically generated and results from the loading at the source and unloading at the sink.

prevacuolar compartment A membrane compartment equivalent to the late endosome in animal cells where sorting occurs before cargo is delivered to the lytic vacuole.

prickles Pointed plant structures that physically deter herbivory and are derived from epidermal cells.

primary active transport The direct coupling of a metabolic energy source such as ATP hydrolysis, oxidation-reduction reaction, or light absorption to active transport by a carrier protein.

primary cell walls The thin (less than 1 μm), unspecialized cell walls that are characteristic of young, growing cells. About 85% polysaccharide and 10% protein by dry weight.

primary cross-regulation Involves distinct signaling pathways regulating a shared transduction component in a positive or a negative manner.

primary dormancy The failure of newly dispersed, mature seeds to germinate under normal conditions, typically induced by abscisic acid (ABA) during seed maturation.

primary endosperm cell The triploid endosperm cell produced by the fusion of the second sperm cell with the two polar nuclei, or the diploid nucleus, of the central cell.

primary growth The phase of plant development that gives rise to new organs and to the basic plant form. It results from cell proliferation in apical meristems, followed by cell elongation and differentiation.

primary inflorescence meristem The meristem that produces stem-bearing flowers; it is formed from the shoot apical meristem.

primary response genes Genes whose expression is necessary for plant morphogenesis and that are expressed rapidly following exposure to a light signal. Often regulated by phytochrome-linked activation of transcription factors. Genes whose expression does not require protein synthesis. *See* secondary response genes.

primary plant axis The longitudinal axis of the plant defined by the positions of the shoot and root apical meristems.

primary plant body The part of the plant directly derived from the shoot and root apical meristems and primary meristems. It is composed of tissues resulting from primary, as opposed to secondary, growth.

primary plasmodesmata Tubular extensions of the plasma membrane, 40 to 50 nm in diameter, that traverse the cell wall and form cytoplasmic connections between cells derived from each other by mitosis.

primary root Root generated directly by growth of the embryonic root or radicle.

primary thickening meristem A specialized meristem located in the apical bud below the leaf primordia in certain monocots, such as palms. It serves to increase the width of the stem behind the apex, enabling the trunk to achieve a considerable height.

primary vein *See* midvein.

primary walls *See* primary cell walls.

proanthocyanidins A group of condensed tannins that are present in many plants that serve as defensive chemicals against plant pathogens and herbivores.

procambium Primary meristematic tissue that differentiates into xylem, phloem, and cambium.

profilins Actin binding proteins that keep the depolymerized globular G-actin monomers charged with ATP, so that they can be rapidly re-integrated into F-actin. They also bind formins and thereby accelerate the formation of F-actin from formins.

programmed cell death (PCD) Process whereby individual cells activate an intrinsic senescence program accompanied by a distinct set of morphological and biochemical changes similar to mammalian apoptosis.

prokaryotic pathway In the chloroplast, the series of reactions for the synthesis of glycerolipids. *See also* eukaryotic pathway.

prolamellar bodies Elaborate semicrystalline lattices of membrane tubules that develop in plastids that have not been exposed to light (etioplasts).

prometaphase Early metaphase stage in which the preprophase band disassembles and new microtubules polymerize to form the mitotic spindle.

promoter The region of the gene that binds RNA polymerase.

prophase The first stage of mitosis (and meiosis) prior to disassembly of the nuclear envelope, during which the chromatin condenses to form distinct chromosomes.

proplastid Type of immature, undeveloped plastid found in meristematic tissue that can convert to various specialized plastid types, such as chloroplasts, amyloplasts, and chromoplasts, during development.

protein bodies Protein storage organelles enclosed by a single unit membrane; found mainly in seed tissues.

PROTEIN PHOSPHATASE 1 (PP1) A signaling intermediate of the phototropin pathway during blue light-induced stomatal opening.

protein stability The rate of protein destruction or inactivation; can contribute to posttranslational regulation and also plays an important role in the overall activity of a gene or its product.

protein storage vacuoles Specialized small vacuoles that accumulate storage proteins, typically in seeds.

26S proteasome A large proteolytic complex that degrades intracellular proteins marked for destruction by the attachment of one or more copies of the small protein ubiquitin.

proteome The entire set of proteins expressed by a cell, tissue, or organism at a certain time.

proteomics The study of proteomes, including the relative abundance and modifications of proteins.

prothallus In ferns, the independent photosynthetic gametophyte.

protoderm In the plant embryo, the surface layer one cell thick that covers both halves of the embryo and will generate the epidermis.

protofilaments Polymerized α- and β-tubulin heterodimers.

proton gradient The concentration gradient of protons (hydrogen ions) across a membrane. Used by chloroplasts and mitochondria to drive ATP production.

proton motive force (PMF) The energetic effect of the electrochemical H^+ gradient across a membrane, expressed in units of electrical potential.

protoplast fusion A technique for incorporating foreign genes into plant genomes by fusion of two genetically different cells from which the cell walls have been removed.

proximal promoter *See* regulatory promotor.

proximal–distal polarity Polarity that develops along the length of a leaf.

pseudogenes Stable but nonfunctional genes; apparently derived from active genes by mutation.

pulvinus (plural *pulvini*) A turgor-driven organ found at the junction between the blade and the petiole of the leaf that provides a mechanical force for leaf movements.

pumps Membrane proteins that carry out primary active transport across a biological membrane. Most pumps transport ions, such as H^+ or Ca^{2+}.

PYR/PYL/RCARs A family of soluble ABA receptors identified as proteins that interact with the PP2C protein phosphatases.

pyruvate dehydrogenase An enzyme in the mitochondrial matrix that decarboxylates pyruvate, producing NADH (from NAD^+), CO_2, and acetic acid in the form of acetyl-CoA (acetic acid bound to coenzyme A).

Q

Q cycle A mechanism for oxidation of plastohydroquinone in chloroplasts and ubihydroquinone in mitochondria.

quantum yield The ratio of the yield of a particular product of a photochemical process to the total number of quanta absorbed.

quantum (plural *quanta*) A discrete packet of energy contained in a photon.

Quartet Model A molecular model that accounts for the interactions of the A-, B-, C-, and E-class genes in specifying floral organ identity during flowering. According to the model, MADS box genes dimerize, and two dimers can form a tetramer. These tetramers are hypothesized to bind CArG-boxes on target genes and modify their expression.

quenching The process by which energy stored in light excited chlorophylls is rapidly dissipated mainly by excitation transfer or photochemistry.

quiescent center Central region of root meristem where cells divide more slowly than surrounding cells, or do not divide at all.

quorum sensing A system of coordinated signals and responses by which populations regulate growth and environmental responses. This is a common mechanisms in microbial organisms.

R

Rabl configuration Proposed positioning of chromosomes inside a nucleus in which all centromeres and all telomeres point in opposite direction.

Rabs A class of targeting recognition proteins for the selective fusion and fission of vesicles and tubules within the endomembrane system.

rachis The main axis of a compound leaf to which leaflets are attached; the main axis of an inflorescence to which flowers are attached.

radial axis The pattern of concentric tissues extending from the outside of a root or stem into its center.

radicle The embryonic root. Usually the first organ to emerge on germination.

raphides Needles of calcium oxalate or carbonate that function in plant defense.

rays Tissues of various height and width that radiate through the secondary xylem and phloem, and are formed from ray initials in the vascular cambium.

reaction center complex A group of electron transfer proteins that receive energy from the antenna complex and convert it into chemical energy using oxidation-reduction reactions.

reactive oxygen species (ROS) These include the superoxide anion ($O_2\bullet^-$), hydrogen peroxide (H_2O_2), and the hydroxyl radical ($HO\bullet$) and singlet oxygen. They are generated in several cell compartments and can act as signals or cause damage to cellular components.

recalcitrant seeds Seeds that are released from the plant with a relatively high water content and active metabolism and, as a consequence, deteriorate upon dehydration and do not survive storage.

receptor kinase A protein in a signaling pathway that detects the presence of a ligand, such as a hormone, by phosphorylating itself or another protein.

receptor-like kinases (RLKs) Transmembrane proteins with putative amino-terminal extracellular domains and carboxyl-terminal intracellular kinase domains, which resemble animal receptor tyrosine kinases. Many plant RLKs specifically phosphorylate serine or threonine residues.

recycling The process whereby membrane content that is added through fusion is sorted and removed through fission.

regulatory promoter Sequence within or adjacent to a gene that regulates the activity of the gene via its core promoter.

regulatory promoter sequences Sequence elements that are part of the core promoter.

release phloem Sieve elements of sinks.

repeat-associated silencing RNAs (ra-siRNAs) Repeat regions from which short interfering RNAs originate.

reporter gene A gene whose expression conspicuously reveals the activity of another gene. Gene engineered to share the same promoter as another gene.

repressor A protein that either alone or in concert with other proteins represses expression of a gene.

rescue When used in genetics, restoration of wild type growth and development.

resistance genes (*R* genes) Genes that function in plant defense against fungi, bacteria, and nematodes in some cases by encoding protein receptors that bind to specific pathogen molecules, elicitors.

respiratory burst oxidase homolog D (RBOHD) An enzyme that generates superoxide using NADPH as an electron donor.

respiratory chain *See* electron transport chain.

respiratory quotient (RQ) The ratio of CO_2 evolution to O_2 consumption.

response regulator One component of the two-component regulatory systems that are composed of a histidine kinase sensor protein and a response regulator protein.

reticulons A class of proteins that control the transition between tubular and cisternal forms of the ER by forming tubules from membrane sheets.

retrograde A backward movement in transport or signaling.

retrotransposons In contrast to DNA transposons, these make an RNA copy of themselves, which is then reverse transcribed into DNA before it is inserted elsewhere in the genome.

rhamnogalacturonan I (RG I) An abundant pectic polysaccharide that has a long backbone of alternating rhamnose and galacturonic acid residues.

rhamnogalacturonan II (RG II) A pectic polysaccharide with a complex structure including apiose residues that may be cross linked by borate esters.

rhizobia Collective term for the genera of soil bacteria that form symbiotic (mutualistic) relationships with members of the plant family Leguminosae.

rhizosphere The immediate microenvironment surrounding the root.

rib zone (RZ) Meristematic cells beneath the central zone in shoot apical meristems that give rise to the internal tissues of the stem.

ribosome The site of cellular protein synthesis and consisting of RNA and protein.

ribulose 5-phosphate In the pentose phosphate pathway, the initial five-carbon product of the oxidation of glucose 6-phosphate; in subsequent reactions, it is converted into sugars containing three to seven carbon atoms.

ribulose-1,5-bisphosphate carboxylase/oxygenase *See* rubisco.

Rieske iron–sulfur protein A protein subunit in the cytochrome b_6f complex, in which two iron atoms are bridged by two sulfur atoms, with two histidine and two cysteine ligands.

ripening The process that causes fruits to become more palatable, including softening, increasing sweetness, loss of acidity, and changes in coloration.

RNA interference (RNAi) pathway An RNA-dependent gene silencing process that is controlled by the RNA-induced silencing complex (RISC) and is initiated by short double-stranded RNA molecules in a cell's cytoplasm.

RNA polymerases A class of enzymes that bind to a gene and transcribe it into an RNA complementary to the DNA sequence.

RNA sequencing (RNA-seq) Technique to measure the abundance of all transcripts within a sample of RNA.

RNA-dependent RNA polymerases (RdRPs) A special class of RNA polymerases that convert single-stranded RNA into double-stranded RNA.

RNA-induced silencing complex (RISC) A multiprotein complex that incorporates one strand of a small interfering RNA (siRNA) or microRNA (miRNA). RISC complexes bind to and cleave mRNA, thereby preventing translation.

RNA-induced transcriptional silencing (RITS) Targeted inactivation of messenger RNA when a short interfering RNA sequence hybridizes to form a double-stranded hybrid.

root apical meristem (RAM) A group of cells at the tip of the root that retain the capacity to proliferate and whose ultimate fate remains undetermined.

root cap Cells at the root apex that cover and protect the meristematic cells from mechanical injury as the root moves through the soil. Site for the perception of gravity and signaling for the gravitropic response.

root hairs Microscopic extensions of root epidermal cells that greatly increase the surface area of the root, thus providing greater capacity for absorption of soil ions and, to a lesser extent, soil water.

root knot nematodes Plant parasites from the genus *Meloidogyne* found in tropical and subtropical soils. Root-knot nematode larvae infect plant roots to form root-knot galls and are a major cause of crop losses.

root meristem growth factors Small signaling peptides that participate in root hair formation.

root pressure A positive hydrostatic pressure in the xylem of roots.

root system architecture The overall geometric arrangement of roots comprising the plant's root system.

root The tissues descending, usually below ground, from the root–shoot junction and which anchor the plant and absorb and conduct water and minerals into the plant.

ROP (Rho-like from plants) GTPases A group of GTPases that participate in control of the cytoskeleton and vesicle trafficking.

ROP-interactive CRIB motif–containing proteins (RICs) Proteins that interact with ROP1 to regulate pollen tube growth and polarity.

ROS scavenging Detoxification of reactive oxygen species via interactions with proteins and electron acceptor molecules.

rough ER The endoplasmic reticulum to which ribsomes are attached.

rubisco-containing bodies Rubisco-containing vesicles that are thought to bud off from senescing chloroplasts, and are subsequently engulfed by autophagosomes and delivered to the vacuole for degradation.

rubisco The acronym for the chloroplast enzyme *ribu*lose *bis*phosphate *c*arboxylase/*o*xygenase. In a carboxylase reaction, rubisco uses atmospheric CO_2 and ribulose 1,5-bisphosphate to form two molecules of 3-phosphoglycerate. It also functions as an oxygenase that incorporates O_2 to ribulose 1,5-bisphosphate to yield one molecule of 3-phosphoglycerate and another of 2-phosphoglycolate. The competition between CO_2 and O_2 for ribulose 1,5-bisphosphate limits net CO_2 fixation.

S

S haplotype Multiple genes, inherited as a single segregating unit, that make up the S locus.

S phase In the cell cycle, the stage during which DNA is replicated; it follows G_1 and precedes G_2.

***S*-locus Cysteine-Rich protein (SCR)** A cysteine-rich protein located in the pollen coat that represents the male *S* determinant in the Brassicaceae.

***S*-locus receptor kinase (SRK)** A serine/threonine receptor kinase located in the plasma membrane of stigma cells that represents the female *S*-determinant in the Brassicaceae.

S-PHASE KINASE-ASSOCIATED PROTEIN1 (SKP1)/Cullin/F-box (SCF) complexes Large protein complexes that function as E3 ubiquitin ligases in the signaling pathways of several plant hormones.

salicylic acid A benzoic acid derivative believed to be an endogenous signal for systemic acquired resistance.

salt stress The adverse effects of excess minerals on plants.

salt-tolerant plants Plants that can survive or even thrive in high-salt soils. *See also* halophytes.

sclerenchyma Plant tissue composed of cells, often dead at maturity, with thick, lignified secondary cell walls. It functions in support of nongrowing regions of the plant.

scutellum The single cotyledon of the grass embryo, specialized for nutrient absorption from the endosperm.

seasonal leaf senescence The pattern of leaf senescence in deciduous trees in temperate climates in which all of the leaves undergo senescence and abscission in the autumn.

second messenger Intracellular molecule (e.g., cyclic AMP, cyclic GMP, calcium, IP_3, or diacylglycerol) whose production has been elicited by a systemic hormone (the primary messenger) binding to a receptor (often on the plasma membrane). Diffuses intracellularly to the target enzymes or intracellular receptor to produce and amplify the response.

secondary active transport Active transport that uses energy stored in the proton motive force or other ion gradient, and operates by symport or antiport.

secondary cell wall Cell wall synthesized by nongrowing cells. Often multilayered and containing lignin, it differs in composition and structure from the primary wall. Forms during cell differentiation after cell expansion ceases.

secondary cross-regulation Regulation by the output of one signal pathway of the abundance or perception of a second signal.

secondary dormancy Seeds that have lost their primary dormancy may become dormant again if exposed to unfavorable conditions that inhibit germination over a period of time.

secondary growth The tissue growth that occurs after elongation is complete. It involves the vascular cambium (producing the secondary xylem and phloem) and the cork cambium (producing the periderm).

secondary inflorescence meristems The inflorescence meristems that develop from the axillary buds of stem-borne leaves.

secondary metabolites Plant compounds that have no direct role in plant growth and development, but function as defenses against herbivores and microbial infection by microbial pathogens, attractants for pollinators and seed-dispersing animals, and as agents of plant–plant competition.

secondary phloem Phloem produced by the vascular cambium.

secondary plasmodesmata Plasmodesmata that form and permit symplastic transport between nonclonally related cells.

secondary response genes Genes whose expression requires protein synthesis and follows that of primary response genes.

secondary walls of woody tissues Thickened walls produced inside the primary cell wall; they are often lignified and play a structural role in supporting the weight of the stem.

secondary wall *See* secondary cell wall.

secondary xylem Xylem produced by the vascular cambium.

seed coat *See* testa.

seed dormancy The state in which a living seed will not germinate even if all the necessary environmental conditions for growth are met. Seed dormancy introduces a temporal delay in the germination process, providing additional time for seed dispersal.

seed longevity The length of time a seed can remain dormant without losing viability.

seed plants Plants in which the embryo is protected and nourished within a seed. The gymnosperms and angiosperms.

seedling establishment The stage that follows seed germination that includes the production of the first photosynthetic leaves and a minimal root system.

selective permeability Membrane property that allows diffusion of some molecules across the membrane to a different extent than other molecules.

self-incompatibility (SI) A general term for various genetic mechanisms in angiosperms for preventing self-fertilization and thus promoting outcrossing (not to be confused with temporal or anatomical mechanisms, such as heterostyly, that also reduce self-pollination).

seminal roots Lateral roots that develop from the embryonic root, or radicle.

senescence down-regulated genes (SDGs) Genes whose expression levels decline during leaf senescence.

senescence-associated genes (SAGs) Genes whose expression levels rise during leaf senescence.

senescence-associated proteins Proteins that are synthesized specifically during senescence.

senescence-associated vacuoles Small, protease-rich, acidic vacuoles that increase in number during senescence in the leaf mesophyll and guard cells, but not in the nongreen epidermal cells. Although they are distinct from rubisco-containing bodies, senescence-associated vacuoles contain rubisco and other stromal enzymes, which they are capable of degrading directly, independent of the autophagic machinery.

senescence An active, genetically controlled, developmental process in which cellular structures and macromolecules are broken down and translocated away from the senescing organ (typically leaves) to actively growing regions that serve as nutrient sinks. Initiated by environmental cues, regulated by hormones.

sensor proteins Specialized plant cellular receptor proteins that perceive external or internal signals. They consist of two domains, an *input domain*, which receives the environmental signal, and a *transmitter domain*, which transmits the signal to the response regulator.

separation layer A cell layer within the abscission zone whose cell walls break down, causing the organ (leaf or fruit) to be shed from the plant.

sequential leaf senescence The pattern of leaf senescence in which there is a gradient of senescence from the growing tip of the shoot to the oldest leaves at the base.

sessile leaf A leaf without a petiole, attached directly to the node via the blade.

shoot apex Consists of the shoot apical meristem plus the most recently formed leaf primordia (organs derived from the apical meristem).

shoot apical meristem (SAM) Meristem at the tip of a shoot. Consists of the terminal central zone (CZ), which contains slowly dividing, undetermined initials, and the flanking peripheral zone (PZ) and rib zone (RZ), in which derivatives of the CZ divide more rapidly and then differentiate.

shoots The aboveground tissues above the root–shoot junction and usually including the stem and leaves.

short interfering RNAs (siRNAs) RNAs that are structurally and functionally quite similar to miRNAs and also lead to the initiation of the RNA interference pathway.

short-day plant (SDP) A plant that flowers only in short days (qualitative SDP) or whose flowering is accelerated by short days (quantitative SDP).

short-distance transport Transport over a distance of only two or three cell diameters. Involved in phloem loading, when sugars move from the mesophyll to the vicinity of the smallest veins of the source leaf, and in phloem unloading, when sugars move from the veins to the sink cells.

short–long-day plants (SLDPs) Plants that flower only after a sequence of short days followed by long days.

siderophores Small molecules secreted by non-grass plants and some microbes to chelate iron, which is then taken up into cells at the root surface.

sieve cells The relatively unspecialized sieve elements of gymnosperms. Contrast with sieve tube elements.

sieve effect The penetration of photosynthetically active light through several layers of cells due to the gaps between chloroplasts permitting the passage of light.

sieve elements Cells of the phloem that conduct sugars and other organic materials throughout the plant. Refers to both sieve tube elements (angiosperms) and sieve cells (gymnosperms).

sieve plates Sieve areas found in angiosperm sieve tube elements; they have larger pores (sieve-plate pores) than other sieve areas and are generally found in end walls of sieve tube elements.

sieve tube elements The highly differentiated sieve elements typical of the angiosperms. Contrast with sieve cells.

sieve tube Tube formed by the joining together of individual sieve tube elements at their end walls.

signal peptide A hydrophobic sequence of 18 to 30 amino acid residues at the amino-terminal end of a chain; it is found on all secretory proteins and most integral membrane proteins and permits their transit across the membrane of the rough ER.

signal recognition particle (SRP) A ribonucleoprotein (protein-RNA complex) that recognizes and targets specific proteins to the endoplasmic reticulum in eukaryotes.

signal transduction pathway A sequence of processes by which an extracellular signal (typically light, a hormone, or a neurotransmitter) interacts with a receptor at the cell surface, causing a change in the level of a second messenger and ultimately a change in cell functioning.

silica cells In members of the grass family, specialized leaf epidermal cells that contain silica bodies.

simple leaf A leaf with one blade.

simple plasmodesmata Plasmodesmata that form a single, unbranched connection between two adjacent cells.

simple sequence repeats (SSR) One group of heterochromatic dispersed repeats that consist of sequences as short as two nucleotides repeated hundreds or even thousands of times. Also known as microsatellites.

singlet oxygen ($^1O_2$) An extremely reactive and damaging form of oxygen formed by reaction of excited chlorophyll with molecular oxygen. Causes damage to cellular components, especially lipids.

sink activity The rate of uptake of photosynthate per unit weight of sink tissue.

sink size The total weight of the sink.

sink strength The ability of a sink organ to mobilize assimilates toward itself. Depends on two factors: sink size and sink activity.

sink Any organ that imports photosynthate, including nonphotosynthetic organs and organs that do not produce enough photosynthetic products to support their own growth or storage needs, such as roots, tubers, developing fruits, and immature leaves. Contrast with source.

size exclusion limit (SEL) The restriction on the size of molecules that can be transported via the symplast. It is imposed by the width of the cytoplasmic sleeve that surrounds the desmotubule in the center of the plasmodesma.

skotomorphogenesis The developmental program plants follow when seeds are germinated and grown in the dark.

smooth ER The endoplasmic reticulum lacking attached ribosomes and usually consisting of tubules.

SNAREs A class of targeting recognition proteins for the selective fusion and fission of vesicles and tubules within the endomembrane system.

soil analysis The chemical determination of the nutrient content in a soil sample typically from the root zone.

soil hydraulic conductivity A measure of the ease with which water moves through the soil.

solar tracking The movement of leaf blades throughout the day so that the planar surface of the blade remains perpendicular to the sun's rays.

solute potential *See* osmotic potential.

solution culture The technique of growing plants without soil whereby their roots are immersed in a nutrient solution. *See also* hydroponics.

sorbitol A sugar alcohol formed by reduction of aldehyde glucose.

sorting out *See* vegetative segregation.

source Any exporting organ that is capable of producing photosynthetic products in excess of its own needs, such as a mature leaf or a storage organ. Contrast with sink.

spacing divisions Asymmetric divisions of the stomatal lineage ground cells that can give rise to secondary meristemoids during stomatal patterning.

specific heat Ratio of the heat capacity of a substance to the heat capacity of a reference substance, usually water. Heat capacity is the amount of heat needed to change the temperature of a unit mass by 1° C. The heat capacity of water is 1 calorie (4.184 Joule) per gram per degree Celsius.

spindle attachment checkpoint A point in the cell cycle (metaphase) where progression through the cycle is halted until the chromosomes are correctly attached to the mitotic spindle.

spindle formation The polymerization and alignment of a bipolar array of microtubules across the nucleus, as the nuclear envelope breaks down in plants. Clusters of the minus ends of two aligned arrays form the spindle poles on either side of the nucleus, while the plus ends are found in the mid-plane of the nucleus.

spines Sharp, stiff plant structures that physically deter herbivores and may assist in water conservation and are derived from leaves.

spokes (plasmodesmatal) Rod-like proteins of unknown nature that span the cytoplasmic sleeve in plasmodesmata.

spongy mesophyll Mesophyll cells of very irregular shape located below the palisade cells and surrounded by large air spaces.

spores Reproductive cells formed in plants by meiosis in the sporophyte generation. They give rise, without fusion with another cell, to the gametophyte generation.

sporophyte generation The stage, or generation, in the life cycle of plants that produces spores. It alternates with the gametophyte generation, in a process called alternation of generations.

sporophyte The diploid (2*N*) multicellular structure that produces haploid spores by meiosis.

sporophytic self-incompatibility (SSI) A type of self-incompatibility in which the pollen grain's incompatibility phenotype is determined by the diploid genome of the pollen parent—specifically, the tapetum of the anther.

SRP receptor A receptor protein on the ER membrane that binds to the ribosome-SRP complex, permitting the ribosome to dock with the translocon pore through which the elongating polypeptide will enter the lumen of the ER.

starch sheath A layer of cells that surrounds the vascular tissues of the shoot and coleoptile and is continuous with the root endodermis. Required for gravitropism in Arabidopsis shoots.

starch–statolith hypothesis Proposed mechanism for gravitropism involving sedimentation of statoliths in statocytes.

starch A polyglucan consisting of long chains of 1,4-linked glucose molecules and branch-points where 1,6-linkages are used. Starch is the carbohydrate storage form in most plants.

starchy endosperm A starch-storing triploid endosperm tissue that comprises the bulk of the seeds of cereals and other members of the grass familily.

statocytes Specialized gravity-sensing plant cells that contain statoliths.

statoliths Cellular inclusions such as amyloplasts that act as gravity sensors by having a high density relative to the cytosol and sedimenting to the bottom of the cell.

stele initials In the root, cells immediately above (proximal to) the quiescent center that give rise to the pericycle and vascular tissue.

stele In the root, the tissues located interior to the endodermis. The stele contains the vascular elements of the root: the phloem and the xylem.

stem The primary axis of the plant that is generally above ground, but anatomically similar belowground modified stems, such as rhizomes and corms, are common.

steroleosin A protein of the single lipid leaflet outer membrane of the oil body that has enzymatic activity (dehydrogenase) on sterols.

sterols A group of lipids containing four hydrocarbon rings that reside in plant membranes and modify membrane fluidity. Sitosterol and cholesterol are common sterols.

stigma/style cysteine-rich adhesin (SCA) A protein secreted by the transmitting tract of lily styles that is involved in the growth and adhesion of pollen tubes.

stipules Small, leaf-like appendages at the bases of leaves in many dicot species.

STOMAGEN A peptide produced by the mesophyll and released to the epidermis that acts as a positive regulator of stomatal density.

stomata (singular *stoma*) A microscopic pore in the leaf epidermis surrounded by a pair of guard cells and in some species, also including subsidiary cells. Stomata regulate the gas exchange (water and CO_2) of leaves by controlling the dimension of a stomatal pore.

stomatal complex The guard cells, subsidiary cells, and stomatal pore, which together regulate leaf transpiration.

stomatal lineage ground cell (SLGC) One of the two daughter cells of the dividing meristemoid mother cell during guard cell differentiation.

stomatal resistance A measurement of the limitation to the free diffusion of gases from and into the leaf posed by the stomatal pores. The inverse of stomatal conductance.

stratification In some plants, a cold temperature requirement for seed germination. The term is derived from the former practice of breaking dormancy by allowing seeds to overwinter in small mounds of alternating layers of seeds and soil.

stress relaxation Selective loosening of bonds between primary cell wall polymers, allowing the polymers to slip by each other, simultaneously increasing the wall surface area and reducing the physical stress in the wall.

stress-response regulons DNA regulatory sequences that function coordinately in stress responses.

stress Disadvantageous influences exerted on a plant by external abiotic or biotic factor(s), such as infection, or heat, water, and anoxia. Measured in relation to plant survival, crop yield, biomass accumulation, or CO_2 uptake.

strigolactones Carotenoid-derived plant hormones that inhibit shoot branching. They also play roles in the soil by stimulating the growth of arbuscular mycorrhizas and the germination of parasitic plant seeds, such as those of *Striga*, the source of their name.

stroma lamellae Unstacked thylakoid membranes within the chloroplast.

stroma The fluid component surrounding the thylakoid membranes of a chloroplast.

stromules Protrusions of the inner and outer membranes of the chloroplast.

subfunctionalization The process by which evolution acts on duplicate genes, causing one copy to be either lost or changed in function, while the other retains its original function.

subjective day When an organism is placed in total darkness, the phase of the rhythm that coincides with the light period of a preceding light/dark cycle. *See* subjective night.

subjective night When an organism is placed in total darkness, the phase of the rhythm that coincides with the dark period of a preceding light/dark cycle. *See* subjective day.

subsidiary cells Specialized epidermal cells that flank the guard cells and work with the guard cells in the control of stomatal apertures.

substrate-level phosphorylation A process that involves the direct transfer of a phosphate group from a substrate molecule to ADP to form ATP.

subtelomeric regions Regions of a chromosome just proximal to the telomeres.

Succinate dehydrogenase (complex II) A multi-subunit protein complex in the mitochondrial electron transport chain that catalyzes the oxidation of succinate and the reduction of ubiquinone.

Sucrose Non-Fermenting Related Kinase2 (SnRK2) A family of kinases that includes ABA-activated protein kinases or stress-activated protein kinases.

sucrose A disaccharide consisting of a glucose and a fructose molecule linked via an ether bond between C-1 on the glucosyl subunit and C-2 on the fructosyl unit. The full chemical name is α-D-glucopyranosyl-(1→2)-β-D-fructofuranoside. Sucrose is the carbohydrate transport form, (e.g., in the phloem between source and sink).

sugar–nucleotide polysaccharide glycosyltransferases A group of enzymes that synthesize the backbones of cell wall polysaccharides.

sunflecks Patches of sunlight that pass through openings in a forest canopy to the forest floor. Major source of incident radiation for plants growing under the forest canopy.

supercool The condition in which cellular water remains liquid because of its solute content, even at temperatures several degrees below its theoretical freezing point.

superoxide dismutase An enzyme that converts superoxide radicals to hydrogen peroxide.

surface tension A force exerted by water molecules at the air–water interface, resulting from the cohesion and adhesion properties of water molecules. This force minimizes the surface area of the air–water interface.

suspensor In seed plant embryogenesis, the structure that develops from the basal cell following the first division of the zygote. Supports, but is not part of, the embryo that develops from the apical cell and hypophysis.

symbiosis The close association of two organisms in a relationship that may or may not be mutually beneficial. Often applied to beneficial (mutualistic) relationships. *See* mutualism.

symplast The continuous system of cell protoplasts interconnected by plasmodesmata.

symplastic transport The intercellular transport of water and solutes through plasmodesmata.

symport A type of secondary active transport in which two substances are moved in the same direction across a membrane.

synaptonemal complex A protein structure that forms between homologous chromosomes during prophase I of meiosis.

syncytium A multinucleate cell that can result from multiple cell fusions of uninuclear cells, usually in response to viral infection.

synergid cells Two cells adjacent to the egg cell of the embryo sac, one of which is penetrated by the pollen tube upon entry into the ovule.

systemic acquired acclimation (SAA) A photoprotective system wherein leaves exposed to higher light levels transmit a signal to preacclimate shaded leaves.

systemic acquired resistance (SAR) The increased resistance throughout a plant to a range of pathogens following infection by a pathogen at one site.

systems biology An approach to examining complex living processes that employs mathematical and computational models to simulate nonlinear biological networks and to better predict their operation.

T

tandem repeats Heterochromatic structures that consist of highly repetitive DNA sequences.

tapetum A layer of secretory cells surrounding the locule of the anther that contribute to the formation of the pollen cell wall.

taproot The main single root axis from which lateral roots develop.

TATA box Located about 25 to 35 bp upstream of the transcriptional start site, this short sequence TATAAA(A) serves as the assembly site for the transcription initiation complex.

telomerase An enzyme that repairs the ends of chromosomes after cell division and prevents them from shortening.

telomere shortening The process by which the telomere (a region of DNA at the end of a chromosome that protects the start of the genetic coding sequence against degradation) undergoes shortening.

telomeres Regions of repetitive DNA that form the chromosome ends and protect them from degradation.

telophase Prior to cytokinesis, the final stage of mitosis (or meiosis) during which the chromatin decondenses, the nuclear envelope re-forms, and the cell plate extends.

temperature coefficient (Q_{10}) The increase in the rate of a process (e.g., respiration) for every 10°C increase in temperature.

temperature compensation A characteristic of circadian rhythms, which can maintain their circadian periodicity over a broad range of temperatures within the physiological range.

tensile strength The ability to resist a pulling force. Water has a high tensile strength.

tension wood A type of reaction wood found in arborescent dicots that forms on the upper side of leaning or horizontal stems or branches.

tension Negative hydrostatic pressure.

tertiary cross-regulation Regulation that involves the outputs of two distinct signaling pathways exerting influences on one another.

testa The outer layer of the seed, derived from the integument of the ovule.

tetrad A pair of replicated homologous chromosomes that are synapsed. Consists of four chromatids.

TFIIB recognition element (BRE) A conserved DNA binding sequence within the core promoter that helps in the regulation of the transcriptional activity of a given gene.

thigmotropism Plant growth in response to touch, enabling roots to grow around rocks and shoots of climbing plants to wrap around structures for support.

thioredoxin A small, ubiquitous protein (approximately 12 kDa) whose active site cysteines participate in thiol-disulfide exchange reactions.

thorns Sharp plant structures that physically deter herbivores and are derived from branches.

thylakoid reactions The chemical reactions of photosynthesis that occur in the specialized internal membranes of the chloroplast (called thylakoids). Include photosynthetic electron transport and ATP synthesis.

thylakoids The specialized, internal, chlorophyll containing membranes of the chloroplast where light absorption and the chemical reactions of photosynthesis take place.

tip growth Localized growth at the tip of a plant cell caused by localized secretion of new wall polymers. Occurs in pollen tubes, root hairs, some sclerenchyma fibers, and cotton fibers, as well as moss protonema and fungal hyphae.

toggle model A model for the conformational change of phytochrome from the Pr to the Pfr form in which chromophore rotation causes the β-hairpin to become helical and exert a tug on the helical spine.

tonoplast The vacuolar membrane.

torpedo stage The third stage of embryogenesis. The structure produced by elongation of the axis of the heart stage embryo and further development of the cotyledons. *See also* globular stage, heart stage.

torus A central thickening found in the pit membranes of tracheids in the xylem of most gymnosperms.

toxic zone The range of nutrient concentrations in excess of the adequate zone and where growth or yield declines.

tracheary elements Water-transporting cells of the xylem.

tracheids Spindle-shaped, water-conducting cells with tapered ends and pitted walls without perforations found in the xylem of both angiosperms and gymnosperms.

tracheophyte *See* vascular plant.

***trans* Golgi network (TGN)** The tubulo-vesicular network that derives from the sloughing *trans* cisternae of the Golgi. It is separate from the early recycling endosome, which is also called the partially coated reticulum in plants.

***trans*-acting factors** The transcription factors that bind to the *cis*-acting sequences.

transcription initiation complex A multiprotein complex of transcription factors required for binding RNA polymerase and initiating transcription.

transcription The process by which the base sequence information in DNA is copied into an RNA molecule.

transcriptional regulation The level of regulation that determines if and when RNA is transcribed from DNA.

transcriptome The entire complement of RNA expressed by a cell, tissue, or organism at a certain time. Includes mRNA, tRNA, rRNA, and any other non-coding RNA.

transcriptomics The study of transcriptomes.

transcytosis Redirection of a secreted protein from one membrane domain within a cell to another polarized domain.

transfer cell A type of companion cell similar to an ordinary companion cell, but with fingerlike projections of the cell wall that greatly increase the surface area of the plasma membrane and increase the capacity for solute transport across the membrane from the apoplast.

transgene A foreign or altered gene that has been inserted into a cell or organism.

transgenic A plant expressing a foreign gene introduced by genetic engineering techniques.

transit peptide An N-terminal amino acid sequence that facilitates the passage of a precursor protein through both the outer and the inner membranes of an organelle such as the chloroplast. The transit peptide is then clipped off.

translation The process whereby a specific protein is synthesized according to the sequence information encoded by the mRNA.

translocation velocity The rate of movement of materials in the phloem sieve elements expressed as the linear distance traveled per unit time.

translocation (1) In protein synthesis, the movement of protein from its site of synthesis (cytoplasm) into the membrane or lumen of an organelle. (2) The movement of photosynthate from sources to sinks in the phloem.

translocons Protein-lined channels in the rough endoplasmic reticulum that form associations with SRP receptors and enable proteins synthesized on ribosomes to enter the ER lumen.

transmitting tract The path of pollen tube growth from the stigma to the micropyle of the ovary.

transpiration ratio The ratio of water loss to photosynthetic carbon gain. Measures the effectiveness of plants in moderating water loss while allowing sufficient CO_2 uptake for photosynthesis.

transpiration The evaporation of water from the surface of leaves and stems.

transport phloem Sieve elements of the connecting pathway.

transport proteins Transmembrane proteins that are involved in the movement of molecules or ions from one side of a membrane to the other side.

transport Molecular or ionic movement from one location to another; may involve crossing a diffusion barrier such as one or more membranes.

transposase An enzyme that catalyzes the movement of a DNA sequence from one site to a different site in the DNA molecule.

transposon tagging The technique of inserting a transposon into a gene and thereby marking that gene with a known DNA sequence.

transposons (transposable elements) DNA elements that can move or be copied from one site in the genome to another site.

treadmilling During interphase, a process by which microtubules in the cortical cytoplasm appear to migrate around the cell periphery due to addition of tubulin heterodimers to the plus end at the same rate as their removal from the minus end.

triacylglycerols Three fatty acyl groups in ester linkage to the three hydroxyl groups of glycerol. Fats and oils.

trichoblasts Root epidermal cells that have the capacity to differentiate into root hairs.

trichomes Unicellular or multicellular hairlike structures that differentiate from the epidermal cells of shoots and roots. Trichomes may be structural or glandular and function in biotic or abiotic plant responses.

triglycerides Three fatty acyl groups in ester linkage to three hydroxyl groups of glycerol. Fats and oils.

triose phosphates A group of three-carbon sugar phosphates.

triploid bridge An often unstable, transitory genomic state between a diploid and tetraploid, formed by the fusion of a typical haploid with an unreduced diploid gamete.

trisomy One kind of aneuploidy in which there are three copies of one type of chromosome rather than the normal two.

tropism Oriented plant growth in response to a perceived directional stimulus from light, gravity, or touch.

tryphine A sticky, adhesive substance rich in proteins, fatty acids, waxes, and other hydrocarbons that coats the exine layer of pollen cell walls.

tube cell *See* vegetative cell.

tubulin A family of cytoskeletal GTP-binding proteins with three members, α-, β-, and γ-tubulin. α-tubulin forms heterodimers with β-tubulin which polymerize to form microtubules. β-tubulin is exposed at the growing plus end and undergoes GTP hydrolysis, while GTP is not hydrolyzed in α-tubulin. The initiation of microtubules is mediated by γ-tubulin, which forms a ring-shaped "primer" for the construction of the microtubule at the minus end of the microtubule.

tunica The outer cell layers of the shoot apical meristem. The outermost tunica layer generates the shoot epidermis.

turgor pressure Force per unit area in a liquid. In a plant cell, the turgor pressure pushes the plasma membrane against the rigid cell wall and provides a force for cell expansion.

turnover The balance between the rate of synthesis and the rate of degradation, usually applied to protein or RNA. An increase in turnover typically refers to an increase in degradation.

two-component regulatory systems Signaling pathways common in prokaryotes. They typically involve a membrane-bound histidine kinase sensor protein that senses environmental signals and a response regulator protein that mediates the response. Although rare in eukaryotes, two-component systems are involved in both ethylene and cytokinin signaling.

type-1 arabinogalactan A pectic polysaccharide with a (1,4)-β-D-galactan backbone decorated with single arabinose residues.

type-A *ARR* In Arabidopsis, genes that encode response regulators that are made up solely of a receiver domain.

type-B *ARR* In Arabidopsis, genes that encode response regulators that have an output domain in addition to a receiver domain.

U

ubiquinone A mobile electron carrier of the mitochondrial electron transport chain. Chemically and functionally similar to plastoquinone in the photosynthetic electron transport chain.

ubiquitin ligase (E3) An SCF complex that is part of the ubiquitination pathway. Binds to proteins destined for degradation. Lysine residues on E3 receive ubiquitin from the conjugate of ubiquitin-activating enzyme (E2) and ubiquitin.

ubiquitin-activating enzyme (E1) Part of the ubiquitination pathway. Initiates ubiquitination by catalyzing the ATP-dependent adenylylation of the C terminus of ubiquitin.

ubiquitin-conjugating enzyme (E2) Part of the ubiquitination pathway. A cysteine residue on E2 receives the adenylylated ubiquitin produced by ubiquitin-activating enzyme, E1.

ubiquitin–proteasome pathway Mechanism for the specific degradation of cellular proteins involving two discrete steps: the polyubiquitination of proteins via the E3 ubiqutin ligase and the degradation of the tagged protein by the 26S proteasome.

ubiquitin A small polypeptide that is covalently attached to proteins by the enzyme ubiquitin ligase using energy from ATP, and that serves as a recognition site for a large proteolytic complex, the proteasome.

uncoupler A chemical compound that increases the proton permeability of membranes and thus uncouples the formation of the proton gradient from ATP synthesis.

uncoupling protein A protein that increases the proton permeability of the inner mitochondrial membrane and thereby decreases energy conservation.

uncoupling A process by which coupled reactions are separated in such a way that the free energy released by one reaction is not available to drive the other reaction.

uniparental inheritance Form of inheritance shown by both mitochondria and plastids, meaning that offspring of sexual reproduction (via pollen and eggs) inherit organelles from only one parent.

unreduced gametes Gametes that have the same number of chromosome sets as the parent cell.

UV RESISTANCE LOCUS 8 (UVR8) The protein receptor that mediates various plant responses to UV-B irradiation.

V

vacuolar H^+-ATPase (V-ATPase) A large, multi-subunit enzyme complex, related to F_oF_1-ATPases, present in endomembranes (tonoplast, Golgi). Acidifies the vacuole and provides the proton motive force for the secondary transport of a variety of solutes into the lumen. V-ATPases also function in the regulation of intracellular protein trafficking.

vacuolar sap The fluid contents of a vacuole, which may include water, inorganic ions, sugars, organic acids, and pigments.

vacuolar-type PCD The type of programmed cell death associated with developmental senescence in plant cells in which the vacuole breaks down, releasing various hydrolases into the cytoplasm.

variegation A condition in which leaves show patterns of white and green. Produced by vegetative segregation and may be due to mutations in nuclear, mitochondrial, or chloroplast genes.

vascular cambium A lateral meristem consisting of fusiform and ray stem cells, giving rise to secondary xylem and phloem elements, as well as ray parenchyma.

vascular plant A plant that has xylem and phloem.

vascular tissues Plant tissues specialized for the transport of water (xylem) and photosynthetic products (phloem).

vegetative cell One of two cells produced by the division of the microspore nucleus during microgametogenesis in angiosperm pollen grains. After engulfing the generative cell, the vegetative cell gives rise to the pollen tube following pollination.

vegetative segregation A major consequence of organellar inheritance (choroplasts and mitochondria) is that a vegetative cell (as opposed to a gamete) can give rise to another vegetative cell via mitosis that is genetically different because one daughter cell may receive organelles with one type of genome, while the other receives organelles with different genetic information.

venation pattern The pattern of veins of a leaf.

vernalization In some species, the cold temperature requirement for flowering. The term is derived from the word for "spring."

vessel elements Nonliving water-conducting cells with perforated end walls found only in angiosperms and a small group of gymnosperms.

vessel A stack of two or more vessel elements in the xylem.

villin An actin binding protein that bundles F-actin filaments.

viscoelastic (rheological, flow) properties Properties that are intermediate between those of a solid and those of a liquid and combine viscous and elastic behavior.

vivipary The precocious germination of seeds in the fruit while still attached to the plant.

volicitin A volatile compound produced by beet armyworm (*Spodoptera exigua*) when feeding on host grasses that attracts the generalist parasitoid wasp *Cotesia marginiventris*.

W

wall collars The callose-containing collars around a plasmodesma that can change size and thereby affect the size exclusion limit.

wall extensibility During primary cell wall expansion, the coefficient that relates growth rate to the turgor pressure that is in excess of the yield threshold.

water potential (Ψ) A measure of the free energy associated with water per unit volume (J m^{-3}). These units are equivalent to pressure units such as pascals. Ψ is a function of the solute potential, the pressure potential, and the gravitational potential: $\Psi = \Psi_s + \Psi_p + \Psi_g$. The term Ψ_g is often ignored because it is negligible for heights under five meters.

wavelength (λ) A unit of measurement for charactering light energy.

The distance between successive wave crests. In the visible spectrum, it corresponds to a color.

whole plant senescence The death of the entire plant, as opposed to the death of individual cells, tissues, or organs.

whorl Pertaining to the concentric pattern of a set of organs that are initiated around the flanks of the meristem.

wilting Loss of rigidity, leading to a flaccid state, due to turgor pressure falling to zero.

wound callose Callose deposited in the sieve pores of damaged sieve elements that seals them off from surrounding intact tissue. As sieve elements recover, the callose disappears from pores.

X

xanthophylls Carotenoids involved in nonphotochemical quenching. The xanthophyll zeaxanthin is associated with the quenched state of photosystem II, and violaxanthin is associated with the unquenched state.

xylan A polymer of the five-carbon sugar xylose.

xylem loading The process whereby ions exit the symplast and enter the conducting cells of the xylem.

xylem The vascular tissue that transports water and ions from the root to the other parts of the plant.

xylogen A proteoglycan-like factor that stimulates xylem differentiation in zinnia suspension cultured cells and is concentrated at the apical ends of the cell walls of differentiating zinnia tracheary elements.

xyloglucan endotransglucosylase (XET) *See* xyloglucan endotransglucosylase/hydrolases.

xyloglucan endotransglucosylase/hydrolases (XTHs) A large family of enzymes, including xyloglucan endotransglucosylase (XET), that have the ability to cut the backbone of a xyloglucan in the cell wall and join one end of the cut chain with the free end of an acceptor xyloglucan.

xyloglucan A hemicellulose with a backbone of 1→4-linked β-D-glucose residues and short side chains that contain xylose, galactose, and sometimes fucose. It is the most abundant hemicellulose in the primary walls of most plants (in grasses it is present, but less abundant).

Y

yield threshold The minimum value for turgor pressure at which measurable extension of the cell wall begins.

yielding properties of the cell wall The capacity of the cell wall to loosen and irreversibly stretch in different ways in response to different internal and external factors.

yielding Long-term irreversible stretching that is characteristic of growing (expanding) cell walls. Nearly lacking in nonexpanding walls.

Z

zeitgebers Environmental signals such as light-to-dark or dark-to-light transitions that synchronize the endogenous oscillator to a 24-hour periodicity.

ZEITLUPE (ZTL) A blue light photoreceptor that regulates day length perception (photoperiodism) and circadian rhythms.

zonation Regional cytological differences in cell division in the shoot apical meristems of seed plants.

zygotic stage The single-celled product of the union of an egg and a sperm.

Illustration Credits

CHAPTER 1

Figure 1.4 Robinson-Beers, K., and Evert, R. F. (1991) Fine structure of plasmodesmata in mature leaves of sugar cane. *Planta* 184: 307–318; Bell, K., and Oparka, K. (2011) Imaging plasmodesmata. *Protoplasma* 248: 9–25; Fitzgibbon, J., Beck, M., Zhou, J., Faulkner, C., Robatzek, S., and Oparka, K. (2013) A developmental framework for complex plasmodesmata formation revealed by large-scale imaging of the *Arabidopsis* leaf epidermis. *Plant Cell* 25: 57–70; Ueki, S., and Citovsky, V. (2011) To gate, or not to gate: regulatory mechanisms for intercellular protein transport and virus movement in plants. *Mol. Plant* 4: 782–793. **Figure 1.7** Buchanan, B. B., Gruissem, W., and Jones, R. L., eds. (2000) *Biochemistry and Molecular Biology of Plants.* American Society of Plant Biologists, Rockville, MD. **Figure 1.8** Fiserova, J., Kiseleva, E., and Goldberg, M. W. (2009) Nuclear envelope and nuclear pore complex structure and organization in tobacco BY-2 cells. *Plant J.* 59: 243–255. **Figure 1.9** Alberts, B., Johnson, A., Lewis, J., Raff, M., Roberts, K., and Walter, P. (2002) *Molecular Biology of the Cell.* 4th ed. Garland Science, New York. **Figure 1.10** Idziak, D., Betekhtin, A., Wolny, E., Lesniewska, K., Wrigth, J., Febrer, M., Bevan, M. Jenkins, G., and Hasterok, R. (2011) Painting the chromosomes of *Brachypodium* – current status and future prospects. *Chromosoma* 120: 469–479. **Figures 1.13 & 1.15** Gunning, B. E. S., and Steer, M. W. (1996) *Plant Cell Biology: Structure and Function of Plant Cells.* Jones and Bartlett, Boston. **Figure 1.18** Huang, A. H. C. (1987) Lipases. In *The Biochemistry of Plants: A Comprehensive Treatise,* Vol. 9: *Lipids: Structure and Function,* P. K. Stumpf, ed., Academic Press, New York, pp. 91–119; Buchanan, B. B., Gruissem, W., and Jones, R. L., eds. (2000) *Biochemistry and Molecular Biology of Plants.* American Society of Plant Physiologists, Rockville, MD. **Figures 1.22 & 1.23** Gunning, B. E. S., and Steer, M. W. (1996) *Plant Cell Biology: Structure and Function of Plant Cells.* Jones and Bartlett, Boston. **Figure 1.30** Xu, X. M., Meulia, T., and Meier, I. (2007) Anchorage of plant RanGAP to the nuclear envelope involves novel nuclear-pore-associated proteins. *Curr. Biol.* 17: 1157–1163; Higaki, T., Kutsuna, N., Sano, T., and Hasezawa, S. (2008) Quantitative analysis of changes in actin microfilament contribution to cell plate development in plant cytokinesis. *BMC Plant Biol.* 8: 80. **Figure 1.31** Seguí-Simarro, J. M., Austin, J. R., White, E. A., and Staehelin, L. A. (2004) Electron tomographic analysis of somatic cell plate formation in meristematic cells of *Arabidopsis* preserved by high-pressure freezing. *Plant Cell* 16: 836–856. **Figures 1.32–1.35** Gunning, B.E.S. (2009) *Plant Cell Biology on DVD: Information for students and a resource for teachers.* Springer, New York (www.springer.com/life+sciences/plant+sciences/book/978-3-642-03690-3). **Figures 1.35 & 1.36** St-Pierre, B., Vazquez-Flota, F. A., and De Luca, V. (1999) Multicellular compartmentation of *Catharanthus roseus* alkaloid biosynthesis predicts intercellular translocation of a pathway intermediate. *Plant Cell* 11: 887–900. **Figure 1.36** Leroux, O. (2012) Collenchyma: a versatile mechanical tissue with dynamic cell walls. *Ann. Bot.* 110: 1083–1098; Rudall, P. J. (1987) Laticifers in Euphorbiaceae – a conspectus. *Bot. J. Linn. Soc.* 94: 143–163. **Figure 1.37** Zhang, W., Wang, X.-Q., and Li, Z.-Y. (2011) The protective schell: Schlereids and their mechanical function in corollas of some species of *Camellia* (Theaceae). *Plant Biol.* 13: 688–692; Kaneda, M., Rensing, K., and Samuels, L. (2010) Secondary cell wall deposition in developing secondary xylem of poplar. *J. Integr. Plant Biol.* 52: 234–243. **Figure 1.38** Froelich, D. R., Mullendore, D. L., Jensen, K. H., Ross-Elliott, T. J., Anstead, J. A., Thompson, G. A., Pelissier, H. C., and Knoblach, M. (2011) Phloem ultrastructure and pressure flow: Sieve-element-occlusion-related agglomerations do not affect translocation. *Plant Cell* 23: 4428–4445. **Figure 1.39** Wightman, R. and Turner, S. (2008) The roles of the cytoskeleton during cellulose deposition at the secondary wall. *Plant J.* 54: 794–805; Samuels, A. L., Rensing, K. H., Douglas, C. J., Mansfield, S. D., Dharmawarhana, D. P., and Ellis, B. E. (2002) Cellular machinery of wood production: differentiation of secondary xylem in *Pinus contorta* var. *latifolia*. *Planta* 216: 72–82.

CHAPTER 2

Figure 2.1 Kato, A., Lamb, J. C., and Birchler, J. A. (2004) Chromosome painting using repetitive DNA sequences as probes for somatic chromosome identification in maize. *Proc. Natl. Acad. Sci. USA* 101: 13554–13559. **Figure 2.3** Miura, A., Yonebayashi, S., Watanabe, K., Toyama, T., Shimadak, H., and Kakutani, T. (2001) Mobilization of transposons by a mutation abolishing full DNA methylation in *Arabidopsis*. *Nature* 411: 212–214. **Figure 2.4** Tiang, C. L., He, Y., and Pawlowski, W. P. (2012) Chromosome organization and dynamics during interphase, mitosis, and meiosis in plants. *Plant Physiol.* 158: 26–34. **Figures 2.5 & 2.6** Ma, H. (2005) Molecular genetic analyses of microsporogenesis and microgametogenesis in flowering plants. *Annu. Rev. Plant Biol.* 56: 393–434. **Figure 2.5** Grandont, L., Jenczewski, E., and Lloyd, A. (2013) Meiosis and its deviations in polyploid plants. *Cytogenet. Genome Res.* 140:

171–84. **Figure 2.7** Bomblies, K., and Madlung, A. (2014) Polyploidy in the *Arabidopsis* genus. *Chromosome Res.* 22: 117–134. **Figure 2.10** Comai, L. (2005) The advantages and disadvantages of being polyploid. *Nat. Rev. Genet.* 6: 836–846.

CHAPTER 3

Figure 3.1 Day, W., Legg, B. J., French, B. K., Johnston, A. E., Lawlor, D. W., and Jeffers, W. de C. (1978) A drought experiment using mobile shelters: The effect of drought on barley yield, water use and nutrient uptake. *J. Agric. Sci.* 91: 599–623; Innes, P., and Blackwell, R. D. (1981) The effect of drought on the water use and yield of two spring wheat genotypes. *J. Agric. Sci.* 102: 341–351; Jones, H. G. (1992) *Plants and Microclimate*, 2nd ed., Cambridge University Press, Cambridge. **Figure 3.2** Whittaker, R. H. (1970) *Communities and Ecosystems*. Macmillan, New York. **Figure 3.8** Nobel, P. S. (1999) *Physicochemical and Environmental Plant Physiology*, 2nd ed. Academic Press, San Diego, CA. **Figure 3.11** Hsiao, T. C., and Xu, L. K. (2000) Sensitivity of growth of roots versus leaves to water stress: Biophysical analysis and relation to water transport. *J. Exp. Bot.* 51: 1595–1616. **Figure 3.15** Hsiao, T. C., and Acevedo, E. (1974) Plant responses to water deficits, efficiency, and drought resistance. *Agr. Meteorol.* 14: 59–84.

CHAPTER 4

Table 4.1 Nobel, P. S. (1999) *Physicochemical and Environmental Plant Physiology*, 2nd ed. Academic Press, San Diego, CA. **Figure 4.3** Kramer, P. J., and Boyer, J. S. (1995) *Water Relations of Plants and Soils.* Academic Press, San Diego, CA. **Figure 4.6** Zimmermann, M. H. (1983) *Xylem Structure and the Ascent of Sap*. Springer, Berlin. **Figure 4.8** Gunning, B. S., and Steer, M. M. (1996) *Plant Cell Biology: Structure and Function*. Jones and Bartlett Publishers, Boston. **Figure 4.9** Sperry, J. S. (2000) Hydraulic constraints on plant gas exchange. *Agric. For. Meteorol.* 104: 13–23. **Figure 4.11** Bange, G. G. J. (1953) On the quantitative explanation of stomatal transpiration. *Acta Bot. Neerl.* 2: 255–296. **Figure 4.12** Zeiger, E., and Hepler, P. K. (1976) Production of guard cell protoplasts from onion and tobacco. *Plant Physiol.* 58: 492–498. **Figure 4.13** Palevitz, B. A. (1981) The structure and development of guard cells. In *Stomatal Physiology*, P. G. Jarvis and T. A. Mansfield, eds., Cambridge University Press, Cambridge, pp. 1–23; Sack, F. D. (1987) The development and structure of stomata. In *Stomatal Function*, E. Zeiger, G. Farquhar, and I. Cowan, eds., Stanford University Press, Stanford, CA, pp. 59–90; Meidner, H., and Mansfield, D. (1968) *Stomatal Physiology.* McGraw-Hill, London. **Figure 4.14** Franks, P. J., and Farquhar, G. D. (2007) The mechanical diversity of stomata and its significance in gas-exchange control. *Plant Physiol.* 143: 78–87.

CHAPTER 5

Table 5.1 Epstein, E. (1999) Silicon. *Annu. Rev. Plant Physiol. Plant Mol. Biol.* 50: 641–664; Epstein, E. (1972) *Mineral Nutrition of Plants: Principles and Perspectives.* John Wiley and Sons, New York. **Table 5.2** Evans, H. J., and Sorger, G. J. (1966) Role of mineral elements with emphasis on the univalent cations. *Annu. Rev. Plant Physiol.* 17: 47–76; Mengel, K., and Kirkby, E. A. (2001) *Principles of Plant Nutrition*, 5th ed. Kluwer Academic Publishers, Dordrecht, Netherlands. **Table 5.3** Epstein, E., and Bloom, A. J. (2005) *Mineral Nutrition of Plants: Principles and Perspectives*, 2nd ed. Sinauer Associates, Sunderland, MA. **Table 5.5** Brady, N. C. (1974) *The Nature and Properties of Soils*, 8th ed. Macmillan, New York. **Figure 5.2** Epstein, E., and Bloom, A. J. (2005) *Mineral Nutrition of Plants: Principles and Perspectives*, 2nd ed. Sinauer Associates, Sunderland, MA. **Figure 5.3** Sievers, R. E., and Bailar, J. C., Jr. (1962) Some metal chelates of ethylenediaminetetraacetic acid, diethylenetriaminepentaacetic acid, and triethylenetriaminehexaacetic acid. *Inorg. Chem.* 1: 174–182. **Figure 5.5** Lucas, R. E., and Davis, J. F. (1961) Relationships between pH values of organic soils and availabilities of 12 plant nutrients. *Soil Sci.* 92: 177–182. **Figures 5.7 & 5.8** Weaver, J. E. (1926) *Root Development of Field Crops.* McGraw-Hill, New York. **Figure 5.10** Mengel, K., and Kirkby, E. A. (2001) *Principles of Plant Nutrition*, 5th ed. Kluwer Academic Publishers, Dordrecht, Netherlands. **Figure 5.11** Bloom, A. J., Jackson, L. E., and Smart, D. R. (1993) Root growth as a function of ammonium and nitrate in the root zone. *Plant Cell Environ.* 16: 199–206. **Figure 5.12** Smith, S. E., and Read, D. J. (2008) *Mycorrhizal Symbiosis*, 3rd ed. Academic Press, Amsterdam, Boston. **Figure 5.14** Rovira, A. D., Bowen, C. D., and Foster, R. C. (1983) The significance of rhizosphere microflora and mycorrhizas in plant nutrition. In *Encyclopedia of Plant Physiology*, New Series, Vol. 15B: *Inorganic Plant Nutrition*, A. Läuchli and R. L. Bieleski, eds., Springer, Berlin, pp. 61–93.

CHAPTER 6

Table 6.1 Higinbotham, N., Etherton, B., and Foster, R. J. (1967) Mineral ion contents and cell transmembrane electropotentials of pea and oat seedling tissue. *Plant Physiol.* 42: 37–46. **Figure 6.4** Higinbotham, N., Graves, J. S., and Davis, R. F. (1970) Evidence for an electrogenic ion transport pump in cells of higher plants. *J. Membr. Biol.* 3: 210–222. **Figure 6.7** Buchanan, B. B., Gruissem, W., and Jones, R. L., eds. (2000) *Biochemistry and Molecular Biology of Plants*. American Society of Plant Physiologists, Rockville, MD; Leng, Q., Mercier, R. W., Hua, B. G., Fromm, H., and Berkowitz, G. A. (2002) Electrophysiological analysis of cloned cyclic nucleotide-gated ion channels. *Plant Physiol.* 128: 400–410. **Figure 6.12** Lin, W., Schmitt, M. R., Hitz, W. D., and Giaquinta, R. T. (1984) Sugar transport into protoplasts isolated from developing soybean cotyledons. *Plant Physiol.* 75: 936–940. **Figure 6.14** Lebaudy, A., Véry, A., and Sentenac, H. (2007) K^+ channel activity in plants: Genes, regulations and functions. *FEBS Lett.* 581: 2357–2366; Very, A. A., and Sentenac, H. (2002) Cation channels in the *Arabidopsis* plasma membrane. *Trends Plant Sci.* 7: 168–175. **Figure 6.16** Palmgren, M. G. (2001) Plant plasma membrane H^+-ATPases: Powerhouses for nutrient uptake. *Annu. Rev. Plant Physiol. Plant Mol. Biol.* 52: 817–845. **Figure 6.17** Kluge, C., Lahr, J., Hanitzsch, M., Bolte, S., Golldack, D., and Dietz, K. J. (2003) New insight into the structure and regulation of the plant vacuolar H^+-ATPase. *J. Bioenerg. Biomembr.* 35: 377–388.

CHAPTER 7

Figure 7.15 Becker, W. M. (1986) *The World of the Cell.* Benjamin/Cummings, Menlo Park, CA. **Figure 7.16** Allen, J. F., and Forsberg, J. (2001) Molecular recognition in thylakoid structure and function. *Trends Plant Sci.* 6: 317–326; Nelson, N., and Ben-Shem, A. (2004) The complex architecture of oxygenic photosynthesis. *Nat. Rev. Mol. Cell Biol.* 5: 971–982. **Figure 7.18** Barros, T., and

Kühlbrandt, W. (2009) Crystallisation, structure and function of plant light-harvesting Complex II. *Biochim. Biophys. Acta* 1787: 753–772. **Figures 7.19 & 7.21** Blankenship, R. E., and Prince, R. C. (1985) Excited-state redox potentials and the Z scheme of photosynthesis. *Trends Biochem. Sci.* 10: 382–383. **Figure 7.22** Barber, J., Nield, N., Morris, E. P., and Hankamer, B. (1999) Subunit positioning in photosystem II revisited. *Trends Biochem. Sci.* 24: 43–45. **Figure 7.23** Ferreira, K. N., Iverson, T. M., Maghlaoui, K., Barber, J., and Iwata, S. (2004) Architecture of the photosynthetic oxygen-evolving center. *Science* 303: 1831–1838; Umena, Y., Kawakami, K., Shen, J.-R., and Kamiya, N. (2011) Crystal structure of oxygen-evolving photosystem II at a resolution of 1.9 Å. *Nature* 473: 55–60. **Figure 7.25** Kurisu, G., Zhang, H. M., Smith, J. L., and Cramer, W. A. (2003) Structure of cytochrome b_6f complex of oxygenic photosynthesis: tuning the cavity. *Science* 302: 1009–1014. **Figure 7.27** Buchanan, B. B., Gruissem, W., and Jones, R. L., eds. (2000) *Biochemistry and Molecular Biology of Plants.* American Society of Plant Physiologists, Rockville, MD; Nelson, N., and Ben-Shem, A. (2004) The complex architecture of oxygenic photosynthesis. *Nat. Rev. Mol. Cell Biol.* 5: 971–982. **Figure 7.29** Jagendorf, A. T. (1967) Acid-based transitions and phosphorylation by chloroplasts. *Fed. Proc. Am. Soc. Exp. Biol.* 26: 1361–1369. **Figure 7.32** Asada, K. (1999) The water–water cycle in chloroplasts: scavenging of active oxygens and dissipation of excess photons. *Annu. Rev. Plant Physiol. Plant Mol. Biol.* 50: 601–639.

CHAPTER 8

Figure 8.12 Edwards, G. E., Franceschi, V. R., and Voznesenskaya, E. V. (2004) Single-cell C_4 photosynthesis versus the dual-cell (Kranz) paradigm. *Annu. Rev. Plant Biol.* 55: 173–196. **Figure 8.15** Ridout, M. J., Parker, M. L., Hedley, C. L., Bogracheva, T. Y. and Morris, V. J. (2003) Atomic force microscopy of pea starch granules: Granule architecture of wild-type parent, *r* and *rb* single mutants, and the *rrb* double mutant. *Carbohydr. Res.* 338: 2135–2147.

CHAPTER 9

Figure 9.3 Smith, H. (1986). The perception of light quality. In *Photomorphogenesis in Plants,* R. E. Kendrick and G. H. M. Kronenberg, eds., Nijhoff, Dordrecht, Netherlands, pp. 187–217. **Figure 9.4** Smith, H. (1994) Sensing the light environment: The functions of the phytochrome family. In *Photomorphogenesis in Plants,* 2nd ed., R. E. Kendrick and G. H. M. Kronenberg, eds., Nijhoff, Dordrecht, Netherlands, pp. 377–416. **Figure 9.5** Vogelmann, T. C., and Björn, L. O. (1983) Response to directional light by leaves of a sun-tracking lupine (*Lupinus succulentus*). *Physiol. Plant.* 59: 533–538. **Figure 9.7** Harvey, G. W. (1979) Photosynthetic performance of isolated leaf cells from sun and shade plants. *Carnegie Inst. Wash. Yb.* 79: 161–164. **Figure 9.8** Björkman, O. (1981) Responses to different quantum flux densities. In *Encyclopedia of Plant Physiology,* New Series, Vol. 12A, O. L. Lange, P. S. Nobel, C. B. Osmond, and H. Zeigler, eds., Springer, Berlin, pp. 57–107. **Figure 9.9** Jarvis, P. G., and Leverenz, J. W. (1983) Productivity of temperate, deciduous and evergreen forests. In *Encyclopedia of Plant Physiology,* New Series, Vol. 12D, O. L. Lange, P. S. Nobel, C. B. Osmond, and H. Ziegler, eds., Springer, Berlin, pp. 233–280. **Figure 9.10** Osmond, C. B. (1994) What is photoinhibition? Some insights from comparisons of shade and sun plants. In *Photoinhibition of Photosynthesis: From Molecular Mechanisms to the Field,* N. R. Baker and J. R. Bowyer, eds., BIOS Scientific, Oxford, pp. 1–24. **Figure 9.11** Demmig-Adams, B., and Adams, W. (1996) The role of xanthophyll cycle carotenoids in the protection of photosynthesis. *Trends Plant Sci.* 1: 21–26. **Figure 9.13** Demming-Adams, B., and Adams, W. (2000) Harvesting sunlight safely. *Nature* 403: 371–372. **Figure 9.15** Berry, J., and Björkman, O. (1980) Photosynthetic response and adaptation to temperature in higher plants. *Annu. Rev. Plant Physiol.* 31: 491–543. **Figure 9.16** Ehleringer, J. R., Cerling, T. E., and Helliker, B. R. (1997) C_4 photosynthesis, atmospheric CO_2, and climate. *Oecologia* 112: 285–299. **Figure 9.17** Ehleringer, J. R. (1978) Implications of quantum yield differences on the distributions of C_3 and C_4 grasses. *Oecologia* 31: 255–267. **Figure 9.18** Barnola, J. M., Raynaud, D., Lorius, C., and Korothevich, Y. S. (1994) Historical CO_2 record from the Vostok ice core. In *Trends '93: A Compendium of Data on Global Change,* T. A. Boden, D. P. Kaiser, R. J. Sepanski, and F. W. Stoss, eds., Carbon Dioxide Information Center, Oak Ridge National Laboratory, Oak Ridge, TN, pp. 7–10; Keeling, C. D., and Whorf, T. P. (1994) Atmospheric CO_2 records from sites in the SIO air sampling network. In *Trends '93: A Compendium of Data on Global Change,* T. A. Boden, D. P. Kaiser, R. J. Sepanski, and F. W. Stoss, eds., Carbon Dioxide Information Center, Oak Ridge National Laboratory, Oak Ridge, TN, pp. 16–26; Neftel, A., Friedle, H., Moor, E., Lötscher, H., Oeschger, H., Siegenthaler, U., and Stauffer, B. (1994) Historical CO_2 record from the Siple Station ice core. In *Trends '93: A Compendium of Data on Global Change,* T. A. Boden, D. P. Kaiser, R. J. Sepanski, and F. W. Stoss, eds., Carbon Dioxide Information Center, Oak Ridge National Laboratory, Oak Ridge, TN, pp. 11–15; Keeling, C. D., Whorf, T. P., Wahlen, M., and Van der Plicht, J. (1995) Interannual extremes in the rate of rise of atmospheric carbon dioxide since 1980. *Nature* 375: 666–670. **Figure 9.20** Berry, J. A., and Downton, J. S. (1982) Environmental regulation of photosynthesis. In *Photosynthesis: Development, Carbon Metabolism and Plant Productivity,* Vol. 2, Govindjee, ed., Academic Press, New York, pp. 263–343. **Figure 9.21** Ehleringer, J. R., Cerling, T. E., and Helliker, B. R. (1997) C_4 photosynthesis, atmospheric CO_2, and climate. *Oecologia* 112: 285–299. **Figure 9.22** Gibson, A. C., and Nobel, P. S. (1986) *The Cactus Primer.* Harvard University Press, Cambridge, MA. **Figure 9.23** Long, S. P., Ainsworth, E. A., Leakey, A. D., Nosberger, J., and Ort, D. R. (2006) Food for thought: Lower-than-expected crop stimulation with rising CO_2 concentrations. *Science* 312: 1918–1921. **Figure 9.24** Cerling, T. E., Harris, J. M., MacFadden, B. J., Leakey, M. G., Quade, J., Eisenmann, V., and Ehleringer, J. R. (1997) Global vegetation change through the Miocene–Pliocene boundary. *Nature* 389: 153–158. **Figure 9.25** Stewart, G. R., Turnbull, M. H., Schmidt, S., and Erskine, P. D. (1995) ^{13}C natural abundance in plant communities along a rainfall gradient: A biological integrator of water availability. *Aust. J. Plant Physiol.* 22: 51–55.

CHAPTER 10

Figure 10.2 Srivastava, A., and Zeiger, E. (1995) Guard cell zeaxanthin tracks photosynthetic active radiation and stomatal apertures in *Vicia faba* leaves. *Plant Cell Environ.* 18: 813–817. **Figure 10.3** Schwartz, A., and Zeiger, E. (1984) Metabolic energy for stomatal opening. Roles of photophosphorylation and oxidative phosphorylation. *Planta* 161: 129–136. **Figure 10.4** Karlsson, P. E. (1986) Blue light regulation of stomata in wheat seedlings. II. Action spectrum and search for action dichroism. *Physiol. Plant.* 66: 207–210. **Figure 10.5** Zeiger, E., and Hepler, P. K. (1977) Light and stomatal function: Blue light stimulates swelling of guard cell protoplasts. *Science* 196: 887–889; Amodeo, G., Srivastava, A., and Zeiger, E. (1992) Vanadate inhibits blue light–stimulated swelling of *Vicia* guard cell protoplasts. *Plant Physiol.* 100: 1567–1570. **Figure 10.6** Shimazaki, K., Iino, M., and Zeiger, E. (1986) Blue light–dependent proton extrusion by guard cell protoplasts of *Vicia faba*. *Nature* 319: 324–326. **Figure 10.7** Serrano, E. E., Zeiger, E., and Hagiwara, S. (1988) Red light stimulates an electrogenic proton pump in *Vicia* guard cell protoplasts. *Proc. Natl. Acad. Sci. USA* 85: 436–440; Assmann, S. M., Simoncini, L., and Schroeder, J. I. (1985) Blue light activates electrogenic ion pumping in guard cell protoplasts of *Vicia faba*. *Nature* 318: 285–287. **Figures 10.8 & 10.9** Talbott, L. D., and Zeiger, E. (1998) The role of sucrose in guard cell osmoregulation. *J. Exp. Bot.* 49: 329–337. **Figure 10.10** Talbott, L. D., Zhu, J., Han, S. W., and Zeiger, E. (2002) Phytochrome and blue light-mediated stomatal opening in the orchid, *Paphiopedilum*. *Plant Cell Physiol.* 43: 639–646. **Figure 10.11** Srivastava, A., and Zeiger, E. (1995) Guard cell zeaxanthin tracks photosynthetic active radiation and stomatal apertures in *Vicia faba* leaves. *Plant Cell Environ.* 18: 813–817. **Figure 10.14** Frechilla, S., Talbott, L. D., Bogomolni, R. A., and Zeiger, E. (2000) Reversal of blue light-stimulated stomatal opening by green light. *Plant Cell Physiol.* 41: 171–176. **Figure 10.15** Talbott, L. D., Hammad, J. W., Harn, L. C, Nguyen, V., Patel, J., and Zeiger, E. (2006) Reversal by green light of blue light-stimulated stomatal opening in intact, attached leaves of *Arabidopsis* operates only in the potassium dependent, morning phase of movement. *Plant Cell Physiol.* 47: 333–339. **Figure 10.16** Karlsson, P. E. (1986) Blue light regulation of stomata in wheat seedlings. II. Action spectrum and search for action dichroism. *Physiol. Plant.* 66: 207–210; Frechilla, S., Talbott, L. D., Bogomolni, R. A., and Zeiger, E. (2000) Reversal of blue light-stimulated stomatal opening by green light. *Plant Cell Physiol.* 41: 171–176.

CHAPTER 11

Table 11.2 Hall, S. M., and Baker, D. A. (1972) The chemical composition of *Ricinus* phloem exudate. *Planta* 106: 131–140. **Table 11.3** Gamalei, Y. V. (1985) Features of phloem loading in woody and herbaceous plants. *Fiziologiya Rastenii (Moscow)* 32: 866–875; van Bel, A. J. E. (1992) Different phloem-loading machineries correlated with the climate. *Acta Bot. Neerl.* 41: 121–141; Rennie, E. A., and Turgeon, R. (2009) A comprehensive picture of phloem loading strategies. *Proc. Natl. Acad. Sci. USA* 106: 14163–14167. **Figure 11.4** Warmbrodt, R. D. (1985) Studies on the root of *Hordeum vulgare* L.—Ultrastructure of the seminal root with special reference to the phloem. *Am. J. Bot.* 72: 414–432. **Figure 11.5** Evert, R. F. (1982) Sieve-tube structure in relation to function. *Bioscience* 32: 789–795; Truernit, E., Bauby, H., Dubreucq, B., Grandjean, O., Runions, J., Barthelemy, J., and Palauqui, J.-C. (2008) High-resolution whole-mount imaging of three-dimensional tissue organization and gene expression enables the study of phloem development and structure in *Arabidopsis*. *Plant Cell* 20: 1494–1503. **Figure 11.6** Schulz, A. (1990) Conifers. In *Sieve Elements: Comparative Structure, Induction and Development*, H.-D. Behnke and R. D. Sjolund, eds., Springer-Verlag, Berlin. **Figure 11.7** Brentwood, B., and Cronshaw, J. (1978) Cytochemical localization of adenosine triphosphatase in the phloem of *Pisum sativum* and its relation to the function of transfer cells. *Planta* 140: 111–120; Turgeon, R., Beebe, D. U., and Gowan, E. (1993) The intermediary cell: Minor-vein anatomy and raffinose oligosaccharide synthesis in the Scrophulariaceae. *Planta* 191: 446–456. **Figure 11.8** Joy, K. W. (1964) Translocation in sugar beet. I. Assimilation of $^{14}CO_2$ and distribution of materials from leaves. *J. Exp. Bot.* 15: 485–494. **Figure 11.10** Nobel, P. S. (2005) *Physicochemical and Environmental Plant Physiology*, 3rd ed., Academic Press, San Diego, CA. **Figure 11.11** Geiger, D. R., and Sovonick, S. A. (1975) Effects of temperature, anoxia and other metabolic inhibitors on translocation. In *Transport in Plants*, 1: *Phloem Transport* (Encyclopedia of Plant Physiology, New Series, Vol. 1), M. H. Zimmerman and J. A. Milburn, eds., Springer, New York, pp. 256–286. **Figure 11.12** Froelich, D. R., Mullendore, D. L., Jensen, K. H., Ross-Elliott, T. J., Anstead, J. A., Thompson, G. A., Pelissier, H. C., and Knoblauch, M. (2011) Phloem ultrastructure and pressure flow: Sieve-Element-Occlusion-Related agglomerations do not affect translocation. *Plant Cell* 23: 4428–4445. **Figure 11.13** Evert, R. F., and Mierzwa, R. J. (1985) Pathway(s) of assimilate movement from mesophyll cells to sieve tubes in the *Beta vulgaris* leaf. In *Phloem Transport. Proceedings of an International Conference on Phloem Transport, Asilomar, CA*, J. Cronshaw, W. J. Lucas, and R. T. Giaquinta, eds. Liss, New York, pp. 419–432. **Figure 11.15** Fondy, B. R. (1975) Sugar selectivity of phloem loading in *Beta vulgaris, vulgaris* L. and *Fraxinus americanus, americana* L. Ph.D. diss., University of Dayton, Dayton, OH. **Figure 11.17** van Bel, A. J. E. (1992) Different phloem-loading machineries correlated with the climate. *Acta Bot. Neerl.* 41: 121–141. **Figure 11.19** Turgeon, R., and Webb, J. A. (1973) Leaf development and phloem transport in *Cucurbita pepo*: Transition from import to export. *Planta* 113: 179–191. **Figure 11.20** Turgeon, R. (2006) Phloem loading: How leaves gain their independence. *Bioscience* 56: 15–24. **Figure 11.21** Schneidereit, A., Imlau, A., and Sauer, N. (2008) Conserved *cis*-regulatory elements for DNA-binding-with-one-finger and homeo-domain-leucine-zipper transcription factors regulate companion cell-specific expression of the *Arabidopsis thaliana* SUCROSE TRANSPORTER 2 gene. *Planta* 228: 651–662. **Figure 11.22** Preiss, J. (1982) Regulation of the biosynthesis and degradation of starch. *Annu. Rev. Plant Physiol.* 33: 431–454. **Figure 11.23** Stadler, R., Wright, K. M., Lauterbach, C., Amon, G., Gahrtz, M., Feuerstein, A., Oparka, K. J., and Sauer, N. (2005) Expression of GFP-fusions in *Arabidopsis* companion cells reveals non-specific protein trafficking into sieve elements and identifies a novel post-phloem domain in roots. *Plant J.* 41: 319–331.

CHAPTER 12

Table 12.2 Brand, M. D. (1994) The stoichiometry of proton pumping and ATP synthesis in mitochondria. *Biochem. (Lond)* 16: 20–24. **Figure 12.5** Perkins, G., Renken, C., Martone, M. E., Young, S. J., Ellisman, M., and Frey, T. (1997) Eceltron tomography of neuronal mitochondria: Three-dimensional structure and organization of cristae and membrane contacts. *J. Struct. Biol.* 119: 260–272; Gunning,

B. E. S., and Steer, M. W. (1996) *Plant Cell Biology: Structure and Function of Plant Cells*. Jones and Bartlett, Boston. **Figure 12.9** Shiba, T., Kido, Y., Sakamoto, K., Inaoka, D. K., Tsuge, C., Tatsumi, R., Takahashi, G., Balogun, E. O., Nara, T., Aoki, T., et al. (2013) Structure of the trypanosome cyanide-insensitive alternative oxidase. *Proc. Natl. Acad. Sci. USA* 110: 4580–4585. **Figure 12.10** Douce, R. (1985) *Mitochondria in Higher Plants: Structure, Function, and Biogenesis*. Academic Press, Orlando, FL.

CHAPTER 13

Figure 13.2 Bloom, A. J. (1997) Nitrogen as a limiting factor: Crop acquisition of ammonium and nitrate. In *Ecology in Agriculture*, L. E. Jackson, ed., Academic Press, San Diego, CA, pp. 145–172. **Figure 13.4** Kleinhofs, A., Warner, R. L., Lawrence, J. M., Melzer, J. M., Jeter, J. M., and Kudrna, D. A. (1989) Molecular genetics of nitrate reductase in barley. In *Molecular and Genetic Aspects of Nitrate Assimilation*, J. L. Wray and J. R. Kinghorn, eds., Oxford Science, New York, pp. 197–211. **Figure 13.6** Pate, J. S. (1983) Patterns of nitrogen metabolism in higher plants and their ecological significance. In *Nitrogen as an Ecological Factor: The 22nd Symposium of the British Ecological Society, Oxford 1981*, J. A. Lee, S. McNeill, and I. H. Rorison, eds., Blackwell, Boston, pp. 225–255. **Figure 13.11** Stokkermans, T. J. W., Ikeshita, S., Cohn, J., Carlson, R. W., Stacey, G., Ogawa, T., and Peters, N. K. (1995) Structural requirements of synthetic and natural product lipo-chitin oligosaccharides for induction of nodule primordia on *Glycine soja*. *Plant Physiol.* 108: 1587–1595. **Figure 13.13** Dixon, R. O. D., and Wheeler, C. T. (1986) *Nitrogen Fixation in Plants*. Chapman and Hall, New York; Buchanan, B., Gruissem, W., and Jones, R., eds. (2000) *Biochemistry and Molecular Biology of Plants*. American Society of Plant Physiologists, Rockville, MD. **Figure 13.16** Rees, D. A. (1977) *Polysaccharide Shapes*. Chapman and Hall, London. **Figure 13.18** Guerinot, M. L., and Yi, Y. (1994) Iron: Nutritious, noxious, and not readily available. *Plant Physiol.* 104: 815–820. **Figure 13.21** Searles, P. S., and Bloom, A. J. (2003) Nitrate photoassimilation in tomato leaves under short-term exposure to elevated carbon dioxide and low oxygen. *Plant Cell Environ.* 26: 1247–1255; Bloom, A. J., Rubio-Asensio, J. S., Randall, L., Rachmilevitch, S., Cousins, A. B., and Carlisle, E. A. (2012) CO_2 enrichment inhibits shoot nitrate assimilation in C_3 but not C_4 plants and slows growth under nitrate in C_3 plants. *Ecology* 93: 355–367.

CHAPTER 14

Figure 14.2 McCann, M. C., Wells, B., and Roberts, K. (1990) Direct visualization of cross-links in the primary plant cell wall. *J. Cell Sci.* 96: 323–334; Zhang, T., Mahgsoudy-Louyeh, S., Tittmann, B., and Cosgrove, D. J. (2013) Visualization of the nanoscale pattern of recently-deposited cellulose microfibrils and matrix materials in never-dried primary walls of the onion epidermis. *Cellulose* 21: 853–862. DOI: 10.1007/s10570-013-9996-1; Roland, J. C., Reis, D., Mosiniak, M., and Vian, B. (1982) Cell wall texture along the growth gradient of the mung bean hypocotyl: Ordered assembly and dissipative processes. *J. Cell Sci.* 56: 303–318. **Figure 14.5** Carpita, N. C., and McCann, M. (2000) The cell wall. In *Biochemistry and Molecular Biology of Plants*, B. B. Buchanan, W. Gruissem, and R. L. Jones eds., American Society of Plant Biologists, Rockville, MD, pp. 52–108. **Figure 14.6** Zhang, T., Mahgsoudy-Louyeh, S., Tittmann, B., and Cosgrove, D. J. (2013) Visualization of the nanoscale pattern of recently-deposited cellulose microfibrils and matrix materials in never-dried primary walls of the onion epidermis. *Cellulose* 21: 853–862. DOI: 10.1007/s10570-013-9996-1; Matthews, J. F., Skopec, C. E., Mason, P. E., Zuccato, P., Torget, R. W., Sugiyama, J., Himmel, M. E., and Brady, J. W. (2006) Computer simulation studies of microcrystalline cellulose Iβ. *Carbohydr. Res.* 341: 138–152. **Figure 14.7** Gunning, B. E. S., and Steer, M. W. (1996) *Plant Cell Biology: Structure and Function of Plant Cells*. Jones and Bartlett, Boston; Kimura, S., Laosinchai, W., Itoh, T., Cui, X., Linder, R., and Brown, R. M., Jr. (1999) Immunogold labeling of rosette terminal cellulose-synthesizing complexes in the vascular plant *Vigna angularis*. *Plant Cell* 11: 2075–2085; Morgan, J. L., Strumillo, J., and Zimmer, J. (2013) Crystallographic snapshot of cellulose synthesis and membrane translocation. *Nature* 493: 181–186; Sethaphong, L., Haigler, C. H., Kubicki, J. D., Zimmer, J., Bonetta, D., DeBolt, S., and Yingling, Y. G. (2013) Tertiary model of a plant cellulose synthase. *Proc. Natl. Acad. Sci. USA* 110: 7512–7517. **Figure 14.8** Cosgrove, D. J. (2005) Growth of the plant cell wall. *Nat. Rev. Mol. Cell Biol.* 6: 850–861. **Figure 14.9** Carpita, N. C., and McCann, M. (2000) The cell wall. In *Biochemistry and Molecular Biology of Plants*, B. B. Buchanan, W. Gruissem, and R. L. Jones eds., American Society of Plant Biologists, Rockville, MD, pp. 52–108. **Figure 14.10** Mohnen, D. (2008) Pectin structure and biosynthesis. *Curr. Opin. Plant Biol.* 11: 266–277. **Figures 14.10 & 14.11** Carpita, N. C., and McCann, M. (2000) The cell wall. In *Biochemistry and Molecular Biology of Plants*, B. B. Buchanan, W. Gruissem, and R. L. Jones eds., American Society of Plant Biologists, Rockville, MD, pp. 52–108. **Figure 14.12** Fry, S. C. (2004) Primary cell wall metabolism: tracking the careers of wall polymers in living plant cells. *New Phytol.* 161: 641–675. **Figure 14.15** Fu, Y., Gu, Y., Zheng, Z., Wasteneys, G., and Yang, Z. (2005) *Arabidopsis* interdigitating cell growth requires two antagonistic pathways with opposing action on cell morphogenesis. *Cell* 120: 687–700; Settleman, J. (2005) Intercalating *Arabidopsis* leaf cells: A jigsaw puzzle of lobes, necks, ROPs, and RICs. *Cell* 120: 570–572. **Figure 14.17** Baskin, T. I., Wilson, J. E., Cork, A., and Williamson R. E. (1994) Morphology and microtubule organization in *Arabidopsis* roots exposed to oryzalin or taxol. *Plant Cell Physiol.* 35: 935–942; Gutierrez, R., Lindeboom, J. J., Paredez, A. R., Emons, A. M., and Ehrhardt, D. W. (2009) *Arabidopsis* cortical microtubules position cellulose synthase delivery to the plasma membrane and interact with cellulose synthase trafficking compartments. *Nat. Cell Biol.* 11: 797–806. **Figure 14.18** Durachko, D. M., and Cosgrove, D. J. (2009) Measuring plant cell wall extension (creep) induced by acidic pH and by alpha-expansin. *J. Vis. Exp.* 11: 1263. **Figure 14.19** Cosgrove, D. J. (1997) Relaxation in a high-stress environment: The molecular bases of extensible cell walls and cell enlargement. *Plant Cell* 9: 1031–1041. **Figure 14.20** Park, Y. B., and Cosgrove, D. J. (2012) A revised architecture of primary cell walls based on biomechanical changes induced by substrate-specific endoglucanases. *Plant Physiol.* 158: 1933–1943; Zhao, Z., Crespi, V. H., Kubicki, J. D., Cosgrove, D. J., and Zhong, L. (2013) Molecular dynamics simulation study of xyloglucan adsorption on cellulose surfaces: Effects of surface hydrophobicity and side-chain variation. *Cellulose* 21: 1025–1039. DOI: 10.1007/s10570-013-

0041-1. **Figure 14.21** Terashima, N., Awano, T., Takabe, K., and Yoshida, M. (2004) Formation of macromolecular lignin in ginkgo xylem cell walls as observed by field emission scanning electron microscopy. *C. R. Biol.* 327: 903–910; Terashima, N., Kitano, K., Kojima, M., Yoshida, M., Yamamoto, H., and Westermark, U. (2009) Nanostructural assembly of cellulose, hemicellulose, and lignin in the middle layer of secondary wall of ginkgo tracheid. *J. Wood Sci.* 55: 409–416. **Figure 14.22** Ralph, J., Brunow, G., and Boerjan, W. (2007) Lignins. In *Encyclopedia of Plant Science*, K Roberts, ed., Wiley, Chichester, pp. 1123–1134. **Figure 14.23** Roppolo, D., and Geldner, N. (2012) Membrane and walls: Who is master, who is servant? *Curr. Opin. Plant Biol.* 15: 608–617.

CHAPTER 15

Table 15.1 Suarez-Rodriguez, M. C., Petersen, M., and Mundy, J. (2010) Mitogen-activated protein kinase signaling in plants. Annu. Rev. Plant Biol. 61: 621–649. **Figure 15.3** Santer, A., and Estelle, M. (2009) Recent advances and emerging trends in plant hormone signaling. *Nature* (*Lond.*) 459: 1071–1078. **Figure 15.6** Wang, X. (2004) Lipid signaling. *Curr. Opin. Plant Biol.* 7: 329–336. **Figure 15.11** Riou-Khamlichi, C., Huntley, R., Jacqmard, A., and Murray, J. A. (1999) Cytokinin activation of *Arabidopsis* cell division through a D-type cyclin. *Science* 283: 1541–1544; Aloni, R., Wolf, A., Feigenbaum, P., Avni, A., and Klee, H. J. (1998) The *Never ripe* mutant provides evidence that tumor-induced ethylene controls the morphogenesis of *Agrobacterium tumefaciens*-induced crown galls in tomato stems. *Plant Physiol.* 117: 841–849. **Figure 15.19** Woodward, W., and Bartel, B. (2005) Auxin: Regulation, action, and interaction. *Annals Botany* 95: 707–735. **Figure 15.22** McKeon, T. A., Fernández-Maculet, J. C., and Yang, S. F. (1995) Biosynthesis and metabolism of ethylene. In *Plant Hormones: Physiology, Biochemistry and Molecular Biology*, 2nd ed., P. J. Davies, ed., Kluwer, Dordrecht, Netherlands, pp. 118–139. **Figure 15.27** Braam, J. (2005) In touch: Plant responses to mechanical stimuli. *New Phytol.* 165: 373–389; Escalante-Pérez, M., Krola, E., Stangea, A., Geigera, D., Al-Rasheidb, K. A. S., Hausec, B., Neherd, E., and Hedrich, R. (2011) A special pair of phytohormones controls excitability, slow closure, and external stomach formation in the Venus flytrap. *Proc. Natl. Acad. Sci. USA* 108: 15492–15497. **Figure 15.30** Ju, C., and Chang, C. (2012) Advances in ethylene signalling: Protein complexes at the endoplasmic reticulum membrane. *AoB PLANTS*. DOI: 10.1093/aobpla/pls031. **Figure 15.31** Jiang, J., Zhang, C., and Wang, X. (2013) Ligand perception, activation, and early signaling of plant steroid receptor Brassinosteroid Insensitive 1. *J. Integr. Plant Biol.* 55 : 1198–1211.

CHAPTER 16

Table 16.3 Jenkins, G. I. (2014) The UV-B photoreceptor UVR8: From structure to physiology. *Plant Cell* 26: 21–37. **Figure 16.6** Shropshire, W., Jr., Klein, W. H., and Elstad, V. B. (1961) Action spectra of photomorphogenic induction and photoinactivation of germination in *Arabidopsis thaliana*. *Plant Cell Physiol.* 2: 63–69; Kelly, J. M., and Lagarias, J. C. (1985) Photochemistry of 124-kilodalton Avena phytochrome under constant illumination in vitro. *Biochemistry* 24: 6003–6010. **Figure 16.7** Thimann, K. V., and Curry, G. M. (1960) Phototropism and phototaxis. In *Comparative Biochemistry*, Vol. 1, M. Florkin and H. S. Mason, eds., Academic Press, New York, pp. 243–306; Swartz, T. E., Corchnoy, S. B., Christie, J. M., Lewis, J. W., Szundi, I., Briggs, W. R., and Bogomolni, R. (2001) The photocycle of a flavin-binding domain of the blue light photoreceptor phototropin. *J. Biol. Chem.* 276: 36493–36500. **Figure 16.9** Burgie, E. S., Bussell, A. N., Walker, J. M., Dubiel, K., and Vierstra, R. D. (2014) Crystal structure of the photosensing module from a red/far-red light-absorbing plant phytochrome. *Proc. Natl. Acad. Sci. USA* 111: 10179–10184. **Figure 16.10** Montgomery, B. L., and Lagarias, J. C. (2002) Phytochrome ancestry: Sensors of bilins and light. *Trends Plant Sci.* 7: 357–366. **Figure 16.11** Yamaguchi, R., Nakamura, M., Mochizuki, N., Kay, S. A., and Nagatani, A. (1999) Light-dependent translocation of a phytochrome B-GFP fusion protein to the nucleus in transgenic *Arabidopsis*. *J. Cell Biol.* 145: 437–445. **Figure 16.12** Briggs, W. R., Mandoli, D. F., Shinkle, J. R., Kaufman, L. S., Watson, J. C., and Thompson, W. F. (1984) Phytochrome regulation of plant development at the whole plant, physiological, and molecular levels. In *Sensory Perception and Transduction in Aneural Organisms*, G. Colombetti, F. Lenci, and P.-S. Song, eds., Plenum, New York, pp. 265–280. **Figure 16.13** Leivar, P., and Monte, E. (2014) PIFs: Systems integrators in plant development. *Plant Cell* 26: 56–78. **Figure 16.15** Spalding, E. P., and Cosgrove, D. J. (1989) Large membrane depolarization precedes rapid blue-light induced growth inhibition in cucumber. *Planta* 178: 407–410. **Figure 16.16** Huang, Y., Baxter, R., Smith, B. S., Partch, C. L., Colbert, C. L., and Deisenhofer, J. (2006) Crystal structure of cryptochrome 3 from *Arabidopsis thaliana* and its implications for photolyase activity. *Proc. Natl. Acad. Sci. USA* 103: 17701–17706. **Figure 16.17** Ahmad, M., Jarillo, J. A., Smirnova, O., and Cashmore, A. R. (1998) Cryptochrome blue light photoreceptors of *Arabidopsis* implicated in phototropism. *Nature* 392: 720–723. **Figure 16.19** Parks, B. M., Folta, K. M., and Spalding, E. P. (2001) Photocontrol of stem growth. *Curr. Opin. Plant Biol.* 4: 436–440. **Figure 16.20** Christie, J. M. (2007) Phototropin blue-light receptors. *Annu. Rev. Plant Biol.* 58: 21–45. **Figure 16.22** Inoue, S.-I., Takemiya, A., and Shimazaki, K.-I. (2010) Phototropin signaling and stomatal opening as a model case. *Curr. Opin. Plant Biol.* 13: 587–593. **Figures 16.23 & 16.24** Wada, M. (2013) Chloroplast movement. *Plant Sci.* 210: 177–182. **Figure 16.27** Jenkins, G. I. (2014) The UV-B photoreceptor UVR8: From structure to physiology. *Plant Cell* 26: 21–37.

CHAPTER 17

Figure 17.3 West, M. A. L., and Harada, J. J. (1993) Embryogenesis in higher plants: An overview. *Plant Cell* 5: 1361–1369. **Figure 17.5** Laux, T., Würschum, T., and Breuninger, H. (2004) Genetic regulation of embryonic pattern formation. *Plant Cell* 16: S190–S202. **Figure 17.7** Scheres, B., Wolkenfelt, H., Willemsen, V., Terlouw, M., Lawson, E., Dean, C., and Weisbeek, P. (1994) Embryonic origin of the *Arabidopsis* primary root and root meristem initials. *Development* 120: 2475–2487. **Figure 17.8** Traas, J., Bellini, C., Nacry, P., Kronenberger, J. Bouchez, D., and Caboche, M. (1995) Normal differentiation patterns in plants lacking microtubular preprophase bands. *Nature* 375: 676–677. **Figure 17.9** Kim, I., Kobayashi, K., Cho, E., and Zambryski, P. C. (2005) Subdomains for transport via plasmodesmata corresponding to the apical-basal axis are established during *Arabidopsis* embryogenesis. *Proc. Natl. Acad. Sci. USA* 102: 11945–11950. **Figure 17.10** Mayer, U., Büttner, G., and Jürgens, G. (1993) Apical–basal pattern formation in the *Arabidopsis* embryo: Studies on the role of the *gnom* gene. *Development* 117: 149–162; Berleth, T., and Jürgens, G. (1993) The role of the *MONOPTEROS* gene in organising the basal body region of the *Arabidopsis* embryo. *Development* 118: 575–587; Mayer, U., Torres-Ruiz, R. A., Berleth, T., Misera, S., and Jürgens, G. (1991) Mutations affecting body organisation in the *Arabidopsis* embryo. *Nature* 353: 402–407. **Figure 17.12** Multani, D. S., Briggs, S. P., Chamberlin, M. A., Blakeslee, J. J., Murphy, A. S., and Johal, G. S. (2003) Loss of an MDR transporter in compact stalks of maize *br2* and sorghum *dw3* mutants. *Science* 302: 81–84. **Figure 17.13** Hadfi, K., Speth, V., and Neuhaus, G. (1998) Auxin-induced developmental patterns in *Brassica juncea* embryos. *Development* 125: 879–887.Liu, C., Xu, Z., and Chua, N. H. (1993) Auxin polar transport is essential for the establishment of bilateral symmetry during early plant embryogenesis. *Plant Cell* 5: 621–630. **Figure 17.16** Abe, M., Katsumata, H., Komeda, Y., and Takahashi, T. (2003) Regulation of shoot epidermal cell differentiation by a pair of homeodomain proteins in *Arabidopsis*. *Development* 130: 635–643. **Figure 17.17** Mähönen, A. P., Bonke, M., Kauppinen, L., Riikonen, M., Benfey, P. N., and Helariutta, Y. (2000) A novel two-component hybrid molecule regulates vascular morphogenesis of the *Arabidopsis* root. *Genes Dev.* 14: 2938–2943. **Figure 17.18** Nakajima, K., and Benfey, P. N. (2002) Signaling in and out: Control of cell division and differentiation in the shoot and root. *Plant Cell* 14: S265–S276. **Figure 17.19** Helariutta, Y., Fukaki, H., Wysocka-Diller, J., Nakajima, K., Jung, J., Sena, G., Hauser, M. T., and Benfey, P. N. (2000) The *SHORT-ROOT* gene controls radial patterning of the *Arabidopsis* root through radial signaling. *Cell* 101: 555–567. **Figure 17.21** Schiefelbein, J. W., Masucci, J. D., and Wang, H. (1997) Building a root: The control of patterning and morphogenesis during root development. *Plant Cell* 9: 1089–1098. **Figure 17.22** Aida, M., Beis, D., Heidstra, R., Willemsen, V., Blilou, I., Galinha, C., Nussaume, L., Noh, Y.-S., Amasino, R., and Scheres, B. (2004) The *PLETHORA* genes mediate patterning of the *Arabidopsis* root stem cell niche. *Cell* 119: 109–120. **Figure 17.23** Müller, B., and Sheen, J. (2008) Cytokinin and auxin interaction in root stem-cell specification during early embryogenesis. *Nature* 453: 1094–1097. **Figure 17.24** Kuhlemeier, C., and Reinhardt, D. (2001) Auxin and phyllotaxis. *Trends Plant Sci.* 6: 187–189. **Figure 17.25** Bowman, J. L., and Eshed, Y. (2000) Formation and maintenance of the shoot apical meristem. *Trends Plant Sci.* 5: 110–115. **Figure 17.26** Steeves, T. A., and Sussex, I. M. (1989) *Patterns in Plant Development*. Cambridge University Press, Cambridge, UK. **Figure 17.27** Jenik, P. D., and Barton, M. K. (2005) Surge and destroy: The role of auxin in plant embryogenesis. *Development* 132: 3577–3585. **Figure 17.28** Laux, T., Würschum, T., and Breuninger, H. (2004) Genetic regulation of embryonic pattern formation. *Plant Cell* 16: S190–S202. **Figure 17.29** Leibfried, A., To, J. P. C., Busch, W., Stehling, S., Kehle, A., Demar, M., Kieber, J. J., and Lohmann, J. U. (2005) WUSCHEL controls meristem function by direct regulation of cytokinin-inducible response regulators. *Nature* 438: 1172–1175. **Figure 17.31** Hudson, A. (2005) Plant meristems: Mobile mediators of cell fate. *Curr. Biol.* 15: R803–805. **Figure 17.33** Reinhardt, D., Pesce, E. R., Stieger, P., Mandel, T., Baltensperger, K., Bennett, M., Traas, J., Friml, J., and Kuhlemeier, C. (2003) Regulation of phyllotaxis by polar auxin transport. *Nature* 426: 255–260; Vernoux, T., Kronenberger, J., Grandjean, O., Laufs, P., and Traas, J. (2000) PIN-FORMED 1 regulates cell fate at the periphery of the shoot apical meristem. *Development* 127: 5157–5165. **Figure 17.34** Miyashima, S., Sebastian, J., Lee, J.-Y., and Helariutta, Y. (2013) Stem cell function during plant vascular development. *EMBO J.* 32: 178–193.

CHAPTER 18

Table 18.1 Bewley, J. D., Bradford, K. J., Hilhorst, H. W. M., and Nonogaki, H. (2013) *Seeds: Physiology of Development, Germination and Dormancy*, 3rd edition. Springer, New York. **Table 18.2** Smith, H. (1982) Light quality photoperception and plant strategy. *Annu. Rev. Plant Physiol.* 33: 481–518. **Figure 18.2** Homrichhausen, T. M., Hewitt, J. R., and Nonogaki, H. (2003) Endo-β-mannanase activity is associated with the completion of embryogenesis in imbibed carrot (*Daucus carota* L.) seeds. *Seed Sci. Res.* 13: 219–227. **Figure 18.4** Li, Y.-C., Rena, J.-P., Cho, M.-J., Zhou, S.-M., Kim, Y.-B., Guo, H.-X., Wong, J.-H., Niu, H.-B., Kim, H.-K., Morigasaki, S., et al. (2009) The level of expression of thioredoxin is linked to fundamental properties and applications of wheat seeds. *Mol. Plant* 2: 430–441. **Figure 18.5** Finch-Savage, W. E. and Leubner-Metzger, G. (2006) Seed dormancy and the control of germination. *New Phytol.* 171: 501–523. **Figure 18.6** Visser, T. (1956) Chilling and apple seed dormancy. *Proc. K. Ned. Akad. Wet. C* 59: 314–324; Grappin, P., Bouinot, D., Sotta, B., Migniac, E., and Julien, M. (2000) After-ripening of tobacco seeds. *Planta* 210: 279–285. **Figure 18.7** Liptay, A., and Schopfer, P. (1983) Effect of water stress, seed coat restraint, and abscisic acid upon different germination capabilities of two tomato lines at low temperature. *Plant Physiol.* 73: 935–938. **Figure 18.8** Nonogaki, H., Bassel, G. W., and Bewley, J. D. (2010) Germination—Still a mystery. *Plant Sci.* 179: 574–581. **Figure 18.9** Bethke, P. C., Schuurink, R., and Jones, R. L. (1997) Hormonal signalling in cereal aleurone. *J. Exp. Bot.* 48: 1337–1356. **Figure 18.10** Gubler, F., Kalla, R., Roberts, J. K., and Jacobsen, J. V. (1995) Gibberellin-regulated expression of a *myb* gene in barley aleurone cells: Evidence of myb transactivation of a high-pI alpha-amylase gene promoter. *Plant Cell* 7: 1879–1891. **Figure 18.13** Cleland, R. E. (1995) Auxin and cell elongation. In *Plant Hormones and Their Role in Plant Growth and Development*, 2nd ed., P. J. Davies, ed., Kluwer, Dordrecht, Netherlands, pp. 214–227; Jacobs, M., and Ray, P. M. (1976) Rapid auxin-induced decrease in free space pH and its relationship to auxin-induced growth in maize and pea. *Plant Physiol.* 58: 203–209. **Figure 18.16** Hartmann, H. T., and Kester, D. E. (1983) *Plant Propagation: Principles and Practices*, 4th ed. Prentice-Hall, Inc., NJ. **Figure 18.17** Shaw, S., and Wilkins, M. B. (1973) The source and lateral transport of growth inhibitors in geotropically stimulated roots of *Zea mays* and *Pisum sativum*. *Planta* 109: 11–26. **Figure 18.18** Baldwin, K. L., Strohm, A. K., and Masson, P. H. (2013) Gravity sensing and signal transduction in vascular plant primary roots. *Am. J. Bot.* 100: 126–142. **Figure 18.19** Blilou, I., Xu, J., Wildwater, M., Willemsen, V., Paponov, I., Friml, J., Heidstra, R., Aida, M., Palme, K., and Scheres, B. (2005) The PIN auxin efflux facilitator network controls growth and patterning in

Arabidopsis roots. *Nature* 433: 39–44. **Figure 18.20** Volkmann, D., and Sievers, A. (1979) Graviperception in multicellular organs. In *Encyclopedia of Plant Physiology*, New Series, Vol. 7: *Physiology of Movements*, W. Haupt and M. E. Feinleib eds., Springer-Verlag, New York, pp. 573–600. **Figure 18.22** Fasano, J. M., Swanson, S. J., Blancaflor, E. B., Dowd, P. E., Kao, T. H., and Gilroy, S. (2001) Changes in root cap pH are required for the gravity response of the *Arabidopsis* root. *Plant Cell* 13: 907–921. **Figure 18.23** Iino, M., and Briggs, W. R. (1984) Growth distribution during first positive phototropic curvature of maize coleoptiles. *Plant Cell Environ.* 7: 97–104. **Figure 18.24** Christie, J. M., Yang, H., Richter, G. L., Sullivan, S., Thomson, C. E., Lin, J., Titapiwatanakun, B., Ennis, M., Kaiserli, E., Lee, O. R., et al. (2011) phot1 inhibition of ABCB19 primes lateral auxin fluxes in the shoot apex required for phototropism. *PLoS Biol.* 9: e1001076. DOI: 10.1371/journal.pbio.1001076. **Figure 18.27** Le, J., Vandenbussche, F., De Cnodder, T., Van Der Straeten, D., and Verbelen, J.-P. (2005) Cell elongation and microtubule behavior in the *Arabidopsis* hypocotyl: Responses to ethylene and auxin. *J. Plant Growth Regul.* 24: 166–178. **Figure 18.28** Binder, B. M., O'Malley, R. C., Moore, J. M., Parks, B. M., Spalding, E. P., and Bleecker, A. B. (2004a) *Arabidopsis* seedling growth response and recovery to ethylene: A kinetic analysis. *Plant Physiol.* 136: 2913–2920; Binder, B. M., Mortimore, L. A., Stepanova, A. N., Ecker, J. R., and Bleecker, A. B. (2004b) Short term growth responses to ethylene in *Arabidopsis* seedlings are EIN3/EIL1 independent. *Plant Physiol.* 136: 2921–2927. **Figure 18.29** Morgan, D. C., and Smith, H. (1979) A systematic relationship between phytochrome-controlled development and species habitat, for plants grown in simulated natural irradiation. *Planta* 145: 253–258. **Figure 18.32** Busse, J. S., and Evert, R. F. (1999) Vascular differentiation and transition in the seedling of *Arabidopsis thaliana* (Brassicaceae). *Int. J. Plant Sci.* 160: 241–251. **Figure 18.33** Lacayo, C. I., Malkin, A. J., Holman, H.-Y. N, Chen, L., Ding, S.-Y., Hwang, M. S., and Thelen, M. P. (2010) Imaging cell wall architecture in single *Zinnia elegans* tracheary elements. *Plant Physiol.* 154: 121–133; Novo-Uzal, E., Fernández-Pérez, F., Herrero, J., Gutiérrez, J., Gómez-Ros, L. V., Bernal, M. A., Díaz, J., Cuello, J., Pomar, F., and Pedreño, M. A. (2013) From *Zinnia* to *Arabidopsis*: Approaching the involvement of peroxidases in lignification. J. Exp. Bot. 64: 3499–3518; Motose, H., Sugiyama, M., and Fukuda, H. (2004) A proteoglycan mediates inductive interaction during plant vascular development. *Nature* 429: 873–878. **Figure 18.35** Bibikova, T., and Gilroy, S. (2003) Root hair development. *J. Plant Growth Regul.* 21: 383–415. **Figure 18.36** Abeles, F. B., Morgan, P. W., and Saltveit, M. E., Jr. (1992) *Ethylene in Plant Biology*, 2nd ed. Academic Press, San Diego, CA. **Figure 18.37** Petricka, J. J., Winter, C. M., and Benfey, P. N. (2012) Control of *Arabidopsis* root development. *Annu. Rev. Plant Biol.* 63: 563–590. **Figure 18.38** Van Norman, J. M., Zhang, J., Cazzonelli, C. I., Pogson, B. J., Harrison, P. J., Bugg, T. D. H., Chan, K. X., Thompson, A. J., and Benfey, P. N. (2013) To branch or not to branch: The role of pre-patterning in lateral root formation. *Development* 140: 4301–4310.

CHAPTER 19

Figure 19.2 Besnard, F., Vernoux, T., and Hamant, O. (2011) Organogenesis from stem cells *in planta*: multiple feedback loops integrating molecular and mechanical signals. *Cell. Mol. Life Sci.* 68: 2885–2906. **Figure 19.3** Sussex, I. M. (1951) Experiments on the cause of dorsiventrality in leaves. *Nature* 167: 651–652. **Figure 19.4** Waites, R., and Hudson, A. (1995) *phantastica*: a gene required for dorsoventrality of leaves in *Antirrhinum majus. Development* 121: 2143–2154. **Figure 19.5** Townsley, B. T., and Sinha, N. R. (2012) A new development: Evolving concepts in leaf ontogeny. *Annu. Rev. Plant Biol.* 63: 535–562. **Figures 19.5 & 19.6** Fukushima, K., and Hasebe, M. (2013) Adaxial–abaxial polarity: The developmental basis of leaf shape diversity. *Genesis* 52: 1–18. **Figure 19.7** Kang, J., and Sinha, N. R. (2010) Leaflet initiation is temporally and spatially separated in simple and complex tomato (*Solanum lycopersicum*) leaf mutants: A developmental analysis. *Botany* 88: 710–724. **Figure 19.8** Hasson, A., Blein, T., and Laufs, P. (2010) Leaving the meristem behind: The genetic and molecular control of leaf patterning and morphogenesis. *C. R. Biol.* 333: 350–360. **Figures 19.10 & 19.11** Lau, S., and Bergmann, D. C. (2012) Stomatal development: A plant's perspective on cell polarity, cell fate transitions and intercellular communication. *Development* 139: 3683–3692. **Figure 19.13** Balkunde, R., Pesch, M., and Hülskamp. (2010) Trichome patterning in *Arabidopsis thaliana*: From genetic to molecular models. *Curr. Top. Dev. Biol.* 91: 299–321. **Figure 19.14** Qing, L., and Aoyama, T. (2012) Pathways for epidermal cell differentiation via the homeobox gene *GLABRA2*: Update on the roles of the classic regulator. *J. Integr. Plant Biol.* 54: 729–737. **Figure 19.15** Sack, L., and Scoffoni, C. (2013). Leaf venation: Structure, function, development, evolution, ecology and applications in the past, present and future. *New Phytol.* 198: 983–1000. **Figure 19.16** Lucas, W. J., Groover, A., Lichtenberger, R., Furuta, K., Yadav, S.-R., Helariutta, Y., He, X.-Q., Fukuda, H., Kang, J., Brady, S. M., et al. (2013) The plant vascular system: Evolution, development and functions. *J. Integr. Plant Biol.* 55: 294–388. **Figure 19.17** Esau, K. (1953) *Plant Anatomy*. Wiley, New York. **Figures 19.19 & 19.20** Bayer, E. M., Smith, R. S., Mandel, T., Nakayama, N., Sauer, M., Prusinkiewicz, P., and Kuhlemeier, C. (2009) Integration of transport-based models for phyllotaxis and midvein formation. *Genes Dev.* 23: 373–384. **Figure 19.21** Lucas, W. J., Groover, A., Lichtenberger, R., Furuta, K., Yadav, S.-R., Helariutta, Y., He, X.-Q., Fukuda, H., Kang, J., Brady, S. M., et al. (2013) The plant vascular system: Evolution, development and functions. *J. Integr. Plant Biol.* 55: 294–388. **Figure 19.22** Petrášek, J., and Friml, J. (2009) Auxin transport routes in plant development. *Development* 136: 2675–2688. **Figure 19.23** Sawchuck, M. G., Edgar, A., and Scarpella, E. (2013) Pattering of leaf vein networks by convergent auxin transport pathways. *PLoS Genet.* 9(2): e1003294. DOI: 10.1371/journal.pgen.1003294; Cheng, Y., Dai, X., and Zhao, Y. (2006) Auxin biosynthesis by the YUCCA flavin monooxygenases controls the formation of floral organs and vascular tissues in *Arabidopsis. Genes Dev.* 20: 1790–1799. **Figure 19.24** Aloni, R., Schwalm, K., Langhans, M., and Ullrich, C. I. (2003) Gradual shifts in sites of free-auxin production during leaf-primordium development and their role in vascular differentiation and leaf morphogenesis in *Arabidopsis. Planta* 216: 841–853. **Figure 19.25** Takiguchi, Y., Imaichi, R., and Kato, M. (1997) Cell division patterns in the apices of subterranean axis and aerial shoot of *Psilotum nudum*

(Psilotaceae): Morphological and phylogenetic implications for the subterranean axis. *Am. J. Bot.* 84: 588–596. **Figure 19.28** Greb, T., Clarenz, O., Schäfer, E., Müller, D., Herrero, R., Schmitz, G., and Theres, K. (2003) Molecular analysis of the *LATERAL SUPPRESOR* gene in *Arabidopsis* reveals a conserved control mechanism for axillary meristem formation. *Genes Dev.* 17: 1175–1187. **Figures 19.29 & 19.30** Domagalska, M. A., and Leyser, O. (2011) Signal integration in the control of shoot branching. *Nature* 12: 211–221. **Figure 19.32** El-Showk, S., Ruonala, R., Helariutta, Y. (2013) Crossing paths: Cytokinin signalling and crosstalk. *Development* 140: 1373–1383. **Figure 19.34** Mason, M. G., Ross, J. J., Babst, B. A., Wienclaw, B. N., and Beveridge, C. A. (2014) Sugar demand, not auxin, is the initial regulator of apical dominance. *Proc. Natl. Acad. Sci. USA* 111: 6092–6097. **Figure 19.37** Hochholdinger, F., and Tuberosa, R. (2009) Genetic and genomic dissection of maize root development and architecture. *Curr. Opin. Plant Biol.* 12: 172–177. **Figure 19.39** Lynch, J. P. (2007) Roots of the second Green Revolution. *Aust. J. Bot.* 55: 493–512. **Figure 19.40** Zhang, Z., Liao, H., and Lucas, W. J. (2014) Molecular mechanisms underlying phosphate sensing, signaling, and adaptation in plants. *J. Integr. Plant Biol.* 56: 192–220. **Figure 19.41** Risopatron, J. P. M., Sun, Y., and Jones, B. J. (2010) The vasculat cambium: Molecular control of cellular structure. *Protoplasma* 247: 145–161.

CHAPTER 20

Table 20.1 Clark, J. R. (1983) Age-related changes in trees. *J. Arboriculture* 9: 201–205. **Figure 20.8** Vince-Prue, D. (1975) *Photoperiodism in Plants.* McGraw-Hill, London; Salisbury, F. B. (1963) Biological timing and hormone synthesis in flowering of *Xanthium. Planta* 49: 518–524; Papenfuss, H. D., and Salisbury, F. B. (1967) Aspects of clock resetting in flowering of *Xanthium. Plant Physiol.* 42: 1562–1568. **Figure 20.9** Coulter, M. W., and Hamner, K. C. (1964) Photoperiodic flowering response of Biloxi soybean in 72 hour cycles. *Plant Physiol.* 39: 848–856. **Figure 20.10** Hayama, R., and Coupland, G. (2004) The molecular basis of diversity in the photoperiodic flowering responses of *Arabidopsis* and rice. *Plant Physiol.* 135: 677–684. **Figure 20.12** Hendricks, S. B., and Siegelman, H. W. (1967) Phytochrome and photoperiodism in plants. *Comp. Biochem.* 27: 211–235; Saji, H., Vince-Prue, D., and Furuya, M. (1983) Studies on the photoreceptors for the promotion and inhibition of flowering in dark-grown seedlings of *Pharbitis nil* choisy. *Plant Cell Physiol.* 67: 1183–1189. **Figure 20.13** Deitzer, G. (1984) Photoperiodic induction in long-day plants. In *Light and the Flowering Process,* D. Vince-Prue, B. Thomas, and K. E. Cockshull, eds., Academic Press, New York, pp. 51–63. **Figure 20.15** Purvis, O. N., and Gregory, F. G. (1952) Studies in vernalization of cereals. XII. The reversibility by high temperature of the vernalized condition in Petkus winter rye. *Ann. Bot.* 1: 569–592. **Figure 20.19** Liu, L., Zhu, Y., Shen, L., and Yu, H. (2013) Emerging insights into florigen transport. *Curr. Opin. Plant Biol.* 16: 607–613. **Figure 20.22** Bewley, J. D., Hempel, F. D., McCormick, S., and Zambryski, P. (2000) Reproductive Development. In: *Biochemistry and Molecular Biology of Plants,* B. B. Buchanan, W. Gruissem, and R. L. Jones, eds., American Society of Plant Biologists, Rockville, MD, pp. 988–1034. **Figure 20.23** Meyerowitz, E. M. (2002) Plants compared to animals: The broadest comparative study of development. *Science* 295: 1482–1485; Krizek, B. A., and Fletcher, J. C. (2005) Molecular mechanisms of flower development: An armchair guide. *Nat. Rev. Genet.* 6: 688–698. **Figure 20.27** Pelaz, S., Gustafson-Brown, C., Kohalmi, S. E., Crosby, W. L., and Yanofsky, M. F. (2001) APETALA1 and SEPALLATA3 interact to promote flower development. *Plant J.* 26: 385–394. **Figure 20.31** Busch, A., and Zachgo, S. (2009) Flower symmetry evolution: Towards understanding the abominable mystery of angiosperm radiation. *BioEssays* 31: 1181–1190.

CHAPTER 21

Figure 21.6 Gasser, C. S., and Robinson-Beers, K. (1993) Pistil development. *Plant Cell* 5: 1231–1239. **Figure 21.10** Johnson, M. A., and Lord, E. (2006) Extracellular guidance cues and intracellular signaling pathways that direct pollen tube growth. In *Plant Cell Monographs,* Vol. 3: *The Pollen Tube,* R. Malho, ed., Springer, New York, p. 223–242; Williams, J. H. (2012) Pollen tube growth rates and the diversification of flowering plant reproductive cycles. *Int. J. Plant Sci.* 173: 649–661. **Figure 21.11** Bowman, J. (1994) *Arabidopsis: An atlas of morphology and development.* Springer-Verlag, New York; Edlund, A. F., Swanson, R., and Preuss, D. (2004) Pollen and stigma structure and function: The role of diversity in pollination. *Plant Cell* 16(Suppl. 1): S84–S97. **Figure 21.12** Konrad, K. R., Wudick, M. M., and Feijó, J. A. (2011) Calcium regulation of tip growth: New genes for old mechanisms. *Curr. Opin. Plant Biol.* 14: 721–730. **Figure 21.13** Cheung, A. Y., Nirooman, S., Zou, Y., and Wu, H. M. (2010) A transmembrane formin nucleates subapical actin assembly and controls tip-focused growth in pollen tubes. *Proc. Natl. Acac. Sci. USA* 107: 16390–16395. **Figure 21.16** Higashiyama, T., Kuroiwa, H., Kawano, S., and Kuroiwa, T. (1998) Guidance in vitro of the pollen tube to the naked embryo sac of *Torenia fournieri. Plant Cell* 10: 2019–2031. **Figure 21.22** Debeaujon, I., Nesi, N., Perez, P., Devic, M., Grandjean, O., Caboche, M., and Lepinieca, L. (2003) Proanthocyanidin-accumulating cells in *Arabidopsis* testa: Regulation of differentiation and role in seed development. *Plant Cell* 15: 2514–2531. **Figure 21.23** Cosségal, M., Vernoud, V., Depège, N., and Rogowsky, P. M. (2007) The embryo surrounding region. *Plant Cell Monogr.* 8: 57–71. DOI: 10.1007/7089_2007_109. **Figure 21.24** Olsen, O.-A. (2004) Nuclear Endosperm Development in Cereals and *Arabidopsis thaliana. Plant Cell* 16(Suppl 1): S214-S227. **Figure 21.25** Otegui, M. S. (2007) Endospern cell walls: Formation, composition, and functions. *Plant Cell Monogr.* 8: 159–178. **Figure 21.26** Olsen, O.-A. (2004) Nuclear Endosperm Development in Cereals and *Arabidopsis thaliana. Plant Cell* 16(Suppl 1): S214-S227. **Figure 21.27** Li, J., and Berger, F. (2012) Endosperm: Food for humankind and fodder for scientific discoveries. *New Phytol.* 195: 290–305. **Figure 21.28** Becraft, P. W., and Yi, G. (2011) Regulation of aleurone development in cereal grains. *J. Exp. Bot.* 62: 1669–1675. **Figure 21.29** Haughn, G., and Chaudhury, A. (2005) Genetic analysis of seed coat development in *Arabidopsis. Trends Plant Sci.* 10: 472–477. **Figure 21.30** Verdier, J., Lalanne, D., Pelletier, S., Torres-Jerez, I., Righetti, K., Bandyopadhyay, K., Leprince, O., Chatelain, E., Vu, B. L., Gouzy, J., et al. (2013) A regulatory network-based approach dissects late maturation processes related to the acquisition of desiccation tolerance and longevity of *Medicago truncatula* seeds. *Plant Physiol.* 163: 757–774. **Figure**

21.31 Delahaie, J., Hundertmark, M., Bove, J., Leprince, O., Rogniaux, H., and Buitink, J. (2013) LEA polypeptide profiling of recalcitrant and orthodox legume seeds reveals ABI3-regulated LEA protein abundance linked to desiccation tolerance. *J. Exp. Bot.* 64: 4559–4573. **Figures 21.33 & 21.34** Seymour, G. B., Østergaard, L., Chapman, N. H., Knapp, S., and Martin, C. (2013) Fruit development and ripening. *Annu. Rev. Plant Biol.* 64: 219–241. **Figure 21.34** Pabón-Mora, N., and Litt, A. (2011) Comparative anatomical and developmental analysis of dry and fleshy fruits of Solanaceae. *Am. J. Bot.* 98: 1415–1436. **Figure 21.35** Fray, R. F., and Grierson, D. (1993) Identification and genetic analysis of normal and mutant phytoene synthase genes of tomato by sequencing, complementation and co-suppression. *Plant Mol. Biol.* 22: 589–602. **Figure 21.37** Oeller, P. W., Lu, M. W., Taylor, L. P., Pike, D. A., and Theologis, A. (1991) Reversible inhibition of tomato fruit senescence by antisense RNA. *Science* 254: 437–439; Grierson, D. (2013) Ethylene and the control of fruit ripening. In *The Molecular Biology and Biochemistry of Fruit Ripening*, G. B. Seymour, G. A. Tucker, M. Poole, and J. J. Giovannoni, eds., Wiley-Blackwell, Oxford, UK, p. 216. **Figure 21.38** Dilley, D. R. (1981) Assessing fruit maturity and ripening and techniques to delay ripening in storage. *Proc. Mich. State Hort. Soc.* 11: 132–146, figure modified in Kupferman, E. (1986) The role of ethylene in determining apple harvest and storage life. *Postharvest Pomology Newsletter*, 4(1), http://postharvest.tfrec.wsu.edu/pages/N4I1C (accessed September 2014). **Figure 21.40** Seymour, G. B., Østergaard, L., Chapman, N. H., Knapp, S., and Martin, C. (2013) Fruit development and ripening. *Annu. Rev. Plant Biol.* 64: 219–241.

CHAPTER 22

Table 22.1 Thomas, H. (2013) Senescence, ageing and death of the whole plant. *New Phytol.* 197: 696–711. DOI: 10.1111/nph.12047. **Figure 22.7** Bassham, D. C., Laporte, M., Marty, F., Moriyasu, Y., Ohsumi, Y., Olsen, L. J., and Yoshimoto, K. (2006) Autophagy in development and stress response of plants. *Autophagy* 2: 2–11; Yoshimoto, K., Hanaoka, H., Sato, S., Kato, T., Tabata, S., Noda, T. and Ohsumi, Y. (2004) Processing of ATG8s, ubiquitin-like proteins, and their deconjugation by ATG4s are essential for plant autophagy. *Plant Cell* 16: 2967–2983. **Figure 22.9** Stahl, E. (1909) *Zur biologie des chlorophylls: Laubfarbe und himmelslicht, vergilbung und etiolement.* G. Fisher Verlag, Jena, Germany. **Figures 22.11 & 22.13** Keskitalo, J., Bergquist, G., Gardestrom, P., and Jansson, S. (2005) A cellular timetable of autumn senescence. *Plant Physiol.* 139: 1635–1648. **Figure 22.14** Krupinska, K., Mulisch, M., Hollmann, J., Tokarz, K., Zschiesche, W., Kage, H., Humbeck, K., and Bilger, W. (2012) An alternative strategy of dismantling of the chloroplasts during leaf senescence observed in a high-yield variety of barley. *Physiol. Plant.* 144: 189–200. **Figure 22.15** Wada, S., Ishida, H., Izumi, M., Yoshimoto, K., Ohsumi, Y., Mae, T., and Makino, A. (2009) Autophagy plays a role in chloroplast degradation during senescence in individually darked leaves. *Plant Physiol.* 149: 885–893. **Figure 22.17** Breeze, E., Harrison, E., McHattie, S., Hughes, L., Hickman, R., Hill, C., Kiddle, S., Kim, Y.-S., Penfold, C. A., Jenkins, D., et al. (2011) High-resolution temporal profiling of transcripts during *Arabidopsis* leaf senescence reveals a distinct chronology of processes and regulation. *Plant Cell* 23: 873–894. **Figure 22.19** Uauy, C., Distelfeld, A., Fahima, T., Blechl, A., and Dubcovsky, J. (2006) A NAC gene regulating senescence improves grain protein, zinc, and iron content in wheat. *Science* 314: 1298–1301. **Figure 22.20** Gan, S., and Amasino, R. M. (1995) Inhibition of leaf senescence by autoregulated production of cytokinin. *Science* 270: 1986–1988. **Figure 22.21** Mothes, K., and Schütte, H. (1961) Über die akkumulation von alpha-aminoisobuttersäure in blattgewebe unter dem einfluβ von kinetin. *Physiol. Plant.* 14: 72–75. **Figure 22.23** Vahala, J., Ruonala, R., Keinänen, M., Tuominen, H., and Kangasjärvi, J. (2003) Ethylene insensitivity modulates ozone-induced cell death in birch (*Betula pendula*). *Plant Physiol.* 132: 185–195. **Figure 22.24** Morgan, P. W. (1984) Is ethylene the natural regulator of abscission? In *Ethylene: Biochemical, Physiological and Applied Aspects*, Y. Fuchs and E. Chalutz, eds., Martinus Nijhoff, The Hague, Netherlands, pp. 231–240. **Figure 22.25** Aalen, R. B., Wildhagen, M., Stø, I. M., and Butenko, M. A. (2013) IDA: A peptide ligand regulating cell separation processes in *Arabidopsis*. *J. Exp. Bot.* 64: 5253–5261. **Figure 22.28** Sillett, S. C., van Pelt, R., Koch, G. W., Ambrose, A. R., Carroll, A. L., Antoine, M. E., and Mifsud, B. M. (2010) Increasing wood production through old age in tall trees. *For. Ecol. Manage.* 259: 976–994. **Figure 22.29** Stephenson, N. L., Das, A. J., Condit, R., Russo, S. E., Baker, P. J., Beckman, N. G., Coomes, D. A., Lines, E. R., Morris, W. K., Rüger, N., et al. (2014) Rate of tree carbon accumulation increases continuously with tree size. *Nature* 507: 90–93.

CHAPTER 23

Figure 23.1 van Dam, N. M. (2009) How plants cope with biotic interactions. *Plant Biol.* 11: 1–5. **Figure 23.2** Gough, C., and Cullimore, J. (2011) Lipo-chitooligosaccharide signaling in endosymbiotic plant–microbe interactions. *Mol. Plant Microbe Interact.* 24: 867–878; Markmann, K., and Parniske, M. (2009) Evolution of root endosymbiosis with bacteria: How novel are nodules? *Trends Plant Sci.* 14: 77–86. **Figure 23.3** Oldroyd, G. E. D., and Downie, A. (2004) Calcium, kinases and nodulation signalling in legumes. *Nat. Rev. Mol. Cell Biol.* 5: 566–576. **Figure 23.4** Goh, C.-H., Veliz Vallejos, D. F., Nicotra, A. B., and Mathesius, U. (2013) The impact of beneficial plant-associated microbes on plant phenotypic plasticity. *J. Chem. Ecol.* 39: 826–839. **Figure 23.21** Christmann, A., and Grill, E. (2013) Electric defence. *Nature* 500: 404–405. **Figure 23.24** Goodspeed, D., Chehab, E. W., Min-Venditti, A., Braam, J., and Covington, M. F. (2012) *Arabidopsis* synchronizes jasmonate-mediated defense with insect circadian behavior. *Proc. Natl. Acad. Sci. USA* 109: 4674–4677. **Figure 23.27** Boller, T., and Felix, G. (2009) A Renaissance of elicitors: Perception of microbe-associate molecular patterns and danger signals by pattern-recognition receptors. *Annu. Rev. Plant Biol.* 60: 379–406.

CHAPTER 24

Table 24.2 Jones, R., Ougham, H., Thomas, H., and Waaland, S. (2013) *The Molecular Life of Plants.* Wiley-Blackwell, Chichester, West Sussex, UK, p. 567. **Table 24.3** Lyons, J. M., Wheaton, T. A., and Pratt, H. K. (1964) Relationship between the physical nature of mitochondrial membranes and chilling sensitivity in plants. *Plant Physiol.* 39: 262–268. **Figure 24.5** Boyer, J. S. (1970) Differing sensitivity of photosynthesis to low leaf water potentials in corn and soybean.

Plant Physiol. 46: 236–239. **Figure 24.6** Mittler, R. (2006) Abiotic stress, the field environment and stress combination. *Trends Plant Sci.* 11: 15–19. **Figure 24.7** Mittler, R., and Blumwald, E. (2010) Genetic engineering for modern agriculture: Challenges and perspectives. *Ann. Rev. Plant Biol.* 61: 443–462. **Figure 24.8** Mittler, R., Finka, A., and Goloubinoff, P. (2012) How do plants feel the heat? *Trends Biochem. Sci.* 37: 118–25. **Figure 24.9** Reddy, A. S., Ali, G. S., Celesnik, H., and Day, I. S. (2011) Coping with stresses: Roles of calcium- and calcium/calmodulin-regulated gene expression. *Plant Cell* 23: 2010–2032. **Figure 24.12** Smékalová, V., Doskočilová, A., Komis, G., and Samaj, J. (2013) Crosstalk between secondary messengers, hormones and MAPK modules during abiotic stress signalling in plants. *Biotechnol. Adv.* 32: 2–11. DOI: 10.1016/j.biotechadv.2013.07.009. **Figure 24.13** Lata, C., and Prasad, M. (2011) Role of DREBs in regulation of abiotic stress responses in plants. *J. Exp. Bot.* 62: 4731–4748. **Figure 24.14** Mittler, R., Vanderauwera, S., Suzuki, N., Miller, G., Tognetti, V. B., Vandepoele, K., Gollery, G., Shulaev, V., and Van Breusegem, F. (2011) ROS signaling: The new wave? *Trends Plant Sci.* 16: 300–309; Suzuki, N., Miller, G., Salazar, C., Mondal, H. A., Shulaev, E., Cortes, D. F., Shuman, J. L., Luo, X., Shah, J., Schlauch, K., et al. (2013) Temporal-spatial interaction between reactive oxygen species and abscisic acid regulates rapid systemic acclimation in plants. *Plant Cell* 25: 3553–69. **Figure 24.15** Gutzat, R., and Mittelsten-Scheid, O. (2012) Epigenetic responses to stress: Triple defense? *Curr. Opin. Plant. Biol.* 15: 568–573. **Figure 24.20** Jones, R., Ougham, H., Thomas, H., and Waaland, S. (2013) *The Molecular Life of Plants.* Wiley-Blackwell, Chichester, West Sussex, UK, p. 568. **Figure 24.21** Baneyx, F., and Mujacic, M. (2004) Recombinant protein folding and misfolding in *Escherichia coli. Nat. Biotechnol.* 22: 1399–1408. **Figure 24.24** Beardsell, M. F., and Cohen, D. (1975) Relationships between leaf water status, abscisic acid levels, and stomatal resistance in maize and sorghum. *Plant Physiol.* 56: 207–212. **Figure 24.25** McAinsh, M. R., Brownlee, C., and Hetherington, A. M. (1990) Abscisic acid-induced elevation of guard cell cytosolic Ca^{2+} precedes stomatal closure. *Nature* 343: 186–188. **Figure 24.28** Sánchez-Calderón, L., Ibarra-Cortés, M. E., and Zepeda-Jazo, I. (2013) Root development and abiotic stress adaptation. In *Abiotic Stress—Plant Responses and Applications in Agriculture*, K. Vahdati and C. Leslie, eds., InTech, Rijeka, Croatia, pp. 135–168. DOI: 10.5772/45842. **Figure 24.29** Saab, I. N., Sharp, R. E., Pritchard, J., and Voetberg, G. S. (1990) Increased endogenous abscisic acid maintains primary root growth and inhibits shoot growth of maize seedlings at low water potentials. *Plant Physiol.* 93:1329–1336.

Photo Credits

Part and Chapter Openers

Chapters 1 and 2: © Vilor/istock.

Part I: © AntiMartina/istock.

Part II: © suriyasilsaksom/istock.

Part III: © Petegar/istock.

Chapter 1

1.3: David McIntyre. **1.29:** © Sebastian Kaulitzki/Shutterstock.

Chapter 2

2.8: *oleracea*: © Jim Mills/istock; *carinata*: © C. Schmitt/Shutterstock; *nigra*: © Roger Whiteway/istock; *juncea*: © Suzannah Skelton/istock; *rapa*: © Chun-Tso Lin/Shutterstock; *napus*: © Peter Austin/istock.

Chapter 3

3.12: David McIntyre.

Chapter 4

4.5: David McIntyre. **4.6B**: © Steve Gschmeissner/Science Source. **4.12B**: © age fotostock Spain, S.L./Alamy. **4.12C**: © Ray Simons/Science Source.

Chapter 6

6.19: © Biodisc/Visuals Unlimited/Alamy.

Chapter 7

7.9: © Biophoto Associates/Science Source.

Chapter 11

11.1: © J. N. A. Lott/Biological Photo Service. **11.2**: © P. Gates/Biological Photo Service.

Chapter 13

13.9: David McIntyre. **13.10**: © Dr. Peter Siver/Visuals Unlimited.

Chapter 14

14.1: © Andrew Syred/Science Source. **14.3A**: David McIntyre. **14.3B**: © Biophoto Associates/Science Source. **14.3C**: © Dennis Drenner/Visuals Unlimited, Inc.

Chapter 15

15.1A: © Nigel Cattlin/Alamy. **15.1B**: © blickwinkel/Alamy. **15.1C**: © Steven Sheppardson/Alamy. **15.1D, E**: David McIntyre. **15.10A, C**: © Sylvan Wittwer/Visuals Unlimited. **15.13**: © Ray Simons/Science Source. **15.27**: © blickwinkel/Alamy.

Chapter 16

16.1: © Shosei/Corbis. **16.2**: © Nigel Cattlin/Alamy. **16.5**: David McIntyre.

Chapter 17

17.1A: © David L. Moore/Alamy. **17.1B**: David McIntyre.

Chapter 18

18.3: © Larry Larsen/Alamy.

Chapter 19

19.9A: © Dr. Ken Wagner/Visuals Unlimited, Inc. **19.9B**: © Robert Harding Picture Library Ltd/Alamy. **19.9**C: © Jon Bertsch/Visuals Unlimited, Inc. **19.9D**: © Garry DeLong/Science Source. **19.12**: © Dr. Stanley Flegler/Visuals Unlimited, Inc. **19.15**: David McIntyre. **19.25A**: © William Ormerod/Visuals Unlimited, Inc. **19.33B**: David McIntyre. **19.35**: © C. J. Wheeler/Alamy. **19.36**: © Heidi Natura/Conservation Research Institute. **19.37B**: © B W Hoffmann/AGE Fotostock.

Chapter 20

20.30A: © Artex67/istock.

Chapter 21

21.4A: © Scientifica/RMF/Visuals Unlimited, Inc. **21.5**: Courtesy of David Twell (Department of Biology) and Stefan Hyman (Electron microscopy Laboratory), University of Leicester. **21.16A**: © blickwinkel/Alamy. **21.34A**: © brozova/istock. **21.36**, **21.39A**: David McIntyre. **21.41**: Courtesy of Andy Davis, John Innes Centre.

Chapter 22

22.1: © Chuck Savage/Corbis. **22.8**: © Chris Wildblood/Alamy. **22.22A**: © Biodisc/Visuals Unlimited, Inc. **22.27**: © WildPictures/Alamy.

Chapter 23

23.7: Courtesy of Agong1/Wikipedia. **23.23**: David McIntyre. **23.24A**: Courtesy of Alton N. Sparks, Jr., University of Georgia, Bugwood.org. **23.36**: © Biodisc/Visuals Unlimited, Inc.

Subject Index

B

D

I

J

K

M

Q

R

S

T

U

X

Y

Z